BASIC CURVES

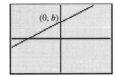

$$y = mx + b$$

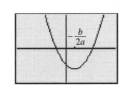

$$y = ax^2 + bx + c \quad (a > 0)$$

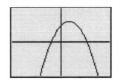

$$y = ax^2 + bx + c \quad (a < 0)$$

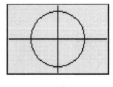

$$x^2 + y^2 = a^2$$

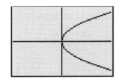

$$y^2 = 4px \quad (p > 0)$$

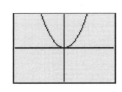

$$x^2 = 4py \quad (p > 0)$$

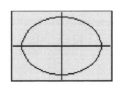

$$\frac{x^2}{a^2} + \frac{y^2}{b^2} = 1$$

$$\frac{y^2}{a^2} + \frac{x^2}{b^2} = 1$$

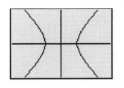

$$\frac{x^2}{a^2} - \frac{y^2}{b^2} = 1$$

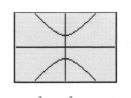

$$\frac{y^2}{a^2} - \frac{x^2}{b^2} = 1$$

$$xy = a \quad (a > 0)$$

$$y = a\sqrt{x} \quad (a > 0)$$

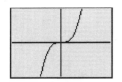

$$y = ax^3 \quad (a > 0)$$

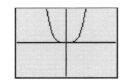

$$y = ax^4 \quad (a > 0)$$

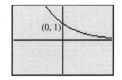

$$y = b^x \quad (b > 1)$$

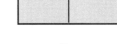

$$y = b^{-x} \quad (b > 1)$$

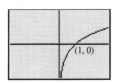

$$y = \log_b x$$

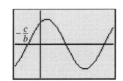

$$y = a \sin(bx + c)$$
$$(a > 0, c > 0)$$

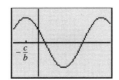

$$y = a \cos(bx + c)$$
$$(a > 0, c > 0)$$

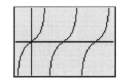

$$y = a \tan x \quad (a > 0)$$

Basic Technical Mathematics with Calculus

SI Version

OTHER TITLES OF RELATED INTEREST

- *Basic Technical Mathematics,* Eighth Edition, by Allyn J. Washington
- *Basic Technical Mathematics with Calculus,* Eighth Edition, by Allyn J. Washington
- *Technical Calculus with Analytic Geometry,* Fourth Edition, by Allyn J. Washington
- *Introduction to Technical Mathematics,* Fourth Edition, by Allyn J. Washington and Mario F. Triola

Basic Technical Mathematics
with Calculus
SI Version

EIGHTH EDITION

ALLYN J. WASHINGTON
Dutchess Community College

PEARSON

Addison
Wesley

Toronto

Library and Archives Canada Cataloguing in Publication

Washington, Allyn, J.
    Basic technical mathematics with calculus: SI version / Allyn J. Washington—8th ed.

    Includes index.

    ISBN 0-321-30689-9

    1. Mathematics—Textbooks.   I. Title
QA37.2.W37   2005   510   C2004-904470-2

ISBN 0-321-30689-9

Vice President, Editorial Director: Michael J. Young

Executive Editor: Dave Ward

Marketing Manager: Toivo Pajo

Supervising Developmental Editor: Suzanne Schaan

Production Editor: Judith Scott

Copy Editor: Julia Cochrane

Production Coordinator: Patricia Ciardullo

Text Designer: Susan Carsten Raymond

Cover Design: Dennis Schaefer

Cover Photos: Daryl Benson/Masterfile; John Kieffer/Index Stock Imagery

About the Cover

The cover shows a silicon wafer set over the backdrop of a circuit board. These thin, round silicon crystal wafers measure between 150 μm and 300 μm, and when placed together, form a 2.5-cm-square integrated circuit (IC). Two decades ago, the equipment needed to perform the function of these ICs would not have fit inside a dorm room.

2 3 4 5   09 08 07 06 05
Printed and bound in the USA.

PEARSON
Addison
Wesley

Toronto

To my loving wife, Millie

Contents

Preface

SCOPE OF THE BOOK

Basic Technical Mathematics with Calculus is intended primarily for students in technical and pre-engineering technology programs or other programs for which coverage of basic mathematics is required. The distinctive feature of this version is that all units of measurement are metric (SI) units, or are acceptable for use with SI units. All other features are the same as those in the regular eighth edition.

Chapters 1 through 20 provide the necessary background for further study with an integrated treatment of algebra and trigonometry. Chapter 21 covers the basic topics of analytic geometry, and Chapter 22 gives an introduction to statistics. Fundamental topics of calculus are covered in Chapters 23 through 30. Numerous applications from many fields of technology are included, primarily to indicate where and how mathematical techniques are used. However, it is not necessary that the student have a specific knowledge of the technical area from which any given problem is taken.

Most students using this text will have a background that includes some algebra and geometry. However, the material is presented in adequate detail for those who may need more study in these areas. The material presented here is sufficient for three to four semesters.

One of the primary reasons for the arrangement of topics in this text is to present material in an order that allows a student to take courses concurrently in allied technical areas, such as physics and electricity. These allied courses normally require a student to know certain mathematical topics by certain definite times; yet the traditional order of topics in mathematics courses makes it difficult to attain this coverage without loss of continuity. However, the material in this book can be rearranged to fit any appropriate sequence of topics. Another feature of this text is that certain topics traditionally included for mathematical completeness have been covered only briefly or have been omitted.

The approach used here is not unduly rigorous mathematically, although all appropriate terms and concepts are introduced as needed and given an intuitive or algebraic foundation. The aim in this book is to help the student develop a feeling for mathematical methods, not simply to provide a collection of formulas. The development of the text material emphasizes that it is essential for the student to have a sound background in algebra and trigonometry in order to understand and succeed in any subsequent work in mathematics.

NEW FEATURES

The eighth edition of *Basic Technical Mathematics with Calculus* includes all the basic features of the earlier editions. However, many sections have been rewritten to some degree to include additional or revised explanatory material, examples, and exercises. Some sections have been extensively revised. Specifically, among the new features of this edition are the following:

NEW CHAPTER INTRODUCTIONS

All 30 of the chapter introductions have been rewritten. Each illustrates specific examples of how the development of technology has been related to the development of mathematics. In these introductions, it is shown that past discoveries in technology led to some of the methods in mathematics, whereas in other cases mathematics topics already known were later very useful in bringing about advances in technology. It is hoped that these introductions will show the student the importance of the role mathematics has had in advances in technology.

EXERCISES DIRECTLY REFERENCED TO TEXT EXAMPLES

The first few exercises in most of the text sections are referenced directly to a specific example of the section. These exercises are worded so that it is necessary for the student to refer to the example in order to complete the required solution. In this way, the student should be able to better review and understand the text material before attempting to solve the exercises that follow.

REVISED COVERAGE

Chapter 3 now includes coverage of shifting a graph. Showing how a graphing calculator is used to check an algebraic operation is shown in Chapter 6. Chapter 13 now has separate sections on exponential functions (13.1) and logarithmic functions (13.2). Section 15.1 now includes both the remainder theorem and synthetic division. Section 17.6 is now a separate section on linear programming. Section 29.7 on Fourier Series now includes half-range expansions. Determinant expansion by minors and the properties of determinants are now covered in the first section of Supplementary Topics, not in Chapter 16.

EXERCISES

There are over 2400 new exercises, including over 500 that illustrate technical applications and over 140 that require some explanation along with the answer. These are in addition to those included in the seventh edition. There are now about 12,000 exercises total in this eighth edition.

FIGURES

There are approximately 90 new figures in this edition, including 20 new calculator screens. There are now over 1300 figures in the text, in addition to those in the answer section.

ADDITIONAL FEATURES

THE GRAPHING CALCULATOR

The graphing calculator is used throughout the text in examples and exercises to help reinforce and develop many topics. There are over 280 graphing calculator screens in this edition. The coverage starts in Section 1.3, where it is used for calculational purposes, and its use for graphing starts in Section 3.5. Additional notes are found in Appendix C.

Appendix C also presents 22 graphing calculator programs that show how the calculator's use can be expanded. Also, a *Graphing Calculator Manual* showing detailed procedures is available as a supplement.

PAGE LAYOUT

Special attention has been given to the page layout. Nearly all examples are started and completed on the same page (there are only five exceptions, and each of these is presented on facing pages). Also, all figures are shown immediately adjacent to the material in which they are discussed.

SPECIAL EXPLANATORY COMMENTS

Throughout the book, special explanatory comments in color have been used in the examples to emphasize and clarify certain important points. Arrows are often used to indicate clearly the part of the example to which reference is made.

PROBLEM-SOLVING TECHNIQUES

Techniques and procedures that summarize the approaches in solving many types of problems have been clearly outlined in color-shaded boxes.

IMPORTANT FORMULAS

Throughout the book, important formulas are set off and displayed so that they can be easily located and used.

SUBHEADS AND KEY TERMS

Many sections include subheads to indicate where the discussion of a new topic starts within the section, and other key terms are noted in the margin for emphasis and easy reference.

SPECIAL CAUTION AND NOTE INDICATORS

NOTE ▶
CAUTION ▶

Two special margin indicators (as shown at the left) are used. The caution indicator identifies errors students commonly make or places where they frequently have difficulty. The note indicator points out text material that is of particular importance in developing or understanding the topic under discussion.

WORD PROBLEMS

There are over 120 examples throughout the text that show the complete solutions of word problems. These are clearly noted by the phrase Solving a Word Problem placed to the left of the example. There are also over 750 exercises in which word problems are to be solved.

WRITING EXERCISES

One specific writing exercise is included at the end of each chapter. These exercises give the student practice in writing explanations of problems solutions. Also, there are over 340 additional exercises throughout the book (at least seven in each chapter) that require at least a sentence or two of explanation as part of the answer. These are noted by the (W) symbol placed to the left of the exercise number. A special Index of Writing Exercises is included at the back of the book.

CHAPTER EQUATIONS, REVIEW EXERCISES, AND PRACTICE TEST

At the end of each chapter, all important equations are listed together for easy reference. Each chapter is also followed by a set of review exercises that covers all the material in the chapter. Following the chapter equations and review exercises is a chapter practice test that students can use to check their understanding of the material. Solutions to all practice test problems are given in the back of the book.

APPLICATIONS AND UNITS OF MEASUREMENT

The examples and exercises illustrate the application of mathematics to all fields of technology. Many relate to modern technology such as computer design, computer-assisted design (CAD), electronics, solar energy, lasers, fiber optics, the environment, and space technology. A special Index of Applications is included near the end of the book.

SUPPLEMENTARY TOPICS

In response to needs of certain programs, eight additional sections of text material are included after Chapter 30. The topics covered are Higher-Order Determinants, Gaussian Elimination, Rotation of Axes, Functions of Two Variables, Curves and Surfaces in Three Dimensions, Partial Derivatives, Double Integrals, and Numerical Solutions of Differential Equations.

EXAMPLES

Over 1400 worked examples are in this text. Of these, over 300 illustrate technical applications.

MARGIN NOTES

Throughout the book, margin notes briefly point out relevant historical events in mathematics and technology. Other margin notes are used to make specific comments related to the text material. Also, where appropriate, equations from earlier material are shown for reference in the margin.

ANSWERS TO EXERCISES

The answers to all odd-numbered exercises (except the end-of-chapter writing exercises) are given at the back of the book. The *Student's Solutions Manual* contains solutions for every other odd-numbered section exercise and the *Instructor's Solutions Manual* contains solutions for all section exercises. The answers to all exercises are given in the *Answer Book*.

FLEXIBILITY OF MATERIAL COVERAGE

The order of material coverage can be changed in many places, and certain sections may be omitted without loss of continuity of coverage. Users of earlier editions have indicated the successful use of numerous variations in coverage. Any changes will depend on the type of course and completeness required. Several possible variations in coverage are included in the *Answer Book*.

SUPPLEMENTS

SUPPLEMENTS FOR THE INSTRUCTOR

Test Bank

The *Test Bank* by Waldo Dyck contains two short-answer test forms for every chapter.

Instructor's Solutions Manual

The *Instructor's Solutions Manual* by Bob Martin contains detailed solutions to every section exercise.

Answer Book

The *Answer Book* by Bob Martin includes all answers to all exercises, including review exercises.

Instructor's Resource CD-ROM (ISBN 0-321-30753-4)

The Test Bank, Instructor's Solutions Manual, Answer Book, and TestGen are all available on the Instructor's Resource CD-ROM. Some of these supplements may also be available to instructors for downloading from a protected location on Pearson Education's online catalogue; see your local sales representative for further information.

Also Available

MathXL

Available on-line with an ID and password, this testing-and-tracking system allows an instructor to administer tests on-line and track students' grades. Instructors can create their own tests and assessments using TestGen software or assign existing tests that correspond with the chapter Practice Tests in the book.

MyMathLab

MyMathLab is a complete on-line course for Addison-Wesley mathematics textbooks that provides multimedia instruction correlated to the textbook content. MyMathLab is easily customizable to suit the needs of students and instructors and provides a comprehensive and efficient on-line course-management system. Instructors can create, copy, edit, assign, and track all tests for their course as well as track student's results. The print supplements are available on-line, side-by-side with the textbook. For more information, visit our Web site at www.mymathlab.com or contact your sales representative for a live demonstration.

TestGen with QuizMaster

TestGen enables instructors to build, edit, print, and administer tests using a computerized bank of questions developed to cover all objectives of the text. Instructors can modify test bank questions or add new questions by using the built-in question editor, which allows users to create graphs, import graphics, insert math notation, and insert variable numbers or text. Tests can be printed or administered online via the Web or other network. TestGen comes packaged with QuizMaster, which allows students to take tests on a local area network. The software is available on a dual-platform Windows/Macintosh CD-ROM.

SUPPLEMENTS FOR THE STUDENT

Student's Solutions Manual (ISBN 0-321-30754-2)

The *Student's Solutions Manual* by Bob Martin includes detailed solutions for every other odd-numbered section exercise.

Graphing Calculator Manual (ISBN 0-321-19740-2)

The *Graphing Calculator Manual* by Robert Seaver presents detailed instructions on the use of various graphing calculators.

Addison-Wesley Math Tutor Center

The Addison-Wesley Math Tutor Center is staffed by qualified mathematics and statistics instructors who provide students with tutoring on examples and odd-numbered exercises from the textbook. Tutoring is available via toll-free telephone, toll-free fax, e-mail, and the Internet. Interactive Web-based technology allows tutors and students to view and work through problems together in real time over the Internet. For more information, please visit our Web site at www.aw-bc.com/tutorcenter or call us at 1-888-777-0463.

QUESTIONS/COMMENTS/INFORMATION

We welcome your comments about this text. You may write to the publisher or the author at:

Pearson Education Canada
26 Prince Andrew Place
Don Mills, Ontario
Canada, M3C2T8

ACKNOWLEDGMENTS

The author gratefully acknowledges the contributions of the following reviewers. Their detailed comments and many suggestions were of great assistance in preparing this eighth edition.

Kathleen M. Acks
Maui Community College

Kathleen L. Almy
Rock Valley College

J. Paul Balog
George Brown College

James Cassidy
Suffolk Community College

Paul Chacon
University of Southern Colorado

Guangxiong Fang
Daniel Webster College

William Ferguson
Columbus State Community College

Maggie Flint
Northeast State Technical Community College

Sarah Flum
Bay de Noc Community College

Parviz Ghavami
Texas State Technical College

James Hardman
Sinclair Community College

Susan J. Hoy
Bristol Community College

Joan E. Jackson
Cincinnati State Technical and Community College

John Jenness
Kwantlen University College

Joe Jordan
John Tyler Community College

John H. Knox
Vermont Technical College

Nestor Komar
Niagara College

Carrie Kyser
Cuyahoga Community College

Bob Malena
Community College of Allegheny County

Kenneth Mann
Catawba Valley Community College

Carol A. McVey
Florence Darlington Technical College

Donald Nevin
Vermont Technical College

Robert C. Opel
Waukesha County Technical College

Beth Osikiewicz
Kent State University—Tuscarawas

Jeff Osikiewicz
Kent State University—Tuscarawas

Jim Rich
Blackhawk Technical College

Joseph Riesen
Fairmont State College

Susan L. Schroeder
University of Houston

Thomas Stark
Cincinnati State Technical and Community College

Emily Sullivan
Bates Technical College

Thomas Sutton
Mohawk College

Jeanne Szarka
Ohio State University—Agricultural Technical Institute

Richard Watkins
Tidewater Community College— Virginia Beach

Daniel Wilshire
Pennsylvania State University— Altoona College

The metrication for this SI edition was prepared by Julia Cochrane, whose assistance with the project was greatly appreciated.

Special thanks go to Bob Martin of Tarrant County Junior College for preparing the *Answer Book,* the *Student's Solutions Manual,* and the *Instructor's Solutions Manual.* Special thanks also to Robert Seaver of Lorain County Community College for preparing the *Graphing Calculator Manual,* and to Waldo Dyck of Red River College for preparing the *Test Bank.* Also, I wish to again thank Thomas Stark of Cincinnati State Technical and Community College for the *RISERS* approach to solving word problems in Appendix A.

My thanks and gratitude go to Jim Bryant who drew all of the chapter-opener drawings; to Martha Ghent and Charles Cox for assisting in the tedious job of reading proof; and to Richard Watkins, Beth Osikiewicz, Russ Baker, and John Esenwa for checking the accuracy of answers. John Jenness of Kwantlen University College conducted a technical check of the SI supplements.

I gratefully acknowledge the cooperation and support of my editors, Maureen O'Connor and Carter Fenton. I am especially grateful to the production editor, Greg Hubit, who has provided very valuable assistance for the past six editions. Also, I wish to acknowledge the very fine work of Elaine Lattanzi of Beacon Publishing Services, who set all of the type for this edition.

Also of great assistance during the production of this edition were Suzanne Alley, Ron Hampton, Kathleen Manley, Susan Carsten Raymond, Dennis Schaefer, Caroline Fell, Dona Kenly, and Lindsay Skay of the Addison-Wesley staff. The team at Pearson Education Canada—Dave Ward, Suzanne Schaan, Judith Scott, and Patricia Ciardullo— made the metric edition possible.

Finally, special mention is due my wife, Millie, to whom I have dedicated this edition. Her help in checking answers and manuscript in earlier editions, but especially her patience, understanding, and support for our over forty-five years together, through the preparation of this edition and all earlier editions, has been invaluable.

A.J.W.

Basic Algebraic Operations

Interest in things such as the land on which they lived, the buildings they constructed, and the motion of the planets led people in early civilizations to keep records and to create methods of counting and ways of measuring. In turn, some of the early ideas of arithmetic, geometry, and trigonometry were developed. From such beginnings, mathematics has played a key role in the great advances in science and technology.

Often, mathematics was developed from studies made in sciences, such as astronomy and physics, to better describe and understand the subject being studied. In some cases, mathematical methods were developed specifically because of the needs in a particular area of application.

Many were interested in the mathematics itself, and they added to what was known at the time. Although this additional knowledge in mathematics may not have been related to any applications at the time it was developed, it often became useful later in applied areas.

In the chapter introductions that follow, examples of the interaction of technology and mathematics are given. From these examples and the text material, it is hoped that you will better understand the important role that math has had and still has in technology. Throughout this text, there are applications from technologies including (but not limited to) aeronautical, business, communications, electricity, electronics, engineering, environmental, heat and air conditioning, mechanical, medical, meteorology, petroleum, product design, solar, and space. To solve the applied problems in this text will require a knowledge of the mathematics presented but will *not* require prior knowledge of the field of application.

It is very important for you to learn and understand the concepts presented in this text in order to have a strong foundation in mathematics to use in your technical field. Developing this understanding will require a serious commitment of time and effort on your part. The author sincerely wishes you the very best success.

A thorough understanding of algebra is essential in order to progress in mathematics, and we begin by reviewing basic concepts that deal with numbers and symbols. These will enable us to develop topics in algebra, which in turn are needed in other areas such as geometry, trigonometry, and calculus.

In the 1500s, 1600s, and 1700s, discoveries in astronomy and the need for more accurate maps and instruments in navigation were very important in leading scientists and mathematicians to develop useful new ideas and methods in mathematics.

Later in the 1800s, scientists were studying the nature of light. This led to a mathematical prediction of the existence of radio waves, now used in many types of communication. Also, in the 1900s and 2000s, mathematics has been vital to the development of electronics and space travel.

1.1 NUMBERS

In technology and science, as well as in everyday life, we use the very familiar **counting numbers** 1, 2, 3, and so on. They are also called **natural numbers** or **positive integers.** The **negative integers** -1, -2, -3, and so on are also very useful in mathematics and its applications. *The **integers** include the positive integers and the negative integers and* **zero,** *which is neither positive nor negative.* This means the integers are the numbers $\ldots, -3, -2, -1, 0, 1, 2, 3$, and so on.

To specify parts of a quantity, *rational numbers* are used. *A **rational number** is any number that can be represented by the division of one integer by another nonzero integer.* Another type of number, an **irrational number,** *cannot be written as the division of one integer by another.*

Irrational numbers were discussed by the Greek mathematician Pythagoras in about 540 B.C.E.

◀ **EXAMPLE 1** The numbers 5 and -19 are integers. They are also rational numbers since they can be written as $\frac{5}{1}$ and $\frac{-19}{1}$, respectively. Normally, we do not write the 1's in the denominators.

The numbers $\frac{5}{8}$ and $\frac{-11}{3}$ are rational numbers because the numerator and the denominator of each are integers.

The numbers $\sqrt{2}$ and π are irrational numbers. It is not possible to find two integers, one divided by the other, to represent either of these numbers. It can be shown that square roots (and other roots) that cannot be expressed exactly in decimal form are irrational. Also, $\frac{22}{7}$ is sometimes used as an *approximation* for π, but it is not equal *exactly* to π. We must remember that $\frac{22}{7}$ is rational and π is irrational.

The decimal number 1.5 is rational since it can be written as $\frac{3}{2}$. Any such *terminating decimal* is rational. The number $0.6666\ldots$, where the 6's continue on indefinitely, is rational since we may write it as $\frac{2}{3}$. In fact, any *repeating decimal* (in decimal form, a specific sequence of digits is repeated indefinitely) is rational. The decimal number $0.673\,273\,273\,2\ldots$ is a repeating decimal where the sequence of digits 732 is repeated indefinitely $\left(0.673\,273\,273\,2\ldots = \frac{1121}{1665}\right)$. ▶

The Real Number System

The integers, the rational numbers, and the irrational numbers, including all such numbers that are positive, negative, or zero, make up the **real number system** *(see Fig. 1.1).* We will use real numbers in this text, with one important exception. In Chapter 12, we will also use **imaginary numbers,** *the name given to square roots of negative numbers.* (The symbol j is used for $\sqrt{-1}$, which is *not* a real number.) However, until Chapter 12, it will be necessary to only *recognize* imaginary numbers when they occur.

Real Numbers

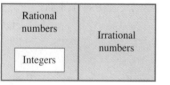

Fig 1.1

◀ **EXAMPLE 2** The number 7 is an integer. It is also rational since $7 = \frac{7}{1}$, and it is a real number since the real numbers include all the rational numbers.

The number 3π is irrational, and it is real since the real numbers include all the irrational numbers.

The numbers $\sqrt{-10}$ and $-\sqrt{-7}$ are imaginary numbers.

The number $\frac{-3}{7}$ is rational and real. The number $-\sqrt{7}$ is irrational and real.

The number $\frac{\pi}{6}$ is irrational and real. The number $\frac{\sqrt{-3}}{2}$ is imaginary. ▶

Fractions were used by early Egyptians and Babylonians. They were used for calculations that involved parts of measurements, property, and possessions.

A **fraction** *may contain any number or symbol representing a number in its numerator or in its denominator.* Therefore, a fraction may be a number that is rational, irrational, or imaginary.

◀ EXAMPLE 3   The numbers $\frac{2}{7}$ and $\frac{-3}{2}$ are fractions, and they are rational.

The numbers $\frac{\sqrt{2}}{9}$ and $\frac{6}{\pi}$ are fractions, but they are not rational numbers. It is not possible to express either as one integer divided by another integer.

The number $\frac{\sqrt{-5}}{6}$ is a fraction, and it is an imaginary number.   ▶

The Number Line

Real numbers may be represented by points on a line. We draw a horizontal line and designate some point on it by O, which we call the **origin** (see Fig. 1.2). The integer *zero* is located at this point. Equal intervals are marked to the right of the origin, and the positive integers are placed at these positions. The other positive rational numbers are located between the integers. The points that cannot be defined as rational numbers represent irrational numbers. We cannot tell whether a given point represents a rational number or an irrational number unless it is specifically marked to indicate its value.

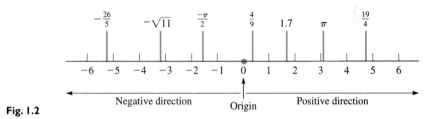

Fig. 1.2

The negative numbers are located on the number line by starting at the origin and marking off intervals *to the left, which is the* **negative direction.** As shown in Fig. 1.2, *the positive numbers are to the right of the origin and the negative numbers are to the left of the origin.* Representing numbers in this way is especially useful for graphical methods.

We next define another important concept of a number. *The* **absolute value** *of a positive number is the number itself, and the absolute value of a negative number is the corresponding positive number.* On the number line, we may interpret the absolute value of a number as the distance (which is always positive) between the origin and the number. Absolute value is denoted by writing the number between vertical lines, as shown in the following example.

◀ EXAMPLE 4   The absolute value of 6 is 6, and the absolute value of −7 is 7. We write these as $|6| = 6$ and $|-7| = 7$. See Fig. 1.3.

Fig. 1.3

Other examples are $\left|\frac{7}{5}\right| = \frac{7}{5}$, $\left|-\sqrt{2}\right| = \sqrt{2}$, $|0| = 0$, $-|\pi| = -\pi$, $|-5.29| = 5.29$, $-|-9| = -9$ since $|-9| = 9$.   ▶

On the number line, *if a first number is to the right of a second number, then the first number is said to be* **greater than** *the second. If the first number is to the left of the second, it is* **less than** *the second number.* The symbol $>$ designates "is greater than," and the symbol $<$ designates "is less than." These are called **signs of inequality.** See Fig. 1.4.

The symbols $=$, $<$, and $>$ were introduced by English mathematicians in the late 1500s.

◖EXAMPLE 5

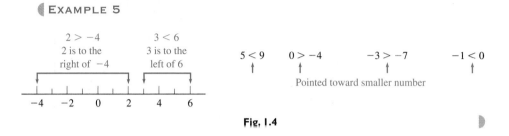

Fig. 1.4

Reciprocal

Every number, except zero, has a **reciprocal.** *The reciprocal of a number is 1 divided by the number.*

◖EXAMPLE 6 The reciprocal of 7 is $\frac{1}{7}$. The reciprocal of $\frac{2}{3}$ is

$$\frac{1}{\frac{2}{3}} = 1 \times \frac{3}{2} = \frac{3}{2} \qquad \text{invert denominator and multiply (from arithmetic)}$$

The reciprocal of 0.5 is $\frac{1}{0.5} = 2$. The reciprocal of $-\pi$ is $-\frac{1}{\pi}$. Note that the negative sign is retained in the reciprocal of a negative number.

For reference, see Appendix B for units of measurement and the symbols used for them.

In applications, *numbers that represent a measurement and are written with units of measurement are called* **denominate numbers.** The next example illustrates the use of units and the symbols that represent them.

◖EXAMPLE 7 To show that a person has a mass of 70 kilograms, we write the person's mass as 70 kg.

To show that the Empire State Building is 381 meters high, we write the height as 381 m.

To show that the speed of a rocket is 1500 meters per second, we write the speed as 1500 m/s. (Note the use of s for second. We use s rather than sec.)

To show that the area of a computer chip is 0.75 square centimeters, we write the area as 0.75 cm^2. (We will not use sq cm.)

To show that the volume of water in a glass tube is 25 cubic centimeters, we write the volume as 25 cm^3. (We will not use cu cm nor cc.)

Literal Numbers

It is usually more convenient to state definitions and operations on numbers in a general form. *To do this, we represent the numbers by letters, called* **literal numbers.** For example, if we want to say "If a first number is to the right of a second number on the number line, then the first number is greater than the second number," we can write "If a is to the right of b on the number line, then $a > b$." Another example of using a literal number is "The reciprocal of n is $1/n$."

Certain literal numbers may take on any allowable value, whereas other literal numbers represent the same value throughout the discussion. *Those literal numbers that may vary in a given problem are called* **variables,** *and those literal numbers that are held fixed are called* **constants.**

◀ EXAMPLE 8 **(a)** The resistance of an electric resistor is R. The current I in the resistor equals the voltage V divided by R, written as $I = V/R$. For this resistor, I and V may take on various values, and R is fixed. This means I and V are variables and R is a constant. For a *different* resistor, the value of R may differ.

(b) The fixed cost for a calculator manufacturer to operate a certain plant is b dollars per day, and it costs a dollars to produce each calculator. The total daily cost C to produce n calculators is

$$C = an + b$$

Here, C and n are variables, and a and b are constants. For *another* plant, the values of a and b would probably differ.

EXERCISES 1.1

In Exercises 1–4, make the given changes in the indicated examples of this section, and then answer the given questions.

1. In the first line of Example 1, change the 5 to −3 and the −19 to 14. What other changes must then be made in the first paragraph?

2. In Example 4, change the 6 to −6. What other changes must then be made in the first paragraph?

3. In the left figure of Example 5, change the 2 to −6. What other changes must then be made?

4. In Example 6, change the $\frac{2}{3}$ to $\frac{3}{2}$. What other changes must then be made?

In Exercises 5–8, designate each of the given numbers as being an integer, rational, irrational, real, or imaginary. (More than one designation may be correct.)

5. $3, -\pi$

6. $\dfrac{5}{4}, \sqrt{-4}$

7. $-\sqrt{-6}, \dfrac{\sqrt{7}}{3}$

8. $-2.33, -\dfrac{\pi}{6}$

In Exercises 9–12, find the absolute value of each number.

9. $3, \dfrac{7}{2}$

10. $-4, \sqrt{2}$

11. $-0.857, -\sqrt{3}$

12. $-\dfrac{\pi}{2}, -\dfrac{19}{4}$

In Exercises 13–20, insert the correct sign of inequality (> or <) between the given numbers.

13. 6   8

14. 7   5

15. π −3.2

16. −4   0

17. −4   −|−3|

18. $-\sqrt{2}$ −1.42

19. $-\dfrac{1}{3}$ $-\dfrac{1}{2}$

20. −0.6   0.2

In Exercises 21–24, find the reciprocal of each number.

21. $3, -\dfrac{1}{3}$

22. $0.25, -\dfrac{4}{\sqrt{3}}$

23. $-\dfrac{5}{\pi}, x$

24. $-\dfrac{8}{3}, \dfrac{y}{b}$

In Exercises 25–28, locate each number on a number line as in Fig. 1.2.

25. $2.5, -\dfrac{1}{2}$

26. $\sqrt{3}, -\dfrac{12}{5}$

27. $-\dfrac{\sqrt{2}}{2}, 2\pi$

28. $\dfrac{123}{19}, -\dfrac{\pi}{6}$

In Exercises 29–52, solve the given problems. Refer to Appendix B for units of measurement and their symbols.

(W) 29. Is an absolute value always positive? Explain.

(W) 30. Is 2.17 rational? Explain.

31. What is the reciprocal of the reciprocal of any positive or negative number?

32. Find a rational number between −0.9 and −1.0 that can be written with a denominator of 11 and an integer in the numerator.

33. Find a rational number between 0.13 and 0.14 that can be written with a numerator of 3 and an integer in the denominator.

34. If $b > a$ and $a > 0$, is $|b - a| < |b| - |a|$?

35. List the following numbers in numerical order, starting with the smallest: $-1, 9, \pi, \sqrt{5}, |-8|, -|-3|, -3.1$.

36. List the following numbers in numerical order, starting with the smallest: $\frac{1}{5}, -\sqrt{10}, -|-6|, -4, 0.25, |-\pi|$.

37. If a and b are positive integers and $b > a$, what type of number is represented by the following?

   (a) $b - a$ **(b)** $a - b$ **(c)** $\dfrac{b - a}{b + a}$

38. If a and b represent positive integers, what kind of number is represented by (a) $a + b$, (b) a/b, and (c) $a \times b$?

39. For any positive or negative integer: (a) Is its absolute value always an integer? (b) Is its reciprocal always a rational number?

40. For any positive or negative rational number: (a) Is its absolute value always a rational number? (b) Is its reciprocal always a rational number?

41. Describe the location of a number x on the number line when (a) $x > 0$ and (b) $x < -4$.

42. Describe the location of a number x on the number line when (a) $|x| < 1$ and (b) $|x| > 2$.

43. For a number $x > 1$, describe the location on the number line of the reciprocal of x.

44. For a number $x < 0$, describe the location on the number line of the number with a value of $|x|$.

45. The heat loss L through a certain type of insulation of thickness t is given by $L = a/t$, where a has a fixed value for this type of insulation. Identify the variables and constants.

46. A sensitive gauge measures the total mass m of a container and the water that forms in it as vapor condenses. It is found that $m = c\sqrt{0.1t + 1}$, where c is the mass of the container and t is the time of condensation. Identify the variables and constants.

47. In an electric circuit, the reciprocal of the total capacitance of two capacitors in series is the sum of the reciprocals of the capacitances. Find the total capacitance of two capacitances of 0.0040 F and 0.0010 F connected in series.

48. Alternating-current (ac) voltages change rapidly between positive and negative values. If a voltage of 100 V decreases to -200 V, (a) which of these voltages is greater, and (b) which is greater in absolute value?

49. The memory of a certain computer has a bits in each byte. Express the number N of bits in n kilobytes in an equation. (A *bit* is a single digit, and bits are grouped in *bytes* in order to represent special characters. Generally, there are 8 bits per byte. If necessary, see Appendix B for the meaning of *kilo*.)

50. A piece y cm long is cut from a board x m long. Give an equation for the length L, in centimeters, of the remaining piece.

51. In a laboratory report, a student wrote "$-20°C > -30°C$." Is this statement correct? Explain.

52. After 5 s, the pressure on a valve is less than 600 kPa. Using t to represent time and p to represent pressure, this statement can be written "for $t > 5$ s, $p < 600$ kPa." In this way, write the statement "when the current I in a circuit is less than 4 A, the voltage V is greater than 12 V."

1.2 FUNDAMENTAL OPERATIONS OF ALGEBRA

The Commutative and Associative Laws

If two numbers are added, it does not matter in which order they are added. (For example, $5 + 3 = 8$ and $3 + 5 = 8$, or $5 + 3 = 3 + 5$.) This statement, generalized and accepted as being correct for all possible combinations of numbers being added, is called the **commutative law** for addition. It states that *the sum of two numbers is the same, regardless of the order in which they are added.* We make no attempt to prove this law in general, but accept that it is true.

In the same way, we have the **associative law** for addition, which states that *the sum of three or more numbers is the same, regardless of the way in which they are grouped for addition.* For example, $3 + (5 + 6) = (3 + 5) + 6$.

The laws just stated for addition are also true for multiplication. Therefore, *the product of two numbers is the same, regardless of the order in which they are multiplied*, and *the product of three or more numbers is the same, regardless of the way in which they are grouped for multiplication.* For example, $2 \times 5 = 5 \times 2$, and $5 \times (4 \times 2) = (5 \times 4) \times 2$.

The Distributive Law

Another very important law is the **distributive law.** It states that *the product of one number and the sum of two or more other numbers is equal to the sum of the products of the first number and each of the other numbers of the sum.* For example,

Note carefully the difference:
associative law: $5 \times (4 \times 2)$
distributive law: $5 \times (4 + 2)$

$$5(4 + 2) = 5 \times 4 + 5 \times 2$$

In this case, it can be seen that the total is 30 on each side.

In practice, these **fundamental laws of algebra** are used naturally without thinking about them, except perhaps for the distributive law.

Not all operations are commutative and associative. For example, division is not commutative, since the order of division of two numbers does matter. For instance, $\frac{6}{5} \neq \frac{5}{6}$ ($\neq$ is read "does not equal)". (Also, see Exercise 50.)

Using literal numbers, the fundamental laws of algebra are as follows:

Commutative law of addition: $a + b = b + a$

Associative law of addition: $a + (b + c) = (a + b) + c$

Commutative law of multiplication: $ab = ba$

Associative law of multiplication: $a(bc) = (ab)c$

Distributive law: $a(b + c) = ab + ac$

Each of these laws is an example of an *identity*, in that the expression to the left of the = sign equals the expression to the right for any value of each of a, b, and c.

OPERATIONS ON POSITIVE AND NEGATIVE NUMBERS

When using the basic operations (addition, subtraction, multiplication, division) on positive and negative numbers, we determine the result to be either positive or negative according to the following rules.

Addition of two numbers of the same sign *Add their absolute values and assign the sum their common sign.*

> From Section 1.1, we recall that a positive number is preceded by no sign. Therefore, in using these rules we show the "sign" of a positive number by simply writing the number itself.

EXAMPLE 1   **(a)** $2 + 6 = 8$ the sum of two positive numbers is positive

(b) $-2 + (-6) = -(2 + 6) = -8$ the sum of two negative numbers is negative

The negative number -6 is placed in parentheses since it is also preceded by a plus sign showing addition. It is not necessary to place the -2 in parentheses.

Addition of two numbers of different signs *Subtract the number of smaller absolute value from the number of larger absolute value and assign to the result the sign of the number of larger absolute value.*

EXAMPLE 2

(a) $2 + (-6) = -(6 - 2) = -4$ ← the negative 6 has the larger absolute value

(b) $-6 + 2 = -(6 - 2) = -4$ ←

(c) $6 + (-2) = 6 - 2 = 4$ ← the positive 6 has the larger absolute value

(d) $-2 + 6 = 6 - 2 = 4$ ←

the subtraction of absolute values

Subtraction of one number from another *Change the sign of the number being subtracted and change the subtraction to addition. Perform the addition.*

EXAMPLE 3   **(a)** $2 - 6 = 2 + (-6) = -(6 - 2) = -4$

Note that after changing the subtraction to addition, and changing the sign of 6 to make it -6, we have precisely the same illustration as Example 2(a).

(b) $-2 - 6 = -2 + (-6) = -(2 + 6) = -8$

Note that after changing the subtraction to addition, and changing the sign of 6 to make it -6, we have precisely the same illustration as Example 1(b).

(c) $-a - (-a) = -a + a = 0$

Subtraction of a Negative Number

This shows that subtracting a number from itself results in zero, even if the number is negative. Therefore, *subtracting a negative number is equivalent to adding a positive number of the same absolute value.*

Multiplication and division of two numbers *The product (or quotient) of two numbers of the same sign is positive. The product (or quotient) of two numbers of different signs is negative.*

▌EXAMPLE 4

(a) $3(12) = 3 \times 12 = 36$ $\qquad \dfrac{12}{3} = 4$ result is positive if both numbers are positive

(b) $-3(-12) = 3 \times 12 = 36$ $\qquad \dfrac{-12}{-3} = 4$ result is positive if both numbers are negative

(c) $3(-12) = -(3 \times 12) = -36$ $\qquad \dfrac{-12}{3} = -\dfrac{12}{3} = -4$ result is negative if one number is positive and the other is negative

(d) $-3(12) = -(3 \times 12) = -36$ $\qquad \dfrac{12}{-3} = -\dfrac{12}{3} = -4$ ▐

ORDER OF OPERATIONS

Often, how we are to combine numbers is clear by grouping the numbers using symbols such as **parentheses,** ( ), and the **bar,** _____ , between the numerator and denominator of a fraction. Otherwise, for an expression in which there are several operations, we use the following order of operations.

ORDER OF OPERATIONS

1. *Operations within specific groupings are done first.*
2. *Perform multiplications and divisions (from left to right).*
3. *Then perform additions and subtractions (from left to right).*

Note that $20 \div (2 + 3) = \frac{20}{2 + 3}$, whereas $20 \div 2 + 3 = \frac{20}{2} + 3$.

▌EXAMPLE 5 **(a)** $20 \div (2 + 3)$ is evaluated by first adding $2 + 3$ and then dividing. The grouping of $2 + 3$ is clearly shown by the parentheses. Therefore, we have $20 \div (2 + 3) = 20 \div 5 = 4$.

(b) $20 \div 2 + 3$ is evaluated by first dividing 20 by 2 and then adding. No specific grouping is shown, and therefore the division is done before the addition. This means $20 \div 2 + 3 = 10 + 3 = 13$.

CAUTION ▶ **(c)** $16 - 2 \times 3$ is evaluated by *first multiplying* 2 *by* 3 and then subtracting. We *do not* *first subtract* 2 *from* 16. Therefore, $16 - 2 \times 3 = 16 - 6 = 10$.

(d) $16 \div 2 \times 4$ is evaluated by first dividing 16 by 2 and then multiplying. From left to right, the division occurs first. Therefore, $16 \div 2 \times 4 = 8 \times 4 = 32$.

(e) $16 \div (2 \times 4)$ is evaluated by first multiplying 2 by 4 and then dividing. The grouping of 2×4 is clearly shown by parentheses. Therefore, this means that $16 \div (2 \times 4) = 16 \div 8 = 2$. ▐

When evaluating expressions, it is generally more convenient to change the operations and numbers so that the result is found by the addition and subtraction of positive numbers. When this is done, we must remember that

$$a + (-b) = a - b \qquad (1.1)$$
$$a - (-b) = a + b \qquad (1.2)$$

◀ EXAMPLE 6   **(a)** $7 + (-3) - 6 = 7 - 3 - 6 = 4 - 6 = -2$ using Eq. (1.1)

(b) $\dfrac{18}{-6} + 5 - (-2) = -3 + 5 + 2 = 2 + 2 = 4$ using Eq. (1.2)

(c) $2(-3) - 2(-4) + \dfrac{25}{-5} = -6 - (-8) + (-5) = -6 + 8 - 5 = -3$

(d) $\dfrac{-12}{2 - 8} + \dfrac{5 - 1}{2(-1)} = \dfrac{-12}{-6} + \dfrac{4}{-2} = 2 + (-2) = 2 - 2 = 0$

In illustrations (b) and (c), we see that the multiplications and divisions were done before the additions and subtractions. In (d) we see that the groupings $(2 - 8$ and $5 - 1)$ were evaluated first. Then we did the divisions and, finally, the addition. ▶

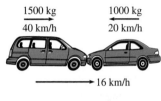

1500 kg → 40 km/h

1000 kg ← 20 km/h

→ 16 km/h

Fig 1.5

◀ EXAMPLE 7   A 1500-kg van going at 40 km/h ran head-on into a 1000-kg car going at 20 km/h. An insurance investigator determined the velocity of the vehicles immediately after the collision from the following calculation. See Fig. 1.5.

$$\frac{1500(40) + (1000)(-20)}{1500 + 1000} = \frac{60\,000 + (-20\,000)}{1500 + 1000} = \frac{60\,000 - 20\,000}{2500}$$

$$= \frac{40\,000}{2500} = 16 \text{ km/h}$$

The numerator and the denominator must be evaluated before the division is performed. The multiplications in the numerator are performed first, followed by the addition in the denominator and the subtraction in the numerator. ▶

OPERATIONS WITH ZERO

Since operations with zero tend to cause some difficulty, we will show them here.

If a is a real number, the operations of addition, subtraction, multiplication, and division with zero are as follows:

$$a + 0 = a$$

$$a - 0 = a \qquad 0 - a = -a$$

$$a \times 0 = 0$$

$$0 \div a = \frac{0}{a} = 0 \qquad (\textbf{if } a \neq 0) \qquad (\neq \text{ means "is not equal to"})$$

◀ EXAMPLE 8   **(a)** $5 + 0 = 5$ **(b)** $-6 - 0 = -6$ **(c)** $0 - 4 = -4$

(d) $\dfrac{0}{6} = 0$ **(e)** $\dfrac{0}{-3} = 0$ **(f)** $\dfrac{5 \times 0}{7} = \dfrac{0}{7} = 0$ ▶

Note that there is no result defined for division by zero. To understand the reason for this, consider the results for $\frac{6}{2}$ and $\frac{6}{0}$.

$$\frac{6}{2} = 3 \quad \text{since} \quad 2 \times 3 = 6$$

If $\frac{6}{0} = b$, then $0 \times b = 6$. This cannot be true because $0 \times b = 0$ for any value of b. Thus,

NOTE ▶   ***division by zero is undefined***

(The special case of $\frac{0}{0}$ is termed *indeterminate*. If $\frac{0}{0} = b$, then $0 = 0 \times b$, which is true for any value of b. Therefore, no specific value of b can be determined.)

◀ EXAMPLE 9

see bottom of page 9 ↓

$\dfrac{2}{5} \div 0$ is undefined $\qquad$ $\dfrac{8}{0}$ is undefined $\qquad$ $\dfrac{7 \times 0}{0 \times 6}$ is indeterminate ▶

The operations with zero will not cause any difficulty if we remember to

CAUTION ▶ *never divide by zero*

Division by zero is the only undefined basic operation. All the other operations with zero may be performed as for any other number.

EXERCISES 1.2

In Exercises 1–4, make the given changes in the indicated examples of this section, and then solve the resulting problems.

1. In Example 5(c), change 3 to (-3) and then evaluate.

2. In Example 6(b), change 18 to -18 and then evaluate.

3. In Example 6(d), interchange the 2 and 8 in the first denominator and then evaluate.

4. In the rightmost illustration in Example 9, interchange the 6 and the 0 above the 6. Is any other change needed?

In Exercises 5–36, evaluate each of the given expressions by performing the indicated operations.

5. $8 + (-4)$ **6.** $-4 + (-7)$ **7.** $-3 + 9$

8. $18 - 21$ **9.** $-19 - (-16)$ **10.** $8 - (-4)$

11. $8(-3)$ **12.** $-9(3)$ **13.** $-7(-5)$

14. $\dfrac{-9}{3}$ **15.** $\dfrac{-6(20 - 10)}{-3}$ **16.** $\dfrac{28}{-7(6 - 5)}$

17. $-2(4)(-5)$ **18.** $3(-4)(6)$ **19.** $2(2 - 7) \div 10$

20. $\dfrac{-64}{-2(8 - 4)}$ **21.** $9 - 0$ **22.** $(7 - 7) \div (5 - 7)$

23. $\dfrac{17 - 7}{7 - 7}$ **24.** $\dfrac{7 - 7}{7 - 7}$

25. $8 - 3(-4)$ **26.** $20 + 8 \div 4$

27. $3 - 2(6) + \left|\dfrac{8}{2}\right|$ **28.** $0 - (-6)(-8) + (-10)$

29. $30(-6)(-2) \div (0 - 40)$ **30.** $\dfrac{7 - |-5|}{-1(-2)}$

31. $\dfrac{24}{3 + (-5)} - 4(-9)$ **32.** $\dfrac{-18}{3} - \dfrac{4 - 6}{-1}$

33. $-7 - \dfrac{|-14|}{2(2 - 3)} - 3(8 - 6)$ **34.** $-7(-3) + \dfrac{6}{-3} - (-9)$

35. $\dfrac{3(-9) - 2(-3)}{3 - 10}$ **36.** $\dfrac{20(-12) - 40(-15)}{98 - |-98|}$

In Exercises 37–44, determine which of the fundamental laws of algebra is demonstrated.

37. $6(7) = 7(6)$ **38.** $6 + 8 = 8 + 6$

39. $6(3 + 1) = 6(3) + 6(1)$ **40.** $4(5 \times \pi) = (4 \times 5)(\pi)$

41. $3 + (5 + 9) = (3 + 5) + 9$

42. $8(3 - 2) = 8(3) - 8(2)$

43. $(\sqrt{5} \times 3) \times 9 = \sqrt{5} \times (3 \times 9)$

44. $(3 \times 6) \times 7 = 7 \times (3 \times 6)$

In Exercises 45–48, for numbers a and b, determine which of the following expressions equals the given expression.
(a) $a + b$ (b) $a - b$ (c) $b - a$ (d) $-a - b$

45. $-a + (-b)$ **46.** $b - (-a)$

47. $-b - (-a)$ **48.** $-a - (-b)$

In Exercises 49–60, answer the given questions. Refer to Appendix B for units of measurement and their symbols.

49. (a) What is the sign of the product of an even number of negative numbers? (b) What is the sign of the product of an odd number of negative numbers?

Ⓦ **50.** Is subtraction commutative? Explain.

Ⓦ **51.** Describe the values of x and y for which (a) $-xy = 1$ and (b) $\frac{x - y}{x - y} = 1$.

Ⓦ **52.** Describe the values of x and y for which (a) $|x + y| = |x| + |y|$ and (b) $|x - y| = |x| + |y|$.

53. Some solar energy systems are used to supplement the utility power company power supplied to a home such that the meter runs backward if the solar energy being generated is greater than the energy being used. With such a system, if the solar power averages 1.5 kW for a 3.0-h period and only 2.1 kW · h is used during this period, what will be the change in the meter reading for this period?

54. A baseball player's batting average (total number of hits divided by total number of at-bats) is expressed in decimal form from 0.000 (no hits for all at-bats) to 1.000 (one hit for each at-bat). A player's batting average is often shown as 0.000 before the first at-bat of the season. Is this a correct batting average? Explain.

55. The daily high temperatures (in °C) in the Falkland Islands in the southern Atlantic Ocean during the first week in July were recorded as 7, 3, −2, −3, −1, 4, and 6. What was the average daily temperature for the week? (Divide the algebraic sum of the readings by the number of readings.)

56. The electric current was measured in a given ac circuit at equal intervals as 0.7 mA, −0.2 mA, −0.9 mA, and −0.6 mA. What was the change in the current between (a) the first two readings, (b) the middle two readings, and (c) the last two readings?

57. One oil-well drilling rig drills 100 m deep the first day and 200 m deeper the second day. A second rig drills 200 m deep the first day and 100 m deeper the second day. In showing that the total depth drilled by each rig was the same, state what fundamental law of algebra is illustrated.

58. An electronics dealer averaged 18 min each when selling 5 computers, and 5 min each when selling 18 cellular phones. In showing that the total time in selling the computers equals the total time in selling the phones, what fundamental law of algebra is illustrated?

59. Each of three delivery trucks is loaded with 50 cases of regular soda and 40 cases of diet soda. Set up the expression for the total number of cases on all three trucks. What fundamental law of algebra is illustrated?

60. A jet travels 600 km/h relative to the air. The wind is blowing at 50 km/h. If the jet travels with the wind for 3 h, set up the expression for the distance traveled. What fundamental law of algebra is illustrated?

1.3 CALCULATORS AND APPROXIMATE NUMBERS

You will be doing many of your calculations on a calculator, and a *graphing calculator* can be used for these calculations and many other operations. In this text, we will restrict our coverage of calculator use to graphing calculators because a *scientific calculator* cannot perform many of the required operations we will cover.

A brief discussion of the graphing calculator appears in Appendix C, and sample calculator screens appear throughout the book. Since there are many models of graphing calculators, *the notation and screen appearance for many operations will differ from one model to another.* You should *practice using your calculator and **review its manual** to be sure how it is used.* Following is an example of a basic calculation done on a graphing calculator.

All calculator screens shown with text material are for a TI-83. They are intended only as an illustration of a calculator screen for the particular operation. Screens for other models may differ.

◀ EXAMPLE 1   Calculate the value of 38.3 − 12.9(−3.58). The numbers are entered as follows. The calculator will perform the multiplication first, following the order of operations shown in Section 1.2. The sign of −3.58 is entered using the $(-)$ key, before 3.58 is entered. The display on the calculator screen is shown in Fig. 1.6.

38.3 $\boxed{-}$ 12.9 $\boxed{\times}$ $\boxed{(-)}$ 3.58 $\boxed{\text{ENTER}}$ keystrokes

This means that 38.3 − 12.9(−3.58) = 84.482.

Note in the display that the negative sign of −3.58 is smaller and a little higher to distinguish it from the minus sign for subtraction. Also note the * shown for multiplication; the asterisk is the standard computer symbol for multiplication.   ▶

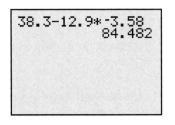

```
38.3-12.9*-3.58
            84.482
```

Fig 1.6

Some calculator keys on different models are labeled differently. For example, on some models, the EXE key is equivalent to the ENTER key.

Looking back into Section 1.2, we see that *the minus sign is used in two different ways:* (1) to indicate subtraction and (2) to designate a negative number. This is clearly shown on a graphing calculator because there is a key for each purpose. The $\boxed{-}$ key is used for subtraction, and the $\boxed{(-)}$ key is used before a number to make it negative.

Calculator keystrokes will generally not be shown, except as they appear in the display screens. They may vary from one model to another.

We will first use a graphing calculator for the purpose of graphing in Section 3.5. Before then we will show some calculational uses of a graphing calculator.

APPROXIMATE NUMBERS AND SIGNIFICANT DIGITS

The final result of a calculation should not be written with any more accuracy than is proper. For example, if the numbers in Example 1 are *approximate*, the result should be written as 84.5, not 84.482. We will see the reason for this later in this section.

Most numbers in technical and scientific work are **approximate numbers,** having been determined by some *measurement*. Certain other numbers are **exact numbers,** having been determined by a *definition* or a *counting* process.

◖ EXAMPLE 2   If a voltage shown on a voltmeter is read as 116 V, the 116 is approximate. Another voltmeter may show the voltage as 115.7 V. However, in reality voltage cannot be determined *exactly*.

If a computer prints out the number of names on a list of 97, this 97 is exact. We know it is not 96 or 98. Since 97 was found from precise counting, it is exact.

By definition, 60 s = 1 min, and the 60 and the 1 are exact.   ◗

Significant Digits

An approximate number may have to include some zeros to properly locate the decimal point. *Except for these zeros, all other digits are called* **significant digits.**

◖ EXAMPLE 3   All numbers in this example are assumed to be approximate.

34.7 has three significant digits.

0.039 has two significant digits. The zeros properly locate the decimal point.

706.1 has four significant digits. The zero is not used for the location of the decimal point. It shows the number of tens in 706.1.

CAUTION ▶

5.90 has three significant digits. *The zero is not necessary as a placeholder* and should not be written unless it is significant.

1400 has two significant digits, unless information is known about the number that makes either or both zeros significant. (A temperature shown as 1400°C has two significant digits. If a price list gives all costs in dollars, a price shown as $1400 has four significant digits.) Without such information, we assume that the zeros are placeholders for proper location of the decimal point.

To show that zeros at the end of a whole number are significant, a notation that can be used is to place a bar over the last significant zero. Using this notation, 78 0̄00 is shown to have four significant digits.

Other approximate numbers with the number of significant digits are 0.0005 (one), 960 000 (two), 0.0709 (three), 1.070 (four), and 700.00 (five).   ◗

From Example 3, we see that *all nonzero digits are significant. Also, zeros not used as placeholders (for location of the decimal point) are significant.*

Accuracy and Precision

In calculations with approximate numbers, the number of significant digits and the position of the decimal point are important. *The* **accuracy** *of a number refers to the number of significant digits it has*, whereas *the* **precision** *of a number refers to the decimal position of the last significant digit.*

◖ EXAMPLE 4   An electric current is measured as 0.31 A on one ammeter and as 0.312 A on another ammeter. Here, 0.312 is more precise since its last digit represents thousandths and 0.31 is expressed only to hundredths. Also, 0.312 is more accurate since it has three significant digits and 0.31 has only two.

A concrete driveway is 130 m long and 0.1 m thick. Here, 130 is more accurate (two significant digits) and 0.1 is more precise (expressed to tenths).   ◗

Rounding Off

The last significant digit of an approximate number is not exact. It has usually been determined by estimating or *rounding off*. However, it is not off by more than one-half of a unit in its place value.

◖ **EXAMPLE 5** When we write the voltage in Example 2 as 115.7 V, we are saying that the voltage is at least 115.65 V and no more than 115.75 V. Any value between these two, rounded off to tenths, would be expressed as 115.7 V.

In changing the fraction $\frac{2}{3}$ to the approximate decimal value 0.667, we are saying that the value is between 0.6665 and 0.6675. ◗

On graphing calculators, it is possible to set the number of decimal places (to the right of the decimal point) to which results will be rounded off.

*To **round off** a number to a specified number of significant digits, discard all digits to the right of the last significant digit (replace them with zeros if needed to properly place the decimal point). If the first digit discarded is 5 or more, increase the last significant digit by 1 (round up). If the first digit discarded is less than 5, do not change the last significant digit (round down).*

◖ **EXAMPLE 6** 70 360 rounded off to three significant digits is 70 400. Here, 3 is the third significant digit and the next digit is 6. Since 6 > 5, we add 1 to 3 and the result, 4, becomes the third significant digit of the approximation. The 6 is then replaced with a zero in order to keep the decimal point in the proper position.

70 430 rounded off to three significant digits, or to the nearest hundred, is 70 400. Here the 3 is replaced with a zero.

CAUTION ◗ 35.003 rounded off to four significant digits is 35.00. *We do not discard the zeros since they are significant* and are not used only to properly place the decimal point.

187.35 rounded off to four significant digits, or to tenths, is 187.4.

187.349 rounded off to four significant digits is 187.3. *We do not round up the 4 and then round up the 3.* ◗

OPERATIONS WITH APPROXIMATE NUMBERS

NOTE ◗ When performing operations on approximate numbers, *we must not express the result to an accuracy or precision that is not valid.* Consider the following examples.

◖ **EXAMPLE 7** A pipe is made in two sections. One is measured as 16.3 m long and the other as 0.927 m long. What is the total length of the two sections together?

It may appear that we simply add the numbers as shown. However, since both numbers are approximate, adding the smallest possible values and the largest possible values, the result differs by 0.1 (17.2 and 17.3) when rounded off to tenths. Rounded off to hundredths (17.18 and 17.28), they do not agree at all since the tenths digit is different. Thus we get a good approximation for the total length if it is rounded off to *tenths*, the precision of the least precise length, and is written as 17.2 m. ◗

```
 16.3   m
 0.927 m
 17.227 m
```

```
smallest values    largest values
 16.25   m          16.35   m
 0.9265 m           0.9275 m
 17.1765 m          17.2775 m
```

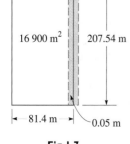

Fig 1.7

◖ **EXAMPLE 8** We find the area of the rectangular piece of land in Fig. 1.7 by multiplying the length, 207.54 m, by the width, 81.4 m. Using a calculator, we find that $(207.54)(81.4) = 16\,893.756$. This apparently means the area is 16 893.756 m^2.

However, the area should not be expressed with this accuracy. Since the length and width are both approximate, we have

$$(207.535 \text{ m})(81.35 \text{ m}) = 16\,882.972\,25 \text{ m}^2 \quad \text{least possible area}$$
$$(207.545 \text{ m})(81.45 \text{ m}) = 16\,904.540\,25 \text{ m}^2 \quad \text{greatest possible area}$$

These values agree when rounded off to three significant digits (16 900 m^2) but do not agree when rounded off to a greater accuracy. Thus, we conclude that the result is accurate only to *three* significant digits, the accuracy of the least accurate measurement, and that the area is written as 16 900 m^2. ◗

Following are the rules used in expressing the result when we perform basic operations on approximate numbers. They are based on reasoning similar to that shown in Examples 7 and 8.

> **OPERATIONS WITH APPROXIMATE NUMBERS**
> 1. *When approximate numbers are added or subtracted, the result is expressed with the precision of the least precise number.*
> 2. *When approximate numbers are multiplied or divided, the result is expressed with the accuracy of the least accurate number.*
> 3. *When the root of an approximate number is found, the result is expressed with the accuracy of the number.*
> 4. *When approximate numbers and exact numbers are involved, the accuracy of the result is limited only by the approximate numbers.*

Always express the result of a calculation with the proper accuracy or precision. **When using a calculator, round off the result if additional digits are displayed.**

CAUTION ▶

When using a calculator, round off only the final result.

EXAMPLE 9 Find the sum of the approximate numbers 73.2, 8.0627, and 93.57. Showing the addition in the standard way and using a calculator, we have

$$73.2 \longleftarrow \text{least precise number (expressed to tenths)}$$
$$8.0627$$
$$\underline{93.57}$$
$$174.8327 \longleftarrow \text{final display must be rounded to tenths}$$

Therefore, the sum of these approximate numbers is 174.8.

When rounding off a number, it may seem difficult to discard the extra digits. However, if you keep those digits, you show a number with too great an accuracy, and it is incorrect to do so.

EXAMPLE 10 In finding the product of the approximate numbers 2.4832 and 30.5 on a calculator, the final display shows 75.7376. However, since 30.5 has only three significant digits, the product is 75.7.

In Example 1, we calculated that $38.3 - 12.9(-3.58) = 84.482$. We know that $38.3 - 12.9(-3.58) = 38.3 + 46.182 = 84.482$. If these numbers are approximate, we must round off the result to tenths, which means the sum is 84.5. We see that *where there is a combination of operations, the final operation determines how the final result is to be rounded off.*

EXAMPLE 11 Using the exact number 600 and the approximate number 2.7, we express the result to tenths if the numbers are added or subtracted. If they are multiplied or divided, we express the result to two significant digits. Since 600 is exact, the accuracy of the result depends only on the approximate number 2.7.

$$600 + 2.7 = 602.7 \qquad 600 - 2.7 = 597.3$$
$$600 \times 2.7 = 1600 \qquad 600 \div 2.7 = 220$$

NOTE ▶

A note regarding the equal sign (=) is in order. We will use it for its defined meaning of "equals exactly" and when the result is an approximate number that has been properly rounded off. Although $\sqrt{27.8} \approx 5.27$, where $\approx$ means "equals approximately," we write $\sqrt{27.8} = 5.27$, since 5.27 has been properly rounded off.

Estimating Results You should *make a rough estimate* of the result when using a calculator. An estimation may prevent accepting an incorrect result after using an incorrect calculator sequence, particularly if the calculator result is far from the estimated value.

◀ EXAMPLE 12 In Example 1 we found that

$$38.3 - 12.9(-3.58) = 84.482 \qquad \text{using exact numbers}$$

When using the calculator, if we forgot to make 3.58 negative, the display would be -7.882, or if we incorrectly entered 38.3 as 83.3, the display would be 129.482.

However, if we estimate the result as

$$40 - 10(-4) = 80$$

we know that a result of -7.882 or 129.482 cannot be correct.

When estimating, we can often use one-significant-digit approximations. If the calculator result is far from the estimate, we should do the calculation again. ▶

EXERCISES 1.3

In Exercises 1–4, make the given changes in the indicated examples of this section, and then solve the given problems.

1. In Example 3, change 0.039 (the second number discussed) to 0.390. Is there any change in the conclusion?

2. In the third paragraph of Example 6, change 35.003 to 35.303 and then find the result.

3. In the first paragraph of Example 10, change 2.4832 to 2.483 and then find the result.

4. In Example 12, change 12.9 to 21.9 and then find the estimated value.

In Exercises 5–8, determine whether the given numbers are approximate or exact.

5. A car with 8 cylinders travels at 90 km/h.

6. A computer chip 0.002 mm thick is priced at $7.50.

7. In 2 h there are 7200 s.

8. A calculator has 50 keys, and its battery lasted for 50 h.

In Exercises 9–14, determine the number of significant digits in each of the given approximate numbers.

9. 107; 3004   **10.** 3600; 730   **11.** 6.80; 6.08

12. 0.8735; 0.0075   **13.** 3000; 3000.1   **14.** 1.00; 0.01

In Exercises 15–20, determine which of the pair of approximate numbers is (a) more precise and (b) more accurate.

15. 30.8; 0.01   **16.** 0.041; 7.673   **17.** 0.1; 78.0

18. 7040; 0.004   **19.** 7000; 0.004   **20.** 50.060; 8.914

In Exercises 21–28, round off the given approximate numbers (a) to three significant digits and (b) to two significant digits.

21. 4.936   **22.** 80.53   **23.** 50 893   **24.** 31 490

25. 9549   **26.** 30.96   **27.** 0.9449   **28.** 0.9999

In Exercises 29–40, assume that all numbers are approximate. (a) Estimate the result and (b) perform the indicated operations on a calculator and compare with the estimate.

29. $3.8 + 0.154 + 47.26$ **30.** $12.78 + 1.0495 - 1.633$

31. $3.64(17.06)$ **32.** $0.49 \div 827$

33. $0.0350 - \dfrac{0.0450}{1.909}$ **34.** $\dfrac{0.3275}{1.096 \times 0.500\,85}$

35. $\dfrac{0.26(-0.4095)}{50.75(0.937)}$ **36.** $\dfrac{326.0}{2.060(3894) - 4008}$

37. $\dfrac{23.962 \times 0.015\,37}{10.965 - 8.249}$ **38.** $\dfrac{0.693\,78 + 0.049\,97}{257.4 \times 3.216}$

39. $\dfrac{3872}{503.1} - \dfrac{2.056 \times 309.6}{395.2}$ **40.** $\dfrac{1}{0.5926} + \dfrac{3.6957}{2.935 - 1.054}$

In Exercises 41–44, perform the indicated operations. The first number is approximate, and the second number is exact.

41. $0.9788 + 14.9$ **42.** $17.311 - 22.98$

43. $3.142(65)$ **44.** $8.62 \div 1728$

In Exercises 45–48, answer the given questions. Refer to Appendix B for units of measurement and their symbols.

45. The manual for a heart monitor lists the frequency of the ultrasound wave as 2.75 MHz. What are the least possible and the greatest possible frequencies?

46. A car manufacturer states that the engine displacement for a certain model is 2400 cm³. What should be the least possible and greatest possible displacements?

Ⓦ**47.** A flash of lightning struck a tower 5.23 km from a person. The thunder was heard 15 s later. The person calculated the speed of sound and reported it as 348.7 m/s. What is wrong with this conclusion?

(W) **48.** A student reports the electric current in a certain experiment as 0.02 A and later notes that the current is 0.023 A. The student states the change in current is 0.003 A. What is wrong with this conclusion?

In Exercises 49–60, perform the indicated calculations on a calculator.

49. Calculate: (a) $2.2 + 3.8 \times 4.5$ (b) $(2.2 + 3.8) \times 4.5$
(Note the use of parentheses for grouping.)

50. Calculate: (a) $6.03 \div 2.25 + 1.77$ (b) $6.03 \div (2.25 + 1.77)$
(Note the use of parentheses for grouping.)

51. (a) Show that π is not exactly equal to 3.1416.
(b) Show that π is not exactly equal to 22/7.

(W) **52.** (a) Note the calculator displays for $2 \div 0.0001$ and $2 \div 0$.
(b) Note the calculator displays for $0.0001 \div 0.0001$ and $0 \div 0$.
(c) In parts (a) and (b), explain why the displays differ.

53. At some point in the decimal equivalent of a rational number, some sequence of digits will start repeating endlessly. An irrational number never has an endlessly repeating sequence of digits. Find the decimal equivalents of (a) 8/33 and (b) π. Note the repetition for 8/33 and that no such repetition occurs for π.

(W) **54.** Following Exercise 53, show that the decimal equivalent of the fraction 124/990 indicates that it is rational. Why is the last digit different?

55. In 3 successive days, a home solar system produced 32.4 MJ, 26.704 MJ, and 36.23 MJ of energy. What was the total energy produced in these 3 days?

56. Two jets flew at 938 km/h and 1450 km/h, respectively. How much faster was the second jet?

57. If 1 K of computer memory has 1024 bytes, how many bytes are there in 256 K of memory? (All numbers are exact.)

58. The power (in W) developed in an electric circuit is the product of the current (in A) and the voltage. What is the power developed in a circuit in which the current is 0.0125 A and the voltage is 12.68 V?

59. The percent of alcohol in a certain car engine coolant is found by performing the calculation $\dfrac{100(40.63 + 52.96)}{105.30 + 52.96}$. Find this percent of alcohol. The number 100 is exact.

60. The tension (in N) in a pulley cable lifting a certain crate was found by calculating the value of $\dfrac{50.45(9.80)}{1 + \dfrac{100.9}{23}}$, where the 1 is exact. Calculate the tension.

1.4 EXPONENTS

In mathematics and its applications, we often have a number multiplied by itself several times. To show this type of product, we use the notation a^n, where a is the number and n is the number of times it appears. *In the expression a^n, the number a is called the* **base,** *and n is called the* **exponent;** in words a^n is read as "the **nth power of a.**"

◀ **EXAMPLE 1** **(a)** $4 \times 4 \times 4 \times 4 \times 4 = 4^5$ the fifth power of 4

(b) $(-2)(-2)(-2)(-2) = (-2)^4$ the fourth power of -2

(c) $a \times a = a^2$ the second power of a, called "a squared"

(d) $\left(\frac{1}{5}\right)\left(\frac{1}{5}\right)\left(\frac{1}{5}\right) = \left(\frac{1}{5}\right)^3$ the third power of $\frac{1}{5}$, called "$\frac{1}{5}$ cubed" ▶

We now state the basic operations with exponents using positive integers as exponents. Therefore, with m and n as positive exponents, we have the following operations.

> *Two forms are shown for Eqs. (1.4) in order that the resulting exponent is a positive integer. We consider negative and zero exponents after the next three examples.*

$$a^m \times a^n = a^{m+n} \tag{1.3}$$

$$\frac{a^m}{a^n} = a^{m-n} \quad (m > n, a \neq 0) \qquad \frac{a^m}{a^n} = \frac{1}{a^{n-m}} \quad (m < n, a \neq 0) \tag{1.4}$$

$$(a^m)^n = a^{mn} \tag{1.5}$$

$$(ab)^n = a^n b^n \qquad \left(\frac{a}{b}\right)^n = \frac{a^n}{b^n} \quad (b \neq 0) \tag{1.6}$$

◀ **EXAMPLE 2** Applying Eq. (1.3), we have

add exponents

$$a^3 \times a^5 = a^{3+5} = a^8$$

We see that this result is correct since we can also write

(3 factors of *a*)(5 factors of *a*) ——————— 8 factors of *a*

$$a^3 \times a^5 = (a \times a \times a)(a \times a \times a \times a \times a) = a^8$$

Applying the first form of Eqs. (1.4), we have

5 > 3

$$\frac{a^5}{a^3} = a^{5-3} = a^2, \qquad \frac{a^5}{a^3} = \frac{\overset{1}{\cancel{a}} \times \overset{1}{\cancel{a}} \times \overset{1}{\cancel{a}} \times a \times a}{\underset{1}{\cancel{a}} \times \underset{1}{\cancel{a}} \times \underset{1}{\cancel{a}}} = a^2$$

Applying the second form of Eqs. (1.4), we have

$$\frac{a^3}{a^5} = \frac{1}{a^{5-3}} = \frac{1}{a^2}, \qquad \frac{a^3}{a^5} = \frac{\overset{1}{\cancel{a}} \times \overset{1}{\cancel{a}} \times \overset{1}{\cancel{a}}}{\underset{1}{\cancel{a}} \times \underset{1}{\cancel{a}} \times \underset{1}{\cancel{a}} \times a \times a} = \frac{1}{a^2}$$

5 > 3

◀ **EXAMPLE 3** Applying Eq. (1.5), we have

multiply exponents

$$(a^5)^3 = a^{5(3)} = a^{15}, \qquad (a^5)^3 = (a^5)(a^5)(a^5) = a^{5+5+5} = a^{15}$$

Applying the first form of Eqs. (1.6), we have

$$(ab)^3 = a^3b^3, \qquad (ab)^3 = (ab)(ab)(ab) = a^3b^3$$

Applying the second form of Eqs. (1.6), we have

$$\left(\frac{a}{b}\right)^3 = \frac{a^3}{b^3}, \qquad \left(\frac{a}{b}\right)^3 = \left(\frac{a}{b}\right)\left(\frac{a}{b}\right)\left(\frac{a}{b}\right) = \frac{a^3}{b^3}$$

CAUTION ▶ When an expression involves a product or a quotient of different bases, ***only exponents of the same base may be combined.*** Consider the following example.

◀ **EXAMPLE 4** Other illustrations using Eqs. (1.3) to (1.6) are as follows:

(a) $(-x^2)^3 = [(-1)x^2]^3 = (-1)^3(x^2)^3 = -x^6$

exponent of 1 — add exponents of *a*

(b) $ax^2(ax)^3 = ax^2(a^3x^3) = a^4x^5$ ← add exponents of *x*

(c) $\dfrac{(3 \times 2)^4}{(3 \times 5)^3} = \dfrac{3^4 2^4}{3^3 5^3} = \dfrac{3 \times 2^4}{5^3}$ **(d)** $\dfrac{(ry^3)^2}{r(y^2)^4} = \dfrac{r^2 y^6}{r y^8} = \dfrac{r}{y^2}$

CAUTION ▶ In illustration (b), note that ***ax^2 means a times the square of x and does not mean a^2x^2,*** whereas $(ax)^3$ *does* mean a^3x^3.

◀ **EXAMPLE 5** In the analysis of the deflection of a beam (the amount it bends), the expression that follows is simplified as shown.

$$\frac{1}{2}\left(\frac{PL}{4EI}\right)\left(\frac{2}{3}\right)\left(\frac{L}{2}\right)^2 = \frac{1}{2}\left(\frac{PL}{4EI}\right)\left(\frac{2}{3}\right)\left(\frac{L^2}{2^2}\right)$$

$$= \frac{\overset{1}{\cancel{2}}PL(L^2)}{\underset{1}{\cancel{2}}(3)(4)(4)EI} = \frac{PL^3}{48EI}$$

L is the length of the beam, and P is the force applied to it. E and I are constants related to the beam. In *simplifying* this expression, we combined exponents of L and divided out the 2 that was in the numerator and in the denominator. ▶

ZERO AND NEGATIVE EXPONENTS

If we let $n = m$ in Eqs. (1.4), we would have $a^m/a^m = a^{m-m} = a^0$. Also, $a^m/a^m = 1$, since any nonzero quantity divided by itself equals 1. Therefore, for Eqs. (1.4) to hold, when $m = n$, we have

$$\boxed{a^0 = 1 \qquad (a \neq 0)} \tag{1.7}$$

Equation (1.7) states that *any nonzero expression raised to the zero power is* 1. Zero exponents can be used with any of the operations for exponents.

◀ **EXAMPLE 6** **(a)** $5^0 = 1$ **(b)** $(2x)^0 = 1$ **(c)** $(ax + b)^0 = 1$

(d) $(a^2b^0c)^2 = a^4b^0c^2 = a^4c^2$ **(e)** $2t^0 = 2(1) = 2$

$b^0 = 1$

CAUTION ▶ We note in illustration (e) that *only t is raised to the zero power*. If the quantity $2t$ were raised to the zero power, it would be written as $(2t)^0$. ▶

If we apply the first form of Eqs. (1.4) to the case where $n > m$, the resulting exponent is negative. This leads to the definition of a negative exponent.

◀ **EXAMPLE 7** Applying both forms of Eqs. (1.4) to a^2/a^7, we have

$$\frac{a^2}{a^7} = a^{2-7} = a^{-5} \quad \text{and} \quad \frac{a^2}{a^7} = \frac{1}{a^{7-2}} = \frac{1}{a^5}$$

If these results are to be consistent, then $a^{-5} = \dfrac{1}{a^5}$. ▶

Although positive exponents are generally preferred in a final result, there are some cases in which zero or negative exponents are to be used. Also, negative exponents are very useful in some operations that we will use later.

Following the reasoning of Example 7, if we define

$$\boxed{a^{-n} = \frac{1}{a^n} \qquad (a \neq 0)} \tag{1.8}$$

then all of the laws of exponents will hold for negative integers.

The use of exponents is taken up in more detail in Chapter 11.

◀ EXAMPLE 8   **(a)** $3^{-1} = \dfrac{1}{3}$ **(b)** $4^{-2} = \dfrac{1}{4^2} = \dfrac{1}{16}$ **(c)** $\dfrac{1}{a^{-3}} = a^3$ change signs of exponents

(d) $\left(\dfrac{a^3 t}{b^2 x}\right)^{-2} = \dfrac{(a^3 t)^{-2}}{(b^2 x)^{-2}} = \dfrac{(b^2 x)^2}{(a^3 t)^2} = \dfrac{b^4 x^2}{a^6 t^2}$ using Eqs. (1.6) and (1.5)

From part (a), where $3^{-1} = 1/3$, we see that the reciprocal of a number x (not 0) is x^{-1}.

ORDER OF OPERATIONS

In Section 1.2, we saw that it is necessary to follow a particular order of operations when performing the basic operations on numbers. Since raising a number to a power is a form of multiplication, this operation is performed before additions and subtractions. In fact, it is performed before multiplications and divisions.

> **ORDER OF OPERATIONS**
> 1. *Operations within specific groupings*
> 2. *Powers*
> 3. *Multiplications and divisions (from left to right)*
> 4. *Additions and subtractions (from left to right)*

◀ EXAMPLE 9 $8 - (-1)^2 - 2(-3)^2 = 8 - 1 - 2(9)$
$$= 8 - 1 - 18 = -11$$

Since there were no specific groupings, we first squared -1 and -3. Next we found the product $2(9)$ in the last term. Finally, the subtractions were performed. Note carefully that *we did not change the sign of* -1 *before we squared it.*

CAUTION ▶

EVALUATING ALGEBRAIC EXPRESSIONS

An algebraic expression is **evaluated** *by* **substituting** *given values of the literal numbers in the expression and calculating the result.* On a calculator, the $\boxed{x^2}$ key is used to square numbers, and the $\boxed{\wedge}$ or $\boxed{x^y}$ key is used for other powers.

On many calculators, there is a specific key or key sequence to evaluate x^3.

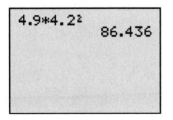

Fig 1.8

◀ EXAMPLE 10   The distance (in m) that an object falls in 4.2 s is found by substituting 4.2 for t in the expression $4.90 t^2$. We show this as

$$t = 4.2 \text{ s} \longleftarrow \text{ substituting}$$
$$4.90(4.2)^2 = 86 \text{ m} \qquad \text{estimation} \longrightarrow 5(4)^2 = 80$$

The result is rounded off to two significant digits (the accuracy of t). The calculator will square 4.2 before multiplying. See Fig. 1.8.

As stated in Section 1.3, in evaluating an expression on a calculator, we should also estimate its value as in Example 10. For an estimate, note that *a negative number raised to an even power gives a positive value and a negative number raised to an odd power gives a negative value.*

◀ EXAMPLE 11 Using the meaning of a power of a number, we have

$$(-2)^2 = (-2)(-2) = 4 \qquad (-2)^3 = (-2)(-2)(-2) = -8$$
$$(-2)^4 = 16 \qquad (-2)^5 = -32 \qquad (-2)^6 = 64 \qquad (-2)^7 = -128$$

◀ EXAMPLE 12 A wire made of a special alloy has an electric resistance R (in Ω) given by $R = a + 0.0115T^3$, where T (in °C) is the temperature (between -4°C and 4°C). Find R for $a = 0.838 \ \Omega$ and $T = -2.87$°C.

Substituting these values, we have

$$R = 0.838 + 0.0115(-2.87)^3 \qquad \text{estimation:}$$
$$= 0.566 \ \Omega \qquad\qquad 0.8 + 0.01(-3)^3 = 0.8 + 0.01(-27) = 0.53$$

Note in the estimation that $(-3)^3 = -27$.

Fig 1.9

Graphing calculators generally use computer symbols in the display for some of the operations to be performed. These symbols are as follows:

Multiplication: * Division: / Powers: ∧

Therefore, to calculate the value of $20 \times 6 + 200/5 - 3^4$, we use the key sequence

$$20 \boxed{\times} 6 \boxed{+} 200 \boxed{\div} 5 \boxed{-} 3 \boxed{\wedge} 4$$

CAUTION ▶ with the result of 79 shown in the display of Fig. 1.9. ***Note carefully that 200 is divided only by 5.*** If it were divided by $5 - 3^4$, then we would use parentheses and show the expression to be evaluated as $20 \times 6 + 200/(5 - 3^4)$.

EXERCISES 1.4

In Exercises 1–4, make the given changes in the indicated examples of this section, and then simplify the resulting expression.

1. In Example 4(a), change $(-x^2)^3$ to $(-x^3)^2$.

2. In Example 6(b), change $(2x)^0$ to $2x^0$.

3. In Example 8(d), interchange the a^3 and b^2.

4. In Example 9, change $(-1)^2$ to $(-1)^3$.

In Exercises 5–52, simplify the given expressions. Express results with positive exponents only.

5. $x^3 x^4$

6. $y^2 y^7$

7. $2b^4 b^2$

8. $3k(k^5)$

9. $\dfrac{m^5}{m^3}$

10. $\dfrac{x^6}{x}$

11. $\dfrac{n^5}{n^9}$

12. $\dfrac{s}{s^4}$

13. $(P^2)^4$

14. $(x^8)^3$

15. $(t^5)^4$

16. $(n^3)^7$

17. $(2n)^3$

18. $(ax)^5$

19. $(nT^2)^{30}$

20. $(3a^2)^3$

21. $\left(\dfrac{2}{b}\right)^3$

22. $\left(\dfrac{F}{t}\right)^{20}$

23. $\left(\dfrac{x^2}{2}\right)^4$

24. $\left(\dfrac{3}{n^3}\right)^3$

25. 7^0

26. $(8a)^0$

27. $-3x^0$

28. $6v^0$

29. 6^{-1}

30. $-w^{-5}$

31. $\dfrac{1}{R^{-2}}$

32. $\dfrac{1}{t^{-48}}$

33. $(-t^2)^7$

34. $(-y^3)^5$

35. $(2x^2)^6$

36. $-(-c^4)^4$

37. $(4xa^{-2})^0$

38. $3(LC^{-1})^0$

39. $-\dfrac{b^{-3}}{b^{-5}}$

40. $2i^{40}i^{-70}$

41. $\dfrac{2v^4}{(2v)^4}$

42. $\dfrac{x^2 x^3}{(x^2)^3}$

43. $\dfrac{(n^2)^4}{(n^4)^2}$

44. $\dfrac{(3t)^{-1}}{3t^0}$

45. $(5^0 x^2 a^{-1})^{-1}$

46. $(3m^{-2}n^4)^{-2}$

47. $\left(\dfrac{4x^{-1}}{a^{-1}}\right)^{-3}$

48. $\left(\dfrac{2b^2}{y^5}\right)^{-2}$

49. $(-8gs^3)^2$

50. $ax^2(-a^2x)^2$

51. $\dfrac{15n^2T^5}{3nT^6}$

52. $\dfrac{(nRT^{-2})^{32}}{R^{-2}T^{32}}$

In Exercises 53–60, evaluate the given expressions. In Exercises 55–60, all numbers are approximate.

53. $7(-4) - (-5)^2$

54. $6 + (-2)^5 - (-2)(8)$

55. $-(-26.5)^2 - (-9.85)^3$

56. $-0.711^2 - (-0.809)^6$

57. $\dfrac{3.07(-1.86)}{(-1.86)^4 + 1.596}$

58. $\dfrac{15.66^2 - (-4.017)^4}{1.044(-3.68)}$

59. $2.38(-60.7)^2 - 2540/1.17^3 + 0.806^5(26.1^3 - 9.88^4)$

60. $0.513(-2.778) - (-3.67)^3 + 0.889^4/(1.89 - 1.09^2)$

In Exercises 61–68, perform the indicated operations.

61. Does $\left(\dfrac{1}{x^{-1}}\right)^{-1}$ represent the reciprocal of x?

(W) **62.** Does $\left(\dfrac{0.2 - 5^{-1}}{10^{-2}}\right)^0$ equal 1? Explain.

63. In developing the "big bang" theory of the origin of the universe, the expression $(kT/(hc))^3(GkThc)^2c$ arises. Simplify this expression.

64. If \$2500 is invested at 4.2% interest, compounded quarterly, the amount in the account after 6 years is $2500(1 + 0.042/4)^{24}$. Calculate this amount (the 1 is exact).

65. In designing a cam for a pump, the expression $\pi\left(\dfrac{r}{2}\right)^3\left(\dfrac{4}{3\pi r^2}\right)$ is used. Simplify this expression.

66. For a certain integrated electric circuit, it is necessary to simplify the expression $\dfrac{gM}{2\pi fC(2\pi fM)^2}$. Perform this simplification.

67. In order to find the electric power (in W) consumed by a flashlight, the expression i^2R must be evaluated, where i is the current (in A) and R is the resistance (in Ω). Find the power consumed if $i = 0.094$ A and $R = 16\ \Omega$.

68. In designing a building, it was determined that the forces acting on an I beam would deflect the beam an amount (in cm), given by $\dfrac{x(1000 - 20x^2 + x^3)}{1850}$, where x is the distance (in m) from one end of the beam. Find the deflection for $x = 6.85$ m. (The 1000 and 20 are exact.)

1.5 SCIENTIFIC NOTATION

In technical and scientific work, we often encounter numbers that are either very large or very small. Such numbers are illustrated in the next example.

Television was invented in the 1920s and first used commercially in the 1940s.

The use of fiber optics was developed in the 1950s.

X rays were discovered by Roentgen in 1895.

◀ **EXAMPLE 1** Television signals travel at about 30 000 000 000 cm/s. The mass of the earth is about 6 000 000 000 000 000 000 000 000 kg. A typical individual fiber in a fiber-optic communications cable has a diameter of 0.000 005 m. Some X rays have a wavelength of about 0.000 000 095 cm. ▶

Writing numbers like those in Example 1 is inconvenient in ordinary notation. Calculators and computers require a more efficient way of expressing such numbers in order to work with them. Therefore, a convenient and useful notation, called *scientific notation*, is used to represent such numbers.

A number in **scientific notation** *is expressed as the product of a number greater than or equal to 1 and less than 10, and a power of 10, and is written as*

$$P \times 10^k$$

where $1 \le P < 10$ and k is an integer. (The symbol $\le$ means "is less than or equal to.")

◀ **EXAMPLE 2** **(a)** $340\,000 = 3.4(100\,000) = 3.4 \times 10^5$

(b) $0.000\,503 = \dfrac{5.03}{10\,000} = \dfrac{5.03}{10^4} = 5.03 \times 10^{-4}$ between 1 and 10

(c) $6.82 = 6.82(1) = 6.82 \times 10^0$ ▶

From Example 2, we see how a number is changed from ordinary notation to scientific notation. *The decimal point is moved so that only one nonzero digit is to its left. The number of places moved is the power of* 10 *(k), which is positive if the decimal point is moved to the left and negative if moved to the right.*

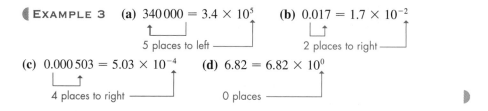

◀ EXAMPLE 3   (a) $340\,000 = 3.4 \times 10^5$ (b) $0.017 = 1.7 \times 10^{-2}$

5 places to left ————   2 places to right ————

(c) $0.000\,503 = 5.03 \times 10^{-4}$ (d) $6.82 = 6.82 \times 10^0$

4 places to right ————   0 places ————   ▶

To change a number from scientific notation to ordinary notation, we reverse the procedure used in Example 3.

◀ EXAMPLE 4   To change 5.83×10^6 to ordinary notation, we move the decimal point six places to the right. Additional zeros must be included to properly locate the decimal point. This means we write

$$5.83 \times 10^6 = 5\,830\,000$$

6 places to right

To change 8.06×10^{-3} to ordinary notation, we must move the decimal point three places to the left. Again, additional zeros must be included. Therefore,

$$8.06 \times 10^{-3} = 0.008\,06$$

3 places to left   ▶

As seen in these examples, scientific notation is an important application of the use of positive and negative exponents. Also, its importance is shown by the metric system use of prefixes to denote powers of 10, as shown in Appendix B.

Scientific notation provides a practical way to handle calculations with very large or very small numbers. First, all numbers are expressed in scientific notation. Then the calculation can be performed on numbers between 1 and 10, using the laws of exponents to find the power of 10 in the final result.

As for large numbers, consider these. A *googol* has been defined as 1 followed by 100 zeros. This means it can be written as 10^{100}. A *googolplex* is 10^{googol}.

◀ EXAMPLE 5   In designing a computer, it was determined that it would be able to process $803\,000$ bits of data in $0.000\,005\,25$ s. (See Exercise 49 of Exercises 1.1 for a brief note on computer data.) The rate of processing the data is

$$5 - (-6) = 11$$

$$\frac{803\,000}{0.000\,005\,25} = \frac{8.03 \times 10^5}{5.25 \times 10^{-6}} = \left(\frac{8.03}{5.25}\right) \times 10^{11} = 1.53 \times 10^{11} \text{ bits/s}$$

As shown, it is proper to leave the result *(rounded off)* in scientific notation. This method is useful when using a calculator and then estimating the result. In this case, the estimate is $(8 \times 10^5) \div (5 \times 10^{-6}) = 1.6 \times 10^{11}$.   ▶

Another advantage of scientific notation is that the precise number of significant digits of a number can be shown directly, even when the final significant digit is 0.

The number 7.50×10^{11} cannot be entered on a scientific calculator in the form 750 000 000 000. It could be entered in this form on a graphing calculator.

Fig 1.10

▸EXAMPLE 6 In evaluating $750\,000\,000\,000^2$, if we know that $750\,000\,000\,000$ has *three* significant digits, we can show the significant digits by writing

$$750\,000\,000\,000^2 = (7.50 \times 10^{11})^2$$
$$= 7.50^2 \times 10^{2\times11} = 56.3 \times 10^{22}$$

If the answer is to be written in scientific notation, we should write it as the product of a number between 1 and 10 and a power of 10. This means we should rewrite it as

$$56.3 \times 10^{22} = (5.63 \times 10)(10^{22}) = 5.63 \times 10^{23}$$ ▸

We can enter numbers in scientific notation on a calculator, as well as have the calculator give results automatically in scientific notation. See the next example.

▸EXAMPLE 7 The wavelength λ (in m) of the light in a red laser beam can be found from the following calculation. Note the significant digits in the numerator.

$$\lambda = \frac{3\,000\,000}{4\,740\,000\,000\,000} = \frac{3.00 \times 10^6}{4.74 \times 10^{12}} = 6.33 \times 10^{-7}\text{ m}$$

The key sequence is 3 $\boxed{\text{EE}}$ 6 $\boxed{\div}$ 4.74 $\boxed{\text{EE}}$ 12 $\boxed{\text{ENTER}}$. See Fig. 1.10. ▸

EXERCISES 1.5

In Exercises 1 and 2, make the given changes in the indicated examples of this section, and then rewrite the number as directed.

1. In the second paragraph of Example 4, change the exponent -3 to 3 and then write the number in ordinary notation.

2. In Example 6, change the exponent 2 to -1 and then write the result in scientific notation.

In Exercises 3–10, change the numbers from scientific notation to ordinary notation.

3. 4.5×10^4 **4.** 6.8×10^7 **5.** 2.01×10^{-3}

6. 9.61×10^{-5} **7.** 3.23×10^0 **8.** 8.40×10^0

9. 1.86×10 **10.** 1×10^{-1}

In Exercises 11–20, change the numbers from ordinary notation to scientific notation.

11. 40 000   **12.** 560 000   **13.** 0.0087   **14.** 0.7

15. 6.09   **16.** 100   **17.** 0.063

18. 0.000 090 8   **19.** 1   **20.** 10

In Exercises 21–24, perform the indicated calculations using a calculator and by first expressing all numbers in scientific notation.

21. 28 000(2 000 000 000)   **22.** 50 000(0.006)

23. $\dfrac{88\,000}{0.0004}$ **24.** $\dfrac{0.000\,03}{6\,000\,000}$

In Exercises 25–28, perform the indicated calculations and then check the result using a calculator. Assume that all numbers are exact.

25. $2 \times 10^{-35} + 3 \times 10^{-34}$ **26.** $5.3 \times 10^{12} - 3.7 \times 10^{10}$

27. $(1.2 \times 10^{29})^3$ **28.** $(2 \times 10^{-16})^{-5}$

In Exercises 29–36, perform the indicated calculations using a calculator. All numbers are approximate.

29. 1280(865 000)(43.8)

30. 0.000 065 9(0.004 86)(3 190 000 000)

31. $\dfrac{0.0732(6710)}{0.001\,34(0.0231)}$ **32.** $\dfrac{0.004\,52}{2430(97\,100)}$

33. $(3.642 \times 10^{-8})(2.736 \times 10^5)$ **34.** $\dfrac{(7.309 \times 10^{-1})^2}{5.9843(2.5036 \times 10^{-20})}$

35. $\dfrac{(3.69 \times 10^{-7})(4.61 \times 10^{21})}{5.004 \times 10^{-9}}$ **36.** $\dfrac{(9.907 \times 10^7)(1.08 \times 10^{12})^2}{(3.603 \times 10^{-5})(2054)}$

In Exercises 37–44, change numbers in ordinary notation to scientific notation or change numbers in scientific notation to ordinary notation. See Appendix B for an explanation of symbols used.

37. The Itaipu Dam on the Parana River between Brazil and Paraguay is the world's largest hydroelectric complex, with an ultimate generating capacity of 12 600 000 kW of power.

38. The speed of light passing through the aqueous humor of the eye is 224 000 000 m/s.

39. A fiber-optic system requires 0.000 003 W of power.

40. A red blood cell measures 0.0075 mm across.

41. To attain an energy density of that in some laser beams, an object would have to be heated to about $10^{30}\,°C$.

42. A unit used in astronomy is the parsec, where 1 parsec is about 3.086×10^{16} m.

43. The power of a radio signal from Galileo, the space probe to Jupiter, is 1.6×10^{-12} W.

44. The electrical force between two electrons is about 2.4×10^{-43} times the gravitational force between them.

In Exercises 45–48, perform the indicated calculations.

45. A computer can do an addition in 7.5×10^{-15} s. How long does it take to perform 5.6×10^{6} additions?

46. Uranium is used in nuclear reactors to generate electricity. About 0.000 000 039% of the uranium disintegrates each day. How much of 0.085 mg of uranium disintegrates in a day?

47. A TV signal travels at 3.00×10^{5} km/s for a total of 7.37×10^{4} km from the station transmitter to a satellite and then to a receiver dish. How long does it take the signal to go from the transmitter to the dish?

48. At $0°C$, the refrigerant Freon is a vapor at a pressure of $P = 1.378 \times 10^{5}$ Pa. If the volume of vapor is $V = 7.865 \times 10^{3}$ cm^3, find the value of PV.

In Exercises 49–52, perform the indicated calculations by first expressing all numbers in scientific notation.

49. One gram of radium produces 37 000 000 000 disintegrations each second. How many disintegrations are there in 1.0 h?

50. The rate of energy radiation (in W) from an object is found by evaluating the expression kT^4, where T is the thermodynamic temperature. Find this value for the human body, for which $k = 0.000\,000\,057$ W/K^4 and $T = 303$ K.

51. In a microwave receiver circuit, the resistance R of a wire 1 m long is given by $R = k/d^2$, where d is the diameter of the wire. Find R if $k = 0.000\,000\,021\,96\ \Omega \cdot$ m^2 and $d = 0.000\,079\,98$ m.

52. The average distance between the sun and earth is about 149 600 000 km, and this distance is called an *astronomical unit* (AU). It takes light about 499.0 s to travel 1 AU. What is the speed of light? Compare this with the speed of the TV signal in Exercise 47.

1.6 ROOTS AND RADICALS

At times we have to find the *square root* of a number, and sometimes we even have to find other roots, such as cube roots. This means we have to find a number that when squared, or cubed and so on equals some given number. For example, to find the square root of 9, we have to find a number that when squared equals 9. Here we see that either 3 or -3 is an answer. Therefore, *either 3 or -3 is a square root of 9 since* $3^2 = 9$ or $(-3)^2 = 9$.

To have a general notation for the square root and have it represent *one* number, *we define the* **principal square root** *of a to be positive if a is positive and represent it by $\sqrt{a}$.* This means $\sqrt{9} = 3$ and not -3.

The general notation for the **principal nth root** of a is $\sqrt[n]{a}$. (When $n = 2$, do not write the 2 for n.) *The $\sqrt{}$ sign is called a* **radical sign.**

Unless we state otherwise, when we refer to the root of a number, it is the principal root.

⟨ **EXAMPLE 1** **(a)** $\sqrt{2}$ (the square root of 2) **(b)** $\sqrt[3]{2}$ (the cube root of 2)

(c) $\sqrt[4]{2}$ (the fourth root of 2) **(d)** $\sqrt[7]{6}$ (the seventh root of 6)

To have a single defined value for all roots (not just square roots) and to consider only real number roots, *we define the* **principal nth root** *of a to be positive if a is positive and to be negative if a is negative and n is odd.* (If a is negative and n is even, the roots are not real.)

⟨ **EXAMPLE 2** **(a)** $\sqrt{169} = 13$ $\left(\sqrt{169} \neq -13\right)$ **(b)** $-\sqrt{64} = -8$

(c) $\sqrt[3]{27} = 3$ since $3^3 = 27$ **(d)** $\sqrt{0.04} = 0.2$ since $0.2^2 = 0.04$

odd

(e) $-\sqrt[4]{256} = -4$ **(f)** $\sqrt[3]{-27} = -3$ **(g)** $-\sqrt[3]{27} = -(+3) = -3$

Another property of square roots is developed by noting illustrations such as $\sqrt{36} = \sqrt{4 \times 9} = \sqrt{4} \times \sqrt{9} = 2 \times 3 = 6$. In general, this property states that *the square root of a product of positive numbers is the product of their square roots.*

$$\sqrt{ab} = \sqrt{a}\sqrt{b} \qquad (a \text{ and } b \text{ positive real numbers}) \qquad (1.9)$$

This property is used in simplifying radicals. It is most useful if either a or b is a **perfect square,** *which is the square of a rational number.*

◖ EXAMPLE 3　(a) $\sqrt{8} = \sqrt{(4)(2)} = \sqrt{4}\sqrt{2} = 2\sqrt{2}$

perfect squares　— simplest form

(b) $\qquad \sqrt{75} = \sqrt{(25)(3)} = \sqrt{25}\sqrt{3} = 5\sqrt{3}$

(c) $\sqrt{4 \times 10^2} = \sqrt{4}\sqrt{10^2} = 2(10) = 20$

(Note that the square root of the square of a positive number is that number.)

In order to represent the square root of a number *exactly*, use Eq. (1.9) to write it in simplest form. However, a decimal *approximation* is often acceptable, and we use the $\boxed{\sqrt{x}}$ key on a calculator. (We will show another way of finding the root of a number in Chapter 11.)

◖ EXAMPLE 4　After reaching its greatest height, the time (in s) for a rocket to fall h m is found by evaluating $0.45\sqrt{h}$. Find the time for the rocket to fall 360 m.

The calculator evaluation is shown in Fig. 1.11. From the display, we see that the rocket takes 8.5 s to fall 360 m. The result is rounded off to two significant digits, the accuracy of 0.45 (an approximate number).

In simplifying a radical, *all operations under a radical sign must be done before finding the root.* The horizontal bar groups the numbers under it.

◖ EXAMPLE 5　(a) $\sqrt{16 + 9} = \sqrt{25} = 5$　first perform the addition $16 + 9$

However, $\sqrt{16 + 9}$ is *not* $\sqrt{16} + \sqrt{9} = 4 + 3 = 7.$

(b) $\sqrt{2^2 + 6^2} = \sqrt{4 + 36} = \sqrt{40} = \sqrt{4}\sqrt{10} = 2\sqrt{10}$, but

$\sqrt{2^2 + 6^2}$ is *not* $\sqrt{2^2} + \sqrt{6^2} = 2 + 6 = 8.$

In defining the principal square root, we did not define the square root of a negative number. However, in Section 1.1, we defined the square root of a negative number to be an **imaginary number.** More generally, *the even root of a negative number is an imaginary number, and the odd root of a negative number is a negative real number.*

◖ EXAMPLE 6

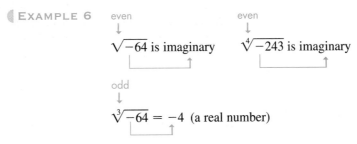

$\qquad\qquad$ even $\qquad\qquad\qquad\qquad$ even
$\qquad\qquad\quad \downarrow \qquad\qquad\qquad\qquad\quad \downarrow$
$\qquad\quad \sqrt{-64}$ is imaginary $\qquad \sqrt[4]{-243}$ is imaginary

$\qquad\quad$ odd
$\qquad\quad \downarrow$
$\qquad\quad \sqrt[3]{-64} = -4$ (a real number)

Side notes:

Try this one on your calculator:
$\sqrt{12\,345\,678\,987\,654\,321}$

The principal *n*th root of a number can be found on a graphing calculator with a specific key sequence.

Fig 1.11

CAUTION ◗

A more detailed discussion of radicals and imaginary numbers is found in Chapters 11 and 12.

EXERCISES 1.6

In Exercises 1–4, make the given changes in the indicated examples of this section and then solve the given problems.

1. In Example 2(b), change the square root to a cube root and then evaluate.

2. In Example 3(b), change $\sqrt{(25)(3)}$ to $\sqrt{(15)(5)}$ and explain whether or not this would be a better expression to use.

3. In Example 5(a), change the $+$ to $\times$ and then evaluate.

4. In the first illustration of Example 6, place a $-$ sign before the radical. Is there any other change in the statement?

In Exercises 5–36, simplify the given expressions. In each of 5–9 and 12–21, the result is an integer.

5. $\sqrt{81}$ 6. $\sqrt{225}$ 7. $-\sqrt{121}$ 8. $-\sqrt{36}$

9. $-\sqrt{49}$ 10. $\sqrt{0.25}$ 11. $\sqrt{0.09}$ 12. $-\sqrt{900}$

13. $\sqrt[3]{125}$ 14. $\sqrt[4]{16}$ 15. $\sqrt[3]{-216}$ 16. $\sqrt[5]{-32}$

17. $\left(\sqrt{5}\right)^2$ 18. $\left(\sqrt{19}\right)^2$ 19. $\left(\sqrt[3]{31}\right)^3$ 20. $\left(\sqrt[4]{53}\right)^4$

21. $\left(-\sqrt{18}\right)^2$ 22. $-\sqrt{32}$ 23. $\sqrt{12}$ 24. $\sqrt{50}$

25. $2\sqrt{84}$ 26. $4\sqrt{108}$ 27. $\sqrt{\dfrac{80}{7-3}}$ 28. $\sqrt{81 \times 10^2}$

29. $\sqrt[3]{8^2}$ 30. $\sqrt[4]{9^2}$ 31. $\dfrac{7^2\sqrt{81}}{3^2\sqrt{49}}$ 32. $\dfrac{2^5\sqrt[3]{243}}{3\sqrt{144}}$

33. $\sqrt{36+64}$ 34. $\sqrt{25+144}$

35. $\sqrt{3^2+9^2}$ 36. $\sqrt{8^2-4^2}$

In Exercises 37–44, find the value of each square root by use of a calculator. Each number is approximate.

37. $\sqrt{85.4}$ 38. $\sqrt{3762}$ 39. $\sqrt{0.4729}$ 40. $\sqrt{0.0627}$

41. (a) $\sqrt{1296+2304}$ (b) $\sqrt{1296}+\sqrt{2304}$

42. (a) $\sqrt{10.6276+2.1609}$ (b) $\sqrt{10.6276}+\sqrt{2.1609}$

43. (a) $\sqrt{0.0429^2-0.0183^2}$ (b) $\sqrt{0.0429^2}-\sqrt{0.0183^2}$

44. (a) $\sqrt{3.625^2+0.614^2}$ (b) $\sqrt{3.625^2}+\sqrt{0.614^2}$

In Exercises 45–52, solve the given problems.

45. The speed (in km/h) of a car that skids to a stop on dry pavement is often estimated by $\sqrt{207s}$, where s is the length (in m) of the skid marks. Estimate the speed if $s = 46$ m.

46. The resistance in an amplifier circuit is found by evaluating $\sqrt{Z^2-X^2}$. Find the resistance for $Z = 5.362\ \Omega$ and $X = 2.875\ \Omega$.

47. The speed (in m/s) of sound in seawater is found by evaluating $\sqrt{B/d}$ for $B = 2.18 \times 10^9$ Pa and $d = 1.03 \times 10^3$ kg/m³. Find this speed, which is important in locating underwater objects using sonar.

48. The terminal speed (in m/s) of a skydiver can be approximated by $\sqrt{40m}$, where m is the mass (in kg) of the skydiver. Calculate the terminal speed (after reaching this speed, the skydiver's speed remains fairly constant before opening the parachute) of a 75-kg skydiver.

49. A TV screen is 38.6 cm wide and 29.0 cm high. The length of a diagonal (the dimension used to describe it—from one corner to the opposite corner) is found by evaluating $\sqrt{w^2+h^2}$, where w is the width and h is the height. Find the diagonal.

50. A car costs $22\,000 new and is worth $15\,000 two years later. The annual rate of depreciation is found by evaluating $100\left(1-\sqrt{V/C}\right)$, where C is the cost and V is the value after 2 years. At what rate did the car depreciate? (100 and 1 are exact.)

51. Is it always true that $\sqrt{a^2} = a$? Explain.

52. For what values of x is (a) $x > \sqrt{x}$, (b) $x = \sqrt{x}$, and (c) $x < \sqrt{x}$?

1.7 ADDITION AND SUBTRACTION OF ALGEBRAIC EXPRESSIONS

Since we use letters to represent numbers, we can see that all operations that can be used on numbers can also be used on literal numbers. In this section, we discuss the methods for adding and subtracting literal numbers.

Addition, subtraction, multiplication, division, and taking of roots are known as **algebraic operations.** *Any combination of numbers and literal symbols that results from algebraic operations is known as an* **algebraic expression.**

When an algebraic expression consists of several parts connected by plus signs and minus signs, each part (along with its sign) is known as a **term** *of the expression. If a given expression is made up of the product of a number of quantities, each of these quantities, or any product of them, is called a* **factor** *of the expression.*

CAUTION ▶   It is important to ***distinguish clearly between terms and factors*** because some operations that are valid for terms are not valid for factors, and conversely.

◀ EXAMPLE 1   In the study of the motion of a rocket, the following algebraic expression may be used.

terms

$$gt^2 - 2vt + 2s$$

factors

This expression has three terms: gt^2, $-2vt$, and $2s$. The first term gt^2 has a factor of g and two factors of t. Any product of these factors is also a factor of gt^2. This means other factors are gt, t^2, and gt^2 itself.   ▶

◀ EXAMPLE 2 $7x(y^2 + x) - \dfrac{x + y}{6x}$ is an algebraic expression with terms $7x(y^2 + x)$ and $-\dfrac{x + y}{6x}$.

The term $7x(y^2 + x)$ has individual factors of 7, x, and $(y^2 + x)$, as well as products of these factors. The factor $y^2 + x$ has two terms, y^2 and x.

The numerator of the term $-\dfrac{x + y}{6x}$ has two terms, and the denominator has factors of 2, 3, and x. The negative sign can be treated as a factor of -1.   ▶

An algebraic expression containing only one term is called a **monomial.** *An expression containing two terms is a* **binomial,** *and one containing three terms is a* **trinomial.** *Any expression containing two or more terms is called a* **multinomial.** Therefore, any binomial or trinomial is also a multinomial.

In any given term, the numbers and literal symbols multiplying any given factor constitute the **coefficient** *of that factor. The product of all the numbers in explicit form is known as the* **numerical coefficient** *of the term. All terms that differ at most in their numerical coefficients are known as* **similar** *or* **like** *terms. That is, similar terms have the same variables with the same exponents.*

◀ EXAMPLE 3   **(a)** $7x^3\sqrt{y}$ is a monomial. It has a numerical coefficient of 7. The coefficient of $\sqrt{y}$ is $7x^3$, and the coefficient of x^3 is $7\sqrt{y}$.

(b) The expression $3a + 2b - 5a$ is a trinomial. The terms $3a$ and $-5a$ are similar since the literal parts are the same.   ▶

like terms   ⎯⎯⎯⎯ unlike other terms due to factor of a

◀ EXAMPLE 4   **(a)** $(4 \times 2b) + 81b - 6ab$ is a multinomial of three terms (a trinomial). The first term has a numerical coefficient of $8 (= 4 \times 2)$, the second has a numerical coefficient of 81, and the third has a numerical coefficient of -6 (the sign is attached to the numerical coefficient). The first two terms are similar, since they differ only in numerical coefficients. The third term is not similar to the others, for it has the factor a.

(b) The commutative law tells us that $x^2y^3 = y^3x^2$. Therefore, the two terms of the expression $3x^2y^3 + 5y^3x^2$ are similar.   ▶

Some calculators can display algebraic expressions and perform algebraic operations.

In adding and subtracting algebraic expressions, we combine similar (or like) terms into a single term. The final ***simplified*** *expression will contain only terms that are not similar.*

◀ **EXAMPLE 5** **(a)** $3x + 2x - 5y = 5x - 5y$

Since there are two similar terms in the original expression, they are added together so that the simplified result has two unlike terms.

(b) $6a^2 - 7a + 8ax$ cannot be simplified since there are no like terms.

(c) $6a + 5c + 2a - c = 6a + 2a + 5c - c$ commutative law

$$= 8a + 4c$$ add like terms

In order to group terms together in an algebraic expression, we use **symbols of grouping.** In this text we use **parentheses** ( ), **brackets** [ ], and **braces** { }. The **bar,** which is used with radicals and fractions, also groups terms. In earlier sections, we used parentheses and the bar.

CAUTION ▶ When adding and subtracting algebraic expressions, it may be necessary to re move symbols of grouping. To do so we must *change the sign of **every term*** within *the symbols if the grouping is preceded by a minus sign. If the symbols of grouping are preceded by a plus sign, each term within the symbols retains its original sign.* This is a result of the distributive law, $a(b + c) = ab + ac$.

◀ **EXAMPLE 6** **(a)** $2(a + 2x) = 2a + 2(2x)$ use distributive law

$$= 2a + 4x$$

(b) $-(+a - 3c) = (-1)(+a - 3c)$ treat $-$ sign as -1

$$= (-1)(+a) + (-1)(-3c)$$

$$= -a + 3c$$ note change of signs

Normally, $+a$ would be written simply as a.

In Example 6(a) we see that $2(a + 2x) = 2a + 4x$. This is true for any values of a and x and is therefore an *identity* (see p. 7). In fact, an expression and any proper form to which it may be changed form an identity when they are shown as equal to each other.

◀ **EXAMPLE 7** $+$ sign before parentheses

(a) $3c + (2b - c) = 3c + 2b - c = 2b + 2c$ use distributive law

$2b = +2b$ signs retained

$-$ sign before parentheses

(b) $3c - (2b - c) = 3c - 2b + c = -2b + 4c$ use distributive law

$2b = +2b$ signs changed

(c) $3c - (-2b + c) = 3c + 2b - c = 2b + 2c$ use distributive law

signs changed

Note in each case that the parentheses are removed and the sign before the parentheses is also removed.

◀ EXAMPLE 8   In designing a certain machine part, it is necessary to perform the following simplification.

$$16(8 - x) - 2(8x - x^2) - (64 - 16x + x^2) = 128 - 16x - 16x + 2x^2 - 64 + 16x - x^2$$
$$= 64 - 16x + x^2$$

NOTE ▶ It is fairly common to have expressions in which more than one symbol of grouping is to be removed in the simplification. Normally, *when several symbols of grouping are to be removed, it is more convenient to remove the innermost symbols first.* This is illustrated in the following example.

◀ EXAMPLE 9   **(a)** $3ax - [ax - (5s - 2ax)] = 3ax - [ax - 5s + 2ax]$ ⟵⌐

$= 3ax - ax + 5s - 2ax$ ⟵   remove parentheses

$= 5s$ remove brackets

(b) $3a^2b - \{[a - (2a^2b - a)] + 2b\} = 3a^2b - \{[a - 2a^2b + a] + 2b\}$ ⟵⌐

$= 3a^2b - \{a - 2a^2b + a + 2b\}$ remove parentheses

$= 3a^2b - a + 2a^2b - a - 2b$ remove brackets

$= 5a^2b - 2a - 2b$ remove braces

Calculators use only parentheses for grouping symbols, and we often need to use one set of parentheses within another set. These are called **nested parentheses.** In the next example, note that the innermost parentheses are removed first.

◀ EXAMPLE 10 $2 - (3x - 2(5 - (7 - x))) = 2 - (3x - 2(5 - 7 + x))$

$= 2 - (3x - 10 + 14 - 2x)$

$= 2 - 3x + 10 - 14 + 2x$

$= -x - 2$

One of the most common errors made by beginning students is changing the sign of only the first term when removing symbols of grouping preceded by a minus sign. CAUTION ▶ *Remember,* **if the symbols are preceded by a minus sign, we must change the sign of all terms.**

EXERCISES 1.7

In Exercises 1–4, make the given changes in the indicated examples of this section, and then solve the resulting problems.

1. In Example 5(a), change $2x$ to $2y$.

2. In Example 7(a), change the sign before $(2b - c)$ from + to −.

3. In Example 9(a), change $[ax - (5s - 2ax)]$ to $[(ax - 5s) - 2ax]$.

4. In Example 9(b), change $\{[a - (2a^2b - a)] + 2b\}$ to $\{a - [2a^2b - (a + 2b)]\}$.

In Exercises 5–52, simplify the given algebraic expressions.

5. $5x + 7x - 4x$

6. $6t - 3t - 4t$

7. $2y - y + 4x$

8. $4C + L - 6C$

9. $2F - 2T - 2 + 3F - T$

10. $x - 2y + 3x - y + z$

11. $a^2b - a^2b^2 - 2a^2b$

12. $xy^2 - 3x^2y^2 + 2xy^2$

13. $s + (4 + 3s)$

14. $5 + (3 - 4n + p)$

15. $v - (4 - 5x + 2v)$

16. $2a - (b - a)$

17. $2 - 3 - (4 - 5a)$

18. $\sqrt{A} + (h - 2\sqrt{A}) - 3\sqrt{A}$

19. $(a - 3) + (5 - 6a)$

20. $(4x - y) - (-2x - 4y)$

21. $-(t - 2u) + (3u - t)$

22. $2(x - 2y) + (5x - y)$

23. $3(2r + s) - (-5s - r)$

24. $3(a - b) - 2(a - 2b)$

25. $-7(6 - 3j) - 2(j + 4)$

26. $-(5t + a^2) - 2(3a^2 - 2st)$

27. $-[(6 - n) - (2n - 3)]$

28. $-[(A - B) - (B - A)]$

29. $2[4 - (t^2 - 5)]$

30. $3[-3 - (a - 4)]$

31. $-2[-x - 2a - (a - x)]$

32. $-2[-3(x - 2y) + 4y]$

33. $a\sqrt{LC} - \left[3 - \left(a\sqrt{LC} + 4\right)\right]$

34. $9v - [6 - (v - 4) + 4v]$

35. $8c - \{5 - [2 - (3 + 4c)]\}$

36. $7y - \{y - [2y - (x - y)]\}$

37. $5p - (q - 2p) - [3q - (p - q)]$

38. $-(4 - x) - [(5x - 7) - (6x + 2)]$

39. $-2\{-(4 - x^2) - [3 + (4 - x^2)]\}$

40. $-\{-[-(x - 2a) - b] - a\}$

41. $5V^2 - (6 - (2V^2 + 3))$

42. $-2x + 2((2x - 1) - 5)$

43. $-(3t - (7 + 2t - (5t - 6)))$

44. $a^2 - 2(x - 5 - (7 - 2(a^2 - 2x) - 3x))$

45. $-4[4R - 2.5(Z - 2R) - 1.5(2R - Z)]$

46. $3\{2.1e - 1.3[f - 2(e - 5f)]\}$

47. In determining the size of a V belt to be used with an engine, the expression $3D - (D - d)$ is used. Simplify this expression.

48. When finding the current in a transistor circuit, the expression $i_1 - (2 - 3i_2) + i_2$ is used. Simplify this expression. (The numbers below the i's are *subscripts*. Different subscripts denote different variables.)

49. Water leaked into a gasoline storage tank at an oil refinery. Finding the pressure in the tank leads to the expression $-4(b - c) - 3(a - b)$. Simplify this expression.

50. Research on a plastic building material leads to $\left[\left(B + \frac{4}{3}\alpha\right) + 2\left(B - \frac{2}{3}\alpha\right)\right] - \left[\left(B + \frac{4}{3}\alpha\right) - \left(B - \frac{2}{3}\alpha\right)\right]$. Simplify this expression.

51. A shipment contains x film cartridges for 15 exposures each and $x + 10$ cartridges for 25 exposures each. What is the total number of photographs that can be taken with the film from this shipment?

52. Each of two stores has $2n + 1$ mouse pads costing \$3 each and $n - 2$ mouse pads costing \$2 each. How much more is the total value of the \$3 mouse pads than the \$2 mouse pads in the two stores?

1.8 MULTIPLICATION OF ALGEBRAIC EXPRESSIONS

To find the product of two or more monomials, we use the laws of exponents as given in Section 1.4 and the laws for multiplying signed numbers as stated in Section 1.2. We first multiply the numerical coefficients to find the numerical coefficient of the product. Then we multiply the literal numbers, remembering that *the exponents may be combined only if the base is the same.*

◀ EXAMPLE 1 (a) $3c^5(-4c^2) = -12c^7$ multiply numerical coefficients and add exponents of c

(b) $(-2b^2y^3)(-9aby^5) = 18ab^3y^8$ add exponents of same base

(c) $2xy(-6cx^2)(3xcy^2) = -36c^2x^4y^3$

If a product contains a monomial that is raised to a power, *we must first raise it to the indicated power* before proceeding with the multiplication.

◀ EXAMPLE 2 (a) $3(2a^2x)^3(-ax) = 3(8a^6x^3)(-ax) = -24a^7x^4$

(b) $2s^3(-st^4)^2(4s^2t) = 2s^3(s^2t^8)(4s^2t) = 8s^7t^9$

We find the product of a monomial and a multinomial by using the distributive law, which states that we *multiply each term of the multinomial by the monomial*. In doing so, we must be careful to give the correct sign to each term of the product.

◀ **EXAMPLE 3** **(a)** $2ax(3ax^2 - 4yz) = 2ax(3ax^2) + (2ax)(-4yz) = 6a^2x^3 - 8axyz$

(b) $5cy^2(-7cx - ac) = (5cy^2)(-7cx) + (5cy^2)(-ac) = -35c^2xy^2 - 5ac^2y^2$ ▶

It is generally not necessary to write out the middle step as it appears in the preceding example. We write the answer directly. For instance, Example 3(a) would appear as $2ax(3ax^2 - 4yz) = 6a^2x^3 - 8axyz$.

We find the product of two multinomials by using the distributive law. The result is that *we multiply each term of one multinomial by each term of the other and add the results*.

Note that, using the distributive law, $(x - 2)(x + 3) = (x - 2)(x) + (x - 2)(3)$ leads to the same result.

◀ **EXAMPLE 4** $(x - 2)(x + 3) = x(x) + x(3) + (-2)(x) + (-2)(3)$

$$= x^2 + 3x - 2x - 6 = x^2 + x - 6$$ ▶

Finding the power of an algebraic expression is equivalent to using the expression as a factor the number of times indicated by the exponent. It is often convenient to write the power of an algebraic expression in this form before multiplying.

two factors

◀ **EXAMPLE 5** **(a)** $(x + 5)^2 = (x + 5)(x + 5) = x^2 + 5x + 5x + 25$

$$= x^2 + 10x + 25$$

(b) $(2a - b)^3 = (2a - b)(2a - b)(2a - b)$ the exponent 3 indicates three factors

$$= (2a - b)(4a^2 - 2ab - 2ab + b^2) = (2a - b)(4a^2 - 4ab + b^2)$$

$$= 8a^3 - 8a^2b + 2ab^2 - 4a^2b + 4ab^2 - b^3$$

$$= 8a^3 - 12a^2b + 6ab^2 - b^3$$

We should note in illustration (a) that

CAUTION ▶ $(x + 5)^2$ *is* **not** *equal to* $x^2 + 25$

since the term $10x$ is not included. We must follow the proper procedure and not simply square each of the terms within the parentheses. ▶

◀ **EXAMPLE 6** An expression used with a lens of a certain telescope is simplified as shown.

$$a(a + b)^2 + a^3 - (a + b)(2a^2 - s^2)$$

$$= a(a + b)(a + b) + a^3 - (2a^3 - as^2 + 2a^2b - bs^2)$$

$$= a(a^2 + ab + ab + b^2) + a^3 - 2a^3 + as^2 - 2a^2b + bs^2$$

$$= a^3 + a^2b + a^2b + ab^2 - a^3 + as^2 - 2a^2b + bs^2$$

$$= ab^2 + as^2 + bs^2$$ ▶

EXERCISES **1.8**

In Exercises 1–4, make the given changes in the indicated examples of this section, and then solve the resulting problems.

1. In Example 2(b), change the factor $(-st^4)^2$ to $(-st^4)^3$.

2. In Example 3(a), change the factor $2ax$ to $-2ax$.

3. In Example 4, change the factor $(x + 3)$ to $(x - 3)$.

4. In Example 5(b), change the exponent 3 to 2.

In Exercises 5–68, perform the indicated multiplications.

5. $(u^7)(ax)$
6. $(2xy)(x^2y^3)$
7. $-ac^2(acx^3)$
8. $-2s^2(-4cs)^2$
9. $(2ax^2)^2(-2ax)$
10. $6pq^3(3pq^2)^2$
11. $a(-a^2x)^3(-2a)$
12. $-2m^2(-3mn)(m^2n)^2$
13. $i^2(R + r)$
14. $2x(p - q)$
15. $-3s(s^2 - 5t)$
16. $-3b(2b^2 - b)$
17. $5m(m^2n + 3mn)$
18. $a^2bc(2ac - 3a^2b)$
19. $3M(-M - N + 2)$
20. $b^2x^2(x^2 - 2x + 1)$
21. $ab^2c^4(ac - bc - ab)$
22. $-4c^2(-9gc - 2c + g^2)$
23. $ax(cx^2)(x + y^3)$
24. $-2(-3st^3)(3s - 4t)$
25. $(x - 3)(x + 5)$
26. $(a + 7)(a + 1)$
27. $(x + 5)(2x - 1)$
28. $(4t_1 + t_2)(2t_1 - 3t_2)$
29. $(2a - b)(3a - 2b)$
30. $(4w^2 - 3)(3w^2 - 1)$
31. $(2s + 7t)(3s - 5t)$
32. $(5p - 2q)(p + 8q)$
33. $(x^2 - 1)(2x + 5)$
34. $(3y^2 + 2)(2y - 9)$
35. $(x^2 - 2x)(x + 4)$
36. $(2ab^2 - 5t)(-ab^2 - 6t)$
37. $(x + 1)(x^2 - 3x + 2)$
38. $(2F + 3)(F^2 - F - 5)$
39. $(4x - x^3)(2 + x - x^2)$
40. $(5a - 3c)(a^2 + ac - c^2)$
41. $2(a + 1)(a - 9)$
42. $-5(y - 3)(y + 6)$
43. $-3(3 - 2T)(3T + 2)$
44. $2n(5 - n)(6n + 5)$
45. $2L(L + 1)(L - 4)$
46. $ax(x + 4)(7 - x^2)$

47. $(2x - 5)^2$
48. $(x - 3)^2$
49. $(x_1 + 3x_2)^2$
50. $(2m + 1)^2$
51. $(xyz - 2)^2$
52. $(b - 2x^2)^2$
53. $2(x + 8)^2$
54. $3(3R + 4)^2$
55. $(2 + x)(3 - x)(x - 1)$
56. $(3x - c^2)^3$
57. $3T(T + 2)(2T - 1)$
58. $[(x - 2)^2(x + 2)]^2$

59. Let $x = 3$ and $y = 4$ to show that (a) $(x + y)^2 \neq x^2 + y^2$ and (b) $(x - y)^2 \neq x^2 - y^2$. ($\neq$ means "does not equal.")

60. Evaluate the product $(98)(102)$ by expressing it as $(100 - 2)(100 + 2)$.

(W) 61. Square an integer between 1 and 9 and subtract 1 from the result. Explain why the result is the product of the integer before and the integer after the one you chose.

(W) 62. Explain how, by appropriate grouping, the product $(x - 2)(x + 3)(x + 2)(x - 3)$ is easier to find. Find the product.

63. In using aircraft radar, the expression $(2R - X)^2 - (R^2 + X^2)$ arises. Simplify this expression.

64. In calculating the temperature variation of an industrial area, the expression $(2T^3 + 3)(T^2 - T - 3)$ arises. Perform the indicated multiplication.

65. In a particular computer design containing n circuit elements, n^2 switches are needed. Find the expression for the number of switches needed for $n + 100$ circuit elements.

66. Simplify the expression $(T^2 - 100)(T - 10)(T + 10)$, which arises when analyzing the energy radiation from an object.

67. In finding the maximum power in part of a microwave transmitter circuit, the expression $(R_1 + R_2)^2 - 2R_2(R_1 + R_2)$ is used. Multiply and simplify.

68. In determining the deflection of a certain steel beam, the expression $27x^2 - 24(x - 6)^2 - (x - 12)^3$ is used. Multiply and simplify.

1.9 DIVISION OF ALGEBRAIC EXPRESSIONS

To find the quotient of one monomial divided by another, we use the laws of exponents and the laws for dividing signed numbers. Again, the *exponents may be combined only if the base is the same.*

◀ EXAMPLE 1 (a) $\dfrac{3c^7}{c^2} = 3c^{7-2} = 3c^5$ (b) $\dfrac{16x^3y^5}{4xy^2} = \dfrac{16}{4}(x^{3-1})(y^{5-2}) = 4x^2y^3$

(c) $\dfrac{-6a^2xy^2}{2axy^4} = -\left(\dfrac{6}{2}\right)\dfrac{a^{2-1}x^{1-1}}{y^{4-2}} = -\dfrac{3a}{y^2}$

divide ↑ subtract
coefficients   exponents

As shown in illustration (c), we use only positive exponents in the final result unless there are specific instructions otherwise.

From arithmetic we may show how a multinomial is to be divided by a monomial. When adding fractions $\left(\text{say } \frac{2}{7} \text{ and } \frac{3}{7}\right)$, we have

$$\frac{2}{7} + \frac{3}{7} = \frac{2+3}{7}$$

Looking at this from *right to left*, we see that *the quotient of a multinomial divided by a monomial is found by dividing each term of the multinomial by the monomial and adding the results*. This can be shown as

$$\frac{a+b}{c} = \frac{a}{c} + \frac{b}{c}$$

CAUTION ▶   Be careful: Although $\dfrac{a+b}{c} = \dfrac{a}{c} + \dfrac{b}{c}$, we must note that $\dfrac{c}{a+b}$ *is not* $\dfrac{c}{a} + \dfrac{c}{b}$.

◀ EXAMPLE 2   **(a)** $\dfrac{4a^2 + 8a}{2a} = \dfrac{4a^2}{2a} + \dfrac{8a}{2a} = 2a + 4$ each term of numerator divided by denominator

(b) $\dfrac{4x^3y - 8x^3y^2 + 2x^2y}{2x^2y} = \dfrac{4x^3y}{2x^2y} - \dfrac{8x^3y^2}{2x^2y} + \dfrac{2x^2y}{2x^2y}$

$$= 2x - 4xy + 1$$

Until you are familiar with the method, it is recommended that you do write out the middle steps.

We usually do not write out the middle step as shown in these illustrations. The divisions of the terms of the numerator by the denominator are usually done by inspection (mentally), and the result is shown as it appears in the next example. ▮

◀ EXAMPLE 3   The expression $\dfrac{2p + v^2d + 2ydg}{2dg}$ is used when analyzing the operation of an irrigation pump. Performing the indicated division, we have

$$\frac{2p + v^2d + 2ydg}{2dg} = \frac{p}{dg} + \frac{v^2}{2g} + y$$

▮

If each term in an algebraic sum is a number or is of the form ax^n, where n is a nonnegative integer, we call the expression a **polynomial** *in x. The greatest value of the exponent n that appears is the* **degree** *of the polynomial.*

A multinomial and a polynomial may differ in that a polynomial cannot have terms like $\sqrt{x}$ or $1/x^2$, but a multinomial may have such terms. Also, a polynomial may have only one term, whereas a multinomial has at least two terms.

◀ EXAMPLE 4   **(a)** $3 + 2x^2 - x^3$ is a polynomial of degree 3.
(b) $x^4 - 3x^2 - \sqrt{x}$ is not a polynomial.

(c) $4x^5$ is a polynomial of degree 5.   **(d)** $\dfrac{1}{4x^5}$ is not a polynomial.

Expressions (a) and (b) are also multinomials since each has more than one term. Expression (b) is not a polynomial due to the $\sqrt{x}$ term. Expression (c) is a polynomial since the exponent is a positive integer. It is also a monomial. Expression (d) is not a polynomial since it can be written as $\frac{1}{4}x^{-5}$, and when written in the form ax^n, n is not positive. ▮

DIVISION OF ONE POLYNOMIAL BY ANOTHER

To divide one polynomial by another, first arrange the dividend (the polynomial to be divided) and the divisor in descending powers of the variable. Then divide the first term of the dividend by the first term of the divisor. The result is the first term of the quotient. Next, multiply the entire divisor by the first term of the quotient and subtract the product from the dividend. Divide the first term of this difference by the first term of the divisor. This gives the second term of the quotient. Multiply this term by the entire divisor and subtract the product from the first difference. Repeat this process until the remainder is zero or a term of lower degree than the divisor. This process is similar to long division of numbers.

EXAMPLE 5 Perform the division $(6x^2 + x - 2) \div (2x - 1)$.

$\left(\text{This division can also be indicated in the fractional form } \dfrac{6x^2 + x - 2}{2x - 1}.\right)$

We set up the division as we would for long division in arithmetic. Then, following the procedure outlined above, we have the following:

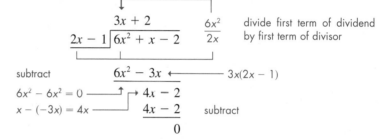

CAUTION ▶

The remainder is zero and the quotient is $3x + 2$. Note that when we subtracted $-3x$ from x, we obtained $4x$.

EXAMPLE 6 Perform the division $\dfrac{4x^3 + 6x^2 + 1}{2x - 1}$. Since there is no x-term in the dividend, we should leave space for any x-terms that might arise.

$$
\begin{array}{r}
2x^2 + 4x + 2 \\
2x - 1 \overline{\smash{)}\ 4x^3 + 6x^2 \qquad + 1} \\
\underline{4x^3 - 2x^2} \\
8x^2 \qquad + 1 \\
\underline{8x^2 - 4x} \\
4x + 1 \\
\underline{4x - 2} \\
3
\end{array}
$$

divisor dividend $\dfrac{4x^3}{2x} = 2x^2$

$6x^2 - (-2x^2) = 8x^2 \longrightarrow$ subtract $\dfrac{8x^2}{2x} = 4x$

$0 - (-4x) = 4x \longrightarrow$ subtract $\dfrac{4x}{2x} = 2$

remainder

The quotient in this case is written as $2x^2 + 4x + 2 + \dfrac{3}{2x - 1}$.

EXERCISES 1.9

In Exercises 1–4, make the given changes in the indicated examples of this section and then perform the indicated divisions.

1. In Example 1(c), change the denominator to $-2a^2xy^5$.

2. In Example 2(b), change the denominator to $2xy^2$.

3. In Example 5, change the dividend to $6x^2 - 7x + 2$.

4. In Example 6, change the sign of the middle term of the numerator from $+$ to $-$.

In Exercises 5–24, perform the indicated divisions.

5. $\dfrac{8x^3y^2}{-2xy}$

6. $\dfrac{-18b^7c^3}{bc^2}$

7. $\dfrac{-16r^3t^5}{-4r^5t}$

8. $\dfrac{51mn^5}{17m^2n^2}$

9. $\dfrac{(15x^2)(4bx)(2y)}{30bxy}$

10. $\dfrac{(5sT)(8s^2T^3)}{10s^3T^2}$

11. $\dfrac{6(ax)^2}{-ax^2}$

12. $\dfrac{12a^2b}{(3ab^2)^2}$

13. $\dfrac{a^2x + 4xy}{x}$

14. $\dfrac{2m^2n - 6mn}{2m}$

15. $\dfrac{3rst - 6r^2st^2}{3rs}$

16. $\dfrac{-5a^2n - 10an^2}{5an}$

17. $\dfrac{4pq^3 + 8p^2q^2 - 16pq^5}{4pq^2}$

18. $\dfrac{a^2x_1x_2^2 + ax_1^3 - ax_1}{ax_1}$

19. $\dfrac{2\pi fL - \pi fR^2}{\pi fR}$

20. $\dfrac{2(ab)^4 - a^3b^4}{3(ab)^3}$

21. $\dfrac{3ab^2 - 6ab^3 + 9a^2b^2}{9a^2b^2}$

22. $\dfrac{2x^2y^2 + 8xy - 12x^2y^4}{2x^2y^2}$

23. $\dfrac{x^{n+2} + ax^n}{x^n}$

24. $\dfrac{3a(F + T)b^2 - (F + T)}{a(F + T)}$

In Exercises 25–42, perform the indicated divisions. Express the answer as shown in Example 6 when applicable.

25. $(2x^2 + 7x + 3) \div (x + 3)$

26. $(3x^2 - 11x - 4) \div (x - 4)$

27. $\dfrac{x^2 - 3x + 2}{x - 2}$

28. $\dfrac{2x^2 - 5x - 7}{x + 1}$

29. $\dfrac{x - 14x^2 + 8x^3}{2x - 3}$

30. $\dfrac{6x^2 + 6 + 7x}{2x + 1}$

31. $(4Z^2 + 23Z + 18) \div (4Z + 3)$

32. $(6x^2 - 20x + 16) \div (3x - 4)$

33. $\dfrac{x^3 + 3x^2 - 4x - 12}{x + 2}$

34. $\dfrac{3x^3 + 19x^2 + 13x - 20}{3x - 2}$

35. $\dfrac{2x^4 + 4x^3 + 2}{x^2 - 1}$

36. $\dfrac{2x^3 - 3x^2 + 8x - 2}{x^2 - x + 2}$

37. $\dfrac{x^3 + 8}{x + 2}$

38. $\dfrac{D^3 - 1}{D - 1}$

39. $\dfrac{x^2 - 2xy + y^2}{x - y}$

40. $\dfrac{3r^2 - 5rR + 2R^2}{r - 3R}$

41. $\dfrac{5E^3 + 8E^2 - 23E - 1}{5E^2 - 7E - 2}$

42. $\dfrac{3x^4 - 2ax^3 - 9a^2x^2 + 2a^4}{3x^2 + ax - 2a^2}$

In Exercises 43–48, perform the indicated divisions.

43. In the optical theory dealing with lasers, the following expression arises: $\dfrac{8A^5 + 4A^3\mu^2E^2 - A\mu^4E^4}{8A^4}$. Perform the indicated division. (μ is the Greek letter mu.)

44. In finding the total resistance of the resistors shown in Fig. 1.12, the expression $\dfrac{6R_1 + 6R_2 + R_1R_2}{6R_1R_2}$ is used. Perform the indicated division.

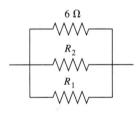

Fig. 1.12

45. When analyzing the potential energy associated with gravitational forces, the expression $\dfrac{GMm[(R + r) - (R - r)]}{2rR}$ arises. Perform the indicated division.

46. A computer model shows that the temperature change T in a certain freezing unit is found by using the expression $\dfrac{3T^3 - 8T^2 + 8}{T - 2}$. Perform the indicated division.

47. In analyzing the displacement of a certain valve, the expression $\dfrac{s^2 - 2s - 2}{s^4 + 4}$ is used. Find the reciprocal of this expression and then perform the indicated division.

48. The voltage and resistance in a certain electric circuit vary with time such that the current is given by the expression $\dfrac{2t^3 + 94t^2 - 290t + 500}{2t + 100}$. By performing the indicated division, find the expression for the current.

1.10 SOLVING EQUATIONS

In this section, we show how algebraic operations are used in solving equations. In the following sections, we show some of the important applications of equations.

An **equation** *is an algebraic statement that two algebraic expressions are equal.* Any value of the literal number representing the **unknown** that produces equality when **substituted** in the equation is said to **satisfy** the equation.

EXAMPLE 1 The equation $3x - 5 = x + 1$ is true only if $x = 3$. For $x = 3$, we have $4 = 4$; for $x = 2$, we have $1 = 3$, which is not correct.

This equation is valid for only one value of the unknown. *An equation valid only for certain values of the unknown is a* **conditional equation.** In this section, nearly all equations we solve will be conditional equations that are satisfied by only one value of the unknown.

EXAMPLE 2 **(a)** The equation $x^2 - 4 = (x - 2)(x + 2)$ is true for all values of x. For example, if $x = 3$, we have $9 - 4 = (3 - 2)(3 + 2)$, or $5 = 5$. If $x = -1$, we have $-3 = -3$. *An equation valid for all values of the unknown is an* **identity.**

We have previously noted *identities* on pages 7 and 28.

(b) The equation $x + 5 = x + 1$ is not true for any value of x. For any value of x we try, we find that the left side is 4 greater than the right side. *Such an equation is called a* **contradiction.**

To **solve** an equation we find the values of the unknown that satisfy it. There is one basic rule to follow when solving an equation:

Perform the same operation on both sides of the equation.

Equations can be solved on many calculators. However, an estimate of the answer may be required in order to find the solution.

We do this to isolate the unknown and thus to find its value.

By performing the same operation on both sides of an equation, the two sides remain equal. Thus,

we may add the same number to both sides, subtract the same number from both sides, multiply both sides by the same number, or divide both sides by the same number (not zero).

Although we may multiply both sides of an equation by zero, this produces $0 = 0$, which is not useful in finding the solution.

EXAMPLE 3 For each of the following equations, we may isolate x, and thereby solve the equation, by performing the indicated operation.

The word *algebra* comes from Arabic and means "a restoration." It refers to the fact that when a number has been added to one side of an equation, the same number must be added to the other side to maintain equality.

$x - 3 = 12$	$x + 3 = 12$	$\dfrac{x}{3} = 12$	$3x = 12$
add 3 to both sides	subtract 3 from both sides	multiply both sides by 3	divide both sides by 3
$x - 3 + 3 = 12 + 3$	$x + 3 - 3 = 12 - 3$	$3\left(\dfrac{x}{3}\right) = 3(12)$	$\dfrac{3x}{3} = \dfrac{12}{3}$
$x = 15$	$x = 9$	$x = 36$	$x = 4$

Each solution should be checked by substitution in the original equation.

The solution of an equation generally requires a combination of the basic operations. Note carefully how these operations are performed in the following examples.

◖ EXAMPLE 4   Solve the equation $2t - 7 = 9$.

We are to perform basic operations to both sides of the equation to finally isolate t on one side. The steps to be followed are suggested by the form of the equation and, in this case, are as follows:

$$2t - 7 = 9 \qquad \text{original equation}$$
$$2t - 7 + 7 = 9 + 7 \qquad \text{add 7 to both sides}$$
$$2t = 16 \qquad \text{combine like terms}$$
$$\frac{2t}{2} = \frac{16}{2} \qquad \text{divide both sides by 2}$$
$$t = 8 \qquad \text{simplify}$$

NOTE ▶   Therefore, we conclude that $t = 8$. Checking *in the original equation*, we have

$$2(8) - 7 \overset{?}{=} 9, \qquad 16 - 7 \overset{?}{=} 9, \qquad 9 = 9$$

Therefore, the solution checks.   ◗

With simpler numbers, the step of adding or subtracting a term, or multiplying or dividing by a factor, is usually done by inspection and not actually written down.

◖ EXAMPLE 5   Solve the equation $3n + 4 = n - 6$.

$$2n + 4 = -6 \qquad \textit{n subtracted from both sides—by inspection}$$
$$2n = -10 \qquad \textit{4 subtracted from both sides—by inspection}$$
$$n = -5 \qquad \textit{both sides divided by 2—by inspection}$$

Checking *in the original equation*, we have $-11 = -11$.   ◗

◖ EXAMPLE 6   Solve the equation $x - 7 = 3x - (6x - 8)$.

$$x - 7 = 3x - 6x + 8 \qquad \text{parentheses removed}$$
$$x - 7 = -3x + 8 \qquad \textit{x-terms combined on right}$$
$$4x - 7 = 8 \qquad \textit{3x added to both sides}$$
$$4x = 15 \qquad \text{7 added to both sides}$$
$$x = \tfrac{15}{4} \qquad \text{both sides divided by 4}$$

Checking in the original equation, we obtain (after simplifying) $-\frac{13}{4} = -\frac{13}{4}$.   ◗

NOTE ▶   Note that we ***always check in the original equation.*** This is done since errors may have been made in finding the later equations.

From these examples we see that the following steps are used in solving the basic equations of this section.

Many other types of equations require more advanced methods for solving. These are considered in later chapters.

PROCEDURE FOR SOLVING EQUATIONS

1. *Remove grouping symbols (distributive law).*

2. *Combine any like terms on each side (also after step 3).*

3. *Perform the same operations on both sides until $x =$ result is obtained.*

4. *Check the solution in the original equation.*

If an equation contains numbers not easily combined by inspection, the best proce-

NOTE ▶   dure is to *first solve for the unknown and then perform the calculation.*

◀ EXAMPLE 7   When finding the current i (in A) in a certain radio circuit, the following equation and solution are used.

$$0.0595 - 0.525i - 8.85(i + 0.003\,16) = 0$$

$$0.0595 - 0.525i - 8.85i - 8.85(0.003\,16) = 0 \qquad \text{note how the above}$$

$$(-0.525 - 8.85)i = 8.85(0.003\,16) - 0.0595 \qquad \text{procedure is followed}$$

$$i = \frac{8.85(0.003\,16) - 0.0595}{-0.525 - 8.85} \qquad \text{evaluate}$$

$$= 0.003\,36 \text{ A}$$

When doing this calculation on a calculator, be careful to properly group the numbers for the division.

In checking, if we use the result above, we find the left side of the original equation to be 3.4×10^{-5}, and if we use the unrounded calculator value, we get zero. Although the *result* should be rounded off, *no rounding off should be done before the final calculation*. An inaccurate result might otherwise be obtained.

NOTE ▶

Ratio and Proportion

The quotient a/b is also called the **ratio** *of a to b. An equation stating that two ratios are equal is called a* **proportion.** Since a proportion is an equation, if one of the numbers is unknown, we can solve for its value as with any equation. Usually this is done by noting the denominators and multiplying each side by a number that will clear the fractions.

◀ EXAMPLE 8   If the ratio of x to 8 equals the ratio of 3 to 4, we have the proportion

$$\frac{x}{8} = \frac{3}{4}$$

We can solve this equation by multiplying both sides by 8. This gives

$$8\left(\frac{x}{8}\right) = 8\left(\frac{3}{4}\right), \quad \text{or} \quad x = 6$$

Substituting $x = 6$ into the original proportion gives the proportion $\frac{6}{8} = \frac{3}{4}$. Since these ratios are equal, the solution checks.

◀ EXAMPLE 9   The ratio of electric current I (in A) to the voltage V across a resistor is constant. If $I = 1.52$ A for $V = 60.0$ V, find I for $V = 82.0$ V.

Since the ratio of I and V is constant, we have the following solution.

$$\frac{1.52}{60.0} = \frac{I}{82.0}$$

$$82.0\left(\frac{1.52}{60.0}\right) = 82.0\left(\frac{I}{82.0}\right)$$

$$2.08 \text{ A} = I$$

$$I = 2.08 \text{ A}$$

Generally, units of measurement will not be shown in intermediate steps. The proper units will be shown with the data and final result.

Checking in the original equation, we have 0.0253 A/V $= 0.0254$ A/V. Since the value of I is rounded up, the solution checks.

We will use the terms *ratio* and *proportion* (particularly ratio) when studying trigonometry in Chapter 4. A detailed discussion of ratio and proportion is found in Chapter 18. A general method of solving equations involving fractions is given in Chapter 6.

EXERCISES 1.10

In Exercises 1–4, make the given changes in the indicated examples of this section and then solve the resulting problems.

1. In Example 3, change 12 to -12 in each of the four illustrations and then solve.

2. In Example 5, change $n - 6$ to $6 - n$ on the right side and then solve.

3. In Example 6, change $(6x - 8)$ to $(8 - 6x)$ and then solve.

4. In Example 9, for the same given values of I and V, find I for $V = 48.0$ V.

In Exercises 5–40, solve the given equations.

5. $x - 2 = 7$

6. $x - 4 = 1$

7. $x + 5 = 4$

8. $s + 6 = -3$

9. $\dfrac{t}{2} = 5$

10. $\dfrac{x}{4} = -2$

11. $4E = -20$

12. $2x = 12$

13. $3t + 5 = -4$

14. $5D - 2 = 13$

15. $5 - 2y = 3$

16. $8 - 5t = 18$

17. $3x + 7 = x$

18. $6 + 4L = 5 - 3L$

19. $2(s - 4) = s$

20. $3(n - 2) = -n$

21. $6 - (r - 4) = 2r$

22. $5 - (x + 2) = 5x$

23. $2(x - 3) - 5x = 7$

24. $4(F + 7) = -7$

25. $0.1x - 0.5(x - 2) = 2$

26. $1.5x - 0.3(x - 4) = 6$

27. $7 - 3(1 - 2p) = 4 + 2p$

28. $3 - 6(2 - 3t) = t - 5$

29. $\dfrac{4x - 2(x - 4)}{3} = 8$

30. $2x = \dfrac{3 - 5(7 - 3x)}{4}$

31. $|x| - 1 = 8$

32. $2 - |x| = 4$

In Exercises 33–40, all numbers are approximate.

33. $5.8 - 0.3(x - 6.0) = 0.5x$

34. $1.9t = 0.5(4.0 - t) - 0.8$

35. $0.15 - 0.24(C - 0.50) = 0.63$

36. $27.5(5.17 - 1.44x) = 73.4$

37. $\dfrac{x}{2.0} = \dfrac{17}{6.0}$

38. $\dfrac{3.0}{7.0} = \dfrac{R}{42}$

39. $\dfrac{165}{223} = \dfrac{13V}{15}$

40. $\dfrac{276x}{17.0} = \dfrac{1360}{46.4}$

In Exercises 41–48, solve the given problems.

41. In finding the maximum operating temperature T (in °C) for a computer integrated circuit, the equation $1.1 = (T - 76)/40$ is used. Find the temperature.

42. To find the voltage V in a circuit in a TV remote-control unit, the equation $1.12V - 0.67(10.5 - V) = 0$ is used. Find V.

43. In blending two gasolines of different octanes, in order to find the number n of liters of one octane needed, the equation $0.14n + 0.06(2000 - n) = 0.09(2000)$ is used. Find n, given that 0.06 and 0.09 are exact and the first zero of 2000 is significant.

44. In order to find the distance x such that the weights are balanced on the lever shown in Fig. 1.13, the equation $210(3x) = 55.3x + 38.5(8.25 - 3x)$ must be solved. Find x. (3 is exact.)

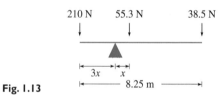

Fig. 1.13

45. A capsule contains two medications in the ratio of 5 to 2. If there are 150 mg of the first medication in the capsule, how many milligrams of the second are there?

46. A person 1.8 m tall is photographed with a 35-mm camera, and the film image is 20 mm. Under the same conditions, how tall is a person whose film image is 16 mm?

W 47. In solving the equation $2(x - 3) + 1 = 2x - 5$, what conclusion can be made?

W 48. In solving the equation $7 - (2 - x) = x + 2$, what conclusion can be made?

1.11 FORMULAS AND LITERAL EQUATIONS

An important application of equations is in the use of *formulas* that are found in geometry and nearly all fields of science and technology. *A* **formula** *is an equation that expresses the relationship between two or more related quantities.* For example, Einstein's famous formula $E = mc^2$ shows the equivalence of energy E to the mass m of an object and the speed of light c.

Einstein published his first paper on relativity in 1905.

NOTE ▶ We can solve a formula for a particular symbol just as we solve any equation. That is, *we isolate the desired symbol by using algebraic operations on the literal numbers.* This is shown in the following examples.

◀ EXAMPLE 1   In Einstein's formula $E = mc^2$, solve for m.

$$E = mc^2 \quad \text{original equation}$$

$$\frac{E}{c^2} = m \quad \text{divide both sides by } c^2$$

$$m = \frac{E}{c^2} \quad \text{switch sides to place } m \text{ at left}$$

The required symbol is usually placed on the left, as shown.   ▶

◀ EXAMPLE 2   A formula relating acceleration a, velocity v, initial velocity v_0, and time t is $v = v_0 + at$. Solve for t.

The subscript $_0$ makes v_0 a different literal symbol from v. (We have used subscripts in a few of the earlier exercises.)

$$v - v_0 = at \quad v_0 \text{ subtracted from both sides}$$

$$t = \frac{v - v_0}{a} \quad \text{both sides divided by } a \text{ and then sides switched}$$ ▶

◀ EXAMPLE 3   In the study of the forces on a certain beam, the equation

$$W = \frac{L(wL + 2P)}{8}$$

Be careful! Just as subscripts can indicate different literal numbers, a capital letter and the same letter in lowercase are different literal numbers. In this example, W and w are different. Also see Exercises 25 and 42 for this section.

is used. Solve for P.

$$8W = \frac{8L(wL + 2P)}{8} \quad \text{multiply both sides by 8}$$

$$8W = L(wL + 2P) \quad \text{simplify right side}$$

$$8W = wL^2 + 2LP \quad \text{remove parentheses}$$

$$8W - wL^2 = 2LP \quad \text{subtract } wL^2 \text{ from both sides}$$

$$P = \frac{8W - wL^2}{2L} \quad \text{divide both sides by } 2L \text{ and switch sides}$$ ▶

◀ EXAMPLE 4   The effect of temperature is important when measurements must be made with great accuracy. The volume V of a special precision container at temperature T in terms of the volume at temperature T_0 is given by

$$V = V_0[1 + b(T - T_0)]$$

where b depends on the material of which the container is made. Solve for T.

Since we are to solve for T, we must isolate the term containing T. This can be done by first removing the grouping symbols and then isolating the term with T.

$$V = V_0[1 + b(T - T_0)] \quad \text{original equation}$$

$$V = V_0[1 + bT - bT_0] \quad \text{remove parentheses}$$

$$V = V_0 + bTV_0 - bT_0V_0 \quad \text{remove brackets}$$

$$V - V_0 + bT_0V_0 = bTV_0 \quad \text{subtract } V_0 \text{ and add } bT_0V_0 \text{ to both sides}$$

$$T = \frac{V - V_0 + bT_0V_0}{bV_0} \quad \text{divide both sides by } bV_0 \text{ and switch sides}$$ ▶

NOTE ▶ If we wish to determine the value of any literal number in an expression for which we know values of the other literal numbers, we should *first solve for the required symbol and then substitute the given values.*

◀ EXAMPLE 5   The electric resistance R (in Ω) of a resistor changes with the temperature T (in °C) according to $R = R_0 + R_0\alpha T$, where R_0 is the resistance at 0°C. For a given resistor, $R_0 = 712\ \Omega$ and $\alpha = 0.004\,55/°C$. Find the value of T for $R = 825\ \Omega$.

We first solve for T and then substitute the given values.

$$R = R_0 + R_0\alpha T$$
$$R - R_0 = R_0\alpha T$$
$$T = \frac{R - R_0}{\alpha R_0}$$

Now substituting, we have

$$T = \frac{825 - 712}{(0.004\,55)(712)}$$

estimation:
$$\frac{800 - 700}{0.005(700)} = \frac{1}{0.035} = 30$$

$$= 34.9°C \quad \text{rounded off}$$

▶

EXERCISES 1.11

In Exercises 1–4, solve for the given letter from the indicated example of this section.

1. For the formula in Example 2, solve for a.

2. For the formula in Example 3, solve for w.

3. For the formula in Example 4, solve for T_0.

4. For the formula in Example 5, solve for α. (Do not evaluate.)

In Exercises 5–38, each of the given formulas arises in the technical or scientific area of study shown. Solve for the indicated letter.

5. $E = IR$, for R (electricity)

6. $PV = nRT$, for T (chemistry)

7. $\theta = kA + \lambda$, for λ (robotics)

8. $W = S_dT - Q$, for Q (air conditioning)

9. $Q = SLd^2$, for L (machine design)

10. $P = 2\pi Tf$, for T (mechanics)

11. $p = p_a + dgh$, for h (hydrodynamics)

12. $2Q = 2I + A + S$, for I (nuclear physics)

13. $A = \dfrac{Rt}{PV}$, for t (jet engine design)

14. $u = -\dfrac{eL}{2m}$, for L (spectroscopy)

15. $ct^2 = 0.3t - ac$, for a (medical technology)

16. $FL = P_1L - P_1d + P_2L$, for d (construction)

17. $T = \dfrac{c + d}{v}$, for d (traffic flow)

18. $L = \dfrac{N\Phi}{i}$, for Φ (electricity)

19. $\dfrac{K_1}{K_2} = \dfrac{m_1 + m_2}{m_1}$, for m_2 (kinetic energy)

20. $f = \dfrac{F}{d - F}$, for d (photography)

21. $a = \dfrac{2mg}{M + 2m}$, for M (pulleys)

22. $v = \dfrac{V(m + M)}{m}$, for M (ballistics)

23. $C_0^2 = C_i^2(1 + 2V)$, for V (electronics)

24. $A_1 = A(M + 1)$, for M (photography)

25. $P = n(p - c)$, for p (economics)

26. $T = 3(T_2 - T_1)$, for T_1 (oil drilling)

27. $F = m\left(g - \dfrac{v^2}{R}\right)$, for v^2 (circular motion)

28. $p_2 = p_1 + rp_1(1 - p_1)$, for r (population growth)

29. $Q_1 = P(Q_2 - Q_1)$, for Q_2 (refrigeration)

30. $p - p_a = dg(y_2 - y_1)$, for y_2 (pressure gauges)

31. $N = N_1T - N_2(1 - T)$, for N_1 (machine design)

32. $t_a = t_c + (1 - h)t_m$, for h (computer access time)

33. $L = \pi(r_1 + r_2) + 2x_1 + x_2$, for r_1 (pulleys)

34. $I = \dfrac{VR_2 + VR_1(1 + \mu)}{R_1 R_2}$, for μ (electronics)

35. $P = \dfrac{V_1(V_2 - V_1)}{gJ}$, for V_2 (jet engine power)

36. $W = T(S_1 - S_2) - Q$, for S_2 (refrigeration)

37. $C = \dfrac{2eAk_1 k_2}{d(k_1 + k_2)}$, for e (electronics)

38. $R = \dfrac{\pi r^4 (p_2 - p_1)}{8nL}$, for p_1 (fluid dynamics)

In Exercises 39–44, find the indicated values.

39. The pressure p (in kPa) at a depth h (in m) below the surface of water is given by the formula $p = p_0 + kh$, where p_0 is the atmospheric pressure. Find h for $p = 205$ kPa, $p_0 = 101$ kPa, and $k = 9.80$ kPa/m.

40. A formula used in determining the total transmitted power P_t in an AM radio signal is $P_t = P_c(1 + 0.500m^2)$. Find P_c if $P_t = 680$ W and $m = 0.925$.

41. A formula relating the Fahrenheit temperature F and the Celsius temperature C is $F = \frac{9}{5}C + 32$. Find the Celsius temperature that corresponds to 90.2°F.

42. In forestry, a formula used to determine the volume V of a log is $V = \frac{1}{2}L(B + b)$, where L is the length of the log and B

and b are the areas of the ends. Find b (in m²) if $V = 1.09$ m³, $L = 4.91$ m, and $B = 0.244$ m². See Fig. 1.14.

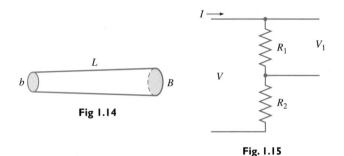

Fig 1.14

Fig. 1.15

43. The voltage V_1 across resistance R_1 (in Ω) is given by

$V_1 = \dfrac{VR_1}{R_1 + R_2}$, where V is the voltage across resistances R_1 and R_2. See Fig. 1.15. Find R_2 (in Ω) if $R_1 = 3.56\ \Omega$, $V_1 = 6.30$ V, and $V = 12.0$ V.

44. The efficiency E of a computer multiprocessor compilation is given by $E = \dfrac{1}{q + p(1 - q)}$, where p is the number of processors and q is the fraction of the compilation that can be performed by the available parallel processors. Find p for $E = 0.66$ and $q = 0.83$.

1.12 APPLIED WORD PROBLEMS

Many applied problems are at first word problems, and we must put them into mathematical terms for solution. Usually the most difficult part in solving a word problem is identifying the information needed for setting up the equation that leads to the solution. To do this, you must read the problem carefully to be sure that you understand all of the terms and expressions used. Following is an approach you should use.

See Appendix A, page A-3, for a variation to the method outlined in these steps. You might find it helpful. Also, some suggested study aids are briefly discussed in Appendix A.

CAUTION ▶

PROCEDURE FOR SOLVING WORD PROBLEMS

1. *Read the statement of the problem.* First, read it quickly for a general overview. Then *reread* slowly and carefully, *listing the information* given.

2. *Clearly identify the unknown quantities* and then *assign an appropriate letter to represent one of them*, stating this choice clearly.

3. *Specify the other unknown quantities* in terms of the one in step 2.

4. *If possible, make a sketch* using the known and unknown quantities.

5. *Analyze the statement* of the problem and *write the necessary equation.* This is often the most difficult step because *some of the information may be implied and not explicitly stated.* Again, a very careful reading of the statement is necessary.

6. *Solve the equation,* clearly stating the solution.

7. *Check the solution* with the original statement of the problem.

Read the following examples very carefully and note just how the outlined procedure is followed.

◀ EXAMPLE 1 A 78-N beam is supported at each end. The supporting force at one end is 10 N more than at the other end. Find the forces.

Since the force at each end is required, we write

$$\text{let } F = \text{the smaller force} \qquad \text{step 2}$$

as a way of establishing the unknown for the equation. Any appropriate letter could be used, and we could have let it represent the larger force.

Also, since the other force is 10 N more, we write

$$F + 10 = \text{the larger force} \qquad \text{step 3}$$

We now draw the sketch in Fig. 1.16. step 4

Since the forces at each end of the beam support the weight of the beam, we have the equation

$$F + (F + 10) = 78 \qquad \text{step 5}$$

This equation can now be solved: $\qquad 2F = 68$

$$F = 34 \text{ N} \qquad \text{step 6}$$

Thus, the smaller force is 34 N, and the larger force is 44 N. This checks with the original statement of the problem. step 7 ▶

◀ EXAMPLE 2 In designing an electric circuit, it is found that 34 resistors with a total resistance of 56 Ω are required. Two different resistances, 1.5 Ω and 2.0 Ω, are used. How many of each are in the circuit?

Since we want to find the number of each resistance, we

$$\text{let } x = \text{number of 1.5-}\Omega \text{ resistors}$$

Also, since there are 34 resistors in all,

$$34 - x = \text{number of 2.0-}\Omega \text{ resistors}$$

We also know that the total resistance of all resistors is 56 Ω. This means that

1.5 Ω number   2.0 Ω
each ⌐ ↓   ⌐ each ⌐ number
   1.5x   + 2.0(34 − x) = 56 ◀──── total resistance of all resistors

total resistance   total resistance
of 1.5-Ω resistors   of 2.0-Ω resistors

$$1.5x + 68 - 2.0x = 56$$
$$-0.5x = -12$$
$$x = 24$$

Therefore, there are 24 1.5-Ω resistors and 10 2.0-Ω resistors. The total resistance of these is $24(1.5) + 10(2.0) = 36 + 20 = 56$ Ω. We see that this checks with the statement of the problem. ▶

Be sure to carefully identify your choice for the unknown. In most problems, there is really a choice. Using the word *let* clearly shows that a specific choice has been made.

The statement after "let *x* (or some other appropriate letter) =" should be clear. It should completely define the chosen unknown.

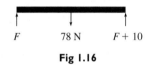

Fig 1.16

"Let *x* = 1.5-Ω resistors" is incomplete. We want to find out how many of them there are.

See Appendix A, page A-3, for a "sketch" that might be used with this example.

◀ EXAMPLE 3   A medical researcher finds that a given sample of an experimental drug can be divided into 4 more slides with 5 mg each than with 6 mg each. How many slides with 5 mg each can be made up?

We are asked to find the number of slides with 5 mg, and therefore we

$$\text{let } x = \text{number of slides with 5 mg}$$

Since the sample may be divided into 4 more slides with 5 mg each than of 6 mg each, we know that

$$x - 4 = \text{number of slides with 6 mg}$$

Since *it is the same sample that is to be divided*, the total mass of the drug on each type is the same. This means

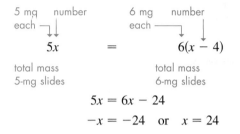

$$5x = 6x - 24$$
$$-x = -24 \quad \text{or} \quad x = 24$$

Therefore, the sample can be divided into 24 slides with 5 mg each, or 20 slides with 6 mg each. Since the total mass, 120 mg, is the same for each set of slides, the solution checks with the statement of the problem.   ◗

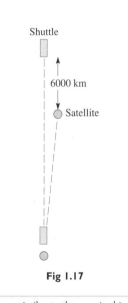

Fig 1.17

A maneuver similar to the one in this example was used to repair the Hubble space telescope in 1994 and 2001.

◀ EXAMPLE 4   A space shuttle maneuvers so that it may "capture" an already orbiting satellite that is 6000 km ahead. If the satellite is moving at 27 000 km/h and the shuttle is moving at 29 500 km/h, how long will it take the shuttle to reach the satellite? (All digits shown are significant.)

First, we let $t =$ the time for the shuttle to reach the satellite. Then, using the fact that the shuttle must go 6000 km farther in the same time, we draw the sketch in Fig. 1.17. Next we use the formula *distance = rate × time* ($d = rt$). This leads to the following equation and solution.

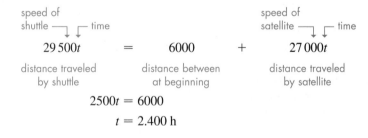

$$2500t = 6000$$
$$t = 2.400 \text{ h}$$

This means that it will take the shuttle 2.400 h to reach the satellite. In 2.400 h, the shuttle will travel 70 800 km, and the satellite will travel 64 800 km. We see that the solution checks with the statement of the problem.   ◗

EXAMPLE 5 A refinery has 7600 L of a gasoline and methanol blend that is 5.00% methanol. How much pure methanol must be added to this blend so that the resulting blend is 10.0% methanol? (All data have three significant digits.)

First, let $x =$ the number of liters of methanol to be added. We want the total volume of methanol to be 10.0% of the volume of the final blend, which is the volume in the original blend plus the volume that is added. See Fig. 1.18. This leads to the following equation and solution.

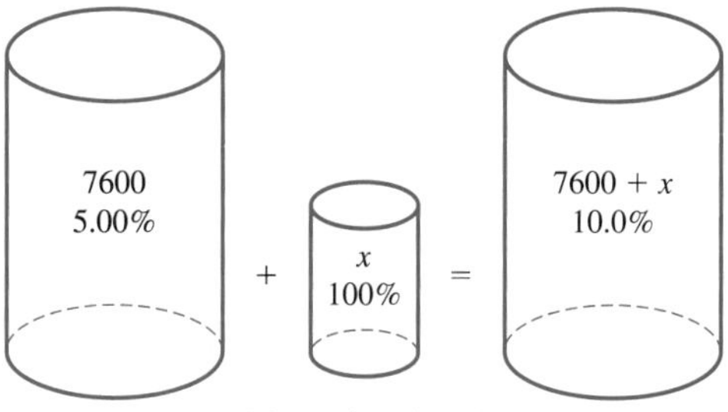

"Let $x =$ methanol" is incomplete. We want to find out the volume (in L) that is to be added.

Liters of methanol

Fig 1.18

5.00% ⟶ original volume   10% ⟶ final volume

$$0.0500(7600) + x = 0.100(7600 + x)$$

methanol in original mixture   methanol added   methanol in final mixture

$$380 + x = 760 + 0.100x$$
$$0.900x = 380$$
$$x = 422 \text{ L}$$

Therefore, 422 L of methanol are to be added. The result checks (to three significant figures) since there would be 802 L of methanol of a total volume of 8020 L.

EXERCISES 1.12

In Exercises 1–4, make the given changes in the indicated examples of this section and then solve the resulting problems.

1. In Example 2, in the second line change 2.0 Ω to 2.5 Ω.

2. In Example 3, in the second line change "4 more slides" to "3 more slides."

3. In Example 4, in the second line change 27 000 km/h to 27 500 km/h.

4. In Example 5, in the second line change "pure methanol" to "of a blend with 50% methanol."

In Exercises 5–32, solve the following problems by first setting up an appropriate equation. Assume all data are accurate to two significant digits unless greater accuracy is given.

5. A certain model new car costs $5000 more today than the same model new car cost 6 years ago. If a person bought a new model today and 6 years ago and spent $49 000 for these cars, what was the cost of each?

6. The flow of one stream into a lake is 45 m³/s more than the flow of a second stream. In 1 h, 4.14 × 10⁵ m³ flow into the lake from the two streams. What is the flow rate of each?

7. Approximately 4.5 million wrecked cars are recycled in two consecutive years. There were 700 000 more recycled the second year than the first year. How many are recycled each year?

8. During a promotion, a business found its Web site was accessed as often on the first day as on the second and third days combined. On the first day, the number was four times that of the third day, on the second day, it was 4000 more than on the third day. How many times was it accessed each day?

9. A developer purchased 70 ha of land for $900 000. If part cost $20 000 per hectare and the remainder cost $10 000 per hectare, how many hectares did the developer buy at each price? (1 hectare = 1 hm².)

10. A vial contains 2000 mg, which is to be used for two dosages. One patient is to be administered 660 mg more than another. How much should be administered to each?

11. After installing a pollution control device, a car's exhaust contained the same amount of pollutant after 5.0 h as it had in 3.0 h. Before the installation the exhaust contained 150 ppm/h (parts per million per hour) of the pollutant. By how much did the device reduce the emission?

12. Three meshed spur gears have a total of 107 teeth. If the second gear has 13 more teeth than the first and the third has 15 more teeth than the second, how many teeth does each have?

13. In the design of a bridge, an engineer determines that four fewer 18-m girders are needed for the span length than 15-m girders. How many 18-m girders are needed?

14. A fuel oil storage depot had an 8-week supply on hand. However, cold weather caused the supply to be used in 6 weeks when 20 000 L extra were used each week. How many liters were in the original supply?

15. The sum of three electric currents that come together at a point in an integrated circuit is zero. If the second current is double the first and the third current is 9.2 μA more than the first, what are the currents? (The sign of a current indicates the direction of flow.)

16. A trucking firm uses one fleet of trucks on round-trip delivery routes of 8 h, and a second fleet, with five more trucks than the first, is used on round-trip delivery routes of 6 h. Budget allotments allow 198 h of delivery time in a week. How many trucks are in each fleet?

17. A primary natural gas pipeline feeds into three smaller pipelines, each of which is 2.6 km longer than the main pipeline. If the total length of the four pipelines is 35.4 km, what is the length of each section of the line?

18. Ten 6.0-V and 12-V batteries have a total voltage of 84 V. How many of each type are there?

19. The admission charge to a theme park is $48 for adults and $30 for children. If the total of admission charges on a certain day was $81 180 and 2002 people entered, how many were adults, and how many were children?

20. Two stock investments cost $15 000. One stock then had a 40% gain and the other a 10% loss. If the net profit is $2000, how much was invested in each stock?

21. A ski lift takes a skier up a slope at 50 m/min. The skier then skis down the slope at 150 m/min. If one round-trip takes 24 min, how long is the slope?

22. The supersonic jet Concorde made a trip averaging 100 km/h less than the speed of sound for 1.00 h and averaging 400 km/h more than the speed of sound for 3.00 h. If the trip covered 5740 km, what is the speed of sound?

23. Trains at each end of the 50.0-km long Eurotunnel under the English Channel start at the same time into the tunnel. Find their speeds if the train from France travels 8.0 km/h faster than the train from England and they pass in 17.0 min. See Fig. 1.19.

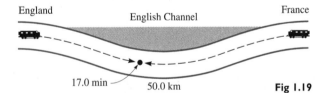

England   English Channel   France

17.0 min   50.0 km   **Fig 1.19**

24. An executive traveling at 60.0 km/h would arrive 10.0 min early for an appointment, whereas at a speed of 45.0 km/h the executive would arrive 5.0 min early. How much time is there until the appointment?

25. One lap at the Canadian Grand Prix in Montreal is 4.36 km. In a race, a car stalls and then starts 30.0 s after a second car. The first car travels at 79.0 m/s, and the second car travels at 73.0 m/s. How long does it take the first car to overtake the second, and which car will be ahead after eight laps?

26. A computer chip manufacturer produces two types of chips. In testing a total of 6100 chips of both types, 0.50% of one type and 0.80% of the other type were defective. If a total of 38 defective chips were found, how many of each type were tested?

27. Two gasoline distributors, A and B, are 367 km apart on a highway. A charges $0.45/L and B charges $0.42/L. Each charges 0.04¢/L per kilometer for delivery. Where on the highway is the cost to the customer the same?

28. An outboard engine uses a gasoline-oil fuel mixture in the ratio of 15 to 1. How much gasoline must be mixed with a gasoline-oil mixture, which is 75% gasoline, to make 8.0 L of the mixture for the outboard engine?

29. A car's radiator contains 12 L of antifreeze at a 25% concentration. How many liters must be drained and then replaced with pure antifreeze to bring the concentration to 50% (the manufacturer's "safe" level)?

30. By mass, a certain roadbed material is 75% crushed rock, and another is 30% crushed rock. How many tonnes of each must be mixed in order to have 250 tonnes of material that is 50% crushed rock?

31. A 5.0-m long car overtakes a 20-m long semitrailer that is traveling at 70 km/h. How fast should the car go to pass the semi in 10 s?

32. An earthquake emits primary waves moving at 8.0 km/s and secondary waves moving at 5.0 km/s. How far from the epicenter of the earthquake is the seismic station if the two waves arrive at the station 2.0 min apart?

CHAPTER 1 EQUATIONS

Commutative law of addition: $a + b = b + a$

Associative law of addition: $a + (b + c) = (a + b) + c$

Distributive law: $a(b + c) = ab + ac$

$$a + (-b) = a - b \tag{1.1}$$
$$a^m \times a^n = a^{m+n} \tag{1.3}$$

Commutative law of multiplication: $ab = ba$

Associative law of multiplication: $a(bc) = (ab)c$

$$a - (-b) = a + b \tag{1.2}$$

$$\frac{a^m}{a^n} = a^{m-n} \quad (m > n, a \neq 0),$$

$$\frac{a^m}{a^n} = \frac{1}{a^{n-m}} \quad (m < n, a \neq 0) \tag{1.4}$$

$$(a^m)^n = a^{mn} \tag{1.5}$$

$$(ab)^n = a^n b^n, \quad \left(\frac{a}{b}\right)^n = \frac{a^n}{b^n} \quad (b \neq 0) \tag{1.6}$$

$$a^0 = 1 \quad (a \neq 0) \tag{1.7}$$

$$a^{-n} = \frac{1}{a^n} \quad (a \neq 0) \tag{1.8}$$

$$\sqrt{ab} = \sqrt{a}\sqrt{b} \quad (a \text{ and } b \text{ positive real numbers}) \tag{1.9}$$

CHAPTER ① REVIEW EXERCISES

In Exercises 1–12, evaluate the given expressions.

1. $(-2) + (-5) - 3$

2. $6 - 8 - (-4)$

3. $\dfrac{(-5)(6)(-4)}{(-2)(3)}$

4. $\dfrac{(-9)(-12)(-4)}{24}$

5. $-5 - |2(-6)| + \dfrac{-15}{3}$

6. $3 - 5(-3 - 2) - \dfrac{12}{-4}$

7. $\dfrac{18}{3 - 5} - (-4)^2$

8. $-(-3)^2 - \dfrac{-8}{(-2) - |-4|}$

9. $\sqrt{16} - \sqrt{64}$

10. $-\sqrt{81 + 144}$

11. $\left(\sqrt{7}\right)^2 - \sqrt[3]{8}$

12. $-\sqrt[4]{16} + \left(\sqrt{6}\right)^2$

In Exercises 13–20, simplify the given expressions. Where appropriate, express results with positive exponents only.

13. $(-2rt^2)^2$

14. $(3a^0 b^{-2})^3$

15. $\dfrac{18m^3 n^4 t}{-3mn^5 t^3}$

16. $\dfrac{15p^4 q^2 r}{5pq^5 r}$

17. $\dfrac{-16N^{-2}(NT^2)}{-2N^0 T^{-1}}$

18. $\dfrac{-35x^{-1}y(x^2 y)}{5xy^{-1}}$

19. $\sqrt{45}$

20. $\sqrt{9 + 36}$

In Exercises 21–24, for each number, (a) determine the number of significant digits and (b) round off each to two significant digits.

21. 8840

22. 21 450

23. 9.040

24. 0.700

In Exercises 25–28, evaluate the given expressions. All numbers are approximate.

25. $37.3 - 16.92(1.067)^2$

26. $\dfrac{8.896 \times 10^{-12}}{3.5954 + 6.0449}$

27. $\dfrac{\sqrt{0.1958 + 2.844}}{3.142(65)^2}$

28. $\dfrac{1}{0.035\,68} + \dfrac{37\,466}{29.63^2}$

In Exercises 29–60, perform the indicated operations.

29. $a - 3ab - 2a + ab$

30. $xy - y - 5y - 4xy$

31. $6LC - (3 - LC)$

32. $-(2x - b) - 3(-x - 5b)$

33. $(2x - 1)(x + 5)$

34. $(x - 4y)(2x + y)$

35. $(x + 8)^2$

36. $(2x + 3y)^2$

37. $\dfrac{2h^3 k^2 - 6h^4 k^5}{2h^2 k}$

38. $\dfrac{4a^2 x^3 - 8ax^4}{2ax^2}$

39. $4R - [2r - (3R - 4r)]$

40. $3b - [2b + 3a - (2a - 3b)] + 4a$

41. $2xy - \{3z - [5xy - (7z - 6xy)]\}$

42. $x^2 + 3b + [(b - y) - 3(2b - y + z)]$

43. $(2x + 1)(x^2 - x - 3)$

44. $(x - 3)(2x^2 - 3x + 1)$

45. $-3y(x - 4y)^2$

46. $-s(4s - 3t)^2$

47. $3p[(q - p) - 2p(1 - 3q)]$

48. $3x[2y - r - 4(s - 2r)]$

49. $\dfrac{12p^3 q^2 - 4p^4 q + 6pq^5}{2p^4 q}$

50. $\dfrac{27s^3 t^2 - 18s^4 t + 9s^2 t}{9s^2 t}$

51. $(2x^2 + 7x - 30) \div (x + 6)$

52. $(4x^2 + 15x - 2) \div (2x + 7)$

53. $\dfrac{3x^3 - 7x^2 + 11x - 3}{3x - 1}$

54. $\dfrac{w^3 - 4w^2 + 7w - 12}{w - 3}$

55. $\dfrac{4x^4 + 10x^3 + 18x - 1}{x + 3}$

56. $\dfrac{8x^3 - 14x + 3}{2x + 3}$

57. $-3\{(r + s - t) - 2[(3r - 2s) - (t - 2s)]\}$

58. $(1 - 2x)(x - 3) - (x + 4)(4 - 3x)$

59. $\dfrac{2y^3 + 9y^2 - 7y + 5}{2y - 1}$

60. $\dfrac{6x^2 + 5xy - 4y^2}{2x - y}$

In Exercises 61–72, solve the given equations.

61. $3x + 1 = x - 8$

62. $4y - 3 = 5y + 7$

63. $\dfrac{5x}{7} = \dfrac{3}{2}$

64. $\dfrac{2(N - 4)}{3} = \dfrac{5}{4}$

65. $6x - 5 = 3(x - 4)$

66. $-2(-4 - y) = 3y$

67. $2s + 4(3 - s) = 6$

68. $2|x| - 1 = 3$

69. $3t - 2(7 - t) = 5(2t + 1)$

70. $6 - 3x - (8 - x) = x - 2(2 - x)$

71. $2.7 + 2.0(2.1x - 3.4) = 0.1$

72. $0.250(6.721 - 2.44x) = 2.08$

In Exercises 73–82, change numbers in ordinary notation to scientific notation or change numbers in scientific notation to ordinary notation. (See Appendix B for an explanation of the symbols that are used.)

73. The maximum pressure exerted by the human heart is about 16 000 Pa.

74. The escape velocity (the velocity required to leave the earth's gravitational field) is about 40 000 km/h.

75. When pictures of the surface of Mars were transmitted to the earth from the Pathfinder mission in 1997, Mars was about 192 000 000 km from earth.

76. Police radar has a frequency of 1.02×10^9 Hz.

77. Among the stars nearest the earth, Centaurus A is about 4.05×10^{13} km away.

78. Before its destruction in 2001, the World Trade Center had nearly 10^6 m^2 of office space. (See Exercise 30, page 131.)

79. A biological cell has a surface area of 0.000 001 2 cm^2.

80. An optical coating on glass to reduce reflections is about 0.000 000 15 m thick.

81. The maximum safe level of radiation in the air of a home due to radon gas is 1.5×10^{-1} Bq/L. (Bq is the symbol for bequerel, the metric unit of radioactivity, where 1 Bq = 1 decay/s.)

82. A computer can execute an instruction in 6.3×10^{-10} s.

In Exercises 83–96, solve for the indicated letter. Where noted, the given formula arises in the technical or scientific area of study.

83. $R = n^2 Z$, for Z (electricity)

84. $R = \dfrac{2GM}{c^2}$, for G (astronomy: black holes)

85. $P = \dfrac{\pi^2 EI}{L^2}$, for E (mechanics)

86. $f = p(c - 1) - c(p - 1)$, for p (thermodynamics)

87. $I = P + Prt$, for t (business)

88. $V = IR + Ir$, for R (electricity)

89. $m = dV(1 - e)$, for e (solar heating)

90. $mu = (m + M)v$, for M (physics: momentum)

91. $N_1 = T(N_2 - N_3) + N_3$, for N_2 (mechanics: gears)

92. $2(J + 1) = \dfrac{f}{B}$, for J (spectroscopy)

93. $R = \dfrac{A(T_2 - T_1)}{H}$, for T_2 (thermal resistance)

94. $Z^2\left(1 - \dfrac{\lambda}{2a}\right) = k$, for λ (radar design)

95. $d = kx^2[3(a + b) - x]$, for a (mechanics: beams)

96. $V = V_0[1 + 3a(T_2 - T_1)]$, for T_2 (thermal expansion)

In Exercises 97–102, perform the indicated calculations.

97. Three adjacent lots have road frontages of 108 m, 89.7 m, and 210 m, respectively. What is the total frontage of these lots?

98. The time (in s) for an object to fall h m is given by the expression $0.45\sqrt{h}$. How long does it take a person to fall 22 m from a sixth-floor window into a net while escaping a fire?

99. The CN Tower in Toronto is 0.553 km high. The Sears Tower in Chicago is 443 m high. How much higher is the CN Tower than the Sears Tower?

100. The time (in s) it takes a computer to check n memory cells is found by evaluating $(n/2650)^2$. Find the time to check 48 cells.

101. The combined electric resistance of two parallel resistors is found by evaluating the expression $\dfrac{R_1 R_2}{R_1 + R_2}$. Evaluate this for $R_1 = 0.0275\ \Omega$ and $R_2 = 0.0590\ \Omega$.

102. The distance (in m) from the earth for which the gravitational force of the earth on a spacecraft equals the gravitational force of the sun on it is found by evaluating $1.5 \times 10^{11}\sqrt{m/M}$, where m and M are the masses of the earth and sun, respectively. Find this distance for $m = 5.98 \times 10^{24}$ kg and $M = 1.99 \times 10^{30}$ kg.

In Exercises 103–106, simplify the given expressions.

103. An analysis of the electric potential of a conductor includes the expression $2V(r - a) - V(b - a)$. Simplify this expression.

104. In finding the value of an annuity, the expression $(Ai - R)(1 + i)^2$ is used. Multiply out this expression.

105. A computer analysis of the velocity of a link in an industrial robot leads to the expression $4(t + h) - 2(t + h)^2$. Simplify this expression.

106. When analyzing the motion of a communications satellite, the expression $\dfrac{k^2 r - 2h^2 k + h^2 r v^2}{k^2 r}$ is used. Perform the indicated division.

In Exercises 107–120, solve the given problems. All data are accurate to two significant digits unless more are given.

107. Two computer software programs cost $190 together. If one costs $72 more than the other, what is the cost of each?

108. One computer hard disk has 1.5 times the memory of another hard disk. If their combined memory is 10.5 GB (gigabytes), what is the memory of each?

109. Three chemical reactions each produce oxygen. If the first produces twice that of the second, the third produces twice that of the first, and the combined total is 560 cm^3, what volume is produced by each?

110. In testing the rate at which a polluted stream flows, a boat that travels at 5.5 km/h in still water took 5.0 h to go downstream between two points, and it took 8.0 h to go upstream between the same two points. What is the rate of flow of the stream?

111. The voltage across a resistor equals the current times the resistance. In a microprocessor circuit, one resistor is 1200 Ω greater than another. The sum of the voltages across them is 12.0 mV. Find the resistances if the current is 2.4 μA in each.

112. An air sample contains 4.0 ppm (parts per million) of two pollutants. The concentration of one is four times the other. What are the concentrations?

113. A 1.5-m piece of tubing weighs 7.6 N. What is the weight of an 8.2-m piece of the same tubing?

114. The fuel for a two-cycle motorboat engine is a mixture of gasoline and oil in the ratio of 15 to 1. How many liters of each are in 6.6 L of mixture?

115. A ship enters the Red Sea from the Suez Canal, moving south at 28.0 km/h. Two hours later, a second ship enters the Red Sea at the southern end, moving north at 35.0 km/h. If the Red Sea is 2230 km long, when will the ships pass?

116. A helicopter used in fighting a forest fire travels at 175 km/h from the fire to a pond and 115 km/h with water from the pond to the fire. If a round-trip takes 30 min, how long does it take from the pond to the fire? See Fig. 1.20.

117. One grade of oil has 0.50% of an additive, and a higher grade has 0.75% of the additive. How many liters of each must be used to have 1000 L of a mixture with 0.65% of the additive?

118. Fifty kilograms of a cement-sand mixture is 40% sand. How many kilograms of sand must be added for the resulting mixture to be 60% sand?

119. An architect plans to have 25% of the floor area of a house in ceramic tile. In all but the kitchen and entry, there are 205 m² of floor area, 15% of which is tile. What area can be planned for the kitchen and entry if each has an all-tile floor?

120. A *karat* equals 1/24 part of gold in an alloy (for example, 9-karat gold is 9/24 gold). How many grams of 9-karat gold must be mixed with 18-karat gold to get 200 g of 14-karat gold?

Writing Exercise

121. A person was asked to explain how to evaluate the expression $\dfrac{1.70}{0.0246(0.0309)}$, where the numbers are approximate. The explanation was: "Divide 1.70 by 0.0246, and then multiply by 0.0309. The calculator display is the answer." Write a short paragraph stating what is wrong with the explanation. (What is the correct result?)

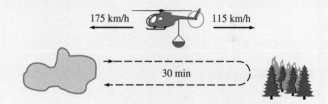

175 km/h   115 km/h

30 min

Fig. 1.20

CHAPTER ① PRACTICE TEST

In Problems 1–5, evaluate the given expressions. In Problems 3 and 5, the numbers are approximate.

1. $\sqrt{9+16}$

2. $\dfrac{(7)(-3)(-2)}{(-6)(0)}$

3. $\dfrac{3.372 \times 10^{-3}}{7.526 \times 10^{12}}$

4. $\dfrac{(+6)(-2) - 3(-1)}{5 - 2}$

5. $\dfrac{346.4 - 23.5}{287.7} - \dfrac{0.944^3}{(3.46)(0.109)}$

In Problems 6–12, perform the indicated operations and simplify. When exponents are used, use only positive exponents in the result.

6. $(2a^0 b^{-2} c^3)^{-3}$

7. $(2x + 3)^2$

8. $3m^2(am - 2m^3)$

9. $\dfrac{8a^3 x^2 - 4a^2 x^4}{-2ax^2}$

10. $\dfrac{6x^2 - 13x + 7}{2x - 1}$

11. $(2x - 3)(x + 7)$

12. $3x - [4x - (3 - 2x)]$

13. Solve for y: $5y - 2(y - 4) = 7$

14. Solve for x: $3(x - 3) = x - (2 - 3d)$

15. Express $0.000\,003\,6$ in scientific notation.

16. List the numbers -3, $|-4|$, $-\pi$, $\sqrt{2}$, and 0.3 in numerical order.

17. What fundamental law is illustrated by $3(5 + 8) = 3(5) + 3(8)$?

18. (a) How many significant digits are in the number 3.0450?
    (b) Round it off to two significant digits.

19. If P dollars are deposited in a bank that compounds interest n times a year, the value of the account after t years is found by evaluating $P(1 + i/n)^{nt}$, where i is the annual interest rate. Find the value of an account for which $P = \$1000$, $i = 5\%$, $n = 2$, and $t = 3$ years (values are exact).

20. In finding the illuminance from a light source, the expression $8(100 - x)^2 + x^2$ is used. Simplify this expression.

21. The equation $L = L_0[1 + \alpha(t_2 - t_1)]$ is used when studying thermal expansion. Solve for t_2.

22. An alloy weighing 20 N is 30% copper. How many newtons of another alloy, which is 80% copper, must be added for the final alloy to be 60% copper?

2 Geometry

In Section 2.5, we see how to find an excellent approximation of the area of an irregular geometric figure, such as a lake. Above is a satellite photograph of Lake Ontario, one of the Great Lakes between the United States and Canada.

When building the pyramids nearly 5000 years ago and today when using MRI (magnetic resonance imaging) to detect a tumor in a human being, the size and shape of an object are measured. Since geometry deals with size and shape, the topics and methods of geometry are important in many of the applications in technology.

Many of the methods of measuring geometric objects were known in ancient times, and most of the geometry used in technology has been known for hundreds of years. In about 300 B.C.E., the Greek mathematician Euclid (who lived and taught in Alexandria, Egypt) organized what was known in geometry. He added many new ideas in a 13-volume set of writings known as the *Elements*. Centuries later it was translated into various languages, and today is second only to the Bible as the most published book in history.

The study of geometry includes the properties and measurements of angles, lines, and surfaces and the basic figures they form. In this chapter, we review the more important methods and formulas for calculating the important geometric measures, such as area and volume. Technical applications are included from areas such as architecture, construction, instrumentation, surveying and civil engineering, mechanical design, and product design of various types, as well as other areas of engineering.

Geometric figures and concepts are also basic to the development of many other areas of mathematics, such as graphing and trigonometry. We will start our study of graphs in Chapter 3 and trigonometry in Chapter 4.

LINES AND ANGLES

It is not possible to define every term we use. In geometry, *the meanings of* **point, line,** and **plane** *are accepted without being defined.* These terms give us a starting point for the definitions of other useful geometric terms.

The amount of rotation of a **ray** *(or* **half-line***) about its endpoint is called an* **angle.** A ray is that part of a line (the word *line* means "straight line") to one side of a fixed point on the line. *The fixed point is the* **vertex** *of the angle. One complete rotation of a ray is an angle with a measure of* 360 **degrees,** *written as* 360°. Some special types of angles are as follows:

Name of angle	Measure of angle
Right angle	*90°*
Straight angle	*180°*
Acute angle	*Between 0° and 90°*
Obtuse angle	*Between 90° and 180°*

EXAMPLE 1 Figure 2.1(a) shows a right angle (marked as ∟). The vertex of the angle is point *B*, and the ray is the half-line *BA*. Figure 2.1(b) shows a straight angle. Figure 2.1(c) shows an acute angle, denoted as ∠*E* (or ∠*DEF* or ∠*FED*). In Fig 2.1(d), ∠*G* is an obtuse angle.

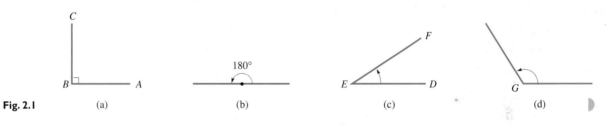

Fig. 2.1 (a)   (b)   (c)   (d)

If two lines intersect such that the angle between them is a right angle, the lines are **perpendicular.** *Lines in the same plane that do not intersect are* **parallel.** These are illustrated in the following example.

EXAMPLE 2 In Fig. 2.2(a), lines *AC* and *DE* are perpendicular (which is shown as $AC \perp DE$) since they meet in a right angle (again, shown as ∟) at *B*.

In Fig. 2.2(b), lines *AB* and *CD* are drawn so they do not meet, even if extended. Therefore, these lines are parallel (which can be shown as $AB \parallel CD$).

In Fig. 2.2(c), $AB \perp BC$, $DC \perp BC$, and $AB \parallel DC$.

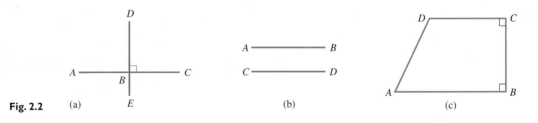

Fig. 2.2   (a)   (b)   (c)

It will be important to recognize perpendicular sides and parallel sides in many of the geometric figures in later sections.

If the sum of the measures of two angles is 180°, then the angles are called **supplementary angles.** *Each angle is the* **supplement** *of the other. If the sum of the measures of two angles is 90°, the angles are called* **complementary angles.** *Each is the* **complement** *of the other.*

EXAMPLE 3 **(a)** In Fig. 2.3(a), $\angle BAC = 55°$, and in Fig. 2.3(b), $\angle DEF = 125°$. Since $55° + 125° = 180°$, $\angle BAC$ and $\angle DEF$ are supplementary angles.

(b) In Fig. 2.4, we see that $\angle POQ$ is a right angle, or $\angle POQ = 90°$. Since $\angle POR + \angle ROQ = \angle POQ = 90°$, $\angle POR$ is the complement of $\angle ROQ$ (or $\angle ROQ$ is the complement of $\angle POR$).

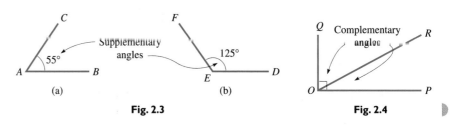

Fig. 2.3   **Fig. 2.4**

It is often necessary to refer to certain special pairs of angles. *Two angles that have a common vertex and a side common between them are known as* **adjacent angles.** *If two lines cross to form equal angles on opposite sides of the point of intersection, which is the common vertex, these angles are called* **vertical angles.**

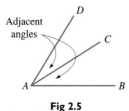

Fig 2.5

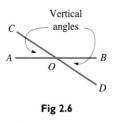

Fig 2.6

EXAMPLE 4 **(a)** In Fig. 2.5, $\angle BAC$ and $\angle CAD$ have a common vertex at A and the common side AC between them so that $\angle BAC$ and $\angle CAD$ are adjacent angles.

(b) In Fig. 2.6, lines AB and CD intersect at point O. Here, $\angle AOC$ and $\angle BOD$ are vertical angles, and they are equal. Also, $\angle BOC$ and $\angle AOD$ are vertical angles and are equal.

We should also be able to identify *the sides of an angle that are adjacent to the angle.* In Fig. 2.5, sides AB and AC are adjacent to $\angle BAC$, and in Fig. 2.6, sides OB and OD are adjacent to $\angle BOD$. Identifying sides adjacent and opposite an angle in a triangle is important in trigonometry.

In a plane, *if a line crosses two or more parallel or nonparallel lines, it is called a* **transversal.** In Fig. 2.7, $AB \parallel CD$, and the transversal of these two parallel lines is the line EF.

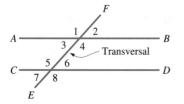

Fig 2.7

When a transversal crosses a pair of parallel lines, certain pairs of equal angles result. In Fig. 2.7, the **corresponding angles** are equal (that is, $\angle 1 = \angle 5$, $\angle 2 = \angle 6$, $\angle 3 = \angle 7$, and $\angle 4 = \angle 8$). Also, the **alternate-interior angles** are equal ($\angle 3 = \angle 6$ and $\angle 4 = \angle 5$), and the **alternate-exterior angles** are equal ($\angle 1 = \angle 8$ and $\angle 2 = \angle 7$).

When more than two parallel lines are crossed by *two* transversals, such as is shown in Fig. 2.8, *the segments of the transversals between the same two parallel lines are called* **corresponding segments.** A useful theorem is that *the ratios of corresponding segments of the transversals are equal.* In Fig. 2.8, this means that

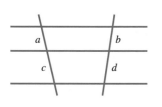

Fig 2.8

$$\frac{a}{b} = \frac{c}{d}$$

(2.1)

◀ EXAMPLE 5　　In Fig. 2.9, part of the beam structure within a building is shown. The vertical beams are parallel. From the distances between beams that are shown, determine the distance x between the middle and right vertical beams.

　　Using Eq. (2.1), we have

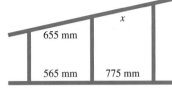

Fig 2.9

$$\frac{655}{565} = \frac{x}{775}$$

$$x = \frac{655(775)}{565}$$

$$= 898 \text{ mm} \quad \text{rounded off} \qquad ▶$$

EXERCISES 2.1

In Exercises 1–4, answer the given questions about the indicated examples of this section.

1. In Example 2, what is the measure of $\angle ABE$ in Fig. 2.2(a)?

2. In Example 3(b), if $\angle POR = 32°$ in Fig. 2.4, what is the measure of $\angle QOR$?

3. In Example 4, how many different pairs of adjacent angles are there in Fig. 2.6?

4. In Example 5, if the segments of 655 mm and 775 mm are interchanged, (a) what is the answer, and (b) is the beam along which x is measured more nearly vertical or more nearly horizontal?

In Exercises 5–12, identify the indicated angles and sides in Fig. 2.10. In Exercises 9 and 10, also find the measures of the indicated angles.

5. Two acute angles

6. Two right angles

7. The straight angle

8. The obtuse angle

9. If $\angle CBD = 65°$, find its complement.

10. If $\angle CBD = 65°$, find its supplement.

11. The sides adjacent to $\angle DBC$

12. The acute angle adjacent to $\angle DBC$

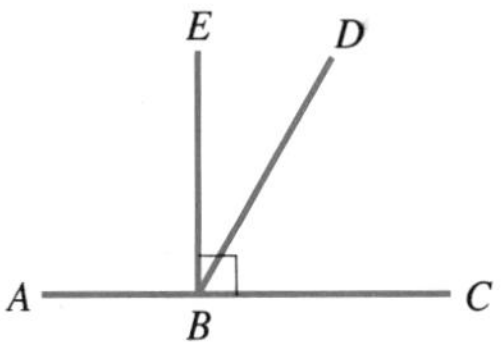

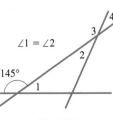

Fig 2.10

In Exercises 13 and 14, use Fig. 2.11. In Exercises 15 and 16, use Fig. 2.12. Find the measures of the indicated angles.

13. $\angle AOB$　　**14.** $\angle AOC$　　**15.** $\angle 3$　　**16.** $\angle 4$

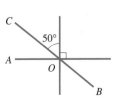

Fig. 2.11

Fig. 2.12

In Exercises 17–20, find the measures of the angles in Fig. 2.13.

17. $\angle 1$　　　**18.** $\angle 2$　　　**19.** $\angle 3$　　　**20.** $\angle 4$

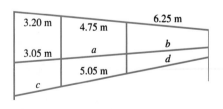

Fig 2.13

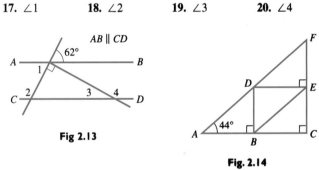

Fig. 2.14

In Exercises 21–24, find the measures of the angles in the truss shown in Fig. 2.14. A truss is a rigid support structure that is used in the construction of buildings and bridges.

21. $\angle BDF$　　**22.** $\angle ABE$　　**23.** $\angle DEB$　　**24.** $\angle DBE$

In Exercises 25–28, find the indicated distances between the straight irrigation ditches shown in Fig. 2.15. The vertical ditches are parallel.

25. a　　　**26.** b　　　**27.** c　　　**28.** d

Fig. 2.15

In Exercises 29–32, solve the given problems.

29. A steam pipe is connected in sections *AB*, *BC*, and *CD*, as shown in Fig. 2.16. What is the angle between sections *BC* and *CD* if *AB* ∥ *CD*?

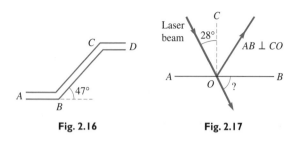

Fig. 2.16   **Fig. 2.17**

30. A laser beam striking a surface is partly reflected, and the remainder of the beam passes straight through the surface, as shown in Fig. 2.17. Find the angle between the surface and the part that passes through.

31. Find the distance on Dundas St. W between Dufferin St. and Ossington Ave. in Toronto, as shown in Fig. 2.18. The north–south streets are parallel.

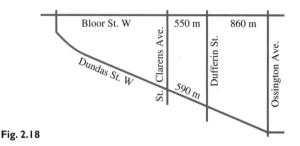

Fig. 2.18

32. An electric circuit board has equally spaced parallel wires with connections at points *A*, *B*, and *C*, as shown in Fig. 2.19. How far is *A* from *C*, if *BC* = 2.15 cm?

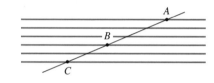

Fig. 2.19

2.2 TRIANGLES

When part of a plane is bounded and closed by straight-line segments, it is called a **polygon,** *and it is named according to the number of sides it has. A* **triangle** *has three sides, a* **quadrilateral** *has four sides, a* **pentagon** *has five sides, a* **hexagon** *has six sides, and so on.* The most important polygons are the triangle, which we consider in this section, and the quadrilateral, which we study in the next section.

TYPES AND PROPERTIES OF TRIANGLES

The properties of the triangle are important in the study of trigonometry, which we start in Chapter 4.

In a **scalene triangle,** no two sides are equal in length. In an **isosceles triangle,** two of the sides are equal in length, and the two *base angles* (the angles opposite the equal sides) are equal. In an **equilateral triangle,** the three sides are equal in length, and each of the three angles is 60°.

The most important triangle in technical applications is the **right triangle.** *In a right triangle, one of the angles is a right angle. The side opposite the right angle is the* **hypotenuse,** *and the other two sides are called* **legs.**

⟨EXAMPLE 1 Figure 2.20(a) shows a scalene triangle. We see that each side is of a different length. Figure 2.20(b) shows an isosceles triangle with two equal sides of 2 m and equal base angles of 40°. Figure 2.20(c) shows an equilateral triangle, each side of which is 5 cm, and each interior angle of which is 60°. Figure 2.20(d) shows a right triangle. The hypotenuse is side *AB*.

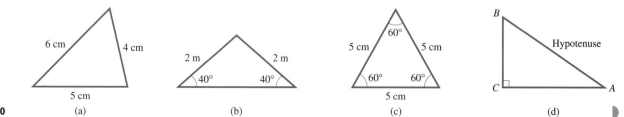

Fig. 2.20   (a)   (b)   (c)   (d)

One very important property of a triangle is that

the sum of the measures of the three angles of a triangle is 180°.

In the next example, we show this property by using material from Section 2.1.

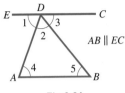

Fig 2.21

◀ **EXAMPLE 2**   In Fig. 2.21, since ∠1, ∠2, and ∠3 constitute a straight angle,

$$\angle 1 + \angle 2 + \angle 3 = 180°$$

Also, by noting alternate-interior angles, we see that ∠1 = ∠4 and ∠3 = ∠5. Therefore, by substitution we have

$$\angle 4 + \angle 2 + \angle 5 = 180°$$

Therefore, if two of the angles of a triangle are known, the third may be found by subtracting the sum of the first two from 180°. ▶

Solving a Word Problem

Fig 2.22

◀ **EXAMPLE 3**   An airplane is flying north and then makes a 90° turn to the west. Later it makes another left turn of 150°. What is the angle of a third left turn that will cause the plane to again fly north? See Fig. 2.22.

From Fig. 2.22, we see that the interior angle of the triangle at *A* is the supplement of 150°, or 30°. Since the sum of the measures of the interior angles of the triangle is 180°, the interior angle at *B* is

$$\angle B = 180° - (90° + 30°) = 60°$$

The required angle is the supplement of 60°, which is 120°. ▶

A line segment drawn from a vertex of a triangle to the *midpoint* of the opposite side is called a **median** of the triangle. A basic property of a triangle is that *the three medians meet at a single point, called the* **centroid** *of the triangle.* See Fig. 2.23. Also, *the three* **angle bisectors** (lines from the vertices that divide the angles in half) *meet at a common point.* See Fig. 2.24.

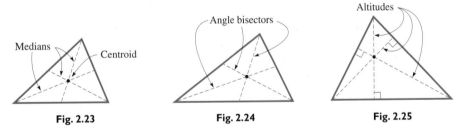

| Fig. 2.23 | Fig. 2.24 | Fig. 2.25 |

An **altitude** *(or* **height**) *of a triangle is the line segment drawn from a vertex perpendicular to the opposite side (or its extension), which is called the* **base** *of the triangle. The three altitudes of a triangle meet at a common point.* See Fig. 2.25. The three common points of the medians, angle bisectors, and altitudes are generally not the same point for a given triangle.

PERIMETER AND AREA OF A TRIANGLE

We now consider two of the most basic measures of a plane geometric figure. The first of these is its **perimeter,** *which is the total distance around it.* In the following example we find the perimeter of a triangle.

Fig 2.26

◀ EXAMPLE 4   Find the perimeter p of a triangle with sides 2.56 m, 3.22 m, and 4.89 m. See Fig. 2.26.

Using the definition of perimeter, for this triangle we have

$$p = 2.56 + 3.22 + 4.89 = 10.67 \text{ m}$$

Therefore, the distance around the triangle is 10.67 m. We express the results to hundredths since each side is given to hundredths.   ◗

The second important measure of a geometric figure is its **area.** Although the concept of area is primarily intuitive, it is easily defined and calculated for the basic geometric figures. *Area gives a measure of the surface of the figure,* just as perimeter gives the measure of the distance around it.

The area A of a triangle of base b and altitude h is

$$A = \tfrac{1}{2}bh \tag{2.2}$$

The following example illustrates the use of Eq. (2.2).

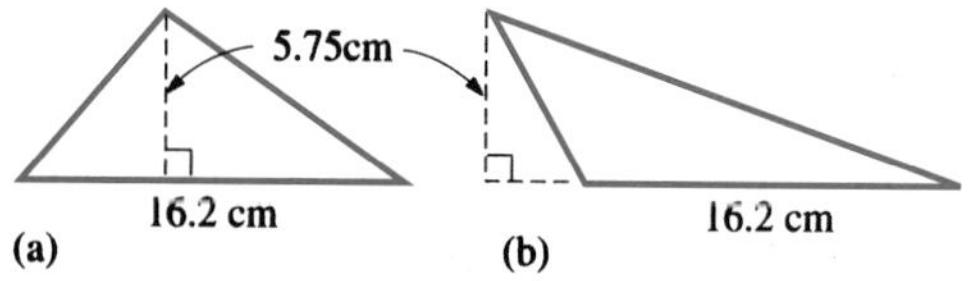

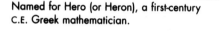

(a)   **(b)**

Fig 2.27

◀ EXAMPLE 5   Find the areas of the triangles in Fig. 2.27(a) and Fig. 2.27(b). Even though the triangles are of different shapes, we see that the base b of each triangle is 16.2 cm and that the altitude h of each is 5.75 cm. Therefore, the area of each triangle is

$$A = \tfrac{1}{2}bh = \tfrac{1}{2}(16.2)(5.75) = 46.6 \text{ cm}^2$$ ◗

Named for Hero (or Heron), a first-century C.E. Greek mathematician.

Another formula for the area of a triangle that is particularly useful when we have *a triangle with three known sides and no right angle is* **Hero's formula,** which is given in Eq. (2.3):

$$A = \sqrt{s(s-a)(s-b)(s-c)}, \tag{2.3}$$
$$\text{where } s = \tfrac{1}{2}(a+b+c)$$

In Eq. (2.3), a, b, and c are the lengths of the sides, and s is one-half of the perimeter.

◀ EXAMPLE 6   A surveyor measures the three sides of a triangular parcel of land between two intersecting straight roads to be 206 m, 293 m, and 187 m, as shown in Fig. 2.28. Find the area of this parcel.

In order to use Eq. (2.3), we first find s:

$$s = \tfrac{1}{2}(206 + 293 + 187) = \tfrac{1}{2}(686) = 343 \text{ m}$$

Now, substituting in Eq. (2.3), we have

$$A = \sqrt{343(343 - 206)(343 - 293)(343 - 187)} = 19\,100 \text{ m}^2$$

The result has been rounded off to three significant digits. In using a calculator, the value of s is stored in memory and then used to find A. It is not necessary to write down anything except the final result.   ◗

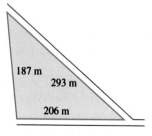

Fig 2.28

THE PYTHAGOREAN THEOREM

As we have noted, one of the most important geometric figures in technical applications is the right triangle. A very important property of a right triangle is given by the **Pythagorean theorem,** which states that

*in a **right triangle,** the square of the length of the hypotenuse equals the sum of the squares of the lengths of the other two sides.*

If c is the length of the hypotenuse and a and b are the lengths of the other two sides (see Fig. 2.29), the Pythagorean theorem is

$$c^2 = a^2 + b^2 \qquad (2.4)$$

Named for the Greek mathematician Pythagoras (sixth century B.C.E.)

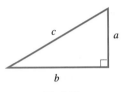

Fig 2.29

Solving a Word Problem

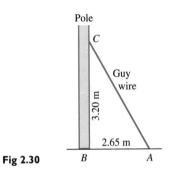

Fig 2.30

◀ **EXAMPLE 7**   A pole is perpendicular to the level ground around it. A guy wire is attached 3.20 m up the pole and at a point on the ground, 2.65 m from the pole. How long is the guy wire?

From the given information, we sketch the pole and guy wire as shown in Fig. 2.30. Using the Pythagorean theorem and then substituting, we have

$$AC^2 = AB^2 + BC^2$$
$$= 2.65^2 + 3.20^2$$
$$AC = \sqrt{2.65^2 + 3.20^2} = 4.15 \text{ m}$$

The guy wire is 4.15 m long. In using the calculator, parentheses are used to group $2.65^2 + 3.20^2$. ▶

SIMILAR TRIANGLES

Calculators can be programmed to perform specific calculations. See Appendix C for a graphing calculator program PYTHAGTH. It can be used to find a side of a right triangle, given the other two sides.

The perimeter and area of a triangle are measures of its *size*. We now consider the shape of triangles.

Two triangles are **similar** *if they have the same shape (but not necessarily the same size).* There are two very important properties of similar triangles.

> **PROPERTIES OF SIMILAR TRIANGLES**
> **1.** *The corresponding angles of similar triangles are equal.*
> **2.** *The corresponding sides of similar triangles are proportional.*

NOTE ▶   For two triangles that are similar, *if one property is true, then the other is also true.* In two similar triangles, the **corresponding sides** *are the sides, one in each triangle, that are between the same pair of equal corresponding angles.*

◀ **EXAMPLE 8**   In Fig. 2.31, a pair of similar triangles are shown. They are similar even though the corresponding parts are not in the same position relative to the page. Using standard symbols, we can write $\triangle ABC \sim \triangle A'B'C'$, where $\triangle$ means "triangle" and $\sim$ means "is similar to."

The pairs of corresponding angles are A and A', B and B', and C and C'. This means $A = A'$, $B = B'$, and $C = C'$.

The pairs of corresponding sides are AB and $A'B'$, BC and $B'C'$, and AC and $A'C'$. In order to show that these corresponding sides are proportional, we write

$$\frac{AB}{A'B'} = \frac{BC}{B'C'} = \frac{AC}{A'C'} \quad \leftarrow \text{sides of } \triangle ABC \\ \leftarrow \text{sides of } \triangle A'B'C'$$

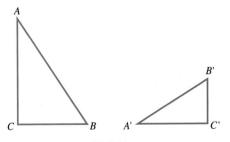

Fig 2.31

If we know that two triangles are similar, we can use the two basic properties of similar triangles to find the unknown parts of one triangle from the known parts of the other triangle. The next example illustrates this in a practical application.

Solving a Word Problem

◖EXAMPLE 9 On level ground a silo casts a shadow 24 m long. At the same time, a pole 4.0 m high casts a shadow 3.0 m long. How tall is the silo? See Fig. 2.32.

The rays of the sun are essentially parallel. The two triangles in Fig. 2.32 are similar since *each has a right angle and the angles at the tops are equal.* The other angles must be equal since the sum of the angles is 180°. The lengths of the hypotenuses are of no importance in this problem, so we use only the other sides in stating the ratios of corresponding sides. Denoting the height of the silo as h, we have

$$\frac{h}{4.0} = \frac{24}{3.0}, \qquad h = 32 \text{ m}$$

We conclude that the silo is 32 m high.

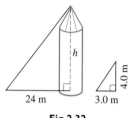

24 m   3.0 m   4.0 m

Fig 2.32

One of the most practical uses of similar geometric figures is that of **scale drawings.** Maps, charts, blueprints, and most drawings that appear in books are familiar examples of scale drawings. Actually, there have been many scale drawings used in this book already.

In any scale drawing, all distances are drawn a certain ratio of the distances they represent, and all angles equal the angles they represent.

◖EXAMPLE 10 In drawing a map of the area shown in Fig. 2.33, a scale of 1 cm = 200 km is used. In measuring the distance between Chicago and Toronto on the map, we find it to be 3.5 cm. The actual distance x between Chicago and Toronto is found from the proportion

scale
↓

actual distance ⟶ $\dfrac{x}{3.5 \text{ cm}} = \dfrac{200 \text{ km}}{1 \text{ cm}}$ or $x = 700$ km
distance on map ⟶

Toronto

3.5 cm   2.7 cm

Chicago   Philadelphia

Fig 2.33

If we did not have the scale but knew that the distance between Chicago and Toronto is 700 km, then by measuring distances on the map between Chicago and Toronto (3.5 cm) and between Toronto and Philadelphia (2.7 cm), we could find the distance between Toronto and Philadelphia. It is found from the following proportion, determined by use of similar triangles:

$$\frac{700 \text{ km}}{3.5 \text{ cm}} = \frac{y}{2.7 \text{ cm}}$$

$$y = \frac{2.7(700)}{3.5} = 540 \text{ km}$$

Similarity requires *equal* angles and *proportional* sides. *If the corresponding angles and the corresponding sides of two triangles are equal, the two triangles are* **congruent.** As a result of this definition, the areas and perimeters of congruent triangles are also equal. Informally, we can say that similar triangles have the same shape, whereas congruent triangles have the same shape and same size.

◖ EXAMPLE 11   A right triangle with legs of 2 cm and 4 cm is congruent to any other right triangle with legs of 2 cm and 4 cm. However, it is similar to any right triangle with legs of 5 cm and 10 cm, since the corresponding sides are proportional. See Fig. 2.34.

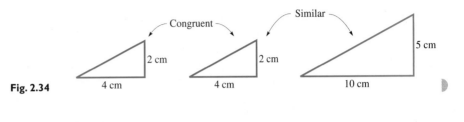

Fig. 2.34

EXERCISES 2.2

In Exercises 1–4, answer the given questions about the indicated examples of this section.

1. In Example 2, if ∠1 = 70° and ∠5 = 45° in Fig. 2.21, what is the measure of ∠2?

2. In Example 5, if 16.2 cm is changed to 61.2 cm, what is the answer?

3. In Example 7, if 2.65 m is changed to 6.25 m, what is the answer?

4. In Example 9, if the 4.0 m is interchanged with the 3.0 m, what is the answer?

In Exercises 5–8, determine ∠A in the indicated figures.

5. Fig. 2.35(a)
6. Fig. 2.35(b)
7. Fig. 2.35(c)
8. Fig. 2.35(d)

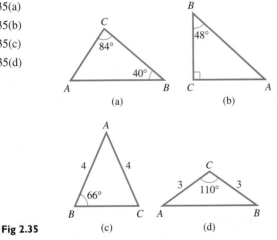

Fig 2.35

In Exercises 9–12, find the perimeter of each triangle.

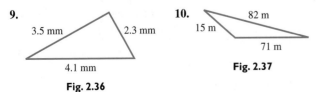

Fig. 2.36

Fig. 2.37

11. An equilateral triangle of side 21.5 cm

12. An isosceles triangle with equal sides of 2.45 dm and third side of 3.22 dm

In Exercises 13–20, find the area of each triangle.

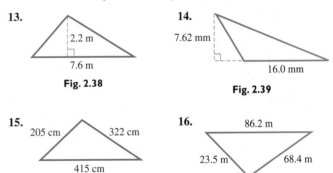

13.

Fig. 2.38

14.

Fig. 2.39

15.

Fig. 2.40

16.

Fig. 2.41

17. Right triangle with legs 3.46 cm and 2.55 cm

18. Right triangle with legs 234 mm and 342 mm

19. An isosceles triangle with equal sides of 0.986 m and third side of 0.884 m

20. An equilateral triangle of side 3.20 dm

In Exercises 21–24, find the third side of the right triangle shown in Fig. 2.42 for the given values.

21. $a = 13.8$ mm, $b = 22.7$ mm
22. $a = 2.48$ m, $b = 1.45$ m
23. $a = 175$ cm, $c = 551$ cm
24. $b = 0.474$ km, $c = 0.836$ km

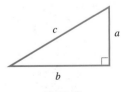

Fig 2.42

In Exercises 25–28, use the right triangle in Fig. 2.43.

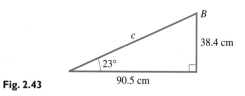

Fig. 2.43

38.4 cm

c

23°

90.5 cm

25. Find ∠*B*.

26. Find side *c*.

27. Find the perimeter.

28. Find the area.

In Exercises 29–48, solve the given problems.

29. In Fig. 2.44, show that △*MKL* ∼ △*MNO*.

30. In Fig. 2.45, show that △*ACB* ∼ △*ADC*.

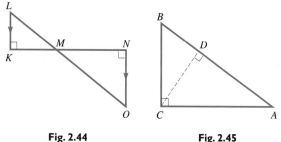

Fig. 2.44   **Fig. 2.45**

31. In Fig. 2.44, if *KN* = 15, *MN* = 9, and *MO* = 12, find *LM*.

32. In Fig. 2.45, if *AD* = 9 and *AC* = 12, find *AB*.

33. For what type of triangle is the centroid the same point as the intersection of altitudes and the intersection of angle bisectors?

(W) **34.** Is it possible that the altitudes of a triangle meet, when extended, outside the triangle? Explain.

35. The angle between the roof sections of an A-frame house is 50°. What is the angle between either roof section and a horizontal rafter?

36. A transmitting tower is supported by a wire that makes an angle of 52° with the level ground. What is the angle between the tower and the wire?

37. A wall pennant is in the shape of an isosceles triangle. If each equal side is 76.6 cm long and the third side is 30.6 cm, what is the area of the pennant?

38. The Bermuda Triangle is sometimes defined as an equilateral triangle 1600 km on a side, with vertices in Bermuda, Puerto Rico, and the Florida coast. Assuming it is flat, what is its approximate area?

39. The sail of a sailboat is in the shape of a right triangle with sides of 3.2 m, 6.0 m, and 6.8 m. What is the area of the sail?

40. An observer is 550 m horizontally from the launch pad of a rocket. After the rocket has ascended 750 m, how far is it from the observer?

41. The base of a 6.0-m ladder is 1.8 m from a wall. How far up on the wall does the ladder reach?

42. The beach shade shown in Fig. 2.46 is made of 30°-60°-90° triangular sections. Find *x*. (In a 30°-60°-90° triangle, the side opposite the 30° angle is one-half the hypotenuse.)

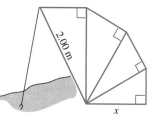

2.00 m

x

Fig 2.46

43. A rectangular room is 18 m long, 12 m wide, and 8.0 m high. What is the length of the longest diagonal from one corner to another corner of the room?

44. On a blueprint, a hallway is 45.6 cm long. The scale is 1.2 cm = 1.0 m. How long is the hallway?

45. Two parallel guy wires are attached to a vertical pole 4.5 m and 5.4 m above the ground. They are secured on the level ground at points 1.2 m apart. How long are the guy wires?

46. To find the width *ED* of a river, a surveyor places markers at *A*, *B*, *C*, and *D*, as shown in Fig. 2.47. The markers are placed such that *AB*∥*ED*, *BC* = 50.0 m, *DC* = 312 m, and *AB* = 80.0 m. How wide is the river?

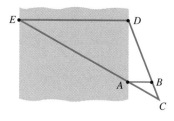

E   *D*

A   *B*

C

Fig. 2.47

47. To find the height of a flagpole, a person places a mirror at *M*, as shown in Fig. 2.48. The person's eyes at *E* are 160 cm above the ground at *A*. From physics, it is known that ∠*AME* = ∠*BMF*. If *AM* = 120 cm and *MB* = 4.5 m, find the height *BF* of the flagpole.

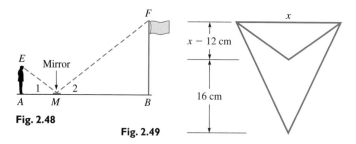

F

x

x − 12 cm

16 cm

E

Mirror

1   2

A   *M*   *B*

Fig. 2.48

Fig. 2.49

48. The cross section of a drainage trough has the shape of an isosceles triangle whose depth is 12 cm less than its width. If the depth is increased by 16 cm and the width remains the same, the area of the cross section is increased by 160 cm². Find the original depth and width. See Fig. 2.49.

Fig 2.50   **Fig 2.51**

QUADRILATERALS

A **quadrilateral** *is a closed plane figure with four sides*, and these four sides form four interior angles. A general quadrilateral is shown in Fig. 2.50.

A **diagonal** *of a polygon is a straight line segment joining any two nonadjacent vertices.* The dashed line is one of two diagonals of the quadrilateral in Fig. 2.51.

TYPES OF QUADRILATERALS

A **parallelogram** *is a quadrilateral in which opposite sides are parallel.* In a parallelogram, opposite sides are equal and opposite angles are equal. A **rhombus** *is a parallelogram with four equal sides.*

A **rectangle** *is a parallelogram in which intersecting sides are perpendicular,* which means that all four interior angles are right angles. In a rectangle, the longer side is usually called the **length,** and the shorter side is called the **width.** *A* **square** *is a rectangle with four equal sides.*

A **trapezoid** *is a quadrilateral in which two sides are parallel.* The parallel sides are called the **bases** of the trapezoid.

◀ EXAMPLE 1   A parallelogram is shown in Fig. 2.52 (a). Opposite sides *a* are equal in length, as are opposite sides *b*. A rhombus with equal sides *s* is shown in Fig. 2.52 (b). A rectangle is shown in Fig. 2.52 (c). The length is labeled *l*, and the width is labeled *w*. A square with equal sides *s* is shown in Fig. 2.52 (d). A trapezoid with bases b_1 and b_2 is shown in Fig. 2.52 (e).

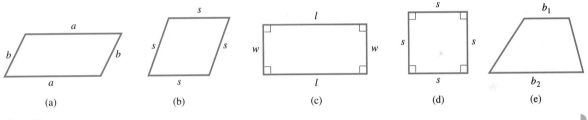

(a)   (b)   (c)   (d)   (e)

Fig. 2.52

PERIMETER AND AREA OF A QUADRILATERAL

The perimeter of a quadrilateral is the sum of the lengths of the four sides.

Solving a Word Problem

540 mm

920 mm

Fig 2.53   540 mm

◀ EXAMPLE 2   An architect designs a room with a rectangular window 920 mm high and 540 mm wide, with another window above in the shape of an equilateral triangle, 540 mm on a side. See Fig. 2.53. How much molding is needed for these windows?

The length of molding is the sum of the perimeters of the windows. For the rectangular window, the opposite sides are equal, which means the perimeter is twice the length *l* plus twice the width *w*. For the equilateral triangle, the perimeter is three times the side *s*. Therefore, the length *L* of molding is

$$L = 2l + 2w + 3s$$
$$= 2(920) + 2(540) + 3(540)$$
$$= 4540 \text{ mm}$$

NOTE ▶ We could write down formulas for the perimeters of the different kinds of triangles and quadrilaterals. However, if we *remember the meaning of perimeter as being the total distance around a geometric figure*, such formulas are not necessary.

For the areas of the square, rectangle, parallelogram, and trapezoid, we have the following formulas.

$A = s^2$	Square of side s (Fig. 2.54)	(2.5)
$A = lw$	Rectangle of length l and width w (Fig. 2.55)	(2.6)
$A = bh$	Parallelogram of base b and height h (Fig. 2.56)	(2.7)
$A = \frac{1}{2}h(b_1 + b_2)$	Trapezoid of bases b_1 and b_2, and height h (Fig. 2.57)	(2.8)

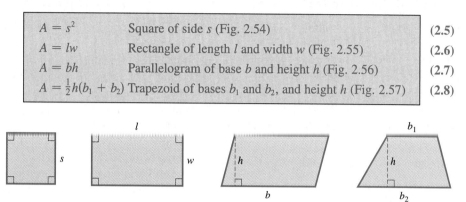

Fig. 2.54   **Fig. 2.55**   **Fig. 2.56**   **Fig. 2.57**

Since a rectangle, a square, and a rhombus are special types of parallelograms, the area of these figures can be found from Eq. (2.7). The area of a trapezoid is of importance when we find areas of irregular geometric figures in Section 2.5.

◀ EXAMPLE 3   A city park is designed with lawn areas in the shape of a right triangle, a parallelogram, and a trapezoid, as shown in Fig. 2.58, with walkways between them. Find the area of each section of lawn and the total lawn area.

$$A_1 = \tfrac{1}{2}bh = \tfrac{1}{2}(72)(45) = 1600 \text{ m}^2 \qquad A_2 = bh = (72)(45) = 3200 \text{ m}^2$$
$$A_3 = \tfrac{1}{2}h(b_1 + b_2) = \tfrac{1}{2}(45)(72 + 35) = 2400 \text{ m}^2$$

The total lawn area is about 7200 m². ▶

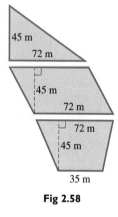

Fig 2.58

Solving a Word Problem

◀ EXAMPLE 4   The length of a rectangular computer chip is 2.0 mm longer than its width. Find the dimensions of the chip if its perimeter is 26.4 mm.

Since the dimensions, the length and the width, are required, let $w =$ the width of the chip. Since the length is 2.0 mm more than the width, we know that $w + 2.0 =$ the length of the chip. See Fig. 2.59.

Since the perimeter of a rectangle is twice the length plus twice the width, we have the equation

$$2(w + 2.0) + 2w = 26.4$$

since the perimeter is given as 26.4 mm. This is the equation we need.

Solving this equation, we have

$$2w + 4.0 + 2w = 26.4$$
$$4w = 22.4$$
$$w = 5.6 \text{ mm} \quad \text{and} \quad w + 2.0 = 7.6 \text{ mm}$$

Therefore, the length is 7.6 mm and the width is 5.6 mm. These values check with the statements of the original problem. ▶

The computer microprocessor chip was first commercially available in 1971.

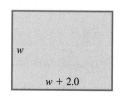
Fig 2.59

EXERCISES 2.3

In Exercises 1–4, make the given changes in the indicated examples of this section and then solve the given problems.

1. In Example 1, interchange the lengths of b_1 and b_2 in Fig. 2.52(e). What type of quadrilateral is the resulting figure?

2. In Example 2, change the equilateral triangle of side 540 mm to a square of side 540 mm and then find the length of molding.

3. In Example 3, change the dimension of 45 m to 55 m in each figure and then find the area.

4. In Example 4, change 2.0 mm to 3.0 mm and then find the dimensions.

In Exercises 5–12, find the perimeter of each figure.

5. Square: side of 65 m

6. Rhombus: side of 2.46 km

7. Rectangle: $l = 0.920$ mm, $w = 0.742$ mm

8. Rectangle: $l = 142$ cm, $w = 126$ cm

9. The parallelogram in Fig. 2.60

10. The parallelogram in Fig. 2.61

11. The trapezoid in Fig. 2.62

12. The trapezoid in Fig. 2.63

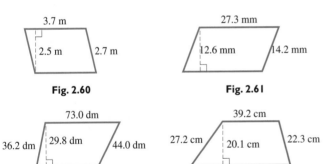

Fig. 2.60   **Fig. 2.61**

Fig. 2.62   **Fig. 2.63**

In Exercises 13–20, find the area of each figure.

13. Square: $s = 2.7$ mm   14. Square: $s = 15.6$ m

15. Rectangle: $l = 0.920$ km, $w = 0.742$ km

16. Rectangle: $l = 142$ cm, $w = 126$ cm

17. The parallelogram in Fig. 2.60

18. The parallelogram in Fig. 2.61

19. The trapezoid in Fig. 2.62

20. The trapezoid in Fig. 2.63

In Exercises 21–24, set up a formula for the indicated perimeter or area. (Do not include dashed lines.)

21. The perimeter of the figure in Fig. 2.64 (a parallelogram and a square attached)

22. The perimeter of the figure in Fig. 2.65 (two trapezoids attached)

23. The area of the figure in Fig. 2.64

24. The area of the figure in Fig. 2.65

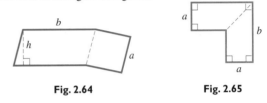

Fig. 2.64   **Fig. 2.65**

In Exercises 25–36, solve the given problems.

(W) 25. If the angle between adjacent sides of a parallelogram is 90°, what conclusion can you make about the parallelogram?

(W) 26. What conclusion can you make about the two triangles formed by the sides and diagonal of a parallelogram? Explain.

(W) 27. Noting how a diagonal of a rhombus divides an interior angle, explain why the automobile jack in Fig. 2.66 is in the shape of a rhombus.

Fig. 2.66   **Fig. 2.67**

28. Part of an electric circuit is wired in the configuration of a rhombus and one of its altitudes as shown in Fig. 2.67. What is the length of wire in this part of the circuit?

29. A walkway 3.0 m wide is constructed along the outside edge of a square courtyard. If the perimeter of the courtyard is 320 m, what is the perimeter of the square formed by the outer edge of the walkway?

30. An architect designs a rectangular window such that the width of the window is 450 mm less than the height. If the perimeter of the window is 4500 mm, what are its dimensions?

31. A designer plans the top of a rectangular workbench to be four times as long as it is wide and then determines that if the width is 1500 mm greater and the length is 3000 mm less, it would be a square. What are its dimensions?

32. A beam support in a building is in the shape of a parallelogram, as shown in Fig. 2.68. Find the area of the side of the beam shown.

Fig. 2.68

33. Each of two walls (with rectangular windows) of an A-frame house has the shape of a trapezoid as shown in Fig. 2.69. If a liter of paint covers 12 m², how much paint is required to paint these walls? (All data are accurate to two significant digits.)

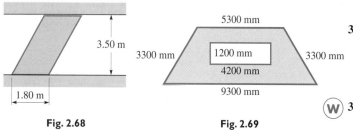

Fig. 2.69

34. A rectangular security area is enclosed on one side by a wall, and the other sides are fenced. The length of the wall is twice the width of the area. The total cost of building the wall and fence is $13 200. If the wall costs $50.00/m and the fence costs $5.00/m, find the dimensions of the area.

W **35.** What is the sum of the measures of the interior angles of a quadrilateral? Explain.

36. Find a formula for the area of a rhombus in terms of its diagonals d_1 and d_2. (See Exercise 27.)

2.4 CIRCLES

The next geometric figure we consider is the circle. *All points on a* **circle** *are at the same distance from a fixed point, the* **center** *of the circle. The distance from the center to a point on the circle is the* **radius** *of the circle. The distance between two points on the circle on a line through the center is the* **diameter.** Therefore, the diameter d is twice the radius r, or $d = 2r$. See Fig. 2.70.

There are also certain special types of lines associated with a circle. *A* **chord** *is a line segment having its endpoints on the circle. A* **tangent** *is a line that touches (does not pass through) the circle at one point. A* **secant** *is a line that passes through two points of the circle.* See Fig. 2.71.

An important property of a tangent is that *a tangent to a circle is perpendicular to the radius drawn to the point of contact.* This is illustrated in the next example.

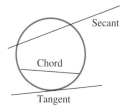

Fig 2.70

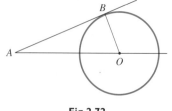

Fig 2.71

◀ **EXAMPLE 1** In Fig. 2.72, O is the center of the circle, and AB is tangent at B. If $\angle OAB = 25°$, find $\angle AOB$.

Since the center is O, OB is a radius of the circle. A tangent is perpendicular to a radius at the point of tangency, which means $\angle ABO = 90°$, so that

$$\angle OAB + \angle OBA = 25° + 90° = 115°$$

Since the sum of the angles of a triangle is 180°, we have

$$\angle AOB = 180° - 115° = 65°$$

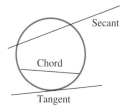

Fig 2.72

CIRCUMFERENCE AND AREA OF A CIRCLE

The perimeter of a circle is called the **circumference.** The formulas for the circumference and area of a circle are as follows:

$c = 2\pi r$	Circumference of a circle of radius r	**(2.9)**
$A = \pi r^2$	Area of a circle of radius r	**(2.10)**

The symbol π (the Greek letter pi), which we use as a number, was first used in this way as a number in the 1700s.

Here, π equals approximately 3.1416. In using a calculator, π can be entered by using the ⬚ π ⬚ key.

EXAMPLE 2 A circular oil spill has a diameter of 2.4 km. This oil spill is to be enclosed within a length of special flexible tubing. What is the area of the spill, and how long must the tubing be? See Fig. 2.73.

We find the area by using Eq. (2.10). Since $d = 2r$, $r = d/2 = 1.2$ km. Therefore, the area is

$$A = \pi r^2 = \pi (1.2)^2$$
$$= 4.5 \text{ km}^2$$

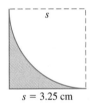

Fig 2.73

The length of the tubing needed to enclose the oil spill is the circumference of the circle. Therefore,

$$c = 2\pi r = 2\pi(1.2) \qquad \text{note that } c = \pi d$$
$$= 7.5 \text{ km}$$

Results have been rounded off to two significant digits, the accuracy of d.

Many applied problems involve a combination of geometric figures. The following example illustrates one such combination.

Solving a Word Problem

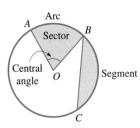

$s = 3.25$ cm

Fig 2.74

EXAMPLE 3 A machine part is a square of side 3.25 cm with a quarter-circle removed (see Fig. 2.74). Find the perimeter and the area of one side of the part.

Setting up a formula for the perimeter, we add the two sides of length s to *one-fourth of the circumference of a circle with radius s*. For the area, we *subtract the area of one-fourth of a circle from the area of the square*. This gives

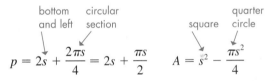

$$p = 2s + \frac{2\pi s}{4} = 2s + \frac{\pi s}{2} \qquad A = s^2 - \frac{\pi s^2}{4}$$

where s is the side of the square and the radius of the circle. Evaluating, we have

$$p = 2(3.25) + \frac{\pi(3.25)}{2} = 11.6 \text{ cm}$$

$$A = 3.25^2 - \frac{\pi(3.25)^2}{4} = 2.27 \text{ cm}^2$$

CIRCULAR ARCS AND ANGLES

*An **arc** is part of a circle, and an angle formed at the center by two radii is a **central angle**.* The measure of an arc is the same as the central angle between the ends of the radii that define the arc. *A **sector** of a circle is the region bounded by two radii and the arc they intercept. A **segment** of a circle is the region bounded by a chord and its arc.* (There are two possible segments for a given chord. The smaller region is a *minor segment*, and the larger region is a *major segment*.). These are illustrated in the following example.

EXAMPLE 4 In Fig. 2.75, a sector of the circle is between radii OA and OB and arc AB (which is denoted by $\overarc{AB}$). If the measure of the central angle at O between the radii is 70°, the measure of $\overarc{AB}$ is also 70°.

In Fig. 2.75, a segment of the circle is the region between chord BC and arc BC ($\overarc{BC}$).

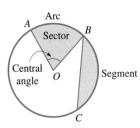

Fig 2.75

An **inscribed angle** *of an arc is one for which the endpoints of the arc are points on the sides of the angle and for which the vertex is a point (not an endpoint) of the arc.* An important property of a circle is that *the measure of an inscribed angle is one-half of its intercepted arc.*

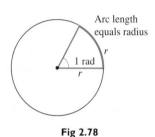

Inscribed angle

Intercepted arc

Fig 2.76

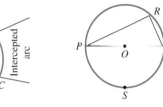

Fig 2.77

▌ EXAMPLE 5 (a) In the circle shown in Fig. 2.76, $\angle ABC$ is inscribed in $\overset{\frown}{ABC}$, and it intercepts $\overset{\frown}{AC}$. If $\overset{\frown}{AC} = 60°$, then $\angle ABC = 30°$.

(b) In the circle shown in Fig. 2.77, PQ is a diameter, and $\angle PRQ$ is inscribed in the semicircular $\overset{\frown}{PRQ}$. Since $\overset{\frown}{PSQ} = 180°$, $\angle PRQ = 90°$. From this we conclude that *an angle inscribed in a semicircle is a right angle.* ▐

RADIAN MEASURE OF AN ANGLE

To this point, we have measured all angles in degrees. There is another measure of an angle, the *radian*, that is defined in terms of an arc of a circle. We will find it of importance when we study trigonometry.

If a central angle of a circle intercepts an arc equal in length to the radius of the circle, the measure of the central angle is defined as 1 **radian.** See Fig. 2.78. Since the radius can be marked off along the circumference 2π times (about 6.283 times), we see that 2π rad $= 360°$ (where rad is the symbol for radian). Therefore,

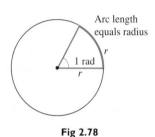

Arc length equals radius

1 rad

Fig 2.78

$$\boxed{\pi \, \text{rad} = 180°} \tag{2.11}$$

is a basic relationship between radians and degrees.

▌ EXAMPLE 6 (a) If we divide each side of Eq. (2.11) by π, we get

$$1 \, \text{rad} = 57.3°$$

where the result has been rounded off.

(b) To change an angle of $118.2°$ to radian measure, we have

$$118.2° = 118.2°\left(\frac{\pi \, \text{rad}}{180°}\right) = 2.06 \, \text{rad}$$

By multiplying $118.2°$ by π rad$/180°$, the unit of measurement that remains is rad, since the degrees "cancel." We will review radian measure again when we study trigonometry. ▐

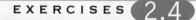

EXERCISES 2.4

In Exercises 1–4, answer the given questions about the indicated examples of this section.

1. In Example 1, if $\angle AOB = 72°$ in Fig. 2.72, then what is the measure of $\angle OAB$?

2. In the first line of Example 2, if "diameter" is changed to "radius," what are the results?

3. In Example 3, if the machine part is the unshaded part (rather than the shaded part) of Fig. 2.74, what are the results?

4. In Example 5(a), if $\angle ABC = 25°$ in Fig. 2.76, then what is the measure of $\overset{\frown}{AC}$?

In Exercises 5–8, refer to the circle with center at O in Fig. 2.79. Identify the following.

5. (a) A secant line
   (b) A tangent line

6. (a) Two chords
   (b) An inscribed angle

7. (a) Two perpendicular lines
   (b) An isosceles triangle

8. (a) A segment
   (b) A sector with an acute central angle

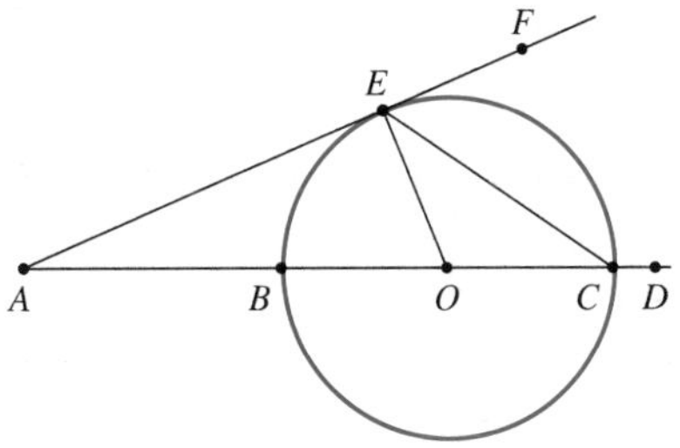

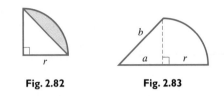

Fig 2.79

In Exercises 9–12, find the circumference of the circle with the given radius or diameter.

9. $r = 275$ cm **10.** $r = 0.563$ m

11. $d = 23.1$ mm **12.** $d = 8.2$ dm

In Exercises 13–16, find the area of the circle with the given radius or diameter.

13. $r = 0.0952$ km **14.** $r = 45.8$ cm

15. $d = 2.33$ m **16.** $d = 125.6$ mm

In Exercises 17–20, refer to Fig. 2.80, where AB is a diameter, TB is a tangent line at B, and $\angle ABC = 65°$. Determine the indicated angles.

17. $\angle CBT$
18. $\angle BCT$
19. $\angle CAB$
20. $\angle BTC$

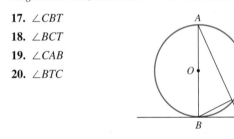

Fig 2.80

In Exercises 21–24, refer to Fig. 2.81. Determine the indicated arcs and angles.

21. $\overset{\frown}{BC}$
22. $\overset{\frown}{AB}$
23. $\angle ABC$
24. $\angle ACB$

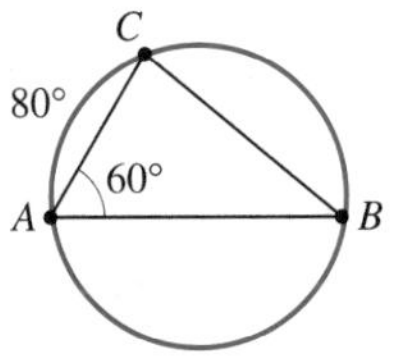

Fig 2.81

In Exercises 25–28, change the given angles to radian measure.

25. $22.5°$ **26.** $60.0°$ **27.** $125.2°$ **28.** $323.0°$

In Exercises 29–32, find a formula for the indicated perimeter or area.

Fig. 2.82   **Fig. 2.83**

29. The perimeter of the quarter-circle in Fig. 2.82

30. The perimeter of the figure in Fig. 2.83. A quarter-circle is attached to a triangle.

31. The area of the segment of the quarter-circle in Fig. 2.82

32. The area of the figure in Fig. 2.83

In Exercises 33–44, solve the given problems.

(**W**) **33.** Describe the location of the midpoints of a set of parallel chords of a circle.

(**W**) **34.** In Fig. 2.84, chords *AB* and *DE* are parallel. What is the relation between $\triangle ABC$ and $\triangle CDE$? Explain.

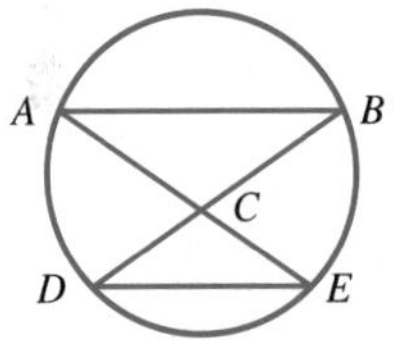

Fig 2.84

35. The radius of the earth's equator is 6370 km. What is the circumference?

36. As a ball bearing rolls along a straight track, it makes 11.0 revolutions while traveling a distance of 109 mm. Find its radius.

37. The rim on a basketball hoop has an inside diameter of 45.7 cm. The largest cross section of a basketball has a diameter of 30.5 cm. What is the ratio of the cross sectional area of the basketball to the area of the hoop?

38. With no change in the rate of flow, by what factor should the diameter of a pipe be increased in order to double the amount of water that flows through the pipe?

39. Using a tape measure, the circumference of a tree is found to be 112 cm. What is the diameter of the tree (assuming a circular cross section)?

40. What is the area of the largest circle that can be cut from a rectangular plate 21.2 cm by 15.8 cm?

41. A window designed between semicircular regions is shown in Fig. 2.85. Find the area of the window.

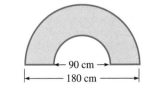

Fig. 2.85

90 cm

180 cm

42. Find the length of the pulley belt shown in Fig. 2.86, if the belt crosses at right angles. The radius of each pulley wheel is 5.50 cm.

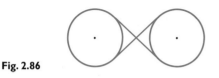

Fig. 2.86

(W) **43.** The velocity of an object moving in a circular path is directed tangent to the circle in which it is moving. A stone on a string moves in a vertical circle, and the string breaks after 5.5 revolutions. If the string was initially in a vertical position, in what direction does it move after the string breaks? Explain.

44. Part of a circular gear with 24 teeth is shown in Fig. 2.87. Find the indicated angle.

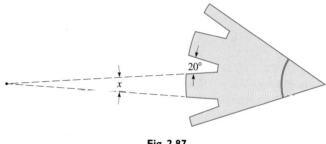

Fig. 2.87

2.5 MEASUREMENT OF IRREGULAR AREAS

To this point, the figures for which we have found areas are well defined, and the areas can be found by direct use of a specific formula. In practice, however, it may be necessary to find the area of a figure with an irregular perimeter or one for which there is no specific formula. In this section, we show two methods of finding a very good *approximation* of such an area. These methods are particularly useful in technical areas such as surveying, architecture, and mechanical design.

THE TRAPEZOIDAL RULE

The first method is based on dividing the required area into trapezoids with equal heights. Considering the area shown in Fig. 2.88, we draw parallel lines at n equal intervals between the edges of the area. We then join the ends of these parallel line segments to form adjacent trapezoids. The sum of the areas of the trapezoids gives a good approximation to the required area.

Calling the lengths of the parallel lines $y_0, y_1, y_2, \ldots, y_n$ and the height of each trapezoid h (the distance between the parallel lines), the total area A is the sum of areas of all the trapezoids. This gives us

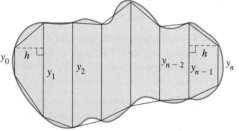

Fig 2.88

| first trapezoid | second trapezoid | third trapezoid | next-to-last trapezoid | last trapezoid |

$$A = \frac{h}{2}(y_0 + y_1) + \frac{h}{2}(y_1 + y_2) + \frac{h}{2}(y_2 + y_3) + \cdots + \frac{h}{2}(y_{n-2} + y_{n-1}) + \frac{h}{2}(y_{n-1} + y_n)$$

$$= \frac{h}{2}(y_0 + y_1 + y_1 + y_2 + y_2 + y_3 + \cdots + y_{n-2} + y_{n-1} + y_{n-1} + y_n)$$

Therefore, the approximate area is

Note carefully that the values of y_0 and y_n are *not* multiplied by 2.

$$A = \frac{h}{2}(y_0 + 2y_1 + 2y_2 + \cdots + 2y_{n-1} + y_n) \qquad (2.12)$$

Equation (2.12) is known as the **trapezoidal rule.** The following examples illustrate its use.

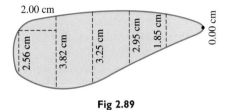

2.00 cm

2.56 cm | 3.82 cm | 3.25 cm | 2.95 cm | 1.85 cm | 0.00 cm

Fig 2.89

EXAMPLE 1   A plate cam for opening and closing a valve is shown in Fig. 2.89. Widths of the face of the cam are shown at 2.00-cm intervals. Find the area of the face of the cam.

From the figure we see that

$$y_0 = 2.56 \text{ cm} \qquad y_1 = 3.82 \text{ cm} \qquad y_2 = 3.25 \text{ cm}$$
$$y_3 = 2.95 \text{ cm} \qquad y_4 = 1.85 \text{ cm} \qquad y_5 = 0.00 \text{ cm}$$

(In making such measurements, often a y-value at one end—or both ends—is zero. In such a case, the end "trapezoid" is actually a triangle.) From the given information in this example, $h = 2.00$ cm. Therefore, using the trapezoidal rule, Eq. (2.12), we have

$$A = \frac{2.00}{2}[2.56 + 2(3.82) + 2(3.25) + 2(2.95) + 2(1.85) + 0.00]$$

$$= 26.3 \text{ cm}^2$$

The area of the face of the cam is approximately 26.3 cm².

When approximating the area with trapezoids, we omit small parts of the area for some trapezoids and include small extra areas for other trapezoids. The omitted areas often approximate the extra areas, which makes the approximation better. Also, the use of smaller intervals improves the approximation since the total omitted area or total extra area is smaller.

See the chapter introduction.

EXAMPLE 2   From a satellite photograph of Lake Ontario (as shown on page 50), one of the Great Lakes between the United States and Canada, measurements of the width of the lake were made along its length, starting at the west end, at 26.0 km intervals. The widths are shown in Fig. 2.90 and are given in the following table.

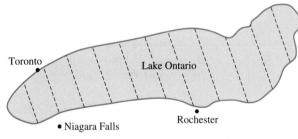

Toronto

Lake Ontario

Rochester

• Niagara Falls

Fig 2.90

Distance from West End (km)	0.0	26.0	52.0	78.0	104	130	156
Width (km)	0.0	46.7	52.1	59.2	60.4	65.7	73.9

Distance from West End (km)	182	208	234	260	286	312
Width (km)	87.0	75.5	66.4	86.1	77.0	0.0

Here we see that $y_0 = 0.0$ km, $y_1 = 46.7$ km, $y_2 = 52.1$ km,..., and $y_n = 0.0$ km. Therefore, using the trapezoidal rule, the approximate area of Lake Ontario is found as follows:

$$A = \frac{26.0}{2}[0.0 + 2(46.7) + 2(52.1) + 2(59.2) + 2(60.4) + 2(65.7) + 2(73.9)$$

$$+ 2(87.0) + 2(75.5) + 2(66.4) + 2(86.1) + 2(77.0) + 0.0] = 19\,500 \text{ km}^2$$

The area of Lake Ontario is actually 19 477 km².

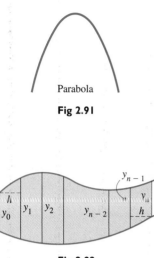

Parabola

Fig 2.91

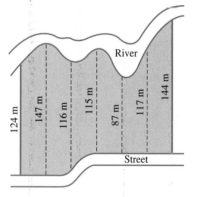

Fig 2.92

Named for the English mathematician Thomas Simpson (1710–1761).

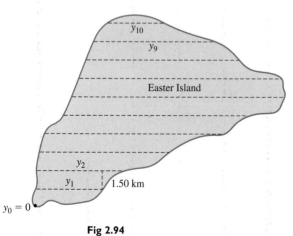

Fig 2.93

SIMPSON'S RULE

For the second method of measuring an irregular area, we also draw parallel lines at equal intervals between the edges of the area. We then join the ends of these parallel lines with curved *arcs*. This takes into account the fact that the perimeters of most figures are curved. The arcs used in this method are not arcs of a circle, but arcs of a *parabola*. A parabola is shown in Fig. 2.91 and is discussed in detail in Chapter 21. (Examples of parabolas are (1) the path of a ball that has been thrown and (2) the cross section of a microwave "dish.")

The development of this method requires advanced mathematics. Therefore, we will simply state the formula to be used. It might be noted that the form of the equation is similar to that of the trapezoidal rule.

The approximate area of the geometric figure shown in Fig. 2.92 is given by

$$A = \frac{h}{3}(y_0 + 4y_1 + 2y_2 + 4y_3 + \cdots + 2y_{n-2} + 4y_{n-1} + y_n) \qquad (2.13)$$

Equation (2.13) is known as **Simpson's rule.** In using Eq. (2.13), *the number n of intervals of width h must be even.*

◀ **EXAMPLE 3** A parking lot is proposed for a riverfront area in a town. The town engineer measured the widths of the area at 30.0-m (three significant digits) intervals, as shown in Fig. 2.93. Find the area available for parking.

First, we see that there are six intervals, which means Eq. (2.13) may be used. With $y_0 = 124$ m, $y_1 = 147$ m,..., $y_6 = 144$ m, and $h = 30.0$ m, we have

$$A = \frac{30.0}{3}[124 + 4(147) + 2(116) + 4(115) + 2(87) + 4(117) + 144]$$

$$= 21\,900 \text{ m}^2$$

For most areas, Simpson's rule gives a somewhat better approximation than the trapezoidal rule. The accuracy of Simpson's rule is also usually improved by using smaller intervals.

◀ **EXAMPLE 4** From an aerial photograph, a cartographer determines the widths of Easter Island at 1.50-km intervals as shown in Fig. 2.94. The widths found are as follows:

Distance from South End (km)	0	1.50	3.00	4.50	6.00	7.50	9.00	10.5	12.0	13.5	15.0
Width (km)	0	4.8	5.7	10.5	15.2	18.5	18.8	17.9	11.3	8.8	3.1

Since there are ten intervals, Simpson's rule may be used. From the table, we have the following values: $y_0 = 0$, $y_1 = 4.8$, $y_2 = 5.7$,..., $y_9 = 8.8$, $y_{10} = 3.1$, and $h = 1.5$. Using Simpson's rule, the cartographer would approximate the area of Easter Island as follows:

$$A = \frac{1.50}{3}(0 + 4(4.8) + 2(5.7) + 4(10.5) + 2(15.2) + 4(18.5)$$

$$+ 2(18.8) + 4(17.9) + 2(11.3) + 4(8.8) + 3.1) = 174 \text{ km}^2$$

Fig 2.94

EXERCISES 2.5

In Exercises 1 and 2, follow the given instructions and solve the given problems related to the indicated examples of this section.

1. In Example 2, use only the distances from the west end of (in km) 0.0, 52.0, 104, 156, 208, 260, and 312. Calculate the area and compare with the answer in the example.

(W) 2. In Example 4, if you use only the data from the south end of (in km) 0, 3.00, 6.00, 9.00, 12.0, and 15.0, would you choose the trapezoidal rule or Simpson's rule to calculate the area? Explain. Do not calculate the area for these data.

In Exercises 3 and 4, answer the given questions related to Fig. 2.95.

(W) 3. Which should be more accurate for finding the area, the trapezoidal rule or Simpson's rule? Explain.

(W) 4. If the trapezoidal rule is used to find the area, will the result probably be too high, about right, or too little? Explain.

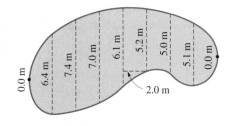

Fig 2.95

In Exercises 5–16, calculate the indicated areas. All data are accurate to at least two significant digits.

5. The widths of a kidney-shaped swimming pool were measured at 2.0-m intervals, as shown in Fig. 2.96. Calculate the surface area of the pool, using the trapezoidal rule.

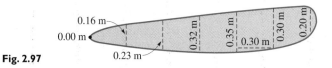

Fig. 2.96

6. Calculate the surface area of the swimming pool in Fig. 2.96, using Simpson's rule.

7. The widths of a cross section of an airplane wing are measured at 0.30-m intervals, as shown in Fig. 2.97. Calculate the area of the cross section, using Simpson's rule.

Fig. 2.97

8. Calculate the area of the cross section of the airplane wing in Fig. 2.97, using the trapezoidal rule.

9. Using aerial photography, the widths of an area burned by a forest fire were measured at 0.5-km intervals, as shown in the following table:

Distance (km)	0.0	0.5	1.0	1.5	2.0	2.5	3.0	3.5	4.0
Width (km)	0.6	2.2	4.7	3.1	3.6	1.6	2.2	1.5	0.8

Determine the area burned by the fire by using the trapezoidal rule.

10. Find the area burned by the forest fire of Exercise 9, using Simpson's rule.

11. A cartographer measured the width of Kruger National Park (and game reserve) in South Africa at 6.0-mm intervals on a map, as shown in Fig. 2.98. The widths are shown in the list that follows. Find the area of the park if the scale of the map is 1.0 mm = 6.0 km.

$y_0 = 7$ mm $y_1 = 15$ mm
$y_2 = 7$ mm $y_3 = 11$ mm
$y_4 = 13$ mm $y_5 = 10$ mm
$y_6 = 9$ mm $y_7 = 12$ mm
$y_8 = 8$ mm $y_9 = 3$ mm

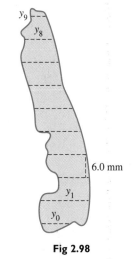

Fig 2.98

12. The widths of an oval-shaped floor were measured at 1.5-m intervals, as shown in the following table:

Distance (m)	0.0	1.5	3.0	4.5	6.0	7.5	9.0	10.5	12.0
Width (m)	0.0	5.0	7.2	8.3	8.6	8.3	7.2	5.0	0.0

Find the area of the floor by using Simpson's rule.

13. The widths of the baseball playing area in Boston's Fenway Park at 14-m intervals are shown in Fig. 2.99. Find the playing area using the trapezoidal rule.

14. Find the playing area of Fenway Park (see Exercise 13) by Simpson's rule.

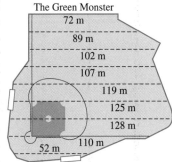

Fig 2.99

15. Soundings taken across a river channel give the following values of distance from one shore with the corresponding depth of the channel.

Distance (m)	0	50	100	150	200	250	300	350	400	450	500
Depth (m)	5	12	17	21	22	25	26	16	10	8	0

Find the area of the cross section of the channel using Simpson's rule.

16. The widths of a bell crank are measured at 2.0-cm intervals, as shown in Fig. 2.100. Find the area of the bell crank if the two connector holes are each 2.50 cm in diameter.

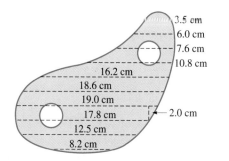

Fig. 2.100

3.5 cm
6.0 cm
7.6 cm
10.8 cm
16.2 cm
18.6 cm
19.0 cm
17.8 cm
12.5 cm
8.2 cm
2.0 cm

In Exercises 17–20, calculate the area of the circle by the indicated method.

The lengths of parallel chords of a circle that are 0.250 cm apart are given in the following table. The diameter of the circle is 2.000 cm. The distance shown is the distance from one end of a diameter.

Distance (cm)	0.000	0.250	0.500	0.750	1.000	1.250	1.500	1.750	2.000
Length (cm)	0.000	1.323	1.732	1.936	2.000	1.936	1.732	1.323	0.000

Using the formula $A = \pi r^2$, the area of the circle is 3.14 cm².

(W) **17.** Find the area of the circle using the trapezoidal rule and only the values of distance of 0.000 cm, 0.500 cm, 1.000 cm, 1.500 cm, and 2.000 cm with the corresponding values of the chord lengths. Explain why the value found is less than 3.14 cm².

(W) **18.** Find the area of the circle using the trapezoidal rule and all values in the table. Explain why the value found is closer to 3.14 cm² than the value found in Exercise 17.

(W) **19.** Find the area of the circle using Simpson's rule and the same table values as in Exercise 17. Explain why the value found is closer to 3.14 cm² than the value found in Exercise 17.

(W) **20.** Find the area of the circle using Simpson's rule and all values in the table. Explain why the value found is closer to 3.14 cm² than the value found in Exercise 19.

2.6 SOLID GEOMETRIC FIGURES

We now review the formulas for the *volume* and *surface area* of some basic solid geometric figures. Just as area is a measure of the surface of a plane geometric figure, **volume** is a measure of the space occupied by a solid geometric figure.

One of the most common solid figures is the **rectangular solid.** This figure has six sides **(faces),** and opposite sides are rectangles. All intersecting sides are perpendicular to each other. The **bases** of the rectangular solid are the top and bottom faces. A **cube** is a rectangular solid with all six faces being equal squares.

A **right circular cylinder** is *generated* by rotating a rectangle about one of its sides. Each **base** is a circle, and the *cylindrical surface* is perpendicular to each of the bases. The **height** is one side of the rectangle, and the **radius** of the base is the other side.

A **right circular cone** is generated by rotating a right triangle about one of its legs. The **base** is a circle, and the **slant height** is the hypotenuse of the right triangle. The **height** is one leg of the right triangle, and the **radius** of the base is the other leg.

The bases of a **right prism** are equal and parallel polygons, and the sides are rectangles. The **height** of a prism is the perpendicular distance between bases. The base of a **pyramid** is a polygon, and the other faces, the **lateral faces,** are triangles that meet at a common point, the **vertex.** A **regular pyramid** has congruent triangles for its lateral faces.

A **sphere** is generated by rotating a circle about a diameter. The **radius** is a line segment joining the center and a point on the sphere. The **diameter** is a line segment through the center and having its endpoints on the sphere.

In the following formulas, V represents the *volume*, A represents the *total surface area*, S represents the *lateral surface area* (bases not included), B represents the *area of the base*, and p represents the *perimeter of the base*.

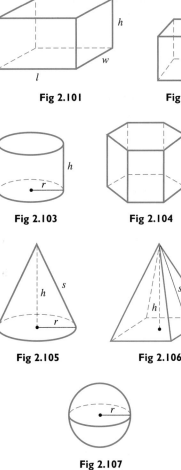

Fig 2.101

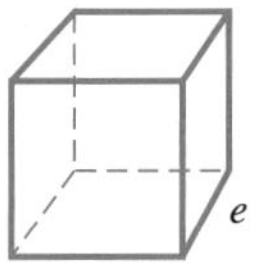

Fig 2.102

Fig 2.103

Fig 2.104

Fig 2.105

Fig 2.106

Fig 2.107

$V = lwh$	Rectangular solid (Fig. 2.101)	(2.14)
$A = 2lw + 2lh + 2wh$		(2.15)
$V = e^3$	Cube (Fig. 2.102)	(2.16)
$A = 6e^2$		(2.17)
$V = \pi r^2 h$	Right circular cylinder (Fig. 2.103)	(2.18)
$A = 2\pi r^2 + 2\pi rh$		(2.19)
$S = 2\pi rh$		(2.20)
$V = Bh$	Right prism (Fig. 2.104)	(2.21)
$S = ph$		(2.22)
$V = \frac{1}{3}\pi r^2 h$	Right circular cone (Fig. 2.105)	(2.23)
$A = \pi r^2 + \pi rs$		(2.24)
$S = \pi rs$		(2.25)
$V = \frac{1}{3}Bh$	Regular pyramid (Fig. 2.106)	(2.26)
$S = \frac{1}{2}ps$		(2.27)
$V = \frac{4}{3}\pi r^3$	Sphere (Fig. 2.107)	(2.28)
$A = 4\pi r^2$		(2.29)

Equation (2.21) is valid for any prism, and Eq. (2.26) is valid for any pyramid. There are other types of cylinders and cones, but we restrict our attention to right circular cylinders and right circular cones, and we will often use "cylinder" or "cone" when referring to them.

The **frustum** of a cone or pyramid is the solid figure that remains after the top is cut off by a plane parallel to the base. Figure 2.108 shows the frustum of a cone.

Fig 2.108

◀ EXAMPLE 1   What volume of concrete is needed for a driveway 25.0 m long, 2.75 m wide, and 0.100 m thick?

The driveway is a rectangular solid for which $l = 25.0$ m, $w = 2.75$ m, and $h = 0.100$ m. Using Eq. (2.14), we have

$$V = (25.0)(2.75)(0.100)$$
$$= 6.88 \text{ m}^3$$

Solving a
Word Problem

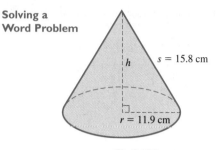

Fig 2.109

‖ **EXAMPLE 2** Calculate the volume of a right circular cone for which the radius $r = 11.9$ cm and the *slant height s* $= 15.8$ cm. See Fig. 2.109.

To find the volume using Eq. (2.23), we need the radius and height of the cone. Therefore, we must first find the height. As noted, the radius and height are the legs of a right triangle, and the slant height is the hypotenuse. To find the height we *use the Pythagorean theorem:*

$$s^2 = r^2 + h^2 \qquad \text{Pythagorean theorem}$$
$$h^2 = s^2 - r^2 \qquad \text{solve for } h$$
$$h = \sqrt{s^2 - r^2}$$
$$= \sqrt{15.8^2 - 11.9^2} = 10.4 \text{ cm}$$

Now, calculating the volume (in using a calculator it is not necessary to record the value of h), we have

$$V = \frac{1}{3}\pi r^2 h \qquad \text{Eq. (2.23)}$$

$$= \frac{1}{3}\pi(11.9^2)(10.4) \qquad \text{substituting}$$

$$= 1540 \text{ cm}^3$$

Solving a
Word Problem

Fig 2.110

‖ **EXAMPLE 3** A grain storage building is in the shape of a cylinder surmounted by a hemisphere (*half a sphere*). See Fig. 2.110. Find the volume of grain that can be stored if the height of the cylinder is 40.0 m and its radius is 12.0 m.

The total volume of the structure is the volume of the cylinder plus the volume of the hemisphere. By the construction we see that the radius of the hemisphere is the same as the radius of the cylinder. Therefore,

$$\overset{\text{cylinder}}{} \quad \overset{\text{hemisphere}}{}$$
$$V = \pi r^2 h + \frac{1}{2}\left(\frac{4}{3}\pi r^3\right) = \pi r^2 h + \frac{2}{3}\pi r^3$$

$$= \pi(12.0)^2(40.0) + \frac{2}{3}\pi(12.0)^3$$

$$= 21\,700 \text{ m}^3$$

E X E R C I S E S **2.6**

In Exercises 1–4, answer the given questions about the indicated examples of this section.

1. In Example 1, if the length is doubled and the thickness is doubled, by what factor is the volume changed?

2. In Example 2, if the value of the height $h = 12.5$ cm is given instead of the slant height, what is the slant height?

3. In Example 2, if the radius is halved and the slant height is doubled, what is the volume?

4. In Example 3, if h is halved, what is the volume?

In Exercises 5–20, find the volume or area of each solid figure for the given values. See Figs. 2.101 to 2.107.

5. Volume of cube: $e = 7.15$ dm

6. Volume of right circular cylinder: $r = 23.5$ cm, $h = 48.4$ cm

7. Total surface area of right circular cylinder: $r = 6.89$ m, $h = 2.33$ m

8. Area of sphere: $r = 67$ mm

9. Volume of sphere: $r = 0.877$ m

10. Volume of right circular cone: $r = 25.1$ mm, $h = 5.66$ mm

11. Lateral area of right circular cone: $r = 78.0$ cm, $s = 83.8$ cm

12. Lateral area of regular pyramid: $p = 3.45$ m, $s = 2.72$ m

13. Volume of regular pyramid: square base of side 16 dm, $h = 13$ dm

14. Volume of right prism: square base of side 29.0 cm, $h = 11.2$ cm

15. Lateral area of regular prism: equilateral triangle base of side 1.092 m, $h = 1.025$ m

16. Lateral area of right circular cylinder: diameter = 25.0 mm, $h = 34.7$ mm

17. Volume of hemisphere: diameter = 0.83 cm

18. Volume of regular pyramid: square base of side 22.4 m, $s = 14.2$ m

19. Total surface area of right circular cone: $r = 3.39$ cm, $h = 0.274$ cm

20. Total surface area of pyramid: All faces and base are equilateral triangles of side 3.67 dm

In Exercises 21–36, solve the given problems.

21. The radius of a cylinder is twice as long as the radius of a cone, and the height of the cylinder is half as long as the height of the cone. What is the ratio of the volume of the cylinder to that of the cone?

22. The base area of a cone is one-fourth of the total area. Find the ratio of the radius to the slant height.

23. In designing a weather balloon, it is decided to double the diameter of the balloon so that it can carry a heavier instrument load. What is the ratio of the final surface area to the original surface area?

24. During a rainfall of 3.00 cm, what weight of water falls on an area of 1.00 km²? Each cubic meter of water weighs 9800 N.

25. A rectangular box is to be used to store radioactive materials. The inside of the box is 12.0 cm long, 9.50 cm wide, and 8.75 cm deep. What is the area of sheet lead that must be used to line the inside of the box?

26. A swimming pool is 15.0 m wide, 24.0 m long, 1.00 m deep at one end, and 2.60 m deep at the other end. How many cubic feet of water can it hold? (The slope on the bottom is constant.) See Fig. 2.111.

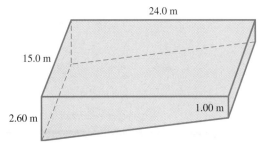

Fig. 2.111

27. The Alaskan oil pipeline is 1200 km long and has a diameter of 1.2 m. What is the maximum volume of the pipeline?

28. A glass prism used in the study of optics has a right triangular base. The legs of the triangle are 3.00 cm and 4.00 cm. The prism is 8.50 cm high. What is the total surface area of the prism? See Fig. 2.112.

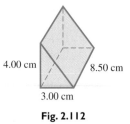

Fig. 2.112

29. The Great Pyramid of Egypt has a square base approximately 230 m on a side. The height of the pyramid is about 150 m. What is its volume? See Fig. 2.113.

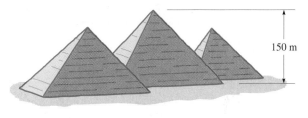

Fig. 2.113

30. A paper cup is in the shape of a cone as shown in Fig. 2.114. What is the surface area of the cup?

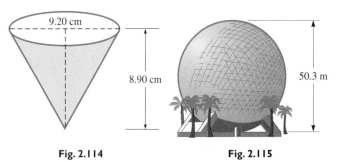

Fig. 2.114   **Fig. 2.115**

31. *Spaceship Earth* (shown in Fig. 2.115) at Epcot Center in Florida is a sphere of 50.3 m in diameter. What is the volume of *Spaceship Earth*?

32. A propane tank is constructed in the shape of a cylinder with a hemisphere at each end, as shown in Fig. 2.116. Find the volume of the tank.

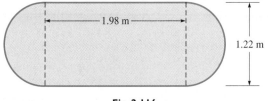

Fig. 2.116

33. A special wedge in the shape of a regular pyramid has a square base 16.0 mm on a side. The height of the wedge is 40.0 mm. What is the total surface area of the wedge?

34. What is the area of a paper label that is to cover the lateral surface of a cylindrical can 8.50 cm in diameter and 11.5 cm high? The ends of the label will overlap 0.50 cm when the label is placed on the can.

35. The side view of a rivet is shown in Fig. 2.117. It is a conical part on a cylindrical part. Find the volume of the rivet.

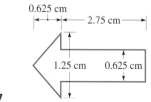

Fig. 2.117

36. A dipstick is made to measure the volume remaining in the conical container shown in Fig. 2.118. How far below the full mark (at the top of the container) on the stick should the mark for half-full be placed?

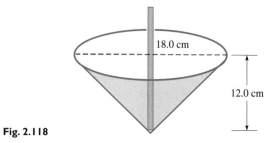

Fig. 2.118

CHAPTER 2 EQUATIONS

Line segments	Fig. 2.8	$\dfrac{a}{b} = \dfrac{c}{d}$	(2.1)
Triangle		$A = \frac{1}{2}bh$	(2.2)
Hero's formula		$A = \sqrt{s(s-a)(s-b)(s-c)}$, where $s = \frac{1}{2}(a+b+c)$	(2.3)
Pythagorean theorem	Fig. 2.29	$c^2 = a^2 + b^2$	(2.4)
Square	Fig. 2.54	$A = s^2$	(2.5)
Rectangle	Fig. 2.55	$A = lw$	(2.6)
Parallelogram	Fig. 2.56	$A = bh$	(2.7)
Trapezoid	Fig. 2.57	$A = \frac{1}{2}h(b_1 + b_2)$	(2.8)
Circle		$c = 2\pi r$	(2.9)
		$A = \pi r^2$	(2.10)
Radians	Fig. 2.78	$\pi \, \text{rad} = 180°$	(2.11)
Trapezoidal rule	Fig. 2.88	$A = \dfrac{h}{2}(y_0 + 2y_1 + 2y_2 + \cdots + 2y_{n-1} + y_n)$	(2.12)
Simpson's rule	Fig. 2.92	$A = \dfrac{h}{3}(y_0 + 4y_1 + 2y_2 + 4y_3 + \cdots + 2y_{n-2} + 4y_{n-1} + y_n)$	(2.13)
Rectangular solid	Fig. 2.101	$V = lwh$	(2.14)
		$A = 2lw + 2lh + 2wh$	(2.15)
Cube	Fig. 2.102	$V = e^3$	(2.16)
		$A = 6e^2$	(2.17)

Right circular cylinder	Fig. 2.103	$V = \pi r^2 h$	(2.18)
		$A = 2\pi r^2 + 2\pi rh$	(2.19)
		$S = 2\pi rh$	(2.20)
Right prism	Fig. 2.104	$V = Bh$	(2.21)
		$S = ph$	(2.22)
Right circular cone	Fig. 2.105	$V = \frac{1}{3}\pi r^2 h$	(2.23)
		$A = \pi r^2 + \pi rs$	(2.24)
		$S = \pi rs$	(2.25)
Regular pyramid	Fig. 2.106	$V = \frac{1}{3}Bh$	(2.26)
		$S = \frac{1}{2}ps$	(2.27)
Sphere	Fig. 2.107	$V = \frac{4}{3}\pi r^3$	(2.28)
		$A = 4\pi r^2$	(2.29)

CHAPTER 2 REVIEW EXERCISES

In Exercises 1–4, use Fig. 2.119. Determine the indicated angles.

1. $\angle CGE$ **2.** $\angle EGF$ **3.** $\angle DGH$ **4.** $\angle EGI$

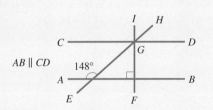

$AB \parallel CD$ 148°

Fig. 2.119

In Exercises 5–12, find the indicated sides of the right triangle shown in Fig. 2.120.

5. $a = 9, b = 40, c = ?$

6. $a = 14, b = 48, c = ?$

7. $a = 400, b = 580, c = ?$

8. $b = 56, c = 65, a = ?$

9. $a = 6.30, b = 3.80, c = ?$

10. $a = 126, b = 25.1, c = ?$

11. $b = 29.3, c = 36.1, a = ?$

12. $a = 0.782, c = 0.885, b = ?$

Fig 2.120

In Exercises 13–20, find the perimeter or area of the indicated figure.

13. Perimeter: equilateral triangle of side 8.5 mm

14. Perimeter: rhombus of side 15.2 cm

15. Area: triangle, $b = 3.25$ m, $h = 1.88$ m

16. Area: triangle of sides 175 cm, 138 cm, 119 cm

17. Circumference of circle: $d = 98.4$ mm

18. Perimeter: rectangle, $l = 2.98$ dm, $w = 1.86$ dm

19. Area: trapezoid, $b_1 = 67.2$ cm, $b_2 = 126.7$ cm, $h = 34.2$ cm

20. Area: circle, $d = 32.8$ m

In Exercises 21–24, find the volume of the indicated solid geometric figure.

21. Prism: base is right triangle with legs 26.0 cm and 34.0 cm, height is 14.0 cm

22. Cylinder: base radius 36.0 cm, height 2.40 cm

23. Pyramid: base area 3850 m², height 125 m

24. Sphere: diameter 22.1 mm

In Exercises 25–28, find the surface area of the indicated solid geometric figure.

25. Total area of cube of edge 5.20 m

26. Total area of cylinder: base diameter 1.20 cm, height 5.80 cm

27. Lateral area of cone: base radius 1.82 mm, height 11.5 mm

28. Total area of sphere: diameter 0.884 m

In Exercises 29–32, use Fig. 2.121. Line CT is tangent to the circle with center at O. Find the indicated angles.

29. $\angle BTA$

30. $\angle TAB$

31. $\angle BTC$

32. $\angle ABT$

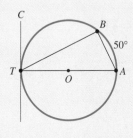

Fig. 2.121

In Exercises 33–36, use Fig. 2.122. Given that AB = 4, BC = 4, CD = 6, and ∠ADC = 53°, find the indicated angle and lengths.

33. ∠ABE

34. AD

35. BE

36. AE

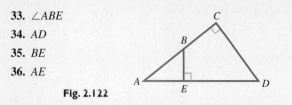

Fig. 2.122

In Exercises 37–40, find the formulas for the indicated perimeters and areas.

37. Perimeter of Fig. 2.123 (a right triangle and semicircle attached)

38. Perimeter of Fig. 2.124 (a square with a quarter circle at each end)

39. Area of Fig. 2.123   **40.** Area of Fig. 2.124

Fig. 2.123   **Fig. 2.124**

In Exercises 41–44, answer the given questions and explain your reasoning in Exercises 43 and 44.

41. Is a square also a rectangle, a parallelogram, and a rhombus?

42. If the measures of two angles of one triangle equal the measures of two angles of a second triangle, are the two triangles similar?

(W) 43. If the dimensions of a plane geometric figure are each multiplied by n, by how much is the area multiplied? Explain, using a circle to illustrate.

(W) 44. If the dimensions of a solid geometric figure are each multiplied by n, by how much is the volume multiplied? Explain, using a cube to illustrate.

In Exercises 45–68, solve the given problems.

45. A machine part is in the shape of a square with equilateral triangles attached to two sides (see Fig. 2.125). Find the perimeter of the machine part.

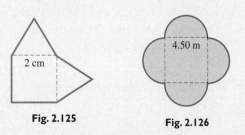

Fig. 2.125   **Fig. 2.126**

46. A patio is designed with semicircular areas attached to a square, as shown in Fig. 2.126. Find the area of the patio.

47. A tooth on a saw is in the shape of an isosceles triangle. If the angle at the point is 38°, find the two base angles.

48. A lead sphere 1.50 cm in diameter is flattened into a circular sheet 14.0 cm in diameter. How thick is the sheet?

49. A ramp for the disabled is designed so that it rises 1.2 m over a horizontal distance of 7.8 m. How long is the ramp?

50. An airplane is 640 m directly above one end of a 3200-m runway. How far is the plane from the glide-slope indicator on the ground at the other end of the runway?

51. A radio transmitting tower is supported by guy wires. The tower and three parallel guy wires are shown in Fig. 2.127. Find the distance AB along the tower.

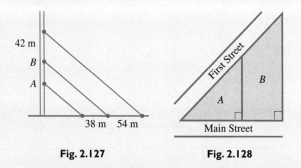

Fig. 2.127   **Fig. 2.128**

52. Find the areas of lots A and B in Fig. 2.128. A has a frontage on Main St. of 140 m, and B has a frontage on Main St. of 84 m. The boundary between lots is 120 m.

53. Find the area of the side of the building shown in Fig. 2.129 (a triangle over a rectangle).

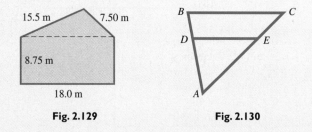

Fig. 2.129   **Fig. 2.130**

54. The metal support in the form of △ABC shown in Fig. 2.130 is strengthened by brace DE, which is parallel to BC. How long is the brace if AB = 24 cm, AD = 16 cm, and BC = 33 cm?

55. A typical scale for an aerial photograph is 1/18 450. In a 20.0- by 25.0-cm photograph with this scale, what is the longest distance (in km) between two locations in the photograph?

56. For a hydraulic press, the mechanical advantage is the ratio of the large piston area to the small piston area. Find the mechanical advantage if the pistons have diameters of 3.10 cm and 2.25 cm.

57. The diameter of the earth is 12 700 km, and a satellite is in orbit at an altitude of 340 km. How far does the satellite travel in one rotation about the earth?

58. The roof of the Louisiana Superdome in New Orleans is supported by a circular steel tension ring 651 m in circumference. Find the area covered by the roof.

59. A rectangular piece of wallboard with two holes cut out for heating ducts is shown in Fig. 2.131. What is the area of the remaining piece?

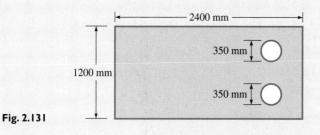

Fig. 2.131

60. The diameter of the sun is 1.38×10^6 km, the diameter of the earth is 1.27×10^4 km, and the distance from the earth to the sun (center to center) is 1.50×10^8 km. What is the distance from the center of the earth to the end of the shadow due to the rays from the sun?

61. Using aerial photography, the width of an oil spill is measured at 250-m intervals, as shown in Fig. 2.132. Using Simpson's rule, find the area of the oil spill.

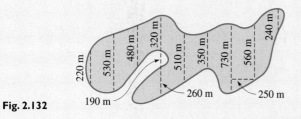

Fig. 2.132

62. To build a highway, it is necessary to cut through a hill. A surveyor measured the cross-sectional areas at 250-m intervals through the cut as shown in the following table. Using the trapezoidal rule, determine the volume of soil to be removed.

Dist. (m)	0	250	500	750	1000	1250	1500	1750
Area (m²)	560	1780	4650	6730	5600	6280	2260	230

63. The Hubble space telescope is within a cylinder 4.3 m in diameter and 13 m long. What is the volume within this cylinder?

64. A horizontal cross section of a concrete bridge pier is a regular hexagon (six sides, all equal in length, and all internal angles are equal), each side of which is 2.50 m long. If the height of the pier is 6.75 m, what is the volume of concrete in the pier?

65. A railroad track 1000.00 m long expands 0.20 m (20 cm) during the afternoon (due to an increase in temperature of about 17°C). Assuming that the track cannot move at either end and that the increase in length causes a bend straight up in the middle of the track, how high is the top of the bend?

66. Two people are talking to each other on cellular phones. If the angle between their signals at the tower is 90°, as shown in Fig. 2.133, how far apart are they?

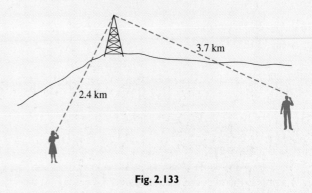

Fig. 2.133

67. A hot-water tank is in the shape of a right circular cylinder surmounted by a hemisphere as shown in Fig. 2.134. How many liters does the tank hold? (1.00 m³ contains 1000 L.)

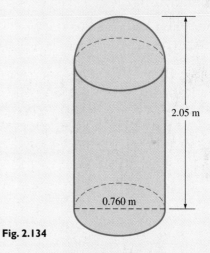

Fig. 2.134

68. A tent is in the shape of a regular pyramid surmounted on a cube. If the edge of the cube is 2.50 m and the total height of the tent is 3.25 m, find the area of the material used in making the tent (not including any floor area).

Writing Exercise

69. The Pentagon, headquarters of the U.S. Department of Defense, is the world's largest office building. It is a regular pentagon (five sides, all equal in length, and all interior angles are equal) 281 m on a side, with a diagonal of length 454 m. Using these data, draw a sketch and write one or two paragraphs to explain how to find the area covered within the outside perimeter of the Pentagon. (What is the area?)

CHAPTER 2 PRACTICE TEST

1. In Fig. 2.135, determine $\angle 1$.

2. In Fig. 2.135, determine $\angle 2$.

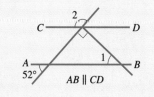

Fig. 2.135

3. A tree is 2.4 m high and casts a shadow 3.0 m long. At the same time, a telephone pole casts a shadow 7.6 m long. How tall is the pole?

4. Find the area of a triangle with sides of 2.46 cm, 3.65 cm, and 4.07 cm.

5. What is the diagonal distance along the floor between corners of a rectangular room 3810 mm wide and 5180 mm long?

6. What is the area of the trapezoid shown in Fig. 2.136?

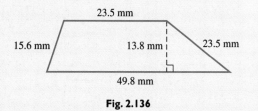

Fig. 2.136

7. The edge of a cube is 4.50 cm. What are (a) the surface area and (b) the volume of the cube?

8. Find the surface area of a tennis ball whose circumference is 21.0 cm.

9. Find the volume of a right circular cone of radius 2.08 m and height 1.78 m.

10. In Fig. 2.137, find $\angle 1$.

11. In Fig. 2.137, find $\angle 2$.

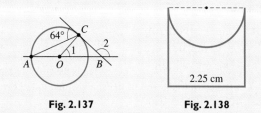

Fig. 2.137   **Fig. 2.138**

12. In Fig. 2.138, find the perimeter of the figure shown. It is a square with a scmicircle removed.

13. In Fig. 2.138, find the area of the figure shown.

14. The width of a marshy area is measured at 50-m intervals, with the results shown in the following table. Using the trapezoidal rule, find the area of the marsh. (All data accurate to two or more significant digits.)

Distance (m)	0	50	100	150	200	250	300
Width (m)	0	90	145	260	205	110	20

By noting, for example, that stones fall faster than leaves, the Greek philosopher Aristotle (about 350 B.C.E.) reasoned that heavier objects fall faster than lighter ones. For about 2000 years, this idea was generally accepted. Then in about 1600, the Italian scientist Galileo showed, by dropping objects from the leaning tower of Pisa, that the distance an object falls in a given time does not depend on its weight.

Galileo is generally credited with first using the *experimental method* by which controlled experiments are used to study natural phenomena. His aim was to find mathematical formulas that could be used to describe these phenomena. He realized that such formulas provided a way of showing a compact and precise relation between the variables.

In technology and science, determining how one quantity depends on others is a primary goal. A rule that shows such a relation is of great importance, and in mathematics such a rule is called a *function*. We start this chapter with a discussion of functions.

Examples of such relations in technology and science, as well as in everyday life, are numerous. Plant growth depends on sunlight and rainfall; traffic flow depends on roadway design; the sales tax on an item depends on the cost of an item; the time to access an Internet site depends on how fast a computer processes data; distance traveled depends on the time and speed of travel; electric voltage depends on the current and resistance. These are but a few of the innumerable possibilities.

The electric power produced in a circuit depends on the resistance R in the circuit. In Section 3.4, we draw a graph of $P = \dfrac{100R}{(0.50 + R)^2}$ to see this type of relationship.

A way of actually seeing how one quantity depends on another is by means of a *graph*. The basic method of graphing was devised by the French mathematicians Descartes and Fermat in the 1630s from their work to combine the methods of algebra and geometry. Their work was very influential in later developments in mathematics and technology, and we will discuss this more in later chapters. In this chapter, we take up these basic methods of drawing and using graphs. In doing so, we will also expand our use of the *graphing calculator*, one of the more recent innovations of technology. As noted in Chapter 1, they were first used in the 1980s.

3.1 INTRODUCTION TO FUNCTIONS

An important method of finding formulas such as those in Section 1.11 is through scientific observation and experimentation, like that of Galileo as mentioned on the previous page. For example, if we were to perform an experiment to determine the relationship between the distance an object falls and the time it falls, we should find (approximately, at least) that $s = 4.9t^2$, where s is the distance in meters and t is the time in seconds. Another example is that electrical measurements of voltage and current through a particular resistor would show that $V = kI$, where V is the voltage, I is the current in the resistor, and k is a constant.

Considerations such as these lead us to one of the basic concepts in mathematics, that of a *function*.

Definition of a Function

Whenever a relationship exists between two variables—such that for every value of the first, there is only one corresponding value of the second—we say that the second variable is a **function** *of the first variable.*

The first variable is called the **independent variable,** *and the second variable is called the* **dependent variable.**

The first variable is termed *independent* since permissible values can be assigned to it arbitrarily, and the second variable is termed *dependent* since its value is determined by the choice of the independent variable. *Values of the independent variable and dependent variable are to be real numbers.* Therefore, there may be restrictions on their possible values. This is discussed in the following section.

EXAMPLE 1   In the equation $y = 2x$, we see that y is a function of x, since for each value of x there is only one value of y. For example, if we substitute $x = 3$, we get $y = 6$ and no other value. By arbitrarily assigning values to x and then substituting, we see that the values of y we obtain *depend* on the values chosen for x. Therefore, x is the independent variable, and y is the dependent variable.

EXAMPLE 2   The power P developed in a certain resistor by a current I is given by $P = 4I^2$. Here, P is a function of I. The dependent variable is P, and the independent variable is I.

EXAMPLE 3   Figure 3.1 shows a cube of edge e. In order to express the volume V as a function of the edge e, we recall from Chapter 2 that $V = e^3$. Here, V is a function of e, since for each value of e there is only one value of V. The dependent variable is V, and the independent variable is e.

If the equation relating the volume and the edge of a cube is written as $e = \sqrt[3]{V}$— that is, if the edge is expressed in terms of the volume—e is a function of V. In this case, e is the dependent variable, and V is the independent variable.

Fig 3.1

There are many ways to express functions. Formulas, tables, charts, and graphs can also define functions. Functions will be of importance throughout the book, and we will use a number of different types of functions in later chapters.

FUNCTIONAL NOTATION

For convenience of notation, the phrase

"function of x" is written as f(x).

This means that "*y* is a function of *x*" may be written as $y = f(x)$. Here, *f* denotes *dependence* and does not represent a quantity or a variable. Therefore, it also follows that $f(x)$ *does not mean f times x.*

CAUTION ▶

◀ **EXAMPLE 4** If $y = 6x^3 - 5x$, we say that *y* is a function of *x*. This function is $6x^3 - 5x$. It is also common to write such a function as $f(x) = 6x^3 - 5x$. However, *y* and $f(x)$ represent the same expression, $6x^3 - 5x$. Using *y*, the quantities are shown, and using $f(x)$, the functional dependence is shown. ▶

One of the most important uses of functional notation is to designate the value of the function for a particular value of the independent variable. That is,

the value of the function f(x) when x = a is written as f(a).

◀ **EXAMPLE 5** For a function $f(x)$, the value of $f(x)$ for $x = 2$ may be expressed as $f(2)$. Thus, substituting 2 for *x* in $f(x) = 3x - 7$, we have

$$f(2) = 3(2) - 7 = -1 \qquad \text{substitute 2 for } x$$

The value of $f(x)$ for $x = -1.4$ is

$$f(-1.4) = 3(-1.4) - 7 = -11.2 \qquad \text{substitute } -1.4 \text{ for } x$$ ▶

We must note that *whatever number a represents, to find f(a), we substitute a for x in f(x).* This is true even if *a* is a literal number, as in the following examples.

◀ **EXAMPLE 6** If $g(t) = \dfrac{t^2}{2t + 1}$, to find $g(a^3)$ we substitute a^3 for *t* in $g(t)$:

$$g(a^3) = \frac{(a^3)^2}{2a^3 + 1} = \frac{a^6}{2a^3 + 1}$$

For the same function, $g(3) = \dfrac{3^2}{2(3) + 1} = \dfrac{9}{7}$. In both cases, to obtain the value of $g(t)$, we simply substitute the value within the parentheses for *t*. ▶

◀ **EXAMPLE 7** The electric resistance *R* of a particular resistor as a function of the temperature *T* (in °C) is given by $R = 10.0 + 0.10T + 0.001T^2$. If a given temperature *T* is increased by 10°C, what is the value of *R* for the increased temperature as a function of the temperature *T*?

We are to determine *R* for a temperature of $T + 10$. Since

$$f(T) = 10.0 + 0.10T + 0.001T^2$$

then

$$f(T + 10) = 10.0 + 0.10(T + 10) + 0.001(T + 10)^2 \qquad \text{substitute } T + 10 \text{ for } T$$
$$= 10.0 + 0.10T + 1.0 + 0.001T^2 + 0.02T + 0.1$$
$$= 11.1 + 0.12T + 0.001T^2$$ ▶

A graphing calculator can be used to evaluate a function in several ways. One is to directly substitute the value into the function as shown in Fig. 3.2(a). A second is to enter the function as Y_1 and evaluate as shown in Fig. 3.2(b). A third way, which is very useful when many values are to be used, is to enter the function as Y_1 and use the *table* feature as shown in Fig. 3.2(c).

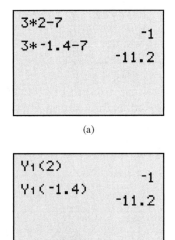

(a)

(b)

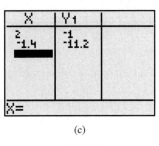

(c)

Fig 3.2

At times we need to define more than one function. We then use different symbols, such as $f(x)$ and $g(x)$, to denote these functions.

◀ EXAMPLE 8   For the functions $f(x) = 5x - 3$ and $g(x) = ax^2 + x$, where a is a constant, we have

$$f(-4) = 5(-4) - 3 = -23 \qquad \text{substitute } -4 \text{ for } x \text{ in } f(x)$$

$$g(-4) = a(-4)^2 + (-4) = 16a - 4 \qquad \text{substitute } -4 \text{ for } x \text{ in } g(x)$$

A function may be looked upon as a set of instructions. These instructions tell us how to obtain the value of the dependent variable for a particular value of the independent variable, even if the instructions are expressed in literal symbols.

◀ EXAMPLE 9   The function $f(x) = x^2 - 3x$ tells us to "square the value of the independent variable, multiply the value of the independent variable by 3, and subtract the second result from the first." An analogy would be a computer that was programmed so that when a number was entered into the program, it would square the number, then multiply the number by 3, and finally subtract the second result from the first. This is represented in diagram form in Fig. 3.3.

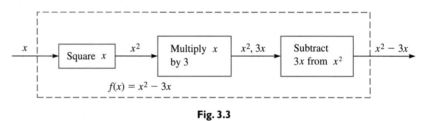

Fig. 3.3

The functions $f(t) = t^2 - 3t$ and $f(n) = n^2 - 3n$ are the same as the function $f(x) = x^2 - 3x$, since the operations performed on the independent variable are the same. ***Although different literal symbols appear, this does not change the function.***

NOTE ▶

EXERCISES 3.1

In Exercises 1–4, solve the given problems related to the indicated examples of this section.

1. In Example 5, find the value of $f(-2)$.

2. In Example 6, evaluate $g(-a^2)$.

3. In Example 7, change "increased" to "decreased" in the third line and then evaluate the function to find the proper expression.

4. In Example 9, change $x^2 - 3x$ to $x^3 + 4x$ and then determine the statements that should be placed in the three boxes in Fig. 3.3.

In Exercises 5–12, find the indicated functions.

5. Express the area A of a circle as a function of (a) its radius r and (b) its diameter d.

6. Express the circumference c of a circle as a function of (a) its radius r and (b) its diameter d.

7. Express the diameter d of a sphere as a function of its volume V.

8. Express the edge e of a cube as a function of its surface area A.

9. Express the area A of a square as a function of its side s; express the side s of a square as a function of its area A.

10. Express the perimeter p of a square as a function of its side s; express the side s of a square as a function of its perimeter p.

11. A circle is inscribed in a square (the circle is tangent to each side of the square). Express the area A of the four corner regions bounded by the circle and the square as a function of the radius r of the circle.

12. Express the area A of an equilateral triangle as a function of its side s.

In Exercises 13–24, evaluate the given functions.

13. $f(x) = 2x + 1$; find $f(1)$ and $f(-1)$.

14. $f(x) = 5x - 9$; find $f(2)$ and $f(-2)$.

15. $f(x) = 5$; find $f(-2)$ and $f(0.4)$.

16. $f(T) = 7.2 - 2.5|T|$; find $f(2.6)$ and $f(-4)$.

17. $\phi(x) = \dfrac{6 - x^2}{2x}$; find $\phi(\pi)$ and $\phi(-2)$.

18. $H(q) = \dfrac{8}{q} + 2\sqrt{q}$; find $H(4)$ and $H(0.16)$.

19. $g(t) = at^2 - a^2t$; find $g\left(-\tfrac{1}{2}\right)$ and $g(a)$.

20. $s(y) = 6\sqrt{y + 1} - 3$; find $s(8)$ and $s(a^2)$.

21. $K(s) = 3s^2 - s + 6$; find $K(-s)$ and $K(2s)$.

22. $T(t) = 5t + 7$; find $T(-2t)$ and $T(t + 1)$.

23. $f(x) = 2x + 4$; find $f(3x) - 3f(x)$.

24. $f(x) = 2x^2 + 1$; find $f(x + 2) - [f(x) + 2]$.

In Exercises 25–28, evaluate the given functions. The values of the independent variable are approximate.

25. Given $f(x) = 5x^2 - 3x$, find $f(3.86)$ and $f(-6.92)$.

26. Given $g(t) = \sqrt{t + 1.0604} - 6t^3$, find $g(0.9261)$.

27. Given $F(H) = \dfrac{2H^2}{H + 0.036\,85}$, find $F(-0.084\,66)$.

28. Given $f(x) = \dfrac{x^4 - 2.0965}{6x}$, find $f(1.9654)$.

 In Exercises 29–36, state the instructions of the function in words as in Example 9.

29. $f(x) = x^2 + 2$

30. $f(x) = 2x - 6$

31. $g(y) = 6y - y^3$

32. $\phi(s) = 8 - 5s + s^2$

33. $R(r) = 3(2r + 5) - 1$

34. $f(z) = \dfrac{4z}{5 - z}$

35. $p(t) = \dfrac{2t - 3}{t + 2}$

36. $Y(y) = 2 + \dfrac{5y}{2(y - 3)}$

In Exercises 37–40, write the equation as given by the statement. Then write the indicated function using functional notation.

37. The surface area A of a cubical open-top aquarium equals 5 times the square of an edge e of the aquarium.

38. A helicopter is at an altitude of 1000 m and is x m horizontally from a fire. Its distance d from the fire is the square root of the sum of 1000 squared and x squared.

39. The present area of the Bering Glacier in Alaska is 8430 km^2, and it is melting at the rate of 140t km^2, where t is the time in centuries. Find the area A at any time in the future.

40. The electrical resistance R of a certain ammeter, in which the resistance of the coil is R_c, is the product of 10 and R_c divided by the sum of 10 and R_c.

In Exercises 41–44, solve the given problems.

41. A demolition ball is used to tear down a building. Its distance s (in m) above the ground as a function of time t (in s) after it is dropped is $s = 17.5 - 4.9t^2$. Since $s = f(t)$, find $f(1.2)$.

42. The stopping distance d (in m) of a car going v km/h is given by $d = 0.2v + 0.008v^2$. Since $d = f(v)$, find $f(30)$, $f(2v)$, and $f(60)$, using both $f(v)$ and $f(2v)$.

43. The electric power P (in W) dissipated in a resistor of resistance R (in Ω) is given by the function $P = \dfrac{200R}{(100 + R)^2}$. Since $P = f(R)$, find $f(R + 10)$.

W 44. (a) Explain the meaning of $f[f(x)]$. (b) Find $f[f(x)]$ for $f(x) = 2x^2$.

3.2 MORE ABOUT FUNCTIONS

DOMAIN AND RANGE

For a given function, *the complete set of possible values of the independent variable is called the **domain** of the function*, and the *complete set of all possible resulting values of the dependent variable is called the **range** of the function.*

As noted earlier, *we will be using only **real numbers** when using functions.* This means there may be restrictions as to the values that may be used since

NOTE ▶ *values that lead to **division by zero** or to **imaginary numbers** may not be included in the domain or the range.*

Note that $f(2) = 6$ and $f(-2) = 6$. This is all right. Two different values of x may give the same value of y, but there may not be two different values of y for one value of x.

EXAMPLE 1 The function $f(x) = x^2 + 2$ is defined for all real values of x. This means its domain is written as *all real numbers*. However, since x^2 is never negative, $x^2 + 2$ is never less than 2. We then write the range as *all real numbers $f(x) \geq 2$*, where the symbol $\geq$ means "is greater than or equal to."

The function $f(t) = \frac{1}{t+2}$ is not defined for $t = -2$, for this value would require division by zero. Also, no matter how large t becomes, $f(t)$ will never exactly equal zero. Therefore, the domain of this function is *all real numbers except -2*, and the range is *all real numbers except* 0.

EXAMPLE 2 The function $g(s) = \sqrt{3 - s}$ is not defined for real numbers greater than 3, since such values make $3 - s$ negative and would result in imaginary values for $g(s)$. This means that the domain of this function is *all real numbers $s \leq 3$*, where the symbol $\leq$ means "is less than or equal to."

Also, since $\sqrt{3 - s}$ means the principal square root of $3 - s$ (see Section 1.6), we know that $g(s)$ cannot be negative. This tells us that the range of the function is *all real numbers $g(s) \geq 0$*.

In Examples 1 and 2, we found the domains by looking for values of the independent variable that cannot be used. The range was found through an inspection of the function. We generally use this procedure, although it may be necessary to use more advanced methods to find the range. Therefore, until we develop other methods, we will look only for the domain of some functions.

Later in this chapter, we will see that the graphing calculator is useful in finding the range of a function.

EXAMPLE 3 Find the domain of the function $f(x) = 16\sqrt{x} + \dfrac{1}{x}$.

From the term $16\sqrt{x}$, we see that x must be greater than or equal to zero in order to have real values. The term $\frac{1}{x}$ indicates that x cannot be zero, because of division by zero. Thus, putting these together, the domain is *all real numbers $x > 0$*.

As for the range, it is *all real numbers $f(x) \geq 12$*. More advanced methods are needed to determine this.

We have seen that the domain may be restricted since we do not use imaginary numbers or divide by zero. The domain may also be restricted by the definition of the function, or by practical considerations in an application.

EXAMPLE 4 A function defined as

$$f(x) = x^2 + 4 \qquad (\text{for } x > 2)$$

has a domain restricted to real numbers greater than 2 by definition. Thus, $f(5) = 29$, but $f(1)$ is not defined, since 1 is not in the domain. Also, the range is all real numbers greater than 8.

The height h (in m) of a certain projectile as a function of the time t (in s) is

$$h = 20t - 4.9t^2$$

Negative values of time have no real meaning in this case. This is generally true in applications. Therefore, the domain is $t \geq 0$. Also, since we know the projectile will not continue in flight indefinitely, there is some upper limit on the value of t. These restrictions are not usually stated unless it affects the solution.

There could be negative values of h if it is possible that the projectile is *below* the launching point at some time (such as a stone thrown from the top of a cliff).

The following example illustrates a function that is defined differently for different intervals of the domain.

▐ EXAMPLE 5 In a certain electric circuit, the current i (in mA) is a function of the time t (in s), which means $i = f(t)$. The function is

$$f(t) = \begin{cases} 8 - 2t & \text{(for } 0 \le t \le 4 \text{ s)} \\ 0 & \text{(for } t > 4 \text{ s)} \end{cases}$$

Since negative values of t are not usually meaningful, $f(t)$ is not defined for $t < 0$. Find the current for $t = 3$ s, $t = 6$ s, and $t = -1$ s.

We are to find $f(3)$, $f(6)$, and $f(-1)$, and we see that values of this function are determined differently depending on the value of t. Since 3 is between 0 and 4,

$$f(3) = 8 - 2(3) = 2 \quad \text{or} \quad i = 2 \text{ mA}$$

Since 6 is greater than 4, $f(6) = 0$, or $i = 0$ mA. We see that $i = 0$ mA for all values of t that are 4 or greater.

Since $f(t)$ is not defined for $t < 0$, $f(-1)$ is not defined. ▌

FUNCTIONS FROM VERBAL STATEMENTS

To find a mathematical function from a verbal statement, we use methods like those for setting up equations in Chapter 1. This is shown in the following examples.

Solving a Word Problem ▐ EXAMPLE 6 The fixed cost for a company to operate a certain plant is $3000 per day. It also costs $4 for each unit produced in the plant. Express the daily cost C of operating the plant as a function of the number n of units produced.

The daily total cost C equals the fixed cost of $3000 plus the cost of producing n units. Since the cost of producing one unit is $4, the cost of producing n units is $4n$. Thus, the total cost C, where $C = f(n)$, is

$$C = 3000 + 4n$$

Here we know that the domain is all values of $n \ge 0$, with some upper limit on n based on the production capacity of the plant. ▌

Solving a Word Problem ▐ EXAMPLE 7 A metallurgist melts and mixes m grams (g) of solder that is 40% tin with n grams of another solder that is 20% tin to get a final solder mixture that contains 200 g of tin. Express n as a function of m. See Fig. 3.4.

The statement leads to the following equation:

tin in first solder		tin in second solder		total amount of tin
$0.40m$	$+$	$0.20n$	$=$	200

Since we want $n = f(m)$, we now solve for n:

$$0.20n = 200 - 0.40m$$
$$n = 1000 - 2m$$

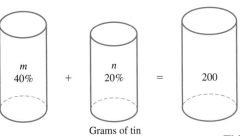

Grams of tin

Fig 3.4

This is the required function. Since neither m nor n can be negative, the domain is all values $0 \le m \le 500$ g, which means that m is greater than or equal to 0 g and less than or equal to 500 g. The range is all values $0 \le n \le 1000$ g. ▌

Solving a Word Problem

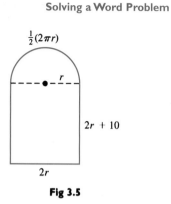

$\frac{1}{2}(2\pi r)$

$2r + 10$

$2r$

Fig 3.5

❨ EXAMPLE 8   An architect designs a window such that it has the shape of a rectangle with a semicircle on top, as shown in Fig. 3.5. The base of the window is 10 cm less than the height of the rectangular part. Express the perimeter p of the window as a function of the radius r of the circular part.

The perimeter is the distance around the window. Since the top part is a semicircle, the length of this top circular part is $\frac{1}{2}(2\pi r)$, and the base of the window is $2r$ because it is equal in length to the dashed line (the diameter of the circle). Finally, the base being 10 cm less than the height of the rectangular part tells us that each vertical side of the rectangle is $2r + 10$. Therefore, the perimeter p, where $p = f(r)$, is

$$p = \tfrac{1}{2}(2\pi r) + 2r + 2(2r + 10)$$
$$= \pi r + 2r + 4r + 20$$
$$= \pi r + 6r + 20$$

We see that the required function is $p = \pi r + 6r + 20$. Since the radius cannot be negative and there would be no window if $r = 0$, the domain of the function is all values $0 < r \le R$, where R is a maximum possible value of r determined by design considerations.   ❩

From the definition of a function we know that any value of the independent variable must yield only one value of the dependent variable. *If a value of the independent variable yields one or more values of the dependent variable, the relationship is called*
Relation   *a* **relation.** A function is a relation for which each value of the independent variable yields only one value of the dependent variable. Therefore, a function is a special type of relation. There are also relations that are not functions.

❨ EXAMPLE 9   For $y^2 = 4x^2$, if $x = 2$, then y can be either 4 or -4. Since a value of x yields more than one single value for y, we see that $y^2 = 4x^2$ is a relation, not a function.   ❩

EXERCISES 3.2

In Exercises 1–4, solve the given problems related to the indicated examples of this section.

1. In Example 1, in the first line change x^2 to $-x^2$. What other changes must be made in the rest of the paragraph?

2. In Example 3, in the first line change $\frac{1}{x}$ to $\frac{1}{x-1}$. What other changes must be made in the first paragraph?

3. In Example 5, find $f(2)$ and $f(5)$.

4. In Example 7, interchange 40% and 20% and then find the function.

In Exercises 5–12, find the domain and range of the given functions. In Exercises 11 and 12, explain your answers.

5. $f(x) = x + 5$

6. $g(u) = 3 - u^2$

7. $G(R) = \dfrac{3.2}{R}$

8. $F(r) = \sqrt{r + 4}$

9. $f(s) = \dfrac{2}{s^2}$

10. $T(t) = 2t^4 + t^2 - 1$

Ⓦ **11.** $H(h) = 2h + \sqrt{h} + 1$ Ⓦ **12.** $f(x) = \dfrac{6}{\sqrt{2 - x}}$

In Exercises 13–16, find the domain of the given functions.

13. $Y(y) = \dfrac{y + 1}{\sqrt{y - 2}}$

14. $f(n) = \dfrac{n}{6 - 2n}$

15. $f(D) = \dfrac{D}{D - 2} + \dfrac{4}{D + 4} - \dfrac{D - 3}{D - 6}$

16. $g(x) = \dfrac{\sqrt{x - 2}}{x - 3}$

In Exercises 17–20, evaluate the indicated functions.

$$F(t) = 3t - t^2 \quad \text{(for } t \leq 2) \qquad h(s) = \begin{cases} 2s & \text{(for } s < -1) \\ s + 1 & \text{(for } s \geq -1) \end{cases}$$

$$f(x) = \begin{cases} x + 1 & \text{(for } x < 1) \\ \sqrt{x + 3} & \text{(for } x \geq 1) \end{cases} \qquad g(x) = \begin{cases} \dfrac{1}{x} & \text{(for } x \neq 0) \\ 0 & \text{(for } x = 0) \end{cases}$$

17. Find $F(2)$ and $F(3)$.

18. Find $h(-8)$ and $h(-0.5)$.

19. Find $f(1)$ and $f(-0.25)$.

20. Find $g(0.2)$ and $g(0)$.

In Exercises 21–32, determine the appropriate functions.

21. A motorist travels at 60 km/h for 2 h and then at 80 km/h for t h. Express the distance d traveled as a function of t.

22. Express the cost C of insulating a cylindrical water tank of height 2 m as a function of its radius r, if the cost of insulation is \$3 per square meter.

23. A rocket burns up at the rate of 2 Mg/min after falling out of orbit into the atmosphere. If the rocket had a mass of 5500 Mg before reentry, express its mass m as a function of the time t, in minutes, of reentry.

24. A computer part costs \$3 to produce and distribute. Express the profit p made by selling 100 of these parts as a function of the price of c dollars each.

25. Upon ascending, a weather balloon ices up at the rate of 0.5 kg/m after reaching an altitude of 1000 m. If the mass of the balloon below 1000 m is 110 kg, express its mass m as a function of its altitude h if $h > 1000$ m.

26. A chemist adds x L of a solution that is 50% alcohol to 100 L of a solution that is 70% alcohol. Express the number n of liters of alcohol in the final solution as a function of x.

27. A company installs underground cable at a cost of \$500 for the first 50 m (or up to 50 m) and \$5 for each meter thereafter. Express the cost C as a function of the length l of underground cable if $l > 50$ m.

28. The *mechanical advantage* of an inclined plane is the ratio of the length of the plane to its height. Express the mechanical advantage M of a plane of length 8 m as a function of its height h.

29. The capacities (in L) of two oil-storage tanks are x and y. The tanks are initially full; 1200 L is removed from them by taking 10% of the contents of the first tank and 40% of the contents of the second tank. (a) Express y as a function of x. (b) Find $f(400)$.

30. A city does not tax the first \$30 000 of a resident's income but taxes any amount over \$30 000 at 5%. (a) Find the tax T as a function of a resident's income I and (b) find $f(25\,000)$ and $f(45\,000)$.

31. In studying the electric current that is induced in wire rotating through a magnetic field, a piece of wire 60 cm long is cut into two pieces. One of these is bent into a circle and the other into a square. Express the total area A of the two figures as a function of the perimeter p of the square.

32. The cross section of an air-conditioning duct is in the shape of a square with semicircles on each side. See Fig. 3.6. Express the area A of this cross section as a function of the diameter d (in cm) of the circular part.

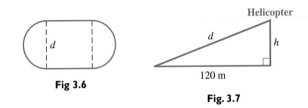

Fig 3.6

Fig. 3.7

In Exercises 33–44, solve the given problems.

33. A helicopter 120 m from a person takes off vertically. Express the distance d from the person to the helicopter as a function of the height h of the helicopter. What are the domain and the range of $d = f(h)$? See Fig. 3.7.

34. A computer program displays a circular image of radius 6 cm. If the radius is decreased by x cm, express the area of the image as a function of x. What are the domain and range of $A = f(x)$?

35. A truck travels 300 km in t h. Express the average speed s of the truck as a function of t. What are the domain and range of $s = f(t)$?

36. A rectangular grazing range with an area of 8 km^2 is to be fenced. Express the length l of the field as a function of its width w. What are the domain and range of $l = f(w)$?

W **37.** The resonant frequency f of a certain electric circuit as a function of the capacitance is $f = \dfrac{1}{2\pi\sqrt{C}}$. Describe the domain.

38. A jet is traveling directly between Calgary, Alberta, and Portland, Oregon, which are 880 km apart. If the jet is x km from Calgary and y km from Portland, find the domain of $y = f(x)$.

39. Express the mass m of the weather balloon in Exercise 25 as a function of any height h in the same manner as the function in Example 5 (and Exercises 17–20) was expressed.

40. Express the cost C of installing any length l of the underground cable in Exercise 27 in the same manner as the function in Example 5 (and Exercises 17–20) was represented.

41. A rectangular piece of cardboard twice as long as wide is to be made into an open box by cutting 5-cm squares from each corner and bending up the sides. (a) Express the volume V of the box as a function of the width w of the piece of cardboard. (b) Find the domain of the function.

42. A spherical buoy 36 cm in diameter is floating in a lake and is more than half above the water. (a) Express the circumference c of the circle of intersection of the buoy and water as a function of the depth d to which the buoy sinks. (b) Find the domain and range of the function.

43. For $f(x - 1) = |x|$, find $f(0)$.

W **44.** For $f(x) = \sqrt{x - 1}$ and $g(x) = x^2$, find the domain of $g[f(x)]$. Explain.

Rectangular (Cartesian) coordinates were developed by the French mathematician René Descartes (1596–1650).

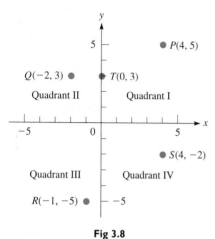

Fig 3.8

NOTE ▶

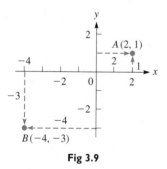

Fig 3.9

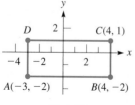

Fig 3.10

3.3 RECTANGULAR COORDINATES

One of the most valuable ways of representing a function is by graphical representation. By using graphs, we can obtain a "picture" of the function; by using this picture, we can learn a great deal about the function.

To make a graphical representation of a function, recall from Chapter 1 that numbers can be represented by points on a line. For a function, we have values of the independent variable as well as the corresponding values of the dependent variable. Therefore, it is necessary to use two different lines to represent the values from each of these sets of numbers. We do this by placing the lines perpendicular to each other.

*Place one line horizontally and label it the **x-axis**.* The values of the independent variable are normally placed on this axis. *The other line is placed vertically and labeled the **y-axis***. Normally, the y-axis is used for values of the dependent variable. *The point of intersection is called the **origin**.* This is the **rectangular coordinate system.**

On the x-axis, positive values are to the right of the origin, and negative values are to the left of the origin. On the y-axis, positive values are above the origin, and negative values are below it. *The four parts into which the plane is divided are called* **quadrants,** which are numbered as in Fig. 3.8.

A point P in the plane is designated by the pair of numbers (x, y), where x is the value of the independent variable and y is the value of the dependent variable. *The x-value is the perpendicular distance of P from the y-axis, and the y-value is the perpendicular distance of P from the x-axis.* The values of x and y, written as (x, y), are the **coordinates** of the point P. Note carefully that the **x-coordinate** *is always written first, and the* **y-coordinate** *is always written second.* (The x-coordinate is also known as the *abscissa,* and the y-coordinate is also known as the *ordinate.*)

◀ EXAMPLE 1   Locate the points $A(2, 1)$ and $B(-4, -3)$ on the rectangular coordinate system.

The coordinates $(2, 1)$ for A mean that the point is 2 units to the *right* of the y-axis and 1 unit *above* the x-axis, as shown in Fig. 3.9. The coordinates $(-4, -3)$ for B mean that the point is 4 units to the *left* of the y-axis and 3 units *below* the x-axis, as shown.

The x-coordinate of point A is 2, and the y-coordinate of A is 1. For point B, its x-coordinate is -4, and its y-coordinate is -3. ▶

◀ EXAMPLE 2   The positions of points $P(4, 5)$, $Q(-2, 3)$, $R(-1, -5)$, $S(4, -2)$, and $T(0, 3)$ are shown in Fig. 3.8. We see that this representation allows for *one point for any pair of values* (x, y). Also note that the point $T(0, 3)$ is on the y-axis. Any such point that is on either axis is not *in* any of the four quadrants. ▶

◀ EXAMPLE 3   Three vertices of the rectangle in Fig. 3.10 are $A(-3, -2)$, $B(4, -2)$, and $C(4, 1)$. What is the fourth vertex?

We use the fact that opposite sides of a rectangle are equal and parallel to find the solution. Since both vertices of the base AB of the rectangle have a y-coordinate of -2, the base is parallel to the x-axis. Therefore, the top of the rectangle must also be parallel to the x-axis. Thus, the vertices of the top must both have a y-coordinate of 1, since one of them has a y-coordinate of 1. In the same way, the x-coordinates of the left side must both be -3. Therefore, the fourth vertex is $D(-3, 1)$. ▶

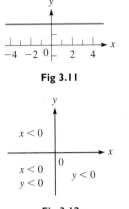

Fig 3.11

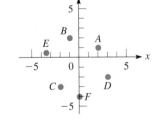

Fig 3.12

❬ EXAMPLE 4 Where are all the points whose *y*-coordinates are 2?

We can see that the question could be stated as: "Where are all the points for which $y = 2$? Since all such points are two units above the *x*-axis, the answer could be stated as "on a line 2 units above the *x*-axis." See Fig. 3.11. ❭

❬ EXAMPLE 5 Where are all points (x, y) for which $x < 0$ and $y < 0$?

Noting that $x < 0$ means "*x* is less than zero," or "*x* is negative," and that $y < 0$ means the same for *y*, we want to determine where both *x* and *y* are negative. Our answer is "in the third quadrant," since both coordinates are negative for all points in the third quadrant, and this is the only quadrant for which this is true. See Fig. 3.12. ❭

EXERCISES 3.3

In Exercises 1 and 2, make the given changes in the indicated examples of this section, and then solve the resulting problems.

1. In Example 3, change $A(-3, -2)$ to $A(-1, -2)$ and then find the fourth vertex.

2. In Example 5, change $y < 0$ to $y > 0$ and then find the location of the points (x, y).

In Exercises 3 and 4, determine (at least approximately) the coordinates of the points shown in Fig. 3.13.

3. *A, B, C*

4. *D, E, F*

Fig. 3.13

In Exercises 5 and 6, plot the given points.

5. $A(2, 7)$, $B(-1, -2)$, $C(-4, 2)$

6. $A\left(3, \frac{1}{2}\right)$, $B(-6, 0)$, $C\left(-\frac{5}{2}, -5\right)$

In Exercises 7–10, plot the given points and then join these points, in the order given, by straight-line segments. Name the geometric figure formed.

7. $A(-1, 4)$, $B(3, 4)$, $C(1, -2)$

8. $A(0, 3)$, $B(0, -1)$, $C(4, -1)$

9. $A(-2, -1)$, $B(3, -1)$, $C(3, 5)$, $D(-2, 5)$

10. $A(-5, -2)$, $B(4, -2)$, $C(6, 3)$, $D(-3, 3)$

In Exercises 11–14, find the indicated coordinates.

11. Three vertices of a rectangle are $(5, 2)$, $(-1, 2)$, and $(-1, 4)$. What are the coordinates of the fourth vertex?

12. Two vertices of an equilateral triangle are $(7, 1)$ and $(2, 1)$. What is the *x*-coordinate of the third vertex?

13. *P* is the point $(3, 2)$. Locate point *Q* such that the *x*-axis is the perpendicular bisector of the line segment joining *P* and *Q*.

14. *P* is the point $(-4, 1)$. Locate point *Q* such that the line segment joining *P* and *Q* is bisected by the origin.

In Exercises 15–32, answer the given questions.

15. Where are all points whose *x*-coordinates are 1?

16. Where are all points whose *y*-coordinates are -3?

17. Where are all points such that $y = 3$?

18. Where are all points such that $|x| = 2$?

19. Where are all points whose *x*-coordinates equal their *y*-coordinates?

20. Where are all points whose *x*-coordinates equal the negative of their *y*-coordinates?

21. What is the *x*-coordinate of all points on the *y*-axis?

22. What is the *y*-coordinate of all points on the *x*-axis?

23. Where are all points for which $x > 0$?

24. Where are all points for which $y < 0$?

25. Where are all points for which $x < -1$?

26. Where are all points for which $y > 4$?

27. Where are all points for which $xy > 0$?

28. Where are all points for which $y/x < 0$?

29. Where are all points for which $xy = 0$?

Ⓦ **30.** Describe the values of *x* and *y* for which $(3, -1)$, $(3, 0)$, and (x, y) are on the same straight line.

31. Find the distance (a) between $(3, -2)$ and $(-5, -2)$ and (b) between $(3, -2)$ and $(3, 4)$.

Ⓦ **32.** Find the distance between $(-5, -2)$ and $(3, 4)$. Explain your method. (*Hint*: See Exercise 31.)

3.4 THE GRAPH OF A FUNCTION

Now that we have introduced the concepts of a function and the rectangular coordinate system, we are in a position to determine the graph of a function. In this way, we will obtain a visual representation of a function.

The graph of a function is the set of all points whose coordinates (x, y) satisfy the functional relationship $y = f(x)$. Since $y = f(x)$, we can write the coordinates of the points on the graph as $(x, f(x))$. Writing the coordinates in this manner tells us exactly how to find them. *We assume a certain value for x and then find the value of the function of x. These two numbers are the coordinates of the point.*

Since there is no limit to the possible number of points that can be chosen, we normally select a few values of x, obtain the corresponding values of the function, plot these points, and then join them. Therefore, we use the following basic procedure in plotting a graph.

As to just what values of x to choose and how many to choose, with a little experience you will usually be able to tell if you have enough points to plot an accurate graph.

PROCEDURE FOR PLOTTING THE GRAPH OF A FUNCTION

1. *Let x take on several values and calculate the corresponding values of y.*
2. *Tabulate these values, arranging the table so that **values of x are increasing**.*
3. *Plot the points and join them from left to right by a **smooth curve** (not short straight-line segments).*

We will show how a graph is displayed on a graphing calculator in the next section.

EXAMPLE 1 Graph the function $f(x) = 3x - 5$.

For purposes of graphing, let $y = f(x)$, or $y = 3x - 5$. Then let x take on various values and determine the corresponding values of y. Note that once we choose a given value of x, we have no choice about the corresponding y-value, as it is determined by evaluating the function. If $x = 0$, we find that $y = -5$. This means that the point $(0, -5)$ is on the graph of the function $3x - 5$. Choosing another value of x—for example, 1—we find that $y = -2$. This means that the point $(1, -2)$ is on the graph of the function $3x - 5$. Continuing to choose a few other values of x, we tabulate the results, as shown in Fig. 3.14. It is best to arrange the table so that the values of x increase; then there is no doubt how they are to be connected, for they are then connected in the order shown. Finally, we connect the points in Fig. 3.14 and see that the graph of the function $3x - 5$ is a straight line.

The table of values for a graph can be seen on a graphing calculator using the *table* feature. Figure 3.15 shows the calculator display of the table for Example 1.

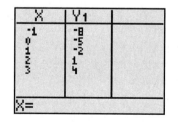

Fig 3.15

Select values of x and calculate corresponding values of y		Tabulate with values of x increasing	
$f(x) = 3x - 5$		x	y
$f(-1) = 3(-1) - 5 = -8$		-1	-8
$f(0) = 3(0) - 5 = -5$		0	-5
$f(1) = 3(1) - 5 = -2$		1	-2
$f(2) = 3(2) - 5 = 1$		2	1
$f(3) = 3(3) - 5 = 4$		3	4

Plot points and join from left to right with smooth curve

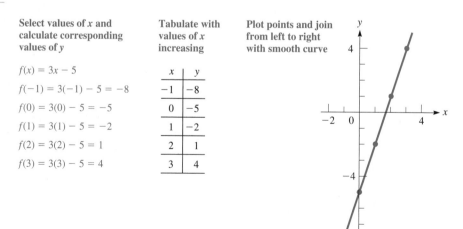

Fig. 3.14

EXAMPLE 2 Graph the function $f(x) = 2x^2 - 4$.

First, let $y = 2x^2 - 4$ and tabulate the values as shown in Fig. 3.16. In determining the values in the table, take particular care to obtain the correct values of y for negative values of x. ***Mistakes are relatively common when dealing with negative numbers.*** We must carefully use the laws for signed numbers. For example, if $x = -2$, we have $y = 2(-2)^2 - 4 = 2(4) - 4 = 8 - 4 = 4$. Once the values are obtained, plot and connect the points with a smooth curve, as shown.

CAUTION ▶

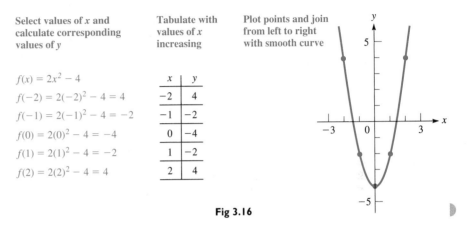

Select values of x and calculate corresponding values of y

$f(x) = 2x^2 - 4$
$f(-2) = 2(-2)^2 - 4 = 4$
$f(-1) = 2(-1)^2 - 4 = -2$
$f(0) = 2(0)^2 - 4 = -4$
$f(1) = 2(1)^2 - 4 = -2$
$f(2) = 2(2)^2 - 4 = 4$

Tabulate with values of x increasing

x	y
-2	4
-1	-2
0	-4
1	-2
2	4

Plot points and join from left to right with smooth curve

Fig 3.16

When graphing a function, we must use extra care with certain parts of the graph. These include the following:

Special Notes on Graphing

1. Since the graphs of most common functions are smooth, *any place that the graph changes in a way that is not expected should be checked* with care. It usually helps to check values of x between which the question arises.

2. The domain of the function may not include all values of x. Remember, ***division by zero is not defined,*** and ***only real values of the variables may be used.***

3. In applications, we must *use only values of the variables that have meaning.* In particular, negative values of some variables, such as time, generally have no meaning.

The following examples illustrate these points.

EXAMPLE 3 Graph the function $y = x - x^2$.

First, we determine the values in the table, as shown with Fig. 3.17. Again, we must be careful when dealing with negative values of x. For the value $x = -1$, we have $y = (-1) - (-1)^2 = -1 - 1 = -2$. Once all values in the table have been found and plotted, note that ***$y = 0$ for both $x = 0$ and $x = 1$.*** The question arises—***what happens between these values?*** Trying $x = \frac{1}{2}$, we find that $y = \frac{1}{4}$. Using this point completes the information needed to complete an accurate graph. Also note that in plotting these graphs we do not stop the graph with the last points determined but indicate that the curve continues by drawing the graph past these points.

x	y
-2	-6
-1	-2
0	0
1	0
2	-2
3	-6

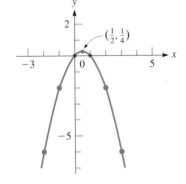

Fig 3.17

◀ **EXAMPLE 4** Graph the function $y = 1 + \dfrac{1}{x}$.

CAUTION ▶

In finding the points on this graph, as shown in Fig. 3.18, note that y is not defined for $x = 0$, due to division by zero. Thus, $x = 0$ is not in the domain, and *we must be careful not to have any part of the curve cross the y-axis* ($x = 0$). Although we cannot let $x = 0$, we can choose other values for x between -1 and 1 that are close to zero. In doing so, we find that as x gets closer to zero, the points get closer and closer to the y-axis, although they do not reach or touch it. In this case, the y-axis is called an **asymptote** of the curve.

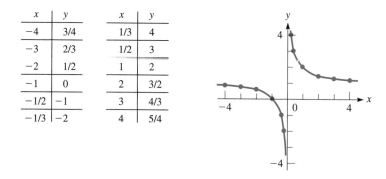

x	y		x	y
-4	3/4		1/3	4
-3	2/3		1/2	3
-2	1/2		1	2
-1	0		2	3/2
$-1/2$	-1		3	4/3
$-1/3$	-2		4	5/4

Fig. 3.18

We see that the curve is smooth, except when $x = 0$ where it is not defined. ▶

◀ **EXAMPLE 5** Graph the function $y = \sqrt{x + 1}$.

NOTE ▶

When finding the points for the graph, we may not let x take on any value less than -1, for *all such values would lead to imaginary values for y* and are not in the domain. Also, since we have the positive square root indicated, the range consists of all values of y that are positive or zero ($y \geq 0$). See Fig. 3.19. Note that the graph starts at $(-1, 0)$.

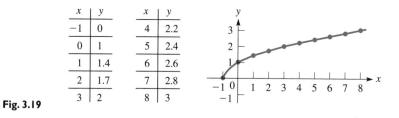

x	y		x	y
-1	0		4	2.2
0	1		5	2.4
1	1.4		6	2.6
2	1.7		7	2.8
3	2		8	3

Fig. 3.19

Although not defined for all values of x, the curve is smooth. ▶

Functions of a particular type have graphs with a certain basic shape, and many have been named. Three examples of this are the *straight line* (Example 1), the *parabola* (Examples 2 and 3), and the *hyperbola* (Example 4). We consider the straight line again in Chapter 5 and the parabola in Chapter 7. All of these graphs and others are studied in detail in Chapter 21. Other types of graphs are found in many of the later chapters.

See the chapter introduction.

The unit of power, the watt (W), is named for James Watt (1736–1819), a British engineer.

EXAMPLE 6 The electric power P (in W) delivered by a certain battery as a function of the resistance R (in Ω) in the circuit is given by $P = \dfrac{100R}{(0.50 + R)^2}$. Plot P as a function of R.

Since negative values for the resistance have no physical significance, we should not plot any values of P for negative values of R. The following table is obtained:

R (Ω)	0	0.25	0.50	1.0	2.0	3.0	4.0	5.0	10.0
P (W)	0.0	44.4	50.0	44.4	32.0	24.5	19.8	16.5	9.1

The values 0.25 and 0.50 are used for R when it is found that P is less for $R = 2$ than for $R = 1$. The sharp change in direction at $R = 1$ should be checked by using these additional points to better see how the curve changes near $R = 1$ and thereby obtain a smoother curve. See Fig. 3.20.

Also note that the scale on the P-axis is different from that on the R-axis. This reflects the different magnitudes and ranges of values used for each of the variables. Different scales are normally used in such cases.

We can make various conclusions from the graph. For example, we see that the maximum power of 50 W occurs for $R = 0.5$ Ω. Also, P decreases as R increases beyond 0.5 Ω. We will consider further the information that can be read from a graph in the next section.

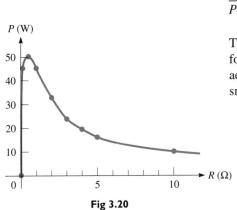

Fig 3.20

The following example illustrates the graph of a function that is defined differently for different intervals of the domain.

EXAMPLE 7 Graph the function $f(x) = \begin{cases} 2x + 1 & \text{(for } x \le 1) \\ 6 - x^2 & \text{(for } x > 1) \end{cases}$

First, let $y = f(x)$ and then tabulate the necessary values. In evaluating $f(x)$, we must be careful to use the proper part of the definition. To see where to start the curve for $x > 1$, we evaluate $6 - x^2$ for $x = 1$, but we must realize that **the curve does not include this point** $(1, 5)$ and starts immediately to its right. To show that it is not part of the curve, draw it as an open circle. See Fig. 3.21.

NOTE ▶

$$f(-2) = 2(-2) + 1 = -3$$
$$f(-1) = 2(-1) + 1 = -1$$
$$f(0) = 2(0) + 1 = 1$$
$$f(1) = 2(1) + 1 = 3$$
$$f(2) = 6 - 2^2 = 2$$
$$f(3) = 6 - 3^2 = -3$$

x	y
-2	-3
-1	-1
0	1
1	3
2	2
3	-3

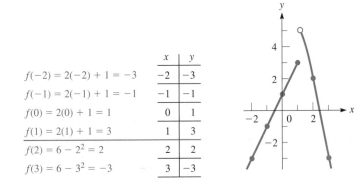

Fig. 3.21

A function such as this one, with a "break" in it, is called *discontinuous*.

A function is defined such that there is only one value of the dependent variable for each value of the independent variable, but a *relation* may have more than one such value of the dependent variable. The following example illustrates how to use a graph to see whether or not a relation is also a function by using the **vertical-line test.**

Vertical-Line Test

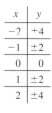

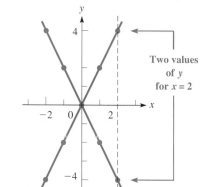

Fig 3.22

◖**EXAMPLE 8**   In Example 9 of Section 3.2, we noted that $y^2 = 4x^2$ is a relation, but not a function. Since $y = 4$ or $y = -4$, normally written as $y = \pm 4$, for $x = 2$, we have two values of y for $x = 2$. Therefore, it is not a function. Making the table as shown at the left, we then draw the graph in Fig. 3.22. When making the table, we also note that there are two values of y for every value of x, except $x = 0$.

If any *vertical line* that crosses the x-axis in the domain intersects the graph in more than one point, it is the graph of a relation that is not a function. Any such vertical line in any of the previous graphs of this section would intersect the graph in only one point. This shows that they are graphs of functions.   ◗

The use of graphs is extensive in mathematics and in applications. For example, in the next section we use graphs to solve equations. Numerous applications can be noted for most types of curves. For example, the parabola has applications in microwave-dish design, suspension bridge design, and the path (trajectory) of a baseball. We will note many more applications of graphs throughout the book.

EXERCISES **3.4**

In Exercises 1–4, make the given changes in the indicated examples of this section and then plot the graphs.

1. In Example 1, change the $-$ sign to $+$.
2. In Example 2, change $2x^2 - 4$ to 4 to $4 - 2x^2$.
3. In Example 4, change the x in the denominator to $x - 1$.
4. In Example 5, change the $+$ sign to $-$.

In Exercises 5–36, graph the given functions.

5. $y = 3x$
6. $y = -2x$
7. $y = 2x - 4$
8. $y = 3x + 5$
9. $y = 7 - 2x$
10. $y = -3$
11. $y = \frac{1}{2}x - 2$
12. $y = 6 - \frac{1}{3}x$
13. $y = x^2$
14. $y = -2x^2$
15. $y = 3 - x^2$
16. $y = x^2 - 3$
17. $y = \frac{1}{2}x^2 + 2$
18. $y = 2x^2 + 1$
19. $y = x^2 + 2x$
20. $h = 20t - 5t^2$
21. $y = x^2 - 3x + 1$
22. $y = 2 + 3x + x^2$
23. $V = e^3$
24. $y = -2x^3$
25. $y = x^3 - x^2$
26. $y = 3x - x^3$
27. $y = x^4 - 4x^2$
28. $y = x^3 - x^4$
29. $P = \frac{1}{V} + 1$
30. $y = \frac{2}{x + 2}$
31. $y = \frac{4}{x^2}$
32. $p = \frac{1}{n^2 + 0.5}$
33. $y = \sqrt{x}$
34. $y = \sqrt{4 - x}$
35. $y = \sqrt{16 - x^2}$
36. $y = \sqrt{x^2 - 16}$

In Exercises 37–60, graph the indicated functions.

37. In blending gasoline, the number of liters n of 85 octane gas to be blended with m L of 92 octane gas is given by the equation $n = 0.40m$. Plot n as a function of m.

38. The consumption of fuel c (in L/h) of a certain engine is determined as a function of the number r of r/min of the engine, to be $c = 0.011r + 4.0$. This formula is valid for 500 r/min to 3000 r/min. Plot c as a function of r. (r is the symbol for revolution.)

39. For a certain model of truck, its resale value V (in dollars) as a function of the total distance m it has been driven is $V = 50\,000 - 0.2m$. Plot V as a function of m for $m \le 100\,000$ km.

40. The resistance R (in Ω) of a resistor as a function of the temperature T (in °C) is given by $R = 250(1 + 0.0032T)$. Plot R as a function of T.

41. The rate H (in W) at which heat is developed in the filament of an electric light bulb as a function of the electric current I (in A) is $H = 240I^2$. Plot H as a function of I.

42. The total annual fraction f of energy supplied by solar energy to a home as a function of the area A (in m^2) of the solar collector is $f = 0.065\sqrt{A}$. Plot f as a function of A.

43. The maximum speed v (in km/h) at which a car can safely travel around a circular turn of radius r (in m) is given by $r = 0.55v^2$. Plot r as a function of v.

44. The height h (in m) of a rocket as a function of the time t (in s) is given by the function $h = 1500t - 4.9t^2$. Plot h as a function of t, assuming level terrain.

45. The power P (in W/h) that a certain windmill generates is given by $P = 0.004v^3$, where v is the wind speed (in km/h). Plot the graph of P vs. v.

46. An astronaut weighs 750 N at sea level. The astronaut's weight at an altitude of x km above sea level is given by $w = 750\left(\dfrac{6400}{6400 + x}\right)^2$. Plot w as a function of x for $x = 0$ to $x = 8000$ km.

47. A formula used to determine the amount of lumber V (in dm^3) that can be cut from a 2-m section of a log of diameter d (in mm) is $V = 0.0013d^2 - 0.043d$. Plot V as a function of d for values of d from 300 mm to 1000 mm.

48. A copper electrode with a mass of 25.0 g is placed in a solution of copper sulfate. An electric current is passed through the solution, and 1.6 g of copper is deposited on the electrode each hour. Express the total mass m on the electrode as a function of the time t and plot the graph.

49. A land developer is considering several options of dividing a large tract into rectangular building lots, many of which would have perimeters of 200 m. For these, the minimum width would be 30 m and the maximum width would be 70 m. Express the areas A of these lots as a function of their widths w and plot the graph.

50. The distance p (in m) from a camera with a 50-mm lens to the object being photographed is a function of the magnification m of the camera, given by $p = \dfrac{0.05(1 + m)}{m}$. Plot the graph for positive values of m up to 0.50.

51. A measure of the light beam that can be passed through an optic fiber is its numerical aperture N. For a particular optic fiber, N is a function of the index of refraction n of the glass in the fiber, given by $N = \sqrt{n^2 - 1.69}$. Plot the graph for $n \leq 2.00$.

52. $F = x^4 - 12x^3 + 46x^2 - 60x + 25$ is the force (in N) exerted by a cam on the arm of a robot, as shown in Fig. 3.23. Noting that x varies from 1 cm to 5 cm, plot the graph.

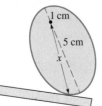

Fig 3.23

53. The number of times S that a certain computer can perform a computation faster with a multiprocessor than with a uniprocessor is given by $S = \dfrac{5n}{4 + n}$, where n is the number of processors. Plot S as a function of n.

54. The voltage V across a capacitor in a certain electric circuit for a 2-s interval is $V = 2t$ during the first second and $V = 4 - 2t$ during the second second. Here, t is the time (in s). Plot V as a function of t.

(W) 55. Plot the graphs of $y = x$ and $y = |x|$ on the same coordinate system. Explain why the graphs differ.

(W) 56. Plot the graphs of $y = 2 - x$ and $y = |2 - x|$ on the same coordinate system. Explain why the graphs differ.

57. Plot the graph of $f(x) = \begin{cases} 3 - x & (\text{for } x < 1) \\ x^2 + 1 & (\text{for } x \geq 1) \end{cases}$.

58. Plot the graph of $f(x) = \begin{cases} \dfrac{1}{x - 1} & (\text{for } x < 0) \\ \sqrt{x + 1} & (\text{for } x \geq 0) \end{cases}$.

(W) 59. Plot the graphs of (a) $y = x + 2$ and (b) $y = \dfrac{x^2 - 4}{x - 2}$. Explain the difference between the graphs.

(W) 60. Plot the graphs of (a) $y = x^2 - x + 1$ and (b) $y = \dfrac{x^3 + 1}{x + 1}$. Explain the difference between the graphs.

In Exercises 61–64, determine whether or not the indicated graph is that of a relation which is a function.

61. Fig. 3.24(a) **62.** Fig. 3.24(b)

63. Fig. 3.24(c) **64.** Fig. 3.24(d)

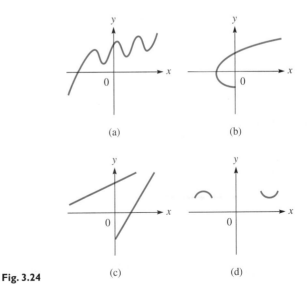

(a) (b)

(c) (d)

Fig. 3.24

3.5 GRAPHS ON THE GRAPHING CALCULATOR

In this section, we will see that a graphing calculator can display a graph quickly and easily. As we have pointed out, to use your calculator effectively, *you must know the sequence of keys* for any operation you intend to use. The manual for any particular model should be used for a detailed coverage of its features.

See Appendix C for a list of graphing calculator features and examples of calculator use (with page references) that are included in this text.

◀ **EXAMPLE 1**   To graph the function $y = 2x + 8$, first display $Y_1 =$ and then enter the $2x + 8$. The display for this is shown in Fig. 3.25(a).

Next use the *window* (or *range*) feature to set the part of the domain and the range that will be seen in the *viewing window*. For this function, set

$$\text{Xmin} = -6, \text{Xmax} = 2, \text{Xscl} = 1, \text{Ymin} = -2, \text{Ymax} = 10, \text{Yscl} = 1$$

The specific detail shown in any nongraphic display depends on the model. A graph with the same *window* settings should appear about the same on all models.

in order to get a good view of the graph. The display for the *window* settings is shown in Fig. 3.25(b).

Then display the graph, using the *graph* (or *exe*) key. The display showing the graph of $y = 2x + 8$ is shown in Fig. 3.25(c).

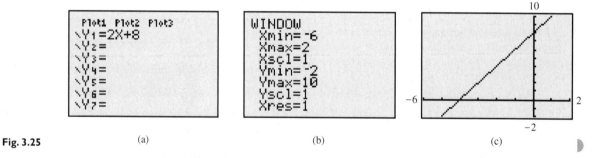

Fig. 3.25   (a)   (b)   (c)

In Example 1, we noted that the *window* feature sets the intervals of the domain and range of the function, which is seen in the viewing window. Unless the settings are appropriate, you may not get a good view of the graph.

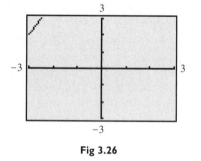

Fig 3.26

In later examples we generally will not show the *window* settings in the text. They will be shown outside the display, as in Fig. 3.26.

◀ **EXAMPLE 2**   In Example 1, the *window* settings were chosen to give a good view of the graph of $y = 2x + 8$. However, if we choose settings of Xmin $= -3$, Xmax $= 3$, Ymin $= -3$, and Ymax $= 3$, we get the view shown in Fig. 3.26. We see that very little of the graph can be seen. With some settings, it is possible that no part of the graph can be seen. Therefore, it is necessary to be careful in choosing the settings. This includes the scale settings (Xscl and Yscl), which should be chosen so that several (not too many or too few) axis markers are used.

Since the settings are easily changed, to get the general location of the graph, it is usually best to first choose intervals between the min and max values that are greater than is probably necessary. They can be reduced as needed.

Also note that the calculator makes the graph just as we have been doing—by plotting points (square dots called *pixels*). It just does it a lot faster. ◀

SOLVING EQUATIONS GRAPHICALLY

An equation can be solved by use of a graph. Most of the time the solution will be approximate, but with a graphing calculator it is possible to get good accuracy in the result. The procedure used is as follows:

> **PROCEDURE FOR SOLVING AN EQUATION GRAPHICALLY**
>
> **1.** *Collect all terms on one side of the equal sign.* This gives us the equation $f(x) = 0$.
> **2.** Set $y = f(x)$ and graph this function.
> **3.** *Find the points where the graph crosses the x-axis. These points are called* the **x-intercepts** *of the graph.* At these points, $y = 0$.
> **4.** *The values of x for which $y = 0$ are the solutions of the equation. (These values are called the **zeros** of the function $f(x)$.)*

The next example is done without a graphing calculator to illustrate the method.

◀ **EXAMPLE 3** Graphically solve the equation $x^2 - 2x = 1$.

Following the above procedure, first rewrite the equation as $x^2 - 2x - 1 = 0$ and then set $y = x^2 - 2x - 1$. Next plot the graph of this function, as shown in Fig. 3.27.

For this curve, we see that it crosses the x-axis between $x = -1$ and $x = 0$ and between $x = 2$ and $x = 3$. This means that there are *two* solutions of the equation, and they can be estimated from the graph as

$$x = -0.5 \quad \text{and} \quad x = 2.5$$

For these values, $y = 0$, which means $x^2 - 2x - 1 = 0$, which in turn means $x^2 - 2x = 1$. ▶

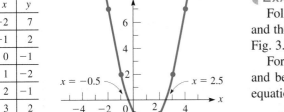

x	y
-2	7
-1	2
0	-1
1	-2
2	-1
3	2
4	7

Fig 3.27

We now show the use of a graphing calculator in solving equations graphically. In the next example, we solve the equation of Example 3.

◀ **EXAMPLE 4** Using a graphing calculator, solve the equation $x^2 - 2x = 1$.

Again, rewrite the equation as $x^2 - 2x - 1 = 0$ and set $y = x^2 - 2x - 1$. Then display this function on the calculator, as shown in Fig. 3.28(a).

To get values of x for which $y = 0$, use the *trace* feature. We move the *cursor* (a blinking pixel) with the arrow keys to find the pixel for which the value of y is closest to zero (it generally will not be *exactly zero*). We then have the display in Fig. 3.28(b), which shows the solutions to be about $x = -0.4$ and $x = 2.4$.

Much more accurate values can be found by using the *zoom* feature, with which any region of the screen can be greatly magnified. Fig. 3.28(c) shows the screen when *zoom* is used near $x = 2.4$. The cursor shows that the solution is about $x = 2.415$. In the same way, the other solution is about $x = -0.415$.

Checking these solutions *in the original equation*, we get $1.002\,225 = 1$ for each. This shows that they check, since we know that they are approximate.

Depending on the calculator model, the equation and cursor coordinates may be displayed when using *trace*.

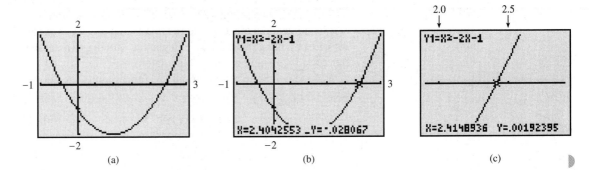

Fig. 3.28   (a)   (b)   (c)

Many calculators have a *zero* (or *root*) feature. The equation in Example 5 can be solved using this feature by finding the zeros on the graph of the function $y = 2x^3 + 3x^2 - 3x$.

Many calculators have an *intersect* feature. The equation in Example 5 can be solved using this feature by finding the points where the curves $y = x^2(2x + 3)$ and $y = 3x$ cross.

◀ **EXAMPLE 5** Solve the equation $t^2(2t + 3) = 3t$ graphically.

Collecting all terms on the left leads to the equation $2t^3 + 3t^2 - 3t = 0$, and we then let $y = 2t^3 + 3t^2 - 3t$. On a calculator, graph $y = 2x^3 + 3x^2 - 3x$ (letting $t = x$), as shown in Fig. 3.29(a). The *window* settings are usually the *default* settings.

This gives us a general idea of the graph, but to get better accuracy change the settings to those shown in Fig. 3.29(b).

Now using the *trace* feature, we get the solutions of $t = -2.2$, $t = 0$, and $t = 0.7$. Here, $t - 0$ is exact (which the calculator may show with $x = 0$, $y = 0$, and no other decimal positions shown). The other solutions are approximate. Checking these solutions, we find that they check (and also that $t = 0$ is an exact solution).

More accurate values can be obtained by using the *zoom* feature.

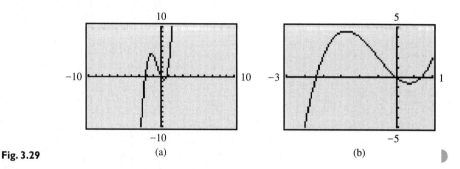

Fig. 3.29　　(a)　　　　　　　(b)

Solving a Word Problem

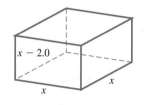

Fig 3.30

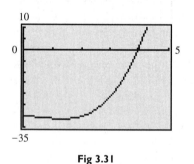

Fig 3.31

◀ **EXAMPLE 6** A rectangular box whose volume is 30 cm³ (two significant digits) is made with a square base and a height that is 2.0 cm less than the length of a side of the base. Find the dimensions of the box by first setting up the necessary equation and then solving it graphically.

Let $x = $ the length of a side of the square base (see Fig. 3.30); the height is then $x - 2.0$. This means that the volume is $(x)(x)(x - 2.0)$, or $x^2(x - 2.0)$. Since the volume is 30 cm³, we have the equation

$$x^2(x - 2.0) = 30$$

To solve it graphically, first rewrite the equation as $x^3 - 2.0x^2 - 30 = 0$ and then set $y = x^3 - 2.0x^2 - 30$. The graphing calculator view is shown in Fig. 3.31. We use only positive values of x since negative values have no meaning. Using the *trace* feature (or *zero* feature or *intersect* feature), we find that $x = 3.9$ is the approximate solution. Therefore, the dimensions are 3.9 cm, 3.9 cm, and 1.9 cm. Checking, these dimensions give a volume of 29 cm³. The difference from 30 cm³ is due to the use of the approximate value of x. A better check can be made by using the unrounded calculator value for x.

FINDING THE RANGE OF A FUNCTION

In Section 3.2, we stated that advanced methods are often necessary to find the range of a function and that we would see that a graphing calculator is useful for this purpose. As with solving equations, we can get very good approximations of the range for most functions by using a graphing calculator.

The method is simply to graph the function and see for what values of y there is a point on the graph. These values of y give us the range of the function.

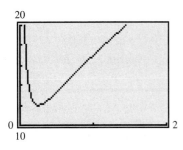

Fig 3.32

EXAMPLE 7 Graphically find the range of the function $y = 16\sqrt{x} + 1/x$.

If we start by using the default *window* settings of Xmin = −10, Xmax = 10, Ymin = −10, Ymax = 10, we see no part of the graph. Looking back to the function, note that x cannot be zero because that would require dividing by zero in the $1/x$ term. Also, $x > 0$ since $\sqrt{x}$ is not defined for $x < 0$. Thus, we should try Xmin = 0 and Ymin = 0.

We also can easily see that $y = 17$ for $x = 1$. This means we should try a value more than 17 for Ymax. Using Ymax = 20, after two or three *window* settings, we use the settings shown in Fig. 3.32. The display shows the range to be all real numbers greater than about 12. (Actually, the range is all real numbers $y \geq 12$.)

SHIFTING A GRAPH

By adding a positive constant to the right side of the function $y = f(x)$, the graph of the function is *shifted* straight up. If a negative constant is added to the right side of the function, the graph is shifted straight down.

EXAMPLE 8 If we add 2 to the right side of $y = x^2$, we get $y = x^2 + 2$. The graphs of these functions are displayed on a calculator as shown in Fig. 3.33. We see that the graph of $y = x^2$ has been shifted up 2 units.

If we add −3 to the right side of $y = x^2$, we get $y = x^2 − 3$. The graphs of these functions are displayed in Fig. 3.34. We see that the graph of $y = x^2$ has been shifted down 3 units.

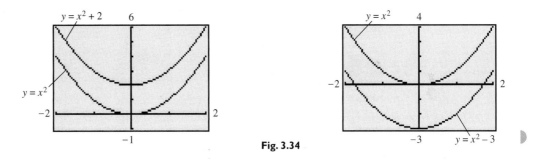

Fig. 3.33

Fig. 3.34

CAUTION ▶ By adding a constant to x in the function $y = f(x)$, the graph of the function is shifted to the right or to the left. *Adding a positive constant to x shifts the graph to the **left**, and adding a negative constant to x shifts the graph to the **right**.*

EXAMPLE 9 For the function $y = x^2$, if we add 2 to x, we get $y = (x + 2)^2$, or if we add −3 to x, we get $y = (x − 3)^2$. The graphs of these three functions are displayed in Fig. 3.35.

We see that the graph of $y = (x + 2)^2$ is 2 units to the left of $y = x^2$ and that the graph of $y = (x − 3)^2$ is 3 units to the right of $y = x^2$.

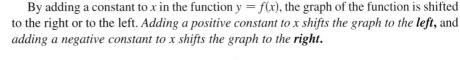

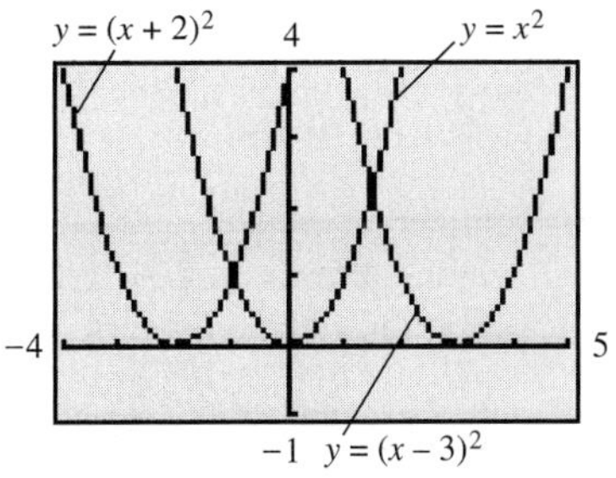

Fig 3.35

Summarizing how a graph is shifted, we have the following:

> **SHIFTING A GRAPH**
>
> **Vertical shifts:** $y = f(x) + k$ shifts the graph of $y = f(x)$
> up k units if $k > 0$ and down k units if $k < 0$.
>
> **Horizontal shifts:** $y = f(x + k)$ shifts the graph of $y = f(x)$
> ***left*** k units if $k > 0$ and ***right*** k units if $k < 0$.

◀ EXAMPLE 10   A function can be shifted both vertically and horizontally. To shift the graph of $y = x^2$ to the right 3 units and down 2 units, add -3 to x and add -2 to the resulting function. In this way, we get $y = (x - 3)^2 - 2$. The graph of this function and the graph of $y = x^2$ are shown in Fig. 3.36.

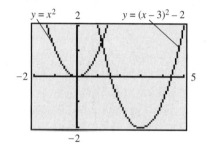

Fig. 3.36

EXERCISES 3.5

In Exercises 1–4, make the indicated changes in the given examples of this section and then solve.

1. In Example 3, change the sign on the left side of the equation from $-$ to $+$.

2. In Example 5, change the $3t$ on the right side of the equation to 3.

3. In Example 7, change the denominator of the second term on the right side of the equation to x^2.

4. In Example 10, in the second line, change "to the right 3 units and down 2 units" to "to the left 2 units and down 3 units."

In Exercises 5–12, display the graphs of the given functions on a graphing calculator. Use appropriate window settings.

5. $y = 3x - 1$

6. $y = 4 - 0.5x$

7. $y = x^2 - 4x$

8. $y = 8 - 2x^2$

9. $y = 6 - x^3$

10. $y = x^4 - 6x^2$

11. $y = \dfrac{2x}{x - 2}$

12. $y = \dfrac{3}{x^2 - 4}$

In Exercises 13–20, use a graphing calculator to solve the given equations to the nearest 0.1.

13. $x^2 + x - 5 = 0$

14. $x^2 - 2x - 4 = 0$

15. $x^3 - 3 = 3x$

16. $x^4 - 2x = 0$

17. $\sqrt{5R + 2} = 3$

18. $\sqrt{x} + 3x = 7$

19. $\dfrac{1}{x^2 + 1} = 0$

20. $T - 2 = \dfrac{1}{T}$

In Exercises 21–28, use a graphing calculator to find the range of the given functions. (The functions of Exercises 25–28 are the same as the functions of Exercises 13–16 of Section 3.2.)

21. $y = \dfrac{4}{x^2 - 4}$

22. $y = \dfrac{x + 1}{x^2}$

23. $y = \dfrac{x^2}{x + 1}$

24. $y = \dfrac{x}{x^2 - 4}$

25. $Y(y) = \dfrac{y + 1}{\sqrt{y - 2}}$

26. $f(n) = \dfrac{n}{6 - 2n}$

27. $f(D) = \dfrac{D}{D - 2} + \dfrac{4}{D + 4} - \dfrac{D - 3}{D - 6}$

28. $g(x) = \dfrac{\sqrt{x - 2}}{x - 3}$

In Exercises 29–36, a function and how it is to be shifted are given. Find the shifted function and then display the given function and the shifted function on the same screen of a graphing calculator.

29. $y = 3x$, up 1

30. $y = x^3$, down 2

31. $y = \sqrt{x}$, right 3

32. $y = \dfrac{2}{x}$, left 4

33. $y = -2x^2$, down 3, left 2

34. $y = -4x$, up 4, right 3

35. $y = \sqrt{2x + 1}$, up 1, left 1

36. $y = \sqrt{x^2 + 4}$, down 2, right 2

In Exercises 37–44, solve the indicated equations graphically. Assume all data are accurate to two significant digits unless greater accuracy is given.

37. In an electric circuit, the current i (in A) as a function of voltage v is given by $i = 0.01v - 0.06$. Find v for $i = 0$.

38. For tax purposes, a corporation assumes that one of its computers depreciates according to the equation $V = 90\,000 - 12\,000t$, where V is the value (in dollars) of the computer after t years. According to this formula, when will the computer be fully depreciated (no value)?

39. Two cubical coolers together hold 40.0 L (40 000 cm³). If the inside edge of one is 5.00 cm greater than the inside edge of the other, what is the inside edge of each?

40. The height h (in m) of a rocket as a function of time t (in s) of flight is given by $h = 15 + 86t - 4.9t^2$. Determine when the rocket is at ground level.

41. The length of a rectangular solar panel is 12 cm more than its width. If its area if 520 cm², find its dimensions.

42. A computer model shows that the cost (in dollars) to remove x percent of a pollutant from a lake is $C = \dfrac{8000x}{100 - x}$. What percent can be removed for $25\,000?

43. In finding the illumination at a point x m from one of two light sources that are 100 m apart, it is necessary to solve the equation $9x^3 - 2400x^2 + 240\,000x - 8\,000\,000 = 0$. Find x.

44. A rectangular storage bin is to be made from a rectangular piece of sheet metal 12 cm by 10 cm, by cutting out equal corners of

side x and bending up the sides. See Fig. 3.37. Find x if the storage bin is to hold 90 cm³.

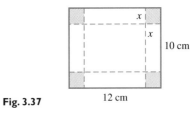

Fig. 3.37

In Exercises 45–52, solve the given problems.

45. Solve the equation of Example 4 to the nearest 0.0001 by using an appropriate feature of a graphing calculator.

46. Solve the equation of Example 5 to the nearest 0.001 by using an appropriate feature of a graphing calculator.

47. The cutting speed s (in cm/s) of a saw in cutting a particular type of metal piece is given by $s = \sqrt{t - 4t^2}$, where t is the time in seconds. What is the maximum cutting speed in this operation (to two significant digits)? (*Hint*: Find the range.)

48. Referring to Exercise 44, explain how to determine the maximum possible capacity for a storage bin constructed in this way. What is the maximum possible capacity (to three significant digits)?

49. A balloon is being blown up at a constant rate. (a) Sketch a reasonable graph of the radius of the balloon as a function of time. (b) Compare to a typical situation that can be described by $r = \sqrt[3]{3t}$, where r is the radius (in cm) and t is the time (in s).

50. A hot-water faucet is turned on. (a) Sketch a reasonable graph of the water temperature as a function of time. (b) Compare to a typical situation described by $T = \dfrac{t^3 + 80}{0.015t^3 + 4}$, where T is the water temperature (in °C) and t is the time (in s).

51. Display the graph of $y = cx^3$ with $c = -2$ and with $c = 2$. Describe the effect of the value of c.

52. Display the graph of $y = cx^4$, with $c = 4$ and with $c = 1/4$. Describe the effect of the value of c.

3.6 GRAPHS OF FUNCTIONS DEFINED BY TABLES OF DATA

As we noted in Section 3.1, there are ways other than formulas to show functions. One important way to show the relationship between variables is by means of a table of values found by observation or from an experiment.

Statistical data often give values that are taken for certain intervals or are averaged over various intervals, and there is no meaning to the intervals *between* the points. Such points should be connected by straight-line segments only to make them stand out better and make the graph easier to read. See Example 1.

◖EXAMPLE 1   The electric energy usage (in MJ) for a certain all-electric house for each month of a year is shown in the following table. Plot these data.

Month	Jan	Feb	Mar	Apr	May	Jun
Energy usage (MJ)	10 504	12 363	10 168	7500	4825	3568

Month	July	Aug	Sep	Oct	Nov	Dec
Energy usage (MJ)	2548	2887	3301	5748	7302	9706

Using statistical plotting features, Fig. 3.39 shows a graphing calculator display of the data of Example 1.

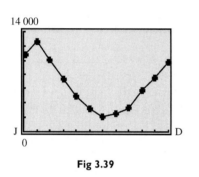

Fig 3.39

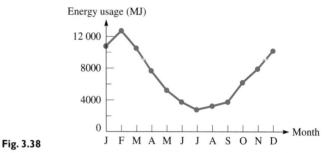

Fig. 3.38

We know that there is no meaning to the intervals between the months, since we have the total number of megajoules for each month. Therefore, we use straight-line segments, but only to make the points stand out better. See Fig. 3.38.   ◗

Data from experiments in science and technology often indicate that the variables could have a formula relating them, although the formula may not be known. In this case, when plotting the graph, the points should be connected by a smooth curve.

◖EXAMPLE 2   Steam in a boiler was heated to 150°C and then allowed to cool. Its temperature T (in °C) was recorded each minute, as shown in the following table. Plot the graph.

The Celsius degree is named for the Swedish astronomer Andres Celsius (1701–1744). He designated 100° as the freezing point of water and 0° as the boiling point. These were later reversed.

Time (min)	0.0	1.0	2.0	3.0	4.0	5.0
Temperature (°C)	150.0	142.8	138.5	135.2	132.7	130.8

Since the temperature changes in a continuous way, there is meaning to the values in the intervals between points. Therefore, these points are joined by a smooth curve, as in Fig. 3.40. Also note that most of the vertical scale was used for the required values, with the indicated break in the scale between 0 and 130.

In Chapter 22, we see how to find an equation that approximates the function relating the variables in a table of values.

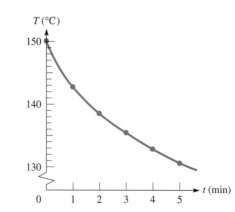

Fig. 3.40

If the graph relating two variables is known, values can be obtained directly from the graph. In finding such a value through the inspection of the graph, we are *reading the graph*.

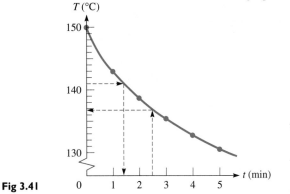

Fig 3.4I

EXAMPLE 3 For the cooling steam in Example 2, we can estimate values of one variable for given values of the other variable.

If we want to know the temperature after 2.5 min, estimate 0.5 of the interval between 2 and 3 on the *t*-axis and mark this point. See Fig. 3.41. Then draw a vertical line from this point to the curve. From the point where it intersects the curve, draw a horizontal line to the *T*-axis. We now estimate that the line crosses at $T = 136.7°C$. Obviously, the number of tenths is a rough estimate.

In the same way, if we want to determine how long the steam took to cool down to 141.0°C, go from 141.0 marker on the *T*-axis to the curve and then to the *t*-axis. This crosses at about $t = 1.4$ min.

LINEAR INTERPOLATION

In Example 3, we see that we can estimate values from a graph. However, unless a very accurate graph is drawn with expanded scales for both variables, only very approximate values can be found. There is a method, called *linear interpolation*, that uses the table itself to get more accurate results.

Linear interpolation *assumes that if a particular value of one variable lies between two of those listed in the table, then the corresponding value of the other variable is at the same proportional distance between the listed values.* On the graph, linear interpolation assumes that two points defined in the table are connected by a straight line. Although this is generally not correct, it is a good approximation if the values in the table are sufficiently close together.

Before the use of electronic calculators, interpolation was used extensively in mathematics textbooks for finding values from mathematics tables. It is still of use when using scientific and technical tables.

EXAMPLE 4 For the cooling steam in Example 2, we can use interpolation to find its temperature after 1.4 min. Since 1.4 min is $\frac{4}{10}$ of the way from 1.0 min to 2.0 min, we will assume that the value of T we want is $\frac{4}{10}$ of the way between 142.8 and 138.5, the values of T for 1.0 min and 2.0 min, respectively. The difference between these values is 4.3, and $\frac{4}{10}$ of 4.3 is 1.7 (rounded off to tenths). Subtracting (the values of T are decreasing) 1.7 from 142.8, we obtain 141.1. Thus, the required value of T is about 141.1°C. (Note that this agrees well with the result in Example 3.)

Another method of indicating the interpolation is shown in Fig. 3.42. From the figure, we have the proportion

$$\frac{0.4}{1.0} = \frac{x}{4.3}$$

$$x = 1.7 \qquad \text{rounded off}$$

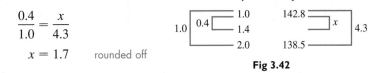

Fig 3.42

Therefore,

$$142.8 - 1.7 = 141.1°C$$

is the required value of T. If the values of T had been increasing, we would have added 1.7 to the value of T for 1.0 min.

EXERCISES **3.6**

In Exercises 1–8, represent the data graphically.

1. The diesel fuel production (in 1000s of liters) at a certain refinery during an 8-week period was as follows:

Week	1	2	3	4	5	6	7	8
Production	765	780	840	850	880	840	760	820

2. The *exchange rate* for the number of Canadian dollars equal to one United States dollar for 1986–2002 is as follows:

Year	1986	1988	1990	1992	1994	1996	1998	2000	2002
Can. Dol.	1.39	1.23	1.17	1.21	1.38	1.37	1.48	1.49	1.57

3. The amount of material necessary to make a cylindrical liter container depends on the diameter, as shown in this table:

Diameter (cm)	6.0	8.0	10	12	14	16
Material (cm²)	723	601	557	560	594	652

4. An oil burner propels air that has been heated to 90°C. The temperature then drops as the distance from the burner increases, as shown in the following table:

Distance (m)	0.0	1.0	2.0	3.0	4.0	5.0	6.0
Temperature (°C)	90	84	76	66	54	46	41

5. A changing electric current in a coil of wire will induce a voltage in a nearby coil. Important in the design of transformers, the effect is called *mutual inductance*. For two coils, the mutual inductance (in H) as a function of the distance between them is given in the following table:

Distance (cm)	0.0	2.0	4.0	6.0	8.0	10.0	12.0
M. ind. (H)	0.77	0.75	0.61	0.49	0.38	0.25	0.17

6. The temperatures felt by the body as a result of the *wind-chill factor* for an outside temperature of 0°C are given in the following table:

Wind speed (km/h)	10	20	30	40	50	60	70
Temp. felt (°C)	−4	−7	−8	−9	−10	−10	−11

7. The time required for a sum of money to double in value, when compounded annually, is given as a function of the interest rate in the following table:

Rate (%)	4	5	6	7	8	9	10
Time (years)	17.7	14.2	11.9	10.2	9.0	8.0	7.3

8. The torque T of an engine, as a function of the frequency f of rotation, was measured as follows:

f (r/min)	500	1000	1500	2000	2500	3000	3500
T (N·m)	175	90	62	45	34	31	27

In Exercises 9 and 10, use the graph in Fig. 3.40, which relates the temperature of cooling steam and the time. Find the indicated values by reading the graph.

9. (a) For $t = 4.3$ min, find T. (b) For $T = 145.0°C$, find t.

10. (a) For $t = 1.8$ min, find T. (b) For $T = 133.5°C$, find t.

In Exercises 11 and 12, use the following table, which gives the voltage produced by a certain thermocouple as a function of the temperature of the thermocouple. Plot the graph. Find the indicated values by reading the graph.

Temperature (°C)	0	10	20	30	40	50
Voltage (V)	0.0	2.9	5.9	9.0	12.3	15.8

11. (a) For $T = 26°C$, find V. (b) For $V = 13.5$ V, find T.

12. (a) For $T = 32°C$, find V. (b) For $V = 4.8$ V, find T.

In Exercises 13–16, find the indicated values by means of linear interpolation.

13. In Exercise 5, find the inductance for $d = 9.2$ cm.

14. In Exercise 6, find the temperature for $s = 17$ km/h.

15. In Exercise 7, find the rate for $t = 10.0$ years.

16. In Exercise 8, find the torque for $t = 2300$ r/min.

In Exercises 17–20, use the following table that gives the rate R of discharge from a tank of water as a function of the height H of water in the tank. For Exercises 17 and 18, plot the graph and find the values from the graph. For Exercises 19 and 20, find the indicated values by linear interpolation.

Height (cm)	0	50	100	200	300	400	600
Rate (m³/s)	0	1.0	1.5	2.2	2.7	3.1	3.5

17. (a) for $R = 2.0$ m³/s, find H. (b) For $H = 240$ cm, find R.

18. (a) For $R = 3.4$ m³/s, find H. (b) For $H = 320$ cm, find R.

19. Find R for $H = 80$ cm. **20.** Find H for $R = 2.5$ m³/s.

In Exercises 21–24, use the following table, which gives the fraction (as a decimal) of the total heating load of a certain system that will be supplied by a solar collector of area A (in m²). Find the indicated values by linear interpolation.

f	0.22	0.30	0.37	0.44	0.50	0.56	0.61
A (m²)	20	30	40	50	60	70	80

21. For $A = 36$ m², find f. **22.** For $A = 52$ m², find f.

23. For $f = 0.59$, find A. **24.** For $f = 0.27$, find A.

In Exercises 25–28, a method of finding values beyond those given is considered. By using a straight-line segment to extend a graph beyond the last known point, we can estimate values from the extension of the graph. The method is known as **linear extrapolation.** *Use this method to estimate the required values from the given graphs.*

25. Using Fig. 3.41, estimate T for $t = 5.3$ min.

26. Using the graph for Exercises 11 and 12, estimate V for $T = 55°C$.

27. Using the graph for Exercises 17–20, estimate R for $H = 700$ cm.

28. Using the graph for Exercises 17–20, estimate R for $H = 800$ cm.

CHAPTER ③ REVIEW EXERCISES

In Exercises 1–4, determine the appropriate function.

1. The radius of a circular water wave increases at the rate of 2 m/s. Express the area of the circle as a function of the time t (in s).

2. A conical sheet-metal hood is to cover an area 6 m in diameter. Find the total surface area A of the hood as a function of its height h.

3. One computer printer prints at the rate of 2000 lines/min for x min, and a second printer prints at the rate of 1800 lines/min for y min. Together they print 50 000 lines. Find y as a function of x.

4. Fencing around a rectangular storage depot area costs twice as much along the front as along the other three sides. The back costs \$10 per meter. Express the cost C of the fencing as a function of the width w if the length (along the front) is 20 m longer than the width.

In Exercises 5–12, evaluate the given functions.

5. $f(x) = 7x - 5$; find $f(3)$ and $f(-6)$.

6. $g(I) = 8 - 3I$; find $g\left(\frac{1}{6}\right)$ and $g(-4)$.

7. $H(h) = \sqrt{1 - 2h}$; find $H(-4)$ and $H(2h)$.

8. $\phi(v) = \dfrac{3v - 2}{|v + 1|}$; find $\phi(-2)$ and $\phi(v + 1)$.

9. $f(x) = 3x^2 - 2x + 4$; find $f(x + h) - f(x)$.

10. $F(x) = x^3 + 2x^2 - 3x$; find $F(3 + h) - F(3)$.

11. $f(x) = 3 - 2x$; find $f(2x) - 2f(x)$.

12. $f(x) = 1 - x^2$; find $[f(x)]^2 - f(x^2)$.

In Exercises 13–16, evaluate the given functions. Values of the independent variable are approximate.

13. $f(x) = 8.07 - 2x$; find $f(5.87)$ and $f(-4.29)$.

14. $g(x) = 7x - x^2$; find $g(45.81)$ and $g(-21.85)$.

15. $G(S) = \dfrac{S - 0.087\,629}{3.0125S}$; find $G(0.174\,27)$ and $G(0.053\,206)$.

16. $h(t) = \dfrac{t^2 - 4t}{t^3 + 564}$; find $h(8.91)$ and $h(-4.91)$.

In Exercises 17–20, determine the domain and the range of the given functions.

17. $f(x) = x^4 + 1$

18. $G(z) = \dfrac{4}{z^3}$

19. $g(t) = \dfrac{2}{\sqrt{t + 4}}$

20. $F(y) = 1 - 2\sqrt{y}$

In Exercises 21–32, plot the graphs of the given functions. Check these graphs by using a graphing calculator.

21. $y = 4x + 2$

22. $y = 5x - 10$

23. $s = 4t - t^2$

24. $y = x^2 - 8x - 5$

25. $y = 3 - x - 2x^2$

26. $y = 6 + 4x + x^2$

27. $y = x^3 - 6x$

28. $V = 3 - 0.5s^3$

29. $y = 2 - x^4$

30. $y = x^4 - 4x$

31. $y = \dfrac{x}{x + 1}$

32. $Z = \sqrt{25 - 2R^2}$

In Exercises 33–40, use a graphing calculator to solve the given equations to the nearest 0.1.

33. $7x - 3 = 0$

34. $3x + 11 = 0$

35. $x^2 + 1 = 6x$

36. $3t - 2 = t^2$

37. $x^3 - x^2 = 2 - x$

38. $5 - x^3 = 2x^2$

39. $\dfrac{1}{x} = 2x$

40. $\sqrt{x} = 2x - 1$

In Exercises 41–44, use a graphing calculator to find the range of the given function.

41. $y = x^4 - 5x^2$

42. $y = x\sqrt{4 - x^2}$

43. $A = w + \dfrac{2}{w}$

44. $y = 2x + \dfrac{3}{\sqrt{x}}$

In Exercises 45–60, solve the given problems.

Ⓦ 45. Explain how $A(a, b)$ and $B(b, a)$ may be in different quadrants.

46. Determine the distance from the origin to the point (a, b).

47. Two vertices of an equilateral triangle are $(0, 0)$ and $(2, 0)$. What is the third vertex?

48. The points $(1, 2)$ and $(1, -3)$ are two adjacent vertices of a square. Find the other vertices.

49. Where are all points for which $|y/x| > 0$?

50. Describe the values of x and y for which $(1, -2)$, $(-1, -2)$, and (x, y) are on the same straight line.

51. Sketch the graph of a function for which the domain is $0 \le x \le 4$ and the range is $1 \le y \le 3$.

52. Sketch the graph of a function for which the domain is all values of x and the range is $-2 < y < 2$.

53. If the function $y = \sqrt{x - 1}$ is shifted left 2 and up 1, what is the resulting function?

54. If the function $y = 3 - 2x$ is shifted right 1 and down 3, what is the resulting function?

(W) **55.** For $f(x) = 2x^3 - 3$, display the graphs of $f(x)$ and $f(-x)$ on a graphing calculator. Describe the graphs in relation to the y-axis.

(W) **56.** For $f(x) = 2x^3 - 3$, display the graphs of $f(x)$ and $-f(x)$ on a graphing calculator. Describe the graphs in relation to the x-axis.

57. An equation used in electronics with a transformer antenna is $I = 12.5\sqrt{1 + 0.5m^2}$. For $I = f(m)$, find $f(0.55)$.

58. The percent p of wood lost in cutting it into boards 38 mm thick due to the thickness t (in mm) of the saw blade is $p = \dfrac{100t}{t + 38}$. Find p if $t = 5.0$ mm. That is, since $p = f(t)$, find $f(5.0)$.

59. The angle A (in degrees) of a robot arm with the horizontal as a function of time t (for 0.0 s to 6.0 s) is given by $A = 8.0 + 12t^2 - 2.0t^3$. What is the greatest value of A to the nearest 0.1°? See Fig. 3.43. (*Hint*: Find the range.)

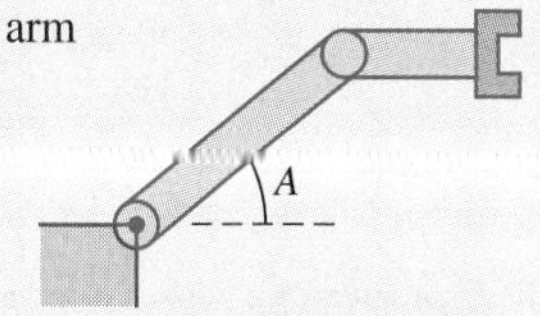

Fig 3.43

60. The electric power P (in W) produced by a certain battery is $P = \dfrac{24R}{R^2 + 1.40R + 0.49}$, where R is the resistance (in Ω) in the circuit. What is the maximum power produced? (*Hint*: Find the range.) See Example 6 on page 95.

In Exercises 61–76, plot the graphs of the indicated functions.

61. When El Niño, a Pacific Ocean current, moves east and warms the water off South America, weather patterns in many parts of the world change significantly. Special buoys along the equator in the Pacific Ocean send data via satellite to monitoring stations. If the temperature T (in °C) at one of these buoys is $T = 28.0 + 0.15t$, where t is the time in weeks between Jan. 1 and Aug. 1 (30 weeks), plot the graph of $T = f(t)$.

62. The change C (in cm) in the length of a 100-m steel bridge girder from its length at 10°C as a function of the temperature T is given by $C = 0.12(T - 10)$. Plot the graph for $T = -20$°C to $T = 40$°C.

63. The profit P (in dollars) a retailer makes in selling 50 cell phones is given by $P = 50(p - 50)$, where p is the selling price. Plot P as a function of p for $p = \$30$ to $p = \$150$.

64. There are 500 L of oil in a tank that has a capacity of 100 000 L. It is filled at a rate of 7000 L/h. Determine the function relating the number of liters N and the time t while the tank is being filled. Plot N as a function of t.

65. The length L (in cm) of a pulley belt is 12 cm longer than the circumference of one of the pulley wheels. Express L as a function of the radius r of the wheel and plot L as a function of r.

66. The pressure loss P (in kPa per 100 m) in a fire hose is given by $P = 0.000\,12Q^2 + 0.0055Q$, where Q is the rate of flow (in L/min). Plot the graph of P as a function of Q.

67. The thermodynamic temperature T (in K) is 273 more than the Celsius reading C. Plot the graph of $T = f(C)$.

68. A company buys a new copier for $1000 and determines that it costs $10 per day to use it (for paper, toner, etc.). Plot the total cost C of the copier as a function of the number n of days of use.

69. For a certain laser device, the laser output power P (in mW) is negligible if the drive current i is less than 80 mA. From 80 mA to 140 mA, $P = 1.5 \times 10^{-6}i^3 - 0.77$. Plot the graph of $P = f(i)$.

70. It is determined that a good approximation for the cost C (in cents/km) of operating a certain car at a constant speed v (in km/h) is given by $C = 0.025v^2 - 1.4v + 35$. Plot C as a function of v for $v = 10$ km/h to $v = 60$ km/h.

71. A medical researcher exposed a virus culture to an experimental vaccine. It was observed that the number of live cells N in the culture as a function of the time t (in h) after exposure was given by $N = \dfrac{1000}{\sqrt{t + 1}}$. Plot the graph of $N = f(t)$.

72. The electric field E (in V/m) from a certain electric charge is given by $E = 25/r^2$, where r is the distance (in m) from the charge. Plot the graph of $E = f(r)$ for values of r up to 10 cm.

73. To draw the approximate shape of an irregular shoreline, a surveyor measured the distances d from a straight wall to the shoreline at 20-m intervals along the wall, as shown in the following table. Plot the graph of distance d as a function of the distance D along the wall.

D (m)	0	20	40	60	80	100	120	140	160
d (m)	15	32	56	33	29	47	68	31	52

74. The percent p of a computer network that is in use during a particular loading cycle as a function of the time t (in s) is given in the following table. Plot the graph of $p = f(t)$.

t (s)	0.0	0.2	0.4	0.6	0.8	1.0	1.2	1.4	1.6
P (%)	0	45	85	90	85	85	60	10	0

75. The vertical sag s (in m) at the middle of an 800-m power line as a function of the temperature T (in °C) is given in the following table. See Fig. 3.44. For the function $s = f(T)$, find $f(14)$ by linear interpolation.

T (°C)	−10	0	10	20
s (m)	3.11	3.23	3.38	3.57

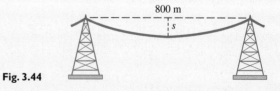

Fig. 3.44

76. In an experiment measuring the pressure p (in kPa) at a given depth d (in m) of seawater, the results in the following table were found. Plot the graph of $p = f(d)$ and from the graph determine $f(10)$.

d (m)	0.0	3.0	6.0	9.0	12	15
p (kPa)	101	131	161	193	225	256

In Exercises 77–84, solve the indicated equations graphically.

77. A person 350 km from home starts toward home and travels at 90 km/h for the first 2.0 h and then slows down to 60 km/h for the rest of the trip. How long does it take the person to be 80 km from home?

78. One industrial cleaner contains 30% of a certain solvent, and another contains 10% of the solvent. To get a mixture containing 50 L of the solvent, 120 L of the first cleaner is used. How much of the second must be used?

79. The solubility s (in kg/m³ of water) of a certain type of fertilizer is given by $s = 135 + 4.9T + 0.19T^2$, where T is the temperature (in °C). Find T for $s = 500$ kg/m³.

80. A 2.00-L (2000-cm³) metal container is to be made in the shape of a right circular cylinder. Express the total area A of metal necessary as a function of the radius r of the base. Then find A for $r = 6.00$ cm, 7.00 cm, and 8.00 cm.

81. In an oil pipeline, the velocity v (in m/s) of the oil as a function of the distance x (in m) from the wall of the pipe is given by $v = 9.6x - 7.5x^2$. Find x for $v = 2.6$ m/s. The diameter of the pipe is 1.20 m.

82. One ball bearing is 1.00 mm more in radius and has twice the volume of another ball bearing. What is the radius of each?

83. A computer, using data from a refrigeration plant, estimates that in the event of a power failure, the temperature (in °C) in the freezers would be given by $T = \dfrac{4t^2}{t + 2} - 20$, where t is the number of hours after the power failure. How long would it take for the temperature to reach 0°C?

84. Two electrical resistors in parallel (see Fig. 3.45) have a combined resistance R_T given by $R_T = \dfrac{R_1 R_2}{R_1 + R_2}$. If $R_2 = R_1 + 2.0$, express R_T as a function of R_1 and find R_1 if $R_T = 6.0\ \Omega$.

Fig. 3.45

Writing Exercise

85. In one or two paragraphs, explain how you would solve the following problem using a graphing calculator: The inner surface area A of a 250.0-cm³ cylindrical cup, as a function of the radius r of the base, is $A = \pi r^2 + \frac{500.0}{r}$. Find r if $A = 175.0$ cm². (What is the answer?) (See if you can derive the formula.)

CHAPTER ③ PRACTICE TEST

1. Given $f(x) = 2x - x^2 + \dfrac{8}{x}$, find $f(-4)$ and $f(2.385)$.

2. A rocket has a mass of 2000 Mg at liftoff. If the first-stage engines burn fuel at the rate of 10 Mg/s, find the mass m of the rocket as a function of the time t (in s) while the first-stage engines operate.

3. Plot the graph of the function $f(x) = 4 - 2x$.

4. Use a graphing calculator to solve the equation $2x^2 - 3 = 3x$ to the nearest 0.1.

5. Plot the graph of the function $y = \sqrt{4 + 2x}$.

6. Locate all points (x, y) for which $x < 0$ and $y = 0$.

7. Find the domain and the range of the function $f(x) = \sqrt{6 - x}$.

8. If the function $y = 2x^2 - 3$ is shifted right 1 and up 3, what is the resulting function?

9. Use a graphing calculator to find the range of the function $y = \dfrac{x^2 + 2}{x + 2}$.

10. A window has the shape of a semicircle over a square, as shown in Fig. 3.46. Express the area of the window as a function of the radius of the circular part.

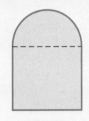

Fig 3.46

11. The voltage V and current i (in mA) for a certain electrical experiment were measured as shown in the following table. Plot the graph of $i = f(V)$ and from the graph find $f(45.0)$.

Voltage (V)	10.0	20.0	30.0	40.0	50.0	60.0
Current (mA)	145	188	220	255	285	315

12. From the table in Problem 11, find the voltage for $i = 200$ mA.

4

The Trigonometric Functions

Using trigonometry, it is often possible to calculate distances that may not be directly measured. In Section 4.5, we show how we can measure the height of Horseshoe Falls on the Canadian side of Niagara Falls.

Triangles are often used in solving applied problems in technology. A short list of such problems includes those in air navigation, surveying, rocket motion, carpentry, structural design, electric circuits, and astronomy. In fact, it was because the Greek astronomer Hipparchus (about 150 B.C.E.) was interested in measuring distances such as that between the earth and the moon that he started the study of **trigonometry.** Using records of earlier works, he organized and developed the real beginnings of this very important field of mathematics in order to make a number of astronomical measurements.

In trigonometry, we develop methods for measuring the sides and angles of triangles, and this in turn allows us to solve related applied problems. Because trigonometry has a great number of applications in many areas of study, it is considered one of the most practical branches of mathematics.

In this chapter, we introduce the basic trigonometric functions and show many applications of right triangles from science and technology. In later chapters, we will discuss other types of triangles and their applications.

As mathematics developed, it became clear, particularly in the 1800s and 1900s, that the trigonometric functions used for solving problems involving triangles were also very valuable in applications in which a triangle is not involved. This important use of the trigonometric functions is now essential in the study of areas such as electronics, mechanical vibrations, acoustics, and optics, and we will study it in later chapters.

4.1 ANGLES

In Chapter 2, we gave a basic definition of an *angle*. In this section, we extend this definition and also give some other important definitions related to angles.

An **angle** *is generated by rotating a ray about its fixed endpoint from an* **initial position** *to a* **terminal position.** *The initial position is called the* **initial side** *of the angle, the terminal position is called the* **terminal side,** *and the fixed endpoint is the* **vertex.** The angle itself is the amount of rotation from the initial side to the terminal side.

If the rotation of the terminal side from the initial side is **counterclockwise,** *the angle is defined as* **positive.** *If the rotation is* **clockwise,** *the angle is* **negative.** In Fig. 4.1, ∠1 is positive and ∠2 is negative.

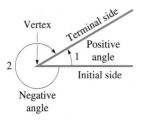

Fig. 4.1

Many symbols are used to designate angles. Among the most widely used are certain Greek letters such as θ (theta), ϕ (phi), α (alpha), and β (beta). Capital letters representing the vertex (e.g., ∠A or simply A) and other literal symbols, such as x and y, are also used commonly.

In Chapter 2, we introduced two measurements of an angle. These are the *degree* and the *radian*. Since degrees and radians are both used on calculators and computers, we will briefly review the relationship between them in this section. However, we will not make use of radians until Chapter 8.

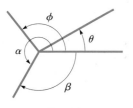

Fig. 4.2

From Section 2.1, we recall that *a* **degree** *is* 1/360 *of one complete rotation.* In Fig. 4.2, $\angle\theta = 30°$, $\angle\phi = 140°$, $\angle\alpha = 240°$, and $\angle\beta = -120°$. Note that β is drawn in a clockwise direction to show that it is negative. The other angles are drawn in a counterclockwise direction to show that they are positive angles.

In Chapter 2, we used degrees and decimal parts of a degree. Most calculators use degrees in this decimal form. Another traditional way is to divide a degree into 60 equal parts called **minutes;** each minute is divided into 60 equal parts called **seconds.** The symbols ′ and ″ are used to designate minutes and seconds, respectively.

In Fig. 4.2, we note that angles α and β have the same initial and terminal sides. *Such angles are called* **coterminal angles.** An understanding of coterminal angles is important in certain concepts of trigonometry.

◀ EXAMPLE 1 Determine the values of two angles that are coterminal with an angle of 145.6°.

Since there are 360° in a complete rotation, we can find a coterminal angle by adding 360° to the given angle to get 505.6°. Another coterminal angle can be found by subtracting 360° from the given angle to get −214.4°. See Fig. 4.3. We could continue to add, or subtract, 360° to get other coterminal angles. ▶

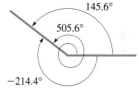

Fig. 4.3

ANGLE CONVERSIONS

NOTE ▶

We will use only degrees as a measure of angles in this chapter. Therefore, when using a calculator, *be sure to use the* **mode** *feature to set the calculator for degrees.* In later chapters, we will use radians. We can change between degrees and radians by using a calculator feature or by the definition (see Section 2.4) of $\pi\,\text{rad} = 180°$.

Before the extensive use of calculators, it was common to use degrees and minutes in tables, whereas calculators use degrees and decimal parts of a degree. Changing from one form to another can be done directly on a calculator by use of the *dms (degree-minute-second)* feature. The following examples illustrate angle conversions by using the definitions and by using the appropriate calculator features.

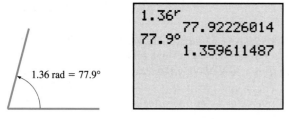

Fig. 4.4

Fig. 4.5

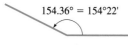

Fig. 4.6

Fig. 4.7

◀ EXAMPLE 2   Express 1.36 rad in degrees.

We know that π rad = 180°, which means 1 rad = 180°/π. Therefore,

$$1.36 \text{ rad} = 1.36\left(\frac{180°}{\pi}\right) = 77.9° \qquad \text{to nearest 0.1°}$$

This angle is shown in Fig. 4.4. We again note that degrees and radians are simply two different ways of measuring an angle.

In Fig. 4.5, a graphing calculator display shows the conversions of 1.36 rad to degrees (calculator in *degree* mode) and 77.9° to radians (calculator in *radian* mode).   ▶

◀ EXAMPLE 3   To change 17°53′ to decimal form, use the fact that 1° = 60′.

Therefore, $53' = \left(\frac{53}{60}\right)^{\circ} = 0.88°$ to nearest 0.01°

This means that 17°53′ = 17.88°. This angle is shown in Fig. 4.6.

To change 154.36° to an angle measured to the nearest minute, we have

$$0.36° = 0.36(60') = 22'$$

This means that 154.36° = 154°22′. See Fig. 4.7.   ▶

STANDARD POSITION OF AN ANGLE

If the initial side of the angle is the positive x-axis and the vertex is the origin, the angle is said to be in **standard position.** The angle is then determined by the position of the terminal side. *If the terminal side is in the first quadrant, the angle is called a* **first-quadrant angle.** Similar terms are used when the terminal side is in the other quadrants. *If the terminal side coincides with one of the axes, the angle is a* **quadrantal angle.** For an angle in standard position, the terminal side can be determined if we know any point, except the origin, on the terminal side.

◀ EXAMPLE 4   A standard position angle of 60° is a first-quadrant angle with its terminal side 60° from the *x*-axis. See Fig. 4.8(a).

A second-quadrant angle of 130° is shown in Fig. 4.8(b).

A third-quadrant angle of 225° is shown in Fig. 4.8(c).

A fourth-quadrant angle of 340° is shown in Fig. 4.8(d).

A standard position angle of −120° is shown in Fig. 4.8(e). Since the terminal side is in the third quadrant, it is a third-quadrant angle.

A standard position angle of 90° is a quadrantal angle since its terminal side is the positive *y*-axis. See Fig. 4.8(f).

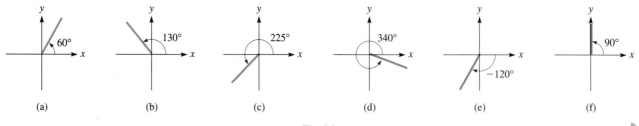

(a)   (b)   (c)   (d)   (e)   (f)

Fig. 4.8

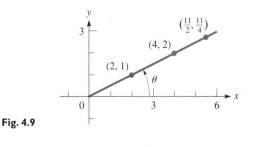

Fig. 4.9

◀ EXAMPLE 5   In Fig. 4.9, θ is in standard position, and the terminal side is uniquely determined by knowing that is passes through the point $(2, 1)$. The same terminal side passes through $(4, 2)$ and $\left(\frac{11}{2}, \frac{11}{4}\right)$, among other points. Knowing that the terminal side passes through any one of these points makes it possible to determine the terminal side of the angle. ▶

EXERCISES 4.1

In Exercises 1–4, find the indicated angles in the given examples of this section.

1. In Example 1, find another angle that is coterminal with the given angle.

2. In Example 3, change $53'$ to $35'$ and then find the decimal form.

3. In Example 4, find another standard position angle that has the same terminal side as the angle in Fig. 4.8(c).

4. In Example 4, find another standard position angle that has the same terminal side as the angle in Fig. 4.8(e).

In Exercises 5–8, draw the given angles.

5. $60°, 120°, -90°$
6. $330°, -150°, 450°$
7. $50°, -360°, -30°$
8. $45°, 245°, -250°$

In Exercises 9–16, determine one positive and one negative coterminal angle for each angle given.

9. $45°$
10. $73°$
11. $-150°$
12. $162°$
13. $70°30'$
14. $153°47'$
15. $278.1°$
16. $-197.6°$

In Exercises 17–20, by means of the definition of a radian, change the given angles in radians to equal angles expressed in degrees to the nearest 0.01°.

17. 0.265 rad
18. 0.838 rad
19. 1.447 rad
20. 3.642 rad

In Exercises 21–24, use a calculator conversion sequence to change the given angles in radians to equal angles expressed in degrees to the nearest 0.01°.

21. 0.329 rad
22. 2.089 rad
23. 4.110 rad
24. 0.067 rad

In Exercises 25–28, use a calculator conversion sequence to change the given angles to equal angles expressed in radians to three significant digits.

25. $56.0°$
26. $137.4°$

27. $284.8°$
28. $-17.5°$

In Exercises 29–32, change the given angles to equal angles expressed to the nearest minute.

29. $47.50°$
30. $315.80°$
31. $-5.62°$
32. $142.87°$

In Exercises 33–36, change the given angles to equal angles expressed in decimal form to the nearest 0.01°.

33. $15°12'$
34. $157°39'$
35. $301°16'$
36. $-4°47'$

In Exercises 37–44, draw angles in standard position such that the terminal side passes through the given point.

37. $(4, 2)$
38. $(-3, 8)$
39. $(-3, -5)$
40. $(6, -1)$
41. $(-7, 5)$
42. $(-4, -2)$
43. $(2, -5)$
44. $(1, 6)$

In Exercises 45–52, the given angles are in standard position. Designate each angle by the quadrant in which the terminal side lies, or as a quadrantal angle.

45. $31°, 310°$
46. $180°, 92°$
47. $175°, -270°$
48. $-5°, 265°$
49. 1 rad, 2 rad
50. 3 rad, π rad
51. 4 rad, $\pi/3$ rad
52. 5 rad, -2 rad

In Exercises 53 and 54, change the given angles to equal angles expressed in decimal form to the nearest 0.001°. In Exercises 55 and 56, change the given angles to equal angles expressed to the nearest second.

53. $21°42'36''$
54. $7°16'23''$
55. $86.274°$
56. $57.019°$

4.2 DEFINING THE TRIGONOMETRIC FUNCTIONS

Important to the definitions and development in this section are the right triangle, the Pythagorean theorem, and the properties of similar triangles. We now briefly review similar triangles and their properties.

As stated in Section 2.2, *two triangles are* **similar** *if they have the same shape (but not necessarily the same size)*. Similar triangles have the following important properties.

PROPERTIES OF SIMILAR TRIANGLES

1. *Corresponding angles are equal.*
2. *Corresponding sides are proportional.*

The **corresponding sides** *are the sides, one in each triangle, that are between the same pair of equal* **corresponding angles.**

◀ EXAMPLE 1　In Fig. 4.10, the triangles are similar and are lettered so that corresponding sides and angles have the same letters. That is, angles A_1 and A_2, angles B_1 and B_2, and angles C_1 and C_2 are pairs of corresponding angles. The pairs of corresponding sides are a_1 and a_2, b_1 and b_2, and c_1 and c_2. From the properties of similar triangles, we know that the corresponding angles are equal, or

$$\angle A_1 = \angle A_2 \qquad \angle B_1 = \angle B_2 \qquad \angle C_1 = \angle C_2$$

Also, the corresponding sides are proportional, which we can show as

$$\frac{a_1}{a_2} = \frac{b_1}{b_2} \qquad \frac{a_1}{a_2} = \frac{c_1}{c_2} \qquad \frac{b_1}{b_2} = \frac{c_1}{c_2}$$

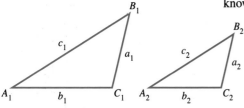

Fig. 4.10

In Example 1, if we multiply both sides of

$$\frac{a_1}{a_2} = \frac{b_1}{b_2} \quad \text{by} \quad \frac{a_2}{b_1}, \quad \text{we get} \quad \frac{a_1}{a_2}\left(\frac{a_2}{b_1}\right) = \frac{b_1}{b_2}\left(\frac{a_2}{b_1}\right)$$

which when simplified gives

$$\frac{a_1}{b_1} = \frac{a_2}{b_2}$$

This shows us that *when two triangles are similar*

the ratio of one side to another side in one triangle is the same as the ratio of the corresponding sides in the other triangle.

Using this we now proceed to the definitions of the trigonometric functions.

We now place an angle θ in standard position and drop perpendicular lines from points on the terminal side to the *x*-axis, as shown in Fig. 4.11. In doing this, we set up similar triangles, each with one vertex at the origin and one side along the *x*-axis.

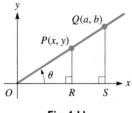

Fig. 4.11

❮ EXAMPLE 2 In Fig. 4.11, we can see that triangles *ORP* and *OSQ* are similar since their corresponding angles are equal (each has the same angle at *O*, a right angle, and therefore equal angles at *P* and *Q*). This means that ratios of the lengths of corresponding sides are equal. For example,

$$\frac{RP}{OR} = \frac{SQ}{OS} \quad \text{which is the same as} \quad \frac{y}{x} = \frac{b}{a}$$

For any position (except at the origin) of *Q* on the terminal side of *θ*, the ratio *b/a* of its ordinate to its abscissa will equal *y/x*. ❯

For an angle in standard position, we may set up six different ratios of the values of *x*, *y*, and *r*, as shown in Fig. 4.12. Because of the properties of similar triangles, for a given angle, any of these ratios has the same value for any point on the terminal side. This means the *values of the ratios depend on the size of the angle, and there is only one value for each ratio.* This means *the ratios are **functions** of the angle, and they are called the* **trigonometric functions.** We now define them, giving their names and the abbreviations that are used when using them.

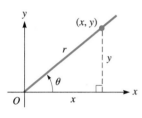

Fig. 4.12

$$
\begin{array}{ll}
\textit{sine of } \theta: \quad \sin \theta = \dfrac{y}{r} & \textit{cosine of } \theta: \quad \cos \theta = \dfrac{x}{r} \\[2mm]
\textit{tangent of } \theta: \quad \tan \theta = \dfrac{y}{x} & \textit{cotangent of } \theta: \quad \cot \theta = \dfrac{x}{y} \\[2mm]
\textit{secant of } \theta: \quad \sec \theta = \dfrac{r}{x} & \textit{cosecant of } \theta: \quad \csc \theta = \dfrac{r}{y}
\end{array}
\tag{4.1}
$$

Here, *the distance r from the origin to the point is called the* **radius vector,** and it is assumed that $r > 0$ (if $r = 0$ there would be no terminal side and therefore no angle).

A given function is not defined when the denominator is zero, and if either $x = 0$ or $y = 0$, this does affect the domain. We will discuss the domains and ranges of these functions in Chapter 10, when we discuss their graphs.

In this chapter, we use the trigonometric functions of acute angles (angles between 0° and 90°). However, *the definitions above are general and may be used with angles of any size.* We will discuss these functions in general in Chapters 8 and 20.

EVALUATING THE TRIGONOMETRIC FUNCTIONS

When evaluating the trigonometric functions, we use the definitions in Eqs. (4.1). We also often use the *Pythagorean theorem*, which we discussed in Section 2.2. For reference, we restate it here. For the right triangle in Fig. 4.13, with hypotenuse *c* and legs *a* and *b*, we have

$$c^2 = a^2 + b^2 \tag{4.2}$$

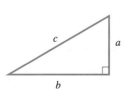

Fig. 4.13

Following are examples of evaluating the trigonometric functions of an angle when a point on the terminal side of the angle is given or can be found.

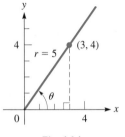

Fig. 4.14

◖EXAMPLE 3   Find the values of the trigonometric functions of the angle θ with its terminal side passing through the point $(3,4)$.

By placing the angle in standard position, as shown in Fig. 4.14, and drawing the terminal side through $(3,4)$, we find by use of the Pythagorean theorem that

$$r = \sqrt{3^2 + 4^2} = \sqrt{25} = 5$$

Using the values $x = 3$, $y = 4$, and $r = 5$, we find that

$$\sin \theta = \frac{4}{5} \qquad \cos \theta = \frac{3}{5} \qquad \tan \theta = \frac{4}{3}$$

$$\cot \theta = \frac{3}{4} \qquad \sec \theta = \frac{5}{3} \qquad \csc \theta = \frac{5}{4}$$

We have left each of these results in the form of a fraction, which is considered to be an *exact form* in that there has been no approximation made. In writing decimal values, we find that $\tan \theta = 1.333$ and $\sec \theta = 1.667$, where these values have been rounded off and are therefore *approximate*.   ◗

◖EXAMPLE 4   Find the values of the trigonometric functions of the angle whose terminal side passes through $(7.27, 4.49)$. The coordinates are approximate.

We show the angle and the given point in Fig. 4.15. From the Pythagorean theorem, we have

$$r = \sqrt{7.27^2 + 4.49^2} = 8.545$$

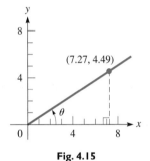

Fig. 4.15

(Here we show a rounded-off value of r. It is not actually necessary to record the value of r since its value can be stored in the memory of a calculator. The reason for recording it here is to show the values used in the calculation of each of the trigonometric functions.) Therefore, we have the following values:

$$\sin \theta = \frac{4.49}{8.545} = 0.525 \qquad \cos \theta = \frac{7.27}{8.545} = 0.851$$

$$\tan \theta = \frac{4.49}{7.27} = 0.618 \qquad \cos \theta = \frac{7.27}{4.49} = 1.62$$

$$\sec \theta = \frac{8.545}{7.27} = 1.18 \qquad \csc \theta = \frac{8.545}{4.49} = 1.90$$

Since the coordinates are approximate, the results are rounded off.

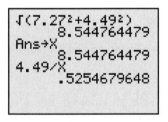

Fig. 4.16

The key sequence for a calculator solution is shown in Fig. 4.16. The result shown in line 2 is for r, which was then stored under x (for convenience of calculator use—it is not the x-coordinate). Storing x as shown is to emphasize the fact that r can be stored for later use (it can be stored simply by using *sto*▶x in line 1). The value of $\sin \theta$ is shown in lines 5 and 6. The values of the other functions can be found by doing the appropriate divisions, using the stored value of r when needed.   ◗

In Example 4, we expressed the result as $\sin \theta = 0.525$. A common error is to omit the angle and give the value as $\sin = 0.525$. This is a meaningless expression, for CAUTION ▶ *we must show the angle* for which we have the value of a function.

If one of the trigonometric functions is known, it is possible to find the values of the other functions. The following example illustrates the method.

◀ EXAMPLE 5   If we know that sin $\theta = 3/7$ and that θ is a first-quadrant angle, we know the ratio of the ordinate to the radius vector (y to r) is 3 to 7. Therefore, the point on the terminal side for which $y = 3$ can be found by use of the Pythagorean theorem. The x-value for this point is

$$x = \sqrt{7^2 - 3^2} = \sqrt{49 - 9} = \sqrt{40} = 2\sqrt{10}$$

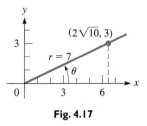

Fig. 4.17

Therefore, the point $(2\sqrt{10}, 3)$ is on the terminal side, as shown in Fig. 4.17.

Therefore, using the values $x = 2\sqrt{10}$, $y = 3$, and $r = 7$, we have the other trigonometric functions of θ. They are

$$\cos \theta = \frac{2\sqrt{10}}{7} \qquad \tan \theta = \frac{3}{2\sqrt{10}} \qquad \cot \theta = \frac{2\sqrt{10}}{3} \qquad \sec \theta = \frac{7}{2\sqrt{10}} \qquad \csc \theta = \frac{7}{3}$$

These values are *exact*. *Approximate* decimal values found on a calculator are

$$\cos \theta = 0.9035 \qquad \tan \theta = 0.4743 \qquad \cot \theta = 2.108$$
$$\sec \theta = 1.107 \qquad \csc \theta = 2.333$$

▶

EXERCISES 4.2

In Exercises 1 and 2, answer the given questions about the indicated examples of this section.

1. In Example 3, if the point $(4, 3)$ replaces the point $(3, 4)$, what are the values?

2. In Example 5, if $4/7$ replaces $3/7$, what are the values?

In Exercises 3–16, find values of the trigonometric functions of the angle (in standard position) whose terminal side passes through the given points. For Exercises 3–14, give answers in exact form. For Exercises 15 and 16, the coordinates are approximate.

3. $(6, 8)$ **4.** $(5, 12)$ **5.** $(15, 8)$ **6.** $(24, 7)$

7. $(0.9, 4.0)$ **8.** $(3.2, 6.0)$ **9.** $(1, \sqrt{15})$ **10.** $(\sqrt{3}, 2)$

11. $(1, 1)$ **12.** $(6, 5)$ **13.** $(5, 2)$ **14.** $(1, \frac{1}{2})$

15. $(0.687, 0.943)$ **16.** $(37.65, 21.87)$

In Exercises 17–24, find the values of the indicated functions. In Exercises 17–20, give answers in exact form. In Exercises 21–24, the values are approximate.

17. Given $\cos \theta = 12/13$, find $\sin \theta$ and $\cot \theta$.

18. Given $\sin \theta = 1/2$, find $\cos \theta$ and $\csc \theta$.

19. Given $\tan \theta = 2$, find $\sin \theta$ and $\sec \theta$.

20. Given $\sec \theta = \sqrt{5}/2$, find $\tan \theta$ and $\cos \theta$.

21. Given $\sin \theta = 0.750$, find $\cot \theta$ and $\csc \theta$.

22. Given $\cos \theta = 0.326$, find $\sin \theta$ and $\tan \theta$.

23. Given $\cot \theta = 0.254$, find $\cos \theta$ and $\tan \theta$.

24. Given $\csc \theta = 1.20$, find $\sec \theta$ and $\cos \theta$.

In Exercises 25–28, each point listed is on the terminal side of an angle. Show that each of the indicated functions is the same for each of the points.

25. $(3, 4)$, $(6, 8)$, $(4.5, 6)$, $\sin \theta$ and $\tan \theta$

26. $(5, 12)$, $(15, 36)$, $(7.5, 18)$, $\cos \theta$ and $\cot \theta$

27. $(0.3, 0.1)$, $(9, 3)$, $(33, 11)$, $\tan \theta$ and $\sec \theta$

28. $(0.4, 0.3)$, $(8, 6)$, $(36, 27)$, $\csc \theta$ and $\cos \theta$

In Exercises 29–36, answer the given questions.

29. If $\tan \theta = 3/4$, what is the value of $\sin^2 \theta + \cos^2 \theta$? [$\sin^2 \theta = (\sin \theta)^2$]

30. If $\sin \theta = 2/3$, what is the value of $\sec^2 \theta - \tan^2 \theta$?

31. If $y = \sin \theta$, what is $\cos \theta$ in terms of y?

32. If $x = \cos \theta$, what is $\tan \theta$ in terms of x?

33. What is x if $(x + 1, 4)$ and $(-2, 6)$ are on the same terminal side of a standard position angle?

34. What is x if $(2, 5)$ and $(7, x)$ are on the same terminal side of a standard position angle?

35. From the definitions of the trigonometric functions, it can be seen that $\csc \theta$ is the reciprocal of $\sin \theta$. What function is the reciprocal of $\cos \theta$?

(W) **36.** Refer to the definitions of the trigonometric functions in Eqs. (4.1). Is the quotient of one of the functions divided by $\cos \theta$ equal to $\tan \theta$? Explain.

4.3 VALUES OF THE TRIGONOMETRIC FUNCTIONS

Using the definitions, we can find values of the trigonometric functions if we know a point on the terminal side of the angle. However, we need to find values of the trigonometric functions for angles measured in degrees, because this is commonly needed in practice.

One way to find the functions of a given angle is to use a scale drawing. That is, we draw the angle in standard position using a protractor and then directly measure the values for x, y, and r for some point on the terminal side. Then, by using the proper ratios, we find the functions of this angle.

We may also use certain facts from geometry to find the functions of certain angles. The next two examples illustrate how this is done.

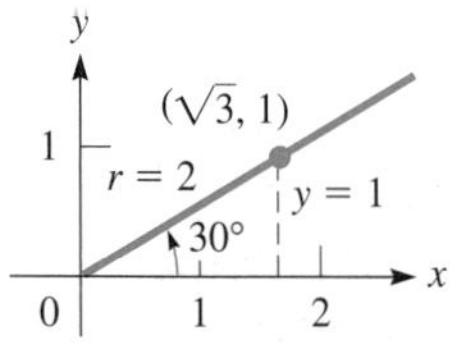

EXAMPLE 1 From geometry, we find that the side opposite a 30° angle in a right triangle is one-half of the hypotenuse. Using this fact and letting $y = 1$ and $r = 2$ (see Fig. 4.18), we calculate that $x = \sqrt{2^2 - 1^2} = \sqrt{3}$ from the Pythagorean theorem. Therefore, with $x = \sqrt{3}$, $y = 1$, and $r = 2$, we have

$$\sin 30° = \frac{1}{2} \qquad \cos 30° = \frac{\sqrt{3}}{2} \qquad \tan 30° = \frac{1}{\sqrt{3}}$$

Using this same method we find the functions of 60° to be

$$\sin 60° = \frac{\sqrt{3}}{2} \qquad \cos 60° = \frac{1}{2} \qquad \tan 60° = \sqrt{3}$$

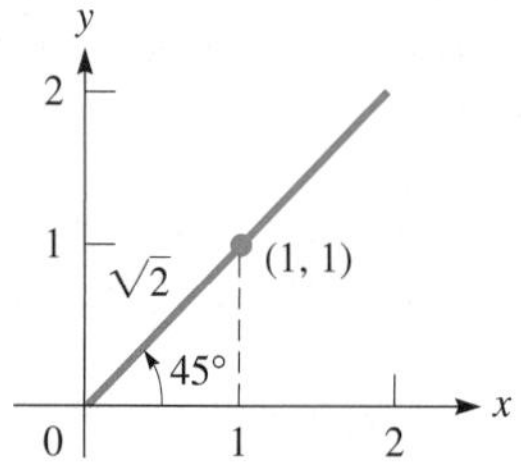

Fig. 4.18

EXAMPLE 2 Find sin 45°, cos 45°, and tan 45°.

From geometry, we know that in an isosceles right triangle the angles are 45°, 45°, and 90°. We know that the sides are in proportion 1, 1, $\sqrt{2}$, respectively. Putting the 45° angle in standard position, we find $x = 1$, $y = 1$, and $r = \sqrt{2}$ (see Fig. 4.19). From this we find

$$\sin 45° = \frac{1}{\sqrt{2}} \qquad \cos 45° = \frac{1}{\sqrt{2}} \qquad \tan 45° = 1$$

Fig. 4.19

As in Example 1, we have given the *exact* values. Decimal approximations are given in the table that follows.

Summarizing the results for 30°, 45°, and 60°, we have:

θ	*(exact values)*			*(decimal approximations)*		
	30°	45°	60°	30°	45°	60°
$\sin\theta$	$\dfrac{1}{2}$	$\dfrac{1}{\sqrt{2}}$	$\dfrac{\sqrt{3}}{2}$	0.500	0.707	0.866
$\cos\theta$	$\dfrac{\sqrt{3}}{2}$	$\dfrac{1}{\sqrt{2}}$	$\dfrac{1}{2}$	0.866	0.707	0.500
$\tan\theta$	$\dfrac{1}{\sqrt{3}}$	1	$\sqrt{3}$	0.577	1.000	1.732

NOTE ▶   It is helpful to be familiar with these values, as they are used in later sections.

The scale-drawing method for finding values of the trigonometric functions gives only approximate results, and geometric methods work only for a limited number of angles. However, it is possible to find these values to any required degree of accuracy for any angle through more advanced methods (using calculus and what are known as *power series*).

The values of the trigonometric functions $\sin \theta$, $\cos \theta$, and $\tan \theta$ are programmed into graphing calculators. For all of our work in the remainder of this chapter, *be sure that your calculator is set for **degrees*** (not radians). The following examples illustrate the use of a calculator in finding trigonometric values.

EXAMPLE 3 Using a graphing calculator to find the value of $\tan 67.36°$, first enter the function and then the angle, just as we have written it. The resulting display is shown in Fig. 4.20.

Therefore, we see that $\tan 67.36° = 2.397\,626\,383$.

Not all calculators will require parentheses around 67.36.

Fig. 4.20

Not only are we able to find values of the trigonometric functions if we know the angle, but we can also find the angle if we know that value of a function. In doing this, we are actually using another important type of mathematical function, an **inverse trigonometric function.** They are discussed in detail in Chapter 20. For the purpose of using a calculator at this point, it is sufficient to recognize and understand the notation that is used.

Inverse Trigonometric Functions

Another notation that is used for $\sin^{-1} x$ is arcsin x.

CAUTION ▶

The notation for "the angle whose sine is x" is $\sin^{-1} x$. This is called the *inverse sine function*. Equivalent meanings are given to $\cos^{-1} x$ (the angle whose cosine is x) and $\tan^{-1} x$ (the angle whose tangent is x). Carefully note that *the -1 used with a trigonometric function in this way shows an inverse trigonometric function and is **not** a negative exponent* ($\sin^{-1} x$ represents an *angle*, not a function of an angle). On a calculator, the $\sin^{-1}$ key is used to find the angle when the sine of that angle is known. The following example illustrates the use of the equivalent $\cos^{-1}$ key.

EXAMPLE 4 If $\cos \theta = 0.3527$, which means that $\theta = \cos^{-1} 0.3527$ (θ is the angle whose cosine is 0.3527), we can use a graphing calculator to find θ. The display for this is shown in Fig. 4.21.

Therefore, we see that $\theta = 69.35°$ (rounded off).

Fig. 4.21

When using the trigonometric functions, the angle is often *approximate*. Angles of 2.3°, 92.3°, and 182.3° are angles with equal accuracy, which shows that *the accuracy of an angle does not depend on the number of digits shown*. The measurement of an angle and the accuracy of its trigonometric functions are shown in the following table:

Angles and Accuracy of Trigonometric Functions

Measurements of Angle to Nearest	Accuracy of Trigonometric Function
1°	2 significant digits
0.1° or 10′	3 significant digits
0.01° or 1′	4 significant digits

We rounded off the result in Example 4 according to this table.

Although we can usually set up the solution of a problem in terms of the sine, cosine, or tangent, there are times when a value of the cotangent, secant, or cosecant is used. We now show how values of these functions are found on a calculator.

Reciprocal Functions

From the definitions, csc $\theta = r/y$ and sin $\theta = y/r$. This means *the value of* csc θ *is the reciprocal of the value of* sin θ. Again, using the definitions, we find that *the value of* sec θ *is the reciprocal of* cos θ and *the value of* cot θ *is the reciprocal of* tan θ. Since the reciprocal of x equals x^{-1}, we use the x^{-1} key along with the sin, cos, and tan keys, to find the values of csc θ, sec θ, and cos θ.

NOTE ▶

Note that here we want the reciprocal $(\cos 27.82°)^{-1}$ and not the angle that would be denoted by using the $\cos^{-1}$ notation.

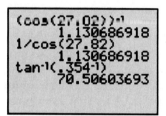

Fig. 4.22

◖EXAMPLE 5 To find the value of sec 27.82°, use the fact that

$$\sec 27.82° = \frac{1}{\cos 27.82°} \quad \text{or} \quad \sec 27.82° = (\cos 27.82°)^{-1}$$

Therefore, we are to find the reciprocal of the value of cos 27.82°. This value can be found using either the first two lines, or the third and fourth lines, of the calculator display shown in Fig. 4.22.

From the display, we see that sec 27.82° = 1.131, with the results rounded off according to the table at the bottom of page 119. ▶

◖EXAMPLE 6 To find the value of θ if cot θ = 0.354, use the fact that

$$\tan \theta = \frac{1}{\cot \theta} = \frac{1}{0.354}$$

The value found by using a graphing calculator is shown in the fifth and sixth lines of Fig. 4.22.

Therefore, $\theta = 70.5°$ (rounded off). ▶

In the following example, we see how to find the value of one function if we know the value of another function of the same angle.

Fig. 4.23

◖EXAMPLE 7 Find sin θ if sec θ = 2.504.

Since the value of sec θ is known, we know that cos θ = 2.504^{-1} (or 1/2.504). This in turn tells us that $\theta = \cos^{-1}(2.504^{-1})$. Since we are to find the value of sin θ, we can see that

$$\sin \theta = \sin(\cos^{-1}(2.504^{-1}))$$

Therefore, we have the calculator display shown in Fig. 4.23.

This means that sin θ = 0.9168 (rounded off). ▶

Calculators have been in common use since the 1970s. Until then the values of the trigonometric functions were generally found by the use of tables (where linear interpolation was used to find values to one more place than was shown in the table). Many standard sources with tables are still available with a precision to at least 10′ or 0.1° (these tables give values with a precision to about five decimal places). However, a calculator is much easier to use than a table, and it can give values to a much greater accuracy. Therefore, we will not use tables in this text.

The following example illustrates the use of the value of a trigonometric function in an applied problem. We consider various types of applications later in the chapter.

◖ **EXAMPLE 8** When a rocket is launched, its horizontal velocity v_x is related to the velocity v with which it is fired by the equation $v_x = v \cos \theta$ (which means $v(\cos \theta)$, but does *not* mean $\cos \theta v$, which is the same as $\cos (\theta v)$). Here, θ is the angle between the horizontal and the direction in which it is fired (see Fig. 4.24). Find v_x if $v = 1250$ m/s and $\theta = 36.0°$.

Substituting the given values of v and θ in $v_x = v \cos \theta$, we have

$$v_x = 1250 \cos 36.0°$$
$$= 1010 \text{ m/s}$$

Therefore, the horizontal velocity is 1010 m/s. ◗

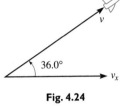

Fig. 4.24

EXERCISES 4.3

In Exercises 1–4, make the given changes in the indicated examples of this section and then find the indicated values.

1. In Example 4, change $\cos \theta$ to $\sin \theta$ and then find the angle.

2. In Example 5, change $\sec 27.82°$ to $\csc 27.82°$ and then find the value.

3. In Example 6, change 0.354 to 0.345 and then find the angle.

4. In Example 7, change $\sin \theta$ to $\tan \theta$ and then find the value.

In Exercises 5–8, use a protractor to draw the given angle. Measure off 10 units (centimeters are convenient) along the radius vector. Then measure the corresponding values of x and y. From these values, determine the trigonometric functions of the angle.

5. 40°   6. 75°   7. 15°   8. 53°

In Exercises 9–24, find the values of the trigonometric functions. Round off results according to the table following Example 4.

9. $\sin 22.4°$
10. $\cos 72.5°$
11. $\tan 57.6°$
12. $\sin 36.0°$
13. $\cos 15.71°$
14. $\tan 8.653°$
15. $\sin 84°$
16. $\cos 47°$
17. $\cot 67.78°$
18. $\csc 22.81°$
19. $\sec 50.4°$
20. $\cot 41.8°$
21. $\csc 49.3°$
22. $\sec 7.8°$
23. $\cot 85.96°$
24. $\csc 76.30°$

In Exercises 25–40, find θ for each of the given trigonometric functions. Round off results according to the table following Example 4.

25. $\cos \theta = 0.3261$
26. $\tan \theta = 2.470$
27. $\sin \theta = 0.9114$
28. $\cos \theta = 0.0427$
29. $\tan \theta = 0.207$
30. $\sin \theta = 0.109$
31. $\cos \theta = 0.650\,07$
32. $\tan \theta = 5.7706$
33. $\csc \theta = 1.245$
34. $\sec \theta = 2.045$
35. $\cot \theta = 0.1443$
36. $\csc \theta = 1.012$
37. $\sec \theta = 3.65$
38. $\cot \theta = 2.08$
39. $\csc \theta = 3.262$
40. $\cot \theta = 0.1519$

In Exercises 41–44, use a calculator to verify the given relationships or statements. [$\sin^2 \theta = (\sin \theta)^2$]

41. $\dfrac{\sin 43.7°}{\cos 43.7°} = \tan 43.7°$

42. $\sin^2 77.5° + \cos^2 77.5° = 1$

43. $\tan 70° = \dfrac{\tan 30° + \tan 40°}{1 - (\tan 30°)(\tan 40°)}$

44. $\sin 78.4° = 2(\sin 39.2°)(\cos 39.2°)$

Ⓦ *In Exercises 45–48, explain why the given statements are true for an acute angle θ.*

45. $\sin \theta$ is always between 0 and 1.

46. $\tan \theta$ can equal any positive real number.

47. $\cos \theta$ decreases in value from 0° to 90°.

48. The value of $\sec \theta$ is never less than 1.

In Exercises 49–52, find the values of the indicated trigonometric functions.

49. Find $\sin \theta$, given $\tan \theta = 1.936$.

50. Find $\cos \theta$, given $\sin \theta = 0.6725$.

51. Find $\tan \theta$, given $\sec \theta = 1.3698$.

52. Find $\csc \theta$, given $\cos \theta = 0.1063$.

In Exercises 53–56, solve the given problems.

53. The sound produced by a jet engine was measured at a distance of 100 m in all directions. The loudness d of the sound (in dB) was found to be $d = 70.0 + 30.0 \cos \theta$, where the 0° line was directed in front of the engine. Calculate d for $\theta = 54.5°$.

54. A brace is used in the structure shown in Fig. 4.25. Its length is $l = a(\sec \theta + \csc \theta)$. Find l if $a = 28.0$ cm and $\theta = 34.5°$.

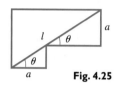

Fig. 4.25

55. The signal from an AM radio station with two antennas d m apart has a wavelength λ (in m). The intensity of the signal depends on the angle θ as shown in Fig. 4.26. An angle of minimum intensity is given by $\sin \theta = 1.50\,\lambda/d$. Find θ if $\lambda = 200$ m and $d = 400$ m.

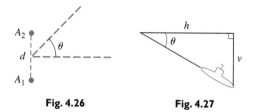

Fig. 4.26   **Fig. 4.27**

56. A submarine dives such that the horizontal distance h it moves and the vertical distance v it dives are related by $v = h \tan \theta$. Here θ is the angle of the dive, as shown in Fig. 4.27. Find θ if $h = 2.35$ km and $v = 1.52$ km.

4.4 THE RIGHT TRIANGLE

We know that a triangle has three sides and three angles. If one side and any other two of these six parts are known, we can find the other three parts. One of the known parts must be a side, for if we know only the three angles, we know only that any triangle with these angles is similar to any other triangle with these angles.

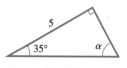

Fig. 4.28

❙ **EXAMPLE 1**   Assume that one side and two angles are known, such as the side of 5 and the angles of 35° and 90° in the triangle in Fig. 4.28. Then we may determine the third angle α by the fact that the sum of the angles of a triangle is always 180°. Of all possible similar triangles having the three angles of 35°, 90°, and 55° (which is α), we have the one with the particular side of 5 between angles of 35° and 90°. Only one triangle with these parts is possible (in the sense that all triangles with the given parts are *congruent* and have equal corresponding angles and sides).   ▶

To **solve a triangle** *means that, when we are given three parts of a triangle (at least one a side), we are to find the other three parts.* In this section, we are going to demonstrate the method of solving a right triangle. *Since one angle of the triangle is 90°, it is necessary to know one side and one other part.* Also, since the sum of the three angles is 180°, we know that *the sum of the other two angles is 90°, and they are acute angles.* It also means they are **complementary angles,** following the definition on page 52.

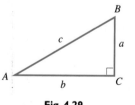

Fig. 4.29

For consistency, when we are labeling the parts of the right triangle *we will use the letters A and B to denote the acute angles and C to denote the right angle. The letters a, b, and c will denote the sides opposite these angles, respectively. Thus, side c is the hypotenuse of the right triangle.* See Fig. 4.29.

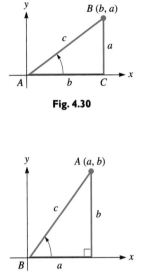

Fig. 4.30

In solving right triangles, we will find it convenient to express the trigonometric functions of the acute angles in terms of the sides. By placing the vertex of angle A at the origin and the vertex of right angle C on the positive x-axis, as shown in Fig. 4.30, we have the following ratios for angle A in terms of the sides of the triangle.

$$\sin A = \frac{a}{c} \qquad \cos A = \frac{b}{c} \qquad \tan A = \frac{a}{b}$$

$$\cot A = \frac{b}{a} \qquad \sec A = \frac{c}{b} \qquad \csc A = \frac{c}{a} \tag{4.3}$$

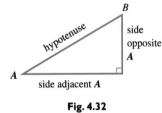

Fig. 4.31

If we placed the vertex of B at the origin, instead of the vertex of angle A, we would obtain the following ratios for the functions of angle B (see Fig. 4.31):

$$\sin B = \frac{b}{c} \qquad \cos B = \frac{a}{c} \qquad \tan B = \frac{b}{a}$$

$$\cot B = \frac{a}{b} \qquad \sec B = \frac{c}{a} \qquad \csc B = \frac{c}{b} \tag{4.4}$$

Equations (4.3) and (4.4) show that we may generalize our definitions of the trigonometric functions of an acute angle of a right triangle (we have chosen $\angle A$ in Fig. 4.32) to be as follows:

Fig. 4.32

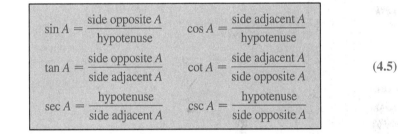

$$\tag{4.5}$$

Which side is adjacent or opposite depends on the angle being considered. In Fig. 4.31, the side opposite A is adjacent to B, and the side adjacent to A is opposite B.

Using the definitions in this form, we can solve right triangles without placing the angle in standard position. The angle need only be a part of any right triangle.

We note from the above discussion that $\sin A = \cos B$, $\tan A = \cot B$, and $\sec A = \csc B$. From this, we conclude that *cofunctions of acute complementary angles are equal.* The sine function and cosine function are cofunctions, the tangent function and cotangent function are cofunctions, and the secant function and cosecant functions are cofunctions.

◀ **EXAMPLE 2** Given $a = 4$, $b = 7$, and $c = \sqrt{65}$ (see Fig. 4.33), find $\sin A$, $\cos A$, $\tan A$, $\sin B$, $\cos B$, and $\tan B$ in exact form and in approximate decimal form (to three significant digits).

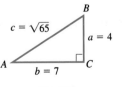

Fig. 4.33

$$\sin A = \frac{\text{side opposite angle } A}{\text{hypotenuse}} = \frac{4}{\sqrt{65}} = 0.496 \qquad \sin B = \frac{\text{side opposite angle } B}{\text{hypotenuse}} = \frac{7}{\sqrt{65}} = 0.868$$

$$\cos A = \frac{\text{side adjacent angle } A}{\text{hypotenuse}} = \frac{7}{\sqrt{65}} = 0.868 \qquad \cos B = \frac{\text{side adjacent angle } B}{\text{hypotenuse}} = \frac{4}{\sqrt{65}} = 0.496$$

$$\tan A = \frac{\text{side opposite angle } A}{\text{side adjacent angle } A} = \frac{4}{7} = 0.571 \qquad \tan B = \frac{\text{side opposite angle } B}{\text{side adjacent angle } B} = \frac{7}{4} = 1.75$$

We see that A and B are complementary angles. Comparing values of the functions of angles A and B, we see that $\sin A = \cos B$ and $\cos A = \sin B$.

We are now ready to solve right triangles. See how the following procedure is used in the examples that follow.

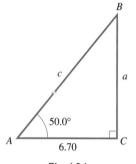

Fig. 4.34

> **PROCEDURE FOR SOLVING A RIGHT TRIANGLE**
>
> 1. *Sketch a right triangle and label the known and unknown sides and angles.*
> 2. *Express each of the three unknown parts in terms of the known parts and solve for the unknown parts.*
> 3. *Check the results.* The sum of the angles should be 180°. If only one side is given, check the computed side with the Pythagorean theorem. If two sides are given, check the angles and computed side by using appropriate trigonometric functions.

◀ **EXAMPLE 3** Solve the right triangle with $A = 50.0°$ and $b = 6.70$.

We first sketch the right triangle shown in Fig. 4.34. (In making the sketch, we should be careful to follow proper labeling of the triangle as outlined on page 122.) We then express unknown side a in terms of known side b and known angle A and solve for a. We will then do the same for unknown side c and unknown angle B.

Finding side a, we know that $\tan A = \dfrac{a}{b}$, which means that $a = b \tan A$. Thus,

$$a = 6.70 \tan 50.0° = 7.98 \qquad \text{lines 1 \& 2 of calculator display in Fig. 4.35}$$

Next, solving for side c, we have $\cos A = \dfrac{b}{c}$, which means $c = \dfrac{b}{\cos A}$.

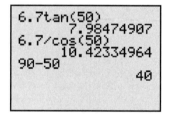

Fig. 4.35

$$c = \frac{6.70}{\cos 50.0°} = 10.4 \qquad \text{lines 3 \& 4 of calculator display in Fig. 4.35}$$

Now solving for B, we know that $A + B = 90°$, or

$$B = 90° - A$$
$$= 90° - 50.0° = 40.0° \qquad \text{lines 5 \& 6 of calculator display in Fig. 4.35}$$

Therefore, $a = 7.98$, $c = 10.4$, and $B = 40.0°$.

Checking the angles: $A + B + C = 50.0° + 40.0° + 90° = 180°$
Checking the sides: $10.4^2 = 108.16$
$$7.98^2 + 6.70^2 = 108.57$$

Since the computed values were rounded off, the values 108.16 and 108.57 show that the values for sides a and c check.

As we calculate the values of the unknown parts, if we store each in the calculator memory, we can get a better check of the solution. In Fig. 4.36, we see the solution for each side, its storage in memory, and the check of these values. *The lines above the display are those that are replaced as the solution proceeds.* From this calculator display we see that the values for sides a and c check very accurately. ▶

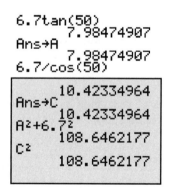

Fig. 4.36

NOTE ▶ In finding the unknown parts, we first expressed them in terms of the known parts. We do this because *it is best to use given values in calculations*. If we use one computed value to find another computed value, any error in the first would be carried to the value of the second. For instance, in Example 3, if we were to find the value of c by using the value of a, any error in a would cause c to be in error as well.

We should also point out that, by inspection, we can make a rough check on the sides and angles of any triangle.

The longest side is always opposite the largest angle, and the shortest side is always opposite the smallest angle.

In a right triangle, *the hypotenuse is always the longest side.* We see that this is true for the sides and angles for the triangle in Example 3, where c is the longest side (opposite the 90° angle) and b is the shortest side and is opposite the angle of 40°.

See Appendix C for a graphing calculator program SLVRTTRI. It can be used to solve a right triangle, given the two legs.

◀ EXAMPLE 4   Solve the right triangle with $b = 56.82$ and $c = 79.55$.

We sketch the right triangle as shown in Fig. 4.37. Since two sides are given, we will use the Pythagorean theorem to find the third side a. Also, we will use the cosine to find $\angle A$.

Since $c^2 = a^2 + b^2$, $a^2 = c^2 - b^2$. Therefore,

$$a = \sqrt{c^2 - b^2} = \sqrt{79.55^2 - 56.82^2}$$
$$= 55.67 \qquad \text{lines 1 \& 2 shown in Fig. 4.38}$$

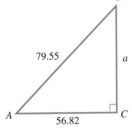

Fig. 4.37

Since $\cos A = \dfrac{b}{c}$, we have

$$\cos A = \frac{56.82}{79.55}$$

$$A = \cos^{-1}\left(\frac{56.82}{79.55}\right) = 44.42° \qquad \text{lines 3–5 shown in Fig. 4.38}$$

It is not necessary to actually calculate the ratio 56.82/79.55. In the same way, we find the value of angle B:

$$\sin B = \frac{56.82}{79.55}$$

$$B = \sin^{-1}\left(\frac{56.82}{79.55}\right) = 45.58° \qquad \text{lines 6–8 shown in Fig. 4.38}$$

Although we used a different function, we did use exactly the same ratio to find B as we used to find A. Therefore, many texts would find B from the fact that $A + B = 90°$, or $B = 90° - A = 90° - 44.42° = 45.58°$. This is also acceptable since any possible error should be discovered when the solution is checked.

We have now found that

$$a = 55.67 \qquad A = 44.42° \qquad B = 45.58°$$

Fig. 4.38

Checking the sides and angles, we first note that side a is the shortest side and is opposite the smallest angle, $\angle A$. Also, the hypotenuse is the longest side. Next, using the sine function (we could use the cosine or tangent) to check the sides, we have

As we noted earlier, the symbol $\approx$ means "equals approximately."

$$\sin 44.42° = \frac{55.67}{79.55}, \text{ or } 0.6999 \approx 0.6998 \qquad \sin 45.58° = \frac{56.82}{79.55}, \text{ or } 0.7142 \approx 0.7143$$

This shows that the values check. As we noted at the end of Example 3, we would get a more accurate check if we save the calculator values as they are found and use them for the check. ▸

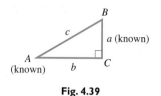

Fig. 4.39

◀ EXAMPLE 5   Given that $\angle A$ and side a are known, express the unknown parts of the right triangle in terms of A and a.

We sketch a right triangle as shown in Fig. 4.39 and set up the required expressions as follows:

Since $\dfrac{a}{b} = \tan A$, we have $a = b \tan A$, or $b = \dfrac{a}{\tan A}$.

Since $\dfrac{a}{c} = \sin A$, we have $a = c \sin A$, or $c = \dfrac{a}{\sin A}$.

Since A is known, $B = 90° - A$.

EXERCISES 4.4

In Exercises 1–4, make the given changes in the indicated examples of this section and then find the indicated values.

1. In Example 2, interchange a and b and then find the values.
2. In Example 3, change 6.70 to 7.60 and then solve the triangle.
3. In Example 4, change 56.82 to 65.82 and then solve the triangle.
4. In Example 5, in lines 1 and 2 change a to b and then find the expressions for the unknown parts.

In Exercises 5–8, draw appropriate figures and verify through observation that only one triangle may contain the given parts (that is, any other which may be drawn will be congruent).

5. A 60° angle included between sides of 3 cm and 6 cm
6. A side of 4 cm included between angles of 40° and 50°
7. A right triangle with a hypotenuse of 5 cm and a leg of 3 cm
8. A right triangle with a 70° angle between the hypotenuse and a leg of 5 cm

In Exercises 9–28, solve the right triangles with the given parts. Round off results. Refer to Fig. 4.40.

9. $A = 77.8°$, $a = 6700$
10. $A = 18.4°$, $c = 0.0897$
11. $a = 150$, $c = 345$
12. $a = 932$, $c = 1240$
13. $B = 32.1°$, $c = 23.8$
14. $B = 64.3°$, $b = 0.652$

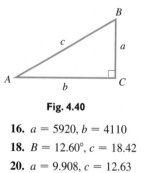

Fig. 4.40

15. $b = 82$, $c = 88$
16. $a = 5920$, $b = 4110$
17. $A = 32.10°$, $c = 56.85$
18. $B = 12.60°$, $c = 18.42$
19. $a = 56.73$, $b = 44.09$
20. $a = 9.908$, $c = 12.63$
21. $B = 37.5°$, $a = 0.862$
22. $A = 52°$, $b = 8.4$
23. $B = 74.18°$, $b = 1.849$
24. $A = 51.36°$, $a = 3692$
25. $a = 591.87$, $b = 264.93$
26. $b = 2.9507$, $c = 5.0864$
27. $A = 12.975°$, $b = 14.592$
28. $B = 84.942°$, $a = 7413.5$

In Exercises 29–32, find the part of the triangle labeled either x or A in the indicated figure.

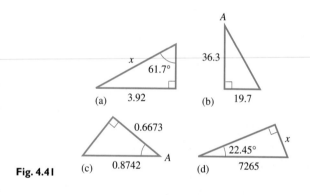

Fig. 4.41

29. Fig. 4.41(a)
30. Fig. 4.41(b)
31. Fig. 4.41(c)
32. Fig. 4.41(d)

In Exercises 33–36, find the indicated part of the right triangle that has the given parts.

33. One leg is 25.6, and the hypotenuse is 37.5. Find the smaller acute angle.
34. One leg is 8.50, and the angle opposite this leg is 52.3°. Find the other leg.
35. The hypotenuse is 827, and one angle is 17.6°. Find the longer leg.
36. The legs are 0.596 and 0.842. Find the larger acute angle.

In Exercises 37–40, refer to Fig. 4.40. In Exercises 37–39, the listed parts are assumed known. Express the other parts in terms of the known parts.

37. A, c
38. a, b
39. B, a

W 40. In Fig. 4.40, is there any combination of two given parts (not including $\angle C$) that does not give a unique solution of the triangle? Explain.

4.5 APPLICATIONS OF RIGHT TRIANGLES

See the chapter introduction.

Solving a Word Problem

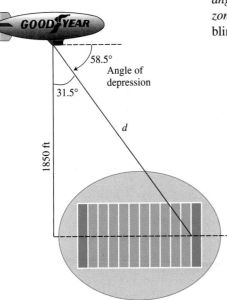

EXAMPLE 1 Horseshoe Falls on the Canadian side of Niagara Falls can be seen from a small boat 2500 ft downstream. The **angle of elevation** *(the angle between the horizontal and the line of sight, when the object is **above** the horizontal)* from the observer to the top of Horseshoe Falls is 4.0°. How high are the Falls?

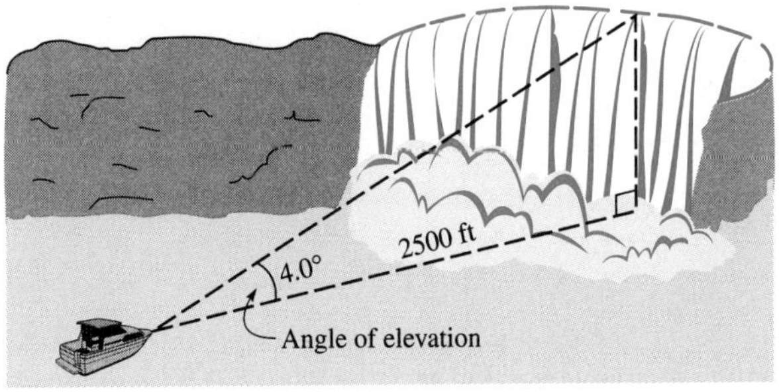

By drawing an appropriate figure, as shown in Fig. 4.42, we show the given information and that we are to find. Here, we let h be the height of the Falls and label the horizontal distance from the boat to the base of the Falls as 2500 ft. From the figure, we see that

$$\frac{h}{2500} = \tan 4.0° \qquad \frac{\text{required opposite side}}{\text{given adjacent side}} = \text{tangent of given angle}$$

$$h = 2500 \tan 4.0°$$

$$= 170 \text{ ft}$$

We have rounded off the result since the data are good only to two significant digits. (Niagara Falls is on the border between the United States and Canada and is divided into the American Falls and the Horseshoe Falls. About 500,000 tonnes of water flow over the Falls each minute.)

Fig. 4.42

Solving a Word Problem

EXAMPLE 2 The Goodyear blimp is 565 m above the ground and south of the Rose Bowl in California during a Super Bowl game. The **angle of depression** *(the angle between the horizontal and the line of sight, when the object is **below** the horizontal)* of the north goal line from the blimp is 58.5°. How far is the observer in the blimp from the goal line?

Again, we sketch a figure as shown in Fig. 4.43. Here, we let d be the distance between the blimp and the north goal line. From the figure, we see that

$$\frac{565}{d} = \cos 31.5° \qquad \frac{\text{given adjacent side}}{\text{required hypotenuse}} = \text{cosine of known angle}$$

$$d = \frac{565}{\cos 31.5°}$$

$$= 663 \text{ m}$$

Here we have rounded off the result to three significant digits, the accuracy of the given information. (The Super Bowl games in 1977, 1980, 1987, and 1993 were played in the Rose Bowl.)

Fig. 4.43

Solving a Word Problem ❙ **EXAMPLE 3** A missile is launched at an angle of 26.55° with respect to the horizontal. If it travels in a straight line over level terrain for 2.000 min and its average speed is 6355 km/h, what is its altitude at this time?

In Fig. 4.44, we let h represent the altitude of the missile after 2.000 min (altitude is measured on a perpendicular line). Also, we determine that in this time the missile has flown 211.8 km in a direct line from the launching site. This is found from the fact that it travels at 6355 km/h for $\dfrac{1}{30.00}$ h (2.000 min), and distance = speed × time. We therefore have $\left(6355\,\dfrac{\text{km}}{\text{h}}\right)\left(\dfrac{1}{30.00}\,\text{h}\right)$ = 211.8 km. This means

$$\frac{h}{211.8} = \sin 26.55° \qquad \frac{\text{required opposite side}}{\text{known hypotenuse}} = \text{sine of given angle}$$

$$h = 211.8(\sin 26.55°)$$

$$= 94.67 \text{ km}$$

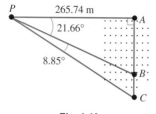

Fig. 4.44

Solving a Word Problem ❙ **EXAMPLE 4** A driver coming to an intersection sees the word STOP in the roadway. From the measurements shown in Fig. 4.45, find the angle θ that the letters make at the driver's eye.

From the figure, we know sides BS and BE in triangle BES and sides BT and BE in triangle BET. This means we can find $\angle TEB$ and $\angle SEB$ by use of the tangent. We then find θ from the fact that $\theta = \angle TEB - \angle SEB$.

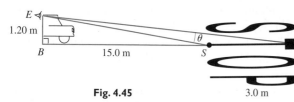

Fig. 4.45

$$\tan \angle TEB = \frac{18.0}{1.20}, \qquad \angle TEB = 86.2°$$

$$\tan \angle SEB = \frac{15.0}{1.20}, \qquad \angle SEB = 85.4°$$

$$\theta = 86.2° - 85.4° = 0.8°$$

❙ **EXAMPLE 5** Using lasers, a surveyor makes the measurements shown in Fig. 4.46, where points B and C are in a marsh. Find the distance between B and C.

Since the distance $BC = AC - AB$, BC is found by finding AC and AB and subtracting:

Fig. 4.46

$$\frac{AB}{265.74} = \tan 21.66°$$

$$AB = 265.74 \tan 21.66°$$

$$\frac{AC}{265.74} = \tan(21.66° + 8.85°)$$

$$AC = 265.74 \tan 30.51°$$

$$BC = AC - AB = 265.74 \tan 30.51° - 265.74 \tan 21.66°$$

$$= 51.06 \text{ m}$$

EXERCISES 4.5

In Exercises 1 and 2, make the given changes in the indicated examples of this section and then find the indicated values.

1. In Example 2, in line 4 change 58.5° to 62.1° and then find the distance.

2. In Example 3, in line 2 change 2.000 min to 3.000 min and then find the altitude.

In Exercises 3–36, solve the given problems. Sketch an appropriate figure, unless the figure is given.

3. A straight 120-m culvert is built down a hillside that makes an angle of 54.0° with the horizontal. Find the height of the hill.

4. In 2000, about 70 tonnes of soil were removed from under the Leaning Tower of Pisa, and the angle the tower made with the ground was increased by about 0.5°. Before that, a point near the top of the tower was 50.5 m from a point at the base (measured along the tower), and this top point was directly above a point on the ground 4.25 m from the same base point. See Fig. 4.47. How much did the point on the ground move toward the base point?

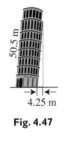

50.5 m

4.25 m

Fig. 4.47

5. A tree has a shadow 22.8 m long when the angle of elevation of the sun is 62.6°. How tall is the tree?

6. The straight arm of a robot is 1.25 m long and makes an angle of 13.0° above a horizontal conveyor belt. How high above the belt is the end of the arm? See Fig. 4.48.

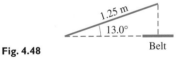

1.25 m

13.0°

Belt

Fig. 4.48

7. The headlights of an automobile are set such that the beam drops 5.10 cm for each 7.50 m in front of the car. What is the angle between the beam and the road?

8. A bullet was fired such that it just grazed the top of a table. It entered a wall, which is 3.84 m from the graze point in the table, at a point 1.41 m above the tabletop. At what angle was the bullet fired above the horizontal? See Fig. 4.49.

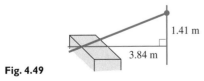

1.41 m

3.84 m

Fig. 4.49

9. A robot is on the surface of Mars. The angle of depression from a camera in the robot to a rock on the surface of Mars is 13.33°. The camera is 196.0 cm above the surface. How far from the camera is the rock?

10. The Sears Tower in Chicago can be seen from a point on the ground known to be 1600 m from the base of the tower. The angle of elevation from the observer to the top of the tower is 16°. How high is the Sears Tower? See Fig. 4.50.

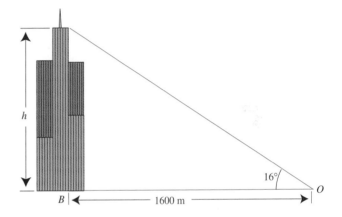

h

B

1600 m

16°

O

Fig. 4.50

11. In the design of a new building, a doorway is 795 mm above the ground. A ramp for the disabled, at an angle of 6.0° with the ground, is to be built to the doorway. How long will the ramp be?

12. On a test flight, during the landing of the space shuttle, the ship was 105 m above the end of the landing strip. It then came in on a constant angle of 7.5° with the landing strip. How far from the end of the landing strip did it first touch ground?

13. From a point on the South Rim of the Grand Canyon, it is found that the angle of elevation of a point on the North Rim is 1.2°. If the horizontal distance between the points is 16 km, how much higher is the point on the North Rim?

14. What is the steepest angle between the surface of a board 3.50 cm thick and a nail 5.00 cm long if the nail is hammered into the board such that it does not go through?

15. A rectangular piece of plywood 1200 mm by 2400 mm is cut from one corner to an opposite corner. What are the angles between edges of the resulting pieces?

16. A guardrail is to be constructed around the top of a circular observation tower. The diameter of the observation area is 12.3 m. If the railing is constructed with 30 equal straight sections, what should be the length of each section?

17. The angle of inclination of a road is often expressed as *percent grade*, which is the vertical rise divided by the horizontal run (expressed as a percent). See Fig. 4.51. A 6.0% grade corresponds to a road that rises 6.0 m for every 100 m along the horizontal. Find the angle of inclination that corresponds to a 6.0% grade.

Fig. 4.51

18. A tabletop is in the shape of a regular octagon (eight sides). What is the greatest distance across the table if one side of the octagon is 0.750 m?

19. To get a good view of a person in front of a teller's window, it is determined that a surveillance camera at a bank should be directed at a point 5.17 m to the right and 2.25 m below the camera. See Fig. 4.52. At what angle of depression should the camera be directed?

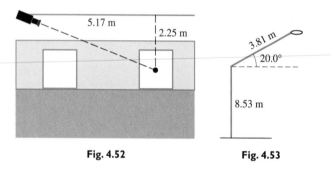

Fig. 4.52   **Fig. 4.53**

20. A street light is designed as shown in Fig. 4.53. How high above the street is the light?

21. A straight driveway is 85.0 m long, and the top is 12.0 m above the bottom. What angle does it make with the horizontal?

22. Part of the Tower Bridge in London is a drawbridge. This part of the bridge is 76.0 m long. When each half is raised, the distance between them is 8.0 m. What angle does each half make with the horizontal? See Fig. 4.54.

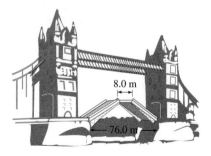

Fig. 4.54

23. A square wire loop is rotating in the magnetic field between two poles of a magnet in order to induce an electric current. The axis of rotation passes through the center of the loop and is midway between the poles, as shown in the side view in Fig. 4.55. How far is the edge of the loop from either pole if the side of the square is 7.30 cm and the poles are 7.66 cm apart when the angle between the loop and the vertical is 78.0°?

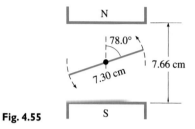

Fig. 4.55

24. From a space probe circling Io, one of Jupiter's moons, at an altitude of 552 km, it was observed that the angle of depression of the horizon was 39.7°. What is the radius of Io?

25. A manufacturing plant is designed to be in the shape of a regular pentagon with 92.5 m on each side. A security fence surrounds the building to form a circle, and each corner of the building is to be 25.0 m from the closest point on the fence. How much fencing is required?

26. A surveyor on the east side of the Rhine River in Germany wishes to find the height of a cliff on the west side. Figure 4.56 shows the measurements that were made. How high is the cliff? (In the figure, the triangle containing the height *h* is vertical and perpendicular to the river.)

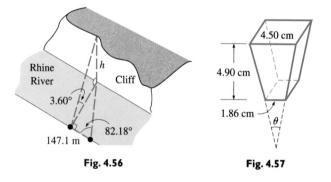

Fig. 4.56   **Fig. 4.57**

27. Find the angle θ in the taper shown in Fig. 4.57. (The front face is an isosceles trapezoid.)

28. What is the circumference of the Arctic Circle (latitude 66°32′N)? The radius of the earth is 6370 km.

29. A stairway 1.0 m wide goes from the bottom of a cylindrical storage tank to the top at a point halfway around the tank. The handrail on the outside of the stairway makes an angle of 31.8° with the horizontal, and the radius of the tank is 11.8 m. Find the length of the handrail. See Fig. 4.58.

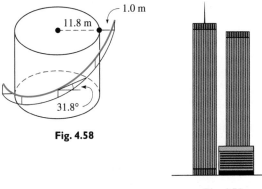

Fig. 4.58

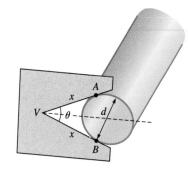

Fig. 4.59

30. An antenna was on the top of the World Trade Center (before it was destroyed in 2001). From a point on the river 2400 m from the Center, the angles of elevation of the top and bottom of the antenna were 12.1° and 9.9°, respectively. How tall was the antenna? (Disregard the small part of the antenna near the base that could not be seen.) The former World Trade Center is shown in Fig. 4.59. (*This problem is included in memory of those who suffered and died as a result of the terrorist attack of September 11, 2001.*)

31. Some of the streets of San Francisco are shown in Fig. 4.60. The distances between intersections A and B, A and D, and C and E are shown. How far is it between intersections C and D?

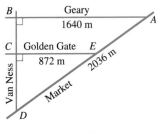

Fig. 4.60

32. A supporting girder structure is shown in Fig. 4.61. Find the length x.

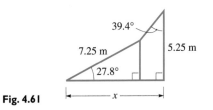

Fig. 4.61

33. The diameter d of a pipe can be determined by noting the distance x on the V-gauge shown in Fig. 4.62. Points A and B indicate where the pipe touches the gauge, and x equals either AV or VB. Find a formula for d in terms of x and θ.

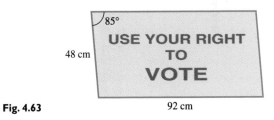

Fig. 4.62

34. The political banner shown in Fig. 4.63 is in the shape of a parallelogram. Find its area.

85°

USE YOUR RIGHT
TO
VOTE

48 cm

92 cm

Fig. 4.63

35. Find a formula for the area of the trapezoidal aqueduct cross section shown in Fig. 4.64.

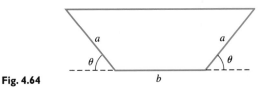

Fig. 4.64

36. A communications satellite is in orbit 35 300 km directly above the earth's equator. What is the greatest latitude from which a signal can travel from the earth's surface to the satellite in a straight line? The radius of the earth is 6370 km.

CHAPTER 4 EQUATIONS

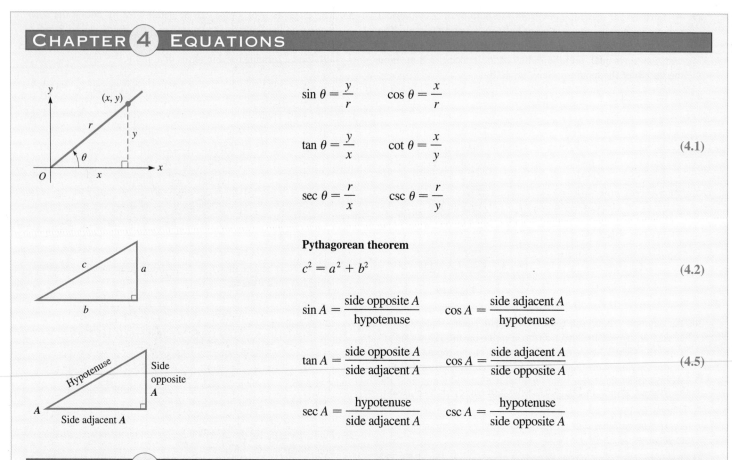

$$\sin \theta = \frac{y}{r} \qquad \cos \theta = \frac{x}{r}$$

$$\tan \theta = \frac{y}{x} \qquad \cot \theta = \frac{x}{y} \tag{4.1}$$

$$\sec \theta = \frac{r}{x} \qquad \csc \theta = \frac{r}{y}$$

Pythagorean theorem

$$c^2 = a^2 + b^2 \tag{4.2}$$

$$\sin A = \frac{\text{side opposite } A}{\text{hypotenuse}} \qquad \cos A = \frac{\text{side adjacent } A}{\text{hypotenuse}}$$

$$\tan A = \frac{\text{side opposite } A}{\text{side adjacent } A} \qquad \cos A = \frac{\text{side adjacent } A}{\text{side opposite } A} \tag{4.5}$$

$$\sec A = \frac{\text{hypotenuse}}{\text{side adjacent } A} \qquad \csc A = \frac{\text{hypotenuse}}{\text{side opposite } A}$$

CHAPTER 4 REVIEW EXERCISES

In Exercises 1–4, find the smallest positive angle and the smallest negative angle (numerically) coterminal with but not equal to the given angle.

1. $17.0°$ **2.** $248.3°$ **3.** $-217.5°$ **4.** $-7.6°$

In Exercises 5–8, express the given angles in decimal form.

5. $31°54'$ **6.** $174°45'$ **7.** $38°6'$ **8.** $321°27'$

In Exercises 9–12, express the given angles to the nearest minute.

9. $17.5°$ **10.** $65.4°$ **11.** $249.7°$ **12.** $126.25°$

In Exercises 13–16, determine the trigonometric functions of the angles (in standard position) whose terminal side passes through the given points. Give answers in exact form.

13. $(24, 7)$ **14.** $(5, 4)$ **15.** $(48, 48)$ **16.** $(1.2, 0.5)$

In Exercises 17–20, find the indicated trigonometric functions. Give answers in decimal form, rounded off to three significant digits.

17. Given $\sin \theta = \frac{5}{13}$, find $\cos \theta$ and $\cot \theta$.

18. Given $\cos \theta = \frac{3}{8}$, find $\sin \theta$ and $\tan \theta$.

19. Given $\tan \theta = 2$, find $\cos \theta$ and $\csc \theta$.

20. Given $\cot \theta = 40$, find $\sin \theta$ and $\sec \theta$.

In Exercises 21–28, find the values of the trigonometric functions. Round off results.

21. $\sin 72.1°$ **22.** $\cos 40.3°$

23. $\tan 61.64°$ **24.** $\sin 49.09°$

25. $\sec 18.4°$ **26.** $\csc 82.4°$

27. $(\cot 7.06°)(\sin 7.06°) - \cos 7.06°$

28. $(\sec 79.36°)(\sin 79.36°) - \tan 79.36°$

In Exercises 29–36, find θ for each of the given trigonometric functions. Round off results.

29. $\cos \theta = 0.950$ **30.** $\sin \theta = 0.630\,52$

31. $\tan \theta = 1.574$ **32.** $\cos \theta = 0.1345$

33. $\csc \theta = 4.713$ **34.** $\cot \theta = 0.7561$

35. $\sec \theta = 25.4$ **36.** $\csc \theta = 1.92$

In Exercises 37–48, solve the right triangles with the given parts. Refer to Fig. 4.65.

37. $A = 17.0°$, $b = 6.00$

38. $B = 68.1°$, $a = 1080$

39. $a = 81.0$, $b = 64.5$

40. $a = 106$, $c = 382$

41. $A = 37.5°$, $a = 12.0$

42. $B = 15.7°$, $c = 12.6$

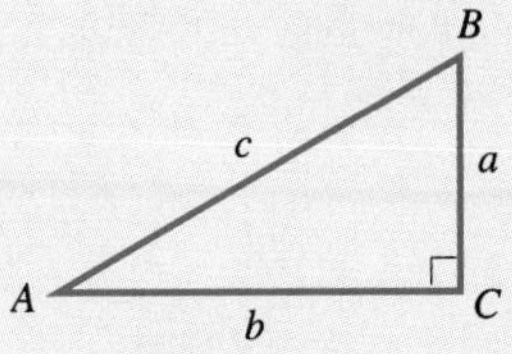

43. $b = 6.508$, $c = 7.642$

44. $a = 72.14$, $b = 14.37$

45. $A = 49.67°$, $c = 0.8253$

46. $B = 4.38°$, $b = 5682$

47. $a = 11.652$, $c = 15.483$

48. $a = 724.39$, $b = 852.44$

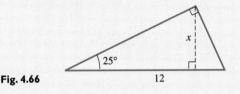

Fig. 4.65

In Exercises 49–84, solve the given problems.

49. Find the value of x for the triangle shown in Fig. 4.66.

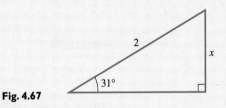

Fig. 4.66

W **50.** Explain three ways in which the value of x can be found for the triangle shown in Fig. 4.67. Which of these methods is the easiest?

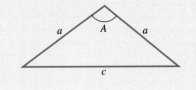

Fig. 4.67

51. Find the perimeter of a regular octagon (eight equal sides with equal interior angles) that is inscribed in a circle (all vertices of the octagon touch the circle) of radius 10.

W **52.** Explain why values of $\sin \theta$ increase as θ increases from $0°$ to $90°$.

53. What is x if $(3, 2)$ and $(x, 7)$ are on the same terminal side of an acute angle?

54. Two legs of a right triangle are 2.607 and 4.517. What is the smaller acute angle?

55. Show that the side c of any triangle ABC is related to the perpendicular h from C to side AB by the equation $c = h \cot A + h \cot B$.

56. For the isosceles triangle shown in Fig. 4.68, show that $c = 2a \sin \frac{A}{2}$.

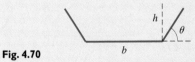

Fig. 4.68

57. The voltage e at any instant in a coil of wire that is turning in a magnetic field is given by $e = E \cos \alpha$, where E is the maximum voltage and α is the angle the coil makes with the field. Find the acute angle α if $e = 56.9$ V and $E = 339$ V.

58. A formula for the area of a quadrilateral is $A = \frac{1}{2}d_1 d_2 \sin \theta$, where d_1 and d_2 are the lengths of the diagonals and θ is the angle between them. Find the area of a four-sided carpet remnant with diagonals 1050 mm and 1330 mm and $\theta = 72.0°$.

59. For a car rounding a curve, the road should be banked at an angle θ according to the equation $\tan \theta = \dfrac{v^2}{gr}$. Here, v is the speed of the car and r is the radius of the curve in the road. See Fig. 4.69. Find θ for $v = 24.2$ m/s, $g = 9.80$ m/s^2, and $r = 282$ m.

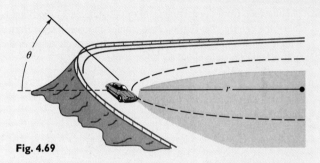

Fig. 4.69

60. The *apparent power S* in an electric circuit in which the power is P and the impedance phase angle is θ is given by $S = P \sec \theta$. Given $P = 12.0$ V$\cdot$A and $\theta = 29.4°$, find S.

61. A surveyor measures two sides and the included angle of a triangular tract of land to be $a = 31.96$ m, $b = 47.25$ m, and $C = 64.09°$. (a) Show that a formula for the area A of the tract is $A = \frac{1}{2}ab \sin C$. (b) Find the area of the tract.

62. A water channel has the cross section of an isosceles trapezoid. See Fig. 4.70. (a) Show that a formula for the area of the cross section is $A = bh + h^2 \cot \theta$. (b) Find A if $b = 12.6$ m, $h = 4.75$ m, and $\theta = 37.2°$.

Fig. 4.70

63. In tracking an airplane on radar, it is found that the plane is 27.5 km on a direct line from the control tower, with an angle of elevation of $10.3°$. What is the altitude of the plane?

64. A straight emergency chute for an airplane is 5.50 m long. In being tested, the end of the chute is 2.9 m above the ground. What angle does the chute make with the ground?

65. The windshield on an automobile is inclined $42.5°$ with respect to the horizontal. Assuming that the windshield is flat and rectangular, what is its area if it is 1.50 m wide and the bottom is 0.48 m in front of the top?

66. A water slide at an amusement park is 25 m long and is inclined at an angle of 52° with the horizontal. How high is the top of the slide above the water level?

67. The window of a house is shaded as shown in Fig. 4.71. What percent of the window is shaded when the angle of elevation θ of the sun is 65°?

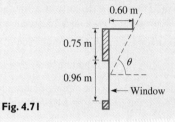

Fig. 4.71

68. The distance from the ground level to the underside of a cloud is called the *ceiling*. See Fig. 4.72. A ground observer 950 m from a searchlight aimed vertically notes that the angle of elevation of the spot of light on a cloud is 76°. What is the ceiling?

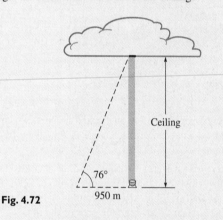

Fig. 4.72

69. The vertical cross section of an attic room in a house is shown in Fig. 4.73. Find the distance *d* across the floor.

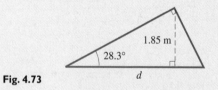

Fig. 4.73

70. The impedance *Z* and resistance *R* in an alternating-current circuit may be represented by letting the impedance be the hypotenuse of a right triangle and the resistance be the side adjacent to the phase angle θ. If $R = 1750 \ \Omega$ and $\theta = 17.38°$, find *Z*.

71. A typical aqueduct built by the Romans dropped on average at an angle of about 0.03° to allow gravity to move the water from the source to the city. For such an aqueduct of 65 km in length, how much higher was the source than the city?

72. A Coast Guard boat 2.75 km from a straight beach can travel at 37.5 km/h. By traveling along a line that is at 69.0° with the beach, how long will it take it to reach the beach? See Fig. 4.74.

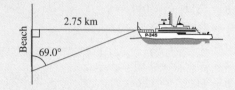

Fig. 4.74

73. In the structural support shown in Fig. 4.75, find *x*.

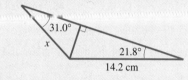

Fig. 4.75

74. A person standing on a level plain hears the sound of a plane, looks in the direction of the sound, but the plane is not there (familiar?). When the sound was heard, it was coming from a point at an angle of elevation of 25°, and the plane was traveling at 720 km/h (201 m/s) at a constant altitude of 850 m along a straight line. If the plane later passes directly over the person, at what angle of elevation should the person have looked directly to see the plane when the sound was heard? (The speed of sound is 340 m/s.) See Fig. 4.76.

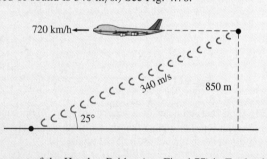

Fig. 4.76

75. The main span of the Humber Bridge (see Fig. 4.77) in England is 1410 m long. The angle *subtended* by the span at the eye of an observer in a helicopter is 2.2°. Show that the distance calculated from the helicopter to the span is about the same if the line of sight is perpendicular to the end or to the middle of the span.

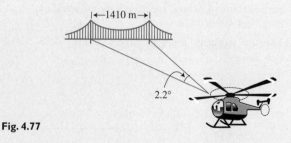

Fig. 4.77

76. Each side piece of the trellis shown in Fig. 4.78 makes an angle of 80.0° with the ground. Find the length of each side piece and the area covered by the trellis.

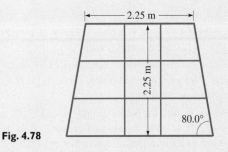

Fig. 4.78

W **77.** The surface of a soccer ball consists of 20 regular hexagons (six sides) interlocked around 12 regular pentagons (five sides). See Fig. 4.79. (a) If the side of each hexagon and pentagon is 45.0 mm, what is the surface area of the soccer ball? (b) Find the surface area, given that the diameter of the ball is 222 mm, (c) Assuming that the given values are accurate, account for the difference in the values found in parts (a) and (b).

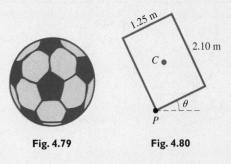

Fig. 4.79   **Fig. 4.80**

78. Through what angle θ must the crate shown in Fig. 4.80 be tipped in order that its center of gravity C be directly above the pivot point P?

79. A laser beam is transmitted with a "width" of 0.002 00°. What is the diameter of a spot of the beam on an object 52 500 km distant? See Fig. 4.81.

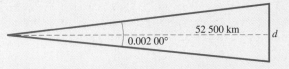

Fig. 4.81

80. Find the gear angle θ in Fig. 4.82 if $t = 0.180$ cm.

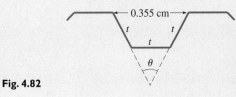

Fig. 4.82

81. A hang glider is directly above the shore of a lake. An observer on a hill is 375 m along a straight line from the shore. From the observer, the angle of elevation of the hang glider is 42.0°, and the angle of depression of the shore is 25.0°. How far above the shore is the hang glider?

82. A ground observer sights a weather balloon to the east at an angle of elevation of 15.0°. A second observer 2.35 km to the east of the first also sights the balloon to the east at an angle of elevation of 24.0°. How high is the balloon? See Fig. 4.83.

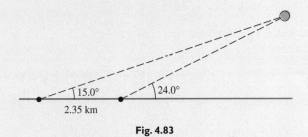

Fig. 4.83

83. A uniform strip of wood 5.0 cm wide frames a trapezoidal window, as shown in Fig. 4.84. Find the left dimension l of the outside of the frame.

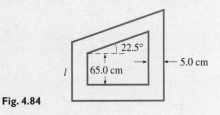

Fig. 4.84

84. A crop-dusting plane flies over a level field at a height of 8.0 m. If the dust leaves the plane through a 30° angle and hits the ground after the plane travels 25 m, how wide a strip is dusted? See Fig. 4.85.

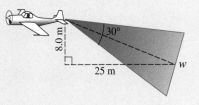

Fig. 4.85

Writing Exercise

85. Two students were discussing a problem in which a distant object was seen through an angle of 2.3°. One found the length of the object (perpendicular to the line of sight) using the tangent of the angle, and the other had the same answer using the sine of the angle. Write a paragraph explaining how this is possible.

CHAPTER ④ PRACTICE TEST

1. Express 37°39′ in decimal form.

2. Find θ to the nearest 0.01° if $\cos \theta = 0.3726$.

3. A ship's captain, desiring to travel due south, discovers that, due to an improperly functioning instrument, the ship has gone 22.62 km in a direction 4.05° east of south. How far from its course (to the east) is the ship?

4. Find $\tan \theta$ in fractional form if $\sin \theta = \dfrac{2}{3}$.

5. Find $\csc \theta$ if $\tan \theta = 1.294$.

6. Solve the right triangle in Fig. 4.86 if $A = 37.4°$ and $b = 52.8$.

7. Solve the right triangle in Fig. 4.86 if $a = 2.49$ and $c = 3.88$.

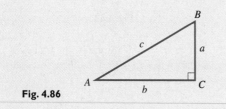

Fig. 4.86

8. The equal sides of an isosceles triangle are each 12.0, and each base angle is 42.0°. What is the length of the third side?

9. If $\tan \theta = 9/40$, find values of $\sin \theta$ and $\cos \theta$. Then evaluate $\sin \theta / \cos \theta$.

10. In finding the wavelength λ (the Greek letter lambda) of light, the equation $\lambda = d \sin \theta$ is used. Find λ if $d = 30.05$ μm and $\theta = 1.167°$. (μ is the prefix for 10^{-6}.)

11. Determine the trigonometric functions of an angle in standard position if its terminal side passes through $(5, 2)$. Give answers in exact and decimal forms.

12. A surveyor sights two points directly ahead. Both are at an elevation 18.525 m lower than the observation point. How far apart are the points if the angles of depression are 13.500° and 21.375°, respectively? See Fig. 4.87.

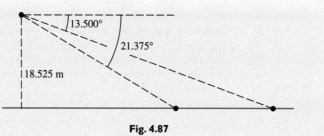

Fig. 4.87

5

Systems of Linear Equations; Determinants

As knowledge about electric circuits was first developing, in 1848 the German physicist Gustav Kirchhoff formulated what are now known as *Kirchhoff's current law* and *Kirchhoff's voltage law*. These laws are still widely used today, and in using them more than one equation is usually set up. To find the needed information about the circuit, it is necessary to find solutions that satisfy all equations at the same time. We will show the solution of circuits using Kirchhoff's laws in some of the exercises later in the chapter. Although best known for these laws of electric circuits, Kirchhoff is also credited in the study of optics as a founder of the modern chemical process known as *spectrum analysis*.

The correct amounts of acid solutions are needed to make a solution of a certain strength. In Section 5.7, we see how a system of equations is used to find these amounts in a certain case.

Methods of solving such *systems* of equations were well known to Kirchhoff, and this allowed the study of electricity to progress rapidly. In fact, just 100 years earlier a book by the English mathematician Colin Maclaurin was published (2 years after his death), in which he covered many topics in algebra in an organized and systematic way. In doing so, he also presented a general method of solving systems of equations. This method is now called *Cramer's rule* (named for the Swiss mathematician Gabriel Cramer, who popularized it in a book he wrote in 1750.) We will explain some of the methods of solution, including Cramer's rule, later in the chapter.

Two or more equations that relate variables are found in many fields of science and technology. These include aeronautics, business, transportation, the analysis of forces on a structure, medical doses, and robotics, as well as electric circuits. These applications often require solutions that satisfy all equations at the same time.

In this chapter, we restrict our attention to *linear* equations (variables occur only to the first power). We will consider systems of two equations with two unknowns and systems of three equations with three unknowns. Systems with other kinds of equations and systems with more unknowns are taken up later in the book.

5.1 LINEAR EQUATIONS

In general, *an equation is termed* **linear** *in a given set of variables if each term contains only one variable, to the first power, or is a constant.*

◀ EXAMPLE 1 $5x - t + 6 = 0$ is linear in x and t, but $5x^2 - t + 6 = 0$ is not linear, due to the presence of x^2.

The equation $4x + y = 8$ is linear in x and y, but $4xy + y = 8$ is not, due to the presence of xy.

The equation $x - 6y + z - 4w = 7$ is linear in x, y, z, and w, but the equation $x - \frac{6}{y} + z - 4w = 7$ is not, due to the presence of $\frac{6}{y}$, where y appears in the denominator.   ▶

An equation that can be written in the form

$$ax + b = 0 \qquad\qquad (5.1)$$

is known as a **linear equation in one unknown.** Here, a and b are constants. We discussed this type of equation in Section 1.10. *The* **solution,** *or* **root,** *of the equation is* $x = -b/a$. We can also see that the solution is the same as the *zero* of the **linear function** $f(x) = ax + b$.

◀ EXAMPLE 2   The equation $2x + 7 = 0$ is a linear equation of the form of Eq. (5.1), with $a = 2$ and $b = 7$.

The solution to the linear equation $2x + 7 = 0$, or the zero of the linear function $f(x) = 2x + 7$, is $-7/2$. From Chapter 3, recall that the *zero* of a function $f(x)$ is the value of x for which $f(x) = 0$.   ▶

In the next example, two specific illustrations are given of applied problems that involve more than one unknown.

See the chapter introduction.

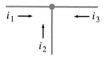

Fig. 5.1

◀ EXAMPLE 3   **(a)** A basic law of direct-current electricity, known as *Kirchhoff's current law,* may be stated as "The algebraic sum of the currents entering any junction in a circuit is zero." If three wires are joined at a junction as in Fig. 5.1, this law leads to the linear equation

$$i_1 + i_2 + i_3 = 0$$

where i_1, i_2, and i_3 are the currents in each of the wires. (Either one or two of these currents must have a negative sign, showing that it is actually leaving the junction.)

(b) Two forces, F_1 and F_2, acting on a beam might be related by the equation

$$2F_1 + 4F_2 = 200$$ ▶

An equation that can be written in the form

$$ax + by = c \qquad\qquad (5.2)$$

NOTE ▶

is a **linear equation in two unknowns.** For such equations, in Chapter 3 we found that for each value of x there is a corresponding value of y. Each of these pairs of numbers is a *solution* to the equation. *A* **solution** *is any set of numbers, one for each variable, that satisfies the equation.*

When we represent the solutions of a linear equation in two unknowns in the form of a graph, we find that the graph is a straight line. Thus, we see the reason for the name *linear*. In the next section, we will further study the graphs of linear equations.

◀ EXAMPLE 4   The equation $2x - y - 4 = 0$ is a linear equation in two unknowns, x and y, since we can write it in the form of Eq. (5.2) as $2x - y = 4$. To graph this equation, we can write it as $y = 2x - 4$. From Fig. 5.2, we see that the graph is a straight line.

The coordinates of any point on the line give us a solution of this equation. For example, the point $(1, -2)$ is on the line. This means that $x = 1$, $y = -2$ is a solution of the equation. We can show that these values are a solution by substituting in the equation $2x - y - 4 = 0$. This gives us

$$2(1) - (-2) - 4 = 0, \qquad 2 + 2 - 4 = 0, \qquad 0 = 0$$

Since we have equality, $x = 1$, $y = -2$ is a solution. In the same way, we can show that $x = 3$, $y = 2$ is a solution.

If we are given the value of one variable, we find the value of the other variable, which gives a solution by substitution. For example, for the equation $2x - y - 4 = 0$, if $x = -1$, we have

$$2(-1) - y - 4 = 0$$
$$-2 - y - 4 = 0$$
$$y = -6$$

Therefore, $x = -1$, $y = -6$ is a solution, and the point $(-1, -6)$ is on the graph.

◀

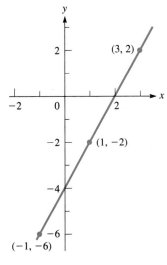

Fig. 5.2

Two linear equations, each containing the same two unknowns,

$$\begin{aligned} a_1 x + b_1 y &= c_1 \\ a_2 x + b_2 y &= c_2 \end{aligned} \qquad (5.3)$$

are said to form a **system of simultaneous linear equations.** *A solution of the system is any pair of values* (x, y) *that satisfies both equations.* Methods of finding the solutions to such systems are the principal concern of this chapter.

◀ EXAMPLE 5   The perimeter of the piece of computer paper shown in Fig. 5.3 is 104 cm. The length is 4 cm more than the width.

From the two statements, we can set up two equations in the two unknown quantities, the length and the width. Letting l = the length and w = the width, we have

$$2l + 2w = 104$$
$$l - w = 4$$

as a system of simultaneous linear equations. The solution of this system is $l = 28$ cm and $w = 24$ cm. These values satisfy both equations since

$$2(28) + 2(24) = 104 \quad \text{and} \quad 28 - 24 = 4$$

This is the only pair of values that satisfies *both* equations. Methods for finding such solutions are taken up in later sections of this chapter.

◀

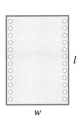

Fig. 5.3

EXERCISES 5.1

In Exercises 1–4, answer the given questions about the indicated examples of this section.

1. In the second line of the third paragraph of Example 1, if $\frac{6}{y}$ is replaced by its reciprocal, is there any change in the conclusion drawn?

2. In the first line of Example 2, if $+$ is changed to $-$, what changes result in the rest of the example?

3. In Example 4, what is the solution if $x = 4$?

4. In the second line of Example 5, if 4 cm is replaced with 6 cm, what is the solution?

In Exercises 5–8, determine whether or not the given pairs of values are solutions of the given linear equations in two unknowns.

5. $2x + 3y = 9$; $(3, 1), \left(5, \frac{1}{3}\right)$

6. $5x + 2y = 1$; $(2, -4), (1, -2)$

7. $-3x + 5y = 13$; $(-1, 2), (4, 5)$

8. $x - 4y = 10$; $(2, -2), (2, 2)$

In Exercises 9–14, for each given value of x, determine the value of y that gives a solution to the given linear equations in two unknowns.

9. $3x - 2y = 12$; $x = 2, x = -3$

10. $-5x + 6y = 60$; $x = -10, x = 8$

11. $x - 4y = 2$; $x = 3, x = -0.4$

12. $3x - 2y = 9$; $x = \frac{2}{3}, x = -3$

13. $4x - 9y = 8$; $x = 2/3, x = -1/2$

14. $2.4y - 4.5x = -3.0$; $x = -0.4, x = 2.0$

In Exercises 15–24, determine whether or not the given pair of values is a solution of the given system of simultaneous linear equations.

15. $x - y = 5$ $x = 4, y = -1$
$2x + y = 7$

16. $2x + y = 8$ $x = -1, y = 10$
$3x - y = -13$

17. $A + 5B = -7$ $A = -2, B = 1$
$3A - 4B = -4$

18. $-3x + y = 1$ $x = \frac{1}{3}, y = 2$
$6x - 3y = -4$

19. $2x - 5y = 0$ $x = \frac{1}{2}, y = -\frac{1}{5}$
$4x + 10y = 4$

20. $6i_1 + i_2 = 5$ $i_1 = 1, i_2 = -1$
$3i_1 - 4i_2 = -1$

21. $3x - 2y = 2.2$ $x = 0.6, y = -0.2$
$5x + y = 2.8$

22. $x - 7y = -3.2$ $x = -1.1, y = 0.3$
$2x + y = 2.5$

23. $30x - 21y = -675$ $x = -12, y = 15$
$42x + 15y = -279$

24. $3.8a + 7.5b = -26.5$ $a = 2.5, b = -4.8$
$2.6a - 12.6b = -51.1$

In Exercises 25–28, answer the given questions.

25. In planning a search pattern from an aircraft carrier, a pilot plans to fly at p km/h relative to a wind that is blowing at w km/h. Traveling with the wind, the ground speed would be 300 km/h, and against the wind the ground speed would be 220 km/h. This leads to two equations:

$p + w = 300$

$p - w = 220$

Are the speeds 260 km/h and 40 km/h?

26. The electric resistance R of a certain resistor is a function of the temperature T given by the equation $R = aT + b$, where a and b are constants. If $R = 1200 \, \Omega$ when $T = 10.0°C$ and $R = 1280 \, \Omega$ when $T = 50.0°C$, we can find the constants a and b by substituting and obtaining the equations

$1200 = 10.0a + b$

$1280 = 50.0a + b$

Are the constants $a = 4.00 \, \Omega/°C$ and $b = 1160 \, \Omega$?

27. The forces acting on part of a structure are shown in Fig. 5.4. An analysis of the forces leads to the equations

$0.80F_1 + 0.50F_2 = 50$

$0.60F_1 - 0.87F_2 = 12$

Are the forces 45 N and 28 N?

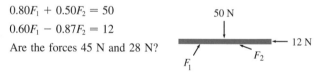

Fig. 5.4

28. A student earned $4000 during the summer and decided to put half into a retirement plan. If the plan was invested in two accounts earning 4.0% and 5.0%, the total income for the first year is $92. The equations to determine the amount x and y are

$x + y = 2000$

$0.040x + 0.050y = 92$

Are the amounts $x = \$1200$ and $y = \$800$?

5.2 GRAPHS OF LINEAR FUNCTIONS

The first method of solving a system of equations that will be studied is a graphical method. Before taking up this method in the next section, we will develop additional ways of graphing a linear equation. We will then be able to analyze and check the graph of a linear equation quickly, often by inspection.

Consider the line that passes through points A, with coordinates (x_1, y_1), and B, with coordinates (x_2, y_2), in Fig. 5.5. Point C is horizontal from A and vertical from B. Thus, C has coordinates (x_2, y_1), and there is a right angle at C. One way of measuring the steepness of this line is to find the ratio of the vertical distance to the horizontal distance between two points. Therefore, *we define the* **slope** *of the line through two points as the difference in the y-coordinates divided by the difference in the x-coordinates.* For points A and B, the slope m is

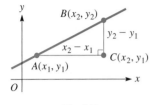

Fig. 5.5

Slope

$$m = \frac{y_2 - y_1}{x_2 - x_1} \qquad (5.4)$$

The slope is often referred to as the *rise* (vertical change) over the *run* (horizontal change). Note that the slope of a vertical line, for which $x_2 = x_1$, is undefined (for $x_2 = x_1$, the denominator of Eq. (5.4) is zero).

◀ **EXAMPLE 1** Find the slope of the line through the points $(2, -3)$ and $(5, 3)$.

In Fig. 5.6, we draw the line through the two given points. By taking $(5, 3)$ as (x_2, y_2), then (x_1, y_1) is $(2, -3)$. We may choose either point as (x_2, y_2), but *once the choice is made the order must be maintained.* Using Eq. (5.4), the slope is

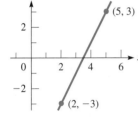

Fig. 5.6

$$m = \frac{3 - (-3)}{5 - 2}$$

$$= \frac{6}{3} = 2$$

The rise is 2 units for each unit (of run) in going from left to right. ▶

◀ **EXAMPLE 2** Find the slope of the line through $(-1, 2)$ and $(3, -1)$.

In Fig. 5.7, we draw the line through these two points. By taking (x_2, y_2) as $(3, -1)$ and (x_1, y_1) as $(-1, 2)$, the slope is

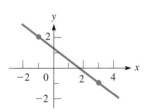

Fig. 5.7

$$m = \frac{-1 - 2}{3 - (-1)}$$

$$= \frac{-3}{3 + 1} = -\frac{3}{4}$$

The line *falls* 3 units for each 4 units in going from left to right. ▶

We note in Example 1 that as *x increases, y increases and that slope is* **positive.** In Example 2, *as x increases, y decreases and that slope is* **negative.** Also, *the larger the absolute value of the slope, the steeper is the line.*

◀ EXAMPLE 3   For each of the following lines shown in Fig. 5.8, we show the difference in the y-coordinates and in the x-coordinates between two points.

In Fig. 5.8(a), a line with a slope of 5 is shown. It rises sharply.

In Fig. 5.8(b), a line with a slope of $\frac{1}{2}$ is shown. It rises slowly.

In Fig. 5.8(c), a line with a slope of -5 is shown. It falls sharply.

In Fig. 5.8(d), a line with a slope of $-\frac{1}{2}$ is shown. It falls slowly.

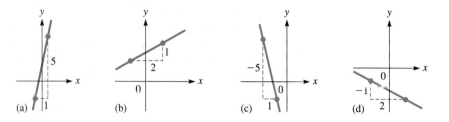

Fig. 5.8   (a)   (b)   (c)   (d)   ▶

SLOPE-INTERCEPT FORM OF THE EQUATION OF A STRAIGHT LINE

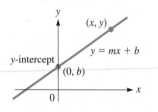

Fig. 5.9

We now show how the slope is related to the equation of a straight line. In Fig. 5.9, if we have two points, $(0, b)$ and a general point (x, y), the slope is

$$m = \frac{y - b}{x - 0}$$

Simplifying this, we have $mx = y - b$, or

$$\boxed{y = mx + b} \tag{5.5}$$

NOTE ▶

In Eq. (5.5), m is the slope, and b is the y-coordinate of the point where the line crosses the y-axis. *This point is the **y-intercept** of the line, and its coordinates are $(0, b)$. Equation (5.5) is the **slope-intercept** form of the equation of a straight line. The coefficient of x is the slope, and the constant is the ordinate of the y-intercept.* The point $(0, b)$ and simply b are both referred to as the y-intercept.

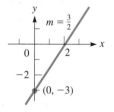

Fig. 5.10

◀ EXAMPLE 4   Find the slope and the y-intercept of the line $y = \frac{3}{2}x - 3$. Since the coefficient of x is $\frac{3}{2}$, this means that the slope is $\frac{3}{2}$. Also, because we can write

$$y = \overset{\text{slope}}{\underset{\downarrow}{\frac{3}{2}}}x + \overset{\text{y-intercept ordinate}}{\underset{\downarrow}{(-3)}}$$

we see that the constant is -3, which means the y-intercept is the point $(0, -3)$. The line is shown in Fig. 5.10.   ▶

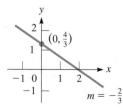

Fig. 5.11

◀ EXAMPLE 5   Find the slope and the y-intercept of the line $2x + 3y = 4$.
We must first write the equation in slope-intercept form. Solving for y, we have

$$y = \overset{\text{slope}}{\underset{\downarrow}{-\frac{2}{3}}}x + \overset{\text{y-intercept ordinate}}{\underset{\downarrow}{\frac{4}{3}}}$$

Therefore, the slope is $-\frac{2}{3}$, and the y-intercept is the point $\left(0, \frac{4}{3}\right)$. See Fig. 5.11.   ▶

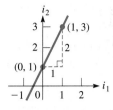

Fig. 5.12

◀ EXAMPLE 6   In analyzing an electric current, two of the currents, i_1 and i_2, were related by $i_2 = 2i_1 + 1$. Graph this equation using the slope-intercept form.

Since the equation is solved for i_2, we treat i_1 as the independent variable and i_2 as the dependent variable. This means that the slope of the line is 2. Also, since the b-term is 1, we can see that the intercept is $(0, 1)$.

We can use this information to sketch the line, as shown in Fig. 5.12. Since the slope is 2, we know that i_2 increases 2 units for each increase of i_1. Thus, starting at the i_2-intercept $(0, 1)$, if i_1 increases by 1, i_2 increases by 2, and we are at the point $(1, 3)$. The line must pass through $(1, 3)$, as well as $(0, 1)$. Therefore, we draw the line through these points. (Negative values may be used for electric currents since the sign shows the direction of flow.) ◗

SKETCHING LINES BY INTERCEPTS

Another way of sketching the graph of a straight line is to find two points on the line and then draw the line through these points. Two points that are easily determined are those where the line crosses the y-axis and the x-axis. We already know that the point where it crosses the y-axis is the y-intercept. In the same way, *the point where it crosses* **NOTE ▶** *the x-axis is called the **x-intercept,*** and the coordinates of the x-intercept are $(a, 0)$.

The intercepts are easily found because one of the coordinates is zero. By setting $x = 0$ and $y = 0$, in turn, and finding the value of the other variable, we get the coordinates of the intercepts. This method works except when both intercepts are at the origin, and we must find one other point, or use the slope-intercept method. In using the intercept method, a third point should be found as a check. The next example shows how a line is sketched by finding its intercepts.

◀ EXAMPLE 7   Sketch the graph of the line $2x - 3y = 6$ by finding its intercepts and one check point. See Fig. 5.13.

First, we let $x = 0$. This gives us $-3y = 6$, or $y = -2$. This gives us the y-intercept, which is the point $(0, -2)$. Next we let $y = 0$, which gives us $2x = 6$, or $x = 3$. This means the x-intercept is the point $(3, 0)$.

The intercepts are enough to sketch the line as shown in Fig. 5.13. To find a check point, we can use any value for x other than 3 or any value of y other than -2. Choosing $x = 1$, we find that $y = -\frac{4}{3}$. This means that the point $\left(1, -\frac{4}{3}\right)$ should be on the line. In Fig. 5.13, we can see that it is on the line.

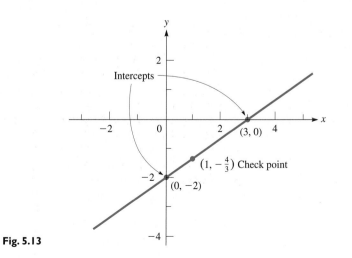

Fig. 5.13

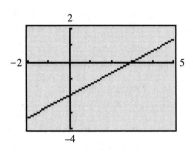

Fig. 5.14

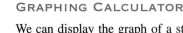

Fig. 5.15

DISPLAYING LINES ON A GRAPHING CALCULATOR

We can display the graph of a straight line on a graphing calculator, just as we did in Section 3.5. In order to enter the equation into the calculator, we must first solve for y. In doing so, the equation will often be in slope-intercept form or be easily written in this form.

Good choices for the *window* settings can be made by using the y-intercept and the slope. The y-intercept tells us a value of y that we usually want displayed, and the slope tells us how the line is directed from that point. Using the y-intercept along with the x-intercept should also lead to good choices for the settings.

◀ EXAMPLE 8   **(a)** Use a graphing calculator to display the graph of the equation in Example 7.

First, solve the equation $2x - 3y = 6$ for y. This gives us $y = \frac{2}{3}x - 2$. Therefore, on the calculator we enter $y_1 = 2x/3 - 2$ and use the *window* settings shown in Fig. 5.14. Note that the graph is the same as that in Fig. 5.13.

(b) Use a graphing calculator to display the graph of the equation $4x + 5y = 100$.

Solving for y, we get $y = -\frac{4}{5}x + 20$, and therefore in the calculator we enter the function $y_1 = -4x/5 + 20$. We also note immediately that the y-intercept is $(0, 20)$, which means we must be careful in choosing the *window* settings. We can also determine that the x-intercept is $(25, 0)$. Therefore, we have the settings and graph shown in Fig. 5.15.   ▶

Further details and discussion of slope and the graphs of linear equations are found in Chapter 21.

EXERCISES 5.2

In Exercises 1–4, answer the given questions about the indicated examples of this section.

1. In Example 2, if the first y-coordinate is changed to -2, what changes result in the example?

2. In the first line of Example 5, if $+$ is changed to $-$, what changes result in the example?

3. In the first line of Example 7, if $-$ is changed to $+$, what changes occur in the graph?

4. In the first line of Example 8(b), if $4x + 5y$ is changed to $5x - 4y$, what changes occur in the graph?

In Exercises 5–12, find the slope of the line that passes through the given points.

5. $(1, 0), (3, 8)$

6. $(3, 1), (2, 7)$

7. $(-1, 2), (-4, 17)$

8. $(-1, -2), (2, 10)$

9. $(5, -3), (-2, -5)$

10. $(3, -4), (-7, -4)$

11. $(0.4, 0.5), (-0.2, 0.2)$

12. $(-2.8, 3.4), (1.2, 4.2)$

In Exercises 13–20, sketch the line with the given slope and y-intercept.

13. $m = 2, (0, -1)$

14. $m = 3, (0, 1)$

15. $m = 0, (0, 2)$

16. $m = -4, (0, -2)$

17. $m = \frac{1}{2}, (0, 0)$

18. $m = \frac{2}{3}, (0, -1)$

19. $m = -9, (0, 20)$

20. $m = -0.3, (0, -1.4)$

In Exercises 21–28, find the slope and the y-intercept of the line with the given equation and sketch the graph using the slope and the y-intercept. A graphing calculator can be used to check your graph.

21. $y = -2x + 1$

22. $y = -4x$

23. $y = x + 4$

24. $y = \frac{4}{5}x + 2$

25. $5x - 2y = 40$

26. $-2y = 7$

27. $24x + 40y = 15$

28. $1.5x - 2.4y = 3.0$

In Exercises 29–36, find the x-intercept and the y-intercept of the line with the given equation. Sketch the line using the intercepts. A graphing calculator can be used to check the graph.

29. $x + 2y = 4$

30. $3x + y = 3$

31. $4x - 3y = 12$

32. $x - 5y = 5$

33. $y = 3x + 6$

34. $y = -2x - 4$

35. $y = -12x + 30$

36. $y = 0.25x + 4.5$

In Exercises 37–40, sketch the indicated lines.

37. The diameter of the large end, d (in cm), of a certain type of machine tool can be found from the equation $d = 0.2l + 1.2$, where l is the length of the tool. Sketch d as a function of l, for values of l to 10 cm. See Fig. 5.16.

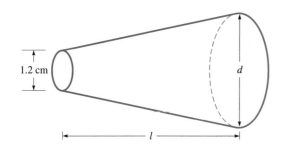

Fig. 5.16

38. In mixing 85 octane gasoline and 93 octane gasoline to produce 91 octane gasoline, the equation $0.85x + 0.93y = 910$ is used. Sketch the graph.

39. Two electric currents, I_1 and I_2 (in mA), in part of a circuit in a computer are related by the equation $4I_1 - 5I_2 = 2$. Sketch I_2 as a function of I_1. These currents can be negative.

40. In 2000, about 9.7×10^9 email messages were sent, and in 2004 about 30.1×10^9 email messages were sent. Assuming that the growth in email messages is linear, set up a function relating the number N of email messages and the time t (in years). Let $t = 0$ be the year 2000. Sketch the graph.

5.3 SOLVING SYSTEMS OF TWO LINEAR EQUATIONS IN TWO UNKNOWNS GRAPHICALLY

Since a solution of a system of simultaneous linear equations in two unknowns is any pair of values (x, y) that satisfies *both* equations, graphically *the solution would be the coordinates of the point of intersection of the two lines.* This must be the case, for the coordinates of this point constitute the only pair of values to satisfy *both* equations. (In some special cases, there may be no solution; in others, there may be many solutions. See Examples 5 and 6.)

Therefore, when we solve two simultaneous linear equations in two unknowns graphically, *we must graph each line and determine the point of intersection.* This may, of course, lead to *approximate results* if the lines cross at points not used to determine the graph.

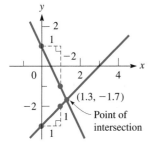

Fig. 5.17

◀ **EXAMPLE 1** Solve the system of equations

$$y = x - 3$$
$$y = -2x + 1$$

Since each of the equations is in slope-intercept form, note that $m = 1$ and $b = -3$ for the first line and that $m = -2$ and $b = 1$ for the second line. Using these values, we sketch the lines, as shown in Fig. 5.17.

From the figure, it can be seen that *the lines cross at about the point* $(1.3, -1.7)$. This means that the solution is approximately

$$x = 1.3 \qquad y = -1.7$$

(The exact solution is $x = \frac{4}{3}$, $y = -\frac{5}{3}$.)

NOTE ▶ In checking the solution, be careful to substitute the values in *both* equations. Making these substitutions gives us

$$-1.7 \overset{?}{=} 1.3 - 3 \quad \text{and} \quad -1.7 \overset{?}{=} -2(1.3) + 1$$
$$= -1.7 \qquad\qquad \approx -1.6$$

These values show that the solution checks. (The point $(1.3, -1.7)$ is *on* the first line and *almost on* the second line. The difference in values when checking the values for the second line is due to the fact that the solution is *approximate*.) ▶

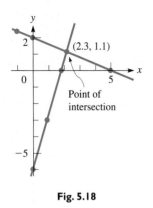

Fig. 5.18

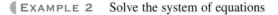

EXAMPLE 2 Solve the system of equations

$$2x + 5y = 10$$
$$3x - y = 6$$

We could write each equation in slope-intercept form in order to sketch the lines. Also, we could use the form in which they are written to find the intercepts. Choosing to find the intercepts and draw lines through them, let $y = 0$; then $x = 0$. Therefore, the intercepts of the first line are the points $(5, 0)$ and $(0, 2)$. A third point is $\left(-1, \frac{12}{5}\right)$. The intercepts of the second line are $(2, 0)$ and $(0, -6)$. A third point is $(1, -3)$. Plotting these points and drawing the proper straight lines, we see that the lines cross at about $(2.3, 1.1)$. $\left[\text{The exact values are } \left(\frac{40}{17}, \frac{18}{17}\right).\right]$ The solution of the system of equations is approximately $x = 2.3$, $y = 1.1$ (see Fig. 5.18).

Checking, we have

$$2(2.3) + 5(1.1) \stackrel{?}{=} 10 \quad \text{and} \quad 3(2.3) - 1.1 \stackrel{?}{=} 6$$
$$10.1 \approx 10 \qquad\qquad\qquad 5.8 \approx 6$$

This shows the solution is correct to the accuracy we can get from the graph.

SOLVING SYSTEMS OF EQUATIONS USING A GRAPHING CALCULATOR

A graphing calculator can be used to find the point of intersection with much greater accuracy than is possible by hand-sketching the lines. Once we locate the point of intersection, the features of the calculator allow us to get the accuracy we need.

EXAMPLE 3 Using a graphing calculator, we now solve the systems of equations in Examples 1 and 2.

Solving the system for Example 1, let $y_1 = x - 3$ and $y_2 = -2x + 1$. Then display the lines as shown in Fig. 5.19(a). Then using the *trace* and *zoom* features, or the *intersect* feature, of a graphing calculator, we find (to the nearest 0.001) that the solution is $x = 1.333$, $y = -1.667$.

Solving the system for Example 2, let $y_1 = -2x/5 + 2$ and $y_2 = 3x - 6$. Then display the lines as shown in Fig. 5.19(b). Then using the *trace* and *zoom* features, or the *intersect* feature, we find (to the nearest 0.001) that the solution is $x = 2.353$, $y = 1.059$.

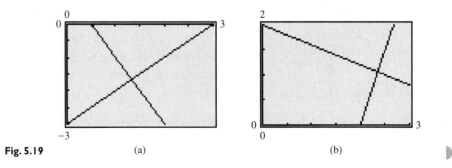

Fig. 5.19   (a)   (b)

WORD PROBLEMS INVOLVING TWO LINEAR EQUATIONS

Linear equations in two unknowns are often useful in solving word problems. Just as in Section 1.12, *we must read the statement carefully in order to identify the unknowns and the information for setting up the equations.*

Solving a Word Problem

EXAMPLE 4 A driver traveled for 1.5 h at a constant speed along a highway. Then, through a construction zone, the driver reduced the car's speed by 40 km/h for 30 min. If 200 km were covered in the 2.0 h, what were the two speeds?

First, let v_h = the highway speed and v_c = the speed in the construction zone. Two equations are found by using

1. distance = rate × time $\left[\text{for units, km} = \left(\tfrac{\text{km}}{\text{h}}\right)\text{h}\right]$, and

2. the fact that "the driver reduced the car's speed by 40 km/h for 30 min."

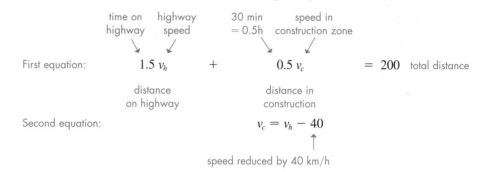

First equation: time on highway (1.5 v_h) + highway speed + 30 min = 0.5h + speed in construction zone (0.5 v_c) = 200 total distance

distance on highway   distance in construction

Second equation: $v_c = v_h - 40$

speed reduced by 40 km/h

Using v_h as the independent variable and v_c as the dependent variable, the sketch is shown in Fig. 5.20(a). With $v_c = y$ and $v_h = x$, we have the calculator display shown in Fig. 5.20(b) with the indicated settings, and $y_1 = -3x + 400$, $y_2 = x - 40$.

We see that the point of intersection is $(110, 70)$, which means that the solution is $v_h = 110$ km/h and $v_c = 70$ km/h. Checking *in the statement of the problem,* we have $(1.5 \text{ h})(110 \text{ km/h}) + (0.5 \text{ h})(70 \text{ km/h}) = 200$ km.

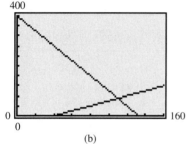

$v_c = -3v_h + 400$
$m = -3, b = 400$

$v_c = v_h - 40$
$m = 1, \ b = -40$

(110, 70)

(a)

(b)

Fig. 5.20

INCONSISTENT AND DEPENDENT SYSTEMS

The lines of each system in the previous examples intersect in a single point, and each system has *one* solution. Such systems are called *consistent* and *independent.* Most systems that we will encounter have just one solution. However, as we now show, *not all systems have just one such solution.* The following examples illustrate a system that has *no solution* and a system that has an *unlimited number of solutions.*

EXAMPLE 5 Solve the system of equations

$$x = 2y + 6$$
$$6y = 3x - 6$$

Writing each of these equations in slope-intercept form (Eq. 5.5), we have for the first equation

$$y = \tfrac{1}{2}x - 3$$

For the second equation, we have

$$y = \tfrac{1}{2}x - 1$$

From these, we see that each line has a slope of $\tfrac{1}{2}$ and that the y-intercepts are $(0, -3)$ and $(0, -1)$. Therefore, we know that the y-intercepts are different, but the *slopes are the same.* Since the slope indicates that each line rises $\tfrac{1}{2}$ unit for y for each unit x increases, *the lines are parallel and do not intersect,* as shown in Fig. 5.21. This means that *there are no solutions* for this system of equations. *Such a system is called* **inconsistent.**

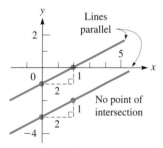

Lines parallel

No point of intersection

Fig. 5.21

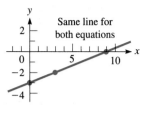

Fig. 5.22

❮ EXAMPLE 6   Solve the system of equations

$$x - 3y = 9$$
$$-2x + 6y = -18$$

We find that the intercepts and a third point for the first line are $(9, 0)$, $(0, -3)$, and $(3, -2)$. In determining the intercepts for the second line, we find that they are $(9, 0)$ and $(0, -3)$, which are also the intercepts of the first line. As a check, we find that the point $(3, -2)$ also satisfies the equation of the second line. This means that the two lines are really the same line. Another check is to write each of these equations in slope-intercept form. This gives us

$$y = \tfrac{1}{3}x - 3$$

for each equation. See Fig. 5.22.

Since the lines are the same, *the coordinates of any point on this common line constitute a solution of the system.* Since **no unique solution can be determined,** the system is called **dependent.**   ❯

Later in the chapter, we will show algebraic ways of finding out whether a given system is consistent, inconsistent, or dependent.

EXERCISES 5.3

In Exercises 1 and 2, answer the given questions about the indicated examples of this section.

1. In the second equation of Example 2, if $-$ is replaced with $+$, what is the solution?

2. In Example 6, by changing what one number in the first equation does the system become (a) inconsistent? (b) Consistent?

In Exercises 3–20, solve each system of equations by sketching the graphs. Use the slope and the y-intercept or both intercepts. Estimate each result to the nearest 0.1 if necessary.

3. $y = -x + 4$
 $y = x - 2$

4. $y = \tfrac{1}{2}x - 1$
 $y = -x + 8$

5. $y = 2x - 6$
 $y = -\tfrac{1}{3}x + 1$

6. $y = \tfrac{1}{2}x - 4$
 $y = 2x + 2$

7. $3x + 2y = 6$
 $x - 3y = 3$

8. $4R - 3V = -8$
 $6R + V = 6$

9. $2x - 5y = 10$
 $3x + 4y = -12$

10. $-5x + 3y = 15$
 $2x + 7y = 14$

11. $s - 4t = 8$
 $2s = t + 4$

12. $y = 4x - 6$
 $y = 2x + 4$

13. $y = -x + 3$
 $y = -2x + 3$

14. $p - 6 = 6v$
 $v = 3 - 3p$

15. $x - 4y = 6$
 $2y = x + 4$

16. $x + y = 3$
 $3x - 2y = 14$

17. $-2r_1 + 2r_2 = 7$
 $4r_1 - 2r_2 = 1$

18. $2x - 3y = -5$
 $3x + 2y = 12$

19. $15y = 20x - 48$
 $32x + 21y = 60$

20. $7A = 8B + 17$
 $5B = 12A - 22$

In Exercises 21–32, solve each system of equations to the nearest 0.1 for each variable by using a graphing calculator.

21. $x = 4y + 2$
 $3y = 2x + 3$

22. $1.2x - 2.4y = 4.8$
 $3.0x = -2.0y + 7.2$

23. $4.0x - 3.5y = 1.5$
 $0.7y + 0.1x = 0.7$

24. $5F - 2T = 7$
 $3F + 4T = 8$

25. $x - 5y = 10$
 $2x - 10y = 20$

26. $18x - 3y = 7$
 $2y = 1 + 12x$

27. $1.9v = 3.2t$
 $1.2t - 2.6v = 6$

28. $3y = 14x - 9$
 $12x + 23y = 0$

29. $5x = y + 3$
 $4x = 2y - 3$

30. $0.75u + 0.67v = 5.9$
 $2.1u - 3.9v = 4.8$

31. $7R = 18V + 13$
 $-1.4R + 3.6V = 2.6$

32. $y = 6x + 2$
 $12x - 2y = -4$

In Exercises 33–36, graphically solve the given problems to the stated accuracy.

33. Chains support a crate, as shown in Fig. 5.23. The equations relating tensions T_1 and T_2 are given below. Determine the tensions to the nearest 1 N from the graph.

$$0.8T_1 - 0.6T_2 = 12$$
$$0.6T_1 + 0.8T_2 = 68$$

Fig. 5.23

34. The equations relating the currents i_1 and i_2 shown in Fig. 5.24 are given below. Find the currents to the nearest 0.1 A.

$$2i_1 + 6(i_1 + i_2) = 12$$
$$4i_2 + 6(i_1 + i_2) = 12$$

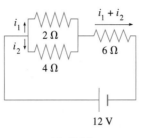

Fig. 5.24

35. An architect designing a parking lot has a row 62.0 m wide to divide into spaces for compact cars and full-size cars. The architect determines that 16 compact car spaces and 6 full-size car spaces use the width, or that 12 compact car spaces and 9 full-size car spaces use all but 0.2 m of the width. What are the widths of the spaces being planned? See Fig. 5.25.

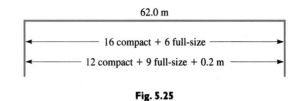

Fig. 5.25

36. A certain car uses 7.2 L/100 km in city driving and 5.4 L/100 km in highway driving. If 36 L of gas are used in traveling 598 km, how many kilometers were driven in the city, and how many were driven on the highway (assuming that only the given rates of usage were actually used)?

5.4 SOLVING SYSTEMS OF TWO LINEAR EQUATIONS IN TWO UNKNOWNS ALGEBRAICALLY

The graphical method of solving two linear equations is good for obtaining a "picture" of the solution. One problem is that graphical methods usually give *approximate* results. If exact solutions are required, we turn to other methods. In this section, we present two algebraic methods of solution.

SOLUTION BY SUBSTITUTION

The first method involves the *elimination of one variable by* **substitution.** The basic idea of this method is to *solve one of the equations for one of the unknowns and substitute this into the other equation.* The result is an equation with only one unknown, and this equation can then be solved for the unknown. Following is the basic procedure to be used.

SOLUTION OF TWO LINEAR EQUATIONS BY SUBSTITUTION

1. *Solve one equation for one of the unknowns.*
2. **Substitute** *this solution into the* **other** *equation.* At this point, we have a linear equation in one unknown.
3. *Solve the resulting equation for the value of the unknown it contains.*
4. *Substitute this value into the equation of step 1 and solve for the other unknown.*
5. *Check the values in* **both original equations.**

In following this procedure, you must first choose an unknown for which to solve. Often it makes little difference, but if it is easier to solve a particular equation for one of the unknowns, that is the one to use.

◀ **EXAMPLE 1**   Solve the following system by substitution.

$$x - 3y = 6$$
$$2x + 3y = 3$$

Here, it is easiest to solve the first equation for x:

step 1
$$x = 3y + 6 \qquad \textbf{(A1)}$$

———— in second equation, x replaced by $3y + 6$

step 2
$$2(3y + 6) + 3y = 3 \qquad \text{substituting}$$

step 3
$$6y + 12 + 3y = 3 \qquad \text{solving for } y$$
$$9y = -9$$
$$y = -1$$

Now, put the value $y = -1$ into the first of the original equations. Since this equation is already solved for x in terms of y, Eq. (A1), we obtain

step 4
$$x = 3(-1) + 6 = 3 \qquad \text{solving for } x$$

step 5
Therefore, the solution of the system is $x = 3$, $y = -1$. As a check, substitute these values into each of the original equations. This gives us $3 - 3(-1) = 6$ and $2(3) + 3(-1) = 3$, which verifies the solution. ▶

◀ **EXAMPLE 2**   Solve the following system by substitution.

$$-5x + 2y = -4$$
$$10x + 6y = 3$$

It makes little different which equation or which unknown is chosen. Therefore,

$$2y = 5x - 4 \qquad \text{solving first equation for } y$$
$$y = \frac{5x - 4}{2} \qquad \textbf{(B1)}$$

———— in second equation, y replaced by $\dfrac{5x - 4}{2}$

$$10x + 6\left(\frac{5x - 4}{2}\right) = 3 \qquad \text{substituting}$$

$$10x + 3(5x - 4) = 3 \qquad \text{solving for } x$$
$$10x + 15x - 12 = 3$$
$$25x = 15$$
$$x = \frac{3}{5}$$

Substituting this value into the expression for y, Eq. (B1), we obtain

$$y = \frac{5(3/5) - 4}{2} = \frac{3 - 4}{2} = -\frac{1}{2} \qquad \text{solving for } y$$

Therefore, the solution of this system is $x = \frac{3}{5}$, $y = -\frac{1}{2}$. Substituting these values in both original equations shows that the solution checks. ▶

SOLUTION BY ADDITION OR SUBTRACTION

The method of substitution is useful if one of the equations can easily be solved for one of the unknowns. However, often a fraction results (such as in Example 2), and this leads to additional algebraic steps to find the solution. Even worse, if the coefficients are themselves decimals or fractions, the algebraic steps are even more involved.

Therefore, we now present another algebraic method of solving a system of equations, *elimination of a variable by* **addition** *or* **subtraction.** Following is the basic procedure to be used.

For reference, Eqs. (5.3) are

$$a_1 x + b_1 y = c_1$$
$$a_2 x + b_2 y = c_2$$

Some prefer always to use addition of the terms of the resulting equations. This method is appropriate because it avoids possible errors that may be caused when subtracting. However, it may require that all signs of one equation be changed.

SOLUTION OF TWO LINEAR EQUATIONS BY ADDITION OR SUBTRACTION

1. *Write the equations in the form of Eqs. (5.3), if they are not already in this form.*
2. *If necessary, multiply all terms of each equation by a constant chosen so that the coefficients of one unknown will be numerically the same in both equations.* (They can have the same or different signs.)
3. (a) *If the numerically equal coefficients have* **different** *signs,* **add** *the terms on each side of the resulting equations.*
   (b) *If the numerically equal coefficients have the* **same** *sign,* **subtract** *the terms on each side of one equation from the terms of the other equation.*
4. *Solve the resulting linear equation in the other unknown.*
5. *Substitute this value into one of the* **original** *equations to find the value of the other unknown.*
6. *Check by substituting both values into both original equations.*

Some graphing calculators have a specific feature for solving simultaneous linear equations. It is necessary only to enter the coefficients and constants to get the solution.

◀ **EXAMPLE 3**   Use the method of elimination by addition or subtraction to solve the system of equations

$$x - 3y = 6 \qquad \text{already in the form of Eqs. (5.3)}$$
$$2x + 3y = 3$$

We look at the coefficients to determine the best way to eliminate one of the unknowns. Since the coefficients of the *y*-terms are *numerically the same and opposite in sign,* we may immediately *add* terms of the two equations together to eliminate *y*. Adding the terms of the left sides and adding the terms of the right sides, we obtain

$$x + 2x - 3y + 3y = 6 + 3$$
$$3x = 9$$
$$x = 3$$

Substituting this value into the first equation, we obtain

$$3 - 3y = 6$$
$$-3y = 3$$
$$y = -1$$

The solution $x = 3$, $y = -1$ agrees with the results obtained for the same problem illustrated in Example 1.

EXAMPLE 4 Use addition or subtraction to solve the system of equations

$$3x - 2y = 4$$
$$x + 3y = 2$$

Looking at the coefficients of x and y, we see that we must multiply the second equation by 3 to make the coefficients of x the same. To make the coefficients of y numerically the same, we must multiply the first equation by 3 and the second equation by 2. Thus, the best method is to multiply the second equation by 3 and eliminate x. Doing

CAUTION ▶ this (be careful to multiply the terms on **both** sides: *a common error is to forget to multiply the value on the* **right**), the coefficients of x have the *same sign*. Therefore, we *subtract* terms of the second equation from those of the first equation:

$$3x - 2y = 4$$
$$3x + 9y = 6 \qquad \text{each term of second equation multiplied by 3}$$

$3x - 3x = 0 \longrightarrow -11y = -2 \qquad \text{subtract}$

$-2y - (+9y) = -11y \longrightarrow y = \dfrac{2}{11} \qquad 4 - 6 = -2$

In order to find the value of x, substitute $y = \frac{2}{11}$ into one of the original equations. Choosing the second equation (its form is somewhat simpler), we have

$$x + 3\left(\frac{2}{11}\right) = 2$$

$$11x + 6 = 22 \qquad \text{multiply each term by 11}$$

$$x = \frac{16}{11}$$

Therefore, the solution is $x = \frac{16}{11}$, $y = \frac{2}{11}$. Substituting these values into both of the original equations shows that the solution checks. ▶

EXAMPLE 5 As noted in Example 4, we can solve the system of equations by first multiplying the first equation by 3 and the second equation by 2, thereby eliminating y. Doing this, we have

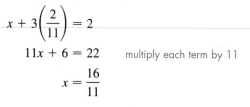

$$9x - 6y = 12 \qquad \text{each term of first equation multiplied by 3}$$
$$2x + 6y = 4 \qquad \text{each term of second equation multiplied by 2}$$

$11x = 16 \qquad \text{add}$

$9x + 2x = 11x \longrightarrow x = \dfrac{16}{11} \qquad 12 + 4 = 16$

$-6y + 6y = 0$

$$3\left(\frac{16}{11}\right) - 2y = 4 \qquad \text{substituting } x = 16/11 \text{ in the first original equation}$$
$$48 - 22y = 44 \qquad \text{multiply each term by 11}$$
$$-22y = -4$$
$$y = \frac{2}{11}$$

Therefore, the solution is $x = \frac{16}{11}$, $y = \frac{2}{11}$, as found in Example 4.

A calculator solution of this system is shown in Fig. 5.26, where $y_1 = 3x/2 - 2$ and $y_2 = -x/3 + 2/3$. The point of intersection is $(1.455, 0.182)$. This solution is the same as the algebraic solutions since $\frac{16}{11} = 1.455$ and $\frac{2}{11} = 0.182$. ▶

We could find y as in Example 4.

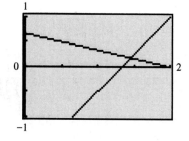

Fig. 5.26

Solving a Word Problem

◀ **EXAMPLE 6** By mass, one alloy is 70% copper and 30% zinc. Another alloy is 40% copper and 60% zinc. How many grams of each of these are required to make 300 g of an alloy that is 60% copper and 40% zinc?

Let A = the required number of grams of the first alloy and B = the required number of grams of the second alloy. Our equations are determined from:

> The second equation is based on the amount of copper. We could have used the amount of zinc, which would have led to the equation $0.30A + 0.60B = 0.40(300)$. Since we need only two equations, we may use any two of these three equations to find the solution.

1. The total mass of the final alloy is 300 g: $A + B = 300$.

2. The final alloy will have 180 g of copper (60% of 300 g), and this comes from 70% of A ($0.70A$) and 40% of B ($0.40B$): $0.70A + 0.40B = 180$.

These two equations can now be solved simultaneously:

$$A + B = 300 \quad \text{sum of masses is 300 g}$$

copper ⟶ $0.70A + 0.40B = 180$ ⟵ 60% of 300 g

70%   mass   40%   mass of
of first   second
alloy   alloy

$$4A + 4B = 1200 \quad \text{multiply each term of first equation by 4}$$
$$7A + 4B = 1800 \quad \text{multiply each term of second equation by 10}$$
$$3A \qquad = 600 \quad \text{subtract first equation from second equation}$$
$$A = 200 \text{ g}$$
$$B = 100 \text{ g} \quad \text{by substituting into first equation}$$

Checking with the statement of the problem, using the percentages of zinc, we have $0.30(200) + 0.60(100) = 0.40(300)$, or $60 \text{ g} + 60 \text{ g} = 120 \text{ g}$. We use zinc here since we used the percentages of copper for the equation. ◗

◀ **EXAMPLE 7** In solving the system of equations

$$4x = 2y + 3$$
$$-y + 2x - 2 = 0$$

first note that the equations are not in the correct form. Therefore, writing them in the form of Eqs. (5.3), we have

$$4x - 2y = 3$$
$$2x - y = 2$$

Now, multiply the second equation by 2 and subtract to get

$$4x - 2y = 3$$
$$4x - 2y = 4$$

$4x - 4x = 0 \longrightarrow 0 = -1$ ⟵ $3 - 4 = -1$
$-2y - (-2y) = 0$ ↗

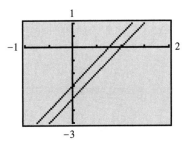

Fig. 5.27

Since 0 does not equal -1, we conclude that there is no solution. When we obtain a result of $0 = a$ ($a \neq 0$), the system of equations is *inconsistent*. As we discussed in the previous section, the lines that represent the equations are parallel. This is shown in the calculator display in Fig. 5.27, where $y_1 = 2x - 3/2$, $y_2 = 2x - 2$. ◗

NOTE ▶ In Example 7, we showed that a result $0 = a$ $(a \neq 0)$ indicates that the system of equations is inconsistent. If we obtain the result $0 = 0$, the system is *dependent*. As shown in the previous section, this means that there is an unlimited number of solutions and the lines that represent the equations are the same line.

EXERCISES 5.4

In Exercises 1–4, make the given changes in the indicated examples of this section and then solve the resulting problems.

1. In Example 1, change the $+$ to $-$ in the second equation and then solve the system of equations.

2. In Example 3, change the $+$ to $-$ in the second equation and then solve the system of equations.

3. In Example 4, change x to $2x$ in the second equation and then solve the system of equations.

4. In Example 7, change 3 to 4 in the first equation and then find if there is any change in the conclusion that is drawn.

In Exercises 5–16, solve the given systems of equations by the method of elimination by substitution.

5. $x = y + 3$
   $x - 2y = 5$

6. $x = 2y + 1$
   $2x - 3y = 4$

7. $p = V - 4$
   $V + p = 10$

8. $y = 2x + 10$
   $2x + y = -2$

9. $x + y = -5$
   $2x - y = 2$

10. $3x + y = 1$
    $3x - 2y = 16$

11. $2x + 3y = 7$
    $6x - y = 1$

12. $2s + 2t = 1$
    $4s - 2t = 17$

13. $33x + 2y = 34$
    $40y = 9x + 11$

14. $3A + 3B = -1$
    $5A = -6B - 1$

15. $0.4p - 0.3n = 0.6$
    $0.2p + 0.4n = -0.5$

16. $6.0x + 4.8y = -8.4$
    $4.8x - 6.5y = -7.8$

In Exercises 17–28, solve the given systems of equations by the method of elimination by addition or subtraction.

17. $x + 2y = 5$
    $x - 2y = 1$

18. $x + 3y = 7$
    $2x + 3y = 5$

19. $2x - 3y = 4$
    $2x + y = -4$

20. $R - 4r = 17$
    $3R + 4r = 3$

21. $12t + 9y = 14$
    $6t = 7y - 16$

22. $3x - y = 3$
    $4x = 3y + 14$

23. $v + 2t = 7$
    $2v + 4t = 9$

24. $3x - y = 5$
    $-9x + 3y = -15$

25. $2x - 3y - 4 = 0$
    $3x + 2 = 2y$

26. $3i_1 + 5 = -4i_2$
    $3i_2 = 5i_1 - 2$

27. $0.3R = 0.7Z + 0.4$
    $0.5Z = 0.7 - 0.2R$

28. $2.50x + 2.25y = 4.00$
    $3.75x - 6.75y = 3.25$

In Exercises 29–40, solve the given systems of equations by either method of this section.

29. $2x - y = 5$
    $6x + 2y = -5$

30. $3x + 2y = 4$
    $6x - 6y = 13$

31. $6x + 3y + 4 = 0$
    $5y = -9x - 6$

32. $1 + 6q = 5p$
    $3p - 4q = 7$

33. $15x + 10y = 11$
    $20x - 25y = 7$

34. $2x + 6y = -3$
    $-6x - 18y = 5$

35. $1.2V + 10.8 = -8.4C$
    $3.6C + 4.8V + 13.2 = 0$

36. $0.66x + 0.66y = -0.77$
    $0.33x - 1.32y = 1.43$

37. $44A = 1 - 15B$
    $5B = 22 + 7A$

38. $12x - 5y = -6$
    $20y + 18x = 13$

39. $2b = 6a - 16$
    $33a = 4b + 39$

40. $30P = 55 - Q$
    $19P + 14Q + 32 = 0$

In Exercises 41–44, solve the given systems of equations by an appropriate algebraic method.

41. Find the voltages V_1 and V_2 of the batteries shown in Fig. 5.28. The terminals are aligned in the same direction in Fig. 5.28(a) and in the opposite directions in Fig. 5.28(b).

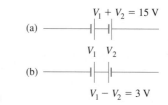

Fig. 5.28

42. A spring of length L is stretched x cm for each newton of weight hung from it. Weights of 3 N and then 5 N are hung from the spring, leading to the equations

   $L + 3x = 18$

   $L + 5x = 22$

   Solve for L and x.

43. Two grades of gasoline are mixed to make a blend with 1.50% of a special additive. Combining x L of a grade with 1.80% of the additive to y L of a grade with 1.00% of the additive gives 10 000 L of the blend. The equations relating x and y are

   $x + y = 10\,000$

   $0.0180x + 0.0100y = 0.0150(10\,000)$

   Find x and y (to three significant digits).

44. A 6.0% solution and a 15.0% solution of a drug are added to 200 mL of a 20.0% solution to make 1200 mL of a 12.0% solution for a proper dosage. The equations relating the number of milliliters of the added solutions are

$$x + y + 200 = 1200$$

$$0.060x + 0.150y + 0.200(200) = 0.120(1200)$$

Find x and y (to three significant digits).

In Exercises 45–54, set up appropriate systems of two linear equations and solve the systems algebraically. All data are accurate to at least two significant digits.

45. The weight W_f supported by the front wheels of a certain car and the weight W_r supported by the rear wheels together equal the weight of the car, 17 700 N. See Fig. 5.29. Also, the ratio of W_r to W_f is 0.847. What are the weights supported by each set of wheels?

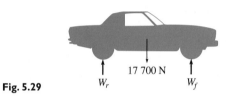

Fig. 5.29 W_r W_f
17 700 N

46. A sprinkler system is used to water two areas. If the total water flow is 980 L/h and the flow through one sprinkler is 65% as much as the other, what is the flow in each?

47. In a test of a heat-seeking rocket, a first rocket is launched at 600 m/s, and the heat-seeking rocket is launched along the same flight path 12 s later at a speed of 960 m/s. Find the times t_1 and t_2 of flight of the rockets until the heat-seeking rocket destroys the first rocket.

48. The *torque* of a force is the product of the force and the perpendicular distance from a specified point. If a lever is supported at only one point and is in balance, the sum of the torques (about the support) of forces acting on one side of the support must equal the sum of the torques of the forces acting on the other side. Find the forces F_1 and F_2 that are in the positions shown in Fig. 5.30(a) and then move to the positions in Fig. 5.30(b). The lever weighs 20 N and is in balance in each case.

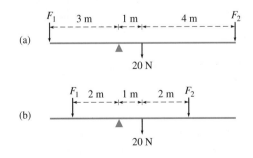

Fig. 5.30

49. A small isolated farm uses a windmill and a gas generator for power. During a 10-day period, they produced 3010 kW·h of power, with the windmill operating at 45.0% of capacity and the generator at capacity. During the following 10-day period, they produced 2900 kW·h with the windmill at 72.0% of capacity and the generator down 60 h for repairs (at capacity otherwise). What is the capacity (in kW) of each?

50. An underwater (but near the surface) explosion is detected by sonar on a ship 30 s before it is heard on the deck. If sound travels at 1500 m/s in water and 330 m/s in air, how far is the ship from the explosion? See Fig. 5.31.

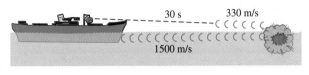
30 s   330 m/s

1500 m/s

Fig. 5.31

51. There are two types of offices in an office building, and a total of 54 offices. One type rents for $900/month and the other type rents for $1250/month. If all offices are occupied and the total rental income is $55 600/month, how many of each type are there?

52. In mixing a weed-killing chemical, a 40% solution of the chemical is mixed with an 85% solution to get 20 L of a 60% solution. How much of each solution is needed?

W)53. What conclusion can you draw from a sales report that states that "sales this month were $8000 more than last month, which means that total sales for both months are $4000 more than twice the sales last month"?

W)54. For an electric circuit, a report stated that current i_1 is twice current i_2 and that twice the sum of the two currents less 6 times i_2 is 6 mA. Explain your conclusion about the values of the currents found from this report.

In Exercises 55 and 56, answer the given questions.

55. What condition(s) must be placed on the constants of the system of equations

$$ax + y = c$$

$$bx + y = d$$

such that there is a unique solution for x and y?

56. What conditions must be placed on the constants of the system of equations in Exercise 55 such that the system is (a) inconsistent? (b) dependent?

5.5 SOLVING SYSTEMS OF TWO LINEAR EQUATIONS IN TWO UNKNOWNS BY DETERMINANTS

Consider two linear equations in two unknowns, as given in Eqs. (5.3):

$$\begin{array}{c} a_1 x + b_1 y = c_1 \\ a_2 x + b_2 y = c_2 \end{array} \tag{5.3}$$

If we multiply the first of these equations by b_2 and the second by b_1, we obtain

$$a_1 b_2 x + b_1 b_2 y = c_1 b_2$$
$$a_2 b_1 x + b_2 b_1 y = c_2 b_1 \tag{5.6}$$

We see that the coefficients of y are the same. Thus, subtracting the second equation from the first, we can solve for x. The solution can be shown to be

$$x = \frac{c_1 b_2 - c_2 b_1}{a_1 b_2 - a_2 b_1} \tag{5.7}$$

In the same manner, we may show that

$$y = \frac{a_1 c_2 - a_2 c_1}{a_1 b_2 - a_2 b_1} \tag{5.8}$$

The expression $a_1 b_2 - a_2 b_1$, which appears in each of the denominators of Eqs. (5.7) and (5.8), is an example of a special kind of expression called a *determinant of the second order*. The determinant $a_1 b_2 - a_2 b_1$ is denoted by

$$\begin{vmatrix} a_1 & b_1 \\ a_2 & b_2 \end{vmatrix}$$

Therefore, by definition, *a* **determinant of the second order** *is*

Second-Order Determinant

$$\begin{vmatrix} a_1 & b_1 \\ a_2 & b_2 \end{vmatrix} = a_1 b_2 - a_2 b_1 \tag{5.9}$$

The numbers a_1 and b_1 are called the **elements** *of the first* **row** *of the determinant. The numbers a_1 and a_2 are the elements of the first* **column** *of the determinant.* In the same manner, the numbers a_2 and b_2 are the elements of the second row, and the numbers b_1 and b_2 are the elements of the second column. *The numbers a_1 and b_2 are the elements of the* **principal diagonal,** *and the numbers a_2 and b_1 are the elements of the* **secondary diagonal.** *Thus, one way of stating the definition indicated in Eq. (5.9) is that the value of a determinant of the second order is found by taking the product of the elements of the principal diagonal and subtracting the product of the elements of the secondary diagonal.*

A diagram that is often helpful for remembering the expansion of a second-order determinant is shown in Fig. 5.32. The following examples illustrate how we carry out the evaluation of determinants.

Determinants were invented by the German mathematician Gottfried Wilhelm Leibniz (1646–1716).

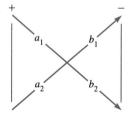

Fig. 5.32

◀ EXAMPLE 1 $\begin{vmatrix} -5 & 8 \\ 3 & 7 \end{vmatrix} = (-5)(7) - 3(8) = -35 - 24 = -59$

◀ EXAMPLE 2   (a) $\begin{vmatrix} 4 & 6 \\ 3 & 17 \end{vmatrix} = 4(17) - (3)(6) = 68 - 18 = 50$

(b) $\begin{vmatrix} 4 & 6 \\ -3 & 17 \end{vmatrix} = 4(17) - (-3)(6) = 68 + 18 = 86$

(c) $\begin{vmatrix} 3.6 & 6.1 \\ -3.2 & -17.2 \end{vmatrix} = 3.6(-17.2) - (-3.2)(6.1) = -42.4$

Note the signs of the terms being combined.

We note that the numerators and denominators of Eqs. (5.7) and (5.8) may be written as determinants. The numerators of the equations are

$$\begin{vmatrix} c_1 & b_1 \\ c_2 & b_2 \end{vmatrix} \quad \text{and} \quad \begin{vmatrix} a_1 & c_1 \\ a_2 & c_2 \end{vmatrix}$$

Therefore, the solutions for x and y of the system of equations

$$\begin{aligned} a_1 x + b_1 y &= c_1 \\ a_2 x + b_2 y &= c_2 \end{aligned} \tag{5.3}$$

may be written directly in terms of determinants, without algebraic operations, as

Note Carefully the Location of c_1 and c_2

$$x = \frac{\begin{vmatrix} c_1 & b_1 \\ c_2 & b_2 \end{vmatrix}}{\begin{vmatrix} a_1 & b_1 \\ a_2 & b_2 \end{vmatrix}} \quad \text{and} \quad y = \frac{\begin{vmatrix} a_1 & c_1 \\ a_2 & c_2 \end{vmatrix}}{\begin{vmatrix} a_1 & b_1 \\ a_2 & b_2 \end{vmatrix}} \tag{5.10}$$

For this reason, determinants provide a quick and easy method of solution of systems of equations. Again, *the denominator of each of Eqs. (5.10) is the same.*

The determinant of the denominator is made up of the coefficients of x and y. Also, *the determinant of the numerator of the solution for x is obtained from the determinant of the denominator by* **replacing the column of a's with the column of c's.** *The determinant of the numerator of the solution for y is obtained from the determinant of the denominator by* **replacing the column of b's with the column of c's.**

Named for the Swiss mathematician Gabriel Cramer (1704–1752).

This result is referred to as **Cramer's rule.** In using Cramer's rule, we must be sure that the equations are written in the form of Eqs. (5.3) before setting up the determinants.

The following examples illustrate the method of solving systems of equations by determinants.

◀ EXAMPLE 3   Solve the following system of equations by determinants:

$$2x + y = 1$$
$$5x - 2y = -11$$

First, note that the equations are in the proper form of Eqs. (5.3) for solution by determinants. Next, set up the determinant for the denominator, which consists of the four coefficients in the system, written as shown. It is

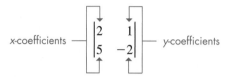

For finding x, the determinant in the numerator is obtained from this determinant by replacing the first column with the constants that appear on the right sides of the equations. Thus, *the numerator for the solution for x is*

For finding y, the determinant in the numerator is obtained from the determinant of the denominator by replacing the second column with the constants that appear on the right sides of the equations. Thus, *the numerator for the solution for y is*

Now set up the solutions for x and y using the determinants above:

Determinants can be evaluated on a graphing calculator. This is shown on page 167.

$$x = \frac{\begin{vmatrix} 1 & 1 \\ -11 & -2 \end{vmatrix}}{\begin{vmatrix} 2 & 1 \\ 5 & -2 \end{vmatrix}} = \frac{1(-2) - (-11)(1)}{2(-2) - (5)(1)} = \frac{-2 + 11}{-4 - 5} = \frac{9}{-9} = -1$$

$$y = \frac{\begin{vmatrix} 2 & 1 \\ 5 & -11 \end{vmatrix}}{\begin{vmatrix} 2 & 1 \\ 5 & -2 \end{vmatrix}} = \frac{2(-11) - (5)(1)}{-9} = \frac{-22 - 5}{-9} = 3$$

Therefore, the solution to the system of equations is $x = -1$, $y = 3$.

Substituting these values into the equations, we have

$$2(-1) + 3 \overset{?}{=} 1 \quad \text{and} \quad 5(-1) - 2(3) \overset{?}{=} -11$$
$$1 = 1 \qquad\qquad\qquad -11 = -11$$

which shows that they check.

NOTE ▶   Since the same determinant appears in each denominator, *it needs to be evaluated only once.* This means that three determinants are to be evaluated in order to solve the system.   ▶

◀ EXAMPLE 4 Solve the following system of equations by determinants. All numbers are approximate.

$$5.3x + 7.2y = 4.5$$
$$3.2x - 6.9y = 5.7$$

constants

$$x = \frac{\begin{vmatrix} 4.5 & 7.2 \\ 5.7 & -6.9 \end{vmatrix}}{\begin{vmatrix} 5.3 & 7.2 \\ 3.2 & -6.9 \end{vmatrix}} = \frac{4.5(-6.9) - 5.7(7.2)}{5.3(-6.9) - 3.2(7.2)} = \frac{-72.09}{-59.61} = 1.2$$

coefficients ⟶

constants

$$y = \frac{\begin{vmatrix} 5.3 & 4.5 \\ 3.2 & 5.7 \end{vmatrix}}{\begin{vmatrix} 5.3 & 7.2 \\ 3.2 & -6.9 \end{vmatrix}} = \frac{5.3(5.7) - 3.2(4.5)}{-59.61} = \frac{15.81}{-59.61} = -0.27$$

coefficients ⟶

The calculations can be done completely on the calculator using the following procedure: (1) Evaluate the denominator and store this value; (2) divide the value of each numerator by the value of the denominator, storing the values of x and y; (3) use the stored values of x and y for checking the solution; (4) round off results (if the numbers are approximate).

Using this procedure, we store the value of the denominator D ($D = -59.61$) in order to calculate x and y, which are stored as X and Y. The check of the solution is shown in the calculator display shown in Fig. 5.33. ◗

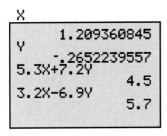

The line with X may be displaced off the screen.

Fig. 5.33

Solving a Word Problem

◀ EXAMPLE 5 Two investments totaling $18 000 yield an annual income of $700. If the first investment has an interest rate of 5.5% and the second a rate of 3.0%, what is the value of each investment?

Let x = the value of the first investment and y = the value of the second investment. We know that the total of the two investments is $18 000. This leads to the equation $x + y = \$18\,000$. The first investment yields $0.055x$ dollars annually, and the second yields $0.030y$ dollars annually. This leads to the equation $0.055x + 0.030y = 700$. These two equations are then solved simultaneously:

$$
\begin{array}{rcll}
x + & y = & 18\,000 & \text{sum of investments} \\
0.055x + & 0.030y = & 700 & \longleftarrow \text{income}
\end{array}
$$

↑ ↑
5.5% value 3.0% value

$$x = \frac{\begin{vmatrix} 18\,000 & 1 \\ 700 & 0.030 \end{vmatrix}}{\begin{vmatrix} 1 & 1 \\ 0.055 & 0.030 \end{vmatrix}} = \frac{540 - 700}{0.030 - 0.055} = \frac{-160}{-0.025} = 6400$$

The value of y can be found most easily by substituting this value of x into the first equation, $y = 18\,000 - x = 18\,000 - 6400 = 11\,600$.

Therefore, the values invested are $6400 and $11 600, respectively. Checking, we see that the total income is $6400(0.055) + \$11\,600(0.030) = \$700$. This agrees with the statement of the problem. ◗

CAUTION ▶   *The equations must be in the form of Eqs. (5.3) before the determinants are set up. The specific positions of the values in the determinants are based on that form of writing the system. If either unknown is missing from an equation, a zero must be placed in the proper position. Also, from Example 4, we see that determinants are easier to use than other algebraic methods when the coefficients are decimals.*

NOTE ▶   If the determinant of the denominator is zero, we do not have a unique solution since this would require division by zero. If the determinant of the denominator is zero and that of the numerator is not zero, the system is *inconsistent*. If the determinants of both numerator and denominator are zero, the system is *dependent*.

EXERCISES 5.5

In Exercises 1–4, make the given changes in the indicated examples of this section and then solve the resulting problems.

1. In Example 2(a), change the 6 to -6 and then evaluate.

2. In Example 2(a), change the 4 to -4 and the 6 to -6 and then evaluate.

3. In Example 3, change the $+$ to $-$ in the first equation and then solve the system of equations.

4. In Example 5, change \$700 to \$830 and then solve for the values of the investments.

In Exercises 5–16, evaluate the given determinants.

5. $\begin{vmatrix} 2 & 4 \\ 3 & 1 \end{vmatrix}$

6. $\begin{vmatrix} -1 & 3 \\ 2 & 6 \end{vmatrix}$

7. $\begin{vmatrix} 3 & -5 \\ 7 & -2 \end{vmatrix}$

8. $\begin{vmatrix} -4 & 7 \\ 1 & -3 \end{vmatrix}$

9. $\begin{vmatrix} 8 & -10 \\ 0 & 4 \end{vmatrix}$

10. $\begin{vmatrix} -4 & -3 \\ -8 & -6 \end{vmatrix}$

11. $\begin{vmatrix} -20 & 110 \\ -70 & -80 \end{vmatrix}$

12. $\begin{vmatrix} -6.5 & 12.2 \\ -15.5 & 34.6 \end{vmatrix}$

13. $\begin{vmatrix} 0.75 & -1.32 \\ 0.15 & 1.18 \end{vmatrix}$

14. $\begin{vmatrix} 0.20 & -0.05 \\ 0.28 & 0.09 \end{vmatrix}$

15. $\begin{vmatrix} 16 & -8 \\ 42 & -15 \end{vmatrix}$

16. $\begin{vmatrix} 43 & -7 \\ -81 & 16 \end{vmatrix}$

In Exercises 17–28, solve the given systems of equations by determinants. (These are the same as those for Exercises 17–28 of Section 5.4.)

17. $x + 2y = 5$
    $x - 2y = 1$

18. $x + 3y = 7$
    $2x + 3y = 5$

19. $2x - 3y = 4$
    $2x + y = -4$

20. $R - 4r = 17$
    $3R + 4r = 3$

21. $12t + 9y = 14$
    $6t = 7y - 16$

22. $3x - y = 3$
    $4x = 3y + 14$

23. $v + 2t = 7$
    $2v + 4t = 9$

24. $3x - y = 5$
    $-9x + 3y = -15$

25. $2x - 3y - 4 = 0$
    $3x + 2 = 2y$

26. $3i_1 + 5 = -4i_2$
    $3i_2 = 5i_1 - 2$

27. $0.3R = 0.7Z + 0.4$
    $0.5Z = 0.7 - 0.2R$

28. $2.50x + 2.25y = 4.00$
    $3.75x - 6.75y = 3.25$

In Exercises 29–36, solve the given systems of equations by determinants. All numbers are approximate.

29. $2.1x - 1.0y = 5.2$
    $5.8x + 1.6y = -5.4$

30. $3.3F + 1.9T = 4.2$
    $5.4F - 6.4T = 13.2$

31. $5.5R = -9.0t - 6.8$
    $6.0t + 3.8R + 4.0 = 0$

32. $12 + 66y = 53x$
    $32x - 39y = 71$

33. $301x - 529y = 1520$
    $385x - 741y = 2540$

34. $0.25d + 0.63n = -0.37$
    $-0.61d - 1.80n = 0.55$

35. $1.2y + 10.8 = -8.4x$
    $3.5x + 4.8y + 12.9 = 0$

36. $6541x + 4397y = -7732$
    $3309x - 8755y = 7622$

In Exercises 37–40, solve the given systems of equations by determinants. All numbers are accurate to at least two significant digits.

37. The forces acting on a link of an industrial robot are shown in Fig. 5.34. The equations for finding forces F_1 and F_2 are

    $F_1 + F_2 = 21$

    $2F_1 = 5F_2$

    Find F_1 and F_2.

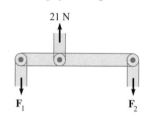

Fig. 5.34

38. The area of a quadrilateral is

$$A = \frac{1}{2}\left(\begin{vmatrix} x_0 & x_1 \\ y_0 & y_1 \end{vmatrix} + \begin{vmatrix} x_1 & x_2 \\ y_1 & y_2 \end{vmatrix} + \begin{vmatrix} x_2 & x_3 \\ y_2 & y_3 \end{vmatrix} + \begin{vmatrix} x_3 & x_0 \\ y_3 & y_0 \end{vmatrix} \right)$$

where (x_0, y_0), (x_1, y_1), (x_2, y_2), and (x_3, y_3) are the rectangular coordinates of the vertices of the quadrilateral, listed counterclockwise. (This *surveyor's formula* can be generalized to find the area of any polygon.)

A surveyor records the locations of the vertices of a quadrilateral building lot on a rectangular coordinate system as $(12.79, 0.00)$, $(67.21, 12.30)$, $(53.05, 47.12)$, and $(10.09, 53.11)$, where distances are in meters. Find the area of the lot.

39. An airplane begins a flight with a total of 144.0 L of fuel stored in two separate wing tanks. During the flight, 25.0% of the fuel in one tank is used, and in the other tank 37.5% of the fuel is used. If the total fuel used is 44.8 L, the amounts x and y used from each tank can be found by solving the system of equations

$x + y = 144.0$

$0.250x + 0.375y = 44.8$

Find x and y.

40. In applying Kirchhoff's laws (see the chapter introduction; for the equations, see, for example, Beiser, *Modern Technical Physics,* 6th ed., p. 550) to the electric circuit shown in Fig. 5.35, the following equations are found. Find the indicated currents I_1 and I_2 (in A).

$52I_1 - 27I_2 = -420$

$-27I_1 + 76I_2 = 210$

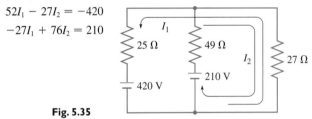

Fig. 5.35

In Exercises 41–48, set up appropriate systems of two linear equations in two unknowns and then solve the systems by determinants. All numbers are accurate to at least two significant digits.

41. A shipment of 320 cell phones and radar detectors was destroyed due to a truck accident. On the insurance claim, the shipper stated that each phone was worth $110, each detector was worth $160, and their total value was $40 700. How many of each were in the shipment?

42. Two types of electromechanical carburetors are being assembled and tested. Each of the first type requires 15 min of assembly time and 2 min of testing time. Each of the second type requires 12 min of assembly time and 3 min of testing time. If 222 min of assembly time and 45 min of testing time are available, how many of each type can be assembled and tested, if all the time is used?

43. A machinery sales representative receives a fixed salary plus a sales commission each month. If $6200 is earned on sales of $70 000 in one month and $4700 is earned on sales of $45 000 in the following month, what are the fixed salary and the commission percent?

44. A moving walkway at an airport is 65.0 m long. A child running at a constant speed takes 20.0 s to run along the walkway in the direction it is moving, and then 52.0 s to run all the way back. What are the speed of the walkway and the speed of the child?

45. A boat carrying illegal drugs leaves a port and travels at 63 km/h. A Coast Guard cutter leaves the port 24 min later and travels at 75 km/h in pursuit of the boat. Find the times each has traveled when the cutter overtakes the boat with drugs. See Fig. 5.36.

Fig. 5.36

46. Sterling silver is 92.5% silver and 7.5% copper. One silver-copper alloy is 94.0% silver, and a second silver-copper alloy is 85.0% silver. How much of each should be used in order to make 100 g of sterling silver?

47. In an experiment, a variable voltage V is in a circuit with a fixed voltage V_0 and a resistance R. The voltage V is related to the current i in the circuit by $V = Ri - V_0$. If $V = 5.8$ V for $i = 2.0$ A and $V = 24.7$ V for $i = 6.2$ A, find V as a function of i.

48. The velocity of sound in steel is 4850 m/s faster than the velocity of sound in air. One end of a long steel bar is struck, and an instrument at the other end measures the time it takes for the sound to reach it. The sound in the bar takes 0.0120 s, and the sound in the air takes 0.180 s. What are the velocities of sound in air and in steel?

5.6 SOLVING SYSTEMS OF THREE LINEAR EQUATIONS IN THREE UNKNOWNS ALGEBRAICALLY

Many technical problems involve systems of linear equations with more than two unknowns. In this section, we solve systems with three unknowns, and later we will show how systems with even more unknowns are solved.

Solving such systems is very similar to solving systems in two unknowns. In this section, we will show the algebraic method, and in the next section we will show how determinants are used. Graphical solutions are not used since a linear equation in three unknowns represents a plane in space. We will, however, briefly show graphical interpretations of systems of three linear equations at the end of this section.

A system of three linear equations in three unknowns written in the form

$$\begin{aligned} a_1x + b_1y + c_1z &= d_1 \\ a_2x + b_2y + c_2z &= d_2 \\ a_3x + b_3y + c_3z &= d_3 \end{aligned}$$

(5.11)

has as its solution the set of values x, y, and z that satisfy all three equations simultaneously. The method of solution involves multiplying *two* of the equations by the proper numbers to eliminate **one** of the unknowns between these equations. We then repeat this process, using a *different pair* of the original equations, being sure that we eliminate the same unknown as we did between the first pair of equations. At this point we have two linear equations in two unknowns that can be solved by any of the methods previously discussed.

◀ **EXAMPLE 1** Solve the following system of equations:

(1)	$4x + y + 3z = 1$	
(2)	$2x - 2y + 6z = 11$	
(3)	$-6x + 3y + 12z = -4$	

We can choose to first eliminate any one of the three unknowns. We have chosen y.

(4)	$8x + 2y + 6z = 2$	(1) multiplied by 2
	$2x - 2y + 6z = 11$	(2)
(5)	$10x \qquad + 12z = 13$	adding

At this point we could have used Eqs. (1) and (3) or Eqs. (2) and (3), but we must set them up to eliminate y.

(6)	$12x + 3y + 9z = 3$	(1) multiplied by 3
	$-6x + 3y + 12z = -4$	(3)
(7)	$18x \qquad - 3z = 7$	subtracting

	$10x + 12z = 13$	(5)
(8)	$72x - 12z = 28$	(7) multiplied by 4
(9)	$82x \qquad = 41$	adding
(10)	$x = \frac{1}{2}$	

Some graphing calculators have a specific feature for solving simultaneous linear equations. It is necessary only to enter the coefficients and constants to get the solution.

(11)	$18\left(\frac{1}{2}\right) - 3z = 7$	substituting (10) in (7)
(12)	$-3z = -2$	
(13)	$z = \frac{2}{3}$	
(14)	$4\left(\frac{1}{2}\right) + y + 3\left(\frac{2}{3}\right) = 1$	substituting (13) and (10) in (1)
(15)	$2 + y + 2 = 1$	
(16)	$y = -3$	

Thus, the solution is $x = \frac{1}{2}$, $y = -3$, $z = \frac{2}{3}$. Substituting in the equations, we have

$$4\left(\tfrac{1}{2}\right) + (-3) + 3\left(\tfrac{2}{3}\right) \stackrel{?}{=} 1 \qquad 2\left(\tfrac{1}{2}\right) - 2(-3) + 6\left(\tfrac{2}{3}\right) \stackrel{?}{=} 11 \qquad -6\left(\tfrac{1}{2}\right) + 3(-3) + 12\left(\tfrac{2}{3}\right) \stackrel{?}{=} -4$$
$$1 = 1 \qquad\qquad\qquad 11 = 11 \qquad\qquad\qquad -4 = -4$$

We see that the solution checks. If we had first eliminated x or z, we would have completed the solution in a similar manner. ▶

◀ EXAMPLE 2   An industrial robot and the forces acting on its main link are shown in Fig. 5.37. An analysis of the forces leads to the following equations relating the forces. Determine the forces.

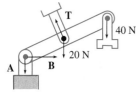

Fig. 5.37

(1)	$A + 60 = 0.8T$		
(2)	$B = 0.6T$		
(3)	$8A + 6B + 80 = 5T$		choose 5 to clear decimals

(4)	$5A \quad\quad - 4T = -300$	(1) multiplied by 5 and rewritten
(5)	$5B - 3T = 0$	(2) multiplied by 5 and rewritten
(6)	$8A + 6B - 5T = -80$	(3) rewritten

(7)	$40A \quad\quad - 32T = -2400$	(4) multiplied by 8
(8)	$40A + 30B - 25T = -400$	(6) multiplied by 5

(9)	$-30B - 7T = -2000$	subtracting
(10)	$30B - 18T = 0$	(5) multiplied by 6

(11)	$-25T = -2000$	adding
(12)	$T = 80 \text{ N}$	
(13)	$A + 60 = 64$	(12) substituted in (1)
(14)	$A = 4 \text{ N}$	
(15)	$B = 48 \text{ N}$	(12) substituted in (2)

Therefore, the forces are $A = 4$ N, $B = 48$ N, and $T = 80$ N. The solution checks when substituted into the original equations. Note that we could have started the solution by substituting Eq. (2) into Eq. (3), thereby first eliminating B.   ▶

Solving a Word Problem

◀ EXAMPLE 3   Three voltages, e_1, e_2, and e_3, where e_3 is three times e_1, are in series with the same polarity (see Fig. 5.38(a)) and have a total voltage of 85 mV. If e_2 is reversed in polarity (see Fig. 5.38(b)), the voltage is 35 mV. Find the voltages.

Since the voltages are in series with the same polarity, $e_1 + e_2 + e_3 = 85$. Then, since e_3 is three times e_1, we have $e_3 = 3e_1$. Then, with the reversed polarity of e_2, we have $e_1 - e_2 + e_3 = 35$. Writing these equations in standard form, we have the following solution:

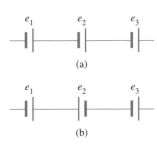

Fig. 5.38

(1)	$e_1 + e_2 + e_3 = 85$	
(2)	$3e_1 \quad\quad - e_3 = 0$	rewriting second equation
(3)	$e_1 - e_2 + e_3 = 35$	

(4)	$2e_1 \quad\quad + 2e_3 = 120$	adding (1) and (3)
(5)	$6e_1 \quad\quad - 2e_3 = 0$	(2) multiplied by 2

(6)	$8e_1 \quad\quad\quad = 120$	adding
(7)	$e_1 = 15 \text{ mV}$	
(8)	$3(15) - e_3 = 0$	substituting (7) in (2)
(9)	$e_3 = 45 \text{ mV}$	
(10)	$15 + e_2 + 45 = 85$	substituting (7) and (9) in (1)
(11)	$e_2 = 25 \text{ mV}$	

Therefore, the three voltages are 15 mV, 25 mV, and 45 mV.

Checking the solution, the sum of the three voltages is 85 mV, e_3 is three times e_1, and the sum of e_1 and e_3 less e_2 is 35 mV.   ▶

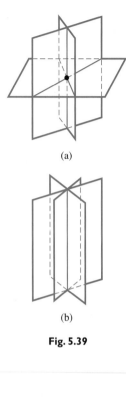

(a)

(b)

Fig. 5.39

We solve systems of equations with more than three unknowns in a manner similar to that used for three unknowns. For example, with four unknowns one unknown is eliminated between three different pairs of equations, and the resulting three equations are solved as with three unknowns.

Linear systems with more than two unknowns may have an unlimited number of solutions or be inconsistent. After eliminating unknowns, if we have $0 = 0$, there is an unlimited number of solutions. If we have $0 = a$ ($a \neq 0$), the system is inconsistent, and there is no solution. (See Exercises 29–32.)

We noted earlier that a linear equation in three unknowns represents a plane in space. For a system of three linear equations in three unknowns, if the three planes intersect at a point, there is a unique solution (Fig. 5.39(a)); if they intersect in a line, there is an unlimited number of solutions (Fig. 5.39(b)). If the planes do not have a common intersection, the system is inconsistent. The planes can be parallel (Fig. 5.40(a)), two of them can be parallel (Fig. 5.40(b)), or they can intersect in three parallel lines (Fig. 5.40(c)). If one plane is coincident with another plane, the system is either inconsistent or has an unlimited number of solutions if they intersect.

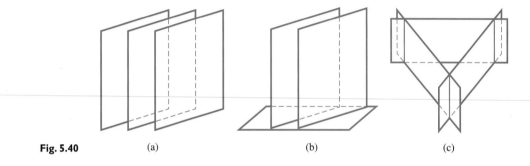

Fig. 5.40   (a)   (b)   (c)

EXERCISES 5.6

In Exercises 1 and 2, make the given changes in Example 1 of this section and then solve the resulting system of equations.

1. In the second equation, change the constant to the right of the = sign from 11 to 12 and in the third equation change the constant to the right of the = sign from -4 to -14.

2. Change the second equation to $8x + 9z = 10$ (no y-term).

In Exercises 3–18, solve the given systems of equations.

3. $x + y + z = 2$
$x - z = 1$
$x + y = 1$

4. $x + y - z = -3$
$x + z = 2$
$2x - y + 2z = 3$

5. $2x + 3y + z = 2$
$-x + 2y + 3z = -1$
$-3x - 3y + z = 0$

6. $2x + y - z = 4$
$4x - 3y - 2z = -2$
$8x - 2y - 3z = 3$

7. $5l + 6w - 3h = 6$
$4l - 7w - 2h = -3$
$3l + w - 7h = 1$

8. $3r + s - t = 2$
$r - 2s + t = 0$
$4r - s + t = 3$

9. $2x - 2y + 3z = 5$
$2x + y - 2z = -1$
$4x - y - 3z = 0$

10. $2u + 2v + 3w = 0$
$3u + v + 4w = 21$
$-u - 3v + 7w = 15$

11. $3x - 7y + 3z = 6$
$3x + 3y + 6z = 1$
$5x - 5y + 2z = 5$

12. $8x + y + z = 1$
$7x - 2y + 9z = -3$
$4x - 6y + 8z = -5$

13. $p + 2q + 2r = 0$
$2p + 6q - 3r = -1$
$4p - 3q + 6r = -8$

14. $9x + 12y + 2z = 23$
$21x + 2y - 5z = -25$
$-18x + 6y - 4z = -11$

15. $2x + 3y - 5z = 7$
$4x - 3y - 2z = 1$
$8x - y + 4z = 3$

16. $2i_1 - 4i_2 - 4i_3 = 3$
$3i_1 + 8i_2 + 2i_3 = -11$
$4i_1 + 6i_2 - i_3 = -8$

17. $r - s - 3t - u = 1$
$2r + 4s - 2u = 2$
$3r + 4s - 2t = 0$
$r + 2t - 3u = 3$

18. $3x + 2y - 4z + 2t = 3$
$5x - 3y - 5z + 6t = 8$
$2x - y + 3z - 2t = 1$
$-2x + 3y + 2z - 3t = -2$

In Exercises 19–28, solve the systems of equations. In Exercises 23–28, it is necessary to set up the appropriate equations. All numbers are accurate to at least three significant digits.

19. A medical supply company has 1150 worker-hours for production, maintenance, and inspection. Using this and other factors, the number of hours used for each operation, P, M, and I, respectively, is found by solving the following system of equations:

$P + M + I = 1150$

$P = 4I - 100$

$P = 6M + 50$

20. Three oil pumps fill three different tanks. The pumping rates of the pumps (in L/h) are r_1, r_2, and r_3, respectively. Because of malfunctions, they do not operate at capacity each time. Their rates can be found by solving the following system of equations:

$r_1 + r_2 + r_3 = 14\,000$

$r_1 + 2r_2 \qquad = 13\,000$

$3r_1 + 3r_2 + 2r_3 = 36\,000$

21. The forces acting on a certain girder, as shown in Fig. 5.41, can be found by solving the following system of equations:

$0.707F_1 - 0.800F_2 \qquad\qquad = 0$

$0.707F_1 + 0.600F_2 - \qquad F_3 = 10.0$

$\qquad 3.00\ F_2 - 3.00\ F_3 = 20.0$

Find the forces, in N.

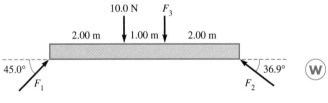

Fig. 5.41

22. Using Kirchhoff's laws (see the chapter introduction; for the equations, see, for example, Beiser, *Modern Technical Physics*, 6th ed., p. 550) for the electric circuit shown in Fig. 5.42, the following equations are found. Find the indicated currents (in A) I_1, I_2, and I_3.

$1.0I_1 + 3.0(I_1 - I_3) = 12$

$2.0I_2 + 4.0(I_2 + I_3) = 12$

$1.0I_1 - 2.0I_2 + 3.0I_3 = 0$

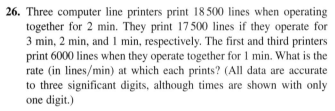

Fig. 5.42

23. Find angles A, B, and C in the roof truss shown in Fig. 5.43.

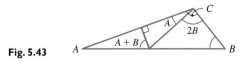

Fig. 5.43

24. Under certain conditions, the cost per kilometer C of operating a car is a function of the speed v (in km/h) of the car, given by $C = av^2 + bv + c$. If $C = 17$ ¢/km for $v = 16$ km/h, $C = 14$ ¢/km for $v = 50$ km/h, and $C = 15$ ¢/km for $v = 80$ km/h, find C as a function of v.

25. The angle θ between two links of a robot arm is given by $\theta = at^3 + bt^2 + ct$, where t is the time during an 11.8-s cycle. If $\theta = 19.0°$ for $t = 1.00$ s, $\theta = 30.9°$ for $t = 3.00$ s, and $\theta = 19.8°$ for $t = 5.00$ s, find the equation $\theta = f(t)$. See Fig. 5.44.

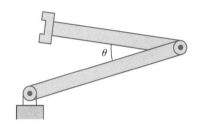

Fig. 5.44

26. Three computer line printers print 18 500 lines when operating together for 2 min. They print 17 500 lines if they operate for 3 min, 2 min, and 1 min, respectively. The first and third printers print 6000 lines when they operate together for 1 min. What is the rate (in lines/min) at which each prints? (All data are accurate to three significant digits, although times are shown with only one digit.)

27. By mass, one fertilizer is 20% potassium, 30% nitrogen, and 50% phosphorus. A second fertilizer has percents of 10, 20, and 70, respectively, and a third fertilizer has percents of 0, 30, and 70, respectively. How much of each must be mixed to get 200 kg of fertilizer with percents of 12, 25, and 63, respectively?

W 28. The average traffic flow (number of vehicles) from noon until 1 P.M. in a certain section of one-way streets in a city is shown in Fig. 5.45. Explain why an analysis of the flow through these intersections is not sufficient to obtain unique values for x, y, and z.

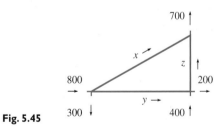

Fig. 5.45

In Exercises 29–32, show that the given systems of equations have either an unlimited number of solutions or no solution. If there is an unlimited number of solutions, find one of them.

29. $x - 2y - 3z = 2$
$x - 4y - 13z = 14$
$-3x + 5y + 4z = 0$

30. $x - 2y - 3z = 2$
$x - 4y - 13z = 14$
$-3x + 5y + 4z = 2$

31. $3x + 3y - 2z = 2$
$2x - y + z = 1$
$x - 5y + 4z = -3$

32. $3x + y - z = -3$
$x + y - 3z = -5$
$-5x - 2y + 3z = -7$

5.7 SOLVING SYSTEMS OF THREE LINEAR EQUATIONS IN THREE UNKNOWNS BY DETERMINANTS

Just as systems of two linear equations in two unknowns can be solved by determinants, so can systems of three linear equations in three unknowns. The system

$$a_1 x + b_1 y + c_1 z = d_1$$
$$a_2 x + b_2 y + c_2 z = d_2 \qquad \text{(5.11)}$$
$$a_3 x + b_3 y + c_3 z = d_3$$

can be solved in general terms by the method of elimination by addition or subtraction. This leads to the following solutions for x, y, and z.

$$x = \frac{d_1 b_2 c_2 + d_3 b_1 c_2 + d_2 b_3 c_1 - d_3 b_2 c_1 - d_1 b_3 c_2 - d_2 b_1 c_3}{a_1 b_2 c_3 + a_3 b_1 c_2 + a_2 b_3 c_1 - a_3 b_2 c_1 - a_1 b_3 c_2 - a_2 b_1 c_3}$$

$$y = \frac{a_1 d_2 c_3 + a_3 d_1 c_2 + a_2 d_3 c_1 - a_3 d_2 c_1 - a_1 d_3 c_2 - a_2 d_1 c_3}{a_1 b_2 c_3 + a_3 b_1 c_2 + a_2 b_3 c_1 - a_3 b_2 c_1 - a_1 b_3 c_2 - a_2 b_1 c_3} \qquad \text{(5.12)}$$

$$z = \frac{a_1 b_2 d_3 + a_3 b_1 d_2 + a_2 b_3 d_1 - a_3 b_2 d_1 - a_1 b_3 d_2 - a_2 b_1 d_3}{a_1 b_2 c_3 + a_3 b_1 c_2 + a_2 b_3 c_1 - a_3 b_2 c_1 - a_1 b_3 c_2 - a_2 b_1 c_3}$$

The expressions that appear in the numerators and denominators of Eqs. (5.12) are examples of a **determinant of the third order.** *This determinant is defined by*

$$\begin{vmatrix} a_1 & b_1 & c_1 \\ a_2 & b_2 & c_2 \\ a_3 & b_3 & c_3 \end{vmatrix} = a_1 b_2 c_3 + a_3 b_1 c_2 + a_2 b_3 c_1 - a_3 b_2 c_1 - a_1 b_3 c_2 - a_2 b_1 c_3 \qquad \text{(5.13)}$$

The elements, rows, columns, and diagonals of a third-order determinant are defined just as are those of a second-order determinant. For example, the principal diagonal is made up of the elements a_1, b_2, and c_3.

Probably the easiest way of remembering the method of finding the value of a third-order determinant is as follows: *Rewrite the first two columns to the right of the determinant. The products of the elements of the principal diagonal and the two parallel diagonals to the right of it are then added. The products of the elements of the secondary diagonal and the two parallel diagonals to the right of it are subtracted from the first sum. The algebraic sum of these six products gives the value of the determinant.* These products are indicated in Fig. 5.46.

> This method is used only for third-order determinants. It does not work for determinants of order higher than three.

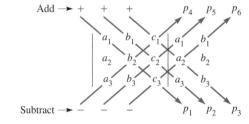

Fig. 5.46

Examples 1 and 2 illustrate this method of evaluating third-order determinants.

Determinants can be evaluated on a calculator using the *matrix* feature. Figure 5.47 shows two windows for the determinant in Example 2. The first shows the window for entering the numbers in the *matrix* (the array of numbers—see page 436). The first four lines of the second window show the matrix displayed, and the last two lines show the evaluation of the determinant.

◀ **EXAMPLE 1**

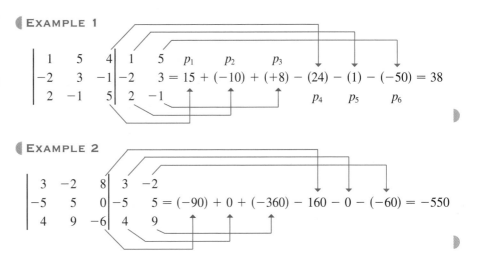

$$\begin{vmatrix} 1 & 5 & 4 \\ -2 & 3 & -1 \\ 2 & -1 & 5 \end{vmatrix} \begin{matrix} 1 & 5 \\ -2 & 3 \\ 2 & -1 \end{matrix} = 15 + (-10) + (+8) - (24) - (1) - (-50) = 38$$

◀ **EXAMPLE 2**

$$\begin{vmatrix} 3 & -2 & 8 \\ -5 & 5 & 0 \\ 4 & 9 & -6 \end{vmatrix} \begin{matrix} 3 & -2 \\ -5 & 5 \\ 4 & 9 \end{matrix} = (-90) + 0 + (-360) - 160 - 0 - (-60) = -550$$

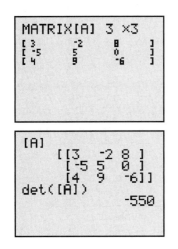

Fig. 5.47

Inspection of Eqs. (5.12) reveals that the numerators of these solutions may also be written in terms of determinants. Thus, we may write the general solution to a system of three equations in three unknowns as

$$x = \frac{\begin{vmatrix} d_1 & b_1 & c_1 \\ d_2 & b_2 & c_2 \\ d_3 & b_3 & c_3 \end{vmatrix}}{\begin{vmatrix} a_1 & b_1 & c_1 \\ a_2 & b_2 & c_2 \\ a_3 & b_3 & c_3 \end{vmatrix}} \qquad y = \frac{\begin{vmatrix} a_1 & d_1 & c_1 \\ a_2 & d_2 & c_2 \\ a_3 & d_3 & c_3 \end{vmatrix}}{\begin{vmatrix} a_1 & b_1 & c_1 \\ a_2 & b_2 & c_2 \\ a_3 & b_3 & c_3 \end{vmatrix}} \qquad z = \frac{\begin{vmatrix} a_1 & b_1 & d_1 \\ a_2 & b_2 & d_2 \\ a_3 & b_3 & d_3 \end{vmatrix}}{\begin{vmatrix} a_1 & b_1 & c_1 \\ a_2 & b_2 & c_2 \\ a_3 & b_3 & c_3 \end{vmatrix}} \qquad (5.14)$$

If the determinant of the denominator is not zero, there is a unique solution to the system of equations. (If all determinants are zero, there is an unlimited number of solutions. If the determinant of the denominator is zero and any of the determinants of the numerators is not zero, the system is inconsistent, and there is no solution.)

An analysis of Eqs. (5.14) shows that the situation is precisely the same as it was when we were using determinants to solve systems of two linear equations. That is, the determinants in the denominators in the expressions for x, y, and z are the same. They consist of elements that are the coefficients of the unknowns. The determinant of the numerator of the solution for x is the same as that of the denominator, except that the column of d's replaces the column of a's. The determinant in the numerator of the solution for y is the same as that of the denominator, except that the column of d's replaces the column of b's. The determinant of the numerator of the solution for z is the same as the determinant of the denominator, except that the column of d's replaces the column of c's. To summarize:

Cramer's Rule

NOTE ▶

The determinant in the denominator of each is made up of the coefficients of x, y, and z. The determinants in the numerators are the same as that in the denominator, except that **the column of d's replaces the column of coefficients of the unknown for which we are solving.**

This, again, is **Cramer's rule.** Remember, the equations must be written in the standard form shown in Eqs. (5.11) before the determinants are formed.

◀ **EXAMPLE 3** Solve the following system by determinants.

$$3x + 2y - 5z = -1$$
$$2x - 3y - z = 11$$
$$5x - 2y + 7x = 9$$

Figure 5.48(a) shows the calculator display for the solutions for x and y in Example 3. Here, det[A] and det[B] are the determinants for the numerators for x and y, respectively, and det[D] is the determinant for the denominator.

Figure 5.48(b) shows the solution for z and the use of the *fraction* feature in changing the result for z into a fraction.

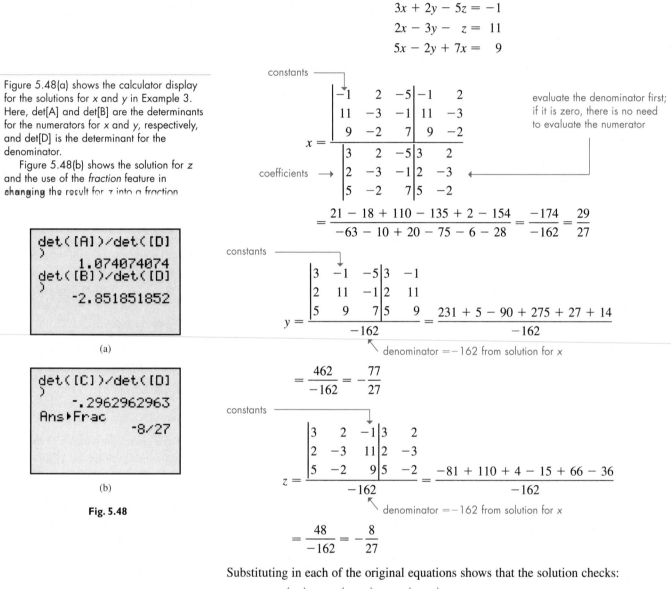

constants ──────

$$x = \frac{\begin{vmatrix} -1 & 2 & -5 \\ 11 & -3 & -1 \\ 9 & -2 & 7 \end{vmatrix} \begin{matrix} -1 & 2 \\ 11 & -3 \\ 9 & -2 \end{matrix}}{\begin{vmatrix} 3 & 2 & -5 \\ 2 & -3 & -1 \\ 5 & -2 & 7 \end{vmatrix} \begin{matrix} 3 & 2 \\ 2 & -3 \\ 5 & -2 \end{matrix}}$$

coefficients ⟶

evaluate the denominator first; if it is zero, there is no need to evaluate the numerator

$$= \frac{21 - 18 + 110 - 135 + 2 - 154}{-63 - 10 + 20 - 75 - 6 - 28} = \frac{-174}{-162} = \frac{29}{27}$$

constants ──────

$$y = \frac{\begin{vmatrix} 3 & -1 & -5 \\ 2 & 11 & -1 \\ 5 & 9 & 7 \end{vmatrix} \begin{matrix} 3 & -1 \\ 2 & 11 \\ 5 & 9 \end{matrix}}{-162} = \frac{231 + 5 - 90 + 275 + 27 + 14}{-162}$$

denominator $= -162$ from solution for x

det([A])/det([D]
)
        1.074074074
det([B])/det([D]
)
        -2.851851852

(a)

$$= \frac{462}{-162} = -\frac{77}{27}$$

constants ──────

$$z = \frac{\begin{vmatrix} 3 & 2 & -1 \\ 2 & -3 & 11 \\ 5 & -2 & 9 \end{vmatrix} \begin{matrix} 3 & 2 \\ 2 & -3 \\ 5 & -2 \end{matrix}}{-162} = \frac{-81 + 110 + 4 - 15 + 66 - 36}{-162}$$

denominator $= -162$ from solution for x

det([C])/det([D]
)
        -.2962962963
Ans▶Frac
        -8/27

(b)

Fig. 5.48

$$= \frac{48}{-162} = -\frac{8}{27}$$

Substituting in each of the original equations shows that the solution checks:

$$3\left(\frac{29}{27}\right) + 2\left(-\frac{77}{27}\right) - 5\left(-\frac{8}{27}\right) = \frac{87 - 154 + 40}{27} = \frac{-27}{27} = -1$$

$$2\left(\frac{29}{27}\right) - 3\left(-\frac{77}{27}\right) - \left(-\frac{8}{27}\right) = \frac{58 + 231 + 8}{27} = \frac{297}{27} = 11$$

$$5\left(\frac{29}{27}\right) - 2\left(-\frac{77}{27}\right) + 7\left(-\frac{8}{27}\right) = \frac{145 + 154 - 56}{27} = \frac{243}{27} = 9$$

After the values of x and y were determined, we could have evaluated z by substituting the values of x and y into one of the original equations.

◗

See the chapter introduction.

Solving a Word Problem

◀ **EXAMPLE 4** An 8.0% solution, an 11% solution, and an 18% solution of nitric acid are to be mixed to get 150 mL of a 12% solution. If the volume of acid from the 8.0% solution equals half the volume of acid from the other two solutions, how much of each is needed?

Let x = volume of 8.0% solution needed, y = volume of 11% solution needed, and z = volume of 18% solution needed.

The fact that the sum of the volumes of the three solutions is 150 mL leads to the equation $x + y + z = 150$. Since there are $0.080x$ mL of pure acid from the first solution, $0.11y$ mL from the second solution, and $0.18z$ mL from the third solution, and $0.12(150)$ mL in the final solution, we are led to the equation $0.080x + 0.11y + 0.18z = 18$. Finally, using the last stated condition, we have the equation $0.080x = 0.5(0.11y + 0.18z)$. These equations are then written in the form of Eqs. (5.11) and solved.

$$
\begin{aligned}
x + \quad y + \quad z &= 150 \qquad \text{sum of volumes}\\
0.080x + \ 0.11y + \ 0.18z &= 18 \qquad \text{volumes of pure acid}\\
0.080x \qquad\qquad\qquad &= 0.055y + 0.090z \quad \text{one-half of acid in others}
\end{aligned}
$$

acid in 8.0% solution

$$
\begin{aligned}
x + \quad y + \quad z &= 150 \qquad \text{standard form of}\\
0.080x + \ 0.11y + \ 0.18z &= 18 \qquad \text{Eqs. (5.11) by rewriting}\\
0.080x - 0.055y - 0.090z &= 0 \qquad \text{third equation}
\end{aligned}
$$

$$
x = \frac{\begin{vmatrix} 150 & 1 & 1 \\ 18 & 0.11 & 0.18 \\ 0 & -0.055 & -0.090 \end{vmatrix} \begin{matrix} 150 & 1 \\ 18 & 0.11 \\ 0 & -0.055 \end{matrix}}{\begin{vmatrix} 1 & 1 & 1 \\ 0.080 & 0.11 & 0.18 \\ 0.080 & -0.055 & -0.090 \end{vmatrix} \begin{matrix} 1 & 1 \\ 0.080 & 0.11 \\ 0.080 & -0.055 \end{matrix}}
$$

$$
= \frac{-1.485 + 0 - 0.990 - 0 + 1.485 + 1.620}{-0.0099 + 0.0144 - 0.0044 - 0.0088 + 0.0099 + 0.0072} = \frac{0.630}{0.0084} = 75
$$

$$
y = \frac{\begin{vmatrix} 1 & 150 & 1 \\ 0.080 & 18 & 0.18 \\ 0.080 & 0 & -0.090 \end{vmatrix} \begin{matrix} 1 & 150 \\ 0.080 & 18 \\ 0.080 & 0 \end{matrix}}{0.0084}
$$

$$
= \frac{-1.620 + 2.160 + 0 - 1.440 - 0 + 1.080}{0.0084} = \frac{0.180}{0.0084} = 21
$$

$$
z = \frac{\begin{vmatrix} 1 & 1 & 150 \\ 0.080 & 0.11 & 18 \\ 0.080 & -0.055 & 0 \end{vmatrix} \begin{matrix} 1 & 1 \\ 0.080 & 0.11 \\ 0.080 & -0.055 \end{matrix}}{0.0084}
$$

the value of z can also be found by substituting $x = 75$ and $y = 21$ into the first equation

$$
= \frac{0 + 1.440 - 0.660 - 1.320 + 0.990 - 0}{0.0084} = \frac{0.450}{0.0084} = 54
$$

Therefore, 75 mL of the 8.0% solution, 21 mL of the 11% solution, and 54 mL of the 18% solution are required to make the 12% solution. Results have been rounded off to two significant digits, the accuracy of the data. Checking with the statement of the problem, we see that these volumes total 150 mL. ▶

As we previously noted, the calculations can be done completely on the calculator using either the *matrix* feature (as illustrated with Examples 2 and 3) or using the method outlined after Example 4 on page 159. Briefly, using this method for three equations is: (1) Evaluate and store the value of the denominator; (2) divide the value of each numerator by the value of the denominator, storing the values of x, y, and z; (3) use the stored values of x, y, and z to check the solution.

EXERCISES 5.7

In Exercises 1 and 2, make the given change in the indicated examples of this section and then solve the resulting probems.

1. In Example 1, interchange the first and second rows of the determinant and then evaluate it.

2. In Example 3, change the constant to the right of the = sign in the first equation from −1 to −3, change the constant to the right of the = sign in the third equation from 9 to 11, and then solve the resulting system of equations.

In Exercises 3–14, evaluate the given third-order determinants.

3. $\begin{vmatrix} 5 & 4 & -1 \\ -2 & -6 & 8 \\ 7 & 1 & 1 \end{vmatrix}$

4. $\begin{vmatrix} -7 & 0 & 0 \\ 2 & 4 & 5 \\ 1 & 4 & 2 \end{vmatrix}$

5. $\begin{vmatrix} 8 & 9 & -6 \\ -3 & 7 & 2 \\ 4 & -2 & 5 \end{vmatrix}$

6. $\begin{vmatrix} -2 & 4 & -1 \\ 5 & -10 & 4 \\ 4 & -8 & 2 \end{vmatrix}$

7. $\begin{vmatrix} -3 & -4 & -8 \\ 5 & -1 & 0 \\ 2 & 10 & -1 \end{vmatrix}$

8. $\begin{vmatrix} 10 & 2 & -7 \\ -2 & -3 & 6 \\ 6 & 5 & -2 \end{vmatrix}$

9. $\begin{vmatrix} 4 & -3 & -11 \\ -9 & 2 & -2 \\ 0 & 1 & -5 \end{vmatrix}$

10. $\begin{vmatrix} 9 & -2 & 0 \\ -1 & 3 & -6 \\ -4 & -6 & -2 \end{vmatrix}$

11. $\begin{vmatrix} 25 & 18 & -50 \\ -30 & 40 & -12 \\ -20 & 55 & -22 \end{vmatrix}$

12. $\begin{vmatrix} 20 & 0 & -15 \\ -4 & 30 & 1 \\ 6 & -1 & 40 \end{vmatrix}$

13. $\begin{vmatrix} 0.1 & -0.2 & 0 \\ -0.5 & 1 & 0.4 \\ -2 & 0.8 & 2 \end{vmatrix}$

14. $\begin{vmatrix} 0.25 & -0.54 & -0.42 \\ 1.20 & 0.35 & 0.28 \\ -0.50 & 0.12 & -0.44 \end{vmatrix}$

In Exercises 15–30, solve the given systems of equations by use of determinants. (Exercises 17–28 are the same as Exercises 3–14 of Section 5.6.)

15. $2x + 3y + z = 4$
$3x - z = -3$
$x - 2y + 2z = -5$

16. $4x + y + z = 2$
$2x - y - z = 4$
$3y + z = 2$

17. $x + y + z = 2$
$x - z = 1$
$x + y = 1$

18. $x + y - z = -3$
$x + z = 2$
$2x - y + 2z = 3$

19. $2x + 3y + z = 2$
$-x + 2y + 3z = -1$
$3x - 3y + z = 0$

20. $2x + y - z = 4$
$4x - 3y - 2z = -2$
$8x - 2y - 3z = 3$

21. $5l + 6w - 3h = 6$
$4l - 7w - 2h = -3$
$3l + w - 7h = 1$

22. $3r + s - t = 2$
$r - 2s + t = 0$
$4r - s + t = 3$

23. $2x - 2y + 3z = 5$
$2x + y - 2z = -1$
$4x - y - 3z = 0$

24. $2u + 2v + 3w = 0$
$3u + v + 4w = 21$
$-u - 3v + 7w = 15$

25. $3x - 7y + 3z = 6$
$3x + 3y + 6z = 1$
$5x - 5y + 2z = 5$

26. $8x + y + z = 1$
$7x - 2y + 9z = -3$
$4x - 6y + 8z = -5$

27. $p + 2q + 2r = 0$
$2p + 6q - 3r = -1$
$4p - 3q + 6r = -8$

28. $9x + 12y + 2z = 23$
$21x + 2y - 5z = -25$
$-18x + 6y - 4z = -11$

29. $3.0x + 4.5y - 7.5z = 10.5$
$4.8x - 3.6y - 2.4z = 1.2$
$4.0x - 0.5y + 2.0z = 1.5$

30. $26L - 52M - 52N = 39$
$45L + 96M + 40N = -80$
$55L + 62M - 11N = -48$

In Exercises 31–40, solve the given problems by determinants. In Exercises 34–40, set up appropriate systems of equations. All numbers are accurate to at least two significant digits.

31. In analyzing the forces on the bell-crank mechanism shown in Fig. 5.49, the following equations are found. Find the forces.

$A - 0.60F = 80$

$B - 0.80F = 0$

$6.0A - 10F = 0$ **Fig. 5.49**

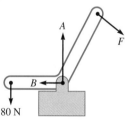

80 N

32. Using Kirchhoff's laws (see the chapter introduction; for the equations, see, for example, Beiser, *Modern Technical Physics*, 6th ed., p. 550) for the electric circuit shown in Fig. 5.50, the following equations are found. Find the indicated currents (in A) I_1, I_2, and I_3.

$19I_1 - 12I_2 = 60$

$12I_1 - 18I_2 + 6.0I_3 = 0$

$6.0I_2 - 18I_3 = 0$

7.0 Ω

I_1 I_2 I_3

60 V   12 Ω   6.0 Ω

12 Ω

Fig. 5.50

33. In a laboratory experiment to measure the acceleration of an object, the distances traveled by the object were recorded for three different time intervals. These data led to the following equations:

$$s_0 + 2v_0 + 2a = 20$$
$$s_0 + 4v_0 + 8a = 54$$
$$s_0 + 6v_0 + 18a = 104$$

Here s_0 is the initial displacement (in m), v_0 is the initial velocity (in m/s), and a is the acceleration (in m/s^2). Find s_0, v_0, and a.

34. A certain 18-hole golf course has par-3, par-4, and par-5 holes, and there are twice as many par-4 holes as par-5 holes. How many holes of each type are there if a golfer has par on every hole for a score of 70?

35. A jar with pennies, nickels, and dimes contains a total of 44 coins worth $2.27. How many of each are there if the number of pennies is 10 less than the number of nickels and dimes together?

36. A mail-order company charges $4 for shipping orders of less than $50, $6 for orders from $50 to $200, and $8 for orders over $200. One day the total shipping charges were $2160 for 384 orders. Find the number of orders shipped at each rate if the number of orders under $50 was 12 more than twice the number of orders over $200.

37. An alloy used in electrical transformers contains nickel (Ni), iron (Fe), and molybdenum (Mo). The percent of Ni is 1% less than five times the percent of Fe. The percent of Fe is 1% more than three times the percent of Mo. Find the percent of each in the alloy.

38. A person invests $20 000, partly at 5.00%, partly at 6.00%, and the remainder at 6.50%. The total annual interest is $1170. Three times the amount invested at 6.00% equals the amount invested at 5.00% and 6.50% combined. How much is invested at each rate?

39. A person spent 1.10 h in a car going to an airport, 1.95 h flying in a jet, and 0.520 h in a taxi to reach the final destination. The jet's speed averaged 12.0 times that of the car, which averaged 15.0 km/h more than the taxi. What was the average speed of each if the trip covered 1140 km?

40. An intravenous aqueous solution is made from three mixtures to get 500 mL with 6.0% of one medication, 8.0% of a second medication, and 86% water. The percents in the mixtures are, respectively, 5.0, 20, 75 (first), 0, 5.0, 95 (second), and 10, 5.0, 85 (third). How much of each is used?

CHAPTER 5 EQUATIONS

Linear equation in one unknown	$ax + b = 0$	(5.1)
Linear equation in two unknowns	$ax + by = c$	(5.2)
System of two linear equations	$a_1 x + b_1 y = c_1$ $a_2 x + b_2 y = c_2$	(5.3)
Definition of slope	$m = \dfrac{y_2 - y_1}{x_2 - x_1}$	(5.4)
Slope-intercept form	$y = mx + b$	(5.5)
Second-order determinant	$\begin{vmatrix} a_1 & b_1 \\ a_2 & b_2 \end{vmatrix} = a_1 b_2 - a_2 b_1$	(5.9)
Cramer's rule	$x = \dfrac{\begin{vmatrix} c_1 & b_1 \\ c_2 & b_2 \end{vmatrix}}{\begin{vmatrix} a_1 & b_1 \\ a_2 & b_2 \end{vmatrix}}$ and $y = \dfrac{\begin{vmatrix} a_1 & c_1 \\ a_2 & c_2 \end{vmatrix}}{\begin{vmatrix} a_1 & b_1 \\ a_2 & b_2 \end{vmatrix}}$	(5.10)
System of three linear equations	$a_1 x + b_1 y + c_1 z = d_1$ $a_2 x + b_2 y + c_2 z = d_2$ $a_3 x + b_3 y + c_3 z = d_3$	(5.11)

Third-order determinant

$$\begin{vmatrix} a_1 & b_1 & c_1 \\ a_2 & b_2 & c_2 \\ a_3 & b_3 & c_3 \end{vmatrix} = a_1b_2c_3 + a_3b_1c_2 + a_2b_3c_1 - a_3b_2c_1 - a_1b_3c_2 - a_2b_1c_3 \quad (5.13)$$

Cramer's rule

$$x = \frac{\begin{vmatrix} d_1 & b_1 & c_1 \\ d_2 & b_2 & c_2 \\ d_3 & b_3 & c_3 \end{vmatrix}}{\begin{vmatrix} a_1 & b_1 & c_1 \\ a_2 & b_2 & c_2 \\ a_3 & b_3 & c_3 \end{vmatrix}} \qquad y = \frac{\begin{vmatrix} a_1 & d_1 & c_1 \\ a_2 & d_2 & c_2 \\ a_3 & d_3 & c_3 \end{vmatrix}}{\begin{vmatrix} a_1 & b_1 & c_1 \\ a_2 & b_2 & c_2 \\ a_3 & b_3 & c_3 \end{vmatrix}} \qquad z = \frac{\begin{vmatrix} a_1 & b_1 & d_1 \\ a_2 & b_2 & d_2 \\ a_3 & b_3 & d_3 \end{vmatrix}}{\begin{vmatrix} a_1 & b_1 & c_1 \\ a_2 & b_2 & c_2 \\ a_3 & b_3 & c_3 \end{vmatrix}} \qquad (5.14)$$

CHAPTER ⑤ REVIEW EXERCISES

In Exercises 1–4, evaluate the given determinants.

1. $\begin{vmatrix} -2 & 5 \\ 3 & 1 \end{vmatrix}$

2. $\begin{vmatrix} 40 & 10 \\ -20 & -60 \end{vmatrix}$

3. $\begin{vmatrix} -18 & -33 \\ -21 & 44 \end{vmatrix}$

4. $\begin{vmatrix} 0.91 & -1.2 \\ 0.73 & -5.0 \end{vmatrix}$

In Exercises 5–8, find the slopes of the lines that pass through the given points.

5. $(2, 0)$, $(4, -8)$

6. $(-1, -5)$, $(-4, 4)$

7. $(40, -20)$, $(-30, -40)$

8. $\left(-6, \frac{1}{2}\right)$, $\left(1, -\frac{7}{2}\right)$

In Exercises 9–12, find the slope and the y-intercepts of the lines with the given equations. Sketch the graphs.

9. $y = -2x + 4$

10. $2y = \frac{2}{3}x - 3$

11. $8x - 2y = 5$

12. $3x = 8 + 3y$

In Exercises 13–20, solve the given systems of equations graphically.

13. $y = 2x - 4$
$y = -\frac{3}{2}x + 3$

14. $y = -3x + 3$
$y = 2x - 6$

15. $4A - B = 6$
$3A + 2B = 12$

16. $2x - 5y = 10$
$3x + y = 6$

17. $7x = 2y + 14$
$y = -4x + 4$

18. $5x = 15 - 3y$
$y = 6x - 12$

19. $3M + 4N = 6$
$2M - 3N = 2$

20. $5x + 2y = 5$
$2x - 4y = 3$

In Exercises 21–30, solve the given systems of equations by using an appropriate algebraic method.

21. $x + 2y = 5$
$x + 3y = 7$

22. $2x - y = 7$
$x + y = 2$

23. $4x + 3y = -4$
$y = 2x - 3$

24. $x = -3y - 2$
$-2x - 9y = 2$

25. $10i - 27v = 29$
$40i + 33v = 69$

26. $3x - 6y = 5$
$7x + 2y = 4$

27. $7x = 2y - 6$
$7y = 12 - 4x$

28. $3R = 8 - 5I$
$6I = 8R + 11$

29. $0.9x - 1.1y = 0.4$
$0.6x - 0.3y = 0.5$

30. $0.42x - 0.56y = 1.26$
$0.98x - 1.40y = -0.28$

In Exercises 31–40, solve the given systems of equations by determinants. (These are the same as for Exercises 21–30.)

31. $x + 2y = 5$
$x + 3y = 7$

32. $2x - y = 7$
$x + y = 2$

33. $4x + 3y = -4$
$y = 2x - 3$

34. $x = -3y - 2$
$-2x - 9y = 2$

35. $10i - 27v = 29$
$40i + 33v = 69$

36. $3x - 6y = 5$
$7x + 2y = 4$

37. $7x = 2y - 6$
$7y = 12 - 4x$

38. $3R = 8 - 5I$
$6I = 8R + 11$

39. $0.9x - 1.1y = 0.4$
$0.6x - 0.3y = 0.5$

40. $0.42x - 0.56y = 1.26$
$0.98x - 1.40y = -0.28$

Ⓦ *In Exercises 41–44, explain your answers. In Exercises 41–43, choose an exercise from among Exercises 31–40 that you think is most easily solved by the indicated method.*

41. Substitution **42.** Addition or subtraction **43.** Determinants

44. Considering the slope m and the y-intercept b, explain how to determine if there is (a) a unique solution, (b) an inconsistent solution, or (c) a dependent solution.

In Exercises 45–48, evaluate the given determinants.

45. $\begin{vmatrix} 4 & -1 & 8 \\ -1 & 6 & -2 \\ 2 & 1 & -1 \end{vmatrix}$

46. $\begin{vmatrix} -500 & 0 & -500 \\ 250 & 300 & -100 \\ -300 & 200 & 200 \end{vmatrix}$

47. $\begin{vmatrix} -2.2 & -4.1 & 7.0 \\ 1.2 & 6.4 & -3.5 \\ -7.2 & 2.4 & -1.0 \end{vmatrix}$ **48.** $\begin{vmatrix} 30 & 22 & -12 \\ 0 & -34 & 44 \\ 35 & -41 & -27 \end{vmatrix}$

In Exercises 49–54, solve the given systems of equations algebraically. In Exercises 53 and 54, the numbers are approximate.

49. $2x + y + z = 4$
$x - 2y - z = 3$
$3x + 3y - 2z = 1$

50. $x + 2y + z = 2$
$3x - 6y + 2z = 2$
$2x - z = 8$

51. $2r + s + 2t = 8$
$3r - 2s - 4t = 5$
$-2r + 3s + 4t = -3$

52. $2u + 2v - w = -2$
$4u - 3v + 2w = -2$
$8u - 4v - 3w = 13$

53. $3.6x + 5.2y - z = -2.2$
$3.2x - 4.8y + 3.9z = 8.1$
$6.4x + 4.1y + 2.3z = 5.1$

54. $32t + 24u + 63v = 32$
$42t - 31u + 19v = 132$
$48t + 12u + 11v = 0$

In Exercises 55–60, solve the given systems of equations by determinants. In Exercises 59 and 60, the numbers are approximate. (These systems are the same as for Exercises 49–54.)

55. $2x + y + z = 4$
$x - 2y - z = 3$
$3x + 3y - 2z = 1$

56. $x + 2y + z = 2$
$3x - 6y + 2z = 2$
$2x - z = 8$

57. $2r + s + 2t = 8$
$3r - 2s - 4t = 5$
$-2r + 3s + 4t = -3$

58. $2u + 2v - w = -2$
$4u - 3v + 2w = -2$
$8u - 4v - 3w = 13$

59. $3.6x + 5.2y - z = -2.2$
$3.2x - 4.8y + 3.9z = 8.1$
$6.4x + 4.1y + 2.3z = 5.1$

60. $32t + 24u + 63v = 32$
$42t - 31u + 19v = 132$
$48t + 12u + 11v = 0$

In Exercises 61–64, solve for x. Here we see that we can solve an equation in which the unknown is an element of a determinant.

61. $\begin{vmatrix} 2 & 5 \\ 1 & x \end{vmatrix} = 3$ **62.** $\begin{vmatrix} -1 & x \\ 3 & 4 \end{vmatrix} = 7$

63. $\begin{vmatrix} x & 1 & 2 \\ 0 & -1 & 3 \\ -2 & 2 & 1 \end{vmatrix} = 5$ **64.** $\begin{vmatrix} 1 & 2 & -1 \\ -2 & 3 & x \\ -1 & 2 & -2 \end{vmatrix} = -3$

In Exercises 65–68, let 1/x = u and 1/y = v. Solve for u and v and then solve for x and y. In this way, we see how to solve systems of equations involving reciprocals.

65. $\dfrac{1}{x} - \dfrac{1}{y} = \dfrac{1}{2}$
$\dfrac{1}{x} + \dfrac{1}{y} = \dfrac{1}{4}$

66. $\dfrac{1}{x} + \dfrac{1}{y} = 3$
$\dfrac{2}{x} + \dfrac{1}{y} = 1$

67. $\dfrac{2}{x} + \dfrac{3}{y} = 3$
$\dfrac{5}{x} - \dfrac{6}{y} = 3$

68. $\dfrac{3}{x} - \dfrac{2}{y} = 4$
$\dfrac{2}{x} + \dfrac{4}{y} = 1$

In Exercises 69 and 70, determine the value of k that makes the system dependent. In Exercises 71 and 72, determine the value of k that makes the system inconsistent.

69. $3x - ky = 6$
$x + 2y = 2$

70. $5x + 20y = 15$
$2x + ky = 6$

71. $kx - 2y = 5$
$4x + 6y = 1$

72. $2x - 5y = 7$
$kx + 10y = 2$

In Exercises 73 and 74, solve the given systems of equations by any appropriate method. All numbers in 73 are accurate to two significant digits, and in 74 they are accurate to three significant digits.

73. A 20-m crane arm with a supporting cable and with a 9000-N box suspended from its end has forces acting on it, as shown in Fig. 5.51. Find the forces (in N) from the following equations.

$F_1 + 2.0F_2 = 26\,000$
$0.87F_1 - F_3 = 0$
$3.0F_1 - 4.0F_2 = 54\,000$

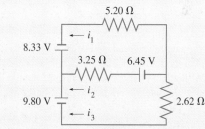

Fig. 5.51

74. In applying Kirchhoff's laws (see Exercise 22 of Section 5.6) to the electric circuit shown in Fig. 5.52, the following equations result. Find the indicated currents (in A).

$i_1 + i_2 + i_3 = 0$
$5.20i_1 - 3.25i_2 = 8.33 - 6.45$
$3.25i_2 - 2.62i_3 = 6.45 - 9.80$

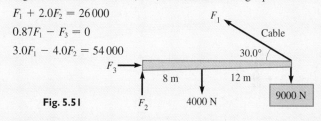

Fig. 5.52

In Exercises 75–88, set up systems of equations and solve by any appropriate method. All numbers are accurate to at least two significant digits.

75. A study found that the fuel consumption for transportation contributes a percent p_1 that is 16% less than the percent p_2 of all other sources combined. Find p_1 and p_2.

76. A certain amount of a fuel contains 150 MJ of potential heat. Part is burned at 80% efficiency, and the rest is burned at 70% efficiency, such that the total amount of heat actually delivered is 114 MJ. Find the amounts burned at each efficiency.

77. A total of 42 tonnes of two types of ore is to be loaded into a smelter. The first type contains 6.0% copper, and the second contains 2.4% copper. Find the necessary amounts of each ore (to the nearest 1 tonne) to produce 2 tonnes of copper.

78. As a 40-m pulley belt makes one revolution, one of the two pulley wheels makes one more revolution than the other. Another wheel of half the radius replaces the smaller wheel and makes six more revolutions than the larger wheel for one revolution of the belt. Find the circumferences of the wheels. See Fig. 5.53.

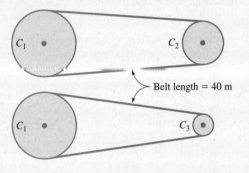

Fig. 5.53

79. A computer analysis showed that the temperature T of the ocean water within 1000 m of a nuclear-plant discharge pipe was given by $T = \dfrac{a}{x + 100} + b$, where x is the distance from the pipe and a and b are constants. If $T = 14°C$ for $x = 0$ and $T = 10°C$ for $x = 900$ m, find a and b.

W 80. In measuring the angles of a triangular parcel, a surveyor noted that one of the angles equaled the sum of the other two angles. What conclusion can be drawn?

81. A satellite is to be launched from a space shuttle. It is calculated that the satellite's speed will be 24 200 km/h if launched directly ahead of the shuttle or 21 400 km/h if launched directly to the rear of the shuttle. What is the speed of the shuttle and the launching speed of the satellite relative to the shuttle? See Fig. 5.54.

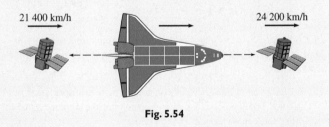

Fig. 5.54

82. The velocity v of sound is a function of the temperature T according to the function $v = aT + b$, where a and b are constants. If $v = 337.5$ m/s for $T = 10.0°C$ and $v = 346.6$ m/s for $T = 25.0°C$, find v as a function of T.

83. The power (in W) dissipated in an electric resistance (in Ω) equals the resistance times the square of the current (in A). If 1.0 A flows through resistance R_1 and 3.0 A flows through resistance R_2, the total power dissipated is 14.0 W. If 3.0 A flows through R_1 and 1.0 A flows through R_2, the total power dissipated is 6.0 W. Find R_1 and R_2.

84. Twelve equal rectangular ceiling panels are placed as shown in Fig. 5.55. If each panel is 150 mm longer than it is wide and a total of 40.0 m of edge and middle strips is used, what are the dimensions of the room?

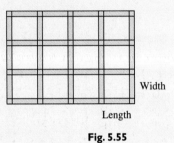

Width

Length

Fig. 5.55

85. The weight of a lever may be considered to be at its center. A 10-m lever of weight w is balanced on a fulcrum 4 m from one end by a load L at that end. If a load of $4L$ is placed at that end, it requires a 20-N weight at the other end to balance the lever. What is the initial load L and the weight w of the lever? (See Exercise 48 of Section 5.4.) See Fig. 5.56.

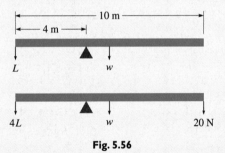

Fig. 5.56

86. Two fuel mixtures, one of 2.0% oil and 98.0% gasoline and another of 8.0% oil and 92.0% gasoline, are to be used to make 10.0 L of a fuel that is 4.0% oil and 96.0% gasoline for use in a chain saw. How much of each mixture is needed?

87. Three computer programs A, B, and C, require a total of 140 MB (megabytes) of hard-disk memory. If three other programs, two requiring the same memory as B and one the same as C, are added to a disk with A, B, and C, a total of 236 MB are required. If three other programs, one requiring the same memory as A and two the same memory as C, are added to a disk with A, B, and C, a total of 304 MB are required. How much memory is required for each of A, B, and C?

88. One ampere of electric current is passed through a solution of sulfuric acid, silver nitrate, and cupric sulfate, releasing hydrogen gas, silver, and copper. A total mass of 1.750 g is released. The mass of silver deposited is 3.40 times the mass of copper deposited, and the mass of copper and 70.0 times the mass of hydrogen combined equals the mass of silver deposited less 0.037 g. How much of each is released?

Writing Exercise

89. Write one or two paragraphs giving reasons for choosing a particular method of solving the following problem. If a first pump is used for 2.2 h and a second pump is used for 2.7 h, 32 m^3 can be removed from a wastewater-holding tank. If the first pump is used for 1.4 h and the second for 2.5 h, 24 m^3 can be removed. How much can each pump remove in 1.0 h? (What is the result to two significant digits?)

CHAPTER ⑤ PRACTICE TEST

1. Find the slope of the line through $(2, -5)$ and $(-1, 4)$.

2. Solve by substitution:

$$x + 2y = 5$$
$$4y = 3 - 2x$$

3. Solve by determinants:

$$3x - 2y = 4$$
$$2x + 5y = -1$$

4. By finding the slope and y-intercept, sketch the graph of $2x + y = 4$.

5. The perimeter of a rectangular ranch is 24 km, and the length is 6.0 km more than the width. Set up equations relating the length l and the width w and then solve for l and w.

6. Solve by addition or subtraction:

$$6N - 2P = 13$$
$$4N + 3P = -13$$

7. In testing an anticholesterol drug, it was found that each milligram of drug administered reduced a person's blood cholesterol level by 2 units. Set up the function relating the cholesterol level C as a function of the dosage d for a person whose cholesterol level is 310 before taking the drug. Sketch the graph.

8. Solve the following system of equations graphically. Determine the values of x and y to the nearest 0.1.

$$2x - 3y = 6$$
$$4x + y = 4$$

9. By volume, one alloy is 60% copper, 30% zinc, and 10% nickel. A second alloy has percents of 50, 30, and 20, respectively. A third alloy is 30% copper and 70% nickel. How much of each alloy is needed to make 100 cm^3 of a resulting alloy with percents of 40, 15, and 45, respectively?

10. Solve for y by determinants:

$$3x + 2y - z = 4$$
$$2x - y + 3z = -2$$
$$x + 4z = 5$$

CHAPTER 6 Factoring and Fractions

The algebraic methods reviewed in Chapter 1 have been enough for our use to now. However, in later chapters additional algebraic methods are needed. In this chapter, we further develop operations with products, quotients, and fractions and show how they are used in solving equations.

The development of the symbols now used in algebra in itself led to advances in mathematics and in science. Until about the year 1500, most problems and their solutions were stated in words, which made them very lengthy and often difficult to follow. For example, to solve the equation $3x + 7 = 8(2x - 5)$, the problem would be stated something like "find a number such that seven added to three times the number is equal to the product of eight and the quantity of five subtracted from twice the number." Then the solution would be written out, word by word.

As time went on, writers abbreviated some of the words in a problem, but they still essentially wrote out the solution in words. Some symbols did start to come into use. For example, the $+$ and $-$ signs first appeared in a published book in the late 1400s, the square root symbol $\sqrt{\ }$ was first used in 1525, and the $=$ sign was introduced in 1557. Then, in the late 1500s, the French lawyer François Viète wrote several articles in which he used symbols, including letters to represent numbers. He so improved the symbolism of algebra that he is often called "the father of algebra." By the mid-1600s, the notation being used was reasonably similar to that we use today.

The use of symbols in algebra made it more useful in all fields of mathematics. It was important in the development of calculus in the 1600s and 1700s. In turn, this gave scientists a powerful tool that allowed for much greater advancement of all areas of science and technology.

Although the primary purpose of this chapter is to develop additional algebraic methods, many technical applications of these operations will be shown. The important applications in optics are noted by the picture of the Hubble telescope above. Other areas of application include electronics and mechanical design.

Great advances in our knowledge of the universe have been made through the use of the Hubble space telescope (shown above) since the mid-1990s. In Section 6.8, we illustrate the use of algebraic operations in the study of planetary motion and in the design of lenses and reflectors in telescopes.

6.1 SPECIAL PRODUCTS

Certain types of algebraic products are used so often that we should be very familiar with them. These are shown in the following equations.

$$a(x + y) = ax + ay \tag{6.1}$$
$$(x + y)(x - y) = x^2 - y^2 \tag{6.2}$$
$$(x + y)^2 = x^2 + 2xy + y^2 \tag{6.3}$$
$$(x - y)^2 = x^2 - 2xy + y^2 \tag{6.4}$$
$$(x + a)(x + b) = x^2 + (a + b)x + ab \tag{6.5}$$
$$(ax + b)(cx + d) = acx^2 + (ad + bc)x + bd \tag{6.6}$$

We see that Eq. (6.1) is the very important *distributive law.* Equations (6.2) to (6.6) are found by using the distributive law along with other basic operations that were developed in Chapter 1.

Equations (6.1) to (6.6) are important **special products** *and should be known* ***thoroughly.*** By carefully studying the products, you will ***see how they are formed,*** and this allows them to be used quickly. Equation (6.6) looks complicated, but when you see how each of the terms on the right side is formed, you will not have to actually memorize it. In using these equations, we must realize that any of the literal numbers may represent any symbol or expression that represents a number.

| Equations (6.1) to (6.6) are identities. (See page 36.) Each is valid for all sets of values of the literal numbers.

◀ EXAMPLE 1   **(a)** Using Eq. (6.1) in the following product, we have

$$6(3r + 2s) = 6(3r) + 6(2s) = 18r + 12s$$

(b) Using Eq. (6.2), we have

Difference of Squares

$$(3r + 2s)(3r - 2s) = (3r)^2 - (2s)^2 = 9r^2 - 4s^2$$

sum of   difference   difference
3r and 2s   of 3r   of squares
        and 2s

Using Eq. (6.1) in (a), we have $a = 6$. In (a) and (b), $3r = x$ and $2s = y$.

◀ EXAMPLE 2   Using Eqs. (6.3) and (6.4) in the following products, we have

$(5a + 2)^2 = (5a + 2)(5a + 2)$

(a) $(5a + 2)^2 = (5a)^2 + 2(5a)(2) + 2^2 = 25a^2 + 20a + 4$ Eq. (6.3)

       square   twice   square
            product

(b) $(5a - 2)^2 = (5a)^2 - 2(5a)(2) + 2^2 = 25a^2 - 20a + 4$ Eq. (6.4)

In these illustrations, we let $x = 5a$ and $y = 2$. *It should be emphasized that*

Avoid a very common error.

CAUTION ▶ $(5a + 2)^2$ is **not** $(5a)^2 + 2^2$, or $25a^2 + 4$

We must very carefully follow the forms of Eqs. (6.3) and (6.4) and be certain to
CAUTION ▶   ***include the middle term,*** 20a. (See Example 5 on page 31.)

◀ EXAMPLE 3   Using Eqs. (6.5) and (6.6) in the following products, we have

(a) $(x + 5)(x - 3) = x^2 + [5 + (-3)]x + (5)(-3) = x^2 + 2x - 15$ Eq. (6.5)

(b) $(4x + 5)(2x - 3) = (4x)(2x) + [(4)(-3) + (5)(2)]x + (5)(-3)$ Eq. (6.6)
$$= 8x^2 - 2x - 15$$

Generally, when we use these special products, we find the middle term mentally and write down the result directly, as shown in the next example.

◀ EXAMPLE 4

(a) $(y - 5)(y + 5) = y^2 - 25$ no middle term—Eq. (6.2)

(b) $(3x - 2)^2 = 9x^2 - 12x + 4$ middle term = $2(3x)(-2)$—Eq. (6.4)

(c) $(x^2 - 4)(x^2 + 7) = x^4 + 3x^2 - 28$ middle term = $(7 - 4)x^2$—Eq. (6.5)
$(x^2)^2 = x^4$

At times, it is necessary to use more than one of the special products to simplify an algebraic expression. When this happens, it may be necessary to write down an intermediate step. This is illustrated in the following examples.

◀ EXAMPLE 5   **(a)** When analyzing the forces on a certain type of beam, the expression $Fa(L - a)(L + a)$ occurs. In expanding this expression, we first multiply $L - a$ by $L + a$ by use of Eq. (6.2). The expansion is completed by using Eq. (6.1), the distributive law.

Eq. (6.2)
$$Fa(L - a)(L + a) = Fa(L^2 - a^2)$$
$$= FaL^2 - Fa^3   \text{Eq. (6.1)}$$

(b) The electrical power delivered to the resistor R in Fig. (6.1) is $R(i_1 + i_2)^2$. Here, i_1 and i_2 are electric currents. To expand this expression, we first perform the square by use of Eq. (6.3) and then complete the expansion by use of Eq. (6.1).

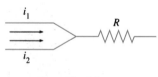

Fig. 6.1

Eq. (6.3)
$$R(i_1 + i_2)^2 = R(i_1^2 + 2i_1 i_2 + i_2^2)$$
$$= Ri_1^2 + 2Ri_1 i_2 + Ri_2^2   \text{Eq. (6.1)}$$

◀ EXAMPLE 6   In determining the product $(x + y - 2)^2$, we may group the quantity $(x + y)$ in an intermediate step. This leads to

$$(x + y - 2)^2 = [(x + y) - 2]^2 = (x + y)^2 - 2(x + y)(2) + 2^2$$
$$= x^2 + 2xy + y^2 - 4x - 4y + 4$$

In this example, we used Eqs. (6.3) and (6.4).

SPECIAL PRODUCTS INVOLVING CUBES

Four other special products occur less frequently. However, they are sufficiently important that they should be readily recognized. They are shown in Eqs. (6.7) to (6.10).

$$(x + y)^3 = x^3 + 3x^2y + 3xy^2 + y^3 \qquad (6.7)$$

$$(x - y)^3 = x^3 - 3x^2y + 3xy^2 - y^3 \qquad (6.8)$$

$$(x + y)(x^2 - xy + y^2) = x^3 + y^3 \qquad (6.9)$$

$$(x - y)(x^2 + xy + y^2) = x^3 - y^3 \qquad (6.10)$$

The following examples illustrate the use of Eqs. (6.7) to (6.10).

EXAMPLE 7 **(a)** $(x + 4)^3 = x^3 + 3(x^2)(4) + 3(x)(4^2) + 4^3$ Eq. (6.7)
$$= x^3 + 12x^2 + 48x + 64$$

(b) $(2x - 5)^3 = (2x)^3 - 3(2x)^2(5) + 3(2x)(5^2) - 5^3$ Eq. (6.8)
$$= 8x^3 - 60x^2 + 150x - 125$$

EXAMPLE 8 **(a)** $(x + 3)(x^2 - 3x + 9) = x^3 + 3^3$ Eq. (6.9)
$$= x^3 + 27$$

(b) $(x - 2)(x^2 + 2x + 4) = x^3 - 2^3$ Eq. (6.10)
$$= x^3 - 8$$

EXERCISES 6.1

In Exercises 1–4, make the given changes in the indicated examples of this section and then solve the resulting problems.

1. In Example 1(b), interchange the + and − signs and then find the product.

2. In Example 3(b), interchange the + and − signs and then find the product.

3. In Example 6, group $y - 2$ instead of $x + y$ and then find the product.

4. In Example 7(a), change + to − in the original expression and then find the product.

In Exercises 5–36, find the indicated products directly by inspection. It should not be necessary to write down intermediate steps [except possibly when using Eq. (6.6)].

5. $40(x - y)$

6. $2x(a - 3)$

7. $2x^2(x - 4)$

8. $3a^2(2a + 7)$

9. $(y + 6)(y - 6)$

10. $(s + 2t)(s - 2t)$

11. $(3v - 2)(3v + 2)$

12. $(ab - c)(ab + c)$

13. $(4x - 5y)(4x + 5y)$

14. $(7s + 2t)(7s - 2t)$

15. $(12 + 5ab)(12 - 5ab)$

16. $(2xy - 11)(2xy + 11)$

17. $(5f + 4)^2$

18. $(i_1 + 3)^2$

19. $(2x + 17)^2$

20. $(9a + 8b)^2$

21. $(L^2 - 1)^2$

22. $(b^2 - 6)^2$

23. $(4a + 7xy)^2$

24. $(3x + 10y)^2$

25. $(0.6s - t)^2$

26. $(0.3p - 0.4q)^2$

27. $(x + 1)(x + 5)$

28. $(y - 8)(y + 5)$

29. $(3 + C^2)(6 + C^2)$

30. $(1 - e^3)(7 - e^3)$

31. $(4x - 5)(5x + 1)$

32. $(2y - 1)(3y - 1)$

33. $(5v - 3)(4v + 5)$

34. $(7s + 6)(2s + 5)$

35. $(3x + 7y)(2x - 9y)$

36. $(8x - y)(3x + 4y)$

Use the special products of this section to determine the products of Exercises 37–62. You may need to write down one or two intermediate steps.

37. $2(x - 2)(x + 2)$

38. $5(n - 5)(n + 5)$

39. $2a(2a - 1)(2a + 1)$

40. $4c(2c - 3)(2c + 3)$

41. $6a(x + 2b)^2$

42. $7r(5r + 2b)^2$

43. $5n^2(2n + 5)^2$

44. $8p(p - 7)^2$

45. $[(2R + 3r)(2R - 3r)]^2$

46. $[(6t - y)(6t + y)]^2$

47. $(x + y + 1)^2$

48. $(x + 2 + 3y)^2$

49. $(3 - x - y)^2$

50. $2(x - y + 1)^2$

51. $(5 - t)^3$

52. $(2s + 3)^3$

53. $(3L + 7R)^3$

54. $(2A - 5B)^3$

55. $(w + h - 1)(w + h + 1)$

56. $(2a - c + 2)(2c - c - 2)$

57. $(x + 2)(x^2 - 2x + 4)$

58. $(s - 3)(s^2 + 3s + 9)$

59. $(4 - 3x)(16 + 12x + 9x^2)$

60. $(2x + 3a)(4x^2 - 6ax + 9a^2)$

61. $(x + y)^2(x - y)^2$

62. $(x - y)(x + y)(x^2 + y^2)$

In Exercises 63–72, use the special products of this section to determine the products. Each comes from the technical area indicated.

63. $P_1(P_0c + G)$ (computers)

64. $h^2L(L + 1)$ (spectroscopy)

65. $4(p + DA)^2$ (photography)

66. $(2J + 3)(2J - 1)$ (lasers)

67. $\frac{1}{2}\pi(R + r)(R - r)$ (architecture)

68. $w(1 - h)(4 - h^2)$ (hydrodynamics)

69. $\frac{L}{6}(x - a)^3$ (mechanics: beams)

70. $(1 - z)^2(1 + z)$ (motion: gyroscope)

71. $L_0[1 + a(T - T_0)]$ (thermal expansion)

72. $(s + 1 + j)(s + 1 - j)$ (electricity)

In Exercises 73–76, solve the given problems.

73. The length of a piece of rectangular floor tile is 3 cm more than twice the side x of a second square piece of tile. The width of the rectangular piece is 3 cm less than twice the side of the square piece. Find the area of the rectangular piece in terms of x (in expanded form).

74. The radius of a circular oil spill is r. It then increases in radius by 40 m before being contained. Find the area of the oil spill at the time it is contained in terms of r (in expanded form). See Fig. 6.2.

Fig. 6.2

75. Find the product $(2x - y - 2)(2x + y + 2)$ by first grouping terms as in Example 6. (*Hint:* This requires grouping the last two terms of each factor.)

(W) **76.** Verify Eqs. (6.8) and (6.9) by multiplication. Then explain why we may refer to Eqs. (6.1) to (6.10) as *identities*.

6.2 FACTORING: COMMON FACTOR AND DIFFERENCE OF SQUARES

At times, we want to determine which expressions can be multiplied together to equal a given algebraic expression. We know from Section 1.7 that when an algebraic expression is the product of two or more quantities, each of these quantities is a ***factor*** of the expression. Therefore, *determining these factors, which is essentially reversing the process of finding a product, is called* **factoring.**

In our work on factoring, we will consider only the factoring of polynomials (see Section 1.9) that have integers as coefficients for all terms. Also, all factors will have integral coefficients. *A polynomial or a factor is called* **prime** *if it contains no factors other than* $+1$ *or* -1 *and plus or minus itself.* Also, we say that an *expression is* **factored completely** *if it is expressed as a product of its prime factors.*

◀ EXAMPLE 1   When we factor the expression $12x + 6x^2$ as

$$12x + 6x^2 = 2(6x + 3x^2)$$

we see that it has not been factored completely. The factor $6x + 3x^2$ is not prime, because it may be factored as

$$6x + 3x^2 = 3x(2 + x)$$

Therefore, the expression $12x + 6x^2$ is factored completely as

$$12x + 6x^2 = 6x(2 + x)$$

The factors x and $(2 + x)$ are prime. We can factor the numerical coefficient, 6, as $2(3)$, but it is standard not to write numerical coefficients in factored form. ▮

To factor expressions easily, we must know how to do algebraic multiplication and really know the special products of the previous section.

NOTE ▶   *The ability to factor algebraic expressions depends heavily on the proper recognition of the special products.*

The special products also give us a way of checking answers and deciding whether a given factor is prime.

COMMON MONOMIAL FACTORS

Often an expression contains a monomial that is common to each term of the expression. Therefore, *the first step in factoring any expression should be to factor out any* **common monomial factor** *that may exist.* To do this, we note the common factor by inspection and then use the reverse of the distributive law, Eq. (6.1), to show the factored form. The following examples illustrate factoring a common monomial factor out of an expression.

For reference, Eq. (6.1) is
$a(x + y) = ax + ay$.

◀ EXAMPLE 2   In factoring $6x - 2y$, we note each term contains a factor of 2:

$$6x - 2y = 2(3x) - 2y = 2(3x - y)$$

Here, 2 is the common monomial factor, and $2(3x - y)$ is the required factored form of $6x - 2y$. Once the common factor has been identified, it is not actually necessary to write a term like $6x$ as $2(3x)$. The result can be written directly.

NOTE ▶   *We check the result by multiplication.* In this case,

$$2(3x - y) = 6x - 2y$$

Since the result of the multiplication gives the original expression, the factored form is correct. ▮

In Example 2, we determined the common factor of 2 by inspection. This is normally the way in which a common factor is found. Once the common factor has been found, the other factor can be determined by dividing the original expression by the common factor.

◀ EXAMPLE 3 Factor: $4ax^2 + 2ax$.

The numerical factor 2 and the literal factors a and x are common to each term. Therefore, the common monomial factor of $4ax^2 + 2ax$ is $2ax$. This means that

$$4ax^2 + 2ax = 2ax(2x) + 2ax(1) = 2ax(2x + 1)$$

Note the presence of the 1 in the factored form. When we divide $4ax^2 + 2ax$ by $2ax$, we get

$$\frac{4ax^2 + 2ax}{2ax} = \frac{4ax^2}{2ax} + \frac{2ax}{2ax}$$
$$= 2x + 1 \hookleftarrow$$

NOTE ▶ noting very carefully that $2ax$ divided by $2ax$ is 1 and *does not simply cancel out leaving nothing*. In cases like this, where the common factor is the same as one of the terms, it is a common error to omit the 1. However,

CAUTION ▶ *we must include the* 1 *in the factor* $2x + 1$.

Without the 1, when the factored form is multiplied out, we would not obtain the original expression.

Usually, the division shown in this example is done by inspection. However, we show it here to emphasize the actual operation that is being performed when we factor out a common factor. ▶

◀ EXAMPLE 4 Factor: $6a^5x^2 - 9a^3x^3 + 3a^3x^2$.

After inspecting each term, we determine that each contains a factor of 3, a^3, and x^2. Thus, the common monomial factor is $3a^3x^2$. This means that

$$6a^5x^2 - 9a^3x^3 + 3a^3x^2 = 3a^3x^2(2a^2 - 3x + 1)$$ ▶

In these examples, note that factoring an expression does not actually change the expression, although it does change the *form* of the expression. In equating the expression to its factored form, we write an *identity*.

It is often necessary to use factoring when solving an equation. This is illustrated in the following example.

◀ EXAMPLE 5 An equation used in the analysis of FM reception is $R_F = \alpha(2R_A + R_F)$. Solve for R_F.

The steps in the solution are as follows:

FM radio was developed in the early 1930s.

$$R_F = \alpha(2R_A + R_F) \qquad \text{original equation}$$
$$R_F = 2\alpha R_A + \alpha R_F \qquad \text{use distributive law}$$
$$R_F - \alpha R_F = 2\alpha R_A \qquad \text{subtract } \alpha R_F \text{ from both sides}$$
$$R_F(1 - \alpha) = 2\alpha R_A \qquad \text{factor out } R_F \text{ on left}$$
$$R_F = \frac{2\alpha R_A}{1 - \alpha} \qquad \text{divide both sides by } 1 - \alpha$$

We see that we collected both terms containing R_F on the left so that we could factor and thereby solve for R_F. ▶

FACTORING THE DIFFERENCE OF TWO SQUARES

For reference, Eq. (6.2) is $(x + y)(x - y) = x^2 - y^2$.

In Eq. (6.2), we see that the product of the sum and the difference of two numbers results in the difference between the squares of the two numbers. Therefore, *factoring the difference of two squares gives factors that are the sum and the difference of the numbers.*

◀ EXAMPLE 6   In factoring $x^2 - 16$, note that x^2 is the square of x and that 16 is the square of 4. Therefore,

Usually in factoring an expression of this type, where it is very clear what numbers are squared, we do not actually write out the middle step as shown. However, if in doubt, write it out.

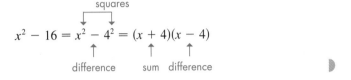

$$x^2 - 16 = x^2 - 4^2 = (x + 4)(x - 4)$$

◀ EXAMPLE 7   **(a)** Since $4x^2$ is the square of $2x$ and 9 is the square of 3, we may factor $4x^2 - 9$ as

$$4x^2 - 9 = (2x)^2 - 3^2 = (2x + 3)(2x - 3)$$

(b) In the same way,

$$(y - 3)^2 - 16x^4 = [(y - 3) + 4x^2][(y - 3) - 4x^2]$$
$$= (y - 3 + 4x^2)(y - 3 - 4x^2)$$

where we note that $16x^4 = (4x^2)^2$.

COMPLETE FACTORING

NOTE ▶   As noted before, ***a common monomial factor should be factored out first.*** However, we must be careful to *see if the other factor can itself be factored.* It is possible, for ex-
CAUTION ▶   ample, that the other factor is the difference of squares. This means that ***complete factoring*** *often requires more than one step.* Be sure to include only prime factors in the final result.

◀ EXAMPLE 8   **(a)** In factoring $20x^2 - 45$, note a common factor of 5 in each term. Therefore, $20x^2 - 45 = 5(4x^2 - 9)$. However, the factor $4x^2 - 9$ itself is the difference of squares. Therefore, $20x^2 - 45$ is completely factored as

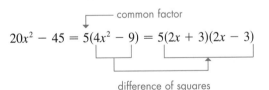

$$20x^2 - 45 = 5(4x^2 - 9) = 5(2x + 3)(2x - 3)$$

difference of squares

(b) In factoring $x^4 - y^4$, note that we have the difference of two squares. Therefore, $x^4 - y^4 = (x^2 + y^2)(x^2 - y^2)$. However, the factor $x^2 - y^2$ is also the difference of squares. This means that

$$x^4 - y^4 = (x^2 + y^2)(x^2 - y^2) = (x^2 + y^2)(x + y)(x - y)$$

CAUTION ▶   ***The factor $x^2 + y^2$ is prime.*** It is **not** equal to $(x + y)^2$. (See Example 2 of Section 6.1.)

FACTORING BY GROUPING

The terms in a polynomial can sometimes be grouped such that the polynomial can then be factored by the methods of this section. The following example illustrates this method of *factoring by grouping.*

◀ EXAMPLE 9   Factor: $2x - 2y + ax - ay$.
  We see that there is no common factor to all four terms, but that each of the first two terms contains a factor of 2, and each of the third and fourth terms contains a factor of a. Grouping terms this way and then factoring each group, we have

$$2x - 2y + ax - ay = (2x - 2y) + (ax - ay)$$
$$= 2(x - y) + a(x - y)$$

NOTE ▶   Each of the two terms we now have contains a ***common binomial factor*** of $x - y$. Since this is a common factor, we have

$$2(x - y) + a(x - y) = (x - y)(2 + a)$$

This means that

$$2x - 2y + ax - ay = (x - y)(2 + a)$$

where the expression on the right is the factored form of the polynomial.

The general method of factoring by grouping can be used with several types of groupings. We will discuss another type in the following section.

EXERCISES 6.2

In Exercises 1–4, make the given changes in the indicated examples of this section and then solve the indicated problems.

1. In Example 3, change the + sign to − and then factor.

2. In Example 3, set the given expression equal to B and then solve for a.

3. In Example 8(a), change the coefficient of the first term from 20 to 5 and then factor.

4. In Example 9, change both − signs to + and then factor.

In Exercises 5–44, factor the given expressions completely.

5. $6x + 6y$
6. $3a - 3b$
7. $5a - 5$
8. $2x^2 + 2$
9. $3x^2 - 9x$
10. $4s^2 + 20s$
11. $7b^2h - 28b$
12. $5a^2 - 20ax$
13. $72n^2 + 6n$
14. $18p^3 - 3p^2$
15. $2x + 4y - 8z$
16. $23a - 46b + 69c$
17. $3ab^2 - 6ab + 12ab^3$
18. $4pq - 14q^2 - 16pq^2$
19. $12pq^2 - 8pq - 28pq^3$
20. $27a^2b - 24ab - 9a$

21. $2a^2 - 2b^2 + 4c^2 - 6d^2$
22. $5a + 10ax - 5ay - 20az$
23. $x^2 - 4$
24. $r^2 - 25$
25. $100 - 9A^2$
26. $49 - Z^4$
27. $36a^4 + 1$
28. $81z^2 - 1$
29. $81s^2 - 25t^2$
30. $36s^2 - 121t^2$
31. $144n^2 - 169p^4$
32. $36a^2b^2 + 169c^2$
33. $(x + y)^2 - 9$
34. $(a - b)^2 - 1$
35. $2x^2 - 8$
36. $5a^2 - 125$
37. $3x^2 - 27z^2$
38. $4x^2 - 100y^2$
39. $2(I - 3)^2 - 8$
40. $a(x + 2)^2 - ay^2$
41. $x^4 - 16$
42. $y^4 - 81$
43. $x^8 - 1$
44. $2x^4 - 8y^4$

In Exercises 45–48, solve for the indicated letter.

45. $2a - b = ab + 3$, for a
46. $n(x + 1) = 5 - x$, for x
47. $3 - 2s = 2(3 - st)$, for s
48. $k(2 - y) = y(2k - 1)$, for y

In Exercises 49–56, factor the given expressions by grouping as illustrated in Example 9.

49. $3x - 3y + bx - by$

50. $am + an + cn + cm$

51. $a^2 + ax - ab - bx$

52. $2y - y^2 - 6y^4 + 12y^3$

53. $x^3 + 3x^2 - 4x - 12$

54. $S^3 - 5S^2 - S + 5$

55. $x^2 - y^2 + x - y$

56. $4p^2 - q^2 + 2p + q$

In Exercises 57 and 58, evaluate the given expressions by using factoring. The results may be checked with a calculator.

57. $\dfrac{8^9 - 8^8}{7}$

58. $\dfrac{5^9 - 5^7}{7^2 - 5^2}$

In Exercises 59–66, factor the expressions completely. In Exercises 65 and 66, it is necessary to set up the proper expression. Each expression comes from the technical area indicated.

59. $2\pi rh + 2\pi r^2$ (container surface area)

60. $4d^2D^2 - 4d^3D - d^4$ (machine design)

61. $Rv + Rv^2 + Rv^3$ (business)

62. $PbL^2 - Pb^3$ (architecture)

63. $rR^2 - r^3$ (pipeline flow)

64. $p_1R^2 - p_1r^2 - p_2R^2 + p_2r^2$ (fluid flow)

65. As large as possible square is cut from a circular metal plate of radius r. Express in factored form the area of the metal pieces that are left.

66. A pipe of outside diameter d is inserted into a pipe of inside radius r. Express in factored form the cross-sectional area within the larger pipe that is outside the smaller pipe.

In Exercises 67–72, solve for the indicated letter. Each equation comes from the technical area indicated.

67. $i_1R_1 = (i_2 - i_1)R_2$, for i_1 (electricity: ammeter)

68. $nV + n_1v = n_1V$, for n_1 (acoustics)

69. $3BY + 5Y = 9BS$, for B (physics: elasticity)

70. $Sq + Sp = Spq + p$, for q (computer design)

71. $ER = AtT_0 - AtT_1$, for t (energy conservation)

72. $R = kT_2^4 - kT_1^4$ (solve for k and factor the resulting denominator) (energy: radiation)

6.3 FACTORING TRINOMIALS

In the previous section, we introduced the concept of factoring and considered factoring based on special products of Eqs. (6.1) and (6.2). We now note that the special products formed from Eqs. (6.3) to (6.6) all result in trinomial (three-term) polynomials. Thus, trinomials of the types formed by these products are important expressions to be factored, and this section is devoted to them.

When factoring an expression based on Eq. (6.5), we start with the expression on the right and then find the factors that are at the left. Therefore, by writing Eq. (6.5) with sides reversed, we have

For reference, Eq. (6.5) is
$(x + a)(x + b) = x^2 + (a + b)x + ab.$

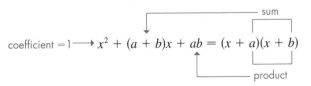

We are to find integers a and b, and they are found by noting that

Factoring a trinomial for which the coefficient of x^2 is 1.

    1. *the coefficient of x^2 is 1,*

    2. *the final constant is the product of the constants a and b in the factors, and*

CAUTION ▶

    3. *the coefficient of x is the sum of a and b.*

As in Section 6.2, we consider only factors in which all terms have integral coefficients.

◀ EXAMPLE 1   In factoring $x^2 + 3x + 2$, we set it up as

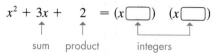

The constant 2 tells us that the product of the required integers is 2. Thus, the only possibilities are 2 and 1 (or 1 and 2). The plus sign before the 2 indicates that the sign before the 1 and 2 in the factors must be the same, either plus or minus. Since the coefficient of x, 3, is the sum of the integers, the plus sign before the 3 tells us that both signs are positive. Therefore,

$$x^2 + 3x + 2 = (x + 2)(x + 1)$$

In factoring $x^2 - 3x + 2$, the analysis is the same until we note that the middle term is negative. This tells us that both signs are negative in this case. Therefore,

$$x^2 - 3x + 2 = (x - 2)(x - 1)$$

For a trinomial containing x^2 and 2 to be factorable, the middle term must be $+3x$ or $-3x$. No other combination of integers gives the proper middle term. Therefore, the expression

$$x^2 + 4x + 2$$

cannot be factored. The integers would have to be 2 and 1, but the middle term would not be $4x$.   ▶

◀ EXAMPLE 2   **(a)** In order to factor $x^2 + 7x - 8$, *we must find two integers whose product is* -8 *and whose sum is* $+7$. The possible factors of -8 are

$$-8 \text{ and } +1 \qquad +8 \text{ and } -1 \qquad -4 \text{ and } +2 \qquad +4 \text{ and } -2$$

Inspecting these, we see that only $+8$ and -1 have the sum of $+7$. Therefore,

$$x^2 + 7x - 8 = (x + 8)(x - 1)$$

> Always multiply the factors together to check to see that you get the **correct middle term.**

In choosing the correct values for the integers, it is usually fairly easy to find a pair for which the product is the final term. However, choosing the pair of integers that correctly fits the middle term is the step that often is not done properly. *Special attention must be given to choosing the integers so that the expansion of the resulting factors has the correct **middle term** of the original expression.*

(b) In the same way, we have

$$x^2 - x - 12 = (x - 4)(x + 3)$$

since -4 and $+3$ is the only pair of integers whose product is -12 and whose sum is -1.

(c) Also,

$$x^2 - 5xy + 6y^2 = (x - 3y)(x - 2y)$$

since -3 and -2 is the only pair of integers whose product is $+6$ and whose sum is -5. Here, we find second terms of each factor with a product of $6y^2$ and sum of $-5xy$, which means that each second term must have a factor of y, as we have shown above.   ▶

For reference, Eqs. (6.3) and (6.4) are
$(x + y)^2 = x^2 + 2xy + y^2$ and
$(x - y)^2 = x^2 - 2xy + y^2$.

In factoring a trinomial in which the first and third terms are perfect squares, we may find that the expression fits the form of Eq. (6.3) or Eq. (6.4), as well as the form of Eq. (6.5). The following example illustrates this case.

◀ EXAMPLE 3 **(a)** To factor $x^2 + 10x + 25$, we must find two integers whose product is $+25$ and whose sum is $+10$. Since $5^2 = 25$, we note that this expression may fit the form of Eq. (6.3). This can be the case only if the first and third terms are perfect squares. Since the sum of $+5$ and $+5$ is $+10$, we have

$$x^2 + 10x + 25 = (x + 5)(x + 5)$$

or

$$x^2 + 10x + 25 = (x + 5)^2$$

(b) To factor $A^4 - 20A^2 + 100$, we note that $A^4 = (A^2)^2$ and $100 = 10^2$. This means that this expression may fit the form of Eq. (6.4). Since the sum of -10 and -10 is -20, we have

$$A^4 - 20A^2 + 100 = (A^2 - 10)(A^2 - 10)$$
$$= (A^2 - 10)^2$$

(c) Just because the first and third terms are perfect squares, we must realize that the expression may be factored but does *not* fit the form of either Eq. (6.3) or (6.4). One example, using x^2 and 100 for the first and third terms, is

$$x^2 + 29x + 100 = (x + 25)(x + 4)$$

▶

FACTORING GENERAL TRINOMIALS

For reference, Eq. (6.6) is
$(ax + b)(cx + d) = acx^2 + (ad + bc)x + bd$.

Factoring expressions based on the special product of Eq. (6.6) often requires some trial and error. However, the amount of trial and error can be kept to a minimum with a careful analysis of the coefficients of x^2 and the constant. Rewriting Eq. (6.6) with sides reversed, we have

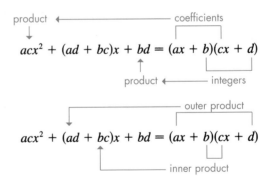

This diagram shows us that

Factoring a trinomial for which the coefficient of x^2 is any positive integer.

1. *the coefficient of x^2 is the product of the coefficients a and c in the factors,*

2. *the final constant is the product of the constants b and d in the factors, and*

CAUTION ▶ 3. *the coefficient of x is the sum of the inner and outer products.*

In finding the factors, we must try possible combinations of a, b, c, and d that give

CAUTION ▶ *the proper inner and outer products for the middle term*

of the given expression.

◖EXAMPLE 4   When factoring $2x^2 + 11x + 5$, we take the factors of 2 to be $+2$ and $+1$ (we use only positive coefficients a and c when the coefficient of x^2 is positive). We now set up the factoring as

$$2x^2 + 11x + 5 = (2x\boxed{})(x\boxed{})$$

NOTE ▶   Since the product of the integers to be found is $+5$, only integers of the same sign need to be considered. Also since ***the sum of the outer and inner products is*** $+11$**,** the integers are positive. The factors of $+5$ are $+1$ and $+5$, and -1 and -5, which means that $+1$ and $+5$ is the only possible pair. Now, trying the factors

$$(2x + 5)(x + 1)$$
$$+5x +2x + 5x = +7x$$
$$+2x$$

we see that $7x$ is not the correct middle term.
  Next, trying

$$(2x + 1)(x + 5)$$
$$+x +x + 10x = +11x$$
$$+10x$$

we have the correct sum of $+11x$. Therefore,

$$2x^2 + 11x + 5 = (2x + 1)(x + 5)$$

For a trinomial with a first term $2x^2$ and a constant $+5$ to be factorable, we can now see that the middle term must be either $+11x$ or $+7x$. This means that $2x^2 + 7x + 5 = (2x + 5)(x + 1)$, but a trinomial such as $2x^2 + 8x + 5$ is not factorable.   ◗

◖EXAMPLE 5   In factoring $4x^2 + 4x - 3$, the coefficient 4 in $4x^2$ shows that the possible coefficients of x in the factors are 4 and 1, or 2 and 2. The 3 shows that the only NOTE ▶ possible constants in the factors are 1 and 3, and *the minus sign with the 3 tells us that these integers have* **different signs.** This gives us the following possible combinations of factors, along with the resulting sum of the outer and inner products:

$$(4x - 3)(x + 1): \ 4x - 3x = +x$$
$$(4x - 1)(x + 3): \ 12x - x = +11x$$
$$(2x + 3)(2x - 1): \ -2x + 6x = +4x$$
$$(2x - 3)(2x + 1): \ 2x - 6x = -4x$$

We see that the factors that have the correct middle term of $+4x$ are $(2x + 3)(2x - 1)$. This means that

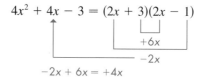

$$4x^2 + 4x - 3 = (2x + 3)(2x - 1)$$
$$+6x$$
$$-2x$$
$$-2x + 6x = +4x$$

Expressing the result with the factors reversed is an equally correct answer.
NOTE ▶   Another hint that is given by the coefficients of the original expression is that *the plus sign with the 4x tells us that the larger of the outer and inner products must be positive.*   ◗

EXAMPLE 6 $\quad 6s^2 + 19st - 20t^2 = (6s - 5t)(s + 4t)$

$$+24st - 5st = +19st$$

CAUTION ▶ There are numerous possibilities for the combinations of 6 and 20. However, we must remember to **check carefully that the** **middle term** *of the expression is the* ***proper result of the factors*** *we have chosen.*

EXAMPLE 7 **(a)** In factoring $9x^2 - 6x + 1$, note that $9x^2$ is the square of $3x$ and 1 is the square of 1. Therefore, we recognize that this expression might fit the perfect square form of Eq. (6.4). This leads us to factor it tentatively as

$$9x^2 - 6x + 1 = (3x - 1)^2 \qquad \text{check middle term: } 2(3x)(-1) = -6x$$

However, before we can be certain that this is correct, we must check to see if the middle term of the expansion of $(3x - 1)^2$ is $-6x$, which is what it must be to fit the form of Eq. (6.4). When we expand $(3x - 1)^2$, we find that the middle term is $-6x$ and therefore that the factorization is correct.

(b) In the same way, we have

$$36x^2 + 84xy + 49y^2 = (6x + 7y)^2 \qquad \text{check middle term: } 2(6x)(7y) = 84xy$$

NOTE ▶ since the middle term of the expansion of $(6x + 7y)^2$ is $84xy$. *We must be careful to include the factors of y in the second terms of the factors.*

Common Monomial Factors
As noted before, we must be careful to factor an expression completely. *We first look for common monomial factors* and *then check each resulting factor to see if it can be factored* when we complete each step.

EXAMPLE 8 When factoring $2x^2 + 6x - 8$, first note the common monomial factor of 2. This leads to

$$2x^2 + 6x - 8 = 2(x^2 + 3x - 4)$$

Now notice that $x^2 + 3x - 4$ is also factorable. Therefore,

$$2x^2 + 6x - 8 = 2(x + 4)(x - 1)$$

Now each factor is prime.

Having noted the common factor of 2 prevents our having to check factors of 2 and 8. If we had not noted the common factor, we might have arrived at

$$(2x + 8)(x - 1) \quad \text{or} \quad (2x - 2)(x + 4)$$

CAUTION ▶ (possibly after a number of trials). Each is correct as far as it goes, but also each of these is ***not complete.*** Since $2x + 8 = 2(x + 4)$, or $2x - 2 = 2(x - 1)$, we can arrive at the proper result shown above. In this way, we would have

$$2x^2 + 6x - 8 = (2x + 8)(x - 1)$$
$$= 2(x + 4)(x - 1)$$

or $\qquad 2x^2 + 6x - 8 = (2x - 2)(x + 4)$
$$= 2(x - 1)(x + 4)$$

Although these factorizations are correct, *it is better to factor out the common factor first.* By doing so, the number of possible factoring combinations is greatly reduced, and the factoring can be done more easily.

Liquid-fuel rockets were designed in the United States in the 1920s but were developed by German engineers. They were first used in the 1940s during the Second World War.

EXAMPLE 9 A study of the path of a certain rocket leads to the expression $16t^2 + 240t - 1600$, where t is the time of flight. Factor this expression.

An inspection shows that there is a common factor of 16. (This might be found by noting successive factors of 2 or 4.) Factoring out 16 leads to

$$16t^2 + 240t - 1600 = 16(t^2 + 15t - 100)$$
$$= 16(t + 20)(t - 5)$$

Here, factors of 100 need to be checked for sums equal to 15. This might take a little time, but it is much simpler than looking for factors of 16 and 1600 with sums equal to 240.

FACTORING BY GROUPING

In the previous section, we introduced the method of factoring by grouping. The following examples use this method for factoring (1) a trinomial and (2) an expression that can be written as the difference of squares.

EXAMPLE 10 Factor the trinomial $6x^2 + 7x - 20$ by grouping.

To factor the trinomial $ax^2 + bx + c$ by grouping, we use the following procedure.

1. Find two numbers whose product is ac and whose sum is b.

2. Write the trinomial with two x-terms having these numbers as coefficients.

3. Complete the factorization by grouping.

For $6x^2 + 7x - 20$, this means we want two numbers with a product of -120 and a sum of 7. Trying products of numbers *with different signs* and a sum of 7, we find the numbers that work are -8 and 15. We then complete steps 2 and 3 of the above procedure as follows:

$$
\begin{aligned}
6x^2 + 7x - 20 &= 6x^2 - 8x + 15x - 20 && 7x = -8x + 15x \\
&= (6x^2 - 8x) + (15x - 20) && \text{group first two terms and last two terms} \\
&= 2x(3x - 4) + 5(3x - 4) && \text{find common factor of each group and} \\
&= (3x - 4)(2x + 5) && \text{note common factor of } 3x - 4
\end{aligned}
$$

Multiplication verifies that these are the correct factors.

EXAMPLE 11 Factor: $x^2 - 4xy + 4y^2 - 9$.

We see that the first three terms of this expression represent $(x - 2y)^2$. Thus, grouping these terms, we have the following solution:

$$
\begin{aligned}
x^2 - 4xy + 4y^2 - 9 &= (x^2 - 4xy + 4y^2) - 9 && \text{group terms} \\
&= (x - 2y)^2 - 9 && \text{factor grouping (note difference of squares)} \\
&= [(x - 2y) + 3][(x - 2y) - 3] && \text{factor difference of squares} \\
&= (x - 2y + 3)(x - 2y - 3)
\end{aligned}
$$

Note that not all groupings work. If we had seen the combination $4y^2 - 9$, which is factorable, and then grouped the first two terms and the last two terms, this would not have led to the factorization.

E X E R C I S E S 6.3

In Exercises 1–6, make the given changes in the indicated examples of this section and then factor.

1. In Example 1, change the 3 to 4 and the 2 to 3.
2. In Example 2(a), change the + before $7x$ to −.
3. In Example 4, change the + before $11x$ to −.
4. In Example 6, change the coefficient 19 to 7.
5. In Example 8, change the 8 to 36.
6. In Example 10, change the + before $7x$ to −.

In Exercises 7–56, factor the given expressions completely.

7. $x^2 + 5x + 4$
8. $x^2 - 5x - 6$
9. $s^2 - s - 42$
10. $a^2 + 14a - 32$
11. $t^2 + 5t - 24$
12. $r^2 - 11r + 18$
13. $x^2 + 2x + 1$
14. $D^2 + 8D + 16$
15. $x^2 - 4xy + 4y^2$
16. $b^2 - 12bc + 36c^2$
17. $3x^2 - 5x - 2$
18. $2n^2 - 13n - 7$
19. $3y^2 - 8y - 3$
20. $5x^2 + 9x - 2$
21. $2s^2 + 13s + 11$
22. $7y^2 - 12y + 5$
23. $3f^4 - 16f^2 + 5$
24. $5R^4 - 3R^2 - 2$
25. $2t^2 + 7t - 15$
26. $3n^2 - 20n + 20$
27. $3t^2 - 7tu + 4u^2$
28. $3x^2 + xy - 14y^2$
29. $4x^2 - 3x - 7$
30. $2z^2 + 13z - 5$
31. $9x^2 + 7xy - 2y^2$
32. $4r^2 + 11rs - 3s^2$
33. $4m^2 + 20m + 25$
34. $16q^2 + 24q + 9$
35. $4x^2 - 12x + 9$
36. $a^2c^2 - 2ac + 1$
37. $9t^2 - 15t + 4$
38. $6t^4 + t^2 - 12$
39. $8b^6 + 31b^3 - 4$
40. $12n^2 + 8n - 15$
41. $4p^2 - 25pq + 6q^2$
42. $12x^2 + 4xy - 5y^2$
43. $12x^2 + 47xy - 4y^2$
44. $8r^2 - 14rs - 9s^2$
45. $2x^2 - 14x + 12$
46. $6y^2 - 33y - 18$
47. $4x^2 + 14x - 8$
48. $12B^2 + 22BH - 4H^2$
49. $ax^3 + 4a^2x^2 - 12a^3x$
50. $6x^4 - 13x^3 + 5x^2$
51. $a^2 + 2ab + b^2 - 4$
52. $x^2 - 6xy + 9y^2 - 4z^2$
53. $25a^2 - 25x^2 - 10xy - y^2$
54. $r^2 - s^2 + 2st - t^2$
55. $4x^{2n} + 13x^n - 12$
56. $12B^{2n} + 19B^nH - 10H^2$

In Exercises 57–68, factor the given expressions completely. Each is from the technical area indicated.

57. $3p^2 + 9p - 54$ (business)
58. $2x^3 - 28x^2 + 98x$ (container design)
59. $4s^2 + 16s + 12$ (electricity)
60. $3e^2 + 18e - 1560$ (fuel efficiency)
61. $200n^2 - 2100n - 3600$ (biology)
62. $bT^2 - 40bT + 400b$ (thermodynamics)
63. $V^2 - 2nBV + n^2B^2$ (chemistry)
64. $a^4 + 8a^2\pi^2f^2 + 16\pi^4f^4$ (periodic motion: energy)
65. $wx^4 - 5wLx^3 + 6wL^2x^2$ (beam design)
66. $1 - 2r^2 + r^4$ (lasers)
67. $3Adu^2 - 4Aduv + Adv^2$ (water power)
68. $k^2A^2 + 2k\lambda A + \lambda^2 - \alpha^2$ (robotics)

In Exercises 69–72, solve the given problems.

69. Often the sum of squares cannot be factored, but $x^4 + 4$ can be factored. Add and subtract $4x^2$ and then use factoring by grouping.
70. Using the method of Exercise 69, factor $x^4 + x^2 + 1$. (First add and subtract x^2.)
71. If the sum of squares $36x^2 + 9$ can be factored, factor it.
(W) 72. Explain why most students would find $24x^2 - 23x - 12$ more difficult to factor than $23x^2 - 18x - 5$.

6.4 THE SUM AND DIFFERENCE OF CUBES

We have seen that the difference of squares can be factored, but that the sum of squares often cannot be factored. We now turn our attention to the sum and difference of cubes, both of which can be factored. Writing Eqs. (6.9) and (6.10) with sides reversed,

Note carefully the positions of the + and − signs.

$$x^3 + y^3 = (x + y)(x^2 - xy + y^2) \qquad (6.9)$$
$$x^3 - y^3 = (x - y)(x^2 + xy + y^2) \qquad (6.10)$$

In these equations, the second factors are prime.

For reference, Eqs. (6.9) and (6.10) are
$$x^3 + y^3 = (x + y)(x^2 - xy + y^2)$$
$$x^3 - y^3 = (x - y)(x^2 + xy + y^2).$$

EXAMPLE 1

(a) $x^3 + 8 = x^3 + 2^3$

$= (x + 2)[(x)^2 - 2x + 2^2]$
$= (x + 2)(x^2 - 2x + 4)$

(b) $x^3 - 1 = x^3 - 1^3$

$= (x - 1)[(x)^2 + (1)(x) + 1^2]$
$= (x - 1)(x^2 + x + 1)$

Table of Cubes

$1^3 =$	1
$2^3 =$	8
$3^3 =$	27
$4^3 =$	64
$5^3 =$	125
$6^3 =$	216

EXAMPLE 2 $\quad 8 - 27x^3 = 2^3 - (3x)^3$ $\qquad$ $8 = 2^3$ and $27x^3 = (3x)^3$

$= (2 - 3x)[2^2 + 2(3x) + (3x)^2]$
$= (2 - 3x)(4 + 6x + 9x^2)$

As we have stated in the previous sections on factoring, *we should start any factoring process by first checking for any common monomial factor* that might be present in each term of the expression. Also, we should check that our result has been factored **completely.**

EXAMPLE 3 $\quad$ In factoring $ax^5 - ax^2$, we first note that each term has a common factor of ax^2. This is factored out to get $ax^2(x^3 - 1)$. However, the expression is not completely factored since $1 = 1^3$, which means that $x^3 - 1$ is the difference of cubes. We complete the factoring by the use of Eq. (6.10). Therefore,

$$ax^5 - ax^2 = ax^2(x^3 - 1)$$
$$= ax^2(x - 1)(x^2 + x + 1)$$

EXAMPLE 4 $\quad$ The volume of material used to make a steel bearing with a hollow core is given by $\frac{4}{3}\pi R^3 - \frac{4}{3}\pi r^3$. Factor this expression.

$$\frac{4}{3}\pi R^3 - \frac{4}{3}\pi r^3 = \frac{4}{3}\pi(R^3 - r^3) \qquad \text{common factor of } \frac{4}{3}\pi$$

$$= \frac{4}{3}\pi(R - r)(R^2 + Rr + r^2) \qquad \text{using Eq. (6.10)}$$

Summary of Methods of Factoring

In our study of factoring, we have seen that we should

first factor out any common monomial factor

and then see if the remaining expression can be further factored as one of the following types:

1. *Difference of squares*

2. *Factorable trinomial*

3. *Sum or difference of cubes*

4. *Factorable by grouping*

Remember, an expression should be factored **completely.**

EXERCISES **6.4**

In Exercises 1 and 2, make the given changes in the indicated examples of this section and then factor.

1. In Example 1(a), change the $+$ before the 8 to $-$.

2. In Example 3, change $-ax^2$ to $+a^4x^2$.

In Exercises 3–26, factor the given expressions completely.

3. $x^3 + 1$ **4.** $R^3 + 27$

5. $8 - t^3$ **6.** $8r^3 - 1$

7. $27x^3 - 8a^3$ **8.** $64x^3 + 125$

9. $2x^3 + 16$ **10.** $3y^3 - 81$

11. $6A^6 - 6A^3$ **12.** $8s^9 - 8$

13. $6x^3y - 6x^3y^4$ **14.** $12a^3 + 96a^3b^3$

15. $x^6y^3 + x^3y^6$ **16.** $16r^3 - 432$

17. $3a^6 - 3a^2$ **18.** $x^6 - 81y^2$

19. $0.001R^3 - 0.064r^3$ **20.** $0.027x^3 + 0.125$

21. $27L^6 + 216L^3$ **22.** $a^3s^5 - 8000a^3s^2$

23. $(a + b)^3 + 64$ **24.** $125 + (2x + y)^3$

25. $64 - x^6$ **26.** $a^6 - 27b^6$

In Exercises 27–32, factor the given expressions completely. Each is from the technical area indicated.

27. $2x^3 + 250$ (computer image)

28. $kT^3 - kT_0^3$ (thermodynamics)

29. $D^4 - d^3D$ (machine design)

30. $(h + 2t)^3 - h^3$ (container design)

31. $QH^4 + Q^4H$ (thermodynamics)

32. $\left(\dfrac{s}{r}\right)^{12} - \left(\dfrac{s}{r}\right)^6$ (molecular interaction)

In Exercises 33 and 34, perform the indicated operations.

33. Perform the division $(x^5 - y^5) \div (x - y)$. Noting the result, determine the quotient $(x^7 - y^7) \div (x - y)$, without dividing. From these results, factor $x^5 - y^5$ and $x^7 - y^7$.

34. Perform the division $(x^5 + y^5) \div (x + y)$. Noting the result, determine the quotient $(x^7 + y^7) \div (x + y)$, without dividing. From these results, factor $x^5 + y^5$ and $x^7 + y^7$.

In Exercises 35 and 36, solve the given problems.

35. Factor $x^6 - y^6$ as the difference of cubes.

W **36.** Factor $x^6 - y^6$ as the difference of squares. Then explain how the result of Exercise 35 can be shown to be the same as the result in this exercise. (*Hint:* See Exercise 70 on page 191.)

6.5 EQUIVALENT FRACTIONS

When we deal with algebraic expressions, we must be able to work effectively with fractions. Since algebraic expressions are representations of numbers, the basic operations on fractions from arithmetic form the basis of our algebraic operations. In this section, we demonstrate a very important property of fractions, and in the following two sections, we establish the basic algebraic operations with fractions.

Fundamental Principle of Fractions

    This important property of fractions, often referred to as the **fundamental principle of fractions,** is that *the value of a fraction is unchanged if both numerator and denominator are multiplied or divided by the same number, provided this number is not zero.* Two fractions are said to be **equivalent** if one can be obtained from the other by use of the fundamental principle.

 EXAMPLE 1   If we multiply the numerator and the denominator of the fraction $\frac{6}{8}$ by 2, we obtain the equivalent fraction $\frac{12}{16}$. If we divide the numerator and the denominator of $\frac{6}{8}$ by 2, we obtain the equivalent fraction $\frac{3}{4}$. Therefore, the fractions $\frac{6}{8}$, $\frac{3}{4}$, and $\frac{12}{16}$ are equivalent.

◖ EXAMPLE 2   We may write

$$\frac{ax}{2} = \frac{3a^2x}{6a}$$

since we get the fraction on the right by multiplying the numerator and the denominator of the fraction on the left by $3a$. This means that the fractions are equivalent. ◗

SIMPLEST FORM, OR LOWEST TERMS, OF A FRACTION

One of the most important operations with a fraction is that of **_reducing_** it to its **simplest form,** or **lowest terms.**

A fraction is said to be in its simplest form if the numerator and the denominator have no common integral factors other than $+1$ or -1.

NOTE ◗

In *reducing* a fraction to its simplest form, use the fundamental principle of fractions by *dividing* both the numerator and the denominator by all factors that are common to each. It will be assumed throughout this text that if any of the literal symbols were to be evaluated, numerical values would be such that none of the denominators would equal zero. Therefore, the undefined operation of division by zero is avoided.

◖ EXAMPLE 3   In order to reduce the fraction

$$\frac{16ab^3c^2}{24ab^2c^5}$$

to its lowest terms, note that both the numerator and the denominator contain the factor $8ab^2c^2$. Therefore,

$$\frac{16ab^3c^2}{24ab^2c^5} = \frac{2b(8ab^2c^2)}{3c^3(8ab^2c^2)} = \frac{2b}{3c^3} \qquad \text{— common factor}$$

Here, we divided out the common factor. The resulting fraction is in lowest terms, since there are no common factors in the numerator and the denominator other than $+1$ or -1. ◗

Cancellation

In simplifying fraction, be very careful in performing the basic step of reducing the fraction to its simplest form. That is,

divide both the numerator and the denominator **_by the common factor._**

This process is called **cancellation.** However, it is a common error to remove any expression that appears in both the numerator and the denominator. If a **_term_** is removed in this way, it is an incorrect application of the cancellation process. We must always

CAUTION ◗   *cancel only* **_factors_** *that are in both the numerator and the denominator.* (Remember— *terms* are separated by $+$ and $-$ signs.) In the following example, we illustrate this very common error in the simplification of fractions. Follow it very carefully.

◀ EXAMPLE 4   When simplifying the expression

$$\frac{x^2(x - 2)}{x^2 - 4}$$ a term, but not a factor, of the denominator

CAUTION ▶   many students would "cancel" the x^2 from the numerator and the denominator. This is incorrect, since **x^2 is a term only** of the denominator.

In order to simplify the above fraction properly, we should factor the denominator. We get

$$\frac{x^2(\overset{1}{\cancel{x - 2}})}{(\underset{1}{\cancel{x - 2}})(x + 2)} = \frac{x^2}{x + 2}$$

Here, the common *factor* $x - 2$ has been divided out.   ▶

The following examples illustrate the proper simplification of fractions.

◀ EXAMPLE 5   **(a)** $\dfrac{2a}{2ax} = \dfrac{1}{x}$ — $2a$ is a factor of the numerator and the denominator

We divide out the common factor of 2a.

(b) $\dfrac{2a}{2a + x}$ $2a$ is a term, but not a factor, of the denominator

CAUTION ▶   ***This cannot be reduced,*** since *there are no common **factors** in the numerator and the denominator.*   ▶

◀ EXAMPLE 6 $\dfrac{2x^2 + 8x}{x + 4} = \dfrac{2x(\overset{1}{\cancel{x + 4}})}{(\underset{1}{\cancel{x + 4}})} = \dfrac{2x}{1}$

$= 2x$

The numerator and the denominator were each divided by $x + 4$ after factoring the numerator. The only remaining factor in the denominator is 1, and it is generally not written in the final result. Another way of writing the denominator is $1(x + 4)$, which shows the ***factor*** of 1 more clearly.   ▶

◀ EXAMPLE 7 $\dfrac{x^2 - 4x + 4}{x^2 - 4} = \dfrac{(x - 2)(\overset{1}{\cancel{x - 2}})}{(x + 2)(\underset{1}{\cancel{x - 2}})}$

$= \dfrac{x - 2}{x + 2}$ — x is a term but not a factor

CAUTION ▶   Here, the numerator and the denominator have each been *factored first and then the common factor $x - 2$ has been divided out.* In the final form, neither the x's nor the 2's may be canceled, since they are not common *factors*.   ▶

◀ EXAMPLE 8   In the mathematical analysis of the vibrations in a certain mechanical system, the following expression and simplification are used:

$$\frac{8s + 12}{4s^2 + 26s + 30} = \frac{4(2s + 3)}{2(2s^2 + 13s + 15)} = \frac{\overset{2}{\cancel{4}}\cancel{(2s + 3)}}{\underset{1}{\cancel{2}}\cancel{(2s + 3)}(s + 5)}$$

$$= \frac{2}{s + 5}$$

In the third fraction, note that the factors common to both the numerator and the denominator are 2 and $(2s + 3)$.   ▷

FACTORS THAT DIFFER ONLY IN SIGN

In simplifying fractions we must be able to distinguish between factors that differ only in *sign.* Since $-(y - x) = -y + x = x - y$, we have

$$\boxed{x - y = -(y - x)}$$ (6.11)

CAUTION ▶   Here, the ***factors $x - y$ and $y - x$ differ only in sign.*** The following examples illustrate the simplification of fractions where a change of signs is necessary.

◀ EXAMPLE 9 $\dfrac{x^2 - 1}{1 - x} = \dfrac{(x - 1)(x + 1)}{-(x - 1)} = \dfrac{x + 1}{-1} = -(x + 1)$

CAUTION ▶   In the second fraction, ***we replaced $1 - x$ with the equal expression $-(x - 1)$.*** In the third fraction, the common factor $x - 1$ was divided out. Finally, we expressed the result in the more convenient form by dividing $x + 1$ by -1. Replacing $1 - x$ with $-(x - 1)$ is the same as factoring -1 from the terms of $1 - x$.   ▷

◀ EXAMPLE 10

$$\frac{2x^4 - 128x}{20 + 7x - 3x^2} = \frac{2x(x^3 - 64)}{(4 - x)(5 + 3x)} = \frac{2x(x - 4)(x^2 + 4x + 16)}{-(x - 4)(3x + 5)}$$

$$= -\frac{2x(x^2 + 4x + 16)}{3x + 5}$$

Again, the factor $4 - x$ has been replaced with the equal expression $-(x - 4)$. This allows us to recognize the common factor of $x - 4$.

Also note that the order of the terms of the factor $5 + 3x$ was changed in writing the third fraction. This was done only to write the terms in the more standard form with the x-term first. However, since both terms are *positive,* it is simply an application of the commutative law of addition, and the factor itself is not actually changed.   ▷

EXERCISES 6.5

In Exercises 1–4, make the given changes in the indicated examples of this section and then solve the resulting problems.

1. In Example 3, change the numerator to $18abc^6$, and then reduce the fraction to lowest terms.

2. In Example 4, change the $-$ sign in the numerator to $+$ and then simplify.

3. In Example 7, change the $-$ sign in the numerator to $+$ and then simplify.

4. In Example 10, change the numerator to $2x^4 - 32x^2$ and then simplify.

In Exercises 5–12, multiply the numerator and the denominator of each fraction by the given factor and obtain an equivalent fraction.

5. $\dfrac{2}{3}$ (by 7)

6. $\dfrac{7}{5}$ (by 9)

7. $\dfrac{ax}{y}$ (by $2x$)

8. $\dfrac{2x^2y}{3n}$ (by $2xn^2$)

9. $\dfrac{2}{x + 3}$ (by $x - 2$)

10. $\dfrac{7}{a - 1}$ (by $a + 2$)

11. $\dfrac{a(x - y)}{x - 2y}$ (by $x + y$)

12. $\dfrac{B - 1}{B + 1}$ (by $B - 1$)

In Exercises 13–20, divide the numerator and the denominator of each fraction by the given factor and obtain an equivalent fraction.

13. $\dfrac{28}{44}$ (by 4)

14. $\dfrac{25}{65}$ (by 5)

15. $\dfrac{4x^2y}{8xy^2}$ (by $2x$)

16. $\dfrac{6a^3b^2}{9a^5b^4}$ (by $3a^2b^2$)

17. $\dfrac{2(R - 1)}{(R - 1)(R + 1)}$ (by $R - 1$)

18. $\dfrac{(x + 5)(x - 3)}{3(x + 5)}$ (by $x + 5$)

19. $\dfrac{s^2 - 3s - 10}{2s^2 + 3s - 2}$ (by $s + 2$)

20. $\dfrac{6x^2 + 13x - 5}{6x^3 - 2x^2}$ (by $3x - 1$)

In Exercises 21–28, replace the A with the proper expression such that the fractions are equivalent.

21. $\dfrac{3x}{2y} = \dfrac{A}{6y^2}$

22. $\dfrac{2R}{R + T} = \dfrac{2R^2T}{A}$

23. $\dfrac{7}{a + 5} = \dfrac{7a - 35}{A}$

24. $\dfrac{a + 1}{5a^2c} = \dfrac{A}{5a^3c - 5a^2c}$

25. $\dfrac{2x^3 + 2x}{x^4 - 1} = \dfrac{A}{x^2 - 1}$

26. $\dfrac{n^2 - 1}{n^3 + 1} = \dfrac{A}{n^2 - n + 1}$

27. $\dfrac{x^2 + 3bx - 4b^2}{x - b} = \dfrac{x + 4b}{A}$

28. $\dfrac{4y^2 - 1}{4y^2 + 6y - 4} = \dfrac{A}{2y + 4}$

In Exercises 29–64, reduce each fraction to simplest form.

29. $\dfrac{2a}{8a}$

30. $\dfrac{6x}{15x}$

31. $\dfrac{18x^2y}{24xy}$

32. $\dfrac{2a^2xy}{6axyz^2}$

33. $\dfrac{a + b}{5a^2 + 5ab}$

34. $\dfrac{t - a}{t^2 - a^2}$

35. $\dfrac{6a - 4b}{4a - 2b}$

36. $\dfrac{5r - 20s}{10r - 5s}$

37. $\dfrac{4x^2 + 1}{4x^2 - 1}$

38. $\dfrac{x^2 - y^2}{x^2 + y^2}$

39. $\dfrac{3x^2 - 6x}{x - 2}$

40. $\dfrac{10T^2 + 15T}{2T + 3}$

41. $\dfrac{2y + 3}{4y^3 + 6y^2}$

42. $\dfrac{3t - 6}{4t^3 - 8t^2}$

43. $\dfrac{x^2 - 8x + 16}{x^2 - 16}$

44. $\dfrac{4a^2 + 12ab + 9b^2}{4a^2 + 6ab}$

45. $\dfrac{2w^4 + 5w^2 - 3}{w^4 + 11w^2 + 24}$

46. $\dfrac{3y^3 + 7y^2 + 4y}{y^2 + 5y + 4}$

47. $\dfrac{5x^2 - 6x - 8}{x^3 + x^2 - 6x}$

48. $\dfrac{4r^2 - 8rs - 5s^2}{6r^2 - 17rs + 5s^2}$

49. $\dfrac{N^4 - 16}{N + 2}$

50. $\dfrac{3 + x(4 + x)}{3 + x}$

51. $\dfrac{t + 4}{(2t + 9)t + 4}$

52. $\dfrac{8A^5 + 8A^4 + 2A^3}{4A + 2}$

53. $\dfrac{(x - 1)(3 + x)}{(3 - x)(1 - x)}$

54. $\dfrac{(2x - 1)(x + 6)}{(x - 3)(1 - 2x)}$

55. $\dfrac{y - x}{2x - 2y}$

56. $\dfrac{x^2 - y^2}{y - x}$

57. $\dfrac{2x^2 - 9x + 4}{4x - x^2}$

58. $\dfrac{3a^2 - 13a - 10}{5 + 4a - a^2}$

59. $\dfrac{(x + 5)(x - 2)(x + 2)(3 - x)}{(2 - x)(5 - x)(3 + x)(2 + x)}$

60. $\dfrac{(2x - 3)(3 - x)(x - 7)(3x + 1)}{(3x + 2)(3 - 2x)(x - 3)(7 + x)}$

61. $\dfrac{x^3 + y^3}{2x + 2y}$

62. $\dfrac{w^3 - 8}{w^2 + 2w + 4}$

63. $\dfrac{6x^2 + 2x}{27x^3 + 1}$

64. $\dfrac{3a^3 - 24}{a^2 - 4a + 4}$

W In Exercises 65–68, after finding the simplest form of each fraction, explain why it cannot be simplified more.

65. (a) $\dfrac{x^2(x + 2)}{x^2 + 4}$ **(b)** $\dfrac{x^4 + 4x^2}{x^4 - 16}$

66. (a) $\dfrac{2x + 3}{2x + 6}$ **(b)** $\dfrac{2(x + 6)}{2x + 6}$

67. (a) $\dfrac{x^2 - x - 2}{x^2 - x}$ **(b)** $\dfrac{x^2 - x - 2}{x^2 + x}$

68. (a) $\dfrac{x^3 - x}{1 - x}$ **(b)** $\dfrac{2x^2 + 4x}{2x^2 + 4}$

In Exercises 69–72, reduce each fraction to simplest form. Each is from the indicated area of application.

69. $\dfrac{mu^2 - mv^2}{mu - mv}$ (nuclear energy)

70. $\dfrac{16(t^2 - 2tt_0 + t_0^2)(t - t_0 - 3)}{3t - 3t_0}$ (rocket motion)

71. $\dfrac{E^2R^2 - E^2r^2}{(R^2 + 2Rr + r^2)^2}$ (electricity)

72. $\dfrac{r_0^3 - r_i^3}{r_0^2 - r_i^2}$ (machine design)

6.6 MULTIPLICATION AND DIVISION OF FRACTIONS

From arithmetic, recall that *the product of two fractions is a fraction whose numerator is the product of the numerators and whose denominator is the product of the denominators of the given fractions.* Also, recall that *we find the quotient of two fractions by inverting the divisor and proceeding as in multiplication.* Symbolically, multiplication of fractions is indicated by

Multiplication of Fractions

$$\frac{a}{b} \times \frac{c}{d} = \frac{ac}{bd}$$

and division is indicated by

Division of Fractions

$$\frac{a}{b} \div \frac{c}{d} = \frac{\dfrac{a}{b}}{\dfrac{c}{d}} = \frac{a}{b} \times \frac{d}{c} = \frac{ad}{bc}$$

The rule for division may be verified by use of the fundamental principle of fractions. By multiplying the numerator and the denominator of the fraction

$$\frac{\dfrac{a}{b}}{\dfrac{c}{d}} \quad \text{by} \quad \frac{d}{c} \quad \text{we obtain} \quad \frac{\dfrac{a}{b} \times \dfrac{d}{c}}{\dfrac{c}{d} \times \dfrac{d}{c}} = \frac{\dfrac{ad}{bc}}{1} = \frac{ad}{bc}$$

The following three examples illustrate the multiplication of fractions.

◖ **EXAMPLE 1**   **(a)** $\dfrac{3}{5} \times \dfrac{2}{7} = \dfrac{(3)(2)}{(5)(7)} = \dfrac{6}{35}$ ← multiply numerators
  ← multiply denominators

(b) $\dfrac{3a}{5b} \times \dfrac{15b^2}{a} = \dfrac{(3a)(15b^2)}{(5b)(a)}$

$= \dfrac{45ab^2}{5ab} = \dfrac{9b}{1}$

$= 9b$

In illustration (b), we divided out the common factor of $5ab$ to reduce the resulting fraction to its lowest terms.

When multiplying fractions, we usually want to express the final result in simplest form, which is generally its most useful form. Since all factors in the numerators and all factors in the denominators are to be multiplied, we should

NOTE ▶

*first only **indicate** the multiplication, but not actually perform it and then factor the numerator and the denominator of the result.*

By *indicate,* we mean to write the factors of the numerator side by side and do the same for the factors of the denominator. If we were to multiply out the numerator and the denominator before factoring, it is very possible that we would not see how to factor the result and therefore would not be able to simplify it. The following example illustrates this point.

◀ EXAMPLE 2   In performing the multiplication

$$\frac{3(x - y)}{(x - y)^2} \times \frac{(x^2 - y^2)}{6x + 9y}$$

if we multiplied out the numerators and the denominators before performing any factoring, we would have to simplify the fraction

$$\frac{3x^3 - 3x^2y - 3xy^2 + 3y^3}{6x^3 - 3x^2y - 12xy^2 + 9y^3}$$

NOTE ▶

It is possible to factor the resulting numerator and denominator, but finding any common factors this way is very difficult. However, as stated above, we should ***first indicate the multiplications*** and not actually do them, and ***the solution is much easier.*** Then we factor the resulting numerator and denominator, divide out (cancel) any common factors, and the solution is then complete. Doing this, we have

$$\frac{3(x - y)}{(x - y)^2} \times \frac{(x^2 - y^2)}{6x + 9y} = \frac{3(x - y)(x^2 - y^2)}{(x - y)^2(6x + 9y)} = \frac{3(x - y)(x + y)(x - y)}{(x - y)^2(3)(2x + 3y)}$$

$$= \frac{\cancel{3}(x - y)^2(x + y)}{\cancel{3}(x - y)^2(2x + 3y)}$$

$$= \frac{x + y}{2x + 3y}$$

The common factor $3(x - y)^2$ is readily recognized using this procedure.   ▶

◀ EXAMPLE 3

It is possible to factor and indicate the product of the factors, showing only a single step, as we have done here.

$$\frac{2x - 4}{4x + 12} \times \frac{2x^2 + x - 15}{3x - 1} = \frac{2(x - 2)(2x - 5)(x + 3)}{4(x + 3)(3x - 1)} \quad \substack{\text{multiplications} \\ \text{indicated}}$$

$$= \frac{(x - 2)(2x - 5)}{2(3x - 1)}$$

Here, the common factor is $2(x + 3)$. It is permissible to multiply out the final form of the numerator and the denominator, but it is often preferable to leave the numerator and the denominator in factored form, as indicated.   ▶

The following examples illustrate the division of fractions.

◀ EXAMPLE 4

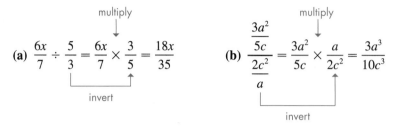

(a) $\dfrac{6x}{7} \div \dfrac{5}{3} = \dfrac{6x}{7} \times \dfrac{3}{5} = \dfrac{18x}{35}$

multiply

invert

(b) $\dfrac{\dfrac{3a^2}{5c}}{\dfrac{2c^2}{a}} = \dfrac{3a^2}{5c} \times \dfrac{a}{2c^2} = \dfrac{3a^3}{10c^3}$

multiply

invert

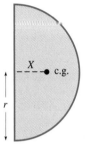

X ---● c.g.

r

Fig. 6.3

◀ EXAMPLE 5   When finding the center of gravity (c.g.) of a uniform flat semi-circular metal plate, the equation $X = \dfrac{4\pi r^3}{3} \div \left(\dfrac{\pi r^2}{2} \times 2\pi \right)$ is derived. Simplify the right side of this equation to find X as a function of r in simplest form. See Fig. 6.3.

The parentheses indicate that we should perform the multiplication first:

$$X = \frac{4\pi r^3}{3} \div \left(\frac{\pi r^2}{2} \times 2\pi \right) = \frac{4\pi r^3}{3} \div \left(\frac{2\pi^2 r^2}{2} \right)$$

$$= \frac{4\pi r^3}{3} \div (\pi^2 r^2) = \frac{4\pi r^3}{3} \times \frac{1}{\pi^2 r^2}$$

$$= \frac{4\pi r^3}{3\pi^2 r^2} = \frac{4r}{3\pi} \qquad \text{divide out the common factor of } \pi r^2$$

This is the exact solution. Approximately, $X = 0.424r$.

◀ EXAMPLE 6

$$\frac{x+y}{3} \div \frac{2x+2y}{6x+15y} = \frac{x+y}{3} \times \frac{6x+15y}{2x+2y} = \frac{(x+y)(3)(2x+5y)}{3(2)(x+y)} \qquad \text{indicate multiplication}$$

invert

$$= \frac{2x+5y}{2} \qquad \text{simplify}$$

◀ EXAMPLE 7

invert

$$\frac{\dfrac{4-x^2}{x^2-3x+2}}{\dfrac{x+2}{x^2-9}} = \frac{4-x^2}{x^2-3x+2} \times \frac{x^2-9}{x+2} = \frac{(2-x)(2+x)(x-3)(x+3)}{(x-2)(x-1)(x+2)} \qquad \text{factor and indicate multiplications}$$

replace $(2-x)$ with $-(x-2)$ and $(2+x)$ with $(x+2)$

$$= \frac{-(x-2)(x+2)(x-3)(x+3)}{(x-2)(x-1)(x+2)}$$

$$= -\frac{(x-3)(x+3)}{x-1} \quad \text{or} \quad \frac{(x-3)(x+3)}{1-x} \qquad \text{simplify}$$

Note the use of Eq. (6.11) when the factor $(2-x)$ was replaced by $-(x-2)$ to get the first form of the answer. As shown, it can also be used to get the alternative form of the answer, although it is not necessary to give this form.

USE OF A GRAPHING CALCULATOR FOR CHECKING ANSWERS

A calculator can be used to check algebraic results. Different numbers can be assigned to each variable to see if the values of the initial expression and the resulting expression are the same. Care must be taken in choosing values (such as not choosing 0 or 1), but in any case the check is valid only for the numbers used.

A graphing calculator gives us a more general way of checking an algebraic result if only one variable is involved. (With more than one variable, values must be assigned to all but one, and the check is less general.) This is shown in the next example.

EXAMPLE 8 To graphically check the result of Example 7, we first let y_1 equal the original expression and y_2 equal the final result. In this case,

$$y_1 = ((4 - x^2)/(x^2 - 3x + 2))/((x + 2)/(x^2 - 9))$$
$$y_2 = -(x - 3)(x + 3)/(x - 1)$$

By graphing y_2 with a heavier curve than y_1 and then watching very carefully, we will see that the graph of y_2 simply goes over the graph of y_1 (see Fig. 6.4(a)). Since the graphs are the same, the results appear to check. However, we cannot consider this as a *proof* that the expressions are equal.

The calculator display shows a check of the solution for all values from $x = -8$ to $x = 8$, except $x = 1$. Actually, using values of x of $-3, -2, 1, 2,$ or 3 is incorrect in that each of these would indicate a division by zero in the original expression. The calculator does show this if we use Xmin and Xmax values such that these x-values come up *exactly*. When we use the *trace* feature for y_1, no y-value shows for x-values of $-3, -2, 2,$ and 3. If you look carefully at the curve in Fig. 6.4(b), you can see missing pixels for these values. This is easier to see with *axes off* as in Fig. 6.4(b).

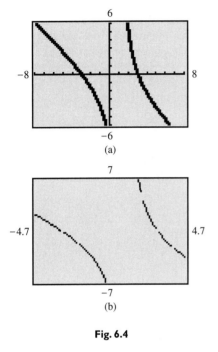

(a)

(b)

Fig. 6.4

EXERCISES 6.6

In Exercises 1–4, make the given changes in the indicated examples of this section and then solve then resulting problems.

1. In Example 2, change the first numerator to $4x + 6y$ and do the multiplication.

2. In Example 3, change the first denominator to $4x - 10$ and do the multiplication.

3. In Example 6, change the denominator $6x + 15y$ to $6x + 12y$ and then do the division.

4. In Example 7, change the numerator $x + 2$ of the divisor to $x + 3$ and then simplify.

In Exercises 5–40, simplify the given expressions involving the indicated multiplications and divisions.

5. $\dfrac{3}{8} \times \dfrac{2}{7}$

6. $\dfrac{11}{5} \times \dfrac{13}{33}$

7. $\dfrac{4x}{3y} \times \dfrac{9y^2}{2}$

8. $\dfrac{18sy^3}{ax^2} \times \dfrac{(ax)^2}{3s}$

9. $\dfrac{2}{9} \div \dfrac{4}{7}$

10. $\dfrac{5}{16} \div \dfrac{25}{13}$

11. $\dfrac{xy}{az} \div \dfrac{bz}{ay}$

12. $\dfrac{sr^2}{2t} \div \dfrac{st}{4}$

13. $\dfrac{4x + 12}{5} \times \dfrac{15t}{3x + 9}$

14. $\dfrac{y^2 + 2y}{6z} \times \dfrac{z^3}{y^2 - 4}$

15. $\dfrac{u^2 - v^2}{u + 2v}(3u + 6v)$

17. $(x - y)\dfrac{x + 2y}{x^2 - y^2}$

17. $\dfrac{2a + 8}{15} \div \dfrac{a^2 + 8a + 16}{25}$

18. $\dfrac{a^2 - a}{3a + 9} \div \dfrac{a^2 - 2a + 1}{a^2 - 9}$

19. $\dfrac{x^4 - 9}{x^2} \div (x^2 + 3)^2$

20. $\dfrac{9B^2 - 16}{B + 1} \div (4 - 3B)$

21. $\dfrac{3ax^2 - 9ax}{10x^2 + 5x} \times \dfrac{2x^2 + x}{a^2x - 3a^2}$

22. $\dfrac{4R^2 - 36}{R^3 - 25R} \times \dfrac{7R - 35}{3R^2 + 9R}$

23. $\dfrac{x^4 - 1}{8x + 16} \times \dfrac{2x^2 - 8x}{x^3 + x}$

24. $\dfrac{2x^2 - 4x - 6}{x^2 - 3x} \times \dfrac{x^3 - 4x^2}{4x^2 - 4x - 8}$

25. $\dfrac{ax + x^2}{2b - cx} \div \dfrac{a^2 + 2ax + x^2}{2bx - cx^2}$

26. $\dfrac{s^4 - 11s^2 + 28}{s^2 + 3} \div \dfrac{s^2 - 4}{2s^2 + 3}$

27. $\dfrac{35a + 25}{12a + 33} \div \dfrac{28a + 20}{36a + 99}$ **28.** $\dfrac{2a^3 + a^2}{2b^3 + b^2} \div \dfrac{2ab + a}{2ab + b}$

29. $\dfrac{x^2 - 6x + 5}{4x^2 - 17x - 15} \times \dfrac{6x + 21}{2x^2 + 5x - 7}$

30. $\dfrac{n^2 + 5n}{3n^2 + 8n + 4} \times \dfrac{2n^2 - 8}{n^3 + 3n^2 - 10n}$

31. $\dfrac{6T^2 - NT - N^2}{2V^2 - 9V - 35} \div \dfrac{8T^2 - 2NT - N^2}{20V^2 + 26V - 60}$

32. $\dfrac{4L^3 - 9L}{8L^2 + 10L - 3} \div \dfrac{2L^3 - 3L^2}{8L^2 + 18L - 5}$

33. $\dfrac{7x^2}{3a} \div \left(\dfrac{a}{x} \times \dfrac{a^2x}{x^2}\right)$

34. $\left(\dfrac{3u}{8v^2} \div \dfrac{9u^2}{2w^2}\right) \times \dfrac{2u^4}{15vw}$

35. $\left(\dfrac{4t^2 - 1}{t - 5} \div \dfrac{2t + 1}{2t}\right) \times \dfrac{2t^2 - 50}{4t^2 + 4t + 1}$

36. $\dfrac{2x^2 - 5x - 3}{x - 4} \div \left(\dfrac{x - 3}{x^2 - 16} \times \dfrac{1}{3 - x}\right)$

37. $\dfrac{x^3 - y^3}{2x^2 - 2y^2} \times \dfrac{x^2 + 2xy + y^2}{x^2 + xy + y^2}$

38. $\dfrac{2M^2 + 4M + 2}{6M - 6} \div \dfrac{5M + 5}{M^2 - 1}$

39. $\dfrac{ax + bx + ay + by}{p - q} \times \dfrac{3p^2 + 4pq - 7q^2}{a + b}$

40. $\dfrac{x^4 + x^5 - 1 - x}{x - 1} \div \dfrac{x - 1}{x}$

In Exercises 41–44, simplify the given expressions and then check your answers with a graphing calculator as in Example 8.

41. $\dfrac{x}{2x + 4} \times \dfrac{x^2 - 4}{3x^2}$ **42.** $\dfrac{4x^2 - 25}{4x^2} \div \dfrac{4x + 10}{8}$

43. $\dfrac{2x^2 + 3x - 2}{2 + 3x - 2x^2} \div \dfrac{5x + 10}{4x + 2}$ **44.** $\dfrac{16x^2 - 8x + 1}{9x} \times \dfrac{12x + 3}{1 - 16x^2}$

In Exercises 45–48, simplify the given expressions. The technical application of each is indicated.

45. $\dfrac{d}{2} \div \dfrac{v_1 d + v_2 d}{4v_1 v_2}$ (average velocity)

46. $\dfrac{c\lambda^2 - c\lambda_0^2}{\lambda_0^2} \div \dfrac{\lambda^2 + \lambda_0^2}{\lambda_0^2}$ (cosmology)

47. $\dfrac{2\pi}{\lambda}\left(\dfrac{a + b}{2ab}\right)\left(\dfrac{ab\lambda}{2a + 2b}\right)$ (optics)

48. $(p_1 - p_2) \div \left(\dfrac{\pi a^4 p_1 - \pi a^4 p_2}{81u}\right)$ (hydrodynamics)

6.7 ADDITION AND SUBTRACTION OF FRACTIONS

From arithmetic, recall that *the sum of a set of fractions that all have the same denominator is the sum of the numerators divided by the common denominator.* Since algebraic expressions represent numbers, this fact is also true in algebra. Addition and subtraction of such fractions are illustrated in the following example.

EXAMPLE 1 **(a)** $\dfrac{5}{9} + \dfrac{2}{9} - \dfrac{4}{9} = \dfrac{5 + 2 - 4}{9}$ ← sum of numerators

← same denominators

$= \dfrac{3}{9} = \dfrac{1}{3}$ ← final result in lowest terms

← use parentheses to show subtraction of both terms

CAUTION ▶

(b) $\dfrac{b}{ax} + \dfrac{1}{ax} - \dfrac{2b - 1}{ax} = \dfrac{b + 1 - (2b - 1)}{ax} = \dfrac{b + 1 - 2b + 1}{ax}$

$= \dfrac{2 - b}{ax}$

LOWEST COMMON DENOMINATOR

NOTE ▶

If the fractions to be combined do not all have the same denominator, we must first change each to an equivalent fraction so that the resulting fractions do have the same denominator. Normally, the denominator that is most convenient and useful is the **lowest common denominator** (abbreviated as **LCD**). *This is the product of all the prime factors that appear in the denominators, with each factor raised to the **highest power** to which it appears **in any one** of the denominators.* This means that the lowest common denominator is the *simplest* algebraic expression into which all given denominators will divide exactly. Following is the procedure for finding the lowest common denominator of a set of fractions.

PROCEDURE FOR FINDING THE LOWEST COMMON DENOMINATOR

1. *Factor each denominator into its prime factors.*
2. *For each different prime factor that appears, note the highest power to which it is raised in any one of the denominators.*
3. *Form the product of all the different prime factors, each raised to the power found in step 2. This product is the lowest common denominator.*

The two examples that follow illustrate the method of finding the LCD.

◀ **EXAMPLE 2**   Find the LCD of the fractions

$$\frac{3}{4a^2b} \qquad \frac{5}{6ab^3} \qquad \frac{1}{4ab^2}$$

We now express each denominator in terms of powers of its prime factors:

highest powers   already seen to be highest power of 2

$$4a^2b = 2^2a^2b \qquad 6ab^3 = 2 \times 3 \times ab^3 \qquad 4ab^2 = 2^2ab^2$$

The prime factors to be considered are 2, 3, a, and b. The largest exponent of 2 that appears is 2. Therefore, 2^2 is a factor of the LCD.

CAUTION ▶

*What matters is that **the highest power of 2 that appears is 2,** not the fact that 2 appears in all three denominators with a total of five factors.*

The largest exponent of 3 that appears is 1 (understood in the second denominator). Therefore, 3 is a factor of the LCD. The largest exponent of a that appears is 2, and the largest exponent of b that appears is 3. Thus, a^2 and b^3 are factors of the LCD. Therefore, the LCD of the fractions is

$$2^2 \times 3 \times a^2b^3 = 12a^2b^3$$

This is the simplest expression into which *each* of the denominators above will divide exactly.

▶

❰ EXAMPLE 3   Find the LCD of the following fractions:

$$\frac{x - 4}{x^2 - 2x + 1} \qquad \frac{1}{x^2 - 1} \qquad \frac{x + 3}{x^2 - x}$$

Factoring each of the denominators, we find that the fractions are

$$\frac{x - 4}{(x - 1)^2} \qquad \frac{1}{(x - 1)(x + 1)} \qquad \frac{x + 3}{x(x - 1)}$$

The factor $(x - 1)$ appears in all the denominators. It is squared in the first fraction and appears only to the first power in the other two fractions. Thus, we must have $(x - 1)^2$ as a factor in the LCD. We do not need a higher power of $(x - 1)$ since, as far as this factor is concerned, each denominator will divide into it evenly. Next, the second denominator has a factor of $(x + 1)$. Therefore, the LCD must also have a factor of $(x + 1)$; otherwise, the second denominator would not divide into it exactly. Finally, the third denominator shows that a factor of x is also needed. The LCD is therefore $x(x + 1)(x - 1)^2$. All three denominators will divide exactly into this expression, and there is no simpler expression for which this is true. ❱

ADDITION AND SUBTRACTION OF FRACTIONS

Once we have found the LCD for the fractions, we multiply the numerator and the denominator of each fraction by the proper quantity to make the resulting denominator in each case the lowest common denominator. After this step, it is necessary only to add the numerators, place this result over the common denominator, and simplify.

❰ EXAMPLE 4   Combine: $\dfrac{2}{3r^2} + \dfrac{4}{rs^3} - \dfrac{5}{3s}$.

By looking at the denominators, notice that the factors necessary in the LCD are 3, r, and s. The 3 appears only to the first power, the largest exponent of r is 2, and the largest exponent of s is 3. Therefore, the LCD is $3r^2s^3$. Now write each fraction with this quantity as the denominator. Since the denominator of the first fraction already contains factors of 3 and r^2, ***it is necessary to introduce the factor of s^3.*** In other words, we must multiply the numerator and the denominator of this fraction by s^3. For similar reasons, we must multiply the numerators and the denominators of the second and third fractions by $3r$ and r^2s^2, respectively. This leads to

NOTE ▶

$$\frac{2}{3r^2} + \frac{4}{rs^3} - \frac{5}{3s} = \frac{2(s^3)}{(3r^2)(s^3)} + \frac{4(3r)}{(rs^3)(3r)} - \frac{5(r^2s^2)}{(3s)(r^2s^2)} \qquad \text{change to equivalent fractions with LCD}$$

$$\text{factors needed in each}$$

$$= \frac{2s^3}{3r^2s^3} + \frac{12r}{3r^2s^3} - \frac{5r^2s^2}{3r^2s^3}$$

$$= \frac{2s^3 + 12r - 5r^2s^2}{3r^2s^3} \qquad \text{combine numerators over LCD}$$

Note that the minus sign for the third term is used in the numerator. ❱

◖ EXAMPLE 5

$$\frac{a}{x-1} + \frac{a}{x+1} = \frac{a(x+1)}{(x-1)(x+1)} + \frac{a(x-1)}{(x+1)(x-1)}$$

change to equivalent fractions with LCD

factors needed

$$= \frac{ax + a + ax - a}{(x+1)(x-1)}$$

combine numerators over LCD

$$= \frac{2ax}{(x+1)(x-1)}$$

simplify

When we multiply each fraction by the quantity required to obtain the proper denominator, we do not actually have to write the common denominator under each numerator. Placing all the products that appear in the numerators over the common denominator is sufficient. Hence, the illustration in this example would appear as

$$\frac{a}{x-1} + \frac{a}{x+1} = \frac{a(x+1) + a(x-1)}{(x-1)(x+1)} = \frac{ax + a + ax - a}{(x-1)(x+1)}$$

$$= \frac{2ax}{(x-1)(x+1)}$$

◖ EXAMPLE 6   The following expression is found in the analysis of the dynamics of missile firing. The indicated addition is performed as shown.

$$\frac{1}{s} - \frac{1}{s+4} + \frac{8}{s^2 + 8s + 16} = \frac{1}{s} - \frac{1}{s+4} + \frac{8}{(s+4)^2}$$

factor third denominator

NOTE ▶

$$= \frac{1(s+4)^2 - 1(s)(s+4) + 8s}{s(s+4)^2}$$

LCD has one factor of s and two factors of $(s+4)$

$$= \frac{s^2 + 8s + 16 - s^2 - 4s + 8s}{s(s+4)^2}$$

expand terms of the numerator

$$= \frac{12s + 16}{s(s+4)^2} = \frac{4(3s+4)}{s(s+4)^2}$$

simplify/factor

We factored the numerator in the final result to see whether or not there were any factors common to the numerator and the denominator. Since there are none, either form of the result is acceptable.

◖ EXAMPLE 7

$$\frac{3x}{x^2 - x - 12} - \frac{x-1}{x^2 - 8x + 16} - \frac{6-x}{2x-8} = \frac{3x}{(x-4)(x+3)} - \frac{x-1}{(x-4)^2} - \frac{6-x}{2(x-4)}$$

factor denominators

$$= \frac{3x(2)(x-4) - (x-1)(2)(x+3) - (6-x)(x-4)(x+3)}{2(x-4)^2(x+3)}$$

change to equivalent fraction with LCD

$$= \frac{[6x^2 - 24x] - [2x^2 + 4x - 6] - [-x^3 + 7x^2 + 6x - 72]}{2(x-4)^2(x+3)}$$

expand in numerator

$$= \frac{6x^2 - 24x - 2x^2 - 4x + 6 + x^3 - 7x^2 - 6x + 72}{2(x-4)^2(x+3)} = \frac{x^3 - 3x^2 - 34x + 78}{2(x-4)^2(x+3)}$$

simplify

In doing this kind of problem, many errors may arise in the use of the minus sign.

CAUTION ▶   Remember, *if a minus sign precedes an expression, the **signs of all terms must be changed*** before they can be combined with other terms.

COMPLEX FRACTIONS

A **complex fraction** *is one in which the numerator, the denominator, or both the numerator and the denominator contain fractions.* The following examples illustrate the simplification of complex fractions.

◀ EXAMPLE 8

The original complex fraction can be written as a division as follows:

$$\frac{2}{x} \div \left(1 - \frac{4}{x}\right)$$

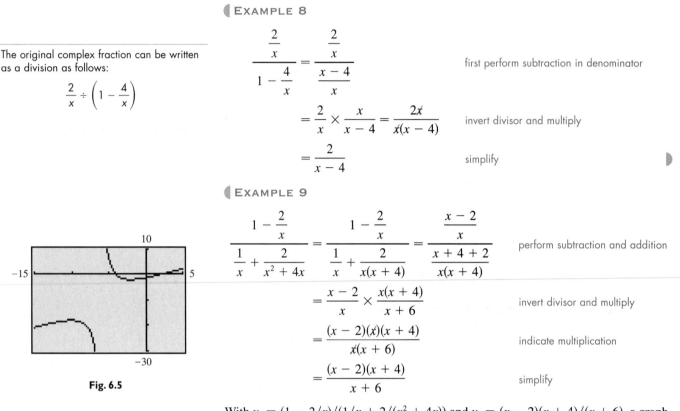

$$\frac{\dfrac{2}{x}}{1 - \dfrac{4}{x}} = \frac{\dfrac{2}{x}}{\dfrac{x - 4}{x}} \qquad \text{first perform subtraction in denominator}$$

$$= \frac{2}{x} \times \frac{x}{x - 4} = \frac{2\cancel{x}}{\cancel{x}(x - 4)} \qquad \text{invert divisor and multiply}$$

$$= \frac{2}{x - 4} \qquad \text{simplify}$$

◀ EXAMPLE 9

Fig. 6.5

$$\frac{1 - \dfrac{2}{x}}{\dfrac{1}{x} + \dfrac{2}{x^2 + 4x}} = \frac{1 - \dfrac{2}{x}}{\dfrac{1}{x} + \dfrac{2}{x(x + 4)}} = \frac{\dfrac{x - 2}{x}}{\dfrac{x + 4 + 2}{x(x + 4)}} \qquad \text{perform subtraction and addition}$$

$$= \frac{x - 2}{x} \times \frac{x(x + 4)}{x + 6} \qquad \text{invert divisor and multiply}$$

$$= \frac{(x - 2)(\cancel{x})(x + 4)}{\cancel{x}(x + 6)} \qquad \text{indicate multiplication}$$

$$= \frac{(x - 2)(x + 4)}{x + 6} \qquad \text{simplify}$$

With $y_1 = (1 - 2/x)/(1/x + 2/(x^2 + 4x))$ and $y_2 = (x - 2)(x + 4)/(x + 6)$, a graphing calculator check of the result (see Example 8 on page 201) is shown in Fig. 6.5. ◀

EXERCISES 6.7

In Exercises 1–4, make the given changes in the indicated examples of this section and then solve the resulting problems.

1. In Example 2, change the denominator of the third fraction to $4a^2b^2$ and then find the LCD.

2. In Example 5, add the fraction $\dfrac{2}{x^2 - 1}$ to those being added and then find the result.

3. In Example 7, change the denominator of the third fraction to $2x + 6$ and then find the result.

4. In Example 8, change the fraction in the numerator to $2/x^2$ and then simplify.

In Exercises 5–46, perform the indicated operations and simplify. For 33, 34, 39, and 40, check the solution with a graphing calculator.

5. $\dfrac{3}{5} + \dfrac{6}{5}$

6. $\dfrac{2}{13} + \dfrac{6}{13}$

7. $\dfrac{1}{x} + \dfrac{7}{x}$

8. $\dfrac{2}{a} + \dfrac{3}{a}$

9. $\dfrac{1}{2} + \dfrac{3}{4}$

10. $\dfrac{5}{9} - \dfrac{1}{3}$

11. $\dfrac{3}{4x} + \dfrac{7a}{4}$

12. $\dfrac{t - 3}{a} - \dfrac{t}{2a}$

13. $\dfrac{a}{x} - \dfrac{b}{x^2}$

14. $\dfrac{2}{s^2} + \dfrac{3}{s}$

15. $\dfrac{6}{5x^3} + \dfrac{a}{25x}$

16. $\dfrac{a}{6y} - \dfrac{2b}{3y^4}$

17. $\dfrac{2}{5a} + \dfrac{1}{a} - \dfrac{a}{10}$

18. $\dfrac{1}{2A} - \dfrac{6}{B} - \dfrac{9}{4C}$

19. $\dfrac{x + 1}{x} - \dfrac{x - 3}{y} - \dfrac{2 - x}{xy}$

20. $5 + \dfrac{1 - x}{2} - \dfrac{3 + x}{4}$

21. $\dfrac{3}{2x - 1} + \dfrac{1}{4x - 2}$

22. $\dfrac{5}{6y + 3} - \dfrac{a}{8y + 4}$

23. $\dfrac{4}{x(x + 1)} - \dfrac{3}{2x}$

24. $\dfrac{3}{ax + ay} - \dfrac{1}{a^2}$

25. $\dfrac{s}{2s-6} + \dfrac{1}{4} - \dfrac{3s}{4s-12}$

26. $\dfrac{2}{x+2} - \dfrac{3-x}{x^2+2x} + \dfrac{1}{x}$

27. $\dfrac{3R}{R^2-9} - \dfrac{2}{3R+9}$

28. $\dfrac{2}{n^2+4n+4} - \dfrac{3}{n+2}$

29. $\dfrac{3}{x^2-8x+16} - \dfrac{2}{4-x}$

30. $\dfrac{2a-b}{c-3d} - \dfrac{b-2a}{3d-c}$

31. $\dfrac{v+4}{v^2+5v+4} - \dfrac{v-2}{v^2-5v+6}$

32. $\dfrac{N-1}{2N^3-4N^2} - \dfrac{5}{2-N}$

33. $\dfrac{x-1}{3x^2-13x+4} - \dfrac{3x+1}{4-x}$

34. $\dfrac{x}{4x^2-12x+5} + \dfrac{2x-1}{4x^2-4x-15}$

35. $\dfrac{t}{t^2-t-6} - \dfrac{2t}{t^2+6t+9} + \dfrac{t}{t^2-9}$

36. $\dfrac{5}{2x^3-3x^2+x} - \dfrac{x}{x^4-x^2} + \dfrac{2-x}{2x^2+x-1}$

37. $\dfrac{1}{w^3+1} + \dfrac{1}{w+1} - 2$

38. $\dfrac{2}{8-x^3} + \dfrac{1}{x^2-x-2}$

39. $\dfrac{\dfrac{1}{x}}{1-\dfrac{1}{x}}$

40. $\dfrac{x-\dfrac{1}{x}}{1-\dfrac{1}{x}}$

41. $\dfrac{\dfrac{x}{y}-\dfrac{y}{x}}{1+\dfrac{y}{x}}$

42. $\dfrac{\dfrac{V^2-9}{V}}{\dfrac{1}{V}-\dfrac{1}{3}}$

43. $\dfrac{2-\dfrac{1}{x}-\dfrac{2}{x+1}}{\dfrac{1}{x^2+2x+1}-1}$

44. $\dfrac{\dfrac{2}{a}-\dfrac{1}{4}-\dfrac{3}{4a-4b}}{\dfrac{1}{4a^2-4b^2}-\dfrac{2}{b}}$

45. $\dfrac{\dfrac{3}{x}+\dfrac{1}{x^2+x}}{\dfrac{1}{x+1}-\dfrac{1}{x-1}}$

46. $\dfrac{\dfrac{1}{u-v}+\dfrac{1}{2u+2v}}{\dfrac{2u}{2u^2-3uv+v^2}+\dfrac{2}{2u-v}}$

The expression $f(x+h) - f(x)$ is frequently used in the study of calculus. (If necessary, refer to Section 3.1 for a review of functional notation.) In Exercises 47–50, determine and then simplify this expression for the given functions.

47. $f(x) = \dfrac{x}{x+1}$

48. $f(x) = \dfrac{3}{2x-1}$

49. $f(x) = \dfrac{1}{x^2}$

50. $f(x) = \dfrac{2}{x^2+4}$

In Exercises 51–57, simplify the given expressions. In Exercise 58, answer the given question.

51. Using the definitions of the trigonometric functions given in Section 4.2, find an expression that is equivalent to
$(\tan\theta)(\cot\theta) + (\sin\theta)^2 - \cos\theta$, in terms of x, y, and r.

52. Using the definitions of the trigonometric functions given in Section 4.2, find an expression that is equivalent to
$\sec\theta - (\cot\theta)^2 + \csc\theta$, in terms of x, y, and r.

53. If $f(x) = 2x - x^2$, find $f\left(\dfrac{1}{a}\right)$.

54. If $f(x) = x^2 + x$, find $f\left(a + \dfrac{1}{a}\right)$.

55. If $f(x) = x - \dfrac{2}{x}$, find $f(a+1)$.

56. If $f(x) = 2x - 3$, find $f[1/f(x)]$.

57. The sum of two numbers a and b is divided by the sum of their reciprocals. Simplify the expression for this quotient.

(W) 58. When adding fractions, explain why it is better to find the lowest common denominator rather than any denominator that is common to the fractions.

In Exercises 59–68, perform the indicated operations. Each expression occurs in the indicated area of application.

59. $\dfrac{3}{4\pi} - \dfrac{3H_0}{4\pi H}$ (transistor theory)

60. $1 + \dfrac{9}{128T} - \dfrac{27P}{64T^3}$ (thermodynamics)

61. $\dfrac{2n^2-n-4}{2n^2+2n-4} + \dfrac{1}{n-1}$ (optics)

62. $\dfrac{b}{x^2+y^2} - \dfrac{2bx^2}{x^4+2x^2y^2+y^4}$ (magnetic field)

63. $\left(\dfrac{3Px}{2L^2}\right)^2 + \left(\dfrac{P}{2L}\right)^2$ (force of a weld)

64. $\dfrac{a}{b^2h} + \dfrac{c}{bh^2} - \dfrac{1}{6bh}$ (strength of materials)

65. $\dfrac{\dfrac{L}{C}+\dfrac{R}{sC}}{sL+R+\dfrac{1}{sC}}$ (electricity)

66. $\dfrac{\dfrac{m}{c}}{1-\dfrac{p^2}{c^2}}$ (airfoil deign)

67. $\dfrac{1}{R^2} + \left(\omega c - \dfrac{1}{\omega L}\right)^2$ (electricity)

68. $\dfrac{\dfrac{x}{h_1}+\dfrac{x-L}{h_2}}{1+\dfrac{x(L-x)}{h_1 h_2}}$ (optics)

6.8 EQUATIONS INVOLVING FRACTIONS

Many important equations in science and technology have fractions in them. Although the solution of these equations will still involve the use of the basic operations stated in Section 1.10, an additional procedure can be used to eliminate the fractions and thereby help lead to the solution. The method is to

NOTE ▶ *multiply each term of the equation by the LCD.*

The resulting equation will not involve fractions and can be solved by methods previously discussed. The following examples illustrate how to solve equations involving fractions.

◀ EXAMPLE 1   Solve for x: $\dfrac{x}{12} - \dfrac{1}{8} = \dfrac{x+2}{6}$.

First, note that the LCD of the terms of the equation is 24. Therefore, multiply each term by 24. This gives

Note carefully that we are multiplying both sides of the equation by the LCD and not combining terms over a common denominator as when adding fractions.

$$\frac{24(x)}{12} - \frac{24(1)}{8} = \frac{24(x+2)}{6}$$ each term multiplied by LCD

Reduce each term to its lowest terms and solve the resulting equation:

$$2x - 3 = 4(x+2) \quad \text{each term reduced}$$
$$2x - 3 = 4x + 8$$
$$-2x = 11$$
$$x = -\frac{11}{2}$$

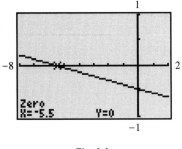

Fig. 6.6

When we check this solution in the original equation, we obtain $-7/12$ on each side of the equal sign. Therefore, the solution is correct.

A graphing calculator solution (see page 98) using $y_1 = x/12 - 1/8 - (x+2)/6$ is shown in Fig. 6.6. The zero of -5.5 agrees with the solution $-11/2$. ▶

◀ EXAMPLE 2   Solve for x: $\dfrac{x}{2} - \dfrac{1}{b^2} = \dfrac{x}{2b}$.

First, determine that the LCD of the terms of the equation is $2b^2$. Then multiply each term by $2b^2$ and continue with the solution:

The procedure for solving equations on page 37 can now be stated as:

1. If the equation has any fractions, first remove the fractions.
2. Combine like terms on each side.
3. Remove grouping symbols.
4. Perform the same operations on both sides, and factor if necessary, until $x =$ result is obtained.
5. Check the solution in the original equation.

$$\frac{2b^2(x)}{2} - \frac{2b^2(1)}{b^2} = \frac{2b^2(x)}{2b} \quad \text{each term multiplied by LCD}$$
$$b^2x - 2 = bx \quad \text{each term reduced}$$
$$b^2x - bx = 2$$
$$x(b^2 - b) = 2 \quad \text{factor}$$
$$x = \frac{2}{b^2 - b}$$

Note the use of factoring in arriving at the final result. Checking shows that each side of the original equation is equal to $\dfrac{1}{b^2(b-1)}$. ▶

See the chapter introduction.

The first telescope was invented by Lippershay, a Dutch lens maker, in about 1608.

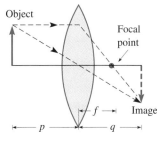

Object

Focal point

f — Image

p — q —

Fig. 6.7

For reference, Eq. (6.11) is $x - y = -(y - x)$.

See the chapter introduction.

In 1609, the Italian scientist Galileo (1564–1642) learned of the invention of the telescope and developed it for astronomical observations. Among his first discoveries were the four largest moons of the planet Jupiter.

◀ **EXAMPLE 3**   An equation relating the focal length f of a lens with the object distance p and the image distance q is given below. See Fig. 6.7. Solve for q.

$$f = \frac{pq}{p + q} \qquad \text{given equation}$$

Since the only denominator is $p + q$, the LCD is also $p + q$. By first multiplying each term by $p + q$, the solution is completed as follows:

$$f(p + q) = \frac{pq(p + q)}{p + q} \qquad \text{each term multiplied by LCD}$$

$$fp + fq = pq \qquad \text{reduce term on right}$$

$$fq - pq = -fp$$

$$q(f - p) = -fp \qquad \text{factor}$$

$$q = \frac{-fp}{f - p} \qquad \text{divide by } f - p$$

$$= -\frac{fp}{-(p - f)} \qquad \text{use Eq. (6.11)}$$

$$= \frac{fp}{p - f}$$

The last form is preferred since there is no minus sign before the fraction. However, either form of the result is correct.   ▶

◀ **EXAMPLE 4**   When developing the equations that describe the motion of the planets, the equation

$$\frac{1}{2}v^2 - \frac{GM}{r} = -\frac{GM}{2a}$$

is found. Solve for M.

First, determine that the LCD of the terms of the equation is $2ar$. Multiplying each term by $2ar$ and proceeding, we have

$$\frac{2ar(v^2)}{2} - \frac{2ar(GM)}{r} = -\frac{2ar(GM)}{2a} \qquad \text{each term multiplied by LCD}$$

$$arv^2 - 2aGM = -rGM \qquad \text{each term reduced}$$

$$rGM - 2aGM = -arv^2$$

$$M(rG - 2aG) = -arv^2 \qquad \text{factor}$$

$$M = -\frac{arv^2}{rG - 2aG}$$

$$= \frac{arv^2}{2aG - rG}$$

The second form of the result is obtained by using Eq. (6.11). Again, note the use of factoring to arrive at the final result.   ▶

◀ EXAMPLE 5   Solve for x: $\dfrac{2}{x+1} - \dfrac{1}{x} = -\dfrac{2}{x^2+x}$.

Multiplying each term by the LCD $x(x+1)$, we have

$$\frac{2(x)(x+1)}{x+1} - \frac{x(x+1)}{x} = -\frac{2x(x+1)}{x(x+1)}$$

Now, simplifying each fraction, we have

$$2x - (x+1) = -2$$

We now complete the solution:

$$2x - x - 1 = -2$$
$$x = -1$$

CAUTION ▶

Extraneous Solutions

Checking this solution in the original equation, notice the zero in the denominators of the first and third terms of the equation. Since division by zero is undefined (see Section 1.2), $x = -1$ cannot be a solution. *Thus, there is* **no solution** *to this equation.* This example points out clearly why it is necessary to check solutions in the original equation. It also shows that *whenever we multiply each term by a common denominator that* **contains the unknown,** *it is possible to obtain a value that is not a solution of the original equation. Such a value is termed an* **extraneous solution.** Only certain equations will lead to extraneous solutions, but we must be careful to identify them when they occur.   ▶

A number of stated problems give rise to equations involving fractions. The following examples illustrate the solution of such problems.

Solving a Word Problem

◀ EXAMPLE 6   An industrial firm uses a computer system that processes and prints out its data for an average day in 20 h. To process the data more rapidly and to handle increased future computer needs, the firm plans to add new components to the system. One set of new components can process the data in 12 h, without the present system. How long would it take the new system, a combination of the present system and the new components, to process the data?

First, let $x =$ the number of hours for the new system to process the data. Next, we know that it takes the present system 20 h to do it. This means that it processes $\frac{1}{20}$ of the data in 1 h, or $\frac{1}{20}x$ of the data in x h. In the same way, the new components can process $\frac{1}{12}x$ of the data in x h. When x h have passed, the new system will have processed all of the data. Therefore,

The first large-scale electronic computer was the ENIAC. It was constructed at the Univ. of Pennsylvania in the mid-1940s and used until 1955. It had 18 000 vacuum tubes and occupied 1400 m² of floor space.

A military programmable computer called *Colossus* was used to break the German codes in the Second World War. It was developed, in secret, before ENIAC.

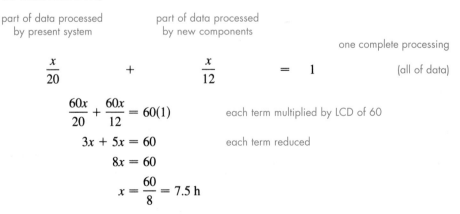

part of data processed   part of data processed
  by present system   by new components

one complete processing

$$\frac{x}{20} \quad + \quad \frac{x}{12} \quad = \quad 1 \qquad \text{(all of data)}$$

$$\frac{60x}{20} + \frac{60x}{12} = 60(1) \qquad \text{each term multiplied by LCD of 60}$$
$$3x + 5x = 60 \qquad \text{each term reduced}$$
$$8x = 60$$
$$x = \frac{60}{8} = 7.5 \text{ h}$$

Therefore, the new system should take about 7.5 h to process the data.   ▶

◖ EXAMPLE 7   A bus averaging 80.0 km/h takes 1.50 h longer to travel from city A to city B than a train that averages 96.0 km/h. How far apart are the cities?

Let d = the distance from city A to city B. Since distance = rate × time, then distance ÷ rate = time. Therefore, the time the bus takes is $d/96.0$ h, and the time the train takes is $d/80.0$ h. This means

$$\frac{d}{80.0} - \frac{d}{96.0} = 1.50$$

$80 = 16(5) = (2^4)(5)$
$96 = 32(3) = (2^5)(3)$
$\text{LCD} = (2^5)(3)(5) = 480$

$$\frac{480d}{80.0} - \frac{480d}{96.0} = 480(1.50)$$

$$6.00d - 5.00d = 720, \qquad d = 720 \text{ km}$$

EXERCISES 6.8

In Exercises 1–4, make the given changes in the indicated examples of this section and then solve for the indicated variable.

1. In Example 2, change the second denominator from b^2 to b and then solve for x.

2. In Example 3, solve for p.

3. In Example 4, solve for G.

4. In Example 5, change the numerator on the right to 1 and then solve for x.

In Exercises 5–32, solve the given equations and check the results.

5. $\dfrac{x}{2} + 6 = 2x$

6. $\dfrac{x}{5} + 2 = \dfrac{15 + x}{10}$

7. $\dfrac{x}{6} - \dfrac{1}{2} = \dfrac{x}{3}$

8. $\dfrac{3N}{8} - \dfrac{3}{4} = \dfrac{N - 4}{2}$

9. $\dfrac{1}{2} - \dfrac{t - 5}{6} = \dfrac{3}{4}$

10. $\dfrac{2x - 7}{3} + 5 = \dfrac{1}{5}$

11. $\dfrac{3x}{7} - \dfrac{5}{21} = \dfrac{2 - x}{14}$

12. $\dfrac{F - 3}{12} - \dfrac{2}{3} = \dfrac{1 - 3F}{2}$

13. $\dfrac{3}{T} + 2 = \dfrac{5}{3}$

14. $\dfrac{1}{2y} - \dfrac{1}{2} = 4$

15. $3 - \dfrac{x - 2}{5x} = \dfrac{1}{5}$

16. $\dfrac{1}{2R} - \dfrac{1}{3} = \dfrac{2}{3R}$

17. $\dfrac{2y}{y - 1} = 5$

18. $\dfrac{x}{2x - 3} = 4$

19. $\dfrac{2}{s} = \dfrac{3}{s - 1}$

20. $\dfrac{5}{n + 2} = \dfrac{3}{2n}$

21. $\dfrac{5}{2x + 4} + \dfrac{3}{x + 2} = 2$

22. $\dfrac{3}{4x - 6} + \dfrac{1}{4} = \dfrac{5}{2x - 3}$

23. $\dfrac{2}{Z - 5} - \dfrac{3}{10 - 2Z} = 3$

24. $\dfrac{4}{4 - x} + 2 - \dfrac{2}{12 - 3x} = \dfrac{1}{3}$

25. $\dfrac{1}{x} + \dfrac{3}{2x} = \dfrac{2}{x + 1}$

26. $\dfrac{3}{t + 3} - \dfrac{1}{t} = \dfrac{5}{2t + 6}$

27. $\dfrac{7}{y} = \dfrac{3}{y - 4} + \dfrac{7}{2y^2 - 8y}$

28. $\dfrac{1}{2x + 3} = \dfrac{5}{2x} - \dfrac{4}{2x^2 + 3x}$

29. $\dfrac{1}{x^2 - x} - \dfrac{1}{x} = \dfrac{1}{x - 1}$

30. $\dfrac{2}{x^2 - 1} - \dfrac{2}{x + 1} = \dfrac{1}{x - 1}$

31. $\dfrac{2}{B^2 - 4} - \dfrac{1}{B - 2} = \dfrac{1}{2B + 4}$

32. $\dfrac{2}{2x^2 + 5x - 3} - \dfrac{1}{4x - 2} + \dfrac{3}{2x + 6} = 0$

In Exercises 33–48, solve for the indicated letter. In Exercises 37–48, each of the given formulas arises in the technical or scientific area of study listed.

33. $2 - \dfrac{1}{b} + \dfrac{3}{c} = 0$, for c

34. $\dfrac{2}{3} - \dfrac{h}{x} = \dfrac{1}{6x}$, for x

35. $\dfrac{t - 3}{b} - \dfrac{t}{2b - 1} = \dfrac{1}{2}$, for t

36. $\dfrac{1}{a^2 + 2a} - \dfrac{y}{2a} = \dfrac{2y}{a + 2}$, for y

37. $\dfrac{s - s_0}{t} = \dfrac{v + v_0}{2}$, for v (velocity of object)

38. $S = \dfrac{P}{A} + \dfrac{Mc}{I}$, for P (machine design)

39. $V = 1.2\left(5.0 + \dfrac{8.0R}{8.0 + R}\right)$, for R
(electric resistance in Fig. 6.8)

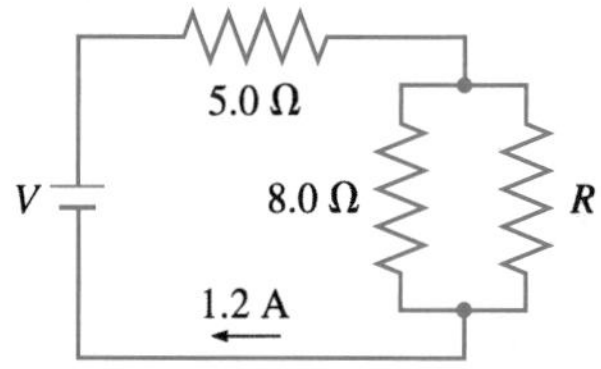

Fig. 6.8

40. $K = \dfrac{ax}{x + b}$, for x (medicine)

41. $z = \dfrac{1}{g_m} - \dfrac{jX}{g_m R}$, for R (FM transmission)

42. $A = \dfrac{1}{2}wp - \dfrac{1}{2}w^2 - \dfrac{\pi}{8}w^2$, for p (architecture)

43. $P = \dfrac{RT}{V - b} - \dfrac{a}{V^2}$, for T (thermodynamics)

44. $\dfrac{1}{x} + \dfrac{1}{nx} = \dfrac{1}{f}$, for n (photography)

45. $\dfrac{1}{R_1} = \dfrac{N_2^2}{N_1^2 R_2} + \dfrac{N_3^2}{N_1^2 R_3}$, for R_1 (transformer resistance)

46. $D = \dfrac{wx^4}{24EI} - \dfrac{wLx^3}{6EI} + \dfrac{wL^2x^2}{4EI}$, for w (beam design)

47. $\dfrac{1}{f} = (n - 1)\left(\dfrac{1}{R_1} + \dfrac{1}{R_2}\right)$, for R_1 (optics)

48. $P = \dfrac{\dfrac{1}{1 + i}}{1 - \dfrac{1}{1 + i}}$, for i (business)

In Exercises 49–58, set up appropriate equations and solve the given stated problems. All numbers are accurate to at least two significant digits.

49. One pipe can fill a certain oil storage tank in 4.0 h, and a second pipe can fill it in 6.0 h. How long will it take to fill the tank if both pipes operate together?

50. One company determines that it will take its crew 450 h to clean up a chemical dump site, and a second company determines that it will take its crew 600 h to clean up the site. How long will it take the two crews working together?

51. One automatic packaging machine can package 100 boxes of machine parts in 12 min, and a second machine can do it in 10 min. A newer model machine can do it in 8.0 min. How long will it take the three machines working together?

52. A painting crew can paint a structure in 12 h, or the crew can paint it in 7.2 h when working with a second crew. How long would it take the second crew to do the job if working alone?

53. An elevator traveled from the first floor to the top floor of a building at an average speed of 2.0 m/s and returned to the first floor at 2.2 m/s. If it was on the top floor for 90 s and the total elapsed time was 5.0 min, how far above the first floor is the top floor?

54. A commuter rapid transit train travels 24 km farther between stops A and B than between stops B and C. If it averages 60 km/h from A to B and 30 km/h between B and C, and an express averages 50 km/h between A and C (not stopping at B), how far apart are stops A and C?

55. A jet takes the same time to travel 2580 km with the wind as it does to travel 1800 km against the wind. If its speed relative to the air is 450 km/h, what is the speed of the wind?

56. An engineer travels from Aberdeen, Scotland, to the Montrose oil field in the North Sea on a ship that averages 28 km/h. After spending 6.0 h at the field, the engineer returns to Aberdeen in a helicopter that averages 140 km/h. If the total trip takes 15.0 h, how far is the Montrose oil field from Aberdeen?

57. The current through each of the resistances R_1 and R_2 in Fig. 6.9 equals the voltage V divided by the resistance. The sum of the currents equals the current i in the rest of the circuit. Find the voltage if $i = 1.2$ A, $R_1 = 2.7\ \Omega$, and $R_2 = 6.0\ \Omega$.

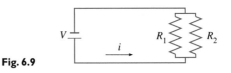

Fig. 6.9

58. A fox, pursued by a greyhound, has a start of 60 leaps. He makes 9 leaps while the greyhound makes but 6; but 3 leaps of the greyhound are equivalent to 7 of the fox. How many leaps must the greyhound make to overcome the fox? (Copied from Davies, Charles, *Elementary Algebra*, New York: A. S. Barnes & Burr, 1852.) (*Hint:* Let the unit of distance be one fox leap.)

In Exercises 59 and 60 find constants A and B such that the equation is true.

59. $\dfrac{x - 12}{x^2 + x - 6} = \dfrac{A}{x + 3} + \dfrac{B}{x - 2}$

60. $\dfrac{23 - x}{2x^2 + 7x - 4} = \dfrac{A}{2x - 1} - \dfrac{B}{x + 4}$

CHAPTER ⑥ EQUATIONS

$a(x + y) = ax + ay$ (6.1)

$(x + y)(x - y) = x^2 - y^2$ (6.2)

$(x + y)^2 = x^2 + 2xy + y^2$ (6.3)

$(x - y)^2 = x^2 - 2xy + y^2$ (6.4)

$(x + a)(x + b) = x^2 + (a + b)x + ab$ (6.5)

$$(ax + b)(cx + d) = acx^2 + (ad + bc)x + bd \tag{6.6}$$

$$(x + y)^3 = x^3 + 3x^2y + 3xy^2 + y^3 \tag{6.7}$$

$$(x - y)^3 = x^3 - 3x^2y + 3xy^2 - y^3 \tag{6.8}$$

$$(x + y)(x^2 - xy + y^2) = x^3 + y^3 \tag{6.9}$$

$$(x - y)(x^2 + xy + y^2) = x^3 - y^3 \tag{6.10}$$

$$x - y = -(y - x) \tag{6.11}$$

CHAPTER 6 REVIEW EXERCISES

In Exercises 1–12, find the products by inspection. No intermediate steps should be necessary.

1. $3a(4x + 5a)$

2. $-7xy(4x^2 - 7y)$

3. $(2a + 7b)(2a - 7b)$

4. $(x - 4z)(x + 4z)$

5. $(2a + 1)^2$

6. $(5C - 2D)^2$

7. $(b - 4)(b + 7)$

8. $(y - 5)(y - 7)$

9. $(2x + 5)(x - 9)$

10. $(4ax - 3)(5ax + 7)$

11. $(2c^a + d)(2c^a - d)$

12. $(3s^n - 2t)(3s^n + 2t)$

In Exercises 13–44, factor the given expressions completely.

13. $3s + 9t$

14. $7x - 28y$

15. $a^2x^2 + a^2$

16. $3ax - 6ax^4 - 9a$

17. $W^2 - 144$

18. $900 - n^2$

19. $16(x + 2)^2 - t^4$

20. $25s^4 - 36t^2$

21. $9t^2 - 6t + 1$

22. $4x^2 - 12x + 9$

23. $25t^2 + 10t + 1$

24. $4c^2 + 36cd + 81d^2$

25. $x^2 + x - 56$

26. $x^2 - 4x - 45$

27. $t^4 - 5t^2 - 36$

28. $N^4 - 11N^2 + 10$

29. $2k^2 - k - 36$

30. $5x^2 + 2x - 3$

31. $4x^2 - 4x - 35$

32. $9F^2 + 7F - 16$

33. $10b^2 + 23b - 5$

34. $12x^2 - 7xy - 12y^2$

35. $4x^2 - 64y^2$

36. $4a^2x^2 + 26a^2x + 36a^2$

37. $250 - 16y^6$

38. $a^4 + 64a$

39. $8x^3 + 27$

40. $R^3 - 125r^3$

41. $ab^2 - 3b^2 + a - 3$

42. $axy - ay + ax - a$

43. $nx + 5n - x^2 + 25$

44. $ty - 4t + y^2 - 16$

In Exercises 45–68, perform the indicated operations and express results in simplest form.

45. $\dfrac{48ax^3y^6}{9a^3xy^6}$

46. $\dfrac{-39r^2s^4t^8}{52rs^5t}$

47. $\dfrac{6x^2 - 7x - 3}{4x^2 - 8x + 3}$

48. $\dfrac{p^4 - 4p^2 - 4}{p^4 - p^2 - 12}$

49. $\dfrac{4x + 4y}{35x^2} \times \dfrac{28x}{x^2 - y^2}$

50. $\left(\dfrac{6x - 3}{x^2}\right)\left(\dfrac{4x^2 - 12x}{12x - 6}\right)$

51. $\dfrac{18 - 6L}{L^2 - 6L + 9} \div \dfrac{L^2 - 2L - 15}{L^2 - 9}$

52. $\dfrac{6x^2 - xy - y^2}{2x^2 + xy - y^2} \div \dfrac{4x^2 - 16y^2}{x^2 + 3xy + 2y^2}$

53. $\dfrac{\dfrac{3x}{7x^2 + 13x - 2}}{\dfrac{6x^2}{x^2 + 4x + 4}}$

54. $\dfrac{\dfrac{3x - 3y}{2x^2 + 3xy - 2y^2}}{\dfrac{3x^2 - 3y^2}{x^2 + 4xy + 4y^2}}$

55. $\dfrac{x + \dfrac{1}{x} + 1}{x^2 - \dfrac{1}{x}}$

56. $\dfrac{\dfrac{4}{y} - 4y}{2 - \dfrac{2}{y}}$

57. $\dfrac{4}{9x} - \dfrac{5}{12x^2}$

58. $\dfrac{3}{10a^2} + \dfrac{1}{4a^3}$

59. $\dfrac{6}{x} - \dfrac{7}{2x} + \dfrac{3}{xy}$

60. $\dfrac{T}{T^2 + 2} - \dfrac{1}{T^3 + 2T}$

61. $\dfrac{a + 1}{a + 2} - \dfrac{a + 3}{a}$

62. $\dfrac{y}{y + 2} - \dfrac{1}{y^2 + 2y}$

63. $\dfrac{2x}{x^2 + 2x - 3} - \dfrac{1}{x^2 + 3x}$

64. $\dfrac{x}{4x^2 + 4x - 3} - \dfrac{3}{4x^2 - 9}$

65. $\dfrac{3x}{2x^2 - 2} - \dfrac{2}{4x^2 - 5x + 1}$

66. $\dfrac{2n - 1}{4 - n} + \dfrac{n + 2}{5n - 20}$

67. $\dfrac{3x}{x^2 + 2x - 3} - \dfrac{2}{x^2 + 3x} + \dfrac{x}{x - 1}$

68. $\dfrac{3}{y^4 - 2y^3 - 8y^2} + \dfrac{y - 1}{y^2 + 2y} - \dfrac{y - 3}{y^2 - 4y}$

In Exercises 69–72, graphically check the results for the indicated exercises of this set on a graphing calculator.

69. Exercise 47

70. Exercise 50

71. Exercise 55

72. Exercise 64

In Exercises 73–76, factor the given expressions. In Exercises 73 and 74, disregard the restriction that coefficients and exponents must be integers. The expressions in Exercises 75 and 76 can be factored if they are first rewritten in a different form.

73. $x^2 - 5$

74. $x - y$

75. $x + y$ (one of two factors is to be x)

76. $x + y$ (one of two factors is to be a)

In Exercises 77–84, solve the given equations.

77. $\dfrac{x}{2} - 3 = \dfrac{x - 10}{4}$

78. $\dfrac{2x}{c} - \dfrac{1}{2c} = \dfrac{3}{c} - x$, for x

79. $\dfrac{2}{t} - \dfrac{1}{at} = 2 + \dfrac{a}{t}$, for t

80. $\dfrac{3}{a^2 y} - \dfrac{1}{ay} = \dfrac{9}{a}$, for y

81. $\dfrac{2x}{2x^2 - 5x} - \dfrac{3}{x} = \dfrac{1}{4x - 10}$

82. $\dfrac{3}{x^2 + 3x} - \dfrac{1}{x} = \dfrac{1}{x + 3}$

83. Given $f(x) = \dfrac{1}{x + 2}$, solve for x if $f(x + 2) = 2f(x)$.

84. Given $f(x) = \dfrac{x}{x + 1}$, solve for x if $3f(x) + f\left(\dfrac{1}{x}\right) = 2$.

In Exercises 85–102, perform the given operations. Where indicated, the expression is found in the stated technical area.

85. Show that $xy = \dfrac{1}{4}[(x + y)^2 - (x - y)^2]$.

86. Show that $x^2 + y^2 = \dfrac{1}{2}[(x + y)^2 + (x - y)^2]$.

87. Multiply: $2zS(S + 1)$ (solid-state physics)

88. Expand: $x^2(4 - x^2)$ (center of mass)

89. Expand: $kr(R - r)$ (blood flow)

90. Expand: $[2b + (n - 1)\lambda]^2$ (optics)

91. Factor: $\pi r_1^2 l - \pi r_2^2 l$ (jet plane fuel supply)

92. Factor: $cT_2 - cT_1 + RT_2 - RT_1$ (pipeline flow)

93. Factor: $4s^3 + 56s^2 + 96s$ (electricity)

94. Factor: $9600t + 8400t^2 - 1200t^3$ (solor energy)

95. Simplify and express in factored form: $(2R - r)^2 - (r^2 + R^2)$ (aircraft radar)

96. Express in factored form: $2R(R + r) - (R + r)^2$ (electricity: power)

97. Expand and simplify: $(n + 1)^3(2n + 1)^3$ (fluid flow in pipes)

98. Expand and simplify: $2(e_1 - e_2)^2 + 2(e_2 - e_3)^2$ (mechanical design)

99. Expand and simplify: $10a(T - t) + a(T - t)^2$ (instrumentation)

100. Expand the third term and then factor by grouping: $pa^2 + (1 - p)b^2 - [pa + (1 - p)b]^2$ (nuclear physics)

101. A metal cube of edge x is heated and each edge increases by 4 mm. Express the increase in volume in factored form.

102. Express the difference in volumes of two ball bearings of radii r mm and 3 mm in factored form. ($r > 3$ mm)

In Exercises 103–114, perform the indicated operations and simplify the given expressions. Each expression is from the indicated technical area of application.

103. $\left(\dfrac{2wtv^2}{Dg}\right)\left(\dfrac{b\pi^2 D^2}{n^2}\right)\left(\dfrac{6}{bt^2}\right)$ (machine design)

104. $\dfrac{m}{c} \div \left[1 - \left(\dfrac{p}{c}\right)^2\right]$ (airfoil design)

105. $\dfrac{\frac{\pi ka}{2}(R^4 - r^4)}{\pi ka(R^2 - r^2)}$ (flywheel rotation)

106. $\dfrac{V}{kp} - \dfrac{RT}{k^2 p^2}$ (electric motors)

107. $1 - \dfrac{d^2}{2} + \dfrac{d^4}{24} - \dfrac{d^6}{120}$ (aircraft emergency locator transmitter)

108. $\dfrac{wx^2}{2T_0} + \dfrac{kx^4}{12T_0}$ (bridge design)

109. $\dfrac{N + n}{2} + \dfrac{(N - n)^2}{4\pi^2 C}$ (machine design)

110. $\dfrac{Am}{k} - \dfrac{g}{2}\left(\dfrac{m}{k}\right)^2 + \dfrac{AML}{k}$ (rocket fuel)

111. $1 - \dfrac{3a}{4r} - \dfrac{a^3}{4r^3}$ (hydrodynamics)

112. $\dfrac{1}{F} + \dfrac{1}{f} - \dfrac{d}{fF}$ (optics)

113. $\dfrac{\frac{u^2}{2g} - x}{\frac{1}{2gc^2} - \frac{u^2}{2g} + x}$ (mechanism design)

114. $\dfrac{V}{\frac{1}{2R} + \frac{1}{2R + 2}}$ (electricity)

In Exercises 115–124, solve for the indicated leter. Each equation is from the indicated technical area of application.

115. $W = mgh_2 - mgh_1$, for m (work done on object)

116. $R_1(V_2 - V_1) + R_2 V_2 = 3R_1 R_2$, for V_2 (electricity)

117. $R = \dfrac{wL}{H(w + L)}$, for L (architecture)

118. $V_0 = \dfrac{V_r A}{1 + \beta A}$, for A (electricity)

119. $E = V_0 + \dfrac{(m + M)V^2}{2} + \dfrac{p^2}{2I}$, for M (nuclear physics)

120. $\dfrac{q_2 - q_1}{d} = \dfrac{f + q_1}{D}$, for q_1 (photography)

121. $s^2 + \dfrac{cs}{m} + \dfrac{kL^2}{mb^2} = 0$, for c (mechnical vibrations)

122. $I = \dfrac{A}{x^2} + \dfrac{B}{(10 - x)^2}$, for A (optics)

123. $V = \dfrac{i}{sC} + \dfrac{V_0}{s}$, for C (electricity)

124. $C = \dfrac{k}{n(1 - k)}$, for k (reinforced concrete design)

In Exercises 125–132, set up appropriate equations and solve the given stated problems. All numbers are accurate to at least two significant digits.

125. If a certain car's lights are left on, the battery will be dead in 4.0 h. If only the radio is left on, the battery will be dead in 24 h. How long will the battery last if both the lights and the radio are left on?

126. Two pumps are being used to fight a fire. One pumps 5000 L in 20 min, and the other pumps 5000 L in 25 min. How long will it take the two pumps together to pump 5000 L?

127. One computer can solve a certain problem in 3.0 s. With the aid of a second computer, the problem is solved in 1.0 s. How long would the second computer take to solve the problem alone?

128. An auto mechanic can do a certain motor job in 3.0 h, and with an assistant he can do it in 2.1 h. How long would it take the assistant to do the job alone?

129. The *relative density* of an object may be defined as its weight in air w_a, divided by the difference of its weight in air and its weight when submerged in water, w_w. For a lead weight, $w_a = 1.097\, w_w$. Find the relative density of lead.

130. A car travels halfway to its destination at 80.0 km/h and the remainder of the distance at 60.0 km/h. What is the average speed of the car for the trip?

131. For electric resistors in parallel, the reciprocal of the combined resistance equals the sum of the reciprocals of the individual resistances. For three resistors of 12 Ω, R ohms, and $2R$ ohms, in parallel, the combined resistance is 6.0 Ω. Find R.

132. An ambulance averaged 36 km/h going to an accident and 48 km/h on its return to the hospital. If the total time for the round-trip was 40 min, including 5 min at the accident scene, how far from the hospital was the accident?

Writing Exercise

133. An architecture student encounters the fraction

$$\frac{2r^2 + 5r - 3}{2r^2 + 7r + 3}$$

Simplify this fraction, and write a paragraph to describe your procedure. When you "cancel," explain what basic operation is being performed.

CHAPTER ⑥ PRACTICE TEST

1. Find the product: $2x(2x - 3)^2$.

2. The following equation is used in elecricity.

Solve for R_1: $\dfrac{1}{R} = \dfrac{1}{R_1 + r} + \dfrac{1}{R_2}$

3. Reduce to simplest form: $\dfrac{2x^2 + 5x - 3}{2x^2 + 12x + 18}$

4. Factor: $4x^2 - 16y^2$

5. Factor: $pb^3 + 8a^3p$ (business application)

6. Factor: $2a - 4T - ba + 2bT$

In Problems 7–9, perform the indicated operations and simplify.

7. $\dfrac{3}{4x^2} - \dfrac{2}{x^2 - x} - \dfrac{x}{2x - 2}$

8. $\dfrac{x^2 + x}{2 - x} \div \dfrac{x^2}{x^2 - 4x + 4}$

9. $\dfrac{1 - \dfrac{3}{2x + 2}}{\dfrac{x}{5} - \dfrac{1}{2}}$

10. If one riveter can do a job in 12 days, and a second riveter can do it in 16 days, how long would it take for them to do it together?

11. Solve for x: $\dfrac{3}{2x^2 - 3x} + \dfrac{1}{x} = \dfrac{3}{2x - 3}$

7

Quadratic Equations

In Section 7.3, we see how the design of a swimming pool area involves the solution of a quadratic equation.

We showed earlier how to solve some basic equations, and in this chapter we develop methods of solving the important *quadratic equation*. In the first three sections, we present algebraic methods of solution, and in Section 7.4 we discuss graphical solutions, including the use of the graphing calculator.

We will see that the graph of the *quadratic function is a parabola*, which has many modern applications in science and technology (Examples 2 and 3 on page 93 illustrate the graph of a parabola). We will discuss the parabola more in this chapter and in detail in Chapter 21.

One of the applications of a parabola is shown by a TV signal from a space satellite to a microwave dish at a home. The signal is reflected off the dish to a receiver at the *focus* of the dish, from which the signal goes to the TV set in the home. The surface of the dish is parabolic, and the fact that parabolic surfaces have this property of reflection was first shown by the Greek mathematician Diocles in about 200 B.C.E. He wanted to find a mirror to reflect the rays of the Sun to a point and cause burning. He proved that this was true for a mirror with a parabolic surface.

Another important application of a parabola (and thereby of quadratic functions) is that of a projectile, examples of which are a baseball, an artillery shell, and a rocket. When Galileo showed that the distance an object falls does not depend on its weight, he discovered that the distance fallen depends on the square of the time of fall. This in turn was shown to mean that the path of a projectile is parabolic (not considering air resistance). To find the location of a projectile for any particular time of flight requires the solution of a quadratic equation.

Other areas of application of quadratic equations include architecture, electric circuits, mechanical systems, forces on structures, and product design. Many are shown in the examples and exercises of this chapter.

7.1 QUADRATIC EQUATIONS; SOLUTION BY FACTORING

Given that a, b, and c are constants ($a \neq 0$), the equation

$$ax^2 + bx + c = 0 \qquad (7.1)$$

is called the **general quadratic equation in x.** The left side of Eq. (7.1) is a polynomial function of degree 2. *This function, $f(x) = ax^2 + bx + c$, is known as the* **quadratic function.** Any equation that can be simplified and then written in the form of Eq. (7.1) is a quadratic equation in one unknown.

Among the applications of quadratic equations and functions are the following examples: In finding the time t of flight of a projectile, we have the equation $s_0 + v_0 t - 4.9t^2 = 0$; in analyzing the electric current i in a circuit, the function $f(i) = Ei - Ri^2$ is found; and in determining the forces at a distance x along a beam, the function $f(x) = ax^2 + bLx + cL^2$ is used.

Since it is the x^2-term in Eq. (7.1) that distinguishes the quadratic equation from other types of equations, the equation is not quadratic if $a = 0$. However, either b or c (or both) may be zero, and the equation is still quadratic. No power of x higher than the second may be present in a quadratic equation. Also, we should be able to properly identify a quadratic equation even when it does not initially appear in the form of Eq. (7.1). The following two examples illustrate how we may recognize quadratic equations.

◀ **EXAMPLE 1** The following are quadratic equations.

$$\underset{\substack{\uparrow \\ a=1}}{x^2} - \underset{\substack{\uparrow \\ b=-4}}{4x} - \underset{\substack{\uparrow \\ c=-5}}{5} = 0$$

To show this equation in the form of Eq. (7.1), it can be written as $1x^2 + (-4)x + (-5) = 0$.

$$\underset{\substack{\uparrow \\ a=3}}{3x^2} - \underset{\substack{\uparrow \\ c=-6}}{6} = 0$$

Since there is no x-term, $b = 0$.

If either (or both) the first power term or the constant is missing, the quadratic is called incomplete.

$$\underset{\substack{\uparrow \\ a=2}}{2x^2} + \underset{\substack{\uparrow \\ b=7}}{7x} = 0$$

Since no constant appears, $c = 0$.

$$(m - 3)x^2 - mx + 7 = 0$$

The constants in Eq. (7.1) may include literal expressions. In this case, $m - 3$ takes the place of a, $-m$ takes the place of b, and $c = 7$.

$$4x^2 - 2x = x^2$$

After all nonzero terms have been collected on the left side, the equation becomes $3x^2 - 2x = 0$.

$$(x + 1)^2 = 4$$

Expanding the left side and collecting all nonzero terms on the left, we have $x^2 + 2x - 3 = 0$. ▶

◀ **EXAMPLE 2** The following are not quadratic equations.

$$bx - 6 = 0$$

There is no x^2-term.

$$x^3 - x^2 - 5 = 0$$

There should be no term of degree higher than 2. Thus, there can be no x^3-term in a quadratic equation.

$$x^2 + x - 7 = x^2$$

When terms are collected, there will be no x^2-term. ▶

SOLUTIONS OF A QUADRATIC EQUATION

Until the 1600s, most mathematicians did not accept negative, irrational, or imaginary roots of an equation. It was also generally accepted that an equation had only one root.

Recall that *the **solution** of an equation consists of all numbers (**roots**) which, when substituted in the equation, give equality.* There are *two* roots for a quadratic equation. At times, these roots are equal (see Example 3), and only one number is actually a solution. Also, the roots can be imaginary, and if this happens, all we wish to do at this point is to recognize that they are imaginary.

EXAMPLE 3 **(a)** The quadratic equation $3x^2 - 7x + 2 = 0$ has roots $x = 1/3$ and $x = 2$. This is seen by substituting these numbers in the equation.

$$3\left(\tfrac{1}{3}\right)^2 - 7\left(\tfrac{1}{3}\right) + 2 = 3\left(\tfrac{1}{9}\right) - \tfrac{7}{3} + 2 = \tfrac{1}{3} - \tfrac{7}{3} + 2 = \tfrac{0}{3} = 0$$
$$3(2)^2 - 7(2) + 2 = 3(4) - 14 + 2 = 14 - 14 = 0$$

(b) The quadratic equation $4x^2 - 4x + 1 = 0$ has a **double root** (*both roots are the same*) of $x = 1/2$. Showing that this number is a solution, we have

$$4\left(\tfrac{1}{2}\right)^2 - 4\left(\tfrac{1}{2}\right) + 1 = 4\left(\tfrac{1}{4}\right) - 2 + 1 = 1 - 2 + 1 = 0$$

(c) The quadratic equation $x^2 + 9 = 0$ has the imaginary roots $x = 3j$ and $x = -3j$, which means $x = 3\sqrt{-1}$ and $x = -3\sqrt{-1}$.

This section deals only with quadratic equations whose quadratic expression is factorable. Therefore, all roots will be rational. Using the fact that

NOTE ▶ *a product is zero if any of its factors is zero*

we have the following steps in solving a quadratic equation.

PROCEDURE FOR SOLVING A QUADRATIC EQUATION BY FACTORING

1. *Collect all terms on the left and simplify* (to the form of Eq. (7.1)).
2. *Factor the quadratic expression.*
3. *Set each factor equal to zero.*
4. *Solve the resulting linear equations. These numbers are the roots of the quadratic equation.*
5. *Check the solutions in the original equation.*

EXAMPLE 4 $x^2 - x - 12 = 0$

$(x - 4)(x + 3) = 0$ factor

$x - 4 = 0 \qquad x + 3 = 0$ set each factor equal to zero

$x = 4 \qquad\quad x = -3$ solve

The roots are $x = 4$ and $x = -3$. We can check them in the original equation by substitution. Therefore,

$$(4)^2 - (4) - 12 \overset{?}{=} 0 \qquad (-3)^2 - (-3) - 12 \overset{?}{=} 0$$
$$0 = 0 \qquad\qquad\qquad 0 = 0$$

Both roots satisfy the original equation.

◀ EXAMPLE 5 $2N^2 + 7N - 4 = 0$

$$(2N - 1)(N + 4) = 0 \qquad \text{factor}$$

$$2N - 1 = 0, \qquad N = \frac{1}{2} \qquad \begin{array}{l}\text{set each factor to}\\\text{zero and solve}\end{array}$$

$$N + 4 = 0, \qquad N = -4$$

Therefore, the roots are $N = \frac{1}{2}$ and $N = -4$. These roots can be checked by the same procedure used in Example 4.   ▶

◀ EXAMPLE 6 $x^2 + 4 = 4x$ equation not in form of Eq. (7.1)

$$x^2 - 4x + 4 = 0 \qquad \text{subtract } 4x \text{ from both sides}$$

$$(x - 2)^2 = 0 \qquad \text{factor}$$

$$x - 2 = 0, \qquad x = 2 \qquad \text{solve}$$

Since $(x - 2)^2 = (x - 2)(x - 2)$, both factors are the same. This means there is a double root of $x = 2$. Substitution shows that $x = 2$ satisfies the original equation.   ▶

NOTE ▶

It is essential for the quadratic expression on the left to be equal to zero (on the right), because if a product equals a nonzero number, it is probable that neither factor will give a correct root. Again, the first step must be to write the equation in the form of Eq. (7.1).

Equations with Fractions

A number of equations involving fractions lead to quadratic equations after the fractions are eliminated. The following two examples, the second being a stated problem, illustrate the process of solving such equations with fractions.

◀ EXAMPLE 7   Solve for x: $\dfrac{1}{x} + 3 = \dfrac{2}{x + 2}$.

$$\frac{x(x + 2)}{x} + 3x(x + 2) = \frac{2x(x + 2)}{x + 2} \qquad \begin{array}{l}\text{multiply each term by}\\\text{the LCD, } x(x + 2)\end{array}$$

$$x + 2 + 3x^2 + 6x = 2x \qquad \text{reduce each term}$$

$$3x^2 + 5x + 2 = 0 \qquad \text{collect terms on left}$$

$$(3x + 2)(x + 1) = 0 \qquad \text{factor}$$

$$3x + 2 = 0, \qquad x = -\frac{2}{3} \qquad \begin{array}{l}\text{set each factor equal}\\\text{to zero and solve}\end{array}$$

$$x + 1 = 0, \qquad x = -1$$

Checking in the original equation, we have

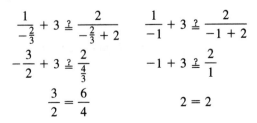

$$\frac{1}{-\frac{2}{3}} + 3 \overset{?}{=} \frac{2}{-\frac{2}{3} + 2} \qquad \frac{1}{-1} + 3 \overset{?}{=} \frac{2}{-1 + 2}$$

$$-\frac{3}{2} + 3 \overset{?}{=} \frac{2}{\frac{4}{3}} \qquad -1 + 3 \overset{?}{=} \frac{2}{1}$$

$$\frac{3}{2} = \frac{6}{4} \qquad 2 = 2$$

We see that the roots check. Remember, if either value gives division by zero, the root is extraneous, and must be excluded from the solution.   ▶

Solving a Word Problem

◀ EXAMPLE 8 A lumber truck travels 60 km from a sawmill to a lumber camp and then back in 7 h travel time. If the truck averages 5 km/h less on the return trip than on the trip to the camp, find its average speed to the camp. See Fig. 7.1.

Let v = the average speed (in km/h) of the truck going to the camp. This means that the average speed of the return trip was $(v - 5)$ km/h.

We also know that $d = vt$ (distance equals speed times time), which tells us that $t = d/v$. Thus, the time for each part of the trip is the distance divided by the speed.

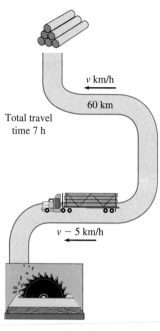

Fig. 7.1

time to camp	time from camp	total time

$$\frac{60}{v} + \frac{60}{v - 5} = 7$$

$60(v - 5) + 60v = 7v(v - 5)$ multiply each term by $v(v - 5)$

$7v^2 - 155v + 300 = 0$ collect terms on the left

$(7v - 15)(v - 20) = 0$ factor

$7v - 15 = 0,$ $v = \dfrac{15}{7}$ set each factor equal to zero and solve

$v - 20 = 0,$ $v = 20$

Only one of these solutions has meaning. The value $v = 15/7$ km/h cannot be the solution since the return speed of 5 km/h less would be negative. Therefore, the solution is $v = 20$ km/h, which means the return speed was 15 km/h. The trip to the camp took 3 h, and the return trip took 4 h, which shows the solution checks. ▶

EXERCISES 7.1

In Exercises 1 and 2, make the given changes in the indicated examples of this section and then solve the resulting quadratic equations.

1. In Example 5, change the + sign before 7N to − and then solve.

2. In Example 7, change the numerator of the first term to 2 and the numerator of the term on the right to 1 and then solve.

In Exercises 3–8, determine whether or not the given equations are quadratic. If the resulting form is quadratic, identify a, b, and c, with $a > 0$. Otherwise, explain why the resulting form is not quadratic.

3. $x(x - 2) = 4$

4. $(3x - 2)^2 = 2$

5. $x^2 = (x + 2)^2$

6. $x(2x^2 + 5) = 7 + 2x^2$

7. $n(n^2 + n - 1) = n^3$

8. $(T - 7)^2 = (2T + 3)^2$

In Exercises 9–42, solve the given quadratic equations by factoring.

9. $x^2 - 4 = 0$

10. $B^2 - 400 = 0$

11. $4y^2 - 9 = 0$

12. $x^2 - 0.16 = 0$

13. $x^2 - 8x - 9 = 0$

14. $s^2 + s - 6 = 0$

15. $R^2 - 7R + 12 = 0$

16. $x^2 - 11x + 30 = 0$

17. $40x - 16x^2 = 0$

18. $15L = 20L^2$

19. $27m^2 = 3$

20. $5p^2 = 80$

21. $3x^2 - 13x + 4 = 0$

22. $7x^2 + 3x - 4 = 0$

23. $A^2 + 8A + 16 = 0$

24. $4x^2 + 25 = 20x$

25. $6x^2 = 13x - 6$

26. $6z^2 = 6 + 5z$

27. $4x(x + 1) = 3$

28. $9t^2 = 9 - t(43 + t)$

29. $x^2 - x - 1 = 1$

30. $2x^2 - 7x + 6 = 3$

31. $x^2 - 4b^2 = 0$

32. $a^2x^2 = 1$

33. $8s^2 + 16s = 90$

34. $18t^2 - 48t + 32 = 0$

35. $(x + 2)^3 = x^3 + 8$

36. $V(V^2 - 4) = V^2(V - 1)$

37. $(x + a)^2 - b^2 = 0$

38. $x^2(a^2 + 2ab + b^2) - x(a + b) = 0$

39. The voltage V across a semiconductor in a computer is given by $V = \alpha I + \beta I^2$, where I is the current (in A). If a 6-V battery is conducted across the semiconductor, find the current if $\alpha = 2\ \Omega$ and $\beta = 0.5\ \Omega/\text{A}$.

40. The mass m (in Mg) of the fuel supply in the first-stage booster of a rocket is $m = 135 - 6t - t^2$, where t is the time (in s) after launch. When does the booster run out of fuel?

41. The power P (in MW) produced between midnight and noon by a nuclear power plant is $P = 4h^2 - 48h + 744$, where h is the hour of the day. At what time is the power 664 MW?

42. In determining the speed s (in km/h) of a car while studying its fuel economy, the equation $s^2 - 16s = 3072$ is used. Find s.

In Exercises 43 and 44, although the equations are not quadratic, factoring will lead to one quadratic factor and the solution can be completed by factoring as with a quadratic equation. Find the three roots of each equation.

43. $x^3 - x = 0$

44. $x^3 - 4x^2 - x + 4 = 0$

In Exercises 45–48, solve the given equations involving fractions.

45. $\dfrac{1}{x - 3} + \dfrac{4}{x} = 2$

46. $2 - \dfrac{1}{x} = \dfrac{3}{x + 2}$

47. $\dfrac{1}{2x} - \dfrac{3}{4} = \dfrac{1}{2x + 3}$

48. $\dfrac{x}{2} + \dfrac{1}{x - 3} = 3$

In Exercises 49–52, set up the appropriate quadratic equations and solve.

49. The spring constant k is the force F divided by the amount x the spring stretches ($k = F/x$). See Fig. 7.2(a). For two springs in series (see Fig. 7.2(b)), the reciprocal of the spring constant k_c for the combination equals the sum of the reciprocals of the individual spring constants. Find the spring constants for each of two springs in series if $k_c = 2$ N/cm and one spring constant is 3 N/cm more than the other.

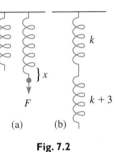

Fig. 7.2

50. The reciprocal of the combined resistance R of two resistances R_1 and R_2 connected in parallel (see Fig. 7.3(a)) is equal to the sum of the reciprocals of the individual resistances. If the two resistances are connected in series (see Fig. 7.3(b)), their combined resistance is the sum of their individual resistances. If two resistances connected in parallel have a combined resistance of 3.0 Ω and the same two resistances have a combined resistance of 16 Ω when connected in series, what are the resistances?

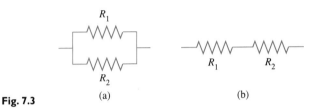

Fig. 7.3 (a) (b)

51. A hydrofoil made the round-trip of 120 km between two islands in 3.5 h of travel time. If the average speed going was 10 km/h less than the average speed returning, find these speeds.

52. A rectangular solar panel is 20 cm by 30 cm. By adding the same amount to each dimension, the area is doubled. How much is added?

7.2 COMPLETING THE SQUARE

Most quadratic equations that arise in applications cannot be solved by factoring. Therefore, we now develop a method, called **completing the square,** that can be used to solve any quadratic equation. In the next section, this method is used to develop a general formula that can be used to solve any quadratic equation.

In the first example, we show the solution of a type of quadratic equation that arises while using the method of completing the square. In the examples that follow it, the method itself is used and described.

The method of completing the square is used later in Section 7.4 and also in Chapter 21.

◀ EXAMPLE 1   In solving $x^2 = 16$, we may write $x^2 - 16 = 0$ and complete the solution by factoring. This gives us $x = 4$ and $x = -4$. Therefore, the principal square root of 16 and its negative both satisfy the original equation. Thus, we may solve $x^2 = 16$ by equating x to the principal square root of 16 and to its negative. The roots of $x^2 = 16$ are 4 and -4.

We may solve $(x - 3)^2 = 16$ in a similar way by equating $x - 3$ to 4 and to -4:

$$x - 3 = 4 \quad \text{or} \quad x - 3 = -4$$

Solving these equations, we obtain the roots 7 and -1.

We may solve $(x - 3)^2 = 17$ in the same way. Thus,

$$x - 3 = \sqrt{17} \quad \text{or} \quad x - 3 = -\sqrt{17}$$

The roots are therefore $3 + \sqrt{17}$ and $3 - \sqrt{17}$. Decimal approximations of these roots are 7.123 and -1.123.

◀EXAMPLE 2 To find the roots of the quadratic equation

$$x^2 - 6x - 8 = 0$$

first note that the left side is not factorable. However,

NOTE ▶ $x^2 - 6x$ *is part of the special product* $(x - 3)^2 = x^2 - 6x + 9$

and this special product is a *perfect square*. By adding 9 to $x^2 - 6x$, we have $(x - 3)^2$. Therefore, we rewrite the original equation as

$$x^2 - 6x = 8$$

NOTE ▶   and then **add 9 to both sides** of the equation. The result is

$$x^2 - 6x + 9 = 17$$

The left side of this equation may be rewritten, giving

$$(x - 3)^2 = 17$$

Now, as in the third illustration of Example 1, we have

$$x - 3 = \pm\sqrt{17}$$

The $\pm$ sign means that $x - 3 = \sqrt{17}$ or $x - 3 = -\sqrt{17}$.

By adding 3 to each side, we obtain

$$x = 3 \pm \sqrt{17}$$

which means $x = 3 + \sqrt{17}$ and $x = 3 - \sqrt{17}$ are the two roots of the equation.

Therefore, by creating an expression that is a perfect square and then using the principal square root and its negative, we were finally able to solve the equation as two linear equations.   ▶

How we determine the number to be added to complete the square is based on the special products in Eqs. (6.3) and (6.4). We rewrite these as

$$(x + a)^2 = x^2 + 2ax + a^2 \tag{7.2}$$
$$(x - a)^2 = x^2 - 2ax + a^2 \tag{7.3}$$

We must be certain that the coefficient of the x^2-term is 1 before we start to complete the square. The coefficient of x in each case is numerically $2a$, and the number added

CAUTION ▶ to complete the square is a^2. Thus, *if we **take half the coefficient of the x-term and square this result,** we have the number that completes the square.* In our example, the numerical coefficient of the x-term was 6, and 9 was added to complete the square. Therefore, the procedure for solving a quadratic equation by completing the square is as follows:

> **SOLVING A QUADRATIC EQUATION BY COMPLETING THE SQUARE**
>
> **1.** *Divide each side by a (the coefficient of x^2).*
> **2.** *Rewrite the equation with the constant on the right side.*
> **3.** *Complete the square: Add the square of one-half of the coefficient of x to both sides.*
> **4.** *Write the left side as a square and simplify the right side.*
> **5.** *Equate the square root of the left side to the principal square root of the right side and to its negative.*
> **6.** *Solve the two resulting linear equations.*

EXAMPLE 3 Solve $2x^2 + 16x - 9 = 0$ by completing the square.

1. Divide each side by 2 to make the coefficient of x^2 equal to 1:

$$x^2 + 8x - \frac{9}{2} = 0$$

2. Put the constant on the right by adding $\frac{9}{2}$ to both sides:

$$x^2 + 8x = \frac{9}{2}$$

CAUTION ▶ **3.** The coefficient of the x-term is 8. Therefore, divide 8 by 2 to get 4. Now, square 4 to get 16 and add 16 to both sides:

$$\frac{1}{2}(8) = 4; \qquad 4^2 = 16$$

$$x^2 + 8x + 16 = \frac{9}{2} + 16 = \frac{41}{2}$$

4. Write the left side as $(x + 4)^2$:

$$(x + 4)^2 = \frac{41}{2} \qquad \text{take the square root of each side}$$

5. Equate $x + 4$ to the principal square root of $\frac{41}{2}$ and its negative:

$$x + 4 = \pm \sqrt{\frac{41}{2}}$$

6. Solve for x:

$$x = -4 \pm \sqrt{\frac{41}{2}}$$

Therefore, the roots are $-4 + \sqrt{\frac{41}{2}}$ and $-4 - \sqrt{\frac{41}{2}}$. Using a calculator to approximate these roots, we get 0.5277 and -8.528. ▶

EXERCISES 7.2

In Exercises 1 and 2, make the given changes in the indicated examples of this section and then solve the resulting quadratic equations by completing the square.

1. In Example 2, change the $-$ sign before $6x$ to $+$.

2. In Example 3, change the coefficient of the second term from 16 to 12.

In Exercises 3–10, solve the given quadratic equations by finding appropriate square roots as in Example 1.

3. $x^2 = 25$ **4.** $x^2 = 100$

5. $x^2 = 7$ **6.** $x^2 = 15$

7. $(x - 2)^2 = 25$ **8.** $(x + 2)^2 = 100$

9. $(x + 3)^2 = 7$ **10.** $(x - 4)^2 = 10$

In Exercises 11–30, solve the given quadratic equations by completing the square. Exercises 11–14 and 17–20 may be checked by factoring.

11. $x^2 + 2x - 8 = 0$ **12.** $x^2 - x - 6 = 0$

13. $D^2 + 3D + 2 = 0$ **14.** $t^2 + 5t - 6 = 0$

15. $n^2 = 4n - 2$ **16.** $(R + 9)(R + 1) = 13$

17. $v(v + 2) = 15$ **18.** $Z^2 + 12 = 8Z$

19. $2s^2 + 5s = 3$ **20.** $4x^2 + x = 3$

21. $3y^2 = 3y + 2$ **22.** $3x^2 = 3 - 4x$

23. $2y^2 - y - 2 = 0$ **24.** $9v^2 - 6v - 2 = 0$

25. $5T^2 - 10T + 4 = 0$ **26.** $4V^2 + 9 = 12V$

27. $9x^2 + 6x + 1 = 0$ **28.** $2x^2 - 3x + 2a = 0$

29. $x^2 + 2bx + c = 0$ **30.** $px^2 + qx + r = 0$

In Exercises 31 and 32, use completing the square to solve the given problems.

31. The voltage V across a certain electronic device is related to the temperature T (in °C) by $V = 4.0T - 0.2T^2$. For what temperature(s) is $V = 15$ V?

32. A rectangular storage area is 8.0 m longer than it is wide. If the area is 28 m², what are its dimensions?

7.3 THE QUADRATIC FORMULA

We now use the method of completing the square to derive a general formula that may be used for the solution of any quadratic equation.

Consider Eq. (7.1), the general quadratic equation:

$$ax^2 + bx + c = 0 \quad (a \neq 0)$$

When we divide through by a, we obtain

$$x^2 + \frac{b}{a}x + \frac{c}{a} = 0$$

Subtracting c/a from each side, we have

$$x^2 + \frac{b}{a}x = -\frac{c}{a}$$

Half of b/a is $b/2a$, which squared is $b^2/4a^2$. Adding $b^2/4a^2$ to each side gives us

$$x^2 + \frac{b}{a}x + \frac{b^2}{4a^2} = -\frac{c}{a} + \frac{b^2}{4a^2}$$

Writing the left side as a perfect square and combining fractions on the right side, we have

$$\left(x + \frac{b}{2a}\right)^2 = \frac{b^2 - 4ac}{4a^2}$$

Equating $x + \dfrac{b}{2a}$ to the principal square root of the right side and its negative,

$$x + \frac{b}{2a} = \frac{\pm\sqrt{b^2 - 4ac}}{2a}$$

When we subtract $b/2a$ from each side and simplify the resulting expression, we obtain the **quadratic formula:**

Quadratic Formula

$$x = \frac{-b \pm \sqrt{b^2 - 4ac}}{2a} \tag{7.4}$$

The quadratic formula, with the $\pm$ sign, means that the solutions to the quadratic equation

$$ax^2 + bx + c = 0$$

are

$$x = \frac{-b + \sqrt{b^2 - 4ac}}{2a}$$

and

$$x = \frac{-b - \sqrt{b^2 - 4ac}}{2a}$$

The quadratic formula gives us a quick general way of solving any quadratic equation. We need only write the equation in the standard form of Eq. (7.1), substitute these numbers into the formula, and simplify.

◀ **EXAMPLE 1** Solve: $x^2 - 5x + 6 = 0$.
$$\underset{a=1}{\uparrow} \quad \underset{b=-5}{\uparrow} \quad \underset{c=6}{\uparrow}$$

Here, using the indicated values of a, b, and c in the quadratic formula, we have

$$x = \frac{-(-5) \pm \sqrt{(-5)^2 - 4(1)(6)}}{2(1)} = \frac{5 \pm \sqrt{25 - 24}}{2} = \frac{5 \pm 1}{2}$$

$$x = \frac{5 + 1}{2} = 3 \quad \text{or} \quad x = \frac{5 - 1}{2} = 2$$

The roots $x = 3$ and $x = 2$ check when substituted in the original equation. ▶

It must be emphasized that, in using the quadratic formula, the entire expression **CAUTION ▶** $-b \pm \sqrt{b^2 - 4ac}$ is divided by $2a$. ***It is a relatively common error to divide only the radical*** $\sqrt{b^2 - 4ac}$.

◀ EXAMPLE 2 Solve: $2x^2 - 7x - 5 = 0$.

$\uparrow$ $\uparrow$ $\uparrow$

$a = 2$ $b = -7$ $c = -5$

Substituting the values for a, b, and c in the quadratic formula, we have

$$x = \frac{-(-7) \pm \sqrt{(-7)^2 - 4(2)(-5)}}{2(2)} = \frac{7 \pm \sqrt{49 + 40}}{4} = \frac{7 \pm \sqrt{89}}{4}$$

$$x = \frac{7 + \sqrt{89}}{4} = 4.108 \quad \text{or} \quad x = \frac{7 - \sqrt{89}}{4} = -0.6085$$

See Appendix C for a graphing calculator program QUADFORM. It can be used to find the roots of a quadratic equation.

The exact roots are $x = \dfrac{7 \pm \sqrt{89}}{4}$ (this form is often used when the roots are irrational).

Approximate decimal values are $x = 4.108$ and $x = -0.6085$.   ◗

◀ EXAMPLE 3 Solve: $9x^2 + 24x + 16 = 0$.
In this example, $a = 9$, $b = 24$, and $c = 16$. Thus,

$$x = \frac{-24 \pm \sqrt{24^2 - 4(9)(16)}}{2(9)} = \frac{-24 \pm \sqrt{576 - 576}}{18} = \frac{-24 \pm 0}{18} = -\frac{4}{3}$$

Here, both roots are $-\frac{4}{3}$, and we write the result as $x = -\frac{4}{3}$ and $x = -\frac{4}{3}$. We will get a double root when $b^2 = 4ac$, as in this case.   ◗

Many calculators have an equation-solving feature. Also, some calculators have a specific feature for solving higher-degree equations.

◀ EXAMPLE 4 Solve: $3x^2 - 5x + 4 = 0$.
In this example, $a = 3$, $b = -5$, and $c = 4$. Therefore,

$$x = \frac{-(-5) \pm \sqrt{(-5)^2 - 4(3)(4)}}{2(3)} = \frac{5 \pm \sqrt{25 - 48}}{6} = \frac{5 \pm \sqrt{-23}}{6}$$

These roots contain imaginary numbers. This happens if $b^2 < 4ac$.   ◗

Examples 1–4 illustrate the *character of the roots* of a quadratic equation. If a, b, and c are rational numbers (see Section 1.1), by noting the value of $b^2 - 4ac$ (called the **discriminant**), we have the following:

If $b^2 - 4ac$ is positive and a perfect square, $ax^2 + bx + c$ is factorable.

CHARACTER OF THE ROOTS OF A QUADRATIC EQUATION

1. If $b^2 - 4ac$ is positive and a perfect square (see Section 1.6), the roots are real, rational, and unequal. (See Example 1, where $b^2 - 4ac = 1$.)

2. If $b^2 - 4ac$ is positive but not a perfect square, the roots are real, irrational, and unequal. (See Example 2, where $b^2 - 4ac = 89$.)

3. If $b^2 - 4ac = 0$, the roots are real, rational, and equal. (See Example 3, where $b^2 - 4ac = 0$.)

4. If $b^2 - 4ac < 0$, the roots contain imaginary numbers and are unequal. (See Example 4, where $b^2 - 4ac = -23$.)

We can use the value of $b^2 - 4ac$ to help in checking the roots or in finding the character of the roots without having to solve the equation completely.

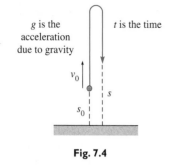

g is the acceleration due to gravity

t is the time

v_0

s_0

s

Fig. 7.4

◀ **EXAMPLE 5** An equation used in the analysis of projectile motion (see Fig. 7.4) is $s = s_0 + v_0t - \frac{1}{2}gt^2$. Solve for t.

$$gt^2 - 2v_0t - 2(s_0 - s) = 0 \qquad \text{multiply by } -2, \text{ put in form of Eq. (7.1)}$$

In this form, we see that $a = g$, $b = -2v_0$, and $c = -2(s_0 - s)$:

$$t = \frac{-(-2v_0) \pm \sqrt{(-2v_0)^2 - 4g(-2)(s_0 - s)}}{2g} = \frac{2v_0 \pm \sqrt{4(v_0^2 + 2gs_0 - 2gs)}}{2g}$$

$$= \frac{2v_0 \pm 2\sqrt{v_0^2 + 2gs_0 - 2gs}}{2g}$$

$$= \frac{v_0 \pm \sqrt{v_0^2 + 2gs_0 - 2gs}}{g}$$

Solving a Word Problem

See the chapter introduction.

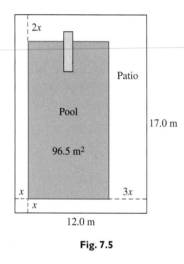

$2x$

Patio

Pool

17.0 m

96.5 m²

x

$3x$

x

12.0 m

Fig. 7.5

◀ **EXAMPLE 6** A rectangular area 17.0 m long and 12.0 m wide is to be used for a patio with a rectangular pool. One end and one side of the patio (the area around the pool for chairs, sunning, etc.) are to be the same width. The other end with the diving board is to be twice as wide, and the other side is to be three times as wide as the narrow side. If the area of the pool is to be 96.5 m², what are the widths of the patio ends and sides and the dimensions of the pool? See Fig. 7.5.

First, let $x =$ the width of the narrow end and side of the patio. This means that the other end is $2x$ in width and the other side is $3x$ in width. Since the area of the pool is 96.5 m², we find x as follows:

pool length   pool width   pool area

$$(17.0 - 3x)(12.0 - 4x) = 96.5$$

$$204 - 68.0x - 36.0x + 12x^2 = 96.5$$

$$12x^2 - 104.0x + 107.5 = 0$$

$$x = \frac{-(-104.0) \pm \sqrt{(-104.0)^2 - 4(12)(107.5)}}{2(12)} = \frac{104.0 \pm \sqrt{5656}}{24}$$

Evaluating the roots, we get $x = 7.5$ m and $x = 1.2$ m. The value $x = 7.5$ m cannot be the required result since the width of the patio would be greater than the width of the entire area. For the value $x = 1.2$ m, the pool would have a length of 13.4 m and a width of 7.2 m. These give an area of 96.5 m², which means it checks. The widths of the patio area are then 1.2 m, 1.2 m, 2.4 m, and 3.6 m.

EXERCISES **7.3**

In Exercises 1–4, make the given changes in the indicated examples of this section and then solve the resulting equations by the quadratic formula.

1. In Example 1, change the − sign before $5x$ to +.

2. In Example 2, change the coefficient of x^2 from 2 to 3.

3. In Example 3, change the + sign before $24x$ to −.

4. In Example 4, change 4 to 3.

In Exercises 5–36, solve the given quadratic equations, using the quadratic formula. Exercises 5–8 are the same as Exercises 11–14 of Section 7.2.

5. $x^2 + 2x - 8 = 0$

6. $x^2 - x - 6 = 0$

7. $D^2 + 3D + 2 = 0$

8. $t^2 + 5t - 6 = 0$

9. $x^2 - 4x + 2 = 0$

10. $x^2 + 10x - 4 = 0$

11. $v^2 + 2v - 15 = 0$

12. $V^2 - 8V + 12 = 0$

13. $2s^2 + 5s = 3$

14. $4x^2 + x = 3$

15. $3y^2 = 3y + 2$

16. $3x^2 = 3 - 4x$

17. $2y^2 - y - 2 = 0$

18. $9v^2 - 6v - 2 = 0$

19. $30y^2 + 23y - 40 = 0$

20. $40x^2 - 62x - 63 = 0$

21. $8t^2 + 61t = -120$

22. $2d(d - 2) = -7$

23. $s^2 = 9 + s(1 - 2s)$

24. $20r^2 = 20r + 1$

25. $25y^2 = 121$

26. $37T = T^2$

27. $15 + 4z = 32z^2$

28. $4x^2 - 12x = 7$

29. $x^2 - 0.20x - 0.40 = 0$

30. $3.2x^2 = 2.5x + 7.6$

31. $0.29Z^2 - 0.18 = 0.63Z$

32. $12.5x^2 + 13.2x = 15.5$

33. $x^2 + 2cx - 1 = 0$

34. $x^2 - 7x + (6 + a) = 0$

35. $b^2x^2 - (b + 1)x + (1 - a) = 0$

36. $c^2x^2 - x - 1 = x^2$

In Exercises 37–40, without solving the given equations, determine the character of the roots.

37. $2x^2 - 7x = -8$

38. $3x^2 + 19x = 14$

39. $3.6t^2 + 2.1 = 7.7t$

40. $0.45s^2 + 0.33 = 0.12s$

In Exercises 41–56, solve the given problems. All numbers are accurate to at least two significant digits.

41. Determine the relationship between a and c in Eq. (7.1) if the roots of a quadratic equation are reciprocals.

42. Without drawing the graph or completely solving the equation, explain how to find the number of x-intercepts of a quadratic function.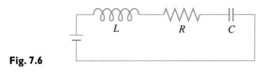

43. In machine design, in finding the outside diameter D_0 of a hollow shaft, the equation $D_0^2 - DD_0 - 0.25D^2 = 0$ is used. Solve for D_0 if $D = 3.625$ cm.

44. A missile is fired vertically into the air. The distance s (in m) above the ground as a function of time t (in s) is given by $s = 100 + 500t - 4.9t^2$. (a) When will the missile hit the ground? (b) When will the missile be 1000 m above the ground?

45. For a rectangle, if the ratio of the length to the width equals the ratio of the length plus the width to the length, the ratio is called the *golden ratio*. Find the value of the golden ratio, which the ancient Greeks thought had the most pleasing properties to look at.

46. When focusing a camera, the distance r the lens must move from the infinity setting is given by $r = f^2/(p - f)$, where p is the distance from the object to the lens, and f is the focal length of the lens. Solve for f.

47. In calculating the current in an electric circuit with an inductance L, a resistance R, and a capacitance C, it is necessary to solve the equation $Lm^2 + Rm + 1/C = 0$. Solve for m in terms of L, R, and C. See Fig. 7.6.

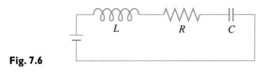
Fig. 7.6

48. In finding the radius r of a circular arch of height h and span b, an architect used the following formula. Solve for h.

$$r = \frac{b^2 + 4h^2}{8h}$$

49. The front of a computer monitor is 36 cm wide and 29 cm high with a uniform edge around the viewing screen. If the edge covers 44% of the monitor front, what is the width of the edge?

50. An investment of $2000 is deposited in a certain annual interest rate. One year later, $3000 is deposited in another account at the same rate. At the end of the second year, the accounts have a total value of $5319.05. What is the interest rate?

51. The length of a tennis court is 12.8 m more than its width. If the area of a tennis court is 262 m², what are its dimensions? See Fig. 7.7.

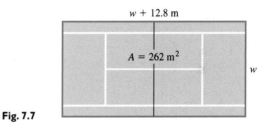

Fig. 7.7

52. Two circular oil spills are tangent to each other. If the distance between centers is 800 m and they cover a combined area of 1.02×10^6 m², what is the radius of each? See Fig. 7.8.

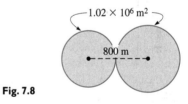

Fig. 7.8

53. In remodeling a house, an architect finds that by adding the same amount to each dimension of a 3.8-m by 5.0-m rectangular room the area would be increased by 11 m². How much must be added to each dimension?

54. Two pipes together drain a wastewater-holding tank in 6.00 h. If used alone to empty the tank, one takes 2.00 h longer than the other. How long does each take to empty the tank if used alone?

55. On some highways, a car can legally travel 20 km/h faster than a truck. Traveling at maximum legal speeds, a car can travel 120 km in 18 min less than a truck. What are the maximum legal speeds for cars and for trucks?

56. For electric capacitors connected in series, the sum of the reciprocals of the capacitances equals the reciprocal of the combined capacitance. If one capacitor has 5.0 µF more capacitance than another capacitor and they are connected in series, what are their capacitances if their combined capacitance is 4.0 µF?

7.4 THE GRAPH OF THE QUADRATIC FUNCTION

In this section, we discuss the graph of the quadratic function $ax^2 + bx + c$ and show the graphical solution of a quadratic equation. By letting $y = ax^2 + bx + c$, we can graph this function, as in Chapter 3. The next example shows the graph of the quadratic function done in this way.

❨ **EXAMPLE 1**   Graph the function $f(x) = x^2 + 2x - 3$.

First, let $y = x^2 + 2x - 3$. We can then set up a table of values and graph the function as shown in Fig. 7.9, or we can display its graph on a graphing calculator as shown in Fig. 7.10.

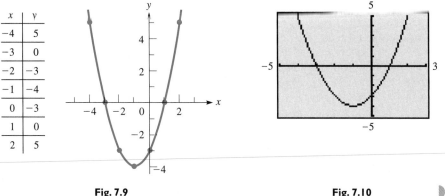

x	y
-4	5
-3	0
-2	-3
-1	-4
0	-3
1	0
2	5

Fig. 7.9   **Fig. 7.10**

The graph of any quadratic function $y = ax^2 + bx + c$ will have the same basic shape as that shown in Fig. 7.9 and is called a **parabola.** (In Section 3.4, we briefly noted that graphs in Examples 2 and 3 were parabolas.) A parabola defined by a quadratic function can open upward (as in Example 1) or downward. The location of the parabola and how it opens depends on the values of a, b, and c, as we will show after Example 2.

NOTE ▶ In Example 1, the parabola has a minimum point at $(-1, -4)$, and the curve opens upward. *All parabolas have an* **extreme point** *of this type. If $a > 0$, the parabola has a* **minimum point,** *and it opens* **upward.** *If $a < 0$, the parabola has a* **maximum point,** *and it opens* **downward.** *The extreme point of the parabola is also known as its* **vertex.**

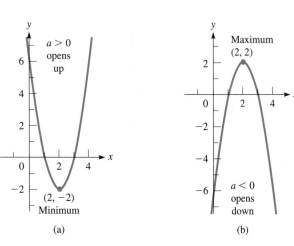

(a)   (b)   **Fig. 7.11**

❨ **EXAMPLE 2**   The graph of $y = 2x^2 - 8x + 6$ is shown in Fig. 7.11(a). For this parabola, $a = 2$ ($a > 0$) and it opens upward. The vertex (a minimum point) is $(2, -2)$.

The graph of $y = -2x^2 + 8x - 6$ is shown in Fig. 7.11(b). For this parabola, $a = -2$ ($a < 0$) and it opens downward. The vertex (a maximum point) is $(2, 2)$.

We can sketch the graph of a parabola by using its basic shape and knowing the location of two or three points, including the vertex. Even when using a graphing calculator, we can get a check on the graph by knowing the vertex and how the parabola opens.

In order to find the coordinates of the extreme point, start with the quadratic function

$$y = ax^2 + bx + c$$

Then, factor a from the two terms containing x, obtaining

$$y = a\left(x^2 + \frac{b}{a}x\right) + c$$

Now, completing the square of the terms within parentheses, we have

$$y = a\left(x^2 + \frac{b}{a}x + \frac{b^2}{4a^2}\right) + c - \frac{b^2}{4a}$$

$$= a\left(x + \frac{b}{2a}\right)^2 + c - \frac{b^2}{4a}$$

Now, look at the factor $\left(x + \frac{b}{2a}\right)^2$. If $x = -b/2a$, the term is zero. If x is any other value,

$\left(x + \frac{b}{2a}\right)^2$ is positive. Thus, if $a > 0$, the value of y increases from the minimum

NOTE ▶ value where $x = -b/2a$, and if $a < 0$, the value of y decreases from the maximum value where $x = -b/2a$. *This means that $x = -b/2a$ is the x-coordinate of the vertex (extreme point). The y-coordinate is found by substituting this x-value in the function.*

Another easily found point is the y-intercept. As with a linear equation, we find the y-intercept where $x = 0$. For $y = ax^2 + bx + c$, if $x = 0$, then $y = c$. *This means that the point $(0, c)$ is the y-intercept.*

◀ **EXAMPLE 3** For the graph of the function $y = 2x^2 - 8x + 6$, find the vertex and y-intercept and sketch the graph. (This function is also used in Example 2.)

First, $a = 2$ and $b = -8$. This means that the x-coordinate of the vertex is

$$\frac{-b}{2a} = \frac{-(-8)}{2(2)} = \frac{8}{4} = 2$$

and the y-coordinate is

$$y = 2(2^2) - 8(2) + 6 = -2$$

Thus, the vertex is $(2, -2)$. Since $a > 0$, it is a minimum point.

Since $c = 6$, the y-intercept is $(0, 6)$.

We can use the minimum point $(2, -2)$ and the y-intercept $(0, 6)$, along with the fact that the graph is a parabola, to get an approximate sketch of the graph. Noting that a parabola increases (or decreases) away from the vertex in the same way on each side of it (it is *symmetric* to a vertical line through the vertex), we sketch the graph in Fig. 7.12. It is the same graph as that shown in Fig. 7.11(a). ▶

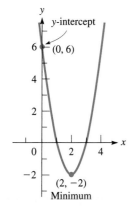

Fig. 7.12

If we are not using a graphing calculator, we may need one or two additional points to get a good sketch of a parabola. This would be true if the y-intercept is close to the vertex. Two points we can find are the x-intercepts, if the parabola crosses the x-axis (one point if the vertex is on the x-axis). They are found by setting $y = 0$ and solving the quadratic equation $ax^2 + bx + c = 0$. Also, we may simply find one or two points other than the vertex and the y-intercept. Sketching a parabola in this way is shown in the following two examples.

◀ **EXAMPLE 4** Sketch the graph of $y = -x^2 + x + 6$.

We first note that $a = -1$ and $b = 1$. Therefore, the x-coordinate of the maximum point $(a < 0)$ is $-\frac{1}{2(-1)} = \frac{1}{2}$. The y-coordinate is $-\left(\frac{1}{2}\right)^2 + \frac{1}{2} + 6 = \frac{25}{4}$. This means that the maximum point is $\left(\frac{1}{2}, \frac{25}{4}\right)$.

The y-intercept is $(0, 6)$.

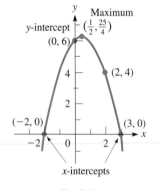

Using these points in Fig. 7.13, since they are close together, does not give a good idea of how wide the parabola opens. Therefore, setting $y = 0$, we solve the equation

$$-x^2 + x + 6 = 0 \qquad \text{multiply each term by } -1$$

or

$$x^2 - x - 6 = 0$$

This equation is factorable. Thus,

$$(x - 3)(x + 2) = 0$$
$$x = 3, -2$$

Fig. 7.13

This means that the x-intercepts are $(3, 0)$ and $(-2, 0)$, as shown in Fig. 7.13.

Also, rather than finding the x-intercepts, we can let $x = 2$ (or some value to the right of the vertex) and then use the point $(2, 4)$. ◗

◀ **EXAMPLE 5** Sketch the graph of $y = x^2 + 1$.

Since there is no x-term, $b = 0$. This means that the x-coordinate of the minimum point $(a > 0)$ is 0 and that the minimum point and the y-intercept are both $(0, 1)$. We know that the graph opens upward, since $a > 0$, which in turn means that it does not cross the x-axis. Now, letting $x = 2$ and $x = -2$, find the points $(2, 5)$ and $(-2, 5)$ on the graph, which is shown in Fig. 7.14. ◗

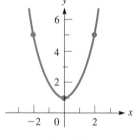

Fig. 7.14

We can see that the domain of the quadratic function is all x. Knowing that the graph has either a maximum point or a minimum point, the range of the quadratic function must be restricted to certain values of y. In Example 3, the range is $f(x) \geq -2$; in Example 4, the range is $f(x) \leq \frac{25}{4}$; and in Example 5, the range is $f(x) \geq 1$.

SOLVING QUADRATIC EQUATIONS GRAPHICALLY

In Chapter 3, we showed how an equation can be solved graphically. Following that method, to solve the equation $ax^2 + bx + c = 0$, let $y = ax^2 + bx + c$ and graph the function. The roots of the equation are the x-coordinates of the points for which $y = 0$ (the x-intercepts). The following examples illustrate solving quadratic equations graphically.

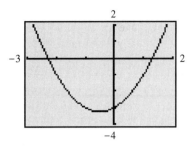

Fig. 7.15

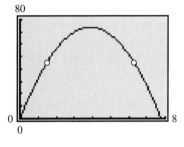

Fig. 7.16

◀ **EXAMPLE 6** Solve the equation $3x = x(2 - x) + 3$ graphically.

First, collect all terms on the left of the equal sign. This gives $x^2 + x - 3 = 0$. We then let $y = x^2 + x - 3$ and graph this function. The minimum point is $\left(-\frac{1}{2}, -\frac{13}{4}\right)$ and the y-intercept is $(0, -3)$.

Using the graphing calculator, these points help us choose the *window* settings shown for the graph of the function in Fig. 7.15. Now, using the *zero* feature (or the *trace* and *zoom* features), we find the roots to be

$$x = -2.30 \quad \text{and} \quad x = 1.30$$

These values (or the stored calculator values) check when substituted in the original equation. ▶

◀ **EXAMPLE 7** A projectile is fired vertically upward from the ground with a velocity of 38 m/s. Its distance above the ground is given by $s = -4.9t^2 + 38t$, where s is the distance (in m) and t is the time (in s). Graph the function and from the graph determine (1) when the projectile will hit the ground and (2) how long it takes to reach 45 m above the ground.

Since $c = 0$, the s-intercept is at the origin. Using $-b/2a$, we find the maximum point at about $(3.9, 74)$. These points help us choose (using x for t, and y for s, on the calculator) the *window* settings for the graph shown in Fig. 7.16.

Using the *trace* and *zoom* features (or the *intersect* feature), we can now determine the required answers.

1. The projectile will hit the ground for $s = 0$, which is shown at the right t-intercept. Thus, it will take about 7.8 s to hit the ground.

2. We find the time for the projectile to get 45 m above the ground by finding the values of t for $s = 45$ m. We see that there are *two* points for which $s = 45$ m. The values of t at these points show that the times are $t = 1.5$ s and $t = 6.3$ s.

From the maximum point, we can also conclude that the maximum height reached by the projectile is about 74 m. This can also be seen on the graph. ▶

EXERCISES 7.4

In Exercises 1 and 2, make the given changes in the indicated examples of this section and then solve the resulting problems.

1. In Example 3, change the $-$ sign before $8x$ to $+$ and then sketch the graph.

2. In Example 6, change the coefficient of x on the left from 3 to 5 and then find the solution graphically.

In Exercises 3–8, sketch the graph of each parabola by using only the vertex and the y-intercept. Check the graph using a graphing calculator.

3. $y = x^2 - 6x + 5$
4. $y = -x^2 - 4x - 3$
5. $y = -3x^2 + 10x - 4$
6. $y = 2x^2 + 8x - 5$
7. $y = x^2 - 4x$
8. $y = -2x^2 - 5x$

In Exercises 9–12, sketch the graph of each parabola by using the vertex, the y-intercept, and the x-intercepts. Check the graph using a graphing calculator.

9. $y = x^2 - 4$
10. $y = x^2 + 3x$
11. $y = -2x^2 - 6x + 8$
12. $y = -3x^2 + 12x - 5$

In Exercises 13–16, sketch the graph of each parabola by using the vertex, the y-intercept, and two other points, not including the x-intercepts. Check the graph using a graphing calculator.

13. $y = 2x^2 + 3$
14. $y = x^2 + 2x + 2$
15. $y = -2x^2 - 2x - 6$
16. $y = -3x^2 - x$

In Exercises 17–24, use a graphing calculator to solve the given equations. If there are no real roots, state this as the answer.

17. $2x^2 - 3 = 0$
18. $5 - x^2 = 0$
19. $-3x^2 + 11x - 5 = 0$
20. $2t^2 = 7t + 4$
21. $x(2x - 1) = -3$
22. $2x - 5 = x^2$
23. $6R^2 = 18 - 7R$
24. $3x^2 - 25 = 20x$

In Exercises 25–30, use a graphing calculator to graph all three parabolas on the same coordinate system. In Exercises 25–28, describe (a) the shifts (see page 100) of $y = x^2$ that occur and (b) how each parabola opens. In Exercises 29 and 30, describe (a) the shifts and (b) the **stretching** and **shrinking**.

25. (a) $y = x^2$ (b) $y = x^2 + 3$ (c) $y = x^2 - 3$

26. (a) $y = x^2$ (b) $y = (x - 3)^2$ (c) $y = (x + 3)^2$

27. (a) $y = x^2$ (b) $y = (x - 2)^2 + 3$ (c) $y = (x + 2)^2 - 3$

28. (a) $y = x^2$ (b) $y = -x^2$ (c) $y = -(x - 2)^2$

29. (a) $y = x^2$ (b) $y = 3x^2$ (c) $y = \frac{1}{3}x^2$

30. (a) $y = x^2$ (b) $y = -3(x - 2)^2$ (c) $y = \frac{1}{3}(x + 2)^2$

In Exercises 31–40, solve the given applied problem.

31. The vertical distance d (in cm) of the end of a robot arm above a conveyor belt in its 8-s cycle is given by $d = 2t^2 - 16t + 47$. Sketch the graph of $d = f(t)$.

32. When mineral deposits form a uniform coating 1 mm thick on the inside of a pipe of radius r (in mm), the cross-sectional area A through which water can flow is $A = \pi(r^2 - 2r + 1)$. Sketch $A = f(r)$.

33. An equipment company determines that the area A (in m²) covered by a rectangular tarpaulin is given by $A = w(8 - w)$, where w is the width of the tarpaulin and its perimeter is 16 m. Sketch the graph of A as a function of w.

34. Under specified conditions, the pressure loss L (in kPa/100 m), in the flow of water through a water line in which the flow is q L/min, is given by $L = 0.0001q^2 + 0.005q$. Sketch the graph of L as a function of q, for $q < 400$ L/min.

35. When analyzing the power P (in W) dissipated in an electric circuit, the equation $P = 50i - 3i^2$ results. Here, i is the current (in A). Sketch the graph of $P = f(i)$.

36. Tests show that the power P (in kW) of an automobile engine as a function of r (in r/min—rpm) is given by $P = -5.0 \times 10^{-6}r^2 + 0.050r - 45$ $(1500 < r < 6000$ r/min). Sketch the graph of P vs. r and find the maximum power that is produced.

37. A missile is fired vertically upward such that its distance s (in m) above the ground is given by $s = 50 + 90t - 4.9t^2$, where t is the time (in s). Sketch the graph and then determine from the graph (a) when the missile will hit the ground (to 0.1 s), (b) how high it will go (to three significant digits), and (c) how long it will take to reach a height of 250 m (to 0.1 s).

38. In a certain electric circuit, the resistance R (in Ω) that gives resonance is found by solving the equation $25R = 3(R^2 + 4)$. Solve this equation graphically (to 0.1 Ω).

39. A security fence is to be built around a rectangular parking area of 2000 m². If the front side of the fence costs $60/m and the other three sides cost $30/m, solve graphically for the dimensions (to 1 m) of the parking area if the fence is to cost $7500. See Fig. 7.17.

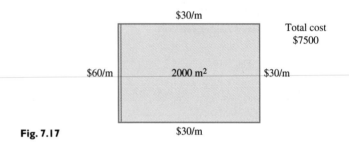

Fig. 7.17

40. An airplane pilot could decrease the time t (in h) needed to travel the 5400 km from Ottawa to London by 60 min if the plane's speed v is increased by 40.0 km/h. Set up the appropriate equation and solve graphically for v (to two significant digits).

CHAPTER 7 EQUATIONS

Quadratic equation	$ax^2 + bx + c = 0$	(7.1)
Quadratic formula	$x = \dfrac{-b \pm \sqrt{b^2 - 4ac}}{2a}$	(7.4)

CHAPTER 7 REVIEW EXERCISES

In Exercises 1–12, solve the given quadratic equations by factoring.

1. $x^2 + 3x - 4 = 0$ **2.** $x^2 + 3x - 10 = 0$

3. $x^2 - 10x + 16 = 0$ **4.** $P^2 - 6P - 27 = 0$

5. $3x^2 + 11x = 4$ **6.** $6y^2 = 11y - 3$

7. $6t^2 = 13t - 5$ **8.** $3x^2 + 5x + 2 = 0$

9. $6s^2 = 25s$ **10.** $6n^2 - 23n - 35 = 0$

11. $4B^2 = 8B + 21$ **12.** $6x^2 = 8 - 47x$

In Exercises 13–24, solve the given quadratic equations by using the quadratic formula.

13. $x^2 - x - 110 = 0$ **14.** $x^2 + 3x - 18 = 0$

15. $m^2 + 2m = 6$ **16.** $D^2 - 7D = 1$

17. $2x^2 - x = 36$ **18.** $3x^2 = 14 - x$

19. $4s^2 - 3s - 2 = 0$

20. $5x^2 + 7x - 2 = 0$

21. $2.1x^2 + 2.3x + 5.5 = 0$

22. $0.30R^2 - 0.42R = 0.15$

23. $6x^2 = 9 - 4x$

24. $24x^2 = 25x + 20$

In Exercises 25–36, solve the given quadratic equations by any appropriate algebraic method.

25. $x^2 + 4x - 4 = 0$

26. $x^2 + 3x + 1 = 0$

27. $3x^2 + 8x + 2 = 0$

28. $3p^2 = 28 - 5p$

29. $4v^2 = v + 5$

30. $6n^2 + 2 = n$

31. $2C^2 + 3C + 7 = 0$

32. $4y^2 - 5y = 8$

33. $a^2x^2 + 2ax + 2 = 0$

34. $16r^2 = 8r - 1$

35. $ay^2 = a - 3y$

36. $2bx = x^2 - 3b$

In Exercises 37–40, solve the given quadratic equations by completing the square.

37. $x^2 - x - 30 = 0$

38. $x^2 = 2x + 5$

39. $2t^2 = t + 4$

40. $4x^2 - 8x = 3$

In Exercises 41–44, solve the given equations.

41. $\dfrac{x-4}{x-1} = \dfrac{2}{x}$

42. $\dfrac{V-1}{3} = \dfrac{5}{V} + 1$

43. $\dfrac{x^2 - 3x}{x - 3} = \dfrac{x^2}{x + 2}$

44. $\dfrac{x-2}{x-5} = \dfrac{15}{x^2 - 5x}$

In Exercises 45–48, sketch the graphs of the given functions by using the vertex, the y-intercept, and one or two other points.

45. $y = 2x^2 - x - 1$

46. $y = -4x^2 - 1$

47. $y = x - 3x^2$

48. $y = 2x^2 + 8x - 10$

In Exercises 49–52, solve the given equations by using a graphing calculator. If there are no real roots, state this as the answer.

49. $2x^2 + x - 4 = 0$

50. $-4x^2 - x - 1 = 0$

51. $3x^2 = -x - 2$

52. $x(15x - 12) = 8$

In Exercises 53–66, solve the given quadratic equations by any appropriate method. All numbers are accurate to at least two significant digits.

53. The bending moment M of a simply supported beam of length L with a uniform load of w kg/m at a distance x from one end is $M = 0.5wLx - 0.5wx^2$. For what values of x is $M = 0$?

54. To find the current in a certain alternating-current circuit, it is necessary to solve the equation $m^2 + 10m + 2000 = 0$. Solve for m.

55. A computer analysis shows that the cost C (in dollars) for a company to make x units of a certain product is given by $C = 0.1x^2 + 0.8x + 7$. How many units can be made for $50?

56. For laminar flow of fluids, the coefficient K used to calculate energy loss due to sudden enlargements is given by $K = 1.00 - 2.67R + R^2$, where R is the ratio of cross-sectional areas. If $K = 0.500$, what is the value of R?

57. At an altitude h (in m) above sea level, the boiling point of water is lower by T °C than the boiling point at sea level, which is 100°C. The difference can be approximated by solving the equation $T^2 + 244T - h = 0$. What is the boiling point in Guadalajara, Mexico (altitude 1500 m)?

58. In a natural gas pipeline, the velocity v (in m/s) of the gas as a function of the distance x (in cm) from the wall of the pipe is given by $v = 5.2x - x^2$. Determine x for $v = 4.8$ m/s.

59. The height h of an object ejected at an angle θ from a vehicle moving with velocity v is given by $h = vt \sin \theta - 4.9t^2$, where t is the time of flight. Find t (to 0.1 s) if $v = 15$ m/s, $\theta = 65°$, and $h = 6.0$ m.

60. In studying the emission of light, in order to determine the angle at which the intensity is a given value, the equation $\sin^2 A - 4 \sin A + 1 = 0$ must be solved. Find angle A (to 0.1°). ($\sin^2 A = (\sin A)^2$.)

61. A computer analysis shows that the number n of electronic components a company should produce for supply to equal demand is found by solving $\dfrac{n^2}{500\,000} = 144 - \dfrac{n}{500}$. Find n.

62. To determine the resistances of two resistors that are to be in parallel in an electric circuit, it is necessary to solve the equation $\dfrac{20}{R} + \dfrac{20}{R + 10} = \dfrac{1}{5}$. Find R (to the nearest 1 Ω).

63. In designing a cylindrical container, the formula $A = 2\pi r^2 + 2\pi rh$ (Eq. (2.19)) is used. Solve for r.

64. In determining the number of bytes b that can be stored on a hard disk, the equation $b = kr(R - r)$ is used. Solve for r.

65. In the study of population growth, the equation $p_2 = p_1 + rp_1(1 - p_1)$ occurs. Solve for p_1.

66. In the study of the velocities of deep-water waves, the equation $v^2 = k^2\left(\dfrac{L}{C} + \dfrac{C}{L}\right)$ occurs. Solve for L.

In Exercises 67–80, set up the necessary equation where appropriate and solve the given problems. All numbers are accurate to at least two significant digits.

67. In testing the effects of a drug, the percent of the drug in the blood was given by $p = 0.090t - 0.015t^2$, where t is the time (in h) after the drug was administered. Sketch the graph of $p = f(t)$.

68. In an electric circuit, the voltage V as a function of the time t (in min) is given by $V = 9.8 - 9.2t + 2.3t^2$. Sketch the graph of $V = f(t)$, for $t \le 5$ min.

69. By adding the same amount to its length and its width, a developer increased the area of a rectangular lot by 3000 m^2 to make it 80 m by 100 m. What were the original dimensions of the lot? See Fig. 7.18.

Fig. 7.18

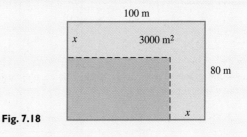

70. A machinery pedestal is made of two concrete cubes, one on top of the other. The pedestal is 2.0 m high and contains 3.5 m³ of concrete. Find the edge of each cube.

71. Concrete contracts as it dries. If the volume of a cubical concrete block is 29 cm³ less and each edge is 0.10 cm less after drying, what was the original length of an edge of the block?

72. A jet flew 1200 km with a tailwind of 50 km/h. The tailwind then changed to 20 km/h for the remaining 570 km of the flight. If the total time of the flight was 3.0 h, find the speed of the jet relative to the air.

73. The length of one rectangular field is 400 m more than the side of a square field. The width is 100 m more than the side of the square field. If the rectangular field has twice the area of the square field, what are the dimensions of each field?

74. A military jet flies directly over and at right angles to the straight course of a commercial jet. The military jet is flying at 200 km/h faster than four times the speed of the commercial jet. How fast is each going if they are 2050 km apart (on a direct line) after 1 h?

75. The width of a rectangular TV screen is 14.5 cm more than the height. If the diagonal is 68.6 cm, find the dimensions of the screen. See Fig. 7.19.

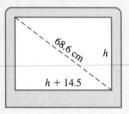

Fig. 7.19

(W) 76. Find the exact value of x that is defined in terms of the *continuing fraction* at the right (the pattern continues endlessly). Explain your method of solution.

$$x = 2 + \cfrac{1}{2 + \cfrac{1}{2 + \cfrac{1}{2 + \cdots}}}$$

77. An electric utility company is placing utility poles along a road. It is determined that five fewer poles per kilometer would be necessary if the distance between poles were increased by 10 m. How many poles are being placed each kilometer?

78. An architect is designing a Norman window (a semicircular part over a rectangular part) as shown in Fig. 7.20. If the area of the window is to be 1.75 m² and the height of the rectangular part is 1.35 m, find the radius of the circular part.

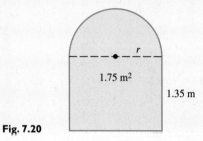

Fig. 7.20

79. A testing station found p parts per million (ppm) of sulfur dioxide in the air as a function of the hour h of the day to be $p = 0.00174(10 + 24h - h^2)$. Sketch the graph of $p = f(h)$ and, from the graph, find the time when $p = 0.205$ ppm.

80. A compact disc (CD) is made such that it is 53.0 mm from the edge of the center hole to the edge of the disc. Find the radius of the hole if 1.36% of the disc is removed in making the hole. See Fig. 7.21.

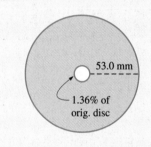

Fig. 7.21

Writing Exercise

81. An electronics student is asked to solve the equation
$$\frac{1}{R_T} = \frac{1}{R} + \frac{1}{R + 1}$$
for R. Write one or two paragraphs explaining your procedure for solving this equation. (What is the solution?)

CHAPTER (7) PRACTICE TEST

In Problems 1–4, algebraically solve for x.

1. $x^2 - 3x - 5 = 0$

2. $2x^2 = 9x - 4$

3. $\dfrac{3}{x} - \dfrac{2}{x + 2} = 1$

4. $2x^2 - x = 6 - 2x(3 - x)$

5. Sketch the graph of $y = 2x^2 + 8x + 5$ using the extreme point and the y-intercept.

6. In electricity the formula $P = EI - RI^2$ is used. Solve for I in terms of E, P, and R.

7. Solve by completing the square: $x^2 - 6x - 9 = 0$.

8. The perimeter of a rectangular window is 8.4 m, and its area is 3.8 m². Find its dimensions.

9. If $y = x^2 - 8x + 8$, sketch the graph using the vertex and any other useful points and find graphically the values of x for which $y = 0$.

8 Trigonometric Functions of Any Angle

When we introduced the trigonometric functions in Chapter 4, we defined them in general but used them only with acute angles. In this chapter, we show how these functions are used with angles of any size and with angles measured in radians.

By the mid-1700s, the trigonometric functions had been used for many years as ratios, as in Chapter 4. It was also known that they are useful in describing periodic functions (functions for which values repeat at specific intervals) without reference to triangles. In about 1750, this led the Swiss mathematician Leonhard Euler to include, for the first time in a textbook, the trigonometric functions of numbers (not angles). As we will see in this chapter, this is equivalent to using these functions of angles of any size, measured in radians.

Euler wrote over 70 volumes in mathematics and applied subjects such as astronomy, mechanics, and music. (Many of these volumes were dictated, as he was blind for the last 17 years of his life.) Although he is noted as one of the great mathematicians of all time, his interest in applied subjects often led him to study and develop topics in mathematics that were used in these applications.

Today, trigonometric functions of numbers are of importance in many areas of application such as electric circuits, mechanical vibrations, and rotational motion. Although electronics were unknown in the 1700s, the trigonometric functions of numbers developed at that time became of great importance to the development of electronics in the 1900s. We will show some of these applications in this chapter and in Chapter 10.

Cruise ships generally measure distances traveled in nautical miles. In Section 8.4, we will see how a nautical mile is defined as a distance along an arc of the earth's surface.

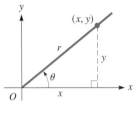

Fig. 8.1

8.1 SIGNS OF THE TRIGONOMETRIC FUNCTIONS

Recall the definitions of the trigonometric functions that were given in Section 4.2. *Here the point (x, y) is a point on the terminal side of angle θ, and r is the radius vector.* See Fig. 8.1.

$$\sin \theta = \frac{y}{r} \qquad \cos \theta = \frac{x}{r} \qquad \tan \theta = \frac{y}{x}$$

$$\cot \theta = \frac{x}{y} \qquad \sec \theta = \frac{r}{x} \qquad \csc \theta = \frac{r}{y} \qquad (8.1)$$

As noted before, these definitions are valid for a standard-position angle of any size. In this section, we determine the *sign* of each function of angles with terminal sides in each of the four quadrants.

We can find the values of the functions if we know the coordinates (x, y) on the terminal side and the radius vector r. *Since r is always taken to be positive, the functions will vary in sign, depending on the values of x and y.* If either x or y is zero in the denominator, the function is undefined. We will consider this in the next section.

Since $\sin \theta = y/r$, *the sign of $\sin \theta$ depends on the sign of y.* Since $y > 0$ in the first and second quadrants and $y < 0$ in the third and four quadrants, *$\sin \theta$ is positive if the terminal side is in the first or second quadrant,* and *$\sin \theta$ is negative if the terminal side is in the third or fourth quadrant.* See Fig. 8.2.

	Quadrant			
	I	II	III	IV
$\sin \theta$	+	+	−	−

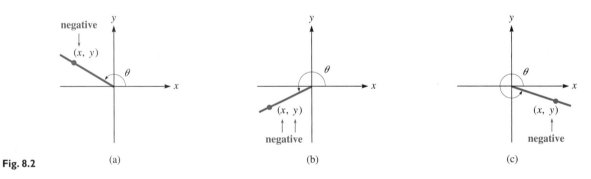

Fig. 8.2   (a)   (b)   (c)

◀ **EXAMPLE 1**   The value of $\sin 20°$ is positive, because the terminal side of $20°$ is in the first quadrant. The value of $\sin 160°$ is positive, because the terminal side of $160°$ is in the second quadrant. The values of $\sin 200°$ and $\sin 340°$ are negative, because the terminal sides are in the third and fourth quadrants, respectively. ▶

Since $\tan \theta = y/x$, *the sign of $\tan \theta$ depends on the ratio of y to x.* Since x and y are both positive in the first quadrant, both negative in the third quadrant, and have different signs in the second and fourth quadrants, *$\tan \theta$ is positive if the terminal side is in the first or third quadrant,* and *$\tan \theta$ is negative if the terminal side is in the second or fourth quadrant.* See Fig. 8.2.

	Quadrant			
	I	II	III	IV
$\tan \theta$	+	−	+	−

◀ **EXAMPLE 2**   The values of $\tan 20°$ and $\tan 200°$ are positive, because the terminal sides of these angles are in the first and third quadrants, respectively. The values of $\tan 160°$ and $\tan 340°$ are negative, because the terminal sides of these angles are in the second and fourth quadrants, respectively. ▶

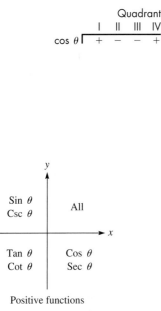

	Quadrant			
cos θ	I	II	III	IV
	+	−	−	+

Since $\cos \theta = x/r$, *the sign of* $\cos \theta$ *depends on the sign of* x. Since $x > 0$ in the first and fourth quadrants and $x < 0$ in the second and third quadrants, $\cos \theta$ *is positive if the terminal side is in the first or fourth quadrant,* and $\cos \theta$ *is negative if the terminal side is in the second or third quadrant.* See Fig. 8.2.

◀ **EXAMPLE 3** The values of $\cos 20°$ and $\cos 340°$ are positive, because the terminal sides of these angles are in the first and fourth quadrants, respectively. The values of $\cos 160°$ and $\cos 200°$ are negative, because the terminal sides of these angles are in the second and third quadrants, respectively. ▶

Since $\csc \theta$ is defined in terms of y and r, as in $\sin \theta$, $\csc \theta$ has the same sign as $\sin \theta$. For similar reasons, $\cot \theta$ has the same sign as $\tan \theta$, and $\sec \theta$ has the same sign as $\cos \theta$. Therefore, as shown in Fig. 8.3,

> **All functions of first-quadrant angles are positive. Sin θ and csc θ are positive for second-quadrant angles. Tan θ and cot θ are positive for third-quadrant angles. Cos θ and sec θ are positive for fourth-quadrant angles. All others are negative.**

y

| Sin θ
Csc θ | All |
| Tan θ
Cot θ | Cos θ
Sec θ |

x

Positive functions

Fig. 8.3

This discussion does not include the *quadrantal angles,* those angles with terminal sides on one of the axes. They will be discussed in the next section.

◀ **EXAMPLE 4** **(a)** The following are *positive:*

$$\sin 150° \quad \cos 290° \quad \tan 190° \quad \cot 260° \quad \sec 350° \quad \csc 100°$$

(b) The following are *negative:*

$$\sin 300° \quad \cos 150° \quad \tan 100° \quad \cot 300° \quad \sec 200° \quad \csc 250°$$ ▶

A calculator will always give the correct sign for a trigonometric function of a given angle. However, as we will see in the next section, a calculator will *not* always give the required angle for a given value of a function.

◀ **EXAMPLE 5** **(a)** If $\sin \theta > 0$, then the terminal side of θ is in either the first or second quadrant.

(b) If $\sec \theta < 0$, then the terminal side of θ is in either the second or third quadrant.

(c) If $\cos \theta > 0$ and $\tan \theta < 0$, then the terminal side of θ is in the fourth quadrant. Only in the fourth quadrant are both signs correct. ▶

◀ **EXAMPLE 6** Determine the trigonometric functions of θ if the terminal side of θ passes through $(-1, \sqrt{3})$. See Fig. 8.4.

We know that $x = -1$, $y = \sqrt{3}$, and from the Pythagorean theorem we find that $r = 2$. Therefore, the trigonometric functions of θ are

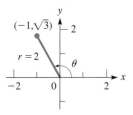

$$\sin \theta = \frac{\sqrt{3}}{2} = 0.8660 \qquad \cos \theta = -\frac{1}{2} = -0.5000 \qquad \tan \theta = -\sqrt{3} = -1.732$$

$$\cot \theta = -\frac{1}{\sqrt{3}} = -0.5774 \qquad \sec \theta = -2 = -2.000 \qquad \csc \theta = \frac{2}{\sqrt{3}} = 1.155$$

Fig. 8.4

The point $(-1, \sqrt{3})$ is in the second quadrant, and the signs of the functions of θ are those of a second-quadrant angle. ▶

When dealing with negative angles, or angles greater than $360°$, the sign of the function is still determined by the location of the terminal side of the angle. For example, $\sin(-120°) < 0$ since the terminal side of $-120°$ is in the third quadrant. Also, for the same reason, $\sin 600° < 0$ (same terminal side as $-120°$ or $240°$).

EXERCISES 8.1

In Exercises 1 and 2, answer the given questions about the indicated examples of this section.

1. In Example 4, if 90° is added to each angle, what is the sign of each resulting function?

2. In Example 6, if the point $(-1, \sqrt{3})$ is replaced with the point $(1, -\sqrt{3})$, what are the resulting values?

In Exercises 3–14, determine the sign of the given trigonometric functions.

3. sin 36°, cos 120°   4. tan 320°, sec 185°

5. csc 98°, cot 82°   6. cos 260°, csc 290°

7. sec 150°, tan 220°   8. sin 335°, cot 265°

9. cos 348°, csc 238°   10. cot 110°, sec 309°

11. tan 460°, sin(−110°)   12. csc(−200°), cos 550°

13. cot(−2°), cos 710°   14. sin 539°, tan(−480°)

In Exercises 15–22, find the trigonometric functions of θ if the terminal side of θ passes through the given point.

15. $(2, 1)$ 16. $(-1, 1)$

17. $(-2, -3)$ 18. $(4, -3)$

19. $(-0.5, 1.2)$ 20. $(-150, -200)$

21. $(50, -20)$ 22. $(0.09, 0.40)$

In Exercises 23–28, for the given values determine the quadrant(s) in which the terminal side of the angle lies.

23. sin θ = 0.5000   24. cos θ = 0.8666

25. tan θ = 1.500   26. sin θ = −0.8666

27. cos θ = −0.5000   28. tan θ = −0.7500

In Exercises 29–40, determine the quadrant in which the terminal side of θ lies, subject to both given conditions.

29. sin θ > 0, cos θ < 0   30. tan θ > 0, cos θ < 0

31. sec θ < 0, cot θ < 0   32. cos θ > 0, csc θ < 0

33. csc θ < 0, tan θ < 0   34. sec θ > 0, csc θ > 0

35. sin θ < 0, tan θ > 0   36. cot θ < 0, sin θ < 0

37. tan θ < 0, cos θ > 0   38. sec θ > 0, csc θ < 0

39. sin θ > 0, cot θ < 0   40. tan θ > 0, csc θ < 0

8.2 TRIGONOMETRIC FUNCTIONS OF ANY ANGLE

In this section, we show how to find the trigonometric functions of an angle of any size. This will be important in Chapter 9, when we discuss vectors and oblique triangles, and in Chapter 10, when we discuss graphs of the trigonometric functions.

Any angle in standard position is coterminal with 0° or a positive angle less than 360°. Since the terminal sides of coterminal angles are the same, the trigonometric functions of coterminal angles are equal. Therefore, we need only be able to find the values of the trigonometric functions of 0° and positive angles less than 360°.

Fig. 8.5

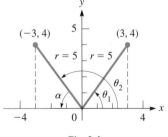

Fig. 8.6

◖EXAMPLE 1   The following pairs of angles are coterminal.

390° and 30°   900° and 180°   −150° and 210° (see Fig. 8.5)

The trigonometric functions of both angles in each pair are equal. For example,

$$\sin 390° = \sin 30° \quad \text{and} \quad \tan(-150°) = \tan 210°$$ ◗

The values of the functions depend only on the values of *x*, *y*, and *r*. Therefore, the absolute value of a function of a second-quadrant angle equals the value of the same function of a first-quadrant angle. For example, in Fig. 8.6, we see that tan $\theta_2 = -4/3$. or $|\tan \theta_2| = 4/3$, and that tan $\theta_1 = 4/3$. This means that $|\tan \theta_2| = \tan \theta_1$. The triangles with θ_1 and α are congruent, which means $\theta_1 = \alpha$. Knowing that the absolute value of a function of θ_2 equals the same function of θ_1 means that

$$|F(\theta_2)| = |F(\theta_1)| = |F(\alpha)| \tag{8.2}$$

where *F* represents any of the trigonometric functions.

Reference Angle

*The angle labeled α is called the **reference angle**. The reference angle of a given angle is the acute angle formed by the terminal side of the angle and the x-axis.*

Using Eq. (8.2) and the fact that $\alpha = 180° - \theta_2$, we may conclude that the value of any trigonometric function of any second-quadrant angle is found from

$$F(\theta_2) = \pm F(180° - \theta_2) = \pm F(\alpha) \qquad (8.3)$$

NOTE ▶ The *sign* used depends on the *sign of the function* in the second quadrant.

◀ EXAMPLE 2 In Fig. 8.6, the trigonometric functions of θ are as follows:

$$\sin \theta_2 = \sin(180° - \theta_2) = \sin \alpha = \sin \theta_1 = \tfrac{4}{5} = 0.8000$$

$$\cos \theta_2 = -\cos \theta_1 = -\tfrac{3}{5} = -0.6000 \qquad \tan \theta_2 = -\tfrac{4}{3} = -1.333$$

$$\cot \theta_2 = -\tfrac{3}{4} = -0.7500 \qquad \sec \theta_2 = -\tfrac{5}{3} = -1.667 \qquad \csc \theta_2 = \tfrac{5}{4} = 1.250 \quad ▶$$

In the same way, we derive the formulas for trigonometric functions of any third- or fourth-quadrant angle. In Fig. 8.7, the reference angle α is found by subtracting $180°$ from θ_3, and the functions of α and θ_1 are numerically equal. In Fig. 8.8, the reference angle α is found by subtracting θ_4 from $360°$.

$$F(\theta_3) = \pm F(\theta_3 - 180°) = \pm F(\alpha) \qquad (8.4)$$

$$F(\theta_4) = \pm F(360° - \theta_4) = \pm F(\alpha) \qquad (8.5)$$

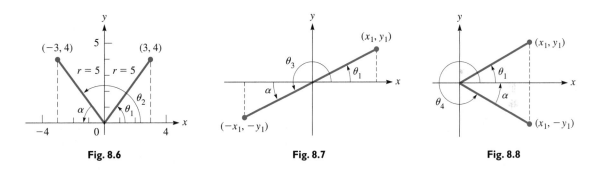

Fig. 8.6   Fig. 8.7   Fig. 8.8

◀ EXAMPLE 3 The trigonometric functions of $\theta_3 = 210°$ can be expressed in terms of the reference angle of $30°$ (see Fig. 8.9), and then evaluated, as follows:

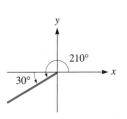

Fig. 8.9

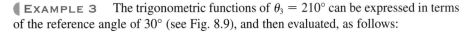

same function   reference angle

$$\sin 210° = -\sin(210° - 180°) = -\sin 30° = -\frac{1}{2} = -0.5000$$

quadrant III $\sin \theta$, $\csc \theta$ — in quadrant III

$$\csc 210° = -\csc 30° = -2.000$$

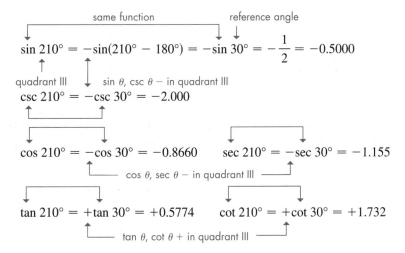

$$\cos 210° = -\cos 30° = -0.8660 \qquad \sec 210° = -\sec 30° = -1.155$$

cos θ, sec θ — in quadrant III

$$\tan 210° = +\tan 30° = +0.5774 \qquad \cot 210° = +\cot 30° = +1.732$$

tan θ, cot θ + in quadrant III   ▶

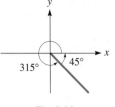

Fig. 8.10

◀ EXAMPLE 4   If $\theta_4 = 315°$, the trigonometric functions of θ_4 are found by using Eq. (8.5) as follows. See Fig. 8.10.

reference angle

$$\sin 315° = -\sin(360° - 315°) = -\sin 45° = -0.7071$$

quadrant IV $\sin \theta$, csc θ − in quadrant IV

$$\csc 315° = -\csc 45° = -1.414$$

$$\cos 315° = +\cos 45° = +0.7071 \qquad \sec 315° = +\sec 45° = +1.414$$

cos θ, sec θ + in quadrant IV

$$\tan 315° = -\tan 45° = -1.000 \qquad \cot 315° = -\cot 45° = -1.000$$

tan θ, cot θ − in quadrant IV

◀ EXAMPLE 5   Other illustrations using Eqs. (8.3), (8.4), and (8.5) follow:

same function   reference angle

$$\sin 160° = +\sin(180° - 160°) = \sin 20° = 0.3420$$
$$\tan 110° = -\tan(180° - 110°) = -\tan 70° = -2.747$$
$$\cos 225° = -\cos(225° - 180°) = -\cos 45° = -0.7071$$
$$\cot 260° = +\cot(260° - 180°) = \cot 80° = 0.1763$$
$$\sec 304° = +\sec(360° - 304°) = \sec 56° = 1.788$$
$$\sin 357° = -\sin(360° - 357°) = -\sin 3° = -0.0523$$

determines   └ proper sign for function in quadrant
quadrant

Fig. 8.11

A calculator will give the values, with the proper signs, of functions like those in Examples 3, 4, and 5. The reciprocal key is used when evaluating cot θ, sec θ, and csc θ, as shown on page 120. A calculator display for tan 110° and sec 304° of Example 5 is shown in Fig. 8.11.

In most examples, we will round off values to four significant digits (as we did in Examples 3, 4, and 5). However, *if the angle is approximate, use the guidelines in Section 4.3 for rounding off values.*

NOTE ▶

◀ EXAMPLE 6   A formula for finding the area of a triangle, knowing sides a and b and the included $\angle C$, is $A = \frac{1}{2}ab \sin C$. A surveyor uses this formula to find the area of a triangular tract of land for which $a = 173.2$ m, $b = 156.3$ m, and $C = 112.51°$. See Fig. 8.12.

To find the area, we substitute into the formula, which gives

$$A = \tfrac{1}{2}(173.2)(156.3) \sin 112.51°$$
$$= 12\,500 \text{ m}^2 \qquad \text{rounded to four significant digits}$$

Fig. 8.12

When these numbers are entered into the calculator, the calculator automatically uses a positive value for sin 112.51°.

Knowing how to use the reference angle is important when using a calculator, because *for a given value of a function of an angle, when finding the angle*

CAUTION ▶ ***the calculator will not necessarily give us directly the required angle.***

It will give an angle we can use, but whether or not it is the required angle for the problem will depend on the problem being solved.

When a value of a trigonometric function is entered into a calculator, it is programmed to give the angle as follows:

For values of sin θ, the calculator displays angles from −90° to 90°:
$$[\sin(-90°) = -1, \sin 0° = 0, \sin 90° = 1]$$

For values of cos θ, the calculator displays angles from 0° to 180°:
$$[\cos 0° = 1, \cos 90° = 0, \cos 180° = -1]$$

*For values of tan θ, the calculator displays angles **between** −90° and 90°:*
$$[\tan 0° = 0, \text{ if } \tan θ < 0, θ \text{ is between } -90° \text{ and } 0°]$$

This means that if sin θ or tan θ is negative, the displayed angle is negative. If cos θ is negative, the displayed angle is greater than 90° and no greater than 180°.

The reason that the calculator displays these angles is shown in Chapter 20, when the inverse trigonometric functions are discussed in detail.

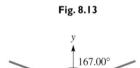

Fig. 8.13

◀ **EXAMPLE 7** For sin θ = 0.2250, we see from the first two lines of the calculator display shown in Fig. 8.13 that θ = 13.00° (rounded off).

This result is correct, but remember that

$$\sin(180° - 13.00°) = \sin 167.00° = 0.2250$$

also. If we need only an acute angle, θ = 13.00° is correct. However, if a second-quadrant angle is required, we see that θ = 167.00° is the angle (see Fig. 8.14). These values can be checked by finding sin 13.00° and sin 167.00°. ▶

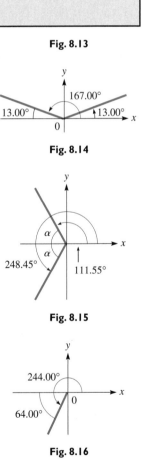

Fig. 8.14

Fig. 8.15

Fig. 8.16

◀ **EXAMPLE 8** For sec θ = −2.722 and 0° ≤ θ < 360° (this means θ may equal 0° or be between 0° and 360°), see from the third and fourth lines of the calculator display shown in Fig. 8.13 that θ = 111.55° (rounded off).

The angle 111.55° is the second-quadrant angle, but sec θ < 0 in the third quadrant as well. The reference angle is α = 180° − 111.55° = 68.45°, and the third-quadrant angle is 180° + 68.45° = 248.45°. Therefore, the two angles between 0° and 360° for which sec θ = −2.722 are 111.55° and 248.45° (see Fig. 8.15). These angles can be checked by finding sec 111.55° and sec 248.45°. ▶

◀ **EXAMPLE 9** Given that tan θ = 2.050 and cos θ < 0, find the angle θ for 0° ≤ θ < 360°.

Since tan θ is positive and cos θ is negative, θ must be a third-quadrant angle. A calculator will display an angle of 64.00° (rounded off) for tan θ = 2.050. However, since we need a third-quadrant angle, *we must add* 64.00° *to* 180°. Thus, the required angle is 244.00° (see Fig. 8.16). Check by finding tan 244.00°.

If tan θ = −2.050 and cos θ < 0, the calculator will display an angle of −64.00° for tan θ = −2.050. We would then have to *recognize that the reference angle is 64.00° and subtract it from 180° to get* 116.00°, the required second-quadrant angle. This can be checked by finding tan 116.00°. ▶

The calculator gives the reference angle (disregarding any minus signs) in all cases except when cos θ is negative. To avoid confusion from the angle displayed by the calculator, *a good procedure is to find the reference angle first.* Then it can be used to determine the angle required by the problem.

NOTE ▶

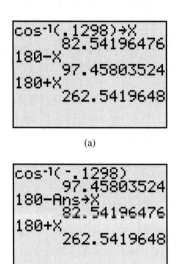

(a)

(b)

Fig. 8.17

We can *find the reference angle by entering the absolute value of the function. The displayed angle will be the reference angle.* The required angle θ is found by using the reference angle as described earlier and as shown in Eqs. (8.6).

$\theta = \alpha$	(first quadrant)
$\theta = 180° - \alpha$	(second quadrant)
$\theta = 180° + \alpha$	(third quadrant)
$\theta = 360° - \alpha$	(fourth quadrant)

(8.6)

◀ **EXAMPLE 10** Given that $\cos \theta = -0.1298$, find θ for $0° \leq \theta < 360°$.

Since $\cos \theta$ is negative, θ is either a second-quadrant angle or a third-quadrant angle. Using 0.1298, the calculator tells us that the reference angle is 82.54°.

For the second-quadrant angle, we subtract 82.54° from 180° to get 97.46°. For the third-quadrant angle, we add 82.54° to 180° to get 262.54°. See Fig. 8.17(a).

Using −0.1298, the calculator displays the second-quadrant angle of 97.46°. For the third-quadrant angle, we must subtract 97.46° from 180° to get the reference angle 82.54°, which is then added to 180° to get 262.54°. Note that it is better to store the reference angle in memory rather than reenter it. See Fig. 8.17(b). ▶

Quadrantal Angles

We use Eqs. (8.3), (8.4), and (8.5) when the terminal side is in one of the quadrants. A **quadrantal angle** *has its terminal side along one of the axes.* Using the definitions of the functions (recalling that $r > 0$), we get the following values for quadrantal angles. These values may be verified by referring to Fig. 8.18.

θ	$\sin \theta$	$\cos \theta$	$\tan \theta$	$\cot \theta$	$\sec \theta$	$\csc \theta$
0°	0.000	1.000	0.000	undef.	1.000	undef.
90°	1.000	0.000	undef.	0.000	undef.	1.000
180°	0.000	−1.000	0.000	undef.	−1.000	undef.
270°	−1.000	0.000	undef.	0.000	undef.	−1.000
360°	Same as the functions of 0° (same terminal side)					

Fig. 8.18

(a) $\theta = 0°$ (b) $\theta = 90°$ (c) $\theta = 180°$ (d) $\theta = 270°$

◀ **EXAMPLE 11** Since $\sin \theta = y/r$, by looking at Fig. 8.18(a) we can see that $\sin 0° = 0/r = 0$.

Since $\tan \theta = y/x$, from Fig. 8.18(b) we see that $\tan 90° = r/0$, which is undefined due to the division by zero. Using a calculator to find $\tan 90°$, the display would indicate an error (due to division by zero).

Since $\cos \theta = x/r$, from Fig. 8.18(c) we see that $\cos 180° = -r/r = -1$.

Since $\cot \theta = x/y$, from Fig. 8.18(d) we see that $\cot 270° = 0/-r = 0$. ▶

Negative Angles

To find the values of the functions of negative angles, we can use functions of corresponding positive angles, if we use the correct *sign.* From Fig. 8.19, note that $\sin \theta = y/r$, and $\sin(-\theta) = -y/r$, and this means $\sin(-\theta) = -\sin \theta$. In the same way, we can get all the relations between the functions of an angle $-\theta$ of any size and the functions of θ. Therefore, using the definitions, we have the following:

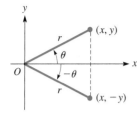

Fig. 8.19

$$\sin(-\theta) = -\sin \theta \qquad \cos(-\theta) = \cos \theta \qquad \tan(-\theta) = -\tan \theta$$
$$\csc(-\theta) = -\csc \theta \qquad \sec(-\theta) = \sec \theta \qquad \cot(-\theta) = -\cot \theta$$

(8.7)

◖ **EXAMPLE 12**   Using Eqs. (8.7), we have the following results:

$$\sin(-60°) = -\sin 60° = -0.8660 \qquad \cos(-60°) = \cos 60° = 0.5000$$
$$\tan(-60°) = -\tan 60° = -1.732 \qquad \cot(-60°) = -\cot 60° = -0.5774$$
$$\sec(-60°) = \sec 60° = 2.000 \qquad \csc(-60°) = -\csc 60° = -1.155$$ ◗

EXERCISES 8.2

In Exercises 1–4, make the given changes in the indicated examples of this section and then solve the resulting problems.

1. In Example 5, add 40° to each angle, express in terms of the same function of a positive acute angle, and then evaluate.

2. In Example 7, make the value of the function negative and then find θ between 0° and 360°.

3. In Example 9, change 2.050 to -2.050 and $<$ to $>$ and then find θ.

4. In Example 10, change the $-$ to $+$ and then find θ.

In Exercises 5–10, express the given trigonometric function in terms of the same function of a positive acute angle.

5. $\sin 160°$, $\cos 220°$

6. $\tan 91°$, $\sec 345°$

7. $\tan 105°$, $\csc 302°$

8. $\cos 190°$, $\cot 290°$

9. $\cos 400°$, $\tan(-400°)$

10. $\tan 920°$, $\csc(-550°)$

In Exercises 11–44, the given angles are approximate. In Exercises 11–18, find the values of the given trigonometric functions by finding the reference angle and attaching the proper sign.

11. $\sin 195°$

12. $\tan 311°$

13. $\cos 106.3°$

14. $\sin 103.4°$

15. $\sec 328.33°$

16. $\cot 516.53°$

17. $\tan(-31.5°)$

18. $\csc(-108.4°)$

In Exercises 19–26, find the values of the given trigonometric functions directly from a calculator.

19. $\tan 152.4°$

20. $\cos 341.4°$

21. $\sin 310.36°$

22. $\tan 242.68°$

23. $\csc 194.82°$

24. $\sec 441.08°$

25. $\cos(-72.61°)$

26. $\sin(-215.5°)$

In Exercises 27–40, find θ for $0° \le \theta < 360°$.

27. $\sin \theta = -0.8480$

28. $\tan \theta = -1.830$

29. $\cos \theta = 0.4003$

30. $\sin \theta = 0.6374$

31. $\cot \theta = -0.212$

32. $\csc \theta = -1.09$

33. $\sin \theta = 0.870$, $\cos \theta < 0$

34. $\tan \theta = 0.932$, $\sin \theta < 0$

35. $\cos \theta = -0.12$, $\tan \theta > 0$

36. $\sin \theta = -0.192$, $\tan \theta < 0$

37. $\tan \theta = -1.366$, $\cos \theta > 0$

38. $\cos \theta = 0.5726$, $\sin \theta < 0$

39. $\sec \theta = 2.047$, $\cot \theta < 0$

40. $\cot \theta = -0.3256$, $\csc \theta > 0$

In Exercises 41–44, determine the function that satisfies the given conditions.

41. Find $\tan \theta$ when $\sin \theta = -0.5736$ and $\cos \theta > 0$.

42. Find $\sin \theta$ when $\cos \theta = 0.422$ and $\tan \theta < 0$.

43. Find $\cos \theta$ when $\tan \theta = -0.809$ and $\csc \theta > 0$.

44. Find $\cot \theta$ when $\sec \theta = 1.122$ and $\sin \theta < 0$.

Ⓦ *In Exercises 45–48, insert the proper sign, $>$ or $<$ or $=$, between the given expressions. Explain your answers.*

45. $\sin 90°$ $2 \sin 45°$

46. $\cos 360°$ $2 \cos 180°$

47. $\tan 180°$ $\tan 0°$

48. $\sin 270°$ $3 \sin 90°$

In Exercises 49–52, evaluate the given expressions.

49. The current i in an alternating-current circuit is given by $i = i_m \sin \theta$, where i_m is the maximum current in the circuit. Find i if $i_m = 0.0259$ A and $\theta = 495.2°$.

50. A force F is related to force F_x directed along the x-axis by $F = F_x \sec \theta$, where θ is the standard position angle for F. See Fig. 8.20. Find F if $F_x = -29.2$ N and $\theta = 127.6°$.

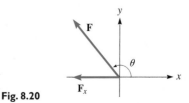

Fig. 8.20

51. For the slider mechanism shown in Fig. 8.21, $y \sin \alpha = x \sin \beta$. Find y if $x = 6.78$ cm, $\alpha = 31.3°$, and $\beta = 104.7°$.

52. A laser follows the path shown in Fig. 8.22. The angle θ is related to the distances a, b, and c by $2ab \cos \theta = a^2 + b^2 - c^2$. Find θ if $a = 12.9$ cm, $b = 15.3$ cm, and $c = 24.5$ cm.

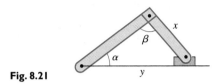

Fig. 8.21

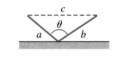

Fig. 8.22

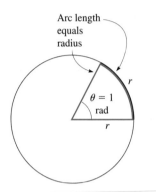

Fig. 8.23

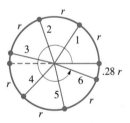

Fig. 8.24

8.3 RADIANS

For many problems in which trigonometric functions are used, particularly those involving the solution of triangles, degree measurements of angles are convenient and quite sufficient. However, division of a circle into 360 equal parts is by definition, and it is arbitrary and artificial (see the margin comment on page 51).

In numerous other types of applications and in more theoretical discussions, the *radian* is a more meaningful measure of an angle. We defined the radian in Chapter 2 and reviewed it briefly in Chapter 4. In this section, we discuss the radian in detail and start by reviewing its definition.

A **radian** *is the measure of an angle with its vertex at the center of a circle and with an intercepted arc on the circle equal in length to the radius of the circle.* See Fig. 8.23.

Since the circumference of any circle in terms of its radius is given by $c = 2\pi r$, the ratio of the circumference to the radius is 2π. This means that the radius may be laid off 2π (about 6.28) times along the circumference, regardless of the length of the radius. Therefore, note that radian measure is independent of the radius of the circle. The definition of a radian is based on an important property of a circle and is therefore a more natural measure of an angle. In Fig. 8.24, the numbers on each of the radii indicate the number of radians in the angle measured in standard position. The circular arrow shows an angle of 6 radians.

Since the radius may be laid off 2π times along the circumference, it follows that there are 2π radians in one complete rotation. Also, there are 360° in one complete rotation. Therefore, 360° is *equivalent* to 2π radians. It then follows that the relation between degrees and radians is 2π rad = 360°, or

Converting Angles

$$\pi \text{ rad} = 180° \tag{8.8}$$

Degrees to Radians

$$1° = \frac{\pi}{180} \text{ rad} = 0.01745 \text{ rad} \tag{8.9}$$

Radians to Degrees

$$1 \text{ rad} = \frac{180°}{\pi} = 57.30° \tag{8.10}$$

From Eqs. (8.8), (8.9), and (8.10), note that we convert angle measurements from degrees to radians or radians to degrees using the following procedure:

> **PROCEDURE FOR CONVERTING ANGLE MEASUREMENTS**
>
> **1.** *To convert an angle measured in degrees to the same angle measured in radians,* **multiply the number of degrees by $\pi/180°$.**
>
> **2.** *To convert an angle measured in radians to the same angle measured in degrees,* **multiply the number of radians by $180°/\pi$.**

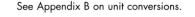

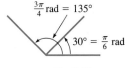

Fig. 8.25

converting degrees to radians

◀ **EXAMPLE 1** **(a)** $18.0° = \left(\dfrac{\pi}{180°}\right)(18.0°) = \dfrac{\pi}{10.0} = 0.314 \text{ rad}$

degrees cancel   (See Fig. 8.25.)

converting radians to degrees

(b) $2.00 \text{ rad} = \left(\dfrac{180°}{\pi}\right)(2.00) = \dfrac{360°}{\pi} = 114.6°$ (See Fig. 8.25.)

Multiplying by $\pi/180°$ or $180°/\pi$ is actually multiplying by 1, because $\pi \text{ rad} = 180°$. The unit of measurement is different, but *the angle is the same.* ▶

See Appendix B on unit conversions.

Because of the definition of the radian, it is common to express radians in terms of π, particularly for angles whose degree measure is a fraction of $180°$.

◀ **EXAMPLE 2** **(a)** Converting $30°$ to radian measure, we have

$$30° = \left(\dfrac{\pi}{180°}\right)(30°) = \dfrac{\pi}{6} \text{ rad} \qquad \text{(See Fig. 8.26)}$$

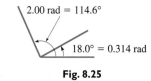

Fig. 8.26

(b) Converting $3\pi/4$ rad to degrees, we have

$$\dfrac{3\pi}{4} \text{ rad} = \left(\dfrac{180°}{\pi}\right)\left(\dfrac{3\pi}{4}\right) = 135° \qquad \text{(See Fig. 8.26)}$$ ▶

NOTE ▶

We wish now to make a very important point. Since π is a number (a little greater than 3) that is the ratio of the circumference of a circle to its diameter, it is the ratio of one length to another. This means *radians have no units,* and *radian measure amounts to measuring angles in terms of real numbers.* It is this property that makes radians useful in many applications. Therefore,

CAUTION ▶

when the angle is measured in radians, it is customary that no units are shown. The radian is understood to be the unit of measurement.

If the symbol rad is used, it is only to emphasize that the angle is in radians.

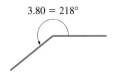

Fig. 8.27

◀ **EXAMPLE 3** **(a)** $60° = \left(\dfrac{\pi}{180°}\right)(60.0°) = \dfrac{\pi}{3.00} = 1.05$ $1.05 = 1.05 \text{ rad}$

no units indicates radian measure

(b) $3.80 = \left(\dfrac{180°}{\pi}\right)(3.80) = 218°$

Because no units are shown for 1.05 and 3.80, they are known to be measured in radians. See Fig. 8.27. ▶

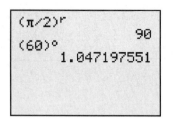

Fig. 8.28

Using the *angle* feature, a calculator can be used directly to change an angle expressed in degrees to an angle expressed in radians, or from an angle expressed in radians to an angle expressed in degrees (as pointed out in Chapter 4 on page 111). In Fig. 8.28, we show the calculator display for changing $\pi/2$ rad to degrees (calculator in degree mode) and for changing $60°$ to radians (calculator in radian mode). How these operations are done depends on the calculator model.

We can use a calculator to find the value of a function of an angle in radians. If the calculator is in radian mode, it then uses values in radians directly and will *consider any angle entered to be in radians*. The mode can be changed as needed, but

CAUTION ▶   *always be careful to have your calculator in the proper mode.*

Check the setting in the *mode* feature. If you are working in degrees, use the degree mode, but if you are working in radians, use the radian mode.

◀ **EXAMPLE 4**   **(a)** To find the value of sin 0.7538, put the calculator in radian mode (note that no units are shown with 0.7538), and the value is found as shown in the first two lines of the calculator display in Fig. 8.29. Therefore,

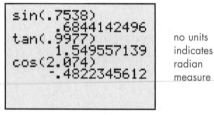

no units
indicates
radian
measure

Fig. 8.29

$$\sin 0.7538 = 0.6844$$

(b) From the third and fourth lines of the display in Fig. 8.29, we see that

$$\tan 0.9977 = 1.550$$

(c) From the last two lines of the display in Fig. 8.29, we see that

$$\cos 2.074 = -0.4822$$

In each case, the calculator was in radian mode.

In the following application, the resulting angle is a unitless number, and it is therefore in radian measure.

◀ **EXAMPLE 5**   The velocity v of an object undergoing simple harmonic motion at the end of a spring is given by

$$v = A\sqrt{\frac{k}{m}} \cos \sqrt{\frac{k}{m}}t \qquad \text{the angle is } \left(\sqrt{\frac{k}{m}}\right)(t)$$

Here, m is the mass of the object (in g), k is a constant depending on the spring, A is the maximum distance the object moves, and t is the time (in s). Find the velocity (in cm/s) after 0.100 s of a 36.0-g object at the end of a spring for which $k = 400 \text{ g/s}^2$, if $A = 5.00$ cm.

Substituting, we have

$$v = 5.00\sqrt{\frac{400}{36.0}} \cos \sqrt{\frac{400}{36.0}}(0.100)$$

Using calculator memory for $\sqrt{\frac{400}{36.0}}$, and with the calculator in radian mode, we have

$$v = 15.7 \text{ cm/s}$$

For certain special situations, we may need to know a reference angle in radians. In order to determine the proper quadrant, we should remember that $\frac{1}{2}\pi = 90°$, $\pi = 180°$, $\frac{3}{2}\pi = 270°$, and $2\pi = 360°$. These are shown in Table 8.1 along with the approximate decimal values for angles in radians. See Fig. 8.30.

Table 8.1

Quadrantal Angles

Degrees	Radians	Radians (decimal)
90°	$\frac{1}{2}\pi$	1.571
180°	π	3.142
270°	$\frac{3}{2}\pi$	4.712
360°	2π	6.283

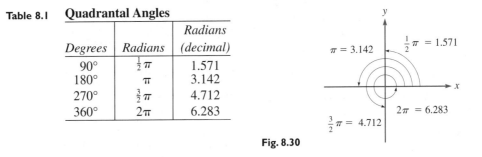

Fig. 8.30

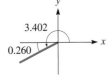

Fig. 8.31

❰ **EXAMPLE 6** An angle of 3.402 is greater than 3.142 but less than 4.712. Thus, it is a third-quadrant angle, and the reference angle is $3.402 - \pi = 0.260$. The ⬚π key can be used. See Fig. 8.31.

An angle of 5.210 is between 4.712 and 6.283. Therefore, it is in the fourth quadrant and the reference angle is $2\pi - 5.210 = 1.073$. ❱

❰ **EXAMPLE 7** Express θ in radians, such that $\cos \theta = 0.8829$ and $0 \le \theta < 2\pi$.

We are to find θ in radians for the given value of $\cos \theta$. Also, since θ is restricted to values between 0 and 2π, we must find a first-quadrant angle and a fourth-quadrant angle ($\cos \theta$ is positive in the first and fourth quadrants). With the calculator in radian mode, we find that

$$\cos^{-1} 0.8829 = 0.4888$$

Therefore, for the fourth-quadrant angle,

$$2\pi - 0.4888 = 5.794$$

This means

$$\theta = 0.4888 \quad \text{or} \quad \theta = 5.794$$

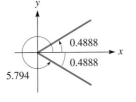

Fig. 8.32

See Fig. 8.32. ❱

When one first encounters radian measure,

CAUTION ❱ *expressions such as sin 1 and sin θ = 1 are often confused.*

The first is equivalent to $\sin 57.30°$, since $57.30° = 1$ (radian). The second means that θ is the angle for which the sine is 1. Since we know that $\sin 90° = 1$, we then can say that $\theta = 90°$ or $\theta = \pi/2$. The following example gives additional illustrations of evaluating expressions involving radians.

❰ **EXAMPLE 8** **(a)** $\sin \dfrac{\pi}{3} = \dfrac{\sqrt{3}}{2}$

(b) $\sin 0.6050 = 0.5688$.

(c) $\tan \theta = 1.709$ means that $\theta = 1.041$ (smallest positive θ).

(d) $\cos \theta = 0.4095$ means that $\theta = 1.149$ (smallest positive θ). ❱

EXERCISES 8.3

In Exercises 1–4, make the given changes in the indicated examples of this section and then solve then resulting problems.

1. In Example 3(b), change 3.80 to 2.80.

2. In Example 4(a), change sin to cos.

3. In Example 7, change cos to sin.

4. In Example 8(a), change $\frac{\pi}{3}$ to $\frac{\pi}{4}$.

In Exercises 5–12, express the given angle measurements in radian measure in terms of π.

5. 15°, 150°   6. 12°, 225°   7. 15°, 330°

8. 36°, 315°   9. 210°, 270°   10. 240°, 300°

11. 720°, 260°   12. 66°, 540°

In Exercises 13–20, the given numbers express angle measure. Express the measure of each angle in terms of degrees.

13. $\frac{2\pi}{5}, \frac{3\pi}{2}$ 14. $\frac{3\pi}{10}, \frac{5\pi}{6}$ 15. $\frac{\pi}{18}, \frac{7\pi}{4}$

16. $\frac{7\pi}{15}, \frac{4\pi}{3}$ 17. $\frac{17\pi}{18}, \frac{5\pi}{3}$ 18. $\frac{11\pi}{36}, \frac{5\pi}{4}$

19. $\frac{\pi}{12}, \frac{3\pi}{20}$ 20. $\frac{7\pi}{30}, \frac{4\pi}{15}$

In Exercises 21–28, express the given angles in radian measure. Round off results to the number of significant digits in the given angle.

21. 23.0°   22. 54.3°   23. 252°   24. 104°

25. 333.5°   26. 168.7°   27. 478.5°   28. −86.1°

In Exercises 29–36, the given numbers express the angle measure. Express the measure of each angle in terms of degrees, with the same accuracy as the given value.

29. 0.750   30. 0.240   31. 3.407   32. 1.703

33. 12.4   34. 34.4   35. −16.42   36. 100.0

In Exercises 37–44, evaluate the given trigonometric functions by first changing the radian measure to degree measure. Round off results to four significant digits.

37. $\sin \frac{\pi}{4}$ 38. $\cos \frac{\pi}{6}$ 39. $\tan \frac{5\pi}{12}$

40. $\sin \frac{7\pi}{18}$ 41. $\cos \frac{5\pi}{6}$ 42. $\tan \frac{7\pi}{3}$

43. $\sec 4.5920$ 44. $\cot 3.2732$

In Exercises 45–52, evaluate the given trigonometric functions directly, without first changing to degree measure.

45. $\tan 0.7359$ 46. $\cos 0.9308$ 47. $\sin 4.24$

48. $\tan 3.47$ 49. $\sec 2.07$ 50. $\sin(-2.34)$

51. $\cot(-4.86)$ 52. $\csc 6.19$

In Exercises 53–60, find θ to four significant digits for $0 \le \theta < 2\pi$.

53. $\sin \theta = 0.3090$ 54. $\cos \theta = -0.9135$

55. $\tan \theta = -0.2126$ 56. $\sin \theta = -0.0436$

57. $\cos \theta = 0.6742$ 58. $\cot \theta = 1.860$

59. $\sec \theta = -1.307$ 60. $\csc \theta = 3.940$

In Exercises 61–68, evaluate the given problems.

61. A unit of angle measurement used in artillery is the *mil*, which is defined as a central angle of a circle that intercepts an arc equal in length to 1/6400 of the circumference. How many mils are in a central angle of 34.4°?

62. Through how many radians does the minute hand of a clock move in 25 min?

63. After the brake was applied, a bicycle wheel went through 1.75 rotations. Through how many radians did a spoke rotate?

64. Through how many radians does a Ferris wheel with 18 seats move when loading passengers on the first 12 seats, assuming the loading process started at seat 1?

65. A flat plate of weight W oscillates as shown in Fig. 8.33. Its potential energy V is given by $V = \frac{1}{2}Wb\theta^2$, where θ is measured in radians. Find V if $W = 8.75$ N, $b = 0.75$ m, and $\theta = 5.5°$.

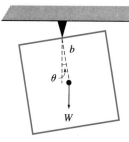

Fig. 8.33

W 66. The charge q (in C) on a capacitor as a function of time is $q = A \sin \omega t$. If t is measured in seconds, in what units is ω measured? Explain.

67. The height h of a rocket launched 1200 m from an observer is found to be $h = 1200 \tan \frac{5t}{3t + 10}$ for $t < 10$ s, where t is the time after launch. Find h for $t = 8.0$ s.

68. The electric intensity I (in W/m²) from the two radio antennas shown in Fig. 8.34 is a function of the angle θ given by $I = 0.023 \cos^2(\pi \sin \theta)$. Find I for $\theta = 40.0°$. ($\cos^2 \alpha = (\cos \alpha)^2$.)

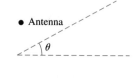

Fig. 8.34   ● Antenna

8.4 APPLICATIONS OF RADIAN MEASURE

Radian measure has numerous applications in mathematics and technology, some of which were illustrated in the last eight exercises of the previous section. In this section, several more applications are shown.

ARC LENGTH

From geometry, we know that *the length of an arc on a circle is proportional to the central angle* and that the length of arc of a complete circle is the circumference. Letting s stand for the length of arc, we may state that $s = 2\pi r$ for a complete circle. Since 2π is the central angle (in radians) of the complete circle, *we have for the length of arc*

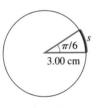

Fig. 8.35

$$s = \theta r \qquad (\theta \text{ in radians}) \tag{8.11}$$

for any circular arc with central angle θ. If we know the central angle in radians and the radius of a circle, we can find the length of a circular arc directly by using Eq. (8.11). See Fig. 8.35.

EXAMPLE 1 In Fig. 8.36, $\theta = \pi/6$ and $r = 3.00$ cm Therefore,

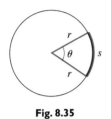

Fig. 8.36

$$s = \left(\frac{\pi}{6}\right)(3.00) = \frac{\pi}{2.00} = 1.57 \text{ cm}$$

with label "θ in radians" pointing to $\frac{\pi}{6}$

Among the important applications of arc length are distances on the earth's surface. For most purposes, the earth may be regarded as a sphere (the diameter at the equator is slightly greater than the distance between the poles). A *great circle* of the earth (or any sphere) is the circle of intersection of the surface of the sphere and a plane that passes through the center.

The equator is a great circle and is designated as $0°$ *latitude*. Other *parallels of latitude* are parallel to the equator with diameters decreasing to zero at the poles, which are $90°$ N and $90°$ S. See Fig. 8.37.

Meridians of longitude are half great circles between the poles. The *prime meridian* through Greenwich, England, is designated as $0°$, with meridians to $180°$ measured east and west from Greenwich. Positions on the surface of the earth are designated by longitude and latitude.

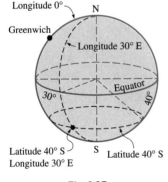

Fig. 8.37

See the chapter introduction.

EXAMPLE 2 The traditional definition of a *nautical mile* is the length of arc along a great circle of the earth for a central angle of $1'$. The modern international definition is a distance of 1852 m. What measurement of the earth's radius does this definition use?

Here, $\theta = 1' = (1/60)°$, and $s = 1852$ m. Solving for r, we have

$$r = \frac{s}{\theta} = \frac{1852}{\left(\frac{1}{60}\right)°\left(\frac{\pi}{180°}\right)} = 6.367 \times 10^6 \text{ m} = 6367 \text{ km}$$

Historically, the fact that the earth is not a perfect sphere has led to many variations in the distance used for a nautical mile.

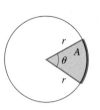

Fig. 8.38

AREA OF A SECTOR OF A CIRCLE

Another application of radians is finding the area of a sector of a circle (see Fig. 8.38). Recall from geometry that areas of sectors of circles are proportional to their central angles. The area of a circle is $A = \pi r^2$, which can be written as $A = \frac{1}{2}(2\pi)r^2$. Since the angle for a complete circle is 2π, *the area of any sector of a circle in terms of the radius and central angle (in radians) is*

$$A = \frac{1}{2}\theta r^2 \quad (\theta \text{ in radians}) \tag{8.12}$$

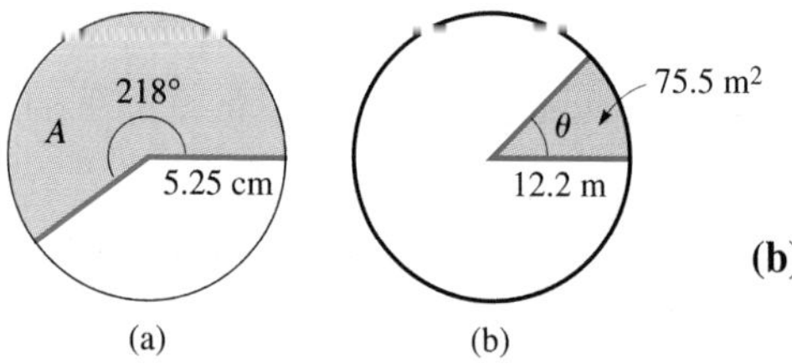

EXAMPLE 3 **(a)** The area of a sector of a circle with central angle 218° and a radius of 5.25 cm (see Fig. 8.39(a)) is

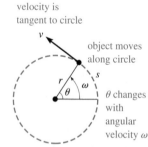

(a)   (b)

Fig. 8.39

$$A = \frac{1}{2}(218)\left(\frac{\pi}{180}\right)(5.25)^2 = 52.4 \text{ cm}^2$$

— θ in radians

(b) Given that the area of a sector is 75.5 m² and the radius is 12.2 m (see Fig. 8.39(b)), we find the central angle by solving for θ and then substituting:

— no units indicates radian measure

$$\theta = \frac{2A}{r^2} = \frac{2(75.5)}{(12.2)^2} = 1.01$$

This means that the central angle is 1.01 rad, or 57.9°.

CAUTION ▶

We should note again that the equations in this section require that the angle θ *be expressed in radians*. A common error is to use θ in degrees.

ANGULAR VELOCITY

velocity is
tangent to circle

object moves
along circle

θ changes
with
angular
velocity ω

Fig. 8.40

The average velocity of a moving object is defined by $v = s/t$, where v is the average velocity, s is the distance traveled, and t is the elapsed time. For an object moving in a circular path with constant speed, the distance traveled is the length of arc through which it moves. Therefore, if we divide both sides of Eq. (8.11) by t, we obtain

$$\frac{s}{t} = \frac{\theta r}{t} = \frac{\theta}{t}r$$

where θ/t is called the *angular velocity* and is designated by ω. Therefore,

$$v = \omega r \tag{8.13}$$

Equation (8.13) expresses the relationship between the **linear velocity** v *and the* **angular velocity** ω *of an object moving around a circle of radius r.* See Fig. 8.40. In the figure, v is shown directed tangent to the circle, for that is its direction for the position shown. The direction of v changes constantly.

The units for ω are radians per unit of time. In this way, the formula can be used directly. However, in practice, ω is often given in revolutions per minute or in some similar unit. In these cases, it is necessary to convert the units of ω to radians per unit of time before substituting in Eq. (8.13).

Some of the typical units used for angular velocity are

rad/s   rad/min   rad/h
°/s   °/min
r/s   r/min

(r represents revolutions.)

(r/min is the same as rpm. This text does not use rpm.)

EXAMPLE 4 A person on a hang glider is moving in a horizontal circular arc of radius 90.0 m with an angular velocity of 0.125 rad/s. The person's linear velocity is

$$v = (0.125 \text{ rad/s})(90.0 \text{ m}) = 11.3 \text{ m/s}$$

(Remember that radians are numbers and are not included in the final set of units.) This means that the person is moving along the circumference of the arc at 11.3 m/s (40.7 km/h).

Solving a Word Problem

The first U.S. communications satellite was launched in July 1962.

EXAMPLE 5 A communication satellite remains at an altitude of 35 920 km above a point on the equator. If the radius of the earth is 6370 km, what is the velocity of the satellite?

In order for the satellite to remain over a point on the equator, it must rotate exactly once each day around the center of the earth (and it must remain at an altitude of 35 920 km). Since there are 2π radians in each revolution, the angular velocity is

$$\omega = \frac{1 \text{ r}}{1 \text{ day}} = \frac{2\pi \text{ rad}}{24 \text{ h}} = 0.2618 \text{ rad/h}$$

The radius of the circle through which the satellite moves is its altitude plus the radius of the earth, or $35\,920 + 6370 = 42\,290$ km. Thus, the velocity is

$$v = 0.2618(42\,290) = 11\,070 \text{ km/h}$$

Solving a Word Problem

EXAMPLE 6 A pulley belt 4.00 m long takes 2.00 s to make one complete revolution. The radius of the pulley is 20.0 cm. What is the angular velocity (in revolutions per minute) of a point on the rim of the pulley? See Fig. 8.41.

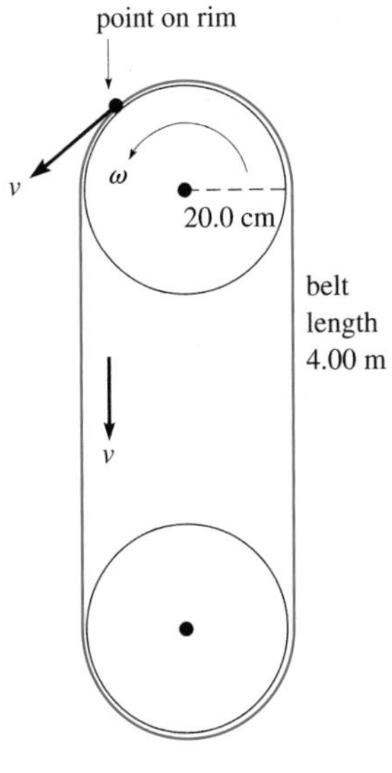

Since the linear velocity of a point on the rim of the pulley is the same as the velocity of the belt, $v = 4.00/2.00 = 2.00$ m/s. The radius of the pulley is $r = 20.0$ cm $= 0.200$ m, and we can find ω by substituting into Eq. (8.13). This gives us

point on rim

20.0 cm

belt length 4.00 m

Fig. 8.41

$$v = \omega r$$
$$2.00 = \omega(0.200)$$
$$\omega = 10.0 \text{ rad/s} \qquad \text{multiply by 60 s/1 min}$$
$$= 600 \text{ rad/min} \qquad \text{multiply by 1 r/}2\pi \text{ rad}$$
$$= 95.5 \text{ r/min} \qquad \text{r is the symbol for revolution}$$

As shown in Appendix B, the change of units can be handled algebraically as

$$10.0 \frac{\text{rad}}{\text{s}} \times 60 \frac{\text{s}}{\text{min}} = 600 \frac{\text{rad}}{\text{min}}$$

$$\frac{600 \text{ rad/min}}{2\pi \text{ rad/r}} = 600 \frac{\text{rad}}{\text{min}} \times \frac{1}{2\pi} \frac{\text{r}}{\text{rad}} = 95.5 \text{ r/min}$$

EXAMPLE 7 The current at any time in a certain alternating-current electric circuit is given by $i = I \sin 120\pi t$, where I is the maximum current and t is the time in seconds. Given that $I = 0.0685$ A, find i for $t = 0.00500$ s.

Substituting, with the calculator in radian mode, we get

$$i = 0.0685 \sin[(120\pi)(0.00500)]$$
$$= 0.0651 \text{ A}$$

EXERCISES 8.4

In Exercises 1–4, make the given changes in the indicated examples of this section, and then solve the resulting problems.

1. In Example 1, change $\pi/6$ to $\pi/4$.

2. In Example 3(a), change 218° to 258°.

3. In Example 4, change 90.0 m to 115 m.

4. In Example 6, change 2.00 s to 2.50 s.

In Exercises 5–16, for an arc length s, area of sector A, and central angle θ of a circle of radius r, find the indicated quantity for the given values.

5. $r = 3.30$ cm, $\theta = \pi/3$, $s = ?$

6. $r = 21.2$ cm, $\theta = 2.65$, $s = ?$

7. $s = 1010$ mm, $\theta = 136.0°$, $r = ?$

8. $s = 0.3456$ m, $\theta = 73.61°$, $A = ?$

9. $s = 0.3913$ km, $r = 0.9449$ km, $A = ?$

10. $s = 3.19$ m, $r = 2.29$ m, $\theta = ?$

11. $r = 4.9$ cm, $\theta = 3.6$, $A = ?$

12. $r = 46.3$ dm, $\theta = 2\pi/5$, $A = ?$

13. $A = 0.0119$ m², $\theta = 326.0°$, $r = ?$

14. $A = 1200$ mm², $\theta = 17°$, $s = ?$

15. $A = 16.5$ m², $r = 4.02$ m, $s = ?$

16. $A = 67.8$ km², $r = 67.8$ km, $\theta = ?$

In Exercises 17–56, solve the given problems.

17. While playing, the left spool of a VCR turns through 820°. For this part of the tape, it is 3.30 cm from the center of the spool to the tape. What length of tape is played?

W 18. The latitude of Manila, Philippines, is 15° N, and the latitude of Shanghai, China, is 31° N. Both are at a longitude of 121° E. What is the distance between Manila and Shanghai? Explain how the angle used in the solution is found. The radius of the earth is 6370 km.

19. Of the estimated natural gas reserves in North America, 4.59×10^9 m³ are in the United States, 2.66×10^9 m³ are in Canada, and 1.99×10^9 m³ are in Mexico. In making a *circle graph* (circular sectors represent percentages of the whole—a *pie chart*) with a radius of 4.00 cm for these data, what are the central angle and area of the sector that represents Canada's reserves?

20. A section of sidewalk is a circular sector of radius 1.25 m and central angle 50.6°. What is the area of this section of sidewalk?

21. When between 12:00 noon and 1:00 P.M. are the minute and hour hands of a clock 180° apart?

22. A cam is in the shape of a circular sector, as shown in Fig. 8.42. What is the perimeter of the cam?

Fig. 8.42   165.58°   1.875 cm

23. A lawn sprinkler can water up to a distance of 25.0 m. It turns through an angle of 115.0°. What area can it water?

24. A spotlight beam sweeps through a horizontal angle of 75.0°. If the range of the spotlight is 110 m, what area can it cover?

25. If a car makes a U-turn in 6.0 s, what is its average angular velocity in the turn?

26. The roller on a computer printer makes 2200 r/min. What is its angular velocity?

27. What is the floor area of the hallway shown in Fig. 8.43? The outside and inside of the hallway are circular arcs.

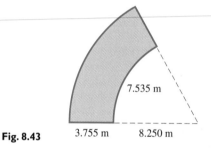

Fig. 8.43   3.755 m   8.250 m   7.535 m

28. The arm of a car windshield wiper is 32.4 cm long and is attached at the middle of a 38.1-cm blade. (Assume that the arm and blade are in line.) What area of the windshield is cleaned by the wiper if it swings through 110.0° arcs?

29. Part of a railroad track follows a circular arc with a central angle of 28.0°. If the radius of the arc of the inner rail is 28.55 m and the rails are 1.44 m apart, how much longer is the outer rail than the inner rail?

30. A wrecking ball is dropped as shown in Fig. 8.44. Its velocity at the bottom of its swing is $v = \sqrt{2gh}$, where g is the acceleration due to gravity. What is its angular velocity at the bottom if $g = 9.80$ m/s² and $h = 4.80$ m?

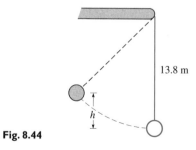

13.8 m

h

Fig. 8.44

31. Part of a security fence is built 2.50 m from a cylindrical storage tank 11.2 m in diameter. What is the area between the tank and this part of the fence if the central angle of the fence is 75.5°? See Fig. 8.45.

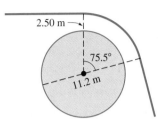

Fig. 8.45

32. Through what angle does the drum in Fig. 8.46 turn in order to lower the crate 10.3 m?

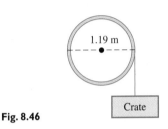

Fig. 8.46

33. A section of road follows a circular arc with a central angle of 15.6°. The radius of the inside of the curve is 285.0 m, and the road is 15.2 m wide. What is the volume of the concrete in the road if it is 0.305 m thick?

34. The propeller of the motor on a motorboat is rotating at 130 rad/s. What is the linear velocity of a point on the tip of a blade if it is 22.5 cm long?

35. A storm causes a pilot to follow a circular-arc route, with a central angle of 12.8°, from city A to city B rather than the straight-line route of 185.0 km. How much farther does the plane fly due to the storm?

36. A highway exit is a circular arc 330 m long with a central angle of 79.4°. What is the radius of curvature of the exit?

37. The paddles of a riverboat have a radius of 2.59 m and revolve at 20.0 r/min. What is the speed of a tip of one of the paddles?

38. The sweep second hand of a watch is 15.0 mm long. What is the linear velocity of the tip?

39. A computer diskette has a diameter of 8.90 cm and rotates at 360.0 r/min. What is the linear velocity of a point on the outer edge?

40. A Ferris wheel 18.0 m in diameter makes one revolution in 3.00 min. Find the speed of a seat on the rim.

41. The sprocket assembly for a 70.0-cm bike is shown in Fig. 8.47. How fast (in r/min) does the rider have to pedal in order to go 25.0 km/h on level ground?

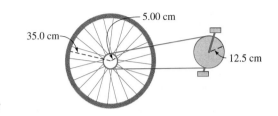

Fig. 8.47

42. The flywheel of a car engine is 0.36 m in diameter. If it is revolving at 750 r/min, through what distance does a point on the rim move in 2.00 s?

43. Two streets meet at an angle of 82.0°. What is the length of the piece of curved curbing at the intersection if it is constructed along the arc of a circle 5.50 m in radius? See Fig. 8.48.

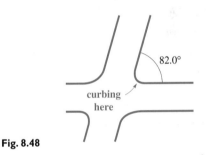

Fig. 8.48

44. An ammeter needle is deflected 52.00° by a current of 0.2500 A. The needle is 3.750 cm long, and a circular scale is used. How long is the scale for a maximum current of 1.500 A?

45. A drill bit 9.53 mm in diameter rotates at 1200 r/min. What is the linear velocity of a point on its circumference?

46. A helicopter blade is 2.75 m long and is rotating at 420 r/min. What is the linear velocity of the tip of the blade?

47. A waterwheel used to generate electricity has paddles 3.75 m long. The speed of the end of a paddle is one-fourth that of the water. If the water is flowing at the rate of 6.50 m/s, what is the angular velocity of the waterwheel?

48. A jet is traveling westward with the sun directly overhead (the jet is on a line between the sun and the center of the earth). How fast must the jet fly in order to keep the sun directly overhead? (Assume that the earth's radius is 6370 km, the altitude of the jet is low, and the earth rotates about its axis once in 24.0 h.)

49. A 1500-kW wind turbine (windmill) rotates at 40.0 r/min. What is the linear velocity of a point on the end of a blade, if the blade is 12.0 m long (from the center of rotation)?

50. What is the linear velocity of a point in Sydney, Australia, which is at a latitude of 33°55′ S? The radius of the earth is 6370 km.

51. Through what total angle does the drive shaft of a car rotate in 1 s when the tachometer reads 2400 r/min?

52. A baseball field is designed such that the outfield fence is along the arc of a circle with its center at second base. If the radius of the circle is 85.0 m, what is the playing area of the field? See Fig. 8.49.

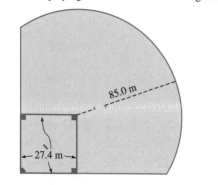

Fig. 8.49

53. The turbine fan blade of a turbojet engine is 1.2 m in diameter and rotates at 250 r/s. How fast is the tip of a blade moving?

54. A patio is in the shape of a circular sector with a central angle of 160.0°. It is enclosed by a railing of which the circular part is 11.6 m long. What is the area of the patio?

55. An oil storage tank 4.25 m long has a flat bottom as shown in Fig. 8.50. The radius of the circular part is 1.10 m. What volume of oil does the tank hold?

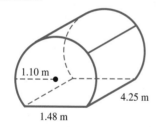

Fig. 8.50

56. Two equal beams of light illuminate the area shown in Fig. 8.51. What area is lit by both beams?

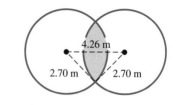

Fig. 8.51

In Exercises 57–60, another use of radians is illustrated.

 57. Use a calculator (in radian mode) to evaluate the ratios $(\sin \theta)/\theta$ and $(\tan \theta)/\theta$ for $\theta = 0.1, 0.01, 0.001$, and 0.0001. From these values explain why it is possible to say that

$$\sin \theta = \tan \theta = \theta \qquad (8.14)$$

approximately for very small angles.

58. Using Eq. (8.14), evaluate tan 0.001°. Compare with a calculator value.

59. An astronomer observes that a star 12.5 light-years away moves through an angle of 0.2″ in 1 year. Assuming it moved in a straight line perpendicular to the initial line of observation, how many km did the star move? (1 light-year = 9.46 km) Use Eq. (8.14).

60. In calculating a back line of a lot, a surveyor discovers an error of 0.05° in an angle measurement. If the lot is 136.0 m deep, by how much is the back line calculation in error? See Fig. 8.52. Use Eq. (8.14).

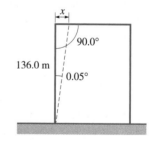

Fig. 8.52

CHAPTER **8** EQUATIONS

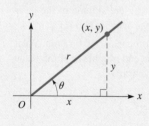

$$\sin \theta = \frac{y}{r} \qquad \cos \theta = \frac{x}{r} \qquad \tan \theta = \frac{y}{x}$$

$$\cot \theta = \frac{x}{y} \qquad \sec \theta = \frac{r}{x} \qquad \csc \theta = \frac{r}{y} \qquad (8.1)$$

$$F(\theta_2) = \pm F(180° - \theta_2) = \pm F(\alpha) \qquad (\alpha \text{ is reference angle}) \qquad (8.3)$$

$$F(\theta_3) = \pm F(\theta_3 - 180°) = \pm F(\alpha) \qquad (8.4)$$

$$F(\theta_4) = \pm F(360° - \theta_4) = \pm F(\alpha) \qquad (8.5)$$

α **is reference angle**

$\theta = \alpha$ (first quadrant)

$\theta = 180° - \alpha$ (second quadrant)

$\theta = 180° + \alpha$ (third quadrant)   (8.6)

$\theta = 360° - \alpha$ (fourth quadrant)

Negative angles

$\sin(-\theta) = -\sin\theta$ $\cos(-\theta) = \cos\theta$ $\tan(-\theta) = -\tan\theta$

$\csc(-\theta) = -\csc\theta$ $\sec(-\theta) = \sec\theta$ $\cot(-\theta) = -\cot\theta$ (8.7)

Radian-degree conversions

$\pi\,\text{rad} = 180°$ (8.8)

$1° = \dfrac{\pi}{180}\,\text{rad} = 0.01745\,\text{rad}$ (8.9)

$1\,\text{rad} = \dfrac{180°}{\pi} = 57.30°$ (8.10)

Circular arc length

$s = \theta r$ (θ in radians)   (8.11)

Circular sector area

$A = \dfrac{1}{2}\theta r^2$ (θ in radians)   (8.12)

Linear and angular velocity

$v = \omega r$ (8.13)

CHAPTER ⑧ REVIEW EXERCISES

In Exercises 1–4, find the trigonometric functions of θ. The terminal side of θ passes through the given point.

1. $(6, 8)$ **2.** $(-12, 5)$

3. $(14, -4)$ **4.** $(-2, -3)$

In Exercises 5–8, express the given trigonometric functions in terms of the same function of a positive acute angle.

5. $\cos 132°$, $\tan 194°$ **6.** $\sin 243°$, $\cot 318°$

7. $\sin 289°$, $\sec(-15°)$ **8.** $\cos 463°$, $\csc(-100°)$

In Exercises 9–12, express the given angle measurements in terms of π.

9. $40°$, $153°$ **10.** $22.5°$, $324°$

11. $408°$, $202.5°$ **12.** $27°$, $-162°$

In Exercises 13–20, the given numbers represent angle measure. Express the measure of each angle in degrees.

13. $\dfrac{7\pi}{5}, \dfrac{13\pi}{18}$ **14.** $\dfrac{3\pi}{8}, \dfrac{7\pi}{20}$

15. $\dfrac{\pi}{15}, \dfrac{11\pi}{6}$ **16.** $\dfrac{17\pi}{10}, \dfrac{5\pi}{4}$

17. 0.560 **18.** 1.354

19. 36.07 **20.** 14.5

In Exercises 21–28, express the given angles in radians (not in terms of π).

21. $102°$ **22.** $305°$

23. $20.25°$ **24.** $148.38°$

25. $262.05°$ **26.** $-18.72°$

27. $136.2°$ **28.** $385.4°$

In Exercises 29–48, determine the values of the given trigonometric functions directly on a calculator. The angles are approximate. Express answers to Exercises 41–44 to four significant digits.

29. $\cos 245.5°$ **30.** $\sin 141.3°$

31. $\cot 295°$ **32.** $\tan 184°$

33. $\csc 247.82°$ **34.** $\sec 96.17°$

35. $\sin 205.24°$ **36.** $\cos 326.72°$

37. $\tan 301.4°$ **38.** $\sin 103.9°$

39. $\tan 436.42°$ **40.** $\cos 162.32°$

41. $\sin \dfrac{9\pi}{5}$ **42.** $\sec \dfrac{5\pi}{8}$

43. $\cos\dfrac{7\pi}{6}$

44. $\tan\dfrac{23\pi}{12}$

45. $\sin 0.5906$

46. $\tan 0.8035$

47. $\csc 2.153$

48. $\cos 7.190$

In Exercises 49–52, find θ in degrees for $0° \le \theta < 360°$.

49. $\tan \theta = 0.1817$

50. $\sin \theta = -0.9323$

51. $\cos \theta = -0.4730$

52. $\cot \theta = 1.196$

In Exercises 53–56, find θ in radians for $0 \le \theta < 2\pi$.

53. $\cos \theta = 0.8387$

54. $\sin \theta = 0.1045$

55. $\sin \theta = -0.8650$

56. $\tan \theta = 2.840$

In Exercises 57–60, find θ in degrees for $0° \le \theta < 360°$.

57. $\cos \theta = -0.7222$, $\sin \theta < 0$

58. $\tan \theta = -1.683$, $\cos \theta < 0$

59. $\cot \theta = 0.4291$, $\cos \theta < 0$

60. $\sin \theta = 0.2626$, $\tan \theta < 0$

In Exercises 61–68, for an arc of length s, area of sector A, and central angle θ of circle of radius r, find the indicated quantity for the given values.

61. $s = 20.3$ cm, $\theta = 107.5°$, $r = ?$

62. $s = 584$ m, $r = 106$ m, $\theta = ?$

63. $A = 265$ mm^2, $r = 12.8$ mm, $\theta = ?$

64. $A = 0.908$ km^2, $\theta = 234.5°$, $r = ?$

65. $r = 4.62$ m, $A = 32.8$ m^2, $s = ?$

66. $\theta = 98.5°$, $A = 0.493$ dm^2, $s = ?$

67. $\theta = 165.4°$, $s = 7.94$ cm, $A = ?$

68. $r = 254$ cm, $s = 76.1$ cm, $A = ?$

In Exercises 69–88, solve the given problems.

69. The instantaneous power p (in W) input to a resistor in an alternating-current circuit is $p = p_m \sin^2 377t$, where p_m is the maximum power input and t is the time (in s). Find p for $p_m = 0.120$ W and $t = 2.00$ ms. $(\sin^2\theta = (\sin\theta)^2.)$

70. The horizontal distance x through which a pendulum moves is given by $x = a(\theta + \sin \theta)$, where a is a constant and θ is the angle between the vertical and the pendulum. Find x for $a = 45.0$ cm and $\theta = 0.175$.

71. A sector gear with a pitch radius of 8.25 cm and a 6.60-cm arc of contact is shown in Fig. 8.53. What is the sector angle θ?

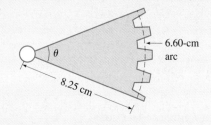

Fig. 8.53

72. Two pulleys have radii of 10.0 cm and 6.00 cm, and their centers are 40.0 cm apart. If the pulley belt is uncrossed, what must be the length of the belt?

73. A special vehicle for traveling on glacial ice has tires that are 1.5 m in diameter. If the vehicle travels at 5.6 km/h, what is the angular velocity (in r/min) of the tire?

74. A rotating circular restaurant at the top of a hotel has a diameter of 32.5 m. If it completes one revolution in 24.0 min, what is the velocity of the outer surface?

75. Find the velocity (in km/h) of the moon as it revolves about the earth. Assume it takes 28 days for one revolution at a distance of 390 000 km from the earth.

76. The stopboard of a shot-put circle is a circular arc 1.22 m in length. The radius of the circle is 1.06 m. What is the central angle?

(W) **77.** The longitude of Anchorage, Alaska, is 150° W, and the longitude of St. Petersburg, Russia, is 30° E. Both cities are at a latitude of 60° N. (a) Find the great circle distance (see page 249) from Anchorage to St. Petersburg over the north pole. (b) Find the distance between them along the 60° N latitude arc. The radius of the earth is 6370 km. What do the results show?

78. A piece of circular filter paper 15.0 cm in diameter is folded such that its effective filtering area is the same as that of a sector with a central angle of 220°. What is the filtering area?

79. To produce an electric current, a circular loop of wire of diameter 25.0 cm is rotating about its diameter at 60.0 r/s in a magnetic field. What is the greatest linear velocity of any point on the loop?

80. Find the area of the decorative glass panel shown in Fig. 8.54. The panel is made of two equal circular sectors and an isosceles triangle.

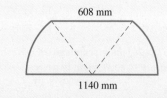

Fig. 8.54

81. A circular hood is to be used over a piece of machinery. It is to be made from a circular piece of sheet metal 1.08 m in radius. A hole 0.25 m in radius and a sector of central angle 80.0° are to be removed to make the hood. What is the area of the top of the hood?

82. The chain on a chain saw is driven by a sprocket 7.50 cm in diameter. If the chain is 108 cm long and makes one revolution in 0.250 s, what is the angular velocity (in r/s) of the sprocket?

83. An *ultracentrifuge*, used to observe the sedimentation of particles such as proteins, may rotate as fast as 80 000 r/min. If it rotates at this rate and is 7.20 cm in diameter, what is the linear velocity of a particle at the outer edge?

84. A computer is programmed to shade in a sector of a pie chart 2.44 cm in radius. If the perimeter of the shaded sector is 7.32 cm, what is the central angle (in degrees) of the sector? See Fig. 8.55.

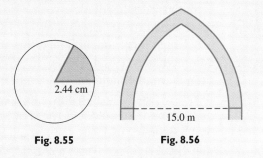

2.44 cm

Fig. 8.55   **Fig. 8.56**

85. A Gothic arch, commonly used in medieval European structures, is formed by two circular arcs. In one type, each arc is one-sixth of a circle, with the center of each at the base on the end of the other arc. See Fig. 8.56. Therefore, the width of the arch equals the radius of each arc. For such an arch, find the area of the opening if the width is 15.0 m.

86. The Trans-Alaska Pipeline was assembled in sections 12.2 m long and 1.22 m in diameter. If the depth of the oil in one horizontal section is 0.305 m, what is the volume of oil in this section?

87. A laser beam is transmitted with a "width" of 0.0008° and makes a circular spot of radius 2.50 km on a distant object. How far is the object from the source of the laser beam? Use Eq. (8.14).

88. The planet Venus subtends an angle of 15″ to an observer on earth. If the distance between Venus and earth is 167 Gm what is the diameter of Venus? Use Eq. (8.14).

Writing Exercise

89. Write a paragraph explaining how you determine the units for the result of the following problem: An astronaut in a spacecraft circles the moon once each 1.95 h. If the altitude of the spacecraft is constant at 113 km, what is its velocity? The radius of the moon is 1740 km. (What is the answer?)

CHAPTER 8 PRACTICE TEST

1. Change 150° to radians in terms of π.

2. Express sin 205° in terms of the sine of a positive acute angle. Do not evaluate.

3. Find sin θ and sec θ if θ is in standard position and the terminal side passes through $(-9, 12)$.

4. An airplane propeller blade is 1.40 m long and rotates at 2200 r/min. What is the linear velocity of a point on the tip of the blade?

5. Given that 3.572 is the measure of an angle, express the angle in degrees.

6. If tan $\theta = 0.2396$, find θ, in degrees, for $0° \le \theta < 360°$.

7. If cos $\theta = -0.8244$ and csc $\theta < 0$, find θ in radians for $0 \le \theta < 2\pi$.

8. The floor of a sunroom is in the shape of a circular sector of arc length 16.0 m and radius 4.25 m. What is the area of the floor?

9. A circular sector has an area of 38.5 cm² and a diameter of 12.2 cm. What is the arc length of the sector?

CHAPTER 9

Vectors and Oblique Triangles

The wind must be considered to find the proper heading for an aircraft. In Section 9.5, we use vectors and oblique triangles to show how this may be done.

In many applications, we often deal with such things as forces and velocities. To study them, both their *magnitudes* **and** *directions* must be known. In general, a quantity for which we must specify *both magnitude and direction* is called a *vector*.

In basic applications, a vector is usually represented by an arrow showing its magnitude and direction, although this was not common before the 1800s. In 1743, the French mathematician d'Alembert published a paper on dynamics in which he used some diagrams, but most of the text was algebraic. In 1788, the French mathematician Lagrange wrote a classic work on *Analytical Mechanics*, but the text was algebraic and included no diagrams.

In the 1800s, calculus was used to greatly advance the use of vectors, and in turn these advancements became very important in further developments in scientific fields such as electromagnetic theory. Also in the 1800s, mathematicians defined a vector more generally and opened up study in new areas of advanced mathematics. In this text, we deal only with the basic meaning of a vector. However, a vector is an excellent example of a math concept that came from a basic physics concept.

Today, vectors are of great importance in many fields of science and technology, including physics, engineering, structural design, and navigation. An aircraft does not necessarily head directly for its destination, for the direction of the jet stream and other winds must be taken into account. All forces on a robot link must be considered to find the resultant force and the direction in which that force moves the link.

After studying vectors, we then show methods of solving triangles that are not right triangles (*oblique* triangles). For such triangles, we use the trig functions of oblique angles. As with right triangles, the applications of oblique triangles are found in many fields of science and technology.

9.1 INTRODUCTION TO VECTORS

We deal with many quantities that may be described only by a number that shows the magnitude. These include lengths, areas, time intervals, monetary amounts, and temperatures. *Quantities such as these, described only by the **magnitude**, are known as* **scalars.**

Scalars

Vectors

As we said on the previous page, *many other quantities, called* **vectors,** *are fully described only when both the **magnitude** and **direction** are specified.* The following example shows the difference between scalars and vectors.

◄ EXAMPLE 1 A jet is traveling at 800 km/h. From this statement alone we know only the *speed* of the jet. *Speed is a scalar quantity,* and it tells us only the *magnitude* of the rate. Knowing only the speed of the jet, we know the rate at which it is moving, but we do not know where it is headed.

If the phrase "in a direction 10° south of west" is added to the sentence about the jet, we specify the direction of travel as well as the speed. We then know the *velocity* of the jet; that is the *direction* of travel as well as the rate at which it is moving. ***Velocity is a vector*** *quantity.* ▶

For an example of the action of two vectors, consider a boat moving in a river. We will assume that the boat is driven by a motor that can move it at 8 km/h in still water and that the river's current is going 6 km/h downstream, as shown in Fig. 9.1. We quickly see that the movement of the boat depends on the direction in which it is headed. If it heads downstream, it moves at 14 km/h, for the water is moving at 6 km/h and the boat moves at 8 km/h with respect to the water. If it heads upstream, however, it moves only at 2 km/h, since the river is acting directly against the motor. If the boat heads directly across the river, the point it reaches on the other side is not directly opposite the point from which it started. This is so because the river is moving the boat downstream *at the same time* the boat moves across the river.

Checking this last case further, assume that the river is 0.4 km wide where the boat is crossing. It takes 0.05 h (0.4 km ÷ 8 km/h = 0.05 h) to cross. In 0.05 h, the river will carry the boat 0.3 km (0.05 h × 6 km/h = 0.3 km) downstream. This means the boat went 0.3 km downstream as it went 0.4 km across the river. From the Pythagorean theorem, we see that it went 0.5 km from its starting point to its finishing point:

$$d^2 = 0.4^2 + 0.3^2 = 0.25$$
$$d = 0.5 \text{ km}$$

Since the 0.5 km was traveled in 0.05 h, the magnitude of the velocity (the *speed*) of the boat was actually

$$v = \frac{d}{t} = \frac{0.5 \text{ km}}{0.05 \text{ h}} = 10 \text{ km/h}$$

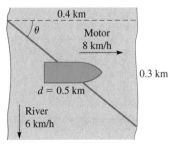

Fig. 9.1

Also, note that the direction of this velocity can be represented along a line that makes an angle θ with the line directed directly across the river, as shown in Fig. 9.1. We can find this angle by noting that

$$\tan \theta = \frac{0.3 \text{ km}}{0.4 \text{ km}} = 0.75$$

$$\theta = \tan^{-1} 0.75 = 37°$$

Therefore, when headed directly across the river, the boat's velocity is 10 km/h directed at an angle of 37° downstream from a line directly across the river.

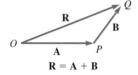

$$R = A + B$$

Fig. 9.2

CAUTION ▶

Polygon Method

ADDITION OF VECTORS

We have just seen two velocity vectors being *added.* Note that these vectors are not added the way numbers are added. We must take into account their directions as well as their magnitudes. Reasoning along these lines, let us now define the sum of two vectors.

We will represent a vector quantity by a letter printed in **boldface** type. The same letter in *italic* (lightface) type represents the magnitude only. Thus, **A** is a vector of magnitude *A*. In handwriting, one usually places an arrow over the letter to represent a vector, such as $\vec{A}$.

Let **A** and **B** represent vectors directed from *O* to *P* and *P* to *Q*, respectively (see Fig. 9.2). *The vector sum* **A** + **B** *is the vector* **R,** *from the* **initial point** *O to the* **terminal point** *Q.* Here, vector **R** is called the **resultant.** *In general, a resultant is a single vector that is the vector sum of any number of other vectors.*

There are two common methods of adding vectors by means of a diagram. The first is illustrated in Fig. 9.3. To add **B** to **A**, shift **B** parallel to itself until its tail touches the head of **A**. *The vector sum* **A** + **B** *is the resultant vector* **R**, *which is drawn from the tail of* **A** *to the head of* **B**. In using this method, we can move a vector for addition as long as *its magnitude and direction remain unchanged.* (Since the magnitude and direction specify a vector, two vectors in different *locations* are considered the same if they have the same magnitude and direction.) When using a diagram to add vectors, it must be drawn with reasonable accuracy.

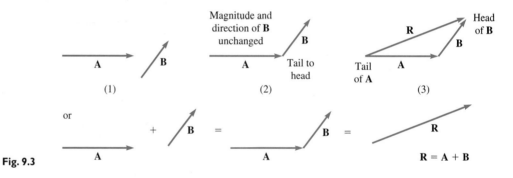

Fig. 9.3

Three or more vectors are added in the same general manner. We place the initial point of the second vector at the terminal point of the first vector, the initial point of the third vector at the terminal point of the second vector, and so on. The resultant is the vector from the initial point of the first vector to the terminal point of the last vector. The order in which they are added does not matter.

◀ **EXAMPLE 2**   The addition of vectors **A, B,** and **C** is shown in Fig. 9.4.

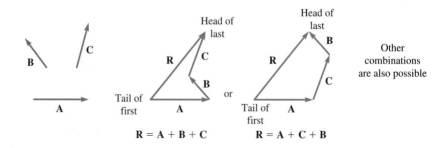

Fig. 9.4

Parallelogram Method   Another method that is convenient when two vectors are being added is to *let the two vectors being added be the sides of a parallelogram. The resultant is then the diagonal of the parallelogram.* The initial point of the resultant is the *common initial point of the two vectors being added.* In using this method, the vectors are first placed tail to tail. This method is illustrated in the following example.

◀ EXAMPLE 3   The addition of vectors **A** and **B** is shown in Fig. 9.5.

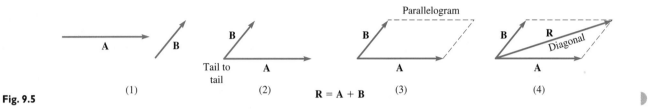

Fig. 9.5

Scalar Multiple of Vector   If vector **C** is in the same direction as vector **A** and **C** has a magnitude *n* times that of **A**, then **C** = *n***A**, where *the vector n***A** *is called the* **scalar multiple** *of vector* **A**. This means that 2**A** is a vector that is twice as long as **A** but is *in the same direction*. Note carefully that only the magnitudes of **A** and 2**A** are different, and their directions are the same. The addition of scalar multiples of vectors is illustrated in the following example.

◀ EXAMPLE 4   For vectors **A** and **B** in Fig. 9.6, find vector 3**A** + 2**B**.

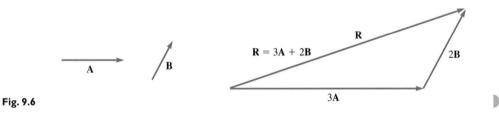

Fig. 9.6

Subtraction of Vectors   Vector **B** is subtracted from vector **A** by reversing the direction of **B** and proceeding as in vector addition. Thus, **A** − **B** = **A** + (−**B**), where *the minus sign indicates that*
NOTE ▶ *vector* −**B** *has the opposite direction of vector* **B.** Vector subtraction is illustrated in the following example.

◀ EXAMPLE 5   For vectors **A** and **B** in Fig. 9.7, find vector 2**A** − **B**.

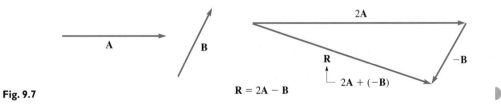

Fig. 9.7

Among the most important applications of vectors is that of the forces acting on a structure or on an object. The next example shows the addition of forces by using the parallelogram method.

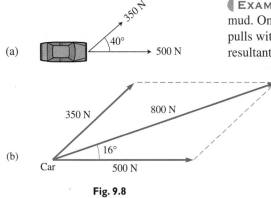

(a)

(b)

Car   500 N

Fig. 9.8

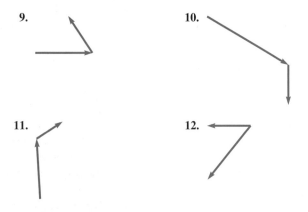

◀ EXAMPLE 6 Two persons pull horizontally on ropes attached to a car mired in mud. One person pulls with a force of 500 N directly to the right, and the other person pulls with a force of 350 N at 40° from the first force, as shown in Fig. 9.8(a). Find the resultant force on the car.

We make a scale drawing of the forces as shown in Fig. 9.8(b), measuring the magnitudes of the forces with a ruler and the angles with a protractor. (The scale drawing of the forces is made larger and with a different scale than that in Fig. 9.8(a) in order to get better accuracy.) We then complete the parallelogram and draw in the diagonal that represents the resultant force. Finally, we find that the resultant force is about 800 N and that it acts at an angle of about 16° from the first force.

▶

Two other important vector quantities are *velocity* and *displacement*. Velocity as a vector is illustrated in Example 1. *The* **displacement** *of an object is the change in its position. Displacement is given by the distance from a reference point and the angle from a reference direction.* The following example illustrates the difference between *distance* and *displacement*.

◀ EXAMPLE 7 A jet travels due east from Belfast, N. Ireland, for 120 km and then turns 70° north of east and travels another 170 km to Edinburgh, Scotland. Find the displacement of Edinburgh from Belfast.

We make a scale drawing in Fig. 9.9 to show the route taken by the jet. Measuring distances with a ruler and angles with a protractor, we find that Edinburgh is about 240 km from Belfast, at an angle of about 42° north of east. By giving both the magnitude and *the direction,* we have given the displacement.

If the jet returned directly from Edinburgh to Belfast, its *displacement* from Belfast would be *zero,* although it traveled a *distance* of 530 km.

▶

Edinburgh

240 km

170 km

42°   70°

Belfast   120 km

Fig. 9.9

EXERCISES **9.1**

In Exercises 1–4, find the resultant vectors if the given changes are made in the indicated examples of this section.

1. In Example 2, what is the resultant of the three vectors if the direction of vector **A** is reversed?

2. In Example 4, for vectors **A** and **B**, what is vector 2**A** + 3**B**?

3. In Example 5, for vectors **A** and **B**, what is vector 2**B** − **A**?

4. In Example 6, if 20° replaces 40°, what is the resultant force?

(W) *In Exercises 5–8, determine whether a scalar or a vector is described in (a) and (b). Explain your answers.*

5. **(a)** A soccer player runs 15 m from the center of the field.
(b) A soccer player runs 15 m from the center of the field toward the opponents' goal.

6. **(a)** A small-craft warning reports winds of 35 km/h.
(b) A small-craft warning reports winds out of the north at 35 km/h.

7. **(a)** An arm of an industrial robot pushes with a 10-N force downward on a part.
(b) A part is being pushed with a 10-N force by an arm of an industrial robot.

8. **(a)** A ballistics test shows that a bullet hit a wall at a speed of 100 m/s.
(b) A ballistics test shows that a bullet hit a wall at a speed of 100 m/s perpendicular to the wall.

In Exercises 9–12, add the given vectors by drawing the appropriate resultant. Use the parallelogram method in Exercises 11 and 12.

9.

10.

11.

12.

In Exercises 13–16, draw the given vectors and find their sum graphically. The magnitude is shown first, followed by the direction as an angle in standard position.

13. 3.6 cm, 0°; 4.3 cm, 90°   **14.** 2.3 cm, 45°; 5.2 cm, 120°

15. 6.0 cm, 150°; 1.8 cm, 315°   **16.** 7.5 cm, 240°; 2.3 cm, 30°

In Exercises 17–36, find the indicated vector sums and differences with the given vectors by means of diagrams. (You might find graph paper to be helpful.)

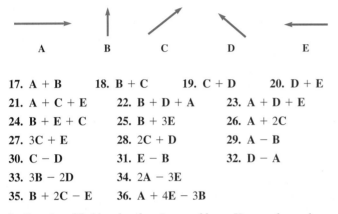

| A | B | C | D | E |

17. A + B   **18.** B + C   **19.** C + D   **20.** D + E

21. A + C + E   **22.** B + D + A   **23.** A + D + E

24. B + E + C   **25.** B + 3E   **26.** A + 2C

27. 3C + E   **28.** 2C + D   **29.** A − B

30. C − D   **31.** E − B   **32.** D − A

33. 3B − 2D   **34.** 2A − 3E

35. B + 2C − E   **36.** A + 4E − 3B

In Exercises 37–44, solve the given problems. Use a ruler and protractor as in Examples 6 and 7.

37. Two forces that act on an airplane wing are called the *lift* and the *drag*. Find the resultant of these forces acting on the airplane wing in Fig. 9.10.

Lift = 4200 N

Drag = 1600 N

Fig. 9.10

38. Two electric charges create an electric field intensity, a vector quantity, at a given point. The field intensity is 30 kN/C to the right and 60 kN/C at an angle of 45° above the horizontal to the right. Find the resultant electric field intensity at this point.

39. A ski tow is moving skiers vertically upward at 24 m/min and horizontally at 44 m/min. What is the velocity of the tow?

40. A small plane travels at 180 km/h in still air. It is headed due south in a wind of 50 km/h from the northeast. What is the resultant velocity of the plane?

41. A driver takes the wrong road at an intersection and travels 4 km north, then 6 km east, and finally 10 km to the southeast to reach the home of a friend. What is the displacement of the friend's home from the intersection?

42. A ship travels 20 km in a direction of 30° south of east and then turns due south for another 40 km. What is the ship's displacement from its initial position?

43. Three ropes hold a helium-filled balloon in place, but two of the ropes break. The remaining rope holds the balloon with a tension of 510 N at an angle of 80° with the ground due to a wind. The weight (a vertical force) of the balloon and contents is 400 N, and the upward buoyant force is 900 N. The wind creates a horizontal force of 90 N on the balloon. What is the resultant force on the balloon?

44. A crate weighing 100 N is suspended by two ropes. The force in one rope is 70 N and is directed to the left at an angle of 60° above the horizontal. What must be the other force in order that the resultant force (including the weight) on the crate is zero?

9.2 COMPONENTS OF VECTORS

Using diagrams is useful in developing an understanding of vectors. However, unless the diagrams are drawn with great care, the results we get are not too accurate. Therefore, other methods are needed in order to get more accurate results.

In this section, we show how a given vector can be made to be the sum of two other vectors, with any required degree of accuracy. In the next section, we show how this lets us add vectors to get their sum with the required accuracy in the result.

Two vectors that, when added together, have a resultant equal to the original vector, are called **components** *of the original vector.* In the illustration of the boat in Section 9.1, the velocities of 8 km/h across the river and 6 km/h downstream are components of the 10 km/h vector directed at the angle θ.

Resolving a Vector into Components

Certain components of a vector are of particular importance. If the initial point of a vector is placed at the origin of a rectangular coordinate system and its direction is given by an angle in standard position, we may find its *x*- and *y*-**components.** *These components are vectors directed along the axes that, when added together, equal the given vector.* The initial points of these components are at the origin, and the terminal points are at the points where perpendicular lines from the terminal point of the given vector cross the axes. *Finding these component vectors is called* **resolving** *the vector into its components.*

◀ EXAMPLE 1   Find the *x*- and *y*-components of the vector **A** shown in Fig. 9.11. The magnitude of **A** is 7.25.

From the figure, we see that A_x, the magnitude of the *x*-component $\mathbf{A}_x$, is related to **A** by

$$\frac{A_x}{A} = \cos 62.0°$$

$$A_x = A \cos 62.0°$$

In the same way, A_y, the magnitude of the *y*-component $\mathbf{A}_y$, is related to **A** ($\mathbf{A}_y$ could be placed along the vertical dashed line) by

$$\frac{A_y}{A} = \sin 62.0°$$

$$A_y = A \sin 62.0°$$

From these relations, knowing that $A = 7.25$, we have

$$A_x = 7.25 \cos 62.0° = 3.40$$
$$A_y = 7.25 \sin 62.0° = 6.40$$

This means that the *x*-component is directed along the *x*-axis to the right and has a magnitude of 3.40. Also, the *y*-component is directed along the *y*-axis upward and its magnitude is 6.40. These two component vectors can replace vector **A,** since the effect they have is the same as **A.**   ▶

◀ EXAMPLE 2   Resolve a vector 14.4 units long and directed at an angle of 126.0° into its *x*- and *y*-components. See Fig. 9.12.

Placing the initial point of the vector at the origin and putting the angle in standard position, note that the vector directed along the *x*-axis, $\mathbf{V}_x$, is related to the vector **V** of magnitude *V* by

┌─ magnitude of vector

$$V_x = V \cos 126.0°$$

└── standard position angle

Since the vector directed along the *y*-axis, $\mathbf{V}_y$, could also be placed along the vertical dashed line, it is related to the vector **V** by

$$V_y = V \sin 126.0°$$

Thus, the vectors $\mathbf{V}_x$ and $\mathbf{V}_y$ have the magnitudes

$$V_x = 14.4 \cos 126.0° = -8.46 \qquad V_y = 14.4 \sin 126.0° = 11.6$$

Therefore, we have resolved the given vector into two components: one, directed along the negative *x*-axis, of magnitude 8.46, and the other, directed along the positive *y*-axis, of magnitude 11.6.

It is also possible to use the reference angle, as long as *the proper sign is attached to each component.* In this case, the reference angle is 54.0°, and therefore

$$V_x = -14.4 \cos 54.0° = -8.46 \qquad V_y = +14.4 \sin 54.0° = 11.6$$

└─ directed along negative *x*-axis   └─ directed along positive *y*-axis

The minus sign shows that the *x*-component is directed to the left.   ▶

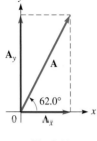

Fig. 9.11

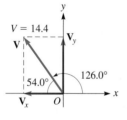

Fig. 9.12

From Examples 1 and 2, we can see that the steps used in finding the x- and y-components of a vector are as follows:

STEPS USED IN FINDING THE x- AND y-COMPONENTS OF A VECTOR

1. *Place vector* **A** *such that* θ *is in standard position.*
2. *Calculate* A_x *and* A_y *from* $A_x = A \cos \theta$ *and* $A_y = A \sin \theta$. We may use the reference angle if we note the direction of the component.
3. *Check the components* to see if each is in the correct direction and has a magnitude that is proper for the reference angle.

◀ **EXAMPLE 3** Resolve vector **A,** of magnitude 375.4 and direction $\theta = 205.32°$, into its x- and y-components. See Fig. 9.13.

By placing **A** such that θ is in standard position, we see that

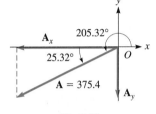

Fig. 9.13

$$A_x = A \cos 205.32° = 375.4 \cos 205.32° = -339.3$$

and

$$A_y = A \sin 205.32° = 375.4 \sin 205.32° = -160.5$$

directed along negative axis

The angle $205.32°$ places the vector in the third quadrant, and each of the components is directed along the negative axis. This must be the case for a third-quadrant angle. Also, the reference angle is $25.32°$, and we see that the magnitude of A_x is greater than the magnitude of A_y, which must be true for a reference angle that is less than $45°$. ▶

◀ **EXAMPLE 4** The tension **T** in a cable supporting the sign shown in Fig. 9.14(a) is 85.0 N. If the cable makes an angle of $53.5°$ with the horizontal, find the horizontal and vertical components of the tension.

The tension is the force the cable exerts on the sign. Showing the tension in Fig. 9.14(b) we see that

$$T_y = T \sin 53.5° = 85.0 \sin 53.5°$$
$$= 68.3 \text{ N}$$
$$T_x = T \cos 53.5° = 85.0 \cos 53.5°$$
$$= 50.6 \text{ N}$$

Note that $T_y > T_x$, which should be the case for an acute angle greater than $45°$. ▶

Fig. 9.14 (a)   (b)

EXERCISES 9.2

In Exercises 1 and 2, find the component vectors if the given changes are made in the indicated examples of this section.

1. In Example 2, find the components if $126.0°$ is changed to $216.0°$.

2. In Example 3, find the components if $205.32°$ is changed to $295.32°$.

In Exercises 3–6, find the horizontal and vertical components of the vectors shown in the given figures. In each, the magnitude of the vector is 750.

3.

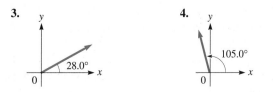

5.

6.

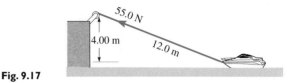

In Exercises 7–16, find the x- and y-components of the given vectors by use of the trigonometric functions. The magnitude is shown first, followed by the direction as an angle in standard position.

7. 8.60 N, $\theta = 68.0°$

8. 9750 N, $\theta = 243.0°$

9. 76.8 m/s, $\theta = 145.0°$

10. 0.0998 dm/s, $\theta = 296.0°$

11. 9040 mm/s², $\theta = 283.3°$

12. 16.4 cm/s², $\theta = 156.5°$

13. 2.65 mN, $\theta = 197.3°$

14. 678 N, $\theta = 22.5°$

15. 0.8734 dm, $\theta = 157.83°$

16. 509.4 m, $\theta = 221.87°$

In Exercises 17–28, find the required horizontal and vertical components of the given vectors.

17. A nuclear submarine approaches the surface of the ocean at 25.0 km/h and at an angle of 17.3° with the surface. What are the components of its velocity? See Fig. 9.15.

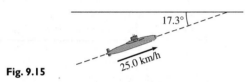

Fig. 9.15

18. Water is flowing downhill at 8.0 m/s through a pipe that is at an angle of 66.4° with the horizontal. What are the components of its velocity?

19. A car is being unloaded from a ship. It is supported by a cable from a crane and guided into position by a horizontal rope. If the tension in the cable is 12 400 N and the cable makes an angle of 3.5° with the vertical, what are the weight W of the car and the tension T in the rope? (The weight of the cable is negligible to that of the car.) See Fig. 9.16.

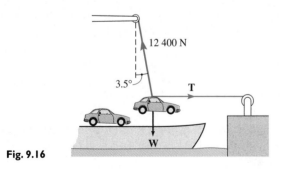

Fig. 9.16

20. A jet is 145 km at a position 37.5° north of east of Tahiti. What are the components of the jet's displacement from Tahiti?

21. The end of a robot arm is 1.20 m on a line 78.6° above the horizontal from the point where it does a weld. What are the components of the displacement from the end of the robot arm to the welding point?

22. The tension in a rope attached to a boat is 55.0 N. The rope is attached to the boat 4.00 m below the level at which it is being drawn in. At the point where there are 12.0 m of rope out, what force tends to bring the boat toward the wharf, and what force tends to raise the boat? See Fig. 9.17.

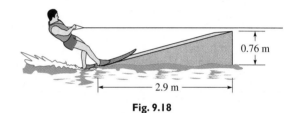

Fig. 9.17

23. A person applies a force of 210 N perpendicular to a jack handle that is at an angle of 25° above the horizontal. What are the horizontal and vertical components of the force?

24. A water skier is pulled up the ramp shown in Fig. 9.18 at 8.5 m/s. How fast is the skier rising when leaving the ramp?

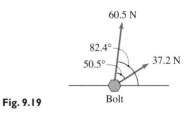

Fig. 9.18

25. A wheelbarrow is being pushed with a force of 310 N directed along the handles, which are at an angle of 20° above the horizontal. What is the effective force for moving the wheelbarrow forward?

26. Vertical wind sheer in the lowest 100 m above the ground is of great importance to aircraft when taking off or landing. It is defined as the rate at which the wind velocity changes per meter above ground. If the vertical wind sheer at 50 m above the ground is 0.75 (km/h)/m directed at angle of 40° above the ground, what are its vertical and horizontal components?

27. At one point, the *Pioneer* space probe was entering the gravitational field of Jupiter at an angle of 2.55° below the horizontal with a velocity of 29 860 km/h. What were the components of its velocity?

28. Two upward forces are acting on a bolt. One force of 60.5 N acts at an angle of 82.4° above the horizontal, and the other force of 37.2 N acts at an angle of 50.5° below the first force. What is the total upward force on the bolt? See Fig. 9.19.

Fig. 9.19

9.3 VECTOR ADDITION BY COMPONENTS

Now that we have developed the meaning of the components of a vector, we are able to add vectors to any degree of required accuracy. To do this, we use the components of the vector, the Pythagorean theorem, and the tangent of the standard-position angle of the resultant. In the following example, two vectors at right angles are added.

EXAMPLE 1   Add vectors **A** and **B**, with $A = 14.5$ and $B = 9.10$. The vectors are at right angles, as shown in Fig. 9.20.

We can find the magnitude R of the resultant vector **R** by use of the Pythagorean theorem. This leads to

$$R = \sqrt{A^2 + B^2} = \sqrt{(14.5)^2 + (9.10)^2}$$
$$= 17.1$$

We now determine the direction of the resultant vector **R** by specifying its direction as the angle θ in Fig. 9.20, that is, the angle that **R** makes with vector **A.** Therefore, we have

$$\tan \theta = \frac{B}{A} = \frac{9.10}{14.5}$$
$$\theta = 32.1°$$

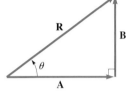

Fig. 9.20

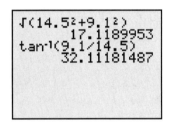

Fig. 9.21

Calculation of the values of R and θ is shown in the calculator window in Fig. 9.21. As we showed in Chapter 4, it is not necessary to record or store the value of the quotient $9.10/14.5$, as we can calculate the value of θ directly on the calculator as $\tan^{-1}(9.10/14.5)$.

Therefore, we see that **R** is a vector of magnitude $R = 17.1$ and in a direction $32.1°$ from vector **A.**

Note that Fig. 9.20 shows vectors **A** and **B** as horizontal and vertical, respectively. This means they are the horizontal and vertical components of the resultant vector **R.** However, we would find the resultant of any two vectors *at right angles* in the same way.

NOTE ▶   If the vectors being added are not at right angles, first *place each vector with its tail at the origin.* Next, *resolve each vector into its x- and y-components.* Then *add the x-components* and *add the y-components to find the x- and y-components of the resultant.* Then, by using the Pythagorean theorem, *find the magnitude of the resultant*, and by use of the tangent, *find the angle that gives the direction of the resultant.*

CAUTION ▶   *Remember, a vector is not completely specified unless both its magnitude and its direction are specified.* A common error is to determine the magnitude, but not to find the angle θ that is used to define its direction.

The following examples illustrate the addition of vectors by first finding the components of the given vectors.

◀ EXAMPLE 2   Find the resultant of two vectors **A** and **B** such that $A = 1200$, $\theta_A = 270.0°$, $B = 1750$, and $\theta_B = 115.0°$.

First place the vectors on a coordinate system with the tail of each at the origin as shown in Fig. 9.22(a). Then resolve each vector into its x- and y-components, as shown in Fig. 9.22(b) and as calculated below. (Note that **A** is vertical and has no horizontal component.) Next, the components are combined, as in Fig. 9.22(c) and as calculated. Finally, the magnitude of the resultant and the angle θ (to determine the direction), as shown in Fig. 9.22(d), are calculated.

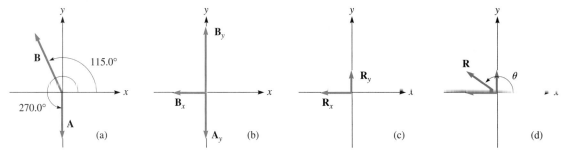

Fig. 9.22

$$A_x = A \cos 270.0° = 1200 \cos 270.0° = 0$$
$$B_x = B \cos 115.0° = 1750 \cos 115.0° = -739.6$$
$$A_y = A \sin 270.0° = 1200 \sin 270.0° = -1200$$
$$B_y = B \sin 115.0° = 1750 \sin 115.0° = 1586$$

⎤ Fig. 9.22(b)

$$R_x = A_x + B_x = 0 - 739.6 = -739.6$$
$$R_y = A_y + B_y = -1200 + 1586 = 386$$

⎤ Fig. 9.22(c)

$$R = \sqrt{R_x^2 + R_y^2} = \sqrt{(-739.6)^2 + 386^2} = 834$$
$$\tan \theta = \frac{R_y}{R_x} = \frac{386}{-739.6} \qquad \theta = 152.4° \longleftarrow 180° - 27.6°$$

⎤ Fig. 9.22(d)

Thus, the resultant has a magnitude of 834 and is directed at a standard-position angle of 152.4°. In finding θ from a calculator,

CAUTION ▶   *the calculator display shows an angle of* $-27.6°$. *However,* θ *is a second-quadrant angle, since* R_x *is negative and* R_y *is positive.*

Therefore, we must use 27.6° as a reference angle. For this reason, it is usually advisable to *find the reference angle first* by disregarding the signs of R_x and R_y when finding θ. Thus,

$$\tan \theta_{\text{ref}} = \left| \frac{R_y}{R_x} \right| = \frac{386}{739.6} \qquad \theta_{\text{ref}} = 27.6°$$

The values shown in this example have been rounded off. In using the calculator, R_x and R_y are each calculated in one step and stored for the calculation of R and θ, and we will show these steps in the next example. However, here we wished to show the individual steps and results to more clearly show the method.

When we found R_y, we saw that we could do a vector addition as a scalar addition. Also, since the magnitude of the components and the resultant are much smaller than either of the original vectors, this vector addition would have been difficult to do accurately by means of a diagram.

◗

◀ EXAMPLE 3   Find the resultant **R** of the two vectors shown in Fig. 9.23(a), **A** of magnitude 8.075 and standard-position angle of 57.26° and **B** of magnitude 5.437 and standard-position angle of 322.15°.

In Fig. 9.23(b), we show the components of vectors **A** and **B,** and then in Fig. 9.23(c), we show the resultant and its components.

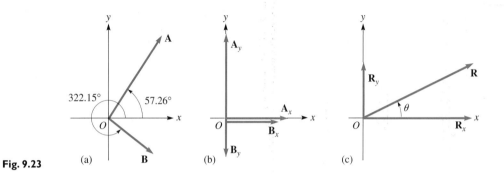

Fig. 9.23   (a)   (b)   (c)

Vector	Magnitude	Angle	x-Component		y-Component	
A	8.075	57.26°	$A_x = 8.075 \cos 57.26°$	$= 4.367$	$A_y = 8.075 \sin 57.26°$	$= 6.792$
B	5.437	322.15°	$B_x = 5.437 \cos 322.15°$	$= 4.293$	$B_y = 5.437 \sin 322.15°$	$= -3.336$
R			$R_x = A_x + B_x$	$= 8.660$	$R_y = A_y + B_y$	$= 3.456$

$$R = \sqrt{R_x^2 + R_y^2} = \sqrt{(8.660)^2 + (3.456)^2} = 9.324$$

$$\theta = \tan^{-1}\frac{R_y}{R_x} = \tan^{-1}\left(\frac{3.456}{8.660}\right) = 21.76° \longleftarrow \text{don't forget the direction}$$

The resultant vector is 9.324 units long and is directed at a standard-position angle of 21.76°, as shown in Fig. 9.23(c). We know that the resultant is in the first quadrant since both R_x and R_y are positive.

In the table above, we have shown rounded-off values for each result. However, when using a calculator, it is necessary only to calculate R_x and R_y in one step each, store these values, and use them to calculate R and θ. This is shown in the calculator window in Fig. 9.24. The lines shown above the window are those that are displaced in proceeding with the solution.

In the calculator solution, $C = R_x$ and $D = R_y$. In finding this solution, we see that both R_x and R_y are positive, which means that θ is a first-quadrant angle.   ▶

```
8.075cos(57.26)+
5.437cos(322.15)
           8.660346589
Ans→C
           8.660346589
8.075sin(57.26)+
5.437sin(322.15)
           3.456028938
Ans→D
           3.456028938
√(C²+D²)
           9.324469908
tan⁻¹(D/C)
           21.75514218
```

Fig. 9.24

Some general formulas can be derived from the previous examples. For a given vector **A,** directed at an angle θ, of magnitude A, and with components A_x and A_y, we have the following relations:

$A_x = A \cos \theta \quad A_y = \sin \theta$	(9.1)
$A = \sqrt{A_x^2 + A_y^2}$	(9.2)
$\theta_{\text{ref}} = \tan^{-1}\dfrac{\lvert A_y \rvert}{\lvert A_x \rvert}$	(9.3)

The value of θ is found by using the reference angle from Eq. (9.3) and the quadrant in which the resultant lies.

From the previous examples, we see that the following procedure is used for adding vectors.

PROCEDURE FOR ADDING VECTORS BY COMPONENTS

1. *Add the x-components of the given vectors to obtain R_x.*

2. *Add the y-components of the given vectors to obtain R_y.*

3. *Find the magnitude of the resultant* **R.** *Use Eq. (9.2) in the form*

$$R = \sqrt{R_x^2 + R_y^2}$$

4. *Find the standard-position angle θ for the resultant* **R.** *First, find the reference angle θ_{ref} for the resultant* **R** *by using Eq. (9.3) in the form*

$$\theta_{ref} = \tan^{-1} \frac{|R_y|}{|R_x|}$$

Some calculators have a specific feature for adding vectors.

See Appendix C for a graphing calculator program ADDVCTR. It can be used to add vectors.

◀ **EXAMPLE 4** Find the resultant of the three given vectors in Fig. 9.25. The magnitudes of these vectors are $T = 422$, $U = 405$, and $V = 210$.

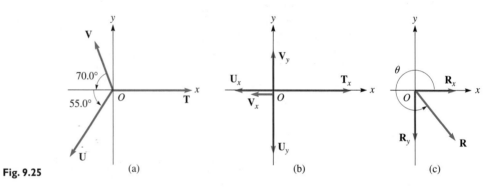

Fig. 9.25   (a)   (b)   (c)

We could change the given angles to standard-position angles. However, we will use the given angles, *being careful to give the proper sign to each component.* In the following table, we show the *x*- and *y*-components of the given vectors, and the sums of these components give us the components of **R.**

```
422cos(0)-405cos
(55)-210cos(70)
       117.8773132
Ans→A
       117.8773132
422sin(0)-405sin
(55)+210sin(70)
      -134.4211276
Ans→B
      -134.4211276
√(A²+B²)
       178.7850679
tan⁻¹(abs(B)/A)
       48.75165091
```

Fig. 9.26

Vector	Magnitude	Ref. Angle	x-Component	y-Component
T	422	0°	422 cos 0° = 422.0	422 sin 0° = 0.0
U	405	55.0°	−405 cos 55.0° = −232.3	−405 sin 55.0° = −331.8
V	210	70.0°	−210 cos 70.0° = −71.8	+210 sin 70.0° = 197.3
R			117.9	note the variation −134.5 due to rounding off

In the calculator solution shown in Fig. 9.26, A = R_x, B = R_y, and abs means absolute value. From this display, we have

$$R = 179 \quad \text{and} \quad \theta_{ref} = 48.8°$$

Since R_x is positive and R_y is negative, we know that θ is **a fourth-quadrant angle.** Therefore, to find θ we subtract θ_{ref} from 360°. This means

$$\theta = 360° - 48.8° = 311.2°$$

EXERCISES 9.3

In Exercises 1 and 2, find the resultant vectors if the given changes are made in the indicated examples of this section.

1. In Example 2, find the resultant if θ_A is changed to $0°$.

2. In Example 3, find the resultant if θ_B is changed to $232.15°$.

*In Exercises 3–6, vectors **A** and **B** are at right angles. Find the magnitude and direction (the angle from vector **A**) of the resultant.*

3. $A = 14.7$
   $B = 19.2$

4. $A = 592$
   $B = 195$

5. $A = 3.086$
   $B = 7.143$

6. $A = 1734$
   $B = 3297$

In Exercises 7–14, with the given sets of components, find R and θ.

7. $R_x = 5.18, R_y = 8.56$

8. $R_x = 89.6, R_y = -52.0$

9. $R_x = -0.982, R_y = 2.56$

10. $R_x = -729, R_y = -209$

11. $R_x = -646, R_y = 2030$

12. $R_x = -31.2, R_y = -41.2$

13. $R_x = 0.6941, R_y = -1.246$

14. $R_x = 7.627, R_y = -6.353$

In Exercises 15–28, add the given vectors by using the trigonometric functions and the Pythagorean theorem.

15. $A = 18.0, \theta_A = 0.0°$
   $B = 12.0, \theta_B = 27.0°$

16. $F = 154, \theta_F = 90.0°$
   $T = 128, \theta_T = 43.0°$

17. $C = 5650, \theta_C = 76.0°$
   $D = 1280, \theta_D = 160.0°$

18. $A = 6.89, \theta_A = 123.0°$
   $B = 29.0, \theta_B = 260.0°$

19. $A = 9.821, \theta_A = 34.27°$
   $B = 17.45, \theta_B = 752.50°$

20. $E = 1.653, \theta_E = 36.37°$
   $F = 0.9807, \theta_F = 253.06°$

21. $A = 21.9, \theta_A = 236.2°$
   $B = 96.7, \theta_B = 11.5°$
   $C = 62.9, \theta_C = 143.4°$

22. $R = 6300, \theta_R = 189.6°$
   $F = 1760, \theta_F = 320.1°$
   $T = 3240, \theta_T = 75.4°$

23. $U = 0.364, \theta_U = 175.7°$
   $V = 0.596, \theta_V = 319.5°$
   $W = 0.129, \theta_W = 100.6°$

24. $A = 6.4, \theta_A = 126°$
   $B = 5.9, \theta_B = 238°$
   $C = 3.2, \theta_C = 72°$

25. The vectors shown in Fig. 9.27

Fig. 9.27

26. The vectors shown in Fig. 9.28

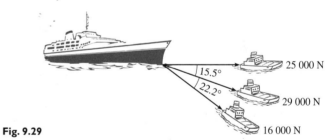

Fig. 9.28

27. In order to move an ocean liner into the channel, three tugboats exert the forces shown in Fig. 9.29. What is the resultant of these forces?

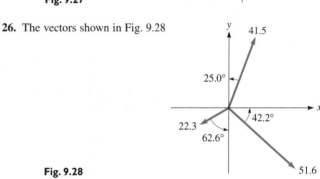

Fig. 9.29

28. A naval cruiser on maneuvers travels 54.0 km at $18.7°$ west of north, then turns and travels 64.5 km at $15.6°$ south of east, and finally turns to travel 72.4 km at $38.1°$ east of south. Find its displacement from its original position. See Fig. 9.30.

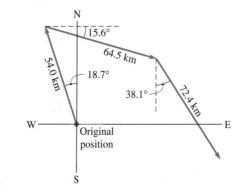

Fig. 9.30

9.4 APPLICATIONS OF VECTORS

In Section 9.1, we introduced the important vector quantities of force, velocity, and displacement, and we found vector sums by use of diagrams. Now we can use the method of Section 9.3 to find sums of these kinds of vectors and others and to use them in various types of applications.

(Top view)

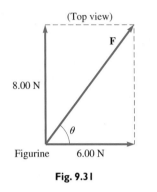

Fig. 9.31

EXAMPLE 1 In centering a figurine on a table, two persons apply forces on it. These forces are at right angles and have magnitudes of 6.00 N and 8.00 N. The angle between their lines of action is 90.0°. What is the resultant of these forces on the figurine?

By means of an appropriate diagram (Fig. 9.31), we may better visualize the actual situation. Note that a good choice of axes (unless specified, it is often convenient to choose the x- and y-axes to fit the problem) is to have the x axis in the direction of the 6.00-N force and the y-axis in the direction of the 8.00-N force. (This is possible since the angle between them is 90°.) With this choice, note that the two given forces will be the x- and y-components of the resultant. Therefore, we arrive at the following results:

$$F_x = 6.00 \text{ N}, \ F_y = 8.00 \text{ N}$$
$$F = \sqrt{(6.00)^2 + (8.00)^2} = 10.0 \text{ N}$$
$$\theta = \tan^{-1} \frac{F_y}{F_x} = \tan^{-1} \frac{8.00}{6.00}$$
$$= 53.1°$$

The metric unit of force, the newton (N), is named for the great English mathematician and physicist Sir Isaac Newton (1642–1727). His name will appear on other pages of this text, as some of his many accomplishments are noted.

The resultant has a magnitude of 10.0 N and acts at an angle of 53.1° from the 6.00-N force.

Solving a Word Problem

EXAMPLE 2 A ship sails 32.50 km due east and then turns 41.25° north of east. After sailing another 16.18 km, where is it with reference to the starting point?

In this problem, we are to find the resultant displacement of the ship from the two given displacements. The problem is diagrammed in Fig. 9.32, where the first displacement is labeled vector **A** and the second as vector **B.**

Since east corresponds to the positive x-direction, note that the x-component of the resultant is **A** + **B**$_x$ and the y-component of the resultant is **B**$_y$. Therefore, we have the following results:

$$R_x = A + B_x = 32.50 + 16.18 \cos 41.25°$$
$$= 32.50 + 12.16$$
$$= 44.66 \text{ km}$$
$$R_y = 16.18 \sin 41.25° = 10.67 \text{ km}$$
$$R = \sqrt{(44.66)^2 + (10.67)^2} = 45.92 \text{ km}$$
$$\theta = \tan^{-1} \frac{10.67}{44.66}$$
$$= 13.44°$$

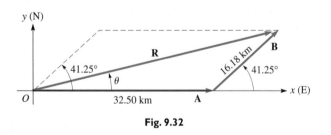

Fig. 9.32

Therefore, the ship is 45.92 km from the starting point, in a direction 13.44° north of east.

Solving a Word Problem

An aircraft's *heading* is the direction in which it is pointed. Its *air speed* is the speed at which it travels through the air surrounding it. Due to the wind, the heading and air speed, and its actual direction and speed relative to the ground, will differ.

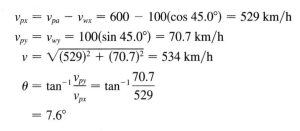

100 km/h

45.0°

θ

600 km/h

Fig. 9.33

EXAMPLE 3 An airplane headed due east is in a wind blowing from the southeast. What is the resultant velocity of the plane with respect to the surface of the earth if the velocity of the plane with respect to the air is 600 km/h and that of the wind is 100 km/h? See Fig. 9.33.

Let $\mathbf{v}_{px}$ be the velocity of the plane in the *x*-direction (east), $\mathbf{v}_{py}$ the velocity of the plane in the *y*-direction, $\mathbf{v}_{wx}$ the *x*-component of the velocity of the wind, $\mathbf{v}_{wy}$ the *y*-component of the velocity of the wind, and $\mathbf{v}_{pa}$ the velocity of the plane with respect to the air. Therefore,

$$v_{px} = v_{pa} - v_{wx} = 600 - 100(\cos 45.0°) = 529 \text{ km/h}$$
$$v_{py} = v_{wy} = 100(\sin 45.0°) = 70.7 \text{ km/h}$$
$$v = \sqrt{(529)^2 + (70.7)^2} = 534 \text{ km/h}$$
$$\theta = \tan^{-1}\frac{v_{py}}{v_{px}} = \tan^{-1}\frac{70.7}{529}$$
$$= 7.6°$$

The plane is traveling 534 km/h and is flying in a direction 7.6° north of east. From this, we observe that a plane does not necessarily head in the direction of its destination.

Equilibrium of Forces

NOTE ▶

As we have seen, an important vector quantity is the force acting on an object. One of the most important applications of vectors involves forces that are in **equilibrium.** *For an object to be in equilibrium, the net force acting on it in any direction must be zero.* This condition is satisfied if the sum of the *x*-components of the force is zero and the sum of the *y*-components of the force is also zero. The following two examples illustrate forces in equilibrium.

Solving a Word Problem

y

$\mathbf{F}_f$

x

80.0 cos 60.0°

60.0°

30.0°

30.0°

30.0°

80.0 N

Fig. 9.34

Newton's *third law of motion* states that when an object exerts a force on another object, the second object exerts on the first object a force of the same magnitude but in the opposite direction. The force exerted by the plank on the block is an illustration of this law. (Sir Isaac Newton, again. See page 272.)

EXAMPLE 4 A cement block is resting on a straight inclined plank that makes an angle of 30.0° with the horizontal. If the block weighs 80.0 N, what is the force of friction between the block and the plank?

The weight of the cement block is the force exerted on the block due to gravity. Therefore, the weight is directed vertically downward. The frictional force tends to oppose the motion of the block and is directed upward along the plank. The frictional force must be sufficient to counterbalance that component of the weight of the block that is directed down the plank for the block to be at rest (not moving). The plank itself "holds up" that component of the weight that is perpendicular to the plank. A convenient set of coordinates (see Fig. 9.34) is one with the origin at the center of the block and with the *x*-axis directed up the plank and the *y*-axis perpendicular to the plank. The magnitude of the frictional force $\mathbf{F}_f$ is given by

$$F_f = 80.0 \cos 60.0° \qquad \text{component of weight down plank equals frictional force}$$

$$= 40.0 \text{ N}$$

We have used the 60.0° angle since it is the reference angle. We could have expressed the frictional force as $F_f = 80.0 \sin 30.0°$.

Here, it is assumed that the block is small enough that we may calculate all forces as though they act at the center of the block (although we know that the frictional force acts at the surface between the block and the plank).

Solving a Word Problem

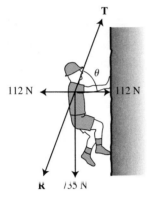

112 N ←   → 112 N

Fig. 9.35

R   735 N

EXAMPLE 5 A 735-N mountain climber suspended by a rope pushes on the side of a cliff with a horizontal force of 112 N. What is the tension **T** in the rope if the climber is in equilibrium? See Fig. 9.35.

For the climber to be in equilibrium, the tension in the rope must be equal and opposite to the resultant of the climber's weight and the force against the cliff. This means that the magnitude of the x-component of the tension is 112 N (the reaction force of the cliff—another illustration of Newton's third law—see Example 4) and the magnitude of the y-component is 735 N (see Fig. 9.35). Therefore,

$$T = \sqrt{112^2 + 735^2}$$
$$= 743 \text{ N}$$
$$\theta = \tan^{-1}\frac{735}{112}$$
$$= 81.3°$$

As with any vector, we must find the direction in which the tension acts (in this case, along the rope).

EXAMPLE 6 For a spacecraft moving in a circular path around the earth, the tangential component $\mathbf{a}_T$ and the centripetal component $\mathbf{a}_R$ of its acceleration are given by the expressions shown in Fig. 9.36. The radius of the circle through which it is moving (from the center of the earth to the spacecraft) is r, its angular velocity is ω, and its angular acceleration is α (the rate at which ω is changing).

Solving a Word Problem

While going into orbit, at one point a spacecraft is moving in a circular path 230 km above the surface of the earth. At this point, $r = 6.60 \times 10^6$ m, $\omega = 1.10 \times 10^{-3}$ rad/s, and $\alpha = 0.420 \times 10^{-6}$ rad/s^2. Calculate the magnitude of the resultant acceleration and the angle it makes with the tangential component.

$$a_R = r\omega^2 = (6.60 \times 10^6)(1.10 \times 10^{-3})^2$$
$$= 7.99 \text{ m/s}^2$$
$$a_T = r\alpha = (6.60 \times 10^6)(0.420 \times 10^{-6})$$
$$= 2.77 \text{ m/s}^2$$

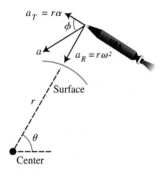

$a_T = r\alpha$
ϕ
a
$a_R = r\omega^2$
Surface
r
θ
Center

Fig. 9.36

Since a tangent line to a circle is perpendicular to the radius at the point of tangency, $\mathbf{a}_T$ is perpendicular to $\mathbf{a}_R$. Thus,

$$a = \sqrt{a_T^2 + a_R^2} = \sqrt{7.99^2 + 2.77^2}$$
$$= 8.45 \text{ m/s}^2$$
$$\phi = \tan^{-1}\frac{a_R}{a_T} = \tan^{-1}\frac{7.99}{2.77}$$
$$= 70.9°$$

The value of $a = 8.45$ m/s^2 is found directly on a calculator, without first rounding off the values of a_R and a_T. If we use the rounded-off values of 7.99 m/s^2 and 2.77 m/s^2 we would get 8.46 m/s^2.

In Exercises 1 and 2, find the necessary quantities if the given changes are made in the indicated examples of this section.

1. In Example 2, find the resultant location if the 41.25° angle is changed to 31.25°.

2. In Example 3, find the resultant velocity if the wind is from the southwest, rather than the southeast.

In Exercises 3–32, solve the given problems.

3. Two hockey players strike the puck at the same time, hitting it with horizontal forces of 34.5 N and 19.5 N that are perpendicular to each other. Find the resultant of these forces.

4. To straighten a small tree, two horizontal ropes perpendicular to each other are attached to the tree. If the tensions in the ropes are 82.3 N and 102 N, what is the resultant force on the tree?

5. In lifting a heavy piece of equipment from the mud, a cable from a crane exerts a vertical force of 6500 N, and a cable from a truck exerts a force of 8300 N at 10.0° above the horizontal. Find the resultant of these forces.

6. At a point in the plane, two electric charges create an electric field (a vector quantity) of 25.9 kN/C at 10.8° above the horizontal to the right and 12.6 kN/C at 83.4 below the horizontal to the right. Find the resultant electric field.

7. A motorboat leaves a dock and travels 1580 m due west, then turns 35.0° to the south and travels another 1640 m to a second dock. What is the displacement of the second dock from the first dock?

8. In Scandinavia, Oslo is 410 km at 8.0° north of west of Stockholm. Copenhagen is 520 km at 53.0° south of west of Stockholm. What is the displacement of Oslo from Copenhagen?

9. From a fixed point, a surveyor locates a pole at 215.6 m due east and a building corner at 358.2 m at 37.72° north of east. What is the displacement of the building from the pole?

10. A rocket is launched with a vertical component of velocity of 2840 km/h and a horizontal component of velocity of 1520 km/h. What is its resultant velocity?

11. A storm front is moving east at 22.0 km/h and south at 12.5 km/h. Find the resultant velocity of the front.

12. To move forward, a helicopter pilot tilts the helicopter forward. If the rotor generates a force of 14 000 N, with a horizontal component (thrust) of 1900 N, what is the vertical component (lift)?

13. The acceleration (a vector quantity) of gravity on a sky diver is 9.8 m/s². If the force of the wind also causes an acceleration of 1.2 m/s² at an angle of 15° above the horizontal, what is the resultant acceleration of the sky diver?

14. In an accident, a truck with momentum (a vector quantity) of 22 100 kg · m/s strikes a car with momentum of 17 800 kg · m/s from the rear. The angle between their directions of motion is 25.0°. What is the resultant momentum?

15. In an automobile safety test, a shoulder and seat belt exerts a force of 425 N directly backward and a force of 368 N backward at an angle of 20.0° below the horizontal on a dummy. If the belt holds the dummy from moving farther forward, what force did the dummy exert on the belt? See Fig. 9.37.

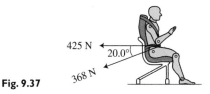

Fig. 9.37

16. Two perpendicular forces act on a ring at the end of a chain that passes over a pulley and holds an automobile engine. If the forces have the values shown in Fig. 9.38, what is the weight of the engine?

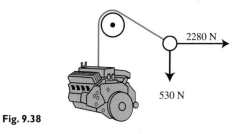

Fig. 9.38

17. A plane flies at 550 km/h into a head wind of 60 km/h at 78° with the direction of the plane. Find the resultant velocity of the plane with respect to the ground. See Fig. 9.39.

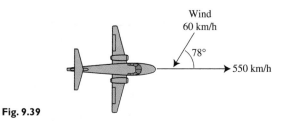

Fig. 9.39

18. A ship's navigator determines that the ship is moving through the water at 17.5 km/h with a heading of 26.3° north of east, but that the ship is actually moving at 19.3 km/h in a direction of 33.7° north of east. What is the velocity of the current?

19. A space shuttle is moving in orbit at 29 370 km/h. A satellite is launched to the rear at 190 km/h at an angle of 5.20° from the direction of the shuttle. Find the velocity of the satellite.

20. A block of ice slides down a (frictionless) ramp with an acceleration of 5.3 m/s². If the ramp makes an angle of 32.7° with the horizontal, find g, the acceleration due to gravity. See Fig. 9.40.

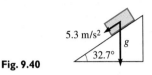

Fig. 9.40

21. A passenger on a cruise ship traveling due east at a speed of 32 km/h notes that the smoke from the ship's funnels makes an angle of 15° with the ship's wake. If the wind is from the southwest, find the speed of the wind.

22. A mine shaft goes due west 75 m from the opening at an angle of 25° below the horizontal surface. It then becomes horizontal and turns 30° north of west and continues for another 45 m. What is the displacement of the end of the tunnel from the opening?

23. A crowbar 1.5 m long is supported underneath 1.2 m from the end by a rock, with the other end under a boulder. If the crowbar is at an angle of 18° with the horizontal and a person pushes down, perpendicular to the crowbar, with a force of 240 N, what vertical force is applied to the boulder? See Fig. 9.41. (The forces are related by $1.2\mathbf{F}_1 = 0.3\mathbf{F}_2$. See Exercise 48 on page 155.)

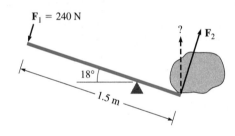

Fig. 9.41

24. A scuba diver's body is directed downstream at 75° to the bank of a river. If the diver swims at 25 m/min, and the water is moving at 5.0 m/min, what is the diver's velocity?

25. While starting up, a circular saw blade 8.20 cm in diameter is rotating at 212 rad/min and has an angular acceleration of 318 rad/min². What is the acceleration of the tip of one of the teeth? (See Example 6.)

26. A boat travels across a river, reaching the opposite bank at a point directly opposite that from which it left. If the boat travels 6.00 km/h in still water and the current of the river flows at 3.00 km/h, what was the velocity of the boat in the water?

27. In searching for a boat lost at sea, a Coast Guard cutter leaves a port and travels 75.0 km due east. It then turns 65° north of east and travels another 75.0 km, and finally turns another 65.0° toward the west and travels another 75.0 km. What is its displacement from the port?

28. A car is held stationary on a ramp by two forces. One is the force of 2140 N by the brakes, which hold it from rolling down the ramp. The other is a reaction force by the ramp of 9850 N, perpendicular to the ramp. This force keeps the car from going through the ramp. See Fig. 9.42. What is the weight of the car, and at what angle with the horizontal is the ramp inclined?

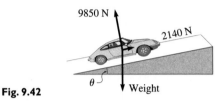

Fig. 9.42

29. A plane is moving at 75.0 m/s, and a package with weather instruments is ejected horizontally from the plane at 15.0 m/s, perpendicular to the direction of the plane. If the vertical velocity v_v (in m/s), as a function of time t (in s) of fall, is given by $v_v = 9.80t$, what is the velocity of the package after 2.00 s (before its parachute opens)?

30. A flat rectangular barge, 48.0 m long and 20.0 m wide, is headed directly across a stream at 4.5 km/h. The stream flows at 3.8 km/h. What is the velocity, relative to the riverbed, of a person walking diagonally across the barge at 5.0 km/h while facing the opposite upstream bank?

31. In Fig. 9.43, a long, straight conductor perpendicular to the plane of the paper carries an electric current i. A bar magnet having poles of strength m lies in the plane of the paper. The vectors $\mathbf{H}_i$, $\mathbf{H}_N$, and $\mathbf{H}_S$ represent the components of the magnetic intensity $\mathbf{H}$ due to the current and to the N and S poles of the magnet, respectively. The magnitudes of the components of $\mathbf{H}$ are given by

$$H_i = \frac{1}{2\pi}\frac{i}{a} \qquad H_N = \frac{1}{4\pi}\frac{m}{b^2} \qquad H_S = \frac{1}{4\pi}\frac{m}{c^2}$$

Given that $a = 0.300$ m, $b = 0.400$ m, $c = 0.300$ m, the length of the magnet is 0.500 m, $i = 4.00$ A, and $m = 2.00$ A · m, calculate the resultant magnetic intensity $\mathbf{H}$. The component $\mathbf{H}_i$ is parallel to the magnet.

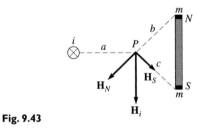

Fig. 9.43

32. Solve the problem of Exercise 31 if $\mathbf{H}_i$ is directed away from the magnet, making an angle of 10.0° with the direction of the magnet.

9.5 OBLIQUE TRIANGLES, THE LAW OF SINES

To this point, we have limited our study of triangle solution to right triangles. However, *many triangles that require solution do not contain a right angle. Such triangles are called* **oblique triangles.** We now discuss solutions of oblique triangles.

In Section 4.4, we stated that we need to know three parts, at least one of them a side, to solve a triangle. There are four possible such combinations of parts, and these combinations are as follows:

Case 1. *Two angles and one side*

Case 2. *Two sides and the angle opposite one of them*

Case 3. *Two sides and the included angle*

Case 4. *Three sides*

There are several ways in which oblique triangles may be solved, but we restrict our attention to the two most useful methods, the **law of sines** and the **law of cosines.** In this section, we discuss the law of sines and show that it may be used to solve Case 1 and Case 2.

Let ABC be an oblique triangle with sides a, b, and c opposite angles A, B, and C, respectively. By drawing a perpendicular h from B to side b, as shown in Fig. 9.44(a), note that $h/c = \sin A$ and $h/a = \sin C$, and therefore

$$h = c \sin A \quad \text{or} \quad h = a \sin C \tag{9.4}$$

This relationship also holds if one of the angles is obtuse. By drawing a perpendicular h to the extension of side b, as shown in Fig. 9.44(b), we see that $h/c = \sin A$.

Also in Fig. 9.44(b), $h/a = \sin(180° - C)$, where $180° - C$ is also the reference angle for the obtuse angle C. For a second quadrant reference angle $180° - C$, we have $\sin[180° - (180° - C)] = \sin C$. Putting this all together, $\sin(180° - C) = \sin C$. This means that

$$h = c \sin A \quad \text{or} \quad h = a \sin(180° - C) = a \sin C \tag{9.5}$$

The results are precisely the same in Eqs. (9.4) and (9.5). Setting the results for h equal to each other, we have

$$c \sin A = a \sin C$$

Dividing each side by $\sin A \sin C$ and reversing sides gives us

$$\frac{a}{\sin A} = \frac{c}{\sin C} \tag{9.6}$$

By dropping a perpendicular from A to a, we also derive the result

$$c \sin B = b \sin C$$

$$\frac{b}{\sin B} = \frac{c}{\sin C} \tag{9.7}$$

If we had dropped a perpendicular to either side a or side c, equations similar to Eqs. (9.6) and (9.7) would have resulted.

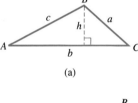

(a)

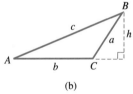

(b)

Fig. 9.44

Combining Eqs. (9.6) and (9.7), *for any triangle with sides a, b, and c, opposite angles A, B, and C, respectively,* such as the one shown in Fig. 9.45, *we have the* **law of sines:**

Law of Sines

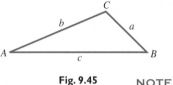

Fig. 9.45   NOTE ▶

$$\frac{a}{\sin A} = \frac{b}{\sin B} = \frac{c}{\sin C} \qquad (9.8)$$

Another form of the law of sines can be obtained by equating the reciprocals of each of the fractions in Eq. (9.8). The law of sines is a statement of proportionality between the sides of a triangle and the sines of the angles opposite them. Note that *there are actually three equations combined in Eq. (9.8).* Of these, we use the one with three known parts of the triangle and we find the fourth part. In finding the complete solution of a triangle, it may be necessary to use two of the three equations

CASE 1: TWO ANGLES AND ONE SIDE

Now we see how the law of sines is used in the solution of a triangle in which two angles and one side are known. If two angles are known, the third may be found from the fact that the sum of the angles in a triangle is 180°. At this point, we must be able to find the ratio between the given side and the sine of the angle opposite it. Then, by use of the law of sines, we may find the other two sides.

◀ **EXAMPLE 1**   Given $c = 6.00$, $A = 60.0°$, and $B = 40.0°$, find a, b, and C.

First, we can see that

$$C = 180.0° - (60.0° + 40.0°) = 80.0°$$

We now know side c and angle C, which allows us to use Eq. (9.8). Therefore, using the equation relating a, A, c, and C, we have

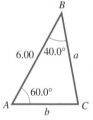

Fig. 9.46

$$\frac{a}{\sin 60.0°} = \frac{6.00}{\sin 80.0°} \quad \text{or} \quad a = \frac{6.00 \sin 60.0°}{\sin 80.0°} = 5.28$$

Now, using the equation relating b, B, c, and C, we have

$$\frac{b}{\sin 40.0°} = \frac{6.00}{\sin 80.0°} \quad \text{or} \quad b = \frac{6.00 \sin 40.0°}{\sin 80.0°} = 3.92$$

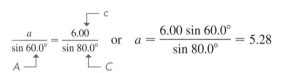

Fig. 9.47

Thus, $a = 5.28$, $b = 3.92$, and $C = 80.0°$. See Fig. 9.46. We could also have used the form of Eq. (9.8) relating a, A, b, and B in order to find b, but any error in calculating a would make b in error as well. Of course, any error in calculating C would make both a and b in error.

The calculator solution is shown in Fig. 9.47.   ▶

NOTE ▶   The solution of a triangle can be checked approximately by noting that *the smallest angle is opposite the shortest side, and the largest angle is opposite the longest side.* Note that this is the case in Example 1, where b (shortest side) is opposite B (smallest angle), and c (longest side) is opposite C (largest angle).

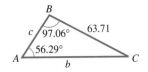

Fig. 9.48

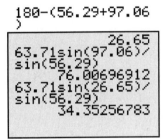

Fig. 9.49

Solving a Word Problem

The first successful helicopter was made in the United States by Igor Sikorsky in 1939.

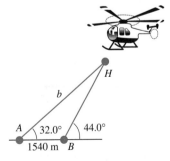

Fig. 9.50

◖ EXAMPLE 2 Solve the triangle with the following given parts: $a = 63.71$, $A = 56.29°$, and $B = 97.06°$. See Fig. 9.48.

From the figure, we see that we are to find angle C and sides b and c. We first determine angle C:

$$C = 180° - (A + B) = 180° - (56.29° + 97.06°)$$
$$= 26.65°$$

Noting the three angles, we know that c is the shortest side (C is the smallest angle) and b is the longest side (B is the largest angle). This means that the length of a is between c and b, or $c < 63.71$ and $b > 63.71$. Now using the ratio $a/\sin A$ of Eq. (9.8) (the law of sines) to find sides b and c, we have

$$\frac{b}{\sin 97.06°} = \frac{63.71}{\sin 56.29°} \quad \text{or} \quad b = \frac{63.71 \sin 97.06°}{\sin 56.29°} = 76.01$$

$$\frac{c}{\sin 26.65°} = \frac{63.71}{\sin 56.29°} \quad \text{or} \quad c = \frac{63.71 \sin 26.65°}{\sin 56.29°} = 34.35$$

Thus, $b = 76.01$, $c = 34.35$, and $C = 26.65°$. Note that $c < a$ and $b > a$, as expected. The calculator solution is shown in Fig. 9.49. ◗

If the given information is appropriate, the law of sines may be used to solve applied problems. The following example illustrates the use of the law of sines in such a problem.

◖ EXAMPLE 3 Two observers A and B sight a helicopter due east. The observers are 1540 m apart, and the angles of elevation they each measure to the helicopter are 32.0° and 44.0°, respectively. How far is observer A from the helicopter? See Fig. 9.50.

Letting H represent the position of the helicopter, we see that angle B within the triangle ABH is $180° - 44.0° = 136.0°$. This means that the angle at H within the triangle is

$$H = 180° - (32.0° + 136.0°) = 12.0°$$

Now, using the law of sines to find required side b, we have

required side → opposite known angle $\dfrac{b}{\sin 136.0°} = \dfrac{1540}{\sin 12.0°}$ ← known side opposite known angle

or

$$b = \frac{1540 \sin 136.0°}{\sin 12.0°} = 5150 \text{ m}$$

Thus, observer A is about 5150 m from the helicopter. ◗

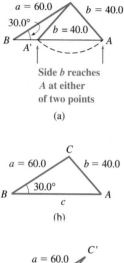

Side b reaches
A at either
of two points

(a)

(h)

(c)

Fig. 9.51

```
sin⁻¹(60sin(30)/4
0)
         48.59037789
Ans→A
         48.59037789
180-(30+A)
         101.4096221
Ans→C
         101.4096221
40sin(C)/sin(30)
         78.41903734
```

```
180-A
         131.4096221
180-(30+Ans)
         18.59037789
40sin(Ans)/sin(3
0)
         25.50401112
```

Fig. 9.52

CASE 2: TWO SIDES AND THE ANGLE OPPOSITE ONE OF THEM

For a triangle in which we know two sides and the angle opposite one of the given sides, the solution will be either *one triangle*, or *two triangles*, or even possibly *no triangle*. The following examples illustrate how each of these results is possible.

EXAMPLE 4 Solve the triangle with the following given parts: $a = 60.0$, $b = 40.0$, and $B = 30.0°$.

First, make a good scale drawing (Fig. 9.51(a)) by drawing angle B and measuring off 60.0 for a. This will more clearly show that side $b = 40.0$ will intersect side c at either position A or A'. This means there are two triangles that satisfy the given values. Using the law of sines, we solve the case for which A is an acute angle:

$$\frac{60.0}{\sin A} = \frac{40.0}{\sin 30.0°} \quad \text{or} \quad \sin A = \frac{60.0 \sin 30.0°}{40.0}$$

$$A = \sin^{-1}\left(\frac{60.0 \sin 30.0°}{40.0}\right) = 48.6°$$

$$C = 180° - (30.0° + 48.6°) = 101.4°$$

Therefore, $A = 48.6°$ and $C = 101.4°$. Using the law of sines again to find c, we have

$$\frac{c}{\sin 101.4°} = \frac{40.0}{\sin 30.0°}$$

$$c = \frac{40.0 \sin 101.4°}{\sin 30.0°} = 78.4$$

Thus, $A = 48.6°$, $C = 101.4°$, and $c = 78.4$. See Fig. 9.51(b).

The other solution is the case in which A', opposite side a, is an obtuse angle. Therefore,

$$A' = 180° - A = 180° - 48.6°$$
$$= 131.4°$$
$$C' = 180° - (30.0° + 131.4°)$$
$$= 18.6°$$

Using the law of sines to find c', we have

$$\frac{c'}{\sin 18.6°} = \frac{40.0}{\sin 30.0°}$$

$$c' = \frac{40.0 \sin 18.6°}{\sin 30.0°} = 25.5$$

This means that the second solution is $A' = 131.4°$, $C' = 18.6°$, and $c' = 25.5$. See Fig. 9.51(c).

The complete sequence for the calculator solution is shown in Fig. 9.52. The upper window shows the completion of the solution for A, C, and c. The lower window shows the solution for A', C', and c'.

◀ **EXAMPLE 5** In Example 4, if $b > 60.0$, only one solution would result. In this case, side b would intercept side c at A. It also intercepts the extension of side c, but this would require that angle B not be included in the triangle (see Fig. 9.53). Thus, only one solution may result if $b > a$.

In Example 4, there would be *no solution* if side b were not at least 30.0. If this were the case, side b would not be long enough to even touch side c. It can be seen that b must at least equal $a \sin B$. If it is just equal to $a \sin B$, there is *one solution*, a right triangle. See Figure 9.54.

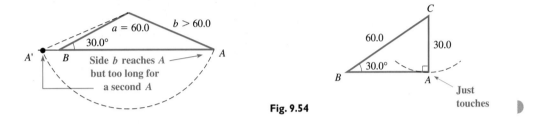

Fig. 9.53

Fig. 9.54

Just touches ▶

Ambiguous Case

Summarizing the results for Case 2 as illustrated in Examples 4 and 5, we make the following conclusions. Given sides a and b and angle A (assuming here that a and A ($A < 90°$) are corresponding parts), we have the following summary of solutions for Case 2.

SUMMARY OF SOLUTIONS:
TWO SIDES AND THE ANGLE OPPOSITE ONE OF THEM

1. **No solution** *if* $a < b \sin A$. See Fig. 9.55(a).
2. **A right triangle solution** *if* $a = b \sin A$. See Fig. 9.55(b).

CAUTION ▶ 3. **Two solutions** *if* $b \sin A < a < b$. See Fig. 9.55(c).
4. **One solution** *if* $a > b$. See Fig. 9.55(d).

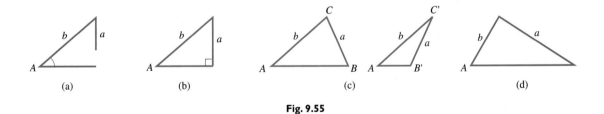

(a) (b) (c) (d)

Fig. 9.55

NOTE ▶ Note that *in order to have two solutions, we must know two sides and the angle opposite one of the sides, and the shorter side must be opposite the known angle.*

If there is *no solution,* the calculator will indicate an *error.* If the solution is a *right triangle,* the calculator will show an angle of *exactly* 90° (no extra decimal digits will be displayed.)

For the reason that two solutions may result from it, Case 2 is called the **ambiguous case.** The following example illustrates Case 2 in an applied problem.

Solving a Word Problem

See the chapter introduction.

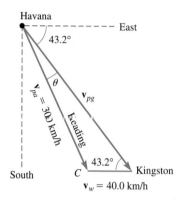

Havana

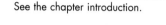

Fig. 9.56

◖ **EXAMPLE 6** Kingston, Jamaica, is 43.2° south of east of Havana, Cuba. What should be the heading of a plane from Havana to Kingston if the wind is from the west at 40.0 km/h and the plane's speed with respect to the air is 300 km/h?

The heading should be set so that the resultant of the plane's velocity with respect to the air $\mathbf{v}_{pa}$ and the velocity of the wind $\mathbf{v}_w$ will be in the direction from Havana to Kingston. This means that the resultant velocity $\mathbf{v}_{pg}$ of the plane with respect to the ground must be at an angle of 43.2° south of east from Havana.

Using the given information we draw the vector triangle shown in Fig. 9.56. In the triangle, we know that the angle at Kingston is 43.2° by noting the alternate interior angles (see page 52). By finding θ, the required heading can be found. There can be only one solution, since $v_{pa} > v_w$. Using the law of sines, we have

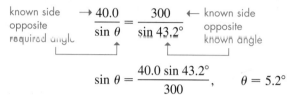

$$\sin \theta = \frac{40.0 \sin 43.2°}{300}, \qquad \theta = 5.2°$$

Therefore, the heading should be 43.2° + 5.2° = 48.4° south of east. Compare this example with Example 3 on page 273.

If we try to use the law of sines for Case 3 or Case 4, we find that we do not have enough information to complete any of the ratios. These cases can, however, be solved by the law of cosines as shown in the next section.

◖ **EXAMPLE 7** Given the three sides of a triangle $a = 5$, $b = 6$, $c = 7$, we would set up the ratios

$$\frac{5}{\sin A} = \frac{6}{\sin B} = \frac{7}{\sin C}$$

The solution cannot be found since each of the three possible equations contains two unknown angles.

EXERCISES 9.5

In Exercises 1 and 2, solve the resulting triangles if the given changes are made in the indicated examples of this section.

1. In Example 2, solve the triangle if the value of B is changed to 82.94°.

2. In Example 4, solve the triangle if the value of a is changed to 70.0.

In Exercises 3–22, solve the triangles with the given parts.

3. $a = 45.7, A = 65.0°, B = 49.0°$

4. $b = 3.07, A = 26.0°, C = 120.0°$

5. $c = 4380, A = 37.4°, B = 34.6°$

6. $a = 93.2, B = 17.9°, C = 82.6°$

7. $a = 4.601, b = 3.107, A = 18.23°$

8. $b = 3.625, c = 2.946, B = 69.37°$

9. $b = 7751, c = 3642, B = 20.73°$

10. $a = 150.4, c = 250.9, C = 76.43°$

11. $b = 0.0742, B = 51.0°, C = 3.4°$

12. $c = 729, B = 121.0°, C = 44.2°$

13. $a = 63.8, B = 58.4°, C = 22.2°$

14. $a = 0.130, A = 55.2°, B = 67.5°$

15. $b = 4384, B = 47.43°, C = 64.56°$

16. $b = 283.2, B = 13.79°, C = 76.38°$

17. $a = 5.240, b = 4.446, B = 48.13°$

18. $a = 89.45, c = 37.36, C = 15.62°$

19. $b = 2880, c = 3650, B = 31.4°$

20. $a = 0.841, b = 0.965, A = 57.1°$

21. $a = 450, b = 1260, A = 64.8°$

22. $a = 20, c = 10, C = 30°$

In Exercises 23–36, use the law of sines to solve the given problems.

23. The base of the dome of the Santa Maria del Fiore Cathedral in Florence, Italy is a regular octagon, 17.4 m on each side. Find the greatest distance across the base of the dome. (Find the length of the longest diagonal.)

24. Two ropes hold a 175-N crate as shown in Fig. 9.57. Find the tensions T_1 and T_2 in the ropes. (*Hint:* Move vectors so that they are tail to head to form a triangle. The vector sum $T_1 + T_2$ must equal 175 N for equilibrium. See page 273.)

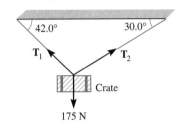

Fig. 9.57

25. Find the tension **T** in the left guy wire attached to the top of the tower shown in Fig. 9.58. (*Hint:* The horizontal components of the tensions must be equal and opposite for equilibrium. See page 273. Thus, move the tension vectors tail to head to form a triangle with a vertical resultant. This resultant equals the upward force at the top of the tower for equilibrium. This last force is not shown and does not have to be calculated.)

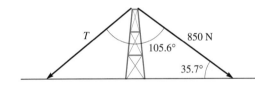

Fig. 9.58

26. Find the distance from Paris to Berlin, from Fig. 9.59.

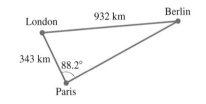

Fig. 9.59

27. Find the length of Cordelia St. in Brisbane, Queensland, Australia, from Fig. 9.60.

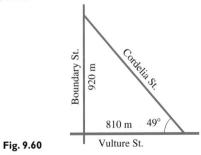

Fig. 9.60

28. When an airplane is landing at a 2510-m runway, the angles of depression to the ends of the runway are 10.0° and 13.5°. How far is the plane from the near end of the runway?

29. Find the total length of the path of the laser beam that is shown in Fig. 9.61.

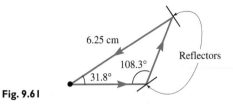

Fig. 9.61

30. In widening a highway, it is necessary for a construction crew to cut into the bank along the highway. The present angle of elevation of the straight slope of the bank is 23.0°, and the new angle is to be 38.5°, leaving the top of the slope at its present position. If the slope of the present bank is 66.0 m long, how far horizontally into the bank at its base must they dig?

31. A communications satellite is directly above the extension of a line between receiving towers A and B. It is determined from radio signals that the angle of elevation of the satellite from tower A is 89.2°, and the angle of elevation from tower B is 86.5°. See Fig. 9.62. If A and B are 1290 km apart, how far is the satellite from A? (Neglect the curvature of the earth.)

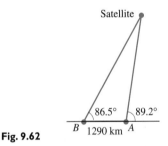

Fig. 9.62

32. Point P on the mechanism shown in Fig. 9.63 is driven back and forth horizontally. If the minimum value of angle θ is 32.0°, what is the distance between extreme positions of P? What is the maximum possible value of angle θ?

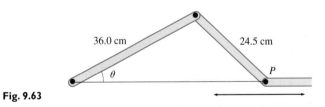

Fig. 9.63

33. A boat owner wishes to cross a river 2.60 km wide and go directly to a point on the opposite side 1.75 km downstream. The boat goes 8.00 km/h in still water, and the stream flows at 3.50 km/h. What should the boat's heading be?

34. An astronaut on the moon drives a lunar rover 16 km in the direction 60.0° north of east from the base. (a) Through what angle must the rover then be turned so that by driving 12 km farther the astronaut can turn again to return to base along a north-south line? (b) How long is the last leg of the trip? (c) Can the astronaut make it back to base if the maximum range of the rover is 40 km?

35. A motorist traveling along a level highway at 75 km/h directly toward a mountain notes that the angle of elevation of the mountain top changes from about 20° to about 30° in a 20-min period. How much closer on a direct line did the mountain top become?

36. A hillside is inclined at 23° with the horizontal. From a given point on the slope, it has been found that a vein of gold is 55 m directly below. At what angle below the hillside slope from another point downhill must a straight 65-m shaft be dug to reach the vein?

9.6 THE LAW OF COSINES

As we noted in the last section, the law of sines cannot be used for Case 3 (two sides and the included angle) and Case 4 (three sides). Therefore, in this section we develop the *law of cosines,* which can be used for Cases 3 and 4. After finding another part of the triangle by the law of cosines, we will find that it is often easier to use the law of sines to complete the solution.

Consider any oblique triangle, for example, either of the triangles shown in Fig. 9.64. For each of these triangles, $h/b = \sin A$, or $h = b \sin A$. Also, by using the Pythagorean theorem, we obtain $a^2 = h^2 + x^2$ for each triangle. Thus (with $(\sin A)^2 = \sin^2 A$),

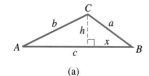
(a)

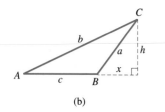
(b)

Fig. 9.64

$$a^2 = b^2 \sin^2 A + x^2 \tag{9.9}$$

In Fig. 9.64(a), note that $(c - x)/b = \cos A$, or $c - x = b \cos A$. Solving for x, we have $x = c - b \cos A$. In Fig. 9.64(b), $c + x = b \cos A$, and solving for x, we have $x = b \cos A - c$. Substituting these relations into Eq. (9.9), we obtain

$$a^2 = b^2 \sin^2 A + (c - b \cos A)^2$$

and

$$a^2 = b^2 \sin^2 A + (b \cos A - c)^2 \tag{9.10}$$

respectively. When expanded, these both give

$$a^2 = b^2 \sin^2 A + b^2 \cos^2 A + c^2 - 2bc \cos A$$
$$= b^2(\sin^2 A + \cos^2 A) + c^2 - 2bc \cos A \tag{9.11}$$

Recalling the definitions of the trigonometric functions, we know that $\sin \theta = y/r$ and $\cos \theta = x/r$. Thus, $\sin^2 \theta + \cos^2 \theta = (y^2 + x^2)/r^2$. However, $x^2 + y^2 = r^2$, which means that

$$\sin^2 \theta + \cos^2 \theta = 1 \tag{9.12}$$

This equation is valid for any angle θ, since we have made no assumptions as to any of the properties of θ. Therefore, by substituting Eq. (9.12) into Eq. (9.11), we arrive at the **law of cosines.**

Law of Cosines

$$\boxed{a^2 = b^2 + c^2 - 2bc \cos A} \tag{9.13}$$

Using the method above, we may also show that

$$\boxed{b^2 = a^2 + c^2 - 2ac \cos B}$$

and

$$\boxed{c^2 = a^2 + b^2 - 2ab \cos C}$$

CASE 3: TWO SIDES AND THE INCLUDED ANGLE

If two sides and the included angle of a triangle are known, the forms of the law of cosines show that we may directly solve for the side opposite the given angle. Then, as noted earlier, the solution may be completed using the law of sines.

◀ **EXAMPLE 1**   Solve the triangle with $a = 45.0$, $b = 67.0$, and $C = 35.0°$. See Fig. 9.65.

Since angle C is known, first solve for side c, using the law of cosines in the form $c^2 = a^2 + b^2 - 2ab \cos C$. Substituting, we have

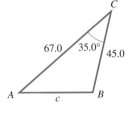

Fig. 9.65

$$c^2 = \underset{\text{unknown side opposite known angle}}{45.0^2 + 67.0^2 - 2(45.0)(67.0) \cos 35.0°}$$

$$c = \sqrt{45.0^2 + 67.0^2 - 2(45.0)(67.0) \cos 35.0°} = 39.7$$

From the law of sines, we now have

$$\frac{45.0}{\sin A} = \frac{67.0}{\sin B} = \frac{39.7}{\sin 35.0°} \quad \begin{matrix}\leftarrow \text{sides} \\ \leftarrow \text{opposite} \\ \leftarrow \text{angles}\end{matrix}$$

$$\sin A = \frac{45.0 \sin 35.0°}{39.7}, \qquad A = 40.6°$$

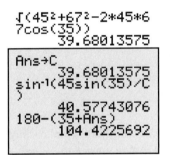

Fig. 9.66

The calculator solution is shown in Fig. 9.66. As shown, we found that $B = 104.4°$ using the fact that the sum of the angles is $180°$, rather than using the relation from the law of sines. Therefore, $c = 39.7$, $A = 40.6°$, and $B = 104.4°$.

After finding side c, solving for angle B rather than angle A, the calculator would show $B = 75.5°$. Then, when subtracting the sum of angles B and C from $180°$, we would get $A = 69.5°$. Although this appears to be correct, it is not. Since the largest angle of a triangle might be greater than $90°$, ***first solve for the smaller unknown angle.*** This smaller angle (opposite the shorter known side) cannot be greater than $90°$, and the larger unknown angle can be found by subtraction as is done in this example.   ◗

CAUTION ◗

◀ **EXAMPLE 2**   Solve the triangle with $a = 0.1762$, $c = 0.5034$, and $B = 129.20°$. See Fig. 9.67.

Again, the given parts are two sides and the included angle, or Case 3. Since B is the known angle, we use the form of the law of cosines that includes angle B. This means that we find b first.

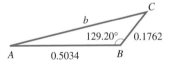

Fig. 9.67

$$b^2 = a^2 + c^2 - 2ac \cos B$$
$$= 0.1762^2 + 0.5034^2 - 2(0.1762)(0.5034) \cos 129.20°$$
$$b = 0.6297$$

From the law of sines, we have

$$\frac{0.1762}{\sin A} = \frac{0.6297}{\sin 129.20°} = \frac{0.5034}{\sin C}$$

$$\sin A = \frac{0.1762 \sin 129.20°}{0.6297}, \qquad A = 12.52°$$

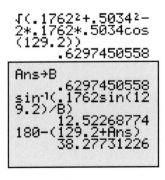

Fig. 9.68

As shown in Fig. 9.68, we complete the solution by using the fact that the sum of the angles is $180°$ to find $C = 38.28°$.

Thus, $b = 0.6297$, $A = 12.52°$, and $C = 38.28°$.   ◗

CASE 4: THREE SIDES

CAUTION ▶

Given the three sides of a triangle, we may solve for the angle opposite any of the sides using the law of cosines. In solving for an angle, *the best procedure is to find the largest angle first.* This avoids the ambiguous case if we switch to the law of sines and the triangle has an obtuse angle. *The largest angle is always opposite the longest side.* Another procedure is to use the law of cosines to find two angles.

◀ **EXAMPLE 3** Solve the triangle given the three sides $a = 49.33$, $b = 21.61$, and $c = 42.57$. See Fig. 9.69.

Since the longest side is $a = 49.33$, first solve for angle A:

$$a^2 = b^2 + c^2 - 2bc \cos A$$

$$\cos A = \frac{b^2 + c^2 - a^2}{2bc} = \frac{21.61^2 + 42.57^2 - 49.33^2}{2(21.61)(42.57)}$$

$$A = 94.81°$$

From the law of sines, we now have

$$\frac{49.33}{\sin A} = \frac{21.61}{\sin B} = \frac{42.57}{\sin C}$$

The complete calculator solution is shown in Fig. 9.70. Therefore,

$$B = 25.88° \quad \text{and} \quad C = 59.31°$$

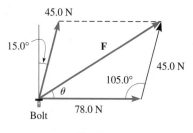

Fig. 9.70

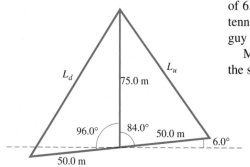

Fig. 9.69

◀ **EXAMPLE 4** Two forces are acting on a bolt. One is a 78.0-N force acting horizontally to the right, and the other is a force of 45.0 N acting upward to the right, 15.0° from the vertical. Find the resultant force **F**. See Fig. 9.71.

Moving the 45.0-N vector to the right and using the lower triangle with the 105.0° angle, the magnitude of **F** is

$$F = \sqrt{78.0^2 + 45.0^2 - 2(78.0)(45.0) \cos 105.0°}$$
$$= 99.6 \text{ N}$$

To find θ, use the law of sines:

$$\frac{45.0}{\sin \theta} = \frac{99.6}{\sin 105.0°}, \qquad \sin \theta = \frac{45.0 \sin 105.0°}{99.6}$$

This gives us $\theta = 25.9°$.

We can also solve this problem using vector components.

Fig. 9.71

Solving a Word Problem

◀ **EXAMPLE 5** A vertical radio antenna is to be built on a hill that makes an angle of 6.0° with the horizontal. Guy wires are to be attached at a point 75.0 m up the antenna and at points 50.0 m from the base of the antenna. What will be the lengths of the guy wires positioned directly up and directly down the hill?

Making an appropriate figure, as in Fig. 9.72, we can set up the equations needed for the solution:

$$L_u^2 = 50.0^2 + 75.0^2 - 2(50.0)(75.0) \cos 84.0°$$

$$L_u = 85.7 \text{ m}$$

$$L_d^2 = 50.0^2 + 75.0^2 - 2(50.0)(75.0) \cos 96.0°$$

$$L_d = 94.4 \text{ m}$$

Fig. 9.72

Following is a summary of solving an oblique triangle by use of the law of sines or the law of cosines.

SOLVING OBLIQUE TRIANGLES

Case 1: Two Angles and One Side
Find the unknown angle by subtracting the sum of the known angles from 180°. Use the **law of sines** to find the unknown sides.

Case 2: Two Sides and the Angle Opposite One of Them
Use the known side and the known angle opposite it to find the angle opposite the other known side. Find the third angle from the fact that the sum of the angles is 180°. Use the **law of sines** to find the third side. **CAUTION:** There may be two solutions. See page 281 for a summary of Case 2 and the *ambiguous case.*

Case 3: Two Sides and the Included Angle
Find the third side by using the **law of cosines.** Find the *smaller* unknown angle (opposite the shorter side) by using the **law of sines.** Complete the solution using the fact that the sum of the angles is 180°.

Case 4: Three Sides
Find the *largest angle* (opposite the longest side) by using the **law of cosines.** Find a second angle by using the **law of sines.** Complete the solution by using the fact that the sum of the angles is 180°.

Other variations in finding the solutions can be used. For example, after finding the third side in Case 3 or finding the largest angle in Case 4, the solution can be completed by using the law of sines. All angles in Case 4 can be found by using the law of cosines. The methods shown above are those normally used.

EXERCISES 9.6

In Exercises 1 and 2, solve the resulting triangles if the given changes are made in the indicated examples of this section.

1. In Example 1, solve the triangle if the value of C is changed to 145°.

2. In Example 3, solve the triangle if the value of a is changed to 29.33.

In Exercises 3–22, solve the triangles with the given parts.

3. $a = 6.00, b = 7.56, C = 54.0°$

4. $b = 87.3, c = 34.0, A = 130.0°$

5. $a = 4530, b = 924, C = 98.0°$

6. $a = 0.0845, c = 0.116, B = 85.0°$

7. $a = 39.53, b = 45.22, c = 67.15$

8. $a = 2.331, b = 2.726, c = 2.917$

9. $a = 385.4, b = 467.7, c = 800.9$

10. $a = 0.2433, b = 0.2635, c = 0.1538$

11. $a = 320, b = 847, C = 158.0°$

12. $b = 18.3, c = 27.1, A = 58.7°$

13. $a = 2140, c = 428, B = 86.3°$

14. $a = 1.13, b = 0.510, C = 77.6°$

15. $b = 103.7, c = 159.1, C = 104.67°$

16. $a = 49.32, b = 54.55, B = 114.36°$

17. $a = 0.4937, b = 0.5956, c = 0.6398$

18. $a = 69.72, b = 49.30, c = 56.29$

19. $a = 723, b = 598, c = 158$

20. $a = 1.78, b = 6.04, c = 4.80$

21. $a = 1500, A = 15°, B = 140°$

22. $a = 17, b = 24, c = 37$

In Exercises 23–36, use the law of cosines to solve the given problems.

23. For a triangle with sides a, b, and c opposite angles A, B, and C, respectively, show that $1 + \cos A = \dfrac{(b + c + a)(b + c - a)}{2bc}$.

W 24. Set up equations (do not solve) to solve the triangle in Fig. 9.73 by the law of cosines. Why is the law of sines easier to use?

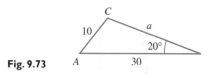

Fig. 9.73

25. A nuclear submarine leaves its base and travels at 23.5 km/h. For 2.00 h, it travels along a course of 32.1° north of west. It then turns an additional 21.5° north of west and travels for another 1.00 h. How far from its base is it?

26. The robot arm shown in Fig. 9.74 places packages on a conveyor belt. What is the distance *x*?

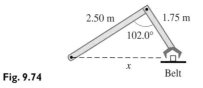

Fig. 9.74

27. Find the angle between the front legs and the back legs of the folding chair shown in Fig. 9.75.

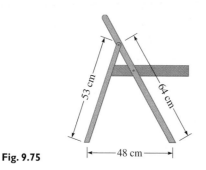

Fig. 9.75

28. In a baseball field, the four bases are at the vertices of a square 90.0 ft on a side. The pitching rubber is 60.5 ft from home plate. See Fig. 9.76. How far is it from the pitching rubber to first base? (1 ft = 0.3048 m.)

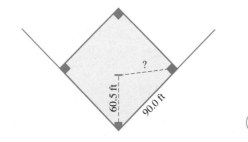

Fig. 9.76

29. A plane leaves an airport and travels 624 km due east. It then turns toward the north and travels another 326 km. It then turns again less than 180° and travels another 846 km directly back to the airport. Through what angles did it turn?

30. The apparent depth of an object submerged in water is less than its actual depth. A coin is actually 5.00 cm from an observer's eye just above the surface, but it appears to be only 4.25 cm. The real light ray from the coin makes an angle with the surface that is 8.1° greater than the angle the apparent ray makes. How much deeper is the coin than it appears to be? See Fig. 9.77.

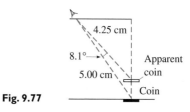

Fig. 9.77

31. A nut is in the shape of a regular hexagon (six sides). If each side is 9.53 mm, what opening on a wrench is necessary to tighten the nut? See Fig. 9.78.

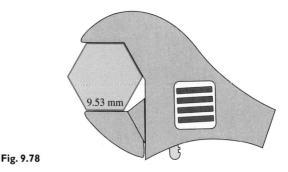

Fig. 9.78

32. Two ropes support a 78.3-N crate from above. The tensions in the ropes are 50.6 N and 37.5 N. What is the angle between the ropes? (See Exercise 24 of Section 9.5.)

33. A ferryboat travels at 11.5 km/h with respect to the water. Because of the river current, it is traveling at 12.7 km/h with respect to the land in the direction of its destination. If the ferryboat's heading is 23.6° from the direction of its destination, what is the velocity of the current?

34. The airline distance from Dublin, Ireland, to Oslo, Norway, is 810 km. It is 490 km from Dublin to Brussels, Belgium, and 680 km from Oslo to Brussels. Find the angle between the routes from Dublin.

35. An air traffic controller sights two planes that are due east from the control tower and headed toward each other. One is 15.8 km from the tower at an angle of elevation of 26.4°, and the other is 32.7 km from the tower at an angle of elevation of 12.4°. How far apart are the planes?

W 36. A triangular machine part has sides of 5 cm and 8 cm. Explain why the law of sines, or the law of cosines, is used to start the solution of the triangle if the third known part is (a) the third side, (b) the angle opposite the 5-cm side, or (c) the angle between the 5-cm and 8-cm sides.

CHAPTER ⑨ EQUATIONS

Vector components $A_x = A \cos \theta \qquad A_y = A \sin \theta$ (9.1)

$$A = \sqrt{A_x^2 + A_y^2}$$ (9.2)

$$\theta_{\text{ref}} = \tan^{-1} \frac{|A_y|}{|A_x|}$$ (9.3)

Law of sines $\dfrac{a}{\sin A} = \dfrac{b}{\sin B} = \dfrac{c}{\sin C}$ (9.8)

Law of cosines $a^2 = b^2 + c^2 - 2bc \cos A$

$$b^2 = a^2 + c^2 - 2ac \cos B$$ (9.13)

$$c^2 = a^2 + b^2 - 2ab \cos C$$

CHAPTER ⑨ REVIEW EXERCISES

In Exercises 1–4, find the x- and y-components of the given vectors by use of the trigonometric functions.

1. $A = 65.0$, $\theta_A = 28.0°$
2. $A = 8.05$, $\theta_A = 149.0°$
3. $A = 0.9204$, $\theta_A = 215.59°$
4. $A = 657.1$, $\theta_A = 343.74°$

*In Exercises 5–8, vectors **A** and **B** are at right angles. Find the magnitude and direction of the resultant.*

5. $A = 327$
$B = 505$
6. $A = 6.8$
$B = 2.9$
7. $A = 4964$
$B = 3298$
8. $A = 26.52$
$B = 89.86$

In Exercises 9–16, add the given vectors by use of the trigonometric functions and the Pythagorean theorem.

9. $A = 780$, $\theta_A = 28.0°$
$B = 346$, $\theta_B = 320.0°$
10. $J = 0.0120$, $\theta_J = 370.5°$
$K = 0.00781$, $\theta_K = 260.0°$
11. $A = 22.51$, $\theta_A = 130.16°$
$B = 7.604$, $\theta_B = 200.09°$
12. $A = 18760$, $\theta_A = 110.43°$
$B = 4835$, $\theta_B = 350.20°$
13. $Y = 51.33$, $\theta_Y = 12.25°$
$Z = 42.61$, $\theta_Z = 291.77°$
14. $A = 703.1$, $\theta_A = 122.54°$
$B = 302.9$, $\theta_B = 214.82°$
15. $A = 75.0$, $\theta_A = 15.0°$
$B = 26.5$, $\theta_B = 192.4°$
$C = 54.8$, $\theta_C = 344.7°$
16. $S = 8120$, $\theta_S = 141.9°$
$T = 1540$, $\theta_T = 165.2°$
$U = 3470$, $\theta_U = 296.0°$

In Exercises 17–36, solve the triangles with the given parts.

17. $A = 48.0°$, $B = 68.0°$, $a = 145$
18. $A = 132.0°$, $b = 7.50$, $C = 32.0°$
19. $a = 22.8$, $B = 33.5°$, $C = 125.3°$
20. $A = 71.0°$, $B = 48.5°$, $c = 8.42$
21. $A = 17.85°$, $B = 154.16°$, $c = 7863$
22. $a = 1.985$, $b = 4.189$, $c = 3.652$
23. $b = 7607$, $c = 4053$, $B = 110.09°$
24. $A = 77.06°$, $a = 12.07$, $c = 5.104$
25. $b = 14.5$, $c = 13.0$, $C = 56.6°$
26. $B = 40.6°$, $b = 7.00$, $c = 18.0$
27. $a = 186$, $B = 130.0°$, $c = 106$
28. $b = 750$, $c = 1100$, $A = 56°$
29. $a = 7.86$, $b = 2.45$, $C = 22.0°$
30. $a = 0.208$, $c = 0.697$, $B = 105.4°$
31. $A = 67.16°$, $B = 96.84°$, $c = 532.9$
32. $A = 43.12°$, $a = 7.893$, $b = 4.113$
33. $a = 17$, $b = 12$, $c = 25$
34. $a = 9064$, $b = 9953$, $c = 1106$
35. $a = 0.530$, $b = 0.875$, $c = 1.25$
36. $a = 47.4$, $b = 40.0$, $c = 45.5$

In Exercises 37–64, solve the given problems.

37. For any triangle ABC show that

$$\frac{a^2 + b^2 + c^2}{2abc} = \frac{\cos A}{a} + \frac{\cos B}{b} + \frac{\cos C}{c}$$

W **38.** In solving a triangle for Case 3 (two sides and the included angle), explain what type of solution is obtained if the included angle is a right angle.

39. An architect determines the two acute angles and one of the legs of a right triangular wall panel. Show that the area A_t is

$$A_t = \frac{a^2 \sin B}{2 \sin A}$$

40. A surveyor determines the three angles and one side of a triangular tract of land. (a) Show that the area A_t can be found from $A_t = \frac{a^2 \sin B \sin C}{2 \sin A}$. (b) For a right triangle, show that this agrees with the formula in Exercise 39.

41. Find the horizontal and vertical components of the force shown in Fig. 9.79.

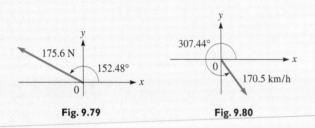

Fig. 9.79   **Fig. 9.80**

42. Find the horizontal and vertical components of the velocity shown in Fig. 9.80.

43. In a ballistics test, a bullet was fired into a block of wood with a velocity of 670 m/s and at an angle of 71.3° with the surface of the block. What was the component of the velocity perpendicular to the surface?

44. A storm cloud is moving at 15 km/h from the northwest. A television tower is 60° south of east of the cloud. What is the component of the cloud's velocity toward the tower?

45. During a 3.00-min period after taking off, the now-retired supersonic jet Concorde traveled at 480 km/h at an angle of 24.0° above the horizontal. What was its gain in altitude during this period?

46. A rocket is launched at an angle of 42.0° with the horizontal and with a speed of 760 m/s. What are its horizontal and vertical components of velocity?

47. Three forces of 3200 N, 1300 N, and 2100 N act on a bolt as shown in Fig. 9.81. Find the resultant force.

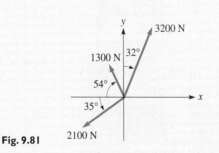

Fig. 9.81

48. In Fig. 9.82, force **F** represents the total surface tension force around the circumference on the liquid in the capillary tube. The vertical component of **F** holds up the liquid in the tube above the liquid surface outside the tube. What is the vertical component of **F**?

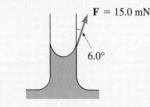

Fig. 9.82

49. A helium-filled balloon rises vertically at 3.3 m/s as the wind carries it horizontally at 5.0 m/s. What is the resultant velocity of the balloon?

50. A crater on the moon is 150 km in diameter. If the distance to the moon (to each side of the crater) from the earth is 390 000 km, what angle is subtended by the crater at an observer's position on the earth?

51. In Fig. 9.83, a damper mechanism in an air-conditioning system is shown. If $\theta = 27.5°$ when the spring is at its shortest and longest lengths, what are these lengths?

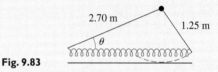

Fig. 9.83

52. A bullet is fired from the ground of a level field at an angle of 39.0° above the horizontal. It travels in a straight line at 670 m/s for 0.20 s when it strikes a target. The sound of the strike is recorded 0.32 s later on the ground. If sound travels at 350 m/s, where is the recording device located?

53. In order to get around an obstruction, an oil pipeline is constructed in two straight sections, one 3.756 km long and the other 4.675 km long, with an angle of 168.85° between the sections where they are joined. How much more pipeline was necessary due to the obstruction?

54. Three pipes of radii 2.50 cm, 3.25 cm, and 4.25 cm are welded together lengthwise. See Fig. 9.84. Find the angles between the center-to-center lines.

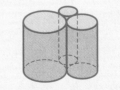

End view

Fig. 9.84

55. Two satellites are being observed at the same observing station. One is 36 200 km from the station, and the other is 30 100 km away. The angle between their lines of observation is 105.4°. How far apart are the satellites?

56. Find the side x in the truss in Fig. 9.85.

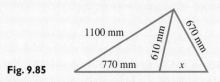

Fig. 9.85

1100 mm, 610 mm, 670 mm, 770 mm, x

57. The angle of depression of a fire noticed west of a fire tower is 6.2°. The angle of depression of a pond, also west of the tower, is 13.5°. If the fire and pond are at the same altitude, and the tower is 2.25 km from the pond on a direct line, how far is the fire from the pond?

58. A surveyor wishes to find the distance between two points between which there is a security-restricted area. The surveyor measures the distance from each of these points to a third point and finds them to be 226.73 m and 185.12 m. If the angle between the lines of sight from the third point to the other points is 126.724°, how far apart are the two points?

59. In Australia, Adelaide is 805 km and 69.0° south of east of Alice Springs. The pilot of an airplane due north of Adelaide radios Alice Springs and finds that the plane is on a line 10.5° south of east from Alice Springs. How far is the plane from Alice Springs?

60. In going around a storm, a plane flies 125 km south, then 140 km at 30.0° south of west, and finally 225 km at 15.0° north of west. What is the displacement of the plane from its original position?

61. A sailboat is headed due north, and its sail is set perpendicular to the wind, which is from the south of west. The component of the

force of the wind in the direction of the heading is 480 N, and the component perpendicular to the heading (the *drift* component) is 650 N. What is the force exerted by the wind, and what is the direction of the wind? See Fig. 9.86.

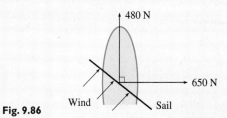

480 N, 650 N, Wind, Sail

Fig. 9.86

62. Boston is 650 km and 21.0° south of west from Halifax, Nova Scotia. Radio signals locate a ship 10.5° east of south from Halifax and 5.6° north of east from Boston. How far is the ship from each city?

63. One end of a 725-m bridge is sighted from a distance of 1630 m. The angle between the lines of sight of the ends of the bridge is 25.2°. From these data, how far is the observer from the other end of the bridge?

64. A plane is traveling horizontally at 400 m/s. A missile is fired horizontally from it 30.0° from the direction in which the plane is traveling. If the missile leaves the plane at 650 m/s, what is its velocity 10.0 s later if the vertical component is given by $v_V = -9.80t$ (in m/s)?

Writing Exercise

65. A laser experiment uses a prism with a triangular base that has sides of 2.00 cm, 3.00 cm, and 4.50 cm. Write two or three paragraphs explaining how to find the angles of the triangle.

CHAPTER 9 PRACTICE TEST

In all triangle solutions, sides a, b, c, are opposite angles A, B, C, respectively.

1. By use of a diagram, find the vector sum $2\mathbf{A} + \mathbf{B}$ for the given sectors.

B, A

2. For the triangle in which $a = 22.5$, $B = 78.6°$, and $c = 30.9$, find b.

3. A surveyor locates a tree 36.50 m to the northeast of a set position. The tree is 21.38 m north of a utility pole. What is the displacement of the utility pole from the set position?

4. For the triangle in which $A = 18.9°$, $B = 104.2°$, and $a = 426$, find c.

5. Solve the triangle in which $a = 9.84$, $b = 3.29$, and $c = 8.44$.

6. Find the horizontal and vertical components of a vector of magnitude 871 that is directed at a standard-position angle of 284.3°.

7. A ship leaves a port and travels due west. At a certain point it turns 31.5° north of west and travels an additional 42.0 km to a point 63.0 km on a direct line from the port. How far from the port is the point where the ship turned?

8. Find the sum of the vectors for which $A = 449$, $\theta_A = 74.2°$, $B = 285$, and $\theta_B = 208.9°$. Use the trigonometric functions and the Pythagorean theorem.

9. Solve the triangle for which $a = 22.3$, $b = 29.6$, and $A = 36.5°$.

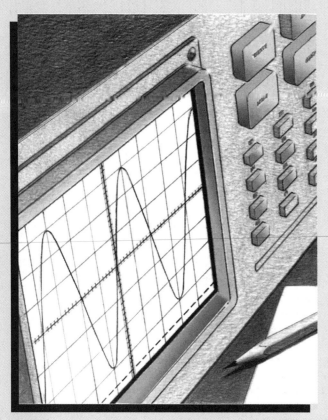

CHAPTER 10

Graphs of the Trigonometric Functions

The electronics era is thought by many to have started in the 1880s with the discovery of the vacuum tube by the American inventor Thomas Edison and the discovery of radio waves by the German physicist Heinrich Hertz. Then in the 1890s, the cathode-ray tube was developed and, as an oscilloscope, has been used since that time to analyze various types of wave forms, such as sound waves and radio waves. Since the mid-1900s, devices similar to a cathode-ray tube have been used in TV picture tubes and computer displays.

What is seen on the screen of an oscilloscope are electric signals that are represented by graphs of trigonometric functions. As noted earlier, the basic method of graphing was developed in the mid-1600s, and using trigonometric functions of numbers has been common since the mid-1700s. Therefore, the graphs of the trigonometric functions were well known in the late 1800s and became very useful in the development of electronics.

The graphs of the trigonometric functions are useful in many areas of application, particularly those that involve wave motion and periodic values. Filtering electronic signals in communications, mixing musical sounds on a tape in a recording studio, studying the seasonal temperatures of an area, and analyzing ocean waves and tides illustrate some of the many applications of periodic motion.

As well as their use in applications, the graphs of the trigonometric functions give us one of the clearest ways of showing the properties of the various functions. Therefore, in this chapter, we show graphs of these functions, with emphasis on the sine and cosine functions.

In Section 10.6, we show the resulting curve when an oscilloscope is used to combine and display electric signals.

10.1 GRAPHS OF $y = a \sin x$ AND $y = a \cos x$

When plotting and sketching the graphs of the trigonometric functions, it is normal to *express the angle in radians.* By using radians, both the independent variable and the dependent variable are real numbers. Therefore,

NOTE ▶ *it is necessary to be able to readily use angles expressed in radians.*

If necessary, review Section 8.3 on radian measure of angles.

In this section, the graphs of the sine and cosine functions are shown. We begin by making a table of values of x and y for the function $y = \sin x$, where x and y are used in the standard way as the *independent variable* and *dependent variable*. We plot the points to obtain the graph in Fig. 10.1.

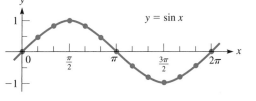

Fig. 10.1

x	0	$\frac{\pi}{6}$	$\frac{\pi}{3}$	$\frac{\pi}{2}$	$\frac{2\pi}{3}$	$\frac{5\pi}{6}$	π
y	0	0.5	0.87	1	0.87	0.5	0

x	$\frac{7\pi}{6}$	$\frac{4\pi}{3}$	$\frac{3\pi}{2}$	$\frac{5\pi}{3}$	$\frac{11\pi}{6}$	2π
y	-0.5	-0.87	-1	-0.87	-0.5	0

The graph of $y = \cos x$ may be drawn in the same way. The next table gives values for plotting the graph of $y = \cos x$, and the graph is shown in Fig. 10.2.

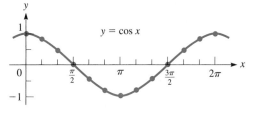

Fig. 10.2

x	0	$\frac{\pi}{6}$	$\frac{\pi}{3}$	$\frac{\pi}{2}$	$\frac{2\pi}{3}$	$\frac{5\pi}{6}$	π
y	1	0.87	0.5	0	-0.5	-0.87	-1

x	$\frac{7\pi}{6}$	$\frac{4\pi}{3}$	$\frac{3\pi}{2}$	$\frac{5\pi}{3}$	$\frac{11\pi}{6}$	2π
y	-0.87	-0.5	0	0.5	0.87	1

The graphs are continued beyond the values shown in the tables to indicate that *they continue indefinitely in each direction.* To show this more clearly, in Figs. 10.3 and 10.4 note the graphs of $y = \sin x$ and $y = \cos x$ from $x = -10$ to $x = 10$.

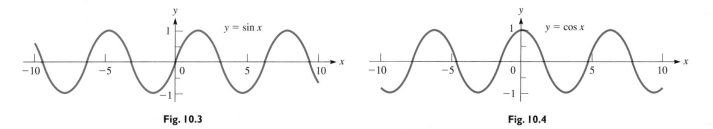

Fig. 10.3   Fig. 10.4

From these tables and graphs, it can be seen that *the graphs of $y = \sin x$ and $y = \cos x$ are of exactly the same shape (called* **sinusoidal***), with the cosine curve displacement $\pi/2$ units to the left of the sine curve.* The shape of these curves should be recognized readily, with special note as to the points at which they cross the axes. This information will be especially valuable in *sketching* similar curves, since the basic sinusoidal shape remains the same. It will not be necessary to plot numerous points every time we wish to sketch such a curve.

To obtain the graph of $y = a \sin x$, note that all the y-values obtained for the graph of $y = \sin x$ are to be multiplied by the number a. In this case, the greatest value of the **Amplitude** sine function is $|a|$. *The number $|a|$ is called the **amplitude** of the curve and represents the greatest y-value of the curve.* Also, the curve will have no value less than $-|a|$. This is true for $y = a \cos x$ as well as $y = a \sin x$.

EXAMPLE 1 Plot the graph of $y = 2 \sin x$.

Since $a = 2$, the amplitude of this curve is $|2| = 2$. This means that the maximum value of y is 2 and the minimum value is $y = -2$. The table of values follows, and the curve is shown in Fig. 10.5.

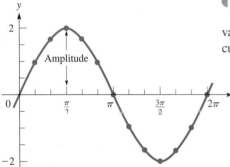

Fig. 10.5

x	0	$\frac{\pi}{6}$	$\frac{\pi}{3}$	$\frac{\pi}{2}$	$\frac{2\pi}{3}$	$\frac{5\pi}{6}$	π
y	0	1	1.73	2	1.73	1	0

x	$\frac{7\pi}{6}$	$\frac{4\pi}{3}$	$\frac{3\pi}{2}$	$\frac{5\pi}{3}$	$\frac{11\pi}{6}$	2π
y	-1	-1.73	-2	-1.73	-1	0

EXAMPLE 2 Plot the graph of $y = -3 \cos x$.

In this case, $a = -3$, and this means that the amplitude is $|-3| = 3$. Therefore, the maximum value of y is 3, and the minimum value of y is -3. The table of values follows, and the curve is shown in Fig. 10.6.

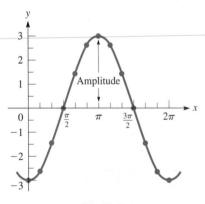

Fig. 10.6

x	0	$\frac{\pi}{6}$	$\frac{\pi}{3}$	$\frac{\pi}{2}$	$\frac{2\pi}{3}$	$\frac{5\pi}{6}$	π
y	-3	-2.6	-1.5	0	1.5	2.6	3

x	$\frac{7\pi}{6}$	$\frac{4\pi}{3}$	$\frac{3\pi}{2}$	$\frac{5\pi}{3}$	$\frac{11\pi}{6}$	2π
y	2.6	1.5	0	-1.5	-2.6	-3

Note from Example 2 that *the effect of the negative sign with the number a is to **invert** the curve about the x-axis.* The effect of the number a can also be seen readily from these examples.

From the previous examples, note that the function $y = a \sin x$ has zeros for $x = 0$, π, 2π and that it has its maximum or minimum values for $x = \pi/2$, $3\pi/2$. The function $y = a \cos x$ has its zeros for $x = \pi/2$, $3\pi/2$ and its maximum or minimum values for $x = 0$, π, 2π. This is summarized in Table 10.1. Therefore, by knowing the general shape of the sine curve, where it has its zeros, and what its amplitude is, *we can rapidly sketch curves of the form $y = a \sin x$ and $y = a \cos x$.*

Since the graphs of $y = a \sin x$ and $y = a \cos x$ can extend indefinitely to the right and to the left, we see that the domain of each is all real numbers. We should note that the key values of $x = 0$, $\pi/2$, π, $3\pi/2$, and 2π are those only for x from 0 to 2π. Corresponding values ($x = 5\pi/2$, 3π, and their negatives) could also be used. Also from the graphs, we can readily see that the range of these functions is $-|a| \leq f(x) \leq |a|$.

Table 10.1

	$x = 0, \pi, 2\pi$	$\frac{\pi}{2}, \frac{3\pi}{2}$
$y = a \sin x$	zeros	max. or min.
$y = a \cos x$	max. or min.	zeros

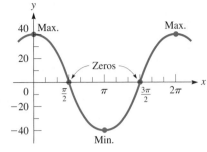

Fig. 10.7

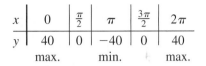

EXAMPLE 3 Sketch the graph of $y = 40 \cos x$.

First, we set up a table of values for the points where the curve has its zeros, maximum points, and minimum points:

x	0	$\frac{\pi}{2}$	π	$\frac{3\pi}{2}$	2π
y	40	0	-40	0	40
	max.		min.		max.

Now, we plot these points and join them, knowing the basic sinusoidal shape of the curve. See Fig. 10.7.

The graphs of $y = a \sin x$ and $y = a \cos x$ can be displayed easily on a graphing calculator. In the next example, we see that the calculator displays the features of the curve we should expect from our previous discussion.

EXAMPLE 4 Display the graph of $y = -2 \sin x$ on a graphing calculator.

Using radian mode on the calculator, Fig. 10.8(a) shows the calculator view for the key values in the following table:

x	0	$\frac{\pi}{2}$	π	$\frac{3\pi}{2}$	2π
y	0	-2	0	2	0
		min.		max.	

We see the amplitude of 2 and the effect of the negative sign in inverting the curve, as expected. Fig. 10.8(b) shows the calculator graph for $x = -10$ to $x = 10$.

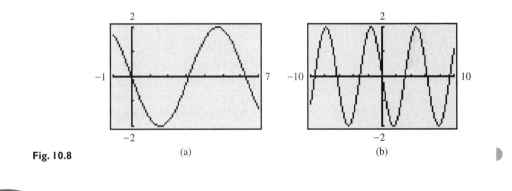

Fig. 10.8   (a)   (b)

EXERCISES 10.1

In Exercises 1 and 2, graph the function if the given changes are made in the indicated examples of this section.

1. In Example 2, if the sign of the coefficient of $\cos x$ is changed, plot the graph of the resulting function.

2. In Example 4, if the sign of the coefficient of $\sin x$ is changed, display the graph of the resulting function.

In Exercises 3–6, complete the following table for the given functions and then plot the resulting graphs.

x	$-\pi$	$-\frac{3\pi}{4}$	$-\frac{\pi}{2}$	$-\frac{\pi}{4}$	0	$\frac{\pi}{4}$	$\frac{\pi}{2}$	$\frac{3\pi}{4}$	π
y									

x	$\frac{5\pi}{4}$	$\frac{3\pi}{2}$	$\frac{7\pi}{4}$	2π	$\frac{9\pi}{4}$	$\frac{5\pi}{2}$	$\frac{11\pi}{4}$	3π
y								

3. $y = \sin x$ **4.** $y = \cos x$

5. $y = 3 \cos x$ **6.** $y = -4 \sin x$

In Exercises 7–22, sketch the graphs of the given functions. Check each using a graphing calculator.

7. $y = 3 \sin x$

8. $y = 5 \sin x$

9. $y = \frac{5}{2} \sin x$

10. $y = 35 \sin x$

11. $y = 200 \cos x$

12. $y = 0.25 \cos x$

13. $y = 0.8 \cos x$

14. $y = \frac{3}{2} \cos x$

15. $y = -\sin x$

16. $y = -3000 \sin x$

17. $y = -1500 \sin x$

18. $y = -0.2 \sin x$

19. $y = -\cos x$

20. $y = -8 \cos x$

21. $y = -50 \cos x$

22. $y = -0.4 \cos x$

Although units of π are convenient, we must remember that π is only a number. Numbers that are not multiples of π may be used. In Exercises 23–26, plot the indicated graphs by finding the values of y that correspond to values of x of 0, 1, 2, 3, 4, 5, 6, and 7 on a calculator. (Remember, the numbers 0, 1, 2, and so on represent radian measure.)

23. $y = \sin x$

24. $y = -30 \sin x$

25. $y = 10 \cos x$

26. $y = 2 \cos x$

In Exercises 27–32, solve the given problems.

27. Find the function and graph it for a function of the form $y = a \sin x$ that passes through $(\pi/2, -2)$.

28. Find the function and graph it for a function of the form $y = a \sin x$ that passes through $(3\pi/2, -2)$.

29. Find the function and graph it for a function of the form $y = a \cos x$ that passes through $(\pi, 2)$.

30. Find the function and graph it for a function of the form $y = a \cos x$ that passes through $(2\pi, -2)$.

31. The graph displayed on an oscilloscope can be represented by $y = -0.05 \sin x$. Display this curve on a graphing calculator.

32. The displacement y (in cm) of the end of the robot arm for welding is $y = 4.75 \cos t$, where t is the time (in s). Display this curve on a graphing calculator.

In Exercises 33–36, the graph of a function of the form $y = a \sin x$ or $y = a \cos x$ is shown. Determine the specific function of each.

33.
34.
35.
36.

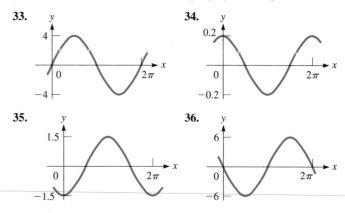

10.2 GRAPHS OF $y = a \sin bx$ AND $y = a \cos bx$

In graphing the function $y = \sin x$, we see that the values of y repeat every 2π units of x. This is because $\sin x = \sin(x + 2\pi) = \sin(x + 4\pi)$, and so forth. For any function F, we say that it has a *period P* if $F(x) = F(x + P)$. For functions that are periodic, such as the sine and cosine, the **period** is the *x-distance between a point and the next corresponding point for which the value of y repeats.*

Period of a Function

Let us now plot the curve $y = \sin 2x$. This means that we choose a value of x, multiply this value by 2, and find the sine of the result. This leads to the following table of values for this function:

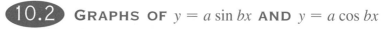

x	0	$\frac{\pi}{8}$	$\frac{\pi}{4}$	$\frac{3\pi}{8}$	$\frac{\pi}{2}$	$\frac{5\pi}{8}$	$\frac{3\pi}{4}$	$\frac{7\pi}{8}$	π	$\frac{9\pi}{8}$	$\frac{5\pi}{4}$
$2x$	0	$\frac{\pi}{4}$	$\frac{\pi}{2}$	$\frac{3\pi}{4}$	π	$\frac{5\pi}{4}$	$\frac{3\pi}{2}$	$\frac{7\pi}{4}$	2π	$\frac{9\pi}{4}$	$\frac{5\pi}{2}$
y	0	0.7	1	0.7	0	−0.7	−1	−0.7	0	0.7	1

Plotting these points, we have the curve shown in Fig. 10.9.

From the table and Fig. 10.9, note that $y = \sin 2x$ repeats after π units of x. The effect of the 2 is that the period of $y = \sin 2x$ is half the period of the curve of $y = \sin x$. We then conclude that if the period of a function $F(x)$ is P, then the period of $F(bx)$ is P/b. Since each of the functions $\sin x$ and $\cos x$ has a period of 2π, *each of the functions $\sin bx$ and $\cos bx$ has a period of $2\pi/b$.*

NOTE ▶

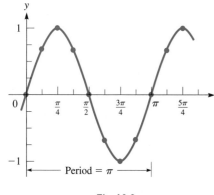

Fig. 10.9

◀ EXAMPLE 1 **(a)** The period of $\sin \overset{\displaystyle \overset{b}{\downarrow}}{3}x$ is $\dfrac{2\pi}{\underset{\longrightarrow}{3}}$, which means that the curve of the

function $y = \sin 3x$ will repeat every $\frac{2\pi}{3}$ (approximately 2.09) units of x.

(b) The period of $\cos \underset{\longrightarrow}{4}x$ is $\dfrac{2\pi}{4} = \dfrac{\pi}{2}$.

(c) The period of $\sin \frac{1}{2}x$ is $\dfrac{2\pi}{\underset{\longrightarrow}{\frac{1}{2}}} = 4\pi$. In this case, the period is longer than that of the

basic sine curve. ▶

◀ EXAMPLE 2 **(a)** The period of $\sin \pi x$ is $2\pi/\pi = 2$. That is, the curve of the
function $\sin \pi x$ repeats every 2 units.

(b) The period of $\cos 3\pi x$ is $\dfrac{2\pi}{3\pi} = \dfrac{2}{3}$.

(c) The period of $\sin \dfrac{\pi}{4}x$ is $\dfrac{2\pi}{\frac{\pi}{4}} = 8$.

(d) The period of $\sin 3x$ is $\dfrac{2\pi}{3} = 2.09$, and the period of $\sin \pi x$ is $\frac{2\pi}{\pi} = 2$. These two

periods differ only sightly since π is only slightly larger than 3. ▶

Combining the value of the period with the value of the amplitude from Section 10.1, we conclude that *the functions $y = a \sin bx$ and $y = a \cos bx$ each has an amplitude of $|a|$ and a period of $2\pi/b$.* These properties are very useful in sketching these functions.

◀ EXAMPLE 3 Sketch the graph of $y = 3 \sin 4x$ for $0 \le x \le \pi$.
Since $a = 3$, the amplitude is 3. Also, the $4x$ tells us that the period is $2\pi/4 = \pi/2$. This means that $y = 0$ for $x = 0$ and for $x = \pi/2$. Since this sine function is zero halfway between $x = 0$ and $x = \pi/2$, we find that $x = 0$ for $x = \pi/4$. Also, the fact that the graph of the sine function reaches its maximum and minimum values halfway between zeros means that $y = 3$ for $x = \pi/8$, and $y = -3$ for $x = 3\pi/8$. Note that the values of x in the following table are those for which $4x = 0, \pi/2, \pi, 3\pi/2, 2\pi$, and so on.

x	0	$\frac{\pi}{8}$	$\frac{\pi}{4}$	$\frac{3\pi}{8}$	$\frac{\pi}{2}$	$\frac{5\pi}{8}$	$\frac{3\pi}{4}$	$\frac{7\pi}{8}$	π
y	0	3	0	-3	0	3	0	-3	0

Using the values from the table and the fact that the curve is sinusoidal in form, we sketch the graph of this function in Fig. 10.10. We see again that knowing the key values and the basic shape of the curve allows us to *sketch* the graph of the curve quickly and easily. ▶

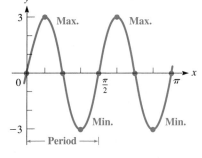

Fig. 10.10

Note from Example 3 that *an important distance in sketching a sine curve or a cosine curve is one-fourth of the period.* For $y = a \sin bx$, it is one-fourth of the period from the origin to the first value of x where y is at its maximum (or minimum) value. Then we proceed another one-fourth period to a zero, another one-fourth period to the next minimum (or maximum) value, another to the next zero (this is where the period is completed), and so on.

Similarly, one-fourth of the period is useful in sketching the graph of $y = \cos bx$. For this function, the maximum (or minimum) value occurs for $x = 0$. At the following one-fourth period values, there is a zero, a minimum (or maximum), a zero, and a maximum (or minimum) at the start of the next period. Therefore,

NOTE ▶ *by finding one-fourth of the period, we can easily find the important values for sketching the curve*

We now summarize the important values for sketching the graphs of $y = a \sin bx$ and $y = a \cos bx$.

IMPORTANT VALUES FOR SKETCHING $y = a \sin bx$ AND $y = a \cos bx$

1. *The amplitude:* $|a|$
2. *The period:* $2\pi/b$
3. *Values of the function for each one-fourth period*

◀ **EXAMPLE 4** Sketch the graph of $y = -2 \cos 3x$ for $0 \le x \le 2\pi$.

Note that the amplitude is 2 and the period is $\frac{2\pi}{3}$. This means that one-fourth of the period is $\frac{1}{4} \times \frac{2\pi}{3} = \frac{\pi}{6}$. Since the cosine curve is at a maximum or minimum for $x = 0$, we find that $y = -2$ for $x = 0$ (the negative value is due to the minus sign before the function), which means it is a minimum point. The curve then has a zero at $x = \frac{\pi}{6}$, a maximum value of 2 at $x = 2\left(\frac{\pi}{6}\right) = \frac{\pi}{3}$, a zero at $x = 3\left(\frac{\pi}{6}\right) = \frac{\pi}{2}$, and its next value of -2 at $x = 4\left(\frac{\pi}{6}\right) = \frac{2\pi}{3}$, and so on. Therefore, we have the following table:

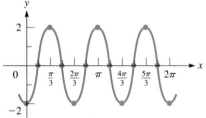

y

2

0   $\frac{\pi}{3}$ $\frac{2\pi}{3}$ π $\frac{4\pi}{3}$ $\frac{5\pi}{3}$ 2π x

-2

Fig. 10.11

x	0	$\frac{\pi}{6}$	$\frac{\pi}{3}$	$\frac{\pi}{2}$	$\frac{2\pi}{3}$	$\frac{5\pi}{6}$	π	$\frac{7\pi}{6}$	$\frac{4\pi}{3}$	$\frac{3\pi}{2}$	$\frac{5\pi}{3}$	$\frac{11\pi}{6}$	2π
y	-2	0	2	0	-2	0	2	0	-2	0	2	0	-2

Using this table and the sinusoidal shape of the cosine curve, we sketch the function in Fig. 10.11. ▶

◀ **EXAMPLE 5** A generator produces a voltage $V = 200 \cos 50\pi t$, where t is the time in seconds (50π has units of rad/s; thus, $50\pi t$ is an angle in radians). Use a graphing calculator to display the graph of V as a function of t for $0 \le t \le 0.06$ s.

The amplitude is 200 V and the period is $2\pi/(50\pi) = 0.04$ s. Since the period is not in terms of π, it is more convenient to use decimal units for t rather than to use units in terms of π as in the previous graphs. Thus, we have the following table of values:

t (s)	0	0.01	0.02	0.03	0.04	0.05	0.06
V (V)	200	0	-200	0	200	0	-200

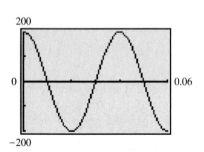

200

0   0.06

-200

Fig. 10.12

For the graphing calculator, use x for t and y for V. This means we graph the function $y_1 = 200 \cos 50\pi x$, as shown in Fig. 10.12. From the amplitude of 200 V and the above table, we choose the *window* values shown in Fig. 10.12. We do not consider negative values of t, for they have no real meaning in this problem. ▶

EXERCISES 10.2

In Exercises 1 and 2, graph the function if the given changes are made in the indicated examples of this section.

1. In Example 3, if the coefficient of x is changed from 4 to 6, sketch the graph of the resulting function.

2. In Example 4, if the coefficient of x is changed from 3 to 4, sketch the graph of the resulting function.

In Exercises 3–22, find the period of each function.

3. $y = 2 \sin 6x$

4. $y = 4 \sin 2x$

5. $y = 3 \cos 8x$

6. $y = 28 \cos 10x$

7. $y = -2 \sin 12x$

8. $y = -\sin 5x$

9. $y = -\cos 16x$

10. $y = -4 \cos 2x$

11. $y = 520 \sin 2\pi x$

12. $y = 2 \sin 3\pi x$

13. $y = 3 \cos 4\pi x$

14. $y = 4 \cos 10\pi x$

15. $y = 15 \sin \frac{1}{3}x$

16. $y = -25 \sin \frac{2}{5}x$

17. $y = -\frac{1}{2} \cos \frac{2}{3}x$

18. $y = \frac{1}{3} \cos \frac{1}{4}x$

19. $y = 0.4 \sin \dfrac{2\pi x}{3}$

20. $y = 1.5 \cos \dfrac{\pi x}{10}$

21. $y = 3.3 \cos \pi^2 x$

22. $y = 2.5 \sin \dfrac{2x}{\pi}$

In Exercises 23–42, sketch the graphs of the given functions. Check each using a graphing calculator. (These are the same functions as in Exercises 3–22.)

23. $y = 2 \sin 6x$

24. $y = 4 \sin 2x$

25. $y = 3 \cos 8x$

26. $y = 28 \cos 10x$

27. $y = -2 \sin 12x$

28. $y = -\sin 5x$

29. $y = -\cos 16x$

30. $y = -4 \cos 2x$

31. $y = 520 \sin 2\pi x$

32. $y = 2 \sin 3\pi x$

33. $y = 3 \cos 4\pi x$

34. $y = 4 \cos 10\pi x$

35. $y = 15 \sin \frac{1}{3}x$

36. $y = -25 \sin \frac{2}{5}x$

37. $y = -\frac{1}{2} \cos \frac{2}{3}x$

38. $y = \frac{1}{3} \cos \frac{1}{4}x$

39. $y = 0.4 \sin \dfrac{2\pi x}{3}$

40. $y = 1.5 \cos \dfrac{\pi x}{10}$

41. $y = 3.3 \cos \pi^2 x$

42. $y = 2.5 \sin \dfrac{2x}{\pi}$

In Exercises 43–46, the period is given for a function of the form $y = \sin bx$. Write the function corresponding to the given period.

43. $\dfrac{\pi}{3}$

44. $\dfrac{2\pi}{5}$

45. 2

46. 6

In Exercises 47–56, solve the given problems.

47. By noting the periods of $\sin 2x$ and $\sin 3x$, find the minimum period of the function $y = \sin 2x + \sin 3x$.

48. By noting the period of $\cos \frac{1}{2}x$ and $\cos \frac{1}{3}x$, find the minimum period of the function $y = \cos \frac{1}{2}x + \cos \frac{1}{3}x$.

49. Find the function and graph it for a function of the form $y = -2 \sin bx$ that passes through $(\pi/4, -2)$ and for which b has the smallest possible positive value.

50. Find the function and graph it for a function of the form $y = 2 \sin bx$ that passes through $(\pi/6, 2)$ and for which b has the smallest possible positive value.

51. Find the function and graph it for a function of the form $y = 2 \cos bx$ that passes through $(\pi, 0)$ and for which b has the smallest possible positive value.

52. Find the function and graph it for a function of the form $y = -2 \cos bx$ that passes through $(\pi/2, 2)$ and for which b has the smallest possible positive value.

53. The standard electric voltage in a 60-Hz alternating-current circuit is given by $V = 170 \sin 120\pi t$, where t is the time in seconds. Sketch the graph of V as a function of t for $0 \leq t \leq 0.05$ s.

54. To tune the instruments of an orchestra before a concert, an A note is struck on a piano. The piano wire vibrates with a displacement y (in mm) given by $y = 3.20 \cos 880\pi t$, where t is in seconds. Sketch the graph of y vs. t for $0 \leq t \leq 0.01$ s.

55. The velocity v (in cm/s) of a piston is $v = 450 \cos 3600t$, where t is in seconds. Sketch the graph of v vs. t for $0 \leq t \leq 0.006$ s.

56. On a Florida beach, the tides have water levels of about 4 m between low and high tides. The period is about 12.5 h. Find a cosine function that describes these tides if high tide is at midnight of a given day. Sketch the graph.

In Exercises 57–60, the graph of a function of the form $y = a \sin bx$ or $y = a \cos bx$ is shown. Determine the specific function of each.

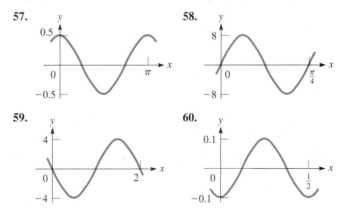

57.

58.

59.

60.

10.3 GRAPHS OF $y = a \sin(bx + c)$ AND $y = a \cos(bx + c)$

Another important quantity in graphing the sine and cosine functions is the *phase angle*. In the function $y = a \sin(bx + c)$, *c represents the* **phase angle.** Its meaning is illustrated in the following example.

EXAMPLE 1 Sketch the graph of $y = \sin\left(2x + \frac{\pi}{4}\right)$.

Note that $c = \pi/4$. Therefore, in order to obtain values for the table, we assume a value for x, multiply it by 2, add $\pi/4$ to this value, and then find the sine of the result. The values that are shown are those for which $2x + \pi/4 = 0, \pi/4, \pi/2, 3\pi/4, \pi$, and so on, which are the important values for $y = \sin 2x$.

x	$-\frac{\pi}{8}$	0	$\frac{\pi}{8}$	$\frac{\pi}{4}$	$\frac{3\pi}{0}$	$\frac{\pi}{2}$	$\frac{5\pi}{8}$	$\frac{3\pi}{4}$	$\frac{7\pi}{8}$	π
y	0	0.7	1	0.7	0	-0.7	-1	-0.7	0	0.7

When solving $2x + \pi/4 = 0$, we get $x = -\pi/8$, and this gives $y = \sin 0 = 0$. The other values for y are found in the same way. The graph is shown in Fig. 10.13.

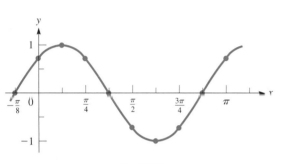

Fig. 10.13

Note from Example 1 that *the graph of* $y = \sin\left(2x + \frac{\pi}{4}\right)$ *is precisely the same as the graph of* $y = \sin 2x$, *except that it is* **shifted** $\pi/8$ *units to the left.* In Fig. 10.14, a graphing calculator view shows the graphs of $y = \sin 2x$ and $y = \sin\left(2x + \frac{\pi}{4}\right)$. We see that the shapes are the same and that the graph of $y = \sin\left(2x + \frac{\pi}{4}\right)$ is about 0.4 unit ($\pi/8 \approx 0.39$) to the left of the graph of $y = \sin 2x$.

In general, the effect of c in the equation $y = a \sin(bx + c)$ is to shift the curve of $y = a \sin bx$ to the left if $c > 0$, or shift the curve to the right if $c < 0$. The amount of this shift is given by $-c/b$. Due to its importance in sketching curves, *the quantity* $-c/b$ *is called the* **displacement** (*or* **phase shift**).

We can see the reason that the displacement is $-c/b$ by noting corresponding points on the graphs of $y = \sin bx$ and $y = \sin(bx + c)$. For $y = \sin bx$, when $x = 0$, then $y = 0$. For $y = \sin(bx + c)$, when $x = -c/b$, then $y = 0$. The point $(-c/b, 0)$ on the graph of $y = \sin(bx + c)$ is $-c/b$ units to the left of the point $(0, 0)$ on the graph of $y = \sin x$. In Fig. 10.14, $-c/b = -\pi/8$.

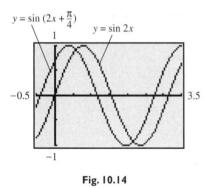

Fig. 10.14

See Appendix C for a graphing calculator program SINECURV. It displays the graphs of $y = \sin x$, $y = 2 \sin x$, $y = \sin 2x$, and $y = 2 \sin(2x - \pi/3)$.

Carefully note the difference between $y = \sin(bx + c)$ and $y = \sin bx + c$. Writing $\sin(bx + c)$ means to find the sine of the quantity of $bx + c$, whereas $\sin bx + c$ means to find the sine of bx and then add the value c.

Therefore, we use the displacement combined with the amplitude and the period along with the other information from the previous sections to sketch curves of the functions $y = a \sin(bx + c)$ and $y = a \cos(bx + c)$, where $b > 0$.

IMPORTANT QUANTITIES TO DETERMINE FOR SKETCHING GRAPHS OF
$y = a \sin(bx + c)$ AND $y = a \cos(bx + c)$

$$\text{Amplitude} = |a|$$

$$\text{Period} = \frac{2\pi}{b}$$

$$\text{Displacement} = -\frac{c}{b}$$

(10.1)

By use of these quantities and the one-fourth period distance, the graphs of the sine and cosine functions can be readily sketched.

A general illustration of the graph of $y = a \sin(bx + c)$ is shown in Fig. 10.15. Note carefully that

CAUTION ▶

the displacement is **negative** *(to the left) for c > 0* (Fig. 10.15(a))

the displacement is **positive** *(to the right) for c < 0* (Fig. 10.15(b))

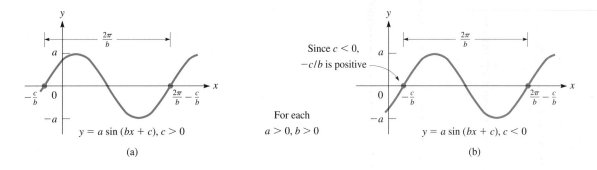

Fig. 10.15

$y = a \sin(bx + c), c > 0$ (a)

Since $c < 0$, $-c/b$ is positive

For each $a > 0, b > 0$

$y = a \sin(bx + c), c < 0$ (b)

NOTE ▶

Note also that we can find the displacement for the graphs of $y = a \sin(bx + c)$ by solving $bx + c = 0$ for x. We see that $x = -c/b$. *This is the same as the horizontal shift for graphs of functions that is shown on page 101.*

◀ **EXAMPLE 2** Sketch the graph of $y = 2 \sin(3x - \pi)$.

First, note that $a = 2$, $b = 3$, and $c = -\pi$. Therefore, the amplitude is 2, the period is $2\pi/3$, and the displacement is $-(-\pi/3) = \pi/3$. (We can also get the displacement from $3x - \pi = 0$, $x = \pi/3$.)

Note that the curve "starts" at $x = \pi/3$ and starts repeating $2\pi/3$ units to the right of this point. Be sure to grasp this point well. *The period tells us the number of units along the x-axis between such corresponding points.* One-fourth of the period is $\frac{1}{4}\left(\frac{2\pi}{3}\right) = \frac{\pi}{6}$.

Important values are at $\frac{\pi}{3}, \frac{\pi}{3} + \frac{\pi}{6} = \frac{\pi}{2}, \frac{\pi}{3} + 2\left(\frac{\pi}{6}\right) = \frac{2\pi}{3}$, and so on. We now make the table of important values and sketch the graph shown in Fig. 10.16.

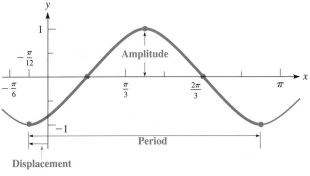

Fig. 10.16

x	0	$\frac{\pi}{6}$	$\frac{\pi}{3}$	$\frac{\pi}{2}$	$\frac{2\pi}{3}$	$\frac{5\pi}{6}$	π
y	0	-2	0	2	0	-2	0

$\leftarrow \frac{1}{4}$ (period) $= \frac{\pi}{6}$

Note that since the period is $2\pi/3$ the curve passes through the origin.

◀ **EXAMPLE 3** Sketch the graph of the function $y = -\cos\left(2x + \frac{\pi}{6}\right)$.

First, we determine that

1. the amplitude is 1
2. the period is $\frac{2\pi}{2} = \pi$
3. the displacement is $-\frac{\pi}{6} \div 2 = -\frac{\pi}{12}$

We now make a table of important values, noting that the curve starts repeating π units to the right of $-\frac{\pi}{12}$.

x	$-\frac{\pi}{12}$	$\frac{\pi}{6}$	$\frac{5\pi}{12}$	$\frac{2\pi}{3}$	$\frac{11\pi}{12}$
y	-1	0	1	0	-1

$\leftarrow \frac{1}{4}$ (period) $= \frac{\pi}{4}$

Fig. 10.17

From this table, we sketch the graph in Fig. 10.17.

Each of the heavy portions of the graphs in Figs. 10.16 and 10.17 is called a *cycle* of the curve. A **cycle** *is any section of the graph that includes exactly one period.*

◀ **EXAMPLE 4**   View the graph of $y = 2\cos\left(\frac{1}{2}x - \frac{\pi}{6}\right)$ on a graphing calculator.
From the values $a = 2$, $b = 1/2$, and $c = -\pi/6$, we determine that

1. the amplitude is 2
2. the period is $2\pi \div \frac{1}{2} = 4\pi$
3. the displacement is $-\left(-\frac{\pi}{6}\right) \div \frac{1}{2} = \frac{\pi}{3}$

We now make a table of important values:

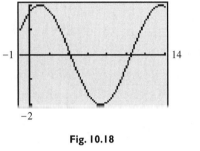

x	$\frac{\pi}{3}$	$\frac{4\pi}{3}$	$\frac{7\pi}{3}$	$\frac{10\pi}{3}$	$\frac{13\pi}{3}$	$\leftarrow \frac{1}{4}$ (period) $= \pi$
y	2	0	-2	0	2	

This table helps us choose the values for the *window* settings in Fig. 10.18. We choose Xmin $= -1$ in order to start to the left of the y-axis and Xmax $= 14$ since $13\pi/3 \approx 13.6$. Also, we choose Ymin $= -2$ and Ymax $= 2$ since the amplitude is 2. We see that the graph in Fig. 10.18 is a little more than one cycle. ▶

Fig. 10.18

◀ **EXAMPLE 5**   The cross section of a certain water wave is $y = 0.7\sin\left(\frac{\pi}{2}x + \frac{\pi}{4}\right)$, where x and y are measured in meters. Display two cycles of y vs. x on a graphing calculator.

From the values $a = 0.7$ m, $b = \pi/2$ m^{-1} (this means 1/m, or per meter), and $c = \pi/4$, we find the amplitude, period, and displacement:

1. amplitude $= 0.7$ m
2. period $= \dfrac{2\pi}{\frac{\pi}{2}} = 4$ m
3. displacement $= -\dfrac{\frac{\pi}{4}}{\frac{\pi}{2}} = -0.5$ m

Using these values, we choose the following values for the *window* settings:

1. Xmin $= -0.5$ (the displacement is -0.5 m)
2. Xmax $= 7.5$ (the period is 4 m, and we want two periods, starting at $x = -0.5$) $(-0.5 + 8 = 7.5)$
3. Ymin $= -0.7$, Ymax $= 0.7$ (the amplitude is 0.7 m)

The graphing calculator view is shown in Fig. 10.19. The negative values of x have the significance of giving points to the wave to the left of the origin. (When *time* is used, no actual physical meaning is generally given to negative values of t.) ▶

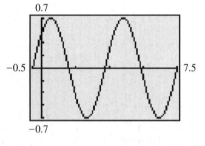

Fig. 10.19

EXERCISES 10.3

In Exercises 1 and 2, graph the function if the given changes are made in the indicated examples of this section.

1. In Example 3, if the sign before $\pi/6$ is changed, sketch the graph of the resulting function.

2. In Example 4, if the sign before $\pi/6$ is changed, sketch the graph of the resulting function.

In Exercises 3–26, determine the amplitude, period, and displacement for each function. Then sketch the graphs of the functions. Check each using a graphing calculator.

3. $y = \sin\left(x - \dfrac{\pi}{6}\right)$

4. $y = 3\sin\left(x + \dfrac{\pi}{4}\right)$

5. $y = \cos\left(x + \dfrac{\pi}{6}\right)$

6. $y = 2\cos\left(x - \dfrac{\pi}{8}\right)$

7. $y = 0.2 \sin\left(2x + \dfrac{\pi}{2}\right)$

8. $y = -\sin\left(3x - \dfrac{\pi}{2}\right)$

9. $y = -\cos(2x - \pi)$

10. $y = 0.4 \sin\left(3x + \dfrac{\pi}{3}\right)$

11. $y = \dfrac{1}{2}\sin\left(\dfrac{1}{2}x - \dfrac{\pi}{4}\right)$

12. $y = 2\sin\left(\dfrac{1}{4}x + \dfrac{\pi}{2}\right)$

13. $y = 30\cos\left(\dfrac{1}{3}x + \dfrac{\pi}{3}\right)$

14. $y = \dfrac{1}{3}\cos\left(\dfrac{1}{2}x - \dfrac{\pi}{8}\right)$

15. $y = \sin\left(\pi x + \dfrac{\pi}{8}\right)$

16. $y = -2\sin(2\pi x - \pi)$

17. $y = \dfrac{3}{4}\cos\left(4\pi x - \dfrac{\pi}{5}\right)$

18. $y = 25\cos\left(3\pi x + \dfrac{\pi}{2}\right)$

19. $y = -0.6\sin(2\pi x - 1)$

20. $y = 1.8\sin\left(\pi x + \dfrac{1}{3}\right)$

21. $y = 40\cos(3\pi x + 2)$

22. $y = 360\cos(6\pi x - 1)$

23. $y = \sin(\pi^2 x - \pi)$

24. $y = -\dfrac{1}{2}\sin\left(2x - \dfrac{1}{\pi}\right)$

25. $y = -\dfrac{3}{2}\cos\left(\pi x + \dfrac{\pi^2}{6}\right)$

26. $y = \pi\cos\left(\dfrac{1}{\pi}x + \dfrac{1}{3}\right)$

In Exercises 27–30, write the equation for the given function with the given amplitude, period, and displacement, respectively.

27. sine, 4, 3π, $-\pi/4$

28. cosine, 8, $2\pi/3$, $\pi/3$

29. cosine, 12, $1/2$, $1/8$

30. sine, 18, 4, -1

In Exercises 31–36, solve the given problems. In Exercises 35 and 36, use a graphing calculator to view the indicated curves.

31. Find the function and graph it for a function of the form $y = 2\sin(2x + c)$ that passes through $(-\pi/8, 0)$ and for which c has the smallest possible positive value.

32. Find the function and graph it for a function of the form $y = 2\cos(2x - c)$ that passes through $(\pi/6, 2)$ and for which c has the smallest possible positive value.

33. A wave traveling in a string may be represented by the equation $y = A\sin 2\pi\left(\dfrac{t}{T} - \dfrac{x}{\lambda}\right)$. Here, A is the amplitude, t is the time the wave has traveled, x is the distance from the origin, T is the time required for the wave to travel one *wave-length* λ (the Greek letter lambda). Sketch three cycles of the wave for which $A = 2.00$ cm, $T = 0.100$ s, $\lambda = 20.0$ cm, and $x = 5.00$ cm.

34. The electric current i (in μA) in a certain circuit is given by $i = 3.8\cos 2\pi(t + 0.20)$, where t is the time in seconds. Sketch three cycles of this function.

35. A certain satellite circles the earth such that its distance y, in kilometers north or south (altitude is not considered) from the equator, is $y = 7200\cos(0.025t - 0.25)$, where t is the time (in min) after launch. View two cycles of the graph.

36. In performing a test on a patient, a medical technician used an ultrasonic signal given by the equation $I = A\sin(\omega t + \theta)$. View two cycles of the graph of I vs. t if $A = 5$ nW/m^2, $\omega = 2 \times 10^5$ rad/s, and $\theta = 0.4$.

(W) *In Exercises 37–40, give the specific form of the equation by evaluating a, b, and c through an inspection of the given curve. Explain how a, b, and c are found.*

37. $y = a\sin(bx + c)$
Fig. 10.20

38. $y = a\cos(bx + c)$
Fig. 10.20

39. $y = a\cos(bx + c)$
Fig. 10.21

40. $y = a\sin(bx + c)$
Fig. 10.21

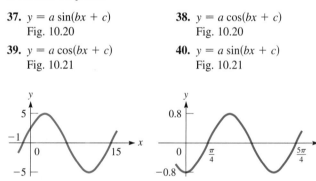

Fig. 10.20   **Fig. 10.21**

10.4 GRAPHS OF $y = \tan x$, $y = \cot x$, $y = \sec x$, $y = \csc x$

We now show the basic form of each curve for the other trigonometric functions. From these, we will then be able to sketch other curves for these functions.

Finding values of the trigonometric functions as in Chapter 8, we set up the following table for $y = \tan x$ and show the graph in Fig. 10.22.

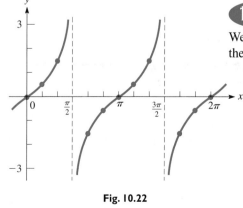

Fig. 10.22

x	0	$\dfrac{\pi}{6}$	$\dfrac{\pi}{3}$	$\dfrac{\pi}{2}$	$\dfrac{2\pi}{3}$	$\dfrac{5\pi}{6}$	π
y	0	0.6	1.7	*	-1.7	-0.6	0

x	$\dfrac{7\pi}{6}$	$\dfrac{4\pi}{3}$	$\dfrac{3\pi}{2}$	$\dfrac{5\pi}{3}$	$\dfrac{11\pi}{6}$	2π
y	0.6	1.7	*	-1.7	-0.6	0

*Undefined

NOTE ▶

The tangent function is not defined for $x = \pm\pi/2$, $x = \pm 3\pi/2$, and so forth. Therefore, using a calculator, note that the value of tan x becomes very large as x gets closer to $\pi/2$, although *there is no point on the curve of* tan x *for* $x = \pi/2$. For example, because $\pi/2 \approx 1.57$, if $x = 1.56$, tan $x = 92.6$. We can also see from the table that *the period of the tangent curve is* π. This differs from the period of the sine and cosine functions.

By knowing the values of sin x, cos x, and tan x, we can find the necessary values of csc x, sec x, and cot x. This is due to the reciprocal relationships among the functions that we showed in Section 4.3. We show these relationships by

$$\csc x = \frac{1}{\sin x} \qquad \sec x = \frac{1}{\cos x} \qquad \cot x = \frac{1}{\tan x} \qquad \textbf{(10.2)}$$

Therefore, to graph $y = \cot x$, $y = \sec x$, and $y = \csc x$, we can obtain the necessary values from the corresponding reciprocal function. Figures 10.23–10.26 are the graphs of these functions as well as a more extensive graph of $y = \tan x$.

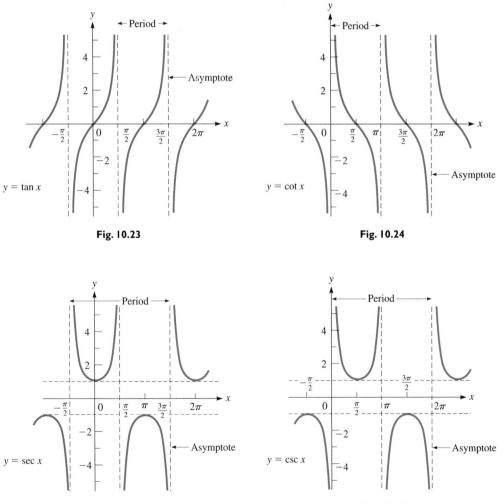

$y = \tan x$ Fig. 10.23

$y = \cot x$ Fig. 10.24

$y = \sec x$ Fig. 10.25

$y = \csc x$ Fig. 10.26

We see from these graphs that the period of $y = \tan x$ and $y = \cot x$ is π and that the period of $y = \sec x$ and $y = \csc x$ is 2π. *The vertical dashed lines in these figures are* **asymptotes** (see Sections 3.4 and 21.6). The curves *approach* these lines, but they never actually touch them.

The functions are not defined for the values of x for which the curve has asymptotes. This means that the domains do not include these values of x. Thus, we see that the domains of $y = \tan x$ and $y = \sec x$ include all real numbers, except the values $x = -\pi/2$, $\pi/2$, $3\pi/2$, and so on. The domains of $y = \cot x$ and $y = \csc x$ include all real numbers except $x = -\pi$, 0, π, 2π, and so on.

From the graphs, we see that the ranges of $y = \tan x$ and $y = \cot x$ are all real numbers, but that the ranges of $y = \sec x$ and $y = \csc x$ do not include the real numbers between -1 and 1.

To sketch functions such as $y = a \sec x$, first sketch $y = \sec x$ and then multiply the y-values by a. Here a is not an amplitude, since the ranges of these functions are not limited in the same way they are for the sine and cosine functions.

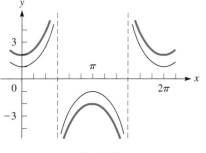

Fig. 10.27

◀ **EXAMPLE 1**   Sketch the graph of $y = 2 \sec x$.

First, we sketch in $y = \sec x$, shown as the light curve in Fig. 10.27. Then we multiply the y-values of this secant function by 2. Although we can only estimate these values and do this approximately, a reasonable graph can be sketched this way. The desired curve is shown in color in Fig. 10.27. ▶

Using a graphing calculator, we can display the graphs of these functions more easily and more accurately than by sketching them. By knowing the general shape and period of the function, the values for the *window* settings can be determined without having to reset them too often.

◀ **EXAMPLE 2**   View at least two cycles of the graph of $y = 0.5 \cot 2x$ on a graphing calculator.

Since the period of $y = \cot x$ is π, the period of $y = \cot 2x$ is $\pi/2$. Therefore, we choose the *window* settings as follows:

Fig. 10.28

Xmin = 0   ($x = 0$ is one asymptote of the curve)

Xmax = 3.2   ($\pi \approx 3.14$; the period is $\pi/2$; two periods is π)

Ymin = -5, Ymax = 5   (the range is all x; this shows enough of the curve)

We must remember to enter the function as $y_1 = 0.5(\tan 2x)^{-1}$, since $\cot x = (\tan x)^{-1}$. The graphing calculator view is shown in Fig. 10.28. We can view many more cycles of the curve with appropriate *window* settings. ▶

◀ **EXAMPLE 3**   View at least two periods of the graph of $y = 2 \sec\left(2x - \frac{\pi}{4}\right)$ on a graphing calculator.

Since the period of $\sec x$ is 2π, the period of $\sec\left(2x - \frac{\pi}{4}\right)$ is $2\pi/2 = \pi$. Recalling that $\sec x = (\cos x)^{-1}$, the curve will have the same displacement as $y = \cos\left(2x - \frac{\pi}{4}\right)$. This displacement is $-\frac{-\pi/4}{2} = \frac{\pi}{8}$. Therefore, we choose the following *window* settings.

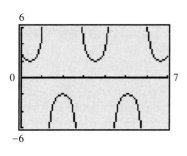

Fig. 10.29

Xmin = 0   (the displacement is positive)

Xmax = 7   (displacement = $\pi/8$; period = π; $\pi/8 + 2\pi = 17\pi/8 \approx 6.7$)

Ymin = -6, Ymax = 6   (there is no curve between $y = -2$ and $y = 2$)

With $y_1 = 2(\cos(2x - \pi/4))^{-1}$, Fig. 10.29 shows the calculator view. ▶

EXERCISES 10.4

In Exercises 1 and 2, view the graphs on a graphing calculator if the given changes are made in the indicated examples of this section.

1. In Example 2, view the graph if 0.5 is changed to 5.

2. In Example 3, view the graph if the sign before $\pi/4$ is changed.

In Exercises 3–6, fill in the following table for each function and plot the graph from these points.

x	$-\frac{\pi}{2}$	$-\frac{\pi}{3}$	$-\frac{\pi}{4}$	$-\frac{\pi}{6}$	0	$\frac{\pi}{6}$	$\frac{\pi}{4}$	$\frac{\pi}{3}$	$\frac{\pi}{2}$	$\frac{2\pi}{3}$	$\frac{3\pi}{4}$	$\frac{5\pi}{6}$	π
y													

3. $y = \tan x$ 4. $y = \cot x$ 5. $y = \sec x$ 6. $y = \csc x$

In Exercises 7–14, sketch the graphs of the given functions by use of the basic curve forms (Figs. 10.23, 10.24, 10.25, and 10.26). See Example 1.

7. $y = 2 \tan x$ 8. $y = 3 \cot x$

9. $y = \frac{1}{2} \sec x$ 10. $y = \frac{3}{2} \csc x$

11. $y = -8 \cot x$ 12. $y = -0.1 \tan x$

13. $y = -3 \csc x$ 14. $y = -6 \sec x$

In Exercises 15–24, view at least two cycles of the graphs of the given functions on a graphing calculator.

15. $y = \tan 2x$ 16. $y = 2 \cot 3x$

17. $y = \frac{1}{2} \sec 3x$ 18. $y = 0.4 \csc 2x$

19. $y = 2 \cot\left(2x + \dfrac{\pi}{6}\right)$ 20. $y = \tan\left(3x - \dfrac{\pi}{2}\right)$

21. $y = 18 \csc\left(3x - \dfrac{\pi}{3}\right)$ 22. $y = 12 \sec\left(2x + \dfrac{\pi}{4}\right)$

23. $y = 3 \tan\left(0.5x - \dfrac{\pi}{8}\right)$ 24. $y = 0.5 \sec\left(0.2x + \dfrac{\pi}{25}\right)$

In Exercises 25–28, sketch the appropriate graphs. Check each on a graphing calculator.

25. Near Antarctica, an iceberg with a vertical face 200 m high is seen from a small boat. At a distance x from the iceberg, the angle of elevation θ of the top of the iceberg can be found from the equation $x = 200 \cot \theta$. Sketch x as a function of θ.

26. In a laser experiment, two mirrors move horizontally in equal and opposite distances from point A. The laser path from and to point B is shown in Fig. 10.30. From the figure we see that $x = a \tan \theta$. Sketch the graph of $x = f(\theta)$ for $a = 5.00$ cm.

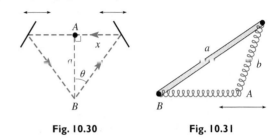

Fig. 10.30   **Fig. 10.31**

27. A mechanism with two springs is shown in Fig. 10.31, where point A is restricted to move horizontally. From the law of sines, we see that $b = (a \sin B)\csc A$. Sketch the graph of b as a function of A for $a = 4.00$ cm and $B = \pi/4$.

28. A cantilever column of length L will buckle if too large a downward force P is applied d units off center. The horizontal deflection x (see Fig. 10.32) is $x = d(\sec(kL) - 1)$, where k is a constant depending on P and $0 < kL < \pi/2$. For a constant d, sketch the graph of x as a function of kL.

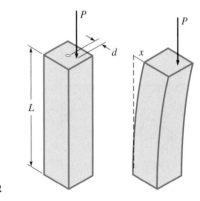

Fig. 10.32

10.5 APPLICATIONS OF THE TRIGONOMETRIC GRAPHS

In Section 8.4, we discussed the velocity of an object moving in a circular path. When the object moves with constant velocity, its *projection* on a diameter moves with what is known as **simple harmonic motion.** For example, if a ball on the end of a string is moving at a constant rate in a circle that is in a plane parallel to rays of light, the shadow of the ball on a wall moves with simple harmonic motion. We now discuss this important physical concept and some of its technical applications.

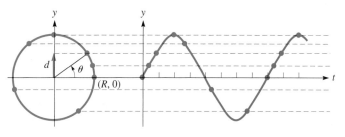

Fig. 10.33

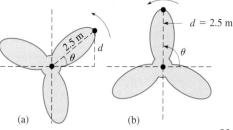

Fig. 10.34

EXAMPLE 1 In Fig. 10.33, assume that a particle starts at the end of the radius at $(R, 0)$ and moves counterclockwise around the circle with constant angular velocity ω. *The displacement of the projection on the y-axis is d* and is given by $d = R \sin \theta$. The displacement is shown for a few different positions of the end of the radius.

Since $\theta/t = \omega$, or $\theta = \omega t$, we have

$$d = R \sin \omega t \qquad (10.3)$$

as the equation for the displacement of this projection, with time t as the independent variable.

For the case where $R = 10.0$ cm and $\omega = 4.00$ rad/s, we have

$$d = 10.0 \sin 4.00t$$

By sketching or viewing the graph of this function, we can find the displacement d of the projection for a given time t. The graph is shown in Fig. 10.34. ▷

In Example 1, note that *time is the independent variable.* This is motion for which the object (the end of the projection) remains at the same horizontal position ($x = 0$) and moves only vertically according to a sinusoidal function. In the previous sections, we dealt with functions in which y is a sinusoidal function of the horizontal displacement x. Think of a water wave. At *one point* of the wave, the motion is only vertical and sinusoidal with time. At *one given time,* a picture would indicate a sinusoidal movement from one horizontal position to the next.

EXAMPLE 2 A windmill is used to pump water. The radius of the blade is 2.5 m, and it is moving with constant angular velocity. If the vertical displacement of the end of the blade is timed from the point it is at an angle of 45° ($\pi/4$ rad) from the horizontal (see Fig. 10.35(a)), the displacement d is given by

$$d = 2.5 \sin\left(\omega t + \frac{\pi}{4}\right)$$

If the blade makes an angle of 90° ($\pi/2$ rad) when $t = 0$ (see Fig. 10.35(b)), the displacement d is given by

$$d = 2.5 \sin\left(\omega t + \frac{\pi}{2}\right) \quad \text{or} \quad d = 2.5 \cos \omega t$$

If timing started at the first maximum for the displacement, the resulting curve for the displacement would be that of the cosine function. ▷

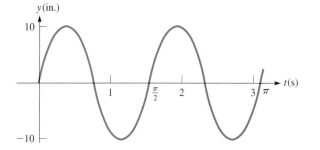

(a)   (b)

Fig. 10.35

Other examples of simple harmonic motion are (1) the movement of a pendulum bob through its arc (a very close approximation to simple harmonic motion), (2) the motion of an object "bobbing" in water, (3) the movement of the end of a vibrating rod (which we hear as sound), and (4) the displacement of a weight moving up and down on a spring. Other phenomena that give rise to equations like those for simple harmonic motion are found in the fields of optics, sound, and electricity. The equations for such phenomena have the same mathematical form because they result from vibratory movement or motion in a circle.

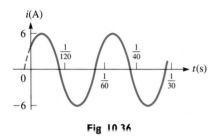

Fig 10.36

Named for the German physicist Heinrich
Hertz (1857–1894).

◀ EXAMPLE 3   A very important use of the trigonometric curves arises in the study
of alternating current, which is caused by the motion of a wire passing through a magnetic field. If the wire is moving in a circular path, with angular velocity ω, the current i
in the wire at time t is given by an equation of the form

$$i = I_m \sin(\omega t + \alpha)$$

where I_m is the maximum current attainable and α is the phase angle.

The current may be represented by a sinusoidal wave. Given that $I_m = 6.00$ A,
$\omega = 120\pi$ rad/s, and $\alpha = \pi/6$, we have the equation

$$i = 6.00 \sin\left(120\pi t + \frac{\pi}{6}\right)$$

From this equation, note that the amplitude is 6.00 A, the period is $\frac{1}{60}$ s, and the displacement is $-\frac{1}{720}$ s. From these values, we draw the graph as shown in Fig. 10.36. Since
the current takes on both positive and negative values, we conclude that it moves alter-
nately in one direction and then the other. ▶

It is a common practice to express the rate of rotation in terms of *the* **frequency** *f,*
the number of cycles per second, rather than directly in terms of the angular velocity ω,
the number of radians per second. *The unit for frequency is the* **hertz** (Hz), *and*
1 Hz = 1 cycle/s. Since there are 2π rad in one cycle, we have

$$\boxed{\omega = 2\pi f} \tag{10.4}$$

It is the frequency f that is referred to in electric current, on radio stations, for musical
tones, and so on.

◀ EXAMPLE 4   For the electric current in Example 3, $\omega = 120\pi$ rad/s. The corresponding frequency f is

$$f = \frac{120\pi}{2\pi} = 60 \text{ Hz}$$

This means that 120π rad/s corresponds to 60 cycles/s. This is the standard frequency
used for alternating current. ▶

EXERCISES 10.5

*In Exercises 1 and 2, answer the given questions about the indicated
examples of this section.*

1. In Example 1, what is the equation relating d and t if the end of
the radius starts at $(0, R)$?

2. In Example 2, if the blade starts at an angle of $-45°$, what is
the equation relating d and t as (a) a sine function? (b) A cosine
function?

A graphing calculator may be used in the following exercises.

*In Exercises 3 and 4, sketch two cycles of the curve of the projection
of Example 1 as a function of time for the given values.*

3. $R = 2.40$ cm, $\omega = 2.00$ rad/s   4. $R = 1.80$ m, $f = 0.250$ Hz

*In Exercises 5 and 6, a point on a cam is 8.30 cm from the center of
rotation. The cam is rotating with a constant angular velocity, and the
vertical displacement $d = 8.30$ cm for $t = 0$ s. See Fig. 10.37. Sketch
two cycles of d as a function of t for the given values.*

5. $f = 3.20$ Hz

6. $\omega = 3.20$ rad/s

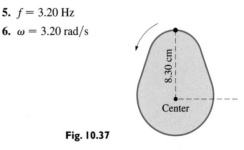

Fig. 10.37

In Exercises 7 and 8, a satellite is orbiting the earth such that its displacement D north of the equator (or south if D < 0) is given by D = A sin(ωt + α). Sketch two cycles of D as a function of t for the given values.

7. $A = 500$ km, $\omega = 3.60$ rad/h, $\alpha = 0$

8. $A = 850$ km, $f = 1.6 \times 10^{-4}$ Hz, $\alpha = \pi/3$

In Exercises 9 and 10, for an alternating-current circuit in which the voltage e is given by e = E cos(ωt + α), sketch two cycles of the voltage as a function of time for the given values.

9. $E = 170$ V, $f = 60.0$ Hz, $\alpha = -\pi/3$

10. $E = 80$ V, $\omega = 377$ rad/s, $\alpha = \pi/2$

In Exercises 11 and 12, refer to the wave in the string described in Exercise 33 of Section 10.3. For a point on the string, the displacement y is given by $y = A \sin 2\pi\left(\dfrac{t}{T} - \dfrac{x}{\lambda}\right)$. *We see that each point on the string moves with simple harmonic motion. Sketch two cycles of y as a function of t for the given values.*

11. $A = 3.20$ cm, $T = 0.050$ s, $\lambda = 40.0$ cm, $x = 5.00$ cm

12. $A = 1.45$ cm, $T = 0.250$ s, $\lambda = 24.0$ cm, $x = 20.0$ cm

In Exercises 13 and 14, the air pressure within a plastic container changes above and below the external atmospheric pressure by $p = p_0 \sin 2\pi ft$. *Sketch two cycles of p = f(t) for the given values.*

13. $p_0 = 280$ kPa, $f = 2.30$ Hz

14. $p_0 = 45.0$ kPa, $f = 0.450$ Hz

In Exercises 15–20, sketch the required curves.

15. Sketch two cycles of the radio signal $e = 0.014 \cos(2\pi ft + \pi/4)$ (e in volts, f in hertz, and t in seconds) for a station broadcasting with $f = 950$ kHz ("95" on the AM radio dial).

16. Sketch two cycles of the acoustical intensity I of the sound wave for which $I = A \cos(2\pi ft - \alpha)$, given that t is in seconds, $A = 0.027$ W/cm^2, $f = 240$ Hz, and $\alpha = 0.80$.

(W)17. The rotating beacon of a parked police car is 12 m from a straight wall. (a) Sketch the graph of the length L of the light beam, where $L = 12 \sec \pi t$, for $0 \le t \le 2.0$ s. (b) Which part(s) of the graph show meaningful values? Explain.

18. The motion of a piston of a car engine approximates simple harmonic motion. Given that the stroke (twice the amplitude) is 0.100 m, the engine runs at 2800 r/min, and the piston starts at the middle of its stroke, find the equation for the displacement d as a function of t. Sketch two cycles.

19. A riverboat's paddle has a 3.7-m radius and rotates at 18 r/min. Find the equation of motion of the vertical displacement (from the center of the wheel) y of the end of a paddle as a function of the time t if the paddle is initially horizontal. Sketch two cycles.

20. The sinusoidal electromagnetic wave emitted by an antenna in a cellular phone system has a frequency of 7.5×10^9 Hz and an amplitude of 0.045 V/m. Find the equation representing the wave if it starts at the origin. Sketch two cycles.

10.6 COMPOSITE TRIGONOMETRIC CURVES

Many applications involve functions that in themselves are a combination of two or more simpler functions. In this section, we discuss methods by which the curve of such a function can be found by combining values from the simpler functions.

◀ EXAMPLE 1 Sketch the graph of $y = 2 + \sin 2x$.

This function is the sum of the simpler functions $y_1 = 2$ and $y_2 = \sin 2x$. We may find values for y by adding 2 to each important value of $y_2 = \sin 2x$.

For $y_2 = \sin 2x$, the amplitude is 1, and the period is $2\pi/2 = \pi$. Therefore, we obtain the values in the following table and sketch the graph in Fig. 10.38.

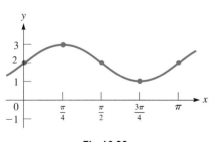

Fig. 10.38

x	0	$\frac{\pi}{4}$	$\frac{\pi}{2}$	$\frac{3\pi}{4}$	π
$\sin 2x$	0	1	0	-1	0
$2 + \sin 2x$	2	3	2	1	2

Note that this is a vertical shift of 2 units of the graph of $y = \sin 2x$, in the same way as discussed on page 101.

▶

ADDITION OF ORDINATES

Another way to sketch the resulting graph is to *first sketch the two simpler curves and then add the y-values graphically. This method is called* **addition of ordinates** *and is illustrated in the following example.*

◀ **EXAMPLE 2** Sketch the graph of $y = 2\cos x + \sin 2x$.

On the same set of coordinate axes, we sketch the curves $y = 2\cos x$ and $y = \sin 2x$. These are shown as dashed and solid light curves in Fig. 10.39. For various values of x, we determine the distance above or below the x-axis of each curve and add these distances, noting that those above the axis are positive and those below the axis are negative. We thereby *graphically* **add** *the y-values* of these curves to get points on the resulting curve, shown in color in Fig. 10.39.

At A, add the two lengths (shown side-by-side for clarity) to get the length for y At B, both lengths are negative, and the value for y is the sum of these negative values. At C, one is positive and the other negative, and we must subtract the lower length from the upper one to get the length for y.

We combine these lengths for enough x-values to get a good curve. Some points are easily found. Where one curve crosses the x-axis, its value is zero, and the resulting curve has its point on the other curve. Here, where $\sin 2x$ is zero, the points for the resulting curve lie on the curve of $2\cos x$.

We should also add values where each curve is at its maximum or its minimum. *Extra care should be taken for those values of x for which one curve is* **positive** *and the other is* **negative.** ▶

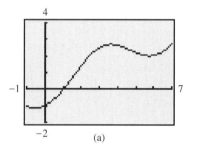

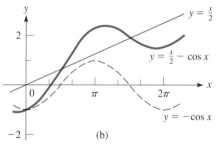

Fig. 10.39

We have seen how a fairly complex curve can be sketched graphically. It is expected that a graphing calculator (or computer *grapher*) will generally be used to view most graphs, particularly ones that are difficult to sketch. A graphing calculator can display such curves much more easily, and with much greater accuracy. Information about the amplitude, period, and displacement is useful in choosing *window* settings. For these graphs it is important that the calculator be in *radian mode.*

◀ **EXAMPLE 3** Use a graphing calculator to display the graph of $y = \frac{x}{2} - \cos x$.

Here, we note that the curve is a combination of the straight line $y = x/2$ and the trigonometric curve $y = \cos x$. There are several good choices for the *window* settings, depending on how much of the curve is to be viewed. To see a little more than one period of $\cos x$, we can make the following choices:

Xmin = −1   (to start to the left of the y-axis)

Xmax = 7   (the period of $\cos x$ is $2\pi \approx 6.3$)

Ymin = −2   (the line passes through $(0,0)$; the amplitude of $y = \cos x$ is 1)

Ymax = 4   (the slope of the line is $1/2$)

The graphing calculator view of the curve is shown in Fig. 10.40(a). The graphs of $y = \frac{x}{2} - \cos x$, $y = \frac{x}{2}$, and $y = -\cos x$ are shown in Fig. 10.40(b).

NOTE ▶ The reason for showing $y = -\cos x$, and not $y = \cos x$, is that if *addition of ordinates* were being used, *it is much easier to add graphic values than to subtract them.* In using the method of addition of ordinates, we could *add* the ordinates of $y = x/2$ and $y = -\cos x$ to get the resulting curve of $y = \frac{x}{2} - \cos x$. ▶

Fig. 10.40

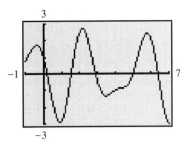

Fig. 10.41

EXAMPLE 4   View the graph of $y = \cos \pi x - 2 \sin 2x$ on a graphing calculator.

The combination of $y = \cos \pi x$ and $y = 2 \sin 2x$ leads to the following choices for the *window* settings:

Xmin $= -1$ (to start to the left of the y-axis)

Xmax $= 7$ (the periods are 2 and π; this shows at least two periods of each)

Ymin $= -3$, Ymax $= 3$ (the sum of the amplitudes is 3)

There are many possible choices for Xmin and Xmax to get a good view of the graph on a calculator. However, since the sum of the amplitudes is 3, note that the curve cannot be below $y = -3$ or above $y = 3$.

The graphing calculator view is shown in Fig. 10.41.

This graph can be constructed by using addition of ordinates, although it is difficult to do very accurately. ◗

LISSAJOUS FIGURES

An important application of trigonometric curves is made when they are added at *right angles*. The methods for doing this are shown in the following examples.

EXAMPLE 5   Plot the graph for which the values of x and y are given by the equations $y = \sin 2\pi t$ and $x = 2 \cos \pi t$. *Equations given in this form, x and y in terms of a third variable, are called* **parametric equations.**

Since both x and y are in terms of t, by assuming values of t we find corresponding values of x and y and use these values to plot the graph. Since the periods of $\sin 2\pi t$ and $2 \cos \pi t$ are $t = 1$ and $t = 2$, respectively, we will use values of $t = 0, 1/4, 1/2, 3/4,$ 1, and so on. These give us convenient values of $0, \pi/4, \pi/2, 3\pi/4, \pi,$ and so on to use in the table. We plot the points in Fig. 10.42.

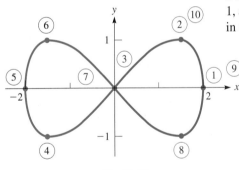

Fig. 10.42

t	0	$\frac{1}{4}$	$\frac{1}{2}$	$\frac{3}{4}$	1	$\frac{5}{4}$	$\frac{3}{2}$	$\frac{7}{4}$	2	$\frac{9}{4}$
x	2	1.4	0	-1.4	-2	-1.4	0	1.4	2	1.4
y	0	1	0	-1	0	1	0	-1	0	1
Point number	1	2	3	4	5	6	7	8	9	10

◗

Since x and y are trigonometric functions of a third variable t and since the x-axis is at right angles to the y-axis, values of x and y obtained in this way result in a combination of two trigonometric curves at right angles. *Figures obtained in this way are called* **Lissajous figures.** Note that the Lissajous figure in Fig. 10.42 *is not a function* since there are *two* values of y for each value of x (except $x = -2, 0, 2$) in the domain.

In practice, Lissajous figures can be shown by applying different voltages to an *oscilloscope* and displaying the electric signals on a screen similar to that on a television set.

Named for the French physicist Jules Lissajous (1822–1880).

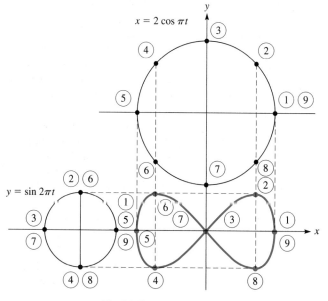

$x = 2 \cos \pi t$

$y = \sin 2\pi t$

Fig. 10.43

◀ **EXAMPLE 6** If a circle is placed on the *x*-axis and another on the *y*-axis, we may represent the coordinates (x, y) for the curve of Example 5 by the lengths of the projections (see Example 1 of Section 10.5) of a point moving around each circle. A careful study of Fig. 10.43 will clarify this. We note that the radius of the circle giving the *x*-values is 2 and that the radius of the circle giving the *y*-values is 1. This is due to the way in which *x* and *y* are defined. Also, due to these definitions, the point revolves around the *y*-circle twice as fast as the corresponding point around the *x*-circle.

See the chapter introduction.

On an oscilloscope, the curve would result when two electric signals are used. The first would have twice the amplitude and one-half the frequency of the other.

Most graphing calculators can be used to display a curve defined by parametric equations. To do this, it is necessary to use the *mode* feature and make the selection for *parametric equations.* Use the manual for the calculator, as there are some differences in how this is done on the various calculators. In the example that follows, we display the graph of parametric equations on a graphing calculator.

◀ **EXAMPLE 7** Use a graphing calculator to display the graph defined by the parametric equations $x = 2 \cos \pi t$ and $y = \sin 2\pi t$. These are the same equations as those used in Examples 5 and 6.

First, select the parametric equation option from the *mode* feature and enter the parametric equations $x_{1T} = 2 \cos \pi t$ and $y_{1T} = \sin 2\pi t$. Then, make the following *window* settings:

Tmin = 0   (standard default settings, and the usual choice)

Tmax = 2   (the periods are 2 and 1; the longer period is 2)

Tstep = .1047   (standard default setting; curve is smoother with 0.01)

Xmin = −2, Xmax = 2   (smallest and largest possible values of *x*), Xscl = 1

Ymin = −1, Ymax = 1   (smallest and largest possible values of *y*), Yscl = 0.5

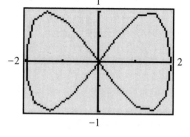

Fig. 10.44

The calculator graph is shown in Fig. 10.44.

EXERCISES 10.6

In Exercises 1–8, sketch the curves of the given functions by addition of ordinates.

1. $y = 1 + \sin x$

2. $y = 3 - 2 \cos x$

3. $y = \frac{1}{3}x + \sin 2x$

4. $y = x - \sin x$

5. $y = \frac{1}{10}x^2 - \sin \pi x$

6. $y = \frac{1}{4}x^2 + \cos 3x$

7. $y = \sin x + \cos x$

8. $y = \sin x + \sin 2x$

In Exercises 9–20, display the graphs of the given functions on a graphing calculator.

9. $y = x^3 + 10 \sin 2x$

10. $y = \dfrac{1}{x^2 + 1} - \cos \pi x$

11. $y = \sin x - \sin 2x$

12. $y = \cos 3x - \sin x$

13. $y = 20 \cos 2x + 30 \sin x$

14. $y = \frac{1}{2} \sin 4x + \cos 2x$

15. $y = 2 \sin x - \cos x$

16. $y = 8 \sin 0.5x - 12 \sin x$

17. $y = \sin \pi x - \cos 2x$

18. $y = 2 \cos 4x - \cos\left(x - \dfrac{\pi}{4}\right)$

19. $y = 2 \sin\left(2x - \dfrac{\pi}{6}\right) + \cos\left(2x + \dfrac{\pi}{3}\right)$

20. $y = 3 \cos 2\pi x + \sin \dfrac{\pi}{2}x$

In Exercises 21–24, plot the Lissajous figures.

21. $x = \sin t, y = \sin t$

22. $x = 2 \cos t, y = \cos (t + 4)$

23. $x = \cos \pi t, y = \sin \pi t$

24. $x = \cos\left(t + \dfrac{\pi}{4}\right), y = \sin 2t$

In Exercises 25–32, use a graphing calculator to display the Lissajous figures.

25. $x = \cos \pi\left(t + \dfrac{1}{6}\right), y = 2 \sin \pi t$

26. $x = \sin^2 \pi t, y = \cos \pi t$

27. $x = 4 \cos 3t, y = \cos 2t$

28. $x = 2 \sin \pi t, y = 3 \sin 3\pi t$

29. $x = \sin t, y = \sin 5t$

30. $x = 5 \cos t, y = 3 \sin 5t$

31. $x = 2 \cos \pi t, y = 3 \sin\left(2\pi t - \dfrac{\pi}{4}\right)$

32. $x = 1.5 \cos 3\pi t, y = 0.5 \cos 5\pi t$

In Exercises 33–40, sketch the appropriate curves. A graphing calculator may be used.

33. An analysis of the temperature records for Montreal indicates that the average daily temperature T (in °C) during the year is approximately $T = 6 - 15 \cos\left[\dfrac{\pi}{6}(x - 0.5)\right]$, where x is measured in months ($x = 0.5$ is Jan. 15, etc.). Sketch the graph of T vs. x for one year.

34. An analysis of data shows that the mean density d (in mg/cm^3) of a calcium compound in the bones of women is given by $d = 139.3 + 48.6 \sin(0.0674x - 0.210)$, where x represents the ages of women ($20 \le x \le 80$ years). (A woman is considered to be osteoporotic if $d < 115$ mg/cm^3.) Sketch the graph.

35. A normal person with a pulse rate of 60 beats/min has a blood pressure of "120 over 80." This means the pressure is oscillating between a high (systolic) of 120 mm of mercury (shown as mm Hg) and a low (diastolic) of 80 mm Hg. Assuming a sinusoidal type of function, find the pressure p as a function of the time t if the initial pressure is 120 mm Hg. Sketch the graph for the first 5 s. (1 mmHg = 133.3 Pa)

36. The strain e (dimensionless) on a cable caused by vibration is $e = 0.0080 - 0.0020 \sin 30t + 0.0040 \cos 10t$, where t is measured in seconds. Sketch two cycles of e as a function of t.

37. The electric current i (in mA) in a certain circuit is given by $i = 0.32 + 0.50 \sin t - 0.20 \cos 2t$, where t is in milliseconds. Sketch two cycles of i as a function of t.

38. The available solar energy depends on the amount of sunlight, and the available time in a day for sunlight depends on the time of the year. An approximate correction factor (in min) to standard time is $C = 10 \sin \dfrac{1}{29}(n - 80) - 7.5 \cos \dfrac{1}{58}(n - 80)$, where n is the number of the day of the year. Sketch C as a function of n.

39. Two signals are seen on an oscilloscope as being at right angles. The equations for the displacements of these signals are $x = 4 \cos \pi t$ and $y = 2 \sin 3\pi t$. Sketch the figure that appears on the oscilloscope.

40. In the study of optics, light is said to be *elliptically polarized* if certain optic vibrations are out of phase. These may be represented by Lissajous figures. Determine the Lissajous figure for two light waves given by $w_1 = \sin \omega t$ and $w_2 = \sin\left(\omega t + \dfrac{\pi}{4}\right)$.

CHAPTER ⑩ EQUATIONS

For the graphs of $y = a \sin(bx + c)$ and $y = a \cos(bx + c)$ Amplitude $= |a|$ Period $= \dfrac{2\pi}{b}$ Displacement $= -\dfrac{c}{b}$ **(10.1)**

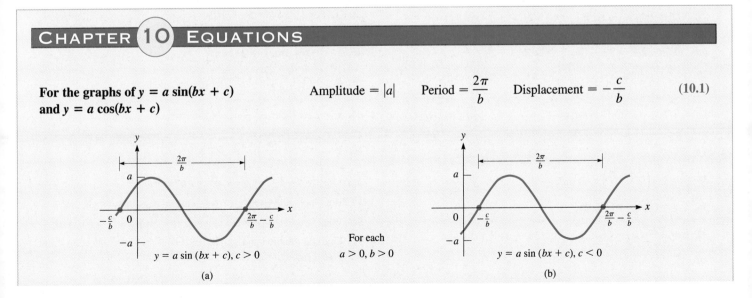

For each $a > 0, b > 0$

$y = a \sin (bx + c), c > 0$

(a)

$y = a \sin (bx + c), c < 0$

(b)

Reciprocal relationships		$\csc x = \dfrac{1}{\sin x}$	$\sec x = \dfrac{1}{\cos x}$	$\cot x = \dfrac{1}{\tan x}$	(10.2)

Simple harmonic motion $d = R \sin \omega t$ (10.3)

Angular velocity and frequency $\omega = 2\pi f$ (10.4)

CHAPTER 10 REVIEW EXERCISES

In Exercises 1–28, sketch the curves of the given trigonometric functions. Check each using a graphing calculator.

1. $y = \frac{2}{3} \sin x$

2. $y = -4 \sin x$

3. $y = -2 \cos x$

4. $y = 2.3 \cos x$

5. $y = 2 \sin 3x$

6. $y = 4.5 \sin 12x$

7. $y = 0.4 \cos 4x$

8. $y = 24 \cos 6x$

9. $y = 3 \cos \frac{1}{3}x$

10. $y = 3 \sin \frac{1}{2}x$

11. $y = \sin \pi x$

12. $y = 36 \sin 4\pi x$

13. $y = 5 \cos 2\pi x$

14. $y = -\cos 6\pi x$

15. $y = -0.5 \sin \frac{\pi}{6}x$

16. $y = 8 \sin \frac{\pi}{4}x$

17. $y = 2 \sin\left(3x - \dfrac{\pi}{2}\right)$

18. $y = 3 \sin\left(\dfrac{x}{2} + \dfrac{\pi}{2}\right)$

19. $y = -2 \cos(4x + \pi)$

20. $y = 0.8 \cos\left(\dfrac{x}{6} - \dfrac{\pi}{2}\right)$

21. $y = -\sin\left(\pi x + \dfrac{\pi}{6}\right)$

22. $y = 250 \sin(3\pi x - \pi)$

23. $y = 8 \cos\left(4\pi x - \dfrac{\pi}{2}\right)$

24. $y = 3 \cos(2\pi x + \pi)$

25. $y = 0.3 \tan 0.5x$

26. $y = \frac{1}{4} \sec x$

27. $y = -\frac{1}{3} \csc x$

28. $y = -5 \cot x$

In Exercises 29–32, sketch the curves of the given functions by addition of ordinates.

29. $y = 2 + \frac{1}{2} \sin 2x$

30. $y = \frac{1}{2}x - \cos \frac{1}{3}x$

31. $y = \sin 2x + 3 \cos x$

32. $y = \sin 3x + 2 \cos 2x$

In Exercises 33–40, display the curves of the given functions on a graphing calculator.

33. $y = 2 \sin x - \cos 2x$

34. $y = 10 \sin 3x - 20 \cos x$

35. $y = \cos\left(x + \dfrac{\pi}{4}\right) - 2 \sin 2x$

36. $y = 2 \cos \pi x + \cos(2\pi x - \pi)$

37. $y = \dfrac{\sin x}{x}$

38. $y = \sqrt{x} \sin 0.5x$

W 39. $y = \sin^2 x + \cos^2 x$ $(\sin^2 x = (\sin x)^2)$
What conclusion can be drawn from the graph?

W 40. $y = \sin\left(x + \dfrac{\pi}{4}\right) - \cos\left(x - \dfrac{\pi}{4}\right) + 1$
What conclusion can be drawn from the graph?

In Exercises 41–44, give the specific form of the indicated equation by evaluating a, b, and c through an inspection of the given curve.

41. $y = a \sin(bx + c)$
(Figure 10.45)

42. $y = a \cos(bx + c)$
(Figure 10.45)

43. $y = a \cos(bx + c)$
(Figure 10.46)

44. $y = a \sin(bx + c)$
(Figure 10.46)

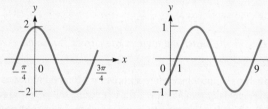

Fig. 10.45   **Fig. 10.46**

In Exercises 45–48, display the Lissajous figures on a graphing calculator.

45. $x = -\cos 2\pi t, \ y = 2 \sin \pi t$

46. $x = \sin\left(t + \dfrac{\pi}{6}\right), \ y = \sin t$

47. $x = 2 \cos\left(2\pi t + \dfrac{\pi}{4}\right), \ y = \cos \pi t$

48. $x = \cos\left(t - \dfrac{\pi}{6}\right), \ y = \cos\left(2t + \dfrac{\pi}{3}\right)$

In Exercises 49–54, solve the given problems.

49. What is the period of the function $y = 2 \cos 0.5x + \sin 3x$?

50. What is the period of the function $y = \sin \pi x + 3 \sin 0.25\pi x$?

51. Find the function and graph it if it is of the form $y = a \sin x$ and passes through $(5\pi/2, 3)$.

52. Find the function and graph it if it is of the form $y = a \cos x$ and passes through $(4\pi, -3)$.

53. Find the function and graph it if it is of the form $y = 3 \cos bx$ and passes through $(\pi/3, -3)$ and b has the smallest possible positive value.

54. Find the function and graph it if it is of the form $y = 3 \sin bx$ and passes through $(\pi/3, 0)$ and b has the smallest possible positive value.

In Exercises 55–76, sketch the appropriate curves. A graphing calculator may be used.

55. The range R of a rocket is given by $R = \dfrac{v_0^2 \sin 2\theta}{g}$. Sketch R as a function of θ for $v_0 = 1000$ m/s and $g = 9.8$ m/s². See Fig. 10.47.

Fig. 10.47

56. The blade of a saber saw moves vertically up and down at 18 strokes per second. The vertical displacement y (in cm) is given by $y = 1.2 \sin 36\pi t$, where t is in seconds. Sketch at least two cycles of the graph of y vs. t.

57. The velocity v (in cm/s) of a piston in a certain engine is given by $v = \omega D \cos \omega t$, where ω is the angular velocity of the crankshaft in radians per second and t is the time in seconds. Sketch the graph of v vs. t if the engine is at 3000 r/min and $D = 3.6$ cm.

58. A light wave for the color yellow can be represented by the equation $y = A \sin 3.4 \times 10^{15} t$. With A as a constant, sketch two cycles of y as a function of t (in s).

59. The electric current i (in A) in a circuit in which there is a *full-wave rectifier* is $i = 10 \, |\sin 120\pi t|$. Sketch the graph of $i = f(t)$ for $0 \le t \le 0.05$ s. What is the period of the current?

60. A circular disk suspended by a thin wire attached to the center of one of its flat faces is twisted through an angle θ. Torsion in the wire tends to turn the disk back in the opposite direction (thus, the name *torsion pendulum* is given to this device). The angular displacement θ (in rad) as a function of time t (in s) is $\theta = \theta_0 \cos(\omega t + \alpha)$, where θ_0 is the maximum angular displacement, ω is a constant that depends on the properties of the disk and wire, and α is the phase angle. Sketch the graph of θ vs. t if $\theta_0 = 0.100$ rad, $\omega = 2.50$ rad/s, and $\alpha = \pi/4$. See Fig. 10.48.

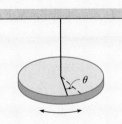

Fig. 10.48

61. The vertical displacement y of a point at the end of a propeller blade of a small boat is $y = 14.0 \sin 40.0\pi t$. Sketch two cycles of y (in cm) as a function of t (in s).

62. In optics, two waves are said to interfere destructively if, when they both pass through a medium, the amplitude of the resulting wave is zero. Sketch the graph of $y = \sin x + \cos(x + \pi/2)$ and find whether or not it would represent destructive interference of two waves.

63. The vertical displacement y (in dm) of a buoy floating in water is given by $y = 3.0 \cos 0.2t + 1.0 \sin 0.4t$, where t is in seconds. Sketch the graph of y as a function of t for the first 40 s.

64. The vertical motion of a rubber raft on a lake approximates simple harmonic motion due to the waves. If the amplitude of the motion is 0.250 m and the period is 3.00 s, find an equation for the vertical displacement y as a function of the time t. Sketch two cycles.

65. A drafting student draws a circle through the three vertices of a right triangle. The hypotenuse of the triangle is the diameter d of the circle, and from Fig. 10.49, we see that $d = a \sec \theta$. Sketch the graph of d as a function of θ for $a = 3.00$ cm.

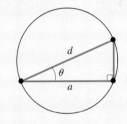

Fig. 10.49

66. The height h (in m) of a certain rocket ascending vertically is given by $h = 800 \tan \theta$, where θ is the angle of elevation from an observer 800 m from the launch pad. Sketch h as a function of θ.

67. At 40° N latitude the number of hours h of daylight each day during the year is approximately $h = 12.2 + 2.8 \sin\left[\frac{\pi}{6}(x - 2.7)\right]$, where x is measured in months ($x = 0.5$ is Jan. 15, etc.) Sketch the graph of h vs. x for one year. (Some of the cities near 40° N are Philadelphia, Madrid, Naples, Ankara, and Beijing.)

W 68. The equation in Exercise 67 can be used for the number of hours of daylight at 40° S latitude with the appropriate change. Explain what change is necessary and determine the proper equation. Sketch the graph. (This would be appropriate for southern Argentina and Wellington, New Zealand.)

69. If the upper end of a spring is not fixed and is being moved with a sinusoidal motion, the motion of the bob at the end of the spring is affected. Sketch the curve if the motion of the upper end of a spring is being moved by an external force and the bob moves according to the equation $y = 4 \sin 2t - 2 \cos 2t$.

70. The loudness L (in dB) of a fire siren as a function of the time t (in s) is approximately $L = 40 - 35 \cos 2t + 60 \sin t$. Sketch this function for $0 \le t \le 10$ s.

71. The path of a roller mechanism used in an assembly-line process is given by $x = \theta - \sin \theta$ and $y = 1 - \cos \theta$. Sketch the path for $0 \le \theta \le 2\pi$.

72. The equations for two voltage signals that give a resulting curve on an oscilloscope are $x = 6 \sin \pi t$ and $y = 4 \cos 4\pi t$. Sketch the graph of the curve displayed on the oscilloscope.

73. The impedance Z (in Ω) and resistance R (in Ω) for an alternating-current circuit are related by $Z = R \sec \theta$, where θ is called the phase angle. Sketch the graph for Z as a function of θ for $-\pi/2 < \theta < \pi/2$.

74. For an object sliding down an inclined plane at constant speed, the coefficient of friction μ between the object and the plane is given by $\mu = \tan \theta$, where θ is the angle between the plane and the horizontal. Sketch the graph of μ vs. θ.

75. The charge q (in C) on a certain capacitor as a function of the time t (in s) is given by $q = 0.0003(3 - 2 \sin 100t \cos 100t)$. Sketch two cycles of q vs. t.

76. The instantaneous power p (in W) in an electric circuit is defined as the product of the instantaneous voltage e and the instantaneous current i (in A). If we have $e = 100 \cos 200t$ and $i = 2 \cos\left(200t + \frac{\pi}{4}\right)$, plot the graph e vs. t and the graph of i vs. t on the same coordinate system. Then sketch the graph of p vs. t by multiplying appropriate values of e and i.

Writing Exercise

77. A wave passing through a string can be described at any instant by the equation $y = a \sin(bx + c)$. Write one or two paragraphs explaining the change in the wave (a) if a is doubled, (b) if b is doubled, and (c) if c is doubled.

CHAPTER 10 PRACTICE TEST

In Problems 1–4, sketch the graphs of the given functions.

1. $y = 0.5 \cos \frac{\pi}{2} x$

2. $y = 2 + 3 \sin x$

3. $y = 3 \sec x$

4. $y = 2 \sin\left(2x - \frac{\pi}{3}\right)$

5. A wave is traveling in a string. The displacement y (in cm) as a function of the time t (in s) from its equilibrium position is given by $y = A \cos(2\pi/T)t$. T is the period (in s) of the motion. If $A = 0.200$ cm and $T = 0.100$ s, sketch two cycles of y vs. t.

6. Sketch the graph of $y = 2 \sin x + \cos 2x$ by addition of ordinates.

7. Use a graphing calculator to display the Lissajous figure for which $x = \sin \pi t$ and $y = 2 \cos 2\pi t$.

8. Sketch two cycles of the curve of a projection of the end of a radius on the y-axis. The radius is of length R and it is rotating counterclockwise about the origin at 2.00 rad/s. It starts at an angle of $\pi/6$ with the positive x-axis.

9. Find the function of the form $y = 2 \sin bx$ if its graph passes through $(\pi/3, 2)$ and b is the smallest possible positive value. Then graph the function.

11 Exponents and Radicals

For use in later chapters, we now further develop the use of exponents and radicals. In previous chapters we have used only exponents that are integers, and by introducing exponents that are fractions we will show the relationship between exponents and radicals $\left(\text{for example, we will show that } \sqrt{x} = x^{1/2}\right)$. In more advanced math and in applications, it is more convenient to use fractional exponents rather than radicals.

As we noted in Chapter 6, the use of symbols led to advances in mathematics and science. As letters became commonly used in the 1600s, it was common to write, for example, x^3 as xxx. For larger powers, this is obviously inconvenient, and the modern use of exponents came into use. The first to use exponents consistently was the French mathematician René Descartes in the 1630s.

The meaning of negative and fractional exponents was first found by the English mathematician Wallis in the 1650s, although he did not write them as we do today. In the 1670s, it was the great English mathematician and physicist Isaac Newton who first used all exponents (positive, negative, and fractional) with modern notation. This improvement in notation made the development of many areas of mathematics, particularly calculus, easier. In this way, it also helped lead to many advances in the applications of mathematics.

As we develop the various operations with exponents and radicals, we will show their uses in some technical areas of application. They are used in a number of formulas in areas such as electronics, hydrodynamics, optics, solar energy, and machine design.

In finding the rate at which solar radiation changes at a solar-energy collector, the following expression is found:

$$\frac{(t^4 + 100)^{1/2} - 2t^3(t + 6)(t^4 + 100)^{-1/2}}{[(t^4 + 100)^{1/2}]^2}$$

In Section 11.2, we show that this can be written in a much simpler form.

11.1 SIMPLIFYING EXPRESSIONS WITH INTEGRAL EXPONENTS

The laws of exponents were given in Section 1.4. For reference, they are

$$a^m \times a^n = a^{m+n} \tag{11.1}$$

$$\frac{a^m}{a^n} = a^{m-n} \quad \text{or} \quad \frac{a^m}{a^n} = \frac{1}{a^{n-m}} \quad (a \neq 0) \tag{11.2}$$

$$(a^m)^n = a^{mn} \tag{11.3}$$

$$(ab)^n = a^n b^n, \quad \left(\frac{a}{b}\right)^n = \frac{a^n}{b^n} \quad (b \neq 0) \tag{11.4}$$

$$a^0 = 1 \quad (a \neq 0) \tag{11.5}$$

$$a^{-n} = \frac{1}{a^n} \quad (a \neq 0) \tag{11.6}$$

Although Eqs. (11.1) to (11.4) were originally defined for positive integers as exponents, we showed in Section 1.4 that, with the definitions given in Eqs. (11.5) and (11.6), they are valid for all integral exponents. Later in this chapter, we will show how fractions may be used as exponents. Since these equations are very important to the development of the topics in this chapter, they should again be reviewed and learned thoroughly.

In this section, we review the use of exponents while using Eqs. (11.1) to (11.6). Then we show how exponents are used and handled in more involved expressions.

◀ EXAMPLE 1 Applying Eq. (11.1), we have

$$a^5 \times a^{-3} = a^{5+(-3)} = a^{5-3} = a^2$$

Applying Eq. (11.1) and then Eq. (11.6), we have

$$a^3 \times a^{-5} = a^{3-5} = a^{-2} = \frac{1}{a^2}$$

NOTE ▶ *Negative exponents are generally not used in the expression of a final result,* unless specified otherwise. However, they are often used in intermediate steps. ▶

◀ EXAMPLE 2 Applying Eq. (11.1), then (11.6), and then (11.4), we have

$$(10^3 \times 10^{-4})^2 = (10^{3-4})^2 = (10^{-1})^2 = \left(\frac{1}{10}\right)^2 = \frac{1}{10^2} = \frac{1}{100}$$

Often, several combinations of the laws can be used to simplify an expression. For example, this expression can be simplified by using Eq. (11.1), then (11.3), and then (11.6), as follows:

$$(10^3 \times 10^{-4})^2 = (10^{3-4})^2 = (10^{-1})^2 = 10^{-2} = \frac{1}{10^2} = \frac{1}{100}$$

The result is in a proper form as either $1/10^2$ or $1/100$. If the exponent is large, then it is common to leave the exponent in the answer. ▶

◀ EXAMPLE 3   Applying Eqs. (11.2) and (11.5), we have

$$\frac{a^2 b^3 c^0}{ab^7} = \frac{a^{2-1}(1)}{b^{7-3}} = \frac{a}{b^4}$$

Applying Eqs. (11.4) and (11.3), we have

$$(x^{-2}y)^3 = (x^{-2})^3(y^3) = x^{-6}y^3 = \frac{y^3}{x^6}$$

Here, the simplification was completed by the use of Eq. (11.6).

◀ EXAMPLE 4 $(x^2 y)^2 \left(\dfrac{2}{x}\right)^{-2} = \dfrac{(x^4 y^2)}{\left(\dfrac{2}{x}\right)^2} = \dfrac{x^4 y^2}{\dfrac{4}{x^2}} = \dfrac{x^4 y^2}{1} \times \dfrac{x^2}{4} = \dfrac{x^6 y^2}{4}$

or

$$(x^2 y)^2 \left(\frac{2}{x}\right)^{-2} = (x^4 y^2)\left(\frac{2^{-2}}{x^{-2}}\right) = (x^4 y^2)\left(\frac{x^2}{2^2}\right) = \frac{x^6 y^2}{4}$$

In the first simplification, we first used Eq. (11.6) and then Eq. (11.4). The simplification was completed by changing the division by a fraction to multiplication and using Eq. (11.1).

In the second simplification, we first used Eq. (11.3), then Eq. (11.6), and finally Eq. (11.1). ▶

Named for the French mathematician and scientist Blaise Pascal (1623–1662).

◀ EXAMPLE 5   When writing a denominate number, if units of measurement appear in the denominator, they can be written using negative exponents. For example, the metric unit for pressure is the *pascal,* where $1\ \text{Pa} = 1\ \text{N/m}^2$. This can be written as

$$1\ \text{Pa} = 1\ \text{N/m}^2 = 1\ \text{N} \cdot \text{m}^{-2}$$

where $1/\text{m}^2 = \text{m}^{-2}$.

Named for the English physicist James Prescott Joule (1818–1899).

The metric unit for energy is the *joule,* where $1\ \text{J} = 1\ \text{kg} \cdot (\text{m} \cdot \text{s}^{-1})^2$, or

$$1\ \text{J} = 1\ \text{kg} \cdot \text{m}^2 \cdot \text{s}^{-2} = 1\ \text{kg} \cdot \text{m}^2/\text{s}^2$$ ▶

Care must be taken to apply the laws of exponents properly. Certain common problems are pointed out in the following examples.

◀ EXAMPLE 6   The expression $(-5x)^0$ equals 1, whereas the expression $-5x^0$ equals -5. For $(-5x)^0$, the parentheses show that the expression $-5x$ is raised to the zero power, whereas for $-5x^0$ only x is raised to the zero power and we have

$$-5x^0 = -5(1) = -5$$

CAUTION ▶   Also, $(-5)^0 = 1$ but $-5^0 = -1$. Again, for $(-5)^0$ parentheses show -5 raised to the zero power, whereas for -5^0 only 5 is raised to the zero power.

Similarly, $(-2)^2 = 4$ and $-2^2 = -4$.

CAUTION ▶   For the same reasons, $2x^{-1} = \dfrac{2}{x}$ whereas $(2x)^{-1} = \dfrac{1}{2x}$. ▶

EXAMPLE 7 $(2a + b^{-1})^{-2} = \dfrac{1}{(2a + b^{-1})^2} = \dfrac{1}{\left(2a + \dfrac{1}{b}\right)^2} = \dfrac{1}{\left(\dfrac{2ab + 1}{b}\right)^2}$

$$= \dfrac{1}{\dfrac{(2ab + 1)^2}{b^2}} = \dfrac{b^2}{(2ab + 1)^2}$$

not necessary to expand the denominator

Another order of operations for simplifying this expression is

$$(2a + b^{-1})^{-2} = \left(2a + \dfrac{1}{b}\right)^{-2} = \left(\dfrac{2ab + 1}{b}\right)^{-2}$$

$$= \dfrac{(2ab + 1)^{-2}}{b^{-2}} = \dfrac{b^2}{(2ab + 1)^2}$$

positive exponents used in the final result

EXAMPLE 8 There is an error that is commonly made in simplifying the type of expression in Example 7. We must be careful to see that

CAUTION ▶ $(2a + b^{-1})^{-2}$ is **not** equal to $(2a)^{-2} + (b^{-1})^{-2}$, or $\dfrac{1}{4a^2} + b^2$

Remember: As noted in Section 6.1, when raising a binomial (or any multinomial) to a power, we cannot simply raise each term to the power to obtain the result.

For reference, Eq. (11.4) is $(ab)^n = a^n b^n$.

However, when raising a product of factors to a power, we use Eq. (11.4). Thus,

$$(2ab^{-1})^{-2} = (2a)^{-2}(b^{-1})^{-2} = \dfrac{b^2}{(2a)^2} = \dfrac{b^2}{4a^2}$$

We see that we must be careful to distinguish between the power of a sum of terms and the power of a product of factors.

CAUTION ▶ From the preceding examples, we see that *when a factor is moved from the denominator to the numerator of a fraction, or conversely, the* **sign** *of the exponent is changed.* We should carefully note the word *factor;* this rule does not apply to moving *terms* in the numerator or the denominator.

EXAMPLE 9 $3L^{-1} - (2L)^{-2} = \dfrac{3}{L} - \dfrac{1}{(2L)^2} = \dfrac{3}{L} - \dfrac{1}{4L^2}$

$$= \dfrac{12L - 1}{4L^2}$$

EXAMPLE 10 $3^{-1}\left(\dfrac{4^{-2}}{3 - 3^{-1}}\right) = \dfrac{1}{3}\left(\dfrac{1}{4^2}\right)\left(\dfrac{1}{3 - \dfrac{1}{3}}\right) = \dfrac{1}{3 \times 4^2}\left(\dfrac{1}{\dfrac{9 - 1}{3}}\right)$

$$= \dfrac{1}{3 \times 4^2}\left(\dfrac{3}{8}\right) = \dfrac{1}{128}$$

◀ EXAMPLE 11

$$\frac{1}{x^{-1}}\left(\frac{x^{-1}-y^{-1}}{x^2-y^2}\right) = \frac{x}{1}\left(\frac{\dfrac{1}{x}-\dfrac{1}{y}}{x^2-y^2}\right) = x\left(\frac{\dfrac{y-x}{xy}}{x^2-y^2}\right)$$

terms (labeled above the numerator $x^{-1}-y^{-1}$)

$$= \frac{\dfrac{x(y-x)}{xy}}{(x-y)(x+y)}$$

$$= \frac{x(y-x)}{xy} \times \frac{1}{(x-y)(x+y)}$$

$$= \frac{x(y-x)}{xy(x-y)(x+y)} = \frac{-(x-y)}{y(x-y)(x+y)}$$

$$= -\frac{1}{y(x+y)}$$

CAUTION ▶ Note that in this example **the x^{-1} and y^{-1} in the numerator could not be moved directly to the denominator with positive exponents** because they are only terms of the original numerator. ◗

◀ EXAMPLE 12 $3(x+4)^2(x-3)^{-2} - 2(x-3)^{-3}(x+4)^3$

$$= \frac{3(x+4)^2}{(x-3)^2} - \frac{2(x+4)^3}{(x-3)^3} = \frac{3(x-3)(x+4)^2 - 2(x+4)^3}{(x-3)^3}$$

$$= \frac{(x+4)^2[3(x-3) - 2(x+4)]}{(x-3)^3}$$

$$= \frac{(x+4)^2(x-17)}{(x-3)^3}$$

Expressions such as the one in this example are commonly found in problems in calculus. ◗

EXERCISES 11.1

In Exercises 1–4, solve the resulting problems if the given changes are made in the indicated examples of this section.

1. In Example 4, change the factor x^2 to x^{-2} and then find the result.

2. In Example 7, change term $2a$ to $2a^{-1}$ and then find the result.

3. In Example 10, change the 3^{-1} in the denominator to 3^{-2} and then find the result.

4. In Example 11, change the sign in the numerator from $-$ to $+$ and then find the result.

In Exercises 5–52, express each of the given expressions in simplest form with only positive exponents.

5. $x^7 x^{-4}$

6. $y^9 y^{-2}$

7. $2a^2 a^{-6}$

8. $5ss^{-5}$

9. 5×5^{-3}

10. $(3^2 \times 4^{-3})^3$

11. $(2\pi x^{-1})^2$

12. $(3xy^{-2})^3$

13. $(5an^{-2})^{-1}$

14. $(6s^2 t^{-1})^{-2}$

15. $(-4)^0$

16. -4^0

17. $-7x^0$

18. $(-7x)^0$

19. $3x^{-2}$

20. $(3x)^{-2}$

21. $(7a^{-1}x)^{-3}$

22. $7a^{-1}x^{-3}$

23. $\left(\dfrac{2}{n^3}\right)^{-3}$

24. $\left(\dfrac{3}{x^3}\right)^{-2}$

25. $\left(\dfrac{a}{b^{-2}}\right)^{-3}$

26. $\left(\dfrac{2n^{-2}}{D^{-1}}\right)^{-2}$

27. $(a+b)^{-1}$

28. $a^{-1} + b^{-1}$

29. $3x^{-2} + 2y^{-2}$

30. $(3x + 2y)^{-2}$

31. $(2a^{-n})^2 \left(\dfrac{3}{2a^n}\right)^{-1}$

32. $(7 \times 3^{-a})\left(\dfrac{3^a}{7}\right)^2$

33. $\left(\dfrac{3a^2}{4b}\right)^{-3}\left(\dfrac{4}{a}\right)^{-5}$

34. $(2np^{-2})^{-2}(4^{-1}p^2)^{-1}$

35. $\left(\dfrac{V^{-1}}{2t}\right)^{-2}\left(\dfrac{t^2}{V^{-2}}\right)^{-3}$

36. $\left(\dfrac{a^{-2}}{b^2}\right)^{-3}\left(\dfrac{a^{-3}}{b^5}\right)^2$

37. $2a^{-2} + (2a^{-2})^4$

38. $3(a^{-1}z^2)^{-3} + c^{-2}z^{-1}$

39. $2 \times 3^{-1} + 4 \times 3^{-2}$

40. $5 \times 2^{-2} - 3^{-1} \times 2^3$

41. $(R_1^{-1} + R_2^{-1})^{-1}$

42. $(2a - b^{-2})^{-1}$

43. $(n^{-2} - 2n^{-1})^2$

44. $(2^{-3} - 4^{-1})^{-2}$

45. $\dfrac{6^{-1}}{4^{-2} + 2}$

46. $\dfrac{x - y^{-1}}{x^{-1} - y}$

47. $\dfrac{x^{-2} - y^{-2}}{x^{-1} - y^{-1}}$

48. $\dfrac{ax^{-2} + a^{-2}x}{a^{-1} + x^{-1}}$

49. $2t^{-2} + t^{-1}(t + 1)$

50. $3x^{-1} - x^{-3}(y + 2)$

51. $(D - 1)^{-1} + (D + 1)^{-1}$

52. $4(2x - 1)(x + 2)^{-1} - (2x - 1)^2(x + 2)^{-2}$

In Exercises 53–68, perform the indicated operations.

53. Express $4^2 \times 64$ (a) as a power of 4 and (b) as a power of 2.

54. Express $1/81$ (a) as a power of 9 and (b) as a power of 3.

55. (a) By use of Eqs. (11.4) and (11.6), show that
$$\left(\dfrac{a}{b}\right)^{-n} = \left(\dfrac{b}{a}\right)^n$$

(b) Verify the equation in part (a) by evaluating each side with $a = 3.576$, $b = 8.091$, and $n = 7$.

(W) **56.** For what integral values of n is $(-3)^{-n} = -3^{-n}$? Explain.

57. For what integral value(s) of n is $n^\pi > \pi^n$?

(W) **58.** Evaluate $(8^{19})^{12}/(8^{16})^{14}$. What happens when you try to evaluate this on a calculator?

59. Solve for x: $2^{5x} = 2^7(2^{2x})^2$.

60. In analyzing the tuning of an electronic circuit, the expression $[\omega\omega_0^{-1} - \omega_0\omega^{-1}]^2$ is used. Expand and simplify this expression.

61. The metric unit of energy, the *joule* (J), can be expressed as $kg \cdot s^{-2} \cdot m^2$. Simplify these units and include *newtons* (see Appendix B) and only positive exponents in the final result.

62. The units for the electric quantity called *permittivity* are $C^2 \cdot N^{-1} \cdot m^{-2}$. Given that $1 F = 1 C^2 \cdot J^{-1}$, show that the units of permittivity are F/m. See Appendix B.

63. When studying a solar energy system, the units encountered are $kg \cdot s^{-1}(m \cdot s^{-2})^2$. Simplify these units and include *joules* (see Example 5) and only positive exponents in the final result.

64. The metric units for the velocity v of an object are $m \cdot s^{-1}$, and the units for the acceleration a of the object are $m \cdot s^{-2}$. What are the units for v/a?

65. Given that $v = a^p t^r$, where v is the velocity of an object, a is its acceleration, and t is the time, use the metric units given in Exercise 64 to show that $p = r = 1$.

66. An expression encountered in finance is
$$\dfrac{p(1 + i)^{-1}[(1 + i)^{-n} - 1]}{(1 + i)^{-1} - 1}$$
where n is an integer. Simplify this expression.

67. An idealized model of the thermodynamic process in a gasoline engine is the *Otto cycle*. The efficiency e of the process is
$$e = \dfrac{\dfrac{T_1 r^\gamma}{r} - \dfrac{T_2 r^\gamma}{r} - T_1 + T_2}{\dfrac{T_1 r^\gamma}{r} - \dfrac{T_2 r^\gamma}{r}}$$
Show that $e = 1 - \dfrac{1}{r^{\gamma-1}}$.

68. In optics, the combined focal length F of two lenses is given by $F = [f_1^{-1} + f_2^{-1} + d(f_1 f_2)^{-1}]^{-1}$, where f_1 and f_2 are the focal lengths of the lenses and d is the distance between them. Simplify the right side of this equation.

11.2 FRACTIONAL EXPONENTS

In Section 11.1, we reviewed the use of integral exponents, including exponents that are negative integers and zero. We now show how rational numbers may be used as exponents. With the appropriate definitions, all the laws of exponents are valid for all rational numbers as exponents.

Equation (11.3) states that $(a^m)^n = a^{mn}$. If we were to let $m = \frac{1}{2}$ and $n = 2$, we would have $(a^{1/2})^2 = a^1$. However, we already have a way of writing a quantity that, when squared, equals a. This is written as $\sqrt{a}$. To be consistent with previous definitions and to allow the laws of exponents to hold, we define

The radical in Eq. (11.7) is the symbol for the nth root of a number. Do not confuse it with $\sqrt{a}$, the square root of a.

$$\boxed{a^{1/n} = \sqrt[n]{a}}$$

(11.7)

In order that Eqs. (11.3) and (11.7) may hold at the same time, we define

$$a^{m/n} = \sqrt[n]{a^m} = \left(\sqrt[n]{a}\right)^m \tag{11.8}$$

These definitions are valid for all the laws of exponents. We must note that Eqs. (11.7) and (11.8) are valid as long as $\sqrt[n]{a}$ does not involve the even root of a negative number. Such numbers are imaginary and are considered in Chapter 12.

For reference, Eq. (11.1) is $a^m \times a^n = a^{m+n}$.

▌ **EXAMPLE 1** We now verify that Eq. (11.1) holds for the above definitions:

$$a^{1/4}a^{1/4}a^{1/4}a^{1/4} = a^{(1/4)+(1/4)+(1/4)+(1/4)} = a^1$$

Now, $a^{1/4} = \sqrt[4]{a}$ by definition. Also, by definition $\sqrt[4]{a}\,\sqrt[4]{a}\,\sqrt[4]{a}\,\sqrt[4]{a} = a$. Equation (11.1) is thereby verified for $n = 4$ in Eq. (11.7).

Equation (11.3) is verified by the following:

Eq. (11.3) is $(a^m)^n = a^{mn}$.

$$(a^{1/4})(a^{1/4})(a^{1/4})(a^{1/4}) = (a^{1/4})^4 = a^1 = \left(\sqrt[4]{a}\right)^4$$

We may interpret $a^{m/n}$ in Eq. (11.8) as the mth power of the nth root of a, as well as the nth root of the mth power of a. This is illustrated in the following example.

▌ **EXAMPLE 2** $\quad 8^{2/3} = \left(\sqrt[3]{8}\right)^2 = (2)^2 = 4 \quad$ or $\quad 8^{2/3} = \sqrt[3]{8^2} = \sqrt[3]{64} = 4$

Although both interpretations of Eq. (11.8) are possible, as indicated in Example 2, in evaluating numerical expressions involving fractional exponents without a calculator, it is almost always best to *find the root first, as indicated by the denominator* of the fractional exponent. This allows us to find the root of the smaller number, which is normally easier to find.

▌ **EXAMPLE 3** To evaluate $(64)^{5/2}$, we should proceed as follows:

$$(64)^{5/2} = [(64)^{1/2}]^5 = 8^5 = 32\,768$$

If we raised 64 to the fifth power first, we would have

$$(64)^{5/2} = (64^5)^{1/2} = (1\,073\,741\,824)^{1/2}$$

We would now have to evaluate the indicated square root. This demonstrates why it is preferable to find the indicated root first.

▌ **EXAMPLE 4** **(a)** $(16)^{3/4} = (16^{1/4})^3 = 2^3 = 8$

(b) $4^{-1/2} = \dfrac{1}{4^{1/2}} = \dfrac{1}{2}$ **(c)** $9^{3/2} = (9^{1/2})^3 = 3^3 = 27$

Eq. (11.6) is $a^{-n} = \dfrac{1}{a^n}$.

We note in (b) that Eq. (11.6) must also hold for negative rational exponents. In writing $4^{-1/2}$ as $1/4^{1/2}$, the *sign* of the exponent is changed.

Fractional exponents allow us to find roots of numbers on a calculator. By use of the appropriate key $\left(\text{on most calculators } \boxed{x^y} \text{ or } \boxed{\wedge}\right)$, we may raise any positive number to any power. For roots, we use the equivalent fractional exponent. Powers that are fractions or decimal in form are entered directly.

Named for the Scottish physicist Lord Kelvin (1824–1907).

Fig. 11.1

EXAMPLE 5 The thermodynamic temperature T (in kelvins (K)) is related to the pressure P (in kPa) of a gas by the equation $T = 80.5P^{2/7}$. Find the value of T for $P = 750$ kPa.

Substituting, we have

$$T = 80.5(750)^{2/7}$$

The calculator display for this calculation is shown in Fig. 11.1. Therefore, $T = 534$ K.

When finding powers of negative numbers, some calculators will show an error. If this is the case, enter the positive value of the number and then enter a negative sign for the result when appropriate. From Section 1.6, we recall that *an even root of a negative number is imaginary and an odd root of a negative number is negative*. If it is an integral power, the basic laws of signs are used.

EXAMPLE 6 Plot the graph of the function $y = 2x^{1/3}$.

In obtaining points for the graph, we use $(1/3)$ as the power when using the calculator. If your calculator does not evaluate powers of negative numbers, enter positive values for the negative values of x and then make the results negative. Since $x^{1/3} = \sqrt[3]{x}$, we know that $x^{1/3}$ is negative for negative values of x because we have an odd root of a negative number. We get the following table of values:

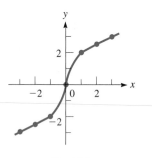

Fig. 11.2

x	-3	-2	-1	0	1	2	3
y	-2.9	-2.5	-2.0	0	2.0	2.5	2.9

The graph is shown in Fig. 11.2. Of course, this curve can easily be shown on a graphing calculator.

Another reason for developing fractional exponents is that they are often easier to use in more complex expressions involving roots. This is true in algebra and in topics from more advanced mathematics. Any expression with radicals can also be expressed with fractional exponents and then simplified. We now show some additional examples with fractional exponents.

For reference, basic forms of Eq. (11.1) to (11.7) are as follows:

$$a^m \times a^n = a^{m+n} \quad (11.1)$$
$$\frac{a^m}{a^n} = a^{m-n} \quad (11.2)$$
$$(a^m)^n = a^{mn} \quad (11.3)$$
$$(ab)^n = a^n b^n \quad (11.4)$$
$$a^0 = 1 \quad (11.5)$$
$$a^{-n} = \frac{1}{a^n} \quad (11.6)$$
$$a^{1/n} = \sqrt[n]{a} \quad (11.7)$$

EXAMPLE 7 **(a)** $(8a^2b^4)^{1/3} = [(8^{1/3})(a^2)^{1/3}(b^4)^{1/3}]$ using Eq. (11.4)

$$= 2a^{2/3}b^{4/3} \quad \text{using Eqs. (11.7) and (11.3)}$$

(b) $a^{3/4}a^{4/5} = a^{3/4+4/5} = a^{31/20}$ using Eq. (11.1)

EXAMPLE 8

$$\left(\frac{4^{-3/2}x^{2/3}y^{-7/4}}{2^{3/2}x^{-1/3}y^{3/4}}\right)^{2/3} = \left(\frac{x^{2/3}x^{1/3}}{2^{3/2}4^{3/2}y^{3/4}y^{7/4}}\right)^{2/3} \quad \text{using Eq. (11.6)}$$

$$= \left(\frac{x^{2/3+1/3}}{2^{3/2}4^{3/2}y^{3/4+7/4}}\right)^{2/3} \quad \text{using Eq. (11.1)}$$

$$= \frac{x^{(1)(2/3)}}{2^{(3/2)(2/3)}4^{(3/2)(2/3)}y^{(10/4)(2/3)}} \quad \text{using Eq. (11.4)}$$

$$= \frac{x^{2/3}}{8y^{5/3}}$$

◀ EXAMPLE 9 $(4x^4)^{-1/2} - 3x^{-3} = \dfrac{1}{(4x^4)^{1/2}} - \dfrac{3}{x^3}$ using Eq. (11.6)

$= \dfrac{1}{2x^2} - \dfrac{3}{x^3}$ using Eq. (11.7)

$= \dfrac{x - 6}{2x^3}$ common denominator ▶

See the chapter introduction.

◀ EXAMPLE 10 The rate R at which solar radiation changes at a solar-energy collector during a day is given by the equation

$$R = \frac{(t^4 + 100)^{1/2} - 2t^3(t + 6)(t^4 + 100)^{-1/2}}{[(t^4 + 100)^{1/2}]^2}$$

Solar collectors supply the power for most space satellites.

Here, R is measured in kW/(m² · h), t is the number of hours from noon, and $-6\,\text{h} \leq t \leq 8\,\text{h}$. Express the right side of this equation in simpler form and find R for $t = 0$ (noon) and for $t = 4\,\text{h}$ (4 P.M.)

Performing the simplification, we have the following steps:

$$R = \frac{(t^4 + 100)^{1/2} - \dfrac{2t^3(t + 6)}{(t^4 + 100)^{1/2}}}{(t^4 + 100)}$$ ← using Eq. (11.6)
← using Eq. (11.3)

$$= \frac{\dfrac{(t^4 + 100)^{1/2}(t^4 + 100)^{1/2} - 2t^3(t + 6)}{(t^4 + 100)^{1/2}}}{(t^4 + 100)}$$ ← common denominator

$$= \frac{(t^4 + 100) - 2t^3(t + 6)}{(t^4 + 100)^{1/2}} \times \frac{1}{t^4 + 100}$$ ← invert divisor and multiply

$$= \frac{100 - 12t^3 - t^4}{(t^4 + 100)^{1/2}(t^4 + 100)} = \frac{100 - 12t^3 - t^4}{(t^4 + 100)^{3/2}}$$ ← using Eq. (11.1)

For $t = 0$: $R = \dfrac{100 - 12(0^3) - 0^4}{(0^4 + 100)^{3/2}} = 0.10\,\text{kW/(m² · h)}$

For $t = 4\,\text{h}$: $R = \dfrac{100 - 12(4^3) - 4^4}{(4^4 + 100)^{3/2}} = -0.14\,\text{kW/(m² · h)}$

We see that the radiation is increasing at noon, and the negative sign tells us that it is decreasing at 4 P.M. ▶

EXERCISES 11.2

In Exercises 1–4, solve the resulting problems if the given changes are made in the indicated examples of this section.

1. In Example 2, change the exponent to 4/3 and then find the result.

2. In Example 4(b), change the exponent to $-3/2$ and then find the result.

3. In Example 6, change the exponent to $-1/3$ and then plot the graph.

4. In Example 9, change the exponent $-1/2$ to $-3/2$ and then find the result.

In Exercises 5–28, evaluate the given expressions.

5. $25^{1/2}$

6. $27^{1/3}$

7. $81^{1/4}$

8. $125^{2/3}$

9. $100^{25/2}$

10. $16^{5/4}$

11. $8^{-1/3}$

12. $16^{-1/4}$

13. $64^{-2/3}$

14. $32^{-4/5}$

15. $5^{1/2}5^{3/2}$

16. $(4^4)^{3/2}$

17. $(3^6)^{2/3}$

18. $\dfrac{121^{-1/2}}{100^{1/2}}$

19. $\dfrac{1000^{1/3}}{400^{-1/2}}$

20. $\dfrac{7^{-1/2}}{6^{-1}7^{1/2}}$

21. $\dfrac{15^{2/3}}{5^2 15^{-1/3}}$

22. $\dfrac{(-27)^{1/3}}{6}$

23. $\dfrac{(-8)^{2/3}}{-2}$

24. $\dfrac{-4}{(-64)^{-2/3}}$

25. $125^{-2/3} - 100^{-3/2}$

26. $32^{0.4} + 25^{-0.5}$

27. $\dfrac{16^{-0.25}}{5} + \dfrac{2^{-0.6}}{2^{0.4}}$

28. $\dfrac{4^{-1}}{36^{-1/2}} - \dfrac{5^{-1/2}}{5^{1/2}}$

In Exercises 29–32, use a calculator to evaluate each expression.

29. $17.98^{1/4}$

30. $750.81^{2/3}$

31. $4.0187^{-4/9}$

32. $0.1863^{-1/6}$

In Exercises 33–56, simplify the given expressions. Express all answers with positive exponents.

33. $B^{7/3}B^{1/7}$

34. $x^{5/6}x^{-1/3}$

35. $\dfrac{y^{-1/2}}{y^{2/5}}$

36. $\dfrac{s^{1/4}s^{2/3}}{s^{-1}}$

37. $\dfrac{x^{3/10}}{x^{-1/5}x^2}$

38. $\dfrac{R^{-2/5}R^2}{R^{-3/10}}$

39. $(8a^3b^6)^{1/3}$

40. $(8b^{-4}c^2)^{2/3}$

41. $(16a^4b^3)^{-3/4}$

42. $(32C^5D^4)^{-2/5}$

43. $\left(\dfrac{a^{5/7}}{a^{2/3}}\right)^{7/4}$

44. $\left(\dfrac{4a^{5/6}b^{-1/5}}{a^{2/3}b^2}\right)^{-1/2}$

45. $\dfrac{1}{2}(4x^2 + 1)^{-1/2}(8x)$

46. $\dfrac{2}{3}(x^3 + 1)^{-1/3}(3x^2)$

47. $\left(\dfrac{6x^{-1/2}y^{2/3}}{18x^{-1}}\right)\left(\dfrac{2y^{1/4}}{x^{1/3}}\right)$

48. $\dfrac{3^{-1}a^{1/2}}{4^{-1/2}b} \div \dfrac{9^{1/2}a^{-1/3}}{2b^{-1/4}}$

49. $(T^{-1} + 2T^{-2})^{-1/2}$

50. $(a^{-2} - a^{-4})^{-1/4}$

51. $(a^3)^{-4/3} + a^{-2}$

52. $(4N^6)^{-1/2} - 2N^{-1}$

53. $[(a^{1/2} - a^{-1/2})^2 + 4]^{1/2}$

54. $4x^{1/2} + \dfrac{1}{2}x^{-1/2}(4x + 1)$

55. $x^2(2x - 1)^{-1/2} + 2x(2x - 1)^{1/2}$

56. $(3n - 1)^{-2/3}(1 - n) - (3n - 1)^{1/3}$

In Exercises 57–60, graph the given functions.

57. $f(x) = 3x^{1/2}$

58. $f(x) = 2x^{2/3}$

59. $f(t) = t^{-4/5}$

60. $f(V) = 4V^{3/2}$

In Exercises 61–68, perform the indicated operations.

61. Express with fractional exponents: (a) $\sqrt[5]{x}$ and (b) $\sqrt[7]{T^3}$.

(W) **62.** (a) Simplify $(x^2 - 4x + 4)^{1/2}$. (b) For what values of x is your answer in part (a) valid? Explain.

(W) **63.** A factor used in determining the performance of a solar-energy storage system is $(A/S)^{-1/4}$, where A is the actual storage capacity and S is a standard storage capacity. If this factor is 0.5, explain how to find the ratio A/S.

64. A factor used in measuring the loudness sensed by the human ear is $(I/I_0)^{0.3}$, where I is the intensity of the sound and I_0 is a reference intensity. Evaluate this factor for $I = 3.2 \times 10^{-6}$ W/m² (ordinary conversation) and $I_0 = 10^{-12}$ W/m².

65. The period T of a satellite circling earth is given by

$$T^2 = kR^3\left(1 + \dfrac{d}{R}\right)^3,$$ where R is the radius of earth, d is the distance of the satellite above earth, and k is a constant. Solve for R, using fractional exponents in the result.

66. The withdrawal resistance R of a nail of diameter d indicates its holding power. One formula for R is $R = ks^{5/2}dh$, where k is a constant, s is the specific gravity of the wood, and h is the depth of the nail in the wood. Solve for s using fractional exponents in the result.

67. The electric current i (in A) in a circuit with a battery of voltage E, a resistance R, and an inductance L, is $i = \dfrac{E}{R}(1 - e^{-Rt/L})$, where t is the time after the circuit is closed. See Fig. 11.3. Find i for $E = 6.20$ V, $R = 1.20$ Ω, $L = 3.24$ H, and $t = 0.001\,00$ s. (The number e is irrational and can be found from the calculator by using the $\boxed{e^x}$ key with $x = 1$.)

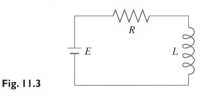

Fig. 11.3

68. For the heat-seeking rocket in pursuit of an aircraft, the distance d (in km) from the rocket to the aircraft is $d = \dfrac{500(\sin \theta)^{1/2}}{(1 - \cos \theta)^{3/2}}$, where θ is shown in Fig. 11.4. Find d for $\theta = 125.0°$.

Fig. 11.4

11.3 SIMPLEST RADICAL FORM

As we have said, any expression with radicals can also be expressed with fractional exponents. For adding or subtracting radicals, there is little advantage to changing form, but with multiplication or division of radicals, fractional exponents have some advantages. Therefore, we now define operations with radicals so that they are consistent with the laws of exponents. This will let us use the form that is more convenient for the operation being performed.

$$\sqrt[n]{a^n} = \left(\sqrt[n]{a}\right)^n = a \tag{11.9}$$

$$\sqrt[n]{a}\sqrt[n]{b} = \sqrt[n]{ab} \tag{11.10}$$

$$\sqrt[m]{\sqrt[n]{a}} = \sqrt[mn]{a} \tag{11.11}$$

$$\frac{\sqrt[n]{a}}{\sqrt[n]{b}} = \sqrt[n]{\frac{a}{b}} \qquad (b \neq 0) \tag{11.12}$$

NOTE ▶ *The number under the radical is called the* **radicand,** *and the number indicating the root being taken is called the* **order** *(or* **index***) of the radical.* To avoid difficulties with imaginary numbers (which are considered in the next chapter), *we will assume that all letters represent positive numbers.*

◀ EXAMPLE 1   Following are illustrations using Eqs. (11.9) to (11.12).

(a) $\sqrt[5]{4^5} = \left(\sqrt[5]{4}\right)^5 = 4$ using Eq. (11.9)

(b) $\sqrt[3]{2}\sqrt[3]{3} = \sqrt[3]{2 \times 3} = \sqrt[3]{6}$ using Eq. (11.10)

(c) $\sqrt[3]{\sqrt{5}} = \sqrt[3 \times 2]{5} = \sqrt[6]{5}$ using Eq. (11.11)

(d) $\dfrac{\sqrt{7}}{\sqrt{3}} = \sqrt{\dfrac{7}{3}}$ using Eq. (11.12)   ▶

◀ EXAMPLE 2   In Example 5 of Section 1.6, we saw that

$$\sqrt{16 + 9} \quad \text{is \textbf{\textit{not}} equal to} \quad \sqrt{16} + \sqrt{9}$$

However, using Eq. (11.10),

$$\sqrt{16 \times 9} = \sqrt{16} \times \sqrt{9}$$
$$= 4 \times 3 = 12$$

NOTE ▶ Therefore, we must *be careful to distinguish between the root of a sum of terms and the root of a product of factors.* This is the same as with powers of sums and powers of products, as shown in Example 8 of Section 11.1. It should be the same, as a root can be interpreted as a fractional exponent.   ▶

◀ EXAMPLE 3   To simplify $\sqrt{75}$, we know that $75 = (25)(3)$ and that $\sqrt{25} = 5$. As in Section 1.6 and now using Eq. (11.10), we write

$$\sqrt{75} = \sqrt{(25)(3)} = \sqrt{25}\sqrt{3} = 5\sqrt{3}$$
$$\underset{\text{perfect square}}{\quad\ \ \uparrow\ \ \quad}$$

This illustrates one step that should always be carried out in simplifying radicals:

NOTE ▶ *Always remove all perfect nth-power factors from the radicand of a radical of order n.*   ▶

◀ EXAMPLE 4 **(a)** $\sqrt{72} = \sqrt{(36)(2)} = \sqrt{36}\sqrt{2} = 6\sqrt{2}$

$\underset{\text{perfect square}}{}$

(b) $\sqrt{a^3b^2} = \sqrt{(a^2)(a)(b^2)} = \sqrt{a^2}\sqrt{a}\sqrt{b^2} = ab\sqrt{a}$

$\underset{\text{perfect squares}}{}$

(c) cube root → $\sqrt[3]{40} = \sqrt[3]{(8)(5)} = \sqrt[3]{8}\sqrt[3]{5} = 2\sqrt[3]{5}$

$\underset{\text{perfect cube}}{}$

(d) fifth root → $\sqrt[5]{64x^8y^{12}} = \sqrt[5]{(32)(2)(x^5)(x^3)(y^{10})(y^2)}$

$\underset{\text{perfect fifth powers}}{}$

$= \sqrt[5]{(32)(x^5)(y^{10})}\sqrt[5]{2x^3y^2}$

$= 2xy^2\sqrt[5]{2x^3y^2}$

NOTE ▶ The next two examples illustrate another procedure used to simplify radicals. It is to *reduce the order of the radical,* when it is possible to do so.

◀ EXAMPLE 5 $\sqrt[6]{8} = \sqrt[6]{2^3} = 2^{3/6} = 2^{1/2} = \sqrt{2}$

Here we started with a sixth root and ended with a square root, thereby reducing the order of the radical. Fractional exponents are often helpful for this.

◀ EXAMPLE 6 **(a)** $\sqrt[8]{16} = \sqrt[8]{2^4} = 2^{4/8} = 2^{1/2} = \sqrt{2}$

(b) $\dfrac{\sqrt[4]{9}}{\sqrt{3}} = \dfrac{\sqrt[4]{3^2}}{\sqrt{3}} = \dfrac{3^{2/4}}{3^{1/2}} = 1$

(c) $\dfrac{\sqrt[6]{8}}{\sqrt{7}} = \dfrac{\sqrt[6]{2^3}}{\sqrt{7}} = \dfrac{2^{1/2}}{7^{1/2}} = \sqrt{\dfrac{2}{7}}$

(d) $\sqrt[9]{27x^6y^{12}} = \sqrt[9]{3^3x^6y^9y^3} = 3^{3/9}x^{6/9}y^{9/9}y^{3/9} = 3^{1/3}x^{2/3}yy^{1/3}$

$= y\sqrt[3]{3x^2y}$

If a radical is to be written in its *simplest form,* the two operations illustrated in the last four examples must be performed. Therefore, we have the following:

STEPS TO REDUCE A RADICAL TO SIMPLEST FORM

1. *Remove all perfect nth-power factors from a radical of order n.*

2. *If possible, reduce the order of the radical.*

When working with fractions, it has traditionally been the practice to write a fraction with radicals in a form in which the denominator contains no radicals. Such a fraction was not considered to be in simplest form unless this was done. This step of simplification was performed primarily for ease of calculation, but with a calculator it does not matter to any extent that there is a radical in the denominator. However, the procedure of writing a radical in this form, called **rationalizing the denominator,** is at times useful for other purposes. Therefore, the following examples show how the process of rationalizing the denominator is carried out.

◀ EXAMPLE 7   To write $\sqrt{\frac{2}{5}}$ in an equivalent form in which the denominator is not included under the radical sign, we *create a perfect square in the denominator* by multiplying the numerator and the denominator under the radical by 5. This gives us $\sqrt{\frac{10}{25}}$, which may be written as $\frac{1}{5}\sqrt{10}$ or $\frac{\sqrt{10}}{5}$. These steps are written as follows:

$$\sqrt{\frac{2}{5}} = \sqrt{\frac{2 \times 5}{5 \times 5}} = \sqrt{\frac{10}{25}} = \frac{\sqrt{10}}{\sqrt{25}} = \frac{\sqrt{10}}{5}$$

perfect square

◀ EXAMPLE 8   **(a)** $\dfrac{5}{\sqrt{18}} = \dfrac{5}{3\sqrt{2}} = \dfrac{5(\sqrt{2})}{3\sqrt{2}(\sqrt{2})} = \dfrac{5\sqrt{2}}{3\sqrt{4}} = \dfrac{5\sqrt{2}}{6}$

perfect square

(b) cube root → $\sqrt[3]{\dfrac{2}{3}} = \sqrt[3]{\dfrac{2 \times 9}{3 \times 9}} = \sqrt[3]{\dfrac{18}{27}} = \dfrac{\sqrt[3]{18}}{\sqrt[3]{27}} = \dfrac{\sqrt[3]{18}}{3}$

perfect cube

In (a), a perfect square was made by multiplying by $\sqrt{2}$. We can indicate this as we have shown or by multiplying the numerator by $\sqrt{2}$ and multiplying the 2 under the radical in the denominator by 2, in which case the denominator would be $3\sqrt{2 \times 2}$. In (b), we want a perfect cube, since a cube root is being found.

◀ EXAMPLE 9   The period T (in s) for one cycle of a simple pendulum is given by $T = 2\pi\sqrt{L/g}$, where L is the length of the pendulum and g is the acceleration due to gravity. Rationalize the denominator on the right side of this equation

Rationalizing this radical, we have

$$T = 2\pi\sqrt{\frac{L}{g}} = 2\pi\sqrt{\frac{gL}{g^2}}$$
$$= \frac{2\pi}{g}\sqrt{gL}$$

See Appendix C for a graphing calculator program TBLROOTS. It displays a table of square roots and cube roots.

◀ EXAMPLE 10   Simplify $\sqrt{\dfrac{1}{2a^2} + 2b^{-2}}$ and rationalize the denominator.

We will write the expression using only positive exponents, and then perform the required operations.

first combine fractions over lowest common denominator →

$$\sqrt{\frac{1}{2a^2} + \frac{2}{b^2}} = \sqrt{\frac{b^2 + 4a^2}{2a^2b^2}} = \frac{\sqrt{b^2 + 4a^2}}{ab\sqrt{2}}$$ ← sum of squares—radical cannot be simplified
$$= \frac{\sqrt{b^2 + 4a^2}\sqrt{2}}{ab\sqrt{2} \times 2}$$
$$= \frac{\sqrt{2(b^2 + 4a^2)}}{2ab}$$

EXERCISES **11.3**

In Exercises 1–4, simplify the resulting expressions if the given changes are made in the indicated examples of this section.

1. In Example 4(b), change the exponent of b to 4 and then find the resulting expression.

2. In Example 6(c), change the 8 to 27 and then find the resulting expression.

3. In Example 8(b), change the root to a fourth root and then find the resulting expression.

4. In Example 10, replace b^{-2} with b^{-4} and then find the resulting expression.

In Exercises 5–62, write each expression in simplest radical form. If a radical appears in the denominator, rationalize the denominator.

5. $\sqrt{24}$
6. $\sqrt{150}$
7. $\sqrt{45}$
8. $\sqrt{98}$
9. $\sqrt{x^2 y^5}$
10. $\sqrt{pq^2 r^7}$
11. $\sqrt{x^2 y^4 z^3}$
12. $\sqrt{12ab^2}$
13. $\sqrt{18R^5 TV^4}$
14. $\sqrt{54M^2 N^3}$
15. $\sqrt[3]{16}$
16. $\sqrt[4]{48}$
17. $\sqrt[5]{96}$
18. $\sqrt[3]{-16}$
19. $\sqrt[3]{8a^2}$
20. $\sqrt[3]{5a^4 b^2}$
21. $\sqrt[4]{64r^3 s^4 t^5}$
22. $\sqrt[5]{16x^5 y^3 z^{11}}$
23. $\sqrt[3]{8}\sqrt[5]{4}$
24. $\sqrt[3]{4}\sqrt[7]{64}$
25. $\sqrt[3]{P}\sqrt[3]{P^2 V}$
26. $\sqrt[6]{3m^5 n^8}\sqrt[6]{9mn}$
27. $\sqrt{\dfrac{3}{2}}$
28. $\sqrt{\dfrac{5}{12}}$
29. $\sqrt[3]{\dfrac{3}{4}}$
30. $\sqrt[4]{\dfrac{2}{25}}$
31. $\sqrt[5]{\dfrac{1}{9}}$
32. $\sqrt[6]{\dfrac{5}{4}}$
33. $\sqrt[4]{400}$
34. $\sqrt[8]{81}$
35. $\sqrt[6]{64}$
36. $\sqrt[9]{27}$
37. $\sqrt{4 \times 10^4}$
38. $\sqrt{4 \times 10^5}$
39. $\sqrt{4 \times 10^6}$
40. $\sqrt[3]{16 \times 10^5}$
41. $\sqrt[4]{4a^2}$
42. $\sqrt[6]{b^2 c^4}$
43. $\sqrt[4]{\dfrac{1}{4}}$
44. $\dfrac{\sqrt[4]{320}}{\sqrt[4]{5}}$
45. $\sqrt[4]{\sqrt[3]{16}}$
46. $\sqrt[5]{\sqrt[4]{9}}$
47. $\sqrt{\sqrt{\sqrt{n}}}$
48. $\sqrt{b^4 \sqrt{a}}$
49. $\sqrt{\dfrac{n}{m^3}}$
50. $\sqrt{\dfrac{2x}{3c^4}}$
51. $\sqrt{28u^3 v^{-5}}$
52. $\sqrt{98x^6 y^{-7}}$
53. $\sqrt{\dfrac{5}{4} - \dfrac{1}{8}}$
54. $\sqrt{\dfrac{1}{a^2} + \dfrac{1}{b}}$
55. $\sqrt{xy^{-1} + x^{-1}y}$
56. $\sqrt{x^2 + 4^{-1}}$
57. $\sqrt{\dfrac{C-2}{C+2}}$
58. $\sqrt{a^2 + 2ab + b^2}$
59. $\sqrt{a^2 + b^2}$
60. $\sqrt{4x^2 - 1}$
61. $\sqrt{9x^2 - 6x + 1}$
62. $\sqrt{\dfrac{1}{?} + 2r + 2r^2}$

In Exercises 63–68, perform the required operation.

63. Display the graphs of $y_1 = \sqrt{x+2}$ and $y_2 = \sqrt{x} + \sqrt{2}$ on a graphing calculator to show graphically that $\sqrt{x+2}$ is not equal to $\sqrt{x} + \sqrt{2}$.

(W) 64. An approximate equation for the efficiency E (in percent) of an engine is $E = 100\left(1 - 1/\sqrt[5]{R^2}\right)$, where R is the compression ratio. Explain how this equation can be written with fractional exponents and then find E for $R = 7.35$.

65. When analyzing the velocity of an object that falls through a very great distance, the expression $a\sqrt{2g/a}$ is derived. Show by rationalizing the denominator that this expression takes on a simpler form.

66. A formula for the angular velocity ω of a disc is $\omega = \sqrt{\dfrac{576EIg}{WL^3}}$. Rationalize the denominator.

67. In analyzing an electronic filter circuit, the expression $\dfrac{8A}{\pi^2 \sqrt{1 + (f_0/f)^2}}$ is used. Rationalize the denominator, expressing the answer without the fraction f_0/f.

68. The expression $\dfrac{1}{2L}\sqrt{R^2 - \dfrac{4L}{C}}$ occurs in the study of electric circuits. Simplify this expression by combining terms under the radical and rationalizing the denominator.

11.4 ADDITION AND SUBTRACTION OF RADICALS

When we add or subtract algebraic expressions, we combine similar terms, those that differ only in numerical coefficients. This is true when adding or subtracting radicals. *The radicals must be similar to perform the addition,* rather than simply indicate the addition. *Radicals are* **similar** *if they differ only in their numerical coefficients. This means they must have the same order and have the same radicand.*

In order to add radicals, we first express each radical in its simplest form, rationalize any denominators, and then combine those that are similar. For those that are not similar, we can only indicate the addition.

◀ EXAMPLE 1   (a) $2\sqrt{7} - 5\sqrt{7} + \sqrt{7} = -2\sqrt{7}$ all similar radicals
This result follows the distributive law, as it should. We can write

$$2\sqrt{7} - 5\sqrt{7} + \sqrt{7} = (2 - 5 + 1)\sqrt{7} = -2\sqrt{7}$$

We can also see that the terms combine just as

$$2x - 5x + x = -2x$$

(b) $\sqrt[5]{6} + 4\sqrt[5]{6} - 2\sqrt[5]{6} = 3\sqrt[5]{6}$ all similar radicals

(c) $\sqrt{5} + 2\sqrt{3} - 5\sqrt{5} = 2\sqrt{3} - 4\sqrt{5}$ answer contains two terms

similar radicals

We note in (c) that we are able only to indicate the final subtraction since the radicals are not similar. ▶

◀ EXAMPLE 2   (a) $\sqrt{2} + \sqrt{8} = \sqrt{2} + \sqrt{4 \times 2} = \sqrt{2} + \sqrt{4}\sqrt{2}$
$$= \sqrt{2} + 2\sqrt{2} = 3\sqrt{2}$$

(b) $\sqrt[3]{24} + \sqrt[3]{81} = \sqrt[3]{8 \times 3} + \sqrt[3]{27 \times 3} = \sqrt[3]{8}\sqrt[3]{3} + \sqrt[3]{27}\sqrt[3]{3}$
$$= 2\sqrt[3]{3} + 3\sqrt[3]{3} = 5\sqrt[3]{3}$$

CAUTION ▶   Notice that $\sqrt{8}$, $\sqrt[3]{24}$, and $\sqrt[3]{81}$ were simplified before performing the additions. We also note that $\sqrt{2} + \sqrt{8}$ **is not** *equal to* $\sqrt{2 + 8}$. ▶

We note in the illustrations of Example 2 that the radicals do not initially appear to be similar. However, after each is simplified we are able to recognize the similar radicals.

◀ EXAMPLE 3   (a) $6\sqrt{7} - \sqrt{28} + 3\sqrt{63} = 6\sqrt{7} - \sqrt{4 \times 7} + 3\sqrt{9 \times 7}$
$$= 6\sqrt{7} - 2\sqrt{7} + 3(3\sqrt{7})$$
$$= 6\sqrt{7} - 2\sqrt{7} + 9\sqrt{7}$$
$$= 13\sqrt{7} \qquad \text{all similar radicals}$$

(b) $3\sqrt{125} - \sqrt{20} + \sqrt{27} = 3\sqrt{25 \times 5} - \sqrt{4 \times 5} + \sqrt{9 \times 3}$
$$= 3(5\sqrt{5}) - 2\sqrt{5} + 3\sqrt{3}$$
$$= 13\sqrt{5} + 3\sqrt{3} \qquad \text{not similar to others}$$ ▶

◀ EXAMPLE 4 $\sqrt{24} + \sqrt{\dfrac{3}{2}} = \sqrt{4 \times 6} + \sqrt{\dfrac{3 \times 2}{2 \times 2}} = \sqrt{4}\sqrt{6} + \sqrt{\dfrac{6}{4}}$
$$= 2\sqrt{6} + \frac{\sqrt{6}}{2} = \frac{4\sqrt{6} + \sqrt{6}}{2} = \frac{5}{2}\sqrt{6}$$

One radical was simplified by removing the perfect square factor, and in the other we rationalized the denominator. Note that we would not be able to combine the radicals if we did not rationalize the denominator of the second radical. ▶

Our main purpose in this section is to add radicals in radical form. However, a decimal value can be obtained by use of a calculator, and in the following example we use decimal values to verify the result of the addition.

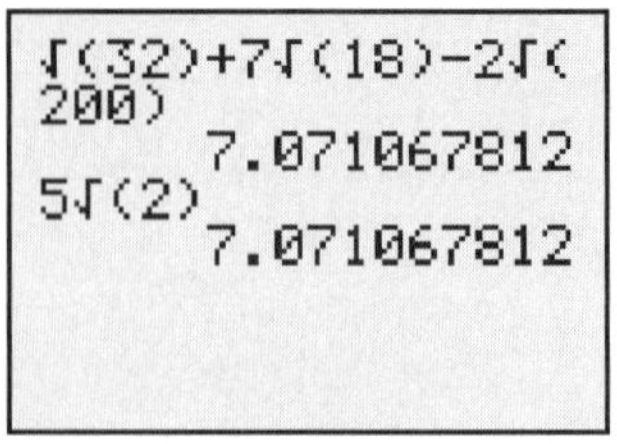

```
√(32)+7√(18)-2√(
200)
        7.071067812
5√(2)
        7.071067812
```

Fig. 11.5

EXAMPLE 5   Use a calculator to verify the result of the following addition:

$$\sqrt{32} + 7\sqrt{18} - 2\sqrt{200} = \sqrt{16 \times 2} + 7\sqrt{9 \times 2} - 2\sqrt{100 \times 2}$$
$$= 4\sqrt{2} + 7(3\sqrt{2}) - 2(10\sqrt{2})$$
$$= 4\sqrt{2} + 21\sqrt{2} - 20\sqrt{2}$$
$$= 5\sqrt{2}$$

The calculator display that verifies this addition is shown in Fig. 11.5.

Next is an example of adding radical expressions that contain literal numbers.

EXAMPLE 6

$$\sqrt{\frac{2}{3a}} - 2\sqrt{\frac{3}{2a}} = \sqrt{\frac{2(3a)}{3a(3a)}} - 2\sqrt{\frac{3(2a)}{2a(2a)}} = \sqrt{\frac{6a}{9a^2}} - 2\sqrt{\frac{6a}{4a^2}}$$

$$= \frac{1}{3a}\sqrt{6a} - \frac{2}{2a}\sqrt{6a} = \frac{1}{3a}\sqrt{6a} - \frac{1}{a}\sqrt{6a}$$

$$= \frac{\sqrt{6a} - 3\sqrt{6a}}{3a} = \frac{-2\sqrt{6a}}{3a}$$

$$= -\frac{2}{3a}\sqrt{6a}$$

EXERCISES 11.4

In Exercises 1 and 2, simplify the resulting expressions if the given changes are made in the indicated examples of this section.

1. In Example 3(b), change $\sqrt{27}$ to $\sqrt{45}$ and then find the resulting simplified expression.

2. In Example 5, change $\sqrt{32}$ to $\sqrt{48}$ and then find the resulting simplified expression.

In Exercises 3–38, express each radical in simplest form, rationalize denominators, and perform the indicated operations.

3. $2\sqrt{3} + 5\sqrt{3}$

4. $8\sqrt{11} - 3\sqrt{11}$

5. $2\sqrt{7} + \sqrt{5} - 3\sqrt{7}$

6. $8\sqrt{6} - 2\sqrt{3} - 5\sqrt{6}$

7. $\sqrt{5} + \sqrt{16 + 4}$

8. $\sqrt{7} + \sqrt{63}$

9. $2\sqrt{3t^2} - 3\sqrt{12t^2}$

10. $4\sqrt{2n^2} - \sqrt{50n^2}$

11. $\sqrt{8a} - \sqrt{32a}$

12. $\sqrt{27x} + 2\sqrt{18x}$

13. $2\sqrt{28} + 3\sqrt{175}$

14. $\sqrt{100 + 25} - 7\sqrt{80}$

15. $2\sqrt{200} - \sqrt{1250} - \sqrt{450}$

16. $2\sqrt{44} - \sqrt{99} + \sqrt{2}\sqrt{88}$

17. $3\sqrt{75R} + 2\sqrt{48R} - 2\sqrt{18R}$

18. $2\sqrt{28} - \sqrt{108} - 2\sqrt{175}$

19. $\sqrt{60} + \sqrt{\frac{5}{3}}$

20. $\sqrt{84} - \sqrt{\frac{3}{7}}$

21. $\sqrt{\frac{1}{2}} + \sqrt{\frac{25}{2}} - \sqrt{18}$

22. $\sqrt{6} - \sqrt{\frac{2}{3}} - \sqrt{18}$

23. $\sqrt[3]{81} + \sqrt[3]{3000}$

24. $\sqrt[3]{-16} + \sqrt[3]{54}$

25. $\sqrt[4]{32} - \sqrt[8]{4}$

26. $\sqrt[6]{\sqrt{2}} - \sqrt[12]{2^{13}}$

27. $\sqrt{a^3b} - \sqrt{4ab^5}$

28. $\sqrt{2R^2I} + \sqrt{8}\sqrt{I^3}$

29. $\sqrt{6}\sqrt{5}\sqrt{3} - \sqrt{40a^2}$

30. $\sqrt{60b^2n} - b\sqrt{135n}$

31. $\sqrt[3]{24a^2b^4} - \sqrt[3]{3a^5b}$

32. $\sqrt[5]{32a^6b^4} + 3a\sqrt[5]{243ab^9}$

33. $\sqrt{\dfrac{a}{c^5}} - \sqrt{\dfrac{c}{a^3}}$

34. $\sqrt{\dfrac{2x}{3y}} + \sqrt{\dfrac{27y}{8x}}$

35. $\sqrt[3]{ab^{-1}} - \sqrt[3]{8a^{-2}b^2}$

36. $\sqrt{\dfrac{2xy^{-1}}{3}} + \sqrt{\dfrac{27x^{-1}y}{8}}$

37. $\sqrt{\dfrac{T-V}{T+V}} - \sqrt{\dfrac{T+V}{T-V}}$

38. $\sqrt{\dfrac{16}{x} + 8 + x} - \sqrt{1 - \dfrac{1}{x}}$

In Exercises 39–42, express each radical in simplest form, rationalize denominators, and perform the indicated operations. Then use a calculator to verify the result.

39. $3\sqrt{45} + 3\sqrt{75} - 2\sqrt{500}$

40. $2\sqrt{40} + 3\sqrt{90} - 5\sqrt{250}$

41. $2\sqrt{\frac{2}{3}} + \sqrt{24} - 5\sqrt{\frac{3}{2}}$

42. $\sqrt{\frac{2}{7}} - 2\sqrt{\frac{7}{2}} + 5\sqrt{56}$

In Exercises 43–48, solve the given problems.

43. Find the exact sum of the positive roots of $x^2 - 2x - 2 = 0$ and $x^2 + 2x - 11 = 0$.

(W) **44.** For the quadratic equation $ax^2 + bx + c = 0$, if a, b, and c are integers, the sum of the roots is a rational number. Explain.

(W) **45.** Without calculating the actual value, determine whether $10\sqrt{11} - \sqrt{1000}$ is positive or negative. Explain.

46. The current I (in A) passing through a resistor R (in Ω) in which P watts of power are dissipated is $I = \sqrt{P/R}$. If the power dissipated in the resistors shown in Fig. 11.6 is W watts, what is the sum of the currents in radical form?

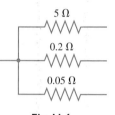

5 Ω

0.2 Ω

0.05 Ω

Fig. 11.6

47. A rectangular piece of plywood 1200 mm by 2400 mm has corners cut from it, as shown in Fig. 11.7. Find the perimeter of the remaining piece in exact form and in decimal form.

2400 mm   300 mm

1200 mm

600 mm

Fig. 11.7

48. Three squares with areas of 150 cm², 54 cm², and 24 cm² are displayed on a computer monitor. What is the sum (in radical form) of the perimeters of these squares?

For reference, Eq. (11.10) is $\sqrt[n]{a}\,\sqrt[n]{b} = \sqrt[n]{ab}$.

11.5 MULTIPLICATION AND DIVISION OF RADICALS

When multiplying expressions containing radicals, we use Eq. (11.10), along with the normal procedures of algebraic multiplication. *Note that the orders of the radicals being multiplied in Eq. (11.10) are the same.* The following examples illustrate the method.

EXAMPLE 1   **(a)** $\sqrt{5}\sqrt{2} = \sqrt{5 \times 2} = \sqrt{10}$

(b) $\sqrt{33}\sqrt{3} = \sqrt{33 \times 3} = \sqrt{99} = \sqrt{9 \times 11} = \sqrt{9}\sqrt{11}$
$$= 3\sqrt{11} \qquad \text{perfect square}$$
or $\sqrt{33}\sqrt{3} = \sqrt{33 \times 3} = \sqrt{11 \times 3 \times 3}$
$$= 3\sqrt{11}$$

Note that we express the resulting radical in simplest form.

EXAMPLE 2   **(a)** $\sqrt[3]{6}\sqrt[3]{4} = \sqrt[3]{6(4)} = \sqrt[3]{24} = \sqrt[3]{8}\sqrt[3]{3}$
$$= 2\sqrt[3]{3} \qquad \text{perfect cube}$$

(b) $\sqrt[5]{8a^3b^4}\sqrt[5]{8a^2b^3} = \sqrt[5]{(8a^3b^4)(8a^2b^3)} = \sqrt[5]{64a^5b^7} = \sqrt[5]{32a^5b^5}\sqrt[5]{2b^2}$
$$= 2ab\sqrt[5]{2b^2} \qquad \text{perfect fifth power}$$

EXAMPLE 3 $\sqrt{2}(3\sqrt{5} - 4\sqrt{2}) = 3\sqrt{2}\sqrt{5} - 4\sqrt{2}\sqrt{2} = 3\sqrt{10} - 4\sqrt{4}$
$$= 3\sqrt{10} - 4(2) = 3\sqrt{10} - 8$$

EXAMPLE 4

$$(5\sqrt{7} - 2\sqrt{3})(4\sqrt{7} + 3\sqrt{3}) = (5\sqrt{7})(4\sqrt{7}) + (5\sqrt{7})(3\sqrt{3}) - (2\sqrt{3})(4\sqrt{7}) - (2\sqrt{3})(3\sqrt{3})$$
$$= (5)(4)\sqrt{7}\sqrt{7} + (5)(3)\sqrt{7}\sqrt{3} - (2)(4)\sqrt{3}\sqrt{7} - (2)(3)\sqrt{3}\sqrt{3}$$
$$= 20(7) + 15\sqrt{21} - 8\sqrt{21} - 6(3)$$
$$= 140 + 7\sqrt{21} - 18$$
$$= 122 + 7\sqrt{21}$$

The calculator display that verifies this multiplication is shown in Fig. 11.8.

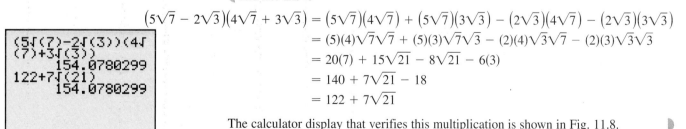

```
(5√(7)-2√(3))(4√
(7)+3√(3))
        154.0780299
122+7√(21)
        154.0780299
```

Fig. 11.8

When raising a single-term radical expression to a power, we use the basic meaning of the power. When raising a binomial to a power, we proceed as with any binomial. We use Eqs. (6.3) and (6.4) along with Eq. (11.10) if the binomial is squared. These are illustrated in the next two examples.

For reference, Eqs. (6.3) and (6.4) are
$(x + y)^2 = x^2 + 2xy + y^2$
$(x - y)^2 = x^2 - 2xy + y^2$

◀ **EXAMPLE 5**   (a) $\left(2\sqrt{7}\right)^2 = 2^2\left(\sqrt{7}\right)^2 = 4(7) = 28$

(b) $\left(2\sqrt{7}\right)^3 = 2^3\left(\sqrt{7}\right)^3 = 8\left(\sqrt{7}\right)^2\left(\sqrt{7}\right) = 8(7)\sqrt{7}$
$$= 56\sqrt{7}$$

◀ **EXAMPLE 6**   (a) $\left(3 + \sqrt{5}\right)^2 = 3^2 + 2(3)\sqrt{5} + \left(\sqrt{5}\right)^2 = 9 + 6\sqrt{5} + 5$
$$= 14 + 6\sqrt{5}$$

(b) $\left(\sqrt{a} - \sqrt{b}\right)^2 = \left(\sqrt{a}\right)^2 - 2\sqrt{a}\sqrt{b} + \left(\sqrt{b}\right)^2$
$$= a + b - 2\sqrt{ab}$$

CAUTION ▶   Again, we note that *to multiply radicals and combine them under one radical sign, it is necessary that the order of the radicals be the same.* If necessary, we can make the order of each radical the same by appropriate operations on each radical separately. Fractional exponents are frequently useful for this purpose.

◀ **EXAMPLE 7**   (a) $\sqrt[3]{2}\sqrt{5} = 2^{1/3}5^{1/2} = 2^{2/6}5^{3/6} = (2^2 5^3)^{1/6} = \sqrt[6]{500}$

(b) $\sqrt[3]{4a^2b}\sqrt[4]{8a^3b^2} = (2^2a^2b)^{1/3}(2^3a^3b^2)^{1/4} = (2^2a^2b)^{4/12}(2^3a^3b^2)^{3/12}$
$$= (2^8a^8b^4)^{1/12}(2^9a^9b^6)^{1/12} = (2^{17}a^{17}b^{10})^{1/12}$$
$$= 2a(2^5a^5b^{10})^{1/12}$$
$$= 2a\sqrt[12]{32a^5b^{10}}$$

DIVISION OF RADICALS

If a fraction involving a radical is to be changed in form, *rationalizing the denominator* or *rationalizing the numerator* is the principal step. Although calculators have made the rationalization of denominators unnecessary for calculation, this process often makes the form of the fraction simpler. Also, rationalizing numerators is useful at times in more advanced mathematics. We now consider rationalizing when the denominator (or numerator) to be rationalized has more than one term.

NOTE ▶   *If the denominator (or numerator) is the sum (or difference) of two terms, at least one of which is a radical, the fraction is rationalized by multiplying both the numerator and the denominator by the difference (or sum) of the same two terms, if the radicals are square roots.*

◀ **EXAMPLE 8**   The fraction $\dfrac{1}{\sqrt{3} - \sqrt{2}}$

can be rationalized by multiplying the numerator and the denominator by $\sqrt{3} + \sqrt{2}$. In this way, the radicals will be removed from the denominator.

$$\frac{1}{\sqrt{3} - \sqrt{2}} \times \frac{\sqrt{3} + \sqrt{2}}{\sqrt{3} + \sqrt{2}} = \frac{\sqrt{3} + \sqrt{2}}{\left(\sqrt{3}\right)^2 - \left(\sqrt{2}\right)^2} = \frac{\sqrt{3} + \sqrt{2}}{3 - 2} = \sqrt{3} + \sqrt{2}$$

change sign

The reason this technique works is that an expression of the form $a^2 - b^2$ is created in the denominator, where a or b (or both) is a radical. We see that the result is a denominator free of radicals.

EXAMPLE 9 Rationalize the denominator of $\dfrac{3\sqrt{6} - \sqrt{2}}{\sqrt{6} + \sqrt{2}}$ and simplify the result.

$$\frac{3\sqrt{6} - \sqrt{2}}{\sqrt{6} + \sqrt{2}}\left(\frac{\sqrt{6} - \sqrt{2}}{\sqrt{6} - \sqrt{2}}\right) = \frac{18 - 4\sqrt{12} + 2}{4}$$

change sign

$$= \frac{20 - 4\left(2\sqrt{3}\right)}{4} = 5 - 2\sqrt{3}$$

We note that, after rationalizing the denominator, the result has a much simpler form than the original expression.

As we noted earlier, in certain types of algebraic operations it may be necessary to rationalize the numerator of an expression. This procedure is illustrated in the following example.

EXAMPLE 10 In studying the properties of a semiconductor, the expression

$$\frac{C_1 + C_2\sqrt{1 + 2V}}{\sqrt{1 + 2V}}$$

is used. Here, C_1 and C_2 are constants and V is the voltage across a junction of the semiconductor. Rationalize the numerator of this expression.

Multiplying numerator and denominator by $C_1 - C_2\sqrt{1 + 2V}$, we have

$$\frac{C_1 + C_2\sqrt{1 + 2V}}{\sqrt{1 + 2V}} = \frac{\left(C_1 + C_2\sqrt{1 + 2V}\right)\left(C_1 - C_2\sqrt{1 + 2V}\right)}{\sqrt{1 + 2V}\left(C_1 - C_2\sqrt{1 + 2V}\right)}$$

$$= \frac{C_1^2 - C_2^2\left(\sqrt{1 + 2V}\right)^2}{C_1\sqrt{1 + 2V} - C_2\sqrt{1 + 2V}\sqrt{1 + 2V}}$$

$$= \frac{C_1^2 - C_2^2(1 + 2V)}{C_1\sqrt{1 + 2V} - C_2(1 + 2V)}$$

EXERCISES 11.5

In Exercises 1–4, perform the indicated operations on the resulting expressions if the given changes are made in the indicated examples of this section.

1. In Example 3, change $4\sqrt{2}$ to $4\sqrt{8}$ and then perform the multiplication.

2. In Example 7(a), change $\sqrt{5}$ to $\sqrt[6]{5}$ and then perform the multiplication.

3. In Example 8, change the sign in the denominator from $-$ to $+$ and then rationalize the denominator.

4. In Example 9, rationalize the numerator of the given expression.

In Exercises 5–46, perform the indicated operations, expressing answers in simplest form with rationalized denominators.

5. $\sqrt{3}\sqrt{10}$ **6.** $\sqrt{2}\sqrt{51}$ **7.** $\sqrt{6}\sqrt{2}$ **8.** $\sqrt{7}\sqrt{14}$

9. $\sqrt[3]{4}\sqrt[3]{2}$ **10.** $\sqrt[5]{4}\sqrt[5]{16}$ **11.** $\left(5\sqrt{2}\right)^2$ **12.** $\left(3\sqrt{5}\right)^3$

13. $\sqrt{8}\sqrt{\frac{5}{2}}$ **14.** $\sqrt[6]{\frac{6}{7}}\sqrt[6]{\frac{2}{3}}$

15. $\sqrt{3}\left(\sqrt{2} - \sqrt{5}\right)$ **16.** $3\sqrt{5}\left(\sqrt{15} - 2\sqrt{5}\right)$

17. $\left(2 - \sqrt{5}\right)\left(2 + \sqrt{5}\right)$ **18.** $\left(2 - \sqrt{5}\right)^2$

19. $\left(3\sqrt{5} - 2\sqrt{3}\right)\left(6\sqrt{5} + 7\sqrt{3}\right)$

20. $\left(3\sqrt{7a} - \sqrt{8}\right)\left(\sqrt{7a} + \sqrt{2}\right)$

21. $\left(3\sqrt{11} - \sqrt{x}\right)\left(2\sqrt{11} + 5\sqrt{x}\right)$

22. $\left(2\sqrt{10} + 3\sqrt{15}\right)\left(\sqrt{10} - 7\sqrt{15}\right)$

23. $\sqrt{a}\left(\sqrt{ab} + \sqrt{c^3}\right)$

24. $\sqrt{3x}\left(\sqrt{3x} - \sqrt{xy}\right)$

25. $\dfrac{\sqrt{6} - 3}{\sqrt{6}}$

26. $\dfrac{5 - \sqrt{10}}{\sqrt{10}}$

27. $\left(\sqrt{2a} - \sqrt{b}\right)\left(\sqrt{2a} + 3\sqrt{b}\right)$

28. $\left(2\sqrt{mn} + 3\sqrt{n}\right)^2$

29. $\sqrt{2}\sqrt[3]{3}$

30. $\sqrt[5]{16}\sqrt[3]{8}$

31. $\dfrac{1}{\sqrt{7} + \sqrt{3}}$

32. $\dfrac{6\sqrt{5}}{5 - 2\sqrt{5}}$

33. $\dfrac{\sqrt{2} - 1}{\sqrt{7} - 3\sqrt{2}}$

34. $\dfrac{2\sqrt{15} - 3}{\sqrt{15} + 4}$

35. $\dfrac{2\sqrt{3} - 5\sqrt{5}}{\sqrt{3} + 2\sqrt{5}}$

36. $\dfrac{\sqrt{15}\ \ 3\sqrt{5}}{2\sqrt{15} - \sqrt{5}}$

37. $\dfrac{2\sqrt{x}}{\sqrt{x} - \sqrt{5}}$

38. $\dfrac{\sqrt{2c} + 3d}{\sqrt{2c} - d}$

39. $\left(\sqrt[5]{\sqrt{6}} - \sqrt{5}\right)\left(\sqrt[5]{\sqrt{6}} + \sqrt{5}\right)$

40. $\sqrt[3]{5} - \sqrt{17}\sqrt[3]{5} + \sqrt{17}$

41. $\left(\sqrt{\dfrac{2}{R}} + \sqrt{\dfrac{R}{2}}\right)\left(\sqrt{\dfrac{2}{R}} - 2\sqrt{\dfrac{R}{2}}\right)$

42. $\left(3 + \sqrt{6 - 2a}\right)\left(2 - \sqrt{6 - 2a}\right)$

43. $\dfrac{\sqrt{x + y}}{\sqrt{x - y} - \sqrt{x}}$

44. $\dfrac{\sqrt{1 + a}}{a - \sqrt{1 - a}}$

45. $\dfrac{\sqrt{a} + \sqrt{a - 2}}{\sqrt{a} - \sqrt{a - 2}}$

46. $\dfrac{\sqrt{T^4 - V^4}}{\sqrt{V^{-2} - T^{-2}}}$

In Exercises 47–50, perform the indicated operations, expressing answers in simplest form with rationalized denominators. Then verify the result with a calculator.

47. $\left(\sqrt{11} + \sqrt{6}\right)\left(\sqrt{11} - 2\sqrt{6}\right)$

48. $\left(2\sqrt{5} - \sqrt{7}\right)\left(3\sqrt{5} + \sqrt{7}\right)$

49. $\dfrac{2\sqrt{6} - \sqrt{5}}{3\sqrt{6} - 4\sqrt{5}}$

50. $\dfrac{\sqrt{7} - 4\sqrt{2}}{5\sqrt{7} - 4\sqrt{2}}$

In Exercises 51–54, combine the terms into a single fraction, but do not rationalize the denominators.

51. $2\sqrt{x} + \dfrac{1}{\sqrt{x}}$

52. $\dfrac{3}{2\sqrt{3x - 4}} - \sqrt{3x - 4}$

53. $\dfrac{x^2}{\sqrt{2x + 1}} + 2x\sqrt{2x + 1}$

54. $4\sqrt{x^2 + 1} - \dfrac{4x}{\sqrt{x^2 + 1}}$

In Exercises 55–58, rationalize the numerator of each fraction.

55. $\dfrac{\sqrt{5} + \sqrt{2}}{3\sqrt{6}}$

56. $\dfrac{\sqrt{19} - 3}{5}$

57. $\dfrac{\sqrt{x + h} - \sqrt{x}}{h}$

58. $\dfrac{\sqrt{3x + 4} + \sqrt{3x}}{8}$

In Exercises 59–68, solve the given problems.

59. By substitution, show that $x = 1 - \sqrt{2}$ is a solution of the equation $x^2 - 2x - 1 = 0$.

W **60.** For the quadratic equation $ax^2 + bx + c = 0$, if a, b, and c are integers, the product of the roots is a rational number. Explain.

61. For an object oscillating at the end of a spring and on which there is a force that retards the motion, the equation $m^2 + bm + k^2 = 0$ must be solved. Here, b is a constant related to the retarding force, and k is the spring constant. By substitution, show that $m = \frac{1}{2}\left(\sqrt{b^2 - 4k^2} - b\right)$ is a solution.

62. Among the products of a specialty furniture company are tables with tops in the shape of a regular octagon (eight sides). Express the area A of a table top as a function of the side s of the octagon.

63. An expression used in determining the characteristics of a spur gear is $\dfrac{50}{50 + \sqrt{V}}$. Rationalize the denominator.

64. In finding the time to drain a tank, the expression $\dfrac{h_2 - h_1}{\sqrt{2g}\left(\sqrt{h_2} - \sqrt{h_1}\right)}$ is used. Rationalize the denominator.

65. In analyzing a tuned amplifier circuit, the expression $\dfrac{2Q}{\sqrt{\sqrt{2} - 1}}$ is used. Rationalize the denominator.

66. When analyzing the ratio of resultant forces when forces **F** and **T** act on a structure, the expression $\sqrt{F^2 + T^2}/\sqrt{F^{-2} + T^{-2}}$ arises. Simplify this expression.

67. The resonant frequency ω of a capacitance C in parallel with a resistance R and inductance L (see Fig. 11.9) is
$$\omega = \dfrac{1}{\sqrt{LC}}\sqrt{1 - \dfrac{R^2 C}{L}}.$$
Combine terms under the radical, rationalize the denominator, and simplify.

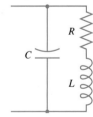

Fig. 11.9

68. In fluid dynamics, the expression $\dfrac{a}{1 + m/\sqrt{r}}$ arises. Combine terms in the denominator, rationalize the denominator, and simplify.

CHAPTER (11) EQUATIONS

Exponents

$$a^m \times a^n = a^{m+n} \tag{11.1}$$

$$\frac{a^m}{a^n} = a^{m-n} \quad \text{or} \quad \frac{a^m}{a^n} = \frac{1}{a^{n-m}} \quad (a \neq 0) \tag{11.2}$$

$$(a^m)^n = a^{mn} \tag{11.3}$$

$$(ab)^n = a^n b^n, \quad \left(\frac{a}{b}\right)^n = \frac{a^n}{b^n} \quad (b \neq 0) \tag{11.4}$$

$$a^0 = 1 \quad (a \neq 0) \tag{11.5}$$

$$a^{-n} = \frac{1}{a^n} \quad (a \neq 0) \tag{11.6}$$

Fractional exponents

$$a^{1/n} = \sqrt[n]{a} \tag{11.7}$$

$$a^{m/n} = \sqrt[n]{a^m} = \left(\sqrt[n]{a}\right)^m \tag{11.8}$$

Radicals

$$\sqrt[n]{a^n} = \left(\sqrt[n]{a}\right)^n = a \tag{11.9}$$

$$\sqrt[n]{a}\sqrt[n]{b} = \sqrt[n]{ab} \tag{11.10}$$

$$\sqrt[m]{\sqrt[n]{a}} = \sqrt[mn]{a} \tag{11.11}$$

$$\frac{\sqrt[n]{a}}{\sqrt[n]{b}} = \sqrt[n]{\frac{a}{b}} \quad (b \neq 0) \tag{11.12}$$

CHAPTER (11) REVIEW EXERCISES

In Exercises 1–28, express each expression in simplest form with only positive exponents.

1. $2a^{-2}b^0$
2. $(2c)^{-1}z^{-2}$
3. $\dfrac{2c^{-1}}{d^{-3}}$
4. $\dfrac{-5x^0}{3y^{-1}}$

5. $3(25)^{3/2}$
6. $32^{2/5}$
7. $400^{-3/2}$
8. $1000^{-2/3}$

9. $\left(\dfrac{3}{t^2}\right)^{-2}$
10. $\left(\dfrac{2x^3}{3}\right)^{-3}$
11. $\dfrac{-8^{2/3}}{49^{-1/2}}$
12. $\dfrac{4^{-45}}{2^{-90}}$

13. $(2a^{1/3}b^{5/6})^6$
14. $(ax^{-1/2}y^{1/4})^8$
15. $(-32m^{15}n^{10})^{3/5}$

16. $(27x^{-6}y^9)^{2/3}$
17. $2L^{-2} - 4C^{-1}$
18. $C^4C^{-1} + (C^4)^{-1}$

19. $\dfrac{2x^{-1}}{2x^{-1} + y^{-1}}$
20. $\dfrac{3a}{(2a)^{-1} - a}$
21. $(a - 3b^{-1})^{-1}$

22. $(2s^{-2} + t)^{-2}$
23. $(x^3 - y^{-3})^{1/3}$
24. $(8a^3)^{2/3}(a^{-2} + 1)^{1/2}$

25. $(W^2 + 2WH + H^2)^{-1/2}$
26. $\left[\dfrac{(9a)^0(4x^2)^{1/3}(3b^{1/2})}{(2b^0)^2}\right]^{-6}$

27. $2x(x - 1)^{-2} - 2(x^2 + 1)(x - 1)^{-3}$
28. $4(1 - T^2)^{1/2} - (1 - T^2)^{-1/2}$

In Exercises 29–74, perform the indicated operations and express the result in simplest radical form with rationalized denominators.

29. $\sqrt{68}$
30. $\sqrt{96}$
31. $\sqrt{ab^5c^2}$
32. $\sqrt{x^3y^4z^6}$

33. $\sqrt{9a^3b^4}$
34. $\sqrt{8x^5y^2}$
35. $\sqrt{84st^3u^{-2}}$
36. $\sqrt{52L^2C^{-5}}$

37. $\dfrac{5}{\sqrt{2s}}$
38. $\dfrac{3a}{\sqrt{5x}}$
39. $\sqrt{\dfrac{11}{27}}$
40. $\sqrt{\dfrac{7}{8V}}$

41. $\sqrt[4]{8m^6n^9}$
42. $\sqrt[3]{9a^7b^{-3}}$
43. $\sqrt[4]{\sqrt[3]{64}}$

44. $\sqrt{a^{-3}\sqrt[5]{b^{12}}}$
45. $\sqrt{36 + 4} - 2\sqrt{10}$
46. $2\sqrt{68x} - \sqrt{153x}$

47. $\sqrt{63} - 2\sqrt{112} - \sqrt{28}$
48. $2\sqrt{20} - \sqrt{80} - 2\sqrt{125}$

49. $a\sqrt{2x^3} + \sqrt{8a^2x^3}$
50. $2\sqrt{m^2n^3} - \sqrt{n^5}$

51. $\sqrt[3]{8a^4} + b\sqrt[3]{a}$
52. $\sqrt[4]{2xy^5} - \sqrt[4]{32xy}$

53. $\sqrt{5}(2\sqrt{5} - \sqrt{11})$
54. $2\sqrt{8}(5\sqrt{2} - \sqrt{6})$

55. $2\sqrt{2}(\sqrt{6} - \sqrt{10})$
56. $3\sqrt{5}(\sqrt{15m^2} + 2\sqrt{35m^2})$

57. $(2 - 3\sqrt{17B})(3 + \sqrt{17B})$
58. $(5\sqrt{6} - 4)(3\sqrt{6} + 5)$

59. $(2\sqrt{7} - 3\sqrt{a})(3\sqrt{7} + \sqrt{a})$

60. $(3\sqrt{2} - \sqrt{13})(5\sqrt{2} + 3\sqrt{13})$

61. $\dfrac{\sqrt{3x}}{2\sqrt{3x} - \sqrt{y}}$
62. $\dfrac{5\sqrt{a}}{2\sqrt{a} - c}$
63. $\dfrac{\sqrt{2}}{\sqrt{3} - 4\sqrt{2}}$

64. $\dfrac{4}{3 - 2\sqrt{7}}$
65. $\dfrac{\sqrt{7} - \sqrt{5}}{\sqrt{5} + 3\sqrt{7}}$
66. $\dfrac{4 - 2\sqrt{6}}{3 + 2\sqrt{6}}$

67. $\dfrac{2\sqrt{x} - a}{3\sqrt{x} + 5a}$
68. $\dfrac{2\sqrt{3y} - \sqrt{2}}{3\sqrt{3y} - 5\sqrt{2}}$
69. $\sqrt{4b^2 + 1}$

70. $\sqrt{a^{-2} + \dfrac{1}{b^2}}$ 71. $\left(\dfrac{2 - \sqrt{15}}{2}\right)^2 - \left(\dfrac{2 - \sqrt{15}}{2}\right)$

72. $\sqrt{2 + a^{-1}b + ab^{-1}} + \sqrt{a^4b^2 + 2a^3b^2 + a^2b^2}$

73. $\sqrt{3 + n}\left(\sqrt{3 + n} - \sqrt{n}\right)^{-1}$

74. $\sqrt{1 + \sqrt{2}}\left(\sqrt{1 + \sqrt{2}} + \sqrt{2}\right)^{-1}$

In Exercises 75 and 76, perform the indicated operations and express the result in simplified radical form with rationalized denominators. In Exercises 77 and 78, without a calculator show that the given equations are true. Then verify each result with a calculator.

75. $\left(\sqrt{7} - 2\sqrt{15}\right)\left(3\sqrt{7} - \sqrt{15}\right)$ 76. $\dfrac{2\sqrt{3} - 7\sqrt{14}}{3\sqrt{3} + 2\sqrt{14}}$

77. $\sqrt{\sqrt{2} - 1}\left(\sqrt{2} + 1\right) = \sqrt{\sqrt{2} + 1}$

78. $\sqrt{\sqrt{3} + 1}\left(\sqrt{3} - 1\right) = \sqrt{2\left(\sqrt{3} - 1\right)}$

In Exercises 79–92, perform the indicated operations.

79. The average annual increase i (in %) of the cost of living over n years is given by $i = 100[(C_2/C_1)^{1/n} - 1]$, where C_1 is the cost of living index for the first year and C_2 is the cost of living index for the last year of the period. Evaluate i if $C_1 = 130.7$ and $C_2 = 172.0$ are the values for 1990 and 2000, respectively.

80. Kepler's third law of planetary motion may be given as $T = kr^{3/2}$, where T is the time for one revolution of a planet around the sun, r is its mean radius from the sun, and $k = 5.46 \times 10^{-13}$ year/km$^{3/2}$. Find the time for one revolution of Venus about the sun if $r = 1.08 \times 10^8$ km.

81. The speed v of a ship of weight W whose engines produce power P is $v = k\sqrt[3]{P/W}$. Express this equation (a) with a fractional exponent and (b) as a radical with the denominator rationalized.

82. In analyzing the orbit of an earth satellite, the expression $\sqrt{1 + \dfrac{2E}{m}\left(\dfrac{h}{GM}\right)^2}$ is used. Combine terms under the radical and express the result with the denominator rationalized.

83. A square plastic sheet of side x is stretched by an amount equal to $\sqrt{x}$ horizontally and vertically. Find the expression for the percent increase in the area of the sheet.

84. In the study of the biological effects of sound, the expression $(2n/\omega r)^{-1/2}$ is found. Express it in simplest rationalized radical form.

85. When studying atomic structure, the expression $\dfrac{v}{n_2^{-2} - n_1^{-2}}$ is used. Write this in simplest form with only positive exponents.

86. In hydrodynamics, the expression $v(4 + 3ar^{-1} - a^3r^{-3})^{-1}$ arises. Express it in simplest form with positive exponents.

87. A square is decreasing in size on a computer screen. To find how fast the side of the square changes, we must rationalize the numerator of $\dfrac{\sqrt{A + h} - \sqrt{A}}{h}$. Perform this operation.

88. A plane takes off 200 km east of its destination, and a 50 km/h wind is from the south. If the plane's velocity is 250 km/h and its heading is always toward its destination, its path can be described as $y = 100[(0.005x)^{0.8} - (0.005x)^{1.2}]$, $0 \le x \le 200$ km. Sketch the path and check using a graphing calculator.

89. In an experiment, a laser beam follows the path shown in Fig. 11.10. Express the length of the path in simplest radical form.

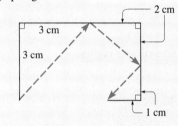

Fig. 11.10

90. The flow rate V (in m/s) through a storm drain pipe is found from $V = 33.5(0.55)^{2/3}(0.0018)^{1/2}$. Find the value of V.

91. The frequency of a certain electric circuit is given by $\dfrac{1}{2\pi\sqrt{\dfrac{LC_1C_2}{C_1 + C_2}}}$. Express this in simplest rationalized radical form.

92. A computer analysis of an experiment showed that the fraction f of viruses surviving X-ray dosages was given by $f = \dfrac{20}{d + \sqrt{3d + 400}}$, where d is the dosage. Express this with the denominator rationalized.

Writing Exercise

93. In calculating the forces on a tower by the wind, it is necessary to evaluate $0.018^{0.13}$. Write a paragraph explaining why this form is preferable to the equivalent radical form.

CHAPTER 11 PRACTICE TEST

In Problems 1–12, simplify the given expressions. For those with exponents, express each result with only positive exponents. For the radicals, rationalize the denominator where applicable.

1. $2\sqrt{20} - \sqrt{125}$

2. $\dfrac{100^{3/2}}{8^{-2/3}}$

3. $(as^{-1/3}t^{3/4})^{12}$

4. $(2x^{-1} + y^{-2})^{-1}$

5. $\left(\sqrt{2x} - 3\sqrt{y}\right)^2$

6. $\sqrt[3]{\sqrt[4]{4}}$

7. $\dfrac{3 - 2\sqrt{2}}{2\sqrt{x}}$

8. $\sqrt{27a^4b^3}$

9. $2\sqrt{2}\left(3\sqrt{10} - \sqrt{6}\right)$

10. $(2x + 3)^{1/2} + (x + 1)(2x + 3)^{-1/2}$

11. $\left(\dfrac{4a^{-1/2}b^{3/4}}{b^{-2}}\right)\left(\dfrac{b^{-1}}{2a}\right)$

12. $\dfrac{2\sqrt{15} + \sqrt{3}}{\sqrt{15} - 2\sqrt{3}}$

13. Express $\dfrac{3^{-1/2}}{2}$ in simplest radical form with a rationalized denominator.

14. In the study of fluid flow in pipes, the expression $0.220N^{-1/6}$ is found. Evaluate this expression for $N = 64 \times 10^6$.

CHAPTER 12   Complex Numbers

Among the important areas of research in the mid-nineteenth century was the study of light, which was determined to be a form of electromagnetic radiation. Today, extensive use of electromagnetic waves is made in a great variety of situations. These include the transmission of signals for radio, television, and cellular phones. Also, radar, microwave ovens, and X-ray machines are among the many other applications of electromagnetic waves.

In the 1860s, James Clerk Maxwell, a Scottish physicist, used the research of others and himself to develop a set of very important equations for electromagnetic radiation. In doing so, he actually predicted *mathematically* the existence of electromagnetic waves, such as radio waves.

It was not until 1887 that Heinrich Hertz, a German physicist, produced and observed radio waves in the laboratory. We see that mathematics was used in predicting the existence of one of the most important devices in use today, over 20 years before it was actually observed.

Important in the study of electromagnetic waves and many areas of electricity and electronics are *complex numbers,* which we study in this chapter. Complex numbers include imaginary numbers as well as real numbers. Other than briefly noting imaginary numbers in Chapters 1 and 7, we have purposely avoided any extended discussion of them until now.

Despite their name, complex numbers and imaginary numbers have very real and useful applications, as we have noted. Besides electricity and electronics, they are used in the studies of mechanical vibrations, optics, and acoustics.

In Section 12.7, we see that tuning in a radio station involves a basic application of complex numbers to electricity.

12.1 BASIC DEFINITIONS

Since the square of a positive number or a negative number is positive, it is not possible to square any real number and have a negative result. Therefore, we must define a number system to include square roots of negative numbers. We will find that such numbers can be used to great advantage in certain applications.

If the radicand in a square root is negative, we can express the indicated root as the product of $\sqrt{-1}$ and the square root of a positive number. *The symbol $\sqrt{-1}$ is defined as the* **imaginary unit** *and is denoted by the symbol j.* Therefore, we have

$$j = \sqrt{-1} \quad \text{and} \quad j^2 = -1 \tag{12.1}$$

Generally, mathematicians use the symbol i for $\sqrt{-1}$, and therefore most nontechnical textbooks use i. However, one of the major technical applications of complex numbers is in electronics, where i represents electric current. Therefore, we will use j for $\sqrt{-1}$, which is also the standard symbol in electronics textbooks.

❰ **EXAMPLE 1**   Express the following square roots in terms of j.

(a) $\sqrt{-9} = \sqrt{(9)(-1)} = \sqrt{9}\sqrt{-1} = 3j$

(b) $\sqrt{-0.25} = \sqrt{0.25}\sqrt{-1} = 0.5j$

(c) $\sqrt{-5} = \sqrt{5}\sqrt{-1} = \sqrt{5}j = j\sqrt{5}$ the form $j\sqrt{5}$ is better than $\sqrt{5}j$

NOTE ❱ When a radical appears, we write the result in a form with the j before the radical as in the last illustration. This clearly shows that the j is not *under* the radical. ❱

❰ **EXAMPLE 2** $\left(\sqrt{-4}\right)^2 = \left(\sqrt{4}\sqrt{-1}\right)^2 = \left(j\sqrt{4}\right)^2 = 4j^2 = -4$

The simplification of this expression does not follow Eq. (11.10), which states that $\sqrt{ab} = \sqrt{a}\sqrt{b}$ for square roots. This is the reason it was noted as being valid only if a and b are not negative. Therefore, we see that *Eq. (11.10) does not always hold for negative values of a and b.*

In simplifying $\left(\sqrt{-4}\right)^2$, we can write it as $\left(\sqrt{-4}\right)\left(\sqrt{-4}\right)$. However, we *cannot now write this product as* $\sqrt{(-4)(-4)}$, for this leads to an incorrect result of $+4$. In fact *we cannot use Eq. (11.10) if both a and b are negative.* It cannot be overemphasized that

CAUTION ❱ $\left(\sqrt{-4}\right)^2$ *is not equal to* $\sqrt{(-4)(-4)}$

Each $\sqrt{-4}$ must be written as $j\sqrt{4}$ before continuing the simplification. ❱

❰ **EXAMPLE 3**   To further illustrate the method of handling square roots of negative numbers, consider the difference between $\sqrt{-3}\sqrt{-12}$ and $\sqrt{(-3)(-12)}$. For these expressions, we have

$$\sqrt{-3}\sqrt{-12} = \left(j\sqrt{3}\right)\left(j\sqrt{12}\right) = \left(\sqrt{3}\sqrt{12}\right)j^2 = \left(\sqrt{36}\right)j^2$$
$$= 6(-1) = -6$$
$$\sqrt{(-3)(-12)} = \sqrt{36} = 6$$

For $\sqrt{-3}\sqrt{-12}$ we have the product of square roots of negative numbers, whereas for $\sqrt{(-3)(-12)}$ we have the product of negative numbers under the radical. We must be careful to note the difference. ❱

NOTE ▶ From Examples 2 and 3 we see that *when we are dealing with square roots of negative numbers, each should be expressed in terms of j before proceeding.* To do this, for any positive real number a we write

$$\sqrt{-a} = j\sqrt{a} \qquad (a > 0)$$

(12.2)

◀ EXAMPLE 4   **(a)** $\sqrt{-6} = \sqrt{(6)(-1)} = \sqrt{6}\sqrt{-1} = j\sqrt{6}$

this step is correct if only
one is negative, as in this case

(b) $-\sqrt{-75} = -\sqrt{(25)(3)(-1)} = -\sqrt{(25)(3)}\sqrt{-1} = -5j\sqrt{3}$

CAUTION ▶ We note that $-\sqrt{-75}$ *is not equal to* $\sqrt{75}$.

At times, we need to raise imaginary numbers to some power. Using the definitions of exponents and of j, we have the following results:

$$j = j \qquad\qquad j^5 = j^4 j = j$$
$$j^2 = -1 \qquad\qquad j^6 = j^4 j^2 = (1)(-1) = -1$$
$$j^3 = j^2 j = -j \qquad\qquad j^7 = j^4 j^3 = (1)(-j) = -j$$
$$j^4 = j^2 j^2 = (-1)(-1) = 1 \qquad j^8 = j^4 j^4 = (1)(1) = 1$$

The powers of j go through the cycle, $j, -1, -j, 1, j, -1, -j, 1$, and so forth. Noting this and the fact that j raised to a power that is a multiple of 4 equals 1 allows us to raise j to any integral power almost on sight.

◀ EXAMPLE 5   **(a)** $j^{10} = j^8 j^2 = (1)(-1) = -1$

(b) $j^{45} = j^{44} j = (1)(j) = j$

(c) $j^{531} = j^{528} j^3 = (1)(-j) = -j$

exponents 8, 44, 528 are multiples of 4

RECTANGULAR FORM OF A COMPLEX NUMBER

Even negative numbers were not widely accepted by mathematicians until late in the sixteenth century.

Imaginary numbers were so named because the French mathematician René Descartes (1596–1650) referred to them as "imaginaries." Most of the mathematical development of them occurred in the eighteenth century.

Complex numbers were named by the German mathematician Karl Friedrich Gauss (1777–1855).

Using real numbers and the imaginary unit j, we define a new kind of number. *A **complex number** is any number that can be written in the form $a + bj$, where a and b are real numbers. If $a = 0$ and $b \ne 0$, we have a number of the form bj, which is a **pure imaginary number**. If $b = 0$, then $a + bj$ is a real number. The form $a + bj$ is known as the **rectangular form** of a complex number, where a is known as the **real part** and b is known as the **imaginary part.** We see that complex numbers include all real numbers and all pure imaginary numbers.

A comment here about the words *imaginary* and *complex* is in order. The choice of the names of these numbers is historical in nature, and unfortunately it leads to some misconceptions about the numbers. The use of *imaginary* does not imply that the numbers do not exist. Imaginary numbers do in fact exist, as they are defined above. In the same way, the use of *complex* does not imply that the numbers are complicated and therefore difficult to understand. With the proper definitions and operations, we can work with complex numbers, just as with any type of number.

A complex number being the sum of a real number and an imaginary number, it is not positive or negative in the usual sense, and it is not equal in the usual way to another complex number. However, the real part is positive or negative (or zero), and the same is true of the imaginary part. Thus, *two complex numbers are equal if and only if their real parts are equal and their imaginary parts are equal.* That is,

NOTE ▶   *two complex numbers, a + bj and x + yj, are equal if a = x and b = y.*

◀ EXAMPLE 6   **(a)** $a + bj = 3 + 4j$ if $a = 3$ and $b = 4$

                (b) $x + yj = 5 - 3j$ if $x = 5$ and $y = -3$

real part ⬑   ⬑ imaginary part   ▶

◀ EXAMPLE 7   What are x and y if $4 - 6j - x = j + jy$?

One way to solve this equation is to arrange the terms so that all the known terms are on the right and all terms containing x and y are on the left. This leads to

$$-x - jy = -4 + 7j$$

From the definition of equality of complex numbers,

$$-x = -4 \quad \text{and} \quad -y = 7$$
$$x = 4 \quad \text{and} \quad y = -7$$

                                                                         ▶

◀ EXAMPLE 8   What values of x and y satisfy the equation

$$x + 3(xj + y) = 5 - j - jy$$

Rearranging the terms so that the known terms are on the right and terms containing x and y are on the left, we have

$$x + 3y + 3jx + jy = 5 - j$$

Next, factoring j from the two terms on the left will put the expression on the left into proper form. This leads to

$$(x + 3y) + (3x + y)j = 5 - j$$

Using the definition of equality, we have

$$x + 3y = 5 \quad \text{and} \quad 3x + y = -1$$

We now solve this system of equations. The solution is $x = -1$ and $y = 2$. Actually, the solution can be obtained at any point by writing each side of the equation in the form $a + bj$ and then equating first the real parts and then the imaginary parts.   ▶

Conjugate   The **conjugate** *of the complex number a + bj is the complex number a − bj.* We see that the sign of the imaginary part is changed to obtain the conjugate.

◀ EXAMPLE 9   **(a)** $3 - 2j$ is the conjugate of $3 + 2j$. We may also say that $3 + 2j$ is the conjugate of $3 - 2j$. Thus each is the conjugate of the other.

(b) $-2 - 5j$ and $-2 + 5j$ are conjugates.

(c) $6j$ and $-6j$ are conjugates.

(d) 3 is the conjugate of 3 (imaginary part is zero).   ▶

EXERCISES 12.1

In Exercises 1–4, perform the indicated operations on the resulting expressions if the given changes are made in the indicated examples of this section.

1. In Example 3, put a − sign before the first radical of the first illustration and then simplify.

2. In Example 4(a), put a *j* in front of the radical and then simplify.

3. In Example 5(a), add 40 to the exponent and then evaluate.

4. In Example 7, change the + on the right side of the equation to − and then solve.

In Exercises 5–16, express each number in terms of j.

5. $\sqrt{-81}$ 6. $\sqrt{-121}$ 7. $-\sqrt{-4}$ 8. $-\sqrt{-49}$

9. $\sqrt{-0.36}$ 10. $-\sqrt{-0.01}$ 11. $\sqrt{-8}$ 12. $\sqrt{-48}$

13. $\sqrt{-\frac{7}{4}}$ 14. $-\sqrt{-\frac{5}{9}}$ 15. $-\sqrt{-4e^2}$ 16. $\sqrt{-\pi^4}$

In Exercises 17–32, simplify each of the given expressions.

17. (a) $\left(\sqrt{-7}\right)^2$ (b) $\sqrt{(-7)^2}$

18. (a) $\sqrt{(-15)^2}$ (b) $\left(\sqrt{-15}\right)^2$

19. (a) $\sqrt{(-2)(-8)}$ (b) $\sqrt{-2}\sqrt{-8}$

20. (a) $\sqrt{-9}\sqrt{-16}$ (b) $\sqrt{(-9)(-16)}$

21. $\sqrt{-\frac{1}{15}}\sqrt{-\frac{27}{5}}$ 22. $-\sqrt{\left(-\frac{4}{7}\right)\left(-\frac{49}{16}\right)}$

23. $-\sqrt{-5}\sqrt{-2}\sqrt{-10}$ 24. $-\sqrt{(-3)(-7)}\sqrt{-21}$

25. (a) j^7 (b) j^{49} 26. (a) $-j^{22}$ (b) j^{408}

27. $j^2 - j^6$ 28. $2j^5 - j^2$

29. $j^{15} - j^{13}$ 30. $3j^{48} + j^{200}$ 31. $-\sqrt{j^2}$ 32. $-\sqrt{-j^2}$

In Exercises 33–44, perform the indicated operations and simplify each complex number to its rectangular form.

33. $2 + \sqrt{-9}$ 34. $-6 + \sqrt{-64}$ 35. $3j - \sqrt{-100}$

36. $-\sqrt{1} - \sqrt{-400}$ 37. $\sqrt{-4j^2} + \sqrt{-4}$ 38. $5 - 2\sqrt{25j^2}$

39. $2j^2 + 3j$ 40. $j^3 - 6$ 41. $\sqrt{18} - \sqrt{-8}$

42. $\sqrt{-27} + \sqrt{12}$ 43. $\left(\sqrt{-2}\right)^2 + j^4$ 44. $\left(2\sqrt{2}\right)^2 - j^6$

In Exercises 45–48, find the conjugate of each complex number.

45. (a) $6 - 7j$ (b) $8 + j$ 46. (a) $-3 + 2j$ (b) $-9 - j$

47. (a) $2j$ (b) -4 48. (a) 6 (b) $-5j$

In Exercises 49–54, find the values of x and y that satisfy the given equations.

49. $7x - 2yj = 14 + 4j$ 50. $2x + 3jy = -6 + 12j$

51. $6j - 7 = 3 - x - yj$ 52. $9 - j = xj + 1 - y$

53. $x - 2j^2 + 7j = yj + 2xj^3$ 54. $2x - 6xj^3 - 3j^2 = yj - y + 7j^5$

In Exercises 55–64, answer the given questions.

55. Are $8j$ and $-8j$ the solutions to the equation $x^2 + 64 = 0$?

56. Are $2j\sqrt{5}$ and $-2j\sqrt{5}$ the solutions to the equation $x^2 + 20 = 0$?

57. Are $2j$ and $-2j$ solutions to the equation $x^4 + 16 = 0$?

58. Are $3j$ and $-3j$ solutions to the equation $x^3 + 27j = 0$?

59. Evaluate $j + j^2 + j^3 + j^4 + j^5 + j^6 + j^7 + j^8$.

60. What is the smallest positive value of *n* for which $j^{-1} = j^n$?

W 61. Is it possible that a given complex number and its conjugate are equal? Explain.

W 62. Is it possible that a given complex number and the negative of its conjugate are equal? Explain.

63. Show that the real part of $x + yj$ equals the imaginary part of $j(x + yj)$.

W 64. Explain why a real number is a complex number, but a complex number may not be a real number.

12.2 BASIC OPERATIONS WITH COMPLEX NUMBERS

The basic operations of addition, subtraction, multiplication, and division of complex numbers are based on the operations for binomials with real coefficients (see Chapters 1 and 6). In performing these operations, we treat *j* as we would any other literal number, although we must properly handle any powers of *j* that might occur. However, *we must* NOTE ▶ *be careful to express all complex numbers in terms of j before performing these operations.* Therefore, we now have the definitions for these operations on complex numbers as shown on the next page.

BASIC OPERATIONS ON COMPLEX NUMBERS

Addition:

$$(a + bj) + (c + dj) = (a + c) + (b + d)j \qquad (12.3)$$

Subtraction:

$$(a + bj) - (c + dj) = (a - c) + (b - d)j \qquad (12.4)$$

Multiplication:

$$(a + bj)(c + dj) = (ac - bd) + (ad + bc)j \qquad (12.5)$$

Division:

$$\frac{a + bj}{c + dj} = \frac{(a + bj)(c - dj)}{(c + dj)(c - dj)} = \frac{(ac + bd) + (bc - ad)j}{c^2 + d^2} \qquad (12.6)$$

Compare Eq. (12.5) with Eq. (6.6).

Eqs. (12.3) and (12.4) show that we *add and subtract complex numbers by combining the real parts and combining the imaginary parts.*

◀ **EXAMPLE 1** (a) $(3 - 2j) + (-5 + 7j) = (3 - 5) + (-2 + 7)j$
$$= -2 + 5j$$

(b) $(7 + 9j) - (6 - 4j) = (7 - 6) + (9 - (-4))j$
$$= 1 + 13j$$

◀ **EXAMPLE 2** $\left(3\sqrt{-4} - 4\right) - \left(6 - 2\sqrt{-25}\right) - \sqrt{-81}$
$$= [3(2j) - 4] - [6 - 2(5j)] - 9j \qquad \text{write in terms of } j$$
$$= [6j - 4] - [6 - 10j] - 9j$$
$$= 6j - 4 - 6 + 10j - 9j$$
$$= -10 + 7j$$

When complex numbers are multiplied, Eq. (12.5) indicates that *we proceed as in any algebraic multiplication,* properly expressing numbers in terms of j and evaluating the power of j.

◀ **EXAMPLE 3** $\left(6 - \sqrt{-4}\right)\left(\sqrt{-9}\right) = (6 - 2j)(3j) \qquad \text{write in terms of } j$
$$= 18j - 6j^2 = 18j - 6(-1)$$
$$= 6 + 18j$$

◀ **EXAMPLE 4**
$(-9.4 - 6.2j)(2.5 + 1.5j) = (-9.4)(2.5) + (-9.4)(1.5j) + (-6.2j)(2.5) + (-6.2j)(1.5j)$
$$= -23.5 - 14.1j - 15.5j - 9.3j^2$$
$$= -23.5 - 29.6j - 9.3(-1)$$
$$= -14.2 - 29.6j$$

The procedure shown by Eq. (12.6) for dividing by a complex number is the same procedure we used for rationalizing the denominator of a fraction with a radical in the denominator. We use this procedure so that we can express any result in the proper form of a complex number. Therefore, *to divide by a complex number, multiply the numerator and the denominator by the conjugate of the denominator.*

Some calculators are programmed to perform operations with complex numbers. In Fig. 12.1 the display of the operations of Examples 2 and 5 are shown for such a calculator. Note the use of i rather than j.

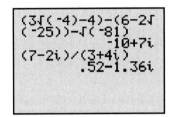

Fig. 12.1

◀ EXAMPLE 5

$$\frac{7 - 2j}{3 + 4j} = \frac{(7 - 2j)(3 - 4j)}{(3 + 4j)(3 - 4j)} \leftarrow \text{multiply by conjugate of denominator}$$

$$= \frac{21 - 28j - 6j + 8j^2}{9 - 16j^2} = \frac{21 - 34j + 8(-1)}{9 - 16(-1)}$$

$$= \frac{13 - 34j}{25}$$

This could be written in the form $a + bj$ as $\frac{13}{25} - \frac{34}{25}j$, but this type of result is generally left as a single fraction. In decimal form, the result would be expressed as $0.52 - 1.36j$.

◀ EXAMPLE 6

(a) $\dfrac{6 - j^3}{2j} = \dfrac{6 - (-j)}{2j} = \dfrac{6 + j}{2j} = \dfrac{6 + j}{2j}\left(\dfrac{-2j}{-2j}\right)$

$$= \frac{-12j - 2j^2}{-4j^2} = \frac{2 - 12j}{4}$$

$$= \frac{1 - 6j}{2}$$

(b) $\dfrac{1}{j} + \dfrac{2}{3 + j} = \dfrac{(3 + j) + 2j}{j(3 + j)} = \dfrac{3 + 3j}{3j - 1} = \dfrac{3 + 3j}{-1 + 3j} \cdot \dfrac{-1 - 3j}{-1 - 3j}$

$$= \frac{-3 - 12j - 9j^2}{1 - 9j^2} = \frac{6 - 12j}{10}$$

$$= \frac{3 - 6j}{5}$$

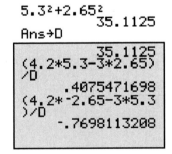

Fig. 12.2

◀ EXAMPLE 7 In an alternating-current circuit, the voltage E is given by $E = IZ$, where I is the current in amperes and Z is the impedance in ohms. Each of these can be represented by complex numbers. Find the complex number representation for I if $E = 4.20 - 3.00j$ volts and $Z = 5.30 + 2.65j$ ohms. (This type of circuit is discussed in more detail in Section 12.7.)

Since $I = E/Z$, we have

$$I = \frac{4.20 - 3.00j}{5.30 + 2.65j} = \frac{(4.20 - 3.00j)(5.30 - 2.65j)}{(5.30 + 2.65j)(5.30 - 2.65j)} \leftarrow \begin{array}{l}\text{multiply by conjugate}\\ \text{of denominator}\end{array}$$

$$= \frac{22.26 - 11.13j - 15.90j + 7.95j^2}{5.30^2 - 2.65^2j^2} = \frac{22.26 - 7.95 - 27.03j}{5.30^2 + 2.65^2}$$

$$= \frac{14.31 - 27.03j}{35.11}$$

$$= 0.408 - 0.770j \text{ amperes}$$

The display for the solution using a calculator that does not operate with complex numbers is shown in Fig. 12.2. Note that the first step was to store the value of the denominator and that we use a negative sign for j^2.

EXERCISES 12.2

In Exercises 1–4, perform the indicated operations on the resulting expressions if the given changes are made in the indicated examples of this section.

1. In Example 1(b), change the sign in the first parentheses from + to − and then perform the addition.

2. In Example 4, change the sign before $6.2j$ from − to +, and then perform the multiplication.

3. In Example 5, change the sign in the denominator from + to − and then simplify.

4. In Example 6(b), change the sign in the second denominator from + to − and then simplify.

In Exercises 5–36, perform the indicated operations, expressing all answers in the form $a + bj$.

5. $(3 - 7j) + (2 - j)$
6. $(-4 - j) + (-7 - 4j)$
7. $(7j - 6) - (3 + j)$
8. $(5.4 - 3.4j) - (2.9j + 5.5)$
9. $0.23 - (0.46 - 0.19j) + 0.67j$
10. $(7 - j) - (4 - 4j) + (6 - j)$
11. $(12j - 21) - (15 - 18j) - 9j$
12. $(0.062j - 0.073) - 0.030j - (0.121 - 0.051j)$
13. $(7 - j)(7j)$
14. $(-2.2j)(1.5j - 4.0)$
15. $(4 - j)(5 + 2j)$
16. $(8j - 5)(7 + 4j)$
17. $(\sqrt{-18}\,\sqrt{-4})(3j)$
18. $\sqrt{-6}\,\sqrt{-12}\,\sqrt{3}$
19. $7j^3 - 7\sqrt{-9}$
20. $6j - 5j^2\sqrt{-63}$
21. $j\sqrt{-7} - j^6\sqrt{112} + 3j$
22. $j^2\sqrt{-7} - \sqrt{-28} + 8$
23. $(3 - 7j)^2$
24. $(4j + 5)^2$
25. $(1 - j)^3$
26. $(1 + j)(1 - j)^2$
27. $\dfrac{6j}{2 - 5j}$
28. $\dfrac{0.25}{3 - \sqrt{-1}}$
29. $\dfrac{1 - j}{3j}$
30. $\dfrac{6 + 5j}{3 - 4j}$
31. $\dfrac{j\sqrt{2} - 5}{j\sqrt{2} + 3}$
32. $\dfrac{j^5 - j^3}{3 + j}$
33. $\dfrac{j^2 - j}{2j - j^8}$
34. $\dfrac{3}{2j} + \dfrac{5}{6 - j}$
35. $\dfrac{4j}{1 - j} - \dfrac{3 + j}{2 + 3j}$
36. $\dfrac{(6j + 5)(2 - 4j)}{(5 - j)(4j + 1)}$

In Exercises 37–44, answer the given problems.

37. Show that $-1 - j$ is a solution to the equation $x^2 + 2x + 2 = 0$.
38. Show that $1 - j\sqrt{3}$ is a solution to the equation $x^2 + 4 = 2x$.
39. Multiply $-3 + j$ by its conjugate.
40. Divide $2 - 3j$ by its conjugate.
41. Write the reciprocal of $3 - j$ in rectangular form.
42. Write the reciprocal of $2 + 5j$ in rectangular form.
43. Write $j^{-2} + j^{-3}$ in rectangular form.
44. When finding the current in a certain electric circuit, the expression $(s + 1 + 4j)(s + 1 - 4j)$ occurs. Simplify this expression.

In Exercises 45–48, solve the given problems. Refer to Example 7.

45. If $I = 0.835 - 0.427j$ amperes and $Z = 250 + 170j$ ohms, find the complex-number representation for E.

46. If $E = 5.70 - 3.65j$ volts and $I = 0.360 - 0.525j$ amperes, find the complex-number representation for Z.

47. If $E = 85 + 74j$ volts and $Z = 2500 - 1200j$ ohms, find the complex-number representation for I.

48. In an alternating-current circuit, two impedances Z_1 and Z_2 have a total impedance Z_T of $Z_T = \dfrac{Z_1 Z_2}{Z_1 + Z_2}$. Find Z_T for $Z_1 = 2 + 3j$ ohms and $Z_2 = 3 - 4j$ ohms.

In Exercises 49–52, answer or explain as indicated.

49. What type of number is the result of (a) adding a complex number to its conjugate and (b) subtracting a complex number from its conjugate?

 50. If the reciprocal of $a + bj$ equals $a - bj$, what condition must a and b satisfy?

51. Explain why the product of a complex number and its conjugate is real and nonnegative.

52. Explain how to show that the reciprocal of the imaginary unit is the negative of the imaginary unit.

12.3 GRAPHICAL REPRESENTATION OF COMPLEX NUMBERS

Since a complex number has an imaginary part as well as a real part, we represent it graphically as a *point*, designated as $a + bj$, in the rectangular coordinate system. *The real part is the x-value* of the point, and *the imaginary part* is the *y-value* of the point. Used in this way, *the coordinate system is called the* **complex plane,** *the horizontal axis is the* **real axis,** and *the vertical axis is the* **imaginary axis.**

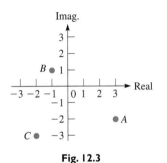

Fig. 12.3

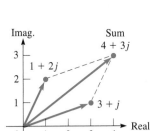

◀ **EXAMPLE 1** In Fig. 12.3, point *A* represents the complex number $3 - 2j$; point *B* represents $-1 + j$; point *C* represents $-2 - 3j$. We note that these complex numbers are represented by the points $(3, -2)$, $(-1, 1)$, and $(-2, -3)$, respectively, of the standard rectangular coordinate system.

We must keep in mind that the meaning given to the points representing complex numbers in the complex plane is different from the meaning given to the points in the standard rectangular coordinate system. A point in the complex plane represents a single complex number, whereas a point in the rectangular coordinate system represents a pair of real numbers. ▶

Let us represent two complex numbers and their sum in the complex plane. Consider, for example, the two complex numbers $1 + 2j$ and $3 + j$. By algebraic addition, the sum is $4 + 3j$. When we draw lines from the origin to these points (see Fig. 12.4), we note that if we think of the complex numbers as being vectors, their sum is the vector sum. Because complex numbers can be used to represent vectors, these numbers are particularly important. *Any complex number can be thought of as representing a vector from the origin to its point in the complex plane.* This leads us to the method used to add complex numbers graphically.

Fig. 12.4

STEPS TO ADD COMPLEX NUMBERS GRAPHICALLY

1. *Find the point corresponding to one of the numbers and draw a line from the origin to this point.*

2. *Repeat step 1 for the second number.*

3. *Complete a parallelogram with the lines drawn as adjacent sides. The resulting fourth vertex is the point representing the sum.*

Note that *this is equivalent to adding vectors by graphical means.*

◀ **EXAMPLE 2** Add the complex numbers $5 - 2j$ and $-2 - j$ graphically.

The solution is indicated in Fig. 12.5. We can see that the fourth vertex of the parallelogram is at $3 - 3j$. This is, of course, the algebraic sum of these two complex numbers. ▶

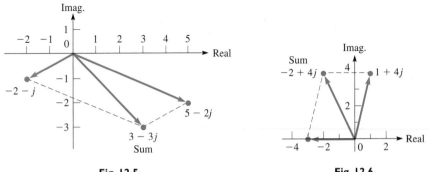

Fig. 12.5   **Fig. 12.6**

◀ **EXAMPLE 3** Add the complex numbers -3 and $1 + 4j$ graphically.

First, note that $-3 = -3 + 0j$, which means that the point representing -3 is on the negative real axis. In Fig. 12.6, we show the numbers -3 and $1 + 4j$ on the graph and complete the parallelogram. From the graph, we see that the sum is $-2 + 4j$. ▶

◀ EXAMPLE 4   Subtract $4 - 2j$ from $2 - 3j$ graphically.

Subtracting $4 - 2j$ is equivalent to adding $-4 + 2j$. Therefore, we complete the solution by adding $-4 + 2j$ and $2 - 3j$, as shown in Fig. 12.7. The result is $-2 - j$.

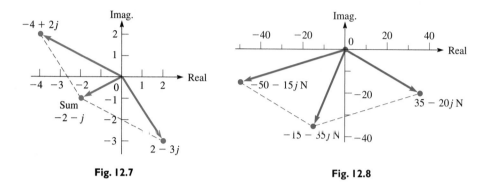

Fig. 12.7   Fig. 12.8

◀ EXAMPLE 5   Two forces acting on an overhead bolt can be represented by $35 - 20j$ N and $-50 - 15j$ N. Find the resultant force graphically.

The forces are shown in Fig. 12.8. From the graph we can see that the sum of the forces, which is the resultant force, is $-15 - 35j$ N.

EXERCISES 12.3

In Exercises 1 and 2, perform the indicated operations for the resulting complex numbers if the given changes are made in the indicated examples of this section.

1. In Example 2, change the sign of the imaginary part of the second complex number and then add the numbers graphically.

2. In Example 4, change the sign of the imaginary part of the second complex number and do the subtraction graphically.

In Exercises 3–8, locate the given numbers in the complex plane.

3. $2 + 6j$ **4.** $-5 + j$ **5.** $-4 - 3j$

6. 4 **7.** $-3j$ **8.** $3 - 4j$

In Exercises 9–28, perform the indicated operations graphically. Check them algebraically.

9. $2 + (3 + 4j)$ **10.** $2j + (-2 + 3j)$

11. $(5 - j) + (3 + 2j)$ **12.** $(3 - 2j) + (-1 - j)$

13. $5j - (1 - 4j)$ **14.** $(2 - j) - 1$

15. $(2 - 4j) + (-2 + j)$ **16.** $(-1 - 2j) + (6 - j)$

17. $(3 - 2j) - (4 - 6j)$ **18.** $(-25 - 40j) - (20 - 55j)$

19. $(80 + 300j) - (260 + 150j)$ **20.** $(-j - 2) - (-1 - 3j)$

21. $(1.5 - 0.5j) + (3.0 + 2.5j)$

22. $(3.5 + 2.0j) - (-4.0 - 1.5j)$

23. $(3 - 6j) - (-1 + 5j)$ **24.** $(-6 - 3j) + (2 - 7j)$

25. $(2j + 1) - 3j - (j + 1)$ **26.** $(6 - j) - 9 - (2j - 3)$

27. $(j - 6) - j + (j - 7)$ **28.** $j - (1 - j) + (3 + 2j)$

In Exercises 29–32, show the given number, its negative, and its conjugate on the same coordinate system.

29. $3 + 2j$ **30.** $-2 + 4j$

31. $-3 - 5j$ **32.** $5 - j$

*In Exercises 33 and 34, show the numbers $a + bj$, $3(a + bj)$, and $-3(a + bj)$ on the same coordinate system. The multiplication of a complex number by a real number is called **scalar multiplication** of the complex number.*

33. $3 - j$ **34.** $-10 - 30j$

In Exercises 35 and 36, perform the indicated vector additions graphically. Check them algebraically.

35. Two ropes hold a boat at a dock. The tensions in the ropes can be represented by $40 + 10j$ N and $50 - 25j$ N. Find the resultant force.

36. Relative to the air, a plane heads north of west with a velocity that can be represented by $-320 + 140j$ km/h. The wind is blowing from south of west with a velocity that can be represented by $40 + 140j$ km/h. Find the resultant velocity of the plane.

12.4 POLAR FORM OF A COMPLEX NUMBER

We have just seen the relationship between complex numbers and vectors. Since we can use one to represent the other, we use this fact to write complex numbers in another way. The new form that we develop has certain advantages when the basic operations of multiplication and division are performed on complex numbers. We discuss these operations later in the chapter.

By drawing a vector from the origin to the point in the complex plane that represents the number $x + yj$, we see the relation between vectors and complex numbers. Further observation indicates that an angle in standard position has been formed. Also, the point $x + yj$ is r units from the origin. In fact, *we can find any point in the complex plane by knowing the angle θ and the value of r.* The necessary equations relating x, y, r, and θ are similar to those we developed for vectors (Eqs. (9.1) to (9.3)). By referring to Fig. 12.9, we can see that

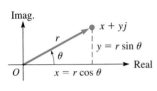

Imag.

$x + yj$

$y = r \sin \theta$

$x = r \cos \theta$

Real

Fig. 12.9

$$x = r \cos \theta \qquad y = r \sin \theta \qquad (12.7)$$

$$r^2 = x^2 + y^2 \qquad \tan \theta = \frac{y}{x} \qquad (12.8)$$

Substituting Eq. (12.7) into the rectangular form $x + yj$ of a complex number, we have

$$x + yj = r \cos \theta + j(r \sin \theta)$$

or

$$x + yj = r(\cos \theta + j \sin \theta) \qquad (12.9)$$

The right side of Eq. (12.9) is called the **polar form** *of a complex number. Sometimes it is referred to as the* **trigonometric form.** An abbreviated form of writing the polar form that is sometimes used is r cis θ. *The length r is called the* **absolute value,** *or the* **modulus,** *and the angle θ is called the* **argument** *of the complex number.* Therefore, Eq. (12.9), along with Eqs. (12.8), defines the polar form of a complex number.

EXAMPLE 1 Represent the complex number $3 + 4j$ graphically and give its polar form.

From the rectangular form $3 + 4j$, we see that $x = 3$ and $y = 4$. Using Eqs. (12.8), we have

$$r = \sqrt{3^2 + 4^2} = 5 \qquad \theta = \tan^{-1} \frac{4}{3} = 53.1°$$

Thus, the polar form is

$$5(\cos 53.1° + j \sin 53.1°)$$

The graphical representation is shown in Fig. 12.10.

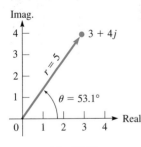

Imag.

4 • $3 + 4j$

3

$r = 5$

2

1    $\theta = 53.1°$

Real

0 1 2 3 4

Fig. 12.10

A note on significant digits is in order here. In writing a complex number as $3 + 4j$ in Example 1, no approximate values are intended. However, in expressing the polar form as $5(\cos 53.1° + j \sin 53.1°)$, we rounded off the angle to the nearest $0.1°$, as it is not possible to express the result exactly in degrees. Therefore, in dealing with nonexact numbers, we will express angles to the nearest $0.1°$. Other results, when approximate, will be expressed to three significant digits, unless a different accuracy is given in the problem. Of course, in applied situations most numbers are approximate, as they are derived through measurement.

Another convenient and widely used notation for the polar form is $r\underline{/\theta}$. We must remember in using this form that it represents a complex number and is simply a shorthand way of writing $r(\cos \theta + j \sin \theta)$. Therefore,

$$r\underline{/\theta} = r(\cos \theta + j \sin \theta) \qquad (12.10)$$

◀ **EXAMPLE 2** (a) $3(\cos 40° + j \sin 40°) = 3\underline{/40°}$

(b) $6.26(\cos 217.3° + j \sin 217.3°) = 6.26\underline{/217.3°}$

(c) $5\underline{/120°} = 5(\cos 120° + j \sin 120°)$

(d) $14.5\underline{/306.2°} = 14.5(\cos 306.2° + j \sin 306.2°)$ ▶

◀ **EXAMPLE 3** Represent the complex number $-2.08 - 3.12j$ graphically and give its polar forms.

The graphical representation is shown in Fig. 12.11. From Eqs. (12.8), we have

$$r = \sqrt{(-2.08)^2 + (-3.12)^2} = 3.75$$

$$\theta_{\text{ref}} = \tan^{-1}\frac{3.12}{2.08} = 56.3° \qquad \theta = 180° + 56.3° = 236.3°$$

Since both the real and imaginary parts are negative, we know that θ is a third-quadrant angle. Therefore, we found the reference angle before finding θ. This means the polar forms are

$$3.75(\cos 236.3° + j \sin 236.3°) = 3.75\underline{/236.3°} \qquad ▶$$

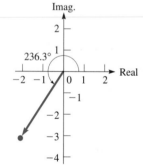

Fig. 12.11

◀ **EXAMPLE 4** The impedance Z (in Ω) in an alternating-current circuit is given by $Z = 3560\underline{/-32.4°}$. Express this in rectangular form.

From the polar form we have $r = 3560 \ \Omega$ and $\theta = -32.4°$ (it is common to use negative angles in this type of application.) This means that we can also write

$$Z = 3560(\cos(-32.4°) + j \sin(-32.4°))$$

See Fig. 12.12. This means that

$$x = 3560 \cos(-32.4°) = 3010$$
$$y = 3560 \sin(-32.4°) = -1910$$

Therefore, the rectangular form is

$$Z = 3010 - 1910j \ \Omega \qquad ▶$$

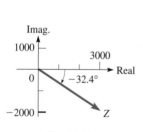

Fig. 12.12

Complex numbers can be converted between rectangular and polar forms on most calculators. This will be illustrated in the next section.

❮ EXAMPLE 5   Represent the numbers 5, −5, 7j, and −7j in polar form.

Since a positive real number is on the positive real axis, its polar form is

$$a = a(\cos 0° + j \sin 0°) = a\underline{/0°}$$

Negative real numbers, being on the negative real axis, are written as

$$a = |a|(\cos 180° + j \sin 180°) = |a|\underline{/180°}$$

Thus,

$$5 = 5(\cos 0° + j \sin 0°) = 5\underline{/0°}$$
$$-5 = 5(\cos 180° + j \sin 180°) = 5\underline{/180°}$$

Positive pure imaginary numbers lie on the positive imaginary axis and are expressed in polar form by

$$bj = b(\cos 90° + j \sin 90°) = b\underline{/90°}$$

Similarly, negative pure imaginary numbers, being on the negative imaginary axis, are written as

$$bj = |b|(\cos 270° + j \sin 270°) = |b|\underline{/270°}$$

Thus,

$$7j = 7(\cos 90° + j \sin 90°) = 7\underline{/90°}$$
$$-7j = 7(\cos 270° + j \sin 270°) = 7\underline{/270°}$$

The *complex numbers* 5, −5, 7j, and −7j are shown graphically in Fig. 12.13.   ❯

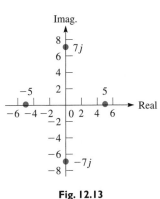

Fig. 12.13

EXERCISES 12.4

In Exercises 1 and 2, change the sign of the real part of the complex number in the indicated example of this section and then perform the indicated operations for the resulting complex number.

1. Example 1   **2.** Example 3

In Exercises 3–18, represent each complex number graphically and give the polar form of each.

3. $8 + 6j$ **4.** $-8 - 15j$ **5.** $3 - 4j$

6. $-5 + 12j$ **7.** $-2.00 + 3.00j$ **8.** $7.00 - 5.00j$

9. $-0.55 - 0.24j$ **10.** $460 - 460j$ **11.** $1 + j\sqrt{3}$

12. $\sqrt{2} - j\sqrt{2}$ **13.** $3.514 - 7.256j$ **14.** $6.231 + 9.527j$

15. -3 **16.** 6 **17.** $9j$ **18.** $-2j$

In Exercises 19–36, represent each complex number graphically and give the rectangular form of each.

19. $5.00(\cos 54.0° + j \sin 54.0°)$

20. $3.00(\cos 232.0° + j \sin 232.0°)$

21. $1.60(\cos 150.0° + j \sin 150.0°)$

22. $2.50(\cos 315.0° + j \sin 315.0°)$

23. $6(\cos 180° + j \sin 180°)$

24. $120(\cos 270° + j \sin 270°)$

25. $0.08(\cos 360° + j \sin 360°)$

26. $15(\cos 0° + j \sin 0°)$

27. $220.8(\cos 155.13° + j \sin 155.13°)$

28. $\cos 600.0° + j \sin 600.0°$

29. $4.75\underline{/172.8°}$ **30.** $1.50\underline{/62.3°}$

31. $0.9326\underline{/229.54°}$ **32.** $277.8\underline{/-342.63°}$

33. $7.32\underline{/-270°}$ **34.** $18.3\underline{/540.0°}$

35. $86.42\underline{/94.62°}$ **36.** $4629\underline{/182.44°}$

In Exercises 37–40, solve the given problems.

37. The voltage of a certain generator is represented by $2.84 - 1.06j$ kV. Write this voltage in polar form.

38. Find the magnitude and direction of a force on a bolt that is represented by $40.5 + 24.5j$ N.

39. The electric field intensity of a light wave can be described by $12.4\underline{/78.3°}$ V/m. Write this in rectangular form.

40. The current in a certain microprocessor circuit is represented by $3.75\underline{/15.0°}$ μA. Write this in rectangular form.

12.5 EXPONENTIAL FORM OF A COMPLEX NUMBER

Another important form of a complex number is the *exponential form*. It is commonly used in electronics, engineering, and physics applications. As we will see in the next section, it is also convenient for multiplication and division of complex numbers, as the rectangular form is for addition and subtraction.

The **exponential form** *of a complex number is written as* $re^{j\theta}$, *where r and θ have the same meanings as given in the previous section, although θ is expressed in radians. The number e is a special irrational number and has an approximate value*

$$e = 2.718\,281\,828\,459\,045\,2$$

This number e is very important in mathematics, and we will see it again in the next chapter. For now it is necessary to accept the value for e, although in calculus its meaning is shown along with the reason it has the above value. We can find its value on a calculator by using the $\boxed{e^x}$ key, with $x = 1$.

In advanced mathematics, it is shown that

$$re^{j\theta} = r(\cos\theta + j\sin\theta) \tag{12.11}$$

By expressing θ in radians, the expression $j\theta$ is an exponent, and it can be shown to obey all the laws of exponents as discussed in Chapter 11. Therefore, we will always

CAUTION ▶ *express θ in radians when using exponential form.*

◀ EXAMPLE 1   Express the number $3 + 4j$ in exponential form.

From Example 1 of Section 12.4, we know that this complex number may be written in polar form as $5(\cos 53.1° + j\sin 53.1°)$. Therefore, we know that $r = 5$. We now express 53.1° in terms of radians as

$$\frac{53.1\pi}{180} = 0.927 \text{ rad}$$

Thus, the exponential form is $5e^{0.927j}$. This means that

$$3 + 4j = 5(\cos 53.1° + j\sin 53.1°) = 5e^{0.927j}$$

degrees to radians

value of r

◀ EXAMPLE 2   Express the number $8.50\underline{/136.3°}$ in exponential form.

Since this complex number is in polar form, we note that $r = 8.50$ and that we must express 136.3° in radians. Changing 136.3° to radians, we have

$$\frac{136.3\pi}{180} = 2.38 \text{ rad}$$

Therefore, the required exponential form is $8.50e^{2.38j}$. This means that

$$8.50\underline{/136.3°} = 8.50e^{2.38j}$$

We see that the principal step in changing from polar form to exponential form is to change θ from degrees to radians.

EXAMPLE 3 Express the number $3.07 - 7.43j$ in exponential form.

From the rectangular form, we have $x = 3.07$ and $y = -7.43$. Therefore,

$$r = \sqrt{(3.07)^2 + (-7.43)^2} = 8.04$$

$$\theta_{\text{ref}} = \tan^{-1}\frac{7.43}{3.07} = 67.6° \qquad \theta = 360° - 67.6° = 292.4°$$

For reference, Eq. (8.8) is π rad $= 180°$.

Since $292.4° = 5.10$ rad, the exponential form is $8.04e^{5.10j}$. (Using radian mode, we could show that $2\pi - \tan^{-1}(7.43/3.07) = 5.10$). This means that

$$3.07 - 7.43j = 8.04e^{5.10j}$$

EXAMPLE 4 Express the complex number $2.00e^{4.80j}$ in polar and rectangular forms.

We first express 4.80 rad as $275.0°$. From the exponential form, we know that $r = 2.00$. Thus, the polar form is

$$2.00(\cos 275.0° + j \sin 275.0°)$$

Using the distributive law, we rewrite the polar form and then evaluate. Thus,

$$2.00e^{4.80j} = 2.00(\cos 275.0° + j \sin 275.0°)$$
$$= 2.00 \cos 275.0° + (2.00 \sin 275.0°)j$$
$$= 0.174 - 1.99j$$

Calculators that operate with complex numbers treat the exponential and polar forms as being the same, since each uses r and θ. Figures 12.14 and 12.15 show the conversions of Examples 5 and 6 on such a calculator.

EXAMPLE 5 Express $(1.846e^{1.229j})^2$ in polar and rectangular forms.

First, we note that $(1.846e^{1.229j})^2 = 1.846^2e^{2.458j}$. Since 2.458 rad $= 140.83°$, the polar form is $1.846^2 \underline{/140.83°} = 3.408 \underline{/140.83°}$.

For the rectangular form, using radian mode we have

$$(1.846e^{1.229j})^2 = 1.846^2e^{2.458j} = 1.846^2(\cos 2.458 + j \sin 2.458)$$
$$= -2.642 + 2.152j$$

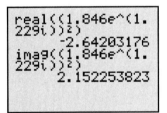

Fig. 12.14

As we noted earlier, an important application of the use of complex numbers is in alternating-current analysis. When an alternating current flows through a given circuit, usually the current and voltage have different phases. That is, they do not reach their peak values at the same time. Therefore, one way of accounting for the magnitude as well as the phase of an electric current or voltage is to write it as a complex number. Here the modulus is the actual magnitude of the current or voltage, and the argument θ is a measure of the phase.

EXAMPLE 6 A current of $2.00 - 4.00j$ amperes flows in a given circuit. Express this current in exponential form and find the magnitude of the current.

From the rectangular form, we have $x = 2.00$ and $y = -4.00$. Therefore,

$$r = \sqrt{(2.00)^2 + (-4.00)^2} = 4.47 \text{ A}$$

$$\theta = \tan^{-1}\frac{-4.00}{2.00} = -63.4°$$

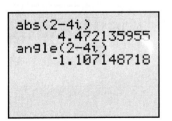

Fig. 12.15

It is normal to express the phase in terms of negative angles when the imaginary part is negative. Changing $63.4°$ to radians, we have $63.4° = 1.11$ rad. This means the exponential form is $4.47e^{-1.11j}$. The modulus is 4.47, which means the magnitude of the current is 4.47 A. Note in the calculator display in Fig. 12.15 that the angle is shown in radians and is negative.

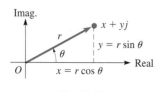

Fig. 12.16

At this point, we summarize the three important forms of a complex number. See Fig. 12.16 for the graphical representation.

> Rectangular: $x + yj$
> Polar: $r(\cos \theta + j \sin \theta) = r\underline{/\theta}$
> Exponential: $re^{j\theta}$

It follows that

$$x + yj = r(\cos \theta + j \sin \theta) = r\underline{/\theta} = re^{j\theta} \qquad (12.12)$$

where

$$r^2 = x^2 + y^2 \qquad \tan \theta = \frac{y}{x} \qquad (12.8)$$

In Eq. (12.12), the argument θ is the same for the exponential and polar forms, although it is expressed in radians in exponential form and is usually expressed in degrees in polar form.

EXERCISES 12.5

In Exercises 1–4, perform the indicated operations for the resulting complex numbers if the given changes are made in the indicated examples of this section.

1. In Example 2, change 136.3° to 226.3° and then find the exponential form.

2. In Example 3, change the sign before 7.43j from − to + and then find the exponential form.

3. In Example 4, change the exponent to 3.80j and then find the polar and rectangular forms.

4. In Example 5, change the exponent 2 to 3 and then find the polar and rectangular forms.

In Exercises 5–24, express the given numbers in exponential form.

5. $3.00(\cos 60.0° + j \sin 60.0°)$

6. $5.00(\cos 135.0° + j \sin 135.0°)$

7. $45.0(\cos 282.3° + j \sin 282.3°)$

8. $2.10(\cos 588.7° + j \sin 588.7°)$

9. $375.5(\cos 95.46° + j \sin 95.46°)$

10. $16.72[\cos(-7.14°) + j \sin(-7.14°)]$

11. $0.515\underline{/198.3°}$

12. $4650\underline{/326.5°}$

13. $4.06\underline{/-61.4°}$

14. $0.0192\underline{/76.7°}$

15. $9245\underline{/296.32°}$

16. $827.6\underline{/470.09°}$

17. $3 - 4j$

18. $-1 - 5j$

19. $-30 + 20j$

20. $600 + 100j$

21. $5.90 + 2.40j$

22. $47.3 - 10.9j$

23. $-634.6 - 528.2j$

24. $-8573 + 5477j$

In Exercises 25–32, express the given complex numbers in polar and rectangular forms.

25. $3.00e^{0.500j}$

26. $2.00e^{1.00j}$

27. $4.64e^{1.85j}$

28. $2.50e^{3.84j}$

29. $3.20e^{-5.41j}$

30. $0.800e^{3.00j}$

31. $0.1724e^{2.391j}$

32. $820.7e^{-3.492j}$

In Exercises 33–36, perform the indicated operations and express results in rectangular and polar forms.

33. $(4.55e^{1.32j})^2$

34. $(0.926e^{0.253j})^3$

35. $(625e^{3.46j})(4.40e^{1.22j})$

36. $(18.0e^{5.13j})(25.5e^{0.77j})$

In Exercises 37–40, perform the indicated operations.

37. The impedance in an antenna circuit is $375 + 110j$ Ω. Write this in exponential form and find the magnitude of the impedance.

38. The intensity of the signal from a radar microwave signal is $37.0[\cos(-65.3°) + j \sin(-65.3°)]$ V/m. Write this in exponential form.

39. The displacement of the end of a vibrating rod is $5.83e^{-1.20j}$. Write this in rectangular form.

40. In an electric circuit, the *admittance* is the reciprocal of the impedance. In a transistor circuit, the impedance is $2800 - 1450j$ Ω. Find the exponential form of the admittance.

12.6 PRODUCTS, QUOTIENTS, POWERS, AND ROOTS OF COMPLEX NUMBERS

We have previously found products and quotients using the rectangular forms of the given complex numbers. However, these operations can also be performed with complex numbers in polar and exponential forms. These operations not only are convenient but also are useful for purposes of finding powers and roots of complex numbers.

We may find the product of two complex numbers by using the exponential form and the laws of exponents. Multiplying $r_1 e^{j\theta_1}$ by $r_2 e^{j\theta_2}$, we have

$$[r_1 e^{j\theta_1}] \times [r_2 e^{j\theta_2}] = r_1 r_2 e^{j\theta_1 + j\theta_2} = r_1 r_2 e^{j(\theta_1 + \theta_2)}$$

We use this equation to express the product of two complex numbers in polar form:

$$[r_1 e^{j\theta_1}] \times [r_2 e^{j\theta_2}] = [r_1(\cos\theta_1 + j\sin\theta_1)] \times [r_2(\cos\theta_2 + j\sin\theta_2)]$$

and

$$r_1 r_2 e^{j(\theta_1 + \theta_2)} = r_1 r_2[\cos(\theta_1 + \theta_2) + j\sin(\theta_1 + \theta_2)]$$

Therefore, the polar expressions are equal, which means that *the product of two complex numbers is*

$$r_1(\cos\theta_1 + j\sin\theta_1)r_2(\cos\theta_2 + j\sin\theta_2)$$
$$= r_1 r_2[\cos(\theta_1 + \theta_2) + j\sin(\theta_1 + \theta_2)] \qquad (12.13)$$
$$(r_1 \underline{/\theta_1})(r_2 \underline{/\theta_2}) = r_1 r_2 \underline{/\theta_1 + \theta_2}$$

NOTE ▶ The *magnitudes are multiplied,* and the *angles are added.*

◀ EXAMPLE 1 Multiply the complex numbers $2 + 3j$ and $1 - j$ by using the polar form of each.

$$\text{For } 2 + 3j: \quad r_1 = \sqrt{2^2 + 3^2} = 3.61 \quad \tan\theta_1 = \frac{3}{2} \quad \theta_1 = 56.3°$$

$$\text{For } 1 - j: \quad r_2 = \sqrt{1^2 + (-1)^2} = 1.41 \quad \tan\theta_2 = \frac{-1}{1} \quad \theta_2 = 315.0°$$

$$(3.61)(\cos 56.3° + j\sin 56.3°)(1.41)(\cos 315.0° + j\sin 315.0°)$$

$$\text{— sum —}$$

$$= (3.61)(1.41)[\cos(56.3° + 315.0°) + j\sin(56.3° + 315.0°)]$$
$$= 5.09(\cos 371.3° + j\sin 371.3°)$$
$$= 5.09(\cos 11.3° + j\sin 11.3°)$$

Note that in the final result the angle is expressed as 11.3°. The angle is usually expressed between 0° and 360°, unless specified otherwise. As we have seen, in some applications it is common to use negative angles. ▶

◀ EXAMPLE 2   When we use the r/θ polar form to multiply the two complex numbers in Example 1, we have

$$r_1 = 3.61 \qquad \theta_1 = 56.3°$$
$$r_2 = 1.41 \qquad \theta_2 = 315.0°$$

$$(3.61\underline{/56.3°})(1.41\underline{/315.0°}) = (3.61)(1.41)\underline{/56.3° + 315.0°}$$

$$= 5.09\underline{/371.3°}$$
$$= 5.09\underline{/11.3°}$$

Again note that the angle in the final result is between $0°$ and $360°$. ▶

If we wish to *divide* one complex number in exponential form by another, we arrive at the following result:

$$r_1 e^{j\theta_1} \div r_2 e^{j\theta_2} = \frac{r_1}{r_2} e^{j(\theta_1 - \theta_2)} \tag{12.14}$$

Therefore, *the result of dividing one complex number in polar form by another is given by*

$$\frac{r_1(\cos \theta_1 + j \sin \theta_1)}{r_2(\cos \theta_2 + j \sin \theta_2)} = \frac{r_1}{r_2}[\cos(\theta_1 - \theta_2) + j \sin(\theta_1 - \theta_2)]$$
$$\frac{r_1\underline{/\theta_1}}{r_2\underline{/\theta_2}} = \frac{r_1}{r_2}\underline{/\theta_1 - \theta_2} \tag{12.15}$$

NOTE ▶   The *magnitudes are divided,* and the *angles are subtracted.*

◀ EXAMPLE 3   Divide the first complex number of Example 1 by the second. Using the polar forms, we have

$$\frac{3.61(\cos 56.3° + j \sin 56.3°)}{1.41(\cos 315.0° + j \sin 315.0°)} = \frac{3.61}{1.41}[\cos(56.3° - 315.0°) + j \sin(56.3° - 315.0°)]$$

$$= 2.56[\cos(-258.7°) + j \sin(-258.7°)]$$
$$= 2.56(\cos 101.3° + j \sin 101.3°)$$ ▶

◀ EXAMPLE 4   Repeating Example 3 using r/θ polar form, we have

$$\frac{3.61\underline{/56.3°}}{1.41\underline{/315.0°}} = \frac{3.61}{1.41}\underline{/56.3° - 315.0°} = 2.56\underline{/-258.7°}$$
$$= 2.56\underline{/101.3°}$$ ▶

CAUTION ▶ We have just seen that multiplying and dividing numbers in polar form can be easily performed. However, if we are to add or subtract numbers in polar form, *we must do the addition or subtraction by using rectangular form.*

◀ EXAMPLE 5   Perform the addition $1.563\underline{/37.56°} + 3.827\underline{/146.23°}$.

In order to do this addition, we must change each number to rectangular form:

$$1.563\underline{/37.56°} + 3.827\underline{/146.23°}$$
$$= 1.563(\cos 37.56° + j \sin 37.56°) + 3.827(\cos 146.23° + j \sin 146.23°)$$
$$= 1.2390 + 0.9528j - 3.1813 + 2.1273j$$
$$= -1.9423 + 3.0801j$$

Now we change this to polar form:

$$r = \sqrt{(-1.9423)^2 + (3.0801)^2} = 3.641$$
$$\tan \theta = \frac{3.0801}{-1.9423} \qquad \theta = 122.24°$$

Therefore,

$$1.563\underline{/37.56°} + 3.827\underline{/146.23°} = 3.641\underline{/122.24°}$$

DEMOIVRE'S THEOREM

For reference, Eq. (11.3) is $(a^m)^n = a^{mn}$.

To raise a complex number to a power, we may use the exponential form of the number in Eq. (11.3). This leads to

$$(re^{j\theta})^n = r^n e^{jn\theta} \tag{12.16}$$

Extending this to polar form, we have

$$\boxed{\begin{array}{c} [r(\cos \theta + j \sin \theta)]^n = r^n(\cos n\theta + j \sin n\theta) \\ (r\underline{/\theta})^n = r^n\underline{/n\theta} \end{array}} \tag{12.17}$$

Named for the mathematician Abraham DeMoivre (1667–1754).

Equation (12.17) is known as **DeMoivre's theorem** *and is valid for all real values of n. It is also used for finding roots of complex numbers if n is a fractional exponent.* We note that *the magnitude is raised to the power,* and *the angle is multiplied by the power.*

◀ EXAMPLE 6   Using DeMoivre's theorem, find $(2 + 3j)^3$.

From Example 1, we know $r = 3.61$ and $\theta = 56.3°$. Therefore,

$$[3.61(\cos 56.3° + j \sin 56.3°)]^3 = (3.61)^3[\cos(3 \times 56.3°) + j \sin(3 \times 56.3°)]$$
$$= 47.0(\cos 168.9° + j \sin 168.9°) = 47.0\underline{/168.9°}$$

Expressing θ in radians, we have $\theta = 56.3° = 0.983$ rad. Therefore,

$$(3.61e^{0.983j})^3 = (3.61)^3 e^{3 \times 0.983j} = 47.0e^{2.95j}$$
$$(2 + 3j)^3 = 47.0(\cos 168.9° + j \sin 168.9°)$$
$$= 47.0\underline{/168.9°} = 47.0e^{2.95j} = -46 + 9j$$

◀ **EXAMPLE 7**   Find the cube root of -1.

Since we know that -1 is a real number, we can find its cube root by means of the definition. That is, $(-1)^3 = -1$. We check this by DeMoivre's theorem. Writing -1 in polar form, we have

$$-1 = 1(\cos 180° + j \sin 180°)$$

Applying DeMoivre's theorem, with $n = \frac{1}{3}$, we obtain

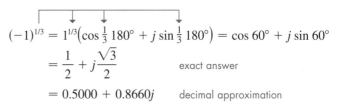

$$(-1)^{1/3} = 1^{1/3}\left(\cos \tfrac{1}{3} 180° + j \sin \tfrac{1}{3} 180°\right) = \cos 60° + j \sin 60°$$

$$= \frac{1}{2} + j\frac{\sqrt{3}}{2} \qquad \text{exact answer}$$

$$= 0.5000 + 0.8660j \qquad \text{decimal approximation}$$

Observe that we did not obtain -1 as the answer. If we check the answer, in the form $\frac{1}{2} + j\frac{\sqrt{3}}{2}$, by actually cubing it, we obtain -1! Therefore, it is a correct answer.

We should note that it is possible to take $\frac{1}{3}$ of any angle up to $1080°$ and still have an angle less than $360°$. Since $180°$ and $540°$ have the same terminal side, let us try writing -1 as $1(\cos 540° + j \sin 540°)$. Using DeMoivre's theorem, we have

$$(-1)^{1/3} = 1^{1/3}\left(\cos \tfrac{1}{3} 540° + j \sin \tfrac{1}{3} 540°\right) = \cos 180° + j \sin 180° = -1$$

We have found the answer we originally anticipated.

Angles of $180°$ and $900°$ also have the same terminal side, so we try

$$(-1)^{1/3} = 1^{1/3}\left(\cos \tfrac{1}{3} 900° + j \sin \tfrac{1}{3} 900°\right) = \cos 300° + j \sin 300°$$

$$= \frac{1}{2} - j\frac{\sqrt{3}}{2} \qquad \text{exact answer}$$

$$= 0.5000 - 0.8660j \qquad \text{decimal approximation}$$

Checking this, we find that it is also a correct root. We may try $1260°$, but $\frac{1}{3}(1260°) = 420°$, which has the same functional values as $60°$, and would give us the answer $0.5000 + 0.8660j$ again.

We have found, therefore, *three cube roots* of -1. They are

$$-1, \quad \frac{1}{2} + j\frac{\sqrt{3}}{2}, \quad \frac{1}{2} - j\frac{\sqrt{3}}{2}$$

These roots are graphed in Fig. 12.17. Note that they are equally spaced on the circumference of a circle of radius 1.   ▶

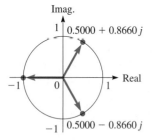

Fig. 12.17

When the results of Example 7 are generalized, it can be proved that *there are n nth roots of a complex number.* When graphed, these roots are on a circle of radius $r^{1/n}$ and are equally spaced $360°/n$ apart. Following is the method for finding these n roots of a complex number.

USING DEMOIVRE'S THEOREM TO FIND THE *n nth* ROOTS OF A COMPLEX NUMBER

1. *Express the number in polar form.*
2. *Express the root as a fractional exponent.*
3. *Use Eq. (12.17) with θ to find one root.*
4. *Use Eq. (12.17) and add 360° to θ, n − 1 times, to find the other roots.*

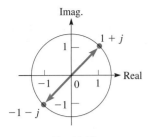

Fig. 12.18

◀ **EXAMPLE 8** Find the two square roots of $2j$.

First we write $2j$ in polar form as $2j = 2(\cos 90° + j \sin 90°)$. To find square roots we use the exponent $1/2$. The first square root is

$$(2j)^{1/2} = 2^{1/2}\left(\cos\frac{90°}{2} + j \sin\frac{90°}{2}\right) = \sqrt{2}(\cos 45° + j \sin 45°) = 1 + j$$

To find the other square root, we add $360°$ to $90°$. This gives us

$$(2j)^{1/2} = 2^{1/2}\left(\cos\frac{450°}{2} + j \sin\frac{450°}{2}\right) = \sqrt{2}(\cos 225° + j \sin 225°) = -1 - j$$

Therefore, the two square roots of $2j$ are $1 + j$ and $-1 - j$. We see in Fig. 12.18 that they are on a circle of radius $\sqrt{2}$ and $180°$ apart. ▶

◀ **EXAMPLE 9** Find all the roots of the equation $x^6 - 64 = 0$.

Solving for x, we have $x^6 = 64$, or $x = \sqrt[6]{64}$. Therefore, we have to find the six sixth roots of 64. Writing 64 in polar form, we have $64 = 64(\cos 0° + j \sin 0°)$. Using the exponent $1/6$ for the sixth root, we have the following solutions:

See Appendix C for a graphing calculator program DEMOIVRE. It can be used to find the roots of a complex number.

Calculators that operate with complex numbers generally find only the first of the n roots.

First root: $\quad 64^{1/6} = 64^{1/6}\left(\cos\dfrac{0°}{6} + j \sin\dfrac{0°}{6}\right) = 2(\cos 0° + j \sin 0°) = 2$

add 360°

Second root: $\quad 64^{1/6} = 64^{1/6}\left(\cos\dfrac{0° + 360°}{6} + j \sin\dfrac{0° + 360°}{6}\right)$
$$= 2(\cos 60° + j \sin 60°) = 1 + j\sqrt{3}$$

add 2 × 360°

Third root: $\quad 64^{1/6} = 64^{1/6}\left(\cos\dfrac{0° + 720°}{6} + j \sin\dfrac{0° + 720°}{6}\right)$
$$= 2(\cos 120° + j \sin 120°) = -1 + j\sqrt{3}$$

add 3 × 360°

Fourth root: $\quad 64^{1/6} = 64^{1/6}\left(\cos\dfrac{0° + 1080°}{6} + j \sin\dfrac{0° + 1080°}{6}\right)$
$$= 2(\cos 180° + j \sin 180°) = -2$$

add 4 × 360°

Fifth root: $\quad 64^{1/6} = 64^{1/6}\left(\cos\dfrac{0° + 1440°}{6} + j \sin\dfrac{0° + 1440°}{6}\right)$
$$= 2(\cos 240° + j \sin 240°) = -1 - j\sqrt{3}$$

add 5 × 360°

Sixth root: $\quad 64^{1/6} = 64^{1/6}\left(\cos\dfrac{0° + 1800°}{6} + j \sin\dfrac{0° + 1800°}{6}\right)$
$$= 2(\cos 300° + j \sin 300°) = 1 - j\sqrt{3}$$

These roots are graphed in Fig. 12.19. Note that they are equally spaced $60°$ apart on the circumference of a circle of radius 2. ▶

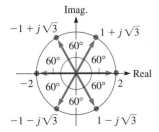

Fig. 12.19

From the text and examples of this and previous sections, we see the advantages for the various forms of writing complex numbers. Rectangular form can be used for all of the basic operations but lends itself best to addition and subtraction. Polar form is used for multiplication, division, raising to powers, and finding roots. Exponential form can be used for multiplication, division, and powers, and also for theoretical purposes (for example, deriving DeMoivre's theorem).

E X E R C I S E S 12.6

In Exercises 1–4, perform the indicated operations for the resulting complex numbers if the given changes are made in the indicated examples of this section.

1. In Example 1, change the sign of the complex part of the second complex number and then perform the multiplication.

2. In Example 2, multiply the same complex numbers as in Exercise 1.

3. In Example 6, change the exponent to 5 and then find the result.

4. In Example 8, replace $2j$ with $-2j$ and then find the roots.

In Exercises 5–20, perform the indicated operations. Leave the result in polar form.

5. $[4(\cos 60° + j \sin 60°)][2(\cos 20° + j \sin 20°)]$

6. $[3(\cos 120° + j \sin 120°)][5(\cos 45° + j \sin 45°)]$

7. $(0.5\underline{/140°})(6\underline{/110°})$

8. $(0.4\underline{/320°})(5.5\underline{/-150°})$

9. $\dfrac{8(\cos 100° + j \sin 100°)}{4(\cos 65° + j \sin 65°)}$

10. $\dfrac{9(\cos 230° + j \sin 230°)}{3(\cos 80° + j \sin 80°)}$

11. $\dfrac{12\underline{/320°}}{5\underline{/-210°}}$

12. $\dfrac{2\underline{/90°}}{4\underline{/75°}}$

13. $[2(\cos 35 + j \sin 35°)]^3$

14. $[3(\cos 120° + j \sin 120°)]^4$

15. $(2\underline{/135°})^8$

16. $(1\underline{/142°})^{10}$

17. $\dfrac{(50\underline{/236°})(2\underline{/84°})}{25\underline{/47°}}$

18. $\dfrac{36\underline{/274°}}{(2\underline{/141°})(6\underline{/195°})}$

19. $\dfrac{(4\underline{/24°})(10\underline{/326°})}{(1\underline{/186°})(8\underline{/77°})}$

20. $\dfrac{(25\underline{/194°})(6\underline{/239'}}{(3\underline{/17°})(10\underline{/29°})}$

In Exercises 21–24, perform the indicated operations. Express results in polar form. See Example 5.

21. $2.78\underline{/56.8°} + 1.37\underline{/207.3°}$

22. $15.9\underline{/142.6°} - 18.5\underline{/71.4°}$

23. $7085\underline{/115.62°} - 4667\underline{/296.34°}$

24. $307.5\underline{/326.54°} + 726.3\underline{/96.41°}$

In Exercises 25–36, change each number to polar form and then perform the indicated operations. Express the result in rectangular and polar forms. Check by performing the same operation in rectangular form.

25. $(3 + 4j)(5 - 12j)$

26. $(-2 + 5j)(-1 - j)$

27. $(7 - 3j)(8 + j)$

28. $(1 + 5j)(4 + 2j)$

29. $\dfrac{7}{1 - 3j}$

30. $\dfrac{8j}{7 + 2j}$

31. $\dfrac{3 + 4j}{5 - 12j}$

32. $\dfrac{-2 + 5j}{-1 - j}$

33. $(3 + 4j)^4$

34. $(-1 - j)^8$

35. $(2 + 3j)^5$

36. $(1 - 2j)^6$

In Exercises 37–42, use DeMoivre's theorem to find all the indicated roots. Be sure to find all roots.

37. The two square roots of $4(\cos 60° + j \sin 60°)$

38. The three cube roots of $27(\cos 120° + j \sin 120°)$

39. The three cube roots of $3 - 4j$

40. The two square roots of $-5 + 12j$

41. The square roots of $1 + j$

42. The cube roots of $\sqrt{3} + j$

In Exercises 43–48, find all of the roots of the given equations.

43. $x^4 - 1 = 0$

44. $x^3 - 8 = 0$

45. $x^3 + 27j = 0$

46. $x^4 - j = 0$

47. $x^5 + 32 = 0$

48. $x^6 + 8 = 0$

In Exercises 49–56, perform the indicated operations.

49. In Example 7, we showed that one cube root of -1 is $\frac{1}{2} - \frac{1}{2}j\sqrt{3}$. Cube this number in rectangular form and show that the result is -1.

(W) 50. Explain why the two square roots of a complex number are negatives of each other.

51. The cube roots of -1 can be found by solving the equation $x^3 + 1 = 0$. Find these roots by factoring $x^3 + 1$ as the sum of cubes and compare with Example 7.

52. The cube roots of 8 can be found by solving the equation $x^3 - 8 = 0$. Find these roots by factoring $x^3 - 8$ as the difference of cubes and compare with Exercise 44.

53. The electric power p (in W) supplied to an element in a circuit is the product of the voltage e and the current i (in A). Find the expression for the power supplied if $e = 6.80 / 56.3°$ volts and $i = 7.05 / -15.8°$ amperes.

54. The displacement d (in cm) of a weight suspended on a system of two springs is $d = 6.03 / 22.5° + 3.26 / 76.0°$ cm. Perform the addition and express the answer in polar form.

55. The voltage across a certain inductor is $V = (8.66 / 90.0°)(50.0 / 135.0°)/(10.0 / 60.0°)$ V. Simplify this expression and find the magnitude of the voltage.

56. In a microprocessor circuit, the current is $I = 3.75 / 15.0°$ μA and the impedance is $Z = 2500 / -35.0°$ Ω. Find the voltage E in rectangular form. (See Example 7 on page 345.)

12.7 AN APPLICATION TO ALTERNATING-CURRENT (AC) CIRCUITS

In the 1880s, it was decided that alternating current (favored by George Westinghouse) would be used to distribute electric power. Thomas Edison had argued for the use of direct current.

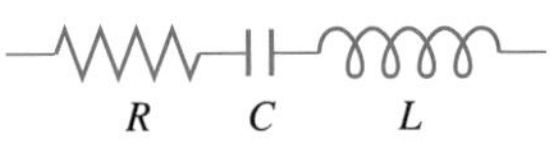

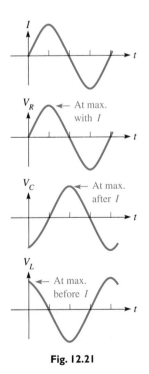

Fig. 12.20

Fig. 12.21

We now show an application of complex numbers in a basic type of alternating-current circuit. We show how the voltage is measured between any two points in a circuit containing a resistance, a capacitance, and an inductance. This circuit is similar to one noted in earlier examples and exercises in this chapter.

A *resistance* is any part of a circuit that tends to obstruct the flow of electric current through the circuit. It is denoted by R (units in ohms, Ω) and in diagrams by —⋀⋀⋀—, as shown in Fig. 12.20. A *capacitance* is two nonconnected plates in a circuit; no current actually flows across the gap between them. In an alternating-current circuit, an electric charge is continually going to and from each plate and, therefore, the current in the circuit is not effectively stopped. It is denoted by C (units in farads, F) and in diagrams by —||— (see Fig. 12.20). An *inductance* is basically a coil of wire in which current is induced because the current is continually changing in the circuit. It is denoted by L (units in henrys, H) and in diagrams by ⟋⟋⟋⟋ (see Fig. 12.20). All these elements affect the voltage in an alternating-current circuit. We state here the relation each has to the voltage and current in the circuit.

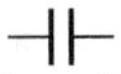

In Chapter 10, we noted that the current and voltage in an alternating-current circuit could be represented by a sine or a cosine curve. Therefore, each reaches peak values periodically. *If they reach their respective peak values at the same time, they are **in phase**. If the voltage reaches its peak before the current, the voltage **leads** the current. If the voltage reaches its peak after the current, the voltage **lags** the current.*

In the study of electricity, it is shown that the voltage across a resistance is in phase with the current. The voltage across a capacitor lags the current by 90°, and the voltage across an inductance leads the current by 90°. This is shown in Fig. 12.21, where, in a given circuit, I represents the current, V_R is the voltage across a resistor, V_C is the voltage across a capacitor, V_L is the voltage across an inductor, and t represents time.

Each element in an alternating-current circuit tends to offer a type of resistance to the flow of current. *The effective resistance of any part of the circuit is called the **reactance**,* and it is denoted by X. The voltage across any part of the circuit whose reactance is X is given by $V = IX$, where I is the current (in A) and V is the voltage (in V). Therefore,

the voltage V_R across a resistor with resistance R,

the voltage V_C across a capacitor with reactance X_C, and

the voltage V_L across an inductor with reactance X_L

are, respectively,

Eqs. (12.18) are based on Ohm's law, which states that the current is proportional to the voltage for a constant resistance. It is named for the German physicist Georg Ohm (1787–1854). The ohm (Ω) is named for him.

$$\boxed{V_R = IR \qquad V_C = IX_C \qquad V_L = IX_L} \tag{12.18}$$

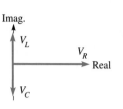

Fig. 12.22

To determine the voltage across a combination of these elements of a circuit, we must account for the reactance, as well as the phase of the voltage across the individual elements. Since the voltage across a resistor is in phase with the current, we represent V_R along the positive real axis as a real number. Since the voltage across an inductance leads the current by 90°, we represent this voltage as a positive, pure imaginary number. In the same way, by representing the voltage across a capacitor as a negative, pure imaginary number, we show that the voltage *lags* the current by 90°. These representations are meaningful since the positive imaginary axis is +90° from the positive real axis and the negative imaginary axis is −90° from the positive real axis. See Fig. 12.22.

The circuit elements shown in Fig. 12.20 are in *series,* and all circuits we consider (except Exercises 22 and 23) are series circuits. The total voltage across a series of all three elements is given by $V_R + V_L + V_C$, which we represent by V_{RLC}. Therefore,

$$V_{RLC} = IR + IX_L j - IX_C j = I[R + j(X_L - X_C)]$$

This expression is also written as

$$\boxed{V_{RLC} = IZ} \tag{12.19}$$

where the symbol Z is called the **impedance** *of the circuit. It is the total effective resistance to the flow of current by a combination of the elements in the circuit,* taking into account the phase of the voltage in each element. From its definition, we see that Z is a complex number.

$$\boxed{Z = R + j(X_L - X_C)} \tag{12.20}$$

with a magnitude

$$\boxed{|Z| = \sqrt{R^2 + (X_L - X_C)^2}} \tag{12.21}$$

Also, as a complex number, it makes an angle θ with the x-axis, given by

$$\boxed{\theta = \tan^{-1}\frac{X_L - X_C}{R}} \tag{12.22}$$

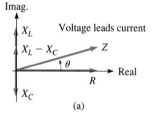

(a)

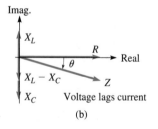

(b)

Fig. 12.23

All these equations are based on phase relations of voltages with respect to the current. Therefore, *the angle θ represents the phase angle between the current and the voltage.* The standard way of expressing θ is to *use a positive angle if the voltage leads the current* and *use a negative angle if the voltage lags the current.* Using Eq. (12.22), a calculator will give the correct angle even when $\tan \theta < 0$.

If the voltage leads the current, then $X_L > X_C$ as shown in Fig. 12.23(a). If the voltage lags the current, then $X_L < X_C$ as shown in Fig. 12.23(b).

In the examples and exercises of this section, the commonly used units and symbols for them are used. For a summary of these units and symbols, including prefixes, see Appendix B.

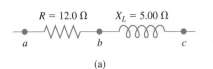

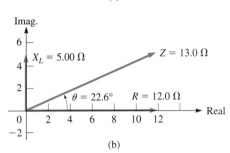

Fig. 12.24

◀ **EXAMPLE 1** In the series circuit shown in Fig. 12.24(a), $R = 12.0 \ \Omega$ and $X_L = 5.00 \ \Omega$. A current of 2.00 A is in the circuit. Find the voltage across each element, the impedance, the voltage across the combination, and the phase angle between the current and the voltage.

The voltage across the resistor (between points a and b) is the product of the current and the resistance ($V = IR$). This means $V_R = (2.00)(12.0) = 24.0$ V. The voltage across the inductor (between points b and c) is the product of the current and the reactance, or $V_L = (2.00)(5.00) = 10.0$ V.

To find the voltage across the combination, between points a and c, we must first find the magnitude of the impedance. Note that *the voltage is not the arithmetic sum of V_R and V_L,* as we must account for the phase.

By Eq. (12.20), the impedance is (there is no capacitor)

$$Z = 12.0 + 5.00j$$

with magnitude

$$|Z| = \sqrt{R^2 + X_L^2} = \sqrt{(12.0)^2 + (5.00)^2} = 13.0 \ \Omega$$

Thus, the magnitude of the voltage across the combination of the resistor and the inductance is

$$|V_{RL}| = (2.00)(13.0) = 26.0 \ \text{V}$$

The phase angle between the voltage and the current is found by Eq. (12.22). This gives

$$\theta = \tan^{-1} \frac{5.00}{12.0} = 22.6°$$

The voltage *leads* the current by 22.6°, and this is shown in Fig. 12.24(b). ▶

See Appendix C for a graphing calculator program IMPEDANC. It can be used to calculate the impedance in a circuit.

◀ **EXAMPLE 2** For a circuit in which $R = 8.00 \ \Omega$, $X_L = 7.00 \ \Omega$, and $X_C = 13.0 \ \Omega$, find the impedance and the phase angle between the current and the voltage.

By the definition of impedance, Eq. (12.20), we have

$$Z = 8.00 + (7.00 - 13.0)j = 8.00 - 6.00j$$

where the magnitude of the impedance is

$$|Z| = \sqrt{(8.00)^2 + (-6.00)^2} = 10.0 \ \Omega$$

The phase angle is found by

$$\theta = \tan^{-1} \frac{-6.00}{8.00} = -36.9°$$

The angle $\theta = -36.9°$ is given directly by the calculator, and it is the angle we want. As we noted after Eq. (12.22), we express θ as a negative angle if the voltage lags the current, as it does in this example. See Fig. 12.25.

From the values above, we write the impedance in polar form as $Z = 10.0 \underline{/-36.9°}$ ohms. ▶

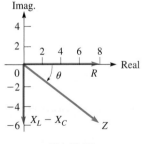

Fig. 12.25

The ampere (A) is named for the French physicist André Ampere (1775–1836).

The volt (V) is named for the Italian physicist Alessandro Volta (1745–1827).

The farad (F) is named for the British physicist Michael Faraday (1791–1867).

The henry (H) is named for the U.S. physicist Joseph Henry (1797–1878).

The coulomb (C) is named for the French physicist Charles Coulomb (1736–1806).

Note that the resistance is represented in the same way as a vector along the positive *x*-axis. Actually, resistance is not a vector quantity but is represented in this manner in order to assign an angle as the phase of the current. The important concept in this analysis is that *the phase **difference** between the current and voltage is constant,* and therefore any direction may be chosen arbitrarily for one of them. Once this choice is made, other phase angles are measured with respect to this direction. A common choice, as above, is to make the phase angle of the current zero. If an arbitrary angle is chosen, it is necessary to treat the current, voltage, and impedance as complex numbers.

◖ EXAMPLE 3   In a particular circuit, the current is $2.00 - 3.00j$ A and the impedance is $6.00 + 2.00j$ Ω. The voltage across this part of the circuit is

$$V = (2.00 - 3.00j)(6.00 + 2.00j) = 12.0 - 14.0j - 6.00j^2$$
$$= 12.0 - 14.0j + 6.00$$
$$= 18.0 - 14.0j \text{ V}$$

The magnitude of the voltage is

$$|V| = \sqrt{(18.0)^2 + (-14.0)^2} = 22.8 \text{ V}$$ ◗

Since the voltage across a resistor is in phase with the current, this voltage can be represented as having a phase difference of zero with respect to the current. Therefore, the resistance is indicated as an arrow in the positive real direction, denoting the fact that the current and the voltage are in phase. *Such a representation is called a **phasor.*** The arrow denoted by *R*, as in Fig. 12.24, is actually the phasor representing the voltage across the resistor. Remember, the positive real axis is arbitrarily chosen as the direction of the phase of the current.

To show properly that the voltage across an inductance leads the current by 90°, its reactance (effective resistance) is multiplied by *j*. We know that there is a positive 90° angle between a positive real number and a positive imaginary number. In the same way, by multiplying the capacitive reactance by $-j$, we show the 90° difference in phase between the voltage and the current in a capacitor, with the current leading. Therefore, jX_L represents the phasor for the voltage across an inductor and $-jX_C$ is the phasor for the voltage across the capacitor. The phasor for the voltage across the combination of the resistance, inductance, and capacitance is *Z*, where the phase difference between the voltage and the current for the combination is the angle θ.

From this, we see that *multiplying a phasor by j means to perform the operation of rotating it through* 90°. For this reason, *j* is also called the *j-operator.*

◖ EXAMPLE 4   Multiplying a positive real number *A* by *j*, we have $A \times j = Aj$, which is a positive imaginary number. In the complex plane, *Aj* is 90° from *A*, which means that by multiplying *A* by *j* we rotated *A* by 90°. Similarly, we see that $Aj \times j = Aj^2 = -A$, which is a negative real number, rotated 90° from *Aj*. Therefore, successive multiplications of *A* by *j* give us

$$A \times j = Aj \qquad \text{positive imaginary number}$$
$$Aj \times j = Aj^2 = -A \qquad \text{negative real number}$$
$$-A \times j = -Aj \qquad \text{negative imaginary number}$$
$$-Aj \times j = -Aj^2 = A \qquad \text{positive real number}$$

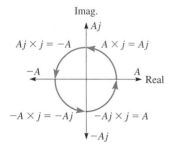

Fig. 12.26

See Fig. 12.26. ◗

An alternating current is produced by a coil of wire rotating through a magnetic field. If the angular velocity of the wire is ω, the capacitive and inductive reactances are given by

$$X_C = \frac{1}{\omega C} \quad \text{and} \quad X_L = \omega L \tag{12.23}$$

Therefore, if ω, C, and L are known, the reactance of the circuit can be found.

EXAMPLE 5 If $R = 12.0\ \Omega$, $L = 0.300$ H, $C = 250\ \mu$F, and $\omega = 80.0$ rad/s, find the impedance and the phase difference between the current and the voltage.

$$X_C = \frac{1}{(80.0)(250 \times 10^{-6})} = 50.0\ \Omega$$

$$X_L = (0.300)(80.0) = 24.0\ \Omega$$

$$Z = 12.0 + (24.0 - 50.0)j = 12.0 - 26.0j$$

$$|Z| = \sqrt{(12.0)^2 + (-26.0)^2} = 28.6\ \Omega$$

$$\theta = \tan^{-1}\frac{-26.0}{12.0} = -65.2°$$

$$Z = 28.6\underline{/-65.2°}\ \Omega$$

The voltage *lags* the current (see Fig. 12.27).

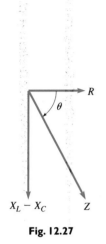

Fig. 12.27

Recall from Section 10.5 that the angular velocity ω is related to the frequency f by the relation $\omega = 2\pi f$. It is very common to use frequency when discussing alternating current.

An important concept in the application of this theory is that of **resonance**. *For resonance, the impedance of any circuit is a minimum, or the total impedance is R.* Thus, $X_L - X_C = 0$. Also, it can be seen that the current and the voltage are in phase under these conditions. Resonance is required for the tuning of radio and television receivers.

See the chapter introduction.

EXAMPLE 6 In the antenna circuit of a radio, the inductance is 4.20 mH, and the capacitance is variable. What range of values of capacitance is necessary for the radio to receive the AM band of radio stations, with frequencies from 530 kHz to 1600 kHz?

For proper tuning, the circuit should be in resonance, or $X_L = X_C$. This means that

$$2\pi f L = \frac{1}{2\pi f C} \quad \text{or} \quad C = \frac{1}{(2\pi f)^2 L}$$

For $f_1 = 530$ kHz $= 5.30 \times 10^5$ Hz and $L = 4.20$ mH $= 4.20 \times 10^{-3}$ H,

$$C_1 = \frac{1}{(2\pi)^2(5.30 \times 10^5)^2(4.20 \times 10^{-3})} = 2.15 \times 10^{-11}\ \text{F} = 21.5\ \text{pF}$$

and for $f_2 = 1600$ kHz $= 1.60 \times 10^6$ Hz and $L = 4.20 \times 10^{-3}$ H, we have

$$C_2 = \frac{1}{(2\pi)^2(1.60 \times 10^6)^2(4.20 \times 10^{-3})} = 2.36 \times 10^{-12}\ \text{F} = 2.36\ \text{pF}$$

The capacitance should be capable of varying from 2.36 pF to 21.6 pF.

From Appendix B, the following prefixes are defined as follows:

Prefix	Factor	Symbol
pico	10^{-12}	p
milli	10^{-3}	m
kilo	10^3	k

EXERCISES 12.7

In Exercises 1 and 2, perform the indicated operations if the given changes are made in the indicated examples of this section.

1. In Example 1, change the value of X_L to 16.0 Ω and then solve the given problem.

2. In Example 5, double the values of L and C and then solve the given problem.

In Exercises 3–6, use the circuit shown in Fig. 12.28. The current in the circuit is 5.75 mA. Determine the indicated quantities.

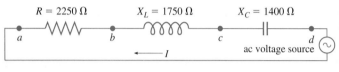

$R = 2250\ \Omega$ $\quad X_L = 1750\ \Omega$ $\quad X_C = 1400\ \Omega$

a $\qquad b$ $\qquad c$ $\qquad d$

$\longleftarrow I$ $\qquad$ ac voltage source

Fig. 12.28

3. The voltage across the resistor (between points a and b).

4. The voltage across the inductor (between points b and c).

5. (a) The magnitude of the impedance across the resistor and the inductor (between points a and c).
   (b) The phase angle between the current and the voltage for this combination.
   (c) The voltage across this combination.

6. (a) The magnitude of the impedance across the resistor, inductor, and capacitor (between points a and d).
   (b) The phase angle between the current and the voltage for this combination.
   (c) The voltage across this combination.

In Exercises 7–10, an alternating-current circuit contains the given combination of circuit elements from among a resistor ($R = 45.0\ \Omega$), a capacitor ($C = 86.2\ \mu\text{F}$), and an inductor ($L = 42.9$ mH). If the frequency in the circuit is $f = 60.0$ Hz, find (a) the magnitude of the impedance and (b) the phase angle between the current and the voltage.

7. The circuit has the inductor and the capacitor (an *LC* circuit).

8. The circuit has the resistor and the capacitor (an *RC* circuit).

9. The circuit has the resistor and the inductor (an *RL* circuit).

10. The circuit has the resistor, the inductor, and the capacitor (an *RLC* circuit).

In Exercises 11–24, solve the given problems.

11. Given that the current in a given circuit is $3.90 - 6.04j$ mA and the impedance is $5.16 + 1.14j$ kΩ, find the magnitude of the voltage.

12. Given that the voltage in a given circuit is $8.375 - 3.140j$ V and the impedance is $2.146 - 1.114j$ Ω, find the magnitude of the current.

13. A resistance ($R = 25.3\ \Omega$) and a capacitance ($C = 2.75$ nF) are in an AM radio circuit. If $f = 1200$ kHz, find the impedance across the resistor and the capacitor.

14. A resistance ($R = 64.5\ \Omega$) and an inductance ($L = 1.08$ mH) are in a telephone circuit. If $f = 8.53$ kHz, find the impedance across the resistor and inductor.

15. The reactance of an inductor is 1200 Ω for $f = 280$ Hz. What is the inductance?

16. A resistor, an inductor, and a capacitor are connected in series across an alternating-current voltage source. A voltmeter measures 12.0 V, 15.5 V, and 10.5 V, respectively, when placed across each element separately. What is the voltage of the source?

17. An inductance of 12.5 μH and a capacitance of 47.0 nF are in series in an amplifier circuit. Find the frequency for resonance.

18. A capacitance ($C = 95.2$ nF) and an inductance are in series in the circuit of a receiver for navigation signals. Find the inductance if the frequency for resonance is 50.0 kHz.

19. In Example 6, what should be the capacitance in order to receive a 680-kHz radio signal?

20. A 220-V source with $f = 60.0$ Hz is connected in series to an inductance ($L = 2.05$ H) and a resistance R in an electric motor circuit. Find R if the current is 0.250 A.

21. The power P (in W) supplied to a series combination of elements in an alternating-current circuit is $P = VI \cos \theta$, where V is the effective voltage, I is the effective current, and θ is the phase angle between the current and voltage. If $V = 225$ mV across the resistor, capacitor, and inductor combination in Exercise 10, determine the power supplied to these elements.

22. For two impedances Z_1 and Z_2 in parallel, the reciprocal of the combined impedance Z_C is the sum of the reciprocals of Z_1 and Z_2. Find the combined impedance for the parallel circuit elements in Fig. 12.29 if the current in the circuit has a frequency of 60.0 Hz.

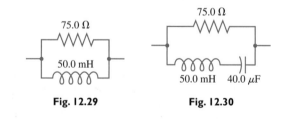

75.0 Ω $\qquad\qquad$ 75.0 Ω

50.0 mH $\qquad\qquad$ 50.0 mH $\quad$ 40.0 μF

Fig. 12.29 $\qquad\qquad$ **Fig. 12.30**

23. Find the combined impedance of the circuit elements in Fig. 12.30. The frequency of the current in the circuit is 60.0 Hz. See Exercise 22.

W 24. Explain why the multiplication of a complex number by -1 may be shown as a rotation of the graph of the number about the origin through 180°.

CHAPTER 12 EQUATIONS

Chapter Equations for Complex Numbers

Imaginary unit

$$j = \sqrt{-1} \quad \text{and} \quad j^2 = -1 \tag{12.1}$$

$$\sqrt{-a} = j\sqrt{a} \qquad (a > 0) \tag{12.2}$$

Basic operations

$$(a + bj) + (c + dj) = (a + c) + (b + d)j \tag{12.3}$$

$$(a + bj) - (c + dj) = (a - c) + (b - d)j \tag{12.4}$$

$$(a + bj)(c + dj) = (ac - bd) + (ad + bc)j \tag{12.5}$$

$$\frac{a + bj}{c + dj} = \frac{(a + bj)(c - dj)}{(c + dj)(c - dj)} = \frac{(ac + bd) + (bc - ad)j}{c^2 + d^2} \tag{12.6}$$

Complex number forms

Rectangular: $x + yj$
Polar: $r(\cos\theta + j\sin\theta) = r\underline{/\theta}$
Exponential: $re^{j\theta}$

$$x = r\cos\theta \qquad y = r\sin\theta \tag{12.7}$$

$$r^2 = x^2 + y^2 \qquad \tan\theta = \frac{y}{x} \tag{12.8}$$

$$x + yj = r(\cos\theta + j\sin\theta) = r\underline{/\theta} = re^{j\theta} \tag{12.12}$$

Product in polar form

$$r_1(\cos\theta_1 + j\sin\theta_1)r_2(\cos\theta_2 + j\sin\theta_2) = r_1r_2[\cos(\theta_1 + \theta_2) + j\sin(\theta_1 + \theta_2)] \tag{12.13}$$

$$(r_1\underline{/\theta_1})(r_2\underline{/\theta_2}) = r_1r_2\underline{/\theta_1 + \theta_2}$$

Quotient in polar form

$$\frac{r_1(\cos\theta_1 + j\sin\theta_1)}{r_2(\cos\theta_2 + j\sin\theta_2)} = \frac{r_1}{r_2}[\cos(\theta_1 - \theta_2) + j\sin(\theta_1 - \theta_2)] \tag{12.15}$$

$$\frac{r_1\underline{/\theta_1}}{r_2\underline{/\theta_2}} = \frac{r_1}{r_2}\underline{/\theta_1 - \theta_2}$$

DeMoivre's theorem

$$[r(\cos\theta + j\sin\theta)]^n = r^n(\cos n\theta + j\sin n\theta) \tag{12.17}$$

$$(r\underline{/\theta})^n = r^n\underline{/n\theta}$$

Chapter Equations for Alternating-Current Circuits

Voltage, current, reactance

$$V_R = IR \qquad V_C = IX_C \qquad V_L = IX_L \tag{12.18}$$

Impedance

$$V_{RLC} = IZ \tag{12.19}$$

$$Z = R + j(X_L - X_C) \tag{12.20}$$

$$|Z| = \sqrt{R^2 + (X_L - X_C)^2} \tag{12.21}$$

Phase angle

$$\theta = \tan^{-1}\frac{X_L - X_C}{R} \tag{12.22}$$

Capacitive reactance and inductive reactance

$$X_C = \frac{1}{\omega C} \quad \text{and} \quad X_L = \omega L \tag{12.23}$$

CHAPTER 12 REVIEW EXERCISES

In Exercises 1–16, perform the indicated operations, expressing all answers in simplest rectangular form.

1. $(6 - 2j) + (4 + j)$ **2.** $(12 + 7j) + (-8 + 6j)$

3. $(18 - 3j) - (12 - 5j)$ **4.** $(-4 - 2j) - \sqrt{-49}$

5. $(2 + j)(4 - j)$ **6.** $(-5 + 3j)(8 - 4j)$

7. $(2j^5)(6 - 3j)(4 + 3j)$ **8.** $j(3 - 2j) - (j^3)(5 + j)$

9. $\dfrac{3}{7 - 6j}$ **10.** $\dfrac{4j}{2 + 9j}$

11. $\dfrac{6 - \sqrt{-16}}{\sqrt{-4}}$ **12.** $\dfrac{3 + \sqrt{-4}}{4 - j}$

13. $\dfrac{5j - (3 - j)}{4 - 2j}$ **14.** $\dfrac{2 + (j - 6)}{1 - 2j}$

15. $\dfrac{j(7 - 3j)}{2 + j}$ **16.** $\dfrac{(2 - j)(3 + 2j)}{4 - 3j}$

In Exercises 17–20, find the values of x and y for which the equations are valid.

17. $3x - 2j = yj - 2$

18. $2xj - 2y = (y + 3)j - 3$

19. $(3 - 2j)(x + jy) = 4 + j$

20. $(x + jy)(7j - 4) = 3j - 16$

In Exercises 21–24, perform the indicated operations graphically. Check them algebraically.

21. $(-1 + 5j) + (4 + 6j)$ **22.** $(7 - 2j) + (-5 + 4j)$

23. $(9 + 2j) - (5 - 6j)$ **24.** $(1 + 4j) - (-3 - 3j)$

In Exercises 25–32, give the polar and exponential forms of each of the complex numbers.

25. $1 - j$ **26.** $4 + 3j$

27. $-2 - 7j$ **28.** $6 - 2j$

29. $1.07 + 4.55j$ **30.** $-327 + 158j$

31. 5000 **32.** $-4j$

In Exercises 33–44, give the rectangular form of each number.

33. $2(\cos 225° + j \sin 225°)$ **34.** $4(\cos 60° + j \sin 60°)$

35. $5.011(\cos 123.82° + j \sin 123.82°)$

36. $2.417(\cos 656.26° + j \sin 656.26°)$

37. $0.62\underline{/-72°}$ **38.** $20\underline{/160°}$

39. $27.08\underline{/346.27°}$ **40.** $1.689\underline{/194.36°}$

41. $2.00e^{0.25j}$ **42.** $e^{3.62j}$

43. $(35.37e^{1.096j})^2$ **44.** $(13.6e^{2.158j})(3.27e^{3.888j})$

In Exercises 45–60, perform the indicated operations. Leave the result in polar form.

45. $[3(\cos 32° + j \sin 32°)][5(\cos 52° + j \sin 52°)]$

46. $[2.5(\cos 162° + j \sin 162°)][8(\cos 115° + j \sin 115°)]$

47. $(40\underline{/18°})(0.5\underline{/245°})$

48. $(0.1254\underline{/172.38°})(27.17\underline{/204.34°})$

49. $\dfrac{24(\cos 165° + j \sin 165°)}{3(\cos 106° + j \sin 106°)}$

50. $\dfrac{18(\cos 403° + j \sin 403°)}{4(\cos 192° + j \sin 192°)}$

51. $\dfrac{245.6\underline{/326.44°}}{17.19\underline{/192.83°}}$ **52.** $\dfrac{100\underline{/206°}}{4\underline{/-320°}}$

53. $0.983\underline{/47.2°} + 0.366\underline{/95.1°}$

54. $17.8\underline{/110.4°} - 14.9\underline{/226.3°}$

55. $7644\underline{/294.36°} - 6871\underline{/17.86°}$

56. $4.944\underline{/327.49°} + 8.009\underline{/7.37°}$

57. $[2(\cos 16° + j \sin 16°)]^{10}$ **58.** $[3(\cos 36° + j \sin 36°)]^6$

59. $(3\underline{/110.5°})^3$ **60.** $(536\underline{/220.3°})^4$

In Exercises 61–64, change each number to polar form and then perform the indicated operations. Express the final result in rectangular and polar forms. Check by performing the same operation in rectangular form.

61. $(1 - j)^{10}$ **62.** $(\sqrt{3} + j)^8(1 + j)^5$

63. $\dfrac{(5 + 5j)^4}{(-1 - j)^6}$ **64.** $(\sqrt{3} - j)^{-8}$

In Exercises 65–68, find all the roots of the given equations.

65. $x^3 + 8 = 0$ **66.** $x^3 - 1 = 0$ **67.** $x^4 + j = 0$ **68.** $x^5 - 32j = 0$

In Exercises 69–72, determine the rectangular form and the polar form of the complex number for which the graphical representation is shown in the given figure.

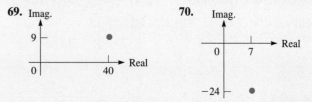

71.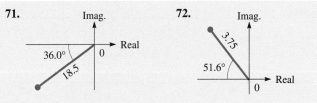

72.

In Exercises 73–80, solve the given problems.

73. Evaluate $x^2 - 2x + 4$ for $x = 5 - 2j$.

74. Evaluate $2x^2 + 5x - 7$ for $x = -8 + 7j$.

75. Using the quadratic formula, solve for x: $x^2 + 3jx - 2 = 0$.

76. Using the quadratic formula, solve for x: $jx^2 - 2x - 3j = 0$.

77. Find a quadratic equation with roots $2 + j$ and $2 - j$.

78. Find a quadratic equation with roots $-3 + 4j$ and $-3 - 4j$.

79. Show that $0.6 + 0.8j$ is the reciprocal of its conjugate.

80. Show that $\frac{1}{2}(1 + j\sqrt{3})$ is the reciprocal of its conjugate.

In Exercises 81–92, find the required quantities.

81. A 60-V alternating-current voltage source is connected in series across a resistor, an inductor, and a capacitor. The voltage across the inductor is 60 V, and the voltage across the capacitor is 60 V. What is the voltage across the resistor?

82. In a series alternating-current circuit with a resistor, an inductor, and a capacitor, $R = 6.50\ \Omega$, $X_C = 3.74\ \Omega$, and $Z = 7.50\ \Omega$. Find X_L.

83. In a series alternating-current circuit with a resistor, an inductor, and a capacitor, $R = 6250\ \Omega$, $Z = 6720\ \Omega$, and $X_L = 1320\ \Omega$. Find the phase angle θ.

84. A coil of wire rotates at 120.0 r/s. If the coil generates a current in a circuit containing a resistance of 12.07 Ω, an inductance of 0.1405 H, and an impedance of 22.35 Ω, what must be the value of a capacitor (in F) in the circuit?

85. What is the frequency f for resonance in a circuit for which $L = 2.65$ H and $C = 18.3\ \mu$F?

86. The displacement of an electromagnetic wave is given by $d = A(\cos \omega t + j \sin \omega t) + B(\cos \omega t - j \sin \omega t)$. Find the expressions for the magnitude and phase angle of d.

87. Two cables lift a crate. The tensions in the cables can be represented by $210 - 120j$ N and $120 + 560j$ N. Express the resultant tension in polar form.

88. A boat is headed across a river with a velocity (relative to the water) that can be represented as $6.5 + 1.7j$ km/h. The velocity of the river current can be represented as $-1.1 - 4.3j$ km/h. Express the resultant velocity of the boat in polar form.

89. In the study of shearing effects in the spinal column, the expression $\dfrac{1}{\mu + j\omega n}$ is found. Express this in rectangular form.

90. In the theory of light reflection on metals, the expression

$$\frac{\mu(1 - kj) - 1}{\mu(1 - kj) + 1}$$

in encountered. Simplify this expression.

91. Show that $e^{j\pi} = -1$.

92. Show that $(e^{j\pi})^{1/2} = j$.

Writing Exercise

93. A computer programmer is writing a program to determine the n nth roots of a real number. Part of the program is to show the number of real roots and the number of pure imaginary roots. Write one or two paragraphs explaining how these numbers of roots can be determined without actually finding the roots.

CHAPTER (12) PRACTICE TEST

1. Add, expressing the resultant in rectangular form:
$(3 - \sqrt{-4}) + (5\sqrt{-9} - 1)$.

2. Multiply, expressing the resultant in polar form: $(2\underline{/130°})(3\underline{/45°})$.

3. Express $2 - 7j$ in polar form.

4. Express in terms of j: (a) $-\sqrt{-64}$, (b) $-j^{15}$.

5. Add graphically: $(4 - 3j) + (-1 + 4j)$.

6. Simplify, expressing the result in rectangular form: $\dfrac{2 - 4j}{5 + 3j}$.

7. Express $2.56(\cos 125.2° + j \sin 125.2°)$ in exponential form.

8. For an alternating-current circuit in which $R = 3.50\ \Omega$, $X_L = 6.20\ \Omega$, and $X_C = 7.35\ \Omega$, find the impedance and the phase angle between the current and the voltage.

9. Express $3.47 - 2.81j$ in exponential form.

10. Find the values of x and y: $x + 2j - y = yj - 3xj$.

11. What is the capacitance of the circuit in a radio that has an inductance of 8.75 mH if it is to receive a station with frequency 600 kHz?

12. Find the cube roots of j.

CHAPTER 13

Exponential and Logarithmic Functions

The growth of population can often be measured using an exponential function. This is illustrated in Section 13.6.

By the early 1600s, astronomy had progressed to the point of finding accurate information about the motion of the heavenly bodies. Also, navigation had led to a more systematic exploration of earth. In making the accurate measurements needed in astronomy and navigation, many lengthy calculations had to be performed, and all such calculations had to be done by hand.

Noting that astronomers' calculations usually involved sines of angles, John Napier (1550–1617), a Scottish mathematician, constructed a table of values that allowed multiplication of these sines by addition of values from the table. These were tables of logarithms, and they first appeared in 1614. Therefore, logarithms were essentially invented in order to make lengthy multiplications by means of addition, thereby making calculations easier.

Napier's logarithms were not in base 10, and the English mathematician Henry Briggs (1561–1631) realized that logarithms in base 10 would make the calculations even easier. He spent many years laboriously developing a table of base 10 logarithms, which was not completed until after his death.

Logarithms were enthusiastically received by mathematicians and scientists as a long-needed tool for lengthy calculations. The great French mathematician Pierre Laplace (1749–1827) stated that logarithms "by shortening the labors doubled the life of the astronomer." Logarithms were commonly used for calculations until the 1970s, when the scientific calculator came into use.

In this chapter, we study the *logarithmic function* and the *exponential function*. Although logarithms are no longer used directly for calculations, they are of great importance in many scientific and technical applications and in advanced mathematics. The basic units used to measure the intensity of sound and those used to measure the intensity of earthquakes are measured in terms of logarithms. In chemistry, the distinction between a base and an acid is defined in terms of logarithms. In electrical transmission lines, power gains and losses are measured in terms of logarithmic units. Exponential functions are used extensively in electronics, mechanical systems, thermodynamics, and nuclear physics. They are also used in biology in studying population growth and in business to calculate compound interest.

13.1 EXPONENTIAL FUNCTIONS

In Chapter 11, we showed that any rational number can be used as an exponent. Now letting *the exponent be a variable,* we define the **exponential function** *as*

$$y = b^x$$ (13.1)

where $b > 0$, $b \neq 1$, and x is any real number. *The number b is called the* **base.**

In using the exponential function, we will use only real numbers. Therefore, the reason that $b > 0$ is that if b were negative any fractional value for x with an even integer for a denominator would lead to an imaginary number. Also, $b \neq 1$ because 1 raised to any real power is 1, which would make y a constant.

◀ **EXAMPLE 1** From the definition, we see that $y = 3^x$ is an exponential function, but $y = (-3)^x$ is not because the base -3 is negative.

However, $y = -3^x$ is an exponential function since it is (-1) times 3^x. Any real number multiple of an exponential function is also an exponential function.

Also, $y = \left(\sqrt{3}\right)^x$ is an exponential function since it can be written as $y = 3^{x/2}$. As long as x is a real number, so is $x/2$. Therefore, the exponent of 3 is real.

The function $y = 3^{-x}$ is an exponential function. If x is real, so is $-x$.

Other illustrations of exponential functions are

$$y = 12^{2.6x} \qquad y = -2(8^{-0.55x}) \qquad y = 35(1.0001)^x$$ ◗

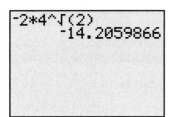

Fig. 13.1

◀ **EXAMPLE 2** Evaluate the function $y = -2(4^x)$ for the given values of x.

(a) If $x = 2$, $y = -2(4^2) = -2(16) = -32$.

(b) If $x = -2$, $y = -2(4^{-2}) = -2/16 = -1/8$.

(c) If $x = 3/2$, $y = -2(4^{3/2}) = -2(8) = -16$.

(d) If $x = \sqrt{2}$, $y = -2(4^{\sqrt{2}}) = -14.206$ (calculator evaluation—see Fig. 13.1) ◗

GRAPHING EXPONENTIAL FUNCTIONS

The graph of a function is useful to show the properties of the function. We now show some representative graphs of the exponential function.

◀ **EXAMPLE 3** Plot the graph of $y = 2^x$.

Assuming values for x and then finding the corresponding values for y, we obtain the following table:

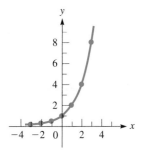

Fig. 13.2

x	-3	-2	-1	0	1	2	3
y	$\frac{1}{8}$	$\frac{1}{4}$	$\frac{1}{2}$	1	2	4	8

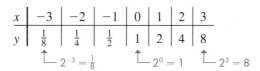

From these values, we plot the curve as shown in Fig. 13.2.

From the graph, we note that the x-axis is an *asymptote* of the curve. As we first noted on page 94, an asymptote is a line that the curve gets closer and closer to as values of x increase (or decrease) without bound, although the curve never actually touches the asymptote. ◗

The bases b of the exponential function of greatest importance in applications are greater than 1. However, in order to understand how the exponential function differs somewhat if $b < 1$, we now use a graphing calculator to display such a graph.

EXAMPLE 4 Display the graph of $y = 3\left(\frac{1}{2}\right)^x$ on a graphing calculator.
Looking at the function, we see that

$$\left(\frac{1}{2}\right)^x = \frac{1}{2^x} = 2^{-x}$$

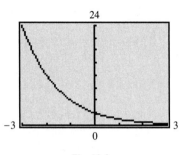

Fig. 13.3

Since x may be any real number, we note that *as x becomes more negative, y will increase rapidly.* Therefore, on a calculator we let $y_1 = 3(1/2)^x$ (or $y_1 = 3(0.5^x)$) and have the display in Fig. 13.3 with the *window* settings shown.

Any exponential curve where $b > 1$ will be similar in shape to that shown in Fig. 13.2, and if $b < 1$ it will be similar to the curve in Fig. 13.3. From these examples, we can see that exponential functions have the following basic features.

BASIC FEATURES OF EXPONENTIAL FUNCTIONS

1. *The domain is all values of x; the range is y > 0.*

2. *The x-axis is an asymptote of the graph of $y = b^x$.*

3. *As x increases, b^x increases if $b > 1$, and b^x decreases if $b < 1$.*

As we have noted, exponential functions are important in many applications. We now illustrate one such application in the next example. Other applications are shown in the exercises.

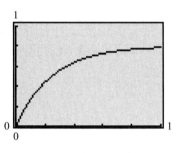

Fig. 13.4

EXAMPLE 5 In an electric circuit in which there is a battery, an inductor, and a resistor, the current i (in A) as a function of the time t (in s) is $i = 0.8(1 - e^{-4t})$. Display the graph of this function on a graphing calculator. Here, e is the same number that we introduced on page 352 and is equal to approximately 2.718.
Here we note that $e^{-4t} = 1$ for $t = 0$, and this means that $i = 0$ for $t = 0$. Also, e^{-4t} becomes very small in a very short time, and this means that i cannot be greater than 0.8 A. With these considerations, we have the settings and curve as shown in Fig. 13.4. Here we used y for i, x for t, and $y_1 = 0.8(1 - e^{-4x})$.

EXERCISES 13.1

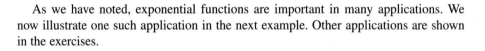

In Exercises 1 and 2, perform the indicated operations if the given changes are made in the indicated examples of this section.

1. In Example 2(c), change the sign of x and then evaluate.

2. In Example 3, change the sign of the exponent and then plot the graph.

In Exercises 3–6, determine if the given functions are exponential functions.

3. (a) $y = 5^x$ (b) $y = 5^{-x}$ **4.** (a) $y = -7^x$ (b) $y = (-7)^{-x}$

5. (a) $y = -7(-5)^{-x}$ (b) $y = -7(5^{-x})$

6. (a) $y = \left(\sqrt{5}\right)^{-x}$ (b) $y = -\left(\sqrt{-5}\right)^x$

In Exercises 7–12, evaluate the exponential function $y = 9^x$ for the given values of x.

7. $x = 0.5$ **8.** $x = 4$ **9.** $x = -2$

10. $x = -0.5$ **11.** $x = -3/2$ **12.** $x = 5/2$

In Exercises 13–18, plot the graphs of the given functions.

13. $y = 4^x$ **14.** $y = 0.25^x$ **15.** $y = 0.2(10^{-x})$

16. $y = 5(1.6^{-x})$ **17.** $y = 0.5\pi^x$ **18.** $y = 2e^x$

In Exercises 19–24, display the graphs of the given functions on a graphing calculator.

19. $y = 0.3(2.55)^x$

20. $y = 1.5(4.15)^x$

21. $y = 0.1(0.25)^x$

22. $y = 0.4(0.95)^x$

23. $i = 1.2(2 + 6^{-t})$

24. $y = 0.5e^{-x}$

In Exercises 25–32, solve the given problems.

25. Use a graphing calculator to find the value(s) for which $x^2 = 2^x$.

26. Use a graphing calculator to find the integral values of x for which $x^3 > 3^x$.

27. The value V of a bank account in which \$250 is invested at 5.00% interest, compounded annually, is $V = 250(1.0500)^t$, where t is the time in years. Find the value of the account after 4 years.

28. The intensity I of an earthquake is given by $I = I_0(10)^R$, where I_0 is a minimum intensity for comparison and R is the Richter scale magnitude of the earthquake. Evaluate I in terms if I_0 if $R = 5.5$.

29. The electric current i (in mA) in the circuit shown in Fig. 13.5 is $i = 2.5(1 - e^{-0.10t})$, where t is the time (in s). Evaluate i for $t = 5.0$ ms.

30. A projection of the annual growth rate p (in %) of the number of users of the Internet is $p = 8.5(1.2^{-t} + 1)$, where t is the number of years after 2005. Display the graph of this function on a graphing calculator from 2005 to 2015.

31. The flash unit on a camera operates by releasing the stored charge on a capacitor. For a particular unit, the charge q (in μC) as a function of the time t (in s) is $q = 100e^{-10t}$. Display the graph on a graphing calculator.

32. The height y (in m) of the Gateway Arch in St. Louis (see Fig. 13.6) is given by $y = 230.9 - 19.5(e^{x/38.9} + e^{-x/38.9})$, where x is the distance (in m) from the point on the ground level directly below the top. Display the graph on a graphing calculator.

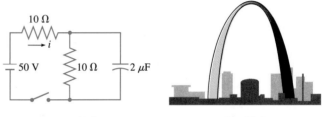

Fig. 13.5   **Fig. 13.6**

13.2 LOGARITHMIC FUNCTIONS

For many uses in mathematics and for many applications, it is necessary to express the exponent x in the exponential function $y = b^x$ in terms of y and the base b. This is done by defining a *logarithm*. Therefore, *if $y = b^x$, the exponent x is the* **logarithm** *of the number y to the base b.* We write this as

$$\text{If } y = b^x, \text{ then } x = \log_b y. \tag{13.2}$$

CAUTION ▶ This means that x is the power to which the base b must be raised in order to equal the number y. That is, x is a logarithm, and *a logarithm is an exponent.* As with the exponential function, for the equation $x = \log_b y$, x may be any real number, b is a positive number other than 1, and y is a positive real number. In Eq. (13.2),

$y = b^x$ *is the* **exponential form,** *and* $x = \log_b y$ *is the* **logarithmic form.**

See Fig. 13.7.

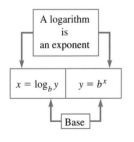

Fig. 13.7

◀ **EXAMPLE 1**   The equation $y = 2^x$ is written as $x = \log_2 y$ when written in logarithmic form. When we choose values of y to find corresponding values of x from this equation, we ask ourselves "2 raised to what power x gives y?"

This means that if $y = 8$, we ask "what power of 2 gives us 8?" Then knowing that $2^3 = 8$, we know that $x = 3$. Therefore, $3 = \log_2 8$. ▶

◀ **EXAMPLE 2**   **(a)** $3^2 = 9$ in logarithmic form is $2 = \log_3 9$.

(b) $4^{-1} = 1/4$ in logarithmic form is $-1 = \log_4(1/4)$.

CAUTION ▶ Remember, *the exponent may be negative.* The *base* must be positive. ▶

◀ **EXAMPLE 3** **(a)** $(64)^{1/3} = 4$ in logarithmic form is $\frac{1}{3} = \log_{64} 4$.

(b) $(32)^{3/5} = 8$ in logarithmic form is $\frac{3}{5} = \log_{32} 8$.

(c) $\log_2 32 = 5$ in exponential form is $32 = 2^5$.

(d) $\log_6 \left(\frac{1}{36}\right) = -2$ in exponential form is $\frac{1}{36} = 6^{-2}$. ▶

◀ **EXAMPLE 4** **(a)** Find b, given that $-4 = \log_b \left(\frac{1}{81}\right)$.

Writing this in exponential form, we have $\frac{1}{81} = b^{-4}$. Thus $\frac{1}{81} = \frac{1}{b^4}$ or $\frac{1}{3^4} = \frac{1}{b^4}$. Therefore, $b = 3$.

(b) Find y, given that $\log_4 y = \frac{1}{2}$.

In exponential form we have $y = 4^{1/2}$, or $y = 2$. ▶

We see that exponential form is very useful for determining values written in logarithmic form. For this reason, it is important that you learn to transform readily from one form to the other.

CAUTION ▶ In order to change a function of the form $y = ab^x$ into logarithmic form, we must first write it as $y/a = b^x$. **The coefficient of b^x must be equal to 1,** which is the form of Eq. (13.1). In the same way, the coefficient of $\log_b y$ must be 1 in order to change it into exponential form.

◀ **EXAMPLE 5** The power supply P (in W) of a certain satellite is given by $P = 75e^{-0.005t}$, where t is the time (in days) after launch. By writing this equation in logarithmic form, solve for t.

In order to have the equation in the exponential form of Eq. (13.1), we must have only $e^{-0.005t}$ on the right. Therefore, by dividing by 75, we have

$$\frac{P}{75} = e^{-0.005t}$$

Writing this in logarithmic form, we have

$$\log_e \left(\frac{P}{75}\right) = -0.005t$$

or

$$t = \frac{\log_e \left(\dfrac{P}{75}\right)}{-0.005} = -200 \log_e \left(\frac{P}{75}\right) \qquad \frac{1}{-0.005} = -200$$

We recall from Section 12.5 that e is the special irrational number equal to about 2.718. It is an important number as a base of logarithms. This is discussed in more detail in Section 13.5. ▶

The Logarithmic Function   When we are working with functions, we must keep in mind that a function is defined by the operation being performed on the independent variable, and not by the letter chosen to represent it. However, for consistency, it is standard practice to let y represent the dependent variable and x represent the independent variable. Therefore, *the* **logarithmic function** *is*

$$\boxed{y = \log_b x} \tag{13.3}$$

As with the exponential function, $b > 0$ and $b \neq 1$.

NOTE ▶

Equations (13.2) and (13.3) do not represent different *functions,* due to the difference in location of the variables, since they represent the *same operation* on the independent variable that appears in each. However, Eq. (13.3) expresses the function with the standard dependent and independent variables.

◀ **EXAMPLE 6**   For the logarithmic function $y = \log_2 x$, we have the standard independent variable x and the standard dependent variable y.

If $x = 16$, $y = \log_2 16$, which means that $y = 4$, since $2^4 = 16$.

If $x = \frac{1}{16}$, $y = \log_2 \left(\frac{1}{16}\right)$, which means that $y = -4$, since $2^{-4} = \frac{1}{16}$.   ▶

GRAPHING LOGARITHMIC FUNCTIONS

We now show some representative graphs for the logarithmic function. From these graphs we can see the basic properties of the logarithmic function.

◀ **EXAMPLE 7**   Plot the graph of $y = \log_2 x$.

We can find the points for this graph more easily if we first put the equation in exponential form: $x = 2^y$. By assuming values for y, we can find the corresponding values for x.

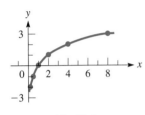

Fig. 13.8

$$\overbrace{}^{2^{-2} = \frac{1}{4}} \qquad \overbrace{}^{2^2 = 4}$$

x	$\frac{1}{4}$	$\frac{1}{2}$	1	2	4	8
y	-2	-1	0	1	2	3

Using these values, we construct the graph seen in Fig. 13.8.   ▶

The log keys on the calculator are [log] (for $\log_{10} x$) and [ln] (for $\log_e x$). Therefore, for now we restrict graphing logarithmic functions to bases 10 and e on the graphing calculator. In Section 13.5, we will see how to use the calculator to graph a logarithmic function with any positive real number base.

◀ **EXAMPLE 8**   Display the graph of $y = 2 \log_{10} x$ on a graphing calculator.

On a graphing calculator, we set $y_1 = 2 \log x$, and the curve is displayed as shown in Fig. 13.9. The settings for Xmin and Xmax were selected since the domain is $x > 0$ and $\log_{10} x$ increases very slowly. The settings for Ymin and Ymax were selected since $\log_{10} 0.1 = -1$, and y does not reach 4 until $x = 100$.   ▶

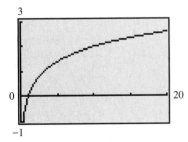

Fig. 13.9

From these graphs, we can see that logarithmic functions have the following features. We consider only the features for which $b > 1$, for these are the bases of importance.

BASIC FEATURES OF LOGARITHMIC FUNCTIONS ($b > 1$)

1. *The domain is $x > 0$; the range is all values of y.*
2. *The negative y-axis is an asymptote of the graph of $y = \log_b x$.*
3. *If $0 < x < 1$, $\log_b x < 0$; if $x = 1$, $\log_b x = 0$; if $x > 1$, $\log_b x > 0$.*
4. *If $x > 1$, x increases more rapidly than $\log_b x$.*

Table I

x	1	4	16	64
$\log_2 x$	0	2	4	6
2^x	2	16	65 536	1.8×10^{19}

We just noted that if $b > 1$ and $x > 1$, x increases more rapidly than $\log_b x$. It is also true that b^x increases more rapidly than x. Actually, as x becomes larger, $\log_b x$ increases very slowly, and b^x increases very rapidly. Using $\log_2 x$ and 2^x and a calculator, we have the table of values shown to the left. This shows that we must choose values of x carefully when graphing these functions.

INVERSE FUNCTIONS

For the exponential function $y = b^x$ and the logarithmic function $y = \log_b x$, if we solve for the independent variable in one of the functions by changing the form, then interchange the variables, we obtain the other function. *Such functions are called* **inverse functions.**

This means that the x- and y-coordinates of inverse functions are interchanged. As a result, the graphs of inverse functions are mirror images of each other across the line $y = x$. This is illustrated in the following example.

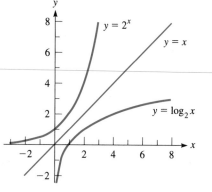

Fig. 13.10

◖ **EXAMPLE 9** The functions $y = 2^x$ and $y = \log_2 x$ are inverse functions. We show this by solving $y = 2^x$ for x and then interchanging x and y.

Writing $y = 2^x$ in logarithmic form gives us $x = \log_2 y$. Then interchanging x and y, we have $y = \log_2 x$, which is the inverse function.

Making a table of values for each function we have

$y = 2^x$:

x	−3	−2	−1	0	1	2	3
y	$\frac{1}{8}$	$\frac{1}{4}$	$\frac{1}{2}$	1	2	4	8

$y = \log_2 x$:

x	$\frac{1}{8}$	$\frac{1}{4}$	$\frac{1}{2}$	1	2	4	8
y	−3	−2	−1	0	1	2	3

We see that the coordinates are interchanged. In Fig. 13.10, note that the graphs of these two functions reflect each other across the line $y = x$. ◗

For a function, there is exactly one value of y in the range for each value of x in the domain. This must also hold for the inverse function. Thus, for a function to have an inverse function, there must be only one x for each y. This is true for $y = b^x$ and $y = \log_b x$, as we have seen earlier in this section.

EXERCISES **13.2**

In Exercises 1–4, perform the indicated operations if the given changes are made in the indicated examples of this section.

1. In Example 3(b), change the exponent to 4/5 and then make any other necessary changes.

2. In Example 4(b), change the 1/2 to 5/2 and then make any other necessary changes.

3. In Example 6, change the logarithm base to 4 and then make any other necessary changes.

4. In Example 7, change the logarithm base to 4 and then plot the graph.

In Exercises 5–16, express the given equations in logarithmic form.

5. $3^3 = 27$

6. $5^2 = 25$

7. $4^4 = 256$

8. $8^2 = 64$

9. $4^{-2} = \frac{1}{16}$

10. $3^{-2} = \frac{1}{9}$

11. $2^{-6} = \frac{1}{64}$

12. $(12)^0 = 1$

13. $8^{1/3} = 2$

14. $(81)^{3/4} = 27$

15. $\left(\frac{1}{4}\right)^2 = \frac{1}{16}$

16. $\left(\frac{1}{2}\right)^{-2} = 4$

In Exercises 17–28, express the given equations in exponential form.

17. $\log_3 81 = 4$

18. $\log_{11} 121 = 2$

19. $\log_9 9 = 1$

20. $\log_{15} 1 = 0$

21. $\log_{25} 5 = \frac{1}{2}$

22. $\log_8 16 = \frac{4}{3}$

23. $\log_{243} 3 = \frac{1}{5}$

24. $\log_{32}\left(\frac{1}{8}\right) = -\frac{3}{5}$

25. $\log_{10} 0.1 = -1$

26. $\log_7\left(\frac{1}{49}\right) = -2$

27. $\log_{0.5} 16 = -4$

28. $\log_{1/3} 3 = -1$

In Exercises 29–44, determine the value of the unknown.

29. $\log_4 16 = x$

30. $\log_5 125 = x$

31. $\log_{10} 0.01 = x$

32. $\log_{16}\left(\frac{1}{4}\right) = x$

33. $\log_7 y = 3$

34. $\log_8(N + 1) = 3$

35. $\log_8(A - 2) = -\frac{2}{3}$

36. $\log_7 y = -2$

37. $\log_b 81 = 2$

38. $\log_b 625 = 4$

39. $\log_b 4 = -\frac{1}{3}$

40. $\log_b 4 = \frac{2}{3}$

41. $\log_{10} 10^{0.2} = x$

42. $\log_5 5^{2.3} = R + 1$

43. $\log_3 27^{-1} = x + 1$

44. $\log_b\left(\frac{1}{4}\right) = -\frac{1}{2}$

In Exercises 45–50, plot the graphs of the given functions.

45. $y = \log_3 x$

46. $y = \log_4 x$

47. $y = \log_{0.5} x$

48. $y = 3 \log_2 x$

49. $N = 0.2 \log_4 v$

50. $A = 2.4 \log_{10}(2r)$

In Exercises 51–54, display the graphs of the given functions on a graphing calculator.

51. $y = 3 \log_e x$

52. $y = 5 \log_{10}|x|$

53. $y = -\log_{10}(-x)$

54. $y = -\log_e(-2x)$

In Exercises 55–64, perform the indicated operations.

55. Evaluate $\log_4 x$ for (a) $x = 1/64$ and (b) $x = -1/2$.

(W) 56. Use a graphing calculator to display the graphs (on the same screen) of $y = \log_e(x^2 + c)$, with $c = -4, 0, 4$. Describe the results.

57. The magnitudes (visual brightness), m_1 and m_2, of two stars are related to their (actual) brightnesses, b_1 and b_2, by the equation $m_1 - m_2 = 2.5 \log_{10}(b_2/b_1)$. Solve for b_2.

58. The velocity v of a rocket at the point at which its fuel is completely burned is given by $v = u \log_e(w_0/w)$, where u is the exhaust velocity, w_0 is the liftoff weight, and w is the burnout weight. Solve for w.

59. An equation relating the number N of atoms of radium at any time t in terms of the number N_0 of atoms at $t = 0$ is $\log_e(N/N_0) = -kt$, where k is a constant. Solve for N.

60. An equation used in measuring the flow of water in a channel is $C = -a \log_{10}(b/R)$. Solve for R.

61. The time t (in ps) required for N calculations by a certain computer design is $t = N + \log_2 N$. Sketch the graph of this function.

62. When a tractor-trailer turns a right angle corner, its rear wheels follow a curve called a *tractrix*, the equation for which is

$$y = \log_e\left(\frac{1 + \sqrt{1 - x^2}}{x}\right) - \sqrt{1 - x^2}.$$ Display the curve on a graphing calculator.

(W) 63. Display the graphs of $y = (1/3)^x$ and $y = 3^{-x}$ and then compare them. Explain what your comparison shows.

(W) 64. In Exercise 47, the graph of $y = \log_{0.5} x$ is plotted. By inspecting the graph and noting the properties of $\log_{0.5} x$, describe some of the differences of logarithms to a base less than 1 from those to a base greater than 1.

In Exercises 65–68, show that the given functions are inverse functions of each other. Then display the graphs of each function and the line $y = x$ on a graphing calculator and note that each is the mirror image of the other across $y = x$.

65. $y = 10^{x/2}$ and $y = 2 \log_{10} x$

66. $y = e^x$ and $y = \log_e x$

67. $y = 3x$ and $y = x/3$

68. $y = 2x + 4$ and $y = 0.5x - 2$

13.3 PROPERTIES OF LOGARITHMS

A logarithm is an exponent. However, a historical curiosity is that logarithms were developed before exponents were used.

Since a logarithm is an exponent, it must follow the laws of exponents. Those laws that we use in this section to derive the very useful properties of logarithms are listed here for reference.

$$b^u b^v = b^{u+v} \tag{13.4}$$

$$\frac{b^u}{b^v} = b^{u-v} \tag{13.5}$$

$$(b^u)^n = b^{nu} \tag{13.6}$$

The next example shows the reasoning used in deriving the properties of logarithms.

◀ **EXAMPLE 1**   We know that $8 \times 16 = 128$. Writing these numbers as powers of 2, we have

$$8 = 2^3 \qquad 16 = 2^4 \qquad 128 = 2^7 = 2^{3+4}$$

The logarithmic forms can be written as

$$3 = \log_2 8 \qquad 4 = \log_2 16 \qquad 3 + 4 = \log_2 128$$

This means that

$$\log_2 8 + \log_2 16 = \log_2 128$$

where

$$8 \times 16 = 128$$

The *sum of the logarithms* of 8 and 16 equals the logarithm of 128, where the *product* of 8 and 16 equals 128.

▶

For reference, Eqs. (13.4) to (13.6) are

$$b^u b^v = b^{u+v}$$

$$\frac{b^u}{b^v} = b^{u-v}$$

$$(b^u)^n = b^{nu}$$

Following Example 1, if we let $u = \log_b x$ and $v = \log_b y$ and write these equations in exponential form, we have $x = b^u$ and $y = b^v$. Therefore, forming the product of x and y, we obtain

$$xy = b^u b^v = b^{u+v} \quad \text{or} \quad xy = b^{u+v}$$

Writing this last equation in logarithmic form yields

$$u + v = \log_b xy$$

or

Logarithm of a Product

$$\log_b xy = \log_b x + \log_b y \tag{13.7}$$

Equation (13.7) states the property that *the logarithm of the product of two numbers is equal to the sum of the logarithms of numbers.*

Using the same definitions of u and v to form the quotient of x and y, we then have

$$\frac{x}{y} = \frac{b^u}{b^v} = b^{u-v} \quad \text{or} \quad \frac{x}{y} = b^{u-v}$$

Writing this last equation in logarithmic form, we have

$$u - v = \log_b\left(\frac{x}{y}\right)$$

or

Logarithm of a Quotient

$$\log_b\left(\frac{x}{y}\right) = \log_b x - \log_b y \tag{13.8}$$

Equation (13.8) states the property that *the logarithm of the quotient of two numbers is equal to the logarithm of the numerator minus the logarithm of the denominator.*

If we again let $u = \log_b x$ and write this in exponential form, we have $x = b^u$. To find the nth power of x, we write

$$x^n = (b^u)^n = b^{nu}$$

Expressing this equation in logarithmic form yields

$$nu = \log_b(x^n)$$

or

Logarithm of a Power

$$\boxed{\log_b(x^n) = n \log_b x} \qquad (13.9)$$

Equation (13.9) states that *the logarithm of the nth power of a number is equal to n times the logarithm of the number.* The exponent n may be integral or fractional.

In Section 13.2, we showed that the base b of logarithms must be a positive number. Since $x = b^u$ and $y = b^v$, this means that x and y are also positive numbers. Therefore, the *properties of logarithms that have just been derived are valid only for positive values of x and y.*

In advanced mathematics, the logarithms of negative and imaginary numbers are defined.

◀ EXAMPLE 2   **(a)** Using Eq. (13.7), we may express $\log_4 15$ as a sum of logarithms:

$$\log_4 15 = \log_4(3 \times 5) = \log_4 3 + \log_4 5$$

logarithm of product
sum of logarithms

(b) Using Eq. (13.8), we may express $\log_4\left(\frac{5}{3}\right)$ as the difference of logarithms:

$$\log_4\left(\frac{5}{3}\right) = \log_4 5 - \log_4 3$$

logarithm of quotient
difference of logarithms

(c) Using Eq. (13.9), we may express $\log_4(t^2)$ as twice $\log_4 t$:

$$\log_4(t^2) = 2 \log_4 t$$

logarithm of power
multiple of logarithm

(d) Using Eq. (13.8) and then Eq. (13.7), we have

$$\log_4\left(\frac{xy}{z}\right) = \log_4(xy) - \log_4 z$$

$$= \log_4 x + \log_4 y - \log_4 z$$

◀ EXAMPLE 3   We may also express a sum or difference of logarithms as the logarithm of a single quantity.

(a) $\log_4 3 + \log_4 x = \log_4(3 \times x) = \log_4 3x$ using Eq. (13.7)

(b) $\log_4 3 - \log_4 x = \log_4\left(\frac{3}{x}\right)$ using Eq. (13.8)

(c) $\log_4 3 + 2 \log_4 x = \log_4 3 + \log_4(x^2) = \log_4 3x^2$ using Eqs. (13.7) and (13.9)

(d) $\log_4 3 + 2 \log_4 x - \log_4 y = \log_4\left(\frac{3x^2}{y}\right)$ using Eqs. (13.7), (13.8), and (13.9)

In Section 13.2, we noted that $\log_b 1 = 0$. Also, since $b = b^1$ in logarithmic form is $\log_b b = 1$, we have $\log_b(b^n) = n \log_b b = n(1) = n$.

Summarizing these properties, we have

$$\boxed{\log_b 1 = 0 \qquad \log_b b = 1} \tag{13.10}$$

$$\boxed{\log_b(b^n) = n} \tag{13.11}$$

These equations may be used to find exact values of certain logarithms.

◀ **EXAMPLE 4** **(a)** We may evaluate $\log_3 9$ using Eq. (13.11):

$$\log_3 9 = \log_3(3^2) = 2$$

We can establish the exact value since the base of logarithms and the number being raised to the power are the same. Of course, this could have been evaluated directly from the definition of a logarithm.

(b) Using Eq. (13.11), we can write $\log_3(3^{0.4}) = 0.4$. Although we did not evaluate $3^{0.4}$, we can evaluate $\log_3(3^{0.4})$. ▬

◀ **EXAMPLE 5** **(a)** $\log_2 6 = \log_2(2 \times 3) = \log_2 2 + \log_2 3 = 1 + \log_2 3$

(b) $\log_5 \frac{1}{5} = \log_5 1 - \log_5 5 = 0 - 1 = -1$

(c) $\log_7 \sqrt{7} = \log_7(7^{1/2}) = \frac{1}{2} \log_7 7 = \frac{1}{2}$ ▬

◀ **EXAMPLE 6** The following illustration shows the evaluation of a logarithm in two different ways. Either method is appropriate.

(a) $\log_5\left(\frac{1}{25}\right) = \log_5 1 - \log_5 25 = 0 - \log_5(5^2) = -2$

(b) $\log_5\left(\frac{1}{25}\right) = \log_5(5^{-2}) = -2$ ▬

◀ **EXAMPLE 7** Use the basic properties of logarithms to solve the following equation for y in terms of x: $\log_b y = 2 \log_b x + \log_b a$.

Using Eq. (13.9) and then Eq. (13.7), we have

$$\log_b y = \log_b(x^2) + \log_b a = \log_b(ax^2)$$

Since we have the logarithm to the base b of different expressions on each side of the resulting equation, the expressions must be equal. Therefore,

$$y = ax^2$$ ▬

◀ **EXAMPLE 8** An equation for the current i and the time t in an electric circuit containing a resistance R and a capacitance C is $\log_e i - \log_e I = -t/RC$ (where R and C are both factors of the denominator). Here, I is the current for $t = 0$. Solve for i as a function of t.

Using Eq. (13.8), we rewrite the left side of this equation, obtaining

$$\log_e\left(\frac{i}{I}\right) = -\frac{t}{RC}$$

Rewriting this in exponential form, we have

$$\frac{i}{I} = e^{-t/RC} \quad \text{or} \quad i = Ie^{-t/RC}$$ ▬

EXERCISES 13.3

In Exercises 1–8, perform the indicated operations on the resulting expressions if the given changes are made in original expressions of the indicated examples of this section.

1. In Example 2(a), change the 15 to 21 and then find the resulting expression.

2. In Example 2(d), change the x to 2 and the z to 3 and then find the resulting expression.

3. In Example 3(b), change the 3 to 5 and then find the resulting expression.

4. In Example 3(c), change the 2 to 3 and then find the resulting expression.

5. In Example 4(a), change the 9 to 27 and then find the result.

6. In Example 5(a), change the 6 to 10 and then find the resulting expression.

7. In Example 5(c), change the 7's to 5's and then find the result.

8. In Example 7, change the 2 to 3 and then find the result.

In Exercises 9–20, express each as a sum, difference, or multiple of logarithms. See Example 2.

9. $\log_5 33$

10. $\log_3 14$

11. $\log_7\left(\frac{5}{3}\right)$

12. $\log_3\left(\frac{2}{11}\right)$

13. $\log_2(a^3)$

14. $\log_8(n^5)$

15. $\log_6 abc$

16. $\log_2\left(\frac{xy}{z^2}\right)$

17. $\log_5 \sqrt[4]{y}$

18. $\log_4 \sqrt[7]{x}$

19. $\log_2\left(\frac{\sqrt{x}}{a^2}\right)$

20. $\log_3\left(\frac{\sqrt[3]{y}}{7}\right)$

In Exercises 21–28, express each as the logarithm of a single quantity. See Example 3.

21. $\log_b a + \log_b c$

22. $\log_2 3 + \log_2 x$

23. $\log_5 9 - \log_5 3$

24. $-\log_8 R + \log_8 V$

25. $-\log_b \sqrt{x} + \log_b x^2$

26. $\log_4 3^3 + \log_4 9$

27. $2 \log_e 2 + 3 \log_e n$

28. $\frac{1}{2} \log_b a - 2 \log_b 5$

In Exercises 29–36, determine the exact value of each of the given logarithms.

29. $\log_2\left(\frac{1}{32}\right)$

30. $\log_3\left(\frac{1}{81}\right)$

31. $\log_2(2^{2.5})$

32. $\log_5(5^{0.1})$

33. $\log_7 \sqrt{7}$

34. $\log_6 \sqrt[3]{6}$

35. $\log_3 \sqrt[4]{27}$

36. $\log_5 \sqrt[3]{25}$

In Exercises 37–44, express each as a sum, difference, or multiple of logarithms. In each case, part of the logarithm may be determined exactly.

37. $\log_3 18$

38. $\log_5 75$

39. $\log_2\left(\frac{1}{6}\right)$

40. $\log_{10}(0.05)$

41. $\log_3 \sqrt{6}$

42. $\log_2 \sqrt[3]{24}$

43. $\log_{10} 3000$

44. $\log_{10}(40^2)$

In Exercises 45–56, solve for y in terms of x.

45. $\log_b y = \log_b 2 + \log_b x$

46. $\log_b y = \log_b 6 - \log_b x$

47. $\log_4 y = \log_4 x - \log_4 5 + \log_4 3$

48. $\log_3 y = -2 \log_3(x + 1) + \log_3 7$

49. $\log_{10} y = 2 \log_{10} 7 - 3 \log_{10} x$

50. $\log_b y = 3 \log_b \sqrt{x} + 2 \log_b 10$

51. $5 \log_2 y - \log_2 x = 3 \log_2 4 + \log_2 a$

52. $4 \log_2 x - 3 \log_2 y = \log_2 27$

53. $\log_2 x + \log_2 y = 1$

54. $3 \log_4 x + \log_4 y = 1$

55. $\frac{2 \log_5 x}{\log_5 3} - \log_5 y = 2$

56. $\log_8 x = 2 \log_8 y + 4$

In Exercises 57–64, solve the given problems.

(W) 57. Explain why $\log_{10}(x + 3)$ is not equal to $\log_{10} x + \log_{10} 3$.

(W) 58. Evaluate $2 \log_2(2x) - \log_2 x^2$. For what values of x is the value of this expression valid? Explain.

(W) 59. Display the graphs of $y = \log_e(e^2 x)$ and $y = 2 + \log_e x$ on a graphing calculator and explain why they are the same.

60. The use of the insecticide DDT was banned in the United States in 1972. A computer analysis shows that an expression relating the amount A still present in an area, the original amount A_0, and the time t (in years) since 1972 is $\log_{10} A = \log_{10} A_0 + 0.1t \log_{10} 0.8$. Solve for A as a function of t.

61. A study of urban density shows that the population density D (in persons/km^2) is related to the distance r (in km) from the city center by $\log_e D = \log_e a - br + cr^2$, where a, b, and c are positive constants. Solve for D as a function of r.

62. When a person ingests a medication capsule, it is found that the rate R (in mg/min) that it enters the bloodstream in time t (in min) is given by $\log_{10} R - \log_{10} 5 = t \log_{10} 0.95$. Solve for R as a function of t.

63. Under certain conditions, the temperature T (in °C) of a cooling object is related to the time t (in min) by the equation $\log_e T = \log_e 65.0 - 0.41t$. Solve for T as a function of t.

64. In analyzing the power gain in an electric circuit, the equation $N = 10(2 \log_{10} I_1 - 2 \log_{10} I_2 + \log_{10} R_1 - \log_{10} R_2)$ is used. Express this with a single logarithm on the right side.

13.4 LOGARITHMS TO THE BASE 10

In Section 13.2, we stated that a base of logarithms must be a positive number, not equal to 1. In the examples and exercises of the previous sections, we used a number of different bases. There are, however, only two bases that are generally used. They are 10 and e, where e is the irrational number approximately equal to 2.718 that we introduced in Section 12.5 and have used in the previous sections of this chapter.

Base 10 logarithms were developed for calculational purposes and were used a great deal for making calculations until the 1970s, when the modern scientific calculator became widely available. Base 10 logarithms are still used in several scientific measurements, and therefore a need still exists for them. Base e logarithms are used extensively in technical and scientific work; we consider them in detail in the next section.

Logarithms to the base 10 are called **common logarithms.** They may be found directly by use of a calculator, and the ⌐log⌐ key is used for this purpose. This, of course, is the same key we have used with logarithmic functions to the base 10 on the graphing calculator. This calculator key indicates the common notation. *When no base is shown, it is assumed to be the base 10.*

NOTE ▶

```
log(426)
       2.629409599
log(.03654)
      -1.437231457
```

Fig. 13.11

◀ EXAMPLE 1   Using a calculator, as shown in the first two lines of the display in Fig. 13.11, we find that

$$\log 426 = 2.629$$

└── no base shown means base is 10

when the result is rounded off. The decimal part of a logarithm is normally expressed to the same accuracy as that of the number of which it is the logarithm, although showing one additional digit in the logarithm is generally acceptable.

Since $10^2 = 100$ and $10^3 = 1000$, and in this case

$$10^{2.629} = 426$$

we see that the 2.629 power of 10 gives a number between 100 and 1000.   ▶

◀ EXAMPLE 2   Finding log 0.036 54, as shown in the third and fourth lines of Fig. 13.11, we obtain

$$\log 0.036\,54 = -1.4372$$

We note that the logarithm here is negative. This should be the case when we recall the meaning of a logarithm. Raising 10 to a negative power gives us a number between 0 and 1, and here we have

$$10^{-1.4372} = 0.036\,54$$ ▶

We may also use a calculator to find a number N if we know log N. *In this case, we refer to N as the* **antilogarithm** *of log N.* On the calculator we use the ⌐10^x⌐ key. We note that it shows the basic definition of a logarithm. (On many scientific calculators, the key sequence ⌐inv⌐ ⌐log⌐ is used. Note that this sequence shows that the exponential and logarithmic functions are inverse functions.)

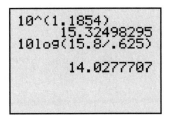

Fig. 13.12

The unit of sound intensity level (used for power gain), the bel (B), is named for the U.S. inventor Alexander Graham Bell (1847–1922). The decibel is the commonly used unit.

◀ **EXAMPLE 3**   Given $\log N = 1.1854$, as shown in the first two lines of the display in Fig. 13.12, we find that

$$N = 15.32$$

where the result has been rounded off. Since $10^1 = 10$ and $10^2 = 100$, we see that $N = 10^{1.1854}$ and is a number between 10 and 100.   ▶

The following example illustrates an application in which a measurement requires the direct use of the value of a logarithm.

◀ **EXAMPLE 4**   The power gain G (in decibels (dB)) of an electronic device is given by $G = 10 \log(P_0/P_i)$, where P_0 is the output power (in W) and P_i is the input power. Determine the power gain for an amplifier for which $P_0 = 15.8$ W and $P_i = 0.625$ W.

Substituting the given values, we have

$$G = 10 \log \frac{15.8}{0.625} = 14.0 \text{ dB}$$

where lines 3 and 4 of the display in Fig. 13.12 show the evaluation on a calculator.   ▶

As noted in the chapter introduction, logarithms were developed for calculational purposes. They were first used in the seventeenth century for making tedious and complicated calculations that arose in astronomy and navigation. These complicated calculations were greatly simplified, since logarithms allowed them to be performed by means of basic additions, subtractions, multiplications, and divisions. Performing calculations in this way provides an opportunity to understand better the meaning and properties of logarithms. Also, certain calculations cannot be done directly on a calculator but can be done by logarithms.

◀ **EXAMPLE 5**   A certain computer design has 64 different sequences of ten binary digits so that the total number of possible states is $(2^{10})^{64} = 1024^{64}$. Evaluate 1024^{64} using logarithms.

Since $\log x^n = n \log x$, we know that $\log 1024^{64} = 64 \log 1024$. Although most calculators will not directly evaluate 1024^{64}, we can use one to find the value of $64 \log 1024$. Since 1024^{64} is *exact,* we will show ten calculator digits until we round off the result. We therefore evaluate 1024^{64} as follows:

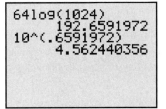

Fig. 13.13

Let $N = 1024^{64}$

$\log N = \log 1024^{64} = 64 \log 1024$ using Eq. (13.9): $\log_b x^n = n \log_b x$

$\qquad = 192.659\,197\,2$

$N = 10^{192.659\,197\,2}$ meaning of logarithm

$\qquad = 10^{192} \times 10^{0.659\,197\,2}$ using Eq. (13.4): $b^u b^v = b^{u+v}$

$\qquad = (10^{192}) \times (4.5624)$ antilogarithm of 0.659 197 2 is 4.5624 (rounded off)

$\qquad = 4.5624 \times 10^{192}$

By using Eq. (13.4), $10^{0.659\,197\,2}$ represents a number between 1 and 10 ($10^0 = 1$ and $10^1 = 10$), and we can write the result immediately in scientific notation.

Although we used a calculator to find 192.659 197 2 and 4.5624 as shown in Fig. 13.13, the calculation was done essentially by logarithms.   ▶

Example 5 shows that calculations using logarithms are based on Eqs. (13.7), (13.8), and (13.9). Multiplication is performed by the addition of logarithms, division is performed by the subtraction of logarithms, and a power is found by a multiple of a logarithm. A root of a number is found by using the fractional exponent form of the power.

E X E R C I S E S **13.4**

In Exercises 1 and 2, find the indicated values if the given changes are made in the indicated examples of this section.

1. In Example 2, change 0.036 54 to 0.3654 and then find the required value.

2. In Example 3, change 1.1854 to 2.1854 and then find the required value.

In Exercises 3–12, find the common logarithm of each of the given numbers by using a calculator.

3. 567

4. 0.0640

5. 9.24×10^6

6. 3.19^3

7. 1.174^{-4}

8. 8.043×10^{-8}

9. $\cos 12.5°$

10. $\tan 50.8°$

11. $\sqrt{274}$

12. $\log_2 16$

In Exercises 13–20, find the antilogarithm of each of the given logarithms by using a calculator.

13. 4.437

14. 0.929

15. −1.3045

16. −6.9788

17. 3.301 12

18. 8.824 36

19. −2.237 46

20. −10.336

In Exercises 21–24, use logarithms to evaluate the given expressions.

21. $(5.98)(14.3)$

22. $\dfrac{895}{73.4^{86}}$

23. $\left(\sqrt[10]{7.32}\right)(2470)^{30}$

24. $\dfrac{126\,000^{20}}{2.63^{2.5}}$

In Exercises 25–28, use a calculator to verify the given values.

25. $\log 5 + \log 7 = \log 35$

26. $\log 500 - \log 20 = \log 25$

27. $\log 81 = 4 \log 3$

28. $\log 6 = 0.5 \log 36$

In Exercises 29–32, find the logarithms of the given numbers.

29. The signal used by some cordless phones is 9.00×10^8 Hz.

30. A large sunspot may be 3.5×10^4 km in diameter.

31. About 1.3×10^{-14}% of carbon atoms are carbon-14, the radioactive isotope used in determining the age of samples from ancient sites.

32. In an air sample taken in an urban area, $5/10^6$ of the air was carbon monoxide.

In Exercises 33–36, find the indicated values.

33. Find T (in K) if $\log T = 8$, where T is the temperature sufficient for nuclear fission.

34. Find v (in m/s) if $\log v = 7.423$, where v is the speed of an electron in a TV picture tube.

35. Find e if $\log e = -0.35$, where e is the efficiency of a certain gasoline engine.

36. Find E (in J) if $\log E = -18.49$, where E is the energy of a photon of visible light.

In Exercises 37 and 38, solve the given problems by finding the appropriate logarithms.

37. A stereo amplifier has an input power of 0.750 W and an output power of 25.0 W. What is the power gain? (See Example 4.)

38. Measured on the Richter scale, the magnitude of an earthquake of intensity I is defined as $R = \log(I/I_0)$, where I_0 is a minimum level for comparison. What is the Richter scale reading for the 1995 Philippine earthquake for which $I = 20\,000\,000I_0$?

In Exercises 39 and 40, use logarithms to perform the indicated calculations.

39. A certain type of optical switch in a fiber-optic system allows a light signal to continue in either of two fibers. How many possible paths could a light signal follow if it passes through 400 such switches?

40. The peak current I_m (in A) in an alternating-current circuit is given by $I_m = \sqrt{\dfrac{2P}{Z \cos \theta}}$, where P is the power developed, Z is the magnitude of the impedance, and θ is the phase angle between the current and voltage. Evaluate I_m for $P = 5.25$ W, $Z = 320\ \Omega$, and $\theta = 35.4°$.

13.5 NATURAL LOGARITHMS

As we have noted, another number important as a base of logarithms is the number e. *Logarithms to the base e are called* **natural logarithms.** Since e is an irrational number equal to about 2.718, it may appear to be a very unnatural choice as a base of logarithms. However, in calculus the reason for its choice and the fact that it is a very natural number for a base of logarithms are shown.

Just as log x refers to logarithms to the base 10, the notation ln x is used to denote logarithms to the base e. We briefly noted this in Section 13.2 in discussing the graphing calculator. Due to the extensive use of natural logarithms, the notation **ln x** is more convenient than **$\log_e x$,** although they mean the same thing.

Since more than one base is important, at times it is useful to change a logarithm from one base to another. If $u = \log_b x$, then $b^u = x$. Taking logarithms of both sides of this last expression to the base a, we have

$$\log_a b^u = \log_a x$$
$$u \log_a b = \log_a x$$
$$u = \frac{\log_a x}{\log_a b}$$

However, $u = \log_b x$, which means that

$$\log_b x = \frac{\log_a x}{\log_a b} \tag{13.12}$$

Equation (13.12) allows us to change a logarithm in one base to a logarithm in another base. The following examples illustrate the method of performing this operation.

◀ **EXAMPLE 1** Change log 20 to a logarithm with base e; that is, find ln 20.
Using Eq. (13.12) with $a = 10$, $b = e$, and $x = 20$, we have

$$\log_e 20 = \frac{\log_{10} 20}{\log_{10} e}$$

or $\quad \ln 20 = \dfrac{\log 20}{\log e} = 2.996 \quad$ see lines 1 and 2 of calculator display in Fig. 13.14

This means that $e^{2.996} = 20$.

◀ **EXAMPLE 2** Find $\log_5 560$.
In Eq. (13.12), if we let $a = 10$ and $b = 5$, we have

$$\log_5 x = \frac{\log x}{\log 5}$$

In this example, $x = 560$. Therefore, we have

$$\log_5 560 = \frac{\log 560}{\log 5} = 3.932 \quad$$ see lines 3 and 4 of calculator display in Fig. 13.14

From the definition of a logarithm, this means that

$$5^{3.932} = 560$$

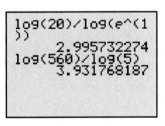

Fig. 13.14

Since natural logarithms are used extensively, it is often convenient to have Eq. (13.12) written specifically for use with logarithms to the base 10 and natural logarithms. First, using $a = 10$ and $b = e$, we have

$$\ln x = \frac{\log x}{\log e} \qquad (13.13)$$

Then, with $a = e$ and $b = 10$, we have

$$\log x = \frac{\ln x}{\ln 10} \qquad (13.14)$$

Note that we really found ln 20 in Example 1 by using Eq. (13.13).

Values of natural logarithms can be found directly on a calculator. The ⎡ ln ⎤ key is used for this purpose. In order to find the antilogarithm of a natural logarithm, we use the ⎡ e^x ⎤ key. The following example illustrates finding a natural logarithm and an antilogarithm on a graphing calculator.

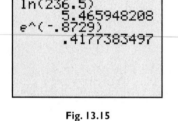

⚞ **EXAMPLE 3** (a) From the first two lines of the calculator display in Fig. 13.15, we find that

$$\ln 236.5 = 5.4659$$

which means that $e^{5.4659} = 236.5$.

(b) Given that $\ln N = -0.8729$, we determine N by finding $e^{-0.8729}$ on the calculator. This gives us (see lines 3 and 4 of the display in Fig. 13.15)

$$N = 0.4177 \qquad ▶$$

Fig. 13.15

Using Eq. (13.12), we can display the graph of a logarithmic function with any base on a graphing calculator, as we show in the next example.

⚞ **EXAMPLE 4** Display the graph of $y = 3 \log_2 x$ on a graphing calculator.
To display this graph, we use the fact that

$$\log_2 x = \frac{\log x}{\log 2} \quad \text{or} \quad \log_2 x = \frac{\ln x}{\ln 2}$$

Therefore, we enter the function

$$y = \frac{3 \log x}{\log 2} \qquad \left(\text{or } y = \frac{3 \ln x}{\ln 2} \right)$$

in the calculator. We can select the *window* settings by considering the domain and range of the logarithmic function and by the fact that $3/\log 2 \approx 10$. This tells us that the range setting should be about ten times that of the curve $y = \log x$. The curve is shown in the calculator display in Fig. 13.16. ▶

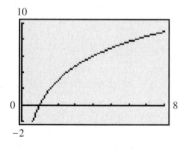

Fig. 13.16

Applications of natural logarithms are found in many fields of technology. One such application is shown in the next example, and others are found in the exercises.

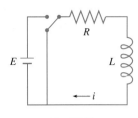

Fig. 13.17

◀ EXAMPLE 5   The electric current i in a circuit containing a resistance and an inductance (see Fig. 13.17) is given by $\ln(i/I) = -Rt/L$, where I is the current at $t = 0$, R is the resistance, t is the time, and L is the inductance. Calculate how long (in s) it takes i to reach 0.430 A, if $I = 0.750$ A, $R = 7.50\ \Omega$, and $L = 1.25$ H.

Solving for t and then evaluating, we have

$$t = -\frac{L\ln(i/I)}{R} = -\frac{L(\ln i - \ln I)}{R} \qquad \text{either form can be used}$$

$$= -\frac{1.25(\ln 0.430 - \ln 0.750)}{7.50} = 0.0927 \text{ s} \qquad \text{evaluating}$$

Therefore, the current changes from 0.750 A to 0.430 A in 0.0927 s.

EXERCISES 13.5

In Exercises 1 and 2, find the indicated values if the given changes are made in the indicated examples of this section.

1. In Example 1, change 20 to 200 and then evaluate.

2. In Example 2, change the base to 4 and then evaluate.

In Exercises 3–8, use logarithms to the base 10 to find the natural logarithms of the given numbers.

3. 26.0 **4.** 631 **5.** 1.562

6. 0.5017 **7.** 0.007 326 7 **8.** 0.000 443 48

In Exercises 9–14, use logarithms to the base 10 to find the indicated logarithms.

9. $\log_7 42$ **10.** $\log_2 86$ **11.** $\log_\pi 245$

12. $\log_{12} 122$ **13.** $\log_{40} 750$ **14.** $\log_{100} 3720$

In Exercises 15–22, find the natural logarithms of the given numbers.

15. 51.4 **16.** 293 **17.** 1.394

18. 65.52 **19.** 0.9917 **20.** 0.002 086

21. $(0.012\,937)^4$ **22.** $\sqrt{0.000\,060\,808}$

In Exercises 23–26, use Eq. (13.14) to find the common logarithms of the given numbers.

23. 45.17 **24.** 8765

25. 0.685 28 **26.** 0.001 429 8

In Exercises 27–34, find the natural antilogarithms of the given logarithms.

27. 2.190 **28.** 3.420 **29.** 0.008 421 0

30. 0.632 **31.** −0.7429 **32.** −2.942 18

33. −23.504 **34.** −0.008 04

In Exercises 35–38, use a graphing calculator to display the indicated graphs.

35. The graph of $y = \log_5 x$ **36.** The graph of $y = 2\log_8 x$

37. Graphically show that $y = 2^x$ and $y = \log_2 x$ are inverse functions. (See Example 9 of Section 13.2.)

(W) **38.** Explain what is meant by the expression $\ln \ln x$. Display the graph of $y = \ln \ln x$ on a calculator.

In Exercises 39–42, use a calculator to verify the given values.

39. $\ln 5 + \ln 8 = \ln 40$ **40.** $2\ln 6 - \ln 3 = \ln 12$

41. $4\ln 3 = \ln 81$ **42.** $\ln 5 - 0.5\ln 25 = \ln 1$

In Exercises 43–52, solve the given problems.

43. Evaluate: $\sqrt{\ln e^9}$.

44. Solve for y in terms of x: $\ln y + 2\ln x = 1 + \ln 5$.

45. Find f (in Hz) if $\ln f = 21.619$, where f is the frequency of the microwaves in a microwave oven.

46. Find k (in 1/Pa) if $\ln k = -21.504$, where k is the compressibility of water.

47. If interest is compounded continuously (daily compounded interest closely approximates this), with an interest rate i, a bank account will double in t years according to $i = (\ln 2)/t$. Find i if the account is to double in 8.5 years.

48. One approximate formula for world population growth is $T = 50.0\ln 2$, where T is the number of years for the population to double. According to this formula, how long does it take for the population to double?

49. For the electric circuit of Example 5, find how long it takes the current to reach 0.1 of the initial value of 0.750 A.

50. The velocity v (in m/s) of a rocket increases as fuel is consumed and ejected. Considering fuel as part of the mass m of a rocket in flight and m_0 is its original mass, the velocity is given by $v = 2500(\ln m_0 - \ln m)$. If 0.75 of the original mass is fuel, find v when all the fuel is used.

51. The distance x traveled by a motorboat in t seconds after the engine is cut off is given by $x = k^{-1}\ln(kv_0t + 1)$, where v_0 is the velocity of the boat at the time the engine is cut and k is a constant. Find how long it takes a boat to go 150 m if $v_0 = 12.0$ m/s and $k = 6.80 \times 10^{-3}$/m.

52. The electric current i (in A) in a circuit containing a 1-H inductor, a 10-Ω resistor, and a 6-V battery is a function of the time t (in s) given by $i = 0.6(1 - e^{-10t})$. Solve for t as a function of i.

13.6 EXPONENTIAL AND LOGARITHMIC EQUATIONS

EXPONENTIAL EQUATIONS

An equation in which the variable occurs in an exponent is called an **exponential equation.** Although some exponential equations may be solved by changing to logarithmic form, they are more generally solved by *taking the logarithm of each side* and then using the basic properties of logarithms.

◀ **EXAMPLE 1** **(a)** We can solve the exponential equation $2^x = 8$ for x by writing it in logarithmic form. This gives us

$$x = \log_2 8 = 3 \qquad 2^3 = 8$$

This method is good if we can directly evaluate the resulting logarithm.

(b) Since 2^x and 8 are equal, the logarithms of 2^x and 8 are also equal. Therefore, we can also solve $2^x = 8$ in a more general way by taking logarithms (to any proper base) of both sides and equating these logarithms. This gives us

$$\log 2^x = \log 8 \qquad \text{or} \qquad \ln 2^x = \ln 8$$

$$x \log 2 = \log 8 \qquad x \ln 2 = \ln 8 \qquad \text{using Eq. (13.9)}$$

$$x = \frac{\log 8}{\log 2} = 3 \qquad x = \frac{\ln 8}{\ln 2} = 3 \qquad \text{using a calculator} \qquad \blacktriangleright$$

For reference, Eqs. (13.7), (13.8), and (13.9) are

$\log_b xy = \log_b x + \log_b y$

$\log_b\left(\dfrac{x}{y}\right) = \log_b x - \log_b y$

$\log_b(x^n) = n \log_b x$

◀ **EXAMPLE 2** Solve the equation $3^{x-2} = 5$.

Taking logarithms of each side and equating them, we have

$$\log 3^{x-2} = \log 5$$

or

$$(x - 2)\log 3 = \log 5 \qquad \text{using Eq. (13.9)}$$

Solving this last equation for x, we have

$$x = 2 + \frac{\log 5}{\log 3} = 3.465$$

This solution means that

$$3^{3.465-2} = 3^{1.465} = 5$$

which can be checked by a calculator. ▶

◀ **EXAMPLE 3** Solve the equation $2(4^{x-1}) = 17^x$.

By taking logarithms of each side, we have the following:

$$\log 2 + (x - 1)\log 4 = x \log 17 \qquad \text{using Eqs. (13.7) and (13.9)}$$

$$x \log 4 - x \log 17 = \log 4 - \log 2$$

$$x(\log 4 - \log 17) = \log 4 - \log 2$$

$$x = \frac{\log 4 - \log 2}{\log 4 - \log 17} = \frac{\log (4/2)}{\log 4 - \log 17} \qquad \text{using Eq. (13.8)}$$

$$= \frac{\log 2}{\log 4 - \log 17}$$

$$= -0.479 \qquad \blacktriangleright$$

EXAMPLE 4 At constant temperature, the atmospheric pressure p (in Pa) at an altitude h (in m) is given by $p = p_0 e^{kh}$, where p_0 is the pressure where $h = 0$ (usually taken as sea level). Given that $p_0 = 101.3$ kPa (atmospheric pressure at sea level) and $p = 68.9$ kPa for $h = 3050$ m, find the value of k.

Since the equation is defined in terms of e, we can solve it most easily by taking natural logarithms of *each side.* By doing this we have the following solution:

$$\ln p = \ln(p_0 e^{kh}) = \ln p_0 + \ln e^{kh} \qquad \text{using Eq. (13.7)}$$
$$= \ln p_0 + kh \ln e = \ln p_0 + kh \qquad \text{using Eq. (13.9), } \ln e = 1$$
$$\ln p - \ln p_0 = kh$$
$$k = \frac{\ln p - \ln p_0}{h}$$

Substituting the given values, we have

$$k = \frac{\ln(68.9 \times 10^3) - \ln(101.3 \times 10^3)}{3050} = -0.000\,126/\text{m}$$

LOGARITHMIC EQUATIONS

Some of the important measurements in scientific and technical work are defined in terms of logarithms. Using these formulas can lead to solving a **logarithmic equation,** *which is an equation with the logarithm of an expression involving the variable.* In solving logarithmic equations, we use the basic properties of logarithms to help change them into a usable form. There is, however, no general algebraic method for solving such equations, and we consider only some special cases.

EXAMPLE 5 The human ear responds to sound on a scale that is approximately proportional to the logarithm of the intensity of the sound. Therefore, the loudness of sound (measured in dB) is defined by the equation $b = 10 \log(I/I_0)$, where I is the intensity of the sound and I_0 is the minimum intensity detectable.

A busy street has a loudness of 70 dB, and riveting has a loudness of 100 dB. To find how many times greater the intensity I_r of the sound of riveting is than the intensity I_c of the sound of the city street, we substitute in the above equation. This gives

$$70 = 10 \log\left(\frac{I_c}{I_0}\right) \quad \text{and} \quad 100 = 10 \log\left(\frac{I_r}{I_0}\right)$$

To solve for I_c and I_r, we divide each side by 10 and then use exponential form:

$$7.0 = \log\left(\frac{I_c}{I_0}\right) \quad \text{and} \quad 10 = \log\left(\frac{I_r}{I_0}\right)$$
$$\frac{I_c}{I_0} = 10^{7.0} \qquad\qquad \frac{I_r}{I_0} = 10^{10}$$
$$I_c = I_0(10^{7.0}) \qquad\qquad I_r = I_0(10^{10})$$

This demonstrates that sound intensity levels change much more than loudness levels. (See Exercise 46.)

Since we want the number of times I_r is greater than I_c, we divide I_r by I_c:

$$\frac{I_r}{I_c} = \frac{I_0(10^{10})}{I_0(10^{7.0})} = \frac{10^{10}}{10^{7.0}} = 10^{3.0} \quad \text{or} \quad I_r = 10^{3.0}I_c = 1000 I_c$$

Thus, the sound of riveting is 1000 times as intense as the sound of the city street.

See the chapter introduction.

For reference, Eqs. (13.7), (13.8), and (13.9) are

$$\log_b xy = \log_b x + \log_b y$$

$$\log_b\left(\frac{x}{y}\right) = \log_b x - \log_b y$$

$$\log_b(x^n) = n \log_b x$$

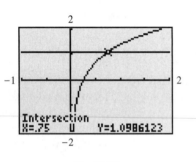

Fig. 13.18

◀ EXAMPLE 6 An analysis of the population of Canada from 1980 to 2000 led to the equation $\log_2 P = \log_2 23.9 + 0.450$, where P is the projected population (in millions) in 2010. Determine this population.

The solution is as follows:

$$\log_2 P - \log_2 23.9 = 0.450$$

$$\log_2(P/23.9) = 0.450 \qquad \text{using Eq. (13.8)}$$

$$P/23.9 = 2^{0.450} \qquad \text{exponential form}$$

$$P = 23.9(2^{0.450})$$

$$= 32.6 \text{ million} \qquad \text{projected 2010 population} \qquad \blacktriangleright$$

◀ EXAMPLE 7 Solve the logarithmic equation $2 \ln 2 + \ln x = \ln 3$.

Using the properties of logarithms, we have the following solution:

$$2 \ln 2 + \ln x = \ln 3$$

$$\ln 2^2 + \ln x - \ln 3 = 0 \qquad \text{using Eq. (13.9)}$$

$$\ln \frac{4x}{3} = 0 \qquad \text{using Eqs. (13.7) and (13.8)}$$

$$\frac{4x}{3} = e^0 = 1 \qquad \text{exponential form}$$

$$4x = 3, \qquad x = 3/4$$

Since $\ln(3/4) = \ln 3 - \ln 4$, this solution checks in the *original equation.*

The graphing calculator solution of this equation is shown in Fig. 13.18 by letting $y_1 = 2 \ln 2 + \ln x$, $y_2 = \ln 3$ and finding the point of intersection. It can also be solved by letting $y_1 = 2 \ln 2 + \ln x - \ln 3$ and finding the zero of the function. $\blacktriangleright$

◀ EXAMPLE 8 Solve the logarithmic equation $2 \log x - 1 = \log(1 - 2x)$.

$$\log x^2 - \log(1 - 2x) = 1$$

$$\log \frac{x^2}{1 - 2x} = 1 \qquad \text{using Eq. (13.8)}$$

$$\frac{x^2}{1 - 2x} = 10^1 \qquad \text{exponential form}$$

$$x^2 = 10 - 20x$$

$$x^2 + 20x - 10 = 0$$

$$x = \frac{-20 \pm \sqrt{400 + 40}}{2} = -10 \pm \sqrt{110}$$

Since logarithms of negative numbers are not defined and $-10 - \sqrt{110}$ is negative and cannot be used in the first term of the original equation, we have

$$x = -10 + \sqrt{110} = 0.488 \qquad \text{use a calculator to check this result} \qquad \blacktriangleright$$

EXERCISES 13.6

In Exercises 1 and 2, find the indicated values if the given changes are made in the indicated examples of this section.

1. In Example 2, change the sign in the exponent from $-$ to $+$ and then solve the equation.

2. In Example 7, change ln 3 to ln 6 and then solve the equation.

In Exercises 3–30, solve the given equations.

3. $2^x = 16$

4. $3^x = \frac{1}{81}$

5. $5^x = 0.3$

6. $\pi^x = 15$

7. $3^{-x} = 0.525$

8. $e^{-x} = 17.54$

9. $6^{x+1} = 10$

10. $5^{x-1} = 2$

11. $3(14^x) = 40$

12. $0.8^x = 0.4$

13. $0.6^x = 2^{x^2}$

14. $15.6^{x+2} = 23^x$

15. $3 \log_8 x = -2$

16. $5 \log_{32} x = -3$

17. $2 \ln x = 1$

18. $4 \ln 2x = \ln e^2$

19. $\log_2 x + \log_2 7 = \log_2 21$

20. $2 \log_2 3 - \log_2 x = \log_2 45$

21. $2 \log(3 - x) = 1$

22. $3 \log(2x - 1) = 1$

23. $\log 12x^2 - \log 3x = 3$

24. $\ln x - \ln\left(\frac{1}{3}\right) = 1$

25. $3 \ln 2 + \ln(x - 1) = \ln 24$

26. $\ln(2x - 1) - 2 \ln 4 = 3 \ln 2$

27. $\frac{1}{2} \log(x + 2) + \log 5 = 1$

28. $2 \log_x 2 + \log_2 x = 3$

29. $\log(2x - 1) + \log(x + 4) = 1$

30. $\log_2 x + \log_2(x + 2) = 3$

In Exercises 31–38, use a graphing calculator to solve the given equations.

31. $15^{-x} = 1.326$

32. $e^{2x} = 3.625$

33. $4(3^x) = 5$

34. $5^{x+2} = e^{2x}$

35. $3 \ln 2x = 2$

36. $\log 4x + \log x = 2$

37. $2 \ln 2 - \ln x = -1$

38. $\log(x - 3) + \log x = \log 4$

In Exercises 39–52, find the indicated quantities.

39. Solve for x: $e^x + e^{-x} = 3$. (*Hint:* Multiply each term by e^x and then it can be treated as a quadratic equation in e^x.)

40. Solve for x: $3^x + 3^{-x} = 4$. See Exercise 39.

41. In computer design, the number N of bits of memory is often expressed as a power of 2, or $N = 2^x$. Find x if $N = 2.68 \times 10^8$ bits.

42. Referring to Exercise 60 on page 381, in what year will the amount of DDT be 25% of the original amount?

43. The temperature T (in °C) of a cooling gold ingot is given by $T = 22 + 98(0.30)^{0.20t}$, where t is the time in minutes. How long will it take the ingot to cool to 35°C?

44. The electric current i (in A) in a circuit containing a resistance R, an inductor L, and a voltage source E is given by $i = \dfrac{E}{R}(1 - e^{-Rt/L})$, where t is the time in seconds. Find t if $i = 0.750$ A, $E = 6.00$ V, $R = 4.50\ \Omega$, and $L = 2.50$ H.

45. In chemistry, the pH value of a solution is a measure of its acidity. The pH value is defined by pH $= -\log(H^+)$, where H^+ is the hydrogen-ion concentration. If the pH of a sample of rainwater is 4.764, find the hydrogen-ion concentration. (If pH < 7, the solution is acid. If pH > 7, the solution is basic.) Acid rain has a pH between 4 and 5, and normal rain is slightly acidic with a pH of about 5.6.

46. Referring to Example 5, show that if the difference in loudness of two sounds is d decibels, the louder sound is $10^{d/10}$ more intense than the quieter sound.

47. Measured on the Richter scale, the magnitude of an earthquake of intensity I is defined as $R = \log(I/I_0)$, where I_o is a minimum level for comparison. How many times I_0 was the 1906 San Francisco earthquake whose magnitude was 8.3 on the Richter scale?

48. How many times more intense was the 1906 San Francisco earthquake than the 1994 Northridge (southern California) earthquake, $R = 6.8$? (See Exercise 47.)

49. Studies have shown that the concentration c (in mg/cm^3 of blood) of aspirin in a typical person is related to the time t (in h) after the aspirin reaches maximum concentration by the equation $\ln c = \ln 15 - 0.20t$. Solve for c as a function of t.

50. In an electric circuit containing a resistor and a capacitor with an initial charge q_0, the charge q on the capacitor at any time t after closing the switch can be found by solving the equation $\ln q = -\dfrac{t}{RC} + \ln q_0$. Here, R is the resistance, and C is the capacitance. Solve for q as a function of t.

51. An earth satellite loses 0.1% of its remaining power each week. An equation relating the power P, the initial power P_0, and the time t (in weeks) is $\ln P = t \ln 0.999 + \ln P_0$. Solve for P as a function of t.

52. In finding the path of a certain plane, the equation $\ln r = \ln a - \ln \cos \theta - (v/w)\ln(\sec \theta + \tan \theta)$ is used. Solve for r. (The conditions of the plane's flight are similar to those described in Exercise 88, p. 338.)

Many exponential and logarithmic equations cannot be solved algebraically as we did in this section. However, they can be solved graphically. For example, such equations can be solved by either of the methods noted at the end of Example 7. In Exercises 53–56, solve the given equations by use of a graphing calculator.

53. $2^x + 3^x = 50$

54. $4^x + x^2 = 25$

55. The curve in which a uniform wire or rope hangs under its own weight is called a *catenary*. An example of a catenary that we see every day is a wire strung between utility poles, as shown in Fig. 13.19. For a particular wire, the equation of the catenary it forms is $y = 2(e^{x/4} + e^{-x/4})$, where (x, y) is a point on the curve. Find x for $y = 5.8$ m.

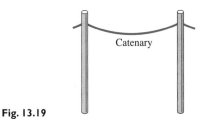

Catenary

Fig. 13.19

56. In finding the current i in a certain electric circuit, the equation $i = 2 \ln(t + 2) - t$ relates the current and the time t (in s). Find t for $i = 0.05$ A.

GRAPHS ON LOGARITHMIC AND SEMILOGARITHMIC PAPER

When constructing the graphs of some functions, one of the variables changes much more rapidly than the other. We saw this in graphing the exponential and logarithmic functions in Sections 13.1 and 13.2. The following example illustrates this point.

EXAMPLE 1 Plot the graph of $y = 4(3^x)$.

Constructing the following table of values,

x	-1	0	1	2	3	4	5
y	1.3	4	12	36	108	324	972

we then plot these values as shown in Fig. 13.20(a).

We see that as x changes from -1 to 5, y changes much more rapidly, from about 1 to nearly 1000. Also, because of the scale that must be used, we see that it is not possible to show accurately the differences in the y-values on the graph.

Even a graphing calculator cannot show the graph accurately for the values near $x = 0$, if we wish to view all of this part of the curve. In fact, the graphing calculator view shows the curve as being on the axis for these values, as we see in Fig. 13.20(b). Note in this figure that we have shown the same values of the range of the function as we did in Fig. 13.20(a).

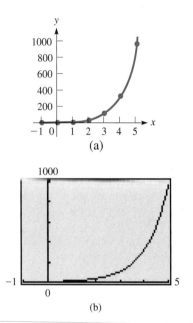

(a)

(b)

Fig. 13.20

It is possible to graph a function with a large change in values, for one or both variables, more accurately than can be done on the standard rectangular coordinate system. This is done by using a scale marked off in distances proportional to the logarithms of the values being represented. *Such a scale is called a* **logarithmic scale.** For example, log 1 = 0, log 2 = 0.301, and log 10 = 1. Thus, on a logarithmic scale, the 2 is placed 0.301 units of distance from the 1 to the 10. Figure 13.21 shows a logarithmic scale with the numbers represented and the distance used for each.

On a logarithmic scale, the distances between the integers are not equal, but this scale does allow for a much greater range of values and much greater accuracy for many of the values. There is another advantage to using logarithmic scales. Many equations that would have more complex curves when graphed on the standard rectangular coordinate system will have simpler curves, often straight lines, when graphed using logarithmic scales. In many cases, this makes the analysis of the curve much easier.

Zero and negative numbers do not appear on the logarithmic scale. In fact, all numbers used on the logarithmic scale must be positive, since the domain of the logarithmic function includes only positive real numbers. Thus, the logarithmic scale must start at some number greater than zero. This number is a power of 10 and can be very small, say, $10^{-6} = 0.000\,001$, but it is positive.

If we wish to use a large range of values for only one of the variables, we use what is known as **semilogarithmic,** or **semilog,** graph paper. On this graph paper, only one axis (usually the y-axis) uses a logarithmic scale. If we wish to use a large range of values for both variables, we use **logarithmic,** or **log-log,** graph paper. Both axes are marked with logarithmic scales.

The following examples illustrate the use of semilog and log-log graph paper.

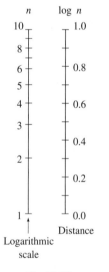

Fig. 13.21

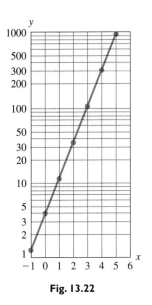

Fig. 13.22

❰ EXAMPLE 2   Construct the graph of $y = 4(3^x)$ on semilogarithmic graph paper. This is the same function as in Example 1, and we repeat the table of values:

x	-1	0	1	2	3	4	5
y	1.3	4	12	36	108	324	972

Again, we see that the range of y-values is large. When we plotted this curve on the rectangular coordinate system in Example 1, we had to use large units along the y-axis. This made the values of 1.3, 4, 12, and 36 appear at practically the same level. However, when we use semilog graph paper, we can label each axis such that all y-values are accurately plotted as well as the x-values.

The logarithmic scale is shown in **cycles,** and we must label the base line of the first cycle as 1 times a power of 10 (0.01, 0.1, 1, 10, 100, and so on) with the following cycle labeled with the next power of 10. The lines between are labeled with 2, 3, 4, and so on, times the proper power of 10. See the vertical scale in Fig. 13.22. We now plot the points in the table on the graph. The resulting graph is a straight line, as we see in Fig. 13.22. Taking logarithms of each side of the equation, we have

$$\log y = \log[4(3^x)] = \log 4 + \log 3^x \qquad \text{using Eq. (13.7)}$$
$$= \log 4 + x \log 3 \qquad \text{using Eq. (13.9)}$$

However, since $\log y$ was plotted automatically (because we used semilogarithmic paper), the graph really represents

$$u = \log 4 + x \log 3$$

where $u = \log y$; $\log 3$ and $\log 4$ are constants, and therefore this equation is of the form $u = mx + b$, which is a straight line (see Section 5.2).

The logarithmic scale in Fig. 13.22 has *three cycles,* since all values of three powers of 10 are represented.   ❱

❰ EXAMPLE 3   Construct the graph of $x^4y^2 = 1$ on logarithmic paper.

First, we solve for y and make a table of values. Considering positive values of x and y, we have

$$y = \sqrt{\frac{1}{x^4}} = \frac{1}{x^2}$$

x	0.5	1	2	8	20
y	4	1	0.25	0.0156	0.0025

We plot these values on log-log paper on which both scales are logarithmic, as shown in Fig. 13.23. We again see that we have a straight line. Taking logarithms of both sides of the equation, we have

$$\log(x^4y^2) = \log 1$$
$$\log x^4 + \log y^2 = 0 \qquad \text{using Eq. (13.7)}$$
$$4 \log x + 2 \log y = 0 \qquad \text{using Eq. (13.9)}$$

If we let $u = \log y$ and $v = \log x$, we then have

$$4v + 2u = 0 \quad \text{or} \quad u = -2v$$

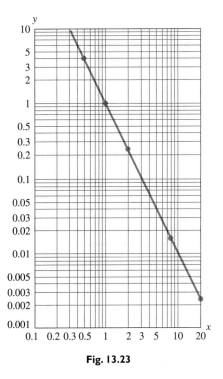

Fig. 13.23

which is the equation of a straight line, as shown in Fig. 13.23. Note, however, that not all graphs on logarithmic paper are straight lines.   ❱

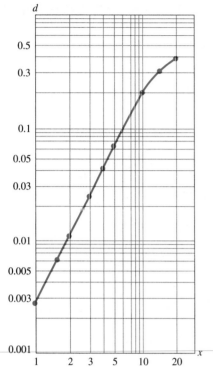

Fig. 13.24

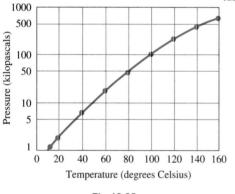

Fig. 13.25

◀ EXAMPLE 4   The deflection (in m) of a certain cantilever beam as a function of the distance x (in m) from one end is

$$d = 0.0001(30x^2 - x^3)$$

If the beam is 20.0 m long, plot a graph of d as a function of x on log-log paper.
   Constructing a table of values, we have

x (m)	1.00	1.50	2.00	3.00	4.00
d (m)	0.002 90	0.006 41	0.0112	0.0243	0.0416

x (m)	5.00	10.0	15.0	20.0
d (m)	0.0625	0.200	0.338	0.400

Since the beam is 20.0 m long, there is no meaning to values of x greater than 20.0 m. The graph is shown in Fig. 13.24.   ◗

   Logarithmic and semilogarithmic paper may be useful for plotting data derived from experimentation. Often the data cover too large a range of values to be plotted on ordinary graph paper. The next example illustrates the use of semilogarithmic paper to plot data.

◀ EXAMPLE 5   The vapor pressure of water depends on the temperature. The following table gives the vapor pressure (in kPa) for the corresponding values of temperature (in °C):

T (°C)	10	20	40	60	80	100	120	140	160
P (kPa)	1.19	2.33	7.34	19.9	47.3	101	199	361	617

These data are then plotted on semilogarithmic paper, as shown in Fig. 13.25. Intermediate values of temperature and pressure can then be read directly from the graph.   ◗

EXERCISES **13.7**

In Exercises 1 and 2, make the given changes in the indicated examples, and then draw the graphs.

  1. In Example 2, change the 4 to 2 and then make the graph.
  2. In Example 3, change the 1 to 4 and then make the graph.

In Exercises 3–10, plot the graphs of the given functions on semilogarithmic paper.

  3. $y = 2^x$
  4. $y = 5^x$
  5. $y = 2(4^x)$
  6. $y = 2^{-x}$
  7. $y = x^3$
  8. $y = 2x^4$
  9. $y = 2x^3 + 4x$
  10. $y = 4x^3 + 2x^2$

In Exercises 11–18, plot the graphs of the given functions on log-log paper.

  11. $y = 0.01x^4$
  12. $y = \sqrt{x}$
  13. $y = x^{2/3}$
  14. $y = x^2 + 2x$
  15. $xy = 4$
  16. $x^2y^3 = 1$
  17. $x^2y^2 = 25$
  18. $x^3y = 8$

In Exercises 19–26, determine the type of graph paper on which the graph of the given function is a straight line. Using the appropriate paper, sketch the graph.

  19. $y = 3^{-x}$
  20. $y = 0.2x^3$
  21. $y = 3x^6$

22. $y = 5(10^{-x})$ **23.** $y = 4^{x/2}$ **24.** $xy^3 = 10$

25. $x\sqrt{y} = 4$ **26.** $y(2^x) = 3$

In Exercises 27–38, plot the indicated graphs.

27. On the moon, the distance s (in m) a rock will fall due to gravity is $s = 0.81t^2$, where t is the time (in s) of fall. Plot the graph of s as a function of t for $0 \le t \le 10$ s on (a) a regular rectangular coordinate system and (b) a semilogarithmic coordinate system.

28. By pumping, the air pressure in a tank is reduced by 18% each second. Thus, the pressure p (in kPa) in the tank is given by $p = 101(0.82)^t$, where t is the time (in s). Plot the graph of p as a function of t for $0 \le t \le 30$ s on (a) a regular rectangular coordinate system and (b) a semilogarithmic coordinate system.

29. Strontium-90 decays according to the equation $N = N_0 e^{-0.028t}$, where N is the amount present after t years and N_0 is the original amount. Plot N as a function of t on semilog paper if $N_0 = 1000$ g.

30. The electric power P (in W) in a certain battery as a function of the resistance R (in Ω) in the circuit is given by $P = \dfrac{100R}{(0.50 + R)^2}$.

Plot P as a function of R on semilog paper, using the logarithmic scale for R and values of R from 0.01 Ω to 10 Ω. Compare the graph with that in Fig. 3.20 on page 95.

31. The acceleration g (in m/s^2) produced by the gravitational force of the earth on a spacecraft is given by $g = 3.99 \times 10^{14}/r^2$, where r is the distance from the center of the earth to the spacecraft. On log-log paper, graph g as a function of r from $r = 6.37 \times 10^6$ m (the earth's surface) to $r = 3.91 \times 10^8$ m (the distance to the moon).

32. In undergoing an adiabatic (no *heat* gained or lost) expansion of a gas, the relation between the pressure p (in kPa) and the volume v (in m^3) is $p^2v^3 = 850$. On log-log paper, graph p as a function of v from $v = 0.10$ m^3 to $v = 10$ m^3.

33. The number of cellular phone subscribers in the United States from 1985 to 2000 is shown in the following table. Plot N as a function of the year on semilog paper.

Year	1985	1988	1991	1994	1997	2000
$N(\times 10^6)$	0.23	2.07	7.56	24.1	55.3	104

34. The period T (in years) and mean distance d (given as a ratio of that of earth) from the sun to the planets (Mercury, Venus, Earth, Mars, Jupiter, Saturn, Uranus, Neptune, Pluto) are given below. Plot T as a function of d on log-log paper.

Planet	M	V	E	M	J	S	U	N	P
d	0.39	0.72	1.00	1.52	5.20	9.54	19.2	30.1	39.5
T	0.24	0.62	1.00	1.88	11.9	29.5	84.0	165	249

35. The intensity level B (in dB) and the frequency (in Hz) for a sound of constant loudness were measured as shown in the table that follows. Plot the data for B as a function of f on semilog paper, using the log scale for f.

f (Hz)	100	200	500	1000	2000	5000	10 000
B (dB)	40	30	22	20	18	24	30

36. The atmospheric pressure p (in kPa) at a given altitude h (in km) is given in the following table. On semilog paper, plot p as a function of h.

h (km)	0	10	20	30	40
p (kPa)	101	25	6.3	2.0	0.53

37. One end of a very hot steel bar is sprayed with a stream of cool water. The rate of cooling R (in °C/s) as a function of the distance d (in cm) from one end of the bar is then measured, with the results shown in the following table. On log-log paper, plot R as a function of d. Such experiments are made to determine the hardness of steel.

d (cm)	0.63	1.3	1.9	2.5
R (°C/s)	600	190	100	72

d (cm)	3.8	5.0	7.5	10	15
R (°C/s)	46	29	17	10	6.0

38. The magnetic intensity H (in A/m) and flux density B (in teslas) of annealed iron are given in the following table. Plot H as a function of B on log-log paper.

B (T)	0.0042	0.043	0.67	1.01
H (A/m)	10	50	100	150

B (T)	1.18	1.44	1.58	1.72
H (A/m)	200	500	1000	10 000

In Exercises 39 and 40, plot the indicated semilogarithmic graphs for the following application.

In a particular electric circuit, called a low-pass filter, the input voltage V_i is across a resistor and a capacitor, and the output voltage V_0 is across the capacitor (see Fig. 13.26). The voltage gain G (in dB) is given by

$$G = 20 \log \frac{1}{\sqrt{1 + (\omega T)^2}}$$

where $\tan \phi = -\omega T.$

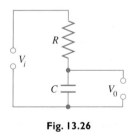

Fig. 13.26

Here, ϕ is the phase angle of V_0/V_i. For values of ωT of 0.01, 0.1, 0.3, 1.0, 3.0, 10.0, 30.0, and 100, plot the indicated graphs. These graphs are called a Bode diagram for the circuit.

39. Calculate values of G for the given values of ωT and plot a semilogarithmic graph of G vs. ωT.

40. Calculate values of ϕ (as negative angles) for the given values of ωT and plot a semilogarithmic graph of ϕ vs. ωT.

CHAPTER ⑬ EQUATIONS

Exponential function	$y = b^x$	(13.1)
Logarithmic form	$x = \log_b y$	(13.2)
Logarithmic function	$y = \log_b x$	(13.3)
Laws of exponents	$b^u b^v = b^{u+v}$	(13.4)
	$\dfrac{b^u}{b^v} = b^{u-v}$	(13.5)
	$(b^u)^n = b^{un}$	(13.6)
Properties of logarithms	$\log_b xy = \log_b x + \log_b y$	(13.7)
	$\log_b\left(\dfrac{x}{y}\right) = \log_b x - \log_b y$	(13.8)
	$\log_b(x^n) = n \log_b x$	(13.9)
	$\log_b 1 = 0 \qquad \log_b b = 1$	(13.10)
	$\log_b(b^n) = n$	(13.11)
Changing base of logarithms	$\log_b x = \dfrac{\log_a x}{\log_a b}$	(13.12)
	$\ln x = \dfrac{\log x}{\log e}$	(13.13)
	$\log x = \dfrac{\ln x}{\ln 10}$	(13.14)

CHAPTER ⑬ REVIEW EXERCISES

In Exercises 1–12, determine the value of x.

1. $\log_{10} x = 4$

2. $\log_9 x = 3$

3. $\log_5 x = -1$

4. $\log_4(\sin x) = -0.5$

5. $\log_2 64 = x$

6. $\log_{12} 144 = x - 3$

7. $\log_8 32 = x$

8. $\log_9 27 = x$

9. $\log_x 36 = 2$

10. $\log_x 243 = 5$

11. $\log_x 10 = \frac{1}{2}$

12. $\log_x 8 = 0$

In Exercises 13–24, express each as a sum, difference, or multiple of logarithms. Wherever possible, evaluate logarithms of the result.

13. $\log_3 2x$

14. $\log_5\left(\dfrac{7}{a}\right)$

15. $\log_3(t^2)$

16. $\log_6 \sqrt{5}$

17. $\log_2 28$

18. $\log_7 98$

19. $\log_3\left(\dfrac{9}{x}\right)$

20. $\log_6\left(\dfrac{5}{36}\right)$

21. $\log_4 \sqrt{48}$

22. $\log_6 \sqrt{72y}$

23. $\log_{10}(1000x^4)$

24. $\log_3(9^2 \times 6^3)$

In Exercises 25–36, solve for y in terms of x.

25. $\log_6 y = \log_6 4 - \log_6 x$

26. $\log_3 y = \frac{1}{2}\log_3 7 + \frac{1}{2}\log_3 x$

27. $(\log_2 3)(\log_2 y) - \log_2 x = 3$

28. $2 \ln y = \ln e^2 - 3 \ln x$

29. $\log_5 x + \log_5 y = \log_5 3 + 1$

30. $\log_7 y = 2 \log_7 5 + \log_7 x + 2$

31. $3 \ln y = 2 + 3 \ln x$

32. $2(\log_9 y + 2 \log_9 x) = 1$

33. $2(\log_4 y - 3 \log_4 x) = 3$

34. $\dfrac{\log_7 x}{\log_7 4} - \log_7 y = 1$

35. $2^y = e^x$

36. $10^y = 3^{x+1}$

In Exercises 37–44, display the graphs of the given functions on a graphing calculator.

37. $y = 0.5(5^x)$

38. $y = 3(2^{-x})$

39. $R = 0.2 \log_4 r$

40. $y = 10 \log_{16} x$

41. $y = \log_{3.15} x$

42. $y = 0.1 \log_{4.05} x$

43. $y = 1 - e^{-|x|}$

44. $s = 2(1 - e^{-0.2t})$

In Exercises 45–48, use logarithms to the base 10 to find the natural logarithms of the given numbers.

45. 8.86

46. 33.0

47. $\sin 2.07$

48. $\sqrt{0.542}$

In Exercises 49–52, use natural logarithms to find logarithms to the base 10 of the given numbers.

49. 65.89

50. 0.0781

51. 0.1197^3

52. $\log 1000$

In Exercises 53–60, solve the given equations.

53. $e^{2x} = 5$

54. $2(5^x) = 15$

55. $3^{x+2} = 5^x$

56. $6^{x+2} = 12^{x-1}$

57. $\log_4 x + \log_4 6 = \log_4 12$

58. $2 \log_3 2 - \log_3(x + 1) = \log_3 5$

59. $\log_8(x + 2) = 2 - \log_8 2$

60. $\log(x + 2) + \log x = 0.4771$

In Exercises 61 and 62, plot the graphs of the given functions on semilogarithmic paper. In Exercises 63 and 64, plot the graphs of the given functions on log-log paper.

61. $y = 6^x$

62. $y = 5x^3$

63. $y = \sqrt[3]{x}$

64. $xy^4 = 16$

If x is eliminated between Eqs. (13.1) and (13.2), we have

$$y = b^{\log_b y} \qquad (13.15)$$

In Exercises 65–68, evaluate the given expressions using Eq. (13.15).

65. $10^{\log 4}$

66. $2e^{\ln 7.5}$

67. $3e^{2\ln 2}$

68. $5(10^{2\log 3})$

In Exercises 69–96, solve the given problems.

69. Use a calculator to verify that $2 \log 3 - \log 6 = \log 1.5$.

70. Use a calculator to verify that $3 \ln 2 + 0.5 \ln 64 = 3 \ln 4$.

71. Evaluate $\sqrt[3]{\ln e^8} - \sqrt{\log 10^4}$.

72. Solve for x: $2^x + 32(2^{-x}) = 12$.

73. If an amount of P dollars is invested at an annual interest rate r (expressed as a decimal), the value V of the investment after

t years is $V = P(1 + r/n)^{nt}$, if interest is compounded n times a year. If \$1000 is invested at an annual interest rate of 6%, compounded semiannually, express V as a function of t and solve for t.

74. The current i (in A) in a certain electric circuit is given by $i = 16(1 - e^{-250t})$, where t is the time (in s). Solve for t.

75. The formula $\ln(I/I_0) = -\beta h$ is used in estimating the thickness of the ozone layer. Here, I_0 is the intensity of a wavelength of sunlight before reaching the earth's atmosphere, I is the intensity of the light after passing through h cm of the ozone layer, and β is a constant. Solve for I.

76. For a radioactive substance with half-life h (the time for half the original amount to decay), an equation relating the time t for N_0 atoms to decay is $h \ln(N_0/N) = t \ln 2$. Solve for N.

77. The bending moment M (in N · m) of a particular concrete column is given by $\log M = 6.663$. What is the value of M?

78. A lottery pays \$500 for a \$1 ticket if a person picks the correct three-digit number determined by the random draw of three numbered balls. The probability of a 50% chance of winning in x drawings is $1 - 0.999^x = 0.5$. For how many drawings does a person have to buy a ticket to have a 50% chance of winning?

79. An approximate formula for the population (in millions) of India since 1995 is $P = 937e^{0.0137t}$, where t is the number of years since 1995. Sketch the graph of P vs. t for 1995 to 2020.

80. A computer analysis of the luminous efficiency E (in lumens/W) of a tungsten lamp as a function of its input power P (in W) is given by $E = 22.0(1 - 0.65e^{-0.008P})$. Sketch the graph of E as a function of P for $0 \le P \le 1000$ W.

81. An original amount of 100 mg of radium radioactively decomposes such that N mg remain after t years. The function relating t and N is $t = 2350(\ln 100 - \ln N)$. Sketch the graph.

82. The time t (in s) to chemically change 5 kg of a certain substance into another is given by $t = -5 \log\left(\dfrac{5 - x}{5}\right)$, where x is the number of kilograms that have been changed at any time. Sketch the graph.

83. An equation that may be used for the angular velocity ω of the slider mechanism in Fig. 13.27 is $2 \ln \omega = \ln 3g + \ln \sin \theta - \ln l$. Solve for $\sin \theta$.

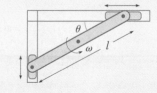

Fig. 13.27

84. Taking into account the weight loss of fuel, the maximum velocity v_m of a rocket is $v_m = u(\ln m_0 - \ln m_s) - gt_f$, where m_0 is the initial mass of the rocket and fuel, m_s is the mass of the rocket shell, t_f is the time during which fuel is expended, u is the velocity of the expelled fuel, and g is the acceleration due to gravity. Solve for m_0.

85. An equation used to calculate the capacity C (in bits/s) of a telephone channel is $C = B \log_2(1 + R)$. Solve for R.

86. An equation used in studying the action of a protein molecule is $\ln A = \ln \theta - \ln(1 - \theta)$. Solve for θ.

87. The magnitudes (visual brightnesses), m_1 and m_2, of two stars are related to their (actual) brightnesses, b_1 and b_2, by the equation $m_1 - m_2 = 2.5 \log(b_2/b_1)$. As a result of this definition, magnitudes may be negative, and *magnitudes decrease as brightnesses increase*. The magnitude of the brightest star, Sirius, is -1.4, and the magnitudes of the faintest stars observable with the naked eye are about 6.0. How much brighter is Sirius than these faintest stars?

88. The power gain of an electronic device such as an amplifier is defined as $n = 10 \log(P_0/P_i)$, where n is measured in decibels, P_0 (in W) is the power output, and P_i (in W) is the power input. If $P_0 = 10.0$ W and $P_i = 0.125$ W, calculate the power gain. (See Example 4 on page 383.)

89. In studying the frictional effects on a flywheel, the revolutions per minute R that it makes as a function of the time t (in min) is given by $R = 4520(0.750)^{2.50t}$. Find t for $R = 1950$ r/min.

90. The efficiency e of a gasoline engine as a function of its compression ratio r is given by $e = 1 - r^{1-\gamma}$, where γ is a constant. Find γ for $e = 0.55$ and $r = 7.5$.

91. The intensity I of light decreases from its value I_0 as it passes a distance x through a medium. Given that $x = k(\ln I_0 - \ln I)$, where k is a constant depending on the medium, find x for $I = 0.850I_0$ and $k = 5.00$ cm.

92. Assume the U.S. national debt D increases at an annual rate of 5% and that the debt in 2000 was \$5.7 trillion. This means $D = 5.7(1.05^t)$, where t is the number of years after 2000. In what year will the debt have doubled?

93. Pure water is running into a brine solution, and the same amount of solution is running out. The number n of kilograms of salt

in the solution after t min is found by solving the equation $\ln n = -0.04t + \ln 20$. Solve for n as a function of t.

94. For the circuit in Fig. 13.28, the current i (in mA) is given by $i = 1.6\, e^{-100t}$. Plot the graph of i as a function of t for the first 0.05 s on semilog paper.

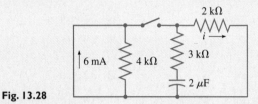

Fig. 13.28

95. For a particular solar-energy system, the collector area A required to supply a fraction F of the total energy is given by $A = 480F^{2.2}$. Plot A (in m^2) as a function of F, from $F = 0.1$ to $F = 0.9$, on semilog paper.

96. The current I (in μA) and resistance R (in Ω) were measured as follows in a certain microcomputer circuit:

$R\ (\Omega)$	100	200	500	1000	2000	5000	10 000
$I\ (\mu\text{A})$	81	41	16	8.2	4.0	1.6	0.8

Plot I as a function of R on log-log paper.

Writing Exercise

97. A machine-design student noted that the edge of a robotic link was shaped like a logarithmic curve. Using a graphing calculator, the student viewed various logarithmic curves, including $y = \log x^2$ and $y = 2 \log x$, for which the student thought the graphs would be identical, but a difference was observed. Write a paragraph explaining what the difference is and why it occurs.

CHAPTER (13) PRACTICE TEST

In Problems 1–4, determine the value of x.

1. $\log_9 x = -\frac{1}{2}$
2. $\log_3 x - \log_3 2 = 2$
3. $\log_x 64 = 3$
4. $3^{3x+1} = 8$
5. Graph the function $y = 2 \log_4 x$.
6. Graph the function $y = 2(3^x)$ on semilog paper.
7. Express $\log_5\left(\dfrac{4a^3}{7}\right)$ as a combination of a sum, difference, and multiple of logarithms, including $\log_5 2$.
8. Solve for y in terms of x: $3 \log_7 x - \log_7 y = 2$.

9. An equation used for a certain electric circuit is $\ln i - \ln I = -t/RC$. Solve for i.
10. Evaluate: $\dfrac{2 \ln 0.9523}{\log 6066}$.
11. Evaluate: $\log_5 732$.
12. If A_0 dollars are invested at 8%, compounded continuously for t years, the value A of the investment is given by $A = A_0 e^{0.08t}$. Determine how long it takes for the investment to double in value.

14 Additional Types of Equations and Systems of Equations

In this chapter, we discuss graphical and algebraic solutions of systems of equations of types different from those of earlier chapters. We also consider solutions of two special types of equations.

One of the methods involves the use of graphs for the solution, similar to that we used in Chapter 3. The intersection of curves was very much a part of a basic method of solution of equations used in the mid-1600s by René Descartes, who developed the coordinate system. Since that time, graphical solutions of equations have been very common and very useful in science and technology.

In the study of optics in the 1800s, it was found that light traveled faster in free space than in other mediums, such as glass and water. Scientists then defined n, the *index of refraction,* to be the ratio of the speed of light in free space to the speed of light in a particular substance. In 1836, the French mathematician Cauchy developed the equation $n = A + B\lambda^{-2} + C\lambda^{-4}$ that related the index of refraction with the wavelength of light λ in a medium. Once the constants A, B, and C were found for a particular medium (by solving three simultaneous equations as in Chapter 5), this equation can be solved for λ for a particular value of n using a method known at the time and is used in this chapter (see Exercise 32 on page 410). Again, we see that an earlier mathematical method was useful in dealing with a new scientific discovery.

The dimensions of a heating system vent are important to its design. In Section 14.1, we discuss a problem in vent design.

Applications of the types of equations and systems of equations of this chapter are found in many fields of science and technology. These include physics, electricity, business, and structural design.

A more complete discussion of the conic sections is found in Chapter 21.

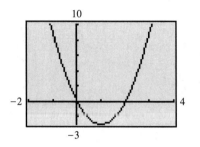

Fig. 14.1

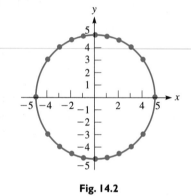

Fig. 14.2

See the *vertical-line test* on page 96.

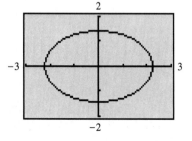

Fig. 14.3

 GRAPHICAL SOLUTION OF SYSTEMS OF EQUATIONS

In this section, we first discuss the graphs of the *circle, parabola, ellipse,* and *hyperbola,* which are known as the **conic sections.** We saw the parabola earlier when discussing quadratic functions in Chapter 7. Then we will find graphical solutions of systems of equations involving these and other nonlinear equations.

◀ **EXAMPLE 1**   Graph the equation $y = 3x^2 - 6x$.

We graphed equations of this form in Section 7.4. Since the general quadratic function is $y = ax^2 + bx + c$, for $y = 3x^2 - 6x$, we have $a = 3$, $b = -6$, and $c = 0$. Therefore, $-b/(2a) = 1$, which means the x-coordinate of the vertex is $-(-6)/6 = 1$. Since $y = -3$ for $x = 1$, the vertex is $(1, -3)$. It is a minimum point since $a > 0$.

Knowing the vertex and the fact that the graph goes through the origin ($c = 0$), we choose appropriate *window* settings and have the display shown in Fig. 14.1.

As we showed in Section 7.4, the curve is a **parabola,** and a parabola always results if the equation is of the form of the quadratic function $y = ax^2 + bx + c$. ◗

◀ **EXAMPLE 2**   Plot the graph of the equation $x^2 + y^2 = 25$.

We first solve this equation for y, and we obtain $y = \sqrt{25 - x^2}$, or $y = -\sqrt{25 - x^2}$, which we write as $y = \pm\sqrt{25 - x^2}$. We now assume values for x and find the corresponding values for y.

x	0	±1	±2	±3	±4	±5
y	±5	±4.9	±4.6	±4	±3	0

If $x > 5$, values of y are imaginary. We cannot plot these because x and y must both be real. When we show $y = \pm3$ for $x = \pm4$, this is a short way of representing four points. These points are $(4, 3)$, $(4, -3)$, $(-4, 3)$, and $(-4, -3)$.

In Fig. 14.2, *the resulting curve is a* **circle.** A circle with its center at the origin results from an equation of the form $x^2 + y^2 = r^2$, where r is the radius. ◗

From the graph of the circle in Fig. 14.2, we see that *the equation of a circle does not represent a function.* There are *two* values of y for most of the values of x in the domain. We must take this into account when displaying the graph of such an equation on a graphing calculator. This is illustrated in the next example.

◀ **EXAMPLE 3**   Display the graph of the equation $2x^2 + 5y^2 = 10$ on a graphing calculator.

First solving for y, we get $y = \pm\sqrt{\dfrac{10 - 2x^2}{5}}$. To display the graph of this equation on a calculator, *we must enter both functions,* one as $y_1 = \sqrt{(10 - 2x^2)/5}$, and the other as $y_2 = -\sqrt{(10 - 2x^2)/5}$.

Trying some *window* settings $\left(\text{or noting that the domain is from } -\sqrt{5} \text{ to } \sqrt{5} \text{ and the range is from } -\sqrt{2} \text{ to } \sqrt{2}\right)$, we get the graphing calculator display that is shown in Fig. 14.3.

The curve is an **ellipse.** An ellipse will be the resulting curve if the equation is of the form $ax^2 + by^2 = c$, where the constants a, b, and c have the same sign, and for which $a \neq b$. ◗

When these types of curves are displayed on the graphing calculator, it is possible that there are small gaps in the curve. The appearance of such gaps depends on the pixel width and the functional values used by the calculator.

◀ EXAMPLE 4 Display the graph of $2x^2 - y^2 = 4$ on a graphing calculator.
Solving for y, we get

$$y = \pm\sqrt{2x^2 - 4}$$

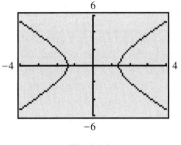

Fig. 14.4

As in Example 3, we enter two functions in the calculator, one with the plus sign and the other with the minus sign, and we have the display shown in Fig. 14.4. The *window* settings are chosen by noting that the values $-\sqrt{2} < x < \sqrt{2}$ are not in the domain of either $y_1 = \sqrt{2x^2 - 4}$ or $y_2 = -\sqrt{2x^2 - 4}$. These values of x would lead to imaginary values of y.

The curve is a **hyperbola,** which results when we have an equation of the form $ax^2 + by^2 = c$, if a and b have *different* signs. ▶

SOLVING SYSTEMS OF EQUATIONS

As in solving systems of linear equations, we solve any system by finding the values of x and y that satisfy both equations at the same time. To solve a system graphically, we graph the equations and find the coordinates of all points of intersection. If the curves do not intersect, the system has no real solutions.

Solving a Word Problem

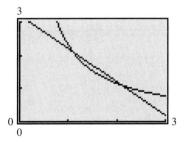

Fig. 14.5

See the chapter introduction.

◀ EXAMPLE 5 For proper ventilation, the vent for a hot-air heating system is to have a rectangular cross-sectional area of 2.3 m² and is to be made from sheet metal 6.4 m wide. Find the dimensions of this cross-sectional area of the vent.

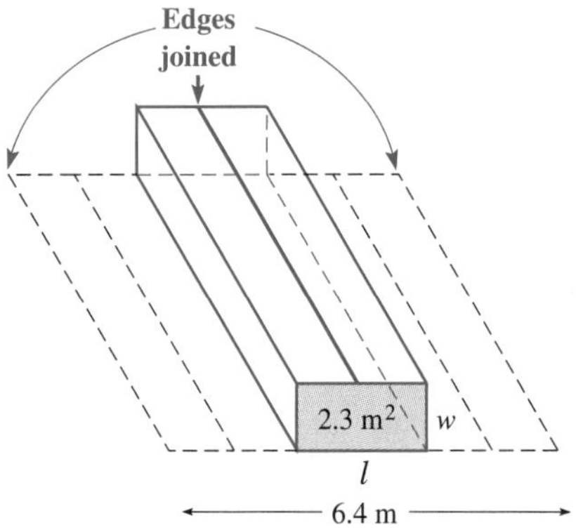

In Fig. 14.5, we have let $l =$ the length and $w =$ the width of the area. Since the area is 2.3 m², we have $lw = 2.3$. Also, since the sheet metal is 6.4 m wide, this is the perimeter of the area. This gives us $2l + 2w = 6.4$, or $l + w = 3.2$. This means that the system of equations to be solved is

$$lw = 2.3$$
$$l + w = 3.2$$

Solving each equation for l, we have

$$l = 2.3/w \quad \text{and} \quad l = 3.2 - w$$

We now display the graphs of these two equations on a graphing calculator, using x for w and y for l. Since negative values of l and w have no meaning to the solution and since the straight line $l = 3.2 - w$ has intercepts of $(3.2, 0)$ and $(0, 3.2)$, we choose the *window* settings as shown in Fig. 14.6.

Using the *intersect* feature (or the *trace* and *zoom* features) with the graph in Fig. 14.6, we find that the solutions are approximately $(1.1, 2.1)$ and $(2.1, 1.1)$. Using the length as the longer dimension, we have the solution of

$$l = 2.1 \text{ m} \quad \text{and} \quad w = 1.1 \text{ m}$$

We see that this checks with the statement of the problem. ▶

In Example 5 we graphed the equation $xy = 2.3$ (having used x for w and y for l). The graph of this equation is also a *hyperbola,* another form of which is $xy = c$. We now show the solutions of two more systems of equations.

Fig. 14.6

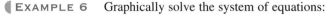

EXAMPLE 6 Graphically solve the system of equations:

$$9x^2 + 4y^2 = 36$$
$$y = 3^x$$

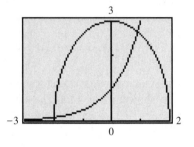

Fig. 14.7

The first equation is of the form represented by an ellipse, as shown in Example 3. The second equation is an exponential function, as discussed in Chapter 13. Solving the first equation for y, we have $y = \pm\frac{1}{2}\sqrt{36 - 9x^2}$. Since this is an ellipse, we note that its domain extends from $x = -2$ to $x = 2$ ($9x^2$ cannot be greater than 36). Since the exponential curve cannot be negative and increases rapidly, we need only graph the upper part of the ellipse (using the $+$ sign). With these considerations, we choose the *window* settings as shown in the calculator display in Fig. 14.7.

The points of intersection give the approximate solutions of $x = -2.0$, $y = 0.1$ and $x = 0.9$, $y = 2.7$.

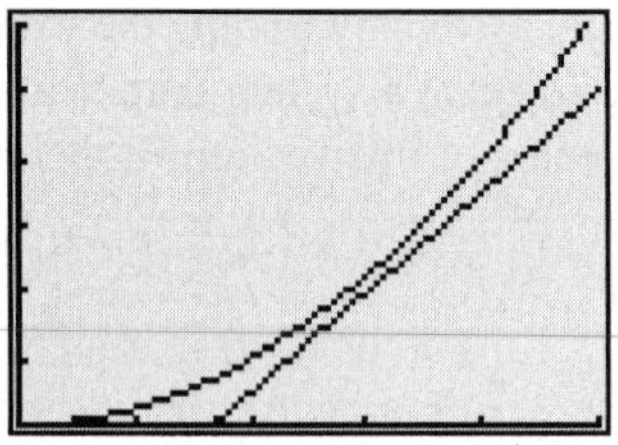

EXAMPLE 7 Graphically solve the system of equations:

$$x^2 = 2y$$
$$3x - y = 5$$

We note that the two curves in this system are a parabola and a straight line. Solving the equation of the parabola for y, we get $y = \frac{1}{2}x^2$. This parabola has its vertex at the origin, and it opens upward since $a > 0$. This means that it is not possible to have a point of intersection for $y < 0$.

The straight line has intercepts of $(0, -5)$ and $(5/3, 0)$. This means that any possible point of intersection must be in the first quadrant. With these considerations, we have the *window* settings as shown in the calculator display in Fig. 14.8.

The curves do not intersect, which means that there are no real solutions to the system of equations.

Fig. 14.8

EXERCISES 14.1

In Exercises 1–4, make the given changes in the indicated examples of this section, and then perform the indicated operations.

1. In Example 1, change the $-$ sign before $6x$ to $+$ and then graph the equation.

2. In Example 2, change the $+$ sign before y^2 to $-$ and then graph the equation.

3. In Example 6, change the coefficient of x^2 to 25 and then solve the system of equations.

4. In Example 7, change the coefficient of y in the first equation to 3 and then solve the system of equations.

In Exercises 5–30, solve the given systems of equations graphically by using a graphing calculator. Find all values to at least the nearest 0.1.

5. $y = 2x$
$x^2 + y^2 = 16$

6. $3x - y = 4$
$y = 6 - 2x^2$

7. $x^2 + 2y^2 = 8$
$x - 2y = 4$

8. $y = 3x - 6$
$xy = 6$

9. $y = x^2 - 2$
$4y = 12x - 17$

10. $4x^2 + 25y^2 = 21$
$10y = 31 - 9x$

11. $8y = 11x^2$
$xy = 30$

12. $y = -2x^2$
$y = x^2 - 6$

13. $y = -x^2 + 4$
$x^2 + y^2 = 9$

14. $y = 2x^2 - 1$
$x^2 + 2y^2 = 16$

15. $x^2 - 4y^2 = 16$
$x^2 + y^2 = 1$

16. $y = 2x^2 - 4x$
$xy = -4$

17. $2x^2 + 3y^2 = 19$
$x^2 + y^2 = 9$

18. $x^2 - y^2 = 4$
$2x^2 + y^2 = 16$

19. $x^2 + y^2 = 1$
$xy = 0.6$

20. $x^2 + y^2 = 0.25$
$x^2 - y^2 = 0.07$

21. $y = x^2$
$y = \sin x$

22. $y = 4x - x^2$
$y = 2 \cos x$

23. $y = e^{-x}$
$x + y = 2$

24. $y = 2^x$
$x^2 + y^2 = 4$

25. $x^2 - y^2 = 1$
$y = \log_2 x$

26. $x^2 + 4y^2 = 16$
$y = 2 \ln x$

27. $y = \ln(x - 1)$
$y = \sin \frac{1}{2}x$

28. $y = \cos x$
$y = \log_3 x$

29. $10^{x+y} = 150$
$y = x^2$

30. $e^{x^2+y^2} = 20$
$xy = 4$

In Exercises 31–36, set up the indicated systems of equations and solve them graphically.

31. A helicopter is located 5.2 km north of east of a radio tower such that it is three times as far north as it is east from the tower. Find the northern and eastern components of the displacement from the tower.

32. A 4.60-m insulating strip is placed completely around a rectangular solar panel with an area of 1.20 m². What are the dimensions of the panel?

33. The power developed in an electric resistor is i^2R, where i is the current. If a first current passes through a 2.0-Ω resistor and a second current passes through a 3.0-Ω resistor, the total power produced is 12 W. If the resistors are reversed, the total power produced is 16 W. Find the currents (in A) if ($i > 0$).

(W) 34. A circular hot tub is located on the square deck of a home. The side of the deck is 7.30 m more than the radius of the hot tub, and there are 72.5 m² of deck around the tub. Find the radius of the hot tub and the length of the side of the deck. Explain your answer.

35. Assume earth is a sphere, with $x^2 + y^2 = 41$ as the equation of a circumference (distance in thousands of km). If a meteorite approaching earth has a path described as $y^2 = 20x + 140$, will the meteorite strike earth? If so, where?

36. Two people meet at the intersection of two perpendicular roads. Each then leaves along a different road with one walking 1.0 km/h faster than the other. If they are 7.0 km apart (on a direct line) after 1.0 h, how fast is each walking?

14.2 ALGEBRAIC SOLUTION OF SYSTEMS OF EQUATIONS

Often the graphical method is the easiest way to solve a system of equations. With a graphing calculator, it is possible to find the result with good accuracy. However, the graphical method does not usually give the *exact* answer. Using algebraic methods to find exact solutions for some systems of equations is either not possible or quite involved. There are systems, however, for which there are relatively simple algebraic solutions. In this section, we consider two useful methods, both of which we discussed before when we were studying systems of linear equations.

SOLUTION BY SUBSTITUTION

The first method is *substitution*. If we can solve one of the equations for one of its variables, we can substitute this solution into the other equation. We then have only one unknown in the resulting equation, and we can then solve this equation by methods discussed in earlier chapters.

◀ **EXAMPLE 1** By substitution, solve the system of equations

$$2x - y = 4$$
$$x^2 - y^2 = 4$$

We solve the first equation for y, obtaining $y = 2x - 4$. We now substitute $2x - 4$ for y in the second equation, getting

$$x^2 - (2x - 4)^2 = 4 \quad \text{in second equation, } y \text{ replaced by } 2x - 4$$

When simplified, this gives a quadratic equation.

$$x^2 - (4x^2 - 16x + 16) = 4$$
$$-3x^2 + 16x - 20 = 0$$
$$x = \frac{-16 \pm \sqrt{256 - 4(-3)(-20)}}{-6} = \frac{-16 \pm \sqrt{16}}{-6} = \frac{-16 \pm 4}{-6} = \frac{10}{3}, 2$$

We now find the corresponding values of y by substituting into $y = 2x - 4$. Thus, we have the solutions $x = \frac{10}{3}$, $y = \frac{8}{3}$, and $x = 2$, $y = 0$. As a check, we find that these values also satisfy the equation $x^2 - y^2 = 4$. Compare these solutions with those that would be obtained from Fig. 14.9.

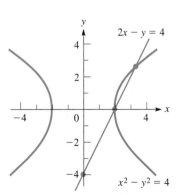

Fig. 14.9

◀ EXAMPLE 2 By substitution, solve the system of equations

$$xy = -2$$
$$2x + y = 2$$

From the first equation we have $y = -2/x$. Substituting this into the second equation, we have

in second equation, y replaced by $-\dfrac{2}{x}$

$$2x + \left(-\frac{2}{x}\right) = 2$$
$$2x^2 - 2 = 2x$$
$$x^2 - x - 1 = 0$$
$$x = \frac{1 \pm \sqrt{1 + 4}}{2} = \frac{1 \pm \sqrt{5}}{2}$$

By substituting these values for x into either of the original equations, we find the corresponding values of y, and we have the solutions

$$x = \frac{1 + \sqrt{5}}{2}, y = 1 - \sqrt{5} \quad \text{and} \quad x = \frac{1 - \sqrt{5}}{2}, y = 1 + \sqrt{5}$$

These can be checked by substituting in the original equations. In decimal form they are

$$x \approx 1.618, y \approx -1.236 \quad \text{and} \quad x \approx -0.618, y \approx 3.236$$

The graphical solutions are shown in Fig. 14.10.

The solutions can also be found by first solving the second equation for y, or either equation for x, and then substituting in the other equation. ▶

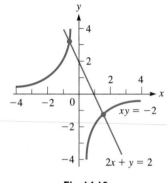

Fig. 14.10

SOLUTION BY ADDITION OR SUBTRACTION

The other algebraic method is that of elimination by *addition or subtraction*. This method is most useful if both equations have only squared terms and constants.

◀ EXAMPLE 3 By addition or subtraction, solve the system of equations

$$2x^2 + y^2 = 9$$
$$x^2 - y^2 = 3$$

We note that if we add the corresponding sides of each equation, y^2 is eliminated. This leads to the solution.

$$2x^2 + y^2 = 9$$
$$\underline{x^2 - y^2 = 3}$$
$$3x^2 \qquad = 12 \qquad \text{add}$$
$$x^2 = 4$$
$$x = \pm 2$$

For $x = 2$, we have two corresponding y-values, $y = \pm 1$. Also, for $x = -2$, we have two corresponding y-values, $y = \pm 1$. Thus we have four solutions:

$$x = 2, y = 1 \qquad x = 2, y = -1 \qquad x = -2, y = 1 \qquad x = -2, y = -1$$

Each solution checks in the original equations. The graphical solutions are shown in Fig. 14.11. ▶

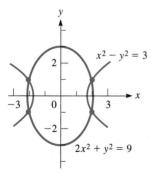

Fig. 14.11

EXAMPLE 4   By addition or subtraction, solve the system of equations

$$3x^2 - 2y^2 = 5$$
$$x^2 + y^2 = 5$$

If we multiply the second equation by 2 and then add the two resulting equations, we get

$$
\begin{aligned}
3x^2 - 2y^2 &= 5 \\
\underline{2x^2 + 2y^2} &= 10 \qquad \text{each term of second equation multiplied by 2} \\
5x^2 \qquad\;\; &= 15 \qquad \text{add} \\
x^2 = 3, \qquad x &= \pm\sqrt{3}
\end{aligned}
$$

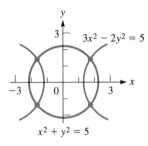

$3x^2 - 2y^2 = 5$

$x^2 + y^2 = 5$

Fig. 14.12

The corresponding values of y for each value of x are $y = \pm\sqrt{2}$. Again, we have four solutions:

$$x = \sqrt{3}, y = \sqrt{2} \qquad x = \sqrt{3}, y = -\sqrt{2}$$
$$x = -\sqrt{3}, y = \sqrt{2} \qquad x = -\sqrt{3}, y = -\sqrt{2}$$

Each solution checks when substituted in the original equations. The graphical solutions are shown in Fig. 14.12.

Solving a Word Problem

EXAMPLE 5   A certain number of machine parts cost \$1000. If they cost \$5 less per part, ten additional parts could be purchased for the same amount of money. What is the cost of each part?

Since the cost of each part is required, we let $c =$ the cost per part. Also, we let $n =$ the number of parts. From the first statement of the problem, we see that $cn = 1000$. Also, from the second statement, we have $(c - 5)(n + 10) = 1000$. Therefore, we are to solve the system of equations

$$cn = 1000$$
$$(c - 5)(n + 10) = 1000$$

Solving the first equation for n and multiplying out the second equation, we have $n = \dfrac{1000}{c}$ and $cn + 10c - 5n - 50 = 1000$. Now, substituting the expression for n into the second equation, we solve for c.

$$c\left(\frac{1000}{c}\right) + 10c - 5\left(\frac{1000}{c}\right) - 50 = 1000$$

$$1000 + 10c - \frac{5000}{c} - 50 = 1000$$

$$10c - \frac{5000}{c} - 50 = 0$$

$$c^2 - 5c - 500 = 0$$

$$(c + 20)(c - 25) = 0$$

$$c = -20, 25$$

Since a negative answer has no significance in this particular situation, we see that the solution is $c = \$25$ per part. Checking with the original statement of the problem, we see that this is correct.

EXERCISES 14.2

In Exercises 1–4, make the given changes in the indicated examples of this section and then solve the resulting systems of equations.

1. In Example 1, change the sign before y in the first equation from $-$ to $+$ and then solve the system.

2. In Example 2, change the right side of the second equation from 2 to 3 and then solve the system.

3. In Example 3, change the coefficient of x^2 in the first equation from 2 to 1 and then solve the system.

4. In Example 4, change the left side of the first equation from $3x^2 - 2y^2$ to $2x^2 - 3y^2$ and then solve the system.

In Exercises 5–28, solve the given systems of equations algebraically.

5. $y = x + 1$
   $y = x^2 + 1$

6. $y = 2x - 1$
   $y = 2x^2 + 2x - 3$

7. $x + 2y = 3$
   $x^2 + y^2 = 26$

8. $s = t + 1$
   $t^2 + s^2 = 25$

9. $x + y = 1$
   $x^2 - y^2 = 1$

10. $x + y = 2$
    $2x^2 - y^2 = 1$

11. $2x - y = 2$
    $2x^2 + 3y^2 = 4$

12. $6y - x = 6$
    $x^2 + 3y^2 = 36$

13. $wh = 1$
    $w + h = 2$

14. $xy = 100$
    $x + y = 20$

15. $xy = 3$
    $3x - 2y = -7$

16. $xy = -4$
    $2x + y = -2$

17. $y = x^2$
    $y = 3x^2 - 50$

18. $M = L^2 - 1$
    $2L^2 - M^2 = 2$

19. $x^2 - y = -1$
    $x^2 + y^2 = 5$

20. $x^2 + y = 5$
    $x^2 + y^2 = 25$

21. $D^2 - 1 = R$
    $D^2 - 2R^2 = 1$

22. $2y^2 - 4x = 7$
    $y^2 + 2x^2 = 3$

23. $x^2 + y^2 = 25$
    $x^2 - 2y^2 = 7$

24. $3x^2 - y^2 = 4$
    $x^2 + 4y^2 = 10$

25. $x^2 + 3y^2 = 37$
    $2x^2 - 9y^2 = 14$

26. $5x^2 - 4y^2 = 15$
    $3y^2 + 4x^2 = 12$

27. $x^2 + y^2 + 4x = 1$
    $x^2 + y^2 - 2y = 9$

28. $x^2 + y^2 - 4x - 2y + 4 = 0$
    $x^2 + y^2 - 2x - 4y + 4 = 0$

(Hint for Exercises 27 and 28: First subtract one equation from the other to get an equation relating x and y. Then substitute this equation in either given equation.)

In Exercises 29–40, solve the indicated systems of equations algebraically. In Exercises 31–40, it is necessary to set up the systems of equations properly.

29. A rocket is fired from behind a ship and follows the path given by $h = 3x - 0.05x^2$, where h is its altitude (in km) and x is the horizontal distance traveled (in km). A missile fired from the ship fol-

lows the path given by $h = 0.8x - 15$. For $h > 0$ and $x > 0$, find where the paths of the rocket and missile cross.

30. A 2-kg block collides with an 8-kg block. Using the physical laws of conservation of energy and conservation of momentum, along with given conditions, the following equations involving the velocities are established:
    $$v_1^2 + 4v_2^2 = 41$$
    $$2v_1 + 8v_2 = 12$$
    Find these velocities (in m/s) if $v_2 > 0$.

31. A rectangular computer chip has a surface area of 2.1 cm^2 and a perimeter of 5.8 cm. Find the length and the width of the chip.

32. The impedance Z in an alternating-current circuit is 2.00 Ω. If the resistance R is numerically equal to the square of the reactance X, find R and X. See Section 12.7.

33. A roof truss is in the shape of a right triangle. If there are 4.60 m of lumber in the truss and the longest side is 2.20 m long, what are the lengths of the other two sides of the truss?

34. In a certain roller mechanism, the radius of one steel ball is 2.00 cm greater than the radius of a second steel ball. If the difference in their masses is 7100 g, find the radii of the balls. The density of steel is 7.70 g/cm^3.

35. A set of equal electrical resistors in series has a total resistance (the sum of the resistances) of 78.0 Ω. Another set of two fewer equal resistors in series also has a total resistance of 78.0 Ω. If each resistor in the second set is 1.3 Ω greater than each of the first set, how many are in each set?

36. Security fencing encloses a rectangular storage area of 1600 m^2 that is divided into two sections by additional fencing parallel to the shorter sides. Find the dimensions of the storage area if 220 m of fencing are used.

37. Squares 4.0 cm on a side are removed from each corner of a rectangular sheet of metal of area 560 cm^2. If the flaps are then bent up to make an open tray with a volume of 960 cm^3, what are the dimensions of the original metal sheet?

38. Two guy wires, one 140 m long and the other 120 m long, are attached at the same point of a TV tower with the longer one secured in the (level) ground 30 m farther from the base of the tower than the shorter one. How high up on the tower are they attached?

39. A jet travels at 990 km/h relative to the air. It takes the jet 1.6 h longer to travel the 5900 km from London to Washington, D.C., against the wind than it takes from Washington to London with the wind. Find the velocity of the wind.

40. In a marketing survey, a company found that the total gross income for selling t tables at a price of p dollars each was \$35 000. It then increased the price of each table by \$100 and found that the total income was only \$27 000 because 40 fewer tables were sold. Find p and t.

14.3 EQUATIONS IN QUADRATIC FORM

Often we encounter equations that can be solved by methods applicable to quadratic equations, even though these equations are not actually quadratic. They do have the property, however, that *with a proper substitution* **they may be written in the form of a quadratic equation.** All that is necessary is that the equation have terms including some variable quantity, its square, and perhaps a constant term. The following example illustrates these types of equations.

NOTE ▶

▌**EXAMPLE 1** **(a)** The equation $x - 2\sqrt{x} - 5 = 0$ is an equation in quadratic form, because if we let $y = \sqrt{x}$ we have $x = \left(\sqrt{x}\right)^2 = y^2$, and the resulting equation is $y^2 - 2y - 5 = 0$.

(b) $t^{-4} - 5t^{-2} + 3 = 0$

$$\overset{\displaystyle \lceil\ (t^{-2})^2}{}$$

By letting $y = t^{-2}$, we have $y^2 - 5y + 3 = 0$.

(c) $t^3 - 3t^{3/2} - 7 = 0$

$$\overset{\displaystyle \lceil\ (t^{3/2})^2}{}$$

By letting $y = t^{3/2}$, we have $y^2 - 3y - 7 = 0$.

(d) $(x + 1)^4 - (x + 1)^2 - 1 = 0$

$$\overset{\displaystyle \lceil\ [(x + 1)^2]^2}{}$$

By letting $y = (x + 1)^2$, we have $y^2 - y - 1 = 0$.

(e) $x^{10} - 2x^5 + 1 = 0$

$$\overset{\displaystyle \lceil\ (x^5)^2}{}$$

By letting $y = x^5$, we have $y^2 - 2y + 1 = 0$. ▶

The following examples illustrate the method of solving equations in quadratic form.

▌**EXAMPLE 2** Solve the equation $2x^4 + 7x^2 = 4$.

We first let $y = x^2$ to write the equation in quadratic form. We will then solve the resulting quadratic equation for y. However, solutions for x are required, so we again let $y = x^2$ to solve for x.

$$2y^2 + 7y - 4 = 0 \qquad \text{let } y = x^2$$
$$(2y - 1)(y + 4) = 0 \qquad \text{factor and solve for } y$$
$$y = \tfrac{1}{2} \quad \text{or} \quad y = -4$$
$$x^2 = \tfrac{1}{2} \quad \text{or} \quad x^2 = -4 \qquad y = x^2 \text{ (to solve for } x\text{)}$$
$$x = \pm\frac{1}{\sqrt{2}} \quad \text{or} \quad x = \pm 2j$$

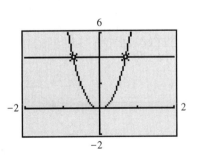

Fig. 14.13

We can let $y_1 = 2x^4 + 7x^2$ and $y_2 = 4$ to solve the system on a graphing calculator. The display is shown in Fig. 14.13, and we see that there are two points of intersection, one for $x = 0.7071$ and the other for $x = -0.7071$. Since $1/\sqrt{2} = 0.7071$, this verifies the real solutions. The imaginary solutions cannot be found graphically. Substitution of each value *in the original equation* shows each value to be a solution. ▶

Two of the solutions in Example 2 are complex numbers. We were able to find these solutions directly from the definition of the square root of a negative number. In some cases (see Exercise 22 of this section), it is necessary to use the method of Section 12.6 to find such complex number solutions.

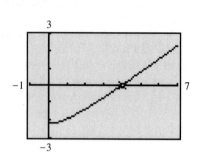

Fig. 14.14

◀ EXAMPLE 3 Solve the equation $x - \sqrt{x} - 2 = 0$.

By letting $y = \sqrt{x}$, we have

$$y^2 - y - 2 = 0$$
$$(y - 2)(y + 1) = 0$$
$$y = 2 \quad \text{or} \quad y = -1$$

Since $y = \sqrt{x}$, we note that y cannot be negative, and this means $y = -1$ cannot lead to a solution. For $y = 2$, we have $x = 4$. Checking, we find that $x = 4$ satisfies the original equation. Therefore, the only solution is $x = 4$.

The graph of $y_1 = x - \sqrt{x} - 2$ is shown in the graphing calculator display in Fig. 14.14. Note that $x = 4$ is the only solution shown. ▶

Extraneous Roots

CAUTION ▶

Example 3 illustrates a very important point. *Whenever an operation involving the unknown is performed on an equation, this operation may introduce roots into a subsequent equation that are not roots of the original equation. Therefore, we must* ***check all answers in the original equation.*** Only operations involving constants—that is, adding, subtracting, multiplying by, or dividing by constants—are certain not to introduce the **extraneous roots.** We first encountered the concept of an extraneous root in Section 6.8, when we discussed equations involving fractions.

◀ EXAMPLE 4 Solve the equation $x^{-2} + 3x^{-1} + 1 = 0$.

By substituting $y = x^{-1}$, we have $y^2 + 3y + 1 = 0$. To solve this equation, we may use the quadratic formula:

$$y = \frac{-3 \pm \sqrt{9 - 4}}{2} = \frac{-3 \pm \sqrt{5}}{2}$$

Since $x = 1/y$, we have

$$x = \frac{2}{-3 + \sqrt{5}} \quad \text{or} \quad x = \frac{2}{-3 - \sqrt{5}}$$

These answers in decimal form are

$$x \approx -2.618 \quad \text{or} \quad x \approx -0.382$$

These results check when substituted in the original equation. In checking these decimal answers, it is more accurate to use the calculator values, before rounding them off. This can be done by storing the calculator values in memory. ▶

◀ EXAMPLE 5 Solve the equation $(x^2 - x)^2 - 8(x^2 - x) + 12 = 0$.

By substituting $y = x^2 - x$, we have

$$y^2 - 8y + 12 = 0$$
$$(y - 2)(y - 6) = 0$$
$$y = 2 \quad \text{or} \quad y = 6$$
$$x^2 - x = 2 \quad \text{or} \quad x^2 - x = 6 \qquad {\scriptstyle y = x^2 - x}$$

Solving each of these equations, we have

$$\begin{array}{ll} x^2 - x - 2 = 0 & x^2 - x - 6 = 0 \\ (x - 2)(x + 1) = 0 & (x - 3)(x + 2) = 0 \\ x = 2 \quad \text{or} \quad x = -1 & x = 3 \quad \text{or} \quad x = -2 \end{array}$$

Each value checks when substituted in the original equation. ▶

Solving a Word Problem ❪ EXAMPLE 6   A rectangular photograph has an area of 120 cm². The diagonal of the photograph is 17 cm. Find its length and width. See Fig. 14.15.

Since the required quantities are the length and width, let l = the length of the photograph and let w = its width. Since the area is 120 cm², $lw = 120$. Also, using the Pythagorean theorem and the fact that the diagonal is 17 cm, we have the equation $l^2 + w^2 = 17^2 = 289$. Therefore, we are to solve the system of equations

$$lw = 120 \qquad l^2 + w^2 = 289$$

Solving the first equation for l, we have $l = 120/w$. Substituting this into the second equation, we have

$$\left(\frac{120}{w}\right)^2 + w^2 = 289$$

$$\frac{14\,400}{w^2} + w^2 = 289$$

$$14\,400 + w^4 = 289w^2$$

Fig. 14.15

Let $x = w^2$.

$$x^2 - 289x + 14\,400 = 0$$

$$x = \frac{-(-289) \pm \sqrt{(-289)^2 - 4(1)(14\,400)}}{2(1)}$$

$$x = 225 \quad \text{or} \quad x = 64$$

The first permanent photograph was taken in 1826 by the French inventor Joseph Niepce (1765–1833).

Therefore, $w^2 = 225$ or $w^2 = 64$.

Solving for w, we get $w = \pm15$ or $w = \pm8.0$. Only the positive values are meaningful in this problem, which means if $w = 8.0$ cm, then $l = 15$ cm. Normally, we designate the longer dimension as the length (although by letting $w = 15$ cm, we get $l = 8.0$ cm). Checking with the statement of the problem, we see that these dimensions for the photograph give an area of 120 cm² and a diagonal of 17 cm.   ◗

EXERCISES ❨14.3❩

In Exercises 1 and 2, make the given changes in the indicated examples of this section and then solve the resulting equations.

1. In Example 2, change the + before the $7x^2$ to − and then solve the equation.

2. In Example 3, change the 2 to 6 and then solve the equation.

In Exercises 3–26, solve the given equations algebraically. In Exercise 10, explain your method.

3. $x^4 - 13x^2 + 36 = 0$ **4.** $4R^4 + 15R^2 = 4$

5. $x^{-2} - 2x^{-1} - 8 = 0$ **6.** $10x^{-2} + 3x^{-1} - 1 = 0$

7. $x^{-4} + 2x^{-2} = 24$ **8.** $x^{-4} + 1 = 2x^{-2}$

9. $2x - 7\sqrt{x} + 5 = 0$ ⓦ **10.** $4x + 3\sqrt{x} = 1$

11. $3\sqrt[3]{x} - 5\sqrt[6]{x} + 2 = 0$ **12.** $\sqrt{x} + 3\sqrt[4]{x} = 28$

13. $x^{2/3} - 2x^{1/3} - 15 = 0$ **14.** $x^3 + 2x^{3/2} - 80 = 0$

15. $2n^{1/2} - 5n^{1/4} = 3$ **16.** $4x^{4/3} + 9 = 13x^{2/3}$

17. $(x - 1) - \sqrt{x - 1} - 2 = 0$

18. $(C + 1)^{-2/3} + 5(C + 1)^{-1/3} - 6 = 0$

19. $(x^2 - 2x)^2 - 11(x^2 - 2x) + 24 = 0$

20. $(x^2 - 1)^2 + (x^2 - 1)^{-2} = 2$

21. $x - 3\sqrt{x - 2} = 6$ $\left(\text{Let } y = \sqrt{x - 2}.\right)$

22. $x^6 + 7x^3 - 8 = 0$

23. $\dfrac{1}{s^2 + 1} + \dfrac{2}{s^2 + 3} = 1$ **24.** $\left(x + \frac{2}{x}\right)^2 - 6x - \frac{12}{x} = -9$

25. $e^{2x} - e^x = 0$ **26.** $10^{2x} - 2(10^x) = 0$

In Exercises 27–30, solve the given equations algebraically and check the solutions with a graphing calculator.

27. $x^4 - 20x^2 + 64 = 0$ **28.** $x^{-2} - x^{-1} - 42 = 0$

29. $x + 2 = 3\sqrt{x}$ **30.** $x^{2/3} - 4x^{1/3} = 12$

In Exercises 31–36, solve the given problems algebraically.

31. The equivalent resistance R_T of two resistors R_1 and R_2 in parallel is given by $R_T^{-1} = R_1^{-1} + R_2^{-1}$. If $R_T = 1.00 \ \Omega$ and $R_2 = \sqrt{R_1}$, find R_1 and R_2.

32. An equation used in the study of the dispersion of light is $\mu = A + B\lambda^{-2} + C\lambda^{-4}$. Solve for λ. (See the chapter introduction.)

33. In the theory dealing with optical interferometers, the equation $\sqrt{F} = 2\sqrt{p}/(1 - p)$ is used. Solve for p if $F = 16$.

34. A special washer is made from a circular disc 3.50 cm in radius by removing a rectangular area of 12.0 cm² from the center. If each corner of the rectangular area is 0.50 cm from the outer edge of the washer, what are the dimensions of the area that is removed?

35. A rectangular TV screen has an area of 2240 cm² and a diagonal of 68.6 cm. Find the dimensions of the screen.

36. A roof truss in the shape of a right triangle has a perimeter of 90 m. If the hypotenuse is 1 m longer than one of the other sides, what are the sides of the truss?

14.4 EQUATIONS WITH RADICALS

Equations with radicals in them are normally solved by squaring both sides of the equation if the radical represents a square root or by a similar operation for the other roots. However, when we do this, we often introduce *extraneous roots*. Thus, it is very important that *all solutions be checked in the original equation.*

NOTE ▶

EXAMPLE 1 Solve the equation $\sqrt{x - 4} = 2$.

By squaring both sides of the equation, we have

$$\left(\sqrt{x - 4}\right)^2 = 2^2$$
$$x - 4 = 4$$
$$x = 8$$

This solution checks when put into the original equation.

EXAMPLE 2 Solve the equation $2\sqrt{3x - 1} = 3x$.

Squaring both sides of the equation gives us

$$\left(2\sqrt{3x - 1}\right)^2 = (3x)^2 \leftarrow \text{don't forget to square the 2}$$
$$4(3x - 1) = 9x^2$$
$$12x - 4 = 9x^2$$
$$9x^2 - 12x + 4 = 0$$
$$(3x - 2)^2 = 0$$
$$x = \frac{2}{3} \qquad \text{(double root)}$$

Checking this solution in the original equation, we have

$$2\sqrt{3\left(\tfrac{2}{3}\right) - 1} \overset{?}{=} 3\left(\tfrac{2}{3}\right), \qquad 2\sqrt{2 - 1} \overset{?}{=} 2, \qquad 2 = 2$$

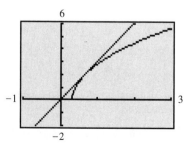

Fig. 14.16

Therefore, the solution $x = \frac{2}{3}$ checks.

We can check this solution graphically by letting $y_1 = 2\sqrt{3x - 1}$ and $y_2 = 3x$. The calculator display is shown in Fig. 14.16. The *intersection* feature shows that the only x-value that the curves have in common is $x = 0.6667$, which agrees with the solution of $x = \frac{2}{3}$. This also means the line y_1 is tangent to the curve of y_2.

◀ EXAMPLE 3   Solve the equation $\sqrt[3]{x - 8} = 2$.

Cubing both sides of the equation, we have

$$x - 8 = 8$$
$$x = 16$$

Checking this solution in the original equation, we get

$$\sqrt[3]{16 - 8} \stackrel{?}{=} 2, \qquad 2 = 2$$

Therefore, the solution checks.

If a radical and other terms are on one side of the equation, we *first isolate the radical.* That is, we rewrite the equation with the radical on one side and all other terms on the other side.

◀ EXAMPLE 4   Solve the equation $\sqrt{x - 1} + 3 = x$.

We first isolate the radical by subtracting 3 from each side. This gives us

$$\sqrt{x - 1} = x - 3$$

We now square both sides and proceed with the solution:

CAUTION ▶

$$\left(\sqrt{x - 1}\right)^2 = (x - 3)^2 \quad \rceil \text{ square the expression on each side,}$$
$$x - 1 = x^2 - 6x + 9 \quad \leftarrow \rfloor \text{ not just the terms separately}$$
$$x^2 - 7x + 10 = 0$$
$$(x - 5)(x - 2) = 0$$
$$x = 5 \quad \text{or} \quad x = 2$$

The solution $x = 5$ checks, but the solution $x = 2$ gives $4 = 2$. Thus, the solution is $x = 5$. The value $x = 2$ is an extraneous root.

◀ EXAMPLE 5   Solve the equation $\sqrt{x + 1} + \sqrt{x - 4} = 5$.

This is most easily solved by first isolating one of the radicals by placing the other radical on the right side of the equation. We then square both sides of the resulting equation.

CAUTION ▶

$$\sqrt{x + 1} = 5 - \sqrt{x - 4} \leftarrow \text{ two terms}$$
$$\left(\sqrt{x + 1}\right)^2 = \left(5 - \sqrt{x - 4}\right)^2$$
$$x + 1 = 25 - 10\sqrt{x - 4} + \left(\sqrt{x - 4}\right)^2 \leftarrow \text{ be careful!}$$
$$= 25 - 10\sqrt{x - 4} + x - 4$$

Now, isolating the radical on one side of the equation and squaring again, we have

$$10\sqrt{x - 4} = 20$$
$$\sqrt{x - 4} = 2 \qquad \text{divide by 10}$$
$$x - 4 = 4 \qquad \text{square both sides}$$
$$x = 8$$

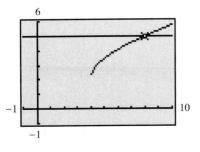

Fig. 14.17

This solution checks. A graphing calculator check is shown in Fig. 14.17, where the $x = 8$ at the point of intersection of $y_1 = \sqrt{x + 1} + \sqrt{x - 4}$ and $y_2 = 5$.

We note again that in squaring $5 - \sqrt{x - 4}$, we do not simply square 5 and $\sqrt{x - 4}$. This is similar to $(5a - 2)^2$ in Example 2 of Section 6.1.

◀ EXAMPLE 6 Solve the equation $\sqrt{x} - \sqrt[4]{x} = 2$.

We can solve this most easily by handling it as an equation in quadratic form. By letting $y = \sqrt[4]{x}$, we have

$$y^2 - y - 2 = 0$$
$$(y - 2)(y + 1) = 0$$
$$y = 2 \quad \text{or} \quad y = -1$$

Since $y = \sqrt[4]{x}$, we know that a negative value of y does not lead to a solution of the original equation. Therefore, we see that $y = -1$ cannot give us a solution. For the other value, $y = 2$, we have

$$\sqrt[4]{x} = 2, \qquad x = 16$$

This checks, because $\sqrt{16} - \sqrt[4]{16} = 4 - 2 = 2$. Therefore, the only solution of the original equation is $x = 16$. ▶

Solving a Word Problem

◀ EXAMPLE 7 Each cross section of a holographic image is in the shape of a right triangle. The perimeter of the cross section is 60 cm, and its area is 120 cm². Find the length of each of the three sides.

If we let the two legs of the triangle be x and y, as shown in Fig. 14.18, from the formulas for the perimeter p and the area A of a triangle, we have

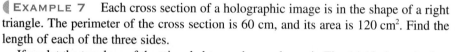

$$p = x + y + \sqrt{x^2 + y^2} \quad \text{and} \quad A = \tfrac{1}{2}xy$$

where the hypotenuse was found by use of the Pythagorean theorem. Using the information given in the statement of the problem, we arrive at the equations

$$x + y + \sqrt{x^2 + y^2} = 60 \quad \text{and} \quad xy = 240$$

Isolating the radical in the first equation and then squaring both sides, we have

$$\sqrt{x^2 + y^2} = 60 - x - y$$
$$x^2 + y^2 = 3600 - 120x - 120y + x^2 + 2xy + y^2$$
$$0 = 3600 - 120x - 120y + 2xy$$

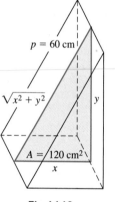

$p = 60\ \text{cm}$

$\sqrt{x^2 + y^2}$

y

$A = 120\ \text{cm}^2$

x

Fig. 14.18

Solving the second of the original equations for y, we have $y = 240/x$. Substituting, we have

$$0 = 3600 - 120x - 120\left(\frac{240}{x}\right) + 2x\left(\frac{240}{x}\right)$$

$$0 = 3600x - 120x^2 - 120(240) + 480x \qquad \text{multiply by } x$$

$$0 = 30x - x^2 - 240 + 4x \qquad \text{divide by } 120$$

$$x^2 - 34x + 240 = 0 \qquad \text{collect terms on left}$$

$$(x - 10)(x - 24) = 0$$

$$x = 10\ \text{cm} \quad \text{or} \quad x = 24\ \text{cm}$$

Holography is a method of producing a three-dimensional image without the use of a lens. The theory of holography was developed in the late 1940s by the British engineer Dennis Gabor (1900–1979). After the invention of lasers, the first holographs were produced in the early 1960s.

If $x = 10$ cm, then $y = 24$ cm, or if $x = 24$ cm, then $y = 10$ cm. Therefore, the legs of the holographic cross section are 10 cm and 24 cm, and the hypotenuse is 26 cm. For these sides, $p = 60$ cm and $A = 120$ cm². We see that these values check with the statement of the problem. ▶

In Exercises 1–4, make the given changes in the indicated examples of this section, and then solve the resulting equations.

1. In Example 2, change the $3x$ on the right to 3 and then solve the equation.

2. In Example 3, change the 8 under the radical to 19 and then solve the equation.

3. In Example 4, change the 3 on the left to 7 and then solve the equation.

4. In Example 5, change the 4 under the second radical to 14 and then solve the equation.

In Exercises 5–34, solve the given equations. In Exercises 19 and 22, explain how the extraneous root is introduced.

5. $\sqrt{x-8} = 2$

6. $\sqrt{x+4} = 3$

7. $\sqrt{8-2x} = x$

8. $2\sqrt{2P+5} = P$

9. $\sqrt{3x+2} = 3x$

10. $\sqrt{5x-1} + 3 = x$

11. $2\sqrt{3-x} - x = 5$

12. $x - 3\sqrt{2x+1} = -5$

13. $\sqrt[3]{y-5} = 3$

14. $\sqrt[4]{5-x} = 2$

15. $5\sqrt{s-6} = s$

16. $\sqrt{x+3} = 4x$

17. $\sqrt{x^2-9} = 4$

18. $t^2 = 3 - \sqrt{2t^2-3}$

(W) 19. $\sqrt{x+4} + 8 = x$

20. $\sqrt{x+15} + 5 = x$

21. $\sqrt{5+\sqrt{x}} = \sqrt{x} - 1$

(W) 22. $\sqrt{13+\sqrt{x}} = \sqrt{x} + 1$

23. $3\sqrt{1-2t} + 1 = 2t$

24. $1 - 2\sqrt{y+4} = y$

25. $2\sqrt{x+2} - \sqrt{3x+4} = 1$

26. $\sqrt{x-1} + \sqrt{x+2} = 3$

27. $\sqrt{5x+1} - 1 = 3\sqrt{x}$

28. $\sqrt{x-7} = \sqrt{x} - 7$

29. $\sqrt{2x-1} - \sqrt{x+11} = -1$

30. $\sqrt{5x-4} - \sqrt{x} = 2$

31. $\sqrt{x-9} = \dfrac{36}{\sqrt{x-9}} - \sqrt{x}$

32. $\sqrt[4]{x+10} = \sqrt{x-2}$

33. $\sqrt{x-2} = \sqrt[4]{x-2} + 12$

34. $\sqrt{3x + \sqrt{3x+4}} = 4$

In Exercises 35–38, solve the given equations algebraically and check the solutions with a graphing calculator.

35. $\sqrt{3x+4} = x$

36. $\sqrt{x-2} + 3 = x$

37. $\sqrt{2x+1} + 3\sqrt{x} = 9$

38. $\sqrt{2x+1} - \sqrt{x+4} = 1$

In Exercises 39–48, solve the given problems.

39. The resonant frequency f in an electric circuit with an inductance L and a capacitance C is given by $f = \dfrac{1}{2\pi\sqrt{LC}}$. Solve for L.

40. A formula used in calculating the range R for radio communication is $R = \sqrt{2rh + h^2}$. Solve for h.

41. An equation used in analyzing a certain type of concrete beam is $k = \sqrt{2np + (np)^2} - np$. Solve for p.

42. In the study of spur gears in contact, the equation $kC = \sqrt{R_1^2 - R_2^2} + \sqrt{r_1^2 - r_2^2} - A$ is used. Solve for r_1^2.

43. The focal length f of a lens, in terms of the image distance q and the object distance p, is given by $\frac{1}{f} = \frac{1}{p} + \frac{1}{q}$. Find p and q if $f = 4$ cm and $p = \sqrt{q}$.

44. The smaller of two cubical boxes is centered on the larger box, and they are taped together with a wide adhesive that just goes around both boxes (see Fig. 14.19). If the edge of the larger box is 1.00 cm greater than that of the smaller box, what are the lengths of the edges of the boxes if 100.0 cm of tape is used?

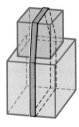

Fig. 14.19

45. A freighter is 5.2 km farther from a Coast Guard station on a straight coast than from the closest point A on the coast. If the station is 8.3 km from A, how far is it from the freighter?

46. The velocity v of an object that falls through a distance h is given by $v = \sqrt{2gh}$, where g is the acceleration due to gravity. Two objects are dropped from heights that differ by 10.0 m such that the sum of their velocities when they strike the ground is 20.0 m/s. Find the heights from which they are dropped if $g = 9.80$ m/s^2.

47. A point D on Denmark's largest island is 3.9 km from the nearest point S on the coast of Sweden (assume the coast is straight, which is nearly the case). A person in a motorboat travels straight from D to a point on the beach x km from S and then travels x km farther along the beach away from S. Find x if the person traveled a total of 7.5 km. See Fig. 14.20.

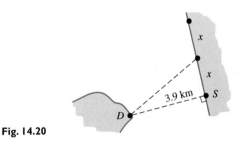

Fig. 14.20

48. The length of the roller belt in Fig. 14.21 is 28.0 m. Find x.

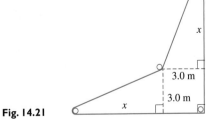

Fig. 14.21

CHAPTER (14) REVIEW EXERCISES

In Exercises 1–10, solve the given systems of equations by use of a graphing calculator.

1. $x + 2y = 6$
$y = 4x^2$

2. $x + y = 3$
$x^2 + y^2 = 25$

3. $3x + 2y = 6$
$x^2 + 4y^2 = 4$

4. $x^2 - 2y = 0$
$y = 3x - 5$

5. $y = x^2 + 1$
$2x^2 + y^2 = 4$

6. $\dfrac{x^2}{4} + y^2 = 1$
$x^2 - y^2 = 1$

7. $y = 11 - x^2$
$y = 2x^2 - 1$

8. $xy = -2$
$y = 1 - 2x^2$

9. $y = x^2 - 2x$
$y = 1 - e^{-x}$

10. $y = \ln x$
$y = \sin x$

In Exercises 11–20, solve each of the given systems of equations algebraically.

11. $y = 4x^2$
$y = 8x$

12. $x + y = 12$
$xy = 20$

13. $2R = L^2$
$R^2 + L^2 = 3$

14. $y = x^2$
$2x^2 - y^2 = 1$

15. $4u^2 + v = 3$
$2u + 3v = 1$

16. $2x^2 + y^2 = 33$
$x^2 + y^2 = 29$

17. $4x^2 - 7y^2 = 21$
$x^2 + 2y^2 = 99$

18. $s - t = 6$
$\sqrt{s} - \sqrt{t} = 1$

19. $4x^2 + 3xy = 4$
$x + 3y = 4$

20. $\dfrac{6}{x} + \dfrac{3}{y} = 4$
$\dfrac{36}{x^2} + \dfrac{36}{y^2} = 13$

In Exercises 21–38, solve the given equations.

21. $x^4 - 20x^2 + 64 = 0$

22. $t^6 - 26t^3 - 27 = 0$

23. $x^{3/2} - 9x^{3/4} + 8 = 0$

24. $x^{1/2} + 3x^{1/4} - 28 = 0$

25. $D^{-2} + 4D^{-1} - 21 = 0$

26. $4x^{-4} + 35x^{-2} = 9$

27. $2x - 3\sqrt{x} - 5 = 0$

28. $e^x + e^{-x} = 2$

29. $\dfrac{4}{r^2 + 1} + \dfrac{7}{2r^2 + 1} = 2$

30. $(x^2 + 5x)^2 - 5(x^2 + 5x) = 6$

31. $3\sqrt{2Z + 4} = 2Z$

32. $\sqrt[3]{x - 2} = 3$

33. $\sqrt{5x + 9} + 1 = x$

34. $2\sqrt{5x - 3} - 1 = 2x$

35. $\sqrt{x + 1} + \sqrt{x} = 2$

36. $\sqrt{3x^2 - 2} - \sqrt{x^2 + 7} = 1$

37. $\sqrt{n + 4} + 2\sqrt{n + 2} = 3$

38. $\sqrt{3x - 2} - \sqrt{x + 7} = 1$

(W) *In Exercises 39 and 40, find the value of x. (In each, the expression on the right is called a* **continued radical***. Also, ... means that the pattern continues indefinitely.) (Hint: Square both sides.) Noting the result, complete the solution and explain your method.*

39. $x = \sqrt{2 + \sqrt{2 + \sqrt{2 + \cdots}}}$

40. $x = \sqrt{6 - \sqrt{6 - \sqrt{6 - \cdots}}}$

In Exercises 41–46, solve the given equations algebraically and check the solutions with a graphing calculator.

41. $x^3 - 2x^{3/2} - 48 = 0$

42. $(x + 1)^4 - 54 = 3(x + 1)^2$

43. $2\sqrt{3x + 1} - \sqrt{x - 1} = 6$

44. $3\sqrt{x} + \sqrt{x - 9} = 11$

45. $\sqrt[3]{x^3 - 7} = x - 1$

46. $\sqrt{x^2 + 7} + \sqrt[4]{x^2 + 7} = 6$

In Exercises 47–54, solve the given problems.

47. Use a graphing calculator to solve the following system of three equations. $x^2 + y^2 = 13$, $y = x - 1$, $xy = 6$.

(W) **48.** Algebraically solve the following system of three equations: $y = -x^2$, $y = x - 1$, $xy = 1$. Explain the results.

49. In the study of atomic structure, the equation
$$L = \frac{h}{2\pi}\sqrt{l(l + 1)} \text{ is used. Solve for } l \ (l > 0).$$

50. The frequency ω of a certain *RLC* circuit is given by
$$\omega = \frac{\sqrt{R^2 + 4(L/C)} + R}{2L}. \text{ Solve for } C.$$

51. In the theory dealing with a suspended cable, the equation $y = \sqrt{s^2 - m^2} - m$ is used. Solve for m.

52. The equation $V = e^2cr^{-2} - e^2Zr^{-1}$ is used in spectroscopy. Solve for r.

53. In an experiment, an object is allowed to fall, stops, and then falls for twice the initial time. The total distance the object falls is 392 cm. The equations relating the times t_1 and t_2 (in s) of fall are $490t_1^2 + 490t_2^2 = 392$ and $t_2 = 2t_1$. Find the times of fall.

54. If two objects collide and the kinetic energy remains constant, the collision is termed perfectly elastic. Under these conditions, if an object of mass m_1 and initial velocity u_1 strikes a second object (initially at rest) of mass m_2, such that the velocities after collision are v_1 and v_2, the following equations are found:

$m_1u_1 = m_1v_1 + m_2v_2$
$\frac{1}{2}m_1u_1^2 = \frac{1}{2}m_1v_1^2 + \frac{1}{2}m_2v_2^2$

Solve these equations for m_2 in terms of u_1, v_1, and m_1.

In Exercises 55–64, set up the appropriate equations and solve them.

55. A wrench is dropped by a worker at a construction site. Four seconds later the worker hears it hit the ground below. How high is the worker above the ground? (The velocity of sound is 331 m/s, and the distance the wrench falls as a function of time is $s = 4.9t^2$.)

56. A rectangular field is enclosed by fencing and a wall along one side and half of an adjacent side. See Fig. 14.22. If the area of the field is 9000 m² and 240 m of fencing are used, what are the dimensions of the field?

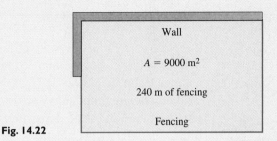

Wall

$A = 9000$ m²

240 m of fencing

Fencing

Fig. 14.22

57. In a certain electric circuit the impedance Z is twice the square of the reactance X, and the resistance R is 0.800 Ω. Find the impedance and the reactance. See Section 12.7.

58. For the plywood piece shown in Fig. 14.23, find x and y.

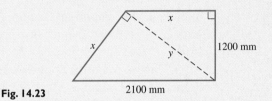

x

x

y

1200 mm

2100 mm

Fig. 14.23

59. The viewing window on a graphing calculator has an area of 1770 mm² and a diagonal of 62 mm. What are the length and width of the rectangle?

60. A trough is made from a piece of sheet metal 12.0 cm wide. The cross section of the trough is shown in Fig. 14.24. Find x.

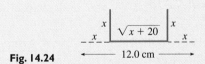

x $\sqrt{x + 20}$ x

x x

Fig. 14.24 ← 12.0 cm →

61. The circular solar cell and square solar cell shown in Fig. 14.25 have a combined surface area of 40.0 cm². Find the radius of the circular cell and the side of the square cell.

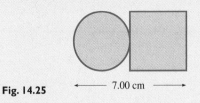

Fig. 14.25 ← 7.00 cm →

62. A plastic band 19.0 cm long is bent into the shape of a triangle with sides $\sqrt{x - 1}$, $\sqrt{5x - 1}$, and 9. Find x.

63. A naval ship travels from Portsmouth, England, to the west end of the Kiel Canal in northern Germany, and later it returns to Portsmouth at a speed that is 9.6 km/h faster. If Portsmouth is 820 km from the Kiel Canal and the total travel time is 35.0 h, find the speed of the ship in each direction.

64. Two trains are approaching the same crossing on tracks that are at right angles to each other. Each is traveling at 60.0 km/h. If one is 6.00 km from the crossing when the other is 3.00 km from it, how much later will they be 4.00 km apart (on a direct line)?

Writing Exercise

65. Using a computer, an engineer designs a triangular support structure with sides (in m) of x, $\sqrt{x - 1}$, and 4.00 m. If the perimeter is to be 9.00 m, the equation to be solved is $x + \sqrt{x - 1} + 4.00 = 9.00$. This equation can be solved by either of two methods used in this chapter. Write one or two paragraphs identifying the methods and explaining how they are used to solve the equation.

CHAPTER ⑭ PRACTICE TEST

1. Solve for x: $x^{1/2} - 2x^{1/4} = 3$.

2. Solve for x: $3\sqrt{x - 2} - \sqrt{x + 1} = 1$.

3. Solve for x: $x^4 - 17x^2 + 16 = 0$.

4. Solve for x and y algebraically:

$x^2 - 2y = 5$
$2x + 6y = 1$

5. Solve for x: $\sqrt[3]{2x + 5} = 5$.

6. The velocity v of an object falling under the influence of gravity in terms of its initial velocity v_0, the acceleration due to gravity g, and the height h fallen is given by $v = \sqrt{v_0^2 + 2gh}$. Solve for h.

7. Solve for x and y graphically: $x^2 - y^2 = 4$
$xy = 2$

8. A rectangular tabletop has a perimeter of 14.0 m and an area of 10.0 m². Find the length and the width of the tabletop.

Equations of Higher Degree

In Section 15.3, we see how the design of a box to hold a product involves the solution of a higher-degree equation.

While solving polynomial equations, Charles Babbage (1792–1871), a British mathematician, found many errors in the handwritten tables of logarithms and other mathematical functions he was using. This led him to use a property of polynomials to design calculating devices to eliminate these errors. Although his devices required too much precision to produce at the time, the design of one included the important features of a modern computer: input, storage, control unit, and output. Because of this design, Babbage is often considered as the father of the computer. (In 1992, his design was used to make the device, and it worked successfully.)

We see that polynomials played an important role in the development of computers. Today, among many other things, computers are used to solve equations, including polynomial equations, which include linear equations (first degree) and quadratic equations (second degree).

Solving polynomial equations of higher degree was of interest to many mathematicians, and special methods were found for some types before 1800. One of the big advances in the solution of these equations was made by Karl Friedrich Gauss, who in 1799 proved the *fundamental theorem of algebra,* which we state later in the chapter. Today, Gauss is considered by many as the greatest mathematician of all time.

In this chapter, we study some of the methods that have been developed for solving higher-degree polynomial equations. The solutions we find include all possible roots, including complex-number roots. Some calculators are programmed to find all such roots, but graphical methods cannot be used to find the complex roots.

Applications of higher-degree equations arise in a number of technical areas. Included in these are the resistance in an electric circuit, finding the dimensions of a container or structure, robotics, and calculating various business production costs.

15.1 THE REMAINDER AND FACTOR THEOREMS; SYNTHETIC DIVISION

In solving higher-degree polynomial equations, the quadratic formula can be used for second-degree equations, and methods have been found for certain third- and fourth-degree equations. It can also be proven that polynomial equations of degree higher than 4 cannot in general be solved algebraically.

In this section, we present two theorems and a simplified method for algebraic division. These will help us in solving polynomial equations later in the chapter.

Any function of the form

$$f(x) = a_0 x^n + a_1 x^{n-1} + \cdots + a_n \tag{15.1}$$

where $a_0 \neq 0$ and n is a positive integer or zero is called a **polynomial function.** We will be considering only polynomials in which the coefficients $a_0, a_1, \ldots, a_n$ are real numbers.

If we divide a polynomial by $x - r$, we find a result of the form

$$f(x) = (x - r)q(x) + R \tag{15.2}$$

where $q(x)$ is the quotient and R is the remainder.

$$
\begin{array}{r}
3x + 11 \\
x - 2 \overline{)3x^2 + 5x - 8} \\
\underline{3x^2 - 6x} \\
11x - 8 \\
\underline{11x - 22} \\
14
\end{array}
$$

◀ EXAMPLE 1   Divide $f(x) = 3x^2 + 5x - 8$ by $x - 2$.

The division is shown at the left, and it shows that

$$3x^2 + 5x - 8 = (x - 2)(3x + 11) + 14$$

where, for this function $f(x)$ with $r = 2$, we identify $q(x)$ and R as

$$q(x) = 3x + 11 \qquad R = 14$$

THE REMAINDER THEOREM

If we now set $x = r$ in Eq. (15.2), we have $f(r) = q(r)(r - r) + R = f(r)(0) + R$, or

$$f(r) = R \tag{15.3}$$

This leads us to the **remainder theorem,** which states that *if a polynomial $f(x)$ is divided by $x - r$ until a constant remainder R is obtained, then $f(r) = R$.* As we can see, this means the remainder equals the value of the function of x at $x = r$.

◀ EXAMPLE 2   In Example 1, $f(x) = 3x^2 + 5x - 8$, $R = 14$, and $r = 2$.

We find that

$$
\begin{aligned}
f(2) &= 3(2^2) + 5(2) - 8 \\
&= 12 + 10 - 8 \\
&= 14
\end{aligned}
$$

Therefore, $f(2) = 14$ verifies that $f(r) = R$ for this example.

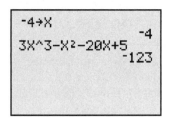

Fig. 15.1

EXAMPLE 3 By using the remainder theorem, determine the remainder when $3x^3 - x^2 - 20x + 5$ is divided by $x + 4$.

In using the remainder theorem, we determine the remainder when the function is divided by $x - r$ by evaluating the function for $x = r$. To have $x + 4$ in the proper form to identify r, we write it as $x - (-4)$. *This means that $r = -4$*, and we therefore evaluate the function $f(x) = 3x^3 - x^2 - 20x + 5$ for $x = -4$, or find $f(-4)$:

$$f(-4) = 3(-4)^3 - (-4)^2 - 20(-4) + 5 = -192 - 16 + 80 + 5$$
$$= -123$$

The remainder is -123 when $3x^3 - x^2 - 20x + 5$ is divided by $x + 4$.

NOTE ▶ To evaluate the function with a calculator, *first store in memory the value to be substituted.* Figure 15.1 shows a calculator display for this example. ▶

THE FACTOR THEOREM

The remainder theorem leads to another important theorem known as the **factor theorem.** It states that *if $f(r) = R = 0$, then $x - r$ is a factor of $f(x)$.* We see in Eq. (15.2) that if the remainder $R = 0$, then $f(x) = (x - r)q(x)$, and this shows that $x - r$ is a factor of $f(x)$. Therefore, we have the following meanings for $f(r) = 0$.

> **ZERO, FACTOR, AND ROOT OF THE FUNCTION $f(x)$**
> If $f(r) = 0$, then $x = r$ is a **zero** of $f(x)$,
> $x - r$ is a **factor** of $f(x)$, and
> $x = r$ is a **root** of the equation $f(x) = 0$

EXAMPLE 4 **(a)** We find that $t + 1$ is a factor of $f(t) = t^3 + 2t^2 - 5t - 6$ because

$$f(-1) = (-1)^3 + 2(-1)^2 - 5(-1) - 6 = -1 + 2 + 5 - 6 = 0$$

(b) However, $t + 2$ is not a factor of $f(t)$ because

$$f(-2) = (-2)^3 + 2(-2)^2 - 5(-2) - 6 = -8 + 8 + 10 - 6 = 4$$ ▶

EXAMPLE 5 Determine if $\frac{2}{3}$ is a zero of $f(x) = 3x^3 + 4x^2 - 16x + 8$.
Evaluating $f\left(\frac{2}{3}\right)$, we have

$$f\left(\frac{2}{3}\right) = 3\left(\frac{2}{3}\right)^3 + 4\left(\frac{2}{3}\right)^2 - 16\left(\frac{2}{3}\right) + 8$$

$$= \frac{8}{9} + \frac{16}{9} - \frac{32}{3} + 8 = \frac{8 + 16 - 96 + 72}{9} = 0$$

Since $f\left(\frac{2}{3}\right) = 0$, $\frac{2}{3}$ is a zero of the function. ▶

SYNTHETIC DIVISION

In the sections that follow, we will find that division of a polynomial by the factor $x - r$ is also useful in solving polynomial equations. Therefore, we now develop a simplified form of long division, known as **synthetic division.** It allows us to easily find the coefficients of the quotient and the remainder. If the degree of the equation is high, it is easier to use synthetic division than to calculate $f(r)$. The method for synthetic division is developed in the following example.

◀ **EXAMPLE 6** Divide $x^4 + 4x^3 - x^2 - 16x - 14$ by $x - 2$.
We first perform this division in the usual manner:

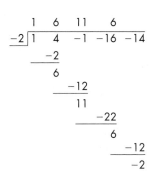

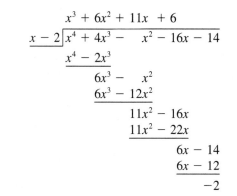

In doing the division, notice that we repeat many terms and that the only important numbers are the coefficients. This means there is no need to write in the powers of x. To the left of the division, we write it without x's and without identical terms.

$$-2 \overline{\begin{array}{rrrrr} 1 & 4 & -1 & -16 & -14 \\ & -2 & -12 & -22 & -12 \\ \hline & 6 & 11 & 6 & -2 \end{array}}$$

All numbers below the dividend may be written in two lines. Then all coefficients of the quotient, except the first, appear in the bottom line. Therefore, the line above the dividend is omitted, and we have the form at the left.

$$\begin{array}{rrrrr|r} 1 & 4 & -1 & -16 & -14 & 2 \\ & -2 & -12 & -22 & -12 & \\ \hline 1 & 6 & 11 & 6 & -2 & \end{array}$$

Now write the first coefficient (in this case, 1) in the bottom line. Also, change the -2 to 2, which is the actual value of r. Then, in the form at the left, write the 2 on the right. *In this form, the 1, 6, 11, and 6 are the coefficients of the x^3, x^2, x, and constant term of the quotient. The -2 is the remainder.*

$$\begin{array}{rrrrr|r} 1 & 4 & -1 & -16 & -14 & 2 \\ & 2 & 12 & 22 & 12 & \\ \hline 1 & 6 & 11 & 6 & -2 & \end{array}$$

Finally, it is easier to use addition rather than subtraction in the process, so we change the signs of the numbers in the middle row. Remember that originally the bottom line was found by subtraction. Therefore, we have the last form on the left.

In the last form at the left, we have 1 (of the bottom row) $\times$ 2 ($= r$) $= 2$, the first number of the middle row. In the second column, $4 + 2 = 6$, the second number in the bottom row. Then, $6 \times 2 (= r) = 12$, the second number of the second row; $-1 + 12 = 11$; $11 \times 2 = 22$; $22 + (-16) = 6$; $6 \times 2 = 12$; and $12 + (-14) = -2$.

We read the bottom line of the last form, the one we use in *synthetic division*, as

$$1x^3 + 6x^2 + 11x + 6 \text{ with a remainder of } -2$$

The method of synthetic division shown in the last form is outlined below. ▶

See Appendix C for a graphing calculator program SYNTHDIV. It can be used to find the coefficients and remainder if a polynomial is divided by $x - R$.

PROCEDURE FOR SYNTHETIC DIVISION

1. *Write the coefficients of $f(x)$. Be certain that the powers are in descending order and that zeros are inserted for missing powers.*

2. *Carry down the left coefficient, then multiply it by r, and place this product under the second coefficient of the top line.*

3. *Add the two numbers in the second column and place the result below. Multiply this sum by r and place the product under the third coefficient of the top line.*

4. *Continue this process until the bottom row has as many numbers as the top row.*

Using synthetic division, the last number in the bottom row is the remainder, and the other numbers are the respective coefficients of the quotient. The first term of the quotient is of degree one less than the first term of the dividend.

◀ **EXAMPLE 7**   Divide $x^5 + 2x^4 - 4x^2 + 3x - 4$ by $x + 3$ using synthetic division.
    Since the powers of x are in descending order, write down the coefficients of $f(x)$. In doing so, we must be certain to include a zero for the missing x^3 term. Next, note that the divisor is $x + 3$, which means that $r = -3$. The -3 is placed to the right. This gives us a top line of

$$\text{coefficients} \rightarrow 1 \quad 2 \quad 0 \quad -4 \quad 3 \quad -4 \quad \boxed{-3} \leftarrow r$$

Next, we carry the left coefficient, 1, to the bottom line and multiply it by r, -3, placing the product, -3, in the middle line under the second coefficient, 2. We then add the 2 and the -3 and place the result, 1, below. This gives

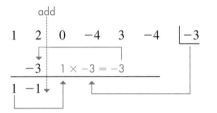

Now multiply the -1 by -3 ($= r$) and place the result, 3, in the middle line under the zero. Now add and continue the process, obtaining the following result:

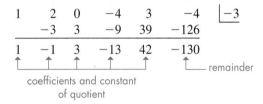

NOTE ▶

Since the degree of the dividend is 5, the degree of the quotient is 4. This means that the quotient is $x^4 - x^3 + 3x^2 - 13x + 42$ and the remainder is -130. In turn, this means that for $f(x) = x^5 + 2x^4 - 4x^2 + 3x - 4$, we have $f(-3) = -130$.   ◗

◀ **EXAMPLE 8**   By synthetic division, divide $3x^4 - 5x + 6$ by $x - 4$.

$$\begin{array}{rrrrr|r} 3 & 0 & 0 & -5 & 6 & \underline{4} \\ & 12 & 48 & 192 & 748 & \\ \hline 3 & 12 & 48 & 187 & 754 & \end{array}$$

The quotient is $3x^3 + 12x^2 + 48x + 187$, and the remainder is 754.   ◗

◀ **EXAMPLE 9**   By synthetic division, determine whether or not $t + 4$ is a factor of $t^4 + 2t^3 - 15t^2 - 32t - 16$.

$$\begin{array}{rrrrr|r} 1 & 2 & -15 & -32 & -16 & \underline{-4} \\ & -4 & 8 & 28 & 16 & \\ \hline 1 & -2 & -7 & -4 & 0 & \end{array}$$

Since the remainder is zero, $t + 4$ is a factor. We may also conclude that

$$f(t) = (t + 4)(t^3 - 2t^2 - 7t - 4)$$ ◗

◀ **EXAMPLE 10** By using synthetic division, determine whether $2x - 3$ is a factor of $2x^3 - 3x^2 + 8x - 12$.

CAUTION ▶   We first note that the coefficient of x in the possible factor is not 1. Thus, *we cannot use $r = 3$, since the factor is not of the form $x - r$.* However, $2x - 3 = 2\left(x - \frac{3}{2}\right)$, which means that if $2\left(x - \frac{3}{2}\right)$ is a factor of the function, $2x - 3$ is a factor. If we use $r = \frac{3}{2}$ and find that the remainder is zero, then $x - \frac{3}{2}$ is a factor.

$$
\begin{array}{rrrr|r}
2 & -3 & 8 & -12 & \frac{3}{2} \\
  & 3 & 0 & 12 & \\
\hline
2 & 0 & 8 & 0 &
\end{array}
$$

Since the remainder is zero, $x - \frac{3}{2}$ is a factor. Also, the quotient is $2x^2 + 8$, which may be factored into $2(x^2 + 4)$. Thus, 2 is also a factor of the function. This means that $2\left(x - \frac{3}{2}\right)$ is a factor of the function, and this in turn means that $2x - 3$ is a factor. This tells us that

$$2x^3 - 3x^2 + 8x - 12 = (2x - 3)(x^2 + 4)$$ ▶

◀ **EXAMPLE 11** Determine whether or not -12.5 is a zero of the function $f(x) = 6x^3 + 61x^2 - 171x + 100$.

If -12.5 is a zero of $f(x)$, then $x - (-12.5)$, or $x + 12.5$, is a factor of $f(x)$, and $f(-12.5) = 0$. We can find the remainder by direct use of the remainder theorem or by synthetic division.

Using synthetic division and a calculator to make the calculations, we have the following setup and calculator sequence:

$$
\begin{array}{rrrr|r}
6 & 61 & -171 & 100 & -12.5 \\
  & -75 & 175 & -50 & \\
\hline
6 & -14 & 4 & 50 &
\end{array}
$$

Since the remainder is 50, and not zero, -12.5 is not a zero of $f(x)$.   ▶

EXERCISES 15.1

In Exercises 1–4, make the given changes in the indicated examples of this section, and then perform the indicated operations.

1. In Example 3, change the $x + 4$ to $x + 3$ and then find the remainder.

2. In Example 4(a), change the $t + 1$ to $t - 1$ and then determine if $t - 1$ is a factor.

3. In Example 7, change the $x + 3$ to $x + 2$ and then perform the synthetic division.

4. In Example 10, change the $2x - 3$ to $2x + 3$ and then determine whether $2x + 3$ is a factor.

In Exercises 5–10, find the remainder by long division.

5. $(x^3 + 2x + 3) \div (x + 1)$

6. $(x^4 - 4x^3 - x^2 + x - 100) \div (x + 3)$

7. $(2x^5 - x^2 + 8x + 44) \div (x + 2)$

8. $(4s^3 - 9s^2 - 24s - 17) \div (s - 5)$

9. $(3x^4 - 9x^3 - x^2 + 5x - 10) \div (x - 3)$

10. $(2x^4 - 10x^2 + 30x - 60) \div (x + 4)$

In Exercises 11–16, find the remainder using the remainder theorem. Do not use synthetic division.

11. $(R^4 + R^3 - 9R^2 + 3) \div (R + 4)$

12. $(4x^4 - x^2 + 5x - 7) \div (x - 3)$

13. $(2x^4 - 7x^3 - x^2 + 8) \div (x - 3)$

14. $(3n^4 - 13n^2 + 10n - 10) \div (n + 4)$

15. $(x^5 - 3x^3 + 5x^2 - 10x + 6) \div (x - 2)$

16. $(3x^4 - 12x^3 - 60x + 4) \div (x - 5)$

In Exercises 17–22, use the factor theorem to determine whether or not the second expression is a factor of the first expression. Do not use synthetic division.

17. $4x^3 + x^2 - 16x - 4, \ x - 2$

18. $3x^3 + 14x^2 + 7x - 4, \ x + 4$

19. $3V^4 - 7V^3 + V + 8, V - 2$

20. $x^5 - 2x^4 + 3x^3 - 6x^2 - 4x + 8, x - 2$

21. $x^6 + 1, x + 1$

22. $x^7 - 128, x + 2$

In Exercises 23–32, perform the indicated divisions by synthetic division.

23. $(x^3 + 2x^2 - x - 2) \div (x - 1)$

24. $(x^3 - 3x^2 - x + 2) \div (x - 2)$

25. $(x^3 + 2x^2 - 3x + 4) \div (x + 1)$

26. $(2x^3 - 4x^2 + x - 1) \div (x + 2)$

27. $(p^6 - 6p^3 - 2p^2 - 6) \div (p - 2)$

28. $(x^5 + 4x^4 - 8) \div (x + 1)$

29. $(x^7 - 128) \div (x - 2)$

30. $(20x^4 + 11x^3 - 89x^2 + 60x - 77) \div (x + 2.75)$

31. $(2x^4 + x^3 + 3x^2 - 1) \div (2x - 1)$

32. $(6t^4 + 5t^3 - 10t + 4) \div (3t - 2)$

In Exercises 33–40, use the factor theorem and synthetic division to determine whether or not the second expression is a factor of the first.

33. $2x^5 - x^3 + 3x^2 - 4;\quad x + 1$

34. $x^5 - 3x^4 - x^2 - 6;\quad x - 3$

35. $4x^3 - 6x^2 + 2x - 2;\quad x - \frac{1}{2}$

36. $3x^3 - 5x^2 + x + 1;\quad x + \frac{1}{3}$

37. $2Z^4 - Z^3 - 4Z^2 + 1;\quad 2Z - 1$

38. $6x^4 + 5x^3 - x^2 + 6x - 2;\quad 3x - 1$

39. $4x^4 + 2x^3 - 8x^2 + 3x + 12;\quad 2x + 3$

40. $3x^4 - 2x^3 + x^2 + 15x + 4;\quad 3x + 4$

In Exercises 41–44, use synthetic division to determine whether or not the given numbers are zeros of the given functions.

41. $x^4 - 5x^3 - 15x^2 + 5x + 14;\quad 7$

42. $r^4 + 5r^3 - 18r - 8;\quad -4$

43. $85x^3 + 348x^2 - 263x + 120;\quad -4.8$

44. $2x^3 + 13x^2 + 10x - 4;\quad \frac{1}{2}$

In Exercises 45–56, solve the given problems.

 **45.** By division, show that $2x - 1$ is a factor of $f(x) = 4x^3 + 8x^2 - x - 2$. May we therefore conclude that $f(1) = 0$? Explain.

46. By division, show that $x^2 + 2$ is a factor of $f(x) = 3x^3 - x^2 + 6x - 2$. May we therefore conclude that $f(-2) = 0$? Explain.

47. For what value of k is $x - 2$ a factor of $f(x) = 2x^3 + kx^2 - x + 14$?

48. For what value of k is $x + 1$ a factor of $f(x) = 3x^4 + 3x^3 + 2x^2 + kx - 1$?

49. Use synthetic division: $(x^3 - 3x^2 + x - 3) \div (x + j)$.

50. Use synthetic division: $(2x^3 - 7x^2 + 10x - 6) \div [x - (1 + j)]$.

51. If $f(x) = -g(x)$, do the functions have the same zeros? Explain.

52. Do the functions $f(x)$ and $f(-x)$ have the same zeros? Explain.

53. In finding the electric current in a certain circuit, it is necessary to factor the denominator of $\dfrac{2s}{s^3 + 5s^2 + 4s + 20}$. Is (a) $(s - 2)$ or (b) $(s + 5)$ a factor?

54. In the theory of the motion of a sphere moving through a fluid, the function $f(r) = 4r^3 - 3ar^2 - a^3$ is used. Is (a) $r = a$ or (b) $r = 2a$ a zero of $f(r)$?

55. In finding the volume V (in cm^3) of a certain gas in equilibrium with a liquid, it is necessary to solve the equation $V^3 - 6V^2 + 12V = 8$. Use synthetic division to determine if $V = 2$ cm^3.

56. An architect is designing a window in the shape of a segment of a circle. An approximate formula for the area is $A = \dfrac{h^3}{2w} + \dfrac{2wh}{3}$, where A is the area, w is the width, and h is the height of the segment. If the width is 1.500 m and the area is 0.5417 m^2, use synthetic division to show that $h = 0.500$ m.

15.2 THE ROOTS OF AN EQUATION

In this section, we present certain theorems that are useful in determining the number of roots in the equation $f(x) = 0$ and the nature of some of these roots. In dealing with polynomial equations of higher degree, it is helpful to have as much of this kind of information as we can find before actually solving for all of the roots, including any possible complex roots. A graph does not show the complex roots an equation may have, but we can verify the real roots, as we show by use of a graphing calculator in some of the examples that follow.

The first of these theorems is so important that it is called the **fundamental theorem of algebra.** It states that

every polynomial equation has at least one (real or complex) root.

This theorem was first proved in 1799 by the German mathematician Karl Gauss (1777–1855) for his doctoral thesis. See the chapter introduction on page 416.

The proof of this theorem is of an advanced nature, and therefore we accept its validity at this time. However, using the fundamental theorem, we can show the validity of other theorems that are useful in solving equations.

Let us now assume that we have a polynomial equation $f(x) = 0$ and that we are looking for its roots. By the fundamental theorem, we know that it has at least one root. Assuming that we can find this root by some means (the factor theorem, for example), we call this root r_1. Thus,

$$f(x) = (x - r_1)f_1(x)$$

where $f_1(x)$ is the polynomial quotient found by dividing $f(x)$ by $(x - r_1)$. However, since the fundamental theorem states that any polynomial equation has at least one root, this must apply to $f_1(x) = 0$ as well. Let us assume that $f_1(x) = 0$ has the root r_2. Therefore, this means that $f(x) = (x - r_1)(x - r_2)f_2(x)$. Continuing this process until one of the quotients is a constant a, we have

$$f(x) = a(x - r_1)(x - r_2) \cdots (x - r_n)$$

Note that one linear factor appears each time a root is found and that the degree of the quotient is one less each time. Thus, there are n linear factors, if the degree of $f(x)$ is n.

Therefore, based on the fundamental theorem of algebra, we have the following two related theorems.

1. *A polynomial of the nth degree can be factored into n linear factors.*

2. *A polynomial equation of degree n has exactly n roots.*

These theorems are illustrated in the following example.

◀ **EXAMPLE 1**   For the equation $f(x) = 2x^4 - 3x^3 - 12x^2 + 7x + 6 = 0$, we are given the factors that we show. In the next section, we will see how to find these factors.

For the function $f(x)$, we have

$$2x^4 - 3x^3 - 12x^2 + 7x + 6 = (x - 3)(2x^3 + 3x^2 - 3x - 2)$$
$$2x^3 + 3x^2 - 3x - 2 = (x + 2)(2x^2 - x - 1)$$
$$2x^2 - x - 1 = (x - 1)(2x + 1)$$
$$2x + 1 = 2\left(x + \tfrac{1}{2}\right)$$

Therefore,

$$2x^4 - 3x^3 - 12x^2 + 7x + 6 = 2(x - 3)(x + 2)(x - 1)\left(x + \tfrac{1}{2}\right) = 0$$

The degree of $f(x)$ is 4. There are four linear factors: $(x - 3)$, $(x + 2)$, $(x - 1)$, and $\left(x + \tfrac{1}{2}\right)$. There are four roots of the equation: 3, -2, 1, and $-\tfrac{1}{2}$. Thus, we have verified each of the theorems above for this example. ▶

It is not necessary for each root of an equation to be different from the other roots. For example, the equation $(x - 1)^2 = 0$ has two roots, both of which are 1. Such roots are referred to as *multiple* (or *repeated*) *roots*.

When we solve the equation $x^2 + 1 = 0$, the roots are j and $-j$. In fact, for any equation (with real coefficients) that has a root of the form $a + bj$ ($b \neq 0$), there is also a root of the form $a - bj$. This is so because we can find the solutions of an equation of the form $ax^2 + bx + c = 0$ from the quadratic formula as

$$\frac{-b + \sqrt{b^2 - 4ac}}{2a} \quad \text{and} \quad \frac{-b - \sqrt{b^2 - 4ac}}{2a}$$

and the only difference between these roots is the sign before the radical. Thus,

NOTE ▶ *if the coefficients of the equation $f(x) = 0$ are real and $a + bj$ ($b \neq 0$) is a complex root, then its conjugate, $a - bj$, is also a root.*

◀ **EXAMPLE 2** Consider the equation $f(x) = (x - 1)^3(x^2 + x + 1) = 0$.

The factor $(x - 1)^3$ shows that there is a triple root of 1, and there is a total of five roots, since the highest-power term would be x^5 if we were to multiply out the function. To find the other two roots, we use the quadratic formula on the *factor* $(x^2 + x + 1)$. This is permissible, since we are finding the values of x for

$$x^2 + x + 1 = 0$$

For this we have

$$x = \frac{-1 \pm \sqrt{1 - 4}}{2}$$

Thus,

$$x = \frac{-1 + j\sqrt{3}}{2} \quad \text{and} \quad x = \frac{-1 - j\sqrt{3}}{2}$$

Therefore, the roots of $f(x) = 0$ are $1, 1, 1, \dfrac{-1 + j\sqrt{3}}{2}$, and $\dfrac{-1 - j\sqrt{3}}{2}$. In the graph of $f(x)$ shown in Fig. 15.2, the *zero* feature gives a value of $x = 1$, but the calculator display does not show that it is a triple root, or that there are complex roots. ▶

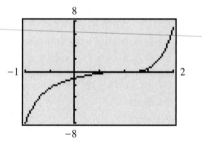

Fig. 15.2

NOTE ▶ From Example 2, we can see that *whenever enough roots are known so that the remaining factor is quadratic, it is possible to find the remaining roots from the quadratic formula.* This is true for finding real or complex roots.

◀ **EXAMPLE 3** Solve the equation $3x^3 + 10x^2 - 16x - 32 = 0$; $-\frac{4}{3}$ is a root.

Using synthetic division and the given root, we have the setup shown at the left. From this, we see that

$$3x^3 + 10x^2 - 16x - 32 = \left(x + \tfrac{4}{3}\right)(3x^2 + 6x - 24)$$

We know that $x + \frac{4}{3}$ is a factor from the given root and that $3x^2 + 6x - 24$ is a factor found from synthetic division. This second factor can be factored as

$$3x^2 + 6x - 24 = 3(x^2 + 2x - 8) = 3(x + 4)(x - 2)$$

Therefore, we have

$$3x^3 + 10x^2 - 16x - 32 = 3\left(x + \tfrac{4}{3}\right)(x + 4)(x - 2)$$

This means the roots are $-\frac{4}{3}$, -4, and 2. Since the three roots are real, they can be found from a calculator display, as shown in Fig. 15.3. ▶

$$\begin{array}{rrrr|l} 3 & 10 & -16 & -32 & \underline{-\tfrac{4}{3}} \\ & -4 & -8 & 32 & \\ \hline 3 & 6 & -24 & 0 & \end{array}$$

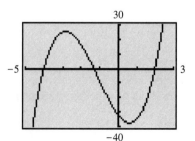

Fig. 15.3

$$
\begin{array}{rrrrr|}
1 & 3 & -4 & -10 & -4 \quad \underline{\,-1} \\
  & -1 & -2 & 6 & 4 \\
\hline
1 & 2 & -6 & -4 & 0
\end{array}
$$

$$
\begin{array}{rrrr|}
1 & 2 & -6 & -4 \quad \underline{\,2} \\
  & 2 & 8 & 4 \\
\hline
1 & 4 & 2 & 0
\end{array}
$$

◖ **EXAMPLE 4** Solve $x^4 + 3x^3 - 4x^2 - 10x - 4 = 0$; -1 and 2 are roots.

Using synthetic division and the root -1, the first setup at the left shows that

$$x^4 + 3x^3 - 4x^2 - 10x - 4 = (x + 1)(x^3 + 2x^2 - 6x - 4)$$

We now know that $x - 2$ must be a factor of $x^3 + 2x^2 - 6x - 4$, since it is a factor of the original function. Again, using synthetic division and this time the root 2, we have the second setup at the left. Thus,

$$x^4 + 3x^3 - 4x^2 - 10x - 4 = (x + 1)(x - 2)(x^2 + 4x + 2)$$

Since the original equation can now be written as

$$(x + 1)(x - 2)(x^2 + 4x + 2) = 0$$

the remaining two roots are found by solving

$$x^2 + 4x + 2 = 0$$

by the quadratic formula. This gives us

$$x = \frac{-4 \pm \sqrt{16 - 8}}{2} = \frac{-4 \pm 2\sqrt{2}}{2} = -2 \pm \sqrt{2}$$

Therefore, the roots are -1, 2, $-2 + \sqrt{2}$, and $-2 - \sqrt{2}$. ◗

◖ **EXAMPLE 5** Solve the equation $3x^4 - 26x^3 + 63x^2 - 36x - 20 = 0$, given that 2 is a double root.

$$
\begin{array}{rrrrr|}
3 & -26 & 63 & -36 & -20 \quad \underline{\,2} \\
  & 6 & -40 & 46 & 20 \\
\hline
3 & -20 & 23 & 10 & 0
\end{array}
$$

$$
\begin{array}{rrrr|}
3 & -20 & 23 & 10 \quad \underline{\,2} \\
  & 6 & -28 & -10 \\
\hline
3 & -14 & -5 & 0
\end{array}
$$

Using synthetic division, we have the first setup at the left. It tells us that

$$3x^4 - 26x^3 + 63x^2 - 36x - 20 = (x - 2)(3x^3 - 20x^2 + 23x + 10)$$

Also, since 2 is a double root, it must be a root of $3x^3 - 20x^2 + 23x + 10 = 0$. Using synthetic division again, we have the second setup at the left. This second quotient $3x^2 - 14x - 5$ factors into $(3x + 1)(x - 5)$. The roots are 2, 2, $-\frac{1}{3}$, and 5.

Since the quotient of the first division is the dividend for the second division, both divisions can be done without rewriting the first quotient as follows:

NOTE ▶

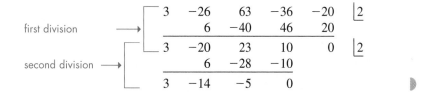

◖ **EXAMPLE 6** Solve the equation $2x^4 - 5x^3 + 11x^2 - 3x - 5 = 0$, given that $1 + 2j$ is a root.

Since $1 + 2j$ is a root, we know that $1 - 2j$ is also a root. Using synthetic division twice, we can then reduce the remaining factor to a quadratic function.

$$
\begin{array}{rrrrr}
2 & -5 & 11 & -3 & -5 \quad \underline{\,1 + 2j} \\
  & 2 + 4j & -11 - 2j & 4 - 2j & 5 \\
\hline
2 & -3 + 4j & -2j & 1 - 2j & 0 \quad \underline{\,1 - 2j} \\
  & 2 - 4j & -1 + 2j & -1 + 2j & \\
\hline
2 & -1 & -1 & 0 &
\end{array}
$$

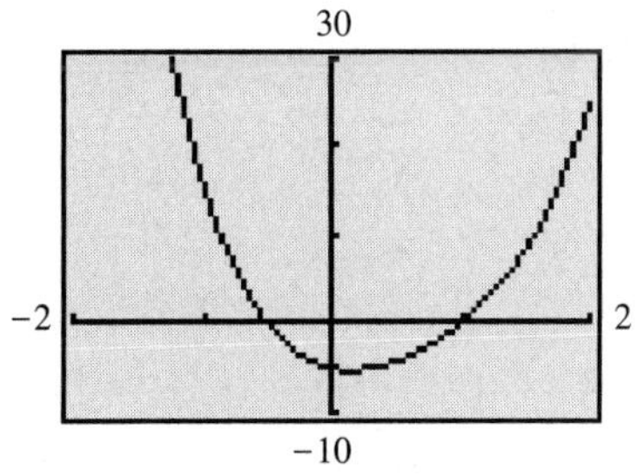

The quadratic factor $2x^2 - x - 1$ factors into $(2x + 1)(x - 1)$. Therefore, the roots of the equation are $1 + 2j$, $1 - 2j$, 1, and $-\frac{1}{2}$. As we can see, the calculator display in Fig. 15.4 shows only the real roots. ◗

Fig. 15.4

$$y = f(x)$$
$$= x^4 - 5x^3 + 5x^2 + 5x - 6$$
$$= (x + 1)(x - 1)(x - 2)(x - 3)$$

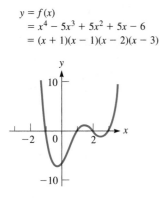

Fig. 15.5

If $f(x)$ is of degree n and the roots of $f(x) = 0$ are real and different, the graph of $f(x)$ crosses the x-axis n times and $f(x)$ changes signs as it crosses. See Fig. 15.5, where $n = 4$, the equation has four changes of sign, and the graph crosses four times.

In Fig. 15.5, we see that for two points on the graph, *the graph must cross the x-axis an odd number of times between points on the different sides of the x-axis and an even number of times between points on the same side of the x-axis, if at all.*

For each pair of complex roots, the number of times the graph crosses the x-axis is reduced by two (Fig. 15.6). For multiple roots, the graph crosses the x-axis once if the multiple is odd, or is tangent to the x-axis if the multiple is even (Fig. 15.7).

The graph must cross the x-axis at least once if the degree of $f(x)$ is odd, because the range includes all real numbers (Figs. 15.6 and 15.7). The curve may not cross the x-axis if the degree of $f(x)$ is even, because the range is bounded at a minimum point or a maximum point (Fig. 15.8), as we have seen for the quadratic function.

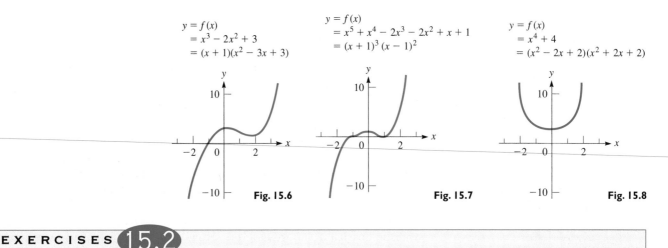

$$y = f(x)$$
$$= x^3 - 2x^2 + 3$$
$$= (x + 1)(x^2 - 3x + 3)$$

Fig. 15.6

$$y = f(x)$$
$$= x^5 + x^4 - 2x^3 - 2x^2 + x + 1$$
$$= (x + 1)^3 (x - 1)^2$$

Fig. 15.7

$$y = f(x)$$
$$= x^4 + 4$$
$$= (x^2 - 2x + 2)(x^2 + 2x + 2)$$

Fig. 15.8

EXERCISES 15.2

In Exercises 1 and 2, make the given changes in the indicated examples of this section and then solve the resulting equation.

1. In Example 2, change the middle term of the second factor to $2x$ and then find the roots.

2. In Example 6, change the middle three terms to the left of the $=$ sign to $-3x^3 + 7x^2 + 7x$ and then solve the equation, given the same root.

In Exercises 3–6, find the roots of the given equations by inspection.

3. $(x - 2)(x^2 - 9) = 0$

4. $x(2x + 5)^2(x^2 - 64) = 0$

5. $(x^2 + 6x + 9)(x^2 + 4) = 0$

6. $(4x^2 + 9)(4x^2 + 4x + 1) = 0$

In Exercises 7–26, solve the given equations using synthetic division, given the roots indicated.

7. $x^3 + x^2 - 8x - 12 = 0$ $(r_1 = -2)$

8. $R^3 + 1 = 0$ $(r_1 = -1)$

9. $2x^3 + 11x^2 + 20x + 12 = 0$ $\left(r_1 = -\frac{3}{2}\right)$

10. $4x^3 - 20x^2 - x + 5 = 0$ $\left(r_1 = \frac{1}{2}\right)$

11. $3x^3 + 2x^2 + 3x + 2 = 0$ $(r_1 = j)$

12. $x^3 + 5x^2 + 9x + 5 = 0$ $(r_1 = -2 + j)$

13. $t^4 + t^3 - 2t^2 + 4t - 24 = 0$ $(r_1 = 2, r_2 = -3)$

14. $x^4 + 2x^3 - 4x^2 - 5x + 6 = 0$ $(r_1 = 1, r_2 = -2)$

15. $x^4 - 9x^2 + 4x + 12 = 0$ (2 is a double root)

16. $4x^4 + 28x^3 + 61x^2 + 42x + 9 = 0$ (-3 is a double root)

17. $6x^4 + 5x^3 - 15x^2 + 4 = 0$ $\left(r_1 = -\frac{1}{2}, r_2 = \frac{2}{3}\right)$

18. $6x^4 - 5x^3 - 14x^2 + 14x - 3 = 0$ $\left(r_1 = \frac{1}{3}, r_2 = \frac{3}{2}\right)$

19. $2x^4 - x^3 - 4x^2 + 10x - 4 = 0$ $(r_1 = 1 + j)$

20. $s^4 - 8s^3 - 72s - 81 = 0$ $(r_1 = 3j)$

21. $x^5 - 3x^4 + 4x^3 - 4x^2 + 3x - 1 = 0$ (1 is a triple root)

22. $12x^5 - 7x^4 + 41x^3 - 26x^2 - 28x + 8 = 0$
$\left(r_1 = 1, r_2 = \frac{1}{4}, r_3 = -\frac{2}{3}\right)$

23. $P^5 - 3P^4 - P + 3 = 0$ $(r_1 = 3, r_2 = j)$

24. $4x^5 + x^3 - 4x^2 - 1 = 0$ $\left(r_1 = 1, r_2 = \frac{1}{2}j\right)$

25. $(x^6 + 2x^5 - 4x^4 - 10x^3 - 41x^2 - 72x - 36 = 0$
(-1 is a double root; $2j$ is a root)

26. $x^6 - x^5 - 2x^3 - 3x^2 - x - 2 = 0$ (j is a double root)

In Exercises 27 and 28, answer the given questions.

Ⓦ **27.** Why cannot a third-degree polynomial function with real coefficients have zeros of 1, 2, and j?

Ⓦ **28.** How can the graph of a fourth-degree polynomial equation have its only x-intercepts as 0, 1, and 2?

15.3 RATIONAL AND IRRATIONAL ROOTS

The product of the factors $(x + 2)(x - 4)(x + 3)$ is $x^3 + x^2 - 14x - 24$. In forming this product, we find that the constant 24 is determined only by the numbers 2, 4, and 3. We see that these numbers represent the roots of the equation if the given function is set equal to zero. In fact, if we find all the integral roots of an equation with integral coefficients and represent the equation in the form

$$f(x) = (x - r_1)(x - r_2) \cdots (x - r_k)f_{k+1}(x) = 0$$

where all the roots indicated are integers, the constant term of $f(x)$ must have factors of $r_1, r_2, \ldots, r_k$. This leads us to the theorem which states that

in a polynomial equation $f(x) = 0$, if the coefficient of the highest power is 1, then any integral roots are factors of the constant term of $f(x)$.

◀ **EXAMPLE 1**   The equation $x^5 - 4x^4 - 7x^3 + 14x^2 - 44x + 120 = 0$ can be written as

$$(x - 5)(x + 3)(x - 2)(x^2 + 4) = 0$$

We now note that $5(3)(2)(4) = 120$. Thus, the roots 5, -3, and 2 are numerical factors of $|120|$. The theorem states nothing about the signs involved.   ▶

If the coefficient a_0 of the highest-power term of $f(x)$ is an integer not equal to 1, the polynomial equation $f(x) = 0$ may have rational roots that are not integers. This coefficient a_0 can be factored from every term of $f(x)$. Thus, any polynomial equation $f(x) = a_0x^n + a_1x^{n-1} + \cdots + a_n = 0$ with integral coefficients can be written in the form

$$f(x) = a_0\left(x^n + \frac{a_1}{a_0}x^{n-1} + \cdots + \frac{a_n}{a_0}\right) = 0$$

Since a_n and a_0 are integers, a_n/a_0 is a rational number. Using the same reasoning as with integral roots applied to the polynomial within the parentheses, we see that any rational roots are factors of a_n/a_0. This leads to the following theorem:

Any rational root of a polynomial equation (with integral coefficients)

$$f(x) = a_0x^n + a_1x^{n-1} + \cdots + a_n = 0$$

is an integral factor of a_n divided by an integral factor of a_0.

We may show this rational root r_r as

$$r_r = \frac{\text{integral factor of } a_n}{\text{integral factor of } a_0} \tag{15.4}$$

◀ **EXAMPLE 2**   If $f(x) = 4x^3 - 3x^2 - 25x - 6 = 0$, any rational roots, if they exist, must be integral factors of 6 divided by integral factors of 4. The integral factors of 6 are 1, 2, 3, and 6, and the integral factors of 4 are 1, 2, and 4. Forming all possible positive and negative quotients, any rational roots that exist will be found in the following list: $\pm 1, \pm\frac{1}{2}, \pm\frac{1}{4}, \pm 2, \pm 3, \pm\frac{3}{2}, \pm\frac{3}{4}, \pm 6$.

The roots of this equation are -2, 3, and $-\frac{1}{4}$.   ▶

Named for the French mathematician René Descartes (1596–1650). (See page 90.)

There are 16 different possible rational roots in Example 2, but we cannot tell which of these are the actual roots. Therefore we now present a rule, known as *Descartes' rule of signs,* which will help us to find these roots.

DESCARTES' RULE OF SIGNS

1. *The number of positive roots of a polynomial equation $f(x) = 0$, cannot exceed the number of changes in sign in $f(x)$ in going from one term to the next in $f(x)$.*
2. *The number of negative roots cannot exceed the number of sign changes in $f(-x)$.*

We can reason this way: If $f(x)$ has all positive terms, then any positive number substituted in $f(x)$ must give a positive value for $f(x)$. This indicates that the number substituted in the function is not a root. Thus, there must be at least one negative and one positive term in the function for any positive number to be a root. This is not a proof, but it does indicate the type of reasoning used in developing the theorem.

◀ **EXAMPLE 3**　By Descartes' rule of signs, determine the maximum number of positive and negative roots of $3x^3 - x^2 - x + 4 = 0$.

Here, $f(x) = 3x^3 - x^2 - x + 4$. The first term is positive, and the second is negative, which indicates a change of sign. The third term is also negative; there is no change of sign from the second to the third term. The fourth term is positive, thus giving us a second change of sign, from the third to the fourth term. Hence, there are two changes in sign, which we can show as follows:

$$f(x) = 3x^3 - x^2 - x + 4$$

$$1 \qquad 2 \quad \longleftarrow \text{ two sign changes}$$

Since there are *two* changes of sign in $f(x)$, there are *no more than two* positive roots of $f(x) = 0$.

To find the maximum possible number of negative roots, we must find the number of sign changes in $f(-x)$. Thus,

$$f(-x) = 3(-x)^3 - (-x)^2 - (-x) + 4$$
$$= -3x^3 - x^2 + x + 4$$

$$\longleftarrow \text{ one sign change}$$

There is only one change of sign in $f(-x)$; therefore, there is one negative root. *When there is just one change of sign in $f(x)$, there is a positive root, and when there is just one change of sign in $f(-x)$, there is a negative root.*

NOTE ▶

◀ **EXAMPLE 4**　For the equation $4x^5 - x^4 - 4x^3 + x^2 - 5x - 6 = 0$, we write

$$f(x) = 4x^5 - x^4 - 4x^3 + x^2 - 5x - 6 \quad \longleftarrow \text{ three sign changes}$$

$$f(-x) = -4x^5 - x^4 + 4x^3 + x^2 + 5x - 6$$

$$\longleftarrow \text{ two sign changes}$$

Thus, there are no more than three positive and two negative roots.

At this point, let us summarize the information we can determine about the roots of a polynomial equation $f(x) = 0$ of degree n and with real coefficients.

> **ROOTS OF A POLYNOMIAL EQUATION OF DEGREE n**
>
> 1. *There are n roots.*
> 2. *Complex roots appear in conjugate pairs.*
> 3. *Any rational roots must be factors of the constant term divided by factors of the coefficient of the highest-power term.*
> 4. *The maximum number of positive roots is the number of sign changes in $f(x)$, and the maximum number of negative roots is the number of sign changes in $f(-x)$.*
> 5. *Once we find $n - 2$ of the roots, we can find the remaining roots by the quadratic formula.*

Some calculators have a specific feature for solving higher-degree equations.

Since synthetic division is easy to perform, it is usually used to try possible roots. When a root is found, the quotient is of one degree less than the degree of the dividend. Each root found makes the ensuing work easier. The following examples show the complete method, as well as two other helpful rules.

◀ **EXAMPLE 5** Find the roots of the equation $2x^3 + x^2 + 5x - 3 = 0$.

Since $n = 3$, there are three roots. If we can find one of these roots, we can use the quadratic formula to find the other two. We have

$$f(x) = 2x^3 + x^2 + 5x - 3 \quad \text{and} \quad f(-x) = -2x^3 + x^2 - 5x - 3$$

which shows there is one positive root and no more than two negative roots, which may or may not be rational. The *possible* rational roots are ± 1, $\pm\frac{1}{2}$, $\pm\frac{3}{2}$, ± 3.

First, trying the root 1 (always a possibility if there are positive roots), we have the synthetic division shown at the left. The remainder of 5 tells us that 1 is not a root, but we have gained some additional information, if we observe closely. If we try any positive number larger than 1, the results in the last row will be larger positive numbers than we now have. The products will be larger, and therefore the sums will also be larger. Thus, there is no positive root larger than 1. This leads to the following rule: *When we are trying a positive root, if the bottom row contains all positive numbers, then there are no roots larger than the value tried.* This rule tells us that there is no reason to try $+\frac{3}{2}$ and $+3$ as roots.

Now let us try $+\frac{1}{2}$, as shown at the left. The zero remainder tells us that $+\frac{1}{2}$ is a root, and the remaining factor is $2x^2 + 2x + 6$, which itself factors to $2(x^2 + x + 3)$. By the quadratic formula we find the remaining roots by solving the equation $x^2 + x + 3 = 0$. This gives us

$$x = \frac{-1 \pm \sqrt{1 - 12}}{2} = \frac{-1 \pm j\sqrt{11}}{2}$$

The three roots are $+\frac{1}{2}$, $\frac{-1 + j\sqrt{11}}{2}$, and $\frac{-1 - j\sqrt{11}}{2}$. There are no negative roots since the nonpositive roots are complex. Proceeding this way, we did not have to try any negative roots. The calculator display in Fig. 15.9 verifies the one real root at $x = 1/2$. ▶

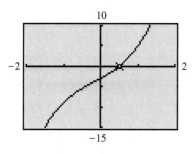

Fig. 15.9

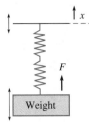

Fig. 15.10

◀ EXAMPLE 6   During a cycle of the movement of the weight on the double spring shown in Fig. 15.10, the force F (in N) on the weight by the spring is

$$F = x^4 - 7x^3 + 12x^2 + 4x$$

where x is the displacement (in cm) of the top of the double spring. For what values of x is $F = 16$ N?

Substituting 16 for F, we see that we are to solve the equation

$$x^4 - 7x^3 + 12x^2 + 4x - 16 = 0$$

To solve this equation, we write

$$f(x) = x^4 - 7x^3 + 12x^2 + 4x - 16$$

$$f(-x) = x^4 + 7x^3 + 12x^2 - 4x - 16$$

We see that there are four roots; there are no more than three positive roots, and there is one negative root. Since the coefficient of x^4 is 1, any possible rational roots must be integers. These possible rational roots are ± 1, ± 2, ± 4, ± 8, and ± 16. Since there is only one negative root, we look for this one first. Trying -2, we have

1	-7	$+12$	$+4$	-16	$\lfloor -2$
	-2	$+18$	-60	$+112$	
1	-9	$+30$	-56	$+96$	

NOTE ▶   If we were to try any negative roots less than -2 (remember, -3 is less than -2), we would find that the numbers would still alternate from term to term in the quotient. Thus, we have this rule: *When we are trying a negative root, if the signs alternate in the bottom row, then there are no roots less than the value tried.* In this case, we know that -4, -8, and -16 cannot be roots.

Next we try -1, as shown at the left. The remainder of zero tells us that -1 is the negative root.

1	-7	$+12$	$+4$	-16	$\lfloor -1$
	-1	8	-20	16	
1	-8	20	-16	0	

Now that we have found the one negative root, we look for the positive roots. Trying 1, we have the setup at the left. The remainder of -3 tells us that 1 is not a root. Next we try 2, as shown at the left. We see that 2 is a root.

1	-8	20	-16	$\lfloor 1$
	1	-7	13	
1	-7	13	-3	

It is not necessary to find any more roots by trial and error. We may now use the quadratic formula or factoring on the equation $x^2 - 6x + 8 = 0$. The remaining roots are 2 and 4. Thus, the roots are -1, 2, 2, and 4. (Note that 2 is a double root.)

1	-8	20	-16	$\lfloor 2$
	2	-12	16	
1	-6	8	0	

These roots now indicate that $F = 16$ N for displacements of -1 cm, 2 cm, and 4 cm.   ▶

By the methods we have presented, we can look for *all roots* of a polynomial equation. These include any possible *complex roots* and *exact values of the rational and irrational roots,* if they exist. These methods allow us to solve a great many polynomial equations for these roots, but there are numerous other equations for which these methods are not sufficient.

When a polynomial equation has more than two irrational roots, we cannot generally find these roots by the methods we have developed. Approximate values can be found on a calculator either graphically or by evaluating the function. These methods are illustrated in the following two examples.

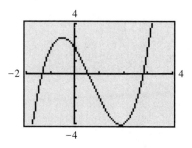

Fig. 15.11

EXAMPLE 7 Find the roots of the equation $x^3 - 2x^2 - 3x + 2 = 0$, using a graphing calculator.

Since the degree of the equation is 3, we know there are three roots, and since the degree is odd, there is at least one real root. Using Descartes' rule of signs we have

$$f(x) = x^3 - 2x^2 - 3x + 2 \qquad \text{two changes of sign}$$
$$f(-x) = -x^3 - 2x^2 + 3x + 2 \qquad \text{one change of sign}$$

This means there is one negative root and no more than two positive roots, if any. Therefore, setting

$$y = x^3 - 2x^2 - 3x + 2$$

and using the *window* settings (after two or three trials) shown in the calculator display in Fig. 15.11, the view shows one negative and two positive roots. Using the *zero* feature (or the *trace* and *zoom* features), these roots are

$$-1.34, 0.53, 2.81 \qquad \text{to the nearest 0.01}$$

Of course, we can use a graphing calculator to help locate and determine the number of real roots of any equation. ▶

Solving a Word Problem

See the chapter introduction.

EXAMPLE 8 The bottom part of a box to hold a jigsaw puzzle is to be made from a rectangular piece of cardboard 37.0 cm by 31.0 cm by cutting out equal squares from the corners, bending up the sides, and taping the corners. See Fig. 15.12. If the volume of the bottom part is to be 2770 cm³, find the side of the square that is to be cut out.

Let $x =$ the side of the square to be cut out. This means

$$2770 = x(37.0 - 2x)(31.0 - 2x) \qquad \text{volume} = 2770 \text{ cm}^3$$
$$= 4x^3 - 136x^2 + 1147x$$
$$4x^3 - 136x^2 + 1147x - 2770 = 0 \qquad \text{simplify with terms on the left}$$

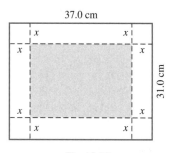

Fig. 15.12

We evaluate $f(x) = 4x^3 - 136x^2 + 1147x - 2770$ for integral values of x and find that $f(4) = -102$ and $f(5) = 65$. This means that $f(x) = 0$ between $x = 4$ and $x = 5$. By entering $f(x)$ as Y_1 in a graphing calculator, we can evaluate the function as shown in the calculator display in Fig. 15.13. This shows the following values:

$$f(4) = -102$$
$$f(5) = 65 \qquad \text{different signs means at least one root between 4 and 5}$$
$$f(4.4) = -15.42 \qquad \text{one root between 4.4 and 4.5}$$
$$f(4.5) = 2$$
$$f(4.48) = -1.353 \qquad \text{root between 4.48 and 4.49, closer to 4.49}$$
$$f(4.49) = 0.3318$$

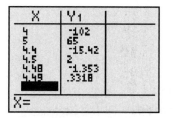

Fig. 15.13

Methods of evaluating functions on a graphing calculator are shown on page 83.

From these values, $x = 4.49$ cm, to three significant digits. This value gives a volume

$$V = 4.49(37.0 - 2(4.49))(31.0 - 2(4.49)) = 2270 \text{ cm}^3$$

which means that the solution checks.

Checking values greater than $x = 5$, we find that there is another root between 6 and 7. Using the same procedure as above, we find that $x = 6.80$ cm is also a solution. (A third root $x = 22.7$ cm is obviously too large.) Therefore, there are two possible squares that can be cut from each corner to get a bottom part with a volume of 2770 cm³. ▶

EXERCISES 15.3

In Exercises 1 and 2, make the given changes in the indicated examples of this section and then perform the indicated operation.

1. In Example 4, change all signs between terms and then find the number of possible positive and negative roots.

2. In Example 5, change the + sign before the $5x$ to − and then find the roots.

In Exercises 3–20, solve the given equations without using a graphing calculator.

3. $x^3 + 2x^2 - x - 2 = 0$

4. $r^3 + x^2 - 5x + 3 = 0$

5. $x^3 + 2x^2 - 5x - 6 = 0$

6. $t^3 - 12t - 16 = 0$

7. $3x^3 - x - 2 = 0$

8. $2t^3 - t^2 - 3t - 1 = 0$

9. $3x^3 + 11x^2 + 5x - 3 = 0$

10. $4x^3 - 5x^2 - 23x + 6 = 0$

11. $x^4 - 11x^2 - 12x + 4 = 0$

12. $8x^4 - 32x^3 - x + 4 = 0$

13. $N^4 - 2N^3 - 13N^2 + 14N + 24 = 0$

14. $4n^4 - 17n^2 + 14n - 3 = 0$

15. $12x^4 + 44x^3 + 21x^2 - 11x = 6$

16. $9x^4 - 3x^3 + 34x^2 - 12x = 8$

17. $D^5 + D^4 - 9D^3 - 5D^2 + 16D + 12 = 0$

18. $x^6 - x^4 - 14x^2 + 24 = 0$

19. $2x^5 - 5x^4 + 6x^3 - 6x^2 + 4x = 1$

20. $2x^5 + 5x^4 - 4x^3 - 19x^2 - 16x = 4$

In Exercises 21–24, use a graphing calculator to solve the given equations to the nearest 0.01.

21. $x^3 - 2x^2 - 5x + 4 = 0$

22. $x^4 - x^3 - 2x^2 - x - 3 = 0$

23. $x^4 + 2x^3 - 3x^2 + 2x - 4 = 0$

24. $2x^5 - 3x^4 + 8x^3 - 4x^2 - 4x + 2 = 0$

In Exercises 25–28, use a calculator to find the irrational root (to the nearest 0.01) that lies between the given values. (See Example 8.)

25. $x^3 - 6x^2 + 10x - 4 = 0$ (0 and 1)

26. $r^4 - r^3 - 3r^2 - r - 4 = 0$ (2 and 3)

27. $3x^3 + 13x^2 + 3x - 4 = 0$ (−1 and 0)

28. $3x^4 - 3x^3 - 11x^2 - x - 4 = 0$ (−2 and −1)

In Exercises 29–44, solve the given problems. Use a graphing calculator in Exercises 36, 37, and 40.

29. Where does the graph of the function
$f(x) = 4x^3 + 3x^2 - 20x - 15$ cross the x-axis?

30. Where does the graph of the function
$f(s) = 2s^4 - s^3 - 5s^2 + 7s - 6$ cross the s-axis?

31. The angular acceleration α (in rad/s^2) of the wheel of a car is given by $\alpha = -0.2t^3 + t^2$, where t is the time (in s). For what values of t is $\alpha = 2.0$ rad/s^2?

32. In finding one of the dimensions d (in cm) of the support columns of a building, the equation $3d^3 + 5d^2 - 400d - 18\,000 = 0$ is found. What is this dimension?

33. The deflection y of a beam at a horizontal distance x from one end is given by $y = k(x^4 - 2Lx^3 - L^3x)$, where L is the length of the beam and k is a constant. For what values of x is the deflection zero?

34. The specific gravity s of a sphere of radius r that sinks to a depth h in water is given by $s = \dfrac{3rh^2 - h^3}{4r^3}$. Find the depth to which a spherical buoy of radius 4.0 cm sinks if $s = 0.50$.

35. A variable electric voltage in a circuit is given by
$V = 0.1t^4 - 1.0t^3 + 3.5t^2 - 5.0t + 2.3$, where t is the time (in s). If the voltage is on for 5.0 s, when is $V = 0$?

36. The pressure difference p (in kPa) at a distance x (in km) from one end of an oil pipeline is given by $p = x^5 - 3x^4 - x^2 + 7x$. If the pipeline is 4 km long, where is $p = 0$?

37. A rectangular tray is made from a square piece of sheet metal 10.0 cm on a side by cutting equal squares from each corner, bending up the sides, and then welding them together. How long is the side of the square that must be cut out if the volume of the tray is 70.0 cm^3?

38. The angle θ of a robot arm with the horizontal as a function of time t (in s) is given by $\theta = 15 + 20t^2 - 4t^3$ for $0 \le t \le 5$ s. Find t for $\theta = 40°$.

39. The radii of four different-sized ball bearings differ by 1.00 mm in radius from one size to the next. If the volume of the largest equals the volumes of the other three combined, find the radii.

40. A rectangular safe is to be made of steel of uniform thickness, including the door. The inside dimensions are 1.20 m, 1.20 m, and 2.00 m. If the volume of steel is 1.25 m^3, find its thickness.

41. For electrical resistors connected in parallel, the reciprocal of the combined resistance equals the sum of the reciprocals of the individual resistances. If three resistors are connected in parallel such that the second resistance is 1 Ω more than the first, and the third is 4 Ω more than the first, find the resistances for a combined resistance of 1 Ω.

42. Each of three revolving doors has a perimeter of 6.60 m and revolves through a volume of 9.50 m^3 in one revolution about their common vertical side. What are the door's dimensions?

43. If a, b, and c are positive integers, find the combinations of the possible positive, negative, and nonreal complex roots if $f(x) = ax^3 - bx^2 + c = 0$.

W 44. An equation $f(x) = 0$ involves only odd powers of x with positive coefficients. Explain why this equation has no real root except $x = 0$.

CHAPTER 15 EQUATIONS

Polynomial function	$f(x) = a_0 x^n + a_1 x^{n-1} + \cdots + a_n$	(15.1)
Remainder theorem	$f(x) = (x - r)q(x) + R$	(15.2)
	$f(r) = R$	(15.3)
Rational roots	$r_r = \dfrac{\text{integral factor of } a_n}{\text{integral factor of } a_0}$	(15.4)

CHAPTER 15 REVIEW EXERCISES

In Exercises 1–4, find the remainder of the indicated division by the remainder theorem.

1. $(2x^3 - 4x^2 - x + 4) \div (x - 1)$
2. $(x^3 - 2x^2 + 9) \div (x + 2)$
3. $(4n^3 + n + 4) \div (n + 3)$
4. $(x^4 - 5x^3 + 8x^2 + 15x - 2) \div (x - 3)$

In Exercises 5–8, use the factor theorem to determine whether or not the second expression is a factor of the first.

5. $x^4 + x^3 + x^2 - 2x - 3; \quad x + 1$
6. $2s^3 - 6s - 4; \quad s - 2$
7. $x^4 + 4x^3 + 5x^2 + 5x - 6; \quad x + 3$
8. $9v^3 + 6v^2 + 4v + 2; \quad 3v + 1$

In Exercises 9–16, use synthetic division to perform the indicated divisions.

9. $(x^3 + 3x^2 + 6x + 1) \div (x - 1)$
10. $(3x^3 - 2x^2 + 7) \div (x - 3)$
11. $(2x^3 - 3x^2 - 4x + 3) \div (x + 2)$
12. $(3D^3 + 8D^2 - 16) \div (D + 4)$
13. $(x^4 - 2x^3 - 3x^2 - 4x - 8) \div (x + 1)$
14. $(x^4 - 6x^3 + x - 8) \div (x - 3)$
15. $(2m^5 - 46m^3 + m^2 - 9) \div (m - 5)$
16. $(x^6 + 63x^3 + 5x^2 - 9x - 8) \div (x + 4)$

In Exercises 17–20, use synthetic division to determine whether or not the given numbers are zeros of the given functions.

17. $y^3 + 5y^2 - 6 = 0; \quad -3$
18. $2x^3 + x^2 - 4x + 4; \quad -2$
19. $2x^4 - x^3 + 2x^2 + x - 1; \quad \frac{1}{2}$
20. $6W^4 - 7W^3 + 2W^2 - 9W - 6; \quad -\frac{2}{3}$

In Exercises 21–32, find all the roots of the given equations, using synthetic division and the given roots.

21. $x^3 + 8x^2 + 17x + 6 = 0 \quad (r_1 = -3)$
22. $3B^3 - B^2 - 24B + 28 = 0 \quad (r_1 = 2)$
23. $3x^4 + 5x^3 + x^2 + x - 10 = 0 \quad (r_1 = 1, r_2 = -2)$
24. $x^4 - x^3 - 5x^2 - x - 6 = 0 \quad (r_1 = 3, r_2 = -2)$
25. $4p^4 - p^2 - 18p + 9 = 0 \quad \left(r_1 = \frac{1}{2}, r_2 = \frac{3}{2}\right)$
26. $x^4 + x^3 - 11x^2 - 9x + 18 = 0 \quad (r_1 = -3, r_2 = 1)$
27. $4x^4 + 4x^3 + x^2 + 4x - 3 = 0 \quad (r_1 = j)$
28. $x^4 + 2x^3 - 4x - 4 = 0 \quad (r_1 = -1 + j)$
29. $x^5 + 3x^4 - x^3 - 11x^2 - 12x - 4 = 0 \quad (-1 \text{ is a triple root})$
30. $24x^5 + 10x^4 + 7x^2 - 6x + 1 = 0 \quad \left(r_1 = -1, r_2 = \frac{1}{4}, r_3 = \frac{1}{3}\right)$
31. $V^5 + 4V^4 + 5V^3 - V^2 - 4V - 5 = 0 \quad (r_1 = 1, r_2 = -2 + j)$
32. $2x^5 - x^4 + 8x - 4 = 0 \quad \left(r_1 = \frac{1}{2}, r_2 = 1 + j\right)$

In Exercises 33–40, solve the given equations.

33. $x^3 + x^2 - 10x + 8 = 0$
34. $x^3 - 8x^2 + 20x = 16$
35. $2r^3 - 3r^2 + 1 = 0$
36. $2x^3 - 3x^2 - 11x + 6 = 0$
37. $6x^3 - x^2 - 12x = 5$
38. $6y^3 + 19y^2 + 2y = 3$
39. $4t^4 - 17t^2 + 14t - 3 = 0$
40. $2x^4 + 5x^3 - 14x^2 - 23x + 30 = 0$

In Exercises 41–60, solve the given problems. Where appropriate, set up the required equations.

41. What are the possible numbers of real zeros (double roots count as two, etc.) for a polynomial with real coefficients and of degree 5?
42. What are the possible combinations of real and nonreal complex zeros (double roots count as two, etc.) of a fourth-degree polynomial?

43. If a graphing calculator shows a real root, how many nonreal complex roots are possible for a sixth-degree polynomial equation $f(x) = 0$?

44. Answer the same question as in Exercise 43 for a seventh-degree polynomial equation.

(W) **45.** Explain how to find k if $x + 2$ is a factor of $f(x) = 3x^3 + kx^2 - 8x - 8$. What is k?

(W) **46.** Explain how to find k if $x - 3$ is a factor of $f(x) = kx^4 - 15x^2 - 5x - 12$. What is k?

47. Where does the graph of the function $f(x) = 6x^4 - 14x^3 + 5x^2 + 5x - 2$ cross the x-axis?

48. Where does the graph of the function $f(x) = 2x^4 - 7x^3 + 11x^2 - 20x + 12$ cross the x-axis?

49. Find the irrational root of the equation $3x^3 - x^2 - 8x - 2 = 0$ that lies between 1 and 2.

50. Find the irrational root of the equation $x^4 + 3x^3 + 6x + 4 = 0$ that lies between −1 and 0.

51. A computer analysis of the number of crimes committed each month in a certain city for the first 10 months of a year showed that $n = x^3 - 9x^2 + 15x + 600$. Here n is the number of monthly crimes and x is the number of the month (as of the last day). In what month were 580 crimes committed?

52. A company determined that the number s (in thousands) of computer chips that it could supply at a price p of less than $5 is given by $s = 4p^2 - 25$, whereas the demand d (in thousands) for the chips is given by $d = p^3 - 22p + 50$. For what price is the supply equal to the demand?

53. In order to find the diameter d (in cm) of a helical spring subject to given forces, it is necessary to solve the equation $64d^3 - 144d^2 + 108d - 27 = 0$. Solve for d.

54. A cubical tablet for purifying water is wrapped in a sheet of foil 0.500 mm thick. The total volume of tablet and foil is 33.1% greater than the volume of the tablet alone. Find the length of the edge of the tablet.

55. For the mirror shown in Fig. 15.14, the reciprocal of the focal distance f equals the sum of the reciprocals of the object distance p (in cm) and image distance q (in cm). If $q = p + 4$ and $f = \dfrac{p+1}{p}$, find p.

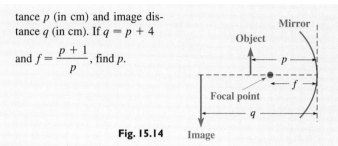

Fig. 15.14

56. Three electric capacitors are connected in series. The capacitance of the second is 1 μF more than that of the first, and the third is 2 μF more than the second. The capacitance of the combination is 1.33 μF. The equation used to determine C, the capacitance of the first capacitor, is

$$\frac{1}{C} + \frac{1}{C+1} + \frac{1}{C+3} = \frac{3}{4}$$

Find the values of the capacitances.

57. The height of a cylindrical oil tank is 3.2 m more than the radius. If the volume of the tank is 680 m³, what are the radius and the height of the tank?

58. A grain storage bin has a square base, each side of which is 5.5 m longer than the height of the bin. If the bin holds 160 m³ of grain, find its dimensions.

59. A rectangular door has a diagonal brace that is 0.300 m longer than the height of the door. If the area of the door is 2.70 m², find its dimensions.

60. The radius of one ball bearing is 1.0 mm greater than the radius of a second ball bearing. If the sum of their volumes is 100 mm³, find the radius of each.

Writing Exercise

61. A computer science student is to write a computer program that will print out the values of n for which $x + r$ is a factor of $x^n + r^n$. Write a paragraph that states which are the values of n and explains how they are found.

CHAPTER (15) PRACTICE TEST

1. Is −3 a zero for the function $2x^3 + 3x^2 + 7x - 6$?

2. Find the remaining roots of the equation $x^4 - 2x^3 - 7x^2 + 20x - 12 = 0$; 2 is a double root.

3. Use synthetic division to perform the division $(x^3 - 5x^2 + 4x - 9) \div (x - 3)$.

4. Use the factor theorem and synthetic division to determine whether or not $2x + 1$ is a factor of $2x^4 + 15x^3 + 23x^2 - 16$.

5. Use the remainder theorem to find the remainder of the division $(x^3 + 4x^2 + 7x - 9) \div (x + 4)$.

6. Solve for x: $2x^4 - x^3 + 5x^2 - 4x - 12 = 0$.

7. The ends of a 10-m beam are supported at different levels. The deflection y of the beam is given by $y = kx^2(x^3 + 436x - 4000)$, where x is the horizontal distance from one end and k is a constant. Find the values of x for which the deflection is zero.

8. A cubical metal block is heated such that its edge increases by 1.0 mm and its volume is doubled. Find the edge of the cube to tenths.

16 Matrices

While working with systems of linear equations and their solutions in the 1850s and 1860s, the English mathematician Arthur Cayley developed the use of a *matrix* (as we will show, a matrix is simply a rectangular array of numbers). His interest was only in the mathematical methods involved, and for many years matrices were used by mathematicians with little or no reference to possible applications. Of course, it was known they could be used whenever simultaneous linear equations were involved.

It was not until the 1920s that one of the first major applications of matrices was made when physicists used them in developing theories about the elementary particles within the atom. Their work has been and is still very important in atomic and nuclear physics.

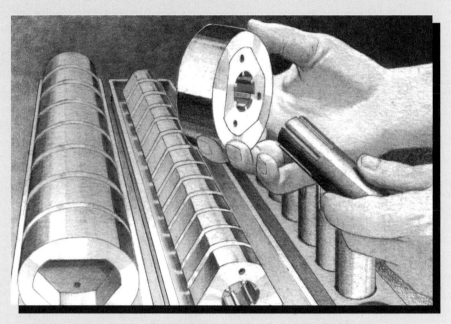

In Section 16.2, we see how to determine the amount of material and worker time required for the production of machine parts.

Since about 1950, matrices have become a very important and useful tool in many other areas of application, such as social science and economics. They are now used extensively in business and industry in making appropriate decisions in research, development, and production.

Matrices are very well adapted for solving systems of equations on a computer. Also, most graphing calculators have a feature for doing matrix operations and using them to solve a system of equations.

Here we have the case of a mathematical method that was developed only for its usefulness in mathematics, but later has been found to be very useful in many areas of application.

16.1 DEFINITIONS AND BASIC OPERATIONS

Systems of linear equations occur in many areas of important technical and scientific applications, and we showed some of these in Chapter 5. The importance of linear systems has led to the development of several methods for their solution.

The method of Arthur Cayley for solving systems of linear equations by the use of *matrices* is readily adaptable for use on a computer and is used on many graphing calculators. For this reason, the method is used much more widely now than before computers were in common use. In this section, we introduce the definitions and some basic operations with matrices. In the sections that follow, we develop additional operations and show how they are used in solving systems of equations. This is only an introduction to the use of matrices, and additional uses and operations with matrices are shown in other sources.

A **matrix** *is an ordered rectangular array of numbers.* To distinguish such an array from a determinant, we enclose it within brackets. As with a determinant, *the individual members are called* **elements** *of the matrix.*

We first introduced the term *matrix* on page 167 when we showed the evaluation of a determinant on a calculator.

◀ EXAMPLE 1   Some examples of matrices are shown here.

$$\begin{bmatrix} 2 & 8 \\ 1 & 0 \end{bmatrix} \qquad \begin{bmatrix} 2 & -4 & 6 \\ -1 & 0 & 5 \end{bmatrix} \qquad \begin{bmatrix} 4 & 6 \\ 0 & -1 \\ -2 & 5 \\ 3 & 0 \end{bmatrix}$$

$$\begin{bmatrix} -1 & 8 & 6 & 7 & 9 \\ 2 & 6 & 0 & 4 & 3 \\ 5 & -1 & 8 & 10 & 2 \end{bmatrix} \qquad \begin{bmatrix} -1 & 2 & 0 & 9 \end{bmatrix}$$

As we can see, it is not necessary for the number of columns and number of rows to be the same, although such is the case for a determinant. However, *if the number of rows does equal the number of columns, the matrix is called a* **square matrix.** We will find that square matrices are of special importance. *If all the elements of a matrix are zero, the matrix is called a* **zero matrix.** It is convenient to designate a given matrix by a capital letter.

CAUTION ▶   We must be careful to distinguish between a matrix and a determinant. *A matrix is simply any* **rectangular array** *of numbers, whereas a determinant is a specific value associated with a* **square** *matrix.*

◀ EXAMPLE 2   Consider the following matrices:

$$A = \begin{bmatrix} 5 & 0 & -1 \\ 1 & 2 & 6 \\ 0 & -4 & -5 \end{bmatrix} \qquad B = \begin{bmatrix} 9 \\ 8 \\ 1 \\ 5 \end{bmatrix} \qquad C = \begin{bmatrix} -1 & 6 & 8 & 9 \end{bmatrix} \qquad O = \begin{bmatrix} 0 & 0 \\ 0 & 0 \end{bmatrix}$$

Matrix A is an example of a square matrix, matrix B is an example of a matrix with four rows and one column, matrix C is an example of a matrix with one row and four columns, and matrix O is an example of a zero matrix. ▮

To be able to refer to specific elements of a matrix and to give a general representation, a double-subscript notation is usually employed. That is,

$$A = \begin{bmatrix} a_{11} & a_{12} & a_{13} \\ a_{21} & a_{22} & a_{23} \\ a_{31} & a_{32} & a_{33} \end{bmatrix}$$

row ⟍⟋⟍ column

We see that the first subscript refers to the row in which the element lies and the second subscript refers to the column in which the element lies.

NOTE ▶ *Two matrices are said to be* **equal** *if and only if they are identical.* That is, they must have the same number of columns, the same number of rows, and the elements must respectively be equal.

◀ EXAMPLE 3   **(a)** $\begin{bmatrix} a_{11} & a_{12} & a_{13} \\ a_{21} & a_{22} & a_{23} \end{bmatrix} = \begin{bmatrix} 1 & -5 & 0 \\ 4 & 6 & -3 \end{bmatrix}$

if and only if $a_{11} = 1$, $a_{12} = -5$, $a_{13} = 0$, $a_{21} = 4$, $a_{22} = 6$, and $a_{23} = -3$.

(b) The matrices

$$\begin{bmatrix} 1 & 2 & 3 \\ -1 & -2 & -5 \end{bmatrix} \quad \text{and} \quad \begin{bmatrix} 1 & 2 & -5 \\ -1 & -2 & 3 \end{bmatrix}$$

are not equal, since the elements in the third column are reversed.

(c) The matrices

$$\begin{bmatrix} 2 & 3 \\ -1 & 5 \end{bmatrix} \quad \text{and} \quad \begin{bmatrix} 2 & 3 & 0 \\ -1 & 5 & 0 \end{bmatrix}$$

are not equal, since the number of columns is different. This is true despite the fact that both elements of the third column are zeros.   ▶

◀ EXAMPLE 4   The forces acting on a bolt are in equilibrium, as shown in Fig. 16.1. Analyzing the horizontal and vertical components as in Section 9.4, we find the following matrix equation. Find forces F_1 and F_2.

$$\begin{bmatrix} 0.98F_1 - 0.88F_2 \\ 0.22F_1 + 0.47F_2 \end{bmatrix} = \begin{bmatrix} 8.0 \\ 3.5 \end{bmatrix}$$

From the equality of matrices, we know that $0.98F_1 - 0.88F_2 = 8.0$ and $0.22F_1 + 0.47F_2 = 3.5$. Therefore, to find the forces F_1 and F_2, we must solve the system of equations

$$0.98F_1 - 0.88F_2 = 8.0$$
$$0.22F_1 + 0.47F_2 = 3.5$$

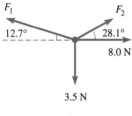

F_1 F_2

12.7°   28.1°

8.0 N

3.5 N

Fig. 16.1

Using determinants, we have

$$F_1 = \frac{\begin{vmatrix} 8.0 & -0.88 \\ 3.5 & 0.47 \end{vmatrix}}{\begin{vmatrix} 0.98 & -0.88 \\ 0.22 & 0.47 \end{vmatrix}} = \frac{8.0(0.47) - 3.5(-0.88)}{0.98(0.47) - 0.22(-0.88)} = 10.5 \text{ N}$$

Using determinants again, or by substituting this value into either equation, we find that $F_2 = 2.6$ N. These values check when substituted into the original matrix equation.   ▶

MATRIX ADDITION AND SUBTRACTION

If two matrices have the same number of rows and the same number of columns, their **sum** *is defined as the matrix consisting of the sums of the corresponding elements.* If the number of rows or the number of columns of the two matrices is not equal, they cannot be added.

◀ **EXAMPLE 5** **(a)**

$$\begin{bmatrix} 8 & 1 & -5 & 9 \\ 0 & -2 & 3 & 7 \end{bmatrix} + \begin{bmatrix} -3 & 4 & 6 & 0 \\ 6 & -2 & 6 & 5 \end{bmatrix} = \begin{bmatrix} 8 + (-3) & 1 + 4 & -5 + 6 & 9 + 0 \\ 0 + 6 & -2 + (-2) & 3 + 6 & 7 + 5 \end{bmatrix}$$

$$= \begin{bmatrix} 5 & 5 & 1 & 9 \\ 6 & -4 & 9 & 12 \end{bmatrix}$$

(b) The matrices

$$\begin{bmatrix} 3 & -5 & 8 \\ 2 & 9 & 0 \\ 4 & -2 & 3 \end{bmatrix} \text{ and } \begin{bmatrix} 3 & -5 & 8 & 0 \\ 2 & 9 & 0 & 0 \\ 4 & -2 & 3 & 0 \end{bmatrix}$$

cannot be added since the second matrix has one more column than the first matrix. This is true even though the extra column contains only zeros. ◗

The product of a number and a matrix (known as **scalar multiplication** *of a matrix) is defined as the matrix whose elements are obtained by multiplying each element of the given matrix by the given number.* Thus, we obtain matrix kA by multiplying the elements of matrix A by k. In this way, $A + A$ and $2A$ are the same matrix.

◀ **EXAMPLE 6** For the matrix A, where

$$A = \begin{bmatrix} -5 & 7 \\ 3 & 0 \end{bmatrix} \text{ we have } 2A = \begin{bmatrix} 2(-5) & 2(7) \\ 2(3) & 2(0) \end{bmatrix} = \begin{bmatrix} -10 & 14 \\ 6 & 0 \end{bmatrix}$$

Figure 16.2 shows a calculator display for A and $2A$. ◗

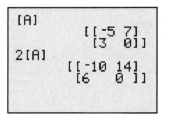

```
[A]
        [[-5 7]
         [3  0]]
2[A]
        [[-10 14]
         [6   0]]
```

Fig. 16.2

By combining the definitions for the addition of matrices and for the scalar multiplication of a matrix, we can define the subtraction of matrices. That is, *the* **difference** *of matrices A and B is given by* $A - B = A + (-B)$. Therefore, we change the sign of each element of B, and proceed as in addition.

The operations of addition, subtraction, and multiplication of a matrix by a number are like those for real numbers. For these operations, the algebra of matrices is like the algebra of real numbers. We see that the following laws hold for matrices.

$A + B = B + A$	(commutative law)	**(16.1)**
$A + (B + C) = (A + B) + C$	(associative law)	**(16.2)**
$k(A + B) = kA + kB$		**(16.3)**
$A + O = A$		**(16.4)**

Here, we have let O represent the zero matrix. We will find in the next section that not all laws for matrix operations are like those for real numbers.

EXERCISES 16.1

In Exercises 1 and 2, make the given changes in the indicated examples of this section and then perform the indicated operations.

1. In Example 5(a), interchange the second and third columns of the second matrix and then add the matrices.

2. In Example 6, find the matrix $-2A$.

In Exercises 3–10, determine the value of the literal numbers in each of the given matrix equalities.

3. $\begin{bmatrix} a & b \\ c & d \end{bmatrix} = \begin{bmatrix} 1 & -3 \\ 4 & 7 \end{bmatrix}$

4. $\begin{bmatrix} x \\ x+y \end{bmatrix} = \begin{bmatrix} 2 \\ 5 \end{bmatrix}$

5. $\begin{bmatrix} x & 2y & z \\ r/4 & -s & -5t \end{bmatrix} = \begin{bmatrix} -2 & 10 & -9 \\ 12 & -4 & 5 \end{bmatrix}$

6. $[a + bj \quad 2c - dj \quad 3e + fj] = [5j \quad a + 6 \quad 3b + c]$
$(j = \sqrt{-1})$

7. $\begin{bmatrix} C + D \\ 2C - D \\ D - 2E \end{bmatrix} = \begin{bmatrix} 5 \\ 4 \\ 6 \end{bmatrix}$

8. $\begin{bmatrix} 2x - 3y \\ x + 4y \end{bmatrix} = \begin{bmatrix} 13 \\ 1 \end{bmatrix}$

9. $\begin{bmatrix} x - 3 & x + y \\ x - z & y + z \\ x + t & y - t \end{bmatrix} = \begin{bmatrix} 5 & 3 \\ 4 & -1 \end{bmatrix}$

10. $\begin{bmatrix} x & y & z \\ x+y & 2x-y & x+2 \end{bmatrix} = \begin{bmatrix} 2 & -3 \\ z & t \end{bmatrix}$

In Exercises 11–14, find the indicated sums of matrices.

11. $\begin{bmatrix} 2 & 3 \\ -5 & 4 \end{bmatrix} + \begin{bmatrix} -1 & 7 \\ 5 & -2 \end{bmatrix}$

12. $\begin{bmatrix} 1 & 0 & 9 \\ 3 & -5 & -2 \end{bmatrix} + \begin{bmatrix} 4 & -1 & 7 \\ 2 & 0 & -3 \end{bmatrix}$

13. $\begin{bmatrix} 50 & -82 \\ -34 & 57 \\ -15 & 62 \end{bmatrix} + \begin{bmatrix} -55 & 82 \\ 45 & 14 \\ 26 & -67 \end{bmatrix}$

14. $\begin{bmatrix} 4.7 & 2.1 & -9.6 \\ -6.8 & 4.8 & 7.4 \\ -1.9 & 0.7 & 5.9 \end{bmatrix} + \begin{bmatrix} -4.9 & -9.6 & -2.1 \\ 3.4 & 0.7 & 0.0 \\ 5.6 & 10.1 & -1.6 \end{bmatrix}$

In Exercises 15–26, use the following matrices to find the indicated matrices.

$$A = \begin{bmatrix} -1 & 4 & -7 & 0 \\ 2 & -6 & -1 & 2 \end{bmatrix} \quad B = \begin{bmatrix} 1 & 5 & -6 & 3 \\ 4 & -1 & 8 & -2 \end{bmatrix}$$

$$C = \begin{bmatrix} 3 & -6 & 9 \\ -4 & 1 & 2 \end{bmatrix}$$

15. $A + B$ **16.** $A - B$ **17.** $A + C$ **18.** $B + C$

19. $2A + B$ **20.** $2B + A$ **21.** $A - 2B$ **22.** $3A - B$

23. $-4A$ **24.** $-3B$ **25.** $-C - A$ **26.** $-\frac{1}{2}A + B$

In Exercises 27–30, use matrices A and B to show that the indicated laws hold for these matrices. In Exercise 27, explain the meaning of the result.

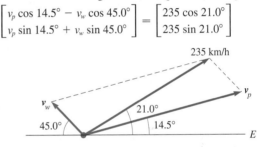

$$A = \begin{bmatrix} -1 & 2 & 3 & 7 \\ 0 & -3 & -1 & 4 \\ 9 & -1 & 0 & -2 \end{bmatrix} \quad B = \begin{bmatrix} 4 & -1 & -3 & 0 \\ 5 & 0 & -1 & 1 \\ 1 & 11 & 8 & 2 \end{bmatrix}$$

(W) **27.** $A + B = B + A$ **28.** $A + O = A$

29. $-(A - B) = B - A$ **30.** $3(A + B) = 3A + 3B$

In Exercises 31 and 32, find the unknown quantities in the given matrix equations.

31. An airplane is flying in a direction $21.0°$ north of east at 235 km/h but is headed $14.5°$ north of east. The wind is from the southeast. Find the speed of the wind v_w and the speed of the plane v_p relative to the wind from the given matrix equation. See Fig. 16.3.

$$\begin{bmatrix} v_p \cos 14.5° - v_w \cos 45.0° \\ v_p \sin 14.5° + v_w \sin 45.0° \end{bmatrix} = \begin{bmatrix} 235 \cos 21.0° \\ 235 \sin 21.0° \end{bmatrix}$$

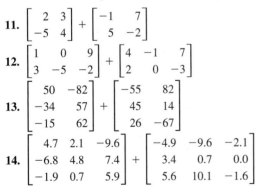

Fig. 16.3

32. Find the electric currents shown in Fig. 16.4 by solving the following matrix equation:

$$\begin{bmatrix} I_1 + I_2 + I_3 \\ -2I_1 + 3I_2 \\ -3I_2 + 6I_3 \end{bmatrix} = \begin{bmatrix} 0 \\ 24 \\ 0 \end{bmatrix}$$

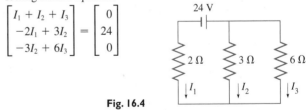

Fig. 16.4

In Exercises 33–36, perform the indicated matrix operations.

33. The contractor of a housing development constructs four different types of houses, with either a carport, a one-car garage, or a two-car garage. The following matrix shows the number of houses of each type and the type of garage.

	Type A	Type B	Type C	Type D
Carport	8	6	0	0
1-car garage	5	4	3	0
2-car garage	0	3	5	6

If the contractor builds two additional identical developments, find the matrix showing the total number of each house-garage type built in the three developments.

34. The inventory of a drug supply company shows that the following numbers of cases of bottles of vitamins C and B_3 (niacin) are in stock: Vitamin C—25 cases of 100-mg bottles, 10 cases of 250-mg bottles, and 32 cases of 500-mg bottles; vitamin B_3—30 cases of 100-mg bottles, 18 cases of 250-mg bottles, and 40 cases of 500-mg bottles. This is represented by matrix A below. After two shipments are sent out, each of which can be represented by matrix B below, find the matrix that represents the remaining inventory.

$$A = \begin{bmatrix} 25 & 10 & 32 \\ 30 & 18 & 40 \end{bmatrix} \qquad B = \begin{bmatrix} 10 & 5 & 6 \\ 12 & 4 & 8 \end{bmatrix}$$

35. One serving of brand K of breakfast cereal provides the given percentages of the given vitamins and minerals: vitamin A, 15%; vitamin C, 25%; calcium, 10%; iron, 25%. One serving of brand G provides: vitamin A, 10%; vitamin C, 10%; calcium, 10%; iron, 45%. One serving of tomato juice provides: vitamin A, 15%; vitamin C, 30%; calcium, 3%; iron, 3%. One serving of orange-pineapple juice provides vitamin A, 0%; vitamin C, 100%; calcium, 2%; iron, 2%. Set up a two-row, four-column matrix B to represent the data for the cereals and a similar matrix J for the juices.

(W) 36. Referring to Exercise 35, find the matrix $B + J$ and explain the meaning of its elements.

16.2 MULTIPLICATION OF MATRICES

The definition for the multiplication of matrices does not have an intuitive basis. However, through the solution of a system of linear equations we can, at least in part, show why multiplication is defined as it is. Consider Example 1.

◀ EXAMPLE 1 If we solve the system of equations

$$2x + y = 1$$
$$7x + 3y = 5$$

we get $x = 2$, $y = -3$. Checking this solution in each of the equations, we get

$$2(2) + 1(-3) = 1$$
$$7(2) + 3(-3) = 5$$

Let us represent the coefficients of the equations by the matrix $\begin{bmatrix} 2 & 1 \\ 7 & 3 \end{bmatrix}$ and the solutions by the matrix $\begin{bmatrix} 2 \\ -3 \end{bmatrix}$. If we now indicate the multiplications of these matrices and perform it as shown

$$\begin{bmatrix} 2 & 1 \\ 7 & 3 \end{bmatrix}\begin{bmatrix} 2 \\ -3 \end{bmatrix} = \begin{bmatrix} 2(2) + 1(-3) \\ 7(2) + 3(-3) \end{bmatrix} = \begin{bmatrix} 1 \\ 5 \end{bmatrix}$$

we note that we obtain a matrix that properly represents the right-side values of the equations. (Note the products and sums in the resulting matrix.) ▶

Following reasons along the lines indicated in Example 1, we now define the **multiplication of matrices.** If the number of columns in a first matrix equals the number of rows in a second matrix, the product of these matrices is formed as follows: *The element in a specified row and a specified column of the product matrix is the sum of the products formed by multiplying each element in the specified row of the first matrix by the corresponding element in the specific column of the second matrix.* The product matrix will have the same number of rows as the first matrix and the same number of columns as the second matrix. Consider the following examples.

◀ EXAMPLE 2   Find the product AB, where

$$A = \begin{bmatrix} 2 & 1 \\ -3 & 0 \\ 1 & 2 \end{bmatrix} \qquad B = \begin{bmatrix} -1 & 6 & 5 & -2 \\ 3 & 0 & 1 & -4 \end{bmatrix}$$

With two columns in matrix A and two rows in matrix B, the product can be formed. The element in the first row and first column of the product is the sum of the products of the corresponding elements of the first row of A and first column of B. The elements in the first row and second column of the product is the sum of the products of corresponding elements of the first row of A and second column of B. We continue until we have three rows (the number in A) and four columns (the number in B).

$$\begin{bmatrix} 2 & 1 \\ -3 & 0 \\ 1 & 2 \end{bmatrix}\begin{bmatrix} -1 & 6 & 5 & -2 \\ 3 & 0 & 1 & -4 \end{bmatrix} = \begin{bmatrix} 2(-1)+1(3) & 2(6)+1(0) & 2(5)+1(1) & 2(-2)+1(-4) \\ -3(-1)+0(3) & -3(6)+0(0) & -3(5)+0(1) & -3(-2)+0(-4) \\ 1(-1)+2(3) & 1(6)+2(0) & 1(5)+2(1) & 1(-2)+2(-4) \end{bmatrix}$$

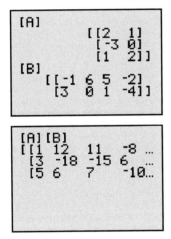

$$= \begin{bmatrix} 1 & 12 & 11 & -8 \\ 3 & -18 & -15 & 6 \\ 5 & 6 & 7 & -10 \end{bmatrix}$$

The elements used to form the element in the first row and first column and the element in the third row and second column of the product are outlined in color. Figure 16.5 shows the calculator display for matrices A and B and their product AB.

In trying to form the product BA, we see that B has four columns and A has three rows. Since these numbers are not the same, the product cannot be formed. Therefore, $AB \neq BA$, which means *matrix multiplication is not commutative (except in special cases)*, and therefore differs from multiplication of real numbers. ▶

Fig. 16.5

◀ EXAMPLE 3   The product of two matrices below may be formed because the first matrix has four columns and the second matrix has four rows. The matrix is formed as shown.

$$\begin{bmatrix} -1 & 9 & 3 & -2 \\ 2 & 0 & -7 & 1 \end{bmatrix}\begin{bmatrix} 6 & -2 \\ 1 & 0 \\ 3 & -5 \\ 3 & 9 \end{bmatrix} = \begin{bmatrix} -1(6)+9(1)+3(3)+(-2)(3) & -1(-2)+9(0)+3(-5)+(-2)(9) \\ 2(6)+0(1)+(-7)(3)+1(3) & 2(-2)+0(0)+(-7)(-5)+1(9) \end{bmatrix}$$

Note that, if the second matrix is to the left of the first, then the product has four rows and four columns.

$$= \begin{bmatrix} -6+9+9-6 & 2+0-15-18 \\ 12+0-21+3 & -4+0+35+9 \end{bmatrix}$$

$$= \begin{bmatrix} 6 & -31 \\ -6 & 40 \end{bmatrix}$$ ▶

IDENTITY MATRIX

There are two special matrices of particular importance in the multiplication of matrices. The first of these is the **identity matrix I,** *which is a square matrix with 1's for elements of the principal diagonal with all other elements zero.* (The principal diagonal starts with the element a_{11}.) It has the property that if it is multiplied by another square matrix with the same number of rows and columns, then the second matrix equals the product matrix.

◀ EXAMPLE 4   Show that $AI = IA = A$ for the matrix

$$A = \begin{bmatrix} 2 & -3 \\ 4 & 1 \end{bmatrix}$$

Since A has two rows and two columns, we choose I with two rows and two columns. Therefore, for this case

$$I = \begin{bmatrix} 1 & 0 \\ 0 & 1 \end{bmatrix} \longleftarrow \text{elements of principal diagonal are 1's}$$

Forming the indicated products, we have results as follows:

$$AI = \begin{bmatrix} 2 & -3 \\ 4 & 1 \end{bmatrix}\begin{bmatrix} 1 & 0 \\ 0 & 1 \end{bmatrix}$$

$$= \begin{bmatrix} 2(1) + (-3)(0) & 2(0) + (-3)(1) \\ 4(1) + 1(0) & 4(0) + 1(1) \end{bmatrix} = \begin{bmatrix} 2 & -3 \\ 4 & 1 \end{bmatrix}$$

$$IA = \begin{bmatrix} 1 & 0 \\ 0 & 1 \end{bmatrix}\begin{bmatrix} 2 & -3 \\ 4 & 1 \end{bmatrix}$$

$$= \begin{bmatrix} 1(2) + 0(4) & 1(-3) + 0(1) \\ 0(2) + 1(4) & 0(-3) + 1(1) \end{bmatrix} = \begin{bmatrix} 2 & -3 \\ 4 & 1 \end{bmatrix}$$

Therefore, we see that $AI = IA = A$.

INVERSE OF A MATRIX

For a given square matrix A, its **inverse** A^{-1} is the other important special matrix. *The matrix A and its inverse A^{-1} have the property that*

$$AA^{-1} = A^{-1}A = I \tag{16.5}$$

If the product of two square matrices equals the identity matrix, the matrices are called inverses of each other. Under certain conditions, the inverse of a given square matrix may not exist, although for most square matrices the inverse does exist. In the next section, we develop the procedure for finding the inverse of a square matrix, and the section that follows shows how the inverse is used in the solution of systems of equations. At this point, we simply show that the product of certain matrices equals the identity matrix and that therefore these matrices are inverses of each other.

◀ EXAMPLE 5   For the given matrices A and B, show that $AB = BA = I$, and therefore that $B = A^{-1}$.

$$A = \begin{bmatrix} 1 & -3 \\ -2 & 7 \end{bmatrix} \qquad B = \begin{bmatrix} 7 & 3 \\ 2 & 1 \end{bmatrix}$$

Forming the products AB and BA, we have the following:

$$AB = \begin{bmatrix} 1 & -3 \\ -2 & 7 \end{bmatrix}\begin{bmatrix} 7 & 3 \\ 2 & 1 \end{bmatrix} = \begin{bmatrix} 7 - 6 & 3 - 3 \\ -14 + 14 & -6 + 7 \end{bmatrix} = \begin{bmatrix} 1 & 0 \\ 0 & 1 \end{bmatrix}$$

$$BA = \begin{bmatrix} 7 & 3 \\ 2 & 1 \end{bmatrix}\begin{bmatrix} 1 & -3 \\ -2 & 7 \end{bmatrix} = \begin{bmatrix} 7 - 6 & -21 + 21 \\ 2 - 2 & -6 + 7 \end{bmatrix} = \begin{bmatrix} 1 & 0 \\ 0 & 1 \end{bmatrix}$$

Since $AB = I$ and $BA = I$, $B = A^{-1}$ and $A = B^{-1}$.

The following example illustrates one kind of application of the multiplication of matrices.

Solving a Word Problem

See the chapter introduction.

❰ EXAMPLE 6   A company makes three types of machine parts. In one day, it produces 40 of type X, 50 of type Y, and 80 of type Z. Each of type X requires 4 units of material and 1 worker-hour to produce; each of type Y requires 5 units of material and 2 worker-hours to produce; each of type Z requires 3 units of material and 2 worker-hours to produce. By representing the number of each type produced as matrix A and the material and time requirements as matrix B, we have

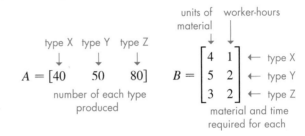

The product AB gives the total number of units of material and the total number of worker-hours needed for the day's production in a one-row, two-column matrix:

$$AB = \begin{bmatrix} 40 & 50 & 80 \end{bmatrix} \begin{bmatrix} 4 & 1 \\ 5 & 2 \\ 3 & 2 \end{bmatrix}$$

$$= \begin{bmatrix} 160 + 250 + 240 & 40 + 100 + 160 \end{bmatrix} = \begin{bmatrix} 650 & 300 \end{bmatrix}$$

Therefore, 650 units of material and 300 worker-hours are required.   ❱

We now have seen how multiplication is defined for matrices. We see that *matrix multiplication is not commutative;* that is, $AB \neq BA$ in general. This is a major difference from the multiplication of real numbers. Another difference is that it is possible that $AB = O$, even though neither A nor B is O. There are, however, some similarities. We have seen, for example, that $AI = A$, where we make I and the number 1 equivalent for the two types of multiplication. Also, *the distributive property $A(B + C) = AB + AC$ holds for matrix multiplication.* This points out additional properties of the algebra of matrices.

EXERCISES 16.2

In Exercises 1 and 2, make the given changes in the indicated examples of this section and then perform the indicated multiplications.

1. In Example 2, interchange columns 1 and 2 in matrix A and then do the multiplication.

2. In Example 5, in A change -2 to 2 and -3 to 3, in B change 2 to -2 and 3 to -3, and then do the multiplications.

In Exercises 3–10, perform the indicated multiplications.

3. $\begin{bmatrix} 4 & -2 \end{bmatrix} \begin{bmatrix} -1 & 0 \\ 2 & 6 \end{bmatrix}$

4. $\begin{bmatrix} -\frac{1}{2} & 6 & -\frac{2}{3} \end{bmatrix} \begin{bmatrix} 4 & 8 \\ \frac{1}{3} & -\frac{1}{2} \\ 0 & 9 \end{bmatrix}$

5. $\begin{bmatrix} 2 & -3 & 1 \\ 0 & 7 & -3 \end{bmatrix} \begin{bmatrix} 9 \\ -2 \\ 5 \end{bmatrix}$

6. $\begin{bmatrix} 0 & -1 & 2 \\ 4 & 11 & 2 \end{bmatrix} \begin{bmatrix} 3 & -1 \\ 1 & 2 \\ 6 & 1 \end{bmatrix}$

7. $\begin{bmatrix} -8 & \frac{3}{4} \\ \frac{7}{2} & -8 \\ -6 & \frac{4}{5} \end{bmatrix} \begin{bmatrix} \frac{1}{4} & -3 \\ 2 & 5 \end{bmatrix}$

8. $\begin{bmatrix} 12 & -47 \\ 43 & -18 \\ 36 & -22 \end{bmatrix} \begin{bmatrix} 25 \\ 66 \end{bmatrix}$

9. $\begin{bmatrix} -1 & 7 \\ 3 & 5 \\ 10 & -1 \\ -5 & 12 \end{bmatrix} \begin{bmatrix} 2 & 1 \\ 5 & -3 \end{bmatrix}$

10. $\begin{bmatrix} 5 & 4 \end{bmatrix} \begin{bmatrix} 4 & -4 \\ -5 & 5 \end{bmatrix}$

In Exercises 11–14, use a graphing calculator to perform the indicated multiplications.

11. $\begin{bmatrix} 2 & -3 \\ 5 & -1 \end{bmatrix}\begin{bmatrix} 3 & 0 & -1 \\ 7 & -5 & 8 \end{bmatrix}$ **12.** $\begin{bmatrix} -7 & 8 \\ 5 & 0 \end{bmatrix}\begin{bmatrix} -9 & 10 \\ 1 & 4 \end{bmatrix}$

13. $\begin{bmatrix} -9.2 & 2.3 & 0.5 \\ -3.8 & -2.4 & 9.2 \end{bmatrix}\begin{bmatrix} 6.5 & -5.2 \\ 4.9 & 1.7 \\ -1.8 & 6.9 \end{bmatrix}$

14. $\begin{bmatrix} 1 & 2 & -6 & 6 & 1 \\ -2 & 4 & 0 & 1 & 2 \end{bmatrix}\begin{bmatrix} 1 \\ -1 \\ 0 \\ 5 \\ 2 \end{bmatrix}$

In Exercises 15–18, find, if possible, AB and BA. If it is not possible, explain why.

15. $A = \begin{bmatrix} 1 & -3 & 8 \end{bmatrix}$ $B = \begin{bmatrix} -1 \\ 5 \\ 7 \end{bmatrix}$

16. $A = \begin{bmatrix} -3 & 2 & 0 \\ 1 & -4 & 5 \end{bmatrix}$ $B = \begin{bmatrix} -2 & 0 \\ 4 & -6 \\ 5 & 1 \end{bmatrix}$

17. $A = \begin{bmatrix} -1 & 2 & 3 \\ 5 & -1 & 0 \end{bmatrix}$ $B = \begin{bmatrix} 1 \\ -5 \\ 2 \end{bmatrix}$

18. $A = \begin{bmatrix} -2 & 1 & 7 \\ 3 & -1 & 0 \\ 0 & 2 & -1 \end{bmatrix}$ $B = \begin{bmatrix} 4 & -1 & 5 \end{bmatrix}$

In Exercises 19–22, show that AI = IA = A.

19. $A = \begin{bmatrix} 1 & 8 \\ -2 & 2 \end{bmatrix}$ **20.** $A = \begin{bmatrix} -3 & 4 \\ 1 & 2 \end{bmatrix}$

21. $A = \begin{bmatrix} 1 & 3 & -5 \\ 2 & 0 & 1 \\ 1 & -2 & 4 \end{bmatrix}$ **22.** $A = \begin{bmatrix} -1 & 2 & 0 \\ 4 & -3 & 1 \\ 2 & 1 & 3 \end{bmatrix}$

In Exercises 23–26, determine whether or not $B = A^{-1}$.

23. $A = \begin{bmatrix} 5 & -2 \\ -2 & 1 \end{bmatrix}$ $B = \begin{bmatrix} 1 & 2 \\ 2 & 5 \end{bmatrix}$

24. $A = \begin{bmatrix} 3 & -4 \\ 5 & -7 \end{bmatrix}$ $B = \begin{bmatrix} 7 & -4 \\ 5 & -2 \end{bmatrix}$

25. $A = \begin{bmatrix} 1 & -2 & 3 \\ 2 & -5 & 7 \\ -1 & 3 & -5 \end{bmatrix}$ $B = \begin{bmatrix} 4 & -1 & 1 \\ 3 & -2 & -1 \\ 1 & -1 & -1 \end{bmatrix}$

26. $A = \begin{bmatrix} 1 & -1 & 3 \\ 3 & -4 & 8 \\ -2 & 3 & -4 \end{bmatrix}$ $B = \begin{bmatrix} 8 & -5 & -4 \\ 4 & -2 & -1 \\ -1 & 1 & 1 \end{bmatrix}$

In Exercises 27–30, determine by matrix multiplication whether or not A is the proper matrix of solution values.

27. $3x - 2y = -1$ $A = \begin{bmatrix} 1 \\ 2 \end{bmatrix}$
  $4x + y = 6$

28. $4x + y = -5$ $A = \begin{bmatrix} -2 \\ 3 \end{bmatrix}$
  $3x + 4y = 6$

29. $3x + y + 2z = 1$
  $x - 3y + 4z = -3$ $A = \begin{bmatrix} -1 \\ 2 \\ 1 \end{bmatrix}$
  $2x + 2y + z = 1$

30. $2x - y + z = 7$
  $x - 3y + 2z = 6$ $A = \begin{bmatrix} 3 \\ -2 \\ -1 \end{bmatrix}$
  $3x + y - z = 8$

In Exercises 31–40, perform the indicated matrix multiplications.

31. Using two rows and columns, show that $(-I)^2 = I$.

32. For $J = \begin{bmatrix} j & 0 \\ 0 & j \end{bmatrix}$, where $j = \sqrt{-1}$, show that $J^2 = -I$, $J^3 = -J$, and $J^4 = I$. Explain the similarity with j^2, j^3, and j^4.

33. Show that $A^2 - I = (A + I)(A - I)$ for $A = \begin{bmatrix} 2 & 4 \\ 3 & 5 \end{bmatrix}$.

34. In the study of polarized light, the matrix product
$\begin{bmatrix} 1 & 0 \\ 0 & -j \end{bmatrix}\begin{bmatrix} 1 & 0 \\ 1 & -j \end{bmatrix}\begin{bmatrix} 1 \\ 1 \end{bmatrix}$ occurs $(j = \sqrt{-1})$. Find this product.

35. In studying the motion of electrons, one of the Pauli spin matrices used is $s_y = \begin{bmatrix} 0 & -j \\ j & 0 \end{bmatrix}$, where $j = \sqrt{-1}$. Show that $s_y^2 = I$.

36. In analyzing the motion of a robotic mechanism, the following matrix multiplication is used. Perform the multiplication and evaluate each element of the result.
$\begin{bmatrix} \cos 60° & -\sin 60° & 0 \\ \sin 60° & \cos 60° & 0 \\ 0 & 0 & 1 \end{bmatrix}\begin{bmatrix} 2 \\ 4 \\ 0 \end{bmatrix}$

37. In an *ammeter,* nearly all the electric current flows through a *shunt,* and the remaining known fraction of current is measured by the meter. See Fig. 16.6. From the given matrix equation, find voltage v_2 and current i_2 in terms of v_1, i_1, and resistance R, whichever may be applicable.

$\begin{bmatrix} v_2 \\ i_2 \end{bmatrix} = \begin{bmatrix} 1 & 0 \\ -\dfrac{1}{R} & 1 \end{bmatrix}\begin{bmatrix} v_1 \\ i_1 \end{bmatrix}$

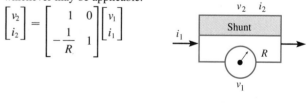

Fig. 16.6

38. In the theory related to the reproduction of color photography, the equation

$$\begin{bmatrix} X \\ Y \\ Z \end{bmatrix} = \begin{bmatrix} 1.0 & 0.1 & 0 \\ 0.5 & 1.0 & 0.1 \\ 0.3 & 0.4 & 1.0 \end{bmatrix} \begin{bmatrix} x \\ y \\ z \end{bmatrix}$$

is found. The X, Y, and Z represent the red, green, and blue densities of the reproductions, respectively, and the x, y, and z represent the red, green, and blue densities, respectively, of the subject. Give the equations relating X, Y, and Z and x, y, and z.

39. The path of an earth satellite can be written as

$$[x \quad y] \begin{bmatrix} 7.10 & -1 \\ 1 & 7.23 \end{bmatrix} \begin{bmatrix} x \\ y \end{bmatrix} = [5.13 \times 10^8]$$

where distances are in km. What type of curve is represented? (See Section 14.1.)

40. Using Kirchhoff's laws on the circuit shown in Fig. 16.7, the following matrix equation is found. By matrix multiplication, find the resulting system of equations.

$$\begin{bmatrix} R_1 + R_2 & -R_2 & 0 \\ -R_2 & R_2 + R_3 + R_4 & -R_4 \\ 0 & -R_4 & R_4 + R_5 \end{bmatrix} \begin{bmatrix} I_1 \\ I_2 \\ I_3 \end{bmatrix} = \begin{bmatrix} V_1 \\ 0 \\ -V_2 \end{bmatrix}$$

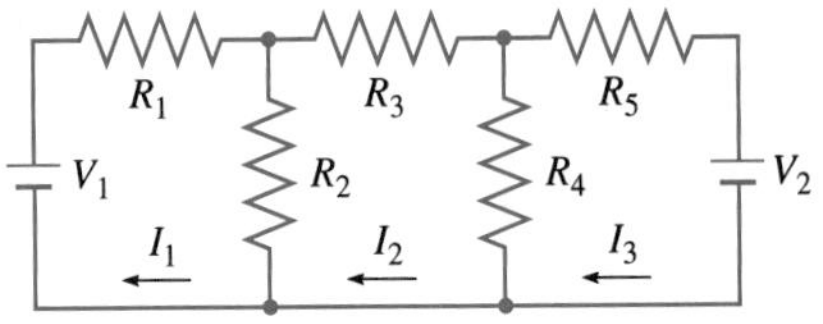

Fig. 16.7

16.3 FINDING THE INVERSE OF A MATRIX

In this section, we show how to find the inverse of a matrix, and in the following section we show how the inverse is used in solving a system of linear equations.

We first show two methods of finding the inverse of a two-row, two-column (2×2) matrix. The first method is as follows:

Inverse of a 2 × 2 Matrix

1. *Interchange the elements on the principal diagonal.*
2. *Change the signs of the off-diagonal elements.*
3. *Divide each resulting element by the determinant of the given matrix.*

This method, which can be used with second-order square matrices *but not with higher-order matrices,* is illustrated in the following example.

◀ **EXAMPLE 1** Find the inverse of the matrix $A = \begin{bmatrix} 2 & -3 \\ 4 & -7 \end{bmatrix}$.

First, we interchange the elements on the principal diagonal and change the signs of the off-diagonal elements. This gives us the matrix

$$\begin{bmatrix} -7 & 3 \\ -4 & 2 \end{bmatrix} \quad \begin{array}{l} \leftarrow \text{signs changed} \\ \leftarrow \text{elements interchanged} \end{array}$$

Now we find the determinant of the original matrix, which means we evaluate

$$\begin{vmatrix} 2 & -3 \\ 4 & -7 \end{vmatrix} = -14 - (-12) = -2$$

If the value of the determinant is zero, the inverse matrix does not exist.

We now divide each element of the second matrix by -2. This gives

$$A^{-1} = \frac{1}{-2} \begin{bmatrix} -7 & 3 \\ -4 & 2 \end{bmatrix} = \begin{bmatrix} \dfrac{-7}{-2} & \dfrac{3}{-2} \\ \dfrac{-4}{-2} & \dfrac{2}{-2} \end{bmatrix} = \begin{bmatrix} \dfrac{7}{2} & -\dfrac{3}{2} \\ 2 & -1 \end{bmatrix} \quad \leftarrow \text{inverse}$$

Checking by multiplication gives

See Exercise 37.

$$AA^{-1} = \begin{bmatrix} 2 & -3 \\ 4 & -7 \end{bmatrix} \begin{bmatrix} \frac{7}{2} & -\frac{3}{2} \\ 2 & -1 \end{bmatrix} = \begin{bmatrix} 7-6 & -3+3 \\ 14-14 & -6+7 \end{bmatrix} = \begin{bmatrix} 1 & 0 \\ 0 & 1 \end{bmatrix} = I$$

Since $AA^{-1} = I$, the matrix A^{-1} is the proper inverse matrix.

Gauss–Jordan Method

The second method, called the *Gauss–Jordan method,* involves *transforming the given matrix into the identity matrix while* **transforming the identity matrix into the inverse.** There are three types of steps allowable in making these transformations:

1. *Any two rows may be interchanged.*

Named for the German mathematician Karl Gauss (1777–1855) and the German geodesist Wilhelm Jordan (1842–1899).

2. *Every element in any row may be multiplied by any number other than zero.*

3. *Any row may be replaced by a row whose elements are the sum of a nonzero multiple of itself and a nonzero multiple of another row.*

NOTE ▶

Note that these are **row operations,** not column operations, and are the operations used in solving a system of equations by addition and subtraction.

◀ EXAMPLE 2 Find the inverse of the matrix

$$A = \begin{bmatrix} 2 & -3 \\ 4 & -7 \end{bmatrix}$$ this is the same matrix as in Example 1

First, we set up the given matrix with the identity matrix as follows:

$$\begin{bmatrix} 2 & -3 & | & 1 & 0 \\ 4 & -7 & | & 0 & 1 \end{bmatrix}$$

The vertical line simply shows the separation of the two matrices.

We wish to transform the left matrix into the identity matrix. Therefore, the first requirement is a 1 for element a_{11}. Therefore, we divide all elements of the first row by 2. This gives the following setup:

$$\begin{bmatrix} 1 & -\frac{3}{2} & | & \frac{1}{2} & 0 \\ 4 & -7 & | & 0 & 1 \end{bmatrix}$$

Next we want to have a zero for element a_{21}. Therefore, we subtract 4 times each element of row 1 from the corresponding element in row 2, replacing the elements of row 2. This gives us the following setup:

As in this example, (1) always work one column at a time, from left to right and (2) never undo the work in a previously completed column.

$$\begin{bmatrix} 1 & -\frac{3}{2} & | & \frac{1}{2} & 0 \\ 4-4(1) & -7-4\left(-\frac{3}{2}\right) & | & 0-4\left(\frac{1}{2}\right) & 1-4(0) \end{bmatrix} \text{ or } \begin{bmatrix} 1 & -\frac{3}{2} & | & \frac{1}{2} & 0 \\ 0 & -1 & | & -2 & 1 \end{bmatrix}$$

Next, we want to have 1, not -1, for element a_{22}. Therefore, we multiply each element of row 2 by -1. This gives

$$\begin{bmatrix} 1 & -\frac{3}{2} & | & \frac{1}{2} & 0 \\ 0 & 1 & | & 2 & -1 \end{bmatrix}$$

Finally, we want zero for element a_{12}. Therefore, we add $\frac{3}{2}$ times each element of row 2 to the corresponding elements of row 1, replacing row 1. This gives

$$\begin{bmatrix} 1+\frac{3}{2}(0) & -\frac{3}{2}+\frac{3}{2}(1) & | & \frac{1}{2}+\frac{3}{2}(2) & 0+\frac{3}{2}(-1) \\ 0 & 1 & | & 2 & -1 \end{bmatrix} \text{ or } \begin{bmatrix} 1 & 0 & | & \frac{7}{2} & -\frac{3}{2} \\ 0 & 1 & | & 2 & -1 \end{bmatrix}$$

At this point, we have transformed the given matrix into the identity matrix, and the identity matrix into the inverse. Therefore, the matrix to the right of the vertical bar in the last setup is the required inverse. Thus,

$$A^{-1} = \begin{bmatrix} \frac{7}{2} & -\frac{3}{2} \\ 2 & -1 \end{bmatrix}$$

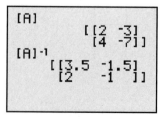

Fig. 16.8

See Fig. 16.8 for a calculator window showing matrix A and its inverse A^{-1}.

In transforming a matrix into the identity matrix, we work on one column at a time, transforming the columns in order from left to right. It is generally best to make the element on the principal diagonal for the column 1 first and then make all other elements in the column 0. This was done in Example 2, and we now illustrate it with another 2×2 matrix, and then we find the inverse of a 3×3 matrix. The method is applicable for any square matrix.

EXAMPLE 3 Find the inverse of the matrix $\begin{bmatrix} -3 & 6 \\ 4 & 5 \end{bmatrix}$.

original setup

$$\begin{bmatrix} -3 & 6 & | & 1 & 0 \\ 4 & 5 & | & 0 & 1 \end{bmatrix} \longrightarrow \begin{bmatrix} 1 & -2 & | & -\frac{1}{3} & 0 \\ 0 & 13 & | & \frac{4}{3} & 1 \end{bmatrix} \longrightarrow \begin{bmatrix} 1 & 0 & | & -\frac{5}{39} & \frac{2}{13} \\ 0 & 1 & | & \frac{4}{39} & \frac{1}{13} \end{bmatrix}$$

row 1 divided by −3   row 2 divided by 13 I A^{-1}

$$\begin{bmatrix} 1 & -2 & | & -\frac{1}{3} & 0 \\ 4 & 5 & | & 0 & 1 \end{bmatrix} \qquad \begin{bmatrix} 1 & -2 & | & -\frac{1}{3} & 0 \\ 0 & 1 & | & \frac{4}{39} & \frac{1}{13} \end{bmatrix}$$

−4 times row 1   2 times row 2
added to row 2   added to row 1

Therefore, $A^{-1} = \begin{bmatrix} -\frac{5}{39} & \frac{2}{13} \\ \frac{4}{39} & \frac{1}{13} \end{bmatrix}$, which can be checked by multiplication.

Inverse of a 3 × 3 Matrix

EXAMPLE 4 Find the inverse of the matrix $\begin{bmatrix} 1 & 2 & -1 \\ 3 & 5 & -1 \\ -2 & -1 & -2 \end{bmatrix}$.

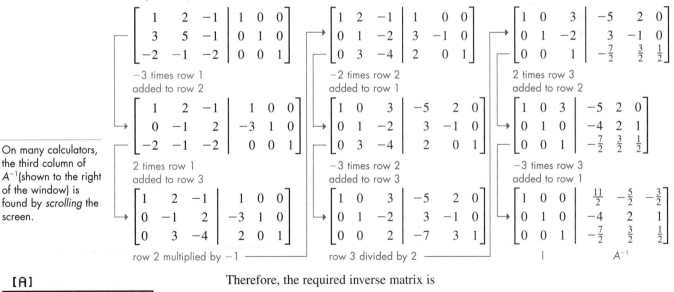

original setup

$$\begin{bmatrix} 1 & 2 & -1 & | & 1 & 0 & 0 \\ 3 & 5 & -1 & | & 0 & 1 & 0 \\ -2 & -1 & -2 & | & 0 & 0 & 1 \end{bmatrix} \qquad \begin{bmatrix} 1 & 2 & -1 & | & 1 & 0 & 0 \\ 0 & 1 & -2 & | & 3 & -1 & 0 \\ 0 & 3 & -4 & | & 2 & 0 & 1 \end{bmatrix} \qquad \begin{bmatrix} 1 & 0 & 3 & | & -5 & 2 & 0 \\ 0 & 1 & -2 & | & 3 & -1 & 0 \\ 0 & 0 & 1 & | & -\frac{7}{2} & \frac{3}{2} & \frac{1}{2} \end{bmatrix}$$

−3 times row 1   −2 times row 2   2 times row 3
added to row 2   added to row 1   added to row 2

$$\begin{bmatrix} 1 & 2 & -1 & | & 1 & 0 & 0 \\ 0 & -1 & 2 & | & -3 & 1 & 0 \\ -2 & -1 & -2 & | & 0 & 0 & 1 \end{bmatrix} \qquad \begin{bmatrix} 1 & 0 & 3 & | & -5 & 2 & 0 \\ 0 & 1 & -2 & | & 3 & -1 & 0 \\ 0 & 3 & -4 & | & 2 & 0 & 1 \end{bmatrix} \qquad \begin{bmatrix} 1 & 0 & 3 & | & -5 & 2 & 0 \\ 0 & 1 & 0 & | & -4 & 2 & 1 \\ 0 & 0 & 1 & | & -\frac{7}{2} & \frac{3}{2} & \frac{1}{2} \end{bmatrix}$$

2 times row 1   −3 times row 2   −3 times row 3
added to row 3   added to row 3   added to row 1

On many calculators, the third column of A^{-1} (shown to the right of the window) is found by *scrolling* the screen.

$$\begin{bmatrix} 1 & 2 & -1 & | & 1 & 0 & 0 \\ 0 & -1 & 2 & | & -3 & 1 & 0 \\ 0 & 3 & -4 & | & 2 & 0 & 1 \end{bmatrix} \qquad \begin{bmatrix} 1 & 0 & 3 & | & -5 & 2 & 0 \\ 0 & 1 & -2 & | & 3 & -1 & 0 \\ 0 & 0 & 2 & | & -7 & 3 & 1 \end{bmatrix} \qquad \begin{bmatrix} 1 & 0 & 0 & | & \frac{11}{2} & -\frac{5}{2} & -\frac{3}{2} \\ 0 & 1 & 0 & | & -4 & 2 & 1 \\ 0 & 0 & 1 & | & -\frac{7}{2} & \frac{3}{2} & \frac{1}{2} \end{bmatrix}$$

row 2 multiplied by −1   row 3 divided by 2 I A^{-1}

Therefore, the required inverse matrix is

$$\begin{bmatrix} \frac{11}{2} & -\frac{5}{2} & -\frac{3}{2} \\ -4 & 2 & 1 \\ -\frac{7}{2} & \frac{3}{2} & \frac{1}{2} \end{bmatrix}$$

which may be checked by multiplication. See Fig. 16.9 for a calculator window showing A and its inverse A^{-1}.

```
[A]
  [[1    2   -1]
   [3    5   -1]
   [-2  -1   -2]]
[A]⁻¹
[[5.5   -2.5 -1.… -1.5]
 [-4    2    1  …  1  ]
 [-3.5 1.5  .5  … .5 ]]
```

Fig. 16.9

In the next example, we review entering a matrix, displaying it, and finding its inverse on a typical calculator. The manual for any particular model should be used to see how the various operations are done on it.

◀ **EXAMPLE 5** By using a graphing calculator, find the inverse of the matrix

$$A = \begin{bmatrix} 2 & -2 & 3 & 2 \\ 3 & 1 & 5 & 2 \\ -2 & 5 & 2 & -3 \\ 4 & -5 & -1 & 4 \end{bmatrix}$$

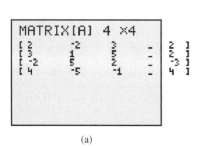

(a)

Using the *matrix* feature, in Fig. 16.10(a) we show the window when the elements of the matrix are entered. Scrolling by use of the arrow key will probably be required to enter all values. The matrix is then displayed as in Fig. 16.10(b).

To get the display for the inverse in Fig. 16.10(c), we enter [A] and then use the $\boxed{x^{-1}}$ key. To avoid long decimal approximations in the display, we have used the *mode* feature to round off values to three decimal places. On most calculators, this still requires scrolling to see all values, but they are easier to read. We have shown those that need scrolling to the right of the window.

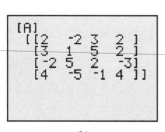

(b)

Fig. 16.10

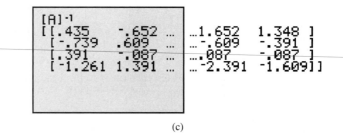

(c)

From these calculator windows, we have found the inverse matrix to be

$$A^{-1} = \begin{bmatrix} 0.435 & -0.652 & 1.652 & 1.348 \\ -0.739 & 0.609 & -0.609 & -0.391 \\ 0.391 & -0.087 & 0.087 & -0.087 \\ -1.261 & 1.391 & -2.391 & -1.609 \end{bmatrix}$$

On a calculator, the *fraction* feature can be used to change decimals to fractions.

Note that calculator entries such as 1E⁻10 mean that it is zero.

The elements of A^{-1} can be checked directly on the calculator by showing that the product $AA^{-1} = I$ (if the rounded values are used, the values of I will be approximate but should be sufficiently close to the necessary 1's and 0's). ▶

EXERCISES **16.3**

In Exercises 1 and 2, make the given changes in the indicated examples of this section and then find the matrix inverses.

1. In Example 1, change the element −7 to −5 and then find the inverse using the same method.
2. In Example 3, change the element −3 to −2 and then find the inverse using the same method.

In Exercises 3–10, find the inverse of each of the given matrices by the method of Example 1 of this section.

3. $\begin{bmatrix} 2 & -5 \\ -2 & 4 \end{bmatrix}$

4. $\begin{bmatrix} -6 & 3 \\ 3 & -2 \end{bmatrix}$

5. $\begin{bmatrix} -1 & 5 \\ 4 & 10 \end{bmatrix}$

6. $\begin{bmatrix} 8 & -1 \\ -4 & -5 \end{bmatrix}$

7. $\begin{bmatrix} 0 & -4 \\ 2 & 6 \end{bmatrix}$

8. $\begin{bmatrix} 7 & -2 \\ -6 & 2 \end{bmatrix}$

9. $\begin{bmatrix} -50 & -45 \\ 26 & 80 \end{bmatrix}$

10. $\begin{bmatrix} 7.2 & -3.6 \\ -1.3 & -5.7 \end{bmatrix}$

34. $\begin{bmatrix} 12.5 & -2.6 & 1.2 & 7.6 \\ -4.6 & 10.0 & -4.7 & -6.8 \\ 5.7 & -3.7 & 7.3 & 11.0 \\ 8.8 & 6.8 & 14.0 & 4.7 \end{bmatrix}$

In Exercises 11–24, find the inverse of each of the given matrices by transforming the identity matrix, as in Examples 2–4.

11. $\begin{bmatrix} 1 & 2 \\ 2 & 3 \end{bmatrix}$

12. $\begin{bmatrix} 1 & 5 \\ -1 & -4 \end{bmatrix}$

13. $\begin{bmatrix} 2 & 4 \\ -1 & -1 \end{bmatrix}$

14. $\begin{bmatrix} -2 & 6 \\ 3 & -4 \end{bmatrix}$

15. $\begin{bmatrix} 2 & 5 \\ -1 & 2 \end{bmatrix}$

16. $\begin{bmatrix} -2 & 3 \\ -3 & 5 \end{bmatrix}$

17. $\begin{bmatrix} 2 & -1 \\ 4 & 6 \end{bmatrix}$

18. $\begin{bmatrix} 1 & -3 \\ 7 & -5 \end{bmatrix}$

19. $\begin{bmatrix} 1 & -3 & -2 \\ -2 & 7 & 3 \\ 1 & -1 & -3 \end{bmatrix}$

20. $\begin{bmatrix} 1 & 2 & -1 \\ 3 & 7 & -5 \\ -1 & -2 & 0 \end{bmatrix}$

21. $\begin{bmatrix} 1 & 3 & 2 \\ -2 & -5 & -1 \\ 2 & 4 & 0 \end{bmatrix}$

22. $\begin{bmatrix} 1 & 3 & 4 \\ -1 & -4 & -2 \\ 4 & 9 & 20 \end{bmatrix}$

23. $\begin{bmatrix} 2 & 4 & 0 \\ 3 & 4 & -2 \\ -1 & 1 & 2 \end{bmatrix}$

24. $\begin{bmatrix} -2 & 6 & 1 \\ 0 & 3 & -3 \\ 4 & -7 & 3 \end{bmatrix}$

In Exercises 25–34, find the inverse of each of the given matrices by using a graphing calculator, as in Example 5. The matrices in Exercises 27–30 are the same as those in Exercises 21–24.

25. $\begin{bmatrix} 2 & 8 \\ -1 & 6 \end{bmatrix}$

26. $\begin{bmatrix} 7 & -3 \\ 6 & -2 \end{bmatrix}$

27. $\begin{bmatrix} 1 & 3 & 2 \\ -2 & -5 & -1 \\ 2 & 4 & 0 \end{bmatrix}$

28. $\begin{bmatrix} 1 & 3 & 4 \\ -1 & -4 & -2 \\ 4 & 9 & 20 \end{bmatrix}$

29. $\begin{bmatrix} 2 & 4 & 0 \\ 3 & 4 & -2 \\ -1 & 1 & 2 \end{bmatrix}$

30. $\begin{bmatrix} -2 & 6 & 1 \\ 0 & 3 & -3 \\ 4 & -7 & 3 \end{bmatrix}$

31. $\begin{bmatrix} 1 & -2 & 1 & 0 \\ 1 & -2 & 2 & -3 \\ 0 & 1 & -1 & 1 \\ -2 & 3 & -2 & 3 \end{bmatrix}$

32. $\begin{bmatrix} 3 & -2 & -1 & 4 \\ 2 & 0 & 5 & 1 \\ -1 & 2 & 1 & -2 \\ 4 & 1 & 3 & 5 \end{bmatrix}$

33. $\begin{bmatrix} 0.2 & 1.2 & -0.8 & -0.5 \\ -0.4 & 3.0 & -1.6 & 0.4 \\ 1.0 & -2.4 & 3.2 & 1.5 \\ -0.1 & 0.4 & 0.0 & 3.0 \end{bmatrix}$

In Exercises 35–40, solve the given problems.

35. Show that the matrix $\begin{bmatrix} 1 & 1 \\ 1 & 1 \end{bmatrix}$ has no inverse.

(W) **36.** Find the determinant of the matrix $\begin{bmatrix} 1 & -2 & 0 \\ -2 & 4 & 8 \\ 3 & -6 & 6 \end{bmatrix}$. Explain what this tells us about its inverse.

37. For the matrix $A = \begin{bmatrix} a & b \\ c & d \end{bmatrix}$, show that

$$\frac{1}{ad - bc}\begin{bmatrix} a & b \\ c & d \end{bmatrix}\begin{bmatrix} d & -b \\ -c & a \end{bmatrix} = \begin{bmatrix} 1 & 0 \\ 0 & 1 \end{bmatrix}$$

This verifies the method of Example 1.

(W) **38.** Describe the relationship between the elements of the matrix $\begin{bmatrix} a & 0 & 0 \\ 0 & b & 0 \\ 0 & 0 & c \end{bmatrix}$ and the elements of its inverse.

39. For the *four-terminal network* shown in Fig. 16.11, it can be shown that the voltage matrix V is related to the coefficient matrix A and the current matrix I by $V = A^{-1}I$, where

$$V = \begin{bmatrix} v_1 \\ v_2 \end{bmatrix} \qquad A = \begin{bmatrix} a_{11} & a_{12} \\ a_{21} & a_{22} \end{bmatrix} \qquad I = \begin{bmatrix} i_1 \\ i_2 \end{bmatrix}$$

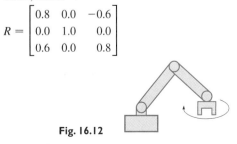

Fig. 16.11

Find the individual equations for v_1 and v_2 that give each in terms of i_1 and i_2.

40. The rotations of a robot arm such as that shown in Fig. 16.12 are often represented by matrices. The values represent trigonometric functions of the angles of rotation. For the following rotation matrix R, find R^{-1}.

$$R = \begin{bmatrix} 0.8 & 0.0 & -0.6 \\ 0.0 & 1.0 & 0.0 \\ 0.6 & 0.0 & 0.8 \end{bmatrix}$$

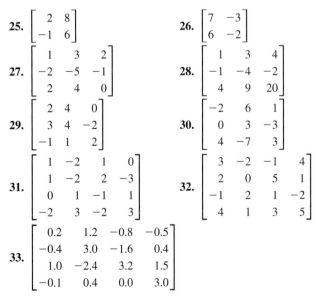

Fig. 16.12

16.4 MATRICES AND LINEAR EQUATIONS

As we stated at the beginning of Section 16.1, matrices can be used to solve systems of equations, and in this section we show one method by which this is done. As we develop this method, it will be apparent that there is a great deal of numerical work involved. However, methods such as this one are easily programmed for use on a computer, which can do the arithmetic work very rapidly. Also, most graphing calculators can perform these operations, and we will show an example at the end of the section in which a calculator is used to solve a system of four equations more readily than with earlier methods. It is the *method* of solving the system of equations that is of primary importance here.

Let us consider the system of equations

$$a_1 x + b_1 y = c_1$$
$$a_2 x + b_2 y = c_2$$

Recalling the definition of equality of matrices, we can write this system as

$$\begin{bmatrix} a_1 x + b_1 y \\ a_2 x + b_2 y \end{bmatrix} = \begin{bmatrix} c_1 \\ c_2 \end{bmatrix} \tag{16.6}$$

If we let

$$A = \begin{bmatrix} a_1 & b_1 \\ a_2 & b_2 \end{bmatrix} \qquad X = \begin{bmatrix} x \\ y \end{bmatrix} \qquad C = \begin{bmatrix} c_1 \\ c_2 \end{bmatrix} \tag{16.7}$$

the left side of Eq. (16.6) can be written as the product of matrices A and X.

$$AX = C \tag{16.8}$$

If we now multiply (on the left) each side of this matrix equation by A^{-1}, we have

$$A^{-1}AX = A^{-1}C$$

Since $A^{-1}A = I$, we have

$$IX = A^{-1}C$$

However, $IX = X$. Therefore,

$$X = A^{-1}C \tag{16.9}$$

NOTE ▶ Equation (16.9) states that *we can solve a system of linear equations by multiplying the one-column matrix of the constants on the right by the inverse of the matrix of the coefficients.* The result is a one-column matrix whose elements are the required values for the solution. Also, note that

CAUTION ▶
$$X = A^{-1}C \quad \text{and } \textbf{\textit{not}} \quad CA^{-1}$$

as the order of matrix multiplication must be carefully followed.

◀ EXAMPLE 1   Use matrices to solve the system of equations

$$2x - y = 7$$
$$5x - 3y = 18$$

We set up the matrix of coefficients and the matrix of constants as

$$A = \begin{bmatrix} 2 & -1 \\ 5 & -3 \end{bmatrix} \quad \text{and} \quad C = \begin{bmatrix} 7 \\ 18 \end{bmatrix}$$

By either of the methods of the previous section, we can determine the inverse of matrix A to be

$$A^{-1} = \begin{bmatrix} 3 & -1 \\ 5 & -2 \end{bmatrix}$$

We now form the matrix product $A^{-1}C$.

$$A^{-1}C = \begin{bmatrix} 3 & -1 \\ 5 & -2 \end{bmatrix}\begin{bmatrix} 7 \\ 18 \end{bmatrix} = \begin{bmatrix} 21 - 18 \\ 35 - 36 \end{bmatrix} = \begin{bmatrix} 3 \\ -1 \end{bmatrix}$$

Since $X = A^{-1}C$, this means that

$$\begin{bmatrix} x \\ y \end{bmatrix} = \begin{bmatrix} 3 \\ -1 \end{bmatrix}$$

Therefore, the required solution is $x = 3$ and $y = -1$, which checks when these values are substituted into the original equations.   ▶

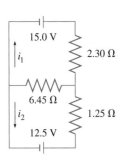

15.0 V

i_1

2.30 Ω

6.45 Ω

i_2

1.25 Ω

12.5 V

Fig. 16.13

◀ EXAMPLE 2   For the electric circuit shown in Fig. 16.13, the equations used to find the currents (in A) i_1 and i_2 are

$$2.30i_1 + 6.45(i_1 + i_2) = 15.0 \qquad 8.75i_1 + 6.45i_2 = 15.0$$
$$1.25i_2 + 6.45(i_1 + i_2) = 12.5 \quad \text{or} \quad 6.45i_1 + 7.70i_2 = 12.5$$

Using matrices to solve this system of equations, we set up the matrix A of coefficients, the matrix C of constants, and the matrix X of currents as

$$A = \begin{bmatrix} 8.75 & 6.45 \\ 6.45 & 7.70 \end{bmatrix} \quad C = \begin{bmatrix} 15.0 \\ 12.5 \end{bmatrix} \quad X = \begin{bmatrix} i_1 \\ i_2 \end{bmatrix}$$

We now find the inverse of A as

$$A^{-1} = \frac{1}{8.75(7.70) - 6.45(6.45)} \begin{bmatrix} 7.70 & -6.45 \\ -6.45 & 8.75 \end{bmatrix} = \begin{bmatrix} 0.2988 & -0.2503 \\ -0.2503 & 0.3395 \end{bmatrix}$$

Therefore,

$$X = A^{-1}C = \begin{bmatrix} 0.2988 & -0.2503 \\ -0.2503 & 0.3395 \end{bmatrix}\begin{bmatrix} 15.0 \\ 12.5 \end{bmatrix}$$
$$= \begin{bmatrix} 0.2988(15.0) - 0.2503(12.5) \\ -0.2503(15.0) + 0.3395(12.5) \end{bmatrix} = \begin{bmatrix} 1.35 \\ 0.49 \end{bmatrix}$$

Therefore, the required currents are $i_1 = 1.35$ A and $i_2 = 0.49$ A. These values check when substituted into the original equations.   ▶

◀ **EXAMPLE 3** Use matrices to solve the system of equations

$$x + 4y - z = 4$$
$$x + 3y + z = 8$$
$$2x + 6y + z = 13$$

Setting up matrices A, C, and X, we have

$$A = \begin{bmatrix} 1 & 4 & -1 \\ 1 & 3 & 1 \\ 2 & 6 & 1 \end{bmatrix} \quad C = \begin{bmatrix} 4 \\ 8 \\ 13 \end{bmatrix} \quad X = \begin{bmatrix} x \\ y \\ z \end{bmatrix}$$

To give another example of finding the inverse of a 3×3 matrix, we briefly show the steps for finding A^{-1}:

$$\begin{bmatrix} 1 & 4 & -1 & 1 & 0 & 0 \\ 1 & 3 & 1 & 0 & 1 & 0 \\ 2 & 6 & 1 & 0 & 0 & 1 \end{bmatrix} \rightarrow \begin{bmatrix} 1 & 4 & -1 & 1 & 0 & 0 \\ 0 & 1 & -2 & 1 & -1 & 0 \\ 0 & -2 & 3 & -2 & 0 & 1 \end{bmatrix} \rightarrow \begin{bmatrix} 1 & 0 & 7 & -3 & 4 & 0 \\ 0 & 1 & -2 & 1 & -1 & 0 \\ 0 & 0 & 1 & 0 & 2 & -1 \end{bmatrix}$$

$$\begin{bmatrix} 1 & 4 & -1 & 1 & 0 & 0 \\ 0 & -1 & 2 & -1 & 1 & 0 \\ 2 & 6 & 1 & 0 & 0 & 1 \end{bmatrix} \quad \begin{bmatrix} 1 & 0 & 7 & -3 & 4 & 0 \\ 0 & 1 & -2 & 1 & -1 & 0 \\ 0 & -2 & 3 & -2 & 0 & 1 \end{bmatrix} \quad \begin{bmatrix} 1 & 0 & 7 & -3 & 4 & 0 \\ 0 & 1 & 0 & 1 & 3 & -2 \\ 0 & 0 & 1 & 0 & 2 & -1 \end{bmatrix}$$

$$\begin{bmatrix} 1 & 4 & -1 & 1 & 0 & 0 \\ 0 & -1 & 2 & -1 & 1 & 0 \\ 0 & -2 & 3 & -2 & 0 & 1 \end{bmatrix} \quad \begin{bmatrix} 1 & 0 & 7 & -3 & 4 & 0 \\ 0 & 1 & -2 & 1 & -1 & 0 \\ 0 & 0 & -1 & 0 & -2 & 1 \end{bmatrix} \quad \begin{bmatrix} 1 & 0 & 0 & -3 & -10 & 7 \\ 0 & 1 & 0 & 1 & 3 & -2 \\ 0 & 0 & 1 & 0 & 2 & -1 \end{bmatrix}$$

Thus, $A^{-1} = \begin{bmatrix} -3 & -10 & 7 \\ 1 & 3 & -2 \\ 0 & 2 & -1 \end{bmatrix}$ and

$$X = A^{-1}C = \begin{bmatrix} -3 & -10 & 7 \\ 1 & 3 & -2 \\ 0 & 2 & -1 \end{bmatrix} \begin{bmatrix} 4 \\ 8 \\ 13 \end{bmatrix} = \begin{bmatrix} -12 - 80 + 91 \\ 4 + 24 - 26 \\ 0 + 16 - 13 \end{bmatrix} = \begin{bmatrix} -1 \\ 2 \\ 3 \end{bmatrix}$$

This means that $x = -1$, $y = 2$, and $z = 3$.

◀ **EXAMPLE 4** Use matrices to solve the system of equations

$$x + 2y - z = -4$$
$$3x + 5y - z = -5$$
$$-2x - y - 2z = -5$$

Setting up matrices A, C, and X, we have

$$A = \begin{bmatrix} 1 & 2 & -1 \\ 3 & 5 & -1 \\ -2 & -1 & -2 \end{bmatrix} \quad C = \begin{bmatrix} -4 \\ -5 \\ -5 \end{bmatrix} \quad X = \begin{bmatrix} x \\ y \\ z \end{bmatrix}$$

Finding A^{-1} (see Example 4 of Section 16.3) and solving for X, we have

$$A^{-1} = \begin{bmatrix} \frac{11}{2} & -\frac{5}{2} & -\frac{3}{2} \\ -4 & 2 & 1 \\ -\frac{7}{2} & \frac{3}{2} & \frac{1}{2} \end{bmatrix}$$

$$X = A^{-1}C = \begin{bmatrix} \frac{11}{2} & -\frac{5}{2} & -\frac{3}{2} \\ -4 & 2 & 1 \\ -\frac{7}{2} & \frac{3}{2} & \frac{1}{2} \end{bmatrix} \begin{bmatrix} -4 \\ -5 \\ -5 \end{bmatrix} = \begin{bmatrix} -2 \\ 1 \\ 4 \end{bmatrix}$$

This means that $x = -2$, $y = 1$, and $z = 4$.

◀ EXAMPLE 5 Use a calculator to perform the necessary matrix operations in solving the following system of equations:

$$2r + 4s - t + u = 5$$
$$r - 2s + 3t - u = -4$$
$$3r + s + 2t - 4u = 8$$
$$4r + 5s - t + 3u = -1$$

First, we set up matrices A, X, and C:

$$A = \begin{bmatrix} 2 & 4 & -1 & 1 \\ 1 & -2 & 3 & -1 \\ 3 & 1 & 2 & -4 \\ 4 & 5 & -1 & 3 \end{bmatrix} \qquad X = \begin{bmatrix} r \\ s \\ t \\ u \end{bmatrix} \qquad C = \begin{bmatrix} 5 \\ -4 \\ 8 \\ -1 \end{bmatrix}$$

It is now necessary only to enter matrices A and C in the calculator and find the matrix product $A^{-1}C$, as shown in the upper window in Fig. 16.14. (There is no need to record or display A^{-1}.) This shows that the solution is

$$r = -2 \qquad s = 3 \qquad t = 0.5 \qquad u = -2.5$$

This solution can be checked on the calculator by storing the resulting matrix X as matrix B and finding the matrix product AB, which should equal matrix C as shown in Fig. 16.14. The product AB is equivalent to substituting each value into the original equations, as shown in Eq. (16.8). ▶

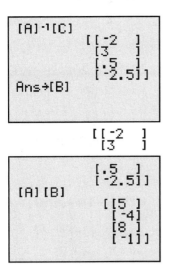

Fig. 16.14

EXERCISES 16.4

In Exercises 1 and 2, make the given changes in the indicated examples of this section and then solve the systems of equations.

1. In Example 1, change the 18 to 19 and then solve the system of equations.

2. In Example 3, change the 8 to 7 and the 13 to 12 and then solve the system of equations.

In Exercises 3–10, solve the given systems of equations by using the inverse of the coefficient matrix. The numbers in parentheses refer to exercises from Section 16.3, where the inverses may be checked.

3. $2x - 5y = -14$ (3)
$-2x + 4y = 11$

4. $-x + 5y = 4$ (5)
$4x + 10y = -4$

5. $x + 2y = 7$ (11)
$2x + 3y = 11$

6. $2x + 4y = -9$ (13)
$-x - y = 2$

7. $2x + 5y = -6$ (15)
$-x + 2y = -6$

8. $x - 3y - 2z = -8$ (19)
$-2x + 7y + 3z = 19$
$x - y - 3z = -3$

9. $x + 3y + 2z = 5$ (21)
$-2x - 5y - z = -1$
$2x + 4y = -2$

10. $2x + 4y = -2$ (23)
$3x + 4y - 2z = -6$
$-x + y + 2z = 5$

In Exercises 11–18, solve the given systems of equations by using the inverse of the coefficient matrix.

11. $2x - 3y = 3$
$4x - 5y = 4$

12. $x + 2y = 3$
$3x + 4y = 11$

13. $2.5x + 2.8y = -3.0$
$3.5x - 1.6y = 9.6$

14. $12x - 5y = -400$
$31x + 25y = 180$

15. $x + 2y + 2z = -4$
$4x + 9y + 10z = -18$
$-x + 3y + 7z = -7$

16. $x - 4y - 2z = -7$
$-x + 5y + 5z = 18$
$3x - 7y + 10z = 38$

17. $2x + 4y + z = 5$
$-2x - 2y - z = -6$
$-x + 2y + z = 0$

18. $4x + y = 2$
$-2x - y + 3z = -18$
$2x + y - z = 8$

In Exercises 19–26, solve the given systems of equations by using the inverse of the coefficient matrix. Use a calculator to perform the necessary matrix operations and display the results and the check. See Example 5.

19. $2x - y - z = 7$
$4x - 3y + 2z = 4$
$3x + 5y + z = -10$

20. $6x + 2y + 9z = 13$
$7x + 6y - 6z = 6$
$5x - 4y + 3z = 15$

21. $u - 3v - 2w = 9$
$3u + 2v + 6w = 20$
$4u - v + 3w = 25$

22. $2x + y - z = 1$
$3x - 2y - 8z = -3$
$x + 3y + z = 10$

23. $x - 5y + 2z - t = -18$
$3x + y - 3z + 2t = 17$
$4x - 2y + z - t = -1$
$-2x + 3y - z + 4t = 11$

24. $2p + q + 5r + s = 5$
$p + q - 3r - 4s = -1$
$3p + 6q - 2r + s = 8$
$2p + 2q + 2r - 3s = 2$

25. $2v + 3w + x - y - 2z = 6$
$6v - 2w - x + 3y - z = 21$
$v + 3w - 4x + 2y + 3z = -9$
$3v - w - x + 7y + 4z = 5$
$v + 6w + 6x - 4y - z = -4$

26. $4x - y + 2z - 2t + u = -15$
$8x + y - z + 4t - 2u = 26$
$2x - 6y - 2z + t - u = 10$
$2x + 5y + z - 3t + 8u = -22$
$4x - 3y + 2z + 4t + 2u = -4$

In Exercises 27–32, solve the indicated systems of equations using the inverse of the coefficient matrix. In Exercises 31 and 32, it is necessary to set up the appropriate equations.

27. For the following system of equations, solve for x^2 and y using the matrix methods of this section, and then solve for x and y.

$x^2 + y = 2$
$2x^2 - y = 10$

28. For the following system of equations, solve for x^2 and y^2 using the matrix methods of this section, and then solve for x and y.

$x^2 - y^2 = 8$
$x^2 + y^2 = 10$

29. Forces **A** and **B** hold up a beam that weighs 254 N, as shown in Fig. 16.15. The equations used to find the forces are

$A \sin 47.2° + B \sin 64.4° = 254$
$A \cos 47.2° - B \cos 64.4° = 0$

Find the magnitude of each force.

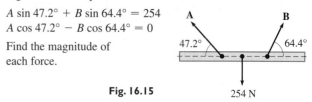

Fig. 16.15

30. In applying Kirchhoff's laws (see Exercise 32 on page 170) to the circuit shown in Fig. 16.16, the following equations are found. Determine the indicated currents (in A).

$I_A + I_B + I_C = 0$
$2I_A - 5I_B = 0$
$5I_B - I_C = -3$

Fig. 16.16

31. A research chemist wants to make 10.0 L of gasoline containing 2.0% of a new experimental additive. Gasoline without additive and two mixtures of gasoline with additive, one with 5.0% and the other with 6.0%, are to be used. If four times as much gasoline without additive as the 5.0% mixture is to be used, how much of each is needed?

32. A river tour boat takes 5.0 h to cruise downstream and 7.0 h for the return upstream. If the river flows at 4.0 km/h, how fast does the boat travel in still water, and how far downstream does the boat go before starting the return trip?

CHAPTER 16 EQUATIONS

Basic laws for matrices	$A + B = B + A$ (commutative law)	(16.1)
	$A + (B + C) = (A + B) + C$ (associative law)	(16.2)
	$k(A + B) = kA + kB$	(16.3)
	$A + O = A$	(16.4)
Inverse matrix	$AA^{-1} = A^{-1}A = I$	(16.5)
Solving systems of equations by matrices	$\begin{bmatrix} a_1x + b_1y \\ a_2x + b_2y \end{bmatrix} = \begin{bmatrix} c_1 \\ c_2 \end{bmatrix}$	(16.6)
	$A = \begin{bmatrix} a_1 & b_1 \\ a_2 & b_2 \end{bmatrix} \quad X = \begin{bmatrix} x \\ y \end{bmatrix} \quad C = \begin{bmatrix} c_1 \\ c_2 \end{bmatrix}$	(16.7)
	$AX = C$	(16.8)
	$X = A^{-1}C$	(16.9)

CHAPTER 16 REVIEW EXERCISES

In Exercises 1–6, determine the values of the literal numbers.

1. $\begin{bmatrix} 2a \\ a - b \end{bmatrix} = \begin{bmatrix} 8 \\ 5 \end{bmatrix}$

2. $\begin{bmatrix} x - y \\ 2x + 2z \\ 4y + z \end{bmatrix} = \begin{bmatrix} 1 \\ 3 \\ -1 \end{bmatrix}$

3. $\begin{bmatrix} 2x & 3y & 2z \\ x + y & 2y + z & z - x \end{bmatrix} = \begin{bmatrix} 4 & -9 & 5 \\ a & b & c \end{bmatrix}$

4. $\begin{bmatrix} a + bj & b \\ aj & b - aj \end{bmatrix} = \begin{bmatrix} 6j & 2d \\ 2cj & ej^2 \end{bmatrix}$ $\left(j = \sqrt{-1} \right)$

5. $\begin{bmatrix} \cos \pi & \sin \frac{\pi}{6} \\ x + y & x - y \end{bmatrix} = \begin{bmatrix} x & y \\ a & b \end{bmatrix}$

6. $\begin{bmatrix} \ln e & \log 100 \\ a^2 & b^2 \end{bmatrix} = \begin{bmatrix} a + b & a - b \\ x & y \end{bmatrix}$

In Exercises 7–12, use the given matrices and perform the indicated operations.

$A = \begin{bmatrix} 2 & -3 \\ 4 & 1 \\ -5 & 0 \\ 2 & -3 \end{bmatrix}$ $B = \begin{bmatrix} -1 & 0 \\ 4 & -6 \\ -3 & -2 \\ 1 & -7 \end{bmatrix}$ $C = \begin{bmatrix} 5 & -6 \\ 2 & 8 \\ 0 & -2 \end{bmatrix}$

7. $A + B$

8. $2C$

9. $B - A$

10. $2C - B$

11. $2A - 3B$

12. $2(A - B)$

In Exercises 13–16, perform the indicated matrix multiplications.

13. $\begin{bmatrix} 2 & -1 \\ -2 & 1 \end{bmatrix} \begin{bmatrix} 1 & -1 \\ 2 & -2 \end{bmatrix}$

14. $\begin{bmatrix} 6 & -4 & 1 & 0 \\ 2 & 0 & -4 & 3 \end{bmatrix} \begin{bmatrix} 7 & -1 & 6 \\ 4 & 0 & 1 \\ 3 & -2 & 5 \\ 9 & 1 & 0 \end{bmatrix}$

15. $\begin{bmatrix} -0.1 & 0.7 \\ 0.2 & 0.0 \\ 0.4 & -0.1 \end{bmatrix} \begin{bmatrix} 0.1 & -0.4 & 0.5 \\ 0.5 & 0.1 & 0.0 \end{bmatrix}$

16. $\begin{bmatrix} 0 & -1 & 6 \\ 8 & 1 & 4 \\ 7 & -2 & -1 \end{bmatrix} \begin{bmatrix} 5 & -1 & 7 & 1 & 5 \\ 0 & 1 & 0 & 4 & 1 \\ 1 & -2 & 3 & 0 & 1 \end{bmatrix}$

In Exercises 17–24, find the inverses of the given matrices. Check each by using a calculator.

17. $\begin{bmatrix} 2 & -5 \\ 2 & -4 \end{bmatrix}$

18. $\begin{bmatrix} -1 & -6 \\ 2 & 10 \end{bmatrix}$

19. $\begin{bmatrix} 0.07 & -0.01 \\ 0.04 & 0.08 \end{bmatrix}$

20. $\begin{bmatrix} 50 & -12 \\ 42 & -80 \end{bmatrix}$

21. $\begin{bmatrix} 1 & 1 & -2 \\ -1 & -2 & 1 \\ 0 & 3 & 4 \end{bmatrix}$

22. $\begin{bmatrix} -1 & -1 & 2 \\ 2 & 3 & 0 \\ 1 & 4 & 1 \end{bmatrix}$

23. $\begin{bmatrix} 2 & -4 & 3 \\ 4 & -6 & 5 \\ -2 & 1 & -1 \end{bmatrix}$

24. $\begin{bmatrix} 3 & 1 & -4 \\ -3 & 1 & -2 \\ -6 & 0 & 3 \end{bmatrix}$

In Exercises 25–32, solve the given systems of equations using the inverse of the coefficient matrix.

25. $2x - 3y = -9$
$4x - y = -13$

26. $5A - 7B = 62$
$6A + 5B = -6$

27. $33x + 52y = -450$
$45x - 62y = 1380$

28. $0.24x - 0.26y = -3.1$
$0.40x + 0.34y = -1.3$

29. $2u - 3v + 2w = 7$
$3u + v - 3w = -6$
$u + 4v + w = -13$

30. $2x + 2y - z = 8$
$x + 4y + 2z = 5$
$3x - 2y + z = 17$

31. $x + 2y + 3z = 1$
$3x - 4y - 3z = 2$
$7x - 6y + 6z = 2$

32. $3x + 2y + z = 2$
$2x + 3y - 6z = 3$
$x + 3y + 3z = 1$

In Exercises 33–40, solve the given systems of equations by using the inverse of the coefficient matrix. Use a calculator to perform the necessary matrix operations and display the results and the check.

33. $3x - 2y + z = 6$
$2x + 3z = 3$
$4x - y + 5z = 6$

34. $7n + p + 2r = 3$
$4n - 2p + 4r = -2$
$2n + 3p - 6r = 3$

35. $2x - 3y + z - t = -8$
$4x + 3z + 2t = -3$
$2y - 3z - t = 12$
$x - y - z + t = 3$

36. $3x + 2y - 2z - 2t = 0$
$5y + 3z + 4t = 3$
$6y - 3z + 4t = 9$
$6x - y + 2z - 2t = -3$

37. $3x - y + 6z - 2t = 8$
$2x + 5y + z + 2t = 7$
$4x - 3y + 8z + 3t = -17$
$3x + 5y - 3z + t = 8$

38. $A + B + 2C - 3D = 15$
$3A + 3B - 8C - 2D = 9$
$6A - 4B + 6C + D = -6$
$2A + 2B - 4C - 2D = 8$

39. $4r - s + 8t - 2u + 4v = -1$
$3r + 2s - 4t + 3u - v = 4$
$3r + 3s + 2t + 5u + 6v = 13$
$6r - s + 2t - 2u + v = 0$
$r - 2s + 4t - 3u + 3v = 1$

40. $2v + 3w + 2x - 2y + 5z = -1$
$7v + 8w + 3x + y - 4z = 3$
$v - 2w - 4x - 4y - 8z = -9$
$3v - w + 7x + 5y - 3z = -18$
$4v + 5w + x + 3y - 6z = 7$

In Exercises 41–44, use matrices A and B.

$$A = \begin{bmatrix} 1 & 0 \\ 3 & 4 \end{bmatrix} \qquad B = \begin{bmatrix} 0 & 1 & 0 \\ 0 & 0 & 1 \\ 1 & 0 & 0 \end{bmatrix}$$

41. Find A^2, A^3, and A^4. **42.** Show that $(A^2)^2 = A^4$.

43. Show that $B^3 = I$. **44.** Show that $B^4 = B$.

In Exercises 45 and 46, use the matrix N.

$$N = \begin{bmatrix} 0 & -1 \\ 1 & 0 \end{bmatrix}$$

45. Show that $N^{-1} = -N$. **46.** Show that $N^2 = -I$.

In Exercises 47 and 48, solve the given problems.

47. For any real number n, show that $\begin{bmatrix} n & 1+n \\ 1-n & -n \end{bmatrix}^2 = I$.

(W) **48.** For the matrix $N = \begin{bmatrix} 1 & 1 \\ 1 & 1 \end{bmatrix}$, find (a) N^2, (b) N^3, (c) N^4. What is N^{20}? Explain.

In Exercises 49–52, use matrices A and B.

$$A = \begin{bmatrix} 1 & -2 \\ 0 & 3 \end{bmatrix} \qquad B = \begin{bmatrix} -3 & 1 \\ 2 & -1 \end{bmatrix}$$

49. Show that $(A + B)(A - B) \neq A^2 - B^2$.

50. Show that $(A + B)^2 \neq A^2 + 2AB + B^2$.

51. Show that the inverse of $2A$ is $A^{-1}/2$.

52. Show that the inverse of $B/2$ is $2B^{-1}$.

In Exercises 53–56, solve the given systems of equations by use of matrices as in Section 16.4.

53. Two electric resistors, R_1 and R_2, are tested with currents and voltages such that the following equations are found:

$$2R_1 + 3R_2 = 26$$
$$3R_1 + 2R_2 = 24$$

Find the resistances R_1 and R_2 (in Ω).

54. A company produces two products, each of which is processed in two departments. Considering the worker time available, the numbers x and y of each product produced each week can be found by solving the system of equations

$$4.0x + 2.5y = 1200$$
$$3.2x + 4.0y = 1200$$

Find x and y.

55. A beam is supported as shown in Fig. 16.17. Find the tension T by solving the following system of equations:

$$0.500F = 0.866T$$
$$0.866F + 0.500T = 350$$

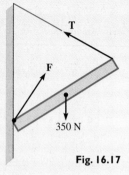

Fig. 16.17

56. To find the electric currents (in A) indicated in Fig. 16.18, it is necessary to solve the following equations.

$$I_A + I_B + I_C = 0$$
$$5I_A - 2I_B = -4$$
$$2I_B - I_C = 0$$

Find I_A, I_B, and I_C.

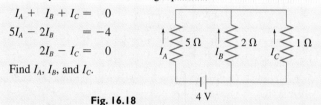

Fig. 16.18

In Exercises 57–60, solve the given problems by setting up the necessary equations and solving them by using matrices.

57. A crime suspect passes an intersection in a car traveling at 180 km/h. The police pass the intersection 3.0 min later in a car traveling at 225 km/h. How long is it before the police overtake the suspect?

58. A contractor needs a backhoe and a generator for two different jobs. Renting the backhoe for 5.0 h and the generator for 6.0 h costs $425 for one job. On the other job, renting the backhoe for 2.0 h and the generator for 8.0 h costs $310. What are the hourly charges for the backhoe and the generator?

59. By mass, three alloys have the following percentages of lead, zinc, and copper.

	Lead	Zinc	Copper
Alloy A	60%	30%	10%
Alloy B	40%	30%	30%
Alloy C	30%	70%	

How many grams of each of alloys A, B, and C must be mixed to get 100 g of an alloy that is 44% lead, 38% zinc, and 18% copper?

60. On a 750-km trip from Paris to Liverpool that took a total of 5.5 h, a person took a limousine to the airport, then a plane, and finally a car to reach the final destination. The limousine took as long as the final car trip and the time for connections. The limousine averaged 55 km/h, the plane averaged 400 km/h, and the car averaged 40 km/h. The plane traveled four times as far as the limousine and car combined. How long did each part of the trip and the connections take?

In Exercises 61–64, perform the indicated matrix operations.

61. An automobile maker has two assembly plants at which cars with either 4, 6, or 8 cylinders and with either standard or automatic transmission are assembled. The annual production at the first plant of cars with the number of cylinders–transmission type (standard, automatic) is as follows:

4: 12 000, 15 000; 6: 24 000, 8000; 8: 4000, 30 000

At the second plant the annual production is

4: 15 000, 20 000; 6: 12 000, 3000; 8: 2000, 22 000

Set up matrices for this production and by matrix addition find the matrix for the total production by the number of cylinders and type of transmission.

62. Set up a matrix representing the information given in Exercise 59. A given shipment contains 500 g of alloy A, 800 g of alloy B, and 700 g of alloy C. Set up a matrix for this information. By multiplying these matrices, obtain a matrix that gives the total weight of lead, zinc, and copper in the shipment.

63. The matrix equation

$$\left[\begin{bmatrix} R_1 & -R_2 \\ -R_2 & R_1 \end{bmatrix} + R_2 \begin{bmatrix} 1 & 0 \\ 0 & 1 \end{bmatrix}\right] \begin{bmatrix} i_1 \\ i_2 \end{bmatrix} = \begin{bmatrix} 6 \\ 0 \end{bmatrix}$$

may be used to represent the system of equations relating the currents and resistances of the circuit in Fig. 16.19. Find this system of equations by performing the indicated matrix operations.

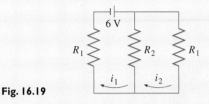

Fig. 16.19

64. A person prepared a meal of the following items, each having the given number of grams of protein, carbohydrates, and fat, respectively. Beef stew: 25, 21, 22; coleslaw: 3, 10, 10; (light) ice cream: 7, 25, 6. If the kilojoule count of each gram of protein, carbohydrate, and fat is 17 kJ/g, 16 kJ/g, and 37 kJ/g, respectively, find the total kJ count of each item by matrix multiplication.

Writing Exercise

65. A hardware company has 60 different retail stores in which 3500 different products are sold. Write a paragraph explaining why matrices provide an efficient method of inventory control for this company.

CHAPTER (16) PRACTICE TEST

1. For matrices A and B, find $A - 2B$.

$$A = \begin{bmatrix} 3 & -1 & 4 \\ 2 & 0 & -2 \end{bmatrix} \quad B = \begin{bmatrix} 1 & 4 & 5 \\ -1 & -2 & 3 \end{bmatrix}$$

2. Evaluate the literal symbols.

$$\begin{bmatrix} 2x & x - y & z \\ x + z & 2y & y + z \end{bmatrix} = \begin{bmatrix} 6 & -2 & 4 \\ a & b & c \end{bmatrix}$$

3. For matrices C and D, find CD and DC.

$$C = \begin{bmatrix} 1 & 0 & 4 \\ 2 & -2 & 1 \\ -1 & 3 & 2 \end{bmatrix} \quad D = \begin{bmatrix} 2 & -2 \\ 4 & -5 \\ 6 & 1 \end{bmatrix}$$

4. Determine whether or not $B = A^{-1}$.

$$A = \begin{bmatrix} 2 & -5 \\ 1 & -2 \end{bmatrix} \quad B = \begin{bmatrix} -2 & 5 \\ -1 & 2 \end{bmatrix}$$

5. For matrix C of Problem 3, find C^{-1}.

6. Solve by using the inverse of the coefficient matrix.

$$\begin{aligned} 2x - 3y &= 11 \\ x + 2y &= 2 \end{aligned}$$

7. Solve the following system of equations by using the inverse of the coefficient matrix. Use a calculator to perform the necessary matrix operations and display the result and check. Write down the display shown in the window with the solutions.

$$\begin{aligned} 7x - 2y + z &= 6 \\ 2x + 3y - 4z &= 6 \\ 4x - 5y + 2z &= 10 \end{aligned}$$

8. Fifty shares of stock A and 30 shares of stock B cost $2600. Thirty shares of stock A and 40 shares of stock B cost $2000. What is the price per share of each stock? Solve by setting up the appropriate equations and then using the inverse of the coefficient matrix.

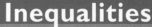

17

Inequalities

Having devoted a great deal of time to the solution of equations and systems of equations, we now turn our attention to solving inequalities and systems of inequalities. In doing so, we will find it necessary to find *all values* of the variable or variables that satisfy the inequality or system of inequalities.

There are numerous technical applications of inequalities. For example, in electricity it might be necessary to find the values of a current that are *greater than* a specified value. In designing a link in a robotic mechanism, it might be necessary to find the forces that are *less than* some specified value. Computers can be programmed to switch from one part of a program to another, based upon a result that is greater than (or less than) some given value.

Systems of linear *equations* have been studied for better than 2000 years, but almost no attention was given to systems of linear *inequalities* until World War II in the 1940s. Problems of deploying personnel and aircraft effectively and allocating supplies efficiently led the U.S. Air Force to have a number of scientists, economists, and mathematicians look for solutions. From this, a procedure for analyzing such problems was devised in 1947 by George Dantzig and his colleagues. Their system involved using systems of linear inequalities and is today called *linear programming.* We introduce the basic method of linear programming in the final section of this chapter.

In Section 17.6, we use inequalities to show how a company can maximize profit in making products such as speaker systems.

Here, we see that a mathematical method was developed as a result of a military need. Today, linear programming is widely used in business and industry in order to set production levels for maximizing profits and minimizing costs.

17.1 PROPERTIES OF INEQUALITIES

In Chapter 1, we first introduced the signs of inequality. To this point, only a basic understanding of their meanings has been necessary to show certain intervals associated with a variable. In this section, we review the meanings and develop certain basic properties of inequalities. We also show the meaning of the solution of an inequality and how it is shown on the number line.

The expression $a < b$ is read as "a is less than b," and the expression $a > b$ is read as "a is greater than b." *These signs define what is known as the **sense** (indicated by the direction of the sign) of the inequality.* Two inequalities are said to have the same sense if the signs of inequality point in the same direction. They are said to have the opposite sense if the signs of inequality point in opposite directions. *The two sides of the inequality are called **members** of the inequality.*

◖ **EXAMPLE 1**   The inequalities $x + 3 > 2$ and $x + 1 > 0$ have the same sense, as do the inequalities $3x - 1 < 4$ and $x^2 - 1 < 3$.

The inequalities $x - 4 < 0$ and $x > -4$ have the opposite sense, as do the inequalities $2x + 4 > 1$ and $3x^2 - 7 < 1$.   ◗

The **solution** *of an inequality consists of all values of the variable that make the inequality a true statement.* Most inequalities with which we deal are **conditional inequalities,** *which are true for some, but not all, values of the variable.* Also, *some inequalities are true for all values of the variable, and they are called **absolute inequalities.*** A solution of an inequality consists of only real numbers, as the terms *greater than* or *less than* have not been defined for complex numbers.

◖ **EXAMPLE 2**   The inequality $x + 1 > 0$ is true for all values of x greater than -1. Therefore, the values of x that satisfy this inequality are written as $x > -1$. This illustrates the difference between the solution of an equation and the solution of an inequality. The solution of an equation normally consists of a few specific numbers,

NOTE ▶ whereas *the solution to an inequality normally consists of an interval of values of the variable.* Any and all values within this interval are termed solutions of the inequality. Since the inequality $x + 1 > 0$ is satisfied only by the values of x in the interval $x > -1$, it is a *conditional inequality.*

The inequality $x^2 + 1 > 0$ is true for all real values of x, since x^2 is never negative. It is an *absolute inequality.*   ◗

There are occasions when it is convenient to combine an inequality with an equality. For such purposes, the symbols $\leq$, meaning *less than or equal to,* and $\geq$, meaning *greater than or equal to,* are used.

◖ **EXAMPLE 3**   If we wish to state that x is positive, we can write $x > 0$. However, the value zero is not included in the solution. If we wish to state that x is not negative, we write $x \geq 0$. Here, zero is part of the solution.

In order to state that x is less than or equal to -5, we write $x \leq -5$.   ◗

In the sections that follow, we will solve inequalities. It is often useful to show the solution on the number line. The next example shows how this is done.

(a)

(b)

Fig. 17.1

Properties of Inequalities

◀ **EXAMPLE 4** **(a)** To graph $x > 2$, we draw a small open circle at 2 on the number line (which is equivalent to the x-axis). Then we draw a solid line to the right of this point with an arrowhead pointing to the right, indicating all values greater than 2. See Fig. 17.1(a). The *open circle* shows that the point is not part of the indicated solution.

(b) To graph $x \leq 1$, we follow the same basic procedure as in part (a), except that we use a solid circle and the arrowhead points to the left. See Fig. 17.1(b). The *solid circle* shows that the point is part of the indicated solution. ▶

We now show the basic operations performed on inequalities. These are the same operations as those performed on equations, but in certain cases the results take on a different form. *The following are the* **properties of inequalities.**

1. *The sense of an inequality is not changed when the same number is added to—or subtracted from—both members of the inequality.* Symbolically, this may be stated as "if $a > b$, then $a + c > b + c$ and $a - c > b - c$."

◀ **EXAMPLE 5** Using Property 1 on the inequality $9 > 6$, we have the following results:

$9 > 6$	$9 > 6$
add 4 to each member	subtract 12 from each member
$9 + 4 > 6 + 4$	$9 - 12 > 6 - 12$
$13 > 10$	$-3 > -6$

Fig. 17.2

In Fig. 17.2, we see that 9 is to the right of 6, 13 is to the right of 10, and -3 is to the right of -6. ▶

2. *The sense of an inequality is not changed if both members are multiplied or divided by the same positive number.* Symbolically, this is stated as "if $a > b$, then $ac > bc$, and $a/c > b/c$, provided that $c > 0$."

◀ **EXAMPLE 6** Using Property 2 on the inequality $8 < 15$, we have the following results:

$8 < 15$	$8 < 15$
multiply both members by 2	divide both members by 2
$2(8) < 2(15)$	$\dfrac{8}{2} < \dfrac{15}{2}$
$16 < 30$	$4 < \dfrac{15}{2}$

▶

CAUTION ▶

3. *The sense of an inequality is reversed if both members are multiplied or divided by the same negative number.* Symbolically, this is stated as "if $a > b$, then $ac < bc$, and $a/c < b/c$, provided that $c < 0$."

NOTE ▶

In using this property of inequalities, be very careful to note that *the inequality sign remains the same if both members are multiplied or divided by a positive number,* but that ***the inequality sign changes if both members are multiplied or divided by a negative number.*** Most of the errors made in dealing with inequalities occur when using this property.

EXAMPLE 7   Using Property 3 on the inequality $4 > -2$, we have the following results:

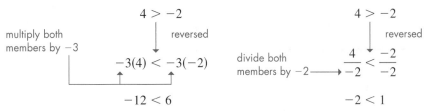

multiply both members by -3 reversed
$$-3(4) < -3(-2)$$
$$-12 < 6$$

divide both members by -2 reversed
$$\frac{4}{-2} < \frac{-2}{-2}$$
$$-2 < 1$$

Fig. 17.3

In Fig. 17.3, we see that 4 is to the right of -2, but that -12 *is to the left of* 6 and that -2 *is to the left of* 1. This is consistent with reversing the sense of the inequality when it is multiplied by -3 and when it is divided by -2.

4. *If both members of an inequality are positive numbers and n is a positive integer, then the inequality formed by taking the nth power of each member, or the nth root of each member, is in the same sense as the given inequality.* Symbolically, this is stated as "if $a > b$, then $a^n > b^n$, and $\sqrt[n]{a} > \sqrt[n]{b}$, provided that $n > 0$, $a > 0$, $b > 0$."

EXAMPLE 8   Using Property 4 on the inequality $16 > 9$, we have

$$16 > 9$$
square both members
$$16^2 > 9^2$$
$$256 > 81$$

$$16 > 9$$
take square root of both members
$$\sqrt{16} > \sqrt{9}$$
$$4 > 3$$

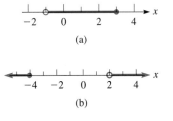

(a)

(b)

Fig. 17.4

Many inequalities have more than two members. In fact, inequalities with three members are common, and care must be used in stating these inequalities.

EXAMPLE 9   **(a)** To state that 5 is less than 6, and also greater than 2, we may write $2 < 5 < 6$, or $6 > 5 > 2$. (Generally the *less than* form is preferred.)

(b) To state that a number x may be greater than -1 *and* also less than or equal to 3, we write $-1 < x \le 3$. (It can also be written as $x > -1$ *and* $x \le 3$.) This is shown in Fig. 17.4(a). Note the use of the open circle and the solid circle.

CAUTION

(c) By writing $x \le -4$ *or* $x > 2$, we state that x is less than or equal to -4, *or* greater than 2. ***It may not be stated as*** $2 < x \le -4$, for this shows x as being less than -4, and also greater than 2, and *no such numbers exist*. See Fig. 17.4(b).

NOTE

Note carefully that ***and*** *is used when the solution consists of values that make* ***both*** *statements true. The word* ***or*** *is used when the solution consists of values that make* ***either*** *statement true.* (In everyday speech, *or* can sometimes mean that one statement is true or another statement is true, but *not* that both are true.)

NOTE

EXAMPLE 10   The inequality $x^2 - 3x + 2 > 0$ is satisfied if x is either greater than 2 *or* less than 1. This is written as $x > 2$ *or* $x < 1$, but it is incorrect to state it as $1 > x > 2$. (If we wrote it this way, we would be saying that the same value of x is less than 1 *and* at the same time greater than 2. Of course, as we noted for this type of situation in Example 9, no such number exists.) Any inequality must be valid for all values satisfying it. However, we could say that the inequality is not satisfied for $1 \le x \le 2$, which means those values of x greater than *or* equal to 1 *and* less than *or* equal to 2 (between or equal to 1 and 2).

Solving a Word Problem

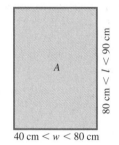

Fig. 17.5

$80 \text{ cm} < l < 90 \text{ cm}$

$40 \text{ cm} < w < 80 \text{ cm}$

A

◀ **EXAMPLE 11** The design of a rectangular solar panel shows that the length l is between 80 cm and 90 cm and the width w between 40 cm and 80 cm. See Fig. 17.5. Find the values of area the panel may have.

Since l is to be less than 90 cm and w less than 80 cm, the area must be less than $(90 \text{ cm})(80 \text{ cm}) = 7200 \text{ cm}^2$. Also, since l is to be greater than 80 cm and w greater than 40 cm, the area must be greater than $(80 \text{ cm})(40 \text{ cm}) = 3200 \text{ cm}^2$. Therefore, the area A may be represented as

$$3200 \text{ cm}^2 < A < 7200 \text{ cm}^2$$

This means the area is greater than 3200 cm² *and* less than 7200 cm². ▶

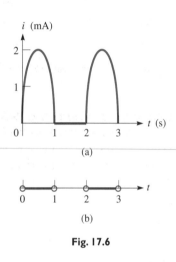

(a)

(b)

Fig. 17.6

◀ **EXAMPLE 12** A semiconductor *diode* has the property that an electric current flows through it in only one direction. If it is an alternating-current circuit, the current in the circuit flows only during the half-cycle when the diode allows it to flow. If a source of current given by $i = 2 \sin \pi t$ (i in mA, t in s) is connected in series with a diode, write the inequalities for the current and the time. Assume the source is on for 3.0 s and a positive current passes through the diode.

We are to find the values of t that correspond to $i > 0$. From the properties of the sine function, we know that $2 \sin \pi t$ has a period of $2\pi/\pi = 2.0$ s. Therefore, the current is zero for $t = 0$, 1.0 s, 2.0 s, and 3.0 s.

The source current is positive for $0 < t < 1.0$ s and for $2.0 \text{ s} < t < 3.0$ s.

The source current is negative for $1.0 \text{ s} < t < 2.0$ s.

Therefore, in the circuit

$$i > 0 \quad \text{for} \quad 0 < t < 1.0 \text{ s} \quad \text{and} \quad 2.0 \text{ s} < t < 3.0 \text{ s}$$
$$i = 0 \quad \text{for} \quad t = 0, 1.0 \text{ s} \le t \le 2.0 \text{ s}$$

A graph of the current in the circuit as a function of time is shown in Fig. 17.6(a). In Fig. 17.6(b) the values of t for which $i > 0$ are shown. ▶

EXERCISES 17.1

In Exercises 1–4, make the given changes in the indicated examples of this section and then perform the indicated operations.

1. In Example 2, in the first paragraph change the $>$ to $<$ and then complete the meaning of the resulting inequality as in the first sentence. Then rewrite the meaning as at the end of the second line.

2. In Example 4(b), change $\le$ to $>$ and then graph the resulting inequality.

3. In Example 7, change the inequality to $-2 > -4$ and then perform the two operations shown in color.

4. In Example 9(b), change the -1 to -3 and the 3 to 1 and then write the two forms in which an inequality represents the statement.

In Exercises 5–12, for the inequality $4 < 9$, state the inequality that results when the given operations are performed on both members.

5. Add 3.

6. Subtract 6.

7. Multiply by 5.

8. Multiply by -2.

9. Divide by -1.

10. Divide by 2.

11. Square both.

12. Take square roots.

In Exercises 13–24, give the inequalities equivalent to the following statements about the number x.

13. Greater than -2

14. Less than 7

15. Less than or equal to 4

16. Greater than or equal to -6

17. Greater than 1 and less than 7

18. Greater than or equal to -2 and less than 6

19. Less than -9, or greater than or equal to -4

20. Less than or equal to 8, or greater than or equal to 12

21. Less than 1, or greater than 3 and less than or equal to 5

22. Greater than or equal to 0 and less than or equal to 2, or greater than 5

23. Greater than -2 and less than 2, or greater than or equal to 3 and less than 4

24. Less than -4, or greater than or equal to 0 and less than or equal to 1, or greater than or equal to 5

(W) *In Exercises 25–28, give verbal statements equivalent to the given inequalities involving the number x.*

25. $0 < x \le 2$ **26.** $x < 5$ or $x > 7$

27. $x < -1$ or $1 \le x < 2$

28. $-1 \le x < 3$ or $5 < x < 7$

In Exercises 29–44, graph the given inequalities on the number line.

29. $x < 3$ **30.** $x \ge -1$

31. $x \le 1$ or $x > 3$ **32.** $x < -3$ or $x \ge 0$

33. $0 \le x < 5$ **34.** $-4 < y < -2$

35. $x \ge -3$ and $x < 5$ **36.** $x > 4$ and $x < 3$

37. $x < -1$ or $1 \le x < 4$

38. $-3 < x < 0$ or $x > 3$

39. $-3 < x < -1$ or $1 < x \le 3$

40. $1 < x \le 2$ or $3 \le x < 4$

41. $t < -3$ or $t > -3$

42. $x < 1$ or $1 < x \le 4$

43. $(x \le 5$ or $x \ge 8)$ and $(3 < x < 10)$

44. $(x < 7$ and $x > 2)$ or $(x > 10$ or $x < 1)$

In Exercises 45–48, answer the given questions about the inequality $0 < a < b$.

45. Is $a^2 < b^2$ a conditional inequality or an absolute inequality?

46. Is $|a - b| < b - a$?

47. If each member of the inequality $2 > 1$ is multiplied by $a - b$, is the result $2(a - b) > (a - b)$?

(W) **48.** What is wrong with the following sequence of steps?
$a < b, ab < b^2, ab - b^2 < 0, b(a - b) < 0, b < 0$

In Exercises 49–56, some applications of inequalities are shown.

49. An electron microscope can magnify an object from 2000 times to 1 000 000 times. Assuming these values are exact, express these magnifications M as an inequality and graph them.

50. A busy person glances at a digital clock that shows 9:36. Another glance a short time later shows the clock at 9:44. Express the amount of time t (in min) that could have elapsed between glances by use of inequalities. Graph these values of t.

51. An earth satellite put into orbit near the earth's surface will have an elliptic orbit if its velocity v is between 29 000 km/h and 40 000 km/h. Write this as an inequality and graph these values of v.

52. Fossils found in Jurassic rocks indicate that dinosaurs flourished during the Jurassic geological period, 140 MY (million years ago) to 200 MY. Write this as an inequality, with t representing past time. Graph the values of t.

53. In executing a program, a computer must perform a set of calculations. Any one of the calculations takes no more than 2565 steps. Express the number n of steps required for a given calculation by an inequality. (Note that n is a positive *integer*.)

54. The velocity v of an ultrasound wave in soft human tissue may be represented as 1550 ± 60 m/s, where the ± 60 m/s gives the possible variation in the velocity. Express the possible velocities by an inequality.

55. The electric intensity E within a charged spherical conductor is zero. The intensity on the surface and outside the sphere equals a constant k divided by the square of the distance r from the center of the sphere. State these relations for a sphere of radius a by using inequalities and graph E as a function of r.

56. If the current from the source in Example 12 is $i = 5 \cos 4\pi t$ and the diode allows only negative current to flow, write the inequalities and draw the graph for the current in the circuit as a function of time for $0 \le t \le 1$ s.

17.2 SOLVING LINEAR INEQUALITIES

Using the properties and definitions discussed in Section 17.1, we can now proceed to solve inequalities. In this section, we solve linear inequalities in one variable. Similar to linear functions as defined in Chapter 5, *a* **linear inequality** *is one in which each term contains only one variable and the exponent of each variable is* 1. We will consider linear inequalities in two variables in Section 17.5.

The procedure for solving a linear inequality in one variable is like that we used in solving basic equations in Chapter 1. We solve the inequality by isolating the variable, and to do this we perform the same operations on each member of the inequality. The operations are based on the properties given in Section 17.1.

◀ EXAMPLE 1   In each of the following inequalities, by performing the indicated operation, we isolate x and thereby solve the inequality.

$x + 2 < 4$	$\dfrac{x}{2} > 4$	$2x \leq 4$
Subtract 2 from each member.	Multiply each member by 2.	Divide each member by 2.
$x < 2$	$x > 8$	$x \leq 2$

Each solution can be checked by substituting any number in the indicated interval into the original inequality. For example, any value less than 2 will satisfy the first inequality, whereas 2 or any number less than 2 will satisfy the third inequality.   ▶

◀ EXAMPLE 2   Solve the following inequality: $3 - 2x \geq 15$.
We have the following solution:

$$3 - 2x \geq 15 \qquad \text{original inequality}$$
$$-2x \geq 12 \qquad \text{subtract 3 from each member}$$

inequality reversed ⟶

$$x \leq -6 \qquad \text{divide each member by } -2$$

CAUTION ▶   Again, carefully note that *the sign of inequality was reversed when each number was divided by -2.* We check the solution by substituting -7 in the original inequality, obtaining $17 \geq 15$.   ▶

◀ EXAMPLE 3   Solve the inequality $2x \leq 3 - x$.
The solution proceeds as follows:

$$2x \leq 3 - x \qquad \text{original inequality}$$
$$3x \leq 3 \qquad \text{add } x \text{ to each member}$$
$$x \leq 1 \qquad \text{divide each member by 3}$$

This solution checks and is represented in Fig. 17.7, as we showed in Section 17.1.

This inequality could have been solved by combining x-terms on the right. In doing so, we would obtain $1 \geq x$. Since this might be misread, it is best to combine the variable terms on the left, as we did above.   ▶

Part of solution

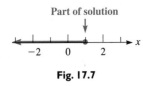

Fig. 17.7

◀ EXAMPLE 4   Solve the inequality $\frac{3}{2}(1 - x) > \frac{1}{4} - x$.

$$\frac{3}{2}(1 - x) > \frac{1}{4} - x \qquad \text{original inequality}$$
$$6(1 - x) > 1 - 4x \qquad \text{multiply each member by 4}$$
$$6 - 6x > 1 - 4x \qquad \text{remove parentheses}$$
$$-6x > -5 - 4x \qquad \text{subtract 6 from each member}$$
$$-2x > -5 \qquad \text{add } 4x \text{ to each member}$$
$$x < \frac{5}{2} \qquad \text{divide each member by } -2$$

Not part of solution

Fig. 17.8

Note that the sense of the inequality was reversed when we divided by -2. This solution is shown in Fig. 17.8. Any value of $x < 5/2$ checks when substituted into the original inequality.   ▶

The following example illustrates an application that involves the solution of an inequality.

◀ EXAMPLE 5   The velocity v (in m/s) of a missile in terms of the time t (in s) is given by $v = 490 - 9.8t$. For how long is the velocity positive? (Since velocity is a vector, this can also be interpreted as asking "How long is the missile moving upward?")

In terms of inequalities, we are asked to find the values of t for which $v > 0$. This means that we must solve the inequality $490 - 9.8t > 0$. The solution is as follows:

$$490 - 9.8t > 0 \qquad \text{original inequality}$$
$$-9.8t > -490 \qquad \text{subtract 490 from each member}$$
$$t < 50 \text{ s} \qquad \text{divide each member by } -9.8$$

Negative values of t have no meaning in this problem. Checking $t = 0$, we find that $v = 490$ m/s. Therefore, the complete solution is $0 \le t < 30$ s.

In Fig. 17.9(a), we show the graph of $v = 490 - 9.8t$, and in Fig. 17.9(b), we show the solution $0 \le t < 50$ s on the number line (which is really the t-axis in this case). Note that the values of v are above the t-axis for those values of t that are part of the solution. This shows the relationship of the graph of v as a function of t, and the solution as graphed on the number line (the t-axis). ▶

INEQUALITIES WITH THREE MEMBERS

◀ EXAMPLE 6   Solve: $-1 < 2x + 3 < 6$.
We have the following solution.

$$-1 < 2x + 3 < 6 \qquad \text{original inequality}$$
$$-4 < 2x < 3 \qquad \text{subtract 3 from each member}$$
$$-2 < x < \frac{3}{2} \qquad \text{divide each member by 2}$$

The solution is shown in Fig. 17.10. ▶

◀ EXAMPLE 7   Solve the inequality $2x < x - 4 \le 3x + 8$.
Since we cannot isolate x in the middle member (or in any member), we rewrite the inequality as

$$2x < x - 4 \quad \text{and} \quad x - 4 \le 3x + 8$$

We then solve each of the inequalities, keeping in mind that the solution must satisfy both of them. Therefore, we have

$$2x < x - 4 \quad \text{and} \quad x - 4 \le 3x + 8$$
$$\qquad\qquad\qquad\qquad -2x \le 12$$
$$x < -4 \qquad\qquad\qquad x \ge -6$$

The solution can be written as $-6 \le x < -4$, and this solution is shown in Fig. 17.11. ▶

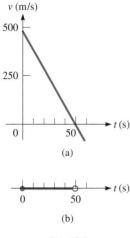

v (m/s)

Fig. 17.9

Fig. 17.10

Fig. 17.11

Solving a Word Problem

◖EXAMPLE 8   In emptying a wastewater tank, one pump can remove no more than 40 L/min. If it operates for 8.0 min and a second pump operates for 5.0 min, what must be the pumping rate of the second pump if 480 L are to be removed?

Let x = the pumping rate of the first pump and y = the pumping rate of the second pump. Since the first operates for 8.0 min and the second for 5.0 min to remove 480 L, we have

first   second
pump   pump   total ⟵ amounts pumped

$$8.0x + 5.0y = 480$$

Since we know that the first pump can remove no more than 40 L/min, which means that $0 \leq x \leq 40$ L/min, we solve for x, then substitute in this inequality:

$x = 60 - 0.625y$	solve for x
$0 \leq 60 - 0.625y \leq 40$	substitute in inequality
$-60 \leq -0.625y \leq -20$	subtract 60 from each member
$96 \geq y \geq 32$	divide each member by -0.625
$32 \leq y \leq 96$ L/min	use $\leq$ symbol (optional step)

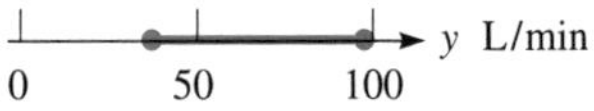

This means that the second pump must be able to pump at least 32 L/min and no more than 96 L/min. See Fig. 17.12.

Although this was a three-member inequality combined with equalities, the solution was done in the same way as with a two-member inequality.   ◗

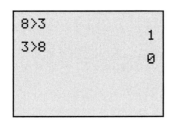

Fig. 17.12

SOLVING AN INEQUALITY WITH A CALCULATOR

A graphing calculator can be used to show the solution of an inequality. On most calculators, a value of 1 is shown if the inequality is satisfied, and a value of 0 is shown if the inequality is not satisfied. This means that if, for example, we enter $8 > 3$ (see the manual to see how $>$ is entered), the calculator displays 1, and if we enter $3 > 8$, it shows 0, as in the display in Fig. 17.13. We can also use an inequality feature to show graphically the solution of an inequality.

Fig. 17.13

◖EXAMPLE 9   Display the solution of the inequality $2x \leq 3 - x$ (see Example 3) on a graphing calculator.

We set y_1 *equal to the inequality,* as shown in Fig. 17.14(a) and then graph y_1. From Fig. 17.14(b), we see that $y_1 = 1$ up to $x = 1$. Thus, the solution that is shown is $x < 1$.

NOTE ▶   We should note that *this method does not distinguish between $<$ and $\leq$ (or $>$ and $\geq$),* so the same solution would be displayed for $2x < 3 - x$. Substituting $x = 1$ into the inequality, we find $2 = 2$, and therefore the complete solution is $x \leq 1$. Compare Fig. 17.14(b) with Fig. 17.7.

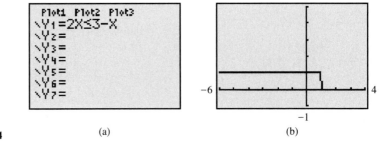

Fig. 17.14   (a)   (b)   ◗

◀ EXAMPLE 10   Display the solution of the inequality $-1 < 2x + 3 < 6$ (see Example 6) on a graphing calculator.

In order to display the solution, we must write the inequality as

$$-1 < 2x + 3 \quad \text{and} \quad 2x + 3 < 6 \qquad \text{see Example 7}$$

Then enter $y_1 = -1 < 2x + 3$ and $2x + 3 < 6$ (consult the manual to determine how "and" is entered) in the calculator, as shown in Fig. 17.15(a). From Fig. 17.15(b), we see that the solution is $-2 < x < 1.5$. The *trace* and *zoom* features can be used to get accurate values. Compare Fig. 17.15(b) with Fig. 17.10.

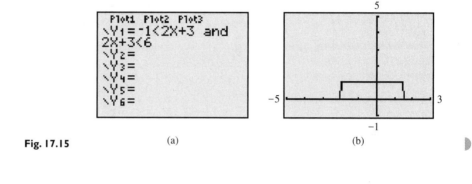

Fig. 17.15   (a)   (b)

E X E R C I S E S   17.2

In Exercises 1–4, make the given changes in the indicated examples of this section and then perform the indicated operations.

1. In Example 2, change the 3 to 21 and then solve the resulting inequality.

2. In Example 4, change the $\frac{1}{4}$ to $\frac{7}{4}$ and then solve and display the resulting inequality.

3. In Example 6, change the $+$ in the middle member to $-$ and then solve the resulting inequality. Graph the solution.

4. In Example 10, change the $+$ in the middle member to $-$ and then display the solution on a graphing calculator.

In Exercises 5–28, solve the given inequalities. Graph each solution.

5. $x - 3 > -4$

6. $x + 2 \le 6$

7. $\frac{1}{2}x < 3$

8. $-4t > 12$

9. $3x - 5 \le -11$

10. $\frac{1}{3}x + 2 \ge 1$

11. $6 - y > 8$

12. $3 - 3x < -1$

13. $\frac{4x - 5}{2} \le x$

14. $1.50 - 5.24x > 3.75 + 2.25x$

15. $180 - 6(x + 12) > 14x + 285$

16. $-2[x - (3 - 2x)] > \frac{1 - 5x}{3} + 2$

17. $2.50(1.50 - 3.40x) < 3.84 - 8.45x$

18. $(2x - 7)(x + 1) \le 4 - x(1 - 2x)$

19. $\frac{1}{3} - \frac{L}{2} < L + \frac{3}{2}$

20. $\frac{x}{5} - 2 > \frac{2}{3}(x + 3)$

21. $-1 < 2x + 1 < 3$

22. $2 < 3R + 1 \le 8$

23. $-4 \le 1 - x < -1$

24. $0 \le 3 - 2x \le 6$

25. $2x < x - 1 \le 3x + 5$

26. $x + 1 \le 7 - x < 2x$

27. $2s - 3 < s - 5 < 3s - 3$

28. $x - 1 < 2x + 2 < 3x + 1$

In Exercises 29–36, solve the inequalities by displaying the solutions on a graphing calculator. See Examples 9 and 10.

29. $3x - 2 < 8 - x$

30. $4(x - 2) > x + 6$

31. $\frac{1}{2}(x + 15) \ge 5 - 2x$

32. $\frac{1}{3}x - 2 \le \frac{1}{2}x + 1$

33. $1 < 5 - 2t < 9$

34. $-3 < 2 - \frac{s}{3} \le -1$

35. $x - 3 < 2x + 5 < 6x + 7$

36. $n - 3 < 2n + 4 \le 1 - n$

In Exercises 37–48, solve the given problems by setting up and solving appropriate inequalities. Graph each solution.

37. Determine the values of x that are in the domain of the function $f(x) = \sqrt{2x - 10}$.

38. Determine the values of x that are in the domain of the function $f(x) = 1/\sqrt{3 - 0.5x}$.

39. The voltage drop V across a resistor is the product of the current i (in A) and the resistance R (in Ω). Find the possible voltage drops across a variable resistor R, if the minimum and maximum resistances are 1.6 kΩ and 3.6 kΩ, respectively, and the current is constant at 2.5 mA.

40. The minimum legal speed on a certain highway is 70 km/h, and the maximum legal speed is 110 km/h. What legal distances can a motorist travel in 4 h on this highway without stopping?

41. A rectangular PV (photovoltaic) solar panel is designed to be 1.42 m long and supply 130 W/m^2 of power. What must the width of the panel be in order to supply between 100 W and 150 W?

42. A beam is supported at each end, as shown in Fig. 17.16. Analyzing the forces leads to the equation $F_1 = 13 - 3d$. For what values of d is F_1 more than 6 N?

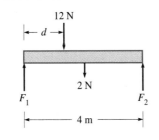

Fig. 17.16

43. The mass m (in g) of silver plate on a dish is increased by electroplating. The mass of silver on the plate is given by $m = 125 + 15.0t$, where t is the time (in h) of electroplating. For what values of t is m between 131 g and 164 g?

44. For a ground temperature of T_0 (in °C), the temperature T (in °C) at a height h (in m) above the ground is given approximately by $T = T_0 - 0.010h$. If the ground temperature is 25°C, for what heights is the temperature above 10°C?

45. During a given rush hour, the numbers of vehicles shown in Fig. 17.17 go in the indicated directions in a one-way street section of a city. By finding the possible values of x and the equation relating x and y, find the possible values of y.

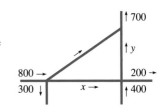

Fig. 17.17

46. One computer printer can print no more than 2500 lines/min. If it operates for 3.0 min and a second printer operates for 2.0 min, what printing rates must the second printer have in order to print 9000 lines?

47. The route of a rapid transit train is 40 km long, and the train makes five stops of equal length. If the train is actually moving for 1 h and each stop must be at least 2 min, what are the lengths of the stops if the train maintains an average speed of at least 30 km/h, including stop times?

48. An oil company plans to install eight storage tanks, each with a capacity of x liters, and five additional tanks, each with a capacity of y liters, such that the total capacity of all tanks is 440 000 L. If capacity y will be at least 40 000 L, what are the possible values of capacity x?

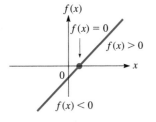

Fig. 17.18

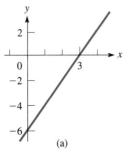

17.3 SOLVING NONLINEAR INEQUALITIES

In this section, we develop methods of solving inequalities with polynomials, expressions with fractions, and nonalgebraic expressions. To develop the method for inequalities with polynomials, we take another look at a linear inequality.

In Fig. 17.18, we see that all values of the linear function $f(x) = ax + b$ ($a \neq 0$) are positive on one side of the point where $f(x) = 0$, and all values of $f(x)$ are negative on the other side. Therefore, *we can solve a linear inequality by expressing it with **zero** on the right and finding the **sign** of the resulting function on either side of the zero.*

❨ **EXAMPLE 1** Solve the inequality $2x - 5 > 1$.

Finding the equivalent inequality with zero on the right, we have $2x - 6 > 0$. Now setting the left member equal to zero, we have

$$2x - 6 = 0 \quad \text{for} \quad x = 3$$

which means $f(x) = 2x - 6$ has one sign for $x > 3$, and the other sign for $x < 3$. Testing values in these intervals, we find, for example, that

$$f(x) = -2 \quad \text{for} \quad x = 2 \quad \text{and} \quad f(x) = +2 \quad \text{for} \quad x = 4$$

This means the solution to the original inequality is $x > 3$. The solution in Fig. 17.19(b) corresponds to the positive values of $f(x) = 2x - 6$ in Fig. 17.19(a).

We could have solved this inequality by methods in the previous section, but the important idea here is to use the *sign* of the function, with zero on the right. ❩

Fig. 17.19

We can extend this method to solving inequalities with polynomials of higher degree. We first find the equivalent inequality with zero on the right and then *factor the function on the left into linear factors and any quadratic factors that lead to complex roots.* As all values of x are considered, each linear factor can change sign at the value for which it is zero. The quadratic factors do not change sign.

This method is especially useful in solving inequalities involving fractions. If a linear factor occurs in the numerator, the function is zero at the value of x for which the linear factor is zero. If such a factor appears in the denominator, the function is undefined where the factor is zero. *The values of x for which a function is zero or undefined are called the* **critical values** *of the function.* As all negative and positive values of x are considered (starting with negative numbers of large absolute value, proceeding through zero, and ending with large positive numbers), a function can change sign only at a critical value.

Critical Values

We now outline the method of solving an inequality by using the critical values of the function. Several examples of the method follow.

USING CRITICAL VALUES TO SOLVE AN INEQUALITY

1. *Determine the equivalent inequality with zero on the right.*
2. *Find all linear factors of the function.*
3. *To find the critical values, set each linear factor equal to zero and solve for x.*
4. *Determine the sign of the function to the left of the leftmost critical value, between critical values, and to the right of the rightmost critical value.*
5. *Those intervals in which the function has the proper sign satisfy the inequality.*

◀ **EXAMPLE 2** Solve the inequality $x^2 - 3 > 2x$.

We first find the equivalent inequality with zero on the right. Therefore, we have $x^2 - 2x - 3 > 0$. We then factor the left member and have

$$(x + 1)(x - 3) > 0$$

Setting each factor equal to zero, we find the left critical value is -1 and the right critical value is 3. All values of x to the left of -1 give the same sign for the function. All values between -1 and 3 give the function the same sign. All values of x to the right of 3 give the same sign to the function. Therefore, *we must determine the sign of $f(x)$ for each of the intervals $x < -1, -1 < x < 3,$ and $x > 3$.*

For $x < -1$, both factors are negative, which means their product is positive, or $(x + 1)(x - 3) > 0$. For $-1 < x < 3$, the left factor is positive, and the right factor is negative, which means their product is negative, or $(x + 1)(x - 3) < 0$. For $x > 3$, both factors are positive, which means their product is positive, or $(x + 1)(x - 3) > 0$.

Summarizing these results for $f(x) = (x + 1)(x - 3)$, we have

$$\text{If } x < -1, f(x) > 0 \qquad \text{if } -1 < x < 3, f(x) < 0 \qquad \text{if } x > 3, f(x) > 0$$

Therefore, the solution to the inequality is $x < -1$ or $x > 3$.

The solution that is shown in Fig. 17.20(b) corresponds to the positive values of the function $f(x) = x^2 - 2x - 3$, shown in Fig. 12.20(a). ▶

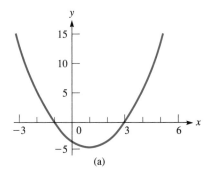

(a)

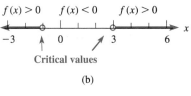

(b)

Fig. 17.20

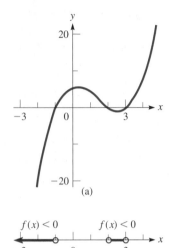

(a)

(b)

Fig. 17.21

EXAMPLE 3 Solve the inequality $x^3 - 4x^2 + x + 6 < 0$.

By methods developed in Chapter 15, we factor the function on the left and obtain $(x + 1)(x - 2)(x - 3) < 0$. The critical values are $-1, 2, 3$. We wish to determine the sign of the left member for the intervals $x < -1$, $-1 < x < 2$, $2 < x < 3$, and $x > 3$. The following table shows each interval, the sign of each factor in each interval, and the resulting sign of the function.

$$f(x) = (x + 1)(x - 2)(x - 3)$$

Interval	$(x + 1)(x - 2)(x - 3)$			Sign of $f(x)$
$x < -1$	$-$	$-$	$-$	$-$
$-1 < x < 2$	$+$	$-$	$-$	$+$
$2 < x < 3$	$+$	$+$	$-$	$-$
$x > 3$	$+$	$+$	$+$	$+$

Since we want values for $f(x) < 0$, the solution is $x < -1$ or $2 < x < 3$. The solution shown in Fig. 17.21(b) corresponds to the values of $f(x) < 0$ in Fig. 17.21(a).

The following example illustrates an applied situation that involves the solution of an inequality.

EXAMPLE 4 The force F (in N) acting on a cam varies according to the time t (in s), and it is given by the function $F = 2t^2 - 12t + 20$. For what values of t, $0 \le t \le 6$ s, is the force at least 4 N?

For a force of at least 4 N, we know that $F \ge 4$ N, or $2t^2 - 12t + 20 \ge 4$. This means we are to solve the inequality $2t^2 - 12t + 16 \ge 0$, and the solution is as follows:

$$2t^2 - 12t + 16 \ge 0$$
$$t^2 - 6t + 8 \ge 0$$
$$(t - 2)(t - 4) \ge 0$$

The critical values are $t = 2$ and $t = 4$, which lead to the following table:

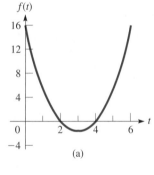

(a)

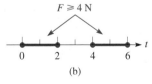

(b)

Fig. 17.22

Interval	$(t - 2)(t - 4)$		Sign of $(t - 2)(t - 4)$
$0 \le t < 2$	$-$	$-$	$+$
$2 < t < 4$	$+$	$-$	$-$
$4 < t \le 6$	$+$	$+$	$+$

We see that the values of t that satisfy the *greater than* part of the problem are $0 \le t < 2$ and $4 < t \le 6$. Since we know that $(t - 2)(t - 4) = 0$ for $t = 2$ and $t = 4$, the solution is

$$0 \le t \le 2 \text{ s} \quad \text{or} \quad 4 \text{ s} \le t \le 6 \text{ s}$$

The graph of $f(t) = 2t^2 - 12t + 16$ is shown in Fig. 17.22(a). The solution, shown in Fig. 17.22(b), corresponds to the values of $f(t)$ that are zero or positive or for which $F \ge 4$ N.

We note here that if the cam rotates in 6-s intervals, the force on the cam is periodic, varying from 2 N to 20 N.

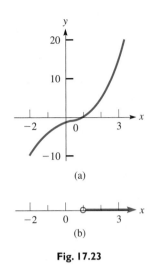

Fig. 17.23

EXAMPLE 5 Solve the inequality $x^3 - x^2 + x - 1 > 0$.

In order to factor this function, we can use the methods of Chapter 15, or we might note that it is factorable by grouping (see Section 6.2). Either method leads to the inequality

$$(x^2 + 1)(x - 1) > 0$$

In this case, we have a quadratic factor $x^2 + 1$ that leads to imaginary roots (j and $-j$) and is never negative. This means we have only one linear factor and therefore only one critical value.

Setting $x - 1 = 0$, we get the critical value $x = 1$. Since $x - 1$ is positive for $x > 1$ and negative for $x < 1$, the solution to the inequality is $x > 1$.

The solution is shown in Fig. 17.23(b), and we see that this solution corresponds to the positive values of $f(x) = x^3 - x^2 + x - 1$ shown in Fig. 17.23(a).

EXAMPLE 6 Find the values of x for which $\sqrt{\dfrac{x - 3}{x + 4}}$ represents a real number.

For the expression to represent a real number, the fraction under the radical must be greater than or equal to zero. This means we must solve the inequality.

$$\frac{x - 3}{x + 4} \geq 0$$

The critical values are found from the factors that are in the numerator or in the denominator. Thus, the critical values are -4 and 3. Considering now the *greater than* part of the $\geq$ sign, we set up the following table:

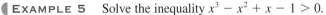

Interval	$\dfrac{x - 3}{x + 4}$	*Sign of* $\dfrac{x - 3}{x + 4}$
$x < -4$	$\dfrac{-}{-}$	$+$
$-4 < x < 3$	$\dfrac{-}{+}$	$-$
$x > 3$	$\dfrac{+}{+}$	$+$

Thus, the values that satisfy the *greater than* part of the problem are those for which $x < -4$ or for which $x > 3$. Now, considering the equality part of the $\geq$ sign, we note that $x = 3$ is valid, for the fraction is zero. However,

if $x = -4$, we have division by zero, and therefore x may not equal -4.

Therefore, the solution of the inequality is $x < -4$ or $x \geq 3$, and this means these are the values for which the original expression represents a real number. The graph of $f(x) = \dfrac{x - 3}{x + 4}$ is shown in Fig. 17.24(a), and the graph of the solution is shown in Fig. 17.24(b).

◀ EXAMPLE 7 Solve the inequality $\dfrac{(x-2)^2(x+3)}{4-x} < 0$.

The critical values are -3, 2, and 4. Thus, we have the following table:

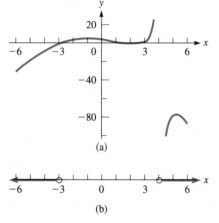

(a)

(b)

Fig. 17.25

Interval	$\dfrac{(x-2)^2(x+3)}{4-x}$	Sign of $\dfrac{(x-2)^2(x+3)}{4-x}$
$x < -3$	$\dfrac{+ \quad -}{+}$	$-$
$-3 < x < 2$	$\dfrac{+ \quad +}{+}$	$+$
$2 < x < 4$	$\dfrac{+ \quad +}{+}$	$+$
$x > 4$	$\dfrac{+ \quad +}{-}$	$-$

Thus, the solution is $x < -3$ or $x > 4$. The graph of the function is shown in Fig. 17.25(a), and the graph of the solution is shown in Fig. 17.25(b). ▶

◀ EXAMPLE 8 Solve the inequality $\dfrac{x+3}{x-1} > 2$.

CAUTION ▶

NOTE ▶

It seems that all we have to do is multiply both members by $x-1$ and solve the resulting linear inequality. However, *this will lead to an incorrect result.* The reason is that we assume $x-1$ is positive when we multiply, but $x-1$ is negative for $x < 1$. To avoid this problem, *first subtract 2 from each member* and combine terms on the left. *This procedure should be used with any inequality involving fractions with the variable in the denominator.* Therefore, the solution is

$$\frac{x+3}{x-1} - 2 > 0 \qquad \text{subtract 2 from both members}$$

$$\frac{x+3-2(x-1)}{x-1} > 0 \qquad \text{combine over the common denominator}$$

$$\frac{5-x}{x-1} > 0 \qquad \text{this is the form to use}$$

The critical values are 1 and 5, and we have the following table of signs:

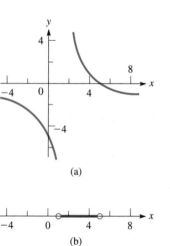

(a)

(b)

Fig. 17.26

Interval	$(5-x)/(x-1)$	Sign of $(5-x)/(x-1)$
$x < 1$	$+ \ / \ -$	$-$
$1 < x < 5$	$+ \ / \ +$	$+$
$x > 5$	$- \ / \ +$	$-$

Thus, the solution is $1 < x < 5$. The graph of $f(x) = (5-x)/(x-1)$ is shown in Fig. 17.26(a), and the graph of the solution is shown in Fig. 17.26(b). ▶

Calculator Solution

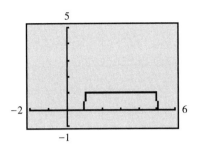

Fig. 17.27

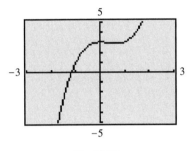

Fig. 17.28

We can use the graphing calculator method of displaying the solution that was developed in Examples 9 and 10 on pages 466 and 467. This method can be used for any of the inequalities of this section, and we illustrate it again in the next example.

◀ **EXAMPLE 9** Solve the inequality of Example 8, $\dfrac{x+3}{x-1} > 2$.

On the calculator, we set $y_1 = (x+3)/(x-1) > 2$, and we then have the display shown in Fig. 17.27. This confirms the solution of $1 < x < 5$. ▶

SOLVING INEQUALITIES GRAPHICALLY

We can get an approximate solution of an inequality from the graph of a function, including functions that are not factorable or not algebraic. The method follows several of the earlier examples. First, write the equivalent inequality with zero on the right and then graph this function. Those values of x corresponding to the proper values of y (either above or below the x-axis) are those that satisfy the inequality.

◀ **EXAMPLE 10** Use a graphing calculator to solve the inequality $x^3 > x^2 - 3$.

Finding the equivalent inequality with zero on the right, we have $x^3 - x^2 + 3 > 0$. On the calculator, we then let $y_1 = x^3 - x^2 + 3$ and graph this function. This gives us the calculator display shown in Fig. 17.28.

We can see that the curve crosses the x-axis only once, between $x = -2$ and $x = -1$. Using the *trace* and *zoom* features (or *zero* feature), we find that this value is about $x = -1.17$. Since we want values of x that correspond to positive values of y, the solution is $x > -1.17$. ▶

EXERCISES  17.3

In Exercises 1–4, make the given changes in the indicated examples of this section and then solve the resulting inequalities.

1. In Example 2, change the $-$ sign before the 3 to $+$ and change $2x$ to $4x$, then solve the resulting inequality, and graph the solution.

2. In Example 5, change both $-$ signs to $+$, then solve the resulting inequality, and graph the solution.

3. In Example 7, change the exponent on $(x-2)$ from 2 to 3, then solve the resulting inequality, and graph the solution.

4. In Example 9, on the right change the 2 to 3, then solve the resulting inequality, and graph the solution.

In Exercises 5–24, solve the given inequalities. Graph each solution. It is suggested that you also graph the function on a graphing calculator as a check.

5. $x^2 - 1 < 0$
6. $x^2 + 3x \geq 0$
7. $2x^2 \leq 4x$
8. $x^2 - 4x > 5$
9. $2x^2 - 12 \leq -5x$
10. $9t^2 + 6t > -1$
11. $x^2 + 4x \leq -4$
12. $6x^2 + 1 < 5x$
13. $R^2 + 4 > 0$
14. $x^4 + 2 < 1$

15. $x^3 + x^2 - 2x > 0$
16. $x^3 - 2x^2 + x \geq 0$
17. $s^3 + 2s^2 - s \geq 2$
18. $n^4 - 2n^3 + 8n + 12 \leq 7n^2$
19. $\dfrac{2x-3}{x+6} \leq 0$
20. $\dfrac{x+5}{x-1} > 0$
21. $\dfrac{x^2 - 6x - 7}{x+5} > 0$
22. $\dfrac{(x-2)^2(5-x)}{(4-x)^3} \leq 0$
23. $\dfrac{x}{x+1} > 1$
24. $\dfrac{2p}{p-1} > 3$

In Exercises 25–32, solve the inequalities by displaying the solutions on a graphing calculator. See Example 9.

25. $3x^2 + 5x \geq 2$
26. $12x^2 + x > 1$
27. $\dfrac{T-8}{3-T} < 0$
28. $\dfrac{3x+1}{x+3} \geq 0$
29. $\dfrac{6-x}{3-x-4x^2} \geq 0$
30. $\dfrac{4-x}{3+2x-x^2} > 0$
31. $\dfrac{x^4(9-x)(x-5)(2-x)}{(4-x)^5} > 0$
32. $\dfrac{2}{x-3} < 4$

In Exercises 33–36, determine the values for x for which the radicals represent real numbers.

33. $\sqrt{(x-1)(x+2)}$

34. $\sqrt{x^2 - 3x}$

35. $\sqrt{-x - x^2}$

36. $\sqrt{\dfrac{x^3 + 6x^2 + 8x}{3 - x}}$

In Exercises 37–44, solve the given inequalities graphically by using a graphing calculator. See Example 10.

37. $x^3 - x > 2$

38. $0.5x^3 < 3 - 2x^2$

39. $x^4 < x^2 - 2x - 1$

40. $3x^4 + x + 1 > 5x^2$

41. $2^x > x + 2$

42. $\log x < 1 - 2x^2$

43. $\sin x < 0.1x^2 - 1$

44. $4 \cos 2x > 2x - 3$

In Exercises 45–50, use inequalities to solve the given problems.

(W) 45. Is $x^2 > x$ for all x? Explain.

(W) 46. Is $x > 1/x$ for all x? Explain.

47. Find an inequality of the form $ax^2 + bx + c > 0$ with $a > 0$ for which the solution is $-1 < x < 4$.

48. Find an inequality of the form $ax^3 + bx < 0$ with $a > 0$ for which the solution is $x < -1$ or $0 < x < 1$.

49. Algebraically find the values of x for which $2^{x+2} > 3^{2x-3}$.

50. Graphically find the values of x for which $2 \log_2 x < \log_3(x + 1)$.

In Exercises 51–60, answer the given questions by solving the appropriate inequalities.

51. The electric power p (in W) delivered to part of a circuit is given by $p = 6i - 4i^2$, where i is the current (in A). For what positive values of i is the power greater than 2 W?

52. The weight w (in Mg) of fuel in a rocket after launch is $w = 2000 - t^2 - 140t$, where t is the time (in min). During what period of time is the weight of fuel greater than 500 Mg?

53. In programming a computer, the formula $63n = 2^x - 1$ may be used. Here, n is the number of items to be added and x is the number of bits needed to represent the sum. Find x if $n < 100$.

54. The object distance p (in cm) and image distance q (in cm) for a camera of focal length 3.00 cm is given by $p = 3.00q/(q - 3.00)$. For what values of q is $p > 12.0$ cm?

55. The weight w (in N) of an object h m above the surface of earth is $w = r^2 w_0/(r + h)^2$, where r is the radius of earth and w_0 is the weight of the object at sea level. Given that $r = 6370$ km, if an object weighs 200 N at sea level, for what altitudes is its weight less than 100 N?

56. One type of machine part costs $10 each, and a second type costs $20 each. How many of the second type can be purchased if at least 30 of the first type are purchased and a total of $1000 is spent?

57. The length of a rectangular microprocessor chip is 2.0 mm more than its width. If its area is less than 35 mm², what values are possible for the width if it must be at least 3.0 mm?

58. A laser source is 2.0 cm from the nearest point P on a flat mirror, and the laser beam is directed at a point Q that is on the mirror and is x cm from P. The beam is then reflected to the receiver, which is x cm from Q. What is x if the total length of the beam is greater than 6.5 cm? See Fig. 17.29.

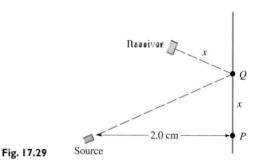

Fig. 17.29

59. A plane takes off from Winnipeg and flies due east at 620 km/h. At the same time, a second plane takes off from the surface of Lake Winnipeg 310 km due north of Winnipeg and flies due north at 560 km/h. For how many hours are the planes less than 1000 km apart?

60. An open box (no top) is formed from a piece of cardboard 8.00 cm square by cutting equal squares from the corners, turning up the resulting sides, and taping the edges together. Find the edges of the squares that are cut out in order that the volume of the box is greater than 32.0 cm³. See Fig. 17.30.

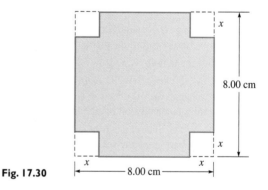

Fig. 17.30

17.4 INEQUALITIES INVOLVING ABSOLUTE VALUES

Inequalities involving absolute values are often useful in later topics in mathematics such as calculus and in applications such as the accuracy of measurements. In this section, we show the meaning of such inequalities and how they are solved.

If we wish to write the inequality $|x| > 1$ without absolute-value signs, we must note that we are considering values of x that are *numerically* larger than 1. Thus, we may write this inequality in the equivalent form $x < -1$ or $x > 1$. We now note that *the original inequality, with an absolute-value sign, can be written in terms of two equivalent inequalities, neither involving absolute values.* If we are asked to write the inequality $|x| < 1$ without absolute-value signs, we write $-1 < x < 1$, since we are considering values of x numerically less than 1.

Following reasoning similar to this, whenever absolute values are involved in inequalities, the following two relations allow us to write equivalent inequalities without absolute values.

$$\text{If } |f(x)| > n, \text{ then } f(x) < -n \text{ or } f(x) > n. \qquad (17.1)$$
$$\text{If } |f(x)| < n, \text{ then } -n < f(x) < n. \qquad (17.2)$$

EXAMPLE 1   Solve the inequality $|x - 3| < 2$.

Here, we want values of x such that $x - 3$ is numerically smaller than 2, or the values of x within 2 units of $x = 3$. These are given by the inequality $1 < x < 5$. Now, using Eq. (17.2), we have

$$-2 < x - 3 < 2$$

By adding 3 to all three members of this inequality, we have

$$1 < x < 5$$

which is the proper interval. See Fig. 17.31.

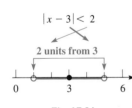

Fig. 17.31

EXAMPLE 2   Solve the inequality $|2x - 1| > 5$.

By using Eq. (17.1), we have

$$2x - 1 < -5 \quad \text{or} \quad 2x - 1 > 5$$

Completing the solution, we have

$$2x < -4 \quad \text{or} \quad 2x > 6 \qquad \text{add 1 to each member}$$
$$x < -2 \qquad x > 3 \qquad \text{divide each member by 2}$$

This means that the given inequality is satisfied for $x < -2$ or for $x > 3$. We must be very careful to remember that *we cannot write this as $3 < x < -2$.* The solution is shown in Fig. 17.32

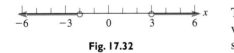

Fig. 17.32

The meaning of this inequality is that the numerical value of $2x - 1$ is greater than 5. By considering values in these intervals, we can see that this is true for values of x less than -2 or greater than 3.

◀ EXAMPLE 3   Solve the inequality $2\left|\dfrac{2x}{3} + 1\right| \geq 4$.

The solution is as follows:

$$2\left|\frac{2x}{3} + 1\right| \geq 4 \qquad \text{original inequality}$$

$$\left|\frac{2x}{3} + 1\right| \geq 2 \qquad \text{divide each member by 2}$$

$$\frac{2x}{3} + 1 \leq -2 \quad \text{or} \quad \frac{2x}{3} + 1 \geq 2 \qquad \text{using Eq. (17.1)}$$

$$2x + 3 \leq -6 \qquad\qquad 2x + 3 \geq 6$$

$$2x \leq -9 \qquad\qquad\quad 2x \geq 3$$

$$x \leq -\frac{9}{2} \qquad\qquad\quad x \geq \frac{3}{2} \qquad \text{solution}$$

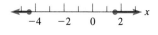

Fig. 17.33

This solution is shown in Fig. 17.33. Note that the sign of equality does not change the method of solution. It simply indicates that $-\frac{9}{2}$ and $\frac{3}{2}$ are included in the solution.   ▶

◀ EXAMPLE 4   Solve the inequality $|3 - 2x| < 3$.
We have the following solution.

$$|3 - 2x| < 3 \qquad \text{original inequality}$$

$$-3 < 3 - 2x < 3 \qquad \text{using Eq. (17.2)}$$

$$-6 < -2x < 0$$

$$3 > x > 0 \qquad \text{divide by } -2 \text{ and reverse signs of inequality}$$

$$0 < x < 3 \qquad \text{solution}$$

Fig. 17.34

The meaning of the inequality is that the numerical value of $3 - 2x$ is less than 3. This is true for values of x between 0 and 3. The solution is shown in Fig. 17.34.   ▶

◀ EXAMPLE 5   A technician measures an electric current and reports that it is 0.036 A with a possible error of ± 0.002 A. Write this result for the current i, using an inequality with absolute values.

The statement of the problem tells us that the current is no less than 0.034 A and no more than 0.038 A. Another way of stating this is that the numerical difference between the true value of i (unknown exactly) and the measured value, 0.036 A, is less than or equal to 0.002 A. Using an absolute-value inequality, this is written as

$$|i - 0.036| \leq 0.002$$

where values are in amperes.

We can see that this inequality is correct by using (Eq. 17.2):

$$-0.002 \leq i - 0.036 \leq 0.002$$

$$0.034 \leq i \leq 0.038 \qquad \text{add 0.036 to each member}$$

This verifies that i should not be less than 0.034 A or no more than 0.038 A. The solution is shown in Fig. 17.35.   ▶

Fig. 17.35

Calculator Solution   Most graphing calculators can be used to display the solution of an inequality that involves absolute values as they can display the solutions to other inequalities (see Example 9 on page 466, Example 10 on page 467, and Example 9 on page 473). Since calculators differ in their operation, check the manual of your calculator to see how an absolute value is entered and displayed. The following example shows the display of the solution of an inequality on a graphing calculator.

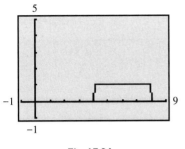

Fig. 17.36

◀ **EXAMPLE 6**   Display the solution to the inequality $\left|\dfrac{x}{2} - 3\right| < 1$ on a graphing calculator.

On the calculator, set $y_1 = \text{abs}(x/2 - 3) < 1$ and obtain the display shown in Fig. 17.36. From this display, we see that the solution is $4 < x < 8$. It is possible that it will take two or three *window* settings to get an accurate solution. ◗

EXERCISES 17.4

In Exercises 1 and 2, make the given changes in the indicated examples of this section and then solve the resulting inequalities.

1. In Example 2, change the $>$ to $<$, solve the resulting inequality, and graph the solution.

2. In Example 4, change the $<$ to $>$, solve the resulting inequality, and graph the solution.

In Exercises 3–24, solve the given inequalities. Graph each solution.

3. $|x - 4| < 1$

4. $|x + 1| < 3$

5. $|5x + 4| > 6$

6. $\left|\dfrac{1}{2}N - 1\right| > 1$

7. $1 + |6x - 5| \le 5$

8. $|5 - 7x| \le 0$

9. $|3 - 4x| > 3$

10. $3 + |3x + 1| \ge 5$

11. $\left|\dfrac{t + 1}{5}\right| < 5$

12. $\left|\dfrac{2x - 9}{4}\right| < 1$

13. $|20x + 85| \le 43$

14. $|2.6x - 9.1| > 10.4$

15. $2|x - 4| > 8$

16. $3|4 - 3x| \le 10$

17. $8 + 3|3 - 2x| < 11$

18. $5 - 4|1 - 7x| > 13$

19. $4|2 - 5x| \ge 6$

20. $2.5|7.1 - 2.0x| \le 6.5$

21. $\left|\dfrac{3R}{5} + 1\right| < 8$

22. $\left|\dfrac{4x}{3} - 5\right| \ge 7$

23. $\left|6.5 - \dfrac{x}{2}\right| \ge 2.3$

24. $\left|27 - \dfrac{2x}{3}\right| > 17$

In Exercises 25–28, solve the given inequalities by displaying the solutions on a graphing calculator. See Example 6.

25. $|2x - 5| < 3$

26. $2|6 - T| > 5$

27. $\left|4 - \dfrac{x}{2}\right| \ge 1$

28. $\left|\dfrac{x}{3} + 2\right| \le 2$

In Exercises 29–32, solve the given quadratic inequalities. Check each by displaying the solution on a graphing calculator.

29. $|x^2 + x - 4| > 2$
   (After using Eq. (17.1), you will have two inequalities. The solution includes the values of x that satisfy *either* of the inequalities.)

30. $|x^2 + 3x - 1| > 3$ (See Exercise 29.)

31. $|x^2 + x - 4| < 2$
   (Use Eq. (17.2), then treat the resulting inequality as two inequalities of the form $f(x) > -n$ and $f(x) < n$. The solution includes the values of x that satisfy *both* of the inequalities.)

32. $|x^2 + 3x - 1| < 3$ (See Exercise 31.)

In Exercises 33–36, solve the given problems.

(W) 33. Solve for x if $|x| < a$ and $a \le 0$. Explain.

34. Solve for x if $|x - 1| < 4$ and $x \ge 0$.

35. The thickness t (in km) of the earth's crust varies and can be described as $|t - 27| \le 23$. What are the minimum and maximum values of the thickness of the earth's crust?

36. The temperature T (in °C) at which a certain integrated-circuit computer chip is designed to operate is given by the inequality $|T - 110| \le 10$. What are the minimum and maximum temperatures of this range?

In Exercises 37–40, use inequalities involving absolute values to solve the given problems.

(W) 37. The production p (in barrels) of oil at a refinery is estimated at $2\,000\,000 \pm 200\,000$. Express p using an inequality with absolute values and describe the production in a verbal statement.

38. The *Mach number M* of a moving object is the ratio of its velocity v to the velocity of sound v_s, and v_s varies with temperature. A jet traveling at 1650 km/h changes its altitude from 500 m to 5500 m. At 500 m (with the temperature at 27°C), $v_s = 1250$ km/h, and at 5500 m (-3°C), $v_s = 1180$ km/h. Express the range of M, using an inequality with absolute values.

39. The voltage V in a certain circuit is given by $V = 6.0 - 200i$, where i is the current (in A). For what values of the current is the absolute value of the voltage less than 2.0 V?

40. A rocket is fired from a plane flying horizontally at 3000 m. The height h (in m) of the rocket above the plane is given by $h = 190t - 4.9t^2$, where t is the time (in s) of flight of the rocket. When is the rocket more than 1300 m above or below the plane? See Fig. 17.37.

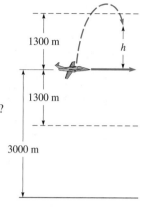

Fig. 17.37

17.5 GRAPHICAL SOLUTION OF INEQUALITIES WITH TWO VARIABLES

To this point, we have considered inequalities with one variable and certain methods of solving them. We may also graphically solve inequalities involving two variables, such as x and y. In this section, we consider the solution of such inequalities.

Let us consider the function $y = f(x)$. We know that the coordinates of points on the graph satisfy the equation $y = f(x)$. However, for points above the graph of the function, we have $y > f(x)$, and for points below the graph of the function, we have $y < f(x)$. Consider the following example.

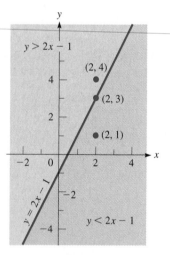

Fig. 17.38

EXAMPLE 1 Consider the linear function $y = 2x - 1$, the graph of which is shown in Fig. 17.38. This equation is satisfied for points on the line. For example, the point $(2, 3)$ is on the line, and we have $3 = 2(2) - 1 = 3$. Therefore, for points on the line, we have $y = 2x - 1$, or $y - 2x + 1 = 0$.

The point $(2, 4)$ is above the line, since we have $4 > 2(2) - 1$, or $4 > 3$. Therefore, for points above the line, we have $y > 2x - 1$, or $y - 2x + 1 > 0$. In the same way, for points below the line, $y < 2x - 1$ or $y - 2x + 1 < 0$. We note this is true for the point $(2, 1)$, since $1 < 2(2) - 1$, or $1 < 3$.

The line for which $y = 2x - 1$ and the regions for which $y > 2x - 1$ and for which $y < 2x - 1$ are shown in Fig. 17.38.

Summarizing,

$$y > 2x - 1 \quad \text{for points } \textit{above} \text{ the line}$$
$$y = 2x - 1 \quad \text{for points } \textit{on} \text{ the line}$$
$$y < 2x - 1 \quad \text{for points } \textit{below} \text{ the line}$$

The illustration in Example 1 leads us to the graphical method of indicating the points that satisfy an inequality with two variables. First, we solve the inequality for y and then determine the graph of the function $y = f(x)$. *If we wish to solve the inequality $y > f(x)$, we indicate the appropriate points by shading in the region above the curve. For the inequality $y < f(x)$, we indicate the appropriate points by shading in the region below the curve.* We note that the complete solution to the inequality consists of all points in an entire region of the plane.

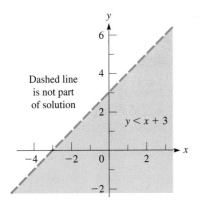

Dashed line is not part of solution

$y < x + 3$

Fig. 17.39

EXAMPLE 2 Draw a sketch of the graph of the inequality $y < x + 3$.

First, we draw the function $y = x + 3$, as shown by the dashed line in Fig. 17.39. Since we wish to find all the points that satisfy the inequality $y < x + 3$, we show these points by shading in the region below the line. The line is shown as a *dashed line* to indicate that points on it do not satisfy the inequality.

Most graphing calculators can be used to display the solution of an inequality involving two variables by shading in an area above a curve, below a curve, or between curves. The manner in which this is done varies according to the model of the calculator. Therefore, the manual should be used to determine how this is done on any particular model of calculator. In Fig. 17.40, such a graphing calculator display is shown for this inequality.

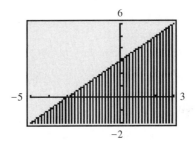

Fig. 17.40

EXAMPLE 3 After a snowstorm, it is estimated that it will take 30 min to plow each kilometer of Route 15 and 45 min to plow each kilometer of Route 80. If no more than 60 plowing-hours are available, what combinations of Route 15 and Route 80 can be plowed?

Let $x =$ kilometers of Route 15 that can be plowed and $y =$ kilometers of Route 80 that can be plowed. The time to plow along each route is the product of the time for each kilometer and the number of kilometers to be plowed. This gives us

time to plow Rt. 15   time to plow Rt. 80   max. available time

$$(0.50 \text{ h/km})(x \text{ km}) + (0.75 \text{ h/km})(y \text{ km}) \leq 60 \text{ h}$$

30 min ⌐   ⌐ 45 min

$$0.50x + 0.75y \leq 60$$
$$y \leq 80 - 0.67x$$

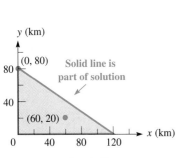

y (km)

(0, 80)   Solid line is part of solution

(60, 20)

Fig. 17.41

Noting that negative values of x and y do not have meaning, we have the graph in Fig. 17.41, shading in the region below the line since we have $y < 80 - 0.67x$ for that region. Any point in the shaded region, or on the axes or the line around the shaded region, gives a solution. The *solid line* indicates that points on it are part of the solution.

The point $(0, 80)$, for example, is a solution and tells us that the 80 km of Route 80 can be plowed if none of Route 15 is plowed. In this case, all 60 h of plowing time are used for Route 80. Another possibility is shown by the point $(60, 20)$, which indicates that 60 km of Route 15 and 20 km of Route 80 can be plowed. In this case, not all of the 60 plowing hours are used.

EXAMPLE 4 Draw a sketch of the graph of the inequality $y > x^2 - 4$.

Although the graph of $y = x^2 - 4$ is not a straight line, the method of solution is the same. We graph the function $y = x^2 - 4$ as a dashed curve, since it is not part of the solution, as shown in Fig. 17.42. We then shade in the region above the curve to indicate the points that satisfy the inequality.

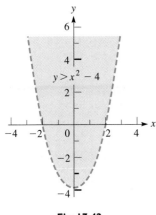

$y > x^2 - 4$

Fig. 17.42

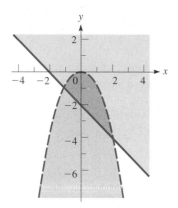

Fig. 17.43

▌**EXAMPLE 5**   Draw a sketch of the region that is defined by the system of inequalities $y \geq -x - 2$ and $y + x^2 < 0$.

Similar to the solution of a system of equations, *the solution of a system of inequalities is any pair of values (x, y) that satisfies both inequalities.* This means we want the region common to both inequalities. In Fig. 17.43, we first shade in the region above the line $y = -x - 2$ and then shade in the region below the parabola $y = -x^2$. The region defined by this system is the darkly shaded region below the parabola that is also above and on the line. The calculator display for this region is shown in Fig. 17.44.

If we are asked to find the region defined by $y \geq -x - 2$ *or* $y + x^2 < 0$, it consists of both shaded regions and all points on the line.

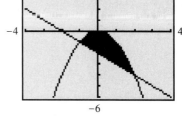

Fig. 17.44

EXERCISES 17.5

In Exercises 1 and 2, make the given changes in the indicated examples of this section and then draw the graph of the resulting inequality.

1. In Example 2, change $x + 3$ to $3 - x$ and then draw the graph of the resulting inequality.

2. In Example 4, change $x^2 - 4$ to $4 - x^2$ and then draw the graph of the resulting inequality.

In Exercises 3–18, draw a sketch of the graph of the given inequality.

3. $y > x - 1$ **4.** $y < 3x - 2$

5. $y \geq 2x + 5$ **6.** $y \leq 3 - x$

7. $3x + 2y + 6 > 0$ **8.** $x + 4y - 8 < 0$

9. $y < x^2$ **10.** $y \leq 2x^2 - 3$

11. $2x^2 - 4x - y > 0$ **12.** $y \leq x^3 - 1$

13. $y < 32x - x^4$ **14.** $y \leq \sqrt{2x + 5}$

15. $y > \dfrac{1}{x^2 + 1}$ **16.** $y < \ln x$

17. $y > 1 + \sin 2x$ **18.** $y > |x| - 3$

In Exercises 19–28, draw a sketch of the graph of the region in which the points satisfy the given system of inequalities.

19. $y > x$
  $y > 1 - x$

20. $y \leq 2x$
  $y \geq x - 1$

21. $y \leq 2x^2$
  $y > x - 2$

22. $y > x^2$
  $y < x + 4$

23. $y > \frac{1}{2}x^2$
  $y \leq 4x - x^2$

24. $y > 4 - x$
  $y < \sqrt{16 - x^2}$

25. $y \geq 0$
  $y \leq \sin x$
  $0 \leq x \leq 3\pi$

26. $y > 0$
  $y > 1 - x$
  $y < e^x$

27. $|y + 2| < 5$
  $|x - 3| \leq 2$

28. $16x + 3y - 12 > 0$
  $y > x^2 - 2x - 3$
  $|2x - 3| < 3$

In Exercises 29–38, use a graphing calculator to display the solution of the given inequality or system of inequalities.

29. $2x + y < 5$ **30.** $4x - y > 1$

31. $y \geq 1 - x^2$ **32.** $y < |4 - 2x|$

33. $y > 2x - 1$
  $y < x^4 - 8$

34. $y < 3 - x$
  $y > 3x - x^3$

35. $y > x^2 + 2x - 8$
  $y < \dfrac{1}{x} - 2$

36. $y > -2x^2$
  $y < 1 - e^{-x}$

37. $y \leq |2x - 3|$
  $y > 1 - 2x^2$

38. $y \geq |4 - x^2|$
  $y < 2 \ln |x|$

In Exercises 39–42, solve the given problems.

39. By an inequality, define the region below the line
$4x - 2y + 5 = 0$.

40. By an inequality, define the region that is bounded by or includes the parabola $x^2 - 2y = 0$, and that contains the point $(1, 0.4)$.

41. Draw a graph of the solution of the system $y \geq 2x^2 - 6$ and $y = x - 3$.

42. Draw a graph of the solution of the system $y < |x + 2|$ and $y = x^2$.

In Exercises 43–48, set up the necessary inequalities and sketch the graph of the region in which the points satisfy the indicated system of inequalities.

43. A telephone company is installing two types of fiber-optic cable in an area. It is estimated that no more than 300 m of type A cable, and at least 200 m but no more than 400 m of type B cable, are needed. Graph the possible lengths of cable that are needed.

44. A refinery can produce gasoline and diesel fuel, in amounts of any combination, except that equipment restricts total production to 7500 L/day. Graph the different possible production combinations of the two fuels.

45. The elements of an electric circuit dissipate p W of power. The power p_R dissipated by a resistor in the circuit is given by $p_R = Ri^2$, where R is the resistance (in Ω) and i is the current (in A). Graph the possible values of p and i for $p > p_R$ and $R = 0.5\ \Omega$.

46. The cross-sectional area A (in m^2) of a certain trapezoid culvert in terms of its depth d (in m) is $A = 2d + d^2$. Graph the possible values of d and A if A is between 1 m^2 and 2 m^2.

47. One pump can remove wastewater at the rate of 250 L/min, and a second pump works at the rate of 150 L/min. Graph the possible values of the time (in min) that each of these pumps operates such that together they pump more than 15 000 L.

48. A rectangular computer chip is being designed such that its perimeter is no more than 30 mm, its width at least 3 mm and its length at least 8 mm. Graph the possible values of the width w and the length l.

17.6 LINEAR PROGRAMMING

This section is an introduction to linear programming. Other methods are developed in a more complete coverage.

An important area in which graphs of inequalities with two or more variables are used is in the branch of mathematics known as **linear programming** (in this context, "programming" does not mean computer programming). This subject, which we mentioned in the chapter introduction, is widely applied in industry, business, economics, and technology. The analysis of many social problems can also be made by use of linear programming.

Linear programming is used to analyze problems such as those related to maximizing profits, minimizing costs, or the use of materials with certain constraints of production. First, we look at a similar type of mathematical problem.

◀ **EXAMPLE 1** Find the maximum value of F, where $F = 2x + 3y$ and x and y are subject to the conditions that

$$x \geq 0,\ y \geq 0$$
$$x + y \leq 6$$
$$x + 2y \leq 8$$

These four inequalities that define the conditions on x and y are known as the **constraints** of the problem, and F is known as the **objective function.**

We now graph this set of inequalities, as shown in Fig. 17.45. Each point in the shaded region (including the line segments on the edges) satisfies all the constraints and is known as a **feasible point.**

The maximum value of F must be found at one of the feasible points. Testing for values at the vertices of the region we have the values in the following table:

Point	$(0,0)$	$(6,0)$	$(4,2)$	$(0,4)$
Value of F	0	12	14	12

If we evaluate F at any other feasible point, we will find that $F < 14$. Therefore, the maximum value of F under the given constraints is 14.

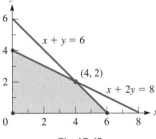

Fig. 17.45

In Example 1, we found the maximum value of a *linear* objective function, subject to *linear* constraints (thus the name *linear programming*). We found the maximum value of *F*, subject to the given constraints, to be at one of the vertices of the region of feasible points. In fact, in the theory of linear programming it is established that

the maximum and minimum values of the objective function will occur at a vertex of the region of feasible points

(or at all points along a line segment connecting two vertices.)

◖ EXAMPLE 2   Find the maximum and minimum values of the objective function $F = 3x + y$, subject to the constraints

$$x \geq 2, y \geq 0$$
$$x + y \leq 8$$
$$2y - x \leq 1$$

The constraints are graphed as shown in Fig. 17.46. We then locate the vertices, and evaluate *F* at these vertices as follows.

Vertex	(2, 0)	(8, 0)	(5, 3)	(2, 1.5)
Value of *F*	6	24	18	7.5

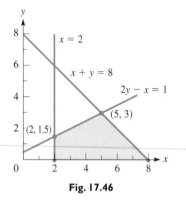

Fig. 17.46

Therefore, we see that the maximum value of *F* is 24, and the minimum value of *F* is 6. If we check the value of *F* at any other feasible point, we will find a value between 6 and 24.   ◗

The following two examples show the use of linear programming in finding a maximum value and a minimum value in applied situations.

Solving a Word Problem

See the chapter introduction.

The loudspeaker was developed in the early 1920s by the U.S. inventor Kellogg Rice.

◖ EXAMPLE 3   A company makes two types of stereo speaker systems, their good-quality system and their highest-quality system. The production of these systems requires assembly of the speaker system itself and the production of the cabinets in which they are installed. The good-quality system requires 3 worker-hours for speaker assembly and 2 worker-hours for cabinet production for each complete system. The highest-quality system requires 4 worker-hours for speaker assembly and 6 worker-hours for cabinet production for each complete system. Available skilled labor allows for a maximum of 480 worker-hours per week for speaker assembly and a maximum of 540 worker-hours per week for cabinet production. It is anticipated that all systems will be sold and that the profit will be $30 for each good-quality system and $75 for each highest-quality system. How many of each system should be produced to provide the greatest profit?

First, let x = the number of good-quality systems and y = the number of highest-quality systems made in one week. Thus, the profit *P* is given by

$$P = 30x + 75y$$

We know that negative numbers are not valid for either x or y, and therefore we have $x \geq 0$ and $y \geq 0$.

The number of available worker-hours per week for each part of the production also restricts the number of systems that can be made. In the speaker-assembly shop, $3x$ worker-hours are needed for each good-quality system and $4y$ worker-hours are needed for each highest-quality system. Because only 480 h are available in the speaker-assembly shop means that $3x + 4y \leq 480$.

In the cabinet shop, it takes $2x$ worker-hours for a good-quality system and $6y$ worker-hours for a highest-quality system. Because only 540 h are available in the cabinet shop means that $2x + 6y \leq 540$.

Therefore, we want to maximize the profit $P = 30x + 75y$, which is the objective function, under the constraints

$x \geq 0, y \geq 0$ number of systems produced cannot be negative

$3x + 4y \leq 480$ worker-hours for speaker assembly

$2x + 6y \leq 540$ worker-hours for cabinet production

See Appendix C for a graphing calculator program LINPROG. It can be used to shade the area for two linear constraints.

The constraints are graphed as shown in Fig. 17.47. We locate the vertices and evaluate the profit P at these points as follows:

Vertex	$(0,0)$	$(160,0)$	$(72,66)$	$(0,90)$
Profit ($)	0	4800	7110	6750

Therefore, we see that the greatest profit of $7110 is made by producing 72 good-quality systems and 66 highest-quality systems.

A way of showing the number of each system to be made for the greatest profit is to assume values of the profit P and graph these lines. For example, for $P = \$3000$ or $P = \$6000$, we have the lines shown in Fig. 17.47. Both amounts of profit are possible with various combinations of speaker systems being produced. However, we note that the line for $P = \$6000$ passes through feasible points farther from the origin. It is also clear that the lines for the profits are parallel and that *the greatest profit attainable is given by the line passing through A,* where $3x + 4y = 40$ and $2x + 6y = 540$ intersect. This also illustrates why the greatest profit is found at one of the vertices of the region of feasible points.

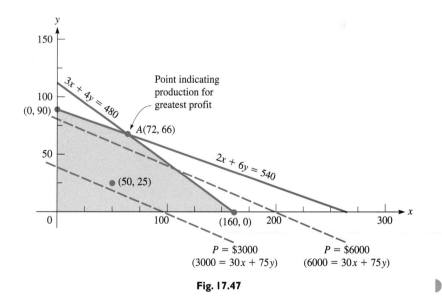

Fig. 17.47

❨EXAMPLE 4   An airline plans to open new routes and use two types of planes, A and B, on these routes. It is expected there would be at least 400 first-class passengers and 2000 economy-class passengers on these routes each day. Plane A costs $18 000/day to operate and has seats for 40 first-class and 80 economy-class passengers. Plane B costs $16 000/day to operate and has seats for 20 first-class and 160 economy-class passengers. How many of each type of plane should be used for these routes to minimize operating costs? (Assume that the planes will be used only on these routes.)

We first let x = the number of A planes and y = the number of B planes to be used. Then the operating cost $C = 18\,000x + 16\,000y$ is the objective function. The constraints are

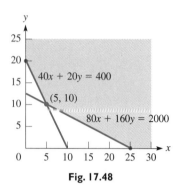

$$x \geq 0, y \geq 0 \qquad \text{number of planes cannot be negative}$$

$$40x + 20y \geq 400 \qquad \text{at least 400 first-class passengers}$$

$$80x + 160y \geq 2000 \qquad \text{at least 2000 economy-class passengers}$$

Fig. 17.48

The constraints are graphed as shown in Fig. 17.48. We see that the region of feasible points is unlimited (there could be more than 2400 passengers), but we still want to evaluate the operating cost C at the vertices. Evaluating C at these points, we have

Vertex	$(25, 0)$	$(5, 10)$	$(0, 20)$
Cost ($)	450 000	250 000	320 000

Therefore, five type-A planes and ten type-B planes should be used on these routes to keep the operating costs at a minimum of $250 000. Obviously, if the number of passengers does not meet the numbers expected, these numbers of planes should be changed accordingly.   ❩

EXERCISES 17.6

In Exercises 1 and 2, make the given changes in the indicated examples of this section and then find the indicated values.

1. In Example 1, change the last constraint to $2x + y \leq 8$. Then graph the feasible points and find the maximum value of F.

2. In Example 2, change the last constraint to $3y - x \leq 4$. Then graph the feasible points and find the maximum and minimum values of F.

In Exercises 3–14, find the indicated maximum and minimum values by the linear programming method of this section. For Exercises 5–14, the constraints are shown below the objective function.

3. Graphing the constraints of a linear programming problem shows the consecutive vertices of the region of feasible points to be $(0, 0)$, $(12, 0)$, $(10, 7)$, $(0, 5)$, and $(0, 0)$. What are the maximum and minimum values of the objective function $F = 3x + 4y$ in this region?

4. Graphing the constraints of a linear programming problem shows the consecutive vertices of the region of feasible points to be $(1, 3)$, $(8, 0)$, $(9, 7)$, $(5, 8)$, $(0, 6)$, and $(1, 3)$. What are the maximum and minimum values of the objective function $F = 2x + 5y$ in this region?

5. Maximum P:
$P = 3x + 5y$
$x \geq 0, y \geq 0$
$2x + y \leq 6$

6. Maximum P:
$P = 2x + 7y$
$x \geq 1, y \geq 0$
$x + 4y \leq 8$

7. Minimum C:
$C = 4x + 6y$
$x \geq 0, y \geq 0$
$x + y \geq 5$
$x + 2y \geq 7$

8. Minimum C:
$C = 6x + 4y$
$x \geq 0, y \geq 0$
$2x + y \geq 6$
$x + y \geq 5$

9. Maximum and minimum F:
$F = x + 3y$
$x \geq 1, y \geq 2$
$y - x \leq 3$
$y + 2x \leq 8$

10. Maximum and minimum F:
$F = 3x - y$
$x \geq 1, y \geq 0$
$x + 4y \leq 8$
$4x + y \leq 8$

11. Maximum P:
$P = 9x + 2y$
$x \geq 0, y \geq 0$
$2x + 5y \leq 10$
$4x + 3y \leq 12$

12. Minimum C:
$C = 3x + 8y$
$x \geq 0, y \geq 0$
$6x + y \geq 6$
$x + 4y \geq 4$

13. Minimum C:
$C = 6x + 4y$
$y \geq 2$
$x + y \leq 12$
$x + 2y \geq 12$
$2x + y \geq 12$

14. Maximum P:
$P = 3x + 4y$
$2x + y \geq 2$
$x + 2y \geq 2$
$x + y \leq 2$

In Exercises 15–20, solve the given linear programming problems.

15. A person wants to invest no more than $9000, part at 6% and part at 5%, with no more than twice as much being invested at 6% than at 5%. How much should be invested at each rate to maximize the income from the investments?

16. An oil refinery refines types A and B of crude oil and can refine as much as 4000 barrels each week. Type A crude has 2 kg of impurities per barrel, type B has 3 kg of impurities per barrel, and the refinery can handle no more than 9000 kg of these impurities each week. How much of each type should be refined in order to maximize profits, if the profit is $4/barrel for type A and $5/barrel for type B?

17. A manufacturer produces a business calculator and a graphing calculator. Each calculator is assembled in two sets of operations, where each operation is in production 8 h during each day. The average time required for a business calculator in the first operation is 3 min, and 6 min is required in the second operation. The graphing calculator averages 6 min in the first operation and 4 min in the second operation. All calculators can be sold; the profit for a business calculator is $8, and the profit for a graphing calculator is $10. How many of each type of calculator should be made each day in order to maximize profit?

(W) **18.** Using the information given in Example 3, with the one change that the profit on each good-quality speaker system is $60 (instead of $30), how many of each system should be made? Explain why this one change in data makes such a change in the solution.

19. Brands A and B of breakfast cereal are both enriched with vitamins P and Q. The necessary information about these cereals is as follows:

	Cereal A	Cereal B	RDA
Vitamin P	0.1 unit/g	0.2 units/g	10 units
Vitamin Q	0.5 units/g	0.3 units/g	30 units
Cost	0.8¢/g	1.2¢/g	

(RDA is the *Recommended Daily Allowance*.) Find the amount of each cereal that together satisfies the RDA of vitamins P and Q at the lowest cost.

20. A computer company makes parts A and B in each of two different plants. It costs $4000 per day to operate the first plant and $5000 per day to operate the second plant. Each day the first plant produces 100 of part A and 200 of part B, while at the second plant 250 of part A and 100 of part B are produced. How many days should each plant operate to produce 2000 of each part and keep operating costs at a minimum?

CHAPTER 17 EQUATIONS

If $|f(x)| > n$, then $f(x) < -n$ or $f(x) > n$. $\hspace{2cm}$ **(17.1)**

If $|f(x)| < n$, then $-n < f(x) < n$. $\hspace{2cm}$ **(17.2)**

CHAPTER 17 REVIEW EXERCISES

In Exercises 1–16, solve each of the given inequalities algebraically. Graph each solution.

1. $2x - 12 > 0$

2. $2.4(T - 4.0) \geq 5.5 - 2.4T$

3. $4 < 2x - 1 < 11$

4. $2x < x + 1 < 4x + 7$

5. $5x^2 + 9x < 2$

6. $x^2 + 2x > 63$

7. $6n^2 - n > 35$

8. $2x^3 + 4 \leq x^2 + 8x$

9. $\dfrac{(2x - 1)(3 - x)}{x + 4} > 0$

10. $x^4 + x^2 \leq 0$

11. $\dfrac{1}{x} < 2$

12. $\dfrac{1}{x - 2} < \dfrac{1}{4}$

13. $|3x + 2| \leq 4$

14. $|4 - 3x| \geq 1$

15. $|3 - 5x| > 7$

16. $\left|2 - \dfrac{y}{2}\right| \leq 0$

In Exercises 17–24, solve the given inequalities on a graphing calculator such that the display is the graph of the solution.

17. $5 - 3x < 0$

18. $6 \leq 4x - 2 < 9$

19. $2 \leq \dfrac{4n - 2}{3} < 3$

20. $3x < 2x + 1 < x - 5$

21. $\dfrac{8 - R}{2R + 1} \leq 0$

22. $\dfrac{(3 - x)^2}{2x + 7} \leq 0$

23. $|x - 2| > 3$

24. $2|2x - 9| < 8$

In Exercises 25–28, use a graphing calculator to solve the given inequalities. Graph the appropriate function and from the graph determine the solution.

25. $x^3 + x + 1 < 0$

26. $\dfrac{2}{R + 2} > 3$

27. $e^{-t} > 0.5$

28. $\sin 2x < 0.8 \quad (0 < x < 4)$

In Exercises 29–40, draw a sketch of the region in which the points satisfy the given inequality or system of inequalities.

29. $y > 4 - x$

30. $y < \dfrac{1}{2}x + 2$

31. $2y - 3x - 4 \le 0$

32. $3y - x + 6 \ge 0$

33. $y > x^2 + 1$

34. $y \le \dfrac{1}{x^2 - 4}$

35. $y - |x + 1| < 0$

36. $2y + 2x^3 + 6x > 3$

37. $y > x + 1$
   $y < 4 - x^2$

38. $y > 2x - x^2$
   $y \ge -2$

39. $y \le \dfrac{1}{x^2 + 1}$
   $y < x - 1$

40. $y < \cos \dfrac{1}{2}x$
   $y > \dfrac{1}{2}e^x$
   $-\pi < x < \pi$

In Exercises 41–48, use a graphing calculator to display the region in which the points satisfy the given inequality or system of inequalities.

41. $y < 3x + 5$

42. $y > 2 - \frac{1}{4}x$

43. $y > 8 + 7x - x^2$

44. $y < x^3 + 4x^2 - x - 4$

45. $y < 32x - x^4$

46. $y > 2x - 1$
   $y < 6 - 3x^2$

47. $y > 1 - x \sin 2x$
   $y < 5 - x^2$

48. $y > |x - 1|$
   $y < 4 + \ln x$

In Exercises 49–52, determine the values of x for which the given radicals represent real numbers.

49. $\sqrt{3 - x}$

50. $\sqrt{x + 5}$

51. $\sqrt{x^2 + 4x}$

52. $\sqrt{\dfrac{x - 1}{x + 2}}$

In Exercises 53–56, find the indicated maximum and minimum values by the method of linear programming. The constraints are shown below the objective function.

53. Maximum P:
   $P = 2x + 9y$
   $x \ge 0, y \ge 0$
   $x + 4y \le 13$
   $3y - x \le 8$

54. Maximum P:
   $P = x + 2y$
   $x \ge 1, y \ge 0$
   $3x + y \le 6$
   $2x + 3y \le 8$

55. Minimum C:
   $C = 3x + 4y$
   $x \ge 0, y \ge 1$
   $2x + 3y \ge 6$
   $4x + 2y \ge 5$

56. Minimum C:
   $C = 2x + 4y$
   $x \ge 0, y \ge 0$
   $x + 3y \ge 6$
   $4x + 7y \ge 18$

In Exercises 57–76, solve the given problems using inequalities. (All data are accurate to at least two significant digits.)

(W) 57. Under what conditions is $|a + b| < |a| + |b|$?

(W) 58. Is $|a - b| < |a| + |b|$ always true? Explain.

59. Find the values for which $f(x) = (x - 2)(x - 3)$ is positive, zero, and negative. Use this information along with $f(0)$ and $f(5)$ to make a rough sketch of the graph of $f(x)$.

60. Follow the same instructions as in Exercise 59 for the function $f(x) = (x - 2)/(x - 3)$.

61. The relationship between Celsius degrees C and Fahrenheit degrees F is $9C = 5F - 160$. For what values of F is $C \ge 37.0°$? (37.0°C is normal body temperature.)

62. The value V (in $) of each building lot in a development is estimated as $V = 65\,000 + 5000t$, where t is the time in years from now. For how long is the value of each lot no more than $90\,000$?

63. The cost C of producing two of one type of calculator and five of a second type is 50. If the cost of producing each of the second type is between 5 and 8, what are the possible costs of producing each of the first type?

64. City A is 600 km from city B. One car starts from A for B 1 h before a second car. The first car averages 60 km/h, and the second car averages 80 km/h for the trip. For what times after the first car starts is the second car ahead of the first car?

65. The pressure p (in kPa) at a depth d (in m) in the ocean is given by $p = 101 + 10.1d$. For what values of d is $p > 500$ kPa?

66. After conducting tests, it was determined that the stopping distance x (in m) of a car traveling 90 km/h was $|x - 95| \le 10$. Express this inequality without absolute values and find the interval of stopping distances that were found in the tests.

67. A heating unit with 80% efficiency and a second unit with 90% efficiency deliver 360 MJ of heat to an office complex. If the first unit consumes an amount of fuel that contains no more than 261 MJ, what is the MJ content of the fuel consumed by the second unit?

68. A rectangular parking lot is to have a perimeter of 100 m and an area no greater than 600 m². What are the possible dimensions of the lot?

69. The electric power p (in W) dissipated in a resistor is given by $p = Ri^2$, where R is the resistance (in Ω) and i is the current (in A). For a given resistor, $R = 12.0\ \Omega$, and the power varies between 2.50 W and 8.00 W. Find the values of the current.

70. The reciprocal of the total resistance of two electric resistances in parallel equals the sum of the reciprocals of the resistances. If a 2.0-Ω resistance is in parallel with a resistance R, with a total resistance greater than 0.5 Ω, find R.

71. The efficiency e (in %) of a certain gasoline engine is given by $e = 100(1 - r^{-0.4})$, where r is the *compression ratio* for the engine. For what values of r is $e > 50\%$?

72. A rocket is fired such that its height h (in km) is given by $h = 41t - t^2$. For what values of t (in min) is the height greater than 400 km?

73. In developing a new product, a company estimates that it will take no more than 1200 min of computer time for research and no more than 1000 min of computer time for development. Graph the possible combinations of the computer times that are needed.

74. A natural-gas supplier has a maximum of 120 worker-hours per week for delivery and for customer service. Graph the possible combinations of times available for these two services.

75. A company produces two types of cameras, the regular model and the deluxe model. For each regular model produced, there is a profit of $8, and for each deluxe model the profit is $15. The same amount of materials is used to make each model, but the supply is sufficient only for 450 cameras per day. The deluxe model requires twice the time to produce as the regular model. If only regular models were made, there would be time enough to produce 600 per day. Assuming all cameras will be sold, how many of each model should be produced if the profit is to be a maximum?

76. A company that manufactures compact disc players gets two different parts, A and B, from two different suppliers. Each package of parts from the first supplier costs $2.00 and contains 6 of each type of part. Each package of parts from the second supplier costs $1.50 and contains 4 of A and 8 of B. How many packages should be bought from each supplier to keep the total cost to a minimum, if production requirements are 600 of A and 900 of B?

Writing Exercises

77. In planning a new city development, an engineer uses a rectangular coordinate system to locate points within the development. A park in the shape of a quadrilateral has corners at $(0, 0)$, $(0, 20)$, $(40, 20)$, and $(20, 40)$ (measurements in meters). Write two or three paragraphs explaining how to describe the park region with inequalities and find these inequalities.

CHAPTER 17 PRACTICE TEST

1. State conditions on x and y in terms of inequalities if the point (x, y) is in the second quadrant.

In Problems 2–7, solve the given inequalities algebraically and graph each solution.

2. $\dfrac{-x}{2} \geq 3$

3. $3x + 1 < -5$

4. $-1 < 1 - 2x < 5$

5. $\dfrac{x^2 + x}{x - 2} \leq 0$

6. $|2x + 1| \geq 3$

7. $|2 - 3x| < 8$

8. Sketch the region in which the points satisfy the following system of inequalities:
$$y < x^2$$
$$y \geq x + 1$$

9. Determine the values of x for which $\sqrt{x^2 - x - 6}$ represents a real number.

10. The length of a rectangular lot is 20 m more than its width. If the area is to be at least 4800 m², what values may the width be?

11. Type A wire costs $0.10 per meter, and type B wire costs $0.20 per meter. Show the possible combinations of lengths of wire that can be purchased for less than $5.00.

12. The range of the visible spectrum in terms of the wavelength λ of light ranges from about $\lambda = 400$ nm (violet) to about $\lambda = 700$ nm (red). Express these values using an inequality with absolute values.

13. Solve the inequality $x^2 > 12 - x$ on a graphing calculator such that the display is the graph of the solution.

14. By using linear programming, find the maximum value of the objective function $P = 5x + 3y$ subject to the following constraints: $x \geq 0$, $y \geq 0$, $2x + 3y \leq 12$, $4x + y \leq 8$.

In Section 18.2, we discuss a space-age application of Newton's universal law of gravitation.

As millions watched on television, men walked on the moon for the first time in 1969. For hundreds of years before that, men on the moon were subjects of dreams and science fiction.

Numerous discoveries in science, engineering, and mathematics helped make it possible for a spacecraft to travel to the moon. These included, for example, discovery and development of rocket flight and numerous developments in electronics in the mid-1900s. However, it can be argued that the beginning came with the formulation of the *universal law of gravitation* by Newton in the 1680s, which was based on earlier discoveries regarding the motion of the planets and the moon. This along with developments in advanced mathematics made it possible to determine what conditions must be met in order to have successful travel in space.

The universal law of gravitation is stated using the terminology of *variation,* the principal topic of this chapter. Using the language of variation, we state how one variable changes as other related variables change. Among the examples we show is that of the universal law of gravitation in Section 18.2.

We begin this chapter by reviewing the meanings of *ratio* and *proportion,* which were first introduced in Chapter 1. Then we see how ratio and proportion lead to variation and setting up numerous relationships. Many of these relationships are based on experimentation and observation, just as in the case of gravitation.

Applications of variation are found in all areas of technology. It is used in acoustics, biology, chemistry, computer technology, economics, electronics, environmental technology, hydrodynamics, mechanics, navigation, optics, physics, space technology, thermodynamics, and other fields.

18.1 RATIO AND PROPORTION

In order to develop the meaning of *variation,* we now review and expand our discussion of *ratio* and *proportion*. First, from Chapter 1 recall that *the quotient a/b is the* **ratio** *of a to b*. Therefore, a fraction is a ratio.

A measurement is the ratio of the measured magnitude to an accepted unit of measurement. For example, measuring the length of an object as 5 cm means it is five times as long as the accepted unit of length, the centimeter. Other examples of ratios are density (weight/volume), relative density (density of object/density of water), and pressure (force/area). Thus, ratios compare quantities of the same kind (for example, the trigonometric ratios) or express the division of magnitudes of different quantities (such a ratio is also called a **rate**).

EXAMPLE 1 The approximate airline distance from Toronto to Los Angeles is 3500 km, and the approximate airline distance from Toronto to Miami is 2000 km. The ratio of these distances is

$$\frac{3500 \text{ km}}{2000 \text{ km}} = \frac{7}{4}$$

The first jet-propelled airplane was flown in Germany in 1928.

Since both units are in kilometers, the resulting ratio is a dimensionless number.

If a jet travels from Toronto to Los Angeles in 5 h, its average speed is

$$\frac{3500 \text{ km}}{5 \text{ h}} = 700 \text{ km/h}$$

In this case we must attach the proper units to the resulting ratio.

As we noted in Example 1, we must be careful to attach the proper units to the resulting ratio. Generally, the ratio of measurements of the same kind should be expressed as a dimensionless number. Consider the following example.

EXAMPLE 2 The length of a certain room is 8 m, and the width of the room is 6 m. Therefore, the ratio of the length to the width is $\frac{8}{6}$, or $\frac{4}{3}$.

If the width of the room is expressed as 6000 mm, we have the ratio 8 m/6000 mm = 1 m/750 mm. However, this does not clearly show the ratio. It is better and more meaningful first to change the units of one of the measurements to the units of the other measurement. Changing the length from 8 m to 8000 mm, we express the ratio as $\frac{4}{3}$, as we saw above. From this ratio we can easily see that the length is $\frac{4}{3}$ as long as the width.

Dimensionless ratios are often used in definitions in mathematics and in technology. For example, the irrational number π is the dimensionless ratio of the circumference of a circle to its diameter. The specific gravity of a substance is the ratio of its density to the density of water. Other illustrations are found in the exercises for this section.

From Chapter 1, also recall that *an equation stating that two ratios are equal is called a* **proportion.** By this definition, a proportion is

$$\frac{a}{b} = \frac{c}{d} \qquad\qquad (18.1)$$

Consider the following example.

◀ EXAMPLE 3   On a certain map, 1 cm represents 10 km. Thus, on this map we have a ratio of 1 cm/10 km. To find the distance represented by 3.5 cm, we can set up the proportion.

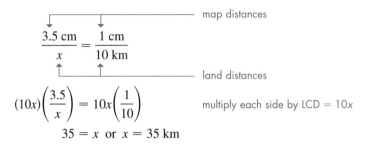

$$\frac{3.5 \text{ cm}}{x} = \frac{1 \text{ cm}}{10 \text{ km}}$$

$$(10x)\left(\frac{3.5}{x}\right) = 10x\left(\frac{1}{10}\right) \qquad \text{multiply each side by LCD} = 10x$$

$$35 = x \text{ or } x = 35 \text{ km}$$

The ratio 1 cm/10 km is the *scale* of the map and has a special meaning, relating map distances in centimeters to land distances in kilometers. In a case like this, we should not change either unit to the other, even though they are both units of length. ▶

Solving a Word Problem

◀ EXAMPLE 4   Given that 1 in. = 2.54 cm (in. is the symbol for the unit of length the *inch*), what is the length in centimeters of the diagonal of a rectangular computer screen that is 10.5 in. long? See Fig. 18.1.

If we equate the ratio of known lengths to the ratio of the given length to the required length, we can find the required length by solving the resulting proportion (which is an equation). This give us

$$\frac{1 \text{ in.}}{2.54 \text{ cm}} = \frac{10.5 \text{ in.}}{x \text{ cm}}$$

$$x = (10.5)(2.54)$$

$$= 26.7 \text{ cm} \qquad \text{rounded off}$$

Fig. 18.1

Therefore, the diagonal of the computer screen is 10.5 in., or 26.7 cm.   ▶

In Appendix B there are additional illustrations of changing units. Also, a listing of all the various units used in the text is included.

◀ EXAMPLE 5   The magnitude of an electric field E is the ratio of the force F on a charge q to the magnitude of q. We can write this as $E = F/q$. If we know the force exerted on a particular charge at some point in the field, we can determine the force that would be exerted on another charge placed at the same point. For example, if we know that a force of 10 nN is exerted on a charge of 4.0 nC, we can then determine the force that would be exerted on a charge of 6.0 nC by the proportion

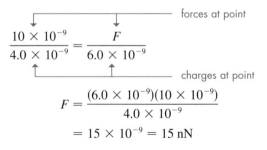

$$\frac{10 \times 10^{-9}}{4.0 \times 10^{-9}} = \frac{F}{6.0 \times 10^{-9}}$$

$$F = \frac{(6.0 \times 10^{-9})(10 \times 10^{-9})}{4.0 \times 10^{-9}}$$

$$= 15 \times 10^{-9} = 15 \text{ nN}$$

▶

Solving a Word Problem ◖ EXAMPLE 6   A certain alloy is 5 parts tin and 3 parts lead. How many grams of each are there in 40 g of the alloy?

First, we let x = the number of grams of tin in the given amount of the alloy. Next, we note that there are 8 total parts of alloy, of which 5 are tin. Thus, 5 is to 8 as x is to 40. This gives the equation

$$\text{parts tin} \longrightarrow \dfrac{5}{8} = \dfrac{x}{40} \longleftarrow \text{grams of tin} \atop \text{total parts} \qquad\qquad \longleftarrow \text{total grams}$$

$$x = 40\left(\dfrac{5}{8}\right) = 25 \text{ g}$$

There are 25 g of tin and 15 g of lead. The ratio 25/15 is the same as 5/3.

EXERCISES 18.1

In Exercises 1 and 2, make the given changes in the indicated examples of this section and then solve the indicated problem.

1. In Example 3, change 10 km to 16 km and then find the required distance.

2. In Example 6, change 5 parts of tin to 7 parts of tin and then find the number of grams of each in the resulting alloy.

In Exercises 3–10, express the ratios in the simplest form.

3. 18 V to 3 V

4. 27 m to 18 m

5. 96 h to 3 days

6. 120 s to 4 min

7. 48 cm to 3 m

8. 6500 cL to 2.6 L

9. 0.14 kg to 3500 mg

10. 2000 μm to 6 mm

In Exercises 11–22, find the required ratios.

11. The *efficiency* of a power amplifier is defined as the ratio of the power output to the power input. Find the efficiency of an amplifier for which the power output is 2.6 W and the power input is 9.6 W.

12. A virus 3.0×10^{-5} cm long appears to be 1.2 cm long through a microscope. What is the *magnification* (ratio of image length to object length) of the microscope?

13. The *coefficient of friction* for two contacting surfaces is the ratio of the frictional force between them to the perpendicular force that presses them together. If it takes 45 N to overcome friction to move a 110-N crate along the floor, what is the coefficient of friction between the crate and the floor? See Fig. 18.2.

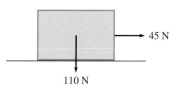

Fig. 18.2

14. The *atomic mass* of an atom of carbon is defined to be 12 u. The ratio of the atomic mass of an atom of oxygen to that of an atom of carbon is $\frac{4}{3}$. What is the atomic mass of an atom of oxygen? (The symbol u represents the *unified atomic mass unit*, where 1 u = 1.66×10^{-27} kg.)

15. An important design feature of an aircraft wing is its *aspect ratio*. It is defined as the ratio of the square of the span of the wing (wingtip to wingtip) to the total area of the wing. If the span of the wing for a certain aircraft is 10.0 m and the area is 18.0 m², find the aspect ratio.

16. For an automobile engine, the ratio of the cylinder volume to compressed volume is the *compression ratio*. If the cylinder volume of 820 cm³ is compressed to 110 cm³, find the compression ratio.

17. The *specific gravity* of a substance is the ratio of its density to the density of water. If the density of steel is 7800 kg/m³ and that of water is 1.0 g/cm³, what is the specific gravity of steel?

18. The *percent grade* of a road is the ratio of vertical rise to the horizontal change in distance (expressed in percent). If a highway rises 75 m for each 1.2 km along the horizontal, what is the percent grade?

19. The *percent error* in a measurement is the ratio of the error in the measurement to the measurement itself, expressed as a percent. When writing a computer program, the memory remaining is determined as 2450 bytes and then it is correctly found to be 2540 bytes. What is the percent error in the first reading?

20. The electric *current* in a given circuit is the ratio of the voltage to the resistance. What is the current (1 V/1 Ω = 1 A) for a circuit where the voltage is 24.0 mV and the resistance is 10.0 Ω?

21. The *mass* of an object is the ratio of its weight to the acceleration g due to gravity. If a space probe weighs 8.46 kN on earth, where $g = 9.80$ m/s², find its mass. (See Appendix B.)

22. *Power* is defined as the ratio of work done to the time required to do the work. If an engine performs 3.65 kJ of work in 15.0 s, find the power developed by the engine. (See Appendix B.)

In Exercises 23–26, find the required quantities from the given proportions.

23. In an electric instrument called a "Wheatstone bridge," electric resistances are related by

$$\frac{R_1}{R_2} = \frac{R_3}{R_4}$$

Find R_2 if $R_1 = 6.00 \ \Omega$,

$R_3 = 62.5 \ \Omega$, and

$R_4 = 15.0 \ \Omega$. See Fig. 18.3.

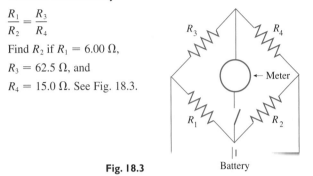

Fig. 18.3

24. For two connected gears, the relation

$$\frac{d_1}{d_2} = \frac{N_1}{N_2}$$

holds, where d is the diameter of the gear and N is the number of teeth. Find N_1 if $d_1 = 2.60$ cm, $d_2 = 11.7$ cm, and $N_2 = 45$. The ratio N_2/N_1 is called the *gear ratio*. See Fig. 18.4.

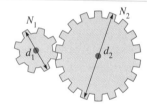

Fig. 18.4

25. According to Boyle's law, the relation

$$\frac{p_1}{p_2} = \frac{V_2}{V_1}$$

holds for pressures p_1 and p_2 and volumes V_1 and V_2 of a gas at constant temperature. Find V_1 if $p_1 = 36.6$ kPa, $p_2 = 84.4$ kPa, and $V_2 = 0.0447$ m^3.

26. In a transformer, an electric current in one coil of wire induces a current in a second coil. For a transformer,

$$\frac{i_1}{i_2} = \frac{t_2}{t_1}$$

where i is the current and t is the number of windings in each coil. In a neon sign amplifier, $i_1 = 1.2$ A and the *turns ratio* $t_2/t_1 = 160$. Find i_2. See Fig. 18.5.

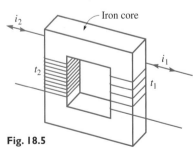

Fig. 18.5

In Exercises 27–44, answer the given questions by setting up and solving the appropriate proportions.

27. Given that 1 kg = 1000 g, what mass in grams is 20.0 kg?

28. Given that 10 000 m^2 = 1 ha, what area in hectares (ha) is 4500 m^2?

29. Given that 1 W · h = 3.6 kJ, what power in kilojoules is 250 W · h?

30. Given that 0.01 d = 864 s, what time in seconds is 2.75 d?

31. Given that $360° = 2\pi$ rad, what angle in degrees is 5.00 rad?

32. Given that 10^4 cm^2 = 10^6 mm^2, what area in square centimeters is 2.50×10^5 mm^2?

33. How many meters per second are equivalent to 45.0 km/h?

34. How many kiloliters per hour are equivalent to 540 L/min?

35. A particular type of automobile engine produces 62 500 cm^3 of carbon monoxide in 2.00 min. How much carbon monoxide is produced in 45.0 s?

36. An airplane consumes 140 L of gasoline in flying 680 km. Under similar conditions, how far can it fly on 240 L?

37. Two separate sections of a roof have the same slope. If the rise and run on one section are, respectively, 3.0 m and 6.3 m, what is the run on the other section if its rise is 4.2 m?

38. When a bullet is fired from a loosely held rifle, the ratio of the mass of the bullet to that of the rifle equals the negative of the reciprocal of the ratio of the velocity of the bullet to that of the rifle. If a 3.0 kg rifle fires a 5.0 g bullet and the velocity of the bullet is 300 m/s, what is the recoil velocity of the rifle?

39. By weight, the ratio of chlorine to sodium in table salt is 35.46 to 23.00. How much sodium is contained in 50.00 kg of salt?

40. An industrial cleaner is diluted 2 parts of cleaner to 5 parts of water. What volume (in mL) of cleaner should be used to get 350 mL of diluted solution?

41. In testing for quality control, it was found that 17 of every 500 computer chips produced by a company in a day were defective. If a total of 595 defective parts were found, what was the total number of chips produced during that day?

42. An electric current of 0.772 mA passes into two wires in which it is divided into currents in the ratio of 2.83 to 1.09. What are the currents in the two wires?

43. One computer line printer can print 2400 lines/min, and a second can print 2800 lines/min. If they print a total of 9100 lines while printing together, how many lines does each print?

44. Of the earth's water area, the Pacific Ocean covers 46.0%, and the Atlantic Ocean covers 23.9%. Together they cover a total of 2.53×10^8 km^2. What is the area of each?

18.2 VARIATION

Named for the French physicist, Jacques Charles (1746–1823).

Scientific laws are often stated in terms of ratios and proportions. For example, Charles' law can be stated as "for a perfect gas under constant pressure, the ratio of any two volumes this gas may occupy equals the ratio of the absolute temperatures." Symbolically, this could be stated as $V_1/V_2 = T_1/T_2$. Thus, if the ratio of the volumes and one of the values of the temperature are known, we can easily find the other temperature.

By multiplying both sides of the proportion of Charles' law by V_2/T_1, we can change the form of the proportion to $V_1/T_1 = V_2/T_2$. This statement says that the ratio of the volume to the temperature (for constant pressure) is constant. Thus, if any pair of values of volume and temperature is known, this ratio of V_1/T_1 can be calculated. This ratio of V_1/T_1 can be called a constant k, which means that Charles' law can be written as $V/T = k$. We now have the statement that the ratio of the volume to temperature is always constant; or, as it is normally stated, "The volume is proportional to the temperature." Therefore, we write $V = kT$, the clearest and most informative statement of Charles' law.

Thus, *for any two quantities always in the same proportion, we say that one is **proportional to** (or **varies directly as**) the second. To show that y is proportional to x (or varies directly as x), we write*

Direct Variation

$$y = kx \tag{18.2}$$

where k is the **constant of proportionality.** This type of relationship is known as **direct variation.**

◀ **EXAMPLE 1** The circumference of a circle is proportional to (varies directly as) the radius r. We write this as $c = 2\pi r$. Since we know that $c = 2\pi r$ for a circle, we know in this case that $k = 2\pi$. ▶

◀ **EXAMPLE 2** The fact that the electric resistance R of a wire varies directly as (is proportional to) its length l is written as $R = kl$. As the length of the wire increases (or decreases), this equation tells us that the resistance increases (or decreases) proportionally. ▶

It is very common that, when two quantities are related, the product of the two quantities remains constant. In such a case, $yx = k$, or

Inverse Variation

$$y = \frac{k}{x} \tag{18.3}$$

This is read as "y **varies inversely as** *x" or "y* **is inversely proportional to** *x."* This type of relationship is known as **inverse variation.**

Named for the English physicist, Robert Boyle (1627–1691).

◀ **EXAMPLE 3** Boyle's law states that "at a given temperature, the pressure p of an ideal gas varies inversely as the volume V." We write this as $p = k/V$. In this case, as the volume of the gas increases, the pressure decreases. ▶

In Fig. 18.6(a), the graph of the equation for direct variation $y = kx$ ($x \geq 0$) is shown. It is a straight line, with slope of k ($k > 0$) and y-intercept of 0. We see that y increases as x increases. In Fig. 18.6(b), the graph of the equation for inverse variation $y = k/x$ ($k > 0, x > 0$) is shown. It is a *hyperbola* (a different form of the equation from that of Example 4 of Section 14.1). As x increases, y decreases.

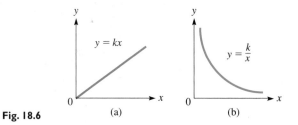

Fig. 18.6

For many relationships, one quantity varies as a specified power of another quantity. The terms *varies directly* and *varies inversely* are used in the following examples with a specified power of the independent variable.

EXAMPLE 4 The statement that the volume V of a sphere varies directly as the cube of its radius r is written as $V = kr^3$. In this case, we know that $k = 4\pi/3$. We see that as the radius increases, the volume increases much more rapidly. For example, if $r = 2.00$ cm, $V = 33.5$ cm³, and if $r = 3.00$ cm, $V = 113$ cm³.

EXAMPLE 5 A company finds that the number n of units of a product that are sold is inversely proportional to the square of the price p of the product. This is written as $n = k/p^2$. As the price of the product is raised, the number of units that are sold decreases much more rapidly.

One quantity may vary as the product of two or more other quantities. Such variation is called **joint variation.** We write

Joint Variation

$$y = kxz \qquad (18.4)$$

to show that y varies jointly as x and z.

EXAMPLE 6 The cost C of a piece of sheet metal varies jointly as the area A of the piece and the cost c per unit area. This we write as $C = kAc$. Here, C increases if the *product Ac* increases.

Direct, inverse, and joint variations may be combined. A given relationship may be a combination of two or all three of these types of variation.

Formulated by the great English mathematician and physicist, Isaac Newton (1642–1727).

EXAMPLE 7 Newton's *universal law of gravitation* can be stated: "The force F of gravitation between two objects varies jointly as the masses m_1 and m_2 of the objects and inversely as the square of the distance r between their centers." We write this as

$$F = \frac{Gm_1m_2}{r^2} \quad \substack{\text{force varies jointly as masses} \\ \text{and} \\ \text{inversely as the square of the distance}}$$

where G is the constant of proportionality.

CAUTION ▶ Note the use of the word *and* in this example. It is used to indicate that F varies in more than one way, but *it is **not** interpreted as addition.*

CALCULATING THE CONSTANT OF
PROPORTIONALITY

Once we have used the given statement to set up a general equation in terms of the variables and the constant of proportionality, we may calculate the value of the constant of proportionality if *one complete set of values* of the variables is known. *This value can then be substituted into the general equation to find the specific equation* relating the variables. We can then find the value of any one of the variables for any set of the others.

◀ EXAMPLE 8   If y varies inversely as x, and $x = 15$ when $y = 4$, find the value of y when $x = 12$.

First, we write

$$y = \frac{k}{x} \qquad \text{general equation from statement}$$

to show that y varies inversely as x. Next, substitute $x = 15$ and $y = 4$ into the equation. This leads to

$$4 = \frac{k}{15}, \quad k = 60 \qquad \text{evaluate } k$$

Thus, for this problem the constant of proportionality is 60, and this may be substituted into $y = k/x$, giving

$$y = \frac{60}{x} \qquad \text{specific equation relating } y \text{ and } x$$

as the equation between y and x. Now, for any given value of x, we may find the value of y. For $x = 12$, we have

$$y = \frac{60}{12} = 5 \qquad \text{evaluating } y \text{ for } x = 12$$

Solving a Word Problem   **◀ EXAMPLE 9**   The frequency f of vibration of a wire varies directly as the square root of the tension T of the wire. If $f = 420 \text{ Hz}$ when $T = 1.14 \text{ N}$, find f when $T = 3.40 \text{ N}$.

The steps in making this evaluation are outlined below:

$$f = k\sqrt{T} \qquad \text{set up general equation: } f \text{ varies directly as } \sqrt{T}$$
$$420 \text{ Hz} = k\sqrt{1.14 \text{ N}} \qquad \text{substitute given set of values and evaluate } k$$
$$k = 393 \text{ Hz/N}^{1/2}$$
$$f = 393 \sqrt{T} \qquad \text{substitute value of } k \text{ to get specific equation}$$
$$f = 393\sqrt{3.40} \qquad \text{evaluate } f \text{ for } T = 3.40 \text{ N}$$
$$= 725 \text{ Hz}$$

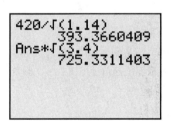

Fig. 18.7

We note that k has a set of units associated with it, and this usually will be the case in applied situations. As long as we do not change the units that are used for any of the variables, the units for the final variable that is evaluated will remain the same. The calculator screen for this calculation is shown in Fig. 18.7.   ▶

Solving a Word Problem ◀ EXAMPLE 1O   The heat H developed in an electric resistor varies jointly as the time t and the square of the current i in the resistor. If H_0 J of heat are developed in t_0 s with i_0 A passing through the resistor, how much heat is developed if both the time and the current are doubled?

$$H = kti^2 \qquad \text{set up general equation}$$

$$H_0 \text{ J} = k(t_0 \text{ s})(i_0 \text{ A})^2 \qquad \text{substitute given values and}$$

$$k = \frac{H_0}{t_0 i_0^2} \text{J}/(\text{s} \cdot \text{A}^2) \qquad \text{evaluate } k$$

$$H = \frac{H_0 t i^2}{t_0 i_0^2} \qquad \text{substitute for } k \text{ to get specific equation}$$

We are asked to determine H when both the time and the current are doubled. This means we are to substitute $t = 2t_0$ and $i = 2i_0$. Making this substitution,

$$H = \frac{H_0(2t_0)(2i_0)^2}{t_0 i_0^2} = \frac{8H_0 t_0 i_0^2}{t_0 i_0^2} = 8H_0$$

The heat developed is eight times that for the original values of i and t.   ▷

Solving a Word Problem ◀ EXAMPLE 11   In Example 7, we stated Newton's universal law of gravitation. This law was formulated in the late seventeenth century, but it has numerous modern space-age applications. Use this law to solve the following problem.

See the chapter introduction.

The first landing on the moon was by the crew of the U.S. spacecraft *Apollo 11* in July 1969.

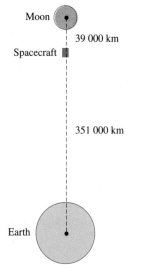

Moon

39 000 km

Spacecraft

351 000 km

Earth

Fig. 18.8

A spacecraft is traveling from the earth to the moon, which are 390 000 km apart. The mass of the moon is 0.0123 that of the earth. How far from the earth is the gravitational force of the earth on the spacecraft equal to the gravitational force of the moon on the spacecraft?

From Example 7, we have the gravitational force between two objects as

$$F = \frac{Gm_1 m_2}{r^2}$$

where the constant of proportionality G is the same for any two objects. Since we want the force between the earth and the spacecraft to equal the force between the moon and the spacecraft, we have

$$\frac{Gm_s m_e}{r^2} = \frac{Gm_s m_m}{(390\,000 - r)^2}$$

where m_s, m_e, and m_m are the masses of the spacecraft, the earth, and the moon, respectively; r is the distance from the earth to the spacecraft; and $390\,000 - r$ is the distance from the moon to the spacecraft. Since $m_m = 0.0123 m_e$, we have

$$\frac{Gm_s m_e}{r^2} = \frac{Gm_s(0.0123 m_e)}{(390\,000 - r\,)^2}$$

$$\frac{1}{r^2} = \frac{0.0123}{(390\,000 - r)^2} \qquad \text{divide each side by } Gm_s m_e$$

$$(390\,000 - r)^2 = 0.0123 r^2 \qquad \text{multiply each side by LCD}$$

$$390\,000 - r = 0.111r \qquad \text{take square roots}$$

$$1.111r = 390\,000$$

$$r = 351\,000 \text{ km}$$

Therefore, the spacecraft is 351 000 km from the earth and 39 000 km from the moon when the gravitational forces are equal. See Fig. 18.8.   ▷

EXERCISES 18.2

In Exercises 1–4, make the given changes in the indicated examples of this section and solve the indicated problems.

1. In Example 1, change radius r to diameter d, write the appropriate equation, and find the value of the constant of proportionality.

2. In Example 8, change "inversely as x" to "inversely as the square of x" and then solve the resulting problem.

3. In Example 9, change 1.14 N to 1.35 N and then solve the resulting problem.

4. In Example 10, change "both the time and current are doubled" to "the time is halved and the current is doubled" and then solve the resulting problem.

In Exercises 5–12, set up the general equations from the given statements.

5. The speed v at which a galaxy is moving away from earth varies directly as its distance r from earth.

6. The demand D for a product varies inversely as its price P.

7. The electric resistance R of a wire varies inversely as the square of its diameter d.

8. The weight w of liquid in a full cubical container is proportional to the cube of an edge e of the container.

9. In a tornado, the pressure P that a roof will withstand is inversely proportional to the square root of the area A of the roof.

10. During an adiabatic (no heat loss or gain) expansion of a gas, the pressure p is inversely proportional to the 3/2 power of the volume V.

11. The stiffness S of a beam varies jointly as its width w and the cube of its depth d.

12. The average electric power P entering a load varies jointly as the resistance R of the load and the square of the effective voltage V, and inversely as the square of the impedance Z.

(W) *In Exercises 13–16, express the meaning of the given equation in a verbal statement, using the language of variation. (k and π are constants.)*

13. $A = \pi r^2$

14. $s = \dfrac{k}{t^{1.2}}$

15. $n = \dfrac{k\sqrt{t}}{u}$

16. $V = \pi r^2 h$

In Exercises 17–20, give the specific equation relating the variables after evaluating the constant of proportionality for the given set of values.

17. V varies directly as the square of H, and $V = 2$ when $H = 64$.

18. n is inversely proportional to the square of p, and $n = \frac{1}{27}$ when $p = 3$.

19. p is proportional to q and inversely proportional to the cube of r, and $p = 6$ when $q = 3$ and $r = 2$.

20. v is proportional to t and the square of s, and $v = 80$ when $s = 2$ and $t = 5$.

In Exercises 21–28, find the required value by setting up the general equation and then evaluating.

21. Find y when $x = 10$ if y varies directly as x, and $y = 20$ when $x = 8$.

22. Find y when $x = 5$ if y varies directly as the square of x, and $y = 6$ when $x = 8$.

23. Find s when $t = 10$ if s is inversely proportional to t, and $s = 100$ when $t = 5$.

24. Find p for $q = 0.8$ if p is inversely proportional to the square of q, and $p = 18$ when $q = 0.2$.

25. Find y for $x = 6$ and $z = 5$ if y varies directly as x and inversely as z, and $y = 60$ when $x = 4$ and $z = 10$.

26. Find r when $n = 16$ if r varies directly as the square root of n and $r = 4$ when $n = 25$.

27. Find f when $p = 2$ and $c = 4$ if f varies jointly as p and the cube of c, and $f = 8$ when $p = 4$ and $c = 0.1$.

28. Find v when $r = 2$, $s = 3$, and $t = 4$ if v varies jointly as r and s and inversely as the square of t, and $v = 8$ when $r = 2$, $s = 6$, and $t = 6$.

In Exercises 29–56, solve the given applied problems involving variation.

29. The volume V of carbon dioxide (CO_2) that is exhausted from a room in a given time varies directly as the initial volume V_0 that is present. If 75 m³ of CO_2 are removed in 1 h from a room with an initial volume of 160 m³, how much is removed in 1 h if the initial volume is 130 m³?

30. The amount of heat H required to melt ice is proportional to the mass m of ice that is melted. If it takes 2.93×10^5 J to melt 875 g of ice, how much heat is required to melt 625 g?

31. In electroplating, the mass m of the material deposited varies directly as the time t during which the electric current is on. Set up the equation for this relationship if 2.50 g are deposited in 5.25 h.

32. Hooke's law states that the force needed to stretch a spring is proportional to the amount the spring is stretched. If 10.0 N stretches a certain spring 4.00 cm, how much will the spring be stretched by a force of 6.00 N?

33. The rate H of heat removal by an air conditioner is proportional to the electric power input P. The constant of proportionality is the *performance coefficient*. Find the performance coefficient of an air conditioner for which $H = 1.8$ kW and $P = 720$ W.

34. The energy E available daily from a solar collector varies directly as the percent p that the sun shines during the day. If a collector provides 1200 kJ for 75% sunshine, how much does it provide for a day during which there is 35% sunshine?

35. The time t required to empty a wastewater-holding tank is inversely proportional to the cross-sectional area A of the drainage pipe. If it takes 2.0 h to empty a tank with a drainage pipe for which $A = 48$ cm^2, how long will it take to empty the tank if $A = 68$ cm^2?

36. The time t required to make a particular trip is inversely proportional to the average speed v. If a jet takes 2.75 h at an average speed of 520 km/h, how long will it take at an average speed of 620 km/h? Explain the meaning of the constant of proportionality.

37. In a physics experiment a given force was applied to three objects. The mass m and the resulting acceleration a were recorded as follows:

m (g)	2.0	3.0	4.0
a (cm/s^2)	30	20	15

(a) Is the relationship $a = f(m)$ one of direct or inverse variation? Explain. (b) Find $a = f(m)$.

38. The lift L of each of three model airplane wings of width w was measured and recorded as follows:

w (cm)	20	40	60
L (N)	10	40	90

(a) Is the relationship $L = f(w)$ one of direct or inverse variation? Explain. (b) Find $L = f(w)$.

39. The power P required to propel a ship varies directly as the cube of the speed s of the ship. If 3.88 MW will propel a ship at 19.3 km/h, what power is required to propel it at 24.2 km/h?

40. The f-number lens setting of a camera varies directly as the square root of the time t that the film is exposed. If the f-number is 8 (written as $f/8$) for $t = 0.0200$ s, find the f-number for $t = 0.0098$ s.

41. The force F on the blade of a wind generator varies jointly as the blade area A and the square of the wind velocity v. Find the equation relating F, A, and v if $F = 76.5$ N when $A = 0.372$ m^2 and $v = 9.42$ m/s.

42. The escape velocity v a spacecraft needs to leave the gravitational field of a planet varies directly as the square root of the product of the planet's radius R and its acceleration due to gravity g. For Mars and earth, $R_M = 0.533R_e$ and $g_M = 0.400g_e$. Find v_M for Mars if $v_e = 11.2$ km/s.

43. The force F between two parallel wires carrying electric currents is inversely proportional to the distance d between the wires. If a force of 0.750 N exists between wires that are 1.25 cm apart, what is the force between them if they are separated by 1.75 cm?

44. The velocity v of a pulse traveling in a string varies directly as the square root of the tension T in the string. If the velocity of a pulse in a string is 150 m/s when the tension is 90.0 N, find the velocity when the tension is 135 N.

45. The average speed s of oxygen molecules in the air is directly proportional to the square root of the absolute temperature T. If the speed of the molecules is 460 m/s at 273 K, what is the speed at 300 K?

46. The time t required to test a computer memory unit varies directly as the square of the number n of memory cells in the unit. If a unit with 4800 memory cells can be tested in 15.0 s, how long does it take to test a unit with 8400 memory cells?

47. The electric resistance R of a wire varies directly as its length l and inversely as its cross-sectional area A. Find the relation between resistance, length, and area for a wire that has a resistance of 0.200 Ω for a length of 60.0 m and cross-sectional area of 0.007 80 cm^2.

48. The general gas law states that the pressure P of an ideal gas varies directly as the thermodynamic temperature T and inversely as the volume V. If $P = 610$ kPa for $V = 10.0$ cm^3 and $T = 290$ K, find V for $P = 400$ kPa and $T = 400$ K.

49. The power P in an electric circuit varies jointly as the resistance R and the square of the current I. If the power is 10.0 W when the current is 0.500 A and the resistance is 40.0 Ω, find the power if the current is 2.00 A and the resistance is 20.0 Ω.

50. The difference $m_1 - m_2$ in magnitudes (visual brightnesses) of two stars varies directly as the base 10 logarithm of the ratio b_2/b_1 of their actual brightnesses. For two particular stars, if $b_2 = 100b_1$ for $m_1 = 7$ and $m_2 = 2$, find the equation relating m_1, m_2, b_1, and b_2.

51. The power gain G by a parabolic microwave dish varies directly as the square of the diameter d of the opening and inversely as the square of the wavelength $λ$ of the wave carrier. Find the equation relating G, d, and $λ$ if $G = 5.5 \times 10^4$ for $d = 2.9$ m and $λ = 3.0$ cm.

52. The intensity I of sound varies directly as the power P of the source and inversely as the square of the distance r from the source. Two sound sources are separated by a distance d, and one has twice the power output of the other. Where should an observer be located on a line between them such that the intensity of each sound is the same?

53. The x-component of the acceleration of an object moving around a circle with constant angular velocity $ω$ varies jointly as cos $ωt$ and the square of $ω$. If the x-component of the acceleration is -11.4 cm/s^2 when $t = 1.00$ s for $ω = 0.524$ rad/s, find the x-component of the acceleration when $t = 2.00$ s.

54. The tangent of the proper banking angle θ of the road for a car making a turn is directly proportional to the square of the car's velocity v and inversely proportional to the radius r of the turn. If $7.75°$ is the proper banking angle for a car traveling at 20.0 m/s around a turn of radius 300 m, what is the proper banking angle for a car traveling at 30.0 m/s around a turn of radius 250 m? See Fig. 18.9.

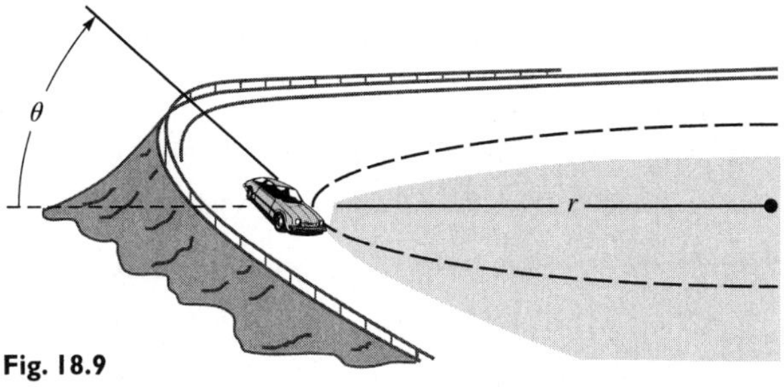

Fig. 18.9

55. The acoustical intensity I of a sound wave is proportional to the square of the pressure amplitude P and inversely proportional to the velocity v of the wave. If $I = 0.474$ W/m^2 for $P = 20.0$ Pa and $v = 346$ m/s, find I if $P = 15.0$ Pa and $v = 320$ m/s.

56. To cook a certain vegetable mix in a microwave oven, the instructions are to cook 120 g for 2.5 min or 240 g for 3.5 min. Assuming the cooking time t is proportional to some power (not necessarily integral) of the weight w, use logarithms to find t as a function of w.

CHAPTER ⑱ EQUATIONS

Proportion	$$\frac{a}{b} = \frac{c}{d}$$	(18.1)
Direct variation	$$y = kx$$	(18.2)
Inverse variation	$$y = \frac{k}{x}$$	(18.3)
Joint variation	$$y = kxz$$	(18.4)

CHAPTER ⑱ REVIEW EXERCISES

In Exercises 1–12, find the indicated ratios.

1. 4 Mg to 20 kg

2. 300 nm to 6 μm

3. 20 mL to 5 cL

4. 12 ks to 2 h

5. The number π equals the ratio of the circumference c of a circle to its diameter d. To check the value of π, a technician used computer simulation to measure the circumference and diameter of a metal cylinder and found the values to be $c = 4.2736$ cm and $d = 1.3603$ cm. What value of π did the technician get?

6. The ratio of the diagonal d of a square to the side s of the square is $\sqrt{2}$. The diagonal and side of the face of a glass cube are found to be $d = 35.375$ mm and $s = 25.014$ mm. What is the value of $\sqrt{2}$ found from these measurements?

7. The mechanical advantage of a lever is the ratio of the output force F_0 to the input force F_i. Find the mechanical advantage if $F_0 = 28$ kN and $F_i = 5000$ N.

8. For an automobile, the ratio of the number n_1 of teeth on the ring gear to the number n_2 of teeth on the pinion gear is the *rear axle ratio* of the car. Find this ratio if $n_1 = 64$ and $n_2 = 20$.

9. The pressure p exerted on a surface is the ratio of the force F on the surface to its area A. Find the pressure on a square patch, 2.25 cm on a side, on a tank if the force on the patch is 37.4 N.

10. The electric resistance R of a resistor is the ratio of the voltage V across the resistor to the current i in the resistor. Find R if $V = 0.632$ V and $i = 2.03$ mA.

11. The *heat of vaporization* of a substance is the amount of heat needed to change a unit amount from liquid to vapor. Experimentation shows 7910 J are needed to change 3.50 g of water to steam. What is the heat of vaporization of water?

12. The electric *current gain G* is the ratio of a constant α to $1 - \alpha$. What is the current gain for $\alpha = 0.95$?

In Exercises 13–28, answer the given questions by setting up and solving the appropriate proportions.

13. On a map of Australia, 37 mm represents 300 km. If the distance on the map between Melbourne and Hobart, Tasmania, is 78 mm, how far is Hobart from Melbourne?

14. Given that 1.000 kg = 1000 g, what is the mass in kilograms of a 14.0-g computer disk?

15. Given that 1.00 kJ = 10^6 mJ, how much heat in kilojoules is produced by a heating element that produces 2660 mJ?

16. Given that 1.00 L = 1000 cm^3, what capacity in liters has a cubical box that is 3.23 cm along an edge?

17. A computer printer can print 3600 characters in 30 s. How many characters can it print in 5.0 min?

18. A solar heater with a collector area of 58.0 m^2 is required to heat 2560 kg of water. Under the same conditions, how much water can be heated by a rectangular solar collector 9.50 m by 8.75 m?

19. The dosage of a certain medicine is 25 mL for each 10 kg of the patient's weight. What is the dosage for a person weighing 56 kg?

20. A woman invests \$50 000 and a man invests \$20 000 in a partnership. If profits are to be shared in the ratio that each invested in the partnership, how much does each receive from \$10 500 in profits?

21. On a certain blueprint, a measurement of 25.0 m is represented by 2.00 mm. What is the actual distance between two points if they are 5.75 mm apart on the blueprint?

22. The chlorine concentration in a water supply is 0.12 part per million. How much chlorine is there in a cylindrical holding tank 4.22 m in radius and 5.82 m high filled from the water supply?

23. One fiber-optic cable carries 60.0% as many messages as another fiber-optic cable. Together they carry 12 000 messages. How many does each carry?

24. Two types of roadbed material, one 50% rock and the other 100% rock, are used in the ratio of 4 to 1 to form a roadbed. If a total of 150 Mg are used, how much rock is in the roadbed?

25. To neutralize 80.0 kg of sodium hydroxide, 98.0 kg of sulfuric acid are needed. How much sodium hydroxide can be neutralized with 37.0 kg of sulfuric acid?

26. A board 15 dm long is cut into two pieces, the lengths of which are in the ratio of 2/3. Find the lengths of the pieces.

27. A total of 322 bolts is in two containers. The ratio of the number of bolts in the first container to the number in the second container is 5/9. How many are in each container?

28. A gasoline company sells octane-87 gas and octane-91 gas in the ratio of 9 to 2. How many of each octane are sold of a total of 16.5 million liters?

In Exercises 29–32, give the specific equation relating the variables after evaluating the constant of proportionality for the given set of values.

29. y varies directly as the square of x, and $y = 27$ when $x = 3$.

30. f varies inversely as l, and $f = 5$ when $l = 8$.

31. v is directly proportional to x and inversely proportional to the cube of y, and $v = 10$ when $x = 5$ and $y = 4$.

32. r varies jointly as u, v, and the square of w, and $r = 8$ when $u = 2$, $v = 4$, and $w = 3$.

In Exercises 33–68, solve the given applied problems.

33. For the lever balanced at the fulcrum, the relation

$$\frac{F_1}{F_2} = \frac{l_0}{L_1}$$

holds, where F_1 and F_2 are forces on opposite sides of the fulcrum at distances L_1 and L_2 (see Fig. 18.10). Find L_2 if $F_1 = 4.50$ N, $F_2 = 6.75$ N, and $L_1 = 17.5$ cm.

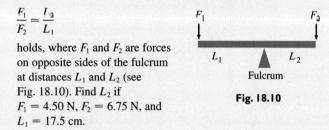
Fig. 18.10

34. A company finds that the volume V of sales of a certain item and the price P of the item are related by

$$\frac{P_1}{P_2} = \frac{V_2}{V_1}$$

Find V_2 if $P_1 = \$8.00$, $P_2 = \$6.00$, and $V_1 = 3000$ per week.

35. The image height h and object height H for the lens shown in Fig. 18.11 are related to the image distance q and object distance p by

$$\frac{h}{H} = \frac{q}{p}$$

Find q if $h = 24$ cm, $H = 84$ cm, and $p = 36$ cm.

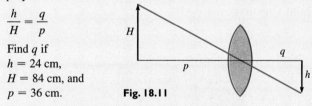
Fig. 18.11

36. For two pulleys connected by a belt, the relation

$$\frac{d_1}{d_2} = \frac{n_2}{n_1}$$

holds, where d is the diameter of the pulley and n is the number of revolutions per unit time it makes. Find n_2 if $d_1 = 4.60$ cm, $d_2 = 8.30$ cm, and $n_1 = 18.0$ r/min.

37. An apartment owner charges rent R proportional to the floor area A of the apartment. Find the equation relating R and A if an apartment of 100 m^2 rents for \$850/month.

38. The number r of aluminum cans that can be made by recycling n used cans is proportional to n. How many cans can be made from 50 000 used cans if $r = 115$ cans for $n = 125$ cans?

39. Under certain conditions, the rate of increase v of bacteria is proportional to the number N of bacteria present. Find v for $N = 7500$ bacteria, if $v = 800$ bacteria/h for $N = 4000$ bacteria.

40. The index of refraction n of a medium varies inversely as the velocity of light v within it. For quartz, $n = 1.46$ and $v = 2.05 \times 10^8$ m/s. What is n for a diamond, in which $v = 1.24 \times 10^8$ m/s?

41. The period T of a pendulum varies directly as the square root of its length L. If $T = \pi/2$ s for $L = 61.0$ cm, find T for $L = 122$ cm.

42. The component of velocity v_x of an object moving in a circle with constant angular velocity ω varies jointly with ω and $\sin \omega t$. If $\omega = \pi/6$ rad/s, and $v_x = -4\pi$ cm/s when $t = 1.00$ s, find v_x when $t = 9.00$ s.

43. The charge C on a capacitor varies directly as the voltage V across it. If the charge is 6.3 μC with a voltage of 220 V across a capacitor, what is the charge on it with a voltage of 150 V across it?

44. The amount of natural gas burned is proportional to the amount of oxygen consumed. If 24.0 kg of oxygen is consumed in burning 15.0 kg of natural gas, how much air, which is 23.2% oxygen by weight, is consumed to burn 50.0 kg of natural gas?

45. The power P of a gas engine is proportional to the area A of the piston. If an engine with a piston area of 50.0 cm^2 can develop 22.5 kW, what power is developed by an engine with a piston area of 40.0 cm^2?

46. The decrease in temperature above a region is directly proportional to the altitude above the region. If the temperature T at the base of the Rock of Gibraltar is 22.0°C and a plane 3.50 km above notes that the temperature is 1.0°C, what is the temperature at the top of Gibraltar, the altitude of which is 430 m? (Assume there are no other temperature effects.)

47. The distance d an object falls under the influence of gravity varies directly as the square of the time t of fall. If an object falls 19.6 m in 2.00 s, how far will it fall is 3.00 s?

48. The kinetic energy E of a moving object varies jointly as the mass m of the object and the square of its velocity v. If a 5.00-kg object, traveling at 10.0 m/s, has a kinetic energy of 250 J, find the kinetic energy of an 8.00-kg object moving at 50.0 m/s.

49. In a particular computer design, N numbers can be sorted in a time proportional to the square of log N. How many times longer does it take to sort 8000 numbers than to sort 2000 numbers?

50. The velocity v of a jet of fluid flowing from an opening in the side of a container is proportional to the square root of the depth d of the opening. If the velocity of the jet from an opening at a depth of 1.22 m is 4.88 m/s, what is the velocity of a jet from an opening at a depth of 7.62 m? See Fig. 18.12.

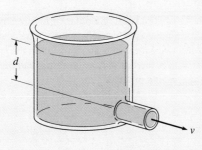

Fig. 18.12

51. In an electric circuit containing an inductance L and a capacitance C, the resonant frequency f is inversely proportional to the square root of the capacitance. If the resonant frequency in a circuit is 25.0 Hz and the capacitance is 95.0 μF, what is the resonant frequency of this circuit if the capacitance is 25.0 μF?

52. The rate of emission R of radiant energy from the surface of a body is proportional to the fourth power of the theromodynamic temperature T. Given that a 25.0-W (the rate of emission) lamp has an operating temperature of 2500 K, what is the operating temperature of a similar 40.0-W lamp?

53. The frequency f of a radio wave is inversely proportional to its wavelength λ. The constant of proportionality is the velocity of the wave, which equals the speed of light. Find this velocity if an FM radio wave has a frequency of 90.9 MHz and a wavelength of 3.29 m.

54. The acceleration of gravity g on a satellite in orbit around the earth varies inversely as the square of its distance r from the center of the earth. If $g = 8.7$ m/s^2 for a satellite at an altitude of 400 km above the surface of the earth, find g if it is 1000 km above the surface. The radius of the earth is 6.4×10^6 m.

55. Using *holography* (a method of producing an image without using a lens), an image of concentric circles is formed. The radius r of each circle varies directly as the square root of the wavelength λ of the light used. If $r = 3.56$ cm for $\lambda = 575$ nm, find r if $\lambda = 483$ nm.

56. A metal circular ring has a circular cross section of radius r. If R is the radius of the ring (measured to the middle of the cross section), the volume V of metal in the ring varies directly as R and the square of r. If $V = 2550$ mm^3 for $r = 2.32$ mm and $R = 24.0$ mm, find V for $r = 3.50$ mm and $R = 32.0$ mm. See Fig. 18.13.

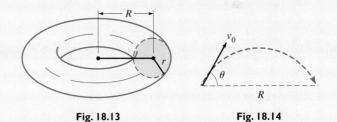

Fig. 18.13   **Fig. 18.14**

57. The range R of a projectile varies jointly as the square of its initial velocity v_0 and the sine of twice the angle θ from the horizontal at which it is fired. See Fig. 18.14. A bullet for which $v_0 = 850$ m/s and $\theta = 22.0°$ has a range of 5.12×10^4 m. Find the range if $v_0 = 750$ m/s and $\theta = 43.2°$.

58. Kepler's third law of planetary motion states that the square of the period of any planet is proportional to the cube of the mean radius (about the sun) of that planet, with the constant of proportionality being the same for all planets. Using the fact that the period of the earth is 1 year and its mean radius is 150×10^6 km, calculate the mean radius for Venus, given that its period is 7.38 months.

59. The stopping distance d of a car varies directly as the square of the velocity v of the car when the brakes are applied. A car moving at 48 km/h can stop in 15 m. What is the stopping distance for the car if it is moving at 85 km/h?

60. The load L that a helical spring can support varies directly as the cube of its wire diameter d and inversely as its coil diameter D. A spring for which $d = 0.120$ cm and $D = 0.953$ cm can support 45.0 N. What is the coil diameter of a similar spring that supports 78.5 N and for which $d = 0.156$ cm?

61. The volume rate of flow R of blood through an artery varies directly as the fourth power of the radius r of the artery and inversely as the distance d along the artery. If an operation is successful in effectively increasing the radius of an artery by 25% and decreasing its length by 2%, by how much is the volume rate of flow increased?

62. The safe, uniformly distributed load L on a horizontal beam, supported at both ends, varies jointly as the width w and the square of the depth d and inversely as the distance D between supports. Given that one beam has double the dimensions of another, how many times heavier is the safe load it can support than the first can support?

63. A bank statement exactly 30 years old is discovered. It states, "This 10-year-old account is now worth $185.03 and pays 4% interest compounded annually." An investment with annual compound interest varies directly as $1 + r$ to the power n, where r is the interest rate expressed as a decimal and n is the number of years of compounding. What was the value of the original investment, and what is it worth now?

64. The distance s that an object falls due to gravity varies jointly as the acceleration g due to gravity and the square of the time t of fall. The acceleration due to gravity on the moon is 0.172 of that on earth. If a rock falls for t_0 seconds on earth, how many times farther would the rock fall on the moon in $3t_0$ seconds?

65. The heat loss L through fiberglass insulation varies directly as the time t and inversely as the thickness d of the fiberglass. If the loss through 20.0 cm of fiberglass is 1.20 MJ in 30 min, what is the loss through 15.0 cm in 1 h 30 min?

66. A quantity important in analyzing the rotation of an object is its *moment of inertia I*. For a ball bearing, the moment of inertia varies directly as its mass m and the square of its radius r. Find the general expression for I if $I = 39.9$ g $\cdot$ cm^2 for $m = 63.8$ g and $r = 1.25$ cm.

67. In the study of polarized light, the intensity I is proportional to the square of the cosine of the angle θ of transmission. If $I = 0.025$ W/m^2 for $\theta = 12.0°$, find I for $\theta = 20.0°$.

68. The force F that acts on a pendulum bob is proportional to the mass m of the bob and the sine of the angle θ the pendulum makes with the vertical. If $F = 0.120$ N for $m = 0.350$ kg and $\theta = 2.00°$, find F for $m = 0.750$ kg and $\theta = 3.50°$.

Writing Exercises

69. A fruit-packing company plans to reduce the size of its fruit juice can (a right circular cylinder) by 10% and keep the price of each can the same (effectively raising the price). The radius and the height of the new can are to be equally proportional to those of the old can. Write one or two paragraphs explaining how to determine the percent decrease in the radius and the height of the old can that is required to make the new can.

CHAPTER 18 PRACTICE TEST

1. Express the ratio of 180 s to 4 min in simplest form.

2. A person 1.8 m tall is photographed, and the film image is 20.0 mm high. Under the same conditions, how tall is a person whose film image is 14.5 mm high?

3. The change L in length of a copper rod varies directly as the change T in temperature of the rod. Set up an equation for this relationship if $L = 2.7$ cm for $T = 150°C$.

4. Given that 1 cm $= 10^4$ μm, what length in centimeters is 7.24 μm?

5. The perimeter of a rectangular solar panel is 210.0 cm. The ratio of the length to the width is 7 to 3. What are the dimensions of the panel?

6. The difference p in pressure in a fluid between that at the surface and that at a point below varies jointly as the density d of the fluid and the depth h of the point. The density of water is 1000 kg/m^3, and the density of alcohol is 800 kg/m^3. This difference in pressure at a point 0.200 m below the surface of water is 1.96 kPa. What is the difference in pressure at a point 0.300 m below the surface of alcohol? (All data are accurate to three significant digits.)

7. The crushing load L of a pillar varies directly as the fourth power of its radius r and inversely as the square of its length l. If one pillar has twice the radius and three times the length of a second pillar, what is the ratio of the crushing load of the first pillar to that of the second pillar?

CHAPTER 19 Sequences and the Binomial Theorem

If a person is saving for the future and invests $1000 at 5%, compounded annually, the value of this investment 40 years later would be about $7040. Even better, if the interest is compounded daily, which is a common method today, it would be worth about $7390 in 40 years. A person saving for retirement would do much better by putting aside $1000 each year and letting the interest accumulate. If $1000 is invested each year at 5%, compounded annually, the total investment would be worth about $126,840 after 40 years. If the interest is compounded daily, it would be worth about $131,000 in 40 years. Obviously, if more is invested, or if the interest rate is higher, the values of these investments would be higher.

Each of these values can be found quickly since the values of the annual investments form what is called a *geometric sequence,* and formulas can be formed for such sums. Such formulas involving compound interest are widely used in calculating values such as monthly car payments, home mortgages, and annuities.

A *sequence* is a set of numbers arranged in some specific way and usually follows a pattern. Sequences have been of interest to people for centuries. There are records that date back to at least 1700 B.C.E. showing calculations involving sequences. Euclid (see page 50) in his *Elements* dealt with sequences in about 300 B.C.E. More advanced forms of sequences were used extensively in the study of advanced mathematics in the 1700s and 1800s. These advances in mathematics have been very important in many areas of science and technology.

Of the many types of sequences, we study certain basic ones in this chapter. Included are those used in the expansion of a binomial to a power. We show applications in areas such as physics and chemistry in studying radioactivity, biology in studying population growth, and, of course, in business when calculating interest.

Sequences are basic to many calculations in business, including compound interest. In Section 19.2, we show such a calculation.

19.1 ARITHMETIC SEQUENCES

A *sequence* of numbers may consist of numbers chosen in any way we may wish to select. This could include a random selection of numbers, although the sequences that are useful follow a pattern. We consider only those sequences that include real numbers or literal numbers that represent real numbers.

An **arithmetic sequence** (*or* **arithmetic progression**) *is a set of numbers in which each number after the first can be obtained from the preceding one by adding to it a fixed number called the* **common difference.** This definition can be expressed in terms of the *recursion formula*

$$a_n = a_{n-1} + d \tag{19.1}$$

where a_n is any term, a_{n-1} is the preceding term, and d is the common difference.

◖ EXAMPLE 1 **(a)** The sequence $2, 5, 8, 11, 14, \ldots$, is an arithmetic sequence with a common difference $d = 3$. We can obtain any term by adding 3 to the previous term. We see that the fifth term is $a_5 = a_4 + d$, or $14 = 11 + 3$.

(b) The sequence $7, 2, -3, -8, \ldots$, is an arithmetic sequence with $d = -5$. We can get any term after the first by adding -5 to the previous term.

The three dots after the 14 in part (a) and after -8 in part (b) mean that the sequences continue. ◗

If we know the first term of an arithmetic sequence, we can find any other term by adding the common difference enough times to get the desired term. This, however, is very inefficient, and there is a general way of finding a particular term.

If a_1 is the first term and d is the common difference, the second term is $a_1 + d$, the third term is $a_1 + 2d$, and so on. For the nth term, we need to add d to the first term $n - 1$ times. Therefore, *the nth term, a_n, of the arithmetic sequence is given by*

nth term

$$a_n = a_1 + (n - 1)d \tag{19.2}$$

Equation (19.2) can be used to find any given term in any arithmetic sequence. We can refer to a_n as the *last term* of an arithmetic sequence if no terms beyond it are included in the sequence. Such a sequence is called a *finite* sequence. If the terms in a sequence continue without end, the sequence is called an *infinite* sequence.

◖ EXAMPLE 2 Find the tenth term of the arithmetic sequence $2, 5, 8, \ldots$.

By subtracting any term from the following term, we find the common difference $d = 3$, and see that the first term $a_1 = 2$. Therefore, the tenth term, a_{10}, is

$$
\begin{array}{ccc}
a_1 & n & d \\
\downarrow & \downarrow & \downarrow
\end{array}
$$
$$a_{10} = 2 + (10 - 1)3 = 2 + (9)(3)$$
$$= 29$$

The three dots after the 8 show that the sequence continues. With no additional information given, this indicates that it is an infinite arithmetic sequence. ◗

❘ EXAMPLE 3   Find the common difference between successive terms of the arithmetic sequence for which the first term is 5 and the 32nd term is −119.

We are to find d given that $a_1 = 5$, $a_{32} = -119$, and $n = 32$. Substitution in Eq. (19.2) gives

$$-119 = 5 + (32 - 1)d$$
$$31d = -124$$
$$d = -4$$

There is no information as to whether this is a finite or an infinite sequence. The solution is the same in either case.   ▶

Solving a Word Problem   ❘ EXAMPLE 4   How many numbers between 10 and 1000 are divisible by 6?

We must first find the smallest and the largest numbers in this range that are divisible by 6. These numbers are 12 and 996. Obviously, the common difference between one multiple of 6 and the next is 6. Thus, we can solve this as an arithmetic sequence with $a_1 = 12$, $a_n = 996$, and $d = 6$. Substituting these values in Eq. (19.2), we have

$$996 = 12 + (n - 1)6$$
$$6n = 990$$
$$n = 165$$

Thus, 165 numbers between 10 and 1000 are divisible by 6.

All the positive multiples of 6 are included in the infinite arithmetic sequence 6, 12, 18, . . . , whereas those between 10 and 1000 are included in the finite arithmetic sequence 12, 18, 24, . . . , 996.   ▶

Solving a Word Problem   ❘ EXAMPLE 5   A package delivery company uses a metal (low-friction) ramp to slide packages from the sorting area to the loading area. If a package is pushed to start it down the ramp at 25 cm/s and the package accelerates as it slides such that it gains 35 cm/s during each second, after how many seconds is the velocity 305 cm/s? See Fig. 19.1.

Here we see that the velocity (in cm/s) of the package after each second is

$$60, 95, 130, \ldots, 305, \ldots$$

Therefore, $a_1 = 60$ (the 25 cm/s was at the beginning, that is, after 0 s), $d = 35$, $a_n = 305$, and we are to find n.

$$305 = 60 + (n - 1)(35)$$
$$245 = 35n - 35$$
$$35n = 280$$
$$n = 8.0$$

This means that the velocity of a package sliding down the ramp is 305 cm/s after 8.0 s.   ▶

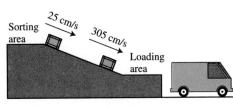

Fig. 19.1

SUM OF n TERMS

Another important quantity related to an arithmetic sequence is the sum of the first n terms. We can indicate this sum by starting the sum with either the first term or with the last term, as shown by these two equations:

$$S_n = a_1 + (a_1 + d) + (a_1 + 2d) + \cdots + (a_n - d) + a_n$$

or

$$S_n = a_n + (a_n - d) + (a_n - 2d) + \cdots + (a_1 + d) + a_1$$

If we now add the corresponding members of these two equations, we obtain the result

$$2S_n = (a_1 + a_n) + (a_1 + a_n) + (a_1 + a_n) + \cdots + (a_1 + a_n) + (a_1 + a_n)$$

Each term on the right in parentheses has the same expression $(a_1 + a_n)$, and there are n such terms. This tells us that *the sum of the first n terms is given by*

$$S_n = \frac{n}{2}(a_1 + a_n) \tag{19.3}$$

The use of Eq. (19.3) is illustrated in the following examples.

❨ EXAMPLE 6   Find the sum of the first 1000 positive integers.
    The first 1000 integers form a finite arithmetic sequence for which $a_1 = 1$, $a_{1000} = 1000$, $n = 1000$, and $d = 1$. Substituting into Eq. (19.3) (in which we do not use the value of d), we have

$$S_{1000} = \frac{1000}{2}(1 + 1000) = 500(1001)$$

$$= 500\,500$$

❨ EXAMPLE 7   Find the sum of the first ten terms of the arithmetic sequence in which the first term is 4 and the common difference is -5.
    We are to find S_n, given that $n = 10$, $a_1 = 4$, and $d = -5$. Since Eq. (19.3) uses the value of a_n but not the value of d, we first find a_{10} by using Eq. (19.2). This gives us

$$a_{10} = 4 + (10 - 1)(-5) = 4 - 45$$

$$= -41$$

Now we can solve for S_{10} by using Eq. (19.3):

$$S_{10} = \frac{10}{2}(4 - 41) = 5(-37)$$

$$= -185$$

If any three of the five values a_1, a_n, n, d, and S_n are given for a particular arithmetic sequence, the other two may be found from Eqs. (19.2) and (19.3). Consider the following example.

◀ EXAMPLE 8   For an arithmetic sequence, given that $a_1 = 2$, $d = \frac{3}{2}$, and $S_n = 72$, find n and a_n.

First, we substitute the given values in Eqs. (19.2) and (19.3) in order to identify what is known and how we may proceed. Substituting $a_1 = 2$ and $d = \frac{3}{2}$ in Eq. (19.2), we obtain

$$a_n = 2 + (n - 1)\left(\frac{3}{2}\right)$$

Substituting $S_n = 72$ and $a_1 = 2$ in Eq. (19.3), we obtain

$$72 = \frac{n}{2}(2 + a_n)$$

We note that n and a_n appear in both equations, which means that we must solve them simultaneously. Substituting the expression for a_n from the first equation into the second equation, we proceed with the solution:

$$72 = \frac{n}{2}\left[2 + 2 + (n - 1)\left(\frac{3}{2}\right)\right]$$
$$72 = 2n + \frac{3n(n - 1)}{4}$$
$$288 = 8n + 3n^2 - 3n$$
$$3n^2 + 5n - 288 = 0$$
$$n = \frac{-5 \pm \sqrt{25 - 4(3)(-288)}}{6} = \frac{-5 \pm \sqrt{3481}}{6} = \frac{-5 \pm 59}{6}$$

Since n must be a positive integer, we find that $n = \dfrac{-5 + 59}{6} = 9$. Using this value in the expression for a_n, we find

$$a_9 = 2 + (9 - 1)\left(\frac{3}{2}\right) = 14$$

Therefore, $n = 9$ and $a_9 = 14$.   ▶

Solving a Word Problem   ◀ EXAMPLE 9   The voltage across a resistor increases such that during each second the increase is 0.002 mV less than during the previous second. Given that the increase during the first second is 0.350 mV, what is the total voltage increase during the first 10.0 s?

We are asked to find the sum of the voltage increases 0.350 mV, 0.348 mV, 0.346 mV, ..., so as to include ten increases. This means we want the sum of an arithmetic sequence for which $a_1 = 0.350$, $d = -0.002$, and $n = 10$. Since we need a_n to use Eq. (19.3), we first calculate it using Eq. (19.2):

$$a_{10} = 0.350 + (10 - 1)(-0.002) = 0.332 \text{ mV}$$

Now we use Eq. (19.3) to find the sum with $a_1 = 0.350$, $a_{10} = 0.332$, and $n = 10$:

$$S_{10} = \frac{10}{2}(0.350 + 0.332) = 3.410 \text{ mV}$$

Thus, the total voltage increase is 3.410 mV.   ▶

EXERCISES 19.1

In Exercises 1–4, make the given changes in the indicated examples of this section and then solve the resulting problems.

1. In Example 3, change -119 to -88 and then find the common difference.

2. In Example 4, change 6 to 5 and then find out how many numbers there are.

3. In Example 6, change 1000 to 500 and then find the sum.

4. In Example 7, change -5 to 5 and then find the sum.

In Exercises 5–8, write the first five terms of the arithmetic sequence with the given values.

5. $a_1 = 4, d = 2$

6. $a_1 = 6, d = -\frac{1}{2}$

7. $a_3 = \frac{5}{2}, a_5 = -\frac{3}{2}$

8. $a_2 = -2, a_5 = 43$

In Exercises 9–16, find the nth term of the arithmetic sequence with the given values.

9. $1, 4, 7, \ldots; n = 8$

10. $-6, -4, -2, \ldots; n = 10$

11. $\frac{7\pi}{4}, \frac{3\pi}{2}, \frac{5\pi}{4}, \ldots; n = 17$

12. $2, \frac{1}{2}, -1, \ldots; n = 25$

13. $a_1 = -0.7, d = 0.4, n = 80$

14. $a_1 = \frac{3}{2}, d = \frac{1}{6}, n = 601$

15. $a_1 = b, d = 2b, n = 25$

16. $a_1 = -c, d = 3c, n = 30$

In Exercises 17–20, find the sum of the n terms of the indicated arithmetic sequence.

17. $n = 20, a_1 = 4, a_{20} = 40$

18. $n = 8, a_1 = -12, d = -2$

19. $-2, -\frac{5}{2}, -3, \ldots; n = 10$

20. $3k, \frac{10}{3}k, \frac{11}{3}k, \ldots; n = 40$

In Exercises 21–32, find any of the values of a_1, d, a_n, n, or S_n that are missing for an arithmetic sequence.

21. $a_1 = 5, d = 8, a_n = 45$

22. $a_1 = -2, n = 60, a_n = 28$

23. $a_1 = \frac{5}{3}, n = 20, S_{20} = \frac{40}{3}$

24. $a_1 = 0.1, a_n = -5.9, S_n = -8.7$

25. $d = 3, n = 30, S_{30} = 1875$

26. $d = 9, a_n = 86, S_n = 455$

27. $a_1 = 7.4, d = -0.5, a_n = 23.1$

28. $a_1 = -\frac{9}{7}, n = 19, a_{19} = -\frac{36}{7}$

29. $a_1 = -5k, d = \frac{1}{2}k, S_n = \frac{23}{2}k$

30. $d = -2c, n = 50, S_{50} = 0$

31. $a_1 = -c, a_n = \frac{b}{2}, S_n = 2b - 4c$

32. $a_1 = 3b, n = 7, d = \frac{b}{3}$

In Exercises 33–56, find the indicated quantities for the appropriate arithmetic sequence.

33. $a_6 = 56, a_{10} = 72$ (find a_1, d, S_n for $n = 10$)

34. $a_{17} = -91, a_2 = -73$ (find a_1, d, S_n for $n = 40$)

Ⓦ 35. Is ln 3, ln 6, ln 12, ... an arithmetic sequence? Explain. If it is, what is the fifth term?

Ⓦ 36. Is sin 2°, sin 4°, sin 6°, ... an arithmetic sequence? Explain. If it is, what is the fifth term?

37. Show that $a, \frac{a+b}{2}, b$ is an arithmetic sequence.

38. If a, b, and c are the first three terms of an arithmetic sequence, find their sum in terms of b only.

39. Find the sum of the first 100 positive integers. (See the margin note on page 506.)

40. Find the number of multiples of 8 between 99 and 999.

41. Find x if $3 - x, -x$, and $\sqrt{9 - 2x}$ are the first three terms of an arithmetic sequence.

42. Is $x, x + 2y, 2x + 3y, \ldots$ an arithmetic sequence? If it is, find the sum of the first 100 terms.

43. A beach now has an area of 9500 m² but is eroding such that it loses 100 m² more of its area each year than during the previous year. If it lost 400 m² during the last year, what will be its area 8 years from now?

44. During a period of heavy rains, on a given day 4500 m³/s of water was being released from a dam. In order to minimize downstream flooding, engineers then reduced the releases by 500 m³/s each day thereafter. How much water was released during the first week of these releases?

45. At a logging camp, 15 layers of logs are so piled that there are 20 logs in the bottom layer, and each layer has 1 fewer log than the layer below it. How many logs are in the pile?

46. In order to prevent an electric current surge in a circuit, the resistance R in the circuit is stepped down by 4.0 Ω after each 0.1 s. If the voltage V is constant at 120 V, do the resulting currents I (in A) form an arithmetic sequence if $V = IR$?

47. There are 12 seats in the first row around a semicircular stage. Each row behind the first has 4 more seats than the row in front of it. How many rows of seats are there if there is a total of 300 seats?

48. A bank loan of $8000 is repaid in annual payments of $1000 plus 10% interest on the unpaid balance. What is the total amount of interest paid?

49. A car depreciates $1800 during the first year after it is bought. Each year thereafter it depreciates $150 less than the year before. How many years after it was bought will it be considered to have no value, and what was the original cost?

50. The sequence of ships' bells is as follows: 12:30 A.M. one bell is rung, and each half hour later one more bell is rung than the previous time until eight bells are rung. The sequence is then repeated starting at 4:30 A.M., again until eight bells are rung. This pattern is followed throughout the day. How many bells are rung in one day?

51. If a tool dropped from a helicopter falls 4.9 m during the first second, 14.7 m during the second second, 24.5 m during the third second, and so on, how high was the helicopter if the tool takes 10.0 s to reach the ground?

52. In preparing a bid for constructing a new building, a contractor determines that the foundation and basement will cost $605 000 and the first floor will cost $360 000. Each floor above the first will cost $15 000 more than the one below it. How much will the building cost if it is to be 18 floors high?

53. Derive a formula for S_n in terms of n, a_1, and d.

W54. A *harmonic sequence* is a sequence of numbers whose reciprocals form an arithmetic sequence. Is a harmonic sequence also an arithmetic sequence? Explain.

55. Show that the sum of the first n positive integers is $\frac{1}{2}n(n + 1)$.

56. Show that the sum of the first n positive odd integers is n^2.

19.2 GEOMETRIC SEQUENCES

A second type of important sequence of numbers is the **geometric sequence** (or **geometric progression**). *In a geometric sequence, each number after the first can be obtained from the preceding one by multiplying it by a fixed number, called the* **common ratio.** We can express this definition in terms of the *recursion formula*

$$a_n = ra_{n-1} \qquad (19.4)$$

where a_n is any term, a_{n-1} is the preceding term, and r is the common ratio. One important application of geometric sequences is in computing compound interest on savings accounts. Other applications are found in areas such as biology and physics.

◀ **EXAMPLE 1** **(a)** The sequence 2, 4, 8, 16,..., is a geometric sequence with a common ratio of 2. Any term after the first can be obtained by multiplying the previous term by 2. We see that the fourth term $a_4 = ra_3$, or $16 = 2(8)$.

(b) The sequence 9, -3, 1, $-1/3$,..., is a geometric sequence with a common ratio of $-1/3$. We can obtain any term after the first by multiplying the previous term by $-1/3$. ▶

If we know the first term, we can find any other term by multiplying by the common ratio a sufficient number of times. When we do this for a general geometric sequence, we can find the nth term in terms of the first term a_1, the common ratio r, and n. Thus, the second term is a_1r, the third term is a_1r^2, and so forth. In general, the expression for the nth term is

nth term

$$a_n = a_1 r^{n-1} \qquad (19.5)$$

◀ **EXAMPLE 2** Find the eighth term of the geometric sequence 8, 4, 2,.... By dividing any term by the previous term, we find the common ratio to be $\frac{1}{2}$. From the terms given, we see that $a_1 = 8$. From the statement of the problem, we know that $n = 8$. Thus, we substitute into Eq. (19.5) to find a_8:

$$a_8 = 8\left(\frac{1}{2}\right)^{8-1} = \frac{8}{2^7} = \frac{1}{16}$$

◀ **EXAMPLE 3** Find the tenth term of the geometric sequence for which $a_1 = \frac{8}{625}$ and $r = -\frac{5}{2}$.

Using Eq. (19.5) to find a_{10}, we have

$$a_{10} = \frac{8}{625}\left(-\frac{5}{2}\right)^{10-1} = \frac{8}{625}\left(-\frac{5^9}{2^9}\right) = -\left(\frac{2^3}{5^4}\right)\left(\frac{5^9}{2^9}\right) = -\frac{5^5}{2^6}$$

$$= -\frac{3125}{64}$$

◀ **EXAMPLE 4** Find the seventh term of a geometric sequence for which the second term is 3, the fourth term is 9, and $r > 0$.

We can find r if we let $a_1 = 3$, $a_3 = 9$, and $n = 3$. (At this point, we are considering a sequence made up of 3, the next number, and 9. These are the second, third, and fourth terms of the original sequence.) Thus,

$$9 = 3r^2, \qquad r = \sqrt{3} \qquad \text{(since } r > 0\text{)}$$

We can now find a_1 of the original sequence by considering just the first two terms of the sequence, a_1 and $a_2 = 3$:

$$3 = a_1\left(\sqrt{3}\right)^{2-1}, \qquad a_1 = \sqrt{3}$$

We can now find the seventh term, using $a_1 = \sqrt{3}$, $r = \sqrt{3}$, and $n = 7$:

$$a_7 = \sqrt{3}\left(\sqrt{3}\right)^{7-1} = \sqrt{3}(3^3) = 27\sqrt{3}$$

We could have shortened this procedure one step by letting the second term be the first term of a new sequence of six terms. If the first term is of no importance in itself, this is acceptable.

Solving a Word Problem

◀ **EXAMPLE 5** In an experiment, 22.0% of a substance changes chemically each 10.0 min. If there is originally 120 g of the substance, how much of it will remain after 45.0 min?

Let $P =$ the portion of the substance remaining after each minute. From the statement of the problem, we know that $r = 0.780$, since 78.0% remains after each 10.0-min period. We also know that $a_1 = 120$ g, and we let n represent the number of minutes of elapsed time. This means that $P = 120(0.780)^{n/10.0}$. It is necessary to divide by 10.0 because the ratio is given for a 10.0-min period. In order to find P when $n = 45.0$ min, we write

$$P = 120(0.780)^{45.0/10.0} = 120(0.780)^{4.50}$$

$$= 39.2 \text{ g}$$

This means that 39.2 g remain after 45.0 min. Note that the power 4.50 represents 4.50 ten-minute periods.

SUM OF n TERMS

A general expression for the sum S_n of the first n terms of a geometric sequence may be found by directly forming the sum and multiplying this equation by r:

$$S_n = a_1 + a_1r + a_1r^2 + \cdots + a_1r^{n-1}$$
$$rS_n = a_1r + a_1r^2 + a_1r^3 + \cdots + a_1r^n$$

If we now subtract the second of these equations from the first, we get $S_n - rS_n = a_1 - a_1 r^n$. All other terms cancel by subtraction. Now, factoring S_n from the terms on the left and a_1 from the terms on the right, we solve for S_n. Thus, *the sum* S_n *of the first n terms of a geometric sequence is*

$$S_n = \frac{a_1(1 - r^n)}{1 - r} \qquad (r \neq 1) \tag{19.6}$$

◀ **EXAMPLE 6**   Find the sum of the first seven terms of the geometric sequence in which the first term is 2 and the common ratio is $\frac{1}{2}$.

We are to find S_n given that $a_1 = 2$, $r = \frac{1}{2}$, and $n = 7$. Using Eq. (19.6), we have

$$S_7 = \frac{2\left(1 - \left(\frac{1}{2}\right)^7\right)}{1 - \frac{1}{2}} = \frac{2\left(1 - \frac{1}{128}\right)}{\frac{1}{2}} = 4\left(\frac{127}{128}\right) = \frac{127}{32}$$

Solving a Word Problem

◀ **EXAMPLE 7**   If \$100 is invested each year at 5% interest compounded annually, what would be the total amount of the investment after 10 years (before the 11th deposit is made)?

After 1 year, the amount invested will have added to it the interest for the year. Therefore, for the last (10th) \$100 invested, its value will become

$$\$100(1 + 0.05) = \$100(1.05) = \$105$$

The next to last \$100 will have interest added twice. After 1 year, its value becomes \$100(1.05), and after 2 years it is $\$100(1.05)(1.05) = \$100(1.05)^2$. In the same way, the value of the first \$100 becomes $\$100(1.05)^{10}$, since it will have interest added 10 times. This means that we are to find the sum of the sequence

$$100(1.05) + 100(1.05)^2 + 100(1.05)^3 + \cdots + 100(1.05)^{10}$$

| 1 year | 2 years | 3 years | 10 years |
| in account | in account | in account | in account |

or

$$100[1.05 + (1.05)^2 + (1.05)^3 + \cdots + (1.05)^{10}]$$

For the sequence in the brackets, we have $a_1 = 1.05$, $r = 1.05$, and $n = 10$. Thus,

$$S_{10} = \frac{1.05[1 - (1.05)^{10}]}{1 - 1.05} = 13.2068$$

The total value of the \$100 investments is $100(13.2068) = \$1320.68$. We see that \$320.68 in interest has been earned. See Fig. 19.2. ▶

See the chapter introduction.

It is reported that Albert Einstein was once asked what was the greatest discovery ever made, and his reply was "compound interest."

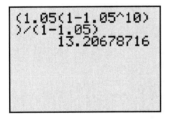

Fig. 19.2

In Exercises 1 and 2, make the given changes in the indicated examples of this section and then solve the given problems.

1. In Example 3, change "tenth" to "seventh" and then solve the given problem.

2. In Example 6, change $\dfrac{1}{2}$ to $\dfrac{1}{3}$ and then find the sum.

In Exercises 3–6, write down the first five terms of the geometric sequence with the given values.

3. $a_1 = 45$, $r = \frac{1}{3}$

4. $a_1 = 0.09$, $r = -\frac{2}{3}$

5. $a_1 = \frac{1}{6}$, $r = 3$

6. $a_1 = -3$, $r = 2$

In Exercises 7–14, find the nth term of the geometric sequence with the given values.

7. $\frac{1}{2}, 1, 2, \ldots; n = 6$

8. $10, 1, 0.1, \ldots; n = 8$

9. $125, -25, 5, \ldots; n = 7$

10. $0.1, 0.3, 0.9, \ldots; n = 5$

11. $a = -2700, r = -\frac{1}{3}, n = 10$

12. $a_1 = 48, r = \frac{1}{2}, n = 12$

13. $10^{100}, -10^{98}, 10^{96}, \ldots; n = 51$

14. $-2, 4k, -8k^2, \ldots; n = 6$

In Exercises 15–20, find the sum of the first n terms of the indicated geometric sequence with the given values.

15. $a_1 = \frac{1}{8}, r = 4, n = 5$

16. $162, -54, 18, \ldots; n = 6$

17. $192, 96, 48, \ldots; n = 6$

18. $a_1 = 9, a_n = 243, n = 4$

19. $a_1 = 96, r = -\frac{k}{2}, n = 10$

20. $\log 2, \log 4, \log 16, \ldots; n = 6$

In Exercises 21–28, find any of the values of a_1, r, a_n, n, or S_n that are missing.

21. $a_1 = \frac{1}{16}, r = 4, n = 6$

22. $r = 0.2, a_n = 0.000\,32, n = 7$

23. $r = \frac{3}{2}, n = 5, S_5 = 211$

24. $r = -\frac{1}{2}, a_n = \frac{1}{8}, n = 7$

25. $a_n = 27, n = 4, S_4 = 40$

26. $a_1 = 3, n = 7, a_7 = 192$

27. $a_1 = 75, r = \frac{1}{5}, a_n = \frac{3}{625}$

28. $r = -2, n = 6, S_6 = 42$

In Exercises 29–52, find the indicated quantities.

(W) 29. Is $3, 3^{x+1}, 3^{2x+1}, \ldots$ a geometric sequence? Explain. If it is, find a_{20}.

30. Find the sum of the first eight terms of the geometric sequence for which the fifth term is 5, the seventh term is 10, and $r > 0$.

31. Show that $a, \sqrt{ab}, b$ are three successive terms of a geometric sequence $(a > 0, b > 0)$.

32. Find x, if $\sqrt{x}, 4, \sqrt{15x + 4}$ are the first three terms of a geometric sequence.

33. The sum of the first three terms of a geometric sequence equals seven times the first term. Find the common ratio.

34. The sum of the first three terms of an arithmetic sequence is 3. What are the numbers if their squares form a geometric sequence?

35. Each stroke of a pump removes 8.2% of the remaining air from a container. What percent of the air remains after 50 strokes?

36. In 1995 the population of Ireland was 3 550 000 and it was estimated that the population would increase at an average annual rate of 0.54% until 2010. What was the estimated 2010 population?

37. An electric current decreases by 12.5% each 1.00 μs. If the initial current is 3.27 mA, what is the current after 8.20 μs?

38. A copying machine is set to reduce the dimensions of material copied by 10%. A drawing 12.0 cm wide is reduced, and then the copies are in turn reduced. What is the width of the drawing on the sixth reduction?

39. How much is an investment of $250 worth after 8 years if it earns annual interest of 7.2% compounded monthly? (7.2% annual interest compounded monthly means that 0.6% (7.2%/12) interest is added each month.)

40. A chemical spill pollutes a stream. A monitoring device finds 620 ppm (parts per million) of the chemical 1.0 km below the spill, and the readings decrease by 12.5% for each kilometer farther downstream. How far downstream is the reading 100 ppm?

41. Measurements show that the temperature of a distant star is presently 9800°C and is decreasing by 10% every 800 years. What will its temperature be in 4000 years?

42. The chlorine in a swimming pool was measured (in ppm, parts per million) to be 1.80, 1.53, 1.30, and 1.10 on four successive days. Noting that these values approximate a geometric sequence, what would be the reading three days after the last reading?

43. The strength of a signal in a fiber-optic cable decreases 12% for every 15 km along the cable. What percent of the signal remains after 100 km?

44. A series of deposits, each of value A and made at equal time intervals, earns an interest rate of i for the time interval. The deposits have a total value of

$$A(1 + i) + A(1 + i)^2 + A(1 + i)^3 + \cdots + A(1 + i)^n$$

after n time intervals (just before the next deposit). Find a formula for this sum.

45. A thermometer is removed from hot water at 100.0°C into a room at 20.0°C. The temperature difference D between the thermometer and the air decreases by 35.0% each minute. What is the temperature reading on the thermometer 10.0 min later?

46. The power on a space satellite is supplied by a radioactive isotope. On a given satellite, the power decreases by 0.2% each day. What percent of the initial power remains after 1 year?

47. If you decided to save money by putting away 1¢ on a given day, 2¢ one week later, 4¢ a week later, and so on, how much would you have to put away 6 months (26 weeks) after putting away the 1¢?

48. How many direct ancestors (parents, grandparents, and so on) does a person have in the ten generations that preceded him or her (assuming that no ancestor appears in more than one line of descent)?

49. Derive a formula for S_n in terms of a_1, r, and a_n.

(W) 50. Write down several terms of a general geometric sequence. Then take the logarithm of each term. Explain why the resulting sequence is an arithmetic sequence.

(W) 51. Do the squares of the terms of a geometric sequence also form a geometric sequence? Explain.

(W) 52. If $a_1, a_2, a_3, \ldots$ is an arithmetic sequence, explain why $2^{a_1}, 2^{a_2}, 2^{a_3}, \ldots$ is a geometric sequence.

19.3 INFINITE GEOMETRIC SERIES

In the previous sections, we developed formulas for the sum of the first n terms of an arithmetic sequence and of a geometric sequence. *The indicated sum of the terms of a sequence is called a* **series.**

◀ EXAMPLE 1   **(a)** The indicated sum of the terms of the arithmetic sequence 2, 5, 8, 11, 14,... is the series $2 + 5 + 8 + 11 + 14 + \dots$.

(b) The indicated sum of terms of the geometric sequence $1, \frac{1}{2}, \frac{1}{4}, \frac{1}{8}, \dots$ is the series $1 + \frac{1}{2} + \frac{1}{4} + \frac{1}{8} + \dots$. ▶

The series associated with a finite sequence will sum up to a real number. The series associated with an infinite arithmetic sequence will not sum up to a real number, as the terms being added become larger and larger numerically. The sum is unbounded, as we can see in Example 1(a). The series associated with an infinite geometric sequence may or may not sum up to a real number, as we now show.

Let us now consider the sum of the first n terms of the infinite geometric sequence $1, \frac{1}{2}, \frac{1}{4}, \dots$. This is the sum of the n terms of the associated geometric series

$$1 + \frac{1}{2} + \frac{1}{4} + \cdots + \frac{1}{2^{n-1}}$$

Here, $a_1 = 1$ and $r = \frac{1}{2}$, and we find that we get the values of S_n for the given values of n in the following table:

n	2	3	4	5	6	7	8	9	10
S_n	$\frac{3}{2}$	$\frac{7}{4}$	$\frac{15}{8}$	$\frac{31}{16}$	$\frac{63}{32}$	$\frac{127}{64}$	$\frac{255}{128}$	$\frac{511}{256}$	$\frac{1023}{512}$

$1 + \frac{1}{2} + \frac{1}{4} + \frac{1}{8}$ ⤴   ⤷ $1 + \frac{1}{2} + \frac{1}{4} + \frac{1}{8} + \frac{1}{16} + \frac{1}{32} + \frac{1}{64}$

The series for $n = 4$ and $n = 7$ are shown. We see that as n gets larger, the numerator of each fraction becomes more nearly twice the denominator. In fact, we find that if we continue to compute S_n as n becomes larger, S_n can be found as close to the value 2 as desired, although it will never actually reach the value 2. For example, if $n = 100$, $S_{100} = 2 - 1.6 \times 10^{-30}$, which could be written as

$$1.999\,999\,999\,999\,999\,999\,999\,999\,999\,998\,4$$

to 32 significant digits. For the sum of the first n terms of a geometric sequence

$$S_n = a_1 \frac{1 - r^n}{1 - r}$$

the term r^n becomes exceedingly small if $|r| < 1$, and if n is sufficiently large, this term is effectively zero. *If this term were exactly zero,* the sum would be

$$S_n = 1 \frac{1 - 0}{1 - \frac{1}{2}} = 2$$

The symbol ∞ for infinity was first used by the English mathematician, John Wallis (1616–1703).

The only problem is that we cannot find any number large enough for n to make $\left(\frac{1}{2}\right)^n$ zero. There is, however, an accepted notation for this. This notation is

$$\lim_{n \to \infty} r^n = 0 \qquad (\text{if } |r| < 1)$$

and it is read as "the limit, as n *approaches* infinity, of r to the nth power is zero."

CAUTION ▶ The symbol ∞ is read as **infinity,** *but it must not be thought of as a number.* It is simply a symbol that stands for a *process* of considering numbers that become large without bound. The number called the **limit** of the sums is simply the number the sums get closer and closer to, as *n* is considered to approach infinity. This notation and terminology are of particular importance in the calculus.

If we consider values of *r* such that $|r| < 1$ and let the values of *n* become unbounded, we find that $\lim_{n \to \infty} r^n = 0$. The formula for *the sum of the terms of an infinite geometric series then becomes*

$$S = \frac{a_1}{1 - r} \qquad (|r| < 1) \tag{19.7}$$

where a_1 is the first term and *r* is the common ratio. If $|r| \geq 1$, *S* is unbounded in value.

◀ **EXAMPLE 2** Find the sum of the infinite geometric series

$$4 - \frac{1}{2} + \frac{1}{16} - \frac{1}{128} + \cdots$$

Here, we see that $a_1 = 4$. We find *r* by dividing any term by the previous term, and we find that $r = -\frac{1}{8}$. We then find the sum by substituting in Eq. (19.7). This gives us

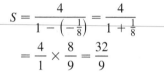

$$S = \frac{4}{1 - \left(-\frac{1}{8}\right)} = \frac{4}{1 + \frac{1}{8}}$$

$$= \frac{4}{1} \times \frac{8}{9} = \frac{32}{9}$$

◀ **EXAMPLE 3** Find the fraction that has as its decimal form $0.121\,212\ldots$.

This decimal form can be considered as being

$$0.12 + 0.0012 + 0.000\,012 + \cdots$$

which means that we have an infinite geometric series in which $a_1 = 0.12$ and $r = 0.01$. Thus,

$$S = \frac{0.12}{1 - 0.01} = \frac{0.12}{0.99}$$

$$= \frac{4}{33}$$

Therefore, the decimal $0.121\,212\ldots$ and the fraction $\frac{4}{33}$ represent the same number. ▶

The decimal in Example 3 is called a **repeating decimal,** because *a particular sequence of digits in the decimal form repeats endlessly.* This example verifies the theorem that any repeating decimal represents a rational number. However, not all repeating decimals start repeating immediately. If the numbers never do repeat, the decimal represents an irrational number. For example, there are no repeating decimals that represent π, $\sqrt{2}$, or *e*. The decimal form of the number *e* does repeat at one point, but the repetition stops. As we noted in Section 12.5, the decimal form of *e* to 16 decimal places is $2.7\,1828\,1828\,4590\,452$. We see that the sequence of digits 1828 repeats only once.

◀ EXAMPLE 4   Find the fraction that has as its decimal form the repeating decimal
$0.503\,453\,453\,45\ldots$.

We first separate the decimal into the beginning, nonrepeating part, and the infinite repeating decimal, which follows. Thus, we have

$$0.503\,453\,453\,45\ldots = 0.50 + 0.003\,453\,453\,45\ldots$$

This means that we are to add $\frac{50}{100}$ to the fraction that represents the sum of the terms of the infinite geometric series $0.003\,45 + 0.000\,003\,45 + \cdots$. For this series, $a_1 = 0.003\,45$ and $r = 0.001$. We find this sum to be

$$S = \frac{0.003\,45}{1 - 0.001} = \frac{0.003\,45}{0.999} = \frac{115}{33\,300} = \frac{23}{6660}$$

Therefore,

$$0.503\,453\,45\ldots = \frac{5}{10} + \frac{23}{6660} = \frac{5(666) + 23}{6660} = \frac{3353}{6660}$$ ▶

Solving a Word Problem   ◀ EXAMPLE 5   Each swing of a certain pendulum bob is 95% as long as the preceding swing. How far does the bob travel in coming to rest if the first swing is 40.0 cm long?

We are to find the sum of the terms of an infinite geometric series for which $a_1 = 40.0$ and $r = 95\% = \frac{19}{20}$. Substituting these values into Eq. (19.7), we obtain

$$S = \frac{40.0}{1 - \frac{19}{20}} = \frac{40.0}{\frac{1}{20}} = (40.0)(20) = 800 \text{ cm}$$

The pendulum bob travels 800 cm (8 m) in coming to rest.   ▶

EXERCISES 19.3

In Exercises 1 and 2, make the given changes in the indicated examples of this section, and then solve the resulting problems.

1. In Example 2, change all − signs to + in the series and then find the sum.

2. In Example 3, change the decimal form to $0.012\,012\,012\ldots$. and then find the fraction.

In Exercises 3–6, find the indicated quantity for an infinite geometric series.

3. $a_1 = 4, r = \frac{1}{2}, S = ?$ 4. $a_1 = 6, r = -\frac{1}{3}, S = ?$

5. $a_1 = 0.5, S = 0.625, r = ?$

6. $S = 4 + 2\sqrt{2}, r = \dfrac{1}{\sqrt{2}}, a_1 = ?$

In Exercises 7–14, find the sums of the given infinite geometric series.

7. $20 - 1 + 0.05 - \cdots$ 8. $9 + 8.1 + 7.29 + \cdots$

9. $1 + \frac{7}{8} + \frac{49}{64} + \cdots$ 10. $6 - 4 + \frac{8}{3} - \cdots$

11. $1 + 0.0001 + 0.000\,000\,01 + \cdots$

12. $300 - 90 + 27 - \cdots$

13. $\left(2 + \sqrt{3}\right) + 1 + \left(2 - \sqrt{3}\right) + \cdots$

14. $\left(1 + \sqrt{2}\right) - 1 + \left(\sqrt{2} - 1\right) - \cdots$

In Exercises 15–28, find the fractions equal to the given decimals.

15. $0.333\,33\ldots$ 16. $0.555\,55\ldots$

17. $0.499\,999\ldots$ 18. $0.999\,999\ldots$

19. $0.404\,040\ldots$ 20. $0.070\,707\ldots$

21. $0.181\,818\ldots$ 22. $0.336\,336\,336\ldots$

23. $0.273\,273\,273\ldots$ 24. $0.822\,22\ldots$

25. $0.366\,666\ldots$ 26. $0.664\,242\,42\ldots$

27. $0.100\,841\,841\,841\ldots$ 28. $0.184\,561\,845\,618\,456\ldots$

In Exercises 29–36, solve the given problems by use of the sum of an infinite geometric series.

29. Liquid is continuously collected in a wastewater-holding tank such that during a given hour only 92.0% as much liquid is collected as in the previous hour. If 28.0 L are collected in the first hour, what must be the minimum capacity of the tank?

30. If 75% of all aluminum cans are recycled, what is the total number of recycled cans that can be made from 400 000 cans that are recycled over and over until all the aluminum from these cans is used up? (Assume no aluminum is lost in the recycling process.)

31. The amounts of plutonium-237 that decay each day because of radioactivity form a geometric sequence. Given that the amounts that decay during each of the first 4 days are 5.882 g, 5.782 g, 5.684 g, and 5.587 g, respectively, what total amount will decay?

32. A helium-filled balloon rose 36.0 m in 1.0 min. Each minute after that, it rose 75% as much as in the previous minute. What was its maximum height?

33. A bicyclist traveling at 10 m/s then coasts to a stop as the bicycle travels 0.90 as far each second as in the previous second. How far does the bicycle travel in coasting to a stop?

(W)**34.** A square has sides of 20 cm. Another square is inscribed in the first square by joining the midpoints of the sides. Assuming that such inscribed squares can be formed endlessly, find the sum of the areas of all the squares and explain how the sum is found. See Fig. 19.3.

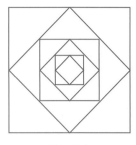

Fig. 19.3

35. Find x if the sum of the terms of the infinite geometric series $1 + 2x + 4x^2 + \ldots$ is 2/3.

36. Find the sum of the terms of the infinite series $1 + 2x + 3x^2 + 4x^3 + \ldots$ for $|x| < 1$. (*Hint:* Use $S - xS$ in the solution.)

19.4 THE BINOMIAL THEOREM

To expand $(x + 2)^5$ would require a number of multiplications that would be a tedious operation. We now develop *the **binomial theorem**, by which it is possible to expand binomials to any given power without direct multiplication.* Using this type of expansion, it is also possible to expand some expressions for which direct multiplication is not possible. The binomial theorem is used to develop expressions needed in certain mathematics topics and in technical applications.

By direct multiplication, we may obtain the following expansions of the binomial $a + b$:

$$(a + b)^0 = 1$$
$$(a + b)^1 = a + b$$
$$(a + b)^2 = a^2 + 2ab + b^2$$
$$(a + b)^3 = a^3 + 3a^2b + 3ab^2 + b^3$$
$$(a + b)^4 = a^4 + 4a^3b + 6a^2b^2 + 4ab^3 + b^4$$
$$(a + b)^5 = a^5 + 5a^4b + 10a^3b^2 + 10a^2b^3 + 5ab^4 + b^5$$

Inspection shows these expansions have certain properties, and we assume that these properties are valid for the expansion of $(a + b)^n$, where n is any positive integer.

PROPERTIES OF THE BINOMIAL $(a + b)^n$

1. *There are $n + 1$ terms.*

2. *The first term is a^n, and the final term is b^n.*

3. *Progressing from the first term to the last, the exponent of a decreases by 1 from term to term, the exponent of b increases by 1 from term to term, and the sum of the exponents of a and b in each term is n.*

4. *If the coefficient of any term is multiplied by the exponent of a in that term and this product is divided by the number of that term, we obtain the coefficient of the next term.*

5. *The coefficients of terms equidistant from the ends are equal.*

EXAMPLE 1   Using the basic properties, develop the expansion for $(a + b)^5$.

Since the exponent of the binomial is 5, we have $n = 5$.

From Property 1, we know that there are six terms.

From Property 2, we know that the first term is a^5 and the final term is b^5.

From Property 3, we know that the factors of a and b in terms 2, 3, 4, and 5 are a^4b, a^3b^2, a^2b^3, and ab^4, respectively.

From Property 4, we obtain the coefficients of terms 2, 3, 4, and 5. In the first term, a^5, the coefficient is 1. Multipling by 5, the power of a, and dividing by 1, the number of the term, we obtain 5, which is the coefficient of the second term. Thus, the second term is $5a^4b$. Again using Property 4, we obtain the coefficient of the third term. The coefficient of the second term is 5. Multiplying by 4, and dividing by 2, we obtain 10. This means that the third term is $10a^3b^2$.

From Property 5, we know that the coefficient of the fifth term is the same as the second and that the coefficient of the fourth term is the same as the third. These properties are illustrated in the following diagram:

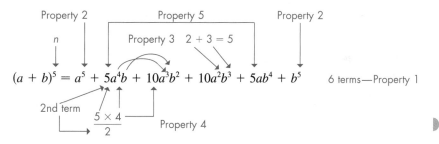

It is not necessary to use the above properties directly to expand a given binomial. If they are applied to $(a + b)^n$, a general formula for the expansion of a binomial may be obtained. In developing and stating the general formula, it is convenient to use the **factorial notation $n!$**, where

$$n! = n(n - 1)(n - 2) \cdots (2)(1) \tag{19.8}$$

We see that $n!$, read "n factorial," represents the product of the first n positive integers.

EXAMPLE 2   **(a)** $3! = (3)(2)(1) = 6$

(b) $5! = (5)(4)(3)(2)(1) = 120$

(c) $8! = (8)(7)(6)(5)(4)(3)(2)(1) = 40\,320$

(d) $3! + 5! = 6 + 120 = 126$

(e) $\dfrac{4!}{2!} = \dfrac{(4)(3)(2)(1)}{(2)(1)} = 12$

In evaluating factorials, we must remember that they represent products of numbers. In part (d), we see that $3! + 5!$ is **not** $(3 + 5)!$. Also, in part (e), we see that $4!/2!$ is **not** $2!$.

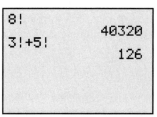

Fig. 19.4

CAUTION ▶

The evaluation of factorials can be done on a calculator. The manual should be consulted as to how to evaluate a factorial on any particular calculator. The evaluations for Example 2(c) and (d) are shown in Fig. 19.4.

See Appendix C for a graphing calculator program BINEXPAN. It gives the first K coefficients of $(AX + B)^N$.

THE BINOMIAL FORMULA

Based on the binomial properties, *the* **binomial theorem** *states that the following* **binomial formula** *is valid for all positive integer values of n* (the binomial theorem is proven through advanced methods).

$$(a + b)^n = a^n + na^{n-1}b + \frac{n(n-1)}{2!}a^{n-2}b^2 + \frac{n(n-1)(n-2)}{3!}a^{n-3}b^3 + \cdots + b^n \qquad (19.9)$$

◀ **EXAMPLE 3** Using the binomial formula, expand $(2x + 3)^6$.

In using the binomial formula for $(2x + 3)^6$, we use $2x$ for a, 3 for b, and 6 for n. Thus,

$$
(2x + 3)^6 = (2x)^6 + 6(2x)^5(3) + \frac{(6)(5)}{2}(2x)^4(3^2) + \frac{(6)(5)(4)}{(2)(3)}(2x)^3(3^3) + \frac{(6)(5)(4)(3)}{(2)(3)(4)}(2x)^2(3^4) + \frac{(6)(5)(4)(3)(2)}{(2)(3)(4)(5)}(2x)(3^5) + 3^6
$$

$$= 64x^6 + 576x^5 + 2160x^4 + 4320x^3 + 4860x^2 + 2916x + 729$$

For the first few integral powers of a binomial $a + b$, the coefficients can be obtained by setting them up in the following pattern, known as **Pascal's triangle.**

Pascal's Triangle

Named for the French scientist and mathematician Blaise Pascal (1623–1662).

$n = 0$						1						
$n = 1$					1		1					
$n = 2$				1		2		1				
$n = 3$			1		3		3		1			
$n = 4$		1		4		6		4		1		
$n = 5$	1		5		10		10		5		1	
$n = 6$	1	6		15		20		15		6		1

see expansions on page 516

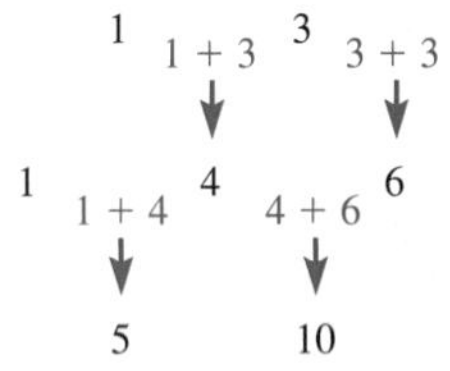

We note that the first and last coefficients shown in each row are 1, and the second and next-to-last coefficients are equal to n. Other coefficients are obtained by adding the two nearest coefficients in the row above, as illustrated in Fig. 19.5 for the indicated section of Pascal's triangle. This pattern may be continued indefinitely, although use of Pascal's triangle is cumbersome for high values of n.

Fig. 19.5

◀ **EXAMPLE 4** Using Pascal's triangle, expand $(5s - 2t)^4$.

Here, we note that $n = 4$. Thus, the coefficients of the five terms are 1, 4, 6, 4, and 1, respectively. Also, here we use $5s$ for a and $-2t$ for b. We are expanding this expression as $[(5s) + (-2t)]^4$. Therefore,

from Pascal's triangle for $n = 4$

$$(5s - 2t)^4 = (5s)^4 + 4(5s)^3(-2t) + 6(5s)^2(-2t)^2 + 4(5s)(-2t)^3 + (-2t)^4$$

$$= 625s^4 - 1000s^3t + 600s^2t^2 - 160st^3 + 16t^4$$

In certain uses of a binomial expansion, it is not necessary to obtain all terms. Only the first few terms are required. The following example illustrates finding the first four terms of an expansion.

◀ **EXAMPLE 5** Find the first four terms of the expansion of $(x + 7)^{12}$.

Here, we use x for a, 7 for b, and 12 for n. Thus, from the binomial formula we have

$$(x + 7)^{12} = x^{12} + 12x^{11}(7) + \frac{(12)(11)}{2}x^{10}(7^2) + \frac{(12)(11)(10)}{(2)(3)}x^9(7^3) + \cdots$$

$$= x^{12} + 84x^{11} + 3234x^{10} + 75\,460x^9 + \cdots$$

If we let $a = 1$ and $b = x$ in the binomial formula, we obtain the **binomial series**

Binomial Series

$$(1 + x)^n = 1 + nx + \frac{n(n-1)}{2!}x^2 + \frac{n(n-1)(n-2)}{3!}x^3 + \cdots \qquad (19.10)$$

which through advanced methods can be shown to be valid for any real number n if $|x| < 1$. When n is either negative or a fraction, we obtain an infinite series. In such a case, we calculate as many terms as may be needed, although such a series is not obtainable through direct multiplication. The binomial series may be used to develop important expressions that are used in applications and more advanced mathematics topics.

◀ **EXAMPLE 6** In the analysis of forces on beams, the expression $1/(1 + m^2)^{3/2}$ is used. Use the binomial series to find the first four terms of the expansion.

Using negative exponents, we have

$$1/(1 + m^2)^{3/2} = (1 + m^2)^{-3/2}$$

Now, in using Eq. (19.10), we have $n = -3/2$ and $x = m^2$:

$$(1 + m^2)^{-3/2} = 1 + \left(-\frac{3}{2}\right)(m^2) + \frac{\left(-\frac{3}{2}\right)\left(-\frac{3}{2} - 1\right)}{2!}(m^2)^2 + \frac{\left(-\frac{3}{2}\right)\left(-\frac{3}{2} - 1\right)\left(-\frac{3}{2} - 2\right)}{3!}(m^2)^3 + \cdots$$

Therefore,

$$\frac{1}{(1 + m^2)^{3/2}} = 1 - \frac{3}{2}m^2 + \frac{15}{8}m^4 - \frac{35}{16}m^6 + \cdots$$

◀ **EXAMPLE 7** Approximate the value of 0.97^7 by use of the binomial series.

We note that $0.97 = 1 - 0.03$, which means $0.97^7 = [1 + (-0.03)]^7$. Using four terms of the binomial series, we have

$$0.97^7 = [1 + (-0.03)]^7$$

$$= 1 + 7(-0.03) + \frac{7(6)}{2!}(-0.03)^2 + \frac{7(6)(5)}{3!}(-0.03)^3$$

$$= 1 - 0.21 + 0.0189 - 0.000\,945 = 0.807\,955$$

From a calculator, we find that $0.97^7 = 0.807\,983$ (to six decimal places), which means these values agree to four significant digits with a value 0.8080. Greater accuracy can be found using the binomial series if more terms are used.

EXERCISES 19.4

In Exercises 1 and 2, make the given changes in the indicated examples of this section and then solve the given problems.

1. In Example 3, change the exponent from 6 to 5 and then perform the expansion.

2. In Example 7, change 0.97 to 0.98 and then approximate the value.

In Exercises 3–12, expand and simplify the given expressions by use of the binomial formula.

3. $(t + 1)^3$

4. $(x - 2)^3$

5. $(2x - 1)^4$

6. $(x^2 + 3)^4$

7. $(2 + 0.1)^5$

8. $(xy - z)^5$

9. $(n + 2\pi)^5$

10. $(1 - j)^6 \quad (j = \sqrt{-1})$

11. $(2a - b^2)^6$

12. $\left(\dfrac{a}{x} + x\right)^6$

In Exercises 13–16, expand and simplify the given expressions by use of Pascal's triangle.

13. $(5x - 3)^4$

14. $(b + 4)^5$

15. $(2a + 1)^6$

16. $(x - 3)^7$

In Exercises 17–24, find the first four terms of the indicated expansions.

17. $(x + 2)^{10}$

18. $(x - 3)^8$

19. $(2a - 1)^7$

20. $(3b + 2)^9$

21. $(x^{1/2} - y)^{12}$

22. $(2a - x^{-1})^{11}$

23. $\left(b^2 + \dfrac{1}{2b}\right)^{20}$

24. $\left(2x^2 + \dfrac{y}{3}\right)^{15}$

In Exercises 25–28, approximate the value of the given expression to three decimal places by using three terms of the appropriate binomial series. Check using a calculator.

25. 1.05^6

26. $\sqrt{0.927}$

27. $\sqrt[3]{1.045}$

28. 0.98^{-7}

In Exercises 29–36, find the first four terms of the indicated expansions by use of the binomial series.

29. $(1 + x)^8$

30. $(1 + x)^{-1/3}$

31. $(1 - x)^{-2}$

32. $\left(1 - 2\sqrt{x}\right)^9$

33. $\sqrt{1 + x}$

34. $\dfrac{1}{\sqrt{1 + x}}$

35. $\dfrac{1}{\sqrt{9 - 9x}}$

36. $\sqrt{4 + x^2}$

In Exercises 37–40, solve the given problems involving factorial notation.

37. Using a calculator, evaluate (a) $17! + 4!$, (b) $21!$, (c) $17! \times 4!$, and (d) $68!$.

38. Using a calculator, evaluate (a) $8! - 7!$, (b) $8!/7!$, (c) $8! \times 7!$ and (d) $56!$.

39. Show that $n! = n \times (n - 1)!$ for $n \geq 2$. To use this equation for $n = 1$, explain why it is necessary to define $0! = 1$ (this is a standard definition of $0!$).

40. Show that $\dfrac{(n + 1)!}{(n - 2)!} = n^3 - n$ for $n \geq 2$. See Exercise 39.

In Exercises 41–44, find the indicated terms by use of the following information. The $r + 1$ term of the expansion of $(a + b)^n$ is given by

$$\frac{n(n - 1)(n - 2) \cdots (n - r + 1)}{r!} a^{n-r} b^r$$

41. The term involving b^5 in $(a + b)^8$

42. The term involving y^6 in $(x + y)^{10}$

43. The fifth term of $(2x - 3b)^{12}$

44. The sixth term of $\left(\sqrt{a} - \sqrt{b}\right)^{14}$

In Exercises 45–52, solve the given problems.

45. Approximate $\sqrt{6}$ to hundredths by noting that $\sqrt{6} = \sqrt{4(1.5)} = 2\sqrt{1 + 0.5}$ and using four terms of the appropriate binomial series.

46. Approximate $\sqrt[3]{10}$ by using the method of Exercise 45.

47. A company purchases a piece of equipment for A dollars, and the equipment depreciates at a rate of r each year. Its value V after n years is $V = A(1 - r)^n$. Expand this expression for $n = 5$.

48. In finding the rate of change of emission of energy from the surface of a body at temperature T, the expression $(T + h)^4$ is used. Expand this expression.

49. In the theory associated with the magnetic field due to an electric current, the expression $1 - \dfrac{x}{\sqrt{a^2 + x^2}}$ is found. By expanding $(a^2 + x^2)^{-1/2}$, find the first three nonzero terms that could be used to approximate the given expression.

50. In the theory related to the dispersion of light, the expression $1 + \dfrac{A}{1 - \lambda_0^2/\lambda^2}$ arises. (a) Let $x = \lambda_0^2/\lambda^2$ and find the first four terms of the expansion of $(1 - x)^{-1}$. (b) Find the same expansion by using long division. (c) Write the original expression in expanded form, using the results of (a) and (b).

51. In finding the rate at which the resistance of a wire changes with its radius, the expression $\dfrac{k}{(r + h)^2} - \dfrac{k}{r^2}$ is used. Use three terms of the binomial expansion of $k(r + h)^{-2}$ to approximate this expression.

52. Find the first four terms of the expansion of $(1 + x)^{-1}$ and then divide $1 + x$ into 1. Compare the results.

CHAPTER ⑲ EQUATIONS

Arithmetic sequences	Recursion formula $\quad a_n = a_{n-1} + d$	(19.1)
	nth term $\qquad\qquad a_n = a_1 + (n-1)d$	(19.2)
	Sum of n terms $\qquad S_n = \dfrac{n}{2}(a_1 + a_n)$	(19.3)
Geometric sequences	Recursion formula $\quad a_n = ra_{n-1}$	(19.4)
	nth term $\qquad\qquad a_n = a_1 r^{n-1}$	(19.5)
	Sum of n terms $\qquad S_n = \dfrac{a_1(1 - r^n)}{1 - r} \quad (r \neq 1)$	(19.6)
Sum of geometric series	$S = \dfrac{a_1}{1 - r} \qquad (|r| < 1)$	(19.7)
Factorial notation	$n! = n(n-1)(n-2) \cdots (2)(1)$	(19.8)
Binomial formula	$(a + b)^n = a^n + na^{n-1}b + \dfrac{n(n-1)}{2!}a^{n-2}b^2 + \dfrac{n(n-1)(n-2)}{3!}a^{n-3}b^3 + \cdots + b^n$	(19.9)
Binomial series	$(1 + x)^n = 1 + nx + \dfrac{n(n-1)}{2!}x^2 + \dfrac{n(n-1)(n-2)}{3!}x^3 + \cdots$	(19.10)

CHAPTER ⑲ REVIEW EXERCISES

In Exercises 1–8, find the indicated term of each sequence.

1. $1, 6, 11, \ldots$ (17th)
2. $1, -3, -7, \ldots$ (21st)

3. $500, 100, 20, \ldots$ (9th)

4. $0.025, 0.01, 0.004, \ldots$ (7th)

5. $8, \frac{7}{2}, -1, \ldots$ (16th)
6. $-1, -\frac{5}{3}, -\frac{7}{3}, \ldots$ (25th)

7. $\frac{3}{4}, \frac{1}{2}, \frac{1}{3}, \ldots$ (7th)
8. $5^{-2}, 5^0, 5^2, \ldots$ (7th)

In Exercises 9–12, find the sum of each sequence with the indicated values.

9. $a_1 = -4, n = 15, a_{15} = 17$ (arith.)

10. $a_1 = 300, d = -\frac{20}{3}, n = 10$

11. $a_1 = 16, r = -\frac{1}{2}, n = 14$

12. $a_1 = 64, a_n = 729, n = 7$ (geom., $r > 0$)

In Exercises 13–24, find the indicated quantities for the appropriate sequences.

13. $a_1 = 17, d = -2, n = 9, S_9 = ?$

14. $d = \frac{4}{3}, a_1 = -3, a_n = 17, n = ?$

15. $a_1 = 4, r = \sqrt{2}, n = 7, a_7 = ?$

16. $a_n = \frac{49}{8}, r = -\frac{2}{7}, S_n = \frac{1911}{32}, a_1 = ?$

17. $a_1 = 80, a_n = -25, S_n = 220, d = ?$

18. $a_1 = 2, d = 0.2, n = 11, S_{11} = ?$

19. $n = 6, r = -0.25, S_6 = 204.75, a_6 = ?$

20. $a_1 = 10, r = 0.1, S_n = 11.111, n = ?$

21. $a_1 = -1, a_n = 32, n = 12, S_{12} = ?$ (arith.)

22. $a_1 = 100, a_n = 6400, S_n = 32\,500, n = ?$ (arith.)

23. $a_1 = 1, n = 7, a_7 = 64, S_7 = ?$

24. $a_1 = \frac{1}{4}, n = 6, a_6 = 8, S_6 = ?$

In Exercises 25–28, find the sums of the given infinite geometric series.

25. $0.9 + 0.6 + 0.4 + \cdots$

26. $80 - 20 + 5 - \cdots$

27. $1 + 1.02^{-1} + 1.02^{-2} + \cdots$

28. $3 - \sqrt{3} + 1 - \cdots$

In Exercises 29–32, find the fractions equal to the given decimals.

29. $0.030303\ldots$
30. $0.363363\ldots$

31. $0.0727272\ldots$
32. $0.25399399399\ldots$

In Exercises 33–36, expand and simplify the given expression. In Exercises 37–40, find the first four terms of the appropriate expansion.

33. $(x - 2)^4$

34. $(3 + 0.1)^4$

35. $(x^2 + 1)^5$

36. $(3n^{1/2} - a)^6$

37. $(a + 2e)^{10}$

38. $\left(\dfrac{x}{4} - y\right)^{12}$

39. $\left(p^2 - \dfrac{q}{6}\right)^9$

40. $\left(2s^2 - \dfrac{3}{2}t^{-1}\right)^{14}$

In Exercises 41–48, find the first four terms of the indicated expansions by use of the binomial series.

41. $(1 + x)^{12}$

42. $(1 - x)^{10}$

43. $\sqrt{1 + x^2}$

44. $\left(4 - 4\sqrt{x}\right)^1$

45. $\sqrt{1 - a^2}$

46. $\sqrt{1 + b^4}$

47. $(2 - 4x)^{-3}$

48. $(1 + 4x)^{-1/4}$

In Exercises 49–84, solve the given problems by use of an appropriate sequence or expansion. All numbers are accurate to at least two significant digits.

49. Find the sum of the first 1000 positive even integers.

50. How many integers divisible by 4 lie between 23 and 121?

51. What is the fifth term of the arithmetic sequence in which the first term is a and the second term is b?

52. Find three consecutive numbers in an arithmetic sequence such that their sum is 15 and the sum of their squares is 77.

53. For a geometric sequence, is it possible that $a_3 = 6$, $a_5 = 9$, and $a_7 = 12$?

54. Find three consecutive numbers in a geometric sequence such that their product is 64 and the sum of their squares is 84.

55. Approximate the value of $(1.06)^{-6}$ by using three terms of the appropriate binomial series. Check using a calculator.

56. Approximate the value of $\sqrt[4]{0.94}$ by using three terms of the appropriate binomial series. Check using a calculator.

57. Approximate the value of $\sqrt{30}$ by noting that $\sqrt{30} = \sqrt{25(1.2)} = 5\sqrt{1 + 0.2}$ and using three terms of the appropriate binomial series.

58. Approximate $\sqrt[3]{29.7}$ by using the *method* of Exercise 57.

59. Each stroke of a pile driver moves a post 2 cm less than the previous stroke. If the first stroke moves the post 24 cm, which stroke moves the post 4 cm?

60. During each hour, an exhaust fan removes 15.0% of the carbon dioxide present in the air in a room at the beginning of the hour. What percent of the carbon dioxide remains after 10.0 h?

61. Each 1.0 mm of a filter through which light passes reduces the intensity of the light by 12%. How thick should the filter be to reduce the intensity of the light to 20%?

62. A pile of dirt and ten holes are in a straight line. It is 20 m from the dirt pile to the nearest hole, and the holes are 8 m apart. If a backhoe takes two trips to fill each hole, how far must it travel in filling all the holes if it starts and ends at the dirt pile?

63. A roof support with equally spaced vertical pieces is shown in Fig. 19.6. Find the total length of the vertical pieces if the shortest one is 254 mm long.

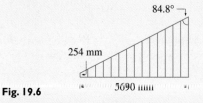

84.8°

254 mm

3690 mm

Fig. 19.6

64. During each microsecond, the current in an electric circuit decreases by 9.3%. If the initial current is 2.45 mA, how long does it take to reach 0.50 mA?

65. A machine that costs $8600 depreciates 1.0% in value each month. What is its value 5 years after it was purchased?

66. The level of chemical pollution in a lake is 4.50 ppb (parts per billion). If the level increases by 0.20 ppb in the following month and by 5.0% less each month thereafter, what will be the maximum level?

67. A piece of paper 0.015 cm thick is cut in half. These two pieces are then placed one on the other and cut in half. If this is repeated such that the paper is cut in half 40 times, how high will the pile be?

68. After the power is turned off, an object on a nearly frictionless surface slows down such that it travels 99.9% as far during 1 s as during the previous second. If it travels 100 cm during the first second after the power is turned off, how far does it travel while stopping?

69. Under gravity, an object falls 4.9 m during the first second, 14.7 m during the second second, 24.5 m during the third second, and so on. How far will it fall during the 20th second?

70. For the object in Exercise 69, what is the total distance fallen during the first 20 s?

71. An object suspended on a spring is oscillating up and down. If the first oscillation is 10.0 cm and each oscillation thereafter is 9/10 of the preceding one, find the total distance the object travels in coming to rest.

72. During each oscillation, a pendulum swings through 85% of the distance of the previous oscillation. If the pendulum swings through 80.8 cm in the first oscillation, through what total distance does it move in 12 oscillations?

73. A person invests $1000 each year at the beginning of the year. What is the total value of these investments after 20 years if they earn 7.5% annual interest, compounded semiannually?

74. In testing a type of insulation, the temperature in a room was made to fall to 2/3 of the initial temperature after 1.0 h, to 2/5 of the initial temperature after 2.0 h, to 2/7 of the initial temperature after 3.0 h, and so on. If the initial temperature was 50.0°C, what was the temperature after 12.0 h? (This is an illustration of a *harmonic sequence*.)

75. Two competing businesses make the same item and sell it initially for $100. One increases the price by $8 each year for 5 years, and the other increases the price by 8% each year for 5 years. What is the difference in price after 5 years?

76. A well driller charges $10.00 for drilling the first meter of a well and for each meter thereafter charges 0.20% more than for the preceding meter. How much is charged for drilling a 150-m well?

77. In hydrodynamics, while studying compressible fluid flow, the expression $\left(1 + \dfrac{a-1}{2} m^2\right)^{a/(a-1)}$ arises. Find the first three terms of the expansion of this expression.

78. In finding the partial pressure P_F of the fluorine gas under certain conditions, the equation

$$P_F = \frac{(1 + 2 \times 10^{-10}) - \sqrt{1 + 4 \times 10^{-10}}}{2} \text{ atm}$$

is found. By using three terms of the expansion for $\sqrt{1+x}$, approximate the value of this expression. (1 atm = 101.3 kPa)

79. During 1 year a beach eroded 1.2 m to a line 48.3 m from the wall of a building. If the erosion is 0.1 m more each year than the previous year, when will the waterline reach the wall?

80. On a highway with a steep incline, the runaway truck ramp is constructed so that a vehicle that has lost its brakes can stop. The ramp is designed to slow a truck in succeeding 20-m distances by 10 km/h, 12 km/h, 14 km/h, ... If the ramp is 160 m long, will it stop a truck moving at 120 km/h when it reaches the ramp?

81. Each application of an insecticide destroys 75% of a certain insect. How many applications are needed to destroy at least 99.9% of the insects?

82. A wire hung between two poles is parabolic in shape. To find the length of wire between two points on the wire, the expression $\sqrt{1 + 0.08x^2}$ is used. Find the first three terms of the binomial expansion of this expression.

(W) 83. Do the reciprocals of the terms of a geometric sequence form a geometric sequence? Explain.

84. The terms a, $a + 12$, $a + 24$ form an arithmetic sequence, and the terms a, $a + 24$, $a + 12$ form a geometric sequence. Find these sequences.

Writing Exercise

85. Derive a formula for the value V after 1 year of an amount A invested at $r\%$ (as a decimal) annual interest, compounded n times during the year. If $A = \$1000$ and $r = 0.10$ (10%), write two or three paragraphs explaining why the amount of interest increases as n increases and stating your approach to finding the maximum possible amount of interest.

CHAPTER (19) PRACTICE TEST

1. Find the sum of the first seven terms of the sequence $6, -2, \frac{2}{3}, \ldots$.

2. For a given sequence $a_1 = 6$, $d = 4$, and $S_n = 126$. Find n.

3. Find the fraction equal to the decimal $0.454\,545\ldots$.

4. Find the first three terms of the expansion of $\sqrt{1 - 4x}$.

5. Expand and simply the expression $(2x - y)^5$.

6. What is the value after 20 years of an investment of $2500 if it draws 5% annual interest compounded annually?

7. Find the sum of the first 100 even integers.

8. A ball is dropped from a height of 8.00 m, and on each rebound it rises to 1/2 of the height it last fell. If it bounces indefinitely, through what total distance will it move?

20

Additional Topics in Trigonometry

In Section 20.2, we show how trigonometric relationships are used in analyzing the voltage produced by a three-phase generator, the most widely used type of polyphase generator of alternating current.

As the use of electricity became widespread in the 1880s, there was a serious debate over the best way to distribute electric power. The American inventor Thomas Edison favored the use of the direct current because it was safer and did not vary with time. Another American inventor and engineer, George Westinghouse, favored alternating current because the voltage could be stepped up and down with transformers during transmission.

Also favoring the use of alternating current was Nikola Tesla, an American (born in Croatia) electrical engineer, and he had a strong influence in the fact that alternating current came to be used for transmission. Tesla developed many electrical devices, among them the *polyphase generator* that allowed alternating current to be transmitted with constant instantaneous power. Using this type of generator, power losses are greatly reduced in transmission lines, which allows for smaller conductors, and the power can be generated far from where it is used.

Three-phase systems are used in most commercial electric generators, and in Section 20.2 we show how a relationship involving trigonometric functions can be used to show a basic property of the current produced by a three-phase generator. Many such relationships among the trigonometric functions can be found from the definitions and other known relationships.

The trigonometric relationships that we develop in this chapter are important for a number of reasons. In fact, we already made use of some of them in Section 10.4 when we graphed certain trigonometric functions, and in Chapter 9 in deriving the law of cosines. In calculus, certain problems use trigonometric relationships for a solution, even including some in which these functions do not appear in the initial problem or final answer. Also, they are useful in a number of technical applications in areas such as electronics, optics, solar energy, and robotics.

Later in the chapter, we see that various trigonometric relationships are used in solving equations with trigonometric functions. Also, we develop the concept of the inverse trigonometric functions that were introduced in Chapter 4.

20.1 FUNDAMENTAL TRIGONOMETRIC IDENTITIES

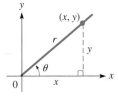

Fig. 20.1

From Chapters 4 and 8, recall that the definition of the sine of an angle θ is $\sin \theta = y/r$ and that the definition of the cosecant of an angle θ is $\csc \theta = r/y$ (see Fig. 20.1). Since $y/r = 1/(r/y)$, we see that $\sin \theta = 1/\csc \theta$. The definitions hold true for *any* angle, which means this relation between $\sin \theta$ and $\csc \theta$ is true for *any* angle. *This type of relation, which is true for any value of the variable, is called an* **identity.** Of course, values where division by zero would be indicated are excluded.

In this section, we develop several important identities among the trigonometric functions. We also show how the basic identities are used to verify other identities.

From the definitions, we have

$$\sin \theta \csc \theta = \frac{y}{r} \times \frac{r}{y} = 1 \quad \text{or} \quad \sin \theta = \frac{1}{\csc \theta} \quad \text{or} \quad \csc \theta = \frac{1}{\sin \theta}$$

$$\cos \theta \sec \theta = \frac{x}{r} \times \frac{r}{x} = 1 \quad \text{or} \quad \cos \theta = \frac{1}{\sec \theta} \quad \text{or} \quad \sec \theta = \frac{1}{\cos \theta}$$

$$\tan \theta \cot \theta = \frac{y}{x} \times \frac{x}{y} = 1 \quad \text{or} \quad \tan \theta = \frac{1}{\cot \theta} \quad \text{or} \quad \cot \theta = \frac{1}{\tan \theta}$$

$$\frac{\sin \theta}{\cos \theta} = \frac{y/r}{x/r} = \frac{y}{x} = \tan \theta \qquad \frac{\cos \theta}{\sin \theta} = \frac{x/r}{y/r} = \frac{x}{y} = \cot \theta$$

Also, from the definitions and the Pythagorean theorem in the form of $x^2 + y^2 = r^2$, we arrive at the following identities.

By dividing the Pythagorean relation through by r^2, we have

$$\left(\frac{x}{r}\right)^2 + \left(\frac{y}{r}\right)^2 = 1 \quad \text{which leads us to} \quad \cos^2 \theta + \sin^2 \theta = 1$$

By dividing the Pythagorean relation by x^2, we have

$$1 + \left(\frac{y}{x}\right)^2 = \left(\frac{r}{x}\right)^2 \quad \text{which leads us to} \quad 1 + \tan^2 \theta = \sec^2 \theta$$

By dividing the Pythagorean relation by y^2, we have

$$\left(\frac{x}{y}\right)^2 + 1 = \left(\frac{r}{y}\right)^2 \quad \text{which leads us to} \quad \cot^2 \theta + 1 = \csc^2 \theta$$

Basic Identities

The term $\cos^2 \theta$ is the common way of writing $(\cos \theta)^2$, and it means to square the value of the cosine of the angle. Obviously, the same holds true for the other functions. Summarizing these results, we have the following important identities.

$\sin \theta = \dfrac{1}{\csc \theta}$ (20.1)	$\tan \theta = \dfrac{\sin \theta}{\cos \theta}$ (20.4)	$\sin^2 \theta + \cos^2 \theta = 1$ (20.6)
$\cos \theta = \dfrac{1}{\sec \theta}$ (20.2)	$\cot \theta = \dfrac{\cos \theta}{\sin \theta}$ (20.5)	$1 + \tan^2 \theta = \sec^2 \theta$ (20.7)
$\tan \theta = \dfrac{1}{\cot \theta}$ (20.3)		$1 + \cot^2 \theta = \csc^2 \theta$ (20.8)

In using these basic identities, θ may stand for any angle or number or expression representing an angle or a number.

◀ **EXAMPLE 1** **(a)** $\sin(x + 1) = \dfrac{1}{\csc(x + 1)}$ using Eq. (20.1)

(b) $\tan 157° = \dfrac{\sin 157°}{\cos 157°}$ using Eq. (20.4)

(c) $\sin^2\left(\dfrac{\pi}{4}\right) + \cos^2\left(\dfrac{\pi}{4}\right) = 1$ using Eq. (20.6)   ▶

◀ **EXAMPLE 2** Let us check the last two illustrations of Example 1 for the particular values of θ that are used.

(a) Using a calculator, we find that

$$\sin 157° = 0.390\,731\,128\,5 \quad \text{and} \quad \cos 157° = -0.920\,504\,853\,5$$

Considering Eq. (20.4) and Example 1(b), by dividing we find that

$$\frac{\sin 157°}{\cos 157°} = \frac{0.390\,731\,128\,5}{-0.920\,504\,853\,5} = -0.424\,474\,816\,2$$

We also find that $\tan 157° = -0.424\,474\,816\,2$, which shows that

$$\tan 157° = \frac{\sin 157°}{\cos 157°}$$

(b) To check Example 1(c), we refer to the values found in Example 2 of Section 4.3, or to Fig. 20.2. These tell us that

$$\sin 45° = \frac{1}{\sqrt{2}} = \frac{\sqrt{2}}{2} \quad \text{and} \quad \cos 45° = \frac{\sqrt{2}}{2}$$

Since $\frac{\pi}{4} = 45°$, by adding the squares of $\sin 45°$ and $\cos 45°$, we have

$$\sin^2\left(\frac{\pi}{4}\right) + \cos^2\left(\frac{\pi}{4}\right) = \left(\frac{\sqrt{2}}{2}\right)^2 + \left(\frac{\sqrt{2}}{2}\right)^2 = \frac{1}{2} + \frac{1}{2} = 1$$

We see that this checks with Eq. (20.6) for these values.   ▶

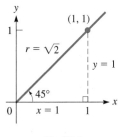

Fig. 20.2

PROVING TRIGONOMETRIC IDENTITIES

A great many identities exist among the trigonometric functions. We are going to use the basic identities that have been developed in Eqs. (20.1) through (20.8), along with a few additional ones developed in later sections to prove the validity of still other identities.

CAUTION ▶ *The ability to prove trigonometric identities depends to a large extent on being very familiar with the basic identities* so that you can *recognize them in somewhat different forms.*

If you do not learn these basic identities and learn them well, you will have difficulty in following the examples and doing the exercises. The more readily you recognize these forms, the more easily you will be able to prove such identities.

In proving identities, look for combinations that appear in, or are very similar to, those in the basic identities. This is illustrated in the following examples.

◖ EXAMPLE 3   In proving the identity

$$\sin x = \frac{\cos x}{\cot x}$$

we know that $\cot x = \dfrac{\cos x}{\sin x}$. Since $\sin x$ appears on the left, substituting for $\cot x$ on the right will eliminate $\cot x$ and introduce $\sin x$. This should help us proceed in proving the identity. Thus,

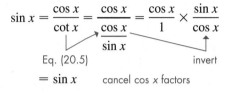

$$\sin x = \frac{\cos x}{\cot x} = \frac{\cos x}{\dfrac{\cos x}{\sin x}} = \frac{\cos x}{1} \times \frac{\sin x}{\cos x}$$

Eq. (20.5)   invert

$$= \sin x \qquad \text{cancel cos } x \text{ factors}$$

By showing that the right side may be changed exactly to $\sin x$, the expression on the left side, we have proved the identity.   ◗

As we point out here, performing the

algebraic operations

carefully and correctly is very important when working with trigonometric expressions. Operations such as substituting, factoring, and simplifying fractions are frequently used.

Some important points should be made in relation to the proof of the identity of Example 3. We must recognize what basic identities may be useful. The proof of an identity requires the use of the basic algebraic operations, and these must be done carefully and correctly. Although in Example 3 we changed the right side to the form on the left, we could have changed the left to the form on the right. From this and the fact that various substitutions are possible, we see that a variety of procedures can be used to prove any given identity.

◖ EXAMPLE 4   Prove that $\tan \theta \csc \theta = \sec \theta$.

In proving this identity, we know that $\tan \theta = \dfrac{\sin \theta}{\cos \theta}$ and also that $\dfrac{1}{\cos \theta} = \sec \theta$. Thus, by substituting for $\tan \theta$, we introduce $\cos \theta$ in the denominator, which is equivalent to introducing $\sec \theta$ in the numerator. Therefore, changing only the left side, we have

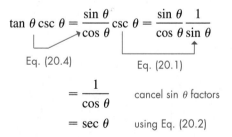

$$\tan \theta \csc \theta = \frac{\sin \theta}{\cos \theta} \csc \theta = \frac{\sin \theta}{\cos \theta} \frac{1}{\sin \theta}$$

Eq. (20.4)   Eq. (20.1)

$$= \frac{1}{\cos \theta} \qquad \text{cancel sin } \theta \text{ factors}$$

$$= \sec \theta \qquad \text{using Eq. (20.2)}$$

Having changed the left side into the form on the right side, we have proven the identity.

NOTE ◗   Many variations of the preceding steps are possible. Also, we could have changed the right side to obtain the form on the left. For example,

$$\tan \theta \csc \theta = \sec \theta = \frac{1}{\cos \theta} \qquad \text{using Eq. (20.2)}$$

$$= \frac{\sin \theta}{\cos \theta \sin \theta} = \frac{\sin \theta}{\cos \theta} \frac{1}{\sin \theta} \qquad \begin{array}{l}\text{multiply numerator and} \\ \text{denominator by sin } \theta \text{ and rewrite}\end{array}$$

$$= \tan \theta \csc \theta \qquad \text{using Eqs. (20.4) and (20.1)}$$ ◗

In proving the identities of Examples 3 and 4, we have shown that the expression on one side of the equal sign can be changed into the expression on the other side. Although making the restriction that we change only one side is not entirely necessary, *we will restrict the method of proof to changing only one side into the same form as the other side.* In this way, we know the form we are to obtain, and by looking ahead we are better able to make the proper changes.

There is no set procedure for working with identities. The most important factors are to (1) *recognize the proper forms,* (2) *see what effect a change may have* before performing it, and (3) *perform it correctly.* Normally, *it is easier to change the form of the more complicated side to the same form as the less complicated side.* If the forms are about the same, a close look often suggests possible steps to use.

CAUTION ▶

◀ EXAMPLE 5   Prove the identity $\dfrac{\cos x \csc x}{\cot^2 x} = \tan x$.

First, we note that the left-hand side has several factors and the right-hand side has only one. Therefore, let us transform the left-hand side. Next, we note that we want $\tan x$ as the final result. We know that $\cot x = 1/\tan x$. Thus,

$$\frac{\cos x \csc x}{\cot^2 x} = \frac{\cos x \csc x}{\dfrac{1}{\tan^2 x}} = \cos x \csc x \tan^2 x$$

At this point, we have two factors of $\tan x$ on the left. Since we want only one, let us factor out one. Therefore,

$$\cos x \csc x \tan^2 x = \tan x(\cos x \csc x \tan x)$$

Now, replacing $\tan x$ within the parentheses by $\sin x/\cos x$, we have

$$\tan x(\cos x \csc x \tan x) = \frac{\tan x(\cos x \csc x \sin x)}{\cos x}$$

Now we may cancel $\cos x$. Also, $\csc x \sin x = 1$ from Eq. (20.1). Finally,

$$\frac{\tan x(\cos x \csc x \sin x)}{\cos x} = \tan x\left(\frac{\cos x}{\cos x}\right)(\csc x \sin x)$$

$$= \tan x(1)(1) = \tan x$$

Since we have transformed the left-hand side into $\tan x$, we have proven the identity. Of course, it is not necessary to rewrite expressions as we did in this example. This was done here only to include the explanations. ▶

◀ EXAMPLE 6   In finding the radiation rate of an accelerated electric charge, it is necessary to show that $\sin^3 \theta = \sin \theta - \sin \theta \cos^2 \theta$. Show this by changing the left side.

Since each term on the right has a factor of $\sin \theta$, we see that we can proceed by writing $\sin^3 \theta$ as $\sin \theta(\sin^2 \theta)$. Then the factor $\sin^2 \theta$ and the $\cos^2 \theta$ on the right suggest the use of Eq. (20.6). Thus, we have

$$\sin^3 \theta = \sin \theta(\sin^2 \theta) = \sin \theta(1 - \cos^2 \theta)$$

$$= \sin \theta - \sin \theta \cos^2 \theta \qquad \text{multiplying}$$

Since we wanted to substitute for $\sin^2 \theta$, we used Eq. (20.6) in the form

$$\sin^2 \theta = 1 - \cos^2 \theta$$

▶

◀ EXAMPLE 7   Prove the identity $\dfrac{\sec^2 y}{\cot y} - \tan^3 y = \tan y$.

Here, we simplify the left side. We can remove $\cot y$ from the denominator, since $\cot y = 1/\tan y$. Also, the presence of $\sec^2 y$ suggests the use of Eq. (20.7). Therefore, we have

$$\frac{\sec^2 y}{\cot y} - \tan^3 y = \frac{\sec^2 y}{\dfrac{1}{\tan y}} - \tan^3 y = \sec^2 y \tan y - \tan^3 y$$

$$= \tan y(\sec^2 y - \tan^2 y) = \tan y(1)$$

$$= \tan y$$

Here, we have used Eq. (20.7) in the form $\sec^2 y - \tan^2 y = 1$.   ▶

◀ EXAMPLE 8   Prove the identity $\dfrac{1 - \sin x}{\sin x \cot x} = \dfrac{\cos x}{1 + \sin x}$.

The combination $1 - \sin x$ also suggests $1 - \sin^2 x$, since multiplying $(1 - \sin x)$ by $(1 + \sin x)$ gives $1 - \sin^2 x$, which can then be replaced by $\cos^2 x$. Thus, changing only the left side, we have

$$\frac{1 - \sin x}{\sin x \cot x} = \frac{(1 - \sin x)(1 + \sin x)}{\sin x \cot x(1 + \sin x)} \quad \text{multiply numerator and denominator by } 1 + \sin x$$

$$= \frac{1 - \sin^2 x}{\sin x\left(\dfrac{\cos x}{\sin x}\right)(1 + \sin x)} = \frac{\cos^2 x}{\cos x(1 + \sin x)} \quad \longleftarrow \text{Eq. (20.6)}$$

cancel sin x

$$= \frac{\cos x}{1 + \sin x} \quad \text{cancel cos x}$$
▶

◀ EXAMPLE 9   Prove the identity $\sec^2 x + \csc^2 x = \sec^2 x \csc^2 x$.

Here, we note the presence of $\sec^2 x$ and $\csc^2 x$ on each side. This suggests the possible use of the square relationships. By replacing the $\sec^2 x$ on the right-hand side by $1 + \tan^2 x$, we can create $\csc^2 x$ plus another term. The left-hand side is the $\csc^2 x$ plus another term, so this procedure should help. Thus, changing only the right side

$$\sec^2 x + \csc^2 x = \sec^2 x \csc^2 x$$

$$= (1 + \tan^2 x)(\csc^2 x) \qquad \text{using Eq. (20.7)}$$

$$= \csc^2 x + \tan^2 x \csc^2 x \qquad \text{multiplying}$$

$$= \csc^2 x + \left(\frac{\sin^2 x}{\cos^2 x}\right)\left(\frac{1}{\sin^2 x}\right) \qquad \text{using Eqs. (20.4) and (20.1)}$$

$$= \csc^2 x + \frac{1}{\cos^2 x} \qquad \text{cancel } \sin^2 x$$

$$= \csc^2 x + \sec^2 x \qquad \text{using Eq. (20.2)}$$

We could have used many other variations of this procedure, and they would have been perfectly valid.   ▶

EXAMPLE 10 Simplify the expression $\dfrac{\csc x}{\tan x + \cot x}$.

We proceed with a simplification such as this in a manner similar to proving an identity, although we do not know just what the result should be. One procedure for this simplification is shown as follows:

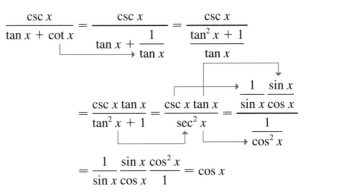

A graphing calculator can be used to check an identity or a simplification. This is done by graphing the function on each side of an identity, or the initial expression and the final expression for simplification. If the two graphs are the same, the identity or simplification is probably shown to be correct, although this is not strictly a proof.

EXAMPLE 11 **(a)** Use a graphing calculator to verify the identity of Example 8. Noting this identity as $\dfrac{1 - \sin x}{\sin x \cot x} = \dfrac{\cos x}{1 + \sin x}$, on a graphing calculator, we let

$$y_1 = (1 - \sin x)/(\sin x/\tan x) \quad \text{(noting that } \cot x = 1/\tan x\text{)}$$
$$y_2 = \cos x/(1 + \sin x)$$

We then graph these two functions as shown in Fig. 20.3. We choose a domain that obviously includes more than one period of each function. Also, after the first curve is plotted, we must watch the screen carefully to see if any new points are plotted for the second curve (use a heavier curve if possible). Since these curves are the same, the identity appears to be verified (although it has not been *proven*).

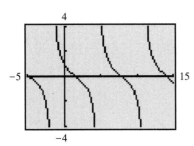

Fig. 20.3

EXERCISES 20.1

In Exercises 1 and 2, make the given changes in the indicated examples of this section and then prove the resulting identities.

1. In Example 3, change the right side to $\frac{\tan x}{\sec x}$ and then prove the resulting identity.

2. In Example 7, change the first term on the left to $\frac{\sin y}{\cos^3 y}$ and then prove the resulting identity.

In Exercises 3–6, use a calculator to check the indicated basic identities for the given angles.

3. Eq. (20.3) for $\theta = 56°$

4. Eq. (20.5) for $\theta = 280°$

5. Eq. (20.6) for $\theta = \dfrac{4\pi}{3}$

6. Eq. (20.7) for $\theta = \dfrac{5\pi}{6}$

In Exercises 7–38, prove the given identities.

7. $\dfrac{\cot \theta}{\cos \theta} = \csc \theta$

8. $\dfrac{\tan y}{\sin y} = \sec y$

9. $\dfrac{\sin x}{\tan x} = \cos x$

10. $\dfrac{\csc \theta}{\sec \theta} = \cot \theta$

11. $\sin y \cot y = \cos y$

12. $\cos x \tan x = \sin x$

13. $\sin x \sec x = \tan x$

14. $\cot \theta \sec \theta = \csc \theta$

15. $\csc^2 x(1 - \cos^2 x) = 1$

16. $\cos^2 x(1 + \tan^2 x) = 1$

17. $\sin x(1 + \cot^2 x) = \csc x$

18. $\sec \theta(1 - \sin^2 \theta) = \cos \theta$

19. $\tan y(\cot y + \tan y) = \sec^2 y$

20. $\csc x(\csc x - \sin x) = \cot^2 x$

21. $\cos \theta \cot \theta + \sin \theta = \csc \theta$

22. $\csc x \sec x - \tan x = \cot x$

23. $\cot \theta \sec^2 \theta - \cot \theta = \tan \theta$

24. $\sin y + \sin y \cot^2 y = \csc y$

25. $\tan x + \cot x = \sec x \csc x$

26. $\tan x + \cot x = \tan x \csc^2 x$

27. $\cos^2 x - \sin^2 x = 1 - 2 \sin^2 x$

28. $\tan^2 y \sec^2 y - \tan^4 y = \tan^2 y$

29. $\dfrac{\sin x}{1 - \cos x} = \csc x + \cot x$

30. $\dfrac{1 + \cos x}{\sin x} = \dfrac{\sin x}{1 - \cos x}$

31. $\dfrac{\sin \theta}{\csc \theta} + \dfrac{\cos \theta}{\sec \theta} = 1$

32. $\dfrac{\sec \theta}{\cos \theta} - \dfrac{\tan \theta}{\cot \theta} = 1$

33. $2 \sin^4 x - 3 \sin^2 x + 1 = \cos^2 x(1 - 2 \sin^2 x)$

34. $\dfrac{\sin^2 \theta + 2 \cos \theta - 1}{\sin^2 \theta + 3 \cos \theta - 3} = \dfrac{1}{1 - \sec \theta}$

35. $\dfrac{1}{2} \sin \pi t \left(\dfrac{\sin \pi t}{1 - \cos \pi t} + \dfrac{1 - \cos \pi t}{\sin \pi t} \right) = 1$

36. $\dfrac{\cot \omega t}{\sec \omega t - \tan \omega t} - \dfrac{\cos \omega t}{\sec \omega t + \tan \omega t} = \sin \omega t + \csc \omega t$

37. $1 + \sin^2 x + \sin^4 x + \cdots = \sec^2 x$

38. $1 - \tan^2 x + \tan^4 x - \cdots = \cos^2 x \quad \left(-\dfrac{\pi}{4} < x < \dfrac{\pi}{4} \right)$

In Exercises 39–46, simplify the given expressions. The result will be one of $\sin x$, $\cos x$, $\tan x$, $\cot x$, $\sec x$, *or* $\csc x$.

39. $\dfrac{\tan x \csc^2 x}{1 + \tan^2 x}$

40. $\dfrac{\cos x - \cos^3 x}{\sin x - \sin^3 x}$

41. $\cot x(\sec x - \cos x)$

42. $\sin x(\tan x + \cot x)$

43. $\dfrac{\tan x + \cot x}{\csc x}$

44. $\dfrac{1 + \tan x}{\sin x} - \sec x$

45. $\dfrac{\cos x + \sin x}{1 + \tan x}$

46. $\dfrac{\sec x - \cos x}{\tan x}$

In Exercises 47–50, use a graphing calculator to verify the given identities by comparing the graphs of each side.

47. $\sin x(\csc x - \sin x) = \cos^2 x$

48. $\cos y(\sec y - \cos y) = \sin^2 y$

49. $\dfrac{\sec x + \csc x}{1 + \tan x} = \csc x$

50. $\dfrac{\cot x + 1}{\cot x} = 1 + \tan x$

In Exercises 51–54, use a graphing calculator to determine whether the given equations are identities.

51. $\sec \theta \tan \theta \csc \theta = \tan^2 \theta + 1$

52. $\sin x \cos x \tan x = \cos^2 x - 1$

53. $\dfrac{2 \cos^2 x - 1}{\sin x \cos x} = \tan x - \cot x$

54. $\cos^3 x \csc^3 x \tan^3 x = \csc^2 x - \cot^2 x$

In Exercises 55–58, solve the given problems involving trigonometric identities.

55. When designing a solar-energy collector, it is necessary to account for the latitude and longitude of the location, the angle of the sun, and the angle of the collector. In doing this, the equation $\cos \theta = \cos A \cos B \cos C + \sin A \sin B$ is used. If $\theta = 90°$, show that $\cos C = -\tan A \tan B$.

56. The path of a point on the circumference of a circle, such as a point on the rim of a bicycle wheel as it rolls along, traces out a curve called a *cycloid*. To find the distance through which the point moves, it is necessary to simplify the expression $(1 - \cos \theta)^2 + \sin^2 \theta$. Perform this simplification.

57. Show that the length l of the straight brace shown in Fig. 20.4 can be found from the equation
$$l = \dfrac{a(1 + \tan \theta)}{\sin \theta}$$

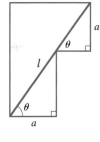

Fig. 20.4

58. In determining the path of least time between two points under certain conditions, it is necessary to show that
$$\sqrt{\dfrac{1 + \cos \theta}{1 - \cos \theta}} \sin \theta = 1 + \cos \theta$$
Show this by transforming the left-hand side.

In Exercises 59–64, solve the given problems.

 59. Explain how to transform $\sin \theta \tan \theta + \cos \theta$ into $\sec \theta$.

(W) 60. Explain how to transform $\tan^2 \theta \cos^2 \theta + \cot^2 \theta \sin^2 \theta$ into 1.

61. Show that $\sin^2 x(1 - \sec^2 x) + \cos^2 x(1 + \sec^4 x)$ has a constant value.

62. Show that $\cot y \csc y \sec y - \csc y \cos y \cot y$ has a constant value.

63. Prove that $\sec^2 \theta + \csc^2 \theta = \sec^2 \theta \csc^2 \theta$ by expressing each function in terms of its x, y, and r definition. See Example 9.

64. Prove that $\dfrac{\csc \theta}{\tan \theta + \cot \theta} = \cos \theta$ by expressing each function in terms of its x, y, and r definition. See Example 10.

In Exercises 65–68, use the given substitutions to show that the given equations are valid. In each, $0 < \theta < \pi/2$.

65. If $x = \cos \theta$, show that $\sqrt{1 - x^2} = \sin \theta$.

66. If $x = 3 \sin \theta$, show that $\sqrt{9 - x^2} = 3 \cos \theta$.

67. If $x = 2 \tan \theta$, show that $\sqrt{4 + x^2} = 2 \sec \theta$.

68. If $x = 4 \sec \theta$, show that $\sqrt{x^2 - 16} = 4 \tan \theta$.

20.2 THE SUM AND DIFFERENCE FORMULAS

For reference, Eq. (12.13) is

$$r_1(\cos\theta_1 + j\sin\theta_1)r_2(\cos\theta_2 + j\sin\theta_2)$$
$$= r_1r_2[\cos(\theta_1 + \theta_2) + j\sin(\theta_1 + \theta_2)]$$

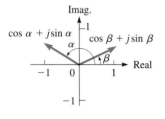

Fig. 20.5

There are other important relations among the trigonometric functions. The most important and useful relations are those that involve twice an angle and half an angle. To obtain these relations, in this section we derive the expressions for the sine and cosine of the sum and difference of two angles. These expressions will lead directly to the desired relations of double and half angles that we will derive in the following section.

Equation (12.13) gives the polar (or trigonometric) form of the product of two complex numbers. We can use this formula to derive the expressions for the sine and cosine of the sum and difference of two angles.

Using Eq. (12.13) to find the product of the complex numbers $\cos\alpha + j\sin\alpha$ and $\cos\beta + j\sin\beta$, which are represented in Fig. 20.5, we have

$$(\cos\alpha + j\sin\alpha)(\cos\beta + j\sin\beta) = \cos(\alpha + \beta) + j\sin(\alpha + \beta)$$

Expanding the left side, and then switching sides, we have

$$\cos(\alpha + \beta) + j\sin(\alpha + \beta) = (\cos\alpha\cos\beta - \sin\alpha\sin\beta) + j(\sin\alpha\cos\beta + \cos\alpha\sin\beta)$$

Since two complex numbers are equal if their real parts are equal and their imaginary parts are equal, we have

$$\sin(\alpha + \beta) = \sin\alpha\cos\beta + \cos\alpha\sin\beta \qquad (20.9)$$

and

$$\cos(\alpha + \beta) = \cos\alpha\cos\beta - \sin\alpha\sin\beta \qquad (20.10)$$

◖ EXAMPLE 1 Verify that $\sin 90° = 1$, by finding $\sin(60° + 30°)$.

$$\sin 90° = \sin(60° + 30°) = \sin 60°\cos 30° + \cos 60°\sin 30° \qquad \text{using Eq. (20.9)}$$

$$= \frac{\sqrt{3}}{2} \times \frac{\sqrt{3}}{2} + \frac{1}{2} \times \frac{1}{2} \qquad \text{for values, see Section 4.3}$$

$$= \frac{3}{4} + \frac{1}{4} = 1$$

It should be obvious from this example that

CAUTION ◗   **$\sin(\alpha + \beta)$ *is* not *equal to* $\sin\alpha + \sin\beta$**

which is something that many students simply assume before they become familiar with the formulas and ideas of this section. If we used such a formula, we would get $\sin 90° = \frac{1}{2}\sqrt{3} + \frac{1}{2} = 1.366$ for the combination $(60° + 30°)$. This is not possible, since the values of the sine never exceed 1 in value. Also, if we used the combination $(45° + 45°)$, we would get 1.414, a different value for the same number, $\sin 90°$. ◗

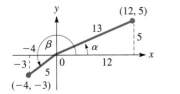

Fig. 20.6

◀ **EXAMPLE 2**   Given that $\sin \alpha = \frac{5}{13}$ (α in the first quadrant) and $\sin \beta = -\frac{3}{5}$ (for β in the third quadrant), find $\cos(\alpha + \beta)$.

Since $\sin \alpha = \frac{5}{13}$ for α in the first quadrant, from Fig. 20.6 we have $\cos \alpha = \frac{12}{13}$. Also, since $\sin \beta = -\frac{3}{5}$ for β in the third quadrant, from Fig. 20.6 we also have $\cos \beta = -\frac{4}{5}$. Then, by using Eq. (20.10), we have

$$\cos(\alpha + \beta) = \cos \alpha \cos \beta - \sin \alpha \sin \beta$$
$$= \frac{12}{13}\left(-\frac{4}{5}\right) - \frac{5}{13}\left(-\frac{3}{5}\right)$$
$$= -\frac{48}{65} + \frac{15}{65} = -\frac{33}{65}$$

From Eqs. (20.9) and (20.10), we can easily find expressions for $\sin(\alpha - \beta)$ and $\cos(\alpha - \beta)$. This is done by finding $\sin[\alpha + (-\beta)]$ and $\cos[\alpha + (-\beta)]$. Thus, we have

$$\sin(\alpha - \beta) = \sin[\alpha + (-\beta)] = \sin \alpha \cos(-\beta) + \cos \alpha \sin(-\beta)$$

Since $\cos(-\beta) = \cos \beta$ and $\sin(-\beta) = -\sin \beta$ (see Eq. (8.7) on page 243), we have

$$\sin(\alpha - \beta) = \sin \alpha \cos \beta - \cos \alpha \sin \beta \qquad \textbf{(20.11)}$$

In the same manner, we find that

$$\cos(\alpha - \beta) = \cos \alpha \cos \beta + \sin \alpha \sin \beta \qquad \textbf{(20.12)}$$

◀ **EXAMPLE 3**   Find $\cos 15°$ from $\cos(45° - 30°)$.

$$\cos 15° = \cos(45° - 30°) = \cos 45° \cos 30° + \sin 45° \sin 30° \qquad \text{using Eq. (20.12)}$$
$$= \frac{\sqrt{2}}{2} \times \frac{\sqrt{3}}{2} + \frac{\sqrt{2}}{2} \times \frac{1}{2} = \frac{\sqrt{6} + \sqrt{2}}{4} \qquad \text{(exact)}$$
$$= 0.9659$$

◀ **EXAMPLE 4**   In analyzing the motion of an object at the end of a spring, the expression $\sin(\omega t + \alpha)\cos \alpha - \cos(\omega t + \alpha)\sin \alpha$ occurs. Simplify this expression.

If we let $x = \omega t + \alpha$, the expression becomes $\sin x \cos \alpha - \cos x \sin \alpha$, which is the form for $\sin(x - \alpha)$. Therefore,

$$\sin(\omega t + \alpha)\cos \alpha - \cos(\omega t + \alpha)\sin \alpha = \sin x \cos \alpha - \cos x \sin \alpha = \sin(x - \alpha)$$
$$= \sin(\omega t + \alpha - \alpha) = \sin \omega t$$

◀ **EXAMPLE 5**   Evaluate $\cos 23° \cos 67° - \sin 23° \sin 67°$.

We note that this expression fits the form of the right side of Eq. (20.10), so

$$\cos 23° \cos 67° - \sin 23° \sin 67° = \cos(23° + 67°)$$
$$= \cos 90°$$
$$= 0$$

NOTE ▶   Again, we are able to evaluate this expression by ***recognizing the form*** of the given expression. Evaluation by a calculator will verify the result.

By dividing the right side of Eq. (20.9) by that of Eq. (20.10), we can determine expressions for $\tan(\alpha + \beta)$, and by dividing the right side of Eq. (20.11) by that of Eq. (20.12), we can determine an expression for $\tan(\alpha - \beta)$. The derivation of these formulas is Exercise 39 of this section. These formulas can be written together, as

$$\tan(\alpha \pm \beta) = \frac{\tan \alpha \pm \tan \beta}{1 \mp \tan \alpha \tan \beta} \qquad (20.13)$$

The formula for $\tan(\alpha + \beta)$ uses the upper signs, and the formula for $\tan(\alpha - \beta)$ uses the lower signs.

Certain trigonometric identities can be proven by the formulas derived in this section. The following examples illustrate this use of these formulas.

◀ EXAMPLE 6 Show that $\tan(\alpha + \beta) \tan(\alpha - \beta) = \dfrac{\tan^2 \alpha - \tan^2 \beta}{1 - \tan^2 \alpha \tan^2 \beta}$.

Using both of Eqs. (20.13), we have

$$\tan(\alpha + \beta) \tan(\alpha - \beta) = \left(\frac{\tan \alpha + \tan \beta}{1 - \tan \alpha \tan \beta} \right) \left(\frac{\tan \alpha - \tan \beta}{1 + \tan \alpha \tan \beta} \right)$$

$$= \frac{\tan^2 \alpha - \tan^2 \beta}{1 - \tan^2 \alpha \tan^2 \beta} \qquad ▶$$

◀ EXAMPLE 7 Prove that $\sin(180° + x) = -\sin x$.

By using Eq. (20.9), we have

$$\sin(180° + x) = \sin 180° \cos x + \cos 180° \sin x$$

Since $\sin 180° = 0$ and $\cos 180° = -1$, we have

$$\sin 180° \cos x + \cos 180° \sin x = (0)\cos x + (-1)\sin x$$

or

$$\sin(180° + x) = -\sin x$$

Although x may or may not be an acute angle, we see that this agrees with the results in Section 8.2 for the sine of the third-quadrant angle if x is acute. See Fig. 20.7.

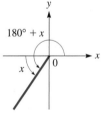

Fig. 20.7

◀ EXAMPLE 8 Show that $\dfrac{\sin(\alpha - \beta)}{\sin \alpha \sin \beta} = \cot \beta - \cot \alpha$.

$$\frac{\sin(\alpha - \beta)}{\sin \alpha \sin \beta} = \frac{\sin \alpha \cos \beta - \cos \alpha \sin \beta}{\sin \alpha \sin \beta} \qquad \text{using Eq. (20.11)}$$

$$= \frac{\sin \alpha \cos \beta}{\sin \alpha \sin \beta} - \frac{\cos \alpha \sin \beta}{\sin \alpha \sin \beta}$$

$$= \frac{\cos \beta}{\sin \beta} - \frac{\cos \alpha}{\sin \alpha}$$

$$= \cot \beta - \cot \alpha \qquad \text{using Eq. (20.5)} \qquad ▶$$

EXAMPLE 9 Show that $\sin\left(\dfrac{\pi}{4} + x\right)\cos\left(\dfrac{\pi}{4} + x\right) = \dfrac{1}{2}(\cos^2 x - \sin^2 x)$. The solution is as follows:

$$\sin\left(\frac{\pi}{4} + x\right)\cos\left(\frac{\pi}{4} + x\right) = \left(\sin\frac{\pi}{4}\cos x + \cos\frac{\pi}{4}\sin x\right)\left(\cos\frac{\pi}{4}\cos x - \sin\frac{\pi}{4}\sin x\right) \quad \text{using Eqs. (20.9) and (20.10)}$$

$$= \sin\frac{\pi}{4}\cos\frac{\pi}{4}\cos^2 x - \sin^2\frac{\pi}{4}\sin x\cos x + \cos^2\frac{\pi}{4}\sin x\cos x - \sin^2 x\sin\frac{\pi}{4}\cos\frac{\pi}{4} \quad \text{expanding}$$

$$= \frac{\sqrt{2}}{2}\frac{\sqrt{2}}{2}\cos^2 x - \left(\frac{\sqrt{2}}{2}\right)^2\sin x\cos x + \left(\frac{\sqrt{2}}{2}\right)^2\sin x\cos x - \frac{\sqrt{2}}{2}\frac{\sqrt{2}}{2}\sin^2 x \quad \text{evaluating}$$

$$= \frac{1}{2}\cos^2 x - \frac{1}{2}\sin^2 x = \frac{1}{2}(\cos^2 x - \sin^2 x)$$

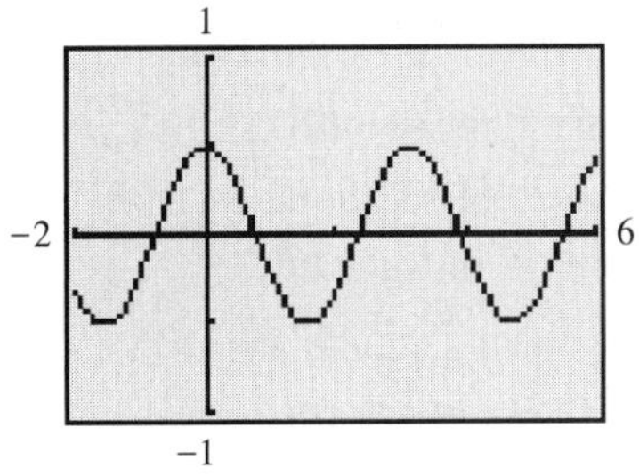

Fig. 20.8

If we check this identity by comparing the graphs of

$$y_1 = \sin(\pi/4 + x)\cos(\pi/4 + x) \quad \text{and} \quad y_2 = (1/2)[(\cos x)^2 - (\sin x)^2]$$

on a graphing calculator, both curves are shown by the graph in Fig. 20.8.

EXAMPLE 10 Alternating electric current is produced essentially by a coil of wire rotating in a magnetic field, and this is the basis for designing generators of alternating current. A three-phase generator uses three coils of wire and thereby produces three electric currents at the same time. This is the most widely used type of *polyphase generator* as mentioned in the chapter introduction on page 524.

The voltages induced in a three-phase generator can be represented as

$$E_1 = E_0 \sin \omega t \qquad E_2 = E_0 \sin\left(\omega t - \frac{2\pi}{3}\right) \qquad E_3 = E_0 \sin\left(\omega t - \frac{4\pi}{3}\right)$$

See the chapter introduction

where E_0 is the maximum voltage and ω is the angular velocity of rotation. Show that the sum of these voltages at any time t is zero.

Setting up the sum $E_1 + E_2 + E_3$ and using Eq. (20.11), we have

$$E_1 + E_2 + E_3 = E_0\left[\sin \omega t + \sin\left(\omega t - \frac{2\pi}{3}\right) + \sin\left(\omega t - \frac{4\pi}{3}\right)\right]$$

$$= E_0\left(\sin \omega t + \sin \omega t\cos\frac{2\pi}{3} - \cos \omega t\sin\frac{2\pi}{3} + \sin \omega t\cos\frac{4\pi}{3} - \cos \omega t\sin\frac{4\pi}{3}\right)$$

$$= E_0\left[\sin \omega t + (\sin \omega t)\left(-\frac{1}{2}\right) - (\cos \omega t)\left(\frac{1}{2}\sqrt{3}\right) + (\sin \omega t)\left(-\frac{1}{2}\right) - (\cos \omega t)\left(-\frac{1}{2}\sqrt{3}\right)\right]$$

$$= E_0\left[(\sin \omega t)\left(1 - \frac{1}{2} - \frac{1}{2}\right) + (\cos \omega t)\left(\frac{1}{2}\sqrt{3} - \frac{1}{2}\sqrt{3}\right)\right] = 0$$

EXERCISES 20.2

In Exercises 1 and 2, make the given changes in the indicated examples of this section and then solve the given problems.

1. In Example 2, change $\frac{5}{13}$ to $\frac{12}{13}$ and then find the value of $\cos(\alpha + \beta)$.

2. In Example 7, change $180° + x$ to $180° - x$ and then determine what other changes result.

In Exercises 3–6, determine the values of the given functions as indicated.

3. Find $\sin 105°$ by using $105° = 60° + 45°$.

4. Find $\tan 75°$ by using $75° = 30° + 45°$.

5. Find $\cos 15°$ by using $15° = 60° - 45°$.

6. Find $\sin 15°$ by using $15° = 45° - 30°$.

In Exercises 7–10, evaluate the given functions with the following information: $\sin \alpha = 4/5$ (α in first quadrant) and $\cos \beta = -12/13$ (β in second quadrant).

7. $\sin(\alpha + \beta)$ **8.** $\tan(\beta - \alpha)$

9. $\cos(\alpha + \beta)$ **10.** $\sin(\alpha - \beta)$

In Exercises 11–18, reduce each of the given expressions to a single term. Expansion of any term is not necessary; proper identification of the form of the expression leads to the proper result.

11. $\sin x \cos 2x + \sin 2x \cos x$

12. $\sin 3x \cos x - \sin x \cos 3x$

13. $\cos(x + y)\cos y + \sin(x + y)\sin y$

14. $\cos(2x - y)\cos y - \sin(2x - y)\sin y$

15. $\dfrac{\tan(x - y) + \tan y}{1 - \tan(x - y)\tan y}$

16. $\sin x \cos(x + 1) + \cos x \sin(x + 1)$

17. $\sin 3x \cos(3x - \pi) - \cos 3x \sin(3x - \pi)$

18. $\cos(x + \pi)\cos(x - \pi) + \sin(x + \pi)\sin(x - \pi)$

In Exercises 19–22, evaluate each of the given expressions. Proper recognition of the given form leads to the result. Verify each result by use of a calculator.

19. $\sin 122° \cos 32° - \cos 122° \sin 32°$

20. $\cos 250° \cos 70° + \sin 250° \sin 70°$

21. $\cos \frac{\pi}{5} \cos \frac{3\pi}{10} - \sin \frac{\pi}{5} \sin \frac{3\pi}{10}$

22. $\dfrac{\tan 18° + \tan 27°}{1 - \tan 18° \tan 27°}$

In Exercises 23–34, prove the given identities.

23. $\sin(180° - x) = \sin x$ **24.** $\cos(3\pi - x) = -\cos x$

25. $\cos(-x) = \cos x$ (*Hint:* $-x = 0 - x$.)

26. $\tan(-x) = -\tan x$

27. $\tan(180° + x) = \tan x$

28. $\sin\left(\frac{\pi}{2} + x\right) = \cos x$

29. $\cos\left(\frac{\pi}{3} + x\right) = \dfrac{\cos x - \sqrt{3} \sin x}{2}$

(W) 30. $\tan(90° + x) = -\cot x$ (Explain why Eq. (20.13) cannot be used for this, but Eqs. (20.9) and (20.10) can be used.)

31. $\sin(x + y)\sin(x - y) = \sin^2 x - \sin^2 y$

32. $\cos(x + y)\cos(x - y) = \cos^2 x - \sin^2 y$

33. $\cos(\alpha + \beta) + \cos(\alpha - \beta) = 2 \cos \alpha \cos \beta$

34. $\cos(x - y) + \sin(x + y) = (\cos x + \sin x)(\cos y + \sin y)$

In Exercises 35–38, verify each identity by comparing the graph of the left side with the graph of the right side on a graphing calculator.

35. $\cos(30° + x) = \dfrac{\sqrt{3} \cos x - \sin x}{2}$

36. $\sin(120° - x) = \dfrac{\sqrt{3} \cos x + \sin x}{2}$

37. $\tan\left(\frac{\pi}{4} + x\right) = \dfrac{1 + \tan x}{1 - \tan x}$ **38.** $\cos\left(\frac{\pi}{2} - x\right) = \sin x$

In Exercises 39–42, derive the given equations in the indicated manner. Equations (20.14), (20.15), and (20.16) are known as the product formulas.

39. By dividing the right side of Eq. (20.9) by that of Eq. (20.10), and dividing the right side of Eq. (20.11) by that of Eq. (20.12), derive Eq. (20.13).

$$\tan(\alpha \pm \beta) = \frac{\tan \alpha \pm \tan \beta}{1 \mp \tan \alpha \tan \beta} \qquad (20.13)$$

(*Hint:* Divide numerator and denominator by $\cos \alpha \cos \beta$.)

40. By adding Eqs. (20.9) and (20.11), derive the equation

$$\sin \alpha \cos \beta = \tfrac{1}{2}[\sin(\alpha + \beta) + \sin(\alpha - \beta)] \qquad (20.14)$$

41. By adding Eqs. (20.10) and (20.12), derive the equation

$$\cos \alpha \cos \beta = \tfrac{1}{2}[\cos(\alpha + \beta) + \cos(\alpha - \beta)] \qquad (20.15)$$

42. By subtracting Eq. (20.10) from Eq. (20.12), derive

$$\sin \alpha \sin \beta = \tfrac{1}{2}[\cos(\alpha - \beta) - \cos(\alpha + \beta)] \qquad (20.16)$$

In Exercises 43–46, additional trigonometric identities are shown. Derive them by letting $\alpha + \beta = x$ and $\alpha - \beta = y$, which leads to $\alpha = \tfrac{1}{2}(x + y)$ and $\beta = \tfrac{1}{2}(x - y)$. The resulting equations are known as the factor formulas.

43. Use Eq. (20.14) and the substitutions above to derive the equation

$$\sin x + \sin y = 2 \sin \tfrac{1}{2}(x + y)\cos \tfrac{1}{2}(x - y) \qquad (20.17)$$

44. Use Eqs. (20.9) and (20.11) and the substitutions above to derive the equation

$$\sin x - \sin y = 2 \sin \tfrac{1}{2}(x - y)\cos \tfrac{1}{2}(x + y) \qquad (20.18)$$

45. Use Eq. (20.15) and the substitutions above to derive the equation

$$\cos x + \cos y = 2 \cos \tfrac{1}{2}(x + y)\cos \tfrac{1}{2}(x - y) \qquad (20.19)$$

46. Use Eq. (20.16) and the substitutions above to derive the equation

$$\cos x - \cos y = -2 \sin \tfrac{1}{2}(x + y)\sin \tfrac{1}{2}(x - y) \qquad (20.20)$$

In Exercises 47–56, use the equations of this section to solve the given problems.

47. Show that $\frac{\sin 2x}{\sin x} = 2 \cos x$. (*Hint:* $\sin 2x = \sin(x + x)$.)

48. Using graphs displayed on a graphing calculator, verify the identity in Exercise 47.

(W) 49. Explain how the exact value of $\sin 75°$ can be found using either Eq. (20.9) or Eq. (20.11).

50. Express $\cos(A + B + C)$ in terms of $\sin A$, $\sin B$, $\sin C$, $\cos A$, $\cos B$, and $\cos C$.

51. The design of a certain three-phase alternating-current generator uses the fact that the sum of the currents $I\cos(\theta + 30°)$, $I\cos(\theta + 150°)$, and $I\cos(\theta + 270°)$ is zero. Show that this is true.

52. The displacements y_1 and y_2 of two waves traveling through the same medium are given by $y_1 = A\sin 2\pi(t/T - x/\lambda)$ and $y_2 = A\sin 2\pi(t/T + x/\lambda)$. Find an expression for the displacement $y_1 + y_2$ of the combination of the waves.

53. An alternating electric current i is given by the equation $i = i_0 \sin(\omega t + \alpha)$. Show that this can be written as $i = i_1 \sin \omega t + i_2 \cos \omega t$, where $i_1 = i_0 \cos \alpha$ and $i_2 = i_0 \sin \alpha$.

54. A weight **w** is held in equilibrium by forces **F** and **T** as shown in Fig. 20.9. Equations relating w, F, and T are

$$F\cos\theta = T\sin\alpha$$

$$w + F\sin\theta = T\cos\alpha$$

Show that $w = \dfrac{T\cos(\theta + \alpha)}{\cos\theta}$.

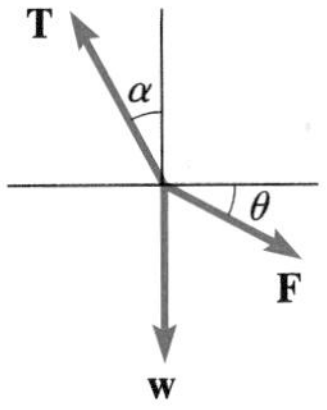

Fig. 20.9

55. For the two bevel gears shown in Fig. 20.10, the equation

$$\tan\alpha = \frac{\sin\beta}{R + \cos\beta} \text{ is used.}$$

Here, R is the ratio of gear 1 to gear 2. Show that

$$R = \frac{\sin(\beta - \alpha)}{\sin\alpha}.$$

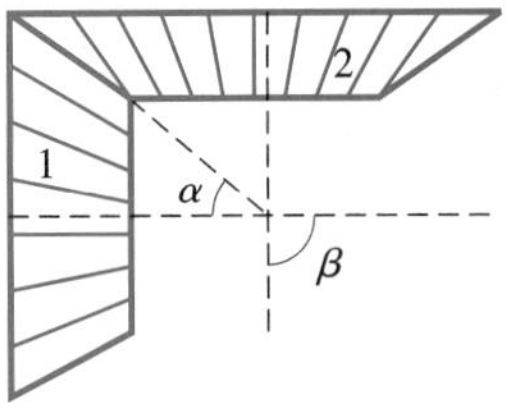

Fig. 20.10

56. In the analysis of the angles of incidence i and reflection r of a light ray subject to certain conditions, the following expression is found:

$$E_2\left(\frac{\tan r}{\tan i} + 1\right) = E_1\left(\frac{\tan r}{\tan i} - 1\right)$$

Show that $E_2 = E_1 \dfrac{\sin(r - i)}{\sin(r + i)}$.

20.3 DOUBLE-ANGLE FORMULAS

If we let $\beta = \alpha$ in Eqs. (20.9) and (20.10), we can derive the important double-angle formulas. By making this substitution in Eq. (20.9), we have

$$\sin(\alpha + \alpha) = \sin(2\alpha) = \sin\alpha\cos\alpha + \cos\alpha\sin\alpha = 2\sin\alpha\cos\alpha$$

Using the same substitution in Eq. (20.10), we have

$$\cos(\alpha + \alpha) = \cos\alpha\cos\alpha - \sin\alpha\sin\alpha = \cos^2\alpha - \sin^2\alpha$$

Again using this substitution in the $\tan(\alpha + \beta)$ form of Eq. (20.13), we have

$$\tan(\alpha + \alpha) = \frac{\tan\alpha + \tan\alpha}{1 - \tan\alpha\tan\alpha} = \frac{2\tan\alpha}{1 - \tan^2\alpha}$$

Then using the basic identity Eq. (20.6), other forms of the equation for $\cos 2\alpha$ may be derived. Summarizing these forms, we have

$$\sin 2\alpha = 2\sin\alpha\cos\alpha \tag{20.21}$$

$$\cos 2\alpha = \cos^2\alpha - \sin^2\alpha \tag{20.22}$$

$$= 2\cos^2\alpha - 1 \tag{20.23}$$

$$= 1 - 2\sin^2\alpha \tag{20.24}$$

$$\tan 2\alpha = \frac{2\tan\alpha}{1 - \tan^2\alpha} \tag{20.25}$$

These double-angle formulas are widely used in applications of trigonometry, especially in calculus. They should be recognized quickly in any of the above forms.

◀ **EXAMPLE 1** **(a)** If $\alpha = 30°$, we have

$$\cos 60° = \cos 2(30°) = \cos^2 30° - \sin^2 30° = \left(\frac{\sqrt{3}}{2}\right)^2 - \left(\frac{1}{2}\right)^2 = \frac{1}{2} \qquad \begin{array}{l}\text{using}\\\text{Eq. (20.22)}\end{array}$$

(b) If $\alpha = 3x$, we have

$$\sin 6x = \sin 2(3x) = 2 \sin 3x \cos 3x \qquad \text{using Eq. (20.21)}$$

(c) If $2\alpha = x$, we may write $\alpha = x/2$, which means that

$$\sin x = \sin 2\left(\frac{x}{2}\right) = 2 \sin \frac{x}{2} \cos \frac{x}{2} \qquad \text{using Eq. (20.21)}$$

(d) If $\alpha = \frac{\pi}{6}$, we have

$$\tan \frac{\pi}{3} = \tan 2\left(\frac{\pi}{6}\right) = \frac{2 \tan \frac{\pi}{6}}{1 - \tan^2\left(\frac{\pi}{6}\right)} = \frac{2\left(\sqrt{3}/3\right)}{1 - \left(\sqrt{3}/3\right)^2} = \sqrt{3} \qquad \text{using Eq. (20.25)} \ \blacktriangleright$$

◀ **EXAMPLE 2** Simplify the expression $\cos^2 2x - \sin^2 2x$.

Since this is the difference of the square of the cosine of an angle and the square of the sine of the same angle, it fits the right side of Eq. (20.22). Therefore, letting $\alpha = 2x$, we have

$$\cos^2 2x - \sin^2 2x = \cos 2(2x) = \cos 4x \qquad\qquad\qquad \blacktriangleright$$

◀ **EXAMPLE 3** To find the area A of a right triangular piece of land, a surveyor may use the formula $A = \frac{1}{4}c^2 \sin 2\theta$, where c is the hypotenuse and θ is *either* of the acute angles. Derive this formula.

In Fig. 20.11, we see that $\sin \theta = a/c$ and $\cos \theta = b/c$, which gives us

$$a = c \sin \theta \quad \text{and} \quad b = c \cos \theta$$

The area is given by $A = \frac{1}{2}ab$, which leads to the solution

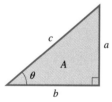

Fig. 20.11

$$A = \frac{1}{2}ab = \frac{1}{2}(c \sin \theta)(c \cos \theta)$$

$$= \frac{1}{2}c^2 \sin \theta \cos \theta = \frac{1}{2}c^2\left(\frac{1}{2} \sin 2\theta\right) \qquad \text{using Eq. (20.21)}$$

$$= \frac{1}{4}c^2 \sin 2\theta$$

In using Eq. (20.21), we divided both sides by 2 to get $\sin \theta \cos \theta = \frac{1}{2} \sin 2\theta$.

If we had labeled the upper acute angle in Fig. 20.11 as θ, we would then have $a = c \cos \theta$ and $b = c \sin \theta$. Substituting these values in the formula for the area gives the same solution. ▶

◀ **EXAMPLE 4** **(a)** Verifying the values of $\sin 90°$, using the functions of $45°$, we have

$$\sin 90° = \sin 2(45°) = 2 \sin 45° \cos 45° = 2\left(\frac{\sqrt{2}}{2}\right)\left(\frac{\sqrt{2}}{2}\right) = 1 \qquad \text{using Eq. (20.21)}$$

(b) Using Eq. (20.25), $\tan 142° = \dfrac{2 \tan 71°}{1 - \tan^2 71°}$. Using a calculator to verify this, we have

$$\tan 142° = -0.781\,285\,626\,5 \quad \text{and} \quad \frac{2 \tan 71°}{1 - \tan^2 71°} = -0.781\,285\,626\,5 \qquad \blacktriangleright$$

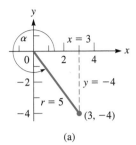

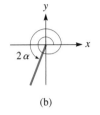

(b)

Fig. 20.12

EXAMPLE 5 Knowing that $\cos \alpha = 3/5$ for an angle in the fourth quadrant, we then see from Fig. 20.12(a) that $\sin \alpha = -4/5$. Therefore, we have

$$\sin 2\alpha = 2 \sin \alpha \cos \alpha \qquad \text{Eq. (20.21)}$$

$$= 2\left(-\frac{4}{5}\right)\left(\frac{3}{5}\right) = -\frac{24}{25}$$

In Fig. 20.12(b), the angle 2α is shown. It is a third-quadrant angle, which verifies the sign of the result. (Since $\cos \alpha = 3/5$, $\alpha \approx 307°$ and $2\alpha = 614°$, which is a third-quadrant angle.)

EXAMPLE 6 Prove the identity $\dfrac{2}{1 + \cos 2x} = \sec^2 x$.

$$\frac{2}{1 + \cos 2x} = \frac{2}{1 + (2 \cos^2 x - 1)} \qquad \text{using Eq. (20.23)}$$

$$= \frac{2}{2 \cos^2 x} = \sec^2 x \qquad \text{using Eq. (20.2)}$$

EXAMPLE 7 Show that $\dfrac{\sin 3x}{\sin x} + \dfrac{\cos 3x}{\cos x} = 4 \cos 2x$.

Since the left side is the more complex side, we change it to the form on the right:

$$\frac{\sin 3x}{\sin x} + \frac{\cos 3x}{\cos x} = \frac{\sin 3x \cos x + \cos 3x \sin x}{\sin x \cos x} \qquad \text{combining fractions}$$

$$= \frac{\sin(3x + x)}{\frac{1}{2} \sin 2x} \qquad \begin{array}{l}\longleftarrow \text{using Eq. (20.9)} \\ \longleftarrow \text{using Eq. (20.21)}\end{array}$$

$$= \frac{2 \sin 4x}{\sin 2x} = \frac{2(2 \sin 2x \cos 2x)}{\sin 2x} \qquad \text{using Eq. (20.21)}$$

$$= 4 \cos 2x$$

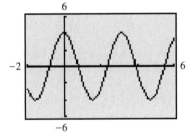

Fig. 20.13

Checking this identity by comparing the graphs of

$$y_1 = (\sin 3x)/(\sin x) + (\cos 3x)/(\cos x) \quad \text{and} \quad y_2 = 4 \cos 2x$$

on a graphing calculator, both curves are shown by the graph in Fig. 20.13.

EXERCISES 20.3

In Exercises 1–4, make the given changes in the indicated examples of this section and then solve the resulting problems.

1. In Example 1(d), change $\frac{\pi}{6}$ to $\frac{\pi}{3}$ and then evaluate $\tan \frac{2\pi}{3}$.

2. In Example 2, change $2x$ to $3x$ and then simplify.

3. In Example 5, change $3/5$ to $4/5$ and then evaluate $\sin 2\alpha$.

4. In Example 6, change the $+$ in the denominator to $-$ and then simplify the expression on the left.

In Exercises 5–8, determine the values of the indicated functions in the given manner.

5. Find $\sin 60°$ by using the functions of $30°$.

6. Find $\sin 120°$ by using the functions of $60°$.

7. Find $\tan 120°$ by using the functions of $60°$.

8. Find $\cos 60°$ by using the functions of $30°$.

In Exercises 9–14, use a calculator to verify the values found by using the double-angle formulas.

9. Find sin 258° directly and by using functions of 129°.

10. Find tan 84° directly and by using functions of 42°.

11. Find cos 96° directly and by using functions of 48°.

12. Find cos 276° directly and by using functions of 138°.

13. Find tan $\frac{2\pi}{5}$ directly and by using functions of $\frac{\pi}{5}$.

14. Find sin(0.2π) directly and by using functions of 0.1π.

In Exercises 15–18, evaluate the indicated functions with the given information.

15. Find sin $2x$ if cos $x = \frac{4}{5}$ (in first quadrant).

16. Find cos $2x$ if sin $x = -\frac{12}{13}$ (in third quadrant).

17. Find tan $2x$ if sin $x = 0.5$ (in second quadrant).

18. Find sin $4x$ if sin $x = 0.6$ (in first quadrant).

In Exercises 19–26, simplify the given expressions. Expansion of any term is not necessary; proper recognition of the form of the expression leads to the proper result.

19. $4 \sin 4x \cos 4x$

20. $4 \sin^2 x \cos^2 x$

21. $1 - 2 \sin^2 4x$

22. $\dfrac{4 \tan 4\theta}{1 - \tan^2 4\theta}$

23. $2 \cos^2 \frac{1}{2}x - 1$

24. $2 \sin \frac{1}{2}x \cos \frac{1}{2}x$

25. $4 \sin^2 2x - 2$

26. $\cos 3x \sin 3x$

In Exercises 27–40, prove the given identities.

27. $\cos^2 \alpha - \sin^2 \alpha = 2 \cos^2 \alpha - 1$

28. $\cos^2 \alpha - \sin^2 \alpha = 1 - 2 \sin^2 \alpha$

29. $\dfrac{\cos x - \tan x \sin x}{\sec x} = \cos 2x$

30. $\cos^4 x - \sin^4 x = \cos 2x$

31. $\dfrac{\sin 4\theta}{\sin 2\theta} = 2 \cos 2\theta$

32. $2 + \dfrac{\cos 2\theta}{\sin^2 \theta} = \csc^2 \theta$

33. $\dfrac{\sin 2\theta}{1 + \cos 2\theta} = \tan \theta$

34. $\dfrac{2 \tan \alpha}{1 + \tan^2 \alpha} = \sin 2\alpha$

35. $1 - \cos 2\theta = \dfrac{2}{1 + \cot^2 \theta}$

36. $\dfrac{\cos^3 \theta + \sin^3 \theta}{\cos \theta + \sin \theta} = 1 - \dfrac{1}{2} \sin 2\theta$

37. $\dfrac{\sin 3x}{\sin x} - \dfrac{\cos 3x}{\cos x} = 2$

38. $\dfrac{\cos 3x}{\sin x} + \dfrac{\sin 3x}{\cos x} = 2 \cot 2x$

39. $\ln(1 - \cos 2x) - \ln(1 + \cos 2x) = 2 \ln \tan x$

40. $\log(20 \sin^2 \theta + 10 \cos 2\theta) = 1$

In Exercises 41–44, verify each identity by comparing the graph of the left side with the graph of the right side on a graphing calculator.

41. $\tan 2\theta = \dfrac{2}{\cot \theta - \tan \theta}$

42. $\dfrac{1 - \tan^2 x}{\sec^2 x} = \cos 2x$

43. $(\sin x + \cos x)^2 = 1 + \sin 2x$

44. $2 \csc 2x \tan x = \sec^2 x$

In Exercises 45–56, solve the given problems.

45. Express sin $3x$ in terms of sin x only.

46. Express cos $3x$ in terms of cos x only.

47. Express cos $4x$ in terms of cos x only.

48. Express sin $4x$ in terms of sin x and cos x.

(W) 49. Without graphing, determine the amplitude and period of the function $y = 4 \sin x \cos x$. Explain.

50. Without graphing, determine the amplitude and period of the function $y = \cos^2 x - \sin^2 x$.

51. The CN Tower in Toronto is 553 m high, and it has an observation deck at the 335 m level. How far from the top of the CN Tower must a 553-m-high helicopter be in order that the angle subtended at the helicopter by the part of the tower above the deck equals the angle subtended by the part of the tower below the deck?

52. The cross section of a radio-wave reflector is defined by $x = \cos 2\theta$, $y = \sin \theta$. Find the relation between x and y by eliminating θ.

53. To find the horizontal range R of a projectile, the equation $R = vt \cos \alpha$ is used, where α is the angle between the line of fire and the horizontal, v is the initial velocity of the projectile, and t is the time of flight. It can be shown that $t = (2v \sin \alpha)/g$, where g is the acceleration due to gravity. Show that $R = (v^2 \sin 2\alpha)/g$. See Fig. 20.14.

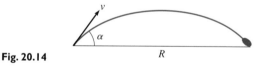

Fig. 20.14

54. In analyzing light reflection from a cylinder onto a flat surface, the expression $3 \cos \theta - \cos 3\theta$ arises. Show that this equals $2 \cos \theta \cos 2\theta + 4 \sin \theta \sin 2\theta$.

55. The instantaneous electric power p in an inductor is given by the equation $p = vi \sin \omega t \sin(\omega t - \pi/2)$. Show that this equation can be written as $p = -\frac{1}{2}vi \sin 2\omega t$.

56. In the study of the stress at a point in a bar, the equation $s = a \cos^2 \theta + b \sin^2 \theta - 2t \sin \theta \cos \theta$ arises. Show that this equation can be written as $s = \frac{1}{2}(a + b) + \frac{1}{2}(a - b)\cos 2\theta - t \sin 2\theta$.

20.4 HALF-ANGLE FORMULAS

If we let $\theta = \alpha/2$ in the identity $\cos 2\theta = 1 - 2 \sin^2 \theta$ and then solve for $\sin(\alpha/2)$,

$$\sin \frac{\alpha}{2} = \pm \sqrt{\frac{1 - \cos \alpha}{2}} \qquad (20.26)$$

Also, with the same substitution in the identity $\cos 2\theta = 2 \cos^2 \theta - 1$, which is then solved for $\cos(\alpha/2)$, we have

$$\cos \frac{\alpha}{2} = \pm \sqrt{\frac{1 + \cos \alpha}{2}} \qquad (20.27)$$

CAUTION ▶ In each of Eqs. (20.26) and (20.27), *the sign chosen depends on the quadrant in which $\frac{\alpha}{2}$ lies.*

◀ EXAMPLE 1   We can find $\sin 15°$ by using the relation

$$\sin 15° = \sqrt{\frac{1 - \cos 30°}{2}} \qquad \text{using Eq. (20.26)}$$

$$= \sqrt{\frac{1 - 0.8660}{2}} = 0.2588$$

Here, the plus sign is used, since $15°$ is in the first quadrant. ▶

◀ EXAMPLE 2   We can find $\cos 165°$ by using the relation

$$\cos 165° = -\sqrt{\frac{1 + \cos 330°}{2}} \qquad \text{using Eq. (20.27)}$$

$$= -\sqrt{\frac{1 + 0.8660}{2}} = -0.9659$$

Here, the minus sign is used, since $165°$ is in the second quadrant, and the cosine of a second-quadrant angle is negative. ▶

◀ EXAMPLE 3   Simplify $\sqrt{\dfrac{1 - \cos 114°}{2}}$ by expressing the result in terms of one-half the given angle. Then, using a calculator, show that the values are equal.

We note that the given expression fits the form of the right side of Eq. (20.26), which means that

$$\sqrt{\frac{1 - \cos 114°}{2}} = \sin \tfrac{1}{2}(114°) = \sin 57°$$

Using a calculator shows that

$$\sqrt{\frac{1 - \cos 114°}{2}} = 0.838\,670\,567\,9 \quad \text{and} \quad \sin 57° = 0.838\,670\,567\,9$$

which verifies the equation for these values. ▶

◀ **EXAMPLE 4** Simplify the expression $\sqrt{\dfrac{9 + 9\cos 6x}{2}}$.

$$\sqrt{\frac{9 + 9\cos 6x}{2}} = \sqrt{\frac{9(1 + \cos 6x)}{2}} = 3\sqrt{\frac{1 + \cos 6x}{2}}$$

$$= 3\cos\tfrac{1}{2}(6x) \qquad \text{using Eq. (20.27) with } \alpha = 6x$$

$$= 3\cos 3x$$

Noting the original expression, we see that $\cos 3x$ cannot be negative. ▶

◀ **EXAMPLE 5** In the kinetic theory of gases, the expression $\sqrt{(1 - \cos\alpha)^2 + \sin^2\alpha}$ is found. Show that this expression equals $2\sin\tfrac{1}{2}\alpha$:

$$\sqrt{(1 - \cos\alpha)^2 + \sin^2\alpha} = \sqrt{1 - 2\cos\alpha + \cos^2\alpha + \sin^2\alpha} \qquad \text{expanding}$$

$$= \sqrt{1 - 2\cos\alpha + 1} \qquad \text{using Eq. (20.26)}$$

$$= \sqrt{2 - 2\cos\alpha}$$

$$= \sqrt{2(1 - \cos\alpha)} \qquad \text{factoring}$$

This last expression is very similar to that for $\sin\tfrac{1}{2}\alpha$, except that no 2 appears in the denominator. Therefore, multiplying the numerator and the denominator under the radical by 2 leads to the solution:

$$\sqrt{2(1 - \cos\alpha)} = \sqrt{\frac{4(1 - \cos\alpha)}{2}} = 2\sqrt{\frac{1 - \cos\alpha}{2}}$$

$$= 2\sin\tfrac{1}{2}\alpha \qquad \text{using Eq. (20.26)}$$

Noting the original expression, we see that $\sin\tfrac{1}{2}\alpha$ cannot be negative. ▶

◀ **EXAMPLE 6** Given that $\tan\alpha = \tfrac{8}{15}$ ($180° < \alpha < 270°$), find $\cos\left(\tfrac{\alpha}{2}\right)$.

Knowing that $\tan\alpha = \tfrac{8}{15}$ for a third-quadrant angle, we determine from Fig. 20.15 that $\cos\alpha = -\tfrac{15}{17}$. This means

$$\cos\frac{\alpha}{2} = -\sqrt{\frac{1 + (-15/17)}{2}} = -\sqrt{\frac{2}{34}} \qquad \text{using Eq. (20.27)}$$

$$= -\frac{1}{17}\sqrt{17} = -0.2425$$

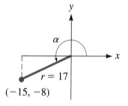

Fig. 20.15

Since $180° < \alpha < 270°$, we know that $90° < \tfrac{\alpha}{2} < 135°$ and therefore $\tfrac{\alpha}{2}$ is in the second quadrant. Since the cosine is negative for second-quadrant angles, we use the negative value of the radical. ▶

◀ **EXAMPLE 7** Show that $2\cos^2\dfrac{x}{2} - \cos x = 1$.

The first step is to substitute for $\cos\tfrac{x}{2}$, which will result in each term on the left being in terms of x and no $\tfrac{x}{2}$ terms will exist. This might allow us to combine terms. So we perform this operation, and we have for the left side

$$2\cos^2\frac{x}{2} - \cos x = 2\left(\frac{1 + \cos x}{2}\right) - \cos x \qquad \text{using Eq. (20.27) with both sides squared}$$

$$= 1 + \cos x - \cos x = 1$$

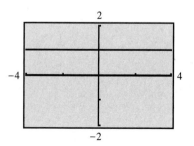

Fig. 20.16

From Fig. 20.16, we verify that the graph of $y_1 = 2[\cos(x/2)]^2 - \cos x$ is the same as the graph of $y_2 = 1$. ▶

◀ EXAMPLE 8 Prove the identity $\sec \dfrac{\alpha}{2} + \csc \dfrac{\alpha}{2} = \dfrac{2\left(\sin \frac{\alpha}{2} + \cos \frac{\alpha}{2}\right)}{\sin \alpha}$.

By expressing $\sin \alpha$ as $2 \sin \dfrac{\alpha}{2} \cos \dfrac{\alpha}{2}$, we have

$$\frac{2\left(\sin \frac{\alpha}{2} + \cos \frac{\alpha}{2}\right)}{\sin \alpha} = \frac{2\left(\sin \frac{\alpha}{2} + \cos \frac{\alpha}{2}\right)}{2 \sin \frac{\alpha}{2} \cos \frac{\alpha}{2}} = \frac{\sin \frac{\alpha}{2}}{\sin \frac{\alpha}{2} \cos \frac{\alpha}{2}} + \frac{\cos \frac{\alpha}{2}}{\sin \frac{\alpha}{2} \cos \frac{\alpha}{2}}$$

See Example 1(c) of Section 20.3

$$= \frac{1}{\cos \frac{\alpha}{2}} + \frac{1}{\sin \frac{\alpha}{2}} = \sec \frac{\alpha}{2} + \csc \frac{\alpha}{2} \qquad \text{using Eqs. (20.2) and (20.1)}$$

◀ EXAMPLE 9 We can find relations for other functions of $\frac{\alpha}{2}$ by expressing these functions in terms of $\sin\left(\frac{\alpha}{2}\right)$ and $\cos\left(\frac{\alpha}{2}\right)$. For example,

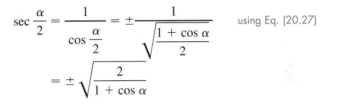

$$\sec \frac{\alpha}{2} = \frac{1}{\cos \dfrac{\alpha}{2}} = \pm \frac{1}{\sqrt{\dfrac{1 + \cos \alpha}{2}}} \qquad \text{using Eq. (20.27)}$$

$$= \pm \sqrt{\frac{2}{1 + \cos \alpha}}$$

EXERCISES 20.4

In Exercises 1 and 2, make the given changes in the indicated examples of this section and then solve the given problem.

1. In Example 3, change the $-$ sign in the numerator to $+$ and then simplify the resulting expression.

2. In Example 6, change $\frac{8}{15}$ $(180° < \alpha < 270°)$ to $-\frac{8}{15}$ $(270° < \alpha < 360°)$ and then solve the resulting problem.

In Exercises 3–8, use the half-angle formulas to evaluate the given functions.

3. $\cos 15°$ **4.** $\sin 22.5°$ **5.** $\sin 75°$

6. $\cos 112.5°$ **7.** $\cos \frac{7\pi}{8}$ **8.** $\sin \frac{11\pi}{12}$

In Exercises 9–12, simplify the given expressions by giving the results in terms of one-half the given angle. Then use a calculator to verify the result.

9. $\sqrt{\dfrac{1 - \cos 236°}{2}}$ **10.** $\sqrt{\dfrac{1 + \cos 98°}{2}}$

11. $\sqrt{1 + \cos 164°}$ **12.** $\sqrt{2 - 2\cos 328°}$

In Exercises 13–18, use the half-angle formulas to simplify the given expressions.

13. $\sqrt{\dfrac{1 - \cos 6x}{2}}$ **14.** $\sqrt{\dfrac{4 + 4\cos 8\beta}{2}}$

15. $\sqrt{8 + 8\cos 4x}$ **16.** $\sqrt{2 - 2\cos 16x}$

17. $\sqrt{4 - 4\cos 10\theta}$ **18.** $\sqrt{18 + 18\cos 1.4x}$

In Exercises 19–22, evaluate the indicated functions with the given information.

19. Find the value of $\sin\left(\frac{\alpha}{2}\right)$ if $\cos \alpha = \frac{12}{13}$ $(0° < \alpha < 90°)$.

20. Find the value of $\cos\left(\frac{\alpha}{2}\right)$ if $\sin \alpha = -\frac{4}{5}$ $(180° < \alpha < 270°)$.

21. Find the value of $\cos\left(\frac{\alpha}{2}\right)$ if $\tan \alpha = -0.2917$ $(90° < \alpha < 180°)$.

22. Find the value of $\sin\left(\frac{\alpha}{2}\right)$ if $\cos \alpha = 0.4706$ $(270° < \alpha < 360°)$.

In Exercises 23–26, derive the required expressions.

23. Derive an expression for $\csc\left(\frac{\alpha}{2}\right)$ in terms of $\cos \alpha$.

24. Derive an expression for $\sec\left(\frac{\alpha}{2}\right)$ in terms of $\sec \alpha$.

25. Derive an expression for $\tan\left(\frac{\alpha}{2}\right)$ in terms of $\sin \alpha$ and $\cos \alpha$.

26. Derive an expression for $\cot\left(\frac{\alpha}{2}\right)$ in terms of $\sin \alpha$ and $\cos \alpha$.

In Exercises 27–32, prove the given identities.

27. $\sin \dfrac{\alpha}{2} = \dfrac{1 - \cos \alpha}{2 \sin \frac{\alpha}{2}}$

28. $2 \cos \dfrac{x}{2} = (1 + \cos x)\sec \dfrac{x}{2}$

29. $2 \sin^2 \frac{x}{2} + \cos x = 1$

30. $2 \cos^2 \frac{\theta}{2} \sec \theta = \sec \theta + 1$

31. $\cos \frac{\theta}{2} = \frac{\sin \theta}{2 \sin \frac{\theta}{2}}$

32. $\cos^2 \frac{x}{2} \left[1 + \left(\frac{\sin x}{1 + \cos x} \right)^2 \right] = 1$

In Exercises 33–36, verify each identity by comparing the graph of the left side with the graph of the right side on a graphing calculator.

33. $2 \sin^2 \frac{\alpha}{2} - \cos^2 \frac{\alpha}{2} = \frac{1 - 3 \cos \alpha}{2}$

34. $\tan \frac{\alpha}{2} = \frac{\sin \alpha}{1 + \cos \alpha}$

35. $\cos^2 \frac{A}{2} - \sin^2 \frac{A}{2} = \frac{\sin 2A}{2 \sin A}$

36. $2 \sin^2 \frac{\theta}{2} = \frac{\sin^2 \theta}{1 + \cos \theta}$

In Exercises 37–44, use the half-angle formulas to solve the given problems.

37. Find $\tan \theta$ if $\sin(\theta/2) = 3/5$.

38. In a right triangle with sides and angles as shown in Fig. 20.17, show that $\sin^2 \frac{A}{2} = \frac{c - b}{2c}$.

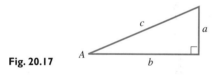

Fig. 20.17

39. In finding the path of a sliding particle, the expression $\sqrt{8 - 8 \cos \theta}$ is used. Simplify this expression.

40. In designing track for a railway system, the equation $d = 4r \sin^2 \frac{A}{2}$ is used. Solve for d in terms of $\cos A$.

41. In electronics, in order to find the *root-mean-square current* in a circuit, it is necessary to express $\sin^2 \omega t$ in terms of $\cos 2\omega t$. Show how this is done.

42. In studying interference patterns of radio signals, the expression $2E^2 - 2E^2 \cos(\pi - \theta)$ arises. Show that this can be written as $4E^2 \cos^2(\theta/2)$.

43. The index of refraction n, the angle A of a prism, and the minimum angle of deflection ϕ are related by

$$n = \frac{\sin \frac{1}{2}(A + \phi)}{\sin \frac{1}{2}A}$$

See Fig. 20.18. Show that an equivalent expression is

$$n = \sqrt{\frac{1 - \cos A \cos \phi + \sin A \sin \phi}{1 - \cos A}}$$

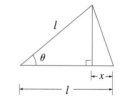

Fig. 20.18

44. For the structure shown in Fig. 20.19, show that $x = 2l \sin^2 \frac{1}{2} \theta$.

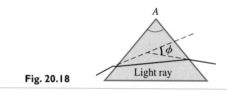

Fig. 20.19

20.5 SOLVING TRIGONOMETRIC EQUATIONS

One of the most important uses of the trigonometric identities is in the solution of equations involving trigonometric functions. The solution of this type of equation consists of the angles that satisfy the equation. When solving for the angle, we generally first solve for a value of a function of the angle and then find the angle from this value of the function.

When equations are written in terms of more than one function, the identities provide a way of changing many of them to equations or factors involving only one function of the same angle. Thus, *the solution is found by using algebraic methods and trigonometric identities and values.* From Chapter 8, recall that we must be careful regarding the sign of the value of a trigonometric function in finding the angle. Figure 20.20 shows again the quadrants in which the functions are positive. Functions not listed are negative.

NOTE ▸

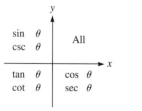

Positive functions

Fig. 20.20

◀ **EXAMPLE 1** Solve the equation $2\cos\theta - 1 = 0$ for all values of θ such that $0 \le \theta \le 2\pi$.

Solving the equation for $\cos\theta$, we obtain $\cos\theta = \frac{1}{2}$. The problem asks for all values of θ from 0 to 2π that satisfy the equation. We know that the cosines of angles in the first and fourth quadrants are positive. Also, we know that $\cos\frac{\pi}{3} = \frac{1}{2}$, which means that $\frac{\pi}{3}$ is the reference angle. Therefore, the solution proceeds as follows:

$$2\cos\theta - 1 = 0$$

$$2\cos\theta = 1 \qquad \text{solve for } \cos\theta$$

$$\cos\theta = \frac{1}{2}$$

$$\theta = \frac{\pi}{3}, \frac{5\pi}{3} \qquad \theta \text{ in quadrants I and IV}$$ ▶

◀ **EXAMPLE 2** Solve the equation $2\cos^2 x - \sin x - 1 = 0$ $(0 \le x < 2\pi)$.

By use of the identity $\sin^2 x + \cos^2 x = 1$, this equation may be put in terms of $\sin x$ only. Thus, we have

$$2(1 - \sin^2 x) - \sin x - 1 = 0 \qquad \text{use identity}$$

$$-2\sin^2 x - \sin x + 1 = 0 \qquad \text{solve for } \sin x$$

$$2\sin^2 x + \sin x - 1 = 0$$

$$(2\sin x - 1)(\sin x + 1) = 0 \qquad \text{factor}$$

Setting each factor equal to zero, we find $\sin x = 1/2$, or $\sin x = -1$. For the domain 0 to 2π, $\sin x = 1/2$ gives $x = \pi/6, 5\pi/6$, and $\sin x = -1$ gives $x = 3\pi/2$. Therefore,

$$x = \frac{\pi}{6}, \frac{5\pi}{6}, \frac{3\pi}{2}$$

These values check when substituted in the original equation. ▶

Graphical Solutions

As with algebraic equations, graphical solutions of trigonometric equations are approximate, whereas algebraic solutions often give exact solutions. As before, *we collect all terms on the left of the equal sign, with zero on the right.* We then *graph the function on the left to find its zeros by finding the values of x where the graph crosses (or is tangent to) the x-axis.*

◀ **EXAMPLE 3** Graphically solve the equation $2\cos^2 x - \sin x - 1 = 0$ such that $0 \le x < 2\pi$ by using a graphing calculator. (This is the same equation as in Example 2.)

Since all the terms of the equation are on the left, with zero on the right, we now set $y = 2\cos^2 x - \sin x - 1$. We then enter this function in the graphing calculator as Y_1, and the graph is displayed in Fig. 20.21. Since angles are expressed in radians and $2\pi = 6.3$, Xmax was chosen as 6.4. Using the *trace* and *zoom* features (or the *zero* feature, or by letting $y_1 = 2\cos^2 x - \sin x - 1$, $y_2 = 0$ and using the *intersect* feature) of a graphing calculator, we find that $y = 0$ for

$$x = 0.52, 2.62, 4.71$$

These values are the same as in Example 2. We note that for $x = 4.71$, the curve *touches* the x-axis but does not cross it. This means it is *tangent* to the x-axis. ▶

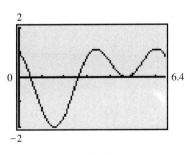

Fig. 20.21

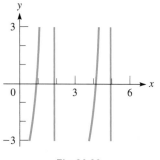

Fig. 20.22

◀ EXAMPLE 4   Solve the equation $\sec^2 x + 2 \tan x - 6 = 0$ $(0 \le x < 2\pi)$.

By use of the identity $1 + \tan^2 x = \sec^2 x$, we may express this equation in terms of $\tan x$ only. Therefore,

$$1 + \tan^2 x + 2 \tan x - 6 = 0 \qquad \text{use identity}$$
$$\tan^2 x + 2 \tan x - 5 = 0$$

Since this is not factorable, we use the quadratic formula to solve for $\tan x$:

$$\tan x = \frac{-2 \pm \sqrt{4 + 20}}{2} = -1 \pm 2.4495$$

Therefore, $\tan x = 1.4495$ and $\tan x = -3.4495$. In radians, we find that $\tan x = 1.4495$ for $x = 0.9669$. Since $\tan x$ is also positive in the third quadrant, we have $x = 4.108$ as well. In the same way, using $\tan x = -3.4495$, we obtain $x = 1.853$ and $x = 4.995$. Therefore, the correction solutions are

$$x = 0.9669, 1.853, 4.108, 4.995$$

These values check in the original equation. Also, they agree with the values of x for which the graph of $y = \sec^2 x + 2 \tan x - 6$ crosses the x-axis in Fig. 20.22 (which can be viewed on a graphing calculator). ▶

◀ EXAMPLE 5   Solve the equation $\cos \frac{x}{2} = 1 + \cos x$ $(0 \le x < 2\pi)$.

By using the half-angle formula for $\cos(x/2)$ and then squaring both sides of the resulting equation, this equation can be solved:

$$\pm \sqrt{\frac{1 + \cos x}{2}} = 1 + \cos x \qquad \text{using identity}$$

$$\frac{1 + \cos x}{2} = 1 + 2 \cos x + \cos^2 x \qquad \text{squaring both sides}$$

$$2 \cos^2 x + 3 \cos x + 1 = 0 \qquad \text{simplifying}$$

$$(2 \cos x + 1)(\cos x + 1) = 0 \qquad \text{factoring}$$

$$\cos x = -\frac{1}{2}, -1$$

$$x = \frac{2\pi}{3}, \frac{4\pi}{3}, \pi$$

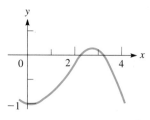

Fig. 20.23

In finding this solution, we squared both sides of the original equation. In doing this, we may have introduced extraneous solutions (see Section 14.3). Thus, we must check each solution in the original equation to see if it is valid. Hence

$$\cos \frac{\pi}{3} \overset{?}{=} 1 + \cos \frac{2\pi}{3} \quad \text{or} \quad \frac{1}{2} \overset{?}{=} 1 + \left(-\frac{1}{2}\right) \quad \text{or} \quad \frac{1}{2} = \frac{1}{2}$$

$$\cos \frac{2\pi}{3} \overset{?}{=} 1 + \cos \frac{4\pi}{3} \quad \text{or} \quad -\frac{1}{2} \overset{?}{=} 1 + \left(-\frac{1}{2}\right) \quad \text{or} \quad -\frac{1}{2} \neq \frac{1}{2}$$

$$\cos \frac{\pi}{2} \overset{?}{=} 1 + \cos \pi \quad \text{or} \quad 0 \overset{?}{=} 1 - 1 \quad \text{or} \quad 0 = 0$$

Thus, the apparent solution $x = \frac{4\pi}{3}$ is not a solution of the original equation. The correct solutions are $x = \frac{2\pi}{3}$ and $x = \pi$.

We can see that these values agree with the values of x for which the graph of $y = \cos(x/2) - 1 - \cos x$ crosses the x-axis in Fig. 20.23. In using a graphing calculator, be sure to enter the function as we have shown it here. ▶

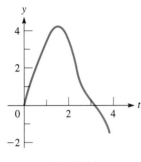

Fig. 20.24

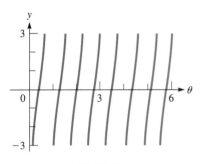

Fig. 20.25

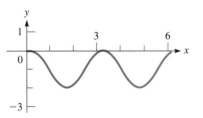

Fig. 20.26

▌ EXAMPLE 6   The vertical displacement y of an object at the end of a spring, which itself is being moved up and down, is given by $y = 3.50 \sin t + 1.20 \sin 2t$. Find the first two values of t (in seconds) for which $y = 0$.

Using the double-angle formula for $\sin 2t$ leads to the solution.

$$3.50 \sin t + 1.20 \sin 2t = 0 \qquad \text{setting } y = 0$$
$$3.50 \sin t + 2.40 \sin t \cos t = 0 \qquad \text{using identities}$$
$$\sin t(3.50 + 2.40 \cos t) = 0 \qquad \text{factoring}$$
$$\sin t = 0 \quad \text{or} \quad \cos t = -1.46$$
$$t = 0.00, 3.14, \dots$$

Since $\cos t$ cannot be numerically larger than 1, there are no values of t for which $\cos t = -1.46$. Thus, the required times are $t = 0.00$ s, 3.14 s.

We can see that these values agree with the values of t for which the graph of $y = 3.50 \sin t + 1.20 \sin 2t$ crosses the t-axis in Fig. 20.24. (In using a graphing calculator, use x for t.) ▮

▌ EXAMPLE 7   Solve the equation $\tan 2\theta - \cot 2\theta = 0$ $(0 \le \theta < 2\pi)$.

$$\tan 2\theta - \frac{1}{\tan 2\theta} = 0 \qquad \text{using } \cot 2\theta = \frac{1}{\tan 2\theta}$$
$$\tan^2 2\theta = 1 \qquad \text{multiplying by } \tan 2\theta \text{ and adding 1 to each side}$$
$$\tan 2\theta = \pm 1 \qquad \text{taking square roots}$$

For $0 \le \theta < 2\pi$, we must have values of 2θ such that $0 \le 2\theta < 4\pi$. Therefore,

$$2\theta = \frac{\pi}{4}, \frac{3\pi}{4}, \frac{5\pi}{4}, \frac{7\pi}{4}, \frac{9\pi}{4}, \frac{11\pi}{4}, \frac{13\pi}{4}, \frac{15\pi}{4}$$

This means that the solutions are

$$\theta = \frac{\pi}{8}, \frac{3\pi}{8}, \frac{5\pi}{8}, \frac{7\pi}{8}, \frac{9\pi}{8}, \frac{11\pi}{8}, \frac{13\pi}{8}, \frac{15\pi}{8}$$

These values satisfy the original equation. Since we multiplied through by $\tan 2\theta$ in the solution, any value of θ that leads to $\tan 2\theta = 0$ would not be valid, since this would indicate division by zero in the original equation.

We see that these solutions agree with the values of θ for which the graph of $y = \tan 2\theta - \cot 2\theta$ crosses the θ-axis in Fig. 20.25. (In using a graphing calculator, use x for θ.) ▮

▌ EXAMPLE 8   Solve the equation $\cos 3x \cos x + \sin 3x \sin x = 1$ $(0 \le x < 2\pi)$.

The left side of this equation is of the general form $\cos(A - x)$, where $A = 3x$. Therefore,

$$\cos 3x \cos x + \sin 3x \sin x = \cos(3x - x) = \cos 2x$$

The original equation becomes

$$\cos 2x = 1$$

This equation is satisfied if $2x = 0$ or $2x = 2\pi$. The solutions are $x = 0$ and $x = \pi$. Only through recognition of the proper trigonometric form can we readily solve this equation.

We see that these solutions agree with the two values of x for which the graph of $y = \cos 3x \cos x + \sin 3x \sin x - 1$ touches the x-axis in Fig. 20.26. ▮

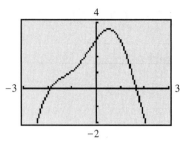

Fig. 20.27

◀ EXAMPLE 9   Solve the equation $\sin 2x + 3 = x^2$.

Although we can substitute $2 \sin x \cos x$ for $\sin 2x$, we cannot express x^2 in terms of a trigonometric function. However, we can find an approximate graphical solution to this equation.

First, write the equation with all terms on the left and zero on the right. This gives us

$$y = \sin 2x + 3 - x^2$$

Then enter this function in a graphing calculator as Y_1, and the graph is displayed in Fig. 20.27. Using the *trace* and *zoom* (or *zero* feature) features, we find that

$$x = -1.90, 1.67$$

These are the only values for which $y = 0$. (The solution can also be found by letting $y_1 = \sin 2x + 3$, $y_2 = x^2$ and using the *intersect* feature of a graphing calculator.) ▶

EXERCISES 20.5

In Exercises 1–4, make the given changes in the indicated examples of this section and then solve the resulting problems.

1. In Example 2, change $2 \cos^2 x$ to $2 \sin^2 x$ and then solve the resulting equation.

2. In Example 4, after $2 \tan x$ change -6 to -4 and then solve the resulting equation.

3. In Example 5, change $\cos \frac{1}{2}x$ to $\sin \frac{1}{2}x$ and on the right of the equal sign change the $+$ to $-$ and then solve the resulting equation.

4. In Example 8, change the $+$ to $-$ and then solve the resulting equation.

In Exercises 5–20, solve the given trigonometric equations analytically (using identities when necessary for exact values when possible) for values of x for $0 \le x < 2\pi$.

5. $\sin x - 1 = 0$ **6.** $2 \cos x + 1 = 0$

7. $\sin\left(x - \frac{\pi}{4}\right) = \cos\left(x - \frac{\pi}{4}\right)$

8. $4 \tan x + 2 = 3(1 + \tan x)$

9. $4 \cos^2 x - 1 = 0$ **10.** $3 \tan^2 x - 1 = 0$

11. $2 \sin^2 x - \sin x = 0$ **12.** $\sin 4x - \sin 2x = 0$

13. $\sin 2x \sin x + \cos x = 0$

14. $\sin x - \sin \dfrac{x}{2} = 0$

15. $2 \cos^2 x - 2 \cos 2x - 1 = 0$

16. $\tan^2 x + 6 = 5 \tan x$

17. $4 \tan x - \sec^2 x = 0$ **18.** $\sin x \sin \frac{1}{2}x = 1 - \cos x$

19. $\sin 2x \cos x - \cos 2x \sin x = 0$

20. $\cos 3x \cos x - \sin 3x \sin x = 0$

In Exercises 21–36, solve the given trigonometric equations analytically and by use of a graphing calculator. Compare results. Use values of x for $0 \le x < 2\pi$.

21. $\tan x + 1 = 0$ **22.** $2 \sin x + 1 = 0$

23. $3 - 4 \cos x = 7 - (2 - \cos x)$

24. $7 \sin x - 2 = 3(2 - \sin x)$

25. $4 \sin^2 x - 3 = 0$ **26.** $|\sin x| = \frac{1}{2}$

27. $\sin 4x - \cos 2x = 0$ **28.** $3 \cos x - 4 \cos^2 x = 0$

29. $2 \sin x = \tan x$ **30.** $\cos 2x + \sin^2 x = 0$

31. $\sin^2 x - 2 \sin x = 1$ **32.** $2 \cos^2 2x + 1 = 3 \cos 2x$

33. $\tan x + 3 \cot x = 4$ **34.** $\tan^2 x + 4 = 2 \sec^2 x$

35. $\sin 2x + \cos 2x = 0$ **36.** $2 \sin 4x + \csc 4x = 3$

In Exercises 37–44, solve the indicated equations analytically.

37. $\sin 3x + \sin x = 0$ (*Hint*: See Eq. (20.17).)

38. $\cos 3x - \cos x = 0$ (*Hint*: See Eq. (20.20).)

39. The acceleration due to gravity g (in m/s^2) varies with latitude, approximately given by $g = 9.7805(1 + 0.0053 \sin^2 \theta)$, where θ is the latitude in degrees. Find θ for $g = 9.8000$.

40. Under certain conditions, the electric current i (in A) in the circuit shown in Fig. 20.28 is
$i = -e^{-100t}(32.0 \sin 624.5t + 0.200 \cos 624.5t)$. For what value of t (in s) is the current first equal to zero?

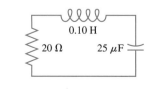

Fig. 20.28

41. To find the angle θ subtended by a certain object on a camera film, it is necessary to solve the equation

$$\frac{p^2 \tan \theta}{0.0063 + p \tan \theta} = 1.6$$

where p is the distance from the camera to the object. Find θ if $p = 4.8$ m.

42. In finding the maximum illuminance from a point source of light, it is necessary to solve the equation $\cos \theta \sin 2\theta - \sin^3 \theta = 0$. Find θ if $0 < \theta < 90°$.

43. The vertical displacement y (in m) of the end of a robot arm is given by $y = 2.30 \cos 0.1t - 1.35 \sin 0.2t$. Find the first four values of t (in s) for which $y = 0$.

44. Looking for a lost ship in the North Atlantic Ocean, a plane flew from Reykjavik, Iceland, 160 km west. It then turned and flew due north and then made a final turn to fly directly back to Reykjavik. If the total distance flown was 480 km, how long were the final two legs of the flight? Solve by setting up and solving an appropriate trigonometric equation. (The Pythagorean theorem may be used only as a check.)

In Exercises 45–52, solve the given equations graphically.

45. $3 \sin x - x = 0$

46. $4 \cos x + 3x = 0$

47. $2 \sin 2x = x^2 + 1$

48. $\sqrt{x} - \sin 3x = 1$

49. $2 \ln x = 1 - \cos 2x$

50. $e^x = 1 + \sin x$

51. In finding the frequencies of vibration of a vibrating wire, the equation $x \tan x = 2.00$ occurs. Find x if $0 < x < \pi/2$.

52. An equation used in astronomy is $\theta - e \sin \theta = M$. Solve for θ for $e = 0.25$ and $M = 0.75$.

20.6 THE INVERSE TRIGONOMETRIC FUNCTIONS

When we studied the exponential and logarithmic functions, we often found it useful to change an expression from one form to the other. We showed that the exponential function $y = b^x$, written in logarithmic form with x as a function of y, is $x = \log_b y$. Then we wrote the logarithmic function as $y = \log_b x$, since it is standard practice to use y as the dependent variable and x as the independent variable.

These two functions, the exponential function $y = b^x$ and the logarithmic function $y = \log_b x$, are called **inverse functions.** *This means that if we solve for the independent variable in terms of the dependent variable in one function, we will arrive at the functional relationship expressed by the other.* It also means that, for every value of x, there is only one corresponding value of y.

Just as we are able to solve $y = b^x$ for the exponent by writing it in logarithmic form, at times it is necessary to solve for the independent variable (the angle) in trigonometric functions. Therefore, we define the **inverse sine function**

$$y = \sin^{-1} x \qquad \left(-\frac{\pi}{2} \le y \le \frac{\pi}{2} \right) \tag{20.28}$$

where y is the angle whose sine is x. This means that x is the value of the sine of the angle y, or $x = \sin y$. (It is necessary to show the range as $-\pi/2 \le y \le \pi/2$, as we will see shortly.)

CAUTION ▶ *In Eq. (20.28), the -1 is **not** an exponent.*

The -1 in $\sin^{-1} x$ is the notation showing the inverse function. We first introduced this notation in Chapter 4 when we were finding the angle with a known value of one of the functions.

The notations Arcsin x, arcsin x, Sin$^{-1} x$ are also used to designate the inverse sine. Some calculators use the $\boxed{\text{inv}}$ key and then the $\boxed{\text{sin}}$ key to find values of the inverse sine. However, since most calculators use the notation $\sin^{-1}$ as a second function on the $\boxed{\text{sin}}$ key, we will continue to use $\sin^{-1} x$ for the inverse sine.

Similar definitions are used for the other inverse trigonometric functions. They also have meanings similar to that of Eq. (20.28).

◀ **EXAMPLE 1** **(a)** $y = \cos^{-1} x$ is read as "y is the angle whose cosine is x." In this case, $x = \cos y$.

(b) $y = \tan^{-1} 2x$ is read as "y is the angle whose tangent is $2x$." In this case, $2x = \tan y$.

(c) $y = \csc^{-1}(1 - x)$ is read as "y is the angle whose cosecant is $1 - x$." In this case, $1 - x = \csc y$, or $x = 1 - \csc y$. ▶

We have seen that $y = \sin^{-1} x$ means that $x = \sin y$. From our previous work with the trigonometric functions, we know that there is an unlimited number of possible values of y for a given value of x in $x = \sin y$. Consider the following example.

◀ **EXAMPLE 2** **(a)** For $x = \sin y$, we know that

$$\sin \frac{\pi}{6} = \frac{1}{2} \quad \text{and} \quad \sin \frac{5\pi}{6} = \frac{1}{2}$$

In fact, $x = \frac{1}{2}$ also for values of y of $-\frac{7\pi}{6}, \frac{13\pi}{6}, \frac{17\pi}{6}$, and so on.

(b) For $x = \cos y$, we know that

$$\cos 0 = 1 \quad \text{and} \quad \cos 2\pi = 1$$

In fact, $\cos y = 1$ for y equal to any even multiple of π. ▶

From Chapter 3, we know that *to have a properly defined **function**, there must be only one value of the dependent variable for a given value of the independent variable.* (A *relation*, on the other hand, may have more than one such value.) Therefore, as in Eq. (20.28), in order to have only one value of y for each value of x in the domain of the inverse trigonometric functions, it is not possible to include all values of y in the range. For this reason, *the range of each of the* **inverse trigonometric functions** *is defined as follows:*

$$-\frac{\pi}{2} \le \sin^{-1} x \le \frac{\pi}{2} \qquad 0 \le \cos^{-1} x \le \pi \qquad -\frac{\pi}{2} < \tan^{-1} x < \frac{\pi}{2} \qquad 0 < \cot^{-1} x < \pi$$

$$0 \le \sec^{-1} x \le \pi \qquad \left(\sec^{-1} x \ne \frac{\pi}{2}\right) \qquad -\frac{\pi}{2} \le \csc^{-1} x \le \frac{\pi}{2} \qquad (\csc^{-1} x \ne 0) \tag{20.29}$$

We must choose a value of y in the range as defined in Eqs. (20.29) that corresponds to a given value of x in the domain. We will discuss the domains and the reasons for these definitions following the next two examples.

◀ **EXAMPLE 3** **(a)** $\sin^{-1}\left(\frac{1}{2}\right) = \frac{\pi}{6}$ first-quadrant angle

This is the only value of the function that lies within the defined range. The value $\frac{5\pi}{6}$ is not correct, even though $\sin\left(\frac{5\pi}{6}\right) = \frac{1}{2}$, since $\frac{5\pi}{6}$ lies outside the defined range.

(b) $\cos^{-1}\left(-\frac{1}{2}\right) = \frac{2\pi}{3}$ second–quadrant angle

Other values such as $\frac{4\pi}{3}$ and $-\frac{2\pi}{3}$ are not correct, since they are not within the defined range for the function $\cos^{-1} x$. ▶

◀ **EXAMPLE 4** $\tan^{-1}(-1) = -\dfrac{\pi}{4}$ *fourth–quadrant angle*

CAUTION ▶

This is the only value within the defined range for the function $\tan^{-1} x$. We must remember that *when x is negative for* $\sin^{-1} x$ *and* $\tan^{-1} x$*, the value of y is a fourth-quadrant angle, expressed as a* **negative angle.** This is a direct result of the definition. (The single exception is $\sin^{-1}(-1) = -\pi/2$, which is a quadrantal angle and is not *in* the fourth quadrant.)

In choosing these values to be the ranges of the inverse trigonometric functions, we first note that the *domain* of $y = \sin^{-1} x$ and $y = \cos^{-1} x$ are each $-1 \le x \le 1$, since the sine and cosine functions take on only these values. Therefore, for each value in this domain, we use only one value of y in the range of the function. Although the domain of $y = \tan^{-1} x$ is all real numbers, we still use only one value of y in the range.

The ranges of the inverse trigonometric functions are chosen so that if x is positive, the resulting value is an angle in the first quadrant. However, care must be taken in choosing the range for negative values of x.

Since the sine of a second-quadrant angle is positive, we cannot choose these angles for $\sin^{-1} x$ for negative values of x. Therefore, we chose fourth-quadrant angles in the form of negative angles in order to have a continuous range of values for $\sin^{-1} x$. The range for $\tan^{-1} x$ is chosen in the same way for similar reasons. However, since the cosine of a fourth-quadrant angle is positive, the range for $\cos^{-1} x$ cannot be the same. To keep a continuous range of values for $\cos^{-1} x$, second-quadrant angles are used.

Values for the other functions are chosen such that the result is also an angle in the first quadrant if x is positive. As for negative values of x, it rarely makes any difference, since either positive values of x arise, or we can use one of the other functions. Our definitions, however, are those that are generally used.

The graphs of the inverse trigonometric functions can be used to show the domains and ranges. We can obtain the graph of the inverse sine function by first sketching the sine curve $x = \sin y$ *along the y-axis.* We then mark the specific part of this curve for which $-\frac{\pi}{2} \le y \le \frac{\pi}{2}$ as the graph of the inverse sine function. The graphs of the other inverse trigonometric functions are found in the same manner. In Figs. 20.29, 20.30, and 20.31, the graphs $x = \sin y$, $x = \cos y$, and $x = \tan y$, respectively, are shown. The heavier, colored portions indicate the graphs of the respective inverse trigonometric functions.

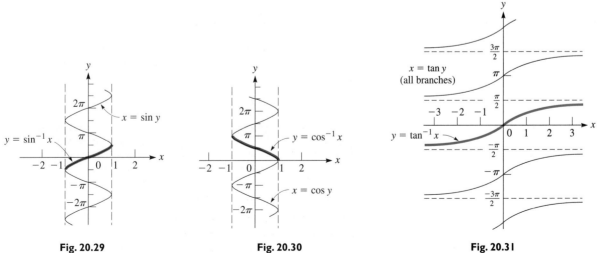

Fig. 20.29   Fig. 20.30   Fig. 20.31

The following examples further illustrate the values and meanings of the inverse trigonometric functions.

▶ **EXAMPLE 5** (a) $\sin^{-1}(-\sqrt{3}/2) = -\pi/3$ (b) $\cos^{-1}(-1) = \pi$
(c) $\tan^{-1} 0 = 0$ (d) $\tan^{-1}(\sqrt{3}) = \pi/3$

Using a calculator in radian mode, we find the following values:

(e) $\sin^{-1} 0.6294 = 0.6808$ (f) $\sin^{-1}(-0.1568) = -0.1574$
(g) $\cos^{-1}(-0.8026) = 2.5024$ (h) $\tan^{-1}(-1.9268) = -1.0921$

We note that the calculator gives values that are in the defined range for each function.

▶

▶ **EXAMPLE 6** Given that $y = \pi - \sec^{-1} 2x$, solve for x.
   We first find the expression for $\sec^{-1} 2x$ and then use the meaning of the inverse secant. The solution follows:

$$y = \pi - \sec^{-1} 2x$$

$$\sec^{-1} 2x = \pi - y \qquad \text{solve for } \sec^{-1} 2x$$

$$2x = \sec(\pi - y) \qquad \text{use meaning of inverse secant}$$

$$x = -\frac{1}{2} \sec y \qquad \sec(\pi - y) = -\sec y$$

As $\sec 2x$ and $2 \sec x$ are different functions, $\sec^{-1} 2x$ and $2 \sec^{-1} x$ are also different functions. Since the values of $\sec^{-1} 2x$ are restricted, so are the resulting values of y.

▶

▶ **EXAMPLE 7** The instantaneous power p in an electric inductor is given by the equation $p = vi \sin \omega t \cos \omega t$. Solve for t.
   Noting the product $\sin \omega t \cos \omega t$ suggests using $\sin 2\alpha = 2 \sin \alpha \cos \alpha$. Then, using the meaning of the inverse sine, we can complete the solution:

$$p = vi \sin \omega t \cos \omega t$$

$$= \frac{1}{2} vi \sin 2\omega t \qquad \text{using double-angle formula}$$

$$\sin 2\omega t = \frac{2p}{vi}$$

$$2\omega t = \sin^{-1}\left(\frac{2p}{vi}\right) \qquad \text{using meaning of inverse sine}$$

$$t = \frac{1}{2\omega} \sin^{-1}\left(\frac{2p}{vi}\right)$$

▶

If we know the value of one of the inverse functions, we can find the trigonometric functions of the angle. If general relations are desired, a representative triangle is very useful. The following examples illustrate these methods.

▶ **EXAMPLE 8** Find $\cos(\sin^{-1} 0.5)$.
   Knowing that the values of inverse trigonometric functions are *angles,* we see that $\sin^{-1} 0.5$ is a first-quadrant angle. Thus, we find $\sin^{-1} 0.5 = \pi/6$. The problem is now to find $\cos(\pi/6)$. This is, of course, $\sqrt{3}/2$, or 0.8660. Thus,

$$\cos(\sin^{-1} 0.5) = \cos(\pi/6) = 0.8660.$$

▶

◀ EXAMPLE 9   **(a)** $\sin(\cot^{-1} 1) = \sin(\pi/4)$ first-quadrant angle

$$= \frac{\sqrt{2}}{2} = 0.7071$$

(b) $\tan[\cos^{-1}(-1)] = \tan \pi$ quadrantal angle
$$= 0$$

(c) $\cos[\sin^{-1}(-0.2395)] = 0.9709$ using a calculator   ▷

◀ EXAMPLE 10   Find $\sin(\tan^{-1} x)$.

We know that $\tan^{-1} x$ is another say of stating "the angle whose tangent is x." Thus, let us draw a right triangle (as in Fig. 20.32) and label one of the acute angles as θ, the side opposite θ as x, and the side adjacent to θ as 1. In this way, we see that, by definition, $\tan \theta = \frac{x}{1}$, or $\theta = \tan^{-1} x$, which means θ is the desired angle. By the Pythagorean theorem, the hypotenuse of this triangle is $\sqrt{x^2 + 1}$. Now we find that $\sin \theta$, which is the same as $\sin(\tan^{-1} x)$, is $x/\sqrt{x^2 + 1}$. Thus,

$$\sin(\tan^{-1} x) = \frac{x}{\sqrt{x^2 + 1}}$$ ▷

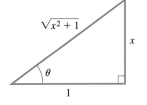

Fig. 20.32

◀ EXAMPLE 11   Find $\cos(2 \sin^{-1} x)$.

From Fig. 20.33, we see that $\theta = \sin^{-1} x$. From the double-angle formulas, we have

$$\cos 2\theta = 1 - 2 \sin^2 \theta$$

Thus, since $\sin \theta = x$, we have

$$\cos(2 \sin^{-1} x) = 1 - 2x^2$$ ▷

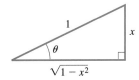

Fig. 20.33

◀ EXAMPLE 12   A triangular brace of sides a, b, and c supports a shelf, as shown in Fig. 20.34. Find the expression for the angle between sides b and c.

The law of cosines leads to the solution:

$$a^2 = b^2 + c^2 - 2bc \cos A$$ law of cosines
$$2bc \cos A = b^2 + c^2 - a^2$$ solving for $\cos A$
$$\cos A = \frac{b^2 + c^2 - a^2}{2bc}$$
$$A = \cos^{-1}\left(\frac{b^2 + c^2 - a^2}{2bc}\right)$$ using meaning of inverse cosine   ▷

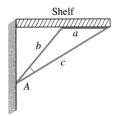

Fig. 20.34

EXERCISES 20.6

In Exercises 1–4, make the given changes in the indicated examples of this section and then solve the resulting problems.

Ⓦ **1.** In Example 1(b), change $2x$ to $3A$ and then write the appropriate statement of meaning.

2. In Example 3(a), change $1/2$ to -1 and then find the appropriate value.

3. In Example 8, change 0.5 to 1 and then find the appropriate value.

4. In Example 10, change sin to cos and then find the appropriate expression.

In Exercises 5–12, write down the meaning of each of the given equations. See Example 1.

5. $y = \tan^{-1} x$

6. $y = \sec^{-1} x$

7. $y = \cot^{-1} 3x$

8. $y = \csc^{-1} 4x$

9. $y = 2 \sin^{-1} x$

10. $y = 3 \tan^{-1} x$

11. $y = 5 \cos^{-1}(2x - 1)$

12. $y = 4 \sin^{-1}(3x + 2)$

In Exercises 13–28, evaluate the given expressions.

13. $\cos^{-1} 0.5$

14. $\sin^{-1} 1$

15. $\tan^{-1} 1$

16. $\cos^{-1} 2$

17. $\tan^{-1}\left(-\sqrt{3}\right)$

18. $\sin^{-1}(-0.5)$

19. $\sec^{-1} 0.5$

20. $\cot^{-1} \sqrt{3}$

21. $\sin^{-1}\left(-\sqrt{2}/2\right)$

22. $\cos^{-1}\left(-\sqrt{3}/2\right)$

23. $\sin\left(\tan^{-1} \sqrt{3}\right)$

24. $\tan\left[\sin^{-1}\left(-\sqrt{2}/2\right)\right]$

25. $\cos[\tan^{-1}(-1)]$

26. $\sec[\cos^{-1}(-0.5)]$

27. $\cos(2 \sin^{-1} 1)$

28. $\sin(2 \tan^{-1} 2)$

In Exercises 29–32, find the exact value of x.

29. $\tan^{-1} x = \sin^{-1} \frac{2}{5}$

30. $\cot^{-1} x = \cos^{-1} \frac{1}{3}$

31. $\sec^{-1} x = -\sin^{-1}\left(-\frac{1}{2}\right)$

32. $\sin^{-1} x = -\tan^{-1}(-1)$

In Exercises 33–40, use a calculator to evaluate the given expressions.

33. $\tan^{-1}(-3.7321)$

34. $\cos^{-1}(-0.6561)$

35. $\sin^{-1} 0.2119$

36. $\tan^{-1} 0.2846$

37. $\tan[\cos^{-1}(-0.6281)]$

38. $\cos[\tan^{-1}(-1.2256)]$

39. $\sin[\tan^{-1}(-0.2297)]$

40. $\tan[\sin^{-1}(-0.3019)]$

In Exercises 41–46, solve the given equations for x.

41. $y = \sin 3x$

42. $y = \cos(x - \pi)$

43. $y = \tan^{-1}(x/4)$

43. $y = 2 \sin^{-1}(x/6)$

45. $1 - y = \cos^{-1}(1 - x)$

46. $2y = \cot^{-1} 3x - 5$

In Exercises 47–54, find an algebraic expression for each of the given expressions.

47. $\tan(\sin^{-1} x)$

48. $\sin(\cos^{-1} x)$

49. $\sin(\sin^{-1} x + \cos^{-1} y)$

50. $\cos(\sin^{-1} x - \cos^{-1} y)$

51. $\sec(\csc^{-1} 3x)$

52. $\tan(\sin^{-1} 2x)$

53. $\sin(2 \sin^{-1} x)$

54. $\cos(2 \tan^{-1} x)$

In Exercises 55–58, solve the given problems with the use of the inverse trigonometric functions.

55. In the analysis of ocean tides, the equation $y = A \cos 2(\omega t + \phi)$ is used. Solve for t.

56. For an object of weight w on an inclined plane that is at an angle θ to the horizontal, the equation relating w and θ is $\mu w \cos \theta = w \sin \theta$, where μ is the coefficient of friction between the surfaces in contact. Solve for θ.

57. The electric current in a certain circuit is given by $i = I_m[\sin(\omega t + \alpha)\cos \phi + \cos(\omega t + \alpha)\sin \phi]$. Solve for t.

58. The time t as a function of the displacement d of a piston is given by $t = \dfrac{1}{2\pi f} \cos^{-1} \dfrac{d}{A}$. Solve for d.

In Exercises 59 and 60, prove that the given expressions are equal. Use the relation for $\sin(\alpha + \beta)$ and show that the sine of the sum of the angles on the left equals the sine of the angle on the right.

59. $\sin^{-1} \dfrac{3}{5} + \sin^{-1} \dfrac{5}{13} = \sin^{-1} \dfrac{56}{65}$

60. $\tan^{-1} \dfrac{1}{3} + \tan^{-1} \dfrac{1}{2} = \dfrac{\pi}{4}$

In Exercises 61–64, evaluate the given expressions.

61. $\sin^{-1} 0.5 + \cos^{-1} 0.5$

62. $\tan^{-1} \sqrt{3} + \cot^{-1} \sqrt{3}$

63. $\sin^{-1} x + \sin^{-1}(-x)$

64. $\sin^{-1} x + \cos^{-1} x$

In Exercises 65–68, solve for the angle A for the given triangles in the given figures in terms of the given sides and angles. For Exercises 65 and 66, explain your method.

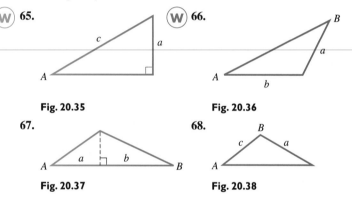

W **65.**

Fig. 20.35

W **66.**

Fig. 20.36

67.

Fig. 20.37

68.

Fig. 20.38

In Exercises 69–72, solve the given problems.

69. The height of the Statue of Liberty is 46.0 m. See Fig. 20.39. From the deck of a boat at a horizontal distance d from the statue, the angles of elevation of the top of the statue and the top of its pedestal are α and β, respectively. Show that

$$\alpha = \tan^{-1}\left(\frac{46.0}{d} + \tan \beta\right).$$

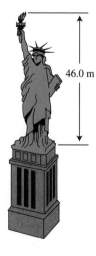

Fig. 20.39

70. Show that the length L of the pulley belt shown in Fig. 20.40 is

$$L = 24 + 11\pi + 10\sin^{-1}\frac{5}{13}.$$

71. If a TV camera is x m from a launch pad of a 50-m rocket that is y m above the ground, find an expression for θ, the angle subtended by the rocket at the camera lens.

(W) **72.** Explain why $\sin^{-1}2x$ is not equal to $2\sin^{-1}x$.

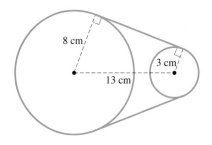

8 cm

3 cm

13 cm

Fig. 20.40

CHAPTER 20 EQUATIONS

Basic trigonometric identities

$$\sin\theta = \frac{1}{\csc\theta} \qquad (20.1)$$

$$\cos\theta = \frac{1}{\sec\theta} \qquad (20.2)$$

$$\tan\theta = \frac{1}{\cot\theta} \qquad (20.3)$$

$$\tan\theta = \frac{\sin\theta}{\cos\theta} \qquad (20.4)$$

$$\cot\theta = \frac{\cos\theta}{\sin\theta} \qquad (20.5)$$

$$\sin^2\theta + \cos^2\theta = 1 \qquad (20.6)$$

$$1 + \tan^2\theta = \sec^2\theta \qquad (20.7)$$

$$1 + \cot^2\theta = \csc^2\theta \qquad (20.8)$$

Sum and difference identities

$$\sin(\alpha + \beta) = \sin\alpha\cos\beta + \cos\alpha\sin\beta \qquad (20.9)$$

$$\cos(\alpha + \beta) = \cos\alpha\cos\beta - \sin\alpha\sin\beta \qquad (20.10)$$

$$\sin(\alpha - \beta) = \sin\alpha\cos\beta - \cos\alpha\sin\beta \qquad (20.11)$$

$$\cos(\alpha - \beta) = \cos\alpha\cos\beta + \sin\alpha\sin\beta \qquad (20.12)$$

$$\tan(\alpha \pm \beta) = \frac{\tan\alpha \pm \tan\beta}{1 \mp \tan\alpha\tan\beta} \qquad (20.13)$$

Double-angle formulas

$$\sin 2\alpha = 2\sin\alpha\cos\alpha \qquad (20.21)$$

$$\cos 2\alpha = \cos^2\alpha - \sin^2\alpha \qquad (20.22)$$

$$= 2\cos^2\alpha - 1 \qquad (20.23)$$

$$= 1 - 2\sin^2\alpha \qquad (20.24)$$

$$\tan 2\alpha = \frac{2\tan\alpha}{1 - \tan^2\alpha} \qquad (20.25)$$

Half-angle formulas

$$\sin\frac{\alpha}{2} = \pm\sqrt{\frac{1 - \cos\alpha}{2}} \qquad (20.26)$$

$$\cos\frac{\alpha}{2} = \pm\sqrt{\frac{1 + \cos\alpha}{2}} \qquad (20.27)$$

Inverse trigonometric functions

$$y = \sin^{-1}x \qquad \left(-\frac{\pi}{2} \le y \le \frac{\pi}{2}\right) \qquad (20.28)$$

$$-\frac{\pi}{2} \le \sin^{-1}x \le \frac{\pi}{2} \qquad 0 \le \cos^{-1}x \le \pi \qquad -\frac{\pi}{2} < \tan^{-1}x < \frac{\pi}{2} \qquad 0 < \cot^{-1}x < \pi$$

$$0 \le \sec^{-1}x \le \pi \qquad \left(\sec^{-1}x \ne \frac{\pi}{2}\right) \qquad -\frac{\pi}{2} \le \csc^{-1}x \le \frac{\pi}{2} \qquad (\csc^{-1}x \ne 0) \qquad (20.29)$$

CHAPTER 20 REVIEW EXERCISES

In Exercises 1–8, determine the values of the indicated functions in the given manner.

1. Find $\sin 120°$ by using $120° = 90° + 30°$.

2. Find $\cos 30°$ by using $30° = 90° - 60°$.

3. Find $\sin 135°$ by using $135° = 180° - 45°$.

4. Find $\tan \frac{5\pi}{4}$ by using $\frac{5\pi}{4} = \pi + \frac{\pi}{4}$.

5. Find $\cos \pi$ by using $\pi = 2\left(\frac{\pi}{2}\right)$.

6. Find $\sin 180°$ by using $180° = 2(90°)$.

7. Find $\tan 60°$ by using $60° = 2(30°)$.

8. Find $\cos 45°$ by using $45° = \frac{1}{2}(90°)$.

In Exercises 9–16, simplify the given expressions by using one of the basic formulas of the chapter. Then use a calculator to verify the result by finding the value of the original expression and the value of the simplified expression.

9. $\sin 14° \cos 38° + \cos 14° \sin 38°$

10. $\cos^2 148° - \sin^2 148°$

11. $2 \sin \frac{\pi}{12} \cos \frac{\pi}{12}$

12. $1 - 2 \sin^2 \frac{\pi}{8}$

13. $\cos 73° \cos 142° + \sin 73° \sin 142°$

14. $\cos 3° \cos 215° - \sin 3° \sin 215°$

15. $\dfrac{4 \tan 12°}{1 - \tan^2 12°}$

16. $\sqrt{\dfrac{1 - \cos 166°}{2}}$

In Exercises 17–24, simplify each of the given expressions. Expansion of any term is not necessary; recognition of the proper form leads to the proper result.

17. $\sin 2x \cos 3x + \cos 2x \sin 3x$

18. $\cos 7x \cos 3x + \sin 7x \sin 3x$

19. $8 \sin 6x \cos 6x$

20. $\dfrac{\tan x + \tan 2x}{1 - \tan x \tan 2x}$

21. $2 - 4 \sin^2 6x$

22. $\cos^2 2x - \sin^2 2x$

23. $\sqrt{2 + 2 \cos 2x}$

24. $\sqrt{32 - 32 \cos 4x}$

In Exercises 25–32, evaluate the given expressions.

25. $\sin^{-1}(-1)$

26. $\sec^{-1} \sqrt{2}$

27. $\cos^{-1} 0.9659$

28. $\tan^{-1}(-0.6249)$

29. $\tan[\sin^{-1}(-0.5)]$

30. $\cos[\tan^{-1}(-\sqrt{3})]$

31. $\sin^{-1}(\tan \pi)$

32. $\cos^{-1}[\tan(-\pi/4)]$

In Exercises 33–44, prove the given identities.

33. $\dfrac{\sec y}{\csc y} = \tan y$

34. $\cos \theta \csc \theta = \cot \theta$

35. $\sin x(\csc x - \sin x) = \cos^2 x$

36. $\cos y(\sec y - \cos y) = \sin^2 y$

37. $\dfrac{\sec^4 x - 1}{\tan^2 x} = 2 + \tan^2 x$

38. $\cos^2 y - \sin^2 y = \dfrac{1 - \tan^2 y}{1 + \tan^2 y}$

39. $2 \csc 2x \cot x = 1 + \cot^2 x$

40. $\sin x \cot^2 x = \csc x - \sin x$

41. $\dfrac{1 - \sin^2 \theta}{1 - \cos^2 \theta} - \cot^2 \theta$

42. $\dfrac{\cos 2\theta}{\cos^2 \theta} = 1 - \tan^2 \theta$

43. $\sin \dfrac{\theta}{2} \cos \dfrac{\theta}{2} = \dfrac{\sin \theta}{2}$

44. $\cos(x - y)\cos y - \sin(x - y)\sin y = \cos x$

In Exercises 45–52, simplify the given expressions. The result will be one of $\sin x$, $\cos x$, $\tan x$, $\cot x$, $\sec x$, or $\csc x$.

45. $\dfrac{\sec x}{\sin x} - \sec x \sin x$

46. $\cos x \cot x + \sin x$

47. $\sin x \tan x + \cos x$

48. $\dfrac{\tan x \csc x}{\sin x} - \cot x$

49. $\dfrac{\sin x \cot x + \cos x}{2 \cot x}$

50. $\dfrac{1 + \cos 2x}{2 \cos x}$

51. $\dfrac{\sin 2x \sec x}{2}$

52. $(\sec x + \tan x)(1 - \sin x)$

In Exercises 53–60, verify each identity by comparing the graph of the left side with the graph of the right side on a graphing calculator.

53. $\dfrac{\cos \theta - \sin \theta}{\cos \theta + \sin \theta} = \dfrac{\cot \theta - 1}{\cot \theta + 1}$

54. $\sin 3y \cos 2y - \cos 3y \sin 2y = \sin y$

55. $\sin 4x(\cos^2 2x - \sin^2 2x) = \dfrac{\sin 8x}{2}$

56. $\csc 2x + \cot 2x = \cot x$

57. $\dfrac{\sin x}{\csc x - \cot x} = 1 + \cos x$

58. $\cos x - \sin \dfrac{x}{2} = \left(1 - 2 \sin \dfrac{x}{2}\right)\left(1 + \sin \dfrac{x}{2}\right)$

59. $\tan \dfrac{\alpha}{2} = \csc \alpha - \cot \alpha$

60. $\sec \dfrac{x}{2} + \csc \dfrac{x}{2} = \dfrac{2\left(\sin \frac{x}{2} + \cos \frac{x}{2}\right)}{\sin x}$

In Exercises 61–64, solve for x.

61. $y = 2 \cos 2x$

62. $y - 2 = 2 \tan\left(x - \dfrac{\pi}{2}\right)$

63. $y = \dfrac{\pi}{4} - 3 \sin^{-1} 5x$

64. $2y = \sec^{-1} 4x - 2$

In Exercises 65–76, solve the given equations for x $(0 \le x < 2\pi)$.

65. $3(\tan x - 2) = 1 + \tan x$

66. $5 \sin x = 3 - (\sin x + 2)$

67. $2(1 - 2 \sin^2 x) = 1$

68. $\sec^2 x = 2 \tan x$

69. $2 \sin^2 \theta + 3 \cos \theta - 3 = 0$

70. $2 \sin 2x + 1 = 0$

71. $\sin x = \sin \dfrac{x}{2}$

71. $\cos 2x = \sin x$

73. $\sin 2x = \cos 3x$

74. $\cos 3x \cos x + \sin 3x \sin x = 0$

75. $\sin^2\left(\dfrac{x}{2}\right) - \cos x + 1 = 0$

76. $\sin x + \cos x = 1$

In Exercises 77–80, solve the given equations graphically.

77. $x + \ln x - 3 \cos^2 x = 2$

78. $e^{\sin x} - 2 = x \cos^2 x$

79. $2 \tan^{-1} x + x^2 = 3$

80. $3 \sin^{-1} x = 6 \sin x + 1$

In Exercises 81–86, find an algebraic expression for each of the given expressions.

81. $\tan(\cot^{-1} x)$

82. $\cos(\csc^{-1} x)$

83. $\sin(2 \cos^{-1} x)$

84. $\cos(\pi - \tan^{-1} x)$

85. $\cos(\sin^{-1} x + \tan^{-1} y)$

86. $\sin(\cos^{-1} x - \tan^{-1} y)$

In Exercises 87–90, use the given substitutions to show that the equations are valid for $0 \le \theta < \pi/2$.

87. If $x = 2 \cos \theta$, show that $\sqrt{4 - x^2} = 2 \sin \theta$.

88. If $x = 2 \sec \theta$, show that $\sqrt{x^2 - 4} = 2 \tan \theta$.

89. If $x = \tan \theta$, show that $\dfrac{x}{\sqrt{1 + x^2}} = \sin \theta$.

90. If $x = \cos \theta$, show that $\dfrac{\sqrt{1 - x^2}}{x} = \tan \theta$.

In Exercises 91–108, use the methods and formulas of this chapter to solve the given problems.

91. Prove: $(\cos \theta + j \sin \theta)^2 = \cos 2\theta + j \sin 2\theta$ $\left(j = \sqrt{-1}\right)$

92. Prove: $(\cos \theta + j \sin \theta)^3 = \cos 3\theta + j \sin 3\theta$ $\left(j = \sqrt{-1}\right)$

93. Solve the inequality $\sin 2x > 2 \sin x$ for $0 \le x < 2\pi$.

94. Show that $(\cos 2\alpha + \sin^2 \alpha)\sec^2 \alpha$ has a constant value.

95. Show that $y = A \sin 2t + B \cos 2t$ may be written as $y = C \sin(2t + \alpha)$, where $C = \sqrt{A^2 + B^2}$ and $\tan \alpha = B/A$. (*Hint:* Let $A/C = \cos \alpha$ and $B/C = \sin \alpha$.)

96. In a right triangle with sides a and b and hypotenuse c, show that $\sin(A/2) = \sqrt{(c - b)/2c}$, where angle A is opposite a.

97. Forces **A** and **B** act on a bolt such that **A** makes an angle θ with the x-axis and **B** makes an angle θ with the y-axis as shown in Fig. 20.41. The resultant **R** has components $R_x = A \cos \theta - B \sin \theta$ and $R_y = A \sin \theta + B \cos \theta$. Using these components, show that $R = \sqrt{A^2 + B^2}$.

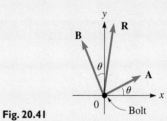

Fig. 20.41

98. For a certain alternating-current generator, the expression $I \cos \theta + I \cos(\theta + 2\pi/3) + I \cos(\theta + 4\pi/3)$ arises. Simplify this expression.

99. Some comets follow a parabolic path that can be described by the equation $r = (k/2)\csc^2(\theta/2)$, where r is the distance to the sun and k is a constant. Show that this equation can be written as $r = k/(1 - \cos \theta)$.

100. In studying the interference of light waves, the identity $\dfrac{\sin \frac{3}{2}x}{\sin \frac{1}{2}x} \sin x = \sin x + \sin 2x$ is used. Prove this identity. $\left(\textit{Hint: } \sin \frac{3}{2}x = \sin\left(x + \frac{1}{2}x\right).\right)$

101. In the study of chemical spectroscopy, the equation $\omega t = \sin^{-1} \dfrac{\theta - \alpha}{R}$ arises. Solve for θ.

102. The power p in a certain electric circuit is given by $p = 2.5[\cos \alpha \sin(\omega t + \phi) - \sin \alpha \cos(\omega t + \phi)]$. Solve for t.

103. In surveying, when determining an azimuth (a measure used for reference purposes), it might be necessary to simplify the expression $(2 \cos \alpha \cos \beta)^{-1} - \tan \alpha \tan \beta$. Perform this operation by expressing it in the simplest possible form when $\alpha = \beta$.

104. In analyzing the motion of an automobile universal joint, the equation $\sec^2 A - \sin^2 B \tan^2 A = \sec^2 C$ is used. Show that this equation is true if $\tan A \cos B = \tan C$.

105. A roof truss is in the shape of an isosceles triangle of height 3.2 m. If the total length of the three members is 25.6 m, what is the length of a rafter? Solve by setting up and solving an appropriate trigonometric equation. (The Pythagorean theorem may be used only as a check.)

106. The angle of elevation of the top of the Eiffel Tower in Paris from a point on level ground 125 m from the center of its base is twice the angle of elevation of the top from a point 325 m farther away. Find the height of the tower.

107. If a plane surface inclined at angle θ moves horizontally, the angle for which the lifting force of the air is a maximum is found by solving the equation $2 \sin \theta \cos^2 \theta - \sin^3 \theta = 0$, where $0 < \theta < 90°$. Solve for θ.

108. To determine the angle between two sections of a certain robot arm, the equation $1.20 \cos \theta + 0.135 \cos 2\theta = 0$ is to be solved. Find the required angle θ if $0° < \theta < 180°$.

Writing Exercise

109. Analyzing the forces on a bridge support, an engineer finds the equation $\tan \theta = 0.4250$. Write a paragraph explaining the difference in solving this equation and evaluating the expression $\tan^{-1} 0.4250$.

CHAPTER 20 PRACTICE TEST

1. Prove that $\sec \theta - \dfrac{\tan \theta}{\csc \theta} = \cos \theta$.

2. Solve for x ($0 \leq x < 2\pi$) analytically, using trigonometric relations where necessary: $\sin 2x + \sin x = 0$.

3. Find an algebraic expression for $\cos(\sin^{-1} x)$.

4. The electric current as a function of the time for a particular circuit is given by $i = 8.00e^{-20t}(1.73 \cos 10.0t - \sin 10.0t)$. Find the time (in s) when the current is first zero.

5. Prove that $\dfrac{\tan \alpha + \tan \beta}{\tan \alpha - \tan \beta} = \dfrac{\sin(\alpha + \beta)}{\sin(\alpha - \beta)}$.

6. Prove that $\cot^2 x - \cos^2 x = \cot^2 x \cos^2 x$.

7. Find $\cos \frac{1}{2}x$ if $\sin x = -\frac{3}{5}$ and $270° < x < 360°$.

8. The intensity of a certain type of polarized light is given by $I = I_0 \sin 2\theta \cos 2\theta$. Solve for θ.

9. Solve graphically: $x - 2 \cos x = 5$.

10. Find the exact value of x: $\cos^{-1} x = -\tan^{-1}(-1)$

21 Plane Analytic Geometry

The development of geometry made little progress from the time of the ancient Greeks until the 1600s. Then in the 1630s, two French mathematicians, René Descartes and Pierre de Fermat, independently introduced algebra into the study of geometry. In doing so, each used algebraic notation and a coordinate system, and this allowed them to analyze the properties of many kinds of curves. Fermat briefly indicated his methods in a letter he wrote in 1636. However, in 1637 Descartes included a much more complete work as an appendix to his *Discourse on Method,* and he is therefore now considered as the founder of *analytic geometry.*

At the time, it was known, for example, that a projectile follows a parabolic path and that the orbits of the planets are ellipses, and there was a renewed interest in curves of various kinds because of their applications. However, Descartes was primarily interested in studying the relationship of algebra to geometry. In doing so, he

In Section 21.4, we show an important application of analytic geometry in the design of a spotlight.

started the development of analytic geometry, which has many applications and also was very important in the invention of calculus that came soon thereafter. In turn, through these advances in mathematics, many more areas of science and technology were able to be developed.

The underlying principle of analytic geometry is the relationship of an algebraic equation and the geometric properties of the curve that represents the equation. In this chapter, we develop equations for a number of important curves and find their properties through an analysis of their equations. Most important among these curves are the *conic sections,* which we briefly introduced in Chapter 14.

As we have noted, analytic geometry has applications in the study of projectile motion and planetary orbits. Other important applications range from the design of gears, airplane wings, and automobile headlights to the construction of bridges and nuclear towers.

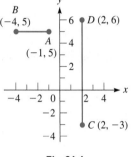

B
$(-4, 5)$
A
$(-1, 5)$
D $(2, 6)$
C $(2, -3)$

Fig. 21.1

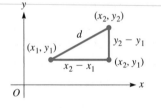

(x_2, y_2)
d
(x_1, y_1)
$y_2 - y_1$
$x_2 - x_1$
(x_2, y_1)

Fig. 21.2

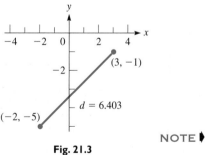

$(3, -1)$
$d = 6.403$
$(-2, -5)$

Fig. 21.3

21.1 BASIC DEFINITIONS

As we noted above, analytic geometry deals with the relationship between an algebraic equation and the geometric curve it represents. In this section, we develop certain basic concepts that will be needed for future use in establishing the proper relationships between an equation and a curve.

THE DISTANCE FORMULA

The first of these concepts involves the distance between any two points in the coordinate plane. If these points lie on a line parallel to the x-axis, the distance from the first point (x_1, y) to the second point (x_2, y) is $|x_2 - x_1|$. The absolute value is used since we are only interested in the magnitude of the distance. Therefore, we could also denote the distance as $|x_1 - x_2|$. Similarly, the distance between two points (x, y_1) and (x, y_2) that lie on a line parallel to the y-axis is $|y_2 - y_1|$ or $|y_1 - y_2|$.

◀ **EXAMPLE 1**   The line segment joining $A(-1, 5)$ and $B(-4, 5)$ in Fig. 21.1 is parallel to the x-axis. Therefore, the distance d between these points is

$$d = |-4 - (-1)| = 3 \quad \text{or} \quad d = |-1 - (-4)| = 3$$

Also in Fig. 21.1, the line segment joining $C(2, -3)$ and $D(2, 6)$ is parallel to the y-axis. The distance d between these points is

$$d = |6 - (-3)| = 9 \quad \text{or} \quad d = |-3 - 6| = 9$$

We now wish to find the length of a line segment joining any two points in the plane. If these points are on a line that is not parallel to either axis (see Fig. 21.2), we use the Pythagorean theorem to find the distance between them. By making a right triangle with the line segment joining the points as the hypotenuse and line segments parallel to the axes as legs, we have *the* **distance formula,** *which gives the distance between any two points in the plane.* This formula is

$$d = \sqrt{(x_2 - x_1)^2 + (y_2 - y_1)^2} \tag{21.1}$$

Here, we choose the positive square root since we are concerned only with the magnitude of the length of the line segment.

◀ **EXAMPLE 2**   The distance between $(3, -1)$ and $(-2, -5)$ is given by

$$d = \sqrt{[(-2) - 3]^2 + [(-5) - (-1)]^2}$$
$$= \sqrt{(-5)^2 + (-4)^2} = \sqrt{25 + 16}$$
$$= \sqrt{41} = 6.403$$

See Fig. 21.3.

NOTE ▶   *It makes no difference which point is chosen as* (x_1, y_1) *and which is chosen as* (x_2, y_2), since the differences in the x-coordinates and the y-coordinates are squared. We obtain the same value for the distance when we calculate it as

$$d = \sqrt{[3 - (-2)]^2 + [(-1) - (-5)]^2}$$
$$= \sqrt{5^2 + 4^2} = \sqrt{41} = 6.403$$

The Slope of a Line

Another important quantity for a line is its *slope,* which we defined in Chapter 5. Here we give it a somewhat more general definition and develop its meaning in more detail, as well as review the basic meaning.

The **slope** *gives a measure of the direction of a line and is defined as the vertical directed distance from one point to another on the same straight line, divided by the horizontal directed distance from the first point to the second.* Thus, the slope, m, is given by

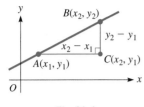

Fig. 21.4

$$m = \frac{y_2 - y_1}{x_2 - x_1} \qquad (21.2)$$

See Fig. 21.4. When the line is horizontal, $y_2 = y_1$ and $m = 0$. When the line is vertical, $x_2 = x_1$ and the slope is undefined.

◀ **Example 3**   The slope of a line joining $(3, -5)$ and $(-2, -6)$ is

$$m = \frac{-6 - (-5)}{-2 - 3} = \frac{-6 + 5}{-5} = \frac{1}{5}$$

See Fig. 21.5. Again we may interpret either of the points as (x_1, y_1) and the other as (x_2, y_2). We can also obtain the slope of this same line from

$$m = \frac{-5 - (-6)}{3 - (-2)} = \frac{1}{5} \qquad ▶$$

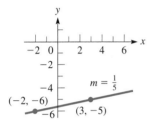

Fig. 21.5

The larger the numerical value of the slope of a line, the more nearly vertical is the line. Also, *a line rising to the right has a positive slope, and a line falling to the right has a negative slope.*

◀ **Example 4**   **(a)** The line in Example 3 has a positive slope, which is numerically small. From Fig. 21.5, it can be seen that the line rises slightly to the right.

(b) The line joining $(3, 4)$ and $(4, -6)$ has a slope of

$$m = \frac{4 - (-6)}{3 - 4} = -10 \quad \text{or} \quad m = \frac{-6 - 4}{4 - 3} = -10$$

This line falls sharply to the right, as shown in Fig. 21.6.   ▶

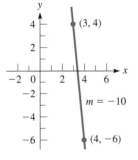

Fig. 21.6

If a given line is extended indefinitely in either direction, it must cross the x-axis at some point unless it is parallel to the x-axis. *The angle measured from the x-axis in a positive direction to the line is called the* **inclination** *of the line* (see Fig. 21.7). The inclination of a line parallel to the x-axis is defined to be zero. *An alternate definition of slope, in terms of the inclination α, is*

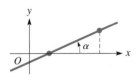

Fig. 21.7

$$m = \tan \alpha \quad (0° \le \alpha < 180°) \qquad (21.3)$$

Since the slope can be defined in terms of any two points on the line, we can choose the x-intercept and any other point. Therefore, from the definition of the tangent of an angle, we see that Eq. (21.3) is in agreement with Eq. (21.2).

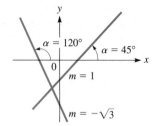

Fig. 21.8

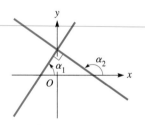

Fig. 21.9

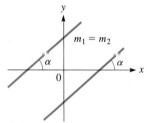

Fig. 21.10

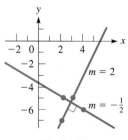

Fig. 21.11

◀ EXAMPLE 5 **(a)** The slope of a line with an inclination of 45° is

$$m = \tan 45° = 1.000$$

(b) If a line has a slope of -1.732, we know that $\tan \alpha = -1.732$. Since $\tan \alpha$ is negative, α must be a second-quadrant angle. Therefore, using a calculator to show that $\tan 60° = 1.732$, we find that $\alpha = 180° - 60° = 120°$. See Fig. 21.8.

We see that if the inclination is an acute angle, the slope is positive and the line rises to the right. If the inclination is obtuse, the slope is negative and the line falls to the right. ◗

Any two parallel lines crossing the x-axis have the same inclination. Therefore, as shown in Fig. 21.9, the *slopes of parallel lines are equal.* This can be stated as

$$\boxed{m_1 = m_2} \qquad \text{(for \| lines)} \qquad \textbf{(21.4)}$$

If two lines are perpendicular, this means that there must be 90° between their inclinations (Fig. 21.10). The relation between their inclinations is

$$\alpha_2 = \alpha_1 + 90°$$

which can be written as

$$90° - \alpha_2 = -\alpha_1$$

If neither line is vertical (the slope of a vertical line is undefined) and we take the tangent in this last relation, we have

$$\tan (90° - \alpha_2) = \tan (-\alpha_1)$$

or

$$\cot \alpha_2 = -\tan \alpha_1$$

since a function of the complement of an angle equals the cofunction of that angle (see page 123) and since $\tan(-\alpha) = -\tan \alpha$ (see page 243). But $\cot \alpha = 1/\tan \alpha$, which means $1/\tan \alpha_2 = -\tan \alpha_1$. Using the inclination definition of slope, we have *as the relation between slopes of perpendicular lines.*

$$\boxed{m_2 = -\frac{1}{m_1} \text{ or } m_1 m_2 = -1} \qquad \text{(for } \perp \text{ lines)} \qquad \textbf{(21.5)}$$

◀ EXAMPLE 6 The line through $(3, -5)$ and $(2, -7)$ has a slope of

$$m_1 = \frac{-5 + 7}{3 - 2} = 2$$

The line through $(4, -6)$ and $(2, -5)$ has a slope of

$$m_2 = \frac{-6 - (-5)}{4 - 2} = -\frac{1}{2}$$

Since the slopes of the two lines are negative reciprocals, we know that the lines are perpendicular. See Fig. 21.11. ◗

See Appendix C for a graphing calculator program SLOPEDIS. It calculates the slope of a line through two points and the distance between the points.

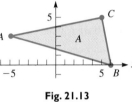

Fig. 21.12

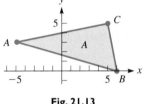

Fig. 21.13

Using the formulas for distance and slope, we can show certain basic geometric relationships. The following examples illustrate the use of the formulas and thereby show the use of algebra in solving problems that are basically geometric. This illustrates the methods of analytic geometry.

EXAMPLE 7 Show that the line segments joining $A(-5,3)$, $B(6,0)$, and $C(5,5)$ form a right triangle. See Fig. 21.12.

If these points are vertices of a right triangle, the slopes of two of the sides must be negative reciprocals. This would show perpendicularity. Thus, we find the slopes of the three lines to be

$$m_{AB} = \frac{3-0}{-5-6} = -\frac{3}{11} \qquad m_{AC} = \frac{3-5}{-5-5} = \frac{1}{5} \qquad m_{BC} = \frac{0-5}{6-5} = -5$$

We see that the slopes of AC and BC are negative reciprocals, which means that $AC \perp BC$. From this we can conclude that the triangle is a right triangle.

EXAMPLE 8 Find the area of the triangle in Example 7. See Fig. 21.13.

Since the right angle is at C, the legs of the triangle are AC and BC. The area is one-half the product of the lengths of the legs of a right triangle. The lengths of the legs are

$$d_{AC} = \sqrt{(-5-5)^2 + (3-5)^2} = \sqrt{104} = 2\sqrt{26}$$
$$d_{BC} = \sqrt{(6-5)^2 + (0-5)^2} = \sqrt{26}$$

Therefore, the area is $A = \frac{1}{2}(2\sqrt{26})(\sqrt{26}) = 26$.

EXERCISES 21.1

In Exercises 1–4, make the given changes in the indicated examples of this section and then solve the resulting problems.

1. In Example 2, change $(-2,-5)$ to $(-2,5)$ and then find the distance.
2. In Example 3, change $(3,-5)$ to $(-3,-5)$ and then find the slope.
3. In Example 5(b), change -1.732 to -0.5774 and then find the inclination.
4. In Example 7, change $B(6,0)$ to $B(-4,-2)$ and then solve the resulting problem.

In Exercises 5–14, find the distance between the given pairs of points.

5. $(3,8)$ and $(-1,-2)$
6. $(-1,3)$ and $(-8,-4)$
7. $(4,-5)$ and $(4,-8)$
8. $(-3,7)$ and $(2,10)$
9. $(-12,20)$ and $(32,-13)$
10. $(23,-9)$ and $(-25,11)$
11. $(\sqrt{32},-\sqrt{18})$ and $(-\sqrt{50},\sqrt{8})$
12. $(e,-\pi)$ and $(-2e,-\pi)$
13. $(1.22,-3.45)$ and $(-1.07,-5.16)$
14. (a,h^2) and $(a+h,(a+h)^2)$

In Exercises 15–24, find the slopes of the lines through the points in Exercises 5–14.

In Exercises 25–28, find the slopes of the lines with the given inclinations.

25. $30°$ 26. $62.5°$ 27. $132.7°$ 28. $135°$

In Exercises 29–32, find the inclinations of the lines with the given slopes.

29. 0.364 30. 0.824 31. -6.691 32. -1.428

In Exercises 33–36, determine whether or not the lines through the two pairs of points are parallel or perpendicular.

33. $(6,-1)$ and $(4,3)$; $(-5,2)$ and $(-7,6)$
34. $(-3,9)$ and $(4,4)$; $(9,-1)$ and $(4,-8)$
35. $(-1,-4)$ and $(2,3)$; $(-5,2)$ and $(-19,8)$
36. $(-a,-2b)$ and $(3a,6b)$; $(2a,-6b)$ and $(5a,0)$

In Exercises 37–40, determine the value of k.

37. The distance between $(-1,3)$ and $(11,k)$ is 13.
38. The distance between $(k,0)$ and $(0,2k)$ is 10.

39. Points $(6, -1)$, $(3, k)$, and $(-3, -7)$ are on the same line.

40. The points in Exercise 39 are the vertices of a right triangle, with the right angle at $(3, k)$.

In Exercises 41–44, show that the given points are vertices of the given geometric figures.

41. $(2, 3)$, $(4, 9)$, and $(-2, 7)$ are vertices of an isosceles triangle.

42. $(-1, 3)$, $(3, 5)$, and $(5, 1)$ are the vertices of a right triangle.

43. $(-5, -4)$, $(7, 1)$, $(10, 5)$, and $(-2, 0)$ are the vertices of a parallelogram.

44. $(-5, 6)$, $(0, 8)$, $(-3, 1)$, and $(2, 3)$ are the vertices of a square.

In Exercises 45–48, find the indicated areas and perimeters.

45. Find the area of the triangle in Exercise 42.

46. Find the area of the square in Exercise 44.

47. Find the perimeter of the triangle in Exercise 41.

48. Find the perimeter of the parallelogram in Exercise 43.

In Exercises 49–52, use the following definition to find the midpoints between the given points on a straight line.

The *midpoint* between points (x_1, y_1) and (x_2, y_2) on a straight line is the point.

$$\left(\frac{x_1 + x_2}{2}, \frac{y_1 + y_2}{2} \right)$$

49. $(-4, 9)$ and $(6, 1)$

50. $(-1, 6)$ and $(-13, -8)$

51. $(-12.4, 25.7)$ and $(6.8, -17.3)$

52. $(2.6, 5.3)$ and $(-4.2, -2.7)$

In Exercises 53–56, solve the given problems.

53. Find the relation between x and y such that (x, y) is always 3 units from the origin.

54. Find the relation between x and y such that (x, y) is always equidistant from the y-axis and $(2, 0)$.

55. Show that the diagonals of a square are perpendicular to each other. (*Hint:* Use $(0, 0)$, $(a, 0)$, and $(0, a)$ as three of the vertices.)

56. The center of a circle is $(2, -3)$, and one end of a diameter is $(-1, 2)$. What are the coordinates of the other end of the diameter?

21.2 THE STRAIGHT LINE

In Chapter 5, we derived the *slope–intercept form* of the equation of the straight line. Here, we extend the development to include other forms of the equation of a straight line. Also, other methods of finding and applying these equations are shown. For completeness, we review some of the material in Chapter 5.

Using the definition of slope, we can derive the general type of equation that represents a straight line. This is another basic method of analytic geometry. That is, equations of a particular form can be shown to represent a particular type of curve. When we recognize the form of the equation, we know the kind of curve it represents. As we have seen, this is of great assistance in sketching the graph.

A straight line can be defined as a *curve* with a constant slope. This means that the value for the slope is the same for any two different points on the line that might be chosen. Thus, considering point (x_1, y_1) on a line to be fixed (Fig. 21.14) and another point $P(x, y)$ that *represents* any other point on the line, we have

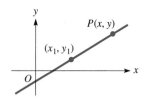

Fig. 21.14

$$m = \frac{y - y_1}{x - x_1}$$

which can be written as

$$\boxed{y - y_1 = m(x - x_1)} \tag{21.6}$$

*Equation (21.6) is the **point-slope form** of the equation of a straight line.* It is useful when we know the slope of a line and some point through which the line passes.

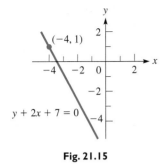

Fig. 21.15

EXAMPLE 1 Find the equation of the line that passes through $(-4, 1)$ with a slope of -2. See Fig. 21.15.

Substituting in Eq. (21.6), we find that

slope

$$y - 1 = (-2)[x - (-4)]$$

coordinates

which can be simplified to $y + 2x + 7 = 0$.

EXAMPLE 2 Find the equation of the line through $(2, -1)$ and $(6, 2)$.

We first find the slope of the line through these points:

$$m = \frac{2 + 1}{6 - 2} = \frac{3}{4}$$

See Appendix C for a graphing calculator program GRAPHLIN. It displays the graph of a line through two given points.

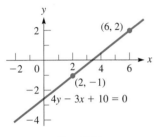

Then, by using either of the two known points and Eq. (21.6), we can find the equation of the line (see Fig. 21.16):

$$y - (-1) = \frac{3}{4}(x - 2)$$

$$4y + 4 = 3x - 6$$

$$4y - 3x + 10 = 0$$

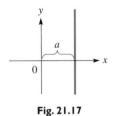

Fig. 21.16

Equation (21.6) can be used for any line except for one parallel to the y-axis. Such a line has an undefined slope. However, it does have the property that all points have the same x-coordinate, regardless of the y-coordinate. *We represent a line parallel to the y-axis (see Fig. 21.17) as*

$$\boxed{x = a} \qquad (21.7)$$

A line parallel to the x-axis has a slope of zero. From Eq. (21.6), we find its equation to be $y = y_1$. To keep the same form as Eq. (21.7), we write this as

Fig. 21.17

$$\boxed{y = b} \qquad (21.8)$$

See Fig. 21.18.

EXAMPLE 3 **(a)** The line $x = 2$ is a line parallel to the y-axis and 2 units to the right of it. This line is shown in Fig. 21.19.

Fig. 21.18

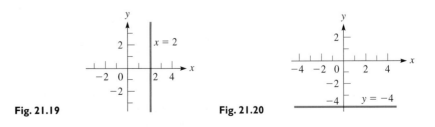

Fig. 21.19   **Fig. 21.20**

(b) The line $y = -4$ is a line parallel to the x-axis and 4 units below it. This line is shown in Fig. 21.20.

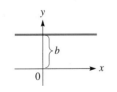

If we choose the special point $(0, b)$, which is the y-intercept of the line, as the point to use in Eq. (21.6), we have $y - b = m(x - 0)$, or

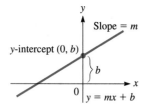

Fig. 21.21

$$y = mx + b \qquad (21.9)$$

Equation (21.9) is the **slope-intercept form** *of the equation of a straight line,* and we first derived it in Chapter 5. Its primary usefulness lies in the fact that once we find the equation of a line and then write it in slope-intercept form, we know that the slope of the line is the coefficient of the x-term and that it crosses the y-axis at the coordinate indicated by the constant term. See Fig. 21.21.

EXAMPLE 4 Find the slope and the y-intercept of the straight line whose equation is $2y + 4x - 5 = 0$.

We write this equation in slope-intercept form:

$$2y = -4x + 5$$

slope ⎤   ⎡ y-coordinate of intercept

$$y = -2x + \frac{5}{2}$$

Since the coefficient of x in this form is -2, the slope is -2. The constant on the right is $5/2$, which means that the y-intercept is $(0, 5/2)$. See Fig. 21.22.

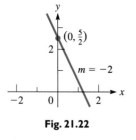

Fig. 21.22

Solving a Word Problem

EXAMPLE 5 The pressure p_0 at the surface of a body of water (due to the atmosphere) is 101 kPa. The pressure p at a depth of 10.0 m is 199 kPa. In general, the pressure difference $p - p_0$ varies directly as the depth h. Sketch a graph of p as a function of h.

The solution is as follows:

$$
\begin{aligned}
p - p_0 &= kh & \text{direct variation} \\
199 - 101 &= k(10.0) & \text{substitute given values} \\
k &= 9.80 \text{ kPa/m} \\
p - 101 &= 9.80h & \text{substitute in first equation} \\
p &= 9.80h + 101
\end{aligned}
$$

We see that this is the equation of a straight line. The slope is 9.80, and the p-intercept is $(0, 101)$. Negative values do not have any physical meaning. The graph is shown in Fig. 21.23.

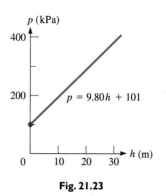

Fig. 21.23

From Eqs. (21.6) and (21.9) and from the examples of this section, we see that the equation of the straight line has certain characteristics: We have a term in y, a term in x, and a constant term if we simplify as much as possible. *This form is represented by the equation*

In Eq. (21.10), A and B cannot both be zero.

$$Ax + By + C = 0 \qquad (21.10)$$

which is known as the **general form** *of the equation of the straight line.* We saw this form before in Chapter 5. Now we have shown why it represents a straight line.

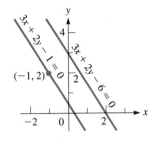

Fig. 21.24

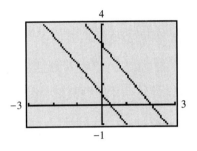

Fig. 21.25

◀ EXAMPLE 6   Find the general form of the equation of the line parallel to the line $3x + 2y - 6 = 0$ and that passes through the point $(-1, 2)$.

Since the line whose equation we want is parallel to the line $3x + 2y - 6 = 0$, it has the same slope. Thus, writing $3x + 2y - 6 = 0$ in slope-intercept form,

$$2y = -3x + 6 \qquad \text{solving for } y$$

$$y = -\frac{3}{2}x + 3$$

Since the slope of $3x + 2y - 6 = 0$ is $-3/2$, the slope of the required line is also $-3/2$. Using $m = -3/2$, the point $(-1, 2)$, and the point-slope form, we have

$$y - 2 = -\frac{3}{2}(x + 1)$$

$$2y - 4 = -3(x + 1)$$

$$3x + 2y - 1 = 0$$

This is the general form of the equation. Both lines are shown in Fig. 21.24.

With $y_1 = -3x/2 + 3$ and $y_2 = -3x/2 + 1/2$, these lines are shown in the calculator display in Fig. 21.25.   ▶

In many physical situations, a linear relationship exists between variables. A few examples of this are (1) the distance traveled by an object and the elapsed time, when the velocity is constant, (2) the amount a spring stretches and the force applied, (3) the change in electric resistance and the change in temperature, (4) the force applied to an object and the resulting acceleration, and (5) the pressure at a certain point within a liquid and the depth of the point.

Solving a Word Problem

◀ EXAMPLE 7   For a period of 6.0 s, the velocity v of a rocket varies linearly with the elapsed time t. If $v = 40$ m/s when $t = 1.0$ s and $v = 55$ m/s when $t = 4.0$ s, find the equation relating v and t and graph the function. From the graph, find the initial velocity and the velocity after 6.0 s. What is the meaning of the slope of the line?

With v as the dependent variable and t as the independent variable, the slope is

$$m = \frac{v_2 - v_1}{t_2 - t_1}$$

Using the information given in the statement of the problem, we have

$$m = \frac{55 - 40}{4.0 - 1.0} = 5.0$$

Then, using the point-slope form of the equation of a straight line, we have

$$v - 40 = 5.0(t - 1.0)$$

$$v = 5.0t + 35$$

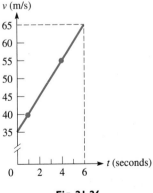

Fig. 21.26

The given values are sufficient to graph the line in Fig. 21.26. There is no need to include negative values of t, since they have no physical meaning. We see that the line crosses the v-axis at 35. This means that the initial velocity (for $t = 0$) is 35 m/s. Also, when $t = 6.0$ s, we see that $v = 65$ m/s.

The slope is the ratio of the change in velocity to the change in time. This is the rocket's *acceleration*. Here the speed of the car increases 5.0 m/s each second. We can express this acceleration as 5.0 (m/s)/s = 5.0 m/s².   ▶

EXERCISES **21.2**

In Exercises 1–4, make the given changes in the indicated examples of this section and then solve the resulting problems.

1. In Example 1, change $(-4, 1)$ to $(4, -1)$ and then find the equation of the line.

2. In Example 2, change $(2, -1)$ to $(-2, 1)$ and then find the equation of the line.

3. In Example 4, change the $+$ before $4x$ to $-$ and then find the slope.

4. In Example 6, change the $+$ before $2y$ to $-$ and then find the general form of the equation.

In Exercises 5–20, find the equation of each of the lines with the given properties. Sketch the graph of each line.

5. Passes through $(-3, 8)$ with a slope of 4.

6. Passes through $(-2, -1)$ with a slope of -2.

7. Passes through $(2, -5)$ and $(4, 2)$.

8. Has an x-intercept $(4, 0)$ and a y-intercept of $(0, -6)$.

9. Passes through $(1, 3)$ and has an inclination of $45°$.

10. Has a y-intercept $(0, -2)$ and an inclination of $120°$.

11. Passes through $(5.3, -2.7)$ and is parallel to the x-axis.

12. Passes through $(-4, -2)$ and is perpendicular to the x-axis.

13. Is parallel to the y-axis and is 3 units to the left of it.

14. Is parallel to the x-axis and is 4.1 units below it.

15. Is perpendicular to a line with a slope of 3 and passes through $(1, -2)$.

16. Is parallel to a line through $(-1, 2)$ and $(3, 1)$ and passes through $(1, 2)$.

17. Is parallel to a line through $(7, -1)$ and $(4, 3)$ and has a y-intercept of $(0, -2)$.

18. Is perpendicular to the line $6.0x - 2.4y - 3.9 = 0$ and passes through $(7.5, -4.7)$.

19. Has a slope of -3 and passes through the intersection of the lines $5x - y = 6$ and $x + y = 12$.

20. Passes through the point of intersection of $2x + y - 3 = 0$ and $x - y - 3 = 0$ and through the point $(4, -3)$.

In Exercises 21–28, reduce the equations to slope-intercept form and find the slope and the y-intercept. Sketch each line.

21. $4x - y = 8$
22. $2x - 3y - 6 = 0$
23. $3x + 5y - 10 = 0$
24. $4y = 6x - 9$
25. $3x - 2y - 1 = 0$
26. $4x + 2y - 5 = 0$
27. $11.2x + 1.6 = 3.2y$
28. $11.5x + 4.60y = 5.98$

In Exercises 29–36, determine whether the given lines are parallel, perpendicular, or neither.

29. $3x - 2y + 5 = 0$ and $4y = 6x - 1$

30. $8x - 4y + 1 = 0$ and $4x + 2y - 3 = 0$

31. $6x - 3y - 2 = 0$ and $x + 2y - 4 = 0$

32. $3y - 2x = 4$ and $6x - 9y = 5$

33. $5x + 2y - 3 = 0$ and $10y = 7 - 4x$

34. $48y - 36x = 71$ and $52x = 17 - 39y$

35. $4.5x - 1.8y = 1.7$ and $2.4x + 6.0y = 0.3$

36. $3.5y = 4.3 - 1.5x$ and $3.6x + 8.4y = 1.7$

In Exercises 37–56, solve the given problems. Exercises 45–56 show some applications of straight lines.

37. Find k if the lines $4x - ky = 6$ and $6x + 3y + 2 = 0$ are parallel.

38. Find k if the lines given in Exercise 37 are perpendicular.

(W) 39. Find k if the lines $3x - y = 9$ and $kx + 3y = 5$ are perpendicular. Explain how this value is found.

(W) 40. Find k such that the line through $(k, 2)$ and $(3, 1 - k)$ is perpendicular to the line $x - 2y = 5$. Explain your method.

41. Find the slope of the line joining points on the graph of $y = x^2$ that have x-coordinates of $-a$ and b $(a > 0, b > 0)$.

42. Show that the *intercept form* $\frac{x}{a} + \frac{y}{b} = 1$ is the equation of a line with x-intercept $(a, 0)$ and y-intercept $(0, b)$.

43. Find the distance from $(4, 1)$ to the line $4x - 3y + 12 = 0$.

44. Find the acute angle between the lines $x + y = 3$ and $2x - 5y = 4$.

45. The velocity v of a box sliding down a long ramp is given by $v = v_0 + at$, where v_0 is the initial velocity, a is the acceleration, and t is the time. If $v_0 = 3.35$ m/s and $v = 9.87$ m/s when $t = 4.50$ s, find v as a function of t. Sketch the graph.

46. The voltage V across part of an electric circuit is given by $V = E - iR$, where E is a battery voltage, i is the current, and R is the resistance. If $E = 6.00$ V and $V = 4.35$ V for $i = 9.17$ mA, find V as a function of i. Sketch the graph (i and V may be negative).

47. The velocity of sound v increases 0.607 m/s for each increase in temperature T of $1.00°C$. If $v = 343$ m/s for $T = 20.0°C$, express v as a function of T.

48. An acid solution is made from x L of a 20% solution and y L of a 30% solution. If the final solution contains 20 L of acid, find the equation relating x and y.

49. The power output P (in W) of a computer chip operating at $120°C$ is proportional to $120 - T_S$, where T_S is the temperature of the surroundings. If $P = 1.0$ W for $T_S = 80°C$, find the equation relating P and T_S.

50. An oil-storage tank is emptied at a constant rate. At 10 A.M., 12 000 L remain, and at 2 P.M., 4000 L remain. If pumping started at 8 A.M., find the equation relating the number of liters n at time t (in h) from 8 A.M. When will the tank be empty?

W **51.** A wall is 15 cm thick. At the outside, the temperature is 3°C, and at the inside, it is 23°C. If the temperature changes at a constant rate through the wall, write an equation of the temperature T in the wall as a function of the distance x from the outside to the inside of the wall. What is the meaning of the slope of the line?

W **52.** The length of a rectangular solar cell is 10 cm more than the width w. Express the perimeter p of the cell as a function of w. What is the meaning of the slope of the line?

53. A light beam is reflected off the edge of an optic fiber at an angle of 0.0032°. The diameter of the fiber is 48 μm. Find the equation of the reflected beam with the x-axis (at the center of the fiber) and the y-axis as shown in Fig. 21.27.

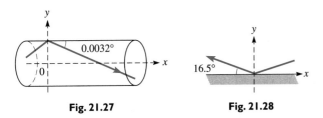

Fig. 21.27   **Fig. 21.28**

54. A police report stated that a bullet caromed upward off a floor at an angle of 16.5° with the floor, as shown in Fig. 21.28. What is the equation of the bullet's path after impact?

W **55.** A survey of the traffic on a particular highway showed that the number of cars passing a particular point each minute varied linearly from 6:30 A.M. to 8:30 A.M. on workday mornings. The study showed that an average of 45 cars passed the point in 1 min at 7 A.M. and that 115 cars passed in 1 min at 8 A.M. If n is the number of cars passing the point in 1 min, and t is the number of minutes after 6:30 A.M., find the equation relating n and t, and graph the equation. From the graph, determine n at 6:30 A.M. and at 8:30 A.M. What is the meaning of the slope of the line?

56. In a research project on cancer, a tumor was determined to weigh 30 mg when first discovered. While being treated, it grew smaller by 2 mg each month. Find the equation relating the weight w of the tumor as a function of the time t in months. Graph the equation.

In Exercises 57–60, treat the given nonlinear functions as linear functions in order to sketch their graphs. At times, this can be useful in showing certain values of a function. For example, $y = 2 + 3x^2$ can be shown as a straight line by graphing y as a function of x^2. A table of values for this graph is shown along with the corresponding graph in Fig. 21.29.

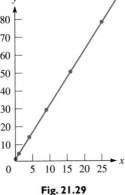

x	0	1	2	3	4	5
x^2	0	1	4	9	16	25
y	2	5	14	29	50	77

Fig. 21.29

57. The number n of memory cells of a certain computer that can be tested in t seconds is given by $n = 1200\sqrt{t}$. Sketch n as a function of $\sqrt{t}$.

58. The force F (in N) applied to a lever to balance a certain weight on the opposite side of the fulcrum is given by $F = 40/d$, where d is the distance (in m) of the force from the fulcrum. Sketch F as a function of $1/d$.

59. A spacecraft is launched such that its altitude h (in km) is given by $h = 300 + 2t^{3/2}$ for $0 \le t < 100$ s. Sketch this as a linear function.

60. The current i (in A) in a certain electric circuit is given by $i = 6(1 - e^{-t})$. Sketch this as a linear function.

In Exercises 61–64, show that the given nonlinear functions are linear when plotted on semilogarithmic or logarithmic paper. In Section 13.7, we noted that graphs on this paper often become straight lines.

61. A function of the form $y = ax^n$ is straight when plotted on logarithmic paper, since $\log y = \log a + n \log x$ is in the form of a straight line. The variables are $\log y$ and $\log x$; the slope can be found from $(\log y - \log a)/\log x = n$, and the intercept is a. (To get the slope from the graph, it is necessary to measure vertical and horizontal distances between two points. The log y-intercept is found where $\log x = 0$, and this occurs when $x = 1$.) Plot $y = 3x^4$ on logarithmic paper to verify this analysis.

62. A function of the form $y = a(b^x)$ is a straight line on semilogarithmic paper, since $\log y = \log a + x \log b$ is in the form of a straight line. The variables are $\log y$ and x, the slope is $\log b$, and the intercept is a. (To get the slope from the graph, we calculate $(\log y - \log a)/x$ for some set of values x and y. The intercept is read directly off the graph where $x = 0$.) Plot $y = 3(2^x)$ on semilogarithmic paper to verify this analysis.

63. If experimental data are plotted on logarithmic paper and the points lie on a straight line, it is possible to determine the function (see Exercise 61). The following data come from an experiment to determine the functional relationship between the pressure p and the volume V of a gas undergoing an adiabatic (no heat loss) change. From the graph on logarithmic paper, determine p as a function of V.

V (m^3)	0.100	0.500	2.00	5.00	10.0
p (kPa)	20.1	2.11	0.303	0.0840	0.0318

64. If experimental data are plotted on semilogarithmic paper, and the points lie on a straight line, it is possible to determine the function (see Exercise 62). The following data come from an experiment designed to determine the relationship between the voltage across an inductor and the time, after the switch is opened. Determine v as a function of t.

v (V)	40	15	5.6	2.2	0.8
t (ms)	0.0	20	40	60	80

21.3 THE CIRCLE

We have found that we can obtain a general equation that represents a straight line by considering a fixed point on the line and then a general point $P(x, y)$ which can represent any other point on the same line. Mathematically, we can state this as "the line is the **locus** of a point $P(x, y)$ that *moves* from a fixed point with constant slope along the line." That is, the point $P(x, y)$ can be considered as a variable point that moves along the line.

In this way, we can define a number of important curves. *A* **circle** *is defined as the locus of a point $P(x, y)$ that moves so that it is always equidistant from a fixed point. We call this fixed distance the* **radius,** *and we call the fixed point the* **center** *of the circle.* Thus, using this definition, calling the fixed point (h, k) and the radius r, we have

$$\sqrt{(x - h)^2 + (y - k)^2} = r$$

or, by squaring both sides, we have

$$(x - h)^2 + (y - k)^2 = r^2 \tag{21.11}$$

Equation (21.11) is called the **standard equation** *of a circle with center at (h, k) and radius r.* See Fig. 21.30.

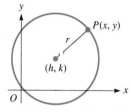

Fig. 21.30

⟨ EXAMPLE 1   The equation $(x - 1)^2 + (y + 2)^2 = 16$ represents a circle with center at $(1, -2)$ and a radius of 4. We determine these values by considering the equation of the circle to be in the form of Eq. (21.11) as

$$(x - 1)^2 + [y - (-2)]^2 = 4^2$$

form requires − signs
coordinates of center
radius

Note carefully the way in which we found the y-coordinate of the center. *We must have a minus sign before each of the coordinates.* Here, to get the y-coordinate, we had to write $+2$ as $-(-2)$. This circle is shown in Fig. 21.31. ⟩

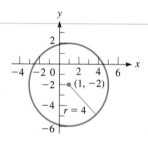

Fig. 21.31

⟨ EXAMPLE 2   Find the equation of the circle with center at $(2, 1)$ and that passes through $(4, 8)$.

In Eq. (21.11), we can determine the equation if we can find h, k, and r for this circle. From the given information, $h = 2$ and $k = 1$. To find r, we use the fact that *all points on the circle must satisfy the equation of the circle.* The point $(4, 8)$ must satisfy Eq. (21.11), with $h = 2$ and $k = 1$. Thus,

$$(4 - 2)^2 + (8 - 1)^2 = r^2 \quad \text{or} \quad r^2 = 53$$

Therefore, the equation of the circle is

$$(x - 2)^2 + (y - 1)^2 = 53$$

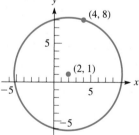

Fig. 21.32

This circle is shown in Fig. 21.32. ⟩

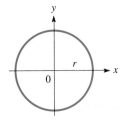

Fig. 21.33

Solving a Word Problem

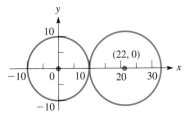

Fig. 21.34

If the center of the circle is at the origin, which means that the coordinates of the center are $(0, 0)$, the equation of the circle (see Fig. 21.33) becomes

$$x^2 + y^2 = r^2 \qquad (21.12)$$

The following example illustrates an application using this type of circle and one with its center not at the origin.

EXAMPLE 3 A student is drawing a friction drive in which two circular disks are in contact with each other. They are represented by circles in the drawing. The first has a radius of 10.0 cm, and the second has a radius of 12.0 cm. What is the equation of each circle if the origin is at the center of the first circle and the positive x-axis passes through the center of the second circle? See Fig. 21.34.

Since the center of the smaller circle is at the origin, we can use Eq. (21.12). Given that the radius is 10.0 cm, we have as its equation

$$x^2 + y^2 = 100$$

The fact that the two disks are in contact tells us that they meet at the point $(10.0, 0)$. Knowing that the radius of the larger circle is 12.0 cm tells us that its center is at $(22.0, 0)$. Thus, using Eq. (21.11) with $h = 22.0$, $k = 0$, and $r = 12.0$,

$$(x - 22.0)^2 + (y - 0)^2 = 12.0^2$$

or

$$(x - 22.0)^2 + y^2 = 144$$

as the equation of the larger circle.

Symmetry

A circle with its center at the origin exhibits an important property of the graphs of many equations. *It is **symmetric** to the x-axis and also to the y-axis.* Symmetry to the x-axis can be thought of as meaning that the lower half of the curve is a reflection of the upper half, and conversely. It can be shown that *if $-y$ can replace y in an equation without changing the equation, the graph of the equation is **symmetric to the x-axis.** Symmetry to the y-axis is similar. If $-x$ can replace x in the equation without changing the equation, the graph is symmetric to the y-axis.*

NOTE ▶

This type of circle is also symmetric to the origin as well as being symmetric to both axes. The meaning of symmetry to the origin is that the origin is the midpoint of any two points (x, y) and $(-x, -y)$ that are on the curve. Thus, *if $-x$ can replace x and $-y$ can replace y at the same time, without changing the equation, the graph of the equation is symmetric to the origin.*

NOTE ▶

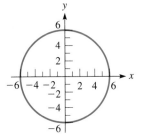

Fig. 21.35

EXAMPLE 4 The equation of the circle with its center at the origin and with a radius of 6 is $x^2 + y^2 = 36$.

The symmetry of this circle can be shown analytically by the substitutions mentioned above. Replacing x with $-x$, we obtain $(-x)^2 + y^2 = 36$. Since $(-x)^2 = x^2$, this equation can be rewritten as $x^2 + y^2 = 36$. Since this substitution did not change the equation, the graph is symmetric to the y-axis.

Replacing y with $-y$, we obtain $x^2 + (-y)^2 = 36$, which is the same as $x^2 + y^2 = 36$. This means that the curve is symmetric to the x-axis.

Replacing x with $-x$ and simultaneously replacing y with $-y$, we obtain $(-x)^2 + (-y)^2 = 36$, which is the same as $x^2 + y^2 = 36$. This means that the curve is symmetric to the origin. This circle is shown in Fig. 21.35.

If we multiply out each of the terms in Eq. (21.11), we may combine the resulting terms to obtain

$$x^2 - 2hx + h^2 + y^2 - 2ky + k^2 = r^2$$

$$\boldsymbol{x^2 + y^2 - 2hx - 2ky + (h^2 + k^2 - r^2) = 0} \qquad (21.13)$$

Since each of h, k, and r is constant for any given circle, the coefficients of x and y and the term within parentheses in Eq. (21.13) are constants. Equation (21.13) can then be written as

$$\boxed{x^2 + y^2 + Dx + Ey + F = 0} \qquad (21.14)$$

Equation (21.14) is called the **general equation** *of the circle.* It tells us that any equation that can be written in that form will represent a circle.

◀ EXAMPLE 5   Find the center and radius of the circle.

$$x^2 + y^2 - 6x + 8y - 24 = 0$$

CAUTION ▶

We can find this information if we write the given equation in standard form. To do so, *we must complete the square in the x-terms and also in the y-terms.* This is done by first writing the equation in the form

$$(x^2 - 6x \quad) + (y^2 + 8y \quad) = 24$$

To complete the square of the x-terms, we take half of -6, which is -3, square it, and add the result, 9, to each side of the equation. In the same way, we complete the square of the y-terms by adding 16 to each side of the equation, which gives

$$\left(\frac{-6}{2}\right)^2 \qquad \left(\frac{8}{2}\right)^2 \qquad \text{add to both sides}$$

$$(x^2 - 6x + 9) + (y^2 + 8y + 16) = 24 + 9 + 16$$

$$(x - 3)^2 + (y + 4)^2 = 49$$

$$\underbrace{(x - 3)^2 + (y - (-4))^2}_{\text{coordinates of center}} = \underbrace{7^2}_{\text{radius}}$$

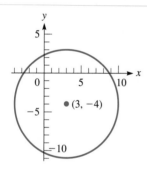

Fig. 21.36

Thus, the center is $(3, -4)$, and the radius is 7 (see Fig. 21.36). ▶

Solving a Word Problem

◀ EXAMPLE 6   A certain pendulum is found to swing through an arc of the circle $3x^2 + 3y^2 - 9.60y - 2.80 = 0$. What is the length (in m) of the pendulum, and from what point is it swinging?

We see that this equation represents a circle by dividing through by 3. This gives us $x^2 + y^2 - 3.20y - 2.80/3 = 0$. The length of the pendulum is the radius of the circle, and the point from which it swings is the center. These are found as follows:

$$x^2 + (y^2 - 3.20y + 1.60^2) = 1.60^2 + 2.80/3 \qquad \text{complete squaes in both } x\text{- and } y\text{-terms}$$

$$x^2 + (y - 1.60)^2 = 3.493 \qquad \text{standard form}$$

Since $\sqrt{3.493} = 1.87$, the length of the pendulum is 1.87 m. The point from which it is swinging is $(0, 1.60)$. See Fig. 21.37.

Replacing x with $-x$, the equation does not change. Replacing y with $-y$, the equation does change (the $3.20y$ term changes sign). Thus, the circle is symmetric only to the y-axis. ▶

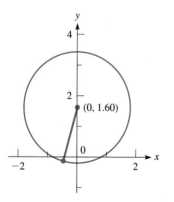

Fig. 21.37

NOTE ▶ In Section 14.1, we noted that the equation of a circle does not represent a *function* since there are two values of y for most values of x in the domain. In fact, *it might be necessary to use the quadratic formula to find the two functions to enter into a graphing calculator* in order to view the curve. This is illustrated in the following example.

◀ EXAMPLE 7 Display the graph of the circle $3x^2 + 3y^2 + 6y - 20 = 0$ on a graphing calculator.

To fit the form of a quadratic equation in y, we write

$$3y^2 + 6y + (3x^2 - 20) = 0$$

Now, using the quadratic formula to solve for y, we let

$$a = 3 \quad b = 6 \quad c = 3x^2 - 20$$

Therefore,

$$y = \frac{-6 \pm \sqrt{6^2 - 4(3)(3x^2 - 20)}}{2(3)}$$

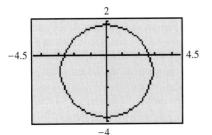

Fig. 21.38

which means we get the *two functions*

$$y_1 = \frac{-6 + \sqrt{276 - 36x^2}}{6} \quad \text{and} \quad y_2 = \frac{-6 - \sqrt{276 - 36x^2}}{6}$$

which are entered into the calculator to get the view shown in Fig. 21.38.

The *window* values were chosen so that the length along the x-axis is about 1.5 times that along the y-axis so as to have less distortion in the circle. There may be gaps at the left and right sides of the circle. ▶

EXERCISES **21.3**

In Exercises 1–4, make the given changes in the indicated examples of this section and then solve the resulting problems.

1. In Example 1, change $(y + 2)^2$ to $(y + 1)^2$ and then find the center and radius. Sketch the circle.

2. In Example 2, change $(2, 1)$ to $(-2, 1)$ and then find the equation of the circle. Sketch the circle.

3. In Example 5, change the $+$ before $8y$ to $-$ and then find the center and radius. Sketch the circle.

4. In Example 7, change the $+$ before $6y$ to $-$ and then display the graph on a graphing calculator.

In Exercises 5–8, determine the center and the radius of each circle.

5. $(x - 2)^2 + (y - 1)^2 = 25$
6. $(x - 3)^2 + (y + 4)^2 = 49$
7. $4(x + 1)^2 + 4y^2 = 9$
8. $9x^2 + 9(y - 6)^2 = 64$

In Exercises 9–24, find the equation of each of the circles from the given information.

9. Center at $(0, 0)$, radius 3
10. Center at $(0, 0)$, radius 1
11. Center at $(2, 2)$, radius 4
12. Center at $\left(\frac{3}{2}, -2\right)$, radius $\frac{5}{2}$
13. Center at $(12, -15)$, radius 18

14. Center at $(-3, -5)$, radius $2\sqrt{3}$
15. The origin and $(-6, 8)$ are ends of a diameter
16. The points $(3, 8)$ and $(-3, 0)$ are the ends of a diameter.
17. Concentric with the circle $(x - 2)^2 + (y - 1)^2 = 4$ and passes through $(4, -1)$
18. Concentric with the circle $(x + 1)^2 + (y - 4)^2 = 9$ and passes through $(-2, 3)$
19. Center at $(-3, 5)$, tangent to the x-axis
20. Center at $(2, -4)$, tangent to the y-axis
21. Tangent to both axes and the lines $y = 4$ and $x = 4$
22. Tangent to both axes, radius 4, in the second quadrant
23. Center at the origin, tangent to the line $x + y = 2$
24. Center at $(5, 12)$, tangent to the line $y = 2x - 3$

In Exercises 25–36, determine the center and radius of each circle. Sketch each circle.

25. $x^2 + (y - 3)^2 = 4$
26. $(x - 2)^2 + (y + 3)^2 = 49$
27. $4(x + 1)^2 + 4(y - 5)^2 = 81$

28. $2(x + 4)^2 + 2(y + 3)^2 = 25$

29. $x^2 + y^2 - 2x - 8 = 0$

30. $x^2 + y^2 - 4x - 6y - 12 = 0$

31. $x^2 + y^2 + 4.20x - 2.60y = 3.51$

32. $x^2 + y^2 + 22x + 14y = 26$

33. $4x^2 + 4y^2 - 16y = 9$

34. $9x^2 + 9y^2 + 18y = 7$

35. $2x^2 + 2y^2 - 4x - 8y - 1 = 0$

36. $3x^2 + 3y^2 - 12x + 4 = 0$

In Exercises 37–40, determine whether the circles with the given equations are symmetric to either axis or to the origin.

37. $x^2 + y^2 = 100$

38. $x^2 + y^2 - 4x - 5 = 0$

39. $3x^2 + 3y^2 + 24y = 8$

40. $5x^2 + 5y^2 - 10x + 20y = 3$

In Exercises 41–56, solve the given problems.

41. Determine whether the circle $x^2 - 6x + y^2 - 7 = 0$ crosses the x-axis.

42. Find the points of intersection of the circle $x^2 + y^2 - x - 3y = 0$ and the line $y = x - 1$.

(W) **43.** Find the locus of a point $P(x, y)$ that moves so that its distance from $(2, 4)$ is twice its distance from $(0, 0)$. Describe the locus.

(W) **44.** Find the equation of the locus of a point $P(x, y)$ that moves so that the line joining it and $(2, 0)$ is always perpendicular to the line joining it and $(-2, 0)$. Describe the locus.

45. Use a graphing calculator to view the circle $x^2 + y^2 + 5y - 4 = 0$.

46. Use a graphing calculator to view the circle $2x^2 + 2y^2 + 2y - x - 1 = 0$.

(W) **47.** What type of graph is represented by the equations
(a) $y = \sqrt{9 - (x - 2)^2}$?   (b) $y = -\sqrt{9 - (x - 2)^2}$?
(c) Are the equations in parts (a) and (b) functions? Explain.

48. What type of graph is represented by the equations
(a) $x^2 + (y - 1)^2 = 0$?   (b) $x^2 + (y - 1)^2 = -1$?

49. In a hoisting device, two of the pulley wheels may be represented by $x^2 + y^2 = 14.5$ and $x^2 + y^2 - 19.6y + 86.0 = 0$. How far apart (in cm) are the wheels?

50. The design of a machine part shows it as a circle represented by the equation $x^2 + y^2 = 42.5$, with a circular hole represented by $x^2 + y^2 + 3.06y - 1.24 = 0$ cut out. What is the least distance (in cm) from the edge of the hole to the edge of the machine part?

51. A wire is rotating in a circular path through a magnetic field to induce an electric current in the wire. The wire is rotating at 60.0 Hz with a constant velocity of 37.7 m/s. Taking the origin at the center of the circle of rotation, find the equation of the path of the wire.

52. A communications satellite remains stationary at an altitude of 36 200 km over a point on the earth's equator. It therefore rotates once each day about the earth's center. Its velocity is constant, but the horizontal and vertical components, v_H and v_V, of the velocity constantly change. Show that the equation relating v_H and v_V (in km/h) is that of a circle. The radius of the earth is 6370 km.

53. Find the equation describing the rim of a circular porthole 0.80 m in diameter if the top is 2.0 m below the surface of the water. Take the origin at the water surface directly above the center of the porthole.

54. An earthquake occurred 37° north of east of a seismic recording station. If the tremors travel at 4.8 km/s and were recorded 25 s later at the station, find the equation of the circle that represents the tremor recorded at the station. Take the station to be at the center of the coordinate system.

55. In analyzing the strain on a beam, *Mohr's circle* is often used. To form it, normal strain is plotted as the x-coordinate and shear strain is plotted as the y-coordinate. The center of the circle is midway between the minimum and maximum values of normal strain on the x-axis. Find the equation of Mohr's circle if the minimum normal strain is 100×10^{-6} and the maximum normal strain is 900×10^{-6} (strain is unitless). Sketch the graph.

56. An architect designs a Norman window, which has the form of a semicircle surmounted on a rectangle, as in Fig. 21.39. Find the area (in m²) of the window if the circular part is on the circle $x^2 + y^2 - 3.00y + 1.25 = 0$.

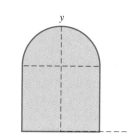

Fig. 21.39

21.4 THE PARABOLA

In Chapter 7, we showed that the graph of a quadratic function is a *parabola*. We now define the parabola more generally and find the general form of its equation.

A **parabola** *is defined as the locus of a point $P(x, y)$ that moves so that it is always equidistant from a given line (the **directrix**) and a given point (the **focus**). The line through the focus that is perpendicular to the directrix is the **axis** of the parabola. The point midway between the focus and directrix is the **vertex**.*

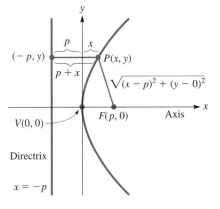

Fig. 21.40

Using the definition, we now find the equation of the parabola with the focus at $(p, 0)$ and the directrix $x = -p$. With these choices, we find a general equation of a parabola with its vertex at the origin.

From the definition, the distance from $P(x, y)$ on the parabola to the focus $(p, 0)$ must equal the distance from $P(x, y)$ to the directrix $x = -p$. The distance from P to the focus is found from the distance formula. The distance from P to the directrix is the perpendicular distance and is along a line parallel to the x-axis. These distances are shown in Fig. 21.40.

Thus, we have

$$\sqrt{(x - p)^2 + (y - 0)^2} = x + p$$

Squaring both sides of this equation, we have

$$(x - p)^2 + y^2 = (x + p)^2$$

or

$$x^2 - 2px + p^2 + y^2 = x^2 + 2px + p^2$$

Simplifying, we obtain

$$\boxed{y^2 = 4px} \tag{21.15}$$

Equation (21.15) is called the **standard form** *of the equation of a parabola with its axis along the x-axis and the vertex at the origin.* Its symmetry to the x-axis can be proven since $(-y)^2 = 4px$ is the same as $y^2 = 4px$.

◀ **EXAMPLE 1**   Find the coordinates of the focus and the equation of the directrix and sketch the graph of the parabola $y^2 = 12x$.

Since the equation of this parabola fits the form of Eq. (21.15), we know that the vertex is at the origin. The coefficient of 12 tells us that

$$4p = 12, \qquad p = 3$$

Since $p = 3$, the focus is the point $(3, 0)$, and the directrix is the line $x = -3$, as shown in Fig. 21.41.   ▶

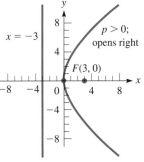

Fig. 21.41

◀ **EXAMPLE 2**   If the focus is to the left of the origin, with the directrix an equal distance to the right, the coefficient of the x-term is negative. This tells us that the parabola opens to the left, rather than to the right, as is the case when the focus is to the right of the origin. For example, the parabola $y^2 = -8x$ has its vertex at the origin, its focus at $(-2, 0)$, and the line $x = 2$ as its directrix. We determine this from the equation as follows:

$$y^2 = -8x \qquad 4p = -8, \qquad p = -2$$

Since $p = -2$, we find

the focus is $(-2, 0)$

the directrix is the line $x = -(-2)$, or $x = 2$

The parabola opens to the left, as shown in Fig. 21.42.   ▶

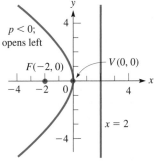

Fig. 21.42

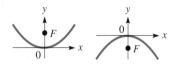

Fig. 21.43

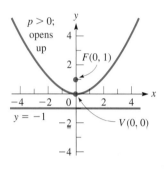

Fig. 21.44

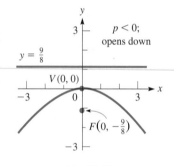

Fig. 21.45

Solving a Word Problem

See the chapter introduction.

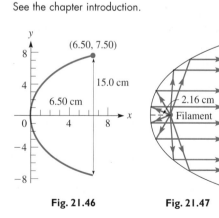

Fig. 21.46   Fig. 21.47

If we chose the focus as the point $(0, p)$ and the directrix as the line $y = -p$ (see Fig. 21.43), we would find that the resulting equation is

$$x^2 = 4py \qquad (21.16)$$

This is the standard form of the equation of a parabola with the y-axis as its axis and the vertex at the origin. Its symmetry to the y-axis can be proved, since $(-x)^2 = 4py$ is the same as $x^2 = 4py$. We note that **the difference between this equation and Eq. (21.15) is that x is squared and y appears to the first power in Eq. (21.16), rather than the reverse, as in Eq. (21.15).**

◀ **EXAMPLE 3**   The parabola $x^2 = 4y$ fits the form of Eq. (21.16). Therefore, its axis is along the y-axis and its vertex is at the origin. From the equation, we find the value of p, which in turn tells us the location of the vertex and the directrix.

$$x^2 = 4y \qquad 4p = 4, \qquad p = 1$$

Focus $(0, p)$ is $(0, 1)$; directrix $y = -p$ is $y = -1$. The parabola is shown in Fig. 21.44, and we see in this case that it opens upward. ▶

◀ **EXAMPLE 4**   The parabola $2x^2 = -9y$ fits the form of Eq. (21.16) if we write it in the form

$$x^2 = -\frac{9}{2}y$$

Here, we see that $4p = -9/2$. Therefore, its axis is along the y-axis, and its vertex is at the origin. Since $4p = -9/2$, we have

$$p = -\frac{9}{8} \qquad \text{focus}\left(0, -\frac{9}{8}\right) \qquad \text{directrix } y = \frac{9}{8}$$

The parabola opens downward, as shown in Fig. 21.45. ▶

◀ **EXAMPLE 5**   In calculus, it can be shown that a light ray coming from the focus of a parabola is reflected off the parabolic surface parallel to the axis of the parabola. This property of a parabolic surface has many applications, including the design of searchlights and spotlights (it was commonly used in automobile headlight design until about 1990, but since then other designs have also been used).

A spotlight reflector is designed with cross sections of equal parabolas. It has an opening of 15.0 cm and is 6.50 cm deep, as shown in Fig. 21.46. Find where the filament of the bulb should be located to produce a beam of light.

Since the parabolic opening is 15.0 cm and 6.50 cm deep, the point $(6.50, 7.50)$ will be on any of the parabolic cross sections. Placing the vertex at the origin, and the axis along the x-axis, the general form of the parabola is $y^2 = 4px$. We can find p by noting that $(6.50, 7.50)$ is on the parabola. This means

$$7.50^2 = 4p(6.50), \qquad p = 2.16$$

and the equation of the parabola is $y^2 = 8.64x$. The filament should be located at the focus, on the axis, 2.16 cm from the vertex, as shown in Fig. 21.47. ▶

Equations (21.15) and (21.16) give us the general form of the equation of a parabola with its vertex at the origin and its focus on one of the coordinate axes. The next example shows the use of the definition to find the equation of a parabola that has its vertex at a point other than the origin.

◀ EXAMPLE 6   Using the definition of the parabola, find the equation of the parabola with its focus at $(2, 3)$ and its directrix the line $y = -1$. See Fig. 21.48.

Choosing a general point $P(x, y)$ on the parabola and equating the distances from this point to $(2, 3)$ and to the line $y = -1$, we have

$$\sqrt{(x - 2)^2 + (y - 3)^2} = y + 1$$

distance P to F = distance P to $y = -1$

Squaring both sides of this equation and simplifying, we have

$$(x - 2)^2 + (y - 3)^2 = (y + 1)^2$$
$$x^2 - 4x + 4 + y^2 - 6y + 9 = y^2 + 2y + 1$$

or

$$8y = 12 - 4x + x^2$$

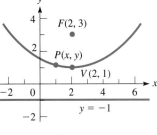

Fig. 21.48

We note that this type of equation has appeared frequently in earlier chapters. The x-term and the constant (12 in this case) are characteristic of a parabola that does not have its vertex at the origin if the directrix is parallel to the x-axis. ▶

We can readily view a parabola on a graphing calculator. If the axis is along the y-axis, or parallel to it, such as in Examples 3, 4, and 6, we simply solve for y and use this function. However, if the axis is along the x-axis, or parallel to it, as in Examples 1, 2, and 5, we get *two* functions to graph. This is similar to Example 7 on page 573. These cases are shown in the following example.

◀ EXAMPLE 7   To display the graph of the parabola in Example 6 on a graphing calculator, we solve for y and enter this function in the calculator. Therefore, we enter the function

$$y_1 = (12 - 4x + x^2)/8$$

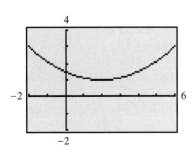

Fig. 21.49

in the calculator, and we get the display shown in Fig. 21.49.

To display the graph of the parabola in Example 1, where $y^2 = 12x$, when we solve for y, we get $y = \pm\sqrt{12x}$. Therefore, we enter the two functions

$$y_1 = \sqrt{12x} \quad \text{and} \quad y_2 = -\sqrt{12x}$$

in the calculator, and we get the display shown in Fig. 21.50. ▶

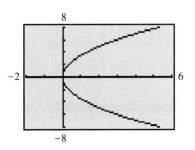

Fig. 21.50

We can conclude that *the equation of a parabola is characterized by the presence of the square of either (but not both) x or y and a first-power term in the other.* We will consider further the equation of the parabola in Sections 21.7 and 21.8.

The parabola has numerous technical applications. The reflection property illustrated in Example 5 has other important applications, such as the design of a radar antenna. The path of a projectile is parabolic. The cables of a suspension bridge are parabolic. These and other applications are illustrated in the exercises.

EXERCISES 21.4

In Exercises 1–4, make the given changes in the indicated examples of this section and then solve the resulting problems. In each, find the focus and directrix, and sketch the parabola.

1. In Example 1, change $12x$ to $20x$.

2. In Example 2, change $-8x$ to $-20x$.

3. In Example 3, change $4y$ to $-6y$.

4. In Example 4, change $-9y$ to $7y$.

In Exercises 5–16, determine the coordinates of the focus and the equation of the directrix of the given parabolas. Sketch each curve.

5. $y^2 = 4x$

6. $y^2 = 16x$

7. $y^2 = -4x$

8. $y^2 = -16x$

9. $x^2 = 8y$

10. $x^2 = y$

11. $x^2 = -4y$

12. $x^2 + 12y = 0$

13. $2y^2 - 5x = 0$

14. $3x^2 = 8y$

15. $y = 0.48x^2$

16. $x = 7.6y^2$

In Exercises 17–28, find the equations of the parabolas satisfying the given conditions. The vertex of each is at the origin.

17. Focus $(3, 0)$

18. Focus $(0, 0.4)$

19. Focus $(0, -0.5)$

20. Focus $(2.5, 0)$

21. Directrix $y = -0.16$

22. Directrix $x = 2$

23. Directrix $x = -84$

24. Directrix $y = 2.3$

25. Axis $x = 0$, passes through $(-1, 8)$

26. Symmetric to x-axis, passes through $(2, -1)$

27. Passes through $(3, 5)$ and $(3, -5)$

28. Passes through $(6, -1)$ and $(-6, -1)$

In Exercises 29–52, solve the given problems.

29. Find the equation of the parabola with focus $(6, 1)$ and directrix $x = 0$ by use of the definition. Sketch the curve.

30. Find the equation of the parabola with focus $(1, 1)$ and directrix $y = 5$ by use of the definition. Sketch the curve.

31. Use a graphing calculator to view the parabola $y^2 + 2x + 8y + 13 = 0$.

32. Use a graphing calculator to view the parabola $y^2 - 2x - 6y + 19 = 0$.

33. The equation of a parabola with vertex (h, k) and axis parallel to the x-axis is $(y - k)^2 = 4p(x - h)$. (This is shown in Section 21.7.) Sketch the parabola for which (h, k) is $(2, -3)$ and $p = 2$.

34. The equation of a parabola with vertex (h, k) and axis parallel to the y-axis is $(x - h)^2 = 4p(y - k)$. (This is shown in Section 21.7.) Sketch the parabola for which (h, k) is $(-1, 2)$ and $p = -3$.

35. The chord of a parabola that passes through the focus and is parallel to the directrix is called the *latus rectum* of the parabola. Find the length of the latus rectum of the parabola $y^2 = 4px$.

36. Find the equation of the circle that has the focus and the vertex of the parabola $x^2 = 8y$ as the ends of a diameter.

37. Find the standard equation of the parabola with vertex $(0, 0)$ and passes through $(2, 2)$ and $(8, 4)$.

W 38. For either standard form of the equation of a parabola, describe what happens to the shape of the parabola as $|p|$ increases.

39. The Golden Gate Bridge at San Francisco Bay is a suspension bridge, and its supporting cables are parabolic. See Fig. 21.51. With the origin at the low point of the cable, what equation represents the cable if the towers are 1280 m apart and the maximum sag is 90 m?

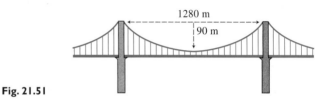

1280 m
90 m

Fig. 21.51

40. The entrance to a building is a parabolic arch 5.6 m high at the center and 7.4 m wide at the base. What equation represents the arch if the vertex is at the top of the arch?

41. The rate of development of heat H (in W) in a resistor of resistance R (in Ω) of an electric circuit is given by $H = Ri^2$, where i is the current (in A) in the resistor. Sketch the graph of H vs. i, if $R = 6.0 \ \Omega$.

42. What is the length of the horizontal bar across the parabolically shaped window shown in Fig. 21.52?

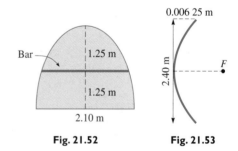

Bar — 1.25 m
1.25 m
2.10 m

0.006 25 m
2.40 m
F

Fig. 21.52   **Fig. 21.53**

43. The primary mirror in the Hubble space telescope has a parabolic cross section, which is shown in Fig. 21.53. What is the focal length (vertex to focus) of the mirror?

44. A rocket is fired horizontally from a plane. Its horizontal distance x and vertical distance y from the point at which it was fired are given by $x = v_0 t$ and $y = \frac{1}{2} g t^2$, where v_0 is the initial velocity of the rocket, t is the time, and g is the acceleration due to gravity. Express y as a function of x and show that it is the equation of a parabola.

45. To launch a spacecraft to the moon, it is first put into orbit around earth and then into a parabolic path toward the moon. Assume the parabolic path is represented by $x^2 = 4py$ and the spacecraft is later observed at $(10, 3)$ (units in thousands of km) after launch. Will this path lead directly to the moon at $(110, 340)$?

46. Under certain load conditions, a beam fixed at both ends is approximately parabolic in shape. If a beam is 4.0 m long and the deflection in the middle is 2.0 cm, find an equation to represent the shape of the beam.

47. A wave entering parallel to the axis of a radio wave antenna with a parabolic cross section is reflected through the focus. What is the equation of the parabola for the antenna with the reflected wave shown in Fig. 21.54?

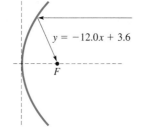

$y = -12.0x + 3.6$

F

Fig. 21.54

48. A wire is fastened 12.0 m up on each of two telephone poles that are 60.0 m apart. Halfway between the poles the wire is 10.0 m above the ground. Assuming the wire is parabolic, find the height of the wire 15.0 m from either pole.

49. The total annual fraction f of energy supplied by solar energy to a home is given by $f = 0.065\sqrt{A}$, where A is the area of the solar collector. Sketch the graph of f as a function of A $(0 < A \le 200 \text{ m}^2)$.

50. The velocity v (in m/s) of a jet of water flowing from an opening in the side of a certain container is given by $v = 4.4\sqrt{h}$, where h is the depth (in m) of the opening. Sketch a graph of v vs. h.

51. A small island is 4 km from a straight shoreline. A ship channel is equidistant between the island and the shoreline. Write an equation for the channel.

52. Under certain circumstances, the maximum power P (in W) in an electric circuit varies as the square of the voltage of the source E_0 and inversely as the internal resistance R_i (in Ω) of the source. If 10 W is the maximum power for a source of 2.0 V and internal resistance of 0.10 Ω, sketch the graph of P vs. E_0 if R_i remains constant.

21.5 THE ELLIPSE

The next important curve is the ellipse. *An **ellipse** is defined as the locus of a point $P(x, y)$ that moves so that the sum of its distances from two fixed points is constant. These fixed points are the **foci** of the ellipse.* Letting this sum of distances be $2a$ and the foci be the points $(-c, 0)$ and $(c, 0)$, we have

$$\sqrt{(x - c)^2 + y^2} + \sqrt{(x + c)^2 + y^2} = 2a$$

See Fig. 21.55. The ellipse has its center at the origin such that c is the length of the line segment from the center to a focus. We will also see that a has a special meaning. Now, from Section 14.4, we see that we should move one radical to the right and then square each side. This leads to the following steps:

$$\sqrt{(x + c)^2 + y^2} = 2a - \sqrt{(x - c)^2 + y^2}$$
$$(x + c)^2 + y^2 = 4a^2 - 4a\sqrt{(x - c)^2 + y^2} + \left(\sqrt{(x - c)^2 + y^2}\right)^2$$
$$x^2 + 2cx + c^2 + y^2 = 4a^2 - 4a\sqrt{(x - c)^2 + y^2} + x^2 - 2cx + c^2 + y^2$$
$$4a\sqrt{(x - c)^2 + y^2} = 4a^2 - 4cx$$
$$a\sqrt{(x - c)^2 + y^2} = a^2 - cx$$
$$a^2(x^2 - 2cx + c^2 + y^2) = a^4 - 2a^2cx + c^2x^2$$
$$(a^2 - c^2)x^2 + a^2y^2 = a^2(a^2 - c^2)$$

We now define $a^2 - c^2 = b^2$ (this reason will be shown presently). Therefore,

$$b^2x^2 + a^2y^2 = a^2b^2$$

Dividing through by a^2b^2, we have

$$\frac{x^2}{a^2} + \frac{y^2}{b^2} = 1 \tag{21.17}$$

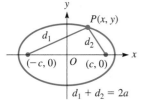

Fig. 21.55

A graphical analysis of this equation is found on the next page.

For reference, Eq. (21.17) is
$$\frac{x^2}{a^2} + \frac{y^2}{b^2} = 1$$

The x-intercepts are $(-a, 0)$ and $(a, 0)$. This means that $2a$ (the sum of distances used in the derivation) is also the distance between the x-intercepts. *The points $(a, 0)$ and $(-a, 0)$ are the* **vertices** *of the ellipse, and the line between them is the* **major axis** (see Fig. 21.56(a)). Thus, *a is the length of the* **semimajor axis.**

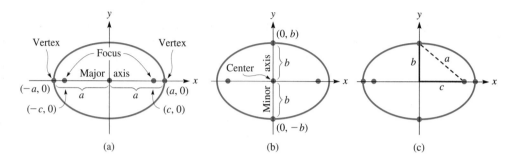

Fig. 21.56

(a)   (b)   (c)

We can now state that *Eq. (21.17) is called the* **standard equation** *of the ellipse with its major axis along the x-axis and its center at the origin.*

The y-intercepts of this ellipse are $(0, -b)$ and $(0, b)$. *The line joining these intercepts is called the* **minor axis** *of the ellipse* (Fig. 21.56(b)), *which means b is the length of the* **semiminor** *axis.* The intercept $(0, b)$ is equidistant from $(-c, 0)$ and $(c, 0)$. Since the sum of the distances from these points to $(0, b)$ is $2a$, the distance $(c, 0)$ to $(0, b)$ must be a. Thus, we have a right triangle with line segments of lengths a, b, and c, with a as hypotenuse (Fig. 21.56(c)). Therefore,

$$a^2 = b^2 + c^2 \tag{21.18}$$

is the relation between distances a, b, and c. This also shows why b was defined as it was in the derivation of Eq. (21.17).

If we choose points on the y-axis as the foci, *the standard equation of the ellipse, with its center at the origin and its major axis along the y-axis, is*

$$\frac{y^2}{a^2} + \frac{x^2}{b^2} = 1 \tag{21.19}$$

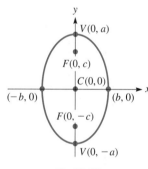

Fig. 21.57

In this case, the vertices are $(0, a)$ and $(0, -a)$, the foci are $(0, c)$ and $(0, -c)$, and the ends of the minor axis are $(b, 0)$ and $(-b, 0)$. See Fig. 21.57.

The ellipses represented by Eqs. (21.17) and (21.19) are both symmetric to both axes and to the origin.

CAUTION

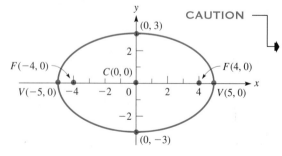

Fig. 21.58

◖ **EXAMPLE 1**   The ellipse $\dfrac{x^2}{25} + \dfrac{y^2}{9} = 1$ seems to fit the form of either Eq. (21.17) or Eq. (21.19). Since $a^2 = b^2 + c^2$, we know that *a is always larger than b.* Since the square of the larger number appears under x^2, we know the equation is in the form of Eq. (21.17). Therefore, $a^2 = 25$ and $b^2 = 9$, or $a = 5$ and $b = 3$. This means that the vertices are $(5, 0)$ and $(-5, 0)$ and the minor axis extends from $(0, -3)$ to $(0, 3)$. See Fig. 21.58.

We find c from the relation $c^2 = a^2 - b^2$. This means that $c^2 = 16$ and the foci are $(4, 0)$ and $(-4, 0)$.   ◗

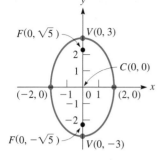

Fig. 21.59

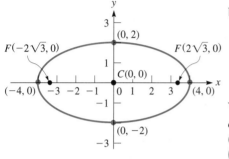

Fig. 21.60

Solving a Word Problem

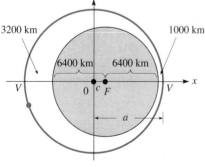

Fig. 21.61

◀ **EXAMPLE 2**   The ellipse

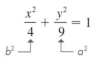

$$\frac{x^2}{4} + \frac{y^2}{9} = 1$$

has vertices $(0, 3)$ and $(0, -3)$. The minor axis extends from $(-2, 0)$ to $(2, 0)$. The equation fits the form of Eq. (21.19) since the larger number appears under y^2. Therefore, $a^2 = 9$, $b^2 = 4$, and $c^2 = 5$. The foci are $\left(0, \sqrt{5}\right)$ and $\left(0, -\sqrt{5}\right)$. This ellipse is shown in Fig. 21.59. ▶

◀ **EXAMPLE 3**   Find the coordinates of the vertices, the ends of the minor axis, and the foci of the ellipse $4x^2 + 16y^2 = 64$.

This equation must be put in standard form first, which we do by dividing through by 64. When this is done, we obtain

$$\frac{x^2}{16} + \frac{y^2}{4} = 1$$

form requires + and 1

We see that $a^2 = 16$ and $b^2 = 4$, which tells us that $a = 4$ and $b = 2$. Then, $c = \sqrt{16 - 4} = \sqrt{12} = 2\sqrt{3}$. Since a^2 appears under x^2, the vertices are $(4, 0)$ and $(-4, 0)$. The ends of the minor axis are $(0, 2)$ and $(0, -2)$, and the foci are $\left(2\sqrt{3}, 0\right)$ and $\left(-2\sqrt{3}, 0\right)$. See Fig. 21.60. ▶

◀ **EXAMPLE 4**   A satellite to study the earth's atmosphere has a minimum altitude of 1000 km and a maximum altitude of 3200 km. If the path of the satellite about the earth is an ellipse with the center of the earth at one focus, what is the equation of its path? Assume that the radius of the earth is 6400 km.

We set up the coordinate system such that the center of the ellipse is at the origin and the center of the earth is at the right focus, as shown in Fig. 21.61. We know that the distance between vertices is

$$2a = 3200 + 6400 + 6400 + 1000 = 17\,000 \text{ km}$$
$$a = 8500 \text{ km}$$

From the right focus to the right vertex is 7400 km. This tells us

$$c = a - 7400 = 8500 - 7400 = 1100 \text{ km}$$

We can now calculate b^2 as

$$b^2 = a^2 - c^2 = 8500^2 - 1100^2 = 7.10 \times 10^7 \text{ km}^2$$

Since $a^2 = 8500^2 = 7.23 \times 10^7 \text{ km}^2$, the equation is

$$\frac{x^2}{7.23 \times 10^7} + \frac{y^2}{7.10 \times 10^7} = 1$$

or

$$7.10x^2 + 7.23y^2 = 5.13 \times 10^8$$

▶

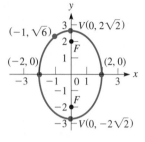

Fig. 21.62

◀ **EXAMPLE 5** Find the equation of the ellipse with its center at the origin and an end of its minor axis at $(2, 0)$ and which passes through $\left(-1, \sqrt{6}\right)$.

Since the center is at the origin and an end of the minor axis is at $(2, 0)$, we know that the ellipse is of the form of Eq. (21.19) and that $b = 2$. Thus, we have

$$\frac{y^2}{a^2} + \frac{x^2}{2^2} = 1$$

In order to find a^2, we use the fact that the ellipse passes through $\left(-1, \sqrt{6}\right)$. This means that these coordinates satisfy the equation of the ellipse. This gives

$$\frac{\left(\sqrt{6}\right)^2}{a^2} + \frac{(-1)^2}{4} = 1, \qquad \frac{6}{a^2} = \frac{3}{4}, \qquad a^2 = 8$$

Therefore, the equation of the ellipse, shown in Fig. 21.62, is

$$\frac{y^2}{8} + \frac{x^2}{4} = 1$$

The following example illustrates the use of the definition of the ellipse to find the equation of an ellipse with its center at a point other than the origin.

◀ **EXAMPLE 6** Using the definition, find the equation of the ellipse with foci at $(1, 3)$ and $(9, 3)$, with major axis of 10.

Recalling that the sum of distances in the definition equals the length of the major axis, we now use the same method as in the derivation of Eq. (21.17).

$\sqrt{(x-1)^2 + (y-3)^2} + \sqrt{(x-9)^2 + (y-3)^2} = 10$	use definition of ellipse
$\sqrt{(x-1)^2 + (y-3)^2} = 10 - \sqrt{(x-9)^2 + (y-3)^2}$	isolate a radical
$x^2 - 2x + 1 + y^2 - 6y + 9 = 100 - 20\sqrt{(x-9)^2 + (y-3)^2}$ $\qquad + x^2 - 18x + 81 + y^2 - 6y + 9$	square both sides and simplify
$20\sqrt{(x-9)^2 + (y-3)^2} = 180 - 16x$	isolate radical
$5\sqrt{(x-9)^2 + (y-3)^2} = 45 - 4x$	divide by 4
$25(x^2 - 18x + 81 + y^2 - 6y + 9) = 2025 - 360x + 16x^2$	square both sides
$9x^2 - 90x + 25y^2 - 150y + 225 = 0$	simplify

The additional x- and y-terms are characteristic of the equation of an ellipse whose center is not at the origin (see Fig. 21.63).

To view this ellipse on a graphing calculator as shown in Fig. 21.64, we solve for y to get the two functions needed. The solutions are $y = \dfrac{15 \pm 3\sqrt{10x - x^2}}{5}$.

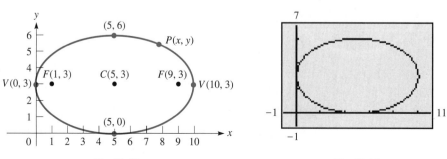

Fig. 21.63　　　　　　　　　　　　　　　　　　　　**Fig. 21.64**

The Polish astronomer Nicolaus Copernicus (1473–1543) is credited as being the first to suggest that the earth revolved about the sun, rather than the previously held belief that the earth was the center of the universe.

We can conclude that *the equation of an ellipse is characterized by the presence of both an x^2-term and a y^2-term, having different coefficients (in value but not in sign).* The difference between the equation of an ellipse and that of a circle is that the coefficients of the squared terms in the equation of the circle are the same, whereas those of the ellipse differ. We will consider the equation of the ellipse further in Sections 21.7 and 21.8.

The ellipse has many applications. The orbits of the planets about the sun are elliptical. Gears, cams, and springs are often elliptical in shape. Arches are often constructed in the form of a semiellipse. These and other applications are illustrated in the exercises.

EXERCISES 21.5

In Exercises 1 and 2, make the given changes in the indicated examples of this section and then solve the given problems.

1. In Example 1, change the 9 to 36; find the vertices, ends of the minor axis, and foci; and sketch the ellipse.

2. In Example 2, interchange the 4 and 9; find the vertices, ends of the minor axis, and foci; and sketch the ellipse.

In Exercises 3–16, find the coordinates of the vertices and foci of the given ellipses. Sketch each curve.

3. $\dfrac{x^2}{4} + \dfrac{y^2}{1} = 1$ 4. $\dfrac{x^2}{100} + \dfrac{y^2}{64} = 1$ 5. $\dfrac{x^2}{25} + \dfrac{y^2}{144} = 1$

6. $\dfrac{x^2}{49} + \dfrac{y^2}{81} = 1$ 7. $\dfrac{4x^2}{25} + \dfrac{y^2}{4} = 1$ 8. $x^2 + \dfrac{9y^2}{25} = 1$

9. $4x^2 + 9y^2 = 324$ 10. $x^2 + 36y^2 = 144$

11. $49x^2 + 4y^2 = 196$ 12. $y^2 = 25(1 - x^2)$

13. $y^2 = 8(2 - x^2)$ 14. $2x^2 + 3y^2 = 600$

15. $4x^2 + 25y^2 = 0.25$ 16. $9x^2 + 4y^2 = 0.09$

In Exercises 17–28, find the equations of the ellipses satisfying the given conditions. The center of each is at the origin.

17. Vertex $(15, 0)$, focus $(9, 0)$

18. Minor axis 8, vertex $(0, -5)$

19. End of minor axis $(0, 3)$, focus $(2, 0)$

20. Vertex $(0, 5)$, focus $(0, -2)$

21. Focus $(0, 2)$, major axis 6

22. Sum of lengths of major and minor axes 18, focus $(3, 0)$

23. Vertex $(8, 0)$, passes through $(2, 3)$

24. Focus $(0, 2)$, passes through $\left(-1, \sqrt{3}\right)$

25. Passes through $(2, 2)$ and $(1, 4)$

26. Passes through $(-2, 2)$ and $\left(1, \sqrt{6}\right)$

27. The sum of distances from (x, y) to $(6, 0)$ and $(-6, 0)$ is 20.

28. The sum of distances from (x, y) to $(0, 2)$ and $(0, -2)$ is 5.

In Exercises 29–52, solve the given problems.

29. Find the equation of the ellipse with foci $(-2, 1)$ and $(4, 1)$ and a major axis of 10 by use of the definition. Sketch the curve.

30. Find the equation of the ellipse with foci $(1, 4)$ and $(1, 0)$ that passes through $(4, 4)$ by use of the definition. Sketch the curve.

31. Use a graphing calculator to view the ellipse
$4x^2 + 3y^2 + 16x - 18y + 31 = 0$.

32. Use a graphing calculator to view the ellipse
$4x^2 + 8y^2 + 4x - 24y + 1 = 0$.

33. The equation of an ellipse with center (h, k) and major axis parallel to the x-axis is $\dfrac{(x - h)^2}{a^2} + \dfrac{(y - k)^2}{b^2} = 1$. (This is shown in Section 21.7.) Sketch the ellipse that has a major axis of 6, a minor axis of 4, and for which (h, k) is $(2, -1)$.

34. The equation of an ellipse with center (h, k) and major axis parallel to the y-axis is $\dfrac{(y - k)^2}{a^2} + \dfrac{(x - h)^2}{b^2} = 1$. (This is shown in Section 21.7.) Sketch the ellipse that has a major axis of 8, a minor axis of 6, and for which (h, k) is $(1, 3)$.

(W) 35. For what values of k does the ellipse $x^2 + ky^2 = 1$ have its vertices on the y-axis? Explain how these values are found.

(W) 36. For what value of k does the ellipse $x^2 + k^2y^2 = 25$ have a focus at $(3, 0)$? Explain how this value is found.

37. Show that the ellipse $2x^2 + 3y^2 - 8x - 4 = 0$ is symmetric to the x-axis.

38. Show that the ellipse $5x^2 + y^2 - 3y - 7 = 0$ is symmetric to the y-axis.

39. Graph the inequality $100x^2 + 49y^2 \le 4900$.

40. For what values of k does $y^2 = 1 + kx^2$ represent an ellipse with foci on the y-axis?

41. The electric power P (in W) dissipated in a resistance R (in Ω) is given by $P = Ri^2$, where i is the current (in A) in the resistor. Find the equation for the total power of 64 W dissipated in two resistors, with resistances 2.0 Ω and 8.0 Ω, respectively, and with currents i_1 and i_2, respectively. Sketch the graph, assuming that negative values of current are meaningful.

42. The *eccentricity e* of an ellipse is defined as $e = c/a$. A cam in the shape of an ellipse can be described by the equation $x^2 + 9y^2 = 81$. Find the eccentricity of this elliptical cam.

43. The planet Pluto moves about the sun in an elliptical orbit, with the sun at one focus. The closest that Pluto approaches the sun is 4.5 Tm, and the farthest it gets from the sun is 7.4 Tm. Find the eccentricity of Pluto's orbit. (See Exercise 42.)

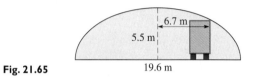 **44.** Halley's Comet has an elliptical orbit with $a = 17.94$ AU (AU is astronomical unit; 1 AU $= 1.5 \times 10^8$ km) and $b = 4.552$ AU, with the sun at one focus. What is the closest that the comet comes to the sun? Explain your method.

45. A draftsman draws a series of triangles with a base from $(-3, 0)$ to $(3, 0)$ and a perimeter of 14 cm (all measurements in centimeters). Find the equation of the curve on which all of the third vertices of the triangles are located.

46. A lithotripter is used to break up a kidney stone by placing the kidney stone at one focus of an ellipsoid end-section and a source of shock waves at the focus of the other end-section. If the vertices of the end-sections are 30.0 cm apart and a minor axis of the ellipsoid is 6.0 cm, how far apart are the foci? In a lithotripter, the shock waves are reflected as the sound waves noted in Exercise 47.

47. An ellipse has a focal property such that a light ray or sound wave emanating from one focus will be reflected through the other focus. Many buildings, such as Statuary Hall in the U.S. Capitol and the Taj Mahal, are built with elliptical ceilings with the property that a sound from one focus is easily heard at the other focus. If a building has a ceiling whose cross sections are part of an ellipse that can be described by the equation $36x^2 + 225y^2 = 8100$ (measurements in meters), how far apart must two persons stand in order to whisper to each other using this focal property?

48. An airplane wing is designed such that a certain cross section is an ellipse 2.80 m wide and 0.40 m thick. Find the equation that can be used to describe the perimeter of this cross section.

49. A road passes through a tunnel with a semielliptical cross section 19.6 m wide and 5.5 m high at the center. What is the height of the tallest vehicle that can pass through the tunnel at a point 6.7 m from the center? See Fig. 21.65.

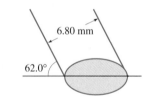

Fig. 21.65

50. An architect designs a window in the shape of an ellipse 1.50 m wide and 1.10 m high. Find the perimeter of the window from the formula $p = \pi(a + b)$. This formula gives a good *approximation* for the perimeter when a and b are nearly equal.

51. The ends of a horizontal tank 20.0 m long are ellipses, which can be described by the equation $9x^2 + 20y^2 = 180$, where x and y are measured in meters. The area of an ellipse is $A = \pi ab$. Find the volume of the tank.

52. A laser beam 6.80 mm in diameter is incident on a plane surface at an angle of $62.0°$, as shown in Fig. 21.66. What is the elliptical area that the laser covers on the surface? (See Exercise 51.)

Fig. 21.66

21.6 THE HYPERBOLA

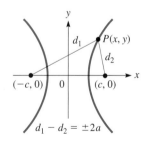

Fig. 21.67

A **hyperbola** *is defined as the locus of a point $P(x, y)$ that moves so that the difference of the distances from two fixed points (the* **foci***) is constant.* We choose the foci to be $(-c, 0)$ and $(c, 0)$ (see Fig. 21.67), and the constant difference to be $2a$. As with the ellipse, c is the length of the line segment from the center to a focus and a (as we will see) the length of the line segment from the center to a vertex. Therefore,

$$\sqrt{(x + c)^2 + y^2} - \sqrt{(x - c)^2 + y^2} = 2a$$

Following the same procedure as with the ellipse, the equation of the hyperbola is

$$\frac{x^2}{a^2} - \frac{y^2}{b^2} = 1 \qquad (21.20)$$

CAUTION ▶ When we derive this equation, *we have a definition of the relation between a, b, and c that is different from that for the ellipse.* This relation is

$$c^2 = a^2 + b^2 \qquad (21.21)$$

In Eq. (21.20), if we let $y = 0$, we find that the x-intercepts are $(-a, 0)$ and $(a, 0)$, just as they are for the ellipse. *These are the* **vertices** *of the hyperbola.* For $x = 0$, we find that we have imaginary solutions for y, which means there are no points on the curve that correspond to a value of $x = 0$.

To find the meaning of b, we solve Eq. (21.20) for y in the special form:

$$\frac{y^2}{b^2} = \frac{x^2}{a^2} - 1$$

$$= \frac{x^2}{a^2} - \frac{a^2 x^2}{a^2 x^2} = \frac{x^2}{a^2}\left(1 - \frac{a^2}{x^2}\right)$$

$$y^2 = \frac{b^2 x^2}{a^2}\left(1 - \frac{a^2}{x^2}\right) \qquad \text{multiply through by } b^2 \text{ and} \\ \text{take square root of each side}$$

$$y = \pm \frac{bx}{a}\sqrt{1 - \frac{a^2}{x^2}} \tag{21.22}$$

We note that, if large values of x are assumed in Eq. (21.22), the quantity under the radical becomes approximately 1. In fact, the larger x becomes, the nearer 1 this expression becomes, since the x^2 in the denominator of a^2/x^2 makes this term nearly zero. Thus, for large values of x, Eq. (21.22) is approximately

$$\boxed{y = \pm \frac{bx}{a}} \tag{21.23}$$

NOTE ▶

Equation (21.23) is seen to represent two straight lines, each of which passes through the origin. One has a slope of b/a, and the other has a slope of $-b/a$. *These lines are called the* **asymptotes** *of the hyperbola.* An **asymptote** *is a line that the curve approaches as one of the variables approaches some particular value.* The graph of the tangent function also has asymptotes, as we saw in Fig. 10.23. We can designate this limiting procedure with notation introduced in Chapter 19 by

$$y \to \frac{bx}{a} \quad \text{as} \quad x \to \pm\infty$$

The easiest way to sketch a hyperbola is to draw its asymptotes and then draw the hyperbola out from each vertex so that it comes closer and closer to each asymptote as x becomes numerically larger. To draw the asymptotes, first draw a small rectangle $2a$ by $2b$ with the origin at the center, as shown in Fig. 21.68. Then straight lines, the asymptotes, are drawn through opposite vertices of the rectangle. This shows us that the significance of the value of b is in the slopes of the asymptotes.

Equation (21.20) is called the **standard equation** *of the hyperbola with its center at the origin. It has a* **transverse axis** *of length $2a$ along the x-axis and a* **conjugate axis** *of length $2b$ along the y-axis.* This means that a represents the length of the semitransverse axis and b represents the length of the semiconjugate axis. See Fig. 21.69. From the definition of c, it is the length of the line segment from the center to a focus. Also, c is the length of the semidiagonal of the rectangle, as shown in Fig. 21.69. This shows us the geometric meaning of the relationship among a, b, and c given in Eq. (21.21).

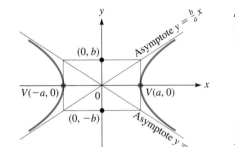

Fig. 21.68

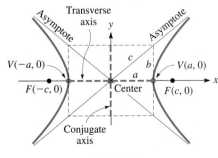

Fig. 21.69

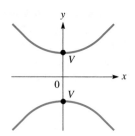

Fig. 21.70

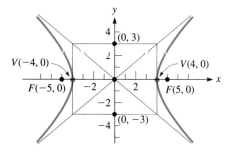

Fig. 21.71

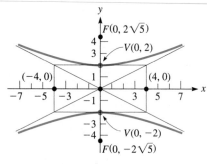

Fig. 21.72

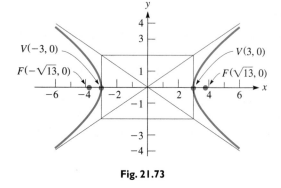

Fig. 21.73

If the transverse axis is along the y-axis and the conjugate axis is along the x-axis, the equation of the hyperbola with its center at the origin (see Fig. 21.70) is

$$\frac{y^2}{a^2} - \frac{x^2}{b^2} = 1 \qquad (21.24)$$

The hyperbolas represented by Eqs. (21.20) and (21.24) are both symmetric to both axes and to the origin.

◀ EXAMPLE 1   The hyperbola $\dfrac{x^2}{16} - \dfrac{y^2}{9} = 1$

a^2 b^2

fits the form of Eq. (21.20). We know that it fits Eq. (21.20) and not Eq. (21.24) since the x^2-term is the positive term with 1 on the right. From the equation, we see that $a^2 = 16$ and $b^2 = 9$, or $a = 4$ and $b = 3$. In turn, this means the vertices are $(4, 0)$ and $(-4, 0)$ and the conjugate axis extends from $(0, -3)$ to $(0, 3)$.

Since $c^2 = a^2 + b^2$, we find that $c^2 = 25$, or $c = 5$. The foci are $(-5, 0)$ and $(5, 0)$.

Drawing the rectangle and the asymptotes in Fig. 21.71, we then sketch in the hyperbola from each vertex toward each asymptote. ◗

◀ EXAMPLE 2   The hyperbola $\dfrac{y^2}{4} - \dfrac{x^2}{16} = 1$

a^2 b^2

has vertices at $(0, -2)$ and $(0, 2)$. Its conjugate axis extends from $(-4, 0)$ to $(4, 0)$. The foci are $\left(0, -2\sqrt{5}\right)$ and $\left(0, 2\sqrt{5}\right)$. We find this directly from the equation since the y^2-term is the positive term with 1 on the right. This means the equation fits the form of Eq. (21.24) with $a^2 = 4$ and $b^2 = 16$. Also, $c^2 = 20$, which means that $c = \sqrt{20} = 2\sqrt{5}$.

Since $2a$ extends along the y-axis, we see that the equations of the asymptotes are $y = \pm (a/b)x$. This is not a contradiction of Eq. (21.23) but the extension of it for a hyperbola with its transverse axis along the y-axis. The ratio a/b gives the slope of the asymptote. The hyperbola is shown in Fig. 21.72. ◗

◀ EXAMPLE 3   Determine the coordinates of the vertices of the hyperbola

$$4x^2 - 9y^2 = 36$$

First, by dividing through by 36, we have

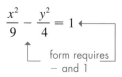

$$\frac{x^2}{9} - \frac{y^2}{4} = 1 \longleftarrow$$
form requires
$-$ and 1

From this form, we see that $a^2 = 9$ and $b^2 = 4$. In turn, this tells us that $a = 3$, $b = 2$, and $c = \sqrt{9 + 4} = \sqrt{13}$. Since a^2 appears under x^2, the equation fits the form of Eq. (21.20). Therefore, the vertices are $(-3, 0)$ and $(3, 0)$ and the foci are $\left(-\sqrt{13}, 0\right)$ and $\left(\sqrt{13}, 0\right)$. The hyperbola is shown in Fig. 21.73. ◗

Solving a Word Problem

◀ EXAMPLE 4   In physics, it is shown that where the velocity of a fluid is greatest, the pressure is the least. In designing an experiment to study this effect in the flow of water, a pipe is constructed such that its lengthwise cross section is hyperbolic. The pipe is 1.0 m long, 0.2 m in diameter at the narrowest point in the middle, and 0.4 m in diameter at each end. What is the equation that represents the cross section of the pipe as shown in Fig. 21.74?

As shown, the hyperbola has its transverse axis along the y-axis and its center at the origin. This means the general equation is given by Eq. (21.24). Since the radius at the middle of the pipe is 0.1 m, we know that $a = 0.1$ m. Also, since it is 1.0 m long and the radius at the end is 0.2 m, we know the point $(0.5, 0.2)$ is on the hyperbola. This point must satisfy the equation.

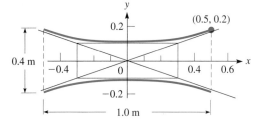

Fig. 21.74

$$\frac{y^2}{a^2} - \frac{x^2}{b^2} = 1 \qquad \text{Eq. (21.24)}$$

point $(0.5, 0.2)$ satisfies equation

$$a = 0.1 \rightarrow \quad \frac{0.2^2}{0.1^2} - \frac{0.5^2}{b^2} = 1$$

$$4 - \frac{0.25}{b^2} = 1, \qquad 3b^2 = 0.25, \qquad b^2 = 0.083$$

$$\frac{y^2}{0.1^2} - \frac{x^2}{0.083} = 1 \qquad \text{substituting } a = 0.1,\ b^2 = 0.083 \text{ in Eq. (21.24)}$$

$$100y^2 - 12x^2 = 1 \qquad \text{equation of cross section}$$ ◗

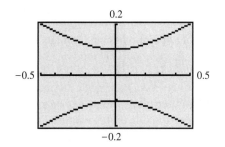

Fig. 21.75

If we use a graphing calculator to display a hyperbola represented by either Eq. (21.20) or (21.24), we have two functions when we solve for y. One represents the upper half of the hyperbola, and the other represents the lower half. For the hyperbola in Example 3, the functions are $y_1 = \sqrt{(4x^2 - 36)/9}$ and $y_2 = -\sqrt{(4x^2 - 36)/9}$. For the hyperbola in Example 4, they are $y_1 = \sqrt{(12x^2 + 1)/100}$ and $y_2 = -\sqrt{(12x^2 + 1)/100}$. A graphing calculator display for this hyperbola is shown in Fig. 21.75.

Equations (21.20) and (21.24) give us the standard forms of the equation of the hyperbola with its center at the origin and its foci on one of the coordinate axes. There is another important equation form that represents a hyperbola, and it is

$$\boxed{xy = c} \tag{21.25}$$

The asymptotes of this hyperbola are the coordinate axes, and the foci are on the line $y = x$ if c is positive or on the line $y = -x$ if c is negative.

The hyperbola represented by Eq. (21.25) is symmetric to the origin, for if $-x$ replaces x and $-y$ replaces y at the same time, we obtain $(-x)(-y) = c$, or $xy = c$. The equation is unchanged. However, if $-x$ replaces x or $-y$ replaces y, but not both, the sign on the left is changed. This means it is not symmetric to either axis. Here, c represents a constant and is not related to the focus. Two examples of this type of hyperbola are shown on the next page.

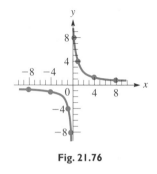

Fig. 21.76

◀ **EXAMPLE 5** Plot the graph of the equation $xy = 4$.

We find the values in the table below and then plot the appropriate points. Here, it is permissible to use a limited number of points, since we know the equation represents a hyperbola. Therefore, using $y = 4/x$, we obtain the values

x	-8	-4	-1	$-\frac{1}{2}$	$\frac{1}{2}$	1	4	8
y	$-\frac{1}{2}$	-1	-4	-8	8	4	1	$\frac{1}{2}$

Note that neither x nor y may equal zero. The hyperbola is shown in Fig. 21.76.

If the constant on the right is negative (for example, if $xy = -4$), then the two branches of the hyperbola are in the second and fourth quadrants. ▶

The first reasonable measurment of the speed of light was made by the Danish astronomer Olof Roemer (1644–1710). He measured the time required for light to come from the moons of Jupiter across the earth's orbit.

◀ **EXAMPLE 6** For a light wave, the product of its frequency f of vibration and its wavelength λ is a constant, and this constant is the speed of light c. For green light, for which $f = 600$ THz, $\lambda = 500$ nm. Graph λ as a function of f for any light wave.

From the statement above, we know that $f\lambda = c$, and from the given values we have

$$(600 \text{ THz})(500 \text{ nm}) = (6.0 \times 10^{14} \text{ Hz})(5.0 \times 10^{-7} \text{ m}) = 3.0 \times 10^8 \text{ m/s}$$

which means $c = 3.0 \times 10^8$ m/s. We are to sketch $f\lambda = 3.0 \times 10^8$. Solving for λ as $\lambda = 3.0 \times 10^8/f$, we have the following table (only positive values have meaning):

f (THz)	750	600	500	430
λ (nm)	400	500	600	700

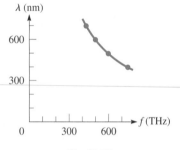

Fig. 21.77

See Fig. 21.77. (Violet light has wavelengths of about 400 nm, orange light has wavelengths of about 600 nm, and red light has wavelengths of about 700 nm.) ▶

We can conclude that *the equation of a hyperbola is characterized by the presence of both an x^2-term and a y^2-term, having different signs, or by the presence of an xy-term with no squared terms.* We will consider the equation of the hyperbola further in Sections 21.7 and 21.8.

The hyperbola has some very useful applications. The LORAN radio navigation system is based on the use of hyperbolic paths. Some reflecting telescopes use hyperbolic mirrors. The paths of comets that never return to pass by the sun are hyperbolic. Some applications are illustrated in the exercises.

EXERCISES 21.6

In Exercises 1 and 2, make the given changes in the indicated examples of this section and then solve the resulting problems.

1. In Example 2, interchange the denominators of 4 and 16; find the vertices, ends of the conjugate axis, and foci; and sketch the hyperbola.

2. In Example 3, change $-9y^2$ to $-y^2$ and then follow the same instructions as in Exercise 1.

In Exercises 3–16, find the coordinates of the vertices and the foci of the given hyperbolas. Sketch each curve.

3. $\dfrac{x^2}{25} - \dfrac{y^2}{144} = 1$

4. $\dfrac{x^2}{16} - \dfrac{y^2}{4} = 1$

5. $\dfrac{y^2}{9} - \dfrac{x^2}{1} = 1$

6. $\dfrac{y^2}{2} - \dfrac{x^2}{2} = 1$

7. $\dfrac{4x^2}{25} - \dfrac{y^2}{4} = 1$

8. $\dfrac{9y^2}{25} - x^2 = 1$

9. $4x^2 - y^2 = 4$

10. $x^2 - 9y^2 = 81$

11. $2y^2 - 5x^2 = 10$

12. $3y^2 - 2x^2 = 300$

13. $y^2 = 4(x^2 + 1)$

14. $y^2 = 9(x^2 - 1)$

15. $4x^2 - y^2 = 0.64$

16. $9y^2 - x^2 = 0.36$

In Exercises 17–28, find the equations of the hyperbolas satisfying the given conditions. The center of each is at the origin.

17. Vertex $(3, 0)$, focus $(5, 0)$

18. Vertex $(0, 1)$, focus $\left(0, \sqrt{3}\right)$

19. Conjugate axis = 12, vertex $(0, 10)$

20. Sum of lengths of transverse and conjugate axes 28, focus $(10, 0)$

21. Passes through $(2, 3)$, focus $(2, 0)$

22. Passes through $\left(8, \sqrt{3}\right)$, vertex $(4, 0)$

23. Passes through $(5, 4)$ and $\left(3, \frac{4}{5}\sqrt{5}\right)$

24. Passes through $(1, 2)$ and $\left(2, 2\sqrt{2}\right)$

25. Asymptote $y = 2x$, vertex $(1, 0)$

26. Asymptote $y = -4x$, vertex $(0, 4)$

27. The difference of distances to (x, y) from $(10, 0)$ and $(-10, 0)$ is 12.

28. The difference of distances to (x, y) from $(0, 4)$ and $(0, -4)$ is 6.

In Exercises 29–48, solve the given problems.

29. Sketch the graph of the hyperbola $xy = 2$.

30. Sketch the graph of the hyperbola $xy = -4$.

31. Find the equation of the hyperbola with foci $(1, 2)$ and $(11, 2)$, and a transverse axis of 8, by use of the definition. Sketch the curve.

32. Find the equation of the hyperbola with vertices $(-2, 4)$ and $(-2, -2)$, and a conjugate axis of 4, by use of the definition. Sketch the curve.

33. Use a graphing calculator to view the hyperbola $x^2 - 4y^2 + 4x + 32y - 64 = 0$.

34. Use a graphing calculator to view the hyperbola $5y^2 - 4x^2 + 8x + 40y + 56 = 0$.

35. The equation of a hyperbola with center (h, k) and transverse axis parallel to the x-axis is $\dfrac{(x - h)^2}{a^2} - \dfrac{(y - k)^2}{b^2} = 1$. (This is shown in Section 21.7). Sketch the hyperbola that has a transverse axis of 4, a conjugate axis of 6, and for which (h, k) is $(-3, 2)$.

36. The equation of a hyperbola with center (h, k) and transverse axis parallel to the y-axis is $\dfrac{(y - k)^2}{a^2} - \dfrac{(x - h)^2}{b^2} = 1$. (This is shown in Section 21.7). Sketch the hyperbola that has a transverse axis of 2, a conjugate axis of 8, and for which (h, k) is $(5, 0)$.

37. Two concentric (same center) hyperbolas are called conjugate hyperbolas if the transverse and conjugate axes of one are, respectively, the conjugate and transverse axes of the other. What is the equation of the hyperbola conjugate to the hyperbola in Exercise 18?

38. As with an ellipse, the *eccentricity* e of a hyperbola is defined as $e = c/a$. Find the eccentricity of the hyperbola $2x^2 - 3y^2 = 24$.

39. Graph the inequality $100x^2 - 49y^2 \le 4900$.

(W) 40. Explain how a branch of a hyperbola differs from a parabola.

41. The cross section of the roof of a storage building shown in Fig. 21.78 is hyperbolic with the horizontal beam passing through the focus. Find the equation of the hyperbola such that its center is at the origin.

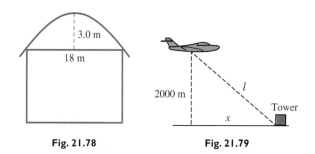

Fig. 21.78   **Fig. 21.79**

42. A plane is flying at a constant altitude of 2000 m. Show that the equation relating the horizontal distance x and the direct line distance l from a control tower to the plane is that of a hyperbola. Sketch the graph of l as a function of x. See Fig. 21.79.

43. A jet travels 600 km at a speed of v km/h for t hours. Graph the equation relating v as a function of t.

44. A drain pipe 100 m long has an inside diameter d (in m) and an outside diameter D (in m). If the volume of material of the pipe itself is 0.50 m³, what is the equation relating d and D? Graph the function of D as a function of d.

45. Ohm's law in electricity states that the product of the current i and the resistance R equals the voltage V across the resistance. If a battery of 6.00 V is placed across a variable resistor R, find the equation relating i and R and sketch the graph of i as a function of R.

46. A ray of light directed at one focus of a hyperbolic mirror is reflected toward the other focus. Find the equation that represents the hyperbolic mirror shown in Fig. 21.80.

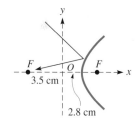

Fig. 21.80

47. A radio signal is sent simultaneously from stations A and B 600 km apart on the Carolina coast. A ship receives the signal from A 1.20 ms before it receives the signal from B. Given that radio signals travel at 300 km/ms, draw a graph showing the possible locations of the ship. This problem illustrates the basis of LORAN.

48. Maximum intensity for monochromatic (single-color) light from two sources occurs where the difference in distances from the sources is an integral number of wavelengths. Find the equation of the curves of maximum intensity in a thin film between the sources where the difference in paths is two wavelengths and the sources are four wavelengths apart. Let the sources be on the x-axis and the origin midway between them. Use units of one wavelength for both x and y.

21.7 TRANSLATION OF AXES

The equations we have considered for the parabola, the ellipse, and the hyperbola are those for which the center of the ellipse or hyperbola, or vertex of the parabola, is at the origin. In this section, we consider, without specific use of the definition, the equations of these curves for the cases in which the axis of the curve is parallel to one of the coordinate axes. This is done by **translation of axes.**

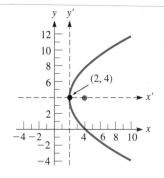

Fig. 21.81

In Fig. 21.81, we choose a point (h, k) in the xy-coordinate plane as the origin of another coordinate system, the $x'y'$-coordinate system. The x'-axis is parallel to the x-axis and the y'-axis is parallel to the y-axis. Every point now has two sets of coordinates (x, y) and (x', y'). We see that

$$x = x' + h \quad \text{and} \quad y = y' + k \tag{21.26}$$

Equation (21.26) can also be written in the form

$$x' = x - h \quad \text{and} \quad y' = y - k \tag{21.27}$$

EXAMPLE 1 Find the equation of the parabola with vertex $(2, 4)$ and focus $(4, 4)$.

If we let the origin of the $x'y'$-coordinate system be the point $(2, 4)$, then the point $(4, 4)$ is the point $(2, 0)$ in the $x'y'$-system. This means $p = 2$ and $4p = 8$. See Fig. 21.82. In the $x'y'$-system, the equation is

$$(y')^2 = 8(x')$$

Using Eqs. (21.27), we have

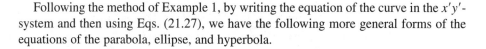

as the equation of the parabola in the xy-coordinate system.

Fig. 21.82

Following the method of Example 1, by writing the equation of the curve in the $x'y'$-system and then using Eqs. (21.27), we have the following more general forms of the equations of the parabola, ellipse, and hyperbola.

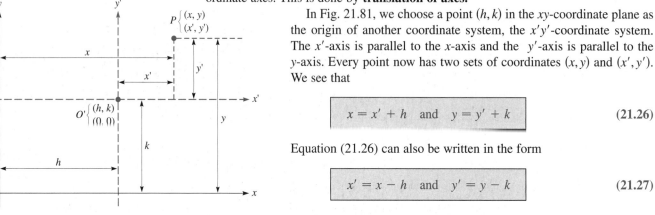

Parabola, vertex (h, k):	$(y - k)^2 = 4p(x - h)$	(axis parallel to x-axis)	(21.28)
	$(x - h)^2 = 4p(y - k)$	(axis parallel to y-axis)	(21.29)
Ellipse, center (h, k):	$\dfrac{(x - h)^2}{a^2} + \dfrac{(y - k)^2}{b^2} = 1$	(major axis parallel to x-axis)	(21.30)
	$\dfrac{(y - k)^2}{a^2} + \dfrac{(x - h)^2}{b^2} = 1$	(major axis parallel to y-axis)	(21.31)
Hyperbola, center (h, k):	$\dfrac{(x - h)^2}{a^2} - \dfrac{(y - k)^2}{b^2} = 1$	(transverse axis parallel to x-axis)	(21.32)
	$\dfrac{(y - k)^2}{a^2} - \dfrac{(x - h)^2}{b^2} = 1$	(transverse axis parallel to y-axis)	(21.33)

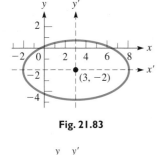

Fig. 21.83

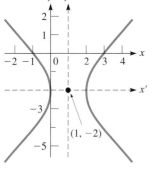

Fig. 21.84

CAUTION ▶

Solving a Word Problem

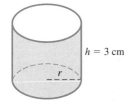

$h = 3$ cm

r

Fig. 21.85

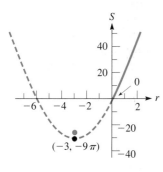

$(-3, -9\pi)$

Fig. 21.86

◀ EXAMPLE 2 Describe the curve of the equation

$$\frac{(x - 3)^2}{25} + \frac{(y + 2)^2}{9} = 1$$

We see that this equation fits the form of Eq. (21.30) with $h = 3$ and $k = -2$. It is the equation of an ellipse with its center at $(3, -2)$ and its major axis parallel to the x-axis. The semimajor axis is $a = 5$, and the semiminor axis is $b = 3$. The ellipse is shown in Fig. 21.83.

◀ EXAMPLE 3 Find the center of the hyperbola $2x^2 - y^2 - 4x - 4y - 4 = 0$.

To analyze this curve, we first complete the square in the x-terms and in the y-terms. This will allow us to recognize properly the choice of h and k.

$$2x^2 - 4x - y^2 - 4y = 4$$

$$2(x^2 - 2x \quad) - (y^2 + 4y \quad) = 4$$

$$2(x^2 - 2x + 1) - (y^2 + 4y + 4) = 4 + 2 - 4$$

We note here that when we added 1 to complete the square of the x-terms within the parentheses, **we were actually adding 2 to the left side.** Thus, we added 2 to the right side. Similarly, when we added 4 to the y-terms within the parentheses, **we were actually subtracting 4 from the left side.** Continuing, we have

$$2(x - 1)^2 - (y + 2)^2 = 2$$

coordinates of center $(1, -2)$

$$\frac{(x - 1)^2}{1} - \frac{(y + 2)^2}{2} = 1$$

Therefore, the center of the hyperbola is $(1, -2)$. See Fig. 21.84. ▶

◀ EXAMPLE 4 Cylindrical glass beakers are to be made with a height of 3 cm. Express the surface area in terms of the radius of the base and sketch the curve.

The total surface area S of a beaker is the sum of the area of the base and the lateral surface area of the side. In general, S in terms of the radius r of the base and height h of the side is $S = \pi r^2 + 2\pi rh$. Since $h = 3$ cm, we have

$$S = \pi r^2 + 6\pi r$$

which is the desired relationship. See Fig. 21.85.

To analyze the equation relating S and r, we complete the square of the r terms:

$$S = \pi(r^2 + 6r)$$

$$S + 9\pi = \pi(r^2 + 6r + 9) \qquad \text{complete the square}$$

$$S + 9\pi = \pi(r + 3)^2$$

vertex $(-3, -9\pi)$

$$(r + 3)^2 = \frac{1}{\pi}(S + 9\pi)$$

This represents a parabola with vertex $(-3, -9\pi)$. Since $4p = 1/\pi$, $p = 1/(4\pi)$, the focus is $\left(-3, \frac{1}{4\pi} - 9\pi\right)$ as shown in Fig. 21.86. The part of the graph for negative r is dashed since only positive values have meaning. ▶

EXERCISES 21.7

In Exercises 1 and 2, make the given changes in the indicated examples of this section and then solve the resulting problem.

1. In Example 2, change $y + 2$ to $y - 2$ and change the sign before the second term from $+$ to $-$. Then describe and sketch the curve.

2. In Example 3, change the fourth and fifth terms from $-4y - 4$ to $+ 6y - 9$ and then find the center.

In Exercises 3–10, describe the curve represented by each equation. Identify the type of curve and its center (or vertex if it is a parabola). Sketch each curve.

3. $(y - 2)^2 = 4(x + 1)$

4. $\dfrac{(x + 4)^2}{4} + \dfrac{(y - 1)^2}{1} = 1$

5. $\dfrac{(x - 1)^2}{4} - \dfrac{(y - 2)^2}{9} = 1$

6. $(y + 5)^2 = -8(x - 2)$

7. $\dfrac{(x + 1)^2}{1} + \dfrac{y^2}{9} = 1$

8. $\dfrac{(y - 4)^2}{16} - \dfrac{(x + 2)^2}{4} = 1$

9. $(x + 3)^2 = -12(y - 1)$

10. $\dfrac{x^2}{0.16} + \dfrac{(y + 1)^2}{0.25} = 1$

In Exercises 11–22, find the equation of each of the curves described by the given information.

11. Parabola: vertex $(-1, 3)$, focus $(3, 3)$

12. Parabola: focus $(2, -5)$, directrix $y = 3$

13. Parabola: axis, directrix are coordinate axes, focus $(12, 0)$

14. Parabola: vertex $(4, 4)$, vertical directrix, passes through $(0, 1)$

15. Ellipse: center $(-2, 2)$, focus $(-5, 2)$, vertex $(-7, 2)$

16. Ellipse: center $(0, 3)$, focus $(12, 3)$, major axis 26 units

17. Ellipse: center $(-2, 1)$, vertex $(-2, 5)$, passes through $(0, 1)$

18. Ellipse: foci $(1, -2)$ and $(1, 10)$, minor axis 5 units

19. Hyperbola: vertex $(-1, 1)$, focus $(-1, 4)$, center $(-1, 2)$

20. Hyperbola: foci $(2, 1)$ and $(8, 1)$, conjugate axis 6 units

21. Hyperbola: vertices $(2, 1)$ and $(-4, 1)$, focus $(-6, 1)$

22. Hyperbola: center $(1, -4)$, focus $(1, 1)$, transverse axis 8 units

In Exercises 23–40, determine the center (or vertex if the curve is a parabola) of the given curve. Sketch each curve.

23. $x^2 + 2x - 4y - 3 = 0$

24. $y^2 - 2x - 2y - 9 = 0$

25. $x^2 + 4y = 24$

26. $x^2 + 4y^2 = 32y$

27. $4x^2 + 9y^2 + 24x = 0$

28. $2x^2 + y^2 + 8x = 8y$

29. $9x^2 - y^2 + 8y = 7$

30. $2x^2 - 4x = 9y - 2$

31. $5x^2 - 4y^2 + 20x + 8y = 4$

32. $0.04x^2 + 0.16y^2 = 0.01y$

33. $4x^2 - y^2 + 32x + 10y + 35 = 0$

34. $2x^2 + 2y^2 - 24x + 16y + 95 = 0$

35. $9x^2 + 4y^2 - 12x + 16y + 16 = 0$

36. $5x^2 - 3y^2 - 40x + 95 = 0$

37. $7x^2 - y^2 - 14x - 16y - 64 = 0$

38. $5x^2 - 2y^2 + 12y + 18 = 0$

39. $9x^2 + 9y^2 - 6x - 24y + 14 = 0$

40. $4y^2 - 15x - 12y + 29 = 0$

In Exercises 41–52, solve the given problems.

41. Find the equation of the hyperbola with asymptotes $x - y = -1$ and $x + y = -3$ and vertex $(3, -1)$.

42. The circle $x^2 + y^2 + 4x - 5 = 0$ passes through the foci and the ends of the minor axis of an ellipse that has its major axis along the x axis. Find the equation of the ellipse.

43. The vertex and focus of one parabola are, respectively, the focus and vertex of a second parabola. Find the equation of the first parabola, if $y^2 = 4x$ is the equation of the second.

44. Identify the curve represented by $4y^2 - x^2 - 6x - 2y = 14$ and view it on a graphing calculator.

45. What is the general form of the equation of a *family* of parabolas if each vertex and focus is on the x-axis?

46. What is the general form of the equation of a *family* of ellipses with foci on the y-axis and if each passes through the origin?

47. An electric current (in A) is $i = 2 + \sin\left(2\pi t - \frac{\pi}{3}\right)$. What is the equation for the current if the origin of the (t', i') system is taken as $\left(\frac{1}{6}, 2\right)$ of the (t, i) system?

48. The stopping distance d (in m) of a car traveling at v km/h is represented by $d = 0.005v^2 + 0.2v$. Where is the vertex of the parabola that represents d?

49. The stream from a fire hose follows a parabolic curve and reaches a maximum height of 18 m at a horizontal distance of 28 m from the nozzle. Find the equation that represents the stream, with the origin at the nozzle. Sketch the graph.

50. For a constant capacitive reactance and a constant resistance, sketch the graph of the impedance and inductive reactance (as abscissas) for an alternating-current circuit. (See Section 12.7.)

51. Two wheels in a friction drive assembly are equal ellipses, as shown in Fig. 21.87. They are always in contact, with the left wheel fixed in position and the right wheel able to move horizontally. Find the equation that can be used to represent the circumference of each wheel in the position shown.

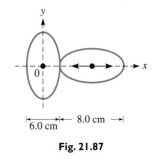

Fig. 21.87

52. An agricultural test station is to be divided into rectangular sections, each with a perimeter of 480 m. Express the area A of each section in terms of its width w and identify the type of curve represented. Sketch the graph of A as a function of w. For what value of w is A the greatest?

21.8 THE SECOND-DEGREE EQUATION

The equations of the circle, parabola, ellipse, and hyperbola are all special cases of the same general equation. In this section, we discuss this equation and how to identify the particular form it takes when it represents a specific type of curve.

Each of these curves can be represented by a **second-degree equation** *of the form*

$$Ax^2 + Bxy + Cy^2 + Dx + Ey + F = 0 \qquad \text{(21.34)}$$

The coefficients of the second-degree equation terms determine the type of curve that results. Recalling the discussions of the general forms of the equations of the circle, parabola, ellipse, and hyperbola from the previous sections of this chapter, Eq. (21.34) represents the indicated curve for given conditions of A, B, and C, as follows:

1. If $A = C$, $B = 0$, a circle.
2. If $A \neq C$ (but they have the same sign), $B = 0$, an ellipse.
3. If A and C have different signs, $B = 0$, a hyperbola.
4. If $A = 0$, $C = 0$, $B \neq 0$, a hyperbola.
5. If either $A = 0$ or $C = 0$ (both not both), $B = 0$, a parabola. (Special cases, such as a single point or no real locus, can also result.)

Another conclusion about Eq. (21.34) is that, if either $D \neq 0$ or $E \neq 0$ (or both), the center of the curve (or the vertex of a parabola) is not at the origin. If $B \neq 0$, the axis of the curve has been rotated. We have considered only one such case (the hyperbola $xy = c$) in this chapter. Rotation of axes is covered as one of the supplementary topics following the final chapter.

❙ **EXAMPLE 1** The equation $2x^2 = 3 - 2y^2$ represents a circle. This can be seen by putting the equation in the form of Eq. (21.34). This form is

$$2x^2 + 2y^2 - 3 = 0$$
$$A = 2 \qquad \qquad C = 2$$

We see that $A = C$. Also, since there is no xy-term, we know that $B = 0$. This means that the equation represents a circle. If we write it as $x^2 + y^2 = \frac{3}{2}$, we see that it fits the form of Eq. (21.12). The circle is shown in Fig. 21.88. ❙

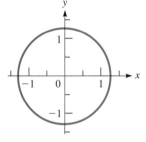

Fig. 21.88

❙ **EXAMPLE 2** The equation $3x^2 = 6x - y^2 + 3$ represents an ellipse. Before we analyze the equation, we should put it in the form of Eq. (21.34). For this equation, this form is

$$3x^2 + y^2 - 6x - 3 = 0$$
$$A = 3 \qquad \qquad C = 1$$

Here, we see that $B = 0$, A and C have the same sign, and $A \neq C$. Therefore, it is an ellipse. The $-6x$ term indicates that the center of the ellipse is not at the origin. The ellipse is shown in Fig. 21.89. ❙

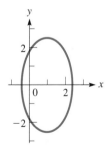

Fig. 21.89

◀ EXAMPLE 3 Identify the curve represented by $2x^2 + 12x = y^2 - 14$. Determine the appropriate quantities for the curve, and sketch the graph.

Writing this equation in the form of Eq. (21.34), we have

$$2x^2 - y^2 + 12x + 14 = 0$$

$$A = 2 \quad\quad\quad C = -1$$

We identify this equation as representing a hyperbola, since A and C have different signs and $B = 0$. We now write it in the standard form of a hyperbola:

$$2x^2 + 12x - y^2 = -14$$

$$2(x^2 + 6x \quad\quad) - y^2 = -14 \qquad \text{complete the square}$$

$$2(x^2 + 6x + 9) - y^2 = -14 + 18$$

$$2(x + 3)^2 - y^2 = 4$$

center is (0, 0)

$$\frac{(x + 3)^2}{2} - \frac{y^2}{4} = 1 \qquad \frac{x'^2}{2} - \frac{y'^2}{4} = 1$$

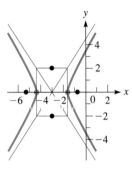

Fig. 21.90

Thus, we see that the center (h, k) of the hyperbola is the point $(-3, 0)$. Also, $a = \sqrt{2}$ and $b = 2$. This means that the vertices are $\left(-3 + \sqrt{2}, 0\right)$ and $\left(-3 - \sqrt{2}, 0\right)$, and the conjugate axis extends from $(-3, 2)$ to $(-3, -2)$. Also, $c^2 = 2 + 4 = 6$, which means that $c = \sqrt{6}$. The foci are $\left(-3 + \sqrt{6}, 0\right)$ and $\left(-3 - \sqrt{6}, 0\right)$. The graph is shown in Fig. 21.90. ▶

◀ EXAMPLE 4 Identify the curve represented by $4y^2 - 23 = 4(4x + 3y)$ and find the appropriate important quantities. Then view it on a graphing calculator.

Writing the equation in the form of Eq. (21.34), we have

$$4y^2 - 16x - 12y - 23 = 0$$

Therefore, we recognize the equation as representing a parabola, since $A = 0$ and $B = 0$. Now, writing the equation in the standard form of a parabola, we have

$$4y^2 - 12y = 16x + 23$$

$$4(y^2 - 3y \quad\quad) = 16x + 23 \qquad \text{complete the square}$$

$$4\left(y^2 - 3y + \frac{9}{4}\right) = 16x + 23 + 9$$

$$4\left(y - \frac{3}{2}\right)^2 = 16(x + 2)$$

vertex $\left(-2, \frac{3}{2}\right)$

$$\left(y - \frac{3}{2}\right)^2 = 4(x + 2) \quad \text{or} \quad y'^2 = 4x'$$

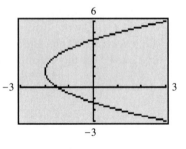

Fig. 21.91

We now note that the vertex is $(-2, 3/2)$ and that $p = 1$. This means that the focus is $(-1, 3/2)$ and the directrix is $x = -3$.

To view the graph of this equation on a graphing calculator, we first solve the equation for y and get $y = \left(3 \pm 4\sqrt{x + 2}\right)/2$. Entering these two functions in the calculator, we get the view shown in Fig. 21.91. ▶

See Appendix C for a graphing calculator program GRAPHCON. It displays the graph of the conic $Ax^2 + Cy^2 + Dx + Ey + F = 0$.

In Chapter 14, when these curves were first introduced, they were referred to as **conic sections.** If a plane is passed through a cone, the intersection of the plane and the cone results in one of these curves; the curve formed depends on the angle of the plane with respect to the axis of the cone. This is shown in Fig. 21.92.

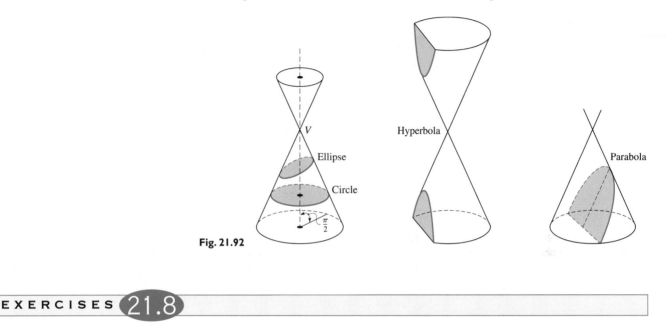

Fig. 21.92

EXERCISES 21.8

In Exercises 1 and 2, make the given changes in the indicated examples of this section and then solve the indicated problem.

1. In Example 1, change the $-$ before the $2y^2$ to $+$ and then determine what type of curve is represented.

2. In Example 3, change $y^2 - 14$ to $14 - y^2$ and then determine what type of curve is represented.

In Exercises 3–24, identify each of the equations as representing a circle, a parabola, an ellipse, a hyperbola, or none of these.

3. $x^2 + 2y^2 - 2 = 0$
4. $x^2 - y = 0$
5. $2x^2 - y^2 - 1 = 0$
6. $y(y + x^2) = 4$
7. $2x^2 + 2y^2 - 3y = 1$
8. $x(x - 3) = y(1 - 2y^2)$
9. $2.2x^2 - x - y = 1.6$
10. $2x^2 + 4y^2 = y + 2x$
11. $x^2 = (y - 1)(y + 1)$
12. $3.2x^2 = 2.1y(1 - 2y)$
13. $36x^2 = 12y(1 - 3y) + 1$
14. $y = 3(1 - 2x)(1 + 2x)$
15. $y(3 - 2x) = x(5 - 2y)$
16. $x(13 - 5x) = 5y^2$
17. $2xy + x - 3y = 6$
18. $(y + 1)^2 = x^2 + y^2 - 1$
19. $2x(x - y) = y(3 - y - 2x)$
20. $2x^2 = x(x - 1) + 4y^2$
21. $(x + 1)^2 + (y + 1)^2 = 2(x + y + 1)$
22. $(2x + y)^2 = 4x(y - 2) - 16$
23. $x(y + 3x) = x^2 + xy - y^2 + 1$
24. $4x(x - 1) = 2x^2 - 2y^2 + 3$

In Exercises 25–32, identify the curve represented by each of the given equations. Determine the appropriate important quantities for the curve and sketch the graph.

25. $x^2 = 8(y - x - 2)$
26. $x^2 = 6x - 4y^2 - 1$
27. $y^2 = 2(x^2 - 2x - 2y)$
28. $4x^2 + 4 = 9 - 8x - 4y^2$
29. $y^2 + 42 = 2x(10 - x)$
30. $x^2 - 4y = y^2 + 4(1 - x)$
31. $4(y^2 - 4x - 2) = 5(4y - 5)$
32. $2(2x^2 - y) = 8 - y^2$

In Exercises 33–36, view the curve for each equation on a graphing calculator. In Exercises 33 and 34, identify the type of curve before viewing it. In Exercises 35 and 36, the axis of each curve has been rotated.

33. $x^2 + 2y^2 - 4x + 12y + 14 = 0$
34. $4y^2 - x^2 + 40y - 4x + 60 = 0$
35. $x^2 + 6xy + 9y^2 - 2x + 14y - 10 = 0$
36. $x^2 - xy + y^2 - 6 = 0$

In Exercises 37–42, use the given values to determine the type of curve represented.

37. For the equation $x^2 + ky^2 = a^2$, what type of curve is represented if (a) $k = 1$, (b) $k < 0$, and (c) $k > 0$ $(k \neq 1)$?

38. For the equation $\dfrac{x^2}{4 - C} - \dfrac{y^2}{C} = 1$, what type of curve is represented if (a) $C < 0$ and (b) $0 < C < 4$? (For $C > 4$, see Exercise 40.)

(W) **39.** In Eq. (21.34), if $A > C > 0$ and $B = D = E = F = 0$, describe the locus of the equation.

(W) **40.** For the equation in Exercise 38, describe the locus of the equation if $C > 4$.

(W) **41.** In Eq. (21.34), if $A = B = C = 0, D \neq 0, E \neq 0$, and $F \neq 0$, describe the locus of the equation.

(W) **42.** In Eq. (21.34), if $A = -C \neq 0, B = D = E = 0$ and $F = C$, describe the locus of the equation if $C > 0$.

In Exercises 43–48, determine the type of curve from the given information.

43. The diagonal brace in a rectangular metal frame is 3.0 cm longer than the length of one of the sides. Determine the type of curve represented by the equation relating the lengths of the sides of the frame.

44. One circular solar cell has a radius that is 2.0 cm less than the radius r of a second circular solar cell. Determine the type of curve represented by the equation relating the total area A of both cells and r.

45. A flashlight emits a cone of light onto the floor. What type of curve is the perimeter of the lighted area on the floor, if the floor cuts completely through the cone of light?

46. What type of curve is the perimeter of the lighted area on the floor of the cone of light in Exercise 45, if the upper edge of the cone is parallel to the floor?

47. A supersonic jet creates a conical shock wave behind it. What type of curve is outlined on the surface of a lake by the shock wave if the jet is flying horizontally?

48. In Fig. 21.92, if the plane cutting the cones passes through the intersection of the upper and lower cones, what type of curve is the intersection of the plane and cones?

21.9 POLAR COORDINATES

The Swiss mathematician Jakob Bernoulli (1654–1705) was among the first to make significant use of polar coordinates.

To this point, we have graphed all curves in the rectangular coordinate system. However, for certain types of curves, other coordinate systems are better adapted. We discuss one of these systems here.

Instead of designating a point by its *x*- and *y*-coordinates, we can specify its location by its radius vector and the angle the radius vector makes with the *x*-axis. Thus, the *r* and θ that are used in the definitions of the trigonometric functions can also be used as the coordinates of points in the plane. The important aspect of choosing coordinates is that, for each set of values, there must be only one point that corresponds to this set. We can see that this condition is satisfied by the use of *r* and θ as coordinates. *In* **polar coordinates,** *the origin is called the* **pole,** *and the half-line for which the angle is zero (equivalent to the positive x-axis) is called the* **polar axis.** The coordinates of a point are designated as (r, θ). We will use radians when measuring the value of θ. See Fig. 21.93.

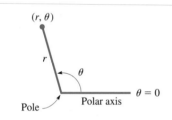

Fig. 21.93

When using polar coordinates, we generally label the lines for some of the values of θ; namely, those for $\theta = 0$ (the polar axis), $\theta = \pi/2$ (equivalent to the positive *y*-axis), $\theta = \pi$ (equivalent to the negative *x*-axis), $\theta = 3\pi/2$ (equivalent to the negative *y*-axis), and possibly others. In Fig. 21.94, these lines and those for multiples of $\pi/6$ are shown. Also, the circles for $r = 1$, $r = 2$, and $r = 3$ are shown in this figure.

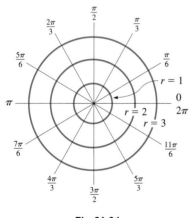

Fig. 21.94

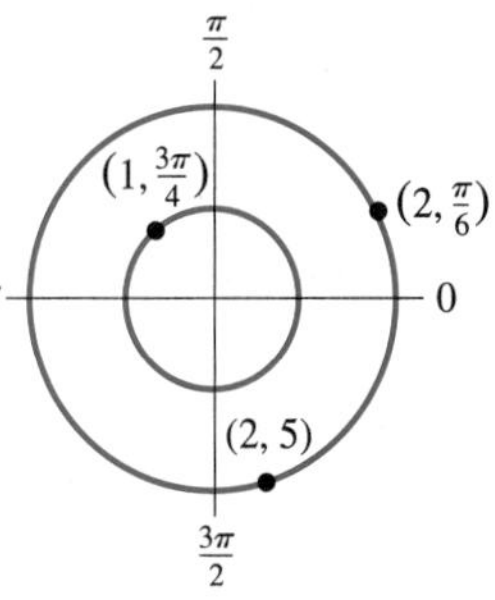

◀ **EXAMPLE 1** **(a)** If $r = 2$ and $\theta = \pi/6$, we have the point shown in Fig. 21.95. The polar coordinates (r, θ) are written as $(2, \pi/6)$. This point corresponds to $\left(\sqrt{3}, 1\right)$ in rectangular coordinates.

(b) The polar coordinate point $(1, 3\pi/4)$ is also shown. It is equivalent to $\left(-\sqrt{2}/2, \sqrt{2}/2\right)$ in rectangular coordinates.

(c) The polar coordinate point $(2, 5)$ is also shown. It is equivalent approximately to $(0.6, -1.9)$ in rectangular coordinates. Remember, the 5 is an angle in radians. ▶

Fig. 21.95

One difference between rectangular coordinates and polar coordinates is that, for each point in the plane, there are limitless possibilities for the polar coordinates of that point. For example, the point $\left(2, \frac{\pi}{6}\right)$ can also be represented by $\left(2, \frac{13\pi}{6}\right)$ since the angles $\frac{\pi}{6}$ and $\frac{13\pi}{6}$ are coterminal. We also remove one restriction on r that we imposed in the definition of the trigonometric functions. That is, r is allowed to take on positive and negative values. If r is negative, θ is located as before, but **the point is found r units from the pole but on the opposite side** from that on which it is positive.

EXAMPLE 2   The coordinates $(3, 2\pi/3)$ and $(3, -4\pi/3)$ represent the same point. However, the point $(-3, 2\pi/3)$ is on the opposite side of the pole, 3 units from the pole. Another possible set of coordinates for the point $(-3, 2\pi/3)$ is $(3, 5\pi/3)$. See Fig. 21.96.

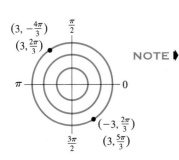

$\left(3, -\frac{4\pi}{3}\right)$
$\left(3, \frac{2\pi}{3}\right)$
$\left(-3, \frac{2\pi}{3}\right)$
$\left(3, \frac{5\pi}{3}\right)$

Fig. 21.96

When plotting a point in polar coordinates, it is generally easier to *first locate the terminal side of θ and then measure r along this terminal side.* This is illustrated in the following example.

EXAMPLE 3   Plot the points $A(2, 5\pi/6)$ and $B(-3.2, -2.4)$ in the polar coordinate system.

To locate A, we determine the terminal side of $\theta = 5\pi/6$ and then determine $r = 2$. See Fig. 21.97.

To locate B, we find the terminal side of $\theta = -2.4$, measuring clockwise from the polar axis (and recalling that $\pi = 3.14 = 180°$). Then we locate $r = -3.2$ on the opposite side of the pole. See Fig. 21.97.

We will find that points with negative values of r occur frequently when plotting curves in polar coordinates.

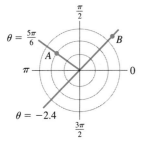

$\theta = \frac{5\pi}{6}$

$\theta = -2.4$

Fig. 21.97

POLAR AND RECTANGULAR COORDINATES

The relationships between the polar coordinates of a point and the rectangular coordinates of the same point come from the definitions of the trigonometric functions. Those most commonly used are (see Fig. 21.98)

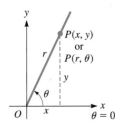

$$x = r\cos\theta \quad y = r\sin\theta \tag{21.35}$$

$$\tan\theta = \frac{y}{x} \quad r = \sqrt{x^2 + y^2} \tag{21.36}$$

Fig. 21.98

The following examples show the use of Eqs. (21.35) and (21.36) in changing coordinates in one system to coordinates in the other system. Also, these equations are used to transform equations from one system to the other.

EXAMPLE 4   Using Eqs. (21.35), we can transform the polar coordinates of $(4, \pi/4)$ into the rectangular coordinates $\left(2\sqrt{2}, 2\sqrt{2}\right)$, since

$$x = 4\cos\frac{\pi}{4} = 4\left(\frac{\sqrt{2}}{2}\right) = 2\sqrt{2} \quad \text{and} \quad y = 4\sin\frac{\pi}{4} = 4\left(\frac{\sqrt{2}}{2}\right) = 2\sqrt{2}$$

See Fig. 21.99.

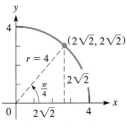

$\left(2\sqrt{2}, 2\sqrt{2}\right)$
$r = 4$

Fig. 21.99

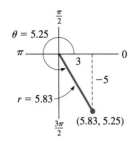

Fig. 21.100

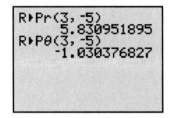

Fig. 21.101

The cyclotron was invented in 1931 at the University of California. It was the first accelerator to deflect particles into circular paths.

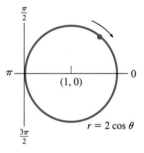

Fig. 21.102

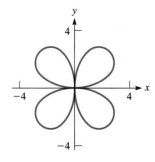

Fig. 21.103

◀ **EXAMPLE 5** Using Eqs. (21.36), we can transform the rectangular coordinates $(3, -5)$ into polar coordinates:

$$\tan \theta = -\frac{5}{3}, \quad \theta = 5.25 \quad (\text{or } -1.03)$$
$$r = \sqrt{3^2 + (-5)^2} = 5.83$$

We know that θ is a fourth-quadrant angle since x is positive and y is negative. Therefore, the point $(3, -5)$ in rectangular coordinates can be expressed as the point $(5.83, 5.25)$ in polar coordinates (see Fig. 21.100). Other polar coordinates for the point are also possible. ▶

Calculators are programmed to make conversions between rectangular coordinates and polar coordinates. The manual for any model should be consulted to determine how any particular model is used for these conversions. For a calculator that uses the *angle* feature, the display for the conversions of Example 5 is shown in Fig. 21.101.

◀ **EXAMPLE 6** If an electrically charged particle enters a magnetic field at right angles to the field, the particle follows a circular path. This fact is used in the design of nuclear particle accelerators.

A proton (positively charged) enters a magnetic field such that its path may be described by the rectangular equation $x^2 + y^2 = 2x$, where measurements are in meters. Find the polar equation of this circle.

We change this equation expressed in the rectangular coordinates x and y into an equation expressed in the polar coordinates r and θ by using the relations $r^2 = x^2 + y^2$ and $x = r \cos \theta$ as follows:

$$\begin{array}{ll} x^2 + y^2 = 2x & \text{rectangular equation} \\ r^2 = 2r \cos \theta & \text{substitute} \\ r = 2 \cos \theta & \text{divided by } r \end{array}$$

This is the polar equation of the circle, which is shown in Fig. 21.102. ▶

◀ **EXAMPLE 7** Find the rectangular equation of the *rose* $r = 4 \sin 2\theta$.

Using the trigonometric identity $\sin 2\theta = 2 \sin \theta \cos \theta$ and Eqs. (21.35) and (21.36) leads to the solution:

$$\begin{array}{ll} r = 4 \sin 2\theta & \text{polar equation} \\ = 4(2 \sin \theta \cos \theta) = 8 \sin \theta \cos \theta & \text{using identity} \\ \sqrt{x^2 + y^2} = 8\left(\frac{y}{r}\right)\left(\frac{x}{r}\right) = \frac{8xy}{r^2} = \frac{8xy}{x^2 + y^2} & \text{using Eqs. (21.35) and (21.36)} \\ x^2 + y^2 = \frac{64x^2y^2}{(x^2 + y^2)^2} & \text{squaring both sides} \\ (x^2 + y^2)^3 = 64x^2y^2 & \text{simplifying} \end{array}$$

Plotting the graph of this equation from the rectangular equation would be complicated. However, as we will see in the next section, plotting this graph in polar coordinates is quite simple. The curve is shown in Fig. 21.103. ▶

EXERCISES 21.9

In Exercises 1–4, make the given changes in the indicated examples of this section and then solve the indicated problems.

1. In Example 2, change $2\pi/3$ to $\pi/3$ and then find another set of coordinates for each point, similar to those shown for the points in the example.

2. In Example 4, change $\pi/4$ to $\pi/6$ and then find the rectangular coordinates.

3. In Example 5, change 3 to -3 and -5 to 5 and then find the polar coordinates.

4. In Example 7, change sin to cos and then find the rectangular equation.

In Exercises 5–16, plot the given polar coordinate points on polar coordinate paper.

5. $\left(3, \dfrac{\pi}{6}\right)$ 6. $(2, \pi)$ 7. $\left(\dfrac{5}{2}, -\dfrac{2\pi}{5}\right)$

8. $\left(5, -\dfrac{\pi}{3}\right)$ 9. $\left(-2, \dfrac{7\pi}{6}\right)$ 10. $\left(-5, \dfrac{\pi}{4}\right)$

11. $\left(-3, -\dfrac{5\pi}{4}\right)$ 12. $\left(-4, -\dfrac{5\pi}{3}\right)$ 13. $(2, 2)$

14. $(-1, -1)$ 15. $(0.5, -8.4)$ 16. $(2.2, -18.8)$

In Exercises 17–20, find a set of polar coordinates for each of the points for which the rectangular coordinates are given.

17. $\left(\sqrt{3}, 1\right)$ 18. $(-1, -1)$

19. $\left(-\dfrac{\sqrt{3}}{2}, -\dfrac{1}{2}\right)$ 20. $(-5, 4)$

In Exercises 21–24, find the rectangular coordinates for each of the points for which the polar coordinates are given.

21. $\left(8, \dfrac{4\pi}{3}\right)$ 22. $(-4, -\pi)$

23. $(3.0, -0.40)$ 24. $(-1.0, 1.0)$

In Exercises 25–36, find the polar equation of each of the given rectangular equations.

25. $x = 3$ 26. $y = x$

27. $x + 2y = 3$ 28. $x^2 + y^2 = 0.81$

29. $x^2 + (y - 2)^2 = 4$ 30. $x^2 - y^2 = 0.01$

31. $x^2 + 4y^2 = 4$ 32. $y^2 = 4x$

33. $x^2 + y^2 = 6y$ 34. $xy = 9$

35. $x^3 + y^3 - 4xy = 0$ 36. $y = \dfrac{2x}{x^2 + 1}$

In Exercises 37–48, find the rectangular equation of each of the given polar equations. In Exercises 37–44, identify the curve that is represented by the equation.

37. $r = \sin \theta$ 38. $r = 4 \cos \theta$

39. $r \cos \theta = 4$ 40. $r \sin \theta = -2$

41. $r = \dfrac{2}{\cos \theta - 3 \sin \theta}$ 42. $r = e^{r \cos \theta} \csc \theta$

43. $r = 4 \cos \theta + 2 \sin \theta$ 44. $r \sin(\theta + \pi/6) = 3$

45. $r = 2(1 + \cos \theta)$ 46. $r = 1 - \sin \theta$

47. $r^2 = \sin 2\theta$ 48. $r^2 = 16 \cos 2\theta$

In Exercises 49–56, solve the given problems. All coordinates given are polar coordinates.

49. The center of a regular hexagon is at the pole with one vertex at $(2, \pi)$. What are the coordinates of the other vertices?

50. Find the distance between the points $(4, \pi/6)$ and $(5, 5\pi/3)$.

51. Under certain conditions, the x- and y-components of a magnetic field B are given by the equations

$$B_x = \dfrac{-ky}{x^2 + y^2} \quad \text{and} \quad B_y = \dfrac{kx}{x^2 + y^2}$$

Write these equations in terms of polar coordinates.

52. In designing a domed roof for a building, an architect uses the equation $x^2 + \dfrac{y^2}{k^2} = 1$, where k is a constant. Write this equation in polar form.

53. The shape of a cam can be described by the polar equation $r = 3 - \sin \theta$. Find the rectangular equation for the shape of the cam.

54. The polar equation of the path of a weather satellite of the earth is $r = \dfrac{7600}{1 + 0.14 \cos \theta}$, where r is measured in kilometers. Find the rectangular equation of the path of this satellite. The path is an ellipse, with the earth at one of the foci.

55. The control tower of an airport is taken to be at the pole, and the polar axis is taken as due east in a polar coordinate graph. How far apart (in km) are planes, at the same altitude, if their positions on the graph are $(6.10, 1.25)$ and $(8.45, 3.74)$?

W 56. The perimeter of a certain type of machine part can be described by the equation $r = a \sin \theta + b \cos \theta$ $(a > 0, b > 0)$. Explain why all such machine parts are circular.

21.10 CURVES IN POLAR COORDINATES

The basic method for finding a curve in polar coordinates is the same as in rectangular coordinates. We assume values of the independent variable—in this case, θ—and then find the corresponding values of the dependent variable r. These points are then plotted and joined, thereby forming the curve that represents the relation in polar coordinates.

Before using the basic method, it is useful to point out that certain basic curves can be sketched directly from the equation. This is done by noting the meaning of each of the polar coordinate variables, r and θ. This is illustrated in the following example.

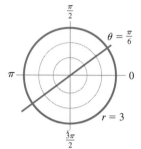

Fig. 21.104

EXAMPLE 1 **(a)** The graph of the polar equation $r = 3$ is a circle of radius 3, with center at the pole. This can be seen to be the case, since $r = 3$ for all possible values of θ. It is not necessary to find specific points for this circle, which is shown in Fig. 21.104.

(b) The graph of $\theta = \pi/6$ is a straight line through the pole. It represents all points for which $\theta = \pi/6$ for all possible values of r, positive or negative. This line is also shown in Fig. 21.104.

EXAMPLE 2 Plot the graph of $r = 1 + \cos \theta$.

We find the following values of r corresponding to the chosen values of θ:

θ	0	$\frac{\pi}{4}$	$\frac{\pi}{2}$	$\frac{3\pi}{4}$	π	$\frac{5\pi}{4}$	$\frac{3\pi}{2}$	$\frac{7\pi}{4}$	2π
r	2	1.7	1	0.3	0	0.3	1	1.7	2
Point Number	1	2	3	4	5	6	7	8	9

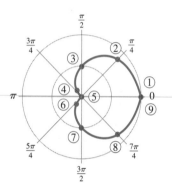

Fig. 21.105

We now see that the points are repeating, and it is unnecessary to find additional points. The curve is called a **cardioid** and is shown in Fig. 21.105.

EXAMPLE 3 Plot the graph of $r = 1 - 2 \sin \theta$.

Choosing values of θ and then finding the corresponding values of r, we find the following table of values:

θ	0	$\frac{\pi}{4}$	$\frac{\pi}{2}$	$\frac{3\pi}{4}$	π	$\frac{5\pi}{4}$	$\frac{3\pi}{2}$	$\frac{7\pi}{4}$	2π
r	1	−0.4	−1	−0.4	1	2.4	3	2.4	1
Point Number	1	2	3	4	5	6	7	8	9

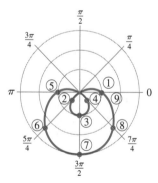

Fig. 21.106

Particular care should be taken in plotting the points for which r is negative. This curve is known as a **limaçon** and is shown in Fig. 21.106.

EXAMPLE 4 A cam is shaped such that the edge of the upper "half" is represented by the equation $r = 2.0 + \cos \theta$ and the lower "half" by the equation $r = \dfrac{3.0}{2.0 - \cos \theta}$, where measurements are in centimeters. Plot the curve that represents the shape of the cam.

We get the points for the edge of the cam by using values of θ from 0 to π for the upper "half" and from π to 2π for the lower half. The table of values follows:

$r = 2.0 + \cos \theta$

θ	0	$\frac{\pi}{4}$	$\frac{\pi}{2}$	$\frac{3\pi}{4}$	π
r	3.0	2.7	2.0	1.3	1.0
Point Number	1	2	3	4	5

$r = \dfrac{3.0}{2.0 - \cos \theta}$

θ	π	$\frac{5\pi}{4}$	$\frac{3\pi}{2}$	$\frac{7\pi}{4}$	2π
r	1.0	1.1	1.5	2.3	3.0
Point Number	6	7	8	9	10

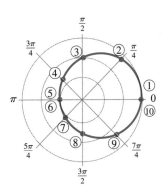

Fig. 21.107

The upper "half" is part of a limaçon and the lower "half" is a semiellipse. The cam is shown in Fig. 21.107.

EXAMPLE 5 Plot the graph of $r = 2 \cos 2\theta$.

In finding values of r, we must be careful first to multiply the values of θ by 2 before finding the cosine of the angle. Also, for this reason, we take values of θ as multiples of $\pi/12$, so as to get enough useful points. The table of values follows:

θ	0	$\frac{\pi}{12}$	$\frac{\pi}{6}$	$\frac{\pi}{4}$	$\frac{\pi}{3}$	$\frac{5\pi}{12}$	$\frac{\pi}{2}$
r	2	1.7	1	0	-1	-1.7	-2

θ	$\frac{7\pi}{12}$	$\frac{2\pi}{3}$	$\frac{3\pi}{4}$	$\frac{5\pi}{6}$	$\frac{11\pi}{12}$	π
r	-1.7	-1	0	1	1.7	2

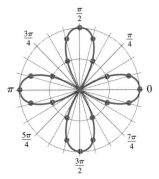

Fig. 21.108

For values of θ starting with π, the values of θ repeat. We have a four-leaf **rose,** as shown in Fig. 21.108.

EXAMPLE 6 Plot the graph of $r^2 = 9 \cos 2\theta$.

Choosing the indicated values of θ, we get the values of r as shown in the following table of values:

θ	0	$\frac{\pi}{8}$	$\frac{\pi}{4}$	$\cdots$	$\frac{3\pi}{4}$	$\frac{7\pi}{8}$	π
r	± 3	± 2.5	0		0	± 2.5	± 3

There are no values of r corresponding to values of θ in the range $\pi/4 < \theta < 3\pi/4$, since twice these angles are in the second and third quadrants and the cosine is negative for such angles. The value of r^2 cannot be negative. Also, the values of r repeat for $\theta > \pi$. The figure is called a **lemniscate** and is shown in Fig. 21.109.

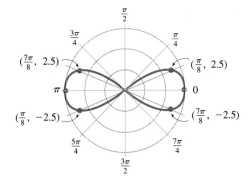

Fig. 21.109

◀ EXAMPLE 7   View the graph of $r = 1 - 2 \cos \theta$ on a graphing calculator.

Using the *mode* feature, a polar equation is displayed using the polar graph option or the parametric graph option, depending on the calculator. (Review the manual for the calculator.)

With the polar graph option, the function is entered directly. The values for the viewing window are determined by settings for x, y and the angles θ that will be used. These values are set in a manner similar to those used for parametric equations. (See page 312 for an example of graphing parametric equations.)

With the parametric graph option, to graph $r = f(\theta)$, we note that $x = r \cos \theta$ and $y = r \sin \theta$. This tells us that

$$x = f(\theta)\cos \theta$$
$$y = f(\theta)\sin \theta$$

Thus, for $r = 1 - 2 \cos \theta$, by using

$$x = (1 - 2 \cos \theta)\cos \theta$$
$$y = (1 - 2 \cos \theta)\sin \theta$$

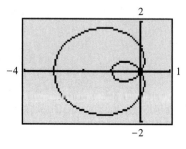

Fig. 21.110

the graph can be displayed, as shown in Fig. 21.110.

▶

EXERCISES 21.10

In Exercises 1–4, make the given changes in the indicated examples of this section and then make the indicated graphs.

1. In Example 1(b), change $\pi/6$ to $5\pi/6$ and then draw the graph.

2. In Example 3, change the $-$ before the $2 \sin \theta$ to $+$ and then plot the graph.

3. In Example 5, change cos to sin and then plot the graph.

4. In Example 7, change the $-$ before the $2 \cos \theta$ to $+$ and then view the graph on a graphing calculator.

In Exercises 5–32, plot the curves of the given polar equations in polar coordinates.

5. $r = 4$ 6. $r = 2$ 7. $\theta = 3\pi/4$

8. $\theta = -1.5$ 9. $r = 4 \sec \theta$ 10. $r = 4 \csc \theta$

11. $r = 2 \sin \theta$ 12. $r = 3 \cos \theta$

13. $1 - r = \cos \theta$ (cardioid)

14. $r + 1 = \sin \theta$ (cardioid)

15. $r = 2 - \cos \theta$ (limaçon)

16. $r = 2 + 3 \sin \theta$ (limaçon)

17. $r = 4 \sin 2\theta$ (rose)   18. $r = 2 \sin 3\theta$ (rose)

19. $r^2 = 4 \sin 2\theta$ (lemniscate)

20. $r^2 = 2 \sin \theta$

21. $r = 2^\theta$ (spiral)   22. $r = 1.5^{-\theta}$ (spiral)

23. $r = 4|\sin 3\theta|$ 24. $r = 2 \sin \theta \tan \theta$ (cissoid)

25. $r = \dfrac{1}{2 - \cos \theta}$ (ellipse)   26. $r = \dfrac{1}{1 - \cos \theta}$ (parabola)

27. $r - 2r \cos \theta = 6$ (hyperbola)

28. $3r - 2r \sin \theta = 6$ (ellipse)

29. $r = 4 \cos \frac{1}{2}\theta$ 30. $r = 2 + \cos 3\theta$

31. $r = 2[1 - \sin(\theta - \pi/4)]$ 32. $r = 4 \tan \theta$

In Exercises 33–40, view the curves of the given polar equations on a graphing calculator.

33. $r = \theta$ $(-20 \le \theta \le 20)$ 34. $r = 0.5^{\sin \theta}$

35. $r = 2 \sec \theta + 1$ 36. $r = 2 \cos(\cos 2\theta)$

37. $r = 3 \cos 4\theta$ 38. $r = 3 \sin 5\theta$

39. $r + 2 = \cos 2\theta$ 40. $2r \cos \theta + r \sin \theta = 2$

In Exercises 41–48, sketch the indicated graphs.

41. An architect designs a patio shaped such that it can be described as the area within the polar curve $r = 4.0 - \sin \theta$, where measurements are in meters. Sketch the curve that represents the perimeter of the patio.

42. The radiation pattern of a certain television transmitting antenna can be represented by $r = 120(1 + \cos \theta)$, where distances (in km) are measured from the antenna. Sketch the radiation pattern.

43. The joint between two links of a robot arm moves in an elliptical path (in cm), given by $r = \frac{25}{10 + 4 \cos \theta}$. Sketch the path.

44. A missile is fired at an airplane and is always directed toward the airplane. The missile is traveling at twice the speed of the airplane. An equation that describes the distance r between the missile and the airplane is $r = \dfrac{70 \sin \theta}{(1 - \cos \theta)^2}$, where θ is the angle between their directions at all times. See Fig. 21.111. This is a *relative pursuit curve*. Sketch the graph of this equation for $\pi/4 \le \theta \le \pi$.

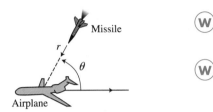

Fig. 21.111

45. In studying the photoelectric effect, an equation used for the rate R at which photoelectrons are ejected at various angles θ is $R = \dfrac{\sin^2 \theta}{(1 - 0.5 \cos \theta)^2}$. Sketch the graph.

46. Sketch the graph of the rectangular equation
$4(x^6 + 3x^4y^2 + 3x^2y^4 + y^6 - x^4 - 2x^2y^2 - y^4) + y^2 = 0$
(*Hint*: The equation can be written as
$4(x^2 + y^2)^3 - 4(x^2 + y^2)^2 + y^2 = 0$.) Transform this equation to polar coordinates and then sketch the curve.

(W) **47.** Noting the graphs in Exercises 17, 18, 37, and 38, what conclusion do you draw about the value of n and the graph of $y = a \sin n\theta$ or $y = a \cos n\theta$?

(W) **48.** (a) Solve the equations $r = 2 \sin \theta$ and $r = 1 + \sin \theta$ simultaneously to find a point of intersection. (b) Display the graphs of the equations on a graphing calculator and note another common point, the coordinates of which do not satisfy both equations. Explain why this occurs.

CHAPTER (21) EQUATIONS

Distance formula	Fig. 21.2	$d = \sqrt{(x_2 - x_1)^2 + (y_2 - y_1)^2}$	(21.1)
Slope	Fig. 21.4	$m = \dfrac{y_2 - y_1}{x_2 - x_1}$	(21.2)
	Fig. 21.7	$m = \tan \alpha \quad (0° \le \alpha < 180°)$	(21.3)
	Fig. 21.9	$m_1 = m_2 \quad$ (for ∥ lines)	(21.4)
	Fig. 21.10	$m_2 = -\dfrac{1}{m_1}$ or $m_1 m_2 = -1 \quad$ (for ⊥ lines)	(21.5)
Straight line	Fig. 21.14	$y - y_1 = m(x - x_1)$	(21.6)
	Fig. 21.17	$x = a$	(21.7)
	Fig. 21.18	$y = b$	(21.8)
	Fig. 21.21	$y = mx + b$	(21.9)
		$Ax + By + C = 0$	(21.10)
Circle	Fig. 21.30	$(x - h)^2 + (y - k)^2 = r^2$	(21.11)
	Fig. 21.33	$x^2 + y^2 = r^2$	(21.12)
		$x^2 + y^2 + Dx + Ey + F = 0$	(21.14)
Parabola	Fig. 21.40	$y^2 = 4px$	(21.15)
	Fig. 21.43	$x^2 = 4py$	(21.16)
Ellipse	Fig. 21.56	$\dfrac{x^2}{a^2} + \dfrac{y^2}{b^2} = 1$	(21.17)
	Fig. 21.56	$a^2 = b^2 + c^2$	(21.18)
	Fig. 21.57	$\dfrac{y^2}{a^2} + \dfrac{x^2}{b^2} = 1$	(21.19)

Hyperbola	Fig. 21.69	$\dfrac{x^2}{a^2} - \dfrac{y^2}{b^2} = 1$		(21.20)
	Fig. 21.69	$c^2 = a^2 + b^2$		(21.21)
	Fig. 21.68	$y = \pm\dfrac{bx}{a}$	(asymptotes)	(21.23)
	Fig. 21.70	$\dfrac{y^2}{a^2} - \dfrac{x^2}{b^2} = 1$		(21.24)
	Fig. 21.76	$xy = c$		(21.25)
Translation of axes	Fig. 21.81	$x = x' + h$ and $y = y' + k$		(21.26)
		$x' = x - h$ and $y' = y - k$		(21.27)
Parabola, vertex (h,k)		$(y - k)^2 = 4p(x - h)$	(axis parallel to x-axis)	(21.28)
		$(x - h)^2 = 4p(y - k)$	(axis parallel to y-axis)	(21.29)
Ellipse, center (h,k)		$\dfrac{(x - h)^2}{a^2} + \dfrac{(y - k)^2}{b^2} = 1$	(major axis parallel to x-axis)	(21.30)
		$\dfrac{(y - k)^2}{a^2} + \dfrac{(x - h)^2}{b^2} = 1$	(major axis parallel to y-axis)	(21.31)
Hyperbola, center (h,k)		$\dfrac{(x - h)^2}{a^2} - \dfrac{(y - k)^2}{b^2} = 1$	(transverse axis parallel to x-axis)	(21.32)
		$\dfrac{(y - k)^2}{a^2} - \dfrac{(x - h)^2}{b^2} = 1$	(transverse axis parallel to y-axis)	(21.33)
Second-degree equation		$Ax^2 + Bxy + Cy^2 + Dx + Ey + F = 0$		(21.34)
Polar coordinates	Fig. 21.98	$x = r\cos\theta \quad y = r\sin\theta$		(21.35)
		$\tan\theta = \dfrac{y}{x} \quad r = \sqrt{x^2 + y^2}$		(21.36)

CHAPTER 21 REVIEW EXERCISES

In Exercises 1–12, find the equation of the indicated curve, subject to the given conditions. Sketch each curve.

1. Straight line: passes through $(1, -7)$ with a slope of 4

2. Straight line: passes through $(-1, 5)$ and $(-2, -3)$

3. Straight line: perpendicular to $3x - 2y + 8 = 0$ and has a y-intercept of $(0, -1)$

4. Straight line: parallel to $2x - 5y + 1 = 0$ and has an x-intercept of $(2, 0)$

5. Circle: concentric with $x^2 + y^2 = 6x$, passes through $(4, -3)$

6. Circle: tangent to the line $x = 3$, center at $(5, 1)$

7. Parabola: focus $(3, 0)$, vertex $(0, 0)$

8. Parabola: vertex $(0, 0)$, passes through $(1, 1)$ and $(-2, 4)$

9. Ellipse: vertex $(10, 0)$, focus $(8, 0)$, tangent to $x = -10$

10. Ellipse: center $(0, 0)$, passes through $(0, 3)$ and $(2, 1)$

11. Hyperbola: $V(0, 13)$, $C(0, 0)$, conj. axis of 24

12. Hyperbola: vertex $(0, 8)$, asymptotes $y = 2x$, $y = -2x$

In Exercises 13–24, find the indicated quantities for each of the given equations. Sketch each curve.

13. $x^2 + y^2 + 6x - 7 = 0$, center and radius

14. $x^2 + y^2 - 4x + 2y - 20 = 0$, center and radius

15. $x^2 = -20y$, focus and directrix

16. $y^2 = 24x$, focus and directrix

17. $16x^2 + y^2 = 16$, vertices and foci

18. $2y^2 - 9x^2 = 18$, vertices and foci

19. $2x^2 - 5y^2 = 0.25$, vertices and foci

20. $2x^2 + 25y^2 = 800$, vertices and foci

21. $x^2 - 8x - 4y - 16 = 0$, vertex and focus

22. $y^2 - 4x + 4y + 24 = 0$, vertex and directrix

23. $4x^2 + y^2 - 16x + 2y + 13 = 0$, center

24. $x^2 - 2y^2 + 4x + 4y + 6 = 0$, center

In Exercises 25–32, plot the given curves in polar coordinates.

25. $r = 4(1 + \sin \theta)$ **26.** $r = 1 - 3\cos \theta$

27. $r = 4\cos 3\theta$ **28.** $r = 3\sin \theta - 4\cos \theta$

29. $r = \dfrac{3}{\sin \theta + 2\cos \theta}$ **30.** $r = \dfrac{1}{2(\sin \theta - 1)}$

31. $r = 2\sin\left(\dfrac{\theta}{2}\right)$ **32.** $r = 1 - \cos 2\theta$

In Exercises 33–36, find the polar equation of each of the given rectangular equations.

33. $y = 2x$ **34.** $2xy = 1$

35. $x^2 + xy + y^2 = 2$ **36.** $x^2 + (y + 3)^2 = 16$

In Exercises 37–40, find the rectangular equation of each of the given polar equations.

37. $r = 2\sin 2\theta$ **38.** $r^2 = \sin \theta$

39. $r = \dfrac{4}{2 - \cos \theta}$ **40.** $r = 4\tan \theta \sec \theta$

In Exercises 41–44, determine the number of real solutions of the given systems of equations by sketching the indicated curves. (See Section 14.1.)

41. $x^2 + y^2 = 9$ **42.** $y = e^x$
  $4x^2 + y^2 = 16$ $x^2 - y^2 = 1$

43. $x^2 + y^2 - 4y - 5 = 0$ **44.** $x^2 - 4y^2 + 2x - 3 = 0$
  $y^2 - 4x^2 - 4 = 0$ $y^2 - 4x - 4 = 0$

In Exercises 45–54, view the curves of the given equations on a graphing calculator.

45. $x^2 + 3y + 2 - (1 + x)^2 = 0$ **46.** $y^2 = 4x + 6$

47. $2x^2 + 2y^2 + 4y - 3 = 0$ **48.** $2x^2 + (y - 3)^2 - 5 = 0$

49. $x^2 - 4y^2 + 4x + 24y - 48 = 0$

50. $x^2 + 2xy + y^2 - 3x + 8y = 0$

51. $r = 3\cos(3\theta/2)$ **52.** $r = 5 - 2\sin 4\theta$

53. $r = 2 - 3\csc \theta$ **54.** $r = 2\sin(\cos 3\theta)$

In Exercises 55–58, find the equation of the locus of a point $P(x, y)$ that moves as stated.

55. Always 4 units from $(3, -4)$

56. Passes through $(7, -5)$ with a constant slope of -2

57. The sum of its distances from $(1, -3)$ and $(7, -3)$ is 8.

58. The difference of its distances from $(3, -1)$ and $(3, -7)$ is 4.

In Exercises 59–96, solve the given problems.

59. In two ways show that the line segments joining $(-3, 11)$, $(2, -1)$, and $(14, 4)$ form a right triangle.

60. Find the equation of the circle that passes through $(3, -2)$, $(-1, -4)$, and $(2, -5)$.

61. Graph the inequality $y > 4(x + 2)^2$.

62. Graph the inequality $4x^2 + 9(y - 2)^2 < 36$.

63. What type of curve is represented by $(x + jy)^2 + (x - jy)^2 = 2$? $\left(j = \sqrt{-1}\right)$

64. For the ellipse in Fig. 21.112, show that the product of the slopes PA and PB is $-b^2/a^2$.

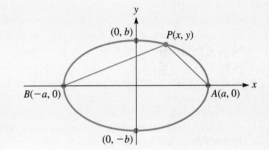

Fig. 21.112

65. Find the area of the square that can be inscribed in the ellipse $7x^2 + 2y^2 = 18$.

66. Using a graphing calculator, determine the number of points of intersection of the polar curves $r = 4|\cos 2\theta|$ and $r = 6\sin[\cos(\cos 3\theta)]$.

67. By means of the definition of a parabola, find the equation of the parabola with focus at $(3, 1)$ and directrix the line $y = -3$. Find the same equation by the method of translation of axes.

68. For what value(s) of k does $x^2 - ky^2 = 1$ represent an ellipse with vertices on the y-axis?

69. The total resistance R_T of two resistances in series in an electric circuit is the sum of the resistances. If a variable resistor R is in series with a 2.5-Ω resistor, express R_T as a function of R and sketch the graph.

70. The acceleration of an object is defined as the change in velocity v divided by the corresponding change in time t. Find the equation relating the velocity v and time t for an object for which the acceleration is 6.0 m/s² and $v = 5.0$ m/s when $t = 0$ s.

71. One computer printer prints 2500 lines/min for x min, and a second printer prints 1500 lines/min for y min. If they print a total of 37 500 lines together, express y as a function of x and sketch the graph.

72. An airplane touches down when landing at 150 km/h. Its velocity v while coming to a stop is given by $v = 150 - 20\,000t$, where t is the time in hours. Sketch the graph of v vs. t.

73. It takes 2.010 kJ of heat to raise the temperature of 1.000 kg of steam by 1.000°C. In a steam generator, a total of y kJ is used to raise the temperature of 50.00 kg of steam from 100°C to T°C. Express y as a function of T and sketch the graph.

74. The temperature in a certain region is 27°C, and at an altitude of 2500 m above the region it is 12°C. If the equation relating the temperature T and the altitude h is linear, find the equation.

75. The radar gun on a police helicopter 170 m above a multilane highway is directed vertically down onto the highway. If the radar gun signal is cone-shaped with a vertex angle of 14°, what area of the highway is covered by the signal?

76. One of the most famous Ferris wheels is Vienna's (Austria) *Prata*. Find the equation representing its circumference, given that $c = 191$ m. Place the origin of the coordinate system 2.0 m below the bottom of the wheel, and the center of the wheel on the y-axis.

77. The arch of a bridge across a river is parabolic. If, at water level, the span of the arch is 80 m and the maximum height above water level is 20 m, what is the equation that represents the arch? Choose the most convenient point for the origin of the coordinate system.

78. A laser source is 2.00 cm from a spherical surface of radius 3.00 cm, and the laser beam is tangent to the surface. By placing the center of the sphere at the origin, and the source on the positive x-axis, find the equation of the line along which the beam shown in Fig. 21.113 is directed.

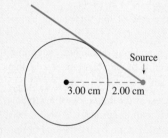

Source

3.00 cm / 2.00 cm

Fig. 21.113

79. The top horizontal cross section of a dam is parabolic. The open area within this cross section is 80 m across and 50 m from front to back. Find the equation of the edge of the open area with the vertex at the origin of the coordinate system and the axis along the x-axis.

80. The *quality factor* Q of a series resonant electric circuit with resistance R, inductance L, and capacitance C is given by $Q = \dfrac{1}{R}\sqrt{\dfrac{L}{C}}$. Sketch the graph of Q and L for a circuit in which $R = 1000\ \Omega$ and $C = 4.00\ \mu$F.

81. At very low temperatures, certain metals have an electric resistance of zero. This phenomenon is called *superconductivity*. A magnetic field also affects the superconductivity. A certain level of magnetic field H_T, the threshold field, is related to the thermodynamic temperature T by $H_T/H_0 = 1 - (T/T_0)^2$, where H_0 and T_0 are specifically defined values of magnetic field and temperature. Sketch the graph of H_T/H_0 vs. T/T_0.

82. A rectangular parking lot is to have a perimeter of 600 m. Express the area A in terms of the width w and sketch the graph.

83. The electric power P (in W) supplied by a battery is given by $P = 12.0i = 0.500i^2$, where i is the current (in A). Sketch the graph of P vs. i.

84. The Colosseum in Rome is in the shape of an ellipse 188 m long and 156 m wide. Find the area of the Colosseum. ($A = \pi ab$ for an ellipse.)

85. A specialty electronics company makes an ultrasonic device to repel animals. It emits a 20–25 kHz sound (above those heard by people), which is unpleasant to animals. The sound covers an elliptical area starting at the device, with the longest dimension extending 36 m from the device and the focus of the area 5 m from the device. Find the area covered by the signal. ($A = \pi ab$.)

86. A study indicated that the fraction f of cells destroyed by various dosages d of X rays is given by the graph in Fig. 21.114. Assuming that the curve is a quarter-ellipse, find the equation relating f and d for $0 \le f \le 1$ and $0 < d \le 10$ units.

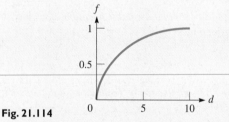

Fig. 21.114

87. A machine-part designer wishes to make a model for an elliptical cam by placing two pins in a design board, putting a loop of string over the pins, and marking off the outline by keeping the string taut. (Note that the definition of the ellipse is being used.) If the cam is to measure 10 cm by 6 cm, how long should the loop of string be and how far apart should the pins be?

88. Soon after reaching the vicinity of the moon, *Apollo 11* (the first spacecraft to land a man on the moon) went into an elliptical lunar orbit. The closest the craft was to the moon in this orbit was 110 km, and the farthest it was from the moon was 310 km. What was the equation of the path if the center of the moon was at one of the foci of the ellipse? Assume that the major axis is along the x-axis and that the center of the ellipse is at the origin. The radius of the moon is 1740 km.

89. The vertical cross section of the cooling tower of a nuclear power plant is hyperbolic, as shown in Fig. 21.115. Find the radius r of the smallest circular horizontal cross section.

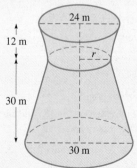

24 m

12 m

r

30 m

30 m

Fig. 21.115

90. In New Zealand, tremors from an earthquake are recorded at Auckland 36 s before they are recorded at Wellington. If the seismographs are 510 km apart and the shock waves from the tremors travel at 5.0 km/s, what is the curve on which lies the point where the earthquake occurred?

91. An electronic instrument located at point P records the sound of a rifle shot and the impact of the bullet striking the target at the same instant. Show that P lies on a branch of a hyperbola.

92. A 20-m rope passes over a pulley 4 m above the ground, and a crate on the ground is attached at one end. The other end of the rope is held at a level of 1 m above the ground and is drawn away from the pulley. Express the height of the crate over the ground in terms of the distance the person is from directly below the crate. Sketch the graph of distance and height. See Fig. 21.116. (Neglect the thickness of the crate.)

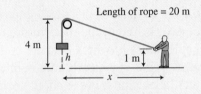

Fig. 21.116

93. A satellite at an altitude proper to make one revolution per day around the center of the earth will have for an excellent approximation of its projection on the earth of its path the curve $r^2 = R^2 \cos 2\left(\theta + \frac{\pi}{2}\right)$, where R is the radius of the earth. Sketch the path of the projection.

94. The vertical cross sections of two pipes as drawn on a drawing board are shown in Fig. 21.117. Find the polar equation of each.

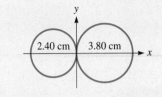

Fig. 21.117

95. The path of a certain plane is $r = 200(\sec \theta + \tan \theta)^{-5}/\cos \theta$, $0 < \theta < \pi/2$. Sketch the path and check it on a graphing calculator. (This path is the same as for the plane in Exercise 88 on page 338.)

96. The sound produced by a jet engine was measured at a distance of 100 m in all directions. The loudness of the sound d (in decibels) was found to be $d = 115 + 10 \cos \theta$, where the 0° line for the angle θ is directed in front of the engine. Sketch the graph of d vs. θ in polar coordinates (use d as r).

Writing Exercise

97. Under a force that varies inversely as the square of the distance from an attracting object (such as the sun exerts on the earth), it can be shown that the equation of the path an object follows is given in general by

$$\frac{1}{r} = a + b \cos \theta$$

where a and b are constants for a particular path. First, transform this equation into rectangular coordinates. Then, write one or two paragraphs explaining why this equation represents one of the conic sections, depending on the values of a and b. It is through this kind of analysis that we know that the paths of the planets and comets are conic sections.

CHAPTER 21 PRACTICE TEST

1. Identify the type of curve represented by the equation $2(x^2 + x) = 1 - y^2$.

2. Sketch the graph of the straight line $4x - 2y + 5 = 0$ by finding its slope and y-intercept.

3. Find the polar equation of the curve whose rectangular equation is $x^2 = 2x - y^2$.

4. Find the vertex and the focus of the parabola $x^2 = -12y$. Sketch the graph.

5. Find the equation of the circle with center at $(-1, 2)$ and that passes through $(2, 3)$.

6. Find the equation of the straight line that passes through $(-4, 1)$ and $(2, -2)$.

7. Where is the focus of a parabolic reflector that is 12.0 cm across and 4.00 cm deep?

8. A hallway 6.0 m wide has a ceiling whose cross section is a semi-ellipse. The ceiling is 3.0 m high at the walls and 4.0 m high at the center. Find the height of the ceiling 1.0 m from each wall.

9. Plot the polar curve $r = 3 + \cos \theta$.

10. Find the center and vertices of the conic section $4y^2 - x^2 - 4x - 8y - 4 = 0$. Show completely the sketch of the curve.

22

Introduction to Statistics

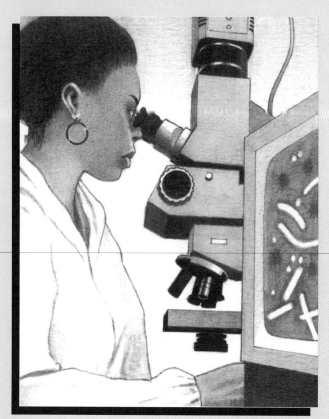

Statistical analysis is used extensively in medical research. In Section 22.6, a method of finding the equation of a line that "fits" given data is shown.

After the invention of the steam engine in the late 1700s by the Scottish engineer James Watt, the production of machine-made goods became widespread during the 1800s. However, it was not until the 1920s that much attention was paid to the quality control of the goods being produced. In 1924, Walter Shewhart of Bell Telephone Laboratories used a statistical chart for controlling product variables; in the 1940s, quality control was used in much of the wartime production.

Quality control is one of the modern uses of *statistics,* the branch of mathematics in which data are collected, analyzed, and interpreted. Today it is nearly impossible to read a newspaper or watch television news without seeing some type of study, in areas such as medicine or politics, that involves statistics. Other fields in which statistical methods are used include biology, physics, psychology, sociology, economics, business, education, electronics, and others.

The first significant use of statistics was made in the 1660s by John Graunt and in the 1690s by Edmund Halley (Halley's Comet), when each published some conclusions about the population in England based on mortality tables. There was little development of statistics until the 1800s, when statistical measures became more widely used. Examples are the scientist Francis Galton who used statistics in the study of human heredity and the nurse Florence Nightingale who used statistical graphs to show that more soldiers died in the Crimean War (in the 1850s) from unsanitary conditions than from combat wounds.

This chapter is an introduction to some of the basic concepts and uses of statistics. We first consider certain basic statistical measures. Then a section is devoted to very large sets of statistical data, followed by a section on *control charts* that are used in statistical process control in industry. The final sections show how to start with a set of points on a graph and find an equation that best "fits" the data. This equation, which shows a basic relationship between the variables, can be useful in research.

22.1 FREQUENCY DISTRIBUTIONS

In statistics we first collect data and thereby form a set of numbers. These numbers could be measurements of electric current, test scores, lengths of objects, or numerous other possibilities. *Data that have been collected but not yet organized are called* **raw data.** In order to obtain useful information from the data, it is necessary to organize it in some way. Normally, a first step in organizing the data is to arrange the numerical values in ascending (or descending) order, called an **array.** An example of raw data and an array is shown in the following example.

◀ **EXAMPLE 1** A survey of 50 users of home computers with Internet access asked each user to estimate carefully the number of hours they spent each week on the Internet. Following are the estimates:

> 12, 20, 15, 14, 7, 10, 12, 25, 18, 5, 10, 24, 16, 3, 12, 14, 28, 8, 13, 18,
>
> 15, 8, 11, 15, 14, 22, 14, 19, 6, 10, 18, 4, 16, 24, 18, 5, 13, 20, 12, 12,
>
> 25, 11, 8, 12, 20, 5, 10, 15, 13, 8

As we can see, no clear pattern can be seen from this raw data. Arranging these in numerical order to form an array, we can summarize the array by showing the number of persons reporting each estimate as follows:

> (hours-persons) 3-1, 4-1, 5-3, 6-1, 7-1, 8-4, 10-4, 11-2, 12-6
>
> 13-3, 14-4, 15-4, 16-2, 18-4, 19-1, 20-3, 22-1, 24-2, 25-2, 28-1 ▶

In Example 1, although a pattern is somewhat clearer from the array than from the raw data, a still clearer pattern is found by *grouping the data.* In the process of grouping, the detail of the raw data is lost, but the advantage is that a much clearer overall pattern of the data can be obtained.

The grouping of data is done by first defining what values are to be included in each group and then tabulating the number of all values that are within each group. *Each group is called a* **class,** and *the number of values in the class is called the* **frequency.** *The table is called a* **frequency distribution table.** This is illustrated in the following example.

◀ **EXAMPLE 2** In Example 1, we note that the estimates vary from 3 h to 28 h. We see that if we form *classes* of 0–4 h, 5–9 h, etc., we will have five possible estimates in each class and that there will be six classes. This gives us the following table of values showing the number of persons (frequency) reporting the indicated estimate of hours on the Internet:

Estimate (hours)	0–4	5–9	10–14	15–19	20–24	25–29
Frequency (persons)	2	9	19	11	6	3

This table shows us the frequency distribution. The 0, 5, 10, and so on are the *lower class limits,* and the 4, 9, 14, and so on are the *upper class limits.* Each class includes five values, which is the *class width.* As with the class limits we have chosen, it is generally preferable to have the same width for each class.

We can see from this frequency distribution table that the pattern of hours on the Internet by the persons responding to the survey is clearer. ▶

At times it is also helpful to know the **relative frequency** *of the class, which is the frequency of the class divided by the total frequency of all classes.* The relative frequency can be expressed as a fraction, decimal, or percent.

❰ EXAMPLE 3   The relative frequency of each class for the data in Example 2 can be shown as in the following table:

Estimated Hours on Internet	Frequency	Relative Frequency (%)	
0–4	2	4	2/50 = 0.04 = 4%
5–9	9	18	
10–14	19	38	
15–19	11	22	
20–24	6	12	
25–29	3	6	
Total	50	100	

Just as graphs are useful in representing algebraic functions, so are they a very convenient method of representing frequency distributions. There are several useful types of graphs for such distributions. *Among the most important of these are the* **histogram** *and the* **frequency polygon.** These graphical representations are generally constructed to represent the data in a frequency distribution table.

In order to represent *grouped* data, where the raw data values are generally not all the same within a given class, we find it necessary to use a representative value for each class. For this we use the **class mark,** *which is found by dividing the sum of the lower and upper class limits by* 2. The following examples illustrate the use of a class mark with a histogram and a frequency polygon.

❰ EXAMPLE 4   *A histogram represents a particular set of data by displaying each class of the data as a rectangle.* Each rectangle is labeled at the center of its base by the *class mark.* The width of each rectangle represents the *class width,* and the height of the rectangle represents the *frequency* of the class.

For the data in Example 2 on estimated hours on the Internet, the class marks are $(0 + 4)/2 = 2$, $(5 + 9)/2 = 7$, and so on. Therefore, a histogram representing these data is shown in Fig. 22.1.

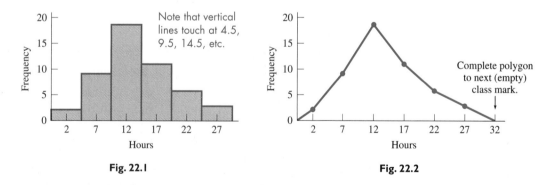

Fig. 22.1   **Fig. 22.2**

❰ EXAMPLE 5   *A frequency polygon is used to represent a set of data by plotting the class marks as abscissas (x-values) and the frequencies as ordinates (y-values).* The resulting points are joined by straight-line segments.

A frequency polygon representing the data in Example 2 on estimated hours on the Internet is shown in Fig. 22.2.

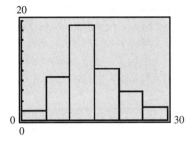

Fig. 22.3

Computer spreadsheets are very useful for this type of analysis.

A graphing calculator may be used to display histograms and frequency polygons. The manual should be reviewed to see how this is done on any particular model. On a calculator, it is necessary to enter both sets of numbers, choose proper *window* settings, and select the correct type of statistical graph. A calculator display for the histogram in Fig. 22.1 is shown in Fig. 22.3.

When dividing data into classes, we must decide how many classes to use. A general guideline is that *the number of classes should be no less than* 5 *nor greater than* 15 (this might go to 20 if we have thousands of values to analyze). The principal consideration is that we get a good pattern of the distribution. This is illustrated in the following example.

◀ EXAMPLE 6   If we divided the data of estimated Internet hours of Example 1 into nine classes, the frequency table would be as follows:

Estimate (hours)	3–5	6–8	9–11	12–14	15–17	18–20	21–23	24–26	27–29
Freq. (respondents)	5	6	6	13	6	8	1	4	1

We can see that this does not give nearly as good an idea of the pattern as the six classes shown in Example 1. ▶

Another way of analyzing data is to use *cumulative* totals. The way this is generally done is to change the frequency into a "less than" or a "more than" **cumulative frequency.** To do this, we *add the class frequencies,* starting either at the greatest class boundary or the lowest class boundary. In our discussion, we will restrict our attention to a "less than" cumulative frequency. The graphical display that is generally used for cumulative frequency is called an **ogive** (pronounced oh-jive).

◀ EXAMPLE 7   For the data on estimated hours on the Internet in Examples 2 and 3, the cumulative frequency is shown in the following table:

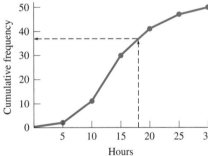

Fig. 22.4

Estimated Hours on Internet	*Cumulative Frequency*
Less than 5	2
Less than 10	11
Less than 15	30
Less than 20	41
Less than 25	47
Less than 30	50

The ogive showing the cumulative frequency for the values in this table is shown in Fig. 22.4. The vertical scale shows the *frequency,* and the horizontal scale shows the *class boundaries.*

One important use of an ogive is to determine the number of values above or below a certain value. For example, to *approximate* the number of respondents that use the Internet *less than* 18 hours per week, we draw a line from the horizontal axis to the ogive and then to the vertical axis as shown in Fig. 22.4. From this, we see that *about* 37 respondents use the Internet *less than* 18 hours per week. ▶

If the data with which we are dealing has only a limited number of values and we do not divide it into classes, we can use the methods we have developed. We use the specific values rather than *class* values. Consider the following example.

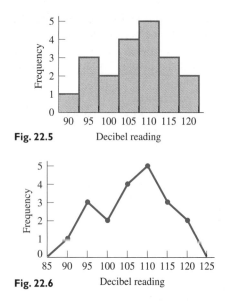

Fig. 22.5   Decibel reading

Fig. 22.6   Decibel reading

❮ **EXAMPLE 8**   A test station measured the loudness of the sound of jet aircraft taking off from a certain airport. The decibel (dB) readings were measured to the nearest 5 dB, and the readings for the first 20 jets were as follows:

$$110, 95, 100, 115, 105, 110, 120, 110, 115, 105,$$
$$90, 95, 105, 110, 100, 115, 105, 120, 95, 110$$

Since there are only seven different values for the 20 readings, the best idea of the pattern of readings is found by using these seven values. Therefore, we have the following frequency distribution table:

Decibel Reading	90	95	100	105	110	115	120
Frequency	1	3	2	4	5	3	2

The histogram for this table is shown in Fig. 22.5, and the frequency polygon is shown in Fig. 22.6. ❯

EXERCISES 22.1

In Exercises 1–4, divide the data in Example 1 into five classes of hours (0–5, 6–11, etc.) on the Internet and then do the following:

1. Form a frequency distribution table.
2. Find the relative frequencies.
3. Draw a histogram.   4. Draw an ogive.

In Exercises 5–12, use the following set of numbers as the raw data.

103, 108, 106, 104, 109, 104, 110, 108, 108, 104, 113, 106, 107, 106, 107, 109, 105, 111, 109, 108

5. Form a frequency distribution table for these numbers.
6. Find the relative frequencies of these numbers.
7. Form a frequency distribution table with five classes for these numbers.
8. Find the relative frequencies for the data in the frequency distribution table in Exercise 7.
9. Draw a histogram for the data of Exercise 7.
10. Draw a frequency polygon for the data of Exercise 7.
11. Form a cumulative frequency table for the data of Exercise 7.
12. Draw an ogive for the data of Exercise 7.

In Exercises 13–32, find the indicated quantities.

13. In testing a computer system, the number of instructions it could perform in 1 ns was measured at different points in a program. The numbers of instructions were recorded as follows:

19, 21, 22, 25, 22, 20, 18, 21, 20, 19, 22, 21, 19, 23, 21

Form a frequency distribution table for these values.

14. For the data of Exercise 13, draw a histogram.

15. For the data of Exercise 13, draw a frequency polygon.
16. For the data of Exercise 13, form a relative frequency distribution table.
17. A strobe light is designed to flash every 2.25 s at a certain setting. Sample bulbs were tested with the following results:

Time (s) between Flashes (class mark)	2.21	2.22	2.23	2.24
Number of Bulbs	2	7	18	41

Time (s)	2.25	2.26	2.27	2.28	2.29
No. Bulbs	56	32	8	3	3

Draw a histogram for these data.

18. For the data of Exercise 17, draw a frequency polygon.
19. For the data of Exercise 17, form a cumulative frequency distribution table.
20. For the data of Exercise 17, draw an ogive.
21. In testing a braking system, the distance required to stop a car from 110 km/h was measured in 120 trials. The results are shown in the following distribution table:

Stopping Distance (m)	47–49	50–52	53–55	56–58
Times Car Stopped	2	15	32	36

Stopping Distance (m)	59–61	62–64	65–67
Times Car Stopped	24	10	1

Form a relative frequency distribution table for these data.

22. For the data in Exercise 21, form a cumulative frequency distribution table.

23. For the data of Exercise 21, draw an ogive.

24. From the ogive in Exercise 23, estimate the number of cars that stopped in less than 57 m.

25. The dosage, in millisieverts (mSv), given by a particular X-ray machine was measured 20 times, with the following readings:

0.425, 0.436, 0.396, 0.421, 0.444, 0.383, 0.437, 0.427, 0.433, 0.434, 0.415, 0.390, 0.441, 0.451, 0.418, 0.426, 0.429, 0.409, 0.436, 0.423

Form a histogram with six classes and the lowest class mark of 0.380 mSv.

26. For the data used for the histogram in Exercise 25, draw a frequency polygon.

27. The life of a certain type of battery was measured for a sample of batteries with the following results (in number of hours):

34, 30, 32, 35, 31, 28, 29, 30, 32, 25, 31, 30, 28, 36, 33, 34, 30, 33, 31, 34, 29, 30, 32

Draw a frequency polygon using six classes.

28. For the data in Exercise 27, draw a cumulative frequency distribution table using six classes.

29. The diameters of a sample of fiber-optic cables were measured with the following results (diameters are class marks):

Diam. (mm)	0.0055	0.0056	0.0057	0.0058	0.0059	0.0060
No. Cables	4	15	32	36	59	64

Diam. (mm)	0.0061	0.0062	0.0063	0.0064	0.0065	0.0066
No. Cables	22	18	10	12	4	4

Draw a histogram for these data.

30. For the data of Exercise 29, draw a histogram with six classes. Compare the pattern of distribution with that of the histogram in Exercise 29.

W 31. Toss four coins 50 times and tabulate the number of heads that appear for each toss. Draw a frequency polygon showing the number of tosses for which 0, 1, 2, 3, or 4 heads appeared. Describe the distribution. (Is it about what should be expected?)

W 32. Most calculators can generate random numbers (between 0 and 1). On a calculator, display 50 random numbers and record the first digit. Draw a histogram showing the number of times for which each first digit (0, 1, 2, ..., 9) appeared. Describe the distribution. (Is it about what should be expected?)

22.2 MEASURES OF CENTRAL TENDENCY

Tables and graphical representations give a general description of data. However, it is often useful and convenient to find representative values for the location of the center of the distribution, and other numbers to give a measure of the deviation from this central value. In this way, we can obtain an arithmetical description of the data. We now discuss the values commonly used to measure the location of the center of the distribution. These are referred to as *measures of central tendency.*

Median   The first of these measures of central tendency is the **median.** *The median is the middle number, that number for which there are as many above it as below it in the distribution.* If there is no middle number, the median is that number halfway between the two numbers nearest the middle of the distribution.

◀ EXAMPLE 1   Given the numbers 5, 2, 6, 4, 7, 4, 7, 2, 8, 9, 4, 11, 9, 1, 3, we first arrange them in numerical order. This arrangement is

┌─ middle number
↓
1, 2, 2, 3, 4, 4, 4, 5, 6, 7, 7, 8, 9, 9, 11

Since there are 15 numbers, the middle number is the eighth. Since the eighth number is 5, the median is 5.

If the number 11 is not included in this set of numbers and there are only 14 numbers in all, the median is that number halfway between the seventh and eighth numbers. Since the seventh is 4 and the eighth is 5, the median is 4.5.   ▶

◀ EXAMPLE 2   In the distribution of Internet hours in Example 1 of Section 22.1, the median is 13 h. There are 50 values in all, and when listed in order, the 24th to 26th values are each 13 h. The number halfway between the 25th and 26th value is the median. Since both are 13 h, the median is 13 h.   ▶

Arithmetic Mean Another very widely applied measure of central tendency is the **arithmetic mean.** *The mean is calculated by finding the sum of all the values and then dividing by the number of values.* (The arithmetic mean is the number most people call the "average." However, in statistics the word *average* has the more general meaning of a measure of central tendency.)

◀ **EXAMPLE 3**   The arithmetic mean of the numbers given in Example 1 is determined by finding the sum of all the numbers and dividing by 15. Therefore, by letting $\bar{x}$ (read as "*x* bar") represent the mean, we have

$$\bar{x} = \frac{5 + 2 + 6 + 4 + 7 + 4 + 7 + 2 + 8 + 9 + 4 + 11 + 9 + 1 + 3}{15}$$

sum of values

$$= \frac{82}{15} \quad 5.5$$

number of values

Thus, the mean is 5.5. (The mean is usually calculated to one more significant digit than was present in the original data.) ▮

If we wish to find the arithmetic mean of a large number of values and if some of them appear more than once, the calculation can be simplified. The mean can be calculated by multiplying each value by its frequency, adding these results, and then dividing by the total number of values (the sum of the frequencies). Letting $\bar{x}$ represent the mean of the values $x_1, x_2, \ldots, x_n$, which occur with frequencies $f_1, f_2, \ldots, f_n$, respectively, we have

This is called a *weighted mean* since each value is given a weighting based on the number of times it occurs.

$$\bar{x} = \frac{x_1 f_1 + x_2 f_2 + \cdots + x_n f_n}{f_1 + f_2 + \cdots + f_n} \tag{22.1}$$

◀ **EXAMPLE 4**   Using Eq. (22.1) to find the arithmetic mean of the numbers of Example 1, we first set up a table of values and their respective frequencies, as follows:

Value	1	2	3	4	5	6	7	8	9	11
Frequency	1	2	1	3	1	1	2	1	2	1

We now calculate the arithmetic mean $\bar{x}$ by using Eq. (22.1):

multiply each value by its frequency and add results

$$\bar{x} = \frac{1(1) + 2(2) + 3(1) + 4(3) + 5(1) + 6(1) + 7(2) + 8(1) + 9(2) + 11(1)}{1 + 2 + 1 + 3 + 1 + 1 + 2 + 1 + 2 + 1}$$

sum of frequencies

$$= \frac{82}{15} = 5.5$$

We see that this agrees with the result of Example 3. ▮

Summations such as those in Eq. (22.1) occur frequently in statistics and other branches of mathematics. In order to simplify writing these sums, the symbol Σ is used to indicate the process of summation. (Σ is the Greek capital letter sigma.) Σx means the sum of the x's.

◀ **EXAMPLE 5**   We can show the sum of the numbers $x_1, x_2, x_3, \ldots, x_n$ as

$$\Sigma x = x_1 + x_2 + x_3 + \cdots + x_n$$

If these numbers are 3, 7, 2, 6, 8, 4, and 9, we have

$$\Sigma x = 3 + 7 + 2 + 6 + 8 + 4 + 9 = 39$$ ▶

Using the summation symbol Σ, we can write Eq. (22.1) for the arithmetic mean as

$$\bar{x} = \frac{x_1 f_1 + x_2 f_2 + x_3 f_3 + \cdots + x_n f_n}{f_1 + f_2 + f_3 + \cdots + f_n} = \frac{\Sigma xf}{\Sigma f} \qquad (22.1)$$

The summation notation Σx is an abbreviated form of the more general notation $\sum_{i=1}^{n} x_i$. This more general form can be used to indicate the sum of the first n numbers of a sequence or to indicate the sum of a certain set within the sequence. For example, for a set of at least 5 numbers, $\sum_{i=3}^{5} x_i$ indicates the sum of the third through the fifth of these numbers (in Example 5, $\sum_{i=3}^{5} x_i = 16$). We will use the abbreviated form Σx to indicate the sum of all the numbers being considered.

◀ **EXAMPLE 6**   We find the arithmetic mean of the Internet hours in Example 1 of Section 22.1 (page 609) by

$$\bar{x} = \frac{\Sigma xf}{\Sigma f} = \frac{3(1) + 4(1) + 5(3) + \cdots + 12(6) + \cdots + 28(1)}{50}$$

$$= \frac{687}{50} = 13.7 \text{ h} \quad \text{(rounded off to tenths)}$$ ▶

The arithmetic mean is one of a number of statistical measures that can be found on a calculator. The use of a calculator will be shown in the next section.

Mode   Another measure of central tendency is *the* **mode,** *which is the value that appears most frequently.* If two or more values appear with the same greatest frequency, each is a mode. If no value is repeated, there is no mode.

◀ **EXAMPLE 7**   **(a)** The mode of the numbers in Example 1 is 4, since it appears three times and no other value appears more than twice.

(b) The modes of the numbers

$$1, 2, 2, 4, 5, 5, 6, 7$$

are 2 and 5, since each appears twice and no other number is repeated.

(c) There is no mode for the values

$$1, 2, 5, 6, 7, 9$$

since none of the values is repeated. ▶

◖ EXAMPLE 8   To find the frictional force between two specially designed surfaces, the force to move a block with one surface along an inclined plane with the other surface is measured ten times. The results, with forces in newtons, are

$$2.2, 2.4, 2.1, 2.2, 2.5, 2.2, 2.4, 2.7, 2.1, 2.5$$

Find the mean, median, and mode of these forces.

To find the mean, we sum the values of the forces and divide this total by 10. This gives

> The mean is useful with many statistical methods and is used extensively. The median is also commonly used and is a good choice if there are some extreme values. The mode is occasionally used (if there is only one) for an estimate of data that is approximately symmetric.

$$\overline{F} = \frac{\sum F}{10} = \frac{2.2 + 2.4 + 2.1 + 2.2 + 2.5 + 2.2 + 2.4 + 2.7 + 2.1 + 2.5}{10}$$

$$= \frac{23.3}{10} = 2.33 \text{ N}$$

The median is found by arranging the values in order and finding the middle value. The values in order are

$$2.1, 2.1, 2.2, 2.2, 2.2, 2.4, 2.4, 2.5, 2.5, 2.7$$

Since there are ten values, we see that the fifth value is 2.2 and the sixth is 2.4. The value midway between these is 2.3, which is the median. Therefore, the median force is 2.3 N.

The mode is 2.2 N, since this value appears three times, which is more than any other value.   ◗

EXERCISES 22.2

In Exercises 1–4, delete the 5 from the data numbers given for Example 1 and then do the following with the resulting data.

1. Find the median.

2. Find the arithmetic mean using the definition, as in Example 3.

3. Find the arithmetic mean using Eq. (22.1), as in Example 4.

4. Find the mode, as in Example 7.

In Exercises 5–16, use the following sets of numbers.

   A: 3, 6, 4, 2, 5, 4, 7, 6, 3, 4, 6, 4, 5, 7, 3
   B: 25, 26, 23, 24, 25, 28, 26, 27, 23, 28, 25
   C: 0.48, 0.53, 0.49, 0.45, 0.55, 0.49, 0.47, 0.55, 0.48, 0.57, 0.51, 0.46
   D: 105, 108, 103, 108, 106, 104, 109, 104, 110, 108, 108, 104, 113, 106, 107, 106, 107, 109, 105, 111, 109, 108

In Exercises 5–8, determine the median of the numbers of the given set.

5. Set *A*

6. Set *B*

7. Set *C*

8. Set *D*

In Exercises 9–12, determine the arithmetic mean of the numbers of the given set.

9. Set *A*

10. Set *B*

11. Set *C*

12. Set *D*

In Exercises 13–16, determine the mode of the numbers of the given set.

13. Set *A*

14. Set *B*

15. Set *C*

16. Set *D*

In Exercises 17–32, the required sets of numbers are those in Exercises 22.1. Find the indicated measures of central tendency.

17. Median of computer instructions in Exercise 13

18. Mean of computer instructions in Exercise 13

19. Mode of computer instructions in Exercise 13

20. Median of times in Exercise 17

21. Mean of times in Exercise 17

22. Mode of times in Exercise 17

23. Median of stopping distances in Exercise 21. (Use the class mark for each class.)

24. Mean of stopping distances in Exercise 21. (Use the class mark for each class.)

25. Mean of X-ray dosages in Exercise 25

26. Median of X-ray dosages in Exercise 25

27. Mode of X-ray dosages in Exercise 25

28. Mean of battery lives in Exercise 27

29. Median of battery lives in Exercise 27

30. Mode of battery lives in Exercise 27

31. Mean of cable diameters in Exercise 29

32. Median of cable diameters in Exercise 29

In Exercises 33–44, find the indicated measures of central tendency.

33. The weekly salaries (in dollars) for the workers in a small factory are as follows:

450, 550, 475, 425, 375, 500, 400,
550, 475, 600, 500, 425, 450, 500

Find the median and the mode of the salaries.

34. Find the mean salary for the salaries in Exercise 33.

35. In a particular month, the electrical usage, rounded to the nearest 400 MJ, of 1000 homes in a certain city was summarized as follows:

Usage	2000	2400	2800	3200	3600	4000	4400	4800
No. Homes	22	80	106	185	380	122	90	15

Find the mean of the electrical usage.

36. Find the median and mode of electrical usage in Exercise 35.

37. A test of air pollution in a city gave the following readings of the concentration of sulfur dioxide (in parts per million) for 18 consecutive days:

0.14, 0.18, 0.27, 0.19, 0.15, 0.22, 0.20, 0.18, 0.15,
0.17, 0.24, 0.23, 0.22, 0.18, 0.32, 0.26, 0.17, 0.23

Find the median and the mode of these readings.

38. Find the mean of the readings in Exercise 37.

39. The *midrange*, another measure of central tendency, is found by finding the sum of the lowest and the highest values and dividing this sum by 2. Find the midrange of the salaries in Exercise 33.

40. Find the midrange of the sulfur dioxide readings in Exercise 37. (See Exercise 39.)

41. Add $100 to each of the salaries in Exercise 33. Then find the median, mean, and mode of the resulting salaries. State any conclusion that might be drawn from the results.

W 42. Multiply each of the salaries in Exercise 33 by 2. Then find the median, mean, and mode of the resulting salaries. State any conclusion that might be drawn from the results.

W 43. Change the final salary in Exercise 33 to $4000, with all other salaries being the same. Then find the mean of these salaries. State any conclusion that might be drawn from the result. (The $4000 here is called an *outlier*, which is an extreme value.)

W 44. Find the median and mode of the salaries indicated in Exercise 43. State any conclusion that might be drawn from the results.

22.3 STANDARD DEVIATION

In using statistics, we generally collect a sample of data and draw certain conclusions about the complete collection of possible values. In statistics, *the complete collection of values (measurements, people, scores, etc.) is called the* **population,** *and a* **sample** *is a subset of the population.* In the applied examples and exercises of the previous two sections, we were using samples of larger populations. For example, in Example 1 on page 609 we used estimates of the time on the Internet from a *sample* of 50 persons, as it would be impractical to get estimates from the total *population* of *all* users of the Internet.

In the preceding section, we discussed measures of central tendency of the data in the various samples. However, regardless of the measure that may be used, it does not tell us whether the values of the population tend to be grouped closely together or spread out over a large range of values. Therefore, we also need some measure of deviation, or spread, of the values from the median or the mean. If the spread is small and the numbers are grouped closely together, the measure of central tendency is more reliable and descriptive of the data than in the case in which the spread is greater.

In statistics, there are several measures of spread that may be defined. In this section, we discuss one that is very widely used: the *standard deviation*. It is defined on the next page.

The **standard deviation** *of a set of* **sample** *values is defined by the equation*

$$s = \sqrt{\frac{\sum (x - \bar{x})^2}{n - 1}}$$

(22.2)

The definition of *s* shows that the following steps are used in computing its value.

STEPS FOR CALCULATING STANDARD DEVIATION

1. *Find the arithmetic mean $\bar{x}$ of the numbers of the set.*

2. *Subtract the mean from each number of the set.*

3. *Square these differences.*

4. *Find the sum of these squares.*

5. *Divide this sum by $n - 1$.*

6. *Find the square root of this result.*

The standard deviation *s* is a positive number. It is a *deviation from the mean*, regardless of whether the individual numbers are greater than or less than the mean. Numbers close together will have a small standard deviation, whereas numbers further apart have a larger standard deviation. Therefore, *the standard deviation becomes larger as the spread of data increases.*

NOTE ▶ Following the steps shown above, we use Eq. (22.2) for the calculation of standard deviation in the following examples.

◀ EXAMPLE 1   Find the standard deviation of the following numbers: 1, 5, 4, 2, 6, 2, 1, 1, 5, 3.

A table of the necessary values is shown below, and steps 1–6 are indicated:

	step 2	step 3
x	$x - \bar{x}$	$(x - \bar{x})^2$
1	-2	4
5	2	4
4	1	1
2	-1	1
6	3	9
2	-1	1
1	-2	4
1	-2	4
5	2	4
3	0	0
30		32 step 4

$$\bar{x} = \frac{30}{10} = 3 \qquad \text{step 1}$$

$$\frac{\sum (x - \bar{x})^2}{n - 1} = \frac{32}{10 - 1} = \frac{32}{9} \qquad \text{step 5}$$

$$s = \sqrt{\frac{32}{9}} = 1.9 \qquad \text{step 6}$$

NOTE ▶ In calculating the standard deviation, it is usually rounded off to one more significant digit than was present in the original data. ▸

◀ EXAMPLE 2   Find the standard deviation of the numbers in Example 1 of Section 22.2.

Since several of the numbers appear more than once, it is helpful to use the frequency of each number in the table, as follows:

x	f	fx	$x - \bar{x}$	$(x - \bar{x})^2$	$f(x - \bar{x})^2$
1	1	1	−4.5	20.25	20.25
2	2	4	−3.5	12.25	24.50
3	1	3	−2.5	6.25	6.25
4	3	12	−1.5	2.25	6.75
5	1	5	−0.5	0.25	0.25
6	1	6	0.5	0.25	0.25
7	2	14	1.5	2.25	4.50
8	1	8	2.5	6.25	6.25
9	2	18	3.5	12.25	24.50
11	1	11	5.5	30.25	30.25
	15	82			123.75

The columns are labeled: step 2 ($x - \bar{x}$), step 3 ($(x - \bar{x})^2$), step 1.

$$\bar{x} = \frac{82}{15} = 5.5$$

$$\frac{\sum f(x - \bar{x})^2}{n - 1} = \frac{123.75}{15 - 1} = \frac{123.75}{14} \quad \text{step 5}$$

$$s = \sqrt{\frac{123.75}{14}} = 3.0 \quad \text{step 6}$$

step 4

In some sources, standard deviation is defined such that n, rather than $n - 1$, is used in the denominator of Eq. (22.2). Using n gives us the *population standard deviation*. However, we generally will be using samples of the population, and the use of $n - 1$ gives better estimates of a population standard deviation when using samples. Therefore, in this text we will use Eq. (22.2) and refer to the sample standard deviation simply as the standard deviation.

It is possible to reduce the computational work required to find the standard deviation. Algebraically, it can be shown (although we will not do so here) that the following equation is another form of Eq. (22.2) and therefore gives the same results.

$$s = \sqrt{\frac{n(\Sigma x^2) - (\Sigma x)^2}{n(n - 1)}} \tag{22.3}$$

Although the form of this equation appears more involved, it does reduce the amount of calculation that is necessary. Consider the following examples.

◀ EXAMPLE 3   Using Eq. (22.3), find s for the numbers in Example 1.

x	x^2
1	1
5	25
4	16
2	4
6	36
2	4
1	1
1	1
5	25
3	9
30	122

$$n = 10$$
$$\Sigma x^2 = 122$$
$$(\Sigma x)^2 = 30^2 = 900$$
$$s = \sqrt{\frac{10(122) - 900}{10(9)}} = 1.9$$

◀ **EXAMPLE 4**   An ammeter measures the electric current in a circuit. In an ammeter, two resistances are connected in parallel, with most of the current passing through a very low resistance called the *shunt*. The resistance of each shunt in a sample of 100 shunts was measured. The results were grouped, and the class mark and frequency for each class are shown in the following table. Calculate the arithmetic mean and the standard deviation of the resistances of the shunts.

R (ohms)	f	fR	fR²
0.200	1	0.200	0.0400
0.210	3	0.630	0.1323
0.220	5	1.100	0.2420
0.230	10	2.300	0.5290
0.240	17	4.080	0.9792
0.250	40	10.000	2.5000
0.260	13	3.380	0.8788
0.270	6	1.620	0.4374
0.280	3	0.840	0.2352
0.290	2	0.580	0.1682
	100	24.730	6.1421

$$\overline{R} = \frac{24.730}{100} = 0.247 \ \Omega$$

$$n = 100$$

$$\Sigma R^2 = 6.1421$$

$$(\Sigma R)^2 = 24.730^2$$

$$s = \sqrt{\frac{100(6.1421) - 24.730^2}{100(99)}} = 0.016$$

The arithmetic mean of the resistances is $0.247 \ \Omega$, with a standard deviation of $0.016 \ \Omega$. ▶

The statistical measures x, Σx, Σx^2, s_x, σ_x (the population standard deviation), and n (and possibly others) can be displayed on a graphing calculator as shown in Fig. 22.7 for the data of Example 5.

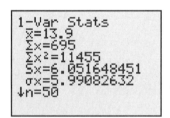

```
1-Var Stats
 x̄=13.9
 Σx=695
 Σx²=11455
Sx=6.051648451
σx=5.99082632
↓n=50
```

Fig. 22.7

◀ **EXAMPLE 5**   Find the standard deviation of the estimated hours on the Internet as grouped in Example 2 of Section 22.1 (page 609). In doing this, we assume that each value in the class is the same as the class mark. The method is not exact, but with a large set of numbers it provides a good approximation with less arithmetic work.

Interval	x	f	fx	fx²
0–4	2	2	4	8
5–9	7	9	63	441
10–14	12	19	228	2 736
15–19	17	11	187	3 179
20–24	22	6	132	2 904
25–29	27	3	81	2 187
		50	695	11 455

$$n = 50$$

$$\Sigma x^2 = 11,455$$

$$(\Sigma x)^2 = 695^2$$

$$s = \sqrt{\frac{50(11\,455) - 695^2}{50(49)}} = 6.1$$

Thus, $s = 6.1$ h. Compare the mean calculated in Example 6 on page 615 (not using class marks) with that shown in the calculator display to the left (using class marks). ▶

In using the statistical measures we have discussed, we must be careful in using and interpreting such measures. Consider the following example.

◀ **EXAMPLE 6**   (a) The numbers 1, 2, 3, 4, 5 have a mean of 3, a median of 3, and a standard deviation of 1.6. These values fairly well describe the center and distribution of the numbers in the set.

(b) The numbers 1, 2, 3, 4, 100 have a mean of 22, a median of 3, and a standard deviation of 44. The large difference between the median and the mean and the very large range of values within one standard deviation of the mean (-22 to 66) indicate that this set of measures does not describe this set of numbers well. In a case like this, the 100 should be checked to see if it is in error. ▶

Example 6 illustrates that the statistical measures can be misleading if the numbers in a set are unevenly distributed. Misleading statistics can also come from the source of the data. Consider the probable results of a survey to find the percent of persons in favor of raising income taxes for the wealthy if the survey is taken at the entrance to a welfare office or if it is taken at the entrance to a stock brokerage firm. There are many other considerations in the proper use and interpretation of statistical measures.

EXERCISES 22.3

In Exercises 1 and 2, in Example 1, change the first 1 to 6 and the first 2 to 7 and then find the standard deviation of the resulting data as directed.

1. Find *s* from the definition, as in Example 1.

2. Find *s* using Eq. 22.3, as in Example 3.

In Exercises 3–14, use the following sets of numbers. They are the same as those used in Exercise 22.2.

A: 3, 6, 4, 2, 5, 4, 7, 6, 3, 4, 6, 4, 5, 7, 3
B: 25, 26, 23, 24, 25, 28, 26, 27, 23, 28, 25
C: 0.48, 0.53, 0.49, 0.45, 0.55, 0.49, 0.47, 0.55, 0.48, 0.57, 0.51, 0.46
D: 105, 108, 103, 108, 106, 104, 109, 104, 110, 108, 108, 104, 113, 106, 107, 106, 107, 109, 105, 111, 109, 108

In Exercises 3–6, use Eq. (22.2) to find the standard deviation s for the indicated sets of numbers.

3. Set *A*   **4.** Set *B*   **5.** Set *C*   **6.** Set *D*

In Exercises 7–10, use Eq. (22.3) to find the standard deviation s for the indicated sets of numbers.

7. Set *A*   **8.** Set *B*   **9.** Set *C*   **10.** Set *D*

In Exercises 11–14, use the statistical feature of a calculator to find the arithmetic mean and the standard deviation s for the indicated sets of numbers.

11. Set *A*   **12.** Set *B*   **13.** Set *C*   **14.** Set *D*

In Exercises 15–20, find the standard deviation s for the indicated sets of numbers from Exercises 22.1 and 22.2.

15. The computer instructions in Exercise 13 of Section 22.1

16. The X-ray dosages in Exercise 25 of Section 22.1

17. The battery lives in Exercise 27 of Section 22.1

18. The salaries of Exercise 33 of Section 22.2

19. The strobe light times in Exercise 17 of Section 22.1

20. The stopping distances in Exercise 21 of Section 22.1

In Exercises 21–24, find the standard deviation s for the indicated sets of numbers using the statistical feature on a graphing calculator.

21. The air pollution data in Exercise 37 of Section 22.2

22. The following data giving the mean number of days of rain for Vancouver, B.C., for the 12 months of the year.

20, 17, 17, 14, 12, 11, 7, 8, 9, 16, 19, 22

23. The fiber-optic cable diameters in Exercise 29 of Section 22.1

24. The electric power usages in Exercise 35 of Section 22.2

22.4 NORMAL DISTRIBUTIONS

The distributions in the previous sections have been for a limited number of values. Let us now consider a very large population, such as the useable lifetime of all of the AA batteries sold in the world in a year. It could have a large number of classes with many values within each class.

We would expect a frequency polygon for this very large population to have its maximum frequency very near the mean and taper off to smaller frequencies on either side. It would probably be shaped very close to the curve shown in Fig. 22.8.

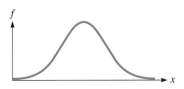

Fig. 22.8

Fig. 22.8

The smooth bell-shaped curve in Fig. 22.8 shows the **normal distribution** of a population large enough that the distribution is considered to be *continuous*. (We can think of this as the points on the frequency distribution curve are so close that they can be considered to be touching.) Using advanced methods, its equation is found to be

$$y = \frac{e^{-(x-\mu)^2/2\sigma^2}}{\sigma\sqrt{2\pi}} \tag{22.4}$$

Here, μ is the *population mean* and σ is the *population standard deviation,* and π and e are the familiar numbers first used in Chapters 2 and 12, respectively.

From Eq. (22.4), we can see that any particular normal distribution for a large population depends on the values of μ and σ. The horizontal location of the curve depends on μ, and the shape (how spread out the curve is) depends on σ, but the bell shape remains. This is illustrated in general in the following example.

◀ **EXAMPLE 1**   In Fig. 22.9, for the left curve $\mu = 10$ and $\sigma = 5$, whereas for the right curve $\mu = 20$ and $\sigma = 10$.

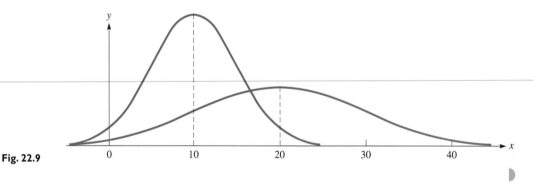

Fig. 22.9

STANDARD NORMAL DISTRIBUTION

As we have just seen, there are innumerable possible normal distributions. However, there is one of particular interest. *The **standard normal distribution** is the normal distribution for which the mean is* 0 *and the standard deviation is* 1. Making these substitutions in Eq. (22.4), we have

$$y = \frac{1}{\sqrt{2\pi}}e^{-x^2/2} \tag{22.5}$$

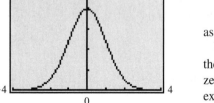

Fig. 22.10

as the equation of the standard normal distribution curve.

From Eq. (22.5), we see that the standard normal distribution curve is symmetric to the y-axis (if we replace x by $-x$, the equation remains unchanged). Since the mean is zero, this means that the curve is symmetric about the mean. We can also see from the exponent of e that y becomes smaller, but is always positive, as x becomes larger numerically, and this in turn means that the x-axis is an asymptote. The graph of the standard normal distribution curve is shown in the graphing calculator display in Fig. 22.10.

Using advanced mathematics, it can be shown that the total area under the standard normal distribution curve (above the *x*-axis) is exactly one unit. As shown in Fig. 22.11, it is found that the area between $x = 0$ and $x = 1$ is 0.3413, the area between $x = 1$ and $x = 2$ is 0.1359, and the area between $x = 2$ and $x = 3$ is 0.0215. Since the curve is symmetric to the *y*-axis, areas for negative values of *x* are equivalent. This means the total area between $x = -3$ and $x = 3$ is about 0.9974 and nearly all the area lies between $x = -3$ and $x = 3$.

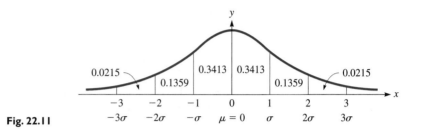

Fig. 22.11

From Fig. 22.11, we can see that *about 68% of the area is within one standard deviation of the mean (in the interval $\mu - \sigma$ to $\mu + \sigma$), and about 95% of the area is within two standard deviations of the mean (in the interval $\mu - 2\sigma$ to $\mu + 2\sigma$).* These percentages are often useful in data analysis.

Since a normal distribution curve gives a measure of the frequency of a particular value within the distribution, *the area under any part of the standard normal curve gives the relative frequency of those values of the distribution.* This is illustrated in the following example.

◀ **EXAMPLE 2** For a normal distribution of values, 0.3413 (34.13%) of the values are between $x = 0$ and $x = 1$, and 68.26% of the values are between $x = -1$ and $x = 1$. For this distribution, 81.85% of the values lie between $x = -2$ and $x = 1$. $(0.1359 + 0.3413 + 0.3413 = 0.8185)$. ▶

We can find the relative frequency of values for any normal distribution by use of the **standard score** z (or *z*-score), which is defined as

$$z = \frac{x - \mu}{\sigma} \tag{22.6}$$

For the normal standard distribution, where $\mu = 0$, if we let $x = \sigma$, then $z = 1$. If we let $x = 2\sigma$, $z = 2$. Therefore, we can see that *a value of z tells us the number of standard deviations the given value of x is above or below the mean.* From the discussion above, we can see that the value of *z* can tell us the area under the curve between the mean and the value of *x* corresponding to that value of *z*. In turn, *this tells us the relative frequency of all values between the mean and the value of x.*

On the next page, Table 22.1 gives the area under the standard normal distribution curve for the given values of *z*. The table includes values only to $z = 3$ since nearly all of the area is between $z = -3$ and $z = 3$. Since the curve is symmetric to the *y*-axis, the values shown are also valid for negative values of *z*.

Table 22.1   Standard Normal (z) Distribution

z	Area	z	Area	z	Area
0.0	0.0000	1.0	0.3413	2.0	0.4772
0.1	0.0398	1.1	0.3643	2.1	0.4821
0.2	0.0793	1.2	0.3849	2.2	0.4861
0.3	0.1179	1.3	0.4032	2.3	0.4893
0.4	0.1554	1.4	0.4192	2.4	0.4918
0.5	0.1915	1.5	0.4332	2.5	0.4938
0.6	0.2257	1.6	0.4452	2.6	0.4953
0.7	0.2580	1.7	0.4554	2.7	0.4965
0.8	0.2881	1.8	0.4641	2.8	0.4974
0.9	0.3159	1.9	0.4713	2.9	0.4981
1.0	0.3413	2.0	0.4772	3.0	0.4987

The following examples illustrate the use of Eq. (22.6) and z-scores.

EXAMPLE 3   For a normal distribution curve based on values of $\mu = 20$ and $\sigma = 5$, find the area between $x = 24$ and $x = 32$. To find this area we use Eq. (22.6) to find the corresponding values of z and then find the difference between these z-scores. These z-scores are

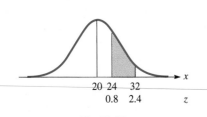

Fig. 22.12

$$z = \frac{24 - 20}{5} = 0.8 \quad \text{and} \quad z = \frac{32 - 20}{5} = 2.4$$

For $z = 0.8$, the area is 0.2881, and for $z = 2.4$, the area is 0.4918. Therefore, the area between $x = 24$ and $x = 32$ (see Fig. 22.12) is

$$0.4918 - 0.2881 = 0.2037$$

This means that the relative frequency of the values between $x = 24$ and $x = 32$ is 20.37%. If we have a large set of measured values with $\mu = 20$ and $\sigma = 5$, we should expect that about 20% of them are between $x = 24$ and $x = 32$.

Solving a Word Problem

EXAMPLE 4   The lifetimes of a certain type of watch battery are normally distributed. The mean lifetime is 400 days, and the standard deviation is 50 days. For a sample of 5000 new batteries, determine how many batteries will last **(a)** between 360 days and 460 days, **(b)** more than 320 days, and **(c)** less than 280 days.

(a) For this distribution, $\mu = 400$ days and $\sigma = 50$ days. Using Eq. (22.6), we find the z-scores for $x = 360$ days and $x = 460$ days. They are

$$z = \frac{360 - 400}{50} = -0.8 \quad \text{and} \quad z = \frac{460 - 400}{50} = 1.2$$

For $z = -0.8$, the area is to the left of the mean, and since the curve is symmetric about the mean, we use the $z = 0.8$ value of the area and add it to the area for $z = 1.2$. Therefore, the area is

$$0.2881 + 0.3849 = 0.6730$$

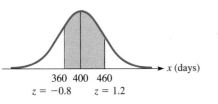

Fig. 22.13

See Fig. 22.13. This means that 67.30% of the 5000 batteries, or 3365 of the batteries, will last between 360 and 460 days.

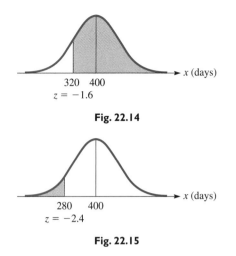

Fig. 22.14

Fig. 22.15

A more complete development of statistics, which usually requires at least an entire book, would also include topics in probability. Here we are pointing out that these standard errors should be considered when dealing with normal distributions.

(b) To determine the number of batteries that will last more than 320 days, we first find the z-score for $x = 320$. It is $z = (320 - 400)/50 = -1.6$. This means we want the total area to the right of $z = -1.6$. In this case, we add the area for $z = 1.6$ to the total area to the right of the mean. Since the total area under the curve is 1.0000, the total area on either side of the mean is 0.5000. Therefore, the area to the right of $z = -1.6$ is $0.4452 + 0.5000 = 0.9452$. See Fig. 22.14. This means that $0.9452 \times 5000 = 4726$ batteries will last more than 320 days.

(c) To find the number of batteries that will last less than 280 days, we first find that $z = (280 - 400)/50 = -2.4$ for $x = 280$. Since we want the total area to the left of $z = -2.4$, we subtract the area for $z = 2.4$ from 0.5000, the total area to the left of the mean. Since the area for $z = 2.4$ is 0.4918, the total area to the left of $z = -2.4$ is $0.5000 - 0.4918 = 0.0082$. See Fig. 22.15. Therefore, $0.0082 \times 5000 = 41$ batteries will last less than 280 days.

STANDARD ERRORS

In Example 4, we assumed that the lifetimes of the batteries were normally distributed. Of course, for any set of 5000 batteries, or any number of batteries for that matter, the lifetimes that actually occur will not follow normal distribution *exactly*. There will be some variation from normal distribution, but for a large sample this variation should be small. The mean and the standard deviation for any sample would probably vary somewhat from that of the population.

In the study of probability, it is shown that if we select all possible samples of size n from a population with a mean μ and standard deviation σ, the mean of the sample means is also μ. Also, *the standard deviation of the sample means*, denoted by $\sigma_{\bar{x}}$, and called **the standard error of the mean,** is

$$\sigma_{\bar{x}} = \frac{\sigma}{\sqrt{n}} \tag{22.7}$$

Also, the *standard error of the standard deviation s of the sample,* denoted by σ_s, and called **the standard error of s,** is

$$\sigma_s = \frac{\sigma}{\sqrt{2n}} \tag{22.8}$$

We can see from these formulas that as the sample size gets larger (*large* is usually considered to be over 30), the less variation there should be in the mean and standard deviation of the sample. This agrees with what we should expect.

EXAMPLE 5 For the sample of 5000 watch batteries in Example 4, we know that $\sigma = 50$ days. Therefore, the standard error in the mean $\bar{x}$ is $50/\sqrt{5000} = 0.7$ day. This means that of all samples of 5000 batteries, about 68% should have a mean lifetime of 400 ± 0.7 day (between 399.3 days and 400.7 days).

The standard error in the standard deviation s is $50/\sqrt{2(5000)} = 0.5$ day. This means that of all samples of 5000 batteries, about 68% should have a standard deviation of 50 ± 0.5 day (between 49.5 days and 50.5 days).

(Considering the significant digits of these values, 68% (within one standard deviation) or even 95% (within two standard deviations) of the sample values of $\bar{x}$ and s should not vary by more than 1 day.)

EXERCISES 22.4

In Exercises 1–4, make the given changes in the indicated examples of this section and then solve the indicated problem.

(W) 1. In Example 1, change the second σ from 10 to 5 and then describe the curve that would result in terms of either or both curves shown in Fig. 22.9.

2. In Example 2, change the final $x = 1$ to $x = 2$ and then find the resulting percent of values for that part of the example.

3. In Example 3, change $x = 32$ to $x = 33$ and then find the resulting area.

4. In Example 4(b), change 320 to 360 and then find the resulting number of batteries.

In Exercises 5–8, use a graphing calculator to display the indicated graph of Eq. (22.4).

5. Display the graph of the normal distribution of values for which $\mu = 10$ and $\sigma = 5$. Compare with the graph shown in Fig. 22.9.

6. Display the graph of the normal distribution of values for which $\mu = 20$ and $\sigma = 10$. Compare with the graph shown in Fig. 22.9.

7. Sketch a graph of a normal distribution of values for which $\mu = 100$ and $\sigma = 10$. Then compare with the graph displayed by a graphing calculator.

(W) 8. Sketch a graph of a normal distribution of values for which $\mu = 100$ and $\sigma = 30$. Then compare with the graph displayed by a graphing calculator. How does this graph differ from that of Exercise 7?

In Exercises 9–12, use the following data and refer to Fig. 22.11. A sample of 200 bags of cement are weighed as a quality check. Over a long period, it has been found that the mean value and standard deviation for this size bag are known and that the weights are normally distributed. Determine how many bags within this sample should have weights that satisfy the following conditions.

9. Within one standard deviation of the mean

10. Within two standard deviations of the mean

11. Between the mean and two standard deviations above the mean

12. Between one standard deviation below the mean and three standard deviations above the mean

In Exercises 13–16, use the following data. Each AA battery in a sample of 500 batteries is checked for its voltage. It has been previously established for this type of battery (when newly produced) that the voltages are distributed normally with $\mu = 1.50$ V and $\sigma = 0.05$ V.

13. How many batteries have voltages between 1.45 V and 1.55 V?

14. How many batteries have voltages between 1.52 V and 1.58 V?

15. What percent of the batteries have voltages below 1.54 V?

16. What percent of the batteries have voltages above 1.64 V?

In Exercises 17–24, use the following data. The lifetimes of a certain type of automobile tire have been found to be distributed normally with a mean lifetime of 100 000 km and a standard deviation of 10 000 km. Answer the following questions for a sample of 5000 of these tires.

17. How many tires will last between 85 000 km and 100 000 km?

18. How many tires will last between 95 000 km and 115 000 km?

19. How many tires will last more than 118 000 km?

20. If the manufacturer guarantees to replace all tires that do not last 75 000 km, what percent of the tires may have to be replaced under this guarantee?

(W) 21. What is the standard error in the mean for all samples of 5000 of these tires? Explain the meaning of this result.

(W) 22. What is the standard error in the standard deviation of all samples of 5000 of these tires? Explain the meaning of this result.

23. What percent of the samples of 5000 of these tires should have a mean lifetime of more than 100 282 km?

24. What percent of the samples of 5000 of these tires should have a mean standard deviation of between 9900 km and 10 100 km?

In Exercises 25–28, solve the given problems.

25. For the strobe light times in Exercise 17 of Section 22.1, find the percent of times within 1 standard deviation of the mean. From Exercise 21 of Section 22.2, we find that $\bar{x} = 2.248$ s, and from Exercise 19 of Section 22.3, we find that $s = 0.014$ s. Compare the results with that of a normal distribution.

26. Follow the same instructions as in Exercise 25 for the fiber-optic diameters in Exercise 29 of Section 22.1. From Exercise 31 of Section 22.2, $\bar{x} = 0.00595$ mm, and from Exercise 23 of Section 22.3, $s = 0.00022$ mm.

27. Follow the same instructions as in Exercise 25 for the hours estimated on the Internet in Example 1 of Section 22.1. From Example 6 of Section 22.2 we find that $\bar{x} = 13.7$ h, and from Example 5 of Section 22.3, we find that $s = 6.1$ h.

(W) 28. Discuss the results found in Exercises 25 and 27, considering the methods used to find the mean and the standard deviation.

22.5 STATISTICAL PROCESS CONTROL

One of the most important uses of statistics in industry is Statistical Process Control (SPC), which is used to maintain and improve product quality. Samples are tested during the production at specified intervals to determine whether the production process needs adjustment to meet quality requirements.

A particular industrial process is considered to be *in control* if it is stable and predictable, and sample measurements fall within upper and lower control limits. The process is *out of control* if it has an unpredictable amount of variation and there are sample measurements outside the control limits due to special causes.

> This is intended only as a brief introduction to this topic. A more complete development requires at least a chapter in a statistics book.

EXAMPLE 1 The manufacturer of 1.5-V batteries states that the voltage of its batteries is no less than 1.45 V or greater than 1.55 V and has designed the manufacturing process to meet these specifications..

If all samples of batteries that are tested have voltages in the proper range with only expected minor variations, the production process is *in control*.

However, if some samples have batteries with voltages out of the proper range, the process is *out of control*. This would indicate some special cause for the problem, such as an improperly operating machine or an impurity getting into the process. The process would probably be halted until the cause is determined.

> Minor variations may be expected, for example, from very small fluctuations in voltage, temperature, or material composition. Special causes resulting in an out-of-control process could include line stoppage, material defect, or an incorrect applied pressure.

CONTROL CHARTS

An important device used in SPC is the *control chart*. It is used to show a trend of a production characteristic over time. Samples are measured at specified intervals of time to see if the measurements are within acceptable limits. The measurements are plotted on a chart to check for trends and abnormalities in the production process.

In making a control chart, we must determine what the mean should be. For a stable process for which previous data are known, it can be based on a production specification or on previous data. For a new or recently modified process, it may be necessary to use present data, although the value may have to be revised for future charts. On a control chart, *this value is used as the population mean, μ.*

It is also necessary to establish the upper and lower control limits. The standard generally used is that 99.7% of the sample measurements should fall within these control limits. This assumes a normal distribution, and we note that this is within three sample standard deviations of the population mean. We will establish these limits by use of a table or a formula that has been made using statistical measures developed in a more complete coverage of quality control. This does follow the normal practice of using a formula or a more complete table in setting up the control limits.

In Fig. 22.16, we show a sample control chart, and on the following pages, we illustrate how control charts are made.

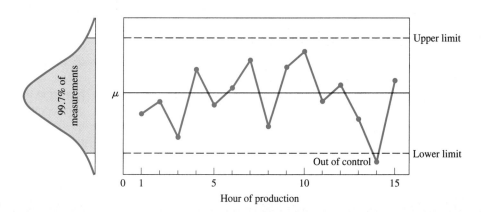

Fig. 22.16

❨EXAMPLE 2 A pharmaceutical company makes a capsule of a prescription drug that contains 500 mg of the drug, according to the label. In a newly modified process of making the capsule, five capsules are tested every 15 min to check the amount of the drug in each capsule. Testing over a 5-h period gave the following results for the 20 subgroups of samples.

Subgroup	*Amount of Drug (in mg) of Five Capsules*					Mean x	Range R
1	503	501	498	507	502	502.2	9
2	497	499	500	495	502	498.6	7
3	496	500	507	503	502	501.6	11
4	512	503	488	500	497	500.0	24
5	504	505	500	508	502	503.8	8
6	495	495	501	497	497	497.0	6
7	503	500	507	499	498	501.4	9
8	494	498	497	501	496	497.2	7
9	502	504	505	500	502	502.6	5
10	500	502	500	496	497	499.0	6
11	502	498	510	503	497	502.0	13
12	497	498	496	502	500	498.6	6
13	504	500	495	498	501	499.6	9
14	500	499	498	501	494	498.4	7
15	498	496	502	501	505	500.4	9
16	500	503	504	499	505	502.2	6
17	487	496	499	498	494	494.8	12
18	498	497	497	502	497	498.2	5
19	503	501	500	498	504	501.2	6
20	496	494	503	502	501	499.2	9
					Sums	9998.0	174
					Means	499.9	8.7

As we noted from the table, *the range R of each sample is the difference between the highest value and the lowest value of the sample.*

From this table of values, we can make an $\bar{x}$ control chart and an R control chart. The $\bar{x}$ chart maintains a check on the average quality level, whereas the R chart maintains a check on the dispersion of the production process. These two control charts are often plotted together and referred to as the $\bar{x}$–R chart.

In order to define the **central line** of the $\bar{x}$ chart, which ideally is equivalent to the value of the population mean μ, we use the mean of the sample means $\bar{\bar{x}}$. For the central line of the R chart, we use $\bar{R}$. From the table, we see that

$$\bar{\bar{x}} = 499.9 \text{ mg} \quad \text{and} \quad \bar{R} = 8.7 \text{ mg}$$

Table 22.2 Control Chart Factors

n	d_2	A	A_2	D_1	D_2	D_3	D_4
5	2.326	1.342	0.577	0.000	4.918	0.000	2.115
6	2.534	1.225	0.483	0.000	5.078	0.000	2.004
7	2.704	1.134	0.419	0.205	5.203	0.076	1.924

The **upper control limit** (UCL) and the **lower control limit** (LCL) for each chart are defined in terms of the mean range $\bar{R}$ and an appropriate constant taken from a table of control chart factors. These factors, which are related to the sample size n, are determined by statistical considerations found in a more complete coverage of quality control. At the left is a brief table of control chart factors (Table 22.2).

The UCL and LCL for the $\bar{x}$ chart are found as follows:

$$\text{UCL}(\bar{x}) = \bar{\bar{x}} + A_2\bar{R} = 499.9 + 0.577(8.7) = 504.9 \text{ mg}$$
$$\text{LCL}(\bar{x}) = \bar{\bar{x}} - A_2\bar{R} = 499.9 - 0.577(8.7) = 494.9 \text{ mg}$$

The UCL and LCL for the R chart are found as follows:

$$\text{LCL}(R) = D_3\bar{R} = 0.000(8.7) = 0.0 \text{ mg}$$
$$\text{UCL}(R) = D_4\bar{R} = 2.115(8.7) = 18.4 \text{ mg}$$

Using these central lines and control limit lines, we now plot the $\bar{x}$ control chart in Fig. 22.17 and the R control chart in Fig. 22.18.

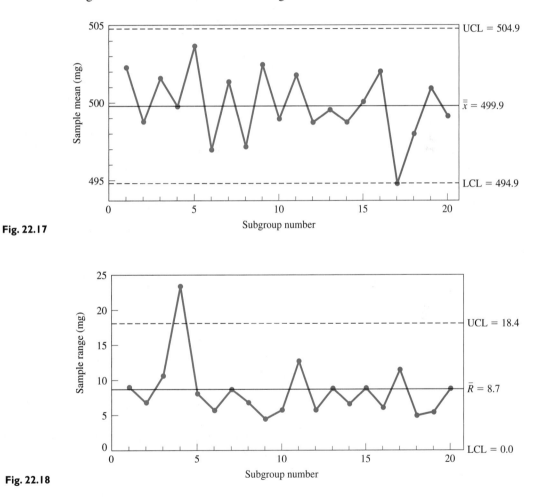

Fig. 22.17

Fig. 22.18

This would be considered a *well-centered process* since $\bar{\bar{x}} = 499.9$ mg, which is very near the target value of 500.0 mg. We do note, however, that subgroup 17 was at a control limit and this might have been due to some special cause, such as the use of a substandard mixture of ingredients. We also note that the process was *out of control* due to some special cause of subgroup 4 since the range was above the upper control limit. We should keep in mind that there are numerous considerations, including human factors, that should be taken into account when making and interpreting control charts and that this is only a very brief introduction to this important industrial use of statistics.

Day	Defective CDs	Proportion Defective
1	22	0.022
2	16	0.016
3	14	0.014
4	18	0.018
5	12	0.012
6	25	0.025
7	36	0.036
8	16	0.016
9	14	0.014
10	22	0.022
11	20	0.020
12	17	0.017
13	26	0.026
14	20	0.020
15	22	0.022
16	28	0.028
17	17	0.017
18	15	0.015
19	25	0.025
20	12	0.012
21	16	0.016
22	22	0.022
23	19	0.019
24	16	0.016
25	20	0.020
Sum	490	

In Example 2, the weight (in mg) of a prescription drug was tested. In quality control, weight is a **variable,** *which is a characteristic that can be* **measured.** Other examples of variables are length, voltage, and pressure.

A characteristic that can be **counted** *is an* **attribute,** which is determined to be either acceptable or not acceptable in testing. Examples of attributes are color (acceptable or not), entries on a customer account (correct or incorrect), and defects (not acceptable) in a product. To monitor an attribute in a production process, we get the *proportion* of defective parts by *dividing the number of defective parts in a sample by the total number of parts in the sample,* and then make a *p control chart.* This is illustrated in the following example.

◀ **EXAMPLE 3** The manufacturer of compact discs has 1000 CDs checked each day for defects (surface scratches, for example). The data for this procedure for 25 days are shown in the table at the left.

For the *p* control chart, the central line is the value of $\bar{p}$, which in this case is

$$\bar{p} = \frac{490}{25\,000} = 0.0196$$

The control limits are each three standard deviations from $\bar{p}$. If *n* is the size of the subgroup, the standard deviation σ_p of a proportion is given by

$$\sigma_p = \sqrt{\frac{\bar{p}(1 - \bar{p})}{n}} = \sqrt{\frac{0.0196(1 - 0.0196)}{1000}} = 0.004\,38$$

Therefore, the control limits are

$$\text{UCL}(p) = 0.0196 + 3(0.004\,38) = 0.0327$$
$$\text{LCL}(p) = 0.0196 - 3(0.004\,38) = 0.0065$$

Using this central line and these control limit lines, we now plot the *p* control chart in Fig. 22.19.

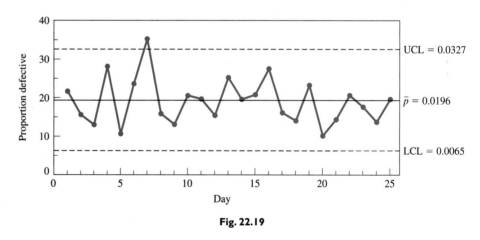

Fig. 22.19

According to the proportion mean of 0.0196, the process produces about 2% defective CDs. We note that the process was out of control on day 7. An adjustment to the production process was probably made to remove the special cause of the additional defective CDs. ◗

EXERCISES 22.5

In Exercises 1–4, in Example 2, change the first subgroup to 497, 499, 502, 493, and 498 and then proceed as directed.

1. Find UCL ($\bar{x}$) and LCL ($\bar{x}$).

2. Find LCL (R) and UCL (R).

(W) 3. How would the $\bar{x}$ control chart differ from Fig. 22.17?

(W) 4. How would the R control chart differ from Fig. 22.18?

In Exercise 5–8, use the following data.

Five automobile engines are taken from the production line each hour and tested for their torque (in N · m) when rotating at a constant frequency. The measurements of the sample torques for 20 h of testing are as follows:

Hour	Torques (in N · m) of Five Engines				
1	366	352	354	360	362
2	370	374	362	366	356
3	358	357	365	372	361
4	360	368	367	359	363
5	352	356	354	348	350
6	366	361	372	370	363
7	365	366	361	370	362
8	354	363	360	361	364
9	361	358	356	364	364
10	368	366	368	358	360
11	355	360	359	362	353
12	365	364	357	367	370
13	360	364	372	358	365
14	348	360	352	360	354
15	358	364	362	372	361
16	360	361	371	366	346
17	354	359	358	366	366
18	362	366	367	361	357
19	363	373	364	360	358
20	372	362	360	365	367

5. Find the central line, UCL, and LCL for the mean.

6. Find the central line, UCL, and LCL for the range.

7. Plot an $\bar{x}$ chart.   8. Plot an R chart.

In Exercise 9–12, use the following data.

Five AC adaptors that are used to charge batteries of a cellular phone are taken from the production line every 15 min and tested for their direct-current output voltage. The output voltages for 24 sample subgroups are as follows:

Subgroup	Output Voltages of Five Adaptors				
1	9.03	9.08	8.85	8.92	8.90
2	9.05	8.98	9.20	9.04	9.12
3	8.93	8.96	9.14	9.06	9.00
4	9.16	9.08	9.04	9.07	8.97
5	9.03	9.08	8.93	8.88	8.95
6	8.92	9.07	8.86	8.96	9.04
7	9.00	9.05	8.90	8.94	8.93
8	8.87	8.99	8.96	9.02	9.03
9	8.89	8.92	9.05	9.10	8.93
10	9.01	9.00	9.09	8.96	8.98
11	8.90	8.97	8.92	8.98	9.03
12	9.04	9.06	8.94	8.93	8.92
13	8.94	8.99	8.93	9.05	9.10
14	9.07	9.01	9.05	8.96	9.02
15	9.01	8.82	8.95	8.99	9.04
16	8.93	8.91	9.04	9.05	8.90
17	9.08	9.03	8.91	8.92	8.96
18	8.94	8.90	9.05	8.93	9.01
19	8.88	8.82	8.89	8.94	8.88
20	9.04	9.00	8.98	8.93	9.05
21	9.00	9.03	8.94	8.92	9.05
22	8.95	8.95	8.91	8.90	9.03
23	9.12	9.04	9.01	8.94	9.02
24	8.94	8.99	8.93	9.05	9.07

9. Find the central line, UCL, and LCL for the mean.

10. Find the central line, UCL, and LCL for the range.

11. Plot an $\bar{x}$ chart.

12. Plot an R chart.

In Exercises 13–16, use the following information.

For a production process for which there is a great deal of data since its last modification, the population mean μ and population standard deviation σ are assumed known. For such a process, we have the following values (using additional statistical analysis):

$\bar{x}$ chart: central line = μ, UCL = $\mu + A\sigma$, LCL = $\mu - A\sigma$

R chart: central line = $d_2\sigma$, UCL = $D_2\sigma$, LCL = $D_1\sigma$

The values of A, d_2, D_2, and D_1 are found in the table of control chart factors in Example 2 (Table 22.2).

13. In the production of robot links and tests for their lenghs, it has been found that $\mu = 2.725$ cm and $\sigma = 0.0032$ cm. Find the central length, UCL, and LCL for the mean if the sample subgroup size is 5.

14. For the robot link samples of Exercise 13, find the central line, UCL, and LCL for the range.

15. After bottling, the volume of soft drink in six sample bottles is checked each 10 minutes. For this process $\mu = 750.0$ mL and $\sigma = 2.2$ mL. Find the central line, UCL, and LCL for the range.

16. For the bottling process of Exercise 15, find the central line, UCL, and LCL for the mean.

In Exercises 17 and 18, use the following data.

A telephone company rechecks the entries for 1000 of its new customers each week for name, address, and phone number. The data collected regarding the number of new accounts with errors, along with the proportion of these accounts with errors, is given in the following table for a 20-week period:

Week	Accounts with Errors	Proportion with Errors
1	52	0.052
2	36	0.036
3	27	0.027
4	58	0.058
5	44	0.044
6	21	0.021
7	48	0.048
8	63	0.063
9	32	0.032
10	38	0.038
11	27	0.027
12	43	0.043
13	22	0.022
14	35	0.035
15	41	0.041
16	20	0.020
17	28	0.028
18	37	0.037
19	24	0.024
20	42	0.042
Total	738	

17. For a p chart, find the values for the central line, UCL, and LCL.

18. Plot a p chart.

In Exercises 19 and 20, use the following data.

The maker of electric fuses checks 500 fuses each day for defects. The number of defective fuses, along with the proportion of defective fuses for 24 days, is shown in the following table.

Day	Number Defective	Proportion Defective
1	26	0.052
2	32	0.064
3	37	0.074
4	16	0.032
5	28	0.056
6	31	0.062
7	42	0.084
8	22	0.044
9	31	0.062
10	28	0.056
11	24	0.048
12	35	0.070
13	30	0.060
14	34	0.068
15	39	0.078
16	26	0.052
17	23	0.046
18	33	0.066
19	25	0.050
20	25	0.050
21	32	0.064
22	23	0.046
23	34	0.068
24	20	0.040
Total	696	

19. For a p chart, find the values for the central line, UCL, and LCL.

20. Plot a p chart.

22.6 LINEAR REGRESSION

We have considered statistical methods for dealing with one variable. We have discussed methods of tabulating, graphing, and measuring the central tendency and the deviations from this value for one variable and have shown an important industrial application in the area of quality control. We now discuss how to find an equation relating two variables for which a set of points is known.

In this section, we show a method of finding the equation of a straight line that passes through a set of data points, and in this way we *fit* the line to the points. In general, *the fitting of a curve to a set of points is called* **regression.** Fitting a straight line to a set of points is *linear regression,* and fitting some other type of curve is called *nonlinear regression.* We consider nonlinear regression in the next section.

Some of the reasons for using regression to find the equation of a curve that passes through a set of points, and thereby "fit" the curve to the points are **(1)** to express a concise relationship between the variables, **(2)** to use the equation to predict certain fundamental results, **(3)** to determine the reliability of certain sets of data, and **(4)** to use the data for testing certain theoretical concepts.

For a given set of several (at least 5 or 6) points for representing pairs of data values, we cannot reasonably expect that the curve of any given equation will pass through all of the points *exactly.* Therefore, when we fit the curve of an equation to the points, we are finding the curve that best approximates passing through the points. It is possible that the curve that best fits the data will not actually pass directly through any of the points, although it should come reasonably close to most of them. Consider the following example.

◖EXAMPLE 1   All the students enrolled in a mathematics course took an entrance test. To study the reliability of this test as an indicator of future success, an instructor tabulated the test scores of ten students (selected at random), along with their course averages at the end of the course, and made a graph of the data. See the table below and Fig. 22.20.

Student	Entrance Test Score, Based on 40	Course Average, Based on 100
A	29	63
B	33	88
C	22	77
D	17	67
E	26	70
F	37	93
G	30	72
H	32	81
I	23	47
J	30	74

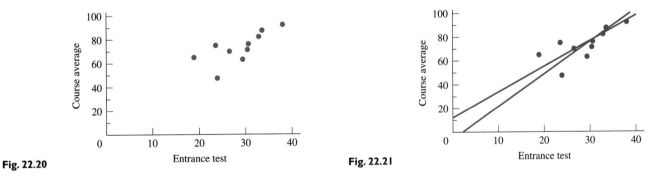

Fig. 22.20   **Fig. 22.21**

We now ask whether there is a functional relationship between the test scores and the course grades. Certainly, no clear-cut relationship exists, but in general we see that the higher the test score, the higher the course grade. This leads to the possibility that there might be some straight line, from which none of the points would vary too significantly. If such a line could be found, then it could be the basis of predictions as to the possible success a student might have in the course, on the basis of his or her grade on the entrance test. Assuming that such a straight line exists, the problem is to find the equation of this line. Figure 22.21 shows two such possible lines.   ▶

There are a number of different methods of determining the straight line that best fits the given data points. We employ the method that is most widely used: the **method of least squares.** *The basic principle of this method is that the sum of the squares of the deviations of all data points from the best line (in accordance with this method) has the least value possible. By* **deviation,** *we mean the difference between the y-value of the line and the y-value for the point (of original data) for a particular value of x.*

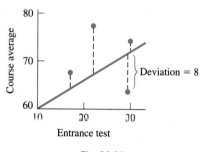

Fig. 22.22

⟨ EXAMPLE 2   In Fig. 22.22, the deviations of some of the points of Example 1 are shown. The point (29, 63) (student A of Example 1) has a deviation of 8 from the indicated line in the figure. Thus, we square the value of this deviation to obtain 64. In order to find the equation of the straight line that best fits the given points, the method of least squares requires that the sum of all such squares be a minimum.

Therefore, in applying this method of least squares, it is necessary to use the equation of a straight line and the coordinates of the points of the data. The deviations of all of these data points are determined, and these values are then squared. It is then necessary to determine the constants for the slope m and the y-intercept b in the equation of a straight line $y = mx + b$ for which the sum of the squared values is a minimum. To do this requires certain methods of advanced mathematics.   ▶

Using the methods that are required from advanced mathematics, we show that *the equation of the* **least-squares line**

$$y = mx + b \qquad (22.9)$$

can be found by calculating the values of the slope m and the y-intercept b by using the formulas

$$m = \frac{n \sum xy - \left(\sum x\right)\left(\sum y\right)}{n \sum x^2 - \left(\sum x\right)^2} \qquad (22.10)$$

and

$$b = \frac{\left(\sum x^2\right)\left(\sum y\right) - \left(\sum xy\right)\left(\sum x\right)}{n \sum x^2 - \left(\sum x\right)^2} \qquad (22.11)$$

In Eqs. (22.10) and (22.11), *the x's and y's are the values of the coordinates of the points in the given data,* and n is the number of points of data. We can reduce the calculational work in finding the values of m and b by noting that the denominators in Eqs. (22.10) and (22.11) are the same. Therefore, in using a calculator, the value of this denominator can be stored in memory.

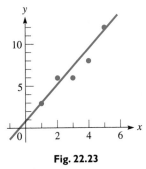

Fig. 22.23

◀ EXAMPLE 3   Find the equation of the least-squares line for the points indicated in the following table. Graph the line and data points on the same graph.

x	1	2	3	4	5
y	3	6	6	8	12

We see from Eqs. (22.10) and (22.11) that we need the sums of x, y, xy, and x^2 in order to find m and b. Thus, we set up a table for these values, along with the necessary calculations, as follows:

x	y	xy	x^2
1	3	3	1
2	6	12	4
3	6	18	9
4	8	32	16
5	12	60	25
sums → 15	35	125	55
↑ Σx	↑ Σy	↑ Σxy	↑ Σx^2

$n = 5$ (5 points)

$$m = \frac{5(125) - (15)(35)}{5(55) - (15)^2} = \frac{100}{50} = 2$$

$$b = \frac{(55)(35) - (125)(15)}{50} = \frac{50}{50} = 1$$

This means that the equation of the least-squares line is $y = 2x + 1$. This line and the data points are shown in Fig. 22.23.

◀ EXAMPLE 4   Find the least-squares line for the data of Example 1.

Here, the x-values will be the entrance-test scores and the y-values are the course averages.

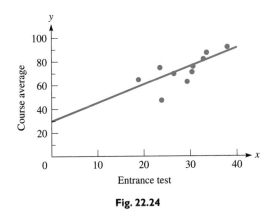

Fig. 22.24

x	y	xy	x^2
29	63	1 827	841
33	88	2 904	1089
22	77	1 694	484
17	67	1 139	289
26	70	1 820	676
37	93	3 441	1369
30	72	2 160	900
32	81	2 592	1024
23	47	1 081	529
30	74	2 220	900
279	732	20 878	8101

$n = 10$

$$m = \frac{10(20\,878) - 279(732)}{10(8101) - 279^2} = 1.44$$

$$b = \frac{8101(732) - 20\,878(279)}{10(8101) - 279^2} = 33.1$$

Thus, the equation of the least-squares line is $y = 1.44x + 33.1$. The line and data points are shown in Fig. 22.24. This line best fits the data, although the fit is obviously approximate. It can be used to predict the approximate course average that a student might be expected to attain, based on the entrance test.

See the chapter introduction.

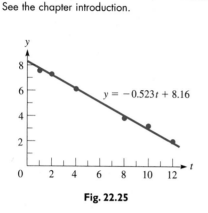

Fig. 22.25

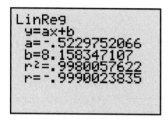

(a)

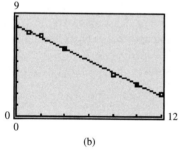

(b)

Fig. 22.26

A graphing calculator can be used to find the slope and the intercept and to display the least-squares line. In the following example, we compare the values using Eqs. (22.9), (22.10), and (22.11) with the corresponding calculator displays.

◀ **EXAMPLE 5** In a research project to determine the amount of a drug that remains in the bloodstream after a given dosage, the amounts y (in mg of drug/dL of blood) were recorded after t h:

t (h)	1.0	2.0	4.0	8.0	10.0	12.0
y (mg/dL)	7.6	7.2	6.1	3.8	2.9	2.0

Find the least-squares line for these data, expressing y as a function of t. Sketch the graph of the line and data points.

The table and calculations are shown below:

t	y	ty	t^2
1.0	7.6	7.6	1.0
2.0	7.2	14.4	4.0
4.0	6.1	24.4	16.0
8.0	3.8	30.4	64.0
10.0	2.9	29.0	100
12.0	2.0	24.0	144
37.0	29.6	129.8	329

$n = 6$

$$m = \frac{6(129.8) - 37.0(29.6)}{6(329) - 37.0^2} = -0.523$$

$$b = \frac{(329)(29.6) - (129.8)(37.0)}{6(329) - 37.0^2} = 8.16$$

The equation of the least-squares line is $y = -0.523t + 8.16$. The line and the data points are shown in Fig. 22.25. This line is useful in determining the effectiveness of the drug. It can also be used to determine when additional medication may be administered.

Using the linear regression feature of a calculator, Fig. 22.26(a) shows the display of the coefficients of the line $y = ax + b$ (note that the slope is a). The value of r is the *coefficient of correlation* (see Exercises 13–16). Figure 22.26(b) shows a calculator display of the points and the line. We see that they agree with Fig. 22.25. ▶

E X E R C I S E S 22.6

In Exercises 1–12, find the equation of the least-squares line for the given data. Graph the line and data points on the same graph.

1. In Example 3, replace the y-values with 3, 7, 9, 9, and 12. Then follow the instructions above.

2.
x	1	2	3	4	5	6	7
y	10	17	28	37	49	56	72

3.
x	20	26	30	38	48	60
y	160	145	135	120	100	90

4.
x	1	3	6	5	8	10	4	7	3	8
y	15	12	10	8	9	2	11	9	11	7

5. In an electrical experiment, the following data were found for the values of current and voltage for a particular element of the circuit. Find the voltage V as a function of the current i.

Current (mA)	15.0	10.8	9.30	3.55	4.60
Voltage (V)	3.00	4.10	5.60	8.00	10.50

6. A particular muscle was tested for its speed of shortening as a function of the force applied to it. The results appear below. Find the speed as a function of the force.

Force (N)	60.0	44.2	37.3	24.2	19.5
Speed (m/s)	1.25	1.67	1.96	2.56	3.05

7. The altitude h (in m) of a rocket was measured at several positions at a horizontal distance x (in m) from the launch site, shown in the table. Find the least-squares line for h as a function of x.

x (m)	0	500	1000	1500	2000	2500
h (m)	0	1130	2250	3360	4500	5600

8. In testing an air-conditioning system, the temperature T in a building was measured during the afternoon hours with the results shown in the table. Find the least-squares line for T as a function of the time t from noon.

t (h)	0.0	1.0	2.0	3.0	4.0	5.0
T (°C)	20.5	20.6	20.9	21.3	21.7	22.0

9. The pressure p was measured along an oil pipeline at different distances from a reference point, with results as shown. Find the least-squares line for p as a function of x. Check the values and line with a graphing calculator.

x (m)	0	50	100	150	200
p (kPa)	4370	4240	4070	3970	3840

10. The heat loss L per hour through various thicknesses of a particular type of insulation was measured as shown in the table. Find the least-squares line for L as a function of t. Check the values and line with a graphing calculator.

t (cm)	3.0	4.0	5.0	6.0	7.0
L (MJ)	5.90	4.80	3.90	3.10	2.45

11. In an experiment on the photoelectric effect, the frequency of light being used was measured as a function of the stopping potential (the voltage just sufficient to stop the photoelectric effect) with the results given below. Find the least-squares line for V as a function of f. The frequency for $V = 0$ is known as the *threshold frequency*. From the graph determine the threshold frequency. Check the values and curve with a graphing calculator.

f (PHz)	0.550	0.605	0.660	0.735	0.805	0.880
V (V)	0.350	0.600	0.850	1.10	1.45	1.80

12. If gas is cooled under conditions of constant volume, it is noted that the pressure falls nearly proportionally as the temperature. If this were to happen until there was no pressure, the theoretical temperature for this case is referred to as *absolute zero*. In an elementary experiment, the following data were found for pressure and temperature under constant volume.

T (°C)	0.0	20	40	60	80	100
P (kPa)	133	143	153	162	172	183

Find the least-squares line for P as a function of T, and from the graph determine the value of absolute zero found in this experiment. Check the values and curve with a graphing calculator.

The linear coefficient of correlation, a measure of the relatedness of two variables, is defined by $r = m(s_x/s_y)$, where s_x and s_y are the standard deviations of the x-values and y-values, respectively. Due to its definition, the values of r lie in the range $-1 \le r \le 1$. If r is near 1, the correlation is considered good. For the values of r between -0.5 and $+0.5$, the correlation is poor. If r is near -1, the variables are said to be negatively correlated; that is, one increases as the other decreases. In Exercises 13–16, compute r for the given data.

13. Exercise 1 **14.** Exercise 2

15. Exercise 4 **16.** Example 1

 NONLINEAR REGRESSION

If the experimental points do not appear to be on a straight line but we recognize them as being approximately on some other type of curve, the method of least squares can be extended to use on these other curves. For example, if the points are apparently on a parabola, we could use the function $y = a + bx^2$. To use the above method, we extend the least-squares line to

$$y = m[f(x)] + b \tag{22.12}$$

Here, $f(x)$ must be calculated first, and then the problem can be treated as a least-squares line to find the values of m and b. Some of the functions $f(x)$ that may be considered for use are x^2, $1/x$, and 10^x.

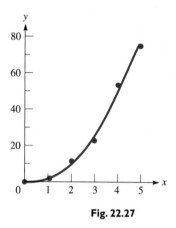

Fig. 22.27

◀ **EXAMPLE 1** Find the least-squares curve $y = mx^2 + b$ for the following points:

x	0	1	2	3	4	5
y	1	5	12	24	53	76

In using Eq. (22.12), $f(x) = x^2$. Our first step is to calculate values of x^2, and then we use x^2 as we used x in finding the equation of the least-squares line.

x	$f(x) = x^2$	y	x^2y	$(x^2)^2$
0	0	1	0	0
1	1	5	5	1
2	4	12	48	16
3	9	24	216	81
4	16	53	848	256
5	25	76	1900	625
	55	171	3017	979

$n = 6$

$$m = \frac{6(3017) - 55(171)}{6(979) - 55^2} = 3.05$$

$$b = \frac{(979)(171) - (3017)(55)}{6(979) - 55^2} = 0.52$$

Therefore, the required equation is $y = 3.05x^2 + 0.52$. The graph of this equation and the data points are shown in Fig. 22.27. ▶

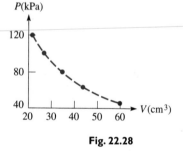

Fig. 22.28

◀ **EXAMPLE 2** In a physics experiment, the pressure p and volume V of a gas were measured at constant temperature. When the points were plotted, they were seen to approximate the hyperbola $y = c/x$. Find the least-squares approximation to the hyperbola $y = m(1/x) + b$ for the given data. See Fig. 22.28.

P (kPa)	V (cm³)	$x (= V)$	$f(x) = \frac{1}{x}$	$y (= P)$	$(\frac{1}{x})y$	$(\frac{1}{x})^2$
120.0	21.0	21.0	0.047 619 0	120.0	5.714 285 7	0.002 267 6
99.2	25.0	25.0	0.040 000 0	99.2	3.968 000 0	0.001 600 0
81.3	31.8	31.8	0.031 446 5	81.3	2.556 603 8	0.000 988 9
60.6	41.1	41.1	0.024 330 9	60.6	1.474 452 6	0.000 592 0
42.7	60.1	60.1	0.016 638 9	42.7	0.710 482 5	0.000 276 9
			0.160 035 3	403.8	14.423 824 6	0.005 725 4

(*Calculator note:* The final digits for the values shown may vary depending on the calculator and how the values are used. Here, all individual values are shown with eight digits (rounded off), although more digits were used. The value of $1/x$ was found from the value of x, with the eight digits shown. However, the values of $(1/x)y$ and $(1/x)^2$ were found from the value of $1/x$, using the extra digits. The sums were found using the rounded-off values shown. However, since the data contain only three digits, any variation in the final digits for $1/x$, $(1/x)y$, or $(1/x)^2$, will not matter.)

$$m = \frac{5(14.423\,824\,6) - 0.160\,035\,3(403.8)}{5(0.005\,725\,4) - 0.160\,035\,3^2} = 2490$$

$$b = \frac{(0.005\,725\,4)(403.8) - (14.423\,824\,6)(0.160\,035\,3)}{5(0.005\,725\,4) - 0.160\,035\,3^2} = 1.2$$

The equation of the hyperbola $y = m(1/x) + b$ is

$$y = \frac{2490}{x} + 1.2$$

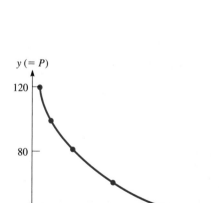

Fig. 22.29

This hyperbola and data points are shown in Fig. 22.29. ▶

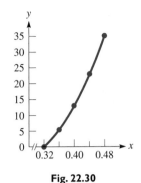

Fig. 22.30

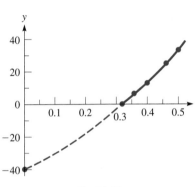

Fig. 22.31

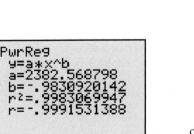

Fig. 22.32

EXAMPLE 3 It has been found experimentally that the tensile strength of brass (a copper–zinc alloy) increases (within certain limits) with the percent of zinc. The following table shows the values that have been found. See Fig. 22.30.

Tensile Strength (GPa)	0.32	0.36	0.40	0.44	0.48
Percent of Zinc	0	5	13	22	34

Fit a curve of the form $y = m(10^x) + b$ to the data. Let x = tensile strength ($\times 10^5$) and y = percent of zinc.

x	$f(x) = 10^x$	y	$(10^x)y$	$(10^x)^2$
0.32	2.089 296 1	0	0.000 000	4.365 158 3
0.36	2.290 867 7	5	11.454 338	5.248 074 6
0.40	2.511 886 4	13	32.654 524	6.309 573 4
0.44	2.754 228 7	22	60.593 031	7.585 775 8
0.48	3.019 951 7	34	102.678 36	9.120 108 4
	12.666 230 6	74	207.380 25	32.628 690 5

(See the note on calculator use in Example 2.)

$$m = \frac{5(207.380\,25) - 12.666\,230\,6(74)}{5(32.628\,690\,5) - 12.666\,230\,6^2} = 36.8$$

$$b = \frac{32.628\,690\,5(74) - 207.380\,25(12.666\,230\,6)}{5(32.628\,690\,5) - 12.666\,230\,6^2} = -78.3$$

The equation of the curve is $y = 36.8(10^x) - 78.3$. It must be remembered that for practical purposes, y must be positive. The graph of the equation is shown in Fig. 22.31, with the solid portion denoting the meaningful part of the curve. The points of the data are also shown.

As we noted and illustrated in the previous section, a graphing calculator can be used to determine the equation of the regression curve, and to display its graph. A typical calculator can fit a number of different types of equations to data, and the regression equations available on a typical calculator are as follows:

Linear:	$y = ax + b$
Quadratic:	$y = ax^2 + bx + c$
Cubic:	$y = ax^3 + bx^2 + cx + d$
Quartic:	$y = ax^4 + bx^3 + cx^2 + dx + e$
Logarithmic:	$y = a + b \ln x$
Exponential:	$y = ab^x$
Power:	$y = ax^b$
Logistic:	$y = \dfrac{c}{1 - ae^{-bx}}$
Sinusoidal:	$y = a \sin(bx + c) + d$

In comparing an equation obtained using Eq. (22.12) with that obtained using a calculator, we must note the difference in the form of the equation. For example, using the power form $y = ax^b$ on a calculator for Example 2, we get the calculator display shown in Fig. 22.32. The differences are due to (**1**) a power of -1 is assumed in our use of Eq. (22.12), and (**2**) there is a b-term in Eq. (22.12) and no b-term in the power form on the calculator.

In the following exercises, find the equation of the indicated least-squares curve. Sketch the curve and plot the data points on the same graph.

1. In Example 1, replace the *y*-values with 2, 3, 10, 25, 44, and 65. Then follow the instructions above.

2. For the points in the following table, find the least-squares curve $y = m\sqrt{x} + b$.

x	0	4	8	12	16
y	1	9	11	14	15

3. For the points in the following table, find the least-squares curve $y = m(1/x) + b$.

x	1.10	2.45	4.04	5.86	6.90	8.54
y	9.85	4.50	2.90	1.75	1.48	1.30

4. For the points in the following table, find the least-squares curve $y = m(10^x) + b$.

x	0.00	0.200	0.500	0.950	1.325
y	6.00	6.60	8.20	14.0	26.0

5. The following data were found for the distance *y* that an object rolled down an inclined plane in time *t*. Determine the least-squares curve $y = mt^2 + b$. Compare the equation with that using the *quadratic regression* feature on a graphing calculator.

t (s)	1.0	2.0	3.0	4.0	5.0
y (cm)	6.0	23	55	98	148

6. The increase in length *y* of a certain metallic rod was measured in relation to particular increases *x* in temperature. Find the least-squares curve $y = mx^2 + b$. Compare the equation with that using the *quadratic regression* feature of a graphing calculator.

x (°C)	50.0	100	150	200	250
y (cm)	1.00	4.40	9.40	16.4	24.0

7. The pressure *p* at which Freon, a refrigerant, vaporizes for temperature *T* is given in the following table. Find the least-squares curve $p = mT^2 + b$. Compare the equation with that using the *quadratic regression* feature of a graphing calculator.

T (°C)	0	10	20	30	40
p (kPa)	480	600	830	1040	1400

8. A fraction *f* of annual hot-water loads at a certain facility are heated by solar energy. The fractions *f* for certain values of the collector area *A* are given in the following table. Find the least-squares curve $f = m\sqrt{A} + b$. Compare the equation with that using the *power regression* feature of a graphing calculator.

A (m²)	0	12	27	56	90
f	0.0	0.2	0.4	0.6	0.8

9. The makers of a special blend of coffee found that the demand for the coffee depended on the price charged. The price *P* per pound and the monthly sales *S* are shown in the following table. Find the least-squares curve $P = m(1/S) + b$.

S (thousands)	240	305	420	480	560
P (dollars)	5.60	4.40	3.20	2.80	2.40

10. The resonant frequency *f* of an electric circuit containing a 4-μF capacitor was measured as a function of an inductance *L* in the circuit. The following data were found. Find the least-squares curve $f = m(1/\sqrt{L}) + b$.

L (H)	1.0	2.0	4.0	6.0	9.0
f (Hz)	490	360	250	200	170

11. The displacement *y* of an object at the end of a spring at given times *t* is shown in the following table. Find the least-squares curve $y = me^{-t} + b$.

t (s)	0.0	0.5	1.0	1.5	2.0	3.0
y (cm)	6.1	3.8	2.3	1.3	0.7	0.3

(W) 12. The average daily temperatures *T* (in °C) for each month in Minneapolis (National Weather Service records) are given in the following table.

t	J	F	M	A	M	J	J	A	S	O	N	D
T (°C)	−12	−8	−2	8	14	20	23	22	16	10	1	−7

Find the least-squares curve $T = m\cos\left[\frac{\pi}{6}(t - 0.5)\right] + b$. Assume the average temperature is for the 15th of each month. Then the values of *t* (in months) are 0.5, 1.5, ..., 11.5. (The fit is fairly good.) Compare the equation using the *sinusoidal regression* feature of a graphing calculator. What are the main reasons for the differences in the equations?

CHAPTER 22 EQUATIONS

Arithmetic mean	$\bar{x} = \dfrac{x_1 f_1 + x_2 f_2 + \cdots + x_n f_n}{f_1 + f_2 + \cdots + f_n}$	(22.1)
Standard deviation	$s = \sqrt{\dfrac{\sum (x - \bar{x})^2}{n - 1}}$	(22.2)
	$s = \sqrt{\dfrac{n(\sum x^2) - (\sum x)^2}{n(n - 1)}}$	(22.3)
Normal distribution	$y = \dfrac{e^{-(x-\mu)^2/2\sigma^2}}{\sigma \sqrt{2\pi}}$	(22.4)
Standard normal distribution	$y = \dfrac{1}{\sqrt{2\pi}} e^{-x^2/2}$	(22.5)
Standard (z) score	$z = \dfrac{x - \mu}{\sigma}$	(22.6)
Standard error of $\bar{x}$	$\sigma_{\bar{x}} = \dfrac{\sigma}{\sqrt{n}}$	(22.7)
Standard error of s	$\sigma_s = \dfrac{\sigma}{\sqrt{2n}}$	(22.8)
Least-squares lines	$y = mx + b$	(22.9)
	$m = \dfrac{n \sum xy - \left(\sum x\right)\left(\sum y\right)}{n \sum x^2 - \left(\sum x\right)^2}$	(22.10)
	$b = \dfrac{\left(\sum x^2\right)\left(\sum y\right) - \left(\sum xy\right)\left(\sum x\right)}{n \sum x^2 - \left(\sum x\right)^2}$	(22.11)
Nonlinear curves	$y = m[f(x)] + b$	(22.12)

CHAPTER 22 REVIEW EXERCISES

In Exercises 1–10, use the following set of numbers.

1098, 1102, 1101, 1095, 1104, 1097, 1107, 1099, 1104, 1093, 1095, 1102, 1101, 1098, 1106, 1098

1. Determine the median.
2. Determine the mode.
3. Determine the mean.
4. Determine the standard deviation.

5. Construct a frequency distribution table with five classes and a lowest class limit of 1093.
6. Draw a frequency polygon for the data in Exercise 5.
7. Draw a histogram for the data in Exercise 5.
8. Construct a relative frequency table for the data in Exercise 5.
9. Construct a cumulative frequency table for the data of Exercise 5.
10. Draw an ogive for the data of Exercise 5.

In Exercises 11–16, use the following data: An important property of oil is its coefficient of viscosity, which gives a measure of how well it flows. In order to determine the viscosity of a certain motor oil, a refinery took samples from 12 different storage tanks and tested them at 50°C. The results (in pascal-seconds) were 0.24, 0.28, 0.29, 0.26, 0.27, 0.26, 0.25, 0.27, 0.28, 0.26, 0.26, 0.25.

11. Find the mean. **12.** Find the median.

13. Find the standard deviation. **14.** Draw a histogram.

15. Draw a frequency polygon. **16.** Determine the mode.

In Exercises 17–24, use the following data: A sample of wind generators was tested for power output when the wind speed was 30 km/h. The following table gives the class marks of the powers produced and the number of generators in each class.

Power (W)	650	660	670	680	690
No. Generators	3	2	7	12	27

Power (W)	700	710	720	730
No. Generators	34	15	16	5

17. Find the median. **18.** Find the mean.

19. Find the mode. **20.** Draw a histogram.

21. Find the standard deviation. **22.** Draw a frequency polygon.

23. Make a cumulative frequency table.

24. Draw an ogive.

In Exercises 25–28, use the following data: A Geiger counter records the presence of high-energy nuclear particles. Even though no apparent radioactive source is present, some particles will be recorded. These are primarily cosmic rays, which are caused by very high–energy particles from outer space. In an experiment to measure the amount of cosmic radiation, the number of counts was recorded during 200 5-s intervals. The following table gives the number of counts and the number of 5-s intervals having this number of counts. Draw a frequency curve for these data.

Counts	0	1	2	3	4	5	6	7	8	9	10
Intervals	3	10	25	45	29	39	26	11	7	2	3

25. Find the median. **26.** Find the mean.

27. Draw a histogram.

28. Make a relative frequency table.

In Exercises 29–32, use the following data: Police radar on a city street recorded the speeds of 110 cars in a 65 km/h zone. The following table shows the class marks of the speeds recorded and the number of cars in each class.

Speed (km/h)	40	45	50	55	60	65	70	75	80	85
No. cars	3	4	4	5	8	22	48	10	4	2

29. Find the mean. **30.** Find the median.

31. Find the standard deviation. **32.** Draw an ogive.

In Exercises 33 and 34, use the following information: A company that makes electric light bulbs tests 500 bulbs each day for defects. The number of defective bulbs, along with the proportion of defective bulbs for 20 days, is shown in the following table.

Day	Number Defective	Proportion Defective
1	23	0.046
2	31	0.062
3	19	0.038
4	27	0.054
5	29	0.058
6	39	0.078
7	26	0.052
8	17	0.034
9	28	0.056
10	33	0.066
11	22	0.044
12	29	0.058
13	20	0.040
14	35	0.070
15	21	0.042
16	32	0.064
17	25	0.050
18	23	0.046
19	29	0.058
20	32	0.064
Total	540	

33. For a p chart, find the values of the central line, UCL, and LCL.

34. Plot a p chart.

In Exercises 35 and 36, use the following information: Five ball bearings are taken from the production line every 15 min and their diameters are measured. The diameters of the sample ball bearings for 16 successive subgroups are given in the following table.

Subgroup	Diameters (mm) of Five Ball Bearings				
1	4.98	4.92	5.02	4.91	4.93
2	5.03	5.01	4.94	5.06	5.07
3	5.05	5.03	5.00	5.02	4.96
4	5.01	4.92	4.91	4.99	5.03
5	4.92	4.97	5.02	4.95	4.94
6	5.02	4.95	5.01	5.07	5.15
7	4.93	5.03	5.02	4.96	4.99
8	4.85	4.91	4.88	4.92	4.90
9	5.02	4.95	5.06	5.04	5.06
10	4.98	4.98	4.93	5.01	5.00
11	4.90	4.97	4.93	5.05	5.02
12	5.03	5.05	4.92	5.03	4.98
13	4.90	4.96	5.00	5.02	4.97
14	5.09	5.04	5.05	5.02	4.97
15	4.88	5.00	5.02	4.97	4.94
16	5.02	5.09	5.03	4.99	5.03

35. Plot an $\bar{x}$ chart. **36.** Plot an R chart.

In Exercises 37–40, use the following data: After analyzing data for a long period of time, it was determined that samples of 500 readings of an organic pollutant for an area are distributed normally. For this pollutant, $\mu = 2.20 \; \mu g/m^3$ and $\sigma = 0.50 \; \mu g/m^3$.

37. In a sample, how many readings are between $1.50 \; \mu g/m^3$ and $2.50 \; \mu g/m^3$?

38. In a sample, how many readings are between $2.50 \; \mu g/m^3$ and $3.50 \; \mu g/m^3$?

39. In a sample, how many readings are above $1.00 \; \mu g/m^3$?

40. In a sample, how many readings are below $2.00 \; \mu g/m^3$?

In Exercises 41–48, find the indicated least-squares curve. Sketch the curve and data points on the same graph.

41. In a certain experiment, the resistance R of a certain resistor was measured as a function of the temperature T. The data found are shown in the following table. Find the least-squares line, expressing R as a function of T.

T (°C)	0.0	20.0	40.0	60.0	80.0	100
R (Ω)	25.0	26.8	28.9	31.2	32.8	34.7

42. An air-pollution monitoring station took samples of air each hour during the later morning hours and tested each sample for the number n of parts per million (ppm) of carbon monoxide. The results are shown in the table, where t is the number of hours after 6 A.M. Find the least-squares line for n as a function of t.

t (h)	0.0	1.0	2.0	3.0	4.0	5.0	6.0
n (ppm)	8.0	8.2	8.8	9.5	9.7	10.0	10.7

43. The *Mach number* of a moving object is the ratio of its speed to the speed of sound (1200 km/h). The following table shows the speed s of a jet aircraft, in terms of Mach numbers, and the time t after it starts to accelerate. Find the least-squares line of s as a function of t. Compare the equation with that using the *linear regression* feature of a graphing calculator.

t (min)	0.00	0.60	1.20	1.80	2.40	3.00
s (Mach number)	0.88	0.97	1.03	1.11	1.19	1.25

44. In an experiment to determine the relation between the load y on a spring and the length x of the spring, the following data were found. Find the least-squares line that expresses y as a function of x. Compare the equation with that using the *linear regression* feature of a graphing calculator.

Load (kg)	0.0	1.0	2.0	3.0	4.0	5.0
Length (cm)	10.0	11.2	12.3	13.4	14.6	15.9

45. The distance s of a missile above the ground at time t after being released from a plane is given by the following table. Find the least-squares curve of the form $s = mt^2 + b$ for these data. Compare the equation with that using the *quadratic regression* feature of a graphing calculator.

t (s)	0.0	3.0	6.0	9.0	12.0	15.0	18.0
s (m)	3000	2960	2820	2600	2290	1900	1410

46. In an elementary experiment that measured the wavelength L of sound as a function of the frequency f, the following results were obtained.

Frequency (Hz)	240	320	400	480	560
Wavelength (cm)	140	107	81.0	70.0	60.0

Find the least-squares curve of the form $L = m(1/f) + b$ for these data.

47. After being heated, the temperature T of an insulated liquid is measured at times t as follows:

t (h)	0	2	4	6	8	10
T (°C)	100	85	72	63	54	48

Plot these points and choose an appropriate function $f(x)$ for $y = m[f(x)] + b$. Then find the equation of the least-squares curve.

48. The vertical distance y of the cable of a suspension bridge above the surface of the bridge is measured at a horizontal distance x along the bridge from its center. See Fig. 22.33. The results are as follows:

x (m)	0	100	200	300	400	500
y (m)	15	17	23	33	47	65

Plot these points and choose an appropriate function $f(x)$ for $y = m[f(x)] + b$. Then find the equation of the least-squares curve.

Fig. 22.33

In Exercises 49–52, use the regression feature of a graphing calculator to solve the given problems related to the following data. Using aerial photography, the area A (in km²) of an oil spill as a function of the time t (in h) after the spill was found to be as follows:

A (km²)	1.4	2.5	4.7	6.8	8.8	10.2
t (h)	1.0	2.0	4.0	6.0	8.0	10.0

49. Find the linear equation $y = ax + b$ to fit this data.

50. Find the quadratic equation $y = ax^2 + bx + c$ to fit this data.

51. Find the power equation $y = ax^b$ to fit this data.

52. Compare the values of the coefficient of correlation r, to determine whether the linear equation or power equation seems to fit the data best.

In Exercises 53–56, solve the given problems.

53. The nth root of the product of n positive numbers is the *geometric mean* of the numbers. Find the geometric mean of the carbon monoxide readings in Exercise 42.

54. One use of the geometric mean (see Exercise 53) is to find an average ratio. By finding the geometric mean, find the average Mach number for the jet in Exercise 43.

55. Show that Eqs. (22.10) and (22.11) satisfy the equation $\bar{y} = m\bar{x} + b$.

56. Given that $\Sigma(x - \bar{x})^2 = \Sigma x^2 - n\bar{x}^2$, derive Eq. (22.3) from Eq. (22.2).

Writing Exercise

57. A study is to be made of the effectiveness of a treatment for glaucoma (a severe eye disorder), depending on the age of the person treated. Write two or three paragraphs explaining what data could be found and how they can be analyzed and then used to predict the effects of the treatment on future patients.

CHAPTER 22 PRACTICE TEST

In Problems 1–3, use the following set of numbers.

5, 6, 1, 4, 9, 5, 7, 3, 8, 10, 5, 8, 4, 9, 6

1. Find the median.

2. Find the mode.

3. Draw a histogram with five classes and the lowest class limit at 1.

In Problems 4–8, use the following data: Two machine parts are considered satisfactorily assembled if their total thickness (to the nearest 0.01 cm) is between or equal to 0.92 cm and 0.94 cm. One hundred assemblies are tested, and the class mark of the thicknesses and the number of assemblies in each class are given in the following table.

Total Thickness (cm)	0.90	0.91	0.92	0.93	0.94	0.95	0.96
Number		3	9	31	38	12	5	2

4. Find the mean.

5. Find the standard deviation.

6. Draw a frequency polygon.

7. Make a relative frequency table.

8. Draw an ogive (less than).

9. For a set of values that are normally distributed, what percent of them is below the value (greater than the mean) for which the z-score is 0.2257?

10. The machine-part assemblies in Problems 4–8 were tested in groups of five each hour for 20 h. Explain, in general, how to use the data from the test subgroups to plot an R chart.

11. Find the equation of the least-squares line for the points indicated in the following table. Graph the line and data points on the same graph.

x	1	3	5	7	9
y	5	11	17	20	27

12. The velocity (in m/s) of an object moving down an inclined plane was measured as a function of the distance (in m) it moved, with the following results:

Distance (m)	1.00	3.00	5.00	7.00	9.00
Velocity (m/s)	1.10	1.90	2.50	2.90	3.30

Find the equation of the least-squares curve of the form $y = m\sqrt{x} + b$, which expresses the velocity as a function of the distance.

CHAPTER 23 The Derivative

Findings related to the motion of the planets and of projectiles in the early 1600s created interest in the mid-1600s as to the motion of objects. Noting, for example, that the velocity of a falling object changes from one instant to the next, just how fast an object is moving at a given instant, its *instantaneous velocity,* was of particular interest. It was also noted that the geometric problem of finding the slope of a curve *at a specific point* was really equivalent to finding the instantaneous velocity of an object, since each involved an *instantaneous rate of change.*

It was well known at the time how to find an *average velocity* (distance traveled divided by time taken) and a slope (difference in *y*-values divided by difference in *x*-values). However, these methods do not work in finding an instantaneous velocity or slope since it means dividing by zero. Therefore, a method for finding the slope of a tangent line to a curve at a given point was a major point of interest in mathematics in the 1600s.

This interest in the motion of objects and the slope of a tangent line led to the development of the area of mathematics known as *calculus.* In this chapter, we start developing the methods of *differential calculus,* which deals with finding the instantaneous rate of change of one quantity with respect to another. Other examples of an instantaneous rate of change are electric current, which is the rate of change of electric charge with respect to time, and the rate of change of light intensity with respect to the distance from the source. In Chapter 25, we will study *integral calculus,* which involves finding the function for which the rate of change is known.

Isaac Newton, the English mathematician and physicist, and Gottfried Wilhelm Leibniz, a German mathematician and philosopher, are credited with the creation of the basic methods of the calculus in the 1660s and 1670s. Others, including the French mathematician Pierre de Fermat, are known to have developed some of the topics related to the calculus in the mid-1600s. In the 1700s and 1800s, many mathematicians further developed and refined the concepts of calculus.

The topic of this chapter, the *derivative,* is the basic concept of differential calculus that is used to measure an instantaneous rate of change. We will show some of the applications of the derivative in fields such as the physical sciences and engineering in this chapter, and we will develop several important types of applications in the next chapter.

Determining how fast an object is moving is important in physics and other areas of technology. In Section 23.4, we develop the method of finding the instantaneous velocity of a moving object.

23.1 LIMITS

Before dealing with the rate of change of a function, we first take up the concept of a *limit*. We encountered a limit with infinite geometric series and with the asymptotes of a hyperbola. It is necessary to develop this concept further.

Continuity

To help develop the concept of a limit, we first consider briefly the **continuity** of a function. *For a function to be **continuous at a point,** the function must exist at the point, and any small change in x produces only a small change in f(x).* In fact, the change in $f(x)$ can be made as small as we wish by restricting the change in x sufficiently, if the function is continuous. Also, *a function is said to be **continuous over an interval** if it is continuous at each point in the interval.*

NOTE ▶

If the domain of a function includes a point and values on only one side of the point, it is *continuous at the point* if the definition of continuity holds for that part of the domain.

◀ EXAMPLE 1   The function $f(x) = 3x^2$ is continuous for all values of x. That is, $f(x)$ is defined for all values of x, and a small change in x for any given value of x produces only a small change in $f(x)$. If we choose $x = 2$ and then let x change by 0.1, 0.01, and so on, we obtain the values in the following table:

x	2	2.1	2.01	2.001
$f(x)$	12	13.23	12.1203	12.012003
Change in x		0.1	0.01	0.001
Change in $f(x)$		1.23	0.1203	0.012003

We can see that the change in $f(x)$ is made smaller by the smaller changes in x. This shows that $f(x)$ is continuous at $x = 2$. Since this type of result would be obtained for any other x we may choose, we see that $f(x)$ is continuous for all values, and therefore it is continuous over the interval of all values of x.   ◗

◀ EXAMPLE 2   The function $f(x) = \dfrac{1}{x - 2}$ is not continuous at $x = 2$. When we substitute 2 for x, we have division by zero. This means the function is not defined. The condition that the function must exist is not satisfied.   ◗

From a graphical point of view, a function that is continuous over an interval has no "breaks" in its graph over that interval. The function is continuous over the interval if we can draw its graph without lifting the marker from the paper. If the function is *discontinuous,* a break occurs because the function is not defined or the definition of the function leads to an instantaneous "jump" in its values.

◀ EXAMPLE 3   **(a)** The graph of the function $f(x) = 3x^2$, which we determined to be continuous for all values of x in Example 1, is shown in Figure 23.1. We see that there are no breaks in the curve.

(b) The graph $f(x) = \dfrac{1}{x - 2}$, which we determined not to be continuous at $x = 2$ in Example 2, is shown in Fig. 23.2. There is a break in the curve for $x = 2$, and this shows that $f(x)$ does not exist at $x = 2$. It is a hyperbola with an asymptote $x = 2$.   ◗

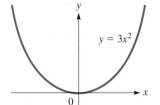

Fig. 23.1

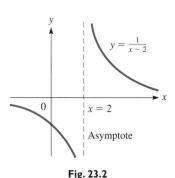

Fig. 23.2

◀ **EXAMPLE 4**   **(a)** For the function represented in Fig. 23.3, the solid circle at $x = 1$ shows that the point is on the graph. Since it is continuous to the right of the point, it is also continuous at the point. Thus, the function is continuous for $x \geq 1$.

(b) The function represented by the graph in Fig. 23.4 is not continuous at $x = 1$. The function is defined (by the solid circle point) for $x = 1$. However, a small change from $x = 1$ may result in a change of at least 1.5 in $f(x)$, regardless of how small a change in x is made. The small change condition is not satisfied.

(c) The function represented by the graph in Fig. 23.5 is not continuous for $x = -2$. The open circle shows that the point is not part of the graph, and therefore $f(x)$ is not defined for $x = -2$. ▶

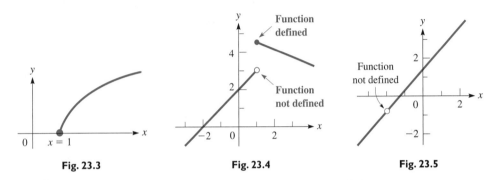

Fig. 23.3   Fig. 23.4   Fig. 23.5

◀ **EXAMPLE 5**   **(a)** We can define the function in Fig. 23.4 as

$$f(x) = \begin{cases} x + 2 & \text{for } x < 1 \\ -\frac{1}{2}x + 5 & \text{for } x \geq 1 \end{cases}$$

where we note that the equation differs for different parts of the domain.

(b) The graph of the function

$$g(x) = \begin{cases} 2x - 1 & \text{for } x \leq 2 \\ -x + 5 & \text{for } x > 2 \end{cases}$$

is shown in Fig. 23.6. We see that it is a continuous function even though the equation for $x \leq 2$ is different from that for $x > 2$. ▶

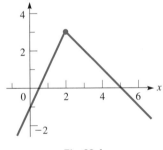

Fig. 23.6

In our earlier discussions of infinite geometric series and the asymptotes of a hyperbola, we used the symbol $\rightarrow$, which means "approaches." *When we say that $x \rightarrow 2$, we mean that x may take on any value as close to 2 as desired, but it also distinctly means* CAUTION ▶ *that **x cannot be set equal to 2.***

◀ **EXAMPLE 6**   Consider the behavior of $f(x) = 2x + 1$ as $x \rightarrow 2$.

Since we are not to use $x = 2$, we use a calculator to set up tables in order to determine values of $f(x)$, as x gets close to 2:

x	1.000	1.500	1.900	1.990	1.999
$f(x)$	3.000	4.000	4.800	4.980	4.998

x	3.000	2.500	2.100	2.010	2.001
$f(x)$	7.000	6.000	5.200	5.020	5.002

values approach 5

We can see that $f(x)$ approaches 5, as x approaches 2, from above 2 and from below 2. ▶

LIMIT OF A FUNCTION

In Example 6, since $f(x) \to 5$ as $x \to 2$, the number 5 is called the limit of $f(x)$ as $x \to 2$. This leads to the meaning of the limit of a function. In general, *the* **limit of a function $f(x)$** *is that value which the function approaches as x approaches the given value a.* This is written as

$$\lim_{x \to a} f(x) = L \tag{23.1}$$

CAUTION ▶

where L is the value of the limit of the function. Remember, in approaching a, x may come as arbitrarily close as desired to a, **but x may not equal a.**

An important conclusion can be drawn from the limit in Example 6. The function $f(x)$ is a continuous function, and $f(2)$ equals the value of the limit as $x \to 2$. In general, it is true that

if $f(x)$ is continuous at $x = a$, then the limit as $x \to a$ equals $f(a)$.

In fact, looking back at our definition of continuity, we see that this is what the definition means. That is, a function $f(x)$ is continuous at $x = a$ if *all three* of the following conditions are satisfied:

1. $f(a)$ exists   **2.** $\lim_{x \to a} f(x)$ exists   **3.** $\lim_{x \to a} f(x) = f(a)$

Although we can evaluate the limit for a continuous function as $x \to a$ by evaluating $f(a)$, it is possible that a function is not continuous at $x = a$ and that the limit exists and can be determined. Thus, we must be able to determine the value of a limit without finding $f(a)$. The following example illustrates the evaluation of such a limit.

◀ EXAMPLE 7   Find $\displaystyle\lim_{x \to 2} \frac{2x^2 - 3x - 2}{x - 2}$.

We note immediately that the function is not continuous at $x = 2$, for division by zero is indicated. Thus, we cannot evaluate the limit by substituting $x = 2$ into the function. Using a calculator to set up tables (see the margin note), we determine the value that $f(x)$ approaches, as x approaches 2:

x	1.000	1.500	1.900	1.990	1.999
$f(x)$	3.000	4.000	4.800	4.980	4.998

x	3.000	2.500	2.100	2.010	2.001
$f(x)$	7.000	6.000	5.200	5.020	5.002

values approach 5

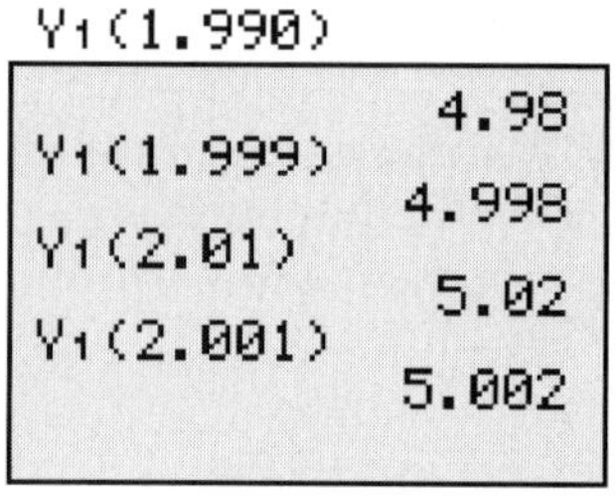

We see that the values obtained are identical to those in Example 6. Since $f(x) \to 5$ as $x \to 2$, we have

$$\lim_{x \to 2} \frac{2x^2 - 3x - 2}{x - 2} = 5$$

Therefore, we see that the limit exists at $x \to 2$, although the function does not exist at $x = 2$.

▶

See the margin note on page 83 for evaluating functions on a calculator. By entering the function as Y_1, it is necessary to enter the function only once for evaluation. See Fig. 23.7.

```
Y₁(1.990)
                    4.98
Y₁(1.999)
                   4.998
Y₁(2.01)
                    5.02
Y₁(2.001)
                   5.002
```

Fig. 23.7

The reason that the functions in Examples 6 and 7 have the same limit is shown in the following example.

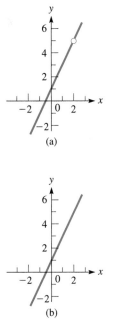

(a)

(b)

Fig. 23.8

EXAMPLE 8 The function $\dfrac{2x^2 - 3x - 2}{x - 2}$ in Example 7 is the same as the function $2x + 1$ in Example 6, except when $x = 2$. By factoring the numerator of the function of Example 7, we have

$$\frac{2x^2 - 3x - 2}{x - 2} = \frac{(2x + 1)(x - 2)}{x - 2} = 2x + 1$$

The cancellation here is valid, as long as x does not equal 2, for we have division by zero at $x = 2$. Also, in finding the limit as $x \to 2$, we do not use the value $x = 2$. Therefore,

$$\lim_{x \to 2} \frac{2x^2 - 3x - 2}{x - 2} = \lim_{x \to 2} (2x + 1) = 5$$

The limits of the two functions are equal, since, again, in finding the limit, we do not let $x = 2$. The graphs of the two functions are shown in Fig. 23.8(a) and (b). We can see from the graphs that the limits are the same, although one of the functions is not continuous.

If $f(x) = 5$ for $x = 2$ is added to the definition of the function in Example 7, it is then the same as $2x + 1$, and its graph is that in Fig. 23.8(b).

The limit of the function in Example 7 was determined by calculating values near $x = 2$ and by means of an algebraic change in the function. This illustrates that limits may be found through the meaning and definition and through other procedures when the function is not continuous. The following example illustrates a function for which the limit does not exist as x approaches the indicated value.

EXAMPLE 9 In trying to find

$$\lim_{x \to 2} \frac{1}{x - 2}$$

we note that $f(x)$ is not defined for $x = 2$, since we would have division by zero. Therefore, we set up the following table to see how $f(x)$ behaves as $x \to 2$:

See Appendix C for the graphing calculator program LIMFUNC. It evaluates $f(x)$ as x approaches a (from values $x > a$).

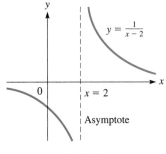

Fig. 23.2

x	3	2.5	2.1	2.01	2.001
$f(x)$	1	2	10	100	1000

$f(x) \to +\infty$

x	1	1.5	1.9	1.99	1.999
$f(x)$	-1	-2	-10	-100	-1000

$f(x) \to -\infty$

We see that $f(x)$ gets larger as $x \to 2$ from above 2 and $f(x)$ gets smaller (large negative values) as $x \to 2$ from below 2. This may be written as $f(x) \to +\infty$ as $x \to 2^+$ and $f(x) \to -\infty$ as $x \to 2^-$, but we must remember that ∞ is not a real number. Therefore, the limit as $x \to 2$ does not exist. The graph of this function is shown in Fig. 23.2, which is shown again for reference.

The following examples further illustrate the evaluation of limits.

◖ EXAMPLE 10   Find $\lim\limits_{x \to 4}(x^2 - 7)$.

Since the function $x^2 - 7$ is continuous at $x = 4$, we may evaluate this limit by substitution. For $f(x) = x^2 - 7$, we have $f(4) = 9$. This means that

$$\lim_{x \to 4}(x^2 - 7) = 9$$

◖ EXAMPLE 11   Find

$$\lim_{t \to 2}\left(\frac{t^2 - 4}{t - 2}\right)$$

Since

$$\frac{t^2 - 4}{t - 2} = \frac{(t - 2)(t + 2)}{t - 2} = t + 2$$

is valid as long as $t \neq 2$, we find that

$$\lim_{t \to 2}\left(\frac{t^2 - 4}{t - 2}\right) = \lim_{t \to 2}(t + 2) = 4$$

Again, we do not have to be concerned with the fact that the cancellation is not valid for $t = 2$. In finding the limit, we do not consider the value of $f(t)$ at $t = 2$.

◖ EXAMPLE 12   Find $\lim\limits_{x \to 0}\left(x\sqrt{x - 3}\right)$.

We see that $x\sqrt{x - 3} = 0$ if $x = 0$, but this function does not have real values for values of x less than 3 other than $x = 0$. *Since x cannot approach 0, f(x) does not approach 0 and the limit does not exist.* The point of this example is that even if $f(a)$ exists, we cannot evaluate the limit by finding $f(a)$, unless $f(x)$ is continuous at $x = a$. Here, $f(a)$ exists but the limit does not exist.

Returning briefly to the discussion of continuity, we again see the need for all three conditions of continuity given on page 648. If $f(a)$ does not exist, the function is discontinuous at $x = a$, and if $f(a)$ does not equal $\lim_{x \to a}f(x)$, a small change in x will not result in a small change in $f(x)$. Also, Example 12 shows another reason to carefully consider the domain of the function in finding the limit.

LIMITS AS x APPROACHES INFINITY

Limits as x approaches infinity are also of importance. However, when dealing with these limits, we must remember that

CAUTION ▶   ∞ *does not represent a real number and that algebraic operations may not be performed on it.*

Therefore, when we write $x \to \infty$, we know we are to consider values of x that are becoming large without bound. We first encountered this concept in Chapter 19 when we discussed infinite geometric series. The following examples illustrate the evaluation of this type of limit.

◀ EXAMPLE 13 The efficiency E of an engine is given by $E = 1 - Q_2/Q_1$, where Q_1 is the heat taken in and Q_2 is the heat ejected by the engine. ($Q_1 - Q_2$ is the work done by the engine.) If, in an engine cycle, $Q_2 = 500$ kJ, determine E as Q_1 becomes large without bound.

We are to find

$$\lim_{Q_1 \to \infty} \left(1 - \frac{500}{Q_1} \right)$$

As Q_1 becomes larger and larger, $500/Q_1$ becomes smaller and smaller and approaches zero. This means $f(Q_1) \to 1$ as $Q_1 \to \infty$. Thus,

$$\lim_{Q_1 \to \infty} \left(1 - \frac{500}{Q_1} \right) = 1$$

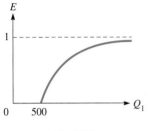

We can verify our reasoning and the value of the limit by making a table of values for Q_1 and E as Q_1 becomes large:

Q_1	500	5000	50 000	500 000	
E	0	0.9	0.99	0.999	values approach 1

Fig. 23.9

Again, we see that $E \to 1$ as $Q_1 \to \infty$. See Fig. 23.9.

This is primarily a theoretical consideration, as there are obvious practical limitations as to how much heat can be supplied to an engine. An engine for which $E = 1$ would operate at 100% efficiency. ▶

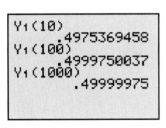

Fig. 23.10

◀ EXAMPLE 14 Find $\lim_{x \to \infty} \dfrac{x^2 + 1}{2x^2 + 3}$.

We note that as $x \to \infty$, both the numerator and the denominator become large without bound. Therefore, we use a calculator (see Fig. 23.10) to make a table to see how $f(x)$ behaves as x becomes very large:

x	1	10	100	1000	
$f(x)$	0.4	0.497 536 945 8	0.499 975 003 7	0.499 999 75	values approach 0.5

From this table, we see that $f(x) \to 0.5$ as $x \to \infty$. See Fig. 23.11.

This limit can also be found through algebraic operations and an examination of the resulting algebraic form. If we divide both the numerator and the denominator of the

NOTE ▶ function by x^2, **which is the largest power of x that appears in either the numerator or the denominator,** we have

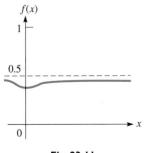

$$\frac{x^2 + 1}{2x^2 + 3} = \frac{1 + \dfrac{1}{x^2}}{2 + \dfrac{3}{x^2}} \quad \begin{array}{l} \leftarrow \text{terms} \to 0 \\ \text{as } x \to \infty \end{array}$$

Fig. 23.11

Here we see that $1/x^2$ and $3/x^2$ both approach zero as $x \to \infty$. This means that the numerator approaches 1 and the denominator approaches 2. Therefore, we get the same result as above since

$$\lim_{x \to \infty} \frac{x^2 + 1}{2x^2 + 3} = \lim_{x \to \infty} \frac{1 + \dfrac{1}{x^2}}{2 + \dfrac{3}{x^2}} = \frac{1}{2}$$

Calculus was not on a sound mathematical basis until limits were properly developed by the French mathematician Augustin-Louis Cauchy (1789–1857) and others in the mid-1800s.

The definitions and development of continuity and of a limit presented in this section are not mathematically rigorous. However, the development is consistent with a more rigorous development, and the concept of a limit is the principal concern.

EXERCISES 23.1

In Exercises 1–4, make the given changes in the indicated examples of this section. Then solve the resulting problems.

1. In Example 2, change the denominator to $x + 2$ and then determine the continuity.

2. In Example 8, change the numerator to $3x^2 - 5x - 2$ and find the resulting limit. Disregard references to Examples 6 and 7.

3. In Example 11, change the denominator to $t + 2$ and then find the limit as $t \to -2$.

4. In Example 14, change the numerator to $4x^2 + 1$ and find the resulting limit.

In Exercises 5–10, determine the values of x for which the function is continuous. If the function is not continuous, determine the reason.

5. $f(x) = 3x - 2$

6. $f(x) = 9 - x^2$

7. $f(x) = \dfrac{2}{x^2 - x}$

8. $f(x) = \dfrac{1}{\sqrt{x}}$

9. $f(x) = \sqrt{\dfrac{x}{x - 2}}$

10. $f(x) = \dfrac{\sqrt{x + 2}}{x}$

In Exercises 11–16, determine the values of x for which the function, as represented by the graphs in Fig. 23.12, is continuous. If the function is not continuous, determine the reason.

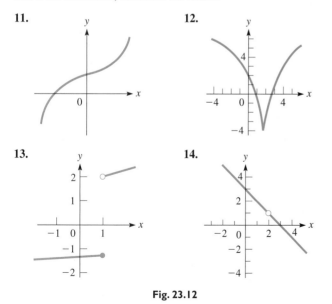

11.

12.

13.

14.

Fig. 23.12

15.

16.

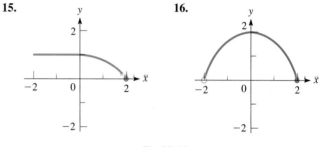

Fig. 23.12

W *In Exercises 17–20, graph the function and determine the values of x for which the functions are continuous. Explain.*

17. $f(x) = \begin{cases} x^2 & \text{for } x < 2 \\ 2 & \text{for } x \geq 2 \end{cases}$

18. $f(x) = \begin{cases} \dfrac{x^3 - x^2}{x - 1} & \text{for } x \neq 1 \\ 1 & \text{for } x = 1 \end{cases}$

19. $f(x) = \begin{cases} \dfrac{2x^2 - 18}{x - 3} & \text{for } x < 3 \text{ or } x > 3 \\ 12 & \text{for } x = 3 \end{cases}$

20. $f(x) = \begin{cases} \dfrac{x + 2}{x^2 - 4} & \text{for } x < -2 \\ \dfrac{x}{8} & \text{for } x > -2 \end{cases}$

In Exercises 21–26, evaluate the indicated limits by evaluating the function for values shown in the table and observing the values that are obtained. Do not change the form of the function.

21. Find $\displaystyle\lim_{x \to 1} \dfrac{x^3 - x}{x - 1}$.

x	0.900	0.990	0.999	1.001	1.010	1.100
$f(x)$						

22. Find $\displaystyle\lim_{x \to -3} \dfrac{x^3 + 2x^2 - 2x + 3}{x + 3}$.

x	-3.100	-3.010	-3.001	-2.999	-2.990	-2.900
$f(x)$						

23. Find $\lim\limits_{x \to 2} \dfrac{2 - \sqrt{x + 2}}{x - 2}$.

x	1.900	1.990	1.999	2.001	2.010	2.100
$f(x)$						

24. Find $\lim\limits_{x \to 0} \dfrac{e^x - 1}{x}$.

x	−0.1	−0.01	−0.001	0.001	0.01	.01
$f(x)$						

25. Find $\lim\limits_{x \to \infty} \dfrac{2x + 1}{5x - 3}$.

x	10	100	1000
$f(x)$			

26. Find $\lim\limits_{x \to \infty} \dfrac{1 - x^2}{8x^2 + 5}$.

x	10	100	1000
$f(x)$			

In Exercises 27–42, evaluate the indicated limits by direct evaluation as in Examples 10–14. Change the form of the function where necessary.

27. $\lim\limits_{x \to 3} (3x - 2)$

28. $\lim\limits_{x \to 4} \sqrt{x^2 - 7}$

29. $\lim\limits_{x \to 0} \dfrac{x^2 + x}{x}$

30. $\lim\limits_{v \to 2} \dfrac{4v^2 - 8v}{v - 2}$

31. $\lim\limits_{x \to -1} \dfrac{x^2 - 1}{3x + 3}$

32. $\lim\limits_{x \to 3} \dfrac{x^2 - 2x - 3}{3 - x}$

33. $\lim\limits_{h \to 3} \dfrac{h^3 - 27}{h - 3}$

34. $\lim\limits_{x \to 1/3} \dfrac{3x - 1}{3x^2 + 5x - 2}$

35. $\lim\limits_{x \to 1} \dfrac{(2x - 1)^2 - 1}{2x - 2}$

36. $\lim\limits_{x \to 4} \dfrac{|x - 4|}{x - 4}$

37. $\lim\limits_{p \to -1} \sqrt{p}\,(p + 1.3)$

38. $\lim\limits_{x \to 1} (x - 1)\sqrt{x^2 - 4}$

39. $\lim\limits_{x \to \infty} \dfrac{3x^2 + 4.5}{x^2 - 1.5}$

40. $\lim\limits_{x \to \infty} \dfrac{x - 1}{7x + 4}$

41. $\lim\limits_{t \to \infty} \dfrac{\sqrt{t^2 + 16}}{t + 1}$

42. $\lim\limits_{x \to \infty} \dfrac{1 - 2x^2}{(4x + 3)^2}$

In Exercises 43 and 44, evaluate the function at 0.1, 0.01, and 0.001 from both sides of the value it approaches. In Exercises 45 and 46, evaluate the function for values of x of 10, 100, and 1000. From these values, determine the limit. Then, by using an appropriate change of algebraic form, evaluate the limit directly and compare values.

43. $\lim\limits_{x \to 0} \dfrac{x^2 - 3x}{x}$

44. $\lim\limits_{x \to 3} \dfrac{2x^2 - 6x}{x - 3}$

45. $\lim\limits_{x \to \infty} \dfrac{2x^2 + x}{x^2 - 3}$

46. $\lim\limits_{x \to \infty} \dfrac{x^2 + 5}{\sqrt{64x^4 + 1}}$

In Exercises 47–52, solve the given problems involving limits.

47. Velocity can be found by dividing the displacement s of an object by the elapsed time t in moving through the displacement. In a certain experiment, the following values were measured for the displacements and elapsed times for the motion of an object. Determine the limiting value of the velocity.

s (cm)	0.480 000	0.280 000	0.029 800	0.002 998 0	0.000 299 98
t (s)	0.200 000	0.100 000	0.010 000	0.001 000 0	0.000 100 00

48. The area A (in mm^2) of the pupil of a certain person's eye is given by $A = \dfrac{36 + 24b^3}{1 + 4b^3}$, where b is the brightness (in lumens) of the light source. Between what values does A vary?

49. A certain object, after being heated, cools at such a rate that its temperature T (in °C) decreases 10% each minute. If the object is originally heated to 100°C, find $\lim\limits_{t \to 10} T$ and $\lim\limits_{t \to \infty} T$, where t is the time (in min).

50. A 5-Ω resistor and a variable resistor of resistance R are placed in parallel. The expression for the resulting resistance R_T is given by $R_T = \dfrac{5R}{5 + R}$. Determine the limiting value of R_T as $R \to \infty$.

51. Using a calculator, find $\lim\limits_{x \to 0} (1 + x)^{1/x}$. Do you recognize the limiting value?

52. Using a calculator in radian mode, find $\lim\limits_{x \to 0} \dfrac{\sin x}{x}$.

*In Exercises 53–56, $\lim\limits_{x \to a^-} f(x)$ means to find the limit as x approaches a from the left only, and $\lim\limits_{x \to a^+} f(x)$ means to find the limit as x approaches a from the right only. These are called **one-sided** limits. Solve the following problems.*

53. Find $\lim\limits_{x \to 4^-} x\sqrt{16 - x^2}$

W **54.** Explain why $\lim\limits_{x \to 0^+} 2^{1/x} \neq \lim\limits_{x \to 0^-} 2^{1/x}$.

W **55.** For $f(x) = \dfrac{x}{|x|}$, find $\lim\limits_{x \to 0^-} f(x)$ and $\lim\limits_{x \to 0^+} f(x)$. Is $f(x)$ continuous at $x = 0$? Explain.

W **56.** In Einstein's theory of relativity, the length L of an object moving at a velocity v is $L = L_0 \sqrt{1 - \dfrac{v^2}{c^2}}$, where c is the speed of light and L_0 is the length of the object at rest. Find $\lim\limits_{v \to c^-} L$ and explain why a limit from the left is used.

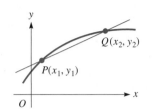

Fig. 23.13

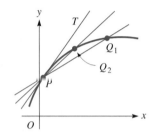

Fig. 23.14

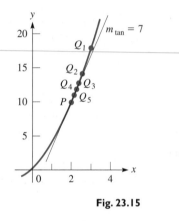

Fig. 23.15

See Appendix C for the graphing calculator program SECTOTAN. It displays the graph of a function and four secant lines as the secant line approaches the tangent line.

23.2 THE SLOPE OF A TANGENT TO A CURVE

Having developed the basic operations with functions and the concept of a limit, we now turn our attention to a graphical interpretation of the rate of change of a function. This interpretation, basic to an understanding of the calculus, deals with the slope of a line tangent to the curve of a function.

Consider the points $P(x_1, y_1)$ and $Q(x_2, y_2)$ in Fig. 23.13. From Chapter 21, we know that the slope of the line through these points is given by

$$m = \frac{y_2 - y_1}{x_2 - x_1}$$

This, however, represents the slope of the line through P and Q and no other line. If we now allow Q to be a point closer to P, the slope of PQ will more closely approximate the slope of a line drawn tangent to the curve at P (see Fig. 23.14). In fact, the closer Q is to P, the better this approximation becomes. It is not possible to allow Q to coincide with P, for then it would not be possible to define the slope of PQ in terms of two points. *The slope of the tangent line, often referred to as the slope of the curve, is the limiting value of the slope of PQ as Q approaches P.*

◀ **EXAMPLE 1**   Find the slope of a line tangent to the curve $y = x^2 + 3x$ at the point $P(2, 10)$ by finding the limit of slopes of the secant lines PQ as Q approaches P.

Let point Q have the x-values of 3.0, 2.5, 2.1, 2.01, and 2.001. Then, using a calculator, we tabulate the necessary values. Since P is the point $(2, 10)$, $x_1 = 2$ and $y_1 = 10$. Thus, using the values of x_2, we tabulate the values of y_2, $y_2 - 10$, $x_2 - 2$ and thereby the values of the slope m:

Point	Q_1	Q_2	Q_3	Q_4	Q_5	P
x_2	3.0	2.5	2.1	2.01	2.001	2
y_2	18.0	13.75	10.71	10.0701	10.007 001	10
$y_2 - 10$	8.0	3.75	0.71	0.0701	0.007 001	
$x_2 - 2$	1.0	0.5	0.1	0.01	0.001	
$m = \dfrac{y_2 - 10}{x_2 - 2}$	8.0	7.5	7.1	7.01	7.001	

We see that the slope of PQ approaches the value of 7 as Q approaches P. Therefore, the slope of the tangent line at $(2, 10)$ is 7. See Fig. 23.15.   ▶

With the proper notation, it is possible to express the coordinates of Q in terms of the coordinates of P. By defining h as $x_2 - x_1$, we have

$$h = x_2 - x_1 \tag{23.2}$$
$$x_2 = x_1 + h \tag{23.3}$$

For the function $y = f(x)$, the point $P(x_1, y_1)$ can be written as $P(x_1, f(x_1))$ and the point $Q(x_2, y_2)$ can be written as $Q(x_1 + h, f(x_1 + h))$.

Using Eq. (23.3), along with the definition of slope, we can express the slope of PQ as

$$m_{PQ} = \frac{f(x_1 + h) - f(x_1)}{(x_1 + h) - x_1} = \frac{f(x_1 + h) - f(x_1)}{h} \qquad (23.4)$$

By our previous discussion, as Q approaches P, the slope of the tangent line is more nearly *approximated* by Eq. (23.4).

◀ **EXAMPLE 2** Find the slope of a line tangent to the curve of $y = x^2 + 3x$ at the point $(2, 10)$. (This is the same slope as calculated in Example 1.)

As in Example 1, point P has the coordinates $(2, 10)$. Thus, the coordinates of any other point Q on the curve can be expressed as $(2 + h, f(2 + h))$. See Fig. 23.16. The slope of PQ then becomes

$$
\begin{aligned}
m_{PQ} &= \frac{f(2 + h) - f(2)}{h} = \frac{[(2 + h)^2 + 3(2 + h)] - [2^2 + 3(2)]}{h} \\
&= \frac{(4 + 4h + h^2 + 6 + 3h) - (4 + 6)}{h} \\
&= \frac{7h + h^2}{h} = 7 + h
\end{aligned}
$$

From this expression, we can see that $m_{PQ} \to 7$ as $h \to 0$. Therefore, we can see that the slope of the tangent line is

$$m_{\text{tan}} = \lim_{h \to 0} m_{PQ} = 7$$

We see that this result agrees with that found in Example 1. ▶

◀ **EXAMPLE 3** Find the slope of a line tangent to the curve of $y = 4x - x^2$ at the point (x_1, y_1).

The points P and Q are $P(x_1, f(x_1))$ and $Q(x_1 + h, f(x_1 + h))$. Therefore,

$$
\begin{aligned}
m_{PQ} &= \frac{f(x_1 + h) - f(x_1)}{h} = \frac{[4(x_1 + h) - (x_1 + h)^2] - (4x_1 - x_1^2)}{h} \\
&= \frac{(4x_1 + 4h - x_1^2 - 2hx_1 - h^2) - (4x_1 - x_1^2)}{h} \\
&= \frac{4h - 2hx_1 - h^2}{h} = 4 - 2x_1 - h
\end{aligned}
$$

Here we see that $m_{PQ} \to 4 - 2x_1$ as $h \to 0$. Therefore,

$$m_{\text{tan}} = 4 - 2x_1$$

This method has an advantage over that used in Example 2. We now have a general expression for the slope of a tangent line for any value x_1. If $x_1 = -1$, $m_{\text{tan}} = 6$ and if $x_1 = 3$, $m_{\text{tan}} = -2$. The tangent lines are shown in Fig. 23.17. ▶

Most calculators have a *tangent* feature. By entering the function and the value of x for which a tangent line is to be drawn, the calculator displays the curve and the tangent line. See Fig. 23.18 for $y = 4x - x^2$ and the tangent line for $x = 3$.

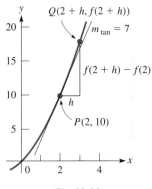

Fig. 23.16

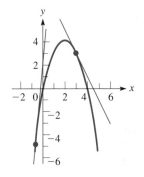

Fig. 23.17

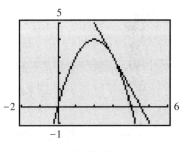

Fig. 23.18

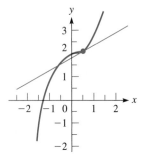

Fig. 23.19

◀ **EXAMPLE 4** Find the expression for the slope of a line tangent to the curve of $y = x^3 + 2$ at the general point (x_1, y_1), and use this expression to find the slope when $x = 1/2$.

For $y = f(x)$, and using the points $P(x_1, f(x_1))$ and $Q(x_1 + h, f(x_1 + h))$, we have the following steps:

$$m_{PQ} = \frac{f(x_1 + h) - f(x_1)}{h} = \frac{[(x_1 + h)^3 + 2] - (x_1^3 + 2)}{h} \quad \text{using Eq. (23.4)}$$

$$= \frac{(x_1^3 + 3x_1^2 h + 3x_1 h^2 + h^3 + 2) - (x_1^3 + 2)}{h} = \frac{3x_1^2 h + 3x_1 h^2 + h^3}{h}$$

$$= 3x_1^2 + 3x_1 h + h^2$$

As $h \to 0$, the expression on the right approaches the value $3x_1^2$. This means that

$$m_{\text{tan}} = 3x_1^2$$

When $x_1 = 1/2$, we find that the slope of the tangent is $3(1/4) = 3/4$. The curve and this tangent line are shown in Fig. 23.19. ▶

In interpreting this slope of a tangent line as the rate of change of a function, we see that if $h = 1$ and the corresponding value of $f(x + h) - f(x)$ is found, it can be said that $f(x)$ changes by the amount $f(x + h) - f(x)$ as x changes by 1 unit. If x changes by a lesser amount (h is less than 1), we can still calculate the ratio of the amount of change in $f(x)$ for the given value of h. Therefore, as long as x changes at all, there will be a corresponding change in $f(x)$. This means that

NOTE ▶ *the ratio of the change in $f(x)$ to the change in x is the* **average rate of change of $f(x)$ with respect to x**.

As $h \to 0$,

NOTE ▶ *the limit of the ratio of the change in $f(x)$ to the change in x is the* **instantaneous rate of change of $f(x)$ with respect to x**.

◀ **EXAMPLE 5** In Example 1, consider points $P(2, 10)$ and $Q(2.5, 13.75)$. From P to Q_2, x changes by 0.5 unit and $f(x)$ changes by 3.75 units. This means *the average change in $f(x)$ for a 1-unit change in x is* $3.75/0.5 = 7.5$ units. However, this is not the rate at which $f(x)$ is changing with respect to x *at* most points within this interval.

At point P, the slope of 7 of the tangent line tells us that $f(x)$ is changing 7 units for a 1-unit change in x. However, *this is an instantaneous rate of change at point P* and tells the rate at which $f(x)$ is changing with respect to x at P. ▶

EXERCISES 23.2

In Exercises 1 and 2, make the given changes in the indicated examples of this section and then find the indicated slopes.

1. In Example 2, change the point $(2, 10)$ to $(3, 18)$ and then find the slope of the tangent line.

2. In Example 3, change the $4x$ to $3x$ and then find the slope of the tangent line.

In Exercises 3–6, use the method of Example 1 to calculate the slope of the line tangent to the curve of each of the given functions. Let Q_1, Q_2, Q_3, and Q_4 have the indicated x-values. Sketch the curve and tangent lines.

3. $y = x^2$; P is $(2, 4)$; let Q have x-values of 1.5, 1.9, 1.99, 1.999.

4. $y = 1 - \frac{1}{2}x^2$; P is $(2, -1)$; let Q have x-values of 1.5, 1.9, 1.99, 1.999.

5. $y = 2x^2 + 5x$; P is $(-2, -2)$; let Q have x-values of $-1.5, -1.9,$ $-1.99, -1.999$.

6. $y = x^3 + 1$; P is $(-1, 0)$; let Q have x-values of $-0.5, -0.9,$ $-0.99, -0.999$.

In Exercises 7–10, use the method of Example 2 to calculate the slope of a line tangent to the curve of each of the functions $y = f(x)$ for the given point P. (These are the same functions and points as in Exercises 3–6.)

7. $y = x^2$; P is $(2, 4)$.

8. $y = 1 - \frac{1}{2}x^2$; P is $(2, -1)$.

9. $y = 2x^2 + 5x$; P is $(-2, -2)$.

10. $y = x^3 + 1$; P is $(-1, 0)$.

In Exercises 11–22, use the method of Example 3 to find a general expression for the slope of a tangent line to each of the indicated curves. Then find the slopes for the given values of x. Sketch the curves and tangent lines.

11. $y = x^2$; $x = 2, x = -1$

12. $y = 1 - \frac{1}{2}x^2$; $x = 2, x = -2$

13. $y = 2x^2 + 5x$; $x = -2, x = 0.5$

14. $y = 4.5 - 3x^2$; $x = 0, x = 2$

15. $y = x^2 + 4x + 2\pi$; $x = -3, x = 2$

16. $y = 2x^2 - 4x$; $x = 1, x = 1.5$

17. $y = 6x - x^2$; $x = -2, x = 3$

18. $y = x^3 - 2x$; $x = -1, x = 0, x = 1$

19. $y = 1.5x^4$; $x = 0, x = 0.5, x = 1$

20. $y = 1 - x^4$; $x = 0, x = 1, x = 2$

21. $y = x^5$; $x = 0, x = 0.5, x = 1$

22. $y = \dfrac{1}{x}$; $x = 0.5, x = 1, x = 2$

In Exercises 23–26, use the tangent feature of a graphing calculator to display the curve and the tangent line for the given values of x.

23. $y = 2x - 3x^2$; $x = 0, x = 0.5$

24. $y = 3x - x^3$; $x = -2, x = 0, x = 2$

25. $y = x^6$; $x = 0, x = 0.5, x = 1$

26. $y = \dfrac{2}{x + 1}$; $x = -0.5, x = 0, x = 1$

In Exercises 27–30, find the average rate of change of y with respect to x from P to Q. Then compare this with the instantaneous rate of change of y with respect to x at P by finding m_{tan} at P.

27. $y = x^2 + 2$; $P(2, 6), Q(2.1, 6.41)$

28. $y = 1 - 2x^2$; $P(1, -1), Q(1.1, -1.42)$

29. $y = 9 - x^3$; $P(2, 1), Q(2.1, -0.261)$

30. $y = x^3 - 6x$; $P(3, 9), Q(3.1, 11.191)$

In Exercises 31 and 32, find the point(s) where the slope of a tangent line to the given curve has the given value.

31. $y = 2x^2, m_{tan} = -4$ **32.** $y = x^3 + 3x^2, m_{tan} = 9$

23.3 THE DERIVATIVE

In the preceding section, we found that we could find the slope of a line tangent to a curve at a point $(x_1, f(x_1))$ by calculating the limit (if it exists) of the difference $f(x_1 + h) - f(x_1)$ divided by h as $h \to 0$. We can write this as

$$m_{tan} = \lim_{h \to 0} \frac{f(x_1 + h) - f(x_1)}{h} \tag{23.5}$$

The limit on the right is defined as *the* **derivative** *of $f(x)$ at x_1*. This is one of the fundamental definitions of calculus.

Considering the limit at each point in the domain of $f(x)$, by letting $x_1 = x$ in Eq. (23.5), we have *the* **derivative** *of the function $f(x)$*

$$f'(x) = \lim_{h \to 0} \frac{f(x + h) - f(x)}{h} \tag{23.6}$$

The process of finding a derivative is called **differentiation.**

When finding the derivative by use of the definition, it is usually easier when the procedure is broken down into steps. A four-step procedure is outlined on the following page.

> **PROCEDURE FOR FINDING THE DERIVATIVE OF A FUNCTION**
>
> **1.** Find $f(x + h)$.
> **2.** Subtract $f(x)$ from $f(x + h)$.
> **3.** Divide the result of step 2 by h.
> **4.** For the result of step 3, find the limit (if it exists) as $h \to 0$.

◀ **EXAMPLE 1** Find the derivative of $y = 2x^2 + 3x$ by using the definition.

With $y = f(x)$, using the above procedure to find the derivative $f'(x)$, we have the following:

$$f(x + h) = 2(x + h)^2 + 3(x + h) \qquad \text{step 1}$$

$$f(x + h) - f(x) = 2(x + h)^2 + 3(x + h) - (2x^2 + 3x) \qquad \text{step 2}$$

$$= 2x^2 + 4xh + 2h^2 + 3x + 3h - 2x^2 - 3x$$

$$= 4xh + 3h + 2h^2$$

$$\frac{f(x + h) - f(x)}{h} = \frac{4hx + 3h + 2h^2}{h} = 4x + 3 + 2h \qquad \text{step 3}$$

$$\lim_{h \to 0} \frac{f(x + h) - f(x)}{h} = \lim_{h \to 0}(4x + 3 + 2h) = 4x + 3 \qquad \text{step 4}$$

$$f'(x) = 4x + 3$$

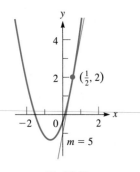

Fig. 23.20

We see that the derivative of the function $2x^2 + 3x$ is the function $4x + 3$. From the meanings of a slope of a tangent line and the derivative, this means we can find the slope of a tangent line for any point on the curve of $y = 2x^2 + 3x$ by substituting the x-coordinate into the expression $4x + 3$. For example, the slope of a tangent line is 5 if $x = 1/2$ (at the point $(1/2, 2)$). See Fig. 23.20. ▶

Since the derivative of a function is itself a function, it is possible that it may not be defined for all values of x. *If the value x_0 is in the domain of the derivative, then the function is said to be differentiable at x_0.* The examples that follow illustrate functions that are not differentiable for all values of x.

◀ **EXAMPLE 2** Find the derivative of $y = \dfrac{3}{x + 2}$ by using the definition.

$$f(x + h) = \frac{3}{x + h + 2} \qquad \text{step 1}$$

$$f(x + h) - f(x) = \frac{3}{x + h + 2} - \frac{3}{x + 2} \qquad \text{step 2}$$

CAUTION ▶

$$= \frac{3(x + 2) - 3(x + h + 2)}{(x + h + 2)(x + 2)} = \frac{-3h}{(x + h + 2)(x + 2)}$$

$$\frac{f(x + h) - f(x)}{h} = \frac{-3h}{h(x + h + 2)(x + 2)} = \frac{-3}{(x + h + 2)(x + 2)} \qquad \text{step 3}$$

$$\lim_{h \to 0} \frac{f(x + h) - f(x)}{h} = \lim_{h \to 0} \frac{-3}{(x + h + 2)(x + 2)} = \frac{-3}{(x + 2)^2} \qquad \text{step 4}$$

$$f'(x) = \frac{-3}{(x + 2)^2}$$

Note that neither the function nor the derivative is defined for $x = -2$. This means the function is not differentiable at $x = -2$. ▶

NOTE ▶ In Example 2, it was necessary to combine fractions in the process of finding the derivative. Such algebraic operations must be done with care. *One of the more common sources of errors is the improper handling of fractions.*

◀ EXAMPLE 3   Find the derivative of $y = 4x^3 + \dfrac{5}{x}$ by using the definition.

$$f(x + h) = 4(x + h)^3 + \frac{5}{x + h}$$

$$f(x + h) - f(x) = 4(x + h)^3 + \frac{5}{x + h} - \left(4x^3 + \frac{5}{x}\right)$$

$$= 4(x^3 + 3x^2h + 3xh^2 + h^3) - 4x^3 + \frac{5}{x + h} - \frac{5}{x}$$

$$= 4x^3 + 12x^2h + 12xh^2 + 4h^3 - 4x^3 + \frac{5x - 5(x + h)}{x(x + h)}$$

$$= 12x^2h + 12xh^2 + 4h^3 - \frac{5h}{x(x + h)}$$

$$\frac{f(x + h) - f(x)}{h} = \frac{12x^2h + 12xh^2 + 4h^3}{h} - \frac{5h}{hx(x + h)}$$

$$\lim_{h \to 0} \frac{f(x + h) - f(x)}{h} = \lim_{h \to 0}\left(12x^2 + 12xh + 4h^2 - \frac{5}{x(x + h)}\right) = 12x^2 - \frac{5}{x^2}$$

$$f'(x) = 12x^2 - \frac{5}{x^2}$$

The algebra is handled most easily if the fractions are combined separately from the other terms.

Note that this function is not differentiable for $x = 0$.   ▶

There are notations other than $f'(x)$ that are used for the derivative. These other notations include y', $D_x y$, and $\dfrac{dy}{dx}$.

◀ EXAMPLE 4   In Example 2, $y = \dfrac{3}{x + 2}$, and we found that the derivative is $\dfrac{-3}{(x + 2)^2}$. Therefore, we may write

$$y' = \frac{-3}{(x + 2)^2} \quad \text{or} \quad \frac{dy}{dx} = \frac{-3}{(x + 2)^2}$$

instead of $f'(x) = \dfrac{-3}{(x + 2)^2}$ as we did in Example 2.

If we wish to find the value of the derivative at some point, such as $(-1, 3)$, we write

$$\frac{dy}{dx} = \frac{-3}{(x + 2)^2}$$

$$\left.\frac{dy}{dx}\right|_{x=-1} = \frac{-3}{(-1 + 2)^2} = -3$$

Note that only the x-coordinate was needed to evaluate the derivative.   ▶

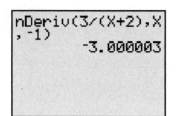

Fig. 23.21

Most calculators have a feature for evaluating derivatives. Since the evaluation is done by numerical approximations, it is called a *numerical derivative*. The display for the evaluation of Example 4 is shown in Fig. 23.21. The result shown is to the default accuracy, but the accuracy can be reset using an additional entry of the derivative feature.

◀ EXAMPLE 5   Find dy/dx for the function $y = \sqrt{x}$ by using the definition.

When finding this derivative, we will find that we have to rationalize the numerator of an expression involving radicals. See page 335 regarding the rationalization of numerators.

$$f(x + h) = \sqrt{x + h}$$

$$f(x + h) - f(x) = \sqrt{x + h} - \sqrt{x}$$

$$\frac{f(x + h) - f(x)}{h} = \frac{\sqrt{x + h} - \sqrt{x}}{h} = \frac{\left(\sqrt{x + h} - \sqrt{x}\right)\left(\sqrt{x + h} + \sqrt{x}\right)}{h\left(\sqrt{x + h} + \sqrt{x}\right)} \quad \text{rationalizing the numerator}$$

$$= \frac{\left(\sqrt{x + h}\right)^2 - \left(\sqrt{x}\right)^2}{h\left(\sqrt{x + h} + \sqrt{x}\right)} = \frac{x + h - x}{h\left(\sqrt{x + h} + \sqrt{x}\right)} = \frac{1}{\sqrt{x + h} + \sqrt{x}}$$

$$\lim_{h \to 0} \frac{f(x + h) - f(x)}{h} = \lim_{h \to 0} \frac{1}{\sqrt{x + h} + \sqrt{x}} = \frac{1}{2\sqrt{x}}$$

$$\frac{dy}{dx} = \frac{1}{2\sqrt{x}}$$

The domain of the function is $x \geq 0$. However, since x appears in the denominator of the derivative, the domain of the derivative is $x > 0$. This means that the function is differentiable for $x > 0$. ▶

CAUTION ▶

One might ask why, when we are finding a derivative, we take the limit as h approaches zero and not simply let h equal zero. If we did this, we would find that the ratio of $(f(x + h) - f(x))/h$ is exactly $0/0$, which requires division by zero. As we know, this is an undefined operation, and therefore h **cannot** **equal zero**. However, it can equal any value as near zero as necessary. This idea is basic in the meaning of the word *limit*.

EXERCISES 23.3

In Exercises 1 and 2, make the given changes in the indicated examples of this section and then find the derivative by using the definition.

1. In Example 1, change $2x^2$ to $4x^2$ and then find the derivative.

2. In Example 2, change $x + 2$ in the denominator to $x - 2$ and then find the derivative.

In Exercises 3–26, find the derivative of each of the functions by using the definition.

3. $y = 3x - 1$

4. $y = 6x + 3$

5. $y = 1 - 2x$

6. $y = 2.3 - 5x$

7. $y = x^2 - 1$

8. $y = 4 - x^2$

9. $y = \pi x^2$

10. $y = -6x^2$

11. $y = x^2 - 7x$

12. $y = x^2 + 4ex$

13. $y = 8x - 2x^2$

14. $y = 3x - \frac{1}{2}x^2$

15. $y = x^3 + 4x - 6$

16. $y = 2x - 4x^3$

17. $y = \dfrac{\sqrt{3}}{x + 2}$

18. $y = \dfrac{3}{2x + 1}$

19. $y = x + \dfrac{4}{3x}$

20. $y = \dfrac{x}{x - 1}$

21. $y = \dfrac{2}{x^2}$

22. $y = \dfrac{2e^2}{x^2 + 4}$

23. $y = x^4 + x^3 + x^2 + x$

24. $y = \frac{1}{3}x^3 + \frac{1}{2}x^2 + x$

25. $y = x^4 - \dfrac{2}{x}$

26. $y = \dfrac{1}{x} + \dfrac{1}{x^2}$

In Exercises 27–30, find the derivative of each function by using the definition. Then evaluate the derivative at the given point. In Exercises 29 and 30, check your result using the derivative evaluation feature of a graphing calculator.

27. $y = 3x^2 - 2x;\ (-1, 5)$

28. $y = 9x - x^3;\ (2, 10)$

29. $y = \dfrac{11}{3x + 2};\ (3, 1)$

30. $y = x^2 - \dfrac{2}{x};\ (-2, 5)$

In Exercises 31–34, find the derivative of each function by using the definition. Then determine the values for which the function is differentiable.

31. $y = 1 + \dfrac{2}{x}$

32. $y = \dfrac{5x}{x - 4}$

33. $y = \dfrac{3}{x^2 - 1}$

34. $y = \dfrac{2}{x^2 + 1}$

In Exercises 35–40, solve the given problems.

35. Find the point(s) on the curve of $y = x^2 - 4x$ for which the slope of a tangent line is 6.

36. Find the point(s) on the curve of $y = 1/(x + 1)$ for which the slope of the tangent line is -1.

(W) 37. Find dy/dx for $y = \sqrt{x + 1}$ by the method of Example 5. For what values of x is the function differentiable? Explain.

38. Find dy/dx for $y = \sqrt{x^2 + 3}$ by the method of Example 5.

39. By noting the derivative for the function in Exercise 23, guess a formula for the derivative of $y = x^n$, where n is a positive integer.

40. By noting the derivative for the function in Exercise 26, guess a formula for the derivative of $y = x^n$, where n is a negative integer. Compare the result with that of Exercise 39.

23.4 THE DERIVATIVE AS AN INSTANTANEOUS RATE OF CHANGE

In Section 23.2, we saw that the slope of a line tangent to a curve at point P was the limiting value of the slope of the line through points P and Q as Q approaches P. In Section 23.3, we defined the limit of the ratio $(f(x + h) - f(x))/h$ as $h \to 0$ as the derivative. Therefore, the first meaning we have given to *the derivative is the slope of a line tangent to a curve,* as we noted in Example 1 of Section 23.3. The following example further illustrates this meaning of the derivative.

◀ EXAMPLE 1 Find the slope of the line tangent to the curve of $y = 4x - x^2$ at the point $(1, 3)$.

We first find the derivative and then evaluate it at the given point:

$$f(x + h) = 4(x + h) - (x + h)^2$$
$$f(x + h) - f(x) = 4(x + h) - (x + h)^2 - (4x - x^2)$$
$$= 4x + 4h - x^2 - 2xh - h^2 - 4x + x^2 = 4h - 2xh - h^2$$
$$\frac{f(x + h) - f(x)}{h} = \frac{4h - 2xh - h^2}{h} = 4 - 2x - h$$
$$\lim_{h \to 0} \frac{f(x + h) - f(x)}{h} = \lim_{h \to 0} (4 - 2x - h) = 4 - 2x$$
$$\frac{dy}{dx} = 4 - 2x$$
$$\left.\frac{dy}{dx}\right|_{(1,3)} = 4 - 2(1) = 2$$

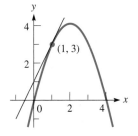

Fig. 23.22

The slope of the tangent line at $(1, 3)$ is 2. Note that only $x = 1$ was needed for the evaluation. The curve and tangent line are shown in Fig. 23.22. ▶

At the end of Section 3.2, we discussed the idea that the ratio $(f(x + h) - f(x))/h$ gives the rate of change of $f(x)$ with respect to x. In defining the derivative as the limit of this ratio as $h \to 0$, it is a measure of the rate of change $f(x)$ with respect to x at point P. However, P may represent any point, which means the value of the derivative changes from one point on a curve to another point.

NOTE ▶ *Therefore, the derivative gives the **instantaneous rate of change** of $f(x)$ with respect to x.*

◀ EXAMPLE 2   In Examples 1 and 2 of Section 23.2, $f(x)$ is changing at the rate of 7 units for a 1-unit change in x, **when x equals exactly 2.** In Example 3 of Section 23.2, $f(x)$ is increasing 6 units for a 1-unit change in x, **when x equals exactly −1,** and $f(x)$ is decreasing 2 units for a 1-unit increase in x, **when x equals exactly 3.** ▶

This gives us a more general meaning of the derivative. If a functional relationship exists between any two variables, then one can be taken to be varying with respect to the other, and the derivative gives us the instantaneous rate of change. There are many applications of this principle, one of which is the velocity of an object. We consider now the case of motion along a straight line, called *rectilinear motion*.

The **average velocity** of an object is found by dividing the change in displacement by the time interval required for this change. As the time interval approaches zero, the limiting value of the average velocity gives the value of the **instantaneous velocity.** Using symbols for the derivative, *the instantaneous velocity of an object moving in rectilinear motion at a specified time t is given by*

Instantaneous Velocity

$$v = \lim_{h \to 0} \frac{s(t + h) - s(t)}{h} \tag{23.7}$$

where $s(t)$ is the displacement as a function of the time t, and h is the time interval that approaches zero. In this case, the derivative has units of displacement divided by units of time, and we can denote it as ds/dt.

See the chapter introduction.

◀ EXAMPLE 3   Find the instantaneous velocity, when $t = 4$ s (exactly), of a falling object for which the distance s (in cm) fallen is the displacement, is given by $s = 490t^2$, by calculating average velocities between $t = 3.5$ s, 3.9 s, 3.99 s, 3.999 s and $t = 4$ s, and then noting the apparent limiting value as $h \to 0$.

The values of h are found by subtracting the given times from 4 s. Also, since $s = 7840$ cm for $t = 4$ s, the differences in distance are found by subtracting the values of s for the given times from 7840 cm. Therefore, we have

t (s)	3.5	3.9	3.99	3.999
s (cm)	6002.5	7452.9	7800.849	7836.0805
$7840 - s$ (cm)	1837.5	387.1	39.15	3.9195
$h = 4 - t$ (s)	0.5	0.1	0.01	0.001
$v = \dfrac{7840 - s}{h}$ (cm/s)	3675.0	3871.0	3915.0	3919.5

We can see that the value of v is approaching 3920 cm/s, which is therefore the instantaneous velocity when $t = 4$ s. ▶

See the chapter introduction.

◀ EXAMPLE 4   Find the expression for the instantaneous velocity of the object of Example 3, for which $s = 490t^2$, where s is the displacement (in cm) and t is the time (in s). Determine the instantaneous velocity for $t = 2$ s and $t = 4$ s.

The required expression is the derivative of s with respect to t.

$$f(t + h) = 490(t + h)^2$$

$$f(t + h) - f(t) = 490(t + h)^2 - 490t^2 = 980th + 490h^2$$

$$\frac{f(t + h) - f(t)}{h} = \frac{980th + 490h^2}{h} = 980t + 490h \qquad \text{expression for instantaneous velocity}$$

$$v = \lim_{h \to 0} \frac{f(t + h) - f(t)}{h} = \lim_{h \to 0} (980t + 490h) = 980t \quad \llcorner$$

$$\frac{ds}{dt}\bigg|_{t=2} = 980(2) = 1960 \text{ cm/s} \quad \text{and} \quad \frac{ds}{dt}\bigg|_{t=4} = 980(4) = 3920 \text{ cm/s}$$

We see that the second result agrees with that found in Example 3.   ▶

By finding $\lim_{h \to 0} (f(x + h) - f(x))/h$, we can find the instantaneous rate of change of $f(x)$ with respect to x. The expression $\lim_{h \to 0} (f(t + h) - f(t))/h$ gives the instantaneous velocity, or instantaneous rate of change of displacement with respect to time. Generalizing, we can say that

NOTE ▶   *the derivative can be interpreted as the instantaneous rate of change of the dependent variable with respect to the independent variable.*

This is true for a differentiable function, no matter what the variables represent.

Solving a Word Problem   ◀ EXAMPLE 5   A spherical balloon is being inflated. Find the expression for the instantaneous rate of change of the volume with respect to the radius. Evaluate this rate of change for a radius of 2.00 m:

$$V = \tfrac{4}{3} \pi r^3 \qquad \text{volume of sphere}$$

$$f(r + h) = \tfrac{4}{3} \pi (r + h)^3 \qquad \text{find derivative}$$

$$f(r + h) - f(r) = \tfrac{4}{3} \pi (r + h)^3 - \tfrac{4}{3} \pi r^3$$

$$= \tfrac{4}{3} \pi (r^3 + 3r^2h + 3rh^2 + h^3 - r^3) = \tfrac{4}{3} \pi (3r^2h + 3rh^2 + h^3)$$

$$\frac{f(r + h) - f(r)}{h} = \frac{4\pi}{3}\left(\frac{3r^2h + 3rh^2 + h^3}{h}\right) = \frac{4\pi}{3}(3r^2 + 3rh + h^2)$$

$$\frac{dV}{dr} = \lim_{h \to 0} \frac{f(r + h) - f(r)}{h} = \lim_{h \to 0}\left(\frac{4\pi}{3}(3r^2 + 3rh + h^2)\right) = 4\pi r^2$$

$$\frac{dV}{dr}\bigg|_{r=2.00\,\text{m}} = 4\pi(2.00)^2 = 16.0\pi = 50.3 \text{ m}^2 \qquad \text{instantaneous rate of change when } r = 2.00 \text{ m}$$

The instantaneous rate of change of the volume with respect to the radius (dV/dr) for $r = 2.00$ m is 50.3 m³/m (this way of showing the units is more meaningful).

As r increases, dV/dr also increases. This should be expected as the volume of a sphere varies directly as the cube of the radius.   ▶

Solving a Word Problem

EXAMPLE 6 The power P produced by an electric current i in a resistor varies directly as the square of the current. Given that 1.2 W of power are produced by a current of 0.50 A in a certain resistor, find an expression for the instantaneous rate of change of power with respect to current. Evaluate this rate of change for $i = 2.5$ A.

We must first find the functional relationship between power and current, by solving the indicated problem in variation:

$$P = ki^2 \qquad 1.2 = k(0.50)^2 \qquad k = 4.8 \text{ W/A}^2 \qquad P = 4.8i^2$$

Now, knowing the function, we may determine the expression for the instantaneous rate of change of P with respect to i by finding the derivative.

$$f(i + h) = 4.8(i + h)^2$$
$$f(i + h) - f(i) = 4.8(i + h)^2 - 4.8i^2 = 4.8(2ih + h^2)$$
$$\frac{f(i + h) - f(i)}{h} = \frac{4.8(2ih + h^2)}{h} = 4.8(2i + h) \qquad \text{expression for instantaneous rate of change}$$
$$\frac{dP}{di} = \lim_{h \to 0} \frac{f(i + h) - f(i)}{h} = \lim_{h \to 0} [4.8(2i + h)] = 9.6i$$
$$\frac{dP}{di}\bigg|_{i=2.5\,\text{A}} = 9.6(2.5) = 24 \text{ W/A} \qquad \text{instantaneous rate of change when } i = 2.5 \text{ A}$$

This tells us that when $i = 2.5$ A, the rate of change of power with respect to current is 24 W/A. Also, we see that the larger the current is, the greater is the increase in power. This should be expected, since the power varies directly as the square of the current.

EXERCISES  23.4

In Exercises 1 and 2, make the given changes in the indicated examples of this section and then solve the resulting problems.

1. In Example 4, change $490t^2$ to $1400t - 490t^2$ and then evaluate the instantaneous velocity at the indicated times.

2. In Example 5, change "volume" in the second line to "surface area" and then evaluate the rate of change for $r = 2.00$ m.

In Exercises 3–6, find the slope of a line tangent to the curve of the given equation at the given point. Sketch the curve and the tangent line.

3. $y = x^2 - 1$; $(2, 3)$

4. $y = 2x - x^2$; $(-1, -3)$

5. $y = \dfrac{16}{3x + 1}$; $(-3, -2)$

6. $y = 3 - \dfrac{16}{x^2}$; $(2, -1)$

In Exercises 7–10, calculate the instantaneous velocity for the indicated value of the time (in s) of an object for which the displacement (in m) is given by the indicated function. Use the method of Example 3 and calculate values of the average velocity for the given values of t and note the apparent limit as the time interval approaches zero.

7. $s = 4t + 10$; when $t = 3$; use values of t of 2.0, 2.5, 2.9, 2.99, 2.999

8. $s = 6 - 3t$; when $t = 4$; use values of t of 3.0, 3.5, 3.9, 3.99, 3.999

9. $s = 3t^2 - 4t$; when $t = 2$; use values of t of 1.0, 1.5, 1.9, 1.99, 1.999

10. $s = 40t - 4.9t^2$; when $t = 0.5$; use values of t of 0.4, 0.45, 0.49, 0.499, 0.4999

In Exercises 11–14, use the definition to find an expression for the instantaneous velocity of an object moving with rectilinear motion according to the given functions (the same as those for Exercises 7–10) relating s (in m) and t (in s). Then calculate the instantaneous velocity for the given value of t.

11. $s = 4t + 10$; $t = 3$

12. $s = 6 - 3t$; $t = 4$

13. $s = 3t^2 - 4t$; $t = 2$

14. $s = 40t - 4.9t^2$; $t = 0.5$

In Exercises 15–20, use the definition to find an expression for the instantaneous velocity of an object moving with rectilinear motion according to the given functions relating s and t.

15. $s = 48t + 12$

16. $s = 3t^2 - 2t^3$

17. $s = 4t^2 - t^4$

18. $s = s_0 + v_0 t + \dfrac{1}{2} at^2$ (s_0, v_0, and a are constants.)

19. $s = 3t - \dfrac{2}{5t}$

20. $s = \dfrac{2t}{t + 2}$

In Exercises 21–24, use the definition to find an expression for the instantaneous acceleration of an object moving with rectilinear motion according to the given functions. The instantaneous acceleration of an object is defined as the instantaneous rate of change of the velocity with respect to time. Here, v is the velocity, s is the displacement, and t is the time.

21. $v = 6t^2 - 4t + 2$

22. $v = \sqrt{2t + 1}$

23. $s = t^3 + 2t$ (Find v, then find a.)

24. $s = s_0 + v_0 t - \frac{1}{2} a t^2$ (s_0, v_0, and a are constants.)
(Find v, then find a.)

In Exercises 25–40, find the indicated instantaneous rates of change.

25. The distance s (in m) above the ground for a projectile fired vertically upward with a velocity of 44 m/s as a function of time t (in s) is given by $s = 44t - 4.9t^2$. Find t for $v = 0$.

(W) 26. For the projectile in Exercise 25, find v for $t = 4.0$ s and for $t = 5.0$ s. What conclusion can be drawn?

27. The electric current i at a point in an electric circuit is the instantaneous rate of change of the electric charge q that passes the point, with respect to the time t. Find i in a circuit for which $q = 30 - 2t$.

28. A load L (in N) is distributed along a beam 10 m long such that $L = 5x - 0.5x^2$, where x is the distance from one end of the beam. Find the expression for the instantaneous rate of change of L with respect to x.

29. A rectangular metal plate contracts while cooling. Find the expression for the instantaneous rate of change of the area A of the plate with respect to its width w, if the length of the plate is constantly three times as long as the width.

30. A circular oil spill is increasing in size. Find the instantaneous rate of change of the area A of the spill with respect to its radius r for $r = 240$ m.

31. The total power P (in W) transmitted by an AM radio station is given by $P = 500 + 250m^2$, where m is the modulation index. Find the instantaneous rate of change of P with respect to m for $m = 0.92$.

32. The bottom of a soft-drink can is being designed as an inverted spherical segment, the volume of which is $V = \frac{1}{6}\pi h^3 + 2.00\pi h$,

where h is the depth (in cm) of the segment. Find the instantaneous rate of change of V with respect to h for $h = 0.60$ cm.

33. The total solar radiation H (in W/m^2) on a particular surface during an average clear day is given by $H = \dfrac{5000}{t^2 + 10}$, where t ($-6 \le t \le 6$) is the number of hours from noon (6 A.M. is equivalent to $t = -6$ h). Find the instantaneous rate of change of H with respect to t at 3 P.M.

34. For the solar radiator in Exercise 33, find the average rate of change of H between 2 P.M. and 4 P.M. Compare with the instantaneous rate of change at 3 P.M.

35. The value (in thousands of dollars) of a certain car is given by the function $V = \dfrac{48}{t + 3}$, where t is measured in years. Find a general expression for the instantaneous rate of change of V with respect to t and evaluate this expression when $t = 3$ years.

36. For the car in Exercise 35, find the average rate of change of V between $t = 2$ years and $t = 4$ years. Compare with the instantaneous rate of change for $t = 3$ years.

37. Oil in a certain machine is stored in a conical reservoir, for which the radius and height are both 4 cm (see Fig. 23.23). Find the instantaneous rate of change of the volume V of oil in the reservoir with respect to the depth d of the oil.

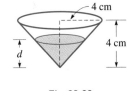

Fig. 23.23

38. The time t required to test a computer memory unit is directly proportional to the square of the number n of memory cells in the unit. For a particular type of unit, $n = 6400$ for $t = 25.0$ s. Find the instantaneous rate of change of t with respect to n for this type of unit for $n = 8000$.

39. A *holograph* (an image formed without using a lens) of concentric circles is formed. The radius r of each circle varies directly as the square root of the wavelength λ of the light used. If $r = 3.72$ cm for $\lambda = 592$ nm, find the expression for the instantaneous rate of change of r with respect to λ.

40. The force F between two electric charges varies inversely as the square of the distance r between them. For two charged particles, $F = 0.12$ N for $r = 0.060$ m. Find the instantaneous rate of change of F with respect to r for $r = 0.120$ m.

23.5 DERIVATIVES OF POLYNOMIALS

The task of finding a derivative can be considerably shortened by using the definition to derive certain basic formulas for derivatives. In this section, we derive the formulas that are used for finding the derivatives of polynomial functions of the form
$$f(x) = a_0 x^n + a_1 x^{n-1} + \cdots + a_n.$$

First, we find the derivative of a constant. Letting $f(x) = c$ and then using the definition of a derivative, we find that $f(x + h) - f(x) = c - c = 0$. This means that

$$\lim_{h \to 0} \frac{f(x + h) - f(x)}{h} = \lim_{h \to 0} \left(\frac{0}{h} \right) = 0$$

From this, we conclude that *the derivative of a constant is zero*. This result holds for all constants. Therefore, if $y = c$, $dy/dx = 0$, or

Derivative of a Constant

$$\frac{dc}{dx} = 0 \tag{23.8}$$

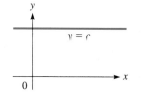

Fig. 23.24

Graphically, this means that for any function of the type $y = c$, the slope is always zero. We know that $y = c$ represents a straight line parallel to the x-axis. From the definition of slope, we know that any line parallel to the x-axis has a slope of zero. See Fig. 23.24. We see that the two results are consistent.

Next, we find the derivative of a positive integral power of x. If $f(x) = x^n$, where n is a positive integer, by using the binomial theorem we have

$$f(x + h) - f(x) = (x + h)^n - x^n = x^n + nx^{n-1}h + \frac{n(n-1)}{2}x^{n-2}h^2 + \cdots + h^n - x^n$$

$$= nx^{n-1}h + \frac{n(n-1)}{2}x^{n-2}h^2 + \cdots + h^n$$

$$\frac{f(x + h) - f(x)}{h} = nx^{n-1} + \frac{n(n-1)}{2}x^{n-2}h + \cdots + h^{n-1}$$

$$\lim_{h \to 0} \frac{f(x + h) - f(x)}{h} = nx^{n-1}$$

Thus, *the derivative of the nth power of x is*

Derivative of the Power of x

$$\frac{dx^n}{dx} = nx^{n-1} \tag{23.9}$$

◖ EXAMPLE 1   Find the derivative of the function $y = -5$.
Since -5 is a constant, applying Eq. (23.8), we have

$$\frac{dy}{dx} = \frac{d(-5)}{dx} = 0$$

◖ EXAMPLE 2   Find the derivative of $y = x^3$.
Using Eq. (23.9), we have

$$\frac{dy}{dx} = \frac{d(x^3)}{dx} = 3x^{3-1} = 3x^2$$

This result is consistent with those found previously in this chapter.

◀ EXAMPLE 3   Find the derivative of the function $y = x$.
In using Eq. (23.9), we have $n = 1$ since $x = x^1$. This means

$$\frac{dy}{dx} = \frac{d(x)}{dx} = (1)x^{1-1} = (1)(x^0)$$

Since $x^0 = 1$, we have

$$\frac{dy}{dx} = 1$$

Thus, the derivative of $y = x$ is 1, which means that the slope of the line $y = x$ is always 1. This is consistent with our previous discussion of the slope of a straight line. ▶

◀ EXAMPLE 4   Find the derivative of the function $v = r^{10}$.
Here, the dependent variable is v, and the independent variable is r. Therefore,

$$\frac{dv}{dr} = \frac{d(r^{10})}{dr} = 10r^{10-1}$$

$$= 10r^9$$ ▶

Next, we find the derivative of a constant times a function of x. We denote this function as u, or to show directly that it is a function of x, as $u(x)$. In finding the derivative of cu with respect to x, we have

> From here on, we will be using functions that are combinations of simpler functions. Therefore, we must denote some functions by symbols other than $f(x)$. We will often be using u and v.

$$\frac{d}{dx}(cu) = \lim_{h \to 0} \frac{cu(x + h) - cu(x)}{h}$$

$$= c \lim_{h \to 0} \frac{u(x + h) - u(x)}{h} = c\frac{du}{dx}$$

Therefore, *the derivative of the product of a constant and a differentiable function of x is the product of the constant and the derivative of the function of x.* This is written as

Derivative of a Constant Times a Function

$$\boxed{\frac{d(cu)}{dx} = c\frac{du}{dx}}$$ (23.10)

◀ EXAMPLE 5   Find the derivative of $y = 3x^2$.
In this case, $c = 3$ and $u = x^2$. Thus, $du/dx = 2x$. Therefore,

$$\frac{dx}{dy} = \frac{d(3x^2)}{dx} = 3\frac{d(x^2)}{dx} = 3(2x)$$

$$= 6x$$ ▶

Occasionally, the derivative of a constant times a function of x is confused with the derivative of a constant that stands alone. It is necessary to clearly distinguish between a constant that multiplies a function and an isolated constant.

If the types of functions for which we have found derivatives are added, the result is a polynomial function with more than one term. The derivative is found by letting $y = u + v$, where u and v are functions of x. Using the definition, we have

$$y = u(x) + v(x)$$

$$\frac{dy}{dx} = \frac{d}{dx}[u(x) + v(x)] = \lim_{h \to 0} \frac{[u(x + h) + v(x + h)] - [u(x) - v(x)]}{h}$$

$$= \lim_{h \to 0}\left[\frac{u(x + h) - u(x)}{h} + \frac{v(x + h) - v(x)}{h}\right]$$

$$= \lim_{h \to 0}\left[\frac{u(x + h) - u(x)}{h}\right] + \lim_{h \to 0}\left[\frac{v(x + h) - v(x)}{h}\right] = \frac{du}{dx} + \frac{dv}{dx}$$

This tells us that *the derivative of the sum of differentiable functions of x is the sum of the derivatives of the functions.* This is written as

Derivative of a Sum

$$\frac{d(u + v)}{dx} = \frac{du}{dx} + \frac{dv}{dx}$$

(23.11)

◀ **EXAMPLE 6** Evaluate the derivative of $f(x) = 2x^4 - 6x^2 - 8x - 9$ at $(-2, 15)$.

First, finding the derivative, we have

$$f'(x) = \frac{d(2x^4)}{dx} - \frac{d(6x^2)}{dx} - \frac{d(8x)}{dx} - \frac{d(9)}{dx}$$

$$= 8x^3 - 12x - 8$$

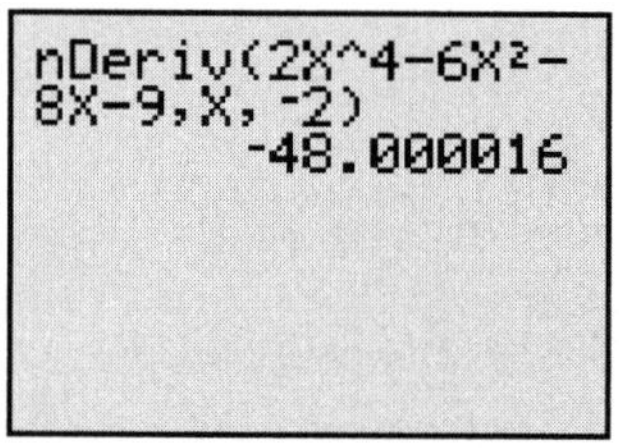

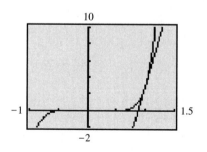

nDeriv(2X^4-6X²-8X-9,X,-2)

  -48.000016

Fig. 23.25

Since the derivative is a function only of x, we now evaluate it for $x = -2$:

$$f'(x)\bigg|_{x=-2} = 8(-2)^3 - 12(-2) - 8 = -48$$

Using the *derivative evaluation* feature of a graphing calculator, we have the display in Fig. 23.25 for this evaluation. We see that the values agree. ▶

◀ **EXAMPLE 7** Find the slope of a line tangent to the curve of $y = 4x^7 - x^4$ at the point $(1, 3)$.

We must find and then evaluate the derivative for the value $x = 1$:

$$\frac{dy}{dx} = 28x^6 - 4x^3 \qquad \text{find derivative}$$

$$\frac{dy}{dx}\bigg|_{x=1} = 28(1) - 4(1) \qquad \text{evaluate derivative}$$

$$= 24$$

Thus, the slope of the tangent line is 24. Again, we note that *the substitution x = 1 must be made after the differentiation has been performed.*

Using the *tangent* feature of a graphing calculator, we have the display for the curve and tangent line in Fig. 23.26. Although it is difficult to approximate a slope as large as 24, we can see that the line is very steep, which tends to verify the result. ▶

Fig. 23.26

◀ EXAMPLE 8   For each 4.0-s cycle, the displacement s (in cm) of a piston is given by the equation $s = t^3 - 6t^2 + 8t$, where t is the time. Find the instantaneous velocity of the piston for $t = 2.6$ s.

$$s = t^3 - 6t^2 + 8t$$

$$\frac{ds}{dt} = 3t^2 - 12t + 8 \qquad \text{find derivative}$$

$$\left.\frac{ds}{dt}\right|_{t=2.6} = 3(2.6)^2 - 12(2.6) + 8 \qquad \text{evaluate derivative}$$

$$= -2.9 \text{ cm/s}$$

This tells us that the piston is moving at -2.9 cm/s (it is moving in a negative direction) when $t = 2.6$ s. ◗

EXERCISES 23.5

In Exercises 1–4, make the given changes in the indicated examples of this section, and then solve the resulting problem.

1. In Example 4, change the exponent 10 to 9 and then find the derivative.

2. In Example 5, change the coefficient 3 to 4 and then find the derivative.

3. In Example 6, change $2x^4$ to $2x^3$ and $-8x$ to $+8x$ and then find and evaluate the derivative for $x = -2$.

4. In Example 7, change $4x^7 - x^4$ to $4x^4 - x^7$ and then display the curve and tangent line for $x = 1$ on a graphing calculator.

In Exercises 5–20, find the derivative of each of the given functions

5. $y = x^5$
6. $y = x^{12}$
7. $f(x) = -4x^9$
8. $y = -7x^6$
9. $y = x^4 - 3\pi$
10. $s = 3t^5 + 4$
11. $y = x^2 + 2x$
12. $y = x^3 - 1.5x^2$
13. $p = 5r^3 - 2r + 1$
14. $y = 6x^2 - 6x + 5$
15. $y = 25x^8 - 34x^5 - x$
16. $y = 4x^4 - 2x + 9$
17. $f(x) = -6x^7 + 5x^3 + \pi^2$
18. $y = 13x^4 - 6x^3 - x - 1$
19. $y = \frac{1}{3}x^3 + \frac{1}{2}x^2$
20. $f(z) = -\frac{1}{4}z^8 + \frac{1}{2}z^4 - 2^3$

In Exercises 21–24, evaluate the derivative of each of the given functions at the given point. In Exercises 23 and 24, check your result using the derivative evaluation feature of a graphing calculator.

21. $y = 6x^2 - 8x + 1$ (2, 9)
22. $s = 2t^3 - 5t^2 + 4$ (-1, -3)
23. $y = 2x^3 + 9x - 7$ (-2, -41)
24. $y = x^4 - 9x^2 - 5x$ (3, -15)

In Exercises 25–28, find the slope of a line tangent to the curve of each of the given functions for the given values of x. Use the tangent feature of a graphing calculator to display the curve and the tangent line to see if the slope is reasonable for the given value of x.

25. $y = 2x^6 - 4x^2$ $(x = -1)$
26. $y = 3x^3 - 9x$ $(x = 1)$
27. $y = 35x - 2x^4$ $(x = 2)$
28. $y = x^4 - \frac{1}{2}x^2 + 2$ $(x = -2)$

In Exercises 29–32, determine an expression for the instantaneous velocity of objects moving with rectilinear motion according to the functions given, if s represents displacement in terms of time t.

29. $s = 6t^5 - 5t + 2$
30. $s = 20 + 60t - 4.9t^2$
31. $s = 2 - 6t - 2t^3$
32. $s = s_0 + v_0 t + \frac{1}{2}at^2$

In Exercises 33–36, s represents the displacement, and t represents the time for objects moving with rectilinear motion, according to the given functions. Find the instantaneous velocity for the given times.

33. $s = 2t^3 - 4t^2$; $t = 4$
34. $s = 120 + 80t - 16t^2$; $t = 2.5$
35. $s = 0.5t^4 - 1.5t^2 + 2.5$; $t = 3$
36. $s = 8t^2 - 10t + 6$; $t = 5$

In Exercises 37–56, solve the given problems by finding the appropriate derivative.

37. For what value(s) of x is the tangent to the curve of $y = 3x^2 - 6x$ parallel to the x-axis? (That is, where is the slope zero?)

38. Find the value of a if the tangent to the curve of $y = ax^2 + 2x$ has a slope of -4 for $x = 2$.

39. For what point(s) on the curve of $y = 3x^2 - 4x$ is the slope of a tangent line equal to 8?

Ⓦ 40. Explain why the curve $y = 5x^3 + 4x - 3$ does not have a tangent line with a slope less than 4.

41. Find the point at which a tangent line to the parabola $y = 2x^2 - 7x$ is perpendicular to the line $x - 3y = 16$.

W **42.** Display the graphs of $y = x^2$ and its derivative on a graphing calculator. State any conclusions you can draw from the relationship of the two graphs.

43. For what value(s) of x is the slope of a line tangent to the curve of $y = 4x^2 + 3x$ equal to the slope of a line tangent to the curve of $y = 5 - 2x^2$?

44. For what value(s) of t is the instantaneous velocity of an object moving according to $s = 5t - 2t^2$ equal to the instantaneous velocity of an object moving according to $s = 3t^2 + 4$?

45. A cylindrical metal container is heated and then allowed to cool. If the radius always equals the height, find an expression for the instantaneous rate of change of the volume V with respect to the radius r.

46. As an ice cube melts uniformly, find the expression for the instantaneous rate of change of the surface area A of the cube with respect to the edge e.

47. The electric power P (in W) as a function of the current i (in A) in a certain circuit is given by $P = 16i^2 + 60i$. Find the instantaneous rate of change of P with respect to i for $i = 0.75$ A.

48. The torque T on the arm of a robotic control mechanism varies directly as the cube of the diameter d of the arm. If $T = 850$ N · m for $d = 0.925$ cm, find the expression for the instantaneous rate of change of T with respect to d.

49. The electric polarization P of a light wave for high values of the electric field E is given by $P = a(c_1E + c_2E^2 + c_3E^3)$, where a, c_1, c_2, and c_3 are constants. Find the expression for the instantaneous rate of change of P with respect to E.

50. The deflection d of a diving board x m from the fixed end at the pool side is given by $d = kx^2(3L - x)$, where L is the length of the diving board and k is a positive constant. Find the expression for the instantaneous rate of change of d with respect to x.

51. The tensile strength S (in N) of a certain material as a function of the temperature T (in °C) is $S = 1600 - 0.000\,022T^2$. Find the instantaneous rate of change of S with respect to T for $T = 65$°C.

52. A tank containing 6000 L of water drains out in 30 min. The volume V of water in the tank after t min of draining is $V = 6000(1 - t/30)^2$. Find the instantaneous time rate of change of V after 15 min of draining.

53. The altitude h (in m) of a jet as a function of the horizontal distance x (in km) it has traveled is given by $h = 0.000\,104x^4 - 0.0417x^3 + 4.21x^2 - 8.33x$. Find the instantaneous rate of change of h with respect to x for $x = 120$ km.

54. The force F (in N) exerted by a cam on a lever is given by $F = x^4 - 12x^3 + 46x^2 - 60x + 25$, where x $(1 \le x \le 5)$ is the distance (in cm) from the center of rotation of the cam to the edge of the cam in contact with the lever (see Fig. 23.27). Find the instantaneous rate of change of F with respect to x when $x = 4.0$ cm.

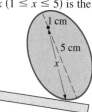

Fig. 23.27

55. Two ball bearings wear down such that the radius r of one is constantly 1.20 mm less than the radius of the other. Find the instantaneous rate of change of the total volume V_T of the two ball bearings with respect to r for $r = 3.30$ mm.

56. An open-top container is to be made from a rectangular piece of cardboard 6.00 cm by 8.00 cm. Equal squares of side x are to be cut from each corner, then the sides are to be bent up and taped together. Find the instantaneous rate of change of the volume V of the container with respect to x for $x = 1.75$ cm.

23.6 DERIVATIVES OF PRODUCTS AND QUOTIENTS OF FUNCTIONS

The formulas developed in the previous section are valid for polynomial functions. However, many functions are not polynomial in form. Some functions can best be expressed as the product of two or more simpler functions, others are the quotient of two simpler functions, and some are expressed as powers of a function. In this section, we develop the formula for the derivative of a product of functions and the formula for the derivative of the quotient of two functions.

◀ **EXAMPLE 1** The functions $f(x) = x^2 + 2$ and $g(x) = 3 - 2x$ can be combined to form functions of the types mentioned above, as we now illustrate:

$$p(x) = f(x)g(x) = (x^2 + 2)(3 - 2x) \quad \text{product of functions}$$

$$q(x) = \frac{g(x)}{f(x)} = \frac{3 - 2x}{x^2 + 2} \qquad \text{quotient of functions}$$

$$F(x) = [g(x)]^2 = (3 - 2x)^2 \qquad \text{power of a function}$$

If u and v are differentiable functions of x, the derivative of the function uv is found by letting $y = uv$, and applying the definition as follows:

$$y = u(x)v(x)$$

$$\frac{dy}{dx} = \frac{d}{dx}[u(x)v(x)] = \lim_{h \to 0} \frac{u(x + h)v(x + h) - u(x)v(x)}{h}$$

By adding and subtracting $u(x + h)v(x)$ in the numerator of the last fraction, we can put it in a form that includes the derivatives of $u(x)$ and $v(x)$. Therefore,

$$\frac{d}{dx}[u(x)v(x)] = \lim_{h \to 0} \frac{u(x + h)v(x + h) - u(x + h)v(x) + u(x + h)v(x) - u(x)v(x)}{h}$$

$$= \lim_{h \to 0} \left[u(x + h)\frac{v(x + h) - v(x)}{h} + v(x)\frac{u(x + h) - u(x)}{h} \right]$$

$$= u(x)\frac{dv(x)}{dx} + v(x)\frac{du(x)}{dx}$$

We conclude that *the derivative of the product of two differentiable functions equals the first function times the derivative of the second function plus the second function times the derivative of the first function.* This is written as

Derivative of a Product

$$\boxed{\frac{d(uv)}{dx} = u\frac{dv}{dx} + v\frac{du}{dx}} \tag{23.12}$$

◀ **EXAMPLE 2** Find the derivative of the product function in Example 1.

$$p(x) = (x^2 + 2)(3 - 2x) \qquad u = x^2 + 2 \qquad v = 3 - 2x$$

$$\frac{d(uv)}{dx} = \quad u \quad \frac{dv}{dx} \quad + \quad v \quad \frac{du}{dx}$$

$$\downarrow \quad \downarrow \qquad \downarrow \quad \downarrow$$

$$p'(x) = (x^2 + 2)(-2) + (3 - 2x)(2x) = -2x^2 - 4 + 6x - 4x^2$$

$$= -6x^2 + 6x - 4 \qquad \qquad \blacktriangleright$$

◀ **EXAMPLE 3** Find the derivative of the function $y = (3 - x - 2x^2)(x^4 - x)$.

In this problem, $u = 3 - x - 2x^2$ and $v = x^4 - x$. Hence,

$$\frac{dy}{dx} = (3 - x - 2x^2)(4x^3 - 1) + (x^4 - x)(-1 - 4x)$$

$$= 12x^3 - 3 - 4x^4 + x - 8x^5 + 2x^2 - x^4 - 4x^5 + x + 4x^2$$

$$= -12x^5 - 5x^4 + 12x^3 + 6x^2 + 2x - 3 \qquad \qquad \blacktriangleright$$

In both of these examples, we could have multiplied the functions first and then taken the derivative as a polynomial. However, we will soon meet functions for which this latter method would not be applicable.

We will now find the derivative of the quotient of two differentiable functions by applying the definition to the function $y = u/v$, as shown on the next page.

$$\frac{dy}{dx} = \frac{d}{dx}\left(\frac{u(x)}{v(x)}\right) = \lim_{h\to 0}\frac{\dfrac{u(x+h)}{v(x+h)} - \dfrac{u(x)}{v(x)}}{h} = \lim_{h\to 0}\frac{v(x)u(x+h) - u(x)v(x+h)}{hv(x+h)v(x)}$$

We can put the last fraction in a form that includes the derivatives of $u(x)$ and $v(x)$ by subtracting and adding $u(x)v(x)$ in the numerator. Therefore,

$$\frac{d}{dx}\left[\frac{u(x)}{v(x)}\right] = \lim_{h\to 0}\frac{v(x)u(x+h) - v(x)u(x) + v(x)u(x) - u(x)v(x+h)}{hv(x+h)v(x)}$$

$$= \lim_{h\to 0}\frac{v(x)\dfrac{u(x+h) - u(x)}{h} - u(x)\dfrac{v(x+h) - v(x)}{h}}{v(x+h)v(x)}$$

$$= \frac{v(x)\dfrac{du(x)}{dx} - u(x)\dfrac{dv(x)}{dx}}{[v(x)]^2}$$

Therefore, *the derivative of the quotient of two differentiable functions equals the denominator times the derivative of the numerator minus the numerator times the derivative of the denominator, all divided by the square of the denominator.*

Derivative of a Quotient

$$\frac{d\left(\dfrac{u}{v}\right)}{dx} = \frac{v\dfrac{du}{dx} - u\dfrac{dv}{dx}}{v^2} \qquad (23.13)$$

EXAMPLE 4 Find the derivative of the quotient indicated in Example 1.

$$q(x) = \frac{3 - 2x}{x^2 + 2} \qquad u = 3 - 2x \qquad v = x^2 + 2$$

$$q'(x) = \frac{\overset{v}{(x^2 + 2)}\overset{\frac{du}{dx}}{(-2)} - \overset{u}{(3 - 2x)}\overset{\frac{dv}{dx}}{(2x)}}{\underset{v^2}{(x^2 + 2)^2}} = \frac{-2x^2 - 4 - 6x + 4x^2}{(x^2 + 2)^2}$$

$$= \frac{2(x^2 - 3x - 2)}{(x^2 + 2)^2}$$

In the first expression for $q'(x)$, be careful not to cancel the factor of $(x^2 + 2)$, as it is not a factor of both terms of the numerator.

EXAMPLE 5 The stress S on a hollow tube is given by

$$S = \frac{16DT}{\pi(D^4 - d^4)}$$

where T is the tension, D is the outer diameter, and d is the inner diameter of the tube. Find the expression for the instantaneous rate of change of S with respect to D, with the other values being constant.

We are to find the derivative of S with respect to D, and it is found as follows:

$$\frac{dS}{dD} = \frac{\pi(D^4 - d^4)(16T) - 16DT(\pi)(4D^3)}{\pi^2(D^4 - d^4)^2} = \frac{16\pi T(D^4 - d^4 - 4D^4)}{\pi^2(D^4 - d^4)^2}$$

$$= \frac{-16T(3D^4 + d^4)}{\pi(D^4 - d^4)^2}$$

EXAMPLE 6 Evaluate the derivative of $y = \dfrac{3x^2 + x}{1 - 4x}$ at $(2, -2)$.

$$\frac{dy}{dx} = \frac{(1 - 4x)(6x + 1) - (3x^2 + x)(-4)}{(1 - 4x)^2}$$

$$= \frac{6x + 1 - 24x^2 - 4x + 12x^2 + 4x}{(1 - 4x)^2}$$

$$= \frac{-12x^2 + 6x + 1}{(1 - 4x)^2}$$

$$\frac{dy}{dx}\bigg|_{x=2} = \frac{-12(2^2) + 6(2) + 1}{[1 - 4(2)]^2} = \frac{-48 + 12 + 1}{49} = \frac{-35}{49}$$

$$= -\frac{5}{7}$$

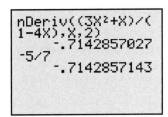

Fig. 23.28

Checking this result by using the *numerical derivative* feature of a graphing calculator, we have the display shown in Fig. 23.28. We see that the calculator evaluation and the decimal value of $-5/7$ agree. (The difference in the last three decimal places is due to the fact that the numerical derivative evaluation is an approximation.)

EXERCISES 23.6

In Exercises 1 and 2, make the given changes in the indicated examples of this section and then solve the resulting problem.

1. In Example 3, change u to $x - 2x^2$ and then find the derivative.

2. In Example 6, change the numerator to $3x^2 - x$ and then find and evaluate the derivative at $x = 2$.

In Exercises 3–8, find the derivative of each function by using Eq. (23.12). Do not find the product before finding the derivative.

3. $y = 6x(3x^2 - 5x)$

4. $y = 2x^3(3x^4 + x)$

5. $s = (3t + 2)(2t - 5)$

6. $f(x) = (3x - 2)(4x^2 + 3)$

7. $y = (x^4 - 3x^2 + 3)(1 - 2x^3)$

8. $y = (x^3 - 6x)(2 - 4x^3)$

In Exercises 9–12, find the derivative of each function by using Eq. (23.12). Then multiply out each function and find the derivative by treating it as a polynomial. Compare the results.

9. $y = (2x - 7)(5 - 2x)$

10. $f(s) = (5s^2 + 2)(2s^2 - 1)$

11. $y = (x^3 - 1)(2x^2 - x - 1)$

12. $y = (3x^2 - 4x + 1)(5 - 6x^2)$

In Exercises 13–24, find the derivative of each function by using Eq. (23.13).

13. $y = \dfrac{x}{2x + 3}$

14. $y = \dfrac{4}{x^3}$

15. $y = \dfrac{\pi}{2x^2 + 1}$

16. $R = \dfrac{5i + 2}{2i + 3}$

17. $y = \dfrac{x^2}{3 - 2x}$

18. $y = \dfrac{e^2}{3x^2 - 5x}$

19. $y = \dfrac{2x - 1}{3x^2 + 2}$

20. $y = \dfrac{2x^3}{4 - x}$

21. $f(x) = \dfrac{3x + 8}{x^2 + 4x + 2}$

22. $y = \dfrac{33x}{4x^5 - 3x - 4}$

23. $y = \dfrac{2x^2 - x - 1}{x^3 + 2x^2}$

24. $y = \dfrac{3x^3 - x}{2x^2 - 5x + 4}$

In Exercises 25–32, evaluate the derivatives of the given functions for the given values of x. In Exercises 25–28, use Eq. (23.12). In Exercises 27, 28, 31, and 32, check your results using the derivative evaluation feature of a graphing calculator.

25. $y = (3x - 1)(4 - 7x)$, $x = 3$

26. $y = (3x^2 - 5)(2x^2 - 1)$, $x = -1$

27. $y = (2x^2 - x + 1)(4 - 2x - x^2)$, $x = -3$

28. $y = (4x^4 + 0.5x^2 + 1)(3x - 2x^2)$, $x = 0.5$

29. $y = \dfrac{3x - 5}{2x + 3}$, $x = -2$

30. $y = \dfrac{2x^2 - 5x}{3x + 2}$, $x = 2$

31. $S = \dfrac{2n^3 - 3n + 8}{2n - 3n^4}$, $n = -1$

32. $y = \dfrac{2x^3 - x^2 - 2}{4x + 3}$, $x = 0.5$

In Exercises 33–52, solve the given problems by finding the appropriate derivatives.

33. What text equation from Section 23.5 is equivalent to the product rule if one of the functions u and v is a constant?

34. By use of the quotient rule, derive a formula for the derivative of the function $1/v(x)$.

35. Using the product rule, find the point(s) on the curve of $y = (2x^2 - 1)(1 - 4x)$ for which the tangent line is $y = 4x - 1$.

W 36. Do the curves of $y = x^2$ and $y = 1/x^2$ cross at right angles? Explain.

37. Find the derivative of $y = \dfrac{x^2(1 - 2x)}{3x - 7}$ in each of the following two ways. (1) Do not multiply out the numerator before finding the derivative. (2) Multiply out the numerator before finding the derivative. Compare the results.

38. Find the derivative of $y = 4x^2 - \dfrac{1}{x - 1}$ in each of the following two ways. (1) Do not combine the terms over a common denominator before finding the derivative. (2) Combine the terms over a common denominator before finding the derivative. Compare the results.

39. Find the slope of a line tangent to the curve of the function $y = (4x + 1)(x^4 - 1)$ at the point $(-1, 0)$. Do not multiply the factors together before taking the derivative. Use the derivative evaluation feature of a graphing calculator to check your result.

40. Find the slope of a line tangent to the curve of the function $y = (3x + 4)(1 - 4x)$ at the point $(2, -70)$. Do not multiply the factors together before taking the derivative. Use the derivative evaluation feature of a graphing calculator to check your result.

41. For what value(s) of x is the slope of a tangent to the curve of $y = \dfrac{x}{x^2 + 1}$ equal to zero? View the graph on a graphing calculator to verify the values found.

42. Determine the sign of the derivative of the function $y = \dfrac{2x - 1}{1 - x^2}$ for the following values of x: $-2, -1, 0, 1, 2$. Is the slope of a tangent line to this curve ever negative? View the graph on a graphing calculator to verify your conclusion.

43. During each cycle, the vertical displacement s of the end of a robot arm is given by $s = (t^2 - 8t)(2t^2 + t + 1)$, where t is the time. Find the expression for the instantaneous velocity of the end of the robot arm. See Fig. 23.29.

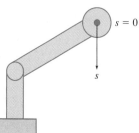

Fig. 23.29

44. The sales S of a product as a function of the time t (in weeks) is given by $S = \dfrac{2300(5t + 1)}{2t + 5}$. Find the instantaneous rate of change of S with respect to t for $t = 9$ weeks.

45. If a constant current of 2 A passes through the *current divider* parallel resistors shown in Fig. 23.30, the current i is given by $i = 8R/(7R + 12)$, where R is a variable resistor. Find di/dR.

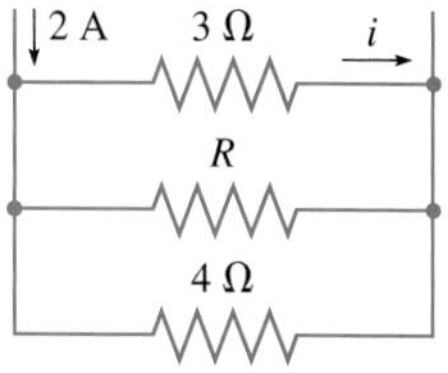

Fig. 23.30

46. The concentration c (in mg/L) of a certain drug in the blood-stream is found to be $c = 25t/(t^2 + 5)$, where t is the time (in h) after the drug is taken. Find di/dt

47. A computer, using data from a refrigeration plant, estimated that in the event of a power failure the temperature T (in °C) in the freezers would be given by $T = \dfrac{2t}{0.05t + 1} - 20$, where t is the number of hours after the power failure. Find the time rate of change of temperature after 6.0 h.

48. The voltage V across a resistor in an electric circuit is the product of the resistance and the current. If the current I (in A) varies with time t (in s) according to the relation $I = 5.00 + 0.01t^2$ and the resistance varies with time according to the relation $R = 15.00 - 0.10t$, find the time rate of change of the voltage when $t = 5.00$ s.

49. The frictional radius r_f of a disc clutch is given by the equation $r_f = \dfrac{2(R^2 + Rr + r^2)}{3(R + r)}$, where R and r are the outer radius and the inner radius of the clutch, respectively. Find the derivative of r_f with respect to R with r constant.

50. In thermodynamics, an equation relating the thermodynamic temperature T, the pressure p, and the volume V of a gas is $T = \left(p + \dfrac{a}{V^2}\right)\left(\dfrac{V - b}{R}\right)$, where a, b, and R are constants. Find the derivative of T with respect to V, assuming p is constant.

51. The electric power P produced by a certain source is given by $P = \dfrac{E^2r}{R^2 + 2Rr + r^2}$, where E is the voltage of the source, R is a resistance of the source, and r is the resistance in the circuit. Find the derivative of P with respect to r, assuming that the other quantities remain constant.

52. In the theory of lasers, the power P radiated is given by the equation $P = \dfrac{kf^2}{\omega^2 - 2\omega f + f^2 + a^2}$, where f is the field frequency and a, k, and ω are constants. Find the derivative of P with respect to f.

23.7 THE DERIVATIVE OF A POWER OF A FUNCTION

In Example 1 of Section 23.6, we illustrated $y = (3 - 2x)^2$ as the power of a function of x, where $3 - 2x$ is the function. If we let $u = 3 - 2x$, we can write $y = u^2$, and in this way y is a function of u, and u is a function of x. This means that y *is a function of a function of x, which is called a* **composite function.**

Since we will often need to find the derivative of a power of a function, which is a composite function, let us look at the derivative of $y = (3 - 2x)^2$.

◀ EXAMPLE 1   Find the derivative of $y = (3 - 2x)^2$.

We will use the product rule by treating $(3 - 2x)^2$ as $(3 - 2x)(3 - 2x)$. The reason for doing it this way will be shown.

$$y = (3 - 2x)^2 = (3 - 2x)(3 - 2x)$$

$$\frac{dy}{dx} = (3 - 2x)(-2) + (3 - 2x)(-2)$$

$$= 2(3 - 2x)(-2)$$

We want to leave the answer in this form in order to compare with a result we can get by letting $y = u^2$ and $u = 3 - 2x$:

$$y = u^2, \qquad \frac{dy}{du} = 2u$$

$$u = 3 - 2x, \qquad \frac{du}{dx} = -2$$

$$\left(\frac{dy}{du}\right)\left(\frac{du}{dx}\right) = 2u(-2) = 2(3 - 2x)(-2)$$

We see that this result is the same as the first result, and therefore for this function we see that $\frac{dy}{dx} = \left(\frac{dy}{du}\right)\left(\frac{du}{dx}\right)$. ▶

As for the function in Example 1, it can be shown that for a differentiable composite function for which y is a function of u and u is a function of x, that

Chain Rule

$$\frac{dy}{dx} = \frac{dy}{du}\frac{du}{dx} \tag{23.14}$$

Equation (23.14) is known as the **chain rule** *for derivatives.*

Using Eq. (23.14) for $y = u^n$, where u is a differentiable function of x, we have

$$\frac{dy}{dx} = \frac{d(u^n)}{du}\frac{du}{dx} \qquad \text{or}$$

Derivative of a Power of a Function of x

$$\frac{du^n}{dx} = nu^{n-1}\left(\frac{du}{dx}\right) \tag{23.15}$$

We use Eq. (23.15) to find the derivative of a power of a differentiable function of x.

◀ **EXAMPLE 2** Find the derivative of $y = (3 - 2x)^3$.

For this function, $n = 3$ and $u = 3 - 2x$. Therefore, $du/dx = -2$. This means

$$\frac{du^n}{dx} = n \quad u \quad ^{n-1} \quad \left(\frac{du}{dx}\right)$$

$$\frac{dy}{dx} = 3(3 - 2x)^2(-2)$$

$$= -6(3 - 2x)^2$$

CAUTION ▶ A common type of error in finding this type of derivative is to omit the du/dx factor; in this case, it is the -2. **The derivative is incomplete and therefore incorrect without this factor.** ▶

◀ **EXAMPLE 3** Find the derivative of $p(x) = (1 - 3x^2)^4$.

In this example, $n = 4$ and $u = 1 - 3x^2$. Hence,

$$p'(x) = 4(1 - 3x^2)^3(-6x) = -24x(1 - 3x^2)^3$$

CAUTION ▶ *(We must not forget the $-6x$.)* ▶

◀ **EXAMPLE 4** Find the derivative of $y = 2x^3(3 - x^3)^4$.

NOTE ▶ Here, we must **use the product rule in combination with the power rule.**

$$\frac{dy}{dx} = 2x^3[4(3 - x^3)^3(-3x^2)] + (3 - x^3)^4[2(3x^2)]$$

$$= -24x^5(3 - x^3)^3 + 6x^2(3 - x^3)^4 = 6x^2(3 - x^3)^3[-4x^3 + (3 - x^3)]$$

$$= 6x^2(3 - 5x^3)(3 - x^3)^3$$ ▶

NOTE ▶ It is better to express the derivative in a factored, simplified form, since this is the form from which useful information may be found. In the next chapter, we will see that an analysis of the derivative has many uses. Therefore, *all derivatives should be in simplest algebraic form.*

To now, we have derived formulas for derivatives of differentiable functions of x raised to positive integral powers. We now show that these formulas are also valid for any rational number used as an exponent. If we raise each side of $y = u^{p/q}$ to the qth power, we have $y^q = u^p$. Applying the power rule, we have

$$qy^{q-1}\left(\frac{dy}{dx}\right) = pu^{p-1}\left(\frac{du}{dx}\right)$$

$$\frac{dy}{dx} = \frac{pu^{p-1}(du/dx)}{qy^{q-1}} = \frac{p}{q}\frac{u^{p-1}}{(u^{p/q})^{q-1}}\frac{du}{dx} = \frac{p}{q}\frac{u^{p-1}}{u^{p-p/q}}\frac{du}{dx}$$

$$= \frac{p}{q}u^{p-1-p+(p/q)}\frac{du}{dx}$$

Thus,

$$\frac{du^{p/q}}{dx} = \frac{p}{q}u^{(p/q)-1}\frac{du}{dx} \tag{23.16}$$

NOTE ▶ We see that in finding the derivative we multiply the function by the rational exponent and subtract 1 from it to find the exponent of the function in the derivative. *This is the same rule as derived for positive integral exponents in Eq. (23.15).*

For reference, Eq. (23.9) is $\dfrac{dx^n}{dx} = nx^{n-1}$.

In deriving Eqs. (23.15) and (23.16), we used Eq. (23.9), and we noted it was valid for positive integral exponents. We can show that Eq. (23.9) is also valid for negative exponents by using the quotient rule on $1/x^n$, which is the same as x^{-n}. Therefore,

the power rule for derivatives, Eq. (23.15), can be extended to include all rational exponents, positive or negative.

This of course includes all integral exponents, positive and negative. Also, we note that Eq. (23.9) is equivalent to Eq. (23.15) with $u = x$ (since $du/dx = 1$).

◖ EXAMPLE 5   We can now find the derivative of $y = \sqrt{x^2 + 1}$.

By using Eq. (23.16) or Eq. (23.15) and writing the square root as the fractional exponent $1/2$, we can derive the result:

$$y = (x^2 + 1)^{1/2}$$

$$\frac{dy}{dx} = \frac{1}{2}(x^2 + 1)^{-1/2}(2x)$$

$$= \frac{x}{(x^2 + 1)^{1/2}}$$

Note that we first rewrite the function in a different, more useful form. This is often an important step before taking the derivative.

To avoid introducing apparently significant factors into the numerator, we do not usually rationalize such fractions. ◗

Having shown that we may use fractional exponents to find derivatives of roots of functions of x, we may also use them to find derivatives of roots of x itself. Consider the following example.

◖ EXAMPLE 6   Find the derivative of $y = 6\sqrt[3]{x^2}$.

We can write this function as $y = 6x^{2/3}$. In finding the derivative, we may use Eq. (23.9) with $n = \frac{2}{3}$. This gives us

$$y = 6x^{2/3}$$

$$\frac{dy}{dx} = 6\left(\frac{2}{3}\right)x^{-1/3} = \frac{4}{x^{1/3}} \qquad \left[\frac{2}{3} - 1\right.$$

We could also use Eq. (23.15) with $u = x$ and $n = \frac{2}{3}$. This give us

$$\frac{dy}{dx} = 6\left(\frac{2}{3}\right)x^{-1/3}(1) = \frac{4}{x^{1/3}} \qquad \left[\frac{du}{dx} = \frac{dx}{dx} = 1\right.$$

This shows us why Eq. (23.9) is equivalent to Eq. (23.15) with $u = x$.

Note that the domain of the function is all real numbers, but the function is not differentiable for $x = 0$. ◗

In the following examples, we illustrate the use of Eqs. (23.9) and (23.15) for the case in which n is a negative exponent. *Special care must be taken in the case of a negative exponent,* so carefully note the caution in each example.

Solving a Word Problem

◀ EXAMPLE 7   The electric resistance R of a wire varies inversely as the square of its radius r. For a given wire, $R = 4.66\ \Omega$ for $r = 0.150$ mm. Find the derivative of R with respect to r for this wire.

Since R varies inversely as the square of r, we have $R = k/r^2$. Then, using the fact that $R = 4.66\ \Omega$ for $r = 0.150$ mm, we have

$$4.66 = \frac{k}{(0.150)^2}, \qquad k = 0.150\ \Omega \cdot mm^2$$

which means that

$$R = \frac{0.105}{r^2}$$

We could find the derivative by the quotient rule. However, when the numerator is constant, the derivative is easily found by using negative exponents:

$$R = \frac{0.105}{r^2} = 0.105 r^{-2}$$

CAUTION ▶

$$\frac{dR}{dr} = 0.105(-2)r^{-3} \longleftarrow -2 - 1 = -3$$

$$= -\frac{0.210}{r^3}$$

Here we used Eq. (23.9) directly. ▶

◀ EXAMPLE 8   Find the derivative of $y = \dfrac{1}{(1 - 4x)^5}$.

The derivative is found as follows:

$$y = \frac{1}{(1 - 4x)^5} = (1 - 4x)^{-5} \qquad \text{use negative exponent}$$

$$\frac{dy}{dx} = (-5)(1 - 4x)^{-6}(-4) \qquad \text{use Eq. (23.15)}$$

$$= \frac{20}{(1 - 4x)^6} \qquad \text{express result with positive exponent}$$

CAUTION ▶   Remember: *Subtracting 1 from −5 gives −6.* ▶

We now see the value of fractional exponents in calculus. They are useful in many algebraic operations, but they are almost essential in calculus. Without fractional exponents, it would be necessary to develop additional formulas to find the derivatives of radical expressions. In order to find the derivative of an algebraic function, we need only those formulas we have already developed. Often it is necessary to combine these formulas, as we saw in Example 4. Actually, most derivatives are combinations. The

NOTE ▶   problem in finding the derivative is *recognizing the form of the function* with which you are dealing. When you have recognized the form, completing the problem is only a matter of mechanics and algebra. You should now see the importance of being able to handle algebraic operations with ease.

EXAMPLE 9   Evaluate the derivative of $y = \dfrac{x}{\sqrt{1-4x}}$ for $x = -2$.

Here, we have a quotient, and in order to find the derivative of this quotient, we must also use the power rule (and a derivative of a polynomial form). With sufficient practice in taking derivatives, we can recognize the rule to use almost automatically. Thus, we find the derivative:

$$\frac{dy}{dx} = \frac{(1-4x)^{1/2}(1) - x\left(\frac{1}{2}\right)(1-4x)^{-1/2}(-4)}{1-4x}$$

$$= \frac{(1-4x)^{1/2} + \dfrac{2x}{(1-4x)^{1/2}}}{1-4x} = \frac{\dfrac{(1-4x)^{1/2}(1-4x)^{1/2} + 2x}{(1-4x)^{1/2}}}{1-4x}$$

$$= \frac{(1-4x) + 2x}{(1-4x)^{1/2}(1-4x)}$$

$$= \frac{1-2x}{(1-4x)^{3/2}}$$

Now, evaluating the derivative for $x = -2$, we have

$$\left.\frac{dy}{dx}\right|_{x=-2} = \frac{1-2(-2)}{[1-4(-2)]^{3/2}} = \frac{1+4}{(1+8)^{3/2}} = \frac{5}{9^{3/2}}$$

$$= \frac{5}{27}$$

EXERCISES 23.7

In Exercises 1–4, make the given changes in the indicated examples of this section and then find the derivatives.

1. In Example 3, change $1 - 3x^2$ to $2 + 3x^3$.

2. In Example 4, change $3 - x^3$ to $2 + x^5$.

3. In Example 5, change $x^2 + 1$ to $2 - 3x^2$.

4. In Example 8, change the exponent 5 to 3.

In Exercises 5–32, find the derivative of each of the given functions.

5. $y = \sqrt{x}$

6. $y = \sqrt[4]{x^3}$

7. $v = \dfrac{3}{t^2}$

8. $y = \dfrac{2}{x^4}$

9. $y = \dfrac{3}{\sqrt[3]{x}}$

10. $y = \dfrac{55}{\sqrt[5]{x^2}}$

11. $y = x\sqrt{x} - \dfrac{1}{x}$

12. $f(x) = 2x^{-3} - 3x^{-2}$

13. $y = (x^2 + 1)^5$

14. $y = (1 - 2x)^4$

15. $y = 2.25(7 - 4x^3)^8$

16. $y = 3(8x^2 - 1)^6$

17. $y = (2x^3 - 3)^{1/3}$

18. $y = (1 - 6x)^{1.5}$

19. $f(y) = \dfrac{3}{(4 - y^2)^4}$

20. $y = \dfrac{\pi^3}{\sqrt{1 - 3x}}$

21. $y = 4(2x^4 - 5)^{0.75}$

22. $r = 5(3\theta^6 - 4)^{2/3}$

23. $y = \sqrt[4]{1 - 8x^2}$

24. $y = \sqrt[3]{4x^6 + 2}$

25. $y = x\sqrt{8x + 5}$

26. $y = x^2(1 - 3x)^5$

27. $y = \dfrac{2\sqrt{1 - 6x}}{x^3}$

28. $R = \dfrac{2T^2}{\sqrt[3]{1 + 4T}}$

29. $y = \dfrac{2x\sqrt{x + 2}}{x + 4}$

30. $y = 8\sqrt{1 + \sqrt{x}}$

31. $f(R) = \sqrt{\dfrac{2R + 1}{4R + 1}}$

32. $y = \left(\dfrac{2x + 1}{3x - 2}\right)^2$

In Exercises 33–36, evaluate the derivatives of the given functions for the given values of x. In Exercises 35 and 36, check your results, using the derivative evaluation feature of a graphing calculator.

33. $y = \sqrt{3x + 4}$, $x = 7$

34. $y = (4 - x^2)^{-1}$, $x = -1$

35. $y = \dfrac{\sqrt{x}}{1 - x}$, $x = 4$

36. $y = x^2\sqrt[3]{3x + 2}$, $x = 2$

In Exercises 37–56, solve the given problems by finding the appropriate derivatives.

37. Find the derivative of $y = 1/x^3$ as (a) a quotient and (b) a negative power of x and show that the results are the same.

38. Let $y = [u(x)]^2$ and find dy/dx, treating $[u(x)]^2$ as the product $u(x)u(x)$. (See Example 1.)

39. Find any values of x for which the derivative of $y = \dfrac{x^2}{\sqrt{x^2 + 1}}$ is zero. View the curve of the function on a graphing calculator to verify the values found.

40. Find any values of x for which the derivative of $y = \dfrac{x}{\sqrt{4x - 1}}$ is zero. View the curve of the function on a graphing calculator to verify the values found.

41. Is the line $x + 3y - 12 = 0$ ever perpendicular to a tangent to the graph of $y = \sqrt{2x + 3}$?

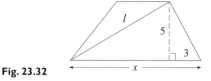**42.** Explain why the graph of $y_1 = \sqrt{x} + a$ cannot be tangent to the graph of $y_2 = \sqrt{4 - 2x}$, regardless of the value of a.

43. Find the slope of a line tangent to the parabola $y^2 = 4x$ at the point $(1, 2)$. Use the derivative evaluation feature of a graphing calculator to check your result.

44. Find the slope of a line tangent to the circle $x^2 + y^2 = 25$ at the point $(4, 3)$. Use the derivative evaluation feature of a graphing calculator to check your result.

45. The displacement s (in cm) of a linkage joint of a robot is given by $s = (8t - t^2)^{2/3}$, where t is the time (in s). Find the velocity of the joint for $t = 6.25$ s.

46. Water is slowly rising in a horizontal drainage pipe. The width w of the water as a function of the depth h is $w = \sqrt{2rh - h^2}$, where r is the radius of the pipe. Find dw/dh for $h = 225$ mm and $r = 600$ mm.

47. When the volume of a gas changes very rapidly, an approximate relation is that the pressure P varies inversely as the $3/2$ power of the volume. If P is 300 kPa when $V = 100$ cm³, find the derivative of P with respect to V. Evaluate this derivative for $V = 100$ cm³.

48. The power gain G of a certain antenna is inversely proportional to the square of the wavelength λ (in m) of the carrier wave. If $G = 5.0 \times 10^4$ for $\lambda = 0.11$ m, find the derivative of G with respect to λ for $\lambda = 0.11$ m.

49. In deep water, the velocity of a wave is $v = k\sqrt{\dfrac{l}{a} + \dfrac{a}{l}}$, where a and k are constants and l is the length of the wave. For what value of l is $dv/dl = 0$?

50. Due to air friction, the drag F on a plane is $F = c_1 v^2 + c_2 v^{-2}$, where v is the plane's velocity and c_1 and c_2 are positive constants. For what values of v is $dF/dv = 0$?

51. The total solar radiation H (in W/m²) on a certain surface during an average clear day is given by
$$H = \frac{4000}{\sqrt{t^6 + 100}} \qquad (-6 < t < 6)$$
where t is the number of hours from noon. Find the rate at which H is changing with time at 4 P.M.

52. In determining the time for a laser beam to go from S to P (see Fig. 23.31), which are in different mediums, it is necessary to find the derivative of the time
$$t = \frac{\sqrt{a^2 + x^2}}{v_1} + \frac{\sqrt{b^2 + (c - x)^2}}{v_2}$$
with respect to x, where a, b, c, v_1, and v_2 are constants. Here, v_1, and v_2 are the velocities of the laser in each medium. Find this derivative.

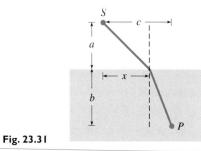

Fig. 23.31

53. The radio waveguide wavelength λ_r is related to its free-space wavelength λ by
$$\lambda_r = \frac{2a\lambda}{\sqrt{4a^2 - \lambda^2}}$$
where a is a constant. Find $d\lambda_r/d\lambda$.

54. The current I in a circuit containing a resistance R and an inductance L is found from the expression
$$I = \frac{V}{\sqrt{R^2 + (\omega L)^2}}$$
Find the expression for the instantaneous rate of change of current with respect to L, assuming that the other quantities remain constant.

55. The length l of a rectangular microprocessor chip is 2 mm longer than its width w. Find the derivative of the length of the diagonal D with respect to w.

56. The trapezoidal structure shown in Fig. 23.32 has an internal support of length l. Find the derivative of l with respect to x.

Fig. 23.32

23.8 DIFFERENTIATION OF IMPLICIT FUNCTIONS

To this point, the functions we have differentiated have been of the form $y = f(x)$. There are, however, occasions when we need to find the derivative of a function determined by an equation that does not express the dependent variable explicitly in terms of the independent variable.

An equation in which y is not expressed explicitly in terms of x may determine one or more functions. *Any such function, where y is defined implicitly as a function of x, is called an* **implicit function.** Some equations defining implicit functions may be solved to determine the explicit functions, and for others it is not possible to solve for the explicit functions. Also, not all such equations define y as a function of x for real values of x.

⟪ EXAMPLE 1   (a) The equation $3x + 4y = 5$ is an equation that defines a function, although it is not in explicit form. In solving for y as $y = -\frac{3}{4}x + \frac{5}{4}$, we have the explicit form of the function.

(b) The equation $y^2 + x = 3$ is an equation that defines two functions, although we do not have the explicit forms. When we solve for y, we obtain the explicit functions $y = \sqrt{3 - x}$ and $y = -\sqrt{3 - x}$.

(c) The equation $y^5 + xy^2 + 3x^2 = 5$ defines y as a function of x, although we cannot actually solve for the explicit algebraic form of the function.

(d) The equation $x^2 + y^2 + 4 = 0$ is not satisfied by any pair of real values of x and y. ▶

Even when it is possible to determine the explicit form of a function given in implicit form, it is not always desirable to do so. In some cases, the implicit form is more convenient than the explicit form.

The derivative of an implicit function may be found directly without having to solve

NOTE ▶ for the explicit function. Thus, *to find dy/dx when y is defined as an implicit function of x, we differentiate each term of the equation with respect to x, regarding y as a differentiable function of x. We then solve for dy/dx, which will usually be in terms of x and y.*

⟪ EXAMPLE 2   Find dy/dx if $y^2 + 2x^2 = 5$.

Here, we find the derivative of each term and then solve for dy/dx. Thus,

$$\frac{d(y^2)}{dx} + \frac{d(2x^2)}{dx} = \frac{d(5)}{dx}$$

$$2y^{2-1}\frac{dy}{dx} + 2\left(2x^{2-1}\frac{dx}{dx}\right) = 0$$

$$2y\frac{dy}{dx} + 4x = 0$$

$$\frac{dy}{dx} = -\frac{2x}{y}$$

For reference, Eq. (23.15) is
$$\frac{du^n}{dx} = nu^{n-1}\frac{du}{dx}.$$

CAUTION ▶ *The factor dy/dx arises from the derivative of the first term as a result of using the derivative of a power of a function of x (Eq. 23.15). The factor dy/dx corresponds to the du/dx of the formula.* In the second term, no factor of dy/dx appears, since there are no y factors in the term. ▶

◀ EXAMPLE 3   Find dy/dx if $3y^4 + xy^2 + 2x^3 - 6 = 0$.

In finding the derivative, we note that the second term is a product, and we must use the product rule for derivatives on it. Thus, we have

$$\frac{d(3y^4)}{dx} + \frac{d(xy^2)}{dx} + \frac{d(2x^3)}{dx} - \frac{d(6)}{dx} = \frac{d(0)}{dx}$$

using product rule

$$12y^3\frac{dy}{dx} + \left[x\left(2y\frac{dy}{dx}\right) + y^2(1)\right] + 6x^2 - \quad 0 = 0$$

$$12y^3\frac{dy}{dx} + 2xy\frac{dy}{dx} + y^2 + 6x^2 = 0 \qquad \text{solve for } \frac{dy}{dx}$$

$$(12y^3 + 2xy)\frac{dy}{dx} = -y^2 - 6x^2$$

$$\frac{dy}{dx} = \frac{-y^2 - 6x^2}{12y^3 + 2xy}$$

◀ EXAMPLE 4   Find dy/dx if $2x^3y + (y^2 + x)^3 = x^4$.

In this case, we use the product rule on the first term and the power rule on the second term:

$$\frac{d(2x^3y)}{dx} + \frac{d(y^2 + x)^3}{dx} = \frac{d(x^4)}{dx}$$

product   power

$$2x^3\left(\frac{dy}{dx}\right) + y(6x^2) + 3(y^2 + x)^2\left(2y\frac{dy}{dx} + 1\right) = 4x^3$$

$$2x^3\frac{dy}{dx} + 6x^2y + 3(y^2 + x)^2\left(2y\frac{dy}{dx}\right) + 3(y^2 + x)^2 = 4x^3$$

$$[2x^3 + 6y(y^2 + x)^2]\frac{dy}{dx} = 4x^3 - 6x^2y - 3(y^2 + x)^2$$

$$\frac{dy}{dx} = \frac{4x^3 - 6x^2y - 3(y^2 + x)^2}{2x^3 + 6y(y^2 + x)^2}$$

◀ EXAMPLE 5   Find the slope of a line tangent to the curve of $2y^3 + xy + 1 = 0$ at the point $(-3, 1)$.

Here, we just find dy/dx and evaluate it for $x = -3$ and $y = 1$:

$$\frac{d(2y^3)}{dx} + \frac{d(xy)}{dx} + \frac{d(1)}{dx} = \frac{d(0)}{dx}$$

$$6y^2\frac{dy}{dx} + x\frac{dy}{dx} + y + 0 = 0$$

$$\frac{dy}{dx} = \frac{-y}{6y^2 + x}$$

$$\frac{dy}{dx}\bigg|_{(-3, 1)} = \frac{-1}{6(1^2) - 3} = \frac{-1}{6 - 3} = -\frac{1}{3}$$

Thus, the slope is $-\frac{1}{3}$.

In Exercises 1 and 2, make the given changes in the indicated examples of this section and then find dy/dx.

1. In Example 2, change y^2 to y^3.

2. In Example 3, change xy^2 to x^2y.

In Exercises 3–22, find dy/dx by differentiating implicitly. When applicable, express the result in terms of x and y.

3. $3x + 2y = 5$

4. $6x - 3y = 4$

5. $4y - 3x^2 = x$

6. $x^5 - 5y = 6 - x$

7. $x^2 - 4y^2 - 9 = 0$

8. $x^2 + 2y^2 - 11 = 0$

9. $y^5 = x^2 - 1$

10. $y^4 = 3x^3 - x$

11. $y^2 + y = x^2 - 4$

12. $2y^3 - y = 7 - x^4$

13. $y + 3xy - 4 = 0$

14. $8y - xy - 7 = 0$

15. $xy^3 + 3y + x^2 = 2\pi^2$

16. $y^2x - \dfrac{5y}{x+1} + 3x = 4$

17. $\dfrac{3x^2}{y^2+1} + y = 3x + 1$

18. $2x - x^3y^2 = y - x^2 - 1$

19. $(2y - x)^4 + x^2 = y + 3$

20. $(y^2 + 2)^3 = x^4y + e^2$

21. $2(x^2 + 1)^3 + (y^2 + 1)^2 = 17$

22. $(2x + 1)(1 - 3y) + y^2 = 13$

In Exercises 23–28, evaluate the derivatives of the given functions at the given points.

23. $3x^3y^2 - 2y^3 = -4;\quad (1, 2)$

24. $2y + 5 - x^2 - y^3 = 0;\quad (2, -1)$

25. $5y^4 + 7 = x^4 - 3y;\quad (3, -2)$

26. $(xy - y^2)^3 = 5y^2 + 22;\quad (4, 1)$

27. $xy^2 + 3x^2 - y^2 + 15 = 0;\quad (-1, 3)$

28. $2(x + y)^3 - y^2/x = 15;\quad (4, -2)$

In Exercises 29–40, solve the given problems by using implicit differentiation.

29. At what point(s) does the graph of $x^2 + y^2 = 4x$ have a horizontal tangent?

30. Show that if $P(x, y)$ is any point on the circle $x^2 + y^2 = a^2$, then a tangent line at P is perpendicular to a line through P and the origin.

31. Find the slope of a line tangent to the curve of the implicit function $xy + y^2 + 2 = 0$ at the point $(-3, 1)$. Use the derivative evaluation feature of a graphing calculator to check your result.

32. Show that the graphs of $2x^2 + y^2 = 24$ and $y^2 = 8x$ are perpendicular at the point $(2, 4)$. Display the graphs on a graphing calculator.

33. The pressure P, volume V, and temperature T of a gas are related by $PV = n(RT + aP - bP/T)$, where a, b, n, and R are constants. For constant V, find dP/dT.

34. Oil moves through a pipeline such that the distance s it moves and the time t are related by $s^3 - t^2 = 7t$. Find the velocity of the oil for $s = 4.01$ m and $t = 5.25$ s.

35. The shelf support shown in Fig. 23.33 is 0.75 m long. Find the expression for dy/dx in terms of x and y.

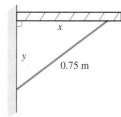

36. An open (no top) right circular cylindrical container of radius r and height h has a total surface area of 940 cm². Find dr/dh in terms of r and h.

37. Two resistors, with resistances r and $r + 2$, are connected in parallel. Their combined resistance R is related to r by the equation $r^2 = 2rR + 2R - 2r$. Find dR/dr.

38. The polar moment of inertia I of a rectangular slab of concrete is given by $I = \frac{1}{12}(b^3h + bh^3)$, where b and h are the base and the height, respectively, of the slab. If I is constant, find the expression for db/dh.

39. A formula relating the length L and radius of gyration r of a steel column is $24C^3Sr^3 = 40C^3r^3 + 9LC^2r^2 - 3L^3$, where C and S are constants. Find dL/dr.

40. A computer is programmed to draw the graph of the implicit function $(x^2 + y^2)^3 = 64x^2y^2$ (see Fig. 23.34 and Example 7 on page 598.) Find the slope of a line tangent to this curve at $(2.00, 0.56)$ and at $(2.00, 3.07)$.

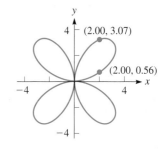

Fig. 23.34

Fig. 23.33

23.9 HIGHER DERIVATIVES

Earlier we noted that the derivative of a function is itself a function. Therefore, we may take its derivative. In this section, we develop the concept and notation for the derivatives of a derivative, as well as show some of the applications.

The Second Derivative

*The derivative of a function is called the **first derivative** of the function. The derivative of the first derivative is called the **second derivative.** Since the second derivative is a function, we may find its derivative, which is called the third derivative. We may continue to find the fourth derivative, fifth derivative, and so on (provided each derivative is defined). The second derivative, third derivative, and so on, are known as **higher derivatives.***

The notations used for higher derivatives follow closely those used for the first derivative. As shown in Section 23.3, the notations for the first derivative are y', $D_x y$, $f'(x)$, and dy/dx. The notations for the second derivative are y'', $D_x^2 y$, $f''(x)$, and d^2y/dx^2. Similar notations are used for other higher derivatives.

◖ EXAMPLE 1   Find the higher derivatives of $y = 5x^3 - 2x$.
We find the first derivative as

$$\frac{dy}{dx} = 15x^2 - 2 \quad \text{or} \quad y' = 15x^2 - 2$$

Next, we obtain the second derivative by finding the derivative of the first derivative:

$$\frac{d^2y}{dx^2} = 30x \quad \text{or} \quad y'' = 30x$$

Continuing to find the successive derivatives, we have

$$\frac{d^3y}{dx^3} = 30 \quad \text{or} \quad y''' = 30$$

$$\frac{d^4y}{dx^4} = 0 \quad \text{or} \quad y^{(4)} = 0$$

Since the third derivative is a constant, the fourth derivative and all successive derivatives will be zero. This can be shown as $d^n y/dx^n = 0$ for $n \geq 4$. ◗

For reference, Eq. (23.12) is
$$\frac{d(uv)}{dx} = u\frac{dv}{dx} + v\frac{du}{dx}.$$

◖ EXAMPLE 2   Find the higher derivatives of $f(x) = x(x^2 - 1)^2$.
Using the product rule, Eq. (23.12), to find the first derivative, we have

$$f'(x) = x(2)(x^2 - 1)(2x) + (x^2 - 1)^2(1)$$
$$= (x^2 - 1)(4x^2 + x^2 - 1) = (x^2 - 1)(5x^2 - 1)$$
$$= 5x^4 - 6x^2 + 1$$

Continuing to find the higher derivatives, we have

$$f''(x) = 20x^3 - 12x$$
$$f'''(x) = 60x^2 - 12$$
$$f^{(4)}(x) = 120x$$
$$f^{(5)}(x) = 120$$
$$f^{(n)}(x) = 0 \qquad \text{for } n \geq 6$$

Note that when using the prime ($f'(x)$) notation the nth derivative may be shown as $f^{(n)}(x)$.

All derivatives after the fifth derivative are equal to zero. ◗

EXAMPLE 3 Evaluate the second derivative of $y = \dfrac{2}{1 - x}$ for $x = -2$.

We write the function as $y = 2(1 - x)^{-1}$ and then find the derivatives:

$$y = 2(1 - x)^{-1}$$

$$\frac{dy}{dx} = 2(-1)(1 - x)^{-2}(-1) = 2(1 - x)^{-2}$$

$$\frac{d^2y}{dx^2} = 2(-2)(1 - x)^{-3}(-1) = 4(1 - x)^{-3} = \frac{4}{(1 - x)^3}$$

Evaluating the second derivative for $x = -2$, we have

$$\left.\frac{d^2y}{dx^2}\right|_{x=-2} = \frac{4}{(1 + 2)^3} = \frac{4}{27}$$

The function is not differentiable for $x = 1$. Also, if we continue to find higher derivatives, the expressions will not become zero, as in Examples 1 and 2.

EXAMPLE 4 Find y'' for the implicit function defined by $2x^2 + 3y^2 = 6$.

Differentiating with respect to x, we have

$$2(2x) + 3(2yy') = 0$$

$$4x + 6yy' = 0 \quad \text{or} \quad 2x + 3yy' = 0 \tag{1}$$

CAUTION ▶ Before differentiating again, we see that **3yy' is a product,** and we note that the derivative of y' is y''. Thus, differentiating again, we have

differentiation of $3yy'$

$$2 + \overline{3yy'' + 3y'(y')} = 0$$

$$2 + 3yy'' + 3(y')^2 = 0 \tag{2}$$

Now, solving Eq. (1) for y' and substituting this into Eq. (2), we have

$$y' = -\frac{2x}{3y}$$

$$2 + 3yy'' + 3\left(-\frac{2x}{3y}\right)^2 = 0$$

$$2 + 3yy'' + \frac{4x^2}{3y^2} = 0$$

$$6y^2 + 9y^3y'' + 4x^2 = 0$$

$$y'' = \frac{-4x^2 - 6y^2}{9y^3} = \frac{-2(2x^2 + 3y^2)}{9y^3}$$

Since $2x^2 + 3y^2 = 6$, we have

$$y'' = \frac{-2(6)}{9y^3} = -\frac{4}{3y^3}$$

As mentioned earlier, higher derivatives are useful in certain applications. This is particularly true of the second derivative. The first and second derivatives are used in the next chapter for several types of applications, and higher derivatives are used when we discuss infinite series in Chapter 29. An important technical application of the second derivative is shown in the example that follows.

In Section 23.4, we briefly discussed the instantaneous velocity of an object, and in the exercises we mentioned acceleration. From that discussion, recall that the instantaneous velocity is the time rate of change of the displacement, and *the* **instantaneous acceleration** *is the time rate of change of the instantaneous velocity.* Therefore, we see that the *acceleration is found from the second derivative of the displacement with respect to time.* Consider the following example.

Solving a Word Problem

❰ **EXAMPLE 5** For the first 12 s after launch, the height s (in m) of a certain rocket is given by $s = 10\sqrt{t^4 + 25} - 50$. Find the vertical acceleration of the rocket when $t = 10.0$ s.

Since the velocity is found from the first derivative and the acceleration is found from the second derivative, we must find the second derivative and then evaluate it for $t = 10.0$ s.

$$s = 10\sqrt{t^4 + 25} - 50$$

$$v = \frac{ds}{dt} = 10\left(\frac{1}{2}\right)(t^4 + 25)^{-1/2}(4t^3) = \frac{20t^3}{(t^4 + 25)^{1/2}}$$

$$a = \frac{dv}{dt} = \frac{d^2s}{dt^2} = \frac{(t^4 + 25)^{1/2}(60t^2) - 20t^3(\frac{1}{2})(t^4 + 25)^{-1/2}(4t^3)}{t^4 + 25} \longleftarrow$$

multiply numerator and denominator by $(t^4 + 25)^{1/2}$

$$= \frac{(t^4 + 25)(60t^2) - 40t^6}{(t^4 + 25)^{3/2}} = \frac{20t^6 + 1500t^2}{(t^4 + 25)^{3/2}}$$

$$= \frac{20t^2(t^4 + 75)}{(t^4 + 25)^{3/2}}$$

Finding the value of the acceleration when $t = 10.0$ s, we have

$$a\big|_{t=10.0} = \frac{20(10.0)^2(10.0^4 + 75)}{(10.0^4 + 25)^{3/2}} = 20.1 \text{ m/s}^2$$

❱

EXERCISES 23.9

In Exercises 1 and 2, make the given changes in the indicated examples of this section and then solve the resulting problem.

1. In Example 1, change $2x$ to $2x^2$.

2. In Example 3, in the denominator change $1 - x$ to $1 + 2x$.

In Exercises 3–10, find all the higher derivatives of the given functions.

3. $y = x^3 + x^2$

4. $f(x) = 3x - x^4$

5. $f(x) = x^3 - 6x^4$

6. $s = 2t^5 + 5t^4$

7. $y = (1 - 2x)^4$

8. $f(x) = (3x + 2)^3$

9. $f(r) = r(4r + 1)^3$

10. $y = x(x - 1)^3$

In Exercises 11–30, find the second derivative of each of the given functions.

11. $y = 2x^7 - x^6 - 3x$

12. $y = 6x - 2x^5$

13. $y = 2x + \sqrt{x}$

14. $r = 3\theta^2 - \dfrac{1}{2\sqrt{\theta}}$

15. $f(x) = \sqrt[4]{8x - 3}$

16. $f(x) = \sqrt[3]{6x + 5}$

17. $f(p) = \dfrac{4.8\pi}{\sqrt{1 + 2p}}$

18. $f(x) = \dfrac{7.5}{\sqrt{3 - 4x}}$

19. $y = 2(2 - 5x)^4$

20. $y = (4x + 1)^6$

21. $y = (3x^2 - 1)^5$

22. $y = 3(2x^3 + 3)^4$

23. $f(x) = \dfrac{2\pi^2}{1 - x}$

24. $f(R) = \dfrac{1 - 3R}{1 + 3R}$

25. $y = \dfrac{x^2}{x + 1}$

26. $y = \dfrac{x}{\sqrt{1 - x^2}}$

27. $x^2 - y^2 = 9$

28. $xy + y^2 = 4$

29. $x^2 - xy = 1 - y^2$

30. $xy = y^2 + 2e^3$

In Exercises 31–36, evaluate the second derivative of the given function for the given value of x.

31. $f(x) = \sqrt{x^2 + 9}, x = 4$

32. $f(x) = x - \dfrac{2}{x^3}, x = -1$

33. $y = 3x^{2/3} - \dfrac{2}{x}, x = -8$　　**34.** $y = 3(1 + 2x)^4, x = \dfrac{1}{2}$

35. $y = x(1 - x)^5, x = 2$　　**36.** $y = \dfrac{x}{2 - 3x}, x = -\dfrac{1}{3}$

In Exercises 37–40, find the acceleration of an object for which the displacement s (in m) is given as a function of the time t (in s) for the given value of t.

37. $s = 26t - 4.9t^2, t = 3.0$ s

38. $s = 3(1 + 2t)^4, t = 0.500$ s

39. $s = \dfrac{16}{0.5t + 1}, t = 2$ s

40. $s = 250\sqrt{6t + 1}, t = 4.0$ s

In Exercises 41–48, solve the given problems by finding the appropriate derivatives.

41. What is the instantaneous rate of change of the first derivative of y with respect to x for $y = (1 - 2x)^4$ for $x = 1$?

42. What is the instantaneous rate of change of the first derivative of y with respect to x for $2xy + y = 1$ for $x = 0.5$?

43. A bullet is fired vertically upward. Its distance s (in m) above the ground is given by $s = 670t - 4.9t^2$, where t is the time (in s). Find the acceleration of the bullet.

44. In testing the brakes on a new model automobile, it was found that the distance s (in m) that it traveled under specified conditions after the brakes were applied was given by $s = 19.2t - 0.40t^3$. What were the velocity and the acceleration of the automobile for $t = 4.00$ s?

45. The voltage V induced in an inductor in an electric circuit is given by $V = L(d^2q/dt^2)$, where L is the inductance (in H). Find the expression for the voltage induced in a 1.60-H inductor if $q = \sqrt{2t + 1} - 1$.

46. How fast is the rate of change of solar radiation changing on the surface in Exercise 33 of Section 23.4 at 3 P.M.?

47. The deflection y (in m) of a 5.00-m beam as a function of the distance x (in m) from one end is $y = 0.0001(x^5 - 25x^2)$. Find the value of d^2y/dx^2 (the rate of change at which the slope of the beam changes) where $x = 3.00$ m.

48. The force F (in N) acting on an object is given by $F = 12\, dv/dt + 2.0v + 5.0$, where v is the velocity (in m/s) and t is the time (in s). If the displacement is given by $s = 25t^{0.60}$, find F for $t = 3.5$ s.

CHAPTER 23 EQUATIONS

Limit of function	$\displaystyle \lim_{x \to a} f(x) = L$	(23.1)
Difference in *x*-coordinates	$h = x_2 - x_1$	(23.2)
	$x_2 = x_1 + h$	(23.3)
Slope	$m_{PQ} = \dfrac{f(x_1 + h) - f(x_1)}{(x_1 + h) - x_1} = \dfrac{f(x_1 + h) - f(x_1)}{h}$	(23.4)
	$m_{\tan} = \displaystyle\lim_{h \to 0} \dfrac{f(x_1 + h) - f(x_1)}{h}$	(23.5)
Definition of derivative	$f'(x) = \displaystyle\lim_{h \to 0} \dfrac{f(x + h) - f(x)}{h}$	(23.6)
Instantaneous velocity	$v = \displaystyle\lim_{h \to 0} \dfrac{s(t + h) - s(t)}{h}$	(23.7)
Derivatives of polynomials	$\dfrac{dc}{dx} = 0$	(23.8)
	$\dfrac{dx^n}{dx} = nx^{n-1}$	(23.9)
	$\dfrac{d(cu)}{dx} = c\dfrac{du}{dx}$	(23.10)
	$\dfrac{d(u + v)}{dx} = \dfrac{du}{dx} + \dfrac{dv}{dx}$	(23.11)

Derivative of product
$$\frac{d(uv)}{dx} = u\frac{dv}{dx} + v\frac{du}{dx}$$
(23.12)

Derivative of quotient
$$\frac{d\frac{u}{v}}{dx} = \frac{v\frac{du}{dx} - u\frac{dv}{dx}}{v^2}$$
(23.13)

Chain rule
$$\frac{dy}{dx} = \frac{dy}{du}\frac{du}{dx}$$
(23.14)

Derivative of power
$$\frac{du^n}{dx} = nu^{n-1}\left(\frac{du}{dx}\right)$$
(23.15)

$$\frac{du^{p/q}}{dx} = \frac{p}{q}u^{(p/q)-1}\frac{du}{dx}$$
(23.16)

CHAPTER 23 REVIEW EXERCISES

In Exercises 1–12, evaluate the given limits.

1. $\lim\limits_{x \to 4}(8 - 3x)$

2. $\lim\limits_{x \to 3}(2x^2 - 10)$

3. $\lim\limits_{x \to -2}\dfrac{|x + 2|}{x + 2}$

4. $\lim\limits_{x \to 1}(x - 1)\sqrt{x^2 + 9}$

5. $\lim\limits_{x \to 2}\dfrac{4x - 8}{x^2 - 4}$

6. $\lim\limits_{x \to 5}\dfrac{x^2 - 25}{3x - 15}$

7. $\lim\limits_{x \to 2}\dfrac{x^2 + 3x - 10}{x^2 - x - 2}$

8. $\lim\limits_{x \to 0}\dfrac{(x - 3)^2 - 9}{x}$

9. $\lim\limits_{x \to \infty}\dfrac{2 + \dfrac{1}{x + 4}}{3 - \dfrac{1}{x^2}}$

10. $\lim\limits_{x \to \infty}\left(7 - \dfrac{1}{x + 1}\right)$

11. $\lim\limits_{x \to \infty}\dfrac{x - 2x^3}{(1 + x)^3}$

12. $\lim\limits_{x \to \infty}\dfrac{\sqrt{4x^2 + 3}}{x + 5}$

In Exercises 13–20, use the definition to find the derivative of each of the given functions.

13. $y = 7 + 5x$

14. $y = 6x - 2$

15. $y = 6 - 2x^2$

16. $y = 2x^2 - x^3$

17. $y = \dfrac{2}{x^2}$

18. $y = \dfrac{x}{1 - 4x}$

19. $y = \sqrt{x + 5}$

20. $y = \dfrac{1}{\sqrt{x}}$

In Exercises 21–36, find the derivative of each of the given functions.

21. $y = 2x^7 - 3x^2 + 5$

22. $y = 8x^7 - 2^5 - x$

23. $y = 4\sqrt{x} - \dfrac{3}{x} + \sqrt{3}$

24. $R = \dfrac{3}{T^2} - 8\sqrt[4]{T}$

25. $f(y) = \dfrac{3y}{1 - 5y}$

26. $y = \dfrac{2x - 1}{x^2 + 1}$

27. $y = (2 - 3x)^4$

28. $y = (2x^2 - 3)^6$

29. $y = \dfrac{3\pi}{(5 - 2x^2)^{3/4}}$

30. $f(Q) = \dfrac{70}{(3Q + 1)^3}$

31. $v = \sqrt{1 + \sqrt{1 + \sqrt{1 + 8s}}}$

32. $y = (x - 1)^3(x^2 - 2)^2$

33. $y = \dfrac{\sqrt{4x + 3}}{2x}$

34. $R = \dfrac{\sqrt{t} + 1}{\sqrt{t} - 1}$

35. $(2x - 3y)^3 = x^2 - y$

36. $x^2y^2 = x^2 + y^2$

In Exercises 37–40, evaluate the derivatives of the given functions for the given values of x. Check your results, using the derivative evaluation feature of a graphing calculator.

37. $y = \dfrac{4}{x} + 2\sqrt[3]{x},\ x = 8$

38. $y = (3x - 5)^4,\ x = -2$

39. $y = 2x\sqrt{4x + 1},\ x = 6$

40. $y = \dfrac{\sqrt{2x^2 + 1}}{3x},\ x = 2$

In Exercises 41–44, find the second derivative of each of the given functions.

41. $y = 3x^4 - \dfrac{1}{x}$

42. $y = \sqrt{1 - 8x}$

43. $s = \dfrac{1 - 3t}{1 + 4t}$

44. $y = 2x(6x + 5)^4$

In Exercises 45–80, solve the given problems.

W **45.** View the graph of $y = \dfrac{2(x^2 - 4)}{x - 2}$ on a graphing calculator with *window* values such that y can be evaluated exactly for $x = 2$. (Xmin = −1 (or 0), Xmax = 4, Ymin = 0, Ymax = 10 will probably work.) Using the *trace* feature, determine the value of y for $x = 2$. Comment on the accuracy of the view and the value found.

(W) 46. A continuous function $f(x)$ is positive at $x = 0$ and negative for $x = 1$. How many solutions does $f(x) = 0$ have between $x = 0$ and $x = 1$? Explain.

47. The velocity v (in m/s) of a weight falling in water is given by $v = \dfrac{6(t + 5)}{t + 1}$, where t is the time (in s). What are (a) the initial velocity and (b) the terminal velocity (as $t \to \infty$)?

48. Two lenses of focal lengths f_1 and f_2, separated by a distance d, are used in the study of lasers. The combined focal length f of this lens combination is $f = \dfrac{f_1 f_2}{f_1 + f_2 - d}$. If f_2 and d remain constant, find the limiting value of f as f_1 continues to increase in value.

49. Find the slope of a line tangent to the curve of $y = 7x^4 - x^3$ at $(-1, 8)$. Use the derivative evaluation feature of a graphing calculator to check your result.

50. Find the slope of a line tangent to the curve of $y = \sqrt[3]{3 - 8x}$ at $(-3, 3)$. Use the derivative evaluation feature of a graphing calculator to check your result.

51. Find the point(s) at which a tangent line to the graph of $y = 1/\sqrt{3x^2 + 3}$ is parallel to the x-axis.

52. Find the point(s) on the graph of $y = 2(1 - 3x)^2$ at which a tangent line is parallel to the line $y = -2x + 5$.

53. Find the equations for (a) the velocity and (b) the acceleration if the displacement s (in m) of an object as a function of the time t (in s) is given by $s = \sqrt{1 + 8t}$.

54. Find the values of the velocity and acceleration for the object in Exercise 53 for $t = 3$ s.

55. The cable of a 200-m suspension bridge can be represented by $y = 0.0015x^2 + C$. At one point, the tension is directed along the line $y = 0.3x - 10$. Find the value of C.

56. The displacement s (in cm) of a piston during each 8-s cycle is given by $s = 8t - t^2$, where t is the time (in s). For what value(s) of t is the velocity of the piston 4 cm/s?

57. The reliability R of a computer system measures the probability that the system will be operating properly after t hours. For one system, $R = 1 - kt + \dfrac{k^2 t^2}{2} - \dfrac{k^3 t^3}{6}$, where k is a constant. Find the expression for the instantaneous rate of change of R with respect to t.

58. The distance s (in m) traveled by a subway train after the brakes are applied is given by $s = 20t - 2t^2$, where t is the time (in s). How far does it travel, after the brakes are applied, in coming to a stop?

59. The electric field E at a distance r from a point charge is $E = k/r^2$, where k is a constant. Find an expression for the instantaneous rate of change of the electric field with respect to r.

60. The velocity of an object moving with constant acceleration can be found from the equation $v = \sqrt{v_0^2 + 2as}$, where v_0 is the initial velocity, a is the acceleration, and s is the distance traveled. Find dv/ds.

61. The voltage induced in an inductor L is given by $E = L(dI/dt)$, where I is the current in the circuit and t is the time. Find the voltage induced in a 0.4-H inductor if the current I (in A) is related to the time (in s) by $I = t(0.01t + 1)^3$.

62. In studying the energy used by a mechanical robotic device, the equation $v = \dfrac{z}{\alpha(1 - z^2) - \beta}$ is used. If α and β are constants, find dv/dz.

63. The frictional radius r_f of a collar used in a braking system is given by $r_f = \dfrac{2(R^3 - r^3)}{3(R^2 - r^2)}$, where R is the outer radius and r is the inner radius. Find dr_f/dR if r is constant.

64. Water is being drained from a pond such that the volume V (in m³) of water in the pond after t hours is given by $V = 5000(60 - t)^2$. Find the rate at which the pond is being drained after 4.00 h.

65. The energy output E of an electric heater is a function of the time t (in s) given by $E = t(1 + 2t)^2$ for $t < 10$ s. Find the power dE/dt (in W) generated by the heater for $t = 8.0$ s.

(W) 66. The amount n (in g) of a compound formed during a chemical change is $n = \dfrac{8t}{2t^2 + 3}$, where t is the time (in s). Find dn/dt for $t = 4.0$ s. What is the meaning of the result?

67. The deflection y of a 10-m beam is $y = kx(x^4 + 450x^2 - 950)$, where k is a constant and x is the horizontal distance from one end. Find the expression for the instantaneous rate of change of y with respect to x.

68. The kinetic energy K (in J) of a rotating flywheel varies directly as the square of its angular velocity ω (in rad/s). If $K = 120$ J for $\omega = 75$ rad/s, find $dK/d\omega$ for $\omega = 140$ rad/s.

69. The frequency f of a certain electronic oscillator is given by $f = \dfrac{1}{2\pi\sqrt{C(L + 2)}}$, where C is a capacitance and L is an inductance. If C is constant, find df/dL.

70. The volume V of fluid produced in the retina of the eye in reaction to exposure to light of intensity I is given by $V = \dfrac{aI^2}{b - I}$, where a and b are constants. Find dV/dI.

71. The temperature T (in °C) in a freezer as a function of the time t (in h) is given by $T = \dfrac{10(1 - t)}{0.5t + 1}$. Find dT/dt.

72. Under certain conditions, the efficiency e (in %) of an internal combustion engine is given by
$$e = 100\left(1 - \dfrac{1}{(V_1/V_2)^{0.4}}\right)$$
where V_1 and V_2 are the maximum and minimum volumes of air in a cylinder, respectively. Assuming that V_2 is kept constant, find the expression for the instantaneous rate of change of efficiency with respect to V_1.

73. The deflection y of a cantilever beam (clamped at one end and free at the other end) is $y = \dfrac{w}{24EI}(6L^2x^2 - 4Lx^3 + x^4)$. Here, L is the length of the beam, and w, E, and I are constants. Find the first four derivatives of y with respect to x. (Each of these derivatives is useful in analyzing the properties of the beam.)

74. The number n of grams of a compound formed during a certain chemical reaction is given by $n = \dfrac{2t}{t + 1}$, where t is the time (in min). Evaluate d^2n/dt^2 (the rate of increase of the amount of the compound being formed) when $t = 4.00$ min.

75. The area of a rectangular patio is to be 75 m². Express the perimeter p of the patio as a function of its width w and find dp/dw.

76. A water tank is being designed in the shape of a right circular cylinder with a volume of 100 m³. Find the expression for the instantaneous rate of change of the total surface area A of the tank with respect to the radius r of the base.

77. An arch over a walkway can be described by the first-quadrant part of the parabola $y = 4 - x^2$. In order to determine the size and shape of rectangular objects that can pass under the arch, express the area A of a rectangle inscribed under the parabola in terms of x. Find dA/dx.

W **78.** A computer analysis showed that a specialized piece of machinery has a value (in dollars) given by $V = 1\,500\,000/(2t + 10)$, where t is the number of years after the purchase. Calculate the value of dV/dt and d^2V/dt^2 for $t = 5$ years. What is the meaning of these values?

79. An airplane flies over an observer with a velocity of 400 km/h and at an altitude of 500 m. If the plane flies horizontally in a straight line, find the rate at which the distance x from the observer

to the plane is changing 0.600 min after the plane passes over the observer. See Fig. 23.35.

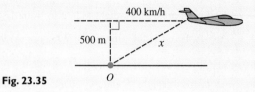

Fig. 23.35

80. The *radius of curvature* of $y = f(x)$ at the point (x, y) on the curve of $y = f(x)$ is given by

$$R = \frac{[1 + (y')^2]^{3/2}}{|y''|}$$

A certain roadway follows the parabola $y = 1.2x - x^2$ for $0 < x < 1.2$, where x is measured in kilometers. Find R for $x = 0.2$ km and $x = 0.6$ km. See Fig. 23.36.

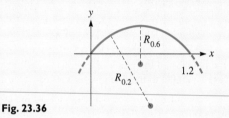

Fig. 23.36

Writing Exercise

81. An engineer designing military rockets uses computer simulation to find the path of a rocket as $y = f(x)$ and the path of an aircraft to be $y = g(x)$. Write two or three paragraphs explaining how the engineer can determine the angle at which the path of the rocket crosses the path of the aircraft.

CHAPTER 23 PRACTICE TEST

1. Find $\lim\limits_{x \to 1} \dfrac{x^2 - x}{x^2 - 1}$.

2. Find $\lim\limits_{x \to \infty} \dfrac{1 - 4x^2}{x + 2x^2}$.

3. Find the slope of a line tangent to the curve of $y = 3x^2 - \dfrac{4}{x^2}$ at $(2, 11)$. Check your result using the derivative evaluation feature of a graphing calculator. Write down the complete value shown on the calculator.

4. The displacement s (in cm) of a pumping machine piston in each cycle is given by $s = t\sqrt{10 - 2t}$, where t is the time (in s). Find the velocity of the piston for $t = 4.00$ s.

5. Find dy/dx: $y = 4x^6 - 2x^4 + \pi^3$

6. Find dy/dx: $y = 2x(5 - 3x)^4$

7. Find dy/dx: $(1 + y^2)^3 - x^2y = 7x$.

8. Under certain conditions, due to the presence of a charge q, the electric potential V along a line is given by

$$V = \frac{kq}{\sqrt{x^2 + b^2}}$$

where k is a constant and b is the minimum distance from the charge to the line. Find the expression for the instantaneous rate of change of V with respect to x.

9. Find the second derivative of $y = \dfrac{2x}{3x + 2}$.

10. By using the definition, find the derivative of $y = 5x - 2x^2$ with respect to x.

Applications of the Derivative

Following the work of Newton and Leibniz, the development of the calculus proceeded rapidly but in a rather disorganized way. Much of the progress in the late 1600s and early 1700s was due to a desire to solve applied problems, particularly in some areas of physics. These included problems such as finding velocities in more complex types of motion, accurately measuring time by use of a pendulum, and finding the equation of a uniform cable hanging under its own weight.

A number of mathematicians, most of whom also studied in various areas of physics, contributed to these advances in calculus. Among them was the Swiss mathematician Leonhard Euler, the most prolific mathematician of all time. Throughout the mid-1700s to later 1700s, he used the idea of a function to much

In Section 24.7, we see how to use the derivative in the design of cylindrical containers such as storage tanks.

better organize the study of algebra, trigonometry, and calculus. In doing so, he fully developed the use of calculus on problems from physics in areas such as planetary motion, mechanics, and optics.

Euler had a nearly unbelievable memory and ability to calculate. At an early age, he memorized the entire *Aeneid* by the Roman poet Virgil and was able to recite it from memory at age 70. In his head, he solved major problems related to the motion of the moon that Newton had not been able to solve. At one time, he was given two solutions to a problem that differed in the 15th decimal place, and he determined, in his head, which was correct. Although blind for the last 17 years of his life, it was one of his most productive periods. From memory, he dictated many of his articles (he wrote a total of over 70 volumes) until his sudden death in 1783.

We have noted some of the problems in technology in which the derivative plays a key role in the solution. Another important type is finding the maximum values or minimum values of functions. Such values are useful, for example, in finding the maximum possible income from production or the least amount of material needed in making a product. In this chapter, we consider several of these kinds of applications of the derivative.

24.1 TANGENTS AND NORMALS

The first application of the derivative we consider involves finding the equation of a line that is *tangent* to a given curve and the equation of a line that is *normal* (perpendicular) to a given curve.

TANGENT LINE

To find the equation of a line tangent to a curve at a given point, we first find the derivative of the function. The derivative is then evaluated at the point, and this gives us the slope of a line tangent to the curve at the point. Then, by using the point-slope form of the equation of a straight line, we find the equation of the tangent line. The following examples illustrate the method.

EXAMPLE 1 Find the equation of the line tangent to the parabola $y = x^2 - 1$ at the point $(-2, 3)$.

Finding the derivative and evaluating it at $x = -2$, we have

$$\frac{dy}{dx} = 2x$$

$$\left.\frac{dy}{dx}\right|_{x=-2} = -4$$

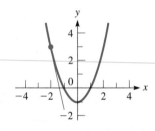

which means that the slope of the tangent line at $(-2, 3)$ is -4. Thus, by using the point-slope form of the equation of the straight line, we obtain the desired equation.

$$y - 3 = -4(x + 2)$$
$$y = -4x - 5$$

Fig. 24.1

The parabola and the tangent line $y = -4x - 5$ are shown in Fig. 24.1.

EXAMPLE 2 Find the equation of the line tangent to the ellipse $4x^2 + 9y^2 = 40$ at the point $(1, 2)$.

The easiest method of finding the derivative of this equation is to treat the equation as an implicit function. In this way, we have the following solution:

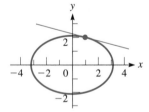

Fig. 24.2

$$8x + 18yy' = 0 \qquad \text{find derivative}$$

$$y' = -\frac{4x}{9y}$$

$$y'|_{(1,2)} = -\frac{4}{18} = -\frac{2}{9} \qquad \begin{array}{l}\text{evaluate derivative to find slope}\\ \text{of tangent line}\end{array}$$

$$y - 2 = -\frac{2}{9}(x - 1) \qquad \text{point-slope form of tangent line}$$

$$9y - 18 = -2x + 2$$

$$2x + 9y - 20 = 0 \qquad \text{standard form of tangent line}$$

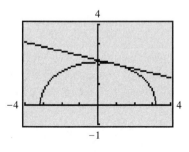

Fig. 24.3

The ellipse and the tangent line $2x + 9y - 20 = 0$ are shown in Fig. 24.2.

To display the function and the tangent line by using the *tangent* feature of a graphing calculator, we would solve for y. Since the point $(1, 2)$ is on the upper half of the ellipse, we would use $y = \frac{2}{3}\sqrt{10 - x^2}$. See Fig. 24.3 for the display. (We could also use the derivative of this function to find the equation of the tangent line.)

NORMAL LINE

About 1700, the word *normal* was adapted from the Latin word *normalis,* which was being used for *perpendicular.*

NOTE ▶

If we wish to obtain the equation of a line normal (perpendicular to a tangent) to a curve, recall that the slopes of perpendicular lines are negative reciprocals. Thus, the derivative is found and evaluated at the specified point. Since this gives the slope of a tangent line, ***we take the negative reciprocal of this number to find the slope of the normal line.*** Then, by using the point-slope form of the equation of a straight line, we find the equation of the normal. The following examples illustrate the method.

◀ **EXAMPLE 3** Find the equation of the line normal to the hyperbola $y = 2/x$ at the point $(2, 1)$.

Taking the derivative of this function and evaluating it for $x = 2$, we have

$$\frac{dy}{dx} = -\frac{2}{x^2}, \qquad \frac{dy}{dx}\bigg|_{x=2} = -\frac{1}{2}$$

Therefore, the slope of a line normal to the curve at $(2, 1)$ is 2. The equation of the normal line is then

$$y - 1 = 2(x - 2)$$

or

$$y = 2x - 3$$

The hyperbola and the normal line are shown in the graphing calculator display shown in Fig. 24.4.

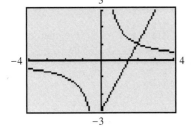

Fig. 24.4

◀ **EXAMPLE 4** Find the y-intercept of the line that is normal to the curve of $y = 2x - \frac{1}{3}x^3$, where $x = 3$.

First, we find the equation of the normal line. We find the y-intercept by writing the equation in slope-intercept form. The solution proceeds as follows:

$$\frac{dy}{dx} = 2 - x^2 \qquad \text{find derivative}$$

$$\frac{dy}{dx}\bigg|_{x=3} = 2 - 3^2 = -7 \qquad \text{evaluate derivative}$$

$$m_{\text{norm}} = \tfrac{1}{7} \qquad \text{negative reciprocal}$$

$$y\big|_{x=3} = 2(3) - \tfrac{1}{3}(3^3) = -3 \qquad \text{find y-coordinate of point}$$

$$y - (-3) = \tfrac{1}{7}(x - 3) \qquad \text{point-slope form of normal line}$$

$$7y + 21 = x - 3$$

$$y = \tfrac{1}{7}x - \tfrac{24}{7} \qquad \text{slope-intercept form}$$

This tells us that the y-intercept is $\left(0, -\frac{24}{7}\right)$. The curve, normal line, and intercept are shown in Fig. 24.5.

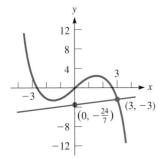

Fig. 24.5

Many of the applications of tangents and normals are geometric. However, there are certain applications in technology, and one of these is shown in the following example. Others are shown in the exercises.

Solving a Word Problem

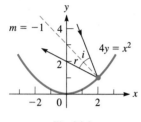

Fig. 24.6

EXAMPLE 5 In Fig. 24.6, the cross section of a parabolic solar reflector is shown, along with an incident ray of light and the reflected ray. The angle of incidence i is equal to the angle of reflection r where both angles are measured with respect to the normal to the surface. If the incident ray strikes at the point where the slope of the normal is -1 and the equation of the parabola is $4y = x^2$, what is the equation of the normal line?

If the slope of the normal line is -1, then the slope of a tangent line is $-\left(\frac{1}{-1}\right) = 1$. Therefore, we know that the value of the derivative at the point of reflection is 1. This allows us to find the coordinates of the point:

$$4y = x^2$$

$$4\frac{dy}{dx} = 2x, \qquad \frac{dy}{dx} = \frac{1}{2}x \qquad \text{find derivative}$$

$$1 = \frac{1}{2}x \qquad \text{substitute } \frac{dy}{dx} = 1$$

$$x = 2$$

This means that the x-coordinate of the point of reflection is 2. We can find the y-coordinate by substituting $x = 2$ into the equation of the parabola. Thus, the point is $(2, 1)$. Since the slope is -1, the equation is

$$y - 1 = (-1)(x - 2)$$
$$y = -x + 3$$

We might note that if the incident ray is vertical, for which $i = 45°$, the reflected ray passes through $(0, 1)$, which is the focus of the parabola. This shows the important reflection property of a parabola that *any incident ray parallel to the axis of a parabola passes through the focus*. We first noted this property in our discussion of the parabola in Example 5 of Section 21.4.

EXERCISES 24.1

In Exercises 1 and 2, make the given changes in the indicated examples of this section and then solve the resulting problems.

1. In Example 2, change $4x^2 + 9y^2$ to $x^2 + 4y^2$, change 40 to 17, and then find the equation of the tangent line.

2. In Example 3, change $2/x$ to $3/(x + 1)$ and then find the equation of the normal line.

In Exercises 3–6, find the equations of the lines tangent to the indicated curves at the given points. In Exercises 3 and 6, sketch the curve and tangent line. In Exercises 4 and 5, use the tangent feature of a graphing calculator to view the curve and tangent line.

3. $y = x^2 + 2$ at $(2, 6)$

4. $y = \frac{1}{3}x^3 - 5x$ at $(3, -6)$

5. $y = \frac{1}{x^2 + 1}$ at $\left(1, \frac{1}{2}\right)$

6. $x^2 + y^2 = 25$ at $(3, 4)$

In Exercises 7–10, find the equations of the lines normal to the indicated curves at the given points. In Exercises 7 and 10, sketch the curve and normal line. In Exercises 8 and 9, use a graphing calculator to view the curve and normal line.

7. $y = 6x - 2x^2$ at $(2, 4)$

8. $y = 8 - x^3$ at $(-1, 9)$

9. $y = \frac{6}{(x^2 + 1)^2}$ at $\left(1, \frac{3}{2}\right)$

10. $x^2 - y^2 = 8$ at $(3, 1)$

In Exercises 11–14, find the equations of the lines tangent or normal to the given curves and with the given slopes. View the curves and lines on a graphing calculator.

11. $y = x^2 - 2x$, tangent line with slope 2

12. $y = \sqrt{2x - 9}$, tangent line with slope 1

13. $y = (2x - 1)^3$, normal line with slope $-\frac{1}{24}$, $x > 0$

14. $y = \frac{1}{2}x^4 + 1$, normal line with slope 4

In Exercises 15–28, solve the given problems involving tangent and normal lines.

15. Find the equations of the tangent and normal lines to the parabola with vertex at $(0, 3)$ and focus at $(0, 0)$, where $x = -1$. Use a graphing calculator to view the curve and lines.

16. Find the equations of the tangent and normal lines to the ellipse with focus at $(4, 0)$, vertex at $(5, 0)$, and center at the origin, where $x = 2$. Use a graphing calculator to view the curve and lines.

17. Show that the line tangent to the graph of $y = x + 2x^2 - x^4$ at $(1, 2)$ is also tangent at $(-1, 0)$.

18. Show that the graphs of $y^2 = 4x + 4$ and $y^2 = 4 - 4x$ cross at right angles.

 19. Without actually finding the points of intersection, explain why the parabola $y^2 = 4x$ and the ellipse $2x^2 + y^2 = 6$ intersect at right angles. (*Hint*: Call a point of intersection (a, b).)

20. Find the y-intercept of the line normal to the curve $y = x^{3/4}$, where $x = 16$.

21. Heat flows normal to isotherms, curves along which the temperature is constant. Find the line along which heat flows through the point $(2, 1)$ and the isotherm is along the graph of $2x^2 + y^2 = 9$.

22. The sparks from an emery wheel to sharpen blades fly off tangent to the wheel. Find the equation along which sparks fly from a wheel described by $x^2 + y^2 = 25$, at $(3, 4)$.

23. A certain suspension cable with supports on the same level is closely approximated as being parabolic in shape. If the supports are 80 m apart and the sag at the center is 10 m, what is the equation of the line along which the tension acts (tangentially) at the right support? (Choose the origin of the coordinate system at the lowest point of the cable.)

24. In a video game, airplanes move from left to right along the path described by $y = 2 + 1/x$. They can shoot rockets tangent to the direction of flight at targets on the x-axis located at $x = 1, 2, 3$, and 4. Will a rocket fired from $(1, 3)$ hit a target?

25. In an electric field, the lines of force are perpendicular to the curves of equal electric potential. In a certain electric field, a curve of equal potential is $y = \sqrt{2x^2 + 8}$. If the line along which the force acts on an electron has an inclination of 135°, find its equation.

26. A radio wave reflects from a reflecting surface in the same way as a light wave (see Example 5). A certain horizontal radio wave reflects off a parabolic reflector such that the reflected wave is 43.60° below the horizontal, as shown in Fig. 24.7. If the equation of the parabola is $y^2 = 8x$, what is the equation of the normal line through the point of reflection?

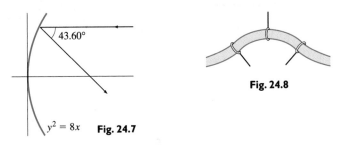

Fig. 24.8

$y^2 = 8x$ **Fig. 24.7**

27. In designing a flexible tubing system, the supports for the tubing must be perpendicular to the tubing. If a section of the tubing follows the curve $y = \dfrac{4}{x^2 + 1}$ ($-2\,\text{dm} < x < 2\,\text{dm}$), along which lines must the supports be directed if they are located at $x = -1$, $x = 0$, and $x = 1$? See Fig. 24.8.

28. On a particular drawing, a pulley wheel can be described by the equation $x^2 + y^2 = 100$ (units in cm). The pulley belt is directed along the lines $y = -10$ and $4y - 3x - 50 = 0$ when first and last making contact with the wheel. What are the first and last points on the wheel where the belt makes contact?

Another mathematical development by the English mathematician and physicist Isaac Newton (1642–1727).

24.2 NEWTON'S METHOD FOR SOLVING EQUATIONS

As we know, finding the roots of an equation $f(x) = 0$ is very important in mathematics and in many types of applications, and we have developed methods of solving many types of equations in the previous chapters. However, for a great many algebraic and nonalgebraic equations, there is no method for finding the roots exactly.

We have shown how equations can be solved graphically, and by using a graphing calculator the roots can be found with great accuracy. In this section, we show a method, known as **Newton's method,** that uses the derivative to locate approximately, but very accurately, the real roots of many kinds of equations. It can be used with polynomial equations of any degree and with other algebraic and nonalgebraic equations.

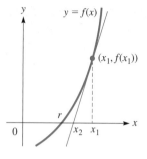

Fig. 24.9

Newton's method is an example of an **iterative method.** In using this type of method, we start with a reasonable guess for a root of the equation. By using the method, we obtain a new value, which is a better approximation. This in turn gives a still better approximation. Continuing in this way, using a calculator we can obtain an approximate answer with the required accuracy. Iterative methods in general are easily programmable for use on a computer.

Let us consider a section of the curve of $y = f(x)$ that (a) crosses the x-axis, (b) always has either a positive slope or a negative slope, and (c) has a slope that either becomes greater or becomes less as x increases. See Fig. 24.9. The curve in the figure crosses the x-axis at $x = r$, which means that $x = r$ is a root of the equation $f(x) = 0$. If x_1 is sufficiently close to r, a line tangent to the curve at $[x_1, f(x_1)]$ will cross the x-axis at a point $(x_2, 0)$, which is closer to r than is x_1.

We know that the slope of the tangent line is the value of the derivative at x_1, or $m_{tan} = f'(x_1)$. Therefore, the equation of the tangent line is

$$y - f(x_1) = f'(x_1)(x - x_1)$$

For the point $(x_2, 0)$ on this line, we have

$$-f(x_1) = f'(x_1)(x_2 - x_1)$$

Solving for x_2, we have the formula

$$x_2 = x_1 - \frac{f(x_1)}{f'(x_1)} \tag{24.1}$$

NOTE ▶ *Here, x_2 is a second approximation to the root. We can then replace x_1 in Eq. (24.1) by x_2 and find a closer approximation, x_3. This process can be repeated as many times as needed to find the root to the required accuracy. This method lends itself well to the use of a calculator or a computer for finding the root.*

◀ **EXAMPLE 1** Find the root of $x^2 - 3x + 1 = 0$ between $x = 0$ and $x = 1$.

Here, $f(x) = x^2 - 3x + 1$. Therefore, $f(0) = 1$ and $f(1) = -1$, which indicates that the root may be near the middle of the interval. Since x_1 must be within the interval, we choose $x_1 = 0.5$.

The derivative is

$$f'(x) = 2x - 3$$

Therefore, $f(0.5) = -0.25$ and $f'(0.5) = -2$, which gives us

$$x_2 = 0.5 - \frac{-0.25}{-2} = 0.375$$

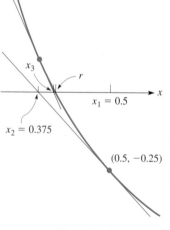

Fig. 24.10

This is a second approximation, which is closer to the actual value of the root. See Fig. 24.10. We can get an even better approximation, x_3, by using the method again with $x_2 = 0.375$, $f(0.375) = 0.015625$, and $f'(0.375) = -2.25$. This gives us

$$x_3 = 0.375 - \frac{0.015625}{-2.25} = 0.3819444$$

Since this is a quadratic equation, we can check this result by using the quadratic formula. Using this formula, we find the root is $x = 0.3819660$. Our result using Newton's method is good to three decimal places. Additional accuracy may be obtained by using the method again as many times as needed. ▶

Solving a Word Problem | **◀ EXAMPLE 2** A spherical water-storage tank holds 500.0 m³. If the outside diameter is 10.0000 m, what is the thickness of the metal of which the tank is made?

Let x = the thickness of the metal. We know that the outside radius of the tank is 5.0000 m. Therefore, using the formula for the volume of a sphere, we have

$$\frac{4\pi}{3}(5.0000 - x)^3 = 500.0$$

$$125.0 - 75.00x + 15.00x^2 - x^3 = 119.366$$

$$x^3 - 15.00x^2 + 75.00x - 5.634 = 0$$

$$f(x) = x^3 - 15.00x^2 + 75.00x - 5.634$$

$$f'(x) = 3x^2 - 30.00x + 75.00$$

Since $f(0) = -5.634$ and $f(0.1) = 1.717$, the root may be closer to 0.1 than to 0.0. Therefore, we let $x_1 = 0.07$. Setting up a table, we have these values:

n	x_n	$f(x_n)$	$f'(x_n)$	$x_n - \dfrac{f(x_n)}{f'(x_n)}$
1	0.07	$-0.457\,157$	72.914 7	0.076 269 750 8
2	0.076 269 750 8	$-0.000\,581\,145$	72.729 358 7	0.076 277 741 3

Since $x_2 = x_3 = 0.0763$ to four decimal places, the thickness is 0.0763 m. This means the inside radius of the tank is 4.9237 m, and this value gives an inside volume of 500.0 m³. In using a calculator, the values in the table are more easily found if the values of x_n, $f(x_n)$, and $f'(x_n)$ are stored in memory for each step. ▶

See Appendix C for the graphing calculator program NEWTON. It finds the *n*th approximation of a root of $f(x) = 0$, using Newton's method.

◀ EXAMPLE 3 Solve the equation $x^2 - 1 = \sqrt{4x - 1}$.

We can see approximately where the root is by sketching the graphs of $y_1 = x^2 - 1$ and $y_2 = \sqrt{4x - 1}$ or by viewing the graphs on a graphing calculator, as shown in Fig. 24.11. From this view, we see that they intersect between $x = 1$ and $x = 2$. Therefore, we choose $x_1 = 1.5$. With

$$f(x) = x^2 - 1 - \sqrt{4x - 1}$$

$$f'(x) = 2x - \frac{2}{\sqrt{4x - 1}}$$

we now find the values in the following table:

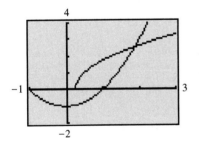

Fig. 24.11

n	x_n	$f(x_n)$	$f'(x_n)$	$x_n - \dfrac{f(x_n)}{f'(x_n)}$
1	1.5	$-0.986\,067\,98$	2.105 572 8	1.968 313 4
2	1.968 313 4	0.252 568 59	3.173 759 8	1.888 733 2
3	1.888 733 2	0.007 052 69	2.996 295 7	1.886 379 4
4	1.886 379 4	0.000 006 20	2.991 026 5	1.886 377 3

Since $x_5 = x_4 = 1.886\,38$ to five decimal places, this is the required solution. (Here, rounded-off values of x_n are shown, although additional digits were carried and used.) This value can be verified on the graphing calculator by using the *intersect* (or *zero*) feature. ▶

E X E R C I S E S **24.2**

In Exercises 1–4, find the indicated roots of the given quadratic equations by finding x_3 from Newton's method. Compare this root with that obtained by using the quadratic formula.

1. In Example 1, change the middle term from $-3x$ to $-5x$ and use the same x_1.

2. $2x^2 - x - 2 = 0$ (between 1 and 2)

3. $3x^2 - 5x - 1 = 0$ (between -1 and 0)

4. $x^2 + 4x + 2 = 0$ (between -4 and -3)

In Exercises 5–16, find the indicated roots of the given equations to at least four decimal places by using Newton's method. Compare with the value of the root found using a graphing calculator.

5. $x^3 - 6x^2 + 10x - 4 = 0$ (between 0 and 1)

6. $x^3 - 3x^2 - 2x + 3 = 0$ (between 0 and 1)

7. $x^3 + 5x^2 + x - 1 = 0$ (the positive root)

8. $2x^3 + 2x^2 - 11x + 3 = 0$ (the larger positive root)

9. $x^4 - x^3 - 3x^2 - x - 4 = 0$ (between 2 and 3)

10. $2x^4 - 2x^3 - 5x^2 - x - 3 = 0$ (between -2 and -1)

11. $x^4 - 2x^3 - 8x - 16 = 0$ (the negative root)

12. $3x^4 - 3x^3 - 11x^2 - x - 4 = 0$ (the negative root)

13. $2x^2 = \sqrt{2x + 1}$ (the positive real solution)

14. $x^3 = \sqrt{x + 1}$ (the real solution)

15. $x = \dfrac{1}{\sqrt{x + 2}}$ (the real solution)

16. $x^{3/2} = \dfrac{1}{2x + 1}$ (the real solution)

In Exercises 17–28, determine the required values by using Newton's method.

17. Find all the real roots of $x^3 - 2x^2 - 5x + 4 = 0$.

18. Find all the real roots of $x^3 - 2x^2 - 2x - 7 = 0$.

(W) 19. Explain how to find $\sqrt[3]{4}$ by using Newton's method.

(W) 20. Explain why Newton's method does not work for finding the root of $x^3 - 3x = 5$ if x_1 is chosen as 1.

21. Use Newton's method to find an expression for x_{n+1}, in terms of x_n and a, for the equation $x^2 - a = 0$. Such an equation can be used to find $\sqrt{a}$.

(W) 22. In Appendix D, page A.18, there is an explanation and example of Newton's method, which was copied directly from *Essays on Several Curious and Useful Subjects in Speculative and Mix'd Mathematicks* by Thomas Simpson (of Simpson's rule). It was published in London in 1740. Explain where the numerical error is in the example and what you think caused the error.

23. The altitude h (in m) of a rocket is given by $h = -2t^3 + 84t^2 + 480t + 10$, where t is the time (in s) of flight. When does the rocket hit the ground?

24. A solid sphere of specific gravity s sinks in water to a depth h (in cm) given by $0.009\,26h^3 - 0.0833h^2 + s = 0$. Find h for $s = 0.786$, if the diameter of the sphere is 6.00 cm.

25. A dome in the shape of a spherical segment is to be placed over the top of a sports stadium. If the radius r of the dome is to be 60.0 m and the volume V within the dome is 180 000 m^3, find the height h of the dome. See Fig. 24.12. $\left(V = \frac{1}{6}\pi h(h^2 + 3r^2).\right)$

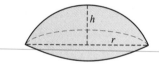

Fig. 24.12

26. The capacitances (in μF) of three capacitors in series are C, $C + 1.00$, and $C + 2.00$. If their combined capacitance is $1.00\ \mu$F, their individual values can be found by solving the equation
$$\frac{1}{C} + \frac{1}{C + 1.00} + \frac{1}{C + 2.00} = 1.00$$
Find these capacitances.

27. An oil-storage tank has the shape of a right circular cylinder with a hemisphere at each end. See Fig. 24.13. If the volume of the tank is 50.0 m^3 and the length l is 4.00 m, find the radius r.

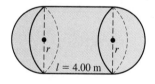

Fig. 24.13

28. A rectangular block of plastic with edges 2.00 cm, 2.00 cm, and 4.00 cm is heated until its volume doubles. By how much does each edge increase if each increases by the same amount?

24.3 CURVILINEAR MOTION

When velocity was introduced in Section 23.4, the discussion was limited to rectilinear motion, or motion along a straight line. A more general discussion of velocity is necessary when we discuss the motion of an object in a plane. There are many important applications of motion in a plane, a principal one being the motion of a projectile.

An important concept in developing this topic is that of a vector. The necessary fundamentals related to vectors are taken up in Chapter 9. Although vectors can be used to represent many physical quantities, we will restrict our attention to their use in describing the velocity and acceleration of an object moving in a plane along a specified path. Such motion is called **curvilinear motion.**

In describing an object undergoing curvilinear motion, it is common to express the x- and y-coordinates of its position separately as functions of time. Equations given in this form—that is, *x and y both given in terms of a third variable (in this case, t)—are said to be in* **parametric form,** which we encountered in Section 10.6. *The third variable, t, is called the* **parameter.**

To find the velocity of an object whose coordinates are given in parametric form, we find its x-component of velocity v_x by determining dx/dt and its y-component of velocity v_y by determining dy/dt. These are then evaluated, and the resultant velocity is found from $v = \sqrt{v_x^2 + v_y^2}$. The direction in which the object is moving is found from $\tan \theta = v_y/v_x$.

◀ EXAMPLE 1   If the horizontal distance x that an object has moved is given by $x = 3t^2$ and the vertical distance y is given by $y = 1 - t^2$, find the resultant velocity when $t = 2$.

To find the resultant velocity, we must find v and θ, by first finding v_x and v_y. After the derivatives are found, they are evaluated for $t = 2$. Therefore,

$$v_x = \frac{dx}{dt} = 6t \qquad v_x|_{t=2} = 12 \qquad \text{find velocity components}$$

$$v_y = \frac{dy}{dt} = -2t \qquad v_y|_{t=2} = -4$$

$$v = \sqrt{12^2 + (-4)^2} = 12.6 \qquad \text{magnitude of velocity}$$

$$\tan \theta = \frac{-4}{12} \qquad\qquad \theta = -18.4° \qquad \text{direction of motion}$$

The path and velocity vectors are shown in Fig. 24.14.   ▶

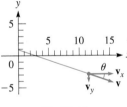

Fig. 24.14

◀ EXAMPLE 2   Find the velocity and direction of motion when $t = 2$ of an object moving such that its x- and y-coordinates of position are given by $x = 1 + 2t$ and $y = t^2 - 3t$.

$$v_x = \frac{dx}{dt} = 2 \qquad\qquad v_x|_{t=2} = 2 \qquad \text{find velocity components}$$

$$v_y = \frac{dy}{dt} = 2t - 3 \qquad v_y|_{t=2} = 1$$

$$v|_{t=2} = \sqrt{2^2 + 1^2} = 2.24 \qquad \text{magnitude of velocity}$$

$$\tan \theta = \frac{1}{2} \qquad\qquad \theta = 26.6° \qquad \text{direction of motion}$$

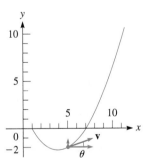

Fig. 24.15

These quantities are shown in Fig. 24.15.   ▶

CAUTION ▶   In these examples, note that *we first find the necessary derivatives and then we evaluate them.* This procedure should always be followed. When a derivative is to be found, it is incorrect to take the derivative of the expression that is the evaluated function.

Acceleration *is the time rate of change of velocity.* Therefore, if the velocity, or its components, is known as a function of time, the acceleration of an object can be found by taking the derivative of the velocity with respect to time. If the displacement is known, the acceleration is found by finding the second derivative with respect to time. Finding the acceleration of an object is illustrated in the following example.

◀ EXAMPLE 3   Find the magnitude and direction of the acceleration when $t = 2$ for an object that is moving such that its x- and y-coordinates of position are given by $x = t^3$ and $y = 1 - t^2$.

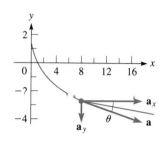

$$v_x = \frac{dx}{dt} = 3t^2 \qquad a_x = \frac{dv_x}{dt} = \frac{d^2x}{dt^2} = 6t \qquad a_x\big|_{t=2} = 12$$

take second derivatives to find acceleration components

$$v_y = \frac{dy}{dt} = -2t \qquad a_y = \frac{dv_y}{dt} = \frac{d^2y}{dt^2} = -2 \qquad a_y\big|_{t=2} = -2$$

$$a\big|_{t=2} = \sqrt{12^2 + (-2)^2} = 12.2 \qquad \text{magnitude of acceleration}$$

$$\tan\theta = \frac{a_y}{a_x} = -\frac{2}{12} \qquad \theta = -9.5° \qquad \text{direction of acceleration}$$

Fig. 24.16

CAUTION ▶   The quadrant in which θ lies is determined from the fact that a_y *is negative and* a_x *is positive.* Thus, θ must be a fourth-quadrant angle (see Fig. 24.16). We see from this example that the magnitude and direction of acceleration are found from its components just as with velocity. ▶

We now summarize the equations used to find the velocity and acceleration of an object for which the displacement is a function of time. They indicate how to find the components, as well as the magnitude and direction, of each.

$v_x = \dfrac{dx}{dt}$	$v_y = \dfrac{dy}{dt}$	velocity components (24.2)
$a_x = \dfrac{dv_x}{dt} = \dfrac{d^2x}{dt^2}$	$a_y = \dfrac{dv_y}{dt} = \dfrac{d^2y}{dt^2}$	acceleration components (24.3)
$v = \sqrt{v_x^2 + v_y^2}$	$a = \sqrt{a_x^2 + a_y^2}$	magnitude (24.4)
$\tan\theta_v = \dfrac{v_y}{v_x}$	$\tan\theta_a = \dfrac{a_y}{a_x}$	direction (24.5)

For reference, Eq. (23.15) is $\dfrac{du^n}{dx} = nu^{n-1}\dfrac{du}{dx}$.

CAUTION ▶   If the curvilinear path an object follows is given with y as a function of x, *the velocity (and acceleration) is found by taking derivatives of each term of the equation with respect to time.* It is assumed that both x and y are functions of time, although these functions are not stated. When finding derivatives we must be careful in using the power rule, Eq. (23.15), so that the factor du/dx is not neglected. In the following examples, we illustrate the use of Eqs. (24.2) to (24.5) in applied situations for which we know the equation of the path of the motion. Again, we must be careful to find the direction of the vector as well as its magnitude in order to have a complete solution.

Solving a Word Problem

EXAMPLE 4 In a physics experiment, a small sphere is constrained to move along a parabolic path described by $y = \frac{1}{3}x^2$. If the horizontal velocity v_x is constant at 6.00 cm/s, find the velocity at the point $(2.00, 1.33)$. See Fig. 24.17.

Since both y and x change with time, both can be considered to be functions of time. Therefore, we can take derivatives of $y = \frac{1}{3}x^2$ with respect to time.

CAUTION ▶

$$\frac{dy}{dt} = \frac{1}{3}\left(2x\frac{dx}{dt}\right) \longleftarrow \frac{dx^2}{dt} = 2x\frac{dx}{dt}$$

$$v_y = \frac{2}{3}xv_x \qquad \text{using Eqs. (24.2)}$$

$$v_y = \frac{2}{3}(2.00)(6.00) = 8.00 \text{ cm/s} \qquad \text{substituting}$$

$$v = \sqrt{6.00^2 + 8.00^2} = 10.0 \text{ cm/s} \qquad \text{magnitude [Eqs. (24.4)]}$$

$$\tan \theta = \frac{8.00}{6.00}, \qquad \theta = 53.1° \qquad \text{direction [Eqs. (24.5)]}$$

Fig. 24.17

Solving a Word Problem

EXAMPLE 5 A helicopter is flying at 18.0 m/s and at an altitude of 120 m when a rescue marker is released from it. The marker maintains a horizontal velocity and follows a path given by $y = 120 - 0.0151x^2$, as shown in Fig. 24.18. Find the magnitude and direction of the velocity and of the acceleration of the marker 3.00 s after release. This is a typical problem in projectile motion.

From the given information, we know that $v_x = dx/dt = 18.0$ m/s. Taking derivatives with respect to time leads to this solution:

$$y = 120 - 0.0151x^2$$

$$\frac{dy}{dt} = -0.0302x\frac{dx}{dt} \qquad \text{taking derivatives}$$

$$v_y = -0.0302xv_x \qquad \text{using Eqs. (24.2)}$$

$$x = (3.00)(18.0) = 54.0 \text{ m} \qquad \text{evaluating at } t = 3.00 \text{ s}$$

$$v_y = -0.0302(54.0)(18.0) = -29.35 \text{ m/s}$$

$$v = \sqrt{18.0^2 + (-29.35)^2} = 34.4 \text{ m/s} \qquad \text{magnitude}$$

$$\tan \theta = \frac{-29.35}{18.0}, \qquad \theta = -58.5° \qquad \text{direction}$$

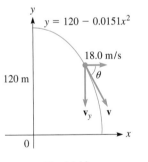

Fig. 24.18

The velocity is 34.4 m/s and is directed at an angle of 58.5° below the horizontal.

To find the acceleration, we return to the equation $v_y = -0.0302xv_x$. Since v_x is constant, we can substitute 18.0 for v_x to get

$$v_y = -0.5436x$$

Again taking derivatives with respect to time, we have

$$\frac{dv_y}{dt} = -0.5436\frac{dx}{dt}$$

$$a_y = -0.5436v_x \qquad \text{using Eqs. (24.3) and (24.2)}$$

$$a_y = -0.5436(18.0) = -9.78 \text{ m/s}^2 \qquad \text{evaluating}$$

We know that v_x is constant, which means that $a_x = 0$. Therefore, the acceleration is 9.78 m/s² and is directed vertically downward.

EXERCISES 24.3

In Exercises 1 and 2, make the given changes in the indicated examples of this section and then solve the resulting problems.

1. In Example 1, change $x = 3t^2$ to $x = 4t^2$ and then find the resultant velocity when $t = 2$.

2. In Example 3, change $y = 1 - t^2$ to $y = 1 - 2t^2$ and then find the acceleration when $t = 2$.

In Exercises 3–6, given that the x- and y-coordinates of a moving particle are given by the indicated parametric equations, find the magnitude and direction of the velocity for the specific value of t. Sketch the curves and show the velocity and its components.

3. $x = 3t, y = 1 - t, t = 4$

4. $x = \dfrac{5t}{2t + 1}, y = 0.1(t^2 + t), t = 2$

5. $x = t(2t + 1)^2, y = \dfrac{6}{\sqrt{4t + 3}}, t = 0.5$

6. $x = \sqrt{1 + 2t}, y = t - t^2, t = 4$

In Exercises 7–10, use the parametric equations and values of t of Exercises 3–6, to find the magnitude and direction of the acceleration in each case.

In Exercises 11–28, find the indicated velocities and accelerations.

11. The water from a valve at the bottom of a water tank follows a path described by $y = 4.0 - 0.20x^2$, where units are in meters. If the velocity v_x is constant at 5.0 m/s, find the resultant velocity at the point $(4.0, 0.80)$.

12. A roller mechanism follows a path described by $y = \sqrt{4x + 1}$, where units are in meters. If $v_x = 2x$, find the resultant velocity (in m/s) at the point $(2.0, 3.0)$.

13. A float is used to test the flow pattern of a stream. It follows a path described by $x = 0.2t^2, y = -0.1t^3$ (x and y in m, t in min). Find the acceleration of the float after 2.0 min.

14. A radio-controlled model car is operated in a parking lot. The coordinates (in m) of the car are given by $x = 3.5 + 2.0t^2$ and $y = 8.5 + 0.25t^3$, where t is the time (in s). Find the acceleration of the car after 2.5 s.

15. A golf ball moves according to the equations $x = 32t$ and $y = 42t - 4.9t^2$, where distances are in meters and time is in seconds. Find the resultant velocity and acceleration of the golf ball for $t = 6.0$ s.

16. A package of relief supplies is dropped and moves according to the parametric equations $x = 45t$ and $y = -4.9t^2$ (x and y in m, t in s). Find the velocity and acceleration when $t = 3.0$ s.

17. A spacecraft moves along a path described by the parametric equations $x = 10(\sqrt{1 + t^4} - 1), y = 40t^{3/2}$ for the first 100 s after launch. Here, x and y are measured in meters, and t is measured in seconds. Find the magnitude and direction of the velocity of the spacecraft 10.0 s and 100 s after launch.

18. An electron moves in an electric field according to the equations $x = 20/\sqrt{1 + t^2}$ and $y = 20t/\sqrt{1 + t^2}$ (x and y in m and t in s). Find the velocity when $t = 1.0$ s.

19. In a computer game, an airplane starts at $(1.00, 4.00)$ (in cm) on the curve $y = 3.00 + x^{-1.50}$ and moves with a constant horizontal velocity of 1.20 cm/s. What is the plane's velocity after 0.500 s?

20. In an aerobic exercise machine, weights are lifted and a person's hands are constrained to move along arcs of the ellipse $16x^2 + 9y^2 = 9$ (in m). If the person's hands move upward at 0.100 m/s, and start at $y = 0$, at what velocity is each moving after 1.50 s?

21. Find the resultant acceleration of the spacecraft in Exercise 17 for the specified times.

22. A ski jump is designed to follow the path given by the equations $x = 3.50t^2$ and $y = 20.0 + 0.120t^4 - 3.00\sqrt{t^4 + 1}$ $(0 \le t \le 4.00$ s$)$ (x and y in m, t in s). Find the velocity and acceleration of a skier when $t = 4.00$ s. See Fig. 24.19.

Fig. 24.19

23. A rocket follows a path given by $y = x - \frac{1}{90}x^3$ (distances in km). If the horizontal velocity is given by $v_x = x$, find the magnitude and direction of the velocity when the rocket hits the ground (assume level terrain) if time is in minutes.

24. A ship is moving around an island on a route described by $y = 3x^2 - 0.2x^3$. If $v_x = 1.2$ km/h, find the velocity of the ship where $x = 3.5$ km.

25. A computer's hard disk is 88.9 mm in diameter and rotates at 3600 r/min. With the center of the disk at the origin, find the velocity components of a point on the rim for $x = 30.5$ mm, if $y > 0$ and $v_x > 0$.

26. A robot arm joint moves in an elliptical path (horizontal major axis 8.0 cm, minor axis 4.0 cm, center at origin). For $y > 0$ and -2 cm $< x < 2$ cm, the joint moves such that $v_x = 2.5$ cm/s. Find its velocity for $x = -1.5$ cm.

27. An airplane ascends such that its gain h in altitude is proportional to the square root of the change x in horizontal distance traveled. If $h = 280$ m for $x = 400$ m and v_x is constant at 350 m/s, find the velocity at this point.

(W) 28. A meteor traveling toward the earth has a velocity inversely proportional to the square root of the distance from the earth's center. State how its acceleration is related to its distance from the center of the earth.

24.4 RELATED RATES

Any two variables that vary with respect to time and between which a relation is known to exist can have the time rate of change of one expressed in terms of the time rate of change of the other. We do this by taking the derivative with respect to the time of the expression that relates the variables, as we did in Examples 4 and 5 of Section 24.3. Since the rates of change are related, this type of problem is referred to as a **related-rate** problem. The following examples illustrate the basic method of solution.

◀ EXAMPLE 1   The voltage of a certain thermocouple as a function of the temperature is given by $E = 2.800T + 0.006T^2$. If the temperature is increasing at the rate of 1.00 °C/min, how fast is the voltage increasing when $T = 100°C$?

Since we are asked to find the time rate of change of voltage, we first take derivatives with respect to time. This gives us

$$\frac{dE}{dt} = 2.800\frac{dT}{dt} + 0.012T\frac{dT}{dt} \longleftarrow \frac{d}{dt}(0.006T^2) = 0.006\left(2T\frac{dT}{dt}\right)$$

CAUTION ▶   ***again being careful to include the factor dT/dt.*** From the given information, we know that $dT/dt = 1.00°C/min$ and that we wish to know dE/dt when $T = 100°C$. Thus,

$$\left.\frac{dE}{dt}\right|_{T=100} = 2.800(1.00) + 0.012(100)(1.00) = 4.00 \text{ V/min}$$

NOTE ▶   *The derivative must be taken before values are substituted.* In this problem, we are finding the time rate of change of the voltage for a specified value of T. For other values of T, dE/dt would have different values.   ▮

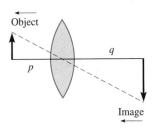

Object

q

p

Image

Fig. 24.20

◀ EXAMPLE 2   The distance q that an image is from a certain lens in terms of p, the distance of the object from the lens, is given by

$$q = \frac{10p}{p - 10}$$

If the object distance is increasing at the rate of 0.200 cm/s, how fast is the image distance changing when $p = 15.0$ cm? See Fig. 24.20.

Taking derivatives with respect to time, we have

don't forget the $\dfrac{dp}{dt}$

$$\frac{dq}{dt} = \frac{(p - 10)\left(10\dfrac{dp}{dt}\right) - 10p\left(\dfrac{dp}{dt}\right)}{(p - 10)^2} = \frac{-100\dfrac{dp}{dt}}{(p - 10)^2}$$

Now, substituting $p = 15.0$ and $dp/dt = 0.200$, we have

$$\left.\frac{dq}{dt}\right|_{p=15} = \frac{-100(0.200)}{(15.0 - 10)^2}$$

$$= -0.800 \text{ cm/s}$$

Thus, the image distance is decreasing (the significance of the minus sign) at the rate of 0.800 cm/s when $p = 15.0$ cm.   ▮

In many related-rate problems, the function is not given but must be set up according to the statement of the problem. The following examples illustrate this type of problem.

Solving a Word Problem

◀ **EXAMPLE 3** A spherical balloon is being blown up such that its volume increases at the constant rate of 2.00 m³/min. Find the rate at which the radius is increasing when it is 3.00 m. See Fig. 24.21.

We are asked to find the relation between the rate of change of the volume of a sphere with respect to time and the corresponding rate of change of the radius with respect to time. Therefore, we are to **take derivatives of the expression for the volume of a sphere with respect to time:**

CAUTION ▶

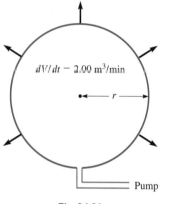

$dV/dt = 2.00$ m³/min

r

Pump

Fig. 24.21

$$V = \frac{4}{3}\pi r^3 \qquad \text{volume of sphere}$$

$$\frac{dV}{dt} = 4\pi r^2 \left(\frac{dr}{dt}\right) \qquad \text{take derivatives with respect to time}$$

$$2.00 = 4\pi(3.00)^2 \left(\frac{dr}{dt}\right) \qquad \text{substitute } \frac{dV}{dt} = 2.00 \text{ m}^3/\text{min and } r = 3.00 \text{ m}$$

$$\frac{dr}{dt}\bigg|_{r=3} = \frac{1}{18.0\pi} \qquad \text{solve for } \frac{dr}{dt}$$

$$= 0.0177 \text{ m/min}$$

Solving a Word Problem

◀ **EXAMPLE 4** The force F of gravity of the earth on a spacecraft varies inversely as the square of the distance r of the spacecraft from the center of the earth. A particular spacecraft weighs 4500 N on the launchpad ($F = 4500$ N for $r = 6370$ km). Find the rate at which F changes later as the spacecraft moves away from the earth at the rate of 12 km/s, where $r = 8500$ km.

First setting up the equation, we have the following solution:

$$F = \frac{k}{r^2} \qquad \text{inverse variation}$$

$$4500 = \frac{k}{6370^2} \qquad \text{substitute } F = 4500 \text{ N}, r = 6370 \text{ km}$$

$$k = 1.83 \times 10^{11} \text{ N} \cdot \text{km}^2 \qquad \text{solve for } k$$

$$F = \frac{1.83 \times 10^{11}}{r^2} \qquad \text{substitute for } k \text{ in equation}$$

$$\frac{dF}{dt} = (1.83 \times 10^{11})(-2)(r^{-3})\frac{dr}{dt} \qquad \text{take derivatives with respect to time}$$

$$= \frac{-3.66 \times 10^{11}}{r^3}\frac{dr}{dt}$$

$$\frac{dF}{dt}\bigg|_{t=8500 \text{ km}} = \frac{-3.66 \times 10^{11}}{8500^3}(12) \qquad \text{evaluate derivative for } r = 8500 \text{ km}, dr/dt = 12 \text{ km/s}$$

$$= -7.2 \text{ N/s}$$

Therefore, the gravitational force is decreasing at the rate of 7.2 N/s. ▶

Solving a Word Problem ◀ **EXAMPLE 5** Two cruise ships leave Vancouver, British Columbia, at noon. Ship *A* travels west at 12.0 km/h (before turning toward Alaska), and ship *B* travels south at 16.0 km/h (toward Seattle). How fast are they separating at 2 P.M.?

In Fig. 24.22, we let $x =$ the distance traveled by *A* and $y =$ the distance traveled by *B*. We can find the distance between them, z, from the Pythagorean theorem. Therefore, we are to find dz/dt for $t = 2.00$ h. Even though there are three variables, each is a function of time. This means we can find dz/dt by taking derivatives of each term with respect to time. This gives us

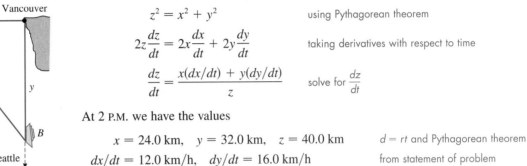

$$z^2 = x^2 + y^2 \qquad \text{using Pythagorean theorem}$$

$$2z\frac{dz}{dt} = 2x\frac{dx}{dt} + 2y\frac{dy}{dt} \qquad \text{taking derivatives with respect to time}$$

$$\frac{dz}{dt} = \frac{x(dx/dt) + y(dy/dt)}{z} \qquad \text{solve for } \frac{dz}{dt}$$

At 2 P.M. we have the values

$$x = 24.0 \text{ km}, \quad y = 32.0 \text{ km}, \quad z = 40.0 \text{ km} \qquad d = rt \text{ and Pythagorean theorem}$$

$$dx/dt = 12.0 \text{ km/h}, \quad dy/dt = 16.0 \text{ km/h} \qquad \text{from statement of problem}$$

$$\left.\frac{dz}{dt}\right|_{z=40} = \frac{(24.0)(12.0) + (32.0)(16.0)}{40.0} = 20.0 \text{ km/h} \qquad \text{substitute values}$$

Fig. 24.22

To Alaska

W ←--- A   x   Vancouver

z   y

B

To Seattle

S

From these examples, we see that we have the following method of solving a related-rates problem.

STEPS FOR SOLVING RELATED-RATES PROBLEMS

1. *Identify the variables and rates* in the problem.
2. If possible, *make a sketch* showing the variables.
3. *Determine the equation* relating the variables.
4. *Differentiate with respect to time.*
5. *Solve for the required rate.*

EXERCISES 24.4

Solve the following problems in related rates.

1. In Example 1, change 0.006 to 0.012 and then find how fast the voltage is increasing when $T = 100°C$.

2. In Example 3, change "volume" to "surface area," change m³/min to m²/min, and then find the rate at which the radius is increasing when it is 3.00 m.

3. How fast is the slope of a tangent to the curve $y = 2/(x + 1)$ changing where $x = 3.0$ if $dx/dt = 0.50$ unit/s?

4. How fast is the slope of a tangent to the curve $y = 2(1 - 2x)^2$ changing where $x = 1.5$ if $dx/dt = 0.75$ unit/s?

5. The electric resistance R (in Ω) of a certain resistor as a function of the temperature T (in °C) is $R = 4.000 + 0.003T^2$. If the temperature is increasing at the rate of 0.100°C/s, find how fast the resistance changes when $T = 150°C$.

6. The kinetic energy K (in J) of an object is given by $K = \frac{1}{2}mv^2$, where m is the mass (in kg) of the object and v is its velocity. If a 250-kg wrecking ball accelerates at 5.00 m/s², how fast is the kinetic energy changing when $v = 30.0$ m/s?

7. The length L (in cm) of a pendulum is slowly decreasing at the rate of 0.100 cm/s. What is the time rate of change of the period T (in s) of the pendulum when $L = 16.0$ cm, if the equation relating the period and length is $T = \pi\sqrt{L/245}$?

8. The voltage V that produces a current I (in A) in a wire of radius r (in cm) is $V = 0.030I/r^2$. If the current increases at 0.020 A/s in a wire of 0.400 mm radius, find the rate at which the voltage is increasing.

9. A plane flying at an altitude of 2.0 km is at a direct distance $D = \sqrt{4.0 + x^2}$ from an airport control tower, where x is the horizontal distance to the tower. If the plane's speed is 350 km/h, how fast is D changing when $x = 6.2$ km?

10. A variable resistor R and an 8-Ω resistor in parallel have a combined resistance R_T given by $R_T = \dfrac{8R}{8 + R}$. If R is changing at 0.30 Ω/min, find the rate at which R_T is changing when $R = 6.0\ \Omega$.

11. The radius r of a ring of a certain holograph (an image produced without using a lens) is given by $r = \sqrt{0.4\lambda}$, where λ is the wavelength of the light being used. If λ is changing at the rate of 0.10×10^{-7} m/s when $\lambda = 6.0 \times 10^{-7}$ m, find the rate at which r is changing.

12. An earth satellite moves in a path that can be described by $\dfrac{x^2}{72.5} + \dfrac{y^2}{71.5} = 1$, where x and y are in thousands of kilometers. If $dx/dt = 12\,900$ km/h for $x = 3200$ km and $y > 0$, find dy/dt.

13. The magnetic field B due to a magnet of length l at a distance r is given by $B = \dfrac{k}{[r^2 + (l/2)^2]^{3/2}}$, where k is a constant for a given magnet. Find the expression for the time rate of change of B in terms of the time rate of change of r.

14. An approximate relationship between the pressure p and volume v of the vapor in a diesel engine cylinder is $pv^{1.4} = k$, where k is a constant. At a certain instant, $p = 4200$ kPa, $v = 75$ cm^3, and the volume is increasing at the rate of 850 cm^3/s. What is the time rate of change of the pressure at this instant?

15. A rectangular swimming pool 16 m by 10 m is being filled at the rate of 0.8 m^3/min. How fast is the height h of the water rising?

16. An engine cylinder 15.0 cm deep is being bored such that the radius increases by 0.100 mm/min. How fast is the volume V of the cylinder changing when the diameter is 9.50 cm?

17. Fatty deposits have decreased the circular cross-sectional opening of a person's artery. A test drug reduces these deposits such that the radius of the opening increases at the rate of 0.020 mm/month. Find the rate at which the area of the opening increases when $r = 1.2$ mm.

18. A computer program increases the side of a square image on the screen at the rate of 0.25 cm/s. Find the rate at which the area of the image increases when the edge is 6.50 cm.

19. A metal cube dissolves in acid such that an edge of the cube decreases by 0.50 mm/min. How fast is the volume of the cube changing when the edge is 8.20 mm?

20. A metal sphere is placed in seawater to study the corrosive effect of seawater. If the surface area decreases at 35 cm^2/year due to corrosion, how fast is the radius changing when it is 12 cm?

21. A uniform layer of ice covers a spherical water-storage tank. As the ice melts, the volume V of ice decreases at a rate that varies directly as the surface area A. Show that the outside radius decreases at a constant rate.

22. A light in a garage is 2.90 m above the floor and 3.65 m behind the door. If the garage door descends vertically at 0.45 m/s, how fast is the door's shadow moving toward the garage when the door is 0.60 m above the floor?

23. One statement of Boyle's law is that the pressure of a gas varies inversely as the volume for constant temperature. If a certain gas occupies 650 cm^3 when the pressure is 230 kPa and the volume is increasing at the rate of 20.0 cm^3/min, how fast is the pressure changing when the volume is 810 cm^3?

24. The tuning frequency f of an electronic tuner is inversely proportional to the square root of the capacitance C in the circuit. If $f = 920$ kHz for $C = 3.5$ pF, find how fast f is changing at this frequency if $dC/dt = 0.3$ pF/s.

25. A spherical metal object is ejected from an earth satellite and reenters the atmosphere. It heats up (until it burns) so that the radius increases at the rate of 5.00 mm/s. What is the time rate of change of volume when the radius is 225 mm?

26. The acceleration due to the gravity g on a spacecraft is inversely proportional to its distance from the center of the earth. At the surface of the earth, $g = 9.80$ m/s^2. Given that the radius of the earth is 6370 km, how fast is g changing on a spacecraft approaching the earth at 1400 m/s at a distance of 41 000 km from the surface?

27. The intensity I of heat varies directly as the strength of the source and inversely as the square of the distance from the source. If an object approaches a heated object of strength 8.00 units at the rate of 50.0 cm/s, how fast is the intensity changing when it is 100 cm from the source?

28. The speed of sound v (in m/s) is $v = 331\sqrt{T/273}$, where T is the temperature (in K). If the temperature is 303 K (30°C) and is rising at 2.0°C/h, how fast is the speed of sound rising?

29. A tank in the shape of an inverted cone has the height of 3.60 m and a radius at the top of 1.15 m. Water is flowing into the tank at the rate of 0.50 m^3/min. How fast is the level rising when it is 1.80 m deep?

30. A ladder is slipping down along a vertical wall. If the ladder is 4.00 m long and the top of it is slipping at the constant rate of 3.00 m/s, how fast is the bottom of the ladder moving along the ground when the bottom is 2.00 m from the wall?

31. A supersonic jet leaves an airfield traveling due east at 1600 km/h. A second jet leaves the same airfield at the same time and travels 1800 km/h along a line north of east such that it remains due north of the first jet. After a half-hour, how fast are the jets separating?

32. A car passes over a bridge at 15.0 m/s at the same time a boat passes under the bridge at a point 10.5 m directly below the car. If the boat is moving perpendicularly to the bridge at 4.0 m/s, how fast are the car and the boat separating 5.0 s later?

33. A rope attached to a boat is being pulled in at a rate of 2.50 m/s. If the water is 5.00 m below the level at which the rope is being drawn in, how fast is the boat approaching the wharf when 13.0 m of rope are yet to be pulled in? See Fig. 24.23.

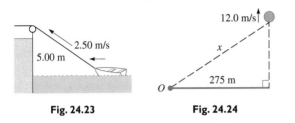

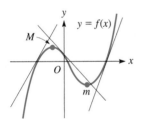

Fig. 24.23   **Fig. 24.24**

34. A weather balloon leaves the ground 275 m from an observer and rises vertically at 12.0 m/s. How fast is the line of sight from the observer to the balloon increasing when the balloon is 450 m high? See Fig. 24.24.

35. A man 1.80 m tall approaches a street light 4.50 m above the ground at the rate of 1.50 m/s. How fast is the end of the man's shadow moving when he is 3.00 m from the base of the light? See Fig. 24.25.

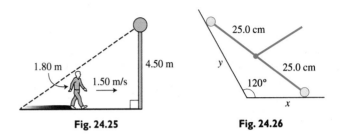

Fig. 24.25   **Fig. 24.26**

36. A roller mechanism, as shown in Fig. 24.26, moves such that the right roller is always in contact with the bottom surface and the left roller is always in contact with the left surface. If the right roller is moving to the right at 1.50 cm/s when $x = 10.0$ cm, how fast is the left roller moving?

24.5 USING DERIVATIVES IN CURVE SKETCHING

For the function in Fig. 24.27, we see that as x increases (from left to right), y also increases until point M is reached. From M to m, y decreases. To the right of m, y again increases. Also, any tangent line left of M or right of m has a positive slope, and any tangent line between M and m has a negative slope. Since the derivative gives us the slope of a tangent line, we see that *as x increases, y increases if the derivative is positive and decreases if the derivative is negative.* This can be stated as

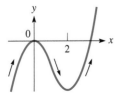

Fig. 24.27

$$f(x) \text{ increases if } f'(x) > 0 \quad \text{and} \quad f(x) \text{ decreases if } f'(x) < 0$$

***It is always assumed that x is increasing.** Also, we assume in our present analysis that $f(x)$ and its derivatives are continuous over the indicated interval.*

◀ EXAMPLE 1   Find those values of x for which the function $f(x) = x^3 - 3x^2$ is increasing and those values for which it is decreasing.

We find these values by finding the derivative and then finding the values of x for which it is positive or negative by solving inequalities. Therefore,

$$f'(x) = 3x^2 - 6x = 3x(x - 2)$$
$$f(x) \text{ is increasing if } 3x(x - 2) > 0, \text{ and decreasing if } 3x(x - 2) < 0$$

We now solve these inequalities by methods in Chapter 17 (see page 469). Finding the **critical values** of $x = 0$ and $x = 2$, we have the following:

If $x < 0, 3x(x - 2) > 0$ or $f'(x) > 0$.   f(x) increasing
If $0 < x < 2, 3x(x - 2) < 0$ or $f'(x) < 0$.   f(x) decreasing
If $x > 2, 3x(x - 2) > 0$ or $f'(x) > 0$.   f(x) increasing

The solution of the inequality $3x(x - 2) > 0$ is $x < 0$ or $x > 2$, which means that $f(x)$ is increasing for these values. Also, $3x(x - 2) < 0$ for $0 < x < 2$, which means that $f(x)$ is decreasing for these values. See Fig. 24.28 for the graph of $f(x) = x^3 - 3x^2$.

Fig. 24.28

Maximum Points and
Minimum Points

The points M and m in Fig. 24.27 are called a **relative maximum point** and a **relative minimum point,** respectively. *This means that M has a greater y-value than any other point near it and that m has a smaller y-value than any point near it.* This does not necessarily mean that M has the greatest y-value of any point on the curve or that m has the least y-value of any point on the curve. However, the points M and m have the greatest or least values of y for that part of the curve (that is why we use the word *relative*). Examination of Fig. 24.27 verifies this point. *The characteristic of both M and m is that the derivative is zero at each point.* (We see that this is so since a tangent line would have a slope of zero at each.) *This is how relative maximum and relative minimum points are located. The derivative is found and then set equal to zero. The solutions of the resulting equation give the x-coordinates of the maximum and minimum points.*

It remains now to determine whether a given value of x, for which the derivative is zero, is the coordinate of a maximum or a minimum point (or neither, which is also possible). From the discussion of increasing and decreasing values for y, we see that *the derivative changes sign from plus to minus when passing through a relative maximum point and from minus to plus when passing through a relative minimum point.* Thus, we find maximum and minimum points by determining those values of x for which the derivative is zero and by properly analyzing the sign change of the derivative. If the sign of the derivative does not change, it is neither a maximum nor a minimum point. This is known as the **first-derivative test for maxima and minima.**

In Fig. 24.29, a diagram for the first-derivative test is shown. The test for a relative maximum is shown in Fig. 24.29(a), and that for a relative minimum is shown in Fig. 24.29(b). For the curves shown in Fig. 24.29, $f(x)$ and $f'(x)$ are continuous throughout the interval shown. (Although $f(x)$ must be continuous, $f'(x)$ may be discontinuous at the maximum point or the minimum point, and the sign changes of the first-derivative test remain valid.)

First-Derivative Test
for Maxima and Minima

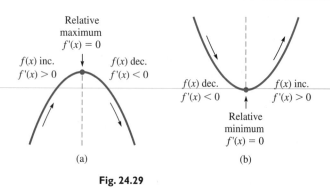

Relative
maximum
$f'(x) = 0$

$f(x)$ inc.
$f'(x) > 0$

$f(x)$ dec.
$f'(x) < 0$

(a)

$f(x)$ dec.
$f'(x) < 0$

$f(x)$ inc.
$f'(x) > 0$

Relative
minimum
$f'(x) = 0$

(b)

Fig. 24.29

◖ **EXAMPLE 2** Find any relative maximum points and relative minimum points on the graph of the function

$$y = 3x^5 - 5x^3$$

Finding the derivative and setting it equal to zero, we have

$$y' = 15x^4 - 15x^2 = 15x^2(x^2 - 1) = 15x^2(x - 1)(x + 1)$$

Therefore,

$$15x^2(x - 1)(x + 1) = 0 \quad \text{for } x = 0, \quad x = 1, \quad x = -1$$

Thus, the sign of the derivative is the same for all points to the left of $x = -1$. For these values, $y' > 0$ (thus, y is increasing). For values of x between -1 and 0, $y' < 0$. For values of x between 0 and 1, $y' < 0$. For values of x greater than 1, $y' > 0$. Thus, the curve has a maximum at $(-1, 2)$ and a minimum at $(1, -2)$. The point $(0, 0)$ is neither a maximum nor a minimum, since the sign of the derivative did not change at this value of x. The graph of $y = 3x^5 - 5x^3$ is shown in Fig. 24.30. ◗

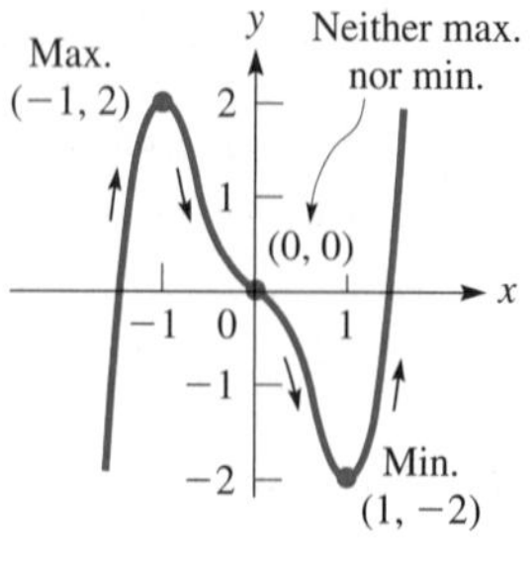

Max.
$(-1, 2)$

Neither max.
nor min.

$(0, 0)$

Min.
$(1, -2)$

Fig. 24.30

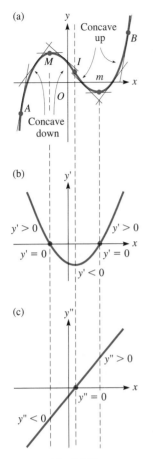

Fig. 24.31

We now look again at the slope of a tangent drawn to a curve. In Fig. 24.31(a), consider the *change* in the values of the slope of a tangent at a point as the point moves from *A* to *B*. At *A* the slope is positive, and as the point moves toward *M*, the slope remains positive but becomes smaller until it becomes zero at *M*. To the right of *M*, the slope is negative and becomes more negative until it reaches *I*. Therefore, *from A to I, the slope continually decreases.* To the right of *I*, the slope remains negative but increases until it becomes zero again at *m*. To the right of *m*, the slope becomes positive and increases to point *B*. Therefore, *from I to B, the slope continually increases.* We say that t*he curve is* **concave down** *from A to I and* **concave up** *from I to B.*

The curve in Fig. 24.31(b) is that of the derivative, and it therefore indicates the values of the slope of *f(x)*. If the slope changes, we are dealing with the rate of change of slope or the rate of change of the derivative. This function is the second derivative. The curve in Fig. 24.31(c) is that of the second derivative. We see that *where the second derivative of a function is* **negative,** *the slope is decreasing, or the curve is* **concave down** *(opens down). Where the second derivative is* **positive,** *the slope is increasing, or the curve is* **concave up** *(opens up).* This may be summarized as follows:

If $f''(x) > 0$, the curve is concave up.

If $f''(x) < 0$, the curve is concave down.

We can also now use this information in the determination of maximum and minimum points. By the nature of the definition of maximum and minimum points and of concavity, it is apparent that *a curve is concave down at a maximum point and concave up at a minimum point.* We can see these properties when we make a close analysis of the curve in Fig. 24.31. Therefore, at $x = a$,

if $f'(a) = 0$ and $f''(a) < 0$,

then *f(x)* has a relative maximum at $x = a$, or

if $f'(a) = 0$ and $f''(a) > 0$,

then *f(x)* has a relative minimum at $x = a$.

Second-Derivative Test for Maxima and Minima

These statements comprise what is known as the **second-derivative test for maxima and minima.** This test is often easier to use than the first-derivative test. However, it can happen that $y'' = 0$ at a maximum or minimum point, and in such cases it is necessary that we use the first-derivative test.

In using the second-derivative test, we should note that $f''(x)$ is *negative* at a *maximum* point and *positive* at a *minimum* point. This is contrary to a natural inclination to think of "maximum" and "positive" together or "minimum" and "negative" together.

Points of Inflection

The points at which the curve changes from concave up to concave down, or from concave down to concave up, are known as **points of inflection.** Thus, point *I* in Fig. 24.31 is a point of inflection. Inflection points are found by determining those values of *x* for which the second derivative changes sign. This is analogous to finding maximum and minimum points by the first-derivative test. In Fig. 24.32, various types of points of inflection are illustrated.

To show that it is necessary for *f(x)* and its derivatives to be continuous, see Exercise 52 of this section.

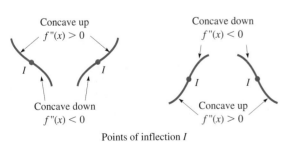

Points of inflection *I*

Fig. 24.32

◀ EXAMPLE 3   Determine the concavity and find any points of inflection of the function $y = x^3 - 3x$.

This requires an inspection and analysis of the second derivative. Therefore, we find the first two derivatives.

$$y' = 3x^2 - 3$$
$$y'' = 6x$$

The second derivative is positive where the function is concave up, and this occurs if $x > 0$. The curve is concave down for $x < 0$, since y'' is negative. Thus, $(0, 0)$ is a point of inflection, since the concavity changes there. The graph of $y = x^3 - 3x$ is shown in Fig. 24.33. ▶

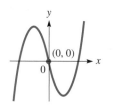

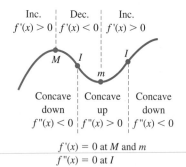

Fig. 24.33

At this point we summarize the information found from the derivatives of a function $f(x)$. See Fig. 24.34.

$f'(x) > 0$ where $f(x)$ increases; $f'(x) < 0$ where $f(x)$ decreases.

$f''(x) > 0$ where the graph of $f(x)$ is concave up; $f''(x) < 0$ where the graph of $f(x)$ is concave down.

If $f'(x) = 0$ at $x = a$, there is a relative maximum point if $f'(x)$ changes from $+$ to $-$ or if $f''(a) < 0$.

If $f'(x) = 0$ at $x = a$, there is a relative minimum point if $f'(x)$ changes from $-$ to $+$ or if $f''(a) > 0$.

If $f'(x) = 0$ at $x = a$, there is a point of inflection if $f''(x)$ changes from $+$ to $-$ or from $-$ to $+$.

Fig. 24.34

The following examples illustrate how the above information is put together to obtain the graph of a function.

◀ EXAMPLE 4   Sketch the graph of $y = 6x - x^2$.

Finding the first two derivatives, we have

$$y' = 6 - 2x = 2(3 - x)$$
$$y'' = -2$$

We now note that $y' = 0$ for $x = 3$. For $x < 3$, we see that $y' > 0$, which means that y is increasing over this interval. Also, for $x > 3$, we note that $y' < 0$, which means that y is decreasing over this interval.

Since y' changes from positive on the left of $x = 3$ to negative on the right of $x = 3$, the curve has a maximum point where $x = 3$. Since $y = 9$ for $x = 3$, this maximum point is $(3, 9)$.

Since $y'' = -2$, this means that its value remains constant for all values of x. Therefore, there are no points of inflection, and the curve is concave down for all values of x. This also shows that the point $(3, 9)$ is a maximum point.

Summarizing, we know that y is increasing for $x < 3$, y is decreasing for $x > 3$, there is a maximum point at $(3, 9)$, and the curve is always concave down. Using this information, we sketch the curve shown in Fig. 24.35.

From the equation, we know this curve is a parabola. We could also find the maximum point from the material of Section 7.4 or Section 21.7. However, using derivatives we can find this kind of important information about the graphs of a great many types of functions. ▶

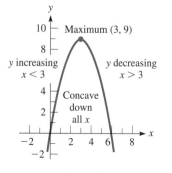

Fig. 24.35

◀ EXAMPLE 5   Sketch the graph of $y = 2x^3 + 3x^2 - 12x$.

Finding the first two derivatives, we have

$$y' = 6x^2 + 6x - 12 = 6(x + 2)(x - 1)$$
$$y'' = 12x + 6 = 6(2x + 1)$$

We note that $y' = 0$ when $x = -2$ and $x = 1$. Using these values in the second derivative, we find that y'' is negative (-18) for $x = -2$ and y'' is positive $(+18)$ when $x = 1$. When $x = -2$, $y = 20$; and when $x = 1$, $y = -7$. Therefore, $(-2, 20)$ is a relative maximum point, and $(1, -7)$ is a relative minimum point.

Next we see that $y' > 0$ if $x < -2$ or $x > 1$. Also, $y' < 0$ for the interval $-2 < x < 1$. Therefore, y is increasing if $x < -2$ or $x > 1$, and y is decreasing if $-2 < x < 1$.

Now we note that $y'' = 0$ when $x = -\frac{1}{2}$, $y'' < 0$ when $x < -\frac{1}{2}$, and $y'' > 0$ when $x > -\frac{1}{2}$. When $x = -\frac{1}{2}$, $y = \frac{13}{2}$. Therefore, there is a point of inflection at $\left(-\frac{1}{2}, \frac{13}{2}\right)$, the curve is concave down if $x < -\frac{1}{2}$, and the curve is concave up if $x > -\frac{1}{2}$.

Finally, by locating the points $(-2, 20)$, $\left(-\frac{1}{2}, \frac{13}{2}\right)$, and $(1, -7)$, we draw the curve *up* to $(-2, 20)$ and then *down* to $\left(-\frac{1}{2}, \frac{13}{2}\right)$, with the curve *concave down.* **Continuing down,** ***but* concave up,** we draw the curve to $(1, -7)$, at which point we start *up* and continue up. We now know the key points and the shape of the curve. See Fig. 24.36. For more precision, additional points may be used.   ◗

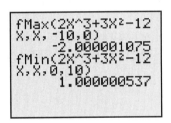

Fig. 24.36

CAUTION ▶

Most graphing calculators have a feature by which we can find the x-value for which a function is maximum (or minimum) over a specified interval. To find a *relative* maximum (or minimum) value, care must be used in choosing the lower and upper values of the interval so as not to include values for which the function may be greater (or less) than the value at the *relative* maximum (or minimum). A typical calculator display (showing the function, the variable, the lower interval value chosen, and the upper interval value chosen) to find the values of x for the maximum point and minimum point of Example 5 is shown in Fig. 24.37.

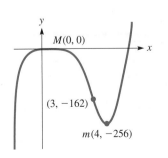

Fig. 24.37

◀ EXAMPLE 6   Sketch the graph of $y = x^5 - 5x^4$.

The first two derivatives are

$$y' = 5x^4 - 20x^3 = 5x^3(x - 4)$$
$$y'' = 20x^3 - 60x^2 = 20x^2(x - 3)$$

We now see that $y' = 0$ when $x = 0$ and $x = 4$. For $x = 0$, $y'' = 0$ also, which means ***we cannot use the second-derivative test*** for maximum and minimum points for $x = 0$ in this case. For $x = 4$, $y'' > 0$ $(+320)$, which means that $(4, -256)$ is a relative minimum point.

Next we note that

$$y' > 0 \text{ for } x < 0 \text{ or } x > 4 \qquad y' < 0 \text{ for } 0 < x < 4$$

Thus, by the first-derivative test, there is a relative maximum point at $(0, 0)$. Also, y is increasing for $x < 0$ or $x > 4$ and decreasing for $0 < x < 4$.

The second derivative indicates that there is a point of inflection at $(3, -162)$. It also indicates that the curve is concave down for $x < 3$ $(x \neq 0)$ and concave up for $x > 3$. There is no point of inflection at $(0, 0)$ since the second derivative does not change sign at $x = 0$.

CAUTION ▶

From this information, we sketch the curve in Fig. 24.38.   ◗

Fig. 24.38

EXERCISES 24.5

In Exercises 1–4, make the given changes in the indicated examples of this section and then solve the resulting problems.

1. In Example 1, after the $-$ sign change $3x^2$ to $6x^2$ and then find the values of x for which $f(x)$ is increasing and those for which it is decreasing.

2. In Example 2, change the $5x^3$ to $15x$ and then find the relative maximum and minimum points.

3. In Example 3, change the $-$ sign to $+$ and then determine the concavity and find any points of inflection.

4. In Example 4, change the $6x$ to $8x$ and then sketch the graph as in the example.

In Exercises 5–8, find those values of x for which the given functions are increasing and those values of x for which they are decreasing.

5. $y = x^2 + 2x$

6. $y = 2 + 6x - 3x^2$

7. $y = 12x - x^3$

8. $y = x^4 - 6x^2$

In Exercises 9–12, find any relative maximum or minimum points of the given functions. (These are the same functions as in Exercises 5–8.)

9. $y = x^2 + 2x$

10. $y = 2 + 6x - 3x^2$

11. $y = 12x - x^3$

12. $y = x^4 - 6x^2$

In Exercises 13–16, find the values of x for which the given function is concave up, the values of x for which it is concave down, and any points of inflection. (These are the same functions as in Exercises 5–8.)

13. $y = x^2 + 2x$

14. $y = 2 + 6x - 3x^2$

15. $y = 12x - x^3$

16. $y = x^4 - 6x^2$

In Exercises 17–20, sketch the graphs of the given functions by determining the appropriate information and points from the first and second derivatives (see Exercises 5–16). Use a graphing calculator to check the graph.

17. $y = x^2 + 2x$

18. $y = 2 + 6x - 3x^2$

19. $y = 12x - x^3$

20. $y = x^4 - 6x^2$

In Exercises 21–32, sketch the graphs of the given functions by determining the appropriate information and points from the first and second derivatives. Use a graphing calculator to check the graph. In Exercises 27–32, use the function maximum-minimum feature to check the relative maximum and minimum points.

21. $y = 12x - 2x^2$

22. $y = 4x^2 - 9$

23. $y = 2x^3 + 6x^2$

24. $y = x^3 - 9x^2 + 15x + 1$

25. $y = x^3 + 3x^2 + 3x + 2$

26. $y = x^3 - 12x + 12$

27. $y = 4x^3 - 24x^2 + 36x$

28. $y = x(x - 4)^3$

29. $y = 4x^3 - 3x^4$

30. $y = x^5 - 20x^2$

(W) 31. $y = x^5 - 5x$

32. $y = x^4 + 32x + 2$

In Exercises 33 and 34, view the graphs of y, y', y'' together on a graphing calculator. State how the graphs of y' and y'' are related to the graph of y.

(W) 33. $y = x^3 - 12x$

34. $y = 24x - 9x^2 - 2x^3$

In Exercises 35–38, describe the indicated features of the given graphs.

35. Display the graph of $y = x^3 + cx$ for $c = -3, -1, 1, 3$ on a graphing calculator. Describe how the graph changes as c varies.

36. Follow the instructions of Exercise 35 for the function $y = x^4 + cx^2$.

37. Display the graph of $y = 2x^5 - 7x^3 + 8x$ on a graphing calculator. Describe the relative locations of the left relative maximum point and the relative minimum points.

38. Describe the following features of the graph in Fig. 24.39 between or at the points A, B, C, D, E, F, and G. (a) Increasing and decreasing, (b) relative maximum and minimum points, (c) concavity, and (d) points of inflection.

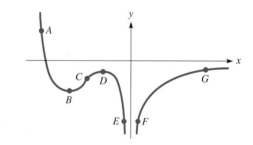

Fig. 24.39

In Exercises 39–48, sketch the indicated curves by the methods of this section. You may check the graphs by using a graphing calculator.

39. A batter hits a baseball that follows a path given by $y = x - 0.025x^2$, where distances are in meters. Sketch the graph of the path of the baseball.

40. The angle θ (in degrees) of a robot arm with the horizontal as a function of the time t (in s) is given by $\theta = 10 + 12t^2 - 2t^3$. Sketch the graph for $0 \le t \le 6$ s.

41. The power P (in W) in a certain electric circuit is given by $P = 4i - 0.5i^2$. Sketch the graph of P vs. i.

42. A computer analysis shows that the thrust T (in kN) of an experimental rocket motor is $T = 20 + 9t - 4t^3$, where t is the time (in min) after the motor is activated. Sketch the graph for the first 2 min.

43. For the force F exerted on the cam on the lever in Exercise 54 on page 670, sketch the graph of F vs. x. (*Hint:* Use methods of Chapter 15 to analyze the first derivative.)

44. The solar-energy power P (in W) produced by a certain solar system does not rise and fall uniformly during a cloudless day because of the system's location. An analysis of records shows that $P = -0.45(2t^5 - 45t^4 + 350t^3 - 1000t^2)$, where t is the time (in h) during which power is produced. Show that, during the solar-power production, the production flattens (inflection) in the middle and then peaks before shutting down. (*Hint:* The solutions are integral.)

45. An electric circuit is designed such that the resistance R (in Ω) is a function of the current i (in mA) according to $R = 75 - 18i^2 + 8i^3 - i^4$. Sketch the graph if $R \geq 0$ and i can be positive or negative.

46. An analysis of data showed that the mean density d (in mg/cm³) of a calcium compound in the bones of women was given by $d = 0.001\,81x^3 - 0.289x^2 + 12.2x + 30.4$, where x represents the ages of women ($20 < x < 80$ years). (A woman probably has osteoporosis if $d < 115$ mg/cm³.) Sketch the graph.

47. A rectangular box is made from a piece of cardboard 8 cm by 12 cm by cutting equal squares from each corner and bending up the sides. See Fig. 24.40. Express the volume of the box as a function of the side of the square that is cut out and then sketch the curve of the resulting equation.

48. A rectangular planter with a square end is to be made from 8.0 m² of redwood. Express the volume of soil the planter can hold as a function of the side of the square of the end. Sketch the curve of the resulting function.

In Exercises 49–51, sketch a continuous curve that has the given characteristics.

49. $f(1) = 0$; $f'(x) > 0$ for all x; $f''(x) < 0$ for all x

50. $f(0) = 1$; $f'(x) < 0$ for all x; $f''(x) < 0$ for $x < 0$; $f''(x) > 0$ for $x > 0$

51. $f(-1) = 0$; $f(2) = 2$; $f'(x) < 0$ for $x < -1$; $f'(x) > 0$ for $x > -1$; $f''(x) < 0$ for $0 < x < 2$; $f''(x) > 0$ for $x < 0$ or $x > 2$

(W) 52. Display the graph of $f(x) = x^{2/3}$ for $-2 < x < 2$ on a graphing calculator. Determine the continuity of $f(x)$, $f'(x)$, and $f''(x)$. Discuss the concavity of the curve in relation to the minimum point. (See the last paragraph on page 709.)

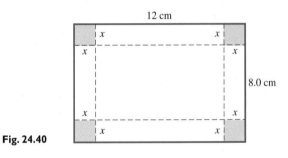

Fig. 24.40

24.6 MORE ON CURVE SKETCHING

When graphing functions in earlier chapters, we often used information that was obtainable from the function itself. We will now use this type of information, along with that found from the derivatives, to graph functions. We will find in graphing any particular function some types of information are of more value than others. The features we will consider in this section are as follows:

FEATURES TO BE USED IN GRAPHING FUNCTIONS

1. *Intercepts* Points for which the graph crosses (or is tangent to) each axis.

2. *Symmetry* For a review of symmetry, see Example 4 and the text before it on page 571.

3. *Behavior as x Becomes Large* We will find what happens to the function as $x \to +\infty$ and as $x \to -\infty$ in a way similar to that when we discussed the asymptotes of a hyperbola on page 585.

4. *Vertical Asymptotes* We will find that we can find vertical asymptotes similar to those noted on pages 94 and 305 by finding values that make factors in the denominator zero.

5. *Domain and Range* For a review of domain and range, see pages 85–87.

6. *Derivatives* We will find information from the first two derivatives as we did in the previous section (pages 707–711).

◀ EXAMPLE 1   Sketch the graph of $y = \dfrac{8}{x^2 + 4}$.

Intercepts If $x = 0$, $y = 2$, which means $(0, 2)$ is an intercept. If $y = 0$, there is no corresponding value of x, since $8/(x^2 + 4)$ is a fraction greater than zero for all x. This also indicates that all points on the curve are above the x-axis.

Symmetry The curve is symmetric to the y-axis since

$$y = \frac{8}{(-x)^2 + 4} \text{ is the same as } y = \frac{8}{x^2 + 4}.$$

The curve is not symmetric to the x-axis since

$$-y = \frac{8}{x^2 + 4} \text{ is not the same as } y = \frac{8}{x^2 + 4}.$$

The curve is not symmetric to the origin since

$$-y = \frac{8}{(-x)^2 + 4} \text{ is not the same as } y = \frac{8}{x^2 + 4}.$$

The value in knowing the symmetry is that we should find those portions of the curve on either side of the y-axis reflections of the other. It is possible to use this fact directly or to use it as a check.

Behavior as x Becomes Large We note that as $x \to \infty$, $y \to 0$ since $8/(x^2 + 4)$ is always a fraction that is greater than zero but which becomes smaller as x becomes larger. Therefore, we see that $y = 0$ is an asymptote. From either the symmetry or the function, we also see that $y \to 0$ as $x \to -\infty$.

Vertical Asymptotes From the discussion of the hyperbola, recall that an asymptote is a line that a curve approaches. We have already noted that $y = 0$ is an asymptote for this curve. This asymptote, the x-axis, is a horizontal line. *Vertical asymptotes, if any exist, are found by determining those values of x for which the denominator of any term is zero.* Such a value of x makes y undefined. Since $x^2 + 4$ cannot be zero, this curve has no vertical asymptotes. The next example illustrates a curve that has a vertical asymptote.

Domain and Range Since the denominator $x^2 + 4$ cannot be zero, x can take on any value. This means the domain of the function is all values of x. Also, we have noted that $8/(x^2 + 4)$ is a fraction greater than zero. Since $x^2 + 4$ is 4 or greater, y is 2 or less. This tells us that the range of the function is $0 < y \leq 2$.

Derivatives Since $y = 8/(x^2 + 4) = 8(x^2 + 4)^{-1}$

$$y' = -8(x^2 + 4)^{-2}(2x)$$
$$= \frac{-16x}{(x^2 + 4)^2}$$

Since $(x^2 + 4)^2$ is positive for all values of x, the sign of y' is determined by the numerator. Thus, we note that $y' = 0$ for $x = 0$ and that $y' > 0$ for $x < 0$ and $y' < 0$ for $x > 0$. The curve, therefore, is increasing for $x < 0$, is decreasing for $x > 0$, and has a maximum point at $(0, 2)$.

The curve for Example 1 is a special case (with $a = 1$) of the curve known as the *witch of Agnesi*. Its general form is

$$y = \frac{8a^3}{x^2 + 4a^2}$$

It is named for the Italian mathematician Maria Gaetana Agnesi (1718–1799). She wrote the first text that contained analytic geometry, differential and integral calculus, series (see Chapter 29), and differential equations (see Chapter 30). The word *witch* was used due to a mistranslation from Italian to English.

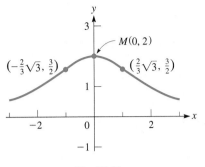

Fig. 24.41

Now finding the second derivative, we have

$$y'' = \frac{(x^2 + 4)^2(-16) + 16x(2)(x^2 + 4)(2x)}{(x^2 + 4)^4} = \frac{-16(x^2 + 4) + 64x^2}{(x^2 + 4)^3}$$

$$= \frac{48x^2 - 64}{(x^2 + 4)^3} = \frac{16(3x^2 - 4)}{(x^2 + 4)^3}$$

We note that y'' is negative for $x = 0$, which confirms that $(0, 2)$ is a maximum point. Also, points of inflection are found for the values of x satisfying $3x^2 - 4 = 0$. Thus, $\left(-\frac{2}{3}\sqrt{3}, \frac{3}{2}\right)$ and $\left(\frac{2}{3}\sqrt{3}, \frac{3}{2}\right)$ are points of inflection. The curve is concave up if $x < -\frac{2}{3}\sqrt{3}$ or $x > \frac{2}{3}\sqrt{3}$, and the curve is concave down if $-\frac{2}{3}\sqrt{3} < x < \frac{2}{3}\sqrt{3}$.

Putting this information together, we sketch the curve shown in Fig. 24.41. Note that this curve could have been sketched primarily by use of the fact that $y \rightarrow 0$ as $x \rightarrow +\infty$ and as $x \rightarrow -\infty$ and the fact that a maximum point exists at $(0, 2)$. However, the other parts of the analysis, such as symmetry and concavity, serve as checks and make the curve more accurate.

EXAMPLE 2 Sketch the graph of $y = x + \dfrac{4}{x}$.

Intercepts If we set $x = 0$, y is undefined. This means that the curve is not *continuous* at $x = 0$ and there are no y-intercepts. If we set $y = 0$, $x + 4/x = (x^2 + 4)/x$ cannot be zero since $x^2 + 4$ cannot be zero. Therefore, there are no intercepts. This may seem to be of little value, but we must realize *this curve does not cross either axis.* This will be of value when we sketch the curve in Fig. 24.42.

Symmetry In testing for symmetry, we find that the curve is not symmetric to either axis. However, this curve does possess symmetry to the origin. This is determined by the fact that when $-x$ replaces x and at the same time $-y$ replaces y, the equation does not change.

Behavior as x Becomes Large As $x \rightarrow +\infty$ and as $x \rightarrow -\infty$, $y \rightarrow x$ since $4/x \rightarrow 0$. Thus, $y = x$ is an asymptote of the curve.

Vertical Asymptotes As we noted in Example 1, vertical asymptotes exist for values of x for which y is undefined. In this equation, $x = 0$ makes the second term on the right undefined, and therefore y is undefined. In fact, as $x \rightarrow 0$ from the positive side, $y \rightarrow +\infty$, and as $x \rightarrow 0$ from the negative side, $y \rightarrow -\infty$. This is derived from the sign of $4/x$ in each case.

Domain and Range Since x cannot be zero, the domain of the function is all x except zero. As for the range, the analysis from the derivatives will show it to be $y \le -4$, $y \ge 4$.

Derivatives Finding the first derivative, we have

$$y' = 1 - \frac{4}{x^2} = \frac{x^2 - 4}{x^2}$$

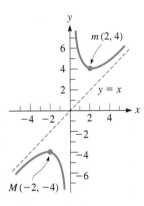

Fig. 24.42

The x^2 in the denominator indicates that the sign of the first derivative is the same as its numerator. The numerator is zero if $x = -2$ or $x = 2$. If $x < -2$ or $x > 2$, then $y' > 0$; and if $-2 < x < 2$, $x \ne 0$, $y' < 0$. Thus, y is increasing if $x < -2$ or $x > 2$, and also y is decreasing if $-2 < x < 2$, except at $x = 0$ (y is undefined). Also, $(-2, -4)$ is a relative maximum point, and $(2, 4)$ is a relative minimum point. The second derivative is $y'' = 8/x^3$. This cannot be zero, but it is negative if $x < 0$ and positive if $x > 0$. Thus, the curve is concave down if $x < 0$ and concave up if $x > 0$. Using this information, we have the curve shown in Fig. 24.42.

◀ **EXAMPLE 3** Sketch the graph of $y = \dfrac{1}{\sqrt{1 - x^2}}$.

Intercepts If $x = 0$, $y = 1$. If $y = 0$, $1/\sqrt{1 - x^2}$ would have to be zero, but it cannot since it is a fraction with 1 as the numerator for all values of x. Thus, $(0, 1)$ is an intercept.

Symmetry The curve is symmetric to the y-axis.

Behavior as x Becomes Large The values of x cannot be considered beyond 1 or -1, for any value of $x < -1$ or $x > 1$ gives imaginary values for y. Thus, the curve does not exist for values of $x < -1$ or $x > 1$.

Vertical Asymptotes If $x = 1$ or $x = -1$, y is undefined. In each case, as $x \to 1$ and as $x \to -1$, $y \to +\infty$.

Domain and Range From the analysis of x becoming large and of the vertical asymptotes, we see that the domain is $-1 < x < 1$. Also, since $\sqrt{1 - x^2}$ is 1 or less, $1/\sqrt{1 - x^2}$ is 1 or more, which means the range is $y \geq 1$.

Derivatives

$$y' = -\frac{1}{2}(1 - x^2)^{-3/2}(-2x) = \frac{x}{(1 - x^2)^{3/2}}$$

We see that $y' = 0$ if $x = 0$. If $-1 < x < 0$, $y' < 0$, and also if $0 < x < 1$, $y' > 0$. Thus, the curve is decreasing if $-1 < x < 0$ and increasing if $0 < x < 1$. There is a minimum point at $(0, 1)$.

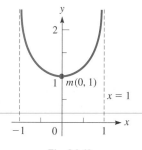

Fig. 24.43

$$y'' = \frac{(1 - x^2)^{3/2} - x(\frac{3}{2})(1 - x^2)^{1/2}(-2x)}{(1 - x^2)^3} = \frac{(1 - x^2) + 3x^2}{(1 - x^2)^{5/2}}$$

$$= \frac{2x^2 + 1}{(1 - x^2)^{5/2}}$$

The second derivative cannot be zero since $2x^2 + 1$ is positive for all values of x. The second derivative is also positive for all permissible values of x, which means the curve is concave up for these values.

Using this information, we sketch the graph in Fig. 24.43. ▶

◀ **EXAMPLE 4** Sketch the graph of $y = \dfrac{x}{x^2 - 4}$.

Intercepts If $x = 0$, $y = 0$, and if $y = 0$, $x = 0$. The only intercept is $(0, 0)$.

Symmetry The curve is not symmetric to either axis. However, since $-y = -x/[(-x)^2 - 4]$ is the same as $y = x/(x^2 - 4)$, it is symmetric to the origin.

Behavior as x Becomes Large As $x \to +\infty$ and as $x \to -\infty$, $y \to 0$. This means that $y = 0$ is an asymptote.

Vertical Asymptotes If $x = -2$ or $x = 2$, y is undefined. As $x \to -2$, $y \to -\infty$ if $x < -2$ since $x^2 - 4$ is positive, and $y \to +\infty$ if $x > -2$ since $x^2 - 4$ is negative. As $x \to 2$, $y \to -\infty$ if $x < 2$, and $y \to +\infty$ if $x > 2$.

Domain and Range The domain is all real values of x except -2 and 2. As for the range, if $x < -2$, $y < 0$ (the numerator is negative and the denominator is positive). If $x > 2$, $y > 0$ (both numerator and denominator are positive). Since $(0, 0)$ is an intercept, we see that the range is all values of y.

Derivatives

$$y' = \frac{(x^2 - 4)(1) - x(2x)}{(x^2 - 4)^2} = -\frac{x^2 + 4}{(x^2 - 4)^2}$$

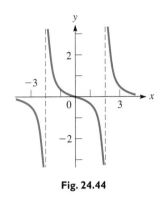

Fig. 24.44

Since $y' < 0$ for all values of x except -2 and 2, the curve is decreasing for all values in the domain.

$$y'' = -\frac{(x^2 - 4)^2(2x) - (x^2 + 4)(2)(x^2 - 4)(2x)}{(x^2 - 4)^4}$$

$$= -\frac{2x(x^2 - 4) - 4x(x^2 + 4)}{(x^2 - 4)^3} = \frac{2x^3 + 24x}{(x^2 - 4)^3} = \frac{2x(x^2 + 12)}{(x^2 - 4)^3}$$

The sign of y'' depends on x and $(x^2 - 4)^3$. If $x < -2$, $y'' < 0$. If $-2 < x < 0$, $y'' > 0$. If $0 < x < 2$, $y'' < 0$. If $x > 2$, $y'' > 0$. This means the curve is concave down for $x < -2$ or $0 < x < 2$ and is concave up for $-2 < x < 0$ or $x > 2$. The curve is sketched in Fig. 24.44.

EXERCISES 24.6

In Exercises 1–18, use the method of the examples of this section to sketch the indicated curves. Use a graphing calculator to check the graph.

1. In Example 2, change the $+$ to $-$ before $4/x$ and then proceed.

2. $y = \dfrac{4}{x^2}$ **3.** $y = \dfrac{2}{x + 1}$ **4.** $y = \dfrac{x}{x - 2}$

5. $y = x^2 + \dfrac{2}{x}$ **6.** $y = x + \dfrac{4}{x^2}$

7. $y = x - \dfrac{1}{x}$ **8.** $y = 3x + \dfrac{1}{x^3}$

9. $y = \dfrac{x^2}{x + 1}$ **10.** $y = \dfrac{9x}{x^2 + 9}$

11. $y = \dfrac{1}{x^2 - 1}$ **12.** $y = \dfrac{x^2 - 1}{x^3}$

13. $y = \dfrac{4}{x} - \dfrac{4}{x^2}$ **14.** $y = 4x + \dfrac{1}{\sqrt{x}}$

15. $y = x\sqrt{1 - x^2}$ **16.** $y = \dfrac{x - 1}{x^2 - 2x}$

17. $y = \dfrac{9x}{9 - x^2}$ **18.** $y = \dfrac{x^2 - 4}{x^2 + 4}$

In Exercises 19–28, solve the given problems.

(W) 19. Display the graph of $y = \dfrac{cx}{1 + c^2x^2}$ on a graphing calculator for $c = -3, -1, 1, 3$. Describe how the graph changes as c varies.

(W) 20. Display the graph of $y = x + \dfrac{4}{x^n}$ on a graphing calculator for $n = 1, 2, 3, 4$. Describe how the graph changes as n varies.

21. The combined capacitance C_T (in μF) of a 6-μF capacitance and a variable capacitance C in series is given by $C_T = \dfrac{6C}{6 + C}$. Sketch the graph.

22. Sketch a graph of P vs. V from van der Waal's equation (see Exercise 50 of Section 23.6), assuming the following values: $R = T = a = 1$ and $b = 0$. For many gases the value of a is much greater than that of b. Even though the values of $R = T = 1$ are not realistic, the *shape* of the curve will be correct for the assumed value of $b = 0$.

23. The reliability R of a computer model is found to be $R = \dfrac{200}{\sqrt{t^2 + 40\,000}}$, where t is the time of operation in hours. ($R = 1$ is perfect reliability, and $R = 0.5$ means there is a 50% chance of a malfunction.) Sketch the graph.

24. The electric power P (in W) produced by a source is given by $P = \dfrac{36R}{R^2 + 2R + 1}$, where R is the resistance in the circuit. Sketch the graph.

25. Assuming that a raindrop is always spherical and as it falls its radius increases from 1 mm to r mm, its velocity v (in mm/s) is $v = k\left(r - \dfrac{1}{r^3}\right)$. With $k = 1$, sketch the graph.

26. If a positive electric charge of $+q$ is placed between two negative charges of $-q$ that are two units apart, Coulomb's law states that the force F on the positive charge is $F = -\dfrac{kq^2}{x^2} + \dfrac{kq^2}{(x - 2)^2}$, where x is the distance from one of the negative charges. Let $kq^2 = 1$ and sketch the graph for $0 < x < 2$.

27. A cylindrical oil drum is to be made such that it will contain 20 kL. Sketch the area of sheet metal required for construction as a function of the radius of the drum.

28. A fence is to be constructed to enclose a rectangular area of $20\,000$ m^2. A previously constructed wall is to be used for one side. Sketch the length of fence to be built as a function of the side of the fence parallel to the wall. See Fig. 24.45.

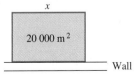

Fig. 24.45

24.7 APPLIED MAXIMUM AND MINIMUM PROBLEMS

Problems from various applied situations frequently occur that require finding a maximum or minimum value of some function. If the function is known, the methods we have already discussed can be used directly. This is discussed in the following example.

The first gasoline-engine automobile was built by the German engineer Karl Benz (1844–1929) in the 1880s.

EXAMPLE 1 An automobile manufacturer, in testing a new engine on one of its new models, found that the efficiency e of the engine as a function of the speed s of the car was given by $e = 0.768s - 0.00004s^3$. Here, e is measured in percent and s is measured in km/h. What is the maximum efficiency of the engine?

In order to find a maximum value, we find the derivative of e with respect to s:

$$\frac{de}{ds} = 0.768 - 0.00012s^2$$

We then set the derivative equal to zero in order to find the value of s for which a maximum may occur:

$$0.768 - 0.00012s^2 = 0$$
$$0.00012s^2 = 0.768$$
$$s^2 = 6400$$
$$s = 80.0 \text{ km/h}$$

We know that s must be positive to have meaning in this problem. Therefore, the apparent solution of $s = -80$ is discarded. The second derivative is

$$\frac{d^2e}{ds^2} = -0.00024s$$

which is negative for any positive value of s. Therefore, we have a maximum for $s = 80.0$. Substituting $s = 80.0$ in the function for e, we obtain

$$e = 0.768(80.0) - 0.00004(80.0^3) = 61.44 - 20.48 = 40.96$$

The maximum efficiency is about 41.0%, which occurs for $s = 80.0$ km/h.

In many problems for which a maximum or minimum value is to be found, the function is not given. To solve such a problem, we use these steps:

STEPS IN SOLVING APPLIED MAXIMUM AND MINIMUM PROBLEMS

1. *Determine the quantity Q to be maximized or minimized.*
2. If possible, *draw a figure illustrating the problem.*
3. *Write an equation for Q in terms of another variable of the problem.*
4. *Take the derivative of the function in step 3.*
5. *Set the derivative equal to zero, and solve the resulting equation.*
6. *Check as to whether the value found in step 5 makes Q a maximum or a minimum.* This might be clear from the statement of the problem, or it might require one of the derivative tests.
7. *Be sure the stated answer is the one the problem required.* Some problems require the maximum or minimum value, and others require values of other variables that give the maximum or minimum value.

CAUTION ▶ *The principal difficulty that arises in these problems is finding the proper function.* We must carefully read the problem to find the information needed to set up the function. The following examples illustrate several types of stated problems involving maximum and minimum values.

Solving a Word Problem ◀ EXAMPLE 2   Find the number that exceeds its square by the greatest amount.

The quantity to be maximized is the difference D between a number x and its square x^2. Therefore, the required function is

$$D = x - x^2$$

Since we want D to be a maximum, we find dD/dx, which is

$$\frac{dD}{dx} = 1 - 2x$$

Setting the derivative equal to zero and solving for x, we have

$$0 = 1 - 2x, \qquad x = \tfrac{1}{2}$$

The second derivative gives $d^2D/dx^2 = -2$, which tells us that the second derivative is always negative. This means that whenever the first derivative is zero it represents a maximum. In many problems, it is not necessary to test for maximum or minimum, since the nature of the problem will indicate which must be the case. For example, in this problem we know that numbers greater than 1 do not exceed their squares at all. The same is true for all negative numbers. Thus, the answer must be between 0 and 1; in this case it is $x = 1/2$. ◗

Solving a Word Problem ◀ EXAMPLE 3   A rectangular corral is to be enclosed with 1600 m of fencing. Find the maximum possible area of the corral.

There are limitless possibilities for rectangles of a perimeter of 1600 m and differing areas. See Fig. 24.46. For example, if the sides are 700 m and 100 m, the area is 70 000 m², or if the sides are 600 m and 200 m, the area is 120 000 m². Therefore, we set up a function for the area of a rectangle in terms of its sides x and y:

$$A = xy$$

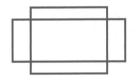

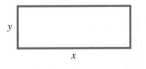

Fig. 24.46

Another important fact is that the perimeter of the corral is 1600 m. Therefore, $2x + 2y = 1600$. Solving for y, we have $y = 800 - x$. By using this expression for y, we can express the area in terms of x only. This gives us

$$A = x(800 - x) = 800x - x^2$$

We complete the solution as follows:

$$\frac{dA}{dx} = 800 - 2x \qquad \text{take derivative}$$

$$800 - 2x = 0 \qquad \text{set derivative equal to zero}$$

$$x = 400 \text{ m}$$

By checking values of the derivative near 400 or by finding the second derivative, we can show that we have a maximum for $x = 400$. This means that $x = 400$ m and $y = 400$ m give the maximum area of 160 000 m² for the corral. ◗

Solving a Word Problem

EXAMPLE 4 The strength S of a beam with a rectangular cross section is directly proportional to the product of its width w and the square of its depth d. Find the dimensions of the strongest beam that can be cut from a log with a circular cross section that is 16.0 cm in diameter. See Fig. 24.47.

Fig. 24.47

The solution proceeds as follows:

$$S = kwd^2 \qquad \text{direct variation}$$
$$d^2 = 256 - w^2 \qquad \text{Pythagorean theorem}$$
$$S = kw(256 - w^2) \qquad \text{substituting}$$
$$= k(256w - w^3) \qquad S = f(w)$$
$$\frac{dS}{dw} = k(256 - 3w^2) \qquad \text{take derivative}$$
$$0 = k(256 - 3w^2) \qquad \text{set derivative equal to zero}$$
$$3w^2 = 256 \qquad \text{solve for } w$$
$$w = \frac{16}{\sqrt{3}} = 9.24 \text{ cm}$$
$$d = \sqrt{256 - \frac{256}{3}} \qquad \text{solve for } d$$
$$= 13.1 \text{ cm}$$

This means that the strongest beam is about 9.24 cm wide and 13.1 cm deep. Since $d^2S/dw^2 = -6kw$ and is negative for $w > 0$ (the only values with meaning in this problem), these dimensions give the maximum strength for the beam.

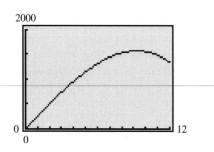

Fig. 24.48

The solution can also be checked on a graphing calculator. By graphing the equation $S = 256w - w^3$ (using y for S, x for w, and $k = 1$ (*the value of k does not affect the solution*)), we see in Fig. 24.48 that S is a maximum for w between 9 and 10. (The value $w = 9.24$ can be found by using the *trace* and *zoom* (or *maximum*) features.)

Solving a Word Problem

EXAMPLE 5 Find the point on the parabola $y = x^2$ that is nearest to the point $(6, 3)$.

In this example we must set up a function for the distance between a general point (x, y) on the parabola and the point $(6, 3)$. The relation is

$$D = \sqrt{(x - 6)^2 + (y - 3)^2}$$

However, to make it easier to take derivatives, we will square both sides of this expression. If a function is a minimum, then so is its square. We will also use the fact that the point (x, y) is on $y = x^2$ by replacing y by x^2. Thus, we have

$$D^2 = (x - 6)^2 + (x^2 - 3)^2 = x^2 - 12x + 36 + x^4 - 6x^2 + 9$$
$$= x^4 - 5x^2 - 12x + 45$$
$$\frac{dD^2}{dx} = 4x^3 - 10x - 12 \qquad \text{take derivative}$$
$$0 = 2x^3 - 5x - 6 \qquad \text{set derivative equal to zero}$$

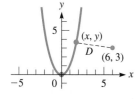

Fig. 24.49

Using synthetic division or some similar method, we find that the solution to this equation is $x = 2$. Thus, the required point on the parabola is $(2, 4)$. (See Fig. 24.49.) We can show that we have a minimum by analyzing the first derivative, by analyzing the second derivative, or by noting that points at much greater distances exist (therefore, it cannot be a maximum).

Solving a Word Problem

◀ EXAMPLE 6   A company determines that it can sell 1000 units of a product per month if the price is $5 for each unit. It also estimates that for each 1¢ reduction in unit price, 10 more units can be sold. Under these conditions, what is the maximum possible income and what price per unit gives this income?

If we let x = the number of units over 1000 sold, the total number of units sold is $1000 + x$. The price for each unit is $5 less 1¢ ($0.01) for each block of 10 units over 1000 that are sold. Thus, the price for each unit is

$$5 - 0.01\left(\frac{x}{10}\right) \quad \text{or} \quad 5 - 0.001x \text{ dollars}$$

The income I is the number of units sold times the price of each unit. Therefore,

$$I = (1000 + x)(5 - 0.001x)$$

Multiplying and finding the first derivative, we have

$$I = 5000 + 4x - 0.001x^2$$

$$\frac{dI}{dx} = 4 - 0.002x \qquad \text{take derivative}$$

$$0 = 4 - 0.002x \qquad \text{set derivative equal to zero}$$

$$x = 2000$$

We note that if $x < 2000$, the derivative is positive, and if $x > 2000$, the derivative is negative. Therefore, if $x = 2000$, I is at a maximum. This means that the maximum income is derived if 2000 units over 1000 are sold, or 3000 units in all. This in turn means that the maximum income is $9000 and the price per unit is $3. These values are found by substituting $x = 2000$ into the expression for I and for the price. ▶

Solving a Word Problem

See the chapter introduction.

◀ EXAMPLE 7   Find the dimensions of a 700-kL cylindrical oil-storage tank that can be made with the least cost of sheet metal, assuming that there is no wasted sheet metal.

Analyzing the wording of the problem carefully, we see that we are to minimize the surface area of a right circular cylinder with a volume of 700 kL. Therefore, we set up expressions for the surface area and the volume (700 kL = 700 m³):

$$A = 2\pi r^2 + 2\pi rh \qquad V = 700 = \pi r^2 h$$

We can express the equation for the area in terms of r only by solving the second equation for h and substituting in the first equation. This gives us

$$h = \frac{700}{\pi r^2}, \qquad A = 2\pi r^2 + 2\pi r\left(\frac{700}{\pi r^2}\right) = 2\pi r^2 + \frac{1400}{r} \qquad \text{(see Fig. 24.50)}$$

In order to find the minimum value of A, we find dA/dr and set it equal to zero:

$$\frac{dA}{dr} = 4\pi r - \frac{1400}{r^2}, \qquad 4\pi r - \frac{1400}{r^2} = 0, \qquad \frac{4\pi r^3 - 1400}{r^2} = 0$$

$$4\pi r^3 - 1400 = 0, \qquad r^3 = \frac{1400}{4\pi} \qquad \text{numerator must} = 0$$

$$r = \sqrt[3]{\frac{1400}{4\pi}} = 4.81 \text{ m} \qquad h = \frac{700}{\pi r^2} = 9.62 \text{ m}$$

Since dA/dr changes sign from negative to positive at (about) $r = 4.81$ m, A is a minimum. This is also verified by the calculator graph of A vs. r in Fig. 24.51. ▶

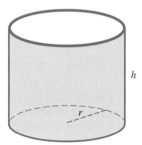

Fig. 24.50

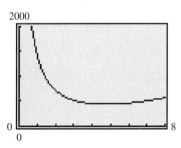

Fig. 24.51

Solving a Word Problem ◖EXAMPLE 8 The illuminance of a light source at any point equals the strength of the source divided by the square of the distance from the source. Two sources, of strengths 8 units and 1 unit, respectively, are 100 m apart. Determine at what point between them the illuminance is the least, assuming that the illuminance at any point is the sum of the illuminances of the two sources.

Let I = the sum of the illuminances and

x = the distance from the source of strength 8

Then, we find that

$$I = \frac{8}{x^2} + \frac{1}{(100 - x)^2}$$

is the function between the illuminance and the distance from the source of strength 8. We must now take a derivative of I with respect to x, set it equal to zero, and solve for x to find the point at which the illuminance is a minimum:

$$\frac{dI}{dx} = -\frac{16}{x^3} + \frac{2}{(100 - x)^3} = \frac{-16(100 - x)^3 + 2x^3}{x^3(100 - x)^3}$$

This function will be zero if the numerator is zero. Therefore, we have

$$2x^3 - 16(100 - x)^3 = 0 \quad \text{or} \quad x^3 = 8(100 - x)^3$$

Taking cube roots of each side, we have

$$x = 2(100 - x) \quad \text{or} \quad x = 66.7 \text{ m}$$

The illuminance is a minimum 66.7 m from the 8-unit source. ▶

EXERCISES 24.7

In the following exercises, solve the given maximum and minimum problems.

1. In Example 3, change 1600 m to 2400 m and then proceed.

2. In Example 7, change 700 kL to 800 kL and then proceed.

3. The height (in m) of a flare shot upward from the ground is given by $s = 34.3t - 4.9t^2$, where t is the time (in s). What is the greatest height to which the flare goes?

4. A small oil refinery estimates that its daily profit P (in dollars) from refining x barrels of oil is $P = 8x - 0.02x^2$. How many barrels should be refined for maximum daily profit, and what is the maximum profit?

5. The power output P of a battery of voltage E and internal resistance R is $P = EI - RI^2$, where I is the current. Find the current for which the power is a maximum.

6. In 2002 the projected U.S. Social Security Fund assets S (in dollars $\times 10^{12}$) was given by $S = -0.000\,74t^3 + 0.020t^2 + 1.1$, where t is the number of years after 2002. What is the maximum projected fund value, and in what year will it occur? Using a graphing calculator, determine when the fund will run out of money.

7. A company projects that its total savings S (in dollars) by converting to a solar-heating system with a solar-collector area A (in m²) will be $S = 360A - 0.10A^3$. Find the area that should give the maximum savings and find the amount of the maximum savings.

8. The altitude h (in m) of a jet that goes into a dive and then again turns upward is given by $h = 16t^3 - 240t^2 + 10\,000$, where t is the time (in s) of the dive and turn. What is the altitude of the jet when it turns up out of the dive?

9. The impedance Z (in Ω) in an electric circuit is given by $Z = \sqrt{R^2 + (X_L - X_C)^2}$. If $R = 2500 \ \Omega$ and $X_L = 1500 \ \Omega$, what value of X_C makes the impedance a minimum?

10. Test results show that, when coughing, the velocity v of the wind in a person's windpipe is $v = kr^2(a - r)$, where a is the radius of the windpipe when not coughing and k is a constant. Find r for the maximum value of v.

11. An alpha particle moves through a magnetic field along the parabolic path $y = x^2 - 4$. Determine the closest that the particle comes to the origin.

12. The electric potential V on the line $3x + 2y = 6$ is given by $V = 3x^2 + 2y^2$. At what point on this line is the potential a minimum?

13. A rectangular hole is to be cut in a wall for a vent. If the perimeter of the hole is 48 cm and the length of the diagonal is a minimum, what are the dimensions of the hole?

14. When two electric resistors R_1 and R_2 are in series, their total resistance (the sum) is 32 Ω. If the same resistors are in parallel, their total resistance (the reciprocal of which equals the sum of the reciprocals of the individual resistances) is the maximum possible for two such resistors. What is the resistance of each?

15. A rectangular microprocessor chip is designed to have an area of 25 mm^2. What must be its dimensions if its perimeter is to be a minimum?

16. The rectangular animal display area in a zoo is enclosed by chain-link fencing and divided into two areas by internal fencing parallel to one of the sides. What dimensions will give the maximum area for the display if a total of 240 m of fencing are used?

17. For the microprocessor chip of Exercise 15, show that the chip will always be a square for a given value of the area A.

18. A rectangular storage area is to be constructed along the side of a tall building. A security fence is required along the remaining three sides of the area. What is the maximum area that can be enclosed with 800 m of fencing?

19. Ship A is traveling due east at 18.0 km/h as it passes a point 40.0 km due south of ship B, which is traveling due south at 16.0 km/h. How much later are the ships nearest each other?

20. An architect is designing a rectangular building in which the front wall costs twice as much per linear meter as the other three walls. The building is to cover 1350 m^2. What dimensions must it have such that the cost of the walls is a minimum?

21. A computer is programmed to display a slowly changing right triangle with its hypotenuse always equal to 12.0 cm. What are the legs of the triangle when it has its maximum area?

22. Canadian postal regulations require that the sum of the three dimensions of a rectangular package not exceed 3 m. What are the dimensions of the largest rectangular box with square ends that can be mailed?

23. A culvert designed with a semicircular cross section of diameter 2.40 m is redesigned to have an isosceles trapezoidal cross section by inscribing the trapezoid in the semicircle. See Fig. 24.52. What is the length of the bottom base b of the trapezoid if its area is to be maximum?

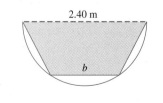

2.40 m

b

Fig. 24.52

24. A 36-cm wide sheet of metal is bent into a rectangular trough as shown in Fig. 24.53. What dimensions give the maximum water flow?

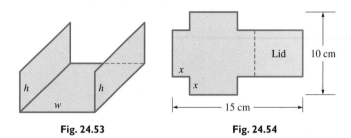

| **Fig. 24.53** | **Fig. 24.54** |

25. A box with a lid is to be made from a rectangular piece of cardboard 10 cm by 15 cm, as shown in Fig. 24.54. Two equal squares of side x are to be removed from one end, and two equal rectangles are to be removed from the other end so that the tabs can be folded to form the box with a lid. Find x such that the volume of the box is a maximum.

26. A lap pool (a pool for swimming laps) is designed to be seven times as long as it is wide. If the area of the sides and bottom is 90.0 m^2, what are the dimensions of the pool if the volume of water it can hold is at maximum?

27. What is the maximum slope of the curve $y = 6x^2 - x^3$?

28. What is the minimum slope of the curve $y = x^5 - 10x^2$?

29. The deflection y of a beam of length L at a horizontal distance x from one end is given by $y = k(2x^4 - 5Lx^3 + 3L^2x^2)$, where k is a constant. For what value of x does the maximum deflection occur?

30. The electric power P (in W) produced by a certain battery is given by $P = \dfrac{144r}{(r + 0.6)^2}$, where r is the resistance in the circuit. For what value of r is the power a maximum?

31. For raising a load, the efficiency E (in %) of a screw with square threads is $E = \dfrac{100T(1 - fT)}{T + f}$, where f is the coefficient of friction and T is the tangent of the pitch angle of the screw. If $f = 0.25$, what acute angle makes E the greatest?

32. Computer simulation shows that the drag F (in N) on a certain airplane is $F = 0.00500v^2 + 3.00 \times 10^8/v^2$, where v is the velocity (in km/h) of the plane. For what velocity is the drag the least?

33. Factories A and B are 8.0 km apart, with factory B emitting eight times the pollutants into the air as factory A. If the number n of particles of pollutants is inversely proportional to the square of the distance from a factory, at what point between A and B is the pollution the least?

34. The potential energy E of an electric charge q due to another charge q_1 at a distance of r_1 is proportional to q_1 and inversely proportional to r_1. If charge q is placed directly between two charges of 2.00 nC and 1.00 nC that are separated by 10.0 mm, find the point at which the total potential energy (the sum due to the other two charges) of q is a minimum.

35. An open box is to be made from a square piece of cardboard whose sides are 20.0 cm long, by cutting equal squares from the corners and bending up the sides. Determine the side of the square that is to be cut out so that the volume of the box may be a maximum. See Fig. 24.55.

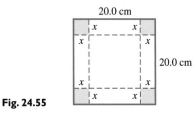

Fig. 24.55

36. A cone-shaped paper cup is to hold 100 cm³ of water. Find the height and radius of the cup that can be made from the least amount of paper.

37. A race track 400 m long is to be built around an area that is a rectangle with a semicircle at each end. Find the open side of the rectangle if the area of the rectangle is to be a maximum. See Fig. 24.56.

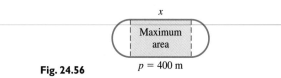

Fig. 24.56

38. A company finds that there is a net profit of $10 for each of the first 1000 units produced each week. For each unit over 1000 produced, there is 2 cents less profit per unit. How many units should be produced each week to net the greatest profit?

39. A beam of rectangular cross section is to be cut from a log 1.00 m in diameter. The stiffness of the beam varies directly as the width and the cube of the depth. What dimensions will give the beam maximum stiffness? See Fig. 24.57.

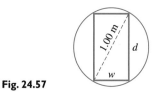

Fig. 24.57

40. On a computer simulation, a target plane is projected to be at $(1.20, 7.00)$ (distances in km), and a rocket is fired along the path $y = 8.00 - 2.00x^2$. How far from the target does the rocket pass from the target plane's projected position?

41. An oil pipeline is to be built from a refinery to a tanker loading area. The loading area is 10.0 km downstream from the refinery and on the opposite side of a river 2.5 km wide. The pipeline is to run along the river and then cross to the loading area. If the pipeline costs $50 000 per km alongside the river and $80 000 per km across the river, find the point P (see Fig. 24.58) at which the pipeline should be turned to cross the river if construction costs are to be a minimum.

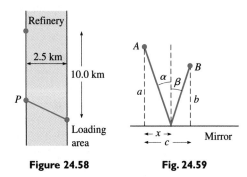

Figure 24.58   **Fig. 24.59**

42. A light ray follows a path of least time. If a ray starts at point A (see Fig. 24.59) and is reflected off a plane mirror to point B, show that the angle of incidence α equals the angle of reflection β. (*Hint*: Set up the expression in terms of x, which will lead to $\sin \alpha = \sin \beta$.)

43. A rectangular building covering 7000 m² is to be built on a rectangular lot as shown in Fig. 24.60. If the building is to be 10.0 m from the lot boundary on each side and 20.0 m from the boundary in front and back, find the dimensions of the building if the area of the lot is a minimum.

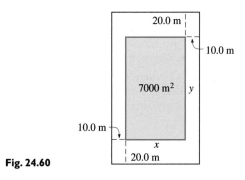

Fig. 24.60

44. A cylindrical cup (no top) is designed to hold 375 cm³ (375 mL). There is no waste in the material used for the sides. However, there is waste in that the bottom is made from a square $2r$ on a side. What are the most economical dimensions for a cup made under these conditions?

24.8 DIFFERENTIALS AND LINEAR APPROXIMATIONS

The symbol dy/dx for the derivatives was first used by Leibniz (see page 645).

To this point, we have used the dy/dx notation for the derivative of y with respect to x, but we have not considered it to be a ratio. In this section, we define the quantities dy and dx, called *differentials,* such that their ratio is equal to the derivative. Then we show that differentials have applications in errors in measurement and in approximating values of functions. Also, in the next chapter, we use the differential notation in the development of integration, which is the inverse process of differentiation.

DIFFERENTIALS

We define the **differential** *of a function* $y = f(x)$ as

$$dy = f'(x)\, dx \qquad\qquad (24.6)$$

In Eq. (24.6), the quantity *dy is the differential of y, and dx is the differential of x.* In this way, we can interpret the derivative as the ratio of the differential of y to the differential of x.

◀ **EXAMPLE 1**   Find the differential of $y = 3x^5 - x$.
    Since $f(x) = 3x^5 - x$, we find $f'(x) = 15x^4 - 1$. This means that

$$dy = \overset{\displaystyle\overset{f'(x)}{\big\downarrow}}{(15x^4 - 1)}\, dx$$

the differential of x

◀ **EXAMPLE 2**   Find the differential of $s = (2t^3 - 1)^4$.

$$ds = 4(2t^3 - 1)^3(6t^2)\, dt$$
$$= 24t^2(2t^3 - 1)^3\, dt$$

◀ **EXAMPLE 3**   Find the differential of $y = \dfrac{4x}{x^2 + 4}$.

$$dy = \frac{(x^2 + 4)(4) - (4x)(2x)}{(x^2 + 4)^2}\, dx \qquad \text{using derivative quotient rule}$$

$$= \frac{4x^2 + 16 - 8x^2}{(x^2 + 4)^2}\, dx = \frac{-4x^2 + 16}{(x^2 + 4)^2}\, dx$$

$$= \frac{-4(x^2 - 4)}{(x^2 + 4)^2}\, dx$$

CAUTION ▶ don't forget the dx

To understand more about differentials and their applications, we now introduce some useful notation. If a variable x changes value from x_1 to x_2, *the difference $x_2 - x_1$* Increments *is called the* **increment** *in x.* Traditionally, in calculus this increment is denoted by Δx ("delta x") which means $\Delta x = x_2 - x_1$. We must be careful to note that Δx is not a product of Δ and x, but a *single symbol* that represents the *change* in x.

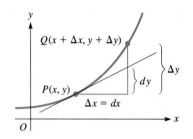

Fig. 24.61

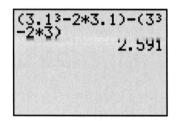

Fig. 24.62

By choosing $\Delta x = dx$, if dx is small, then the differential in y, dy, closely approximates the increment in y, Δy. This is the basis of the applications of the differential, and to understand this better, let us look at Fig. 24.61. We see that points $P(x, y)$ and $Q(x + \Delta x, y + \Delta y)$ lie on the graph of $f(x)$ and that the increments of x and y between the points are Δx and Δy. At point P, $f'(x) = dy/dx$, which means that the slope of a tangent line at P is indicated by dy/dx. With $dx = \Delta x$, we see that as dx becomes smaller, dy more nearly approximates Δy, which is the actual difference in the y-values. Therefore, *for small values of Δx, dy can be used to approximate Δy.*

◀ EXAMPLE 4   Calculate Δy and dy for $y = x^3 - 2x$ for $x = 3$ and $\Delta x = 0.1$.
First, we find Δy by calculating $f(3.1) - f(3)$. Therefore,

$$\Delta y = f(3.1) - f(3) = [3.1^3 - 2(3.1)] - [3^3 - 2(3)] = 2.591$$

See the calculator display in Fig. 24.62.
The differential of y is

$$dy = (3x^2 - 2)\, dx$$

Since $dx = \Delta x$, we have

$$dy = [3(9) - 2](0.1) = 2.5$$

Thus, $\Delta y = 2.591$ and $dy = 2.5$. In this case, dy is very nearly equal to Δy. ▶

ESTIMATING ERRORS IN MEASUREMENT

The fact that dy can be used to approximate Δy is useful in finding the error in a result from a measurement, if the data are in error, or the equivalent problem of finding the change in the result if a change is made in the data. Even though such changes can be found by using a calculator, the differential can be used to set up a general expression for the change of a particular function.

Solving a Word Problem   ◀ EXAMPLE 5   The edge of a cube of gold was measured to be 3.850 cm. From this value, the volume was found. Later it was discovered that the value of the edge was 0.020 cm too small. By approximately how much was the volume in error?
The volume V of a cube, in terms of an edge e, is $V = e^3$. Since we wish to find the change in V for a given change in e, we want the value of dV for $e = 3.850$ cm and $de = 0.020$ cm.
First, finding the general expression for dV, we have

$$dV = 3e^2\, de$$

Now, evaluating this expression for the given values, we have

$$dV = 3(3.850)^2(0.020) = 0.89 \text{ cm}^3$$

In this case, the volume was in error by about 0.89 cm³. As long as de is small compared with e, we can calculate an error or change in the volume of a cube by calculating the value of $3e^2\, de$. ▶

Often when considering the error of a given value or result, *the actual numerical value of the error, the* **absolute error,** *is not as important as its size in relation to the size of the quantity itself. The ratio of the absolute error to the size of the quantity itself is known as the* **relative error**, which is commonly expressed as a percent.

◀ EXAMPLE 6   Referring to Example 5, we see that the absolute error in the edge was 0.020 cm. The relative error in the edge was

$$\frac{de}{e} = \frac{0.020}{3.85} = 0.0052 = 0.52\%$$

The absolute error in the volume was 0.89 cm³, and the original value of the volume was $3.850^3 = 57.07$ cm³. This means the relative error in the volume was

$$\frac{dV}{V} = \frac{0.89}{57.07} = 0.016 = 1.6\%$$

▶

LINEAR APPROXIMATIONS

For reference, Eq. (21.6) is $y - y_1 = m(x - x_1)$.

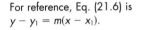

Continuing our discussion related to Fig. 24.61, on the curve of the function $f(x)$ at the point $(a, f(a))$, the slope of a tangent line is $f'(a)$. See Fig. 24.63. For a point (x, y) on the tangent line, by using the point-slope form for the equation of a straight line, Eq. (21.6), we have

$$f(x) = f(a) + f'(a)(x - a)$$

For the points on $y = f(x)$ near $x = a$, we can use the tangent line to approximate the function, as shown in Fig. 24.63. Therefore, the approximation $f(x) \approx L(x)$ *is the* **linear approximation** *of* $f(x)$ *near* $x = a$, *where the function*

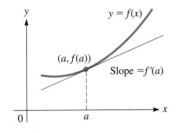

Fig. 24.63

$$\boxed{L(x) = f(a) + f'(a)(x - a)} \tag{24.7}$$

is called the **linearization** *of* $f(x)$ *at* $x = a$.

◀ EXAMPLE 7   Find the linearization of the function $f(x) = \sqrt{2x + 1}$ at $x = 4$. Use it to approximate $\sqrt{9.06}$.

The solution is as follows:

$$f(x) = \sqrt{2x + 1}$$

$$f'(x) = \tfrac{1}{2}(2x + 1)^{-1/2}(2) = \frac{1}{\sqrt{2x + 1}} \qquad \text{find the derivative}$$

$$f(4) = \sqrt{2(4) + 1} = 3 \qquad f'(4) = \frac{1}{\sqrt{2(4) + 1}} = \frac{1}{3} \qquad \text{evaluate } f(4) \text{ and } f'(4)$$

$$L(x) = 3 + \tfrac{1}{3}(x - 4) = \frac{x + 5}{3} \qquad \text{Use Eq. (24.7)}$$

$$\sqrt{2x + 1} \approx \frac{x + 5}{3} \qquad \begin{array}{l}2x + 1 = 9.06, \\ x = 4.03\end{array}$$

$$\sqrt{9.06} \approx \frac{4.03 + 5}{3} = \frac{9.03}{3} = 3.01 \qquad \begin{array}{l}\text{by calculator,} \\ \sqrt{9.06} = 3.009\,983\,389\end{array}$$

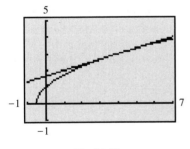

Fig. 24.64

In Fig. 24.64, a graphing calculator display showing $f(x) = \sqrt{2x + 1}$ and the tangent line (using the *tangent* feature) is shown. We see that the tangent line gives a good approximation of the function when x is near 4.

▶

EXERCISES 24.8

In Exercises 1–4, make the given changes in the indicated examples of this section and then solve the resulting problems.

1. In Example 3, change x^2 to x^3 and then find dy.

2. In Example 4, change the $-$ before $2x$ to $+$ and then calculate Δy and dy.

3. In Example 6, change 0.020 to 0.025 and then find de/e and dV/V.

4. In Example 7, change $x = 4$ to $x = 12$ and then approximate $\sqrt{25.06}$.

In Exercises 5–16, find the differential of each of the given functions.

5. $y = x^5 + x$

6. $y = 3x^2 + 6$

7. $V = \dfrac{2}{r^5} + 3\pi^2$

8. $y = 2\sqrt{x} - \dfrac{1}{x}$

9. $s = 2(3t^2 - 5)^4$

10. $y = 5(4 + 3x)^{1/3}$

11. $y = \dfrac{2}{3x^2 + 1}$

12. $R = \sqrt{\dfrac{u}{1 + 2u}}$

13. $y = x^2(1 - x)^3$

14. $y = 6x\sqrt{1 - 4x}$

15. $y = \dfrac{x}{5x + 2}$

16. $y = \dfrac{3x + 1}{\sqrt{2x - 1}}$

In Exercises 17–20, find the values of Δy and dy for the given values of x and dx.

17. $y = 7x^2 + 4x, x = 4, \Delta x = 0.2$

18. $y = (x^2 + 2x)^3, x = 7, \Delta x = 0.02$

19. $y = x\sqrt{1 + 4x}, x = 12, \Delta x = 0.06$

20. $y = \dfrac{x}{\sqrt{6x - 1}}, x = 3.5, \Delta x = 0.025$

In Exercises 21–24, find the linearization $L(x)$ of the given functions for the given values of a. Display $f(x)$ and $L(x)$ on the same calculator screen.

21. $f(x) = x^2 + 2x, a = 0$

22. $f(x) = 2\sqrt[3]{x}, a = 8$

23. $f(x) = \dfrac{1}{2x + 1}, a = -1$

24. $g(x) = x\sqrt{2x + 8}, a = -2$

In Exercises 25–36, solve the given problems by finding the appropriate differential.

25. If a spacecraft circles the earth at an altitude of 250 km, how much farther does it travel in one orbit than an airplane that circles the earth at a low altitude? The radius of the earth is 6370 km.

26. Approximate the amount of paint needed to apply one coat of paint 0.50 mm thick on a hemispherical dome 55 m in diameter.

27. The radius of a circular manhole cover is measured to be 40.6 ± 0.05 cm (this means the possible error in the radius is 0.05 cm). Estimate the possible relative error in the area of the top of the cover.

28. The side of a square microprocessor chip is measured as 0.950 cm, and later it is measured as 0.952 cm. What is the difference in the calculations of the area due to the difference in the measurements of the side? See Fig. 24.65.

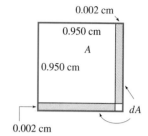

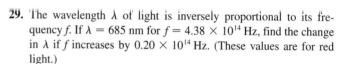

Fig. 24.65

29. The wavelength λ of light is inversely proportional to its frequency f. If $\lambda = 685$ nm for $f = 4.38 \times 10^{14}$ Hz, find the change in λ if f increases by 0.20×10^{14} Hz. (These values are for red light.)

30. The velocity of an object rolling down a certain inclined plane is given by $v = \sqrt{40 + 4.9h}$, where h is the distance (in m) traveled along the plane by the object. What is the increase in velocity (in m/s) of an object in moving from 20.0 m to 20.5 m along the plane? What is the relative change in the velocity?

31. The radius r of a holograph is directly proportional to the square root of the wavelength λ of the light used. Show that $dr/r = \frac{1}{2}d\lambda/\lambda$.

32. The gravitational force F of the earth on an object is inversely proportional to the square of the distance r of the object from the center of the earth. Show that $dF/F = -2dr/r$.

33. Show that an error of 2% in the measurement of the side of a square results in an error of approximately 4% in the calculation of the area.

34. Show that an error of 2% in the measurement of the radius of a sphere results in an error of approximately 6% in the calculation of the volume.

35. Calculate $\sqrt{4.05}$, using differentials.

W 36. Explain how to evaluate 2.03^4.

In Exercises 37–40, solve the linearization problems.

37. Linearize $f(x) = \sqrt{2 - x}$ for $a = 1$ and use it to approximate the value of $\sqrt{1.9}$.

W 38. Explain how to evaluate $\sqrt[3]{8.03}$, using linearization.

39. The capacitance C (in μF) in an element of an electronic tuner is given by $C = \dfrac{3.6}{\sqrt{1 + 2V}}$, where V is the voltage. Linearize C for $V = 4.0$ V.

40. A 16-Ω resistor is put in parallel with a variable resistor of resistance R. The combined resistance of the two resistors is $R_T = \dfrac{16R}{16 + R}$. Linearize R_T for $R = 4.0\ \Omega$.

CHAPTER 24 EQUATIONS

Newton's method

$$x_2 = x_1 - \frac{f(x_1)}{f'(x_1)} \qquad (24.1)$$

Curvilinear motion

$$v_x = \frac{dx}{dt} \qquad v_y = \frac{dy}{dt} \qquad (24.2)$$

$$a_x = \frac{dv_x}{dt} = \frac{d^2x}{dt^2} \qquad a_y = \frac{dv_y}{dt} = \frac{d^2y}{dt^2} \qquad (24.3)$$

$$v = \sqrt{v_x^2 + v_y^2} \qquad a = \sqrt{a_x^2 + a_y^2} \qquad (24.4)$$

$$\tan \theta_v = \frac{v_y}{v_x} \qquad \tan \theta_a = \frac{a_y}{a_x} \qquad (24.5)$$

Curve sketching and maximum and minimum values

$f'(x) > 0$ where $f(x)$ increases; $f'(x) < 0$ where $f(x)$ decreases.

$f''(x) > 0$ where the graph of $f(x)$ is concave up; $f''(x) < 0$ where the graph of $f(x)$ is concave down.

If $f'(x) = 0$ at $x = a$, there is a relative maximum point if $f'(x)$ changes from $+$ to $-$ or if $f''(a) < 0$.

If $f'(x) = 0$ at $x = a$, there is a relative minimum point if $f'(x)$ changes from $-$ to $+$ or if $f''(a) > 0$.

If $f''(x) = 0$ at $x = a$, there is a point of inflection if $f''(x)$ changes from $+$ to $-$ or from $-$ to $+$.

Differential

$$dy = f'(x)\, dx \qquad (24.6)$$

Linearization

$$L(x) = f(a) + f'(a)(x - a) \qquad (24.7)$$

CHAPTER 24 REVIEW EXERCISES

In Exercises 1–6, find the equations of the tangent and normal lines. Use a graphing calculator to view the curve and the line.

1. Find the equation of the line tangent to the parabola $y = 3x - x^2$ at the point $(-1, -4)$.

2. Find the equation of the line tangent to the curve $y = x^2 - \dfrac{6}{x}$ at the point $(2, 1)$.

3. Find the equation of the line normal to $x^2 - 4y^2 = 9$ at the point $(5, 2)$.

4. Find the equation of the line normal to $y = \dfrac{1}{\sqrt{x-2}}$ at the point $\left(6, \frac{1}{2}\right)$.

5. Find the equation of the line tangent to the curve $y = \sqrt{x^2 + 3}$ and that has a slope of $\frac{1}{2}$.

6. Find the equation of the line normal to the curve $y = \dfrac{1}{2x+1}$ and that has a slope of $\frac{1}{2}$ if $x \geq 0$.

In Exercises 7–12, find the indicated velocities and accelerations.

7. Given that the x- and y-coordinates of a moving particle are given as a function of time t by the parametric equations $x = \sqrt{t} + t$, $y = \frac{1}{12}t^3$, find the magnitude and direction of the velocity when $t = 4$.

8. If the x- and y-coordinates of a moving object as functions of time t are given by $x = 0.1t^2 + 1$, $y = \sqrt{4t + 1}$, find the magnitude and direction of the velocity when $t = 6$.

9. An object moves along the curve $y = 0.5x^2 + x$ such that $v_x = 0.5\sqrt{x}$. Find v_y at $(2, 4)$.

10. A particle moves along the curve of $y = \dfrac{1}{x+2}$ with a constant velocity in the x-direction of 4 cm/s. Find v_y at $\left(2, \frac{1}{4}\right)$.

11. Find the magnitude and direction of the acceleration for the particle in Exercise 7.

12. Find the magnitude and direction of the acceleration for the particle in Exercise 10.

In Exercises 13–16, find the indicated roots of the given equations to at least four decimal places by use of Newton's method.

13. $x^3 - 3x^2 - x + 2 = 0$ (between 0 and 1)

14. $2x^3 - 4x^2 - 9 = 0$ (between 2 and 3)

15. $x^2 = \frac{3}{x}$ (the real solution)

16. $\sqrt{x-2} = \frac{1}{(x-6)^2}$ (the real solution)

In Exercises 17–24, sketch the graphs of the given functions by information obtained from the function as well as information obtained from the derivatives. Use a graphing calculator to check the graph.

17. $y = 4x^2 + 16x$

18. $y = x^3 + 2x^2 + x + 1$

19. $y = 27x - x^3$

20. $y = x(6-x)^3$

21. $y = x^4 - 32x$

22. $y = x^4 + 4x^3 + 16x$

23. $y = \dfrac{x^2}{\sqrt{x^2-1}}$

24. $y = x^3 + \dfrac{3}{x}$

In Exercises 25–28, find the differential of each of the given functions.

25. $y = 4x^3 + \dfrac{1}{x}$

26. $y = \dfrac{1}{(2x-1)^2}$

27. $y = x\sqrt[3]{1-3x}$

28. $s = \sqrt{\dfrac{2+t}{2-t}}$

In Exercises 29 and 30, evaluate $\Delta y - dy$ for the given functions and values.

29. $y = x^3$, $x = 2$, $\Delta x = 0.1$

30. $y = 6x^2 - x$, $x = 3$, $\Delta x = 0.2$

In Exercises 31 and 32, find the linearization of the given functions for the given values of a.

31. $f(x) = \sqrt{x^4 + 3x^2 + 8}$, $a = 2$

32. $f(x) = x^2(x+1)^4$, $a = -2$

In Exercises 33–40, solve the given problems by finding the appropriate differentials.

33. A weather balloon 3.500 m in radius becomes covered with a uniform layer of ice 1.2 cm thick. What is the volume of the ice?

34. The total power P (in W) transmitted by an AM radio transmitter is $P = 460 + 230m^2$, where m is the modulation index. What is the change in power if m changes from 0.86 to 0.89?

35. A cylindrical silo with a flat top is 12.0 m in diameter and is 12.0 m high. By how much is the volume changed if the radius is increased by 1.0 m and the height is unchanged?

36. A ski slope follows a path that can be represented by $y = 0.010x^2 - 0.86x + 24$. What is the change in the slope of the path when x changes from 26 m to 28 m?

37. The impedance Z of an electric circuit as a function of the resistance R and the reactance X is given by $Z = \sqrt{R^2 + X^2}$. Derive an expression of the relative error in impedance for an error in R and a given value of X.

38. Evaluate $\sqrt{8.94}$.

39. Evaluate 3.02^5.

40. Show that the relative error in the calculation of the volume of a sphere is approximately three times the relative error in the measurement of the radius.

In Exercises 41–76, solve the given problems.

41. The parabolas $y = x^2 + 2$ and $y = 4x - x^2$ are tangent to each other. Find the equation of the line tangent to them at the point of tangency.

42. Find the equation of the line tangent to the curve of $y = x^4 - 8x$ and perpendicular to the line $4y - x + 5 = 0$.

43. The deflection y (in m) of a beam at a horizontal distance x (in m) from one end is given by $y = k(x^4 - 30x^3 + 1000x)$, where k is a constant. Observing the equation and using Newton's method, find the values of x where the deflection is zero, if the beam is 10.000 m long.

44. The edges of a rectangular water tank are 3.00 m, 5.00 m, and 8.00 m. By Newton's method, determine by how much each edge should be increased equally to double the volume of the tank.

45. A parachutist descends (after the parachute opens) in a path that can be described by $x = 8t$ and $y = -0.15t^2$, where distances are in meters and time is in seconds. Find the parachutist's velocity upon landing if the landing occurs when $t = 12$ s.

46. One of the curves on an automobile test track can be described by $y = 225/x$, where dimensions are in meters. For a car approaching the curve at a constant velocity of 120 km/h, find the x- and y-components of the velocity at $(10.0, 22.5)$.

47. In Fig. 24.66, the tension T supports the 40.0-N weight. The relation between the tension T and the deflection d is

$d = \dfrac{1000}{\sqrt{T^2 - 400}}$. If the tension is increasing at 2.00 N/s when $T = 28.0$ N, how fast is the deflection (in cm) changing?

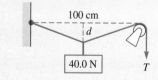

Fig. 24.66

48. The impedance Z (in Ω) in a particular electric circuit is given by $Z = \sqrt{48 + R^2}$, were R is the resistance. If R is increasing at a rate of 0.45 Ω/min for $R = 6.5\ \Omega$, find the rate at which Z is changing.

49. By using the methods of this chapter to graph $y = x^2 - \frac{2}{x}$, graph the solution of the inequality $x^2 > \frac{2}{x}$.

50. Display the graphs of $y_1 = x^2 - \frac{2}{x}$ and $y_2 = x^2$ on the same screen of a graphing calculator, and explain why y_2 can be considered to be a nonlinear asymptote of y_1.

51. An analysis of the power output P (in kW/m³) of a certain turbine showed that it depended on the flow rate r (in m³/s) of water to the turbine according to the equation $P = 0.030r^3 - 2.6r^2 + 71r - 200$ ($6 \le r \le 30$ m³/s). Determine the rate for which P is a maximum.

52. The altitude h (in m) of a certain rocket as a function of the time t (in s) after launching is given by $h = 550t - 4.9t^2$. What is the maximum altitude the rocket attains?

53. Sketch the continuous curve having these characteristics:

$f(0) = 2$ $f'(x) < 0$ for $x < 0$ $f''(x) > 0$ for all x
$f'(x) > 0$ for $x > 0$

54. Sketch a continuous curve having these characteristics:

$f(0) = 1$ $f'(0) = 0$ $f''(x) < 0$ for $x < 0$
$f'(x) > 0$ for $|x| > 0$ $f''(x) > 0$ for $x > 0$

55. A horizontal cylindrical oil tank (the length is parallel to the ground) of radius 2.00 m is being emptied. Find how fast the width w of the oil surface is changing when the depth h is 0.500 m and changing at the rate of 0.0500 m/min.

56. The current I (in A) in a circuit with a resistance R (in Ω) and a battery whose voltage is E and whose internal resistance is r (in Ω) is given by $I = E/(R + r)$. If R changes at the rate of 0.250 Ω/min, how fast is the current changing when $R = 6.25\ \Omega$, if $E = 3.10$ V and $r = 0.230\ \Omega$?

57. The radius of a circular oil spill is increasing at the rate of 15 m/min. How fast is the area of the spill changing when the radius is 400 m?

58. A baseball diamond is a square 90.0 ft on a side. See Fig. 24.67. As a player runs from first base toward second base at 18.0 ft/s, at what rate is the player's distance from home plate increasing when the player is 40.0 ft from first base? (1 ft = 0.3048 m.)

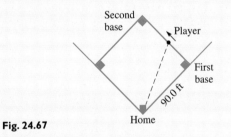

Fig. 24.67

59. A special insulation strip is to be sealed completely around three edges of a rectangular solar panel. If 200 cm of the strip are used, what is the maximum area of the panel?

60. A swimming pool with a rectangular surface of 130 m² is to have a cement border area that is 4.00 m wide at each end and 2.75 m wide at the sides. Find the surface dimensions of the pool if the total area covered is to be a minimum.

61. A study showed that the percent y of persons surviving burns to x percent of the body is given by $y = \dfrac{300}{0.0005x^2 + 2} - 50$.

Linearize this function with $a = 50$ and sketch the graphs of y and $L(x)$.

62. A company estimates that the sales S (in dollars) of a new product will be $S = 5000t/(t + 4)^2$, where t is the time (in months) after it is put into production. Sketch the graph of S vs. t.

63. An airplane flying horizontally at 2400 m is moving toward a radar installation at 1110 km/h. If the plane is directly over a point on the ground 8.00 km from the radar installation, what is its actual speed? See Fig. 24.68.

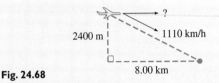

Fig. 24.68

64. The base of a conical machine part is being milled such that the height is decreasing at the rate of 0.050 cm/min. If the part originally had a radius of 1.0 cm and a height of 3.0 cm, how fast is the volume changing when the height is 2.8 cm?

65. The reciprocal of the total capacitance C_T of electrical capacitances in series equals the sum of the reciprocals of the individual capacitances. If the sum of two capacitances is 12 μF, find their values if their total capacitance in series is a maximum.

66. A cable is to be from point A to point B on a wall and then to point C. See Fig. 24.69. Where is B located if the total length of cable is a minimum?

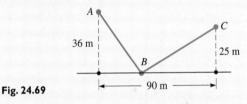

Fig. 24.69

67. A box with a square base and an open top is to be made of 27 dm² of cardboard. What is the maximum volume that can be contained within the box?

68. The strength of a rectangular beam is proportional to the product of its width w and the square of its depth d. Find the dimensions of the strongest beam that can be cut from a circular log 16 cm in diameter.

69. A person in a boat 4 km from the nearest point P on a straight shoreline wants to go to point A on the shoreline, 5 km from P. If the person can row at 3 km/h and walk at 5 km/h, at what point on the shoreline should the boat land in order that point A can be reached in the least time? See Fig. 24.70.

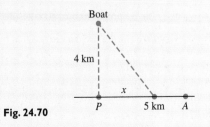

Fig. 24.70

70. A machine part is to be in the shape of a circular sector of radius r and central angle θ. Find r and θ if the area is one unit and the perimeter is a minimum. See Fig. 24.71.

Fig. 24.71

71. An open drawer for small tools is to be made from a rectangular piece of heavy sheet metal 36.0 cm by 30.0 cm, by cutting out equal squares from two corners and bending up the three sides, as shown in Fig. 24.72. Find the side of the square that should be cut out so that the volume of the drawer is a maximum.

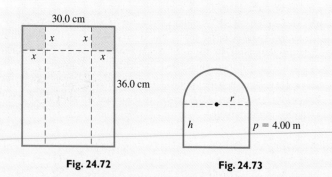

Fig. 24.72   **Fig. 24.73**

72. A Norman window has the form of a rectangle surmounted by a semicircle. Find the dimensions (radius of circular part and height of rectangular part) of the window that will admit the most light if the perimeter of the window is 4.00 m. See Fig. 24.73.

73. A pile of sand in the shape of a cone has a radius that always equals the altitude. If 3.00 m³ of sand are poured onto the pile each minute, how fast is the radius increasing when the pile is 2.50 m high?

74. A book is designed such that its (rectangular) pages have 2.5-cm margins at the top and bottom, 1.5-cm margins on the sides, and a total area of 320 cm². What are the page dimensions that give the maximum printed area?

75. A specially made cylindrical container is made of stainless steel sides and bottom and a silver top. If silver is ten times as expensive as stainless steel, what are the most economical dimensions of the container if it is to hold 314 cm³?

76. An object is moving in a horizontal circle of a radius 2.00 m at the rate of 2.00 rad/s. If the object is on the end of a string and the string breaks after 1.05 s, causing the object to travel along a line tangent to the circle, what is the equation of the path of the object after the string breaks? (Choose the origin of the coordinate system at the center of the circle and assume the object started on a positive x-axis moving counterclockwise.)

Writing Exercise

77. A container manufacturer makes various sizes of closed cylindrical plastic containers for shipping liquid products. Write two or three paragraphs explaining how to determine the ratio of the height to radius of the container such that the least amount of plastic is used for each size. Include the reason why it is not necessary to specify the volume of the container in finding this ratio.

CHAPTER 24 PRACTICE TEST

1. Find the equation of the line tangent to the curve $y = x^4 - 3x^2$ at the point $(1, -2)$.

2. For $y = 3x^2 - x$, evaluate (a) Δy, (b) dy, and (c) $\Delta y - dy$ for $x = 3$ and $\Delta x = 0.1$.

3. If the x- and y-coordinates of a moving object as functions of time are given by the parametric equations $x = 3t^2$, $y = 2t^3 - t^2$, find the magnitude and direction of the acceleration when $t = 2$.

4. The electric power (in W) produced by a certain source is given by $P = \dfrac{144r}{(r + 0.6)^2}$, where r is the resistance (in Ω) in the circuit. For what value of r is the power a maximum?

5. Find the root of the equation $x^2 - \sqrt{4x + 1} = 0$ between 1 and 2 to four decimal places by use of Newton's method. Use $x_1 = 1.5$ and find x_3.

6. Linearize the function $y = \sqrt{2x + 4}$ for $a = 6$.

7. Sketch the graph of $y = x^3 + 6x^2$ by finding the values of x for which the function is increasing, decreasing, concave up, and concave down and by finding any maximum points, minimum points, and points of inflection.

8. Sketch the graph of $y = \dfrac{4}{x^2} - x$ by finding the same information as required in Problem 7, as well as intercepts, symmetry, behavior as x becomes large, vertical asymptotes, and the domain and range.

9. Trash is being compacted into a cubical volume. The edge of the cube is decreasing at the rate of 0.10 m/s. When an edge of the cube is 1.25 m, how fast is the volume changing?

10. A rectangular field is to be fenced and then divided in half by a fence parallel to two opposite sides. If a total of 6000 m of fencing is used, what is the maximum area that can be fenced?

CHAPTER 25 Integration

Finding areas of geometric figures had been studied by the ancient Greeks, and they had found how to find the area of any polygon. Also, they had found that they could find areas of a curved figure by inscribing polygons in the figure and then letting the number of sides of the polygon increase. There was little more progress in finding such areas until the 1600s when analytic geometry was developed.

Several mathematicians of the 1600s studied both the area problem and the tangent problem that was discussed in Chapter 23. These included the French mathematician Pierre de Fermat and the English mathematician Isaac Barrow, both of whom developed a few formulas by which tangents and areas could be found. However, Newton and Leibniz found that these two problems were related and determined general methods of finding them. For these reasons, Newton and Leibniz are credited with the creation of calculus.

In Section 25.4, integration is used to find the rate of flow of water over an obstacle.

Finding tangents and finding areas appear to be very different, but they are closely connected. As we will show, finding an area uses the inverse process of finding the slope of a tangent line, which we have shown can be interpreted as an instantaneous rate of change.

In physical and technical applications, we often find information that gives us the instantaneous rate of change of a variable. With such information, we have to reverse the process of differentiation in order to find the function that relates the variables. This leads to the problem of finding the function when we know its derivative. This procedure is known as *integration*, which is the inverse process of differentiation.

This means that areas are found by integration. There are also many applications of integration in science and technology. A few of these applications will be illustrated in this chapter, and several specific applications will be developed in the next chapter.

25.1 ANTIDERIVATIVES

We now show how to reverse the process of finding a derivative or a differential. *This reverse process is known as* **antidifferentiation.** In the next section, we formalize the process, and it is only the basic idea that is the topic of this section.

◖EXAMPLE 1 Find a function for which the derivative is $8x^3$. That is, find an antiderivative of $8x^3$.

When finding the derivative of a constant times a power of x, we multiply the constant coefficient by the power of x and reduce the power by 1. Therefore, in this case, the power of x must have been 4 before the differentiation was performed.

If we let the derivative function be $f(x) = 8x^3$ and then let its antiderivative function be $F(x) = ax^4$ (by increasing the power of x in $f(x)$ by 1), we can find the value of a by equating the derivative of $F(x)$ to $f(x)$. This gives us

$$F'(x) = 4ax^3 = 8x^3, \quad 4a = 8, \quad a = 2$$

This means that $F(x) = 2x^4$. ◗

◖EXAMPLE 2 Find an antiderivative of $v^2 + 2v$.

As for the v^2, we know that the power of v required in an antiderivative is 3. Also, to make the coefficient correct, we must multiply by $\frac{1}{3}$. The $2v$ should be recognized as the derivative of v^2. Therefore, we have as an antiderivative $\frac{1}{3}v^3 + v^2$. ◗

In Examples 1 and 2, we note that we could add any constant to the antiderivative given as the result and still have a correct antiderivative. This is due to the fact that the derivative of a constant is zero. This is considered further in the following section. For the examples and exercises in this section, we will not include any such constants in the results.

NOTE ▶ We note that when we find an antiderivative of a given function, we obtain another function. Thus, *we can define an* **antiderivative** *of the function $f(x)$ to be a function $F(x)$ such that $F'(x) = f(x)$.*

◖EXAMPLE 3 Find an antiderivative of the function $f(x) = \sqrt{x} - \dfrac{2}{x^3}$.

Since we wish to find an antiderivative of $f(x)$, we know that $f(x)$ is the derivative of the required function.

Considering the term $\sqrt{x}$, we first write it as $x^{1/2}$. To have x to the $\frac{1}{2}$ power in the derivative, we must have x to the $\frac{3}{2}$ power in the antiderivative. Knowing that the derivative of $x^{3/2}$ is $\frac{3}{2}x^{1/2}$, we write $x^{1/2}$ as $\frac{2}{3}\left(\frac{3}{2}x^{1/2}\right)$. Thus, the first term of the antiderivative is $\frac{2}{3}x^{3/2}$.

As for the term $-2/x^3$, we write it as $-2x^{-3}$. This we recognize as the derivative of x^{-2}, or $1/x^2$.

This means that an antiderivative of the function

$$\text{function} \qquad f(x) = \sqrt{x} - \frac{2}{x^3} \quad \longleftarrow \text{ derivative}$$

$$\text{is the function}$$

$$\text{antiderivative} \longrightarrow F(x) = \frac{2}{3}x^{3/2} + \frac{1}{x^2} \qquad \text{function}$$

◗

A great many functions of which we must find an antiderivative are not polynomials or simple powers of x. It is these functions that may cause more difficulty in the general process of antidifferentiation. Pay special attention to the following examples, for they illustrate a type of problem that you will find to be very important.

◀ **EXAMPLE 4** Find an antiderivative of the function $f(x) = 3(x^3 - 1)^2(3x^2)$.

Noting that we have a power of $x^3 - 1$ in the derivative, it is reasonable that the antiderivative may include a power of $x^3 - 1$. Since, in the derivative, $x^3 - 1$ is raised to the power 2, the antiderivative would then have $x^3 - 1$ raised to the power 3. Noting that the derivative of $(x^3 - 1)^3$ is $3(x^3 - 1)^2(3x^2)$, the desired antiderivative is

$$F(x) = (x^3 - 1)^3$$

CAUTION ▶ We note that ***the factor $3x^2$ does not appear in the antiderivative.*** It is included when finding the derivative. Therefore it must be present for $(x^3 - 1)^3$ to be a proper antiderivative, but must be excluded when finding an antiderivative. ▶

◀ **EXAMPLE 5** Find an antiderivative of the function $f(x) = (2x + 1)^{1/2}$.

Here, we note a power of $2x + 1$ in the derivative, which infers that the antiderivative has a power of $2x + 1$. Since, in finding a derivative, 1 is subtracted from the power of $2x + 1$, we should add 1 in finding the antiderivative. Thus, we should have $(2x + 1)^{3/2}$ as part of the antiderivative. Finding a derivative of $(2x + 1)^{3/2}$, we obtain $\frac{3}{2}(2x + 1)^{1/2}(2) = 3(2x + 1)^{1/2}$. This differs from the given derivative by the factor of 3. Thus, if we write $(2x + 1)^{1/2} = \frac{1}{3}[3(2x + 1)^{1/2}]$, we have the required antiderivative as

$$F(x) = \frac{1}{3}(2x + 1)^{3/2}$$

Checking, the derivative of $\frac{1}{3}(2x + 1)^{3/2}$ is $\frac{1}{3}(\frac{3}{2})(2x + 1)^{1/2}(2) = (2x + 1)^{1/2}$. ▶

EXERCISES 25.1

In Exercises 1–4, make the given changes in the indicated examples of this section and then solve the resulting problems.

1. In Example 1, change the coefficient 8 to 12.

2. In Example 2, change v^2 to v^3.

3. In Example 3, change $2/x^3$ to $3/x^4$.

4. In Example 5, change $2x$ to $4x$.

In Exercises 5–12, determine the value of a that makes $F(x)$ an antiderivative of $f(x)$.

5. $f(x) = 3x^2$, $F(x) = ax^3$

6. $f(x) = 5x^4$, $F(a) = ax^5$

7. $f(x) = 18x^5$, $F(x) = ax^6$

8. $f(x) = 40x^7$, $F(x) = ax^8$

9. $f(x) = 9\sqrt{x}$, $F(x) = ax^{3/2}$

10. $f(x) = 10x^{1/4}$, $F(x) = ax^{5/4}$

11. $f(x) = \dfrac{1}{x^2}$, $F(x) = \dfrac{a}{x}$

12. $f(x) = \dfrac{6}{x^4}$, $F(x) = \dfrac{a}{x^3}$

In Exercises 13–36, find antiderivatives of the given functions.

13. $f(x) = \frac{5}{2}x^{3/2}$

14. $f(x) = \frac{4}{3}x^{1/3}$

15. $f(t) = 6t^3 + 2$

16. $f(x) = 12x^5 + 2x$

17. $f(x) = 2x^2 - x$

18. $f(x) = x^2 - 5$

19. $f(x) = 2\sqrt{x} + 3$

20. $f(s) = 9\sqrt[3]{s} - 3$

21. $f(x) = -\dfrac{7}{x^6}$

22. $f(x) = \dfrac{8}{x^5}$

23. $f(v) = 4v + 3\pi^2$

24. $f(x) = \dfrac{1}{2\sqrt{x}} + \sqrt{3}$

25. $f(x) = x^2 + 2 + x^{-2}$

26. $f(x) = x\sqrt{x} - x^{-3}$

27. $f(x) = 6(2x + 1)^5(2)$

28. $f(R) = 3(R^2 + 1)^2(2R)$

29. $f(p) = 4(p^2 - 1)^3(2p)$

30. $f(x) = 5(2x^4 + 1)^4(8x^3)$

31. $f(x) = x^3(2x^4 + 1)^4$

32. $f(x) = x(1 - x^2)^7$

33. $f(x) = \frac{3}{2}(6x + 1)^{1/2}(6)$

34. $f(y) = \frac{5}{4}(1 - y)^{1/4}(-1)$

35. $f(x) = (3x + 1)^{1/3}$

36. $f(x) = (4x + 3)^{1/2}$

25.2 THE INDEFINITE INTEGRAL

In the previous section, in developing the basic technique of finding an antiderivative, we noted that the results given are not unique. That is, we could have added any constant to the answers and the result would still have been correct. Again, this is the case since the derivative of a constant is zero.

EXAMPLE 1 The derivatives of x^3, $x^3 + 4$, $x^3 - 7$, and $x^3 + 4\pi$ are all $3x^2$. This means that any of the functions listed, as well as others, would be a proper answer to the problem of finding an antiderivative of $3x^2$.

From Section 24.8, we know that the differential of a function $F(x)$ can be written as $d[F(x)] = F'(x)\,dx$. Therefore, since finding a differential of a function is closely related to finding the derivative, so is the antiderivative closely related to the process of finding the function for which the differential is known.

The notation used for finding the general form of the antiderivative, the **indefinite integral,** *is written in terms of the differential. Thus, the indefinite integral of a function $f(x)$, for which $dF(x)/dx = f(x)$, or $dF(x) = f(x)\,dx$, is defined as*

$$\int f(x)\,dx = F(x) + C \qquad (25.1)$$

Here, $f(x)$ is called the **integrand,** $F(x) + C$ is the indefinite integral, and C is an arbitrary constant, called the **constant of integration.** It represents any of the constants that may be attached to an antiderivative to have a proper result. We must have additional information beyond a knowledge of the differential to assign a specific value to C. The symbol $\int$ is the **integral sign,** and it indicates that the inverse of the differential is to be found. *Determining the indefinite integral is called* **integration,** which we can see is essentially the same as finding an antiderivative.

EXAMPLE 2 In performing the integration

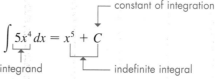

$$\int 5x^4\,dx = x^5 + C$$

we might think that the inclusion of this constant C would affect the derivative of the function x^5. However, the only effect of the C is to raise or lower the curve. The slope of $x^5 + 2$, $x^5 - 2$, or any function of the form $x^5 + C$ is the same for any given value of x. As Fig. 25.1 shows, tangents drawn to the curves are all parallel for the same value of x.

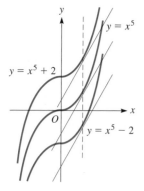

Fig. 25.1

At this point we shall derive some basic formulas for integration. Since we know $d(cu)/dx = c(du/dx)$, where u is a function of x and c is a constant, we can write

$$\int c\,du = c\int du = cu + C \qquad (25.2)$$

Also, since the derivative of a sum of functions equals the sum of the derivatives, we write

$$\int (du + dv) = u + v + C \tag{25.3}$$

To find the differential of a power of a function, we multiply by the power, subtract 1 from it, and multiply by the differential of the function. *To find the integral, we reverse this procedure to get the power formula for integration:*

$$\int u^n \, du = \frac{u^{n+1}}{n + 1} + C \qquad (n \neq -1) \tag{25.4}$$

We must be able to recognize the proper form and the component parts to use these formulas. Unless you do this and have a good knowledge of differentiation, you will have trouble using Eq. (25.4). Most of the difficulty, if it exists, arises from an improper identification of *du*.

◀ **EXAMPLE 3**   Integrate: $\int 6x \, dx$.

We must identify u, n, du, and any multiplying constants. Noting that 6 is a multiplying constant, we identify x as u, which means dx must be du and $n = 1$.

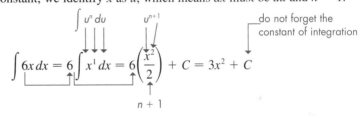

We see that our result checks since the differential of $3x^2 + C$ is $6x \, dx$.   ▶

◀ **EXAMPLE 4**   Integrate: $\int (5x^3 - 6x^2 + 1) \, dx$.

Here, we must use a combination of Eqs. (25.2), (25.3), and (25.4). Therefore,

$$\int (5x^3 - 6x^2 + 1) \, dx = \int 5x^3 \, dx + \int (-6x^2) \, dx + \int dx$$

$$= 5\int x^3 \, dx - 6\int x^2 \, dx + \int dx$$

In the first integral, $u = x$, $n = 3$, and $du = dx$. In the second, $u = x$, $n = 2$, and $du = dx$. The third uses Eq. (25.2) directly, with $c = 1$ and $du = dx$. This means

$$5\int x^3 \, dx - 6\int x^2 \, dx + \int dx = 5\left(\frac{x^4}{4}\right) - 6\left(\frac{x^3}{3}\right) + x + C$$

$$= \frac{5}{4}x^4 - 2x^3 + x + C$$ ▶

◀ EXAMPLE 5 Integrate: $\int \left(\sqrt{r} - \dfrac{1}{r^3} \right) dr$.

In order to use Eq. (25.4), we must first write $\sqrt{r} = r^{1/2}$ and $1/r^3 = r^{-3}$:

$$\int \left(\sqrt{r} - \frac{1}{r^3} \right) dr = \int r^{1/2}\, dr - \int r^{-3}\, dr = \frac{1}{\frac{3}{2}}r^{3/2} - \frac{1}{-2}r^{-2} + C$$

$$= \frac{2}{3}r^{3/2} + \frac{1}{2}r^{-2} + C = \frac{2}{3}r^{3/2} + \frac{1}{2r^2} + C$$

Again we note that a good knowledge of differential forms is essential for the proper recognition of u and du.

◀ EXAMPLE 6 Integrate: $\int (x^2 + 1)^3 (2x\, dx)$.

We first note that $n = 3$, for this is the power involved in the function being integrated. If $n = 3$, then $x^2 + 1$ must be u. If $u = x^2 + 1$, then $du = 2x\, dx$. Thus, the integral is in proper form for integration *as it stands*. Using the power formula,

$$\int (x^2 + 1)^3 (2x\, dx) = \frac{(x^2 + 1)^4}{4} + C$$

CAUTION ▶

It must be emphasized that the entire quantity **(2x dx) *must be equated to du.*** Normally, u and n are recognized first, and then du is derived from u.

Showing the use of u directly, we can write the integration as

$$\int (x^2 + 1)^3 (2x\, dx) = \int u^3\, du = \frac{1}{4}u^4 + C = \frac{(x^2 + 1)^4}{4} + C$$

◀ EXAMPLE 7 Integrate: $\int x^2 \sqrt{x^3 + 2}\, dx$.

We first note that $n = \frac{1}{2}$ and u is then $x^3 + 2$. Since $u = x^3 + 2$, $du = 3x^2\, dx$. Now we group $3x^2\, dx$ as du. Since there is no 3 under the integral sign, we introduce one. In order not to change the numerical value, we also introduce $\frac{1}{3}$, normally before the integral sign. In this way we take advantage of the fact that *a constant (and only a constant) factor may be moved across the integral sign.*

CAUTION ▶

$$\int x^2 \sqrt{x^3 + 2}\, dx = \frac{1}{3} \int 3x^2 \sqrt{x^3 + 2}\, dx = \frac{1}{3} \int \sqrt{x^3 + 2}\,(3x^2\, dx)$$

Here we indicate the proper grouping to have the proper form of Eq. (25.5):

$$\int x^2 \sqrt{x^3 + 2}\, dx = \frac{1}{3} \int \sqrt{x^3 + 2}\,(3x^2\, dx) = \frac{1}{3}\left(\frac{2}{3}\right)(x^3 + 2)^{3/2} + C$$

$$= \frac{2}{9}(x^3 + 2)^{3/2} + C$$

We have integrated certain basic functions. Other methods are used to integrate other types of functions, and some of these are discussed in Chapter 28. Also, many functions cannot be integrated.

The $1/\frac{3}{2}$ was written as $\frac{2}{3}$, since this form is more convenient with fractions.

With $u = x^3 + 2$ and using u directly in the integration, we can write

$$\int x^2 \sqrt{x^3 + 2}\, dx = \int (x^3 + 2)^{1/2}(x^2\, dx)$$

$$= \int u^{1/2}\left(\frac{1}{3}du\right) = \frac{1}{3}\int u^{1/2}\, du \qquad \text{integrating in terms of } u$$

$$= \frac{1}{3}\left(\frac{2}{3}\right)u^{3/2} + C = \frac{2}{9}u^{3/2} + C$$

$$= \frac{2}{9}(x^3 + 2)^{3/2} + C \qquad \text{substituting } x^3 + 2 = u$$

EVALUATING THE CONSTANT OF INTEGRATION

To find the constant of integration, we need information such as a set of values that satisfy the function. A point through which the curve passes would provide the necessary information. This is illustrated in the following examples.

◀ **EXAMPLE 8** Find y in terms of x, given that $dy/dx = 3x - 1$ and the curve passes through $(1, 4)$.

We write the equation as

$$dy = (3x - 1)\, dx$$

and then indicate and perform the integration:

$$\int dy = \int (3x - 1)\, dx$$

$$y = \frac{3}{2}x^2 - x + C$$

Since the curve passes through $(1, 4)$, the coordinates must satisfy the equation. Thus,

$$4 = \frac{3}{2} - 1 + C \quad \text{or} \quad C = \frac{7}{2}$$

This means that the solution is

$$y = \frac{3}{2}x^2 - x + \frac{7}{2} \quad \text{or} \quad 2y = 3x^2 - 2x + 7$$

The graph of this parabola is shown in Fig. 25.2. ▶

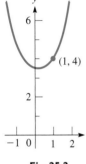

Fig. 25.2

◀ **EXAMPLE 9** The time rate of change of the displacement (velocity) of a robot arm is $ds/dt = t\sqrt{9 - t^2}$. Find the expression for the displacement as a function of time if $s = 0$ cm when $t = 0$ s.

We start the solution by writing

$$ds = t\sqrt{9 - t^2}\, dt$$

$$\int ds = \int t(9 - t^2)^{1/2}\, dt$$

To integrate the expression on the right, we recognize that $n = \frac{1}{2}$ and $u = 9 - t^2$, and therefore $du = -2t\, dt$. This means we need a -2 with the t and dt to form the proper du, which in turn means we place a $-\frac{1}{2}$ before the integral sign:

$$\int ds = \int t(9 - t^2)^{1/2}\, dt = -\frac{1}{2} \int (9 - t^2)^{1/2}(-2t\, dt)$$

$$s = -\frac{1}{2}\left(\frac{2}{3}\right)(9 - t^2)^{3/2} + C = -\frac{1}{3}(9 - t^2)^{3/2} + C$$

Since $s = 0$ cm when $t = 0$ s, we have

$$0 = -\frac{1}{3}(9)^{3/2} + C \quad \text{or} \quad C = 9$$

This means that the expression for the displacement is

$$s = -\frac{1}{3}(9 - t^2)^{3/2} + 9$$

EXERCISES 25.2

In Exercises 1–4, make the given changes in the indicated examples of this section, and then solve the resulting problems.

1. In Example 3, change the coefficient 6 to 8.

2. In Example 5, change $\sqrt{r}$ to $\sqrt[3]{r}$.

3. In Example 6, change the power 3 to 4.

4. In Example 8, change $(1, 4)$ to $(2, 3)$.

In Exercises 5–36, integrate each of the given expressions.

5. $\displaystyle\int 2x\,dx$

6. $\displaystyle\int 5x^4\,dx$

7. $\displaystyle\int x^7\,dx$

8. $\displaystyle\int 0.6y^5\,dy$

9. $\displaystyle\int 2x^{3/2}\,dx$

10. $\displaystyle\int 6\sqrt[3]{x}\,dx$

11. $\displaystyle\int 9R^{-4}\,dR$

12. $\displaystyle\int \frac{4}{\sqrt{x}}\,dx$

13. $\displaystyle\int (x^2 - x^5)\,dx$

14. $\displaystyle\int (1 - 3x)\,dx$

15. $\displaystyle\int (9x^2 + x + 3)\,dx$ 16. $\displaystyle\int x(x - 2)^2\,dx$

17. $\displaystyle\int \left(\frac{t^2}{2} - \frac{2}{t^2}\right)dt$

18. $\displaystyle\int \frac{3x^2 - 4}{x^2}\,dx$

19. $\displaystyle\int \sqrt{x}\,(x^2 - x)\,dx$

20. $\displaystyle\int (3R\sqrt{R} - 5R^2)\,dR$

21. $\displaystyle\int (2x^{-2/3} + 3^{-2})\,dx$

22. $\displaystyle\int (x^{1/3} + x^{1/5} + x^{-1/7})\,dx$

23. $\displaystyle\int (1 + 2s^2)^2\,ds$

24. $\displaystyle\int (x^2 + 4x + 4)^{1/3}\,dx$

25. $\displaystyle\int (x^2 - 1)^5(2x\,dx)$

26. $\displaystyle\int (x^3 - 2)^6(3x^2\,dx)$

27. $\displaystyle\int (x^4 + 3)^4(4x^3\,dx)$

28. $\displaystyle\int (1 - 2x)^{1/3}(-2\,dx)$

29. $\displaystyle\int (2\theta^5 + 5)^7\theta^4\,d\theta$

30. $\displaystyle\int 6x^2(1 - x^3)^{4/3}\,dx$

31. $\displaystyle\int \sqrt{8x + 1}\,dx$

32. $\displaystyle\int \frac{dV}{(0.3 + 2V)^3}$

33. $\displaystyle\int \frac{x\,dx}{\sqrt{6x^2 + 1}}$

34. $\displaystyle\int \frac{2x^2\,dx}{\sqrt{2x^3 + 1}}$

35. $\displaystyle\int \frac{x - 1}{\sqrt{x^2 - 2x}}\,dx$

36. $\displaystyle\int (x^2 - x)\left(x^3 - \frac{3}{2}x^2\right)^8\,dx$

In Exercises 37–40, find y in terms of x.

37. $\dfrac{dy}{dx} = 6x^2$, curve passes through $(0, 2)$

38. $\dfrac{dy}{dx} = 8x + 1$, curve passes through $(-1, 4)$

39. $\dfrac{dy}{dx} = x^2(1 - x^3)^5$, curve passes through $(1, 5)$

40. $\dfrac{dy}{dx} = 2x^3(x^4 - 6)^4$, curve passes through $(2, 10)$

Ⓦ In Exercises 41–56, solve the given problems. In Exercises 41–46, explain your answers.

41. Is $\displaystyle\int 3x^2\,dx = x^3$?

42. Can $\displaystyle\int (x^2 - 1)^2\,dx$ be integrated with $u = x^2 - 1$?

43. Can $\displaystyle\int (4x^3 + 3)^5 x^4\,dx$ be integrated with $u = 4x^3 + 3$ and $du = x^4\,dx$?

44. Is $\displaystyle\int \sqrt{2x + 1}\,dx = \frac{2}{3}(2x + 1)^{1/2}$?

45. Is $\displaystyle\int 3(2x + 1)^2\,dx = (2x + 1)^3 + C$?

46. Is $\displaystyle\int x^{-2}\,dx = -\frac{1}{3}x^{-3} + C$?

47. Find the equation of the curve whose slope is $-x\sqrt{1 - 4x^2}$ and that passes through $(0, 7)$.

48. Find the equation of the curve whose slope is $\sqrt{6x - 3}$ and that passes through $(2, -1)$.

49. The time rate of change of electric current in a circuit is given by $di/dt = 4t - 0.6t^2$. Find the expression for the current as a function of time if $i = 2$ A when $t = 0$ s.

50. The rate of change of the frequency f of an electronic oscillator with respect to the inductance L is $df/dL = 80(4 + L)^{-3/2}$. Find f as a function of L if $f = 80$ Hz for $L = 0$ H.

51. The rate of change of the temperature T (in °C) from the center of a blast furnace to a distance r (in m) from the center is given by $dT/dr = -4500(r + 1)^{-3}$. Express T as a function of r if $T = 2500$°C for $r = 0$.

52. The rate of change of current i (in mA) in a circuit with a variable inductance is given by $di/dt = 300(5.0 - t)^{-2}$, where t (in ms) is the time the circuit is closed. Find i as a function of t if $i = 300$ mA for $t - 2.0$ ms.

53. At a given site, the rate of change of the annual fraction f of energy supplied by solar energy with respect to the solar-collector area A (in m²) is $\dfrac{df}{dA} = \dfrac{0.005}{\sqrt{0.01A + 1}}$. Find f as a function of A if $f = 0$ for $A = 0$ m².

54. An analysis of a company's records shows that in a day the rate of change of profit p (in dollars) in producing x generators is $\dfrac{dp}{dx} = \dfrac{600(30 - x)}{\sqrt{60x - x^2}}$. Find the profit in producing x generators if a loss of $5000 is incurred if none are produced.

55. Find the equation of the curve for which the second derivative is 6. The curve passes through $(1, 2)$ with a slope of 8.

Ⓦ 56. The second derivative of a function is $12x^2$. Explain how to find the function if its curve passes through the points $(1, 6)$ and $(2, 21)$. Find the function.

25.3 THE AREA UNDER A CURVE

In geometry, there are formulas and methods for finding the areas of regular figures. By means of integration, it is possible to find the area between curves for which we know the equations. The next example illustrates the basic idea behind the method.

◀ EXAMPLE 1 Approximate the area in the first quadrant to the left of the line $x = 4$ and under the parabola $y = x^2 + 1$. First, make this approximation by inscribing two rectangles of equal width under the parabola and finding the sum of the areas of these rectangles. Then, improve the approximation by repeating the process with eight rectangles.

The area to be approximated is shown in Fig. 25.3(a). The area with two rectangles inscribed under the curve is shown in Fig. 25.3(b). The first approximation, admittedly small, of the area can be found by adding the areas of the two rectangles. Both rectangles have a width of 2. The left rectangle is 1 unit high, and the right rectangle is 5 units high. Thus, the area of the two rectangles is

$$A = 2(1 + 5) = 12$$

See Appendix C for the graphing calculator program AREAUNCV. It approximates the area under a curve by summing the areas of n rectangles.

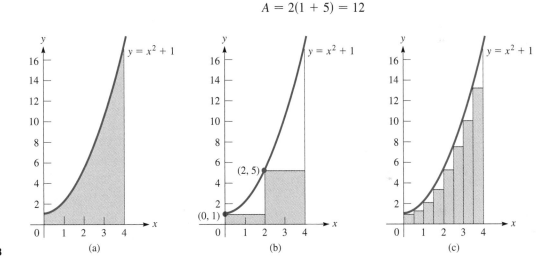

Fig. 25.3

(a)   (b)   (c)

A much better approximation is found by inscribing the eight rectangles as shown in Fig. 25.3(c). Each of these rectangles has a width of $\frac{1}{2}$. The leftmost rectangle has a height of 1. The next has a height of $\frac{5}{4}$, which is determined by finding y for $x = \frac{1}{2}$. The next rectangle has a height of 2, which is found by evaluating y for $x = 1$. Finding the heights of all rectangles and multiplying their sum by $\frac{1}{2}$ gives the area of the eight rectangles as

$$A = \frac{1}{2}\left(1 + \frac{5}{4} + 2 + \frac{13}{4} + 5 + \frac{29}{4} + 10 + \frac{53}{4}\right) = \frac{43}{2} = 21.5$$

Table 25.1

Number of Rectangles n	Total Area of Rectangles
8	21.5
100	25.014 4
1 000	25.301 344
10 000	25.330 134

An even better approximation could be obtained by inscribing more rectangles under the curve. The greater the number of rectangles, the more nearly the sum of their areas equals the area under the curve. See Table 25.1. By using integration later in this section, we determine the *exact* area to be $\frac{76}{3} = 25\frac{1}{3}$. ▶

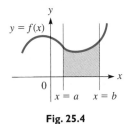

Fig. 25.4

We now develop the basic method used to find the area under a curve, which is the area bounded by the curve, the x-axis, and the lines $x = a$ and $x = b$. See Fig. 25.4. We assume here that $f(x)$ is never negative in the interval $a < x < b$. In Chapter 26, we will extend the method such that $f(x)$ may be negative.

In finding the area under a curve, we consider the sum of the areas of inscribed rectangles, as the number of rectangles is assumed to increase without bound. The reason for this last condition is that, as we saw in Example 1, as the number of rectangles increases, the approximation of the area is better.

◀ **EXAMPLE 2**　Find the area under the straight line $y = 2x$, above the x-axis, and to the left of the line $x = 4$.

Since this figure is a right triangle, the area can easily be found. However, the **NOTE ▶** *method* we use here is the important concept. We first subdivide the interval from $x = 0$ to $x = 4$ into n inscribed rectangles of Δx in width. The extremities of the intervals are labeled $a, x_1, x_2, \ldots, b\,(= x_n)$, as shown in Fig. 25.5, where

$$x_1 = \Delta x \qquad x_2 = 2\,\Delta x, \ldots \qquad x_{n-1} = (n-1)\,\Delta x \qquad b = n\,\Delta x$$

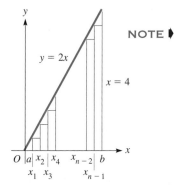

The area of each of these n rectangles is as follows:

First:　$f(a)\,\Delta x$, where $f(a) = f(0) = 2(0) = 0$ is the height.
Second:　$f(x_1)\,\Delta x$, where $f(x_1) = 2(\Delta x) = 2\,\Delta x$ is the height.
Third:　$f(x_2)\,\Delta x$, where $f(x_2) = 2(2\,\Delta x) = 4\,\Delta x$ is the height.
Fourth:　$f(x_3)\,\Delta x$, where $f(x_3) = 2(3\,\Delta x) = 6\,\Delta x$ is the height.

$\vdots$

Last:　$f(x_{x-1})\,\Delta x$, where $f[(n-1)\,\Delta x] = 2(n-1)\,\Delta x$ is the height.

Fig. 25.5

These areas are summed up as follows:

NOTE ▶
$$A_n = \boxed{f(a)\,\Delta x + f(x_1)\,\Delta x + f(x_2)\,\Delta x + \cdots + f(x_{n-1})\,\Delta x} \tag{25.5}$$

$$= 0 + 2\,\Delta x(\Delta x) + 4\,\Delta x(\Delta x) + \cdots + 2[n-1]\,\Delta x]\,\Delta x$$
$$= 2(\Delta x)^2[1 + 2 + 3 + \cdots + (n-1)]$$

Now, $b = n\,\Delta x$, or $4 = n\,\Delta x$, or $\Delta x = 4/n$. Thus,

$$A_n = 2\left(\frac{4}{n}\right)^2[1 + 2 + 3 + \cdots + (n-1)]$$

The sum of the arithmetic sequence $1 + 2 + 3 + \cdots + n - 1$ is

$$s = \frac{n-1}{2}(1 + n - 1) = \frac{n(n-1)}{2} = \frac{n^2 - n}{2}$$

Now the expression for the sum of the areas can be written as

$$A_n = \frac{32}{n^2}\left(\frac{n^2 - n}{2}\right) = 16\left(1 - \frac{1}{n}\right)$$

This expression is an approximation of the actual area under consideration. The larger n becomes, the better the approximation. If we let $n \to \infty$ (which is equivalent to letting $\Delta x \to 0$), the limit of this sum will equal the area in question.

This checks with the geometric result.

$$A = \lim_{n\to\infty} 16\left(1 - \frac{1}{n}\right) = 16 \qquad 1/n \to 0 \text{ as } n \to \infty$$

NOTE ▶　*The area under the curve is the limit of the sum of the areas of the inscribed rectangles, as the number of rectangles approaches infinity.*

▶

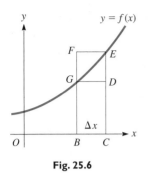

Fig. 25.6

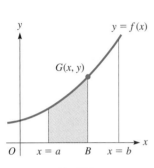

Fig. 25.7

The method indicated in Example 2 illustrates the interpretation of finding an area as a summation process, although it should not be considered as a proof. However, we will find that integration proves to be a much more useful method for finding an area. Let us now see how integration can be used directly.

Let ΔA represent the area *BCEG* under the curve, as indicated in Fig. 25.6. We see that the following inequality is true for the indicated areas:

$$A_{BCDG} < \Delta A < A_{BCEF}$$

If the point G is now designated as (x, y) and E as $(x + \Delta x, y + \Delta y)$, we have $y\,\Delta x < \Delta A < (y + \Delta y)\,\Delta x$. Dividing through by Δx, we have

$$y < \frac{\Delta A}{\Delta x} < y + \Delta y$$

Now we take the limit as $\Delta x \to 0$ (Δy then approaches zero). This results in

$$\frac{dA}{dx} = y \tag{25.6}$$

This is true since the left member of the inequality is y and the right member approaches y. Also, in the definition of the derivative, Eq. (23.6), $f(x + h) - f(x)$ is equivalent to ΔA, and h is equivalent to Δx, which means

$$\lim_{\Delta x \to 0} \frac{\Delta A}{\Delta x} = \frac{dA}{dx}$$

We shall now use Eq. (25.6) to show the method of finding the complete area under a curve. We now let $x = a$ be the left boundary of the desired area and $x = b$ be the right boundary (Fig. 25.7). The area under the curve to the right of $x = a$ and bounded on the right by the line *GB* is now designated as A_{ax}. From Eq. (25.6) we have

$$dA_{ax} = \left[y\,dx \right]_a^x \quad \text{or} \quad A_{ax} = \left[\int y\,dx \right]_a^x$$

where $\left[\;\;\right]_a^x$ is the notation used to indicate the boundaries of the area. Thus,

$$A_{ax} = \left[\int f(x)\,dx \right]_a^x = [F(x) + C]_a^x \tag{25.7}$$

But we know that if $x = a$, then $A_{aa} = 0$. Thus, $0 = F(a) + C$, or $C = -F(a)$. Therefore,

$$A_{ax} = \left[\int f(x)\,dx \right]_a^x = F(x) - F(a) \tag{25.8}$$

Now, to find the area under the curve that reaches from a to b, we write

$$A_{ab} = F(b) - F(a) \tag{25.9}$$

Thus, the area under the curve that reaches from a to b is given by

$$A_{ab} = \left[\int f(x)\,dx \right]_a^b = F(b) - F(a) \tag{25.10}$$

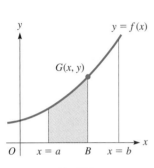

Fig. 25.8

NOTE ▶

This shows that the area under the curve may be found by integrating the function $f(x)$ to find the function $F(x)$, which is then evaluated at each boundary value. The area is the difference between these values of $F(x)$. See Fig. 25.8.

Integration as Summation

In Example 2, we found an area under a curve by finding the limit of the sum of the areas of the inscribed rectangles as the number of rectangles approaches infinity. Equation (25.10) expresses the area under a curve in terms of integration. We can now see that we have obtained an area by summation and also expressed it in terms of integration. Therefore, we conclude that

summations can be evaluated by integration.

Also, we have seen the connection between the problem of finding the slope of a tangent to a curve (differentiation) and the problem of finding an area under a curve (integration). We would not normally suspect that these two problems would have solutions that lead to reverse processes. We have also seen that the definition of integration has much more application than originally anticipated.

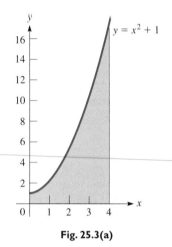

Fig. 25.3(a)

◀ **EXAMPLE 3** Find the area under the curve $y = x^2 + 1$ between the y-axis and the line $x = 4$. This is the area of Example 1 shown in Fig. 25.3(a), which is shown again here for reference.

Using Eq. (25.10), we note that $f(x) = x^2 + 1$. This means that

$$\int (x^2 + 1)\,dx = \frac{1}{3}x^3 + x + C$$

Therefore, with $F(x) = \frac{1}{3}x^3 + x$, the area is given by

$$A_{0,4} = F(4) - F(0) \qquad \text{using Eq. (25.10)}$$

$$= \left[\frac{1}{3}(4^3) + 4\right] - \left[\frac{1}{3}(0^3) + 0\right] \qquad \text{evaluating } F(x) \text{ at } x = 4 \text{ and } x = 0$$

$$= \frac{1}{3}(64) + 4 = \frac{76}{3}$$

We note that this is about 4 square units greater than the value obtained by the approximation in Example 1. This result means that the exact area is $25\frac{1}{3}$, as stated at the end of Example 1. ▶

◀ **EXAMPLE 4** Find the area under the curve $y = x^3$ that is between the lines $x = 1$ and $x = 2$.

In Eq. (25.10), $f(x) = x^3$. Therefore,

$$\overset{\overset{\displaystyle F(x)}{\big|}}{\int x^3\,dx = \frac{1}{4}x^4 + C}$$

$$A_{1,2} = F(2) - F(1) = \left[\frac{1}{4}(2^4)\right] - \left[\frac{1}{4}(1^4)\right] \qquad \text{using Eq. (25.10) and evaluating}$$

$$= 4 - \frac{1}{4} = \frac{15}{4}$$

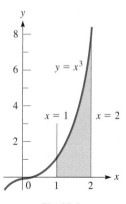

Fig. 25.9

Therefore, we see that the required area is $15/4$. Again, as in Example 3, this is an exact area, not an approximation. The curve and the area are shown in Fig. 25.9. ▶

Note that we do not have to include the constant of integration when finding areas. Any constant added to $F(x)$ cancels out when $F(a)$ is subtracted from $F(b)$.

EXERCISES 25.3

In Exercises 1–4, make the given changes in the indicated examples of this section and then solve the resulting problems.

1. In Example 1, change $x = 4$ to $x = 2$ and find the area of (a) two inscribed rectangles and (b) four inscribed rectangles.

2. In Example 3, change $x = 4$ to $x = 2$ and compare the results with those of Exercise 1.

3. In Example 4, change $x = 2$ to $x = 3$.

4. In Example 4, change $x = 1$ to $x = 2$ and $x = 2$ to $x = 3$. Note that the result added to the result of Example 4 is the same as the result for Exercise 3.

In Exercises 5–14, find the approximate area under the curves of the given equations by dividing the indicated intervals into n subintervals and then add up the areas of the inscribed rectangles. There are two values of n for each exercise and therefore two approximations for each area. The height of each rectangle may be found by evaluating the function for the proper value of x. See Example 1.

5. $y = 3x$, between $x = 0$ and $x = 3$, for
   (a) $n = 3$ ($\Delta x = 1$), (b) $n = 10$ ($\Delta x = 0.3$)

6. $y = 2x + 1$, between $x = 0$ and $x = 2$, for
   (a) $n = 4$ ($\Delta x = 0.5$), (b) $n = 10$ ($\Delta x = 0.2$)

7. $y = x^2$, between $x = 0$ and $x = 2$, for
   (a) $n = 5$ ($\Delta x = 0.4$), (b) $n = 10$ ($\Delta x = 0.2$)

8. $y = x^2 + 2$, between $x = 0$ and $x = 3$, for
   (a) $n = 3$ ($\Delta x = 1$), (b) $n = 10$ ($\Delta x = 0.3$)

9. $y = 4x - x^2$, between $x = 1$ and $x = 4$, for
   (a) $n = 6$, (b) $n = 10$

10. $y = 1 - x^2$, between $x = 0.5$ and $x = 1$, for
    (a) $n = 5$, (b) $n = 10$

11. $y = \dfrac{1}{x^2}$, between $x = 1$ and $x = 5$, for (a) $n = 4$, (b) $n = 8$

12. $y = \sqrt{x}$, between $x = 1$ and $x = 4$, for (a) $n = 3$, (b) $n = 12$

13. $y = \dfrac{1}{\sqrt{x + 1}}$, between $x = 3$ and $x = 8$, for
    (a) $n = 5$, (b) $n = 10$

14. $y = 2x\sqrt{x^2 + 1}$, between $x = 0$ and $x = 6$, for
    (a) $n = 6$, (b) $n = 12$

In Exercises 15–24, find the exact area under the given curves between the indicated values of x. The functions are the same as those for which approximate areas were found in Exercises 5–14.

15. $y = 3x$, between $x = 0$ and $x = 3$

16. $y = 2x + 1$, between $x = 0$ and $x = 2$

17. $y = x^2$, between $x = 0$ and $x = 2$

18. $y = x^2 + 2$, between $x = 0$ and $x = 3$

19. $y = 4x - x^2$, between $x = 1$ and $x = 4$

20. $y = 1 - x^2$, between $x = 0.5$ and $x = 1$

21. $y = \dfrac{1}{x^2}$, between $x = 1$ and $x = 5$

22. $y = \sqrt{x}$, between $x = 1$ and $x = 4$

23. $y = \dfrac{1}{\sqrt{x + 1}}$, between $x = 3$ and $x = 8$

(W) 24. $y = 2x\sqrt{x^2 + 1}$, between $x = 0$ and $x = 6$
   Explain the reason for the difference between this result and the two values found in Exercise 14.

*In developing the concept of the area under a curve, we first (in Examples 1 and 2) considered rectangles **inscribed** under the curve. A more complete development also considers rectangles **circumscribed** above the curve and shows that the limiting area of the circumscribed rectangles equals the limiting area of the inscribed rectangles as the number of rectangles increases without bound. See Fig. 25.10 for an illustration of inscribed and circumscribed rectangles.*

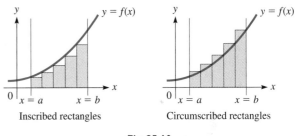

Inscribed rectangles   Circumscribed rectangles

Fig. 25.10

(W) *In Exercises 25–28, find the sum of the areas of 10 circumscribed rectangles for each curve and show that the exact area (as shown in Exercises 15–18) is between the sum of the areas of the circumscribed rectangles and the inscribed rectangles (as found in Exercises 5(b)–8(b)). Also note that the mean of the two sums is close to the exact value.*

25. $y = 3x$ between $x = 0$ and $x = 3$ (compare with Exercises 5(b) and 15). Why is the mean of the sums of the inscribed rectangles and circumscribed rectangles equal to the exact value?

26. $y = 2x + 1$ between $x = 0$ and $x = 2$ (compare with Exercises 6(b) and 16). Why is the mean of the sums of the inscribed rectangles and circumscribed rectangles equal to the exact value?

27. $y = x^2$ between $x = 0$ and $x = 2$ (compare with Exercises 7(b) and 17). Why is the mean of the sums of the inscribed rectangles and circumscribed rectangles greater than the exact value?

28. $y = x^2 + 2$ between $x = 0$ and $x = 3$ (compare with Exercises 8(b) and 18). Why is the mean of the sums of the inscribed rectangles and circumscribed rectangles greater than the exact value?

25.4 THE DEFINITE INTEGRAL

Using reasoning similar to that in the preceding section, *we define the* **definite integral** *of a function f(x) as*

$$\int_a^b f(x)\,dx = F(b) - F(a) \tag{25.11}$$

where $F'(x) = f(x)$. **We call this a** **definite integral** *because the final result of integrating and evaluating is a* **number.** (The *indefinite* integral had an arbitrary constant in the result.) *The numbers a and b are called the* **lower limit** *and the* **upper limit,** *respectively. We can see that the value of a definite integral is found by evaluating the function (found by integration) at the upper limit and subtracting the value of this function at the lower limit.*

Limits of Integration

From Section 25.3, we know that this definite integral can be interpreted as the area under the curve of $y = f(x)$ from $x = a$ to $x = b$, and in general as a summation. *This summation interpretation will be applied to many kinds of applied problems.*

That integration is equivalent to the limit of a sum is the reason that Leibniz (see page 645) used an elongated S for the integral sign. It stands for the Latin word for *sum.*

◀ EXAMPLE 1   Evaluate the integral $\int_0^2 x^4\,dx$.

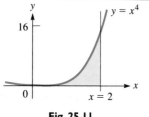

Fig. 25.11

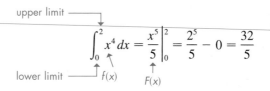

$$\int_0^2 x^4\,dx = \left.\frac{x^5}{5}\right|_0^2 = \frac{2^5}{5} - 0 = \frac{32}{5}$$

Note that a vertical line—with the limits written at the top and the bottom—is the way the value is indicated after integration, but before evaluation.

The area that this definite integral can represent is shown in Fig. 25.11.   ▶

◀ EXAMPLE 2   Evaluate $\int_1^3 (x^{-2} - 1)\,dx$.

$$\int_1^3 (x^{-2} - 1)\,dx = \left. -\frac{1}{x} - x\right|_1^3 = \left(-\frac{1}{3} - 3\right) - (-1 - 1)$$

$$= -\frac{10}{3} + 2 = -\frac{4}{3}$$

◀ EXAMPLE 3   Evaluate $\int_0^1 5z(z^2 + 1)^5\,dz$.

For purposes of integration, $n = 5$, $u = z^2 + 1$, and $du = 2z\,dz$. Hence,

$$\int_0^1 5z(z^2 + 1)^5\,dz = \frac{5}{2} \int_0^1 (z^2 + 1)^5(2z\,dz)$$

$$= \frac{5}{2}\left(\frac{1}{6}\right)(z^2 + 1)^6\Big|_0^1$$

$$= \frac{5}{12}(2^6 - 1^6) = \frac{5(63)}{12} = \frac{105}{4}$$

◀ EXAMPLE 4   Evaluate $\displaystyle\int_{0.1}^{2.7} \frac{dx}{\sqrt{4x+1}}$.

In order to integrate, we have $n = -\frac{1}{2}$, $u = 4x + 1$, and $du = 4\,dx$. Therefore,

$$\int_{0.1}^{2.7} \frac{dx}{\sqrt{4x+1}} = \int_{0.1}^{2.7} (4x+1)^{-1/2}\,dx = \frac{1}{4}\int_{0.1}^{2.7} (4x+1)^{-1/2}(4\,dx)$$

$$= \frac{1}{4}\left(\frac{1}{\frac{1}{2}}\right)(4x+1)^{1/2}\Big|_{0.1}^{2.7} = \frac{1}{2}(4x+1)^{1/2}\Big|_{0.1}^{2.7} \qquad \text{integrate}$$

$$= \frac{1}{2}\left(\sqrt{11.8} - \sqrt{1.4}\right) = 1.126 \qquad \text{evaluate} \qquad ▶$$

◀ EXAMPLE 5   Evaluate $\displaystyle\int_{0}^{4} \frac{x+1}{(x^2+2x+2)^3}\,dx$.

For integrating, $n = -3$, $u = x^2 + 2x + 2$, and $du = (2x+2)\,dx$.

$$\int_{0}^{4} (x^2+2x+2)^{-3}(x+1)\,dx = \frac{1}{2}\int_{0}^{4} (x^2+2x+2)^{-3}[2(x+1)\,dx]$$

$$= \frac{1}{2}\left(\frac{1}{-2}\right)(x^2+2x+2)^{-2}\Big|_{0}^{4} \qquad \text{integrate}$$

$$= -\frac{1}{4}(16+8+2)^{-2} + \frac{1}{4}(0+0+2)^{-2} \qquad \text{evaluate}$$

$$= \frac{1}{4}\left(-\frac{1}{26^2} + \frac{1}{2^2}\right) = \frac{1}{4}\left(\frac{1}{4} - \frac{1}{676}\right)$$

$$= \frac{1}{4}\left(\frac{168}{676}\right) = \frac{21}{338}$$

Fig. 25.12

The value of a definite integral can be found on a calculator. Figure 25.12 shows a typical display using the *numerical integral* feature to evaluate the integral of this example. Also shown is the check that the fraction above gives the same result. ▶

The following example illustrates an application of the definite integral. In Chapter 26, we will see that the definite integral has applications in many areas of science and technology.

See the chapter introduction.

◀ EXAMPLE 6   The rate of flow Q (in m^3/s) of water over a certain dam is found by evaluating the definite integral in the equation $Q = \int_{0}^{1.25} 240\sqrt{1.50-y}\,dy$. See Fig. 25.13. Find Q.

The solution is as follows:

$$Q = \int_{0}^{1.25} 240\sqrt{1.50-y}\,dy = -240\int_{0}^{1.25} (1.50-y)^{1/2}(-dy)$$

$$= -240\left(\frac{2}{3}\right)(1.50-y)^{3/2}\Big|_{0}^{1.25} \qquad \text{integrate}$$

$$= -160[(1.50-1.25)^{3/2} - (1.50-0)^{3/2}] \qquad \text{evaluate}$$

$$= -160(0.25^{3/2} - 1.50^{3/2}) = 274 \text{ m}^3/s \qquad ▶$$

y

1.25 m   1.50 m

Fig. 25.13

The definition of the definite integral is valid regardless of the source of $f(x)$. That is, we may apply the definite integral whenever we want to sum a function in a manner similar to that which we use to find an area.

EXERCISES 25.4

In Exercises 1 and 2, make the given changes in the indicated examples of this section and then solve the resulting problems.

1. In Example 2, change the upper limit from 3 to 4.

2. In Example 4, change $4x$ to $2x$.

In Exercises 3–34, evaluate the given definite integrals.

3. $\int_0^1 2x\,dx$

4. $\int_0^2 3x^2\,dx$

5. $\int_1^4 x^{5/2}\,dx$

6. $\int_4^9 (p^{3/2} - 3)\,dp$

7. $\int_3^6 \left(\frac{1}{\sqrt{x}} + 2\right) dx$

8. $\int_{1.2}^{1.6} \left(5 + \frac{6}{x^4}\right) dx$

9. $\int_{-1.6}^{0.7} (1 - x)^{1/3}\,dx$

10. $\int_1^5 \sqrt{2x - 1}\,dx$

11. $\int_{-2}^2 (T - 2)(T + 2)\,dT$

12. $\int_1^2 (3x^5 - 2x^3)\,dx$

13. $\int_{0.5}^{2.2} (\sqrt[3]{x} - 2)\,dx$

14. $\int_{2.7}^{5.3} \left(\frac{1}{x\sqrt{x}} + 4\right) dx$

15. $\int_0^4 (1 - \sqrt{x})^2\,dx$

16. $\int_1^4 \frac{y + 4}{\sqrt{y}}\,dy$

17. $\int_{-2}^{-1} 2x(4 - x^2)^3\,dx$

18. $\int_0^1 x(3x^2 - 1)^3\,dx$

19. $\int_0^4 \frac{x\,dx}{\sqrt{x^2 + 9}}$

20. $\int_{0.2}^{0.7} x^2(x^3 + 2)^{3/2}\,dx$

21. $\int_{2.75}^{3.25} \frac{dx}{\sqrt[3]{6x + 1}}$

22. $\int_2^6 \frac{2\,dx}{\sqrt{4x + 1}}$

23. $\int_1^3 \frac{2x\,dx}{(2x^2 + 1)^3}$

24. $\int_{12.6}^{17.2} \frac{3\,dx}{(6x - 1)^2}$

25. $\int_3^7 \sqrt{16t^2 + 8t + 1}\,dt$

26. $\int_{-5}^1 \sqrt{6 - 2x}\,dx$

27. $\int_0^2 2x(9 - 2x^2)^2\,dx$

28. $\int_{-1}^2 V(V^3 + 1)\,dV$

29. $\int_0^1 (x^2 + 3)(x^3 + 9x + 6)^2\,dx$

30. $\int_2^3 \frac{x^2 + 1}{(x^3 + 3x)^2}\,dx$

31. $\int_{-1}^2 \frac{8x - 2}{(2x^2 - x + 1)^3}\,dx$

32. $\int_{-3}^{-2} (3x^2 - 2)\sqrt[3]{2x^3 - 4x + 1}\,dx$

33. $\int_{\sqrt{5}}^3 2x\sqrt[4]{x^4 + 8x^2 + 16}\,dx$

34. $\int_{-2}^0 \left(\sqrt{2x + 4} - \sqrt[3]{3x + 8}\right) dx$

In Exercises 35–44, solve the given problems.

W 35. Show that $\int_0^1 x^3\,dx + \int_1^2 x^3\,dx = \int_0^2 x^3\,dx$. In terms of area, explain the result.

W 36. Show that $\int_0^1 x\,dx > \int_0^1 x^2\,dx$ and $\int_1^2 x\,dx < \int_1^2 x^2\,dx$. In terms of area, explain the result.

37. Given that $\int_0^9 \sqrt{x}\,dx = 18$, evaluate $\int_0^9 2\sqrt{t}\,dt$.

38. Evaluate $\int_{-1}^1 t^{2k}\,dt$, where k is a positive integer.

39. The work W (in N · m) in winding up an 80-m cable is $W = \int_0^{80} (1000 - 5x)\,dx$. Evaluate W.

W 40. The total volume V of liquid flowing through a certain pipe of radius R is $V = k(R^2 \int_0^R r\,dr - \int_0^R r^3\,dr)$, where k is a constant. Evaluate V and explain why R, but not r, can be to the left of the integral sign.

41. The surface area A (in m²) of a certain parabolic radio-wave reflector is $A = 4\pi \int_0^2 \sqrt{3x + 9}\,dx$. Evaluate A.

42. The total force (in N) on the circular end of a water tank is $F = 19\,600 \int_0^5 y\sqrt{25 - y^2}\,dy$. Evaluate F.

43. In finding the average electron energy in a metal at very low temperatures, the integral $\frac{3N}{2E_F^{3/2}} \int_0^{E_F} E^{3/2}\,dE$ is used. Evaluate this integral.

44. In finding the electric field E caused by a surface electric charge on a disk, the equation $E = k \int_0^R \frac{r\,dr}{(x^2 + r^2)^{3/2}}$ is used. Evaluate the integral.

NUMERICAL INTEGRATION: THE TRAPEZOIDAL RULE

For data and functions that cannot be directly integrated by available methods, it is possible to develop numerical methods of integration. These numerical methods are of greater importance today since they are readily adaptable for use on a calculator or computer. There are a great many such numerical techniques for approximating the value of an integral. In this section, we develop one of these, the *trapezoidal rule.* In the following section, another numerical method is discussed.

We know from Sections 25.3 and 25.4 that we can interpret a definite integral as the area under a curve. We will therefore show how to approximate the value of the integral by approximating the appropriate area by a set of inscribed trapezoids. The basic idea here is very similar to that used when rectangles were inscribed under a curve. However, the use of trapezoids reduces the error and provides a better approximation.

The area to be found is subdivided into n intervals of equal width. Perpendicular lines are then dropped from the curve (or points, if only a given set of numbers is available). If the points on the curve are joined by straight-line segments, the area of successive parts under the curve is approximated by finding the areas of the trapezoids formed. However, if these points are not too far apart, the approximation will be very good (see Fig. 25.14). From geometry, recall that the area of a trapezoid equals one-half the product of the sum of the bases times the altitude. For these trapezoids, the bases are the y-coordinates, and the altitudes are h. Therefore, when we indicate the sum of these trapezoidal areas, we have

$$A_T = \frac{1}{2}(y_0 + y_1)h + \frac{1}{2}(y_1 + y_2)h + \frac{1}{2}(y_2 + y_3)h + \cdots$$

$$+ \frac{1}{2}(y_{n-2} + y_{n-1})h + \frac{1}{2}(y_{n-1} + y_n)h$$

We note, when this addition is performed, that the result is

$$A_T = h\left(\frac{1}{2}y_0 + y_1 + y_2 + \cdots + y_{n-1} + \frac{1}{2}y_n\right) \tag{25.12}$$

The y-values to be used either are derived from the function $y = f(x)$ or are the y-coordinates of a set of data.

Since A_T approximates the area under the curve, it also approximates the value of the definite integral, or

$$\int_a^b f(x)\,dx \approx \frac{h}{2}(y_0 + 2y_1 + 2y_2 + \cdots + 2y_{n-1} + y_n) \tag{25.13}$$

Equation (25.13) is known as the **trapezoidal rule.** This is the same rule as Eq. (2.12), which we used to measure irregular areas (see page 68). We can now use it to find the approximate value of an integral.

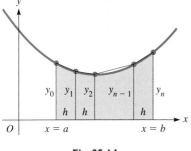

Fig. 25.14

See Appendix C for the graphing calculator program TRAPRULE. It evaluates an integral using the trapezoidal rule with n intervals.

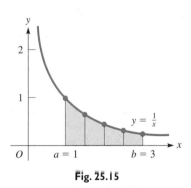

Fig. 25.15

Table 25.2

Number of Trapezoids n	Total Area of Trapezoids
4	1.116 666 7
100	1.098 641 9
1 000	1.098 612 6
10 000	1.098 612 3

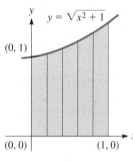

Fig. 25.16

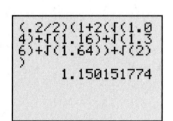

Fig. 25.17

◖ **EXAMPLE 1** Approximate the value of $\int_1^3 \dfrac{1}{x}\,dx$ by the trapezoidal rule. Let $n = 4$.

We are to approximate the area under $y = 1/x$ from $x = 1$ to $x = 3$ by dividing the area into four trapezoids. This area is found by applying Eq. (25.12), which is the approximate value of the integral, as shown in Eq. (25.13). Figure 25.15 shows the graph. In this example, $f(x) = 1/x$, and

$$h = \frac{3-1}{4} = \frac{1}{2} \qquad y_0 = f(a) = f(1) = 1$$

$$y_1 = f\left(\frac{3}{2}\right) = \frac{2}{3} \qquad y_2 = f(2) = \frac{1}{2}$$

$$y_3 = f\left(\frac{5}{2}\right) = \frac{2}{5} \qquad y_n = y_4 = f(b) = f(3) = \frac{1}{3}$$

$$A_T = \frac{1/2}{2}\left[1 + 2\left(\frac{2}{3}\right) + 2\left(\frac{1}{2}\right) + 2\left(\frac{2}{5}\right) + \frac{1}{3}\right]$$

$$= \frac{1}{4}\left(\frac{15 + 20 + 15 + 12 + 5}{15}\right) = \frac{1}{4}\left(\frac{67}{15}\right) = \frac{67}{60}$$

Therefore,

$$\int_1^3 \frac{1}{x}\,dx \approx \frac{67}{60}$$

We cannot perform this integration directly by methods developed up to this point. As we increase the number of trapezoids, the value becomes more accurate. See Table 25.2. The actual value to seven decimal places is 1.098 612 3. ◗

◖ **EXAMPLE 2** Approximate the value of $\int_0^1 \sqrt{x^2 + 1}\,dx$ by the trapezoidal rule. Let $n = 5$.

Figure 25.16 shows the graph. In this example,

$$h = \frac{1-0}{5} = 0.2$$

$$y_0 = f(0) = 1 \qquad\qquad y_1 = f(0.2) = \sqrt{1.04} = 1.019\,803\,9$$
$$y_2 = f(0.4) = \sqrt{1.16} = 1.077\,033\,0 \qquad y_3 = f(0.6) = \sqrt{1.36} = 1.166\,190\,4$$
$$y_4 = f(0.8) = \sqrt{1.64} = 1.280\,624\,8 \qquad y_5 = f(1) = \sqrt{2.00} = 1.414\,213\,6$$

Hence, we have

$$A_T = \frac{0.2}{2}[1 + 2(1.109\,803\,9) + 2(1.077\,033\,0) + 2(1.166\,190\,4)$$

$$+ 2(1.280\,624\,8) + 1.414\,213\,6]$$

$$= 1.150 \qquad \text{(rounded off)}$$

This means that

$$\int_0^1 \sqrt{x^2 + 1}\,dx \approx 1.150 \qquad \text{the actual value is 1.148 to three decimal places}$$

The entire calculation can be done on a calculator without tabulating values, as shown in the display shown in Fig. 25.17. ◗

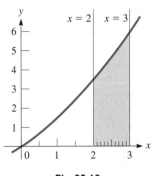

Fig. 25.18

Fig. 25.19

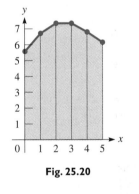

Fig. 25.20

❮ **EXAMPLE 3**   Approximate the value of the integral $\int_2^3 x\sqrt{x+1}\,dx$ by using the trapezoidal rule. Use $n = 10$.

In Fig. 25.18, the graph of the function and the area used in the trapezoidal rule are shown. From the given values, we have $h = \frac{3-2}{10} = 0.1$. Therefore,

$$y_0 = f(2) = 2\sqrt{3} = 3.464\,1016 \qquad y_1 = f(2.1) = 2.1\sqrt{3.1} = 3.697\,4315$$
$$y_2 = f(2.2) = 2.2\sqrt{3.2} = 3.935\,4796 \qquad y_3 = f(2.3) = 2.3\sqrt{3.3} = 4.178\,1575$$
$$y_4 = f(2.4) = 2.4\sqrt{3.4} = 4.425\,3813 \qquad y_5 = f(2.5) = 2.5\sqrt{3.5} = 4.677\,0717$$
$$y_6 = f(2.6) = 2.6\sqrt{3.6} = 4.933\,1532 \qquad y_7 = f(2.7) = 2.7\sqrt{3.7} = 5.193\,5537$$
$$y_8 = f(2.8) = 2.8\sqrt{3.8} = 5.458\,2048 \qquad y_9 = f(2.9) = 2.9\sqrt{3.9} = 5.727\,0411$$
$$y_{10} = f(3) = 3\sqrt{4} = 6.000\,0000$$

$$A_T = \frac{0.1}{2}[3.464\,1016 + 2(3.697\,4315) + \cdots + 2(5.727\,0411) + 6.000\,0000]$$

$$= 4.6958$$

Therefore, $\int_2^3 x\sqrt{x+1}\,dx \approx 4.6958$. We see from the calculator display shown in Fig. 25.19 that the value is 4.6954 to four decimal places. ❯

❮ **EXAMPLE 4**   The following points were found empirically:

x	0	1	2	3	4	5
y	5.68	6.75	7.32	7.35	6.88	6.24

Approximate the value of the integral of the function defined by these points between $x = 0$ and $x = 5$ by the trapezoidal rule.

In order to find A_T, we use the values of y_0, y_1, and so on, directly from the table. We also note that $h = 1$. The graph is shown in Fig. 25.20. Therefore, we have

$$A_T = \frac{1}{2}[5.68 + 2(6.75) + 2(7.32) + 2(7.35) + 2(6.88) + 6.24] = 34.26$$

Although we do not know the algebraic form of the function, we can state that

$$\int_0^5 f(x)\,dx \approx 34.26$$ ❯

EXERCISES 25.5

In Exercises 1 and 2, make the given changes in the indicated examples of this section and then solve the resulting problems.

1. In Example 1, change n from 4 to 2.

2. In Example 3, change n from 10 to 5.

In Exercises 3–6, (a) approximate the value of each of the given integrals by use of the trapezoidal rule, using the given value of n, and (b) check by direct integration.

3. $\int_0^2 2x^2\,dx,\ n = 4$

4. $\int_0^1 (1 - x^2)\,dx,\ n = 3$

5. $\int_1^4 \left(1 + \sqrt{x}\right)dx,\ n = 6$

6. $\int_3^8 \sqrt{1 + x}\,dx,\ n = 5$

In Exercises 7–14, approximate the value of each of the given integrals by use of the trapezoidal rule, using the given value of n.

7. $\int_2^3 \frac{1}{2x}\,dx,\ n = 2$

8. $\int_2^6 \frac{dx}{x + 3},\ n = 4$

9. $\int_0^5 \sqrt{25 - x^2}\,dx,\ n = 5$

10. $\int_0^2 \sqrt{x^3 + 1}\,dx,\ n = 4$

11. $\int_1^5 \frac{1}{x^2 + x}\,dx,\ n = 10$

12. $\int_2^4 \frac{1}{x^2 + 1}\,dx,\ n = 10$

13. $\int_0^4 2^x\,dx,\ n = 12$

14. $\int_0^{1.5} 10^x\,dx,\ n = 15$

In Exercises 15 and 16, approximate the values of the integrals defined by the given sets of points.

15. $\int_2^{14} y \, dx$

x	2	4	6	8	10	12	14
y	0.67	2.34	4.56	3.67	3.56	4.78	6.87

16. $\int_{1.4}^{3.2} y \, dx$

x	1.4	1.7	2.0	2.3	2.6	2.9	3.2
y	0.18	7.87	18.23	23.53	24.62	20.93	20.76

In Exercises 17–20, solve the given problems by using the trapezoidal rule.

W **17.** Explain why the approximate value of the integral in Exercise 5 is less than the exact value.

18. A force F that a distributed electric charge has on a point charge is $F = k \int_0^2 \dfrac{dx}{(4 + x^2)^{3/2}}$, where x is the distance along the distributed charge and k is a constant. With $n = 8$, evaluate F in terms of k.

19. The length L (in m) of telephone wire needed (considering the sag) between two poles exactly 100 m apart is $L = 2 \int_0^{50} \sqrt{6.4 \times 10^{-7} x^2 + 1} \, dx$. With $n = 10$, evaluate L (to six significant digits).

20. The amount A (in standard pollution index) of a pollutant in the air in a city is measured to be $A = \dfrac{150}{1 + 0.25(t - 4.0)^2} + 25$, where t is the time (in h) after 6 A.M. With $n = 6$, find the total value of A between 6 A.M. and noon.

25.6 SIMPSON'S RULE

The numerical method of integration developed in this section is also readily programmable for use on a computer or easily usable with the necessary calculations done on a calculator. It is obtained by interpreting the definite integral as the area under a curve, as we did in developing the trapezoidal rule, and by approximating the curve by a set of parabolic arcs. The use of parabolic arcs, rather than chords as with the trapezoidal rule, usually gives a better approximation.

Since we will be using parabolic arcs, we first derive a formula for the area that is under a parabolic arc. The curve shown in Fig. 25.21 represents the parabola $y = ax^2 + bx + c$. The points shown on this curve are $(-h, y_0)$, $(0, y_1)$, and (h, y_2). The area under the parabola is given by

$$A = \int_{-h}^{h} y \, dx = \int_{-h}^{h} (ax^2 + bx + c) \, dx = \frac{ax^3}{3} + \frac{bx^2}{2} + cx \Big|_{-h}^{h}$$

$$= \frac{2}{3} ah^3 + 2ch$$

$$A = \frac{h}{3}(2ah^2 + 6c) \tag{25.14}$$

The coordinates of the three points also satisfy the equation $y = ax^2 + bx + c$. This means that

$$y_0 = ah^2 - bh + c$$
$$y_1 = c$$
$$y_2 = ah^2 + bh + c$$

By finding the sum of $y_0 + 4y_1 + y_2$, we have

$$y_0 + 4y_1 + y_2 = 2ah^2 + 6c \tag{25.15}$$

Substituting Eq. (25.15) into Eq. (25.14), we have

$$A = \frac{h}{3}(y_0 + 4y_1 + y_2) \tag{25.16}$$

We note that the area depends only on the distance h and the three y-coordinates.

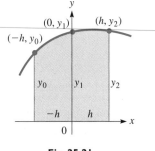

Fig. 25.21

Now let us consider the area under the curve in Fig. 25.22. If a parabolic arc is passed through the points (x_0, y_0), (x_1, y_1), and (x_2, y_2), we may use Eq. (25.16) to approximate the area under the curve between x_0 and x_2. We again note that the distance h is the difference in the x-coordinates. Therefore, the area under the curve between x_0 and x_2 is

$$A_1 = \frac{h}{3}(y_0 + 4y_1 + y_2)$$

Similarly, if a parabolic arc is passed through the three points starting with (x_2, y_2), the area between x_2 and x_4 is

$$A_2 = \frac{h}{3}(y_2 + 4y_3 + y_4)$$

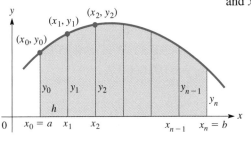

Fig. 25.22

The sum of these areas is

$$A_1 + A_2 = \frac{h}{3}(y_0 + 4y_1 + 2y_2 + 4y_3 + y_4) \tag{25.17}$$

CAUTION ▶ We can continue this procedure until the approximate value of the entire area has been found. We must note, however, that *the number of intervals n of width h must be even.* Therefore, generalizing on Eq. (25.17) and recalling again that the value of the definite integral is the area under the curve, we have

$$\int_a^b f(x)\, dx \approx \frac{h}{3}(y_0 + 4y_1 + 2y_2 + 4y_3 + 2y_4 + \cdots + 4y_{n-1} + y_n) \tag{25.18}$$

Although named for the English mathematician Thomas Simpson (1710–1761), he did not discover the rule. It was well known when he included it in some of his many books on mathematics.

Equation (25.18) is known as **Simpson's rule.** As with the trapezoidal rule, we used Simpson's rule in Chapter 2 to measure irregular areas (see page 70). We now see how it is derived. Simpson's rule is also used in many calculator models for the evaluation of definite integrals.

◀ **EXAMPLE 1** Approximate the value of the integral $\int_0^1 \dfrac{dx}{x+1}$ by Simpson's rule. Let $n = 2$.

In Fig. 25.23, the graph of the function and the area used are shown. We are to approximate the integral by using Eq. (25.18). We therefore note that $f(x) = 1/(x + 1)$. Also, $x_0 = a = 0$, $x_1 = 0.5$, and $x_2 = b = 1$. This is due to the fact that $n = 2$ and $h = 0.5$ since the total interval is 1 unit (from $x = 0$ to $x = 1$). Therefore,

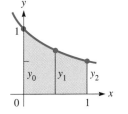

Fig. 25.23

$$y_0 = \frac{1}{0 + 1} = 1.0000 \qquad y_1 = \frac{1}{0.5 + 1} = 0.6667 \qquad y_2 = \frac{1}{1 + 1} = 0.5000$$

Substituting, we have

$$\int_0^1 \frac{dx}{x+1} = \frac{0.5}{3}[1.0000 + 4(0.6667) + 0.5000]$$

$$= 0.694$$

To three decimal places, the actual value of the integral is 0.693. We will consider the method of integrating this function in a later chapter. ▶

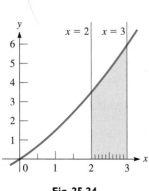

Fig. 25.24

◀ EXAMPLE 2 Approximate the value of $\int_2^3 x\sqrt{x + 1}\,dx$ by Simpson's rule. Use $n = 10$.

Since the necessary values for this function are shown in Example 3 of Section 25.5, we shall simply tabulate them here. ($h = 0.1$.) See Fig. 25.24.

$$y_0 = 3.464\,101\,6 \qquad y_1 = 3.697\,431\,5 \qquad y_2 = 3.935\,479\,6 \qquad y_3 = 4.178\,157\,5$$

$$y_4 = 4.425\,381\,3 \qquad y_5 = 4.677\,071\,7 \qquad y_6 = 4.933\,153\,2 \qquad y_7 = 5.193\,553\,7$$

$$y_8 = 5.458\,204\,8 \qquad y_9 = 5.727\,041\,1 \qquad y_{10} = 6.000\,000\,0$$

Therefore, we evaluate the integral as follows:

$$\int_2^3 x\sqrt{x + 1}\,dx = \frac{0.1}{3}[3.464\,101\,6 + 4(3.697\,431\,5) + 2(3.935\,479\,6)$$

$$+ 4(4.178\,157\,5) + 2(4.425\,381\,3) + 4(4.677\,071\,7)$$

$$+ 2(4.933\,153\,2) + 4(5.193\,553\,7) + 2(5.458\,204\,8)$$

$$+ 4(5.727\,041\,1) + 6.000\,000\,0]$$

$$= \frac{0.1}{3}(140.861\,56) = 4.695\,385\,4$$

This result agrees with the actual value to the eight significant digits shown. The value we obtained with the trapezoidal rule was 4.6958.

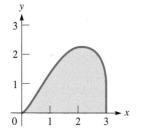

Fig. 25.25

◀ EXAMPLE 3 The rear stabilizer of a certain aircraft is shown in Fig. 25.25. The area A (in m^2) of one side of the stabilizer is $A = \int_0^3 (3x^2 - x^3)^{0.6}\,dx$. Find this area, using Simpson's rule with $n = 6$.

Here, we note that $f(x) = (3x^2 - x^3)^{0.6}$, $a = 0$, $b = 3$, and $h = \dfrac{3 - 0}{6} = 0.5$. Therefore,

$$y_0 = f(0) = [3(0)^2 - 0^3]^{0.6} = 0$$

$$y_1 = f(0.5) = [3(0.5)^2 - 0.5^3]^{0.6} = 0.754\,272\,0$$

$$y_2 = f(1) = [3(1)^2 - 1^3]^{0.6} = 1.515\,716\,6$$

$$y_3 = f(1.5) = [3(1.5)^2 - 1.5^3]^{0.6} = 2.074\,742\,8$$

$$y_4 = f(2) = [3(2)^2 - 2^3]^{0.6} = 2.297\,396\,7$$

$$y_5 = f(2.5) = [3(2.5)^2 - 2.5^3]^{0.6} = 1.981\,116\,5$$

$$y_6 = f(3) = [3(3)^2 - 3^3]^{0.6} = 0$$

$$A = \int_0^3 (3x^2 - x^3)^{0.6}\,dx = \frac{0.5}{3}[0 + 4(0.754\,272\,0) + 2(1.515\,716\,6)$$

$$+ 4(2.074\,742\,8) + 2(2.297\,396\,7) + 4(1.981\,116\,5) + 0]$$

$$= 4.477\,792\,0 \text{ m}^2$$

Thus, the area of one side of the stabilizer is 4.478 m^2 (rounded off).

As with the trapezoidal rule, since the calculation can be done completely on a calculator, it is not necessary to record the above values. Using $y = (3x^2 - x^3)^{0.6}$, the display in Fig. 25.26 shows the calculator evaluation of the area.

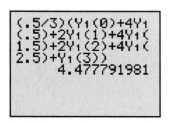

Fig. 25.26

EXERCISES 25.6

In Exercises 1 and 2, make the given changes in the indicated examples of this section, and then solve the indicated problems.

1. In Example 1, change the denominator $x + 1$ to $x + 2$ and then find the approximate value of the integral.

2. In Example 2, change n such that $h = 0.2$ and explain why Simpson's rule cannot be used.

In Exercises 3–6, (a) approximate the value of each of the given integrals by use of Simpson's rule, using the given value of n, and (b) check by direct integration.

3. $\int_0^2 (1 + x^3)\, dx$, $n = 2$

4. $\int_0^8 x^{1/3}\, dx$, $n = 2$

5. $\int_1^4 \left(2x + \sqrt{x}\right) dx$, $n = 6$

6. $\int_0^2 x\sqrt{x^2 + 1}\, dx$, $n = 4$

In Exercises 7–12, approximate the value of each of the given integrals by use of Simpson's rule, using the given values of n. Exercises 8–10 are the same as Exercises 10–12 of Section 25.5.

7. $\int_0^5 \sqrt{25 - x^2}\, dx$, $n = 4$

8. $\int_0^2 \sqrt{x^3 + 1}\, dx$, $n = 4$

9. $\int_1^5 \dfrac{1}{x^2 + x}\, dx$, $n = 10$

10. $\int_2^4 \dfrac{1}{x^2 + 1}\, dx$, $n = 10$

11. $\int_{-4}^5 (2x^4 + 1)^{0.1}\, dx$, $n = 6$

12. $\int_0^{2.4} \dfrac{dx}{\left(4 + \sqrt{x}\right)^{3/2}}$, $n = 8$

In Exercises 13 and 14, approximate the values of the integrals defined by the given sets of point by using Simpson's rule. These are the same as Exercises 15 and 16 of Section 25.5.

13. $\int_2^{14} y\, dx$

x	2	4	6	8	10	12	14
y	0.67	2.34	4.56	3.67	3.56	4.78	6.87

14. $\int_{1.4}^{3.2} y\, dx$

x	1.4	1.7	2.0	2.3	2.6	2.9	3.2
y	0.18	7.87	18.23	23.53	24.62	20.93	20.76

In Exercises 15 and 16, solve the given problems, using Simpson's rule.

15. The distance $\bar{x}$ (in cm) from one end of a barrel plug (with vertical cross section) to its center of mass, as shown in Fig. 25.27, is $\bar{x} = 0.9129 \int_0^3 x\sqrt{0.3 - 0.1x}\, dx$. Find $\bar{x}$ with $n = 12$.

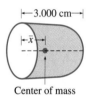

Fig. 25.27　　Center of mass

16. The average value of the electric current i_{av} (in A) in a circuit for the first 4 s is $i_{av} = \frac{1}{4} \int_0^4 (4t - t^2)^{0.2}\, dt$. Find i_{av} with $n = 10$.

CHAPTER 25 EQUATIONS

Indefinite integral	$\displaystyle\int f(x)\, dx = F(x) + C$	(25.1)
Integrals	$\displaystyle\int c\, du = c\int du = cu + C$	(25.2)
	$\displaystyle\int (du + dv) = u + v + C$	(25.3)
Power formula	$\displaystyle\int u^n\, du = \dfrac{u^{n+1}}{n + 1} + C \quad (n \neq -1)$	(25.4)
Area under a curve	$A_{ab} = \left[\displaystyle\int f(x)\, dx\right]_a^b = F(b) - F(a)$	(25.10)
Definite integral	$\displaystyle\int_a^b f(x)\, dx = F(b) - F(a)$	(25.11)
Trapezoidal rule	$\displaystyle\int_a^b f(x)\, dx \approx \dfrac{h}{2}(y_0 + 2y_1 + 2y_2 + \cdots + 2y_{n-1} + y_n)$	(25.13)
Simpson's rule	$\displaystyle\int_a^b f(x)\, dx \approx \dfrac{h}{3}(y_0 + 4y_1 + 2y_2 + 4y_3 + 2y_4 + \cdots + 4y_{n-1} + y_n)$	(25.18)

CHAPTER 25 REVIEW EXERCISES

In Exercises 1–24, evaluate the given integrals.

1. $\int (4x^3 - x)\,dx$

2. $\int (5 + 3t^2)\,dt$

3. $\int \sqrt{u}\,(u^2 + 2)\,du$

4. $\int x(x - 3x^4)\,dx$

5. $\int_1^4 \left(\dfrac{\sqrt{x}}{2} + \dfrac{2}{\sqrt{x}}\right)dx$

6. $\int_1^2 \left(x + \dfrac{1}{x^2}\right)dx$

7. $\int_0^2 x(4 - x)\,dx$

8. $\int_0^1 2t(2t + 1)^2\,dt$

9. $\int \left(3 + \dfrac{2}{x^3}\right)dx$

10. $\int \left(3\sqrt{x} + \dfrac{1}{2\sqrt{x}} - \dfrac{1}{4}\right)dx$

11. $\int_{-2}^5 \dfrac{dx}{\sqrt[3]{x^2 + 6x + 9}}$

12. $\int_{0.35}^{0.85} x\left(\sqrt{1 - x^2} + 1\right)dx$

13. $\int \dfrac{dn}{(2 - 5n)^3}$

14. $\int \dfrac{1}{x^2}\sqrt{1 + \dfrac{1}{x}}\,dx$

15. $\int 3(7 - 2x)^{3/4}\,dx$

16. $\int (y^3 + 3y^2 + 3y + 1)^{2/3}\,dy$

17. $\int_0^2 \dfrac{3x\,dx}{\sqrt[3]{1 + 2x^2}}$

18. $\int_1^6 \dfrac{2\,dx}{(3x - 2)^{3/4}}$

19. $\int x^2(1 - 2x^3)^4\,dx$

20. $\int 3R^3(1 - 5R^4)\,dR$

21. $\int \dfrac{(2 - 3x^2)\,dx}{(2x - x^3)^2}$

22. $\int \dfrac{x^2 - 3}{\sqrt{6 + 9x - x^3}}\,dx$

23. $\int_1^3 (x^2 + x + 2)(2x^3 + 3x^2 + 12x)\,dx$

24. $\int_0^2 (4x + 18x^2)(x^2 + 3x^3)^2\,dx$

In Exercises 25–34, solve the given problems.

25. Find the equation of the curve that passes through $(-1, 3)$ for which the slope is given by $3 - x^2$.

26. Find the equation of the curve that passes through $(1, -2)$ for which the slope is $x(x^2 + 1)^2$.

(W) 27. Perform the integration $\int (1 - 2x)\,dx$ (a) term by term, labeling the constant of integration as C_1, and then (b) by letting $u = 1 - 2x$, using the general power rule and labeling the constant of integration as C_2. Is $C_1 = C_2$? Explain.

(W) 28. Following the methods (a) and (b) in Exercise 27, perform the integration $\int (3x + 2)\,dx$. In (b) let $u = 3x + 2$. Is $C_1 = C_2$? Explain.

(W) 29. Show that $\int_0^1 x^3\,dx = \int_1^2 (x - 1)^3\,dx$. In terms of area, explain this result.

30. If $\int_0^1 [f(x) - g(x)]\,dx = 3$ and $\int_0^1 g(x)\,dx = -1$, find the value of $\int_0^1 2f(x)\,dx$.

31. Use Eq. (25.10) to find the area under $y = 6x - 1$ between $x = 1$ and $x = 3$.

32. Use Eq. (25.10) to find the first-quadrant area under $y = 8x - x^4$.

(W) 33. Given that $f(x)$ is continuous, $f(x) > 0$, and $f''(x) < 0$ for $a \le x \le b$, explain why the exact value of $\int_a^b f(x)\,dx$ is greater than the approximate value found by use of the trapezoidal rule.

34. It is shown in more advanced works that when evaluating $\int_a^b f(x)\,dx$, the maximum error in using Simpson's rule is $\dfrac{M(b - a)^5}{180n^4}$, where M is the greatest absolute value of the fourth derivative of $f(x)$ for $a \le x \le b$. Evaluate the maximum error for the integral $\int_2^3 \dfrac{dx}{2x}$ with $n = 4$.

In Exercises 35 and 36, solve the given problems by using the trapezoidal rule. In Exercises 37 and 38, solve the given problems using Simpson's rule.

35. Approximate $\int_1^3 \dfrac{dx}{2x - 1}$ with $n = 4$.

36. The streamflow F (in m³/s) passing through a cross section of a stream is found by evaluating $F = \int_0^L h(x)\,dx$, where L is the width of the stream cross section and $h(x)$ is the product of the depth and velocity of the stream x m from the bank. From the following table, estimate F with $L = 24$ m.

x (m)	0.0	3.0	6.0	9.0	12.0	15.0	18.0	21.0	24.0
$h(x)$	0.00	0.59	0.13	0.34	0.76	0.65	0.29	0.07	0.00

37. Approximate $\int_1^3 \dfrac{dx}{2x - 1}$ with $n = 4$ (see Exercise 35).

38. The velocity v (in km/h) of a car was recorded at 1-min intervals as shown. Estimate the distance traveled by the car.

t (min)	0	1	2	3	4	5	6	7	8	9	10
v (km/h)	60	62	65	69	72	74	76	77	77	75	76

In Exercises 39–44, use the function $y = x\sqrt[3]{2x^2 + 1}$ and approximate the area under the curve between $x = 1$ and $x = 4$ by the indicated method. See Fig. 25.28.

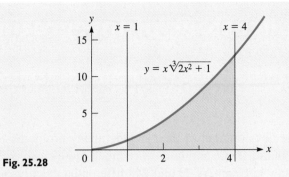

Fig. 25.28

39. Find the sum of the areas of three inscribed rectangles.

40. Find the sum of the areas of six inscribed rectangles.

41. Use the trapezoidal rule with $n = 3$.

42. Use the trapezoidal rule with $n = 6$.

43. Use Simpson's rule with $n = 6$.

44. Use integration (for the exact area).

In Exercises 45–48, find the area of the archway, as shown in Fig. 25.29, by the indicated method. The archway can be described as the area bounded by the elliptical arc $y = 4 + \sqrt{1 + 8x - 2x^2}$, $x = 0$, $x = 4$, and $y = 0$, where dimensions are in meters.

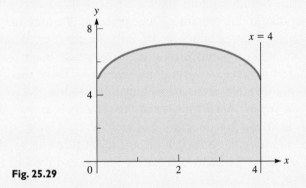

Fig. 25.29

45. Find the sum of the areas of eight inscribed rectangles.

46. Use the trapezoidal rule with $n = 8$.

47. Use Simpson's rule with $n = 8$.

48. Use the integration evaluation feature on a calculator.

In Exercises 49–52, solve the given problems by integration.

49. The deflection y of a certain beam at a distance x from one end is given by $dy/dx = k(2L^3 - 12Lx + 2x^4)$, where k is a constant and L is the length of the beam. Find y as a function of x if $y = 0$ for $x = 0$.

50. The total electric charge Q on a charged sphere is given by $Q = k \int \left(r^2 - \dfrac{r^3}{R} \right) dr$, where k is a constant, r is the distance from the center of the sphere, and R is the radius of the sphere. Find Q as a function of r if $Q = Q_0$ for $r = R$.

51. Part of the deck of a boat is the parabolic area shown in Fig. 25.30. The area A (in m²) is $A = 2 \int_0^5 \sqrt{5 - y}\, dy$. Evaluate A.

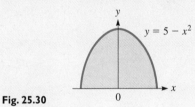

Fig. 25.30

52. The distance s (in cm) through which a cam follower moves in 4 s is $s = \int_0^4 t\sqrt{4 + 9t^2}\, dt$. Evaluate s.

Writing Exercise

53. A computer science student is writing a program to find a good approximation for the value of π by using the formula $A = \pi r^2$ for a circle. The value of π is to be found by approximating the area of a circle with a given radius. Write two or three paragraphs explaining how the value of π can be approximated in this way. Include any equations and values that may be used, but do not actually make the calculations.

CHAPTER 25 PRACTICE TEST

1. Find an antiderivative of $f(x) = 2x - (1 - x)^4$.

2. Integrate: $\int x\sqrt{1 - 2x^2}\, dx$.

3. Find y in terms of x if $dy/dx = (6 - x)^4$ and the curve passes through $(5, 2)$.

4. Approximate the area under $y = \dfrac{1}{x + 2}$ between $x = 1$ and $x = 4$ (above the x-axis) by inscribing six rectangles and finding the sum of their areas.

5. Evaluate $\displaystyle\int_1^4 \dfrac{dx}{x + 2}$ by using the trapezoidal rule with $n = 6$.

6. Evaluate the definite integral of Problem 5 by using Simpson's rule with $n = 6$.

7. The total electric current i (in A) to pass a point in the circuit between $t = 1$ s and $t = 3$ s is $i = \displaystyle\int_1^3 \left(t^2 + \dfrac{1}{t^2} \right) dt$. Evaluate i.

26

Applications of Integration

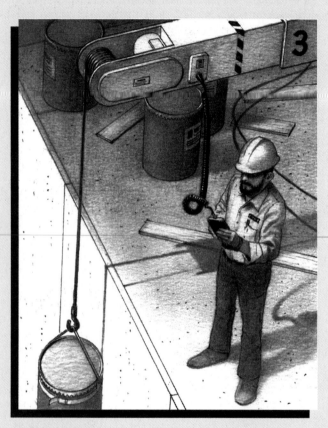

In Section 26.6, integration is used to find the work done in winding up a cable.

With the development of the calculus, many problems being studied in the 1600s and later were much more easily solved. As we saw in Chapters 23 and 24, differential calculus led to the solution of problems such as finding velocities, maximum and minimum values, and various types of rates of change.

Integral calculus also led to the solution of many types of problems that were being studied. As Newton and Leibniz developed the methods of integral calculus, they were interested in finding areas and the problems that could be solved by finding these areas. They were also very interested in applications in which the rate of change was known and therefore led to the relation between the variables being studied.

One of the problems being studied in the 1600s by the French mathematician and physicist Blaise Pascal was that of the pressure within a liquid and the force on the walls of the container due to this pressure. Since pressure is a measure of the force on an area, finding the force became essentially solving an area problem. Also at the time, many mathematicians and physicists were studying various kinds of motion, such as motion along a curved path and the motion of a rotating object. When information about the velocity is known, the solution is found by integrating. Later in the 1800s, since electric current is the time rate change of electric charge, the current could be found by integrating known expressions for the charge.

As it turns out, integration is useful in many areas of science, engineering, and technology. It has important applications in areas such as electricity, mechanics, architecture, machine design, and business, as well as other areas of physics and geometry.

In the first section of this chapter, we present some important applications of the indefinite integral, with emphasis on the motion of an object and the voltage across a capacitor. In the remaining sections, we show uses of the definite integral related to geometry, mechanics, work by a variable force, and force due to liquid pressure.

APPLICATIONS OF THE INDEFINITE INTEGRAL

VELOCITY AND DISPLACEMENT

Velocity as a first derivative and acceleration as a second derivative were introduced in Chapter 23.

We first apply integration to the problem of finding the displacement and velocity as functions of time, when we know the relationship between acceleration and time, and certain values of displacement and velocity. As shown in Section 25.2, these values are needed for finding the constants of integration that are introduced.

Recalling that the acceleration a of an object is given by $a = dv/dt$, we can find the expression for the velocity v in terms of a, the time t, and the constant of integration. Therefore, we write $dv = a\,dt$, or

$$v = \int a\,dt \qquad (26.1)$$

If the acceleration is constant, we have

$$v = at + C_1 \qquad (26.2)$$

In general, Eq. (26.1) is used to find the velocity as a function of time when we know the acceleration as a function of time. Since the case of constant acceleration is often encountered, Eq. (26.2) can often be used.

EXAMPLE 1   Find the expression for the velocity if $a = 12t$, given that $v = 8$ when $t = 1$.

Using Eq. (26.1), we have

$$v = \int (12t)\,dt = 6t^2 + C_1$$

Substituting the known values, we have $8 = 6 + C_1$, or $C_1 = 2$. This means that

$$v = 6t^2 + 2$$

EXAMPLE 2   For an object falling under the influence of gravity, the acceleration due to gravity is essentially constant. Its value is $-9.8\ \text{m/s}^2$. (The negative sign is chosen so that *all quantities directed up are positive* and **all quantities directed down are negative.**) Find the expression for the velocity of an object under the influence of gravity if $v = v_0$ when $t = 0$.

NOTE ▶

We write

$$v = \int (-9.8)\,dt \qquad \text{substitute } a = -9.8 \text{ into Eq. (26.1)}$$

$$= -9.8t + C_1 \qquad \text{integrate}$$

$$v_0 = -9.8(0) + C_1 \qquad \text{substitute given values}$$

$$C_1 = v_0 \qquad \text{solve for } C_1$$

$$v = v_0 - 9.8t \qquad \text{substitute}$$

The velocity v_0 is called the *initial velocity*. If the object is given an initial upward velocity of 40 m/s, $v_0 = 40$ m/s. If the object is dropped, $v_0 = 0$. If the object is given an initial downward velocity of 40 m/s, $v_0 = -40$ m/s.

Once we have the expression for velocity, we can then integrate to find the expression for displacement s in terms of the time. Since $v = ds/dt$, we can write $ds = v\,dt$, or

$$s = \int v\,dt \qquad (26.3)$$

Consider the following examples.

Solving a Word Problem

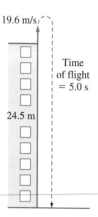

19.6 m/s

Time of flight = 5.0 s

24.5 m

Fig. 26.1

◀ EXAMPLE 3   A ball is thrown vertically from the top of a building 24.5 m high and hits the ground 5.0 s later. What initial velocity was the ball given?

Measuring vertical distances from the ground, we know that $s = 24.5$ m when $t = 0$ and that $v = v_0 - 9.8t$. Thus,

$$s = \int (v_0 - 9.8t)\,dt = v_0 t - 4.9t^2 + C \qquad \text{integrate}$$

$$24.5 = v_0(0) - 4.9(0) + C, \qquad C = 24.5 \qquad \text{evaluate } C$$

$$s = v_0 t - 4.9t^2 + 24.5$$

We also know that $s = 0$ when $t = 5.0$ s. Thus,

$$0 = v_0(5.0) - 4.9(5.0)^2 + 24.5 \qquad \text{substitute given values}$$

$$5.0v_0 = 98.0$$

$$v_0 = 19.6 \text{ m/s}$$

This means that the initial velocity was 19.6 m/s upward. See Fig. 26.1.

◀ EXAMPLE 4   During the initial stage of launching a spacecraft vertically, the acceleration a (in m/s^2) of the spacecraft is $a = 6t^2$. Find the height s of the spacecraft after 6.0 s if $s = 12$ m for $t = 0.0$ s and $v = 16$ m/s for $t = 2.0$ s.

First, we use Eq. (26.1) to get an expression for the velocity:

$$v = \int 6t^2\,dt = 2t^3 + C_1 \qquad \text{integrate}$$

$$16 = 2(2.0)^3 + C_1, \qquad C_1 = 0 \qquad \text{evaluate } C_1$$

$$v = 2t^3$$

We now use Eq. (26.3) to get an expression for the displacement:

$$s = \int 2t^3\,dt = \tfrac{1}{2}t^4 + C_2 \qquad \text{integrate}$$

$$12 = \tfrac{1}{2}(0.0)^4 + C_2, \qquad C_2 = 12 \qquad \text{evaluate } C_2$$

$$s = \tfrac{1}{2}t^4 + 12$$

Now, finding s for $t = 6.0$ s, we have

$$s = \tfrac{1}{2}(6.0)^4 + 12 = 660 \text{ m}$$

VOLTAGE ACROSS A CAPACITOR

The second basic application of the indefinite integral we will discuss comes from the field of electricity. By definition, *the current i in an electric circuit equals the time rate of change of the charge q (in coulombs) that passes a given point in the circuit, or*

$$i = \frac{dq}{dt} \qquad (26.4)$$

Rewriting this expression in differential notation as $dq = i\,dt$ and integrating both sides of the equation, we have

$$q = \int i\,dt \qquad (26.5)$$

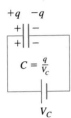

Fig. 26.2

Now, the voltage V_C across a capacitor with capacitance C (see Fig. 26.2) is given by $V_C = q/C$. By combining equations, the voltage V_C is given by

$$V_C = \frac{1}{C} \int i\,dt \qquad (26.6)$$

Here, V_C is measured in volts, C in farads, i in amperes, and t in seconds.

◀ EXAMPLE 5   The current in a certain electric circuit as a function of time is given by $i = 6t^2 + 4$. Find an expression for the amount of charge q that passes a point in the circuit as a function of time. Assuming that $q = 0$ when $t = 0$, determine the total charge that passes the point in 2 s.

Since $q = \int i\,dt$, we have

$$q = \int (6t^2 + 4)\,dt \qquad \text{substitute into Eq. (26.5)}$$

$$= 2t^3 + 4t + C \qquad \text{integrate}$$

The last expression is the desired expression giving charge as a function of time. We note that when $t = 0$, then $q = C$, which means that the constant of integration represents the initial charge, or the charge that passed a given point before we started timing. Using q_0 to represent this charge, we have

$$q = 2t^3 + 4t + q_0$$

Now, returning to the second part of the problem, we see that $q_0 = 0$. Therefore, evaluating q for $t = 2$ s, we have

$$q = 2(8) + 4(2) = 24 \text{ C}$$

(Here, the symbol C represents coulombs and is not the C for capacitance of Eq. (26.6) or the constant of integration.) This is the charge that passes any specified point in the circuit in 2 s.   ▶

Solving a Word Problem **EXAMPLE 6** The voltage across a 5.0-μF capacitor is zero. What is the voltage after 20 ms if a current of 75 mA charges the capacitor?

Since the current is 75 mA, we know that $i = 0.075$ A $= 7.5 \times 10^{-2}$ A. We see that **we must use the proper power of 10 that corresponds to each prefix.** Since $5.0 \ \mu$F $= 5.0 \times 10^{-6}$ F, we have

CAUTION ▶

$$V_C = \frac{1}{5.0 \times 10^{-6}} \int 7.5 \times 10^{-2} \, dt \qquad \text{substituting into Eq. (26.6)}$$

$$= (1.5 \times 10^4) \int dt$$

$$= (1.5 \times 10^4)t + C_1 \qquad \text{integrate}$$

From the given information we know that $V_C = 0$ when $t = 0$. Thus,

$$0 = (1.5 \times 10^4)(0) + C_1 \quad \text{or} \quad C_1 = 0 \qquad \text{evaluate } C_1$$

This means that

$$V_C = (1.5 \times 10^4)t$$

Evaluating this expression for $t = 20 \times 10^{-3}$ s, we have

$$V_C = (1.5 \times 10^4)(20 \times 10^{-3})$$

$$= 30 \times 10 = 300 \text{ V}$$

◗

Solving a Word Problem **EXAMPLE 7** A certain capacitor is measured to have a voltage of 100 V across it. At this instant a current as a function of time given by $i = 0.06\sqrt{t}$ is sent through the circuit. After 0.25 s, the voltage is measured to be 140 V. What is the capacitance of the capacitor?

Substituting $i = 0.06\sqrt{t}$, we find that

$$V_C = \frac{1}{C} \int \left(0.06\sqrt{t} \, dt\right) = \frac{0.06}{C} \int t^{1/2} \, dt \qquad \text{using Eq. (26.6)}$$

$$= \frac{0.04}{C} t^{3/2} + C_1 \qquad \text{integrate}$$

From the given information we know that $V_C = 100$ V when $t = 0$. Thus,

$$100 = \frac{0.04}{C}(0) + C_1 \quad \text{or} \quad C_1 = 100 \text{ V} \qquad \text{evaluate } C_1$$

This means that

$$V_C = \frac{0.04}{C} t^{3/2} + 100$$

We also know that $V_C = 140$ V when $t = 0.25$ s. Therefore,

$$140 = \frac{0.04}{C}(0.25)^{3/2} + 100$$

$$40 = \frac{0.04}{C}(0.125)$$

or

$$C = 1.25 \times 10^{-4} \text{ F} = 125 \ \mu\text{F}$$

◗

EXERCISES 26.1

1. In Example 3, change 5.0 s to 1.0 s and then solve the resulting problem.

2. In Example 7, change $0.06\sqrt{t}$ to $0.06t$ and then solve the resulting problem.

3. What is the velocity (in m/s) of a sandbag 1.5 s after it is released from a hot-air balloon that is stationary in the air?

4. A hoop is started upward along an inclined plane at 5.0 m/s. If the acceleration of the hoop is 2.0 m/s² downward along the plane, find the velocity of the hoop after 6.0 s.

5. A conveyor belt 8.00 m long moves at 0.25 m/s. If a package is placed at one end, find its displacement from the other end as a function of time.

6. During each cycle, the velocity v (in mm/s) of a piston is $v = 6t - 6t^2$, where t is the time (in s). Find the displacement s of the piston after 0.75 s if the initial displacement is zero.

7. While in the barrel of a tennis ball machine, the acceleration a (in m/s²) of a ball is $a = 30\sqrt{1 - 4t}$, where t is the time (in s). If $v = 0$ for $t = 0$, find the velocity of the ball as it leaves the barrel at $t = 0.25$ s.

8. A person skis down a slope with an acceleration (in m/s²) given by $a = \dfrac{600t}{(60 + 0.5t^2)^2}$, where t is the time (in s). Find the skier's velocity as a function of time if $v = 0$ when $t = 0$.

9. If a car decelerates at 250 m/s² (about the maximum a human body can survive) during an accident, and the car was going at 96 km/h at impact, over what distance must an airbag stop a person in order to survive the crash?

10. The engine of a lunar lander is cut off when the lander is 5.0 m above the surface of the moon and descending at 2.0 m/s. If the acceleration due to gravity on the moon is 1.6 m/s², what is the speed of the lander just before it touches the surface?

11. If an aircraft is to attain a take-off velocity of 75 m/s after traveling 240 m along the flight deck of an aircraft carrier, find the aircraft's acceleration (assumed constant).

12. A ball is thrown straight up from the edge of the roof of a 60.0-m-tall building. If a second ball is dropped from the roof 2.00 s later, what must be the initial velocity of the first ball if both are to reach the ground at the same time?

13. What must be the nozzle velocity of the water from a fire hose if it is to reach a point 30 m directly above the nozzle?

14. An arrow is shot upward with a vertical velocity of 40.0 m/s from the edge of a cliff. If it hits the ground below after 9.0 s, how high is the cliff?

15. In coming to a stop, the acceleration of a car is $-4t$. If it is traveling at 32.0 m/s when the brakes are applied, how far does it travel while stopping?

16. A hoist mechanism raises a crate with an acceleration (in m/s²) $a = \sqrt{1 + 0.2t}$, where t is the time in seconds. Find the dis-

placement of the crate as a function of time if $v = 0$ m/s and $s = 2$ m for $t = 0$ s.

17. The electric current in a microprocessor circuit is 0.230 μA. How many coulombs pass a given point in the circuit in 1.50 ms?

18. The electric current (in mA) in a computer circuit as a function of time (in s) is $i = 0.3 - 0.2t$. What total charge passes a point in the circuit in 0.050 s?

19. In an amplifier circuit, the current i (in A) changes with time t (in s) according to $i = 0.06t\sqrt{1 + t^2}$. If 0.015 C of charge has passed a point in the circuit at $t = 0$, find the total charge to have passed the point at $t = 0.25$ s.

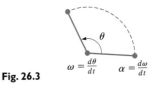 20. The current i (in μA) in a certain microprocessor circuit is given by $i = 8 - t$, where t is the time (in μs) and $0 \le t \le 20$ μs. If $q_0 = 0$, for what value of t, greater than zero, is $q = 0$? What interpretation can be given to this result?

21. The voltage across a 2.5-μF capacitor in a copying machine is zero. What is the voltage after 12 ms if a current of 25 mA charges the capacitor?

22. The voltage across an 8.50-nF capacitor in an FM receiver circuit is zero. Find the voltage after 2.00 μs if a current (in mA) $i = 0.042t$ charges the capacitor.

23. The voltage across a 3.75-μF capacitor in a television circuit is 4.50 mV. Find the voltage after 0.565 ms if a current (in μA) $i = \sqrt[3]{1 + 6t}$ further charges the capacitor.

24. A current $i = t/\sqrt{t^2 + 1}$ (in A) is sent through an electric dryer circuit containing a previously uncharged 2.0-μF capacitor. How long does it take for the capacitor voltage to reach 120 V?

25. The angular velocity ω is the time rate of change of the angular displacement θ of a rotating object. See Fig. 26.3. In testing the shaft of an engine, its angular velocity is $\omega = 16t + 0.50t^2$, where t is the time (in s) of rotation. Find the angular displacement through which the shaft goes in 10.0 s.

$$\omega = \frac{d\theta}{dt} \qquad \alpha = \frac{d\omega}{dt}$$

Fig. 26.3

26. The angular acceleration α is the time rate of change of angular velocity ω of a rotating object. See Fig. 26.3. When starting up, the angular acceleration of a helicopter blade is $\alpha = \sqrt{8t + 1}$. Find the expression for θ if $\omega = 0$ and $\theta = 0$ for $t = 0$.

27. An inductor in an electric circuit is essentially a coil of wire in which the voltage is affected by a changing current. By definition, the voltage caused by the changing current is given by $V_L = L(di/dt)$, where L is the inductance (in H). If $V_L = 12.0 - 0.2t$ for a 3.0-H inductor, find the current in the circuit after 20 s if the initial current was zero.

28. If the inner and outer walls of a container are at different temperatures, the rate of change of temperature with respect to the distance from one wall is a function of the distance from the wall. Symbolically, this is stated as $dT/dx = f(x)$, where T is the temperature. If x is measured from the outer wall, at 20°C, and $f(x) = 72x^2$, find the temperature at the inner wall if the container walls are 0.5 cm thick.

29. Surrounding an electrically charged particle is an electric field. The rate of change of electric potential with respect to the distance from the particle creating the field equals the negative of the value of the electric field. That is, $dV/dx = -E$, where E is the electric field. If $E = k/x^2$, where k is a constant, find the electric potential at a distance x_1 from the particle, if $V \to 0$ as $x \to \infty$.

30. The rate of change of the vertical deflection y with respect to the horizontal distance x from one end of a beam is a function of x.

For a particular beam, the function is $k(x^5 + 1350x^3 - 7000x^2)$, where k is a constant. Find y as a function of x.

31. Fresh water is flowing into a brine solution, with an equal volume of mixed solution flowing out. The amount of salt in the solution decreases, but more slowly as time increases. Under certain conditions, the time rate of change of mass of salt (in g/min) is given by $-1/\sqrt{t + 1}$. Find the mass m of salt as a function of time if 1000 g were originally present. Under these conditions, how long would it take for all the salt to be removed?

32. A holograph of a circle is formed. The rate of change of the radius r of the circle with respect to the wavelength λ of the light used is inversely proportional to the square root of λ. If $dr/d\lambda = 3.55 \times 10^4$ and $r = 4.08$ cm for $\lambda = 574$ nm, find r as a function of λ.

26.2 AREAS BY INTEGRATION

In Section 25.3, we introduced the method of finding the area under a curve by integration. We also showed that the area can be found by a summation process on the rectangles inscribed under the curve, which means that integration can be interpreted as a summation process. *The applications of the definite integral use this summation interpretation of the integral.* We now develop a general procedure for finding the area for which the bounding curves are known by summing the areas of inscribed rectangles and using integration for the summation.

The first step is to make a sketch of the area. Next, a representative **element of area** dA (a typical rectangle) is drawn. In Fig. 26.4, the width of the element is dx. The length of the element is determined by the y-coordinate (of the vertex of the element) of the point on the curve. Thus, the length is y. The area of this element is $y\,dx$, which in turn means that $dA = y\,dx$, or

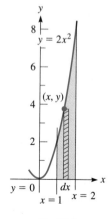

Fig. 26.4

$$A = \int_a^b y\,dx = \int_a^b f(x)\,dx \tag{26.7}$$

This equation states that the elements are to be summed (this is the meaning of the integral sign) from a (the left boundary) to b (the right boundary).

◀ **EXAMPLE 1** Find the area bounded by $y = 2x^2$, $y = 0$, $x = 1$, and $x = 2$.

This area is shown in Fig. 26.5. The rectangle shown is the representative element. Its area is $y\,dx$. The elements are to be summed from $x = 1$ to $x = 2$.

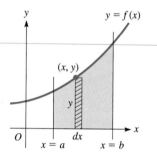

Fig. 26.5

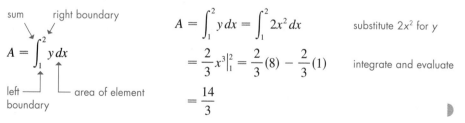

$$A = \int_1^2 y\,dx = \int_1^2 2x^2\,dx \qquad \text{substitute } 2x^2 \text{ for } y$$

$$= \frac{2}{3}x^3\Big|_1^2 = \frac{2}{3}(8) - \frac{2}{3}(1) \qquad \text{integrate and evaluate}$$

$$= \frac{14}{3}$$

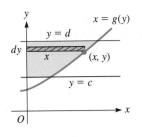

Fig. 26.6

In Figs. 26.4 and 26.5, the elements are vertical. It is also possible to use horizontal elements, and many problems are simplified by using them. In using horizontal elements, the length (longest dimension) is measured in terms of the x-coordinate of the point on the curve, and the width becomes dy. In Fig. 26.6, the area of the element is $x\,dy$, which means $dA = x\,dy$, or

$$A = \int_c^d x\,dy = \int_c^d g(y)\,dy \qquad (26.8)$$

In using Eq. (26.8), the elements are summed from c (the lower boundary) to d (the upper boundary). In the following example, the area is found by use of both vertical and horizontal elements of area.

◀ EXAMPLE 2   Find the area in the first quadrant bounded by $y = 9 - x^2$.

The area to be found is shown in Fig. 26.7. First, using the vertical element of length y and width dx, we have

$$A = \int_0^3 y\,dx \qquad \text{sum of areas of elements}$$

$$= \int_0^3 (9 - x^2)\,dx \qquad \text{substitute } 9 - x^2 \text{ for } y$$

$$= \left(9x - \frac{x^3}{3}\right)\bigg|_0^3 \qquad \text{integrate}$$

$$= (27 - 9) - 0 = 18 \qquad \text{evaluate}$$

Now, using the horizontal element of length x and width dy, we have

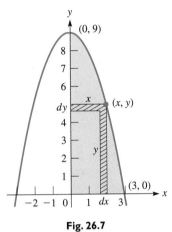

Fig. 26.7

$$A = \int_0^9 x\,dy \qquad \text{sum of areas of elements}$$

$$= \int_0^9 \sqrt{9 - y}\,dy = -\int_0^9 (9 - y)^{1/2}(-dy) \qquad \text{substitute } \sqrt{9 - y} \text{ for } x$$

$$= -\frac{2}{3}(9 - y)^{3/2}\bigg|_0^9 \qquad \text{integrate}$$

$$= -\frac{2}{3}(9 - 9)^{3/2} + \frac{2}{3}(9 - 0)^{3/2} \qquad \text{evaluate}$$

$$= \frac{2}{3}(27) = 18$$

CAUTION ▶

Note that the limits for the vertical elements are 0 and 3, whereas those for the horizontal elements are 0 and 9. These limits are determined by the direction in which the elements are summed. As we have noted, ***vertical elements are summed from left to right, and horizontal elements are summed from bottom to top.*** Doing it in this way means that the summation will be done in a positive direction.   ▶

The choice of vertical or horizontal elements is determined by (1) which one leads to the simplest solution or (2) the form of the resulting integral. In some problems, it makes little difference which is chosen. However, our present methods of integration do not include many types of integrals.

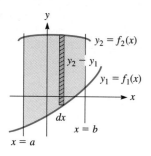

Fig. 26.8

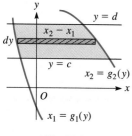

Fig. 26.9

AREA BETWEEN TWO CURVES

It is also possible to find the area between two curves when one of the curves is not an axis. In such a case, the length of the element of area becomes the difference in the y- or x-coordinates, depending on whether a vertical element or a horizontal element is used.

In Fig. 26.8, by using vertical elements, the element of area is bounded on the bottom by $y_1 = f_1(x)$ and on the top by $y_2 = f_2(x)$. The length of the element is $y_2 - y_1$, and its width is dx. Thus, the area is

$$A = \int_a^b (y_2 - y_1)\, dx \qquad (26.9)$$

In Fig. 26.9, by using horizontal elements, the element of area is bounded on the left by $x_1 = g_1(y)$ and on the right by $x_2 = g_2(y)$. The length of the element is $x_2 - x_1$, and its width is dy. Thus, the area is

$$A = \int_c^d (x_2 - x_1)\, dy \qquad (26.10)$$

The following examples show the use of Eqs. (26.9) and (26.10) to find the indicated areas.

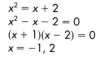

EXAMPLE 3 Find the area bounded by $y = x^2$ and $y = x + 2$.

First, by sketching each curve, we see that the area to be found is that shown in Fig. 26.10. The points of intersection of these curves are found by solving the equations simultaneously. The solution for the x-values is shown at the left. We then find the y-coordinates by substituting into either equation. The substitution shows the points of intersection to be $(-1, 1)$ and $(2, 4)$.

Here, we choose vertical elements, since they are all bounded at the top by the line $y = x + 2$ and at the bottom by the parabola $y = x^2$. If we were to choose horizontal elements, the bounding curves are different above $(-1, 1)$ from below this point. Choosing horizontal elements would then require two separate integrals for solution. Therefore, using vertical elements, we have

$$A = \int_{-1}^{2} (y_{\text{line}} - y_{\text{parabola}})\, dx \qquad \text{using Eq. (26.9)}$$

$$= \int_{-1}^{2} (x + 2 - x^2)\, dx = \left(\frac{x^2}{2} + 2x - \frac{x^3}{3} \right) \Big|_{-1}^{2}$$

$$= \left(2 + 4 - \frac{8}{3} \right) - \left(\frac{1}{2} - 2 + \frac{1}{3} \right)$$

$$= \frac{10}{3} + \frac{7}{6} = \frac{27}{6}$$

$$= \frac{9}{2}$$

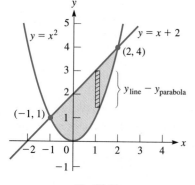

Fig. 26.10

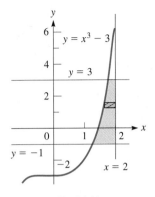

Fig. 26.11

◀ **EXAMPLE 4** Find the area bounded by the curve $y = x^3 - 3$ and the lines $x = 2$, $y = -1$, and $y = 3$.

Sketching the curve and lines, we show the area in Fig. 26.11. Horizontal elements are better, since they avoid having to evaluate the area in two parts. Therefore, we have

$$A = \int_{-1}^{3} (x_{\text{line}} - x_{\text{cubic}}) \, dy = \int_{-1}^{3} \left(2 - \sqrt[3]{y + 3}\right) dy \qquad \text{using Eq. (26.10)}$$

$$= 2y - \frac{3}{4}(y + 3)^{4/3} \Big|_{-1}^{3} = \left[6 - \frac{3}{4}(6^{4/3})\right] - \left[-2 - \frac{3}{4}(2^{4/3})\right]$$

$$= 8 - \frac{9}{2}\sqrt[3]{6} + \frac{3}{2}\sqrt[3]{2} = 1.713$$

As we see, the choice of horizontal elements leads to limits of -1 and 3. If we had chosen vertical elements, the limits would have been $\sqrt[3]{2}$ and $\sqrt[3]{6}$ for the area to the left of $\left(\sqrt[3]{6}, 3\right)$, and $\sqrt[3]{6}$ and 2 to the right of this point. ▶

CAUTION ▶ It is important to set up the element of area so that its length is positive. If the difference is taken incorrectly, the result will show a negative area. ***Getting positive lengths can be ensured for vertical elements if we subtract y of the lower curve from y of the upper curve. For horizontal elements, we should subtract x of the left curve from x of the right curve.*** This important point is illustrated in the following example.

◀ **EXAMPLE 5** Find the area bounded by $y = x^3 - 3x - 2$ and the x-axis.

Sketching the graph, we find that $y = x^3 - 3x - 2$ has a maximum point at $(-1, 0)$, a minimum point at $(1, -4)$, and an intercept at $(2, 0)$. The graph is shown in Fig. 26.12, and we see that the area is *below* the x-axis. In using vertical elements, we see that the top is the x-axis ($y = 0$) and the bottom is the curve of $y = x^3 - 3x - 2$. Therefore, we have

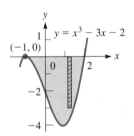

Fig. 26.12

$$A = \int_{-1}^{2} [0 - (x^3 - 3x - 2)] \, dx = \int_{-1}^{2} (-x^3 + 3x + 2) \, dx$$

$$= -\frac{1}{4}x^4 + \frac{3}{2}x^2 + 2x \Big|_{-1}^{2}$$

$$= \left[-\frac{1}{4}(2^4) + \frac{3}{2}(2^2) + 2(2)\right] - \left[-\frac{1}{4}(-1)^4 + \frac{3}{2}(-1)^2 + 2(-1)\right]$$

$$= \frac{27}{4} = 6.75$$

If we had simply set up the area as $A = \int_{-1}^{2} (x^3 - 3x - 2) \, dx$, we would have found $A = -6.75$. The negative sign shows that the area is below the x-axis. Again, we avoid any complications with negative areas by making the length of the element positive.

Also note that since

$$0 - (x^3 - 3x - 2) = -(x^3 - 3x - 2)$$

an area bounded on top by the x-axis can be found by setting up the area as being "under" the curve and using the negative of the function. ▶

NOTE ▶ We must *be very careful if the bounding curves of an area cross.* In such a case, for part of the area one curve is above the area, and for a different part of the area this same curve is below the area. When this happens, *two integrals must be used* to find the area. The following example illustrates the necessity of using this procedure.

◀ **EXAMPLE 6** Find the area between $y = x^3 - x$ and the x-axis.

We note from Fig. 26.13 that the area to the left of the origin is above the axis and the area to the right is below. If we find the area from

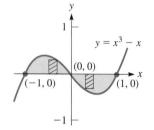

Fig. 26.13

$$A = \int_{-1}^{1} (x^3 - x)\, dx = \frac{x^4}{4} - \frac{x^2}{2} \Big|_{-1}^{1}$$

$$= \left(\frac{1}{4} - \frac{1}{2}\right) - \left(\frac{1}{4} - \frac{1}{2}\right) = 0$$

we see that the apparent area is zero. From the figure, we know this is not correct. Noting that the y-values (of the area) are negative to the right of the origin, we set up the integrals

$$A = \int_{-1}^{0} (x^3 - x)\, dx + \int_{0}^{1} [0 - (x^3 - x)]\, dx$$

$$= \left(\frac{x^4}{4} - \frac{x^2}{2}\right) \Big|_{-1}^{0} - \left(\frac{x^4}{4} - \frac{x^2}{2}\right) \Big|_{0}^{1}$$

$$= 0 - \left(\frac{1}{4} - \frac{1}{2}\right) - \left(\frac{1}{4} - \frac{1}{2}\right) + 0 = \frac{1}{2}$$

The area under a curve can be applied to various kinds of functions. This is illustrated in the following example and in some of the exercises that follow.

◀ **EXAMPLE 7** Measurements of solar radiation on a particular surface indicated that the rate r, in joules per hour, at which solar energy is received during the day is given by the equation $r = 3600(12t^2 - t^3)$, where t is the time in hours. Since r is a rate, we may write $r = dE/dt$, where E is the energy, in joules, received at the surface. This means that $dE = 3600(12t^2 - t^3)\, dt$, and we can find the total energy by evaluating the definite integral

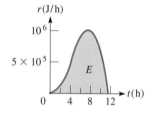

Fig. 26.14

$$E = 3600 \int_{0}^{12} (12t^2 - t^3)\, dt$$

This integral can be interpreted as being the area under $f(t) = 3600(12t^2 - t^3)$ from $t = 0$ to $t = 12$, as shown in Fig. 26.14 (the limits of integration are found from the t-intercepts of the curve). Evaluating this integral, we have

$$E = 3600 \int_{0}^{12} (12t^2 - t^3)\, dt = 3600\left(4t^3 - \frac{1}{4}t^4\right) \Big|_{0}^{12}$$

$$= 3600\left[4(12^3) - \frac{1}{4}(12^4) - 0\right]$$

$$= 6.22 \times 10^6\ \text{J}$$

Therefore, 6.22 MJ of energy were received in 12 h.

The display in Fig. 26.15 shows the calculator evaluation for this integral, where x is used for t in using the calculator. It can be seen that the values check. ◀

```
fnInt(3600(12X²-
X^3),X,0,12)
          6220800
```

Fig. 26.15

EXERCISES 26.2

In Exercises 1 and 2, make the given changes in the indicated examples of this section and then find the resulting areas.

1. In Example 1, change $x = 2$ to $x = 3$.

2. In Example 3, change $y = x + 2$ to $y = 2x$.

In Exercises 3–28, find the areas bounded by the indicated curves.

3. $y = 4x$, $y = 0$, $x = 1$ **4.** $y = 3x^2$, $y = 0$, $x = 3$

5. $y = 6 - 4x$, $x = 0$, $y = 0$, $y = 3$

6. $y = x^2 + 2$, $x = 0$, $y = 4$ $(x > 0)$

7. $y = x^2 - 4$, $y = 0$, $x = 4$

8. $y = x^2 - 2x$, $y = 0$

9. $y = x^{-2}$, $y = 0$, $x = 2$, $x = 3$

10. $y = 16 - x^2$, $y = 0$, $x = 1$, $x = 2$

11. $y = \sqrt{x}$, $x = 0$, $y = 1$, $y = 3$

12. $y = 2\sqrt{x + 1}$, $x = 0$, $y = 4$

13. $y = 2/\sqrt{x}$, $x = 0$, $y = 1$, $y = 4$

14. $x = y^2 - y$, $x = 0$

15. $y = 4 - 2x$, $x = 0$, $y = 0$, $y = 3$

16. $y = x$, $y = 2 - x$, $x = 0$

17. $y = x - 2\sqrt{x}$, $y = 0$

18. $y = x^4 - 2x^3$, $y = 0$

19. $y = x^2$, $y = 2 - x$, $x = 0$ $(x \geq 0)$

20. $y = x^2$, $y = 2 - x$, $y = 1$

21. $y = x^4 - 8x^2 + 16$, $y = 16 - x^4$

22. $y = \sqrt{x - 1}$, $y = 3 - x$, $y = 0$

23. $y = x^2 + 5x$, $y = 3 - x^2$

24. $y = x^3$, $y = x^2 + 4$, $x = -1$

25. $y = x^5$, $x = -1$, $x = 2$, $y = 0$

26. $y = x^2 + 2x - 8$, $y = x + 4$

27. $x + 2y = 0$, $y = x^2$, $y = x + 6$

28. $y = x^2$, $y = x^{1/3}$, between $x = -1$ and $x = 1$

In Exercises 29–32, solve the given problems.

(W) **29.** Describe a region for which the area is found by evaluating the integral $\int_1^2 (2x^2 - x^3)\, dx$.

(W) **30.** Why can the integral $\int_a^2 (2 + x - x^2)\, dx$ be used to find the area bounded by $x = a$, $y = 0$, and $y = 2 + x - x^2$ if $a = -1$, but not if $a = -2$?

31. Find the value of c such that the region bounded by $y = x^2$ and $y = 4$ is divided by $y = c$ into two regions of equal area.

32. Find the value(s) of c such that the region bounded by $y = x^2 - c^2$ and $y = x^2 + c^2$ has an area of 576.

In Exercises 33–36, find the areas bounded by the indicated curves, using (a) vertical elements and (b) horizontal elements.

33. $y = 8x$, $x = 0$, $y = 4$ **34.** $y = x^3$, $x = 0$, $y = 3$

35. $y = x^4$, $y = 8x$ **36.** $y = 4x$, $y = x^3$

In Exercises 37–44, some applications of areas are shown.

37. Certain physical quantities are often represented as an area under a curve. By definition, power is the time rate of change of performing work. Thus, $p = dw/dt$, or $dw = p\, dt$. Therefore, if $p = 12t - 4t^2$, find the work (in J) performed in 3 s by finding the area under the curve of p vs. t. See Fig. 26.16.

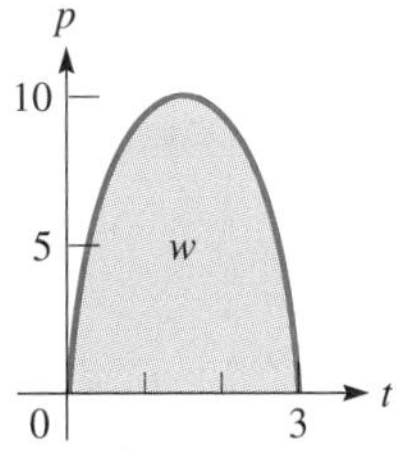

Fig. 26.16

38. The total electric charge Q (in C) to pass a point in the circuit from time t_1 to t_2 is $Q = \int_{t_1}^{t_2} i\, dt$, where i is the current (in A). Find Q if $t_1 = 1$ s, $t_2 = 4$ s, and $i = 0.0032t\sqrt{t^2 + 1}$.

39. Since the displacement s, velocity v, and time t of a moving object are related by $s = \int v\, dt$, it is possible to represent the change in displacement as an area. A rocket is launched such that its vertical velocity v (in km/s) as a function of time t (in s) is $v = 1 - 0.01\sqrt{2t + 1}$. Find the change in vertical displacement from $t = 10$ s to $t = 100$ s.

40. The total cost C (in dollars) of production can be interpreted as an area. If the cost per unit C' (in dollars per unit) of producing x units is given by $100/(0.01x + 1)^2$, find the total cost of producing 100 units by finding the area under the curve of C' vs x.

41. A cam is designed such that one face of it is described as being the area between the curves $y = x^3 - 2x^2 - x + 2$ and $y = x^2 - 1$ (units in cm). Show that this description does not uniquely describe the face of the cam. Find the area of the face of the cam, if a complete description requires that $x \leq 1$.

42. Using CAD (computer-assisted design), an architect programs a computer to sketch the shape of a swimming pool designed between the curves

$$y = \frac{800x}{(x^2 + 10)^2} \qquad y = 0.5x^2 - 4x \qquad x = 8$$

(dimensions in m). Find the area of the surface of the pool.

43. A coffee-table top is designed to be the region between $y = 0.25x^4$ and $y = 12 - 0.25x^4$. What is the area (in dm^2) of the table top?

44. A window is designed to be the area between a parabolic section and a straight base, as shown in Fig. 26.17. What is the area of the window?

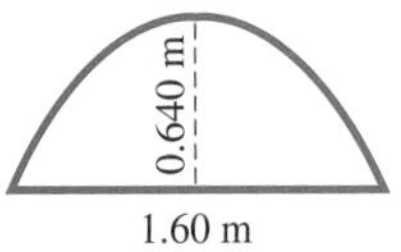

0.640 m

1.60 m

Fig. 26.17

26.3 VOLUMES BY INTEGRATION

Consider a region in the *xy*-plane and its representative element of area, as shown in Fig. 26.18(a). When the region is revolved about the *x*-axis, it is said to generate a **solid of revolution,** which is also shown in the figure. We now show methods of finding volumes of solids that are generated in this way.

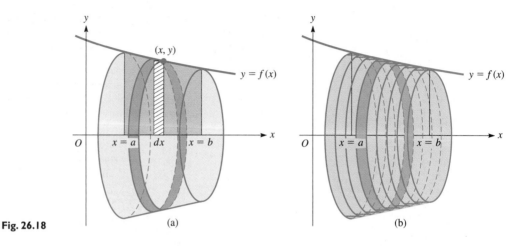

Fig. 26.18

(a)   (b)

As the region revolves about the *x*-axis, so does its representative element, which generates a solid for which the volume is known—an infinitesimally thin cylindrical **disk.** The volume of a right circular cylinder is π times its radius squared times its height (in this case, the thickness) of the cylinder. Since the element is revolved about the *x*-axis, the *y*-coordinate of the point on the curve that touches the element is the radius. Also, the thickness is dx. This disk, the representative **element of volume,** has a volume of $dV = \pi y^2 \, dx$. Summing these elements of volume from left to right, as shown in Fig. 26.18(b), we have for the total volume

> This is the sum of the volumes of the disks whose thicknesses approach zero as the number of disks approaches infinity.

$$V = \pi \int_a^b y^2 \, dx = \pi \int_a^b [f(x)]^2 \, dx \tag{26.11}$$

The element of volume is a **disk,** *and by use of Eq. (26.11) we can find the volume of the solid generated by a region bounded by the x-axis, which is revolved about the x-axis.*

◀ **EXAMPLE 1**   Find the volume of the solid generated by revolving the region bounded by $y = x^2$, $x = 2$, and $y = 0$ about the *x*-axis. See Fig. 26.19.

From the figure, we see that the radius of the disk is y and its thickness is dx. The elements are summed from left ($x = 0$) to right ($x = 2$):

$$V = \pi \int_0^2 y^2 \, dx \qquad \text{using Eq. (26.11)}$$

$$= \pi \int_0^2 (x^2)^2 \, dx = \pi \int_0^2 x^4 \, dx \qquad \text{substitute } x^2 \text{ for } y$$

$$= \frac{\pi}{5} x^5 \Big|_0^2 = \frac{32\pi}{5} \qquad \text{integrate and evaluate}$$

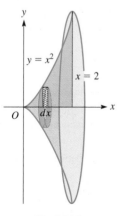

Fig. 26.19

Since π is used in Eq. (26.11), it is common to leave results in terms of π. In applied problems, a decimal result would normally be given.

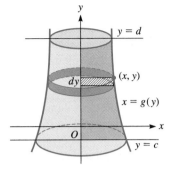

Fig. 26.20

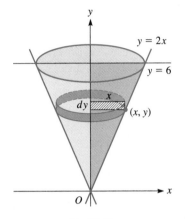

Fig. 26.21

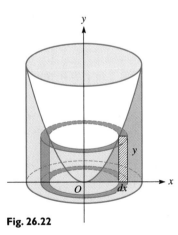

Fig. 26.22

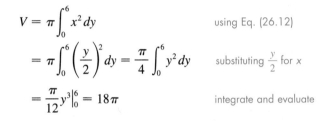

Fig. 26.23

If a region bounded by the y-axis is revolved about the y-axis, the volume of the solid generated is given by

$$V = \pi \int_c^d x^2 \, dy \qquad (26.12)$$

In this case, the radius of the element of volume, a *disk*, is the *x*-coordinate of the point on the curve, and the thickness of the disk is *dy*, as shown in Fig. 26.20. One should always be careful to identify the radius and the thickness properly.

◀ EXAMPLE 2   Find the volume of the solid generated by revolving the region bounded by $y = 2x$, $y = 6$, and $x = 0$ about the *y*-axis.

Figure 26.21 shows the volume to be found. Note that the radius of the disk is *x* and its thickness is *dy*.

$$
\begin{aligned}
V &= \pi \int_0^6 x^2 \, dy & \text{using Eq. (26.12)} \\
&= \pi \int_0^6 \left(\frac{y}{2}\right)^2 dy = \frac{\pi}{4}\int_0^6 y^2 \, dy & \text{substituting } \tfrac{y}{2} \text{ for } x \\
&= \frac{\pi}{12} y^3 \Big|_0^6 = 18\pi & \text{integrate and evaluate}
\end{aligned}
$$

Since this volume is a right circular cone, it is possible to check the result:

$$V = \frac{1}{3}\pi r^2 h = \frac{1}{3}\pi(3^2)(6) = 18\pi$$

If the region in Fig. 26.22 is revolved about the *y*-axis, the element of the area *y dx* generates a different element of volume from that when it is revolved about the *x*-axis. In Fig. 26.22, this element of volume is a **cylindrical shell.** *The total volume is made up of an infinite number of concentric shells.* When the volumes of these shells are summed, we have the total volume generated. Thus, we must now find the approximate volume *dV* of the representative shell. By finding the circumference of the base and multiplying this by the height, we obtain an expression for the surface area of the shell. Then, by multiplying this by the thickness of the shell, we find its volume. The volume of the representative **shell** shown in Fig. 26.23(a) is

Shell
$$dV = 2\pi(\text{radius}) \times (\text{height}) \times (\text{thickness}) \qquad (26.13)$$

Similarly, the volume of a *disk* is (see Fig. 26.23(b))

Disk
$$dV = \pi(\text{radius})^2 \times (\text{thickness}) \qquad (26.14)$$

It is generally better to remember the formulas for the elements of volume in the general forms given in Eqs. (26.13) and (26.14), and not in the specific forms such as Eqs. (26.11) and (26.12) (both of these use *disks*). If we remember the formulas in this way, we can readily apply these methods to finding any such volume of a solid of revolution.

◀ **EXAMPLE 3**   Use the method of cylindrical shells to find the volume of the solid generated by revolving the first-quadrant region bounded by $y = 4 - x^2$, $x = 0$, and $y = 0$ about the y-axis.

From Fig. 26.24, we identify the radius, the height, and the thickness of the shell:

$$\text{radius} = x \qquad \text{height} = y \qquad \text{thickness} = dx$$

CAUTION ▶ *The fact that the elements of area that generate the shells go from $x = 0$ to $x = 2$ determines the limits of integration as **0** and **2**.* Therefore,

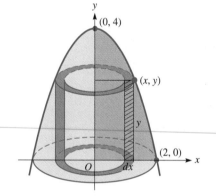

$$V = 2\pi \int_0^2 xy\, dx \longleftarrow \text{thickness} \qquad\qquad \text{using Eq. (26.13)}$$
$$\text{radius} \overset{\uparrow\uparrow}{} \text{height}$$

$$= 2\pi \int_0^2 x(4 - x^2)\, dx = 2\pi \int_0^2 (4x - x^3)\, dx \qquad \text{substitute } 4 - x^2 \text{ for } y$$

$$= 2\pi \left(2x^2 - \frac{1}{4}x^4 \right) \Big|_0^2 \qquad\qquad \text{integrate}$$

$$= 8\pi \qquad\qquad \text{evaluate}$$

Fig. 26.24

We can find the volume shown in Example 3 by using disks, as we show in the following example.

◀ **EXAMPLE 4**   Use the method of disks to find the volume indicated in Example 3.
From Fig. 26.25, we identify the radius and the thickness of the disk:

$$\text{radius} = x \qquad \text{thickness} = dy$$

CAUTION ▶ *Since the elements of area that generate the disks go from $y = 0$ to $y = 4$, the limits of integration are **0** and **4**.* Thus,

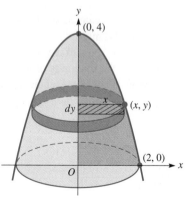

$$V = \pi \int_0^4 x^2\, dy \qquad\qquad \text{using Eq. (26.14)}$$
$$\text{radius} \overset{\uparrow\uparrow}{} \text{thickness}$$

$$= \pi \int_0^4 (4 - y)\, dy \qquad \text{substitute } \sqrt{4 - y} \text{ for } x$$

$$= \pi \left(4y - \frac{1}{2}y^2 \right) \Big|_0^4 \qquad \text{integrate}$$

$$= 8\pi \qquad\qquad \text{evaluate}$$

We see that the volume of 8π using disks agrees with the result we obtained using shells in Example 3.

Fig. 26.25

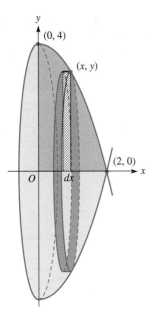

Fig. 26.26

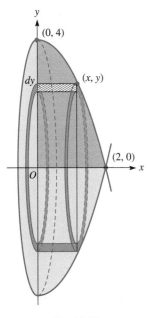

Fig. 26.27

◀ EXAMPLE 5   By using disks, find the volume of the solid generated if the first-quadrant region bounded by $y = 4 - x^2$, $x = 0$, and $y = 0$ is revolved about the x-axis. (This is the same region as used in Examples 3 and 4.)

For the disk in Fig. 26.26, we have

$$\text{radius} = y \qquad \text{thickness} = dx$$

and the limits of integration are $x = 0$ and $x = 2$. This gives us

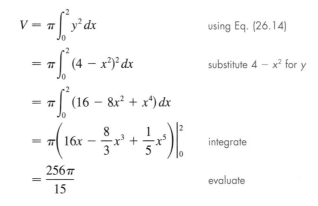

$$V = \pi \int_0^2 y^2 \, dx \qquad\qquad \text{using Eq. (26.14)}$$

$$= \pi \int_0^2 (4 - x^2)^2 \, dx \qquad \text{substitute } 4 - x^2 \text{ for } y$$

$$= \pi \int_0^2 (16 - 8x^2 + x^4) \, dx$$

$$= \pi \left(16x - \frac{8}{3}x^3 + \frac{1}{5}x^5 \right) \Big|_0^2 \qquad \text{integrate}$$

$$= \frac{256\pi}{15} \qquad\qquad\qquad \text{evaluate} \qquad\qquad ▶$$

We now show how to set up the integral to find the volume of the solid shown in Example 5 by using cylindrical shells. As it turns out, we are not able at this point to integrate the expression that arises, but we are still able to set up the proper integral.

◀ EXAMPLE 6   Use the method of cylindrical shells to find the volume indicated in Example 5.

From Fig. 26.27, we see for the shell we have

$$\text{radius} = y \qquad \text{height} = x \qquad \text{thickness} = dy$$

Since the elements go from $y = 0$ to $y = 4$, the limits of integration are 0 and 4. Hence,

$$V = 2\pi \int_0^4 xy \, dy \qquad\qquad \text{using Eq. (26.13)}$$

$$= 2\pi \int_0^4 \sqrt{4 - y}\,(y \, dy) \qquad \text{substitute } \sqrt{4 - y} \text{ for } x$$

$$= \frac{256\pi}{15}$$

The method of performing the integration $\int \sqrt{4 - y}\,(y \, dy)$ has not yet been discussed. We present the answer here for the reader's information to show that the volume found in this example is the same as that found in Example 5. ▶

In the next example, we show how to find the volume of the solid generated if a region is revolved about a line other than one of the axes. We will see that a proper choice of the radius, height, and thickness for Eq. (26.13) leads to the result.

◀ **EXAMPLE 7**   Find the volume of the solid generated if the region in Example 3 is revolved about the line $x = 2$.

Shells are convenient, since the volume of a shell can be expressed as a single integral. We can find the radius, height, and thickness of the shell from Fig. 26.28. We carefully note that ***the radius is not x but 2 − x,*** since the region is revolved about $x = 2$. We see that

CAUTION ▶

$$\text{radius} = 2 - x \qquad \text{height} = y \qquad \text{thickness} = dx$$

Since the elements that generate the shells go from $x = 0$ to $x = 2$, the limits of integration are 0 and 2. This means we have

$$V = 2\pi \int_0^2 (2 - x)y\,dx \longleftarrow \text{thickness} \qquad \text{using Eq. (25.13)}$$

$$= 2\pi \int_0^2 (2 - x)(4 - x^2)\,dx \qquad \text{substitute } 4 - x^2 \text{ for } y$$

$$= 2\pi \int_0^2 (8 - 2x^2 - 4x + x^3)\,dx$$

$$= 2\pi \left(8x - \frac{2}{3}x^3 - 2x^2 + \frac{1}{4}x^4 \right)\Big|_0^2 \qquad \text{integrate}$$

$$= \frac{40\pi}{3} \qquad \text{evaluate}$$

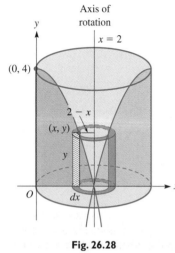

Axis of rotation

$x = 2$

$(0, 4)$

$2 - x$

(x, y)

y

dx

Fig. 26.28

```
fnInt(2π(2-X)(4-
X²),X,0,2)
        41.88790205
40π/3
        41.88790205
```

Fig. 26.29

(If the region had been revolved about the line $x = 3$, the only difference in the integral would have been that $r = 3 - x$. Everything else, including the limits, would have remained the same.)

The display in Fig. 26.29 shows the calculator evaluation for this integral. The first form of the integral was used to avoid any possible errors that might be made in the simplification. We see that the values agree.

▶

EXERCISES 26.3

In Exercises 1 and 2, make the given changes in the indicated examples of this section and then find the indicated volumes.

1. In Example 1, change $y = x^2$ to $y = x^3$.

2. In Example 3, change $y = 4 - x^2$ to $y = 4 - x$.

In Exercises 3–6, find the volume generated by revolving the region bounded by $y = 2 - x$, $x = 0$, and $y = 0$ about the indicated axis, using the indicated element of volume.

3. x-axis (disks)   **4.** y-axis (disks)

5. y-axis (shells)   **6.** x-axis (shells)

In Exercises 7–16, find the volume generated by revolving the regions bounded by the given curves about the x-axis. Use the indicated method in each case.

7. $y = x$, $y = 0$, $x = 2$ (disks)

8. $y = \sqrt{x}$, $x = 0$, $y = 2$ (shells)

9. $y = 3\sqrt{x}$, $y = 0$, $x = 4$ (disks)

10. $y = 2x - x^2$, $y = 0$ (disks)

11. $y = x^3$, $y = 8$, $x = 0$ (shells)

12. $y = x^2$, $y = x$ (shells)

13. $y = x^2 + 1$, $x = 0$, $x = 3$, $y = 0$ (disks)

14. $y = 6 - x - x^2$, $x = 0$, $y = 0$ (quadrant I), (disks)

15. $x = 4y - y^2 - 3$, $x = 0$ (shells)

16. $y = x^4$, $x = 0$, $y = 1$, $y = 2$ (shells)

In Exercises 17–26, find the volume generated by revolving the regions bounded by the given curves about the y-axis. Use the indicated method in each case.

17. $y = x^{1/3}$, $x = 0$, $y = 2$ (disks)

18. $y = \sqrt{x^2 - 1}$, $y = 0$, $x = 3$ (shells)

19. $y = 2\sqrt{x}$, $x = 0$, $y = 2$ (disks)

20. $y^2 = x$, $y = 4$, $x = 0$ (disks)

21. $x^2 - 4y^2 = 4$, $x = 3$ (shells)

22. $y = 3x^2 - x^3$, $y = 0$ (shells)

23. $x = 6y - y^2$, $x = 0$ (disks)

24. $x^2 + 4y^2 = 4$ (quadrant I), (disks)

25. $y = \sqrt{4 - x^2}$ (quadrant I), (shells)

26. $y = 8 - x^3$, $x = 0$, $y = 0$ (shells)

In Exercises 27–36, find the indicated volumes by integration.

W **27.** Describe a region that is revolved about the *x*-axis to generate a volume found by evaluating the integral $\pi \int_1^2 x^3 \, dx$.

W **28.** Describe a region that is revolved about the *y*-axis to generate a volume found by the integral in Exercise 27.

29. Find the volume generated if the region of Exercise 10 is revolved about the line $x = 2$.

30. Find the volume generated if the region bounded by $y = \sqrt{x}$ and $y = x/2$ is revolved about the line $y = 4$.

31. Derive the formula for the volume of a right circular cone of radius *r* and height *h* by revolving the area bounded by $y = (r/h)x$, $y = 0$, and $x = h$ about the *x*-axis.

W **32.** Explain how to derive the formula for the volume of a sphere by using the disk method.

33. A *drumlin* is an oval hill composed of relatively soft soil that was deposited beneath glacial ice (the campus of Dutchess Community College in Poughkeepsie, New York, is on a drumlin). Computer analysis showed that the surface of a certain drumlin can be approximated by $y = 10(1 - 0.0001x^2)$ revolved 180° about the *x*-axis from $x = -100$ to $x = 100$ (see Fig. 26.30). Find the volume (in m³) of this drumlin.

34. A commercial dirigible used for outdoor advertising has a helium-filled balloon in the shape of an ellipse revolved about its major axis. If the balloon is 41.3 m long and 12.0 m in diameter, what volume of helium is required to fill it? See Fig. 26.31.

<-----41.3 m----->   12.0 m

Fig. 26.31

35. A hole 2.00 cm in diameter is drilled through the center of a spherical lead weight 6.00 cm in diameter. How much lead is removed? See Fig. 26.32.

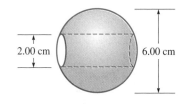

2.00 cm   6.00 cm

Fig. 26.32

36. All horizontal cross sections of a keg 1.20 m tall are circular, and the sides of the keg are parabolic. The diameter at the top and bottom is 0.80 m, and the diameter in the middle is 1.0 m. Find the volume that the keg holds.

$y = 10 (1 - 0.0001x^2)$

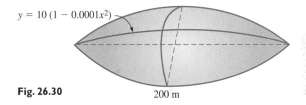

Fig. 26.30   200 m

26.4 CENTROIDS

In the study of mechanics, a very important property of an object is its center of mass. In this section, we explain the meaning of center of mass and then show how integration is used to determine the center of mass for regions and solids of revolution.

If a mass m is at a distance d from a specified point O, the **moment** *of the mass about O is defined as md. If several masses* $m_1, m_2, \ldots, m_n$ *are at distances* $d_1, d_2, \ldots, d_n$, *respectively, from point O, the total moment (as a group) about O is defined as* $m_1d_1 + m_2d_2 + \cdots + m_nd_n$. *The* **center of mass** *is that point* $\overline{d}$ *units from O at which all the masses could be concentrated to get the same total moment. Therefore* $\overline{d}$ *is defined by the equation*

$$m_1d_1 + m_2d_2 + \cdots + m_nd_n = (m_1 + m_2 + \cdots + m_n)\overline{d} \qquad (26.15)$$

The moment of a mass is a measure of its tendency to rotate about a point. A weight far from the point of balance of a long rod is more likely to make the rod turn than if the same weight were placed near the point of balance. It is easier to open a door if you push near the doorknob than if you push near the hinges. This is the type of physical property that the moment of mass measures.

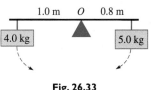

1.0 m   *O*   0.8 m

4.0 kg   5.0 kg

Fig. 26.33

◀ **EXAMPLE 1**   One of the simplest and most basic illustrations of moments and center of mass is seen in balancing a long rod with masses of different sizes, one on either side of the balance point.

In Fig. 26.33, a mass of 5.0 kg is hung from the rod 0.8 m to the right of point *O*. We see that this 5.0-kg mass tends to turn the rod clockwise. A mass placed on the opposite side of *O* will tend to turn the rod counterclockwise. Neglecting the mass of the rod, in order to balance the rod at *O*, the moments must be equal in magnitude but opposite in sign. Therefore, a 4.0-kg mass would have to be placed 1.0 m to the left.

Thus, with $d_1 = 0.8$ m, $d_2 = -1.0$ m, we see that

$$(5.0 + 4.0)\bar{d} = 5.0(0.8) + 4.0(-1.0) = 4.0 - 4.0$$
$$\bar{d} = 0.0 \text{ m}$$

The center of mass of the combination of the 5.0-kg mass and the 4.0-kg mass is at *O*.

CAUTION ▶ Also, note that *we must use* **directed distances** *in finding moments*. ▮

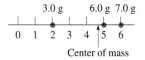

3.0 g   6.0 g 7.0 g

0   1   2   3   4   5   6

Center of mass

Fig. 26.34

◀ **EXAMPLE 2**   A mass of 3.0 g is placed at $(2.0, 0)$ on the *x*-axis (distances in cm). Another mass of 6.0 g is placed at $(5.0, 0)$, and a third mass of 7.0 g is placed at $(6.0, 0)$. See Fig. 26.34. Find the center of mass of these three masses.

Taking the reference point as the origin, we find $d_1 = 2.0$ cm, $d_2 = 5.0$ cm, and $d_3 = 6.0$ cm. Thus, $m_1 d_1 + m_2 d_2 + m_3 d_3 = (m_1 + m_2 + m_3)\bar{d}$ becomes

$$3.0(2.0) + 6.0(5.0) + 7.0(6.0) = (3.0 + 6.0 + 7.0)\bar{d} \quad \text{or} \quad \bar{d} = 4.9 \text{ cm}$$

This means that the center of mass of the three masses is at $(4.9, 0)$. Therefore, a mass of 16.0 g placed at this point has the same moment as the three masses as a unit. ▮

Solving a Word Problem

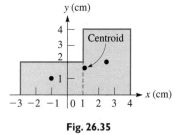

y (cm)

4
3   Centroid
2
1

−3 −2 −1   0 1   2   3   4   → *x* (cm)

Fig. 26.35

◀ **EXAMPLE 3**   Find the center of mass of the flat metal plate that is shown in Fig. 26.35.

We first note that the center of mass is not *on* either axis. This can be seen from the fact that the major portion of the area is in the first quadrant. *We will therefore measure the moments with respect to each axis to find the point that is the center of mass. This point is also called the* **centroid** *of the plate.*

The easiest method of finding this centroid is to divide the plate into rectangles, as indicated by the dashed line in Fig. 26.35, and assume that we may consider the mass of each rectangle to be concentrated at its center. In this way, the center of the left rectangle is at $(-1.0, 1.0)$ (distances in cm), and the center of the right rectangle is at $(2.5, 2.0)$. The mass of each rectangle area, assumed uniform, is proportional to its area. The area of the left rectangle is 8.0 cm^2, and that of the right rectangle is 12.0 cm^2. Thus, taking moments with respect to the *y*-axis, we have

$$8.0(-1.0) + 12.0(2.5) = (8.0 + 12.0)\bar{x}$$

where $\bar{x}$ is the *x*-coordinate of the centroid. Solving for $\bar{x}$, we have $\bar{x} = 1.1$ cm.

Now taking moments with respect to the *x*-axis, we have

$$8.0(1.0) + 12.0(2.0) = (8.0 + 12.0)\bar{y}$$

Since the center of mass does not depend on the density of the metal, we have assumed the constant of proportionality to be 1.

where $\bar{y}$ is the *y*-coordinate of the centroid. Thus, $\bar{y} = 1.6$ cm. This means that the coordinates of the centroid, the center of mass, are $(1.1, 1.6)$. This may be interpreted as meaning that a plate of this shape would balance on a single support under this point. As an approximate check, we note from the figure that this point appears to be a reasonable balance point for the plate. ▮

CENTROID OF A THIN, FLAT PLATE BY INTEGRATION

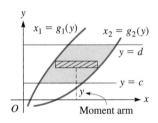

Fig. 26.36

If a thin, flat plate covers the region bounded by $y_1 = f_1(x)$, $y_2 = f_2(x)$, $x = a$, and $x = b$, as shown in Fig. 26.36, the moment of the mass of the element of area about the y-axis is given by $(k\,dA)x$, where k is the mass per unit area. In this expression, $k\,dA$ is the mass of the element, and x is its distance (moment arm) from the y-axis. The element dA may be written as $(y_2 - y_1)\,dx$, which means that the moment may be written as $kx(y_2 - y_1)\,dx$. If we then sum up the moments of all the elements and express this as an integral (which, of course, means sum), we have $k\int_a^b x(y_2 - y_1)\,dx$. If we consider all the mass of the plate to be concentrated at one point $\bar{x}$ units from the y-axis, the moment would be $(kA)\bar{x}$, where kA is the mass of the entire plate and $\bar{x}$ is the distance the center of mass is from the y-axis. By the previous discussion, these two expressions should be equal. This means $k\int_a^b x(y_2 - y_1)\,dx = kA\bar{x}$. Since k appears on each side of the equation, we divide it out (we are assuming that the mass per unit area is constant). The area A is found by the integral $\int_a^b (y_2 - y_1)\,dx$. Therefore, the x-coordinate of the centroid of the plate is given by

$$\bar{x} = \frac{\displaystyle\int_a^b x(y_2 - y_1)\,dx}{\displaystyle\int_a^b (y_2 - y_1)\,dx} \qquad (26.16)$$

Equation (26.16) gives us the x-coordinate of the centroid of the plate if vertical elements are used. Note that

CAUTION ▶   *the two integrals in Eq. (26.16) must be evaluated separately.*

We cannot cancel out the apparent common factor $y_2 - y_1$, and we cannot combine quantities and perform only one integration. The two integrals must be evaluated separately first. Then any possible cancellations of factors common to the numerator and the denominator may be made.

Following the same reasoning that we used in developing Eq. (26.16), if a thin plate covering the region bounded by the functions $x_1 = g_1(y)$, $x_2 = g_2(y)$, $y = c$, and $y = d$, as shown in Fig. 26.37, the y-coordinate of the centroid of the plate is given by the equation

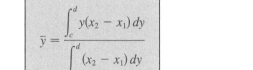

$$\bar{y} = \frac{\displaystyle\int_c^d y(x_2 - x_1)\,dy}{\displaystyle\int_c^d (x_2 - x_1)\,dy} \qquad (26.17)$$

Fig. 26.37

In this equation, horizontal elements are used.

In applying Eqs. (26.16) and (26.17), we should keep in mind that each denominator of the right-hand sides gives the area of the plate and that, once we have found this area, we may use it for both $\bar{x}$ and $\bar{y}$. In this way, we can avoid having to set up and perform one of the indicated integrations. Also, in finding the coordinates of the centroid, we should look for and utilize any symmetry the region may have.

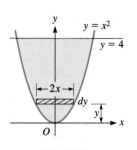

Fig. 26.38

◀ EXAMPLE 4   Find the coordinates of the centroid of a thin plate covering the region bounded by the parabola $y = x^2$ and the line $y = 4$.

We sketch a graph indicating the region and an element of area (see Fig. 26.38). The curve is a parabola whose axis is the y-axis. Since the region is symmetric to the y-axis, the centroid must be on this axis. This means that the x-coordinate of the centroid is zero, or $\bar{x} = 0$. To find the y-coordinate of the centroid, we have

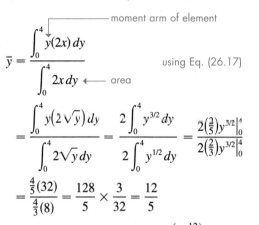

$$\bar{y} = \frac{\displaystyle\int_0^4 y(2x)\,dy}{\displaystyle\int_0^4 2x\,dy} \qquad \text{using Eq. (26.17)}$$

$$= \frac{\displaystyle\int_0^4 y\left(2\sqrt{y}\right)dy}{\displaystyle\int_0^4 2\sqrt{y}\,dy} = \frac{2\displaystyle\int_0^4 y^{3/2}\,dy}{2\displaystyle\int_0^4 y^{1/2}\,dy} = \frac{2\left(\frac{2}{5}\right)y^{5/2}\big|_0^4}{2\left(\frac{2}{3}\right)y^{3/2}\big|_0^4} \qquad \begin{array}{l}\text{integrate and evaluate numerator}\\\text{and denominator separately}\end{array}$$

$$= \frac{\frac{4}{5}(32)}{\frac{4}{3}(8)} = \frac{128}{5} \times \frac{3}{32} = \frac{12}{5}$$

The coordinates of the centroid are $\left(0, \frac{12}{5}\right)$. This plate would balance if a single pointed support were to be put under this point.   ◗

Solving a Word Problem

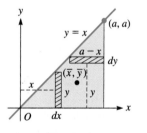

Fig. 26.39

◀ EXAMPLE 5   Find the coordinates of the centroid of an isosceles right triangular plate with side a.

We must first set up the region in the xy-plane. One choice is to place the triangle with one vertex at the origin and the right angle on the x-axis (see Fig. 26.39). Since each side is a, the hypotenuse passes through the point (a, a). The equation of the hypotenuse is $y = x$. The x-coordinate of the centroid is found by using Eq. (26.16):

$$\bar{x} = \frac{\displaystyle\int_0^a xy\,dx}{\displaystyle\int_0^a y\,dx} = \frac{\displaystyle\int_0^a x(x)\,dx}{\displaystyle\int_0^a x\,dx} = \frac{\displaystyle\int_0^a x^2\,dx}{\displaystyle\int_0^a x\,dx} = \frac{\frac{1}{3}x^3\big|_0^a}{\frac{1}{2}x^2\big|_0^a} = \frac{\frac{a^3}{3}}{\frac{a^2}{2}} = \frac{2a}{3}$$

The y-coordinate of the centroid is found by using Eq. (26.17):

$$\bar{y} = \frac{\displaystyle\int_0^a y(a - x)\,dy}{\frac{a^2}{2}} = \frac{\displaystyle\int_0^a y(a - y)\,dy}{\frac{a^2}{2}} = \frac{\displaystyle\int_0^a (ay - y^2)\,dy}{\frac{a^2}{2}}$$

$$= \frac{\frac{ay^2}{2} - \frac{y^3}{3}\big|_0^a}{\frac{a^2}{2}} = \frac{\frac{a^3}{6}}{\frac{a^2}{2}} = \frac{a}{3}$$

Thus, the coordinates of the centroid are $\left(\frac{2}{3}a, \frac{1}{3}a\right)$. The results indicate that the center of mass is $\frac{1}{3}a$ units from each of the equal sides.   ◗

CENTROID OF A SOLID OF REVOLUTION

Another figure for which we wish to find the centroid is a solid of revolution. If the density of the solid is constant, the centroid is on the axis of revolution. The problem that remains is to find just where on the axis the centroid is located.

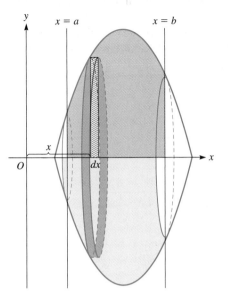

Fig. 26.40

If a region bounded by the x-axis, as shown in Fig. 26.40, is revolved about the x-axis, a vertical element of area generates a disk element of volume. The center of mass of the disk is at its center, and we may consider its mass concentrated there. The moment about the y-axis of a typical element is $x(k)(\pi y^2\,dx)$, where x is the moment arm, k is the density, and $\pi y^2\,dx$ is the volume. The sum of the moments of the elements can be expressed as an integral; it equals the volume times the density times the x-coordinate of the centroid of the volume. Since π and the density k would appear on each side of the equation, they cancel and need not be written. Therefore,

$$\bar{x} = \frac{\displaystyle\int_a^b xy^2\,dx}{\displaystyle\int_a^b y^2\,dx} \tag{26.18}$$

is the equation for the x-coordinate of the centroid of a solid of revolution about the x-axis.

In the same manner, we may find that *the y-coordinate of the centroid of a solid of revolution about the y-axis is*

$$\bar{y} = \frac{\displaystyle\int_c^d yx^2\,dy}{\displaystyle\int_c^d x^2\,dy} \tag{26.19}$$

◀ **EXAMPLE 6** Find the coordinates of the centroid of the solid generated by revolving the first-quadrant region under the curve $y = 4 - x^2$ about the x-axis.

Since the curve (see Fig. 26.41) is rotated about the x-axis, the centroid is on the x-axis, which means that $\bar{y} = 0$. We find the x-coordinate as follows:

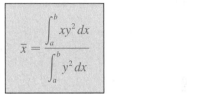

$$\bar{x} = \frac{\displaystyle\int_0^2 \overset{\text{moment arm}}{xy^2}\,dx}{\displaystyle\int_0^2 y^2\,dx} \qquad \text{using Eq. (26.18)}$$

$$= \frac{\displaystyle\int_0^2 x(4 - x^2)^2\,dx}{\displaystyle\int_0^2 (4 - x^2)^2\,dx} = \frac{\displaystyle\int_0^2 (16x - 8x^3 + x^5)\,dx}{\displaystyle\int_0^2 (16 - 8x^2 + x^4)\,dx}$$

$$= \frac{8x^2 - 2x^4 + \frac{1}{6}x^6 \big|_0^2}{16x - \frac{8}{3}x^3 + \frac{1}{5}x^5 \big|_0^2} = \frac{32 - 32 + \frac{64}{6}}{32 - \frac{64}{3} + \frac{32}{5}} = \frac{5}{8}$$

The coordinates of the centroid are $\left(\frac{5}{8}, 0\right)$.

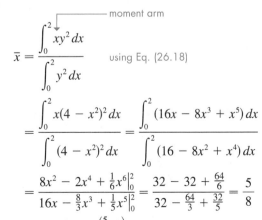

Fig. 26.41

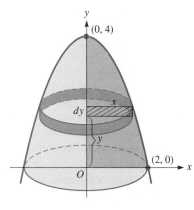

Fig. 26.42

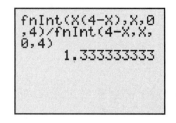

Fig. 26.43

EXAMPLE 7 Find the coordinates of the centroid of the volume generated by revolving the first-quadrant region under the curve $y = 4 - x^2$ about the y-axis as shown in Fig. 26.42. (This is the same region as in Example 6.)

Since the curve is rotated about the y-axis, $\bar{x} = 0$. The y-coordinate is

$$
\bar{y} = \frac{\overset{\text{moment arm}}{\displaystyle\int_0^4 y x^2 \, dy}}{\displaystyle\int_0^4 x^2 \, dy} \qquad \text{using Eq. (26.19)}
$$

$$
= \frac{\displaystyle\int_0^4 y(4 - y) \, dy}{\displaystyle\int_0^4 (4 - y) \, dy} = \frac{\displaystyle\int_0^4 (4y - y^2) \, dy}{\displaystyle\int_0^4 (4 - y) \, dy} = \frac{2y^2 - \frac{1}{3}y^3 \big|_0^4}{4y - \frac{1}{2}y^2 \big|_0^4}
$$

$$
= \frac{32 - \frac{64}{3}}{16 - 8} = \frac{4}{3}
$$

The coordinates of the centroid are $\left(0, \frac{4}{3}\right)$.

Figure 26.43 shows the calculator evaluation for $\bar{y}$. Note that the division is performed in one calculation, although each integral is evaluated separately.

Solving a Word Problem

EXAMPLE 8 Find the centroid of a solid right circular cone of radius a and altitude h.

To generate a right circular cone, we may revolve a right triangle about one of its legs (Figs. 26.44). Placing a leg of length h along the x-axis, we rotate the right triangle whose hypotenuse is given by $y = (a/h)x$ about the x-axis. Therefore,

$$
\bar{x} = \frac{\overset{\text{moment arm}}{\displaystyle\int_0^h x y^2 \, dx}}{\displaystyle\int_0^h y^2 \, dx} \qquad \text{using Eq. (26.18)}
$$

$$
= \frac{\displaystyle\int_0^h x\left[\left(\frac{a}{h}\right)x\right]^2 dx}{\displaystyle\int_0^h \left[\left(\frac{a}{h}\right)x\right]^2 dx} = \frac{\left(\frac{a^2}{h^2}\right)\left(\frac{1}{4}x^4\right)\Big|_0^h}{\left(\frac{a^2}{h^2}\right)\left(\frac{1}{3}x^3\right)\Big|_0^h} = \frac{3}{4}h
$$

Therefore, the centroid is located along the altitude $\frac{3}{4}$ of the way from the vertex, or $\frac{1}{4}$ of the way from the base.

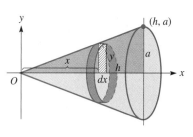

Fig. 26.44

EXERCISES 26.4

In Exercises 1 and 2, make the given changes in the indicated examples of this section and then find the coordinates of the centroid.

1. In Example 4, change $y = x^2$ to $y = |x|$ ($y = x$ for $x \geq 0$, and $y = -x$ for $x < 0$).

2. In Example 6, change $y = 4 - x^2$ to $y = 4 - x$.

In Exercises 3–6, find the center of mass (in cm) of the particles with the given masses located at the given points on the x-axis.

3. 5.0 g at $(1.0, 0)$, 8.5 g at $(4.2, 0)$, 3.6 g at $(2.5, 0)$

4. 2.3 g at $(1.3, 0)$, 6.5 g at $(5.8, 0)$, 1.2 g at $(9.5, 0)$

5. 42 g at $(-3.5, 0)$, 24 g at $(0, 0)$, 15 g at $(2.6, 0)$, 84 g at $(3.7, 0)$

6. 550 g at $(-42, 0)$, 230 g at $(-27, 0)$, 470 g at $(16, 0)$, 120 g at $(22, 0)$

In Exercises 7–10, find the coordinates (to 0.01 in.) of the centroids of the uniform flat-plate machine parts shown.

7. Fig. 26.45(a)   **8.** Fig. 26.45(b)

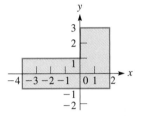 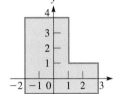

9. Fig. 26.45(c)   **10.** Fig. 26.45(d)

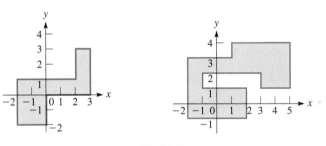

Fig. 26.45

In Exercises 11–32, find the coordinates of the centroids of the given figures. In Exercises 11–20, each region is covered by a thin, flat plate.

11. The region bounded by $y = x^2$ and $y = 2$

12. The semicircular region in Fig. 26.46

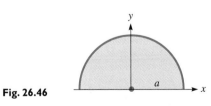

Fig. 26.46

13. The region bounded by $y = 4 - x$ and the axes

14. The region bounded by $y = x^3$, $x = 2$, and the x-axis

15. The region bounded by $y = x^2$ and $y = x^3$

16. The region bounded by $y^2 = x$, $y = 2$, and $x = 0$

17. The region bounded by $y = 2x$, $y = 3x$, and $y = 6$

18. The region bounded by $y = x^{2/3}$, $x = 8$, and $y = 0$

19. The region bounded by $x^2 = 4py$ and $y = a$ if $p > 0$ and $a > 0$

20. The region above the x-axis, bounded by the ellipse with vertices $(a, 0)$ and $(-a, 0)$, and minor axis $2b$ (The area of an ellipse is πab.)

21. The solid generated by revolving the region bounded by $y = x^3$, $y = 0$, and $x = 1$ about the x-axis

22. The solid generated by revolving the region bounded by $y = 2 - 2x$, $x = 0$, and $y = 0$ about the y-axis

23. The solid generated by revolving the region in the first quadrant bounded by $y^2 = 4x$, $y = 0$, and $x = 1$ about the y-axis

24. The solid generated by revolving the region bounded by $y = x^2$, $x = 2$, and the x-axis about the x-axis

25. The solid generated by revolving the region bounded by $y^2 = 4x$ and $x = 1$ about the x-axis

26. The solid generated by revolving the region bounded by $x^2 - y^2 = 9$, $y = 4$, and the x-axis about the y-axis

(W) 27. Explain how to find the centroid of a right triangular plate with legs a and b. Find the location of the centroid.

28. Find the location of the centroid of a hemisphere of radius a.

29. A lens with semielliptical vertical cross sections and circular horizontal cross sections is shown in Fig. 26.47. For proper installation in an optical device, its centroid must be known. Locate its centroid.

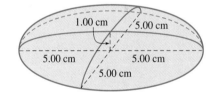

Fig. 26.47

30. A sanding machine disc can be described as the solid generated by rotating the region bounded by $y^2 = 4/x$, $y = 1$, $y = 2$, and the y-axis about the y-axis (measurements in cm). Locate the centroid of the disc.

31. A highway marking pylon has the shape of a frustum of a cone. Find its centroid if the radii of its bases are 5.00 cm and 20.0 cm and the height between bases is 60.0 cm.

32. A floodgate is in the shape of an isosceles trapezoid. Find the location of the centroid of the floodgate if the upper base is 20 m, the lower base is 12 m, and the height between bases is 6.0 m. See Fig. 26.48.

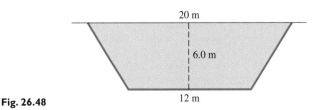

Fig. 26.48

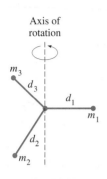

Axis of rotation

Fig. 26.49

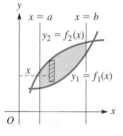

Radius of Gyration

Fig. 26.50

26.5 MOMENTS OF INERTIA

In the discussion of rotational motion in physics, an important quantity is the **moment of inertia** of an object. The moment of inertia of an object rotating about an axis is analogous to the mass of a moving object. In each case, *the moment of inertia or mass is the measure of the tendency of the object to resist a change in motion.*

Suppose that a particle of mass m is rotating about some point: We define its moment of inertia as md^2, where d is the distance from the particle to the point. If a group of particles of masses $m_1, m_2, \ldots, m_n$ are rotating about an axis, as shown in Fig. 26.49, the moment of inertia I with respect to the axis of the group is

$$I = m_1 d_1^2 + m_2 d_2^2 + \cdots + m_n d_n^2$$

where the d's are the respective distances of the particles from the axis. If all the masses were at the same distance R from the axis of rotation, so that the total moment of inertia were the same, we would have

$$m_1 d_1^2 + m_2 d_2^2 + \cdots + m_n d_n^2 = (m_1 + m_2 + \cdots + m_n)R^2 \qquad (26.20)$$

where R is called the **radius of gyration.**

EXAMPLE 1 Find the moment of inertia and the radius of gyration of the array of three masses, one of 3.0 g at $(-2.0, 0)$, another of 5.0 g at $(1.0, 0)$, and the third of 4.0 g at $(4.0, 0)$, with respect to the origin (distances in cm). See Fig. 26.50.

The moment of inertia of the array is

$$I = 3.0(-2.0)^2 + 5.0(1.0)^2 + 4.0(4.0)^2 = 81 \text{ g} \cdot \text{cm}^2$$

The radius of gyration is found from $I = (m_1 + m_2 + m_3)R^2$. Thus,

$$81 = (3.0 + 5.0 + 4.0)R^2, \qquad R^2 = \frac{81}{12}, \qquad R = 2.6 \text{ cm}$$

Therefore, a mass of 12.0 g placed at $(2.6, 0)$ (or $(-2.6, 0)$) has the same rotational inertia about the origin as the array of masses as a unit.

MOMENT OF INERTIA OF A THIN, FLAT PLATE

If a thin, flat plate covering the region is bounded by the curves of the functions $y_1 = f_1(x)$, $y_2 = f_2(x)$ and the lines $x = a$ and $x = b$, as shown in Fig. 26.51, the moment of inertia of this plate with respect to the y-axis, I_y, is given by the sum of the moments of inertia of the individual elements. The mass of each element is $k(y_2 - y_1)\,dx$, where k is the mass per unit area and $(y_2 - y_1)\,dx$ is the area of the element. The distance of the element from the y-axis is x. Representing this sum as an integral, we have

$$I_y = k \int_a^b x^2 (y_2 - y_1)\,dx \qquad (26.21)$$

Fig. 26.51

To find the radius of gyration of the plate with respect to the y-axis, R_y, we would first find the moment of inertia, divide this by the mass of the plate, and take the square root of this result.

In the same manner, the moment of inertia of a thin plate, with respect to the x-axis, bounded by $x_1 = g_1(y)$ and $x_2 = g_2(y)$ is given by

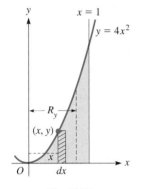

Fig. 26.52

$$I_x = k \int_c^d y^2 (x_2 - x_1) \, dy \qquad (26.22)$$

We find the radius of gyration of the plate with respect to the x-axis, R_x, in the same manner as we find it with respect to the y-axis (see Fig. 26.52).

◀ EXAMPLE 2 Find the moment of inertia and the radius of gyration of the plate covering the region bounded by $y = 4x^2$, $x = 1$, and the x-axis with respect to the y-axis.
We find the moment of inertia of this plate (see Fig. 26.53) as follows:

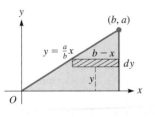

Fig. 26.53

┌ distance from element to axis

$$I_y = k \int_0^1 x^2 y \, dx \qquad \text{using Eq. (26.21)}$$

$$= k \int_0^1 x^2 (4x^2) \, dx = 4k \int_0^1 x^4 \, dx$$

$$= 4k \left(\frac{1}{5} x^5 \right) \Big|_0^1 = \frac{4k}{5}$$

To find the radius of gyration, we first determine the mass of the plate:

$$m = k \int_0^1 y \, dx = k \int_0^1 (4x^2) \, dx \qquad m = kA$$

$$= 4k \left(\frac{1}{3} x^3 \right) \Big|_0^1 = \frac{4k}{3}$$

$$R_y^2 = \frac{I_y}{m} = \frac{4k}{5} \times \frac{3}{4k} = \frac{3}{5} \qquad R_y^2 = I_y/m$$

$$R_y = \sqrt{\frac{3}{5}} = \frac{\sqrt{15}}{5}$$

Solving a Word Problem

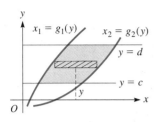

Fig. 26.54

◀ EXAMPLE 3 Find the moment of inertia of a right triangular plate with sides a and b with respect to side b. Assume that $k = 1$.
Placing the triangle as shown in Fig. 26.54, we see that the equation of the hypotenuse is $y = (a/b)x$. The moment of inertia is

┌ distance from element to axis

$$I_x = \int_0^a y^2 (b - x) \, dy \qquad \text{using Eq. (26.22)}$$

$$= \int_0^a y^2 \left(b - \frac{b}{a} y \right) dy = b \int_0^a \left(y^2 - \frac{1}{a} y^3 \right) dy$$

$$= b \left(\frac{1}{3} y^3 - \frac{1}{4a} y^4 \right) \Big|_0^a$$

$$= b \left(\frac{a^3}{3} - \frac{a^3}{4} \right) = \frac{ba^3}{12}$$

MOMENT OF INERTIA OF A SOLID

In applications, among the most important moments of inertia are those of solids of revolution. Since **all parts of an element of mass should be at the same distance from the axis,** the most convenient element of volume to use is the cylindrical shell. In Fig. 26.55, if the region bounded by the curves $y_1 = f_1(x)$, $y_2 = f_2(x)$, $x = a$, and $x = b$ is revolved about the y-axis, the moment of inertia of the element of volume is $k[2\pi x(y_2 - y_1)\,dx](x^2)$, where k is the density, $2\pi x(y_2 - y_1)\,dx$ is the volume of the element, and x^2 is the square of the distance from the y-axis. Expressing the sum of the elements as an integral, *the moment of inertia of the solid with respect to the y-axis, I_y, is*

> Note carefully that Eq. (26.23) gives the moment of inertia with respect to the y-axis and that $(y_2 - y_1)$ is the height of the shell (see Fig. 26.23).

$$I_y = 2\pi k \int_a^b (y_2 - y_1)x^3\,dx \qquad (26.23)$$

The radius of gyration of the solid with respect to the y-axis, R_y, is found by determining (1) the moment of inertia, (2) the mass of the solid, and (3) the square root of the quotient of the moment of inertia divided by the mass.

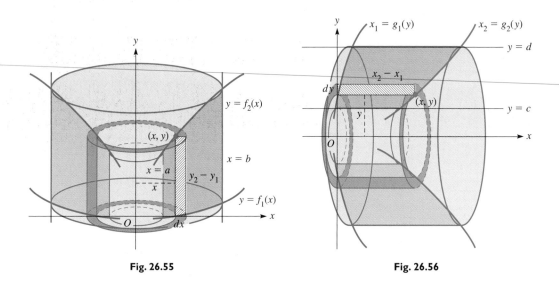

Fig. 26.55　　　　　　**Fig. 26.56**

The moment of inertia of the solid (see Fig. 26.56) generated by revolving the region bounded by $x_1 = g_1(y)$, $x_2 = g_2(y)$, $y = c$, and $y = d$ about the x-axis, I_x, is given by

> Note carefully that Eq. (26.24) gives the moment of inertia with respect to the x-axis and that $(x_2 - x_1)$ is the height of the shell (see Fig. 26.23).

$$I_x = 2\pi k \int_c^d (x_2 - x_1)y^3\,dy \qquad (26.24)$$

The radius of gyration of the solid with respect to the x-axis, R_x, is found in the same manner as R_y.

EXAMPLE 4 Find the moment of inertia and the radius of gyration with respect to the *x*-axis of the solid generated by revolving the region bounded by the curves of $y^3 = x$, $y = 2$, and the *y*-axis about the *x*-axis. See Fig. 26.57.

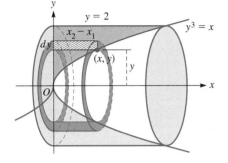

Fig. 26.57

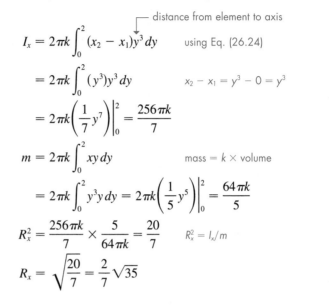

$$\text{distance from element to axis}$$

$$I_x = 2\pi k \int_0^2 (x_2 - x_1) y^3 \, dy \qquad \text{using Eq. (26.24)}$$

$$= 2\pi k \int_0^2 (y^3) y^3 \, dy \qquad x_2 - x_1 = y^3 - 0 = y^3$$

$$= 2\pi k \left(\frac{1}{7} y^7\right)\Big|_0^2 = \frac{256\pi k}{7}$$

$$m = 2\pi k \int_0^2 xy \, dy \qquad \text{mass} = k \times \text{volume}$$

$$= 2\pi k \int_0^2 y^3 y \, dy = 2\pi k \left(\frac{1}{5} y^5\right)\Big|_0^2 = \frac{64\pi k}{5}$$

$$R_x^2 = \frac{256\pi k}{7} \times \frac{5}{64\pi k} = \frac{20}{7} \qquad R_x^2 = I_x/m$$

$$R_x = \sqrt{\frac{20}{7}} = \frac{2}{7}\sqrt{35}$$

Solving a Word Problem

EXAMPLE 5 As noted at the beginning of this section, the moment of inertia is important when studying the rotational motion of an object. For this reason, the moments of inertia of various objects are calculated, and the formulas tabulated. Such formulas are usually expressed in terms of the mass of the object.

Among the objects for which the moment of inertia is important is a solid disk. Find the moment of inertia of a disk with respect to its axis and in terms of its mass.

To generate a disk (see Fig. 26.58), we rotate the region bounded by the axes, $x = r$ and $y = b$, about the *y*-axis. We then have

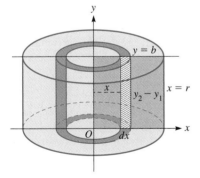

Fig. 26.58

$$\text{distance from element to axis}$$

$$I_y = 2\pi k \int_0^r (y_2 - y_1) x^3 \, dx \qquad \text{using Eq. (26.23)}$$

$$= 2\pi k \int_0^r (b) x^3 \, dx = 2\pi k b \int_0^r x^3 \, dx \qquad y_2 - y_1 = b - 0 = b$$

$$= 2\pi k b \left(\frac{1}{4} x^4\right)\Big|_0^r = \frac{\pi k b r^4}{2}$$

The mass of the disk is $k(\pi r^2)b$. Rewriting the expression for I_y, we have

$$I_y = \frac{(\pi k b r^2) r^2}{2} = \frac{mr^2}{2}$$

Due to the limited methods of integration available at this point, we cannot integrate the expressions for the moments of inertia of circular areas or of a sphere. These will be introduced in Section 28.8 in the exercises, by which point the proper method of integration will have been developed.

E X E R C I S E S 26.5

In Exercises 1 and 2, make the given changes in the indicated examples of this section, and then solve the resulting problems.

1. In Example 2, change $y = 4x^2$ to $y = 4x$.

2. In Example 4, change $y^3 = x$ to $y^2 = x$.

In Exercises 3–6, find the moment of inertia (in g · cm²) and the radius of gyration (in cm) with respect to the origin of each of the given arrays of masses located at the given points on the x-axis.

3. 5.0 g at $(2.4, 0)$, 3.2 g at $(3.5, 0)$

4. 3.4 g at $(-1.5, 0)$, 6.0 g at $(2.1, 0)$, 2.6 g at $(3.8, 0)$

5. 45.0 g at $(-3.80, 0)$, 90.0 g at $(0.00, 0)$, 62.0 g at $(5.50, 0)$

6. 564 g at $(-45.0, 0)$, 326 g at $(-22.5, 0)$, 720 g at $(15.4, 0)$, 205 g at $(64.0, 0)$

In Exercises 7–28, find the indicated moment of inertia or radius of gyration.

7. Find the moment of inertia of a plate covering the region bounded by $x = -1$, $x = 1$, $y = 0$, and $y = 1$ with respect to the x-axis.

8. Find the radius of gyration of a plate covering the region bounded by $x = 2$, $x = 4$, $y = 0$, and $y = 4$, with respect to the y-axis.

9. Find the moment of inertia of a plate covering the first-quadrant region bounded by $y^2 = x$, $x = 4$, and the x-axis with respect to the x-axis.

10. Find the moment of inertia of a plate covering the region bounded by $y = 2x$, $x = 1$, $x = 2$, and the x-axis with respect to the y-axis.

11. Find the radius of gyration of a plate covering the region bounded by $y = x^3$, $x = 2$, and the x-axis with respect to the y-axis.

12. Find the radius of gyration of a plate covering the first-quadrant region bounded by $y^2 = 1 - x$ with respect to the x-axis.

13. Find the moment of inertia of a right triangular plate with sides a and b with respect to side a in terms of the mass of the plate.

14. Find the moment of inertia of a rectangular plate of sides a and b with respect to side a. Express the result in terms of the mass of the plate.

15. Find the radius of gyration of a plate covering the region bounded by $y = x^2$, $x = 2$, and the x-axis with respect to the x-axis.

16. Find the radius of gyration of a plate covering the region bounded by $y^2 = x^3$, $y = 8$, and the y-axis with respect to the y-axis.

17. Find the radius of gyration of the plate of Exercise 16 with respect to the x-axis.

18. Find the radius of gyration of a plate covering the first-quadrant region bounded by $x = 1$, $y = 2 - x$, and the y-axis with respect to the y-axis.

19. Find the moment of inertia with respect to its axis of the solid generated by revolving the region bounded by $y^2 = x$, $y = 2$, and the y-axis about the x-axis.

20. Find the radius of gyration with respect to its axis of the solid generated by revolving the first-quadrant region under the curve $y = 4 - x^2$ about the y-axis.

21. Find the radius of gyration with respect to its axis of the solid generated by revolving the region bounded by $y = 2x - x^2$ and the x-axis about the y-axis.

22. Find the radius of gyration with respect to its axis of the solid generated by revolving the region bounded by $y = 2x$ and $y = x^2$ about the y-axis.

23. Find the moment of inertia in terms of its mass of a right circular cone of radius r and height h with respect to its axis. See Fig. 26.59.

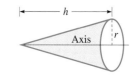

Fig. 26.59

24. Find the moment of inertia in terms of its mass of a circular hoop of radius r and of negligible thickness with respect to its center.

25. A rotating drill head is in the shape of a right circular cone. Find the moment of inertia of the drill head with respect to its axis if its radius is 0.600 cm, its height is 0.800 cm, and its mass is 3.00 g. (See Exercise 23.)

26. Find the moment of inertia (in kg · m²) of a rectangular door 2 m high and 1 m wide with respect to its hinges if $k = 3$ kg/m². (See Exercise 14.)

27. Find the moment of inertia of a flywheel with respect to its axis if its inner radius is 4.0 cm, its outer radius is 6.0 cm, and its mass is 1.2 kg. See Fig. 26.60.

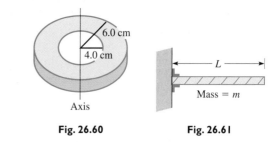

Fig. 26.60   **Fig. 26.61**

(W) 28. A cantilever beam is supported only at its left end, as shown in Fig. 26.61. Explain how to find the formula for the moment of inertia of this beam with respect to a vertical axis through its left end if its length is L and its mass is m. (Consider the mass to be distributed evenly along the beam. This is not an area or volume type of problem.) Find the formula for the moment of inertia.

26.6 OTHER APPLICATIONS

We have seen that the definite integral is used to find the exact measure of the sum of products in which one factor is an increment that approaches a limit of zero. This makes the definite integral a powerful mathematical tool in that a great many applications can be expressed in this form. The following examples show three more applications of the definite integral, and others are shown in the exercises.

WORK BY A VARIABLE FORCE

In physics, **work** *is defined as the product of a constant force times the distance through which it acts.* When we consider the work done in stretching a spring, the first thing we recognize is that the more the spring is stretched, the greater is the force necessary to stretch it. Thus the force varies. However, if we are stretching the spring a distance Δx, where we are considering the limit as $\Delta x \to 0$, the force can be considered as approaching a constant over Δx. Adding the product of force$_1$ times Δx_1, force$_2$ times Δx_2, and so forth, we see that the total is the sum of these products. Thus, the work can be expressed as a definite integral in the form

$$W = \int_a^b f(x)\, dx \qquad (26.25)$$

CAUTION

where $f(x)$ is the force as a function of the distance the spring is stretched. The *limits a and b refer to the initial and final distances the spring is stretched* **from its normal length.**

One problem remains: We must find the function $f(x)$. From physics, we learn that the force required to stretch a spring is proportional to the amount it is stretched (Hooke's law). If a spring is stretched x units from its normal length, then $f(x) = kx$. From conditions stated for a particular spring, the value of k may be determined. Thus, $W = \int_a^b kx\, dx$ is the formula for finding the total work done in stretching a spring. Here, a and b are the initial and final amounts the spring is stretched from its natural length.

Named for the English physicist Robert Hooke (1635–1703).

Solving a Word Problem  **EXAMPLE 1** A spring of natural length 12 cm requires a force of 6.0 N to stretch it 2.0 cm. See Fig. 26.62. Find the work done in stretching it 6.0 cm.

From Hooke's law, we find the constant k for the spring as

$$f(x) = kx, \qquad 6.0 = k(2.0), \qquad k = 3.0 \text{ N/cm}$$

Since the spring is to be stretched 6.0 cm, $a = 0$ (it starts unstretched) and $b = 6.0$ (it is 6.0 cm longer than its normal length). Therefore, the work done in stretching it is

$$W = \int_0^{6.0} 3.0x\, dx = 1.5x^2 \Big|_0^{6.0} \qquad \text{using Eq. (26.25)}$$

$$= 54 \text{ N} \cdot \text{cm}$$

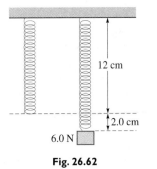

Fig. 26.62

Problems involving work by a variable force arise in many fields of technology. On the following page is an illustration from electricity that deals with the motion of an electric charge through an electric field created by another electric charge.

Electric charges are of two types, designated as positive and negative. A basic law is that charges of the same sign repel each other and charges of opposite signs attract each other. *The force between charges is proportional to the product of their charges, and inversely proportional to the square of the distance between them.*

The force $f(x)$ between electric charges is therefore given by

$$f(x) = \frac{kq_1q_2}{x^2}$$

(26.26)

when q_1 and q_2 are the charges (in coulombs), x is the distance (in meters), the force is in newtons, and $k = 9.0 \times 10^9 \, \text{N} \cdot \text{m}^2/\text{C}^2$. For other systems of units, the numerical value of k is different. We can find the work done when electric charges move toward each other or when they separate by use of Eq. (26.26) in Eq. (26.25).

Solving a Word Problem

❙ EXAMPLE 2 Find the work done when two α-particles, $q = 0.32$ aC each, move until they are 10 nm apart, if they were originally separated by 1.0 m.

From the given information, we have for each α-particle

$$q = 0.32 \text{ aC} = 0.32 \times 10^{-18} \, \text{C} = 3.2 \times 10^{-19} \, \text{C}$$

Since the particles start 1.0 m apart and are moved to 10 nm apart, $a = 1.0$ m and $b = 10 \times 10^{-9}$ m $= 10^{-8}$ m. The work done is

f(x) from Eq. (26.26)

$$W = \int_{1.0}^{10^{-8}} \frac{9.0 \times 10^9 (3.2 \times 10^{-19})^2}{x^2} \, dx \qquad \text{using Eq. (26.25)}$$

$$= 9.2 \times 10^{-28} \int_{1.0}^{10^{-8}} \frac{dx}{x^2} = 9.2 \times 10^{-28} \left(-\frac{1}{x} \right) \Big|_{1.0}^{10^{-8}}$$

$$= -9.2 \times 10^{-28}(10^8 - 1) = -9.2 \times 10^{-20} \, \text{J}$$

Since $10^8 \gg 1$, where $\gg$ means "much greater than," the 1 may be neglected in the calculation. The minus sign in the result means that work must be done *on* the system to move the particles toward each other. If free to move, they tend to separate. ❙

The following is another type of problem involving work by a variable force.

Solving a Word Problem

❙ EXAMPLE 3 Find the work done in winding up 60.0 m of a 100-m cable that weighs 4.00 N/m. See Fig. 26.63.

See the chapter introduction.

First, we let x denote the length of cable that has been wound up at any time. Then the force required to raise the remaining cable equals the weight of the cable that has not yet been wound up. This weight is the product of the unwound cable length, $100 - x$, and its weight per unit length, 4.00 N/m, or

$$f(x) = 4.00(100 - x)$$

Since 60.0 m of cable are to be wound up, $a = 0$ (none is initially wound up) and $b = 60.0$ m. The work done is

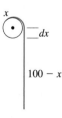

Fig. 26.63

$$W = \int_0^{60.0} 4.00(100 - x) \, dx \qquad \text{using Eq. (26.25)}$$

$$= \int_0^{60.0} (400 - 4.00x) \, dx = 400x - 2.00x^2 \Big|_0^{60.0} = 16\,800 \, \text{N} \cdot \text{m}$$

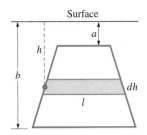

Fig. 26.64

FORCE DUE TO LIQUID PRESSURE

The second application of integration in this section deals with the force due to liquid pressure. The force F on an area A at the depth h in a liquid of density w is $F = whA$. Let us assume that the plate shown in Fig. 26.64 is submerged vertically in water. Using integration to sum the forces on the elements of area, *the total force on the plate is given by*

$$F = w \int_a^b lh\,dh \qquad (26.27)$$

Here, l is the length of the element of area, h is the depth of the element of area, w is the weight per unit volume of the liquid, a is the depth of the top, and b is the depth of the bottom of the area on which the force is exerted.

Solving a Word Problem

◀ EXAMPLE 4   A vertical floodgate of a dam is 3.00 m wide and 2.00 m high. Find the force on the floodgate if its upper edge is 1.00 m below the surface of the water. See Fig. 26.65.

Each element of area of the floodgate has a length of 3.00 m, which means that $l = 3.00$ m. Since the top of the gate is 1.00 m below the surface, $a = 1.00$ m, and since the gate is 2.00 m high, $b = 3.00$ m. Using $w = 9800$ N/m³, we have the force on the gate as

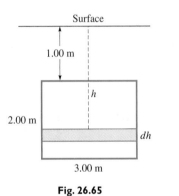

Fig. 26.65

$$F = 9800 \int_{1.00}^{3.00} 3.00h\,dh \qquad \text{using Eq. (26.27)}$$

$$= 29\,400 \int_{1.00}^{3.00} h\,dh$$

$$= 14\,700h^2 \Big|_{1.00}^{3.00} = 14\,700(9.00 - 1.00)$$

$$= 118\,000 \text{ N} = 118 \text{ kN}$$

Solving a Word Problem

NOTE▶

◀ EXAMPLE 5   The vertical end of a tank of water is in the shape of a right triangle as shown in Fig. 26.66. What is the force on the end of the tank?

In setting up the figure, it is convenient to use coordinate axes. It is also convenient to have the y-axis directed downward, since we integrate from the top to the bottom of the area. The equation of the line OA is $y = \frac{1}{2}x$. Thus, we see that the length of an element of area of the end of the tank is $4.0 - x$, the depth of the element of area is y, the top of the tank is $y = 0$, and the bottom is $y = 2.0$ m. Therefore, the force on the end of the tank is ($w = 9800$ N/m³)

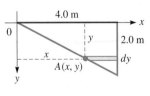

Fig. 26.66

$$F = 9800 \int_0^{2.0} \overset{\text{length}}{\overbrace{(4.0 - x)}} \overset{\text{depth}}{\overbrace{(y)}}(dy) \qquad \text{using Eq. (26.27)}$$

$$= 9800 \int_0^{2.0} (4.0 - 2y)(y\,dy)$$

$$= 19\,600 \int_0^{2.0} (2.0y - y^2)\,dy = 19\,600 \left(1.0y^2 - \frac{1}{3}y^3 \right) \Big|_0^{2.0}$$

$$= 26\,100 \text{ N}$$

AVERAGE VALUE OF A FUNCTION

The third application of integration shown in this section is that of the *average value* of a function. In general, *an average is found by summing up the quantities to be averaged and then dividing by the total number of them.* Generalizing on this and using integration for the summation, *the* **average value** *of a function y with respect to x from x = a to x = b is given by*

$$y_{av} = \frac{\int_a^b y \, dx}{b - a} \tag{26.28}$$

The following examples illustrate applications of the average value of a function.

Solving a Word Problem   ◀ EXAMPLE 6   The velocity v (in m/s) of an object falling under the influence of gravity as a function of the time t (in s) is given by $v = 9.80t$. What is the average velocity of the object with respect to time for the first 3.0 s?

In this case, we want the average value of the function v from $t = 0$ to $t = 3.0$ s. This gives us

$$v_{av} = \frac{\int_0^{3.0} v \, dt}{3.0 - 0} \quad \text{using Eq. (26.28)}$$

$$= \frac{\int_0^{3.0} 9.80t \, dt}{3.0} = \frac{4.90t^2}{3.0} \Big|_0^{3.0}$$

$$= 14.7 \text{ m/s}$$

This result can be interpreted as meaning that an average velocity of 14.7 m/s for 3.0 s would result in the same distance, 44.1 m, being traveled by the object as that with the variable velocity. Since $s = \int v \, dt$, the numerator represents the distance traveled. ◗

Solving a Word Problem   ◀ EXAMPLE 7   The power P developed in a certain resistor as a function of the current i is $P = 6i^2$. What is the average power (in W) with respect to the current as the current changes from 2.0 A to 5.0 A?

In this case, we are to find the average value of the function P from $i = 2.0$ A to $i = 5.0$ A. This average value of P is

$$P_{av} = \frac{\int_{2.0}^{5.0} P \, di}{5.0 - 2.0} \quad \text{using Eq. (26.28)}$$

$$= \frac{6 \int_{2.0}^{5.0} i^2 \, di}{3.0} = \frac{2i^3}{3.0} \Big|_{2.0}^{5.0} = \frac{2(125 - 8.0)}{3.0} = 78 \text{ W} \quad ◗$$

In general, it might be noted that the average value of y with respect to x is that value of y which, when multiplied by the length of the interval for x, gives the same area as that under the curve of y as a function of x.

EXERCISES 26.6

1. In Example 1, find the work done in stretching the spring from a length of 15.0 cm to a length of 18.0 cm.

2. In Example 3, find the work done in winding up all the cable.

3. In Example 4, find the force on the floodgate if the upper edge is 2.00 m below the surface.

4. In Example 6, find the average velocity of the object with respect to time between 3.00 s and 6.00 s.

5. The spring of a spring balance is 8.0 cm long when there is no weight on the balance, and it is 9.5 cm long with 6.0 N hung from the balance. How much work is done in stretching it from 8.0 cm to a length of 10.0 cm?

6. How much work is done in stretching the spring of Exercise 5 from a length of 10.0 cm to 12.0 cm?

7. A force of 1200 N compresses a spring from its natural length of 18 cm to a length of 16 cm. How much work is done in compressing it from 16 cm to 14 cm?

8. A force F of 25 N on the spring in the lever-spring mechanism shown in Fig. 26.67 stretches the spring by 16 mm. How much work is done by the 25-N force in stretching the spring?

9. An electron has a 1.6×10^{-19} C negative charge. How much work is done in separating two electrons from 1.0 pm to 4.0 pm?

Fig. 26.67

10. How much work is done in separating an electron (see Exercise 9) and an oxygen nucleus, which has a positive charge of 1.3×10^{-18} C, from a distance of 2.0 μm to a distance of 1.0 m?

11. The gravitational force (in N) of attraction between two objects is given by $F = k/x^2$, where x is the distance between the objects. If the objects are 10 m apart, find the work required to separate them until they are 100 m apart. Express the result in terms of k.

12. Find the work done by winding up 20 m of a 25-m rope on which the force of gravity is 6.0 N/m.

13. Find the work done in winding up a 50.0-m cable that weighs 40.0 N.

14. A chain is being unwound from a winch. The force of gravity on it is 12.0 N/m. When 20 m have been unwound, how much work is done by gravity in unwinding another 30 m?

15. At liftoff, a rocket weighs 32.5 tonnes, including the weight of its fuel. It is fired vertically, and, during the first stage of ascent, the fuel is consumed at the rate of 1.25 tonnes per 1000 m of ascent. How much work is done in lifting the rocket to an altitude of 12 000 m?

16. While descending, a 550-N weather balloon enters a zone of freezing rain in which ice forms on the balloon at the rate of

7.50 N per 100 m of descent. Find the work done on the balloon during the first 1000 m of descent through the freezing rain.

17. Find the work done in pumping the water out of the top of a cylindrical tank 2.0 m in radius and 4.0 m high, given that the tank is initially full and water weighs 9.8 kN/m^3. (*Hint:* If horizontal slices dx m thick are used, each element weighs $9800(\pi)(2.00^2)\,dx$ N, and each element must be raised $4.0 - x$ m, if x is the distance from the base to the element (see Fig. 26.68). In this way the force, which is the weight of the slice, and the distance through which the force acts are determined. Thus, the products of force and distance are summed by integration.)

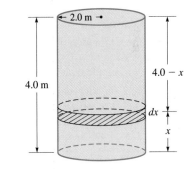

Fig. 26.68

18. A hemispherical tank of radius 3.0 m is full of water. Find the work done in pumping the water out of the top of the tank. (See Exercise 17. This problem is similar, except that the weight of each element is $9800\pi(\text{radius})^2(\text{thickness})$, where the radius of each element is different. If we let x be the radius of an element and y be the distance the element must be raised, we have $9800\pi x^2\,dy$, with $x^2 + y^2 = 9.0$.)

19. One end of a spa is a vertical rectangular wall 4.00 m wide. What is the force exerted on this wall by the water if it is 0.80 m deep?

20. Find the force on one side of a cubical container 6.0 cm on an edge if the container is filled with mercury. The density of mercury is 133 kN/m^3.

21. A rectangular sea aquarium observation window is 3.00 m wide and 2.00 m high. What is the force on this window if the upper edge is 1.50 m below the surface of the water? The density of sea-water is 10.1 kN/m^3.

22. A horizontal tank has circular ends, each with a radius of 2.00 m. It is filled to a depth of 2.00 m with oil of density 9400 N/m^3. Find the force on one end of the tank.

23. A right triangular plate of base 2.0 m and height 1.0 m is submerged vertically in water, as shown in Fig. 26.69. Find the force on one side of the plate.

Fig. 26.69

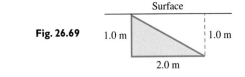

24. Find the force on the plate in Exercise 23 if the top vertex is 3.0 m below the surface. (W)

25. A dam is in the shape of the area bounded by $y = x^2$ and $y = 20$ (distances in m). Find the force on the area below $y = 4$ if the surface of the water is at the top of the dam.

26. A horizontal cylindrical tank has vertical elliptical ends with the minor axis horizontal. The major axis is 2.0 m and the minor axis is 1.3 m. Find the force on one end if the tank is half-filled with fuel oil of density 7.8 kN/m³.

(W) **27.** A watertight cubical box with an edge of 2.00 m is suspended in water such that the top surface is 1.00 m below water level. Find the total force on the top of the box and the total force on the bottom of the box. What meaning can you give to the difference of these two forces?

28. Find the force on the region bounded by $x = 2y - y^2$ and the y-axis if the upper point of the area is at the surface of the water. All distances are in meters.

29. The electric current i (in A) as a function of the time t (in s) for a certain circuit is given by $i = 4t - t^2$. Find the average value of the current with respect to time for the first 4.0 s.

30. The temperature T (in °C) recorded in a city during a given day approximately followed the curve of $T = 0.001\,00t^4 - 0.280t^2 + 25.0$, where t is the number of hours from noon ($-12\,\text{h} \le t \le 12\,\text{h}$). What was the average temperature during the day?

31. The efficiency e (in %) of an automobile engine is given by $e = 0.768s - 0.000\,04s^3$, where s is the speed (in km/h) of the car. Find the average efficiency with respect to the speed for $s = 30.0\,\text{km/h}$ to $s = 90.0\,\text{km/h}$. (See Example 1 of Section 24.7.)

32. Find the average value of the volume of a sphere with respect to the radius. Explain the meaning of the result. (W)

33. The length of arc s of a curve from $x = a$ to $x = b$ is
$$s = \int_a^b \sqrt{1 + \left(\frac{dy}{dx}\right)^2}\, dx$$
The cable of a bridge can be described by the equation $y = 0.04x^{3/2}$ from $x = 0$ to $x = 100$ m. Find the length of the cable. See Fig. 26.70.

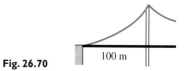

Fig. 26.70

34. A rocket takes off in a path described by the equation $y = \frac{2}{3}(x^2 - 1)^{3/2}$. Find the distance traveled by the rocket for $x = 1.0$ km to $x = 3.0$ km. (See Exercise 33.)

35. The area of a surface of revolution from $x = a$ to $x = b$ is
$$S = 2\pi \int_a^b y \sqrt{1 + \left(\frac{dy}{dx}\right)^2}\, dx$$
Find the formula for the lateral surface area of a right circular cone of radius r and height h.

36. The grinding surface of a grinding machine can be described as the surface generated by rotating the curve $y = 0.2x^3$ from $x = 0$ to $x = 2.0$ cm about the x axis. Find the grinding surface area. (See Exercise 35.)

CHAPTER 26 EQUATIONS

Velocity	$v = \int a\, dt$	(26.1)
	$v = at + C_1$	(26.2)
Displacement	$s = \int v\, dt$	(26.3)
Electric current	$i = \dfrac{dq}{dt}$	(26.4)
Electric charge	$q = \int i\, dt$	(26.5)
Voltage across capacitor	$V_C = \dfrac{1}{C} \int i\, dt$	(26.6)

Area

$$A = \int_a^b y\,dx = \int_a^b f(x)\,dx \tag{26.7}$$

$$A = \int_c^d x\,dy = \int_c^d g(y)\,dy \tag{26.8}$$

$$A = \int_a^b (y_2 - y_1)\,dx \tag{26.9}$$

$$A = \int_c^d (x_2 - x_1)\,dy \tag{26.10}$$

Volume

$$V = \pi \int_a^b y^2\,dx = \pi \int_a^b [f(x)]^2\,dx \tag{26.11}$$

$$V = \pi \int_c^d x^2\,dy \tag{26.12}$$

Shell

$$dV = 2\pi(\text{radius}) \times (\text{height}) \times (\text{thickness}) \tag{26.13}$$

Disk

$$dV = \pi(\text{radius})^2 \times (\text{thickness}) \tag{26.14}$$

Center of mass

$$m_1 d_1 + m_2 d_2 + \cdots + m_n d_n = (m_1 + m_2 + \cdots + m_n)\overline{d} \tag{26.15}$$

Centroid of flat plate

$$\overline{x} = \frac{\displaystyle\int_a^b x(y_2 - y_1)\,dx}{\displaystyle\int_a^b (y_2 - y_1)\,dx} \tag{26.16}$$

$$\overline{y} = \frac{\displaystyle\int_c^d y(x_2 - x_1)\,dy}{\displaystyle\int_c^d (x_2 - x_1)\,dy} \tag{26.17}$$

Centroid of solid of revolution

$$\overline{x} = \frac{\displaystyle\int_a^b xy^2\,dx}{\displaystyle\int_a^b y^2\,dx} \tag{26.18}$$

$$\overline{y} = \frac{\displaystyle\int_c^d yx^2\,dy}{\displaystyle\int_c^d x^2\,dy} \tag{26.19}$$

Radius of gyration

$$m_1 d_1^2 + m_2 d_2^2 + \cdots + m_n d_n^2 = (m_1 + m_2 + \cdots + m_n)R^2 \tag{26.20}$$

Moment of inertia of flat plate

$$I_y = k \int_a^b x^2(y_2 - y_1)\,dx \tag{26.21}$$

$$I_x = k \int_c^d y^2(x_2 - x_1)\,dy \tag{26.22}$$

Moment of inertia of solid of revolution

$$I_y = 2\pi k \int_a^b (y_2 - y_1)x^3\,dx \tag{26.23}$$

$$I_x = 2\pi k \int_c^d (x_2 - x_1)y^3\,dy \tag{26.24}$$

Work	$W = \int_a^b f(x)\,dx$	(26.25)
Force between electric charges	$f(x) = \dfrac{kq_1q_2}{x^2}$	(26.26)
Force due to liquid pressure	$F = w \int_a^b lh\,dh$	(26.27)
Average value	$y_{\text{av}} = \dfrac{\displaystyle\int_a^b y\,dx}{b-a}$	(26.28)

CHAPTER 26 REVIEW EXERCISES

1. How long after it is dropped does a baseball reach a speed of 42 m/s (the speed of a good fastball)?

2. If the velocity v (m/s) of a subway train after the brakes are applied can be expressed as $v = \sqrt{400 - 20t}$, where t is the time in seconds, how far does it travel in coming to a stop?

3. A weather balloon is rising at the rate of 10.0 m/s when a small metal part drops off. If the balloon is 60.0 m high at this instant, when will the part hit the ground?

4. A float is dropped into a river at a point where it is flowing at 1.5 m/s. How far does the float travel in 30 s if it accelerates downstream at 0.010 m/s²?

5. The acceleration of an object rolling down an inclined plane is 4.0 m/s². If it starts from rest at the top of the plane, find an equation for the distance the object travels from the top as a function of time.

6. What is the initial vertical velocity of a baseball that just reaches the ceiling of an indoor stadium that is 65 m high?

7. The electric current i (in A) in a circuit as a function of the time t (in s) is $i = 0.25(2\sqrt{t} - t)$. Find the total charge to pass a point in the circuit in 2.0 s.

8. The current i (in A) in a certain electric circuit is given by $i = \sqrt{1 + 4t}$, where t is the time (in s). Find the charge that passes a given point from $t = 1.0$ s to $t = 3.0$ s if $q_0 = 0$.

9. The voltage across a 5.5-nF capacitor in an FM radio receiver is zero. What is the voltage after 25 μs if a current of 12 mA charges the capacitor?

10. The initial voltage across a capacitor is zero, and $V_C = 2.50$ V after 8.00 ms. If a current $i = t/\sqrt{t^2 + 1}$, where i is the current (in A) and t is the time (in s), charges the capacitor, find the capacitance C of the capacitor.

11. The distribution of weight on a cable is not uniform. If the slope of the cable at any point is given by $dy/dx = 20 + 0.025x^2$ and if the origin of the coordinate system is at the lowest point, find the equation that gives the curve described by the cable.

12. The time rate of change of the reliability R (in %) of a computer system is $dR/dt = -2.5(0.05t + 1)^{-1.5}$, where t is the time (in h). If $R = 100$ for $t = 0$, find R for $t = 100$ h.

13. Find the area between $y = \sqrt{1 - x}$ and the coordinate axes.

14. Find the area bounded by $y = 3x^2 - x^3$ and the x-axis.

15. Find the area bounded by $y^2 = 2x$ and $y = x - 4$.

16. Find the area bounded by $y = 1/(2x + 1)^2$, $y = 0$, $x = 1$, and $x = 2$.

17. Find the area between $y = x^2$ and $y = x^3 - 2x^2$.

18. Find the area between $y = x^2 + 2$ and $y = 3x^2$.

19. Show that the curve $y = x^n$ ($n > 0$) divides the square bounded by $x = 0$, $y = 0$, $x = 1$, and $y = 1$ into two regions, the areas of which are in the ratio $n/1$.

20. Find the value of a such that the line $x = a$ bisects the area under the curve $y = 1/x^2$ between $x = 1$ and $x = 4$.

21. Find the volume generated by revolving the region bounded by $y = 3 + x^2$ and the line $y = 4$ about the x-axis.

22. Find the volume generated by revolving the region bounded by $y = 8x - x^4$ and the x-axis about the x-axis.

23. Find the volume generated by revolving the region bounded by $y = x^3 - 4x^2$ and the x-axis about the y-axis.

24. Find the volume generated by revolving the region bounded by $y = x$ and $y = 3x - x^2$ about the y-axis.

25. Find the volume generated by revolving an ellipse about its major axis.

26. A hole of radius 1.00 cm is bored along the diameter of a sphere of radius 4.00 cm. Find the volume of the material that is removed from the sphere.

27. Find the center of mass of the following array of four masses in the xy-plane (distances in cm): 60 g at $(4, 4)$, 160 g at $(-3, 6)$, 70 g at $(-5, -4)$, 130 g at $(3, -5)$.

28. Find the centroid of the flat-plate machine part shown in Fig. 26.71. Each section is uniform, and the mass of the section to the right of the y-axis is twice that of the section to the left.

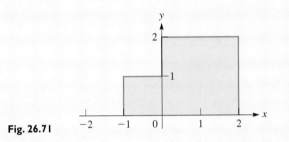

Fig. 26.71

29. Find the centroid of a flat plate covering the region bounded by $y^2 = x^3$ and $y = 2x$.

30. Find the centroid of a flat plate covering the region bounded by $y = 2x - 4$, $x = 1$, and $y = 0$.

31. Find the centroid of the volume generated by revolving the region bounded by $y = \sqrt{x}$, $x = 1$, $x = 4$, and $y = 0$ about the x-axis.

32. Find the centroid of the volume generated by revolving the region bounded by $yx^4 = 1$, $y = 1$, and $y = 4$ about the y-axis.

33. Find the moment of inertia of a flat plate covering the region bounded by $y = 3x - x^2$ and $y = x$ with respect to the y-axis.

34. Find the radius of gyration of a flat plate covering the region bounded by $y = 8 - x^3$, and the axes, with respect to the y-axis.

35. Find the moment of inertia with respect to its axis of a lead bullet that is defined by revolving the region bounded by $y = 3.00x^{0.10}$, $x = 0$, $x = 20.0$, and $y = 0$ about the x-axis (all measurements in mm). The density of lead is 0.0114 g/mm³.

36. Find the radius of gyration with respect to its axis of a rotating machine part that can be defined by revolving the region bounded by $y = 1/x$, $x = 1.00$, $x = 4.00$, and $y = 0.25$ (all measurements in cm) about the x-axis.

37. A pail and its contents weigh 80 N. The pail is attached to the end of a 30-m rope that weighs 20 N and is hanging vertically. How much work is done in winding up the rope with the pail attached?

38. The gravitational force (in N) of the earth on a satellite (the weight of the satellite) is given by $F = 10^{11}/x^2$, where x is the vertical distance (in km) form the center of the earth to the satellite. How much work is done in moving the satellite from the earth's surface to an altitude of 3000 km? The radius of the earth is 6370 km.

39. The rear stabilizer of a certain aircraft can be described as the region under the curve $y = 3x^2 - x^3$, as shown in Fig. 26.72. Find the x-coordinate (in m) of the centroid of the stabilizer.

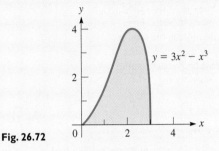

Fig. 26.72

40. The water in a spherical tank 20.0 m in radius is 15.0 m deep at the deepest point. How much water is in the tank?

41. The nose cone of a rocket has the shape of a semiellipse revolved about its major axis, as shown in Fig. 26.73. What is the volume of the nose cone?

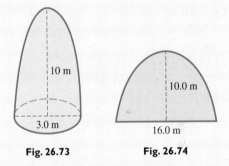

Fig. 26.73   **Fig. 26.74**

42. The deck area of a boat is a parabolic section as shown in Fig. 26.74. What is the area of the deck?

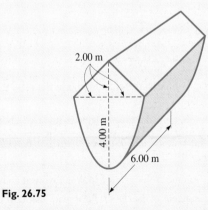

Fig. 26.75

43. The vertical ends of a fuel storage tank have a parabolic bottom section and a triangular top section, as shown in Fig. 26.75. What volume does the tank hold?

44. The capillary tube shown in Fig. 26.76 has circular horizontal cross sections of inner radius 1.1 mm. What is the volume of the liquid in the tube above the level of liquid outside the tube if the top of the liquid in the center vertical cross section is described by the equation $y = x^4 + 1.5$, as shown?

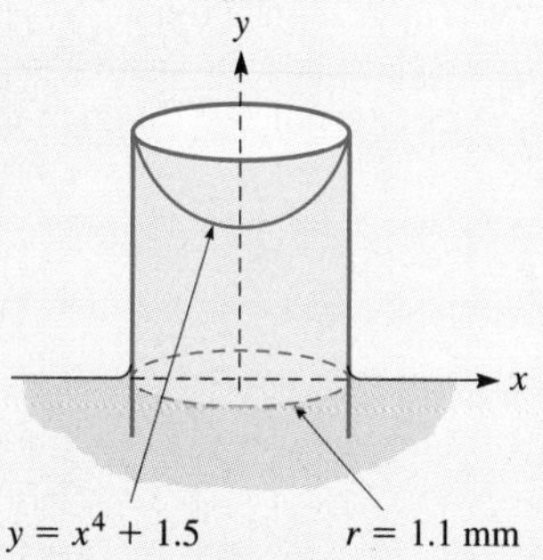

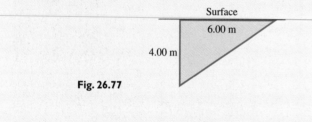

$y = x^4 + 1.5$ $r = 1.1$ mm

Fig. 26.76

45. A cylindrical chemical waste-holding tank 4.50 m in radius has a depth of 3.25 m. Find the total force on the circular side of the tank when it is filled with liquid with a density of 10.6 kN/m³.

46. A section of a dam is in the shape of a right triangle. The base of the triangle is 6.00 m and is in the surface of the water. If the triangular section goes to a depth of 4.00 m, find the force on it. See Fig. 26.77.

Surface

6.00 m

4.00 m

Fig. 26.77

47. The electric resistance of a wire is inversely proportional to the square of its radius. If a certain wire has a resistance of 0.30 Ω when its radius is 2.0 mm, find the average value of the resistance with respect to the radius if the radius changes from 2.0 mm to 2.1 mm.

48. A horizontal straight section of pipe is supported at its center by a vertical wire as shown in Fig. 26.78. Find the formula for the moment of inertia of the pipe with respect to an axis along the wire if the pipe is of length L and mass m.

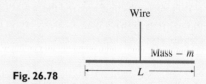

Wire

Mass — m

Fig. 26.78

Writing Exercise

49. A tub for holding liquids has a circular top of radius a. All cross sections of the tub that are perpendicular to a fixed diameter of the top are squares. Write one or two paragraphs explaining how to derive the formula that gives the volume of the tub. What is the formula?

CHAPTER 26 PRACTICE TEST

In Problems 1–3, use the region bounded by $y = \frac{1}{4}x^2$, $y = 0$, and $x = 2$.

1. Find the area.

2. Find the coordinates of the centroid of a flat plate that covers the region.

3. Find the volume if the given region is revolved about the x-axis.

In Problems 4 and 5, use the first-quadrant region bounded by $y = x^2$, $x = 0$, and $y = 9$.

4. Find the volume if the given region is revolved about the x-axis.

5. Find the moment of inertia of a flat plate that covers the region, with respect to the y-axis.

6. The velocity v of an object as a function of the time t is $v = 60 - 4t$. Find the expression for the displacement s if $s = 10$ for $t = 0$.

7. The natural length of a spring is 8.0 cm. A force of 12 N stretches it to a length of 10.0 cm. How much work is done in stretching it from a length of 10.0 cm to a length of 14.0 cm?

8. A vertical rectangular floodgate is 6.00 m wide and 2.00 m high. Find the force on the gate if its upper edge is 1.00 m below the surface of the water ($w = 9.80$ kN/m³).

Differentiation of Transcendental Functions

While studying vibrations in a rod in the late 1730s, the Swiss mathematician Euler noted that trigonometric functions arose naturally as solutions to equations in which derivatives appeared. This was the first treatment of the trigonometric functions as functions of numbers essentially as we do today. Later, in 1755, Euler wrote a textbook on differential calculus in which he included differentiation of the trigonometric, inverse trigonometric, logarithmic, and exponential functions. These are the most important of the *transcendental* (nonalgebraic) *functions,* as named by Euler.

The calculus of the transcendental functions was formulated mostly in the 1700s, by Euler and a number of other mathematicians, and has helped lead to the rapid progress made later in many technical and scientific areas. For example, the transcendental functions, along with their derivatives and integrals, have been of great importance in the development of the fields of electricity and electronics in the 1800s and 1900s, particularly with respect to alternating current.

In this chapter, we develop formulas for the derivatives of these transcendental functions, and in the following chapter we will take up integration that involves these functions. Other areas in which we will show important applications include harmonic motion, rocket motion, monetary interest calculations, population growth, acoustics, and optics.

In Sections 27.4 and 27.7, we use derivatives of transcendental functions in analyzing the motion of a rocket.

27.1 DERIVATIVES OF THE SINE AND COSINE FUNCTIONS

If we find the derivative of the sine function, we can use it to find the derivatives of the other trigonometric and inverse trigonometric functions. Therefore, we now find the derivative of the sine function.

Let $y = \sin x$, where x is expressed in radians. If x changes by an amount h, from the definition of the derivative, we have

$$\frac{dy}{dx} = \lim_{h \to 0} \frac{\sin(x + h) - \sin x}{h}$$

Referring now to Eq. (20.18), we have

$$\frac{dy}{dx} = \lim_{h \to 0} \frac{2 \sin \frac{1}{2}(x + h - x) \cos \frac{1}{2}(x + h + x)}{h}$$

$$= \lim_{h \to 0} \frac{\sin(h/2) \cos(x + h/2)}{h/2}$$

Looking ahead to the next step of letting $h \to 0$, we see that the numerator and denominator both approach zero. This situation is precisely the same as that in which we were finding the derivatives of the algebraic functions. To find the limit, we must find

$$\lim_{h \to 0} \frac{\sin(h/2)}{h/2}$$

since these are the factors that cause the numerator and the denominator to approach zero.

In finding this limit, we let $\theta = h/2$ for convenience of notation. This means that we are to determine $\lim_{\theta \to 0} \frac{\sin \theta}{\theta}$. Of course, it would be convenient to know before proceeding if this limit does actually exist. Therefore, by using a calculator, we can develop a table of values of $\frac{\sin \theta}{\theta}$ as θ becomes very small:

θ (radians)	0.5	0.1	0.05	0.01	0.001
$\dfrac{\sin \theta}{\theta}$	0.958 851 1	0.998 334 2	0.999 583 4	0.999 983 3	0.999 999 8

We see from this table that the limit of $\frac{\sin \theta}{\theta}$, as $\theta \to 0$, appears to be 1.

In order to prove that $\lim_{\theta \to 0} \frac{\sin \theta}{\theta} = 1$, we use a geometric approach. Considering Fig. 27.1, we see that the following inequality is true:

Area triangle OBD < area sector OBD < area triangle OBC

$$\frac{1}{2} r(r \sin \theta) < \frac{1}{2} r^2 \theta < \frac{1}{2} r(r \tan \theta) \quad \text{or} \quad \sin \theta < \theta < \tan \theta$$

For reference, Eq. (20.18) is $\sin x - \sin y = 2 \sin \frac{1}{2}(x - y) \cos \frac{1}{2}(x + y)$.

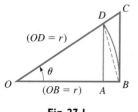

$(OD = r)$

θ

O $(OB = r)$ A B

Fig. 27.1

Remembering that we want to find the limit of $(\sin \theta)/\theta$, we next divide through by $\sin \theta$ and then take reciprocals:

$$1 < \frac{\theta}{\sin \theta} < \frac{1}{\cos \theta} \quad \text{or} \quad 1 > \frac{\sin \theta}{\theta} > \cos \theta$$

When we consider the limit as $\theta \to 0$, we see that the left member remains 1 and the right member approaches 1. Thus, $(\sin \theta)/\theta$ must approach 1. This means

$$\lim_{\theta \to 0} \frac{\sin \theta}{\theta} = \lim_{h \to 0} \frac{\sin(h/2)}{h/2} = 1 \tag{27.1}$$

Using the result in Eq. (27.1) in the expression for dy/dx, we have

$$\lim_{h \to 0} \left[\cos(x + h/2) \frac{\sin(h/2)}{h/2} \right] = \cos x$$

or

$$\frac{dy}{dx} = \cos x \tag{27.2}$$

To find the derivative of $y = \sin u$, where u is a function of x, we use the chain rule, Eq. (23.14), which we repeat here for reference:

$$\frac{dy}{dx} = \frac{dy}{du} \frac{du}{dx} \tag{27.3}$$

Therefore, for $y = \sin u$, $dy/du = \cos u$, we have

$$\frac{d(\sin u)}{dx} = \cos u \frac{du}{dx} \tag{27.4}$$

◀ **EXAMPLE 1**   Find the derivative of $y = \sin 2x$.
In this example, $u = 2x$. Thus,

$$\frac{dy}{dx} = \frac{d(\sin 2x)}{dx} = \cos 2x \frac{d(2x)}{dx} = (\cos 2x)(2) \qquad \text{using Eq. (27.4)}$$

$$= 2 \cos 2x$$

◀ **EXAMPLE 2**   Find the derivative of $y = 2 \sin(x^2)$.
In this example, $u = x^2$, which means that $du/dx = 2x$. Hence,

$$\frac{dy}{dx} = 2[\cos(x^2)](2x) \qquad \text{using Eq. (27.4)}$$

$$= 4x \cos(x^2)$$

CAUTION ▶   It is important here, just as it is *in finding the derivatives of powers of all functions, to remember to include the factor du/dx.*

◀ EXAMPLE 3   Find the derivative of $r = \sin^2 \theta$.

This example is a combination of the use of the power rule, Eq. (23.15) and the derivative of the sine function Eq. (27.4). Since $\sin^2 \theta$ means $(\sin \theta)^2$, in using the power rule we have $u = \sin \theta$. Thus,

For reference, Eq. (23.15) is
$\dfrac{du^n}{dx} = nu^{n-1}\left(\dfrac{du}{dx}\right)$.

$$\frac{dr}{d\theta} = 2(\sin \theta)\frac{d \sin \theta}{d\theta} \qquad \text{using Eq. (23.15)}$$

$$= 2 \sin \theta \cos \theta \qquad \text{using Eq. (27.4)}$$

$$= \sin 2\theta \qquad \text{using identity (Eq. (20.21))}$$

In order to find the derivative of the cosine function, we write it in the form $\cos u = \sin\left(\frac{\pi}{2} - u\right)$. Thus, if $y = \sin\left(\frac{\pi}{2} - u\right)$, we have

$$\frac{dy}{dx} = \cos\left(\frac{\pi}{2} - u\right)\frac{d\left(\frac{\pi}{2} - u\right)}{dx} = \cos\left(\frac{\pi}{2} - u\right)\left(-\frac{du}{dx}\right)$$

$$= -\cos\left(\frac{\pi}{2} - u\right)\frac{du}{dx}$$

Since $\cos\left(\frac{\pi}{2} - u\right) = \sin u$, we have

$$\boxed{\frac{d(\cos u)}{dx} = -\sin u \frac{du}{dx}} \tag{27.5}$$

◀ EXAMPLE 4   The electric power p developed in a resistor of an amplifier circuit is $p = 25 \cos^2 120\pi t$, where t is the time. Find the expression for the time rate of change of power.

From Chapter 23, we know that we are to find the derivative dp/dt. Therefore,

$$p = 25 \cos^2 120\pi t$$

$$\frac{dp}{dt} = 25(2 \cos 120\pi t)\frac{d \cos 120\pi t}{dt} \qquad \text{using Eq. (23.15)}$$

$$= 50 \cos 120\pi t(-\sin 120\pi t)\frac{d(120\pi t)}{dt} \qquad \text{using Eq. (27.5)}$$

$$= (-50 \cos\ 120\pi t \sin 120\pi t)(120\pi)$$

$$= -6000\pi \cos 120\pi t \sin 120\pi t$$

$$= -3000\pi \sin 240\pi t \qquad \text{using Eq. (20.21)}$$

For reference, Eq. (20.21) is
$\sin 2\alpha = 2 \sin \alpha \cos \alpha$.

◀ EXAMPLE 5   Find the derivative of $y = \sqrt{1 + \cos 2x}$.

$$y = (1 + \cos 2x)^{1/2}$$

$$\frac{dy}{dx} = \frac{1}{2}(1 + \cos 2x)^{-1/2}\frac{d(1 + \cos 2x)}{dx} \qquad \text{using Eq. (23.15)}$$

$$= \frac{1}{2}(1 + \cos 2x)^{-1/2}(-\sin 2x)(2) \qquad \text{using Eq. (27.5)}$$

$$= -\frac{\sin 2x}{\sqrt{1 + \cos 2x}}$$

◖ **EXAMPLE 6** Find the differential of $y = \sin 2x \cos x^2$.

From Section 24.8, recall that the differential of a function $y = f(x)$ is $dy = f'(x)\,dx$. Thus, using the derivative product rule and the derivatives of the sine and cosine functions, we arrive at the following result:

$$y = \sin 2x \cos x^2 \qquad\qquad y = (\sin 2x)(\cos x^2)$$
$$dy = [\sin 2x(-\sin x^2)(2x) + \cos x^2(\cos 2x)(2)]\,dx$$
$$= (-2x \sin 2x \sin x^2 + 2 \cos 2x \cos x^2)\,dx \qquad\qquad ◗$$

◖ **EXAMPLE 7** Find the slope of a line tangent to the curve of $y = 5 \sin 3x$, where $x = 0.2$.

Here we are to find the derivative of $y = 5 \sin 3x$ and then evaluate the derivative for $x = 0.2$. Therefore, we have the following:

$$y = 5 \sin 3x$$
$$\frac{dy}{dx} = 5(\cos 3x)(3) = 15 \cos 3x \qquad \text{find derivative}$$
$$\left.\frac{dy}{dx}\right|_{x=0.2} = 15 \cos 3(0.2) = 15 \cos 0.6 \qquad \text{evaluate}$$
$$= 12.38$$

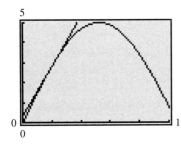

Fig. 27.2

CAUTION ▶ In evaluating the slope, we must remember that $x = 0.2$ means the ***values are in radians***. Therefore, the slope is 12.38.

As we showed in Chapter 23, the *tangent* feature of a graphing calculator can be used to display a curve and the line tangent at a specific point. In Fig. 27.2, the calculator display for this function and its tangent at $x = 0.2$ is shown.

Also, reviewing the use of the calculator, the evaluation of this derivative using the *numerical derivative* feature is shown in the calculator display in Fig. 27.3. ◗

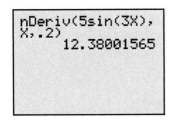

```
nDeriv(5sin(3X),
X,.2)
      12.38001565
```

Fig. 27.3

EXERCISES 27.1

In Exercises 1 and 2, make the given changes in the indicated examples of this section and then find the derivatives.

1. In Example 3, in the given function change θ to $2\theta^2$.

2. In Example 5, in the given function change $2x$ to x^2.

In Exercises 3–34, find the derivatives of the given functions.

3. $y = \sin(x + 2)$

4. $y = 3 \sin 4x$

5. $y = 2 \sin(2x^3 - 1)$

6. $y = 5 \sin(2 - 3t)$

7. $y = 6 \cos \frac{1}{2}x$

8. $y = \cos(1 - x)$

9. $y = 2 \cos(3x - \pi)$

10. $y = 4 \cos(6x^2 + 5)$

11. $r = \sin^2 3\pi\theta$

12. $y = 3 \sin^3(2x^4 + 1)$

13. $y = 3 \cos^3(5x + 2)$

14. $y = 4 \cos^2 \sqrt{x}$

15. $y = x \sin 3x$

16. $y = x^2 \sin \pi x$

17. $y = 3x^3 \cos 5x$

18. $y = 0.5\theta \cos(2\theta + \pi/4)$

19. $y = \sin x^2 \cos 2x$

20. $y = 6 \sin x \cos 4x$

21. $y = \sqrt{1 + \sin 4x}$

22. $y = (x - \cos^2 x)^4$

23. $r = \dfrac{\sin(3t - \pi/3)}{2t}$

24. $y = \dfrac{2x + 3}{\sin 2\pi x}$

25. $y = \dfrac{2 \cos x^2}{3x - 1}$

26. $y = \dfrac{\cos^2 3x}{1 + 2 \sin^2 2x}$

27. $y = 2 \sin^2 3x \cos 2x$

28. $y = \cos^3 4x \sin^2 2x$

29. $s = \sin(\sin 2t)$

30. $z = 0.2 \cos(\sin 3\phi)$

31. $y = \sin^3 x - \cos 2x$

32. $y = x \sin x + \cos x$

33. $p = \dfrac{1}{\sin s} + \dfrac{1}{\cos s}$

34. $y = 2x \sin x + 2 \cos x - x^2 \cos x$

In Exercises 35–56, solve the given problems.

35. Using a graphing calculator: (a) display the graph of $y = (\sin x)/x$ to verify that $(\sin \theta)/\theta \to 1$ as $\theta \to 0$, and (b) verify the values for $(\sin \theta)/\theta$ in the table on page 798.

36. Evaluate $\lim_{\theta \to 0} (\tan \theta)/\theta$. (Use the fact that $\lim_{\theta \to 0} (\sin \theta)/\theta = 1$.)

W 37. On a calculator, find the values of (a) cos 1.0000 and (b) $(\sin 1.0001 - \sin 1.0000)/0.0001$. Compare the values and give the meaning of each in relation to the derivative of the sine function where $x = 1$.

W 38. On a calculator, find the values of (a) $-\sin 1.0000$ and (b) $(\cos 1.0001 - \cos 1.0000)/0.0001$. Compare the values and give the meaning of each in relation to the derivative of the cosine function where $x = 1$.

39. On the graph of $y = \sin x$ in Fig. 27.4, draw tangent lines at the indicated points and determine the slopes of these tangent lines. Then plot the values of these slopes for the same values of x and join the points with a smooth curve. Compare the resulting curve with $y = \cos x$. (Note the meaning of the derivative as the slope of a tangent line.)

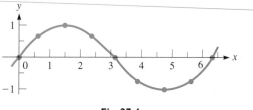

Fig. 27.4

40. Repeat the instructions given in Exercise 39 for the graph of $y = \cos x$ in Fig. 27.5. Compare the resulting curve with $y = \sin x$. (Be careful in this comparison and remember the difference between $y = \sin x$ and the derivative of $y = \cos x$. As in Exercise 39, note the meaning of the derivative as the slope of a tangent line.)

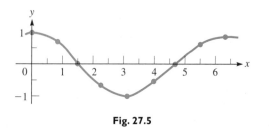

Fig. 27.5

41. Find the derivative of the implicit function $\sin(xy) + \cos 2y = x^2$.

42. Find the derivative of the implicit function $x \cos 2y + \sin x \cos y = 1$.

43. Show that $\dfrac{d^4 \sin x}{dx^4} = \sin x$.

44. If $y = \cos 2x$, show that $\dfrac{d^2 y}{dx^2} = -4y$.

45. Find the derivative of each member of the identity $\cos 2x = 2 \cos^2 x - 1$ and thereby obtain another trigonometric identity.

46. Find the derivative of each member of the identity $2 \sin^2 x = 1 - \cos 2x$ and thereby obtain another trigonometric identity.

47. Use differentials to estimate the value of sin 31°.

48. Find the linearization $L(x)$ of the function $f(x) = \sin(\cos x)$ for $a = \pi/2$.

49. Find the slope of a line tangent to the curve of $y = x \cos 2x$ where $x = 1.20$. Verify the result by using the *derivative-evaluating* feature of a graphing calculator.

50. Find the slope of a line tangent to the curve of $y = \dfrac{2 \sin 3x}{x}$, where $x = 0.15$. Verify the result by using the *derivative-evaluating* feature of a graphing calculator.

51. The voltage V in a certain electric circuit as a function of the time t (in s) is given by $V = 3.0 \sin 188t \cos 188t$. How fast is the voltage changing when $t = 2.0$ ms?

52. A water slide at an amusement park follows the curve (y in m) $y = 2.0 + 2.0 \cos (0.53x + 0.40)$ for $0 \le x \le 5.0$ m. Find the angle with the horizontal of the slide for $x = 2.5$ m.

53. The blade of a saber saw moves vertically up and down, and its displacement y (in cm) is given by $y = 1.85 \sin 36\pi t$, where t is the time (in s). Find the velocity of the blade for $t = 0.0250$ s.

54. The current i (in A) in an amplifier circuit as a function of the time t (in s) is given by $i = 0.10 \cos(120\pi t + \pi/6)$. Find the expression for the voltage across a 2.0-mH inductor in the circuit. (See Exercise 27 of Section 26.1 on page 763.)

55. In testing a heat-seeking rocket, it is always moving directly toward a remote-controlled aircraft. At a certain instant the distance r (in km) from the rocket to the aircraft is $r = \dfrac{100}{1 - \cos \theta}$, where θ is the angle between their directions of flight. Find $dr/d\theta$ for $\theta = 120°$. See Fig. 27.6.

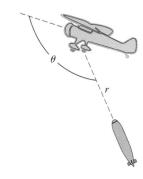

Fig. 27.6

56. The number N of reflections of a light ray passing through an optic fiber of length L and diameter d is $N = \dfrac{L \sin \theta}{d\sqrt{n^2 - \sin^2 \theta}}$. Here, n is the index of refraction of the fiber, and θ is the angle between the light ray and the fiber's axis. Find $dN/d\theta$.

27.2 DERIVATIVES OF THE OTHER TRIGONOMETRIC FUNCTIONS

We can find the derivatives of the other trigonometric functions by expressing the functions in terms of the sine and cosine. After we perform the differentiation, we use trigonometric relations to put the derivative in a convenient form.

We obtain the derivative of tan u by expressing tan u as sin u/cos u. Therefore, letting $y = \sin u / \cos u$, by employing the quotient rule, we have

$$\frac{dy}{dx} = \frac{\cos u[\cos u(du/dx)] - \sin u[-\sin u(du/dx)]}{\cos^2 u}$$

$$= \frac{\cos^2 u + \sin^2 u}{\cos^2 u}\frac{du}{dx} = \frac{1}{\cos^2 u}\frac{du}{dx} = \sec^2 u\frac{du}{dx}$$

$$\boxed{\frac{d(\tan u)}{dx} = \sec^2 u\frac{du}{dx}} \tag{27.6}$$

We find the derivative of cot u by letting $y = \cos u / \sin u$ and again using the quotient rule:

$$\frac{dy}{dx} = \frac{\sin u[-\sin u(du/dx)] - \cos u[\cos u(du/dx)]}{\sin^2 u}$$

$$= \frac{-\sin^2 u - \cos^2 u}{\sin^2 u}\frac{du}{dx}$$

$$\boxed{\frac{d(\cot u)}{dx} = -\csc^2 u\frac{du}{dx}} \tag{27.7}$$

To obtain the derivative of sec u, we let $y = 1/\cos u$. Then,

$$\frac{dy}{dx} = -(\cos u)^{-2}\left[(-\sin u)\left(\frac{du}{dx}\right)\right] = \frac{1}{\cos u}\frac{\sin u}{\cos u}\frac{du}{dx}$$

$$\boxed{\frac{d(\sec u)}{dx} = \sec u \tan u\frac{du}{dx}} \tag{27.8}$$

We obtain the derivative of csc u by letting $y = 1/\sin u$. And so,

$$\frac{dy}{dx} = -(\sin u)^{-2}\left(\cos u\frac{du}{dx}\right) = -\frac{1}{\sin u}\frac{\cos u}{\sin u}\frac{du}{dx}$$

$$\boxed{\frac{d(\csc u)}{dx} = -\csc u \cot u\frac{du}{dx}} \tag{27.9}$$

Note the convenient forms of these derivatives that are obtained by using basic trigonometric identities.

EXAMPLE 1   Find the derivative of $y = 2 \tan 8x$.
The derivative is

$$\frac{dy}{dx} = 2(\sec^2 8x)(8) \qquad \text{using Eq. (27.6)}$$

$$= 16 \sec^2 8x$$

EXAMPLE 2   Find the derivative of $y = 3 \sec^2 4x$.
Using the power rule and Eq. (27.8), we have

$$\frac{dy}{dx} = 3(2)(\sec 4x)\frac{d(\sec 4x)}{dx} \qquad \text{using } \frac{du^n}{dx} = nu^{n-1}\frac{du}{dx}$$

$$= 6(\sec 4x)(\sec 4x \tan 4x)(4) \qquad \text{using } \frac{d \sec u}{dx} = \sec u \tan u \frac{du}{dx}$$

$$= 24 \sec^2 4x \tan 4x$$

EXAMPLE 3   Find the derivative of $y = t \csc^3 2t$.
Using the power rule, the product rule, and Eq. (27.9), we have

$$\frac{dy}{dt} = t(3 \csc^2 2t)(-\csc 2t \cot 2t)(2) + (\csc^3 2t)(1)$$

$$= \csc^3 2t(-6t \cot 2t + 1)$$

EXAMPLE 4   Find the derivative of $y = (\tan 2x + \sec 2x)^3$.
Using the power rule and Eqs. (27.6) and (27.8), we have

$$\frac{dy}{dx} = 3(\tan 2x + \sec 2x)^2[\sec^2 2x(2) + \sec 2x \tan 2x(2)]$$

$$= 3(\tan 2x + \sec 2x)^2(2 \sec 2x)(\sec 2x + \tan 2x)$$

$$= 6 \sec 2x(\tan 2x + \sec 2x)^3$$

EXAMPLE 5   Find the differential of $r = \sin 2\theta \tan \theta^2$.
Here we are to find the derivative of the given function and multiply by $d\theta$. Therefore, using the product rule along with Eqs. (27.4) and (27.6), we have

$$dr = [(\sin 2\theta)(\sec^2 \theta^2)(2\theta) + (\tan \theta^2)(\cos 2\theta)(2)]d\theta$$

$$= (2\theta \sin 2\theta \sec^2 \theta^2 + 2 \cos 2\theta \tan \theta^2) d\theta \qquad \text{don't forget the } d\theta.$$

EXAMPLE 6   Find dy/dx if $\cot 2x - 3 \csc xy = y^2$.
In finding the derivative of this implicit function, we must be careful not to forget the factor dy/dx when it occurs. The derivative is found as follows:

$$\cot 2x - 3 \csc xy = y^2$$

$$(-\csc^2 2x)(2) - 3(-\csc xy \cot xy)\left(x\frac{dy}{dx} + y\right) = 2y\frac{dy}{dx}$$

$$3x \csc xy \cot xy \frac{dy}{dx} - 2y\frac{dy}{dx} = 2 \csc^2 2x - 3y \csc xy \cot xy$$

$$\frac{dy}{dx} = \frac{2 \csc^2 2x - 3y \csc xy \cot xy}{3x \csc xy \cot xy - 2y}$$

❨ **EXAMPLE 7** Evaluate the derivative of $y = \dfrac{2x}{1 - \cot 3x}$, where $x = 0.25$.

Finding the derivative, we have

$$\frac{dy}{dx} = \frac{(1 - \cot 3x)(2) - 2x(\csc^2 3x)(3)}{(1 - \cot 3x)^2}$$

$$= \frac{2 - 2\cot 3x - 6x\csc^2 3x}{(1 - \cot 3x)^2}$$

Now, substituting $x = 0.25$, we have

$$\left.\frac{dy}{dx}\right|_{x=0.25} = \frac{2 - 2\cot 0.75 - 6(0.25)\csc^2 0.75}{(1 - \cot 0.75)^2}$$

$$= -626.0$$

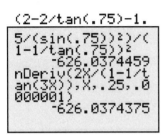

Fig. 27.7

In using the calculator, we note again that we must have it in **radian mode.** To evaluate the above expression for dy/dx, we must use the reciprocals of tan 0.75 and sin 0.75 for cot 0.75 and csc 0.75, respectively, as shown in the first four lines of the display in Fig. 27.7. The last four lines of Fig. 27.7 show a check of our result using the *numerical derivative* feature of the calculator.   ❩

EXERCISES 27.2

In Exercises 1 and 2, make the given changes in the indicated examples of this section and then find the derivatives.

1. In Example 2, in the given function change $4x$ to x^2.

2. In Example 5, in the given function change θ^2 to 3θ.

In Exercises 3–34, find the derivatives of the given functions.

3. $y = \tan 5x$

4. $y = 3\tan(3x + 2)$

5. $y = 5\cot(0.25\pi - \theta)$

6. $y = 3\cot 6x$

7. $y = 3\sec 2x$

8. $y = \sec\sqrt{1 - x}$

9. $y = -3\csc\sqrt{2x + 3}$

10. $h = 0.5\csc(1 - 2\pi t)$

11. $y = 5\tan^2 \pi x$

12. $y = 2\tan^2(x^2)$

13. $y = 2\cot^4 \frac{1}{2}x$

14. $y = \cot^2(1 - x^2)$

15. $y = \sqrt{\sec 4x}$

16. $y = 0.8\sec^3 5u$

17. $y = 3\csc^4 7x$

18. $y = \csc^2(2x^2)$

19. $r = t^2 \tan 0.5t$

20. $y = 3x\sec 2\pi x$

21. $y = 4\cos x\csc x^2$

22. $y = \frac{1}{2}\sin 2x\sec x$

23. $y = \dfrac{\csc x}{x}$

24. $u = \dfrac{\cot 0.25z}{2z}$

25. $y = \dfrac{2\cos 4x}{1 + \cot 3x}$

26. $y = \dfrac{\tan^2 3x}{2 + \sin x^2}$

27. $y = \frac{1}{3}\tan^3 x - \tan x$

28. $y = \csc 2x - 2\cot 2x$

29. $r = \tan(\sin 2\pi\theta)$

30. $y = x\tan x + \sec^2 2x$

31. $y = \sqrt{2x + \tan 4x}$

32. $y = (1 - \csc^2 3x)^3$

33. $x\sec y - 2y = \sin 2x$

34. $3\cot(x + y) = \cos y^2$

In Exercises 35–38, find the differentials of the given functions.

35. $y = 4\tan^2 3x$

36. $y = 2.5\sec^3 2t$

37. $y = \tan 4x\sec 4x$

38. $y = 2x\cot 3x$

In Exercises 39–52, solve the given problems.

Ⓦ **39.** On a calculator, find the values of (a) $\sec^2 1.0000$ and (b) $(\tan 1.0001 - \tan 1.0000)/0.0001$. Compare the values and give the meaning of each in relation to the derivative of tan x where $x = 1$.

Ⓦ **40.** On a calculator, find the values of (a) $\sec 1.0000 \tan 1.0000$ and (b) $(\sec 1.0001 - \sec 1.0000)/0.0001$. Compare the values and give the meaning of each in relation to the derivative of sec x where $x = 1$.

41. (a) Display the graph of $y = \tan x$ on a graphing calculator, and using the *derivative* feature, evaluate dy/dx for $x = 1$. (b) Display the graph of $y = \sec^2 x$ and evaluate y for $x = 1$. (Compare the values in parts (a) and (b).)

42. Follow the instructions in Exercise 41, using the graphs of $y = \sec x$ and $y = \sec x\tan x$.

43. Find the derivative of each member of the identity $1 + \tan^2 x = \sec^2 x$ and show that the results are equal.

44. Find the derivative of each member of the identity $1 + \cot^2 x = \csc^2 x$ and show that the results are equal.

45. Find the slope of a line tangent to the curve of $y = 2\cot 3x$ where $x = \pi/12$. Verify the result by using the *numerical derivative* feature of a graphing calculator.

46. Find the slope of a line normal to the curve of $y = \csc\sqrt{2x + 1}$ where $x = 0.45$. Verify the result by using the *numerical derivative* feature of a graphing calculator.

47. Show that $y = 2 \tan x - \sec x$ satisfies $\dfrac{dy}{dx} = \dfrac{2 - \sin x}{\cos^2 x}$.

48. Show that $y = \cos^3 x \tan x$ satisfies

$$\cos x \frac{dy}{dx} + 3y \sin x - \cos^2 x = 0.$$

49. The vertical displacement y (in cm) of the end of an industrial robot arm for each cycle is $y = 2t^{1.5} - \tan 0.1t$, where t is the time (in s). Find its vertical velocity for $t = 15$ s.

50. The electric charge q (in C) passing a given point in a circuit is given by $q = t \sec\sqrt{0.2t^2 + 1}$, where t is the time (in s). Find the current i (in A) for $t = 0.80$ s. ($i = dq/dt$.)

51. An observer to a rocket launch was 1000 m from the takeoff position. The observer found the angle of elevation of the rocket as a function of time to be $\theta = 3t/(2t + 10)$. Therefore, the height h (in m) of the rocket was $h = 1000 \tan\dfrac{3t}{2t + 10}$. Find the time rate of change of height after 5.0 s. See Fig. 27.8.

52. A surveyor measures the distance between two markers to be 378.00 m. Then, moving along a line equidistant from the markers, the distance d from the surveyor to each marker is $d = 189.00 \csc\frac{1}{2}\theta$, where θ is the angle between the lines of sight to the markers. See Fig. 27.9. By using differentials, find the change in d if θ changes from 98.20° to 98.45°.

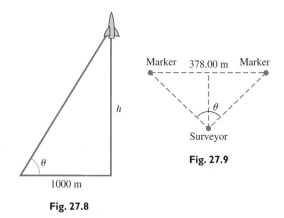

Marker   378.00 m   Marker

h

Surveyor

Fig. 27.9

θ

1000 m

Fig. 27.8

27.3 DERIVATIVES OF THE INVERSE TRIGONOMETRIC FUNCTIONS

To obtain the derivative of $y = \sin^{-1} u$, we first solve for u in the form $u = \sin y$ and then take derivatives with respect to x:

$$\frac{du}{dx} = \cos y \frac{dy}{dx}$$

Solving this equation for dy/dx, we obtain

$$\frac{dy}{dx} = \frac{1}{\cos y}\frac{du}{dx} = \frac{1}{\sqrt{1 - \sin^2 y}}\frac{du}{dx} = \frac{1}{\sqrt{1 - u^2}}\frac{du}{dx}$$

We choose the positive square root since $\cos y > 0$ for $-\frac{\pi}{2} < y < \frac{\pi}{2}$, which is the range of the defined values of $\sin^{-1} u$. Therefore, we obtain the following result:

$$\boxed{\frac{d(\sin^{-1} u)}{dx} = \frac{1}{\sqrt{1 - u^2}}\frac{du}{dx}} \qquad (27.10)$$

Note that the derivative of the inverse sine function is an algebraic function.

◀ **EXAMPLE 1** Find the derivative of $y = \sin^{-1} 4x$.

$$\frac{dy}{dx} = \frac{1}{\sqrt{1 - (4x)^2}}\overbrace{(4)}^{\frac{du}{dx}} = \frac{4}{\sqrt{1 - 16x^2}} \qquad \text{using Eq. (27.10)}$$

We find the derivative of the inverse cosine function by letting $y = \cos^{-1} u$ and by following the same procedure as that used in finding the derivative of $\sin^{-1} u$:

$$u = \cos y, \qquad \frac{du}{dx} = -\sin y \frac{dy}{dx}$$

$$\frac{dy}{dx} = -\frac{1}{\sin y}\frac{du}{dx} = -\frac{1}{\sqrt{1 - \cos^2 y}}\frac{du}{dx}$$

Therefore,

$$\boxed{\frac{d(\cos^{-1} u)}{dx} = -\frac{1}{\sqrt{1 - u^2}}\frac{du}{dx}} \qquad (27.11)$$

The positive square root is chosen here since $\sin y > 0$ for $0 < y < \pi$, which is the range of the defined values of $\cos^{-1} u$. We note that the derivative of the inverse cosine is the negative of the derivative of the inverse sine.

By letting $y = \tan^{-1} u$, solving for u, and taking derivatives, we find the derivative of the inverse tangent function:

$$u = \tan y, \qquad \frac{du}{dx} = \sec^2 y \frac{dy}{dx}, \qquad \frac{dy}{dx} = \frac{1}{\sec^2 y}\frac{du}{dx} = \frac{1}{1 + \tan^2 y}\frac{du}{dx}$$

Therefore,

$$\boxed{\frac{d(\tan^{-1} u)}{dx} = \frac{1}{1 + u^2}\frac{du}{dx}} \qquad (27.12)$$

We can see that the derivative of the inverse tangent is an algebraic function also.

The inverse sine, inverse cosine, and inverse tangent prove to be of the greatest importance in applications and in further development of mathematics. Therefore, the formulas for the derivatives of the other inverse functions are not presented here, although they are included in the exercises.

Solving a Word Problem

◀ **EXAMPLE 2** A 20-N force acts on a sign as shown in Fig. 27.10. Express the angle θ as a function of the x-component F_x of the force, and then find the expression for the instantaneous rate of change of θ with respect to F_x.

From the figure, we see that $F_x = 20 \cos \theta$. Solving for θ, we have $\theta = \cos^{-1}(F_x/20)$. To find the instantaneous rate of change of θ with respect to F_x, we are to take the derivative $d\theta/dF_x$:

$$\theta = \cos^{-1}\frac{F_x}{20} = \cos^{-1} 0.05F_x$$

$$\frac{d\theta}{dF_x} = -\frac{1}{\sqrt{1 - (0.05F_x)^2}}(0.05) \qquad \text{using Eq. (27.11)}$$

$$= \frac{-0.05}{\sqrt{1 - 0.0025F_x^2}}$$

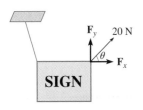

$\mathbf{F}_y$ 20 N

$\mathbf{F}_x$

SIGN

Fig. 27.10

◀ **EXAMPLE 3** Find the derivative of $y = (x^2 + 1)\tan^{-1} x - x$.

Using the product rule along with Eq. (27.12) on the first term, we have

$$\frac{dy}{dx} = (x^2 + 1)\left(\frac{1}{1 + x^2}\right)(1) + (\tan^{-1} x)(2x) - 1$$

$$\underline{\hspace{2cm}}\text{using Eq. (27.12)}$$

$$= 2x \tan^{-1} x$$

◀ **EXAMPLE 4** Find the differential of $y = x(\tan^{-1} 2x)^2$.

Using the product rule, the power rule Eq. (23.15), and Eq. (27.12), along with the meaning of the differential, we have

$$\text{using Eq. (23.15)}$$

$$dy = \left[x(2)(\tan^{-1} 2x)\left(\frac{1}{1 + (2x)^2}\right)(2) + (\tan^{-1} 2x)^2(1)\right] dx$$

$$\underline{\hspace{2cm}}\text{using Eq. (27.12)}$$

$$= \tan^{-1} 2x\left(\frac{4x}{1 + 4x^2} + \tan^{-1} 2x\right) dx$$

For reference, Eq. (23.15) is
$$\frac{du^n}{dx} = nu^{n-1}\left(\frac{du}{dx}\right).$$

◀ **EXAMPLE 5** Find the derivative of $y = x \sin^{-1} 2x + \frac{1}{2}\sqrt{1 - 4x^2}$.

We write

$$\text{using Eq. (27.10)}\qquad\qquad\text{using Eq. (23.15)}$$

$$\frac{dy}{dx} = x\left(\frac{2}{\sqrt{1 - 4x^2}}\right) + \sin^{-1} 2x + \frac{1}{2}\left(\frac{1}{2}\right)(1 - 4x^2)^{-1/2}(-8x)$$

$$= \frac{2x}{\sqrt{1 - 4x^2}} + \sin^{-1} 2x - \frac{2x}{\sqrt{1 - 4x^2}}$$

$$= \sin^{-1} 2x$$

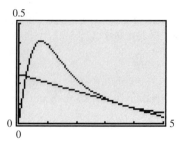

0.5

0

0   5

0

Fig. 27.11

◀ **EXAMPLE 6** Find the slope of a tangent to the curve of $y = \dfrac{\tan^{-1} x}{x^2 + 1}$, where $x = 3.60$. In Fig. 27.11, the calculator display of the function and the tangent line is shown.

Here we are to find the derivative and then evaluate it for $x = 3.60$.

$$\frac{dy}{dx} = \frac{(x^2 + 1)\left(\dfrac{1}{1 + x^2}\right)(1) - (\tan^{-1} x)(2x)}{(x^2 + 1)^2}\qquad\text{take derivative}$$

$$= \frac{1 - 2x \tan^{-1} x}{(x^2 + 1)^2}$$

$$\left.\frac{dy}{dx}\right|_{x=3.60} = \frac{1 - 2(3.60)(\tan^{-1} 3.60)}{(3.60^2 + 1)^2}\qquad\text{evaluate}$$

$$= -0.0429$$

This result can be checked by using the *numerical derivative* feature of a graphing calculator.

EXERCISES 27.3

In Exercises 1 and 2, make the given changes in the indicated examples of this section and then find the derivatives.

1. In Example 1, in the given function change $4x$ to x^2.

2. In Example 3, in the given function change $(x^2 + 1)\tan^{-1} x$ to $(4x^2 + 1)\tan^{-1} 2x$.

In Exercises 3–34, find the derivatives of the given functions.

3. $y = \sin^{-1} 7x$

4. $y = \sin^{-1}(1 - x^2)$

5. $y = 2 \sin^{-1} 3x^3$

6. $y = \sin^{-1} \sqrt{1 - 2x}$

7. $y = 3.6 \cos^{-1} 0.5s$

8. $\theta = 0.2 \cos^{-1} 5t$

9. $y = 2 \cos^{-1} \sqrt{2 - x}$

10. $y = 3 \cos^{-1}(x^2 + 0.5)$

11. $y = \tan^{-1} \sqrt{x}$

12. $y = \tan^{-1}(1 - x)$

13. $y = 6 \tan^{-1}(1/x)$

14. $y = 4 \tan^{-1} \pi x^4$

15. $y = 5x \sin^{-1} x$

16. $y = x^2 \cos^{-1} x$

17. $v = 0.4u \tan^{-1} 2u$

18. $y = (x^2 + 1)\sin^{-1} 4x$

19. $T = \dfrac{3R - 1}{\sin^{-1} 2R}$

20. $\theta = \dfrac{\tan^{-1} 2r}{\pi r}$

21. $y = \dfrac{\sin^{-1} 2x}{\cos^{-1} 2x}$

22. $y = \dfrac{x^2 + 1}{\tan^{-1} x}$

23. $y = 2(\cos^{-1} 4x)^3$

24. $r = 0.5(\sin^{-1} 3t)^4$

25. $y = [\sin^{-1}(4x + 1)]^2$

26. $y = \sqrt{\sin^{-1}(x - 1)}$

27. $y = \tan^{-1}\left(\dfrac{1 - t}{1 + t}\right)$

28. $y = \dfrac{1}{\cos^{-1} 2x}$

29. $y = \dfrac{1}{1 + 4x^2} - \tan^{-1} 2x$

30. $y = \sin^{-1} x - \sqrt{1 - x^2}$

31. $y = 3(4 - \cos^{-1} 2x)^3$

32. $\sin^{-1}(x + y) + y = x^2$

33. $2 \tan^{-1} xy + x = 3$

34. $y = \sqrt{2\pi - \sin^{-1} 4x}$

In Exercises 35–52, solve the given problems.

(W) 35. On a calculator, find the values of (a) $1/\sqrt{1 - 0.5^2}$ and (b) $(\sin^{-1} 0.5001 - \sin^{-1} 0.5000)/0.0001$. Compare the values and give the meaning of each in relation to the derivative of $\sin^{-1} x$ where $x = 0.5$.

(W) 36. On a calculator, find the values of (a) $1/(1 + 0.5^2)$ and (b) $(\tan^{-1} 0.5001 - \tan^{-1} 0.5000)/0.0001$. Compare the values and give the meaning of each in relation to the derivative of $\tan^{-1} x$ where $x = 0.5$.

37. Find the differential of the function $y = (\sin^{-1} x)^3$.

38. Find the linearization $L(x)$ of the function $f(x) = 2x \cos^{-1} x$ for $a = 0$.

39. Find the slope of a line tangent to the curve of $y = x/\tan^{-1} x$ at $x = 0.80$. Verify the result by using the *numerical derivative* feature of a graphing calculator.

(W) 40. Explain what is wrong with a problem that requires finding the derivative of $y = \sin^{-1}(x^2 + 1)$.

41. Use a graphing calculator to display the graphs of $y = \sin^{-1} x$ and $y = 1/\sqrt{1 - x^2}$. By roughly estimating slopes of tangent lines of $y = \sin^{-1} x$, note that $y = 1/\sqrt{1 - x^2}$ gives reasonable values for the derivative of $y = \sin^{-1} x$.

42. Use a graphing calculator to display the graphs of $y = \tan^{-1} x$ and $y = 1/(1 + x^2)$. By roughly estimating slopes of tangent lines of $y = \tan^{-1} x$, note that $y = 1/(1 + x^2)$ gives reasonable values for the derivative of $y = \tan^{-1} x$.

43. Find the second derivative of the function $y = \tan^{-1} 2x$.

44. Show that $\dfrac{d(\cot^{-1} u)}{dx} = -\dfrac{1}{1 + u^2}\dfrac{du}{dx}$.

45. Show that $\dfrac{d(\sec^{-1} u)}{dx} = \dfrac{1}{\sqrt{u^2(u^2 - 1)}}\dfrac{du}{dx}$.

46. Show that $\dfrac{d(\csc^{-1} u)}{dx} = -\dfrac{1}{\sqrt{u^2(u^2 - 1)}}\dfrac{du}{dx}$.

47. In the analysis of the waveform of an AM radio wave, the equation $t = \dfrac{1}{\omega} \sin^{-1}\dfrac{A - E}{mE}$ arises. Find dt/dm, assuming that the other quantities are constant.

48. An equation that arises in the theory of solar collectors is $\alpha = \cos^{-1}\dfrac{2f - r}{r}$. Find the expression for $d\alpha/dr$ if f is constant.

49. When an alternating current passes through a series *RLC* circuit, the voltage and current are out of phase by angle θ (see Section 12.7). Here $\theta = \tan^{-1}[(X_L - X_C)/R]$, where X_L and X_C are the reactances of the inductor and capacitor, respectively, and R is the resistance. Find $d\theta/dX_C$ for constant X_L and R.

50. When passing through glass, a light ray is refracted (bent) such that the angle of refraction r is given by $r = \sin^{-1}[(\sin i)/\mu]$. Here, i is the angle of incidence, and μ is the index of refraction of the glass (see Fig. 27.12). For different types of glass, μ differs. Find the expression for dr for a constant value of i.

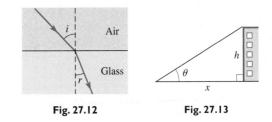

Fig. 27.12   **Fig. 27.13**

51. As a person approaches a building of height h, the angle of elevation of the top of the building is a function of the person's distance from the building. Express the angle of elevation θ in terms of h and the distance x from the building and then find $d\theta/dx$. Assume the person's height is negligible to that of the building. See Fig. 27.13.

52. A triangular metal frame is designed as shown in Fig. 27.14. Express angle A as a function of x and evaluate dA/dx for $x = 6$ cm.

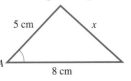

Fig. 27.14

27.4 APPLICATIONS

With our development of the formulas for the derivatives of the trigonometric and inverse trigonometric functions, it is now possible for us to apply these derivatives in the same manner as we applied the derivatives of algebraic functions. We can now use trigonometric and inverse trigonometric functions to solve tangent and normal, Newton's method, time rate of change, curve tracing, maximum and minimum, and differential application problems. The following examples illustrate the use of these functions in these types of problems.

EXAMPLE 1 Sketch the curve $y = \sin^2 x - \dfrac{x}{2} \, (0 \le x \le 2\pi)$.

First, by setting $x = 0$, we see that the only easily obtainable intercept is $(0, 0)$. Replacing x by $-x$ and y by $-y$, we find that the curve is not symmetric to either axis or to the origin. Also, since x does not appear in a denominator, there are no vertical asymptotes. We are considering only the restricted domain $0 \le x \le 2\pi$. (Without this restriction, the domain is all x and the range is all y.)

In order to find the information from the derivatives, we write

$$\frac{dy}{dx} = 2 \sin x \cos x - \frac{1}{2} = \sin 2x - \frac{1}{2}$$

$$\frac{d^2y}{dx^2} = 2 \cos 2x$$

Relative maximum and minimum points will occur for $\sin 2x = \frac{1}{2}$. Thus, we have possible relative maximum and minimum points for

$$2x = \frac{\pi}{6}, \frac{5\pi}{6}, \frac{13\pi}{6}, \frac{17\pi}{6}, \quad \text{or} \quad x = \frac{\pi}{12}, \frac{5\pi}{12}, \frac{13\pi}{12}, \frac{17\pi}{12}$$

Now, using the second derivative, we find that d^2y/dx^2 is positive for $x = \frac{\pi}{12}$ and $x = \frac{13\pi}{12}$ and is negative for $x = \frac{5\pi}{12}$ and $x = \frac{17\pi}{12}$. Thus, the maximum points are $\left(\frac{5\pi}{12}, 0.279\right)$ and $\left(\frac{17\pi}{12}, -1.29\right)$. Minimum points are $\left(\frac{\pi}{12}, -0.064\right)$ and $\left(\frac{13\pi}{12}, -1.63\right)$. Inflection points occur for $\cos 2x = 0$, or

$$2x = \frac{\pi}{2}, \frac{3\pi}{2}, \frac{5\pi}{2}, \frac{7\pi}{2}, \quad \text{or} \quad x = \frac{\pi}{4}, \frac{3\pi}{4}, \frac{5\pi}{4}, \frac{7\pi}{4}$$

Therefore, the points of inflection are $\left(\frac{\pi}{4}, 0.11\right)$, $\left(\frac{3\pi}{4}, -0.68\right)$, $\left(\frac{5\pi}{4}, -1.46\right)$, and $\left(\frac{7\pi}{4}, -2.25\right)$. Using this information, we sketch the curve in Fig. 27.15.

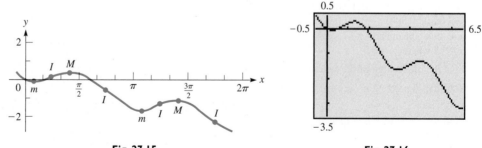

Fig. 27.15   Fig. 27.16

See Fig. 27.16 for a graphing calculator display of this graph.

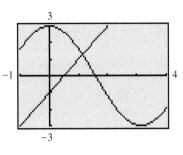

Fig. 27.17

EXAMPLE 2   By using Newton's method, solve the equation $2x - 1 = 3 \cos x$.

First we locate the required root approximately by sketching $y_1 = 2x - 1$ and $y_2 = 3 \cos x$. Using the graphing calculator view shown in Fig. 27.17, we see that they intersect between $x = 1$ and $x = 2$, near $x = 1.2$. Therefore, using $x_1 = 1.2$, with

$$f(x) = 2x - 1 - 3 \cos x$$
$$f'(x) = 2 + 3 \sin x$$

we use Eq. (24.1), which is

$$x_2 = x_1 - \frac{f(x_1)}{f'(x_1)}$$

To find x_2, we have

$$f(x_1) = 2(1.2) - 1 - 3 \cos 1.2 = 0.312\,926\,7$$
$$f'(x_1) = 2 + 3 \sin 1.2 = 4.796\,117\,3$$
$$x_2 = 1.2 - \frac{0.312\,926\,7}{4.796\,117\,3} = 1.134\,754\,2$$

Finding the next approximation, we find $x_3 = 1.134\,236\,6$, which is accurate to the value shown. Again, when using the calculator it is not necessary to list the values of $f(x_1)$ and $f'(x_1)$, as the complete calculation can be done directly on the calculator.

If we use the *intersect* feature of a graphing calculator, we find that we obtain the value of x_3 given above.

Solving a Word Problem

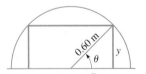

Fig. 27.18

EXAMPLE 3   Logs with a circular cross section 1.20 m in diameter are cut in half lengthwise. Find the largest rectangular cross-sectional area that can then be cut from one of the halves.

First, we draw Fig. 27.18 to help set up the necessary equation. From the figure we see that $x = 0.60 \cos \theta$ and $y = 0.60 \sin \theta$, which means that the area of the rectangle inscribed within the semicircular area is

$$A = (2x)y = 2(0.60 \cos \theta)(0.60 \sin \theta) = 0.72 \cos \theta \sin \theta$$
$$= 0.36 \sin 2\theta \qquad \text{using trigonometric identity (Eq. (20.21))}$$

Now, taking the derivative and setting it equal to zero, we have

$$\frac{dA}{d\theta} = (0.36 \cos 2\theta)(2) = 0.72 \cos 2\theta$$

$$0.72 \cos 2\theta = 0, \qquad 2\theta = \frac{\pi}{2}, \qquad \theta = \frac{\pi}{4} \qquad \cos \frac{\pi}{2} = 0$$

(Using $2\theta = \frac{3\pi}{2}$, $\theta = \frac{3\pi}{4}$ leads to the same solution.) Since the minimum area is zero, we have the maximum area when $\theta = \frac{\pi}{4}$, and this maximum area is

$$A = 0.36 \sin 2\left(\frac{\pi}{4}\right) = 0.36 \sin \frac{\pi}{2} = 0.36 \text{ m}^2 \qquad \sin \frac{\pi}{2} = 1$$

Therefore, the largest rectangular cross-sectional area is 0.36 m^2.

Solving a Word Problem

See the chapter introduction.

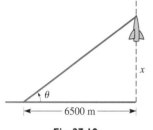

Fig. 27.19

◀ EXAMPLE 4　A rocket is taking off vertically at a distance of 6500 m from an observer. If, when the angle of elevation is 38.4°, it is changing at the rate of 5.0°/s, how fast is the rocket ascending?

From Fig. 27.19, we see that

$$\tan \theta = \frac{x}{6500}, \quad \text{or} \quad \theta = \tan^{-1} \frac{x}{6500}$$

Taking derivatives with respect to time, we have

$$\frac{d\theta}{dt} = \frac{1}{1 + (x/6500)^2} \frac{dx/dt}{6500}$$

$$= \frac{6500 \, dx/dt}{6500^2 + x^2}$$

When evaluating this expression, we must remember to express angles in radians. Therefore, this means that $d\theta/dt = 5.0°/s = 0.0873$ rad/s. Substituting this value and $x = 6500 \tan 38.4° = 5150$ m, we have

$$0.0873 = \frac{6500 \, dx/dt}{6500^2 + 5150^2}$$

$$\frac{dx}{dt} = 924 \text{ m/s}$$

◀ EXAMPLE 5　A particle is moving so that its x- and y-coordinates are given by $x = \cos 2t$ and $y = \sin 2t$. Find the magnitude and direction of its velocity when $t = \pi/8$.

Taking derivatives with respect to time, we have

$$v_x = \frac{dx}{dt} = -2 \sin 2t \qquad v_y = \frac{dy}{dt} = 2 \cos 2t$$

$$v_x|_{t=\pi/8} = -2 \sin 2\left(\frac{\pi}{8}\right) = -2\left(\frac{\sqrt{2}}{2}\right) = -\sqrt{2} \qquad \text{evaluating}$$

$$v_y|_{t=\pi/8} = 2 \cos 2\left(\frac{\pi}{8}\right) = 2\left(\frac{\sqrt{2}}{2}\right) = \sqrt{2}$$

$$v = \sqrt{v_x^2 + v_y^2} = \sqrt{2 + 2} = 2 \qquad \text{magnitude}$$

$$\tan \theta = \frac{v_y}{v_x} = \frac{\sqrt{2}}{-\sqrt{2}} = -1$$

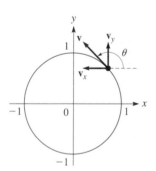

Fig. 27.20

Since v_x is negative and v_y is positive, $\theta = 135°$. Note that in this example θ is the angle, in standard position, between the horizontal and the resultant velocity.

By plotting the curve we note that it is a circle (see Fig. 27.20). Thus, the object is moving about a circle in a counterclockwise direction. It can be determined in another way that the curve is a circle. If we square each of the expressions defining x and y and then add these, we have the equation $x^2 + y^2 = \cos^2 2t + \sin^2 2t = 1$. This is a circle of radius 1.

Solving a Word Problem

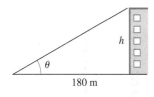

Fig. 27.21

◖ **EXAMPLE 6** At a point 180 m from the base of a building on level ground, the angle of elevation of the top of the building is 30.00°. What would be the error in calculating the height of the building due to an error 0.25° in the measurement of the angle?

From Fig. 27.21, we see that $h = 180 \tan \theta$. To find the error in h, we must find the differential dh. Therefore, we have

$$dh = 180 \sec^2 \theta \, d\theta$$

The possible error in θ is 0.25°, which in *radian measure* is $0.25\pi/180$, which is the value we should use in the calculation of dh. We can use the value $\theta = 30.00°$ if we have the calculator set in degree mode. Calculating dh, we have

$$dh = \frac{180(0.25\pi/180)}{\cos^2 30.00°} = 1.05 \text{ m}$$

In using the calculator, we divide by $\cos^2 30.00°$ since $\sec \theta = 1/\cos \theta$.

We see that an error of 0.25° in the angle can result in an error of over 1 m in the calculated value of the height. ◗

EXERCISES 27.4

In Exercises 1 and 2, make the given changes in the indicated examples of this section and then solve the resulting problems.

1. In Example 1, change $\sin^2 x$ to $\sin x$.

2. In Example 5, change $\cos 2t$ to $3 \cos 2t$, and $\sin 2t$ to $2 \sin 2t$.

In Exercises 3–36, solve the given problems.

3. Show that the slopes of the sine and cosine curves are negatives of each other at the points of intersection.

4. Show that the tangent curve is always increasing (when the tangent is defined.)

5. Show that the curve of $y = \tan^{-1} x$ is always increasing.

6. Sketch the graph of $y = \sin x + \cos x$ ($0 \le x \le 2\pi$). Check the graph on a graphing calculator.

7. Sketch the graph of $y = x - \tan x$ $\left(-\frac{\pi}{2} < x < \frac{\pi}{2}\right)$. Check the graph on a graphing calculator.

8. Sketch the graph of $y = 2 \sin x + \sin 2x$ ($0 \le x \le 2\pi$). Check the graph on a graphing calculator.

9. Find the equation of the line tangent to the curve of $y = \sin 2x$ at $x = \frac{5\pi}{8}$.

10. Find the equation of a line normal to the curve of $y = \tan^{-1}(x/2)$ at $x = 3$.

11. By Newton's method, find the positive root of the equation $x^2 - 4 \sin x = 0$. Verify the result by using the *intersect* (or *zero*) feature of a graphing calculator.

12. By Newton's method, find the smallest positive root of the equation $\tan x = 2x$. Verify the result by using the *intersect* (or *zero*) feature of a graphing calculator.

13. Find the minimum value of the function $y = 6 \cos x - 8 \sin x$.

14. Find the maximum value of the function
$y = \tan^{-1}(1 + x) + \tan^{-1}(1 - x)$.

15. In studying water waves, the vertical displacement y (in m) of a wave was determined to be $y = 0.50 \sin 2t + 0.30 \cos t$, where t is the time (in s). Find the velocity and the acceleration for $t = 0.40$ s.

16. At 30°N latitude, the number of hours h of daylight each day during the year is given approximately by the equation
$h = 12.1 + 2.0 \sin\left[\frac{\pi}{6}(x - 2.7)\right]$, where x is measured in months ($x = 0.5$ is Jan. 15, etc.). Find the date of the longest day and the date of the shortest day. (Cities near 30°N are Houston, Texas, and Cairo, Egypt.)

17. Find the time rate of change of the horizontal component T_x of the constant 46.6-N tension shown in Fig. 27.22 if $d\theta/dt = 0.36°/s$ for $\theta = 14.2°$.

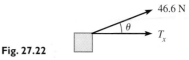

Fig. 27.22

18. The *apparent power* P_a (in W) in an electric circuit whose power is P and whose impedance phase angle is θ is given by $P_a = P \sec \theta$. Given that P is constant at 12 W, find the time rate of change of P_a if θ is changing at the rate of 0.050 rad/min, when $\theta = 40.0°$.

19. A point on the rim of a 133.3-mm computer floppy disk can be described by the equations $x = 66.7 \cos 12\pi t$ and $y = 66.7 \sin 12\pi t$. Find the velocity of the point for $t = 1.250$ s.

20. A machine is programmed to move an etching tool such that the position (in cm) of the tool is given by $x = 2 \cos 3t$ and $y = \cos 2t$, where t is the time (in s). Find the velocity of the tool for $t = 4.1$ s.

21. Find the acceleration of the tool of Exercise 20 for $t = 4.1$ s.

22. The volume V (in m³) of water used each day by a community during the summer is found to be $V = 2500 + 480 \sin(\pi t/90)$, where t is the number of the summer day, and $t = 0$ is the first day of summer. On what summer day is the water usage the greatest?

23. A person observes an object dropped from the top of a building 40.0 m away. If the top of the building is 60.0 m above the person's eye level, how fast is the angle of elevation of the object changing after 1.0 s? (The distance the object drops is given by $s = 4.9t^2$.) See Fig. 27.23.

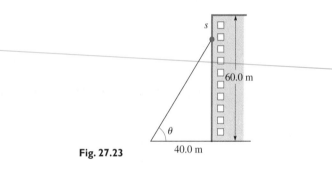

Fig. 27.23

24. A car passes directly under a police helicopter 150 m above a straight and level highway. After the car has traveled another 20.0 m, the angle of depression of the car from the helicopter is decreasing at the rate of 0.215 rad/s. What is the speed of the car?

25. A searchlight is 225 m from a straight wall. As the beam moves along the wall, the angle between the beam and the perpendicular to the wall is increasing at the rate of 1.5°/s. How fast is the length of the beam increasing when it is 315 m long? See Fig. 27.24.

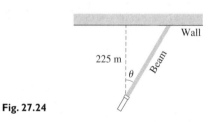

Fig. 27.24

26. In a modern hotel, where the elevators are directly observable from the lobby area (and a person can see from the elevators), a person in the lobby observes one of the elevators rising at the rate of 4.00 m/s. If the person was 16.0 m from the elevator when it left the lobby, how fast is the angle of elevation of the line of sight to the elevator increasing 10.0 s later?

27. If a block is placed on a plane inclined with the horizontal at an angle θ such that the block just moves down the plane, the coefficient of friction μ is given by $\mu = \tan \theta$. Use differentials to find the change in μ if θ changes from 20° to 21°.

28. The electric power p (in W) developed in a resistor in an FM receiver circuit is $p = 0.0307 \cos^2 120\pi t$, where t is the time (in s). Linearize p for $t = 0.0010$ s.

29. When an astronaut views the horizon of earth from a spacecraft at an altitude of 610 km, the angle θ in Fig. 27.25 is found to be $65.8° \pm 0.5°$. Use differentials to approximate the possible error in the astronaut's calculation of earth's radius.

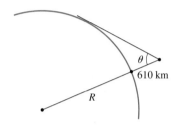

Fig. 27.25

30. A surveyor measures two sides and the included angle of a triangular parcel of land to be 82.04 m, 75.37 m, and 38.38°. What error is caused in the calculation of the third side by an error of 0.15° in the angle?

31. The volume V (in L) of air in a person's lungs during one normal cycle of inhaling and exhaling at any time t is $V = 0.48(1.2 - \cos 1.26t)$. What is the maximum flow rate (in L/s) of air?

32. To connect the four vertices of a square with the minimum amount of electric wire requires using the wiring pattern shown in Fig. 27.26. Find θ for the total length of wire ($L = 4x + y$) to be a minimum.

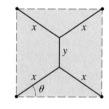

Fig. 27.26

33. The strength S of a rectangular beam is directly proportional to the product of its width w and the square of its depth d. Use trigonometric functions to find the dimensions of the strongest beam that can be cut from a circular log 16.0 cm in diameter. (See Example 4 on page 720.)

34. An architect is designing a window in the shape of an isosceles triangle with a perimeter of 150 cm. What is the vertex angle of the window of greatest area?

35. A wall is 1.8 m high and 1.2 m from a building. What is the length of the shortest pole that can touch the building and the ground beyond the wall? (*Hint:* From Fig. 27.27, it can be shown that $y = 1.8 \csc \theta + 1.2 \sec \theta$.)

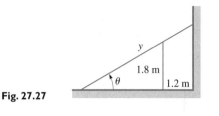

Fig. 27.27

36. The television screen at a sports arena is vertical and 2.4 m high. The lower edge is 8.5 m above an observer's eye level. If the best view of the screen is obtained when the angle subtended by the screen at eye level is a maximum, how far from directly below the screen must the observer's eye be? See Fig. 27.28.

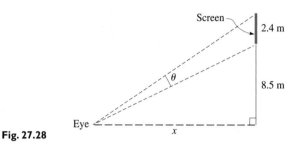

Fig. 27.28

27.5 DERIVATIVE OF THE LOGARITHMIC FUNCTION

The next function for which we will find the derivative is the logarithmic function by using the definition of a derivative.

If we let $y = \log_b x$, we have

$$\frac{dy}{dx} = \lim_{h \to 0} \frac{\log_b(x + h) - \log_b x}{h} = \lim_{h \to 0} \frac{\log_b \frac{x + h}{x}}{h}$$

$$= \lim_{h \to 0} \frac{1}{x} \frac{x}{h} \log_b \left(1 + \frac{h}{x}\right)$$

$$= \frac{1}{x} \lim_{h \to 0} \log_b \left(1 + \frac{h}{x}\right)^{x/h}$$

(We multiplied and divided by x for purposes of evaluating the limit, as we now will show.)

In order to find dy/dx, we must determine

$$\lim_{h \to 0} \left(1 + \frac{h}{x}\right)^{x/h}$$

We can see that the exponent becomes unbounded, but the number being raised to this exponent approaches 1. Therefore, we will investigate this limiting value.

To find an approximate value, let us graph the function $y = (1 + t)^{1/t}$ (for purposes of graphing, we let $h/x = t$). We construct a table of values and then graph the function (see Fig. 27.29).

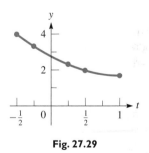

Fig. 27.29

t	-0.5	-0.25	$+0.25$	$+0.50$	$+1.00$
y	4.00	3.16	2.44	2.25	2.00

Only these values are shown, since we are interested in the y-value corresponding to $t = 0$. We see from the graph that this value is approximately 2.7. Choosing very small values of t, we may obtain these values:

t	0.1	0.01	0.001	0.000 1
y	2.5937	2.7048	2.7169	2.718 15

NOTE ▸ By methods developed in Chapter 29, it can be shown that this value is about 2.718 281 8. ***The limiting value is the irrational number e.*** This is the same number used in the exponential form of a complex number in Chapter 12 and as the base of natural logarithms in Chapter 13.

Returning to the derivative of the logarithmic function, we have

$$\frac{dy}{dx} = \lim_{h \to 0} \left[\frac{1}{x} \log_b \left(1 + \frac{h}{x} \right)^{x/h} \right] = \frac{1}{x} \log_b e$$

Therefore,

$$\frac{dy}{dx} = \frac{1}{x} \log_b e$$

For reference, Eq. (27.3) is
$$\frac{dy}{dx} = \frac{dy}{du}\frac{du}{dx}.$$

Now for $y = \log_b u$, where u is a function of x, using Eq. (27.3), we have

$$\frac{d(\log_b u)}{dx} = \frac{1}{u} \log_b e \frac{du}{dx} \tag{27.13}$$

At this point, we see that if we choose e as the base of a system of logarithms, the above formula becomes

$$\frac{d(\ln u)}{dx} = \frac{1}{u}\frac{du}{dx} \tag{27.14}$$

The choice of e as the base b makes $\log_e e = 1$; thus, this factor does not appear in Eq. (27.14). We now see why the number e is chosen as the base for a system of logarithms, the natural logarithms. The notation $\ln u$ is the same as that used in Chapter 13 for natural logarithms.

◖ **EXAMPLE 1** Find the derivative of $y = \log 4x$.
Using Eq. (27.13), we find

$$\frac{dy}{dx} = \frac{1}{4x} (\log e)(4) \qquad \overset{\frac{du}{dx}}{}$$

$$= \frac{1}{x} \log e \qquad \qquad \log e = 0.4343 \qquad \qquad ▹$$

◖ **EXAMPLE 2** Find the derivative of $s = \ln 3t^4$.
Using Eq. (27.14), we have (with $u = 3t^4$)

$$\frac{ds}{dt} = \frac{1}{3t^4}(12t^3) \qquad \overset{}{\underset{\frac{du}{dt}}{}}$$

$$= \frac{4}{t} \qquad \qquad \qquad \qquad ▹$$

◖ EXAMPLE 3 Find the derivative of $y = \ln \tan 4x$.

Using Eq. (27.14), along with the derivative of the tangent, we have

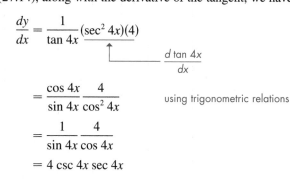

$$\frac{dy}{dx} = \frac{1}{\tan 4x}(\sec^2 4x)(4)$$

$$\frac{d \tan 4x}{dx}$$

$$= \frac{\cos 4x}{\sin 4x}\frac{4}{\cos^2 4x} \qquad \text{using trigonometric relations}$$

$$= \frac{1}{\sin 4x}\frac{4}{\cos 4x}$$

$$= 4 \csc 4x \sec 4x$$

Frequently, we can find the derivative of a logarithmic function more simply if we **NOTE ▸** *use the properties of logarithms to simplify the expression before the derivative is found.* The following examples illustrate this.

For reference, Eqs. (13.7), (13.8), and (13.9) are:

$\log_b xy = \log_b x + \log_b y$

$\log_b \left(\dfrac{x}{y}\right) = \log_b x - \log_b y$

$\log_b x^n = n \log_b x$

◖ EXAMPLE 4 Find the derivative of $y = \ln \dfrac{x-1}{x+1}$.

In this example, it is easier to find the derivative if we write y in the form

$$y = \ln(x - 1) - \ln(x + 1)$$

by using the properties of logarithms (Eq. (13.8)). Hence,

$$\frac{dy}{dx} = \frac{1}{x-1} - \frac{1}{x+1} = \frac{x+1-x+1}{(x-1)(x+1)}$$

$$= \frac{2}{x^2 - 1}$$

◖ EXAMPLE 5 **(a)** Find the derivative of $y = \ln(1 - 2x)^3$.

First, using Eq. (13.9), we rewrite the equation as $y = 3\ln(1 - 2x)$. Then we have

$$\frac{dy}{dx} = 3\left(\frac{1}{1-2x}\right)(-2) = \frac{-6}{1-2x}$$

(b) Find the derivative of $y = \ln^3(1 - 2x)$.

First, we note that

$$y = \ln^3(1 - 2x) = [\ln(1 - 2x)]^3$$

where $\ln^3(1 - 2x)$ is usually the preferred notation.

NOTE ▸   Next, we must be careful to distinguish this function from that in part (a). For $y = \ln^3(1 - 2x)$, it is the logarithm of $1 - 2x$ that is being cubed, whereas for $y = \ln(1 - 2x)^3$, it is $1 - 2x$ that is being cubed.

Now, finding the derivative of $y = \ln^3(1 - 2x)$, we have

$$\frac{dy}{dx} = 3[\ln^2(1 - 2x)]\left(\frac{1}{1-2x}\right)(-2)$$

$$= -\frac{6\ln^2(1 - 2x)}{1-2x} \qquad \frac{d \ln(1-2x)}{dx}$$

◀ EXAMPLE 6   Evaluate the derivative of $y = \ln\left[(\sin 2x)\left(\sqrt{x^2 + 1}\right)\right]$ for $x = 0.375$.
First, using the properties of logarithms, we rewrite the function as

$$y = \ln \sin 2x + \frac{1}{2} \ln(x^2 + 1)$$

Now we have

$$\frac{dy}{dx} = \frac{1}{\sin 2x}(\cos 2x)(2) + \frac{1}{2}\left(\frac{1}{x^2 + 1}\right)(2x) \qquad \text{take the derivative}$$

$$= 2 \cot 2x + \frac{x}{x^2 + 1}$$

$$\left.\frac{dy}{dx}\right|_{x=0.375} = 2 \cot 0.750 + \frac{0.375}{0.375^2 + 1} = 2.48 \qquad \text{evaluate}$$

▶

EXERCISES 27.5

In Exercises 1 and 2, make the given changes in the indicated examples of this section and then find the derivatives.

1. In Example 3, in the given function change tan to cos.

2. In Example 4, in the given function change $x - 1$ to x^2.

In Exercises 3–34, find the derivatives of the given functions.

3. $y = \log x^2$

4. $y = \log_2 6x$

5. $y = 2 \log_5(3x + 1)$

6. $y = 3 \log_7(x^2 + 1)$

7. $u = 0.2 \ln(1 - 3x)$

8. $y = 2 \ln(3x^2 - 1)$

9. $y = 2 \ln \tan 2x$

10. $s = \ln \sin^2 t$

11. $R = \ln \sqrt{T}$

12. $y = 5 \ln \sqrt{4x - 3}$

13. $y = \ln(x^2 + 2x)^3$

14. $s = [\ln(2t^3 - t)]^2$

15. $v = 3[t + \ln t^2]^2$

16. $y = x^2 \ln 2x$

17. $y = \dfrac{3x}{\ln(2x + 1)}$

18. $y = \dfrac{8 \ln x}{x}$

19. $y = \ln(\ln x)$

20. $r = 0.5 \ln \cos(\pi\theta^2)$

21. $y = \ln \dfrac{2x}{1 + x}$

22. $y = \ln\left(x\sqrt{x + 1}\right)$

23. $y = \sin \ln x$

24. $y = \tan^{-1} \ln 2x$

25. $u = 3v \ln^2 2v$

26. $h = 0.1s \ln^4 s$

27. $y = \ln(x \tan x)$

28. $y = \ln\left(x + \sqrt{x^2 - 1}\right)$

29. $r = \ln \dfrac{v^2}{v + 2}$

30. $y = \sqrt{x + \ln 3x}$

31. $y = \sqrt{x^2 + 1} - \ln \dfrac{1 + \sqrt{x^2 + 1}}{x}$

32. $3 \ln xy + \sin y = x^2$

33. $y = x - \ln^2(x + y)$

34. $y = \ln(x + \ln x)$

In Exercises 35–56, solve the given problems.

Ⓦ **35.** On a calculator find the value of $(\ln 2.0001 - \ln 2.0000)/0.0001$ and compare it with 0.5. Give the meanings of the value found and 0.5 in relation to the derivative of $\ln x$, where $x = 2$.

Ⓦ **36.** On a calculator find the value of $(\ln 0.5001 - \ln 0.5000)/0.0001$ and compare it with 2. Give the meanings of the value found and 2 in relation to the derivative of $\ln x$, where $x = 0.5$.

37. Using a graphing calculator, (a) display the graph of $y = (1 + x)^{1/x}$ to verify that $(1 + x)^{1/x} \to 2.718$ as $x \to 0$ and (b) verify the values for $(1 + x)^{1/x}$ in the tables on page 815.

38. (a) Display the graph of $y = \ln x$ on a graphing calculator, and using the derivative feature, evaluate dy/dx for $x = 2$. (b) Display the graph of $y = 1/x$, and evaluate y for $x = 2$. (c) Compare the values in parts (a) and (b).

39. Given that $\ln \sin 45° = -0.3466$, use differentials to approximate $\ln \sin 44°$.

40. Find the second derivative of the function $y = x^2 \ln x$.

41. Evaluate the derivative of $y = \sin^{-1} 2x + \sqrt{1 - 4x^2}$, where $x = 0.250$.

42. Evaluate the derivative of $y = \ln \sqrt{\dfrac{2x + 1}{3x + 1}}$, where $x = 2.75$.

43. Find the linearization $L(x)$ for the function $f(x) = 2 \ln \tan x$ for $a = \pi/4$.

44. Find the differential of the function $y = 6 \log_x 2$.

45. Find the slope of a line tangent to the curve of $y = \tan^{-1} 2x + \ln(4x^2 + 1)$,g where $x = 0.625$. Verify the result by using the *numerical derivative* feature of a graphing calculator.

46. Find the slope of a line tangent to the curve of $y = x \ln 2x$ at $x = 2$. Verify the result by using the *numerical derivative* feature of a graphing calculator.

47. Find the derivative of $y = x^x$ by first taking logarithms of each side of the equation. Explain why Eq. (23.15) cannot be used to find the derivative of this function.

48. Find the derivative of $y = (\sin x)^x$ by first taking logarithms of each side of the equation. Explain why Eq. (23.15) cannot be used to find the derivative of this function.

49. Find the derivatives of $y_1 = \ln(x^2)$ and $y_2 = 2 \ln x$, and evaluate these derivatives for $x = -1$. Explain your results.

50. The inductance L (in μH) of a coaxial cable is given by $L = 0.032 + 0.15 \log (a/x)$, where a and x are the radii of the outer and inner conductors, respectively. For constant a, find dL/dx.

51. If the loudness b (in dB) of a sound of intensity I is given by $b = 10 \log(I/I_0)$, where I_0 is a constant, find the expression for db/dt in terms of dI/dt.

52. The time t for a particular computer system to process N bits of data is directly proportional to $N \ln N$. Find the expression for dt/dN.

53. When a tractor-trailer turns a right-angle corner, the rear wheels follow a curve known as a *tractrix*, the equation for which is
$$y = \ln\left(\frac{1 + \sqrt{1 + x^2}}{x}\right) - \sqrt{1 - x^2}. \text{ Find } dy/dx.$$

54. When designing a computer to sort files on a hard disk, the equation $y = xA \log_x A$ arises. If A is constant, find dy/dx.

55. When air friction is considered, the time t (in s) it takes a certain falling object to attain a velocity v (in m/s) is given by
$$t = 5 \ln \frac{5}{5 - 0.1v}. \text{ Find } dt/dv \text{ for } v = 10.0 \text{ m/s}.$$

56. The electric potential V at a point P at a distance x from an electric charge distributed along a wire of length $2a$ (see Fig. 27.30) is
$$V = k \ln \frac{\sqrt{a^2 + x^2} + a}{\sqrt{a^2 + x^2} - a}, \text{ where } k \text{ is a constant. Find the expression for the electric field } E, \text{ which is defined as } E = -dV/dx.$$

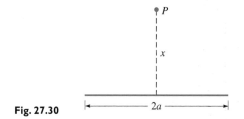

Fig. 27.30

27.6 DERIVATIVE OF THE EXPONENTIAL FUNCTION

To obtain the derivative of the exponential function, we let $y = b^u$ and then take natural logarithms of both sides:

$$\ln y = \ln b^u = u \ln b$$

$$\frac{1}{y}\frac{dy}{dx} = \ln b \frac{du}{dx}$$

$$\frac{dy}{dx} = y \ln b \frac{du}{dx}$$

Thus,

$$\frac{d(b^u)}{dx} = b^u \ln b \left(\frac{du}{dx}\right) \tag{27.15}$$

If we let $b = e$, Eq. (27.15) becomes

$$\frac{d(e^u)}{dx} = e^u \left(\frac{du}{dx}\right) \tag{27.16}$$

The simplicity of Eq. (27.16) compared with Eq. (27.15) again shows the advantage of choosing e as the base of natural logarithms. It is for this reason that e appears so often in applications of calculus.

◀ **EXAMPLE 1**   Find the derivative of $y = e^x$.

Using Eq. (27.16), we have

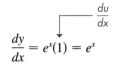

$$\frac{dy}{dx} = e^x(1) = e^x$$

We see that the derivative of the function e^x equals itself. This exponential function is widely used in applications of calculus. ▶

For reference, Eq. (23.15) is
$$\frac{du^n}{dx} = nu^{n-1}\left(\frac{du}{dx}\right).$$

We should note carefully that Eq. (23.15) is used with a variable raised to a constant exponent, whereas with Eqs. (27.15) and (27.16) we are finding the derivative of a constant raised to a variable exponent.

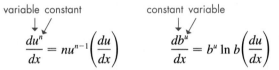

In the following example, we must note carefully this difference in the type of function that leads us to use either Eq. (23.15) or Eq. (27.15) in order to find the derivative.

◀ **EXAMPLE 2**   Find the derivatives of $y = (4x)^2$ and $y = 2^{4x}$.

Using Eq. (23.15), we have   Using Eq. (27.15), we have

$$y = (4x)^2$$ $$y = 2^{4x}$$

$$\frac{dy}{dx} = 2(4x)^1(4)$$ $$\frac{dy}{dx} = 2^{4x}(\ln 2)(4)$$

$$= 32x$$ $$= (4 \ln 2)(2^{4x})$$ ▶

We continue now with additional examples of the use of Eq. (27.16).

◀ **EXAMPLE 3**   Find the derivative of $y = \ln \cos e^{2x}$.

Using Eq. (27.16), along with the derivatives of the logarithmic and cosine functions, we have

$$\frac{dy}{dx} = \frac{1}{\cos e^{2x}}\frac{d \cos e^{2x}}{dx} \qquad \text{using } \frac{d \ln u}{dx} = \frac{1}{u}\frac{du}{dx}$$

$$= \frac{1}{\cos e^{2x}}(-\sin e^{2x})\frac{de^{2x}}{dx} \qquad \text{using } \frac{d \cos u}{dx} = -\sin u \frac{du}{dx}$$

$$= -\frac{\sin e^{2x}}{\cos e^{2x}}(e^{2x})(2) \qquad \text{using } \frac{de^u}{dx} = e^u \frac{du}{dx}$$

$$= -2e^{2x}\tan e^{2x} \qquad \text{using } \frac{\sin \theta}{\cos \theta} = \tan \theta$$ ▶

◀ **EXAMPLE 4**   Find the derivative of $r = \theta e^{\tan \theta}$.

Here we use Eq. (27.16) with the derivatives of a product and the tangent:

$$\frac{dr}{d\theta} = \theta e^{\tan \theta}(\sec^2 \theta) + e^{\tan \theta}(1)$$

$$= e^{\tan \theta}(\theta \sec^2 \theta + 1)$$ ▶

◀ **EXAMPLE 5** Find the derivative of $y = (e^{1/x})^2$.

In this example, we use Eqs. (23.15) and (27.16):

$$\frac{dy}{dx} = 2(e^{1/x})(e^{1/x})\left(-\frac{1}{x^2}\right)$$

using Eqs. (27.16) and (23.15) to find $\frac{du}{dx}$ of Eq. (23.15)

$$= \frac{-2(e^{1/x})^2}{x^2} = \frac{-2e^{2/x}}{x^2}$$

This problem could have also been solved by first writing the function as $y = e^{2/x}$, which is an equivalent form determined by the laws of exponents. When we use this form, the derivative becomes

using Eq. (23.15) to find $\frac{du}{dx}$ of Eq. (27.16)

$$\frac{dy}{dx} = e^{2/x}\left(-\frac{2}{x^2}\right) = \frac{-2e^{2/x}}{x^2}$$

This change in form of the function simplifies the steps necessary for finding the derivative. ▶

◀ **EXAMPLE 6** Find the derivative of $y = (3e^{4x} + 4x^2 \ln x)^3$.

Using the general power rule (Eq. (23.15)) for the derivatives, the derivative of the exponential function (Eq. (27.16)), the derivative of a product (Eq. (23.12)), and the derivative of a logarithm (Eq. (27.14)), we have

$$\frac{dy}{dx} = 3(3e^{4x} + 4x^2 \ln x)^2\left[12e^{4x} + 4x^2\left(\frac{1}{x}\right) + (\ln x)(8x)\right]$$

$$= 3(3e^{4x} + 4x^2 \ln x)^2(12e^{4x} + 4x + 8x \ln x)$$ ▶

◀ **EXAMPLE 7** Find the slope of a line tangent to the curve of $y = \dfrac{3e^{2x}}{x^2 + 1}$, where $x = 1.275$.

Here we are to find the derivative and then evaluate it for $x = 1.275$. The solution is as follows:

$$\frac{dy}{dx} = \frac{(x^2 + 1)(3e^{2x})(2) - 3e^{2x}(2x)}{(x^2 + 1)^2}$$ take the derivative

$$= \frac{6e^{2x}(x^2 - x + 1)}{(x^2 + 1)^2}$$

$$\frac{dy}{dx}\bigg|_{x=1.275} = \frac{6e^{2(1.275)}(1.275^2 - 1.275 + 1)}{(1.275^2 + 1)^2}$$ evaluate

$$= 15.05$$

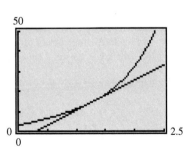

Fig. 27.31

Using the *tangent* feature of a graphing calculator, the function and the tangent line are shown in Fig. 27.31. The value of the derivative can be checked by using the *numerical derivative* feature. ▶

EXERCISES 27.6

In Exercises 1 and 2, make the given changes in the indicated examples of this section and then solve the resulting problems.

1. In Example 3, in the given function change cos to sin.

2. In Example 7, in the given function change $x^2 + 1$ to $x + 1$.

In Exercises 3–32, find the derivatives of the given functions.

3. $y = 4^{6x}$

4. $y = 10^{x^2}$

5. $y = e^{\sqrt{x}}$

6. $r = 0.3e^{\theta^2}$

7. $y = 4e^t(e^{2t} - e^t)$

8. $y = 0.2 \ln(e^{5x} + 1)$

9. $R = Te^{-T}$

10. $y = 5x^2e^{2x}$

11. $y = xe^{\sin x}$

12. $y = 4e^x \sin \frac{1}{2}x$

13. $r = \dfrac{2(e^{2s} - e^{-2s})}{e^{2s}}$

14. $u = \dfrac{e^{0.5v}}{2v}$

15. $y = e^{-3x} \sin 4x$

16. $y = (\cos 2x)(e^{x^2-1})$

17. $y = \dfrac{2e^{3x}}{4x + 3}$

18. $y = \dfrac{7 \ln 2x}{e^{2x} + 2}$

19. $y = \ln(e^{x^2} + 4)$

20. $p = (3e^{2n} + e^2)^3$

21. $y = (2e^{2x})^3 \sin x^2$

22. $y = (e^{3/x} \cos x)^2$

23. $u = 4\sqrt{\ln 2t + e^{2t}}$

24. $y = (2e^{x^2} + x^2)^3$

25. $y = xe^{xy} + \sin y$

26. $y = 4e^{-2/x} \ln y + 1$

27. $y = 3e^{2x} \ln x$

28. $r = 0.4e^{2\theta} \ln \cos \theta$

29. $y = \ln \sin 2e^{6x}$

30. $y = 6 \tan e^{x+1}$

31. $y = 2 \sin^{-1} e^{2x}$

32. $W = \tan^{-1} e^{3s}$

In Exercises 33–52, solve the given problems.

(W) 33. On a calculator, find the values of (a) e and (b) $(e^{1.0001} - e^{1.0000})/0.0001$. Compare the values and give the meaning of each in relation to the derivative of e^x, where $x = 1$.

(W) 34. On a calculator, find the values of (a) e^2 and (b) $(e^{2.0001} - e^{2.0000})/0.0001$. Compare the values and give the meaning of each in relation to the derivative of e^x, where $x = 2$.

35. Display the graph of $y = e^x$ on a graphing calculator. Using the *derivative* feature, evaluate dy/dx for $x = 2$ and compare with the value of y for $x = 2$.

36. Find a formula for the nth derivative of $y = ae^{bx}$.

37. Find the slope of a line tangent to the curve of $y = e^{-2x} \cos 2x$ for $x = 0.625$. Verify the result by using the *numerical derivative* feature of a graphing calculator.

38. Find the slope of a line tangent to the curve of $y = \dfrac{e^{-x}}{1 + \ln 4x}$ for $x = 1.842$. Verify the result by using the *numerical derivative* feature of a graphing calculator.

39. Find the differential of the function $y = \dfrac{2e^{4x}}{x + 2}$.

40. Find the linearization of the function $f(x) = \dfrac{6e^{4x}}{2x + 3}$ for $a = 0$.

41. Use a graphing calculator to display the graph of $y = e^x$. By roughly estimating slopes of tangent lines, note that it is reasonable that these values are equal to the y-coordinates of the points at which these estimates are made. (*Remember:* For $y = e^x$, $dy/dx = e^x$ also.)

42. Use a graphing calculator to display the graphs of $y = e^{-x}$ and $y = -e^{-x}$. By roughly estimating slopes of tangent lines of $y = e^{-x}$, note that $y = -e^{-x}$ gives reasonable values for the derivative of $y = e^{-x}$.

43. Show that $y = xe^{-x}$ satisfies the equation $(dy/dx) + y = e^{-x}$.

44. Show that $y = e^{-x} \sin x$ satisfies the equation
$$\frac{d^2y}{dx^2} + 2\frac{dy}{dx} + 2y = 0.$$

45. For $y = \dfrac{e^{2x} - 1}{e^{2x} + 1}$, show that $\dfrac{dy}{dx} = 1 - y^2$.

46. If $e^x + e^y = e^{x+y}$, show that $dy/dx = -e^{y-x}$.

47. For what values of m does the function $y = ae^{mx}$ satisfy the equation $y'' + y' - 6y = 0$?

48. The average energy consumption C (in MJ/year) of a certain model of refrigerator-freezer is approximately $C = 5350e^{-0.0748t} + 1800$, where t is measured in years, with $t = 0$ corresponding to 1990, and a newer model is produced each year. Assuming the function is continuous, use differentials to estimate the reduction of the 2006 model from that of the 2005 model.

49. The reliability R $(0 \le R \le 1)$ of a certain computer system is given by $R = e^{-0.002t}$, where t is the time of operation (in h). Find dR/dt for $t = 100$ h.

50. A thermometer is taken from a freezer at $-16°C$ and placed in a room at $24°C$. The temperature T of the thermometer as a function of the time t (in min) after removal is given by $T = 8.0(3.0 - 5.0e^{-0.50t})$. How fast is the temperature changing when $t = 6.0$ min?

51. For the electric circuit shown in Fig. 27.32, the current i (in A) is given by $i = 4.42e^{-66.7t} \sin 226t$, where t is the time (in s). Find the expression for di/dt.

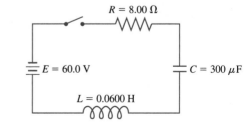

Fig. 27.32

52. Under certain assumptions of limitations to population growth, the population P (in billions) of the world is given by the *logistic equation* $P = \dfrac{10}{1 + 0.65e^{-0.060t}}$, where t is the number of years after the year 2000. Find the expression for dP/dt.

In Exercises 53–56, use the following information.

The **hyperbolic sine** *of u is defined as*

$$\sinh u = \frac{1}{2}(e^u - e^{-u})$$

Figure 27.33 shows the graph of $y = \sinh x$.

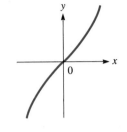

Fig. 27.33

The **hyperbolic cosine** *of u is defined as*

$$\cosh u = \frac{1}{2}(e^u + e^{-u})$$

Figure 27.34 shows the graph of $y = \cosh x$.

Fig. 27.34

These functions are called hyperbolic *functions since, if* $x = \cosh u$ *and* $y = \sinh u$, x *and* y *satisfy the equation of the hyperbola* $x^2 - y^2 = 1$.

53. Verify the fact that the exponential expressions for the hyperbolic sine and hyperbolic cosine given above satisfy the equation of the hyperbola.

54. Show that $\sinh u$ and $\cosh u$ satisfy the identity $\cosh^2 u - \sinh^2 u = 1$.

55. Show that

$$\frac{d}{dx} \sinh u = \cosh u \frac{du}{dx} \quad \text{and}$$

$$\frac{d}{dx} \cosh u = \sinh u \frac{du}{dx}$$

where u is a function of x.

56. Show that

$$\frac{d^2 \sinh x}{dx^2} = \sinh x \quad \text{and} \quad \frac{d^2 \cosh x}{dx^2} = \cosh x$$

27.7 APPLICATIONS

The following examples show applications of the logarithmic and exponential functions to curve tracing, Newton's method, and time-rate-of-change problems. Certain other applications are indicated in the exercises.

EXAMPLE 1 Sketch the graph of the function $y = x \ln x$.

First, we note that x cannot be zero since $\ln x$ is not defined at $x = 0$. Since $\ln 1 = 0$, we have an intercept at $(1, 0)$. There is no symmetry to the axes or origin, and there are no vertical asymptotes. Also, because $\ln x$ is defined only for $x > 0$, the domain is $x > 0$.

Finding the first two derivatives, we have

$$\frac{dy}{dx} = x\left(\frac{1}{x}\right) + \ln x = 1 + \ln x \qquad \frac{d^2y}{dx^2} = \frac{1}{x}$$

The first derivative is zero if $\ln x = -1$, or $x = e^{-1}$. The second derivative is positive for this value of x. Thus, there is a minimum point at $(1/e, -1/e)$. Since the domain is $x > 0$, the second derivative indicates that the curve is always concave up. In turn, we now see that the range of the function is $y \geq -1/e$. The graph is shown in Fig. 27.35.

Although the curve approaches the origin as x approaches zero, the origin is not included on the graph of the function.

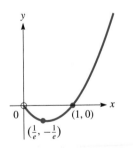

Fig. 27.35

EXAMPLE 2 Sketch the graph of the function $y = e^{-x} \cos x$ $(0 \le x \le 2\pi)$.

This curve has intercepts for all values for which $\cos x$ is zero. Those values in the domain $0 \le x \le 2\pi$ for which $\cos x = 0$ are $x = \frac{\pi}{2}$ and $x = \frac{3\pi}{2}$. The factor e^{-x} is always positive, and $e^{-x} = 1$ for $x = 0$, which means $(0, 1)$ is also an intercept. There is no symmetry to the axes or the origin, and there are no vertical asymptotes.

Next, finding the first derivative, we have

$$\frac{dy}{dx} = -e^{-x} \sin x - e^{-x} \cos x = -e^{-x}(\sin x + \cos x)$$

Setting the derivative equal to zero, since e^{-x} is always positive, we have

$$\sin x + \cos x = 0, \qquad \tan x = -1, \qquad x = \frac{3\pi}{4}, \frac{7\pi}{4}$$

Now, finding the second derivative, we have

$$\frac{d^2y}{dx^2} = -e^{-x}(\cos x - \sin x) - e^{-x}(-1)(\sin x + \cos x) = 2e^{-x} \sin x$$

The sign of the second derivative depends only on $\sin x$. Therefore,

$$\frac{d^2y}{dx^2} > 0 \quad \text{for } x = \frac{3\pi}{4} \quad \text{and} \quad \frac{d^2y}{dx^2} < 0 \quad \text{for } x = \frac{7\pi}{4}$$

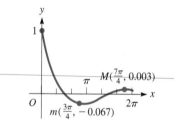

This means that $\left(\frac{3\pi}{4}, -0.067\right)$ is a minimum and $\left(\frac{7\pi}{4}, 0.003\right)$ is a maximum.

Also from the second derivative, points of inflection occur for $x = 0$, π, and 2π since $\sin x = 0$ for these values. The graph is shown in Fig. 27.36.

Fig. 27.36

EXAMPLE 3 Find the root of the equation $e^{2x} - 4 \cos x = 0$, which lies between 0 and 1, by using Newton's method.

Here,

$$f(x) = e^{2x} - 4 \cos x$$
$$f'(x) = 2e^{2x} + 4 \sin x$$

This means that $f(0) = -3$ and $f(1) = 5.2$. Therefore, we choose $x_1 = 0.5$. Using Eq. (24.1), which is

$$x_2 = x_1 - \frac{f(x_1)}{f'(x_1)}$$

we have these values:

$$f(x_1) = e^{2(0.5)} - 4 \cos 0.5 = -0.792\,048\,4$$
$$f'(x_1) = 2e^{2(0.5)} + 4 \sin 0.5 = 7.354\,265\,8$$
$$x_2 = 0.5 - \frac{-0.792\,048\,4}{7.354\,265\,8} = 0.607\,699\,2$$

Using the method again, we find $x_3 = 0.597\,975\,1$, which is correct to three decimal places.

Solving a Word Problem ❙ **EXAMPLE 4**   One model for population growth is that the population P at time t is given by $P = P_0 e^{kt}$, where P_0 is the *initial* population ($t = 0$, when timing starts for the population being considered) and k is a constant. Show that the instantaneous time rate of change of population is directly proportional to the population present at time t.

To find the time rate of change, we find the derivative dP/dt:

$$\frac{dP}{dt} = (P_0 e^{kt})(k) = kP_0 e^{kt}$$

$$= kP \qquad \text{since } P = P_0 e^{kt}$$

Thus, we see that population growth increases as population increases. ❙

See the chapter introduction.

❙ **EXAMPLE 5**   A rocket is moving such that the only force on it is due to gravity and its mass is decreasing at a constant rate r. If it moves vertically, its velocity v as a function of the time t is given by

$$v = v_0 - gt - k \ln\left(1 - \frac{rt}{m_0}\right)$$

where v_0 is the initial velocity, g is the acceleration due to gravity, t is the time, m_0 is the initial mass, and k is a constant. Determine the expression for the acceleration.

Since the acceleration is the time rate of change of the velocity, we must find dv/dt. Therefore,

$$\frac{dv}{dt} = -g - k\frac{1}{1 - \dfrac{rt}{m_0}}\left(\frac{-r}{m_0}\right) = -g + \frac{km_0}{m_0 - rt}\left(\frac{r}{m_0}\right)$$

$$= \frac{kr}{m_0 - rt} - g$$ ❙

EXERCISES 27.7

In Exercises 1 and 2, make the given changes in the indicated examples of this section and then solve the resulting problems.

1. In Example 2, in the given function change cos to sin.

2. In Example 3, in the given function change e^{2x} to $2e^x$.

In Exercises 3–14, sketch the graphs of the given functions. Check each by displaying the graph on a graphing calculator.

3. $y = \ln \cos x$

4. $y = \dfrac{2 \ln x}{x}$

5. $y = xe^{-x}$

6. $y = \dfrac{e^x}{x}$

7. $y = \ln \dfrac{1}{x^2 + 1}$

8. $y = \ln \dfrac{e}{x}$

9. $y = 4e^{-x^2}$

10. $y = x - e^x$

11. $y = \ln x - x$

12. $y = e^{-x} \sin x$

13. $y = \frac{1}{2}(e^x - e^{-x})$ (See Exercise 53 of Section 27.6.)

14. $y = \frac{1}{2}(e^x + e^{-x})$ (See Exercise 53 of Section 27.6.)

In Exercises 15–40, solve the given problems by finding the appropriate derivative.

(**W**) **15.** Find the values of x for which the graphs of $y_1 = e^{2/x^2}$ and $y_2 = e^{-2/x^2}$ are increasing and decreasing. Explain why they differ as they do.

(**W**) **16.** Find the values of x for which the graphs of $y_1 = \ln(x^2 + 1)$ and $y_2 = \ln(x^2 - 1)$ are concave up and concave down. Explain why they differ as they do.

17. Find the equation of the line tangent to the curve of $y = x^2 \ln x$ at the point $(1, 0)$.

18. Find the equation of the line tangent to the curve of $y = \tan^{-1} 2x$, where $x = 1$.

19. Find the equation of the line normal to the curve of $y = 2 \sin \frac{1}{2}x$, where $x = 3\pi/2$.

20. Find the equation of the line normal to the curve of $y = e^{2x}/x$, at $x = 1$.

21. By Newton's method, solve the equation $x^2 - 2 + \ln x = 0$. Check the result by using the *intersect* (or *zero*) feature of a graphing calculator.

22. By Newton's method, find the value of x for which $y = e^{\cos x}$ is minimum for $0 < x < 2\pi$. Check the result by displaying the graph on a graphing calculator and then finding the minimum point.

23. The electric current i (in A) through an inductor of 0.50 H as a function of time t (in s) is $i = e^{-5.0t} \sin 120\pi t$. Find the voltage across the inductor for $t = 1.0$ ms. (See Exercise 27 on page 763.)

24. A computer analysis showed that the population density D (in persons/km²) at a distance r (in km) from the center of a city is approximately $D = 200(1 + 5e^{-0.01r^2})$ if $r < 20$ km. At what distance from the city center does the decrease in population density (dD/dr) itself start to decrease?

25. The power supply P (in W) in a satellite is $P = 100e^{-0.005t}$, where t is measured in days. Find the time rate of change of power after 100 days.

26. The number N of atoms of radium at any time t is given in terms of the number at $t = 0$, N_0, by $N = N_0 e^{-kt}$. Show that the time rate of change of N is proportional to N.

27. The vapor pressure p and thermodynamic temperature T of a gas are related by the equation $\ln p = \dfrac{a}{T} + b \ln T + c$, where a, b, and c are constants. Find the expression for dp/dT.

28. The charge q on a capacitor in a circuit containing a capacitor of capacitance C, a resistance R, and a source of voltage E is given by $q = CE(1 - e^{-t/RC})$. Show that this equation satisfies the equation $R\dfrac{dq}{dt} + \dfrac{q}{C} = E$.

29. Assuming that force is proportional to acceleration, show that a particle moving along the x-axis, so that its displacement $x = ae^{kt} + be^{-kt}$, has a force acting on it which is proportional to its displacement.

30. The radius of curvature at a point on a curve is given by
$$R = \frac{[1 + (dy/dx)^2]^{3/2}}{d^2y/dx^2}$$
A roller mechanism moves along the path defined by $y = \ln \sec x$ $(-1.5 \text{ dm} \le x \le 1.5 \text{ dm})$. Find the radius of curvature of this path for $x = 0.85$ dm.

31. Sketch the graph of $y = \ln \sec x$, marking that part which is the path of the roller mechanism of Exercise 30.

32. In an electronic device, the maximum current density i_m as a function of the temperature T is given by $i_m = AT^2e^{k/T}$, where A and k are constants. Find the expression for a small change in i_m for a small change in T.

33. A meteorologist sketched the path of the jet stream on a map of the northern United States and southern Canada on which all latitudes were parallel and all longitudes were parallel and equally spaced. A computer analysis showed this path to be $y = 6.0e^{-0.020x} \sin 0.20x$ $(0 \le x \le 60)$, where the origin is 125.0°W, 45.0°N and $(60, 0)$ is 65.0°W, 45.0°N. Find the locations of the maximum and minimum latitudes of the jet stream between 65°W and 125°W for that day.

34. The reliability R $(0 \le R \le 1)$ of a certain computer system after t hours of operation is found from $R = 3e^{-0.004t} - 2e^{-0.006t}$. Use Newton's method to find how long the system operates to have a reliability of 0.8 (80% probability that there will be no system failure).

35. An object on the end of a spring is moving so that its displacement (in cm) from the equilibrium position is given by $y = e^{-0.5t}(0.4 \cos 6t - 0.2 \sin 6t)$. Find the expression for the velocity of the object. What is the velocity when $t = 0.26$ s? The motion described by this equation is called *damped harmonic motion*.

36. A package of weather instruments is propelled into the air to an altitude of about 7 km. A parachute then opens, and the package returns to the surface. The altitude y of the package as a function of the time t (in min) is given by $y = \dfrac{10t}{e^{0.4t} + 1}$. Find the vertical velocity of the package for $t = 8.0$ min.

37. The speed s of signaling by use of a certain communications cable is directly proportional to $x^2 \ln x^{-1}$, where x is the ratio of the radius of the core of the cable to the thickness of the surrounding insulation. For what value of x is s a maximum?

38. A computer is programmed to inscribe a series of rectangles in the first quadrant under the curve of $y = e^{-x}$. What is the area of the largest rectangle that can be inscribed?

39. The relative number N of gas molecules in a container that are moving at a velocity v can be shown to be $N = av^2e^{-bv^2}$, where a and b are constants. Find v for the maximum N.

40. A missile is launched and travels along a path that can be represented by $y = \sqrt{x}$. A radar tracking station is located 2.00 km directly behind the launch pad. Placing the launch pad at the origin and the radar station at $(-2.00, 0)$, find the largest angle of elevation required of the radar to track the missile.

CHAPTER (27) EQUATIONS

Limit of $\dfrac{\sin\theta}{\theta}$ as $\theta \to 0$

$$\lim_{\theta\to 0}\frac{\sin\theta}{\theta} = \lim_{h\to 0}\frac{\sin(h/2)}{h/2} = 1 \tag{27.1}$$

Chain rule

$$\frac{dy}{dx} = \frac{dy}{du}\frac{du}{dx} \tag{27.3}$$

Derivatives

$$\frac{d(\sin u)}{dx} = \cos u \frac{du}{dx} \tag{27.4}$$

$$\frac{d(\cos u)}{dx} = -\sin u \frac{du}{dx} \tag{27.5}$$

$$\frac{d(\tan u)}{dx} = \sec^2 u \frac{du}{dx} \tag{27.6}$$

$$\frac{d(\cot u)}{dx} = -\csc^2 u \frac{du}{dx} \tag{27.7}$$

$$\frac{d(\sec u)}{dx} = \sec u \tan u \frac{du}{dx} \tag{27.8}$$

$$\frac{d(\csc u)}{dx} = -\csc u \cot u \frac{du}{dx} \tag{27.9}$$

$$\frac{d(\sin^{-1} u)}{dx} = \frac{1}{\sqrt{1-u^2}}\frac{du}{dx} \tag{27.10}$$

$$\frac{d(\cos^{-1} u)}{dx} = -\frac{1}{\sqrt{1-u^2}}\frac{du}{dx} \tag{27.11}$$

$$\frac{d(\tan^{-1} u)}{dx} = \frac{1}{1+u^2}\frac{du}{dx} \tag{27.12}$$

$$\frac{d(\log_b u)}{dx} = \frac{1}{u}\log_b e \frac{du}{dx} \tag{27.13}$$

$$\frac{d(\ln u)}{dx} = \frac{1}{u}\frac{du}{dx} \tag{27.14}$$

$$\frac{d(b^u)}{dx} = b^u \ln b \frac{du}{dx} \tag{27.15}$$

$$\frac{d(e^u)}{dx} = e^u \frac{du}{dx} \tag{27.16}$$

CHAPTER (27) REVIEW EXERCISES

In Exercises 1–40, find the derivative of the given functions.

1. $y = 3\cos(4x - 1)$

2. $y = 4\sec(1 - x^3)$

3. $u = 0.2\tan\sqrt{3 - 2v}$

4. $y = 5\sin(1 - 6x)$

5. $y = \csc^2(3x + 2)$

6. $r = \cot^2 5\pi\theta$

7. $y = 3\cos^4 x^2$

8. $y = 2\sin^3\sqrt{x}$

9. $y = (e^{x-3})^2$

10. $y = 0.5e^{\sin 2x}$

11. $y = 3\ln(x^2 + 1)$

12. $R = \ln(3 + \sin T^2)$

13. $y = 3\tan^{-1}\left(\dfrac{x}{3}\right)$

14. $y = 0.4\cos^{-1}(2\pi t + 1)$

15. $\theta = \ln\sin^{-1} 0.1t$

16. $y = \sin(\tan^{-1} x)$

17. $y = \sqrt{\csc 4x + \cot 4x}$

18. $y = \cos^2(\tan x)$

19. $y = 7\ln(x - e^{-x})^2$

20. $h = \ln\sqrt[3]{\sin 6\theta}$

21. $y = \dfrac{\cos^2 x}{e^{3x} + 1}$

22. $y = \sqrt{\dfrac{1 + \cos 2x}{2}}$

23. $v = \dfrac{u^2}{\tan^{-1} 2u}$

24. $y = \dfrac{\sin^{-1} x}{4x}$

25. $y = \ln(\csc x^2)$

26. $u = 0.5\ln\tan e^x$

27. $y = \ln^2(3 + \sin x)$

28. $y = \ln(3 + \sin x)^2$

29. $L = 0.1e^{-2t}\sec \pi t$

30. $y = 5\,e^{3x}\ln x$

31. $y = \sqrt{\sin 2x + e^{4x}}$

32. $x + y\ln 2x = y^2$

33. $\tan^{-1}\dfrac{y}{x} = x^2 e^y$

34. $3y + \ln xy = 2 + x^2$

35. $r = 0.5t(e^{2t} + 1)(e^{-2t} - 1)$

36. $y = (\ln 4x - \tan 4x)^3$

37. $\ln xy + ye^{-x} = 1$

38. $y = x(\sin^{-1} x)^2 + 2\sqrt{1 - x^2}\,\sin^{-1} x - 2x$

39. $y = x\cos^{-1} x - \sqrt{1 - x^2}$

40. $W = \ln(4s^2 + 1) + \tan^{-1} 2s$

In Exercises 41–44, sketch the graphs of the given functions. Check each by displaying the graph on a graphing calculator.

41. $y = x - \cos x$

42. $y = 4\sin x + \cos 2x$

43. $y = x(\ln x)^2$

44. $y = \ln(1 + x)$

In Exercises 45–48, find the equations of the indicated tangent or normal lines.

45. Find the equation of the line tangent to the curve of $y = 4\cos^2(x^2)$ at $x = 1$.

46. Find the equation of the line tangent to the curve of $y = \ln\cos x$ at $x = \frac{\pi}{6}$.

47. Find the equation of the line normal to the curve of $y = e^{x^2}$ at $x = \frac{1}{2}$.

48. Find the equation of the line normal to the curve of $y = \tan^{-1} x$ at $x = 1$.

In Exercises 49–88, solve the given problems.

49. Find the derivative of each member of the identity $\sin^2 x + \cos^2 x = 1$ and show that the results are equal.

50. Find the derivative of each member of the identity
$$\sin(x + 1) = \sin x\cos 1 + \cos x\sin 1$$
and show that the results are equal.

51. By Newton's method, solve the equation $e^x - x^2 = 0$. Check the solution by displaying the graph on a calculator and using the *intersect* (or *zero*) feature.

52. By Newton's method, solve the equation $x^2 = \tan^{-1} x$. Check the solution by displaying the graph on a calculator and using the *intersect* (or *zero*) feature.

53. Find the values of x for which the graph of $y = e^x - 2e^{-x}$ is concave up.

54. Find the values of x for which the graph of $y = 2e^{\sin(x/2)}$ has maximum or minimum points.

55. If a 200-N crate is dragged along a horizontal floor by a force F (in N) acting along a rope at an angle θ with the floor, the magnitude of F is given by $F = \dfrac{200\,\mu}{\mu\sin\theta + \cos\theta}$, where μ is the coefficient of friction. Evaluate the instantaneous rate of change of F with respect to θ when $\theta = 15°$ if $\mu = 0.20$.

56. Find the date of the maximum number of hours of daylight for cities at 40°N. See Exercise 67 on page 315.

57. Periodically a robot moves a part vertically y cm in an automobile assembly line. If y as a function of the time t (in s) is $y = 0.75\left(\sec\sqrt{0.15t} - 1\right)$, find the velocity at which the part is moved after 5.0 s of each period.

58. The length L (in m) of the shadow of a tree 15 m tall is $L = 15\cot\theta$, where θ is the angle of elevation of the sun. Approximate the change in L if θ changes from 50° to 52°.

59. An analysis of temperature records for Sydney, Australia, indicates that the average daily temperature (in °C) during the year is given approximately by $T = 17.2 + 5.2\cos\left[\frac{\pi}{6}(x - 0.50)\right]$, where x is measured in months ($x = 0.5$ is Jan. 15, etc.). What is the *daily* time rate of change of temperature on March 1? (*Hint:* 12 months/365 days = 0.033 month/day = dx/dt.)

60. An earth-orbiting satellite is launched such that its altitude (in km) is given by $y = 240(1 - e^{-0.05t})$, where t is the time (in min). Find the vertical velocity of the satellite for $t = 10.0$ min.

61. Power can be defined as the time rate of doing work. If work is being done in an electric circuit according to $W = 10\cos 2t$, find P as a function of t.

62. The value V of a bank account in which $1000 is deposited and then earns 6% annual interest, compounded continuously (daily compounding approximates this, and some banks actually use continuous compounding), is $V = 1000e^{0.06t}$ after t years. How fast is the account growing after exactly 2 years?

63. In determining how to divide files on the hard disk of a computer, we can use the equation $n = xN\log_x N$. Sketch the graph of n as a function of x for $1 < x \le 10$ if $N = 8$.

64. Under certain conditions, the potential V (in V) due to a magnet is given by $V = -k\ln\left(1 + \dfrac{L}{x}\right)$, where L is the length of the magnet and x is the distance from the point where the potential is measured. Find the expression for dV/dx.

65. In the theory of making images by holography, an expression used for the light-intensity distribution is $I = kE_0^2\cos^2\frac{1}{2}\theta$, where k and E_0 are constant and θ is the phase angle between two light waves. Find the expression for $dI/d\theta$.

66. If we neglect air resistance, the range R of a bullet fired at an angle θ with the horizontal is $R = \dfrac{v_0^2}{g}\sin 2\theta$, where v_0 is the initial velocity and g is the acceleration due to gravity. Find θ for the maximum range. See Fig. 27.37.

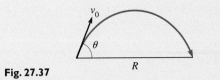

Fig. 27.37

67. In the design of a cone-type clutch, an equation that relates the cone angle θ and the applied force F is $\theta = \sin^{-1}(Ff/R)$, where R is the frictional resistance and f is the coefficient of friction. For constant R and f, find $d\theta/dF$.

68. If inflation makes the dollar worth 5% less each year, then the value of \$100 in t years will be $V = 100(0.95)^t$. What is the approximate change in the value during the fourth year?

69. An object attached to a cord of length l, as shown in Fig. 27.38, moves in a circular path. The angular velocity ω is given by $\omega = \sqrt{g/(l \cos \theta)}$. By use of differentials, find the approximate change in ω if θ changes from 32.50° to 32.75°, given that $g = 9.800 \text{ m/s}^2$ and $l = 0.6375$ m.

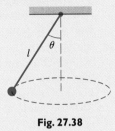

Fig. 27.38

70. An analysis of samples of air for a city showed that the number of parts per million p of sulfur dioxide on a certain day was $p = 0.05 \ln(2 + 24t - t^2)$, where t is the hour of the day. Using differentials, find the approximate change in the amount of sulfur dioxide between 10 A.M. and noon.

71. According to Newton's law of cooling (Isaac Newton, again), the rate at which a body cools is proportional to the difference in temperature between it and the surrounding medium. By use of this law, the temperature T (in °C) of an engine coolant as a function of the time t (in min) is $T = 30 + 60(0.5)^{0.200t}$. The coolant was initially at 90°C, and the air temperature was 30°C. Linearize this function for $t = 5.00$ min and display the graphs of $T = f(t)$ and $L(t)$ on a graphing calculator.

72. The charge q on a certain capacitor in an amplifier circuit as a function of time t is given by $q = e^{-0.1t}(0.2 \sin 120\pi t + 0.8 \cos 120 \pi t)$. The current i in the circuit is the instantaneous time rate of change of the charge. Find the expression for i as a function of t.

73. A projection of the number n (in millions) of users of the Internet is $n = 160 - 140e^{-0.30t}$, where t is the number of years after 2000. What is the projected annual rate of increase in 2010?

74. A football is thrown horizontally (very little arc) at 18 m/s parallel to the sideline. A TV camera is 31 m from the path of the football. Find $d\theta/dt$, the rate at which the camera must turn to follow the ball when $\theta = 15°$. See Fig. 27.39.

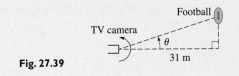

Fig. 27.39

75. An architect designs an arch of height y (in m) over a walkway by the curve of the equation $y = 3.00e^{-0.500x^2}$. What are the dimensions of the largest rectangular passage area under the arch?

76. A force P (in N) at an angle θ above the horizontal drags a 50-N box across a level floor. The coefficient of friction between the floor and the box is constant and equals 0.20. The magnitude of the force P is given by $P = \dfrac{(0.20)(50)}{0.20 \sin \theta + \cos \theta}$. Find θ such that P is a minimum.

77. A jet is flying at 220 m/s directly away from the control tower of an airport. If the jet is at a constant altitude of 1700 m, how fast is the angle of elevation of the jet from the control tower changing when it is 13.0°?

78. The current i in an electric circuit with a resistance R and an inductance L is $i = i_0 e^{-Rt/L}$, where i_0 is the initial current. Show that the time rate of change of the current is directly proportional to the current.

79. A silo constructed as shown in Fig. 27.40 is to hold 2880 m³ of silage when completely full. It can be shown (can you?) that the surface area S (in m²) (not including the base) is $S = 640 + 81\pi\left(\csc \theta - \frac{2}{3} \cot \theta\right)$. Find θ such that S is a minimum.

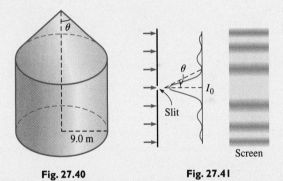

Fig. 27.40   **Fig. 27.41**

80. Light passing through a narrow slit forms patterns of light and dark (see Fig. 27.41). The intensity I of the light at an angle θ is given by $I = I_0 \left[\dfrac{\sin(k \sin \theta)}{k \sin \theta}\right]^2$, where k and I_0 are constants. Show that the maximum and minimum values of I occur for $k \sin \theta = \tan(k \sin \theta)$.

81. When a wheel rolls along a straight line, a point P on the circumference traces a curve called a *cycloid*. See Fig. 27.42. The parametric equations of a cycloid are $x = r(\theta - \sin \theta)$ and $y = r(1 - \cos \theta)$. Find the velocity of the point on the rim of a wheel for which $r = 5.500$ cm and $d\theta/dt = 0.12$ rad/s for $\theta = 35.0°$. (An inverted cycloid is the path of least time of descent (the *brachistochrone*) of an object acted on only by gravity.)

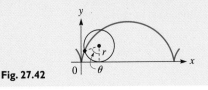

Fig. 27.42

82. In the study of atomic spectra, it is necessary to solve the equation $x = 5(1 - e^{-x})$ for x. Use Newton's method to find the solution.

83. The illuminance from a point source of light varies directly as the cosine of the angle of incidence (measured from the perpendicular) and inversely as the square of the distance r from the source. How high above the center of a circle of radius 10.0 cm should a light be placed so that illuminance at the circumference will be a maximum? See Fig. 27.43.

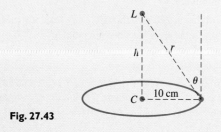

Fig. 27.43

84. A Y-shaped metal bracket is to be made such that its height is 10.0 cm and its width across the top is 6.00 cm. What shape will require the least amount of material? See Fig. 27.44.

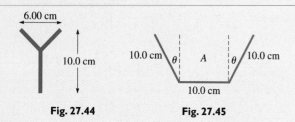

Fig. 27.44   **Fig. 27.45**

85. A gutter is to be made from a sheet of metal 30.0 cm wide by turning up strips of width 10.0 cm along each side to make equal angles θ with the vertical. Sketch a graph of the cross-sectional area A as a function of θ. See Fig. 27.45.

86. The displacement y (in cm) of a weight on a spring in water is given by $y = 3.0te^{-0.20t}$, where t is the time (in s). What is the maximum displacement? (For this type of displacement, the motion is called *critically damped*, as the weight returns to its equilibrium position as quickly as possible without oscillating.)

87. Show that the equation of the hyperbolic cosine function

$$y = \frac{H}{w} \cosh \frac{wx}{H} \qquad (w \text{ and } H \text{ are constants})$$

satisfies the equation

$$\frac{d^2y}{dx^2} = \frac{w}{H}\sqrt{1 + \left(\frac{dy}{dx}\right)^2}$$

(see Exercise 53 on page 823). A *catenary* (see Exercise 55 on page 391) is the curve of a uniform cable hanging under its own weight and is in the shape of a hyperbolic cosine curve. Also, this shape (inverted) was chosen for the St. Louis Gateway Arch (shown in Fig. 27.46) and makes the arch self-supporting.

Fig. 27.46

88. A conical filter is made from a circular piece of wire mesh of radius 24.0 cm by cutting out a sector with central angle θ and then taping the cut edges of the remaining piece together (see Fig. 27.47). What is the maximum possible volume the resulting filter can hold?

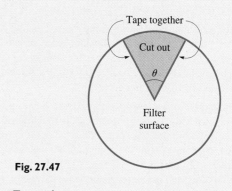

Fig. 27.47

Writing Exercise

89. To find the area of the largest rectangular microprocessor chip with a perimeter of 40 mm, it is possible to use either an algebraic function or a trigonometric function. Write two or three paragraphs to explain how each type of function can be used to find the required area.

CHAPTER 27 PRACTICE TEST

In Problems 1–3, find the derivative of each of the functions.

1. $y = \tan^3 2x + \tan^{-1} 2x$ **2.** $y = 2(3 + \cot 4x)^3$

3. $y \sec 2x = \sin^{-1} 3y$

4. Find the differential of the function $y = \dfrac{\cos^2(3x + 1)}{x}$.

5. Find the slope of a tangent to the curve of $y = \ln \dfrac{2x - 1}{1 + x^2}$ for $x = 2$.

6. Find the expression for the time rate of change of electric current that is given by the equation $i = 8e^{-t} \sin 10t$, where t is the time.

7. Sketch the graph of the function $y = xe^x$.

8. A balloon leaves the ground 250 m from an observer and rises at the rate of 5.0 m/s. How fast is the angle of elevation of the balloon increasing after 8.0 s?

Methods of Integration

In developing calculus, mathematicians saw that integration and differentiation were inverse processes, and therefore many integrals could be formed by finding the antiderivative. However, many of the integrals that arose mathematically and from the study of mechanical systems did not fit a form from which the antiderivative could be found directly. This led to the creation of numerous methods to integrate various types of functions.

Some of these methods of integration had been developed by the early 1700s and were used by many mathematicians, including Newton and Leibniz. By the mid-1700s, many of these methods were included in textbooks. One of these texts was written by the Italian mathematician Maria Agnesi. Her text included analytic geometry, differential calculus, integral calculus, and some advanced topics and was noted for clear and organized explanations with many examples. Another important set of textbooks was written by Euler, who wrote separate texts on precalculus topics, differential calculus, and integral calculus. These texts were also noted for clear and organized presentations and were widely used until the early 1800s. Euler's integral calculus text included most of the methods of integration presented in this chapter.

Being able to integrate functions by using special methods, as well as by directly using antiderivatives, made integral calculus much more useful and helped in developing many areas of geometry, science, and technology in the 1800s and 1900s. In earlier chapters, we have noted a number of applications of integration, and additional examples are found in the examples and exercises of this chapter.

In this chapter, we expand the use of the general power formula for integration for use with integrands that include transcendental functions. We then develop additional standard forms for transcendental functions and several special methods of integration for integrands that do not directly fit standard forms. In using all forms of integration, *recognition of the integral form* is of great importance.

In Section 28.5, we show an application of integration that is important in the design of electric appliances, such as an electric heater.

28.1 THE GENERAL POWER FORMULA

The first formula for integration that we will discuss is the general power formula, and we will expand its use to include transcendental integrands. It was first introduced with the integration of basic algebraic forms in Chapter 25 and is repeated here for reference.

$$\int u^n \, du = \frac{u^{n+1}}{n+1} + C \qquad (n \neq -1) \tag{28.1}$$

CAUTION ▶ In applying Eq. (28.1) to transcendental integrands, as well as with algebraic integrands, *we must properly recognize the quantities u, n, and du.* This requires familiarity with the differential forms of Chapters 23 and 27.

◀ EXAMPLE 1   Integrate: $\int \sin^3 x \cos x \, dx$.

Since $d(\sin x) = \cos x \, dx$, we note that this integral fits the form of Eq. (28.1) for $u = \sin x$. Thus, with $u = \sin x$, we have $du = \cos x \, dx$, which means that this integral is of the form $\int u^3 \, du$. Therefore, the integration can now be completed:

$$\int \sin^3 x \cos x \, dx = \int \sin^3 x \overset{du}{(\cos x \, dx)}$$

$$= \frac{1}{4} \sin^4 x + C \quad \longleftarrow \quad \text{do not forget the constant of integration}$$

CAUTION ▶ We note here that *the factor* cos *x is a necessary part of the du* in order to have the proper form of integration *and therefore does not appear in the final result.*

We check our result by finding the derivative of $\frac{1}{4} \sin^4 x + C$, which is

$$\frac{d}{dx} \left(\frac{1}{4} \sin^4 x + C \right) = \frac{1}{4} (4) \sin^3 x \cos x$$

$$= \sin^3 x \cos x$$

◀ EXAMPLE 2   Integrate: $\int 2\sqrt{1 + \tan \theta} \sec^2 \theta \, d\theta$.

Here we note that $d(\tan \theta) = \sec^2 \theta \, d\theta$, which means that the integral fits the form of Eq. (28.1) with

$$u = 1 + \tan \theta \qquad du = \sec^2 \theta \, d\theta \qquad n = \tfrac{1}{2}$$

The integral is of the form $\int u^{1/2} \, du$. Thus,

$$\int 2\sqrt{1 + \tan \theta} \, (\sec^2 \theta \, d\theta) = 2 \int \underset{u}{(1 + \tan \theta)^{1/2}} \overset{du}{(\sec^2 \theta \, d\theta)}$$

$$= 2 \left(\frac{2}{3} \right) (1 + \tan \theta)^{3/2} + C$$

$$= \frac{4}{3} (1 + \tan \theta)^{3/2} + C$$

◀ **EXAMPLE 3** Integrate: $\int \ln x \left(\dfrac{dx}{x} \right)$.

By noting that $d(\ln x) = \dfrac{dx}{x}$, we have

$$u = \ln x \qquad du = \frac{dx}{x} \qquad n = 1$$

This means that the integral is of the form $\int u \, du$. Thus,

$$\int \ln x \left(\frac{dx}{x} \right) = \frac{1}{2} (\ln x)^2 + C = \frac{1}{2} \ln^2 x + C$$

◀ **EXAMPLE 4** Find the value of $\displaystyle\int_0^{0.5} \frac{\sin^{-1} x}{\sqrt{1 - x^2}} \, dx$.

For purposes of integrating, we see that

$$u = \sin^{-1} x \qquad du = \frac{dx}{\sqrt{1 - x^2}} \qquad n = 1$$

$$\int_0^{0.5} \frac{\sin^{-1} x}{\sqrt{1 - x^2}} \, dx = \int_0^{0.5} \sin^{-1} x \left(\frac{dx}{\sqrt{1 - x^2}} \right) \qquad \int u \, du$$

$$= \frac{(\sin^{-1} x)^2}{2} \Big|_0^{0.5} \qquad\qquad \text{integrate}$$

$$= \frac{\left(\frac{\pi}{6} \right)^2}{2} - 0 = \frac{\pi^2}{72} \qquad\qquad \text{evaluate}$$

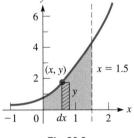

Fig. 28.1

The display in Fig. 28.1 shows the calculator check for this result.

◀ **EXAMPLE 5** Find the first-quadrant area bounded by $y = \dfrac{e^{2x}}{\sqrt{e^{2x} + 1}}$ and $x = 1.5$.

The area to be found is shown in Fig. 28.2. The area of the representative element is $y \, dx$. Therefore, the area is found by evaluating the integral

$$\int_0^{1.5} \frac{e^{2x} \, dx}{\sqrt{e^{2x} + 1}}$$

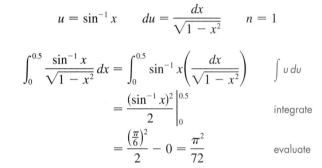

Fig. 28.2

For the purpose of integration, $n = -\frac{1}{2}$, $u = e^{2x} + 1$, and $du = 2e^{2x} \, dx$. Therefore,

$$\int_0^{1.5} (e^{2x} + 1)^{-1/2} e^{2x} \, dx = \frac{1}{2} \int_0^{1.5} (e^{2x} + 1)^{-1/2} (2e^{2x} \, dx)$$

$$= \frac{1}{2} (2)(e^{2x} + 1)^{1/2} \Big|_0^{1.5} = (e^{2x} + 1)^{1/2} \Big|_0^{1.5}$$

$$= \sqrt{e^3 + 1} - \sqrt{2} = 3.178$$

This means that the area is 3.178 square units.

EXERCISES **28.1**

In Exercises 1 and 2, make the given changes in the indicated examples of this section and then solve the given problems.

1. In Example 1, change $\sin^3 x$ to $\cos^3 x$. What other change must be made in the integrand to have a result of $\frac{1}{4} \cos^4 x + C$?

2. In Example 3, change $\ln x$ to $\ln x^2$ and then integrate.

In Exercises 3–26, integrate each of the functions.

3. $\int \sin^4 x \cos x \, dx$

4. $\int \cos^5 x (-\sin x \, dx)$

5. $\int 0.4\sqrt{\cos \theta} \sin \theta \, d\theta$

6. $\int 8 \sin^{1/3} x \cos x \, dx$

7. $\int 4 \tan^2 x \sec^2 x \, dx$

8. $\int \sec^3 x (\sec x \tan x) \, dx$

9. $\int_0^{\pi/8} \cos 2x \sin 2x \, dx$

10. $\int_{\pi/6}^{\pi/4} 3\sqrt{\cot x} \csc^2 x \, dx$

11. $\int (\sin^{-1} x)^3 \left(\dfrac{dx}{\sqrt{1 - x^2}} \right)$

12. $\int \dfrac{20(\cos^{-1} 2t)^4 \, dt}{\sqrt{1 - 4t^2}}$

13. $\int \dfrac{5 \tan^{-1} 5x}{1 + 25x^2} \, dx$

14. $\int \dfrac{\sin^{-1} 4x \, dx}{\sqrt{1 - 16x^2}}$

15. $\int [\ln(x + 1)]^2 \dfrac{dx}{x + 1}$

16. $\int 0.8(3 + 2 \ln u)^3 \dfrac{du}{u}$

17. $\int_0^{1/2} \dfrac{\ln(2x + 3)}{2x + 3} \, dx$

18. $\int_1^e \dfrac{(1 - 2 \ln x) \, dx}{x}$

19. $\int (4 + e^x)^3 e^x \, dx$

20. $\int 2\sqrt{1 - e^{-x}} (e^{-x} \, dx)$

21. $\int \dfrac{e^{2t} \, dt}{(1 - e^{2t})^3}$

22. $\int \dfrac{(1 + 3e^{-2x})^4 \, dx}{e^{2x}}$

23. $\int (1 + \sec^2 x)^4 (\sec^2 x \tan x \, dx)$

24. $\int (e^x + e^{-x})^{1/4} (e^x - e^{-x}) \, dx$

25. $\int_{\pi/6}^{\pi/4} (1 + \cot x)^2 \csc^2 x \, dx$

26. $\int_{\pi/3}^{\pi/2} \dfrac{\sin \theta \, d\theta}{\sqrt{1 + \cos \theta}}$

In Exercises 27–36, solve the given problems by integration.

27. Find the area under the curve $y = \dfrac{1 + \tan^{-1} 2x}{1 + 4x^2}$ from $x = 0$ to $x = 2$.

28. Find the first-quadrant area bounded by $y = \dfrac{\ln(4x + 1)}{4x + 1}$ and $x = 2$.

29. Rewrite the integral $\int \dfrac{\sqrt{(1 + e^{-r})(1 - e^{-r})}}{e^{2r}} \, dr$ so that it fits the form $\int u^n \, du$ and identify u, n, and du.

30. Rewrite the integral $\int \dfrac{\tan^3 x}{\cos^2 x} \, dx$ so that it fits the form $\int u^n \, du$ and identify u, n, and du.

31. The general expression for the slope of a given curve is $(\ln x)^2/x$. If the curve passes through $(1, 2)$, find its equation.

32. Find the equation of the curve for which $dy/dx = (1 + \tan 2x)^2 \sec^2 2x$ if the curve passes through $(2, 1)$.

33. In the development of the expression for the total pressure P on a wall due to molecules with mass m and velocity v striking the wall, the equation $P = mnv^2 \int_0^{\pi/2} \sin \theta \cos^2 \theta \, d\theta$ is found. The symbol n represents the number of molecules per unit volume, and θ represents the angle between a perpendicular to the wall and the direction of the molecule. Find the expression for P.

34. The solar energy E passing through a hemispherical surface per unit time, per unit area, is $E = 2\pi I \int_0^{\pi/2} \cos \theta \sin \theta \, d\theta$, where I is the solar intensity and θ is the angle at which it is directed (from the perpendicular). Evaluate this integral.

35. After an electric power interruption, the current i in a circuit is given by $i = 3(1 - e^{-t})^2(e^{-t})$, where t is the time. Find the expression for the total electric charge q to pass a point in the circuit if $q = 0$ for $t = 0$.

36. A space vehicle is launched vertically from the ground such that its velocity v (in km/s) is given by $v = [\ln^2(t^3 + 1)] \dfrac{t^2}{t^3 + 1}$, where t is the time (in s). Find the altitude of the vehicle after 10.0 s.

28.2 THE BASIC LOGARITHMIC FORM

The general power formula for integration, Eq. (28.1), is valid for all values of n except $n = -1$. If n were set equal to -1, this would cause the result to be undefined. When we obtained the derivative of the logarithmic function, we found

$$\frac{d(\ln u)}{dx} = \frac{1}{u} \frac{du}{dx}$$

This means the differential of the logarithmic form is $d(\ln u) = du/u$. Reversing the process, we then determine that $\int du/u = \ln u + C$. In other words, when the exponent of the expression being integrated is -1, the expression is a logarithmic form.

Logarithms are defined only for positive numbers. Thus, $\int du/u = \ln u + C$ is valid if $u > 0$. If $u < 0$, then $-u > 0$. In this case, $d(-u) = -du$, or $\int (-du)/(-u) = \ln(-u) + C$. However, $\int du/u = \int(-du)/(-u)$. These results can be combined into a single form using the absolute value of u. Therefore,

$$\int \frac{du}{u} = \ln|u| + C \qquad (28.2)$$

◀ EXAMPLE 1   Integrate: $\int \dfrac{dx}{x + 1}$.

Since $d(x + 1) = dx$, this integral fits the form of Eq. (28.2) with $u = x + 1$ and $du = dx$. Therefore, we have

$$\int \frac{dx}{x + 1} = \ln|x + 1| + C$$

◀ EXAMPLE 2   Newton's law of cooling states that the rate at which an object cools is directly proportional to the difference in its temperature T and the temperature of the surrounding medium. By use of this law, the time t (in min) a certain object takes to cool from 80°C to 50°C in air at 20°C is found to be

$$t = -9.8 \int_{80}^{50} \frac{dT}{T - 20}$$

Find the value of t.

We see that the integral fits Eq. (28.2) with $u = T - 20$ and $du = dT$. Thus,

$$t = -9.8 \int_{80}^{50} \frac{dT}{T - 20} \quad \longleftarrow du \quad \longleftarrow u$$

$$= -9.8 \ln|T - 20|_{80}^{50} \qquad \text{integrate}$$

$$= -9.8(\ln 30 - \ln 60) \qquad \text{evaluate}$$

$$= -9.8 \ln \frac{30}{60} = -9.8 \ln(0.50) \qquad \ln x - \ln y = \ln \frac{x}{y}$$

$$= 6.8 \text{ min}$$

◀ EXAMPLE 3   Integrate: $\int \dfrac{\cos x}{\sin x}\, dx$.

We note that $d(\sin x) = \cos x\, dx$. This means that this integral fits the form of Eq. (28.2) with $u = \sin x$ and $du = \cos x\, dx$. Thus,

$$\int \frac{\cos x}{\sin x}\, dx = \int \frac{\cos x\, dx}{\sin x} \quad \longleftarrow du \quad \longleftarrow u$$

$$= \ln|\sin x| + C$$

◀ EXAMPLE 4   Integrate: $\int \dfrac{x\,dx}{4 - x^2}$.

This integral fits the form of Eq. (28.2) with $u = 4 - x^2$ and $du = -2x\,dx$. This means that we must introduce a factor of -2 into the numerator and a factor of $-\frac{1}{2}$ before the integral. Therefore,

$$\int \frac{x\,dx}{4 - x^2} = -\frac{1}{2}\int \frac{-2x\,dx}{4 - x^2} \begin{matrix} \longleftarrow du \\ \longleftarrow u \end{matrix}$$

$$= -\frac{1}{2}\ln|4 - x^2| + C$$

CAUTION ▶   We should note that if the quantity $4 - x^2$ were raised to any power other than that in the example, we would have to employ the general power formula for integration. For example,

$$\int \frac{x\,dx}{(4 - x^2)^2} = -\frac{1}{2}\int \frac{-2x\,dx}{(4 - x^2)^2} \begin{matrix} \longleftarrow du \\ \\ \longleftarrow u^2 \end{matrix}$$

$$= -\frac{1}{2}\frac{(4 - x^2)^{-1}}{-1} + C = \frac{1}{2(4 - x^2)} + C \qquad ❭$$

◀ EXAMPLE 5   Integrate: $\int \dfrac{e^{4x}\,dx}{1 + 3e^{4x}}$.

Since $d(1 + 3e^{4x})/dx = 12e^{4x}$, we see that we can use Eq. (28.2) with $u = 1 + 3e^{4x}$ and $du = 12e^{4x}\,dx$. Therefore, we write

$$\int \frac{e^{4x}\,dx}{1 + 3e^{4x}} = \frac{1}{12}\int \frac{12e^{4x}\,dx}{1 + 3e^{4x}} \qquad \text{introduce factors of 12}$$

$$= \frac{1}{12}\ln|1 + 3e^{4x}| + C \qquad \text{integrate}$$

$$= \frac{1}{12}\ln(1 + 3e^{4x}) + C \qquad 1 + 3e^{4x} > 0 \text{ for all } x \qquad ❭$$

◀ EXAMPLE 6   Evaluate: $\int_0^{\pi/8} \dfrac{\sec^2 2\theta}{1 + \tan 2\theta}\,d\theta$.

Since $d(1 + \tan 2\theta) = 2\sec^2 2\theta\,d\theta$, we see that we can use Eq. (28.2) with $u = 1 + \tan 2\theta$ and $du = 2\sec^2 2\theta\,d\theta$. Therefore, we have

$$\int_0^{\pi/8} \frac{\sec^2 2\theta}{1 + \tan 2\theta}\,d\theta = \frac{1}{2}\int_0^{\pi/8} \frac{2\sec^2 2\theta\,d\theta}{1 + \tan 2\theta} \qquad \text{introduce factors of 2}$$

$$= \frac{1}{2}\ln|1 + \tan 2\theta|\big|_0^{\pi/8} \qquad \text{integrate}$$

$$= \frac{1}{2}(\ln|1 + 1| - \ln|1 + 0|) \qquad \text{evaluate}$$

$$= \frac{1}{2}(\ln 2 - \ln 1) = \frac{1}{2}(\ln 2 - 0)$$

$$= \frac{1}{2}\ln 2 \qquad\qquad\qquad\qquad ❭$$

❨ EXAMPLE 7   Find the volume within the piece of tapered tubing shown in Fig. 28.3, which can be described as the volume generated by revolving the region bounded by the curve of $y = \dfrac{3}{\sqrt{4x + 3}}$, $x = 2.50$ cm, and the axes about the x-axis.

The volume can be found by setting up only one integral by using a disk element of volume, as shown. The volume is found as follows:

$$V = \pi \int_0^{2.50} y^2 \, dx = \pi \int_0^{2.50} \left(\frac{3}{\sqrt{4x + 3}} \right)^2 dx$$

$$= \pi \int_0^{2.50} \frac{9 \, dx}{4x + 3} = \frac{9\pi}{4} \int_0^{2.50} \frac{4 \, dx}{4x + 3}$$

$$= \frac{9\pi}{4} \ln(4x + 3) \Big|_0^{2.50} = \frac{9\pi}{4} (\ln 13.0 - \ln 3)$$

$$= \frac{9\pi}{4} \ln \frac{13.0}{3} = 10.4 \text{ cm}^3$$

Therefore, the volume within this piece of tubing is about 10.4 cm³.

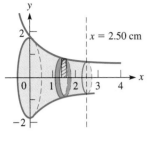

Fig. 28.3

EXERCISES 28.2

In Exercises 1 and 2, make the given changes in the indicated examples of this section and then solve the given problems.

1. In Example 1, change $x + 1$ to $x^2 + 1$. What other change must be made in the integrand to have a result of $\ln|x^2 + 1| + C$?

2. In Example 3, interchange $\cos x$ and $\sin x$ and then integrate.

In Exercises 3–30, integrate each of the given functions.

3. $\displaystyle\int \frac{dx}{1 + 4x}$

4. $\displaystyle\int \frac{dx}{1 - 4x}$

5. $\displaystyle\int \frac{2x \, dx}{4 - 3x^2}$

6. $\displaystyle\int \frac{4\sqrt{u} \, du}{1 + u\sqrt{u}}$

7. $\displaystyle\int_0^2 \frac{dx}{8 - 3x}$

8. $\displaystyle\int_{-1}^3 \frac{2x^3 \, dx}{x^4 + 1}$

9. $\displaystyle\int \frac{0.4 \csc^2 2\theta \, d\theta}{\cot 2\theta}$

10. $\displaystyle\int \frac{\sin 3x}{\cos 3x} \, dx$

11. $\displaystyle\int_0^{\pi/2} \frac{\cos x \, dx}{1 + \sin x}$

12. $\displaystyle\int_0^{\pi/4} \frac{\sec^2 x \, dx}{4 + \tan x}$

13. $\displaystyle\int \frac{e^{-x}}{1 - e^{-x}} \, dx$

14. $\displaystyle\int \frac{5e^{3x}}{1 - e^{3x}} \, dx$

15. $\displaystyle\int \frac{1 + e^x}{x + e^x} \, dx$

16. $\displaystyle\int \frac{3e^t \, dt}{\sqrt{e^{2t} + 4e^t + 4}}$

17. $\displaystyle\int \frac{\sec x \tan x \, dx}{1 + 4 \sec x}$

18. $\displaystyle\int \frac{\sin 2x}{1 - \cos^2 x} \, dx$

19. $\displaystyle\int_1^3 \frac{1 + x}{4x + 2x^2} \, dx$

20. $\displaystyle\int_1^2 \frac{4x + 6x^2}{x^2 + x^3} \, dx$

21. $\displaystyle\int \frac{0.5 \, dr}{r \ln r}$

22. $\displaystyle\int \frac{dx}{x(1 + 2 \ln x)}$

23. $\displaystyle\int \frac{2 + \sec^2 x}{2x + \tan x} \, dx$

24. $\displaystyle\int \frac{x + \cos 2x}{x^2 + \sin 2x} \, dx$

25. $\displaystyle\int \frac{2 \, dx}{\sqrt{1 - 2x}}$

26. $\displaystyle\int \frac{4x \, dx}{(1 + x^2)^2}$

27. $\displaystyle\int \frac{x + 2}{x^2} \, dx$

28. $\displaystyle\int \frac{3v^2 - 2v}{v^2} \, dv$

29. $\displaystyle\int_0^{\pi/12} \frac{\sec^2 3x}{4 + \tan 3x} \, dx$

30. $\displaystyle\int_1^2 \frac{x^2 + 1}{x^3 + 3x} \, dx$

In Exercises 31–44, solve the given problems by integration.

31. Find the area bounded by $y(x + 1) = 1$, $x = 0$, $y = 0$, and $x = 2$. See Fig. 28.4.

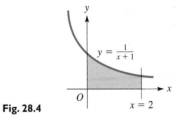

Fig. 28.4

Ⓦ **32.** Evaluate $\displaystyle\int_1^2 x^{-1} \, dx$ and $\displaystyle\int_2^4 x^{-1} \, dx$. Give a geometric interpretation of these two results.

33. Integrate $\displaystyle\int \frac{x - 4}{x + 4} \, dx$ by first using algebraic division to change the form of the integrand.

34. Integrate $\int \sec x \, dx$ by first multiplying the integrand by

$\dfrac{\sec x + \tan x}{\sec x + \tan x}$ (a special form of 1).

35. Find the volume generated by revolving the area bounded by $y = 1/(x^2 + 1)$, $x = 0$, $x = 1$, and $y = 0$ about the y-axis. Use shells.

36. Find the volume of the solid generated by revolving the region bounded by $y = \dfrac{2}{\sqrt{3x + 1}}$, $x = 0$, $x = 3.5$, and $y = 0$ about the x-axis.

37. The general expression for the slope of a curve is $\dfrac{\sin x}{3 + \cos x}$. If the curve passes through the point $(\pi/3, 2)$, find its equation.

38. Find the average value of the function $xy = 4$ from $x = 1$ to $x = 2$.

39. Under ideal conditions, the natural law of population growth is that population increases at a rate proportional to the population P present at any time t. This leads to the equation $t = \dfrac{1}{k} \int \dfrac{dP}{P}$. Assuming ideal conditions for the United States, if $P = 249$ million in 1990 ($t = 0$) and $P = 275$ million in 2000 ($t = 10$ years), find the population that is projected in 2020 ($t = 30$ years).

40. In determining the temperature that is absolute zero (0 K, or about $-273°$C), the equation $\ln T = -\int \dfrac{dr}{r - 1}$ is used. Here T is the thermodynamic temperature and r is the ratio between certain specific vapor pressures. If $T = 273.16$ K for $r = 1.3361$, find T as a function of r (if $r > 1$ for all T).

41. The time t and electric current i for a certain circuit with a voltage E, a resistance R, and an inductance L is given by

$t = L \int \dfrac{di}{E - iR}$. If $t = 0$ for $i = 0$, integrate and express i as a function of t.

42. Conditions are often such that a force proportional to the velocity tends to retard the motion of an object moving through a resisting medium. Under such conditions, the acceleration of a certain object moving down an inclined plane is given by $20 - v$. This leads to the equation $t = \int \dfrac{dv}{20 - v}$. If the object starts from rest, find the expression for the velocity as a function of time.

43. An architect designs a wall panel that can be described as the first-quadrant area bounded by $y = \dfrac{50}{x^2 + 20}$ and $x = 3.00$. If the area of the panel is 6.61 m², find the x-coordinate (in m) of the centroid of the panel.

44. The electric power p developed in a certain resistor is given by $p = 3 \int \dfrac{\sin \pi t}{2 + \cos \pi t} dt$, where t is the time. Express p as a function of t.

28.3 THE EXPONENTIAL FORM

In deriving the derivative for the exponential function, we obtained the result $de^u/dx = e^u(du/dx)$. This means that the differential of the exponential form is $d(e^u) = e^u \, du$. Reversing this form to find the proper form of the integral for the exponential function, we have

$$\int e^u \, du = e^u + C \tag{28.3}$$

◀ **EXAMPLE 1** Integrate: $\int xe^{x^2} \, dx$.

Since $d(x^2) = 2x \, dx$, we can write this integral in the form of Eq. (28.3) with $u = x^2$ and $du = 2x \, dx$. Thus,

$$\int xe^{x^2} \, dx = \frac{1}{2} \int e^{x^2}(2x \, dx)$$

$$= \frac{1}{2} e^{x^2} + C$$

◀ EXAMPLE 2   For an electric circuit containing a direct voltage source E, a resistance R, and an inductance L, the current i and time t are related by $ie^{Rt/L} = \dfrac{E}{L} \int e^{Rt/L}\, dt$. See Fig. 28.5. If $i = 0$ for $t = 0$, perform the integration and then solve for i as a function of t.

For this integral, we see that $u = \dfrac{Rt}{L}$, which means that $du = \dfrac{R\,dt}{L}$. The solution is then as follows:

$$ie^{Rt/L} = \frac{E}{L} \int e^{Rt/L}\, dt = \frac{E}{L}\left(\frac{L}{R}\right) \int e^{Rt/L}\left(\frac{R\,dt}{L}\right) \qquad \text{introduce factor } \frac{R}{L}$$

$$= \frac{E}{R} e^{Rt/L} + C \qquad \text{integrate}$$

$$0(e^0) = \frac{E}{R} e^0 + C, \qquad C = -\frac{E}{R} \qquad i = 0 \text{ for } t = 0; \text{ evaluate } C$$

$$ie^{Rt/L} = \frac{E}{R} e^{Rt/L} - \frac{E}{R} \qquad \text{substitute for } C$$

$$i = \frac{E}{R} - \frac{E}{R} e^{-Rt/L} = \frac{E}{R}\left(1 - e^{-Rt/L}\right) \qquad \text{solve for } i$$

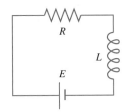

Fig. 28.5

◀ EXAMPLE 3   Integrate: $\displaystyle \int \frac{dx}{e^{3x}}$.

This integral can be put in proper form by writing it as $\int e^{-3x}\, dx$. In this form, $u = -3x$, $du = -3\, dx$. Thus,

$$\int \frac{dx}{e^{3x}} = \int e^{-3x}\, dx = -\frac{1}{3} \int e^{-3x}(-3\, dx)$$

$$= -\frac{1}{3} e^{-3x} + C$$

◀ EXAMPLE 4   Integrate: $\displaystyle \int \frac{4e^{3x} - 3e^x}{e^{x+1}}\, dx$.

This can be put in the proper form for integration and then integrated, as follows:

$$\int \frac{4e^{3x} - 3e^x}{e^{x+1}}\, dx = \int \frac{4e^{3x}}{e^{x+1}}\, dx - \int \frac{3e^x}{e^{x+1}}\, dx$$

$$= 4 \int e^{3x-(x+1)}\, dx - 3 \int e^{x-(x+1)}\, dx \quad \text{using Eq. (11.2)}$$

$$= 4 \int e^{2x-1}\, dx - 3 \int e^{-1}\, dx$$

$$= \frac{4}{2} \int e^{2x-1}(2\, dx) - \frac{3}{e} \int dx$$

$$= 2e^{2x-1} - \frac{3}{e} x + C$$

For reference, Eq. (11.2) is $\dfrac{a^m}{a^n} = a^{m-n}$.

◀ **EXAMPLE 5** Evaluate: $\int_0^{\pi/2} (\sin 2\theta)(e^{\cos 2\theta}) \, d\theta$.

With $u = \cos 2\theta$, $du = -2 \sin 2\theta \, d\theta$, we have

$$\int_0^{\pi/2} (\sin 2\theta)(e^{\cos 2\theta}) \, d\theta = -\frac{1}{2} \int_0^{\pi/2} (e^{\cos 2\theta})(-2 \sin 2\theta \, d\theta)$$

$$= -\frac{1}{2} e^{\cos 2\theta} \Big|_0^{\pi/2} \qquad \text{integrate}$$

$$= -\frac{1}{2}\left(\frac{1}{e} - e\right) = 1.175 \qquad \text{evaluate}$$

The calculator check for this evaluation is shown in Fig. 28.6.

```
fnInt(sin(2θ)e^(
cos(2θ)),θ,0,π/2
)
         1.175201194
-.5(e^(-1)-e^(1)
)
         1.175201194
```

Fig. 28.6

◀ **EXAMPLE 6** Find the equation of the curve for which $\dfrac{dy}{dx} = \dfrac{e^{\sqrt{x+1}}}{\sqrt{x+1}}$ if the curve passes through $(0, 1)$.

The solution of this problem requires that we integrate the given function and then evaluate the constant of integration. Hence,

$$dy = \frac{e^{\sqrt{x+1}}}{\sqrt{x+1}} dx \qquad \int dy = \int \frac{e^{\sqrt{x+1}}}{\sqrt{x+1}} dx$$

For purposes of integrating the right-hand side,

$$u = \sqrt{x+1} \quad \text{and} \quad du = \frac{1}{2\sqrt{x+1}} dx$$

$$y = 2 \int e^{\sqrt{x+1}}\left(\frac{1}{2\sqrt{x+1}} dx\right)$$

$$= 2e^{\sqrt{x+1}} + C$$

Letting $x = 0$ and $y = 1$, we have $1 = 2e + C$, or $C = 1 - 2e$. This means that the equation is

$$y = 2e^{\sqrt{x+1}} + 1 - 2e$$

The graph of this function is shown in Fig. 28.7.

Fig. 28.7

EXERCISES 28.3

In Exercises 1 and 2, make the given changes in the indicated examples of this section and then solve the given problems.

1. In Example 1, change e^{x^2} to e^{x^3}. What other change must be made in the integrand to have a result of $e^{x^3} + C$?

2. In Example 4, change e^{x+1} to e^{x-1} and then integrate.

In Exercises 3–26, integrate each of the given functions.

3. $\displaystyle\int e^{7x}(7 \, dx)$

4. $\displaystyle\int e^{x^4}(4x^3 \, dx)$

5. $\displaystyle\int e^{2x+5} \, dx$

6. $\displaystyle\int 2e^{-4x} \, dx$

7. $\displaystyle\int_{-2}^{2} 6e^{s/2} \, ds$

8. $\displaystyle\int_{1}^{2} 3e^{4x} \, dx$

9. $\displaystyle\int 6x^2 e^{x^3} \, dx$

10. $\displaystyle\int xe^{-x^2} \, dx$

11. $\displaystyle\int_{1}^{4} \frac{e^{\sqrt{x}}}{\sqrt{x}} dx$

12. $\displaystyle\int_{0}^{1} 4(\ln e^u)e^{-2u^2} \, du$

13. $\displaystyle\int 4(\sec \theta \tan \theta)e^{2 \sec \theta} \, d\theta$

14. $\displaystyle\int (\sec^2 x)e^{\tan x} \, dx$

15. $\displaystyle\int \sqrt{e^{2y} + e^{3y}} \, dy$

16. $\displaystyle\int \frac{4 \, dx}{x^2 e^{1/x}}$

17. $\displaystyle\int_{1}^{3} 3e^{2x}(e^{-2x} - 1) \, dx$

18. $\displaystyle\int_{0}^{0.5} \frac{3e^{3x+1}}{e^x} dx$

19. $\displaystyle\int \frac{2 \, dx}{\sqrt{x}e^{\sqrt{x}}}$

20. $\displaystyle\int \frac{4 \, dx}{e^{\sin x} \sec x}$

21. $\int \dfrac{e^{\tan^{-1}x}}{x^2+1}\,dx$

22. $\int \dfrac{e^{\sin^{-1}2x}\,dx}{\sqrt{1-4x^2}}$

23. $\int \dfrac{e^{\cos 3x}\,dx}{\csc 3x}$

24. $\int (e^x - e^{-x})^2\,dx$

25. $\int_0^{\pi} (\sin 2x)e^{\cos^2 x}\,dx$

26. $\int_0^2 \dfrac{e^{2t}\,dt}{\sqrt{e^{2t}+4}}$

In Exercises 27–40, solve the given problems by integration.

27. Find the area bounded by $y = 3e^x$, $x = 0$, $y = 0$, and $x = 2$.

28. Find the area bounded by $x = a$, $x = b$, $y = 0$, and $y = e^x$. Explain the meaning of the result.

29. Integrate $\int 2e^{x^2+\ln x}\,dx$ by first showing that $e^{\ln x} = x$. (*Hint:* Let $y = e^{\ln x}$ and then take natural logarithms of both sides.)

30. Integrate $\int \dfrac{e^{2x}\,dx}{1+e^x}$ by first changing the form of the integrand by algebraic division.

31. Find the volume generated by revolving the region bounded by $y = e^{x^2}$, $x = 1$, $y = 0$, and $x = 2$ about the y-axis. See Fig. 28.8.

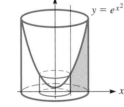

Fig. 28.8

32. Find the equation of the curve for which $dy/dx = \sqrt{e^{x+3}}$ if the curve passes through $(1, 0)$.

33. Find the average value of the function $y = e^{2x}$ from $x = 0$ to $x = 4$.

34. Find the moment of inertia with respect to the y-axis of a flat plate that covers the first-quadrant region bounded by $y = e^{x^3}$, $x = 1$, and the axes.

35. Using Eq. (27.15), show that $\int b^u\,du = \dfrac{b^u}{\ln b} + C$ $(b > 0, b \neq 1)$.

36. Find the first-quadrant area bounded by $y = 2^x$ and $x = 3$. See Exercise 35.

37. For an electric circuit containing a voltage source E, a resistance R, and a capacitance C, an equation relating the charge q on the capacitor and the time t is $qe^{t/RC} = \dfrac{E}{R}\int e^{t/RC}\,dt$. See Fig. 28.9. If $q = 0$ for $t = 0$, perform the integration and then solve for q as a function of t.

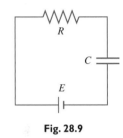

Fig. 28.9

38. In the theory dealing with energy propagation of lasers, the equation $E = a\int_0^{I_0} e^{-Tx}\,dx$ is used. Here a, I_0, and T are constants. Evaluate this integral.

39. The St. Louis Gateway Arch (see Fig. 27.46) has a shape that is given approximately by (measurements in m) $y = -19.46(e^{x/38.92} + e^{-x/38.92}) + 230.9$. What is the maximum height of the Arch?

40. The force F (in N) exerted by a robot programmed to staple carton sections together is given by $F = 6\int e^{\sin \pi t}\cos \pi t\,dt$, where t is the time (in s). Find F as a function of t if $F = 0$ for $t = 1.5$ s.

28.4 BASIC TRIGONOMETRIC FORMS

In this section, we discuss the integrals of the six trigonometric functions and the trigonometric integrals that arise directly from reversing the formulas for differentiation. Other trigonometric forms will be discussed later.

By reversing differentiation formulas, these integral formulas are obtained:

$$\int \sin u\,du = -\cos u + C \qquad (28.4)$$

$$\int \cos u\,du = \sin u + C \qquad (28.5)$$

$$\int \sec^2 u\,du = \tan u + C \qquad (28.6)$$

$$\int \csc^2 u\,du = -\cot u + C \qquad (28.7)$$

$$\int \sec u \tan u\,du = \sec u + C \qquad (28.8)$$

$$\int \csc u \cot u\,du = -\csc u + C \qquad (28.9)$$

◀ EXAMPLE 1   Integrate: $\int x \sec^2 x^2 \, dx$.

With $u = x^2$, $du = 2x \, dx$, we have

$$\int x \sec^2 x^2 \, dx = \frac{1}{2} \int (\sec^2 x^2)(2x \, dx) \quad\quad\quad du \;/\; u$$

$$= \frac{1}{2} \tan x^2 + C \quad\quad\quad \text{using Eq. (28.6)}$$

◀ EXAMPLE 2   Integrate: $\int \dfrac{\tan 2x}{\cos 2x} \, dx$.

By using the basic identity $\sec \theta = 1/\cos \theta$, we can transform this integral into the form $\int \sec 2x \tan 2x \, dx$. In this form, $u = 2x$, $du = 2 \, dx$. Therefore,

$$\int \frac{\tan 2x}{\cos 2x} \, dx = \int \sec 2x \tan 2x \, dx = \frac{1}{2} \int \sec 2x \tan 2x (2 \, dx) \quad\quad du \;/\; u$$

$$= \frac{1}{2} \sec 2x + C \quad\quad \text{using Eq. (28.8)}$$

◀ EXAMPLE 3   The vertical velocity v (in cm/s) of the end of a vibrating rod is given by $v = 80 \cos 20\pi t$, where t is the time in seconds. Find the vertical displacement y (in cm) as a function of t if $y = 0$ for $t = 0$.

Since $v = dy/dt$, we have the following solution:

$$\frac{dy}{dt} = 80 \cos 20\pi t$$

$$\int dy = \int 80 \cos 20\pi t \, dt = \frac{80}{20\pi} \int (\cos 20\pi t)(20\pi \, dt) \quad\quad \text{set up integration}$$

$$y = \frac{4}{\pi} \sin 20\pi t + C \quad\quad \text{using Eq. (28.5)}$$

$$0 = 1.27 \sin 0 + C, \quad C = 0 \quad\quad \text{evaluate } C$$

$$y = 1.27 \sin 20\pi t \quad\quad \text{solution}$$

To find the integrals for the other trigonometric functions, we must change them to a form for which the integral can be determined by methods previously discussed. We can accomplish this by using the basic trigonometric relations.

The formula for $\int \tan u \, du$ is found by expressing the integral in the form $\int (\sin u / \cos u) \, du$. We recognize this as being a logarithmic form, where the u of the logarithmic form is $\cos u$ in this integral. The differential of $\cos u$ is $-\sin u \, du$. Therefore, we have

$$\int \tan u \, du = \int \frac{\sin u}{\cos u} \, du = -\int \frac{-\sin u \, du}{\cos u} = -\ln|\cos u| + C$$

The formula for $\int \cot u \, du$ is found by writing it in the form $\int (\cos u / \sin u) \, du$. In this manner, we obtain the result

$$\int \cot u \, du = \int \frac{\cos u}{\sin u} \, du = \int \frac{\cos u \, du}{\sin u} = \ln|\sin u| + C$$

The formula for $\int \sec u \, du$ is found by writing it in the form

$$\int \frac{\sec u(\sec u + \tan u)}{\sec u + \tan u} \, du$$

We see that this form is also a logarithmic form, since

$$d(\sec u + \tan u) = (\sec u \tan u + \sec^2 u) \, du$$

The right side of this equation is the expression appearing in the numerator of the integral. Thus,

$$\int \sec u \, du = \int \frac{\sec u(\sec u + \tan u) \, du}{\sec u + \tan u} = \int \frac{\sec u \tan u + \sec^2 u}{\sec u + \tan u} \, du$$

$$= \ln|\sec u + \tan u| + C$$

To obtain the formula for $\int \csc u \, du$, we write it in the form

$$\int \frac{\csc u(\csc u - \cot u) \, du}{\csc u - \cot u}$$

Thus, we have

$$\int \csc u \, du = \int \frac{\csc u(\csc u - \cot u)}{\csc u - \cot u} \, du = \int \frac{(-\csc u \cot u + \csc^2 u) \, du}{\csc u - \cot u}$$

$$= \ln|\csc u - \cot u| + C$$

Summarizing these results, we have the following integrals:

$$\int \tan u \, du = -\ln|\cos u| + C \qquad (28.10)$$

$$\int \cot u \, du = \ln|\sin u| + C \qquad (28.11)$$

$$\int \sec u \, du = \ln|\sec u + \tan u| + C \qquad (28.12)$$

$$\int \csc u \, du = \ln|\csc u - \cot u| + C \qquad (28.13)$$

◀ EXAMPLE 4   Integrate: $\int \tan 4\theta \, d\theta$.
Noting that $u = 4\theta$, $du = 4 \, d\theta$, we have

$$\int \tan 4\theta \, d\theta = \frac{1}{4} \int \tan 4\theta(4 \, d\theta) \qquad \text{introducing factors of 4}$$

$$= -\frac{1}{4} \ln|\cos 4\theta| + C \qquad \text{using Eq. (28.10)}$$

◀ EXAMPLE 5   Integrate: $\int \dfrac{\sec e^{-x} \, dx}{e^x}$.
In this integral, $u = e^{-x}$, $du = -e^{-x} \, dx$. Therefore,

$$\int \frac{\sec e^{-x} \, dx}{e^x} = -\int (\sec e^{-x})(-e^{-x} \, dx) \qquad \text{introducing } - \text{sign}$$

$$= -\ln|\sec e^{-x} + \tan e^{-x}| + C \qquad \text{using Eq. (28.12)}$$

EXAMPLE 6 Evaluate: $\displaystyle\int_{\pi/6}^{\pi/4} \frac{1 + \cos x}{\sin x} dx$.

The solution is as follows:

$$\int_{\pi/6}^{\pi/4} \frac{1 + \cos x}{\sin x} dx = \int_{\pi/6}^{\pi/4} \csc x\, dx + \int_{\pi/6}^{\pi/4} \cot x\, dx \qquad \text{using Eqs. (20.1) and (20.5)}$$

$$= \ln|\csc x - \cot x|\Big|_{\pi/6}^{\pi/4} + \ln|\sin x|\Big|_{\pi/6}^{\pi/4} \qquad \text{integrating}$$

$$= \ln\left|\sqrt{2} - 1\right| - \ln\left|2 - \sqrt{3}\right| + \ln\left|\tfrac{1}{2}\sqrt{2}\right| - \ln\left|\tfrac{1}{2}\right| \qquad \text{evaluating}$$

$$= \ln \frac{\left(\tfrac{1}{2}\sqrt{2}\right)\left(\sqrt{2} - 1\right)}{\left(\tfrac{1}{2}\right)\left(2 - \sqrt{3}\right)} = \ln \frac{2 - \sqrt{2}}{2 - \sqrt{3}} = 0.782$$

EXERCISES 28.4

In Exercises 1 and 2, make the given changes in the indicated examples of this section and then solve the given problems.

1. In Example 1, change $\sec^2 x^2$ to $\sec^2 x^3$. What other change must be made in the integrand to have a result of $\tan x^3 + C$?

2. In Example 4, change tan to cot and then integrate.

In Exercises 3–26, integrate each of the given functions.

3. $\displaystyle\int \cos 2x\, dx$

4. $\displaystyle\int 4 \sin(2 - x)\, dx$

5. $\displaystyle\int 0.3 \sec^2 3\theta\, d\theta$

6. $\displaystyle\int \csc 2x \cot 2x\, dx$

7. $\displaystyle\int \sec \tfrac{1}{2}x \tan \tfrac{1}{2}x\, dx$

8. $\displaystyle\int e^x \csc^2(e^x)\, dx$

9. $\displaystyle\int_{0.5}^{1} x^2 \cot x^3\, dx$

10. $\displaystyle\int_{0}^{1} 6 \sin \tfrac{1}{2}t \sec \tfrac{1}{2}t\, dt$

11. $\displaystyle\int 3\phi \sec^2 \phi^2 \cos \phi^2\, d\phi$

12. $\displaystyle\int 2 \csc 3x\, dx$

13. $\displaystyle\int \frac{\sin(1/x)}{x^2}\, dx$

14. $\displaystyle\int \frac{3\, dx}{\sin 4x}$

15. $\displaystyle\int_{0}^{\pi/6} \frac{dx}{\cos^2 2x}$

16. $\displaystyle\int_{0}^{1} \frac{2e^s\, ds}{\sec e^s}$

17. $\displaystyle\int \frac{\sec 5x}{\cot 5x}\, dx$

18. $\displaystyle\int \frac{\sin 2x}{\cos^2 x}\, dx$

19. $\displaystyle\int \sqrt{\tan^2 2x + 1}\, dx$

20. $\displaystyle\int 5(\tan u)(\ln \cos u)\, du$

21. $\displaystyle\int \frac{1 + \sin 2T}{\tan 2T}\, dT$

22. $\displaystyle\int \frac{1 - \cot^2 x}{\cos^2 x}\, dx$

23. $\displaystyle\int \frac{1 - \sin x}{1 + \cos x}\, dx$

24. $\displaystyle\int \frac{1 + \sec^2 x}{x + \tan x}\, dx$

25. $\displaystyle\int_{0}^{\pi/9} \sin 3x(\csc 3x + \sec 3x)\, dx$

26. $\displaystyle\int_{\pi/4}^{\pi/3} (1 + \sec x)^2\, dx$

In Exercises 27–36, solve the given problems by integration.

27. Find the area bounded by $y = 2 \tan x$, $x = \frac{\pi}{4}$, and $y = 0$.

28. Find the area under the curve $y = \sin x$ from $x = 0$ to $x = \pi$.

29. Although $\displaystyle\int \frac{dx}{1 + \sin x}$ does not appear to fit a form for integration, show that it can be integrated by multiplying the numerator and the denominator by $1 - \sin x$.

30. Evaluate $\displaystyle\int_{a}^{a+2\pi} \sin x\, dx$ for any real value of a. Show an interpretation of the result in terms of the area under a curve.

31. Find the volume generated by revolving the region bounded by $y = \sec x$, $x = 0$, $x = \frac{\pi}{3}$, and $y = 0$ about the x-axis.

32. Find the volume generated by revolving the region bounded by $y = \cos x^2$, $x = 0$, $y = 0$, and $x = 1$ about the y-axis.

33. The angular velocity ω (in rad/s) of a pendulum is $\omega = -0.25 \sin 2.5t$. Find the angular displacement θ as a function of t if $\theta = 0.10$ for $t = 0$.

34. If the current i (in A) in a certain electric circuit is given by $i = 110 \cos 377t$, find the expression for the voltage across a 500-μF capacitor as a function of time. The initial voltage is zero. Show that the voltage across the capacitor is 90° out of phase with the current.

35. A fin on a wind-direction indicator has a shape that can be described as the region bounded by $y = \tan x^2$, $y = 0$, and $x = 1$. Find the x-coordinate (in m) of the centroid of the fin if its area is 0.3984 m².

36. A force is given as a function of the distance from the origin as $F = \dfrac{2 + \tan x}{\cos x}$. Express the work done by this force as a function of x if $W = 0$ for $x = 0$.

28.5 OTHER TRIGONOMETRIC FORMS

By use of trigonometric relations developed in Chapter 20, it is possible to transform many integrals involving powers of the trigonometric functions into integrable form. We now show the relationships that are useful for these integrals.

$$\cos^2 x + \sin^2 x = 1 \qquad \textbf{(28.14)}$$

$$1 + \tan^2 x = \sec^2 x \qquad \textbf{(28.15)}$$

$$1 + \cot^2 x = \csc^2 x \qquad \textbf{(28.16)}$$

$$2\cos^2 x = 1 + \cos 2x \qquad \textbf{(28.17)}$$

$$2\sin^2 x = 1 - \cos 2x \qquad \textbf{(28.18)}$$

CAUTION ▶ *To integrate a product of powers of the sine and cosine, we use Eq. (28.14)* ***if at least one of the powers is odd.*** The method is based on transforming the integral so that it is made up of powers of either the sine or cosine and the first power of the other. In this way this first power becomes a factor of *du*.

◀ **EXAMPLE 1** Integrate: $\int \sin^3 x \cos^2 x \, dx$.

Since $\sin^3 x = \sin^2 x \sin x = (1 - \cos^2 x)\sin x$, we can write this integral with powers of $\cos x$ along with $-\sin x$, which is the necessary *du* for this integral. Therefore,

$$\int \sin^3 x \cos^2 x \, dx = \int (1 - \cos^2 x)(\sin x)(\cos^2 x) \, dx \qquad \text{using Eq. (28.14)}$$

$$= \int (\cos^2 x - \cos^4 x)(\sin x \, dx)$$

$$= -\int \cos^2 x(-\sin x \, dx) + \int \cos^4 x(-\sin x \, dx)$$

$$= -\frac{1}{3}\cos^3 x + \frac{1}{5}\cos^5 x + C$$

◀ **EXAMPLE 2** Integrate: $\int \cos^5 2x \, dx$.

Since $\cos^5 2x = \cos^4 2x \cos 2x = (1 - \sin^2 2x)^2 \cos 2x$, it is possible to write this integral with powers of $\sin 2x$ along with $\cos 2x \, dx$. Thus, with the introduction of a factor of 2, $(\cos 2x)(2 \, dx)$ is the necessary *du* for this integral. Thus,

$$\int \cos^5 2x \, dx = \int (1 - \sin^2 2x)^2 \cos 2x \, dx \qquad \text{using Eq. (28.14)}$$

$$= \int (1 - 2\sin^2 2x + \sin^4 2x) \cos 2x \, dx$$

$$= \int \cos 2x \, dx - \int 2\sin^2 2x \cos 2x \, dx + \int \sin^4 2x \cos 2x \, dx$$

$$= \frac{1}{2}\int \cos 2x(2 \, dx) - \int \sin^2 2x(2 \cos 2x \, dx) + \frac{1}{2}\int \sin^4 2x(2 \cos 2x \, dx)$$

$$= \frac{1}{2}\sin 2x - \frac{1}{3}\sin^3 2x + \frac{1}{10}\sin^5 2x + C$$

CAUTION ▶ *In products of powers of the sine and cosine, **if the powers to be integrated are even,** **we use Eqs. (28.17) and (28.18) to transform the integral.*** Those most commonly met are $\int \cos^2 u\, du$ and $\int \sin^2 u\, du$. Consider the following examples.

◀ EXAMPLE 3   Integrate: $\int \sin^2 2x\, dx$.

Using Eq. (28.18) in the form $\sin^2 2x = \frac{1}{2}(1 - \cos 4x)$, this integral can be transformed

CAUTION ▶ into a form that can be integrated. (Here we note ***the x of Eq. (28.18) is treated as 2x*** for this integral.) Therefore, we write

$$\int \sin^2 2x\, dx = \int \left[\frac{1}{2}(1 - \cos 4x) \right] dx \qquad \text{using Eq. (28.18)}$$

$$= \frac{1}{2} \int dx - \frac{1}{8} \int \cos 4x(4\, dx)$$

$$= \frac{x}{2} - \frac{1}{8} \sin 4x + C$$

To integrate even powers of the secant, powers of the tangent, or products of the secant and tangent, we use Eq. (28.15) to transform the integral. In transforming, the

NOTE ▶ forms we look for are ***powers of the tangent with*** $\sec^2 x$***,*** which becomes part of du, or ***powers of the secant along with*** $\sec x \tan x$***,*** which becomes part of du in this case. Similar transformations are made when we integrate powers of the cotangent and cosecant, with the use of Eq. (28.16).

◀ EXAMPLE 4   Integrate: $\int \sec^3 t \tan t\, dt$.

By writing $\sec^3 t \tan t$ as $\sec^2 t(\sec t \tan t)$, we can use the $\sec t \tan t\, dt$ as the du of the integral. Thus,

$$\int \sec^3 t \tan t\, dt = \int (\sec^2 t)(\overset{\displaystyle du}{\overline{\sec t \tan t\, dt}})$$

$$= \frac{1}{3} \sec^3 t + C$$

◀ EXAMPLE 5   Integrate: $\int \tan^5 x\, dx$.

Since $\tan^5 x = \tan^3 x \tan^2 x = \tan^3 x(\sec^2 x - 1)$, we can write this integral with powers of $\tan x$ along with $\sec^2 x\, dx$. Thus, $\sec^2 x\, dx$ becomes the necessary du of the integral. It is necessary to replace $\tan^2 x$ with $\sec^2 x - 1$ twice during the integration. Therefore,

$$\int \tan^5 x\, dx = \int \tan^3 x(\sec^2 x - 1)\, dx \qquad \text{using Eq. (28.15)}$$

$$= \int \tan^3 x(\sec^2 x\, dx) - \int \tan^3 x\, dx$$

$$= \frac{1}{4} \tan^4 x - \int \tan x(\sec^2 x - 1)\, dx \qquad \text{using Eq. (28.15) again}$$

$$= \frac{1}{4} \tan^4 x - \int \tan x(\sec^2 x\, dx) + \int \tan x\, dx$$

$$= \frac{1}{4} \tan^4 x - \frac{1}{2} \tan^2 x - \ln|\cos x| + C$$

◀ EXAMPLE 6   Integrate: $\int_0^{\pi/4} \dfrac{\tan^3 x}{\sec^3 x}\, dx$.

This integral requires the use of several trigonometric relationships to obtain integrable forms.

$$\int_0^{\pi/4} \frac{\tan^3 x}{\sec^3 x}\, dx = \int_0^{\pi/4} \frac{(\sec^2 x - 1)\tan x}{\sec^3 x}\, dx \qquad \text{using Eq. (28.15)}$$

$$= \int_0^{\pi/4} \frac{\tan x}{\sec x}\, dx - \int_0^{\pi/4} \frac{\tan x\, dx}{\sec^3 x} \qquad \frac{\tan x}{\sec x} = \frac{\sin x}{\cos x \sec x} = \sin x$$

$$= \int_0^{\pi/4} \sin x\, dx - \int_0^{\pi/4} \cos^2 x \sin x\, dx \qquad \frac{\tan x}{\sec^3 x} = \frac{\tan x}{\sec x \sec^2 x} = \sin x \cos^2 x$$

$$= -\cos x + \frac{1}{3}\cos^3 x \Big|_0^{\pi/4} \qquad \text{integrate}$$

$$= -\frac{\sqrt{2}}{2} + \frac{1}{3}\left(\frac{\sqrt{2}}{2}\right)^3 - \left(-1 + \frac{1}{3}\right) \qquad \text{evaluate}$$

$$= \frac{8 - 5\sqrt{2}}{12} = 0.0774$$

```
fnInt((tan(X))^3
(cos(X)^3),X,0,π
/4)
         .0774110157
(8-5√(2))/12
         .0774110157
```

Fig. 28.10

See Fig. 28.10 for a calculator check of this result.   ▶

See the chapter introduction.

◀ EXAMPLE 7   The *root-mean-square value of a function* with respect to x is defined by

$$\boxed{y_{\text{rms}} = \sqrt{\frac{1}{T}\int_0^T y^2\, dx}} \qquad (28.19)$$

Usually the value of T that is of importance is the period of the function. Find the root-mean-square value of the electric current i (in A) used in a home electric heater, for which $i = 17.7\cos 120\pi t$, for one period.

In most countries, the rms voltage is 240 V. In the United States and Canada, it is 120 V.

The period is $\frac{2\pi}{120\pi} = \frac{1}{60.0}$ s. Thus, we must find the square root of the integral

$$\frac{1}{1/60.0}\int_0^{1/60.0} (17.7\cos 120\pi t)^2\, dt = 18\,800 \int_0^{1/60.0} \cos^2 120\pi t\, dt$$

Evaluating this integral, we have

$$18\,800 \int_0^{1/60.0} \cos^2 120\pi t\, dt = 9400 \int_0^{1/60.0} (1 + \cos 240\pi t)\, dt$$

$$= 9400 t \Big|_0^{1/60.0} + \frac{9400}{240\pi}\int_0^{1/60.0} \cos 240\pi t (240\pi\, dt)$$

$$= 156.7 + \frac{235}{6\pi}\sin 240\pi t \Big|_0^{1/60.0} = 156.7$$

This means the root-mean-square current is

$$i_{\text{rms}} = \sqrt{156.7} = 12.5 \text{ A}$$

This value of the current, often referred to as the *effective current,* is the value of direct current that would produce the same quantity of heat energy in the same time. It is important in the design of electric heaters and other electric appliances.   ▶

EXERCISES 28.5

In Exercises 1 and 2, answer the given questions related to the indicated examples of this section.

1. In Example 3, what change must be made in the integrand in order to have a result of $\frac{1}{6}\sin^3 2x + C$?

2. In Example 5, what change must be made in the integrand in order to have a result of $\frac{1}{6}\tan^6 x + C$?

In Exercises 3–30, integrate each of the given functions.

3. $\displaystyle\int \sin^2 x \cos x\, dx$

4. $\displaystyle\int \sin x \cos^5 x\, dx$

5. $\displaystyle\int \sin^3 2x\, dx$

6. $\displaystyle\int 3\cos^3 T\, dT$

7. $\displaystyle\int 4(\cos^4\theta - \sin^4\theta)\, d\theta$

8. $\displaystyle\int \sin^3 x \cos^6 x\, dx$

9. $\displaystyle\int_0^{\pi/4} 5\sin^5 x\, dx$

10. $\displaystyle\int_{\pi/3}^{\pi/2} 10\sin t(1 - \cos 2t)^2\, dt$

11. $\displaystyle\int \sin^2 x\, dx$

12. $\displaystyle\int \cos^2 2x\, dx$

13. $\displaystyle\int 2(1 + \cos 3\phi)^2\, d\phi$

14. $\displaystyle\int_0^1 \sin^2 4x\, dx$

15. $\displaystyle\int \tan^3 x\, dx$

16. $\displaystyle\int \frac{6\cot^2 y}{\tan y}\, dy$

17. $\displaystyle\int_0^{\pi/4} \tan x \sec^4 x\, dx$

18. $\displaystyle\int \cot 4x \csc^4 4x\, dx$

19. $\displaystyle\int \tan^4 2x\, dx$

20. $\displaystyle\int 4\cot^4 x\, dx$

21. $\displaystyle\int 0.5\sin s \sin 2s\, ds$

22. $\displaystyle\int \sqrt{\tan x}\, \sec^4 x\, dx$

23. $\displaystyle\int (\sin x + \cos x)^2\, dx$

24. $\displaystyle\int (\tan 2x + \cot 2x)^2\, dx$

25. $\displaystyle\int \frac{1 - \cot x}{\sin^4 x}\, dx$

26. $\displaystyle\int \frac{(\sin u + \sin^2 u)^2}{\sec u}\, du$

27. $\displaystyle\int_{\pi/6}^{\pi/4} \cot^5 p\, dp$

28. $\displaystyle\int_{\pi/6}^{\pi/3} \frac{2\, dx}{1 + \sin x}$

29. $\displaystyle\int \sec^6 x\, dx$

30. $\displaystyle\int \tan^7 x\, dx$

In Exercises 31–44, solve the given problems by integration.

31. Find the volume generated by revolving the region bounded by $y = \sin x$ and $y = 0$, from $x = 0$ to $x = \pi$, about the x-axis.

32. Find the volume generated by revolving the region bounded by $y = \tan^3(x^2)$, $y = 0$, and $x = \frac{\pi}{4}$ about the y-axis.

33. Find the area bounded by $y = \sin x$, $y = \cos x$, and $x = 0$ in the first quadrant.

34. Find the length of the curve $y = \ln\cos x$ from $x = 0$ to $x = \frac{\pi}{3}$. (See Exercise 33 of Section 26.6).

(W) 35. Show that $\int \sin x \cos x\, dx$ can be integrated in two ways. Explain the difference in the answers.

(W) 36. Show that $\int \sec^2 x \tan x\, dx$ can be integrated in two ways. Explain the difference in the answers.

37. Show that $\displaystyle\int_0^\pi \sin^2 nx\, dx = \frac{1}{2}\pi$, where n is any positive integer.

38. Find the root-mean-square current in a circuit from $t = 0$ s to $t = 0.50$ s if $i = i_0 \sin t\sqrt{\cos t}$.

39. In the study of the rate of radiation by an accelerated charge, the following integral must be evaluated: $\int_0^\pi \sin^3\theta\, d\theta$. Find the value of the integral.

40. In finding the volume of a special O-ring for a space vehicle, the integral $\displaystyle\int \frac{\sin^2\theta}{\cos^2\theta}\, d\theta$ must be evaluated. Perform this integration.

41. For a voltage $V = 340\sin 120\pi t$, show that the root-mean-square voltage for one period is 240 V.

42. For a current $i = i_0 \sin\omega t$, show that the root-mean-square current for one period is $i_0/\sqrt{2}$.

43. In the analysis of the intensity of light from a certain source, the equation $I = A\int_{-a/2}^{a/2} \cos^2[b\pi(c - x)]\, dx$ is used. Here, A, a, b, and c are constants. Evaluate this integral. (The simplification is quite lengthy.)

44. In the study of the lifting force L due to a stream of fluid passing around a cylinder, the equation $L = k\int_0^{2\pi}(a\sin\theta + b\sin^2\theta - b\sin^3\theta)\, d\theta$ is used. Here, k, a, and b are constants and θ is the angle from the direction of flow. Evaluate the integral.

28.6 INVERSE TRIGONOMETRIC FORMS

For reference, Eq. (27.10) is
$$\frac{d(\sin^{-1} u)}{dx} = \frac{1}{\sqrt{1 - u^2}} \frac{du}{dx}.$$

Using Eq. (27.10), we can find the differential of $\sin^{-1}(u/a)$, where a is constant:

$$d\left(\sin^{-1} \frac{u}{a}\right) = \frac{1}{\sqrt{1 - (u/a)^2}} \frac{du}{a} = \frac{a}{\sqrt{a^2 - u^2}} \frac{du}{a} = \frac{du}{\sqrt{a^2 - u^2}}$$

Reversing this differentiation formula, we have the important integration formula

$$\int \frac{du}{\sqrt{a^2 - u^2}} = \sin^{-1} \frac{u}{a} + C \tag{28.20}$$

By finding the differential of $\tan^{-1}(u/a)$, we have

$$d\left(\tan^{-1} \frac{u}{a}\right) = \frac{1}{1 + (u/a)^2} \frac{du}{a} = \frac{a^2}{a^2 + u^2} \frac{du}{a} = \frac{a\,du}{a^2 + u^2}$$

Now, reversing this differential, we have

$$\int \frac{du}{a^2 + u^2} = \frac{1}{a} \tan^{-1} \frac{u}{a} + C \tag{28.21}$$

This shows one of the principal uses of the inverse trigonometric functions: They provide a solution to the integration of important algebraic functions.

◀ **EXAMPLE 1** Integrate: $\int \frac{dx}{\sqrt{9 - x^2}}$.

This integral fits the form of Eq. (28.20) with $u = x$, $du = dx$, and $a = 3$. Thus,

$$\int \frac{dx}{\sqrt{9 - x^2}} = \int \frac{dx}{\sqrt{3^2 - x^2}}$$

$$= \sin^{-1} \frac{x}{3} + C \qquad \blacktriangleright$$

◀ **EXAMPLE 2** The volume flow rate Q (in m³/s) of a constantly flowing liquid is given by $Q = 24 \int_0^2 \frac{dx}{6 + x^2}$, where x is the distance from the center of flow. Find the value of Q.

For the integral, we see that it fits Eq. (28.21) with $u = x$, $du = dx$, and $a = \sqrt{6}$.

$$Q = 24 \int_0^2 \frac{dx}{6 + x^2} = 24 \int_0^2 \frac{dx}{\left(\sqrt{6}\right)^2 + x^2}$$

$$= \frac{24}{\sqrt{6}} \tan^{-1} \frac{x}{\sqrt{6}} \Big|_0^2$$

$$= \frac{24}{\sqrt{6}} \left(\tan^{-1} \frac{2}{\sqrt{6}} - \tan^{-1} 0\right)$$

$$= 6.71 \text{ m}^3/\text{s} \qquad \blacktriangleright$$

◀ **EXAMPLE 3**   Integrate: $\displaystyle\int \frac{dx}{\sqrt{25 - 4x^2}}$.

This integral fits the form of Eq. (28.20) with $u = 2x$, $du = 2\,dx$, and $a = 5$. Thus, in order to have the proper du, we must include a factor of 2 in the numerator, and therefore we also place a $\frac{1}{2}$ before the integral. This leads to

$$\int \frac{dx}{\sqrt{25 - 4x^2}} = \frac{1}{2} \int \frac{2\,dx}{\sqrt{5^2 - (2x)^2}} \longleftarrow du$$

$$= \frac{1}{2} \sin^{-1} \frac{2x}{5} + C$$

◀ **EXAMPLE 4**   Integrate: $\displaystyle\int_{-1}^{3} \frac{dx}{x^2 + 6x + 13}$.

At first glance, it does not appear that this integral fits any of the forms presented up to this point. However, by writing the denominator in the form

CAUTION ▶ $(x^2 + 6x + 9) + 4 = (x + 3)^2 + 2^2$, *we recognize that* $u = x + 3$, $du = dx$, and $a = 2$. Thus,

$$\int_{-1}^{3} \frac{dx}{x^2 + 6x + 13} = \int_{-1}^{3} \frac{dx}{(x + 3)^2 + 2^2} \longleftarrow du$$

$$= \frac{1}{2} \tan^{-1} \frac{x + 3}{2} \Big|_{-1}^{3} \qquad \text{integrate}$$

$$= \frac{1}{2} (\tan^{-1} 3 - \tan^{-1} 1) \qquad \text{evaluate}$$

$$= 0.2318$$

Now we can see the use of completing the square when we are transforming integrals into proper form.

◀ **EXAMPLE 5**   Integrate: $\displaystyle\int \frac{2r + 5}{r^2 + 9}\,dr$.

By writing this integral as the sum of two integrals, we may integrate each of these separately:

$$\int \frac{2r + 5}{r^2 + 9}\,dr = \int \frac{2r\,dr}{r^2 + 9} + \int \frac{5\,dr}{r^2 + 9}$$

The first integral is a logarithmic form, and the second is an inverse tangent form. For the first, $u = r^2 + 9$, $du = 2r\,dr$. For the second, $u = r$, $du = dr$, $a = 3$.

$$\int \frac{2r\,dr}{r^2 + 9} + 5 \int \frac{dr}{r^2 + 9} = \ln|r^2 + 9| + \frac{5}{3} \tan^{-1} \frac{r}{3} + C$$

CAUTION ▶   The inverse trigonometric integral forms show very well the importance of ***proper recognition of the form of the integral.*** It is important that these forms are not confused with those of the general power rule or the logarithmic form.

❰ EXAMPLE 6   The integral $\int \dfrac{dx}{\sqrt{1-x^2}}$ is of the inverse sine form with $u = x$, $du = dx$, and $a = 1$. Thus,

$$\int \frac{dx}{\sqrt{1-x^2}} = \sin^{-1} x + C$$

The integral $\int \dfrac{x\,dx}{\sqrt{1-x^2}}$ is not of the inverse sine form due to the factor of x in the numerator. It is integrated by use of the general power rule, with $u = 1 - x^2$, $du = -2x\,dx$, and $n = -\frac{1}{2}$. Thus,

$$\int \frac{x\,dx}{\sqrt{1-x^2}} = -\sqrt{1-x^2} + C$$

The integral $\int \dfrac{x\,dx}{1-x^2}$ is of the basic logarithmic form with $u = 1 - x^2$ and $du = -2x\,dx$. If $1 - x^2$ is raised to any power other than 1 in the denominator, we would use the general power rule. To be of the inverse sine form, we would have the square root of $1 - x^2$ and no factor of x, as in the first illustration, Thus,

$$\int \frac{x\,dx}{1-x^2} = -\frac{1}{2}\ln|1-x^2| + C$$

❰ EXAMPLE 7   The following integrals are of the form indicated.

$\int \dfrac{dx}{1+x^2}$	Inverse tangent form	$u = x,\ du = dx$
$\int \dfrac{x\,dx}{1+x^2}$	Logarithmic form	$u = 1 + x^2,\ du = 2x\,dx$
$\int \dfrac{x\,dx}{\sqrt{1+x^2}}$	General power form	$u = 1 + x^2,\ du = 2x\,dx$
$\int \dfrac{dx}{1+x}$	Logarithmic form	$u = 1 + x,\ du = dx$
$\int \dfrac{x\,dx}{\sqrt{1-x^4}}$	Inverse sine form	$u = x^2,\ du = 2x\,dx$
$\int \dfrac{x\,dx}{1+x^4}$	Inverse tangent form	$u = x^2,\ du = 2x\,dx$

There are a number of integrals whose forms appear to be similar to those in Examples 6 and 7, but which do not fit the forms we have discussed. They include

$$\int \frac{dx}{\sqrt{x^2-1}} \qquad \int \frac{dx}{\sqrt{1+x^2}} \qquad \int \frac{dx}{1-x^2} \qquad \int \frac{dx}{x\sqrt{1+x^2}}$$

We will develop methods to integrate some of these forms, and all of them can be integrated by the tables discussed in Section 28.11.

EXERCISES 28.6

In Exercises 1 and 2, answer the given questions related to the indicated examples of this section.

1. In Example 1, what change must be made in the integrand in order to have a result of $\sqrt{9 - x^2} + C$?

2. In Example 2, what change must be made in the integrand in order that the integration would lead to a result of $\ln(6 + x^2) + C$?

In Exercises 3–26, integrate each of the given functions.

3. $\displaystyle\int \frac{dx}{\sqrt{4 - x^2}}$

4. $\displaystyle\int \frac{dx}{\sqrt{49 - x^2}}$

5. $\displaystyle\int \frac{dx}{64 + x^2}$

6. $\displaystyle\int \frac{6p^2\,dp}{4 + p^6}$

7. $\displaystyle\int \frac{x\,dx}{\sqrt{1 - 16x^4}}$

8. $\displaystyle\int_0^1 \frac{2\,dx}{\sqrt{9 - 4x^2}}$

9. $\displaystyle\int_0^2 \frac{3e^{-t}\,dt}{1 + 9e^{-2t}}$

10. $\displaystyle\int_1^3 \frac{dx}{49 + 4x^2}$

11. $\displaystyle\int_0^{0.4} \frac{2\,dx}{\sqrt{4 - 5x^2}}$

12. $\displaystyle\int \frac{dx}{2\sqrt{x}\sqrt{1 - x}}$

13. $\displaystyle\int \frac{8x\,dx}{9x^2 + 16}$

14. $\displaystyle\int \frac{4y\,dy}{\sqrt{25 - 16y^2}}$

15. $\displaystyle\int_1^e \frac{3\,du}{u[1 + (\ln u)^2]}$

16. $\displaystyle\int_0^1 \frac{4x\,dx}{1 + x^4}$

17. $\displaystyle\int \frac{e^x\,dx}{\sqrt{1 - e^{2x}}}$

18. $\displaystyle\int \frac{\sec^2 x\,dx}{\sqrt{1 - \tan^2 x}}$

19. $\displaystyle\int \frac{dT}{T^2 + 2T + 2}$

20. $\displaystyle\int \frac{2\,dx}{x^2 + 8x + 17}$

21. $\displaystyle\int \frac{4\,dx}{\sqrt{-4x - x^2}}$

22. $\displaystyle\int \frac{0.3\,ds}{\sqrt{2s - s^2}}$

23. $\displaystyle\int_{\pi/6}^{\pi/2} \frac{2\cos 2\theta\,d\theta}{1 + \sin^2 2\theta}$

24. $\displaystyle\int_{-4}^0 \frac{dx}{x^2 + 4x + 5}$

25. $\displaystyle\int \frac{2 - x}{\sqrt{4 - x^2}}\,dx$

26. $\displaystyle\int \frac{3 - 2x}{1 + 4x^2}\,dx$

In Exercises 27–30, identify the form of each integral as being inverse sine, inverse tangent, logarithmic, or general power, as in Examples 6 and 7. Do not integrate. In each part (a), explain how the choice was made.

27. (a) $\displaystyle\int \frac{2\,dx}{4 + 9x^2}$ **(b)** $\displaystyle\int \frac{2\,dx}{4 + 9x}$ **(c)** $\displaystyle\int \frac{2x\,dx}{\sqrt{4 + 9x^2}}$

28. (a) $\displaystyle\int \frac{2x\,dx}{4 - 9x^2}$ **(b)** $\displaystyle\int \frac{2\,dx}{\sqrt{4 - 9x}}$ **(c)** $\displaystyle\int \frac{2x\,dx}{4 + 9x^2}$

29. (a) $\displaystyle\int \frac{2x\,dx}{\sqrt{4 - 9x^2}}$ **(b)** $\displaystyle\int \frac{2\,dx}{\sqrt{4 - 9x^2}}$ **(c)** $\displaystyle\int \frac{2\,dx}{4 - 9x}$

30. (a) $\displaystyle\int \frac{2\,dx}{9x^2 + 4}$ **(b)** $\displaystyle\int \frac{2x\,dx}{\sqrt{9x^2 - 4}}$ **(c)** $\displaystyle\int \frac{2x\,dx}{9x^2 - 4}$

In Exercises 31–40, solve the given problems by integration.

31. Explain how to integrate $\displaystyle\int \frac{dx}{\sqrt{x}(1 + x)}$. What is the result?

32. Integrate $\displaystyle\int \sqrt{\frac{1 + x}{1 - x}}\,dx$ by first multiplying the numerator and denominator of the fraction under the radical by $1 + x$.

33. Find the area bounded by $y(1 + x^2) = 1$, $x = 0$, $y = 0$, and $x = 2$.

34. Find the area bounded by $y\sqrt{4 - x^2} = 1$, $x = 0$, $y = 0$, and $x = 1$. See Fig. 28.11.

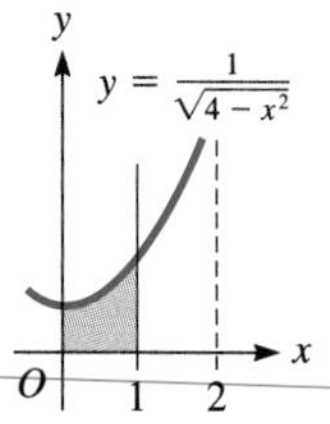

Fig. 28.11

35. To find the electric field E from an electric charge distributed uniformly over the entire xy-plane at a distance d from the plane, it is necessary to evaluate the integral $kd \displaystyle\int \frac{dx}{d^2 + x^2}$. Here x is the distance from the origin to the element of charge. Perform the indicated integration.

36. An oil-storage tank can be described as the volume generated by revolving the region bounded by $y = 24/\sqrt{16 + x^2}$, $x = 0$, $y = 0$, and $x = 3$ about the x-axis. Find the volume (in m³) of the tank.

37. In dealing with the theory for simple harmonic motion, it is necessary to solve the equation $\displaystyle\frac{dx}{\sqrt{A^2 - x^2}} = \sqrt{\frac{k}{m}}\,dt$ (k, m, and A are constants). Determine the solution if $x = x_0$ when $t = 0$.

38. During each cycle, the velocity v (in m/s) of a robotic welding device is given by $v = 2t - \dfrac{12}{2 + t^2}$, where t is the time (in s). Find the expression for the displacement s (in m) as a function of t if $s = 0$ for $t = 0$.

39. Find the moment of inertia with respect to the y-axis for a flat plate covering the region bounded by $y = 1/(1 + x^6)$, the x-axis, $x = 1$, and $x = 2$.

40. Find the length of arc along the curve $y = \sqrt{1 - x^2}$ between $x = 0$ and $x = 1$. (See Exercise 33 of Section 26.6.)

28.7 INTEGRATION BY PARTS

There are many methods of transforming integrals into forms that can be integrated by one of the basic formulas. In the preceding sections we saw that completing the square and trigonometric identities can be used for this purpose. In this section and the following one, we develop two general methods. The method of *integration by parts* is discussed in this section.

Since the derivative of a product of functions is found by use of the formula

$$\frac{d(uv)}{dx} = u\frac{dv}{dx} + v\frac{du}{dx}$$

the differential of a product of functions is given by $d(uv) = u\,dv + v\,du$. Integrating both sides of this equation, we have $uv = \int u\,dv + \int v\,dv$. Solving for $\int u\,dv$, we obtain

$$\int u\,dv = uv - \int v\,du \qquad (28.22)$$

Integration by use of Eq. (28.22) is called **integration by parts.**

◀ EXAMPLE 1   Integrate: $\int x \sin x\,dx$.

This integral does not fit any of the previous forms we have discussed, since neither x nor $\sin x$ can be made a factor of a proper du. However, by choosing $u = x$ and $dv = \sin x\,dx$, integration by parts may be used. Thus,

$$u = x \qquad dv = \sin x\,dx$$

By finding the differential of u and integrating dv, we find du and v. This gives us

$$du = dx \qquad v = -\cos x + C_1$$

Now, substituting in Eq. (28.22), we have

$$\int \overset{u}{\downarrow} \ \overset{dv}{\downarrow} \ = \ \overset{u}{\downarrow} \ \overset{v}{\downarrow} \ - \int \ \overset{v}{\downarrow} \ \overset{du}{\downarrow}$$

$$\int (x)(\sin x\,dx) = (x)(-\cos x + C_1) - \int (-\cos x + C_1)(dx)$$

$$= -x\cos x + C_1 x + \int \cos x\,dx - \int C_1\,dx$$

$$= -x\cos x + C_1 x + \sin x - C_1 x + C$$

$$= -x\cos x + \sin x + C$$

Other choices of u and dv may be made, but they are not useful. For example, if we choose $u = \sin x$ and $dv = x\,dx$, then $du = \cos x\,dx$ and $v = \frac{1}{2}x^2 + C_2$. This makes $\int v\,du = \int \left(\frac{1}{2}x^2 + C_2\right)(\cos x\,dx)$, which is more complex than the integrand of the original problem.

NOTE ▶   We also note that the constant C_1 that was introduced when we integrated dv does not appear in the final result. This constant will always cancel out, and therefore *we will not show any constant of integration when finding v.*   ◗

As in Example 1, there is often more than one choice as to the part of the integrand that is selected to be u and the part that is selected to be dv. There are no set rules that may be stated for the best choice of u and dv, but two guidelines may be stated.

> **GUIDELINES FOR CHOOSING u AND dv**
>
> CAUTION ▶
>
> **1.** *The quantity u is normally chosen such that du/dx is of simpler form than u.*
> **2.** *The differential dv is normally chosen such that $\int dv$ is easily integrated.*

Working examples, and thereby gaining experience in methods of integration, is the best way to determine when this method should be used and how to use it.

◀ **EXAMPLE 2** Integrate: $\int x\sqrt{1-x}\,dx$.

We see that this form does not fit the general power rule, for $x\,dx$ is not a factor of the differential of $1-x$. By choosing $u = x$ and $dv = \sqrt{1-x}\,dx$, we have $du/dx = 1$, and v can readily be determined. Thus,

$$u = x \qquad dv = \sqrt{1-x}\,dx = (1-x)^{1/2}\,dx$$

$$du = dx \qquad v = -\frac{2}{3}(1-x)^{3/2}$$

Substituting in Eq. (28.22), we have

$$\int \underset{u}{x}[\underset{dv}{(1-x)^{1/2}\,dx}] = \underset{u}{x}\left[\underset{v}{-\frac{2}{3}(1-x)^{3/2}}\right] - \int\left[\underset{v}{-\frac{2}{3}(1-x)^{3/2}}\right]\underset{du}{dx}$$

At this point, we see that we can complete the integration. Thus,

$$\int x(1-x)^{1/2}\,dx = -\frac{2x}{3}(1-x)^{3/2} + \frac{2}{3}\int(1-x)^{3/2}\,dx$$

$$= -\frac{2x}{3}(1-x)^{3/2} + \frac{2}{3}\left(-\frac{2}{5}\right)(1-x)^{5/2} + C$$

$$= -\frac{2}{3}(1-x)^{3/2}\left[x + \frac{2}{5}(1-x)\right] + C$$

$$= -\frac{2}{15}(1-x)^{3/2}(2+3x) + C$$

◀ **EXAMPLE 3** Integrate: $\int \sqrt{x}\ln x\,dx$.

For this integral we have

$$u = \ln x \qquad dv = x^{1/2}\,dx$$

$$du = \frac{1}{x}\,dx \qquad v = \frac{2}{3}x^{3/2}$$

$$\int \sqrt{x}\ln x\,dx = \frac{2}{3}x^{3/2}\ln x - \frac{2}{3}\int x^{1/2}\,dx$$

$$= \frac{2}{3}x^{3/2}\ln x - \frac{4}{9}x^{3/2} + C$$

◀ EXAMPLE 4   Integrate: $\int \sin^{-1} x\, dx$.
    We write

$$u = \sin^{-1} x \qquad dv = dx$$

$$du = \frac{dx}{\sqrt{1 - x^2}} \qquad v = x$$

$$\int \sin^{-1} x\, dx = x \sin^{-1} x - \int \frac{x\, dx}{\sqrt{1 - x^2}} = x \sin^{-1} x + \frac{1}{2} \int \frac{-2x\, dx}{\sqrt{1 - x^2}}$$

$$= x \sin^{-1} x + \sqrt{1 - x^2} + C$$

◀ EXAMPLE 5   In a certain electric circuit, the current i (in A) is given by the equation $i = te^{-t}$, where t is the time (in s). Find the charge q to pass a point in the circuit between $t = 0$ s and $t = 1.0$ s.

Since $i = \dfrac{dq}{dt}$, we have $\dfrac{dq}{dt} = te^{-t}$. Thus, with $q = \displaystyle\int_0^{1.0} te^{-t}\, dt$, we are to solve for q. For this integral,

$$u = t \qquad dv = e^{-t}\, dt$$

$$du = dt \qquad v = -e^{-t}$$

$$q = \int_0^{1.0} te^{-t}\, dt = -te^{-t}\Big|_0^{1.0} + \int_0^{1.0} e^{-t}\, dt = -te^{-t} - e^{-t}\Big|_0^{1.0}$$

$$= -e^{-1.0} - e^{-1.0} + 1.0 = 0.26\ \text{C}$$

Therefore, 0.26 C pass a given point in the circuit.

◀ EXAMPLE 6   Integrate: $\int e^x \sin x\, dx$.
    Let $u = \sin x$, $dv = e^x\, dx$, $du = \cos x\, dx$, $v = e^x$.

$$\int e^x \sin x\, dx = e^x \sin x - \int e^x \cos x\, dx$$

NOTE ▶   At first glance, it appears that we have made no progress in applying the method of integration by parts. We note, however, that when we integrated $\int e^x \sin x\, dx$, part of the result was a term of $\int e^x \cos x\, dx$. This implies that *if $\int e^x \cos x\, dx$ were integrated, a term of $\int e^x \sin x\, dx$ might result.* Thus, the method of integration by parts is now applied to the integral $\int e^x \cos x\, dx$:

$$u = \cos x \qquad dv = e^x\, dx \qquad du = -\sin x\, dx \qquad v = e^x$$

And so $\int e^x \cos x\, dx = e^x \cos x + \int e^x \sin x\, dx$. Substituting this expression into the expression for $\int e^x \sin x\, dx$, we obtain

$$\int e^x \sin x\, dx = e^x \sin x - \left(e^x \cos x + \int e^x \sin x\, dx \right)$$

$$= e^x \sin x - e^x \cos x - \int e^x \sin x\, dx$$

$$2\int e^x \sin x\, dx = e^x(\sin x - \cos x) + 2C$$

$$\int e^x \sin x\, dx = \frac{e^x}{2}(\sin x - \cos x) + C$$

Thus, by combining integrals of like form, we obtain the desired result.

EXERCISES 28.7

In Exercises 1 and 2, answer the given questions related to the indicated examples of this section.

(W) **1.** In Example 2, do the choices $u = \sqrt{1-x}$ and $dv = x\,dx$ work for this integral? Explain.

(W) **2.** In Example 6, do the choices $u = e^x$ and $dv = \sin x\,dx$ work for this integral? Explain.

In Exercises 3–18, integrate each of the given functions.

3. $\displaystyle\int \theta \cos \theta\,d\theta$ **4.** $\displaystyle\int x \sin 2x\,dx$ **5.** $\displaystyle\int xe^{2x}\,dx$

6. $\displaystyle\int 3xe^x\,dx$ **7.** $\displaystyle\int x \sec^2 x\,dx$ **8.** $\displaystyle\int_0^{\pi/4} x \sec x \tan x\,dx$

9. $\displaystyle\int 2 \tan^{-1} x\,dx$ **10.** $\displaystyle\int \ln s\,ds$ **11.** $\displaystyle\int_{-3}^0 \frac{4t\,dt}{\sqrt{1-t}}$

12. $\displaystyle\int x\sqrt{x+1}\,dx$ **13.** $\displaystyle\int x \ln x\,dx$ **14.** $\displaystyle\int x^2 \ln 4x\,dx$

15. $\displaystyle\int 2\phi^2 \sin \phi \cos \phi\,d\phi$ **16.** $\displaystyle\int_0^1 r^2 e^{2r}\,dr$

17. $\displaystyle\int_0^{\pi/2} e^x \cos x\,dx$ **18.** $\displaystyle\int e^{-x} \sin 2x\,dx$

In Exercises 19–32, solve the given problems by integration.

19. To integrate $\displaystyle\int e^{-\sqrt{x}}\,dx$, the only choices possible of $u = e^{-\sqrt{x}}$ and $dv = dx$ do not work. However, if we first let $t = \sqrt{x}$, $dt = dx/2\sqrt{x}$, the integration can be done by parts. Perform this integration.

20. To integrate $\displaystyle\int x \ln(x+1)\,dx$, the substitution $t = x + 1$, $dt = dx$ leads to an integral that can be done readily by parts. Perform this integration in this way.

21. Find the area bounded by $y = xe^{-x}$, $y = 0$, and $x = 2$. See Fig. 28.12.

22. Find the area bounded by $y = 2(\ln x)/x^2$, $y = 0$, and $x = 3$.

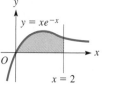

$y = xe^{-x}$

$x = 2$

Fig. 28.12

23. Find the volume generated by revolving the region bounded by $y = \tan^2 x$, $y = 0$, and $x = 0.5$ about the y-axis.

24. Find the volume generated by revolving the region bounded by $y = \sin x$ and $y = 0$ from $x = 0$ to $x = \pi$ about the y-axis.

25. Find the x-coordinate of the centroid of a flat plate covering the region bounded by $y = \cos x$ and $y = 0$ for $0 \le x \le \pi/2$.

26. Find the moment of inertia with respect to its axis of the solid generated by revolving the region bounded by $y = e^x$, $x = 1$, and the coordinate axes about the y-axis.

27. Find the root-mean-square value of the function $y = \sqrt{\sin^{-1} x}$ between $x = 0$ and $x = 1$. (See Example 7 of Section 28.5.)

28. The general expression for the slope of a curve is $dy/dx = x^3\sqrt{1 + x^2}$. Find the equation of the curve if it passes through the origin.

29. Computer simulation shows that the velocity v (in m/s) of a test car is $v = t^3/\sqrt{t^2 + 1}$ from $t = 0$ to $t = 8.0$ s. Find the expression for the distance traveled by the car in t s.

30. The nose cone of a rocket has the shape of the solid that is generated by revolving the region bounded by $y = \ln x$, $y = 0$, and $x = 9.5$ about the x-axis. Find the volume (in m³) of the nose cone. See Fig. 28.13.

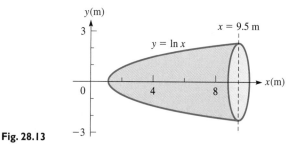

Fig. 28.13

31. The current in a given circuit is given by $i = e^{-2t} \cos t$. Find an expression for the amount of charge that passes a given point in the circuit as a function of the time, if $q_0 = 0$.

32. In finding the average length $\bar{x}$ (in nm) of a certain type of large molecule, we use the equation $\bar{x} = \lim_{b \to \infty} [0.1 \int_0^b x^3 e^{-x^2/8}\,dx]$. Evaluate the integral and then use a calculator to show that $\bar{x} \to 3.2$ nm as $b \to \infty$.

28.8 INTEGRATION BY TRIGONOMETRIC SUBSTITUTION

We have seen that trigonometric relations provide a means of transforming trigonometric integrals into forms that can be integrated. As we will show in this section, they can also be useful for integrating certain types of algebraic integrals. Substitutions based on Eqs. (28.14), (28.15), and (28.16), which are referenced in the margin on the next page, prove to be particularly useful for integrating expressions that involve radicals. The following examples illustrate the method.

◀ **EXAMPLE 1** Integrate: $\displaystyle\int \frac{dx}{x^2\sqrt{1-x^2}}$.

If we let $x = \sin \theta$, the radical becomes $\sqrt{1 - \sin^2 \theta} = \cos \theta$. Therefore, by making this substitution, the integral can be transformed into a trigonometric integral. We

CAUTION ▶ must be careful to ***replace all factors of the integral by proper expressions in terms of*** θ. With $x = \sin \theta$, by finding the differential of x, we have $dx = \cos \theta \, d\theta$.

Now, by substituting $x = \sin \theta$ and $dx = \cos \theta \, d\theta$ into the integral, we have

For reference, Eqs. (28.14), (28.15), and (28.16) are
$\cos^2 x + \sin^2 x = 1$
$1 + \tan^2 x = \sec^2 x$
$1 + \cot^2 x = \csc^2 x$

$$\int \frac{dx}{x^2\sqrt{1-x^2}} = \int \frac{\cos \theta \, d\theta}{\sin^2 \theta \sqrt{1 - \sin^2 \theta}} \qquad \text{substituting}$$

$$= \int \frac{\cos \theta \, d\theta}{\sin^2 \theta \cos \theta} = \int \csc^2 \theta \, d\theta \qquad \text{using trigonometric relations}$$

This last integral can be integrated by using Eq. (28.7). This leads to

$$\int \csc^2 \theta \, d\theta = -\cot \theta + C$$

CAUTION ▶ We have now performed the integration, but *the answer we now have is in terms of* θ, *and we must express the result in terms of* x. Making a triangle with an angle θ such that $\sin \theta = x/1$ (see Fig. 28.14), we may express any of the trigonometric functions in terms of x. (This is the method used with inverse trigonometric functions.) Thus,

$$\cot \theta = \frac{\sqrt{1 - x^2}}{x}$$

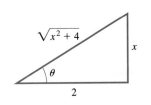

Fig. 28.14

Therefore, the result of the integration becomes

$$\int \frac{dx}{x^2\sqrt{1-x^2}} = -\cot \theta + C = -\frac{\sqrt{1 - x^2}}{x} + C$$

◀ **EXAMPLE 2** Integrate: $\displaystyle\int \frac{dx}{\sqrt{x^2 + 4}}$.

If we let $x = 2 \tan \theta$, the radical in this integral becomes

$$\sqrt{x^2 + 4} = \sqrt{4 \tan^2 \theta + 4} = 2\sqrt{\tan^2 \theta + 1} = 2\sqrt{\sec^2 \theta} = 2 \sec \theta$$

Therefore, with $x = 2 \tan \theta$ and $dx = 2 \sec^2 \theta \, d\theta$, we have

$$\int \frac{dx}{\sqrt{x^2 + 4}} = \int \frac{2 \sec^2 \theta \, d\theta}{\sqrt{4 \tan^2 \theta + 4}} = \int \frac{2 \sec^2 \theta \, d\theta}{2 \sec \theta} \qquad \text{substituting}$$

$$= \int \sec \theta \, d\theta = \ln|\sec \theta + \tan \theta| + C \qquad \text{using Eq. (28.12)}$$

$$= \ln\left|\frac{\sqrt{x^2 + 4}}{2} + \frac{x}{2}\right| + C = \ln\left|\frac{\sqrt{x^2 + 4} + x}{2}\right| + C \qquad \text{see Fig. (28.15)}$$

Fig. 28.15

This answer is acceptable, but by using the properties of logarithms, we have

$$\ln\left|\frac{\sqrt{x^2 + 4} + x}{2}\right| + C = \ln\left|\sqrt{x^2 + 4} + x\right| + (C - \ln 2)$$

$$= \ln\left|\sqrt{x^2 + 4} + x\right| + C'$$

Combining constants, as we did in writing $C' = C - \ln 2$, is a common practice in integration problems.

◀ EXAMPLE 3   Integrate: $\int \dfrac{2\,dx}{x\sqrt{x^2 - 9}}$.

If we let $x = 3 \sec \theta$, the radical in this integral becomes

$$\sqrt{x^2 - 9} = \sqrt{9 \sec^2 \theta - 9} = 3\sqrt{\sec^2 \theta - 1} = 3\sqrt{\tan^2 \theta} = 3 \tan \theta$$

Therefore, with $x = 3 \sec \theta$ and $dx = 3 \sec \theta \tan \theta\, d\theta$, we have

$$\int \frac{2\,dx}{x\sqrt{x^2 - 9}} = 2\int \frac{3 \sec \theta \tan \theta\, d\theta}{3 \sec \theta \sqrt{9 \sec^2 \theta - 9}} = 2\int \frac{\tan \theta\, d\theta}{3 \tan \theta}$$

$$= \frac{2}{3}\int d\theta = \frac{2}{3}\theta + C = \frac{2}{3}\sec^{-1}\frac{x}{3} + C$$

It is not necessary to refer to a triangle to express the result in terms of x. The solution is found by solving $x = 3 \sec \theta$ for θ, as indicated.   ▶

These examples show that by making the proper substitution, we can integrate algebraic functions by using the equivalent trigonometric forms. In summary, for the indicated radical form, the following trigonometric substitutions are used:

Basic Trigonometric Substitutions

For	$\sqrt{a^2 - x^2}$	use	$x = a \sin \theta$
For	$\sqrt{a^2 + x^2}$	use	$x = a \tan \theta$
For	$\sqrt{x^2 - a^2}$	use	$x = a \sec \theta$

(28.23)

◀ EXAMPLE 4   The joint between two links of a robot arm moves back and forth along the curve defined by $y = 3.0 \ln x$ from $x = 1.0$ cm to $x = 4.0$ cm. Find the distance the joint moves in one cycle.

For reference, the equation in Exercise 33 on page 792 is

$$s = \int_a^b \sqrt{1 + \left(\frac{dy}{dx}\right)^2}\, dx$$

The required distance is found by use of the equation for the length of arc, shown in Exercise 33 on page 792. Therefore, we find the derivative as $dy/dx = 3.0/x$, which means the total distance s moved in one cycle is

$$s = 2\int_{1.0}^{4.0} \sqrt{\left[1 + \left(\frac{3.0}{x}\right)^2\right]}\, dx = 2\int_{1.0}^{4.0} \frac{\sqrt{x^2 + 9.0}}{x}\, dx$$

To integrate, we make the substitution $x = 3.0 \tan \theta$ and $dx = 3.0 \sec^2 \theta\, d\theta$. Thus,

$$\int \frac{\sqrt{x^2 + 9.0}}{x}\, dx = \int \frac{\sqrt{9.0 \tan^2 \theta + 9.0}}{3.0 \tan \theta}(3.0 \sec^2 \theta\, d\theta) = 3.0\int \frac{\sec \theta}{\tan \theta}(1 + \tan^2 \theta)\, d\theta$$

$$= 3.0\left(\int \frac{\sec \theta}{\tan \theta}\, d\theta + \int \tan \theta \sec \theta\, d\theta\right) = 3.0\left(\int \csc \theta\, d\theta + \int \tan \theta \sec \theta\, d\theta\right) \quad \text{see Fig. 28.16}$$

$$= 3.0(\ln|\csc \theta - \cot \theta| + \sec \theta) + C = 3.0\left[\ln\left|\frac{\sqrt{x^2 + 9.0}}{x} - \frac{3.0}{x}\right| + \frac{\sqrt{x^2 + 9.0}}{3.0}\right] + C$$

Limits have not been included, due to the change in variables. Evaluating, we have

$$s = 2\int_{1.0}^{4.0} \frac{\sqrt{x^2 + 9.0}}{x}\, dx = 6.0\left[\ln\left|\frac{\sqrt{x^2 + 9.0} - 3.0}{x}\right| + \frac{\sqrt{x^2 + 9.0}}{3.0}\right]\Bigg|_{1.0}^{4.0}$$

$$= 6.0\left[\left(\ln 0.50 - \ln\frac{\sqrt{10.0} - 3.0}{1.0}\right) + \frac{5.0}{3.0} - \frac{\sqrt{10.0}}{3.0}\right] = 10.4 \text{ cm}$$ ▶

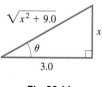

Fig. 28.16

EXERCISES 28.8

In Exercises 1 and 2, answer the given equations related to the indicated examples of this section.

 1. In Example 1, how must the integrand be changed in order to have a result of $\sin^{-1} x + C$?

 2. In Example 2, how must the integrand be changed in order to have a result of $\tan^{-1}(x/2) + C$?

In Exercises 3–8, give the proper trigonometric substitution and find the transformed integral, but do not integrate.

3. $\int \dfrac{\sqrt{9 - x^2}}{x^2}\, dx$ **4.** $\int \dfrac{dx}{\sqrt{x^2 - 16}}$ **5.** $\int \dfrac{dx}{x^2\sqrt{x^2 + 1}}$

6. $\int \dfrac{dx}{(1 - x^2)^{3/2}}$ **7.** $\int \dfrac{dx}{x\sqrt{x^2 - 1}}$ **8.** $\int \sqrt{25 + x^2}\, dx$

In Exercises 9–24, integrate each of the given functions.

9. $\int \dfrac{\sqrt{1 - x^2}}{x^2}\, dx$ **10.** $\int_0^4 \dfrac{dt}{(t^2 + 9)^{3/2}}$

11. $\int \dfrac{2\, dx}{\sqrt{x^2 - 4}}$ **12.** $\int \dfrac{\sqrt{x^2 - 25}}{x}\, dx$

13. $\int \dfrac{6\, dz}{z^2\sqrt{z^2 + 9}}$ **14.** $\int \dfrac{3\, dx}{x\sqrt{4 - x^2}}$

15. $\int \dfrac{4\, dx}{(4 - x^2)^{3/2}}$ **16.** $\int \dfrac{6p^3\, dp}{\sqrt{9 + p^2}}$

17. $\int_0^{0.5} \dfrac{x^3\, dx}{\sqrt{1 - x^2}}$ **18.** $\int_4^5 \dfrac{\sqrt{x^2 - 16}}{x^2}\, dx$

19. $\int \dfrac{5\, dx}{\sqrt{x^2 + 2x + 2}}$ **20.** $\int \dfrac{dx}{\sqrt{x^2 + 2x}}$

21. $\int_{2.5}^3 \dfrac{dy}{y\sqrt{4y^2 - 9}}$ **22.** $\int \sqrt{16 - x^2}\, dx$

23. $\int \dfrac{2\, dx}{\sqrt{e^{2x} - 1}}$ **24.** $\int \dfrac{12 \sec^2 u\, du}{(4 - \tan^2 u)^{3/2}}$

In Exercises 25–32, solve the given problems by integration.

25. Find the area of a circle of radius 1.

26. Find the area bounded by $y = \dfrac{1}{x^2\sqrt{x^2 - 1}}$, $x = \sqrt{2}$, $x = \sqrt{5}$, and $y = 0$. See Fig. 28.17.

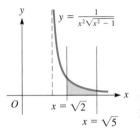

Fig. 28.17

27. Find the moment of inertia with respect to the y-axis of a flat plate covering the first-quadrant region under the circle $x^2 + y^2 = a^2$ in terms of its mass.

28. Find the moment of inertia of a sphere of radius a with respect to its axis in terms of its mass.

29. Find the volume generated by revolving the region bounded by $y = \dfrac{\sqrt{x^2 - 16}}{x^2}$, $y = 0$, and $x = 5$ about the y-axis.

30. The vertical cross section of a highway culvert is defined by the region within the ellipse $1.00x^2 + 9.00y^2 = 9.00$, where dimensions are in meters. Find the area of the cross section of the culvert.

31. If an electric charge Q is distributed along a straight wire of length $2a$, the electric potential V at a point P, which is at a distance b from the center of the wire, is $V = kQ\displaystyle\int_{-a}^{a} \dfrac{dx}{\sqrt{b^2 + x^2}}$. Here k is a constant and x is the distance along the wire. Evaluate the integral.

32. An electric insulating ring for a machine part can be described as the volume generated by revolving the region bounded by $y = x^2\sqrt{x^2 - 4}$, $y = 0$, and $x = 2.5$ cm about the y-axis. Find the volume (in cm^3) of material in the ring.

28.9 **INTEGRATION BY PARTIAL FRACTIONS: NONREPEATED LINEAR FACTORS**

In the earlier sections of this chapter, we saw how *reversing differential forms* allows us to integrate various types of functions. Then, in the previous two sections, we have seen how the appropriate use of a *derivative of a product* and *trigonometric identities* are used to put integrals into a form that can be integrated. Now, in this section and the next section, we show an *algebraic method* by which integrands that are fractions can be changed into a form that can be integrated.

In algebra, we combine fractions into a single fraction by means of addition. However, if we wish to integrate an expression that contains a rational fraction, in which both the numerator and the denominator are polynomials, it is often advantageous to reverse the operation of addition and express the rational fraction as the sum of simpler fractions.

◀ EXAMPLE 1　In attempting to integrate $\int \dfrac{7 - x}{x^2 + x - 2}\, dx$, we find that it does not fit any of the standard forms in this chapter. However, we can show that

$$\frac{7 - x}{x^2 + x - 2} = \frac{2}{x - 1} - \frac{3}{x + 2}$$

This means that

$$\int \frac{7 - x}{x^2 + x - 2}\, dx = \int \frac{2\, dx}{x - 1} - \int \frac{3\, dx}{x + 2}$$
$$= 2 \ln|x - 1| - 3 \ln|x + 2| + C$$

We see that by writing the fraction in the original integrand as the sum of the simpler fractions, each resulting integrand can be integrated.　▶

In Example 1, we saw that the integral is readily determined once the rational fraction $(7 - x)/(x^2 + x - 2)$ is replaced by the simpler fractions. In this section and the next, we describe how certain rational fractions can be expressed in terms of simpler fractions and thereby be integrated. *This technique is called the* **method of partial fractions.**

In order to express the rational fraction $f(x)/g(x)$ in terms of simpler partial fractions, *the degree of the numerator $f(x)$ must be less than that of the denominator $g(x)$.* If this is not the case, we divide numerator by denominator until the remainder is of the proper form. Then the denominator $g(x)$ is factored into a product of linear and quadratic factors. The method of determining the partial fractions depends on the factors that are obtained. In advanced algebra, the form of the partial fractions is shown (but we shall not show the proof here).

There are four cases for the types of factors of the denominator. They are (1) *nonrepeated linear factors*, (2) *repeated linear factors*, (3) *nonrepeated quadratic factors*, and (4) *repeated quadratic factors*. In this section, we consider the case of nonrepeated linear factors, and the other cases are discussed in the next section.

NOTE ▶

NONREPEATED LINEAR FACTORS

For the case of nonrepeated linear factors, we use the fact that *corresponding to each linear factor $ax + b$, occurring once in the denominator, there will be a partial fraction of the form*

$$\frac{A}{ax + b}$$

where A is a constant to be determined. The following examples illustrate the method.

◀ **EXAMPLE 2**   Integrate: $\int \dfrac{7 - x}{x^2 + x - 2} \, dx$. (This is the same integral as in Example 1. Here we see how the partial fractions are found.)

First, we note that the degree of the numerator is 1 (the highest power term is x) and that of the denominator is 2 (the highest power term is x^2). Since the degree of the denominator is higher, we may proceed to factoring it. Thus,

$$\frac{7 - x}{x^2 + x - 2} = \frac{7 - x}{(x - 1)(x + 2)}$$

The method of partial fractions essentially reverses the process of combining fractions over a common denominator. By determining that

$$\frac{7 - x}{x^2 + x - 2} = \frac{2}{x - 1} - \frac{3}{x + 2}$$

reverses the process by which

$$\frac{2}{x - 1} - \frac{3}{x + 2} = \frac{2(x + 2) - 3(x - 1)}{(x - 1)(x + 2)}$$
$$= \frac{7 - x}{x^2 + x - 2}$$

There are two linear factors, $(x - 1)$ and $(x + 2)$, in the denominator, and they are different. This means that there are two partial fractions. Therefore, we write

$$\frac{7 - x}{(x - 1)(x + 2)} = \frac{A}{x - 1} + \frac{B}{x + 2} \qquad (1)$$

We are to determine constants A and B so that Eq. (1) is an identity. In finding A and B, we clear Eq. (1) of fractions by multiplying both sides by $(x - 1)(x + 2)$.

$$7 - x = A(x + 2) + B(x - 1) \qquad (2)$$

Equation (2) is also an identity, which means that there are two ways of determining the values of A and B.

NOTE ▶   *Solution by substitution:* Since Eq. (2) is an identity, *it is true for any value of x.* Thus, in turn we pick $x = -2$ and $x = 1$, for each of these values makes a factor on the right equal to zero, and the values of B and A are easily found. Therefore,

$$\text{For } x = -2: \ \ 7 - (-2) = A(-2 + 2) + B(-2 - 1)$$
$$9 = -3B \qquad B = -3$$
$$\text{For } x = 1: \qquad 7 - 1 = A(1 + 2) + B(1 - 1)$$
$$6 = 3A, \qquad A = 2$$

NOTE ▶   *Solution by equating coefficients:* Since Eq. (2) is an identity, another way of finding the constants A and B is to *equate coefficients of like powers of x from each side.* Thus, writing Eq. (2) as

$$7 - x = (2A - B) + (A + B)x$$

we have

$$2A - B = 7 \qquad \text{equating constants: } x^0 \text{ terms}$$
$$A + B = -1 \qquad \text{equating coefficients of } x$$

$$
\begin{aligned}
2A - B &= 7 \\
\underline{A + B} &= \underline{-1} \\
3A &= 6 \\
A &= 2 \\
2(2) - B &= 7 \\
-B &= 3 \\
B &= -3
\end{aligned}
$$

Now, using the values $A = 2$ and $B = -3$ (as found at the left), we have

$$\frac{7 - x}{(x - 1)(x + 2)} = \frac{2}{x - 1} - \frac{3}{x + 2}$$

Therefore, the integral is found as in Example 1.

$$\int \frac{7 - x}{x^2 + x - 2} \, dx = \int \frac{7 - x}{(x - 1)(x + 2)} \, dx = \int \frac{2 \, dx}{x - 1} - \int \frac{3 \, dx}{x + 2}$$
$$= 2 \ln|x - 1| - 3 \ln|x + 2| + C$$

Using the properties of logarithms, we may write this as

$$\int \frac{7 - x}{x^2 + x - 2} \, dx = \ln\left|\frac{(x - 1)^2}{(x + 2)^3}\right| + C$$

◀ EXAMPLE 3   Integrate: $\displaystyle\int \frac{6x^2 - 14x - 11}{(x + 1)(x - 2)(2x + 1)}\,dx.$

The denominator is factored and is of degree 3 (when multiplied out, the highest power term of x is x^3). This means we have three nonrepeated linear factors.

$$\frac{6x^2 - 14x - 11}{(x + 1)(x - 2)(2x + 1)} = \frac{A}{x + 1} + \frac{B}{x - 2} + \frac{C}{2x + 1}$$

Multiplying through by $(x + 1)(x - 2)(2x + 1)$, we have

$$6x^2 - 14x - 11 = A(x - 2)(2x + 1) + B(x + 1)(2x + 1) + C(x + 1)(x - 2)$$

We now substitute the values of 2, $-\frac{1}{2}$, and -1 for x. Again, these are chosen because they make factors of the coefficients of A, B, or C equal to zero, although any values may be chosen. Therefore,

For $x = 2$: $6(4) - 14(2) - 11 = A(0)(5) + B(3)(5) + C(3)(0),$ $B = -1$

For $x = -\dfrac{1}{2}$: $6\left(\dfrac{1}{4}\right) - 14\left(-\dfrac{1}{2}\right) - 11 = A\left(-\dfrac{5}{2}\right)(0) + B\left(\dfrac{1}{2}\right)(0) + C\left(\dfrac{1}{2}\right)\left(-\dfrac{5}{2}\right),$ $C = 2$

For $x = -1$: $6(1) - 14(-1) - 11 = A(-3)(-1) + B(0)(-1) + C(0)(-3),$ $A = 3$

Therefore,

$$\int \frac{6x^2 - 14x - 11}{(x + 1)(x - 2)(2x + 1)}\,dx = \int \frac{3\,dx}{x + 1} - \int \frac{dx}{x - 2} + \int \frac{2\,dx}{2x + 1}$$

$$= 3\ln|x + 1| - \ln|x - 2| + \ln|2x + 1| + C_1 = \ln\left|\frac{(2x + 1)(x + 1)^3}{x - 2}\right| + C_1$$

Here we have let the constant of integration be C_1 since we used C as the numerator of the third partial fraction.   ▶

◀ EXAMPLE 4   Integrate: $\displaystyle\int \frac{2x^4 - x^3 - 9x^2 + x - 12}{x^3 - x^2 - 6x}\,dx.$

Since the numerator is of a higher degree than the denominator, we must first divide the numerator by the denominator. This gives

$$\frac{2x^4 - x^3 - 9x^2 + x - 12}{x^3 - x^2 - 6x} = 2x + 1 + \frac{4x^2 + 7x - 12}{x^3 - x^2 - 6x}$$

We must now express this rational fraction in terms of its partial fractions.

$$\frac{4x^2 + 7x - 12}{x^3 - x^2 - 6x} = \frac{4x^2 + 7x - 12}{x(x + 2)(x - 3)} = \frac{A}{x} + \frac{B}{x + 2} + \frac{C}{x - 3}$$

Clearing fractions, we have

$$4x^2 + 7x - 12 = A(x + 2)(x - 3) + Bx(x - 3) + Cx(x + 2)$$

Now, using values of x of -2, 3, and 0 for substitution, we obtain the values of $B = -1$, $C = 3$, and $A = 2$, respectively. Therefore,

$$\int \frac{2x^4 - x^3 - 9x^2 + x - 12}{x^3 - x^2 - 6x}\,dx = \int\left(2x + 1 + \frac{2}{x} - \frac{1}{x + 2} + \frac{3}{x - 3}\right)dx$$

$$= x^2 + x + 2\ln|x| - \ln|x + 2| + 3\ln|x - 3| + C_1$$

$$= x^2 + x + \ln\left|\frac{x^2(x - 3)^3}{x + 2}\right| + C_1$$ ▶

In Example 2, we showed two ways of finding the values of A and B. The method of substitution is generally easier to use with linear factors, and we used it in Examples 3 and 4. However, as we will see in the next section, the method of equating coefficients can be very useful for other cases with partial fractions.

EXERCISES 28.9

In Exercises 1 and 2, make the given changes in the integrands of the indicated examples of this section, and then find the resulting fractions to be used in the integration. Do not integrate.

1. In Example 2, change the numerator to $10 - x$.

2. In Example 3, change the numerator to $x^2 - 12x - 10$.

In Exercises 3–6, write out the form of the partial fractions, similar to that shown in Eq. (1) of Example 2, that would be used to perform the indicated integrations. Do not evaluate the constants.

3. $\int \dfrac{3x + 2}{x^2 + x}\,dx$

4. $\int \dfrac{9 - x}{x^2 + 2x - 3}\,dx$

5. $\int \dfrac{x^2 - 6x - 8}{x^3 - 4x}\,dx$

6. $\int \dfrac{2x^2 - 5x - 7}{x^3 + 2x^2 - x - 2}\,dx$

In Exercises 7–22, integrate each of the given functions.

7. $\int \dfrac{x + 3}{(x + 1)(x + 2)}\,dx$

8. $\int \dfrac{x + 2}{x(x + 1)}\,dx$

9. $\int \dfrac{dx}{x^2 - 4}$

10. $\int \dfrac{p - 9}{2p^2 - 3p + 1}\,dp$

11. $\int \dfrac{x^2 + 3}{x^2 + 3x}\,dx$

12. $\int \dfrac{x^3}{x^2 + 3x + 2}\,dx$

13. $\int_0^1 \dfrac{2t + 4}{3t^2 + 5t + 2}\,dt$

14. $\int_1^3 \dfrac{x - 1}{4x^2 + x}\,dx$

15. $\int \dfrac{4x^2 - 10}{x(x + 1)(x - 5)}\,dx$

16. $\int \dfrac{4x^2 + 21x + 6}{(x + 2)(x - 3)(x + 4)}\,dx$

17. $\int \dfrac{6x^2 - 2x - 1}{4x^3 - x}\,dx$

18. $\int_2^3 \dfrac{dR}{R^3 - R}$

19. $\int_1^2 \dfrac{x^3 + 7x^2 + 9x + 2}{x(x^2 + 3x + 2)}\,dx$

20. $\int \dfrac{2x^3 + x - 1}{x^3 + x^2 - 4x - 4}\,dx$

21. $\int \dfrac{dV}{(V^2 - 4)(V^2 - 9)}$

22. $\int \dfrac{5x^3 - 2x^2 - 15x + 24}{x^4 - 2x^3 - 11x^2 + 12x}\,dx$

In Exercises 23–32, solve the given problems by integration.

23. Derive the general formula $\int \dfrac{du}{u(a + bu)} = -\dfrac{1}{a}\ln\dfrac{a + bu}{u} + C$.

24. Derive the general formula $\int \dfrac{du}{u^2 - a^2} = \dfrac{1}{2a}\ln\dfrac{u - a}{u + a} + C$.

25. Find the area bounded by $y = (x - 16)/(x^2 - 5x - 14)$, $y = 0$, $x = 2$, and $x = 4$.

26. Find the first-quadrant area bounded by $y = 1/(x^3 + 3x^2 + 2x)$, $x = 1$, and $x = 3$.

27. Find the volume generated if the region of Exercise 26 is revolved about the y-axis.

28. Find the x-coordinate of the centroid of a flat plate that covers the region bounded by $y(x^2 - 1) = 1$, $y = 0$, $x = 2$, and $x = 4$.

29. The general expression for the slope of a curve is $(3x + 5)/(x^2 + 5x)$. Find the equation of the curve if it passes through $(1, 0)$.

30. The current i (in A) as a function of the time t (in s) in a certain electric circuit is given by $i = (4t + 3)/(2t^2 + 3t + 1)$. Find the total charge that passes a given point in the circuit during the first second.

31. The force F (in N) applied by a stamping machine in making a certain computer part is $F = 4x/(x^2 + 3x + 2)$, where x is the distance (in cm) through which the force acts. Find the work done by the force from $x = 0$ to $x = 0.500$ cm.

32. Under specified conditions, the time t (in min) required to form x g of a substance during a chemical reaction is given by $t = \int dx/[(4 - x)(2 - x)]$. Find the equation relating t and x if $x = 0$ g when $t = 0$ min.

 ## 28.10 INTEGRATION BY PARTIAL FRACTIONS: OTHER CASES

In the previous section, we introduced the method of partial fractions and considered the case of nonrepeated linear factors. In this section, we develop the use of partial fractions for the cases of repeated linear factors and nonrepeated quadratic factors. We also briefly discuss the case of repeated quadratic factors.

REPEATED LINEAR FACTORS

For repeated linear factors, we use the fact that *corresponding to each linear factor ax + b that occurs n times in the denominator there will be n partial fractions*

$$\frac{A_1}{ax + b} + \frac{A_2}{(ax + b)^2} + \cdots + \frac{A_n}{(ax + b)^n}$$

where $A_1, A_2, \ldots, A_n$ are constants to be determined.

◀ **EXAMPLE 1**   Integrate: $\displaystyle\int \frac{dx}{x(x + 3)^2}$.

Here we see that the denominator has a factor of x and two factors of $x + 3$. For the factor of x, we use a partial fraction as in the previous section, for it is a nonrepeated factor. For the factor $x + 3$, we need two partial fractions, one with a denominator of $x + 3$ and the other with a denominator of $(x + 3)^2$. Thus, we write

NOTE ▶

$$\frac{1}{x(x + 3)^2} = \boxed{\frac{A}{x} + \frac{B}{x + 3} + \frac{C}{(x + 3)^2}}$$

Multiplying each side by $x(x + 3)^2$, we have

$$1 = A(x + 3)^2 + Bx(x + 3) + Cx \tag{1}$$

Using the values of x of -3 and 0, we have

For $x = -3$: $1 = A(0^2) + B(-3)(0) + (-3)C,$ $C = -\dfrac{1}{3}$

For $x = 0$: $1 = A(3^2) + B(0)(3) + C(0),$ $A = \dfrac{1}{9}$

Since no other numbers make a factor in Eq. (1) equal to zero, we either choose some other value of x or equate coefficients of some power of x in Eq. (1). Since Eq. (1) is an identity, we may choose any value of x. With $x = 1$, we have

$$1 = A(4^2) + B(1)(4) + C(1)$$
$$1 = 16A + 4B + C$$

Using the known values of A and C, we have

$$1 = 16\left(\frac{1}{9}\right) + 4B - \frac{1}{3}, \qquad B = -\frac{1}{9}$$

This means that

$$\frac{1}{x(x + 3)^2} = \frac{\frac{1}{9}}{x} + \frac{-\frac{1}{9}}{x + 3} + \frac{-\frac{1}{3}}{(x + 3)^2}$$

or

$$\int \frac{dx}{x(x + 3)^2} = \frac{1}{9}\int \frac{dx}{x} - \frac{1}{9}\int \frac{dx}{x + 3} - \frac{1}{3}\int \frac{dx}{(x + 3)^2}$$

$$= \frac{1}{9}\ln|x| - \frac{1}{9}\ln|x + 3| - \frac{1}{3}\left(\frac{1}{-1}\right)(x + 3)^{-1} + C_1$$

$$= \frac{1}{9}\ln\left|\frac{x}{x + 3}\right| + \frac{1}{3(x + 3)} + C_1$$

The properties of logarithms are used to write the final form of the answer.

▶

◀ **EXAMPLE 2** Integrate: $\int \dfrac{3x^3 + 15x^2 + 21x + 15}{(x - 1)(x + 2)^3} dx.$

First, we set up the partial fractions as

NOTE ▶

$$\frac{3x^3 + 15x^2 + 21x + 15}{(x - 1)(x + 2)^3} = \boxed{\frac{A}{x - 1} + \frac{B}{x + 2} + \frac{C}{(x + 2)^2} + \frac{D}{(x + 2)^3}}$$

Next we clear fractions:

$$3x^3 + 15x^2 + 21x + 15 = A(x + 2)^3 + B(x - 1)(x + 2)^2$$
$$+ C(x - 1)(x + 2) + D(x - 1) \qquad (1)$$

For $x = 1$: $\qquad\qquad 3 + 15 + 21 + 15 = 27A, \qquad 54 = 27A, \qquad A = 2$

For $x = -2$: $\quad 3(-8) + 15(4) + 21(-2) + 15 = -3D, \qquad 9 = -3D, \qquad D = -3$

To find B and C, we equate coefficients of powers of x. Therefore, we write Eq. (1) as

$$3x^3 + 15x^2 + 21x + 15 = (A + B)x^3 + (6A + 3B + C)x^2$$
$$+ (12A + C + D)x + (8A - 4B - 2C - D)$$

Coefficients of x^3: $\quad 3 = A + B, \qquad\qquad 3 = 2 + B, \qquad\qquad B = 1$

Coefficients of x^2: $15 = 6A + 3B + C, \qquad 15 = 12 + 3 + C, \qquad C = 0$

$$\frac{3x^3 + 15x^2 + 21x + 15}{(x - 1)(x + 2)^3} = \frac{2}{x - 1} + \frac{1}{x + 2} + \frac{0}{(x + 2)^2} + \frac{-3}{(x + 2)^3}$$

$$\int \frac{3x^3 + 15x^2 + 21x + 15}{(x - 1)(x + 2)^3} dx = 2 \int \frac{dx}{x - 1} + \int \frac{dx}{x + 2} - 3 \int \frac{dx}{(x + 2)^3}$$

$$= 2 \ln|x - 1| + \ln|x + 2| - 3\left(\frac{1}{-2}\right)(x + 2)^{-2} + C_1$$

$$= \ln|(x - 1)^2(x + 2)| + \frac{3}{2(x + 2)^2} + C_1 \qquad \blacktriangleright$$

If there is one repeated factor in the denominator and it is the only factor present in the denominator, a substitution is easier and more convenient than using partial fractions. This is illustrated in the following example.

◀ **EXAMPLE 3** Integrate: $\int \dfrac{x\,dx}{(x - 2)^3}.$

This could be integrated by first setting up the appropriate partial fractions. However, the solution is more easily found by using the substitution $u = x - 2$. Using this, we have

$$u = x - 2 \qquad x = u + 2 \qquad dx = du$$

$$\int \frac{x\,dx}{(x - 2)^3} = \int \frac{(u + 2)(du)}{u^3} = \int \frac{du}{u^2} + 2 \int \frac{du}{u^3}$$

$$= \int u^{-2}\,du + 2 \int u^{-3}\,du = \frac{1}{-u} + \frac{2}{-2}u^{-2} + C$$

$$= -\frac{1}{u} - \frac{1}{u^2} + C = -\frac{u + 1}{u^2} + C$$

$$= -\frac{x - 2 + 1}{(x - 2)^2} + C = \frac{1 - x}{(x - 2)^2} + C \qquad \blacktriangleright$$

NONREPEATED QUADRATIC FACTORS

For the case of nonrepeated quadratic factors, we use the fact that *corresponding to each irreducible quadratic factor $ax^2 + bx + c$ that occurs once in the denominator there is a partial fraction of the form*

$$\frac{Ax + B}{ax^2 + bx + c}$$

where A and B are constants to be determined. (Here, an *irreducible quadratic factor* is one that cannot be further factored into linear factors involving only real numbers.) This case is illustrated in the following examples.

◀ EXAMPLE 4   Integrate: $\displaystyle\int \frac{4x + 4}{x^3 + 4x}\,dx$.

In setting up the partial fractions, we note that the denominator factors as $x^3 + 4x = x(x^2 + 4)$. Here the factor $x^2 + 4$ cannot be further factored. This means we have

NOTE ▶

$$\frac{4x + 4}{x^3 + 4x} = \frac{4x + 4}{x(x^2 + 4)} = \boxed{\frac{A}{x} + \frac{Bx + C}{x^2 + 4}}$$

Clearing fractions, we have

$$4x + 4 = A(x^2 + 4) + Bx^2 + Cx$$
$$= (A + B)x^2 + Cx + 4A$$

Equating coefficients of powers of x gives us

For x^2: $0 = A + B$
For x: $4 = C$
For constants: $4 = 4A, \quad A = 1$

Therefore, we easily find that $B = -1$ from the first equation. This means that

$$\frac{4x + 4}{x^3 + 4x} = \frac{1}{x} + \frac{-x + 4}{x^2 + 4}$$

and

$$\int \frac{4x + 4}{x^3 + 4x}\,dx = \int \frac{1}{x}\,dx + \int \frac{-x + 4}{x^2 + 4}\,dx$$
$$= \int \frac{1}{x}\,dx - \int \frac{x\,dx}{x^2 + 4} + \int \frac{4\,dx}{x^2 + 4}$$
$$= \ln|x| - \frac{1}{2}\ln|x^2 + 4| + 2\tan^{-1}\frac{x}{2} + C_1$$

We could use the properties of logarithms to combine the first two terms of the result. ▶

◀ EXAMPLE 5   Integrate: $\int \dfrac{x^3 + 3x^2 + 2x + 4}{x^2(x^2 + 2x + 2)}\,dx$.

In the denominator, we have a repeated linear factor, x^2, and a quadratic factor. Therefore,

NOTE ▶

$$\frac{x^3 + 3x^2 + 2x + 4}{x^2(x^2 + 2x + 2)} = \boxed{\frac{A}{x} + \frac{B}{x^2} + \frac{Cx + D}{x^2 + 2x + 2}}$$

$$x^3 + 3x^2 + 2x + 4 = Ax(x^2 + 2x + 2) + B(x^2 + 2x + 2) + Cx^3 + Dx^2$$
$$= (A + C)x^3 + (2A + B + D)x^2 + (2A + 2B)x + 2B$$

Equating coefficients, we find that

For constants:	$2B = 4,$	$B = 2$	
For x:	$2A + 2B = 2,$	$A + B = 1,$	$A = -1$
For x^2:	$2A + B + D = 3,$	$-2 + 2 + D = 3,$	$D = 3$
For x^3:	$A + C = 1,$	$-1 + C = 1,$	$C = 2$

$$\frac{x^3 + 3x^2 + 2x + 4}{x^2(x^2 + 2x + 2)} = -\frac{1}{x} + \frac{2}{x^2} + \frac{2x + 3}{x^2 + 2x + 2}$$

$$\int \frac{x^3 + 3x^2 + 2x + 4}{x^2(x^2 + 2x + 2)}\,dx = -\int \frac{dx}{x} + 2\int \frac{dx}{x^2} + \int \frac{2x + 3}{x^2 + 2x + 2}\,dx$$

$$= -\ln|x| - 2\left(\frac{1}{x}\right) + \int \frac{2x + 2 + 1}{x^2 + 2x + 2}\,dx$$

$$= -\ln|x| - \frac{2}{x} + \int \frac{2x + 2}{x^2 + 2x + 2}\,dx + \int \frac{dx}{(x^2 + 2x + 1) + 1}$$

$$= -\ln|x| - \frac{2}{x} + \ln|x^2 + 2x + 2| + \tan^{-1}(x + 1) + C_1$$

CAUTION ▶   *Note the manner in which the integral with the quadratic denominator was handled for the purpose of integration.* First, the numerator, $2x + 3$, was written in the form $(2x + 2) + 1$ so that we could fit the logarithmic form with the $2x + 2$. Then we completed the square in the denominator of the final integral so that it then fit an inverse tangent form. ▶

REPEATED QUADRATIC FACTORS

Finally, considering the case of repeated quadratic factors, we use the fact that *corresponding to each irreducible quadratic factor $ax^2 + bx + c$ that occurs n times in the denominator there will be n partial fractions*

$$\frac{A_1x + B_1}{ax^2 + bx + c} + \frac{A_2x + B_2}{(ax^2 + bx + c)^2} + \cdots + \frac{A_nx + B_n}{(ax^2 + bx + c)^n}$$

where $A_1, A_2, \ldots, A_n, B_1, B_2, \ldots, B_n$ are constants to be determined. The procedures that lead to the solution are the same as those for the other cases. Exercises 19 and 20 in the following set are solved by using these partial fractions for repeated quadratic factors.

EXERCISES **28.10**

In Exercises 1–4, make the given changes in the integrands of the indicated examples of this section. Then write out the equation that shows the partial fractions (like those opposite NOTE) that would be used for the integration.

1. In Example 1, change the numerator from 1 to 2.
2. In Example 2, change the denominator to $(x - 1)^2(x + 2)^2$.
3. In Example 4, change the denominator to $x^3 - 9x^2$.
4. In Example 5, change the denominator to $x^2(x^2 + 3x + 2)$.

In Exercises 5–20, integrate each of the given functions.

5. $\displaystyle \int \frac{x - 8}{x^3 - 4x^2 + 4x} dx$

6. $\displaystyle \int \frac{dT}{T^3 - T^2}$

7. $\displaystyle \int \frac{2\,dx}{x^2(x^2 - 1)}$

8. $\displaystyle \int_1^3 \frac{3x^3 + 8x^2 + 10x + 2}{x(x + 1)^3} dx$

9. $\displaystyle \int_1^2 \frac{2s\,ds}{(s - 3)^3}$

10. $\displaystyle \int \frac{x\,dx}{(x + 2)^4}$

11. $\displaystyle \int \frac{x^3 - 2x^2 - 7x + 28}{(x + 1)^2(x - 3)^2} dx$

12. $\displaystyle \int \frac{4\,dx}{(x + 1)^2(x - 1)^2}$

13. $\displaystyle \int_0^2 \frac{x^2 + x + 5}{(x + 1)(x^2 + 4)} dx$

14. $\displaystyle \int \frac{v^2 + v - 1}{(v^2 + 1)(v - 2)} dv$

15. $\displaystyle \int \frac{5x^2 + 8x + 16}{x^2(x^2 + 4x + 8)} dx$

16. $\displaystyle \int \frac{2x^2 + x + 3}{(x^2 + 2)(x - 1)} dx$

17. $\displaystyle \int \frac{10x^3 + 40x^2 + 22x + 7}{(4x^2 + 1)(x^2 + 6x + 10)} dx$

18. $\displaystyle \int_3^4 \frac{5x^3 - 4x}{x^4 - 16} dx$

19. $\displaystyle \int \frac{2r^3}{(r^2 + 1)^2} dr$

20. $\displaystyle \int \frac{-x^3 + x^2 + x + 3}{(x + 1)(x^2 + 1)^2} dx$

In Exercises 21–28, solve the given problems by integration.

21. Find the area bounded by $y = \dfrac{x - 3}{x^3 + x^2}$, $y = 0$, and $x = 1$.

22. Find the first quadrant area bounded by $y = \dfrac{3x^2 + 2x + 9}{(x^2 + 9)(x + 1)}$ and $x = 2$.

23. Find the volume generated by revolving the first-quadrant region bounded by $y = 4/(x^4 + 6x^2 + 5)$ and $x = 2$ about the y-axis.

24. Find the volume generated by revolving the first-quadrant region bounded by $y = x/(x + 3)^2$ and $x = 3$ about the x-axis.

25. Under certain conditions, the velocity v (in m/s) of an object moving along a straight line as a function of the time t (in s) is given by $v = \dfrac{t^2 + 14t + 27}{(2t + 1)(t + 5)^2}$. Find the distance traveled by the object during the first 2.00 s.

26. By a computer analysis, the electric current i (in A) in a certain circuit is given by $i = \dfrac{0.0010(7t^2 + 16t + 48)}{(t + 4)(t^2 + 16)}$, where t is the time (in s). Find the total charge that passes a point in the circuit in the first 0.250 s.

27. Find the x-coordinate of the centroid of a flat plate covering the region bounded by $y = 4/(x^3 + x)$, $x = 1$, $x = 2$, and $y = 0$.

28. The slope of a curve is given by $\dfrac{dy}{dx} = \dfrac{29x^2 + 36}{4x^4 + 9x^2}$. Find the equation of the curve if it passes through $(1, 5)$.

28.11 INTEGRATION BY USE OF TABLES

In this chapter, we have introduced certain basic integrals and have also brought in some methods of reducing other integrals to these basic forms. Often this transformation and integration requires a number of steps to be performed, and therefore integrals are tabulated for reference. The integrals found in tables have been derived by using the methods introduced thus far, as well as many other methods that can be used. Therefore, an understanding of the basic forms and some of the basic methods is very useful in finding integrals from tables. Such an understanding forms a basis for proper recognition of the forms that are used in the tables, as well as the types of results that may be expected. Therefore, **the use of the tables depends on proper recognition of the form and the variables and constants of the integral.** The following examples illustrate the use of the table of integrals found in Appendix E. More extensive tables are available in other sources.

CAUTION ▶

EXAMPLE 1 Integrate: $\displaystyle\int \frac{x\,dx}{\sqrt{2+3x}}$.

We first note that this integral fits the form of Formula 6 of Appendix E, with $u = x$, $a = 2$, and $b = 3$. Therefore,

$$\int \frac{x\,dx}{\sqrt{2+3x}} = -\frac{2(4-3x)\sqrt{2+3x}}{27} + C$$

EXAMPLE 2 Integrate: $\displaystyle\int \frac{\sqrt{4-9x^2}}{x}\,dx$.

This fits the form of Formula 18, with proper identification of constants; $u = 3x$, $du = 3\,dx$, $a = 2$. Hence,

$$\int \frac{\sqrt{4-9x^2}}{x}\,dx = \int \frac{\sqrt{4-9x^2}}{3x}\,3\,dx$$

$$= \sqrt{4-9x^2} - 2\ln\left(\frac{2+\sqrt{4-9x^2}}{3x}\right) + C$$

For reference, Formula 18 is

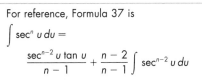

$$\int \frac{\sqrt{a^2-u^2}}{u}\,du =$$
$$\sqrt{a^2-u^2} - a\ln\left(\frac{a+\sqrt{a^2-u^2}}{u}\right)$$

EXAMPLE 3 Integrate: $\int 5\sec^3 2x\,dx$.

This fits the form of Formula 37; $n = 3$, $u = 2x$, $du = 2\,dx$. And so,

$$\int 5\sec^3 2x\,dx = 5\left(\frac{1}{2}\right)\int \sec^3 2x(2\,dx)$$

$$= \frac{5}{2}\frac{\sec 2x\tan 2x}{2} + \frac{5}{2}\left(\frac{1}{2}\right)\int \sec 2x(2\,dx)$$

For reference, Formula 37 is

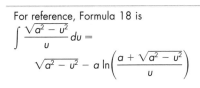

$$\int \sec^n u\,du =$$
$$\frac{\sec^{n-2} u\tan u}{n-1} + \frac{n-2}{n-1}\int \sec^{n-2} u\,du$$

To complete this integral, we must use the basic form of Eq. (28.12). Thus, we complete it by

$$\int 5\sec^3 2x\,dx = \frac{5\sec 2x\tan 2x}{4} + \frac{5}{4}\ln|\sec 2x + \tan 2x| + C$$

For reference, Eq. (28.12) is

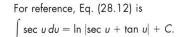

$$\int \sec u\,du = \ln|\sec u + \tan u| + C.$$

EXAMPLE 4 Find the area bounded by $y = x^2\ln 2x$, $y = 0$, and $x = e$.

From Fig. 28.18, we see that the area is

$$A = \int_{0.5}^{e} x^2\ln 2x\,dx$$

This integral fits the form of Formula 46 if $u = 2x$. Thus, we have

$$A = \frac{1}{8}\int_{0.5}^{e}(2x)^2\ln 2x(2\,dx) = \frac{1}{8}(2x)^3\left[\frac{\ln 2x}{3} - \frac{1}{9}\right]_{0.5}^{e}$$

$$= e^3\left(\frac{\ln 2e}{3} - \frac{1}{9}\right) - \frac{1}{8}\left(\frac{\ln 1}{3} - \frac{1}{9}\right) = e^3\left(\frac{3\ln 2e - 1}{9}\right) + \frac{1}{72}$$

$$= 9.118$$

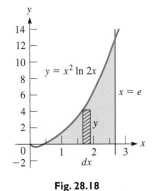

Fig. 28.18

NOTE ▶

The proper identification of u and du is the key step in the use of tables. Therefore, for the integrals in the following example the proper u and du, along with the appropriate formula from the table, are identified, but the integrations are not performed.

◀ **EXAMPLE 5** (a) $\int x\sqrt{1 - x^4}\,dx$ $u = x^2,$ $du = 2x\,dx$ Formula 15

(b) $\int \dfrac{(4x^6 - 9)^{3/2}}{x}\,dx$ $u = 2x^3,$ $du = 6x^2\,dx$ Formula 22

introduce a factor of x^2 into numerator and denominator

(c) $\int x^3 \sin x^2\,dx$ $u = x^2,$ $du = 2x\,dx$ Formula 47   ▶

Following is a brief summary of the approach to integration we have used to obtain the exact result. Also, definite integrals may be approximated by methods such as the trapezoidal rule or Simpson's rule.

BASIC APPROACH TO INTEGRATING A FUNCTION

1. *Write the integral such that it fits an integral form. Either a basic form as developed in this chapter or a form from a table of integrals may be used.*
2. *Use a method of transforming the integral such that an integral form may be used. Appropriate methods are covered in this chapter or in other sources.*

EXERCISES 28.11

In Exercises 1 and 2, make the given changes in the indicated examples of this section, and then state which formula from Appendix E would be used to complete the integration.

1. In Example 1, change the denominator to $(2 + 3x)^2$.

2. In Example 2, in the numerator change $-$ to $+$.

In Exercises 3 and 4, identify u, du, and the formula from Appendix E that would be used to complete the integration.

3. $\int \dfrac{x\,dx}{(4 - x^4)^{3/2}}$

4. $\int x^5 \ln x^3\,dx$

In Exercises 5–44, integrate each function by using the table in Appendix E.

5. $\int \dfrac{3x\,dx}{2 + 5x}$

6. $\int \dfrac{4x\,dx}{(1 + x)^2}$

7. $\int_2^7 4x\sqrt{2 + x}\,dx$

8. $\int \dfrac{dx}{x^2 - 4}$

9. $\int \dfrac{dy}{(y^2 + 4)^{3/2}}$

10. $\int_0^{\pi/3} \sin^3 x\,dx$

11. $\int \sin 2x \sin 3x\,dx$

12. $\int 6 \sin^{-1} 3x\,dx$

13. $\int \dfrac{\sqrt{4x^2 - 9}}{x}\,dx$

14. $\int \dfrac{(9x^2 + 16)^{3/2}}{x}\,dx$

15. $\int \cos^5 4x\,dx$

16. $\int 0.2 \tan^2 2\phi\,d\phi$

17. $\int 6r \tan^{-1} r^2\,dr$

18. $\int 5xe^{4x}\,dx$

19. $\int_1^2 (4 - x^2)^{3/2}\,dx$

20. $\int \dfrac{3\,dx}{9 - 16x^2}$

21. $\int \dfrac{dx}{x\sqrt{4x^2 + 1}}$

22. $\int \dfrac{\sqrt{4 + x^2}}{x}\,dx$

23. $\int \dfrac{8\,dx}{x\sqrt{1 - 4x^2}}$

24. $\int \dfrac{dx}{x(1 + 4x)^2}$

25. $\int_0^{\pi/12} \sin\theta \cos 5\theta\,d\theta$

26. $\int_0^2 x^2 e^{3x}\,dx$

27. $\int x^5 \cos x^3\,dx$

28. $\int 5 \sin^3 t \cos^2 t\,dt$

29. $\int \dfrac{2x\,dx}{(1 - x^4)^{3/2}}$

30. $\int \dfrac{dx}{x(1 - 4x)}$

31. $\int_1^3 \dfrac{\sqrt{3 + 5x^2}}{x}\,dx$

32. $\int_0^1 \dfrac{\sqrt{9 - 4x^2}}{x}\,dx$

33. $\int x^3 \ln x^2\,dx$

34. $\int \dfrac{1.2u\,du}{u^2\sqrt{u^4 - 9}}$

35. $\int \dfrac{9x^2\,dx}{(x^6 - 1)^{3/2}}$

36. $\int x^7\sqrt{x^4 + 4}\,dx$

37. Find the length of arc of the curve $y = x^2$ from $x = 0$ to $x = 1$. (See Exercise 33 of Section 26.6.)

38. Find the moment of inertia with respect to its axis of the solid generated by revolving the region bounded by $y = 3 \ln x$, $x = e$, and the x-axis about the y-axis.

39. Find the area of an ellipse with a major axis $2a$ and a minor axis $2b$.

40. The voltage across a 5.0-μF capacitor in an electric circuit is zero. What is the voltage after 5.00 μs if a current i (in mA) as a function of the time t (in s) given by $i = \tan^{-1} 2t$ charges the capacitor?

41. Find the force (in N) on the region bounded by $x = 1/\sqrt{1 + y}$, $y = 0$, $y = 3$, and the y-axis, if the surface of the water is at the upper edge of the region.

42. If 6.00 g of a chemical are placed in water, the time t (in min) it takes to dissolve half of the chemical is given by

$$t = 560 \int_3^6 \frac{dx}{x(x + 4)},$$ where x is the amount of undissolved chemical at any time. Evaluate t.

43. The dome of a sports arena is the surface generated by revolving $y = 20.0 \cos 0.0196x$ $(0 \le x \le 80.0 \text{ m})$ about the y-axis. Find the volume within the dome.

44. If an electric charge Q is distributed along a wire of length $2a$, the force F exerted on an electric charge q placed at point P is

$$F = kqQ \int \frac{b \, dx}{(b^2 + x^2)^{3/2}}.$$ Integrate to find F as a function of x.

CHAPTER ㉘ EQUATIONS

Integrals

$$\int u^n \, du = \frac{u^{n+1}}{n + 1} + C \quad (n \ne -1) \tag{28.1}$$

$$\int \frac{du}{u} = \ln|u| + C \tag{28.2}$$

$$\int e^u \, du = e^u + C \tag{28.3}$$

$$\int \sin u \, du = -\cos u + C \tag{28.4}$$

$$\int \cos u \, du = \sin u + C \tag{28.5}$$

$$\int \sec^2 u \, du = \tan u + C \tag{28.6}$$

$$\int \csc^2 u \, du = -\cot u + C \tag{28.7}$$

$$\int \sec u \tan u \, du = \sec u + C \tag{28.8}$$

$$\int \csc u \cot u \, du = -\csc u + C \tag{28.9}$$

$$\int \tan u \, du = -\ln|\cos u| + C \tag{28.10}$$

$$\int \cot u \, du = \ln|\sin u| + C \tag{28.11}$$

$$\int \sec u \, du = \ln|\sec u + \tan u| + C \tag{28.12}$$

$$\int \csc u \, du = \ln|\csc u - \cot u| + C \tag{28.13}$$

Trigonometric relations	$\cos^2 x + \sin^2 x = 1$	(28.14)
	$1 + \tan^2 x = \sec^2 x$	(28.15)
	$1 + \cot^2 x = \csc^2 x$	(28.16)
	$2 \cos^2 x = 1 + \cos 2x$	(28.17)
	$2 \sin^2 x = 1 - \cos 2x$	(28.18)
Root-mean-square value	$y_{\text{rms}} = \sqrt{\dfrac{1}{T} \displaystyle\int_0^T y^2\, dx}$	(28.19)
Integrals	$\displaystyle\int \dfrac{du}{\sqrt{a^2 - u^2}} = \sin^{-1} \dfrac{u}{a} + C$	(28.20)
	$\displaystyle\int \dfrac{du}{a^2 + u^2} = \dfrac{1}{a} \tan^{-1} \dfrac{u}{a} + C$	(28.21)
	$\displaystyle\int u\, dv = uv - \int v\, du$	(28.22)
Trigonometric substitutions	For $\sqrt{a^2 - x^2}$ use $x = a \sin \theta$	
	For $\sqrt{a^2 + x^2}$ use $x = a \tan \theta$	(28.23)
	For $\sqrt{x^2 - a^2}$ use $x = a \sec \theta$	

CHAPTER 28 REVIEW EXERCISES

In Exercises 1–40, integrate the given functions without using a table of integrals.

1. $\displaystyle\int e^{-2x}\, dx$

2. $\displaystyle\int e^{\cos 2x} \sin x \cos x\, dx$

3. $\displaystyle\int \dfrac{dx}{x(\ln 2x)^2}$

4. $\displaystyle\int_1^8 y^{1/3} \sqrt{y^{4/3} + 9}\, dy$

5. $\displaystyle\int_0^{\pi/2} \dfrac{4 \cos \theta\, d\theta}{1 + \sin \theta}$

6. $\displaystyle\int \dfrac{\sec^2 x\, dx}{2 + \tan x}$

7. $\displaystyle\int \dfrac{2\, dx}{25 + 49x^2}$

8. $\displaystyle\int \dfrac{dx}{\sqrt{1 - 4x^2}}$

9. $\displaystyle\int_0^{\pi/2} \cos^3 2\theta\, d\theta$

10. $\displaystyle\int_0^{\pi/8} \sec^3 2x \tan 2x\, dx$

11. $\displaystyle\int_0^2 \dfrac{x\, dx}{4 + x^2}$

12. $\displaystyle\int_1^e \dfrac{\ln v^2\, dv}{\ln e^v}$

13. $\displaystyle\int (\sin t + \cos t)^2 \sin t\, dt$

14. $\displaystyle\int \dfrac{\sin^3 x\, dx}{\sqrt{\cos x}}$

15. $\displaystyle\int \dfrac{e^x\, dx}{1 + e^{2x}}$

16. $\displaystyle\int \dfrac{p + 25}{p^2 - 25}\, dp$

17. $\displaystyle\int \sec^4 3x\, dx$

18. $\displaystyle\int \dfrac{(1 - \cos^2 \theta)\, d\theta}{1 + \cos 2\theta}$

19. $\displaystyle\int \dfrac{2x^2 + 6x + 1}{2x^3 - x^2 - x}\, dx$

20. $\displaystyle\int \dfrac{4 - e^{\sqrt{x}}}{\sqrt{x}\, e^{\sqrt{x}}}\, dx$

21. $\displaystyle\int \dfrac{3x\, dx}{4 + x^4}$

22. $\displaystyle\int_1^3 \dfrac{2\, dR}{\sqrt{R}\,(1 + R)}$

23. $\displaystyle\int \dfrac{4\, dx}{\sqrt{4x^2 - 9}}$

24. $\displaystyle\int \dfrac{x^2\, dx}{\sqrt{9 - x^2}}$

25. $\displaystyle\int \dfrac{e^{2x}\, dx}{\sqrt{e^{2x} + 1}}$

26. $\displaystyle\int \dfrac{x^2 - 2x + 3}{(x - 1)^3}\, dx$

27. $\displaystyle\int \dfrac{2x^2 + 3x + 18}{x^3 + 9x}\, dx$

28. $\displaystyle\int_{1/2}^{e/2} \dfrac{(4 + \ln 2u)^3\, du}{u}$

29. $\displaystyle\int_0^{\pi/6} 3 \sin^2 3\phi\, d\phi$

30. $\displaystyle\int \sin^4 x\, dx$

31. $\displaystyle\int x \csc^2 2x\, dx$

32. $\displaystyle\int x \tan^{-1} x\, dx$

33. $\displaystyle\int \dfrac{3u^2 - 6u - 2}{3u^3 + u^2}\, du$

34. $\displaystyle\int \dfrac{R^2 + 3}{R^4 + 3R^2 + 2}\, dR$

35. $\int e^{2x} \cos e^{2x}\, dx$

36. $\int \dfrac{3\, dx}{x^2 + 6x + 10}$

37. $\int_1^e \dfrac{3 \cos(\ln x)\, dx}{x}$

38. $\int_1^3 \dfrac{2\, dx}{x^2 - 2x + 5}$

39. $\int \dfrac{u^2 - 1}{u + 2}\, du$

40. $\int \dfrac{\log_x 2\, dx}{x \ln x}$

In Exercises 41–76, solve the given problem by integration.

41. For the integral $\int \dfrac{dx}{x\sqrt{2 - x^2}}$, using a trigonometric substitution, find the transformed integral but do not integrate.

42. For the integral $\int \dfrac{dx}{\sqrt{x^2 + 4x + 3}}$, using a trigonometric substitution, find the transformed integral but do not integrate.

(W) 43. Show that $\int e^x(e^x + 1)^2\, dx$ can be integrated in two ways. Explain the difference in the answers.

(W) 44. Show that $\int \frac{1}{x}(1 + \ln x)\, dx$ can be integrated in two ways. Explain the difference in the answers.

45. Evaluate $\int \sin^2 x\, dx$ (a) by using Eq. (28.18) and (b) by using Eq. (28.22). Show that the results are equivalent.

46. Integrate $\int \dfrac{dx}{1 + e^x}$ by first rewriting the integrand. (*Hint:* It is possible to multiply the numerator and the denominator by an appropriate expression.)

(W) 47. The integral $\int \dfrac{x}{\sqrt{x^2 + 4}}\, dx$ can be integrated in more than one way. Explain what methods can be used and which is simpler.

48. Find the equation of the curve for which $dy/dx = e^x(2 - e^x)^2$, if the curve passes through $(0, 4)$.

49. Find the equation of the curve for which $dy/dx = \sec^4 x$, if the curve passes through the origin.

50. Find the equation of the curve for which $\dfrac{dy}{dx} = \dfrac{\sqrt{4 + x^2}}{x^4}$, if the curve passes through $(2, 1)$.

51. Find the area bounded by $y = 4e^{2x}$, $x = 1.5$, and the axes.

52. Find the area bounded by $y = x/(1 + x)^2$, the x-axis, and the line $x = 4$.

53. Find the area inside the circle $x^2 + y^2 = 25$ and to the right of the line $x = 3$.

54. Find the area bounded by $y = x\sqrt{x + 4}$, $y = 0$, and $x = 5$.

55. Find the area bounded by $y = \tan^{-1} 2x$, $x = 2$, and the x-axis.

56. In polar coordinates, the area A bounded by the curve $r = f(\theta)$, $\theta = \alpha$, and $\theta = \beta$ is found by evaluating the integral $A = \dfrac{1}{2}\int_\alpha^\beta r^2\, d\theta$. Find the area bounded by $\alpha = 0$, $\beta = \pi/2$, and $y = e^\theta$.

57. Find the volume generated by revolving the region bounded by $y = xe^x$, $y = 0$, and $x = 2$ about the y-axis.

58. Find the volume generated by revolving about the y-axis the region bounded by $y = x + \sqrt{x + 1}$, $x = 3$, and the axes.

59. Find the volume of the solid generated by revolving the region bounded by $y = e^x \sin x$ and the x-axis between $x = 0$ and $x = \pi$ about the x-axis.

60. Find the centroid of a flat plate that covers the region bounded by $y = \ln x$, $x = 2$, and the x-axis.

61. Find the length of arc along the curve of $y = \ln \sin x$ from $x = \pi/3$ to $x = 2\pi/3$. See Exercise 33 of Section 26.6.

62. Find the area of the surface generated by revolving the curve of $y = \sqrt{4 - x^2}$ from $x = -2$ to $x = 2$ about the x-axis. See Exercise 35 of Section 26.6.

63. The change in the thermodynamic entity of entropy ΔS may be expressed as $\Delta S = \int (c_v/T)\, dT$, where c_v is the heat capacity at constant volume and T is the temperature. For increased accuracy, c_v is often given by the equation $c_v = a + bT + cT^2$, where a, b, and c are constants. Express ΔS as a function of temperature.

64. A certain type of chemical reaction leads to the equation $dt = \dfrac{dx}{k(a - x)(b - x)}$, where a, b, and k are constants. Solve for t as a function of x.

65. An electric transmission line between two towers has a shape given by $y = 16.0(e^{x/32} + e^{-x/32})$. Find the length of transmission line if the towers are 50.0 m apart (from $x = -25.0$ m to $x = 25.0$ m). See Exercise 33 on page 792 for the length of arc formula.

66. An object at the end of a spring is immersed in liquid. Its velocity (in cm/s) is then described by the equation $v = 2e^{-2t} + 3e^{-5t}$, where t is the time (in s). Such motion is called *overdamped*. Find the displacement s as a function of t if $s = -1.6$ cm for $t = 0$.

67. When we consider the resisting force of the air, the velocity v (in m/s) of a falling brick in terms of the time t (in s) is given by $dv/(9.8 - 0.1v) = dt$. If $v = 0$ when $t = 0$, find v as a function of t.

68. The power delivered to an electric circuit is given by $P = ei$, where e and i are the instantaneous voltage and the instantaneous current in the circuit, respectively. The mean power, averaged over a period $2\pi/\omega$, is given by $P_{av} = \dfrac{\omega}{2\pi}\int_0^{2\pi/\omega} ei\, dt$. If $e = 20 \cos 2t$ and $i = 3 \sin 2t$, find the average power over a period of $\pi/4$.

69. Find the root-mean-square value for one period of the electric current i if $i = 2 \sin t$.

70. In atomic theory, when finding the number n of atoms per unit volume of a substance, we use the equation $n = A \int_0^\pi e^{a \cos \theta} \sin \theta\, d\theta$. Perform the indicated integration.

71. In the study of the effects of an electric field on molecular orientation, the integral $\int_0^\pi (1 + k \cos \theta) \cos \theta \sin \theta\, d\theta$ is used. Evaluate this integral.

72. In finding the lift of the air flowing around an airplane wing, we use the integral $\int_{-\pi/2}^{\pi/2} \theta^2 \cos \theta\, d\theta$. Evaluate this integral.

73. Find the volume within the piece of tubing in an oil distribution line shown in Fig. 28.19. All cross sections are circular.

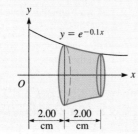

$y = e^{-0.1x}$

| 2.00 cm | 2.00 cm |

Fig. 28.19

74. A metal plate has a shape shown in Fig. 28.20. Find the x-coordinate of the centroid of the plate.

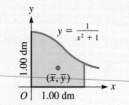

$y = \dfrac{1}{x^2 + 1}$

1.00 dm

$(\bar{x}, \bar{y})$

Fig. 28.20   O | 1.00 dm

75. The nose cone of a space vehicle is to be covered with a heat shield. The cone is designed such that a cross section x m from the tip and perpendicular to its axis is a circle of radius $1.5x^{2/3}$ m. Find the surface area of the heat shield if the nose cone is 4.00 m long. See Fig. 28.21. (See Exercise 35 of Section 26.6.)

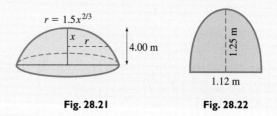

$r = 1.5x^{2/3}$

4.00 m

1.25 m

1.12 m

Fig. 28.21   **Fig. 28.22**

76. A window has a shape of a semiellipse, as shown in Fig. 28.22. What is the area of the window?

Writing Exercise

77. The side of a cutting blade designed using CAD (computer-assisted design) can be described as the region bounded by $y = 4/(1 + e^x)$, $x = 2.8$, and the axes. Write two or three paragraphs explaining how this area may be found by *algebraically* changing the form of the appropriate integral in either one of two ways. (The evaluation of the integral requires only the use of basic forms of this chapter.)

CHAPTER 28 PRACTICE TEST

In Problems 1–7, evaluate the given integrals.

1. $\displaystyle\int (\sec x - \sec^3 x \tan x)\, dx$

2. $\displaystyle\int \sin^3 x\, dx$

3. $\displaystyle\int \tan^3 2x\, dx$

4. $\displaystyle\int \cos^2 4\theta\, d\theta$

5. $\displaystyle\int \frac{dx}{x^2\sqrt{4 - x^2}}$

6. $\displaystyle\int x e^{-2x}\, dx$

7. $\displaystyle\int \frac{x^3 + 5x^2 + x + 2}{x^4 + x^2}\, dx$

8. The electric current in a certain circuit is given by $i = \displaystyle\int \frac{6t + 1}{4t^2 + 9}\, dt$, where t is the time. Integrate and find the resulting function if $i = 0$ for $t = 0$.

9. Find the first-quadrant area bounded by $y = \dfrac{1}{\sqrt{16 - x^2}}$ and $x = 3$.

CHAPTER 29

Expansion of Functions in Series

In the mid-1600s, mathematicians found that transcendental functions can be represented by polynomials and that by using these polynomials it was possible to more easily calculate the values of these functions. In this chapter, we show how a given function may be expressed in terms of a polynomial and how this polynomial is used to evaluate the function.

The polynomials that we will develop are known as *power series,* and they can be expressed with an unlimited number of terms. Although first noted for their usefulness in calculating values of transcendental functions, many mathematicians, including Newton, used power series extensively to further develop various areas of mathematics. In fact, the French mathematician Joseph-Louis Lagrange (1736–1813) attempted to make power series the basis for the development of all methods in calculus.

Many types of electronic devices are used to control the current in a circuit. In Section 29.6, we see how series are used to analyze one such device.

Another type of series was developed by the French physicist and mathematician Jean Baptiste Joseph Fourier (1768–1830) in the study of heat conduction. In 1822, he showed that a function can be expressed in a series of sine and cosine terms. Today, these series are very important in the study of electricity and electronics. They are also useful in the study of mechanical vibrations and other applications that are periodic in nature. We study these series in the last two sections of this chapter.

We see again that a concept, first used to ease calculation, became very important in the later development of mathematics. Also, a concept developed for the study of heat, long before the advent of electronics, has become important in electronics.

In Chapter 19, we discussed arithmetic and geometric sequences and the concept of an infinite geometric sequence. In the first section of this chapter, we further develop these topics for use in the sections that follow.

29.1 INFINITE SERIES

As well as arithmetic sequences and geometric sequences, there are many other ways of generating sequences of numbers. The squares of the integers 1, 4, 9, 16, 25 ... form a sequence. Also, the successive approximations $x_1, x_2, x_3, \ldots$ found by using Newton's method in solving a particular equation form a sequence.

In general, *a* **sequence** *(or* **infinite sequence***) is an infinite succession of numbers. Each of the numbers is a* **term** *of the sequence.* Each term of the sequence is associated with a positive integer, although at times it is convenient to associate the first term with zero (or some specified positive integer). We shall use a_n to designate the term of the sequence corresponding to the integer n.

◀ **EXAMPLE 1** Find the first three terms of the sequence for which $a_n = 2n + 1$, $n = 1, 2, 3, \ldots$.

Substituting the values of n, we obtain the values

$$a_1 = 2(1) + 1 = 3 \qquad a_2 = 2(2) + 1 = 5 \qquad a_3 = 2(3) + 1 = 7, \ldots$$

Therefore, we have the sequence $3, 5, 7, \ldots$.

Given $a_n = 2n + 1$ for $n = 0, 1, 2, \ldots$, the sequence is $1, 3, 5, \ldots$. ▶

Many calculators can display sequences. In Fig. 29.1, a calculator display of the sequence for Example 1 is shown.

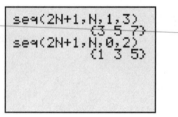

Fig. 29.1

As we stated in Chapter 19, *the indicated sum of the terms of a sequence is called an* **infinite series.** Thus, for the sequence

$$a_1, a_2, a_3, \ldots, a_n, \ldots$$

the associated infinite series is

$$a_1 + a_2 + a_3 + \cdots + a_n + \cdots$$

Using the summation sign Σ (see Section 22.2) to indicate the sum, we have

Infinite Series

$$\sum_{n=1}^{\infty} a_n = a_1 + a_2 + a_3 + \cdots + a_n + \cdots \tag{29.1}$$

Since it is not possible to actually find the sum of infinitely many terms, we define the sum for an infinite series in terms of a limit. For the infinite series of Eq. (29.1), we let S_n represent the sum of the first n terms. Therefore,

$$S_1 = a_1$$
$$S_2 = a_1 + a_2$$
$$S_3 = a_1 + a_2 + a_3$$
$$S_n = a_1 + a_2 + a_3 + \cdots + a_n$$

Partial Sum

The numbers $S_1, S_2, S_3, \ldots, S_n, \ldots$ form a sequence. *Each term of this sequence is called a* **partial sum.** *We say that the infinite series, Eq. (29.1), is* **convergent** *and has the sum S given by*

$$S = \lim_{n \to \infty} S_n = \lim_{n \to \infty} \sum_{i=1}^{n} a_i \tag{29.2}$$

if this limit exists. If the limit does not exist, the series is **divergent.**

◖ EXAMPLE 2   For the infinite series

$$\sum_{n=0}^{\infty} \frac{1}{5^n} = \frac{1}{5^0} + \frac{1}{5^1} + \frac{1}{5^2} + \cdots + \frac{1}{5^n} + \cdots$$

the first six partial sums are

$$S_0 = 1 \qquad\qquad\qquad \text{first term}$$

$$S_1 = 1 + \frac{1}{5} = 1.2 \qquad\qquad \text{sum of first two terms}$$

$$S_2 = 1 + \frac{1}{5} + \frac{1}{25} = 1.24 \qquad \text{sum of first three terms}$$

$$S_3 = 1 + \frac{1}{5} + \frac{1}{25} + \frac{1}{125} = 1.248$$

$$S_4 = 1 + \frac{1}{5} + \frac{1}{25} + \frac{1}{125} + \frac{1}{625} = 1.2496$$

$$S_5 = 1 + \frac{1}{5} + \frac{1}{25} + \frac{1}{125} + \frac{1}{625} + \frac{1}{3125} = 1.249\,92$$

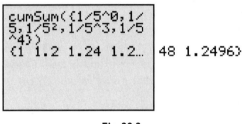

Fig. 29.2

These values are found using the standard calculational features of a calculator or by the *cumulative sum* feature as shown in Fig. 29.2. Here it appears that the sequence of partial sums approaches the value 1.25. We therefore conclude that this infinite series converges and that its sum is approximately 1.25. (In Example 4 of this section, we show that this infinite series does in fact converge and have a sum of 1.25.) ◗

◖ EXAMPLE 3   **(a)** The infinite series

$$\sum_{n=1}^{\infty} 5^n = 5 + 5^2 + 5^3 + \cdots + 5^n + \cdots$$

is a divergent series. The first four partial sums are

$$S_1 = 5 \qquad S_2 = 30 \qquad S_3 = 155 \qquad S_4 = 780$$

Obviously they are increasing without bound.

(b) The infinite series

$$\sum_{n=0}^{\infty} (-1)^n = 1 + (-1) + 1 + (-1) + \cdots + (-1)^n + \cdots$$

has as its first five partial sums

$$S_0 = 1 \qquad S_1 = 0 \qquad S_2 = 1 \qquad S_3 = 0 \qquad S_4 = 1$$

The values of these partial sums do not approach a limiting value, and therefore the series diverges. ◗

See Appendix C for a graphing calculator program PARTSUMS. It evaluates the first *n* partial sums of an infinite series.

Since convergent series are those that have a value associated with them, they are the ones that are of primary use to us. However, generally it is not easy to determine whether a given series is convergent, and many types of tests have been developed for this purpose. These tests for convergence may be found in most textbooks that include the more advanced topics in calculus.

One important series for which we are able to determine the convergence, and its sum if convergent, is the geometric series. For this series the nth partial sum is

$$S_n = a_1 + a_1 r + a_1 r^2 + \cdots + a_1 r^{n-1}$$

where r is the fixed number by which we multiply a given term to get the next term. In Chapter 19 we determined that if $|r| < 1$, the sum S of the infinite geometric series is

$$S = \lim_{n \to \infty} S_n = \frac{a_1}{1 - r} \qquad (29.3)$$

If $r = 1$, we see that the series is $a_1 + a_1 + a_1 + \cdots + a_1 + \cdots$ and is therefore divergent. If $r = -1$, the series is $a_1 - a_1 + a_1 - a_1 + \cdots$ and is also divergent. If $|r| > 1$, $\lim_{n \to \infty} r^n$ is unbounded. Therefore, *the geometric series is convergent only if* $|r| < 1$ *and has the value given by Eq. (29.3).*

CAUTION ▶

◀ **EXAMPLE 4**   Show that the infinite series

$$\sum_{n=0}^{\infty} \frac{1}{5^n} = \frac{1}{5^0} + \frac{1}{5^1} + \frac{1}{5^2} + \cdots + \frac{1}{5^n} + \cdots$$

is convergent and find its sum. This is the same series as in Example 2.

We see that this is a geometric series with $r = \frac{1}{5}$. Since $|r| < 1$, the series is convergent. We find the sum to be

$$S = \frac{1}{1 - \frac{1}{5}} = \frac{1}{\frac{4}{5}} = \frac{5}{4} = 1.25 \qquad \text{using Eq. (29.3)}$$

This value agrees with the conclusion in Example 2.   ▶

EXERCISES 29.1

In Exercises 1 and 2, make the given changes in the indicated examples of this section and then solve the given problems.

1. In Example 3(a), change 5^n to 0.5^n. What other changes occur?

2. In Example 4, change $n = 0$ to $n = 1$. What is the value of S?

In Exercises 3–6, give the first four terms of the sequences for which a_n is given.

3. $a_n = n^2, \quad n = 1, 2, 3, \ldots$

4. $a_n = \frac{2}{3^n}, \quad n = 1, 2, 3, \ldots$

5. $a_n = \frac{1}{n + 2}, \quad n = 0, 1, 2, \ldots$

6. $a_n = \frac{n^2 + 1}{2n + 1}, \quad n = 0, 1, 2, \ldots$

In Exercises 7–10, give (a) the first four terms of the sequence for which a_n is given and (b) the first four terms of the infinite series associated with the sequence.

7. $a_n = \left(-\frac{2}{5}\right)^n, \quad n = 1, 2, 3, \ldots$

8. $a_n = \frac{1}{n} + \frac{1}{n + 1}, \quad n = 1, 2, 3, \ldots$

9. $a_n = \cos \frac{n\pi}{2}, \quad n = 0, 1, 2, \ldots$

10. $a_n = \frac{1}{n(n + 1)}, \quad n = 2, 3, 4, \ldots$

In Exercises 11–14, find the nth term of the given infinite series for which $n = 1, 2, 3, \ldots$.

11. $\frac{1}{2} + \frac{1}{3} + \frac{1}{4} + \frac{1}{5} + \cdots$

12. $\frac{1}{2} + \frac{1}{4} + \frac{1}{8} + \frac{1}{16} + \cdots$

13. $\frac{1}{2 \times 3} + \frac{1}{3 \times 4} + \frac{1}{4 \times 5} + \frac{1}{5 \times 6} + \cdots$

14. $-\frac{2}{3} + \frac{4}{9} - \frac{8}{27} + \frac{16}{81} - \cdots$

In Exercises 15–22, find the first five partial sums of the given series and determine whether the series appears to be convergent or divergent. If it is convergent, find its approximate sum.

15. $1 + \dfrac{1}{8} + \dfrac{1}{27} + \dfrac{1}{64} + \dfrac{1}{125} + \cdots$

16. $1 + 2 + 5 + 10 + 17 + \cdots$

17. $1 + \dfrac{1}{2} + \dfrac{2}{3} + \dfrac{3}{4} + \dfrac{4}{5} + \cdots$

18. $\dfrac{1}{3} - \dfrac{1}{9} + \dfrac{1}{27} - \dfrac{1}{81} + \dfrac{1}{243} - \cdots$

19. $\displaystyle\sum_{n=0}^{\infty} \sqrt{n}$ **20.** $\displaystyle\sum_{n=1}^{\infty} \dfrac{2}{n(n+1)}$

21. $\displaystyle\sum_{n=1}^{\infty} \dfrac{2n+1}{n^2(n+1)^2}$ **22.** $\displaystyle\sum_{n=1}^{\infty} \dfrac{n}{2n+1}$

In Exercises 23–30, test each of the given geometric series for convergence or divergence. Find the sum of each series that is convergent.

23. $1 + 2 + 4 + \cdots + 2^n + \cdots$

24. $1 + \dfrac{1}{2} + \dfrac{1}{4} + \cdots + \dfrac{1}{2^n} + \cdots$

25. $1 - \dfrac{1}{3} + \dfrac{1}{9} - \cdots + \left(-\dfrac{1}{3}\right)^n + \cdots$

26. $1 - \dfrac{3}{2} + \dfrac{9}{4} - \cdots + \left(-\dfrac{3}{2}\right)^n + \cdots$

27. $10 + 9 + 8.1 + 7.29 + 6.561 + \cdots$

28. $4 + 1 + \dfrac{1}{4} + \dfrac{1}{16} + \dfrac{1}{64} + \cdots$

29. $512 - 64 + 8 - 1 + \dfrac{1}{8} - \cdots$

30. $16 + 12 + 9 + \dfrac{27}{4} + \dfrac{81}{16} + \cdots$

In Exercises 31–40, solve the given problems as indicated.

31. Using a calculator, take successive square roots of 2 and find at least 20 approximate values for the terms of the sequence $2^{1/2}, 2^{1/4}, 2^{1/8}, 2^{1/16}, \ldots$. From the values that are obtained, (a) what do you observe about the value of $\lim\limits_{n\to\infty} 2^{1/2^n}$? (b) Determine whether the infinite series for this sequence converges or diverges.

(W) 32. Using a calculator, (a) take successive square roots of 0.01 and then (b) take successive square roots of 100. From these sequences of square roots, state any general conclusions that might be drawn.

33. Referring to Chapter 19, we see that the sum of the first n terms of a geometric sequence is

$$S_n = \dfrac{a_1(1 - r^n)}{1 - r} \quad (r \neq 1) \qquad \text{Eq. (19.6)}$$

where a_1 is the first term and r is the common ratio. We can visualize the corresponding infinite series by graphing the function

$f(x) = a_1(1 - r^x)/(1 - r)$ $(r \neq 1)$ (or using a calculator that can graph a sequence.) The graph represents the sequence of partial sums for values where $x = n$, since $f(n) = S_n$.

Use a graphing calculator to visualize the first five partial sums of the series

$$\dfrac{1}{2} + \dfrac{1}{4} + \dfrac{1}{8} + \cdots$$

What value does the infinite series approach? (*Remember:* Only points for which x is an integer have real meaning.)

34. Following Exercise 33, use a graphing calculator to show that the sum of the infinite series of Example 4 is 1.25. (*Be careful:* Because of the definition of the series, $x = 1$ corresponds to $n = 0$.)

35. The value V (in dollars) of a certain investment after n years can be expressed as

$V = 100(1.05 + 1.05^2 + 1.05^3 + \cdots + 1.05^n)$

(a) By finding partial sums, determine whether this series converges or diverges. (b) Following Exercise 33, use a graphing calculator to visualize the first ten partial sums. (See Example 7 of Section 19.2.)

36. If an electric discharge is passed through hydrogen gas, a spectrum of isolated parallel lines, called the Balmer series, is formed. See Fig. 29.3. The wavelengths λ (in nm) of the light for these lines is given by the formula

$$\dfrac{1}{\lambda} = 1.097 \times 10^{-2}\left(\dfrac{1}{2^2} - \dfrac{1}{n^2}\right) \quad (n = 3, 4, 5, \ldots)$$

Find the wavelengths of the first three lines and the shortest wavelength of all the lines of the series.

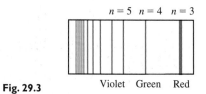

Fig. 29.3 Violet Green Red

37. Use geometric series to show that $\displaystyle\sum_{n=0}^{\infty} x^n = \dfrac{1}{1-x}$ for $|x| < 1$.

38. Use geometric series to show that $\displaystyle\sum_{n=0}^{\infty} (-1)^n x^n = \dfrac{1}{1+x}$ for $|x| < 1$.

39. If term a_1 is given along with a rule to find term a_{n+1} from term a_1, the sequence is said to be defined *recursively*. If $a_1 = 2$ and $a_{n+1} = (n + 1)a_n$, find the first five terms of the sequence.

40. A sequence is defined recursively (see Exercise 39) by $x_1 = \dfrac{N}{2}$, $x_{n+1} = \dfrac{1}{2}\left(x_n + \dfrac{N}{x_n}\right)$. With $N = 10$, find x_6 and compare the value with $\sqrt{10}$. It can be seen that $\sqrt{N}$ can be approximated using this recursion sequence.

29.2 MACLAURIN SERIES

In this section, we develop a very important basic polynomial form of a function. Before developing the method using calculus, we will review how this can be done for some functions algebraically.

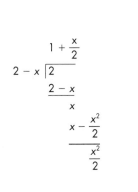

◀ **EXAMPLE 1** By using long division (as started at the left), we have

$$\frac{2}{2-x} = 1 + \frac{1}{2}x + \frac{1}{4}x^2 + \cdots + \left(\frac{1}{2}x\right)^{n-1} + \cdots \qquad (1)$$

where n is the number of the term of the expression on the right. Since x represents a number, the right-hand side of Eq. (1) becomes a geometric series.

From Eq. (29.3) we know that the sum of a geometric series with first term a_1 and common ratio r is

$$S = \frac{a_1}{1 - r}$$

where $|r| < 1$ and the series converges.

If $x = 1$, the right-hand side of Eq. (1) is

$$1 + \frac{1}{2} + \frac{1}{4} + \cdots + \left(\frac{1}{2}\right)^{n-1} + \cdots$$

For this series, $r = \frac{1}{2}$ and $a_1 = 1$, which means that the series converges and $S = 2$. If $x = 3$, the right-hand side of Eq. (1) is

$$1 + \frac{3}{2} + \frac{9}{4} + \cdots + \left(\frac{3}{2}\right)^{n-1} + \cdots$$

which diverges since $r > 1$. Referring to the left side of Eq. (1), we see that it also equals 2 when $x = 1$. Thus, we see that the two sides agree for $x = 1$, but that the series diverges for $x = 3$. In fact, as long as $|x| < 2$, the series will converge to the value of the function on the left. From this we conclude that the series on the right properly represents the function on the left, as long as $|x| < 2$. ▶

From Example 1, we see that an algebraic function may be properly represented by a function of the form

Power Series

$$f(x) = a_0 + a_1x + a_2x^2 + \cdots + a_nx^n + \cdots \qquad (29.4)$$

Equation (29.4) is known as a **power-series expansion** *of the function* $f(x)$. The problem now arises as to whether or not functions in general may be represented in this form. If such a representation were possible, it would provide a means of evaluating the transcendental functions for the purpose of making tables of values. Also, since a power-series expansion is in the form of a polynomial, it makes algebraic operations much simpler due to the properties of polynomials. A further study of calculus shows many other uses of power series.

In Example 1, we saw that the function could be represented by a power series as long as $|x| < 2$. That is, if we substitute any value of x in this interval into the series and also into the function, the series will converge to the value of the function. *This interval of values for which the series converges is called the* **interval of convergence.**

◀ EXAMPLE 2 In Example 1, the interval of convergence for the series

$$1 + \frac{1}{2}x + \frac{1}{4}x^2 + \cdots + \left(\frac{1}{2}x\right)^{n-1} + \cdots$$

is $|x| < 2$. We saw that the series converges for $x = 1$, with $S = 2$, and that the value of the function is 2 for $x = 1$. This verifies that $x = 1$ is in the interval of convergence.

Also, we saw that the series diverges for $x = 3$, which verifies that $x = 3$ is not in the interval of convergence. ▶

At this point, we will assume that unless otherwise noted, the functions with which we will be dealing may be properly represented by a power-series expansion (it takes more advanced methods to prove that this is generally possible), for appropriate intervals of convergence. We will find that the methods of calculus are very useful in developing the method of general representation. Thus, writing a general power series, along with the first few derivatives, we have

$$f(x) = a_0 + a_1x + a_2x^2 + a_3x^3 + a_4x^4 + a_5x^5 + \cdots + a_nx^n + \cdots$$
$$f'(x) = a_1 + 2a_2x + 3a_3x^2 + 4a_4x^3 + 5a_5x^4 + \cdots + na_nx^{n-1} + \cdots$$
$$f''(x) = 2a_2 + 2(3)a_3x + 3(4)a_4x^2 + 4(5)a_5x^3 + \cdots + (n-1)na_nx^{n-2} + \cdots$$
$$f'''(x) = 2(3)a_3 + 2(3)(4)a_4x + 3(4)(5)a_5x^2 + \cdots + (n-2)(n-1)na_nx^{n-3} + \cdots$$
$$f^{iv}(x) = 2(3)(4)a_4 + 2(3)(4)(5)a_5x + \cdots + (n-3)(n-2)(n-1)na_nx^{n-4} + \cdots$$

NOTE ▶ Regardless of the values of the constants a_n for any power series, **if $x = 0$, the left and right sides must be equal,** and all the terms on the right are zero except the first. Thus, setting $x = 0$ in each of the above equations, we have

$$f(0) = a_0 \qquad f'(0) = a_1 \qquad f''(0) = 2a_2$$
$$f'''(0) = 2(3)a_3 \qquad f^{iv}(0) = 2(3)(4)a_4$$

Solving each of these for the constants a_n, we have

$$a_0 = f(0) \qquad a_1 = f'(0) \qquad a_2 = \frac{f''(0)}{2!} \qquad a_3 = \frac{f'''(0)}{3!} \qquad a_4 = \frac{f^{iv}(0)}{4!}$$

Substituting these into the expression for $f(x)$, we have

Maclaurin Series

$$f(x) = f(0) + f'(0)x + \frac{f''(0)x^2}{2!} + \frac{f'''(0)x^3}{3!} + \cdots + \frac{f^n(0)x^n}{n!} + \cdots \tag{29.5}$$

Named for the Scottish mathematician Colin Maclaurin (1698–1746).

Equation (29.5) is known as the **Maclaurin series expansion** *of a function.* For a function to be represented by a Maclaurin expansion, the function and all of its derivatives must exist at $x = 0$. Also, we note that the factorial notation introduced in Section 19.4 is used in writing the Maclaurin series expansion.

As we mentioned earlier, one of the uses we will make of series expansions is that of determining the values of functions for particular values of x. If x is sufficiently small, successive terms become smaller and smaller and the series will converge rapidly. This is considered in the sections that follow.

◀ EXAMPLE 3   Find the first four terms of the Maclaurin series expansion of $f(x) = \dfrac{2}{2-x}$.

This is written as

$$f(x) = \frac{2}{2-x} \qquad f(0) = 1$$

$$f'(x) = \frac{2}{(2-x)^2} \qquad f'(0) = \frac{1}{2}$$

find derivatives and evaluate each at $x = 0$

$$f''(x) = \frac{4}{(2-x)^3} \qquad f''(0) = \frac{1}{2}$$

$$f'''(x) = \frac{12}{(2-x)^4} \qquad f'''(0) = \frac{3}{4}$$

$$f(x) = 1 + \frac{1}{2}x + \frac{1}{2}\left(\frac{x^2}{2!}\right) + \frac{3}{4}\left(\frac{x^3}{3!}\right) + \cdots \qquad \text{using Eq. (29.5)}$$

or

$$\frac{2}{2-x} = 1 + \frac{1}{2}x + \frac{1}{4}x^2 + \frac{1}{8}x^3 + \cdots$$

We see that this result agrees with that obtained by direct division.

◀ EXAMPLE 4   Find the first four terms of the Maclaurin series expansion of $f(x) = e^{-x}$.

We write

$$f(x) = e^{-x} \qquad f(0) = 1$$
$$f'(x) = -e^{-x} \qquad f'(0) = -1$$
$$f''(x) = e^{-x} \qquad f''(0) = 1$$
$$f'''(x) = -e^{-x} \qquad f''''(0) = -1$$

find derivatives and evaluate each at $x = 0$

$$f(x) = 1 + (-1)x + 1\left(\frac{x^2}{2!}\right) + (-1)\left(\frac{x^3}{3!}\right) + \cdots \qquad \text{using Eq. (29.5)}$$

or

$$e^{-x} = 1 - x + \frac{x^2}{2!} - \frac{x^3}{3!} + \cdots$$

◀ EXAMPLE 5   Find the first three nonzero terms of the Maclaurin series expansion of $f(x) = \sin 2x$.

We write

$f(x) = \sin 2x$	$f(0) = 0$	$f'''(x) = -8\cos 2x$	$f'''(0) = -8$
$f'(x) = 2\cos 2x$	$f'(0) = 2$	$f^{iv}(x) = 16\sin 2x$	$f^{iv}(0) = 0$
$f''(x) = -4\sin 2x$	$f''(0) = 0$	$f^{v}(x) = 32\cos 2x$	$f^{v}(0) = 32$

$$f(x) = 0 + 2x + 0 + (-8)\frac{x^3}{3!} + 0 + 32\frac{x^5}{5!} + \cdots$$

$$\sin 2x = 2x - \frac{4}{3}x^3 + \frac{4}{15}x^5 - \cdots$$

This series is called an **alternating series,** *since every other term is negative.*

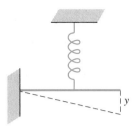

Fig. 29.4

❮ EXAMPLE 6 A lever is attached to a spring as shown in Fig. 29.4. Frictional forces in the spring are just sufficient so that the lever does not oscillate after being depressed. Such motion is called *critically damped*. The displacement y of the lever as a function of the time t for one such case is $y = (1 + t)e^{-t}$. In order to study the motion for small values of t, a polynomial form of $y = f(t)$ is to be used. Find the first four nonzero terms of the expansion.

$$\begin{aligned} f(t) &= (1 + t)e^{-t} & f(0) &= 1 \\ f'(t) &= (1 + t)e^{-t}(-1) + e^{-t} = -te^{-t} & f'(0) &= 0 \\ f''(t) &= te^{-t} - e^{-t} & f''(0) &= -1 \\ f'''(t) &= -te^{-t} + e^{-t} + e^{-t} = 2e^{-t} - te^{-t} & f'''(0) &= 2 \\ f^{iv}(t) &= -2e^{-t} + te^{-t} - e^{-t} = te^{-t} - 3e^{-t} & f^{iv}(0) &= -3 \end{aligned}$$

$$f(t) = 1 + 0 + (-1)\frac{t^2}{2!} + 2\frac{t^3}{3!} + (-3)\frac{t^4}{4!} + \cdots$$

$$(1 + t)e^{-t} = 1 - \frac{t^2}{2} + \frac{t^3}{3} - \frac{t^4}{8} + \cdots$$

❯

EXERCISES 29.2

In Exercises 1 and 2, make the given changes in the indicated examples of this section and then find the resulting series.

1. In Example 3, in $f(x)$ change the denominator to $2 + x$.

2. In Example 5, in $f(x)$ change $2x$ to $(-2x)$.

In Exercises 3–16, find the first three nonzero terms of the Maclaurin expansion of the given functions.

3. $f(x) = e^x$

4. $f(x) = \sin x$

5. $f(x) = \cos x$

6. $f(x) = \ln(1 + x)$

7. $f(x) = \sqrt{1 + x}$

8. $f(x) = e^{-2x}$

9. $f(x) = \cos 4\pi x$

10. $f(x) = e^x \sin x$

11. $f(x) = \dfrac{1}{1 - x}$

12. $f(x) = \dfrac{1}{(1 + x)^2}$

13. $f(x) = \ln(1 - 2x)$

14. $f(x) = (1 + x)^{3/2}$

15. $f(x) = \cos^2 x$

16. $f(x) = \ln(1 + 4x)$

In Exercises 17–22, find the first two nonzero terms of the Maclaurin expansion of the given functions.

17. $f(x) = \tan^{-1} x$

18. $f(x) = \cos x^2$

19. $f(x) = \tan x$

20. $f(x) = \sec x$

21. $f(x) = \ln \cos x$

22. $f(x) = xe^{\sin x}$

In Exercises 23–32, solve the given problems.

Ⓦ **23.** Is it possible to find a Maclaurin expansion for (a) $f(x) = \csc x$ or (b) $f(x) = \ln x$? Explain.

Ⓦ **24.** Is it possible to find a Maclaurin expansion for (a) $f(x) = \sqrt{x}$ or (b) $f(x) = \sqrt{1 + x}$? Explain.

25. Find the first three nonzero terms of the Maclaurin expansion for (a) $f(x) = e^x$ and (b) $f(x) = e^{x^2}$. Compare these expansions.

26. By finding the Maclaurin expansion of $f(x) = (1 + x)^n$, derive the first four terms of the binomial series, which is Eq. (19.10). Its interval of convergence is $|x| < 1$ for all values of n.

27. If $f(x) = e^{3x}$, compare the Maclaurin expansion with the linearization for $a = 0$.

28. Find the Maclaurin series for $f(x) = \frac{1}{2}(e^x + e^{-x})$. (See Exercises 53–56 on page 823.)

29. If $f(x) = x^2$, show that this function is obtained when a Maclaurin expansion is found.

30. If $f(x) = x^4 + 2x^2$, show that this function is obtained when a Maclaurin expansion is found.

31. The reliability R $(0 \le R \le 1)$ of a certain computer system is $R = e^{-0.001t}$, where t is the time of operation (in min). Express $R = f(t)$ in polynomial form by using the first three terms of the Maclaurin expansion.

32. In the analysis of the optical paths of light from a narrow slit S to a point P as shown in Fig. 29.5, the law of cosines is used to obtain the equation

$$c^2 = a^2 + (a + b)^2 - 2a(a + b)\cos\frac{s}{a}$$

where s is part of the circular arc AB. By using two nonzero terms of the Maclaurin expansion of $\cos\frac{s}{a}$, simplify the right side of the equation. (In finding the expansion, let $x = \frac{s}{a}$ and then substitute back into the expansion.)

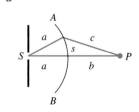

Fig. 29.5

29.3 CERTAIN OPERATIONS WITH SERIES

The series found in Exercises 3 to 6 and 26 (the binomial series) of Section 29.2 are of particular importance. They are used to evaluate exponential functions, trigonometric functions, logarithms, powers, and roots, as well as develop other series. For reference, we give them here with their intervals of convergence.

$$e^x = 1 + x + \frac{x^2}{2!} + \frac{x^3}{3!} + \cdots \qquad \text{(all } x) \qquad (29.6)$$

$$\sin x = x - \frac{x^3}{3!} + \frac{x^5}{5!} - \cdots \qquad \text{(all } x) \qquad (29.7)$$

$$\cos x = 1 - \frac{x^2}{2!} + \frac{x^4}{4!} - \cdots \qquad \text{(all } x) \qquad (29.8)$$

$$\ln(1 + x) = x - \frac{x^2}{2} + \frac{x^3}{3} - \frac{x^4}{4} + \cdots \qquad (|x| < 1) \qquad (29.9)$$

$$(1 + x)^n = 1 + nx + \frac{n(n - 1)}{2!}x^2 + \cdots \qquad (|x| < 1) \qquad (29.10)$$

In the next section, we will see how to use these series in finding values of functions. In this section, we see how new series are developed by using the above basic series, and we also show other uses of series.

When we discussed functions in Chapter 3, we mentioned functions such as $f(2x)$

NOTE ▶ and $f(-x)$. *By using functional notation and the preceding series, we can find the series expansions of many other series without using direct expansion.* This can often save time in finding a desired series.

◀ **EXAMPLE 1** Find the Maclaurin expansion of e^{2x}.

From Eq. (29.6), we know the expansion of e^x. Hence,

$$f(x) = 1 + x + \frac{x^2}{2!} + \frac{x^3}{3!} + \cdots$$

Since $e^{2x} = f(2x)$, we have

$$f(2x) = 1 + (2x) + \frac{(2x)^2}{2!} + \frac{(2x)^3}{3!} + \cdots \qquad \text{in } f(x), \text{ replace } x \text{ by } 2x$$

$$e^{2x} = 1 + 2x + 2x^2 + \frac{4x^3}{3} + \cdots$$

◀ **EXAMPLE 2** Find the Maclaurin expansion of $\sin x^2$.

From Eq. (29.7), we know the expansion of $\sin x$. Therefore,

$$f(x) = x - \frac{x^3}{3!} + \frac{x^5}{5!} - \cdots$$

$$f(x^2) = (x^2) - \frac{(x^2)^3}{3!} + \frac{(x^2)^5}{5!} - \cdots \qquad \text{in } f(x), \text{ replace } x \text{ by } x^2$$

$$\sin x^2 = x^2 - \frac{x^6}{3!} + \frac{x^{10}}{5!} - \cdots$$

Direct expansion of this series is quite lengthy.

The basic algebraic operations may be applied to series in the same manner they are applied to polynomials. That is, we may add, subtract, multiply, or divide series in order to obtain other series. The interval of convergence for the resulting series is that which is common to those of the series being used. The multiplication of series is illustrated in the following example.

◀ EXAMPLE 3   Multiply the series for e^x and $\cos x$ in order to obtain the series expansion for $e^x \cos x$.

Using the series expansion for e^x and $\cos x$ as shown in Eqs. (29.6) and (29.8), we have the following indicated multiplication:

$$e^x \cos x = \left(1 + x + \frac{x^2}{2!} + \frac{x^3}{3!} + \frac{x^4}{4!} + \cdots\right)\left(1 - \frac{x^2}{2!} + \frac{x^4}{4!} - \cdots\right)$$

By multiplying the series on the right, we have the following result, considering through the x^4 terms in the product.

$$1\left(1 - \frac{x^2}{2!} + \frac{x^4}{4!}\right) \quad x\left(1 - \frac{x^2}{2!}\right) \quad \frac{x^2}{2!}\left(1 - \frac{x^2}{2!}\right) \quad \left(\frac{x^3}{3!} + \frac{x^4}{4!}\right)(1)$$

$$e^x \cos x = 1 - \frac{x^2}{2} + \frac{x^4}{24} + x - \frac{x^3}{2} + \frac{x^2}{2} - \frac{x^4}{4} + \frac{x^3}{6} + \frac{x^4}{24} + \cdots$$

$$= 1 + x - \frac{1}{3}x^3 - \frac{1}{6}x^4 + \cdots$$

It is also possible to use the operations of differentiation and integration to obtain series expansions, although the proof of this is found in more advanced texts. Consider the following example.

◀ EXAMPLE 4   Show that by differentiating the expansion for $\ln(1 + x)$ term by term, the result is the same as the expansion for $\dfrac{1}{1 + x}$.

The series for $\ln(1 + x)$ is shown in Eq. (29.9) as

$$\ln(1 + x) = x - \frac{x^2}{2} + \frac{x^3}{3} - \frac{x^4}{4} + \cdots$$

Differentiating, we have

$$\frac{1}{1 + x} = 1 - \frac{2x}{2} + \frac{3x^2}{3} - \frac{4x^3}{4} + \cdots$$

$$= 1 - x + x^2 - x^3 + \cdots$$

Using the binomial expansion for $\dfrac{1}{1 + x} = (1 + x)^{-1}$, we have

$$(1 + x)^{-1} = 1 + (-1)x + \frac{(-1)(-2)}{2!}x^2 + \frac{(-1)(-2)(-3)}{3!}x^3 + \cdots \qquad \text{using Eq. (29.10)}$$
$$\text{with } n = -1$$

$$= 1 - x + x^2 - x^3 + \cdots$$

We see that the results are the same.

For reference, Eq. (12.11) is
$re^{j\theta} = r(\cos\theta + j\sin\theta)$.

We can use algebraic operations on series to verify that the definition of the exponential form of a complex number, as shown in Eq. (12.11), is consistent with other definitions. The only assumption required here is that the Maclaurin expansions for e^x, $\sin x$, and $\cos x$ are also valid for complex numbers. This is shown in advanced calculus. Thus,

$$e^{j\theta} = 1 + j\theta + \frac{(j\theta)^2}{2!} + \frac{(j\theta)^3}{3!} + \cdots = 1 + j\theta - \frac{\theta^2}{2!} - j\frac{\theta^3}{3!} + \cdots \quad (29.11)$$

$$j\sin\theta = j\theta - j\frac{\theta^3}{3!} + \cdots \quad (29.12)$$

$$\cos\theta = 1 - \frac{\theta^2}{2!} + \cdots \quad (29.13)$$

When we add the terms of Eq. (29.12) to those of Eq. (29.13), the result is the series given in Eq. (29.11). Thus,

$$\boxed{e^{j\theta} = \cos\theta + j\sin\theta} \quad (29.14)$$

A comparison of Eqs. (12.11) and (29.14) indicates the reason for the choice of the definition of the exponential form of a complex number.

An additional use of power series is now shown. Many integrals that occur in practice cannot be integrated by methods given in the preceding chapters. However, power series can be very useful in giving excellent approximations to some definite integrals.

❲ **EXAMPLE 5** Find the first-quadrant area bounded by $y = \sqrt{1 + x^3}$ and $x = 0.5$. From Fig. 29.6, we see that the area is

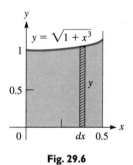

Fig. 29.6

$$A = \int_0^{0.5} \sqrt{1 + x^3}\, dx$$

This integral does not fit any form we have used. However, its value can be closely approximated by using the binomial expansion for $\sqrt{1 + x^3}$ and then integrating.

Using the binomial expansion to find the first three terms of the expansion for $\sqrt{1 + x^3}$, we have

$$\sqrt{1 + x^3} = (1 + x^3)^{0.5} = 1 + 0.5x^3 + \frac{0.5(-0.5)}{2}(x^3)^2 + \cdots$$

$$= 1 + 0.5x^3 - 0.125x^6 + \cdots$$

Substituting in the integral, we have

$$A = \int_0^{0.5} (1 + 0.5x^3 - 0.125x^6 + \cdots)\, dx$$

$$= x + \frac{0.5}{4}x^4 - \frac{0.125}{7}x^7 + \cdots \Big|_0^{0.5}$$

$$= 0.5 + 0.007\,812\,5 - 0.000\,139\,5 + \cdots = 0.507\,673 + \cdots$$

We can see that each of the terms omitted was very small. The result shown is correct to four decimal places, or $A = 0.5077$. Additional accuracy can be obtained by using more terms of the expansion. ❩

EXAMPLE 6 Evaluate: $\int_0^{0.1} e^{-x^2} dx$.

We write

$$e^{-x^2} = 1 + (-x^2) + \frac{(-x^2)^2}{2!} + \cdots \qquad \text{using Eq. (29.6)}$$

Thus,

$$\int_0^{0.1} e^{-x^2} dx = \int_0^{0.1} \left(1 - x^2 + \frac{x^4}{2} - \cdots \right) dx \qquad \text{substitute}$$

$$= \left(x - \frac{x^3}{3} + \frac{x^5}{10} - \cdots \right)\Bigg|_0^{0.1} \qquad \text{integrate}$$

$$= 0.1 - \frac{0.001}{3} + \frac{0.00001}{10} = 0.0996677 \qquad \text{evaluate}$$

This answer is correct to the indicated accuracy.

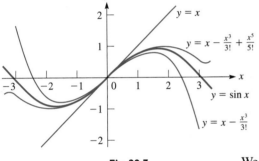

Fig. 29.7

The question of accuracy now arises. The integrals just evaluated indicate that the more terms used, the greater the accuracy of the result. To graphically show the accuracy involved, Fig. 29.7 depicts the graphs of $y = \sin x$ and the graphs of

$$y = x \qquad y = x - \frac{x^3}{3!} \qquad y = x - \frac{x^3}{3!} + \frac{x^5}{5!}$$

which are the first three approximations of $y = \sin x$. We can see that each term added gives a better fit to the curve of $y = \sin x$. Also, this gives a graphical representation of the meaning of a series expansion.

We have just shown that the more terms included, the more accurate the result. For small values of x, a Maclaurin series gives good accuracy with a very few terms. In this case the series *converges* rapidly, as we mentioned earlier. For this reason, a Maclaurin series is of particular use for small values of x. For larger values of x, usually a function is expanded in a Taylor series (see Section 29.5). Of course, if we omit any term in a series, there is some error in the calculation.

EXERCISES 29.3

In Exercises 1 and 2, make the given changes in the indicated examples of this section, and then find the resulting series.

1. In Example 1, change e^{2x} to e^{2x^2}.

2. In Example 3, change e^x to e^{-x}.

In Exercises 3–10, find the first four nonzero terms of the Maclaurin expansions of the given functions by using Eqs. (29.6) to (29.10).

3. $f(x) = e^{3x}$

4. $f(x) = e^{-2x}$

5. $f(x) = \sin \frac{1}{2} x$

6. $f(x) = \sin x^4$

7. $f(x) = x \cos 4x$

8. $f(x) = \sqrt{1 - x^4}$

9. $f(x) = \ln(1 + x^2)$

10. $f(x) = x^2 \ln(1 - x)$

In Exercises 11–14, evaluate the given integrals by using three terms of the appropriate series.

11. $\int_0^1 \sin x^2 \, dx$

12. $\int_0^{0.4} \sqrt[4]{1 - 2x^2} \, dx$

13. $\int_0^{0.2} \cos \sqrt{x} \, dx$

14. $\int_{0.1}^{0.2} \frac{\cos x - 1}{x} \, dx$

In Exercises 15–24, find the indicated series by the given operation.

15. Find the first four terms of the Maclaurin expansion of the function $f(x) = \frac{2}{1 - x^2}$ by adding the terms of the series for the functions $\frac{1}{1 - x}$ and $\frac{1}{1 + x}$.

16. Find the first four nonzero terms of the expansion of the function $f(x) = \frac{1}{2}(e^x - e^{-x})$ by subtracting the terms of the appropriate series. The result is the series for sinh x. (See Exercise 53 of Section 27.6.)

17. Find the first three terms of the expansion for $e^x \sin x$ by multiplying the proper expansions together, term by term.

18. Find the first three nonzero terms of the expansion for $f(x) = \tan x$ by dividing the series for $\sin x$ by that for $\cos x$.

19. By using the properties of logarithms and the series for $\ln(1 + x)$, find the series for $x^2 \ln(1 - x)^2$.

20. By using the properties of logarithms and the series for $\ln(1 + x)$, find the series for $\ln \dfrac{1 + x}{1 - x}$.

21. Show that by differentiating term by term the expansion for $\sin x$, the result is the expansion for $\cos x$.

22. Show that by differentiating term by term the expansion for e^x, the result is also the expansion for e^x.

23. Show that by integrating term by term the expansion for $\cos x$, the result is the expansion for $\sin x$.

24. Show that by integrating term by term the expansion for $-1/(1 - x)$ (see Exercise 11 of Section 29.2), the result is the expansion for $\ln(1 - x)$.

In Exercises 25–32, solve the given problems.

25. Evaluate $\int_0^1 e^x \, dx$ directly and compare the result obtained by using four terms of the series for e^x and then integrating.

26. Evaluate $\displaystyle\lim_{x \to 0} \frac{\sin x}{x}$ by using the series expansion for $\sin x$. Compare the result with Eq. (27.1).

27. Find the approximate value of the area bounded by $y = x^2 e^x$, $x = 0.2$, and the x-axis by using three terms of the appropriate Maclaurin series.

28. Find the approximate area under the graph of $y = \dfrac{1}{\sqrt{2\pi}} e^{-x^2/2}$ from $x = -1$ to $x = 1$ by using three terms of the appropriate series. See Fig. 29.8 and Fig. 22.11.

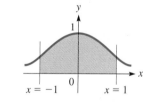

Fig. 29.8

29. The *Fresnel integral* $\displaystyle\int_0^x \cos t^2 \, dt$ is used in the analysis of beam displacements (and in optics). Evaluate this integral for $x = 0.2$ by using two terms of the appropriate series.

30. The dome of a sports arena is designed as the surface generated by revolving the curve of $y = 20.0 \cos 0.0196x$ $(0 \le x \le 80.0 \text{ m})$ about the y-axis. Find the volume within the dome by using three terms of the appropriate series. (This is the same dome as in Exercise 43, page 871. Compare the results).

31. In the theory of relativity, when studying the kinetic (moving) energy of an object, the equation $K = \left[\left(1 - \dfrac{v^2}{c^2} \right)^{-1/2} - 1 \right] mc^2$ is used. Here, for a given object, K is the kinetic energy, v is its velocity, and c is the velocity of light. If v is much smaller than c, show that $K = \frac{1}{2}mv^2$, which is the classical expression for K.

32. The charge q on a capacitor in a certain electric circuit is given by $q = ce^{-at} \sin 6at$, where t is the time. By multiplication of series, find the first four nonzero terms of the expansion for q.

In Exercises 33–36, use a graphing calculator to display (a) the given function and (b) the first three series approximations of the function in the same display. Each display will be similar to that in Fig. 29.7 for the function $y = \sin x$ and its first three approximations. Be careful in choosing the appropriate window values.

33. $y = e^x$

34. $y = \cos x$

35. $y = \ln(1 + x)$ $(|x| < 1)$

36. $y = \sqrt{1 + x}$ $(|x| < 1)$

29.4 COMPUTATIONS BY USE OF SERIES EXPANSIONS

As we mentioned at the beginning of the previous section, power-series expansions can be used to compute numerical values of exponential functions, trigonometric functions, logarithms, powers, and roots. By including a sufficient number of terms in the expansion, we can calculate these values to any degree of accuracy that may be required.

It is through such calculations that tables of values can be made, and decimal approximations of numbers such as e and π can be found. Also, many of the values found on a calculator or a computer are calculated by using series expansions that have been programmed into the chip which is in the calculator or computer.

◀ EXAMPLE 1   Calculate the value of $e^{0.1}$.

In order to evaluate $e^{0.1}$, we substitute 0.1 for x in the expansion for e^x. The more terms that are used, the more accurate a value we can obtain. The limit of the partial sums would be the actual value. However, since $e^{0.1}$ is irrational, we cannot express the exact value in decimal form.

Therefore, the value is found as follows:

$$e^x = 1 + x + \frac{x^2}{2!} + \cdots \qquad \text{Eq. (29.6)}$$

$$e^{0.1} = 1 + 0.1 + \frac{(0.1)^2}{2} + \cdots \qquad \text{substitute 0.1 for } x$$

$$= 1.105 \qquad \text{using 3 terms}$$

Using a calculator, we find that $e^{0.1} = 1.105\,170\,918$, which shows that our answer is valid to the accuracy shown. ▶

CAUTION ▶

◀ EXAMPLE 2   Calculate the value of sin 2°.

In finding trigonometric values, we must be careful to *express the angle in radians*. Thus, the value of sin 2° is found as follows:

$$\sin x = x - \frac{x^3}{3!} + \cdots \qquad \text{Eq. (29.7)}$$

$$\sin 2° = \left(\frac{\pi}{90}\right) - \frac{(\pi/90)^3}{6} + \cdots \qquad 2° = \frac{\pi}{90} \text{ rad}$$

$$= 0.034\,899\,496\,3 \qquad \text{using 2 terms}$$

A calculator gives the value $0.034\,899\,496\,7$. Here we note that the second term is much smaller than the first. In fact, a good approximation of 0.0349 can be found by using just one term. We now see that $\sin \theta \approx \theta$ for small values of θ, as we noted in Section 8.4. ▶

◀ EXAMPLE 3   Calculate the value of cos 0.5429.

Since the angle is expressed in radians, we have

$$\cos 0.5429 = 1 - \frac{0.5429^2}{2} + \frac{0.5429^4}{4!} - \cdots \qquad \text{using Eq. (29.8)}$$

$$= 0.856\,249\,5 \qquad \text{using 3 terms}$$

A calculator shows that $\cos 0.5429 = 0.856\,214\,082\,4$. Since the angle is not small, additional terms are needed to obtain this accuracy. With one more term, the value $0.856\,213\,9$ is obtained. ▶

◀ EXAMPLE 4   Calculate the value of ln 1.2.

$$\ln(1 + x) = x - \frac{x^2}{2} + \frac{x^3}{3} - \cdots \qquad \text{Eq. (29.9)}$$

$$\ln 1.2 = \ln(1 + 0.2)$$

$$= 0.2 - \frac{(0.2)^2}{2} + \frac{(0.2)^3}{3} - \cdots = 0.1827$$

To four significant digits, $\ln(1.2) = 0.1823$. One more term is required to obtain this accuracy. ▶

We now illustrate the use of series in error calculations. We also discussed this as an application of differentials. A series solution allows as close a value of the calculated error as needed, whereas only one term can be found using differentials.

◀ EXAMPLE 5   The velocity v of an object that has fallen h m is $v = 4.43\sqrt{h}$. Find the approximate error in calculating the velocity of an object that has fallen 100.0 m, with a possible error of 2.0 m.

NOTE ▶   If we **let** $v = 4.43\sqrt{100.0 + x}$, **where x is the error in h,** we may express v as a Maclaurin expansion in x:

$$f(x) = 4.43(100.0 + x)^{1/2} \qquad f(0) = 44.3$$
$$f'(x) = 2.22(100.0 + x)^{-1/2} \qquad f'(0) = 0.222$$
$$f''(x) = -1.11(100.0 + x)^{-3/2} \qquad f''(0) = -0.00111$$

Therefore,

$$v = 4.43\sqrt{100.0 + x} = 4.43 + 0.222x - 0.00056x^2 + \cdots$$

Since the calculated value of v for $x = 0$ is 44.3, the error e in the value of v is

$$e = 0.222x - 0.00056x^2 + \cdots$$

Calculating, the error for $x = 2.0$ is

$$e = 0.222(2.0) - 0.00056(4.0) = 0.444 - 0.002 = 0.442 \text{ m/s}$$

The value 0.444 is that which is found using differentials. The additional terms are corrections to this term. The additional term in this case shows that the first term is a good approximation to the error. Although this problem can be done numerically, a series solution allows us to find the error for any value of x.   ▶

Solving a Word Problem   ◀ EXAMPLE 6   From a point on the surface of the earth, a laser beam is aimed tangentially toward a vertical rod 2 km distant. How far up on the rod does the beam touch? (Assume the earth is a perfect sphere of radius 6400 km.)

From Fig. 29.9, we see that

$$x = 6400 \sec \theta - 6400$$

Finding the series for $\sec \theta$, we have

$$f(\theta) = \sec \theta \qquad f(0) = 1$$
$$f'(\theta) = \sec \theta \tan \theta \qquad f'(0) = 0$$
$$f''(\theta) = \sec^3 \theta + \sec \theta \tan^2 \theta \qquad f''(0) = 1$$

Thus, the first two nonzero terms are $\sec \theta = 1 + (\theta^2/2)$. Therefore,

$$x = 6400(\sec \theta - 1)$$
$$= 6400\left(1 + \frac{\theta^2}{2} - 1\right) = 3200\,\theta^2$$

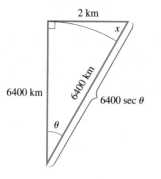

Fig. 29.9

The first two terms of the expansion for $\tan \theta$ are $\theta + \theta^3/3$, which means that $\tan \theta \approx \theta$, since θ is small (see Section 8.4). From Fig. 29.9, $\tan \theta = 2/6400$, and therefore $\theta = 1/3200$. Therefore, we have

$$x = 3200\left(\frac{1}{3200}\right)^2 = \frac{1}{3200} = 0.0003 \text{ km}$$

This means the 2-km-long beam touches the rod only 30 cm above the surface!   ▶

EXERCISES 29.4

In Exercises 1 and 2, make the given changes in the indicated examples of this section, and then solve the resulting problems.

1. In Example 1, change $e^{0.1}$ to $e^{-0.1}$.

2. In Example 4, change ln 1.2 to ln 0.8.

In Exercises 3–18, calculate the value of each of the given functions. Use the indicated number of terms of the appropriate series. Compare with the value found directly on a calculator.

3. $e^{0.2}$ (3)

4. 1.01^{-1} (4)

5. $\sin 0.1$ (2)

6. $\cos 0.05$ (2)

7. e (7)

8. $1/\sqrt{e}$ (5)

9. $\cos 3°$ (2)

10. $\sin 4°$ (2)

11. $\ln(1.4)$ (4)

12. $\ln(0.95)$ (4)

13. $\sin 0.3625$ (3)

14. $\cos 0.4072$ (3)

15. $\ln 0.9861$ (3)

16. $\ln 1.0534$ (3)

17. 1.032^6 (3)

18. 0.9982^8 (3)

In Exercises 19–22, calculate the value of each of the given functions. In Exercises 19 and 20, use the expansion for $\sqrt{1 + x}$, and in Exercises 21 and 22 use the expansion for $\sqrt[3]{1 + x}$. Use three terms of the appropriate series.

19. $\sqrt{1.1076}$

20. $\sqrt{0.7915}$

21. $\sqrt[3]{0.9628}$

22. $\sqrt[3]{1.1392}$

In Exercises 23–26, calculate the maximum error of the values calculated in the indicated exercises. If a series is alternating (every other term is negative), the maximum possible error in the calculated value is the value of the first term omitted.

23. Exercise 5

24. Exercise 4

25. Exercise 9

26. Exercise 11

In Exercises 27–36, solve the given problems by using series expansions.

27. We can evaluate π by use of $\frac{1}{4}\pi = \tan^{-1}\frac{1}{2} + \tan^{-1}\frac{1}{3}$ (see Exercise 60 of Section 20.6), along with the series for $\tan^{-1} x$. The first three terms are $\tan^{-1} x = x - \frac{1}{3}x^3 + \frac{1}{5}x^5$. Using these terms, expand $\tan^{-1}\frac{1}{2}$ and $\tan^{-1}\frac{1}{3}$ and approximate the value of π.

28. Use the fact that $\frac{1}{4}\pi = \tan^{-1}\frac{1}{7} + 2\tan^{-1}\frac{1}{3}$ to approximate the value of π. (See Exercise 27.)

W 29. Explain why $e^x > 1 + x + \frac{1}{2}x^2$ for $x > 0$.

30. Using a calculator, determine how many terms of the expansion for $\ln(1 + x)$ are needed to give the value of ln 1.3 accurate to five decimal places.

31. The time t (in years) for an investment to increase by 10% when the interest rate is 6% is given by $t = \dfrac{\ln 1.1}{0.06}$. Evaluate this expression by using the first four terms of the appropriate series.

32. The period T of a pendulum of length L is given by

$$T = 2\pi\sqrt{\frac{L}{g}}\left(1 + \frac{1}{4}\sin^2\frac{\theta}{2} + \frac{9}{64}\sin^4\frac{\theta}{2} + \cdots\right)$$

where g is the acceleration due to gravity and θ is the maximum angular displacement. If $L = 1.000$ m and $g = 9.800$ m/s², calculate T for $\theta = 10.0°$ (a) if only one term (the 1) of the series is used and (b) if two terms of the indicated series are used. In the second term, substitute one term of the series for $\sin^2(\theta/2)$.

W 33. The current in a circuit containing a resistance R, an inductance L, and a battery whose voltage is E is given by the equation $i = \dfrac{E}{R}(1 - e^{-Rt/L})$, where t is the time. Approximate this expression by using the first three terms of the appropriate exponential series. Under what conditions will this approximation be valid?

34. The image distance q from a certain lens as a function of the object distance p is given by $q = 20p/(p - 20)$. Find the first three nonzero terms of the expansion of the right side. From this expression, calculate q for $p = 2.00$ cm and compare it with the value found by substituting 2.00 in the original expression.

35. At what height above the shoreline on the French side of Lake Geneva must an observer be in order to see a point 14 km distant on the shoreline of the Swiss side of the lake? (This is the widest part of the lake.) The radius of the earth is 6400 km.

36. The efficiency E (in %) of an internal combustion engine in terms of its compression ratio e is given by $E = 100(1 - e^{-0.40})$. Determine the possible approximate error in the efficiency for a compression ratio measured to be 6.00 with a possible error of 0.50. (*Hint:* Set up a series for $(6 + x)^{-0.40}$.)

29.5 TAYLOR SERIES

To obtain accurate values of a function for values of x that are not close to zero, it is usually necessary to use many terms of a Maclaurin expansion. However, we can use another type of series, called a **Taylor series,** *which is a more general expansion than a Maclaurin expansion.* Also, functions for which a Maclaurin series may not be found may have a Taylor series.

The basic assumption in formulating a Taylor expansion is that a function may be expanded in a polynomial of the form

$$f(x) = c_0 + c_1(x - a) + c_2(x - a)^2 + \cdots \qquad (29.15)$$

Following the same line of reasoning as in deriving the Maclaurin expansion, we may find the constants $c_0, c_1, c_2, \ldots$ That is, derivatives of Eq. (29.15) are taken, and the function and its derivatives are evaluated at $x = a$. This leads to

$$f(x) = f(a) + f'(a)(x - a) + \frac{f''(a)(x - a)^2}{2!} + \cdots \qquad (29.16)$$

Named for the English mathematician Brook Taylor (1685–1737).

*Equation (29.16) is the **Taylor series expansion** of a function.* It converges rapidly for values of x that are close to a, and this is illustrated in Examples 3 and 4.

◀ EXAMPLE 1   Expand $f(x) = e^x$ in a Taylor series with $a = 1$.

$$\begin{aligned}
f(x) &= e^x & f(1) &= e & \text{find derivatives and evaluate each at } x = 1 \\
f'(x) &= e^x & f'(1) &= e \\
f''(x) &= e^x & f''(1) &= e \\
f'''(x) &= e^x & f'''(1) &= e
\end{aligned}$$

$$f(x) = e + e(x - 1) + e\frac{(x - 1)^2}{2!} + e\frac{(x - 1)^3}{3!} + \cdots \qquad \text{using Eq. (29.16)}$$

$$e^x = e\left[1 + (x - 1) + \frac{(x - 1)^2}{2} + \frac{(x - 1)^3}{6} + \cdots \right]$$

This series can be used in evaluating e^x for values of x near 1. ▶

◀ EXAMPLE 2   Expand $f(x) = \sqrt{x}$ in powers of $(x - 4)$.

Another way of stating this is to find the Taylor series for $f(x) = \sqrt{x}$, with $a = 4$. Thus,

$$\begin{aligned}
f(x) &= x^{1/2} & f(4) &= 2 & \text{find derivatives and evaluate each at } x = 4 \\
f'(x) &= \frac{1}{2x^{1/2}} & f'(4) &= \frac{1}{4} \\
f''(x) &= -\frac{1}{4x^{3/2}} & f''(4) &= -\frac{1}{32} \\
f'''(x) &= \frac{3}{8x^{5/2}} & f'''(4) &= \frac{3}{256}
\end{aligned}$$

$$f(x) = 2 + \frac{1}{4}(x - 4) - \frac{1}{32}\frac{(x - 4)^2}{2!} + \frac{3}{256}\frac{(x - 4)^3}{3!} - \cdots \qquad \text{using Eq. (29.16)}$$

$$\sqrt{x} = 2 + \frac{(x - 4)}{4} - \frac{(x - 4)^2}{64} + \frac{(x - 4)^3}{512} - \cdots$$

This series would be used to evaluate square roots of numbers near 4.

In Fig. 29.10, we show a graphing calculator view of $y = \sqrt{x}$ and $y = 1 + x/4$, which are the first two terms of the Taylor series, as well as being the linearization of $f(x)$ at $x = 4$. Each passes through $(4, 2)$, and they have nearly equal values of y for values of x near 4. ▶

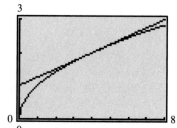

3

0
0   8

Fig. 29.10

In the last section, we evaluated functions by using Maclaurin series. In the following examples, we use Taylor series to evaluate functions.

◀ EXAMPLE 3   By using Taylor series, evaluate $\sqrt{4.5}$.
Using the four terms of the series found in Example 2, we have

$$\sqrt{4.5} = 2 + \frac{(4.5-4)}{4} - \frac{(4.5-4)^2}{64} + \frac{(4.5-4)^3}{512} \qquad \text{substitute 4.5 for } x$$

$$= 2 + \frac{(0.5)}{4} - \frac{(0.5)^2}{64} + \frac{(0.5)^3}{512}$$

$$= 2.121\,337\,891$$

The value found directly on a calculator is $2.121\,320\,344$. Therefore, the value found by these terms of the series expansion is correct to four decimal places. ◗

In Example 3, we saw that successive terms become small rapidly. If a value of x is chosen such that $x - a$ is larger, the successive terms may not become small rapidly, and many terms may be required. Therefore, ***we should choose the value of a as conveniently close as possible to the x-values that will be used.*** Also, we should note that a Maclaurin expansion for $\sqrt{x}$ cannot be used since the derivatives of $\sqrt{x}$ are not defined for $x = 0$.

CAUTION ▶

◀ EXAMPLE 4   Calculate the approximate value of $\sin 29°$ by using three terms of the appropriate Taylor expansion.
Since the value of $\sin 30°$ is known to be $\frac{1}{2}$, we let $a = \frac{\pi}{6}$ (remember, we must use values expressed in radians), when we evaluate the expansion for $x = 29°$ (when expressed in radians) the quantity $(x - a)$ is $-\frac{\pi}{180}$ (equivalent to $-1°$). This means that its numerical values are small and become smaller when it is raised to higher powers. Therefore,

$$f(x) = \sin x \qquad f\left(\frac{\pi}{6}\right) = \frac{1}{2} \qquad \text{find derivatives and evaluate each at } x = \frac{\pi}{6}$$

$$f'(x) = \cos x \qquad f'\left(\frac{\pi}{6}\right) = \frac{\sqrt{3}}{2}$$

$$f''(x) = -\sin x \qquad f''\left(\frac{\pi}{6}\right) = -\frac{1}{2}$$

$$f(x) = \frac{1}{2} + \frac{\sqrt{3}}{2}\left(x - \frac{\pi}{6}\right) - \frac{1}{4}\left(x - \frac{\pi}{6}\right)^2 - \cdots \qquad \text{using Eq. (29.16)}$$

$$\sin x = \frac{1}{2} + \frac{\sqrt{3}}{2}\left(x - \frac{\pi}{6}\right) - \frac{1}{4}\left(x - \frac{\pi}{6}\right)^2 - \cdots \qquad f(x) = \sin x$$

$$\sin 29° = \sin\left(\frac{\pi}{6} - \frac{\pi}{180}\right) \qquad 29° = 30° - 1° = \frac{\pi}{6} - \frac{\pi}{180}$$

$$= \frac{1}{2} + \frac{\sqrt{3}}{2}\left(\frac{\pi}{6} - \frac{\pi}{180} - \frac{\pi}{6}\right) - \frac{1}{4}\left(\frac{\pi}{6} - \frac{\pi}{180} - \frac{\pi}{6}\right)^2 - \cdots \qquad \text{substitute } \frac{\pi}{6} - \frac{\pi}{180} \text{ for } x$$

$$= \frac{1}{2} + \frac{\sqrt{3}}{2}\left(-\frac{\pi}{180}\right) - \frac{1}{4}\left(-\frac{\pi}{180}\right)^2 - \cdots$$

$$= 0.484\,808\,850\,9$$

The value found directly on a calculator is $0.484\,809\,620\,2$. ◗

EXERCISES 29.5

In Exercises 1 and 2, make the given changes in the indicated examples of this section, and then solve the resulting problems.

1. In Example 2, change $(x - 4)$ to $(x - 1)$.

2. In Example 4, change sin 29° to sin 31°.

In Exercises 3–10, evaluate the given functions by using the series developed in the examples of this section.

3. $e^{1.2}$

4. $e^{0.7}$

5. $\sqrt{4.2}$

6. $\sqrt{3.5}$

7. sin 32°

8. sin 28°

9. sin 29.53°

10. $\sqrt{3.8527}$

In Exercises 11–18, find the first three nonzero terms of the Taylor expansion for the given function and given value of a.

11. e^{-x} $(a = 2)$

12. $\cos x$ $\left(a = \frac{\pi}{4}\right)$

13. $\sin x$ $\left(a = \frac{\pi}{3}\right)$

14. $\ln x$ $(a = 3)$

15. $\sqrt[3]{x}$ $(a = 8)$

16. $\dfrac{1}{x}$ $(a = 2)$

17. $\tan x$ $\left(a = \frac{\pi}{4}\right)$

18. $\ln \sin x$ $\left(a = \frac{\pi}{2}\right)$

In Exercises 19–26, evaluate the given functions by using three terms of the appropriate Taylor series.

19. e^{π}

20. $\ln 3.1$

21. $\sqrt{9.3}$

22. 2.056^{-1}

23. $\sqrt[3]{8.3}$

24. tan 46°

25. sin 61°

26. cos 42°

In Exercises 27–32, solve the given problems.

27. By completing the steps indicated before Eq. (29.16) in the text, complete the derivation of Eq. (29.16).

28. Find the first three terms of the Taylor expansion of $f(x) = \ln x$ with $a = 1$. Compare this Taylor expansion with the linearization $L(x)$ of $f(x)$ with $a = 1$. Compare the graphs of $f(x)$, $L(x)$, and the Taylor expansion on a graphing calculator.

29. Show that the polynomial $2x^3 + x^2 - 3x + 5$ can be written as $2(x - 1)^3 + 7(x - 1)^2 + 5(x - 1) + 5$.

30. Calculate $\sqrt{3}$ using the series in Example 2 and compare with the value using the series in Exercise 1. Which is the better approximation?

31. Calculate sin 31° by using three terms of the Maclaurin expansion for sin x. Also calculate sin 31° by using three terms of the Taylor expansion in Example 4 (see Exercise 2). Compare the accuracy of the values obtained with that found directly on a calculator.

32. In the analysis of the electric potential of an electric charge distributed along a straight wire of length L, the expression $\ln \dfrac{x + L}{x}$ is used. Find three terms of the Taylor expansion of this expression in powers of $(x - L)$.

(W) *In Exercises 33–36, use a graphing calculator to display (a) the function in the indicated exercise of this set and (b) the first two terms of the Taylor series found for that exercise in the same display. Describe how closely the graph in part (b) fits the graph in part (a). Use the given values of x for Xmin and Xmax.*

33. Exercise 13 (sin x), $x = 0$ to $x = 2$

34. Exercise 15 $\left(\sqrt[3]{x}\right)$, $x = 0$ to $x = 16$

35. Exercise 16 $(1/x)$, $x = 0$ to $x = 4$

36. Exercise 17 (tan x), $x = 0$ to $x = 1.5$

29.6 INTRODUCTION TO FOURIER SERIES

Many problems encountered in the various fields of science and technology involve functions that are periodic. *A periodic function is one for which $F(x + P) = F(x)$, where P is the period.* We noted that the trigonometric functions are periodic when we discussed their graphs in Chapter 10. Illustrations of applied problems that involve periodic functions are alternating-current voltages and mechanical oscillations.

Therefore, in this section we use a series made of terms of sines and cosines. This allows us to represent complicated periodic functions in terms of the simpler sines and cosines. It also provides us a good approximation over a greater interval than Maclaurin and Taylor series, which give good approximations with a few terms only near a specific value. Illustrations of applications of this type of series are given in Example 3 and in the exercises.

We will assume that a function $f(x)$ may be represented by the series of sines and cosines as indicated:

$$f(x) = a_0 + a_1 \cos x + a_2 \cos 2x + \cdots + a_n \cos nx + \cdots$$
$$+ b_1 \sin x + b_2 \sin 2x + \cdots + b_n \sin nx + \cdots \quad (29.17)$$

Since all the sines and cosines indicated in this expansion have a period of 2π (the period of any given term may be less than 2π, but all do repeat every 2π units—for example, $\sin 2x$ has a period of π, but it also repeats every 2π), the series expansion indicated in Eq. (29.17) will also have a period of 2π. *This series is called a* **Fourier series.**

The principal problem to be solved is that of finding the coefficients a_n and b_n. Derivatives proved to be useful in finding the coefficients for a Maclaurin expansion. We use the properties of certain integrals to find the coefficients of a Fourier series. To utilize these properties, we multiply all terms of Eq. (29.17) by $\cos mx$ and then evaluate from $-\pi$ to π (in this way we take advantage of the period 2π). Thus, we have

$$\int_{-\pi}^{\pi} f(x) \cos mx \, dx = \int_{-\pi}^{\pi} (a_0 + a_1 \cos x + a_2 \cos 2x + \cdots)(\cos mx) \, dx$$
$$+ \int_{-\pi}^{\pi} (b_1 \sin x + b_2 \sin 2x + \cdots)(\cos mx) \, dx \quad (29.18)$$

Using the methods of integration of Chapter 28, we now find the values of the coefficients a_n and b_n. For the coefficients a_n we find that the values differ depending on whether or not $n = m$. Therefore, first considering the case for which $n \neq m$, we have

$$\int_{-\pi}^{\pi} a_0 \cos mx \, dx = \frac{a_0}{m} \sin mx \Big|_{-\pi}^{\pi}$$
$$= \frac{a_0}{m}(0 - 0) = 0 \quad (29.19)$$

$$\int_{-\pi}^{\pi} a_n \cos nx \cos mx \, dx$$
$$= a_n \left[\frac{\sin(n-m)x}{2(n-m)} + \frac{\sin(n+m)x}{2(n+m)} \right]\Big|_{-\pi}^{\pi} = 0 \qquad (n \neq m) \quad (29.20)$$

These values are all equal to zero since the sine of any multiple of π is zero.

Now, considering the case for which $n = m$, we have

$$\int_{-\pi}^{\pi} a_n \cos nx \cos nx \, dx = \int_{-\pi}^{\pi} a_n \cos^2 nx \, dx$$
$$= \left(\frac{a_n x}{2} + \frac{a_n}{2n} \sin nx \cos nx \right)\Big|_{-\pi}^{\pi}$$
$$= \frac{a_n x}{2}\Big|_{-\pi}^{\pi} = \pi a_n \quad (29.21)$$

On the next page, we continue by finding the values of the coefficient b_n in Eq. (29.18).

Named for the French mathematician and physicist Jean Baptiste Joseph Fourier (1768–1830).

Now, finding the values of the coefficient b_n, we have

$$\int_{-\pi}^{\pi} b_n \sin nx \cos mx \, dx = b_n \left[-\frac{\cos(n-m)x}{2(n-m)} - \frac{\cos(n+m)x}{2(n+m)} \right]\Big|_{-\pi}^{\pi}$$

$$= b_n \left[-\frac{\cos(n-m)\pi}{2(n-m)} - \frac{\cos(n+m)\pi}{2(n+m)} \right.$$

$$\left. + \frac{\cos(n-m)(-\pi)}{2(n-m)} + \frac{\cos(n+m)(-\pi)}{2(n+m)} \right]$$

$$= 0 \text{ [since } \cos\theta = \cos(-\theta)] \qquad (n \neq m) \qquad (29.22)$$

$$\int_{-\pi}^{\pi} b_n \sin nx \cos nx \, dx = \frac{b_n}{2n} \sin^2 nx \Big|_{-\pi}^{\pi} = 0 \qquad (29.23)$$

These integrals are seen to be zero, except for the one specific case of $\int_{-\pi}^{\pi} a_n \cos nx \cos mx \, dx$ when $n = m$, for which the result is indicated in Eq. (29.21). Using these results in Eq. (29.18), we have

$$\int_{-\pi}^{\pi} f(x) \cos nx \, dx = a_n \int_{-\pi}^{\pi} \cos^2 nx \, dx = \pi a_n$$

$$a_n = \frac{1}{\pi} \int_{-\pi}^{\pi} f(x) \cos nx \, dx \qquad (29.24)$$

This equation allows us to find the coefficients a_n, except a_0. We find the term a_0 by direct integration of Eq. (29.17) from $-\pi$ to π. When we perform this integration, all the sine and cosine terms integrate to zero, thereby giving the result

$$\int_{-\pi}^{\pi} f(x) \, dx = \int_{-\pi}^{\pi} a_0 \, dx = a_0 x \Big|_{-\pi}^{\pi} = 2\pi a_0$$

$$a_0 = \frac{1}{2\pi} \int_{-\pi}^{\pi} f(x) \, dx \qquad (29.25)$$

By multiplying all terms of Eq. (29.17) by $\sin mx$ and then integrating from $-\pi$ to π, we find the coefficients b_n. We obtain the result

$$b_n = \frac{1}{\pi} \int_{-\pi}^{\pi} f(x) \sin nx \, dx \qquad (29.26)$$

We can restate our equations for the Fourier series of a function $f(x)$:

$$f(x) = a_0 + a_1 \cos x + a_2 \cos 2x + \cdots + a_n \cos nx + \cdots$$
$$+ b_1 \sin x + b_2 \sin 2x + \cdots + b_n \sin nx + \cdots \qquad (29.17)$$

where the coefficients are found by

$$a_0 = \frac{1}{2\pi} \int_{-\pi}^{\pi} f(x) \, dx \qquad (29.25)$$

$$a_n = \frac{1}{\pi} \int_{-\pi}^{\pi} f(x) \cos nx \, dx \qquad (29.24)$$

$$b_n = \frac{1}{\pi} \int_{-\pi}^{\pi} f(x) \sin nx \, dx \qquad (29.26)$$

◖ EXAMPLE 1   Find the Fourier series for the square wave function

$$f(x) = \begin{cases} -1 & -\pi \le x < 0 \\ 1 & 0 \le x < \pi \end{cases}$$

(Many of the functions we shall expand in Fourier series are discontinuous (not continuous) like this one. See Section 23.1 for a discussion of continuity.)

CAUTION ◗   Since $f(x)$ is defined differently for the intervals of x indicated, *it requires two integrals for each coefficient:*

using Eq. (29.25) $a_0 = \dfrac{1}{2\pi} \displaystyle\int_{-\pi}^{0} (-1)\, dx + \dfrac{1}{2\pi} \int_{0}^{\pi} (1)\, dx = -\dfrac{x}{2\pi}\Big|_{-\pi}^{0} + \dfrac{x}{2\pi}\Big|_{0}^{\pi} = -\dfrac{1}{2} + \dfrac{1}{2} = 0$

using Eq. (29.24) $a_n = \dfrac{1}{\pi} \displaystyle\int_{-\pi}^{0} (-1)\cos nx\, dx + \dfrac{1}{\pi} \int_{0}^{\pi} (1)\cos nx\, dx = -\dfrac{1}{n\pi}\sin nx\Big|_{-\pi}^{0} + \dfrac{1}{n\pi}\sin nx\Big|_{0}^{\pi} = 0 + 0 = 0$

for all values of n, since $\sin n\pi = 0$;

using Eq. (29.26) with $n = 1$ $b_1 = \dfrac{1}{\pi} \displaystyle\int_{-\pi}^{0} (-1)\sin x\, dx + \dfrac{1}{\pi} \int_{0}^{\pi} (1)\sin x\, dx = \dfrac{1}{\pi}\cos x\Big|_{-\pi}^{0} - \dfrac{1}{\pi}\cos x\Big|_{0}^{\pi}$

$= \dfrac{1}{\pi}(1 + 1) - \dfrac{1}{\pi}(-1 - 1) = \dfrac{4}{\pi}$

using Eq. (29.26) with $n = 2$ $b_2 = \dfrac{1}{\pi} \displaystyle\int_{-\pi}^{0} (-1)\sin 2x\, dx + \dfrac{1}{\pi} \int_{0}^{\pi} (1)\sin 2x\, dx = \dfrac{1}{2\pi}\cos 2x\Big|_{-\pi}^{0} - \dfrac{1}{2\pi}\cos 2x\Big|_{0}^{\pi}$

$= \dfrac{1}{2\pi}(1 - 1) - \dfrac{1}{2\pi}(1 - 1) = 0$

using Eq. (29.26) with $n = 3$ $b_3 = \dfrac{1}{\pi} \displaystyle\int_{-\pi}^{0} (-1)\sin 3x\, dx + \dfrac{1}{\pi} \int_{0}^{\pi} (1)\sin 3x\, dx = \dfrac{1}{3\pi}\cos 3x\Big|_{-\pi}^{0} - \dfrac{1}{3\pi}\cos 3x\Big|_{0}^{\pi}$

$= \dfrac{1}{3\pi}(1 + 1) - \dfrac{1}{3\pi}(-1 - 1) = \dfrac{4}{3\pi}$

In general, if n is even, $b_n = 0$, and if n is odd, then $b_n = 4/n\pi$. Therefore,

$$f(x) = \frac{4}{\pi}\sin x + \frac{4}{3\pi}\sin 3x + \frac{4}{5\pi}\sin 5x + \cdots = \frac{4}{\pi}\left(\sin x + \frac{1}{3}\sin 3x + \frac{1}{5}\sin 5x + \cdots\right)$$

A graph of the function as defined, and the curve found by using the first three terms of the Fourier series, are shown in Fig. 29.11.

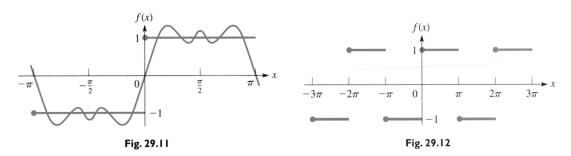

Fig. 29.11   **Fig. 29.12**

Since functions found by Fourier series have a period of 2π, they can represent functions with this period. If the function $f(x)$ were defined to be periodic with period 2π, with the same definitions as originally indicated, we would graph the function as shown in Fig. 29.12. The Fourier series representation would follow it as in Fig. 29.11. If more terms were used, the fit would be closer.

◀ **EXAMPLE 2** Find the Fourier series for the function

$$f(x) = \begin{cases} 1 & -\pi \le x < 0 \\ x & 0 \le x < \pi \end{cases}$$

For the periodic function, let $f(x + 2\pi) = f(x)$ for all x.

A graph of three periods of this function is shown in Fig. 29.13.

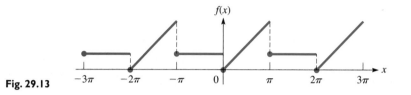

Fig. 29.13

Now, finding the coefficients, we have

$$a_0 = \frac{1}{2\pi} \int_{-\pi}^{0} dx + \frac{1}{2\pi} \int_{0}^{\pi} x\,dx = \frac{x}{2\pi}\Big|_{-\pi}^{0} + \frac{x^2}{4\pi}\Big|_{0}^{\pi} \qquad \text{using Eq. (29.25)}$$

$$= \frac{1}{2} + \frac{\pi}{4} = \frac{2 + \pi}{4}$$

$$a_1 = \frac{1}{\pi} \int_{-\pi}^{0} \cos x\,dx + \frac{1}{\pi} \int_{0}^{\pi} x \cos x\,dx \qquad \text{using Eq. (29.24) with } n = 1$$

$$= \frac{1}{\pi} \sin x\Big|_{-\pi}^{0} + \frac{1}{\pi}(\cos x + x \sin x)\Big|_{0}^{\pi} = -\frac{2}{\pi}$$

$$a_2 = \frac{1}{\pi} \int_{-\pi}^{0} \cos 2x\,dx + \frac{1}{\pi} \int_{0}^{\pi} x \cos 2x\,dx \qquad \text{using Eq. (29.24) with } n = 2$$

$$= \frac{1}{2\pi} \sin 2x\Big|_{-\pi}^{0} + \frac{1}{4\pi}(\cos 2x + 2x \sin 2x)\Big|_{0}^{\pi} = 0$$

$$a_3 = \frac{1}{\pi} \int_{-\pi}^{0} \cos 3x\,dx + \frac{1}{\pi} \int_{0}^{\pi} x \cos 3x\,dx \qquad \text{using Eq. (29.24) with } n = 3$$

$$= \frac{1}{3\pi} \sin 3x\Big|_{-\pi}^{0} + \frac{1}{9\pi}(\cos 3x + 3x \sin 3x)\Big|_{0}^{\pi} = -\frac{2}{9\pi}$$

$$b_1 = \frac{1}{\pi} \int_{-\pi}^{0} \sin x\,dx + \frac{1}{\pi} \int_{0}^{\pi} x \sin x\,dx \qquad \text{using Eq. (29.26) with } n = 1$$

$$= -\frac{1}{\pi} \cos x\Big|_{-\pi}^{0} + \frac{1}{\pi}(\sin x - x \cos x)\Big|_{0}^{\pi} = \frac{\pi - 2}{\pi}$$

$$b_2 = \frac{1}{\pi} \int_{-\pi}^{0} \sin 2x\,dx + \frac{1}{\pi} \int_{0}^{\pi} x \sin 2x\,dx \qquad \text{using Eq. (29.26) with } n = 2$$

$$= -\frac{\cos 2x}{2\pi}\Big|_{-\pi}^{0} + \frac{\sin 2x - 2x \cos 2x}{4\pi}\Big|_{0}^{\pi} = -\frac{1}{2}$$

Therefore, the first few terms of the Fourier series are

$$f(x) = \frac{2 + \pi}{4} - \frac{2}{\pi} \cos x - \frac{2}{9\pi} \cos 3x - \cdots + \left(\frac{\pi - 2}{\pi}\right) \sin x - \frac{1}{2} \sin 2x + \cdots$$

See the chapter introduction.

◖ EXAMPLE 3 Certain electronic devices allow an electric current to pass through in only one direction. When an alternating current is applied to the circuit, the current exists for only half the cycle. Figure 29.14 is a representation of such a current as a function of time. This type of electronic device is called a *half-wave rectifier.* Derive the Fourier series for a rectified wave for which half is defined by $f(t) = \sin t \; (0 \le t \le \pi)$ and for which the other half is defined by $f(t) = 0$.

In finding the Fourier coefficients, we first find a_0 as

$$a_0 = \frac{1}{2\pi} \int_0^\pi \sin t \, dt = \frac{1}{2\pi}(-\cos t)\Big|_0^\pi = \frac{1}{2\pi}(1+1) = \frac{1}{\pi}$$

In the previous example, we evaluated each of the coefficients individually. Here we show how to set up a general expression for a_n and another for b_n. Once we have determined these, we can substitute values of n in the formula to obtain the individual coefficients:

$$a_n = \frac{1}{\pi} \int_0^\pi \sin t \cos nt \, dt = -\frac{1}{2\pi}\left[\frac{\cos(1-n)t}{1-n} + \frac{\cos(1+n)t}{1+n}\right]_0^\pi$$

$$= -\frac{1}{2\pi}\left[\frac{\cos(1-n)\pi}{1-n} + \frac{\cos(1+n)\pi}{1+n} - \frac{1}{1-n} - \frac{1}{1+n}\right]$$

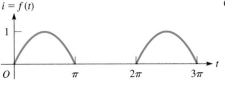

$i = f(t)$

Fig. 29.14

See Formula 40 in the table of integrals in Appendix E. It is valid for all values of n except $n = 1$. Now we write

$$a_1 = \frac{1}{\pi}\int_0^\pi \sin t \cos t \, dt = \frac{1}{2\pi}\sin^2 t \Big|_0^\pi = 0$$

$$a_2 = -\frac{1}{2\pi}\left(\frac{-1}{-1} + \frac{-1}{3} - \frac{1}{-1} - \frac{1}{3}\right) = -\frac{2}{3\pi}$$

$$a_3 = -\frac{1}{2\pi}\left(\frac{1}{-2} + \frac{1}{4} - \frac{1}{-2} - \frac{1}{4}\right) = 0$$

$$a_4 = -\frac{1}{2\pi}\left(\frac{-1}{-3} + \frac{-1}{5} - \frac{1}{-3} - \frac{1}{5}\right) = -\frac{2}{15\pi}$$

$$b_n = \frac{1}{\pi}\int_0^\pi \sin t \sin nt \, dt = \frac{1}{2\pi}\left[\frac{\sin(1-n)t}{1-n} - \frac{\sin(1+n)t}{1+n}\right]_0^\pi$$

$$= -\frac{1}{2\pi}\left[\frac{\sin(1-n)\pi}{1-n} - \frac{\sin(1+n)\pi}{1+n}\right]$$

See Formula 39 in Appendix E. It is valid for all values of n except $n = 1$.

Therefore, we have

$$b_1 = \frac{1}{\pi}\int_0^\pi \sin t \sin t \, dt = \frac{1}{\pi}\int_0^\pi \sin^2 t \, dt = \frac{1}{2\pi}(t - \sin t \cos t)\Big|_0^\pi = \frac{1}{2}$$

We see that $b_n = 0$ if $n > 1$, since each is evaluated in terms of the sine of a multiple of π.

Therefore, the Fourier series for the rectified wave is

$$f(t) = \frac{1}{\pi} + \frac{1}{2}\sin t - \frac{2}{\pi}\left(\frac{1}{3}\cos 2t + \frac{1}{15}\cos 4t + \cdots\right)$$

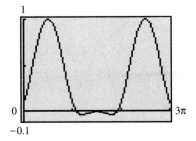

Fig. 29.15

The graph of these terms of the Fourier series is shown in the graphing calculator display in Fig. 29.15.

◗

All the types of periodic functions included in this section (as well as many others) may actually be seen on an oscilloscope when the proper signal is set into it. In this way the oscilloscope may be used to analyze the periodic nature of such phenomena as sound waves and electric currents.

EXERCISES 29.6

In Exercises 1 and 2, make the given changes in Example 1 of this section and then find the resulting Fourier series.

1. Change the -1 to -2, and the 1 to 2.

2. Change the -1 to 0.

In Exercises 3–12, find at least three nonzero terms (including a_0 and at least two cosine terms and two sine terms if they are not all zero) of the Fourier series for the given functions, and sketch at least three periods of the function.

3. $f(x) = \begin{cases} 1 & -\pi \le x < 0 \\ 0 & 0 \le x < \pi \end{cases}$

4. $f(x) = \begin{cases} 0 & -\pi \le x < -\frac{\pi}{2}, \frac{\pi}{2} \le x < \pi \\ 2 & -\frac{\pi}{2} \le x < \frac{\pi}{2} \end{cases}$

5. $f(x) = \begin{cases} 1 & -\pi \le x < 0 \\ 2 & 0 \le x < \pi \end{cases}$

6. $f(x) = \begin{cases} 0 & -\pi \le x < 0, \frac{\pi}{2} < x < \pi \\ 1 & 0 \le x \le \frac{\pi}{2} \end{cases}$

7. $f(x) = \begin{cases} 0 & -\pi \le x < 0 \\ x & 0 \le x < \pi \end{cases}$

8. $f(x) = x \quad -\pi \le x < \pi$

9. $f(x) = \begin{cases} -1 & -\pi \le x < 0 \\ 0 & 0 \le x < \frac{\pi}{2} \\ 1 & \frac{\pi}{2} \le x < \pi \end{cases}$

10. $f(x) = x^2 \quad -\pi \le x < \pi$

11. $f(x) = \begin{cases} -x & -\pi \le x < 0 \\ x & 0 \le x < \pi \end{cases}$

12. $f(x) = \begin{cases} 0 & -\pi \le x < 0 \\ x^2 & 0 \le x < \pi \end{cases}$

In Exercises 13–18, use a graphing calculator to display the terms of the Fourier series given in the indicated example or answer for the indicated exercise. Compare with the sketch of the function. For each calculator display use Xmin = -8 and Xmax = 8.

13. Example 1   **14.** Example 2   **15.** Exercise 5

16. Exercise 7   **17.** Exercise 11   **18.** Exercise 10

In Exercises 19 and 20, solve the given problems.

19. Find the Fourier expansion of the electronic device known as a *full-wave rectifier.* This is found by using as the function for the current $f(t) = -\sin t$ for $-\pi \le t \le 0$ and $f(t) = \sin t$ for $0 < t \le \pi$. The graph of this function is shown in Fig. 29.16. The portion of the curve to the left of the origin is dashed because from a physical point of view we can give no significance to this part of the wave, although mathematically we can derive the proper form of the Fourier expansion by using it.

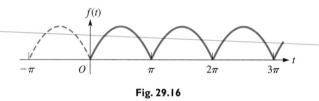

Fig. 29.16

20. The loudness L (in decibels) of a certain siren as a function of time t (in s) can be described by the function

$$\begin{aligned} L &= 0 & -\pi \le t < 0 \\ L &= 100t & 0 \le t < \pi/2 \\ L &= 100(\pi - t) & \pi/2 \le t < \pi \end{aligned}$$

with a period of 2π seconds (where only positive values of t have physical significance). Find a_0, the first nonzero cosine term, and the first two nonzero sine terms of the Fourier expansion for the loudness of the siren. See Fig. 29.17.

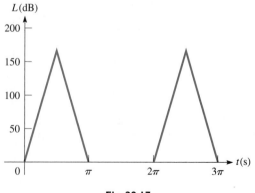

Fig. 29.17

When finding the Fourier expansion of some functions, it may turn out that all the sine terms evaluate to be zero or that all the cosine terms evaluate to be zero. In fact, in Example 1 on page 897 we see that all of the cosine terms were zero and that the expansion contained only sine terms. We now show how to quickly determine if an expansion will contain only sine terms, or only cosine terms.

EVEN FUNCTIONS AND ODD FUNCTIONS

In Chapter 21 (page 571), we showed that when $-x$ replaces x in a function $f(x)$, and the function does not change, the curve of the function is symmetric to the y-axis. *Such a function is called an* **even function.**

◀ EXAMPLE 1   We can show that the function $y = \cos x$ is an even function by using the Maclaurin expansions for $\cos x$ and $\cos(-x)$. These are

$$\cos x = 1 - \frac{x^2}{2} + \frac{x^4}{24} - \cdots$$

$$\cos(-x) = 1 - \frac{(-x)^2}{2} + \frac{(-x)^4}{24} \cdots = 1 - \frac{x^2}{2} + \frac{x^4}{24} - \cdots$$

Since the expansions are the same, $\cos x$ is an even function. ◗

NOTE ▶ Since $\cos x$ is an even function and all of its terms are even functions, it follows that *an **even function** will have a Fourier series that contains only **cosine** terms (and possibly a constant term).*

◀ EXAMPLE 2   The Fourier series for the function

$$f(x) = \begin{cases} 0 & -\pi \le x < -\pi/2, \ \pi/2 \le x < \pi \\ 1 & -\pi/2 \le x < \pi/2 \end{cases}$$

is $f(x) = \dfrac{1}{2} + \dfrac{2}{\pi}\left(\cos x - \dfrac{1}{3}\cos 3x + \dfrac{1}{5}\cos 5x - \cdots\right)$. We see that $f(x) = f(-x)$, which means it is an even function. We also see that its Fourier series expansion contains only cosine terms (and a constant). Thus, *when finding the Fourier series we do not have to find any sine terms.* The graph of $f(x)$ in Fig. 29.18 shows its symmetry to the y-axis. ◗

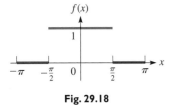

$f(x)$

Fig. 29.18

Again referring to Chapter 21 (page 571), we recall that if $-x$ replaces x and $-y$ replaces y at the same time, and the function does not change, then the function is symmetric to the origin. *Such a function is called an* **odd function.**

◀ EXAMPLE 3   We can show that the function $y = \sin x$ is an odd function by using the Maclaurin expansions for $\sin x$ and $-\sin(-x)$ (the $-$ sign before $\sin(-x)$ is equivalent to making y negative). These are

$$\sin x = x - \frac{x^3}{6} + \frac{x^5}{120} - \cdots$$

$$-\sin(-x) = -\left[(-x) - \frac{(-x)^3}{6} + \frac{(-x)^5}{120} - \cdots\right] = x - \frac{x^3}{6} + \frac{x^5}{120} - \cdots$$

Since $\sin x = -\sin(-x)$, $\sin x$ is an odd function. ◗

Since $\sin x$ is an odd function and all its terms are odd functions, it follows that an **odd function** will have a Fourier series that contains only **sine** terms (and no constant term).

◀ **EXAMPLE 4** As we showed in Example 1 on page 897, the Fourier series for the function

$$f(x) = \begin{cases} -1 & -\pi \le x < 0 \\ 1 & 0 \le x < \pi \end{cases}$$

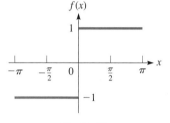

$f(x)$

Fig. 29.19

is $f(x) = \dfrac{4}{\pi}\left(\sin x + \dfrac{1}{3}\sin 3x + \dfrac{1}{5}\sin 5x + \cdots\right)$. We can see that $f(-x) = -f(-x)$, which means $f(x)$ is an odd function. We also see that its Fourier series expansion contains only sine terms. Therefore, *when finding the Fourier series, we do not have to find any cosine terms.* The graph of $f(x)$ in Fig. 29.19 shows its symmetry to the origin. ▶

If a constant k is added to a function $f_1(x)$, the resulting function $f(x)$ is

$$f(x) = k + f_1(x)$$

NOTE ▶ Therefore, if we know the Fourier series expansion for $f_1(x)$, *the Fourier series expansion of $f(x)$ is found by adding k to the Fourier series expansion of $f_1(x)$.*

◀ **EXAMPLE 5** The values of the function

$$f(x) = \begin{cases} 1 & -\pi \le x < -\pi/2,\ \pi/2 \le x < \pi \\ 2 & -\pi/2 \le x < \pi/2 \end{cases}$$

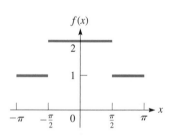

$f(x)$

Fig. 29.20

are all 1 greater than those of the function of Example 2. Therefore, denoting the function of Example 2 as $f_1(x)$, we have $f(x) = 1 + f_1(x)$. This means that the Fourier series for $f(x)$ is

$$f(x) = 1 + \left[\frac{1}{2} + \frac{2}{\pi}\left(\cos x - \frac{1}{3}\cos 3x + \frac{1}{5}\cos 5x - \cdots\right)\right]$$

$$= \frac{3}{2} + \frac{2}{\pi}\left(\cos x - \frac{1}{3}\cos 3x + \frac{1}{5}\cos 5x - \cdots\right)$$

In Fig. 29.20, we see that the graph of $f(x)$ is shifted up vertically by 1 unit from the graph of $f_1(x)$ in Fig. 29.18. This is equivalent to a vertical translation of axes. We also note that $f(x)$ is an even function. ▶

◀ **EXAMPLE 6** The values of the function

$$f(x) = \begin{cases} -\frac{3}{2} & -\pi \le x < 0 \\ \frac{1}{2} & 0 \le x < \pi \end{cases}$$

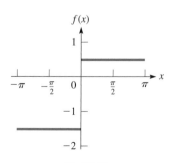

$f(x)$

Fig. 29.21

are all $\frac{1}{2}$ less than those of the function of Example 4. Therefore, denoting the function of Example 4 as $f_1(x)$, we have $f(x) = -\frac{1}{2} + f_1(x)$. This means that the Fourier series for $f(x)$ is

$$f(x) = -\frac{1}{2} + \frac{4}{\pi}\left(\sin x - \frac{1}{3}\sin 3x + \frac{1}{5}\sin 5x - \cdots\right)$$

In Fig. 29.21, we see that the graph of $f(x)$ is shifted vertically down by $\frac{1}{2}$ unit from the graph of $f_1(x)$ in Fig. 29.19. Although $f(x)$ is not an odd function, it would be an odd function if its origin were translated to $\left(0, -\frac{1}{2}\right)$. ▶

FOURIER SERIES WITH PERIOD 2L

The standard form of a Fourier series we have considered to this point is defined over the interval from $x = -\pi$ to $x = \pi$. At times it is preferable to have a series that is defined over a different interval.

Noting that

$$\sin\frac{n\pi}{L}(x + 2L) = \sin n\left(\frac{\pi x}{L} + 2\pi\right) = \sin\frac{n\pi x}{L}$$

we see that $\sin(n\pi x/L)$ has a period of $2L$. Thus, by using $\sin(n\pi x/L)$ and $\cos(n\pi x/L)$ and the same method of derivation, the following equations are found for the coefficients for the Fourier series for the interval from $x = -L$ to $x = L$.

$$a_0 = \frac{1}{2L}\int_{-L}^{L} f(x)\,dx \tag{29.27}$$

$$a_n = \frac{1}{L}\int_{-L}^{L} f(x)\cos\frac{n\pi x}{L}\,dx \tag{29.28}$$

$$b_n = \frac{1}{L}\int_{-L}^{L} f(x)\sin\frac{n\pi x}{L}\,dx \tag{29.29}$$

◀ EXAMPLE 7   Find the Fourier series for the function

$$f(x) = \begin{cases} 0 & -4 \le x < 0 \\ 2 & 0 \le x < 4 \end{cases}$$

and for which the period is 8. See Fig. 29.22.

Since the period is 8, $L = 4$. Next we note that $f(x) = 1 + f_1(x)$, where $f_1(x)$ is an odd function (from the definition of $f(x)$ and from Fig. 29.22 we can see the symmetry to the point $(0, 1)$). Therefore, *the constant is 1 and there are no cosine terms* in the Fourier series for $f(x)$. Now, finding the sine terms, we have

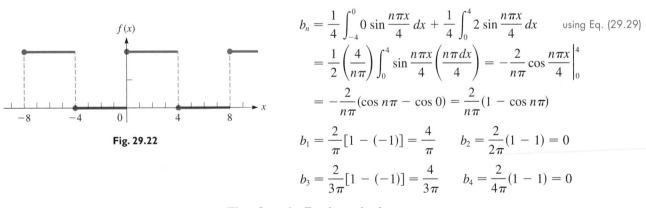

Fig. 29.22

$$b_n = \frac{1}{4}\int_{-4}^{0} 0\sin\frac{n\pi x}{4}\,dx + \frac{1}{4}\int_{0}^{4} 2\sin\frac{n\pi x}{4}\,dx \quad \text{using Eq. (29.29)}$$

$$= \frac{1}{2}\left(\frac{4}{n\pi}\right)\int_{0}^{4}\sin\frac{n\pi x}{4}\left(\frac{n\pi\,dx}{4}\right) = -\frac{2}{n\pi}\cos\frac{n\pi x}{4}\Big|_{0}^{4}$$

$$= -\frac{2}{n\pi}(\cos n\pi - \cos 0) = \frac{2}{n\pi}(1 - \cos n\pi)$$

$$b_1 = \frac{2}{\pi}[1 - (-1)] = \frac{4}{\pi} \qquad b_2 = \frac{2}{2\pi}(1 - 1) = 0$$

$$b_3 = \frac{2}{3\pi}[1 - (-1)] = \frac{4}{3\pi} \qquad b_4 = \frac{2}{4\pi}(1 - 1) = 0$$

Therefore, the Fourier series is

$$f(x) = 1 + \frac{4}{\pi}\sin\frac{\pi x}{4} + \frac{4}{3\pi}\sin\frac{3\pi x}{4} + \cdots$$

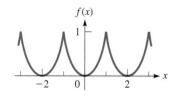

Fig. 29.23

◀ **EXAMPLE 8** Find the Fourier series for the function

$$f(x) = x^2 \quad -1 \le x < 1$$

for which the period is 2. See Fig. 29.23.

Since the period is 2, $L = 1$. Next we note that $f(x) = f(-x)$, which means it is an even function. Therefore, there are no sine terms in the Fourier series. Finding the constant and the cosine terms, we have

$$a_0 = \frac{1}{2(1)} \int_{-1}^{1} x^2 \, dx = \frac{1}{6} x^3 \Big|_{-1}^{1} = \frac{1}{6}(1 + 1) = \frac{1}{3}$$

$$a_n = \frac{1}{1} \int_{-1}^{1} x^2 \cos \frac{n\pi x}{1} \, dx = \int_{-1}^{1} x^2 \cos n\pi x \, dx \qquad \text{integrating by parts: } u = x^2, \ du = 2x \, dx, \\ dv = \cos n\pi x \, dx, \ v = (1/n\pi)\sin n\pi x$$

$$= x^2 \left(\frac{1}{n\pi} \sin n\pi x \right) \Big|_{-1}^{1} - \frac{2}{n\pi} \int_{-1}^{1} x \sin n\pi x \, dx \qquad \text{integrating by parts: } u = x, \ du = dx, \\ dv = \sin n\pi x \, dx, \ v = (-1/n\pi)\cos n\pi x$$

$\sin n\pi = 0 \qquad = \frac{1}{n\pi} \sin n\pi - \frac{1}{n\pi} \sin(-n\pi) - \frac{2}{n\pi} \left[x \left(-\frac{1}{n\pi} \cos n\pi x \right) \Big|_{-1}^{1} - \left(-\frac{1}{n\pi} \int_{-1}^{1} \cos n\pi x \, dx \right) \right]$

$\sin n\pi = 0 \qquad = \frac{2}{n^2\pi^2}[\cos n\pi + \cos(-n\pi)] + \frac{1}{n^2\pi^2} \sin n\pi x \Big|_{-1}^{1} = \frac{2}{n^2\pi^2}(2 \cos n\pi) = \frac{4}{n^2\pi^2} \cos n\pi$

$$a_1 = \frac{4}{\pi^2} \cos \pi = -\frac{4}{\pi^2} \qquad a_2 = \frac{4}{4\pi^2} \cos 2\pi = \frac{4}{4\pi^2} \qquad a_3 = \frac{4}{9\pi^2} \cos 3\pi = -\frac{4}{9\pi^2}$$

Therefore, the Fourier series is

$$f(x) = \frac{1}{3} - \frac{4}{\pi^2} \left(\cos \pi x - \frac{1}{4} \cos 2\pi x + \frac{1}{9} \cos 3\pi x - \cdots \right)$$

▶

HALF-RANGE EXPANSIONS

We have seen that the Fourier series expansion for an even function contains only cosine terms (and possibly a constant), and the expansion of an odd function contains only sine terms. It is also possible to specify a function to be even or odd, such that the expansion will contain only cosine terms or only sine terms.

Considering the symmetry of an even function, the area under the curve from $-L$ to 0 is the same as the area under the curve from 0 to L (see Fig. 29.24). This means the value of the integral from $-L$ to 0 equals the value of the integral from 0 to L. Therefore, the value of the integral from $-L$ to L equals twice the value of the integral from 0 to L, or

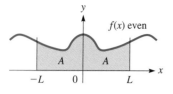

Fig. 29.24

$$\int_{-L}^{L} f(x) \, dx = 2 \int_{0}^{L} f(x) \, dx \qquad f(x) \text{ even}$$

Therefore, to obtain the Fourier coefficients for an expression from $-L$ to L for an even function, we can multiply the coefficients obtained using Eqs. (29.27) and (29.28) from 0 to L by 2. Similar reasoning shows that the Fourier coefficients for an expansion from $-L$ to L for an odd function may be found by multiplying the coefficients obtained using Eq. (29.29) from 0 to L by 2.

A **half-range Fourier cosine series** *is a series that contains only cosine terms,* and *a* **half-range Fourier sine series** *is a series that contains only sine terms.* To find the half-range expansion for a function $f(x)$, it is defined for interval 0 to L (*half* of the interval from $-L$ to L) and then specified as odd or even, thereby clearly defining the function in the interval from $-L$ to 0. This means that *the Fourier coefficients for a half-range cosine series are given by*

$$a_0 = \frac{1}{L}\int_0^L f(x)\,dx \quad \text{and} \quad a_n = \frac{2}{L}\int_0^L f(x)\cos\frac{n\pi x}{L}\,dx \qquad (n = 1, 2, \ldots) \qquad (29.30)$$

Similarly, *the Fourier coefficients for a half-range sine series are given by*

$$b_n = \frac{2}{L}\int_0^L f(x)\sin\frac{n\pi x}{L}\,dx \qquad (n = 1, 2, \ldots) \qquad (29.31)$$

◀ **EXAMPLE 9** Find $f(x) = x$ in a half-range cosine series for $0 \le x < 2$.

Since we are to have a cosine series, we extend the function to be an even function with its graph as shown in Fig. 29.25. The dark portion between $x = 0$ and $x = L$ shows the given function as defined, and light portions show the extension that makes it an even function. Now, by use of Eqs. (29.30) we find the Fourier expansion coefficients, with $L = 2$.

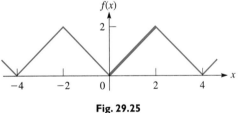

$$a_0 = \frac{1}{2}\int_0^2 x\,dx = \frac{1}{4}x^2\Big|_0^2 = 1$$

$$a_n = \frac{2}{2}\int_0^2 x\cos\frac{n\pi x}{2}\,dx = x\left(\frac{2}{n\pi}\sin\frac{n\pi x}{2}\right) - \left(\frac{-4}{n^2\pi^2}\cos\frac{n\pi x}{2}\right)\Big|_0^2$$

$$= \frac{4}{n^2\pi^2}(\cos n\pi - 1) \qquad (n \ne 0)$$

Fig. 29.25

If n is even, $\cos n\pi - 1 = 0$. Therefore, we evaluate a_n for the odd values of n, and find the expansion is

$$f(x) = 1 - \frac{8}{\pi^2}\left(\cos\frac{\pi x}{2} + \frac{1}{9}\cos\frac{3\pi x}{2} + \frac{1}{25}\cos\frac{5\pi x}{2} + \cdots\right)$$

◀ **EXAMPLE 10** Expand $f(x) = x$ in a half-range sine series for $0 \le x < 2$.

Since we are to have a sine series, we extend the function to be an odd function with its graph as shown in Fig. 29.26. Again, the dark portion shows the given function as defined, and light portions show the extension that makes it an odd function. By using Eq. (29.31) we find the Fourier expansion coefficients, with $L = 2$.

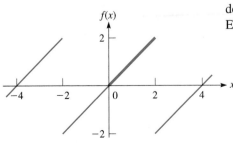

$$b_n = \frac{2}{2}\int_0^2 x\sin\frac{n\pi x}{2}\,dx$$

$$= x\left(\frac{-2}{n\pi}\cos\frac{n\pi x}{2}\right) - \left(\frac{-4}{n^2\pi^2}\sin\frac{n\pi x}{2}\right)\Big|_0^2 = -\frac{4}{n\pi}\cos n\pi$$

$$f(x) = \frac{4}{\pi}\left(\sin\frac{\pi x}{2} - \frac{1}{2}\sin\pi x + \frac{1}{3}\sin\frac{3\pi x}{2} - \cdots\right)$$

Fig. 29.26

EXERCISES 29.7

In Exercises 1–4, write the Fourier series for each function by comparing it to an appropriate function given in an example of this section. Do not use any of the formulas for a_0, a_n, or b_n.

1. $f(x) = \begin{cases} 2 & -\pi \le x < -\pi/2, \; \pi/2 \le x < \pi \\ 3 & -\pi/2 \le x < \pi/2 \end{cases}$

2. $f(x) = \begin{cases} -\frac{1}{2} & -\pi \le x < 0 \\ \frac{3}{2} & 0 \le x < \pi \end{cases}$

3. $f(x) = \begin{cases} -2 & -4 \le x < 0 \\ 0 & 0 \le x < 4 \end{cases}$

4. $f(x) = \begin{cases} -\frac{1}{3} & -\pi \le x < -\pi/2, \; \pi/2 \le x < \pi \\ \frac{2}{3} & -\pi/2 \le x < \pi/2 \end{cases}$

In Exercises 5–12, determine whether the given function is even, or odd, or neither. One period is defined for each function.

5. $f(x) = \begin{cases} 5 & -3 \le x < 0 \\ 0 & 0 \le x < 3 \end{cases}$ **6.** $f(x) = \begin{cases} -1 & -2 \le x < 0 \\ 1 & 0 \le x < 2 \end{cases}$

7. $f(x) = \begin{cases} 2 & -1 \le x < 1 \\ 0 & -2 \le x < -1, \; 1 \le x < 2 \end{cases}$

8. $f(x) = \begin{cases} 0 & -2 \le x < 0, \; 1 \le x < 2 \\ 1 & 0 \le x < 1 \end{cases}$

9. $f(x) = |x| \quad -4 \le x < 4$ **10.** $f(x) = \begin{cases} 0 & -1 \le x < 0 \\ e^x & 0 \le x < 1 \end{cases}$

11. $f(x) = -x \cos 3x \quad -3 \le x < 3$

12. $f(x) = x \sin 2x \cos x \quad -4 \le x < 4$

In Exercises 13–18, find at least three nonzero terms (including a_0 and at least two cosine terms and two sine terms if they are not all zero) of the Fourier series for the function from the indicated exercise of this section. Sketch at least three periods of the function.

13. Exercise 5   **14.** Exercise 6   **15.** Exercise 7

16. Exercise 8   **17.** Exercise 9   **18.** Exercise 10

In Exercises 19–24, solve the given problems.

19. Expand $f(x) = 1$ in a half-range sine series for $0 \le x < 4$.

20. Expand $f(x) = 1 \; (0 \le x < 2)$, $f(x) = 0 \; (2 \le x < 4)$ in a half-range cosine series for $0 \le x < 4$.

21. Expand $f(x) = x^2$ in a half-range cosine series for $0 \le x < 2$.

22. Expand $f(x) = x^2$ in a half-range sine series for $0 \le x < 2$.

23. Each pulse of a pulsating force F of a pressing machine is 8 N. The force lasts for 1 s, followed by a 3-s pause. Thus, it can be represented by $F = 0$ for $-2 \le t < 0$ and $1 \le t < 2$, and $F = 8$ for $0 \le t < 1$, with a period of 4 s (only positive values of t have physical significance). Find the Fourier series for the force.

24. A pulsating electric current i (in mA) with a period of 2 s can be described by $i = e^{-t}$ for $-1 \le t < 1$ s for one period (only positive values of t have physical significance). Find the Fourier series that represents this current.

CHAPTER 29 EQUATIONS

Infinite series	$\displaystyle\sum_{n=1}^{\infty} a_n = a_1 + a_2 + a_3 + \cdots + a_n + \cdots$	(29.1)
Sum of series	$S = \lim_{n \to \infty} S_n = \lim_{n \to \infty} \displaystyle\sum_{i=1}^{n} a_i$	(29.2)
Sum of geometric series	$S = \lim_{n \to \infty} S_n = \dfrac{a_1}{1-r}$	(29.3)
Power series	$f(x) = a_0 + a_1 x + a_2 x^2 + \cdots + a_n x^n + \cdots$	(29.4)
Maclaurin series	$f(x) = f(0) + f'(0)x + \dfrac{f''(0)x^2}{2!} + \dfrac{f'''(0)x^3}{3!} + \cdots + \dfrac{f^n(0)x^n}{n!} + \cdots$	(29.5)

Special series

$$e^x = 1 + x + \frac{x^2}{2!} + \frac{x^3}{3!} + \cdots \qquad \text{(all } x\text{)}$$ (29.6)

$$\sin x = x - \frac{x^3}{3!} + \frac{x^5}{5!} - \cdots \qquad \text{(all } x\text{)}$$ (29.7)

$$\cos x = 1 - \frac{x^2}{2!} + \frac{x^4}{4!} - \cdots \qquad \text{(all } x\text{)}$$ (29.8)

$$\ln(1 + x) = x - \frac{x^2}{2} + \frac{x^3}{3} - \frac{x^4}{4} + \cdots \qquad (|x| < 1)$$ (29.9)

$$(1 + x)^n = 1 + nx + \frac{n(n - 1)}{2!}x^2 + \cdots \qquad (|x| < 1)$$ (29.10)

Taylor series

$$f(x) = f(a) + f'(a)(x - a) + \frac{f''(a)(x - a)^2}{2!} + \cdots$$ (29.16)

Fourier series

$$f(x) = a_0 + a_1 \cos x + a_2 \cos 2x + \cdots + a_n \cos nx + \cdots$$
$$+ b_1 \sin x + b_2 \sin 2x + \cdots + b_n \sin nx + \cdots$$ (29.17)

Period = 2π

$$a_0 = \frac{1}{2\pi} \int_{-\pi}^{\pi} f(x)\, dx$$ (29.25)

$$a_n = \frac{1}{\pi} \int_{-\pi}^{\pi} f(x)\cos nx\, dx$$ (29.24)

$$b_n = \frac{1}{\pi} \int_{-\pi}^{\pi} f(x)\sin nx\, dx$$ (29.26)

Period = $2L$

$$a_0 = \frac{1}{2L} \int_{-L}^{L} f(x)\, dx$$ (29.27)

$$a_n = \frac{1}{L} \int_{-L}^{L} f(x)\cos \frac{n\pi x}{L}\, dx$$ (29.28)

$$b_n = \frac{1}{L} \int_{-L}^{L} f(x)\sin \frac{n\pi x}{L}\, dx$$ (29.29)

Half-range expansions

$$a_0 = \frac{1}{L} \int_0^L f(x)\, dx \quad \text{and} \quad a_n = \frac{2}{L} \int_0^L f(x)\cos \frac{n\pi x}{L}\, dx \quad (n = 1, 2, \ldots)$$ (29.30)

$$b_n = \frac{2}{L} \int_0^L f(x)\sin \frac{n\pi x}{L}\, dx \quad (n = 1, 2, \ldots)$$ (29.31)

CHAPTER ㉙ REVIEW EXERCISES

In Exercises 1–8, find the first three nonzero terms of the Maclaurin expansion of the given functions.

1. $f(x) = \dfrac{1}{1 + e^x}$

2. $f(x) = e^{\cos x}$

3. $f(x) = \sin 2x^2$

4. $f(x) = \dfrac{1}{(1 - x)^2}$

5. $f(x) = (x + 1)^{1/3}$

6. $f(x) = \dfrac{x^2}{1 + x^2}$

7. $f(x) = \sin^{-1} x$

8. $f(x) = \dfrac{1}{1 - \sin x}$

In Exercises 9–20, calculate the value of each of the given functions. Use three terms of the appropriate series.

9. $e^{-0.2}$

10. $\ln(1.10)$

11. $\sqrt[3]{1.3}$

12. $\sin 3.5°$

13. 1.086^{-1}

14. 0.9839^{10}

15. $\ln 0.8172$

16. $\cos 0.1376$

17. $\tan 43.62°$

18. $\sqrt[4]{260}$

19. $\sqrt{148}$

20. $\cos 47°$

In Exercises 21 and 22, evaluate the given integrals by using three terms of the appropriate series.

21. $\int_{0.1}^{0.2} \dfrac{\cos x}{\sqrt{x}}\, dx$

22. $\int_{0}^{0.1} \sqrt[3]{1 + x^2}\, dx$

In Exercises 23 and 24, find the first three terms of the Taylor expansion for the given function and value of a.

23. $\cos x \quad (a = \pi/3)$

24. $\ln \cos x \quad (a = \pi/4)$

In Exercises 25–28, write the Fourier series for each function by comparing it to an appropriate function in an example of either Section 29.6 or 29.7. One period is given for each function. Do not use any formulas for a_0, a_n, or b_n.

25. $f(x) = \begin{cases} 0 & -\pi \le x < 0 \\ x - 1 & 0 \le x < \pi \end{cases}$

26. $f(x) = x^2 - 1 \quad -1 \le x < 1$

27. $f(x) = \begin{cases} \pi - 1 & -4 \le x < 0 \\ \pi + 1 & 0 \le x < 4 \end{cases}$

28. $f(x) = \begin{cases} 1 & -\pi \le x < 0 \\ 1 + \sin x & 0 \le x < \pi \end{cases}$

In Exercises 29–32, find at least three nonzero terms (including a_0 and at least two cosine terms and two sine terms if they are not all zero) of the Fourier series for the given function. One period is given for each function.

29. $f(x) = \begin{cases} 0 & -\pi \le x < -\pi/2,\ \pi/2 \le x < \pi \\ 1 & -\pi/2 \le x < \pi/2 \end{cases}$ (See Example 2, page 901.)

30. $f(x) = \begin{cases} -x & -\pi \le x < 0 \\ 0 & 0 \le x < \pi \end{cases}$

31. $f(x) = x \quad -2 \le x < 2$

32. $f(x) = \begin{cases} -2 & -3 \le x < 0 \\ 2 & 0 \le x < 3 \end{cases}$

In Exercises 33–64, solve the given problems.

33. Test the series $1000 + 800 + 640 + 512 + \ldots$ for convergence or divergence. If convergent, find its sum.

34. Test the series $1 + 1.1 + 1.21 + 1.331 + \ldots$ for convergence or divergence. If convergent, find its sum.

35. Find the sum of the series $64 + 48 + 36 + 27 + \cdots$.

36. Find the first five partial sums of the series $\sum\limits_{n=1}^{\infty} \dfrac{n}{3n + 1}$ and determine whether it appears to be convergent or divergent.

37. If h is small, show that $\sin(x + h) - \sin(x - h) = 2h \cos x$.

38. Find the first three nonzero terms of the Maclaurin expansion of the function $\sin x + x \cos x$ by differentiating the expansion term by term for $x \sin x$.

39. Using the properties of logarithms and Eq. (29.9), find four terms of the Maclaurin expansion of $\ln(1 + x)^4$.

40. By multiplication of series, show that the first two terms of the Maclaurin series for $2 \sin x \cos x$ are the same as those of the series for $\sin 2x$.

41. Find the first four nonzero terms of the expansion for $\cos^2 x$ by using the identity $\cos^2 x = \frac{1}{2}(1 + \cos 2x)$ and the series for $\cos x$.

42. Evaluate the integral $\int_0^1 x \sin x\, dx$ (a) by methods of Chapter 28 and (b) using three terms of the series for $\sin x$. Compare results.

43. Find the first three terms of the Maclaurin expansion for $\sec x$ by finding the reciprocal of the series for $\cos x$.

44. By simplifying the sum of the squares for the Maclaurin series for $\sin x$ and $\cos x$, verify (to this extent) that $\sin^2 x + \cos^2 x = 1$.

45. From the Maclaurin series for $f(x) = 1/(1 - x)$ (see Exercise 11 of Section 29.2), find the series for $1/(1 + x)$.

46. Show that the Maclaurin expansions for $\cos x$ and $\cos(-x)$ are the same.

47. Expand $f(x) = x^2$ in a half-range cosine series for $0 < x \le 1$.

48. Expand $f(x) = 2 - x$ in a half-range sine series for $0 < x \le 2$.

49. Calculate $e^{0.9}$ by using four terms of the Maclaurin expansion for e^x. Also calculate $e^{0.9}$ by using the first three terms of the Taylor expansion in Example 1 on page 892. Compare the accuracy of the values obtained with that found directly on a calculator.

50. Find the volume generated by revolving the region bounded by $y = e^{-x}$, $y = 0$, $x = 0$, and $x = 0.1$ about the y-axis by using three terms of the appropriate series.

51. Find the approximate area between the curve of $y = \dfrac{x - \sin x}{x^2}$ and the x-axis between $x = 0.1$ and $x = 0.2$.

52. Find the approximate value of the moment of inertia with respect to its axis of the solid generated by revolving the smaller region bounded by $y = \sin x$, $x = 0.3$, and the x-axis about the y-axis. Use two terms of the appropriate series.

53. Find three terms of the Maclaurin series for $\tan^{-1} x$ by integrating the series for $1/(1 + x^2)$, term by term.

54. The displacement y (in m) of a water wave as a function of the time t (in s) is $y = 0.5 \sin 0.5t - 0.2 \sin 0.4t$. Find the first three terms of the Maclaurin series for the displacement.

55. The number N of radioactive nuclei in a radioactive sample is $N = N_0 e^{-\lambda t}$. Here t is the time, N_0 is the number at $t = 0$, and λ is the *decay constant*. By using four terms of the appropriate series, express the right side of this equation as a polynomial.

56. The length of Lake Erie is a great circle arc of 390 km. If the lake is assumed to be flat, use series to find the error in calculating the distance from the center of the lake to the center of the earth. The radius of the earth is 6400 km.

57. From what height can a person see a point 10 km distant on earth's surface?

58. The vertical displacement y of a mass at the end of a spring is given by $y = \sin 3t - \cos 2t$, where t is the time. By subtraction of series, find the first four nonzero terms of the series for y.

59. The electric potential V at a distance x along a certain surface is given by $V = \ln \dfrac{1 + x}{1 - x}$. Find the first four terms of the Maclaurin series for V.

60. If a mass M is hung from a spring of mass m, the ratio of the masses is $m/M = k\omega \tan k\omega$, where k is a constant and ω is a measure of the frequency of vibration. By using two terms of the appropriate series, express m/M as a polynomial in terms of ω.

61. In the study of electromagnetic radiation, the expression $\dfrac{N_0}{1 - e^{-k/T}}$ is used. Here T is the thermodynamic temperature, and N_0 and k are constants. Show that this expression can be written as $N_0(1 + e^{-k/T} + e^{-2k/T} + \cdots)$. (*Hint:* Let $x = e^{-k/T}$.)

62. In the analysis of reflection from a spherical mirror, it is necessary to express the x-coordinate on the surface shown in Fig. 29.27 in terms of the y-coordinate and the radius R. Using the equation of the semicircle shown, solve for x (note that $x \le R$). Then express the result as a series. (Note that the first approximation gives a parabolic surface.)

63. A certain electric current is pulsating so that the current as a function of time is given by $f(t) = 0$ if $-\pi \le t < 0$ and $\pi/2 < t < \pi$. If $0 < t < \pi/2$, $f(t) = \sin t$. Find the Fourier expansion for this pulsating current and sketch three periods

64. The force F applied to a spring system as a function of the time t is given by $F = t/\pi$ if $0 \le t \le \pi$ and $F = 0$ if $\pi < t < 2\pi$. If the period of the force is 2π, find the first few terms of the Fourier series that represents the force.

Writing Exercise

65. A computer science class is assigned to write a program to make a table of values of the sine, cosine, and tangent of an angle in degrees to the nearest $0.1°$. Write a few paragraphs explaining how these values may be found using the known values for $0°$, $30°$, and $45°$ (see page 118), trigonometric relations in Chapter 20, and series from this chapter, without using many terms of any series.

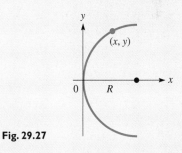

Fig. 29.27

CHAPTER 29 PRACTICE TEST

1. By direct expansion, find the first four nonzero terms of the Maclaurin expansion for $f(x) = (1 + e^x)^2$.

2. Find the first three nonzero terms of the Taylor expansion for $f(x) = \cos x$, with $a = \pi/3$.

3. Evaluate $\ln 0.96$ by using four terms of the expansion for $\ln(1 + x)$.

4. Find the first three nonzero terms of the expansion for $f(x) = \dfrac{1}{\sqrt{1 - 2x}}$ by using the binomial series.

5. Evaluate $\int_0^1 x \cos x \, dx$ by using three terms of the appropriate series.

6. An electric current is pulsating such that it is a function of the time with a period of 2π. If $f(t) = 2$ for $0 \le t < \pi$ and $f(t) = 0$ for the other half-cycle, find the first three nonzero terms of the Fourier series for this current.

7. $f(x) = x^2 + 2$ for $-2 \le x < 2$ (period $= 4$). Is $f(x)$ an even function, an odd function, or neither? Expressing the Fourier series as $F(x)$, what is the Fourier series for the function $g(x) = x^2 - 1$ for $-2 \le x < 2$ (period $= 4$)? (Do *not* use integration to derive specific terms for either series.)

Differential Equations

In Section 30.5, we show how differential equations are used in the study of historical events by using the method of carbon dating.

Many of the physical problems being studied in the 1700s, such as velocity and light, led to equations that involved derivatives or differentials. These equations are called *differential equations*. Therefore, solving these differential equations became a very important topic of mathematical development during the eighteenth century.

In this chapter, we study some of the basic methods of solving differential equations. Many of these methods were first developed by the famous Swiss mathematician Leonhard Euler (1707–1783). He is undoubtedly the most prolific mathematician of all time, in that his work in mathematics and other fields filled over 70 large volumes.

The final topic covered in this chapter is the solution of differential equations by *Laplace transforms*. They are named for the French mathematician Pierre Laplace (1749–1827). Actually, Laplace had devised a mathematical method in the late 1700s that the English electric engineer Oliver Heaviside (1850–1925) refined and developed into its present useful form in the late 1800s. The Laplace transform is particularly useful for solving problems involving electrical circuits and mechanical systems.

Here we see again that a field of mathematics was developed in response to the need for solving real-life physical problems. Also, we see that a method developed for purely mathematical reasons was then very usefully applied 100 years later in electricity, an area of study that did not exist when the method was first devised.

We actually solved a few simple differential equations in earlier chapters when we started a solution with the expression for the slope of a tangent line or the velocity of an object. Also, we have noted the applications of differential equations in electrical circuits, mechanical systems, and the study of light. Other areas of application include chemical reactions, interest calculations, changes in pressure and temperature, population growth, forces on beams and structures, and nuclear energy.

30.1 Solutions of Differential Equations

A **differential equation** *is an equation that contains derivatives or differentials.* Most differential equations we shall consider contain first and/or second derivatives, although some will have higher derivatives. *An equation that contains only first derivatives is called a* **first-order** *differential equation. An equation that contains second derivatives, and possibly first derivatives, is called a* **second-order** *differential equation. In general, the* **order** *of the differential equation is that of the highest derivative in the equation, and the* **degree** *is the highest power of that derivative.*

◀ EXAMPLE 1 **(a)** The equation $dy/dx + x = y$ is a first-order differential equation since it contains only a first derivative.

(b) The equations $\dfrac{d^2y}{dx^2} + y = 3x^2$ and $\dfrac{d^2y}{dx^2} + 2\dfrac{dy}{dx} = x$ are second-order differential

equations since each contains a second derivative and no higher derivatives. The dy/dx in the second equation does not affect the order.

(c) The equation $\dfrac{d^2y}{dx^2} + \left(\dfrac{dy}{dx}\right)^4 - y = 6$ is a differential equation of the second order

and first degree. That is, the highest derivative that appears is the second, and it is raised to the first power. Since the second derivative appears, the fourth power of the first derivative does not affect the degree. ▶

In our discussion of differential equations, we will restrict our attention to equations of the first degree.

A **solution** *of a differential equation is a relation between the variables that satisfies the differential equation.* That is, when this relation is substituted into the differential equation, an algebraic identity results. *A solution containing a number of independent*

General Solution *arbitrary constants equal to the order of the differential equation is called the* **general solution** *of the equation. When specific values are given to at least one of these con-*

Particular Solution *stants, the solution is called a* **particular solution.**

◀ EXAMPLE 2 Any coefficients that are not specified numerically after like terms have been combined are independent arbitrary constants. In the expression $c_1x + c_2 + c_3x$, there are only two arbitrary constants since the x-terms may be combined; $c_2 + c_4x$ is an equivalent expression with $c_4 = c_1 + c_3$. ▶

◀ EXAMPLE 3 $y = c_1e^{-x} + c_2e^{2x}$ is the general solution of the differential equation

$$\frac{d^2y}{dx^2} - \frac{dy}{dx} = 2y$$

The order of this differential equation is 2, and there are two independent arbitrary constants in the solution. The equation $y = 4e^{-x}$ is a particular solution. It can be derived from the general solution by letting $c_1 = 4$ and $c_2 = 0$. Each of these solutions can be shown to satisfy the differential equation by taking two derivatives and substituting. ▶

To solve a differential equation, we have to find some method of transforming the equation so that each term may be integrated. Some of these methods will be considered after this section. The purpose here is to show that a given equation is a solution

NOTE ▶ of the differential equation *by taking the required derivatives and to show that an identity results after substitution.*

EXAMPLE 4 Show that $y = c_1 \sin x + c_2 \cos x$ is the general solution of the differential equation $y'' + y = 0$.

The function and its first two derivatives are

$$y = c_1 \sin x + c_2 \cos x$$
$$y' = c_1 \cos x - c_2 \sin x$$
$$y'' = -c_1 \sin x - c_2 \cos x$$

Substituting these into the differential equation, we have

$$\begin{array}{ccc} y'' & + & y & = 0 \\ \downarrow & & \downarrow & \end{array}$$

$$(-c_1 \sin x - c_2 \cos x) + (c_1 \sin x + c_2 \cos x) = 0 \quad \text{or} \quad 0 = 0$$

We know that this must be the general solution, since there are two independent arbitrary constants and the order of the differential equation is 2.

EXAMPLE 5 Show that $y = cx + x^2$ is a solution of the differential equation $xy' - y = x^2$.

Taking one derivative of the function and substituting into the differential equation, we have

$$y = cx + x^2$$

$$xy' - y = x^2$$

$$y' = c + 2x$$

$$x(c + 2x) - (cx + x^2) = x^2 \quad \text{or} \quad x^2 = x^2$$

EXERCISES 30.1

In Exercises 1 and 2, show that the indicated solutions are in fact solutions of the differential equations in the indicated examples.

1. In Example 3, two solutions are shown for the given differential equation. Show that each is a solution.

2. In Example 4, show that $y_1 = c \sin x + 5 \cos x$ and $y_2 = 2 \sin x - 3 \cos x$ are solutions of the given differential equation.

In Exercises 3–6, determine whether the given equation is the general solution or a particular solution of the given differential equation.

3. $\dfrac{dy}{dx} + 2xy = 0, \quad y = e^{-x^2}$

4. $y' \ln x - \dfrac{y}{x} = 0, \quad y = c \ln x$

5. $y'' + 3y' - 4y = 3e^x, \quad y = c_1 e^x + c_2 e^{-4x} + \frac{3}{5}xe^x$

6. $\dfrac{d^2y}{dx^2} + 4y = 0, \quad y = c_1 \sin 2x + 3 \cos 2x$

In Exercises 7–10, show that each function $y = f(x)$ is a solution of the given differential equation.

7. $\dfrac{dy}{dx} - y = 1; \quad y = e^x - 1, \quad y = 5e^x - 1$

8. $\dfrac{dy}{dx} = 2xy^2; \quad y = -\dfrac{1}{x^2}, \quad y = -\dfrac{1}{x^2 + c}$

9. $y'' + 4y = 0; \quad y = 3 \cos 2x, \quad y = c_1 \sin 2x + c_2 \cos 2x$

10. $y'' = 2y'; \quad y = 3e^{2x}, \quad y = 2e^{2x} - 5$

In Exercises 11–32, show that the given equation is a solution of the given differential equation.

11. $\dfrac{dy}{dx} = 2x, \quad y = x^2 + 1$

12. $\dfrac{dy}{dx} = 1 - 3x^2, \quad y = 2 + x - x^3$

13. $\dfrac{dy}{dx} - 3 = 2x, \quad y = x^2 + 3x$

14. $xy' = 2y, \quad y = cx^2$

15. $y' + 2y = 2x, \quad y = ce^{-2x} + x - \frac{1}{2}$

16. $y' - 3x^2 = 1, \quad y = x^3 + x + c$

17. $y'' + 9y = 4 \cos x, \quad 2y = \cos x$

18. $y'' - 4y' + 4y = e^{2x}, \quad y = e^{2x}\left(c_1 + c_2 x + \dfrac{x^2}{2}\right)$

19. $x^2y' + y^2 = 0$, $xy = cx + cy$

20. $xy' - 3y = x^2$, $y = cx^3 - x^2$

21. $x\dfrac{d^2y}{dx^2} + \dfrac{dy}{dx} = 0$, $y = c_1 \ln x + c_2$

22. $y'' + 4y = 10e^x$, $y = c_1 \sin 2x + c_2 \cos 2x + 2e^x$

23. $y' + y = 2 \cos x$, $y = \sin x + \cos x - e^{-x}$

24. $(x + y) - xy' = 0$, $y = x \ln x - cx$

25. $y'' + y' = 6 \sin 2x$, $y = e^{-x} - \frac{3}{5} \cos 2x - \frac{6}{5} \sin 2x$

26. $xy'' + y' = 16x^3$, $y = x^4 + c_1 + c_2 \ln x$

27. $\cos x \dfrac{dy}{dx} + \sin x = 1 - y$, $y = \dfrac{x + c}{\sec x + \tan x}$

28. $2xyy' + x^2 = y^2$, $x^2 + y^2 = cx$

29. $(y')^2 + xy' = y$, $y = cx + c^2$

30. $x^4(y')^2 - xy' = y$, $y = c^2 + \dfrac{c}{x}$

31. $\dfrac{d^3y}{dx^3} = \dfrac{d^2y}{dx^2}$, $y = c_1 + c_2x + c_3e^x$

32. $\dfrac{d^3y}{dx^3} + 4\dfrac{d^2y}{dx^2} + 4\dfrac{dy}{dx} = 0$, $y = c_1 + c_2e^{-2x} + xe^{-2x}$

Separation of Variables

We will now solve differential equations of the first order and first degree. Of the many methods for solving such equations, a few are presented in this and the next two sections. The first of these is *the method of* **separation of variables.**

A differential equation of the first order and first degree contains the first derivative to the first power. That is, it may be written as $dy/dx = f(x, y)$. This type of equation is more commonly expressed in its differential form,

$$M(x, y)\, dx + N(x, y)\, dy = 0 \tag{30.1}$$

where $M(x, y)$ and $N(x, y)$ may represent constants, functions of either x or y, or functions of x and y.

To solve an equation of the form of Eq. (30.1), we must integrate. However, if $M(x, y)$ contains y, the first term cannot be integrated. Also, if $N(x, y)$ contains x, the second term cannot be integrated. If it is possible to rewrite Eq. (30.1) as

$$A(x)\, dx + B(y)\, dy = 0 \tag{30.2}$$

NOTE ▶ where $A(x)$ does not contain y and $B(y)$ does not contain x, then *we may find the solution by integrating each term and adding the constant of integration.* (In rewriting Eq. (30.1), if division is used, the solution is not valid for values that make the divisor zero.) Many differential equations can be solved in this way.

◀ EXAMPLE 1 Solve the differential equation $dx - 4xy^3\, dy = 0$.

We can write this equation as

$$(1)\ dx + (-4xy^3)\, dy = 0$$

which means that $M(x, y) = 1$ and $N(x, y) = -4xy^3$.

CAUTION ▶ We must remove the x from the coefficient of dy *without introducing y into the coefficient of dx.* We do this by dividing each term by x, which gives us

$$dx/x - 4y^3\, dy = 0$$

It is now possible to integrate each term. Performing this integration, we have

$$\ln |x| - y^4 = c$$

The constant of integration c becomes the arbitrary constant of the solution. ▶

NOTE ▶

In Example 1, we showed the integration of dx/x as $\ln|x|$, which follows our discussion in Section 28.2. We know $\ln|x| = \ln x$ if $x > 0$ and $\ln|x| = \ln(-x)$ if $x < 0$. Since we know the values being used when we find a particular solution, *we generally will not use the absolute value notation when integrating logarithmic forms.* We would show the integration of dx/x as $\ln x$, with the understanding that we know $x > 0$. When using negative values of x, we would express it as $\ln(-x)$.

◀ **EXAMPLE 2** Solve the differential equation $xy\,dx + (x^2 + 1)\,dy = 0$.

In order to integrate each term, it is necessary to divide each term by $y(x^2 + 1)$. When this is done, we have

$$\frac{x\,dx}{x^2 + 1} + \frac{dy}{y} = 0$$

Integrating, we obtain the solution

$$\frac{1}{2}\ln(x^2 + 1) + \ln y = c$$

For reference, Eq. (13.9) is $\log_b x^n = n \log_b x$ and Eq. (13.7) is $\log_b xy = \log_b x + \log_b y$.

It is possible to make use of the properties of logarithms to make the form of this solution neater. If we write the constant of integration as $\ln c_1$, rather than c, we have $\frac{1}{2}\ln(x^2 + 1) + \ln y = \ln c_1$. Multiplying through by 2 and using the property of logarithms given by Eq. (13.9), we have $\ln(x^2 + 1) + \ln y^2 = \ln c_1^2$. Next, using the property of logarithms given by Eq. (13.7), we then have $\ln(x^2 + 1)y^2 = \ln c_1^2$, which means

$$(x^2 + 1)y^2 = c_1^2$$

CAUTION ▶

NOTE ▶

This form of the solution is more compact and generally would be preferred. However, *any expression that represents a constant may be chosen as the constant of integration* and leads to a correct solution. In checking answers, we must remember that a different choice of constant will lead to a different form of the solution. Thus, two different-appearing answers may both be correct. *Often there is more than one reasonable choice of a constant, and different forms of the solution may be expected.*

◀ **EXAMPLE 3** Solve the differential equation $\dfrac{d\theta}{dt} = \dfrac{\theta}{t^2 + 4}$.

The solution proceeds as follows:

$$\frac{d\theta}{\theta} = \frac{dt}{t^2 + 4} \qquad \text{separate variables by multiplying by } dt \text{ and dividing by } \theta$$

$$\ln\theta = \frac{1}{2}\tan^{-1}\frac{t}{2} + \frac{c}{2} \qquad \text{integrate}$$

$$2\ln\theta = \tan^{-1}\frac{t}{2} + c$$

$$\ln\theta^2 = \tan^{-1}\frac{t}{2} + c$$

Note the different forms of the result using $c/2$ as the constant of integration. These forms would differ somewhat had we chosen c as the constant.

The choice of $\ln c$ as the constant of integration (on the left) is also reasonable. It would lead to the result $2\ln c\theta = \tan^{-1}(t/2)$.

In order to separate the variables of a differential equation that contains exponential functions, it may be necessary to use the properties of exponents. Also, when trigonometric functions are involved, the basic trigonometric identities may be needed.

◀ EXAMPLE 4   Solve the differential equation $2e^{3x} \sin y\, dx + e^x \csc y\, dy = 0$.

In the dx-term we want only a function of x, which means that we must divide by $\sin y$. Also, the dy-term indicates that we must divide by e^x. Thus,

$$\frac{2e^{3x} \sin y\, dx}{e^x \sin y} + \frac{e^x \csc y\, dy}{e^x \sin y} = 0 \qquad \text{divide by } e^x \sin y$$

$$2e^{2x}\, dx + \csc^2 y\, dy = 0 \qquad \text{variables separated}$$

$$e^{2x}(2\, dx) + \csc^2 y\, dy = 0 \qquad \text{form for integrating}$$

$$e^{2x} - \cot y = c \qquad \text{integrate} \qquad ▶$$

FINDING PARTICULAR SOLUTIONS

In order to find a particular solution of a differential equation, we must have information that allows us to evaluate the constant of integration. The following examples show how particular solutions of differential equations are found. Also, we will show graphically the difference between the general solution and the particular solution.

◀ EXAMPLE 5   Solve the differential equation $(x^2 + 1)^2\, dy + 4x\, dx = 0$, subject to the condition that $x = 1$ when $y = 3$.

Separating variables, we have

$$dy + \frac{4x\, dx}{(x^2 + 1)^2} = 0 \qquad \text{dividing by } (x^2 + 1)^2$$

$$y - \frac{2}{x^2 + 1} = c \qquad \text{integrating}$$

$$y = \frac{2}{x^2 + 1} + c \qquad \text{general solution}$$

Since a specific set of values is given, we can evaluate the constant of integration and thereby get a particular solution. Using the values $x = 1$ and $y = 3$, we have

$$3 = \frac{2}{1 + 1} + c, \qquad c = 2 \qquad \text{evaluate } c$$

which gives us

$$y = \frac{2}{x^2 + 1} + 2 \qquad \text{particular solution}$$

The general solution defines a *family* of curves, one member of the family for each value of c that may be considered. A few of these curves are shown in Fig. 30.1. When c is specified as in the particular solution, we have the specific (darker) curve shown in Fig. 30.1. ▶

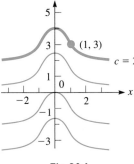

Fig. 30.1

See pages 991–992 for numerical methods of solving differential equations that can be written in the form $dy/dx = f(x, y)$.

◀ EXAMPLE 6 Find the particular solution in Example 2 if the function is subject to the condition that $x = 0$ when $y = e$.

Using the solution $\frac{1}{2} \ln(x^2 + 1) + \ln y = c$, we have

$$\frac{1}{2} \ln(0 + 1) + \ln e = c, \qquad \frac{1}{2} \ln 1 + 1 = c, \qquad c = 1$$

The particular solution is then

$$\frac{1}{2} \ln(x^2 + 1) + \ln y = 1 \qquad \text{substitute } c = 1$$

$$\ln(x^2 + 1) + 2 \ln y = 2$$

$$\ln y^2(x^2 + 1) = 2 \qquad \text{using properties of logarithms}$$

$$y^2(x^2 + 1) = e^2 \qquad \text{exponential form}$$

Using the general solution $(x^2 + 1)y^2 = c_1^2$, we have

$$(0 + 1)e^2 = c_1^2, \qquad c_1^2 = e^2$$

$$y^2(x^2 + 1) = e^2$$

NOTE ▶ which is precisely the same solution as above. This shows that *the choice of the form of the constant does not affect the final result, and the constant is truly arbitrary.* ▶

EXERCISES 30.2

In Exercises 1 and 2, make the given changes in the indicated examples of this section and then solve the resulting differential equations.

1. In Example 2, change the first term to $2xy\,dx$.

2. In Example 4, change the second term to $e^{2x} \csc y\,dy$.

In Exercises 3–32, solve the given differential equations. Explain your method of solution for Exercise 15.

3. $2x\,dx + dy = 0$

4. $y^2\,dy + x^3\,dx = 0$

5. $y^2\,dx + dy = 0$

6. $y\,dt + t\,dy = 0$

7. $\dfrac{dV}{dP} = -\dfrac{V}{P^2}$

8. $\dfrac{2\,dy}{dx} = \dfrac{y(x + 1)}{x}$

9. $x^2 + (x^3 + 5)y' = 0$

10. $xyy' + \sqrt{1 + y^2} = 0$

11. $dy + \ln xy\,dx = (4x + \ln y)\,dx$

12. $r\sqrt{1 - \theta^2}\,\dfrac{dr}{d\theta} = \theta + 4$

13. $e^{x^2}\,dy = x\sqrt{1 - y}\,dx$

14. $\sqrt{1 + 4x^2}\,dy = y^3x\,dx$

(W) **15.** $e^{x+y}\,dx + dy = 0$

16. $e^{2x}\,dy + e^x\,dx = 0$

17. $y' - y = 4$

18. $ds - s^2\,dt = 9\,dt$

19. $x\dfrac{dy}{dx} = y^2 + y^2 \ln x$

20. $(yx^2 + y)\dfrac{dy}{dx} = \tan^{-1} x$

21. $y \tan x\,dx + \cos^2 x\,dy = 0$

22. $\sin x \sec y\,dx = dy$

23. $yx^2\,dx = y\,dx - x^2\,dy$

24. $e^{\cos \theta} \tan \theta\,d\theta + \sec \theta\,dy = 0$

25. $y\sqrt{1 - x^2}\,dy + 2\,dx = 0$

26. $(x^3 + x^2)\,dx + (x + 1)y\,dy = 0$

27. $xe^{x^2 - y}\,dx = dy$

28. $\cos^2 \phi + y \csc \phi\dfrac{dy}{d\phi} = 0$

29. $2 \ln t\,dt + t\,di = 0$

30. $2y(x^3 + 1)\,dy + 3x^2(y^2 - 1)\,dx = 0$

31. $y^2e^x + (e^x + 1)\dfrac{dy}{dx} = 0$

32. $V + 1 + (1 + \sin T) \sec T\dfrac{dV}{dT} = 0$

In Exercises 33–40, find the particular solution of the given differential equation for the indicated values.

33. $\dfrac{dy}{dx} + yx^2 = 0; \quad x = 0$ when $y = 1$

34. $\dfrac{dy}{dx} + 2y = 6; \quad x = 0$ when $y = 1$

35. $(xy^2 + x)\dfrac{dy}{dx} = \ln x; \quad x = 1$ when $y = 0$

36. $\dfrac{ds}{dt} = \sec s; \quad t = 0$ when $s = 0$

37. $y' = (1 - y)\cos x; \quad x = \pi/6$ when $y = 0$

38. $x\,dy = y \ln y\,dx; \quad x = 2$ when $y = e$

39. $y^2e^x\,dx + e^{-x}\,dy = y^2\,dx; \quad x = 0$ when $y = 2$

40. $2y \cos y\,dy - \sin y\,dy = y \sin y\,dx; \quad x = 0$ when $y = \pi/2$

30.3 INTEGRATING COMBINATIONS

NOTE ▶

Many differential equations cannot be solved by the method of separation of variables. Many other methods have been developed for solving such equations. One of these methods is based on the fact that *certain combinations of basic differentials can be integrated as a unit.* The following differentials suggest some of these combinations that may occur:

$$d(xy) = x\,dy + y\,dx \tag{30.3}$$

$$d(x^2 + y^2) = 2(x\,dx + y\,dy) \tag{30.4}$$

$$d\left(\frac{y}{x}\right) = \frac{x\,dy - y\,dx}{x^2} \tag{30.5}$$

$$d\left(\frac{x}{y}\right) = \frac{y\,dx - x\,dy}{y^2} \tag{30.6}$$

Equation (30.3) suggests that if the combination $x\,dy + y\,dx$ occurs in a differential equation, we should look for a function of xy as a solution. Equation (30.4) suggests that if the combination $x\,dx + y\,dy$ occurs, we should look for a function of $x^2 + y^2$. Equations (30.5) and (30.6) suggest that if either of the combinations $x\,dy - y\,dx$ or $y\,dx - x\,dy$ occurs, we should look for a function of y/x or x/y.

EXAMPLE 1 Solve the differential equation $x\,dy + y\,dx + xy\,dy = 0$.

By dividing through by xy, we have

$$\frac{x\,dy + y\,dx}{xy} + dy = 0$$

The left term is the differential of xy divided by xy. Thus, it integrates to $\ln xy$.

$$\frac{d(xy)}{xy} + dy = 0$$

for which the solution is

$$\ln xy + y = c$$

EXAMPLE 2 Solve the differential equation $y\,dx - x\,dy + x\,dx = 0$.

The combination of $y\,dx - x\,dy$ suggests that this equation might make use of either Eq. (30.5) or (30.6). This would require dividing through by x^2 or y^2. If we divide by

CAUTION ▶

y^2, the last term cannot be integrated, but *division by x^2 still allows integration of the last term.* Performing this division, we obtain

$$\frac{y\,dx - x\,dy}{x^2} + \frac{dx}{x} = 0$$

This left combination is the negative of Eq. (30.5). Thus, we have

$$-d\left(\frac{y}{x}\right) + \frac{dx}{x} = 0$$

for which the solution is $-\dfrac{y}{x} + \ln x = c$.

◀ **EXAMPLE 3** Solve the differential equation $(x^2 + y^2 + x)\, dx + y\, dy = 0$.

Regrouping the terms of this equation, we have

$$(x^2 + y^2)\, dx + (x\, dx + y\, dy) = 0 \qquad \text{divide each term by } x^2 + y^2$$

$$dx + \frac{x\, dx + y\, dy}{x^2 + y^2} = 0$$

The right term now can be put in the form of du/u (with $u = x^2 + y^2$) by multiplying each of the terms of the numerator by 2. This leads to

$$dx + \left(\frac{1}{2}\right) \frac{2x\, dx + 2y\, dy}{x^2 + y^2} = 0 \qquad d(x^2 + y^2) = 2x\, dx + 2y\, dy$$

$$x + \frac{1}{2}\ln(x^2 + y^2) = \frac{c}{2} \quad \text{or} \quad 2x + \ln(x^2 + y^2) = c$$

◀ **EXAMPLE 4** Find the particular solution of the differential equation

$$(x^3 + xy^2 + 2y)\, dx + (y^3 + x^2y + 2x)\, dy = 0$$

which satisfies the condition that $x = 1$ when $y = 0$.

Regrouping the terms of the equation, we have

$$x(x^2 + y^2)\, dx + y(x^2 + y^2)\, dy + 2(y\, dx + x\, dy) = 0$$

Factoring $x^2 + y^2$ from each of the first two terms gives

$$(x^2 + y^2)(x\, dx + y\, dy) + 2(y\, dx + x\, dy) = 0$$

$$\frac{1}{2}(x^2 + y^2)\overset{d(x^2+y^2)}{(2x\, dx + 2y\, dy)} + 2\overset{d(xy)}{(y\, dx + x\, dy)} = 0$$

$$\frac{1}{2}\left(\frac{1}{2}\right)(x^2 + y^2)^2 + 2xy + \frac{c}{4} = 0 \qquad \text{integrating}$$

$$(x^2 + y^2)^2 + 8xy + c = 0$$

Using the given condition gives $(1 + 0)^2 + 0 + c = 0$, or $c = -1$. The particular solution is then $(x^2 + y^2)^2 + 8xy = 1$.

NOTE ▶ The use of integrating combinations depends on proper recognition of the forms. It may take two or three arrangements to find the combination that leads to the solution. Of course, many equations cannot be arranged so as to give integrable combinations in all terms.

EXERCISES 30.3

In Exercises 1 and 2, make the given changes in the indicated examples of this section and then solve the resulting differential equations.

1. In Example 1, change the third term to $2xy^2\, dy$.

2. In Example 2, change the third term to $2\, dx$.

In Exercises 3–18, solve the given differential equations.

3. $x\, dy + y\, dx + x\, dx = 0$

4. $(2y + t)\, dy + y\, dt = 0$

5. $y\, dx - x\, dy + x^3\, dx = 2\, dx$

6. $x\, dy - y\, dx + y^2\, dx = 0$

7. $A^3\, dr + A^2 r\, dA + r\, dA = A\, dr$

8. $\sec(xy)\, dx + (x\, dy + y\, dx) = 0$

9. $x^3 y^4 (x\, dy + y\, dx) = 3\, dy$

10. $x\, dy + y\, dx + 4xy^3\, dy = 0$

11. $\sqrt{x^2 + y^2}\, dx - 2y\, dy = 2x\, dx$

12. $R\,dR + (R^2 + T^2 + T)\,dT = 0$

13. $\tan(x^2 + y^2)\,dy + x\,dx + y\,dy = 0$

14. $(x^2 + y^3)^2\,dy + 2x\,dx + 3y^2\,dy = 0$

(W) **15.** $y\,dy + (y^2 - x^2)\,dx = x\,dx$ (Explain your solution.)

16. $e^{x+y}(dx + dy) + 4x\,dx = 0$

17. $10x\,dy + 5y\,dx + 3y\,dy = 0$

18. $2(u\,dv + v\,du)\ln uv + 3u^3v\,du = 0$

In Exercises 19–24, find the particular solutions to the given differential equations that satisfy the given conditions.

19. $2(x\,dy + y\,dx) + 3x^2\,dx = 0;$ $x = 1$ when $y = 2$

20. $t\,dt + s\,ds = 2(t^2 + s^2)\,dt;$ $t = 1$ when $s = 0$

21. $y\,dx - x\,dy = y^3\,dx + y^2x\,dy;$ $x = 2$ when $y = 4$

22. $e^{x/y}(x\,dy - y\,dx) = y^4\,dy;$ $x = 0$ when $y = 2$

23. $2\csc(xy)\,dx + x\,dy + y\,dx = 0;$ $x = 0$ when $y = \pi/2$

24. $\sqrt[3]{x^2 + y^2}\,dy = 3(x\,dx + y\,dy);$ $x = 0$ when $y = 8$

30.4 THE LINEAR DIFFERENTIAL EQUATION OF THE FIRST ORDER

There is one type of differential equation of the first order and first degree for which an integrable combination can always be found. *It is the* **linear differential equation** *of the first order and is of the form*

$$dy + Py\,dx = Q\,dx \tag{30.7}$$

NOTE ▶ *where P and Q are functions of x only.* This type of equation occurs widely in applications.

If each side of Eq. (30.7) is multiplied by $e^{\int P\,dx}$, it becomes integrable, since the left side becomes of the form du with $u = ye^{\int P\,dx}$ and the right side is a function of x only. This is shown by finding the differential of $ye^{\int P\,dx}$. Thus,

$$d(ye^{\int P\,dx}) = e^{\int P\,dx}(dy + Py\,dx)$$

In finding the differential of $\int P\,dx$ we use the fact that, by definition, these are reverse processes. Thus, $d(\int P\,dx) = P\,dx$. Therefore, if each side is multiplied by $e^{\int P\,dx}$, the left side may be immediately integrated to $ye^{\int P\,dx}$, and the right-side integration may be indicated. The solution becomes

$$ye^{\int P\,dx} = \int Qe^{\int P\,dx}\,dx + c \tag{30.8}$$

◀ **EXAMPLE 1** Solve the differential equation $dy + \left(\dfrac{2}{x}\right)y\,dx = 4x\,dx$.

This equation fits the form of Eq. (30.7) with $P = 2/x$ and $Q = 4x$. The first expression to find is $e^{\int P\,dx}$. In this case this is

$$e^{\int (2/x)\,dx} = e^{2\ln x} = e^{\ln x^2} = x^2 \qquad \text{see text comments following example}$$

The left side integrates to yx^2, while the right side becomes $\int 4x(x^2)\,dx$. Thus,

$$ye^{\int P\,dx} = \int Qe^{\int P\,dx}\,dx + c$$

$$y(x^2) = \int (4x)(x^2)\,dx + c \qquad \text{using Eq. (30.8)}$$

$$yx^2 = \int 4x^3\,dx + c = x^4 + c \qquad \text{integrating}$$

$$y = x^2 + cx^{-2}$$

NOTE ▶ As in Example 1, in finding the factor $e^{\int P\,dx}$ we often obtain an expression of the form $e^{\ln u}$. Using the properties of logarithms, we now show that $e^{\ln u} = u$:

$$\text{Let } y = e^{\ln u}$$

$$\ln y = \ln e^{\ln u} = \ln u(\ln e) = \ln u$$

$$y = u \qquad \text{or} \qquad e^{\ln u} = u$$

NOTE ▶ Also, in finding $e^{\int P\,dx}$, the constant of integration in the exponent $\int P\,dx$ can always be taken as zero, as we did in Example 1. To show why this is so, let $P = 2/x$ as in Example 1:

$$e^{\int (2/x)\,dx} = e^{\ln x^2 + c} = (e^{\ln x^2})(e^c) = x^2 e^c$$

The solution to the differential equation, as given in Eq. (30.8), is then

$$y(x^2)(e^c) = \int 4x(x^2)(e^c)\,dx + c_1 e^c$$

Regardless of the value of c, the factor e^c can be divided out. Therefore, it is convenient to let $c = 0$ and have $e^c = 1$.

◀ **EXAMPLE 2** Solve the differential equation $x\,dy - 3y\,dx = x^3\,dx$.

Putting this equation in the form of Eq. (30.7) by dividing through by x gives $dy - (3/x)y\,dx = x^2\,dx$. Here, $P = -3/x$, $Q = x^2$, and the factor $e^{\int P\,dx}$ becomes

$$e^{\int (-3/x)\,dx} = e^{-3\ln x} = e^{\ln x^{-3}} = x^{-3}$$

Therefore,

$$y e^{\int P\,dx} = \int Q e^{\int P\,dx}\,dx + c$$

$$y x^{-3} = \int x^2(x^{-3})\,dx + c \qquad \text{using Eq. (30.8)}$$

$$= \int x^{-1}\,dx + c = \ln x + c$$

$$y = x^3(\ln x + c)$$

◀ **EXAMPLE 3** Solve the differential equation $dy + y\,dx = x\,dx$.

Here, $P = 1$, $Q = x$, and $e^{\int P\,dx} = e^{\int (1)\,dx} = e^x$. Therefore,

$$y e^x = \int x e^x\,dx + c = e^x(x - 1) + c \qquad \begin{array}{l}\text{using Eq. (30.8) and integrating}\\ \text{by parts or tables}\end{array}$$

$$y = x - 1 + c e^{-x}$$

◀ **EXAMPLE 4** Solve the differential equation $\cos x \dfrac{dy}{dx} = 1 - y \sin x$.

Writing this in the form of Eq. (30.7), we have

$$dy + y \tan x\,dx = \sec x\,dx \qquad \text{dividing by } \cos x$$

Thus, with $P = \tan x$, we have

$$e^{\int P\,dx} = e^{\int \tan x\,dx} = e^{-\ln \cos x} = \sec x \qquad \text{see the first NOTE after Example 1}$$

$$y \sec x = \int \sec^2 x\,dx = \tan x + c \qquad \text{using Eq. (30.8)}$$

$$y = \sin x + c \cos x$$

◀ EXAMPLE 5   For the differential equation $dy = (1 - 2y)x\,dx$, find the particular solution such that $x = 0$ when $y = 2$.

The solution proceeds as follows:

$$dy + 2xy\,dx = x\,dx \qquad \text{form of Eq. (30.7)}$$

$$e^{\int P\,dx} = e^{\int 2x\,dx} = e^{x^2} \qquad \text{find } e^{\int P\,dx}$$

$$ye^{x^2} = \int xe^{x^2}\,dx \qquad \text{using Eq. (30.8)}$$

$$= \frac{1}{2}e^{x^2} + c \qquad \text{general solution}$$

$$(2)(e^0) = \frac{1}{2}(e^0) + c, \qquad 2 = \frac{1}{2} + c, \qquad c = \frac{3}{2} \qquad \begin{array}{l}x = 0,\ y = 2;\\ \text{evaluate } c\end{array}$$

$$ye^{x^2} = \frac{1}{2}e^{x^2} + \frac{3}{2} \qquad \text{substitute } c = \frac{3}{2}$$

$$y = \frac{1}{2}(1 + 3e^{-x^2}) \qquad \text{particular solution} \quad \blacktriangleright$$

EXERCISES 30.4

In Exercises 1 and 2, make the given changes in the indicated examples of this section and then solve the resulting differential equations.

1. In Example 1, change the right side to $3\,dx$.

2. In Example 3, change the right side to $2\,dx$.

In Exercises 3–28, solve the given differential equations.

3. $dy + y\,dx = e^{-x}\,dx$

4. $dy + 3y\,dx = e^{-3x}\,dx$

5. $dy + 2y\,dx = e^{-4x}\,dx$

6. $di + i\,dt = e^{-t}\cos t\,dt$

7. $\dfrac{dy}{dx} - 2y = 4$

8. $2\dfrac{dy}{dx} = 5 - 6y$

9. $x\,dy - y\,dx = 3x\,dx$

10. $x\,dy + 3y\,dx = dx$

11. $2x\,dy + y\,dx = 8x^3\,dx$

12. $3x\,dy - y\,dx = 9x\,dx$

13. $dr + r\cot\theta\,d\theta = d\theta$

14. $y' = x^2 y + 3x^2$

15. $\sin x\dfrac{dy}{dx} = 1 - y\cos x$

16. $\dfrac{dv}{dt} - \dfrac{v}{t} = \ln t$

17. $y' + y = 3$

18. $y' + 2y = \sin x$

19. $ds = (te^{4t} + 4s)\,dt$

20. $y' - 2y = 2e^{2x}$

21. $y' = x^3(1 - 4y)$

22. $y' + y\tan x = -\sin x$

23. $x\dfrac{dy}{dx} = y + (x^2 - 1)^2$

24. $dy = dt - \dfrac{y\,dt}{(1 + t^2)\tan^{-1} t}$

25. $x\,dy + (1 - 3x)y\,dx = 3x^2 e^{3x}\,dx$

26. $(1 + x^2)\,dy + xy\,dx = x\,dx$

27. $\tan\theta\dfrac{dr}{d\theta} - r = \tan^2\theta$

28. $y' + y = y^2$ (Solve by letting $y = 1/u$ and solving the resulting linear equation for u.)

Ⓦ *In Exercises 29 and 30, solve the given differential equations. Explain how each can be solved using either of two different methods.*

29. $y' = 2(1 - y)$

30. $x\,dy = (2x - y)\,dx$

In Exercises 31–36, find the indicated particular solutions of the given differential equations.

31. $\dfrac{dy}{dx} + 2y = e^{-x};\quad x = 0$ when $y = 1$

32. $dq - 4q\,du = 2\,du;\quad q = 2$ when $u = 0$

33. $y' + 2y\cot x = 4\cos x;\quad x = \pi/2$ when $y = 1/3$

34. $y'\sqrt{x} + \tfrac{1}{2}y = e^{\sqrt{x}};\quad x = 1$ when $y = 3$

35. $(\sin x)y' + y = \tan x;\quad x = \pi/4$ when $y = 0$

36. $f(x)\,dy + 2yf'(x)\,dx = f(x)f'(x)\,dx;\ f(x) = -1$ when $y = 3$

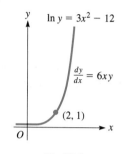

Fig. 30.2

30.5 ELEMENTARY APPLICATIONS

The differential equations of the first order and first degree we have discussed thus far have numerous applications in geometry and the various fields of technology. In this section we illustrate some of these applications.

EXAMPLE 1 The slope of a given curve is given by the expression $6xy$. Find the equation of the curve if it passes through the point $(2, 1)$.

Since the slope is $6xy$, the differential equation for the curve is

$$\frac{dy}{dx} = 6xy$$

We now want to find the particular solution of this equation for which $x = 2$ when $y = 1$. The solution follows:

$$\frac{dy}{y} = 6x\, dx \qquad \text{separate variables}$$

$$\ln y = 3x^2 + c \qquad \text{general solution}$$

$$\ln 1 = 3(2^2) + c \qquad \text{evaluate } c$$

$$0 = 12 + c, \qquad c = -12$$

$$\ln y = 3x^2 - 12 \qquad \text{particular solution}$$

The graph of this solution is shown in Fig. 30.2.

EXAMPLE 2 *A curve that intersects all members of a family of curves at right angles is called an* **orthogonal trajectory** *of the family.* Find the equations of the orthogonal trajectories of the parabolas $x^2 = cy$. As before, each value of c gives us a particular member of the family.

The derivative of the given equation is $dy/dx = 2x/c$. This equation contains the constant c, which depends on the point (x, y) on the parabola. ***Eliminating this constant between the equations of the parabolas and the derivative,*** we have

CAUTION ▶

$$c = \frac{x^2}{y} \qquad \frac{dy}{dx} = \frac{2x}{c} = \frac{2x}{x^2/y} \quad \text{or} \quad \frac{dy}{dx} = \frac{2y}{x}$$

This equation gives a general expression for the slope of any of the members of the family. For a curve to be perpendicular, its slope must equal the negative reciprocal of this expression, or the slope of the orthogonal trajectories must be

$$\left.\frac{dy}{dx}\right|_{\text{OT}} = -\frac{x}{2y} \quad \longleftarrow \text{ this equation must not contain the constant } c$$

Solving this differential equation gives the family of orthogonal trajectories.

$$2y\, dy = -x\, dx$$

$$y^2 = -\frac{x^2}{2} + \frac{c}{2}$$

$$2y^2 + x^2 = c \qquad \text{orthogonal trajectories}$$

Thus the orthogonal trajectories are ellipses. Note in Fig. 30.3 that each parabola intersects each ellipse at right angles.

Some calculators have a feature by which a family of curves can be graphed.

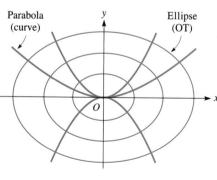

Fig. 30.3

Solving a Word Problem

See the chapter introduction.

Radioactivity was discovered in 1898 by the French physicist Henri Becquerel (1852–1908). Carbon dating was developed in 1947 by the U.S. chemist Willard Libby (1908–1980).

EXAMPLE 3 Radioactive elements decay at rates that are proportional to the amount of the element present. Carbon-14 decays such that one-half of an original amount decays into other forms in about 5730 years. By measuring the proportion of carbon-14 in remains at an ancient site, the approximate age of the remains can be determined. This method, called *carbon dating,* is used to determine the dates of prehistoric events.

The analysis of some wood at the site of the ancient city of Troy showed that the concentration of carbon-14 was 67.8% of the concentration that new wood would have. Determine the equation relating the amount of carbon-14 present with the time and then determine the age of the wood at the site of Troy.

Let N_0 be the original amount and N be the amount present at any time t (in years). The rate of decay can be expressed as a derivative. Therefore, since the rate of change is proportional to N, we have the equation

$$\frac{dN}{dt} = kN$$

Solving this differential equation, we have

$\dfrac{dN}{N} = k\,dt$	separate variables
$\ln N = kt + \ln c$	general solution
$\ln N_0 = k(0) + \ln c$	$N = N_0$ for $t = 0$
$c = N_0$	solve for c
$\ln N = kt + \ln N_0$	substitute N_0 for c
$\ln N - \ln N_0 = kt$	use properties of logarithms
$\ln \dfrac{N}{N_0} = kt$	
$N = N_0 e^{kt}$	exponential form

Now, using the condition that one-half of carbon-14 decays in 5730 years, we have $N = N_0/2$ when $t = 5730$ years. This gives

$$\frac{N_0}{2} = N_0 e^{5730k} \quad \text{or} \quad \frac{1}{2} = (e^k)^{5730}$$

CAUTION

$$e^k = 0.5^{1/5730}$$

Therefore, the equation relating N and t is

$$N = N_0(0.5)^{t/5730}$$

Since the present concentration is $N = 0.678N_0$, we can solve for t:

$$0.678N_0 = N_0(0.5)^{t/5730} \quad \text{or} \quad 0.678 = 0.5^{t/5730}$$

$$\ln 0.678 = \ln 0.5^{t/5730}, \quad \ln 0.678 = \frac{t}{5730}\ln 0.5 \quad \text{solving an exponential equation}$$

$$t = 5730\frac{\ln 0.678}{\ln 0.5} = 3210 \text{ years}$$

Therefore, the wood at the Troy site is about 3210 years old.

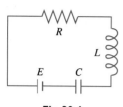

Fig. 30.4

◀ **EXAMPLE 4** The general equation relating the current i, voltage E, inductance L, capacitance C, and resistance R of a simple electric circuit (see Fig. 30.4) is

$$L\frac{di}{dt} + Ri + \frac{q}{C} = E \qquad (30.9)$$

where q is the charge on the capacitor. Find the general expression for the current in a circuit containing an inductance, a resistance, and a voltage source if $i = 0$ when $t = 0$.

The differential equation for this circuit is

$$L\frac{di}{dt} + Ri = E$$

Using the method of the linear differential equation of the first order, we have the equation

$$di + \frac{R}{L}i\,dt = \frac{E}{L}\,dt$$

The factor $e^{\int P\,dt}$ is $e^{\int (R/L)\,dt} = e^{(R/L)t}$. This gives

$$ie^{(R/L)t} = \frac{E}{L}\int e^{(R/L)t}\,dt = \frac{E}{R}e^{(R/L)t} + c$$

Letting the current be zero for $t = 0$, we have $c = -E/R$. The result is

$$ie^{(R/L)t} = \frac{E}{R}e^{(R/L)t} - \frac{E}{R}$$

$$i = \frac{E}{R}\left(1 - e^{-(R/L)t}\right)$$

(We can see that $i \to E/R$ as $t \to \infty$. In practice, the exponential term becomes negligible very quickly.) ◗

Solving a Word Problem

◀ **EXAMPLE 5** Fifty liters of brine originally containing 3.00 kg of salt are in a tank into which 2.00 L of water run each minute with the same amount of mixture running out each minute. How much salt is in the tank after 10.0 min?

Let $x =$ the number of kilograms of salt in the tank after t min. Each liter of brine contains $x/50$ kg of salt, and in time dt, $2\,dt$ L *of mixture leave the tank* **with** $(\boldsymbol{x/50})(\boldsymbol{2\,dt})$ **kg** *of salt*. The amount of salt that is leaving may also be written as $-dx$ (the minus sign is included to show that x is decreasing). Thus,

CAUTION ◗

$$-dx = \frac{2x\,dt}{50} \quad \text{or} \quad \frac{dx}{x} = -\frac{dt}{25}$$

This leads to $\ln x = -(t/25) + \ln c$. Using the fact that $x = 3.00$ kg when $t = 0$, we find that $\ln 3.00 = \ln c$, or $c = 3.00$. Therefore,

$$x = 3.00e^{-t/25}$$

is the general expression for the amount of salt in the tank at time t. Therefore, when $t = 10.0$ min, we have

$$x = 3.00e^{-10/25} = 3.00e^{-0.4} = 3.00(0.670) = 2.01 \text{ kg}$$

There are 2.01 kg of salt in the tank after 10.0 min. (Although the data were given with three significant digits, we did not use all significant digits in writing the equations that were used.) ◗

Solving a Word Problem

EXAMPLE 6 An object moving through (or across) a resisting medium often experiences a retarding force that is approximately proportional to the velocity as well as the force that causes the motion. An example of this is a ball that falls due to the force of gravity and air resistance produces a retarding force. Applying Newton's laws of motion (from physics) to this ball leads to the equation

$$m\frac{dv}{dt} = F - kv \tag{30.10}$$

where m is the mass of the object, v is the velocity of the object, t is the time, F is the force causing the motion, and k ($k > 0$) is a constant. The quantity kv is the retarding force.

> There is at least some resistance to the motion of any object moving through a medium (not a vacuum). The exact nature of the resistance is not usually known.

We assume that these conditions hold for a falling object whose mass is 5.00 kg and experiences a force (its own weight) of 49.0 N. The object starts from rest, and the air causes a retarding force numerically equal to 0.200 times the velocity.

Substituting in Eq. (30.10) and then solving the differential equations, we have

$$5\frac{dv}{dt} = 49 - 0.2\,v$$

$$\frac{5dv}{49 - 0.2v} = dt \qquad \text{separate variables}$$

$$-25\ln(49 - 0.2v) = t - 25\ln c \qquad \text{integrate}$$

$$\ln(49 - 0.2v) = -\frac{t}{25} + \ln c \qquad \text{solve for } v$$

$$\ln\frac{49 - 0.2v}{c} = -\frac{t}{25}$$

$$49 - 0.2v = ce^{-t/25}$$

$$0.2v = 49 - ce^{-t/25}$$

$$v = 5(49 - ce^{-t/25}) \qquad \text{general solution}$$

Since the object started from rest, $v = 0$ when $t = 0$. Thus,

> The data were given to three significant digits, but not all significant digits were used in writing the equations.

$$0 = 5(49 - c) \quad \text{or} \quad c = 49 \qquad \text{evaluate } c$$

$$v = 245(1 - e^{-t/25}) \qquad \text{particular solution}$$

$$= 245(1 - e^{-0.200}) = 245(1 - 0.819) \qquad \text{evaluating } v \text{ for } t = 5.00 \text{ s}$$

$$= 44.4 \text{ m/s}$$

After 5.00 s the velocity is 44.4 m/s. Without the air resistance, the velocity would be about 49.0 m/s.

EXERCISES 30.5

In Exercises 1–4, make the given changes in the indicated examples of this section, and then solve the resulting problems.

1. In Example 2, change $x^2 = cy$ to $y^2 = cx$.

2. In Example 4, if the term $L\frac{di}{dt}$ is deleted (no inductance) in Eq. (30.9), find the expression for the charge in the circuit. (There is a capacitance C, and $i = dq/dt$.) The initial charge is zero.

3. In Example 5, change 2.00 L to 1.00 L.

4. In Example 6, change 0.200 to 0.100 (for the retarding force).

In Exercises 5–8, find the equation of the curve for the given slope and point through which it passes. Use a graphing calculator to display the curve.

5. Slope given by $2x/y$; passes through $(2, 3)$

6. Slope given by $-y/(x + y)$; passes through $(-1, 3)$

7. Slope given by $y + x$; passes through $(0, 1)$

8. Slope given by $-2y + e^{-x}$; passes through $(0, 2)$

In Exercises 9–12, find the equation of the orthogonal trajectories of the curves for the given equations. Use a graphing calculator to display at least two members of the family of curves and at least two of the orthogonal trajectories.

9. The exponential curves $y = ce^x$

10. The cubic curves $y = cx^3$

11. The curves $y = c(\sec x + \tan x)$

12. The family of circles, all with centers at the origin

In Exercises 13–48, solve the given problems by solving the appropriate differential equation.

13. The isotope neon-23 decays such that half of an original amount disintegrates in 40.0 s. Find the relation between the amount present and the time and then find the percent remaining after 60.0 s.

14. Radium-226 decays such that 10% of the original amount disintegrates in 246 years. Find the half-life (the time for one-half of the original amount to disintegrate) of radium-226.

15. A possible health hazard in the home is radon gas. It is radioactive and about 90.0% of an original amount disintegrates in 12.7 days. Find the half-life of radon gas. (The problem with radon is that it is a gas and is being continually produced by the radioactive decay of minute amounts of radioactive radium found in the soil and rocks of an area.)

16. Noting Example 3, another element used to date more recent events is helium-3, which has a half-life of 12.3 years. If a building has 5.0% of helium-3 that a new building would have, about how old is the building?

17. A radioactive element leaks from a nuclear power plant at a constant rate r, and it decays at a rate that is proportional to the amount present. Find the relation between the amount N present in the environment in which it leaks and the time t, if $N = 0$ when $t = 0$.

18. Most use of the pesticide DDT was banned in the United States in 1972. It has been found that 19% of an initial amount of DDT is degraded into harmless products in 10 years. If no DDT has been used in an area since 1972, in what year will the concentration become only 20% of the 1972 amount?

19. The growth of the population P of a nation with a constant immigration rate I may be expressed as $\dfrac{dP}{dt} = kP + I$, where t is in years. If the population of Canada in 2001 was 31.5 million and about 0.1 million immigrants enter Canada each year, what will the population of Canada be in 2015, given that the growth rate k is about 1.0% (0.010) annually?

20. Assuming that the natural environment of the earth is limited and that the maximum population it can sustain is M, the rate of growth of the population P is given by the *logistic* differential equation $\dfrac{dP}{dt} = kP(M - P)$. Using this equation for the earth, if $P = 6.0$ billion in 2000, $k = 0.0038$, $M = 50$ billion, what will be the population of the earth in 2025?

21. The rate of change of the radial stress S on the walls of a pipe with respect to the distance r from the axis of the pipe is given by $r\dfrac{dS}{dr} = 2(a - S)$, where a is a constant. Solve for S as a function of r.

22. The velocity v of a meteor approaching the earth is given by $v\dfrac{dv}{dr} = -\dfrac{GM}{r^2}$, where r is the distance from the center of the earth, M is the mass of the earth, and G is a universal gravitational constant. If $v = 0$ for $r = r_0$, solve for v as a function of r.

23. Assume that the rate at which highway construction increases is directly proportional to the total length M of all highways already completed at time t (in years). Solve for M as a function of t if $M = 5250$ km for a certain region when $t = 0$ and $M = 5460$ km for $t = 2.00$ years.

24. The marginal profit function gives the change in the total profit P of a business due to a change in the business, such as adding new machinery or reducing the size of the sales staff. A company determines that the marginal profit dP/dx is $e^{-x^2} - 2Px$, where x is the amount invested in new machinery. Determine the total profit (in thousands of dollars) as a function of x, if $P = 0$ for $x = 0$.

25. According to Newton's law of cooling, the rate at which a body cools is proportional to the difference in temperature between it and the surrounding medium. Assuming Newton's law holds, how long will it take a cup of hot water, initially at 90°C, to cool to 40°C if the room temperature is 25°C, if it cools to 60°C in 5.0 min?

26. An object whose temperature is 100°C is placed in a medium whose temperature is 20°C. The temperature of the object falls to 50°C in 10 min. Express the temperature T of the object as a function of time t (in min). (Assume it cools according to Newton's law of cooling as stated in Exercise 25.)

27. If interest in a bank account is compounded continuously, the amount grows at a rate that is proportional to the amount present in the account. Interest that is compounded daily very closely approximates this situation. Determine the amount in an account after one year if $1000 is placed in the account and it pays 4% interest per year, compounded continuously.

28. The rate of change in the intensity I of light below the surface of the ocean with respect to the depth y is proportional to I. If the intensity at 5.0 m is 50% of the intensity I_0 at the surface, at what depth is the intensity 15% of I_0?

29. If the current in an RL circuit with a voltage source E is zero when $t = 0$ (see Example 4), show that $\lim\limits_{t \to \infty} i = E/R$. See Fig. 30.5.

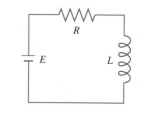

Fig. 30.5

30. If a circuit contains only an inductance and a resistance, with $L = 2.0$ H and $R = 30\ \Omega$, find the current i as a function of time t if $i = 0.020$ A when $t = 0$. See Fig. 30.6.

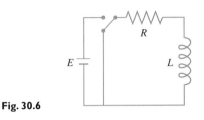

Fig. 30.6

31. An amplifier circuit contains a resistance R, an inductance L, and a voltage source $E \sin \omega t$. Express the current in the circuit as a function of the time t if the initial current is zero.

32. A radio transmitter circuit contains a resistance of $2.0\ \Omega$, a variable inductor of $100 - t$ henrys, and a voltage source of 4.0 V. Find the current i in the circuit as a function of the time t for $0 \le t \le 100$ s if the initial current is zero.

33. If a circuit contains only a resistance R and a capacitance C, find the expression relating the charge on the capacitor in terms of the time if $i = dq/dt$ and $q = q_0$ when $t = 0$. See Fig. 30.7.

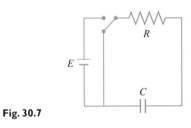

Fig. 30.7

34. A circuit contains a 4.0-μF capacitor, a 450-Ω resistor, and a 20.0-mV voltage source. If the charge on the capacitor is 20.0 nC when $t = 0$, find the charge after 0.010 s. ($i = dq/dt$.)

35. One hundred liters of brine originally containing 4.0 kg of salt are in a tank into which 5.0 L of water run each minute. The same amount of mixture from the tank leaves each minute. How much salt is in the tank after 20 min?

36. Repeat Exercise 35 with the change that the water entering the tank contains 0.10 kg of salt per liter.

37. An object falling under the influence of gravity has a variable acceleration given by $9.8 - v$, where v represents the velocity. If the object starts from rest, find an expression for the velocity in terms of the time. Also find the limiting value of the velocity (find $\lim_{t\to\infty} v$).

38. In a ballistics test, a bullet is fired into a sandbag. The acceleration of the bullet within the sandbag is $-15\sqrt{v}$, where v is the velocity (in m/s). When will the bullet stop if it enters the sandbag at 300 m/s?

39. A boat with a mass of 150 kg is being towed at 8.0 km/h. The tow rope is then cut, and a motor that exerts a force of 80 N on the boat is started. If the water exerts a retarding force that numerically equals twice the velocity, what is the velocity of the boat 3.0 min later?

40. A parachutist is falling at a rate of 60.0 m/s when her parachute opens. If the air resists the fall with a force equal to $5v^2$, find the velocity as a function of time. The person and equipment have a combined mass of 100 kg (weight is 980 N).

41. For each cycle, a roller mechanism follows a path described by $y = 2x - x^2$, $y \ge 0$, such that $dx/dt = 6t - 3t^2$. Find x and y (in cm) in terms of the time t (in s) if x and y are zero for $t = 0$.

42. In studying the flow of water in a stream, it is found that an object follows the hyperbolic path $y(x + 1) = 10$ such that $(t + 1)\,dx = (x - 2)\,dt$. Find x and y (in m) in terms of time t (in s) if $x = 4$ m and $y = 2$ m for $t = 0$.

43. The rate of change of air pressure p (in kPa) with respect to height h (in m) is approximately proportional to the pressure. If the pressure is 100 kPa when $h = 0$ and $p = 80$ kPa when $h = 2000$ m, find the expression relating pressure and height.

44. Water flows from a vertical cylindrical storage tank through a hole of area A at the bottom of the tank. The rate of flow is $2.6A\sqrt{h}$, where h is the distance (in m) from the surface of the water to the hole. If h changes from 9.0 m to 8.0 m in 16 min, how long will it take the tank to empty? See Fig. 30.8.

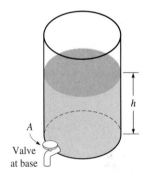

Fig. 30.8

45. Assume that the rate of depreciation of an object is proportional to its value at any time t. If a car costs $\$16\,500$ new and its value 3 years later is $\$9850$, what is its value 11 years after it was purchased?

46. Assume that sugar dissolves at a rate proportional to the undissolved amount. If there are initially 525 g of sugar and 225 g remain after 4.00 min, how long does it take to dissolve 375 g?

47. Fresh air is being circulated into a room whose volume is 120 m^3. Under specified conditions the number of cubic meters x of carbon dioxide present at any time t (in min) is found by solving the differential equation $dx/dt = 1 - 5.0\,x$. Find x as a function of t if $x = 0.35$ m^3 when $t = 0$.

48. The lines of equal potential in a field of force are all at right angles to the lines of force. In an electric field of force caused by charged particles, the lines of force are given by $x^2 + y^2 = cx$. Find the equation of the lines of equal potential. Use a graphing calculator to view a few members of the lines of force and those of equal potential.

30.6 HIGHER-ORDER HOMOGENEOUS EQUATIONS

Another important type of differential equation is the linear differential equation of higher order with constant coefficients. First, we shall briefly describe the general higher-order equation and the notation we shall use with this type of equation.

The general **linear differential equation of the *n*th order** *is of the form*

$$a_0 \frac{d^n y}{dx^n} + a_1 \frac{d^{n-1} y}{dx^{n-1}} + \cdots + a_{n-1} \frac{dy}{dx} + a_n y = b \qquad (30.11)$$

where the a's and b are either functions of x or constants.

For convenience of notation, the *n*th derivative with respect to the independent variable will be denoted by D^n. Here, D is called the **operator,** since it denotes the *operation* of differentiation. Using this notation with x as the independent variable, Eq. (30.11) becomes

$$a_0 D^n y + a_1 D^{n-1} y + \cdots + a_{n-1} Dy + a_n y = b \qquad (30.12)$$

If b = 0, the general linear equation is called **homogeneous,** *and if b ≠ 0, it is called* **nonhomogeneous.** Both types of equations have important applications.

◄ EXAMPLE 1 Using the operator form of Eq. (30.12), the differential equation

$$\frac{d^3 y}{dx^3} - 3 \frac{d^2 y}{dx^2} + 4 \frac{dy}{dx} - 2y = e^x \sec x$$

is written as

$$D^3 y - 3D^2 y + 4Dy - 2y = e^x \sec x$$

This equation is nonhomogeneous since $b = e^x \sec x$. ▶

Although the a's may be functions of x, we shall restrict our attention to the cases in which they are constants. We shall, however, consider both homogeneous equations and nonhomogeneous equations. Also, since second-order linear equations are the most commonly found in elementary applications, we shall devote most of our attention to them. The methods used to solve second-order equations may be applied to equations of higher order, and we shall consider certain of these higher-order equations.

SECOND-ORDER HOMOGENEOUS EQUATIONS WITH CONSTANT COEFFICIENTS

Using the operator notation, *a second-order, linear, homogeneous differential equation with constant coefficients is one of the form*

$$a_0 D^2 y + a_1 Dy + a_2 y = 0 \qquad (30.13)$$

where the a's are constants. The following example indicates the kind of solution we should expect for this type of equation.

◀ **EXAMPLE 2** Solve the differential equation $D^2y - Dy - 2y = 0$.

First, we put this equation in the form $(D^2 - D - 2)y = 0$. This is another way of saying that we are to take the second derivative of y, subtract the first derivative, and finally subtract twice the function. This expression may now be factored as $(D - 2)(D + 1)y = 0$. (We will not develop the algebra of the operator D. However, most such algebraic operations can be shown to be valid.) This formula tells us to find the first derivative of the function and add this to the function. Then twice this result is to be subtracted from the derivative of this result. If we let $z = (D + 1)y$, which is valid since $(D + 1)y$ is a function of x, we have $(D - 2)z = 0$. This equation is easily solved by separation of variables. Thus,

$$\frac{dz}{dx} - 2z = 0 \qquad \frac{dz}{z} - 2\,dx = 0 \qquad \ln z - 2x = \ln c_1$$

$$\ln \frac{z}{c_1} = 2x \quad \text{or} \quad z = c_1 e^{2x}$$

Replacing z by $(D + 1)y$, we have

$$(D + 1)y = c_1 e^{2x}$$

This is a linear equation of the first order. Then,

$$dy + y\,dx = c_1 e^{2x}\,dx$$

The factor $e^{\int P\,dx}$ is $e^{\int dx} = e^x$. And so,

$$ye^x = \int c_1 e^{3x}\,dx = \frac{c_1}{3}e^{3x} + c_2 \qquad \text{using Eq. (30.8)}$$

$$y = c_1' e^{2x} + c_2 e^{-x}$$

where $c_1' = \frac{1}{3}c_1$. This example indicates that solutions of the form e^{mx} result for this equation. ▶

Based on the result of Example 2, assume that an equation of the form of Eq. (30.13) has a particular solution ce^{mx}. Substituting this into Eq. (30.13) gives

$$a_0 cm^2 e^{mx} + a_1 cm e^{mx} + a_2 ce^{mx} = 0$$

Since the exponential function $e^{mx} > 0$ for all real x, this equation will be satisfied if m is a root of the equation

Auxiliary Equation

$$a_0 m^2 + a_1 m + a_2 = 0 \tag{30.14}$$

Equation (30.14) is called the **auxiliary equation** *of Eq. (30.13). Note that it may be formed directly by inspection from Eq. (30.13).*

There are two roots of the auxiliary Eq. (30.14), and there are two arbitrary constants in the solution of Eq. (30.13). These factors lead us to *the general solution of Eq. (30.13), which is*

General Solution

$$y = c_1 e^{m_1 x} + c_2 e^{m_2 x} \tag{30.15}$$

where m_1 and m_2 are the solutions of Eq. (30.14). We see that this is in agreement with the results of Example 2.

◀ EXAMPLE 3   Solve the differential equation $D^2y - 5Dy + 6y = 0$.

From this operator form of the differential equation, we write the auxiliary equation

$$m^2 - 5m + 6 = 0$$

Solving the auxiliary equation, we have

$$(m - 3)(m - 2) = 0$$
$$m_1 = 3 \qquad m_2 = 2$$

Now, using Eq. (30.15), we write the solution of the differential equation as

$$y = c_1e^{3x} + c_2e^{2x}$$

It makes no difference which constant is written with each exponential term.   ▶

◀ EXAMPLE 4   Solve the differential equation $y'' = 6y'$.

We first rewrite this equation using the D notation for derivatives. Also, we want to write it in the proper form of a homogeneous equation. This gives us

$$D^2y - 6Dy = 0$$

Proceeding with the solution, we have

$$m^2 - 6m = 0 \qquad \text{auxiliary equation}$$
$$m(m - 6) = 0 \qquad \text{solve for } m$$
$$m_1 = 0 \qquad m_2 = 6$$
$$y = c_1e^{0x} + c_2e^{6x} \qquad \text{using Eq. (30.15)}$$

Since $e^{0x} = 1$, we have

$$y = c_1 + c_2e^{6x} \quad \text{general solution}$$ ▶

Third-Order Equation   ◀ EXAMPLE 5   Solve the differential equation $2\dfrac{d^3y}{dx^3} + \dfrac{d^2y}{dx^2} - 7\dfrac{dy}{dx} = 0$.

Although this is a third-order equation, the *method* of solution is the same as in the previous examples. Using the D-notation for derivatives, we have

$$2D^3y + D^2y - 7Dy = 0$$

This means the auxiliary equation is

$$2m^3 + m^2 - 7m = 0, \qquad m(2m^2 + m - 7) = 0$$

We can see that one root is $m = 0$. The quadratic factor is not factorable, but we can find the roots from it by using the quadratic formula. This gives

$$m = \frac{-1 \pm \sqrt{1 + 56}}{4} = \frac{-1 \pm \sqrt{57}}{4}$$

Since there are *three* roots, there are *three* arbitrary constants. Following the solution indicated by Eq. (30.15), we have

$$y = c_1e^{0x} + c_2e^{(-1+\sqrt{57})x/4} + c_3e^{(-1-\sqrt{57})x/4}$$

Again, $e^{0x} = 1$. Also, factoring $e^{-x/4}$ from the second and third terms, we have

$$y = c_1 + e^{-x/4}(c_2e^{x\sqrt{57}/4} + c_3e^{-x\sqrt{57}/4})$$ ▶

◀ EXAMPLE 6   Solve the differential equation $D^2y - 2Dy - 15y = 0$ and find the particular solution that satisfies the conditions $Dy = 2$ and $y = -1$, when $x = 0$. (It is necessary to give two conditions since there are two constants to evaluate.)

We have

$$m^2 - 2m - 15 = 0, \qquad (m - 5)(m + 3) = 0$$
$$m_1 = 5 \qquad m_2 = -3$$
$$y = c_1 e^{5x} + c_2 e^{-3x}$$

CAUTION ▶   This equation is the general solution. In order to evaluate the constants c_1 and c_2, **we use the given conditions to find two simultaneous equations in c_1 and c_2.** These are then solved to determine the particular solution. Thus,

$$y' = 5c_1 e^{5x} - 3c_2 e^{-3x}$$

Using the given conditions in the general solution and its derivative, we have

$$c_1 + c_2 = -1 \qquad y = -1 \text{ when } x = 0$$
$$5c_1 - 3c_2 = 2 \qquad Dy = 2 \text{ when } x = 0$$

The solution to this system of equations is $c_1 = -\frac{1}{8}$ and $c_2 = -\frac{7}{8}$. The particular solution becomes

$$y = -\frac{1}{8}e^{5x} - \frac{7}{8}e^{-3x} \quad \text{or} \quad 8y + e^{5x} + 7e^{-3x} = 0$$

In all examples and exercises of this section, all the roots of the auxiliary equation are different, and they do not include complex numbers. For such roots, the solutions have a different form. They are the topic of the following section.

EXERCISES 30.6

In Exercises 1 and 2, make the given changes in the indicated examples of this section and then solve the resulting differential equations.

1. In Example 3, delete the $+6y$ term.

2. In Example 4, add $16y$ to the right side.

In Exercises 3–24, solve the given differential equations.

3. $\dfrac{d^2y}{dx^2} - \dfrac{dy}{dx} - 6y = 0$

4. $\dfrac{d^2y}{dx^2} + \dfrac{dy}{dx} = 0$

5. $3\dfrac{d^2y}{dx^2} + 4\dfrac{dy}{dx} + y = 0$

6. $\dfrac{d^2y}{dx^2} - 2\dfrac{dy}{dx} - 8y = 0$

7. $D^2y - 3Dy = 0$

8. $D^2y = y$

9. $2D^2y - 3y = Dy$

10. $D^2y + 7Dy + 6y = 0$

11. $3D^2y + 12y = 20Dy$

12. $4D^2y + 12Dy = 7y$

13. $3y'' + 8y' - 3y = 0$

14. $8y'' + 6y' - 9y = 0$

15. $3y'' + 2y' - y = 0$

16. $2y'' - 7y' + 6y = 0$

17. $2\dfrac{d^2y}{dx^2} - 4\dfrac{dy}{dx} + y = 0$

18. $\dfrac{d^2y}{dx^2} + \dfrac{dy}{dx} = 5y$

19. $4D^2y - 3Dy - 2y = 0$

20. $2D^2y - 3Dy - y = 0$

21. $y'' = 3y' + y$

22. $5y'' - y' = 3y$

23. $y'' + y' = 8y$

24. $8y'' = y' + y$

In Exercises 25–28, find the particular solutions of the given differential equations that satisfy the given conditions.

25. $D^2y - 4Dy - 21y = 0$; $Dy = 0$ and $y = 2$ when $x = 0$

26. $4D^2y - Dy = 0$; $Dy = 2$ and $y = 4$ when $x = 0$

27. $D^2y - Dy = 12y$; $y = 0$ when $x = 0$, and $y = 1$ when $x = 1$

28. $2D^2y + 5Dy = 0$; $y = 0$ when $x = 0$, and $y = 2$ when $x = 1$

In Exercises 29–32, solve the given third- and fourth-order differential equations.

29. $y''' - 2y'' - 3y' = 0$

30. $D^3y - 6D^2y + 11Dy - 6y = 0$

31. $D^4y - 5D^2y + 4y = 0$

32. $D^4y - D^3y - 9D^2y + 9Dy = 0$

30.7 AUXILIARY EQUATION WITH REPEATED OR COMPLEX ROOTS

In solving higher-order homogeneous differential equations in the previous section, we purposely avoided repeated or complex roots of the auxiliary equation. In this section we develop the solutions for such equations. The following example indicates the type of solution that results from the case of repeated roots.

EXAMPLE 1 Solve the differential equation $D^2y - 4Dy + 4y = 0$.

Using the method of Example 2 of the previous section, we have the following steps:

$$(D^2 - 4D + 4)y = 0, \qquad (D - 2)(D - 2)y = 0, \qquad (D - 2)z = 0$$

where $z = (D - 2)y$. The solution to $(D - 2)z = 0$ is found by separation of variables. And so

$$\frac{dz}{dx} - 2z = 0 \qquad \frac{dz}{z} - 2\,dx = 0$$

$$\ln z - 2x = \ln c_1 \quad \text{or} \quad z = c_1 e^{2x}$$

Substituting back, we have $(D - 2)y = c_1 e^{2x}$, which is a linear equation of the first order. Then

$$dy - 2y\,dx = c_1 e^{2x}\,dx \qquad e^{\int -2\,dx} = e^{-2x}$$

This leads to

$$ye^{-2x} = c_1 \int dx = c_1 x + c_2 \quad \text{or} \quad y = c_1 xe^{2x} + c_2 e^{2x}$$

This example indicates the type of solution that results when the auxiliary equation has repeated roots. If the method of the previous section were to be used, the solution of the above example would be $y = c_1 e^{2x} + c_2 e^{2x}$. This would not be the general solution, since both terms are similar, which means that there is only one independent constant. The constants can be combined to give a solution of the form $y = ce^{2x}$, where $c = c_1 + c_2$. This solution would contain only one constant for a second-order equation. ▸

For reference, Eq. (30.13) is $a_0 D^2y + a_1 Dy + a_2 y = 0$ and Eq. (30.14) is $a_0 m^2 + a_1 m + a_2 = 0$.

Based on the above example, *the solution to Eq. (30.13) when the auxiliary Eq. (30.14) has repeated roots is*

Solution with Repeated Roots

$$\boxed{y = e^{mx}(c_1 + c_2 x)} \tag{30.16}$$

where m is the double root. (In Example 1, this double root is 2.)

EXAMPLE 2 Solve the differential equation $(D + 2)^2 y = 0$.

The auxiliary equation is $(m + 2)^2 = 0$, for which the solutions are $m = -2, -2$. Since we have repeated roots, the solution of the differential equation is

$$y = e^{-2x}(c_1 + c_2 x) \qquad \text{using Eq. (30.16)}$$ ▸

◀ **EXAMPLE 3** Solve the differential equation $\dfrac{d^2y}{dx^2} - 10\dfrac{dy}{dx} + 25y = 0$.

The solution is as follows:

$$D^2y - 10\,Dy + 25y = 0 \qquad \text{using operator } D \text{ notation}$$
$$m^2 - 10m + 25 = 0 \qquad \text{auxiliary equation}$$
$$(m - 5)^2 = 0 \qquad \text{solve for } m$$
$$m = 5, 5 \qquad \text{double root}$$
$$y = e^{5x}(c_1 + c_2 x) \qquad \text{using Eq. (30.16)}$$

When the auxiliary equation has complex roots, it can be solved by the method of the previous section and the solution can be put in a more useful form. For complex roots of the auxiliary equation $m = \alpha \pm j\beta$, the solution is of the form

$$y = c_1 e^{(\alpha + j\beta)x} + c_2 e^{(\alpha - j\beta)x} = e^{\alpha x}(c_1 e^{j\beta x} + c_2 e^{-j\beta x})$$

Using the exponential form of a complex number, Eq. (12.11), we have

$$y = e^{\alpha x}[c_1 \cos \beta x + jc_1 \sin \beta x + c_2 \cos(-\beta x) + jc_2 \sin(-\beta x)]$$
$$= e^{\alpha x}(c_1 \cos \beta x + c_2 \cos \beta x + jc_1 \sin \beta x - jc_2 \sin \beta x)$$
$$= e^{\alpha x}(c_3 \cos \beta x + c_4 \sin \beta x)$$

where $c_3 = c_1 + c_2$ and $c_4 = jc_1 - jc_2$.

Therefore, *if the auxiliary equation has complex roots of the form $\alpha \pm j\beta$,*

Solution with Complex Roots

$$\boxed{y = e^{\alpha x}(c_1 \sin \beta x + c_2 \cos \beta x)} \qquad (30.17)$$

is the solution to Eq. (30.13). The c_1 and c_2 here are not the same as those above. They are simply the two arbitrary constants of the solution.

◀ **EXAMPLE 4** Solve the differential equation $D^2y - Dy + y = 0$.

We have the following solution:

$$m^2 - m + 1 = 0 \qquad \text{auxiliary equation}$$
$$m = \frac{1 \pm j\sqrt{3}}{2} \qquad \text{complex roots}$$
$$\alpha = \frac{1}{2} \quad \beta = \frac{\sqrt{3}}{2} \qquad \text{identify } \alpha \text{ and } \beta$$
$$y = e^{x/2}\left(c_1 \sin \frac{\sqrt{3}}{2}x + c_2 \cos \frac{\sqrt{3}}{2}x\right) \qquad \text{using Eq. (30.17)}$$

◀ **EXAMPLE 5** Solve the differential equation $D^3y + 4\,Dy = 0$.

This is a third-order equation, which means there are three arbitrary constants in the general solution.

$$m^3 + 4m = 0, \qquad m(m^2 + 4) = 0 \qquad \text{auxiliary equation}$$
$$m_1 = 0 \quad m_2 = 2j \quad m_3 = -2j \qquad \text{three roots, two of them complex}$$
$$\alpha = 0 \quad \beta = 2 \qquad \text{identify } \alpha \text{ and } \beta \text{ for complex roots}$$
$$y = c_1 e^{0x} + e^{0x}(c_2 \sin 2x + c_3 \cos 2x) \qquad \text{using Eqs. (30.15) and (30.17)}$$
$$= c_1 + c_2 \sin 2x + c_3 \cos 2x \qquad e^0 = 1$$

▌**EXAMPLE 6**   Solve the differential equation $y'' - 2y' + 12y = 0$, if $y' = 2$ and $y = 1$ when $x = 0$.

$$D^2y - 2\,Dy + 12y = 0 \qquad \text{using operator } D \text{ notation}$$
$$m^2 - 2m + 12 = 0 \qquad \text{auxiliary equation}$$
$$m = \frac{2 \pm \sqrt{4 - 48}}{2} = 1 \pm j\sqrt{11} \qquad \text{complex roots: } \alpha = 1, \beta = \sqrt{11}$$
$$y = e^x\!\left(c_1 \cos \sqrt{11}x + c_2 \sin \sqrt{11}x\right) \qquad \text{general solution}$$

Using the condition that $y = 1$ when $x = 0$, we have

$$1 = e^0(c_1 \cos 0 + c_2 \sin 0) \quad \text{or} \quad c_1 = 1$$

Since $y' = 2$ when $x = 0$, we find the derivative and then evaluate c_2.

$$y' = e^x\!\left(c_1 \cos \sqrt{11}x + c_2 \sin \sqrt{11}x - \sqrt{11}c_1 \sin \sqrt{11}x + \sqrt{11}c_2 \cos \sqrt{11}x\right)$$
$$2 = e^0\!\left(\cos 0 + c_2 \sin 0 - \sqrt{11} \sin 0 + \sqrt{11}c_2 \cos 0\right) \qquad y' = 2 \text{ when } x = 0$$
$$2 = 1 + \sqrt{11}c_2, \qquad c_2 = \tfrac{1}{11}\sqrt{11} \qquad \text{solve for } c_2$$
$$y = e^x\!\left(\cos \sqrt{11}x + \tfrac{1}{11}\sqrt{11} \sin \sqrt{11}x\right) \qquad \text{particular solution} \qquad ▐$$

If the root of the auxiliary equation is repeated more than once—for example, a *triple root*—an additional term with another arbitrary constant and the next higher power of x is added to the solution for each additional root. Also, if a pair of complex roots is repeated, an additional term with a factor of x and another arbitrary constant is added for each root of the pair. These are illustrated in the following example.

Roots Repeated More Than Once   ▌**EXAMPLE 7**   **(a)** For the differential equation $D^3y + 3\,D^2y + 3\,Dy + y = 0$, the auxiliary equation is

$$m^3 + 3m^2 + 3m + 1 = 0, \qquad (m + 1)^3 = 0$$

Each of the three roots is $m = -1$. The equation is a third-order equation, which means there are three arbitrary constants. Therefore, the general solution is

$$y = e^{-x}(c_1 + c_2x + c_3x^2)$$

Repeated Complex Roots   **(b)** For the differential equation $D^4y + 8\,D^2y + 16y = 0$, the auxiliary equation is

$$m^4 + 8m^2 + 16 = 0, \qquad (m^2 + 4)^2 = 0$$

With two factors of $m^2 + 4$, the roots are $2j$, $2j$, $-2j$, and $-2j$. The fourth-order equation, and four roots, indicate four arbitrary constants. Since $e^{0x} = 1$, the general solution is

$$y = (c_1 + c_2x)\sin 2x + (c_3 + c_4x)\cos 2x \qquad ▐$$

Knowing the various types of possible solutions, it is possible to determine the differential equation if the solution is known. Consider the following example.

▌**EXAMPLE 8**   **(a)** A solution of $y = c_1e^x + c_2e^{2x}$ indicates an auxiliary equation with roots of $m_1 = 1$ and $m_2 = 2$. Thus, the auxiliary equation is $(m - 1)(m - 2) = 0$, and the simplest form of the differential equation is $D^2y - 3\,Dy + 2y = 0$.

(b) A solution of $y = e^{2x}(c_1 + c_2x)$ indicates repeated roots $m_1 = m_2 = 2$ of the auxiliary equation $(m - 2)^2 = 0$, and a differential equation $D^2y - 4\,Dy + 4y = 0$. ▐

In Exercises 1–4, make the given changes in the indicated examples of this section, and then solve the resulting differential equations.

1. In Example 3, change the $-$ sign to $+$.

2. In Example 3, delete the second term $\left(-10\frac{dy}{dx}\right)$.

3. In Example 3, delete the third term $(+25y)$.

4. In Example 4, change the $-$ sign to $+$.

In Exercises 5–32, solve the given differential equations.

5. $\dfrac{d^2y}{dx^2} - 2\dfrac{dy}{dx} + y = 0$ 　　 **6.** $\dfrac{d^2y}{dx^2} - 6\dfrac{dy}{dx} + 9y = 0$

7. $D^2y + 12Dy + 36y = 0$ 　　 **8.** $16D^2y + 8Dy + y = 0$

9. $\dfrac{d^2y}{dx^2} + 9y = 0$ 　　 **10.** $\dfrac{d^2y}{dx^2} + y = 0$

11. $D^2y + Dy + 2y = 0$ 　　 **12.** $D^2y + 4y = 2Dy$

13. $D^4y - y = 0$ 　　 **14.** $4D^2y = 12Dy - 9y$

15. $4D^2y + y = 0$ 　　 **16.** $9D^2y + 4y = 0$

17. $16y'' - 24y' + 9y = 0$ 　　 **18.** $9y'' - 24y' + 16y = 0$

19. $25y'' + 2y = 0$ 　　 **20.** $y'' - 4y' + 5y = 0$

21. $2D^2y + 5y = 4Dy$ 　　 **22.** $D^2y + 4Dy + 6y = 0$

23. $25y'' + 16y = 40y'$ 　　 **24.** $9y''' + 0.6y'' + 0.01y' = 0$

25. $2D^2y - 3Dy - y = 0$ 　　 **26.** $D^2y - 5Dy - 4y = 0$

27. $3D^2y + 12Dy = 2y$ 　　 **28.** $36D^2y = 25y$

29. $D^3y - 6D^2y + 12Dy - 8y = 0$

30. $D^4y - 2D^3y + 2D^2y - 2Dy + y = 0$

31. $D^4y + 2D^2y + y = 0$ 　　 **32.** $2D^4y - y = 0$

In Exercises 33–36, find the particular solutions of the given differential equations that satisfy the given conditions.

33. $y'' + 2y' + 10y = 0$; 　$y = 0$ when $x = 0$ and $y = e^{-\pi/6}$ when $x = \pi/6$

34. $9D^2y + 16y = 0$; 　$Dy = 0$ and $y = 2$ when $x = \pi/2$

35. $D^2y + 16y = 8Dy$; 　$Dy = 2$ and $y = 4$ when $x = 0$

36. $D^4y + 3D^3y + 2D^2y = 0$; 　$y = 0$ and $Dy = 4$ and $D^2y = -8$ and $D^3y = 16$ when $x = 0$

In Exercises 37–40, find the simplest form of the second-order differential homogeneous linear differential equation that has the given solution. In Exercises 38 and 39, explain how the equation is found.

37. $y = c_1e^{3x} + c_2e^{-3x}$ 　　 Ⓦ **38.** $y = c_1e^{3x} + c_2xe^{3x}$

Ⓦ **39.** $y = c_1\cos 3x + c_2\sin 3x$

40. $y = c_1e^{2x}\cos x + c_2e^{2x}\sin x$

30.8 SOLUTIONS OF NONHOMOGENEOUS EQUATIONS

We now consider the solution of a nonhomogeneous linear equation of the form

$$a_0D^2y + a_1Dy + a_2y = b \qquad (30.18)$$

where b is a function of x or is a constant. When the solution is substituted into the left side, we must obtain b. Solutions found from the methods of Sections 30.6 and 30.7 give zero when substituted into the left side, but they do contain the arbitrary constants necessary in the solution. If we could find a particular solution that when substituted into the left side produced b, it could be added to the solution containing the arbitrary constants. Therefore, *the solution is of the form*

$$y = y_c + y_p \qquad (30.19)$$

where y_c, called the **complementary solution,** *is obtained by solving the corresponding homogeneous equation and where y_p is the* **particular solution** *necessary to produce the expression b of Eq. (30.18). It should be noted that y_p satisfies the differential equation, but it has no arbitrary constants and therefore cannot be the general solution. The arbitrary constants are part of y_c.*

◀ **EXAMPLE 1**   The differential equation $D^2y - Dy - 6y = e^x$ has the solution

$$y = c_1e^{3x} + c_2e^{-2x} - \frac{1}{6}e^x$$

where the complementary solution y_c and particular solution y_p are

$$y_c = c_1e^{3x} + c_2e^{-2x} \qquad y_p = -\frac{1}{6}e^x$$

The complementary solution y_c is obtained by solving the corresponding homogeneous equation $D^2y - Dy - 6y = 0$, and we shall discuss below the method of finding y_p. Again we note that y_c contains the arbitrary constants, y_p contains the expression needed to produce the e^x on the right, and therefore both are needed to have the complete general solution.   ▶

By inspecting the form of b on the right side of the equation, we can find the form that the particular solution must have. Since a combination of the particular solution and its derivatives must form the function b,

CAUTION ▶   *y_p is an expression that contains all possible forms of b and its derivatives.*

The method that is used to find the exact form of y_p is called the **method of undetermined coefficients.**

◀ **EXAMPLE 2**   **(a)** If the function b is $4x$, we choose the particular solution y_p to be of the form $y_p = A + Bx$. The Bx-term is included to account for the $4x$. Since the derivative of Bx is a constant, the A-term is included to account for any first derivative of the Bx-term that may be present. Since the derivative of A is zero, no other terms are needed to account for higher-derivative terms of the Bx-term.

(b) If the function b is e^{2x}, we choose the form of the particular solution to be

NOTE ▶ $y_p = Ce^{2x}$. *Since all derivatives of Ce^{2x} are a constant times e^{2x}, no other forms appear in the derivatives, and no other forms are needed in y_p.*

(c) If the function b is $4x + e^{2x}$, we choose the form of the particular solution to be $y_p = A + Bx + Ce^{2x}$.   ▶

◀ **EXAMPLE 3**   **(a)** If b is of the form $x^2 + e^{-x}$, we choose the particular solution to be of the form $y_p = A + Bx + Cx^2 + Ee^{-x}$.

(b) If b is of the form $xe^{-2x} - 5$, we choose the form of the particular solution to be $y_p = Ae^{-2x} + Bxe^{-2x} + C$.

(c) If b is of the form $x \sin x$, we choose the form of the particular solution to be $y_p = A \sin x + B \cos x + Cx \sin x + Ex \cos x$. All these types of terms occur in the derivatives of $x \sin x$.   ▶

◀ **EXAMPLE 4**   **(a)** If b is of the form $e^x + xe^x$, we would then choose y_p to be of CAUTION ▶   the form $y_p = Ae^x + Bxe^x$. These terms occur for xe^x and its derivatives. *Since the form of the e^x-term of b is already included in Ae^x, we do not include another e^x- term in y_p.*

(b) In the same way, if b is of the form $2x + 4x^2$, we choose the form of y_p to be NOTE ▶ $y_p = A + Bx + Cx^2$. *There are only forms that occur in either $2x$ or $4x^2$ and their derivatives.*   ▶

Once we have determined the form of y_p, we have to find the numerical values of the coefficients $A, B, \ldots$. The *method of undetermined coefficients is to*

NOTE ▶   *substitute the chosen form of y_p into the differential equation and equate the coefficients of like terms.*

◖ EXAMPLE 5   Solve the differential equation $D^2y - Dy - 6y = e^x$.

In this case the solution of the auxiliary equation $m^2 - m - 6 = 0$ gives us the roots $m_1 = 3$ and $m_2 = -2$. Thus,

$$y_c = c_1e^{3x} + c_2e^{-2x}$$

The proper form of y_c is $y_p = Ae^x$. This means that $Dy_p = Ae^x$ and $D^2y_p = Ae^x$. Substituting y_p and its derivatives into the differential equation, we have

$$Ae^x - Ae^x - 6Ae^x = e^x$$

To produce equality, the coefficients of e^x must be the same on each side of the equation. Thus,

$$-6A = 1 \quad \text{or} \quad A = -1/6$$

Therefore, $y_p = -\frac{1}{6}e^x$. This gives the complete solution $y = y_c + y_p$,

$$y = c_1e^{3x} + c_2e^{-2x} - \frac{1}{6}e^x \qquad \text{see Example 1}$$

This solution checks when substituted into the original differential equation.

◖ EXAMPLE 6   Solve the differential equation $D^2y + 4y = x - 4e^{-x}$.

In this case we have $m^2 + 4 = 0$, which gives us $m_1 = 2j$ and $m_2 = -2j$. Therefore, $y_c = c_1 \sin 2x + c_2 \cos 2x$.

The proper form of the particular solution is $y_p = A + Bx + Ce^{-x}$. Finding two derivatives and then substituting into the differential equation gives

$$y_p = A + Bx + Ce^{-x} \qquad Dy_p = B - Ce^{-x} \qquad D^2y_p = Ce^{-x}$$

$$\underline{\hspace{3cm}} \quad D^2y + 4y = x - 4e^{-x} \qquad \text{differential equation}$$

$$(Ce^{-x}) + 4(A + Bx + Ce^{-x}) = x - 4e^{-x} \qquad \text{substituting}$$

$$Ce^{-x} + 4A + 4Bx + 4Ce^{-x} = x - 4e^{-x}$$

$$(4A) + (4B)x + (5C)e^{-x} = 0 + (1)x + (-4)e^{-x} \qquad \text{note coefficients}$$

CAUTION ▶   *Equating the constants, and the coefficients* of x and e^{-x} on either side gives

$$4A = 0 \qquad 4B = 1 \qquad 5C = -4$$
$$A = 0 \qquad B = 1/4 \qquad C = -4/5$$

This means that the particular solution is

$$y_p = \frac{1}{4}x - \frac{4}{5}e^{-x}$$

In turn this tells us that the complete solution is

$$y = c_1 \sin 2x + c_2 \cos 2x + \frac{1}{4}x - \frac{4}{5}e^{-x}$$

Substitution into the original differential equation verifies this solution.

◀ EXAMPLE 7   Solve the differential equation $D^3y - 3D^2y + 2Dy = 10 \sin x$.

$m^3 - 3m^2 + 2m = 0$ $m(m - 1)(m - 2) = 0$ $m_1 = 0$ $m_2 = 1$ $m_3 = 2$ auxiliary equation

$y_c = c_1 + c_2e^x + c_3e^{2x}$ complementary solution

We now find the particular solution:

$y_p = A \sin x + B \cos x$ particular solution form

$Dy_p = A \cos x - B \sin x$ $D^2y_p = -A \sin x - B \cos x$ $D^3y_p = -A \cos x + B \sin x$ find three derivatives

$(-A \cos x + B \sin x) - 3(-A \sin x - B \cos x) + 2(A \cos x - B \sin x) = 10 \sin x$ substitute into differential equation

$(3A - B)\sin x + (A + 3B)\cos x = 10 \sin x$

$3A - B = 10$ $A + 3B = 0$ equate coefficients

The solution of this system is $A = 3$, $B = -1$.

$y_p = 3 \sin x - \cos x$ particular solution

$y = c_1 + c_2e^x + c_3e^{2x} + 3 \sin x - \cos x$ complete general solution

Check this solution in the differential equation.

◀ EXAMPLE 8   Find the particular solution of $y'' + 16y = 2e^{-x}$ if $Dy = -2$ and $y = 1$ when $x = 0$.

In this case we must not only find y_c and y_p, but we must also evaluate the constants of y_c from the given conditions. The solution is as follows:

$D^2y + 16y = 2e^{-x}$ operator D form

$m^2 + 16 = 0$, $m = \pm 4j$ auxiliary equation

$y_c = c_1 \sin 4x + c_2 \cos 4x$ complementary solution

$y_p = Ae^{-x}$ particular solution form

$Dy_p = -Ae^{-x}$ $D^2y_p = Ae^{-x}$

$Ae^{-x} + 16Ae^{-x} = 2e^{-x}$ substituting

$17Ae^{-x} = 2e^{-x}$, $A = \dfrac{2}{17}$ equate coefficients

$y_p = \dfrac{2}{17}e^{-x}$

$y = c_1 \sin 4x + c_2 \cos 4x + \dfrac{2}{17}e^{-x}$ complete general solution

We now evaluate c_1 and c_2 from the given conditions:

$Dy = 4c_1 \cos 4x - 4c_2 \sin 4x - \dfrac{2}{17}e^{-x}$

$1 = c_1(0) + c_2(1) + \dfrac{2}{17}(1)$, $c_2 = \dfrac{15}{17}$ $y = 1$ when $x = 0$

$-2 = 4c_1(1) - 4c_2(0) - \dfrac{2}{17}(1)$, $c_1 = -\dfrac{8}{17}$ $Dy = -2$ when $x = 0$

$y = -\dfrac{8}{17} \sin 4x + \dfrac{15}{17} \cos 4x + \dfrac{2}{17}e^{-x}$ required particular solution

This solution checks when substituted into the differential equation.

A SPECIAL CASE

It may happen that a term of the proposed y_p is similar to a term of y_c. Since any term of y_c gives zero when substituted in the differential equation, so will that term of the proposed y_p. This means the proposed y_p must be modified. Therefore,

> From Eq. (30.18), we note that the function b is the function on the right side of the differential equation.

if a term of the proposed y_p is similar to a term of y_c, any term of the proposed y_p, included to account for the similar term of the function b, must be multiplied by the smallest possible integral power of x such that any resulting term y_p is not similar to the term of y_c.

The following example shows that this is not as involved as it sounds.

EXAMPLE 9 Solve the differential equation $D^2y - 2Dy + y = x + e^x$.

We find that the auxiliary equation and complementary solution are

$$m^2 - 2m + 1 = 0, \qquad (m-1)^2 = 0 \qquad m_1 = 1 \qquad m_2 = 1$$

$$y_c = e^x(c_1 + c_2x)$$

Based on the function b on the right side, the *proposed* form of y_p is

$$y_p = A + Bx + Ce^x \qquad \text{proposed form}$$

We now note that the term Ce^x is similar to the term c_1e^x of y_c. Therefore, we must multiply the term Ce^x by the smallest power of x such that it is not similar to any term of y_c. If we multiply by x, the term becomes similar to c_2xe^x. Therefore, we must multiply Ce^x by x^2 such that y_p is

> Note that the $A + Bx$ are not multiplied by x^2 since they are not included in y_p to account for the e^x on the right.

$$y_p = A + Bx + Cx^2e^x \qquad \text{correct modified form}$$

Using this form of y_p, we now complete the solution.

$$Dy_p = B + Cx^2e^x + 2Cxe^x \qquad D^2y_p = Cx^2e^x + 2Cxe^x + 2Ce^x + 2Cxe^x = 2Ce^x + 4Cxe^x + Cx^2e^x$$

$$(2Ce^x + 4Cxe^x + Cx^2e^x) - 2(B + Cx^2e^x + 2Cxe^x) + (A + Bx + Cx^2e^x) = x + e^x$$

$$(A - 2B) + Bx + 2Ce^x = x + e^x$$

$$A - 2B = 0 \qquad B = 1 \qquad 2C = 1 \qquad A = 2 \qquad B = 1 \qquad C = 1/2$$

$$y_p = 2 + x + \tfrac{1}{2}x^2e^x$$

$$y = e^x(c_1 + c_2x) + 2 + x + \tfrac{1}{2}x^2e^x$$

EXERCISES 30.8

In Exercises 1–4, make the given changes in the indicated examples of this section and then solve the given problems.

1. In Example 3(a), add $2x$ to the form b and then determine the form of y_p.

2. In Example 5, change e^x to e^{2x} and then find the solution.

3. In Example 6, on the right side change x to $\sin x$, and then find the proper form of y_c and y_p.

4. In Example 8, change the right side to $x^2 + xe^x$ and then find the proper form for y_p.

In Exercises 5–16, solve the given differential equations. The form of y_p is given.

5. $D^2y - Dy - 2y = 4$ (Let $y_p = A$.)

6. $D^2y - Dy - 6y = 4x$ (Let $y_p = A + Bx$.)

7. $D^2y - y = 2 + x^2$ (Let $y_p = A + Bx + Cx^2$.)

8. $D^2y + 4Dy + 3y = 2 + e^x$ (Let $y_p = A + Be^x$.)

9. $y'' - 3y' = 2e^x + xe^x$ (Let $y_p = Ae^x + Bxe^x$.)

10. $y'' + y' - 2y = 8 + 4x + 2xe^{2x}$
(Let $y_p = A + Bx + Ce^{2x} + Exe^{2x}$.)

11. $9D^2y - y = \sin x$ (Let $y_p = A \sin x + B \cos x$.)

12. $D^2y + 4y = \sin x + 4$ (Let $y_p = A + B \sin x + C \cos x$.)

13. $\dfrac{d^2y}{dx^2} - 2\dfrac{dy}{dx} + y = 2x + x^2 + \sin 3x$
(Let $y_p = A + Bx + Cx^2 + E \sin 3x + F \cos 3x$.)

14. $D^2y - y = e^{-x}$ (Let $y_p = Axe^{-x}$.)
15. $D^2y + 4y = -12 \sin 2x$ (Let $y_p = Ax \sin 2x + Bx \cos 2x$.)
16. $y'' - 2y' + y = 3 + e^x$ (Let $y_p = A + Bx^2e^x$.)

In Exercises 17–32, solve the given differential equations.

17. $\dfrac{d^2y}{dx^2} - \dfrac{dy}{dx} - 30y = 10$ **18.** $2\dfrac{d^2y}{dx^2} + 11\dfrac{dy}{dx} - 6y = 8x$

19. $3\dfrac{d^2y}{dx^2} + 13\dfrac{dy}{dx} - 10y = 14e^{3x}$

20. $\dfrac{d^2y}{dx^2} + 4y = 2 \sin 3x$

21. $D^2y - 4y = \sin x + 2 \cos x$

22. $2D^2y + 5Dy - 3y = e^x + 4e^{2x}$

23. $D^2y + y = 4 + \sin 2x$

24. $D^2y - Dy + y = x + \sin x$
25. $D^2y + 5Dy + 4y = xe^x + 4$
26. $3D^2y + Dy - 2y = 4 + 2x + e^x$
27. $y''' - y' = \sin 2x$ **28.** $D^4y - y = x$
29. $D^2y + y = \cos x$ **30.** $4y'' - 4y' + y = 4e^{x/2}$
31. $D^2y + 2Dy = 8x + e^{-2x}$ **32.** $D^3y - Dy = 4e^{-x} + 3e^{2x}$

In Exercises 33–36, find the particular solution of each differential equation for the given conditions.

33. $D^2y - Dy - 6y = 5 - e^x$; $Dy = 4$ and $y = 2$ when $x = 0$
34. $3y'' - 10y' + 3y = xe^{-2x}$; $y' = -\frac{9}{35}$ and $y = -\frac{13}{35}$ when $x = 0$
35. $y'' + y = x + \sin 2x$; $y' = 1$ and $y = 0$ when $x = \pi$
36. $D^2y - 2Dy + y = xe^{2x} - e^{2x}$; $Dy = 4$ and $y = -2$ when $x = 0$

30.9 APPLICATIONS OF HIGHER-ORDER EQUATIONS

We now show important applications of second-order differential equations to simple harmonic motion and simple electric circuits. Also, we will show an application of a fourth-order differential equation to the deflection of a beam.

Simple Harmonic Motion

EXAMPLE 1   Simple harmonic motion may be defined as motion in a straight line for which the acceleration is proportional to the displacement and in the opposite direction. Examples of this type of motion are a weight on a spring, a simple pendulum, and an object bobbing in water. If x represents the displacement, d^2x/dt^2 is the acceleration.

Using the definition of simple harmonic motion, we have

$$\frac{d^2x}{dt^2} = -k^2x$$

(We chose k^2 for convenience of notation in the solution.) We write this equation in the form

$$D^2x + k^2x = 0 \qquad \text{here, } D = d/dt$$

The roots of the auxiliary equation are kj and $-kj$, and the solution is

$$x = c_1 \sin kt + c_2 \cos kt$$

This solution indicates an oscillating motion, which is known to be the case. If, for example, $k = 4$ and we know that $x = 2$ and $Dx = 0$ (which means the velocity is zero) for $t = 0$, we have

$$Dx = 4c_1 \cos 4t - 4c_2 \sin 4t$$
$$2 = c_1(0) + c_2(1) \qquad x = 2 \text{ for } t = 0$$
$$0 = 4c_1(1) - 4c_2(0) \qquad Dx = 0 \text{ for } t = 0$$

which gives $c_1 = 0$ and $c_2 = 2$. Therefore,

$$x = 2 \cos 4t$$

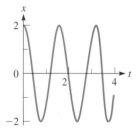

Fig. 30.9

is the equation relating the displacement and time; Dx is the velocity and D^2x is the acceleration. See Fig. 30.9.

EXAMPLE 2 In practice, an object moving with simple harmonic motion will in time cease to move due to unavoidable frictional forces. A "freely" oscillating object has a retarding force that is approximately proportional to the velocity. The differential equation for this case is $D^2x = -k^2x - bDx$. This results from applying (from physics) Newton's second law of motion (see the margin note at the left). Again, using the operator $D^2x = d^2x/dt^2$, the term D^2x represents the acceleration of the object, the term $-k^2x$ is a measure of the restoring force (of the spring, for example), and the term $-bDx$ represents the retarding (damping) force. This equation can be written as

$$D^2x + bDx + k^2x = 0$$

The auxiliary equation is $m^2 + bm + k^2 = 0$, for which the roots are

$$m = \frac{-b \pm \sqrt{b^2 - 4k^2}}{2}$$

If $k = 3$ and $b = 4$, $m = -2 \pm j\sqrt{5}$, which means the solution is

$$x = e^{-2t}\left(c_1 \sin \sqrt{5}t + c_2 \cos \sqrt{5}t\right) \tag{1}$$

Here, $4k^2 > b^2$, and this case is called **underdamped.** In this case the object oscillates as the amplitude becomes smaller.

If $k = 2$ and $b = 5$, $m = -1, -4$, which means the solution is

$$x = c_1e^{-t} + c_2e^{-4t} \tag{2}$$

Here, $4k^2 < b^2$, and the case is called **overdamped.** It will be noted that the motion is not oscillatory, since no sine or cosine terms appear. In this case the object returns slowly to equilibrium without oscillating.

If $k = 2$ and $b = 4$, $m = -2, -2$, which means the solution is

$$x = e^{-2t}(c_1 + c_2t) \tag{3}$$

Here, $4k^2 = b^2$, and the case is called **critically damped.** Again the motion is not oscillatory. In this case there is just enough damping to prevent any oscillations. The object returns to equilibrium in the minimum time.

See Fig. 30.10, in which Eqs. (1), (2), and (3) are represented in general. Of course, the actual values depend on c_1 and c_2, which in turn depend on the conditions imposed on the motion.

Newton's second law states that the net force acting on an object is equal to its mass times its acceleration. (This is one of Newton's best-known contributions to physics.)

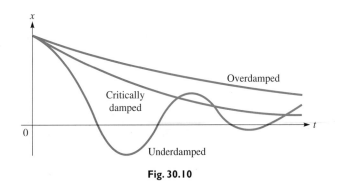

Fig. 30.10

Solving a Word Problem ❙ EXAMPLE 3   In testing the characteristics of a particular type of spring, it is found that a weight of 4.90 N stretches the spring 0.490 m when the weight and spring are placed in a fluid that resists the motion with a force equal to twice the velocity. If the weight is brought to rest and then given a velocity of 12.0 m/s, find the equation of motion. See Fig. 30.11.

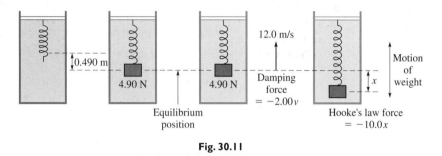

Fig. 30.11

In order to find the equation of motion, we use Newton's second law of motion (see Example 2). The weight (one force) at the end of the spring is offset by the equilibrium position force exerted by the spring, in accordance with Hooke's law (see Section 26.6). Therefore, the net force acting on the weight is the sum of the Hooke's law force due to the displacement from the equilibrium position and the resisting force. Using Newton's second law, we have

mass × acceleration = resisting force + Hooke's law force

$$m D^2 x = -2.00 Dx - kx$$

The mass of an object is its weight divided by the acceleration due to gravity. The weight is 4.90 N, and the acceleration due to gravity is 9.80 m/s². Thus, the mass m is

$$m = \frac{4.90 \text{ N}}{9.80 \text{ m/s}^2} = 0.500 \text{ kg.}$$

where the kilogram is the unit of mass.

The constant k for the Hooke's law force is found from the fact that the spring stretches 0.490 m for a force of 4.90 N. Thus, using Hooke's law,

$$4.90 = k(0.490), \qquad k = 10.0 \text{ N/m}$$

This means that the differential equation to be solved is

$$0.500 D^2 x + 2.00 Dx + 10.0x = 0$$

or

$$1.00 D^2 x + 4.00 Dx + 20.0x = 0$$

Solving this equation, we have

$$1.00 m^2 + 4.00 m + 20.0 = 0 \qquad \text{auxiliary equation}$$

$$m = \frac{-4.00 \pm \sqrt{16.0 - 4(20.0)(1.00)}}{2.00}$$

$$= -2.00 \pm 4.00 j \qquad \text{complex roots}$$

$$x = e^{-2.00t}(c_1 \cos 4.00t + c_2 \sin 4.00t) \qquad \text{general solution}$$

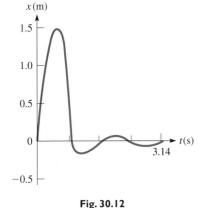

Fig. 30.12

Since the weight started from the equilibrium position with a velocity of 12.0 m/s, we know that $x = 0$ and $Dx = 12.0$ for $t = 0$. Thus,

$$0 = e^0(c_1 + 0c_2) \quad \text{or} \quad c_1 = 0 \qquad x = 0 \text{ for } t = 0$$

Thus, since $c_1 = 0$, we have

$$x = c_2 e^{-2.00t} \sin 4.00t$$

$$Dx = c_2 e^{-2.00t}(\cos 4.00t)(4.00) + c_2 \sin 4.00t(e^{-2.00t})(-2.00)$$

$$12.0 = c_2 e^0(1)(4.00) + c_2(0)(e^0)(-2.00) \qquad Dx = 12.0 \text{ for } t = 0$$

$$c_2 = 3.00$$

This means that the equation of motion is

$$x = 3.00e^{-2.00t} \sin 4.00t$$

The motion is underdamped; the graph is shown in Fig. 30.12.

It is possible to have an additional force acting on a weight such as the one in Example 3. For example, a vibratory force may be applied to the support of the spring. In such a case, called *forced vibrations,* this additional external force is added to the other net force. This means that the added force $F(t)$ becomes a nonzero function on the right side of the differential equation, and we must then solve a nonhomogeneous equation.

Electric Circuits

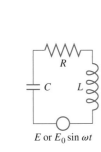

Fig. 30.13

◀ EXAMPLE 4 The impressed voltage in the electric circuit equals the sum of the voltages across the components of the circuit. For a circuit with a resistance R, an inductance L, a capacitance C, and a voltage source E (see Fig. 30.13), we have

$$L\frac{d^2q}{dt^2} + R\frac{dq}{dt} + \frac{q}{C} = E \tag{30.20}$$

By definition, q represents the electric charge, $dq/dt = i$ is the current, and d^2q/dt^2 is the time rate of change of current. This equation may be written as

$$LD^2q + RDq + q/C = E$$

The auxiliary equation is $Lm^2 + Rm + 1/C = 0$. The roots are

$$m = \frac{-R \pm \sqrt{R^2 - 4L/C}}{2L} = -\frac{R}{2L} \pm \sqrt{\frac{R^2}{4L^2} - \frac{1}{LC}}$$

If we let $a = R/2L$ and $\omega = \sqrt{1/LC - R^2/4L^2}$, we have (assuming complex roots, which corresponds to realistic values of R, L, and C)

$$q_c = e^{-at}(c_1 \sin \omega t + c_2 \cos \omega t)$$

This indicates an oscillating charge, or an alternating current. However, the exponential term usually is such that the current dies out rapidly unless there is a source of voltage in the circuit.

If there is no source of voltage in the circuit of Example 4, we have a homogeneous differential equation to solve. If we have a constant voltage source, the particular solution is of the form $q_p = A$. If there is an alternating voltage source, the particular solution is of the form $q_p = A \sin \omega_1 t + B \cos \omega_1 t$, where ω_1 is the angular velocity of the source. After a very short time, the exponential factor in the complementary solution makes it negligible. For this reason it is referred to as the **transient** term, and *the particular solution is the* **steady-state** *solution*. Therefore, to find the steady-state solution, we need find only the particular solution.

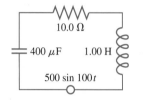

Fig. 30.14

◀ **EXAMPLE 5**   Find the steady-state solution for the current in a circuit containing the following elements: $C = 400 \ \mu\text{F}$, $L = 1.00 \text{ H}$, $R = 10.0 \ \Omega$, and a voltage source of 500 sin 100t. See Fig. 30.14.

This means the differential equation to be solved is

$$\frac{d^2q}{dt^2} + 10\frac{dq}{dt} + \frac{10^4}{4}q = 500 \sin 100t$$

Since we wish to find the steady-state solution, we must find q_p, from which we may find i_p by finding a derivative. The solution now follows:

$$q_p = A \sin 100t + B \cos 100t \qquad \text{particular solution form}$$

$$\frac{dq_p}{dt} = 100A \cos 100t - 100B \sin 100t$$

$$\frac{d^2q_p}{dt} = -10^4A \sin 100t - 10^4B \cos 100t$$

$$-10^4A \sin 100t - 10^4B \cos 100t + 10^3A \cos 100t - 10^3B \sin 100t \qquad \text{substitute into differential equation}$$

$$+ \frac{10^4}{4}A \sin 100t + \frac{10^4}{4}B \cos 100t = 500 \sin 100t$$

$$(-0.75 \times 10^4A - 10^3B)\sin 100t + (-0.75 \times 10^4B + 10^3A)\cos 100t = 500 \sin 100t$$

$$-7.5 \times 10^3A - 10^3B = 500 \qquad \text{equate coefficients of sin 100}t$$

$$10^3A - 7.5 \times 10^3B = 0 \qquad \text{equate coefficients of cos 100}t$$

Solving these equations, we obtain

$$B = -8.73 \times 10^{-3} \quad \text{and} \quad A = -65.5 \times 10^{-3}$$

Therefore,

$$q_p = -65.5 \times 10^{-3} \sin 100t - 8.73 \times 10^{-3} \cos 100t$$

$$i_p = \frac{dq_p}{dt} = -6.55 \cos 100t + 0.87 \sin 100t$$

which is the required solution. (We assumed three significant digits for the data but did not use all of them in most equations of the solution.)   ◗

It should be noted that the complementary solutions of the mechanical and electric cases are of identical form. There is also an equivalent mechanical case to that of an impressed sinusoidal voltage source in the electric case. This arises in the case of forced vibrations, when an external force affecting the vibrations is applied to the system. Thus, we may have transient and steady-state solutions to mechanical and other nonelectric situations.

Deflection of Beams

In the study of the strength of materials and elasticity it is shown that the deflection y of a beam of length L satisfies the differential equation $EI\,d^4y/dx^4 = w(x)$, where EI is a measure of the stiffness of the beam and $w(x)$ is the weight distribution along the beam. See Fig. 30.15. Since this is a fourth-order equation, it is necessary to specify four conditions to obtain a solution. These conditions are determined by the way in which the ends, where $x = 0$ and where $x = L$, are held. For an end held in the specified manner, these conditions are: *clamped:* $y = 0$ and $y' = 0$; *hinged:* $y = 0$ and $y'' = 0$; *free:* $y'' = 0$ and $y''' = 0$ ($y' = 0$ indicates no change in alignment; $y'' = 0$ indicates no curvature; $y''' = 0$ indicates no shearing force). Since the conditions are given for specific positions, this kind of problem is called a *boundary value problem*. Consider the following example.

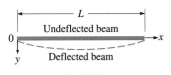

L
Undeflected beam
0 $\rightarrow x$
y Deflected beam

Fig. 30.15

Solving a Word Problem

❙ EXAMPLE 6   A uniform beam of length L is hinged at both ends and has a constant load distribution of w due to its own weight. Find the deflection y of the beam in terms of the distance x from one end.

Using the differential equation given above, we have $EI\,d^4y/dx^4 = w$. For convenience in the solution, let $k = w/EI$. Thus, the solution is as follows:

$$D^4y = k$$
$$m^4 = 0 \qquad m_1 = m_2 = m_3 = m_4 = 0$$

Since the four roots of the auxiliary equation are equal,

$$y_c = c_1 + c_2x + c_3x^2 + c_4x^3$$

The form of y_c indicates that we must multiply the *proposed* $y_p = A$ by x^4 so that it is not similar to any of the terms of y_c. This gives us $y_p = Ax^4$. Therefore,

$$y_p = Ax^4 \qquad Dy_p = 4Ax^3 \qquad D^2y_p = 12Ax^2 \qquad D^3y_p = 24Ax \qquad D^4y_p = 24A$$
$$24A = k, \qquad A = k/24$$

$$y = c_1 + c_2x + c_3x^2 + c_4x^3 + \frac{k}{24}x^4 \qquad \text{general solution}$$

We now find four known conditions to evaluate the four constants from the discussion above the example. For a beam hinged at both ends, at both $x = 0$ and $x = L$, $y = 0$ and $D^2y = 0$. We now find four derivatives and use these conditions.

$$Dy = c_2 + 2c_3x + 3c_4x^2 + \frac{k}{6}x^3 \qquad D^2y = 2c_3 + 6c_4x + \frac{k}{2}x^2 \qquad \text{find derivatives}$$

$$D^3y = 6c_4 + kx \qquad D^4y = k$$

At $x = 0$, $y = 0$: $c_1 = 0$;   At $x = 0$, $D^2y = 0$: $c_3 = 0$ use conditions to evaluate constants

At $x = L$, $y = 0$: $0 = c_2L + c_4L^3 + \dfrac{kL^4}{24}$

At $x = L$, $D^2y = 0$: $0 = 6c_4L + \dfrac{kL^2}{2}$

$$c_4 = -\frac{kL}{12} \qquad 0 = c_2L + \left(-\frac{kL}{12}\right)L^3 + \frac{kL^4}{24} \qquad c_2 = \frac{kL^3}{24}$$

$$y = \frac{kL^3}{24}x - \frac{kL}{12}x^3 + \frac{kx^4}{24} = \frac{k}{24}(L^3x - 2Lx^3 + x^4) \qquad \text{particular solution}$$

$$= \frac{w}{24EI}(L^3x - 2Lx^3 + x^4) \qquad k = w/EI$$

❙

EXERCISES 30.9

In Exercises 1 and 2, make the given changes in the indicated examples of this section and then solve the resulting problems.

1. In Example 1, change the conditions that $x = 2$ and $Dx = 0$ for $t = 0$ to $x = 0$ and $Dx = 2$ for $t = 0$.

2. In Example 5, change the voltage source to $500 \cos 100t$.

In Exercises 3–24, solve the given problems.

3. An object moves with simple harmonic motion according to $D^2x + 0.2Dx + 100x = 0$, $D = d/dt$. Find the displacement as a function of time, subject to the conditions $x = 4$ and $Dx = 0$ when $t = 0$.

4. What must be the value of b so that the motion of an object given by the equation $D^2x + bDx + 100x = 0$ is critically damped?

5. When the angular displacement θ of a pendulum is small (less than about 6°), the pendulum moves with simple harmonic motion closely approximated by $D^2\theta + \dfrac{g}{l}\theta = 0$. Here, $D = d/dt$, g is the acceleration due to gravity, and l is the length of the pendulum. Find θ as a function of time (in s) if $g = 9.8$ m/s², $l = 1.0$ m, $\theta = 0.1$, and $D\theta = 0$ when $t = 0$. Sketch the curve.

6. A block of wood floating in oil is depressed from its equilibrium position such that its equation of motion is $D^2y + 8Dy + 3y = 0$, where y is the displacement (in cm) and $D = d/dt$. Find its displacement after 12 s if $y = 6.0$ cm and $Dy = 0$ when $t = 0$.

7. A car suspension is depressed from its equilibrium position such that its equation of motion is $D^2y + bDy + 25y = 0$, where y is the displacement and $D = d/dt$. What must be the value of b if the motion is critically damped?

8. A mass of 0.820 kg stretches a given spring by 0.250 m. The mass is pulled down 0.150 m below the equilibrium position and released. Find the equation of motion of the mass if there is no damping.

9. A 4.00-N weight stretches a certain spring 5.00 cm. With this weight attached, the spring is pulled 10.0 cm longer than its equilibrium length and released. Find the equation of the resulting motion, assuming no damping.

10. Find the solution for the spring of Exercise 9 if a damping force numerically equal to the velocity is present.

11. Find the solution for the spring of Exercise 9 if no damping is present but an external force of $4 \sin 2t$ is acting on the spring.

12. Find the solution for the spring of Exercise 9 if the damping force of Exercise 10 and the impressed force of Exercise 11 are both acting.

13. Find the equation relating the charge and the time in an electric circuit with the following elements: $L = 0.200$ H, $R = 8.00\ \Omega$, $C = 1.00\ \mu$F, and $E = 0$. In this circuit, $q = 0$ and $i = 0.500$ A when $t = 0$.

14. For a given electric circuit, $L = 2$ mH, $R = 0$, $C = 50$ nF, and $E = 0$. Find the equation relating the charge and the time if $q = 10^5$ C and $i = 0$ when $t = 0$.

15. For a given circuit, $L = 0.100$ H, $R = 0$, $C = 100\ \mu$F, and $E = 100$ V. Find the equation relating the charge and the time if $q = 0$ and $i = 0$ when $t = 0$.

16. Find the relation between the current and the time for the circuit of Exercise 15.

17. For a radio tuning circuit, $L = 0.500$ H, $R = 10.0\ \Omega$, $C = 200\ \mu$F, and $E = 120 \sin 120\pi t$. Find the equation relating the charge and time.

18. Find the steady-state current for the circuit of Exercise 17.

19. In a given electric circuit $L = 8.00$ mH, $R = 0$, $C = 0.500\ \mu$F, and $E = 20.0e^{-200t}$ mV. Find the relation between the current and the time if $q = 0$ and $i = 0$ for $t = 0$.

20. Find the current as a function of time for a circuit in which $L = 0.400$ H, $R = 60.0\ \Omega$, $C = 0.200\ \mu$F, and $E = 0.800e^{-100t}$ V, if $q = 0$ and $i = 5.00$ mA for $t = 0$.

21. Find the steady-state current for a circuit with $L = 1.00$ H, $R = 5.00\ \Omega$, $C = 150\ \mu$F, and $E = 120 \sin 100t$ V.

22. Find the steady-state solution for the current in an electric circuit containing the following elements: $C = 20.0\ \mu$F, $L = 2.00$ H, $R = 20.0\ \Omega$, and $E = 200 \sin 10t$.

23. A *cantilever* beam is clamped at the end $x = 0$ and is free at the end $x = L$. Find the equation for the deflection y of the beam in terms of the distance x from one end if it has a constant load distribution of w due to its own weight. See Fig. 30.16.

Fig. 30.16

24. A beam 10 m in length is hinged at both ends and has a variable load distribution of $w = kEIx$, where $k = 7.2 \times 10^{-4}$/m and x is the distance from one end. Find the equation of the deflection y in terms of x.

Named for the French mathematician and astronomer Pierre Laplace (1749–1827).

30.10 LAPLACE TRANSFORMS

Laplace transforms *provide an algebraic method of obtaining a **particular** solution of a differential equation from stated initial conditions.* Since this is frequently what is wanted, Laplace transforms are often used in engineering and electronics. The treatment in this text is intended only an as introduction to Laplace transforms.

The Laplace transform of a function $f(t)$ is defined as the function $F(s)$ as

$$F(s) = \int_0^\infty e^{-st} f(t)\, dt \qquad (30.21)$$

By writing the transform as $F(s)$, we show that the result of integrating and evaluating is a function of s. To denote that we are dealing with "the Laplace transform of the function $f(t)$," the notation $\mathcal{L}(f)$ is used. Thus,

$$F(s) = \mathcal{L}(f) = \int_0^\infty e^{-st} f(t)\, dt \qquad (30.22)$$

We shall see that both notations are quite useful.

In Eqs. (30.21) and (30.22), we note that the upper limit is ∞, which means it is unbounded. This integral is one type of what is known as an **improper integral.** In evaluating at the upper limit, it is necessary to find the limit of the resulting function as the upper limit approaches infinity. This may be shown as

$$\lim_{c \to \infty} \int_0^c e^{-st} f(t)\, dt$$

where we substitute c for t in the resulting function and determine the limit as $c \to \infty$ to determine the result for the upper limit. This also means that the Laplace transform, $F(s)$, is defined only for those values of s for which the limit is defined.

◀ **EXAMPLE 1**   Find the Laplace transform of the function $f(t) = t, t > 0$.
By the definition of the Laplace transform

$$\mathcal{L}(f) = \mathcal{L}(t) = \int_0^\infty e^{-st} t\, dt$$

This may be integrated by parts or by Formula (44) in Appendix E. Using the formula, we have

$$\mathcal{L}(t) = \int_0^\infty t e^{-st}\, dt = \lim_{c \to \infty} \int_0^c t e^{-st}\, dt = \lim_{c \to \infty} \left. \frac{e^{-st}(-st-1)}{s^2} \right|_0^c$$

$$= \lim_{c \to \infty} \left[\frac{e^{-sc}(-sc-1)}{s^2} \right] + \frac{1}{s^2}$$

Now, for $s > 0$, as $c \to \infty$, $e^{-sc} \to 0$ and $sc \to \infty$. However, although we cannot prove it here, $e^{-sc} \to 0$ much faster than $sc \to \infty$. We can see that this is reasonable, for $ce^{-c} = 4.5 \times 10^{-4}$ for $c = 10$ and $ce^{-c} = 3.7 \times 10^{-42}$ for $c = 100$. Thus, the value at the upper limit approaches zero, which means the limit is zero. Thus,

$$\mathcal{L}(t) = \frac{1}{s^2} \qquad \text{defined for } s > 0$$

◖ EXAMPLE 2   Find the Laplace transform of the function $f(t) = \cos at$.
By definition,

$$\mathcal{L}(f) = \mathcal{L}(\cos at) = \int_0^\infty e^{-st} \cos at \, dt$$

Using Formula (50) in Appendix E, we have

$$\mathcal{L}(\cos at) = \int_0^\infty e^{-st} \cos at \, dt = \lim_{c \to \infty} \int_0^c e^{-st} \cos at \, dt$$

$$= \lim_{c \to \infty} \frac{e^{-st}(-s \cos at + a \sin at)}{s^2 + a^2} \Big|_0^c$$

$$= \lim_{c \to \infty} \frac{e^{-sc}(-s \cos ac + a \sin ac)}{s^2 + a^2} - \left(-\frac{s}{s^2 + a^2} \right)$$

$$= 0 + \frac{s}{s^2 + a^2} = \frac{s}{s^2 + a^2} \qquad (s > 0)$$

Therefore, the Laplace transform of the function $\cos at$ is

$$\mathcal{L}(\cos at) = \frac{s}{s^2 + a^2}$$

In both examples the resulting transform was an algebraic function of s.   ◗

We now present a short table of Laplace transforms. They are sufficient for our work in this chapter. More complete tables are available in many references.

Table of Laplace Transforms

	$f(t) = \mathcal{L}^{-1}(F)$	$\mathcal{L}(f) = F(s)$		$f(t) = \mathcal{L}^{-1}(F)$	$\mathcal{L}(f) = F(s)$
1.	1	$\dfrac{1}{s}$	11.	te^{-at}	$\dfrac{1}{(s+a)^2}$
2.	$\dfrac{t^{n-1}}{(n-1)!}$	$\dfrac{1}{s^n}\,(n = 1,2,3,\ldots)$	12.	$t^{n-1}e^{-at}$	$\dfrac{(n-1)!}{(s+a)^n}$
3.	e^{-at}	$\dfrac{1}{s+a}$	13.	$e^{-at}(1 - at)$	$\dfrac{s}{(s+a)^2}$
4.	$1 - e^{-at}$	$\dfrac{a}{s(s+a)}$	14.	$[(b-a)t + 1]e^{-at}$	$\dfrac{s+b}{(s+a)^2}$
5.	$\cos at$	$\dfrac{s}{s^2 + a^2}$	15.	$\sin at - at \cos at$	$\dfrac{2a^3}{(s^2 + a^2)^2}$
6.	$\sin at$	$\dfrac{a}{s^2 + a^2}$	16.	$t \sin at$	$\dfrac{2as}{(s^2 + a^2)^2}$
7.	$1 - \cos at$	$\dfrac{a^2}{s(s^2 + a^2)}$	17.	$\sin at + at \cos at$	$\dfrac{2as^2}{(s^2 + a^2)^2}$
8.	$at - \sin at$	$\dfrac{a^3}{s^2(s^2 + a^2)}$	18.	$t \cos at$	$\dfrac{s^2 - a^2}{(s^2 + a^2)^2}$
9.	$e^{-at} - e^{-bt}$	$\dfrac{b - a}{(s+a)(s+b)}$	19.	$e^{-at} \sin bt$	$\dfrac{b}{(s+a)^2 + b^2}$
10.	$ae^{-at} - be^{-bt}$	$\dfrac{s(a - b)}{(s+a)(s+b)}$	20.	$e^{-at} \cos bt$	$\dfrac{s+a}{(s+a)^2 + b^2}$

An important property of transforms is the **linearity property,**

$$\mathscr{L}[af(t) + bg(t)] = a\mathscr{L}(f) + b\mathscr{L}(g) \tag{30.23}$$

We state this property here since it determines that the transform of a sum of functions is the sum of the transforms. This is of definite importance when dealing with a sum of functions. This property is a direct result of the definition of the Laplace transform.

Another Laplace transform important to the solution of a differential equation is the transform of the derivative of a function. Let us first find the Laplace transform of the first derivative of a function.

By definition,

$$\mathscr{L}(f') = \int_0^\infty e^{-st} f'(t) \, dt$$

To integrate by parts, let $u = e^{-st}$ and $dv = f'(t) \, dt$, so $du = -se^{-st} \, dt$ and $v = f(t)$ (the integral of the derivative of a function is the function). Therefore,

$$\mathscr{L}(f') = e^{-st} f(t) \Big|_0^\infty + s \int_0^\infty e^{-st} f(t) \, dt$$

$$= 0 - f(0) + s\mathscr{L}(f)$$

It is noted that the integral in the second term on the right is the Laplace transform of $f(t)$ by definition. Therefore, *the Laplace transform of the first derivative of a function is*

$$\mathscr{L}(f') = s\mathscr{L}(f) - f(0) \tag{30.24}$$

Applying the same analysis, we may find *the Laplace transform of the second derivative of a function. It is*

$$\mathscr{L}(f'') = s^2\mathscr{L}(f) - sf(0) - f'(0) \tag{30.25}$$

Here it is necessary to integrate by parts twice to derive the result. The transforms of higher derivatives are found in a similar manner.

Equations (30.24) and (30.25) allow us to express the transform of each derivative in terms of s and the transform itself. This is illustrated in the following example.

◀ **EXAMPLE 3** Given that $f(0) = 0$ and $f'(0) = 1$, express the transform of $f''(t) - 2f'(t)$ in terms of s and the transform of $f(t)$.

By using the linearity property and the transforms of the derivatives, we have

$$\mathscr{L}[f''(t) - 2f'(t)] = \mathscr{L}(f'') - 2\mathscr{L}(f') \qquad \text{using Eq. (30.23)}$$

$$= [s^2\mathscr{L}(f) - sf(0) - f'(0)] - 2[s\mathscr{L}(f) - f(0)] \qquad \text{using Eqs. (30.25) and (30.24)}$$

$$= [s^2\mathscr{L}(f) - s(0) - 1] - 2[s\mathscr{L}(f) - 0] \qquad \text{substitute given values}$$

$$= (s^2 - 2s)\mathscr{L}(f) - 1 \qquad\qquad\qquad\qquad ▷$$

INVERSE TRANSFORMS

If the Laplace transform of a function is known, it is then possible to find the function by finding the **inverse transform,**

$$\mathcal{L}^{-1}(F) = f(t) \tag{30.26}$$

where $\mathcal{L}^{-1}$ denotes the inverse transform.

◀ EXAMPLE 4   If $F(s) = \dfrac{s}{s^2 + a^2}$, from Transform (5) of the table we see that

$$\mathcal{L}^{-1}(F) = \mathcal{L}^{-1}\left(\frac{s}{s^2 + a^2}\right) = \cos at$$

$$f(t) = \cos at$$

◀ EXAMPLE 5   If $(s^2 - 2s)\mathcal{L}(f) - 1 = 0$, then

$$\mathcal{L}(f) = \frac{1}{s^2 - 2s} \quad \text{or} \quad F(s) = \frac{1}{s(s - 2)}$$

Therefore, we have

$$f(t) = \mathcal{L}^{-1}(F) = \mathcal{L}^{-1}\left[\frac{1}{s(s - 2)}\right] \qquad \text{inverse transform}$$

$$= -\frac{1}{2}\mathcal{L}^{-1}\left[\frac{-2}{s(s - 2)}\right] \qquad \text{fit form of Transform (4)}$$

$$= -\frac{1}{2}(1 - e^{2t}) \qquad \text{use Transform (4)}$$

The introduction of the factor -2 in Example 5 illustrates that it often takes some algebra to get $F(s)$ to match the proper form in the table. The following examples show that completing the square (see Section 7.2) and partial fractions (see Sections 28.9 and 28.10) can be used to assist in making $F(s)$ fit a form in the table.

Completing the Square   ◀ EXAMPLE 6   If $F(s) = \dfrac{s + 5}{s^2 + 6s + 10}$, then

$$\mathcal{L}^{-1}(F) = \mathcal{L}^{-1}\left[\frac{s + 5}{s^2 + 6s + 10}\right]$$

It appears that this function does not fit any of the forms given. However,

$$s^2 + 6s + 10 = (s^2 + 6s + 9) + 1 = (s + 3)^2 + 1$$

By writing $F(s)$ as

$$F(s) = \frac{(s + 3) + 2}{(s + 3)^2 + 1} = \frac{s + 3}{(s + 3)^2 + 1} + \frac{2}{(s + 3)^2 + 1}$$

we can find the inverse of each term. Therefore,

$$\mathcal{L}^{-1}(F) = e^{-3t}\cos t + 2e^{-3t}\sin t \qquad \text{using Transforms (20) and (19)}$$

$$f(t) = e^{-3t}(\cos t + 2\sin t)$$

Partial Fractions   ❰ EXAMPLE 7   If $F(s) = \dfrac{5s^2 - 17s + 32}{s^3 - 8s^2 + 16s}$, then

$$\mathcal{L}^{-1}(F) = \mathcal{L}^{-1}\left[\frac{5s^2 - 17s + 32}{s^3 - 8s^2 + 16s}\right]$$

To fit forms in the table, we will now use partial fractions.

$$\frac{5s^2 - 17s + 32}{s^3 - 8s^2 + 16s} = \frac{5s^2 - 17s + 32}{s(s-4)^2} = \frac{A}{s} + \frac{B}{s-4} + \frac{C}{(s-4)^2}$$ factor of s, repeated factor $s - 4$

$$5s^2 - 17s + 32 = A(s-4)^2 + Bs(s-4) + Cs$$ multiply each side by $s(s-4)^2$

$s = 0$: $32 = 16A$, $A = 2$

$s = 4$: $5(4^2) - 17(4) + 32 = 4C$, $C = 11$

s^2 terms: $5 = A + B$, $5 = 2 + B$, $B = 3$

$$\mathcal{L}^{-1}(F) = \mathcal{L}^{-1}\left[\frac{2}{s} + \frac{3}{s-4} + \frac{11}{(s-4)^2}\right]$$ substitute in $F(s)$

$$\mathcal{L}^{-1}(F) = f(t) = 2 + 3e^{4t} + 11te^{4t}$$ using Transforms (1), (3), (11) ❱

EXERCISES 30.10

In Exercises 1–4, make the given changes in the indicated examples of this section, and then solve the resulting problems.

1. In Example 1, change the function $f(t)$. Let $f(t) = 1$.

2. In Example 2, change the function $f(t)$. Let $f(t) = \sin at$.

3. In Example 3, interchange the values of $f(0)$ and $f'(0)$.

4. In Example 4, in the function $F(s)$, change the numerator to a.

In Exercises 5–12, find the transforms of the given functions by use of the table.

5. $f(t) = e^{3t}$

6. $f(t) = 1 - \cos 2t$

7. $f(t) = 5t^3e^{-2t}$

8. $f(t) = 2e^{-3t}\sin 4t$

9. $f(t) = \cos 2t - \sin 2t$

10. $f(t) = 2t\sin 3t + e^{-3t}\cos t$

11. $f(t) = 3 + 2t\cos 3t$

12. $f(t) = t^3 - 3te^{-t}$

In Exercises 13–16, express the transforms of the given expressions in terms of s and $\mathcal{L}(f)$.

13. $y'' + y'$, $f(0) = 0$, $f'(0) = 0$

14. $y'' - 3y'$, $f(0) = 2$, $f'(0) = -1$

15. $2y'' - y' + y$, $f(0) = 1$, $f'(0) = 0$

16. $y'' - 3y' + 2y$, $f(0) = -1$, $f'(0) = 2$

In Exercises 17–28, find the inverse transforms of the given functions of s.

17. $F(s) = \dfrac{2}{s^3}$

18. $F(s) = \dfrac{3}{s^2 + 4}$

19. $F(s) = \dfrac{1}{2s + 6}$

20. $F(s) = \dfrac{3}{s^4 + 4s^2}$

21. $F(s) = \dfrac{1}{s^3 + 3s^2 + 3s + 1}$

22. $F(s) = \dfrac{s^2 - 1}{s^4 + 2s^2 + 1}$

23. $F(s) = \dfrac{s + 2}{(s^2 + 9)^2}$

24. $F(s) = \dfrac{s + 3}{s^2 + 4s + 13}$

25. $F(s) = \dfrac{4s^2 - 8}{(s + 1)(s - 2)(s - 3)}$

26. $F(s) = \dfrac{3s + 1}{(s - 1)(s^2 + 1)}$

27. $F(s) = \dfrac{2s + 3}{s^2 - 2s + 5}$

Ⓦ **28.** $F(s) = \dfrac{3s^4 + 3s^3 + 6s^2 + s + 1}{s^5 + s^3}$ (Explain your method of solution.)

30.11 **SOLVING DIFFERENTIAL EQUATIONS BY LAPLACE TRANSFORMS**

We will now show how certain differential equations can be solved by using Laplace transforms. *It must be remembered that these solutions are the **particular** solutions of the equations subject to the given conditions.* The necessary operations were developed in the preceding section. The following examples illustrate the method.

EXAMPLE 1   Solve the differential equation $2y' - y = 0$, if $y(0) = 1$. (Note that we are using y to denote the function.)

Taking transforms of each term in the equation, we have

$$\mathcal{L}(2y') - \mathcal{L}(y) = \mathcal{L}(0)$$
$$2\mathcal{L}(y') - \mathcal{L}(y) = 0$$

$\mathcal{L}(0) = 0$ by direct use of the definition of the transform. Now, using Eq. (30.24), $\mathcal{L}(y') = s\mathcal{L}(y) - y(0)$, we have

$$2[s\mathcal{L}(y) - 1] - \mathcal{L}(y) = 0 \qquad y(0) = 1$$

Solving for $\mathcal{L}(y)$, we obtain

$$2s\mathcal{L}(y) - \mathcal{L}(y) = 2$$
$$\mathcal{L}(y) = \frac{2}{2s - 1} = \frac{1}{s - \frac{1}{2}}$$

Finding the inverse transform, we have

$$y = e^{t/2} \qquad \text{using Transform (3)}$$

You should check this solution with that obtained by methods developed earlier. Also, it should be noted that the solution was essentially an algebraic one. This points out the power and usefulness of Laplace transforms. ***We are able to translate a differential equation into an algebraic form,*** which can in turn be translated into the solution of the differential equation. Thus, we can solve a differential equation by using algebra and specific algebraic forms.

NOTE ▶

EXAMPLE 2   Solve the differential equation $y'' + 2y' + 2y = 0$, if $y(0) = 0$ and $y'(0) = 1$.

Using the same steps as outlined in Example 1, we have

$$\mathcal{L}(y'') + 2\mathcal{L}(y') + 2\mathcal{L}(y) = 0 \qquad \text{take transforms}$$
$$[s^2\mathcal{L}(y) - sy(0) - y'(0)] + 2[s\mathcal{L}(y) - y(0)] + 2\mathcal{L}(y) = 0 \qquad \text{using Eqs. (30.25) and (30.24)}$$
$$[s^2\mathcal{L}(y) - s(0) - 1] + 2[s\mathcal{L}(y) - 0] + 2\mathcal{L}(y) = 0 \qquad \text{substitute given values}$$
$$s^2\mathcal{L}(y) - 1 + 2s\mathcal{L}(y) + 2\mathcal{L}(y) = 0$$
$$(s^2 + 2s + 2)\mathcal{L}(y) = 1 \qquad \text{solve for } \mathcal{L}(y)$$
$$\mathcal{L}(y) = \frac{1}{s^2 + 2s + 2} = \frac{1}{(s + 1)^2 + 1} \qquad \text{take inverse transform}$$
$$y = e^{-t} \sin t \qquad \text{using Transform (19)}$$

EXAMPLE 3 Using Laplace transforms, solve the differential equation $y'' + y = \cos t$, if $y(0) = 1$ and $y'(0) = 2$.

$$\mathscr{L}(y'') + \mathscr{L}(y) = \mathscr{L}(\cos t) \qquad \text{take transforms}$$

$$[s^2\mathscr{L}(y) - s(1) - 2] + \mathscr{L}(y) = \frac{s}{s^2 + 1} \qquad \text{using Eq. (30.25) and Transform (5)}$$

$$(s^2 + 1)\mathscr{L}(y) = \frac{s}{s^2 + 1} + s + 2$$

$$\mathscr{L}(y) = \frac{s}{(s^2 + 1)^2} + \frac{s}{s^2 + 1} + \frac{2}{s^2 + 1}$$

$$y = \frac{t}{2}\sin t + \cos t + 2\sin t \qquad \text{using Transforms (16), (5), (6)}$$

Solving a Word Problem

EXAMPLE 4 A spring is stretched 0.31 m by a weight of 4.9 N (mass of 0.5 kg). The medium resists the motion of the object with a force of $4v$, where v is the velocity of motion. The differential equation describing the displacement y is

$$\frac{1}{2}\frac{d^2y}{dt^2} + 4\frac{dy}{dt} + 16y = 0 \qquad \text{see Example 3, page 942}$$

Find y as a function of time t, if $y(0) = 1$ and $dy/dx = 0$ for $t = 0$.

Clearing fractions and denoting derivatives by y'' and y', we have the following differential equation and solution.

$$y'' + 8y' + 32y = 0$$

$$\mathscr{L}(y'') + 8\mathscr{L}(y') + 32\mathscr{L}(y) = 0 \qquad \text{take transforms}$$

$$[s^2\mathscr{L}(y) - s(1) - 0] + 8[s\mathscr{L}(y) - 1] + 32\mathscr{L}(y) = 0 \qquad \text{substitute given values}$$

$$(s^2 + 8s + 32)\mathscr{L}(y) = s + 8 \qquad \text{solve for } \mathscr{L}(y)$$

$$\mathscr{L}(y) = \frac{s + 8}{(s + 4)^2 + 4^2} = \frac{s + 4}{(s + 4)^2 + 4^2} + \frac{4}{(s + 4)^2 + 4^2} \qquad \text{fit transform forms}$$

$$y = e^{-4t}\cos 4t + e^{-4t}\sin 4t = e^{-4t}(\cos 4t + \sin 4t) \qquad \text{take inverse transforms}$$

The graph of this solution is shown in Fig. 30.17.

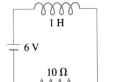

Fig. 30.17

EXAMPLE 5 The initial current in the circuit shown in Fig. 30.18 is zero. Find the current i as a function of the time t.

The differential equation for this circuit is

$$\frac{di}{dt} + 10i = 6 \qquad \text{using Eq. (30.20)}$$

Fig. 30.18

Following the procedures outlined in the previous examples, the solution is found.

$$\mathscr{L}\left(\frac{di}{dt}\right) + 10\mathscr{L}(i) = \mathscr{L}(6) \qquad \text{take transforms}$$

$$[s\mathscr{L}(i) - 0] + 10\mathscr{L}(i) = \frac{6}{s} \qquad \text{substitute given values and find transform on right}$$

$$\mathscr{L}(i) = \frac{6}{s(s + 10)} \qquad \text{solve for } \mathscr{L}(i)$$

$$i = 0.6(1 - e^{-10t}) \qquad \text{take inverse transform}$$

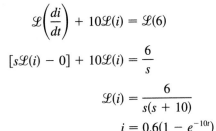

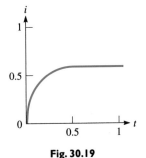

Fig. 30.19

The graph of this solution is shown in Fig. 30.19.

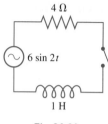

Fig. 30.20

EXAMPLE 6 An electric circuit in an FM radio transmitter contains a 1-H inductor and a 4-Ω resistor. It is being tested using a voltage source of 6 sin 2t. If the initial current is zero, find the current i as a function of time t. See Fig. 30.20.

The solution is as follows:

$$(1)Di + 4i = 6 \sin 2t \qquad \text{differential equation, } D = d/dt$$

$$\mathcal{L}(Di) + 4\mathcal{L}(i) = 6\mathcal{L}(\sin 2t) \qquad \text{take transforms}$$

$$[s\mathcal{L}(i) - 0] + 4\mathcal{L}(i) = \frac{6(2)}{s^2 + 4} \qquad i(0) = 0$$

$$\mathcal{L}(i) = \frac{12}{(s + 4)(s^2 + 4)} = \frac{A}{s + 4} + \frac{Bs + C}{s^2 + 4} \qquad \text{use partial fractions}$$

$$12 = A(s^2 + 4) + B(s^2 + 4s) + C(s + 4)$$

$s = 0$: $12 = 4A + 4C$, $3 = A + C$
s terms: $0 = 4B + C$
s^2 terms: $0 = A + B$

The equations that give us the following values are shown in the margin at the left.

$$A = 0.6 \qquad B = -0.6 \qquad C = 2.4$$

$$\mathcal{L}(i) = 0.6\left(\frac{1}{s + 4}\right) - 0.6\left(\frac{s}{s^2 + 4}\right) + 1.2\left(\frac{2}{s^2 + 4}\right) \qquad \text{take inverse transforms}$$

$$i = 0.6e^{-4t} - 0.6 \cos 2t + 1.2 \sin 2t$$

This is checked by showing that $i(0) = 0$ and that it satisfies the original equation.

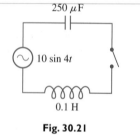

Fig. 30.21

EXAMPLE 7 An electric circuit contains a 0.1-H inductor, a 250-μF capacitor, a voltage source of 10 sin 4t, and negligible resistance (R = 0). See Fig. 30.21. If the initial charge on the capacitor and initial current are zero, find the current as a function of time.

The solution is as follows:

$$0.1 D^2q + \frac{1}{250 \times 10^{-6}}q = 10 \sin 4t \qquad \text{differential equation, } D = d/dt$$

$$D^2q + 40\,000q = 100 \sin 4t$$

$$\mathcal{L}(D^2q) + 40\,000\mathcal{L}(q) = 100\mathcal{L}(\sin 4t) \qquad \text{take transforms}$$

$s = 0$: $400 = 16B + 200^2E$
s terms: $0 = 16A + 200^2C$
s^2 terms: $0 = B + E$
s^3 terms: $0 = A + C$

$$[s^2\mathcal{L}(q) - sq(0) - Dq(0)] + 40\,000\mathcal{L}(q) = \frac{400}{s^2 + 16} \qquad q(0) = 0, D(q) = 0$$

$$\mathcal{L}(q) = \frac{400}{(s^2 + 200^2)(s^2 + 16)} = \frac{As + B}{s^2 + 200^2} + \frac{Cs + E}{s^2 + 16} \qquad \text{use partial fractions}$$

$$400 = (As + B)(s^2 + 16) + (Cs + E)(s^2 + 200^2)$$

The equations that give us the following values are shown in the margin at the left.

$$A = 0 \qquad B = -0.010 \qquad C = 0 \qquad E = 0.010$$

$$\mathcal{L}(q) = \frac{0.010}{s^2 + 16} - \frac{0.010}{s^2 + 200^2} = \frac{0.010}{4}\left(\frac{4}{s^2 + 16}\right) - \frac{0.010}{200}\left(\frac{200}{s^2 + 200^2}\right)$$

$$q = 0.0025 \sin 4t - 5.0 \times 10^{-5} \sin 200t \qquad \text{take inverse transforms}$$

$$i = 0.010 \cos 4t - 0.010 \cos 200t$$

EXERCISES 30.11

In Exercises 1–4, make the given changes in the indicated examples of this section, and then solve the resulting problems.

 1. In Example 1, change the function $y(0)$ from 1 to 2.

 2. In Example 2, interchange the values of $y(0)$ and $y'(0)$.

 3. In Example 3, interchange the values of $y(0)$ and $y'(0)$.

 4. In Example 5, change the initial current to 1 A.

In Exercises 5–36, solve the given differential equations by Laplace transforms. The function is subject to the given conditions.

 5. $y' + y = 0$, $y(0) = 1$

 6. $y' - 2y = 0$, $y(0) = 2$

 7. $2y' - 3y = 0$, $y(0) = -1$

 8. $y' + 2y = 1$, $y(0) = 0$

 9. $y' + 3y = e^{-3t}$, $y(0) = 1$

 10. $y' + 2y = te^{-2t}$, $y(0) = 0$

 11. $y'' + 4y = 0$, $y(0) = 0$, $y'(0) = 1$

 12. $9y'' - 4y = 0$, $y(0) = 1$, $y'(0) = 0$

 13. $y'' + 2y' = 0$, $y(0) = 0$, $y'(0) = 2$

 14. $y'' + 2y' + y = 0$, $y(0) = 0$, $y'(0) = -2$

 15. $y'' - 4y' + 5y = 0$, $y(0) = 1$, $y'(0) = 2$

 16. $4y'' + 4y' + y = 0$, $y(0) = 1$, $y'(0) = 0$

 17. $y'' + y = 1$, $y(0) = 1$, $y'(0) = 1$

 18. $y'' + 4y = 2t$, $y(0) = 0$, $y'(0) = 0$

 19. $y'' + 2y' + y = e^{-t}$, $y(0) = 1$, $y'(0) = 2$

 20. $2y'' + 8y = 3 \sin 2t$, $y(0) = 0$, $y'(0) = 0$

 21. $y'' - 4y = 10e^{3t}$, $y(0) = 5$, $y'(0) = 0$

 22. $y'' - 2y' + y = e^{2t}$, $y(0) = 1$, $y'(0) = 3$

 23. $y'' - y = 5 \sin 2t$, $y(0) = 0$, $y'(0) = 1$

 24. $y'' + y' - 2y = \sin 3t$, $y(0) = 0$, $y'(0) = 0$

 25. A constant force of 6 N moves a 2-kg mass through a medium that resists the motion with a force equal to the velocity v. The equation relating the velocity and the time is $2\dfrac{dv}{dt} = 6 - v$. Find v as a function of t if the object starts from rest.

 26. A pendulum moves with simple harmonic motion according to the differential equation $D^2\theta + 20\theta = 0$, where θ is the angular displacement and $D = d/dt$. Find θ as a function of t if $\theta = 0$ and $D\theta = 0.40$ rad/s when $t = 0$.

 27. A 50-Ω resistor, a 4.0-μF capacitor, and a 40-V battery are connected in series. Find the charge on the capacitor as a function of time t if the initial charge is zero.

 28. A 2-H inductor, an 80-Ω resistor, and an 8-V battery are connected in series. Find the current in the circuit as a function of time if the initial current is zero.

 29. A 10-H inductor, a 40-μF capacitor, and a voltage supply whose voltage is given by $100 \sin 50t$ are connected in series in an electric circuit. Find the current as a function of the time if the initial charge on the capacitor is zero and the initial current is zero.

 30. A 20-mH inductor, a 40-Ω resistor, a 50-μF capacitor, and a voltage source of $100e^{-1000t}$ are connected in series in an electric circuit. Find the charge on the capacitor as a function of time t, if $q = 0$ and $i = 0$ when $t = 0$.

 31. The weight on a spring undergoes forced vibrations according to the equation $D^2y + 9y = 18 \sin 3t$. Find its displacement y as a function of the time t, if $y = 0$ and $Dy = 0$ when $t = 0$.

 32. A spring is stretched 1 m by a 20-N weight. The spring is stretched 0.5 m below the equilibrium position with the weight attached and then released. If it is in a medium that resists the motion with a force equal to $12v$, where v is the velocity, find the displacement y of the weight as a function of the time.

 33. For the electric circuit shown in Fig. 30.22, find the current as a function of the time t if the initial current is zero.

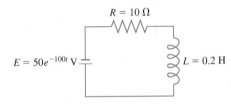

Fig. 30.22

 34. For the electric circuit shown in Fig. 30.23, find the current as a function of time t if the initial charge on the capacitor is zero and the initial current is zero.

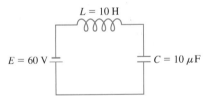

Fig. 30.23

 35. For the electric circuit shown in Fig. 30.24, find the current as a function of the time t, if the initial charge on the capacitor is zero and the initial current is zero.

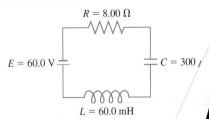

Fig. 30.24

 36. For the beam in Example 6 on page 945, find the de function of x using Laplace transforms. The Lapl the fourth derivative y^{iv} is given by
$$\mathcal{L}(f^{iv}) = s^4\mathcal{L}(f) - s^3f(0) - s^2f'(0) - sf''(0)$$
Also, since $y'(0)$ and $y''(0)$ are not given, $\flat$ sume $y'(0) = a$, and $y'''(0) = b$. It is then and b to obtain the solution.

CHAPTER 30 EQUATIONS

Separation of variables	$M(x, y)\,dx + N(x, y)\,dy = 0$	(30.1)
	$A(x)\,dx + B(y)\,dy = 0$	(30.2)
Integrating combinations	$d(xy) = x\,dy + y\,dx$	(30.3)
	$d(x^2 + y^2) = 2(x\,dx + y\,dy)$	(30.4)
	$d\left(\dfrac{y}{x}\right) = \dfrac{x\,dy - y\,dx}{x^2}$	(30.5)
	$d\left(\dfrac{x}{y}\right) = \dfrac{y\,dx - x\,dy}{y^2}$	(30.6)
Linear differential equation of first order	$dy + Py\,dx = Q\,dx$	(30.7)
	$ye^{\int P\,dx} = \displaystyle\int Qe^{\int P\,dx}\,dx + c$	(30.8)
Electric circuit	$L\dfrac{di}{dt} + Ri + \dfrac{q}{C} = E$	(30.9)
Motion in resisting medium	$m\dfrac{dv}{dt} = F - kv$	(30.10)
General linear differential equation	$a_0\dfrac{d^ny}{dx^n} + a_1\dfrac{d^{n-1}y}{dx^{n-1}} + \cdots + a_{n-1}\dfrac{dy}{dx} + a_n y = b$	(30.11)
	$a_0 D^n y + a_1 D^{n-1} y + \cdots + a_{n-1} Dy + a_n y = b$	(30.12)
Homogeneous linear differential equation	$a_0 D^2 y + a_1 Dy + a_2 y = 0$	(30.13)
Auxiliary equation	$a_0 m^2 + a_1 m + a_2 = 0$	(30.14)
Distinct roots	$y = c_1 e^{m_1 x} + c_2 e^{m_2 x}$	(30.15)
Repeated roots	$y = e^{mx}(c_1 + c_2 x)$	(30.16)
Complex roots	$y = e^{\alpha x}(c_1 \sin \beta x + c_2 \cos \beta x)$	(30.17)
Nonhomogeneous linear differential equation	$a_0 D^2 y + a_1 Dy + a_2 y = b$	(30.18)
	$y = y_c + y_p$	(30.19)
Circuit	$L\dfrac{d^2q}{dt^2} + R\dfrac{dq}{dt} + \dfrac{q}{C} = E$	(30.20)
Transforms	$F(s) = \displaystyle\int_0^\infty e^{-st}f(t)\,dt$	(30.21)
	$F(s) = \mathcal{L}(f) = \displaystyle\int_0^\infty e^{-st}f(t)\,dt$	(30.22)
	$\mathcal{L}[af(t) + bg(t)] = a\mathcal{L}(f) + b\mathcal{L}(g)$	(30.23)
	$\mathcal{L}(f') = s\mathcal{L}(f) - f(0)$	(30.24)
	$\mathcal{L}(f'') = s^2\mathcal{L}(f) - sf(0) - f'(0)$	(30.25)
Inverse transform	$\mathcal{L}^{-1}(F) = f(t)$	(30.26)

CHAPTER 30 REVIEW EXERCISES

In Exercises 1–32, find the general solution to the given differential equations.

1. $4xy^3\,dx + (x^2 + 1)\,dy = 0$ **2.** $\dfrac{dy}{dx} = e^{x-y}$

3. $\sin 2x\,dx + y\sin x\,dy = \sin x\,dx$

4. $x\,dy + y\,dx = y\,dy$

5. $2D^2y + Dy = 0$ **6.** $2D^2y - 5Dy + 2y = 0$

7. $y'' + 2y' + y = 0$ **8.** $y'' + 2y' + 2y = 0$

9. $(x + y)\,dx + (x + y^3)\,dy = 0$ **10.** $R\ln L\,dL = L\,dR$

11. $x\dfrac{dy}{dx} - 3y = x^2$ **12.** $dy - 2y\,dx = (x - 2)e^x\,dx$

13. $dy = 2y\,dx + y^2\,dx$ **14.** $x^2y\,dy = (1 + x)\csc y\,dx$

15. $D^2y + 2Dy + 6y = 0$ **16.** $4D^2y - 4Dy + y = 0$

17. $y' + 4y = 2e^{-2x}$ **18.** $2uv\,du = (2v - \ln v)\,dv$

19. $\sin x\dfrac{dy}{dx} + y\cos x + x = 0$ **20.** $y\,dy = (x^2 + y^2 - x)\,dx$

21. $2\dfrac{d^2s}{dt^2} + \dfrac{ds}{dt} - 3s = 6$ **22.** $\dfrac{d^2y}{dx^2} + 6\dfrac{dy}{dx} + 9y = 3x$

23. $y'' + y' - y = 2e^x$ **24.** $4D^3y + 9Dy = xe^x$

25. $9D^2y - 18Dy + 8y = 16 + 4x$

26. $y'' + y = 4\cos 2x$

27. $D^3y - D^2y + 9Dy - 9y = \sin x$

28. $y'' + y' = e^x + \cos 2x$

29. $y'' - 7y' - 8y = 2e^{-x}$ **30.** $3y'' - 6y' = 4 + xe^x$

31. $D^2y + 25y = 50\cos 5x$ **32.** $D^2y + 4y = 8x\sin 2x$

In Exercises 33–40, find the indicated particular solution of the given differential equations.

33. $3y' = 2y\cot x;\quad x = \dfrac{\pi}{2}$ when $y = 2$

34. $T\,dV - V\,dT = V^3\,dV;\quad T = 1$ when $V = 3$

35. $y' = 4x - 2y;\quad x = 0$ when $y = -2$

36. $xy^2\,dx + e^x\,dy = 0;\quad x = 0$ when $y = 2$

37. $\dfrac{d^2y}{dx^2} + \dfrac{dy}{dx} + 4y = 0;\quad Dy = \sqrt{15},\ y = 0$ when $x = 0$

38. $5y'' + 7y' - 6y = 0;\quad y' = 10,\ y = 2$ when $x = 0$

39. $D^2y + 4Dy + 4y = 4\cos x;\quad Dy = 1,\ y = 0$ when $x = 0$

40. $y'' - 2y' + y = e^x + x;\quad y = 0,\ y' = 0$ when $x = 0$

In Exercises 41–48, solve the given differential equations by using Laplace transforms, where the function is subject to the given conditions.

41. $4y' - y = 0,\ y(0) = 1$ **42.** $2y' - y = 4,\ y(0) = 1$

43. $y' - 3y = e^t,\ y(0) = 0$ **44.** $y' + 2y = e^{-2t},\ y(0) = 2$

45. $y'' + y = 0,\ y(0) = 0,\ y'(0) = -4$

46. $y'' + 4y' + 5y = 0,\ y(0) = 1,\ y'(0) = 1$

47. $y'' + 9y = 3e^t,\ y(0) = 0,\ y'(0) = 0$

48. $y'' - 2y' + y = e^x + x,\ y(0) = 0,\ y'(0) = 1$

In Exercises 49–84, solve the given problems.

49. An object moves along a hyperbolic path described by $xy = 1$, such that $dx/dt = 2t$. Express x and y in terms of t if $x = 1$, $y = 1$ when $t = 0$.

50. An object moves along a parabolic path described by $y = x^2 + x$, such that $dx/dt = 4t + 1$. Express x and y in terms of t, if both x and y are zero when $t = 0$.

51. The time rate of change of volume of an evaporating substance is proportional to the surface area. Express the radius of an evaporating sphere of ice as a function of time. Let $r = r_0$ when $t = 0$. (*Hint:* Express both V and A in terms of the radius r.)

52. An insulated tank is filled with a solution containing radioactive cobalt. Due to the radioactivity, energy is released and the temperature T (in °C) of the solution rises with the time t (in h). The following equation expresses the relation between temperature and time for a specific case:

$$56\,600 = 262(T - 70) + 20\,200\dfrac{dT}{dt}$$

If the initial temperature is 70°C, what is the temperature 24 h later?

53. In a certain chemical reaction, the velocity of the reaction is proportional to the mass m of the chemical that remains unchanged. If m_0 is the initial mass and dm/dt is the velocity of the reaction, find m as a function of the time t.

54. Under proper conditions, bacteria grow at a rate proportional to the number present. In a certain culture there were 10^4 bacteria present at a given time, and there were 3.0×10^5 bacteria present after 10 h. How many were present after 5.0 h?

55. An object with a mass of 1.00 kg slides down a long inclined plane. The effective force of gravity is 200 N, and the motion is retarded by a force numerically equal to the velocity. If it starts from rest, what is the velocity (in m/s) 4.00 s later?

56. A 760-N object falls from rest under the influence of gravity. Find the equation for the velocity at any time t (in s) if the air resists the motion with a force numerically equal to twice the velocity.

57. After 1.0 h, is it noted that 10% of a certain radioactive material has decayed. Find the half-life of the material.

58. There are initially 100 mg of a certain radioactive material, and after 2.0 years there are 95 mg remaining. Find the expression for the mass m at any time t.

59. Radioactive potassium 40 with a half-life of 1.28×10^9 years is used for dating rock samples. If a given rock sample has 75% of its original amount of potassium 40, how old is the rock?

60. When a gas undergoes an adiabatic change (no gain or loss of heat), the rate of change of pressure with respect to volume is directly proportional to the pressure and inversely proportional to the volume. Express the pressure in terms of the volume.

61. Under ideal conditions, the natural law of population change is that the population increases at a rate proportional to the population at any time. Under these conditions, project the population of the world in 2010 if it reached 5.0 billion in 1987 and 6.0 billion in 1999.

62. A spherical balloon is being blown up such that its volume V increases at a rate proportional to its surface area. Show that this leads to the differential equation $dV/dt = kV^{2/3}$ and solve for V as a function of t.

63. Find the orthogonal trajectories of the family of curves $y = cx^5$.

64. Find the equation of the curves for which their normals at all points are in the direction with the lines connecting the points and the origin.

65. Find the temperature after 1.0 h of an object originally 100°C, if it cools to 90° in 5.0 min in air that is at 20°C. (See Exercise 25 on page 926.)

66. If a circuit contains a resistance R, a capacitance C, and a source of voltage E, express the charge q on the capacitor as a function of time.

?-H inductor, a 40-Ω resistor, and a 20-V battery are connected ~ies. Find the current in the circuit as a function of time if the ~urrent is zero.

~ moved according to the equation $D^2y + 400y = 0$, ~ d/dt. Find the displacement y as a function of the 10 and $Dy = 0$ when $t = 0$.

~ is stretched 0.50 m by a 40-N weight. With this ~ on it, the spring is stretched 0.50 m beyond the ~ and released. Find the equation of the result-~edium in which the weight is suspended ~ a force equal to 16 times the velocity. ~derdamped, critically damped, or over-~e.

70. The end of a vibrating rod moves according to the equation $D^2y + 0.2\,Dy + 4000y = 0$, where y is the displacement and $D = d/dt$. Find y as a function of t if $y = 3.00$ cm and $Dy = -0.300$ cm/s when $t = 0$.

71. A 0.5-H inductor, a 6-Ω resistor, and a 20-mF capacitor are connected in series with a generator for which $E = 24 \sin 10t$. Find the charge on the capacitor as a function of time if the initial charge and initial current are zero.

72. A 5.00-mH inductor and a 10.0-μF capacitor are connected in series with a voltage source of $0.200e^{-200t}$ V. Find the charge on the capacitor as a function of time if $q = 0$ and $i = 4.00$ mA when $t = 0$.

73. Find the equation for the current as a function of time if a resistor of 20 Ω, an inductor of 4 H, a capacitor of 100 μF, and a battery of 100 V are in series. The initial charge on the capacitor is 10 mC, and the initial current is zero.

74. If an electric circuit contains an inductance L, a capacitor with a capacitance C, and a sinusoidal source of voltage $E_0 \sin \omega t$, express the charge q on the capacitor as a function of the time. Assume $q = 0$, $i = 0$ when $t = 0$.

75. The differential equation relating the current and time for a certain electric circuit is $2\,di/dt + i = 12$. Solve this equation by use of Laplace transforms given that the initial current is zero. Evaluate the current for $t = 0.300$ s.

76. A 6-H inductor and a 30-Ω resistor are connected in series with a voltage source of $10 \sin 20t$. Find the current as a function of time if the initial current is zero. Use Laplace transforms.

77. A 0.25-H inductor, a 4.0-Ω resistor, and a 100-μF capacitor are connected in series. If the initial charge on the capacitor is 400 μC and the initial current is zero, find the charge on the capacitor as a function of time. Use Laplace transforms.

78. An inductor of 0.5 H, a resistor of 6 Ω, and a capacitor of 200 μF are connected in series. If the initial charge on the capacitor is 10 mC and the initial current is zero, find the charge on the capacitor as a function of time after the switch is closed. Use Laplace transforms.

79. A mass of 0.25 kg stretches a spring for which $k = 16$ N/m. An external force of $\cos 8t$ is applied to the spring. Express the displacement y of the object as a function of time if the initial displacement and velocity are zero. Use Laplace transforms.

80. A spring is stretched 1.00 m by a mass of 5.00 kg (assume the weight to be 50.0 N). Find the displacement y of the object as a function of time if $y(0) = 1$ m and $dy/dt = 0$ when $t = 0$. Use Laplace transforms.

81. Air containing 20% oxygen passes into a 5.00-L container initially filled with 100% oxygen. A uniform mixture of the air and oxygen then passes from the container at the same rate. What volume of oxygen is in the container after 5.00 L of air have passed into it?

82. The approximate differential equation relating the displacement y of a beam at a horizontal distance x from one end is $EI\dfrac{d^2y}{dx^2} = M$, where E is the modulus of elasticity, I is the moment of inertia of the cross section of the beam perpendicular to its axis, and M is the bending moment at the cross section. If $M = 2000x - 40x^2$ for a particular beam of length L for which $y = 0$ when $x = 0$ and when $x = L$, express y in terms of x. Consider E and I as constants.

83. When a circular disk of mass m and radius r is suspended by a wire at the center of one of its flat faces and the disk is twisted through an angle θ, torsion in the wire tends to turn the disk back in the opposite direction. The differential equation for this case is $\dfrac{1}{2}mr^2\dfrac{d^2\theta}{dt^2} = -k\theta$, where k is a constant. Determine the equation of motion if $\theta = \theta_0$ and $d\theta/dt = \omega_0$ when $t = 0$. See Fig. 30.25.

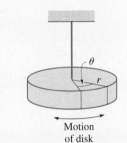

Fig. 30.25

Motion
of disk

84. The gravitational acceleration of an object is inversely proportional to the square of its distance r from the center of the earth. Use the chain rule, Eq. (23.14), to show that the acceleration is $\dfrac{dv}{dt} = v\dfrac{dv}{dr}$, where $v = \dfrac{dr}{dt}$ is the velocity of the object. Then solve for v as a function of r if $dv/dt = -g$ and $v = v_0$ for $r = R$, where R is the radius of the earth. Finally, show that a spacecraft must have a velocity of at least $v_0 = \sqrt{2gR}$ ($= 11$ km/s) in order to escape from the earth's gravitation. (Note the expression for v^2 as $r \to \infty$.)

Writing Exercise

85. An electric circuit contains an inductor L, a resistor R, and a battery of voltage E. The initial current in the circuit is zero. Write three or four paragraphs explaining how the differential equation for the current in the circuit is solved using (a) separation of variables, (b) the linear differential equation of the first order, and (c) Laplace transforms.

CHAPTER 30 PRACTICE TEST

In Problems 1–6, find the general solution of each of the given differential equations.

1. $x\dfrac{dy}{dx} + 2y = 4$

2. $y'' + 2y' + 5y = 0$

3. $x\,dx + y\,dy = x^2\,dx + y^2\,dx$

4. $2D^2y - Dy = 2\cos x$

5. $\dfrac{d^2y}{dx^2} - 4\dfrac{dy}{dx} + 4y = 3x$

6. $D^2y - 2Dy - 8y = 4e^{-2x}$

7. Find the particular solution of the differential equation

$(xy + y)\dfrac{dy}{dx} = 2$, if $y = 2$ when $x = 0$.

8. If interest in a bank account is compounded continuously, the amount grows at a rate that is proportional to the amount present. Derive the equation for the amount A in an account with continuous compounding in which the initial amount is A_0 and the interest rate is r as a function of the time t after A_0 is deposited.

9. Using Laplace transforms, solve the differential equation $y'' + 9y = 9$, if $y(0) = 0$ and $y'(0) = 1$.

10. Using Laplace transforms, solve the differential equation $D^2y - Dy - 2y = 12$, if $y(0) = 0$ and $y'(0) = 0$.

11. Find the equation for the current as a function of the time in a circuit containing a 2-H inductance, an 8-Ω resistor, 6-V battery in series, if $i = 0$ when $t = 0$.

12. A mass of 0.5 kg stretches a spring for which $k = 32$ this weight attached, the spring is pulled 0.3 m lo equilibrium length and released. Find the equation motion, assuming no damping. (The acceleration 9.8 m/s^2.)

ion

e (in s)

r, and a

N/m. With

ger than its

f the resulting

ue to gravity is

Supplementary Topics

HIGHER-ORDER DETERMINANTS

When we introduced determinants in Chapter 5, we limited our discussion to second- and third-order determinants. In this section, we show some basic methods of evaluating higher-order determinants.

From Section 5.7, recall that a *third-order determinant is defined as*

$$\begin{vmatrix} a_1 & b_1 & c_1 \\ a_2 & b_2 & c_2 \\ a_3 & b_3 & c_3 \end{vmatrix} = a_1b_2c_3 + a_3b_1c_2 + a_2b_3c_1 - a_3b_2c_1 - a_1b_3c_2 - a_2b_1c_3 \qquad \text{(S1.1)}$$

Rearranging terms and factoring a_1, $-a_2$, and a_3, we have

$$\begin{vmatrix} a_1 & b_1 & c_1 \\ a_2 & b_2 & c_2 \\ a_3 & b_3 & c_3 \end{vmatrix} = a_1(b_2c_3 - b_3c_2) - a_2(b_1c_3 - b_3c_1) + a_3(b_1c_2 - b_2c_1) \qquad \text{(S1.2)}$$

$$= a_1\begin{vmatrix} b_2 & c_2 \\ b_3 & c_3 \end{vmatrix} - a_2\begin{vmatrix} b_1 & c_1 \\ b_3 & c_3 \end{vmatrix} + a_3\begin{vmatrix} b_1 & c_1 \\ b_2 & c_2 \end{vmatrix} \qquad \text{(S1.3)}$$

In Eq. (S1.3) we see that the third-order determinant is expanded as products of the elements of the first column and second-order determinants whose elements are in neither the same row nor same column as the first-column element. These determinants are called *minors*. In general, *a **minor** of an element of a determinant is the determinant that results by deleting the row and column in which the element lies.*

⟨ EXAMPLE 1

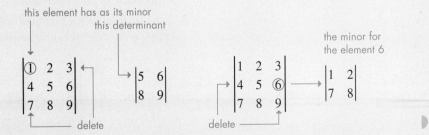

We now see that Eq. (S1.3) expresses the expansion of a third-order determinant as the sum of the products of the elements of the first column and their minors, with the second term assigned a minus sign. Actually this is only one of several ways of expressing the expansion. However, it does lead to a general theorem regarding the expansion of a determinant of any order. The foregoing provides a basis for this theorem, although it cannot be considered as a proof. The theorem is given on the next page.

EXPANSION OF A DETERMINANT BY MINORS

The value of a determinant of order n may be found by forming the n products of the elements of any column (or row) and their minors. A product is given a plus sign if the sum of the number of the column and the number of the row in which the element lies is even, and a minus sign if this sum is odd. The algebraic sum of the terms thus obtained is the value of the determinant.

EXAMPLE 2   In evaluating the following determinant, note that the third column has two zeros. Therefore, expanding by the third column will require less numerical work. This evaluation is performed as follows:

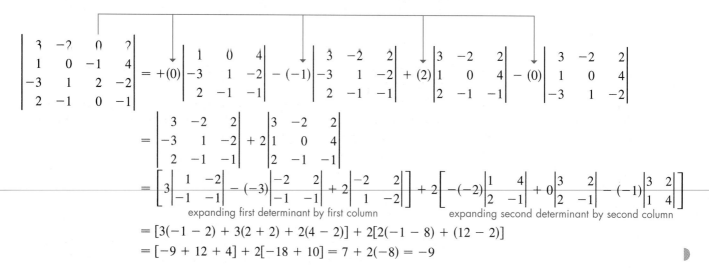

$$= [3(-1-2) + 3(2+2) + 2(4-2)] + 2[2(-1-8) + (12-2)]$$

$$= [-9 + 12 + 4] + 2[-18 + 10] = 7 + 2(-8) = -9$$

PROPERTIES OF DETERMINANTS

We can expand a determinant of any order by minors, but even a fourth-order determinant usually requires a great deal of calculational work. We now present some basic properties of determinants with which they can be evaluated, often with less work than with minors, and not requiring a calculator or computer.

1. *If each element above or each element below the principal diagonal of a determinant is zero, then the product of the elements of the principal diagonal is the value of the determinant.*

2. *If **all** corresponding rows and columns of a determinant are interchanged, the value of the determinant is unchanged.*

EXAMPLE 3   **(a)**
$$\begin{vmatrix} 2 & 1 & 5 & 8 \\ 0 & -5 & 7 & 9 \\ 0 & 0 & 4 & -6 \\ 0 & 0 & 0 & 3 \end{vmatrix} = 2(-5)(4)(3) = -120 \quad \text{property 1}$$

(b)
$$\begin{vmatrix} 1 & 3 & -1 \\ 2 & 0 & 4 \\ -2 & 5 & -6 \end{vmatrix} \quad \text{we obtain the determinant} \quad \begin{vmatrix} 1 & 2 & -2 \\ 3 & 0 & 5 \\ -1 & 4 & -6 \end{vmatrix} \quad \begin{array}{l}\text{property 2} \\ \text{the value of each} \\ \text{determinant is } -18\end{array}$$

3. *If two columns (or rows) of a determinant are identical, the value of the determinant is zero.*

4. *If two columns (or rows) of a determinant are interchanged, the value of the determinant is changed in sign.*

◖ EXAMPLE 4   **(a)**

$$
\begin{vmatrix} 3 & 5 & 2 \\ -4 & 6 & 9 \\ -4 & 6 & 9 \end{vmatrix} = 0 \quad \text{property 3}
$$

identical

(b)
$$
\begin{vmatrix} 3 & 0 & 2 \\ 1 & 1 & 5 \\ 2 & 1 & 3 \end{vmatrix} \quad \text{and} \quad \begin{vmatrix} 2 & 0 & 3 \\ 5 & 1 & 1 \\ 3 & 1 & 2 \end{vmatrix}
$$

property 4
the value of the first is −8, and
the value of the second is 8

5. *If all elements of a column (or row) are multiplied by the same number k, the value of the determinant is multiplied by k.*

6. *If all the elements of any column (or row) are multiplied by the same number k, and the resulting numbers are added to the corresponding elements of another column (or row), the value of the determinant is unchanged.*

◖ EXAMPLE 5   **(a)**
$$
\begin{vmatrix} -1 & 0 & 6 \\ 6 & 3 & -6 \\ 0 & 5 & 3 \end{vmatrix} = 3 \begin{vmatrix} -1 & 0 & 6 \\ 2 & 1 & -2 \\ 0 & 5 & 3 \end{vmatrix}
$$

property 5
the value of the first is 141,
and the value of the second
is 47
$141 = 3(47)$

(b) The value of the following determinant is unchanged if we multiply each element of the first row by 2 and add these numbers to the corresponding elements of the second row. The value of each determinant is −37. The great value in using Property 6 is that we can purposely place zeros in the resulting determinant.

$4 \times 2 = 8 \quad -1 \times 2 = -2 \quad 3 \times 2 = 6$

$$
\begin{vmatrix} 4 & -1 & 3 \\ 2 & 2 & 1 \\ 1 & 0 & -3 \end{vmatrix} = \begin{vmatrix} 4 & -1 & 3 \\ 2+8 & 2+(-2) & 1+6 \\ 1 & 0 & -3 \end{vmatrix} = \begin{vmatrix} 4 & -1 & 3 \\ 10 & 0 & 7 \\ 1 & 0 & -3 \end{vmatrix} \quad \text{or} \quad \begin{vmatrix} 4 & -1 & 3 \\ 10 & 0 & 7 \\ 1 & 0 & -3 \end{vmatrix} = \begin{vmatrix} 4 & -1 & 3 \\ 2 & 2 & 1 \\ 1 & 0 & -3 \end{vmatrix}
$$

With the use of these six properties, determinants of higher order can be evaluated much more easily. The technique is to

obtain zeros in a given column (or row) in all positions except one.

We can then expand by this column (or row), thereby reducing the order of the determinant. Property 6 is probably the most valuable for obtaining the zeros.

▌EXAMPLE 6   Using these properties, we evaluate the following determinant.

$$\begin{vmatrix} 3 & 2 & -1 & 1 \\ -1 & 1 & 2 & 3 \\ 2 & 2 & 1 & 4 \\ 0 & -1 & -2 & 2 \end{vmatrix} = \begin{vmatrix} 0 & 5 & 5 & 10 \\ -1 & 1 & 2 & 3 \\ 2 & 2 & 1 & 4 \\ 0 & -1 & -2 & 2 \end{vmatrix}$$

Each element of the second row is multiplied by 3, and the resulting numbers are added to the corresponding elements of the first row. Here we have used Property 6. In this way a zero has been placed in column 1, row 1.

$$= \begin{vmatrix} 0 & 5 & 5 & 10 \\ -1 & 1 & 2 & 3 \\ 0 & 4 & 5 & 10 \\ 0 & -1 & -2 & 2 \end{vmatrix}$$

Each element of the second row is multiplied by 2, and the resulting numbers are added to the corresponding elements of the third row. Again, we have used Property 6. Also, a zero has been placed in the first column, third row. We now have three zeros in the first column.

$$= -(-1) \begin{vmatrix} 5 & 5 & 10 \\ 4 & 5 & 10 \\ -1 & -2 & 2 \end{vmatrix}$$

Expand the determinant by the first column. We have now reduced the determinant to a third-order determinant.

$$= 5 \begin{vmatrix} 1 & 1 & 2 \\ 4 & 5 & 10 \\ -1 & -2 & 2 \end{vmatrix}$$

Factor 5 from each element of the first row. Here we are using Property 5.

$$= 5(2) \begin{vmatrix} 1 & 1 & 1 \\ 4 & 5 & 5 \\ -1 & -2 & 1 \end{vmatrix}$$

Factor 2 from each element of the third column. Again we are using Property 5. Also, by doing this we have reduced the size of the numbers, and the resulting numbers are somewhat easier to work with.

$$= 10 \begin{vmatrix} 1 & 1 & 1 \\ 0 & 1 & 1 \\ -1 & -2 & 1 \end{vmatrix}$$

Each element of the first row is multiplied by -4, and the resulting numbers are added to the corresponding elements of the second row. Here we are using Property 6. We have placed a zero in the first column, second row.

$$= 10 \begin{vmatrix} 1 & 1 & 1 \\ 0 & 1 & 1 \\ 0 & -1 & 2 \end{vmatrix}$$

Each element of the first row is added to the corresponding element of the third row. Again, we have used Property 6. A zero has been placed in the first column, third row. We now have two zeros in the first column.

$$= 10(1) \begin{vmatrix} 1 & 1 \\ -1 & 2 \end{vmatrix}$$

Expand the determinant by the first column.

$$= 10(2 + 1) = 30$$

Expand the second-order determinant.

A somewhat more systematic method is to place zeros below the principal diagonal and then use Property 1.   ▌

To display and evaluate a determinant on a calculator, see page 167.

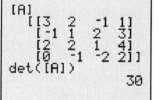

Fig. S1.1

The methods of this section and that of expansion by minors illustrate the ways in which determinants were evaluated before the extensive use of computers and calculators. As we noted in Chapter 5, most graphing calculators can be used to quickly evaluate determinants. In fact, most can evaluate determinants up to the sixth- (or possibly higher) order determinants. Figure S1.1 shows the calculator evaluation of the fourth-order determinant of Example 6.

SOLVING SYSTEMS OF LINEAR EQUATIONS BY DETERMINANTS

We can use the expansion of determinants by minors to solve systems of linear equations. **Cramer's rule** *for solving systems of linear equations, as stated in Section 5.7, is valid for any system of n equations in n unknowns.*

EXAMPLE 7 Solve the following system of equations:

$$
\begin{aligned}
x + 2y + z \quad\;\; &= 5 \\
2x \quad\;\; + z + 2t &= 1 \\
x - y + 3z + 4t &= -6 \\
4x - y \quad\;\;\; - 2t &= 0
\end{aligned}
$$

constants

expanding by fourth row

$$
x = \frac{\begin{vmatrix} 5 & 2 & 1 & 0 \\ 1 & 0 & 1 & 2 \\ -6 & -1 & 3 & 4 \\ 0 & -1 & 0 & -2 \end{vmatrix}}{\begin{vmatrix} 1 & 2 & 1 & 0 \\ 2 & 0 & 1 & 2 \\ 1 & -1 & 3 & 4 \\ 4 & -1 & 0 & -2 \end{vmatrix}} = \frac{-(0)\begin{vmatrix} 2 & 1 & 0 \\ 0 & 1 & 2 \\ -1 & 3 & 4 \end{vmatrix} + (-1)\begin{vmatrix} 5 & 1 & 0 \\ 1 & 1 & 2 \\ -6 & 3 & 4 \end{vmatrix} - (0)\begin{vmatrix} 5 & 2 & 0 \\ 1 & 0 & 2 \\ -6 & -1 & 4 \end{vmatrix} + (-2)\begin{vmatrix} 5 & 2 & 1 \\ 1 & 0 & 1 \\ -6 & -1 & 3 \end{vmatrix}}{(1)\begin{vmatrix} 0 & 1 & 2 \\ -1 & 3 & 4 \\ -1 & 0 & -2 \end{vmatrix} - 2\begin{vmatrix} 2 & 1 & 2 \\ 1 & 3 & 4 \\ 4 & 0 & -2 \end{vmatrix} + (1)\begin{vmatrix} 2 & 0 & 2 \\ 1 & -1 & 4 \\ 4 & -1 & -2 \end{vmatrix} - (0)\begin{vmatrix} 2 & 0 & 1 \\ 1 & -1 & 3 \\ 4 & -1 & 0 \end{vmatrix}}
$$

expanding by first row

$$
= \frac{-(-26) - 2(-14)}{1(0) - 2(-18) + 1(18)} = \frac{26 + 28}{36 + 18} = \frac{54}{54} = 1
$$

We could use minors or the properties of determinants for the evaluations. Noting the two zeros in the fourth row of the numerator, we expanded by minors. For the denominator, which we need to evaluate only once, we used minors of the first row, although we could have used the properties to create zeros in the first column.

In solving for y, we again note the two zeros in the fourth row:

expanding by fourth row

$$
y = \frac{\begin{vmatrix} 1 & 5 & 1 & 0 \\ 2 & 1 & 1 & 2 \\ 1 & -6 & 3 & 4 \\ 4 & 0 & 0 & -2 \end{vmatrix}}{54} = \frac{-4\begin{vmatrix} 5 & 1 & 0 \\ 1 & 1 & 2 \\ -6 & 3 & 4 \end{vmatrix} + (-2)\begin{vmatrix} 1 & 5 & 1 \\ 2 & 1 & 1 \\ 1 & -6 & 3 \end{vmatrix}}{54}
$$

$$
= \frac{-4(-26) - 2(-29)}{54} = \frac{104 + 58}{54} = \frac{162}{54} = 3
$$

Substituting $x = 1$ and $y = 3$ in the first equation gives us $z = -2$. Then substituting $x = 1$ and $y = 3$ in the fourth equation gives us $t = 1/2$. Therefore, the required solution is $x = 1$, $y = 3$, $z = -2$, $t = 1/2$. We can check the solution by substituting in the second or third equation (we used the first and fourth to *find* values of z and t).

EXERCISES S.1

In Exercises 1–4, evaluate each determinant by inspection. Observation will allow evaluation by using the properties of this section.

1. $\begin{vmatrix} 4 & -5 & 8 \\ 0 & 3 & -8 \\ 0 & 0 & -5 \end{vmatrix}$

2. $\begin{vmatrix} 3 & 0 & 0 \\ 0 & 10 & 0 \\ -9 & -1 & -5 \end{vmatrix}$

3. $\begin{vmatrix} 3 & -2 & 4 & 2 \\ 5 & -1 & 2 & -1 \\ 3 & -2 & 4 & 2 \\ 0 & 3 & -6 & 0 \end{vmatrix}$

4. $\begin{vmatrix} -12 & -24 & -24 & 15 \\ 12 & 32 & 32 & -35 \\ -22 & 18 & 18 & 18 \\ 44 & 0 & 0 & -26 \end{vmatrix}$

In Exercises 5–8, use the given value of the determinant at the right and the properties of this section to evaluate the following determinants.

$$\begin{vmatrix} 2 & -3 & 1 \\ -4 & 1 & 3 \\ 1 & -3 & -2 \end{vmatrix} = 40$$

5. $\begin{vmatrix} 2 & 1 & -3 \\ -4 & 3 & 1 \\ 1 & -2 & -3 \end{vmatrix}$

6. $\begin{vmatrix} 2 & -3 & 1 \\ -4 & 1 & 3 \\ 2 & -6 & -4 \end{vmatrix}$

7. $\begin{vmatrix} 2 & -3 & -1 \\ -4 & 1 & -3 \\ 1 & -3 & 2 \end{vmatrix}$

8. $\begin{vmatrix} 2 & -4 & 1 \\ -3 & 1 & -3 \\ 1 & 3 & -2 \end{vmatrix}$

In Exercises 9–18, evaluate the given determinants by expansion by minors.

9. $\begin{vmatrix} 3 & 0 & 0 \\ -2 & 1 & 4 \\ 4 & -2 & 5 \end{vmatrix}$

10. $\begin{vmatrix} 10 & 0 & -3 \\ -2 & -4 & 1 \\ 3 & 0 & 2 \end{vmatrix}$

11. $\begin{vmatrix} 3 & 1 & 0 \\ -2 & 3 & -1 \\ 4 & 2 & 5 \end{vmatrix}$

12. $\begin{vmatrix} -4 & 3 & -2 \\ -2 & 2 & 4 \\ -1 & 5 & -3 \end{vmatrix}$

13. $\begin{vmatrix} 4 & 3 & 6 & 0 \\ 3 & 0 & 0 & 4 \\ 5 & 0 & 1 & 2 \\ 2 & 1 & 1 & 7 \end{vmatrix}$

14. $\begin{vmatrix} 6 & -3 & -6 & 3 \\ -2 & 1 & 2 & -1 \\ 18 & 7 & -1 & 5 \\ 0 & -1 & 10 & 10 \end{vmatrix}$

15. $\begin{vmatrix} 1 & 3 & -3 & 5 \\ 4 & 2 & 1 & 2 \\ 3 & 2 & -2 & 2 \\ 0 & 1 & 2 & -1 \end{vmatrix}$

16. $\begin{vmatrix} -2 & 2 & 1 & 3 \\ 1 & 4 & 3 & 1 \\ 4 & 3 & -2 & -2 \\ 3 & -2 & 1 & 5 \end{vmatrix}$

17. $\begin{vmatrix} 1 & 2 & 0 & 1 & 0 \\ 0 & 2 & 1 & 0 & 1 \\ 1 & 0 & -1 & 1 & -1 \\ -2 & 0 & -1 & 2 & 1 \\ 1 & 0 & 2 & -1 & -2 \end{vmatrix}$

18. $\begin{vmatrix} -1 & 3 & 5 & 0 & -5 \\ 0 & 1 & 7 & 3 & -2 \\ 5 & -2 & -1 & 0 & 3 \\ -3 & 0 & 2 & -1 & 3 \\ 6 & 2 & 1 & -4 & 2 \end{vmatrix}$

In Exercises 19–28, use the determinants for Exercises 9–18 and evaluate each using the properties of determinants. Do not evaluate directly more than one second-order determinant.

In Exercises 29–32, solve the given systems of equations by determinants. Evaluate by expansion by minors.

29. $x + t = 0$
$3x + y + z = -1$
$2y - z + 3t = 1$
$2z - 3t = 1$

30. $2x + y + z = 4$
$2y - 2z - t = 3$
$3y - 3z + 2t = 1$
$6x - y + t = 0$

31. $x + 2y - z = 6$
$y - 2z - 3t = -5$
$3x - 2y + t = 2$
$2x + y + z - t = 0$

32. $2p + 3r + s = 4$
$p - 2r - 3s + 4t = -1$
$3p + r + s - 5t = 3$
$-p + 2r + s + 3t = 2$

In Exercises 33–36, solve the given systems of equations by determinants. Evaluate by using the properties of determinants.

33. $2x + y + z = 2$
$3y - z + 2t = 4$
$y + 2z + t = 0$
$3x + 2z = 4$

34. $2x + y + z = 0$
$x - y + 2t = 2$
$2y + z + 4t = 2$
$5x + 2z + 2t = 4$

35. $D + E + 2F = 1$
$2D - E + G = -2$
$D - E - F - 2G = 4$
$2D - E + 2F - G = 0$

36. $3x + y + t = 0$
$3z + 2t = 8$
$6x + 2y + 2z + t = 3$
$3x - y - z - t = 0$

In Exercises 37–40, solve the given problems by using determinants.

37. In applying Kirchoff's laws (see Exercise 40 on page 161) to the circuit shown in Fig. S1.2, the following equations are found. Determine the indicated currents (in A).

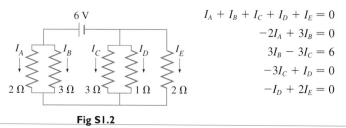

Fig S1.2

$I_A + I_B + I_C + I_D + I_E = 0$
$-2I_A + 3I_B = 0$
$3I_B - 3I_C = 6$
$-3I_C + I_D = 0$
$-I_D + 2I_E = 0$

38. In analyzing the forces A, B, C, and D shown on the beam in Fig. S1.3, the following equations are used. Find these forces.

$A + B = 850$
$A + B + 400 = 0.8C + 0.6D$
$0.6C = 0.8D$
$5A - 5B + 4C - 3D = 0$

Fig S1.3

39. In testing for air pollution, a given air sample contained 6.0 parts per million (ppm) of four pollutants, sulfur dioxide (SO_2), nitric oxide (NO), nitrogen dioxide (NO_2), and carbon monoxide (CO). The ppm of CO was 10 times that of SO_2, which in turn equaled those of NO and NO_2. There was a total of 0.8 ppm of SO_2 and NO. How many ppm of each were present in the air sample?

40. Three computer programs, A, B, and C, use 15% of a computer's 6.0 GB (gigabyte) hard-drive memory. If two additional programs are added, one requiring the same memory as A and the other half the memory of C, 22% of the memory will be used. However, if two other programs are added to A, B, and C, one requiring half the memory of B and the other the same memory as C, 25% of the memory will be used. How many megabytes of memory are required for each of A, B, and C?

S.2 Gaussian Elimination

Named for the German mathematician Karl Gauss (1777–1855).

We now show a general method that can be used to solve a system of linear equations. The procedure is similar to that used in finding the inverse of a matrix in Section 16.3. It is known as **Gaussian elimination** and is commonly used in computer programs. When using this method, we first rewrite the system of equations in a different form, and the next example illustrates this form.

$$x - 3y - z = 1$$
$$y + 2z = 5$$
$$z = 2$$

◀ EXAMPLE 1 In solving the system of equations shown to the left, we see that the third equation directly gives us the value $z = 2$. Since the second equation contains only y and z, we can substitute $z = 2$ into the second equation to get $y = 1$. Then, we can find x by substituting $y = 1$ and $z = 2$ into the first equation, and we get $x = 6$.

Gaussian elimination is based on first writing the system of equations in a form like that shown in Example 1, so that the last equation shows the value of one unknown, and the others are found by substituting back, as in Example 1. The method is based on the use of two operations on the equations of the original system.

1. *Both sides of an equation may be multiplied by a constant.*
2. *A multiple of one equation may be added to another equation.*

Using three linear equations in three unknowns as an example, by using the above operations we can change the system

$$a_1 x + b_1 y + c_1 z = d_1$$
$$a_2 x + b_2 y + c_2 z = d_2 \tag{S2.1}$$
$$a_3 x + b_3 y + c_3 z = d_3$$

into the equivalent system

$$x + b_4 y + c_4 z = d_4$$
$$y + c_5 z = d_5 \tag{S2.2}$$
$$z = d_6$$

The solution is completed by substituting the value of z into the second equation, and then substituting the values of y and z into the first equation, as in Example 1.

◀ EXAMPLE 2 Solve the given system of equations by Gaussian elimination. Given system of equations (equations and solution at left)

This is in the general form of Eq. (S2.1).

$$2x + y = 4$$
$$3x - 2y = 3$$

$$x + \tfrac{1}{2}y = 2$$
$$3x - 2y = 3$$

We want the coeffficient of x in the first equation to be 1. Therefore, divide the first equation by 2, the coefficient of x.

$$x + \tfrac{1}{2}y = 2$$
$$- \tfrac{7}{2}y = -3$$

We next eliminate x in the second equation by subtracting 3 times the first equation from the second equation.

This is in the general form of Eq. (S2.2).

$$x + \tfrac{1}{2}y = 2$$
$$y = \tfrac{6}{7}$$

Now we solve the second equation for y by dividing by $-\tfrac{7}{2}$.

$$x + \tfrac{1}{2}\left(\tfrac{6}{7}\right) = 2$$
$$x = \tfrac{11}{7}$$

To find the value of x, we substitute the value of $y = \tfrac{6}{7}$ into the first equation.

The solution is $x = \tfrac{11}{7}$, $y = \tfrac{6}{7}$, which checks when substituted into the original equations.

◀ EXAMPLE 3 Solve the given system of equations by Gaussian elimination.
Given system of equations (equations and solution at left)

This is in the general form of Eq. (S2.1).

$$x + 3y - 2z = -5$$
$$2x - y + 4z = 7$$
$$-3x + 2y - 3z = -1$$

Since the coeffficient of x in the first equation is 1, proceed to the next step. Subtract 2 times the first equation from the second equation and add 3 times the first equation to the third equation to eliminate x from the second and third equations.

$$x + 3y - 2z = -5$$
$$- 7y + 8z = 17$$
$$11y - 9z = -16$$

To get the coefficient of y equal to 1 in the second equation, divide it by -7.

$$x + 3y - 2z = -5$$
$$y - \tfrac{8}{7}z = -\tfrac{17}{7}$$
$$11y - 9z = -16$$

To eliminate y from the third equation, subtract 11 times the second equation from the third equation.

$$x + 3y - 2z = -5$$
$$y - \tfrac{8}{7}z = -\tfrac{17}{7}$$
$$\tfrac{25}{7}z = \tfrac{75}{7}$$

This is in the form of Eq. (S2.2).

Solve for z by multiplying the third equation by $\tfrac{7}{25}$ (or dividing by $\tfrac{25}{7}$).

$$x + 3y - 2z = -5$$
$$y - \tfrac{8}{7}z = -\tfrac{17}{7}$$
$$z = 3$$

The value of y is found by substituting $z = 3$ into the second equation.

$$y - \tfrac{8}{7}(3) = -\tfrac{17}{7}$$
$$y = 1$$

The value of x is found by substituting $y = 1$ and $z = 3$ back into the first equation.

$$x + 3(1) - 2(3) = -5$$
$$x = -2$$

The solution is $x = -2$, $y = 1$, $z = 3$, which checks when substituted into the original equations.

◀ EXAMPLE 4 Solve the given system of equations by Gaussian elimination.
Given system of equations (equations and solution at left)

This is in the general form of Eq. (S2.1).

$$4y + z = 2$$
$$2x + 6y - 2z = 3$$
$$4x + 8y - 5z = 4$$

Since the first equation does not contain x, which means that $a_1 = 0$, we cannot divide by a_1. Therefore, interchange the first and second equations and then divide the new first equation by 2.

$$x + 3y - z = \tfrac{3}{2}$$
$$4y + z = 2$$
$$4x + 8y - 5z = 4$$

Eliminate x in the third equation by subtracting 4 times the first equation from the third equation.

$$x + 3y - z = \tfrac{3}{2}$$
$$4y + z = 2$$
$$- 4y - z = -2$$

Make the coefficient of y in the second equation equal to 1 by dividing the second equation by 4.

$$x + 3y - z = \tfrac{3}{2}$$
$$y + \tfrac{1}{4}z = \tfrac{1}{2}$$
$$- 4y - z = -2$$

Eliminate y from the third equation by adding 4 times the second equation to the third equation.

This is the form that Eq. (S2.2) takes in this case.

$$x + 3y - z = \tfrac{3}{2}$$
$$y + \tfrac{1}{4}z = \tfrac{1}{2}$$
$$0 = 0$$

Since the third equation, $0 = 0$, is correct, we can continue. Although there is no specific value for z, it is possible to express both x and y in terms of z.

$$x + 3y - z = \tfrac{3}{2}$$
$$y = \tfrac{1}{2} - \tfrac{1}{4}z$$

Solve the second equation for y. In this case, it is expressed in terms of z. We no longer need to include the third equation.

$$x + 3(\tfrac{1}{2} - \tfrac{1}{4}z) - z = \tfrac{3}{2}$$
$$x = \tfrac{7}{4}z$$

Substitute the solution for y in the first equation and solve for x in terms of z.

CAUTION ▶

We have expressed the solution as $x = \tfrac{7}{4}z$ and $y = \tfrac{1}{2} - \tfrac{1}{4}z$. Since both x and y are expressed in terms of z and there is no specific value of z, the value of z can be chosen arbitrarily. This means **there is an unlimited number of solutions.** For example, if $z = 4$, then $x = 7$ and $y = -\tfrac{1}{2}$. If $z = -2$, then $x = -\tfrac{7}{2}$ and $y = 1$.

When we solved systems of linear equations in Chapters 5 and 16, we found that not all systems have unique solutions, as in Examples 2 and 3. In Example 4, we illustrated the use of Gaussian elimination on a system of equations for which the solution is not unique.

In Example 4, one of the equations became $0 = 0$, and there was an unlimited number of solutions. If any of the equations of a system becomes $0 = a$, $a \neq 0$, then the system is inconsistent and there is no solution. Example 5 that follows illustrates such a system.

If a system of equations has more unknowns than equations, or if it can be written in this way, as in Example 3, it usually has an unlimited number of solutions. It is possible, however, that such a system is inconsistent.

If a system of equations has more equations than unknowns, it is inconsistent unless enough equations become $0 = 0$ such that at least one solution is found. The following example illustrates two systems of equations in which there are more equations than unknowns.

First system		Second system	
$x + 2y =$	5	$x + 2y =$	5
$3x - y =$	1	$3x - y =$	1
$4x + y =$	6	$4x + y =$	2
$x + 2y =$	5	$x + 2y =$	5
$-7y =$	-14	$-7y =$	-14
$-7y =$	-14	$-7y =$	-18
$x + 2y =$	5	$x + 2y =$	5
$y =$	2	$y =$	2
$-7y =$	-14	$-7y =$	-18
$x + 2y =$	5	$x + 2y =$	5
$y =$	2	$y =$	2
$0 =$	0	$0 =$	-4
$x =$	1		

EXAMPLE 5 Solve the following systems of equations by Gaussian elimination.

The solutions are shown at the left. We note that each system has three equations and two unknowns.

In the solution of the first system, the third equation becomes $0 = 0$, and only two equations are needed to find the solution $x = 1$, $y = 2$.

In the solution of the second system, the third equation becomes $0 = -4$, which means the system is inconsistent and there is no solution.

The solutions are shown graphically in Figs. S2.1 and S2.2. In Fig. S2.1, each of the three lines passes through the point $(1, 2)$, whereas in Fig. S2.2 there is no point common to the three lines.

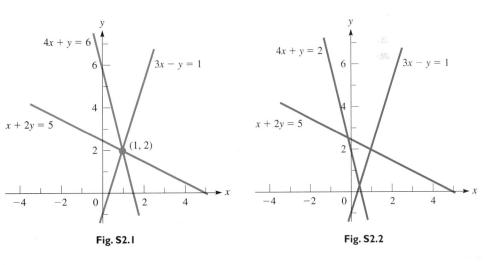

Fig. S2.1   Fig. S2.2

EXERCISES S.2

In Exercises 1–28, solve the given systems of equations by Gaussian elimination. If there is an unlimited number of solutions, find two of them.

1. $x + 2y = 4$
    $3x - y = 5$

2. $2x + y = 1$
    $5x + 2y = 1$

3. $5x - 3y = 2$
    $-2x + 4y = 3$

4. $-3x + 2y = 4$
    $4x + y = -5$

5. $2x + y - z = 0$
    $4x + y + z = 2$
    $-2x - 2y + 3z = 0$

6. $2y + 2z = -1$
    $3x - 4y + 3z = 1$
    $4x + 2y + 5z = 4$

7. $2x - 3y + z = 4$
$6y - 4x - 2z = 9$

8. $3s + 4t - u = -5$
$2u - 6s - 8t = 10$

9. $x + 3y + 3z = -3$
$2x + 2y + z = -5$
$-2x - y + 4z = 6$

10. $3x - y + 2z = 3$
$4x - 2y + z = 3$
$6x + 6y + 3z = 4$

11. $w + 2x - y + 3z = 12$
$2w - 2y - z = 3$
$3x - y - z = -1$
$-w + 2x + y + 2z = 3$

12. $2x - 3y + 2z + 2t = 3$
$4x + 2y - 3z = -4$
$2x - y + 3z + 2t = 3$
$6x + 3y - 2z - t = 2$

13. $x - 4y + z = 2$
$3x - y + 4z = -4$

14. $4x + z = 6$
$2x - y - 2z = -2$

15. $2x - y + z = 5$
$3x + 2y - 2z = 4$
$5x + 8y - 8z = 5$

16. $3u + 6v + 2w = -2$
$u + 3v - 4w = 2$
$2u - 3v - 2w = -2$

17. $x + 3y + z = 4$
$2x - 6y - 3z = 10$
$4x - 9y + 3z = 4$

18. $3x + 2y - z = 3$
$2x - y - 3z = 2$
$-x + 4y + 5z = -1$

19. $2x - 4y = 7$
$3x + 5y = -6$
$9x - 7y = 15$

20. $4x - y = 5$
$2x + 2y = 3$
$6x - 4y = 7$
$2x + y = 4$

21. $3x + 5y = -2$
$24x - 18y = 13$
$15x - 33y = 19$
$6x + 68y = -33$

22. $x + 3y - z = 1$
$3x - y + 4z = 4$
$-2x + 2y + 3z = 17$
$3x + 7y + 5z = 23$

23. $x - 2y - 2z = 3$
$2x + y + 3z = 4$
$-2x - y - z = 5$
$3x + 3y - 2z = 2$

24. $2x - y - 2z - t = 4$
$4x + 2y + 3z + 2t = 3$
$-2x - y + 4z = -2$

25. $s + 2t - 3u = 2$
$6t + 3s - 9u = 6$
$7s + 14t - 21u = 13$

26. $x + 2y - 3z + 2t = 3$
$4y - 6z + 4t = 1$

27. $r - s - 3t - u = 1$
$2r + 4s - 2u = 2$
$r + 5s + 3t - u = 1$
$3r + 4s - 2t = 0$
$r + 2t - 3u = 3$

28. $x + 2y - 3z = 4$
$2x - y - 6z + 2t = 2$
$x + 3y + 3z - t = 1$

In Exercises 29–32, set up systems of equations and solve by Gaussian elimination.

29. One computer can perform x calculations per second, and a second computer can perform y calculations per second. If each operates for 2.00 s, 25.0 million calculations are performed. If the first operates for 4.00 s and the second for 3.00 s, 43.2 million calculations are performed. Find x and y.

30. The voltage across an electric resistor equals the current (in A) times the resistance (in Ω). If a current of 3.00 A passes through each of two resistors, the sum of the voltages is 10.5 V. If 2.00 A passes through the first resistor and 4.00 A passes through the second resistor, the sum of the voltages is 13.0 V. Find the resistances.

31. Three machines together produce 650 parts each hour. Twice the production of the second machine is 10 parts/h more than the sum of the production of the other two machines. If the first operates for 3.00 h and the others operate for 2.00 h, 1550 parts are produced. Find the production rate of each machine.

32. A total of $12 000 is invested, part at 6.5%, part at 6.0%, and part at 5.5%, yielding a total annual interest of $726. The income from the 6.5% part yields $128 less than that for the other two parts combined. How much is invested at each rate?

S.3 ROTATION OF AXES

In Chapter 21, we discussed the circle, parabola, ellipse, and hyperbola and how these curves are represented by the second-degree equation

$$Ax^2 + Bxy + Cy^2 + Dx + Ey + F = 0 \qquad \text{(S3.1)}$$

Our discussion included the properties of the curves and their equations with center (vertex of a parabola) at the origin. However, except for the special case of the hyperbola $xy = c$, we did not cover what happens when the axes are rotated about the origin.

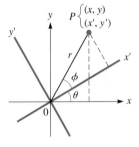

Fig. S3.1

*If a set of axes is rotated about the origin through an angle θ, as shown in Fig. S3.1, we say that there has been a **rotation of axes**.* In this case, each point P in the plane has two sets of coordinates, (x, y) in the original system and (x', y') in the rotated system.

If we now let r equal the distance from the origin O to point P and let ϕ be the angle between the x'-axis and the line OP, we have

$$x' = r \cos \phi \qquad\qquad y' = r \sin \phi \qquad\qquad \text{(S3.2)}$$

$$x = r \cos(\theta + \phi) \qquad\qquad y = r \sin(\theta + \phi) \qquad\qquad \text{(S3.3)}$$

Using the cosine and sine of the sum of two angles, we can write Eqs. (S3.3) as

$$
\begin{aligned}
x &= r \cos \phi \cos \theta - r \sin \phi \sin \theta \\
y &= r \cos \phi \sin \theta + r \sin \phi \cos \theta
\end{aligned}
\qquad\qquad \text{(S3.4)}
$$

Now, using Eqs. (S3.2), we have

$$
\boxed{
\begin{aligned}
x &= x' \cos \theta - y' \sin \theta \\
y &= x' \sin \theta + y' \cos \theta
\end{aligned}
}
\qquad\qquad \text{(S3.5)}
$$

In our derivation, we have used the special case when θ is acute and P is in the first quadrant of both sets of axes. When simplifying equations of curves using Eqs. (S3.5), we find that a rotation through a positive acute angle θ is sufficient. It can be shown, however, that Eqs. (S3.5) hold for any θ and position of P.

◀ **EXAMPLE 1** Transform $x^2 - y^2 + 8 = 0$ by rotating the axes through $45°$.

When $\theta = 45°$, the rotation equations (S3.5) become

$$x = x' \cos 45° - y' \sin 45° = \frac{x'}{\sqrt{2}} - \frac{y'}{\sqrt{2}}$$

$$y = x' \sin 45° + y' \cos 45° = \frac{x'}{\sqrt{2}} + \frac{y'}{\sqrt{2}}$$

Substituting into the equation $x^2 - y^2 + 8 = 0$ gives

$$\left(\frac{x'}{\sqrt{2}} - \frac{y'}{\sqrt{2}}\right)^2 - \left(\frac{x'}{\sqrt{2}} + \frac{y'}{\sqrt{2}}\right)^2 + 8 = 0$$

$$\frac{1}{2}x'^2 - x'y' + \frac{1}{2}y'^2 - \frac{1}{2}x'^2 - x'y' - \frac{1}{2}y'^2 + 8 = 0$$

$$x'y' = 4$$

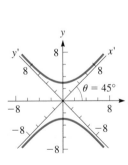

Fig. S3.2

The graph and both sets of axes are shown in Fig. S3.2. The original equation represents a hyperbola. We have the $xy = c$ form with rotation through $45°$. ◗

When we showed the type of curve represented by the second-degree equation Eq. (S3.1) in Section 21.8, the standard forms of the parabola, ellipse, and hyperbola required that $B = 0$. This means that there is no xy-term in the equation. In Section 21.8, we considered only one case $(xy = c)$ for which $B \neq 0$.

If we can remove the xy-term from the second-degree equation, the analysis of the graph is simplified. By a proper rotation of axes we find that Eq. (S3.1) can be transformed into an equation that has no $x'y'$-term.

By substituting Eqs. (S3.5) into Eq. (S3.1) and then simplifying, we have

$$(A \cos^2 \theta + B \sin \theta \cos \theta + C \sin^2 \theta)x'^2 + [B \cos 2\theta - (A - C)\sin 2\theta]x'y'$$
$$+ (A \sin^2 \theta - B \sin \theta \cos \theta + C \cos^2 \theta)y'^2 + (D \cos \theta + E \sin \theta)x' + (E \cos \theta - D \sin \theta)y' + F = 0$$

If there is to be no $x'y'$-term, its coefficient must be zero. This means that $B \cos 2\theta - (A - C)\sin 2\theta = 0$, or

Angle of Rotation

$$\tan 2\theta = \frac{B}{A - C} \qquad (A \neq C) \tag{S3.6}$$

Eq. (S3.6) gives the angle of rotation except when $A = C$. In this case, the coefficient of the $x'y'$-term is $B \cos 2\theta$, which is zero if $2\theta = 90°$. Thus,

$$\theta = 45° \qquad (A = C) \tag{S3.7}$$

Consider the following example.

◀ **EXAMPLE 2** By rotation of axes, transform $8x^2 + 4xy + 5y^2 = 9$ into a form without an xy-term. Identify and sketch the curve.

Here, $A = 8$, $B = 4$, and $C = 5$. Therefore, using Eq. (S3.6), we have

For reference: Eq. (20.2) is

$\cos \theta = \dfrac{1}{\sec \theta}$

Eq. (20.7) is $1 + \tan^2 \theta = \sec^2 \theta$

Eq. (20.26) is $\sin \dfrac{\alpha}{2} = \pm \sqrt{\dfrac{1 - \cos \alpha}{2}}$

Eq. (20.27) is $\cos \dfrac{\alpha}{2} = \pm \sqrt{\dfrac{1 + \cos \alpha}{2}}$

$$\tan 2\theta = \frac{4}{8 - 5} = \frac{4}{3}$$

Since $\tan 2\theta$ is positive, we may take 2θ as an acute angle, which means θ is also acute. For the transformation, we need $\sin \theta$ and $\cos \theta$. We find these values by first finding the value of $\cos 2\theta$ and then using the half-angle formulas:

$$\cos 2\theta = \frac{1}{\sec 2\theta} = \frac{1}{\sqrt{1 + \tan^2 2\theta}} = \frac{1}{\sqrt{1 + \left(\frac{4}{3}\right)^2}} = \frac{3}{5} \qquad \text{using Eqs. (20.2) and (20.7)}$$

Now, using the half-angle formulas, Eqs. (20.26) and (20.27), we have

$$\sin \theta = \sqrt{\frac{1 - \cos 2\theta}{2}} = \sqrt{\frac{1 - \frac{3}{5}}{2}} = \frac{1}{\sqrt{5}} \qquad \cos \theta = \sqrt{\frac{1 + \cos 2\theta}{2}} = \sqrt{\frac{1 + \frac{3}{5}}{2}} = \frac{2}{\sqrt{5}}$$

Here, θ is about 26.6°. Now substituting these values into Eqs. (S3.5), we have

$$x = x'\left(\frac{2}{\sqrt{5}}\right) - y'\left(\frac{1}{\sqrt{5}}\right) = \frac{2x' - y'}{\sqrt{5}} \qquad y = x'\left(\frac{1}{\sqrt{5}}\right) + y'\left(\frac{2}{\sqrt{5}}\right) = \frac{x' + 2y'}{\sqrt{5}}$$

Now, substituting into the equation $8x^2 + 4xy + 5y^2 = 9$ gives

$$8\left(\frac{2x' - y'}{\sqrt{5}}\right)^2 + 4\left(\frac{2x' - y'}{\sqrt{5}}\right)\left(\frac{x' + 2y'}{\sqrt{5}}\right) + 5\left(\frac{x' + 2y'}{\sqrt{5}}\right)^2 = 9$$
$$8(4x'^2 - 4x'y' + y'^2) + 4(2x'^2 + 3x'y' - 2y'^2) + 5(x'^2 + 4x'y' + 4y'^2) = 45$$
$$45x'^2 + 20y'^2 = 45$$
$$\frac{x'^2}{1} + \frac{y'^2}{\frac{9}{4}} = 1$$

This is an ellipse with semimajor axis of 3/2 and semiminor axis of 1. See Fig. S3.3.

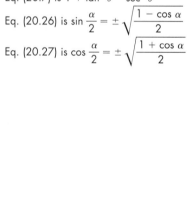

Fig. S3.3

In Example 2, $\tan 2\theta$ was positive, and we made 2θ and θ positive. If, when using Eq. (S3.6), $\tan 2\theta$ is negative, we then make 2θ obtuse ($90° < 2\theta < 180°$). In this case, $\cos 2\theta$ will be negative, but θ will be acute ($45° < \theta < 90°$).

In Section 21.7, we showed the use of translation of axes in writing an equation in standard form if $B = 0$. In this section, we have seen how rotation of axes is used to eliminate the xy-term. It is possible that both a translation of axes and a rotation of axes are needed to write an equation in standard form.

In Section 21.8, we identified a conic section by inspecting the values of A and C when $B = 0$. If $B \neq 0$, these curves are identified as follows:

1. If $B^2 - 4AC = 0$, a parabola
2. If $B^2 - 4AC < 0$, an ellipse
3. If $B^2 - 4AC > 0$, a hyperbola

Special cases such as a point, parallel or intersecting lines, or no curve may result.

◀ **EXAMPLE 3** For the equation $16x^2 - 24xy + 9y^2 + 20x - 140y - 300 = 0$, identify the curve and simplify it to standard form. Sketch the graph and display it on a graphing calculator.

With $A = 16$, $B = -24$, and $C = 9$, using Eq. (S3.6), we have

$$\tan 2\theta = \frac{-24}{16 - 9} = -\frac{24}{7}$$

In this case, $\tan 2\theta$ is negative, and we take 2θ to be an obtuse angle. We then find that $\cos 2\theta = -7/25$. In turn, we find that $\sin \theta = 4/5$ and $\cos \theta = 3/5$. Here, θ is about $53.1°$. Using these values in Eqs. (S3.5), we find that

$$x = \frac{3x' - 4y'}{5} \qquad y = \frac{4x' + 3y'}{5}$$

Substituting these into the original equation and simplifying, we get

$$y'^2 - 4x' - 4y' - 12 = 0$$

This equation represents a parabola with its axis parallel to the x'-axis. The vertex is found by completing the square:

$$(y' - 2)^2 = 4(x' + 4)$$

The vertex is the point $(-4, 2)$ in the $x'y'$-rotated system. Therefore,

$$y''^2 = 4x''$$

is the equation in the $x''y''$-rotated and then translated system. The graph and the coordinate systems are shown in Fig. S3.4.

To display the curve on a graphing calculator, we solve for y by using the quadratic formula. Writing the equation as

$$9y^2 + (-24x - 140)y + (16x^2 + 20x - 300) = 0$$

Therefore, we see that in using the quadratic formula, $a = 9$, $b = -24x - 140$, and $c = 16x^2 + 20x - 300$. Now, solving for y, we have

$$y = \frac{24x + 140 \pm \sqrt{(-24x - 140)^2 - 4(9)(16x^2 + 20x - 300)}}{18}$$

We **enter both functions indicated by the $\pm$ sign** to get the display in Fig. S3.5. ▶

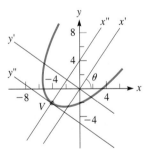

Fig. S3.4

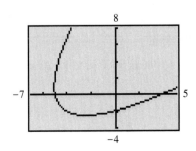

Fig. S3.5

In Exercises 1–4, transform the given equations by rotating the axes through the given angle. Identify and sketch each curve.

1. $x^2 - y^2 = 25$, $\theta = 45°$

2. $x^2 + y^2 = 16$, $\theta = 60°$

3. $8x^2 - 4xy + 5y^2 = 36$, $\theta = \tan^{-1} 2$

4. $2x^2 + 24xy - 5y^2 = 8$, $\theta = \tan^{-1} \frac{3}{4}$

In Exercises 5–10, transform each equation to a form without an xy-term by a rotation of axes. Identify and sketch each curve. Then display each curve on a graphing calculator.

5. $x^2 + 2xy + y^2 - 2x + 2y = 0$

6. $5x^2 - 6xy + 5y^2 = 32$

7. $3x^2 + 4xy = 4$

8. $9x^2 - 24xy + 16y^2 - 320x - 240y = 0$

9. $11x^2 - 6xy + 19y^2 = 20$

10. $x^2 + 4xy - 2y^2 = 6$

In Exercises 11 and 12, transform each equation to a form without an xy-term by a rotation of axes. Then transform the equation to a standard form by a translation of axes. Identify and sketch each curve. Then display each curve on a graphing calculator.

11. $16x^2 - 24xy + 9y^2 - 60x - 80y + 400 = 0$

12. $73x^2 - 72xy + 52y^2 + 100x - 200y + 100 = 0$

S.4 FUNCTIONS OF TWO VARIABLES

Many situations in science, engineering, and technology involve functions with more than one independent variable. Although some of these applications involve three or more independent variables, we shall concern ourselves primarily with functions of two variables.

In this section we establish the meaning of a function of two variables, and in the section that follows we discuss the graph of this type of function.

Many familiar formulas express one variable in terms of two or more other variables. The following example illustrates one from geometry.

◖ **EXAMPLE 1**   The total surface area of a right circular cylinder is a function of the radius and the height of the cylinder. That is, the area will change if either or both of these change. The formula for the total surface area is

$$A = 2\pi r^2 + 2\pi rh$$

We say that A is a function of r and h. See Fig. S4.1.   ◗

Fig. S4.1

There are numerous applications of functions of two variables. For example, the voltage in a simple electric circuit depends on the resistance and current. The moment of a force depends on the magnitude of the force and the distance of the force from the axis. The pressure of a gas depends on its temperature and volume. The temperature at a point on a plate depends on the coordinates of the point.

We define a function of two variables as follows: *If z is uniquely determined for given values of x and y, then z is a function of x and y.* The notation used is similar to that used for one independent variable. It is $z = f(x, y)$, where both x and y are independent variables. Therefore, it follows that $f(a, b)$ means *"the value of the function when x = a and y = b."*

◖ **EXAMPLE 2**   If $f(x, y) = 3x^2 + 2xy - y^3$, find $f(-1, 2)$. Substituting, we have

$$f(-1, 2) = 3(-1)^2 + 2(-1)(2) - (2)^3$$
$$= 3 - 4 - 8 = -9$$ ◗

EXAMPLE 3   If $f(x, y) = 2xy^2 - y$, find $f(x, 2x) - f(x, x^2)$.

We note that in each evaluation the x factor remains as x, but that we are to substitute $2x$ for y and subtract the function for which x^2 is substituted for y.

$$f(x, 2x) - f(x, x^2) = [2x(2x)^2 - (2x)] - [2x(x^2)^2 - x^2]$$
$$= [8x^3 - 2x] - [2x^5 - x^2]$$
$$= 8x^3 - 2x - 2x^5 + x^2$$
$$= -2x^5 + 8x^3 + x^2 - 2x$$

This type of difference of functions is important in Section S.7.

Restricting values of the function to real numbers means that certain restrictions may be placed on the independent variables. *Values of either x or y or both that lead to division by zero or to imaginary values for z are not permissible.*

EXAMPLE 4   If $f(x, y) = \dfrac{3xy}{(x - y)(x + 3)}$, all values of x and y are permissible except those for which $x = y$ and $x = -3$. Each of these would indicate division by zero.

If $f(x, y) = \sqrt{4 - x^2 - y^2}$, neither x nor y may be greater than 2 in absolute value, and the sum of their squares may not exceed 4. Otherwise, imaginary values of the function would result.

One of the primary difficulties students have with functions is setting them up from stated conditions. Although many examples of this have been encountered in our previous work, an example of setting up a function of two variables should prove helpful. Although no general rules can be given for this procedure, a careful analysis of the statement should lead to the desired function.

Solving a Word Problem

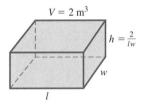

$V = 2\text{ m}^3$

$h = \dfrac{2}{lw}$

w

l

Fig. S4.2

EXAMPLE 5   An open rectangular metal box is to be made to contain 2 m³. The cost of sheet metal is c cents per square meter. Express the cost of the sheet metal needed to construct the box as a function of the length and width of the box.

The cost in question depends on the surface area of the sides of the box. An "open" box is one that has no top. Thus, the surface area of the box is

$$S = lw + 2lh + 2hw$$

where l is the length, w the width, and h the height of the box (see Fig. S4.2). However, this expression contains three independent variables. Using the condition that the volume of the box is 2 m³, we have $lwh = 2$. Since we wish to have only l and w, we solve this expression for h and find that $h = 2/lw$. Substituting for h in the expression for the surface area, we have

$$S = lw + 2l\left(\frac{2}{lw}\right) + 2\left(\frac{2}{lw}\right)w$$
$$= lw + \frac{4}{w} + \frac{4}{l}$$

Since the cost C is given by $C = cS$, we have the expression for the cost as

$$C = c\left(lw + \frac{4}{w} + \frac{4}{l}\right)$$

EXERCISES **S.4**

In Exercises 1–4, determine the indicated functions.

1. Express the volume of a right circular cone as a function of the radius of the base and the height.

2. A cylindrical can is to be made to contain a volume V. Express the total surface area of the can as a function of V and the radius of the can.

3. The angle between two forces $\mathbf{F}_1$ and $\mathbf{F}_2$ is 30°. Express the magnitude of the resultant $\mathbf{R}$ in terms of F_1 and F_2. See Fig. S4.3.

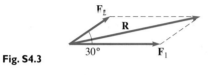

Fig. S4.3

4. A computer-leasing firm charges a monthly fee F based on the length of time a corporation has used the service, plus $100 for every hour the computer is used during the month. Express the total monthly charge T as a function of F and the number of hours h the computer is used.

In Exercises 5–16, evaluate the given functions.

5. $f(x, y) = 2x - 6y$; find $f(0, -4)$

6. $g(r, s) = r - 2rs - r^2 s$; find $g(-2, 1)$

7. $f(r, \theta) = 2r(r \tan \theta - \sin 2\theta)$; find $f(3, \pi/4)$

8. $f(x, t) = \ln(xt + x^2 - 4)$; find $f(2, e)$

9. $Y(y, t) = \dfrac{2 - 3y}{t - 1} + 2y^2 t$; find $Y(y, 2)$

10. $X(x, t) = -6xt + xt^2 - t^3$; find $X(x, -t)$

11. $g(x, z) = z \tan^{-1}(x^2 + xz)$; find $g(-x, z)$

12. $F(r, x) = re^x(2rx + 1)$; find $F(3x, x)$

13. $H(p, q) = p - \dfrac{p - 2q^2 - 5q}{p + q}$; find $H(p, q + k)$

14. $f(x, y) = x^2 - 2xy - 4x$; find $f(x + h, y + k) - f(x, y)$

15. $f(x, y) = xy + x^2 - y^2$; find $f(x, x) - f(x, 0)$

16. $g(y, z) = 3y^3 - y^2 z + 5z^2$; find $g(3z^2, z) - g(z, z)$

In Exercises 17 and 18, determine which values of x and y, if any, are not permissible.

17. $f(x, y) = \dfrac{\sqrt{y}}{2x}$

18. $f(x, y) = \sqrt{x^2 + y^2 - x^2 y - y^3}$

In Exercises 19–24, solve the given problems.

19. For a certain electric circuit, the current i (in A) in terms of the voltage E and resistance R (in Ω) is given by

$i = \dfrac{E}{R + 0.25}$. Find the current for $E = 1.50$ V and $R = 1.20\ \Omega$

and for $E = 1.60$ V and $R = 1.05\ \Omega$.

20. The reciprocal of the image distance q from a lens as a function of the object distance p and the focal length f of the lens is

$$\frac{1}{q} = \frac{1}{f} - \frac{1}{p}$$

Find the image distance of an object 20 cm from a lens whose focal length is 5 cm.

21. A rectangular solar cell panel has a perimeter p and a width w. Express the area A of the panel in terms of p and w and evaluate the area for $p = 250$ cm and $w = 55$ cm. See Fig. S4.4.

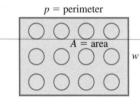

Fig. S4.4

22. A gasoline storage tank is in the shape of a right cylinder with a hemisphere at each end, as shown in Fig. S4.5. Express the volume V of the tank in terms of r and h and then evaluate the volume for $r = 1.25$ m and $h = 4.17$ m.

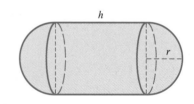

Fig. S4.5

23. The crushing load L of a pillar varies as the fourth power of its radius r and inversely as the square of its length l. Express L as a function of r and l for a pillar 6.0 m tall and 1.0 m in diameter that is crushed by a load of 20 Mg.

24. The resonant frequency f (in Hz) of an electric circuit containing an inductance L and capacitance C is inversely proportional to the square root of the product of the inductance and the capacitance. If the resonant frequency of a circuit containing a 4-H inductor and a 64-μF capacitor is 10 Hz, express f as a function of L and C.

S.5 CURVES AND SURFACES IN THREE DIMENSIONS

We will now undertake a brief description of the graphical representation of a function of two variables. We shall show first a method of representation in the rectangular coordinate system in two dimensions. The following example illustrates the method.

◀ EXAMPLE 1   In order to represent $z = 2x^2 + y^2$, we will assume various values of z and sketch the resulting equation in the xy-plane. For example, if $z = 2$ we have

$$2x^2 + y^2 = 2$$

We recognize this as an ellipse with its major axis along the y-axis and vertices at $(0, \sqrt{2})$ and $(0, -\sqrt{2})$. The ends of the minor axis are at $(1, 0)$ and $(-1, 0)$. However, the ellipse $2x^2 + y^2 = 2$ represents the function $z = 2x^2 + y^2$ only for the value of $z = 2$. If $z = 4$, we have $2x^2 + y^2 = 4$, which is another ellipse. In fact, for all positive values of z, an ellipse is the resulting curve. Negative values of z are not possible, since neither x^2 nor y^2 may be negative. Figure S5.1 shows the ellipses that are obtained by using the indicated values of z. ▶

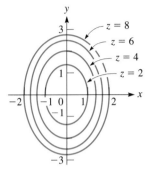

Fig. S5.1

The method of representation illustrated in Example 1 is useful if only a few specific values of z are to be used, or at least if the various curves do not intersect in such a way that they cannot be distinguished. If a general representation of z as a function of x and y is desired, it is necessary to use three coordinate axes, one each for x, y, and z. The most widely applicable system of this kind is to place a third coordinate axis at right angles to each of the x- and y-axes. In this way we employ three dimensions for the representation.

The three mutually perpendicular coordinate axes, the x-axis, the y-axis, and the z-axis, are the basis of the **rectangular coordinate system in three dimensions.** Together they form three mutually perpendicular planes in space, the xy-plane, the yz-plane, and the xz-plane. To every point in space of the coordinate system is associated the set of numbers (x, y, z). The point at which the axes meet is the *origin*. The positive directions of the axes are indicated in Fig. S5.2. *That part of space in which all values of the coordinates are positive is called the first* **octant.** Numbers are not assigned to the other octants.

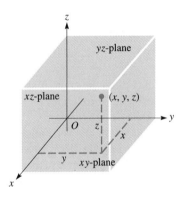

Fig. S5.2

◀ EXAMPLE 2   Represent the point $(2, 4, 3)$ in rectangular coordinates.

We first note that a certain distortion is necessary to represent values of x reasonably, since the x-axis "comes out" of the plane of the page. Units $\sqrt{2}/2$ ($= 0.7$) as long as those used on the other axes give a good representation. With this in mind, we draw a line 4 units long from the point $(2, 0, 0)$ on the x-axis in the xy-plane. This locates the point $(2, 4, 0)$. From this point a line 3 units long is drawn vertically upward. This locates the desired point, $(2, 4, 3)$. The point may be located by starting from $(0, 4, 0)$, proceeding 2 units *parallel* to the x-axis to $(2, 4, 0)$ and then proceeding vertically 3 units to $(2, 4, 3)$. It may also be located by starting from $(0, 0, 3)$ (see Fig. S5.3). ▶

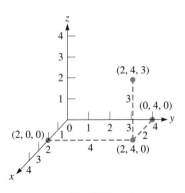

Fig. S5.3

We now show certain basic techniques by which three-dimensional figures may be drawn. We start by showing the general equation of a plane.

In Chapter 21 we showed that the graph of the equation $Ax + By + C = 0$ in two dimensions is a straight line. By the following example, we will verify that *the graph of the equation*

Equation of a Plane

$$Ax + By + Cz + D = 0 \qquad \text{(S5.1)}$$

is a **plane** *in three dimensions.*

◀ **EXAMPLE 3**   Show that the graph of $2x + 3y + z - 6 = 0$ in three dimensions is a plane.

NOTE ▶   *If we let any of the three variables take on a specific value, we obtain a linear equation in the other two variables.* For example, the point $\left(\frac{1}{2}, 1, 2\right)$ satisfies the equation and therefore lies on the graph of the equation. For $x = \frac{1}{2}$, we have

$$3y + z - 5 = 0$$

which is the equation of a straight line. This means that all pairs of values of y and z that satisfy this equation, along with $x = \frac{1}{2}$, satisfy the given equation. Thus, for $x = \frac{1}{2}$, the straight line $3y + z - 5 = 0$ lies on the graph of the equation.

For $z = 2$, we have

$$2x + 3y - 4 = 0$$

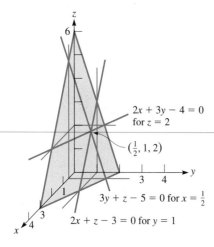

Fig. S5.4

which is also a straight line. By similar reasoning this line lies on the graph of the equation. Since two lines through a point define a plane, these lines through $\left(\frac{1}{2}, 1, 2\right)$ define a plane. This plane is the graph of the equation (see Fig. S5.4).

For any point on the graph, there is a straight line on the graph that is parallel to one of the coordinate planes. Therefore, there are intersecting straight lines through the point. Thus, the graph is a plane. A similar analysis can be made for any equation of the same form. ▶

Since we know that the graph of an equation of the form of Eq. (S5.1) is a plane, its graph can be found by determining its three intercepts, and the plane can then be represented by drawing in the lines between these intercepts. If the plane passes through the origin, by letting two of the variables in turn be zero, two straight lines that define the plane are found.

◀ **EXAMPLE 4**   Sketch the graph of $3x - y + 2z - 4 = 0$.

The intercepts of the graph of an equation are those points where it crosses the respective axes. Thus, by letting two of the variables at a time equal zero, we obtain the intercepts. For the given equation the intercepts are $\left(\frac{4}{3}, 0, 0\right)$, $(0, -4, 0)$, and $(0, 0, 2)$. These points are located (see Fig. S5.5), and lines are drawn between them to represent the plane. ▶

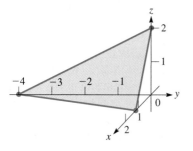

Fig. S5.5

The graph of an equation in three variables, which is essentially equivalent to a function with two independent variables, is a **surface** *in space. This is seen for the plane and will be verified for other equations in the examples that follow.*

The intersection of two surfaces is a **curve** *in space.* This has been seen in Examples 3 and 4, since the intersections of the given planes and the coordinate planes are lines (which in the general sense are curves). *We define the* **traces** *of a surface to be the curves of intersection of the surface and the coordinate planes.* The traces of a plane are those lines drawn between the intercepts to represent the plane. Many surfaces may be sketched by finding their traces and intercepts.

Traces

◀ **EXAMPLE 5** Find the intercepts and traces, and sketch the graph of the equation $z = 4 - x^2 - y^2$.

The intercepts of the graph of the equation are $(2, 0, 0)$, $(-2, 0, 0)$, $(0, 2, 0)$, $(0, -2, 0)$, and $(0, 0, 4)$.

Since the traces of a surface lie within the coordinate planes, for each trace one of the variables is zero. Thus, by *letting each variable in turn be zero, we find the trace of the surface in the plane of the other two variables.* Therefore, the traces of this surface are

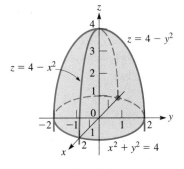

in the yz-plane: $z = 4 - y^2$ (a parabola)
in the xz-plane: $z = 4 - x^2$ (a parabola)
in the xy-plane: $x^2 + y^2 = 4$ (a circle)

Fig. S5.6

Using the intercepts and sketching the traces, we obtain the surface represented by the equation as shown in Fig. S5.6. This figure is called a **circular paraboloid**. ▶

There are numerous techniques for analyzing the equation of a surface in order to obtain its graph. Another which we shall discuss here, which is closely associated with a trace, is that of a **section**. *By assuming a specific value of one of the variables, we obtain an equation in two variables, the graph of which lies in a plane parallel to the coordinate plane of the two variables.* The following example illustrates sketching a surface by use of intercepts, traces, and sections.

Section

◀ **EXAMPLE 6** Sketch the graph of $4x^2 + y^2 - z^2 = 4$.

The intercepts are $(1, 0, 0)$, $(-1, 0, 0)$, $(0, 2, 0)$, and $(0, -2, 0)$. We note that there are no intercepts on the z-axis, for this would necessitate $z^2 = -4$.

The traces are

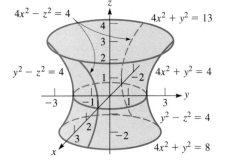

Fig. S5.7

in the yz-plane: $y^2 - z^2 = 4$ (a hyperbola)
in the xz-plane: $4x^2 - z^2 = 4$ (a hyperbola)
in the xy-plane: $4x^2 + y^2 = 4$ (an ellipse)

The surface is reasonably defined by these curves, but by assuming suitable values of z we may indicate its shape better. For example, if $z = 3$, we have $4x^2 + y^2 = 13$, which is an ellipse. In using it we must remember that it is valid for $z = 3$ and therefore should be drawn 3 units above the xy-plane. If $z = -2$, we have $4x^2 + y^2 = 8$, which is also an ellipse. Thus, we have the following sections:

for $z = 3$: $4x^2 + y^2 = 13$ (an ellipse)
for $z = -2$: $4x^2 + y^2 = 8$ (an ellipse)

Other sections could be found, but these are sufficient to obtain a good sketch of the graph (see Fig. S5.7). The figure is called an **elliptic hyperboloid**. ▶

Having developed the rectangular coordinate system in three dimensions, we can compare the graph of a function using two dimensions and three dimensions. The next example shows the surface for the function of Example 1.

◖ EXAMPLE 7   Sketch the graph of $z = 2x^2 + y^2$.
The only intercept is $(0, 0, 0)$. The traces are

in the yz-plane: $z = y^2$ (a parabola)
in the xz-plane: $z = 2x^2$ (a parabola)
in the xy-plane: the origin

The trace in the xy-plane is only the point of origin, since $2x^2 + y^2 = 0$ may be written as $y^2 = -2x^2$, which is true only for $x = 0$ and $y = 0$.

To get a better graph we should use some positive values for z. As we noted in Example 1, negative values of z cannot be used. Since we used $z = 2$, $z = 4$, $z = 6$, and $z = 8$ in Example 1, we shall use these values here. Therefore,

for $z = 2$: $2x^2 + y^2 = 2$ for $z = 4$: $2x^2 + y^2 = 4$
for $z = 6$: $2x^2 + y^2 = 6$ for $z = 8$: $2x^2 + y^2 = 8$

Each of these sections is an ellipse. The surface, called an **elliptic paraboloid,** is shown in Fig. S5.8. Compare with Fig. S5.1. ◗

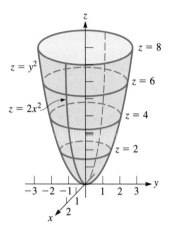

Fig. S5.8

Example 7 illustrates how *topographic maps* may be drawn. These maps represent three-dimensional terrain in two dimensions. For example, if Fig. S5.8 represents an excavation in the surface of the earth, then Fig. S5.1 represents the curves of constant elevation, or *contours,* with equally spaced elevations measured from the bottom of the excavation.

An equation with only two variables may represent a surface in space. Since only two variables are included in the equation, the surface is independent of the other variable. Another interpretation is that all sections, for all values of the variable not included, are the same. That is, ***all sections parallel to the coordinate plane of the included variables are the same as the trace in that plane.***

NOTE ◗

◖ EXAMPLE 8   Sketch the graph of $x + y = 2$ in the rectangular coordinate system in three dimensions and in two dimensions.

Since z does not appear in the equation, we can consider the equation to be $x + y + 0z = 2$. Therefore, we see that *for any value of z* the section is the straight line $x + y = 2$. Thus, the graph is a plane as shown in Fig. S5.9(a). The graph as a straight line in two dimensions is shown in Fig. S5.9(b).

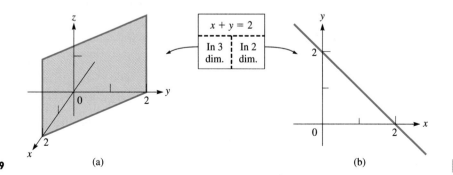

Fig. S5.9   (a)   (b)

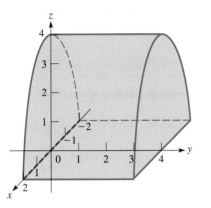

Fig. S5.10

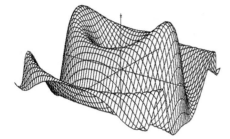

Fig. S5.11

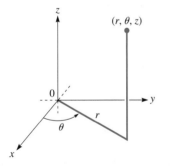

Fig. S5.12

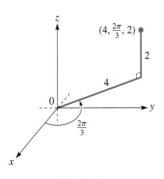

Fig. S5.13

◖ **EXAMPLE 9** The graph of the equation $z = 4 - x^2$ in three dimensions is a surface whose sections, for all values of y, are given by the parabola $z = 4 - x^2$. The surface is shown in Fig. S5.10. ◗

The surface in Example 9 is known as a **cylindrical surface.** In general, a cylindrical surface is one that can be generated by a line moving parallel to a fixed line while passing through a plane curve.

It must be realized that most of the figures shown extend beyond the ranges indicated by the traces and sections. However, these traces and sections are convenient for representing and visualizing these surfaces.

There are various computer programs (and some graphing calculators) that can be used to display three-dimensional surfaces. They generally use sections in planes that are perpendicular to both the x- and y-axes. Such computer-drawn surfaces are of great value for all types of surfaces, especially those of a complex nature. Figure S5.11 shows the graph of $z = \dfrac{\sin(2x^2 + y^2)}{x^2 + 1}$ drawn by a computer program.

CYLINDRICAL COORDINATES

Another set of coordinates that can be used to display a three-dimensional figure are *cylindrical coordinates, which are polar coordinates and the z-axis combined.* In using cylindrical coordinates, every point in space is designated by the coordinates (r, θ, z), as shown in Fig. S5.12. The equations relating the rectangular coordinates (x, y, z) and the cylindrical coordinates (r, θ, z) of a point are

$$x = r \cos \theta \qquad y = r \sin \theta \qquad z = z$$
$$r^2 = x^2 + y^2 \qquad \tan \theta = \frac{y}{x}$$

(S5.2)

The following examples illustrate the use of cylindrical coordinates.

◖ **EXAMPLE 10** **(a)** Plot the point with cylindrical coordinates $(4, 2\pi/3, 2)$ and find its corresponding rectangular coordinates.

The point is shown in Fig. S5.13. From Eqs. (S5.2), we have

$$x = 4 \cos \frac{2\pi}{3} = 4\left(-\frac{1}{2}\right) = -2 \qquad y = 4 \sin \frac{2\pi}{3} = 4\left(\frac{\sqrt{3}}{2}\right) = 2\sqrt{3} \qquad z = 2$$

Therefore, the rectangular coordinates of the point are $\left(-2, 2\sqrt{3}, 2\right)$.

(b) Find the cylindrical coordinates of the point with rectangular coordinates $(2, -2, 6)$.

Using Eqs. (S5.2), we have

$$r = \sqrt{2^2 + (-2)^2} = 2\sqrt{2} \qquad z = 6$$
$$\tan \theta = \frac{-2}{2} = -1 \quad \text{(quadrant IV)} \qquad \theta = 2\pi - \frac{\pi}{4} = \frac{7\pi}{4}$$

The cylindrical coordinates are $\left(2\sqrt{2}, 7\pi/4, 6\right)$. As with polar coordinates, the value of θ can be $7\pi/4 + 2n\pi$, where n is an integer. ◗

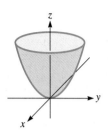

Fig. S5.14

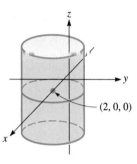

Fig. S5.15

◀ EXAMPLE 11   Find the equation for the surface $4x^2 + 4y^2 = z$ in cylindrical coordinates and sketch the surface.

Factoring the 4 from the terms on the left, we have $4(x^2 + y^2) = z$. We now use the fact that $x^2 + y^2 = r^2$ to write this equation in cylindrical coordinates as $4r^2 = z$. From this equation we see that $z \geq 0$ and that as z increases we have circular sections of increasing radius. Therefore, it is a **circular paraboloid**. See Fig. S5.14.   ▶

◀ EXAMPLE 12   Find the equation for the surface $r = 4 \cos \theta$ in rectangular coordinates and sketch the surface.

If we multiply each side of the equation by r, we obtain $r^2 = 4r \cos \theta$. Now from Eqs. S5.2, we have $r^2 = x^2 + y^2$ and $r \cos \theta = x$, which means that $x^2 + y^2 = 4x$, or

$$x^2 - 4x + 4 + y^2 = 4$$
$$(x - 2)^2 + y^2 = 4$$

We recognize this as a cylinder that has as its trace in the xy-plane a circle with center at $(2, 0, 0)$, as shown in Fig. S5.15.   ▶

If the center of the trace of a circular cylinder in the xy-plane is at the origin, the equation of the cylinder in cylindrical coordinates is simply $r = a$. For this reason, these coordinates are called "cylindrical" coordinates.

EXERCISES   S.5

In Exercises 1 and 2, use the method of Example 1 and draw the graphs of the given equations for the given values of z.

1. $z = x^2 + y^2$, $z = 1$, $z = 4$, $z = 9$

2. $z = y - x^2$, $z = 0$, $z = 2$, $z = 4$

In Exercises 3–16, sketch the graphs of the given equations in the rectangular coordinate system in three dimensions.

3. $x + y + 2z - 4 = 0$ **4.** $4x - 2y + z - 8 = 0$

5. $z = y - 2x - 2$ **6.** $x + 2y = 4$

7. $x^2 + y^2 + z^2 = 4$ **8.** $z = x^2 + y^2$

9. $z = 4 - 4x^2 - y^2$ **10.** $x^2 + y^2 - 4z^2 = 4$

11. $z = 2x^2 + y^2 + 2$ **12.** $x^2 - y^2 - z^2 = 9$

13. $x^2 + y^2 = 16$ **14.** $y^2 + 9z^2 = 9$

15. $z^2 = 9x^2 + 4y^2$ **16.** $4z = x^2$

In Exercises 17–22, perform the indicated operations involving cylindrical coordinates.

17. Find the rectangular coordinates of the points whose cylindrical coordinates are (a) $(3, \pi/4, 5)$, (b) $(2, \pi/2, 3)$, (c) $(4, \pi/3, 2)$.

18. Find the cylindrical coordinates of the points whose rectangular coordinates are (a) $(3, 4, 5)$, (b) $(8, 15, -6)$, (c) $\left(\sqrt{3}, -2, 1\right)$.

Ⓦ **19.** Describe the surface for which the cylindrical coordinate equation is (a) $r = 2$, (b) $\theta = 2$, (c) $z = 2$.

20. Write the equation $x^2 + y^2 + 4z^2 = 4$ in cylindrical coordinates and sketch the surface.

21. Write the equation $r^2 = 4z$ in rectangular coordinates and sketch the surface.

22. Write the equation $r = 4 \cos \theta$ in rectangular coordinates and sketch the surface.

In Exercises 23–28, sketch the indicated curves and surfaces.

23. Curves that represent a constant temperature are called *isotherms*. The temperature at a point (x, y) of a flat plate is t (°C), where $t = 4x - y^2$. In two dimensions, draw the isotherms for $t = -4, 0, 8$.

24. At a point (x, y) in the xy-plane, the electric potential V (in volts) is given by $V = y^2 - x^2$. Draw the lines of equal potential for $V = -9, 0, 9$.

25. An electric charge is so distributed that the electric potential at all points on an imaginary surface is the same. Such a surface is called an *equipotential surface*. Sketch the graph of the equipotential surface whose equation is $2x^2 + 2y^2 + 3z^2 = 6$.

26. The surface of a small hill can be roughly approximated by the equation $z(2x^2 + y^2 + 100) = 1500$, where the units are meters. Draw the surface of the hill and the contours for $z = 3$ m, $z = 6$ m, $z = 9$ m, $z = 12$ m, and $z = 15$ m.

27. Sketch the line in space defined by the intersection of the planes $x + 2y + 3z - 6 = 0$ and $2x + y + z - 4 = 0$.

28. Sketch the graph of $x^2 + y^2 - 2y = 0$ in three dimensions and in two dimensions.

S.6 PARTIAL DERIVATIVES

In Chapter 23, when we showed that the derivative is the instantaneous rate of change of one variable with respect to another, only one independent variable was involved. To extend the derivative to functions of two (or more) variables, we find the derivative of the function with respect to one of the independent variables, while the other is held constant.

If $z = f(x, y)$ and y is held constant, z becomes a function of x alone. The derivative of this function with respect to x is termed the **partial derivative** *of z with respect to x. Similarly, if x is held constant, the derivative of the function with respect to y is the* **partial derivative** *of z with respect to y.*

For the function $z = f(x, y)$, the notations used for the partial derivative of z with respect to x include

The symbol ∂ was introduced by the German mathematician Carl Jacobi (1804–1851).

$$\frac{\partial z}{\partial x} \qquad \frac{\partial f}{\partial x} \qquad f_x \qquad \frac{\partial}{\partial x} f(x, y) \qquad f_x(x, y)$$

Similarly, $\partial z / \partial y$ denotes the partial derivative of z with respect to y. In speaking, this is often shortened to "the partial of z with respect to y."

◀ EXAMPLE 1 If $z = 4x^2 + xy - y^2$, find $\partial z / \partial x$ and $\partial z / \partial y$.

To find the partial derivative of z with respect to x, we treat y as a constant.

treat as constant

$$z = 4x^2 + xy - y^2$$
$$\frac{\partial z}{\partial x} = 8x + y$$

To find the partial derivative of z with respect to y, we treat x as a constant.

treat as constant

$$z = 4x^2 + xy - y^2$$
$$\frac{\partial z}{\partial x} = x - 2y$$

◀ EXAMPLE 2 If $z = \dfrac{x \ln y}{x^2 + 1}$, find $\partial z / \partial x$ and $\partial z / \partial y$.

$$\frac{\partial z}{\partial x} = \frac{(x^2 + 1)(\ln y) - (x \ln y)(2x)}{(x^2 + 1)^2} = \frac{(1 - x^2)\ln y}{(1 + x^2)^2}$$
$$\frac{\partial z}{\partial y} = \left(\frac{x}{x^2 + 1}\right)\left(\frac{1}{y}\right) = \frac{x}{y(x^2 + 1)}$$

We note that in finding $\partial z / \partial x$ it is necessary to use the quotient rule, since x appears in both numerator and denominator. However, when finding $\partial z / \partial y$, the only derivative needed is that of $\ln y$.

EXAMPLE 3 For the function $f(x, y) = x^2 y \sqrt{2 + xy^2}$, find $f_y(2, 1)$.

The notation $f_y(2, 1)$ means the partial derivative of f with respect to y, evaluated for $x = 2$ and $y = 1$. Thus, first finding $f_y(x, y)$, we have

$$f(x, y) = x^2 y (2 + xy^2)^{1/2}$$

$$f_y(x, y) = x^2 y \left(\frac{1}{2}\right)(2 + xy^2)^{-1/2}(2xy) + (2 + xy^2)^{1/2}(x^2)$$

$$= \frac{x^3 y^2}{(2 + xy^2)^{1/2}} + x^2(2 + xy^2)^{1/2}$$

$$= \frac{x^3 y^2 + x^2(2 + xy^2)}{(2 + xy^2)^{1/2}} = \frac{2x^2 + 2x^3 y^2}{(2 + xy^2)^{1/2}}$$

$$f_y(2, 1) = \frac{2(4) + 2(8)(1)}{(2 + 2)^{1/2}} = 12$$

To determine the geometric interpretation of a partial derivative, assume that $z = f(x, y)$ is the surface shown in Fig. S6.1. Choosing a point P on the surface, we then draw a plane through P parallel to the xz-plane. On this plane through P, the value of y is constant. The intersection of this plane and the surface is the curve as indicated. *The partial derivative of z with respect to x represents the slope of a line tangent to this curve.* When the values of the coordinates of point P are substituted into the expression for this partial derivative, it gives the slope of the tangent line at that point. In the same way, the partial derivative of z with respect to y, evaluated at P, gives the slope of the line tangent to the curve that is found from the intersection of the surface and the plane parallel to the yz-plane through P.

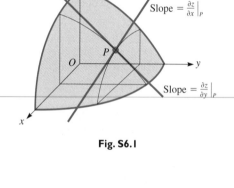

Slope $= \left.\dfrac{\partial z}{\partial x}\right|_P$

Slope $= \left.\dfrac{\partial z}{\partial y}\right|_P$

Fig. S6.1

EXAMPLE 4 Find the slope of a line tangent to the surface $2z = x^2 + 2y^2$ and parallel to the xz-plane at the point $(2, 1, 3)$. Also, find the slope of the line tangent to this surface and parallel to the yz-plane at the same point.

Finding the partial derivative with respect to x, we have

$$2\frac{\partial z}{\partial x} = 2x \quad \text{or} \quad \frac{\partial z}{\partial x} = x$$

This derivative, evaluated at the point (2, 1, 3), will give us the slope of the line tangent that is also parallel to the xz-plane. Therefore, the first required slope is

$$\left.\frac{\partial z}{\partial x}\right|_{(2,1,3)} = 2$$

The partial derivative of z with respect to y, evaluated at $(2, 1, 3)$, will give us the second required slope. Thus,

$$\frac{\partial z}{\partial y} = 2y, \quad \left.\frac{\partial z}{\partial y}\right|_{(2,1,3)} = 2$$

Therefore, both slopes are 2. See Fig. S6.2.

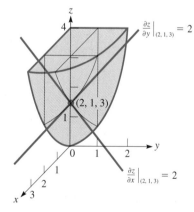

$\left.\dfrac{\partial z}{\partial y}\right|_{(2,1,3)} = 2$

$(2, 1, 3)$

$\left.\dfrac{\partial z}{\partial x}\right|_{(2,1,3)} = 2$

Fig. S6.2

The general interpretation of the partial derivative follows that of a derivative of a function with one independent variable. *The partial derivative $f_x(x_0, y_0)$ is the instantaneous rate of change of the function $f(x, y)$ with respect to x, with y held constant at the value of y_0.* This holds regardless of what the variables represent.

Applications of partial derivatives are found in many fields of technology. We show here an application from electricity, and others are found in the exercises.

◀ EXAMPLE 5   An electric circuit in a microwave transmitter has parallel resistances r and R. The current through r can be found from

$$i = \frac{IR}{r + R}$$

where I is the total current for the two branches. Assuming that I is constant at 85.4 mA, find $\partial i / \partial r$ and evaluate it for $R = 0.150\ \Omega$ and $r = 0.032\ \Omega$.

Substituting for I and finding the partial derivative, we have the following:

$$i = \frac{85.4R}{r + R} = 85.4R(r + R)^{-1}$$

$$\frac{\partial i}{\partial r} = (-1)(85.4R)(r + R)^{-2}(1) = \frac{-85.4R}{(r + R)^2}$$

$$\left.\frac{\partial i}{\partial r}\right|_{\substack{r=0.032 \\ R=0.150}} = \frac{-85.4(0.150)}{(0.032 + 0.150)^2} = -387\ \text{mA}/\Omega$$

This result tells us that the current is decreasing at the rate of 387 mA per ohm of change in the smaller resistor at the instant when $r = 0.032\ \Omega$, for a fixed value of $R = 0.150\ \Omega$. ▶

Since the partial derivatives $\partial f / \partial x$ and $\partial f / \partial y$ are functions of x and y, we can take partial derivatives of each of them. This gives rise to **partial derivatives of higher order,** in a manner similar to the higher derivatives of a function of one independent variable. *The possible* **second-order partial derivatives** *of a function $f(x, y)$ are*

$$\frac{\partial^2 f}{\partial x^2} = \frac{\partial}{\partial x}\left(\frac{\partial f}{\partial x}\right) \qquad \frac{\partial^2 f}{\partial y^2} = \frac{\partial}{\partial y}\left(\frac{\partial f}{\partial y}\right)$$

$$\frac{\partial^2 f}{\partial x\, \partial y} = \frac{\partial}{\partial x}\left(\frac{\partial f}{\partial y}\right) \qquad \frac{\partial^2 f}{\partial y\, \partial x} = \frac{\partial}{\partial y}\left(\frac{\partial f}{\partial x}\right)$$

◀ EXAMPLE 6   Find the second-order partial derivatives of $z = x^3 y^2 - 3xy^3$.
First, we find $\partial z / \partial x$ and $\partial z / \partial y$:

$$\frac{\partial z}{\partial x} = 3x^2 y^2 - 3y^3 \quad \text{or} \quad \frac{\partial z}{\partial y} = 2x^3 y - 9xy^2$$

Therefore, we have the following second-order partial derivatives:

$$\frac{\partial^2 z}{\partial x^2} = \frac{\partial}{\partial x}\left(\frac{\partial z}{\partial x}\right) = 6xy^2 \qquad \frac{\partial^2 z}{\partial y^2} = \frac{\partial}{\partial y}\left(\frac{\partial z}{\partial y}\right) = 2x^3 - 18xy$$

$$\frac{\partial^2 z}{\partial x\, \partial y} = \frac{\partial}{\partial x}\left(\frac{\partial z}{\partial y}\right) = 6x^2 y - 9y^2 \qquad \frac{\partial^2 z}{\partial y\, \partial x} = \frac{\partial}{\partial y}\left(\frac{\partial z}{\partial x}\right) = 6x^2 y - 9y^2 \quad ▶$$

In Example 6 we note that

$$\boxed{\frac{\partial^2 z}{\partial x\,\partial y} = \frac{\partial^2 z}{\partial y\,\partial x}} \qquad \text{(S6.1)}$$

In general, this is true if the function and partial derivatives are continuous.

◀ **EXAMPLE 7**   For $f(x, y) = \tan^{-1}\dfrac{y}{x^2}$, show that $\dfrac{\partial^2 f}{\partial x\,\partial y} = \dfrac{\partial^2 f}{\partial y\,\partial x}$.

Finding $\partial f/\partial x$ and $\partial f/\partial y$, we have

$$\frac{\partial f}{\partial x} = \frac{1}{1 + \left(\dfrac{y}{x^2}\right)^2}\left(\frac{-2y}{x^3}\right) = \frac{x^4}{x^4 + y^2}\left(-\frac{2y}{x^3}\right) = \frac{-2xy}{x^4 + y^2}$$

$$\frac{\partial f}{\partial y} = \frac{1}{1 + \left(\dfrac{y}{x^2}\right)^2}\left(\frac{1}{x^2}\right) = \frac{x^4}{x^4 + y^2}\left(\frac{1}{x^2}\right) = \frac{x^2}{x^4 + y^2}$$

Now, finding $\partial^2 f/\partial x\,\partial y$ and $\partial^2 f/\partial y\,\partial x$, we have

$$\frac{\partial^2 f}{\partial x\,\partial y} = \frac{(x^4 + y^2)(2x) - x^2(4x^3)}{(x^4 + y^2)^2} = \frac{-2x^5 + 2xy^2}{(x^4 + y^2)^2}$$

$$\frac{\partial^2 f}{\partial y\,\partial x} = \frac{(x^4 + y^2)(-2x) - (-2xy)(2y)}{(x^4 + y^2)^2} = \frac{-2x^5 + 2xy^2}{(x^4 + y^2)^2}$$

We see that they are equal. ▶

EXERCISES S.6

In Exercises 1–12, find the partial derivative of the dependent variable or function with respect to each of the independent variables.

1. $z = 5x + 4x^2 y$

2. $z = \dfrac{x^2}{y} - 2xy$

3. $f(x, y) = xe^{2y}$

4. $f(x, y) = 3y^2 \cos 2x$

5. $\phi = r\sqrt{1 + 2rs}$

6. $z = (x^2 + xy^3)^4$

7. $z = \sin xy$

8. $y = \ln(r^2 + s)$

9. $f(x, y) = \dfrac{2\sin^3 2x}{1 - 3y}$

10. $z = \dfrac{\sin^{-1} xy}{3 + x^2}$

11. $z = y \sin x + \cos xy$

12. $u = ye^{x + 2y}$

In Exercises 13–16, evaluate the indicated partial derivatives at the given points.

13. $z = 3xy - x^2$, $\left.\dfrac{\partial z}{\partial x}\right|_{(1, -2, -7)}$

14. $z = x^2 \cos 4y$, $\left.\dfrac{\partial z}{\partial y}\right|_{(2, \frac{\pi}{2}, 4)}$

15. $z = x\sqrt{x^2 - y^2}$, $\left.\dfrac{\partial z}{\partial x}\right|_{(5, 3, 20)}$

16. $z = e^y \ln xy$, $\left.\dfrac{\partial z}{\partial y}\right|_{(e, 1, e)}$

In Exercises 17–20, find all the second partial derivatives.

17. $z = 2xy^3 - 3x^2 y$

18. $F(x, y) = y \ln(x + 2y)$

19. $z = \dfrac{x}{y} + e^x \sin y$

20. $f(x, y) = \dfrac{2 + \cos y}{1 + x^2}$

In Exercises 21–28, solve the given problems.

21. Find the slope of a line tangent to the surface $z = 9 - x^2 - y^2$ and parallel to the yz-plane that passes through $(1, 2, 4)$. Repeat the instructions for the line through $(2, 2, 1)$. Draw an appropriate figure.

22. A metal plate in the shape of a circular segment of radius r expands by being heated. Express the width w (straight dimension) as a function of r and the height h. Then find both $\partial w/\partial r$ and $\partial w/\partial h$.

23. Two resistors R_1 and R_2, placed in parallel, have a combined resistance R_T given by $\dfrac{1}{R_T} = \dfrac{1}{R_1} + \dfrac{1}{R_2}$. Find $\dfrac{\partial R_T}{\partial R_1}$.

24. A metallic machine part contracts while cooling. It is in the shape of a hemisphere attached to a cylinder, as shown in Fig. S6.3. Find the rate of change of volume with respect to r when $r = 2.65$ cm and $h = 4.20$ cm.

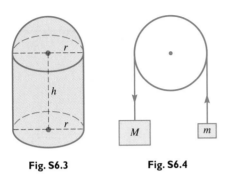

Fig. S6.3   **Fig. S6.4**

25. Two masses M and m are attached as shown in Fig. S6.4. If $M > m$, the downward acceleration a of mass M is given by $a = \dfrac{M - m}{M + m}g$, where g is the acceleration due to gravity. Show that $M\dfrac{\partial a}{\partial M} + m\dfrac{\partial a}{\partial m} = 0$.

26. If an observer and a source of sound are moving toward or away from each other, the observed frequency of sound is different from that emitted. This is known as the *Doppler effect.* The equation relating the frequency f_0 the observer hears and the frequency f_s emitted by the source (a constant) is $f_0 = f_s\left(\dfrac{v + v_0}{v - v_s}\right)$, where v is the velocity of sound in air (a constant), v_0 is the velocity of the observer, and v_s is the velocity of the source. Show that $f_s\dfrac{\partial f_0}{\partial v_s} = f_0\dfrac{\partial f_0}{\partial v_0}$.

27. The *mutual conductance* (in $1/\Omega$) of a certain electronic device is defined as $g_m = \partial i_b/\partial e_c$. Under certain circumstances, the current i_b (in μA) is given by $i_b = 50(e_b + 5e_c)^{1.5}$. Find g_m when $e_b = 200$ V and $e_c = -20$ V.

28. The displacement y at any point in a taut, flexible string depends on the distance x from one end of the string and the time t. Show that $y(x, t) = 2\sin 2x\cos 4t$ satisfies the *wave equation* $\dfrac{\partial^2 y}{\partial t^2} = a^2\dfrac{\partial^2 y}{\partial x^2}$ with $a = 2$.

S.7 DOUBLE INTEGRALS

We now turn our attention to integration in the case of a function of two variables. The analysis has similarities to that of partial differentiation, in that an operation is performed while holding one of the independent variables constant.

If $z = f(x, y)$ and we wish to integrate with respect to x and y, we first consider either x or y constant and integrate with respect to the other. After this integral is evaluated, we then integrate with respect to the variable first held constant. We shall now define this type of integral and then give an appropriate geometric interpretation.

If $z = f(x, y)$ the **double integral** *of the function over x and y is defined as*

$$\int_a^b\left[\int_{g(x)}^{G(x)} f(x, y)\, dy\right] dx$$

NOTE ▶ *It will be noted that the limits on the inner integral are functions of x and those on the outer integral are explicit values of x. In performing the integration,* ***x is held constant while the inner integral is found and evaluated.*** *This results in a function of x only. This function is then integrated and evaluated.*

It is customary not to include the brackets in stating a double integral. Therefore, we write

$$\int_a^b\left[\int_{g(x)}^{G(x)} f(x, y)\, dy\right] dx = \int_a^b\int_{g(x)}^{G(x)} f(x, y)\, dy\, dx \qquad \text{(S7.1)}$$

◀ EXAMPLE 1   Evaluate $\int_0^1 \int_{x^2}^x xy \, dy \, dx$.

First, we integrate the inner integral with y as the variable and x as a constant.

$$\overset{\text{treat as constant}}{\int_{x^2}^x xy \, dy} = \left(x \frac{y^2}{2} \right) \Big|_{x^2}^x = x \left(\frac{x^2}{2} - \frac{x^4}{2} \right) = \frac{1}{2}(x^3 - x^5)$$

This means

$$\int_0^1 \int_{x^2}^x xy \, dy \, dx = \int_0^1 \frac{1}{2}(x^3 - x^5) \, dx = \frac{1}{2} \left(\frac{x^4}{4} - \frac{x^6}{6} \right) \Big|_0^1$$

$$= \frac{1}{2} \left(\frac{1}{4} - \frac{1}{6} \right) - \frac{1}{2}(0) = \frac{1}{24}$$

◀ EXAMPLE 2   Evaluate $\int_0^{\pi/2} \int_0^{\sin y} e^{2x} \cos y \, dx \, dy$.

CAUTION ▶   Since the inner differential is dx (*the inner limits must then be functions of y, which may be constant*), we first integrate with x as the variable and y as a constant. The second integration is with y as the variable.

$$\int_0^{\pi/2} \int_0^{\sin y} e^{2x} \cos y \, dx \, dy = \int_0^{\pi/2} \left[\frac{1}{2} e^{2x} \cos y \right]_0^{\sin y} dy$$

$$= \frac{1}{2} \int_0^{\pi/2} (e^{2\sin y} \cos y - \cos y) \, dy$$

$$= \frac{1}{2} \left[\frac{1}{2} e^{2\sin y} - \sin y \right]_0^{\pi/2} = \frac{1}{2} \left(\frac{1}{2} e^2 - 1 \right) - \frac{1}{2} \left(\frac{1}{2} - 0 \right)$$

$$= \frac{1}{4} e^2 - \frac{1}{2} - \frac{1}{4} = \frac{1}{4}(e^2 - 3) = 1.097$$

For the geometric interpretation of a double integral, consider the surface shown in Fig. S7.1(a). An **element of volume** (dimensions of dx, dy, and z) extends from the xy-plane to the surface. With x a constant, sum (integrate) these elements of volume from the left boundary, $y = g(x)$, to the right boundary, $y = G(x)$. Now the volume of the vertical slice is a function of x, as shown in Fig. S7.1(b). By summing (integrating) the volumes of these slices from $x = a$ ($x = 0$ in the figure) to $x = b$, we have the complete volume as shown in Fig. S7.1(c).

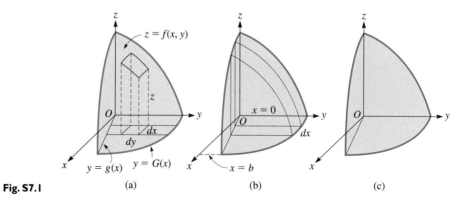

Fig. S7.1   (a)   (b)   (c)

Thus, *we may interpret a double integral as the* **volume under a surface,** in the same way as the integral was interpreted as the area of a plane figure. We may find the volume of a more general figure, as this volume is not necessarily a volume of revolution, as discussed in Section 26.3.

◀ EXAMPLE 3 Find the volume that is in the first octant and under the plane $x + 2y + 4z - 8 = 0$. See Fig. S7.2.

This figure is a tetrahedron, for which $V = \frac{1}{3}Bh$. Assuming the base is in the xy-plane, $B = \frac{1}{2}(4)(8) = 16$, and $h = 2$. Therefore, $V = \frac{1}{3}(16)(2) = \frac{32}{3}$ cubic units. We shall use this value to check the one we find by double integration.

To find $z = f(x, y)$, we solve the given equation for z. Thus,

$$z = \frac{8 - x - 2y}{4}$$

CAUTION ▶

Next, we must find the limits on y and x. Choosing to integrate over y first, we see that y goes from $y = 0$ to $y = (8 - x)/2$. ***This last limit is the trace of the surface in the xy-plane.*** Next, we note that x goes from $x = 0$ to $x = 8$. Therefore, we set up and evaluate the integral:

$$V = \int_0^8 \int_0^{(8-x)/2} \left(\frac{8 - x - 2y}{4}\right) dy\, dx$$

$$= \int_0^8 \left[\frac{1}{4}(8y - xy - y^2)\Big|_0^{(8-x)/2}\right] dx$$

$$= \frac{1}{4}\int_0^8 \left[8\left(\frac{8-x}{2}\right) - x\left(\frac{8-x}{2}\right) - \left(\frac{8-x}{2}\right)^2\right] dx$$

$$= \frac{1}{4}\int_0^8 \left(32 - 4x - 4x + \frac{x^2}{2} - 16 + 4x - \frac{x^2}{4}\right) dx$$

$$= \frac{1}{4}\int_0^8 \left(16 - 4x + \frac{x^2}{4}\right) dx$$

$$= \frac{1}{4}\left(16x - 2x^2 + \frac{x^3}{12}\right)\Big|_0^8 = \frac{1}{4}\left(128 - 128 + \frac{512}{12}\right)$$

$$= \frac{1}{4}\left(\frac{128}{3}\right) = \frac{32}{3} \text{ cubic units}$$

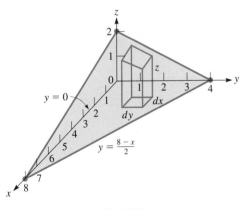

Fig. S7.2

We see that the values obtained by the two different methods agree.

◀ EXAMPLE 4 Find the volume above the xy-plane, below the surface $z = xy$, and enclosed by the cylinder $y = x^2$ and the plane $y = x$.

Constructing the figure, shown in Fig. S7.3, we now note that $z = xy$ is the desired function of x and y. Integrating over y first, the limits on y are $y = x^2$ to $y = x$. The corresponding limits on x are $x = 0$ to $x = 1$. Therefore, the double integral to be evaluated is

$$V = \int_0^1 \int_{x^2}^x xy\, dy\, dx$$

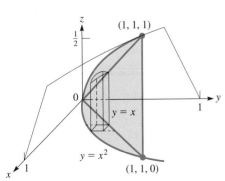

Fig. S7.3

This integral has already been evaluated in Example 1 of this section, and we can now see the geometric interpretation of that integral. Using the result from Example 1, we see that the required volume is 1/24 cubic unit.

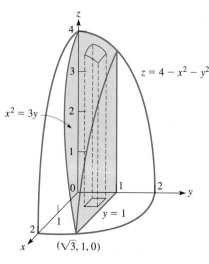

Fig. S7.4

◖ **EXAMPLE 5** Find the volume that is in the first octant under the surface $z = 4 - x^2 - y^2$ and that is between the cylinder $x^2 = 3y$ and the plane $y = 1$. See Fig. S7.4.

Setting up the integration such that we integrate over x first, we have

$$V = \int_0^1 \int_0^{\sqrt{3y}} (4 - x^2 - y^2) \, dx \, dy$$

$$= \int_0^1 \left[4x - \frac{x^3}{3} - y^2 x \right]_0^{\sqrt{3y}} dy$$

$$= \int_0^1 \left(4\sqrt{3y} - \sqrt{3}y^{3/2} - \sqrt{3}y^{5/2} \right) dy$$

$$= \sqrt{3} \left[4 \left(\frac{2}{3} \right) y^{3/2} - \frac{2}{5} y^{5/2} - \frac{2}{7} y^{7/2} \right] \Big|_0^1$$

$$= \sqrt{3} \left(\frac{8}{3} - \frac{2}{5} - \frac{2}{7} \right) = \frac{208\sqrt{3}}{105} = 3.431 \text{ cubic units}$$

If we integrate over y first, the integral is

$$V = \int_0^{\sqrt{3}} \int_{x^2/3}^1 (4 - x^2 - y^2) \, dy \, dx$$

Integrating and evaluating this integral, we arrive at the same result.

EXERCISES S.7

In Exercises 1–8, evaluate the given double integrals.

1. $\displaystyle\int_2^4 \int_0^1 xy^2 \, dx \, dy$

2. $\displaystyle\int_0^4 \int_1^{\sqrt{y}} (x - y) \, dx \, dy$

3. $\displaystyle\int_0^1 \int_0^{\sqrt{1-x^2}} y \, dy \, dx$

4. $\displaystyle\int_4^9 \int_0^x \sqrt{x - y} \, dy \, dx$

5. $\displaystyle\int_1^e \int_1^y \frac{1}{x} \, dx \, dy$

6. $\displaystyle\int_0^{\pi/6} \int_0^1 y \sin x \, dy \, dx$

7. $\displaystyle\int_0^{\ln 3} \int_0^x e^{2x+3y} \, dy \, dx$

8. $\displaystyle\int_0^{1/2} \int_y^{y^2} \frac{dx \, dy}{\sqrt{y^2 - x^2}}$

In Exercises 9–14, find the indicated volumes by double integration.

9. The first-octant volume under the plane $x + y + z - 4 = 0$

10. The volume above the xy-plane and under the surface $z = 4 - x^2 - y^2$

11. The first-octant volume bounded by the xy-plane, the planes $x = y$, $y = 2$, and $z = 2 + x^2 + y^2$

12. The first-octant volume under the plane $z = x + y$ and inside the cylinder $x^2 + y^2 = 9$

13. A wedge is to be made in the shape shown in Fig. S7.5 (all vertical cross sections are equal right triangles). By double integration, find the volume of the wedge.

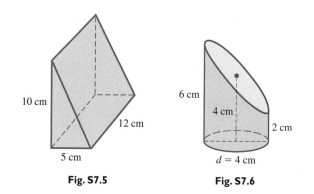

Fig. S7.5   **Fig. S7.6**

14. A circular piece of pipe is cut as shown in Fig. S7.6. Find the volume within the pipe. (*Hint:* Place the z-axis along the axis of the pipe, with the base of the pipe in the xy-plane.)

In Exercises 15 and 16, draw the appropriate figure.

15. Draw the appropriate figure indicating a volume that is found from the integral

$$\int_0^{1/2} \int_{x^2}^1 (4 - x - 2y) \, dy \, dx$$

16. Repeat Exercise 15 for the integral

$$\int_1^2 \int_0^{2-y} \sqrt{1 + x^2 + y^2} \, dx \, dy$$

NUMERICAL SOLUTIONS OF DIFFERENTIAL EQUATIONS

In Chapter 30 we presented some methods of solving differential equations. Some other types of differential equations do not have exact solutions. Therefore, in this section we show how differential equations can be solved numerically. We will demonstrate one basic method and one of the more advanced methods.

EULER'S METHOD

To find an approximate solution to a differential equation of the form $dy/dx = f(x, y)$, that passes through a known point (x_0, y_0), we write the equation as $dy = f(x, y) \, dx$ and then approximate dy as $y_1 - y_0$, and replace dx with Δx. From Section 24.8, we recall that Δy closely approximates dy for a small dx and that $dx = \Delta x$. This gives us

$$y_1 = y_0 + f(x_0, y_0) \, \Delta x \qquad \text{and} \qquad x_1 = x_0 + \Delta x$$

Therefore, we now know another point (x_1, y_1) that is on (or very nearly on) the curve of the solution. We can now repeat this process using (x_1, y_1) as a known point to obtain a next point (x_2, y_2). Continuing this process, we can get a series of points that are approximately on the solution curve. The method is called ***Euler's method.***

Named for the Swiss mathematician Leonhard Euler (1707–1783).

See Appendix C for a graphing calculator program EULRMETH. It gives numerical values of the solution of a differential equation using Euler's method.

◀ **EXAMPLE 1** For the differential equation $dy/dx = x + y$, use Euler's method to find the y-values of the solution for $x = 0$ to $x = 0.5$ with $\Delta x = 0.1$, if the curve of the solution passes through $(0, 1)$.

Using the method outlined above, we have $x_0 = 0$, $y_0 = 1$, and

$$y_1 = 1 + (0 + 1)(0.1) = 1.1 \qquad \text{and} \qquad x_1 = 0 + 0.1 = 0.1$$

This tells us that the curve passes (or nearly passes) through the point $(0.1, 1.1)$. Assuming this point is correct, we use it to find the next point on the curve.

$$y_2 = 1.1 + (0.1 + 1.1)(0.1) = 1.22 \qquad \text{and} \qquad x_2 = 0.1 + 0.1 = 0.2$$

Therefore, the next approximate point is $(0.2, 1.22)$. Continuing this process we find a set of points that would approximately satisfy the function that is the solution of the differential equation. Tabulating results, we have the following table.

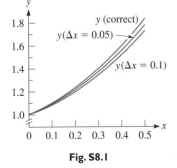

Fig. S8.1

x	y	Correct value of y
0.0	1.0000	1.0000
0.1	1.1000	1.1103
0.2	1.2200	1.2428
0.3	1.3620	1.3997
0.4	1.5282	1.5836
0.5	1.7210	1.7974

the values shown have been rounded off, although more digits were carried in the calculations

In this case we are able to find the correct values since the equation can be written as $dy/dx - y = x$, and the solution is $y = 2e^x - x - 1$. Although numerical methods are generally used with equations that cannot be solved exactly, we chose this equation so that we could compare values obtained with known values.

We can see that as x increases, the error in y increases. More accurate values can be found by using smaller values of Δx. In Fig. S8.1 the solution curves using $\Delta x = 0.1$ and $\Delta x = 0.05$ are shown along with the correct values of y.

Euler's method is easy to use and understand, but it is less accurate than other methods. We will show one of the more accurate methods in the next example.

Named for the German mathematicians Carl Runge (1856–1927) and Martin Kutta (1867–1944).

RUNGE–KUTTA METHOD

For more accurate numerical solutions of a differential equation, the ***Runge–Kutta method*** is often used. Starting at a first point (x_0, y_0), the coordinates of the second point (x_1, y_1) are found by using a weighted average of the slopes calculated at the points where $x = x_0$, $x = x_0 + \frac{1}{2}\Delta x$, and $x = x_0 + \Delta x$. The formulas for y_1 and x_1 are

$$y_1 = y_0 + \frac{1}{6}H(J + 2K + 2L + M) \qquad \text{and} \qquad x_1 = x_0 + H \qquad \text{(for convenience, } H = \Delta x)$$

where

$$J = f(x_0, y_0)$$
$$K = f(x_0 + 0.5H, y_0 + 0.5HJ)$$
$$L = f(x_0 + 0.5H, y_0 + 0.5HK)$$
$$M = f(x_0 + H, y_0 + HL)$$

We have used uppercase letters to correspond to calculator use. Traditional sources normally use h for H and a lowercase letter (such as k) with subscripts for J, K, L, and M, and express 0.5 as 1/2.

As with Euler's method, once (x_1, y_1) is determined we use the formulas again to find (x_2, y_2) by replacing (x_0, y_0) with (x_1, y_1). The following example illustrates the use of the Runge–Kutta method.

See Appendix C for a graphing calculator program RUNGKUTT. It gives numerical values of the solution of a differential equation using the Runge–Kutta method.

❰ **EXAMPLE 2** For the differential equation $dy/dx = x + \sin xy$, use the Runge–Kutta method to find y-values of the solution for $x = 0$ to $x = 0.5$ with $\Delta x = 0.1$, if the curve of the solution passes through $(0, 0)$.

Using the formulas and method outlined above, we have the following solution, with calculator notes to the right of the equations. Also, calculator symbols are used on the right sides of the equations to indicate the way in which they should be entered.

$$x_0 = 0 \qquad \text{store as } X$$
$$y_0 = 0 \qquad \text{store as } Y$$
$$H = 0.1$$
$$J = X + \sin XY = 0 \qquad \text{store as } J$$
$$K = X + 0.5H + \sin[(X + 0.5H)(Y + 0.5HJ)] = 0.05 \qquad \text{store as } K$$
$$L = X + 0.5H + \sin[(X + 0.5H)(Y + 0.5HK)] = 0.050\,125 \qquad \text{store as } L$$
$$M = X + H + \sin[(X + H)(Y + HL)] = 0.100\,501\,25 \qquad \text{store as } M$$
$$y_1 = Y + (H/6)(J + 2K + 2L + M) = 0.005\,012\,520\,8 \qquad \text{store as } Y$$
$$x_1 = X + H = 0.1 \qquad \text{store as } X$$

We now use (x_1, y_1) as we just used (x_0, y_0) to get the next point (x_2, y_2). Following is a table indicating the values that are obtained on the calculator. The graph of these points is shown in Fig. S8.2.

x	y
0.0	0.0
0.1	0.005\,012\,520\,8
0.2	0.020\,201\,339\,5
0.3	0.046\,027\,845\,5
0.4	0.083\,286\,818\,1
0.5	0.133\,146\,006\,2

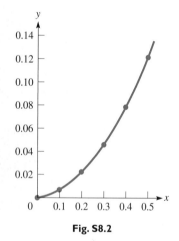

Fig. S8.2

EXERCISES **S.8**

In Exercises 1–8, use Euler's method to find y-values of the solution for the given values of x and Δx, if the curve of the solution passes through the given point. Check the results against known values by solving the differential equations exactly. Plot the graphs of the solutions in Exercises 1–4.

1. $\dfrac{dy}{dx} = x + 1$; $x = 0$ to $x = 1$; $\Delta x = 0.2$; $(0, 1)$

2. $\dfrac{dy}{dx} = \sqrt{2x + 1}$; $x = 0$ to $x = 1.2$; $\Delta x = 0.3$; $(0, 2)$

3. $\dfrac{dy}{dx} = y(0.4x + 1)$; $x = -0.2$ to $x = 0.3$; $\Delta x = 0.1$; $(-0.2, 2)$

4. $\dfrac{dy}{dx} = y + e^x$; $x = 0$ to $x = 0.5$; $\Delta x = 0.1$; $(0, 0)$

5. The differential equation of Exercise 1 with $\Delta x = 0.1$

6. The differential equation of Exercise 2 with $\Delta x = 0.1$

7. The differential equation of Exercise 3 with $\Delta x = 0.05$

8. The differential equation of Exercise 4 with $\Delta x = 0.05$

In Exercises 9–14, use the Runge–Kutta method to find y-values of the solution for the given values of x and Δx, if the curve of the solution passes through the given point.

9. $\dfrac{dy}{dx} = xy + 1$; $x = 0$ to $x = 0.4$; $\Delta x = 0.1$; $(0, 0)$

10. $\dfrac{dy}{dx} = x^2 + y^2$; $x = 0$ to $x = 0.4$; $\Delta x = 0.1$; $(0, 1)$

11. $\dfrac{dy}{dx} = e^{xy}$; $x = 0$ to $x = 1$; $\Delta x = 0.2$; $(0, 0)$

12. $\dfrac{dy}{dx} = \sqrt{1 + xy}$; $x = 0$ to $x = 0.2$; $\Delta x = 0.05$; $(0, 1)$

13. $\dfrac{dy}{dx} = \cos(x + y)$; $x = 0$ to $x = 0.6$; $\Delta x = 0.1$; $(0, \pi/2)$

14. $\dfrac{dy}{dx} = y + \sin x$; $x = 0.5$ to $x = 1.0$; $\Delta x = 0.1$; $(0.5, 0)$

In Exercises 15 and 16, solve the given problems.

15. An electric circuit contains a 1-H inductor, a 2-Ω resistor, and a voltage source of $\sin t$. The resulting differential equation relating the current i and the time t is $di/dt + 2i = \sin t$. Find i after 0.5 s by Euler's method with $\Delta t = 0.1$ s if the initial current is zero. Solve the equation exactly and compare the values.

16. An object is being heated such that the rate of change of the temperature T (in °C) with respect to time t (in min) is $dT/dt = \sqrt[3]{1 + t^3}$. Find T for $t = 5$ min by using the Runge–Kutta method with $\Delta t = 1$ min, if the initial temperature is 0°C.

APPENDIX A Study Aids

A.1 INTRODUCTION

The primary objective of this text is to give you an understanding of mathematics so that you can use it effectively as a tool in your technology. Without an understanding of the basic methods, knowledge is usually short-lived. However, if you do understand, you will find your work much more enjoyable and rewarding. This is true in any course you may take, be it in mathematics or in any other field.

Mathematics is an indispensable tool in almost all scientific fields of study. You will find it used to a greater and greater degree as you work in your chosen field. Generally, in the introductory portions of allied courses, it is enough to have a grasp of elementary concepts in algebra and geometry. However, as you progress in your field, the need for more mathematics will be apparent. This text is designed to develop these necessary tools so that they will be available to you in your technical courses. *You cannot derive the full benefit from your mathematics course unless you devote the necessary amount of time to develop a sound understanding of the subject.*

Your mathematics background probably includes some geometry and algebra. Therefore, some of the topics covered in this book may seem familiar to you, especially in the earlier chapters. However, it is likely that your background in some of these areas is not complete, either because you have not studied mathematics for a while or because you did not fully understand the topics when you first encountered them. *If a topic is familiar, take the opportunity to clarify any points on which you are not certain* and do not reason that there is no sense in studying it again. In almost every topic, you probably will find certain points that can use further study. *If the topic is new to you, use your time effectively to develop an understanding of the methods involved* and do not simply memorize problems of a certain type.

There is only one good way to develop the understanding and working knowledge necessary in any course, and that is to **work with it.** Many students consider mathematics difficult. They say that it is their lack of mathematical ability and the complexity of the material that makes it difficult. Some of the topics in mathematics, especially in the more advanced areas, do require a certain aptitude for full comprehension. *However, a large proportion of poor grades in elementary mathematics courses results from the fact that* **the student is not willing to put in the necessary time to develop a full understanding.** The student takes a quick glance through the material, tries a few exercises, is largely unsuccessful, and then decides that the material is "impossible." A detailed reading of the text, following the illustrative examples carefully, and then solving the exercises would lead to more success and therefore make the work much more enjoyable and rewarding. No matter what text is used, what methods are used in the course, or what other variables may be introduced, *if you do not put in an adequate amount of time studying, you will not derive the proper results.* More detailed suggestions for study are included in the following section.

A.2 SUGGESTIONS FOR STUDY

When you are studying the material presented in this text, the following suggestions may help you to derive full benefit from the time you devote to it.

1. Before attempting to do the exercises, read through the material preceding them.

2. Follow the illustrative examples carefully, being certain that you know how to proceed from step to step. You should then have a good idea of the methods involved.

3. Work through the exercises, spending a reasonable amount of time on each problem. If you cannot solve a certain problem in a reasonable amount of time, leave it and go to the next. Return to this problem later. If you find many problems difficult, you should reread the explanatory material and the examples to determine what point or points you have not understood.

4. When you have completed the exercises, or at least most of them, glance back through the explanatory material to be sure you understand the methods and principles.

5. If you have gone through the first four steps and certain points still elude you, ask to have these points clarified in class. Do not be afraid to ask questions; only be sure that you have made a sincere effort on your own before you ask them.

Some study habits that are useful not only here but in all of your other subjects are the following:

1. Put in the time required to develop the material fully, being certain that you are making effective use of your time. A good place to study helps immeasurably.

2. Learn the *methods and principles* being presented. Memorize as little as possible, for although certain basic facts are more expediently learned by memorization, these should be kept to a minimum.

3. Keep up with the material in all of your courses. Do not let yourself get so behind in your studies that it becomes difficult to make up the time. Usually the time is never really made up. Studying only before tests is a poor way of learning and is usually ineffective.

4. When you are taking examinations, always read each question carefully before attempting the solution. Solve those you find easiest first and do not spend too much time on any one problem. Also, use all the time available for the examination. If you finish early, use the reminder of the time to check your work.

If you consider these suggestions carefully and follow good study habits, you should enjoy a successful learning experience in this course as well as in other courses.

A.3 SOLVING WORD PROBLEMS

Drill-type problems require a working knowledge of the methods presented, and some algebraic steps to change the algebraic form may be required to complete the solution. *Word problems,* however, require a proper interpretation of the statement of the problem before they can be put in a form for solution.

We have to put word problems in symbolic form in order to solve them, and it is this procedure that most students find difficult. Because such problems require more than going through a certain routine, they demand more analysis and appear to be more difficult. Among the reasons for the student's difficulty at solving word problems are (1) unsuccessful previous attempts at solving word problems, leading the student to believe that all word problems are "impossible," (2) a poorly organized approach to the solution, and (3) failure to read the problem carefully, thereby having an improper and incomplete interpretation of the statement given. These can be overcome with proper attitude and care.

NOTE ▶ *A specific procedure for solving word problems is shown on page 42,* when word problems are first covered in our study of algebra. There are over 120 completely worked examples of word problems (as well as numerous other examples that show a similar analysis) throughout this text, illustrating proper interpretations and approaches to these problems. After Chapter 1, these examples are noted in the margin by the phrase *Solving a Word Problem.*

RISERS

The procedure shown on page 42 is similar to that used by most instructors and texts. One of the variations that a number of instructors use is called *RISERS.* This is a word formed from the first letters (an *acronym*) of the words that outline the procedure. These are *Read, Imagine, Sketch, Equate, Relate,* and *Solve.* We now briefly outline this procedure here.

Read the statement of the problem carefully.

Imagine. Take time to get a mental image of the situation described.

Sketch a figure.

Equate, *on the sketch,* the known and unknown quantities.

Relate the known and unknown quantities with an equation.

Solve the equation.

There are problems where a *sketch* may simply be words and numbers placed so that we may properly *equate* the known and unknown quantities. For example, in Example 2 on page 43, the *sketch, equate,* and *relate* steps might look like this:

sketch	34 resistors		56 Ω
	1.5-Ω resistors	2.0-Ω resistors	
equate	x	$34 - x$	
resistance	$1.5x$	$2.0(34 - x)$	56
relate	$1.5x$	$+ 2.0(34 - x) = 56$	

NOTE ▶ If you follow the method on page 42, or this RISERS variation, or any appropriate step-by-step method, and *write out the solution neatly,* you will find that word problems lend themselves to solution more readily than you have previously found.

APPENDIX B

Units of Measurement; The Metric System

B.1 THE METRIC SYSTEM (SI)

Most scientific and technical calculations involve numbers that represent a measurement or count of a specific physical quantity. *Such numbers are called* **denominate numbers,** *and associated with these denominate numbers are* **units of measurement.** For calculations and results to be meaningful, we must know these units. For example, if we measure the length of an object to be 12, we must know whether it is being measured in meters, centimeters, or some other specified unit of length.

Two basic systems of units, the **SI metric system** and the **United States Customary** system, are in use today. The U.S. Customary system traditionally has been known as the *British system.* (The SI metric system is now used in Great Britain, although many measurements in the traditional British system are also still used.) Nearly every country in the world now uses the SI metric system. In the United States both systems are used, and international trade has led most major U.S. industrial firms to convert their products to the metric system. Also, due to world trade, to further promote conversion to the metric system, the U.S. Congress has passed legislation stating that the metric system is the preferred system and requiring Federal agencies to use the metric system in their business-related activities. It should also be noted that the metric system is used worldwide in nearly all scientific work.

The metric system was first adopted in France in the late eighteenth century. Since then several systems using metric units have been developed, each adding some specific units, and some used British units in certain cases. Therefore, a wide variety of metric units was being used.

In order to simplify the metric system, a long series of international discussions were held, and in 1960 a modernized metric system was established. It is the **International System of Units,** called SI, and it is the system that has now replaced all previous systems of units. In this text, the metric system and SI are synonymous, and *all units are SI units or are acceptable for use with SI.*

In the SI system there are seven **base units** that are used to measure fundamental quantities. These base units form the foundation of the system. The base units are: (1) meter (length), (2) kilogram (mass), (3) second (time), (4) ampere (electric current), (5) kelvin (thermodynamic temperature), (6) mole (amount of substance), and (7) candela (luminous intensity).

The SI system also has **supplementary units** (used for measuring plane and solid angles), **derived units** (which are used for numerous quantities), as well as other units that are acceptable for use. The only supplementary unit used in this text is the *radian* (see Section 8.3).

SI Metric System

Many derived units have special names, and many others do not. For example, the volt is defined as a meter²-kilogram/second³-ampere, which is in terms of (a unit of length)² (a unit of mass)/(a unit of time)³ (a unit of electric current). The unit for acceleration has no special name, and is left in terms of the base units: meter per second squared.

Among the units permitted for use with the SI system are the units of time, the *minute, hour, day,* and *year*. Also permitted are the *degree, minute,* and *second* measurements for an angle, the *liter* for capacity, and the *degree Celsius* for temperature.

In designating units of measurement it is convenient to use designated **symbols,** rather than the complete name of the unit each time. In Table B-2 a list of quantities, the units of measurement, and the appropriate symbols are listed.

We note that the kilogram is the base unit of *mass*. The distinction between mass and *force* is very significant in physics, and this distinction causes difficulty when discussing the *weight* of an object. *Weight, which is the force with which an object is attracted to the earth, is different from mass, which is a measure of the amount of material in an object*. Although they are different quantities, mass and weight are very closely related. Near the surface of the earth the weight of an object is directly proportional to its mass. However, at great distance from the earth the weight of an object will be essentially zero, whereas its mass does not change.

The kilogram is commonly used for weight, although it is mass that is actually being denoted. The metric unit of force, the *newton,* is also used for weight. Thus, it is preferable to *specify the mass of an object in kilograms* and *the force of gravity of an object in newtons,* and thereby avoid the use of the term *weight* unless its meaning is completely clear.

As for temperature, the *kelvin* is the base unit for thermodynamic temperature, but *degrees Celsius* (formerly *Centigrade*) is the unit that is used for most general purposes. For temperature intervals, $1°C = 1$ K, although on a scale, $0°C = 273.15$ K.

The base units and derived units are often not a convenient size for many types of measurements. Therefore, other units of more convenient sizes are used. To denote these other units, the SI system employs certain prefixes to denote different orders of magnitude. These prefixes, along with their meanings and symbols, are shown in Table B-1.

The failure to convert units caused the $125 000 000 Mars Climate Orbiter to fly too close to Mars and break up in the Martian atmosphere in September 1999. A spacecraft team submitted force data in pounds, but the mission controllers assumed the data were in newtons. The system for checking data did not note the change in units. (1 lb = 4.448 N)

Table B.I **Metric Prefixes**

Prefix	Factor	Symbol	Prefix	Factor	Symbol
exa	10^{18}	E	deci	10^{-1}	d
peta	10^{15}	P	centi	10^{-2}	c
tera	10^{12}	T	milli	10^{-3}	m
giga	10^{9}	G	micro	10^{-6}	μ
mega	10^{6}	M	nano	10^{-9}	n
kilo	10^{3}	k	pico	10^{-12}	p
hecto	10^{2}	h	femto	10^{-15}	f
deca	10^{1}	da	atto	10^{-18}	a

Table B.2 Quantities and Their Associated Units

Quantity	Quantity Symbol	Metric (SI) Unit		In Terms of Other SI Units
		Name	Symbol	
Length	s	**meter**	m	
Mass	m	**kilogram**	kg	
Force	F	newton	N	$m \cdot kg/s^2$
Time	t	**second**	s	
Area	A		m^2	
Volume	V		m^3	
Capacity	V	liter	L	$(1\ L = 1\ dm^3)$
Velocity	v		m/s	
Acceleration	a		m/s^2	
Density	d, ρ		kg/m^3	
Pressure	p	pascal	Pa	N/m^2
Energy, work	E, W	joule	J	$N \cdot m$
Power	P	watt	W	J/s
Period	T		s	
Frequency	f	hertz	Hz	1/s
Angle	θ	radian	rad	
Electric current	I, i	**ampere**	A	
Electric charge	q	coulomb	C	$A \cdot s$
Electric potential	V, E	volt	V	$J/(A \cdot s)$
Capacitance	C	farad	F	s/Ω
Inductance	L	henry	H	$\Omega \cdot s$
Resistance	R	ohm	Ω	V/A
Thermodynamic temperature	T	**kelvin**	K	
Temperature	T	degrees Celsius	°C	$1°C = 1\ K)$
Quantity of heat	Q	joule	J	
Amount of substance	n	**mole**	mol	
Luminous intensity	I	**candela**	cd	

Special Notes:

1. The SI base units are shown in boldface type.

2. The unit symbols shown above are those that are used in the text. Many of them were adopted with the adoption of the SI system. This means, for example, that we use s rather than sec for seconds, and A rather than amp for amperes. Also, other units, such as volt are not spelled out, a common practice in the past.

3. The liter and degree Celsius are not actually SI units. However, they are recognized for use with the SI system due to their practical importance. Also, the symbol for liter has several variations. Presently L is recognized for use in the United States and Canada, l is recognized by the International Committee of Weights and Measures, and ℓ is also recognized for use in several countries.

4. Other units of time, along with their symbols, which are recognized for use with the SI system and are used in this text, are minute, min; hour, h; day, d.

5. Many additional specialized units are used with the SI system. However, most of those that appear in this text are shown in the table. A few of the specialized units are noted when used in the text. One which is frequently used is that for revolution, r.

6. There are a number of units that were used with the metric system prior to the development of the SI system. However, many of these are not to be used with the SI system. Among those that were commonly used are the dyne, erg, and calorie.

◀ EXAMPLE 1   Some commonly used units that use prefixes in Table B.2, along with their meanings, are shown below.

Unit	Symbol	Meaning	Unit	Symbol	Meaning
megohm	MΩ	10^6 ohms	milligram	mg	10^{-3} gram
kilometer	km	10^3 meters	microfarad	μF	10^{-6} farad
centimeter	cm	10^{-2} meter	nanosecond	ns	10^{-9} second

(Mega is shortened to meg when used with "ohm.")   ▶

It is the relationship among units within the metric system that makes it convenient to use. *Units of different magnitudes differ by a power of ten, and therefore only a shift of the decimal point is often the change necessary when changing units.* No specific arrangement exists within the U.S. system.

There are a number of matters of style that are employed with denominate numbers and their units when using the SI system. Following are some of the basic rules of style, although many others may be found in standard reference sources: (1) Symbols remain unaltered if the unit is in the plural. For example, either volt or volts is designated by the symbol V. (2) When designating square or cubic units, we use exponents in the designation. For example, we use m^2 rather than sq m. (3) The multiplication of unit symbols is designated by a dot of multiplication. For example, m · N means metre newton (whereas mN means millinewton). (4) A solidus (/) or negative exponent is used to designate the division of one unit symbol by another. For example, a meter per second would be designated by m/s or m · s^{-1}. Generally we have used the solidus in this text. (5) In long numbers, the digits are separated into groups of three, to the left and right of the decimal point. A space is not necessary if four digits appear. For example, we write 56 000 rather than 56,000 (that is, use only a space, and not a comma). Also, we write 0.000 000 7 rather than 0.0000007. We may write either 0.1234 or 0.123 4, although we should be consistent. In this text, no space is used with four-digit numbers.

B.2 CHANGING UNITS

When using denominate numbers, it may be necessary to change from one set of units to another. *A change within a given system is called a* **reduction,** and *a change from one system to another is called a* **conversion.** Some calculators are programmed to do conversions.

NOTE ▶   To change a given number of one set of units into another set of units, *we perform algebraic operations with units in the same manner as we do with any algebraic symbol.* Consider the following example.

◀ EXAMPLE 1   If we had a number representing meters per second to be multiplied by another number representing seconds per minute, as far as the units are concerned, we have

$$\frac{m}{s} \times \frac{s}{min} = \frac{m \times s}{s \times min} = \frac{m}{min}$$

This means that the final result would be in meters per minute.

In changing a number of one set of units to another set of units, we use reduction and conversion factors and the principle illustrated in Example 1. The convenient way to use the values in the tables is in the form of fractions. Since the given values are equal to each other, their quotient is 1. For example, since 100 cm = 1 m,

NOTE ▶

$$\frac{100 \text{ cm}}{1 \text{ m}} = 1 \quad \text{or} \quad \frac{1 \text{ m}}{100 \text{ cm}} = 1$$

since each represents the division of a certain length by itself. Multiplying a quantity by 1 does not change its value. The following examples illustrate changing units.

◀ EXAMPLE 2   Change 20 kg to milligrams.

$$20 \text{ kg} = 20 \text{ kg} \left(\frac{10^3 \text{ g}}{1 \text{ kg}}\right)\left(\frac{10^3 \text{ mg}}{1 \text{ g}}\right)$$
$$= 20 \times 10^6 \text{ mg} = 2.0 \times 10^7 \text{ mg}$$

We note that this result is found essentially by moving the decimal point three places when changing from kilograms to grams, and another three places when changing from grams to milligrams.

◀ EXAMPLE 3   Change 30 m/h to centimeters per second:

$$30 \frac{m}{h} = \left(30 \frac{m}{h}\right)\left(\frac{100 \text{ cm}}{1 \text{ m}}\right)\left(\frac{1 \text{ h}}{60 \text{ min}}\right)\left(\frac{1 \text{ min}}{60 \text{ s}}\right)$$
$$= \frac{(30)(100) \text{ cm}}{(60)(60) \text{ s}} = 0.833 \frac{\text{cm}}{\text{s}}$$

The only units remaining after the division are those required.

◀ EXAMPLE 4   Change 575 g/cm³ to kilograms per cubic meter:

$$575 \frac{g}{cm^3} = \left(575 \frac{g}{cm^3}\right)\left(\frac{100 \text{ cm}}{1 \text{ m}}\right)^3\left(\frac{1 \text{ kg}}{1000 \text{ g}}\right)$$
$$= \left(575 \frac{g}{cm^3}\right)\left(\frac{10^6 \text{ cm}^3}{1 \text{ m}^3}\right)\left(\frac{1 \text{ kg}}{10^3 \text{ g}}\right)$$
$$= 575 \times 10^3 \frac{\text{kg}}{\text{m}^3} = 5.75 \times 10^5 \frac{\text{kg}}{\text{m}^3}$$

◀ EXAMPLE 5   Change 62.8 kPa to newtons per square centimeter.

$$62.8 \text{ kPa} = \left(62.8 \times 10^3 \frac{N}{m^2}\right)\left(\frac{1 \text{ m}}{10^2 \text{ cm}}\right)^2$$

$$= \left(62.8 \times 10^3 \frac{N}{m^2}\right)\left(\frac{1 \text{ m}^2}{10^4 \text{ cm}^2}\right)$$

$$= \frac{62.8 \times 10^3}{10^4} \frac{N}{cm^2} = 6.28 \frac{N}{cm^2}$$

EXERCISES   B

In Exercises 1–4, give the symbol and the meaning for the given unit.

1. megahertz

2. kilowatt

3. millimeter

4. picosecond

In Exercises 5–8, give the name and the meaning for the units whose symbols are given.

5. kV

6. GΩ

7. mA

8. pF

In Exercises 9–44, make the indicated changes of units.

9. Change 1 km to centimeters.

10. Change 1 kg to milligrams.

11. Change 20 s to megaseconds.

12. Change 800 Pa to kilopascals.

13. Change 810.15 K to degrees Celsius.

14. Change 31.50°C to kelvins

15. Change 250 mm² to square meters.

16. Change 1.75 m² to square centimeters.

17. Change 80.0 m³ to cubic decimeters.

18. Change 360 μm³ to cubic centimeters.

19. Change 50 mL to liters.

20. Change 0.125 L to milliliters.

21. Change 45.0 m/s to centimeters per second.

22. Change 200 mm/s to meters per second.

23. Change 38.2 cm/s to meters per hour.

24. Change 1.32 km/h to centimeters per minute.

25. Change 9.80 m/s² to centimeters per minute squared.

26. Change 5.10 g/cm³ to kilograms per cubic millimeter.

27. Change 25 h to milliseconds.

28. Change 30 ns to minutes.

29. Change 5.25 mV to watts per ampere.

30. Change 15.0 μF to millicoulombs per volt.

31. A weather satellite orbiting the earth has a mass of 2200 kg. How many megagrams is this?

32. At sea level, atmospheric pressure is about 101 300 Pa. How many kilopascals is this?

33. A car's gasoline tank holds 56 L. What is this capacity in cubic centimeters?

34. A hockey puck has a mass of about 0.160 kg. What is its mass in milligrams?

35. The density of water is 1000 kg/m³. Change this to grams per liter.

36. Water flows from a kitchen faucet at the rate of 8500 mL/min. What is this rate in liters per second?

37. The speed of sound is about 332 m/s. Change this speed to kilometers per hour.

38. The acceleration due to gravity is about 980 cm/s². Convert this to meters per squared hour.

39. Fifteen grams of a medication are to be dissolved in 0.060 L of water. Express this concentration in milligrams per deciliter.

40. The earth's surface receives energy from the sun at the rate of 1.35 kW/m². Reduce this to joules per second square centimeter.

41. The moon travels about 2 400 000 km in about 28 d in one rotation about the earth. Express its velocity in meters per second.

42. The average density of the earth is about 5.52 g/cm³. Change this to kilograms per cubic meter.

43. A typical electric current density in a wire is 1.2×10^6 A/m². Express this in milliamperes per square centimeter.

44. A certain car travels 24 km on 2.0 L of gas. Express the fuel consumption in litres per 100 kilometers.

In Exercises 45–48, use the following information. A commercial jet with 230 passengers on a 2850-km flight from Vancouver to Chicago averaged 765 km/h and used fuel at the rate of 5650 L/h.

45. How many hours long was the flight?

46. How long in seconds did it take to use 1.0 L of fuel?

47. What was the fuel consumption in km/L?

48. What was the fuel consumption in L/passenger?

The Graphing Calculator

C.1 INTRODUCTION

Until the 1970s, the personal calculating device for engineers and scientists was the slide rule. The microprocessor chip became commercially available in 1971, and the scientific calculator became widely used during the 1970s. Then in the 1980s, the graphing calculator was developed and is now used extensively.

The scientific calculator is still used by many when only accurate calculations are required. However, the graphing calculator can perform all the operations of a scientific calculator and numerous other operations. It also has the major advantage that entries can be seen in the viewing *window,* and this allows a visual check of entered data.

For these reasons, the graphing calculator is now used much more than the scientific calculator by engineers and scientists and in the more advanced mathematics courses. Therefore, as we stated on page 11, we restrict our coverage of calculator use in this text to graphing calculators.

This appendix includes a brief discussion of graphing calculator features. We then show its use in the listing of the over 130 examples with sample screen displays throughout the text. A set of exercises that uses the basic calculational and graphing features is also included. The final section of this appendix gives some graphing calculator programs that can be used to perform more extensive operations with fewer steps.

C.2 THE GRAPHING CALCULATOR

Since their introduction in the 1980s, many types and models of graphing calculators have been developed. All models can do all the calculational operations (and more) and display any of the graphs required in this text. Some models can also do symbolic operations. However, regardless of the model you are using, *to determine how the features of a particular model are used, refer to the manual for that model.*

In using a calculator, keep in mind that certain operations do not have defined results. If such an operation is attempted, the calculator will show an error display. Operations that can result in an error display include division by zero, square root of a negative number, logarithm of a negative number, and the inverse trigonometric function of a value outside the defined interval. Also, an error display may result if an improper sequence of keys is used.

On a graphing calculator, negative numbers are entered by using the $\boxed{(-)}$ key, and subtraction is entered by using the $\boxed{-}$ key. There is additional discussion of this on page 11 of Section 1.3. Use of the $\boxed{-}$ key for the $\boxed{(-)}$ key will usually result in an error display.

Most of the keys are used to access two or three different features. These features may also have subroutines that may be used. To activate a second use, the *2nd* (or *shift*) key is used. The *alpha* key is used for entering alphabetical characters when using more than one literal symbol (x is usually available directly) or when entering a calculator program.

GRAPHING CALCULATOR FEATURES

We now list some of the more important features of a graphing calculator with an explanation of their use. (This listing uses the designations on a TI-83 calculator. Your calculator may use a different designation for some features.)

Feature	Use
MODE	Determines how numbers and graphs are displayed. A *menu* is displayed that shows settings that may be used. The *mode* can be set, for example, (1) for the number of decimal places displayed in numbers; (2) to display angles in degrees or in radians; (3) to display the graph of a function or of parametric equations; (4) to display a graph in rectangular or polar coordinates. See the *mode* feature for other possible settings.
Y=	Used to enter functions. These functions are to be graphed or used in some other way.
WINDOW	Used to set the boundaries of the viewed portion of a graph. (On some calculators this is the *range* feature.)
GRAPH	Used to display the graph of a function. (On some models this feature puts the calculator in graphing mode.)
TRACE	Used to move the cursor from pixel to pixel along a displayed graph.
ZOOM	Used to adjust the viewing window, usually to magnify the view of a particular part of a graph.
STO▶	Used to store in memory using variable names.
STAT	Used to access various statistical features.
MATH	Used to access specific mathematical features. These include fractions, *n*th roots, maximum and minimum values, absolute values, and others.
TEST	Used to enter special symbols such as $>$ and $<$ and to enter special logic words such as *and* and *or.*
MATRX	Used to enter matrix values and perform matrix operations such as the evaluation of a determinant.
DRAW	Used to make certain types of drawings, such as shading in an area between curves or a tangent line to a curve.
PRGM	Used to enter and access calculator programs. A set of sample programs is given in the last section of this appendix.

A *Graphing Calculator Lab Manual* supplement for this text is also available. It provides general directions for using all graphing calculator features and specific key steps for some models.

There are many other features on a graphing calculator that are used to perform specific types of calculations and operations. Again, *to determine how the features of a particular model are used, refer to the manual for that model.*

Examples of the use of these features are found in the text examples in the listing that follows on the next page.

EXAMPLES OF CALCULATOR USE

The first calculational use of a graphing calculator is shown in Section 1.3 on page 11. The first graphical use is shown in Section 3.5 on page 98.

The following list shows the pages on which examples of the use of a graphing calculator are given in the text. Calculator *window* screens are shown with each example. The list is divided into calculator uses that are primarily calculational and those that are primarily graphical.

EXERCISES C.2

The following exercises provide an opportunity for practice in performing calculations and a few basic graphs on a graphing calculator. There are many additional exercises throughout the book that require these types of calculations. Many additional exercises that require graphing of various types of curves are found after graphing is introduced in Chapter 3.

In Exercises 1–92, perform the indicated calculations on a graphing calculator.

1. $47.08 + 8.94$

2. $654.1 + 407.7$

3. $4724 - 561.9$

4. $0.9365 - 8.077$

5. 0.0396×471

6. 26.31×0.9393

7. $76.7 \div 194$

8. $52{,}060 \div 75.09$

9. 3.76^2

10. 0.986^2

11. $\sqrt{0.2757}$

12. $\sqrt{60.36}$

13. $\dfrac{1}{0.0749}$

14. $\dfrac{1}{607.9}$

15. $(19.66)^{2.3}$

16. $(8.455)^{1.75}$

17. $\sin 47.3°$

18. $\sin 1.15$

19. $\cos 3.85$

20. $\cos 119.1°$

21. $\tan 306.8°$

22. $\tan 0.537$

23. $\sec 6.11$

24. $\csc 242.0°$

25. $\sin^{-1} 0.6607$ (in degrees)

26. $\cos^{-1}(-0.8311)$ (in radians)

27. $\tan^{-1}(-2.441)$ (in radians)

28. $\sin^{-1} 0.0737$ (in degrees)

29. $\log 3.857$

30. $\log 0.9012$

31. $\ln 808$

32. $\ln 70.5$

33. $10^{0.545}$

34. $10^{-0.0915}$

35. $e^{-5.17}$

36. $e^{1.672}$

37. $(4.38 + 9.07) \div 6.55$

38. $(382 + 964) \div 844$

39. $4.38 + (9.07 \div 6.55)$

40. $382 + (964 \div 844)$

41. $\dfrac{5.73 \times 10^{11}}{20.61 - 7.88}$

42. $\dfrac{7.09 \times 10^{23}}{284 + 839}$

43. $50.38\pi^2$

44. $\dfrac{5\pi}{14.6}$

45. $\sqrt{1.65^2 + 6.44^2}$

46. $\sqrt{0.735^2 + 0.409^2}$

47. $3(3.5)^4 - 4(3.5)^2$

48. $\dfrac{3(-1.86)}{(-1.86)^2 + 1}$

49. $29.4 \cos 72.5°$

50. $\dfrac{477}{\sin 58.7°}$

51. $\dfrac{4 + \sqrt{(-4)^2 - 4(3)(-9)}}{2(3)}$

52. $\dfrac{-5 - \sqrt{5^2 - 4(4)(-7)}}{2(4)}$

53. $\dfrac{0.176(180)}{\pi}$

54. $\dfrac{209.6\pi}{180}$

55. $\dfrac{1}{2}\left(\dfrac{51.4\pi}{180}\right)(7.06)^2$

56. $\dfrac{1}{2}\left(\dfrac{148.2\pi}{180}\right)(49.13)^2$

57. $\sin^{-1}\dfrac{27.3 \sin 36.5°}{46.8}$

58. $\dfrac{0.684 \sin 76.1°}{\sin 39.5°}$

59. $\sqrt{3924^2 + 1762^2 - 2(3924)(1762)\cos 106.2°}$

60. $\cos^{-1}\dfrac{8.09^2 + 4.91^2 - 9.81^2}{2(8.09)(4.91)}$

61. $\sqrt{5.81 \times 10^8} + \sqrt[3]{7.06 + 10^{11}}$

62. $(6.074 \times 10^{-7})^{2/5} - (1.447 \times 10^{-5})^{4/9}$

63. $\dfrac{3}{2\sqrt{7} - \sqrt{6}}$

64. $\dfrac{7\sqrt{5}}{4\sqrt{5} - \sqrt{11}}$

65. $\tan^{-1}\dfrac{7.37}{5.06}$

66. $\tan^{-1}\dfrac{46.3}{-25.5}$

67. $2 + \dfrac{\log 12}{\log 7}$

68. $\dfrac{10^{0.4115}}{\pi}$

69. $\dfrac{26}{2}(-1.450 + 2.075)$

70. $\dfrac{4.55(1 - 1.08^{15})}{1 - 1.08}$

71. $\sin^2\left(\dfrac{\pi}{7}\right) + \cos^2\left(\dfrac{\pi}{7}\right)$

72. $\sec^2\left(\dfrac{2}{9}\pi\right) - \tan^2\left(\dfrac{2}{9}\pi\right)$

73. $\sin 31.6° \cos 58.4° + \sin 58.4° \cos 31.6°$

74. $\cos^2 296.7° - \sin^2 296.7°$

75. $\sqrt{(1.54 - 5.06)^2 + (-4.36 - 8.05)^2}$

76. $\sqrt{(7.03 - 2.94)^2 + (3.51 - 6.44)^2}$

77. $\dfrac{(4.001)^2 - 16}{4.001 - 4}$

78. $\dfrac{(2.001)^2 + 3(2.001) - 10}{2.001 - 2}$

79. $\dfrac{4\pi}{3}(8.01^3 - 8.00^3)$

80. $4\pi(76.3^2 - 76.0^2)$

81. $0.01\left(\dfrac{1}{2}\sqrt{2} + \sqrt{2.01} + \sqrt{2.02} + \dfrac{1}{2}\sqrt{2.03}\right)$

82. $0.2\left[\dfrac{1}{2}(3.5)^2 + 3.7^2 + 3.9^2 + \dfrac{1}{2}(4.1)^2\right]$

83. $\dfrac{e^{0.45} - e^{-0.45}}{e^{0.45} + e^{-0.45}}$

84. $\ln \sin 2e^{-0.055}$

85. $\ln \dfrac{2 - \sqrt{2}}{2 - \sqrt{3}}$

86. $\sqrt{\dfrac{9}{2} + \dfrac{9 \sin 0.2\pi}{8\pi}}$

87. $2 + \dfrac{0.3}{4} - \dfrac{(0.3)^2}{64} + \dfrac{(0.3)^3}{512}$

88. $\dfrac{1}{2} + \dfrac{\pi\sqrt{3}}{360} - \dfrac{1}{4}\left(\dfrac{\pi}{180}\right)^2$

89. $160(1 - e^{-1.50})$

90. $e^{-3.60}(\cos 1.20 + 2 \sin 1.20)$

91. $\dfrac{(10)(9)(8)(7)(6)}{5!}$

92. $\dfrac{20! - 15!}{20! + 15!}$

In Exercises 93–108, display the graphs of the given functions on a graphing calcaulator. Use window settings to properly display the curve.

93. $y = 2x$

94. $y = 6 - x$

95. $y = x^2$

96. $y = 4 - 2x^2$

97. $y = 0.5x^3$

98. $y = 2x^2 - x^4$

99. $y = \sqrt{4 - x}$

100. $y = \sqrt{25 - x^2}$

101. $y = \log x$

102. $y = 2^x$

103. $y = 2e^x$

104. $y = 3 \ln x$

105. $y = \sin 2x$

106. $y = 2 \cos x$

107. $y = \tan^{-1} x$

108. $y = \dfrac{6}{x}$

C.3 GRAPHING CALCULATOR PROGRAMS

Programs like those used on a computer can be stored in the memory of a graphing calculator. As mentioned earlier, such programs are used to perform more extensive operations with fewer steps, often significantly fewer steps. Generally, it is necessary only to enter certain data to obtain the required results.

Following are 14 programs that were written for use on a TI-83 graphing calculator. For other calculator models, it may be necessary to adapt the steps indicated to the format and designated operations of that model.

The chapter, program title, and page on which the program reference appears are shown. A brief description of each program is also given.

Chapter 2   PYTHAGTH   page 57

This program solves for any side of a right triangle, given the other two sides. The input is chosen from the menu.

:Menu("PYTHAGTH", "FINDLEGA",A,"FINDLEGB",B, "FINDHYPC",C)
:Lbl A:Prompt B,C
:Disp "A=", $\sqrt{}$ (C²−B²):Stop
:Lbl B:Prompt A,C
:Disp "B=", $\sqrt{}$ (C²−A²):Stop
:Lbl C:Prompt A,B
:Disp "C=", $\sqrt{}$ (A²+B²)

Chapter 4   SLVRTTRI   page 125

This program solves a right triangle, given both legs.

:Prompt A,B
:Disp "C=", $\sqrt{}$ (A²+B²)
:Disp "θA=", tan⁻¹(A/B)
:Disp "θB=", tan⁻¹(B/A)

Chapter 7   QUADFORM   page 225

This program solves the quadratic equation $Ax^2 + Bx + C = 0$ for real roots.

:Prompt A,B,C
:B²−4AC→D
:If D<0: Then:Disp "IMAGINARY ROOTS":Stop:End
:Disp "ROOTS ARE",(⁻B+ $\sqrt{}$ (D))/(2A),(⁻B− $\sqrt{}$ (D))/(2A)

Chapter 9   ADDVCTR   page 270

This program adds two vectors, given their magnitudes and angles.

:Disp "ENTER MAGNITUDES"
:Input "A=",A:Input"B=",B
:Disp "ENTER ANGLES"
:Input "θA=",P:Input "θB=",Q
:Acos(P)+Bcos(Q)→R
:Asin(P)+Bsin(Q)→S
:tan⁻¹(S/R)→T
:Disp "R=", $\sqrt{}$ (R²+S²)
:If R<0:Goto A:If S<0:Goto B
:Disp "θR=",T:Stop
:Lbl A:Disp "θR=",T+180:Stop
:Lbl B:Disp "θR=",T+360

Chapter 10   SINECURV   page 300

This program displays the graphs of $y_1 = \sin x$, $y_2 = 2 \sin x$, $y_3 = \sin 2x$, and $y_4 = 2 \sin(2x - \pi/3)$. There is no input.

```
:¯2→Xmin:7→Xmax
:¯2→Ymin:2→Ymax
:"sin(X)"→Y₁:"2sin(X)"→Y₂
:"sin(2X)"→Y₃:"2sin(2X−π/3)"→Y₄
:Disp "Y₁=sin(X)"
:Disp "Y₂=2sin(X)"
:Disp "Y₃=sin(2X)"
:Disp "Y₄=2sin(2X−π/3)"
:Pause:DispGraph:Pause:ClrHome
```

Chapter 11   TBLROOTS   page 329

This program displays a table of square roots and cube roots. Input to TblSet should be made before running the program.

```
:"√(X)"→Y₁
:"³√(X)"→Y₂
:DispTable:Pause:ClrHome
```

Chapter 12   DEMOIVRE   page 359

This program finds the nth roots of the complex number $a + bj$. Displayed are the real and imaginary parts of each root.

```
:Prompt A,B,N
:√((A²+B²)^(1/N))→R
:If A<0:(tan⁻¹(B/A)+180)/N→θ
:If A>0 and B<0:(tan⁻¹(B/A)+360)/N→θ
:If A=0 and B>0:90/N→θ
:If A=0 and B<0:270/N→θ
:If A>0 and B≥0:(tan⁻¹(B/A))/N→θ
:For(I,1,N)
:Disp "ROOT",I
:Disp "REAL PART",Rcos(θ)
:Disp "IMAG PART",Rsin(θ)
:θ+360/N→θ:Pause:End
```

Chapter 12   IMPEDANC   page 363

This program calculates the impedance Z and phase angle θ, given a resistance R, inductance L, capacitance C, and frequency F. (Any of R, L, or C can be zero.)

```
:Prompt R,L,F,C
:2πFL→I
:If C=0:Goto A
:(2πFC)⁻¹→Q
:If R=0:Goto B
:Disp "Z=",√(R²+(I−Q)²)
:Disp "θ=",tan⁻¹((I−Q)/R):Stop
:Lbl A:Disp "Z=",√(R²+I²):Disp "θ=",tan⁻¹(I/R):Stop
:Lbl B:Disp "Z=",abs(I−Q)
:If I>Q:Disp "θ=90"
:If I<Q:Disp "θ=−90"
```

Chapter 15   SYNTHDIV   page 419

This program gives the coefficients and the remainder if a polynomial is divided by $x - R$. The coefficients are entered as a {LIST}.

```
:Disp "LIST COEFFICIENTS"
:Input L₁
:Input "R=",R
:L₁(1)→F:dim(L₁)→D
:Disp "QUOTIENT="
:For(N,1,D−1)
:Pause F:RF+L₁(N+1)→F:End
:Disp "REMAINDER=",F
```

Chapter 17   LINPROG   page 483

This program double shades the area of feasible points for two linear constraints. *Trace* is used to find the point of maximum value or minimum value.

```
:0→Xmin:0→Ymin
:Input "Xmax=",Xmax
:Input "Xscl=",Xscl
:Input "Ymax=",Ymax
:Input "Yscl=",Yscl
:Input "Y₁=",Y₁
:Input "Y₂=",Y₂
:Disp "FOR MINIMUM"
:Disp "ENTER A=1"
:Disp "FOR MAXIMUM"
:Disp "ENTER A=2"
:Prompt A
:If A=1: Goto A
:Shade(0,Y₁,0,Xmax,1,3)
:Shade(0,Y₂,0,Xmax,2,3)
:Trace:Pause:ClrHome:Stop
:Lbl A
:Shade(Y₁,Ymax,0,Xmax,1,3)
:Shade(Y₂,Ymax,0,Xmax,2,3)
:Trace:Pause:ClrHome
```

Chapter 19   BINEXPAN   page 518

This program gives the first k coefficients of $(ax + b)^n$.

```
:Prompt K, A,B,N
:A^N→T
:Disp "COEFFICIENTS="
:For (C,1,K): Disp T
:T(N−C+1)B/AC→T
:Pause:End
```

Chapter 21   SLOPEDIS   page 563

This program calculates the slope m of a line through two points (a, c) and (e, f) and the distance between the points. It also displays the line segment on a split screen. Set proper *window* values before running the program.

```
:Prompt A,C,E,F
:Horiz
:Output (2,1,"M=")
:Output (2,4,(F−C)/(E−A))
:Output (3,1,"D=")
:Output (3,4,√((E−A)²+(F−C)²))
:Line (A,C,E,F)
:Pause:Full
```

Chapter 21   GRAPHLIN   page 565

This program displays the graph of a line through two points (a, c) and (d, e). The *window* values should be set appropriately before running the program. Default *window* is $(-9, 9)$ by $(-6, 6)$.

```
:Prompt A,C,D,E
:(E−C)/(D−A)→M:C−MA→B
:"MX+B"→Y₁
:−9→Xmin:9→Xmax:1→Xscl
:−6→Ymin:6→Ymax:1→Yscl
:DispGraph
:Pause:ClrHome
```

Chapter 21   GRAPHCON   page 595

This program displays the graph of a conic of the form $ax^2 + cy^2 + dx + ey + f = 0$. The *window* settings may have to be changed in the program. The default *window* is $(-9, 9)$ by $(-6, 6)$.

```
:Prompt A,C,D,E,F
:If C=0:Goto A
:(−E+√(E²−4C(AX²+DX+F)))/(2C)→Y₁
:(−E−√(E²−4C(AX²+DX+F)))/(2C)→Y₂
:Lbl B
:−9→Xmin:9→Xmax:1→Xscl
:−6→Ymin:6→Ymax:1→Yscl
:DispGraph
:Pause:ClrHome:Stop
:Lbl A: "(AX²+DX+F)/(−E)"→Y₁
:Goto B
```

Chapter 23   LIMFUNC   page 649

This program evaluates the limit of $f(x)$ as $x \to a^+$.

```
:Input "F(X)=",Y₁
:Input "A=",A
:1→D:1→N
:For(I,1,8):A+D→X
:Disp "FOR X=",X
:Disp "F(X)=",Y₁
:.1D→D
:Pause:End
```

Chapter 23   SECTOTAN   page 654

This program displays the graph of a function and four secant lines as the secant line approaches the tangent line at a given point. Input the function Y_1, A (the x-coordinate of the point of tangency), and H.

```
:ClrDraw
:Input "Y₁=",Y₁
:DispGraph:Pause
:Prompt A,H
:A→X:Y₁→B
:For(I,1,4):A+H→X
:(Y₁−B)/H→M
:DrawF M(X−A)+B
:Pause: .7H→H
:End:ClrHome
```

Chapter 24   NEWTON   page 697

This program finds the nth approximation of a root of the equation $f(x) = 0$ using Newton's method. Input the function $f(x)$, its derivative dy/dx, n, and the estimate A.

```
:Input "Y=",Y₁
:Input "DY/DX=", Y₂
:Input "N=", N
:Input "A=",X
:For(I,2,N)
:Disp "ROOT",I:Disp "IS",X−Y₁/Y₂
:X−Y₁/Y₂→X
:Pause:End
```

Chapter 25   AREAUNCV   page 741

This program approximates the area under a curve by summing the areas of n rectangles (inscribed for an increasing curve, circumscribed for a decreasing curve) from $x = a$ to $x = b$.

```
:Input "F(X)=",Y₁
:Prompt A,B,N
:(B−A)/N→D:0→R
:For(I,0,N−1)
:A+ID→X:R+DY₁→R
:End
:Disp "AREA=",R
```

Chapter 25   TRAPZRUL   page 749

This program evaluates the integral of $f(x)\,dx$ from $x = a$ to $x = b$ by using the trapezoidal rule with n intervals.

```
:Input "F(X)=",Y₁
:Prompt A,B,N
:(B−A)/N→D:A→X:Y₁→E
:B→X:Y₁→F:E+F→S
:For(I,1,N−1)
:A+ID→X:S+2Y₁→S:End
:Disp "APPROX VALUE OF" :Disp "INTEGRAL IS"
:Disp SD/2
```

Chapter 29   PARTSUMS   page 877

This program evaluates the first *n* partial sums of an infinite series.

```
:Input "GENERAL TERM=",Y₁
:Input "N=",J:0→S
:For(N,1,J)
:Disp "FOR N=",N
:Disp "SUM=",S+Y₁
:S+Y₁→S
:Pause:End
```

Supplementary Topic S.7   EULRMETH   page 991

This program gives numerical values of y for the solution of the differential equation $dy/dx = f(x, y)$ from $x = a$ to $x = b$ with an initial y-value of y_0 and $dx = H$, using Euler's method.

```
:Input "F(X,Y)=",Y₁
:Prompt A,B,H
:Input "Y0=",Y
:abs(int((B−A)/H+.5))→N
:For(I,1,N+1):A+(I−1)H→X
:Disp "X=",X
:Disp "Y=",Y
:Y+Y₁H→Y
:Pause:End
```

Supplementary Topic S.7   RUNGKUTT   page 992

This program gives numerical values of y for the solution of the differential equation $dy/dx = f(x, y)$ from $x = a$ to $x = b$ with an initial y-value of y_0 and $dx = H$, using the Runge–Kutta method.

```
:Input "F(X,Y)=",Y₁
:Prompt A,B,H
:Input "Y0=",C
:abs(int((B−A)/H+.5))→N:C→Y
:Disp "X=",A
:Disp "Y=",C:Pause
:For(I,1,N)
:A+(I−1)H→X:HY₁→J
:X+H/2→X:C+J/2→Y
:HY₁→K:C+K/2→Y
:HY₁→L:X+H/2→X
:C+L→Y:HY₁→M
:C+(J+2K+2L+M)/6→Y
:Y→C
:Disp "X=", X
:Disp "Y=", Y
:Pause:End
```

EXERCISES C.3

In each of the following exercises, write and test a graphing calculator program to perform the indicated operations for a topic in the indicated chapter.

1. (Chapter 2) To find the total surface area and volume of a right circular cylinder for given values of the radius r and height h.

2. (Chapter 5) To solve the system of equations $ax + by = c$ and $dx + ey = f$ for given values of the constants.

3. (Chapter 6) To find the coefficients e, f, and g for the product of binomials $(ax + b)(cx + d) = ex^2 + fx + g$ for given values of the constants.

4. (Chapter 8) To find the arc length and area of a circular sector for given values of the radius r and angle θ.

5. (Chapter 9) To solve a triangle given angles A, B, and included side C.

6. (Chapter 13) To find the logarithm to any given base b.

7. (Chapter 18) To find the constant of proportionality k for $y = kr^a s^b/t^c$, given the values of y, r, a, s, b, t, and c.

8. (Chapter 19) For a geometric sequence, to calculate the nth term, sum of n terms, and the sum of the infinite series for given values of the first term a, ratio r, and n.

9. (Chapter 23) To evaluate the derivative of $y = cx^n$ at $x = a$ for given values of a, c, and n.

10. (Chapter 25) To evaluate the definite integral $\int_a^b cx^n \, dx$ for given values of a, b, c, and n.

At the right is an explanation and example of Newton's method for solving equations. They were copied directly from *Essays on Several Curious and Useful Subjects in Speculative and Mix'd Mathematicks* by Thomas Simpson (of Simpson's rule). It was published in London in 1740.

See Exercise 22, page 698.

A new Method for the Solution of Equations in Numbers.

CASE I.

When only one Equation is given, and one Quantity (x) to be determined.

TAKE the Fluxion of the given Equation (be it what it will) fuppofing, x, the unknown, to be the variable Quantity; and having divided the whole by $\dot{x}$, let the Quotient be reprefented by A. Eftimate the Value of x pretty near the Truth, fubftituting the fame in the Equation, as alfo in the Value of A, and let the Error, or refulting Number in the former, be divided by this numerical Value of A, and the Quotient be fubtracted from the faid former Value of x; and from thence will arife a new Value of that Quantity much nearer to he Truth than the former, wherewith proceeding as before, another new Value may be had, and fo another, &c. 'till we arrive to any Degree of Accuracy defired.

EXAMPLE I.

LET $300x - x^3 - 1000$ be given $= 0$; to find a Value of x. From $300\dot{x} - 3x^2\dot{x}$, the Fluxion of the given Equation, having expunged $\dot{x}$, ($Cafe$ I.) there will be $300 - 3xx = A$: And, becaufe it appears by Infpection, that the Quantity $300x - x^3$, when x is $= 3$, will be lefs, and when $x = 4$, greater than 1000, I eftimate x at 3.5, and fubftitute inftead thereof, both in the Equation and in the Value of A, finding the Error in the former $= 7.125$, and the Value of the latter $= 263.25$: Wherefore, by taking $\frac{7.125}{263.25} = .027$ from 3.5 there will remain 3.473 for a new Value of x; with which proceeding as before, the next Error, and the next Value of A, will come out .00962518, and 263.815 refpectively; and from thence the third Value of $x = 3.47296351$; which is true, at leaft, to 7 or 8 Places.

E A Table of Integrals

The basic forms of Chapter 28 are not included. The constant of integration is omitted.

Forms containing $a + bu$ and $\sqrt{a + bu}$

1. $\displaystyle\int \frac{u\,du}{a + bu} = \frac{1}{b^2}[(a + bu) - a\ln(a + bu)]$

2. $\displaystyle\int \frac{du}{u(a + bu)} = -\frac{1}{a}\ln\frac{a + bu}{u}$

3. $\displaystyle\int \frac{u\,du}{(a + bu)^2} = \frac{1}{b^2}\left(\frac{a}{a + bu} + \ln(a + bu)\right)$

4. $\displaystyle\int \frac{du}{u(a + bu)^2} = \frac{1}{a(a + bu)} - \frac{1}{a^2}\ln\frac{a + bu}{u}$

5. $\displaystyle\int u\sqrt{a + bu}\,du = -\frac{2(2a - 3bu)(a + bu)^{3/2}}{15b^2}$

6. $\displaystyle\int \frac{u\,du}{\sqrt{a + bu}} = -\frac{2(2a - bu)\sqrt{a + bu}}{3b^2}$

7. $\displaystyle\int \frac{du}{u\sqrt{a + bu}} = \frac{1}{\sqrt{a}}\ln\left(\frac{\sqrt{a + bu} - \sqrt{a}}{\sqrt{a + bu} + \sqrt{a}}\right), \qquad a > 0$

8. $\displaystyle\int \frac{\sqrt{a + bu}}{u}\,du = 2\sqrt{a + bu} + a\int \frac{du}{u\sqrt{a + bu}}$

Forms containing $\sqrt{u^2 \pm a^2}$ and $\sqrt{a^2 - u^2}$

9. $\displaystyle\int \frac{du}{u^2 - a^2} = \frac{1}{2a}\ln\frac{u - a}{u + a}$

10. $\displaystyle\int \frac{du}{\sqrt{u^2 \pm a^2}} = \ln\left(u + \sqrt{u^2 \pm a^2}\right)$

11. $\displaystyle\int \frac{du}{u\sqrt{u^2 + a^2}} = -\frac{1}{a}\ln\left(\frac{a + \sqrt{u^2 + a^2}}{u}\right)$

12. $\displaystyle\int \frac{du}{u\sqrt{u^2 - a^2}} = \frac{1}{a}\sec^{-1}\frac{u}{a}$

13. $\displaystyle\int \frac{du}{u\sqrt{a^2 - u^2}} = -\frac{1}{a}\ln\left(\frac{a + \sqrt{a^2 - u^2}}{u}\right)$

14. $\displaystyle\int \sqrt{u^2 \pm a^2}\,du = \frac{u}{2}\sqrt{u^2 \pm a^2} \pm \frac{a^2}{2}\ln\left(u + \sqrt{u^2 \pm a^2}\right)$

15. $\displaystyle\int \sqrt{a^2 - u^2}\,du = \frac{u}{2}\sqrt{a^2 - u^2} + \frac{a^2}{2}\sin^{-1}\frac{u}{a}$

16. $\displaystyle\int \frac{\sqrt{u^2 + a^2}}{u}\, du = \sqrt{u^2 + a^2} - a \ln\left(\frac{a + \sqrt{u^2 + a^2}}{u}\right)$

17. $\displaystyle\int \frac{\sqrt{u^2 - a^2}}{u}\, du = \sqrt{u^2 - a^2} - a \sec^{-1}\frac{u}{a}$

18. $\displaystyle\int \frac{\sqrt{a^2 - u^2}}{u}\, du = \sqrt{a^2 - u^2} - a \ln\left(\frac{a + \sqrt{a^2 - u^2}}{u}\right)$

19. $\displaystyle\int (u^2 \pm a^2)^{3/2}\, du = \frac{u}{4}(u^2 \pm a^2)^{3/2} \pm \frac{3a^2 u}{8}\sqrt{u^2 \pm a^2} + \frac{3a^4}{8}\ln\left(u + \sqrt{u^2 \pm a^2}\right)$

20. $\displaystyle\int (a^2 - u^2)^{3/2}\, du = \frac{u}{4}(a^2 - u^2)^{3/2} + \frac{3a^2 u}{8}\sqrt{a^2 - u^2} + \frac{3a^4}{8}\sin^{-1}\frac{u}{a}$

21. $\displaystyle\int \frac{(u^2 + a^2)^{3/2}}{u}\, du = \frac{1}{3}(u^2 + a^2)^{3/2} + a^2\sqrt{u^2 + a^2} - a^3 \ln\left(\frac{a + \sqrt{u^2 + a^2}}{u}\right)$

22. $\displaystyle\int \frac{(u^2 - a^2)^{3/2}}{u}\, du = \frac{1}{3}(u^2 - a^2)^{3/2} - a^2\sqrt{u^2 - a^2} + a^3 \sec^{-1}\frac{u}{a}$

23. $\displaystyle\int \frac{(a^2 - u^2)^{3/2}}{u}\, du = \frac{1}{3}(a^2 - u^2)^{3/2} - a^2\sqrt{a^2 - u^2} + a^3 \ln\left(\frac{a + \sqrt{a^2 - u^2}}{u}\right)$

24. $\displaystyle\int \frac{du}{(u^2 \pm a^2)^{3/2}} = \pm \frac{u}{a^2\sqrt{u^2 \pm a^2}}$

25. $\displaystyle\int \frac{du}{(a^2 - u^2)^{3/2}} = \frac{u}{a^2\sqrt{a^2 - u^2}}$

26. $\displaystyle\int \frac{du}{u(u^2 + a^2)^{3/2}} = \frac{1}{a^2\sqrt{u^2 + a^2}} - \frac{1}{a^3}\ln\left(\frac{a + \sqrt{u^2 + a^2}}{u}\right)$

27. $\displaystyle\int \frac{du}{u(u^2 - a^2)^{3/2}} = -\frac{1}{a^2\sqrt{u^2 - a^2}} - \frac{1}{a^3}\sec^{-1}\frac{u}{a}$

28. $\displaystyle\int \frac{du}{u(a^2 - u^2)^{3/2}} = \frac{1}{a^2\sqrt{a^2 - u^2}} - \frac{1}{a^3}\ln\left(\frac{a + \sqrt{a^2 - u^2}}{u}\right)$

Trigonometric forms

29. $\displaystyle\int \sin^2 u\, du = \frac{u}{2} - \frac{1}{2}\sin u \cos u$

30. $\displaystyle\int \sin^3 u\, du = -\cos u + \frac{1}{3}\cos^3 u$

31. $\displaystyle\int \sin^n u\, du = -\frac{1}{n}\sin^{n-1} u \cos u + \frac{n-1}{n}\int \sin^{n-2} u\, du$

32. $\displaystyle\int \cos^2 u\, du = \frac{u}{2} + \frac{1}{2}\sin u \cos u$

33. $\displaystyle\int \cos^3 u\, du = \sin u - \frac{1}{3}\sin^3 u$

34. $\displaystyle\int \cos^n u\, du = \frac{1}{n}\cos^{n-1} u \sin u + \frac{n-1}{n}\int \cos^{n-2} u\, du$

35. $\displaystyle\int \tan^n u\, du = \frac{\tan^{n-1} u}{n-1} - \int \tan^{n-2} u\, du$

36. $\displaystyle\int \cot^n u\, du = -\frac{\cot^{n-1} u}{n-1} - \int \cot^{n-2} u\, du$

37. $\displaystyle\int \sec^n u\, du = \frac{\sec^{n-2} u \tan u}{n-1} + \frac{n-2}{n-1} \int \sec^{n-2} u\, du$

38. $\displaystyle\int \csc^n u\, du = \frac{\csc^{n-2} u \cot u}{n-1} + \frac{n-2}{n-1} \int \csc^{n-2} u\, du$

39. $\displaystyle\int \sin au \sin bu\, du = \frac{\sin(a-b)u}{2(a-b)} - \frac{\sin(a+b)u}{2(a+b)}$

40. $\displaystyle\int \sin au \cos bu\, du = -\frac{\cos(a-b)u}{2(a-b)} - \frac{\cos(a+b)u}{2(a+b)}$

41. $\displaystyle\int \cos au \cos bu\, du = \frac{\sin(a-b)u}{2(a-b)} + \frac{\sin(a+b)u}{2(a+b)}$

42. $\displaystyle\int \sin^m u \cos^n u\, du = \frac{\sin^{m+1} u \cos^{n-1} u}{m+n} + \frac{n-1}{m+n} \int \sin^m u \cos^{n-2} u\, du$

43. $\displaystyle\int \sin^m u \cos^n u\, du = -\frac{\sin^{m-1} u \cos^{n+1} u}{m+n} + \frac{m-1}{m+n} \int \sin^{m-2} u \cos^n u\, du$

Other forms

44. $\displaystyle\int u e^{au}\, du = \frac{e^{au}(au-1)}{a^2}$

45. $\displaystyle\int u^2 e^{au}\, du = \frac{e^{au}}{a^3}(a^2 u^2 - 2au + 2)$

46. $\displaystyle\int u^n \ln u\, du = u^{n+1}\left(\frac{\ln u}{n+1} - \frac{1}{(n+1)^2}\right)$

47. $\displaystyle\int u \sin u\, du = \sin u - u \cos u$

48. $\displaystyle\int u \cos u\, du = \cos u + u \sin u$

49. $\displaystyle\int e^{au} \sin bu\, du = \frac{e^{au}(a \sin bu - b \cos bu)}{a^2 + b^2}$

50. $\displaystyle\int e^{au} \cos bu\, du = \frac{e^{au}(a \cos bu + b \sin bu)}{a^2 + b^2}$

51. $\displaystyle\int \sin^{-1} u\, du = u \sin^{-1} u + \sqrt{1 - u^2}$

52. $\displaystyle\int \tan^{-1} u\, du = u \tan^{-1} u - \frac{1}{2} \ln(1 + u^2)$

Since statements will vary for writing exercises (W), answers here are in abbreviated form. Answers are not included for end-of-chapter writing exercises.

Exercises 1.1, page 5

1. Change $\frac{5}{1}$ to $\frac{-3}{1}$ and $\frac{-19}{1}$ to $\frac{14}{1}$. **3.** Change $2 > -4$, 2 is to the right of -4, to $-6 < -4$, -6 is to the left of -4. Also the figure must be changed to show -6 to left of -4.

5. Integer, rational, real; irrational, real

7. Imaginary; irrational, real **9.** $3, \frac{7}{2}$ **11.** $0.857, \sqrt{3}$

13. $6 < 8$ **15.** $\pi > -3.2$ **17.** $-4 < -|-3|$

19. $-\frac{1}{3} > -\frac{1}{2}$ **21.** $\frac{1}{3}, -3$ **23.** $-\frac{\pi}{5}, \frac{1}{x}$

25.

27.

29. No, $|0| = 0$ **31.** The number itself **33.** $\frac{3}{22}$ or $\frac{3}{23}$

35. $-3.1, -|-3|, -1.9, \sqrt{5}, \pi, |-8|$

37. (a) Positive integer (b) Negative integer (c) Positive rational number less than 1

39. (a) Yes (b) Yes **41.** (a) To right of origin (b) To left of -4

43. Between 0 and 1 **45.** L, t are variables; a is constant

47. 0.0008 F **49.** $N = 1000an$

51. Yes; -20 is to right of -30

Exercises 1.2, page 10

1. 22 **3.** -4 **5.** 4 **7.** 6 **9.** -3 **11.** -24

13. 35 **15.** 20 **17.** 40 **19.** -1 **21.** 9

23. Undefined **25.** 20 **27.** -5 **29.** -9 **31.** 24

33. -6 **35.** 3 **37.** Commutative law of multiplication

39. Distributive law **41.** Associative law of addition

43. Associative law of multiplication **45.** d **47.** b

49. (a) Positive (b) Negative

51. (a) Negative reciprocals of each other (b) They may not be equal.

53. -2.4 kW · h **55.** 2°C

57. 100 m + 200 m = 200 m + 100 m; commutative law of addition

59. $3(50 + 40)$ cases; distributive law

Exercises 1.3, page 15

1. 0.390 has 3 sig. digits; the zero is not needed for proper location of the decimal point.

3. 75.7 **5.** 8 is exact; 90 is approx. **7.** 2 and 7200 are exact

9. 3, 4 **11.** 3, 3 **13.** 1, 5 **15.** (a) 0.01 (b) 30.8

17. (a) Same (b) 78.0 **19.** (a) 0.004 (b) Same

21. (a) 4.94 (b) 4.9 **23.** (a) 50 900 (b) 51 000

25. (a) 9550 (b) 9500 **27.** (a) 0.945 (b) 0.94

29. (a) 51 (b) 51.2 **31.** (a) 68 (b) 62.1

33. (a) 0.015 (b) 0.0114 **35.** (a) -0.002 (b) -0.0022

37. (a) 0.1 (b) 0.1356 **39.** (a) 6.5 (b) 6.086 **41.** 15.8788

43. 204.2 **45.** 2.745 MHz, 2.755 MHz

47. Too many sig. digits; time has only 2 sig. digits

49. (a) 19.3 (b) 27

51. (a) $\pi = 3.141592654$ (b) $\frac{22}{7} = 3.142857143$

53. (a) 0.242 424 242 4 (b) 3.141 592 654 **55.** 95.3 MJ

57. 262 144 bytes **59.** 59.14%

Exercises 1.4, page 20

1. x^6 **3.** $\frac{a^6x^2}{b^4t^2}$ **5.** x^7 **7.** $2b^6$ **9.** m^2 **11.** $\frac{1}{n^4}$

13. P^8 **15.** t^{20} **17.** $8n^3$ **19.** $n^{30}T^{60}$ **21.** $\frac{8}{b^3}$

23. $\frac{x^8}{16}$ **25.** 1 **27.** -3 **29.** $\frac{1}{6}$ **31.** R^2

33. $-t^{14}$ **35.** $64x^{12}$ **37.** 1 **39.** $-b^2$ **41.** $\frac{1}{8}$

43. 1 **45.** $\frac{a}{x^2}$ **47.** $\frac{x^3}{64a^3}$ **49.** $64g^2s^6$ **51.** $\frac{5n}{T}$

53. -53 **55.** 253 **57.** -0.421 **59.** 9990 **61.** Yes

63. $\frac{G^2k^5T^5}{h}$ **65.** $\frac{r}{6}$ **67.** 0.14 W

Exercises 1.5, page 23

1. 8060 **3.** 45 000 **5.** 0.002 01 **7.** 3.23 **9.** 18.6

11. 4×10^4 **13.** 8.7×10^{-3} **15.** 6.09×10^0

17. 6.3×10^{-2} **19.** 1×10^0 **21.** 5.6×10^{13}

23. 2.2×10^8 **25.** 3.2×10^{-34} **27.** 1.728×10^{87}

29. 4.85×10^{10} **31.** 1.59×10^7 **33.** 9.965×10^{-3}

35. 3.40×10^{23} **37.** 1.26×10^7 kW **39.** 3×10^{-6} W

41. $1\,000\,000\,000\,000\,000\,000\,000\,000\,000\,000\,000°\text{C}$
43. $0.000\,000\,000\,001\,6\,\text{W}$ **45.** $4.2 \times 10^{-8}\,\text{s}$
47. $2.46 \times 10^{-1}\,\text{s}$ **49.** 1.3×10^{14} disintegrations
51. $3.433\,\Omega$

Exercises 1.6, page 26

1. -4 **3.** 12 **5.** 9 **7.** -11 **9.** -7 **11.** 0.3
13. 5 **15.** -6 **17.** 5 **19.** 31 **21.** 18 **23.** $2\sqrt{3}$
25. $4\sqrt{21}$ **27.** $2\sqrt{5}$ **29.** 4 **31.** 7 **33.** 10
35. $3\sqrt{10}$ **37.** 9.24 **39.** 0.6877 **41.** (a) 60 (b) 84
43. (a) 0.0388 (b) 0.0246 **45.** $98\,\text{km/h}$ **47.** $1450\,\text{m/s}$
49. $48.3\,\text{cm}$ **51.** no, not true if $a < 0$

Exercises 1.7, page 29

1. $3x - 3y$ **3.** $4ax + 5s$ **5.** $8x$ **7.** $y + 4x$
9. $5F - 3T - 2$ **11.** $-a^2b - a^2b^2$ **13.** $4s + 4$
15. $5x - v - 4$ **17.** $5a - 5$ **19.** $-5a + 2$
21. $-2t + 5u$ **23.** $7r + 8s$ **25.** $-50 + 19j$
27. $-9 + 3n$ **29.** $18 - 2t^2$ **31.** $6a$ **33.** $2a\sqrt{LC} + 1$
35. $4c - 6$ **37.** $8p - 5q$ **39.** $-4x^2 + 22$ **41.** $7V^2 - 3$
43. $-6t + 13$ **45.** $4Z - 24R$ **47.** $2D + d$
49. $-b + 4c - 3a$ **51.** $40x + 250$

Exercises 1.8, page 32

1. $-8s^8t^{13}$ **3.** $x^2 - 5x + 6$ **5.** a^3x **7.** $-a^2c^3x^3$
9. $-8a^3x^5$ **11.** $2a^8x^3$ **13.** $i^2R + i^2r$ **15.** $-3s^3 + 15st$
17. $5m^3n + 15m^2n$ **19.** $-3M^2 - 3MN + 6M$
21. $a^2b^2c^5 - ab^3c^5 - a^2b^3c^4$ **23.** $acx^4 + acx^3y^3$
25. $x^2 + 2x - 15$ **27.** $2x^2 + 9x - 5$ **29.** $6a^2 - 7ab + 2b^2$
31. $6s^2 + 11st - 35t^2$ **33.** $2x^3 + 5x^2 - 2x - 5$
35. $x^3 + 2x^2 - 8x$ **37.** $x^3 - 2x^2 - x + 2$
39. $x^5 - x^4 - 6x^3 + 4x^2 + 8x$ **41.** $2a^2 - 16a - 18$
43. $18T^2 - 15T - 18$ **45.** $2L^3 - 6L^2 - 8L$
47. $4x^2 - 20x + 25$ **49.** $x_1^2 + 6x_1x_2 + 9x_2^2$
51. $x^2y^2z^2 - 4xyz + 4$ **53.** $2x^2 + 32x + 128$
55. $-x^3 + 2x^2 + 5x - 6$ **57.** $6T^3 + 9T^2 - 6T$
59. (a) $49 \neq 9 + 16$ (b) $1 \neq 9 - 16$
61. $n^2 - 1 = (n - 1)(n + 1)$ for any n **63.** $3R^2 - 4RX$
65. $n^2 + 200n + 10\,000$ **67.** $R_1^2 - R_2^2$

Exercises 1.9, page 35

1. $\dfrac{3}{y^3}$ **3.** $3x - 2$ **5.** $-4x^2y$ **7.** $\dfrac{4t^4}{r^2}$ **9.** $4x^2$
11. $-6a$ **13.** $a^2 + 4y$ **15.** $t - 2rt^2$ **17.** $q + 2p - 4q^3$

19. $\dfrac{2L}{R} - R$ **21.** $\dfrac{1}{3a} - \dfrac{2b}{3a} + 1$ **23.** $x^2 + a$ **25.** $2x + 1$
27. $x - 1$ **29.** $4x^2 - x - 1, R = -3$ **31.** $Z + 5 + \dfrac{3}{4Z + 3}$
33. $x^2 + x - 6$ **35.** $2x^2 + 4x + 2, R = 4x + 4$
37. $x^2 - 2x + 4$ **39.** $x - y$ **41.** $E + 3 + \dfrac{5}{5E^2 - 7E - 2}$
43. $A + \dfrac{\mu^2E^2}{2A} - \dfrac{\mu^4E^4}{8A^3}$ **45.** $\dfrac{GMm}{R}$
47. $s^2 + 2s + 6 + \dfrac{16s + 16}{s^2 - 2s - 2}$

Exercises 1.10, page 39

1. $-9, -15, -36, -4$ **3.** $\dfrac{1}{8}$ **5.** 9 **7.** -1 **9.** 10
11. -5 **13.** -3 **15.** 1 **17.** $-\dfrac{7}{2}$ **19.** 8 **21.** $\dfrac{10}{3}$
23. $-\dfrac{13}{3}$ **25.** -2.5 **27.** 0 **29.** 8 **31.** 9 or -9
33. 9.5 **35.** -1.5 **37.** 5.7 **39.** 0.85 **41.** $120°\text{C}$
43. $750\,\text{L}$ **45.** $60\,\text{mg}$ **47.** True for all x

Exercises 1.11, page 41

1. $\dfrac{v - v_0}{t}$ **3.** $\dfrac{V_0 + bTV_0 - V}{bV_0}$ **5.** $\dfrac{E}{I}$ **7.** $\theta - kA$
9. $\dfrac{Q}{Sd^2}$ **11.** $\dfrac{p - p_a}{dg}$ **13.** $\dfrac{APV}{R}$ **15.** $\dfrac{0.3t - ct^2}{c}$
17. $Tv - c$ **19.** $\dfrac{K_1m_1 - K_2m_1}{K_2}$ **21.** $\dfrac{2mg - 2am}{a}$
23. $\dfrac{C_0^2 - C_1^2}{2C_1^2}$ **25.** $\dfrac{P + nc}{n}$ **27.** $\dfrac{mgR - FR}{m}$
29. $\dfrac{Q_1 + PQ_1}{P}$ **31.** $\dfrac{N + N_2 - N_2T}{T}$ **33.** $\dfrac{L - \pi r_2 - 2x_1 - x_2}{\pi}$
35. $\dfrac{gJP + V_1^2}{V_1}$ **37.** $\dfrac{Cd(k_1 + k_2)}{2Ak_1k_2}$ **39.** $10.6\,\text{m}$ **41.** $32.3°\text{C}$
43. $3.22\,\Omega$

Exercises 1.12, page 45

1. 29 1.5-Ω resistors, 5 2.5-Ω resistors   **3.** $3.000\,\text{h}$
5. $\$22\,000, \$27\,000$
7. 1.9 million the first year, 2.6 million the second year
9. 20 ha at $\$20\,000/\text{ha}$, 50 ha at $\$10\,000/\text{ha}$
11. $60\,\text{ppm/h}$ **13.** 20 girders
15. $-2.3\,\mu\text{A}, -4.6\,\mu\text{A}, 6.9\,\mu\text{A}$ **17.** $6.9\,\text{km}, 9.5\,\text{km}$
19. 1190 adults, 812 children   **21.** $900\,\text{m}$
23. $84.2\,\text{km/h}, 92.2\,\text{km/h}$ **25.** $395\,\text{s}$, first car

27. 146 km from A   **29.** 4 L   **31.** 79 km/h

Review Exercises for Chapter 1, page 47

1. -10 **3.** -20 **5.** -22 **7.** -25 **9.** -4 **11.** 5

13. $4r^2t^4$ **15.** $-\dfrac{6m^2}{nt^2}$ **17.** $\dfrac{8T^3}{N}$ **19.** $3\sqrt{5}$

21. (a) 3 (b) 8800   **23.** (a) 4 (b) 9.0   **25.** 18.0

27. 1.3×10^{-4} **29.** $-a - 2ab$ **31.** $7LC - 3$

33. $2x^2 + 9x - 5$ **35.** $x^2 + 16x + 64$ **37.** $hk - 3h^2k^4$

39. $7R - 6r$ **41.** $13xy - 10z$ **43.** $2x^3 - x^2 - 7x - 3$

45. $-3x^2y + 24xy^2 - 48y^3$ **47.** $-9p^2 + 3pq + 18p^2q$

49. $\dfrac{6q}{p} - 2 + \dfrac{3q^4}{p^3}$ **51.** $2x - 5$ **53.** $x^2 - 2x + 3$

55. $4x^3 - 2x^2 + 6x,\ R = -1$ **57.** $15r - 3s - 3t$

59. $y^2 + 5y - 1,\ R = 4$ **61.** $-\dfrac{9}{2}$ **63.** $\dfrac{21}{10}$ **65.** $-\dfrac{7}{3}$

67. 3   **69.** $-\dfrac{19}{5}$ **71.** 1   **73.** 1.6×10^4 Pa

75. 1.92×10^8 km   **77.** 40 500 000 000 000 km

79. 1.2×10^{-6} cm^2 **81.** 0.15 Bq/L   **83.** $\dfrac{R}{n^2}$ **85.** $\dfrac{PL^2}{\pi^2 I}$

87. $\dfrac{I - P}{Pr}$ **89.** $\dfrac{dV - m}{dV}$ **91.** $\dfrac{N_1 + TN_3 - N_3}{T}$

93. $\dfrac{RH + AT_1}{A}$ **95.** $\dfrac{d - 3kbx^2 + kx^3}{3kx^2}$ **97.** 410 m

99. 110 m   **101.** 0.0188 Ω **103.** $2rV - aV - bV$

105. $4t + 4h - 2t^2 - 4th - 2h^2$ **107.** \$59, \$131

109. 160 cm^3, 80 cm^3, 320 cm^3 **111.** 1900 Ω, 3100 Ω

113. 42 N   **115.** 34.6 h after second ship enters Red Sea

117. 400 L, 600 L   **119.** 27 m^2

Exercises 2.1, page 53

1. $90°$ **3.** 4   **5.** $\angle EBD, \angle DBC$ **7.** $\angle ABC$ **9.** $25°$

11. BD, BC **13.** $140°$ **15.** $145°$ **17.** $62°$ **19.** $28°$

21. $134°$ **23.** $44°$ **25.** 4.53 m   **27.** 3.40 m   **29.** $133°$

31. 920 m

Exercises 2.2, page 59

1. $65°$ **3.** 7.02 m   **5.** $56°$ **7.** $48°$ **9.** 9.9 mm

11. 64.5 cm   **13.** 8.4 m^2 **15.** 32,300 cm^2 **17.** 4.41 cm^2

19. 0.390 m^2 **21.** 26.6 mm   **23.** 522 cm   **25.** $67°$

27. 227.2 cm

29. $\angle K = \angle N = 90°; \angle LMK = \angle OMN;$
     $\angle KLM = \angle NOM; \triangle MKL \sim \triangle MNO$

31. 8   **33.** Equilateral triangle   **35.** $65°$ **37.** 1150 cm^2

39. 9.6 m^2 **41.** 5.7 m   **43.** 23 m   **45.** 7.5 m, 9.0 m

47. 6.0 m

Exercises 2.3, page 63

1. Trapezoid   **3.** 8900 m^2 **5.** 260 m   **7.** 3.324 mm

9. 12.8 m   **11.** 214.4 dm   **13.** 7.3 mm^2 **15.** 0.683 km^2

17. 9.3 m^2 **19.** 2000 dm^2 **21.** $p = 4a + 2b$

23. $A = bh + a^2$ **25.** It is a rectangle.

27. The diagonal always divides the rhombus into two congruent triangles. All outer sides are always equal.

29. 344 m   **31.** 1500 mm, 3000 mm   **33.** 2.7 L

35. $360°$. A diagonal divides a quadrilateral into two triangles, and the interior angles of each triangle are $180°$.

Exercises 2.4, page 66

1. $18°$ **3.** $p = 11.6$ cm, $A = 8.30$ cm^2 **5.** (a) AD (b) AF

7. (a) $AF \perp OE$ (b) $\triangle OEC$ **9.** 17.3 cm   **11.** 72.6 mm

13. 0.0285 km^2 **15.** 4.26 m^2 **17.** $25°$ **19.** $25°$

21. $120°$ **23.** $40°$ **25.** 0.393 rad   **27.** 2.185 rad

29. $p = \frac{1}{2}\pi r + 2r$ **31.** $A = \frac{1}{4}\pi r^2 - \frac{1}{2}r^2$

33. All are on the same diameter.   **35.** 40 000 km   **37.** $\frac{0.445}{1}$

39. 35.7 cm   **41.** 9500 cm^2

43. Horizontally and opposite to original direction

Exercises 2.5, page 71

1. 18,100 km^2

3. Simpson's rule should be more accurate in that it accounts better for the arcs between points on the upper curve.

5. 84 m^2 **7.** 0.45 m^2 **9.** 9.8 km^2 **11.** 19,000 km^2

13. 11 800 m^2 **15.** 8100 m^2

17. 2.73 cm^2 The trapezoids are inside the boundary and do not include some of the area.

19. 2.98 cm^2 The ends of the areas are curved so that they can get closer to the boundary.

Exercises 2.6, page 74

1. The volume is four times as much.   **3.** 1150 cm^3 **5.** 366 dm^3

7. 399 m^2 **9.** 2.83 m^3 **11.** 20 500 cm^2 **13.** 1100 dm^3

15. 3.358 m^2 **17.** 0.15 cm^3 **19.** 72.3 cm^2 **21.** $\frac{6}{1}$

23. $\frac{4}{1}$ **25.** 604 cm^2 **27.** 1.4×10^6 m^3

29. 2.6×10^6 m^3 **31.** 66 600 m^3 **33.** 1560 mm^2

35. 1.10 cm^3

Review Exercises for Chapter 2, page 77

1. $32°$ **3.** $32°$ **5.** 41   **7.** 700   **9.** 7.36   **11.** 21.1

13. 25.5 mm   **15.** 3.06 m^2 **17.** 309 mm   **19.** 3320 cm^2

21. 6190 cm^3 **23.** 160 000 m^3 **25.** 162 m^2 **27.** 66.6 mm^2

29. 25°   **31.** 65°   **33.** 53°   **35.** 2.4

37. $p = \pi a + b + \sqrt{4a^2 + b^2}$ **39.** $A = ab + \frac{1}{2}\pi a^2$

41. yes   **43.** n^2; $A = \pi(nr)^2 = n^2(\pi r^2)$ **45.** 12 cm

47. 71°   **49.** 7.9 m   **51.** 30 m   **53.** 215 ft^2

55. 5.91 km   **57.** 42 000 km   **59.** 2.7 × 10^6 mm^2

61. 1.0 × 10^6 m^2 **63.** 190 m^3 **65.** 10 m   **67.** 873 L

Exercises 3.1, page 84

1. -13 **3.** $f(T - 10) = 9.1 + 0.08T + 0.001T^2$

5. (a) $A = \pi r^2$ (b) $A = \frac{1}{4}\pi d^2$ **7.** $d = \sqrt[3]{\dfrac{6V}{\pi}}$

9. $A = s^2, s = \sqrt{A}$ **11.** $A = 4r^2 - \pi r^2$ **13.** 3, -1

15. 5, 5   **17.** $\dfrac{6 - \pi^2}{2\pi}, -\dfrac{1}{2}$ **19.** $\frac{1}{4}a + \frac{1}{2}a^2, 0$

21. $3s^2 + s + 6, 12s^2 - 2s + 6$ **23.** -8 **25.** 62.9, 260

27. -0.2998

29. Square the value of the independent variable and add 2.

31. Cube the value of the independent variable and subtract this from 6 times the value of the independent variable.

33. Multiply by 3 the sum of twice r added to 5 and subtract 1.

35. Subtract 3 from twice t and divide this quantity by the sum of t and 2.

37. $A = 5e^2$, $f(e) = 5e^2$

39. $A = 8430 - 140t$, $f(t) = 8430 - 140t$ **41.** 10.4 m

43. $f(R + 10) = \dfrac{200(R + 10)}{(110 + R)^2}$

Exercises 3.2, page 88

1. $-x^2 + 2$ is never greater than 2. The range is all real numbers $f(x) \le 2$.

3. 4 mA, 0 mA

5. Domain: all real numbers; range: all real numbers

7. Domain: all real numbers except 0; range: all real numbers except 0

9. Domain: all real numbers except 0; range: all real numbers $f(s) > 0$

11. Domain: all real numbers $h \ge 0$; range: all real numbers $H(h) \ge 1$; $\sqrt{h}$ cannot be negative

13. All real numbers $y > 2$

15. All real numbers except -4, 2, and 6   **17.** 2, not defined

19. 2, 0.75   **21.** $d = 120 + 80t$ **23.** $w = 5500 - 2t$

25. $m = 0.5h - 390$ **27.** $C = 5l + 250$

29. (a) $y = 3000 - 0.25x$ (b) 2900 L

31. $A = \dfrac{1}{16}p^2 + \dfrac{(60 - p)^2}{4\pi}$

33. $d = \sqrt{14,400 + h^2}$; domain: $h \ge 0$; range: $d \ge 120\ m$

35. $s = \dfrac{300}{t}$; domain: $t > 0$; range: $s > 0$ (upper limits depend on truck)

37. Domain is all values of $C > 0$, with some upper limit depending on the circuit.

39. $m = \begin{cases} 0.5h - 390 & \text{for } h > 1000 \\ 110 & \text{for } 0 \le h \le 1000 \end{cases}$

41. (a) $V = 10w^2 - 150w + 500$ (b) $w \ge 10$ cm   **43.** 1

Exercises 3.3, page 91

1. $(-1, 1)$ **3.** $(2, 1), (-1, 2), (-2, -3)$

5.

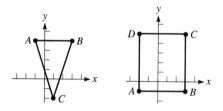

7. Isosceles triangle   **9.** Rectangle   **11.** $(5, 4)$

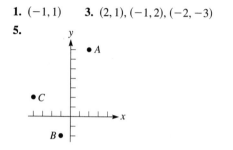

13. $(3, -2)$

15. On a line parallel to the y-axis, 1 unit to the right

17. On a line parallel to the x-axis, 3 units above

19. On a line through the origin that bisects the first and third quadrants

21. 0   **23.** To the right of the y-axis

25. To the left of a line that is parallel to the y-axis, 1 unit to its left

27. In first or third quadrant   **29.** On either axis

31. (a) 8 (b) 6

Exercises 3.4, page 96

1.   **3.**   **5.**

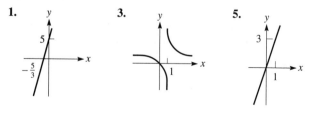

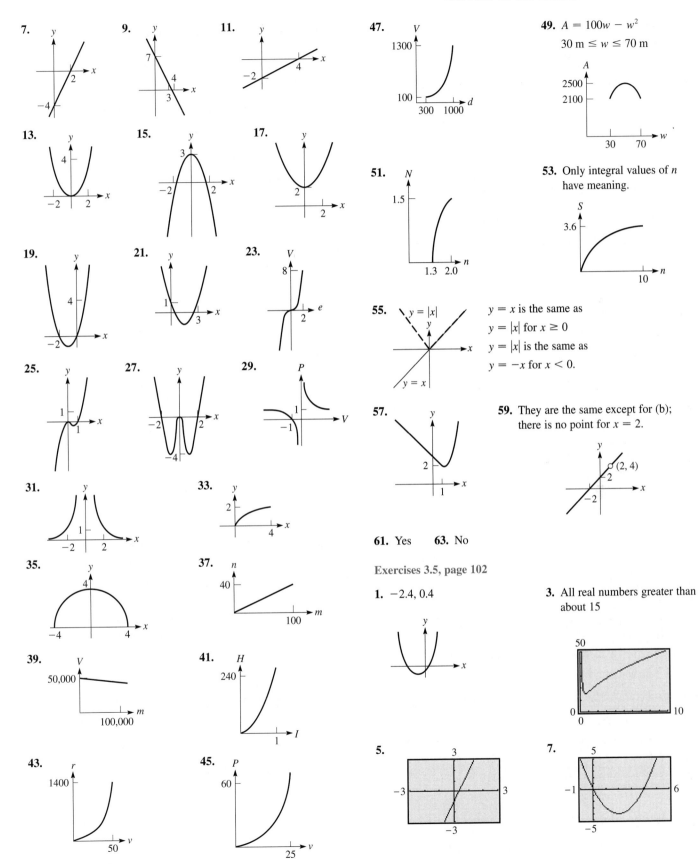

7.

9.

11.

13.

15.

17.

19.

21.

23.

25.

27.

29.

31.

33.

35.

37.

39.

41.

43.

45.

47.

49. $A = 100w - w^2$

30 m $\leq w \leq$ 70 m

51.

53. Only integral values of n have meaning.

55. $y = x$ is the same as $y = |x|$ for $x \geq 0$

$y = |x|$ is the same as $y = -x$ for $x < 0$.

57.

59. They are the same except for (b); there is no point for $x = 2$.

61. Yes   **63.** No

Exercises 3.5, page 102

1. $-2.4, 0.4$

3. All real numbers greater than about 15

5.

7.

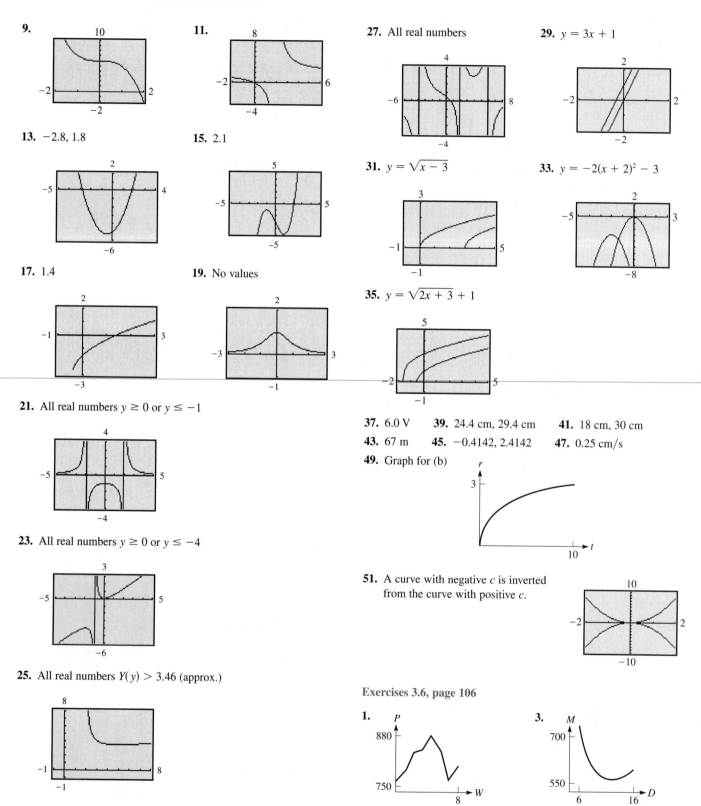

9.

11.

13. −2.8, 1.8

15. 2.1

17. 1.4

19. No values

21. All real numbers $y \geq 0$ or $y \leq -1$

23. All real numbers $y \geq 0$ or $y \leq -4$

25. All real numbers $Y(y) > 3.46$ (approx.)

27. All real numbers

29. $y = 3x + 1$

31. $y = \sqrt{x - 3}$

33. $y = -2(x + 2)^2 - 3$

35. $y = \sqrt{2x + 3} + 1$

37. 6.0 V   **39.** 24.4 cm, 29.4 cm   **41.** 18 cm, 30 cm

43. 67 m   **45.** −0.4142, 2.4142   **47.** 0.25 cm/s

49. Graph for (b)

51. A curve with negative c is inverted from the curve with positive c.

Exercises 3.6, page 106

1.

3.

5.

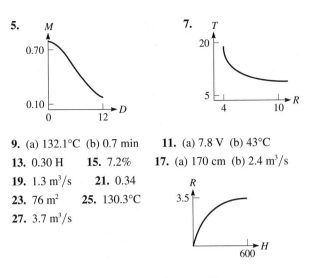

7.

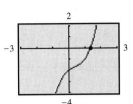

9. (a) 132.1°C (b) 0.7 min   **11.** (a) 7.8 V (b) 43°C

13. 0.30 H   **15.** 7.2%   **17.** (a) 170 cm (b) 2.4 m³/s

19. 1.3 m³/s   **21.** 0.34

23. 76 m²   **25.** 130.3°C

27. 3.7 m³/s

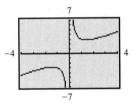

Review Exercises for Chapter 3, page 107

1. $A = 4\pi t^2$ **3.** $y = -\dfrac{10}{9}x + \dfrac{250}{9}$ **5.** 16, −47

7. 3, $\sqrt{1 - 4h}$ **9.** $3h^2 + 6hx - 2h$ **11.** −3

13. −3.67, 16.7   **15.** 0.165 03, −0.214 76

17. Domain: all real numbers; range: all real numbers $f(x) \geq 1$

19. Domain: all real numbers $t > -4$; range: all real numbers $g(t) > 0$

21.

23.

25.

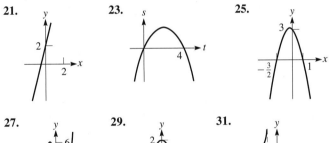

27.

29.

31.

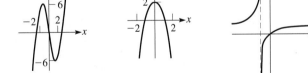

33. 0.4

35. 0.2, 5.8

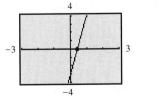

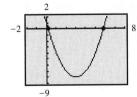

37. 1.4

39. −0.7, 0.7

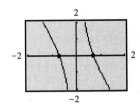

41. All real numbers $y \geq -6.25$

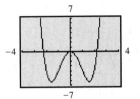

43. All real numbers $y \leq -2.83$ or $y \geq 2.83$

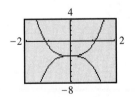

45. Either a or b is positive, the other is negative.

47. $\left(1, \sqrt{3}\right)$ or $\left(1, -\sqrt{3}\right)$ **49.** In any quadrant (not on an axis)

51. Many possibilities (two shown)   **53.** $y = \sqrt{x + 1} + 1$

55. They are reflections of each other across the y-axis.

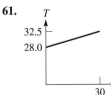

57. 13.4   **59.** 72.0°

61.

63.

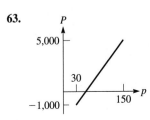

65. $L = 2\pi r + 12$

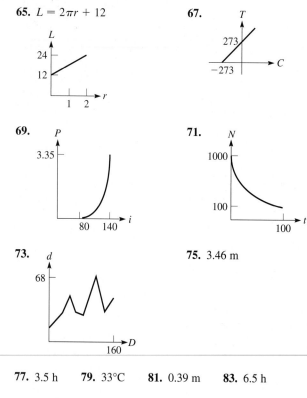

67.

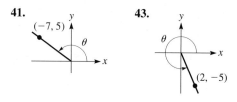

69.

71.

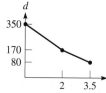

73.

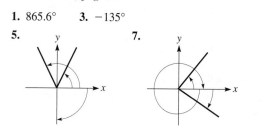

75. 3.46 m

77. 3.5 h   **79.** 33°C   **81.** 0.39 m   **83.** 6.5 h

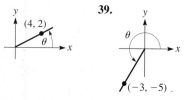

Exercises 4.1, page 113

1. 865.6°   **3.** −135°

5.

7.

9. 405°, −315°   **11.** 210°, −510°   **13.** 430° 30′, −289° 30′
15. 638.1°, −81.9°   **17.** 15.18°   **19.** 82.91°   **21.** 18.85°
23. 235.49°   **25.** 0.977   **27.** 4.97   **29.** 47° 30′
31. −5° 37′   **33.** 15.20°   **35.** 301.27°

37.

39.

41.

43.

45. I, IV   **47.** II, quadrantal   **49.** I, II   **51.** III, I
53. 21.710°   **55.** 86° 16′ 26″

Exercises 4.2, page 117

1. $\sin\theta = \dfrac{3}{5}$, $\cos\theta = \dfrac{4}{5}$, $\tan\theta = \dfrac{3}{4}$, $\cot\theta = \dfrac{4}{3}$,

$\sec\theta = \dfrac{5}{4}$, $\csc\theta = \dfrac{5}{3}$

3. $\sin\theta = \dfrac{4}{5}$, $\cos\theta = \dfrac{3}{5}$, $\tan\theta = \dfrac{4}{3}$, $\cot\theta = \dfrac{3}{4}$,

$\sec\theta = \dfrac{5}{3}$, $\csc\theta = \dfrac{5}{4}$

5. $\sin\theta = \dfrac{8}{17}$, $\cos\theta = \dfrac{15}{17}$, $\tan\theta = \dfrac{8}{15}$, $\cot\theta = \dfrac{15}{8}$,

$\sec\theta = \dfrac{17}{15}$, $\csc\theta = \dfrac{17}{8}$

7. $\sin\theta = \dfrac{40}{41}$, $\cos\theta = \dfrac{9}{41}$, $\tan\theta = \dfrac{40}{9}$, $\cot\theta = \dfrac{9}{40}$,

$\sec\theta = \dfrac{41}{9}$, $\csc\theta = \dfrac{41}{40}$

9. $\sin\theta = \dfrac{\sqrt{15}}{4}$, $\cos\theta = \dfrac{1}{4}$, $\tan\theta = \sqrt{15}$, $\cot\theta = \dfrac{1}{\sqrt{15}}$,

$\sec\theta = 4$, $\csc\theta = \dfrac{4}{\sqrt{15}}$

11. $\sin\theta = \dfrac{1}{\sqrt{2}}$, $\cos\theta = \dfrac{1}{\sqrt{2}}$, $\tan\theta = 1$, $\cot\theta = 1$,

$\sec\theta = \sqrt{2}$, $\csc\theta = \sqrt{2}$

13. $\sin\theta = \dfrac{2}{\sqrt{29}}$, $\cos\theta = \dfrac{5}{\sqrt{29}}$, $\tan\theta = \dfrac{2}{5}$, $\cot\theta = \dfrac{5}{2}$,

$\sec\theta = \dfrac{\sqrt{29}}{5}$, $\csc\theta = \dfrac{\sqrt{29}}{2}$

15. $\sin\theta = 0.808$, $\cos\theta = 0.589$, $\tan\theta = 1.37$, $\cot\theta = 0.729$,

$\sec\theta = 1.70$, $\csc\theta = 1.24$

17. $\dfrac{5}{13}, \dfrac{12}{5}$ **19.** $\dfrac{2}{\sqrt{5}}, \sqrt{5}$ **21.** 0.882, 1.33

23. 0.246, 3.94   **25.** $\sin\theta = \dfrac{4}{5}$, $\tan\theta = \dfrac{4}{3}$

27. $\tan\theta = \dfrac{1}{3}$, $\sec\theta = \dfrac{3}{\sqrt{10}}$ **29.** 1   **31.** $\cos\theta = \sqrt{1 - y^2}$

33. $-\dfrac{7}{3}$ **35.** $\sec\theta$

Exercises 4.3, page 121

1. 20.65°   **3.** 71.0°

5. $\sin 40° = 0.64$, $\cos 40° = 0.77$, $\tan 40° = 0.84$, $\cot 40° = 1.19$,
$\sec 40° = 1.31$, $\csc 40° = 1.56$

7. $\sin 15° = 0.26$, $\cos 15° = 0.97$, $\tan 15° = 0.27$, $\cot 15° = 3.73$,
$\sec 15° = 1.04$, $\csc 15° = 3.86$

9. 0.381   **11.** 1.58   **13.** 0.9626   **15.** 0.99   **17.** 0.4085

19. 1.569   **21.** 1.32   **23.** 0.07063   **25.** 70.97°

27. 65.70°   **29.** 11.7°   **31.** 49.453°   **33.** 53.44°

35. 81.79°   **37.** 74.1°   **39.** 17.85°   **41.** $0.9556 = 0.9556$

43. $2.747 = 2.747$ **45.** y is always less than r.

47. For a given r, x decreases as θ increases from 0° to 90°.

49. 0.8885   **51.** 0.93614   **53.** 87 dB   **55.** 48.6°

Exercises 4.4, page 126

1. $\sin A = 0.868$, $\cos A = 0.496$, $\tan A = 1.75$, $\sin B = 0.496$,
$\cos B = 0.868$, $\tan B = 0.571$

3. $a = 44.68$, $A = 34.17°$, $B = 55.83°$

5. 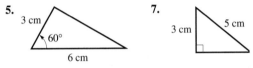 **7.**

9. $B = 12.2°$, $b = 1450$, $c = 6850$

11. $A = 25.8°$, $B = 64.2°$, $b = 311$

13. $A = 57.9°$, $a = 20.2$, $b = 12.6$

15. $A = 21°$, $a = 32$, $B = 69°$

17. $a = 30.21$, $B = 57.90°$, $b = 48.16$

19. $A = 52.15°$, $B = 37.85°$, $c = 71.85$

21. $A = 52.5°$, $b = 0.661$, $c = 1.09$

23. $A = 15.82°$, $a = 0.5239$, $c = 1.922$

25. $A = 65.886°$, $B = 24.114°$, $c = 648.46$

27. $a = 3.3621$, $B = 77.025°$, $c = 14.974$

29. 4.45   **31.** 40.24°   **33.** 43.1°   **35.** 788

37. $a = c \sin A$, $b = c \cos A$, $B = 90° - A$

39. $A = 90° - B$, $b = a \tan B$, $c = \dfrac{a}{\cos B}$

Exercises 4.5, page 129

1. 639 m   **3.** 97 m   **5.** 44.0 m   **7.** 0.4°   **9.** 850.1 cm

11. 7610 mm   **13.** 0.34 km   **15.** 26.6°, 63.4°, 90.0°

17. 3.4°   **19.** 23.5°   **21.** 8.1°   **23.** 3.07 cm   **25.** 651 m

27. 30.2°   **29.** 47.3 m   **31.** 642 m

33. $d = 2x \tan(0.5\theta)$ **35.** $A = a(b + a \cos \theta)\sin \theta$

Review Exercises for Chapter 4, page 132

1. 377.0°, −343.0°   **3.** 142.5°, −577.5°   **5.** 31.9°

7. 38.1°   **9.** 17° 30′   **11.** 249° 42′

13. $\sin \theta = \dfrac{7}{25}$, $\cos \theta = \dfrac{24}{25}$, $\tan \theta = \dfrac{7}{24}$, $\cot \theta = \dfrac{24}{7}$,
$\sec \theta = \dfrac{25}{24}$, $\csc \theta = \dfrac{25}{7}$

15. $\sin \theta = \dfrac{1}{\sqrt{2}}$, $\cos \theta = \dfrac{1}{\sqrt{2}}$, $\tan \theta = 1$, $\cot \theta = 1$,
$\sec \theta = \sqrt{2}$, $\csc \theta = \sqrt{2}$

17. 0.923, 2.40   **19.** 0.447, 1.12   **21.** 0.952   **23.** 1.853

25. 1.05   **27.** 0   **29.** 18.2°   **31.** 57.57°   **33.** 12.25°

35. 87.7°   **37.** $a = 1.83$, $B = 73.0°$, $c = 6.27$

39. $A = 51.5°$, $B = 38.5°$, $c = 104$

41. $B = 52.5°$, $b = 15.6$, $c = 19.7$

43. $A = 31.61°$, $a = 4.006$, $B = 58.39°$

45. $a = 0.6292$, $B = 40.33°$, $b = 0.5341$

47. $A = 48.813°$, $B = 41.187°$, $b = 10.196$

49. 4.6   **51.** 61.2   **53.** 10.5

55. $x/h = \cot A$, $y/h = \cot B$, $c = x + y = h \cot A + h \cot B$

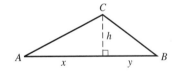

57. 80.3°   **59.** 12.0°

61. (a) $A = \frac{1}{2}bh = \frac{1}{2}b(a \sin C) = \frac{1}{2}ab \sin C$ (b) 679.2 m^2

63. 4.92 km   **65.** 0.977 m^2 **67.** 56%   **69.** 4.43 m

71. 34 m   **73.** 10.2 cm

75. Middle: $\dfrac{705}{\tan 1.1°} = 36\,700$ m; end: $\dfrac{1410}{\tan 2.2°} = 36\,700$ m

77. (a) 147 000 mm^2 (b) 155 000 mm^2

 (c) Curved surfaces covering each pentagon or hexagon would
 have greater area than plane area.

79. 1.83 km   **81.** 464 m   **83.** 73.3 cm

Exercises 5.1, page 140

1. Yes. The equation is then linear.   **3.** $x = 4$, $y = 4$

5. Yes; no   **7.** Yes; yes   **9.** $-3, -\dfrac{21}{2}$ **11.** $\dfrac{1}{4}, -0.6$

13. $-\dfrac{16}{27}, -\dfrac{10}{9}$ **15.** Yes   **17.** No   **19.** No   **21.** Yes

23. Yes   **25.** Yes   **27.** No

Exercises 5.2, page 144

1. $m = \dfrac{1}{4}$. The line rises 1 unit for each 4 units in going from left to right.

3. The y-intercept is $(0, 2)$. The checkpoint is $\left(1, \dfrac{4}{3}\right)$.

5. 4   **7.** -5 **9.** $\dfrac{2}{7}$ **11.** $\dfrac{1}{2}$

13. 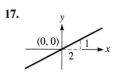   **15.**

17. 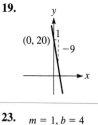   **19.**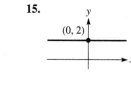

21. $m = -2, b = 1$ **23.** $m = 1, b = 4$

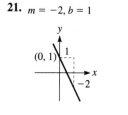

25. $m = \dfrac{5}{2}, b = -20$ **27.** $m = -\dfrac{3}{5}, b = \dfrac{3}{8}$

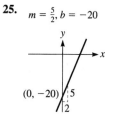

29. 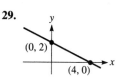   **31.**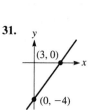

33. 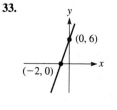   **35.**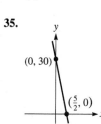

37. 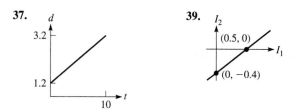   **39.**

Exercises 5.3, page 148

1. $x = 1.5, y = 1.4$ **3.** $x = 3.0, y = 1.0$
5. $x = 3.0, y = 0.0$ **7.** $x = 2.2, y = -0.3$
9. $x = -0.9, y = -2.3$ **11.** $s = 1.1, t = -1.7$
13. $x = 0.0, y = 3.0$ **15.** $x = -14.0, y = -5.0$
17. $r_1 = 4.0, r_2 = 7.5$ **19.** $x = 2.1, y = -0.4$
21. $x = -3.6, y = -1.4$ **23.** $x = 1.1, y = 0.8$
25. Dependent   **27.** $t = -1.9, v = -3.2$
29. $x = 1.5, y = 4.5$ **31.** Inconsistent   **33.** 50 N, 47 N
35. 2.6 m, 3.4 m

Exercises 5.4, page 154

1. $x = -3, y = -3$ **3.** $x = \dfrac{16}{13}, y = -\dfrac{2}{13}$

5. $x = 1, y = -2$ **7.** $V = 7, p = 3$ **9.** $x = -1, y = -4$

11. $x = \dfrac{1}{2}, y = 2$ **13.** $x = 1, y = \dfrac{1}{2}$

15. $p = \dfrac{9}{22}, n = -\dfrac{16}{11}$ **17.** $x = 3, y = 1$

19. $x = -1, y = -2$ **21.** $t = -\dfrac{1}{3}, y = 2$ **23.** Inconsistent

25. $x = -\dfrac{14}{5}, y = -\dfrac{16}{5}$ **27.** $R = 2.38, Z = 0.45$

29. $x = \dfrac{1}{2}, y = -4$ **31.** $t = -\dfrac{2}{3}, y = 0$

33. $x = \dfrac{3}{5}, y = \dfrac{1}{5}$ **35.** $C = -1, V = -2$

37. $A = -1, B = 3$ **39.** $a = \dfrac{1}{3}, b = -7$

41. $V_1 = 9.0$ V, $V_2 = 6.0$ V   **43.** $x = 6250$ L, $y = 3750$ L
45. $W_r = 8120$ N, $W_f = 9580$ N   **47.** $t_1 = 32$ s, $t_2 = 20$ s
49. 7.00 kW, 9.39 kW
51. 34 at \$900/month, 20 at \$1250/month
53. Incorrect conclusion or error in sales figures; system of equations is inconsistent.
55. $a \neq b$

Exercises 5.5, page 160

1. 86 **3.** $x = -13, y = -27$ **5.** -10 **7.** 29 **9.** 32

11. 9300 **13.** 1.083 **15.** 96 **17.** $x = 3, y = 1$

19. $x = -1, y = -2$ **21.** $t = -\dfrac{1}{3}, y = 2$

23. Inconsistent **25.** $x = -\dfrac{14}{5}, y = -\dfrac{16}{5}$

27. $R = 2.38, Z = 0.45$ **29.** $x = 0.32, y = -4.5$

31. $R = 4.0, t = -3.2$ **33.** $x = -11.2, y = -9.26$

35. $x = -1.0, y = -2.0$ **37.** $F_1 = 15$ N, $F_2 = 6.0$ N

39. $x = 73.6$ L, $y = 70.4$ L **41.** 210 phones, 110 detectors

43. \$2000, 6% **45.** 2.5 h, 2.1 h **47.** $V = 4.5i - 3.2$

Exercises 5.6, page 164

1. $x = 1, y = -4, z = \dfrac{1}{3}$ **3.** $x = 2, y = -1, z = 1$

5. $x = 4, y = -3, z = 3$ **7.** $l = \dfrac{1}{2}, w = \dfrac{2}{3}, h = \dfrac{1}{6}$

9. $x = \dfrac{2}{3}, y = -\dfrac{1}{3}, z = 1$ **11.** $x = \dfrac{4}{15}, y = -\dfrac{3}{5}, z = \dfrac{1}{3}$

13. $p = -2, q = \dfrac{2}{3}, r = \dfrac{1}{3}$ **15.** $x = \dfrac{3}{4}, y = 1, z = -\dfrac{1}{2}$

17. $r = 0, s = 0, t = 0, u = -1$

19. $P = 800$ h, $M = 125$ h, $I = 225$ h

21. $F_1 = 9.43$ N, $F_2 = 8.33$ N, $F_3 = 1.67$ N

23. $A = 22.5°, B = 45.0°, C = 112.5°$

25. $\theta = 0.295t^3 - 5.53t^2 + 24.2t$ **27.** 70 kg, 100 kg, 30 kg

29. Unlimited: $x = -10, y = -6, z = 0$ **31.** No solution

Exercises 5.7, page 170

1. -38 **3.** 122 **5.** 651 **7.** -439 **9.** 202

11. 29 440 **13.** 0.128 **15.** $x = -1, y = 2, z = 0$

17. $x = 2, y = -1, z = 1$ **19.** $x = 4, y = -3, z = 3$

21. $l = \dfrac{1}{2}, w = \dfrac{2}{3}, h = \dfrac{1}{6}$ **23.** $x = \dfrac{2}{3}, y = -\dfrac{1}{3}, z = 1$

25. $x = \dfrac{4}{15}, y = -\dfrac{3}{5}, z = \dfrac{1}{3}$ **27.** $p = -2, q = \dfrac{2}{3}, r = \dfrac{1}{3}$

29. $x = \dfrac{3}{4}, y = 1, z = -\dfrac{1}{2}$

31. $A = 125$ N, $B = 60$ N, $F = 75$ N

33. $s_0 = 2$ m, $v_0 = 5$m/s, $a = 4$ m/s^2

35. 17 pennies, 12 nickels, 15 dimes

37. 79% Ni, 16% Fe, 5% Mo

39. 30.9 km/h, 45.9 km/h, 551 km/h

Review Exercises for Chapter 5, page 172

1. -17 **3.** -1485 **5.** -4 **7.** $\dfrac{2}{7}$

9. $m = -2, b = 4$ **11.** $m = 4, b = -\dfrac{5}{2}$

13. $x = 2.0, y = 0.0$ **15.** $A = 2.2, B = 2.7$

17. $x = 1.5, y = -1.9$ **19.** $M = 1.5, N = 0.4$

21. $x = 1, y = 2$ **23.** $x = \dfrac{1}{2}, y = -2$ **25.** $i = 2, v = -\dfrac{1}{3}$

27. $x = -\dfrac{6}{19}, y = \dfrac{36}{19}$ **29.** $x = 1.10, y = 0.54$

31. $x = 1, y = 2$ **33.** $x = \dfrac{1}{2}, y = -2$

35. $i = 2, v = -\dfrac{1}{3}$ **37.** $x = -\dfrac{6}{19}, y = \dfrac{36}{19}$

39. $x = 1.10, y = 0.54$ **41.** Best choices: 33, 34

43. Best choices: 39, 40 **45.** -115 **47.** 230.08

49. $x = 2, y = -1, z = 1$ **51.** $r = 3, s = -1, t = \dfrac{3}{2}$

53. $x = -0.17, y = 0.16, z = 2.4$ **55.** $x = 2, y = -1, z = 1$

57. $r = 3, s = -1, t = \dfrac{3}{2}$ **59.** $x = -0.17, y = 0.16, z = 2.4$

61. 4 **63.** $-\dfrac{15}{7}$ **65.** $x = \dfrac{8}{3}, y = -8$ **67.** $x = 1, y = 3$

69. -6 **71.** $-\dfrac{4}{3}$ **73.** $F_1 = 21\,000$ N, $F_2 = 2400$ N,

$F_3 = 18\,000$ N **75.** $p_1 = 42\%, p_2 = 58\%$

77. 28 tonnes of 6% copper, 14 tonnes of 2.4% copper

79. $a = 440$ m $\cdot$ °C, $b = 9.6$°C **81.** 22 800 km/h, 1400 km/h

83. $R_1 = 0.50\ \Omega, R_2 = 1.5\ \Omega$ **85.** $L = 10$ N, $w = 40$ N

87. 68 MB, 24 MB, 48 MB

Exercises 6.1, page 179

1. $9r^2 - 4s^2$ **3.** $x^2 + 2xy + y^2 - 4x - 4y + 4$

5. $40x - 40y$ **7.** $2x^3 - 8x^2$ **9.** $y^2 - 36$ **11.** $9v^2 - 4$

13. $16x^2 - 25y^2$ **15.** $144 - 25a^2b^2$ **17.** $25f^2 + 40f + 16$

19. $4x^2 + 68x + 289$ **21.** $L^4 - 2L^2 + 1$

23. $16a^2 + 56axy + 49x^2y^2$ **25.** $0.36s^2 - 1.2st + t^2$

27. $x^2 + 6x + 5$ **29.** $C^4 + 9C^2 + 18$ **31.** $20x^2 - 21x - 5$

33. $20v^2 + 13v - 15$ **35.** $6x^2 - 13xy - 63y^2$ **37.** $2x^2 - 8$

39. $8a^3 - 2a$ **41.** $6ax^2 + 24abx + 24ab^2$

43. $20n^4 + 100n^3 + 125n^2$ **45.** $16R^4 - 72R^2r^2 + 81r^4$

47. $x^2 + y^2 + 2xy + 2x + 2y + 1$

49. $x^2 + y^2 + 2xy - 6x - 6y + 9$ **51.** $125 - 75t + 15t^2 - t^3$

53. $27L^3 + 189L^2R + 441LR^2 + 343R^3$

55. $w^2 + 2wh + h^2 - 1$ **57.** $x^3 + 8$ **59.** $64 - 27x^3$

61. $x^4 - 2x^2y^2 + y^4$ **63.** $P_0P_1c + P_1G$

65. $4p^2 + 8pDA + 4D^2A^2$ **67.** $\frac{1}{2}\pi R^2 - \frac{1}{2}\pi r^2$

69. $\frac{L}{6}x^3 - \frac{L}{2}ax^2 + \frac{L}{2}a^2x - \frac{L}{6}a^3$ **71.** $L_0 + aL_0T - aL_0T_0$

73. $4x^2 - 9$ **75.** $4x^2 - y^2 - 4y - 4$

Exercises 6.2, page 184

1. $2ax(2x - 1)$ **3.** $5(x + 3)(x - 3)$ **5.** $6(x + y)$

7. $5(a - 1)$ **9.** $3x(x - 3)$ **11.** $7b(bh - 4)$

13. $6n(12n + 1)$ **15.** $2(x + 2y - 4z)$ **17.** $3ab(b - 2 + 4b^2)$

19. $4pq(3q - 2 - 7q^2)$ **21.** $2(a^2 - b^2 + 2c^2 - 3d^2)$

23. $(x + 2)(x - 2)$ **25.** $(10 + 3A)(10 - 3A)$

27. $36a^4 + 1$ (prime)   **29.** $(9s + 5t)(9s - 5t)$

31. $(12n - 13p^2)(12n + 13p^2)$ **33.** $(x + y + 3)(x + y - 3)$

35. $2(x + 2)(x - 2)$ **37.** $3(x + 3z)(x - 3z)$

39. $2(I - 1)(I - 5)$ **41.** $(x^2 + 4)(x + 2)(x - 2)$

43. $(x^4 + 1)(x^2 + 1)(x + 1)(x - 1)$ **45.** $\frac{3 + b}{2 - b}$

47. $\frac{3}{2(t - 1)}$ **49.** $(3 + b)(x - y)$ **51.** $(a - b)(a + x)$

53. $(x + 2)(x - 2)(x + 3)$ **55.** $(x - y)(x + y + 1)$

57. $8^8 = 16{,}777{,}216$ **59.** $2\pi r(h + r)$ **61.** $Rv(1 + v + v^2)$

63. $r(R - r)(R + r)$ **65.** $r^2(\pi - 2)$ **67.** $\frac{i_2R_2}{R_1 + R_2}$

69. $\frac{5Y}{3(3S - Y)}$ **71.** $\frac{ER}{A(T_0 - T_1)}$

Exercises 6.3, page 191

1. $(x + 3)(x + 1)$ **3.** $(2x - 1)(x - 5)$ **5.** $2(x + 6)(x - 3)$

7. $(x + 1)(x + 4)$ **9.** $(s - 7)(s + 6)$ **11.** $(t + 8)(t - 3)$

13. $(x + 1)^2$ **15.** $(x - 2y)^2$ **17.** $(3x + 1)(x - 2)$

19. $(3y + 1)(y - 3)$ **21.** $(2s + 11)(s + 1)$

23. $(3f^2 - 1)(f^2 - 5)$ **25.** $(2t - 3)(t + 5)$

27. $(3t - 4u)(t - u)$ **29.** $(4x - 7)(x + 1)$

31. $(9x - 2y)(x + y)$ **33.** $(2m + 5)^2$ **35.** $(2x - 3)^2$

37. $(3t - 4)(3t - 1)$ **39.** $(8b^3 - 1)(b^3 + 4)$

41. $(4p - q)(p - 6q)$ **43.** $(12x - y)(x + 4y)$

45. $2(x - 1)(x - 6)$ **47.** $2(2x - 1)(x + 4)$

49. $ax(x + 6a)(x - 2a)$ **51.** $(a + b + 2)(a + b - 2)$

53. $(5a + 5x + y)(5a - 5x - y)$ **55.** $(4x^n - 3)(x^n + 4)$

57. $3(p + 6)(p - 3)$ **59.** $4(s + 1)(s + 3)$

61. $100(2n + 3)(n - 12)$ **63.** $(V - nB)^2$

65. $wx^2(x - 2L)(x - 3L)$ **67.** $Ad(3u - v)(u - v)$

69. $(x^2 + 2 + 2x)(x^2 + 2 - 2x)$ **71.** $9(4x^2 + 1)$

Exercises 6.4, page 193

1. $(x - 2)(x^2 + 2x + 4)$ **3.** $(x + 1)(x^2 - x + 1)$

5. $(2 - t)(4 + 2t + t^2)$ **7.** $(3x - 2a)(9x^2 + 6ax + 4a^2)$

9. $2(x + 2)(x^2 - 2x + 4)$ **11.** $6A^3(A - 1)(A^2 + A + 1)$

13. $6x^3y(1 - y)(1 + y + y^2)$ **15.** $x^3y^3(x + y)(x^2 - xy + y^2)$

17. $3a^2(a^2 + 1)(a + 1)(a - 1)$

19. $0.001(R - 4r)(R^2 + 4Rr + 16r^2)$

21. $27L^3(L + 2)(L^2 - 2L + 4)$

23. $(a + b + 4)(a^2 + 2ab + b^2 - 4a - 4b + 16)$

25. $(2 - x)(2 + x)(16 + 4x^2 + x^4)$

27. $2(x + 5)(x^2 - 5x + 25)$ **29.** $D(D - d)(D^2 + Dd + d^2)$

31. $QH(H + Q)(H^2 - HQ + Q^2)$

33. $x^5 - y^5 = (x - y)(x^4 + x^3y + x^2y^2 + xy^3 + y^4)$

$x^7 - y^7 = (x - y)(x^6 + x^5y + x^4y^2 + x^3y^3 + x^2y^4 + xy^5 + y^6)$

35. $(x + y)(x - y)(x^4 + x^2y^2 + y^4)$

Exercises 6.5, page 197

1. $\frac{3c}{4b}$ **3.** $\frac{x + 2}{x - 2}$ **5.** $\frac{14}{21}$ **7.** $\frac{2ax^2}{2xy}$ **9.** $\frac{2x - 4}{x^2 + x - 6}$

11. $\frac{ax^2 - ay^2}{x^2 - xy - 2y^2}$ **13.** $\frac{7}{11}$ **15.** $\frac{2xy}{4y^2}$ **17.** $\frac{2}{R + 1}$

19. $\frac{s - 5}{2s - 1}$ **21.** $9xy$ **23.** $a^2 - 25$ **25.** $2x$ **27.** 1

29. $\frac{1}{4}$ **31.** $\frac{3x}{4}$ **33.** $\frac{1}{5a}$ **35.** $\frac{3a - 2b}{2a - b}$

37. $\frac{4x^2 + 1}{(2x + 1)(2x - 1)}$ (cannot be reduced)   **39.** $3x$

41. $\frac{1}{2y^2}$ **43.** $\frac{x - 4}{x + 4}$ **45.** $\frac{2w^2 - 1}{w^2 + 8}$ **47.** $\frac{5x + 4}{x(x + 3)}$

49. $(N - 2)(N^2 + 4)$ **51.** $\frac{1}{2t + 1}$ **53.** $\frac{x + 3}{x - 3}$ **55.** $-\frac{1}{2}$

57. $-\frac{2x - 1}{x}$ **59.** $\frac{(x + 5)(x - 3)}{(5 - x)(x + 3)}$ **61.** $\frac{x^2 - xy + y^2}{2}$

63. $\frac{2x}{9x^2 - 3x + 1}$

65. (a) $\frac{x^2(x + 2)}{x^2 + 4}$ (b) $\frac{x^2}{x^2 - 4}$ Numerator and denominator have no common factor. In each, x^2 is not a factor of the denominator.

67. (a) $\dfrac{(x - 2)(x + 1)}{x(x - 1)}$ (b) $\dfrac{x - 2}{x}$ Numerator and denominator have no common factor. In each, x is not a factor of the numerator.

69. $u + v$ **71.** $\dfrac{E^2(R - r)}{(R + r)^3}$

Exercises 6.6, page 201

1. $\dfrac{2(x + y)}{3(x - y)}$ **3.** $x + 2y$ **5.** $\dfrac{3}{28}$ **7.** $6xy$ **9.** $\dfrac{7}{18}$

11. $\dfrac{xy^2}{bz^2}$ **13.** $4t$ **15.** $3(u + v)(u - v)$ **17.** $\dfrac{10}{3(a + 4)}$

19. $\dfrac{x^2 - 3}{x^2(x^2 + 3)}$ **21.** $\dfrac{3x}{5a}$ **23.** $\dfrac{(x + 1)(x - 1)(x - 4)}{4(x + 2)}$

25. $\dfrac{x^2}{a + x}$ **27.** $\dfrac{15}{4}$ **29.** $\dfrac{3}{4x + 3}$ **31.** $\dfrac{2(3T + N)(5V - 6)}{(V - 7)(4T + N)}$

33. $\dfrac{7x^4}{3a^4}$ **35.** $\dfrac{4t(2t - 1)(t + 5)}{(2t + 1)^2}$ **37.** $\dfrac{x + y}{2}$

39. $(x + y)(3p + 7q)$ **41.** $\dfrac{x - 2}{6x}$ **43.** $\dfrac{2(2x - 1)}{5(2 - x)}$

45. $\dfrac{2v_1 v_2}{v_1 + v_2}$ **47.** $\dfrac{\pi}{2}$

Exercises 6.7, page 206

1. $12a^2b^3$ **3.** $\dfrac{x^3 - 10x^2 + 36x - 90}{2(x - 4)^2(x + 3)}$ **5.** $\dfrac{9}{5}$ **7.** $\dfrac{8}{x}$

9. $\dfrac{5}{4}$ **11.** $\dfrac{3 + 7ax}{4x}$ **13.** $\dfrac{ax - b}{x^2}$ **15.** $\dfrac{30 + ax^2}{25x^3}$

17. $\dfrac{14 - a^2}{10a}$ **19.** $\dfrac{-x^2 + 4x + xy + y - 2}{xy}$ **21.** $\dfrac{7}{2(2x - 1)}$

23. $\dfrac{5 - 3x}{2x(x + 1)}$ **25.** $\dfrac{-3}{4(s - 3)}$ **27.** $\dfrac{7R + 6}{3(R + 3)(R - 3)}$

29. $\dfrac{2x - 5}{(x - 4)^2}$ **31.** $\dfrac{-4}{(v + 1)(v - 3)}$ **33.** $\dfrac{9x^2 + x - 2}{(3x - 1)(x - 4)}$

35. $\dfrac{13t^2 + 27t}{(t - 3)(t + 2)(t + 3)^2}$ **37.** $\dfrac{-2w^3 + w^2 - w}{(w + 1)(w^2 - w + 1)}$

39. $\dfrac{1}{x - 1}$ **41.** $\dfrac{x - y}{y}$ **43.** $-\dfrac{(x + 1)(x - 1)(2x + 1)}{x^2(x + 2)}$

45. $-\dfrac{(3x + 4)(x - 1)}{2x}$ **47.** $\dfrac{h}{(x + 1)(x + h + 1)}$

49. $\dfrac{-2hx - h^2}{x^2(x + h)^2}$ **51.** $\dfrac{y^2 - rx + r^2}{r^2}$ **53.** $\dfrac{2a - 1}{a^2}$

55. $\dfrac{a^2 + 2a - 1}{a + 1}$ **57.** $\dfrac{a + b}{\frac{1}{a} + \frac{1}{b}} = ab$ **59.** $\dfrac{3(H - H_0)}{4\pi H}$

61. $\dfrac{n(2n + 1)}{2(n + 2)(n - 1)}$ **63.** $\dfrac{P^2(9x^2 + L^2)}{4L^4}$

65. $\dfrac{sL + R}{s^2LC + sRC + 1}$ **67.** $\dfrac{\omega^2 L^2 + \omega^4 R^2 C^2 L^2 - 2\omega^2 R^2 CL + R^2}{\omega^2 R^2 L^2}$

Exercises 6.8, page 211

1. $\dfrac{2}{b - 1}$ **3.** $\dfrac{arv^2}{2aM - rM}$ **5.** 4 **7.** -3 **9.** $\dfrac{7}{2}$

11. $\dfrac{16}{21}$ **13.** -9 **15.** $-\dfrac{2}{13}$ **17.** $\dfrac{5}{3}$ **19.** -2

21. $\dfrac{3}{4}$ **23.** $\dfrac{37}{6}$ **25.** -5 **27.** $\dfrac{63}{8}$ **29.** No solution

31. $\dfrac{2}{3}$ **33.** $\dfrac{3b}{1 - 2b}$ **35.** $\dfrac{(2b - 1)(b + 6)}{2(b - 1)}$

37. $\dfrac{2s - 2s_0 - v_0 t}{t}$ **39.** $\dfrac{40V - 240}{78 - 5.0V}$ **41.** $\dfrac{jX}{1 - g_m z}$

43. $\dfrac{PV^3 - bPV^2 + aV - ab}{RV^2}$ **45.** $\dfrac{N_1^2 R_2 R_3}{N_2^2 R_3 + N_3^2 R_2}$

47. $\dfrac{fnR_2 - fR_2}{R_2 + f - fn}$ **49.** 2.4 h   **51.** 3.2 min   **53.** 220 m

55. 80 km/h   **57.** 2.2 V   **59.** $A = 3, B = -2$

Review Exercises for Chapter 6, page 213

1. $12ax + 15a^2$ **3.** $4a^2 - 49b^2$ **5.** $4a^2 + 4a + 1$

7. $b^2 + 3b - 28$ **9.** $2x^2 - 13x - 45$ **11.** $4c^{2a} - d^2$

13. $3(s + 3t)$ **15.** $a^2(x^2 + 1)$ **17.** $(W + 12)(W - 12)$

19. $(4x + 8 + t^2)(4x + 8 - t^2)$ **21.** $(3t - 1)^2$ **23.** $(5t + 1)^2$

25. $(x + 8)(x - 7)$ **27.** $(t + 3)(t - 3)(t^2 + 4)$

29. $(2k - 9)(k + 4)$ **31.** $(2x + 5)(2x - 7)$

33. $(5b - 1)(2b + 5)$ **35.** $4(x + 4y)(x - 4y)$

37. $2(5 - 2y^2)(25 + 10y^2 + 4y^4)$ **39.** $(2x + 3)(4x^2 - 6x + 9)$

41. $(a - 3)(b^2 + 1)$ **43.** $(x + 5)(n - x + 5)$

45. $\dfrac{16x^2}{3a^2}$ **47.** $\dfrac{3x + 1}{2x - 1}$ **49.** $\dfrac{16}{5x(x - y)}$ **51.** $-\dfrac{6}{L - 5}$

53. $\dfrac{x + 2}{2x(7x - 1)}$ **55.** $\dfrac{1}{x - 1}$ **57.** $\dfrac{16x - 15}{36x^2}$ **59.** $\dfrac{5y + 6}{2xy}$

61. $\dfrac{-2(2a + 3)}{a(a + 2)}$ **63.** $\dfrac{2x^2 - x + 1}{x(x + 3)(x - 1)}$

65. $\dfrac{12x^2 - 7x - 4}{2(x - 1)(x + 1)(4x - 1)}$ **67.** $\dfrac{x^3 + 6x^2 - 2x + 2}{x(x - 1)(x + 3)}$

69.   **71.**

73. $\left(x + \sqrt{5}\right)\left(x - \sqrt{5}\right)$ **75.** $x\left(1 + \dfrac{y}{x}\right)$ **77.** 2

79. $-\dfrac{(a-1)^2}{2a}$ **81.** $\dfrac{10}{3}$ **83.** -6

85. $\dfrac{1}{4}[(x+y)^2 - (x-y)^2] = \dfrac{1}{4}(x^2 + 2xy + y^2 - x^2 + 2xy - y^2)$

$$= \dfrac{1}{4}(4xy) = xy$$

87. $2zS^2 + 2zS$ **89.** $krR - kr^2$ **91.** $\pi l(r_1 + r_2)(r_1 - r_2)$

93. $4s(s+2)(s+12)$ **95.** $R(3R - 4r)$

97. $8n^6 + 36n^5 + 66n^4 + 63n^3 + 33n^2 + 9n + 1$

99. $10aT - 10at + aT^2 - 2aTt + at^2$ **101.** $4(3x^2 + 12x + 16)$

103. $\dfrac{12\pi^2 wv^2 D}{gn^2 t}$ **105.** $\dfrac{R^2 + r^2}{2}$

107. $\dfrac{120 - 60d^2 + 5d^4 - d^6}{120}$

109. $\dfrac{2\pi^2 CN + 2\pi^2 Cn + N^2 - 2nN + n^2}{4\pi^2 C}$

111. $\dfrac{4r^3 - 3ar^2 - a^3}{4r^3}$ **113.** $\dfrac{c^2 u^2 - 2gc^2 x}{1 - c^2 u^2 + 2gc^2 x}$

115. $\dfrac{W}{g(h_2 - h_1)}$ **117.** $\dfrac{RHw}{w - RH}$

119. $\dfrac{2EI - 2IV_0 - mIV^2 - p^2}{IV^2}$ **121.** $\dfrac{-mb^2 s^2 - kL^2}{b^2 s}$

123. $\dfrac{i}{sV - V_0}$ **125.** 3.4 h   **127.** 1.5 s   **129.** 11.3

131. 18 Ω

Exercises 7.1, page 220

1. $-\frac{1}{2}, 4$ **3.** $a = 1, b = -2, c = -4$

5. Not quadratic, no x^2-term   **7.** $a = 1, b = -1, c = 0$

9. $2, -2$ **11.** $\frac{3}{2}, -\frac{3}{2}$ **13.** $-1, 9$ **15.** $3, 4$ **17.** $0, \frac{5}{2}$

19. $\frac{1}{3}, -\frac{1}{3}$ **21.** $\frac{1}{3}, 4$ **23.** $-4, -4$ **25.** $\frac{2}{3}, \frac{3}{2}$ **27.** $\frac{1}{2}, -\frac{3}{2}$

29. $2, -1$ **31.** $2b, -2b$ **33.** $\frac{5}{2}, -\frac{9}{2}$ **35.** $0, -2$

37. $b - a, -b - a$ **39.** 2 A, -6 A (if current is negative)

41. 2 A.M., 10 A.M.   **43.** $-1, 0, 1$ **45.** $\frac{3}{2}, 4$ **47.** $-2, \frac{1}{2}$

49. 3 N/cm, 6 N/cm   **51.** 30 km/h, 40 km/h

Exercises 7.2, page 223

1. $-3 \pm \sqrt{17}$ **3.** $-5, 5$ **5.** $-\sqrt{7}, \sqrt{7}$ **7.** $-3, 7$

9. $-3 \pm \sqrt{7}$ **11.** $2, -4$ **13.** $-2, -1$ **15.** $2 \pm \sqrt{2}$

17. $-5, 3$ **19.** $-3, \frac{1}{2}$ **21.** $\frac{1}{6}(3 \pm \sqrt{33})$

23. $\frac{1}{4}(1 \pm \sqrt{17})$ **25.** $\frac{1}{5}(5 \pm \sqrt{5})$ **27.** $-\frac{1}{3}, -\frac{1}{3}$

29. $-b \pm \sqrt{b^2 - c}$ **31.** 5°C, 15°C

Exercises 7.3, page 226

1. $-3, -2$ **3.** $\frac{4}{3}$ (double root)   **5.** $2, -4$ **7.** $-2, -1$

9. $2 \pm \sqrt{2}$ **11.** $-5, 3$ **13.** $-3, \frac{1}{2}$ **15.** $\frac{1}{6}(3 \pm \sqrt{33})$

17. $\frac{1}{4}(1 \pm \sqrt{17})$ **19.** $-\frac{8}{5}, \frac{5}{6}$ **21.** $\frac{1}{16}(-61 \pm \sqrt{-119})$

23. $\frac{1}{6}(1 \pm \sqrt{109})$ **25.** $\pm\frac{11}{5}$ **27.** $\frac{3}{4}, -\frac{5}{8}$ **29.** $-0.54, 0.74$

31. $-0.26, 2.43$ **33.** $-c \pm \sqrt{c^2 + 1}$

35. $\dfrac{b + 1 \pm \sqrt{-3b^2 + 2b + 4b^2 a + 1}}{2b^2}$ **37.** Imaginary, unequal

39. Real, irrational, unequal   **41.** $a = c$ **43.** 4.376 cm

45. $\frac{1}{2}(1 + \sqrt{5}) = 1.618$ **47.** $\dfrac{-R \pm \sqrt{R^2 - 4L/C}}{2L}$

49. 4.0 cm   **51.** 11.0 m by 23.8 m   **53.** 1.1 m

55. 100 km/h, 80 km/h

Exercises 7.4, page 231

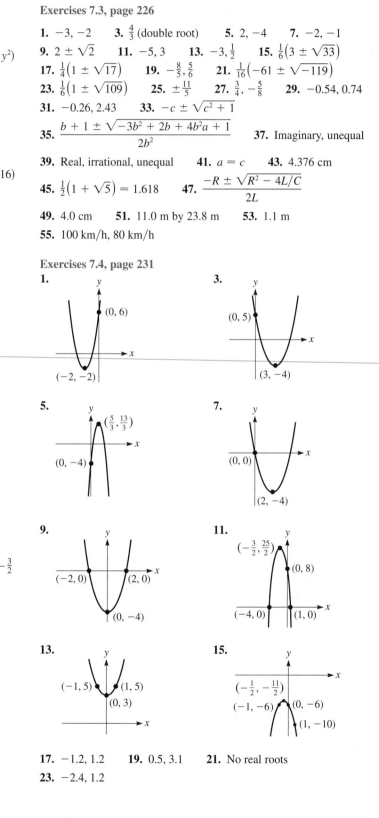

1. $(0, 6)$, $(-2, -2)$

3. $(0, 5)$, $(3, -4)$

5. $\left(\frac{5}{3}, \frac{13}{3}\right)$, $(0, -4)$

7. $(0, 0)$, $(2, -4)$

9. $(-2, 0)$, $(2, 0)$, $(0, -4)$

11. $\left(-\frac{3}{2}, \frac{25}{2}\right)$, $(0, 8)$, $(-4, 0)$, $(1, 0)$

13. $(-1, 5)$, $(1, 5)$, $(0, 3)$

15. $\left(-\frac{1}{2}, -\frac{11}{2}\right)$, $(-1, -6)$, $(0, -6)$, $(1, -10)$

17. $-1.2, 1.2$ **19.** $0.5, 3.1$ **21.** No real roots

23. $-2.4, 1.2$

25.

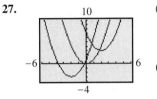

$y = x^2 + 3$
$y = x^2$
$y = x^2 - 3$

(a) The parabola $y = x^2 + 3$ is 3 units above, and the parabola $y = x^3 - 3$ is 3 units below.
(b) All parabolas open up.

27.

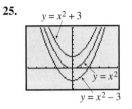

(a) Parabola (b) is 2 units to the right and 3 units up.
Parabola (c) is 2 units to the left and 3 units down.
(b) All parabolas open up.

29.

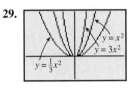

$y = x^2$
$y = 3x^2$
$y = \frac{1}{3}x^2$

The parabola $y = 3x^2$ rises more quickly, and the parabola $y = \frac{1}{3}x^2$ rises more slowly.

31.

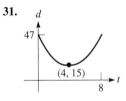

d
47
(4, 15)
8   t

33.

A
16
8   w

35.

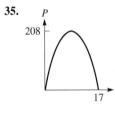

P
208
17   i

37. (a) After 19 s
(b) 460 m
(c) 2.6 s, 16 s

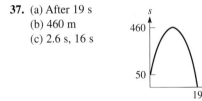

s
460
50
19   t

39. 62 m by 32 m, or 22 m by 93 m

Review Exercises for Chapter 7, page 232

1. $-4, 1$ **3.** $2, 8$ **5.** $\frac{1}{3}, -4$ **7.** $\frac{1}{2}, \frac{5}{3}$ **9.** $0, \frac{25}{6}$
11. $-\frac{3}{2}, \frac{7}{2}$ **13.** $-10, 11$ **15.** $-1 \pm \sqrt{7}$ **17.** $-4, \frac{9}{2}$
19. $\frac{1}{8}(3 \pm \sqrt{41})$ **21.** $\frac{1}{42}(-23 \pm \sqrt{-4091})$
23. $\frac{1}{6}(-2 \pm \sqrt{58})$ **25.** $-2 \pm 2\sqrt{2}$ **27.** $\frac{1}{3}(-4 \pm \sqrt{10})$
29. $-1, \frac{5}{4}$ **31.** $\frac{1}{4}(-3 \pm \sqrt{-47})$ **33.** $\dfrac{-1 \pm \sqrt{-1}}{a}$

35. $\dfrac{-3 \pm \sqrt{9 + 4a^2}}{2a}$ **37.** $-5, 6$ **39.** $\frac{1}{4}(1 \pm \sqrt{33})$
41. $3 \pm \sqrt{7}$ **43.** 0 (3 is not a solution)

45.

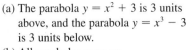

y
x
$(0, -1)$
$(\frac{1}{4}, -\frac{9}{8})$

47.

y
$(\frac{1}{6}, \frac{1}{12})$
$(0, 0)$ x

49. $-1.7, 1.2$ **51.** No real roots   **53.** $0, L$ **55.** 17
57. 94°C   **59.** 0.6 s, 2.2 s   **61.** 8000
63. $\dfrac{-\pi h \pm \sqrt{\pi^2 h^2 + 2\pi A}}{2\pi}$ **65.** $\dfrac{1 + r \pm \sqrt{(1 + r)^2 - 4rp_2}}{2r}$

67.

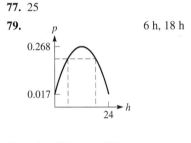

p
0.135
6   t

69. 61 m, 81 m   **71.** 9.9 cm
73. 970 m, 670 m; 570 m, 570 m   **75.** 40.7 cm, 55.2 cm
77. 25

79.

p 6 h, 18 h
0.268
0.017
24   h

Exercises 8.1, page 238

1. (a) $-, +, -, -, +, -$ (b) $+, -, +, +, +, -$ **3.** $+, -$
5. $+, +$ **7.** $-, +$ **9.** $+, -$ **11.** $-, -$ **13.** $-, +$
15. $\sin \theta = \dfrac{1}{\sqrt{5}}$, $\cos \theta = \dfrac{2}{\sqrt{5}}$, $\tan \theta = \dfrac{1}{2}$, $\cot \theta = 2$,

   $\sec \theta = \dfrac{1}{2}\sqrt{5}$, $\csc \theta = \sqrt{5}$

17. $\sin \theta = -\dfrac{3}{\sqrt{13}}$, $\cos \theta = -\dfrac{2}{\sqrt{13}}$, $\tan \theta = \dfrac{3}{2}$, $\cot \theta = \dfrac{2}{3}$,

   $\sec \theta = -\dfrac{1}{2}\sqrt{13}$, $\csc \theta = -\dfrac{1}{3}\sqrt{13}$

19. $\sin \theta = \dfrac{12}{13}$, $\cos \theta = -\dfrac{5}{13}$, $\tan \theta = -\dfrac{12}{5}$, $\cot \theta = -\dfrac{5}{12}$,

   $\sec \theta = -\dfrac{13}{5}$, $\csc \theta = \dfrac{13}{12}$

21. $\sin \theta = -\dfrac{2}{\sqrt{29}}$, $\cos \theta = \dfrac{5}{\sqrt{29}}$, $\tan \theta = -\dfrac{2}{5}$, $\cot \theta = -\dfrac{5}{2}$,

$\sec \theta = \dfrac{1}{5}\sqrt{29}$, $\csc \theta = -\dfrac{1}{2}\sqrt{29}$

23. I, II **25.** I, III **27.** II, III **29.** II **31.** II

33. IV **35.** III **37.** IV **39.** II

Exercises 8.2, page 243

1. $\sin 200° = -\sin 20° = -0.3420$,
$\tan 150° = -\tan 30° = -0.5774$,
$\cos 265° = -\cos 85° = -0.0872$,
$\cot 300° = -\cot 60° = -0.5774$, $\sec 344° = \sec 16° = 1.040$,
$\sin 397° = \sin 37° = 0.6018$

3. $296.00°$ **5.** $\sin 20°$; $-\cos 40°$ **7.** $-\tan 75°$; $-\csc 58°$

9. $\cos 40°$; $-\tan 40°$ **11.** $-\sin 15° = -0.26$

13. $-\cos 73.7° = -0.281$ **15.** $\sec 31.67° = 1.175$

17. $-\tan 31.5° = -0.613$ **19.** -0.523 **21.** -0.7620

23. -3.910 **25.** 0.2989 **27.** $237.99°, 302.01°$

29. $66.40°, 293.60°$ **31.** $102.0°, 282.0°$ **33.** $119.5°$

35. $263°$ **37.** $306.21°$ **39.** $299.24°$ **41.** -0.7003

43. -0.777 **45.** $<$ **47.** $=$ **49.** 0.0183 A

51. 12.6 cm

Exercises 8.3, page 248

1. $160°$ **3.** $1.082, 2.060$ **5.** $\dfrac{\pi}{12}, \dfrac{5\pi}{6}$ **7.** $\dfrac{5\pi}{12}, \dfrac{11\pi}{6}$

9. $\dfrac{7\pi}{6}, \dfrac{3\pi}{2}$ **11.** $4\pi, \dfrac{13\pi}{9}$ **13.** $72°, 270°$ **15.** $10°, 315°$

17. $170°, 300°$ **19.** $15°, 27°$ **21.** 0.401 **23.** 4.40

25. 5.821 **27.** 8.351 **29.** $43.0°$ **31.** $195.2°$ **33.** $710°$

35. $-940.8°$ **37.** 0.7071 **39.** 3.732 **41.** -0.8660

43. -8.327 **45.** 0.9056 **47.** -0.89 **49.** -2.1

51. 0.149 **53.** $0.3141, 2.827$ **55.** $2.932, 6.074$

57. $0.8309, 5.452$ **59.** $2.442, 3.841$ **61.** 612 mil

63. 11.0 rad **65.** 0.030 N · m **67.** 2900 m

Exercises 8.4, page 252

1. 2.36 cm **3.** 14.4 m/s **5.** 3.46 cm **7.** 426 mm

9. 0.1849 km^2 **11.** 43 cm^2 **13.** 0.0647 m **15.** 8.21 m

17. 47.2 cm **19.** $1.81, 14.5$ cm^2 **21.** 32.73 min past noon

23. 627 m^2 **25.** 0.52 rad/s **27.** 34.73 m^2 **29.** 0.704 m

31. 22.6 m^2 **33.** 369 m^3 **35.** 0.4 km **37.** 325 m/min

39. $10\,100$ cm/min **41.** 75.8 r/min **43.** 9.41 m

45. 35.9 m/min **47.** 0.433 rad/s **49.** 3000 m/min

51. 250 rad **53.** 940 m/s **55.** 14.9 m^3

57. Ratios become very close to 1 **59.** 1.15×10^8 km

Review Exercises for Chapter 8, page 255

1. $\sin \theta = \dfrac{4}{5}$, $\cos \theta = \dfrac{3}{5}$, $\tan \theta = \dfrac{4}{3}$, $\cot \theta = \dfrac{3}{4}$, $\sec \theta = \dfrac{5}{3}$,

$\csc \theta = \dfrac{5}{4}$

3. $\sin \theta = -\dfrac{2}{\sqrt{53}}$, $\cos \theta = \dfrac{7}{\sqrt{53}}$, $\tan \theta = -\dfrac{2}{7}$, $\cot \theta = -\dfrac{7}{2}$,

$\sec \theta = \dfrac{\sqrt{53}}{7}$, $\csc \theta = -\dfrac{\sqrt{53}}{2}$

5. $-\cos 48°$, $\tan 14°$ **7.** $-\sin 71°$, $\sec 15°$ **9.** $\dfrac{2\pi}{9}, \dfrac{17\pi}{20}$

11. $\dfrac{34\pi}{15}, \dfrac{9\pi}{8}$ **13.** $252°, 130°$ **15.** $12°, 330°$ **17.** $32.1°$

19. $2067°$ **21.** 1.78 **23.** 0.3534 **25.** 4.5736

27. 2.377 **29.** -0.415 **31.** -0.47 **33.** -1.080

35. -0.4264 **37.** -1.64 **39.** 4.140 **41.** -0.5878

43. -0.8660 **45.** 0.5569 **47.** 1.197 **49.** $10.30°, 190.30°$

51. $118.23°, 241.77°$ **53.** $0.5759, 5.707$ **55.** $4.187, 5.238$

57. $223.76°$ **59.** $246.78°$ **61.** 10.8 cm

63. $3.23 = 185.3°$ **65.** 14.2 m **67.** 10.92 cm^2

69. 0.0562 W **71.** $0.800 = 45.8°$ **73.** 19.8 r/min

75. 3600 km/h

77. (a) 6680 km (b) $10\,000$ km; great circle route shortest of all routes

79. 4710 cm/s **81.** 2.70 m^2 **83.** 1.81×10^6 cm/s

85. 138 m^2 **87.** 3.58×10^5 km

Exercises 9.1, page 262

1. **3.**

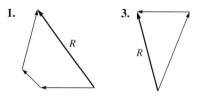

5. (a) Scalar since no direction is given. (b) Vector since direction is given.

7. (a) Vector: magnitude and direction are specified; (b) scalar: only magnitude is specified.

9. **11.**

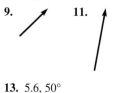

13. $5.6, 50°$ **15.** $4.3, 156°$

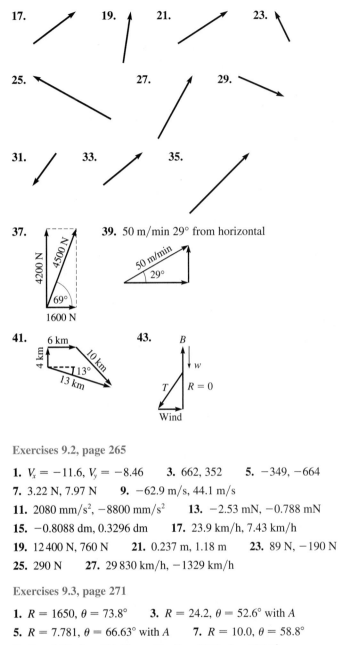

17.   **19.**   **21.**   **23.**

25.   **27.**   **29.**

31.   **33.**   **35.**

37.

4200 N   4500 N   1600 N   69°

39. 50 m/min 29° from horizontal

50 m/min   29°

41. 6 km   4 km   10 km   13°   13 km

43. B w T $R = 0$ Wind

11. 25.3 km/h, 29.6° S of E   **13.** 9.6 m/s^2, 7.0° from vertical

15. 781 N, 9.3° above horizontal

17. 540 km/h, 6° from direction of plane

19. 29 180 km/h, 0.03° from direction of shuttle

21. 12 km/h   **23.** 910 N   **25.** 184 000 cm/min^2, $\phi = 89.6°$

27. 138 km, 65.0° N of E

29. 79.0 m/s, 11.3° from direction of plane, 75.6° from vertical

31. 4.05 A/m, 11.5° with magnet

Exercises 9.5, page 282

1. $b = 76.01, c = 50.01, C = 40.77°$

3. $b = 38.1, C = 66.0°, c = 46.1$

5. $a = 2800, b = 2620, C = 108.0°$

7. $B = 12.20°, C = 149.57°, c = 7.448$

9. $a = 11{,}050, A = 149.70°, C = 9.57°$

11. $A = 125.6°, a = 0.0776, c = 0.00566$

13. $A = 99.4°, b = 55.1, c = 24.4$

15. $A = 68.01°, a = 5520, c = 5376$

17. $A_1 = 61.36°, C_1 = 70.51°, c_1 = 5.628; A_2 = 118.64°, C_2 = 13.23°, c_2 = 1.366$

19. $A_1 = 107.3°, a_1 = 5280, C_1 = 41.3°; A_2 = 9.9°, a_2 = 952, C_2 = 138.7°$

21. No solution   **23.** 45.5 m   **25.** 880 N   **27.** 1220 m

29. 13.94 cm   **31.** 27 300 km

33. 77.3° with bank downstream   **35.** 23 km

Exercises 9.6, page 287

1. $A = 14.0°, B = 21.0°, c = 107$

3. $A = 50.3°, B = 75.7°, c = 6.31$

5. $A = 70.9°, B = 11.1°, c = 4750$

7. $A = 34.72°, B = 40.67°, C = 104.61°$

9. $A = 18.21°, B = 22.28°, C = 139.51°$

11. $A = 6.0°, B = 16.0°, c = 1150$

13. $A = 82.3°, b = 2160, C = 11.4°$

15. $A = 36.24°, B = 39.09°, a = 97.22$

17. $A = 46.94°, B = 61.82°, C = 71.24°$

19. $A = 137.9°, B = 33.7°, C = 8.4°$

21. $b = 3700, C = 25°, c = 2400$

23. (Intermediate steps) $a^2 = b^2 + c^2 - 2bc \cos A$,
$2bc \cos A = (b^2 + c^2 + 2bc) - a^2 - 2bc$,
$2bc(1 + \cos A) = (b + c)^2 - a^2$

25. 69.4 km   **27.** 47.3°   **29.** 57.3°, 141.7°   **31.** 16.5 mm

33. 5.09 km/h   **35.** 17.8 km

Review Exercises for Chapter 9, page 289

1. $A_x = 57.4, A_y = 30.5$ **3.** $A_x = -0.7485, A_y = -0.5357$

5. $R = 602, \theta = 57.1°$ with A **7.** $R = 5960, \theta = 33.60°$ with A

Exercises 9.2, page 265

1. $V_x = -11.6, V_y = -8.46$ **3.** 662, 352   **5.** $-349, -664$

7. 3.22 N, 7.97 N   **9.** -62.9 m/s, 44.1 m/s

11. 2080 mm/s^2, -8800 mm/s^2 **13.** -2.53 mN, -0.788 mN

15. -0.8088 dm, 0.3296 dm   **17.** 23.9 km/h, 7.43 km/h

19. 12 400 N, 760 N   **21.** 0.237 m, 1.18 m   **23.** 89 N, -190 N

25. 290 N   **27.** 29 830 km/h, -1329 km/h

Exercises 9.3, page 271

1. $R = 1650, \theta = 73.8°$ **3.** $R = 24.2, \theta = 52.6°$ with A

5. $R = 7.781, \theta = 66.63°$ with A **7.** $R = 10.0, \theta = 58.8°$

9. $R = 2.74, \theta = 111.0°$ **11.** $R = 2130, \theta = 107.7°$

13. $R = 1.426, \theta = 299.12°$ **15.** $R = 29.2, \theta = 10.8°$

17. $R = 5920, \theta = 88.4°$ **19.** $R = 27.27, \theta = 33.14°$

21. $R = 50.2, \theta = 50.3°$ **23.** $R = 0.242, \theta = 285.9°$

25. $R = 235, \theta = 121.7°$ **27.** 68 000 N, 15° from 25 000-N force

Exercises 9.4, page 275

1. 47.10 km, 10.27° N of E

3. 36.9 N, 29.5° from the 34.5-N force

5. 11 000 N, 44° above horizontal

7. 3070 m, 17.8° S of W   **9.** 229.4 m, 72.82° N of E

9. $R = 965$, $\theta = 8.6°$ **11.** $R = 26.12$, $\theta = 146.03°$

13. $R = 71.93$, $\theta = 336.50°$ **15.** $R = 99.42$, $\theta = 359.57°$

17. $b = 181$, $C = 64.0°$, $c = 175$

19. $A = 21.2°$, $b = 34.8$, $c = 51.5$

21. $a = 17{,}340$, $b = 24{,}660$, $C = 7.99°$

23. $A = 39.88°$, $a = 5194$, $C = 30.03°$

25. $A_1 = 54.8°$, $a_1 = 12.7$, $B_1 = 68.6°$; $A_2 = 12.0°$, $a_2 = 3.24$, $B_2 = 111.4°$

27. $A = 32.3°$, $b = 267$, $C = 17.7°$

29. $A = 148.7°$, $B = 9.3°$, $c = 5.66$

31. $a = 1782$, $b = 1920$, $C = 16.00°$

33. $A = 37°$, $B = 25°$, $C = 118°$

35. $A = 20.6°$, $B = 35.6°$, $C = 123.8°$

37. Add the 3 forms of law of cosines together; simplify; divide by $2abc$.

39. $A_t = \frac{1}{2}ab$; $b = \dfrac{a \sin B}{\sin A}$; substitute   **41.** -155.7 N, 81.14 N

43. 630 m/s   **45.** 9.8 km   **47.** $R = 2700$ N, $\theta = 107°$

49. 6.1 m/s, $35°$ with horizontal   **51.** 2.30 m, 2.49 m

53. 0.039 km   **55.** 52 700 km   **57.** 2.65 km   **59.** 293 km

61. 810 N, $36°$ N of E   **63.** 1280 m or 1680 m (ambiguous)

Exercises 10.1, page 295

1. $y = 3 \cos x$

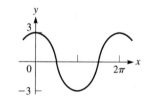

3. 0, -0.7, -1, -0.7, 0, 0.7, 1, 0.7, 0, -0.7, -1, -0.7, 0, 0.7, 1, 0.7, 0

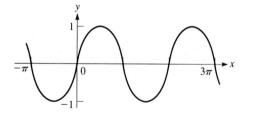

5. -3, -2.1, 0, 2.1, 3, 2.1, 0, -2.1, -3, -2.1, 0, 2.1, 3, 2.1, 0, -2.1, -3

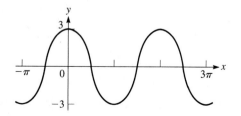

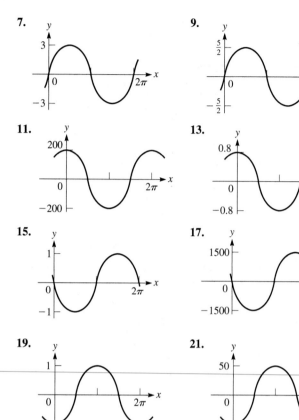

7.

9.

11.

13.

15.

17.

19.

21.

23. 0, 0.84, 0.91, 0.14, -0.76, -0.96, -0.28, 0.66

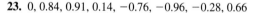

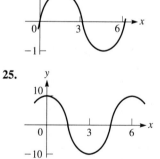

25.

27. $y = -2 \sin x$

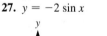

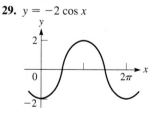

29. $y = -2 \cos x$

31.

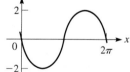

33. $y = 4 \sin x$ **35.** $y = -1.5 \cos x$

Exercises 10.2, page 299

1. $y = 3 \sin 6x$

49. $y = -2 \sin 2x$

51. $y = 2 \cos \dfrac{1}{2}x$

3. $\dfrac{\pi}{3}$ **5.** $\dfrac{\pi}{4}$ **7.** $\dfrac{\pi}{6}$ **9.** $\dfrac{\pi}{8}$ **11.** 1 **13.** $\dfrac{1}{2}$

15. 6π **17.** 3π **19.** 3 **21.** $\dfrac{2}{\pi}$

53.

23. **25.**

55.

27. **29.**

57. $y = \frac{1}{2} \cos 2x$ **59.** $y = -4 \sin \pi x$

Exercises 10.3, page 302

31. **33.**

1. $y = -\cos\!\left(2x - \dfrac{\pi}{6}\right)$

3. $1, 2\pi, \dfrac{\pi}{6}$

35. **37.**

5. $1, 2\pi, -\dfrac{\pi}{6}$

7. $0.2, \pi, -\dfrac{\pi}{4}$

39. **41.**

9. $1, \pi, \dfrac{\pi}{2}$

11. $\dfrac{1}{2}, 4\pi, \dfrac{\pi}{2}$

43. $y = \sin 6x$ **45.** $y = \sin \pi x$ **47.** 2π

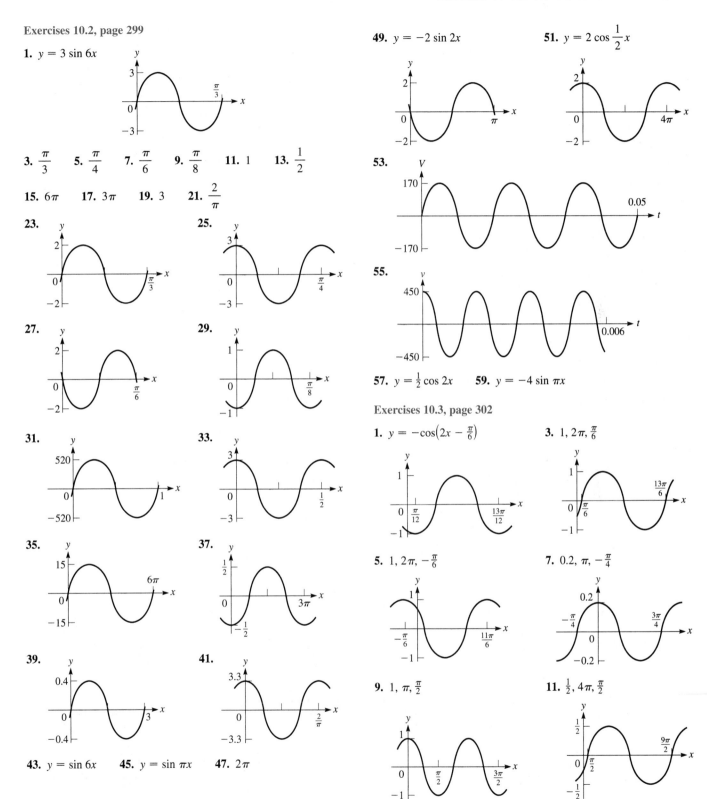

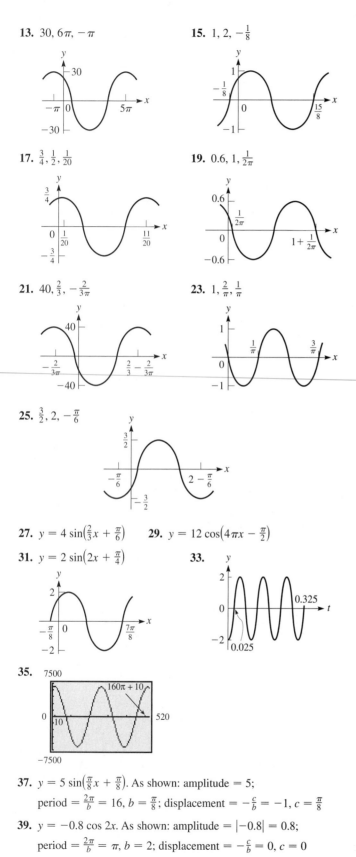

13. $30, 6\pi, -\pi$

15. $1, 2, -\frac{1}{8}$

17. $\frac{3}{4}, \frac{1}{2}, \frac{1}{20}$

19. $0.6, 1, \frac{1}{2\pi}$

21. $40, \frac{2}{3}, -\frac{2}{3\pi}$

23. $1, \frac{2}{\pi}, \frac{1}{\pi}$

25. $\frac{3}{2}, 2, -\frac{\pi}{6}$

27. $y = 4\sin\left(\frac{2}{3}x + \frac{\pi}{6}\right)$

29. $y = 12\cos\left(4\pi x - \frac{\pi}{2}\right)$

31. $y = 2\sin\left(2x + \frac{\pi}{4}\right)$

33.
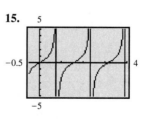

35.
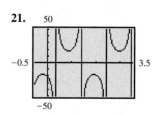

37. $y = 5\sin\left(\frac{\pi}{8}x + \frac{\pi}{8}\right)$. As shown: amplitude $= 5$;
period $= \frac{2\pi}{b} = 16, b = \frac{\pi}{8}$; displacement $= -\frac{c}{b} = -1, c = \frac{\pi}{8}$

39. $y = -0.8\cos 2x$. As shown: amplitude $= |-0.8| = 0.8$;
period $= \frac{2\pi}{b} = \pi, b = 2$; displacement $= -\frac{c}{b} = 0, c = 0$

Exercises 10.4, page 306

1. $y = 5\cot 2x$
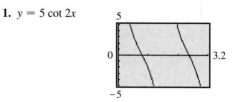

3. Undef., $-1.7, -1, -0.58, 0, 0.58, 1, 1.7$, undef., $-1.7, -1,$
$-0.58, 0$
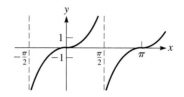

5. Undef., $2, 1.4, 1.2, 1, 1.2, 1.4, 2$, undef., $-2, -1.4, -1.2, -1$
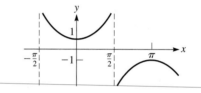

7.
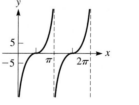

9.
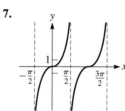

11.
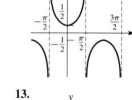

13.
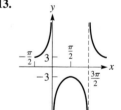

15.
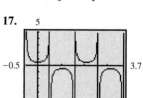

17.

19.
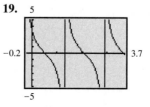

21.

23.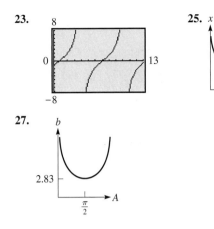

25.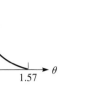

Exercises 10.6, page 312

1.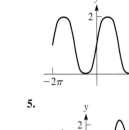

3.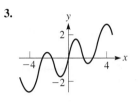

27.

5.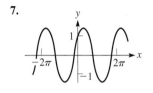

7.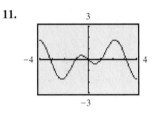

Exercises 10.5, page 308

1. $d = R \cos \omega t$

3.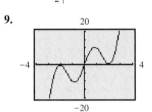

9.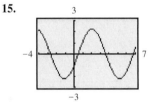

11.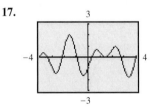

5.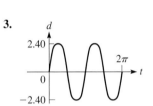

7.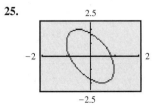

13.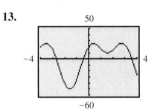

15.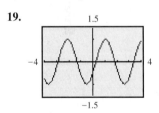

9.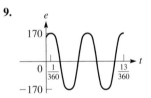

11.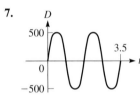

17.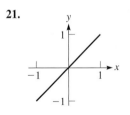

19.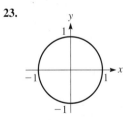

13.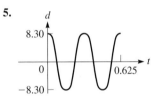

15.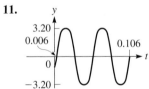

21.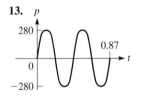

23.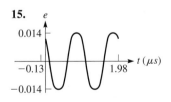

17. (a) 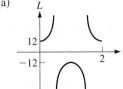   (b) $L > 0$, above x-axis

25.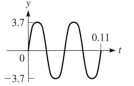

27.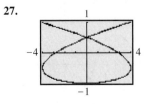

19. $y = 3.7 \sin 36\pi t$

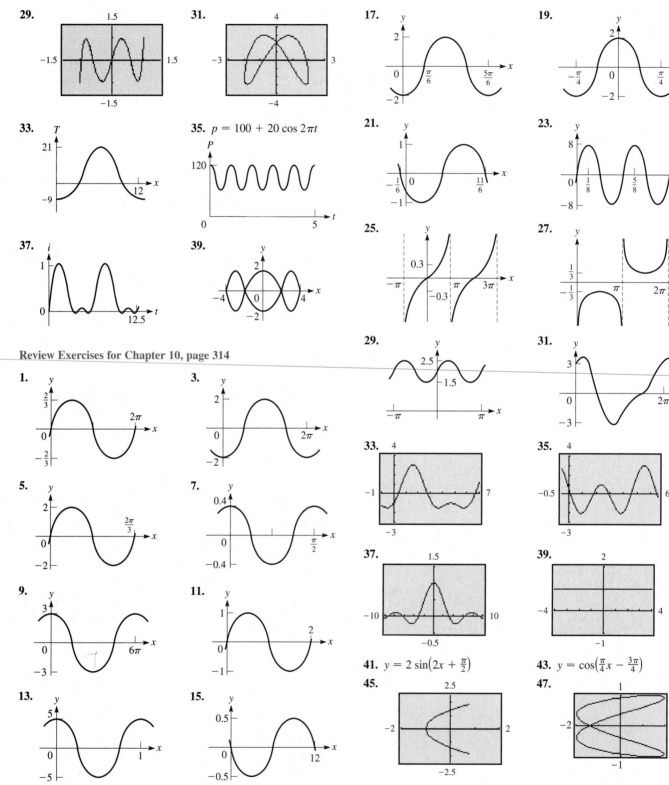

29.

31.

33.

35. $p = 100 + 20 \cos 2\pi t$

37.

39.

17.

19.

21.

23.

25.

27.

Review Exercises for Chapter 10, page 314

1.

3.

5.

7.

9.

11.

13.

15.

29.

31.

33.

35.

37.

39.

41. $y = 2 \sin\left(2x + \frac{\pi}{2}\right)$

43. $y = \cos\left(\frac{\pi}{4}x - \frac{3\pi}{4}\right)$

45.

47.

49. 4π

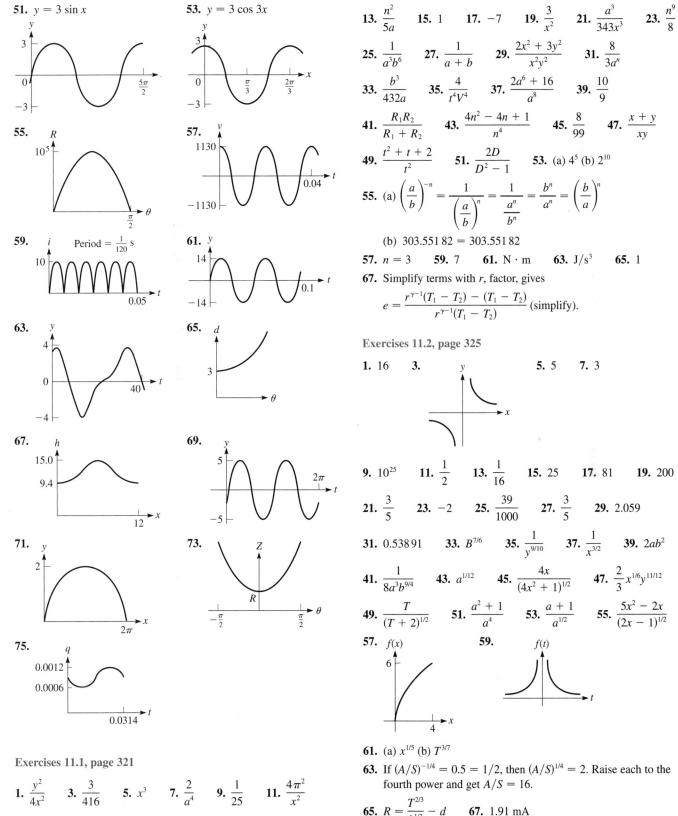

51. $y = 3 \sin x$

53. $y = 3 \cos 3x$

55.

57.

59. Period $= \frac{1}{120}$ s

61.

63.

65.

67.

69.

71.

73.

75.

Exercises 11.1, page 321

1. $\dfrac{y^2}{4x^2}$ **3.** $\dfrac{3}{416}$ **5.** x^3 **7.** $\dfrac{2}{a^4}$ **9.** $\dfrac{1}{25}$ **11.** $\dfrac{4\pi^2}{x^2}$

13. $\dfrac{n^2}{5a}$ **15.** 1   **17.** -7 **19.** $\dfrac{3}{x^2}$ **21.** $\dfrac{a^3}{343x^3}$ **23.** $\dfrac{n^9}{8}$

25. $\dfrac{1}{a^3b^6}$ **27.** $\dfrac{1}{a+b}$ **29.** $\dfrac{2x^2 + 3y^2}{x^2y^2}$ **31.** $\dfrac{8}{3a^n}$

33. $\dfrac{b^3}{432a}$ **35.** $\dfrac{4}{t^4V^4}$ **37.** $\dfrac{2a^6 + 16}{a^8}$ **39.** $\dfrac{10}{9}$

41. $\dfrac{R_1R_2}{R_1 + R_2}$ **43.** $\dfrac{4n^2 - 4n + 1}{n^4}$ **45.** $\dfrac{8}{99}$ **47.** $\dfrac{x+y}{xy}$

49. $\dfrac{t^2 + t + 2}{t^2}$ **51.** $\dfrac{2D}{D^2 - 1}$ **53.** (a) 4^5 (b) 2^{10}

55. (a) $\left(\dfrac{a}{b}\right)^{-n} = \dfrac{1}{\left(\dfrac{a}{b}\right)^n} = \dfrac{1}{\dfrac{a^n}{b^n}} = \dfrac{b^n}{a^n} = \left(\dfrac{b}{a}\right)^n$

(b) $303.551\,82 = 303.551\,82$

57. $n = 3$ **59.** 7   **61.** N $\cdot$ m   **63.** J/s^3 **65.** 1

67. Simplify terms with r, factor, gives

$$e = \dfrac{r^{\gamma-1}(T_1 - T_2) - (T_1 - T_2)}{r^{\gamma-1}(T_1 - T_2)} \text{ (simplify)}.$$

Exercises 11.2, page 325

1. 16   **3.**   **5.** 5   **7.** 3

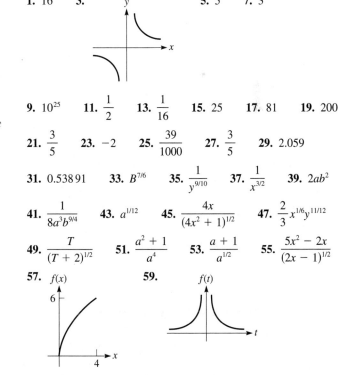

9. 10^{25} **11.** $\dfrac{1}{2}$ **13.** $\dfrac{1}{16}$ **15.** 25   **17.** 81   **19.** 200

21. $\dfrac{3}{5}$ **23.** -2 **25.** $\dfrac{39}{1000}$ **27.** $\dfrac{3}{5}$ **29.** 2.059

31. 0.538 91   **33.** $B^{7/6}$ **35.** $\dfrac{1}{y^{9/10}}$ **37.** $\dfrac{1}{x^{3/2}}$ **39.** $2ab^2$

41. $\dfrac{1}{8a^3b^{9/4}}$ **43.** $a^{1/12}$ **45.** $\dfrac{4x}{(4x^2 + 1)^{1/2}}$ **47.** $\dfrac{2}{3}x^{1/6}y^{11/12}$

49. $\dfrac{T}{(T + 2)^{1/2}}$ **51.** $\dfrac{a^2 + 1}{a^4}$ **53.** $\dfrac{a + 1}{a^{1/2}}$ **55.** $\dfrac{5x^2 - 2x}{(2x - 1)^{1/2}}$

57. $f(x)$ **59.** $f(t)$

61. (a) $x^{1/5}$ (b) $T^{3/7}$

63. If $(A/S)^{-1/4} = 0.5 = 1/2$, then $(A/S)^{1/4} = 2$. Raise each to the fourth power and get $A/S = 16$.

65. $R = \dfrac{T^{2/3}}{k^{1/3}} - d$ **67.** 1.91 mA

Exercises 11.3, page 330

1. $ab^2\sqrt{a}$ **3.** $\dfrac{\sqrt[4]{54}}{3}$ **5.** $2\sqrt{6}$ **7.** $3\sqrt{5}$ **9.** $xy^2\sqrt{y}$

11. $xy^2z\sqrt{z}$ **13.** $3R^2V^2\sqrt{2RT}$ **15.** $2\sqrt[3]{2}$ **17.** $2\sqrt[5]{3}$

19. $2\sqrt[3]{a^2}$ **21.** $2st\sqrt[4]{4r^3t}$ **23.** 2   **25.** $P\sqrt[3]{V}$

27. $\dfrac{1}{2}\sqrt{6}$ **29.** $\dfrac{1}{2}\sqrt[3]{6}$ **31.** $\dfrac{1}{3}\sqrt[5]{27}$ **33.** $2\sqrt{5}$ **35.** 2

37. 200   **39.** 2000   **41.** $\sqrt{2a}$ **43.** $\dfrac{1}{2}\sqrt{2}$ **45.** $\sqrt[3]{2}$

47. $\sqrt[8]{n}$ **49.** $\dfrac{\sqrt{mn}}{m^2}$ **51.** $\dfrac{2u\sqrt{7uv}}{v^3}$ **53.** $\dfrac{3}{4}\sqrt{2}$

55. $\dfrac{\sqrt{xy(x^2+y^2)}}{xy}$ **57.** $\dfrac{\sqrt{C^2-4}}{C+2}$ **59.** $\sqrt{a^2+b^2}$

61. $3x-1$ **63.**

65. $\sqrt{2ag}$

67. $\dfrac{8Af\sqrt{f^2+f_0^2}}{\pi^2(f^2+f_0^2)}$

37. $\dfrac{2x+2\sqrt{5x}}{x-5}$ **39.** 1   **41.** $\dfrac{2-R-R^2}{R}$

43. $-\dfrac{\sqrt{x^2-y^2}+\sqrt{x^2+xy}}{y}$ **45.** $a-1+\sqrt{a(a-2)}$

47. $-1-\sqrt{66}=-9.124\,038\,4$

49. $-\dfrac{16+5\sqrt{30}}{26}=-1.668\,697\,2$ **51.** $\dfrac{2x+1}{\sqrt{x}}$

53. $\dfrac{5x^2+2x}{\sqrt{2x+1}}$ **55.** $\dfrac{1}{\sqrt{30}-2\sqrt{3}}$ **57.** $\dfrac{1}{\sqrt{x+h}+\sqrt{x}}$

59. $(1-\sqrt{2})^2-2(1-\sqrt{2})-1$
$=1-2\sqrt{2}+2-2+2\sqrt{2}-1=0$

61. $\left[\dfrac{1}{2}\left(\sqrt{b^2-4k^2}-b\right)\right]^2+b\left[\dfrac{1}{2}\left(\sqrt{b^2-4k^2}-b\right)\right]+k^2$

$=\dfrac{1}{4}(b^2-4k^2)-\dfrac{b}{2}\sqrt{b^2-4k^2}+\dfrac{1}{4}b^2$

$+\dfrac{b}{2}\sqrt{b^2-4k^2}-\dfrac{1}{2}b^2+k^2=0$

63. $\dfrac{2500-50\sqrt{V}}{2500-V}$ **65.** $2Q\sqrt{\sqrt{2}+1}$ **67.** $\dfrac{\sqrt{LC-R^2C^2}}{LC}$

Exercises 11.4, page 332

1. $16\sqrt{5}$ **3.** $7\sqrt{3}$ **5.** $\sqrt{5}-\sqrt{7}$ **7.** $3\sqrt{5}$

9. $-4t\sqrt{3}$ **11.** $-2\sqrt{2a}$ **13.** $19\sqrt{7}$ **15.** $-20\sqrt{2}$

17. $23\sqrt{3R}-6\sqrt{2R}$ **19.** $\dfrac{7}{3}\sqrt{15}$ **21.** 0   **23.** $13\sqrt[3]{3}$

25. $\sqrt[4]{2}$ **27.** $(a-2b^2)\sqrt{ab}$ **29.** $(3-2a)\sqrt{10}$

31. $(2b-a)\sqrt[3]{3a^2b}$ **33.** $\dfrac{(a^2-c^3)\sqrt{ac}}{a^2c^3}$ **35.** $\dfrac{(a-2b)\sqrt[3]{ab^2}}{ab}$

37. $\dfrac{-2V\sqrt{T^2-V^2}}{T^2-V^2}$ **39.** $15\sqrt{3}-11\sqrt{5}=1.384\,014\,4$

41. $\dfrac{1}{6}\sqrt{6}=0.408\,248\,3$ **43.** $3\sqrt{3}$

45. Positive, $\sqrt{1000}=10\sqrt{10}$, $\sqrt{11}>\sqrt{10}$

47. $5400+900\sqrt{2}=6670$ mm

Exercises 11.5, page 335

1. $3\sqrt{10}-16$ **3.** $\sqrt{3}-\sqrt{2}$ **5.** $\sqrt{30}$ **7.** $2\sqrt{3}$

9. 2   **11.** 50   **13.** $2\sqrt{5}$ **15.** $\sqrt{6}-\sqrt{15}$ **17.** -1

19. $48+9\sqrt{15}$ **21.** $66+13\sqrt{11x}-5x$

23. $a\sqrt{b}+c\sqrt{ac}$ **25.** $\dfrac{2-\sqrt{6}}{2}$ **27.** $2a-3b+2\sqrt{2ab}$

29. $\sqrt[6]{72}$ **31.** $\dfrac{1}{4}\left(\sqrt{7}-\sqrt{3}\right)$

33. $\dfrac{1}{11}\left(\sqrt{7}+3\sqrt{2}-6-\sqrt{14}\right)$ **35.** $\dfrac{1}{17}\left(-56+9\sqrt{15}\right)$

Review Exercises for Chapter 11, page 337

1. $\dfrac{2}{a^2}$ **3.** $\dfrac{2d^3}{c}$ **5.** 375   **7.** $\dfrac{1}{8000}$ **9.** $\dfrac{t^4}{9}$ **11.** -28

13. $64a^2b^5$ **15.** $-8m^9n^6$ **17.** $\dfrac{2C-4L^2}{CL^2}$ **19.** $\dfrac{2y}{x+2y}$

21. $\dfrac{b}{ab-3}$ **23.** $\dfrac{(x^3y^3-1)^{1/3}}{y}$ **25.** $\dfrac{1}{W+H}$

27. $\dfrac{-2(x+1)}{(x-1)^3}$ **29.** $2\sqrt{17}$ **31.** $b^2c\sqrt{ab}$ **33.** $3ab^2\sqrt{a}$

35. $\dfrac{2t\sqrt{21st}}{u}$ **37.** $\dfrac{5\sqrt{2s}}{2s}$ **39.** $\dfrac{1}{9}\sqrt{33}$ **41.** $mn^2\sqrt[4]{8m^2n}$

43. $\sqrt{2}$ **45.** 0   **47.** $-7\sqrt{7}$ **49.** $3ax\sqrt{2x}$

51. $(2a+b)\sqrt[3]{a}$ **53.** $10-\sqrt{55}$ **55.** $4\sqrt{3}-4\sqrt{5}$

57. $6-51B-7\sqrt{17B}$ **59.** $42-7\sqrt{7a}-3a$

61. $\dfrac{6x+\sqrt{3xy}}{12x-y}$ **63.** $-\dfrac{8+\sqrt{6}}{29}$ **65.** $\dfrac{13-2\sqrt{35}}{29}$

67. $\dfrac{6x-13a\sqrt{x}+5a^2}{9x-25a^2}$ **69.** $\sqrt{4b^2+1}$ **71.** $\dfrac{15-2\sqrt{15}}{4}$

73. $\dfrac{3+n+\sqrt{n^2+3n}}{3}$ **75.** $51-7\sqrt{105}=-20.728\,655$

77. $\sqrt{\sqrt{2}-1}\left(\sqrt{2}+1\right)=\sqrt{\left(\sqrt{2}-1\right)\left(\sqrt{2}+1\right)^2}$
$=\sqrt{\left(\sqrt{2}-1\right)\left(3+2\sqrt{2}\right)}=\sqrt{\sqrt{2}+1}=1.553\,774\,0$

79. 2.784%   **81.** (a) $v=k(P/W)^{1/3}$ (b) $v=\dfrac{k\sqrt[3]{PW^2}}{W}$

83. $\dfrac{100(1 + 2\sqrt{x})}{x}$ **85.** $\dfrac{n_1^2 n_2^2 v}{n_1^2 - n_2^2}$ **87.** $\dfrac{1}{\sqrt{A + h} + \sqrt{A}}$

89. $6\sqrt{2}$ cm   **91.** $\dfrac{\sqrt{LC_1 C_2 (C_1 + C_2)}}{2\pi LC_1 C_2}$

Exercises 12.1, page 343

1. 6   **3.** -1 **5.** $9j$ **7.** $-2j$ **9.** $0.6j$ **11.** $2j\sqrt{2}$

13. $\dfrac{1}{2}j\sqrt{7}$ **15.** $-2ej$ **17.** (a) -7 (b) 7   **19.** (a) 4 (b) -4

21. $-\dfrac{3}{5}$ **23.** $10j$ **25.** (a) $-j$ (b) j **27.** 0   **29.** $-2j$

31. $-j$ **33.** $2 + 3j$ **35.** $-7j$ **37.** $2 + 2j$

39. $-2 + 3j$ **41.** $3\sqrt{2} - 2j\sqrt{2}$ **43.** -1

45. (a) $6 + 7j$ (b) $8 - j$ **47.** (a) $-2j$ (b) -4

49. $x = 2, y = -2$ **51.** $x = 10, y = -6$

53. $x = -2, y = 3$ **55.** Yes   **57.** No   **59.** 0

61. Yes; imag. part is zero.   **63.** Each is x.

Exercises 12.2, page 346

1. $1 - 5j$ **3.** $\dfrac{1}{25}(29 + 22j)$ **5.** $5 - 8j$ **7.** $-9 + 6j$

9. $-0.23 + 0.86j$ **11.** $-36 + 21j$ **13.** $7 + 49j$

15. $22 + 3j$ **17.** $-18j\sqrt{2}$ **19.** $-28j$ **21.** $3\sqrt{7} + 3j$

23. $-40 - 42j$ **25.** $-2 - 2j$ **27.** $\dfrac{1}{29}(-30 + 12j)$

29. $-\dfrac{1}{3}(1 + j)$ **31.** $\dfrac{1}{11}(-13 + 8j\sqrt{2})$ **33.** $\dfrac{-1 + 3j}{5}$

35. $\dfrac{-35}{13} + \dfrac{33j}{13}$

37. $(-1 - j)^2 + 2(-1 - j) + 2 = 1 + 2j - 1 - 2 - 2j + 2 = 0$

39. 10   **41.** $\dfrac{3}{10} + \dfrac{1}{10}j$ **43.** $-1 + j$

45. $281 + 35.2j$ volts   **47.** $0.016 + 0.038j$ amperes

49. (a) Real (b) Pure imaginary

51. Product is sum of squares of two real numbers
$(a + bj)(a - bj) = a^2 + b^2$

Exercises 12.3, page 348

1. $3 - j$ **3.**   **5.**

7. $-3j$ **9.** $5 + 4j$ **11.** $8 + j$

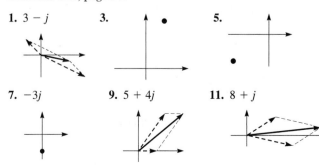

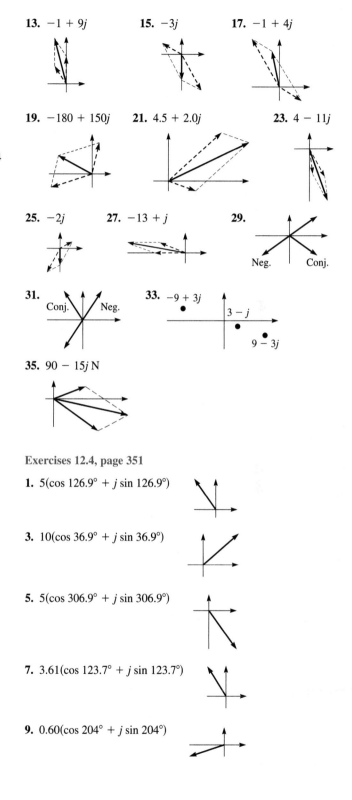

13. $-1 + 9j$ **15.** $-3j$ **17.** $-1 + 4j$

19. $-180 + 150j$ **21.** $4.5 + 2.0j$ **23.** $4 - 11j$

25. $-2j$ **27.** $-13 + j$ **29.**

Neg.   Conj.

31.

Conj.   Neg.

33. $-9 + 3j$
$3 - j$
$9 - 3j$

35. $90 - 15j$ N

Exercises 12.4, page 351

1. $5(\cos 126.9° + j \sin 126.9°)$

3. $10(\cos 36.9° + j \sin 36.9°)$

5. $5(\cos 306.9° + j \sin 306.9°)$

7. $3.61(\cos 123.7° + j \sin 123.7°)$

9. $0.60(\cos 204° + j \sin 204°)$

11. $2(\cos 60° + j \sin 60°)$

13. $8.062(\cos 295.84° + j \sin 295.84°)$

15. $3(\cos 180° + j \sin 180°)$ **17.** $9(\cos 90° + j \sin 90°)$

19. $2.94 + 4.05j$ **21.** $-1.39 + 0.800j$ **23.** -6

25. 0.08 **27.** $-200.3 + 92.86j$

29. $-4.71 + 0.595j$ **31.** $-0.6052 - 0.7096j$ **33.** $7.32j$

35. $-6.961 + 86.14j$ **37.** $3.03\underline{/339.5°}$ kV

39. $2.51 + 12.1j$ V/m

Exercises 12.5, page 354

1. $8.50e^{3.95j}$

3. $2.00(\cos 217.7° + j \sin 217.7°) = -1.58 - 1.22j$

5. $3.00e^{1.05j}$ **7.** $45.0e^{4.93j}$ **9.** $375.5e^{1.666j}$ **11.** $0.515e^{3.46j}$

13. $4.06e^{-1.07j} = 4.06e^{5.21j}$ **15.** $9245e^{5.172j}$ **17.** $5.00e^{5.36j}$

19. $36.1e^{2.55j}$ **21.** $6.37e^{0.386j}$ **23.** $825.7e^{3.836j}$

25. $3.00\underline{/28.6°}$; $2.63 + 1.44j$

27. $4.64\underline{/106.0°}$; $-1.28 + 4.46j$

29. $3.20\underline{/50.0°} = 2.06 + 2.45j$

31. $0.1724\underline{/136.99°}$; $-0.1261 + 0.1176j$

33. $-18.2 + 9.95j$, $20.7\underline{/151.3°}$

35. $-89.1 - 2750j$; $2750\underline{/268.1°}$

37. $391e^{0.285j}$ ohms; $391 \ \Omega$ **39.** $2.11 - 5.43j$

Exercises 12.6, page 360

1. $5.09(\cos 101.3° + j \sin 101.3°)$

3. $613(\cos 281.5° + j \sin 281.5°)$ **5.** $8(\cos 80° + j \sin 80°)$

7. $3(\cos 250° + j \sin 250°)$ **9.** $2(\cos 35° + j \sin 35°)$

11. $2.4\underline{/170°}$ **13.** $8(\cos 105° + j \sin 105°)$

15. $256(\cos 0° + j \sin 0°)$ **17.** $4\underline{/273°}$ **19.** $5\underline{/87°}$

21. $1.73\underline{/79.8°}$ **23.** $11{,}750\underline{/115.91°}$

25. $65.0(\cos 345.7° + j \sin 345.7°) = 63 - 16j$

27. $61.4(\cos 343.9° + j \sin 343.9°)$; $59 - 17j$

29. $2.21(\cos 71.6° + j \sin 71.6°)$; $\dfrac{7}{10} + \dfrac{21}{10}j$

31. $0.385(\cos 120.5° + j \sin 120.5°) = \dfrac{1}{169}(-33 + 56j)$

33. $625(\cos 212.5° + j \sin 212.5°) = -527 - 336j$

35. $609(\cos 281.5° + j \sin 281.5°)$; $122 - 597j$

37. $2(\cos 30° + j \sin 30°)$; $2(\cos 210° + j \sin 210°)$

39. $-0.364 + 1.67j$, $-1.26 - 1.15j$, $1.63 - 0.520j$

41. $1.10 + 0.455j$, $-1.10 - 0.455j$ **43.** $1, -1, j, -j$

45. $3j, -\dfrac{3}{2}(\sqrt{3} + j), \dfrac{3}{2}(\sqrt{3} - j)$

47. $1.62 + 1.18j$, $-0.618 + 1.90j$, -2, $-0.618 - 0.190j$, $1.62 - 1.18j$

49. $\left[\dfrac{1}{2}(1 - j\sqrt{3})\right]^3 = \dfrac{1}{8}\left[1 - 3(j\sqrt{3}) + 3(j\sqrt{3})^2\right.$

$\left. - (j\sqrt{3})^3\right] = \dfrac{1}{8}\left[1 - 3j\sqrt{3} - 9 + 3j\sqrt{3}\right]$

$= \dfrac{1}{8}(-8) = -1$

51. $-1, \dfrac{1}{2} + j\dfrac{\sqrt{3}}{2}, \dfrac{1}{2} - j\dfrac{\sqrt{3}}{2}$ **53.** $p = 47.9\underline{/40.5°}$ watts

55. $43.3\underline{/165°}$, 43.3 V

Exercises 12.7, page 366

1. $V_R = 24.0$ V, $V_L = 32.0$ V, $V_{RL} = 40.0$ V, $\theta = 53.1°$ (voltage leads current)   **3.** 12.9 V   **5.** (a) $2850 \ \Omega$ (b) $37.9°$

(c) 16.4 V   **7.** (a) $14.6 \ \Omega$ (b) $-90.0°$ **9.** (a) $47.8 \ \Omega$ (b) $19.8°$

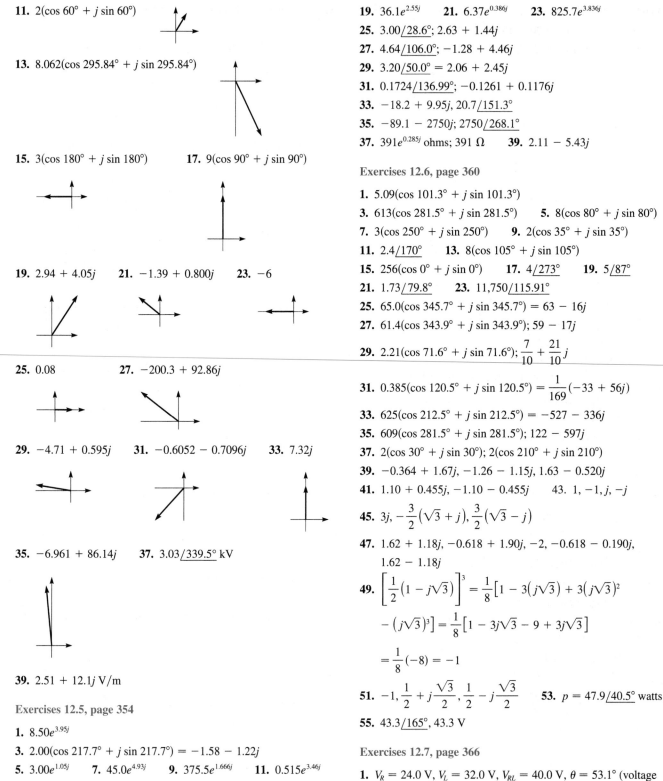

11. 38.0 V   **13.** 54.5 Ω, −62.3°   **15.** 0.682 H

17. 2.08 × 10^5 Hz   **19.** 1.30 × 10^{-11} F = 13.0 pF

21. 1.02 mW   **23.** 21.4 − 33.9*j*

Review Exercises for Chapter 12, page 368

1. 10 − *j*   **3.** 6 + 2*j*   **5.** 9 + 2*j*   **7.** −12 + 66*j*

9. $\frac{1}{85}$(21 + 18*j*)   **11.** −2 − 3*j*   **13.** $\frac{1}{10}$(−12 + 9*j*)

15. $\frac{1}{5}$(13 + 11*j*)   **17.** $x = -\frac{2}{3}, y = -2$

19. $x = \frac{10}{13}, y = \frac{11}{13}$ **21.** 3 + 11*j*   **23.** 4 + 8*j*

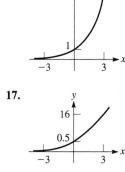

25. 1.41(cos 315° + *j* sin 315°) = $1.41e^{5.50j}$

27. 7.28(cos 254.1° + *j* sin 254.1°) = $7.28e^{4.43j}$

29. 4.67(cos 76.8° + *j* sin 76.8°); $4.67e^{1.34j}$

31. 5000$\underline{/0°}$, $5000e^{0j}$ **33.** −1.41 − 1.41*j*

35. −2.789 + 4.162*j*   **37.** 0.19 − 0.59*j*   **39.** 26.31 − 6.427*j*

41. 1.94 + 0.495*j*   **43.** −728.1 + 1017*j*

45. 15(cos 84° + *j* sin 84°)   **47.** 20$\underline{/263°}$

49. 8(cos 59° + *j* sin 59°)   **51.** 14.29$\underline{/133.61°}$

53. 1.26$\underline{/59.7°}$ **55.** 9682$\underline{/249.52°}$

57. 1024(cos 160° + *j* sin 160°)   **59.** 27$\underline{/331.5°}$

61. 32(cos 270° + *j* sin 270°) = −32*j*

63. $\frac{625}{2}$(cos 270° + *j* sin 270°) = $-\frac{625}{2}j$

65. $1 + j\sqrt{3}, -2, 1 - j\sqrt{3}$

67. 0.383 + 0.924*j*, −0.924 + 0.383*j*, −0.383 − 0.924*j*,
    0.924 − 0.383*j*   **69.** 40 + 9*j*, 41(cos 12.7° + *j* sin 12.7°)

71. −15.0 − 10.9*j*, 18.5(cos 216.0° + *j* sin 216.0°)

73. 15 − 16*j*   **75.** −*j*, −2*j*   **77.** $x^2 - 4x + 5 = 0$

79. $\frac{1}{0.6 - 0.8j} = \frac{0.6 + 0.8j}{(0.6 - 0.8j)(0.6 + 0.8j)} = 0.6 + 0.8j$

81. 60 V   **83.** −21.6°   **85.** 22.9 Hz

87. 550$\underline{/53°}$ N   **89.** $\frac{u - j\omega n}{u^2 + \omega^2 n^2}$

91. $e^{j\pi} = \cos \pi + j \sin \pi = -1$

Exercises 13.1, page 372

1. $-\frac{1}{4}$ **3.** (a) Yes (b) Yes

5. (a) No (base cannot be negative) (b) Yes   **7.** 3

9. $\frac{1}{81}$ **11.** $\frac{1}{27}$

13.
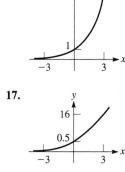

15.

17.
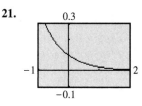

19.
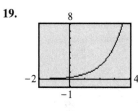

21.
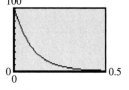

23.
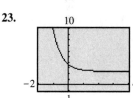

25. −0.7667, 2, 4   **27.** $303.88   **29.** 0.0012 mA

31.

Exercises 13.2, page 376

1. $32^{4/5} = 16$ in logarithmic form is $\frac{4}{5} = \log_{32} 16$.

3. If $x = 16, y = \log_4 16$ means $y = 2$, since $4^2 = 16$. If $x = \frac{1}{16}$, $y = \log_4\left(\frac{1}{16}\right)$ means $y = -2$, since $4^{-2} = \frac{1}{16}$.

5. $\log_3 27 = 3$ **7.** $\log_4 256 = 4$ **9.** $\log_4\left(\frac{1}{16}\right) = -2$

11. $\log_2\left(\frac{1}{64}\right) = -6$ **13.** $\log_8 2 = \frac{1}{3}$

15. $\log_{1/4}\left(\frac{1}{16}\right) = 2$ **17.** $81 = 3^4$ **19.** $9 = 9^1$

21. $5 = 25^{1/2}$ **23.** $3 = 243^{1/5}$ **25.** $0.1 = 10^{-1}$

27. $16 = (0.5)^{-4}$ **29.** 2   **31.** −2   **33.** 343

35. $\dfrac{9}{4}$ **37.** 9   **39.** $\dfrac{1}{64}$ **41.** 0.2   **43.** −4

45.

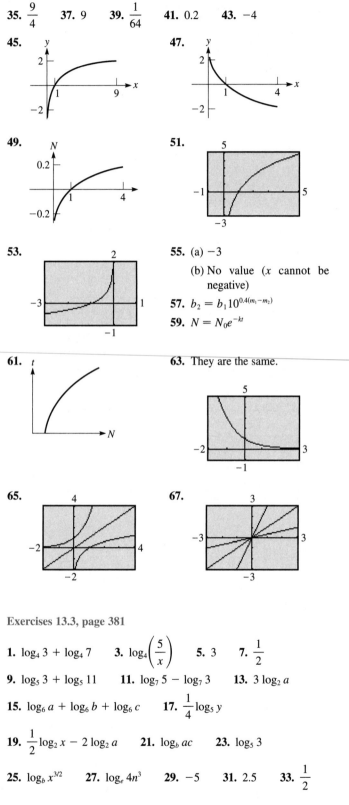

47.

49.

51.

53.

55. (a) −3
(b) No value (x cannot be negative)
57. $b_2 = b_1 10^{0.4(m_1 - m_2)}$
59. $N = N_0 e^{-kt}$

61.

63. They are the same.

65.

67.

Exercises 13.3, page 381

1. $\log_4 3 + \log_4 7$ **3.** $\log_4\left(\dfrac{5}{x}\right)$ **5.** 3   **7.** $\dfrac{1}{2}$

9. $\log_5 3 + \log_5 11$ **11.** $\log_7 5 - \log_7 3$ **13.** $3 \log_2 a$

15. $\log_6 a + \log_6 b + \log_6 c$ **17.** $\dfrac{1}{4}\log_5 y$

19. $\dfrac{1}{2}\log_2 x - 2\log_2 a$ **21.** $\log_b ac$ **23.** $\log_5 3$

25. $\log_b x^{3/2}$ **27.** $\log_e 4n^3$ **29.** −5   **31.** 2.5   **33.** $\dfrac{1}{2}$

35. $\dfrac{3}{4}$ **37.** $2 + \log_3 2$ **39.** $-1 - \log_2 3$

41. $\dfrac{1}{2}(1 + \log_3 2)$ **43.** $3 + \log_{10} 3$ **45.** $y = 2x$

47. $y = \dfrac{3x}{5}$ **49.** $y = \dfrac{49}{x^3}$ **51.** $y = 2(2ax)^{1/5}$ **53.** $y = \dfrac{2}{x}$

55. $y = \dfrac{1}{25}x^{2/\log_5 3}$ **57.** $\log_{10} x + \log_{10} 3 = \log_{10} 3x$

59. $y = \log_e(e^2 x) = 2\log_e e + \log_e x = 2 + \log_e x$

61. $D = ae^{cr^2 - br}$

63. $T = 65e^{-0.41t}$

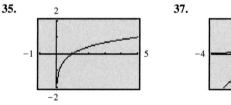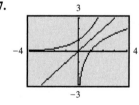

Exercises 13.4, page 384

1. −0.4372   **3.** 2.754   **5.** 6.966   **7.** −0.2787
9. −0.0104   **11.** 1.219   **13.** 27 400   **15.** 0.049 60
17. 2000.4   **19.** 0.005 788 2   **21.** 85.5   **23.** 7.37×10^{101}
25. $0.6990 + 0.8451 = 1.5441$ **27.** $1.9085 = 1.9085$
29. 8.9542   **31.** −13.886   **33.** 10^8 K   **35.** 0.45
37. 15.2 dB   **39.** $2^{400} = 2.58 \times 10^{120}$

Exercises 13.5, page 387

1. 5.298   **3.** 3.258   **5.** 0.4460   **7.** −4.916 23   **9.** 1.92
11. 4.806   **13.** 1.795   **15.** 3.940   **17.** 0.3322
19. −0.008 335   **21.** −17.390 66   **23.** 1.6549
25. −0.164 13   **27.** 8.935   **29.** 1.0085   **31.** 0.4757
33. 6.1993×10^{-11}

35.

37.

39. $1.6094 + 2.0794 = 3.6889$ **41.** $4.3944 = 4.3944$
43. 3   **45.** 2.45×10^9 Hz   **47.** 8.155%   **49.** 0.384 s
51. 21.7 s

Exercises 13.6, page 390

1. −0.535   **3.** 4   **5.** −0.748   **7.** 0.587   **9.** 0.285
11. 0.982   **13.** $0, \dfrac{\ln 0.6}{\ln 2} = -0.737$ **15.** $\dfrac{1}{4}$ **17.** 1.649

19. 3   **21.** −0.162   **23.** 250   **25.** 4   **27.** 2

29. 1.42   **31.** −0.1042   **33.** 0.2031   **35.** 0.9739

37. 10.87   **39.** ±0.9624   **41.** 28.0   **43.** 8.4 min

45. 1.72×10^{-5} **47.** $10^{8.3} = 2.0 \times 10^{8}$ **49.** $c = 15e^{-0.20t}$

51. $P = P_0(0.999)^t$ **53.** $x = 3.353$

55. ±3.7 m

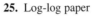

Exercises 13.7, page 394

1.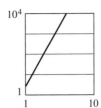

3.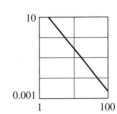

5.

7.

9.

11.

13.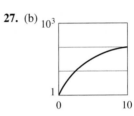

15.

17.

19. Semilog paper

21. Log-log paper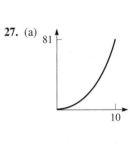

23. Semilog paper

25. Log-log paper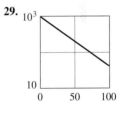

27. (a)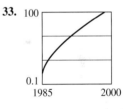

27. (b)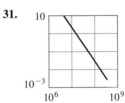

29.

31.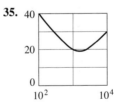

33.

35.

37.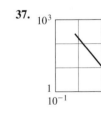

39.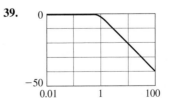

Review Exercises for Chapter 13, page 396

1. 10 000   **3.** $\frac{1}{5}$ **5.** 6   **7.** $\frac{5}{3}$ **9.** 6   **11.** 100

13. $\log_3 2 + \log_3 x$ **15.** $2 \log_3 t$ **17.** $2 + \log_2 7$

19. $2 - \log_3 x$ **21.** $1 + \frac{1}{2} \log_4 3$ **23.** $3 + 4 \log_{10} x$

25. $y = \dfrac{4}{x}$ **27.** $y = (8x)^{1/\log_2 3}$ **29.** $y = \dfrac{15}{x}$ **31.** $y = e^{2/3}x$

33. $y = 8x^3$ **35.** $y = \dfrac{x}{\ln 2} = 1.44x$

37. 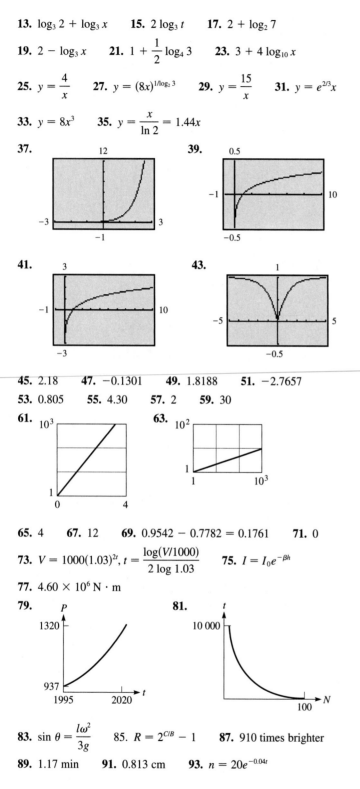 **39.**

41. **43.**

45. 2.18 **47.** -0.1301 **49.** 1.8188 **51.** -2.7657

53. 0.805 **55.** 4.30 **57.** 2 **59.** 30

61. **63.**

65. 4 **67.** 12 **69.** $0.9542 - 0.7782 = 0.1761$ **71.** 0

73. $V = 1000(1.03)^{2t}, \; t = \dfrac{\log(V/1000)}{2 \log 1.03}$ **75.** $I = I_0 e^{-\beta h}$

77. $4.60 \times 10^6 \, \text{N} \cdot \text{m}$

79. **81.**

83. $\sin \theta = \dfrac{l\omega^2}{3g}$ **85.** $R = 2^{C/B} - 1$ **87.** 910 times brighter

89. 1.17 min **91.** 0.813 cm **93.** $n = 20e^{-0.04t}$

95.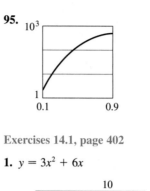

Exercises 14.1, page 402

1. $y = 3x^2 + 6x$ **3.** $x = -1.2, y = 0.3;$
$x = 0.8, y = 2.3$

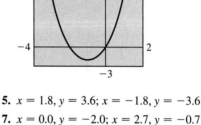

5. $x = 1.8, y = 3.6; x = -1.8, y = -3.6$

7. $x = 0.0, y = -2.0; x = 2.7, y = -0.7$

9. $x = 1.5, y = 0.2$ **11.** $x = 2.8, y = 10.7$

13. $x = 1.1, y = 2.8; x = -1.1, y = 2.8;$
$x = 2.4, y = -1.8; x = -2.4, y = -1.8$

15. No real solutions

17. $x = -2.8, y = -1.0; x = 2.8, y = 1.0;$
$x = 2.8, y = -1.0; x = -2.8, y = 1.0$

19. No real solutions

21. $x = 0.0, y = 0.0; x = 0.9, y = 0.8$

23. $x = -1.1, y = 3.1; x = 1.8, y = 0.2$

25. $x = 1.0, y = 0.0$ **27.** $x = 3.6, y = 1.0$

29. $x = -2.06, y = 4.23; x = 1.06, y = 1.12$

31. 4.9 km N, 1.6 km E **33.** 2.2 A, 0.9 A **35.** No

Exercises 14.2, page 406

1. $x = 2, y = 0; x = \dfrac{10}{3}, y = -\dfrac{8}{3}$

3. $x = -\sqrt{6}, y = -\sqrt{3}; x = -\sqrt{6}, y = \sqrt{3};$
$x = \sqrt{6}, y = -\sqrt{3}; x = \sqrt{6}, y = \sqrt{3}$

5. $x = 0, y = 1; x = 1, y = 2$

7. $x = -\dfrac{19}{5}, y = \dfrac{17}{5}; x = 5, y = -1$ **9.** $x = 1, y = 0$

11. $x = \dfrac{2}{7}(3 + \sqrt{2}), y = \dfrac{2}{7}(-1 + 2\sqrt{2});$

$x = \dfrac{2}{7}(3 - \sqrt{2}), y = \dfrac{2}{7}(-1 - 2\sqrt{2})$

13. $w = 1, h = 1$ **15.** $x = \dfrac{2}{3}, y = \dfrac{9}{2}; x = -3, y = -1$

17. $x = -5, y = 25; x = 5, y = 25$

19. $x = 1, y = 2; x = -1, y = 2$

21. $D = 1, R = 0; D = -1, R = 0; D = \frac{1}{2}\sqrt{6}, R = \frac{1}{2};$

$D = -\frac{1}{2}\sqrt{6}, R = \frac{1}{2}$

23. $x = \sqrt{19}, y = \sqrt{6}; x = \sqrt{19}, y = -\sqrt{6};$
$x = -\sqrt{19}, y = \sqrt{6}; x = -\sqrt{19}, y = -\sqrt{6}$

25. $x = -5, y = -2; x = -5, y = 2; x = 5, y = -2; x = 5, y = 2$

27. $x = -3, y = 2; x = -1, y = -2$

29. $x = 50$ km, $h = 25$ km **31.** 1.5 cm, 1.4 cm

33. 2.19 m, 0.21 m **35.** 12, 10 **37.** 28 cm by 20 cm

39. 130 km/h

Exercises 14.3, page 409

1. $\pm\frac{1}{2}j\sqrt{2}, \pm 2$ **3.** $-3, -2, 2, 3$ **5.** $-\frac{1}{2}, \frac{1}{4}$

7. $-\frac{1}{2}, \frac{1}{2}, \frac{1}{6}j\sqrt{6}, -\frac{1}{6}j\sqrt{6}$ **9.** $1, \frac{25}{4}$ **11.** $\frac{64}{729}, 1$

13. $-27, 125$ **15.** 81 **17.** 5 **19.** $-2, -1, 3, 4$

21. 18 **23.** $1, -1, j\sqrt{2}, -j\sqrt{2}$ **25.** 0 **27.** $\pm 2, \pm 4$

29. 1, 4 **31.** $R_1 = 2.62\ \Omega, R_2 = 1.62\ \Omega$ **33.** 0.610

35. 40.4 cm, 55.4 cm

Exercises 14.4, page 413

1. $\frac{13}{12}$ **3.** 10 **5.** 12 **7.** 2 **9.** $\frac{2}{3}$ **11.** -1

13. 32 **15.** 4, 9 **17.** ± 5

19. 12 (Extraneous root introduced in squaring both sides of $\sqrt{x+4} = x - 8$.)

21. 16 **23.** $\frac{1}{2}$ **25.** 7, -1 **27.** 0 **29.** 5 **31.** 25

33. 258 **35.** 4 **37.** 4 **39.** $L = \frac{1}{4\pi^2 f^2 C}$

41. $\frac{k^2}{2n(1-k)}$ **43.** $p = 4.83$ cm, $q = 23.3$ cm **45.** 9.2 km

47. 2.7 km

Review Exercises for Chapter 14, page 414

1. $x = -0.9, y = 3.5; x = 0.8, y = 2.6$

3. $x = 2.0, y = 0.0; x = 1.6, y = 0.6$

5. $x = 0.8, y = 1.6; x = -0.8, y = 1.6$

7. $x = -2, y = 7; x = 2, y = 7$

9. $x = 0.0, y = 0.0; x = 2.4, y = 0.9$

11. $x = 0, y = 0; x = 2, y = 16$

13. $L = \sqrt{2}, R = 1; L = -\sqrt{2}, R = 1$

15. $u = \frac{1}{12}(1 + \sqrt{97}), v = \frac{1}{18}(5 - \sqrt{97});$
$u = \frac{1}{12}(1 - \sqrt{97}), v = \frac{1}{18}(5 + \sqrt{97})$

17. $x = 7, y = 5; x = 7, y = -5; x = -7, y = 5;$
$x = -7, y = -5$

19. $x = -2, y = 2; x = \frac{2}{3}, y = \frac{10}{9}$

21. $-4, -2, 2, 4$ **23.** 1, 16 **25.** $\frac{1}{3}, -\frac{1}{7}$ **27.** $\frac{25}{4}$

29. $\sqrt{3}, -\sqrt{3}, \frac{1}{2}j\sqrt{3}, -\frac{1}{2}j\sqrt{3}$ **31.** 6 **33.** 8 **35.** $\frac{9}{16}$

37. $\frac{1}{3}(11 - 4\sqrt{15})$ **39.** 2 **41.** 4 **43.** 5 **45.** $-1, 2$

47. $x = -2, y = -3; x = 3, y = 2$

49. $l = \frac{1}{2}\left(-1 + \sqrt{1 + 16\pi^2 L^2/h^2}\right)$

51. $m = \frac{1}{2}\left(-y \pm \sqrt{2s^2 - y^2}\right)$ **53.** 0.40 s, 0.80 s **55.** 70 m

57. $Z = 1.09\ \Omega, X = 0.738\ \Omega$ **59.** 34 mm, 52 mm

61. 0.353 cm, 6.29 cm **63.** 42.5 km/h, 52.1 km/h

Exercises 15.1, page 421

1. -25 **3.** Coefficients $1\ 0\ 0\ -4\ 11, R = -26$ **5.** 0

7. -40 **9.** -4 **11.** 51 **13.** -28 **15.** 14 **17.** Yes

19. No **21.** No **23.** $x^2 + 3x + 2, R = 0$

25. $x^2 + x - 4, R = 8$

27. $p^5 + 2p^4 + 4p^3 + 2p^2 + 2p + 4, R = 2$

29. $x^6 + 2x^5 + 4x^4 + 8x^3 + 16x^2 + 32x + 64, R = 0$

31. $x^3 + x^2 + 2x + 1, R = 0$ **33.** No **35.** No **37.** Yes

39. No **41.** Yes **43.** Yes

45. $(4x^3 + 8x^2 - x - 2) \div (2x - 1) = 2x^2 + 5x + 2$; no, because the coefficient of x in $2x - 1$ is 2, not 1.

47. -7 **49.** $x^2 - (3 + j)x + 3j, R = 0$

51. Yes, because both functions cross the x-axis at the same values.

53. (a) No (b) Yes **55.** Yes

Exercises 15.2, page 426

(Note: Unknown roots listed)

1. $1, 1, 1, -1, -1$ **3.** $-3, 2, 3$ **5.** $-3, -3, 2j, -2j$

7. $-2, 3$ **9.** $-2, -2$ **11.** $-j, -\frac{2}{3}$ **13.** $2j, -2j$

15. $-1, -3$ **17.** $-2, 1$ **19.** $1 - j, \frac{1}{2}, -2$ **21.** $j, -j$

23. $-j, -1, 1$ **25.** $-2j, 3, -3$

27. If j is a root, $-j$ must also be a root, and $-j$ is not one of the roots.

Exercises 15.3, page 432

1. No more than two positive roots and three negative roots

3. $1, -1, -2$ **5.** $2, -1, -3$

7. $1, \dfrac{1}{6}(-3 + j\sqrt{15}), \dfrac{1}{6}(-3 - j\sqrt{15})$ **9.** $\dfrac{1}{3}, -3, -1$

11. $-2, -2, 2 \pm \sqrt{3}$ **13.** $-3, -1, 2, 4$

15. $-3, -\dfrac{2}{3}, -\dfrac{1}{2}, \dfrac{1}{2}$ **17.** $2, 2, -1, -1, -3$

19. $\dfrac{1}{2}, 1, 1, j, -j$ **21.** $-1.86, 0.68, 3.18$ **23.** $-3.24, 1.24$

25. 0.59 **27.** -0.77 **29.** $-\dfrac{3}{4}, \sqrt{5}, -\sqrt{5}$

31. 1.8 s, 4.5 s   **33.** $0, L$ **35.** 0.9 s, 4.1 s

37. 1.23 cm or 2.14 cm   **39.** 3.0 mm, 4.0 mm; 5.0 mm, 6.0 mm

41. $2\,\Omega, 3\,\Omega, 6\,\Omega$ **43.** Two positive, one negative; zero positive, one negative, two nonreal complex

Review Exercises for Chapter 15, page 433

1. 1 **3.** -107 **5.** Yes   **7.** No

9. $x^2 + 4x + 10, R = 11$ **11.** $2x^2 - 7x + 10, R = -17$

13. $x^3 - 3x^2 - 4, R = -4$

15. $2m^4 + 10m^3 + 4m^2 + 21m + 105, R = 516$ **17.** No

19. Yes   **21.** (unlisted roots) $\dfrac{1}{2}(-5 \pm \sqrt{17})$

23. (unlisted roots) $\dfrac{1}{3}(-1 \pm j\sqrt{14})$

25. (unlisted roots) $-1 + j\sqrt{2}, -1 - j\sqrt{2}$

27. (unlisted roots) $-j, \dfrac{1}{2}, -\dfrac{3}{2}$ **29.** (unlisted roots) $2, -2$

31. (unlisted roots) $-2 - j, \dfrac{1}{2}(-1 \pm j\sqrt{3})$ **33.** $1, 2, -4$

35. $-\dfrac{1}{2}, 1, 1$ **37.** $-1, -\dfrac{1}{2}, \dfrac{5}{3}$

39. $\dfrac{1}{2}, \dfrac{3}{2}, -1 + \sqrt{2}, -1 - \sqrt{2}$ **41.** $1, 3, 5$ **43.** 2 or 4

45. Use k in synthetic division, $k = 4$.

47. $1, 0.4, 1.5, -0.6$ (last three are irrational)

49. 1.91 **51.** April   **53.** 0.75 cm   **55.** 2 cm

57. 5.1 m, 8.3 m   **59.** $h = 2.25$ m, $w = 1.20$ m

Exercises 16.1, page 439

1. $\begin{bmatrix} 5 & 7 & -1 & 9 \\ 6 & 4 & 1 & 12 \end{bmatrix}$ **3.** $a = 1, b = -3, c = 4, d = 7$

5. $x = -2, y = 5, z = -9, r = 48, s = 4, t = -1$

7. $C = 3, D = 2, E = -2$ **9.** Elements cannot be equated; different number of rows.

11. $\begin{bmatrix} 1 & 10 \\ 0 & 2 \end{bmatrix}$ **13.** $\begin{bmatrix} -5 & 0 \\ 11 & 71 \\ 11 & -5 \end{bmatrix}$ **15.** $\begin{bmatrix} 0 & 9 & -13 & 3 \\ 6 & -7 & 7 & 0 \end{bmatrix}$

17. Cannot be added   **19.** $\begin{bmatrix} -1 & 13 & -20 & 3 \\ 8 & -13 & 6 & 2 \end{bmatrix}$

21. $\begin{bmatrix} -3 & -6 & 5 & -6 \\ -6 & -4 & -17 & 6 \end{bmatrix}$ **23.** $\begin{bmatrix} 4 & -16 & 28 & 0 \\ -8 & 24 & 4 & -8 \end{bmatrix}$

25. Elements cannot be added; different number of columns.

27. $A + B = B + A = \begin{bmatrix} 3 & 1 & 0 & 7 \\ 5 & -3 & -2 & 5 \\ 10 & 10 & 8 & 0 \end{bmatrix}$

29. $-(A - B) = B - A = \begin{bmatrix} 5 & -3 & -6 & -7 \\ 5 & 3 & 0 & -3 \\ -8 & 12 & 8 & 4 \end{bmatrix}$

31. $v_w = 31.0$ km/h, $v_p = 249$ km/h   **33.** $\begin{bmatrix} 24 & 18 & 0 & 0 \\ 15 & 12 & 9 & 0 \\ 0 & 9 & 15 & 18 \end{bmatrix}$

35. $B = \begin{bmatrix} 15 & 25 & 10 & 25 \\ 10 & 10 & 10 & 45 \end{bmatrix}, J = \begin{bmatrix} 15 & 30 & 3 & 3 \\ 0 & 100 & 2 & 2 \end{bmatrix}$

Exercises 16.2, page 443

1. $\begin{bmatrix} 5 & 6 & 7 & -10 \\ -9 & 0 & -3 & 12 \\ 1 & 12 & 11 & -8 \end{bmatrix}$ **3.** $[-8 \quad -12]$ **5.** $\begin{bmatrix} 29 \\ -29 \end{bmatrix}$

7. $\begin{bmatrix} -\dfrac{1}{2} & \dfrac{111}{4} \\ -\dfrac{121}{8} & -\dfrac{101}{2} \\ \dfrac{1}{10} & 22 \end{bmatrix}$ **9.** $\begin{bmatrix} 33 & -22 \\ 31 & -12 \\ 15 & 13 \\ 50 & -41 \end{bmatrix}$ **11.** $\begin{bmatrix} -15 & 15 & -26 \\ 8 & 5 & -13 \end{bmatrix}$

13. $\begin{bmatrix} -49.43 & 55.20 \\ -53.02 & 79.16 \end{bmatrix}$

15. $AB = [40], BA = \begin{bmatrix} -1 & 3 & -8 \\ 5 & -15 & 40 \\ 7 & -21 & 56 \end{bmatrix}$

17. $AB = \begin{bmatrix} -5 \\ 10 \end{bmatrix}$, BA not defined   **19.** $AI = IA = A$

21. $AI = IA = A$ **23.** $B = A^{-1}$ **25.** $B = A^{-1}$ **27.** Yes

29. No   **31.** $\begin{bmatrix} -1 & 0 \\ 0 & -1 \end{bmatrix}\begin{bmatrix} -1 & 0 \\ 0 & -1 \end{bmatrix} = \begin{bmatrix} 1 & 0 \\ 0 & 1 \end{bmatrix}$

33. $A^2 - I = (A + I)(A - I) = \begin{bmatrix} 15 & 28 \\ 21 & 36 \end{bmatrix}$

35. $\begin{bmatrix} 0 & -j \\ j & 0 \end{bmatrix}\begin{bmatrix} 0 & -j \\ j & 0 \end{bmatrix} = \begin{bmatrix} 1 & 0 \\ 0 & 1 \end{bmatrix}$

37. $v_2 = v_1, i_2 = -v_1/R + i_1$ **39.** Ellipse

Exercises 16.3, page 448

1. $\begin{bmatrix} -\frac{5}{2} & \frac{3}{2} \\ -2 & 1 \end{bmatrix}$
3. $\begin{bmatrix} -2 & -\frac{5}{2} \\ -1 & -1 \end{bmatrix}$
5. $\begin{bmatrix} -\frac{1}{3} & \frac{1}{6} \\ \frac{2}{15} & \frac{1}{30} \end{bmatrix}$

7. $\begin{bmatrix} \frac{3}{4} & \frac{1}{2} \\ -\frac{1}{4} & 0 \end{bmatrix}$
9. $\begin{bmatrix} -\frac{8}{283} & -\frac{9}{566} \\ \frac{13}{1415} & \frac{5}{283} \end{bmatrix}$
11. $\begin{bmatrix} -3 & 2 \\ 2 & -1 \end{bmatrix}$

13. $\begin{bmatrix} -\frac{1}{2} & -2 \\ \frac{1}{2} & 1 \end{bmatrix}$
15. $\begin{bmatrix} \frac{2}{9} & -\frac{5}{9} \\ \frac{1}{9} & \frac{2}{9} \end{bmatrix}$
17. $\begin{bmatrix} \frac{3}{8} & \frac{1}{16} \\ -\frac{1}{4} & \frac{1}{8} \end{bmatrix}$

19. $\begin{bmatrix} -18 & -7 & 5 \\ -3 & -1 & 1 \\ -5 & -2 & 1 \end{bmatrix}$
21. $\begin{bmatrix} 2 & 4 & \frac{7}{2} \\ -1 & -2 & -\frac{3}{2} \\ 1 & 1 & \frac{1}{2} \end{bmatrix}$

23. $\begin{bmatrix} \frac{5}{2} & -2 & -2 \\ -1 & 1 & 1 \\ \frac{7}{4} & -\frac{3}{2} & -1 \end{bmatrix}$
25. $\begin{bmatrix} 0.3 & -0.4 \\ 0.05 & 0.1 \end{bmatrix}$

27. $\begin{bmatrix} 2 & 4 & 3.5 \\ -1 & -2 & -1.5 \\ 1 & 1 & 0.5 \end{bmatrix}$
29. $\begin{bmatrix} 2.5 & -2 & -2 \\ -1 & 1 & 1 \\ 1.75 & -1.5 & -1 \end{bmatrix}$

31. $\begin{bmatrix} 1 & 2 & 3 & 1 \\ 1 & 3 & 3 & 2 \\ 2 & 4 & 3 & 3 \\ 1 & 1 & 1 & 1 \end{bmatrix}$

33. $\begin{bmatrix} 2.537 & -0.950 & 0.159 & 0.470 \\ -0.213 & 0.687 & 0.290 & -0.272 \\ -1.006 & 0.870 & 0.496 & -0.532 \\ 0.113 & -0.123 & -0.033 & 0.385 \end{bmatrix}$

35. $\begin{vmatrix} 1 & 1 \\ 1 & 1 \end{vmatrix} = 0$ means the inverse does not exist.

37. $\dfrac{1}{ad-bc}\begin{bmatrix} ad-bc & -ba+ab \\ cd-dc & -bc+ad \end{bmatrix} = \begin{bmatrix} 1 & 0 \\ 0 & 1 \end{bmatrix}$

39. $v_1 = (a_{22}i_1 - a_{12}i_2)/(a_{11}a_{22} - a_{12}a_{21})$
$v_2 = (-a_{21}i_1 + a_{11}i_2)/(a_{11}a_{22} - a_{12}a_{21})$

Exercises 16.4, page 453

1. $x=2, y=-3$ **3.** $x=\frac{1}{2}, y=3$ **5.** $x=1, y=3$

7. $x=2, y=-2$ **9.** $x=-1, y=0, z=3$

11. $x=-\frac{3}{2}, y=-2$ **13.** $x=1.6, y=-2.5$

15. $x=2, y=-4, z=1$ **17.** $x=2, y=-\frac{1}{2}, z=3$

19. $x=1, y=-2, z=-3$ **21.** $u=2, v=-5, w=4$

23. $x=2, y=3, z=-2, t=1$

25. $v=2, w=-1, x=\frac{1}{2}, y=\frac{3}{2}, z=-3$

27. $x=\pm 2, y=-2$ **29.** $A=118$ N, $B=186$ N

31. 6.4 L, 1.6 L, 2.0 L

Review Exercises for Chapter 16, page 455

1. $a = 4, b = -1$

3. $x=2, y=-3, z=\frac{5}{2}, a=-1, b=-\frac{7}{2}, c=\frac{1}{2}$

5. $x=-1, y=\frac{1}{2}, a=-\frac{1}{2}, b=-\frac{3}{2}$ **7.** $\begin{bmatrix} 1 & -3 \\ 8 & -5 \\ -8 & -2 \\ 3 & -10 \end{bmatrix}$

9. $\begin{bmatrix} -3 & 3 \\ 0 & -7 \\ 2 & -2 \\ -1 & -4 \end{bmatrix}$
11. $\begin{bmatrix} 7 & -6 \\ -4 & 20 \\ -1 & 6 \\ 1 & 15 \end{bmatrix}$
13. $\begin{bmatrix} 0 & 0 \\ 0 & 0 \end{bmatrix}$

15. $\begin{bmatrix} 0.34 & 0.11 & -0.05 \\ 0.02 & -0.08 & 0.10 \\ -0.01 & -0.17 & 0.20 \end{bmatrix}$
17. $\begin{bmatrix} -2 & \frac{5}{2} \\ -1 & 1 \end{bmatrix}$

19. $\begin{bmatrix} \frac{40}{3} & \frac{5}{3} \\ -\frac{20}{3} & \frac{35}{3} \end{bmatrix}$
21. $\begin{bmatrix} 11 & 10 & 3 \\ -4 & -4 & -1 \\ 3 & 3 & 1 \end{bmatrix}$

23. $\begin{bmatrix} \frac{1}{2} & -\frac{1}{2} & -1 \\ -3 & 2 & 1 \\ -4 & 3 & 2 \end{bmatrix}$
25. $x=-3, y=1$

27. $x=10, y=-15$ **29.** $u=-1, v=-3, w=0$

31. $x=1, y=\frac{1}{2}, z=-\frac{1}{3}$ **33.** $x=3, y=1, z=-1$

35. $x=1, y=2, z=-3, t=1$

37. $x=-\frac{1}{3}, y=3, z=\frac{2}{3}, t=-4$

39. $r=\frac{1}{2}, s=-7, t=-\frac{3}{4}, u=5, v=\frac{3}{2}$

41. $\begin{bmatrix} 1 & 0 \\ 15 & 16 \end{bmatrix}, \begin{bmatrix} 1 & 0 \\ 63 & 64 \end{bmatrix}, \begin{bmatrix} 1 & 0 \\ 255 & 256 \end{bmatrix}$

43. $B^3 = \begin{bmatrix} 1 & 0 & 0 \\ 0 & 1 & 0 \\ 0 & 0 & 1 \end{bmatrix}$ **45.** $N^{-1} = -N = \begin{bmatrix} 0 & 1 \\ -1 & 0 \end{bmatrix}$

47. $\begin{bmatrix} n & 1+n \\ 1-n & -n \end{bmatrix}\begin{bmatrix} n & 1+n \\ 1-n & -n \end{bmatrix}$
$= \begin{bmatrix} n^2+(1-n^2) & (n+n^2)-(n+n^2) \\ (n-n^2)-(n-n^2) & (1-n^2)+n^2 \end{bmatrix}$

49. $(A+B)(A-B) = \begin{bmatrix} -6 & 2 \\ 4 & 2 \end{bmatrix}, A^2-B^2 = \begin{bmatrix} -10 & -4 \\ 8 & 6 \end{bmatrix}$

51. $\begin{bmatrix} \frac{1}{2} & \frac{1}{3} \\ 0 & \frac{1}{6} \end{bmatrix} = \frac{1}{2}\begin{bmatrix} 1 & \frac{2}{3} \\ 0 & \frac{1}{3} \end{bmatrix}$ **53.** $R_1 = 4\,\Omega, R_2 = 6\,\Omega$

55. $F = 303$ N, $T = 175$ N

57. 0.20 h after police pass intersection   **59.** 30 g, 50 g, 20 g

61. $\begin{bmatrix} 12\,000 & 24\,000 & 4\,000 \\ 15\,000 & 8\,000 & 30\,000 \end{bmatrix}$

$+ \begin{bmatrix} 15\,000 & 12\,000 & 2\,000 \\ 20\,000 & 3\,000 & 22\,000 \end{bmatrix}$

$= \begin{bmatrix} 27\,000 & 36\,000 & 6\,000 \\ 35\,000 & 11\,000 & 52\,000 \end{bmatrix}$

63. $(R_1 + R_2)i_1 - R_2 i_2 = 6$
$-R_2 i_1 + (R_1 + R_2)i_2 = 0$

Exercises 17.1, page 462

1. $x + 1 < 0$ is true for all values of x less than -1; $x < -1$.

3. $-2 > -4$, multiplied by -3 gives $6 < 12$; divided by -2 gives $1 < 2$.

5. $7 < 12$ **7.** $20 < 45$ **9.** $-4 > -9$ **11.** $16 < 81$

13. $x > -2$ **15.** $x \le 4$ **17.** $1 < x < 7$

19. $x < -9$ or $x \ge -4$ **21.** $x < 1$ or $3 < x \le 5$

23. $-2 < x < 2$ or $3 \le x < 4$

25. x is greater than 0 and less than or equal to 2.

27. x is less than -1, or greater than or equal to 1 and less than 2.

29. 3

31. 1 3

33. 0 5

35. -3 5

37. -1 1 4

39. -3 -1 1 3

41. -3

43. 3 5 8 10

45. Absolute inequality **47.** Yes, if $a - b > 0$; no, if $a - b < 0$

49. $2000 \le M \le 1\,000\,000$ **51.** $29\,000 < v < 40\,000$ km/h

2 000   1 000 000

0   29 000   40 000

53. $0 < n \le 2565$ steps **55.** $E = 0$ for $0 < r < a$
$E = k/r^2$ for $r \ge a$

Exercises 17.2, page 467

1. $x \le 3$ **3.** $1 < x < \dfrac{9}{2}$ **5.** $x > -1$

1 $\frac{9}{2}$ -1

7. $x < 6$ **9.** $x \le -2$

6 -2

11. $y < -2$ **13.** $x \le \dfrac{5}{2}$

-2 $\frac{5}{2}$

15. $x < -\dfrac{177}{20}$ **17.** $x > -1.80$

$-\frac{177}{20}$ -1.80

19. $L > -\dfrac{7}{9}$ **21.** $-1 < x < 1$

$-\frac{7}{9}$ -1 1

23. $2 < x \le 5$ **25.** $-3 \le x < -1$

2 5 -3 -1

27. No values **29.** $x < \dfrac{5}{2}$

0

31. $x \ge -1$ **33.** $-2 < t < 2$

35. $x > -\dfrac{1}{2}$

37. $x \geq 5$

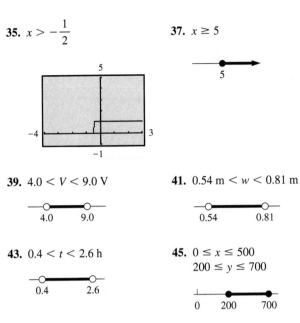

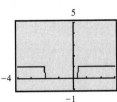

39. $4.0 < V < 9.0$ V

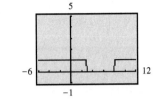

41. 0.54 m $< w < 0.81$ m

43. $0.4 < t < 2.6$ h

45. $0 \leq x \leq 500$
$200 \leq y \leq 700$

47. 2 min $\leq$ stop times ≤ 4 min

Exercises 17.3, page 473

1. $x < 1$ or $x > 3$

3. $-3 < x < 2$ or $x > 4$

5. $-1 < x < 1$

7. $0 \leq x \leq 2$

9. $-4 \leq x \leq \dfrac{3}{2}$

11. $x = -2$

13. All R

15. $-2 < x < 0, x > 1$

17. $-2 \leq s \leq -1, s \geq 1$

19. $-6 < x \leq \dfrac{3}{2}$

21. $-5 < x < -1, x > 7$

23. $x < -1$

25. $x \leq -2, x \geq \dfrac{1}{3}$

27. $T < 3, T > 8$

29. $-1 < x < \dfrac{3}{4}, x \geq 6$

31. $2 < x < 4, 5 < x < 9$

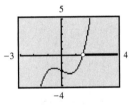

33. $x \leq -2, x \geq 1$

35. $-1 \leq x \leq 0$

37. $x > 1.52$

39. $-1.39 < x < -0.43$

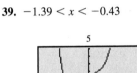

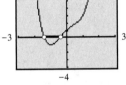

41. $x < -1.69, x > 2.00$

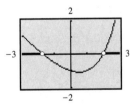

43. $x < -4.43, -3.11 < x < -1.08,$
$x > 3.15$

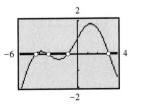

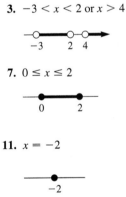

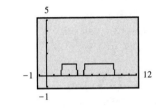

45. No; not true if $0 \leq x \leq 1$ **47.** $x^2 - 3x - 4 > 0$

49. $x < 3.113, x < \dfrac{3\log 3 + 2\log 2}{2\log 3 - \log 2}$ **51.** $0.5 < i < 1$ A

53. $x \leq 12$ **55.** $h > 2640$ km   **57.** $3.0 \leq w < 5.0$ mm

59. $0 \leq t < 0.92$ h

Exercises 17.4, page 477

1. $-2 < x < 3$

3. $3 < x < 5$

5. $x < -2, x > \dfrac{2}{5}$

7. $\dfrac{1}{6} \leq x \leq \dfrac{3}{2}$

9. $x < 0, x > \dfrac{3}{2}$

11. $-26 < x < 24$

13. $-6.4 \leq x \leq -2.1$

15. $x < 0, x > 8$

17. $1 < x < 2$

19. $x \leq \dfrac{1}{10}, x \geq \dfrac{7}{10}$

21. $-15 < R < \dfrac{35}{3}$

23. $x \leq 8.4, x \geq 17.6$

25. $1 < x < 4$

27. $x \leq 6, x \geq 10$

29. $x < -3, -2 < x < 1, x > 2$

31. $-3 < x < -2, 1 < x < 2$

33. $x = 0$ valid only for $a = 0$ **35.** 4 km, 50 km

37. $|p - 2\,000\,000| \leq 200\,000$; production is at least 1 800 000 barrels, but not greater than 2 200 000 barrels.

39. $0.020 < i < 0.040$ A

Exercises 17.5, page 480

1. $y < 3 - x$ **3.**

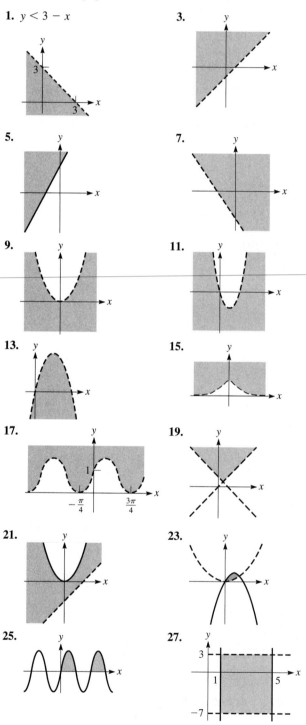

5.   **7.**

9.   **11.**

13.   **15.**

17.   **19.**

21.   **23.**

25.   **27.**

29.

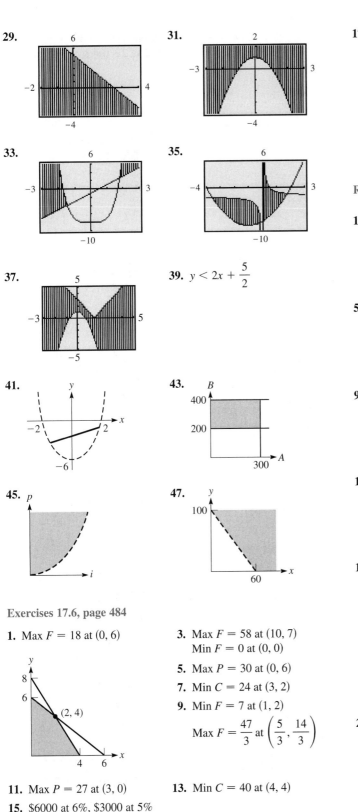

31.

33.

35.

37.

39. $y < 2x + \dfrac{5}{2}$

41.

43.

45.

47.

Exercises 17.6, page 484

1. Max $F = 18$ at $(0, 6)$

3. Max $F = 58$ at $(10, 7)$
Min $F = 0$ at $(0, 0)$

5. Max $P = 30$ at $(0, 6)$

7. Min $C = 24$ at $(3, 2)$

9. Min $F = 7$ at $(1, 2)$
Max $F = \dfrac{47}{3}$ at $\left(\dfrac{5}{3}, \dfrac{14}{3}\right)$

11. Max $P = 27$ at $(3, 0)$

13. Min $C = 40$ at $(4, 4)$

15. \$6000 at 6%, \$3000 at 5%

17. 40 business models
60 graphing models

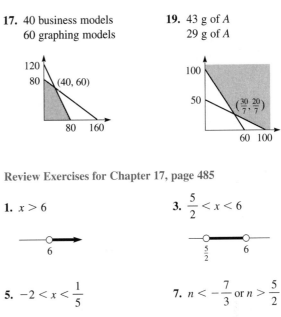

19. 43 g of A
29 g of A

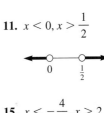

Review Exercises for Chapter 17, page 485

1. $x > 6$

3. $\dfrac{5}{2} < x < 6$

5. $-2 < x < \dfrac{1}{5}$

7. $n < -\dfrac{7}{3}$ or $n > \dfrac{5}{2}$

9. $x < -4, \dfrac{1}{2} < x < 3$

11. $x < 0, x > \dfrac{1}{2}$

13. $-2 \le x \le \dfrac{2}{3}$

15. $x < -\dfrac{4}{5}, x > 2$

17. $x > \dfrac{5}{3}$

19. $2 \le n < \dfrac{11}{4}$

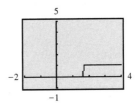

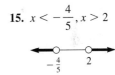

21. $R < -\dfrac{1}{2}, R \ge 8$

23. $x < -1, x > 5$

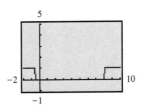

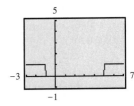

25. $x < -0.68$

27. $x < 0.69$

29.

31.

33.

35.

37.

39.

41.

43.

45.

47.

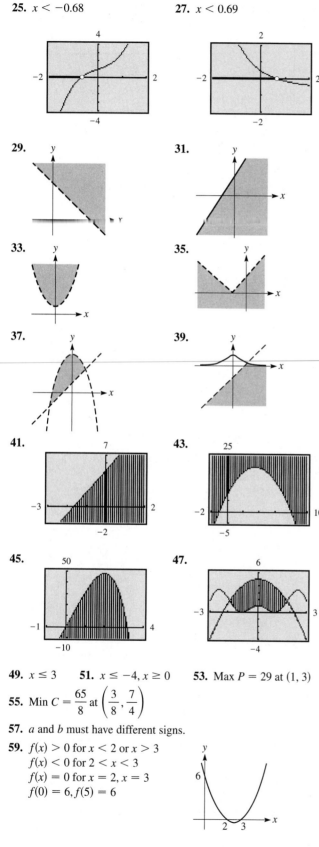

49. $x \le 3$ **51.** $x \le -4, x \ge 0$ **53.** Max $P = 29$ at $(1, 3)$

55. Min $C = \dfrac{65}{8}$ at $\left(\dfrac{3}{8}, \dfrac{7}{4}\right)$

57. a and b must have different signs.

59. $f(x) > 0$ for $x < 2$ or $x > 3$
$f(x) < 0$ for $2 < x < 3$
$f(x) = 0$ for $x = 2, x = 3$
$f(0) = 6, f(5) = 6$

61. $F \ge 98.6°$ **63.** Between \$5 and \$12.50 **65.** $d > 39.5$ m

67. $168 \le B \le 400$ MJ **69.** $0.456 < i < 0.816$ A

71. $r > 5.7$ **73.**

75. 300 regular
150 deluxe

Exercises 18.1, page 491

1. 56 km **3.** 6 **5.** $\dfrac{4}{3}$ **7.** $\dfrac{4}{25}$ **9.** 40

11. $0.27 = 27\%$ **13.** 0.41 **15.** 5.56 **17.** 7.8

19. 3.5% **21.** 863 kg **23.** 1.44 Ω **25.** 0.103 m³

27. 20 000 g **29.** 900 kJ **31.** 286° **33.** 12.5 m/s

35. 23 400 cm³ **37.** 8.8 m **39.** 19.67 kg

41. 17 500 chips **43.** 4200 lines, 4900 lines

Exercises 18.2, page 497

1. $C = kd, k = \pi$ **3.** 667 Hz **5.** $v = kr$ **7.** $R = \dfrac{k}{d^2}$

9. $P = \dfrac{k}{\sqrt{A}}$ **11.** $S = kwd^3$

13. A varies directly as the square of r.

15. n varies as the square root of t and inversely as u.

17. $V = \dfrac{H^2}{2048}$ **19.** $p = \dfrac{16q}{r^3}$ **21.** 25 **23.** 50 **25.** 180

27. 2.56×10^5 **29.** 61 m³ **31.** $m = 0.476t$ **33.** 2.5

35. 1.4 h **37.** (a) Inverse (b) $a = 60/m$ **39.** 7.65 MW

41. $F = 2.32Av^2$ **43.** 0.536 N **45.** 480 m/s

47. $R = \dfrac{2.60 \times 10^{-5}l}{A}$ **49.** 80.0 W **51.** $G = \dfrac{5.9d^2}{\lambda^2}$

53. -6.57 cm/s² **55.** 0.288 W/m²

Review Exercises for Chapter 18, page 499

1. 200 **3.** $\dfrac{2}{5}$ **5.** 3.1417 **7.** 5.6 **9.** 7.39 N/cm²

11. 2260 J/g **13.** 630 km **15.** 2.66×10^{-3} kJ

17. 36 000 characters **19.** 140 mL **21.** 71.9 m

23. 4500, 7500 **25.** 30.2 kg **27.** 115 bolts, 207 bolts

29. $y = 3x^2$ **31.** $v = \dfrac{128x}{y^3}$ **33.** 11.7 cm **35.** 10 cm

37. $R = 8.5\,A$ **39.** 1500 bacteria/h   **41.** 2.22 s

43. 4.3 μC   **45.** 18.0 kW   **47.** 44.1 m   **49.** 1.4

51. 48.7 Hz   **53.** 2.99×10^8 m/s   **55.** 3.26 cm

57. 5.73×10^4 m   **59.** 47 m   **61.** 150%

63. $125.00, $600.13   **65.** 4.80 MJ   **67.** 0.023 W/m^2

Exercises 19.1, page 508

1. -3 **3.** 125 250   **5.** 4, 6, 8, 10, 12

7. $\dfrac{13}{2}, \dfrac{9}{2}, \dfrac{5}{2}, \dfrac{1}{2}, -\dfrac{3}{2}$ **9.** 22   **11.** $-\dfrac{9\pi}{4}$ **13.** 30.9

15. $49b$ **17.** 440   **19.** $-\dfrac{85}{2}$ **21.** $n = 6, S_6 = 150$

23. $d = -\dfrac{2}{19}, a_{20} = -\dfrac{1}{3}$ **25.** $a_1 = 19, a_{30} = 106$

27. No solution (n cannot be negative)   **29.** $n = 23, a_{23} = 6k$

31. $n = 8, d = \dfrac{1}{14}(b + 2c)$ **33.** $a_1 = 36, d = 4, S_{10} = 540$

35. Yes, $d = \ln 2; a_5 = \ln 48$ **37.** $d = \dfrac{b - a}{2}$ **39.** 5050

41. -8 **43.** 2700 m^2 **45.** 195 logs   **47.** 10 rows

49. 13 years, $11 700   **51.** 490 m

53. $S_n = \dfrac{1}{2}n[2a_1 + (n - 1)d]$ **55.** $a_1 = 1, a_n = n$

Exercises 19.2, page 511

1. $\dfrac{25}{8}$ **3.** 45, 15, 5, $\dfrac{5}{3}, \dfrac{5}{9}$ **5.** $\dfrac{1}{6}, \dfrac{1}{2}, \dfrac{3}{2}, \dfrac{9}{2}, \dfrac{27}{2}$

7. 16   **9.** $\dfrac{1}{125}$ **11.** $\dfrac{100}{729}$ **13.** 1   **15.** $\dfrac{341}{8}$ **17.** 378

19. $\dfrac{3(2^{10} - k^{10})}{16(2 + k)}$ **21.** $a_6 = 64, S_6 = \dfrac{1365}{16}$

23. $a_1 = 16, a_5 = 81$ **25.** $a_1 = 1, r = 3$

27. $n = 7, S_n = \dfrac{58\,593}{625}$ **29.** Yes; $r = 3^x; a_{20} = 3^{19x+1}$

31. $r = \sqrt{\dfrac{b}{a}}$ **33.** 2 or -3 **35.** 1.4%   **37.** 1.09 mA

39. $443.96   **41.** 5800°C   **43.** 43%   **45.** 21.1°C

47. $671 088.64   **49.** $S_n = \dfrac{a_1 - ra_n}{1 - r}$

51. Yes, the ratio is squared.

Exercises 19.3, page 515

1. $\dfrac{32}{7}$ **3.** 8   **5.** 0.2   **7.** $\dfrac{400}{21}$ **9.** 8   **11.** $\dfrac{10,000}{9999}$

13. $\dfrac{1}{2}(5 + 3\sqrt{3})$ **15.** $\dfrac{1}{3}$ **17.** 0.5   **19.** $\dfrac{40}{99}$ **21.** $\dfrac{2}{11}$

23. $\dfrac{91}{333}$ **25.** $\dfrac{11}{30}$ **27.** $\dfrac{100\,741}{999\,000}$ **29.** 350 L   **31.** 346 g

33. 100 m   **35.** $-\dfrac{1}{4}$

Exercises 19.4, page 520

1. $32x^5 + 240x^4 + 720x^3 + 1080x^2 + 810x + 243$

3. $t^3 + 3t^2 + 3t + 1$ **5.** $16x^4 - 32x^3 + 24x^2 - 8x + 1$

7. 40.841 01

9. $n^5 + 10\pi n^4 + 40\pi^2 n^3 + 80\pi^3 n^2 + 80\pi^4 n + 32\pi^5$

11. $64a^6 - 192a^5b^2 + 240a^4b^4 - 160a^3b^6 + 60a^2b^8 - 12ab^{10} + b^{12}$

13. $625x^4 - 1500x^3 + 1350x^2 - 540x + 81$

15. $64a^6 + 192a^5 + 240a^4 + 160a^3 + 60a^2 + 12a + 1$

17. $x^{10} + 20x^9 + 180x^8 + 960x^7 + \cdots$

19. $128a^7 - 448a^6 + 672a^5 - 560a^4 + \cdots$

21. $x^6 - 12x^{11/2}y + 66x^5y^2 - 220x^{9/2}y^3 + \cdots$

23. $b^{40} + 10b^{37} + \dfrac{95}{2}b^{34} + \dfrac{285}{2}b^{31} + \cdots$

25. 1.338   **27.** 1.015   **29.** $1 + 8x + 28x^2 + 56x^3 + \cdots$

31. $1 + 2x + 3x^2 + 4x^3 + \cdots$

33. $1 + \dfrac{1}{2}x - \dfrac{1}{8}x^2 + \dfrac{1}{16}x^3 - \cdots$

35. $\dfrac{1}{3}\left[1 + \dfrac{1}{2}x + \dfrac{3}{8}x^2 + \dfrac{5}{16}x^3 + \cdots\right]$

37. (a) 3.557×10^{14} (b) 5.109×10^{19} (c) 8.536×10^{15}
(d) 2.480×10^{96}

39. $n! = n(n - 1)(n - 2)(\cdots)(2)(1) = n \times (n - 1)!$; for
$n = 1, 1! = 1 \times 0!$. Since $1! = 1, 0!$ must $= 1$.

41. $56a^3b^5$ **43.** $10\,264\,320x^8b^4$ **45.** 2.45

47. $V = A(1 - 5r + 10r^2 - 10r^3 + 5r^4 - r^5)$

49. $1 - \dfrac{x}{a} + \dfrac{x^3}{2a^3} - \cdots$ **51.** $-\dfrac{2hk}{r^3} + \dfrac{3kh^2}{r^4}$

Review Exercises for Chapter 19, page 521

1. 81   **3.** 1.28×10^{-3} **5.** $-\dfrac{119}{2}$ **7.** $\dfrac{16}{243}$ **9.** $\dfrac{195}{2}$

11. $\dfrac{16\,383}{1536}$ **13.** 81   **15.** 32   **17.** -15 **19.** -0.25

21. 186   **23.** $\dfrac{455}{2}$ (as), 127 (gs), or 43 (gs)   **25.** 2.7

27. 51   **29.** $\dfrac{1}{33}$ **31.** $\dfrac{4}{55}$

33. $x^4 - 8x^3 + 24x^2 - 32x + 16$

35. $x^{10} + 5x^8 + 10x^6 + 10x^4 + 5x^2 + 1$

37. $a^{10} + 20a^9e + 180a^8e^2 + 960a^7e^3 + \cdots$

39. $p^{18} - \dfrac{3}{2}p^{16}q + p^{14}q^2 - \dfrac{7}{18}p^{12}q^3 + \cdots$

41. $1 + 12x + 66x^2 + 220x^3 + \cdots$

43. $1 + \dfrac{1}{2}x^2 - \dfrac{1}{8}x^4 + \dfrac{1}{16}x^6 - \cdots$

45. $1 - \dfrac{1}{2}a^2 - \dfrac{1}{8}a^4 - \dfrac{1}{16}a^6 - \cdots$

47. $\dfrac{1}{8} + \dfrac{3}{4}x + 3x^2 + 10x^3 + \cdots$ **49.** 1,001,000

51. $4b - 3a$ **53.** No **55.** 0.7156 **57.** 5.475

59. 11th **61.** 12.6 mm **63.** 7690 mm **65.** \$4700

67. 1.65×10^{10} cm $= 165\,000$ km **69.** 191 m **71.** 100 cm

73. \$47,340.80 **75.** \$6.93

77. $1 + \dfrac{1}{2}am^2 + \dfrac{1}{8}am^4$ **79.** 21 years

81. Five applications **83.** Yes; term to term ratios are equal.

Exercises 20.1, page 530

(*Note:* "Answers" to trigonometric identities are intermediate steps of suggested reductions of the left member.)

1. $\sin x = \dfrac{\tan x}{\sec x} = \dfrac{\dfrac{\sin x}{\cos x}}{\dfrac{1}{\cos x}} = \dfrac{\sin x}{\cos x} \cdot \dfrac{\cos x}{1} = \sin x$

3. $1.483 = \dfrac{1}{0.6745}$

5. $\left(-\dfrac{1}{2}\sqrt{3}\right)^2 + \left(-\dfrac{1}{2}\right)^2 = \dfrac{3}{4} + \dfrac{1}{4} = 1$

7. $\dfrac{\cos \theta}{\sin \theta}\left(\dfrac{1}{\cos \theta}\right) = \dfrac{1}{\sin \theta}$

9. $\dfrac{\sin x}{\dfrac{\sin x}{\cos x}} = \dfrac{\sin x}{1}\left(\dfrac{\cos x}{\sin x}\right)$ **11.** $\sin y\left(\dfrac{\cos y}{\sin y}\right)$

13. $\sin x\left(\dfrac{1}{\cos x}\right)$ **15.** $\csc^2 x(\sin^2 x)$

17. $\sin x(\csc^2 x) = (\sin x)(\csc x)(\csc x)$
$= \sin x\left(\dfrac{1}{\sin x}\right)\csc x$

19. $\tan y \cot y + \tan^2 y = 1 + \tan^2 y$

21. $\cos \theta\left(\dfrac{\cos \theta}{\sin \theta}\right) + \sin \theta = \dfrac{\cos^2 \theta + \sin^2 \theta}{\sin \theta} = \dfrac{1}{\sin \theta}$

23. $\cot \theta(\sec^2 \theta - 1) = \cot \theta \tan^2 \theta = (\cot \theta \tan \theta)\tan \theta$

25. $\dfrac{\sin x}{\cos x} + \dfrac{\cos x}{\sin x} = \dfrac{\sin^2 x + \cos^2 x}{\cos x \sin x} = \dfrac{1}{\cos x \sin x}$

27. $(1 - \sin^2 x) - \sin^2 x$

29. $\dfrac{\sin x(1 + \cos x)}{1 - \cos^2 x} = \dfrac{1 + \cos x}{\sin x}$

31. $\dfrac{\sin \theta}{\dfrac{1}{\sin \theta}} + \dfrac{\cos \theta}{\dfrac{1}{\cos \theta}} = \sin^2\theta + \cos^2\theta$

33. $(2 \sin^2 x - 1)(\sin^2 x - 1)$

35. $\dfrac{\sin \pi t}{2}\left(\dfrac{\sin^2 \pi t + (1 - \cos \pi t)^2}{(1 - \cos \pi t)\sin \pi t}\right)$
$= \dfrac{\sin^2 \pi t + 1 - 2\cos \pi t + \cos^2 \pi t}{2(1 - \cos \pi t)} = \dfrac{2(1 - \cos \pi t)}{2(1 - \cos \pi t)}$

37. Infinite series: $\dfrac{1}{1 - \sin^2 x} = \dfrac{1}{\cos^2 x}$ **39.** $\cot x$

41. $\sin x$ **43.** $\sec x$ **45.** $\cos x$

47. **49.**

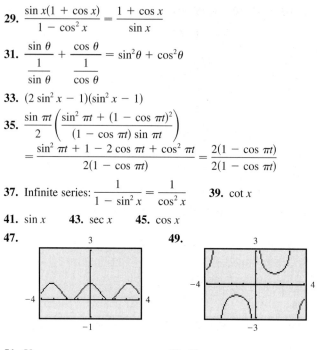

51. Yes **53.** No

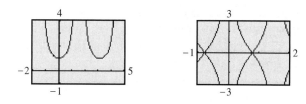

55. $0 = \cos A \cos B \cos C + \sin A \sin B$,
$\cos C = -\dfrac{\sin A \sin B}{\cos A \cos B}$

57. $l = a \csc \theta + a \sec \theta = a\left(\dfrac{1}{\sin \theta} + \dfrac{\tan \theta}{\sin \theta}\right)$

59. Replacing $\tan \theta$ with $\sin \theta/\cos \theta$ gives $\sin^2 \theta/\cos \theta + \cos \theta$. Using LCD of $\cos \theta$ gives $(\sin^2 \theta + \cos^2 \theta)/\cos \theta$ which equals $\sec \theta$.

61. $\sin^2 x - \sin^2 x \sec^2 x + \cos^2 x + \cos^2 x \sec^4 x$
$= \sin^2 x - \tan^2 x + \cos^2 x + \sec^2 x = 2$

63. $\left(\dfrac{r}{x}\right)^2 + \left(\dfrac{r}{y}\right)^2 = \dfrac{r^2(x^2 + y^2)}{x^2y^2}$

65. $\sqrt{1 - \cos^2 \theta} = \sqrt{\sin^2 \theta}$

67. $\sqrt{4 + 4\tan^2 \theta} = 2\sqrt{1 + \tan^2 \theta}$

Exercises 20.2, page 535

1. $\dfrac{16}{65}$

3. $\sin 105° = \sin 60° \cos 45° + \cos 60° \sin 45°$

$= \dfrac{\sqrt{3}}{2} \dfrac{\sqrt{2}}{2} + \dfrac{1}{2} \dfrac{\sqrt{2}}{2} = 0.9659$

5. $\cos 15° = \cos(60° - 45°)$

$= \cos 60° \cos 45° + \sin 60° \sin 45°$

$= \left(\dfrac{1}{2}\right)\left(\dfrac{1}{2}\sqrt{2}\right) + \left(\dfrac{1}{2}\sqrt{3}\right)\left(\dfrac{1}{2}\sqrt{2}\right)$

$= \dfrac{1}{4}\sqrt{2} + \dfrac{1}{4}\sqrt{6} = \dfrac{1}{4}(\sqrt{2} + \sqrt{6}) = 0.9659$

7. $-\dfrac{33}{65}$ **9.** $-\dfrac{56}{65}$ **11.** $\sin 3x$ **13.** $\cos x$

15. $\tan x$ **17.** 0 **19.** 1 **21.** 0

23. $\sin(180° - x) = \sin 180° \cos x - \cos 180° \sin x$

$= (0)\cos x - (-1)\sin x$

25. $\cos(0° - x) = \cos 0° \cos x + \sin 0° \sin x$

$= (1)\cos x + (0)\sin x$

27. $\tan(180° + x) = \dfrac{\tan 180° + \tan x}{1 - \tan 180° \tan x} = \tan x$

29. $\cos\left(\dfrac{\pi}{3} + x\right) = \cos \dfrac{\pi}{3}\cos x - \sin \dfrac{\pi}{3}\sin x$

$= \dfrac{1}{2}\cos x - \dfrac{1}{2}\sqrt{3}\sin x$

31. $(\sin x \cos y + \cos x \sin y)(\sin x \cos y - \cos x \sin y)$

$= \sin^2 x \cos^2 y - \cos^2 x \sin^2 y$

$= \sin^2 x(1 - \sin^2 y) - (1 - \sin^2 x)\sin^2 y$

33. $(\cos \alpha \cos \beta - \sin \alpha \sin \beta)$

$+ (\cos \alpha \cos \beta + \sin \alpha \sin \beta)$

35.

37.

39, 41, 43, 45. Use the indicated method.

47. $\dfrac{\sin(x + x)}{\sin x} = \dfrac{\sin x \cos x + \cos x \sin x}{\sin x}$

$= \dfrac{2 \sin x \cos x}{\sin x}$

49. Using Eq. (20.9), $\sin 75° = \sin(30° + 45°)$. Using Eq. (20.11), $\sin 75° = \sin(135° - 60°)$. (Other angles are also possible.)

51. $I(\cos \theta \cos 30° - \sin \theta \sin 30°)$

$+ I(\cos \theta \cos 150° - \sin \theta \sin 150°)$

$+ I(\cos \theta \cos 270° - \sin \theta \sin 270°)$

$= I\left[\dfrac{1}{2}\sqrt{3}\cos \theta - \dfrac{1}{2}\sin \theta - \dfrac{1}{2}\sqrt{3}\cos \theta - \dfrac{1}{2}\sin \theta\right.$

$\left. + 0 - (-\sin \theta)\right]$

53. $i_0 \sin(\omega t + \alpha) = i_0(\sin \omega t \cos \alpha + \cos \omega t \sin \alpha)$

55. $\tan \alpha(R + \cos \beta) = \sin \beta, R = \dfrac{\sin \beta - \tan \alpha \cos \beta}{\tan \alpha}$

$= \dfrac{\sin \beta \cos \alpha - \cos \beta \sin \alpha}{\cos \alpha \tan \alpha}$

Exercises 20.3, page 539

1. $-\sqrt{3}$ **3.** $-\dfrac{24}{25}$

5. $\sin 60° = \sin 2(30°) = 2 \sin 30° \cos 30°$

$= 2\left(\dfrac{1}{2}\right)\left(\dfrac{1}{2}\sqrt{3}\right) = \dfrac{1}{2}\sqrt{3}$

7. $\tan 120° = \dfrac{2 \tan 60°}{1 - \tan^2 60°} = \dfrac{2\sqrt{3}}{1 - (\sqrt{3})^2} = -\sqrt{3}$

9. $\sin 258° = 2 \sin 129° \cos 129° = -0.978\,147\,6$

11. $\cos 96° = \cos^2 48° - \sin^2 48° = -0.104\,528\,5$ **13.** 3.0777

15. $\dfrac{24}{25}$ **17.** $-\sqrt{3}$ **19.** $2 \sin 8x$ **21.** $\cos 8x$

23. $\cos x$ **25.** $-2 \cos 4x$ **27.** $\cos^2 \alpha - (1 - \cos^2 \alpha)$

29. $\dfrac{\cos x - (\sin x/\cos x)\sin x}{1/\cos x} = \cos^2 x - \sin^2 x$

31. $\dfrac{2 \sin 2\theta \cos 2\theta}{\sin 2\theta}$ **33.** $\dfrac{2 \sin \theta \cos \theta}{1 + 2\cos^2 \theta - 1} = \dfrac{\sin \theta}{\cos \theta}$

35. $1 - (1 - 2\sin^2 2\theta) = \dfrac{2}{\csc^2 \theta}$

37. $\dfrac{\sin 3x \cos x - \cos 3x \sin x}{\sin x \cos x} = \dfrac{\sin 2x}{\frac{1}{2}\sin 2x}$

39. $\ln \dfrac{1 - \cos 2x}{1 + \cos 2x} = \ln \dfrac{2 \sin^2 x}{2 \cos^2 x} = \ln \tan^2 x$

41.

43.

45. $3 \sin x - 4 \sin^3 x$ **47.** $8 \cos^4 x - 8 \cos^2 x + 1$

49. amp. $= 2$, per. $= \pi$; write equation as

$y = 2(2 \sin x \cos x) = 2 \sin 2x$.

51. 474 m   **53.** $R = v\left(\dfrac{2v \sin \alpha}{g}\right)\cos \alpha = \dfrac{v^2(2 \sin \alpha \cos \alpha)}{g}$

55. $vi \sin \omega t \sin\left(\omega t - \dfrac{\pi}{2}\right)$

$= vi \sin \omega t\left(\sin \omega t \cos \dfrac{\pi}{2} - \cos \omega t \sin \dfrac{\pi}{2}\right)$

$= vi \sin \omega t[-(\cos \omega t)(1)] = -\dfrac{1}{2}vi(2 \sin \omega t \cos \omega t)$

Exercises 20.4, page 543

1. $\cos 57°$

3. $\cos 15° = \cos \dfrac{1}{2}(30°) = \sqrt{\dfrac{1 + \cos 30°}{2}} = \sqrt{\dfrac{1.8660}{2}}$

$= 0.9659$

5. $\sin 75° = \sin \dfrac{1}{2}(150°) = \sqrt{\dfrac{1 - \cos 150°}{2}}$

$= \sqrt{\dfrac{1.8660}{2}} = 0.9659$

7. -0.9239 **9.** $\sin 118° = 0.882\,947\,6$

11. $\sqrt{2\left(\dfrac{1 + \cos 164°}{2}\right)} = \sqrt{2}\cos 82° = 0.196\,820\,5$

13. $\sin 3x$ **15.** $4\cos 2x$ **17.** $2\sqrt{2}\sin 5\theta$ **19.** $\dfrac{1}{26}\sqrt{26}$

21. 0.1414 **23.** $\pm\sqrt{\dfrac{2}{1 - \cos \alpha}}$

25. $\tan \dfrac{1}{2}\alpha = \dfrac{1 - \cos \alpha}{\sin \alpha} = \dfrac{\sin \alpha}{1 + \cos \alpha}$

27. $\dfrac{1 - \cos \alpha}{2\sin \frac{1}{2}\alpha} = \dfrac{1 - \cos \alpha}{2\sqrt{\frac{1}{2}(1 - \cos \alpha)}} = \sqrt{\dfrac{1 - \cos \alpha}{2}}$

29. $2\left(\dfrac{1 - \cos x}{2}\right) + \cos x$

31. $\sqrt{\dfrac{(1 + \cos \theta)(1 - \cos \theta)}{2(1 - \cos \theta)}} = \dfrac{\sin \theta}{\sqrt{4\left(\dfrac{1 - \cos \theta}{2}\right)}}$

33.

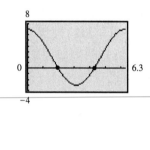

35.

37. $\pm\dfrac{24}{7}$ **39.** $4\sin \dfrac{\theta}{2}$

41. $\sin \omega t = \pm\sqrt{\dfrac{1 - \cos 2\omega t}{2}}$,

$\sin^2 \omega t = \dfrac{1}{2}(1 - \cos 2\omega t)$

43. $\dfrac{\sqrt{\dfrac{1 - \cos(A + \phi)}{2}}}{\sqrt{\dfrac{1 - \cos A}{2}}} = \sqrt{\dfrac{1 - \cos(A + \phi)}{1 - \cos A}}$

Exercises 20.5, page 548

1. $\dfrac{\pi}{2}, \dfrac{7\pi}{6}, \dfrac{11\pi}{6}$ **3.** $0, \dfrac{\pi}{3}, \dfrac{5\pi}{3}$ **5.** $\dfrac{\pi}{2}$ **7.** $\dfrac{\pi}{2}, \dfrac{3\pi}{2}$

9. $\dfrac{\pi}{3}, \dfrac{2\pi}{3}, \dfrac{4\pi}{3}, \dfrac{5\pi}{3}$ **11.** $0, \dfrac{\pi}{6}, \dfrac{5\pi}{6}, \pi$ **13.** $\dfrac{\pi}{2}, \dfrac{3\pi}{2}$

15. $\dfrac{\pi}{4}, \dfrac{3\pi}{4}, \dfrac{5\pi}{4}, \dfrac{7\pi}{4}$ **17.** $0.2618, 1.309, 3.403, 4.451$

19. $0, \pi$ **21.** $\dfrac{3\pi}{4} = 2.36, \dfrac{7\pi}{4} = 5.50$

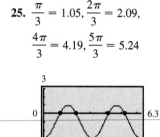

23. $1.9823, 4.3009$ **25.** $\dfrac{\pi}{3} = 1.05, \dfrac{2\pi}{3} = 2.09,$

$\dfrac{4\pi}{3} = 4.19, \dfrac{5\pi}{3} = 5.24$

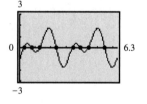

27. $\dfrac{\pi}{12} = 0.26, \dfrac{\pi}{4} = 0.79, \dfrac{5\pi}{12} = 1.31, \dfrac{3\pi}{4} = 2.36,$

$\dfrac{13\pi}{12} = 3.40, \dfrac{5\pi}{4} = 3.93, \dfrac{17\pi}{12} = 4.45, \dfrac{7\pi}{4} = 5.50$

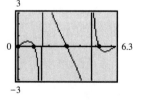

29. $0 = 0.00, \dfrac{\pi}{3} = 1.05, \pi = 3.14, \dfrac{5\pi}{3} = 5.24$

31. 3.569, 5.856

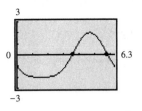

33. 0.7854, 1.249, 3.927, 4.391

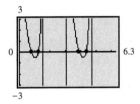

35. $\dfrac{3\pi}{8} = 1.18, \dfrac{7\pi}{8} = 2.75, \dfrac{11\pi}{8} = 4.32, \dfrac{15\pi}{8} = 5.89$

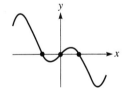

37. $0, \dfrac{\pi}{2}, \pi, \dfrac{3\pi}{2}$ **39.** $37.8°$ **41.** 6.56×10^{-4}

43. 10.2 s, 15.7 s, 21.2 s, 47.1 s
45. $-2.28, 0.00, 2.28$ **47.** 0.29, 0.95

49. 2.10   **51.** 1.08

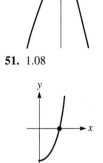

Exercises 20.6, page 553

1. y is the angle whose tangent is $3A$.   **3.** 0
5. y is the angle whose tangent is x.

7. y is the angle whose cotangent is $3x$.
9. y is twice the angle whose sine is x.
11. y is five times the angle whose cosine is $2x - 1$.

13. $\dfrac{\pi}{3}$ **15.** $\dfrac{\pi}{4}$ **17.** $-\dfrac{\pi}{3}$ **19.** No value   **21.** $-\dfrac{\pi}{4}$

23. $\dfrac{1}{2}\sqrt{3}$ **25.** $\dfrac{1}{2}\sqrt{2}$ **27.** -1 **29.** $2/\sqrt{21}$

31. $2/\sqrt{3}$ **33.** -1.3090 **35.** 0.2135   **37.** -1.2389

39. -0.2239 **41.** $x = \dfrac{1}{3}\sin^{-1} y$ **43.** $x = 4\tan y$

45. $x = 1 - \cos(1 - y)$ **47.** $\dfrac{x}{\sqrt{1 - x^2}}$

49. $xy + \sqrt{1 - x^2 - y^2 + x^2 y^2}$ **51.** $\dfrac{3x}{\sqrt{9x^2 - 1}}$

53. $2x\sqrt{1 - x^2}$ **55.** $t = \dfrac{1}{2\omega}\cos^{-1}\dfrac{y}{A} - \dfrac{\phi}{\omega}$

57. $t = \dfrac{1}{\omega}\left(\sin^{-1}\dfrac{i}{I_m} - \alpha - \phi\right)$

59. $\sin\left(\sin^{-1}\dfrac{3}{5} + \sin^{-1}\dfrac{5}{13}\right) = \dfrac{3}{5}\cdot\dfrac{12}{13} + \dfrac{4}{5}\cdot\dfrac{5}{13} = \dfrac{56}{65}$

61. $\dfrac{\pi}{2}$ **63.** 0   **65.** $\sin^{-1}\left(\dfrac{a}{c}\right)$ **67.** $\tan^{-1}\left(\dfrac{b\tan B}{a}\right)$

69. Let y = height to top of pedestal; $\tan\alpha = \dfrac{46.0 + y}{d}$,

$\tan\beta = \dfrac{y}{d}$; $\tan\alpha = \dfrac{46.0 + d\tan\beta}{d}$

71. $\theta = \tan^{-1}\left(\dfrac{y + 50}{x}\right) - \tan^{-1}\left(\dfrac{y}{x}\right)$

Review Exercises for Chapter 20, page 556

1. $\sin(90° + 30°) = \sin 90° \cos 30° + \cos 90° \sin 30°$
$= (1)\left(\dfrac{1}{2}\sqrt{3}\right) + (0)\left(\dfrac{1}{2}\right) = \dfrac{1}{2}\sqrt{3}$

3. $\sin(180° - 45°) = \sin 180° \cos 45° - \cos 180° \sin 45°$
$= 0\left(\dfrac{1}{2}\sqrt{2}\right) - (-1)\left(\dfrac{1}{2}\sqrt{2}\right) = \dfrac{1}{2}\sqrt{2}$

5. $\cos\left(2\dfrac{\pi}{2}\right) = \cos^2\dfrac{\pi}{2} - \sin^2\dfrac{\pi}{2} = 0 - 1^2 = -1$

7. $\tan 2(30°) = \dfrac{2\tan 30°}{1 - \tan^2 30°} = \dfrac{2(\sqrt{3}/3)}{1 - (\sqrt{3}/3)^2} = \sqrt{3}$

9. $\sin 52° = 0.788\,010\,8$ **11.** $\dfrac{1}{2}$ **13.** $\cos(-69°) = 0.358\,367\,9$

15. $2\tan 24° = 0.890\,457\,4$ **17.** $\sin 5x$ **19.** $4\sin 12x$

21. $2\cos 12x$ **23.** $2\cos x$ **25.** $-\dfrac{\pi}{2}$ **27.** 0.2619

29. $-\dfrac{1}{3}\sqrt{3}$ **31.** 0   **33.** $\dfrac{\frac{1}{\cos y}}{\frac{1}{\sin y}} = \dfrac{1}{\cos y}\cdot\dfrac{\sin y}{1}$

35. $\sin x \csc x - \sin^2 x = 1 - \sin^2 x$

37. $\dfrac{(\sec^2 x - 1)(\sec^2 x + 1)}{\tan^2 x} = \sec^2 x + 1$

39. $2\left(\dfrac{1}{\sin 2x}\right)\left(\dfrac{\cos x}{\sin x}\right) = 2\left(\dfrac{1}{2\sin x \cos x}\right)\left(\dfrac{\cos x}{\sin x}\right)$

$= \dfrac{1}{\sin^2 x}$

41. $\dfrac{\cos^2 \theta}{\sin^2 \theta}$　　**43.** $\dfrac{1}{2}\left(2\sin\dfrac{\theta}{2}\cos\dfrac{\theta}{2}\right)$　　**45.** $\cot x$　　**47.** $\sec x$

49. $\sin x$　　**51.** $\sin x$

53.

55.

57.

59.

61. $x = \dfrac{1}{2}\cos^{-1}\dfrac{1}{2}y$　　**63.** $x = \dfrac{1}{5}\sin\dfrac{1}{3}\left(\dfrac{1}{4}\pi - y\right)$

65. 1.2925, 4.4341　　**67.** $\dfrac{\pi}{6}, \dfrac{5\pi}{6}, \dfrac{7\pi}{6}, \dfrac{11\pi}{6}$　　**69.** $0, \dfrac{\pi}{3}, \dfrac{5\pi}{3}$

71. $0, \dfrac{2\pi}{3}$　　**73.** 0.3142, 1.571, 2.827, 4.084, 4.712, 5.341

75. 0　　**77.** 1.56, 2.16, 3.46　　　　**79.** $-2.31, 1.14$

81. $\dfrac{1}{x}$　　**83.** $2x\sqrt{1 - x^2}$　　**85.** $\dfrac{\sqrt{1 - x^2} - xy}{\sqrt{1 + y^2}}$

87. $2\sqrt{1 - \cos^2\theta}$　　**89.** $\dfrac{\tan \theta}{\sqrt{1 + \tan^2\theta}} = \dfrac{\tan \theta}{\sec \theta}$

91. $(\cos \theta + j\sin \theta)^2 = (\cos^2 \theta - \sin^2 \theta) + j(2\sin \theta \cos \theta)$

93. $\pi < x < 2\pi$　　**95.** $C\left(\dfrac{A}{C}\sin 2t + \dfrac{B}{C}\cos 2t\right)$
$= C(\cos \alpha \sin 2t + \sin \alpha \cos 2t)$

97. $R = \sqrt{(A\cos \theta - B\sin \theta)^2 + (A\sin \theta + B\cos \theta)^2}$
$= \sqrt{A^2(\cos^2\theta + \sin^2\theta) + B^2(\sin^2\theta + \cos^2\theta)}$

99. $\dfrac{k}{2}\cdot\dfrac{1}{\sin^2\frac{\theta}{2}} = \dfrac{k}{2}\cdot\dfrac{1}{\frac{1 - \cos\theta}{2}}$　　**101.** $\theta = \alpha + R\sin \omega t$

103. $\dfrac{\cos 2\alpha}{2\cos^2\alpha}$　　**105.** 6.8 m　　**107.** 54.7°

Exercises 21.1, page 563

1. $\sqrt{61}$　　**3.** 150°　　**5.** $2\sqrt{29}$　　**7.** 3　　**9.** 55

11. $2\sqrt{53}$　　**13.** 2.86　　**15.** $\dfrac{5}{2}$　　**17.** Undefined　　**19.** $-\dfrac{3}{4}$

21. $-\dfrac{5}{9}$　　**23.** 0.747　　**25.** $\dfrac{1}{3}\sqrt{3}$　　**27.** -1.084

29. 20.0°　　**31.** 98.50°　　**33.** Parallel　　**35.** Perpendicular

37. $8, -2$　　**39.** -3　　**41.** Two sides equal $2\sqrt{10}$.

43. $m_1 = \dfrac{5}{12}, m_2 = \dfrac{4}{3}$　　**45.** 10　　**47.** $4\sqrt{10} + 4\sqrt{2} = 18.3$

49. (1.5)　　**51.** $(-2.8, 4.2)$　　**53.** $x^2 + y^2 = 9$

55. $m_1 = 1, m_2 = -1$

Exercises 21.2, page 568

1. $y + 2x - 7 = 0$　　**3.** 2

5. $4x - y + 20 = 0$　　　　**7.** $7x - 2y - 24 = 0$

9. $x - y + 2 = 0$　　　　**11.** $y = -2.7$

13. $x = -3$　　　　**15.** $x + 3y + 5 = 0$

17. $4x + 3y + 6 = 0$　　　　**19.** $3x + y - 18 = 0$

21. $y = 4x - 8$; $m = 4$, $(0, -8)$

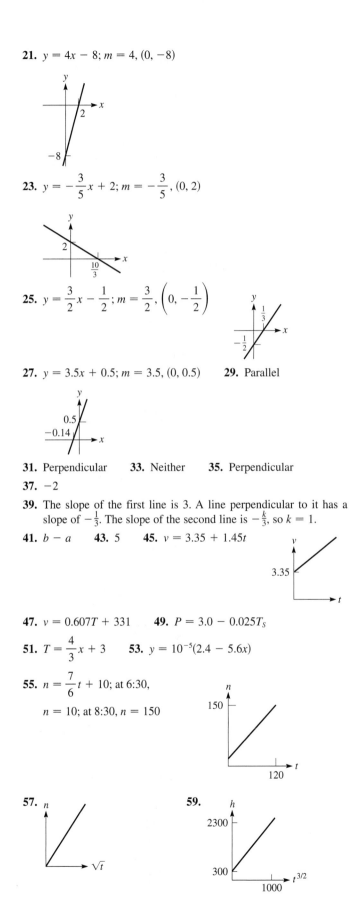

23. $y = -\dfrac{3}{5}x + 2$; $m = -\dfrac{3}{5}$, $(0, 2)$

25. $y = \dfrac{3}{2}x - \dfrac{1}{2}$; $m = \dfrac{3}{2}$, $\left(0, -\dfrac{1}{2}\right)$

27. $y = 3.5x + 0.5$; $m = 3.5$, $(0, 0.5)$ **29.** Parallel

31. Perpendicular   **33.** Neither   **35.** Perpendicular

37. -2

39. The slope of the first line is 3. A line perpendicular to it has a slope of $-\frac{1}{3}$. The slope of the second line is $-\frac{k}{3}$, so $k = 1$.

41. $b - a$ **43.** 5   **45.** $v = 3.35 + 1.45t$

47. $v = 0.607T + 331$ **49.** $P = 3.0 - 0.025T_S$

51. $T = \dfrac{4}{3}x + 3$ **53.** $y = 10^{-5}(2.4 - 5.6x)$

55. $n = \dfrac{7}{6}t + 10$; at 6:30,

$n = 10$; at 8:30, $n = 150$

57.   **59.**

61.   **63.**

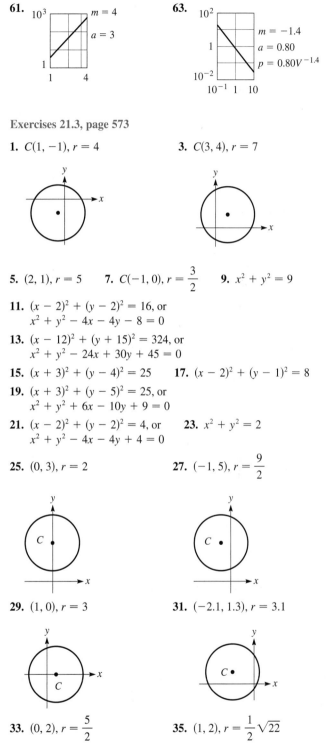

Exercises 21.3, page 573

1. $C(1, -1)$, $r = 4$ **3.** $C(3, 4)$, $r = 7$

5. $(2, 1)$, $r = 5$ **7.** $C(-1, 0)$, $r = \dfrac{3}{2}$ **9.** $x^2 + y^2 = 9$

11. $(x - 2)^2 + (y - 2)^2 = 16$, or
$x^2 + y^2 - 4x - 4y - 8 = 0$

13. $(x - 12)^2 + (y + 15)^2 = 324$, or
$x^2 + y^2 - 24x + 30y + 45 = 0$

15. $(x + 3)^2 + (y - 4)^2 = 25$ **17.** $(x - 2)^2 + (y - 1)^2 = 8$

19. $(x + 3)^2 + (y - 5)^2 = 25$, or
$x^2 + y^2 + 6x - 10y + 9 = 0$

21. $(x - 2)^2 + (y - 2)^2 = 4$, or   **23.** $x^2 + y^2 = 2$
$x^2 + y^2 - 4x - 4y + 4 = 0$

25. $(0, 3)$, $r = 2$ **27.** $(-1, 5)$, $r = \dfrac{9}{2}$

29. $(1, 0)$, $r = 3$ **31.** $(-2.1, 1.3)$, $r = 3.1$

33. $(0, 2)$, $r = \dfrac{5}{2}$ **35.** $(1, 2)$, $r = \dfrac{1}{2}\sqrt{22}$

37. Symmetric to both axes and origin

39. Symmetric to y-axis   **41.** (7, 0), (−1, 0)

43. $3x^2 + 3y^2 + 4x + 8y - 20 = 0$, circle

45.

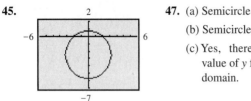

47. (a) Semicircle

(b) Semicircle

(c) Yes, there is only one value of y for each x in the domain.

49. 2.82 in.   **51.** $x^2 + y^2 = 0.0100$

53. $x^2 + (y - 2.4)^2 = 0.16$

55. $(x - 500 \times 10^{-6})^2 + y^2 = 0.16 \times 10^{-6}$

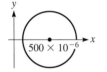

Exercises 21.4, page 578

1. $F(5, 0), x = -5$ **3.** $F\left(0, -\dfrac{3}{2}\right), y = \dfrac{3}{2}$

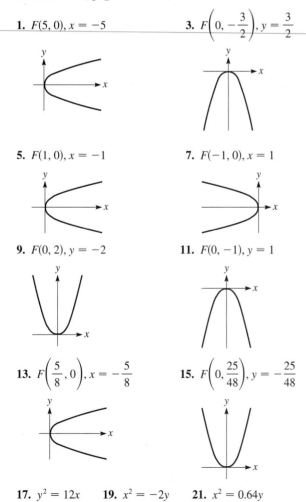

5. $F(1, 0), x = -1$ **7.** $F(-1, 0), x = 1$

9. $F(0, 2), y = -2$ **11.** $F(0, -1), y = 1$

13. $F\left(\dfrac{5}{8}, 0\right), x = -\dfrac{5}{8}$ **15.** $F\left(0, \dfrac{25}{48}\right), y = -\dfrac{25}{48}$

17. $y^2 = 12x$ **19.** $x^2 = -2y$ **21.** $x^2 = 0.64y$

23. $x^2 = 336y$ **25.** $x^2 = \dfrac{1}{8}y$ **27.** $y^2 = \dfrac{25}{3}x$

29. $y^2 - 2y - 12x + 37 = 0$ **31.**

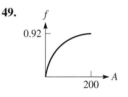

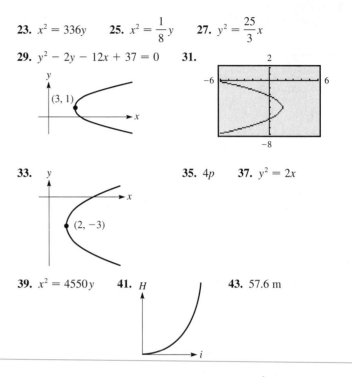

33.

35. $4p$ **37.** $y^2 = 2x$

39. $x^2 = 4550y$ **41.** H   **43.** 57.6 m

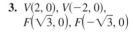

45. No (a course correction is necessary)   **47.** $y^2 = 1.2x$

49.

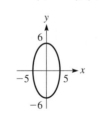

51. $y^2 = 8x$ or $x^2 = 8y$ with vertex midway between island and shore

Exercises 21.5, page 583

1. $V(0, 6), V(0, -6),$
ends minor axis $(5, 0),(-5, 0)$
foci $\left(0, \sqrt{11}\right), \left(0, -\sqrt{11}\right)$

3. $V(2, 0), V(-2, 0),$
$F\left(\sqrt{3}, 0\right), F\left(-\sqrt{3}, 0\right)$

5. $V(0, 12), V(0, -12),$
$F\left(0, \sqrt{119}\right), F\left(0, -\sqrt{119}\right)$

7. $V\left(\dfrac{5}{2}, 0\right), V\left(-\dfrac{5}{2}, 0\right),$
$F\left(\dfrac{3}{2}, 0\right), F\left(-\dfrac{3}{2}, 0\right)$

9. $V(9, 0)$, $V(-9, 0)$,
   $F(3\sqrt{5}, 0)$, $F(-3\sqrt{5}, 0)$

11. $V(0, 7)$, $V(0, -7)$,
   $F(0, \sqrt{45})$, $F(0, -\sqrt{45})$

13. $V(0, 4)$, $V(0, -4)$,
   $F(0, \sqrt{14})$, $F(0, -\sqrt{14})$

15. $V(0.25, 0)$, $V(-0.25, 0)$,
   $F(0.23, 0)$, $F(-0.23, 0)$

17. $\dfrac{x^2}{225} + \dfrac{y^2}{144} = 1$, or $144x^2 + 225y^2 = 32{,}400$

19. $\dfrac{x^2}{13} + \dfrac{y^2}{9} = 1$, or $9x^2 + 13y^2 = 117$

21. $\dfrac{y^2}{9} + \dfrac{x^2}{5} = 1$, or $9x^2 + 5y^2 = 45$

23. $\dfrac{x^2}{64} + \dfrac{15y^2}{144} = 1$, or $3x^2 + 20y^2 = 192$

25. $\dfrac{x^2}{5} + \dfrac{y^2}{20} = 1$, or $4x^2 + y^2 = 20$

27. $\dfrac{x^2}{100} + \dfrac{y^2}{64} = 1$, or $16x^2 + 25y^2 = 1600$

29. $16x^2 + 25y^2 - 32x - 50y - 359 = 0$

31.

33.

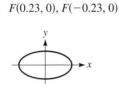

$(-1, -1)$ $(5, -1)$

35. Write equation as $\dfrac{x^2}{1} + \dfrac{y^2}{1/k} = 1$. Thus, $\sqrt{\dfrac{1}{k}} > 1$, or $k < 1$,

37. $2x^2 + 3y^2 - 8x - 4 = 2x^2 + 3(-y)^2 - 8x - 4$

39.

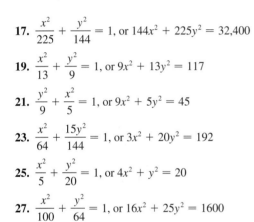

41. $i_1^2 + 4i_2^2 = 32$

43. 0.24   **45.** $7x^2 + 16y^2 = 112$ **47.** 27.5 m

49. 4.0 m   **51.** 843 ft³

Exercises 21.6, page 588

1. $V(0, -4)$, $V(0, 4)$,
   conj. axis $(-2, 0)$, $(2, 0)$
   $F(0, -2\sqrt{5})$, $F(0, 2\sqrt{5})$

3. $V(5, 0)$, $V(-5, 0)$,
   $F(13, 0)$, $F(-13, 0)$

5. $V(0, 3)$, $V(0, -3)$,
   $F(0, \sqrt{10})$, $F(0, -\sqrt{10})$

7. $V\left(-\dfrac{5}{2}, 0\right)$, $V\left(\dfrac{5}{2}, 0\right)$,
   $F\left(-\dfrac{1}{2}\sqrt{41}, 0\right)$, $F\left(\dfrac{1}{2}\sqrt{41}, 0\right)$

9. $V(1, 0)$, $V(-1, 0)$,
   $F(\sqrt{5}, 0)$, $F(-\sqrt{5}, 0)$

11. $V(0, \sqrt{5})$, $V(0, -\sqrt{5})$,
   $F(0, \sqrt{7})$, $F(0, -\sqrt{7})$

13. $V(0, 2)$, $V(0, -2)$,
   $F(0, \sqrt{5})$, $F(0, -\sqrt{5})$

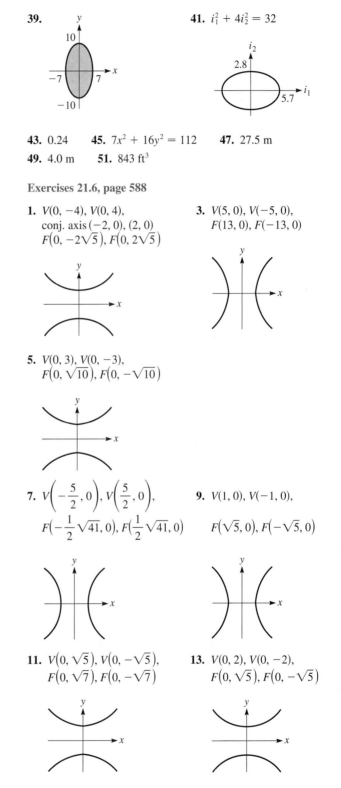

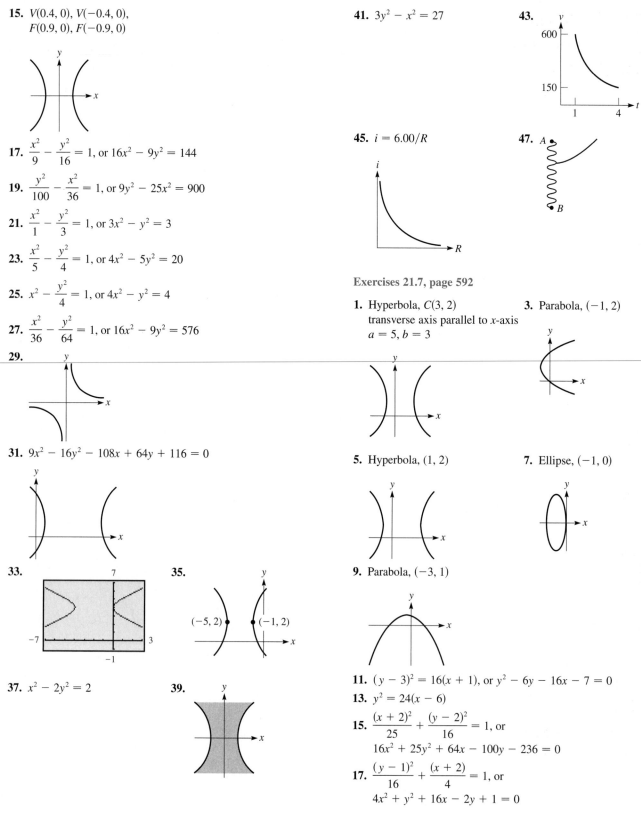

15. $V(0.4, 0)$, $V(-0.4, 0)$,
$F(0.9, 0)$, $F(-0.9, 0)$

17. $\dfrac{x^2}{9} - \dfrac{y^2}{16} = 1$, or $16x^2 - 9y^2 = 144$

19. $\dfrac{y^2}{100} - \dfrac{x^2}{36} = 1$, or $9y^2 - 25x^2 = 900$

21. $\dfrac{x^2}{1} - \dfrac{y^2}{3} = 1$, or $3x^2 - y^2 = 3$

23. $\dfrac{x^2}{5} - \dfrac{y^2}{4} = 1$, or $4x^2 - 5y^2 = 20$

25. $x^2 - \dfrac{y^2}{4} = 1$, or $4x^2 - y^2 = 4$

27. $\dfrac{x^2}{36} - \dfrac{y^2}{64} = 1$, or $16x^2 - 9y^2 = 576$

29.

31. $9x^2 - 16y^2 - 108x + 64y + 116 = 0$

33.

35. $(-5, 2)$ $(-1, 2)$

37. $x^2 - 2y^2 = 2$

39.

41. $3y^2 - x^2 = 27$

43.

45. $i = 6.00/R$

47. A B

Exercises 21.7, page 592

1. Hyperbola, $C(3, 2)$
transverse axis parallel to x-axis
$a = 5$, $b = 3$

3. Parabola, $(-1, 2)$

5. Hyperbola, $(1, 2)$

7. Ellipse, $(-1, 0)$

9. Parabola, $(-3, 1)$

11. $(y - 3)^2 = 16(x + 1)$, or $y^2 - 6y - 16x - 7 = 0$

13. $y^2 = 24(x - 6)$

15. $\dfrac{(x + 2)^2}{25} + \dfrac{(y - 2)^2}{16} = 1$, or
$16x^2 + 25y^2 + 64x - 100y - 236 = 0$

17. $\dfrac{(y - 1)^2}{16} + \dfrac{(x + 2)}{4} = 1$, or
$4x^2 + y^2 + 16x - 2y + 1 = 0$

19. $\dfrac{(y-2)^2}{1} - \dfrac{(x+1)^2}{3} = 1$, or

$x^2 - 3y^2 + 2x + 12y - 8 = 0$

21. $\dfrac{(x+1)^2}{9} - \dfrac{(y-1)^2}{16} = 1$, or

$16x^2 - 9y^2 + 32x + 18y - 137 = 0$

23. Parabola, $(-1, -1)$ **25.** Parabola, $(0, 6)$

27. Ellipse, $(-3, 0)$ **29.** Hyperbola, $(0, 4)$

31. Hyperbola, $(-2, 1)$ **33.** Hyperbola, $(-4, 5)$

35. Ellipse, $\left(\dfrac{2}{3}, -2\right)$ **37.** Hyperbola, $(1, -8)$

39. Circle, $\left(\dfrac{1}{3}, \dfrac{4}{3}\right)$

41. $x^2 - y^2 + 4x - 2y - 22 = 0$

43. $y^2 + 4x - 4 = 0$ **45.** $y^2 = 4p(x - h)$ **47.** $i' = \sin 2\pi t'$

49. $(x - 28)^2 = -\dfrac{28^2}{18}(y - 18)$

51. $\dfrac{x^2}{9.0} + \dfrac{y^2}{16} = 1,\ \dfrac{(x - 7.0)^2}{16} + \dfrac{y^2}{9.0} = 1$

Exercises 21.8, page 595

1. Hyperbola   **3.** Ellipse   **5.** Hyperbola   **7.** Circle
9. Parabola   **11.** Hyperbola   **13.** Circle
15. None (straight line)   **17.** Hyperbola   **19.** Ellipse
21. None (point at origin)   **23.** Ellipse
25. Parabola: $V(-4, 0)$; $F(-4, 2)$

27. Hyperbola; $C(1, -2)$;   **29.** Ellipse; $C(5, 0)$;
$V\left(1, -2 \pm \sqrt{2}\right)$ $V\left(5, \pm 2\sqrt{2}\right)$

31. Parabola; $V\left(-\dfrac{1}{2}, \dfrac{5}{2}\right)$; $F = \left(\dfrac{1}{2}, \dfrac{5}{2}\right)$

33. Ellipse

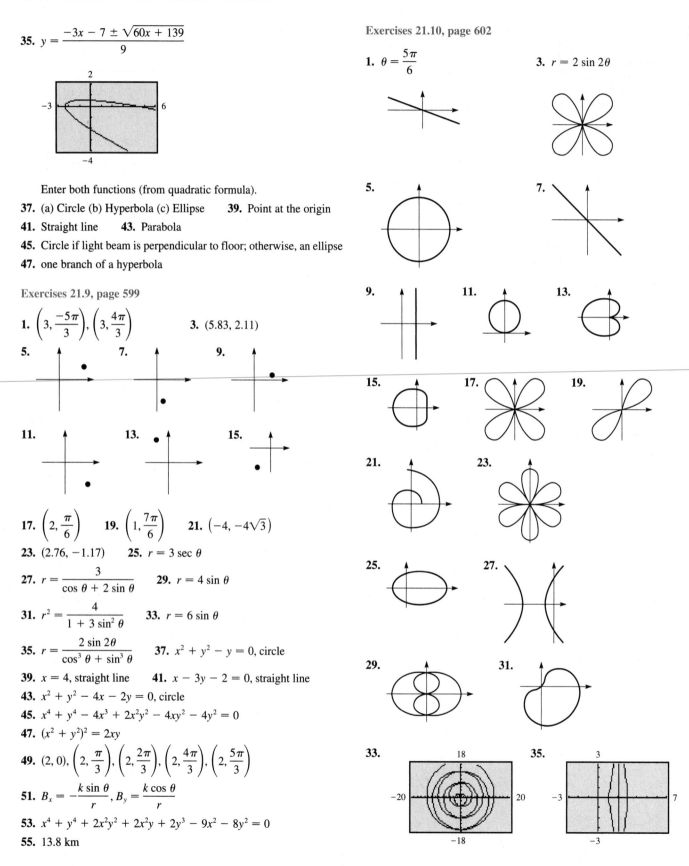

35. $y = \dfrac{-3x - 7 \pm \sqrt{60x + 139}}{9}$

Enter both functions (from quadratic formula).

37. (a) Circle (b) Hyperbola (c) Ellipse   **39.** Point at the origin

41. Straight line   **43.** Parabola

45. Circle if light beam is perpendicular to floor; otherwise, an ellipse

47. one branch of a hyperbola

Exercises 21.9, page 599

1. $\left(3, \dfrac{-5\pi}{3}\right), \left(3, \dfrac{4\pi}{3}\right)$ **3.** (5.83, 2.11)

5.   **7.**   **9.**

11.   **13.**   **15.**

17. $\left(2, \dfrac{\pi}{6}\right)$ **19.** $\left(1, \dfrac{7\pi}{6}\right)$ **21.** $\left(-4, -4\sqrt{3}\right)$

23. (2.76, −1.17)   **25.** $r = 3 \sec \theta$

27. $r = \dfrac{3}{\cos \theta + 2 \sin \theta}$ **29.** $r = 4 \sin \theta$

31. $r^2 = \dfrac{4}{1 + 3 \sin^2 \theta}$ **33.** $r = 6 \sin \theta$

35. $r = \dfrac{2 \sin 2\theta}{\cos^3 \theta + \sin^3 \theta}$ **37.** $x^2 + y^2 - y = 0$, circle

39. $x = 4$, straight line   **41.** $x - 3y - 2 = 0$, straight line

43. $x^2 + y^2 - 4x - 2y = 0$, circle

45. $x^4 + y^4 - 4x^3 + 2x^2y^2 - 4xy^2 - 4y^2 = 0$

47. $(x^2 + y^2)^2 = 2xy$

49. $(2, 0), \left(2, \dfrac{\pi}{3}\right), \left(2, \dfrac{2\pi}{3}\right), \left(2, \dfrac{4\pi}{3}\right), \left(2, \dfrac{5\pi}{3}\right)$

51. $B_x = -\dfrac{k \sin \theta}{r}, B_y = \dfrac{k \cos \theta}{r}$

53. $x^4 + y^4 + 2x^2y^2 + 2x^2y + 2y^3 - 9x^2 - 8y^2 = 0$

55. 13.8 km

Exercises 21.10, page 602

1. $\theta = \dfrac{5\pi}{6}$ **3.** $r = 2 \sin 2\theta$

5.   **7.**

9.   **11.**   **13.**

15.   **17.**   **19.**

21.   **23.**

25.   **27.**

29.   **31.**

33.   **35.**

37.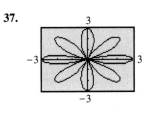

39.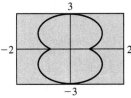

17. $V(0, 4), V(0, -4), F(0, \sqrt{15}), F(0, -\sqrt{15})$

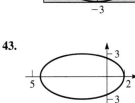

41.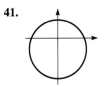

43.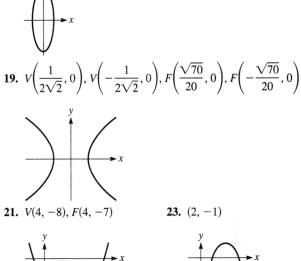

19. $V\left(\dfrac{1}{2\sqrt{2}}, 0\right), V\left(-\dfrac{1}{2\sqrt{2}}, 0\right), F\left(\dfrac{\sqrt{70}}{20}, 0\right), F\left(-\dfrac{\sqrt{70}}{20}, 0\right)$

45.

47. If n is odd, there are n loops. If n is even, there are $2n$ loops.

21. $V(4, -8), F(4, -7)$ **23.** $(2, -1)$

Review Exercises for Chapter 21, page 604

1. $4x - y - 11 = 0$

3. $2x + 3y + 3 = 0$

    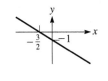

25.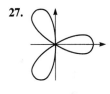

27.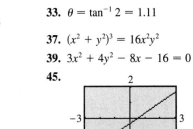

5. $x^2 - 6x + y^2 - 1 = 0$

7. $y^2 = 12x$

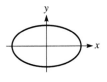   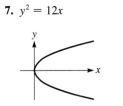

29.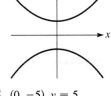

31.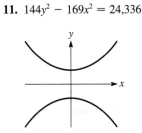

9. $9x^2 + 25y^2 = 900$

11. $144y^2 - 169x^2 = 24{,}336$

33. $\theta = \tan^{-1} 2 = 1.11$ **35.** $r^2 = \dfrac{2}{1 + \sin\theta\cos\theta}$

37. $(x^2 + y^2)^3 = 16x^2y^2$

39. $3x^2 + 4y^2 - 8x - 16 = 0$ **41.** 4   **43.** 2

45.

47.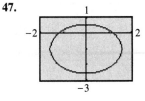

13. $(-3, 0), r = 4$

15. $(0, -5), y = 5$

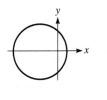

49.

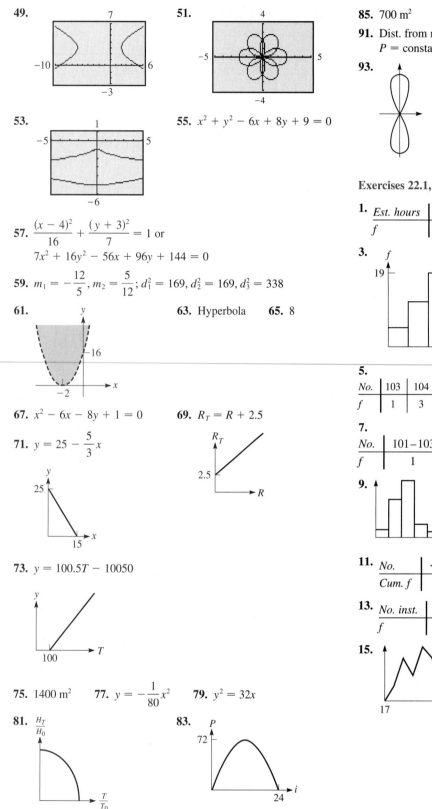

51.

53.

55. $x^2 + y^2 - 6x + 8y + 9 = 0$

57. $\dfrac{(x-4)^2}{16} + \dfrac{(y+3)^2}{7} = 1$ or
$7x^2 + 16y^2 - 56x + 96y + 144 = 0$

59. $m_1 = -\dfrac{12}{5}, m_2 = \dfrac{5}{12}; d_1^2 = 169, d_2^2 = 169, d_3^2 = 338$

61.

63. Hyperbola   **65.** 8

67. $x^2 - 6x - 8y + 1 = 0$ **69.** $R_T = R + 2.5$

71. $y = 25 - \dfrac{5}{3}x$

73. $y = 100.5T - 10050$

75. 1400 m²   **77.** $y = -\dfrac{1}{80}x^2$ **79.** $y^2 = 32x$

81.   **83.**

85. 700 m²   **87.** 18 cm, 8 cm   **89.** 11.3 m

91. Dist. from rifle to P − dist. from target to
P = constant (related to dist. from rifle to target).

93.

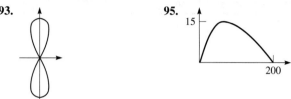

95.

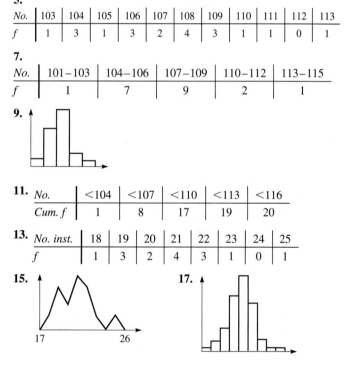

Exercises 22.1, page 612

1.

Est. hours	0–5	6–11	12–17	18–23	24–29
f	5	12	19	9	5

3.

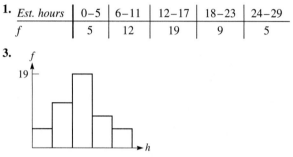

5.

No.	103	104	105	106	107	108	109	110	111	112	113
f	1	3	1	3	2	4	3	1	1	0	1

7.

No.	101–103	104–106	107–109	110–112	113–115
f	1	7	9	2	1

9.

11.

No.	<104	<107	<110	<113	<116
Cum. f	1	8	17	19	20

13.

No. inst.	18	19	20	21	22	23	24	25
f	1	3	2	4	3	1	0	1

15.   **17.**

19.

Time (s)	<2.22	<2.23	<2.24	<2.25	<2.26	<2.27	<2.28	<2.29	<2.30
Cum. f (%)	2	9	27	68	124	156	164	167	170

21.

Dist. (m)	47–49	50–52	53–55	56–58	59–61	62–64	65–67
f (%)	1.7	12.5	26.7	30.0	20.0	8.3	0.8

23. 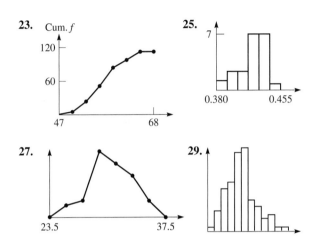 **25.**

5. 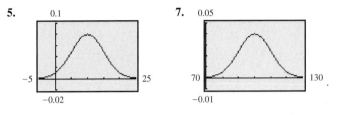 **7.**

9. 136   **11.** 95   **13.** 341   **15.** 78.81%

17. 2166   **19.** 179

21. 141; about 68% of samples have mean lifetime from 99 859 km to 100 141 km.

23. 2.28%   **25.** 76% (normal dist. is 68%)

27. 68% (normal dist. is 68%)

27. **29.**

Exercises 22.5, page 631

1. UCL$(\bar{x})$ = 504.7 mg, LCL$(\bar{x})$ = 494.7 mg

3. The first point would be below the $\bar{x}$ line at 499.7. The UCL and LCL lines would be 0.2 unit lower.

5. $\bar{\bar{x}}$ = 361.8 N · m, UCL = 368.9 N · m, LCL = 354.6 N · m

7.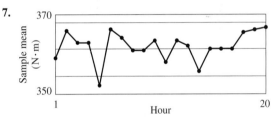

31. The greatest frequency should be at 2 and the least at 0 and 4.

 (Graphs will generally have approximately the shape shown.)

9. $\bar{\bar{x}}$ = 8.986 V, UCL = 9.081 V, LCL = 8.891 V

11.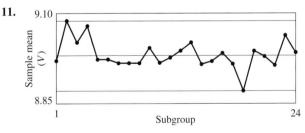

Exercises 22.2, page 616

1. 5   **3.** 5.5   **5.** 4   **7.** 0.49   **9.** 4.6   **11.** 0.503

13. 4   **15.** 0.48, 0.49, 0.55   **17.** 21   **19.** 21

21. 2.248 s   **23.** 57 m   **25.** 0.4237 mSv   **27.** 0.436 mSv

29. 31 h   **31.** 0.005 95 mm   **33.** $475, $500

35. 3450 MJ   **37.** 0.195, 0.18   **39.** $487.50

41. $575, $577, $600; if each value is increased by the same amount, the median, mean, and mode are also increased by this amount.

43. $727; an outlier can make the mean a poor measure of the center of the distribution.

Exercises 22.3, page 621

1. 2.2   **3.** 1.55   **5.** 0.039   **7.** 1.55   **9.** 0.039

11. 4.6, 1.55   **13.** 0.503, 0.039   **15.** 1.8   **17.** 2.6 h

19. 0.014 s   **21.** 0.047   **23.** 0.00022 mm

Exercises 22.4, page 626

1. The peak would be as high as for the left curve, and it would be centered as for the right curve.

3. 0.2072

13. μ = 2.725 cm, UCL = 2.729 cm, LCL = 2.721 cm

15. μ = 5.57 mL, UCL = 11.17 mL, LCL = 0.00 mL

17. $\bar{p}$ = 0.0369, UCL = 0.0548, LCL = 0.0190

19. $\bar{p}$ = 0.0580, UCL = 0.0894, LCL = 0.0266

Exercises 22.6, page 636

1. $y = 2x + 2$

3. $y = -1.77x + 191$

5. $V = -0.590i + 11.3$

7. $h = 2.24x + 5.2$

9. $p = -2.66x + 4364$

11. $V = 4.32 \times 10^{-15}f - 2.03$
$f_0 = 0.470$ PHz

13. 0.953   **15.** -0.901

Exercises 22.7, page 640

1. $y = 2.59x^2 + 1.07$

3. $y = 10.9/x$

5. $y = 5.97t^2 + 0.38$

7. $p = 0.551T^2 + 540$

9. $P = \dfrac{1343}{S}$

11. $y = 6.20e^{-t} - 0.05$

Review Exercises for Chapter 22, page 641

1. 1100   **3.** 1100

5.

No.	1093–1095	1096–1098	1099–1101
f	3	4	3

No.	1102–1104	1105–1107
f	4	2

7.

9.

No.	<1096	<1099	<1102	<1105	<1108
f	3	7	10	14	16

11. 0.264 Pa · s   **13.** 0.014 Pa · s

15.

17. 700 W   **19.** 700 W

21. 17.3 W

23.

Power (W)	<660	<670	<680	<690
Cum. f	3	5	12	24

Power (W)	<700	<710	<720	<730	<740
Cum. f	51	85	100	116	121

25. 4   **27.**

29. 66.2 km/h

31. 9.0 km/h

33. $\bar{p} = 0.0540$, UCL $= 0.0843$, LCL $= 0.0237$

35.

37. 322   **39.** 495   **41.** $R = 0.0983T + 25.0$

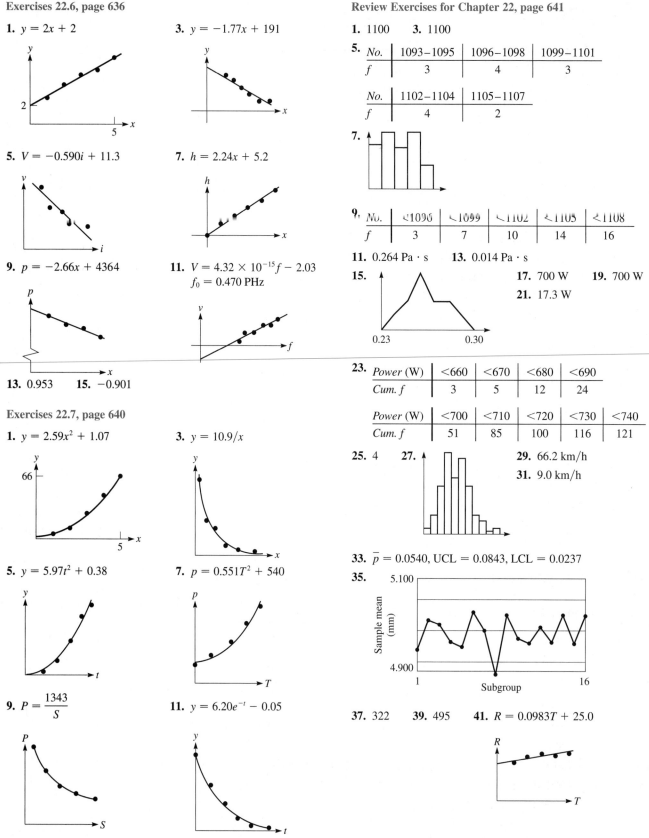

43. $s = 0.123t + 0.887$ **45.** $s = -4.90t^2 + 3000$

47. $y = -5.2x + 96.2$ (Curve has a good fit to a straight line but fits $T = 82.6e^{-0.1t} + 17.3$ better.)

49. $y = 0.997x + 0.581$ **51.** $y = 1.39x^{0.878}$ **53.** 9.2 ppm

55. Substitute expressions for m, b, and $\bar{x}$. Simplify to $\bar{y} = \bar{y}$.

Exercises 23.1, page 652

1. Not cont. at $x = -2$ **3.** -4 **5.** Cont. all x

7. Not cont. $x = 0$ and $x = 1$, div. by zero

9. Cont. $x \le 0$, $x > 2$; function not defined   **11.** Cont. all x

13. Not cont. $x = 1$, small change   **15.** Cont. $x \le 2$

17. Not cont. $x = 2$, small change   **19.** Cont. all x

21.

x	0.900	0.990	0.999	1.001
$f(x)$	1.7100	1.9701	1.9970	2.0030

x	1.010	1.100
$f(x)$	2.0301	2.3100

$\lim\limits_{x \to 1} f(x) = 2$

23.

x	1.900	1.990	1.999	2.001
$f(x)$	-0.2516	-0.2502	-0.25002	-0.24998

x	2.010	2.100
$f(x)$	-0.2498	-0.2485

$\lim\limits_{x \to 2} f(x) = -0.25$

25.

x	10	100	1000
$f(x)$	0.4468	0.4044	0.4004

$\lim\limits_{x \to \infty} f(x) = 0.4$

27. 7   **29.** 1   **31.** $-\dfrac{2}{3}$ **33.** 27   **35.** 2

37. Does not exist   **39.** 3   **41.** 1

43.

x	-0.1	-0.01	-0.001	0.001
$f(x)$	-3.1	-3.01	-3.001	-2.999

x	0.01	0.1
$f(x)$	-2.99	-2.9

$\lim\limits_{x \to 0} f(x) = -3$

45.

x	10	100	1000
$f(x)$	2.1649	2.0106	2.0010

$\lim\limits_{x \to \infty} f(x) = 2$

47. 3 cm/s   **49.** 34.9°C, 0°C   **51.** e **53.** 0

55. -1; $+1$; no, $\lim\limits_{x \to 0^+} f(x) \ne \lim\limits_{x \to 0^-} f(x)$

Exercises 23.2, page 656

1. 9   **3.** (Slopes) 3.5, 3.9, 3.99 3.999; $m = 4$ **5.** (Slopes) -2, -2.8, -2.98, -2.998; $m = -3$

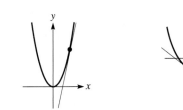

7. 4   **9.** -3 **11.** $m_{\tan} = 2x_1$; 4, -2

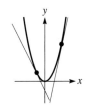

13. $m_{\tan} = 4x_1 + 5$; -3, 7   **15.** $m_{\tan} = 2x_1 + 4$; -2, 8

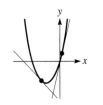
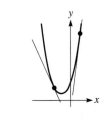

17. $m_{\tan} = 6 - 2x_1$; 10, 0   **19.** $m_{\tan} = 6x_1^3$; 0, 0.75, 6

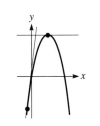
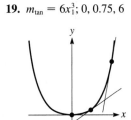

21. $m_{\tan} = 5x_1^4$; 0, 0.31, 5

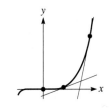

23.   **25.**

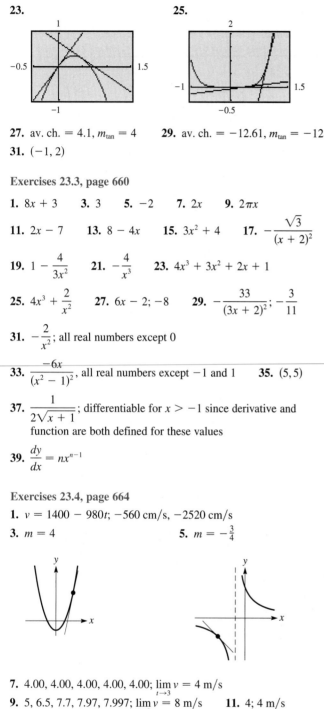

27. av. ch. $= 4.1$, $m_{\tan} = 4$ **29.** av. ch. $= -12.61$, $m_{\tan} = -12$

31. $(-1, 2)$

Exercises 23.3, page 660

1. $8x + 3$ **3.** 3 **5.** -2 **7.** $2x$ **9.** $2\pi x$

11. $2x - 7$ **13.** $8 - 4x$ **15.** $3x^2 + 4$ **17.** $-\dfrac{\sqrt{3}}{(x + 2)^2}$

19. $1 - \dfrac{4}{3x^2}$ **21.** $-\dfrac{4}{x^3}$ **23.** $4x^3 + 3x^2 + 2x + 1$

25. $4x^3 + \dfrac{2}{x^2}$ **27.** $6x - 2$; -8 **29.** $-\dfrac{33}{(3x + 2)^2}$; $-\dfrac{3}{11}$

31. $-\dfrac{2}{x^2}$; all real numbers except 0

33. $\dfrac{-6x}{(x^2 - 1)^2}$, all real numbers except -1 and 1 **35.** $(5, 5)$

37. $\dfrac{1}{2\sqrt{x + 1}}$; differentiable for $x > -1$ since derivative and function are both defined for these values

39. $\dfrac{dy}{dx} = nx^{n-1}$

Exercises 23.4, page 664

1. $v = 1400 - 980t$; -560 cm/s, -2520 cm/s

3. $m = 4$ **5.** $m = -\dfrac{3}{4}$

7. $4.00, 4.00, 4.00, 4.00, 4.00$; $\lim\limits_{t \to 3} v = 4$ m/s

9. $5, 6.5, 7.7, 7.97, 7.997$; $\lim\limits_{t \to 2} v = 8$ m/s   **11.** 4; 4 m/s

13. $6t - 4$; 8 m/s   **15.** 48 **17.** $8t - 4t^3$

19. $3 + \dfrac{2}{5t^2}$ **21.** $12t - 4$ **23.** $6t$ **25.** 4.5 s

27. -2 **29.** $6w$ **31.** 460 W   **33.** -83.1 W/(m² · h)

35. $-\dfrac{48}{(t + 3)^2}$; $-\$1300$/year   **37.** πd^2 **39.** $24.2/\sqrt{\lambda}$

Exercises 23.5, page 669

1. $9r^8$ **3.** $6x^2 - 12x + 8$; 56 **5.** $5x^4$ **7.** $-36x^8$

9. $4x^3$ **11.** $2x + 2$ **13.** $15r^2 - 2$

15. $200x^7 - 170x^4 - 1$ **17.** $-42x^6 + 15x^2$

19. $x^2 + x$ **21.** 16 **23.** 33

25. -4 **27.** -29

29. $30t^4 - 5$ **31.** $-6 - 6t^2$ **33.** 64 **35.** 45 **37.** 1

39. $(2, 4)$ **41.** $(1, -5)$ **43.** $-\dfrac{1}{4}$ **45.** $3\pi r^2$

47. 84 W/A   **49.** $a(c_1 + 2c_2 E + 3c_3 E^2)$

51. $-0.002\,86$ N/°C   **53.** -80.5 m/km   **55.** 391 mm²

Exercises 23.6, page 673

1. $-12x^5 + 5x^4 + 6x^2 - 2x$

3. $6x(6x - 5) + (3x^2 - 5x)(6) = 54x^2 - 60x$

5. $(3t + 2)(2) + (2t - 5)(3) = 12t - 11$

7. $(x^4 - 3x^2 + 3)(-6x^2) + (1 - 2x^3)(4x^3 - 6x)$
$\quad = -14x^6 + 30x^4 + 4x^3 - 18x^2 - 6x$

9. $(2x - 7)(-2) + (5 - 2x)(2) = -8x + 24$

11. $(x^3 - 1)(4x - 1) + (2x^2 - x - 1)(3x^2)$
$\quad = 10x^4 - 4x^3 - 3x^2 - 4x + 1$

13. $\dfrac{3}{(2x + 3)^2}$ **15.** $-\dfrac{4\pi x}{(2x^2 + 1)^2}$ **17.** $\dfrac{6x - 2x^2}{(3 - 2x)^2}$

19. $\dfrac{-6x^2 + 6x + 4}{(3x^2 + 2)^2}$ **21.** $\dfrac{-3x^2 - 16x - 26}{(x^2 + 4x + 2)^2}$

23. $\dfrac{-2x^3 + 2x^2 + 5x + 4}{x^3(x + 2)^2}$ **25.** -107 **27.** 75

29. 19 **31.** -5.64 **33.** Eq. (23.10)   **35.** $(0, -1)$

37. (1) $\dfrac{-12x^3 + 45x^2 - 14x}{(3x - 7)^2}$ (2) $\dfrac{-12x^3 + 45x^2 - 14x}{(3x - 7)^2}$

39. 12 **41.** $1, -1$

43. $8t^3 - 45t^2 - 14t - 8$ **45.** $\dfrac{96}{(7R + 12)^2}$ **47.** 1.2°C/h

49. $\dfrac{2R(R + 2r)}{3(R + r)^2}$ **51.** $\dfrac{E^2(R - r)}{(R + r)^3}$

Exercises 23.7, page 679

1. $36x^2(2 + 3x^3)^3$ **3.** $-\dfrac{3x}{(2 - 3x^2)^{1/2}}$

5. $\dfrac{1}{2x^{1/2}}$ **7.** $-\dfrac{6}{t^3}$ **9.** $-\dfrac{1}{x^{4/3}}$ **11.** $\dfrac{3}{2}x^{1/2} + \dfrac{1}{x^2}$

13. $10x(x^2 + 1)^4$ **15.** $-216x^2(7 - 4x^3)^7$

17. $\dfrac{2x^2}{(2x^3 - 3)^{2/3}}$ **19.** $\dfrac{24y}{(4 - y^2)^5}$ **21.** $\dfrac{24x^3}{(2x^4 - 5)^{0.25}}$

23. $\dfrac{-4x}{(1 - 8x^2)^{3/4}}$ **25.** $\dfrac{12x + 5}{(8x + 5)^{1/2}}$ **27.** $\dfrac{6(5x - 1)}{x^4(1 - 6x)^{1/2}}$

29. $\dfrac{x^2 + 12x + 16}{(x + 4)^2(x + 2)^{1/2}}$ **31.** $\dfrac{-1}{\sqrt{2R + 1}(4R + 1)^{3/2}}$ **33.** $\dfrac{3}{10}$

35. $\dfrac{5}{36}$ **37.** $\dfrac{x^3(0) - 1(3x^2)}{x^6} = -3x^{-4}$

39. $x = 0$ **41.** Yes, at $\left(-\dfrac{1}{5}, \dfrac{61}{15}\right)$

43. 1   **45.** -1.35 cm/s   **47.** $\dfrac{-450\,000}{V^{5/2}}$, -4.50 kPa/cm³

49. $l = a$ **51.** -45.2 W/(m² · h)   **53.** $\dfrac{8a^3}{(4a^2 - \lambda^2)^{3/2}}$

55. $\dfrac{2(w + 1)}{(2w^2 + 4w + 4)^{1/2}}$

Exercises 23.8, page 683

1. $-\dfrac{4x}{3y^2}$ **3.** $-\dfrac{3}{2}$ **5.** $\dfrac{6x + 1}{4}$ **7.** $\dfrac{x}{4y}$ **9.** $\dfrac{2x}{5y^4}$

11. $\dfrac{2x}{2y + 1}$ **13.** $\dfrac{-3y}{3x + 1}$ **15.** $\dfrac{-2x - y^3}{3xy^2 + 3}$

17. $\dfrac{3(y^2 + 1)(y^2 - 2x + 1)}{(y^2 + 1)^2 - 6x^2y}$ **19.** $\dfrac{4(2y - x)^3 - 2x}{8(2y - x)^3 - 1}$

21. $\dfrac{-3x(x^2 + 1)^2}{y(y^2 + 1)}$ **23.** 3   **25.** $-\dfrac{108}{157}$ **27.** $\dfrac{1}{4}$

29. $(2, 2), (2, -2)$ **31.** 1   **33.** $\dfrac{nRT^2 + bnP}{VT^2 - anT^2 + bnT}$

35. $-\dfrac{x}{y}$ **37.** $\dfrac{r - R + 1}{r + 1}$ **39.** $\dfrac{2C^2r(12CSr - 20Cr - 3L)}{3(C^2r^2 - L^2)}$

Exercises 23.9, page 686

1. $y' = 15x^2 - 4x, y'' = 30x - 4, y''' = 30, y^{(n)} = 0\ (n \geq 4)$

3. $y' = 3x^2 + 2x, y'' = 6x + 2, y''' = 6, y^{(n)} = 0\ (n \geq 4)$

5. $f'(x) = 3x^2 - 24x^3, f''(x) = 6x - 72x^2,$
     $f'''(x) = 6 - 144x, f^{(4)}(x) = -144, f^{(n)}(x) = 0\ (n \geq 5)$

7. $y' = -8(1 - 2x)^3, y'' = 48(1 - 2x)^2,$
     $y''' = -192(1 - 2x), y^{(4)} = 384, y^{(n)} = 0\ (n \geq 5)$

9. $f'(r) = (16r + 1)(4r + 1)^2, f''(r) = 24(8r + 1)(4r + 1),$
     $f'''(r) = 96(16r + 3), f^{iv}(r) = 1536, f^{(n)}(r) = 0\ (n \geq 5)$

11. $84x^5 - 30x^4$ **13.** $-\dfrac{1}{4x^{3/2}}$ **15.** $-\dfrac{12}{(8x - 3)^{7/4}}$

17. $\dfrac{14.4\pi}{(1 + 2p)^{5/2}}$ **19.** $600(2 - 5x)^2$

21. $30(27x^2 - 1)(3x^2 - 1)^3$ **23.** $\dfrac{4\pi^2}{(1 - x)^3}$ **25.** $\dfrac{2}{(x + 1)^3}$

27. $-\dfrac{9}{y^3}$ **29.** $-\dfrac{6(x^2 - xy + y^2)}{(2y - x)^3}$ **31.** $\dfrac{9}{125}$ **33.** $-\dfrac{13}{384}$

35. -50 **37.** -9.8 m/s²   **39.** 1 m/s²   **41.** 48

43. -9.8 m/s²   **45.** $-\dfrac{1.60}{(2t + 1)^{3/2}}$ **47.** 0.049

Review Exercises for Chapter 23, page 688

1. -4 **3.** Does not exist.   **5.** 1   **7.** $\dfrac{7}{3}$ **9.** $\dfrac{2}{3}$ **11.** -2

13. 5   **15.** $-4x$ **17.** $-\dfrac{4}{x^3}$ **19.** $\dfrac{1}{2\sqrt{x + 5}}$

21. $14x^6 - 6x$ **23.** $\dfrac{2}{x^{1/2}} + \dfrac{3}{x^2}$ **25.** $\dfrac{3}{(1 - 5y)^2}$

27. $-12(2 - 3x)^3$ **29.** $\dfrac{9\pi x}{(5 - 2x^2)^{7/4}}$

31. $\dfrac{1}{\sqrt{1 + \sqrt{1 + \sqrt{1 + 8s}}}\ \sqrt{1 + \sqrt{1 + 8s}}\ \sqrt{1 + 8s}}$

33. $\dfrac{-2x - 3}{2x^2(4x + 3)^{1/2}}$ **35.** $\dfrac{2x - 6(2x - 3y)^2}{1 - 9(2x - 3y)^2}$ **37.** $\dfrac{5}{48}$

39. $\dfrac{74}{5}$ **41.** $36x^2 - \dfrac{2}{x^3}$ **43.** $\dfrac{56}{(1 + 4t)^3}$

45. It appears to be 8, but using *trace*, there is no value shown for $x = 2$.

Point (2, 8) is missing

47. (a) 30 m/s (b) 6 m/s   **49.** -31 **51.** $\left(0, \dfrac{1}{3}\sqrt{3}\right)$

53. (a) $v = \dfrac{4}{(1 + 8t)^{1/2}}$ (b) $a = \dfrac{-16}{(1 + 8t)^{3/2}}$ **55.** 5

57. $-k + k^2t - \frac{1}{2}k^3t^2$ **59.** $-\frac{2k}{r^3}$

61. $0.4(0.01t + 1)^2(0.04t + 1)$

63. $\frac{2R(R + 2r)}{3(R + r)^2}$ **65.** 830 W   **67.** $5k(x^4 + 270x^2 - 190)$

69. $-\frac{1}{4\pi\sqrt{C}(L + 2)^{3/2}}$ **71.** $\frac{-15}{(0.5t + 1)^2}$

73. $y' = \frac{w}{6EI}(3L^2x - 3Lx^2 + x^3)$

$y'' = \frac{w}{2EI}(L - x)^2$

$y''' = \frac{w}{EI}(x - L)$

$y^{iv} = \frac{w}{EI}$

75. $p = 2w + \frac{150}{w}, \frac{dp}{dw} = 2 - \frac{150}{w^2}$

77. $A = 4x - x^3, \frac{dA}{dx} = 4 - 3x^2$

79. 397 km/h

Exercises 24.1, page 694

1. $x + 8y - 17 = 0$

3. $4x - y - 2 = 0$

5. $2y + x - 2 = 0$

7. $x - 2y + 6 = 0$

9. $2x - 6y + 7 = 0$

11. $y = 2x - 4$

13. $y - 8 = -\frac{1}{24}\left(x - \frac{3}{2}\right)$, or $2x + 48y - 387 = 0$

15. $2x - 12y + 37 = 0$
$72x + 12y + 37 = 0$

17. The line is $y - x - 1 = 0$.

19. Take derivatives; evaluate at (a, b); show that product $m_1m_2 = -4a/b^2 = -1$.

21. $x - 4y + 2 = 0$ **23.** $x - 2y - 20 = 0$

25. $x + y - 6 = 0$

27. $x + 2y - 3 = 0, x = 0, x - 2y + 3 = 0$

Exercises 24.2, page 698

1. $x_3 = 0.2086$; calculator: 0.2087 (to four decimal places)

3. $-0.180\,460\,4$ **5.** $0.585\,786\,4$ **7.** $0.348\,894\,2$

9. $2.561\,552\,8$ **11.** $-1.236\,068\,0$ **13.** $0.917\,543\,3$

15. $0.618\,034\,0$ **17.** $-1.855\,772\,5, 0.678\,362\,8, 3.177\,409\,7$

19. Find the real root of $x^3 - 4 = 0$; $1.587\,401\,1$

21. $x_{n+1} = \frac{1}{2}x_n + \frac{a}{2x_n}$ **23.** 47 s   **25.** 29.5 m   **27.** 1.61 m

Exercises 24.3, page 702

1. $16.5, -14.0°$ **3.** $3.16, 341.6°$ **5.** $8.07, 352.4°$

7. $a = 0$ **9.** $20.0, 3.7°$ **11.** 9.4 m/s, $302°$

13. 1.3 m/min^2, $288°$ **15.** 36 m/s, $332°$; 9.8 m/s^2, $270°$

17. 276 m/s, $43.5°$; 2090 m/s, $16.7°$ **19.** 1.32 cm/s, $-24.9°$

21. 22.1 m/s^2, $25.4°$; 20.2 m/s^2, $8.5°$ **23.** 21.2 km/min, $296.6°$

25. $x^2 + y^2 = 1.75^2$; $v_x = 731$ m/min, $v_y = -690$ m/min

27. 370 m/s, $19°$

Exercises 24.4, page 705

1. 5.20 V/min   **3.** 0.031 /s   **5.** 0.0900 Ω/s   **7.** $0.002\,51$

9. 330 km/h   **11.** 4.1×10^{-6} m/s   **13.** $\frac{dB}{dt} = \frac{-3kr(dr/dt)}{[r^2 + (l/2)^2]^{5/2}}$

15. 0.0050 m/min   **17.** 0.15 mm^2/month

19. -101 mm^3/min   **21.** $\frac{dV}{dt} = kA; \frac{dr}{dt} = k$

23. -4.6 kPa/min   **25.** 3.18×10^6 mm^3/s

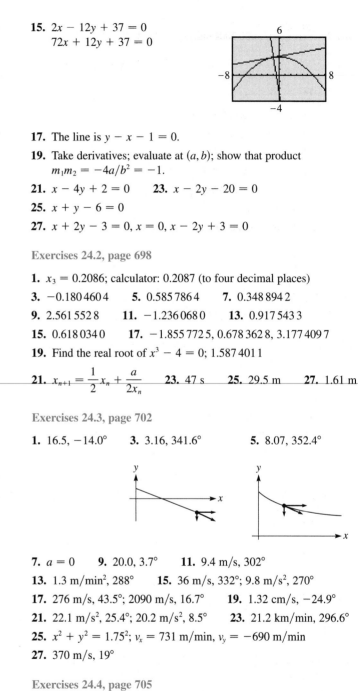

27. $I = \dfrac{8k}{x^2}; \dfrac{dI}{dt} = 0.000\,800k$ unit/s   **29.** 0.48 m/min

31. 820 km/h   **33.** 2.71 m/s   **35.** 2.50 m/s

Exercises 24.5, page 712

1. Inc. $x < 0$, $x > 4$, dec. $0 < x < 4$

3. Conc. down $x < 0$, conc. up $x > 0$, infl. $(0, 0)$

5. Inc. $x > -1$, dec. $x < -1$

7. Inc. $-2 < x < 2$, dec. $x < -2$, $x > 2$

9. Min. $(-1, -1)$ **11.** Min. $(-2, -16)$, Max. $(2, 16)$

13. Conc. up all x

15. Conc. up $x < 0$, conc. down $x > 0$, infl. $(0, 0)$

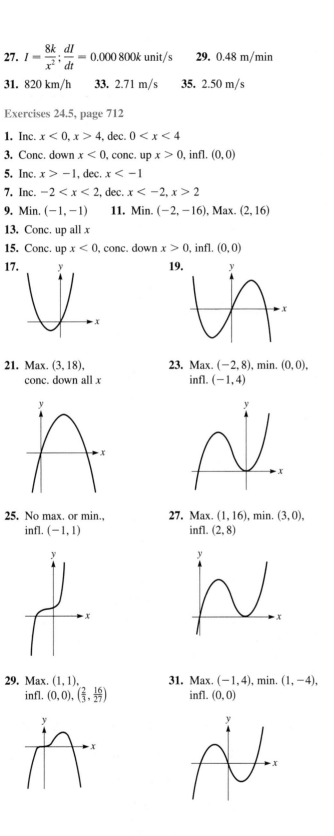

17.

19.

21. Max. $(3, 18)$, conc. down all x

23. Max. $(-2, 8)$, min. $(0, 0)$, infl. $(-1, 4)$

25. No max. or min., infl. $(-1, 1)$

27. Max. $(1, 16)$, min. $(3, 0)$, infl. $(2, 8)$

29. Max. $(1, 1)$, infl. $(0, 0)$, $\left(\frac{2}{3}, \frac{16}{27}\right)$

31. Max. $(-1, 4)$, min. $(1, -4)$, infl. $(0, 0)$

33. Where $y' > 0$, y inc.
$y' = 0$, y has a max. or min.
$y' < 0$, y dec.
$y'' > 0$, y conc. up
$y'' = 0$, y has infl.
$y'' < 0$, y conc. down

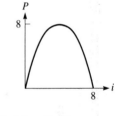

35. Curve has a maximum and a minimum for $c < 0$ but is always increasing for $c > 0$.

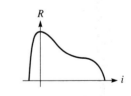

37. The left relative maximum point is above the left relative minimum point but below the right relative minimum point.

39. Max. $(20, 10)$ **41.** Max. $(4, 8)$

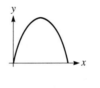

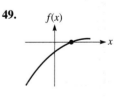

43. Max. $(3, 16)$ **45.** Max. $(0, 75)$, infl. $(1, 64)$, $(3, 48)$

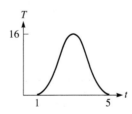

47. $V = 4x^3 - 40x^2 + 96x$, max. $(1.57, 67.6)$ **49.**

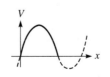

51.

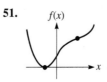

Exercises 24.6, page 717

1. $y = x - \dfrac{4}{x}$
int. $(2, 0)$, $(-2, 0)$;
inc. all x, except $x = 0$;
conc. up $x < 0$,
conc. down $x > 0$

3. Dec. $x < -1, x > -1$,
conc. up $x > -1$,
conc. down $x < -1$,
int. $(0, 2)$,
asym. $x = -1, y = 0$

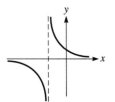

5. Int. $\left(-\sqrt[3]{2}, 0\right)$, min. $(1, 3)$,
infl. $\left(-\sqrt[3]{2}, 0\right)$, asym. $x = 0$

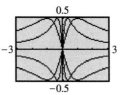

7. Int. $(1, 0)$, $(-1, 0)$,
asym. $x = 0, y = x$,
conc. up $x < 0$,
conc. down $x > 0$

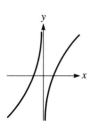

9. Int. $(0, 0)$, max. $(-2, -4)$,
min. $(0, 0)$, asym. $x = -1$

11. Int. $(0, -1)$,
max. $(0, -1)$,
asym. $x = 1$,
$x = -1, y = 0$

13. Int. $(1, 0)$, max. $(2, 1)$,
infl. $\left(3, \frac{8}{9}\right)$,
asym. $x = 0, y = 0$

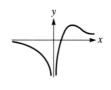

15. Int. $(0, 0)$, $(1, 0)$, $(-1, 0)$,
max. $\left(\frac{1}{2}\sqrt{2}, \frac{1}{2}\right)$,
min. $\left(-\frac{1}{2}\sqrt{2}, -\frac{1}{2}\right)$

17. Int. $(0, 0)$, infl. $(0, 0)$
asym. $x = -3$,
$x = 3, y = 0$

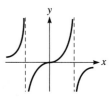

19. As c goes from -3 to $+3$,
the graph goes from second
and fourth quadrants to the
third and first quadrants.

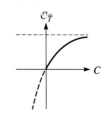

21. Int. $(0, 0)$, asym. $C_T = 6$,
inc. $C \geq 0$,
conc. down $C \geq 0$

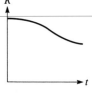

23. Int. $(0, 1)$, max. $(0, 1)$
infl. $(141, 0.82)$,
asym. $R = 0$

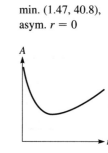

25. Int. $(1, 0)$, $(-1, 0)$;
inc. all x, except $x = 0$;
conc. up $x < 0$,
conc. down $x > 0$

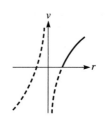

27. $A = 2\pi r^2 + \dfrac{40}{r}$,
min. $(1.47, 40.8)$,
asym. $r = 0$

Exercises 24.7, page 722

1. $360\,000$ m^2 **3.** 60 m **5.** $\dfrac{E}{2R}$ **7.** 35 m^2, \$8300

9. $1500\ \Omega$ **11.** 1.94 units **13.** 12 cm by 12 cm

15. 5 mm, 5 mm

17. $xy = A, p = 2x + 2y; p = 2x + \frac{2A}{x}; \frac{dp}{dx} = 2 - \frac{2A}{x^2};$
$x = \sqrt{A}, y = \sqrt{A}; x = y$ (square)

19. 1.1 h **21.** 8.49 cm, 8.49 cm **23.** 1.20 m

25. 2 cm **27.** 12 **29.** $0.58L$ **31.** $38.0°$

33. 2.7 km from A **35.** 3.3 cm   **37.** 100 m

39. $w = 0.50$ m, $d = 0.87$ m   **41.** 8.0 km from refinery

43. 59.2 m, 118 m

Exercises 24.8, page 728

1. $\dfrac{8(2 - x^3)}{(x^3 + 4)^2}\,dx$ **3.** $\dfrac{de}{e} = 0.0065 = 0.65\%$; $\dfrac{dV}{V} = 0.019 = 1.9\%$

5. $(5x^4 + 1)\,dx$ **7.** $\dfrac{-10\,dr}{r^6}$ **9.** $48t(3t^2 - 5)^3\,dt$

11. $\dfrac{-12x\,dx}{(3x^2 + 1)^2}$ **13.** $x(1 - x)^2(-5x + 2)\,dx$

15. $\dfrac{2\,dx}{(5x + 2)^2}$ **17.** 12.28, 12   **19.** 0.626 490 3, 0.6257

21. $L(x) = 2x$ **23.** $L(x) = -2x - 3$

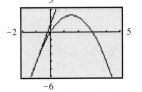

25. 1570 km   **27.** 0.25%   **29.** -31 nm

31. $\dfrac{dr}{r} = \dfrac{kd\lambda/(2\lambda^{1/2})}{k\lambda^{1/2}}$ **33.** $\dfrac{dA}{A} = \dfrac{2\,ds}{s}$ **35.** 2.0125

37. $L(x) = -\tfrac{1}{2}x + \tfrac{3}{2}$; 1.45   **39.** $L(V) = 1.73 - 0.13V$

Review Exercises for Chapter 24, page 729

1. $5x - y + 1 = 0$ **3.** $8x + 5y - 50 = 0$

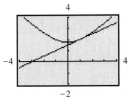

5. $x - 2y + 3 = 0$ **7.** 4.19, 72.6°   **9.** 2.12

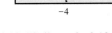

11. 2.00, 90.9°   **13.** 0.745 898 3   **15.** 1.4422

17. Min. $(-2, -16)$,
conc. up all x

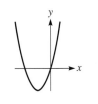

19. Int. $(0, 0)$, $(\pm 3\sqrt{3}, 0)$;
max. $(3, 54)$, min. $(-3, -54)$;
infl. $(0, 0)$

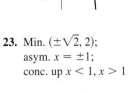

21. Min. $(2, -48)$;
conc. up $x < 0$, $x > 0$

23. Min. $(\pm\sqrt{2}, 2)$;
asym. $x = \pm 1$;
conc. up $x < 1$, $x > 1$

25. $\left(12x^2 - \dfrac{1}{x^2}\right)dx$ **27.** $\dfrac{(1 - 4x)\,dx}{(1 - 3x)^{2/3}}$ **29.** 0.061

31. $L(x) = \dfrac{1}{3}(11x - 4)$ **33.** 1.85 m³   **35.** 3200 m³

37. $\dfrac{R\,dR}{R^2 + X^2}$ **39.** 251.1   **41.** $2x - y + 1 = 0$

43. 0.0 m, 6.527 m   **45.** 8.8 m/s, 336°   **47.** -7.44 cm/s

49. 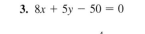   **51.** 22 m³/s

53.    **55.** -0.113 m/min

57. 38 000 m²/min   **59.** 5000 cm²

61. Max. $(0, 100)$; infl. $(37, 63)$;   **63.** 1160 km/h
int. $(0, 100)$, $(89, 0)$
$L(x) = 113 - 1.42x$

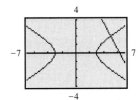

65. 6 μF, 6 μF   **67.** 13.5 dm³   **69.** 3 km   **71.** 6.6 cm

73. 0.153 m/min   **75.** $r = 2.09$ cm; $h = 23.0$ cm

Exercises 25.1, page 735

1. $F(x) = 3x^4$ **3.** $F(x) = \frac{2}{3}x^{3/2} + \frac{1}{x^3}$

5. 1   **7.** 3   **9.** 6   **11.** -1 **13.** $x^{5/2}$

15. $\frac{3}{2}t^4 + 2t$ **17.** $\frac{2}{3}x^3 - \frac{1}{2}x^2$ **19.** $\frac{4}{3}x^{3/2} + 3x$

21. $\frac{7}{5x^5}$ **23.** $2v^2 + 3\pi^2 v$ **25.** $\frac{1}{3}x^3 + 2x - \frac{1}{x}$

27. $(2x + 1)^6$ **29.** $(p^2 - 1)^4$ **31.** $\frac{1}{40}(2x^4 + 1)^5$

33. $(6x + 1)^{3/2}$ **35.** $\frac{1}{4}(3x + 1)^{4/3}$

Exercises 25.2, page 740

1. $4x^2 + C$ **3.** $\frac{(x^2 + 1)^5}{5} + C$ **5.** $x^2 + C$

7. $\frac{1}{8}x^8 + C$ **9.** $\frac{4}{5}x^{5/2} + C$ **11.** $-\frac{3}{R^3} + C$

13. $\frac{1}{3}x^3 - \frac{1}{6}x^6 + C$ **15.** $3x^3 + \frac{1}{2}x^2 + 3x + C$

17. $\frac{1}{6}t^3 + \frac{2}{t} + C$ **19.** $\frac{2}{7}x^{7/2} - \frac{2}{5}x^{5/2} + C$

21. $6x^{1/3} + \frac{1}{9}x + C$ **23.** $s + \frac{4}{3}s^3 + \frac{4}{5}s^5 + C$

25. $\frac{1}{6}(x^2 - 1)^6 + C$ **27.** $\frac{1}{5}(x^4 + 3)^5 + C$

29. $\frac{1}{80}(2\theta^5 + 5)^8 + C$ **31.** $\frac{1}{12}(8x + 1)^{3/2} + C$

33. $\frac{1}{6}\sqrt{6x^2 + 1} + C$ **35.** $\sqrt{x^2 - 2x} + C$

37. $y = 2x^3 + 2$ **39.** $y = 5 - \frac{1}{18}(1 - x^3)^6$

41. No. The result should include the constant of integration.

43. No. With $u = 4x^3 + 3$, $du = 12x^2\,dx$, and the x^4 would have to be x^2.

45. No. There should be a factor of $1/2$ in the result.

47. $12y = 83 + (1 - 4x^2)^{3/2}$ **49.** $i = 2t^2 - 0.2t^3 + 2$

51. $T = 2250(r + 1)^{-2} + 250$ **53.** $f = \sqrt{0.01A + 1} - 1$

55. $y = 3x^2 + 2x - 3$

Exercises 25.3, page 745

1. (a) 3 (b) $\frac{15}{4}$ **3.** 20   **5.** 9, 12.15   **7.** 1.92, 2.28

9. 7.625, 8.208   **11.** 0.464, 0.5995   **13.** 1.92, 1.96

15. 13.5   **17.** $\frac{8}{3}$ **19.** 9   **21.** 0.8   **23.** 2

25. 14.85; the extra area above $y = 3x$ using circumscribed rectangles is the same as the omitted area under $y = 3x$ using inscribed rectangles.

27. 3.08; the extra area above $y = x^2$ using circumscribed rectangles is greater than the omitted area under $y = x^2$ using inscribed rectangles.

Exercises 25.4, page 748

1. $-\frac{9}{4}$ **3.** 1   **5.** $\frac{254}{7}$ **7.** $6 + 2\sqrt{6} - 2\sqrt{3}$ **9.** 2.53

11. $-\frac{32}{3}$ **13.** $\frac{33}{20}\sqrt[3]{\frac{11}{5}} - \frac{3}{8}\sqrt[3]{\frac{1}{2}} - \frac{17}{5} = -1.552$ **15.** $\frac{4}{3}$

17. $-\frac{81}{4}$ **19.** 2   **21.** $\frac{1}{4}(20.5^{2/3} - 17.5^{2/3}) = 0.1875$

23. $\frac{88}{3249} = 0.0271$ **25.** 84   **27.** $\frac{364}{3}$ **29.** $\frac{3880}{9}$

31. $\frac{33}{784} = 0.0421$ **33.** $\frac{2}{3}(13\sqrt{13} - 27) = 13.25$

35. $\frac{1}{4} + \frac{15}{4} = 4$; under $y = x^3$, the area from $x = 0$ to $x = 1$ plus the area from $x = 1$ to $x = 2$ equals the area from $x = 0$ to $x = 2$.

37. 36   **39.** 64 000 N $\cdot$ m   **41.** 86.8 m^2 **43.** $\frac{3NE_F}{5}$

Exercises 25.5, page 751

1. $\frac{7}{6}$ **3.** $\frac{11}{2} = 5.50, \frac{16}{3} = 5.33$ **5.** $7.661, \frac{23}{3} = 7.667$

7. 0.2042   **9.** 18.98   **11.** 0.5205   **13.** 21.74   **15.** 45.36

17. The tops of all trapezoids are below $y = 1 + \sqrt{x}$.

19. 100.027 m

Exercises 25.6, page 755

1. 0.406   **3.** (a) 6 (b) 6   **5.** (a) 19.67 (b) 19.67

7. 19.27   **9.** 0.5114   **11.** 13.147   **13.** 44.63

15. 1.200 cm

Review Exercises for Chapter 25, page 756

1. $x^4 - \frac{1}{2}x^2 + C$ **3.** $\frac{2}{7}u^{7/2} + \frac{4}{3}u^{3/2} + C$ **5.** $\frac{19}{3}$

7. $\frac{16}{3}$ **9.** $3x - \frac{1}{x^2} + C$ **11.** 3

13. $\frac{1}{10(2 - 5n^2)} + C$ **15.** $-\frac{6}{7}(7 - 2x)^{7/4} + C$

17. $\frac{9}{8}(3\sqrt[3]{3} - 1)$ **19.** $-\frac{1}{30}(1 - 2x^3)^5 + C$

21. $-\frac{1}{2x - x^3} + C$ **23.** $\frac{3350}{3}$ **25.** $y = 3x - \frac{1}{3}x^3 + \frac{17}{3}$

27. (a) $x - x^2 + C_1$
(b) $-\frac{1}{4}(1 - 2x)^2 + C_2 = x - x^2 + C_2 - \frac{1}{4}; C_1 = C_2 - \frac{1}{4}$

29. $0.25 = 0.25$; $y = x^3$ shifted 1 unit to the right is $y = (x - 1)^3$. Therefore, areas are the same.

31. 22   **33.** The graph of $f(x)$ is concave down. This means the tops of the trapezoids are below $y = f(x)$.

35. 0.842   **37.** 0.811   **39.** 13.6   **41.** 19.3016

43. 19.0356   **45.** 24.68 m^2 **47.** 25.81 m^2

49. $y = k(2L^3 x - 6Lx^2 + \frac{2}{5}x^5)$

51. 14.9 m^2

Exercises 26.1, page 763

1. -19.6 m/s   **3.** -15 m/s   **5.** $s = 8.00 - 0.25t$

7. 5.0 m/s   **9.** 1.4 m   **11.** 12 m/s^2 **13.** 24 m/s

15. 85.3 m   **17.** 0.345 nC   **19.** 0.017 C   **21.** 120 V

23. 4.65 mV   **25.** 970 rad   **27.** 66.7 A   **29.** $\frac{k}{x_1}$

31. $m = 1002 - 2\sqrt{t + 1}$, 2.51×10^5 min

Exercises 26.2, page 769

1. $\frac{52}{3}$ **3.** 2 **5.** $\frac{27}{8}$ **7.** $\frac{32}{3}$ **9.** $\frac{1}{6}$ **11.** $\frac{26}{3}$

13. 3 **15.** $\frac{15}{4}$ **17.** $\frac{8}{3}$ **19.** $\frac{7}{6}$ **21.** $\frac{256}{15}$ **23.** $\frac{343}{24}$

25. $\frac{65}{6}$ **27.** $\frac{14}{3}$

29. The area bounded by $x = 1$, $y = 2x^2$, and $y = x^3$ or the area bounded by $x = 1$, $y = 0$, and $y = 2x^2 - x^3$.

31. $4^{2/3}$ **33.** 1 **35.** $\frac{48}{5}$ **37.** 18.0 J **39.** 80.8 km

41. 4 cm^2 **43.** 42.5 dm^2

Exercises 26.3, page 774

1. $\frac{128\pi}{7}$ **3.** $\frac{8}{3}\pi$ **5.** $\frac{8}{3}\pi$ **7.** $\frac{8}{3}\pi$ **9.** 72π **11.** $\frac{768}{7}\pi$

13. $\frac{348}{5}\pi$ **15.** $\frac{16}{3}\pi$ **17.** $\frac{128}{7}\pi$ **19.** $\frac{2}{5}\pi$ **21.** $\frac{10\pi}{3}\sqrt{5}$

23. $\frac{1296}{5}\pi$ **25.** $\frac{16}{3}\pi$

27. The region bounded by $y = x^{3/2}$, $y = 0$, $x = 1$, and $x = 2$

29. $\frac{8}{3}\pi$ **31.** $\frac{1}{3}\pi r^2 h$ **33.** 16 800 m^3 **35.** 18.3 cm^3

Exercises 26.4, page 780

1. $\left(0, \frac{8}{3}\right)$ **3.** 2.9 cm **5.** 1.2 cm **7.** (−0.5 in., 0.5 in.)

9. (0.32 in., 0.23 in.) **11.** $\left(0, \frac{6}{5}\right)$ **13.** $\left(\frac{4}{3}, \frac{4}{3}\right)$ **15.** $\left(\frac{3}{5}, \frac{12}{35}\right)$

17. $\left(\frac{5}{3}, 4\right)$ **19.** $\left(0, \frac{3}{5}a\right)$ **21.** $\left(\frac{7}{8}, 0\right)$ **23.** $\left(0, \frac{5}{6}\right)$ **25.** $\left(\frac{2}{3}, 0\right)$

27. $\left(\frac{2}{3}b, \frac{1}{3}a\right)$. Place triangle with a vertex at origin, side b and right angle on x-axis. Equation of hypotenuse is $y = ax/b$. Use Eqs. (26.16) and (26.17).

29. 0.375 cm above center of base

31. 19.3 cm from larger base

Exercises 26.5, page 786

1. $I_y = k, R_y = \frac{1}{2}\sqrt{2}$ **3.** 68 g · cm^2, 2.9 cm

5. 2530 g · cm^2, 3.58 cm **7.** $\frac{2}{3}k$ **9.** $\frac{64}{15}k$ **11.** $\frac{2}{3}\sqrt{6}$

13. $\frac{1}{6}mb^2$ **15.** $\frac{4}{7}\sqrt{7}$ **17.** $\frac{8}{11}\sqrt{55}$ **19.** $\frac{64}{3}\pi k$

21. $\frac{2}{5}\sqrt{10}$ **23.** $\frac{3}{10}mr^2$ **25.** 0.324 g · cm^2

27. 31.2 kg · cm^2

Exercises 26.6, page 791

1. 41 N · cm **3.** 176 kN **5.** 8.0 N · cm **7.** 3600 N · cm

9. 1.7×10^{-16} J **11.** $0.09k$ N · m **13.** 1800 N · m

15. 3.00×10^5 m · tonne **17.** 9.85×10^5 N · m **19.** 12.5 kN

21. 152 kN **23.** 6500 N **25.** 1.84 MN

27. 3.92×10^4 N, 1.18×10^5 N buoyant force **29.** 2.7 A

31. 35.3% **33.** 109 m **35.** $S = \pi r\sqrt{r^2 + h^2}$

Review Exercises for Chapter 26, page 794

1. 4.3 s **3.** 4.7 s **5.** $s = 2t^2$ **7.** 0.44 C **9.** 55 V

11. $y = 20x + \frac{1}{120}x^3$ **13.** $\frac{2}{3}$ **15.** 18 **17.** $\frac{27}{4}$

19. $A_{\text{top}}/A_{\text{bot}} = \left(\dfrac{n}{n + 1}\right)\left(\dfrac{1}{n + 1}\right) = \dfrac{n}{1}$

21. $\frac{48}{5}\pi$ **23.** $\frac{512}{5}\pi$ **25.** $\frac{4}{3}\pi ab^2$ **27.** (−0.5 cm, 0.6 cm)

29. $\left(\frac{40}{21}, \frac{10}{3}\right)$ **31.** $\left(\frac{14}{5}, 0\right)$ **33.** $\frac{8}{5}k$ **35.** 68.7 g · mm^2

37. 2700 N · m **39.** 1.8 m **41.** 47 m^3 **43.** 88.0 m^3

45. 1580 kN **47.** 0.29 Ω

Exercises 27.1, page 801

1. $8\theta \sin 2\theta^2 \cos 2\theta^2 = 4\theta \sin 4\theta^2$ **3.** $\cos(x + 2)$

5. $12x^2 \cos(2x^3 - 1)$ **7.** $-3 \sin \frac{1}{2}x$ **9.** $-6 \sin(3x - \pi)$

11. $6\pi \sin 3\pi\theta \cos 3\pi\theta = 3\pi \sin 6\pi\theta$

13. $-45 \cos^2(5x + 2)\sin(5x + 2)$

15. $\sin 3x + 3x \cos 3x$ **17.** $9x^2 \cos 5x - 15x^3 \sin 5x$

19. $2x \cos x^2 \cos 2x - 2 \sin x^2 \sin 2x$

21. $\dfrac{2 \cos 4x}{\sqrt{1 + \sin 4x}}$ **23.** $\dfrac{3t \cos(3t - \pi/3) - \sin(3t - \pi/3)}{2t^2}$

25. $\dfrac{4x(1 - 3x)\sin x^2 - 6 \cos x^2}{(3x - 1)^2}$

27. $4 \sin 3x(3 \cos 3x \cos 2x - \sin 3x \sin 2x)$

29. $2 \cos 2t \cos(\sin 2t)$ **31.** $3 \sin^2 x \cos x + 2 \sin 2x$

33. $-\dfrac{\cos s}{\sin^2 s} + \dfrac{\sin s}{\cos^2 s}$

35. (a) (b) See the table.

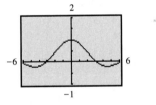

37. (a) 0.540 302 3, value of derivative
(b) 0.540 260 2, slope of secant line

39. Resulting curve is $y = \cos x$.

41. $\dfrac{2x - y \cos xy}{x \cos xy - 2 \sin 2y}$

43. $\dfrac{d \sin x}{dx} = \cos x, \dfrac{d^2 \sin x}{dx^2} = -\sin x,$

$\dfrac{d^3 \sin x}{dx^3} = -\cos x, \dfrac{d^4 \sin x}{dx^4} = \sin x$

45. $\sin 2x = 2 \sin x \cos x$ **47.** 0.515 **49.** −2.36

51. 410 V/s **53.** −199 cm/s **55.** −38.5 km

Exercises 27.2, page 805

1. $12x \sec^2 x^2 \tan x^2$ **3.** $5 \sec^2 5x$ **5.** $5 \csc^2(0.25\pi - \theta)$

7. $6 \sec 2x \tan 2x$ **9.** $\dfrac{3}{\sqrt{2x + 3}} \csc \sqrt{2x + 3} \cot \sqrt{2x + 3}$

11. $10\pi \tan \pi x \sec^2 \pi x$ **13.** $-4\cot^3 \frac{1}{2}x \csc^2 \frac{1}{2}x$

15. $2\tan 4x\sqrt{\sec 4x}$ **17.** $-84\csc^4 7x \cot 7x$

19. $0.5t^2 \sec^2 0.5t + 2t \tan 0.5t$

21. $-4\csc x^2(2x\cos x \cot x^2 + \sin x)$

23. $-\dfrac{\csc x(x\cot x + 1)}{x^2}$

25. $\dfrac{2(-4\sin 4x - 4\sin 4x \cot 3x + 3\cos 4x \csc^2 3x)}{(1 + \cot 3x)^2}$

27. $\sec^2 x(\tan^2 x - 1)$ **29.** $2\pi \cos 2\pi\theta \sec^2(\sin 2\pi\theta)$

31. $\dfrac{1 + 2\sec^2 4x}{\sqrt{2x + \tan 4x}}$ **33.** $\dfrac{2\cos 2x - \sec y}{x \sec y \tan y - 2}$

35. $24\tan 3x \sec^2 3x\, dx$

37. $4\sec 4x(\tan^2 4x + \sec^2 4x)\, dx$

39. (a) 3.4255188, value of derivative
   (b) 3.4260524, slope of secant line

41. (a) (b)

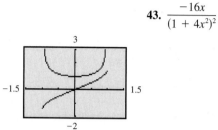

43. $2\tan x \sec^2 x = 2\sec x(\sec x \tan x)$ **45.** -12

47. $2\sec^2 x - \sec x \tan x = \dfrac{2}{\cos^2 x} - \dfrac{\sin x}{\cos^2 x}$

49. -8.4 cm/s **51.** 140 m/s

Exercises 27.3, page 809

1. $\dfrac{2x}{\sqrt{1 - x^4}}$ **3.** $\dfrac{7}{\sqrt{1 - 49x^2}}$ **5.** $\dfrac{18x^2}{\sqrt{1 - 9x^6}}$

7. $\dfrac{-1.8}{\sqrt{1 - 0.25s^2}}$ **9.** $\dfrac{1}{\sqrt{(x - 1)(2 - x)}}$ **11.** $\dfrac{1}{2\sqrt{x}(1 + x)}$

13. $-\dfrac{6}{x^2 + 1}$ **15.** $\dfrac{5x}{\sqrt{1 - x^2}} + 5\sin^{-1}x$

17. $\dfrac{0.8u}{1 + 4u^2} + 0.4\tan^{-1} 2u$ **19.** $\dfrac{3\sqrt{1 - 4R^2}\sin^{-1} 2R - 6R + 2}{\sqrt{1 - 4R^2}(\sin^{-1} 2R)^2}$

21. $\dfrac{2(\cos^{-1} 2x + \sin^{-1} 2x)}{\sqrt{1 - 4x^2}(\cos^{-1} 2x)^2}$ **23.** $\dfrac{-24(\cos^{-1} 4x)^2}{\sqrt{1 - 16x^2}}$

25. $\dfrac{4\sin^{-1}(4x + 1)}{\sqrt{-4x^2 - 2x}}$ **27.** $\dfrac{-1}{t^2 + 1}$ **29.** $\dfrac{-2(2x + 1)^2}{(1 + 4x^2)^2}$

31. $\dfrac{18(4 - \cos^{-1} 2x)^2}{\sqrt{1 - 4x^2}}$ **33.** $-\dfrac{x^2 y^2 + 2y + 1}{2x}$

35. (a) 1.1547005, value of derivative
   (b) 1.1547390, slope of secant line

37. $\dfrac{3(\sin^{-1} x)^2\, dx}{\sqrt{1 - x^2}}$ **39.** 0.41

41. **43.** $\dfrac{-16x}{(1 + 4x^2)^2}$

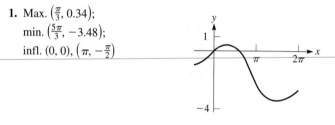

45. Let $y = \sec^{-1} u$; solve for u; take derivatives; substitute.

47. $\dfrac{E - A}{\omega m\sqrt{m^2 E^2 - (A - E)^2}}$ **49.** $-\dfrac{R}{R^2 + (X_L - X_C)^2}$

51. $\theta = \tan^{-1}\dfrac{h}{x}; \dfrac{d\theta}{dx} = \dfrac{-h}{x^2 + h^2}$

Exercises 27.4, page 813

1. Max. $\left(\frac{\pi}{3}, 0.34\right)$;
   min. $\left(\frac{5\pi}{3}, -3.48\right)$;
   infl. $(0, 0), \left(\pi, -\frac{\pi}{2}\right)$

3. $d\sin x/dx = \cos x$ and $d\cos x/dx = -\sin x$, and $\sin x = \cos x$ at points of intersection.

5. $\dfrac{1}{x^2 + 1}$ is always positive.

7. Dec. $x > 0, x < 0$;
   infl. $(0, 0)$;
   asym. $x = \dfrac{\pi}{2}, x = -\dfrac{\pi}{2}$

9. $8\sqrt{2}x + 8y + 4\sqrt{2} - 5\pi\sqrt{2} = 0$ **11.** 1.9337538

13. -10 **15.** 0.58 m/s, -1.7 m/s^2 **17.** -0.072 N/s

19. 2510 mm/s, $270°$ **21.** 17.4 cm/s^2, $176°$

23. -0.085 rad/s **25.** 8.08 m/s **27.** 0.020 **29.** 280 km

31. 0.60 L/s **33.** $w = 9.24$ cm, $d = 13.1$ cm **35.** 4.2 m

Exercises 27.5, page 818

1. $-4\tan 4x$ **3.** $\dfrac{2\log e}{x}$ **5.** $\dfrac{6\log_5 e}{3x + 1}$ **7.** $\dfrac{-0.6}{1 - 3x}$

9. $\dfrac{4\sec^2 2x}{\tan 2x} = 4\sec 2x \csc 2x$ **11.** $\dfrac{1}{2T}$ **13.** $\dfrac{6(x + 1)}{x^2 + 2x}$

15. $\dfrac{6(t + 2)(t + \ln t^2)}{t}$ **17.** $\dfrac{3(2x + 1)\ln(2x + 1) - 6x}{(2x + 1)[\ln(2x + 1)]^2}$

19. $\dfrac{1}{x \ln x}$ **21.** $\dfrac{1}{x^2 + x}$ **23.** $\dfrac{\cos \ln x}{x}$

25. $6 \ln 2v + 3 \ln^2 2v$ **27.** $\dfrac{x \sec^2 x + \tan x}{x \tan x}$ **29.** $\dfrac{v + 4}{v(v + 2)}$

31. $\dfrac{\sqrt{x^2 + 1}}{x}$ **33.** $\dfrac{x + y - 2\ln(x + y)}{x + y + 2\ln(x + y)}$

35. 0.5 is value of derivative; 0.499 987 5 is slope of secant line.

37. (a)   (b) See the table.

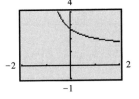

39. -0.3641 **41.** 1.15   **43.** $L(x) = 4x - \pi$ **45.** 2.73

47. $x^x(\ln x + 1)$. In Eq. (23.15) the exponent is constant. For x^x, both the base and the exponent are variables.

49. $\dfrac{dy_1}{dx}\bigg|_{x=-1} = \dfrac{2}{x}\bigg|_{x=-1} = -2$ $[\ln(x^2)$ defined for all $x, x \neq 0]$

$\dfrac{dy_2}{dx}\bigg|_{x=-1} = \dfrac{2}{x}\bigg|_{x=-1}$ does not exist $(\ln x$ not defined for $x \leq 0)$

51. $\dfrac{10 \log e}{I}\dfrac{dI}{dt}$ **53.** $\dfrac{x}{x^2 + 1 + \sqrt{1 + x^2}} - \dfrac{1}{x} + \dfrac{x}{\sqrt{1 - x^2}}$

55. 0.125 s^2/m

Exercises 27.6, page 822

1. $2e^{2x} \cot e^{2x}$ **3.** $(6 \ln 4)4^{6x}$ **5.** $\dfrac{e^{\sqrt{x}}}{2\sqrt{x}}$ **7.** $4e^{2t}(3e^t - 2)$

9. $e^{-T}(1 - T)$ **11.** $e^{\sin x}(x \cos x + 1)$ **13.** $8e^{-4s}$

15. $e^{-3x}(4 \cos 4x - 3 \sin 4x)$ **17.** $\dfrac{2e^{3x}(12x + 5)}{(4x + 3)^2}$

19. $\dfrac{2xe^{x^2}}{e^{x^2} + 4}$ **21.** $16e^{6x}(x \cos x^2 + 3 \sin x^2)$

23. $\dfrac{2(1 + 2te^{2t})}{t\sqrt{\ln 2t + e^{2t}}}$ **25.** $\dfrac{e^{xy}(xy + 1)}{1 - x^2 e^{xy} - \cos y}$

27. $\dfrac{3e^{2x}}{x} + 6e^{2x} \ln x$ **29.** $12e^{6x} \cot 2e^{6x}$ **31.** $\dfrac{4e^{2x}}{\sqrt{1 - e^{4x}}}$

33. (a) 2.718 281 8, value of derivative
(b) 2.718 417 7, slope of secant line

35.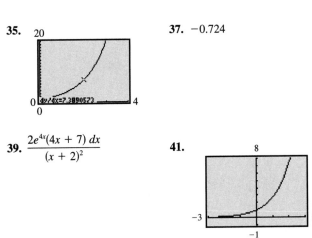

37. -0.724

39. $\dfrac{2e^{4x}(4x + 7)\,dx}{(x + 2)^2}$ **41.**

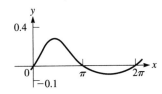

43. $(-xe^{-x} + e^{-x}) + (xe^{-x}) = e^{-x}$

45. $\dfrac{2e^{2x}(e^{2x} + 1) - 2e^{2x}(e^{2x} - 1)}{(e^{2x} + 1)^2} = \dfrac{4e^{2x}}{(e^{2x} + 1)^2}$

$= \dfrac{(e^{2x} + 1)^2 - (e^{2x} - 1)^2}{(e^{2x} + 1)^2}$

47. $-3, 2$ **49.** $-0.001 64$/h

51. $e^{-66.7t}(999 \cos 226t - 295 \sin 226t)$

53. Substitute and simplify.

55. $\dfrac{d}{dx}\left[\dfrac{1}{2}(e^u - e^{-u})\right] = \dfrac{1}{2}(e^u + e^{-u})\dfrac{du}{dx}$;

$\dfrac{d}{dx}\left[\dfrac{1}{2}(e^u + e^{-u})\right] = \dfrac{1}{2}(e^u - e^{-u})\dfrac{du}{dx}$

Exercises 27.7, page 825

1. Int. $(0, 0)$, $(\pi, 0)$, $(2\pi, 0)$;
max. $\left(\dfrac{\pi}{4}, 0.322\right)$;
min. $\left(\dfrac{5\pi}{4}, -0.014\right)$;
infl. $\left(\dfrac{\pi}{2}, 0.208\right)$

3. Int. $(0, 0)$, max. $(0, 0)$,
not defined for $\cos x < 0$,
asym. $x = -\dfrac{1}{2}\pi, \dfrac{1}{2}\pi, \ldots$

5. Int. $(0, 0)$, max. $\left(1, \dfrac{1}{e}\right)$,
infl. $\left(2, \dfrac{2}{e^2}\right)$, asym. $y = 0$

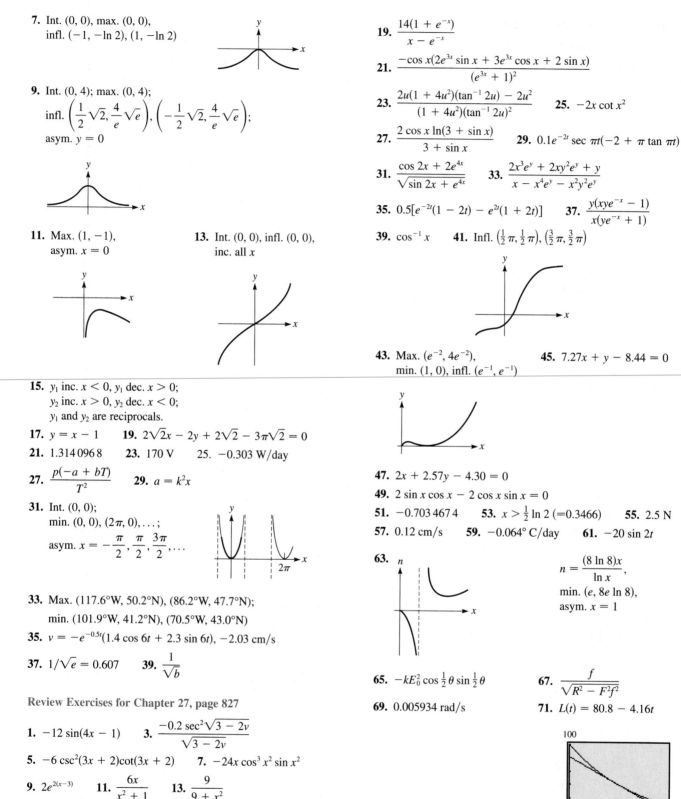

7. Int. (0, 0), max. (0, 0),
 infl. $(-1, -\ln 2)$, $(1, -\ln 2)$

9. Int. (0, 4); max. (0, 4);
 infl. $\left(\frac{1}{2}\sqrt{2}, \frac{4}{e}\sqrt{e}\right)$, $\left(-\frac{1}{2}\sqrt{2}, \frac{4}{e}\sqrt{e}\right)$;
 asym. $y = 0$

11. Max. $(1, -1)$,
 asym. $x = 0$

13. Int. (0, 0), infl. (0, 0),
 inc. all x

15. y_1 inc. $x < 0$, y_1 dec. $x > 0$;
 y_2 inc. $x > 0$, y_2 dec. $x < 0$;
 y_1 and y_2 are reciprocals.

17. $y = x - 1$ **19.** $2\sqrt{2}x - 2y + 2\sqrt{2} - 3\pi\sqrt{2} = 0$

21. $1.314\,096\,8$ **23.** 170 V   **25.** -0.303 W/day

27. $\dfrac{p(-a + bT)}{T^2}$ **29.** $a = k^2 x$

31. Int. (0, 0);
 min. (0, 0), $(2\pi, 0), \ldots$;
 asym. $x = -\dfrac{\pi}{2}, \dfrac{\pi}{2}, \dfrac{3\pi}{2}, \ldots$

33. Max. (117.6°W, 50.2°N), (86.2°W, 47.7°N);
 min. (101.9°W, 41.2°N), (70.5°W, 43.0°N)

35. $v = -e^{-0.5t}(1.4\cos 6t + 2.3\sin 6t)$, -2.03 cm/s

37. $1/\sqrt{e} = 0.607$ **39.** $\dfrac{1}{\sqrt{b}}$

Review Exercises for Chapter 27, page 827

1. $-12\sin(4x - 1)$ **3.** $\dfrac{-0.2\sec^2\sqrt{3 - 2v}}{\sqrt{3 - 2v}}$

5. $-6\csc^2(3x + 2)\cot(3x + 2)$ **7.** $-24x\cos^3 x^2 \sin x^2$

9. $2e^{2(x-3)}$ **11.** $\dfrac{6x}{x^2 + 1}$ **13.** $\dfrac{9}{9 + x^2}$

15. $\dfrac{0.1}{\sin^{-1} 0.1t \sqrt{1 - 0.01t^2}}$ **17.** $(-2\csc 4x)\sqrt{\csc 4x + \cot 4x}$

19. $\dfrac{14(1 + e^{-x})}{x - e^{-x}}$

21. $\dfrac{-\cos x(2e^{3x}\sin x + 3e^{3x}\cos x + 2\sin x)}{(e^{3x} + 1)^2}$

23. $\dfrac{2u(1 + 4u^2)(\tan^{-1} 2u) - 2u^2}{(1 + 4u^2)(\tan^{-1} 2u)^2}$ **25.** $-2x\cot x^2$

27. $\dfrac{2\cos x \ln(3 + \sin x)}{3 + \sin x}$ **29.** $0.1e^{-2t}\sec \pi t(-2 + \pi\tan \pi t)$

31. $\dfrac{\cos 2x + 2e^{4x}}{\sqrt{\sin 2x + e^{4x}}}$ **33.** $\dfrac{2x^3 e^y + 2xy^2 e^y + y}{x - x^4 e^y - x^2 y^2 e^y}$

35. $0.5[e^{-2t}(1 - 2t) - e^{2t}(1 + 2t)]$ **37.** $\dfrac{y(xye^{-x} - 1)}{x(ye^{-x} + 1)}$

39. $\cos^{-1} x$ **41.** Infl. $\left(\frac{1}{2}\pi, \frac{1}{2}\pi\right)$, $\left(\frac{3}{2}\pi, \frac{3}{2}\pi\right)$

43. Max. $(e^{-2}, 4e^{-2})$,
 min. (1, 0), infl. (e^{-1}, e^{-1}) **45.** $7.27x + y - 8.44 = 0$

47. $2x + 2.57y - 4.30 = 0$

49. $2\sin x\cos x - 2\cos x\sin x = 0$

51. $-0.703\,467\,4$ **53.** $x > \frac{1}{2}\ln 2\ (=0.3466)$ **55.** 2.5 N

57. 0.12 cm/s   **59.** $-0.064°$ C/day   **61.** $-20\sin 2t$

63.

$n = \dfrac{(8\ln 8)x}{\ln x}$,
min. $(e, 8e\ln 8)$,
asym. $x = 1$

65. $-kE_0^2 \cos \frac{1}{2}\theta \sin \frac{1}{2}\theta$ **67.** $\dfrac{f}{\sqrt{R^2 - F^2 f^2}}$

69. 0.005934 rad/s   **71.** $L(t) = 80.8 - 4.16t$

73. 2 100 000/year   **75.** 2.00 m, 1.82 m   **77.** -0.0065 rad/s

79. 48.2°   **81.** 0.40 cm/s, 72.5°   **83.** 7.07 cm

85. $A = 100 \cos \theta (1 + \sin \theta)$

Max. $\left(\frac{\pi}{6}, 130\right)$

87. $\dfrac{w}{H} \cosh \dfrac{wx}{H} = \dfrac{w}{H} \sqrt{1 + \sinh^2 \dfrac{wx}{H}}$

Exercises 28.1, page 834

1. Change $\cos x$ to $-\sin x$.   **3.** $\frac{1}{5} \sin^5 x + C$

5. $-\frac{4}{15}(\cos \theta)^{3/2} + C$ **7.** $\frac{4}{3} \tan^3 x + C$ **9.** $\frac{1}{8}$

11. $\frac{1}{4}(\sin^{-1} x)^4 + C$ **13.** $\frac{1}{2}(\tan^{-1} 5x)^2 + C$

15. $\frac{1}{3}[\ln(x + 1)]^3 + C$ **17.** 0.179   **19.** $\frac{1}{4}(4 + e^x)^4 + C$

21. $\dfrac{1}{4(1 - e^{2r})^2} + C$ **23.** $\frac{1}{10}(1 + \sec^2 x)^5 + C$

25. 4.1308   **27.** 1.102

29. $\frac{1}{2}\int (1 - e^{-2r})^{1/2}(2e^{-2r})\, dr; u = 1 - e^{-2r}, n = \frac{1}{2}, du = 2e^{-2r}\, dr$

31. $y = \frac{1}{3}(\ln x)^3 + 2$ **33.** $\frac{1}{3}mnv^2$ **35.** $q = (1 - e^{-t})^3$

Exercises 28.2, page 837

1. Change dx to $2x\, dx$.   **3.** $\frac{1}{4} \ln|1 + 4x| + C$

5. $-\frac{1}{3} \ln|4 - 3x^2| + C$ **7.** $\frac{1}{3} \ln 4 = 0.462$

9. $-0.2 \ln|\cot 2\theta| + C$ **11.** $\ln 2 = 0.693$

13. $\ln|1 - e^{-x}| + C$ **15.** $\ln|x + e^x| + C$

17. $\frac{1}{4} \ln|1 + 4 \sec x| + C$ **19.** $\frac{1}{4} \ln 5 = 0.402$

21. $0.5 \ln|\ln r| + C$ **23.** $\ln|2x + \tan x| + C$

25. $-2\sqrt{1 - 2x} + C$ **27.** $\ln|x| - \frac{2}{x} + C$

29. $\frac{1}{3} \ln\left(\frac{5}{4}\right) = 0.0744$ **31.** 1.10

33. $\displaystyle\int \frac{x - 4}{x + 4}\, dx = \int \left(1 - \frac{8}{x + 4}\right) dx = x - 8 \ln|x + 4| + C$

35. $\pi \ln 2 = 2.18$ **37.** $y = \ln\left(\dfrac{3.5}{3 + \cos x} + 2\right)$

39. 335 million   **41.** $i = \dfrac{E}{R}(1 - e^{-Rt/L})$ **43.** 1.41 m

Exercises 28.3, page 840

1. Change $x\, dx$ to $3x^2\, dx$.   **3.** $e^{7x} + C$ **5.** $\frac{1}{2}e^{2x+5} + C$

7. 28.2   **9.** $2e^{x^3} + C$ **11.** $2(e^2 - e) = 9.34$

13. $2e^{2 \sec \theta} + C$ **15.** $\frac{2}{3}(1 + e^y)^{3/2} + C$

17. $6 - \dfrac{3(e^6 - e^2)}{2} = -588.06$ **19.** $-\dfrac{4}{e^{\sqrt{x}}} + C$

21. $e^{\tan^{-1} x} + C$ **23.** $-\frac{1}{3}e^{\cos 3x} + C$ **25.** 0

27. $3e^2 - 3 = 19.2$ **29.** $y = e^{\ln x}$; $\ln y = \ln e^{\ln x}$; $y = x$; $e^{x^2} + C$

31. $\pi(e^4 - e) = 163$ **33.** $\frac{1}{8}(e^8 - 1) = 372$

35. $\ln b \int b^u\, du = b^u + C_1$ **37.** $q = EC(1 - e^{-t/RC})$

39. 192 m

Exercises 28.4, page 844

1. Change $x\, dx$ to $3x^2\, dx$.   **3.** $\frac{1}{2} \sin 2x + C$

5. $0.1 \tan 3\theta + C$ **7.** $2 \sec \frac{1}{2}x + C$ **9.** 0.6365

11. $\frac{3}{2} \ln|\sec \phi^2 + \tan \phi^2| + C$ **13.** $\cos\left(\dfrac{1}{x}\right) + C$

15. $\frac{1}{2}\sqrt{3}$ **17.** $\frac{1}{5} \sec 5x + C$

19. $\frac{1}{2} \ln|\sec 2x + \tan 2x| + C$ **21.** $\frac{1}{2}(\ln|\sin 2T| + \sin 2T) + C$

23. $\csc x - \cot x - \ln|\csc x - \cot x| + \ln|\sin x| + C$

25. $\frac{1}{9}\pi + \frac{1}{3} \ln 2 = 0.580$ **27.** 0.693

29. Integral changes to $\int (\sec^2 x - \sec x \tan x)\, dx$

31. $\pi\sqrt{3} = 5.44$ **33.** $\theta = 0.10 \cos 2.5t$ **35.** 0.7726 m

Exercises 28.5, page 848

1. Change dx to $\cos 2x\, dx$.   **3.** $\frac{1}{3} \sin^3 x + C$

5. $-\frac{1}{2} \cos 2x + \frac{1}{6} \cos^3 2x + C$ **7.** $2 \sin 2\theta + C$

9. $\frac{1}{24}(64 - 43\sqrt{2}) = 0.1329$ **11.** $\frac{1}{2}x - \frac{1}{4} \sin 2x + C$

13. $\frac{1}{3}(9\phi + 4 \sin 3\phi + \sin 3\phi \cos 3\phi) + C$

15. $\frac{1}{2} \tan^2 x + \ln|\cos x| + C$ **17.** $\frac{3}{4}$

19. $\frac{1}{6} \tan^3 2x - \frac{1}{2} \tan 2x + x + C$ **21.** $\frac{1}{3} \sin^3 s + C$

23. $x - \frac{1}{2} \cos 2x + C$

25. $\frac{1}{4} \cot^4 x - \frac{1}{3} \cot^3 x + \frac{1}{2} \cot^2 x - \cot x + C$

27. $1 + \frac{1}{2} \ln 2 = 1.347$ **29.** $\frac{1}{5} \tan^5 x + \frac{2}{3} \tan^3 x + \tan x + C$

31. $\frac{1}{2}\pi^2 = 4.935$ **33.** $\sqrt{2} - 1 = 0.414$

35. $\int \sin x \cos x\, dx = \frac{1}{2} \sin^2 x + C_1 = -\frac{1}{2} \cos^2 x + C_2$;

$\quad C_2 = C_1 + \frac{1}{2}$

37. $\displaystyle\int_0^\pi \sin^2 nx\, dx = \frac{1}{2} \int_0^\pi (1 - \cos 2nx)\, dx$

$\qquad = \left. \frac{1}{2}x - \sin 2nx \right|_0^\pi = \dfrac{\pi}{2}$

39. $\frac{4}{3}$ **41.** $V = \sqrt{\dfrac{1}{1/60.0} \displaystyle\int_0^{1/60.0} (340 \sin 120\pi t)^2\, dt} = 240$ V

43. $\dfrac{aA}{2} + \dfrac{A}{2b\pi} \sin ab\pi \cos 2bc\pi$

Exercises 28.6, page 852

1. Change dx to $-x\, dx$.   **3.** $\sin^{-1} \frac{1}{2}x + C$

5. $\frac{1}{8} \tan^{-1} \frac{1}{8}x + C$ **7.** $\frac{1}{8} \sin^{-1} 4x^2 + C$ **9.** 0.8634

11. $\frac{2}{5}\sqrt{5} \sin^{-1} \frac{1}{5}\sqrt{5} = 0.415$ **13.** $\frac{4}{9} \ln|9x^2 + 16| + C$

15. 2.356   **17.** $\sin^{-1} e^x + C$ **19.** $\tan^{-1}(T + 1) + C$

21. $4 \sin^{-1} \frac{1}{2}(x + 2) + C$ **23.** -0.714

25. $2 \sin^{-1}(\frac{1}{2}x) + \sqrt{4 - x^2} + C$

27. (a) Inverse tangent, $\int \dfrac{du}{a^2 + u^2}$ where $u = 3x$,

$du = 3\,dx$, $a = 2$; numerator cannot fit du of denominator. Positive $9x^2$ leads to inverse tangent form.

(b) logarithmic, $\int \dfrac{du}{u}$ where $u = 4 + 9x$, $du = 9\,dx$

(c) general power, $\int u^{-1/2}\,du$ where $u = 4 + 9x^2$, $du = 18x\,dx$

29. (a) General power, $\int u^{-1/2}\,du$ where $u = 4 - 9x^2$,

$du = -18x\,dx$; numerator can fit du of denominator. Square root becomes $-1/2$ power. Does not fit inverse

sine form.   (b) Inverse sine, $\int \dfrac{du}{\sqrt{a^2 - u^2}}$ where

$u = 3x$, $du = 3dx$, $a = 2$ (c) Logarithmic, $\int \dfrac{du}{u}$

where $u = 4 - 9x$, $du = -9\,dx$

31. Form fits inverse tangent integral with $u = \sqrt{x}$, $du = \dfrac{dx}{2\sqrt{x}}$.

Result is $2 \tan^{-1} \sqrt{x} + C$.

33. $\tan^{-1} 2 = 1.11$ **35.** $k \tan^{-1} \dfrac{x}{d} + C$

37. $\sin^{-1} \dfrac{x}{A} = \sqrt{\dfrac{k}{m}}\, t + \sin^{-1} \dfrac{x_0}{A}$ **39.** $0.22k$

Exercises 28.7, page 856

1. No. Integral $\int v\,du$ is more complex than the given integral.

3. $\cos \theta + \theta \sin \theta + C$ **5.** $\frac{1}{2}xe^{2x} - \frac{1}{4}e^{2x} + C$

7. $x \tan x + \ln|\cos x| + C$ **9.** $2x \tan^{-1} x - 2 \ln\sqrt{1 + x^2} + C$

11. $-\frac{32}{3}$ **13.** $\frac{1}{2}x^2 \ln x - \frac{1}{4}x^2 + C$

15. $\frac{1}{2}\phi \sin 2\phi - \frac{1}{4}(2\phi^2 - 1)\cos 2\phi + C$

17. $\frac{1}{2}(e^{\pi/2} - 1) = 1.91$ **19.** $-2e^{-\sqrt{x}}(\sqrt{x} + 1) + C$

21. $1 - \dfrac{3}{e^2} = 0.594$ **23.** 0.1104 **25.** $\frac{1}{2}\pi - 1 = 0.571$

27. 0.756 **29.** $s = \frac{1}{3}\big[(t^2 - 2)\sqrt{t^2 + 1} + 2\big]$

31. $q = \frac{1}{5}[e^{-2t}(\sin t - 2 \cos t) + 2]$

Exercises 28.8, page 859

1. Delete the x^2 before the radical in the denominator.

3. $x = 3 \sin \theta$, $\int \cot^2 \theta\,d\theta$ **5.** $x = \tan \theta$, $\int \csc \theta \cot \theta\,d\theta$

7. $x = \sec \theta$, $\int d\theta$ **9.** $-\dfrac{\sqrt{1 - x^2}}{x} - \sin^{-1} x + C$

11. $2 \ln\big|x + \sqrt{x^2 - 4}\big| + C$ **13.** $-\dfrac{2\sqrt{z^2 + 9}}{3z} + C$

15. $\dfrac{x}{\sqrt{4 - x^2}} + C$ **17.** $\dfrac{16 - 9\sqrt{3}}{24} = 0.017$

19. $5 \ln\big|\sqrt{x^2 + 2x + 2} + x + 1\big| + C$ **21.** $0.039\,97$

23. $2 \sec^{-1} e^x + C$ **25.** π **27.** $\frac{1}{4}ma^2$ **29.** 2.68

31. $kQ \ln \dfrac{\sqrt{a^2 + b^2} + a}{\sqrt{a^2 + b^2} - a}$

Exercises 28.9, page 863

1. $\dfrac{3}{x - 1} - \dfrac{4}{x + 2}$ **3.** $\dfrac{A}{x} + \dfrac{B}{x + 1}$

5. $\dfrac{A}{x} + \dfrac{B}{x + 2} + \dfrac{C}{x - 2}$ **7.** $\ln\left|\dfrac{(x + 1)^2}{x + 2}\right| + C$

9. $\dfrac{1}{4} \ln\left|\dfrac{x - 2}{x + 2}\right| + C$ **11.** $x + \ln\left|\dfrac{x}{(x + 3)^4}\right| + C$

13. 1.057 **15.** $\ln\left|\dfrac{x^2(x - 5)^3}{x + 1}\right| + C$

17. $\dfrac{1}{4} \ln\left|\dfrac{x^4(2x + 1)^3}{2x - 1}\right| + C$ **19.** $1 + \ln \dfrac{16}{3} = 2.674$

21. $\dfrac{1}{60} \ln\left|\dfrac{(V + 2)^3(V - 3)^2}{(V - 2)^3(V + 3)^2}\right| + C$

23. $\dfrac{1}{u(a + bu)} = \dfrac{A}{u} + \dfrac{B}{a + bu}$; $1 = A(a + bu) + Bu$;

$A = \dfrac{1}{a}$, $B = -\dfrac{b}{a}$

25. 1.322 **27.** $2\pi \ln \dfrac{6}{5} = 1.146$

29. $y = \ln\left|\dfrac{x(x + 5)^2}{36}\right|$ **31.** 0.1633 N $\cdot$ cm

Exercises 28.10, page 868

1. $\dfrac{2}{x(x + 3)^2} = \dfrac{A}{x} + \dfrac{B}{x + 3} + \dfrac{C}{(x + 3)^2}$

3. $\dfrac{4x + 4}{x^3 - 9x^2} = \dfrac{A}{x} + \dfrac{B}{x^2} + \dfrac{C}{x - 9}$

5. $\dfrac{3}{x - 2} + 2 \ln\left|\dfrac{x - 2}{x}\right| + C$ **7.** $\dfrac{2}{x} + \ln\left|\dfrac{x - 1}{x + 1}\right| + C$

9. $-\dfrac{5}{4}$ **11.** $-\dfrac{2}{x + 1} - \dfrac{1}{x - 3} + \ln|x + 1| + C$

13. $\dfrac{1}{8}\pi + \ln 3 = 1.491$ **15.** $-\dfrac{2}{x} + \dfrac{3}{2} \tan^{-1} \dfrac{x + 2}{2} + C$

17. $\dfrac{1}{4} \ln(4x^2 + 1) + \ln|x^2 + 6x + 10| + \tan^{-1}(x + 3) + C$

19. $\ln(r^2 + 1) + \dfrac{1}{r^2 + 1} + C$ **21.** $2 + 4 \ln \dfrac{2}{3} = 0.3781$

23. $\pi \ln \dfrac{25}{9} = 3.210$ **25.** 0.9190 m   **27.** 1.369

Exercises 28.11, page 870

1. Formula 3 **3.** Formula 25; $u = x^2$, $du = 2x\,dx$

5. $\frac{3}{25}[2 + 5x - 2\ln|2 + 5x|] + C$ **7.** $\frac{3544}{15} = 236.3$

9. $\dfrac{y}{4\sqrt{y^2 + 4}} + C$ **11.** $\frac{1}{2}\sin x - \frac{1}{10}\sin 5x + C$

13. $\sqrt{4x^2 - 9} - 3\sec^{-1}\left(\dfrac{2x}{3}\right) + C$

15. $\frac{1}{20}\cos^4 4x \sin 4x + \frac{1}{5}\sin 4x - \frac{1}{15}\sin^3 4x + C$

17. $3r^2 \tan^{-1} r^2 - \frac{3}{2}\ln(1 + r^4) + C$ **19.** $\frac{1}{4}(8\pi - 9\sqrt{3}) = 2.386$

21. $-\ln\left(\dfrac{1 + \sqrt{4x^2 + 1}}{2x}\right) + C$

23. $-8\ln\left(\dfrac{1 + \sqrt{1 - 4x^2}}{2x}\right) + C$ **25.** 0.0208

27. $\frac{1}{3}(\cos x^3 + x^3 \sin x^3) + C$ **29.** $\dfrac{x^2}{\sqrt{1 - x^4}} + C$

31. 4.892 **33.** $\frac{1}{4}x^4\left(\ln x^2 - \frac{1}{2}\right) + C$ **35.** $-\dfrac{3x^3}{\sqrt{x^6 - 1}} + C$

37. $\frac{1}{4}\left[2\sqrt{5} + \ln\left(2 + \sqrt{5}\right)\right] = 1.479$ **39.** $\pi a b$

41. 32.7 kN **43.** $187\,000\text{ m}^3$

Review Exercises for Chapter 28, page 872

1. $-\frac{1}{2}e^{-2x} + C$ **3.** $-\dfrac{1}{\ln 2x} + C$ **5.** $4\ln 2 = 2.773$

7. $\frac{2}{35}\tan^{-1}\frac{7}{5}x + C$ **9.** 0 **11.** $\frac{1}{2}\ln 2 = 0.3466$

13. $\frac{2}{3}\sin^3 t - \cos t + C$ **15.** $\tan^{-1} e^x + C$

17. $\frac{1}{9}\tan^3 3x + \frac{1}{3}\tan 3x + C$ **19.** $\ln\left|\dfrac{(x - 1)^3}{x(2x + 1)}\right| + C$

21. $\frac{3}{4}\tan^{-1}\dfrac{x^2}{2} + C$ **23.** $2\ln\left|2x + \sqrt{4x^2 - 9}\right| + C$

25. $\sqrt{e^{2x} + 1} + C$ **27.** $2\ln|x| + \tan^{-1}\dfrac{x}{3} + C$

29. $\dfrac{\pi}{4} = 0.7854$ **31.** $-\frac{1}{2}x\cot 2x + \frac{1}{4}\ln|\sin 2x| + C$

33. $\dfrac{2}{u} + \ln|3u + 1| + C$ **35.** $\frac{1}{2}\sin e^{2x} + C$

37. $3\sin 1 = 2.524$ **39.** $\frac{1}{2}u^2 - 2u + 3\ln|u + 2| + C$

41. $x = \sqrt{2}\sin\theta; \dfrac{1}{\sqrt{2}}\displaystyle\int \csc\theta\,d\theta$

43. $\frac{1}{3}(e^x + 1)^3 + C_1 = \frac{1}{3}e^{3x} + e^{2x} + e^x + C_2; C_2 = C_1 + \frac{1}{3}$

45. (a) $\dfrac{1}{2}\displaystyle\int (1 - \cos 2x)\,dx = \dfrac{x}{2} - \dfrac{1}{4}\sin 2x + C_1$

(b) $\displaystyle\int \sin x(\sin x\,dx) = -\sin x \cos x + \int \cos^2 x\,dx$

$= -\sin x \cos x + \displaystyle\int (1 - \sin^2 x)\,dx$

47. Power rule with $u = x^2 + 4$, $du = 2x\,dx$, and $n = -1/2$ is easier to use than a trigonometric substitution with $x = 2\tan\theta$.

49. $y = \frac{1}{3}\tan^3 x + \tan x$ **51.** $2(e^3 - 1) = 38.17$

53. 11.18 **55.** $\frac{1}{4}(8\tan^{-1} 4 - \ln 17) = 1.943$

57. $4\pi(e^2 - 1) = 80.29$ **59.** $\frac{1}{8}\pi(e^{2\pi} - 1) = 209.9$

61. $\ln 3 = 1.10$ **63.** $\Delta S = a\ln T + bT + \frac{1}{2}cT^2 + C$

65. 55.2 m **67.** $v = 98(1 - e^{-0.1t})$ **69.** $\sqrt{2}$ **71.** $\frac{2}{3}k$

73. 3.47 cm^3 **75.** 73.0 m^2

Exercises 29.1, page 878

1. Converges; $S_1 = 0.5$, $S_2 = 0.75$, $S_3 = 0.875$, $S_4 = 0.9375$

3. 1, 4, 9, 16 **5.** $\frac{1}{2}, \frac{1}{3}, \frac{1}{4}, \frac{1}{5}$

7. (a) $-\frac{2}{5}, \frac{4}{25}, -\frac{8}{125}, \frac{16}{625}$ (b) $-\frac{2}{5} + \frac{4}{25} - \frac{8}{125} + \frac{16}{625} - \cdots$

9. (a) $1, 0, -1, 0$ (b) $1 + 0 - 1 + 0 + \cdots$

11. $a_n = \dfrac{1}{n + 1}$ **13.** $a_n = \dfrac{1}{(n + 1)(n + 2)}$

15. $1, 1.125, 1.162\,037\,0, 1.177\,662\,0, 1.185\,662\,0$; convergent; 1.2

17. $1, 1.5, 2.166\,666\,7, 2.916\,666\,7, 3.716\,666\,7$; divergent

19. $0, 1, 2.4142, 4.1463, 6.1463$; divergent

21. $0.75, 0.888\,888\,9, 0.937\,500\,0, 0.960\,000\,0, 0.972\,222\,2$; convergent; 1 **23.** Divergent

25. Convergent, $S = \frac{3}{4}$ **27.** Convergent, $S = 100$

29. Convergent, $S = \frac{4096}{9}$ **31.** (a) 1 (b) Diverges

33. 1 **35.** (a) Diverges

(b) $y = 2100(1.05^x - 1)$

37. $r = x; S = \dfrac{x^0}{1 - x} = \dfrac{1}{1 - x}$ **39.** $2, 4, 12, 48, 240$

Exercises 29.2, page 883

1. $\dfrac{2}{2 + x} = 1 - \dfrac{1}{2}x + \dfrac{1}{4}x^2 - \dfrac{1}{8}x^3 + \cdots$

3. $1 + x + \frac{1}{2}x^2 + \cdots$ **5.** $1 - \frac{1}{2}x^2 + \frac{1}{24}x^4 - \cdots$

7. $1 + \frac{1}{2}x - \frac{1}{8}x^2 + \cdots$ **9.** $1 - 8\pi^2 x^2 + \frac{32}{3}\pi^4 x^4 - \cdots$

11. $1 + x + x^2 + \cdots$ **13.** $-2x - 2x^2 - \frac{8}{3}x^3 - \cdots$

15. $1 - x^2 + \frac{1}{3}x^4 - \cdots$ **17.** $x - \frac{1}{3}x^3 + \cdots$

19. $x + \frac{1}{3}x^3 + \cdots$ **21.** $-\frac{1}{2}x^2 - \frac{1}{12}x^4 - \cdots$

23. No. Functions are not defined at $x = 0$.

25. $e^x = 1 + x + \dfrac{x^2}{2} + \cdots, e^{x^2} = 1 + x^2 + \dfrac{x^4}{2} + \cdots$

27. $f(x) = 1 + 3x + \frac{9}{2}x^2 + \cdots; L(x) = 1 + 3x$

29. $f(x) = x^2$

31. $R = e^{-0.001t} = 1 - 0.001t + (5 \times 10^{-7})t^2 - \cdots$

Exercises 29.3, page 887

1. $e^{2x^2} = 1 + 2x^2 + 2x^4 + \frac{4}{3}x^6 + \cdots$

3. $1 + 3x + \frac{9}{2}x^2 + \frac{9}{2}x^3 + \cdots$

5. $\dfrac{x}{2} - \dfrac{x^3}{2^3 3!} + \dfrac{x^5}{2^5 5!} - \dfrac{x^7}{2^7 7!} + \cdots$

7. $x - 8x^3 + \frac{32}{3}x^5 - \frac{256}{45}x^7 + \cdots$

9. $x^2 - \frac{1}{2}x^4 + \frac{1}{3}x^6 - \frac{1}{4}x^8 + \cdots$ **11.** 0.3103

13. 0.1901 **15.** $2(1 + x^2 + x^4 + x^6 + \cdots)$

17. $x + x^2 + \frac{1}{3}x^3 + \cdots$ **19.** $-2x^3 - x^4 - \dfrac{2x^5}{3} - \dfrac{x^6}{2} - \cdots$

21. $\dfrac{d}{dx}\left(x - \dfrac{1}{6}x^3 + \dfrac{1}{120}x^5 - \cdots\right) = 1 - \dfrac{1}{2}x^2 + \dfrac{1}{24}x^4 - \cdots$

23. $\displaystyle\int \cos x\, dx = x - \dfrac{x^3}{3!} + \cdots$

25. $\displaystyle\int_0^1 e^x\, dx = 1.718\,281\,8,$

$\displaystyle\int_0^1 \left(1 + x + \tfrac{1}{2}x^2 + \tfrac{1}{6}x^3\right) dx = 1.708\,333\,3$

27. $0.003\,099$ **29.** $0.199\,968$

31. $\left[\left(-\dfrac{v^2}{c^2}\right)^{-1/2} - 1\right]mc^2 = \left[1 + \left(-\dfrac{1}{2}\right)\left(-\dfrac{v^2}{c^2}\right) + \dfrac{\left(-\frac{1}{2}\right)\left(-\frac{3}{2}\right)}{2}\right.$

$\left.\left(-\dfrac{v^2}{c^2}\right)^2 + \cdots - 1\right]mc^2$

$= 1 + \dfrac{1}{2}mv^2 + \dfrac{3}{8}m\dfrac{v^4}{c^2} + \cdots - 1 = \dfrac{1}{2}mv^2$, if v is much

smaller than c.

33. 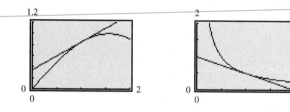   **35.**

Exercises 29.4, page 891

1. 0.905 **3.** $1.22,\ 1.221\,402\,8$ **5.** $0.099\,833\,3,\ 0.099\,833\,4$

7. $2.718\,055\,6,\ 2.718\,281\,8$ **9.** $0.998\,629\,2,\ 0.998\,629\,5$

11. $0.334\,933\,3,\ 0.336\,472\,2$ **13.** $0.354\,613\,0,\ 0.354\,612\,9$

15. $-0.013\,997\,5,\ -0.013\,997\,5$ **17.** $1.207\,36,\ 1.208\,03$

19. $1.052\,352\,8$ **21.** $0.987\,446\,2$ **23.** 8.3×10^{-8}

25. 3.1×10^{-7} **27.** 3.146

29. The terms of the expansion for e^x after those on the right side of the inequality have a positive value.

31. 1.59 years   **33.** $i = \dfrac{E}{L}\left(t - \dfrac{Rt^2}{2L}\right)$; small values of t

35. 15 m

Exercises 29.5, page 894

1. $\sqrt{x} = 1 + \dfrac{1}{2}(x - 1) - \dfrac{1}{8}(x - 1)^2 + \dfrac{1}{16}(x - 1)^3 - \cdots$

3. 3.32 **5.** 2.049 **7.** $0.515\,05$ **9.** $0.492\,88$

11. $e^{-2}\left[1 - (x - 2) + \dfrac{(x - 2)^2}{2!} - \cdots\right]$

13. $\dfrac{1}{2}\left[\sqrt{3} + \left(x - \dfrac{1}{3}\pi\right) - \dfrac{\sqrt{3}}{2!}\left(x - \dfrac{1}{3}\pi\right)^2 - \cdots\right]$

15. $2 + \dfrac{1}{12}(x - 8) - \dfrac{1}{288}(x - 8)^2 + \cdots$

17. $1 + 2\left(x - \dfrac{1}{4}\pi\right) + 2\left(x - \dfrac{1}{4}\pi\right)^2 + \cdots$ **19.** 23.1308

21. 3.0496 **23.** 2.0247 **25.** $0.874\,62$

27. Use the indicated method.

29. $2x^3 + x^2 - 3x + 5 = 5 + 5(x - 1) + 14\dfrac{(x - 1)^2}{2}$

$+ 12\dfrac{(x - 1)^3}{6}$

31. $0.515\,040\,8,\ 0.515\,038\,8,\ 0.515\,038\,1$

33. Graph of part (b) fits   **35.** Graph of part (b) fits
well near $x = \pi/3$.   well near $x = 2$.

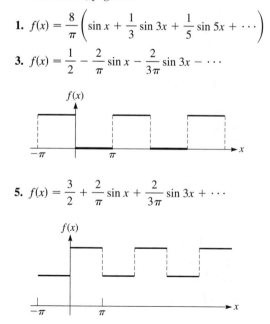

Exercises 29.6, page 900

1. $f(x) = \dfrac{8}{\pi}\left(\sin x + \dfrac{1}{3}\sin 3x + \dfrac{1}{5}\sin 5x + \cdots\right)$

3. $f(x) = \dfrac{1}{2} - \dfrac{2}{\pi}\sin x - \dfrac{2}{3\pi}\sin 3x - \cdots$

5. $f(x) = \dfrac{3}{2} + \dfrac{2}{\pi}\sin x + \dfrac{2}{3\pi}\sin 3x + \cdots$

7. $f(x) = \dfrac{\pi}{4} - \dfrac{2}{\pi}\left(\cos x + \dfrac{1}{9}\cos 3x + \cdots\right)$

$\quad + \left(\sin x - \dfrac{1}{2}\sin 2x + \cdots\right)$

9. $f(x) = -\dfrac{1}{4} - \dfrac{1}{\pi}\cos x + \dfrac{1}{3\pi}\cos 3x - \cdots$

$\quad + \dfrac{3}{\pi}\sin x - \dfrac{1}{\pi}\sin 2x + \dfrac{1}{\pi}\sin 3x - \cdots$

11. $f(x) = \dfrac{\pi}{2} - \dfrac{4}{\pi}\cos x - \dfrac{4}{9\pi}\cos 3x - \cdots$

13. **15.**

17.

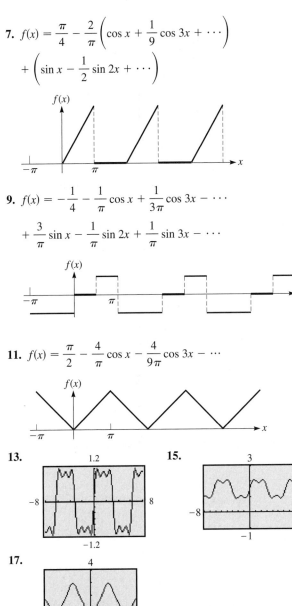

19. $f(t) = \dfrac{2}{\pi} - \dfrac{4}{3\pi}\cos 2t - \dfrac{4}{15\pi}\cos 4t - \cdots$

Exercises 29.7, page 906

1. $\dfrac{5}{2} + \dfrac{2}{\pi}\left(\cos x - \dfrac{1}{3}\cos 3x + \dfrac{1}{5}\cos 5x - \cdots\right)$

3. $-1 + \dfrac{4}{\pi}\sin\dfrac{\pi x}{4} + \dfrac{4}{3\pi}\sin\dfrac{3\pi x}{4} + \cdots$

5. Neither **7.** Even **9.** Even **11.** Odd

13. $f(x) = \dfrac{5}{2} - \dfrac{10}{\pi}\left(\sin\dfrac{\pi x}{3} + \dfrac{1}{3}\sin \pi x - \cdots\right)$

15. $f(x) = 1 + \dfrac{4}{\pi}\cos\dfrac{\pi x}{2} - \dfrac{4}{3\pi}\cos\dfrac{3\pi x}{2} + \cdots$

17. $f(x) = 2 - \dfrac{16}{\pi^2}\left(\cos\dfrac{\pi x}{4} + \dfrac{1}{9}\cos\dfrac{3\pi x}{4} + \cdots\right)$

19. $f(x) = \dfrac{4}{\pi}\left(\sin\dfrac{\pi x}{4} + \dfrac{1}{3}\sin\dfrac{3\pi x}{4} + \dfrac{1}{5}\sin\dfrac{5\pi x}{4} + \cdots\right)$

21. $f(x) = \dfrac{4}{3} - \dfrac{16}{\pi^2}\left(\cos\dfrac{\pi x}{2} - \dfrac{1}{4}\cos \pi x + \dfrac{1}{9}\cos\dfrac{3\pi x}{2} - \cdots\right)$

23. $f(t) = 2 + \dfrac{8}{\pi}\left(\cos\dfrac{\pi}{2}t - \dfrac{1}{3}\cos\dfrac{3\pi}{2}t + \cdots + \sin\dfrac{\pi}{2}t\right.$

$\quad\left. + \sin \pi t + \dfrac{1}{3}\sin\dfrac{3\pi}{2}t + \cdots\right)$

Review Exercises for Chapter 29, page 907

1. $\dfrac{1}{2} - \dfrac{1}{4}x + \dfrac{1}{48}x^3 - \cdots$ **3.** $2x^2 - \dfrac{4}{3}x^6 + \dfrac{4}{15}x^{10} - \cdots$

5. $1 + \dfrac{1}{3}x - \dfrac{1}{9}x^2 + \cdots$ **7.** $x + \dfrac{1}{6}x^3 + \dfrac{3}{40}x^5 + \cdots$

9. 0.82 **11.** 1.09 **13.** 0.9214 **15.** -0.2015

17. 0.95299 **19.** 12.1655 **21.** 0.259

23. $\dfrac{1}{2} - \dfrac{1}{2}\sqrt{3}\left(x - \dfrac{1}{3}\pi\right) - \dfrac{1}{4}\left(x - \dfrac{1}{3}\pi\right)^2 + \cdots$

25. $f(x) = \dfrac{\pi - 2}{4} - \dfrac{2}{\pi}\left(\cos x + \dfrac{1}{9}\cos 3x + \cdots\right)$

$\quad + \left(\dfrac{\pi - 2}{\pi}\right)\sin x - \dfrac{1}{2}\sin 2x + \cdots$

27. $f(x) = \pi + \dfrac{4}{\pi}\left(\sin\dfrac{\pi x}{4} + \dfrac{1}{3}\sin\dfrac{3\pi x}{4} + \cdots\right)$

29. $f(x) = \dfrac{1}{2} + \dfrac{2}{\pi}\left(\cos x - \dfrac{1}{3}\cos 3x + \cdots\right)$

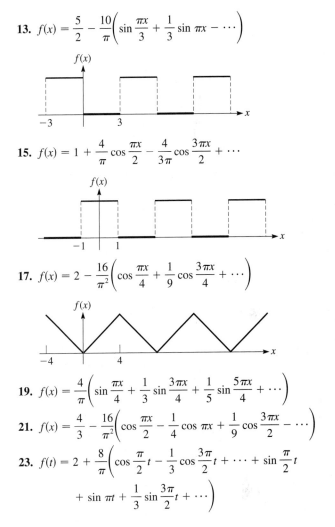

31. $f(x) = \dfrac{4}{\pi}\left(\sin\dfrac{\pi x}{2} - \dfrac{1}{2}\sin\pi x + \dfrac{1}{3}\sin\dfrac{3\pi x}{2} - \cdots\right)$

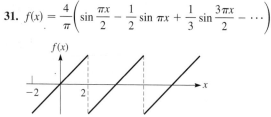

33. Convergent, $S = 5000$ **35.** $S = 256$

37. $(x + h) - \dfrac{(x + h)^3}{3!} + \cdots - (x - h) + \dfrac{(x - h)^3}{3!} - \cdots$

$\qquad = 2h - \dfrac{2hx^2}{2!} + \cdots = 2h\left(1 - \dfrac{x^2}{2!} + \cdots\right)$

39. $4x - 2x^2 + \frac{4}{3}x^3 - x^4 + \cdots$

41. $1 - x^2 + \frac{1}{3}x^4 - \frac{2}{45}x^6 + \cdots$

43. $1 + \dfrac{x^2}{2} + \dfrac{5x^4}{24} + \cdots$ **45.** $1 - x + x^2 - \cdots$

47. $\dfrac{1}{3} + \dfrac{4}{\pi^2}\left(-\cos\pi x + \dfrac{1}{4}\cos 2\pi x - \dfrac{1}{9}\cos 3\pi x + \cdots\right)$

49. $2.4265, 2.4600, 2.459\,603\,1$ **51.** $0.002\,496\,88$

53. $x - \dfrac{x^3}{3} + \dfrac{x^5}{5} - \cdots$ **55.** $N_0\left(1 - \lambda t + \dfrac{\lambda^2 t^2}{2} - \dfrac{\lambda^3 t^3}{6} + \cdots\right)$

57. 7.8 m **59.** $2x + \frac{2}{3}x^3 + \frac{2}{5}x^5 + \frac{2}{7}x^7 + \cdots$

61. $N_0[1 + e^{-k/T} + (e^{-k/T})^2 + \cdots]$

$\qquad = N_0(1 + e^{-k/T} + e^{-2k/T} + \cdots)$

63. $f(t) = \dfrac{1}{2\pi} + \dfrac{1}{\pi}\left(\dfrac{1}{2}\cos t - \dfrac{1}{3}\cos 2t + \cdots\right)$

$\qquad + \dfrac{1}{4}\sin t + \dfrac{2}{3\pi}\sin 2t + \cdots$

f(t) graph with peaks, axis labeled $-\pi$, π, t

Exercises 30.1, page 912

1. $(c_1 e^{-x} + 4c_2 e^{2x}) - (-c_1 e^{-x} + 2c_2 e^{2x}) = 2(c_1 e^{-x} + c_2 e^{2x})$;

$(4e^{-x}) - (-4e^{-x}) = 2(4e^{-x})$ **3.** Particular solution

5. General solution

(The following "answers" are the unsimplified expressions obtained by substituting functions and derivatives.)

7. $e^x - (e^x - 1) = 1; 5e^x - (5e^x - 1) = 1$

9. $-12\cos 2x + 4(3\cos 2x) = 0$;

$(-4c_1\sin 2x - 4c_2\cos 2x) + 4(c_1\sin 2x + c_2\cos 2x) = 0$

11. $2x = 2x$ **13.** $(2x + 3) - 3 = 2x$

15. $(-2ce^{-2x} + 1) + 2\left(ce^{-2x} + x - \frac{1}{2}\right) = 2x$

17. $-\frac{1}{2}\cos x + \frac{9}{2}\cos x = 4\cos x$

19. $x^2\left[-\dfrac{c^2}{(x - c)^2}\right] + \left[\dfrac{cx}{(x - c)}\right]^2 = 0$

21. $x\left(-\dfrac{c_1}{x^2}\right) + \dfrac{c_1}{x} = 0$

23. $(\cos x - \sin x + e^{-x}) + (\sin x + \cos x - e^{-x}) = 2\cos x$

25. $\left(e^{-x} + \frac{12}{5}\cos 2x + \frac{24}{5}\sin 2x\right) +$
$\left(-e^{-x} + \frac{6}{5}\sin 2x - \frac{12}{5}\cos 2x\right) = 6\sin 2x$

27. $\cos x\left[\dfrac{(\sec x + \tan x) - (x + c)(\sec x \tan x + \sec^2 x)}{(\sec x + \tan x)^2}\right]$
$\qquad + \sin x = 1 - \dfrac{x + c}{\sec x + \tan x}$

29. $c^2 + cx = cx + c^2$ **31.** $c_3 e^x = c_3 e^x$

Exercises 30.2, page 916

1. $y(x^2 + 1) = c$ **3.** $y = c - x^2$ **5.** $x - \dfrac{1}{y} = c$

7. $\ln V = \dfrac{1}{P} + c$ **9.** $\ln(x^3 + 5) + 3y = c$

11. $y = 2x^2 + x - x\ln x + c$ **13.** $4\sqrt{1 - y} = e^{-x^2} + c$

15. $e^x - e^{-y} = c$; e^{x+y} is the same as $e^x e^y$. Divide each term by e^y and integrate.

17. $\ln(y + 4) = x + c$ **19.** $y(1 + \ln x)^2 + cy + 2 = 0$

21. $\tan^2 x + 2\ln y = c$ **23.** $x^2 + 1 + x\ln y + cx = 0$

25. $y^2 + 4\sin^{-1} x = c$ **27.** $e^{x^2} = 2e^y + c$

29. $i = c - (\ln t)^2$ **31.** $\ln(e^x + 1) - \dfrac{1}{y} = c$

33. $3\ln y + x^3 = 0$ **35.** $\frac{1}{3}y^3 + y = \frac{1}{2}\ln^2 x$

37. $2\ln(1 - y) = 1 - 2\sin x$ **39.** $e^{2x} - \dfrac{2}{y} = 2(e^x - 1)$

Exercises 30.3, page 918

1. $\ln xy + y^2 = c$ **3.** $2xy + x^2 = c$ **5.** $x^3 - 2y = cx - 4$

7. $A^2 r - r = cA$ **9.** $(xy)^4 = 12\ln y + c$

11. $2\sqrt{x^2 + y^2} = x + c$ **13.** $y = c - \frac{1}{2}\ln\sin(x^2 + y^2)$

15. $\ln(y^2 - x^2) + 2x = c$; subtract $x\,dx$ from each side and divide through by $y^2 - x^2$. **17.** $5xy^2 + y^3 = c$

19. $2xy + x^3 = 5$ **21.** $2x = 2xy^2 - 15y$

23. $2x + 1 = \cos xy$

Exercises 30.4, page 921

1. $y = x + cx^{-2}$ **3.** $y = e^{-x}(x + c)$ **5.** $y = -\frac{1}{2}e^{-4x} + ce^{-2x}$

7. $y = -2 + ce^{2x}$ **9.** $y = x(3\ln x + c)$

11. $y = \dfrac{8}{7}x^3 + \dfrac{c}{\sqrt{x}}$ **13.** $r = -\cot\theta + c\csc\theta$

15. $y = (x + c)\csc x$ **17.** $y = 3 + ce^{-x}$

19. $2s = e^{4t}(t^2 + c)$ **21.** $y = \frac{1}{4} + ce^{-x^4}$

23. $3y = x^4 - 6x^2 - 3 + cx$ **25.** $xy = (x^3 + c)e^{3x}$

27. $r = \sin\theta[\ln(\tan\theta + \sec\theta)] + c\sin\theta$

29. Can solve by separation of variables: $\dfrac{dy}{1-y} = 2\,dx$.

Can also solve as linear differential equation of first order:

$dy + 2y\,dx = 2\,dx$; $y = 1 + ce^{-2x}$

31. $y = e^{-x}$ **33.** $y = \frac{4}{3}\sin x - \csc^2 x$

35. $y(\csc x - \cot x) = \ln\dfrac{(\sqrt{2}-1)(\csc 2x - \cot 2x)}{\csc x - \cot x}$

Exercises 30.5, page 925

1. $y^2 + 2x^2 = c$ **3.** 2.46 kg

5. $y^2 = 2x^2 + 1$ **7.** $y = 2e^x - x - 1$

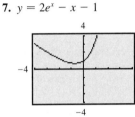

9. $y^2 = c - 2x$ **11.** $y^2 = c - 2\sin x$

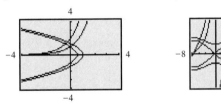

13. $N = N_0(0.5)^{t/40}$, 35.4%   **15.** 3.82 days

17. $N = \dfrac{r}{k}(1 - e^{-kt})$ **19.** 37.7 million   **21.** $S = a + \dfrac{c}{r^2}$

23. $5250e^{0.0196t}$ **25.** 12 min   **27.** \$1040.81

29. $\lim\limits_{t\to\infty}\dfrac{E}{R}(1 - e^{-Rt/L}) = \dfrac{E}{R}$

31. $i = \dfrac{E}{R^2 + \omega^2 L^2}(R\sin\omega t - \omega L\cos\omega t + \omega L e^{-Rt/L})$

33. $q = q_0 e^{-t/RC}$ **35.** 1.5 kg   **37.** $v = 9.8(1 - e^{-t})$, 9.8

39. 20 m/s   **41.** $x = 3t^2 - t^3$, $y = 6t^2 - 2t^3 - 9t^4 + 6t^5 - t^6$

43. $p = 100(0.8)^{h/2000}$ **45.** \$2490   **47.** $x = 0.20 + 0.15^{-2.0t}$

Exercises 30.6, page 931

1. $y = c_1 + c_2 e^{5x}$ **3.** $y = c_1 e^{3x} + c_2 e^{-2x}$

5. $y = c_1 e^{-x} + c_2 e^{-x/3}$ **7.** $y = c_1 + c_2 e^{3x}$

9. $y = c_1 e^{3x/2} + c_2 e^{-x}$ **11.** $y = c_1 e^{6x} + c_2 e^{2x/3}$

13. $y = c_1 e^{x/3} + c_2 e^{-3x}$ **15.** $y = c_1 e^{x/3} + c_2 e^{-x}$

17. $y = e^x(c_1 e^{x\sqrt{2}/2} + c_2 e^{-x\sqrt{2}/2})$

19. $y = e^{3x/8}(c_1 e^{x\sqrt{41}/8} + c_2 e^{-x\sqrt{41}/8})$

21. $y = e^{3x/2}(c_1 e^{x\sqrt{13}/2} + c_2 e^{-x\sqrt{13}/2})$

23. $y = e^{-x/2}(c_1 e^{x\sqrt{33}/2} + c_2 e^{-x\sqrt{33}/2})$

25. $y = \frac{1}{5}(3e^{7x} + 7e^{-3x})$ **27.** $y = \dfrac{e^3}{e^7 - 1}(e^{4x} - e^{-3x})$

29. $y = c_1 + c_2 e^{-x} + c_3 e^{3x}$

31. $y = c_1 e^x + c_2 e^{-x} + c_3 e^{2x} + c_4 e^{-2x}$

Exercises 30.7, page 935

1. $y = e^{-5x}(c_1 + c_2 x)$ **3.** $y = c_1 + c_2 e^{10x}$

5. $y = (c_1 + c_2 x)e^x$ **7.** $y = (c_1 + c_2 x)e^{-6x}$

9. $y = c_1 \sin 3x + c_2 \cos 3x$

11. $y = e^{-x/2}(c_1 \sin\frac{1}{2}\sqrt{7}x + c_2 \cos\frac{1}{2}\sqrt{7}x)$

13. $y = c_1 e^x + c_2 e^{-x} + c_3 \sin x + c_4 \cos x$

15. $y = c_1 \sin\frac{1}{2}x + c_2 \cos\frac{1}{2}x$ **17.** $y = (c_1 + c_2 x)e^{3x/4}$

19. $y = c_1 \sin\frac{1}{5}\sqrt{2}x + c_2 \cos\frac{1}{5}\sqrt{2}x$

21. $y = e^x(c_1 \cos\frac{1}{2}\sqrt{6}x + c_2 \sin\frac{1}{2}\sqrt{6}x)$ **23.** $y = (c_1 + c_2 x)e^{4x/5}$

25. $y = e^{3x/4}(c_1 e^{x\sqrt{17}/4} + c_2 e^{-x\sqrt{17}/4})$

27. $y = c_1 e^{x(-6+\sqrt{42})/3} + c_2 e^{x(-6-\sqrt{42})/3}$

29. $y = e^{2x}(c_1 + c_2 x + c_3 x^2)$

31. $y = (c_1 + c_2 x)\sin x + (c_3 + c_4 x)\cos x$

33. $y = e^{-x}\sin 3x$ **35.** $y = e^{4x}(4 - 14x)$ **37.** $D^2 y - 9y = 0$

39. $D^2 y + 9y = 0$. The sum of $\cos 3x$ and $\sin 3x$ with no exponential factor indicates imaginary roots with $\alpha = 0$ and $\beta = 3$.

Exercises 30.8, page 939

1. $y_p = A + Bx + Cx^2 + Ee^{-x}$

3. $y_c = c_1 \sin 2x + c_2 \cos 2x$; $y_p = A\sin x + B\cos x + Ce^{-x}$

5. $y = c_1 e^{2x} + c_2 e^{-x} - 2$ **7.** $y = c_1 e^{-x} + c_2 e^x - 4 - x^2$

9. $y = c_1 + c_2 e^{3x} - \frac{3}{4}e^x - \frac{1}{2}xe^x$

11. $y = c_1 e^{x/3} + c_2 e^{-x/3} - \frac{1}{10}\sin x$

13. $y = (c_1 + c_2 x)e^x + 10 + 6x + x^2 - \frac{2}{25}\sin 3x + \frac{3}{50}\cos 3x$

15. $y = c_1 \sin 2x + c_2 \cos 2x + 3x\cos 2x$

17. $y = c_1 e^{-5x} + c_2 e^{6x} - \frac{1}{3}$ **19.** $y = c_1 e^{2x/3} + c_2 e^{-5x} + \frac{1}{4}e^{3x}$

21. $y = c_1 e^{2x} + c_2 e^{-2x} - \frac{1}{5}\sin x - \frac{2}{5}\cos x$

23. $y = c_1 \sin x + c_2 \cos x - \frac{1}{3}\sin 2x + 4$

25. $y = c_1 e^{-x} + c_2 e^{-4x} - \frac{7}{100}e^x + \frac{1}{10}xe^x + 1$

27. $y = c_1 + c_2 e^x + c_3 e^{-x} + \frac{1}{10}\cos 2x$

29. $y = c_1 \sin x + c_2 \cos x + \frac{1}{2}x\sin x$

31. $y = c_1 + c_2 e^{-2x} + 2x^2 - 2x - \frac{1}{2}xe^{-2x}$

33. $y = \frac{1}{6}(11e^{3x} + 5e^{-2x} + e^x - 5)$

35. $y = -\frac{2}{3}\sin x + \pi\cos x + x - \frac{1}{3}\sin 2x$

Exercises 30.9, page 946

1. $x = \dfrac{1}{2}\sin 4t$ **3.** $x = e^{-0.1t}(0.04\sin 10t + 4\cos 10t)$

5. $\theta = 0.1 \cos 3.1t$ **7.** 10

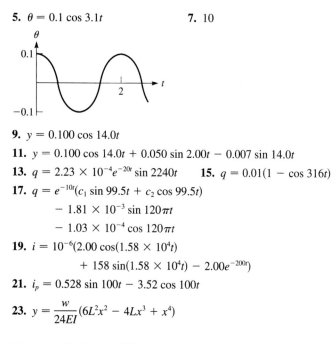

9. $y = 0.100 \cos 14.0t$

11. $y = 0.100 \cos 14.0t + 0.050 \sin 2.00t - 0.007 \sin 14.0t$

13. $q = 2.23 \times 10^{-4}e^{-20t} \sin 2240t$ **15.** $q = 0.01(1 - \cos 316t)$

17. $q = e^{-10t}(c_1 \sin 99.5t + c_2 \cos 99.5t)$
$\quad - 1.81 \times 10^{-3} \sin 120\pi t$
$\quad - 1.03 \times 10^{-4} \cos 120\pi t$

19. $i = 10^{-6}(2.00 \cos(1.58 \times 10^4 t)$
$\quad + 158 \sin(1.58 \times 10^4 t) - 2.00e^{-200t})$

21. $i_p = 0.528 \sin 100t - 3.52 \cos 100t$

23. $y = \dfrac{w}{24EI}(6L^2x^2 - 4Lx^3 + x^4)$

Exercises 30.10, page 951

1. $F(s) = \displaystyle\int_0^\infty e^{-st}\,dt = -\dfrac{1}{s}e^{-st}\Big|_0^\infty = \dfrac{1}{s}$

3. $(s^2 - 2s)\mathcal{L}(f) - s + 2$ **5.** $\dfrac{1}{s - 3}$ **7.** $\dfrac{30}{(s + 2)^4}$

9. $\dfrac{s - 2}{s^2 + 4}$ **11.** $\dfrac{3}{s} + \dfrac{2(s^2 - 9)}{(s^2 + 9)^2}$ **13.** $s^2\mathcal{L}(f) + s\mathcal{L}(f)$

15. $(2s^2 - s + 1)\mathcal{L}(f) - 2s + 1$ **17.** t^2 **19.** $\tfrac{1}{2}e^{-3t}$

21. $\tfrac{1}{2}t^2 e^{-t}$ **23.** $\tfrac{1}{54}(9t \sin 3t + 2 \sin 3t - 6t \cos 3t)$

25. $-\tfrac{1}{3}e^{-t} - \tfrac{8}{3}e^{2t} + 7e^{3t}$ **27.** $\tfrac{1}{2}e^t(4 \cos 2t + 5 \sin 2t)$

Exercises 30.11, page 955

1. $y = 2e^{t/2}$ **3.** $y = \dfrac{t}{2} \sin t + \sin t + 2 \cos t$ **5.** $y = e^{-t}$

7. $y = -e^{3t/2}$ **9.** $y = (1 + t)e^{-3t}$ **11.** $y = \tfrac{1}{2} \sin 2t$

13. $y = 1 - e^{-2t}$ **15.** $y = e^{2t} \cos t$ **17.** $y = 1 + \sin t$

19. $y = e^{-t}(\tfrac{1}{2}t^2 + 3t + 1)$ **21.** $y = 2e^{3t} + 3e^{-2t}$

23. $y = \tfrac{3}{2}e^t - \tfrac{3}{2}e^{-t} - \sin 2t$ **25.** $v = 6(1 - e^{-t/2})$

27. $q = 1.6 \times 10^{-4}(1 - e^{-5000t})$ **29.** $i = 5t \sin 50t$

31. $y = \sin 3t - 3t \cos 3t$ **33.** $i = 5.0e^{-50t} - 5.0e^{-100t}$

35. $i = 4.42e^{-66.7t} \sin 226t$

Review Exercises for Chapter 30, page 957

1. $2 \ln(x^2 + 1) - \dfrac{1}{2y^2} = c$ **3.** $y^2 = 2x - 4 \sin x + c$

5. $y = c_1 + c_2 e^{-x/2}$ **7.** $y = (c_1 + c_2x)e^{-x}$

9. $2x^2 + 4xy + y^4 = c$ **11.** $y = cx^3 - x^2$

13. $y = c(y + 2)e^{2x}$ **15.** $y = e^{-x}(c_1 \sin \sqrt{5}x + c_2 \cos \sqrt{5}x)$

17. $y = e^{-2x} + ce^{-4x}$ **19.** $y = \tfrac{1}{2}(c - x^2) \csc x$

21. $s = c_1 e^t + c_2 e^{-3t/2} - 2$

23. $y = e^{-x/2}(c_1 e^{x\sqrt{5}/2} + c_2 e^{-x\sqrt{5}/2}) + 2e^x$

25. $y = c_1 e^{2x/3} + c_2 e^{4x/3} + \tfrac{1}{2}x + \tfrac{25}{8}$

27. $y = c_1 e^x + c_2 \sin 3x + c_3 \cos 3x - \tfrac{1}{16}(\sin x + \cos x)$

29. $y = c_1 e^{-x} + c_2 e^{8x} - \tfrac{2}{9}xe^{-x}$

31. $y = c_1 \sin 5x + c_2 \cos 5x + 5x \sin 5x$

33. $y^3 = 8 \sin^2 x$ **35.** $y = 2x - 1 - e^{-2x}$

37. $y = 2e^{-x/2}\sin(\tfrac{1}{2}\sqrt{15}x)$

39. $y = \tfrac{1}{25}[16 \sin x + 12 \cos x - 3e^{-2x}(4 + 5x)]$

41. $y = e^{t/4}$ **43.** $y = \tfrac{1}{2}(e^{3t} - e^t)$ **45.** $y = -4 \sin t$

47. $y = \tfrac{1}{10}(3e^t - \sin 3t - 3 \cos 3t)$ **49.** $x = t^2 + 1; y = \dfrac{1}{t^2 + 1}$

51. $r = r_0 + kt$ **53.** $m = m_0(1 - e^{-kt})$ **55.** 3.93 m/s

57. 6.6 h **59.** 5.31×10^8 years **61.** 7.1 billion

63. $5y^2 + x^2 = c$ **65.** 36.1°C **67.** $i = 0.5(1 - e^{-20t})$

69. $y = 0.25e^{-2t}(2 \cos 4t + \sin 4t)$, underdamped

71. $q = e^{-6t}(0.4 \cos 8t + 0.3 \sin 8t) - 0.4 \cos 10t$

73. $i = 0$ **75.** $i = 12(1 - e^{-t/2}); i(0.3) = 1.67$ A

77. $q = 10^{-4}e^{-8t}(4.0 \cos 200t + 0.16 \sin 200t)$

79. $y = 0.25t \sin 8t$ **81.** 2.47 L

83. $\theta = \theta_0 \cos \omega t + \dfrac{\omega_0}{\omega} \sin \omega t; \omega = \sqrt{\dfrac{2k}{mr^2}}$

Exercises S.1, page 965

1. -60 **3.** 0 **5.** -40 **7.** -40 **9.** 39 **11.** 57

13. -13 **15.** -72 **17.** 0 **19.** 39 **21.** 57

23. -13 **25.** -72 **27.** 0

29. $x = -1, y = 0, z = 2, t = 1$

31. $x = 1, y = 2, z = -1, t = 3$

33. $x = 2, y = -1, z = -1, t = 3$

35. $D = 1, E = 2, F = -1, G = -2$

37. $\dfrac{33}{16}$ A, $\dfrac{11}{8}$ A, $-\dfrac{5}{8}$ A, $-\dfrac{15}{8}$ A, $-\dfrac{15}{16}$ A

39. ppm SO_2: 0.5, NO: 0.3, NO_2: 0.2, CO: 5.0

Exercises S.2, page 969

1. $x = 2, y = 1$ **3.** $x = \dfrac{17}{14}, y = \dfrac{19}{14}$

5. $x = -1, y = 4, z = 2$ **7.** Inconsistent

9. $x = -2, y = -\dfrac{2}{3}, z = \dfrac{1}{3}$

11. $w = 1, x = 0, y = -2, z = 3$

13. Unlimited: $x = -3$, $y = -1$, $z = 1$; $x = 12$, $y = 0$, $z = -10$

15. Inconsistent **17.** $x = 4$, $y = \dfrac{2}{3}$, $z = -2$

19. $x = \dfrac{1}{2}$, $y = -\dfrac{3}{2}$ **21.** $x = \dfrac{1}{6}$, $y = -\dfrac{1}{2}$

23. Inconsistent **25.** Inconsistent

27. $r = 0$, $s = 0$, $t = 0$, $u = -1$ **29.** 5.7 calc/s, 6.8 calc/s

31. 250 parts/h, 220 parts/h, 180 parts/h

Exercises S.3, page 974

1. hyperbola;
$2x'y' + 25 = 0$

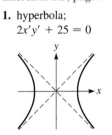

3. ellipse;
$4x'^2 + 9y'^2 = 36$

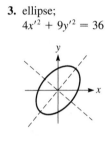

5. parabola;
$x'^2 + \sqrt{2}y' = 0$

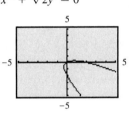

7. hyperbola;
$4x'^2 - y'^2 = 4$

9. ellipse;
$x'^2 + 2y'^2 = 2$

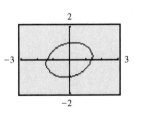

11. parabola
$y'^2 - 4x' + 16 = 0$
$y''^2 = 4x''$

Exercises S.4, page 976

1. $V = \frac{1}{3}\pi r^2 h$ **3.** $R = \sqrt{F_1^2 + F_2^2 + 1.732F_1F_2}$ **5.** 24

7. 12 **9.** $2 - 3y + 4y^2$ **11.** $z\tan^{-1}(x^2 - xz)$

13. $\dfrac{p^2 + pq + kp - p + 2q^2 + 4kq + 2k^2 + 5q + 5k}{p + q + k}$

15. 0 **17.** $x \neq 0$, $y \geq 0$ **19.** 1.03 A, 1.23 A

21. $A = \dfrac{pw - 2w^2}{2}$, 3850 cm² **23.** $L = \dfrac{(1.15 \times 10^4)r^4}{l^2}$

Exercises S.5, page 982

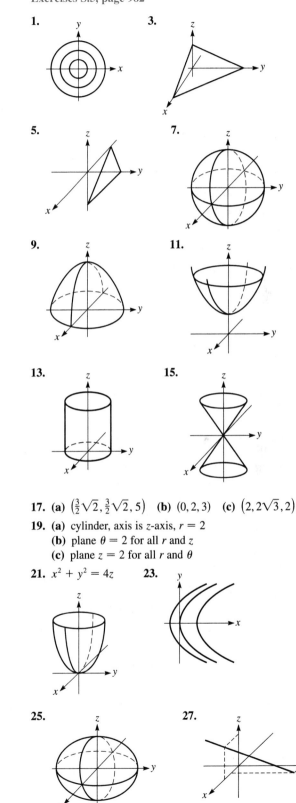

17. (a) $\left(\frac{3}{2}\sqrt{2}, \frac{3}{2}\sqrt{2}, 5\right)$ (b) $(0, 2, 3)$ (c) $\left(2, 2\sqrt{3}, 2\right)$

19. (a) cylinder, axis is z-axis, $r = 2$
(b) plane $\theta = 2$ for all r and z
(c) plane $z = 2$ for all r and θ

21. $x^2 + y^2 = 4z$ **23.**

Exercises S.6, page 986

1. $\dfrac{\partial z}{\partial x} = 5 + 8xy,\ \dfrac{\partial z}{\partial y} = 4x^2$ **3.** $\dfrac{\partial f}{\partial x} = e^{2y},\ \dfrac{\partial f}{\partial y} = 2xe^{2y}$

5. $\dfrac{\partial \phi}{\partial r} = \dfrac{1 + 3rs}{\sqrt{1 + 2rs}},\ \dfrac{\partial \phi}{\partial s} = \dfrac{r^2}{\sqrt{1 + 2rs}}$

7. $\dfrac{\partial z}{\partial x} = y \cos xy,\ \dfrac{\partial z}{\partial y} = x \cos xy$

9. $\dfrac{\partial f}{\partial x} = \dfrac{12 \sin^2 2x \cos 2x}{1 - 3y},\ \dfrac{\partial f}{\partial y} = \dfrac{6 \sin^3 2x}{(1 - 3y)^2}$

11. $\dfrac{\partial z}{\partial x} = y \cos x - y \sin xy,\ \dfrac{\partial z}{\partial y} = \sin x - x \sin xy$

13. -8 **15.** $\dfrac{41}{4}$

17. $\dfrac{\partial^2 z}{\partial x^2} = -6y,\ \dfrac{\partial^2 z}{\partial y^2} = 12xy,\ \dfrac{\partial^2 z}{\partial x \partial y} = 6y^2 - 6x$

19. $\dfrac{\partial^2 z}{\partial x^2} = e^x \sin y,\ \dfrac{\partial^2 z}{\partial y^2} = \dfrac{2x}{y^3} - e^x \sin y,$

$\dfrac{\partial^2 z}{\partial x \partial y} = \dfrac{\partial^2 z}{\partial y \partial x} = -\dfrac{1}{y^2} + e^x \cos y$

21. $-4, -4$ **23.** $\left(\dfrac{R_2}{R_1 + R_2}\right)^2$

25. $M\left(\dfrac{2mg}{(M + m)^2}\right) + m\left(\dfrac{-2Mg}{(M + m)^2}\right) = 0$ **27.** $3.75 \times 10^{-3}\ 1/\Omega$

Exercises S.7, page 990

1. $\dfrac{28}{3}$ **3.** $\dfrac{1}{3}$ **5.** 1 **7.** $\dfrac{74}{5}$ **9.** $\dfrac{32}{3}$ **11.** $\dfrac{28}{3}$

13. $300\ \text{cm}^3$ **15.**

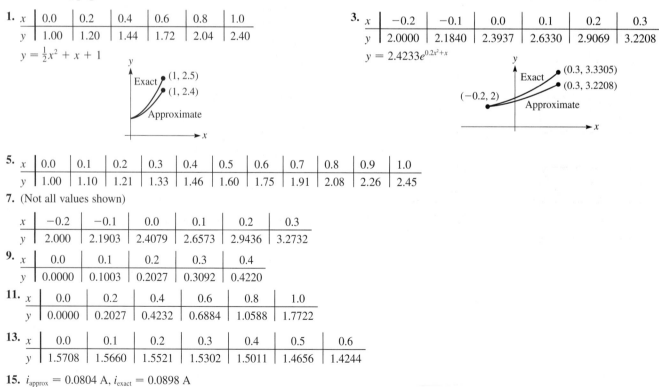

Exercises S.8, page 993

1.

x	0.0	0.2	0.4	0.6	0.8	1.0
y	1.00	1.20	1.44	1.72	2.04	2.40

$y = \frac{1}{2}x^2 + x + 1$

3.

x	−0.2	−0.1	0.0	0.1	0.2	0.3
y	2.0000	2.1840	2.3937	2.6330	2.9069	3.2208

$y = 2.4233e^{0.2x^2 + x}$

5.

x	0.0	0.1	0.2	0.3	0.4	0.5	0.6	0.7	0.8	0.9	1.0
y	1.00	1.10	1.21	1.33	1.46	1.60	1.75	1.91	2.08	2.26	2.45

7. (Not all values shown)

x	−0.2	−0.1	0.0	0.1	0.2	0.3
y	2.000	2.1903	2.4079	2.6573	2.9436	3.2732

9.

x	0.0	0.1	0.2	0.3	0.4
y	0.0000	0.1003	0.2027	0.3092	0.4220

11.

x	0.0	0.2	0.4	0.6	0.8	1.0
y	0.0000	0.2027	0.4232	0.6884	1.0588	1.7722

13.

x	0.0	0.1	0.2	0.3	0.4	0.5	0.6
y	1.5708	1.5660	1.5521	1.5302	1.5011	1.4656	1.4244

15. $i_{\text{approx}} = 0.0804\ \text{A},\ i_{\text{exact}} = 0.0898\ \text{A}$

Exercises for Appendix B, page A.9

1. MHz, 1 MHZ = 10^6 Hz   **3.** mm, 1 mm = 10^{-3} m

5. kilovolt, 1 kV = 10^3 V   **7.** milliampere, 1 mA = 10^{-3} A

9. 10^5 cm   **11.** 2.0×10^{-5} Ms   **13.** 537.00°C

15. 2.50×10^{-4} m²   **17.** 8.00×10^4 dm³   **19.** 0.050 L

21. 4.50×10^3 cm/s   **23.** 1.38×10^3 m/h

25. 3.53 cm/min²   **27.** 9.0×10^7 ms   **29.** 5.25×10^{-3} W/A

31. 2.2 Mg   **33.** 5.6×10^4 cm³   **35.** 1000 g/L

37. 1.20×10^3 km/h   **39.** 25 000 mg/dL   **41.** 992 m/s

43. 1.2×10^5 mA/cm²   **45.** 3.73 h   **47.** 0.135 km/L

Exercises for C.2, page A.13

(Most answers have been rounded off to four significant digits.)

1. 56.02   **3.** 4162.1   **5.** 18.65   **7.** 0.3954   **9.** 14.14

11. 0.5251   **13.** 13.35   **15.** 944.6   **17.** 0.7349

19. −0.7594   **21.** −1.337   **23.** 1.015   **25.** 41.35°

27. −1.182   **29.** 0.5862   **31.** 6.695   **33.** 3.508

35. 0.005 685   **37.** 2.053   **39.** 5.765   **41.** 4.501×10^{10}

43. 497.2   **45.** 6.648   **47.** 401.2   **49.** 8.841

51. 2.523   **53.** 10.08   **55.** 22.36   **57.** 20.3°   **59.** 4729

61. 3.301×10^4 **63.** 1.056   **65.** 55.5°   **67.** 3.277

69. 8.125   **71.** 1.000   **73.** 1.000   **75.** 12.90   **77.** 8.001

79. 8.053   **81.** 0.04259   **83.** 0.4219   **85.** 0.7822

87. 2.073 646 5   **89.** 124.3   **91.** 252

93.
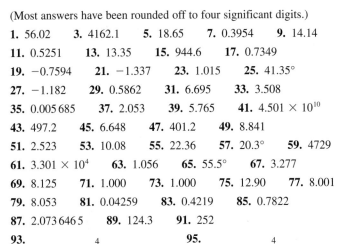

95.

97.
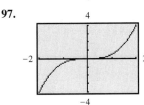

99.

101.
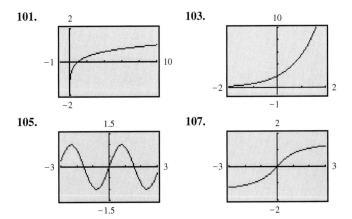

103.

105.

107.

Exercises for C.3, page A.16

(Program names are suggested.)

1. CYLINDER
:Prompt R,H
:Disp "TOTAL AREA=",2πR(R+H)
:Disp "VOLUME=",πR²H

3. PRODBINO
:Prompt A,B,C,D
:Disp "E=",A*C
:Disp "F=",A*D+B*C
:Disp "G=",B*D

5. SLVTRASA
:Input "θA=",A
:Input "θB=",B
:Input "SC=",F
:180−A−B→C
:Disp "θC=",C
:Disp "SA=",Fsin(A)/sin(C)
:Disp "SB=",Fsin(B)/sin(C)

7. CNSTPROP
:Prompt Y,R,A,S,B,T,C
:Disp "K="
:Disp YT^C/((R^A)(S^B))

9. EVALDERV
:Prompt C,N,A
:Disp "DY/DX=",NCA^(N−1)

Chapter 1

1. $\sqrt{9 + 16} = \sqrt{25} = 5$

2. $\dfrac{(7)(-3)(-2)}{(-6)(0)}$ is undefined (division by zero).

3. $\dfrac{3.372 \times 10^{-3}}{7.526 \times 10^{12}} = 4.480 \times 10^{-16}$

```
3.372E-3/7.526E1
2
     4.48046771E-16
```

4. $\dfrac{(+6)(-2) - 3(-1)}{5 - 2} = \dfrac{-12 - (-3)}{3} = \dfrac{-12 + 3}{3}$

$$= \dfrac{-9}{3} = -3$$

5. $\dfrac{346.4 - 23.5}{287.7} - \dfrac{0.944^3}{(3.46)(0.109)} = -1.108$

```
(346.4-23.5)/287
.7-.944^3/(3.46*
.109)
       -1.10820764
```

6. $(2a^0b^{-2}c^3)^{-3} = 2^{-3}a^{0(-3)}b^{(-2)(-3)}c^{3(-3)} = 2^{-3}a^0b^6c^{-9}$

$$= \dfrac{b^6}{8c^9}$$

7. $(2x + 3)^2 = (2x + 3)(2x + 3)$
$= 2x(2x) + 2x(3) + 3(2x) + 3(3)$
$= 4x^2 + 6x + 6x + 9$
$= 4x^2 + 12x + 9$

8. $3m^2(am - 2m^3) = 3m^2(am) + 3m^2(-2m^3)$
$= 3am^3 - 6m^5$

9. $\dfrac{8a^3x^2 - 4a^2x^4}{-2ax^2} = \dfrac{8a^3x^2}{-2ax^2} - \dfrac{4a^2x^4}{-2ax^2} = -4a^2 - (-2ax^2)$

$$= -4a^2 + 2ax^2$$

10.
$$
\begin{array}{r}
3x - 5 \quad \text{(quotient)} \\
2x - 1\overline{\smash{\big)}\,6x^2 - 13x + 7} \\
\underline{6x^2 - 3x} \\
-10x + 7 \\
\underline{-10x + 5} \\
2 \quad \text{(remainder)}
\end{array}
$$

11. $(2x - 3)(x + 7)$
$= 2x(x) + 2x(7) + (-3)(x) + (-3)(7)$
$= 2x^2 + 14x - 3x - 21 = 2x^2 + 11x - 21$

12. $3x - [4x - (3 - 2x)] = 3x - [4x - 3 + 2x]$
$= 3x - [6x - 3]$
$= 3x - 6x + 3 = -3x + 3$

13. $5y - 2(y - 4) = 7$
$5y - 2y + 8 = 7$
$3y + 8 - 8 = 7 - 8$
$3y = -1$
$y = -1/3$

14. $\qquad 3(x - 3) = x - (2 - 3d)$
$3x - 9 = x - 2 + 3d$
$3x - 9 - x + 9 = x - 2 + 3d - x + 9$
$2x = 3d + 7$
$x = \dfrac{3d + 7}{2}$

15. $0.000\,003\,6 = 3.6 \times 10^{-6}$ (six places to right)

16.

$-\pi$	-3	0.3	$\sqrt{2}$	$|-4|$	(order)
-3.14	-3	0.3	1.41	4	(value)

17. $3(5 + 8) = 3(5) + 3(8)$ illustrates distributive law.

18. (a) 5, (b) 3.0 (zero is significant)

19. Evaluation:
$1000(1 + 0.05/2)^{2(3)} = 1000(1.025)^6 = \$1159.69$

20. $8(100 - x)^2 + x^2 = 8(100 - x)(100 - x) + x^2$
$= 8(10\,000 - 200x + x^2) + x^2$
$= 80\,000 - 1600x + 8x^2 + x^2$
$= 80\,000 - 1600x + 9x^2$

21. $L = L_0[1 + \alpha(t_2 - t_1)]$
$= L_0[1 + \alpha t_2 - \alpha t_1]$
$= L_0 + \alpha L_0 t_2 - \alpha L_0 t_1$
$L - L_0 + \alpha L_0 t_1 = \alpha L_0 t_2$
$t_2 = \dfrac{L - L_0 + \alpha L_0 t_1}{\alpha L_0}$

22. Let n = number of newtons of second alloy
$0.3(20) + 0.8n = 0.6(n + 20)$
$6 + 0.8n = 0.6n + 12$
$0.2n = 6$
$n = 30$ N

Chapter 2

1. $\angle 1 + \angle 3 + 90° = 180°$ (sum of angles of a triangle)
$\angle 3 = 52°$ (vertical angles)
$\angle 1 = 180° - 90° - 52° = 38°$

2. $\angle 2 + \angle 4 = 180°$ (straight angle)
$\angle 4 = 52°$ (corresponding angles)
$\angle 2 + 52° = 180°$
$\angle 2 = 128°$

3.

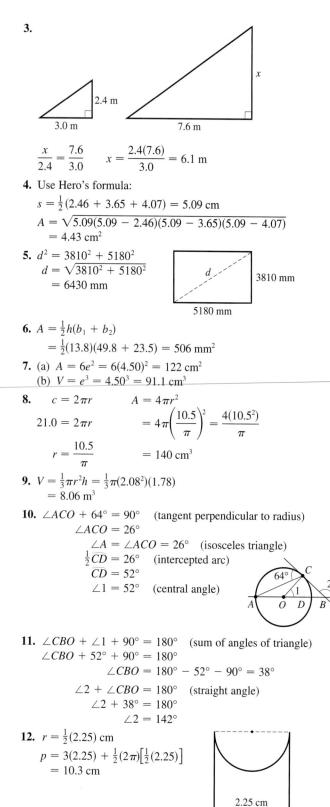

$$\frac{x}{2.4} = \frac{7.6}{3.0} \qquad x = \frac{2.4(7.6)}{3.0} = 6.1 \text{ m}$$

4. Use Hero's formula:

$$s = \tfrac{1}{2}(2.46 + 3.65 + 4.07) = 5.09 \text{ cm}$$
$$A = \sqrt{5.09(5.09 - 2.46)(5.09 - 3.65)(5.09 - 4.07)}$$
$$= 4.43 \text{ cm}^2$$

5. $d^2 = 3810^2 + 5180^2$
$d = \sqrt{3810^2 + 5180^2}$
$= 6430 \text{ mm}$

6. $A = \tfrac{1}{2}h(b_1 + b_2)$
$= \tfrac{1}{2}(13.8)(49.8 + 23.5) = 506 \text{ mm}^2$

7. (a) $A = 6e^2 = 6(4.50)^2 = 122 \text{ cm}^2$
(b) $V = e^3 = 4.50^3 = 91.1 \text{ cm}^3$

8. $c = 2\pi r \qquad A = 4\pi r^2$
$21.0 = 2\pi r \qquad = 4\pi\left(\dfrac{10.5}{\pi}\right)^2 = \dfrac{4(10.5^2)}{\pi}$
$r = \dfrac{10.5}{\pi} \qquad = 140 \text{ cm}^3$

9. $V = \tfrac{1}{3}\pi r^2 h = \tfrac{1}{3}\pi(2.08^2)(1.78)$
$= 8.06 \text{ m}^3$

10. $\angle ACO + 64° = 90°$ (tangent perpendicular to radius)
$\angle ACO = 26°$
$\angle A = \angle ACO = 26°$ (isosceles triangle)
$\tfrac{1}{2}\overset{\frown}{CD} = 26°$ (intercepted arc)
$\overset{\frown}{CD} = 52°$
$\angle 1 = 52°$ (central angle)

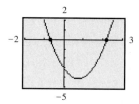

11. $\angle CBO + \angle 1 + 90° = 180°$ (sum of angles of triangle)
$\angle CBO + 52° + 90° = 180°$
$\angle CBO = 180° - 52° - 90° = 38°$
$\angle 2 + \angle CBO = 180°$ (straight angle)
$\angle 2 + 38° = 180°$
$\angle 2 = 142°$

12. $r = \tfrac{1}{2}(2.25) \text{ cm}$
$p = 3(2.25) + \tfrac{1}{2}(2\pi)\left[\tfrac{1}{2}(2.25)\right]$
$= 10.3 \text{ cm}$

2.25 cm

13. $A = 2.25^2 - \tfrac{1}{2}\pi\left[\tfrac{1}{2}(2.25)\right]^2 = 3.07 \text{ cm}^2$

14. $A = \tfrac{1}{2}(50)[0 + 2(90) + 2(145) + 2(260)$
$+ 2(205) + 2(110) + 20]$
$= 41\,000 \text{ m}^2$

Chapter 3

1. $f(x) = 2x - x^2 + \dfrac{8}{x}$

$$f(-4) = 2(-4) - (-4)^2 + \dfrac{8}{-4}$$
$$= -8 - 16 - 2$$
$$= -26$$

$$f(2.385) = 2(2.385) - (2.385)^2 + \dfrac{8}{2.385} = 2.436$$

2. $w = 2000 - 10t$

3. $f(x) = 4 - 2x$

x	y
-1	6
0	4
1	2
2	0
3	-2
4	-4

$y = 4 - 2x$
$y = 4 - 2(-1) = 6$
$y = 4 - 2(0) = 4$
$y = 4 - 2(1) = 2$
$y = 4 - 2(2) = 0$
$y = 4 - 2(3) = -2$
$y = 4 - 2(4) = -4$

4. $y = 2x^2 - 3x - 3$

$x = -0.7$ and $x = 2.2$

5. $y = \sqrt{4 + 2x}$

x	y
-2	0
-1	1.4
0	2
1	2.4
2	2.8
4	3.5

$y = \sqrt{4 + 2(-2)} = 0$
$y = \sqrt{4 + 2(-1)} = 1.4$
$y = \sqrt{4 + 2(0)} = 2$
$y = \sqrt{4 + 2(1)} = 2.4$
$y = \sqrt{4 + 2(2)} = 2.8$
$y = \sqrt{4 + 2(4)} = 3.5$

6. On negative x-axis

7. $f(x) = \sqrt{6 - x}$
Domain: $x \le 6$; x cannot be greater than 6 to have real values of $f(x)$.
Range: $f(x) \ge 0$; $\sqrt{6 - x}$ is the principal square root of $6 - x$ and cannot be negative.

8. Shifting $y = 2x^2 - 3$ to the right 1 and up 3 gives
$y = 2(x - 1)^2 - 3 + 3 = 2(x - 1)^2 = 2x^2 - 4x + 2$.

9. Range: $y < -8.9$ or $y > 0.9$

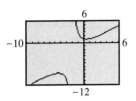

10. Let r = radius of circular part

      square   **semicircle**

$A = (2r)(2r) + \frac{1}{2}(\pi r^2)$
$= 4r^2 + \frac{1}{2}\pi r^2$

11.

Voltage V	10.0	20.0	30.0	40.0	50.0	60.0
Current i	145	188	220	255	285	315

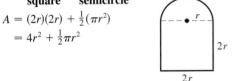

$f(45.0) = 270$ mA

12. $V = 20.0$ V for $i = 188$ mA.
$V = 30.0$ V for $i = 220$ mA.
$\dfrac{x}{10} = \dfrac{12}{32}$, $x = 3.8$ (rounded off)
$V = 20.0 + 3.8 = 23.8$ V
for $i = 200$ mA

$$10\begin{bmatrix} x \begin{bmatrix} -20.0 \\ \\ -30.0 \end{bmatrix} & \begin{array}{c} 188 \\ 200 \\ 220 \end{array} \end{bmatrix} 12 \Bigg] 32$$

Chapter 4

1. $39' = \left(\frac{39}{60}\right)^\circ = 0.65°$
$37°39' = 37.65°$

2. $\cos \theta = 0.3726$; $\theta = 68.12°$

3. Let x = distance from course to east

$\dfrac{x}{22.62} = \sin 4.05°$
$x = 22.62 \sin 4.05°$
$= 1.598$ km

4. $\sin \theta = \dfrac{2}{3}$
$x = \sqrt{3^2 - 2^2} = \sqrt{5}$
$\tan \theta = \dfrac{2}{\sqrt{5}}$

5. $\tan \theta = 1.294$; $\csc \theta = 1.264$

6. $B = 90° - 37.4° = 52.6°$ $\dfrac{52.8}{c} = \cos 37.4°$

$\dfrac{a}{52.8} = \tan 37.4°$ $c = \dfrac{52.8}{\cos 37.4°}$
$a = 52.8 \tan 37.4°$ $= 66.5$
$= 40.4$

7. $2.49^2 + b^2 = 3.88^2$
$b = \sqrt{3.88^2 - 2.49^2} = 2.98$
$\sin A = \dfrac{2.49}{3.88}$
$A = 39.9°$ $B = 50.1°$

8. $\dfrac{s/2}{12.0} = \cos 42.0°$
$s = 24.0 \cos 42.0° = 17.8$

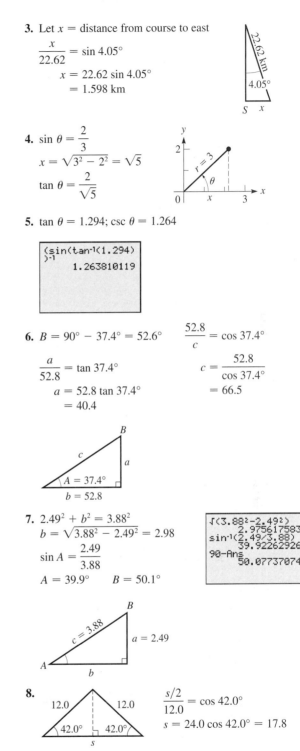

9.

$h = \sqrt{9^2 + 40^2} = 41$

$\sin \theta = \dfrac{9}{41}, \cos \theta = \dfrac{40}{41}$

$\dfrac{\sin \theta}{\cos \theta} = \dfrac{9/41}{40/41} = \dfrac{9}{40}$

10. $\lambda = d \sin \theta$
$= 30.05 \sin 1.167°$
$= 0.6120 \ \mu\text{m}$

11. $r = \sqrt{5^2 + 2^2} = \sqrt{29}$

$\sin \theta = \dfrac{2}{\sqrt{29}} = 0.3714$ $\qquad$ $\csc \theta = \dfrac{\sqrt{29}}{2} = 2.693$

$\cos \theta = \dfrac{5}{\sqrt{29}} = 0.9285$ $\qquad$ $\sec \theta = \dfrac{\sqrt{29}}{5} = 1.077$

$\tan \theta = \dfrac{2}{5} = 0.4000$ $\qquad$ $\cot \theta = \dfrac{5}{2} = 2.500$

12. Distance between points is $x - y$.

$\dfrac{18.525}{x} = \tan 13.500°$ $\qquad$ $\dfrac{18.525}{y} = \tan 21.375°$

$x - y = \dfrac{18.525}{\tan 13.500°} - \dfrac{18.525}{\tan 21.375°} = 29.831 \text{ m}$

Chapter 5

1. Points $(2, -5)$ and $(-1, 4)$

$m = \dfrac{4 - (-5)}{-1 - 2} = \dfrac{9}{-3} = -3$

2. $\begin{aligned} x + 2y &= 5 \\ 4y &= 3 - 2x \\ x &= 5 - 2y \end{aligned}$ $\qquad$ $\begin{aligned} 4y &= 3 - 2(5 - 2y) \\ 4y &= 3 - 10 + 4y \\ 0 &= -7 \\ &\text{Inconsistent} \end{aligned}$

3. $3x - 2y = 4$
$2x + 5y = -1$

$x = \dfrac{\begin{vmatrix} 4 & -2 \\ -1 & 5 \end{vmatrix}}{\begin{vmatrix} 3 & -2 \\ 2 & 5 \end{vmatrix}} = \dfrac{20 - 2}{15 - (-4)} = \dfrac{18}{19}$

$y = \dfrac{\begin{vmatrix} 3 & 4 \\ 2 & -1 \end{vmatrix}}{19} = \dfrac{-3 - 8}{19} = -\dfrac{11}{19}$

4. $2x + y = 4$
$y = -2x + 4$
$m = -2 \qquad b = 4$

5. $2l + 2w = 24$ (perimeter)
$\begin{aligned} l &= w + 6.0 \\ l + w &= 12 \\ \hline l - w &= 6.0 \\ \hline 2l &= 18 \\ l &= 9 \text{ km} \\ 9 &= w + 6 \\ w &= 3 \text{ km} \end{aligned}$

6. $\begin{aligned} 6N - 2P &= 13 \\ 4N + 3P &= -13 \\ \hline 18N - 6P &= 39 \\ 8N + 6P &= -26 \\ \hline 26N &= 13 \\ N &= \dfrac{1}{2} \end{aligned}$

$6\left(\dfrac{1}{2}\right) - 2P = 13$
$-2P = 10$
$P = -5$

7. $C = 310 - 2d$

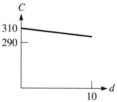

8. $\begin{aligned} 2x - 3y &= 6 \\ x = 0: y &= -2 \\ y = 0: x &= 3 \\ \text{Int: } (0, -2), &(3, 0) \\ x = 1.3 \quad & y = -1.1 \end{aligned}$ $\qquad$ $\begin{aligned} 4x + y &= 4 \\ x = 0: y &= 4 \\ y = 0: x &= 1 \\ \text{Int: } (0, 4), &(1, 0) \end{aligned}$

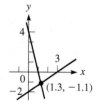

9. Let x = vol. of first alloy
y = vol. of second alloy
z = vol. of third alloy

$$
\begin{array}{llll}
x + & y + & z = 100 & \text{total vol.} \\
0.6x + & 0.5y + & 0.3z = 40 & \text{copper} \\
0.3x + & 0.3y & = 15 & \text{zinc}
\end{array}
$$

$$
\begin{array}{l}
x + y + z = 100 \\
6x + 5y + 3z = 400 \\
\hline
3x + 3y = 150 \\
3x + 2y = 100 \\
\hline
3x + 3y = 150
\end{array}
$$

$$y = 50 \text{ cm}^3$$
$$x = 0 \text{ cm}^3$$
$$z = 50 \text{ cm}^3$$

10.
$$3x + 2y - z = 4$$
$$2x - y + 3z = -2$$
$$x + 4z = 5$$

$$
y = \frac{\begin{vmatrix} 3 & 4 & -1 \\ 2 & -2 & 3 \\ 1 & 5 & 4 \end{vmatrix}}{\begin{vmatrix} 3 & 2 & -1 \\ 2 & -1 & 3 \\ 1 & 0 & 4 \end{vmatrix}}
$$

$$
= \frac{-24 + 12 - 10 - 2 - 45 - 32}{-12 + 6 + 0 - 1 - 0 - 16}
$$

$$
= \frac{-101}{-23} = \frac{101}{23}
$$

Chapter 6

1. $2x(2x - 3)^2 = 2x[(2x)^2 - 2(2x)(3) + 3^2]$
$= 2x(4x^2 - 12x + 9)$
$= 8x^3 - 24x^2 + 18x$

2. $\dfrac{1}{R} = \dfrac{1}{R_1 + r} + \dfrac{1}{R_2}$

$$\frac{RR_2(R_1 + r)}{R} = \frac{RR_2(R_1 + r)}{R_1 + r} + \frac{RR_2(R_1 + r)}{R_2}$$

$$R_2(R_1 + r) = RR_2 + R(R_1 + r)$$
$$R_1R_2 + rR_2 = RR_2 + RR_1 + rR$$
$$R_1R_2 - RR_1 = RR_2 + rR - rR_2$$
$$R_1(R_2 - R) = RR_2 + rR - rR_2$$
$$R_1 = \frac{RR_2 + rR - rR_2}{R_2 - R}$$

3. $\dfrac{2x^2 + 5x - 3}{2x^2 + 12x + 18} = \dfrac{(2x - 1)(x + 3)}{2(x + 3)^2} = \dfrac{2x - 1}{2(x + 3)}$

4. $4x^2 - 16y^2 = 4(x^2 - 4y^2) = 4(x + 2y)(x - 2y)$

5. $pb^3 + 8a^3p = p(b^3 + 8a^3)$
$= p(b + 2a)(b^2 - 2ab + 4a^2)$

6. $2a - 4T - ba + 2bT = 2(a - 2T) - b(a - 2T)$
$= (2 - b)(a - 2T)$

7. $\dfrac{3}{4x^2} - \dfrac{2}{x^2 - x} - \dfrac{x}{2x - 2} = \dfrac{3}{4x^2} - \dfrac{2}{x(x - 1)} - \dfrac{x}{2(x - 1)}$

$$= \frac{3(x - 1) - 2(4x) - x(2x^2)}{4x^2(x - 1)}$$

$$= \frac{3x - 3 - 8x - 2x^3}{4x^2(x - 1)}$$

$$= \frac{-2x^3 - 5x - 3}{4x^2(x - 1)}$$

8. $\dfrac{x^2 + x}{2 - x} \div \dfrac{x^2}{x^2 - 4x + 4} = \dfrac{x^2 + x}{2 - x} \times \dfrac{x^2 - 4x + 4}{x^2}$

$$= \frac{x(x + 1)}{2 - x} \times \frac{(x - 2)^2}{x^2}$$

$$= -\frac{x(x + 1)(x - 2)^2}{(x - 2)(x^2)}$$

$$= -\frac{(x + 1)(x - 2)}{x}$$

9. $\dfrac{1 - \dfrac{3}{2x + 2}}{\dfrac{x}{5} - \dfrac{1}{2}} = \dfrac{\dfrac{2(x + 1) - 3}{2(x + 1)}}{\dfrac{2x - 5}{10}}$

$$= \frac{2x + 2 - 3}{2(x + 1)} \times \frac{10}{2x - 5}$$

$$= \frac{5(2x - 1)}{(x + 1)(2x - 5)}$$

10. Let t = time working together

$$\frac{t}{12} + \frac{t}{16} = 1$$

LCD of 12 and 16 is 48.

$$\frac{48t}{12} + \frac{48t}{16} = 48$$

$$4t + 3t = 48$$

$$t = \frac{48}{7} = 6.9 \text{ days}$$

11. $\dfrac{3}{2x^2 - 3x} + \dfrac{1}{x} = \dfrac{3}{2x - 3}$ $\qquad$ LCD $= x(2x - 3)$

$$\frac{3x(2x - 3)}{x(2x - 3)} + \frac{x(2x - 3)}{x} = \frac{3x(2x - 3)}{2x - 3}$$

$$3 + (2x - 3) = 3x$$

$$x = 0$$

No solution due to division by zero in first two terms

Chapter 7

1. $x^2 - 3x - 5 = 0$
Not factorable
$a = 1, \quad b = -3, \quad c = -5$

$$x = \frac{-(-3) \pm \sqrt{(-3)^2 - 4(1)(-5)}}{2(1)}$$

$$= \frac{3 \pm \sqrt{29}}{2}$$

2. $2x^2 = 9x - 4$
$2x^2 - 9x + 4 = 0$
$(2x - 1)(x - 4) = 0$
$2x - 1 = 0 \quad x - 4 = 0$
$$x = \frac{1}{2} \quad \text{or} \quad x = 4$$

3. $\dfrac{3}{x} - \dfrac{2}{x + 2} = 1$
$$\frac{3x(x + 2)}{x} - \frac{2x(x + 2)}{x + 2} = x(x + 2)$$
$$3(x + 2) - 2x = x^2 + 2x$$
$$3x + 6 - 2x = x^2 + 2x$$
$$0 = x^2 + x - 6$$
$$(x + 3)(x - 2) = 0$$
$$x = -3, 2$$

4. $2x^2 - x = 6 - 2x(3 - x)$
$2x^2 - x = 6 - 6x + 2x^2$
$5x = 6 \quad$ (not quadratic)
$$x = \frac{6}{5}$$

5. $y = 2x^2 + 8x + 5$
$$\frac{-b}{2a} = \frac{-8}{2(2)} = -2$$
$y = 2(-2)^2 + 8(-2) + 5 = -3$
Min. pt. $(a > 0)$ is $(-2, -3)$,
$c = 5$, y-intercept is $(0, 5)$.

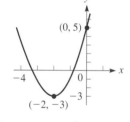

6. $P = EI - RI^2$
$RI^2 - EI + P = 0$
$$I = \frac{-(-E) \pm \sqrt{(-E)^2 - 4RP}}{2R}$$
$$= \frac{E \pm \sqrt{E^2 - 4RP}}{2R}$$

7. $x^2 - 6x - 9 = 0$
$x^2 - 6x \qquad = 9$
$x^2 - 6x + 9 = 9 + 9$
$(x - 3)^2 = 18$
$x - 3 = \pm\sqrt{18}$
$x = 3 \pm 3\sqrt{2}$

8. Let $w =$ width of window
$\quad\quad h =$ height of window
$2w + 2h = 8.4 \quad$ (perimeter)
$w + h = 4.2 \quad h = 4.2 - w$
$\quad\quad w(4.2 - w) = 3.8 \quad$ (area)
$\quad\quad -w^2 + 4.2w = 3.8$
$w^2 - 4.2w + 3.8 = 0$
$$w = \frac{-(-4.2) \pm \sqrt{(-4.2)^2 - 4(3.8)}}{2}$$
$= 2.88, 1.32$
$w = 1.3$ m, $h = 2.9$ m $\quad$ or
$w = 2.9$ m, $h = 1.3$ m

9. $y = x^2 - 8x + 8$
$a = 1, b = -8$
$$\frac{-b}{2a} = \frac{-(-8)}{2(1)} = 4$$
$y = 4^2 - 8(4) + 8 = -8$
$V(4, -8), c = 8, y$-int $(0, 8)$
For $x = 8, y = 8$
$y = 0$ for $x = 1.2, 6.8$

Chapter 8

1. $150° = \left(\dfrac{\pi}{180}\right)(150) = \dfrac{5\pi}{6}$

2. $\sin 205° = -\sin(205° - 180°) = -\sin 25°$

3. $x = -9, \quad y = 12$
$r = \sqrt{(-9)^2 + 12^2} = 15$
$$\sin\theta = \frac{12}{15} = \frac{4}{5}$$
$$\sec\theta = \frac{15}{-9} = -\frac{5}{3}$$

4. $r = 1.40$ m
$\omega = 2200$ r/min $= (2200 \text{ r/min})(2\pi \text{ rad/r})$
$\quad = 4400\pi$ rad/min
$v = \omega r = (4400\pi)(1.40) = 19\,000$ m/min

5. $3.572 = 3.572\left(\dfrac{180°}{\pi}\right) = 204.7°$

6. $\tan\theta = 0.2396$
$\theta_{\text{ref}} = 13.47°$
$\theta = 13.47° \quad$ or
$\theta = 180° + 13.47° = 193.47°$

7. $\cos\theta = -0.8244$, $\csc\theta < 0$;
$\cos\theta$ negative, $\csc\theta$ negative,
θ in third quadrant
$\theta_{\text{ref}} = 0.6017, \qquad \theta = 3.7432$

8. $s = r\theta$, $16.0 = 4.25\theta$
$$\theta = \frac{16.0}{4.25} = 3.76 \text{ rad}$$
$$A = \frac{1}{2}\theta r^2 = \frac{1}{2}(3.76)(4.25)^2$$
$$= 34.0 \text{ m}^2$$

$s = 16.0$ m

$r = 4.25$ m

9. $A = \dfrac{1}{2}\theta r^2$

$A = 38.5$ cm², $d = 12.2$ cm, $r = 6.10$ cm

$38.5 = \dfrac{1}{2}\theta(6.10)^2$, $\theta = \dfrac{2(38.5)}{6.10^2} = 2.07$

$s = r\theta$, $s = 6.10(2.07) = 12.6$ cm

Chapter 9

1.

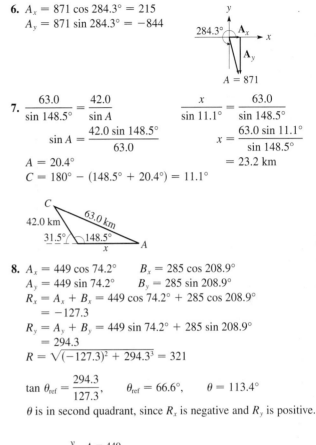

2. $b^2 = a^2 + c^2 - 2ac\cos B$

$b = \sqrt{22.5^2 + 30.9^2 - 2(22.5)(30.9)\cos 78.6°} = 34.4$

3. $x^2 = 36.50^2 + 21.38^2 - 2(36.50)(21.38)\cos 45.00°$

$x = 26.19$ m $\qquad \dfrac{21.38}{\sin\alpha} = \dfrac{26.19}{\sin 45.00°}$

$\sin\alpha = \dfrac{21.38\sin 45.00°}{26.19}, \qquad \alpha = 35.26°$

$\theta = 45.00° - 35.26° = 9.74°$

Displacement is 26.19 m,

9.74° N of E.

4. $C = 180° - (18.9° + 104.2°) = 56.9°$

$\dfrac{c}{\sin C} = \dfrac{a}{\sin A} \qquad c = \dfrac{426\sin 56.9°}{\sin 18.9°} = 1100$

5. Since a is longest side, find A first.

$a^2 = b^2 + c^2 - 2bc\cos A$

$\cos A = \dfrac{b^2 + c^2 - a^2}{2bc} = \dfrac{3.29^2 + 8.44^2 - 9.84^2}{2(3.29)(8.44)}$

$A = 105.4°$

$\dfrac{b}{\sin B} = \dfrac{a}{\sin A}$

$\sin B = \dfrac{b\sin A}{a} = \dfrac{3.29\sin 105.4°}{9.84}$

$B = 18.8°$

$C = 180° - (105.4° + 18.8°) = 55.8°$

6. $A_x = 871\cos 284.3° = 215$

$A_y = 871\sin 284.3° = -844$

7. $\dfrac{63.0}{\sin 148.5°} = \dfrac{42.0}{\sin A} \qquad \dfrac{x}{\sin 11.1°} = \dfrac{63.0}{\sin 148.5°}$

$\qquad \sin A = \dfrac{42.0\sin 148.5°}{63.0} \qquad x = \dfrac{63.0\sin 11.1°}{\sin 148.5°}$

$A = 20.4° \qquad\qquad\qquad\qquad = 23.2$ km

$C = 180° - (148.5° + 20.4°) = 11.1°$

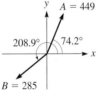

8. $A_x = 449\cos 74.2° \qquad B_x = 285\cos 208.9°$

$A_y = 449\sin 74.2° \qquad B_y = 285\sin 208.9°$

$R_x = A_x + B_x = 449\cos 74.2° + 285\cos 208.9°$

$\qquad = -127.3$

$R_y = A_y + B_y = 449\sin 74.2° + 285\sin 208.9°$

$\qquad = 294.3$

$R = \sqrt{(-127.3)^2 + 294.3^3} = 321$

$\tan\theta_{\text{ref}} = \dfrac{294.3}{127.3}, \qquad \theta_{\text{ref}} = 66.6°, \qquad \theta = 113.4°$

θ is in second quadrant, since R_x is negative and R_y is positive.

9. $29.6\sin 36.5° = 17.6$

$17.6 < 22.3 < 29.6$ means two solutions.

$\dfrac{29.6}{\sin B} = \dfrac{22.3}{\sin 36.5°}, \qquad \sin B = \dfrac{29.6\sin 36.5°}{22.3}$

$B_1 = 52.1° \qquad C_1 = 180° - 36.5° - 52.1° = 91.4°$

$B_2 = 180° - 52.1° = 127.9°,$

$C_2 = 180° - 36.5° - 127.9° = 15.6°$

$\dfrac{c_1}{\sin 91.4°} = \dfrac{22.3}{\sin 36.5°}, \qquad c_1 = \dfrac{22.3\sin 91.4°}{\sin 36.5°} = 37.5$

$\dfrac{c_2}{\sin 15.6°} = \dfrac{22.3}{\sin 36.5°}, \qquad c_2 = \dfrac{22.3\sin 15.6°}{\sin 36.5°} = 10.1$

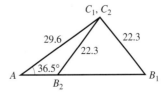

Chapter 10

1. $y = 0.5 \cos \dfrac{\pi}{2} x$

Amp. = 0.5, disp. = 0,

per. $= \dfrac{2\pi}{\pi/2} = 4$

x	0	1	2	3	4
y	0.5	0	−0.5	0	0.5

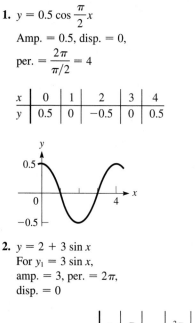

2. $y = 2 + 3 \sin x$
For $y_1 = 3 \sin x$,
amp. = 3, per. = 2π,
disp. = 0

x	0	$\frac{\pi}{2}$	π	$\frac{3\pi}{2}$	2π
$y_1 = 3 \sin x$	0	3	0	−3	0
$y = 2 + 3 \sin x$	2	5	2	−1	2

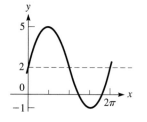

3. $y = 3 \sec x$

$\sec x = \dfrac{1}{\cos x}$

4. $y = 2 \sin\left(2x - \dfrac{\pi}{3}\right)$

Amp. = 2, per. $= \dfrac{2\pi}{2} = \pi$

disp. $= -\dfrac{-\pi/3}{2} = \dfrac{\pi}{6}$

x	$\frac{\pi}{6}$	$\frac{5\pi}{12}$	$\frac{2\pi}{3}$	$\frac{11\pi}{12}$	$\frac{7\pi}{6}$
y	0	2	0	−2	0

$\frac{\pi}{6} + \frac{\pi}{4} = \frac{5\pi}{12}, \frac{\pi}{6} + \frac{\pi}{2} = \frac{2\pi}{3}, \frac{\pi}{6} + \frac{3\pi}{4} = \frac{11\pi}{12}$

5. $y = A \cos \dfrac{2\pi}{T} t$
$A = 0.200$ cm, $T = 0.100$ s,
$y = 0.200 \cos 20\pi t$,
amp. = 0.200 cm, per. = 0.100 s

6. $y = 2 \sin x + \cos 2x$
For $y_1 = 2 \sin x$,
amp. = 2, per. = 2π, disp. = 0
For $y_2 = \cos 2x$,
amp. = 1, per. $= \frac{2\pi}{2} = \pi$, disp. = 0

7. $x = \sin \pi t, y = 2 \cos 2\pi t$

8. $d = R \sin\left(\omega t + \dfrac{\pi}{6}\right)$
$\omega = 2.00$ rad/s
$d = R \sin\left(2.00t + \dfrac{\pi}{6}\right)$

Amp. = R, per. $= \dfrac{2\pi}{2.00} = \pi$ s = 3.14 s,

disp. $= \dfrac{-\pi/6}{2.00} = -\dfrac{\pi}{12}$ s = −0.26 s

9. $y = 2 \sin bx$
$\left(\frac{\pi}{3}, 2\right)$, is first max. with $x > 0$
period $= \frac{2\pi}{b}, \frac{1}{4}\left(\frac{2\pi}{b}\right) = \frac{\pi}{3}, \frac{\pi}{2b} = \frac{\pi}{3}, b = \frac{3}{2}$
$y = 2 \sin \frac{3x}{2}$

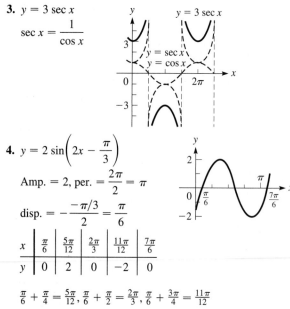

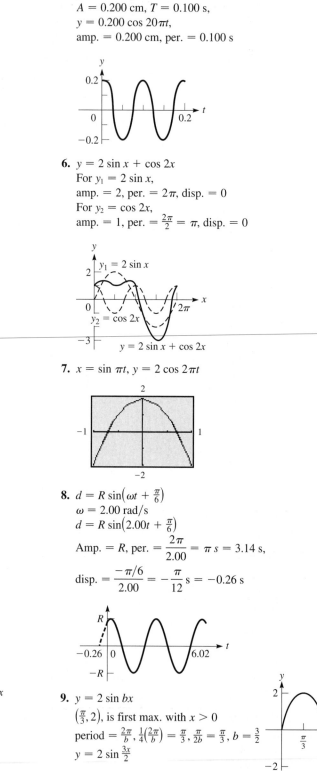

Chapter 11

1. $2\sqrt{20} - \sqrt{125} = 2\sqrt{4 \times 5} - \sqrt{25 \times 5}$
$$= 2(2\sqrt{5}) - 5\sqrt{5}$$
$$= 4\sqrt{5} - 5\sqrt{5} = -\sqrt{5}$$

2. $\dfrac{100^{3/2}}{8^{-2/3}} = (100^{3/2})(8^{2/3}) = [(100^{1/2})^3][(8^{1/3})^2]$
$$= (10^3)(2^2) = 4000$$

3. $(as^{-1/3}t^{3/4})^{12} = a^{12}s^{(-1/3)(12)}t^{(3/4)(12)} = a^{12}s^{-4}t^9 = \dfrac{a^{12}t^9}{s^4}$

4. $(2x^{-1} + y^{-2})^{-1} = \dfrac{1}{2x^{-1} + y^{-2}} = \dfrac{1}{\dfrac{2}{x} + \dfrac{1}{y^2}} = \dfrac{1}{\dfrac{2y^2 + x}{xy^2}}$
$$= \dfrac{xy^2}{2y^2 + x}$$

5. $(\sqrt{2x} - 3\sqrt{y})^2 = (\sqrt{2x})^2 - 2\sqrt{2x}(3\sqrt{y}) + (3\sqrt{y})^2$
$$= 2x - 6\sqrt{2xy} + 9y$$

6. $\sqrt[3]{\sqrt[4]{4}} = \sqrt[12]{4} = \sqrt[12]{2^2} = 2^{2/12} = 2^{1/6} = \sqrt[6]{2}$

7. $\dfrac{3 - 2\sqrt{2}}{2\sqrt{x}} = \dfrac{3 - 2\sqrt{2}}{2\sqrt{x}} \times \dfrac{\sqrt{x}}{\sqrt{x}} = \dfrac{3\sqrt{x} - 2\sqrt{2x}}{2x}$

8. $\sqrt{27a^4b^3} = \sqrt{9 \times 3 \times (a^2)^2(b^2)(b)} = 3a^2b\sqrt{3b}$

9. $2\sqrt{2}(3\sqrt{10} - \sqrt{6}) = 2\sqrt{2}(3\sqrt{10}) - 2\sqrt{2}(\sqrt{6})$
$$= 6\sqrt{20} - 2\sqrt{12}$$
$$= 6(2\sqrt{5}) - 2(2\sqrt{3})$$
$$= 12\sqrt{5} - 4\sqrt{3}$$

10. $(2x + 3)^{1/2} + (x + 1)(2x + 3)^{-1/2}$
$$= (2x + 3)^{1/2} + \dfrac{x + 1}{(2x + 3)^{1/2}}$$
$$= \dfrac{(2x + 3)^{1/2}(2x + 3)^{1/2} + x + 1}{(2x + 3)^{1/2}} = \dfrac{2x + 3 + x + 1}{(2x + 3)^{1/2}}$$
$$= \dfrac{3x + 4}{(2x + 3)^{1/2}}$$

11. $\left(\dfrac{4a^{-1/2}b^{3/4}}{b^{-2}}\right)\left(\dfrac{b^{-1}}{2a}\right) = \dfrac{4a^{-1/2}b^{3/4-1}}{2ab^{-2}} = \dfrac{2b^{2-1/4}}{a^{1+1/2}} = \dfrac{2b^{7/4}}{a^{3/2}}$

12. $\dfrac{2\sqrt{15} + \sqrt{3}}{\sqrt{15} - 2\sqrt{3}} = \dfrac{2\sqrt{15} + \sqrt{3}}{\sqrt{15} - 2\sqrt{3}} \times \dfrac{\sqrt{15} + 2\sqrt{3}}{\sqrt{15} + 2\sqrt{3}}$
$$= \dfrac{2(15) + 4\sqrt{45} + \sqrt{45} + 2(3)}{15 - 4(3)}$$
$$= \dfrac{36 + 15\sqrt{5}}{3}$$
$$= 12 + 5\sqrt{5}$$

13. $\dfrac{3^{-1/2}}{2} = \dfrac{1}{2 \times 3^{1/2}} = \dfrac{1}{2\sqrt{3}}$
$$= \dfrac{\sqrt{3}}{2\sqrt{3}\sqrt{3}} = \dfrac{\sqrt{3}}{6}$$

14. $0.220N^{-1/6} \qquad N = 64 \times 10^6$
$$0.220(64 \times 10^6)^{-1/6} = \dfrac{0.220}{(64 \times 10^6)^{1/6}}$$
$$= \dfrac{0.220}{2 \times 10} = 0.011$$

Chapter 12

1. $(3 - \sqrt{-4}) + (5\sqrt{-9} - 1) = (3 - 2j) + [5(3j) - 1]$
$$= 3 - 2j + 15j - 1$$
$$= 2 + 13j$$

2. $(2\underline{/130°})(3\underline{/45°}) = (2)(3)\underline{/130° + 45°} = 6\underline{/175°}$

3. $2 - 7j$: $r = \sqrt{2^2 + (-7)^2} = \sqrt{53} = 7.28$
$$\tan\theta = \dfrac{-7}{2} = -3.500, \qquad \theta_{\text{ref}} = 74.1°, \qquad \theta = 285.9°$$
$$2 - 7j = 7.28(\cos 285.9° + j\sin 285.9°)$$
$$= 7.28\underline{/285.9°}$$

4. (a) $-\sqrt{-64} = -(8j) = -8j$
(b) $-j^{15} = -j^{12}j^3 = (-1)(-j) = j$

5. $(4 - 3j) + (-1 + 4j) = 3 + j$

6. $\dfrac{2 - 4j}{5 + 3j} = \dfrac{(2 - 4j)(5 - 3j)}{(5 + 3j)(5 - 3j)} = \dfrac{10 - 26j + 12j^2}{25 - 9j^2}$
$$= \dfrac{10 - 26j - 12}{25 - 9(-1)} = \dfrac{-2 - 26j}{34} = -\dfrac{1 + 13j}{17}$$

7. $2.56(\cos 125.2° + j\sin 125.2°) = 2.56e^{2.185j}$
$$125.2° = \dfrac{125.2\pi}{180} = 2.185 \text{ rad}$$

8. $R = 3.50\ \Omega, X_L = 6.20\ \Omega, X_C = 7.35\ \Omega$
$$|Z| = \sqrt{R^2 + (X_L - X_C)^2}$$
$$= \sqrt{3.50^2 + (6.20 - 7.35)^2} = 3.68\ \Omega$$
$$\tan\theta = \dfrac{X_L - X_C}{R} = \dfrac{6.20 - 7.35}{3.50} = \dfrac{-1.15}{3.50}$$
$$\theta = -18.2°$$

9. $3.47 - 2.81j = 4.47e^{5.60j}$
$$R = \sqrt{3.47^2 + (-2.81)^2} = 4.47$$
$$\tan\theta = \dfrac{-2.81}{3.47}, \qquad \theta_{\text{ref}} = 39.0°$$
$$\theta = 321.0° = 5.60 \text{ rad}$$

10. $x + 2j - y = yj - 3xj$
$$x - y + 3xj - yj = -2j$$
$$(x - y) + (3x - y)j = 0 - 2j$$
$$x - y = \quad 0$$
$$\underline{3x - y = -2}$$
$$2x = -2$$
$$x = -1, \qquad y = -1$$

11. $L = 8.75 \text{ mH} = 8.75 \times 10^{-3} \text{ H}$

$f = 600 \text{ kHz} = 6.00 \times 10^5 \text{ Hz}$

$$2\pi f L = \frac{1}{2\pi f C}$$

$$C = \frac{1}{(2\pi f)^2 L} = \frac{1}{(2\pi)^2 (6.00 \times 10^5)^2 (8.75 \times 10^{-3})}$$

$$= 8.04 \times 10^{-12} = 8.04 \text{ pF}$$

12. $j = 1(\cos 90° + j \sin 90°)$

$$j^{1/3} = 1^{1/3}\left(\cos \frac{90°}{3} + j \sin \frac{90°}{3}\right)$$

$$= \cos 30° + j \sin 30° = 0.8660 + 0.5000j$$

$$= 1^{1/3}\left(\cos \frac{90° + 360°}{3} + j \sin \frac{90° + 360°}{3}\right)$$

$$= \cos 150° + j \sin 150° = -0.8660 + 0.5000j$$

$$= 1^{1/3}\left(\cos \frac{90° + 720°}{3} + j \sin \frac{90° + 720°}{3}\right)$$

$$= \cos 270° + j \sin 270° = -j$$

Cube roots of j: $0.8660 + 0.5000j$
$$-0.8660 + 0.5000j$$
$$-j$$

Chapter 13

1. $\log_9 x = -\dfrac{1}{2}$

$$x = 9^{-1/2}$$

$$= \frac{1}{9^{1/2}} = \frac{1}{3}$$

2. $\log_3 x - \log_3 2 = 2$

$$\log_3 \frac{x}{2} = 2$$

$$\frac{x}{2} = 3^2$$

$$x = 2(3^2) = 18$$

3. $\log_x 64 = 3$

$$64 = x^3$$

$$4^3 = x^3$$

$$x = 4$$

4. $3^{3x+1} = 8$

$$(3x + 1)\log 3 = \log 8$$

$$3x + 1 = \frac{\log 8}{\log 3}$$

$$x = \frac{1}{3}\left(\frac{\log 8}{\log 3} - 1\right) = 0.298$$

5. $y = 2\log_4 x$

x	$\frac{1}{4}$	1	4	16
y	-2	0	2	4

$\log_4 \frac{1}{4} = -1$, $\log_4 16 = 2$

6. $y = 2(3^x)$

x	-1	0	1	2
y	0.7	2	6	18

x	3	4	5
y	54	162	486

7. $\log_5\left(\dfrac{4a^3}{7}\right) = \log_5 4a^3 - \log_5 7$

$$= \log_5 4 + \log_5 a^3 - \log_5 7$$

$$= \log_5 4 + 3 \log_5 a - \log_5 7$$

$$= 2 \log_5 2 + 3 \log_5 a - \log_5 7$$

8. $3 \log_7 x - \log_7 y = 2$

$$\log_7 x^3 - \log_7 y = 2$$

$$\log_7 \frac{x^3}{y} = 2 \qquad \frac{x^3}{y} = 7^2$$

$$x^3 = 49y \qquad y = \frac{1}{49}x^3$$

9. $\ln i - \ln I = -t/RC$

$$\ln \frac{i}{I} = -t/RC$$

$$\frac{i}{I} = e^{-t/RC} \qquad i = Ie^{-t/RC}$$

10. $\dfrac{2 \ln 0.9523}{\log 6066} = -0.025\,84$

11. $\log_b x = \dfrac{\log_a x}{\log_a b}$

$$\log_5 732 = \frac{\log 732}{\log 5} = 4.098$$

12. $A = A_0 e^{0.08t} \qquad A = 2A_0$

$$2A_0 = A_0 e^{0.08t} \qquad 2 = e^{0.08t}$$

$$\ln 2 = \ln e^{0.08t} = 0.08t \qquad t = \frac{\ln 2}{0.08} = 8.66 \text{ years}$$

Chapter 14

1. $x^{1/2} - 2x^{1/4} = 3$

Let $y = x^{1/4}$

$$y^2 - 2y - 3 = 0$$

$$(y - 3)(y + 1) = 0$$

$$y = 3, -1$$

$$x^{1/4} \neq -1$$

$$x^{1/4} = 3, \qquad x = 81$$

Check: $81^{1/2} - 2(81^{1/4}) = 3$

$$9 - 6 = 3$$

Solution: $x = 81$

2. $3\sqrt{x-2} - \sqrt{x+1} = 1$

$$3\sqrt{x-2} = 1 + \sqrt{x+1}$$
$$9(x-2) = 1 + 2\sqrt{x+1} + (x+1)$$
$$8x - 20 = 2\sqrt{x+1}$$
$$4x - 10 = \sqrt{x+1}$$
$$16x^2 - 80x + 100 = x + 1$$
$$16x^2 - 81x + 99 = 0$$
$$x = \frac{81 \pm \sqrt{81^2 - 4(16)(99)}}{32}$$
$$= \frac{81 \pm 15}{32} = 3, \frac{33}{16}$$

Check: $x = 3$: $3\sqrt{3-2} - \sqrt{3+1} \stackrel{?}{=} 1$

$$3 - 2 = 1$$
$$x = \tfrac{33}{16}: 3\sqrt{\tfrac{33}{16}-2} - \sqrt{\tfrac{33}{16}+1} \stackrel{?}{=} 1$$
$$\tfrac{3}{4} - \tfrac{7}{4} \neq 1$$

Solution: $x = 3$

3. $x^4 - 17x^2 + 16 = 0$

Let $y = x^2$

$$y^2 - 17y + 16 = 0$$
$$(y - 1)(y - 16) = 0$$
$$y = 1, 16$$
$$x^2 = 1, 16$$
$$x = -1, 1, -4, 4$$

All values check.

4. $x^2 - 2y = 5$

$\underline{2x + 6y = 1}$

$$y = \tfrac{1}{2}(x^2 - 5)$$
$$2x + 6\left(\tfrac{1}{2}\right)(x^2 - 5) = 1$$
$$3x^2 + 2x - 16 = 0$$
$$(3x + 8)(x - 2) = 0$$
$$x = -\tfrac{8}{3}, 2$$
$$x = -\tfrac{8}{3}: y = \tfrac{1}{2}\left(\tfrac{64}{9} - 5\right) = \tfrac{19}{18}$$
$$x = 2: y = \tfrac{1}{2}(4 - 5) = -\tfrac{1}{2}$$
$$x = -\tfrac{8}{3}, y = \tfrac{19}{18}$$

or $x = 2, y = -\tfrac{1}{2}$

5. $\sqrt[3]{2x + 5} = 5$

$$2x + 5 = 125$$
$$2x = 120$$
$$x = 60$$

6. $v = \sqrt{v_0^2 + 2gh}$

$$v^2 = v_0^2 + 2gh$$
$$2gh = v^2 - v_0^2$$
$$h = \frac{v^2 - v_0^2}{2g}$$

7. $x^2 - y^2 = 4$ $\qquad xy = 2$

$$y = \pm\sqrt{x^2 - 4} \qquad y = \frac{2}{x}$$

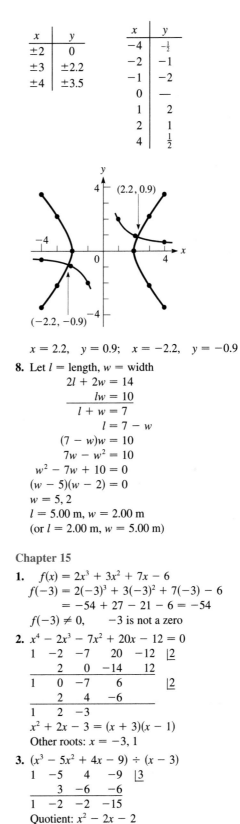

x	y
± 2	0
± 3	± 2.2
± 4	± 3.5

x	y
-4	$-\tfrac{1}{2}$
-2	-1
-1	-2
0	—
1	2
2	1
4	$\tfrac{1}{2}$

$x = 2.2, \quad y = 0.9; \quad x = -2.2, \quad y = -0.9$

8. Let l = length, w = width

$$2l + 2w = 14$$
$$\underline{lw = 10}$$
$$l + w = 7$$
$$l = 7 - w$$
$$(7 - w)w = 10$$
$$7w - w^2 = 10$$
$$w^2 - 7w + 10 = 0$$
$$(w - 5)(w - 2) = 0$$
$$w = 5, 2$$

$l = 5.00$ m, $w = 2.00$ m

(or $l = 2.00$ m, $w = 5.00$ m)

Chapter 15

1. $f(x) = 2x^3 + 3x^2 + 7x - 6$

$$f(-3) = 2(-3)^3 + 3(-3)^2 + 7(-3) - 6$$
$$= -54 + 27 - 21 - 6 = -54$$

$f(-3) \neq 0, \qquad -3$ is not a zero

2. $x^4 - 2x^3 - 7x^2 + 20x - 12 = 0$

```
1   -2   -7    20   -12  |2
      2    0   -14    12
1    0   -7     6        |2
      2    4    -6
1    2   -3
```

$x^2 + 2x - 3 = (x + 3)(x - 1)$

Other roots: $x = -3, 1$

3. $(x^3 - 5x^2 + 4x - 9) \div (x - 3)$

```
1   -5    4    -9   |3
      3   -6    -6
1   -2   -2   -15
```

Quotient: $x^2 - 2x - 2$

Remainder $= -15$

4. $f(x) = 2x^4 + 15x^3 + 23x^2 - 16$

$2x + 1 = 2\left(x + \frac{1}{2}\right)$

$$
\begin{array}{rrrrr|r}
2 & 15 & 23 & 0 & -16 & \underline{-\frac{1}{2}} \\
 & -1 & -7 & -8 & 4 & \\
\hline
2 & 14 & 16 & -8 & -12 &
\end{array}
$$

Remainder is not zero;

$2x + 1$ is not a factor.

5. $(x^3 + 4x^2 + 7x - 9) \div (x + 4)$

$f(x) = x^3 + 4x^2 + 7x - 9$

$f(-4) = (-4)^3 + 4(-4)^2 + 7(-4) - 9$

$\quad = -64 + 64 - 28 - 9 = -37$

Remainder $= -37$

6. $2x^4 - x^3 + 5x^2 - 4x - 12 = 0$

$f(x) = 2x^4 - x^3 + 5x^2 - 4x - 12$

$f(-x) = 2x^4 + x^3 + 5x^2 + 4x - 12$

$n = 4$; 4 roots

No more than 3 positive roots

One negative root

Rational roots: factors of 12 divided by factors by 2

Possible rational roots: $\pm 1, \pm 2, \pm 3, \pm 4, \pm 6, \pm 12, \pm\frac{1}{2}, \pm\frac{3}{2}$

$$
\begin{array}{rrrrr|r}
2 & -1 & 5 & -4 & -12 & \underline{2} \\
 & 4 & 6 & 22 & 36 & \\
\hline
2 & 3 & 11 & 18 & 24 &
\end{array}
$$

2 is too large.

$$
\begin{array}{rrrrr|r}
2 & -1 & 5 & -4 & -12 & \underline{\frac{3}{2}} \\
 & 3 & 3 & 12 & 12 & \\
\hline
2 & 2 & 8 & 8 & 0 & \underline{-1} \\
 & -2 & 0 & -8 & & \\
\hline
2 & 0 & 8 & 0 & &
\end{array}
$$

$2x^2 + 8 = 0$, $\quad x^2 + 4 = 0$, $\quad x = \pm 2j$

roots: $\frac{3}{2}, -1, 2j, -2j$

7. $y = kx^2(x^3 + 436x - 4000)$

$y = 0, kx^2(x^3 + 436x - 4000) = 0$

$x = 0, x^3 + 436x - 4000 = 0$

$$
\begin{array}{rrrr|r}
1 & 0 & 436 & -4000 & \underline{8} \\
 & 8 & 64 & 4000 & \\
\hline
1 & 8 & 500 & 0 &
\end{array}
$$

$y = 0$ for $x = 0$ m, $\quad x = 8$ m

8. Let $x =$ length of edge.

$V = x^3, \quad 2V = (x + 1)^3$

$2x^3 = x^3 + 3x^2 + 3x + 1$

$x^3 - 3x^2 - 3x - 1 = 0$

$y = x^3 - 3x^2 - 3x - 1$

$x = 3.8$ mm

Chapter 16

1. $A = \begin{bmatrix} 3 & -1 & 4 \\ 2 & 0 & -2 \end{bmatrix}$ $\quad B = \begin{bmatrix} 1 & 4 & 5 \\ -1 & -2 & 3 \end{bmatrix}$

$2B = \begin{bmatrix} 2 & 8 & 10 \\ -2 & -4 & 6 \end{bmatrix}$

$A - 2B = \begin{bmatrix} 3 - 2 & -1 - 8 & 4 - 10 \\ 2 + 2 & 0 + 4 & -2 - 6 \end{bmatrix}$

$\quad = \begin{bmatrix} 1 & -9 & -6 \\ 4 & 4 & -8 \end{bmatrix}$

2. $\begin{bmatrix} 2x & x - y & z \\ x + z & 2y & y + z \end{bmatrix} = \begin{bmatrix} 6 & -2 & 4 \\ a & b & c \end{bmatrix}$

$2x = 6 \qquad x - y = -2 \qquad z = 4$

$x = 3 \qquad 3 - y = -2$

$\qquad\qquad y = 5$

$x + z = a \qquad 2y = b \qquad y + z = c$

$3 + 4 = a \qquad 2(5) = b \qquad 5 + 4 = c$

$\quad a = 7 \qquad\quad b = 10 \qquad\quad c = 9$

3. $CD = \begin{bmatrix} 1 & 0 & 4 \\ 2 & -2 & 1 \\ -1 & 3 & 2 \end{bmatrix}\begin{bmatrix} 2 & -2 \\ 4 & -5 \\ 6 & 1 \end{bmatrix}$

$\quad = \begin{bmatrix} 2 + 0 + 24 & -2 + 0 + 4 \\ 4 - 8 + 6 & -4 + 10 + 1 \\ -2 + 12 + 12 & 2 - 15 + 2 \end{bmatrix}$

$\quad = \begin{bmatrix} 26 & 2 \\ 2 & 7 \\ 22 & -11 \end{bmatrix}$

$DC = \begin{bmatrix} 2 & -2 \\ 4 & -5 \\ 6 & 1 \end{bmatrix}\begin{bmatrix} 1 & 0 & 4 \\ 2 & -2 & 1 \\ -1 & 3 & 2 \end{bmatrix}$ not defined, since D has 2 columns and C has 3 rows

4. $A = \begin{bmatrix} 2 & -5 \\ 1 & -2 \end{bmatrix} \qquad B = \begin{bmatrix} -2 & 5 \\ -1 & 2 \end{bmatrix}$

$AB = \begin{bmatrix} 2 & -5 \\ 1 & -2 \end{bmatrix}\begin{bmatrix} -2 & 5 \\ -1 & 2 \end{bmatrix} = \begin{bmatrix} -4 + 5 & 10 - 10 \\ -2 + 2 & 5 - 4 \end{bmatrix}$

$\quad = \begin{bmatrix} 1 & 0 \\ 0 & 1 \end{bmatrix} = I$

$BA = \begin{bmatrix} -2 & 5 \\ -1 & 2 \end{bmatrix}\begin{bmatrix} 2 & -5 \\ 1 & -2 \end{bmatrix} = \begin{bmatrix} -4 + 5 & 10 - 10 \\ -2 + 2 & 5 - 4 \end{bmatrix}$

$\quad = \begin{bmatrix} 1 & 0 \\ 0 & 1 \end{bmatrix} = I$

$AB = BA = I, B = A^{-1}$

5. $\left[\begin{array}{rrr|rrr} 1 & 0 & 4 & 1 & 0 & 0 \\ 2 & -2 & 1 & 0 & 1 & 0 \\ -1 & 3 & 2 & 0 & 0 & 1 \end{array}\right] \rightarrow \left[\begin{array}{rrr|rrr} 1 & 0 & 4 & 1 & 0 & 0 \\ 0 & -2 & -7 & -2 & 1 & 0 \\ 0 & 3 & 6 & 1 & 0 & 1 \end{array}\right] \rightarrow$

$\left[\begin{array}{rrr|rrr} 1 & 0 & 4 & 1 & 0 & 0 \\ 0 & 1 & \frac{7}{2} & 1 & -\frac{1}{2} & 0 \\ 0 & 3 & 6 & 1 & 0 & 1 \end{array}\right] \rightarrow \left[\begin{array}{rrr|rrr} 1 & 0 & 4 & 1 & 0 & 0 \\ 0 & 1 & \frac{7}{2} & 1 & -\frac{1}{2} & 0 \\ 0 & 0 & -\frac{9}{2} & -2 & \frac{3}{2} & 1 \end{array}\right] \rightarrow$

$\left[\begin{array}{rrr|rrr} 1 & 0 & 4 & 1 & 0 & 0 \\ 0 & 1 & \frac{7}{2} & 1 & -\frac{1}{2} & 0 \\ 0 & 0 & 1 & \frac{4}{9} & -\frac{1}{3} & -\frac{2}{9} \end{array}\right] \rightarrow \left[\begin{array}{rrr|rrr} 1 & 0 & 0 & -\frac{7}{9} & \frac{4}{3} & \frac{8}{9} \\ 0 & 1 & 0 & -\frac{5}{9} & \frac{2}{3} & \frac{7}{9} \\ 0 & 0 & 1 & \frac{4}{9} & -\frac{1}{3} & -\frac{2}{9} \end{array}\right]$

$C^{-1} = \begin{bmatrix} -\frac{7}{9} & \frac{4}{3} & \frac{8}{9} \\ -\frac{5}{9} & \frac{2}{3} & \frac{7}{9} \\ \frac{4}{9} & -\frac{1}{3} & -\frac{2}{9} \end{bmatrix}$

6. $2x - 3y = 11$
$\quad\;\; x + 2y = 2$

$A = \begin{bmatrix} 2 & -3 \\ 1 & 2 \end{bmatrix}$ $A^{-1} = \begin{bmatrix} \frac{2}{7} & \frac{3}{7} \\ -\frac{1}{7} & \frac{2}{7} \end{bmatrix}$ $C = \begin{bmatrix} 11 \\ 2 \end{bmatrix}$

$A^{-1}C = \begin{bmatrix} \frac{2}{7} & \frac{3}{7} \\ -\frac{1}{7} & \frac{2}{7} \end{bmatrix}\begin{bmatrix} 11 \\ 2 \end{bmatrix} = \begin{bmatrix} \frac{22}{7} + \frac{6}{7} \\ -\frac{11}{7} + \frac{4}{7} \end{bmatrix} = \begin{bmatrix} 4 \\ -1 \end{bmatrix}$

$x = 4 \qquad y = -1$

7. $A = \begin{bmatrix} 7 & -2 & 1 \\ 2 & 3 & -4 \\ 4 & -5 & 2 \end{bmatrix}$ $C = \begin{bmatrix} 6 \\ 6 \\ 10 \end{bmatrix}$

```
[A]⁻¹[C]
        [[.5  ]
         [-3  ]
         [-3.5]]
```

$x = 0.5 \qquad y = -3 \qquad z = -3.5$

8. Let $A = $ number of shares of stock A
$\qquad B = $ number of shares of stock B
$\quad 50A + 30B = 2600$
$\quad \underline{30A + 40B = 2000}$
$\quad \;\; 5A + \;\; 3B = 260$
$\quad \;\; \underline{3A + \;\; 4B = 200}$

Let $C = $ coefficient matrix

$C = \begin{bmatrix} 5 & 3 \\ 3 & 4 \end{bmatrix}, \quad \begin{vmatrix} 5 & 3 \\ 3 & 4 \end{vmatrix} = 20 - 9 = 11$

$C^{-1} = \frac{1}{11}\begin{bmatrix} 4 & -3 \\ -3 & 5 \end{bmatrix} = \begin{bmatrix} \frac{4}{11} & -\frac{3}{11} \\ -\frac{3}{11} & \frac{5}{11} \end{bmatrix}$

$\begin{bmatrix} A \\ B \end{bmatrix} = \begin{bmatrix} \frac{4}{11} & -\frac{3}{11} \\ -\frac{3}{11} & \frac{5}{11} \end{bmatrix}\begin{bmatrix} 260 \\ 200 \end{bmatrix} = \begin{bmatrix} \frac{4}{11}(260) - \frac{3}{11}(200) \\ -\frac{3}{11}(260) + \frac{5}{11}(200) \end{bmatrix}$

$\qquad = \begin{bmatrix} 40 \\ 20 \end{bmatrix}$

$A = 40$ shares $\qquad B = 20$ shares

Chapter 17

1. $x < 0, y > 0$

2. $\dfrac{-x}{2} \geq 3$
$\quad\;\; -x \geq 6$
$\qquad\; x \leq -6$

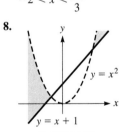

3. $3x + 1 < -5$
$\qquad 3x < -6$
$\qquad\;\; x < -2$

4. $-1 < 1 - 2x < 5$
$\quad\; -2 < -2x < 4$
$\qquad 1 > x > -2$
$\quad\; -2 < x < 1$

5. $\dfrac{x^2 + x}{x - 2} \leq 0, \dfrac{x(x + 1)}{x - 2} \leq 0$

Interval	$\dfrac{x(x + 1)}{x - 2}$	Sign
$x < -1$	$\dfrac{-\;\;-}{-}$	$-$
$-1 < x < 0$	$\dfrac{-\;\;+}{-}$	$+$
$0 < x < 2$	$\dfrac{+\;\;+}{-}$	$-$
$x > 2$	$\dfrac{+\;\;+}{+}$	$+$

Solution: $x \leq -1$ or $0 \leq x < 2$ (x cannot equal 2)

6. $|2x + 1| \geq 3$
$\quad 2x + 1 \geq 3, \qquad 2x + 1 \leq -3$
$\qquad 2x \geq 2, \qquad\quad\; 2x \leq -4$
$\qquad\;\; x \geq 1 \quad$ or $\quad\;\; x \leq -2$

7. $|2 - 3x| < 8$
$\quad -8 < 2 - 3x < 8$
$\; -10 < -3x < 6$
$\quad \dfrac{10}{3} > x > -2$
$\quad -2 < x < \dfrac{10}{3}$

8.

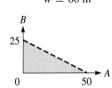

9. If $\sqrt{x^2 - x - 6}$ is real,
then $x^2 - x - 6 \geq 0$.
$(x - 3)(x + 2) \geq 0$
$x \leq -2$ or $x \geq 3$

10. Let $w = $ width, $l = $ length
$\qquad\qquad l = w + 20 \qquad w^2 + 20w - 4800 \geq 0$
$\qquad\qquad wl \geq 4800 \qquad (w + 80)(w - 60) \geq 0$
$\quad w(w + 20) \geq 4800 \qquad\qquad\qquad w \geq 60 \text{ m}$

11. Let $A = $ length of type A wire
$\qquad B = $ length of type B wire
$\quad 0.10A + 0.20B < 5.00$
$\qquad\;\; A + 2B < 50$

12. $|\lambda - 550 \text{ nm}| < 150 \text{ nm}$ (within 150 nm of $\lambda = 550$ nm)

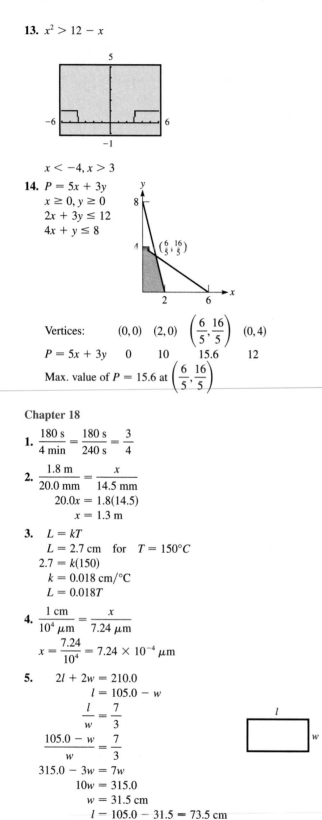

13. $x^2 > 12 - x$

$x < -4, x > 3$

14. $P = 5x + 3y$
$x \geq 0, y \geq 0$
$2x + 3y \leq 12$
$4x + y \leq 8$

Vertices: $(0,0)$ $(2,0)$ $\left(\dfrac{6}{5}, \dfrac{16}{5}\right)$ $(0,4)$

$P = 5x + 3y$ $\quad 0 \quad\quad 10 \quad\quad 15.6 \quad\quad 12$

Max. value of $P = 15.6$ at $\left(\dfrac{6}{5}, \dfrac{16}{5}\right)$

Chapter 18

1. $\dfrac{180 \text{ s}}{4 \text{ min}} = \dfrac{180 \text{ s}}{240 \text{ s}} = \dfrac{3}{4}$

2. $\dfrac{1.8 \text{ m}}{20.0 \text{ mm}} = \dfrac{x}{14.5 \text{ mm}}$
$20.0x = 1.8(14.5)$
$x = 1.3 \text{ m}$

3. $L = kT$
$L = 2.7 \text{ cm} \quad \text{for} \quad T = 150°C$
$2.7 = k(150)$
$k = 0.018 \text{ cm/°C}$
$L = 0.018T$

4. $\dfrac{1 \text{ cm}}{10^4 \text{ } \mu\text{m}} = \dfrac{x}{7.24 \text{ } \mu\text{m}}$

$x = \dfrac{7.24}{10^4} = 7.24 \times 10^{-4} \text{ } \mu\text{m}$

5. $2l + 2w = 210.0$
$l = 105.0 - w$
$\dfrac{l}{w} = \dfrac{7}{3}$
$\dfrac{105.0 - w}{w} = \dfrac{7}{3}$
$315.0 - 3w = 7w$
$10w = 315.0$
$w = 31.5 \text{ cm}$
$l = 105.0 - 31.5 = 73.5 \text{ cm}$

6. $p = kdh$
Using values of water,
$1.96 = k(1000)(0.200)$
$k = 0.009\,80 \text{ kPa} \cdot \text{m}^2/\text{kg}$
For alcohol,
$p = 0.009\,80(800)(0.300)$
$\quad = 2.35 \text{ kPa}$

7. Let $L_1 = $ crushing load of first pillar
$L_2 = $ crushing load of second pillar
$$L_2 = \frac{kr_2^4}{l_2^2} \qquad L_1 = \frac{k(2r_2)^4}{(3l_2)^2}$$

$$\frac{L_1}{L_2} = \frac{\dfrac{k(2r_2)^4}{(3l_2)^2}}{\dfrac{kr_2^4}{l_2^2}} = \frac{16kr_2^4}{9l_2^2} \times \frac{l_2^2}{kr_2^4} = \frac{16}{9}$$

Chapter 19

1. $6, -2, \frac{2}{3}, \ldots$;
geometric sequence

$a_1 = 6 \qquad r = \dfrac{-2}{6} = -\dfrac{1}{3}$

$S_7 = \dfrac{6\left[1 - \left(-\frac{1}{3}\right)^7\right]}{1 - \left(-\frac{1}{3}\right)}$

$\quad = \dfrac{6\left(1 + \dfrac{1}{3^7}\right)}{\dfrac{4}{3}} = \dfrac{1094}{243}$

2. $a_1 = 6, d = 4, s_n = 126$;
arithmetic sequence
$126 = \frac{n}{2}(6 + a_n)$
$a_n = 6 + (n - 1)4 = 2 + 4n$
$126 = \frac{n}{2}[6 + (2 + 4n)]$
$252 = n(8 + 4n) = 4n^2 + 8n$
$n^2 + 2n - 63 = 0$
$(n + 9)(n - 7) = 0$
$\qquad n = 7$

3. $0.454545\ldots = 0.45 + 0.0045 + 0.000045 + \cdots$
$a = 0.45 \qquad r = 0.01$
$S = \dfrac{0.45}{1 - 0.01} = \dfrac{0.45}{0.99} = \dfrac{5}{11}$

4. $\sqrt{1 - 4x} = (1 - 4x)^{1/2}$

$\quad = 1 + \left(\frac{1}{2}\right)(-4x) + \dfrac{\frac{1}{2}\left(\frac{1}{2} - 1\right)}{2}(-4x)^2 + \cdots$

$\quad = 1 - 2x - 2x^2 + \cdots$

5. $(2x - y)^5 = (2x)^5 + 5(2x)^4(-y) + \dfrac{5(4)}{2}(2x)^3(-y)^2$

$\qquad + \dfrac{5(4)(3)}{2(3)}(2x)^2(-y)^3$

$\qquad + \dfrac{5(4)(3)(2)}{2(3)(4)}(2x)(-y)^4 + (-y)^5$

$\quad = 32x^5 - 80x^4y + 80x^3y^2 - 40x^2y^3 + 10xy^4 - y^5$

6. $5\% = 0.05$
Value after 1 year is
$2500 + 2500(0.05) = 2500(1.05)$
$V_{20} = 2500(1.05)^{20}$
$= \$6633.24$

7. $2 + 4 + \cdots + 200$
$a_1 = 2, \quad a_{100} = 200, \quad n = 100$
$S_{100} = \frac{100}{2}(2 + 200)$
$= 10\,100$

8. Ball falls 8.00 m, rises 4.00 m, falls 4.00 m, etc.
Distance $= 8.00 + (4.00 + 4.00)$
$\qquad + (2.00 + 2.00) + \cdots$
$\qquad = 8.00 + 8.00 + 4.00 + 2.00 + \cdots$
$\qquad = 8.00 + \dfrac{8.00}{1 - 0.5} = 8.00 + 16.0 = 24.0$ m

Chapter 20

1. $\sec \theta - \dfrac{\tan \theta}{\csc \theta} = \dfrac{1}{\cos \theta} - \dfrac{\dfrac{\sin \theta}{\cos \theta}}{\dfrac{1}{\sin \theta}} = \dfrac{1}{\cos \theta} - \dfrac{\sin^2 \theta}{\cos \theta}$

$\qquad = \dfrac{1 - \sin^2 \theta}{\cos \theta} = \dfrac{\cos^2 \theta}{\cos \theta} = \cos \theta$

$\qquad \cos \theta = \cos \theta$

2. $\sin 2x + \sin x = 0$
$2 \sin x \cos x + \sin x = 0$
$\sin x(2 \cos x + 1) = 0$

$\sin x = 0 \qquad \cos x = -\dfrac{1}{2}$

$x = 0, \pi, \dfrac{2\pi}{3}, \dfrac{4\pi}{3}$

3. $\theta = \sin^{-1} x$
$\cos \theta = \cos(\sin^{-1} x)$
$\qquad = \sqrt{1 - x^2}$

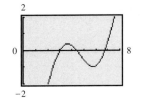

4. $i = 8.00e^{-20t}(1.73 \cos 10.0t - \sin 10.0t)$
Since $8.00e^{-20t}$ will not equal zero, $i = 0$ if
$1.73 \cos 10.0t - \sin 10.0t = 0$
$1.73 = \tan 10.0t$
$10.0t = \tan^{-1} 1.73$
$\quad t = 0.105$ s

5. $\dfrac{\tan \alpha + \tan \beta}{\tan \alpha - \tan \beta} = \dfrac{\dfrac{\sin \alpha}{\cos \alpha} + \dfrac{\sin \beta}{\cos \beta}}{\dfrac{\sin \alpha}{\cos \alpha} - \dfrac{\sin \beta}{\cos \beta}}$

$\qquad = \dfrac{\sin \alpha \cos \beta + \sin \beta \cos \alpha}{\sin \alpha \cos \beta - \sin \beta \cos \alpha}$

$\qquad = \dfrac{\sin (\alpha + \beta)}{\sin (\alpha - \beta)}$

6. $\cot^2 x - \cos^2 x = \dfrac{\cos^2 x}{\sin^2 x} - \cos^2 x =$
$\cos^2 x(\csc^2 x - 1) = \cos^2 x \cot^2 x;$
$\cos^2 x \cot^2 x = \cos^2 x \cot^2 x$

7. $\sin x = -\dfrac{3}{5}$
$\cos x = \dfrac{4}{5}$

$\cos \dfrac{x}{2} = -\sqrt{\dfrac{1 + \left(\dfrac{4}{5}\right)}{2}} = -\sqrt{\dfrac{9}{10}} = -0.9487$

since $270° < x < 360°,\; 135° < \dfrac{x}{2} < 180°;$
$\dfrac{x}{2}$ is in second quadrant, where $\cos \dfrac{x}{2}$ is negative.

8. $I = I_0 \sin 2\theta \cos 2\theta$
$\dfrac{2I}{I_0} = 2 \sin 2\theta \cos 2\theta = \sin 4\theta$

$4\theta = \sin^{-1} \dfrac{2I}{I_0}$

$\theta = \dfrac{1}{4} \sin^{-1} \dfrac{2I}{I_0}$

9. $y = x - 2 \cos x - 5$
$x = 3.02, 4.42, 6.77$

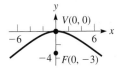

10. $\cos^{-1} x = -\tan^{-1}(-1)$
$\tan^{-1}(-1) = -\dfrac{\pi}{4}$
$\cos^{-1} x = -\left(-\dfrac{\pi}{4}\right) = \dfrac{\pi}{4}$
$x = \cos \dfrac{\pi}{4} = \dfrac{1}{2}\sqrt{2}$

Chapter 21

1. $\qquad 2(x^2 + x) = 1 - y^2$
$2x^2 + y^2 + 2x - 1 = 0$
$A \neq C$ (same sign)
$B = 0$: ellipse

2. $4x - 2y + 5 = 0$
$\qquad y = 2x + \dfrac{5}{2}$
$\qquad m = 2 \qquad b = \dfrac{5}{2}$

3. $\qquad x^2 = 2x - y^2$
$x^2 + y^2 = 2x$
$\qquad r^2 = 2r \cos \theta$
$\qquad r = 2 \cos \theta$

4. $x^2 = -12y$
$4p = -12, \qquad p = -3$
$V(0,0) \qquad F(0,-3)$

5. Center $(-1, 2)$; $h = -1$, $k = 2$
$$r = \sqrt{(2+1)^2 + (3-2)^2}$$
$$= \sqrt{10}$$
$$(x+1)^2 + (y-2)^2 = 10$$
or
$$x^2 + y^2 + 2x - 4y - 5 = 0$$

6. $m = \dfrac{-2-1}{2+4} = -\dfrac{1}{2}$
$$y - 1 = -\tfrac{1}{2}(x+4)$$
$$2y - 2 = -x - 4$$
$$x + 2y + 2 = 0$$

7. $x^2 = 4py$
$(6.00)^2 = 4p(4.00)$
$p = 2.25$
Focus is 2.25 cm
from vertex.

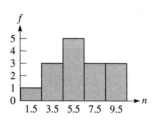

8. $a = 3$ $b = 1$

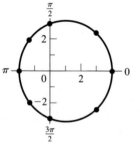

$$\frac{x^2}{9} + \frac{y^2}{1} = 1$$
Find y for $x = 2$ m.
$$\frac{4}{9} + \frac{y^2}{1} = 1$$
$$y^2 = \frac{5}{9} \qquad y = 0.7 \text{ m}$$
$$h = 3.0 + 0.7 = 3.7 \text{ m}$$

9. $r = 3 + \cos\theta$

r	0	$\frac{\pi}{4}$	$\frac{\pi}{2}$	$\frac{3\pi}{4}$	π	$\frac{5\pi}{4}$	$\frac{3\pi}{2}$	$\frac{7\pi}{4}$	2π
θ	4.0	3.7	3.0	2.3	2.0	2.3	3.0	3.7	4.0

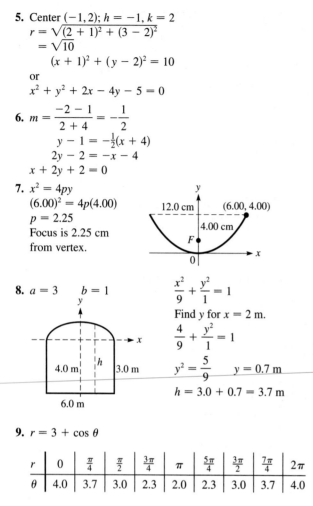

10. $4y^2 - x^2 - 4x - 8y - 4 = 0$
$$4(y^2 - 2y \quad) - (x^2 + 4x \quad) = 4$$
$$4(y^2 - 2y + 1) - (x^2 + 4x + 4) = 4 + 4 - 4$$
$$\frac{(y-1)^2}{1^2} - \frac{(x+2)^2}{2^2} = 1$$
$C(-2, 1)$ $V(-2, 0)$ $V(-2, 2)$

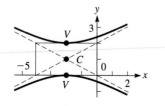

Chapter 22

1.

Number	1	2	3	4	5	6	7	8	9	10
Frequency	1	0	1	2	3	2	1	2	2	1

$\Sigma f = 15$
Median is eighth number; median $= 6$.

2. 5 appears three times, and no other number appears more than twice; mode $= 5$.

3.

Number	1–2	3–4	5–6	7–8	9–10
Frequency	1	3	5	3	3

4. Let $t = $ thickness
$$\bar{t} = \frac{3(0.90) + 9(0.91) + 31(0.92) + 38(0.93) + 12(0.94) + 5(0.95) + 2(0.96)}{3 + 9 + 31 + 38 + 12 + 5 + 2}$$
$$= \frac{92.7}{100} = 0.927 \text{ cm}$$

5. $\Sigma x^2 = 3(0.90)^2 + 9(0.91)^2 + 31(0.92)^2 + 38(0.93)^2$
$\qquad + 12(0.94)^2 + 5(0.95)^2 + 2(0.96)^2 = 85.9464$
$(\Sigma x)^2 = 92.7^2$
$$s = \sqrt{\frac{100(85.9464) - 92.7^2}{100(99)}} = 0.117 \text{ cm}$$

6.

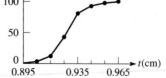

7.

t (cm)	0.90	0.91	0.92	0.93	0.94	0.95	0.96
%	3	9	31	38	12	5	2

8.

t (cm)	<0.905	<0.915	<0.925	<0.935
cum f	3	12	43	81

t (cm)	<0.945	<0.955	<0.965
cum f	93	98	100

9. $0.5000 + 0.2257 = 0.7257$ (total area to the left)
$$= 72.57\%$$

10. Find the range of each subgroup (subtract the lowest value from the largest value). Find the mean R of these ranges (divide their sum by 20). This is the value of the central line. Multiply R by the appropriate control chart factors to obtain the upper control limit (UCL) and the lower control limit (LCL). Draw the central line, the UCL line, and the LCL line on a graph. Plot the points for each R corresponding to its subgroup and join successive points by straight-line segments.

11.

x	y	xy	x^2
1	5	5	1
3	11	33	9
5	17	85	25
7	20	140	49
9	27	243	81
25	80	506	165

$n = 5$

$$m = \frac{5(506) - (25)(80)}{5(165) - 25^2}$$
$$= 2.65$$
$$b = \frac{165(80) - (506)(25)}{5(165) - 25^2}$$
$$= 2.75$$
$$y = 2.65x + 2.75$$

12.

x	$\sqrt{x}$	y	$\sqrt{x}y$	$(\sqrt{x})^2 = x$
1.00	1.000	1.10	1.1000	1.000
3.00	1.732	1.90	3.2908	3.000
5.00	2.236	2.50	5.5900	5.000
7.00	2.646	2.90	7.6734	7.000
9.00	3.000	3.30	9.9000	9.000
	10.614	11.70	27.5542	25.000

$n = 5$

$$m = \frac{5(27.5542) - (10.614)(11.70)}{5(25.000) - (10.614)^2} = 1.10$$

$$b = \frac{(25.000)(11.70) - (27.5542)(10.614)}{5(25.000) - (10.614)^2} = 0.00$$

(rounded off)

$$y = 1.10\sqrt{x}$$

Chapter 23

1. $\lim\limits_{x \to 1} \dfrac{x^2 - x}{x^2 - 1} = \lim\limits_{x \to 1} \dfrac{x(x-1)}{(x+1)(x-1)} = \lim\limits_{x \to 1} \dfrac{x}{x+1} = \dfrac{1}{2}$

2. $\lim\limits_{x \to \infty} \dfrac{1 - 4x^2}{x + 2x^2} = \lim\limits_{x \to \infty} \dfrac{\dfrac{1}{x^2} - 4}{\dfrac{1}{x} + 2} = -2$

3. $y = 3x^2 - \dfrac{4}{x^2}$

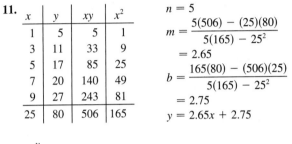

$$\frac{dy}{dx} = 6x + \frac{8}{x^3}$$
$$\left.\frac{dy}{dx}\right|_{x=2} = 6(2) + \frac{8}{2^3}$$
$$= 13$$
$$m_{\tan} = 13$$

4. $s = t\sqrt{10 - 2t}$

$$v = \frac{ds}{dt} = t(\tfrac{1}{2})(10 - 2t)^{-1/2}(-2) + (10 - 2t)^{1/2}(1)$$

$$= \frac{-t}{(10 - 2t)^{1/2}} + (10 - 2t)^{1/2} = \frac{10 - 3t}{(10 - 2t)^{1/2}}$$

$$v|_{t=4.00} = \frac{10 - 3(4.00)}{[10 - 2(4.00)]^{1/2}} = \frac{10 - 12.00}{2.00^{1/2}} = -1.41 \text{ cm/s}$$

5. $y = 4x^6 - 2x^4 + \pi^3$

$$\frac{dy}{dx} = 4(6x^5) - 2(4x^3) \qquad \pi^3 \text{ is constant}$$
$$= 24x^5 - 8x^3$$

6. $y = 2x(5 - 3x)^4$

$$\frac{dy}{dx} = 2x[4(5 - 3x)^3(-3)] + (5 - 3x)^4(2)$$
$$= -24x(5 - 3x)^3 + 2(5 - 3x)^4$$
$$= 2(5 - 3x)^3(-12x + 5 - 3x)$$
$$= 2(5 - 15x)(5 - 3x)^3$$
$$= 10(1 - 3x)(5 - 3x)^3$$

7. $(1 + y^2)^3 - x^2y = 7x$
$$3(1 + y^2)^2(2yy') - x^2y' - y(2x) = 7$$
$$6y(1 + y^2)^2y' - x^2y' = 7 + 2xy$$
$$y' = \frac{7 + 2xy}{6y(1 + y^2)^2 - x^2}$$

8. $V = \dfrac{kq}{\sqrt{x^2 + b^2}} = kq(x^2 + b^2)^{-1/2}$

$$\frac{dV}{dx} = kq\left(-\frac{1}{2}\right)(x^2 + b^2)^{-3/2}(2x) = \frac{-kqx}{(x^2 + b^2)^{3/2}}$$

9. $y = \dfrac{2x}{3x + 2}$

$$\frac{dy}{dx} = \frac{(3x + 2)(2) - 2x(3)}{(3x + 2)^2} = \frac{4}{(3x + 2)^2} = 4(3x + 2)^{-2}$$

$$\frac{d^2y}{dx^2} = -2(4)(3x + 2)^{-3}(3) = \frac{-24}{(3x + 2)^3}$$

10. $y = 5x - 2x^2$

$$f(x + h) = 5(x + h) - 2(x + h)^2$$
$$f(x + h) - f(x) = 5(x + h) - 2(x + h)^2 - (5x - 2x^2)$$
$$= 5h - 4xh - 2h^2$$
$$\frac{f(x + h) - f(x)}{h} = \frac{5h - 4xh - 2h^2}{h} = 5 - 4x - 2h$$
$$\lim_{h \to 0} \frac{f(x + h) - f(x)}{h} = 5 - 4x$$

Chapter 24

1. $y = x^4 - 3x^2$

$$\frac{dy}{dx} = 4x^3 - 6x$$

$$\frac{dy}{dx}\Big|_{x=1} = 4(1^3) - 6(1)$$
$$= -2$$

$$y - (-2) = -2(x - 1)$$
$$y = -2x$$

2. $y = 3x^2 - x$

$$\Delta y = f(3.1) - f(3)$$
$$[3(3.1)^2 - 3.1] - [3(3^2) - 3] = 1.73$$
$$dy = (6x - 1)\,dx$$
$$= [6(3) - 1](0.1) = 1.7$$

$$\Delta y - dy = 0.03$$

3. $x = 3t^2$ $\qquad\qquad y = 2t^3 - t^2$

$$v_x = \frac{dx}{dt} = 6t \qquad\qquad v_y = \frac{dy}{dt} = 6t^2 - 2t$$

$$a_x = \frac{dv_x}{dt} = \frac{d^2x}{dt^2} = 6 \qquad a_y = \frac{dv_y}{dt} = \frac{d^2y}{dt^2} = 12t - 2$$

$$a_x|_{t=2} = 6 \qquad\qquad a_y|_{t=2} = 12(2) - 2 = 22$$

$$a|_{t=2} = \sqrt{6^2 + 22^2} = 22.8 \qquad \tan\theta = \frac{22}{6}, \quad \theta = 74.7°$$

4. $P = \dfrac{144r}{(r + 0.6)^2}$

$$\frac{dP}{dr} = \frac{144[(r + 0.6)^2(1) - r(2)(r + 0.6)(1)]}{(r + 0.6)^4}$$

$$= \frac{144[(r + 0.6) - 2r]}{(r + 0.6)^3} = \frac{144(0.6 - r)}{(r + 0.6)^3}$$

$$\frac{dP}{dr} = 0; \qquad 0.6 - r = 0, \qquad r = 0.6\,\Omega$$

$$\left(r < 0.6, \frac{dP}{dr} > 0; r > 0.6, \frac{dP}{dr} < 0\right)$$

5. $x^2 - \sqrt{4x + 1} = 0; \qquad f(x) = x^2 - \sqrt{4x + 1}$

$$f'(x) = 2x - \tfrac{1}{2}(4x + 1)^{-1/2}(4) = 2x - \frac{2}{(4x + 1)^{1/2}}$$

n	x_n	$f(x_n)$	$f'(x_n)$	$x_n - \dfrac{f(x_n)}{f'(x_n)}$
1	1.5	$-0.395\,751\,3$	2.244\,071\,1	1.676\,354\,2
2	1.676\,354\,2	0.034\,300\,1	2.632\,211\,8	1.663\,323\,3

$$x_3 = 1.6633$$

6. $y = \sqrt{2x + 4}, a = 6$

$$\frac{dy}{dx} = \frac{1}{2}(2x + 4)^{-1/2}(2) = \frac{1}{\sqrt{2x + 4}}$$

$$\frac{dy}{dx}\Big|_{x=6} = \frac{1}{\sqrt{2(6) + 4}} = \frac{1}{4}$$

$$f(6) = \sqrt{2(6) + 4} = 4$$

$$L(x) = 4 + \tfrac{1}{4}(x - 6) = \tfrac{1}{4}x + \tfrac{5}{2}$$

7. $y = x^3 + 6x^2$

$$y' = 3x^2 + 12x = 3x(x + 4)$$
$$y'' = 6x + 12 = 6(x + 2)$$

$x < -4$	y inc.	Max. $(-4, 32)$
$-4 < x < 0$	y dec.	Min. $(0, 0)$
$x > 0$	y inc.	Infl. $(-2, 16)$
$x < -2$	y conc. down	
$x > -2$	y conc. up	

8. $y = \dfrac{4}{x^2} - x$

$$y' = -\frac{8}{x^3} - 1 = -\frac{8 + x^3}{x^3}$$

$$y'' = \frac{24}{x^4}$$

$x < -2$	y dec.
$-2 < x < 0$	y inc.
$x > 0$	y dec., conc. up
$x < 0$	y conc. up

Min. $(-2, 3)$, no infl., int. $(\sqrt[3]{4}, 0)$, sym. none; as $x \to \pm\infty$, $y \to -x$, asym. $y = -x$, $x = 0$. Domain: all real x except 0; range: all real y.

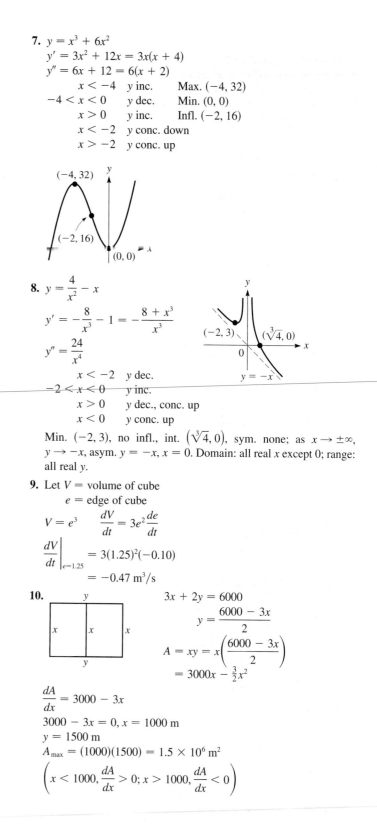

9. Let V = volume of cube
$\qquad e$ = edge of cube

$$V = e^3 \qquad \frac{dV}{dt} = 3e^2\frac{de}{dt}$$

$$\frac{dV}{dt}\Big|_{e=1.25} = 3(1.25)^2(-0.10)$$
$$= -0.47 \text{ m}^3/\text{s}$$

10.

$$3x + 2y = 6000$$

$$y = \frac{6000 - 3x}{2}$$

$$A = xy = x\left(\frac{6000 - 3x}{2}\right)$$
$$= 3000x - \tfrac{3}{2}x^2$$

$$\frac{dA}{dx} = 3000 - 3x$$

$$3000 - 3x = 0, x = 1000 \text{ m}$$
$$y = 1500 \text{ m}$$
$$A_{max} = (1000)(1500) = 1.5 \times 10^6 \text{ m}^2$$

$$\left(x < 1000, \frac{dA}{dx} > 0; x > 1000, \frac{dA}{dx} < 0\right)$$

Chapter 25

1. Power of x required for $2x$ is 2. Therefore, multiply by $1/2$. Antiderivative of $2x = \frac{1}{2}(2x^2) = x^2$. Power of $(1-x)^4$ required is 5. Derivative of $(1-x)^5$ is $5(1-x)^4(-1)$. Writing $-(1-x)^4$ as $\frac{1}{5}[5(1-x)^4(-1)]$, the antiderivative of $-(1-x)^4$ is $\frac{1}{5}(1-x)^5$. Therefore, the antiderivative of $2x - (1-x)^4$ is $x^2 + \frac{1}{5}(1-x)^5$.

2. $\int x\sqrt{1-2x^2}\,dx = \int x(1-2x^2)^{1/2}\,dx$

$u = 1-2x^2 \qquad du = -4x\,dx \qquad n = \frac{1}{2} \qquad n+1 = \frac{3}{2}$

$\int x(1-2x^2)^{1/2}\,dx = -\frac{1}{4}\int(1-2x^2)^{1/2}(-4x\,dx)$

$\qquad = -\frac{1}{4}\left(\frac{2}{3}\right)(1-2x^2)^{3/2} + C$

$\qquad = -\frac{1}{6}(1-2x^2)^{3/2} + C$

3. $\dfrac{dy}{dx} = (6-x)^4,\ dy = (6-x)^4\,dx$

$\int dy = \int(6-x)^4\,dx$

$y = -\int(6-x)^4(-dx) = -\dfrac{1}{5}(6-x)^5 + C$

$2 = -\frac{1}{5}(6-5) + C,\ C = \frac{11}{5}$

$y = -\frac{1}{5}(6-x)^5 + \frac{11}{5}$

4. $y = \dfrac{1}{x+2} \qquad n = 6 \qquad \Delta x = \dfrac{4-1}{6} = \dfrac{1}{2}$

$A = \frac{1}{2}\left(\frac{2}{7} + \frac{1}{4} + \frac{2}{9} + \frac{1}{5} + \frac{2}{11} + \frac{1}{6}\right) = 0.6532$

x	1	$\frac{3}{2}$	2	$\frac{5}{2}$	3	$\frac{7}{2}$	4
y	$\frac{1}{3}$	$\frac{2}{7}$	$\frac{1}{4}$	$\frac{2}{9}$	$\frac{1}{5}$	$\frac{2}{11}$	$\frac{1}{6}$

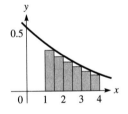

5. (See values for Problem 4.)

$\int_1^4 \dfrac{dx}{x+2} = \dfrac{1}{4}\left[\dfrac{1}{3} + 2\left(\dfrac{2}{7}\right) + 2\left(\dfrac{1}{4}\right) + 2\left(\dfrac{2}{9}\right)\right.$

$\left. \qquad + 2\left(\dfrac{1}{5}\right) + 2\left(\dfrac{2}{11}\right) + \dfrac{1}{6}\right]$

$\qquad = 0.6949$

6. (See values for Problem 4.)

$\int_1^4 \dfrac{dx}{x+2} = \dfrac{1}{6}\left[\dfrac{1}{3} + 4\left(\dfrac{2}{7}\right) + 2\left(\dfrac{1}{4}\right) + 4\left(\dfrac{2}{9}\right)\right.$

$\left. \qquad + 2\left(\dfrac{1}{5}\right) + 4\left(\dfrac{2}{11}\right) + \dfrac{1}{6}\right] = 0.6932$

7. $i = \int_1^3\left(t^2 + \dfrac{1}{t^2}\right)dt = \dfrac{1}{3}t^3 - \dfrac{1}{t}\Big|_1^3$

$\qquad = \frac{1}{3}(27) - \frac{1}{3} - \left(\frac{1}{3} - 1\right) = 9.3\ \text{A}$

Chapter 26

1. $A = \displaystyle\int_0^2 \dfrac{1}{4}x^2\,dx$

$\qquad = \dfrac{1}{12}x^3\Big|_0^2 = \dfrac{2}{3}$

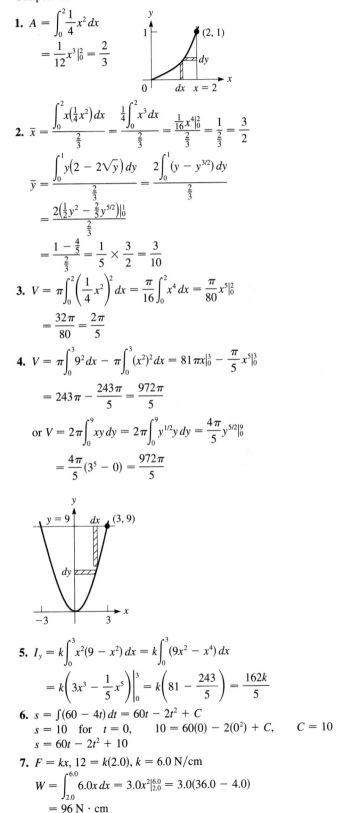

2. $\bar{x} = \dfrac{\displaystyle\int_0^2 x\left(\frac{1}{4}x^2\right)dx}{\frac{2}{3}} = \dfrac{\frac{1}{4}\displaystyle\int_0^2 x^3\,dx}{\frac{2}{3}} = \dfrac{\frac{1}{16}x^4\big|_0^2}{\frac{2}{3}} = \dfrac{1}{\frac{2}{3}} = \dfrac{3}{2}$

$\bar{y} = \dfrac{\displaystyle\int_0^1 y(2 - 2\sqrt{y})\,dy}{\frac{2}{3}} = \dfrac{2\displaystyle\int_0^1 (y - y^{3/2})\,dy}{\frac{2}{3}}$

$\qquad = \dfrac{2\left(\frac{1}{2}y^2 - \frac{2}{5}y^{5/2}\right)\big|_0^1}{\frac{2}{3}}$

$\qquad = \dfrac{1 - \frac{4}{5}}{\frac{2}{3}} = \dfrac{1}{5} \times \dfrac{3}{2} = \dfrac{3}{10}$

3. $V = \pi\displaystyle\int_0^2 \left(\dfrac{1}{4}x^2\right)^2 dx = \dfrac{\pi}{16}\displaystyle\int_0^2 x^4\,dx = \dfrac{\pi}{80}x^5\big|_0^2$

$\qquad = \dfrac{32\pi}{80} = \dfrac{2\pi}{5}$

4. $V = \pi\displaystyle\int_0^3 9^2\,dx - \pi\displaystyle\int_0^3 (x^2)^2\,dx = 81\pi x\big|_0^3 - \dfrac{\pi}{5}x^5\big|_0^3$

$\qquad = 243\pi - \dfrac{243\pi}{5} = \dfrac{972\pi}{5}$

or $V = 2\pi\displaystyle\int_0^9 xy\,dy = 2\pi\displaystyle\int_0^9 y^{1/2}y\,dy = \dfrac{4\pi}{5}y^{5/2}\big|_0^9$

$\qquad = \dfrac{4\pi}{5}(3^5 - 0) = \dfrac{972\pi}{5}$

5. $I_y = k\displaystyle\int_0^3 x^2(9 - x^2)\,dx = k\displaystyle\int_0^3 (9x^2 - x^4)\,dx$

$\qquad = k\left(3x^3 - \dfrac{1}{5}x^5\right)\Big|_0^3 = k\left(81 - \dfrac{243}{5}\right) = \dfrac{162k}{5}$

6. $s = \int(60 - 4t)\,dt = 60t - 2t^2 + C$

$\qquad s = 10$ for $t = 0, \qquad 10 = 60(0) - 2(0^2) + C, \qquad C = 10$

$\qquad s = 60t - 2t^2 + 10$

7. $F = kx,\ 12 = k(2.0),\ k = 6.0\ \text{N/cm}$

$\qquad W = \displaystyle\int_{2.0}^{6.0} 6.0x\,dx = 3.0x^2\big|_{2.0}^{6.0} = 3.0(36.0 - 4.0)$

$\qquad = 96\ \text{N} \cdot \text{cm}$

8. $F = 9.80 \int_{1.00}^{3.00} 6.00h\, dh = (9.80)(6.00)\frac{1}{2}h^2\big|_{1.00}^{3.00}$

$= (9.80)(3.00)(9.00 - 1.00) = 235$ kN

1.00 m

dh 2.00 m

6.00 m

$\theta = \tan^{-1}\dfrac{x}{250}$

$\dfrac{d\theta}{dt} = \dfrac{1}{1 + \dfrac{x^2}{250^2}}\dfrac{dx/dt}{250} = \dfrac{250^2}{250^2 + x^2}\dfrac{dx/dt}{250}$

$\dfrac{d\theta}{dt}\bigg|_{t=8.0} = \dfrac{250}{250^2 + 40^2}(5.0) = 0.020$ rad/s

x

θ

250 m

Chapter 27

1. $y = \tan^3 2x + \tan^{-1} 2x$

$\dfrac{dy}{dx} = 3(\tan^2 2x)(\sec^2 2x)(2) + \dfrac{2}{1 + (2x)^2}$

$= 6\tan^2 2x \sec^2 2x + \dfrac{2}{1 + 4x^2}$

2. $y = 2(3 + \cot 4x)^3$

$\dfrac{dy}{dx} = 2(3)(3 + \cot 4x)^2(-\csc^2 4x)(4)$

$= -24(3 + \cot 4x)^2 \csc^2 4x$

3. $y \sec 2x = \sin^{-1} 3y$

$y(\sec 2x \tan 2x)(2) + (\sec 2x)(y') = \dfrac{3y'}{\sqrt{1 - (3y)^2}}$

$\sqrt{1 - 9y^2} \sec 2x(2y \tan 2x + y') = 3y'$

$y' = \dfrac{2y\sqrt{1 - 9y^2} \sec 2x \tan 2x}{3 - \sqrt{1 - 9y^2} \sec 2x}$

4. $y = \dfrac{\cos^2(3x + 1)}{x}$

$dy = \dfrac{x\{2\cos(3x+1)[-\sin(3x+1)(3)]\} - \cos^2(3x+1)(1)}{x^2}dx$

$= \dfrac{-6x\cos(3x+1)\sin(3x+1) - \cos^2(3x+1)}{x^2}dx$

5. $y = \ln\dfrac{2x - 1}{1 + x^2}$

$= \ln(2x - 1) - \ln(1 + x^2)$

$\dfrac{dy}{dx} = \dfrac{2}{2x - 1} - \dfrac{2x}{1 + x^2}$

$m_{\tan} = \dfrac{dy}{dx}\bigg|_{x=2} = \dfrac{2}{4 - 1} - \dfrac{4}{1 + 4} = \dfrac{2}{3} - \dfrac{4}{5} = -\dfrac{2}{15}$

6. $i = 8e^{-t} \sin 10t$

$\dfrac{di}{dt} = 8[e^{-t}\cos 10t(10) + \sin 10t(-e^{-t})]$

$= 8e^{-t}(10\cos 10t - \sin 10t)$

7. $y = xe^x$

$y' = xe^x + e^x = e^x(x + 1)$

$y'' = xe^x + e^x + e^x = e^x(x + 2)$

$x < -1$ y dec.

$x > -1$ y inc.

$x < -2$ y conc. down

$x > -2$ y conc. up

Int. (0, 0), min. $(-1, -e^{-1})$

infl. $(-2, -2e^{-2})$, asym. $y = 0$

8. For $t = 8.0$ s, $x = 40$ m

y

$(-2, -2e^{-2})$

$(0, 0)$

$(-1, -e^{-1})$

x

Chapter 28

1. $\displaystyle\int (\sec x - \sec^3 x \tan x)\, dx$

$= \displaystyle\int \sec x\, dx - \int \sec^2 x(\sec x \tan x)\, dx$

$= \ln|\sec x + \tan x| - \tfrac{1}{3}\sec^3 x + C$

2. $\displaystyle\int \sin^3 x\, dx = \int \sin^2 x \sin x\, dx$

$= \displaystyle\int (1 - \cos^2 x)\sin x\, dx$

$= \displaystyle\int \sin x\, dx - \int \cos^2 x \sin x\, dx$

$= -\cos x + \tfrac{1}{3}\cos^3 x + C$

3. $\displaystyle\int \tan^3 2x\, dx = \int \tan 2x(\tan^2 2x)\, dx$

$= \displaystyle\int \tan 2x(\sec^2 2x - 1)\, dx$

$= \tfrac{1}{2}\displaystyle\int \tan 2x \sec^2 2x(2\, dx) - \tfrac{1}{2}\int \tan 2x(2\, dx)$

$= \tfrac{1}{4}\tan^2 2x + \tfrac{1}{2}\ln|\cos 2x| + C$

4. $\displaystyle\int \cos^2 4\theta\, d\theta = \tfrac{1}{2}\int (1 + \cos 8\theta)\, d\theta$

$= \tfrac{1}{2}\displaystyle\int d\theta + \tfrac{1}{16}\int \cos 8\theta(8\, d\theta)$

$= \tfrac{1}{2}\theta + \tfrac{1}{16}\sin 8\theta + C$

5. Let $x = 2\sin\theta$,

$dx = 2\cos\theta\, d\theta$.

$\displaystyle\int \dfrac{dx}{x^2\sqrt{4 - x^2}} = \int \dfrac{2\cos\theta\, d\theta}{4\sin^2\theta\sqrt{4 - 4\sin^2\theta}}$

$= \tfrac{1}{4}\displaystyle\int \dfrac{\cos\theta\, d\theta}{\sin^2\theta\sqrt{\cos^2\theta}} = \tfrac{1}{4}\int \csc^2\theta\, d\theta$

$= -\tfrac{1}{4}\cot\theta + C$

$= -\dfrac{\sqrt{4 - x^2}}{4x} + C$

2

x

θ

$\sqrt{4 - x^2}$

6. $\displaystyle\int xe^{-2x}\, dx$; $u = x$, $du = dx$, $dv = e^{-2x}\, dx$, $v = -\tfrac{1}{2}e^{-2x}$

$\displaystyle\int xe^{-2x}\, dx = x\left(-\tfrac{1}{2}e^{-2x}\right) - \int\left(-\tfrac{1}{2}e^{-2x}\right)dx$

$= -\tfrac{1}{2}xe^{-2x} - \tfrac{1}{4}e^{-2x} + C$

7. $\int \dfrac{x^3 + 5x^2 + x + 2}{x^4 + x^2}\, dx$

$\dfrac{x^3 + 5x^2 + x + 2}{x^4 + x^2} = \dfrac{x^3 + 5x^2 + x + 2}{x^2(x^2 + 1)}$

$\qquad\qquad\qquad = \dfrac{A}{x} + \dfrac{B}{x^2} + \dfrac{Cx + D}{x^2 + 1}$

$x^3 + 5x^2 + x + 2 = Ax(x^2 + 1) + B(x^2 + 1) + Cx^3 + Dx^2$

x^3-terms: $1 = A + C \qquad x^2$-terms: $5 = B + D$

x-terms: $1 = A \qquad\qquad\quad x = 0{:}\ 2 = B$

$A = 1, B = 2, C = 0, D = 3$

$\int \dfrac{x^3 + 5x^2 + x + 2}{x^4 + x^2}\, dx = \int \dfrac{dx}{x} + \int \dfrac{2\, dx}{x^2} + \int \dfrac{3\, dx}{x^2 + 1}$

$\qquad\qquad\qquad = \ln|x| - \dfrac{2}{x} + 3 \tan^{-1}x + C$

8. $i = \int \dfrac{6t + 1}{4t^2 + 9}\, dt = \int \dfrac{6t\, dt}{4t^2 + 9} + \int \dfrac{dt}{4t^2 + 9}$

$\quad = \dfrac{6}{8} \int \dfrac{8t\, dt}{4t^2 + 9} + \dfrac{1}{2} \int \dfrac{2\, dt}{9 + (2t)^2}$

$\quad = \dfrac{3}{4} \ln(4t^2 + 9) + \dfrac{1}{2}\left(\dfrac{1}{3}\right)\tan^{-1}\dfrac{2t}{3} + C$

$i = 0$ for $t = 0{:}\ 0 = \dfrac{3}{4}\ln 9 + \dfrac{1}{6}\tan^{-1}0 + C,\ C = -\dfrac{3}{4}\ln 9$

$i = \dfrac{3}{4}\ln(4t^2 + 9) + \dfrac{1}{6}\tan^{-1}\dfrac{2t}{3} - \dfrac{3}{4}\ln 9$

$\quad = \dfrac{3}{4}\ln\dfrac{4t^2 + 9}{9} + \dfrac{1}{6}\tan^{-1}\dfrac{2t}{3}$

9. $y = \dfrac{1}{\sqrt{16 - x^2}};\ A = \int_0^3 y\, dx$

$\qquad = \int_0^3 \dfrac{dx}{\sqrt{16 - x^2}}$

$\qquad = \sin^{-1}\dfrac{x}{4}\Big|_0^3$

$\qquad = \sin^{-1}\dfrac{3}{4} = 0.8481$

Chapter 29

1. $f(x) = (1 + e^x)^2 \qquad f(0) = (1 + 1)^2 = 4$

$f'(x) = 2(1 + e^x)(e^x) \qquad f'(0) = 2(1 + 1)(1) = 4$

$\quad = 2e^x + 2e^{2x}$

$f''(x) = 2e^x + 4e^{2x} \qquad f''(0) = 2(1) + 4(1) = 6$

$f'''(x) = 2e^x + 8e^{2x} \qquad f'''(0) = 2(1) + 8(1) = 10$

$(1 + e^x)^2 = 4 + 4x + \dfrac{6}{2}x^2 + \dfrac{10}{6}x^3 + \cdots$

$\qquad\qquad = 4 + 4x + 3x^2 + \dfrac{5}{3}x^3 + \cdots$

2. $f(x) = \cos x \qquad f\left(\dfrac{\pi}{3}\right) = \dfrac{1}{2}$

$f'(x) = -\sin x \qquad f'\left(\dfrac{\pi}{3}\right) = -\dfrac{\sqrt{3}}{2}$

$f''(x) = -\cos x \qquad f''\left(\dfrac{\pi}{3}\right) = -\dfrac{1}{2}$

$\cos x = \dfrac{1}{2} - \dfrac{\sqrt{3}}{2}\left(x - \dfrac{\pi}{3}\right) - \dfrac{\dfrac{1}{2}\left(x - \dfrac{\pi}{3}\right)^2}{2} + \cdots$

$\quad = \dfrac{1}{2}\left[1 - \sqrt{3}\left(x - \dfrac{\pi}{3}\right) - \dfrac{1}{2}\left(x - \dfrac{\pi}{3}\right)^2 + \cdots\right]$

3. $\ln(1 + x) = x - \dfrac{x^2}{2} + \dfrac{x^3}{3} - \dfrac{x^4}{4} + \cdots$

$\ln 0.96 = \ln(1 - 0.04)$

$\quad = -0.04 - \dfrac{(-0.04)^2}{2} + \dfrac{(-0.04)^3}{3} - \dfrac{(-0.04)^4}{4}$

$\quad = -0.040\,822\,0$

4. $f(x) = \dfrac{1}{\sqrt{1 - 2x}} = (1 - 2x)^{-1/2}$

$(1 + x)^n = 1 + nx + \dfrac{n(n - 1)}{2}x^2 + \cdots$

$(1 - 2x)^{-1/2} = 1 + \left(-\dfrac{1}{2}\right)(-2x) + \dfrac{-\frac{1}{2}\left(-\frac{3}{2}\right)}{2}(-2x)^2 + \cdots$

$\qquad\qquad = 1 + x + \dfrac{3}{2}x^2 + \cdots$

5. $\int_0^1 x \cos x\, dx = \int_0^1 x\left(1 - \dfrac{x^2}{2} + \dfrac{x^4}{24}\right) dx$

$\qquad\qquad = \int_0^1 \left(x - \dfrac{x^3}{2} + \dfrac{x^5}{24}\right) dx$

$\qquad\qquad = \dfrac{1}{2}x^2 - \dfrac{x^4}{8} + \dfrac{x^6}{144}\Big|_0^1$

$\qquad\qquad = \tfrac{1}{2} - \tfrac{1}{8} + \tfrac{1}{144} - 0 = 0.3819$

6. $f(t) = 0 \qquad -\pi \le t < 0$

$f(t) = 2 \qquad\ 0 \le t < \pi$

$a_0 = \dfrac{1}{2\pi}\int_0^\pi 2\, dt = \dfrac{1}{\pi}t\Big|_0^\pi = 1$

$a_n = \dfrac{1}{\pi}\int_0^\pi 2 \cos nt\, dt = \dfrac{2}{n\pi}\sin nt\Big|_0^\pi$

$\quad = 0 \quad$ for all n

$b_n = \dfrac{1}{\pi}\int_0^\pi 2 \sin nt\, dt = \dfrac{2}{n\pi}(-\cos nt)\Big|_0^\pi$

$\quad = \dfrac{2}{n\pi}(1 - \cos n\pi)$

$b_1 = \dfrac{2}{\pi}(1 + 1) = \dfrac{4}{\pi} \qquad b_2 = \dfrac{2}{2\pi}(1 - 1) = 0$

$b_3 = \dfrac{2}{3\pi}(1 + 1) = \dfrac{4}{3\pi}$

$f(t) = 1 + \dfrac{4}{\pi}\sin t + \dfrac{4}{3\pi}\sin 3t + \cdots$

7. $f(x) = x^2 + 2, f(-x) = (-x)^2 + 2 = x^2 + 2, -f(-x) \ne f(x)$

Since $f(x) = f(-x), f(x)$ is an even function.

Since $f(x) \ne -f(-x), f(x)$ is not an odd function.

$g(x) = x^2 - 1, g(x) = f(x) - 3$

Fourier series for $g(x)$ is $F(x) - 3$.

Chapter 30

1. $x\dfrac{dy}{dx} + 2y = 4$

$dy + \dfrac{2}{x}y\,dx = \dfrac{4}{x}\,dx$

$e^{\int \frac{2}{x}dx} = e^{2\ln x} = x^2$

$yx^2 = \int \dfrac{4}{x}x^2\,dx = \int 4x\,dx = 2x^2 + c$

$y = 2 + \dfrac{c}{x^2}$

2. $y'' + 2y' + 5y = 0$

$m^2 + 2m + 5 = 0$

$m = \dfrac{-2 \pm \sqrt{4-20}}{2} = -1 \pm 2j$

$y = e^{-x}(c_1 \sin 2x + c_2 \cos 2x)$

3. $x\,dx + y\,dy = x^2\,dx + y^2\,dx = (x^2 + y^2)\,dx$

$\dfrac{x\,dx + y\,dy}{x^2 + y^2} = dx$

$\tfrac{1}{2}\ln(x^2 + y^2) = x + \tfrac{1}{2}c \quad \ln(x^2 + y^2) = 2x + c$

4. $2D^2y - Dy = 2\cos x$

$2m^2 - m = 0 \qquad m = 0, \tfrac{1}{2}$

$y_c = c_1 + c_2 e^{x/2}$

$y_p = A\sin x + B\cos x$

$Dy_p = A\cos x - B\sin x$

$D^2y_p = -A\sin x - B\cos x$

$2(-A\sin x - B\cos x) - (A\cos x - B\sin x) = 2\cos x$

$-2A + B = 0 \qquad -2B - A = 2 \qquad A = -\tfrac{2}{5} \qquad B = -\tfrac{4}{5}$

$y = c_1 + c_2 e^{x/2} - \tfrac{2}{5}\sin x - \tfrac{4}{5}\cos x$

5. $\dfrac{d^2y}{dx^2} - 4\dfrac{dy}{dx} + 4y = 3x$

$m^2 - 4m + 4 = 0 \qquad m = 2, 2$

$y_c = (c_1 + c_2 x)e^{2x}$

$y_p = A + Bx \qquad Dy_p = B \qquad D^2y_p = 0$

$0 - 4B + 4(A + Bx) = 3x$

$4A - 4B = 0 \qquad 4B = 3$

$B = \tfrac{3}{4} \qquad A = \tfrac{3}{4}$

$y = (c_1 + c_2 x)e^{2x} + \tfrac{3}{4} + \tfrac{3}{4}x$

6. $D^2y - 2Dy - 8y = 4e^{-2x}$

$m^2 - 2m - 8 = 0 \qquad (m+2)(m-4) = 0 \qquad m = -2, 4$

$y_c = c_1 e^{-2x} + c_2 e^{4x}$

$y_p = Axe^{-2x}$ (factor of x necessary due to first term of y_c)

$Dy_p = Ae^{-2x} - 2Axe^{-2x}$

$D^2y_p = -2Ae^{-2x} - 2Ae^{-2x} + 4Axe^{-2x} = -4Ae^{-2x} + 4Axe^{-2x}$

$(-4Ae^{-2x} + 4Axe^{-2x}) - 2(Ae^{-2x} - 2Axe^{-2x})$

$\quad - 8Axe^{-2x} = 4e^{-2x}$

$-6Ae^{-2x} = 4e^{-2x} \qquad A = -\tfrac{2}{3}$

$y = c_1 e^{-2x} + c_2 e^{4x} - \tfrac{2}{3}xe^{-2x}$

7. $(xy + y)\dfrac{dy}{dx} = 2$

$y(x + 1)\,dy = 2\,dx$

$y\,dy = \dfrac{2\,dx}{x + 1}$

$\tfrac{1}{2}y^2 = 2\ln(x + 1) + c$

$y = 2$ when $x = 0$

$\tfrac{1}{2}(4) = 2\ln(1) + c, \quad c = 2$

$\tfrac{1}{2}y^2 = 2\ln(x + 1) + 2$

$y^2 = 4\ln(x + 1) + 4$

8. $\dfrac{dA}{dt} = rA$

$\dfrac{dA}{A} = r\,dt$

$\ln A = rt + \ln c$

$\ln\dfrac{A}{c} = rt$

$A = ce^{rt}$

$A_0 = ce^0,$

$c = A_0$

$A = A_0 e^{rt}$

9. $y'' + 9y = 9$

$y(0) = 0 \qquad y'(0) = 1$

$\mathcal{L}(y'') + 9\mathcal{L}(y) = \mathcal{L}(9)$

$s^2\mathcal{L}(y) - s(0) - 1 + 9\mathcal{L}(y) = \dfrac{9}{s}$

$(s^2 + 9)\mathcal{L}(y) = \dfrac{9}{s} + 1$

$\mathcal{L}(y) = \dfrac{9}{s(s^2 + 9)} + \dfrac{1}{s^2 + 9}$

$y = 1 - \cos 3t + \tfrac{1}{3}\sin 3t$

10. $D^2y - Dy - 2y = 12 \qquad y(0) = 0, y'(0) = 0$

$\mathcal{L}(y'') - \mathcal{L}(y') - 2\mathcal{L}(y) = \mathcal{L}(12)$

$s^2\mathcal{L}(y) - s(0) - 0 - [s\mathcal{L}(y) - 0] - 2\mathcal{L}(y) = \dfrac{12}{s}$

$\mathcal{L}(y)(s^2 - s - 2) = \dfrac{12}{s}$

$\mathcal{L}(y) = \dfrac{12}{s(s^2 - s - 2)} = \dfrac{12}{s(s + 1)(s - 2)}$

$\quad = \dfrac{A}{s} + \dfrac{B}{s + 1} + \dfrac{C}{s - 2}$

$12 = A(s + 1)(s - 2) + Bs(s - 2) + Cs(s + 1)$

$s = 0: \qquad 12 = -2A, A = -6$

$s = -1: \qquad 12 = 3B, B = 4$

$s = 2: \qquad 12 = 6C, C = 2$

$\mathcal{L}(y) = -\dfrac{6}{s} + \dfrac{4}{s + 1} + \dfrac{2}{s - 2}$

$y = -6 + 4e^{-t} + 2e^{2t}$

11. $L\dfrac{d^2q}{dt^2} + R\dfrac{dq}{dt} = E \qquad L\dfrac{di}{dt} + Ri = E$

$L = 2\,\text{H} \qquad R = 8\,\Omega \qquad E = 6\,\text{V}$

$2\dfrac{di}{dt} + 8i = 6 \qquad di + 4i\,dt = 3\,dt$

$e^{\int 4\,dt} = e^{4t} \qquad ie^{4t} = \displaystyle\int 3e^{4t}\,dt = \tfrac{3}{4}e^{4t} + c$

$i = 0 \quad \text{for} \quad t = 0 \qquad 0 = \tfrac{3}{4} + c \qquad c = -\tfrac{3}{4}$

$ie^{4t} = \tfrac{3}{4}e^{4t} - \tfrac{3}{4}$

$i = \tfrac{3}{4}(1 - e^{-4t}) = 0.75(1 - e^{-4t})$

12. $m = 0.5\,\text{kg},\, k = 32\,\text{N/m}$

$mD^2x = -kx, \quad 0.5D^2x + 32x = 0$

$D^2x + 64x = 0$

$m^2 + 64 = 0; \qquad m = \pm 8j$

$x = c_1 \sin 8t + c_2 \cos 8t$

$Dx = 8c_1 \cos 8t - 8c_2 \sin 8t$

$x = 0.3\,\text{m} \qquad Dx = 0 \quad \text{for} \quad t = 0$

$0.3 = c_1 \sin 0 + c_2 \cos 0, \qquad c_2 = 0.3$

$0 = 8c_1 \cos 0 - 8c_2 \sin 0, \qquad c_1 = 0$

$x = 0.3 \cos 8t$

INDEX OF WRITING EXERCISES

The final review exercise in each chapter is a writing exercise. Each of these 30 exercises will require at least a paragraph to provide a good explanation of the problem presented. Also, there are over 340 other exercises throughout the text (at least seven in each chapter) marked with ⓦ before the exercise number or instructions that require at least one complete sentence (up to a paragraph) to provide the explanation needed.

Following is a listing of these writing exercises. The first number shown is the page number, and the numbers in parentheses are the exercise numbers. The * denotes the final review exercise of the chapter.

ALGEBRA

Exponents and Radicals

$a^m \cdot a^n = a^{m+n}$

$\dfrac{a^m}{a^n} = a^{m-n} \quad a \neq 0$

$(a^m)^n = a^{mn}$

$(ab)^n = a^n b^n$

$\left(\dfrac{a}{b}\right)^n = \dfrac{a^n}{b^n} \quad b \neq 0$

$a^0 = 1 \quad a \neq 0$

$a^{-n} = \dfrac{1}{a^n} \quad a \neq 0$

$a^{m/n} = \sqrt[n]{a^m} = (\sqrt[n]{a})^m$

$\sqrt{ab} = \sqrt{a}\,\sqrt{b}$

Special Products

$a(x + y) = ax + ay$

$(x + y)(x - y) = x^2 - y^2$

$(x + y)^2 = x^2 + 2xy + y^2$

$(x - y)^2 = x^2 - 2xy + y^2$

Quadratic Equation and Formula

$ax^2 + bx + c = 0$

$x = \dfrac{-b \pm \sqrt{b^2 - 4ac}}{2a}$

Properties of Logarithms

$\log_b x + \log_b y = \log_b xy$

$\log_b x - \log_b y = \log_b\left(\dfrac{x}{y}\right)$

$n \log_b x = \log_b (x^n)$

Complex Numbers

$\sqrt{-a} = j\sqrt{a} \quad (a > 0)$

$x + yj = r(\cos \theta + j \sin \theta)$

$\quad = re^{j\theta} = r\underline{/\theta}$

Variation

Direct variation: $y = kx$

Inverse variation: $y = k/x$

TRIGONOMETRY

$\sin \theta = \dfrac{y}{r} \quad \cos \theta = \dfrac{x}{r} \quad \tan \theta = \dfrac{y}{x} \quad \cot \theta = \dfrac{x}{y} \quad \sec \theta = \dfrac{r}{x} \quad \csc \theta = \dfrac{r}{y}$

Law of Sines: $\dfrac{a}{\sin A} = \dfrac{b}{\sin B} = \dfrac{c}{\sin C}$

Law of Cosines: $a^2 = b^2 + c^2 - 2bc \cos A$

$\pi \text{ rad} = 180°$

Basic Identities

$\sin \theta = \dfrac{1}{\csc \theta} \quad \cos \theta = \dfrac{1}{\sec \theta} \quad \tan \theta = \dfrac{1}{\cot \theta} \quad \tan \theta = \dfrac{\sin \theta}{\cos \theta} \quad \cot \theta = \dfrac{\cos \theta}{\sin \theta}$

$\sin^2\theta + \cos^2\theta = 1 \quad 1 + \tan^2\theta = \sec^2\theta \quad 1 + \cot^2\theta = \csc^2\theta$

$\sin 2\alpha = 2 \sin \alpha \cos \alpha \quad \cos 2\alpha = \cos^2\alpha - \sin^2\alpha = 2\cos^2\alpha - 1 = 1 - 2\sin^2\alpha$

$\sin \dfrac{\alpha}{2} = \pm\sqrt{\dfrac{1 - \cos \alpha}{2}} \quad \cos \dfrac{\alpha}{2} = \pm\sqrt{\dfrac{1 + \cos \alpha}{2}}$